D0215766

HUMAN PHYSIOLOGY

BRYAN DERRICKSON
VALENCIA COLLEGE

Director, Life Sciences	Kevin Witt
Executive Editor	Bonnie Roesch
Associate Editor	Brittany Cheetham
Developmental Editor	Karen Trost
Senior Marketing Manager	Maria Guarascio
Senior Product Designer	Linda Muriello
Senior Media Specialist	Svetlana Barskaya
Market Solutions Assistant	Lindsey Myers
Senior Production Editor	Trish McFadden
Photo Manager	MaryAnne Price
Illustration Coordinators	Anna Melhorn/Claudia Durell
Senior Designer	Thomas Nery

This book is printed on acid free paper. ♾

Founded in 1807, John Wiley & Sons, Inc. has been a valued source of knowledge and understanding for more than 200 years, helping people around the world meet their needs and fulfill their aspirations. Our company is built on a foundation of principles that include responsibility to the communities we serve and where we live and work. In 2008, we launched a Corporate Citizenship Initiative, a global effort to address the environmental, social, economic, and ethical challenges we face in our business. Among the issues we are addressing are carbon impact, paper specifications and procurement, ethical conduct within our business and among our vendors, and community and charitable support. For more information, please visit our website: www.wiley.com/go/citizenship.

ISBN: 978-0-470-38140-3

Printed in the United States of America 10 9 8 7 6 5 4 3 2 1

The inside back cover will contain printing identification and country of origin if omitted from this page. In addition, if the ISBN on the back cover differs from the ISBN on this page, the one on the back cover is correct.

About the Author

Courtesy of Bryan Derrickson

Bryan Derrickson is Professor of Biology at Valencia College in Orlando, Florida, where he teaches human anatomy and physiology as well as general biology and human sexuality. He received his bachelor's degree in biology from Morehouse College and his Ph.D. in cell biology from Duke University. Bryan's study at Duke was in the Physiology Division within the Department of Cell Biology, so while his degree is in cell biology, his training focused on physiology. At Valencia, he frequently serves on faculty hiring committees. He has served as a member of the Faculty Senate, which is the governing body of the college, and as a member of the Faculty Academy Committee (now called the Teaching and Learning Academy), which sets the standards for the acquisition of tenure by faculty members. Nationally, he is a member of the Human Anatomy and Physiology Society (HAPS) and the National Association of Biology Teachers (NABT). Bryan has always wanted to teach. Inspired by several biology professors while in college, he decided to pursue physiology with an eye to teaching at the college level. He is completely dedicated to the success of his students. He particularly enjoys the challenges of his diverse student population, in terms of their age, ethnicity, and academic ability, and finds being able to reach all of them, despite their differences, a rewarding experience. His students continually recognize Bryan's efforts and care by nominating him for a campus award known as the "Valencia Professor Who Makes Valencia a Better Place to Start." Bryan has received this award three times.

In honor of my mother, **Rosalind Gilmer Derrickson** **B.H.D.**

Courtesy of Bryan Derrickson

Preface

Welcome to your course in human physiology! Many of you are taking this course because you hope to pursue a career in medicine or one of the allied health professions. Others of you may be taking the course because you are simply interested in learning how your own amazing body functions. Whatever your motivations, ***Human Physiology,*** and ***WileyPLUS Learning Space*** have all the content and tools needed to ensure that you receive a solid foundation and the knowledge and skills to reach your desired goals.

I am passionate about the discipline of physiology, and have an equally strong passion for teaching. Physiology is a fascinating subject and I enjoy conveying this information to students and guiding them through the intricacies of the many complex functions of the human body. I wrote this text because of my desire to share the story of physiology with a wider audience.

Human Physiology is a comprehensive text that uses four main underlying principles in physiology as a foundation for specific details of all the systems of the human body. These principles include *homeostasis, mechanisms of action, communication,* and *integration.* As you progress through the text, you will discover these underlying themes supporting your understanding of core physiological concepts.

Most importantly, I endeavored to distinguish this text by uniquely combining three powerful elements: (1) clear, easy-to-follow writing style supported with carefully developed figures; (2) an emphasis on the development of vital critical thinking skills; and (3) a fully integrated digital platform rich in interactive activities and media. Together these elements provide a superior level of coverage that helps prepare you for successful careers in medicine and allied health.

THE NARRATIVE AND VISUALS

Each chapter in ***Human Physiology*** is written in a style that is very straightforward and easy to understand. This type of writing style facilitates comprehension and retention of key physiological concepts. Words can convey a lot when carefully chosen, but if you are like most students today, visual representation of the material is of equal importance to you. You'll find clear visual explanations that bring the words on the page to life. In addition, each chapter is filled with boxed information on relevant clinical connections and other real-life examples that will stir your interest in, and solidify understanding of, the relevant science at hand.

CRITICAL THINKING

Understanding how to think critically about the scientific information presented is vital to your success. ***Human Physiology, 1st edition*** and ***WileyPLUS Learning Space*** have several features that give you the opportunity to hone this essential skill. *Critical Thinking* exercises help you to think logically and critically about real-life scenarios that involve important physiological concepts. *Research to Reality* is a unique feature that provides you with the opportunity to analyze and interpret real scientific data from primary research articles. *Physiological Equation* boxes describe key physiological equations that you can use to analyze and quantify certain physiological parameters. *Ponder This* questions at the end of each chapter provide additional opportunities to think critically about physiological information.

ENGAGING DIGITALLY

I am so excited to have my text fully integrated with ***WileyPLUS Learning Space.*** This platform allows you to create a personalized study plan, assess your progress along the way, and make deeper connections as you interact with the course material and with one another. This collaborative learning environment provides immediate insight into both your strengths and your problem areas through a combination of dynamic course materials—such as 3D animations, game-like exercises, and laboratory simulations—and visual reports so you can act on what's most important to help you master the course material. ***WileyPLUS Learning Space*** also includes ***ORION***—integrated, adaptive practice that helps you build proficiency on particular topics and use your study time most effectively.

Acknowledgments

Human Physiology*, *1st edition and ***WileyPLUS Learning Space*** would not be possible without the help of many, particularly the numerous academic colleagues that collaborated with us along the way. First to thank are three contributors whose work informed and enhanced our focus on critical thinking and connections to real life situations and activities:

David Mallory, Marshall University, was instrumental in the design and execution of the *Critical Thinking Boxes* and step-by-step questions found in each chapter. His choice of topics for these boxes adds great interest and really highlights the connections to the concepts at hand. Thank you so much, David.

Lynn Diener, Mount Mary University, researched and chose most of the primary research papers used in the ***Research to Reality*** activities. She used her expertise and experience in implementing this kind of investigation in her own classes to bring this unique feature to this project. I am very grateful, Lynn.

Heidi Bustamante, University of Colorado, Boulder, excels at writing thoughtful questions that challenge students to critically think about what they have learned. I am so happy that she chose to enhance this text by writing the end-of-chapter *Ponder This* questions. I really appreciate your work, Heidi.

I am also very grateful that Wiley commissioned a board of academic advisors with expertise in teaching physiology to help inform not only individual resources integrated with this text, but also as a sounding board for all of us as we progressed through development.

WILEY ADVISORY BOARD FOR ANATOMY AND PHYSIOLOGY

Heidi Bustamante *University of Colorado, Boulder*
Patrick Cafferty *Emory University*
Janet Casagrand *University of Colorado, Boulder*
Lynn Diener *Mount Mary University*
John Erickson *Ivy Tech State College-SE*
Melaney Farr *Salt Lake Community College*
Geoff Goellner *Minnesota State University*
Michael Griffin *Angelo State University*
Wanda Hargroder *Louisiana State University*
DJ Hennager *Kirkwood Community College*
Sandra Hutchinson *Santa Monica College*
Heather Labbe *University of Montana - Missoula*
Tom Lancraft *St. Petersburg College*
David Mallory *Marshall University*
Russell Nolan *Baton Rouge Community College*
Terry Thompson *Wor-Wic Community College*
Tracy Wagner *Washburn University of Topeka*
Paul Wagner *Washburn University of Topeka*

I thank the following group of academics who have contributed to the creation and integration of this text with ***WileyPLUS Learning Space*** with ***ORION***. Their thoughtful execution of resources and assessment has raised the bar for excellence. Thank you, so very much.

Brian Antonsen *Marshall University*
Heidi Bustamante *University of Colorado, Boulder*
Patrick Cafferty *Emory University*
Janet Casagrand *University of Colorado, Boulder*
Lynn Diener *Mount Mary University*
John Erickson *Ivy Tech State College-SE*
Gregory Loftin *Metropolitan Community College of Kansas City*
Geoff Goellner *Minnesota State University*
Michael Griffin *Angelo State University*
Jill Tall *Youngstown State University*
Tracy Wagner *Washburn University of Topeka*
Paul Wagner *Washburn University of Topeka*
Chad Wayne *University of Houston*

The development of a first edition text and media for a course as complex as human physiology is a long process and would not be possible without the continued involvement of those "in the trenches" teaching the course, who guided and informed our choices all along the way. I am very grateful to my colleagues who have reviewed the drafts of manuscript, participated in focus groups and workshops, or offered suggestions for improvement.

Ateegh Al-Arabi *Johnson County Community College*
Brenda Alston-Mills *Michigan State University*
Gwen Bachman *University of Nebraska, Lincoln*
Ari Berowitz *University of Oklahoma, Norman*
Eric Blough *Marshall University*
Sunny K. Boyd *University of Notre Dame*
Carol Britson *University of Mississippi*
Jackie Brittingham *Simpson College*
Brent Bruot *Kent State University*
Heidi Bustamante *University of Colorado, Boulder*
Sherell Byrd *Fort Lewis College*
Martin Burg *Grand Valley State University*
Phyllis Callahan *Miami University*
Jackie Carnegie *University of Ottawa*
Robert Carroll *East Carolina University*
Janet Casagrand *University of Colorado, Boulder*
Daniel Castellanos *Florida International University*
Chris Dewitt *University of South Carolina Aiken*
Debora Christensen *Drake University*
Ruth Clark *Washington University, St. Louis*
Josefa Cubina *New York Institute of Technology*
Maria de Bellard *California State University, Northridge*
Michael Deschenes *College of William and Mary*

Lynn Diener *Mount Mary University*
Stephen Dodd *University of Florida*
John Erickson *Ivy Tech*
Max G. Ervin *Middle Tennessee State University*
Carol Fassbinder-Orth *Creighton University*
Ralph Ferges *Palomar College*
James Ferraro *Southern Illinois School of Medicine*
Michael Finkler *Indiana University - Kokomo*
John Fishback *Ozarks Tech Community College*
David Flory *University of Central Florida*
Victor Fomin *University of Delaware*
Michelle French *University of Toronto*
Geoff Goellner *Minnesota State University*
Barbara Goodman *University of South Dakota*
Mike Griffin *Angelo State University*
Bryan Hamilton *Waynesburg University*
Janet Haynes *Long Island University*
Susan Heaphy *The Ohio State University at Lima*
Steven Heidemann *Michigan State University*
Stephen Henderson *California State University, Chico*
William Higgins *University of Maryland*
Steven Hobbs *University of Colorado*
James Hoffman *Diablo Valley College*
Kelly Johnson *University of Kansas*
Steven King *Oregon Health & Science University*
Jennifer Kneafsey *Tulsa Community College - Northeast Campus*
Megan Knoch *Indiana University of Pennsylvania*
Joan Lafuze *Indiana University East*
Dean Lauritzen *City College of San Francisco*
John Lepri *University of North Carolina, Greensboro*
Paul Lonquich *California State University, Northridge*
Jennifer Lundmark *California State University - Sacramento*
David Mallory *Marshall University*
Theresa Martin *San Mateo Community College*
Michael Masson *Santa Barbara Community College*
Tamara Mau *University of California, Berkeley*
Eric McElroy *College of Charleston*
Thomas McNeilis *Dixie State College of Utah*
John McReynolds *University of Michigan*
Jamie Melling *Western University*
Jeanne Mitchell *Truman State University*
Diane Morel *University of the Sciences*
Tim Mullican *Dakota Wesleyan University*
Cheryl Neudauer *Minneapolis Community and Technical College*
David Nutting *University of Tennessee, Memphis*
Linda Ogren *University of California - Santa Cruz*
Ok-Kyong Park-Sarge *University of Kentucky*
William Percy *University of South Dakota*
David Pistole *Indiana University of Pennsylvania*
Robert Preston *Illinois State University*
David Petzel *Creighton University*
Steve Price *Virginia Commonwealth University*
Peter Reiser *Ohio State University*
Nick A. Ritucci *Wright State University*
Laurel Roberts *University of Pittsburgh*
Sonia Rocah-Sanchez *Creighton University*
Dean Scherer *Oklahoma State University*
David Schulz *University of Missouri*
Virginia Shea *University of North Carolina, Chapel Hill*
Allison Shearer *Grossmont College*
Rachel Smetanka *Southern Utah University*
James Strauss *Pennsylvania State University*
Erik Swenson *University of Washington*
Jill Tall *Youngstown State University*
Bonnie Tarricone *Ivy Tech Community College at Indianapolis*
Mark Thomas *University of Northern Colorado*
Maureen Tubbiola *St. Cloud State University*
Paul Wagner *Washburn University of Topeka*
Tracy Wagner *Washburn University of Topeka*
Curt Walker *Dixie State College of Utah*
Richard Walker *University of Calgary*
R. Douglass Watson *University of Alabama, Birmingham*
Chad Wayne *University of Houston*

Last, but certainly not least, I owe tremendous gratitude to the team at Wiley. This team of collaborative publishing professionals is dedicated, enthusiastic, and talented and each brings a level of skill to the job that, I believe, is unparalleled. My hat is off to the entire team: Bonnie Roesch, Executive Editor; Karen Trost, Developmental Editor; Brittany Cheetham, Associate Editor; Maria Guarascio, Senior Marketing Manager; Linda Muriello, Senior Product Designer; Lindsey Myers, Market Solutions Assistant; Trish McFadden, Senior Production Editor; Mary Ann Price, Senior Photo Editor; Claudia Volano and Anna Melhorn, Illustration Coordinators; Thomas Nery, Senior Designer. I also want to acknowledge Kevin Witt, Director, for his support of this project. My heartfelt thanks to you all!

Bryan Derrickson
Department of Science, PO Box 3028
Valencia College
Orlando, FL 32802
physioBRYAN@aol.com

Brief Contents

Contents

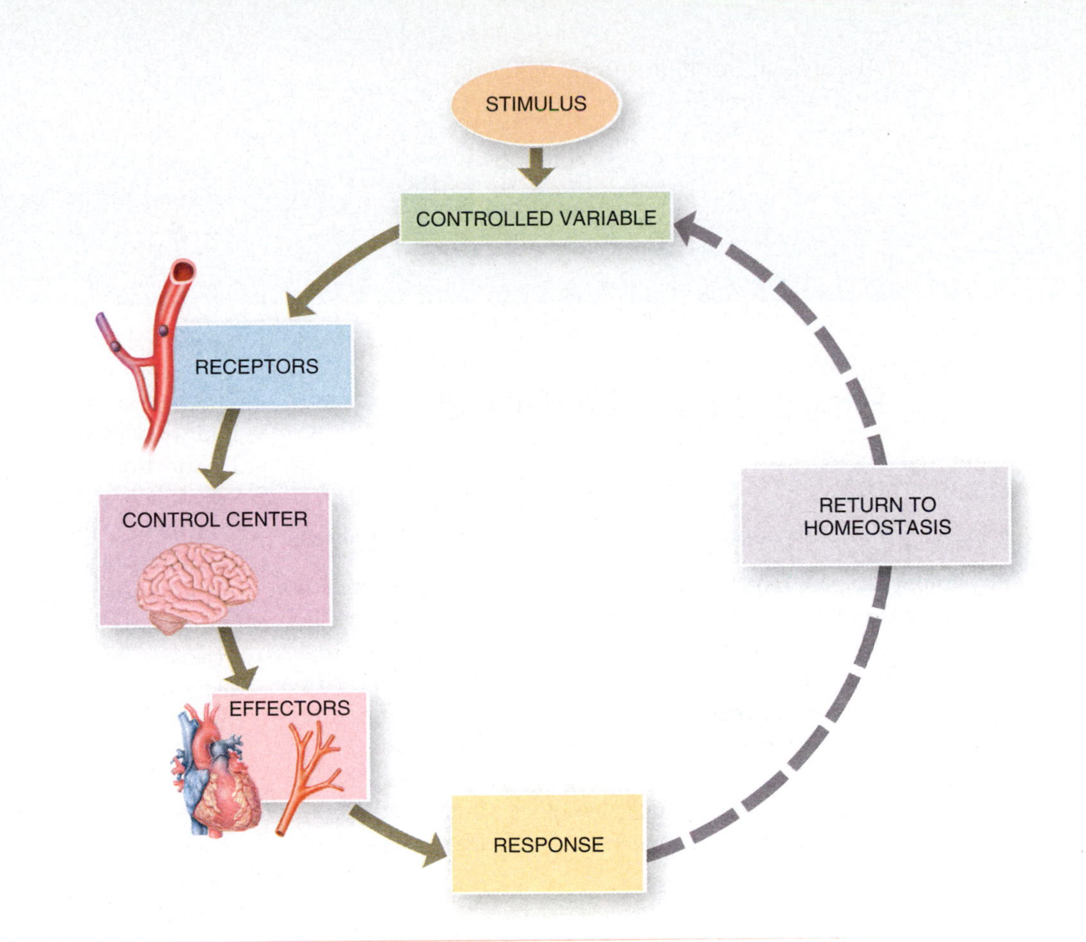

1 An Introduction to Physiology

The Human Body and Homeostasis

Humans have many ways to maintain homeostasis, the state of relative stability of the body's internal environment. Disruptions to homeostasis often set in motion corrective cycles, called feedback systems, that help restore the conditions needed for health and life.

LOOKING BACK TO MOVE AHEAD...

- One of the fundamental principles of biology is the cell theory, which states that (1) the cell is basic unit of life; (2) all organisms are composed of one or more cells; and (3) cells arise from pre-existing cells.
- Organisms are classified into three domains: Bacteria, Archaea, and Eukarya; humans belong to domain Eukarya, kingdom Animalia, phylum Chordata, subphylum Vertebrata, and class Mammalia.
- All organisms have a binomial (a two-part Latin scientific name) that consists of a genus and species; the binomial for humans is *Homo sapiens*, which means "wise man" (*homo-* = *man; sapiens* = wise).
- Compared to other organisms, humans have several distinguishing features: erect posture; bipedal locomotion (ability to walk on two legs); and a large, well-developed brain that allows for analytical skills and complex thought.

Your fascinating journey through the human body begins with an overview of the meaning of physiology, followed by a discussion of the organization of the human body and the properties that it shares with all living things. Next, you will discover how the body regulates its own internal environment through an unceasing process known as homeostasis. You will then explore physiology as a science—its historical background and the general methods of acquiring and organizing physiological information. The chapter concludes with a discussion of the key themes of physiology.

1.1 Physiology Defined

OBJECTIVE

- Define physiology and identify several of its subdisciplines.

Physiology (fiz′-ē-OL-ō-jē) is the study of the functions of an organism and its constituent parts. This text focuses on *human physiology*—how the parts of the human body work. Human physiology considers topics such as the molecular mechanism responsible for muscle contraction; communication between cells using chemical messengers; the parts of the brain involved in language comprehension and expression; and the maintenance of blood pressure by the coordinated efforts of the heart, kidneys, brain, and glands. Because physiology has a broad scope, it is divided into many subdisciplines, several of which are described in TABLE 1.1.

TABLE 1.1 Selected Subdisciplines of Physiology

Subdiscipline	Study of . . .
Molecular physiology	Functions of individual molecules, such as proteins.
Cell physiology	Functions of cells.
Neurophysiology	Functions of the nervous system.
Endocrinology	Hormones (chemical regulators in the blood) and how they control body functions.
Cardiovascular physiology	Functions of the heart, blood vessels, and blood.
Immunology	How the body defends itself against disease-causing agents.
Respiratory physiology	Functions of the air passageways and lungs.
Renal physiology	Functions of the kidneys.
Gastrointestinal physiology	Functions of the stomach and intestines.
Integrative physiology	How different parts of the body work together to accomplish a particular function.
Exercise physiology	Changes in cell and organ functions as a result of muscular activity.
Pathophysiology	Functional changes associated with disease and aging.

Whereas physiology deals with body functions, *anatomy* is the study of body structure. However, the two cannot truly be separated: The function of a body part is a reflection of its structure. For example, the walls of the air sacs in the lungs are very thin, permitting rapid movement of inhaled oxygen into the blood. By contrast, the lining of the urinary bladder is much thicker to prevent the escape of urine into the pelvic cavity, yet its construction allows for considerable stretching as the urinary bladder fills with urine. Because structure and function are so closely related, as the function of a body part is discussed in the text, relevant information about its structure is provided to help clarify your understanding of the topic.

CHECKPOINT

1. Which subdiscipline of physiology would most likely explore how the kidneys and lungs work together to maintain the acid–base balance of your body fluids? (*Hint:* Refer to TABLE 1.1).

1.2 Levels of Organization in the Body

OBJECTIVES

- Describe the levels of organization that comprise the human body.
- Explain the functions of the twelve body systems.

As you study physiology, your exploration of the human body will extend from atoms and molecules to the whole person. From the smallest to the largest, six levels of organization will help you to understand how the body functions: the chemical, cellular, tissue, organ, system, and organismal levels of organization (FIGURE 1.1).

- ***Chemical level.*** This very basic level includes **atoms**, the smallest units of matter that participate in chemical reactions, and **molecules**, two or more atoms joined together. Certain atoms, such as carbon (C), hydrogen (H), oxygen (O), nitrogen (N), phosphorus (P), calcium (Ca), and sulfur (S), are essential for maintaining life. Two familiar molecules found in the body are deoxyribonucleic acid (DNA), the genetic material passed from one generation to the next, and glucose, the main type of sugar in the bloodstream. Chapter 2 focuses on the chemical level of organization.
- ***Cellular level.*** Molecules combine to form **cells**, the basic structural and functional units of an organism. Cells are the smallest units capable of performing all life processes. Among the many kinds of cells in your body are epithelial cells, connective tissue cells, muscle cells, and neurons (nerve cells). FIGURE 1.1 shows a smooth muscle cell, one of the three types of muscle cells in the body.

FIGURE 1.1 Levels of organization in the human body.

The levels of organization in the body are chemical, cellular, tissue, organ, system, and organismal.

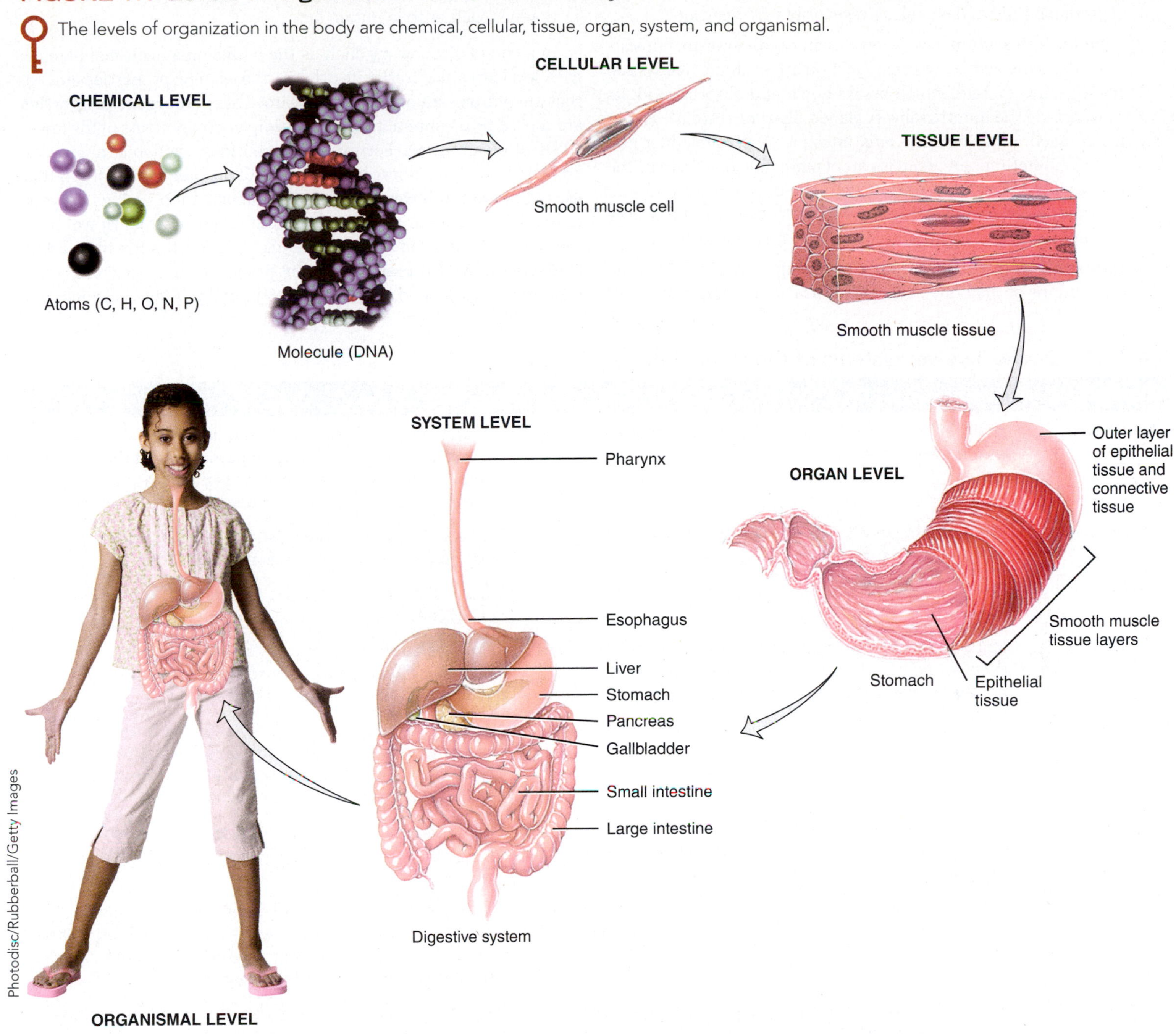

Which level of organization is composed of two or more different types of tissues that work together to perform a specific function?

- ***Tissue level.*** A **tissue** is a group of similar cells that work together to perform a particular function. There are just four basic types of tissue in your body: epithelial tissue, connective tissue, muscle tissue, and nervous tissue. *Epithelial tissue* covers body surfaces, lines hollow organs and ducts, and forms glands. *Connective tissue* supports and protects body organs, stores energy reserves as fat, and helps provide the body with immunity to disease-causing agents. *Muscle tissue* contracts to produce movement, maintain posture, and generate heat. *Nervous tissue* detects and responds to changes in the body's external or internal environment. Shown in FIGURE 1.1 is smooth muscle tissue, which consists of tightly packed smooth muscle cells. The cellular and tissue levels of organization are described in Chapter 3.
- ***Organ level.*** An **organ** is a structure composed of two or more different types of tissues. It has a specific function and usually (but not always) has a recognizable shape. Examples of organs are the stomach, heart, liver, lungs, brain, and skin. FIGURE 1.1 shows how several types of tissues comprise the stomach. The stomach's outer covering is a layer of epithelial tissue and connective tissue that reduces friction when the stomach moves and rubs against other organs. Underneath are smooth muscle tissue layers, which contract to churn and mix food and then push it into the next digestive organ,

the small intestine. The innermost lining is an epithelial tissue layer that produces fluid and chemicals responsible for digestion.

- ***System level.*** A **system,** also known as an **organ system,** consists of related organs with a common function. An example of a system is the digestive system, which breaks down and absorbs food. Its organs include the mouth, salivary glands, pharynx (throat), esophagus, stomach, small intestine, large intestine, liver, gallbladder, and pancreas. Sometimes an organ is part of more than one system. The pancreas, for example, is part of both the digestive system and the hormone-producing endocrine system. TABLE 1.2 introduces the components and functions of the twelve systems of the body.
- ***Organismal level.*** An **organism** is any living individual. All of the organ systems of the body collectively form the organism. In other words, the organism is the totality of all of its organ systems functioning together to maintain life.

In a complex hierarchy such as the body's organizational plan, as each level gives rise to the next highest level, new properties emerge that are not present at the levels below. These **emergent properties** are caused by the interactions of the simpler components of the lower levels of organization. For example, emotions, thoughts, memories, and intelligence are emergent properties of the brain (organ level) that are not present at lower levels of brain organization such as nervous tissue (tissue level) or individual neurons (cellular level). However, as the cells and tissues of the brain interact with each other in a variety of different ways, the brain's emergent properties arise. Because emergent properties depend on the interactions of lower-level components,

TABLE 1.2 The Twelve Systems of the Human Body

System	Components	Functions
Nervous	Brain, spinal cord, nerves, and special sense organs, such as the eyes and ears.	Generates action potentials to regulate body activities; detects changes in the body's external and internal environments, interprets the changes, and responds by causing muscular contractions or glandular secretions.
Muscular	Muscles composed of skeletal muscle tissue, so-called because it is usually attached to bones.	Produces body movements, such as walking; stabilizes body position (posture); generates heat.
Skeletal	Bones, joints, and associated cartilages.	Supports and protects the body; aids body movements; houses cells that produce blood cells.
Endocrine	Hormone-producing glands (pituitary gland, thyroid gland, parathyroid glands, adrenal glands, and pineal gland) and hormone-producing cells in several other organs and tissues.	Regulates body activities by releasing hormones, which are chemical messengers transported in blood from an endocrine gland or tissue to a target organ.
Cardiovascular	Heart, blood vessels, and blood.	The heart pumps blood through blood vessels; blood carries oxygen and nutrients to cells and carries carbon dioxide and other wastes away from cells.
Immune	Lymphocytes (white blood cells), lymph nodes, bone marrow, thymus, spleen, tonsils, and lymphoid tissue of the gut.	Defends body against microbes and other foreign substances.
Lymphatic	Lymphatic vessels, lymph, lymph nodes, bone marrow, thymus, spleen, tonsils, and lymphoid tissue of the gut.	Drains excess interstitial fluid; returns filtered plasma proteins back to the blood; carries out immune responses (part of the lymphatic system also functions as the immune system); transports dietary lipids.
Integumentary	Skin and associated structures, such as hair, nails, sweat glands, and oil glands.	Protects the body from the external environment; helps regulate body temperature; eliminates some wastes.
Respiratory	Nose, pharynx (throat), larynx (voice box), trachea (windpipe), bronchi, and lungs.	Transfers oxygen from inhaled air to blood and carbon dioxide from blood to exhaled air; helps regulate acid–base balance of body fluids.
Urinary	Kidneys, ureters, urinary bladder, and urethra.	Eliminates wastes and excess substances in urine; regulates volume and chemical composition of blood; helps regulate acid–base balance of body fluids.
Digestive	Mouth, pharynx (throat), esophagus, stomach, small and large intestines, salivary glands, liver, gallbladder, and pancreas.	Achieves physical and chemical breakdown of food; absorbs nutrients; eliminates solid wastes.
Reproductive	Gonads (testes in males and ovaries in females) and associated organs (epididymis, vas deferens, and penis in males; fallopian tubes, uterus, and vagina in females).	Gonads produce gametes (sperm or eggs) that unite to form a new organism; gonads also release hormones that regulate reproduction and other body processes; associated organs transport and store gametes.

they are not properties of any single one of these simpler components and cannot be predicted just by knowing that these components exist. The interaction of the components of the body to give rise to emergent properties is an example of integration. **Integration** is the process by which several components work together for a common, unified purpose. Thus, the body is much more than the sum of its parts: It is the result of integrated activities between components at essentially all levels of organization.

CHECKPOINT

2. Define the following terms: atom, molecule, cell, tissue, organ, system, and organism.
3. Refer to TABLE 1.2. Which body systems help eliminate wastes?
4. What is an emergent property?

1.3 Life Processes

OBJECTIVE

- Identify the important life processes of the human body.

Certain processes distinguish organisms, or living things, from nonliving things. Following are the six most important life processes of the human body:

1. **Metabolism** is the sum of all of the chemical reactions that occur in the body. One phase of metabolism is **catabolism**, the breakdown of complex chemical substances into simpler components. The other phase of metabolism is **anabolism**, the formation of complex chemical substances from smaller, simpler components. For example, digestive processes catabolize (break down) proteins in food into amino acids. These amino acids are then used to anabolize (build) new proteins that make up body structures such as muscles and bones.
2. **Responsiveness** is the body's ability to detect and respond to changes. For example, a decrease in body temperature represents a change in the internal environment (within the body), and turning your head toward the sound of squealing brakes is a response to change in the external environment (outside the body). Different cells in the body respond to environmental changes in characteristic ways. Neurons respond by generating electrical signals known as *action potentials*. Muscle cells respond by contracting, which generates force to move body parts.
3. **Movement** includes motion of the whole body, individual organs, single cells, and even tiny structures inside cells. For example, the coordinated action of leg muscles moves your whole body from one place to another when you walk or run. After you eat a meal that contains fats, your gallbladder contracts and squirts bile into the gastrointestinal tract to aid in the digestion of these fats. When a body tissue is damaged or infected, certain leukocytes (white blood cells) move from the blood into the affected tissue to help clean up and repair the area. Inside the cell, various parts move from one position to another to carry out their functions.
4. **Growth** is an increase in body size that results from an increase in the size of existing cells, an increase in the number of cells, or both. In addition, a tissue sometimes increases in size because the amount of material between cells increases. In a growing bone, for example, mineral deposits accumulate between bone cells, causing the bone to grow in length and width.
5. **Differentiation** is the development of a cell from an unspecialized to a specialized state. Each type of cell in the body has a specialized structure and function that differs from that of its precursor (ancestor) cells. For example, erythrocytes (red blood cells) and several types of leukocytes all arise from the same unspecialized precursor cells in bone marrow. Such precursor cells, which can divide and give rise to cells that undergo differentiation, are known as **stem cells**. Also through differentiation, a zygote (fertilized egg) develops into an embryo, and then into a fetus, an infant, a child, and finally an adult.
6. **Reproduction** refers either to (1) the formation of new cells for tissue growth, repair, or replacement, or (2) the production of a new individual. The formation of new cells occurs through cell division. The production of a new individual occurs through the fertilization of an egg by a sperm cell to form a zygote, followed by repeated cell divisions and the differentiation of these cells.

When one or more life processes ceases to occur properly, the result is death of cells and tissues, which may lead to death of the organism. Clinically, loss of the heartbeat, absence of spontaneous breathing, and loss of brain functions indicate death in the human body.

CHECKPOINT

5. The formation of a muscle cell from a precursor cell is an example of which of the six life processes in the human body?

1.4 Homeostasis

OBJECTIVES

- Define homeostasis.
- Distinguish between the body's internal environment and external environment.
- Describe the components of a feedback system.
- Contrast the operation of negative and positive feedback systems.
- Explain feedforward control.

The cells that comprise the human body are able to thrive because they live in the relative constancy of the body's internal environment despite continual changes in the body's external environment. The maintenance of relatively stable conditions in the body's internal environment is known as **homeostasis** (hō′mē-ō-STĀ-sis; *homeo-* = sameness; *-stasis* = standing still). It occurs because of the ceaseless interplay of the body's many regulatory processes. Homeostasis is a dynamic steady state. The term *dynamic* is used to refer to homeostasis because each regulated parameter can change over a narrow range that is compatible with life. For example, the level of glucose is maintained between 70 and 110 milligrams of glucose per 100 milliliters of blood. It normally does not fall too low between meals or rise too high even after eating a high-glucose meal. The term *steady state* is used to refer to homeostasis because energy is needed to keep the regulated parameter at a relatively constant level. Steady state is not the same as equilibrium. In an *equilibrium*, conditions remain constant without the expenditure of energy. Each structure in the body, from the cellular level to the systemic level, contributes in some way to keeping the internal environment of the body within normal limits.

Maintenance of Body Fluid Volume and Composition is Essential to Homeostasis

An important aspect of homeostasis is maintaining the volume and composition of **body fluids**, dilute, watery solutions containing dissolved chemicals that are found inside cells as well as surrounding them (**FIGURE 1.2**). In a lean adult, body fluids make up about 55–60% of total body mass; this percentage is lower in an individual with more adipose tissue (fat) because fat cells contain less water than skeletal muscle. About two-thirds of body fluid is **intracellular fluid (ICF)** (*intra-* = inside), the fluid within cells. The other one-third, called **extracellular fluid (ECF)** (*extra-* = outside), is fluid outside body cells. ECF consists of two components: (1) **interstitial fluid** (*inter-* = between), the fluid that fills the narrow spaces between cells, and (2) **plasma**, the fluid portion of blood. Interstitial fluid constitutes about 80% of the ECF, and plasma comprises the remaining 20%.

The proper functioning of body cells depends on precise regulation of the composition of their surrounding fluid. Because extracellular fluid surrounds the cells of the body, it serves as the body's **internal environment**. By contrast, the **external environment** of the body is the space that surrounds the entire body.

FIGURE 1.3 is a simplified view of the body that shows how a number of organ systems allow substances to be exchanged between the external environment, internal environment, and body cells in order to maintain homeostasis. Note that the integumentary system covers the outer surface of the body. Although this system does not play a major role in the exchange of materials, it protects the internal environment from damaging agents in the external environment. From the external environment, oxygen enters plasma through the respiratory system and nutrients enter plasma through the digestive system. After entering plasma, these substances are transported throughout the body by the cardiovascular system. Oxygen and nutrients eventually leave plasma and enter interstitial fluid by crossing the walls of blood capillaries, the smallest blood vessels of the body. Blood capillaries are specialized to allow the transfer of material between plasma and interstitial fluid. From interstitial fluid, oxygen and nutrients are taken up by cells and metabolized for energy. During this process, the cells produce waste products, which enter interstitial fluid and then move across blood capillary walls into plasma. The cardiovascular system transports these wastes to the appropriate organs for elimination from the body into the external environment. The waste product CO_2 is removed from the body by the respiratory system; nitrogen-containing wastes, such as urea and ammonia, are eliminated from the body by the urinary system.

FIGURE 1.2 Body fluid compartments.

The fluid within cells is intracellular fluid; the fluid outside cells is extracellular fluid, which consists of interstitial fluid and plasma.

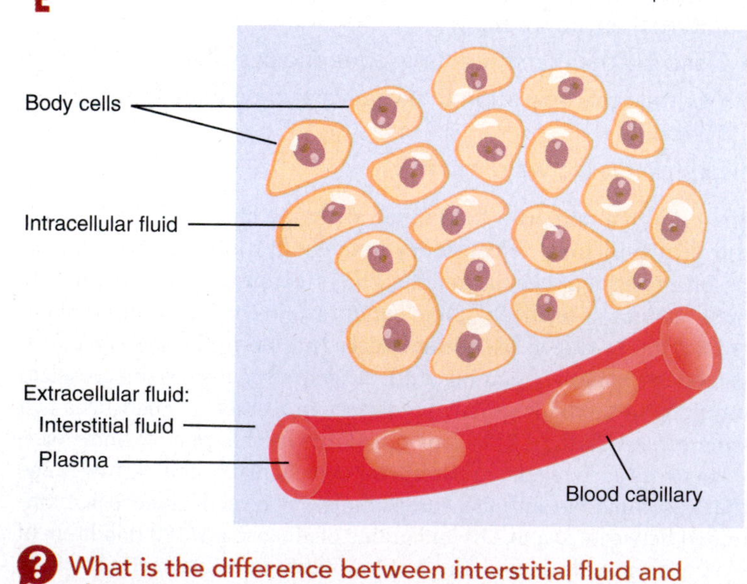

What is the difference between interstitial fluid and plasma?

Homeostasis Is Regulated via Feedback Systems and Feedforward Control

Homeostasis in the human body is continually being disturbed. Some disruptions come from the external environment in the form of physical insults such as the intense heat of a hot summer day or a lack of enough oxygen for that two-mile run. Other disruptions originate in the internal environment, such as a blood glucose level that falls too low when you skip breakfast. Homeostatic imbalances may also occur due to psychological stresses in our social environment—the demands of work and school, for example. In most cases the disruption of homeostasis is mild and temporary, and the responses of body cells quickly restore balance in the internal environment. However, in some cases the disruption of homeostasis may be intense and prolonged, as in poisoning, overexposure to temperature extremes, severe infection, or major surgery.

Fortunately, the body has many regulating systems that can usually bring the internal environment back into balance. Most often, the nervous system and the endocrine system, working together or independently, provide the needed corrective measures. The nervous system regulates homeostasis by sending action potentials (electrical signals) to organs that can counteract changes from the balanced state. The endocrine system includes many glands, organs, and tissues that secrete messenger molecules called *hormones* into the blood. Action potentials typically cause rapid changes, but hormones usually work more slowly. Both means of regulation, however, work toward the same end, usually through negative feedback systems.

FIGURE 1.3 A simplified view of exchanges between the external and internal environments. Note that the linings of the respiratory, digestive, and urinary systems are continuous with the external environment.

The internal environment of the body refers to the extracellular fluid (interstitial fluid and plasma) that surrounds body cells.

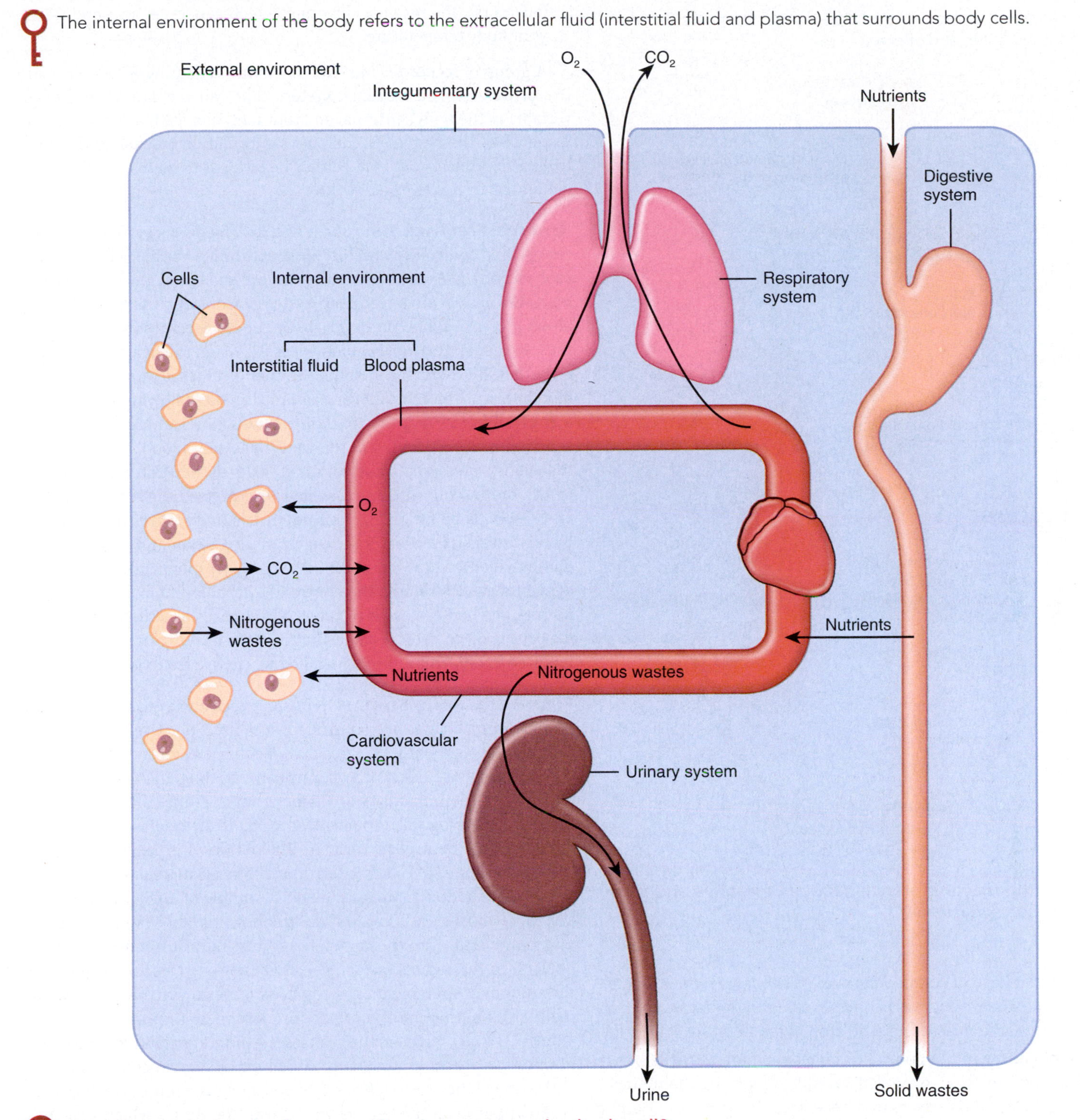

How does a nutrient in the external environment reach a body cell?

Feedback Systems

The body can regulate its internal environment through many feedback systems. A **feedback system** or *feedback loop* is a cycle of events in which a parameter of the internal environment is monitored, evaluated, changed, remonitored, reevaluated, and so on. Each monitored parameter is called a **controlled variable**. Examples of controlled variables include body temperature, blood pressure, blood glucose level, blood pH, and blood oxygen content. Any disruption that changes a controlled variable is called a *stimulus*. A feedback system includes three basic components—a receptor, a control center, and an effector (FIGURE 1.4).

1. A **receptor** is a body structure that monitors changes in a controlled variable and sends input to a control center. Typically, the

FIGURE 1.4 Operation of a feedback system. The solid return arrow symbolizes feedback.

The three basic components of a feedback system are the receptor, control center, and effector.

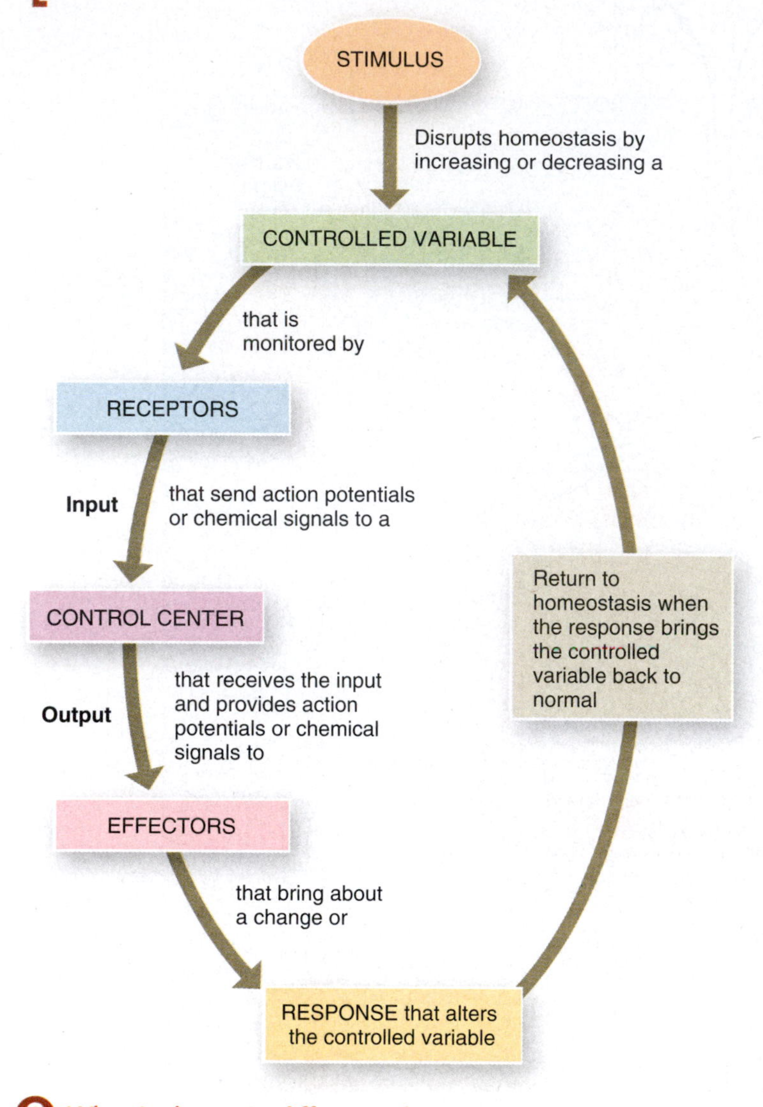

What is the main difference between negative and positive feedback systems?

input is in the form of action potentials or chemical signals. For example, certain nerve endings in the skin sense temperature and can detect changes, such as a dramatic drop in temperature.

2. A **control center** in the body determines the narrow range or **set point** within which a controlled variable should be maintained, evaluates the input it receives from receptors, and generates output commands when they are needed. *Output* from the control center typically occurs as action potentials, or hormones or other chemical signals. In the skin temperature example, the brain acts as the control center, receiving action potentials from the skin receptors and generating action potentials as output.
3. An **effector** is a body structure that receives output from the control center and produces a *response* or effect that changes the controlled variable. Nearly every organ or tissue in the body can behave as an effector; in many cases, the effector is a muscle or a gland. When your body temperature drops sharply, your brain (control center) sends action potentials (output) to your skeletal muscles (effectors). The result is shivering, which generates heat and raises your body temperature.

A group of receptors and effectors communicating with their control center forms a feedback system that can regulate a controlled variable in the body's internal environment. In a feedback system, the response of the system "feeds back" information to change the controlled variable in some way, either negating it (negative feedback) or enhancing it (positive feedback).

NEGATIVE FEEDBACK SYSTEMS A **negative feedback system** *reverses* a change in a controlled variable. Most controlled variables in the body, such as body temperature, blood pressure, and blood glucose level, are regulated by negative feedback systems. Consider the regulation of blood pressure. Blood pressure (BP) is the force exerted by blood as it presses against the walls of blood vessels. When the heart beats faster or harder, BP increases. If some internal or external stimulus causes blood pressure (controlled variable) to rise, the following sequence of events occurs (FIGURE 1.5). *Baroreceptors* (the receptors), pressure-sensitive neurons located in the walls of certain blood vessels, detect the higher pressure. These neurons send action potentials (input) to the brain (control center), which interprets the electrical signals and responds by sending action potentials (output) to the heart and blood vessels (the effectors). Heart rate decreases and blood vessels dilate (widen), which cause BP to decrease (response). This sequence of events quickly returns the controlled variable—blood pressure—to normal, and homeostasis is restored. Notice that the activity of the effector causes BP to drop, a result that negates the original stimulus (an increase in BP). This is why it is called a negative feedback system.

POSITIVE FEEDBACK SYSTEMS A **positive feedback system** *strengthens* or *reinforces* a change in a controlled variable. A positive feedback system operates similarly to a negative feedback system except for the way the response affects the controlled variable. The control center still provides commands to an effector, but this time the effector produces a physiological response that adds to or *reinforces* the initial change in the controlled variable. The action of a positive feedback system continues until it is interrupted by some mechanism.

Normal childbirth provides a good example of a positive feedback system (FIGURE 1.6). The first contractions of labor (stimulus) push part of the fetus into the cervix, the lowest part of the uterus, which opens into the vagina. Stretch-sensitive neurons (receptors) monitor the amount of stretching of the cervix (controlled variable). As stretching increases, they send more action potentials (input) to the brain (control center), which in turn causes the pituitary gland to release the hormone oxytocin (output) into the blood. Oxytocin causes muscles in the wall of the uterus (effector) to contract even more forcefully. The contractions push the fetus farther down the uterus, which stretches the cervix even more. The cycle of stretching, hormone release, and ever-stronger contractions is interrupted only by the birth of the baby. Then stretching of the cervix ceases and oxytocin is no longer released.

This example suggests some important differences between positive and negative feedback systems. Because a positive feedback system continually reinforces a change in a controlled variable, some event outside the system must shut it off. If the action of a positive feedback system is not stopped, it can "run away" and may even produce life-threatening conditions in the body. The action of a negative feedback system, by

FIGURE 1.5 Homeostatic regulation of blood pressure by a negative feedback system. The dashed return arrow with a negative sign surrounded by a circle symbolizes negative feedback.

If the response reverses the stimulus, a system is operating by negative feedback.

FIGURE 1.6 Positive feedback control of labor contractions during birth of a baby. The dashed return arrow with a positive sign surrounded by a circle symbolizes positive feedback.

If the response enhances or intensifies the stimulus, a system is operating by positive feedback.

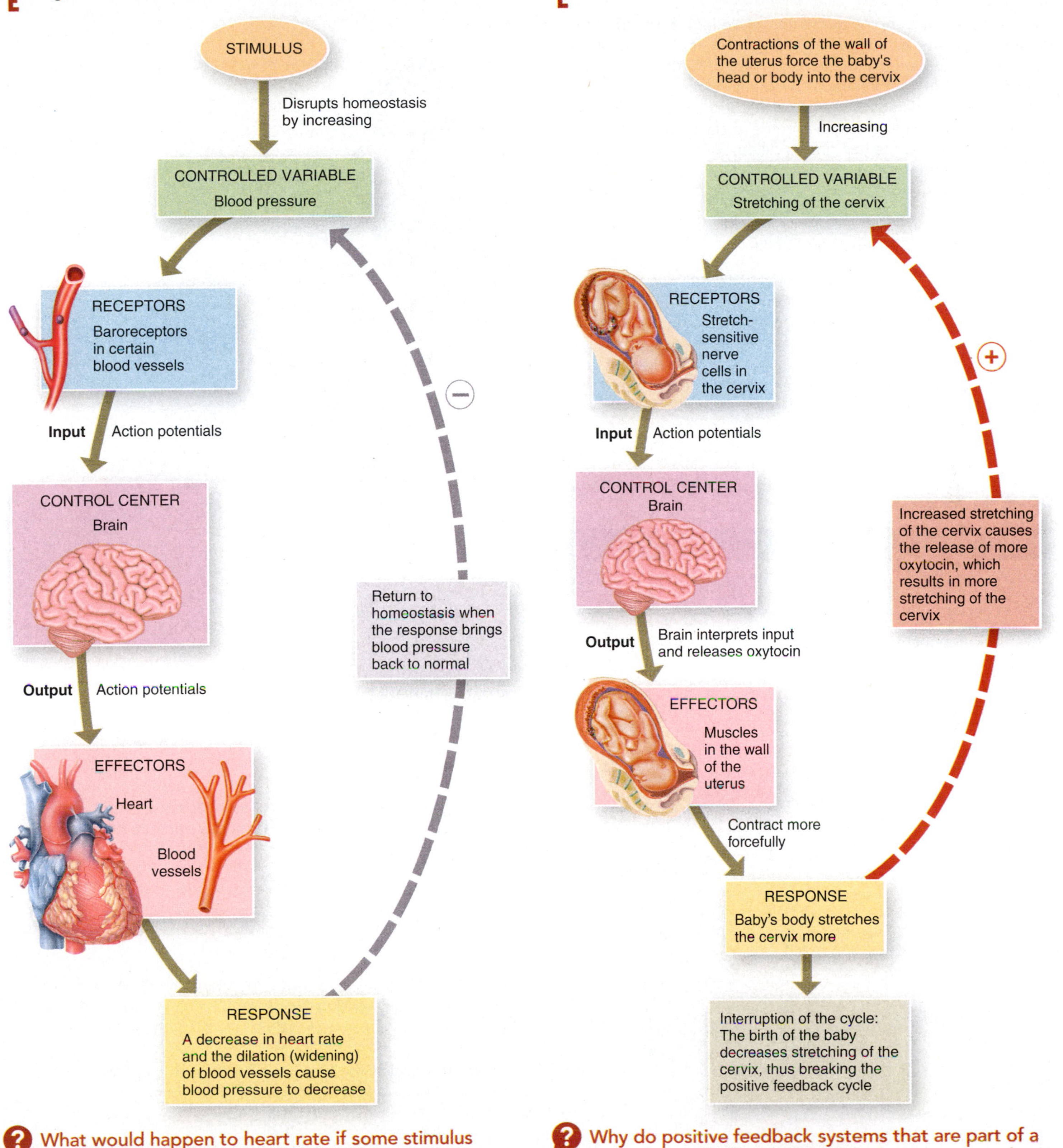

What would happen to heart rate if some stimulus caused blood pressure to decrease? Would this occur by way of positive feedback or negative feedback?

Why do positive feedback systems that are part of a normal physiological response include some mechanism that terminates the system?

Critical Thinking

Working up a Sweat

Julie is the starting point guard for her college basketball team. On the day of one of her games, the gym is hot and crowded. The fans are cheering and excitement is in the air! As the game begins, Julie notices that her forehead is sweating. As the game continues, more areas of her body begin to sweat, and her palms leave wet prints on the ball. Eventually her entire jersey becomes drenched with perspiration. Once half-time is called, Julie is able to rest for a few minutes, cool down, and rehydrate.

SOME THINGS TO KEEP IN MIND:

Various mechanisms are employed to keep our core body temperature in the normal range. Thermoregulation begins with thermoreceptors, which detect changes in body temperature. The thermoreceptors then relay that information to the thermoregulation center in the brain. This center in turn stimulates sweat glands in the skin to secrete sweat onto the skin surface. As the water in the sweat evaporates, large quantities of heat energy leave the skin surface and the body cools down.

SOME INTERESTING FACTS:

Average number of sweat glands in the human body	3 million
Maximum amount of sweat produced in an hour	3 liters
Average amount of sweat lost daily under normal conditions	600 mL

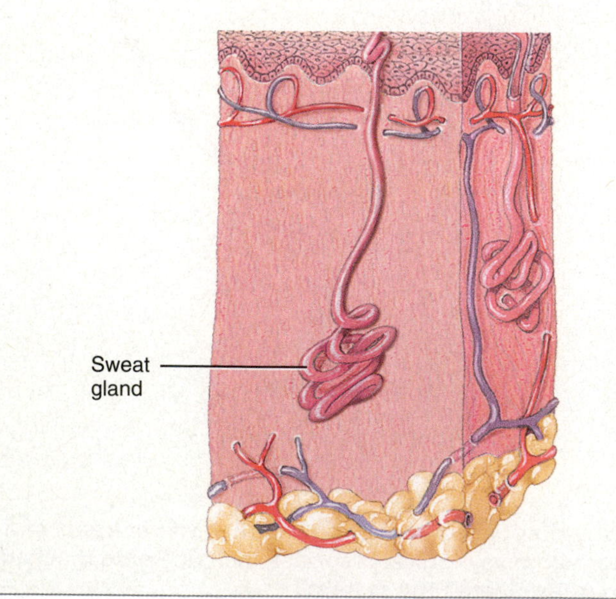

What are the specific components of the feedback system involved in causing Julie to sweat when her body temperature becomes too high?

Why is this feedback system considered to be negative feedback?

Besides body temprature, what other controlled variables are regulated by a negative feedback system?

What would happen to Julie if her increased body temperature were regulated by a positive feedback system?

Why is thermoregulation so essential to maintaining homeostasis?

contrast, slows and then stops as the controlled variable returns to its normal state. Usually, positive feedback systems reinforce conditions that do not happen very often, and negative feedback systems regulate conditions in the body that remain fairly stable over long periods.

Feedforward Control

In a feedback system, events occur *in response to* a change in a controlled variable. In **feedforward control**, events occur *in anticipation of* a change in a controlled variable. To understand how feedforward control works, consider the following example. Suppose that the nutrient concentration in your internal environment becomes too low, which causes you to feel hungry. If you see, smell, or think about food during this time, a feedforward mechanism causes your mouth to salivate and your stomach to secrete gastric juice. This anticipatory response, which is part of the *cephalic phase* (*cephalic* = head) of digestion, prepares the digestive system for food that is about to be eaten, making it easier for the food to be digested and nutrients to enter the internal environment. You will learn more about the cephalic phase of digestion in Chapter 21 (see Section 21.8).

Homeostatic Imbalances Can Lead to Disorders, Diseases, or Even Death

As long as all of the body's controlled variables remain within certain narrow limits, homeostasis is maintained, body cells function efficiently, and the body stays healthy. However, should one or more components of the body lose their ability to contribute to homeostasis, the normal balance among all of the body's processes may be disturbed. If the homeostatic imbalance is moderate, a disorder or disease may occur; if it is severe, death may result.

A **disorder** is any abnormality of structure or function. **Disease** is a more specific term for an illness characterized by a recognizable set of signs and symptoms. **Signs** are *objective* changes that a clinician can observe and measure, such as swelling, fever, high blood pressure, or paralysis. **Symptoms** are *subjective* changes in body functions that are not apparent to an observer, such as headache, nausea, or anxiety. You will learn about specific disorders and diseases as the functions of the various parts of the body are discussed in the chapters to come.

CLINICAL CONNECTION

Diagnosis of Disease

Diagnosis (dī′-ag-NŌ-sis; *dia-* = through; *-gnosis* = knowledge) is the science and skill of distinguishing one disorder or disease from another. The patient's signs and symptoms, his or her medical history, a physical exam, and laboratory tests provide the basis for making a diagnosis. Taking a ***medical history*** consists of collecting information about events that might be related to a patient's illness. These include the chief complaint (primary reason for seeking medical attention), history of present illness, past medical problems, family medical problems, social history, and review of symptoms. A ***physical examination*** is an orderly evaluation of the body and its functions. This process includes **inspection** (the examiner observes the body for any changes that deviate from normal), **palpation** (the examiner feels body surfaces with the hands to detect enlarged or tender organs or abnormal masses), **auscultation** (the examiner listens to body sounds to evaluate the functioning of certain organs, often using a stethoscope to amplify the sounds), and **percussion** (the examiner taps on the body surface with the fingertips and listens to the resulting echo, which can provide information about the size, consistency, and position of an underlying structure). The physical examination also includes measurements of vital signs (temperature, pulse, respiratory rate, and blood pressure), and sometimes laboratory tests.

CHECKPOINT

6. Why is extracellular fluid called the internal environment of the body?
7. What types of disturbances can act as stimuli that initiate a feedback system?
8. How are negative and positive feedback systems similar? How are they different?
9. What is the significance of feedforward control?

1.5 Physiology as a Science

OBJECTIVES

- Describe the history of physiology.
- Identify the steps of the scientific method.
- Explain the importance of scientific literature.
- Discuss the mechanistic approach to explaining body function.
- Define concept mapping.

The History of Physiology Spans Thousands of Years

The term *physiology*, which literally means "study of nature" (*physio-* = nature; *-logy* = study of), is derived from the Greek word *physiologoi* (fiz′-ē-OL-ō-goy), a name that refers to a group of ancient Greek philosophers of the sixth and fifth centuries B.C. who speculated about the existence and purpose of all things (living and nonliving) in the natural world. They searched for rational explanations about the phenomena that they observed around them and rejected traditional supernatural explanations. Prominent among the *physiologoi* were Thales, Anaximander, Pythagoras, and Democritus.

Over the next few centuries, the scope of physiology began to focus on how living things in nature function. The Greek physician Hippocrates (460–375 B.C.), considered the father of medicine, thought that the normal functioning of the body depended on the balance of four types of bodily fluids, or humors (blood, phlegm, yellow bile, and black bile) and that illness resulted when these humors were out of balance. This *humoral theory* influenced Western medicine for the next 2000 years and was eventually disproved in the 18th century. The Greek philosopher Aristotle (384–322 B.C.) proposed that every part of the body is formed for a specific purpose and that the function of a given body part can be deduced from its structure. The Greek physician Erasistratus (304–250 B.C.) accurately described the function of heart valves and distinguished between sensory nerves and motor nerves. Galen (A.D. 130–201), another Greek physician, correctly described functions of the kidneys and spinal nerves and demonstrated that arteries contain blood. However, Galen erroneously believed that air entering the body was transformed by the brain, liver, and heart into "souls" or "spirits" that governed the vital functions of the body. He also incorrectly described the movement of blood through the heart. Despite his errors, Galen became the undisputed authority on medicine in the Western world. His views went unchallenged for about 1400 years—that is, until the Renaissance.

The Renaissance was a cultural movement in Europe from the 14th to the 17th centuries characterized by a revival of classical influences on art and literature. It also marked the beginning of modern science: Experimentation in the physical and life sciences started to flourish; the way that science was performed began to change, with an emphasis on scientific procedure and reproducibility of results (the scientific method); and generally accepted scientific concepts passed down from the ancient Greeks were now being reexamined. This movement when science started to change to its modern form is known as the **Scientific Revolution**, which occurred during the 16th and 17th centuries. A major physiological discovery during the Scientific Revolution was made by the English physician William Harvey (1578–1657). Through dissection and experimentation, he was able to correctly explain the circulation of blood through the body: The heart pumps blood into arteries, arteries carry blood away from the heart, and veins return blood back to the heart. Harvey's discovery is considered to be the beginning of modern experimental physiology.

During the 18th and 19th centuries, there were many noteworthy contributions to physiology. The Italian physiologist Lazzaro Spallanzani (1729–1799) demonstrated that digestion of food in the stomach involves a chemical process. The English physician William Hewson (1739–1774) determined that a substance now known as fibrinogen is necessary for blood clotting to occur. The German physiologist Carl Ludwig (1816–1895) invented the *kymograph*, an instrument designed to measure and record variations in fluid pressure. He was also the first to use isolated, perfused organs for experimentation. By the middle of the 19th century, the cell as the fundamental unit of life (the cell theory) was firmly established, so physiologists began to study cell function. The French physiologist Claude Bernard (1813–1878) first proposed that the cells of a multicellular organism flourish because they live in the relative constancy of *le milieu interieur*—the internal environment—despite continual changes in the organism's external environment. The American physiologist Walter B. Cannon (1871–1945) later coined the term *homeostasis* to describe this internal constancy.

Many advances in physiology were also achieved during the 20th century. In the early 1900s, the German physiologist Otto Frank (1865–1944) and the English physiologist Ernest Starling (1866–1927) described how the strength of the heart's contraction is affected by the degree of stretch of the heart wall—the so-called *Frank-Starling law of the heart*. In the 1950s, the English physiologists Alan Hodgkin (1914–1998) and Andrew Huxley (1917–2012) described the mechanism of the action potential in a giant squid axon. During that same decade, the English biologist Hugh Huxley (1929–2013) along with several others described the sliding filament mechanism of muscle contraction. In the 1970s and 1980s, the American pharmacologist Alfred Gilman (1941–present) and the American biochemist Martin Rodbell (1925–1998) described the function of G proteins in signal transduction pathways.

Today physiology is considered a mature discipline, as most of the functions at the organ and system levels have been elucidated. However, there is still more work to be done. Physiologists are currently focusing on functions at the cell and molecular levels. For example, in the last decade of the twentieth century, the genomes of humans, mice, fruit flies, and more than 50 microbes were sequenced. As a result, research in the field of **genomics**, the study of the relationships between the genome (all of the DNA of an organism) and the biological functions of an organism, has flourished. Another active area of research is **proteomics**, the study of the relationship between the proteome (all of the proteins of an organism) and the biological functions of that organism. Another key area of research during this millennium is integration: Twenty-first-century physiologists are seeking to understand how various parts of the body work together to accomplish a particular function.

The Scientific Method Is a Systematic Way of Acquiring Knowledge About the Natural World

The current knowledge that physiologists have about how the body functions has been obtained through observation and experimentation. The process of acquiring knowledge about some aspect of the natural world in a systematic way is known as the **scientific method**, which typically consists of four steps: (1) make an observation, (2) formulate a hypothesis, (3) design an experiment to test the hypothesis, and (4) interpret the data.

1. ***Make an observation.*** This observation may occur during experimentation or it may be gleaned from researching the scientific literature. For example, a physiologist who is interested in studying high-density lipoproteins (HDLs), also known as "good" cholesterol, might observe from reading a few scientific articles that athletes have higher levels of HDLs in their bloodstream than sedentary individuals.
2. ***Formulate a hypothesis.*** After making an observation, a hypothesis is formulated. A **hypothesis** is a tentative explanation for the observation. The hypothesis must be testable by experimentation and capable of being refuted. A logical hypothesis from the observation about HDLs given above is that exercise increases the HDL level in the bloodstream.
3. ***Design an experiment to test the hypothesis.*** A good experiment alters a single variable that the physiologist believes plays a critical role in the initial observation, while all other variables are kept constant. The variable that is altered is known as the **independent variable**. Any factor that varies in response to changes in the independent variable is known as the **dependent variable**. In the HDL hypothesis, exercise is the independent variable, and the HDL blood concentration is the dependent variable. To test the effect of exercise on the HDL level in the blood, the physiologist designs an experiment in which several groups of people exercise for varying amounts of time each day over a period of three months. One group does not exercise at all, another group exercises for 30 minutes a day, another group for 1 hour a day, and so on. Every week over the course of the experiment, the HDL blood concentration of each individual of each group is measured. A vital component of any experiment is the **control**, which is the part of the experiment in which the independent variable is not altered. Having a control allows the physiologist to know that the changes in the experimental group are due to a change in the independent variable and not some other factor. In the HDL experiment, the control is the group of people who do not exercise at all over the three-month period.

In an experiment such as this one, many variables other than exercise might affect the outcome of the results. The physiologist tries to anticipate these issues and find ways to eliminate them. For example, some people in this experiment might be more physically fit than others. To eliminate this factor, the physiologist would include in the experiment only people who have the same approximate age, body weight, and overall degree of health. The physiologist would also check to make sure that the initial HDL levels in the bloodstream are about the same in all individuals participating in the study. Another possible source of variation is diet. Different foods can affect the cholesterol level in the bloodstream. So the physiologist would need to make sure that all participants in the study have the same type of diet each day over the entire testing period. Still yet another possible source of variation is small population size. The smaller the number of people participating in this study, the more likely there will be errors or outliers in the data. Ideally the physiologist would involve thousands of individuals in this experiment, with each group consisting of hundreds of individuals. However, experiments with such a large number of people are often hard to manage, so limits might have to be set in terms of the number of people that can participate. It is important to point out that some variables might be out of the physiologist's control. A good example is genetics: Some people may be genetically prone to producing more or less HDLs and other types of cholesterol. If this is the case for some participants in this study, their results might not change significantly in response to exercise.

As the HDL experiment proceeds over the three-month period, the physiologist collects data. Scientific data may either be *quantitative* (involves numerical measurements) or *qualitative* (involves descriptions). In this experiment, the data obtained is quantitative because it involves measuring the HDL concentration in the blood.

A final point about the design of this experiment is that humans are used as the subjects because there is no threat to their well-being. However, in physiology experiments that are invasive, **animal models** such as rats or other mammals that resemble humans in structure are often used as the subjects. The results are then extrapolated to humans. Experiments that involve animal models have strict protocols to make sure that the animals are treated humanely.

4. ***Interpret the data.*** After the data is collected from the experiment, the results are interpreted. Data are often displayed on graphs and in tables. Interpretation of the data involves the use of *statistics* to make sure that the results obtained are not by chance but instead are real phenomena. In some experiments, the data may not support the hypothesis, and the physiologist will have to modify the hypothesis and design a new experiment to test that new hypothesis. In other cases, the data may support the hypothesis. Suppose that, for the HDL experiment, the physiologist finds that exercise does indeed increase the HDL level in the bloodstream. Although the data support the hypothesis, the physiologist should repeat the experiment several times to make sure that it is *reproducible.* An experiment must be reproducible to have merit in the scientific community. When a hypothesis is tested over and over again by independent scientists and the data support that hypothesis, the hypothesis becomes a **scientific theory**. Just because a hypothesis becomes a theory does not mean that it cannot ever be refuted. At some point in the future, new evidence may indicate that the theory is no longer true. So it is very possible that some physiological information that is currently accepted as fact by the scientific community may be disproved in years to come.

The scientific method is a way to investigate phenomena in the natural world using a fixed series of steps. However, in a more general process known as **scientific inquiry,** a scientist investigates a problem without a requisite order. The essence of scientific inquiry is essentially the same as the scientific method except that the physiologist is free to investigate the problem in any order that he or she wants. For example, the physiologist may perform an experiment and, based on what she finds, she may then develop a hypothesis.

Scientific Literature Helps Physiologists Conduct Research

A physiologist uses scientific literature as a research tool—to read about current studies in the field or as a means to publish the results of experiments that he or she has performed. The main sources of scientific information are journals, books, and the Internet.

- ***Journals.*** A journal is a collection of scientific articles, either original research articles or review articles, about a particular topic. Journals are usually **peer-reviewed**, meaning that a group of scientific experts makes sure that a given scientific article meets certain criteria before it is published. Examples of peer-reviewed journals include the *American Journal of Physiology, Nature*, and *Proceedings of the National Academy of Sciences.* Journal articles offer the most current scientific research that is available.
- ***Books.*** Books can also be a source of scientific information. A science textbook, such as this one, summarizes the main concepts in a particular scientific field. These concepts have been obtained over the years from numerous scientific experiments. Although books are wonderful sources of scientific information, they take longer to publish, so they are not as current as the latest edition of a journal.
- ***Internet.*** The Internet is another source of scientific information. However, information found on the Internet may not always be reliable. If you are searching for a particular scientific topic, use reputable websites that are written and maintained by scientific professionals. Examples of reputable scientific websites include www.nih.gov, www.mayoclinic.com, and www.webmd.com. Another reputable website is www.pubmed.com, a great search engine for finding journal articles about a particular topic. Many journals have websites that allow you to access scientific information online. For example, the website for the journal *Nature* is www.nature.com.

Physiologists Use the Mechanistic Approach to Explain How the Body Functions

At the core of physiology is the desire to explain the mechanisms responsible for the various functions of the body. This approach to explaining body function is called the **mechanistic approach**: It describes how a particular event in the body occurs using cause-and-effect sequences. For example, consider the following question: "Why does the body sweat when its temperature becomes too hot?" The mechanistic approach to answering this question would be as follows: Thermoreceptors in the skin detect the hot temperature and then send signals to the brain, which in turn sends signals that activate sweat glands. Activation of sweat glands causes the body to sweat, and the body cools off as sweat evaporates from the skin surface.

Another approach to explaining body function is the **teleological approach**, which describes why a particular function occurs without mentioning the mechanism involved. In response to the question, "Why does the body sweat when its temperature becomes too hot?", the teleological approach to answering this question would simply be "To cool off the body." Because the teleological approach does not provide the mechanism of how the physiological event occurs, it is not the approach that physiologists prefer to use.

Concept Mapping Allows Physiologists to Illustrate the Relationships Between Ideas

As a physiologist studies body functions, the information that is learned often needs to be organized. This can be achieved through a process known as **concept mapping**, in which certain ideas or concepts are graphically displayed to illustrate their relationships to one another. A common type of concept map used in physiology is a **process map**, in which pieces of information about a particular process are connected by arrows to show cause-and-effect sequences. Because the information flows from one part of the map to the other, a process map is also known as a **flowchart**. Flowcharts are used throughout this text; examples include FIGURE 13.21 and FIGURE 18.16. Some flowcharts are also called **feedback system** (*feedback loop*) **diagrams** if arrows are used to show that an initial change in a controlled variable causes a sequence of events that either reverses the change in the controlled variable (negative feedback) or enhances the change in the controlled variable (positive feedback). FIGURES 1.4, 1.5, and 1.6 are examples of feedback system diagrams.

CHECKPOINT

10. What are the physiological contributions of Erasistratus? Galen? William Harvey? Claude Bernard?

11. What are the current areas of physiological research?

12. What is the difference between the scientific method and scientific inquiry?

13. What are some examples of reputable scientific websites?

14. How does the mechanistic approach to explaining body function differ from the teleological approach?

15. What is a flowchart?

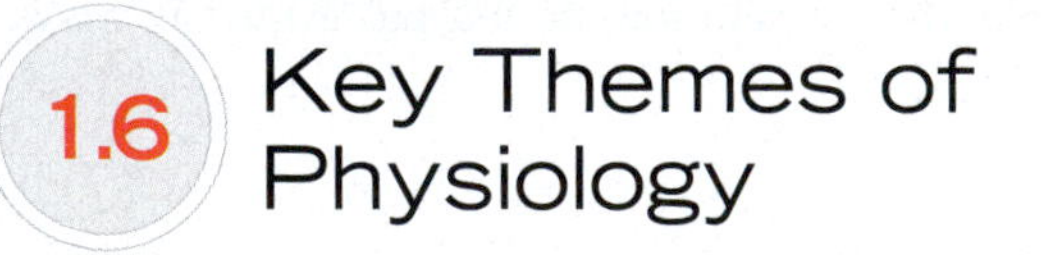

1.6 Key Themes of Physiology

OBJECTIVE

- Identify the key themes of physiology.

Four key themes of physiology appear throughout this book: homeostasis, integration, mechanism of action, and communication.

- ***Homeostasis.*** Earlier in this chapter you learned that homeostasis is the maintenance of relatively stable conditions in the body's internal environment. This occurs through the operation of feedback systems. Each feedback system consists of (1) a receptor that monitors a controlled variable, (2) a control center that determines the set point of the controlled variable and evaluates input from the receptor, and (3) an effector that receives output commands from the control center and produces a response that changes the controlled variable (see FIGURE 1.4). Because homeostasis is vital to the normal functioning of the body, it is the most important theme of physiology. Information relevant to homeostasis is presented in every chapter of this book. Furthermore, in certain chapters you will learn how controlled variables such as blood pressure, body temperature, blood glucose level, blood oxygen level, and blood pH are regulated to maintain homeostasis. For example, in Chapter 13 you will discover how the blood glucose concentration is regulated.
- ***Integration.*** Integration occurs when several components work together to accomplish a particular function. Essentially all of the components of the body—molecules, cells, tissues, organs, and organ systems—work together at various levels to keep the body alive. Because integration occurs throughout the body, it is a major theme of physiology. As you read the chapters of this book, you will encounter numerous examples of integration. For example, in Chapter 18 you will learn that breathing involves several organs of different systems—the lungs (respiratory system), brain (nervous system), and skeletal muscles (muscular system). Other examples of integration that you will explore in this text include control of movement (Chapter 12); growth of bone (Chapter 13); regulation of blood pressure (Chapter 15); acid–base balance (Chapter 20); and the so-called integrative functions of the brain: wakefulness and sleep, language, emotions, motivation, and learning and memory (Chapter 8).
- ***Mechanism of Action.*** Physiologists explain how the body works by indicating the mechanisms that are involved. This mechanistic approach describes how a physiological event occurs using cause-and-effect sequences. Because mechanism of action is the way that body function is explained, it is a prominent theme of physiology. As you read through this book, you will learn the mechanisms of action of a large number of physiological events, ranging from the system level down to the molecular level. Because the current focus of physiology research is at the cell and molecular levels, this text emphasizes cell and molecular mechanisms of function. Beginning with Chapter 2, each chapter will have at least one physiological event described at the cell or molecular level. Some chapters will have multiple physiological events described in this manner. For example, in Chapter 9 you will learn about the cell and molecular mechanisms responsible for each of your sensations.
- ***Communication.*** Another key theme of physiology is communication. The cells of the body must communicate with one another in order for the body to function. Chapter 6 introduces the various ways that cells communicate with each other. For example, a main method of communication between cells is by the release of an *extracellular chemical messenger*. This messenger may be a *hormone* (a chemical that enters the blood to reach a distant target cell), a *neurotransmitter* (a chemical released at the junction between two neurons or between a neuron and a muscle cell), or a *local mediator* (a chemical that communicates with a nearby cell without entering the bloodstream). In several of the remaining chapters of this book, you will learn more details about how these chemical messengers function. For example, in Chapter 13 you will explore the roles of the different types of hormones in the body.

CHECKPOINT

16. Which of the four key themes of physiology is most important?

FROM RESEARCH TO REALITY

Diet and Blood Pressure

Reference
Appel, L. et al. (1997). A clinical trial of the effects of dietary patterns on blood pressure. *New England Journal of Medicine.* 336(16):1117–1124.

How much might your diet impact your blood pressure?

kivoart/Getty Images

The saying goes "You are what you eat." Is this true when you talk about blood pressure? Could a change in your diet change your blood pressure? Could diet be as effective as drugs at decreasing high blood pressure?

Article description:
Researchers gave 459 subjects a control diet for three weeks and then randomized those subjects into three diets: the control diet, a fruits-and-vegetables diet, or a combination diet. The control diet was reflective of the typical American diet; the fruits-and-vegetables diet was similar to the control diet but included more fruits and vegetables and fewer sweets and snacks; and the combination diet was rich in fruits and vegetables, but low in saturated fat, total fat, and cholesterol. The blood pressure of each subject was monitored for six weeks to see if diet could have any effect on blood pressure.

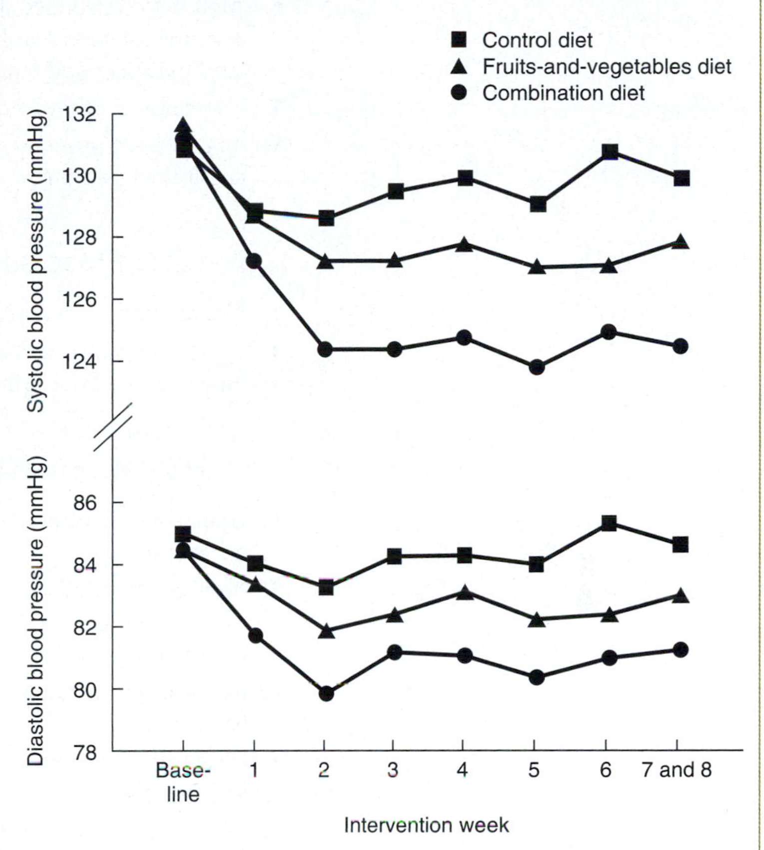

Go to WileyPLUS Learning Space and use the data from this article to answer the questions posed there and to discover more about diet and blood pressure.

CHAPTER REVIEW

1.1 Physiology Defined

1. Physiology is the study of the functions of an organism and its constituent parts.
2. Human physiology describes how the parts of the human body work.

1.2 Levels of Organization in the Body

1. The human body consists of six levels of organization: chemical, cellular, tissue, organ, system, and organismal.
2. Cells are the basic structural and functional units of an organism and the smallest living units in the human body.

3. Tissues are groups of cells and the materials surrounding them that work together to perform a particular function.
4. Organs are composed of two or more different types of tissues; they have specific functions and usually have recognizable shapes.
5. Systems consist of related organs that have a common function.
6. TABLE 1.1 introduces the twelve systems of the human body: the endocrine, nervous, muscular, skeletal, cardiovascular, immune, lymphatic, integumentary, respiratory, urinary, digestive, and reproductive systems.
7. An organism is any living individual.
8. An emergent property is a new property that exists at a given level of organization and not at the levels below.

1.3 Life Processes

1. All organisms carry on certain processes that distinguish them from nonliving things.
2. The most important life processes of the human body are metabolism, responsiveness, movement, growth, differentiation, and reproduction.

1.4 Homeostasis

1. Homeostasis is the maintenance of relatively stable conditions in the body's internal environment.
2. Body fluids are dilute, watery solutions. Intracellular fluid (ICF) is fluid inside cells, and extracellular fluid (ECF) is fluid outside cells. Interstitial fluid is the ECF that fills spaces between cells; plasma is the ECF within blood vessels.
3. Because it surrounds all body cells, extracellular fluid is called the body's internal environment.
4. Disruptions of homeostasis come from external stimuli, internal stimuli, and psychological stresses.
5. When disruption of homeostasis is mild and temporary, responses of body cells quickly restore balance in the internal environment. If disruption is extreme, regulation of homeostasis may fail.
6. Most often, the nervous and endocrine systems, acting together or separately, regulate homeostasis. The nervous system detects body changes and sends action potentials to counteract the changes. The endocrine system regulates homeostasis by secreting hormones.
7. Feedback systems include three components: (1) Receptors monitor changes in a controlled variable and send input to a control center; (2) the control center determines the set point at which a controlled variable should be maintained, evaluates the input it receives from receptors, and generates output commands when they are needed; and (3) effectors receive output from the control center and produce a response (effect) that alters the controlled variable.
8. If a response reverses the original stimulus, the system is operating by negative feedback. One example of negative feedback is the regulation of blood pressure. If a response enhances the original stimulus, the system is operating by positive feedback. One example of positive feedback is uterine contractions during the birth of a baby.
9. In addition to feedback systems, homeostasis may also involve feedforward control. In feedforward control, events occur in anticipation of a change in a controlled variable. An example of feedback control occurs when the sight, smell, or thought of food causes your mouth to salivate and your stomach to produce gastric juice.
10. Disruptions of homeostasis—homeostatic imbalances—can lead to disorders, diseases, and even death.

CHAPTER REVIEW

1.5 Physiology as a Science

1. The term *physiology* is derived from the physiologoi, a group of ancient Greek philosophers who speculated about the existence and purpose of all things in nature.
2. Over time, the scope of physiology began to focus on how living things in nature function. Hippocrates, Aristotle, Erasistratus, and Galen (all Greek) were among the first to study body function.
3. During the Renaissance, old views about science began to change, resulting in a Scientific Revolution. William Harvey's discovery of the circulation of blood is considered to be the beginning of modern experimental physiology.
4. Over the next several centuries, there were many significant discoveries about body function. Currently, physiology is considered a mature science because most of the functions at the organ and systems levels have been elucidated. Physiologists are currently focusing on describing functions at the cell and molecular levels.
5. Because physiology is a science, knowledge about the body is obtained in a systematic way by using the scientific method. There are four steps to the scientific method: (1) make an observation, (2) formulate a hypothesis, (3) design an experiment to test the hypothesis, and (4) interpret the data.
6. As a physiologist studies body function, valuable information can be obtained by researching scientific literature. Sources of scientific information include journals, books, and the Internet.
7. The mechanistic approach to explaining body function describes how a particular event in the body occurs using cause-and-effect sequences.
8. Concept mapping is the process in which pieces of information are graphically displayed to illustrate their relationship to one another.

1.6 Key Themes of Physiology

1. Four key themes of physiology reoccur throughout this textbook: homeostasis, integration, mechanism of action, and communication.
2. Homeostasis is the maintenance of relatively stable conditions in the body's internal environment.
3. Integration is the process by which several components work together for a common purpose.
4. Mechanism of action refers to the cause-and-effect sequence that is used to describe a physiological event.
5. Communication is the process by which different parts of the body communicate with one another.

1. It is a very cold (25°F) day outside and you have left for class without taking your jacket. Once you start walking across campus, the cold receptors in your skin start firing action potentials to the brain. The brain integrates the incoming information and sends an output signal to your skeletal muscles and the blood vessels just underneath your skin. Your skeletal muscles start to produce rapid contractions causing you to shiver, while the blood vessels of your skin constrict, pulling blood away from the cold environmental air. These two compensation mechanism reduce further loss of heat as well as help with the production of new heat. For this scenario, properly identify the stimulus, receptor, control center, effector, and response and explain what type of feedback mechanism this represents.

2. You are a prestigious scientist working in a lab with the hope to create a drug that will burn fat twice as fast when combined with exercise compared to exercise alone. You have gone through many research articles to investigate the effects that others have found on exercise and burning fat. You find that exercise is correlated with the amount of fat that can be burned, and decide to go ahead with your study. You design an experiment that includes 2 study groups, each with 100 subjects. All groups are instructed to adhere to a strict diet and a daily exercise routine of 30 minutes on a treadmill, walking 3.0 mph. Group 1 receives a sugar pill while group 2 receives your new drug. Provide your hypothesis, independent and dependent variables, and the control for your study.

3. If you were to drink a gallon of salt water, assuming that all of this salt water is absorbed into the body, which fluid compartment would be first, and most directly, impacted by this fluid? Defend your answer.

ANSWERS TO FIGURE QUESTIONS

1.1 An organ (the organ level) is composed of two or more different types of tissues that work together to perform a specific function.

1.2 Interstitial fluid is the fluid between cells; plasma is the fluid portion of blood.

1.3 A nutrient moves from the external environment into plasma via the digestive system, then into the interstitial fluid, and then to a body cell.

1.4 In negative feedback systems, the response reverses the original stimulus, but in positive feedback systems, the response enhances the original stimulus.

1.5 When something causes blood pressure to decrease, heart rate increases because of operation of this negative feedback system.

1.6 Because positive feedback systems continually intensify or reinforce the original stimulus, some mechanism is needed to end the response.

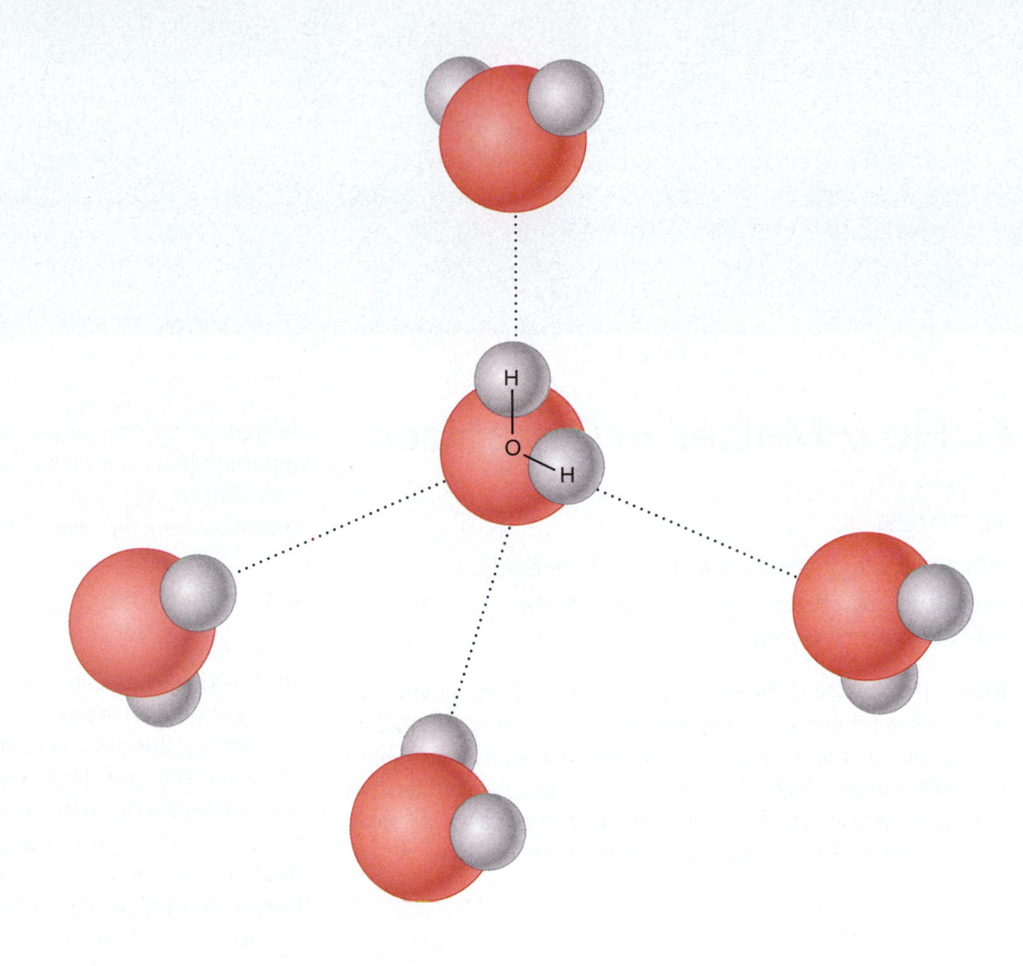

2 Chemical Composition of the Body

Chemistry and Homeostasis

Maintaining the proper assortment and quantity of thousands of different chemicals in your body, and monitoring the interactions of these chemicals with one another, are two important aspects of homeostasis.

LOOKING BACK TO MOVE AHEAD...

- From smallest to largest, the six levels of organization in the human body are the chemical, cellular, tissue, organ, system, and organismal levels (Section 1.2).
- Metabolism is the sum of all of the chemical reactions that occur in the body (Section 1.3).
- Water is present inside cells (intracellular fluid) and outside cells (extracellular fluid) (Section 1.4).
- Homeostasis is the maintenance of relatively stable conditions in the body's internal environment due to the ceaseless interplay of the body's many regulatory processes (Section 1.4).

You learned in Chapter 1 that the chemical level of organization, the lowest level of organization in the body, consists of atoms and molecules. These small particles ultimately form body organs and systems of astonishing size and complexity. In this chapter, you will examine the properties of atoms and how atoms bond together to form molecules. You will also learn about the importance of water, the different types of liquid mixtures (solutions, colloids, and suspensions), and the concept of pH. This chapter concludes with a discussion of five categories of large molecules that are vital to the body: carbohydrates, lipids, proteins, nucleic acids, and adenosine triphosphate (ATP).

2.1 How Matter Is Organized

OBJECTIVES

- Identify the main chemical elements of the body.
- Describe the structures of atoms, ions, molecules, free radicals, and compounds.

Chemistry is the science of the structure and interactions of matter. All living and nonliving things consist of **matter**, which is anything that occupies space and has **mass**. Mass is the amount of matter in any object, which does not change. *Weight*, the force of gravity acting on matter, does change. When objects are farther from Earth, the pull of gravity is weaker; this is why the weight of an astronaut is close to zero in outer space.

All Forms of Matter Are Composed of Atoms

Matter exists in three states: solid, liquid, and gas. *Solids*, such as bones and teeth, are compact and have a definite shape and volume. *Liquids*, such as blood plasma, have a definite volume and assume the shape of their container. *Gases*, like oxygen and carbon dioxide, have neither a definite shape nor volume. All forms of matter—both living and nonliving—are made up of atoms. An **atom** is the smallest stable unit of matter. Atoms are extremely small. Two hundred thousand of the largest atoms would fit on the period at the end of this sentence. Hydrogen atoms, the smallest atoms, have a diameter less than 0.1 nanometer, and the largest atoms are only five times larger.

Matter composed of one kind of atom is called an **element**. Each element is a pure substance that cannot be split into simpler substances by ordinary chemical means. Scientists now recognize 118 elements. Of these, 92 occur naturally on Earth. The rest have been produced from the natural elements using particle accelerators or nuclear reactors. Each element is designated by a **chemical symbol**, one or two letters of the element's name in English, Latin, or another language. Examples of chemical symbols are H for hydrogen, C for carbon, O for oxygen, N for nitrogen, Ca for calcium, and Na for sodium (*natrium* = sodium).*

Twenty-six different chemical elements normally are present in your body. Just four elements, called the *major elements*, constitute 96% of the body's mass: oxygen, carbon, hydrogen, and nitrogen. Eight others, the *lesser elements*, contribute 3.6% to the body's mass: calcium, phosphorus (P), potassium (K), sulfur (S), sodium, chlorine (Cl), magnesium (Mg), and iron (Fe). An additional 14 elements—the *trace elements*—are present in tiny amounts. Together, they account for the remaining 0.4% of the body's mass. Several trace elements have important functions in the body. For example, iodine is needed to make thyroid hormones. The functions of some trace elements are unknown. TABLE 2.1 lists the main chemical elements of the human body.

* The periodic table of elements, which lists all of the known chemical elements, can be found in Appendix B.

An Atom Contains Protons, Neutrons, and Electrons

An individual atom consists of dozens of different **subatomic particles**. However, only three types of subatomic particles are important for understanding the chemical reactions in the human body: protons, neutrons, and electrons (FIGURE 2.1). The dense central core of an atom is its **nucleus**. Within the **nucleus** are positively charged **protons (p^+)** and uncharged (neutral) **neutrons (n^0)**. The tiny, negatively charged **electrons (e^-)** move about in a large space surrounding the nucleus. They do not follow a fixed path or orbit but instead form a negatively charged "cloud" that envelops the nucleus (FIGURE 2.1a).

FIGURE 2.1 Two representations of the structure of an atom. Electrons move about the nucleus, which contains neutrons and protons. (a) In the electron cloud model of an atom, the shading represents the general region where electrons are located. (b) In the electron shell model, individual electrons are grouped into concentric circles according to the shells they occupy. Both models depict a carbon atom, with six protons, six neutrons, and six electrons.

An atom is the smallest unit of matter that retains the properties and characteristics of its element.

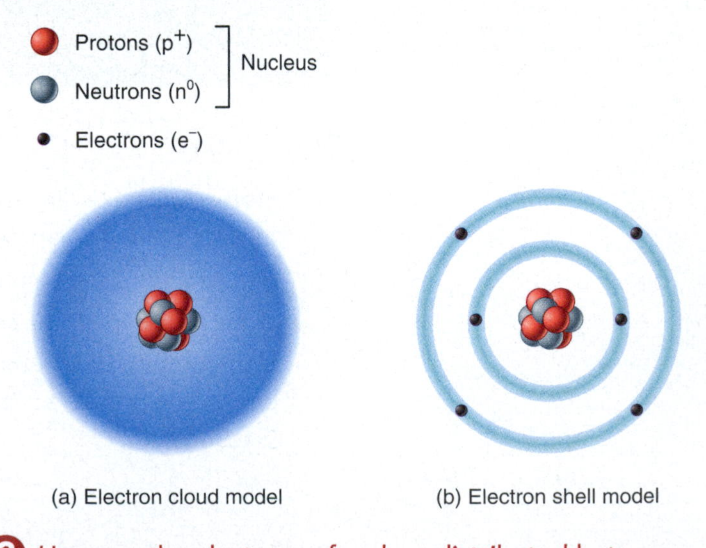

How are the electrons of carbon distributed between the first and second electron shells?

TABLE 2.1 Main Chemical Elements in the Body

Chemical Element (Symbol)	% of Total Body Mass	Significance
MAJOR ELEMENTS	(96)	
Oxygen (O)	65.0	Part of water and many organic (carbon-containing) molecules; used to generate ATP, a molecule used by cells to temporarily store chemical energy.
Carbon (C)	18.5	Forms backbone chains and rings of all organic molecules: carbohydrates, lipids (fats), proteins, and nucleic acids (DNA and RNA).
Hydrogen (H)	9.5	Constituent of water and most organic molecules; ionized form (H^+) makes body fluids more acidic.
Nitrogen (N)	3.2	Component of all proteins and nucleic acids.
LESSER ELEMENTS	(3.6)	
Calcium (Ca)	1.5	Contributes to hardness of bones and teeth; ionized form (Ca^{2+}) needed for blood clotting, release of some hormones, contraction of muscle, and many other processes.
Phosphorus (P)	1.0	Component of nucleic acids and ATP; required for normal bone and tooth structure.
Potassium (K)	0.35	Ionized form (K^+) is the most plentiful cation (positively charged particle) in intracellular fluid; needed to generate action potentials.
Sulfur (S)	0.25	Component of some vitamins and many proteins.
Sodium (Na)	0.2	Ionized form (Na^+) is the most plentiful cation in extracellular fluid; essential for maintaining water balance; needed to generate action potentials.
Chlorine (Cl)	0.2	Ionized form (Cl^-) is the most plentiful anion (negatively charged particle) in extracellular fluid; essential for maintaining water balance.
Magnesium (Mg)	0.1	Ionized form (Mg^{2+}) needed for action of many enzymes, molecules that increase the rate of chemical reactions in organisms.
Iron (Fe)	0.005	Ionized forms (Fe^{2+} and Fe^{3+}) are part of hemoglobin (oxygen-carrying protein in erythrocytes) and some enzymes.
TRACE ELEMENTS	(0.4)	Aluminum (Al), boron (B), chromium (Cr), cobalt (Co), copper (Cu), fluorine (F), iodine (I), manganese (Mn), molybdenum (Mo), selenium (Se), silicon (Si), tin (Sn), vanadium (V), and zinc (Zn).

Even though their exact positions cannot be predicted, specific groups of electrons are most likely to move about within certain regions around the nucleus. These regions, called **electron shells**, are depicted as simple circles around the nucleus. Because each electron shell can hold a specific number of electrons, the electron shell model best conveys this aspect of atomic structure (FIGURE 2.1b). The first electron shell (the one nearest the nucleus) never holds more than 2 electrons. The second shell holds a maximum of 8 electrons, and the third can hold up to 18 electrons. The electron shells fill with electrons in a specific order, beginning with the first shell. For example, notice in FIGURE 2.2 that sodium (Na), which has 11 electrons total, contains 2 electrons in the first shell, 8 electrons in the second shell, and 1 electron in the third shell. The most massive element present in the human body is iodine, which has a total of 53 electrons: 2 in the first shell, 8 in the second shell, 18 in the third shell, 18 in the fourth shell, and 7 in the fifth shell.

The number of electrons in an atom of an element always equals the number of protons. Because each electron and proton carries a single charge, the negatively charged electrons and the positively charged protons balance each other. Thus, each atom is electrically neutral; its total charge is zero.

Each Atom Has an Atomic Number and a Mass Number

The *number of protons* in the nucleus of an atom is an atom's **atomic number.** FIGURE 2.2 shows that atoms of different elements have different atomic numbers because they have different numbers of protons. For example, oxygen has an atomic number of 8 because its nucleus has 8 protons, and sodium has an atomic number of 11 because its nucleus has 11 protons.

The **mass number** of an atom is the sum of its protons and neutrons. Because sodium has 11 protons and 12 neutrons, its mass number is 23 (FIGURE 2.2). Although all atoms of one element have the same number of protons, they may have different numbers of neutrons and thus different mass numbers. **Isotopes** are atoms of an element that have different numbers of neutrons and therefore different mass numbers. In a sample of oxygen, for example, most atoms have 8 neutrons, and a few have 9 or 10, but all have 8 protons and 8 electrons. Most isotopes are stable, which means that their nuclear structure does not change over time. The stable isotopes of oxygen are designated ^{16}O, ^{17}O, and ^{18}O (or O-16, O-17, and O-18). The numbers indicate the mass number of each isotope. As you will discover shortly, the number of electrons

FIGURE 2.2 Atomic structures of several stable atoms.

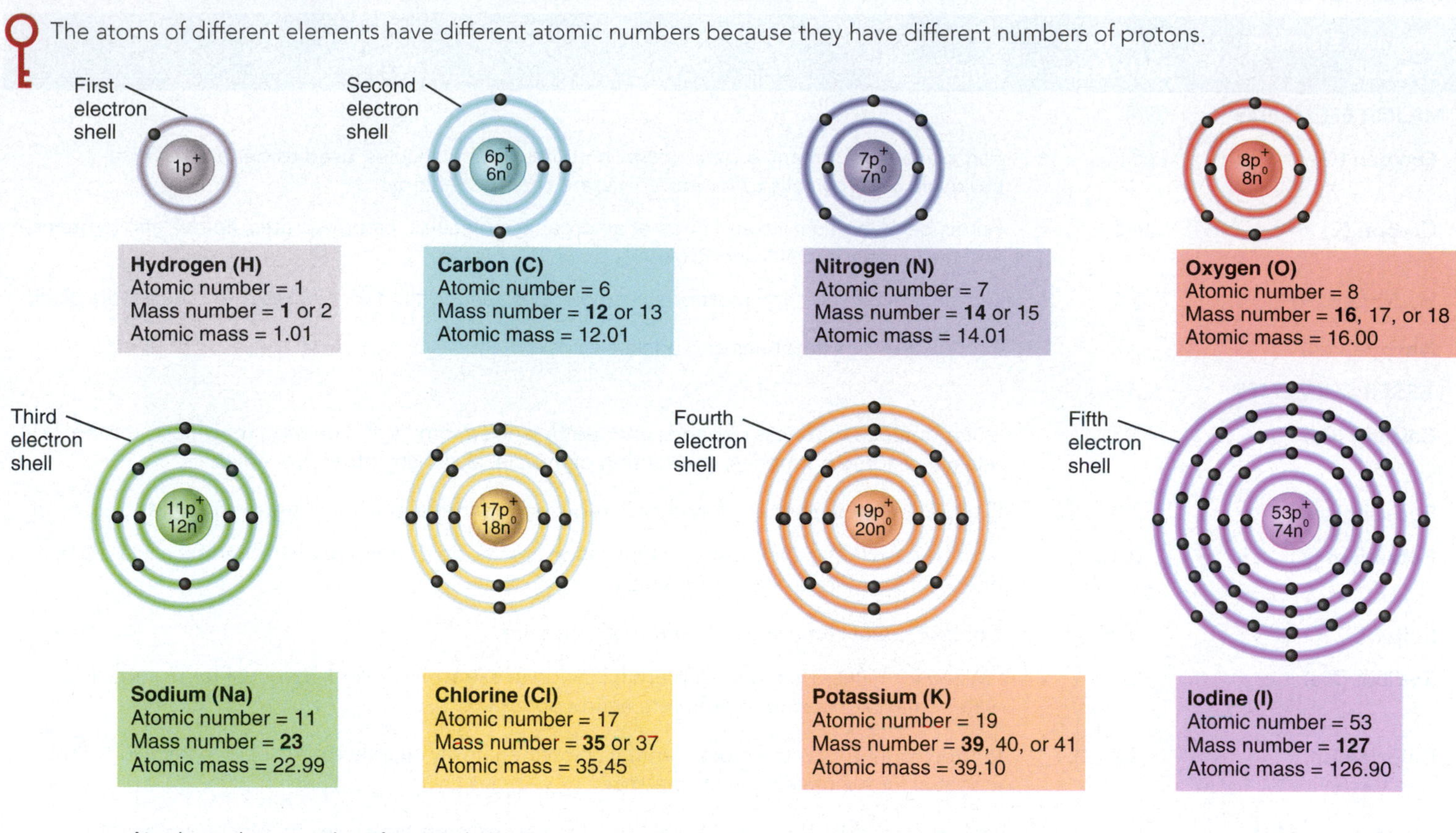

Atomic number = number of protons in an atom
Mass number = number of protons and neutrons in an atom (boldface indicates most common isotope)
Atomic mass = average mass of all stable atoms of a given element in daltons

Which four of these elements are present most abundantly in living organisms?

of an atom determines its chemical properties. Although the isotopes of an element have different numbers of neutrons, they have identical chemical properties because they have the same number of electrons.

Certain isotopes called **radioactive isotopes** are unstable; their nuclei decay (spontaneously change) into a stable configuration. Examples are H-3, C-14, O-15, and O-19. As they decay, these atoms emit radiation—either subatomic particles or packets of energy—and in the process often transform into a different element. For example, the radioactive isotope of carbon, C-14, decays to N-14. The decay of a radioisotope may be as fast as a fraction of a second or as slow as millions of years. The **half-life** of an isotope is the time required for half of the radioactive atoms in a sample of that isotope to decay into a more stable form. The half-life of C-14, which is used to determine the age of organic samples, is 5600 years; the half-life of I-131, an important clinical tool, is 8 days.

Atomic Mass Is Measured In Daltons

The standard unit for measuring the mass of atoms and their subatomic particles is a **dalton**, also known as an *atomic mass unit (amu)*. A neutron has a mass of 1.008 daltons, and a proton has a mass of 1.007 daltons. The mass of an electron, at 0.0005 dalton, is almost 2000 times smaller than the mass of a neutron or proton. The **atomic mass** (also called the *atomic weight*) of an element is the average mass of all of its naturally occurring isotopes. Typically, the atomic mass of an element is close to the mass number of its most abundant isotope.

CLINICAL CONNECTION

Harmful and Beneficial Effects of Radiation

Radioactive isotopes may have either harmful or beneficial effects on the human body. Their radiations can break apart molecules, posing a serious threat to the body by producing tissue damage and/or causing various types of cancer. Although the decay of naturally occurring radioactive isotopes typically releases just a small amount of radiation into the environment, localized accumulations can occur. Radon-222, a colorless and odorless gas that is a naturally occurring radioactive breakdown product of uranium, may seep out of the soil and accumulate in buildings. It is not only associated with many cases of lung cancer in smokers but has also been implicated in many cases of lung cancer in nonsmokers. Beneficial effects of certain radioisotopes include their use in medical imaging procedures to diagnose and treat certain disorders. Some radioisotopes can be used as **tracers** to follow the movement of certain substances through the body. Thallium-201 is used to monitor blood flow through the heart during an exercise stress test. Iodine-131 is used to detect cancer of the thyroid gland and to assess its size and activity, and it may also be used to destroy part of an overactive thyroid gland. Cesium-137 is used to treat advanced cervical cancer, and iridium-192 is used to treat prostate cancer.

Atoms Can Give Rise to Ions, Molecules, and Compounds

As you have just learned, atoms of the same element have the same number of protons. The atoms of each element have a characteristic way of losing, gaining, or sharing their electrons when interacting with other atoms to achieve stability. The way that electrons behave enables atoms in the body to exist in electrically charged forms called ions, or to join with each other into complex combinations called molecules. If an atom either *gives up* or *gains* electrons, it becomes an ion. An **ion** is an atom that has a positive or negative charge because it has unequal numbers of protons and electrons. *Ionization* is the process of giving up or gaining electrons. An ion of an atom is symbolized by writing its chemical symbol followed by the number of its positive (+) or negative (−) charges. Thus, Ca^{2+} stands for a calcium ion that has two positive charges because it has lost two electrons.

When two or more atoms *share* electrons, the resulting combination is called a **molecule**. A *molecular formula* indicates the elements and the number of atoms of each element that make up a molecule. A molecule may consist of two atoms of the same kind, such as an oxygen molecule (**FIGURE 2.3a**). The molecular formula for a molecule of oxygen is O_2. The subscript 2 indicates that the molecule contains two atoms of oxygen. Two or more different kinds of atoms may also form a molecule, as in a water molecule (H_2O). In H_2O, one atom of oxygen shares electrons with two atoms of hydrogen.

A **compound** is a substance that contains atoms of two or more different elements. Most of the atoms in the body are joined into compounds. Water (H_2O) and sodium chloride (NaCl), common table salt, are compounds. However, a molecule of oxygen (O_2) is not a compound because it consists of atoms of only one element.

A **free radical** is an atom or molecule with an unpaired electron in the outermost shell. A common example is superoxide, which is formed by the addition of an electron to an oxygen molecule (**FIGURE 2.3b**). Having an unpaired electron makes a free radical unstable, highly reactive, and destructive to nearby molecules. Free radicals become stable by either giving up their unpaired electron to, or taking on an electron from, another molecule. In so doing, free radicals may break apart important body molecules and disrupt homeostasis.

FIGURE 2.3 Atomic structures of an oxygen molecule and a superoxide free radical.

A free radical has an unpaired electron in its outermost electron shell.

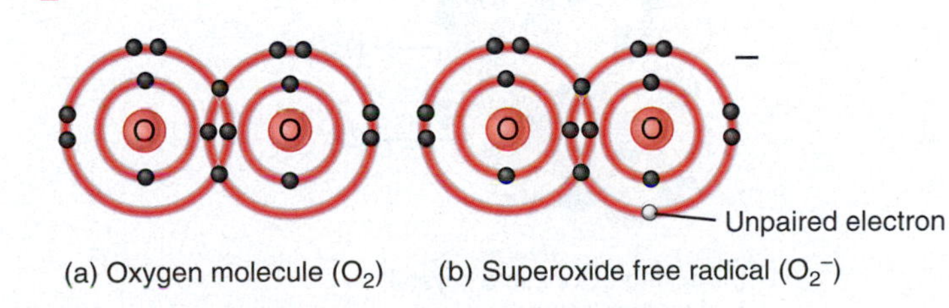

(a) Oxygen molecule (O_2) (b) Superoxide free radical (O_2^-)

What substances in the body can inactivate oxygen-derived free radicals?

CLINICAL CONNECTION

Free Radicals and Their Effects on Health

In our bodies, several processes can generate **free radicals**, including exposure to ultraviolet radiation in sunlight, exposure to X-rays, and some reactions that occur during normal metabolic processes. Certain harmful substances, such as carbon tetrachloride (a solvent used in dry cleaning), also give rise to free radicals when they participate in metabolic reactions in the body. Among the many disorders, diseases, and conditions linked to oxygen-derived free radicals are cancer, atherosclerosis, Alzheimer's disease, emphysema, diabetes mellitus, cataracts, macular degeneration, and deterioration associated with aging. Consuming more **antioxidants**—substances that inactivate oxygen-derived free radicals—is thought to slow the pace of damage caused by free radicals. Important dietary antioxidants include selenium, zinc, beta-carotene, and vitamins C and E.

CHECKPOINT

1. What are the names and chemical symbols of the 12 most abundant chemical elements in the human body?
2. What are the atomic number, mass number, and atomic mass of carbon? How are they related?
3. Is nitric oxide (NO) an example of a molecule, a compound, or both?
4. How can free radicals exert negative effects on the body?

2.2 Chemical Bonds

OBJECTIVES

- Explain the significance of valence electrons.
- Distinguish among the different types of chemical bonds.

The forces that hold together the atoms of a molecule or a compound are **chemical bonds**. The likelihood that an atom will form a chemical bond with another atom depends on the number of electrons in its outermost shell, also called the **valence shell**. An atom with a valence shell holding eight electrons is *chemically stable*, which means it is unlikely to form chemical bonds with other atoms. Neon, for example, has eight electrons in its valence shell, and for this reason it does not bond easily with other atoms. The valence shell of hydrogen and helium is the first electron shell, which holds a maximum of two electrons. Because helium has two valence electrons, it too is stable and seldom bonds with other atoms. Hydrogen, on the other hand, has only one valence electron (see **FIGURE 2.2**), so it binds readily with other atoms.

The atoms of most biologically important elements do not have eight electrons in their valence shells. Under the right conditions, two or more atoms can interact in ways that produce a chemically stable arrangement of eight valence electrons for each atom. This chemical principle, called the **octet rule** (*octet* = set of eight), helps explain why atoms interact in predictable ways. One atom is more likely to interact with another atom if doing so will leave both with eight valence electrons. For this to happen, an atom either empties its partially filled valence shell, fills it with donated electrons, or shares electrons with other atoms. Not all chemical bonds are formed from the loss or gain of valence electrons or the sharing of valence electrons. Some chemical bonds form as a result of the interactions between parts of different molecules or different parts of the same molecule. Let's now consider four types of chemical bonds: ionic bonds, covalent bonds, hydrogen bonds, and van der Waals interactions.

Ionic Bonds Involve the Loss or Gain of Electrons

As you have already learned, when atoms lose or gain one or more valence electrons, ions are formed. Positively and negatively charged ions are attracted to one another—opposites attract. The force of attraction that holds together ions with opposite charges is an **ionic bond**. Consider sodium and chlorine atoms, the components of common table salt. Sodium has one valence electron (FIGURE 2.4a). If sodium *loses* this electron, it is left with the eight electrons in its second shell, which becomes the valence shell. As a result, the total number of protons (11) exceeds the number of electrons (10). Thus, the sodium atom has become a **cation** (KAT-ī-on), or positively charged ion. A sodium ion has a charge of 1+ and is written Na^+. By contrast, chlorine has seven valence electrons (FIGURE 2.4b). If chlorine *gains* an electron from a neighboring atom, it will have a complete octet in its third electron shell. After gaining an electron, the total number of electrons (18) exceeds the number of protons (17), and the chlorine atom has become an **anion** (AN-ī-on), a negatively charged ion. The ionic form of chlorine is called a *chloride* ion. It has a charge of 1− and is written Cl^-. When an atom of sodium donates its sole valence electron to an atom of chlorine, the resulting positive and negative charges pull both ions tightly together, forming an ionic bond (FIGURE 2.4c). The resulting compound is sodium chloride, written NaCl.

In general, ionic compounds exist as solids, with an orderly, repeating arrangement of the ions, as in a crystal of NaCl (FIGURE 2.4d). A crystal of NaCl may be large or small—the total number of ions can vary—but the ratio of Na^+ to Cl^- is always 1:1. An ionic compound that breaks apart into positive and negative ions in solution is called an **electrolyte** (e-LEK-trō-līt). Most ions in the body are dissolved in body fluids as electrolytes, so named because their solutions can conduct an electric current. (The chemistry and importance of electrolytes will be discussed in Chapter 20.) TABLE 2.2 lists the names and symbols of the most common ions in the body.

TABLE 2.2 Common Ions and Ionic Compounds in the Body

Cations		Anions	
Name	**Symbol**	**Name**	**Symbol**
Hydrogen ion	H^+	Fluoride ion	F^-
Sodium ion	Na^+	Chloride ion	Cl^-
Potassium ion	K^+	Iodide ion	I^-
Ammonium ion	NH_4^+	Hydroxide ion	OH^-
Magnesium ion	Mg^{2+}	Bicarbonate ion	HCO_3^-
Calcium ion	Ca^{2+}	Oxide ion	O^{2-}
Iron (II) ion	Fe^{2+}	Sulfate ion	SO_4^{2-}
Iron (III) ion	Fe^{3+}	Phosphate ion	PO_4^{3-}

FIGURE 2.4 Ions and ionic bond formation. (a) A sodium atom can have a complete octet of electrons in its outermost shell by losing one electron. (b) A chlorine atom can have a complete octet by gaining one electron. (c) An ionic bond may form between oppositely charged ions. (d) In a crystal of NaCl, each Na^+ is surrounded by six Cl^-. In (a) to (c), the electron that is lost or accepted is colored red.

An ionic bond is the force of attraction that holds together oppositely charged ions.

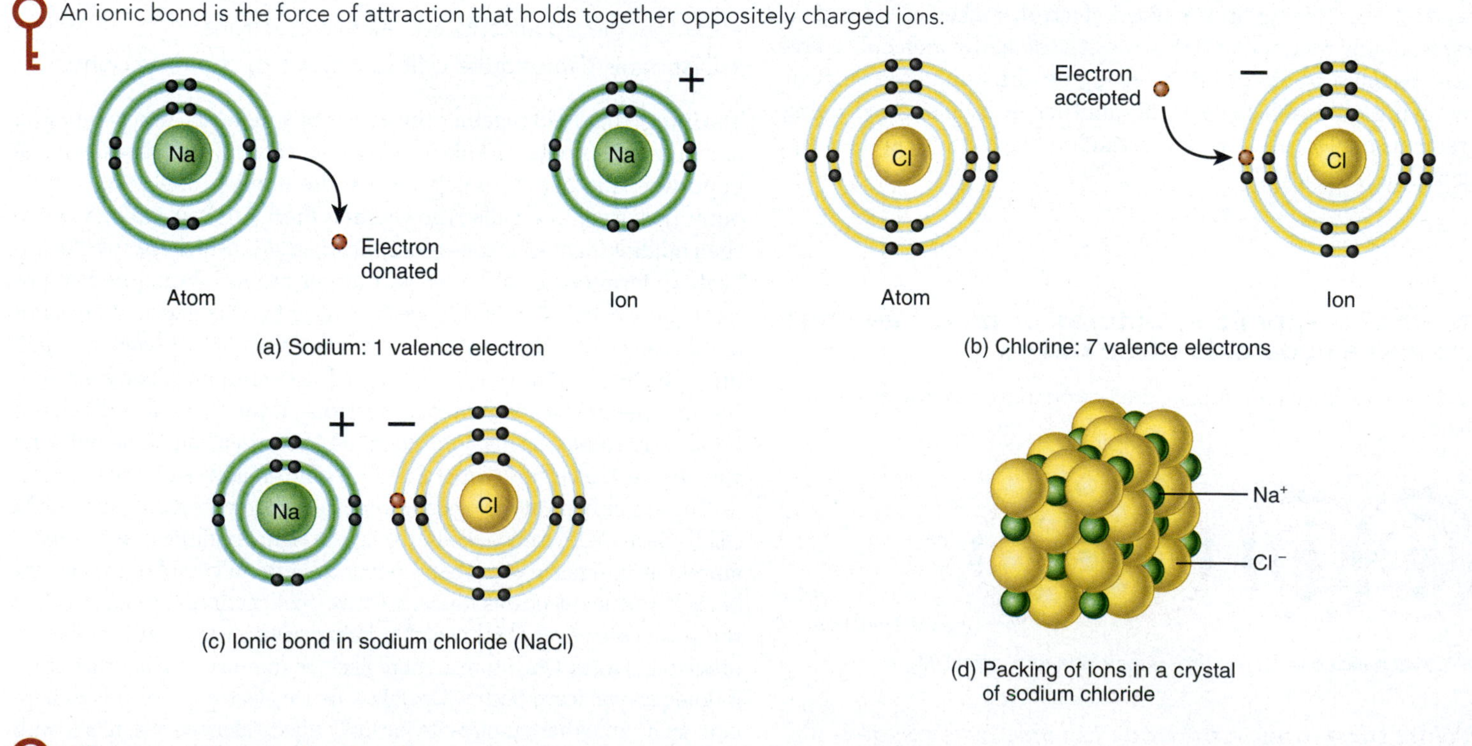

What are cations and anions?

Covalent Bonds Involve Sharing of Electrons

The strongest type of chemical bond is the covalent bond. In a **covalent bond**, two or more atoms *share* electrons rather than gaining or losing them. Covalent bonds may form between atoms of the same element or between atoms of different elements. They are the most common chemical bonds in the body, and the molecules that result from them comprise most of the body's structures.

Atoms form a covalently bonded molecule by sharing one, two, or three pairs of valence electrons. A **single covalent bond** results when two atoms share one electron pair. For example, a molecule of hydrogen forms when two hydrogen atoms share their single valence electrons (FIGURE 2.5a), which allows both atoms to have a full valence shell at least part of the time. A **double covalent bond** results when two atoms share two pairs of electrons, as happens in an oxygen molecule (FIGURE 2.5b). A **triple covalent bond** occurs when two atoms share three pairs of electrons, as in a molecule of nitrogen (FIGURE 2.5c). Notice in the *structural formulas* for covalently bonded molecules in FIGURE 2.5 that the number of lines between the chemical symbols for two atoms indicates whether the bond is a single (—), double (=), or triple (≡) covalent bond.

FIGURE 2.5 Covalent bond formation. The red electrons are shared equally in (a) to (d) and unequally in (e). In writing the structural formula of a covalently bonded molecule, each straight line between the chemical symbols for two atoms denotes a pair of shared electrons. In molecular formulas, the number of atoms in each molecule is noted by subscripts.

In a covalent bond, two atoms share one, two, or three pairs of valence electrons.

DIAGRAMS OF ATOMIC AND MOLECULAR STRUCTURE		STRUCTURAL FORMULA	MOLECULAR FORMULA
(a) Hydrogen atoms	Hydrogen molecule	H — H	H_2
(b) Oxygen atoms	Oxygen molecule	O = O	O_2
(c) Nitrogen atoms	Nitrogen molecule	N ≡ N	N_2
(d) Carbon atom, Hydrogen atoms	Methane molecule	H—C—H with H above and below C	CH_4
(e) Oxygen atom, Hydrogen atoms	Water molecule (δ^-, δ^+, δ^+)	O bonded to two H	H_2O

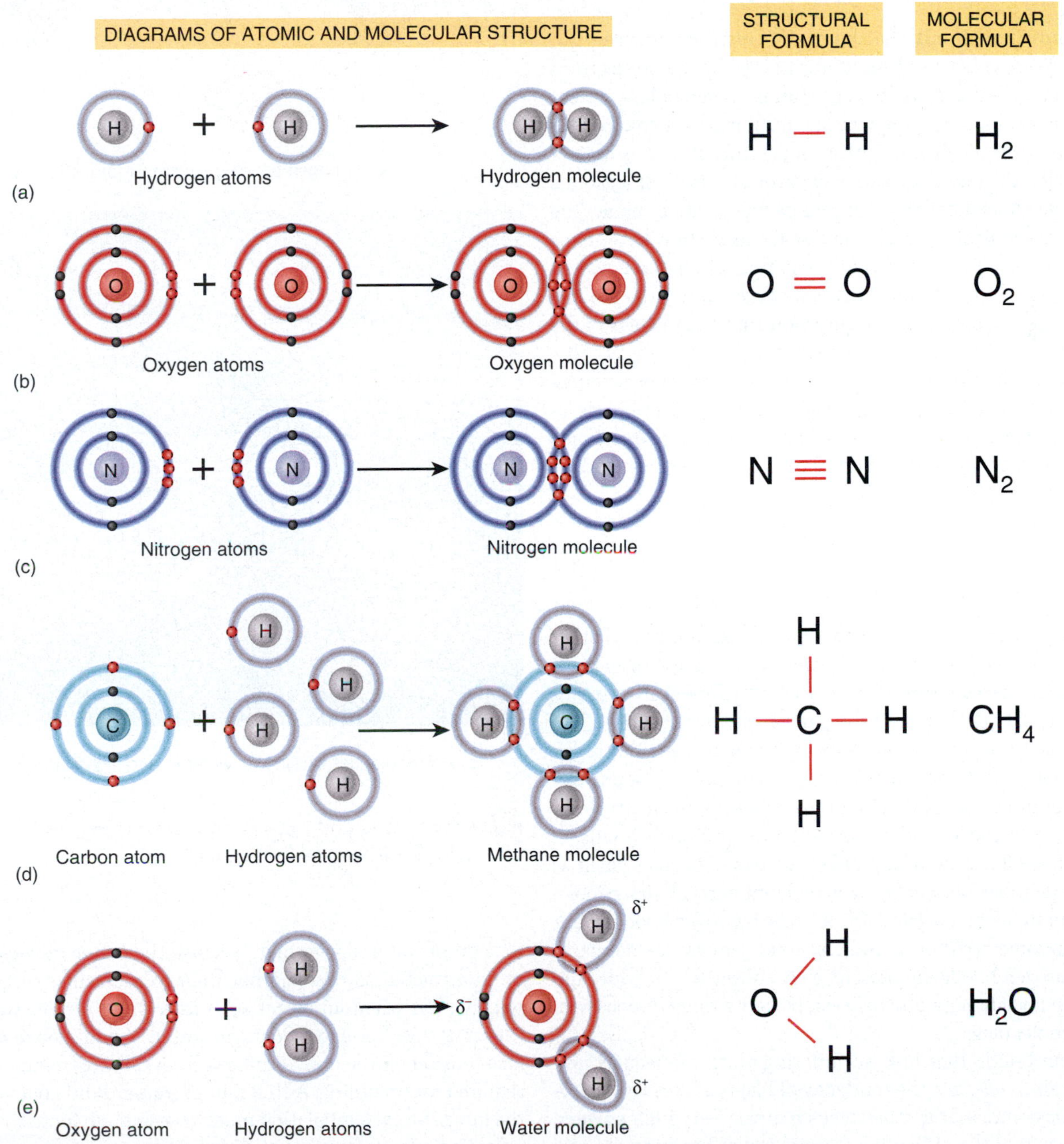

What is the principal difference between an ionic bond and a covalent bond?

The same principles of covalent bonding that apply to atoms of the same element also apply to covalent bonds between atoms of different elements. The gas methane (CH_4) contains covalent bonds formed between the atoms of two different elements, one carbon and four hydrogens (FIGURE 2.5d). The valence shell of the carbon atom can hold eight electrons but has only four of its own. The single electron shell of a hydrogen atom can hold two electrons, but each hydrogen atom has only one of its own. A methane molecule contains four separate single covalent bonds. Each hydrogen atom shares one pair of electrons with the carbon atom.

In some covalent bonds, two atoms share the electrons equally—one atom does not attract the shared electrons more strongly than the other atom. This type of bond is a **nonpolar covalent bond**. The bonds between two identical atoms are always nonpolar covalent bonds (FIGURE 2.5a–c). The bonds between carbon and hydrogen atoms are also nonpolar, such as the four C—H bonds in a methane molecule (FIGURE 2.5d).

In a **polar covalent bond**, the sharing of electrons between two atoms is unequal—the nucleus of one atom attracts the shared electrons more strongly than the nucleus of the other atom. When polar covalent bonds form, the resulting molecule has a partial negative charge near the atom that attracts electrons more strongly. This atom has greater **electronegativity**, the power to attract electrons to itself. At least one other atom in the molecule then will have a partial positive charge. The partial charges are indicated by a lowercase Greek delta with a minus or plus sign: δ^- or δ^+. A very important example of a polar covalent bond in living systems is the bond between oxygen and hydrogen in a molecule of water (FIGURE 2.5e); in this molecule, the nucleus of the oxygen atom attracts the electrons more strongly than the nuclei of the hydrogen atoms, so the oxygen atom is said to have greater electronegativity. Later in the chapter, you will see how polar covalent bonds allow water to dissolve many molecules that are important to life. Bonds between nitrogen and hydrogen and those between oxygen and carbon are also polar covalent bonds.

Hydrogen Bonds Result from Attraction of Oppositely Charged Regions of Molecules

The polar covalent bonds that form between hydrogen atoms and other atoms can give rise to a third type of chemical bond, a hydrogen bond (FIGURE 2.6a). A **hydrogen bond** forms when a hydrogen atom with a partial positive charge (δ^+) attracts the partial negative charge (δ^-) of a neighboring electronegative atom, most often an oxygen or nitrogen atom. Thus, hydrogen bonds result from attraction of oppositely charged parts of molecules rather than from sharing of electrons, as in covalent bonds, or the loss or gain of electrons, as in ionic bonds.

Hydrogen bonds form important links between molecules (such as neighboring water molecules) or between different parts of a large molecule (such as a protein or a nucleic acid). A single hydrogen bond is weak compared to an ionic bond or a covalent bond. However, when large numbers of hydrogen bonds are present, they collectively provide substantial strength and stability, and help establish the three-dimensional shape of large molecules.

The hydrogen bonds that link neighboring water molecules give water considerable *cohesion*, the tendency of like particles to stay together. The cohesion of water molecules creates a very high **surface tension**, a measure of the difficulty of stretching or breaking the surface of a liquid. At the boundary between water and air, the surface tension of water is very high because the water molecules are much more attracted to one another than to molecules in the air. This is readily seen when an insect such as a water strider walks on water (FIGURE 2.6b) or a leaf floats on water. The influence of water's surface tension on the body can be seen in the way it increases the work required for breathing. A thin film of watery fluid coats the air sacs of the lungs. So each inhalation must have enough force to overcome the opposing effect of surface tension as the air sacs stretch and enlarge while taking in air.

FIGURE 2.6 Hydrogen bonding and surface tension. In (a), each water molecule forms hydrogen bonds (indicated by dotted lines) with three to four neighboring water molecules.

Hydrogen bonds occur because hydrogen atoms in one water molecule are attracted to the partial negative charge of the oxygen atom in another water molecule.

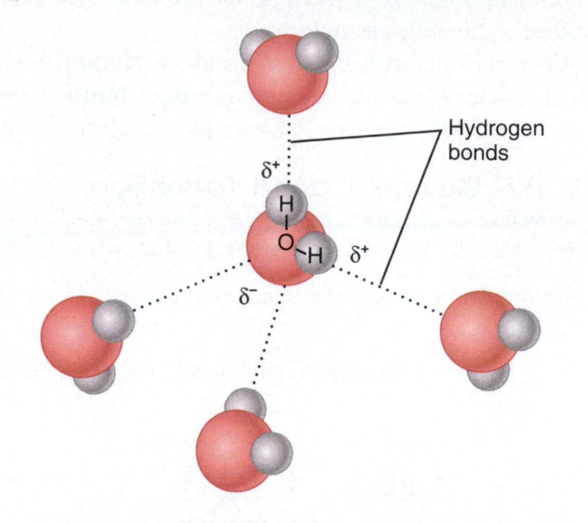

(a) Hydrogen bonding among water molecules

(b) A water strider walking on water due to high surface tension.

Why would you expect ammonia (NH_3) to form hydrogen bonds with water molecules?

Van der Waals Interactions Involve Transient Fluctuations in Electron Distribution

Because electrons are in constant motion, they are not always evenly distributed around the atoms of a molecule. For example, in a nonpolar molecule, the distribution of electrons at any given instant may be greater on one side of the molecule than the other, even though the atoms of the molecule share electrons equally with one another. These transient fluctuations in electron distribution in turn cause transient unequal charge distributions in the molecule: Temporarily, one side of the molecule has a partial negative charge (δ^-) because it has a higher density of electrons, and the other side has a partial positive charge (δ^+) because it has a lower electron density. Which side of the nonpolar molecule is δ^- or δ^+ varies over time as the electrons change their positions in the molecule. If two nonpolar molecules are brought close to one another, a transient unequal charge distribution in one nonpolar molecule can induce a complementary transient unequal charge distribution in the other nonpolar molecule, and a van der Waal interaction forms between the two molecules (FIGURE 2.7a). A **van der Waals interaction** is a weak attraction between nearby molecules due to transient unequal charge distributions in these molecules. This attraction bonds the molecules together for as long as the unequal charge distributions last. A van der Waals interaction can also form between two polar molecules or between a polar molecule and a nonpolar molecule. In polar molecules, the distribution of electrons fluctuates, but on average the electrons stay closer to the side of the molecule that contains the electronegative atom. Thus, polar molecules have relatively permanent unequal electron distributions, meaning that one part of the molecule is essentially always δ^- and the other part δ^+. If two polar molecules with complementary unequal charge distributions are in close proximity and have the proper orientation, a van der Waals interaction will attract the two molecules to each other. Similarly, if a polar molecule is near a nonpolar molecule, the polar molecule induces a complementary unequal charge distribution in the nonpolar molecule and then the two molecules are attracted to each other by a van der Waals interaction. Van der Waals interactions not only bind molecules to one another, but they can also bind different parts of a large molecule to help establish that molecule's three-dimensional shape.

Because it is a weak type of attraction, a van der Waals interaction forms only when molecules are in close proximity to one another. If two molecules are far away from each other, no attraction exists. As the two molecules get closer, the attraction increases. However, there is a limit to this attraction: If the molecules get too close to each other, the electron clouds of the two molecules start to overlap and the electrons repel one another, forcing the molecules apart.

An individual van der Waals interaction is a very weak bond, even weaker than a hydrogen bond. However, multiple van der Waals interactions collectively provide the strength and stability needed to hold molecules together. For example, geckos (small lizards) are able to climb a wall because of van der Waals interactions (FIGURE 2.7b). On the bottom surface of each gecko foot are hundreds of thousands of tiny, thin hairs. As these hairs come in contact with a wall, numerous van der Waals interactions between the molecules in the hairs of the gecko's foot and the molecules in the wall cumulatively provide enough strength to keep the gecko attached to the wall.

FIGURE 2.7 Van der Waals interactions.

Van der Waals interactions are weak attractions between nearby molecules due to unequal charge distributions in these molecules.

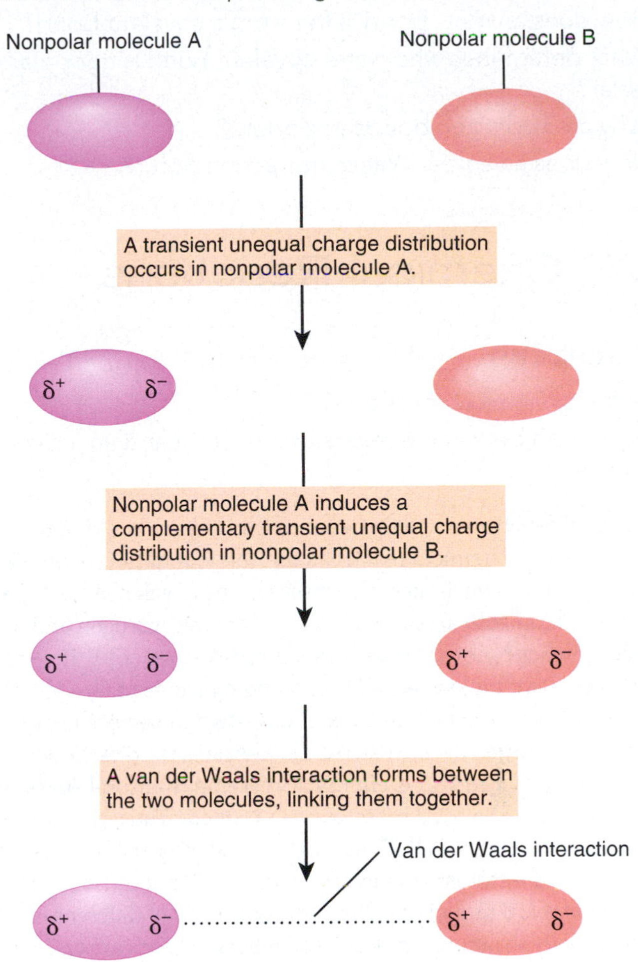

(a) A van der Waals interaction between two nonpolar molecules.

(b) A gecko climbing a wall due to numerous van der Waals interactions

Why does a nonpolar molecule have transient, unequal charge distributions?

CHECKPOINT

5. Which electron shell is the valence shell of an atom, and what is its significance?
6. How does an ionic bond differ from a covalent bond?
7. What determines whether a covalent bond is nonpolar or polar?
8. Why are hydrogen bonds important?
9. How does a van der Waals interaction occur?

2.3 Chemical Reactions

OBJECTIVES

- Define a chemical reaction.
- Distinguish between a reversible reaction and an irreversible reaction.

A **chemical reaction** occurs when new bonds form or old bonds break between atoms. Chemical reactions are the foundation of all life processes and, as you have seen, the interactions of valence electrons are the basis of all chemical reactions. Consider how hydrogen and oxygen molecules react to form water molecules (FIGURE 2.8). The starting substances—two molecules of H_2 and one molecule of O_2—are known as the **reactants**. The ending substances—two molecules of H_2O—are the **products**. The arrow in FIGURE 2.8 indicates the direction in which the reaction proceeds. In a chemical reaction, the total mass of the reactants equals the total mass of the products. Thus, the number of atoms of each element is the same before and after the reaction. However, because the atoms are rearranged, the reactants and products have different chemical properties. Through thousands of different chemical reactions, body structures are built and body functions are carried out.

Some chemical reactions are irreversible. **Irreversible reactions** proceed in only one direction, from reactants to products, as indicated by a single arrow:

$$\underset{\text{Reactants}}{A + B} \longrightarrow \underset{\text{Products}}{C + D}$$

FIGURE 2.8 The chemical reaction between two hydrogen molecules (H_2) and one oxygen molecule (O_2) to form two molecules of water (H_2O). Note that the reaction occurs by breaking old bonds and making new bonds.

The number of atoms of each element is the same before and after making new bonds.

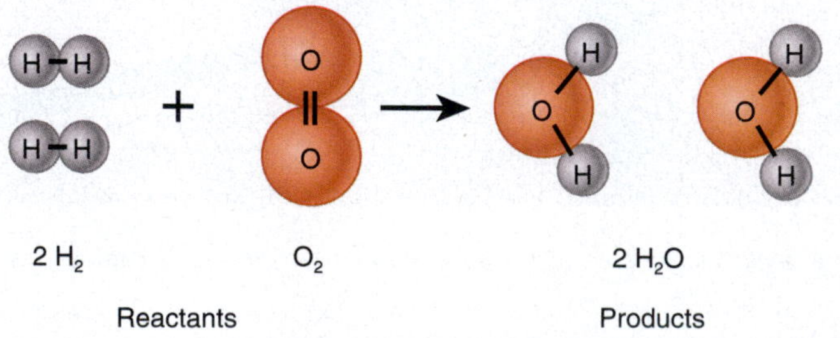

Why does this reaction require two molecules of H_2?

Other chemical reactions may be reversible. In a **reversible reaction**, the products can revert to the original reactants. A reversible reaction is indicated by two half arrows pointing in opposite directions:

$$\underset{\text{Reactants}}{A + B} \rightleftharpoons \underset{\text{Products}}{C + D}$$

In the generalized example above, the *forward reaction* occurs when reactants are converted to products; the *reverse reaction* occurs when products are converted to reactants. When a reaction is at **chemical equilibrium**, the rate of the forward reaction is equal to the rate of the reverse reaction. In other words, at equilibrium, the conversion of reactants to products is occurring at the same rate as the conversion of products to reactants.

An example of a reversible reaction is the reaction between carbon dioxide and water to form carbonic acid:

$$\underset{\substack{\text{Carbon}\\\text{dioxide}}}{CO_2} + \underset{\text{Water}}{H_2O} \rightleftharpoons \underset{\substack{\text{Carbonic}\\\text{acid}}}{H_2CO_3}$$

When the reaction proceeds in the forward direction, carbonic acid is formed; when the reaction proceeds in the reverse direction, carbonic acid breaks down into carbon dioxide and water.

One factor that influences the net direction of a reversible reaction is the concentrations of the reactants and products. Increasing the concentration of the reactants relative to the products drives the reaction in the forward direction; increasing the concentration of the products relative to the reactants drives the reaction in the reverse direction. This relationship between the net direction of a reversible reaction and the concentrations of the reactants and products is known as the **law of mass action**. During your study of physiology, you will become very familiar with the law of mass action because it will be applied to several chemical reactions that are described in this text (see Sections 2.5, 4.3, 6.2, 11.4, and 18.5).

Chemical reactions involve changes in energy. *Energy* is the capacity to do work. Some chemical reactions release more energy than they absorb; other chemical reactions absorb more energy than they release. The energy changes associated with chemical reactions will be described in detail in Chapter 4.

CHECKPOINT

10. What is the relationship between reactants and products in a chemical reaction?
11. What is happening when a reversible reaction is at chemical equilibrium?
12. What is the law of mass action?

2.4 Inorganic Compounds and Solutions

OBJECTIVES

- Describe the properties of water and those of inorganic acids, bases, and salts.
- Distinguish among solutions, colloids, and suspensions.
- Explain the role of buffer systems in maintaining pH.

Most of the chemicals in your body exist in the form of compounds. Biologists and chemists divide these compounds into two principal classes: inorganic compounds and organic compounds. **Inorganic compounds** usually lack carbon and are structurally simple. They include water and many salts, acids, and bases. Inorganic compounds may have either ionic or covalent bonds. Water makes up 55–60% of a lean adult's total body mass; all other inorganic compounds combined add 1–2%. Examples of inorganic compounds that contain carbon are carbon dioxide (CO_2), bicarbonate ion (HCO_3^-), and carbonic acid (H_2CO_3). Even though these substances contain carbon, they are considered inorganic because of their small size and simple structure. **Organic compounds** always contain carbon, usually contain hydrogen, and always have covalent bonds. Most are large molecules, and many are made up of long chains of carbon atoms. Organic compounds, which you will learn more about in Section 2.5, make up the remaining 38–44% of the human body.

Water Is Vital to Life

Water is the most important and abundant inorganic compound in all living systems. Although you might be able to survive for weeks without food, without water you would die in a matter of days. Nearly all of the body's chemical reactions occur in a watery medium. Water has many properties that make it such an indispensable compound for life. You have already learned about the most important property of water, its polarity—the uneven sharing of valence electrons that confers a partial negative charge near the one oxygen atom and two partial positive charges near the two hydrogen atoms in a water molecule (see FIGURE 2.5e). This property alone makes water an excellent solvent for ionic or polar substances, gives water molecules cohesion (the tendency to stick together), and allows water to resist temperature changes.

Water as a Solvent

In medieval times people searched in vain for a "universal solvent," a substance that would dissolve all other materials. They found nothing that worked as well as water. Although it is the most versatile solvent known, water is not the universal solvent sought by medieval alchemists. If it were, no container could hold it because it would dissolve all potential containers! What exactly is a solvent? In a **solution**, a substance called the **solvent** dissolves another substance called the **solute**. Usually there is more solvent than solute in a solution. For example, your sweat is a dilute solution of water (the solvent) plus small amounts of salts (the solutes).

The versatility of water as a solvent for ionized or polar substances is due to its polar covalent bonds and its bent shape, which allows each water molecule to interact with several neighboring ions or molecules. Solutes that are charged or contain polar covalent bonds are **hydrophilic** (*hydro-* = water; *-philic* = loving), which means they dissolve easily in water. Common examples of hydrophilic solutes are sugar and salt. Molecules that contain mainly nonpolar covalent bonds, by contrast, are **hydrophobic** (*-phobic* = fearing). They are not very water soluble. Examples of hydrophobic compounds include animal fats and vegetable oils.

To understand the dissolving power of water, consider what happens when a crystal of a salt such as sodium chloride (NaCl) is placed in water (FIGURE 2.9). The electronegative oxygen atom in water molecules attracts the sodium ions (Na^+), and the electropositive hydrogen atoms in water molecules attract the chloride ions (Cl^-). Soon, water molecules surround and separate Na^+ and Cl^- ions from each other at the surface of the crystal, breaking the ionic bonds that hold NaCl together. The water molecules surrounding the ions also lessen the chance that Na^+ and Cl^- will come together and reform an ionic bond.

FIGURE 2.9 How polar water molecules dissolve salts and polar substances. When a crystal of sodium chloride is placed in water, the slightly negative oxygen end (red) of water molecules is attracted to the positive sodium ions (Na^+), and the slightly positive hydrogen portions (gray) of water molecules are attracted to the negative chloride ions (Cl^-). In addition to dissolving sodium chloride, water also causes it to dissociate, or separate into charged particles.

Water is a versatile solvent because its polar covalent bonds, in which electrons are shared unequally, create positive and negative regions.

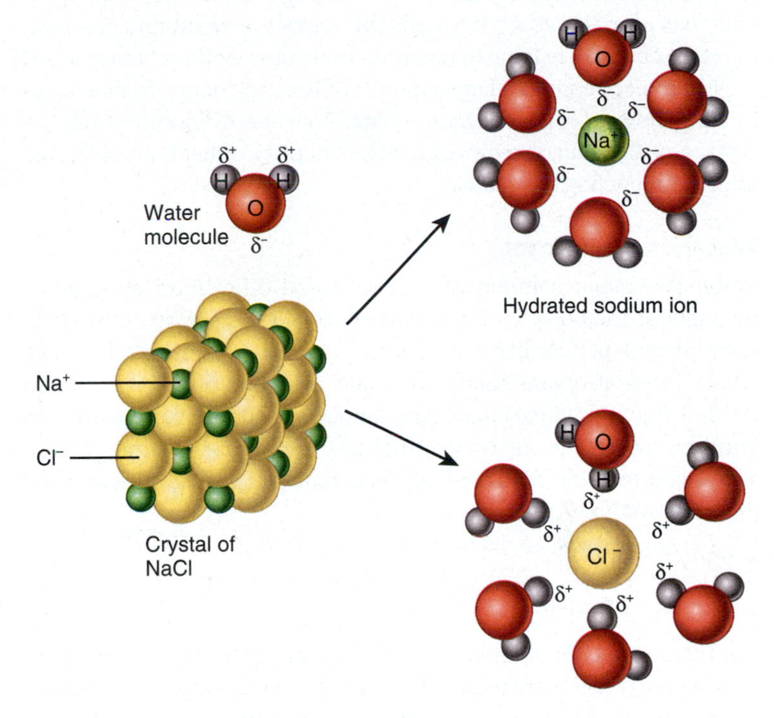

Table sugar (sucrose) easily dissolves in water but is not an electrolyte. Is it likely that all the covalent bonds between atoms in table sugar are nonpolar bonds? Why or why not?

The ability of water to form solutions is essential to health and survival. Because water can dissolve so many different substances, it is an ideal medium for chemical reactions. Water enables dissolved reactants to collide and form products. Water also dissolves waste products, which allows them to be flushed out of the body in the urine.

Because nonpolar molecules are hydrophobic, when they are present in an aqueous environment they aggregate to minimize their exposure to water. This association of nonpolar molecules is referred to as a **hydrophobic interaction**. Such an interaction is important because it can help establish the three-dimensional shape of a large molecule as nonpolar parts of that molecule aggregate to get away from water in the surroundings.

Thermal Properties of Water

In comparison to most substances, water can absorb or release a relatively large amount of heat with only a modest change in its own temperature. For this reason, water is said to have a high *heat capacity.* The reason for this property is the large number of hydrogen bonds in water. As water absorbs heat energy, some of the energy is used to break hydrogen bonds. Less energy is then left over to increase the motion of water molecules, which would increase the water's temperature. The high heat capacity of water is the reason it is used in automobile radiators; it cools the engine by absorbing heat without its own temperature rising to an unacceptably high level. The large amount of water in the body has a similar effect: It lessens the impact of environmental temperature changes, helping to maintain body temperature homeostasis.

Water also requires a large amount of heat to change from a liquid to a gas: Its *heat of vaporization* is high. As water evaporates from the surface of the skin, it removes a large quantity of heat, providing an important cooling mechanism.

Water as a Lubricant

Water is a major component of mucus and other lubricating fluids throughout the body. Lubrication is especially necessary in the chest (pleural and pericardial cavities) and abdomen (peritoneal cavity), where internal organs touch and slide over one another. It is also needed at joints, where bones, ligaments, and tendons rub against one another. Inside the gastrointestinal tract, mucus and other watery secretions moisten foods, aiding their smooth passage through the digestive system.

Solutions, Colloids, and Suspensions Are Types of Mixtures

A **mixture** is a combination of elements or compounds that are physically blended together but not bound by chemical bonds. For example, the air you are breathing is a mixture of gases that includes nitrogen, oxygen, argon, and carbon dioxide. Three common liquid mixtures are solutions, colloids, and suspensions.

Once mixed together, solutes in a solution remain evenly dispersed among the solvent molecules. Because the solute particles in a solution are very small, a solution looks clear and transparent.

A **colloid** differs from a solution mainly because of the size of its particles. The solute particles in a colloid are large enough to scatter light, just as water droplets in fog scatter light from a car's headlight beams. For this reason, colloids usually appear translucent or opaque. Milk is an example of a liquid that is both a colloid and a solution: The large milk proteins make it a colloid, whereas calcium salts, milk sugar (lactose), ions, and other small particles are in solution.

The solutes in both solutions and colloids do not settle out and accumulate on the bottom of the container. In a **suspension**, by contrast, the suspended material may mix with the liquid or suspending medium for some time, but eventually it will settle out. Blood is an example of a suspension. When freshly drawn from the body, blood has an even, reddish color. After blood sits for a while in a test tube, erythrocytes (red blood cells) settle out of the suspension and drift to the bottom of the tube (see **FIGURE 16.1a**). The upper layer, the liquid portion of blood, appears pale yellow and is called plasma. Plasma is both a solution of ions and other small solutes and a colloid due to the presence of larger plasma proteins.

The **concentration** of a solution may be expressed in several ways: molarity, equivalents, and mass per volume.

- ***Molarity.*** The term **molarity** refers to the number of moles of solute per liter of solution. The unit used to report molarity is moles per liter (mol/L) or molar (M). A **mole** is the amount of any substance that has a mass in grams equal to the sum of the atomic masses of all of its atoms. For example, one mole of the element chlorine (atomic mass = 35.45) is 35.45 grams, and one mole of the salt sodium chloride (NaCl) is 58.44 grams (22.99 for Na + 35.45 for Cl). To make a 1 molar (M) solution of NaCl, dissolve 1 mole of NaCl (58.44g) in enough water to make a total of 1 liter of solution. Just as a dozen always means 12 of something, a mole of anything has the same number of particles: 6.023×10^{23}. This huge number is called *Avogadro's number.* Thus, measurements of substances that are stated in moles tell us about the numbers of atoms, ions, or molecules present. This is important when chemical reactions are occurring because each reaction requires a set number of atoms of specific elements. Because body fluids are very dilute, the solute concentrations in these solutions are often expressed as millimoles per liter (mmol/L; 1 mmol = 0.001 mol) or millimolar (mM). For example, the normal concentration of glucose in the blood is about 5 mmol/L. This means that, on average, there are 5 mmol (0.005 mol) of glucose per liter of blood.
- ***Equivalents.*** When the solutes of a solution are ions, it is often important to know the amount of charges that these ions contribute to the solution. This is achieved by expressing the concentrations of ions in units of equivalents per liter (Eq/L). An **equivalent** is equal to one mole of positive or negative charges. For example, in a 1 M solution of sodium ions (Na^+), there is the equivalent of one mole of charges because each Na^+ ion in the mole contributes one unit of charge. Likewise, in a 1 M solution of chloride ions (Cl^-), there is the equivalent of one mole of charges because each Cl^- ion in the mole contributes one unit of charge. As these two examples illustrate, for *monovalent ions* (ions that have a single positive or negative charge), the number of Eq/L is equal to the number of mol/L. Now consider a 1 M solution of magnesium ions (Mg^{2+}). In this solution, there is the equivalent of two moles of charges because each Mg^{2+} ion in the mole contributes two units of charge. This means that a 1 M solution of Mg^{2+} ions is equal to 2 Eq/L. So, for *divalent ions* (ions that have two positive or negative charges), the number of Eq/L is twice the number of mol/L. Here is another scenario to help you understand the concept of an equivalent: Suppose the concentration of Na^+ ions in a solution is 0.140 Eq/L. What does that mean? It means that the Na^+ ions contribute 0.140 equivalent (0.140 mole of charge) per liter of solution. In dilute body fluids, such as those in the body, the concentration of ions are typically expressed in units of milliequivalents per liter (mEq/L; 1 mEq = 0.001 Eq). For example, the Cl^- concentration in blood is about 100 mEq/L. This means that Cl^- ions contribute 100 mEq (100 mmol of charge) per liter of blood. Likewise, the normal Mg^{2+} concentration in the blood is about 2 mEq/L. This means that Mg^{2+} ions contribute 2 mEq (2 mmol of charge) per liter of blood.
- ***Mass per volume.*** Another way to express the concentration of a solution is mass per volume, which indicates the mass of a solute found in a given volume of solution. A unit often used to express mass per volume is grams of solute per liter of solution (g/L). For

example, the normal hemoglobin concentration in the blood is about 150 g/L. This means that there are 150 g of hemoglobin per liter of blood under normal conditions. For solute concentrations in dilute solutions, mass per volume is typically expressed as milligrams per deciliter (mg/dL) (Note that one deciliter is 0.1 L or 100 mL.) For example, the normal blood level of creatinine (a waste product of muscle metabolism) is about 0.9 mg/dL. This means that, on average, there is about 0.9 mg of creatinine per dL (or 100 mL) of blood. Mass per volume can also be expressed as a percentage. For instance, a 10% NaCl solution consists of 10 parts NaCl per 100 parts total solution. Therefore, to make a 10% NaCl solution, take 10 g of NaCl and add enough water to make a total of 100 mL of solution.

Inorganic Acids, Bases, and Salts Dissociate When Dissolved in Water

When inorganic acids, bases, or salts dissolve in water, they **dissociate**; that is, they separate into ions and become surrounded by water molecules. An **acid** (FIGURE 2.10a) is a substance that dissociates into one or more **hydrogen ions (H^+)** and one or more anions. Because H^+ is a single proton with one positive charge, an acid is also referred to as a **proton donor**. A **base**, by contrast (FIGURE 2.10b), removes H^+ from a solution and is therefore a **proton acceptor**. Many bases dissociate into one or more **hydroxide ions (OH^-)** and one or more cations.

A **salt**, when dissolved in water, dissociates into cations and anions, neither of which is H^+ or OH^- (FIGURE 2.10c). In the body, salts such as potassium chloride are electrolytes that are important for carrying electrical currents (ions flowing from one place to another), especially in nerve and muscle tissues. The ions of salts also provide many essential chemical elements in intracellular fluid and in extracellular fluids such as blood, lymph, and the interstitial fluid of tissues.

FIGURE 2.10 Dissociation of inorganic acids, bases, and salts.

Dissociation is the separation of inorganic acids, bases, and salts into ions in a solution.

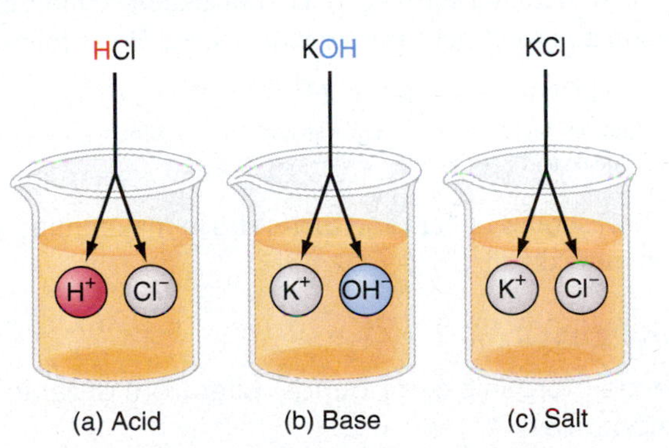

The compound $CaCO_3$ (calcium carbonate) dissociates into a calcium ion (Ca^{2+}) and a carbonate ion (CO_3^{2-}). Is it an acid, base, or a salt? What about H_2SO_4, which dissociates into two H^+ and one SO_4^{2-}?

Acids and bases react with one another to form salts. For example, the reaction of hydrochloric acid (HCl) and potassium hydroxide (KOH), a base, produces the salt potassium chloride (KCl) and water (H_2O). This reaction can be written as follows:

$$\underset{\text{Acid}}{HCl} + \underset{\text{Base}}{KOH} \longrightarrow \underset{\text{Dissociated ions}}{H^+ + Cl^- + K^+ + OH^-} \longrightarrow \underset{\text{Salt}}{KCl} + \underset{\text{Water}}{H_2O}$$

The Body's Chemical Reactions Are Sensitive to Changes in pH

To ensure homeostasis, intracellular and extracellular fluids must contain almost balanced quantities of acids and bases. The more hydrogen ions (H^+) dissolved in a solution, the more acidic the solution; the more hydroxide ions (OH^-), the more basic (alkaline) the solution. The chemical reactions that take place in the body are very sensitive to even small changes in the acidity or alkalinity of the body fluids in which they occur. Any departure from the narrow limits of normal H^+ and OH^- concentrations greatly disrupts body functions.

A solution's acidity or alkalinity is expressed on the **pH scale**, which extends from 0 to 14 (FIGURE 2.11). The **pH** of a solution is defined as the negative logarithm of the hydrogen ion concentration in moles per liter. This relationship is expressed by the following equation, using brackets [ ] to indicate the hydrogen ion concentration:

$$pH = -\log [H^+]$$

Therefore, a solution that has a H^+ concentration of 10^{-4} mol/L has a pH of 4. A solution with a H^+ concentration of 10^{-7} mol/L has a pH of 7; a solution with a H^+ concentration of 10^{-9} mol/L has a pH of 9; and so on. Because the pH scale is logarithmic, each change in one pH unit represents a *tenfold* change in the number of hydrogen ions (H^+). A pH of 6 denotes 10 times more H^+ than a pH of 7, and a pH of 8 indicates 10 times fewer H^+ than a pH of 7 and 100 times fewer H^+ than a pH of 6.

The midpoint of the pH scale is 7, where the concentrations of H^+ and OH^- are equal. A substance with a pH of 7, such as pure water, is neutral. A solution that has more H^+ than OH^- is an **acidic solution** and has a pH below 7. A solution that has more OH^- than H^+ is a **basic (alkaline) solution** and has a pH above 7.

Buffer Systems Convert Strong Acids or Bases into Weak Acids or Bases

Although the pH of body fluids may differ, the normal limits for each fluid are quite narrow. TABLE 2.3 shows the pH values for certain body fluids along with those of some common substances outside the body. Homeostatic mechanisms maintain the pH of blood between 7.35 and 7.45 (average = 7.4), which is slightly more basic than pure water. You will learn in Chapter 20 that if the pH of blood falls below 7.35, a condition called *acidosis* occurs, and if the pH rises above 7.45, it results in a condition called *alkalosis*. Both conditions can seriously compromise homeostasis. Saliva is slightly acidic, and semen is slightly basic. Because the kidneys help remove excess acid from the body, urine can be quite acidic.

Even though strong acids and bases are continually taken into and formed by the body, the pH of fluids inside and outside cells remains almost constant. One important reason is the presence of **buffer systems**, which convert strong acids or bases into weak acids or bases. Strong acids

FIGURE 2.11 The pH scale. A pH below 7 indicates an acidic solution—more H^+ than OH^-. $[H^+]$ = hydrogen ion concentration; $[OH^-]$ = hydroxide ion concentration.

The lower the numerical value of the pH, the more acidic is the solution because the H^+ concentration becomes progressively greater. A pH above 7 indicates a basic (alkaline) solution; that is, there are more OH^- than H^+. The higher the pH, the more basic is the solution.

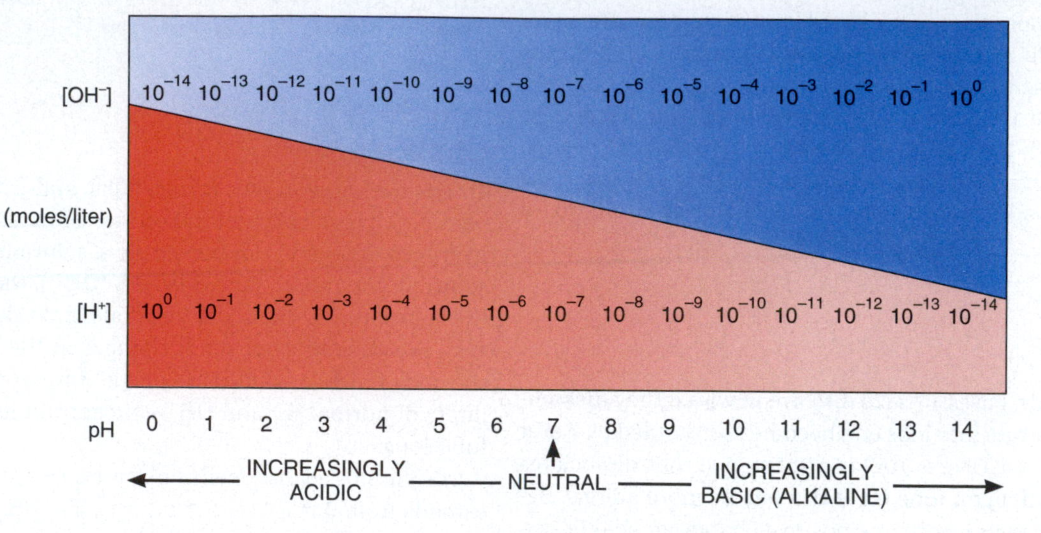

At pH 7 (neutrality), the concentration of H^+ and OH^- are equal (10^{-7} mol/liter). What are the concentrations of H^+ and OH^- at pH 6? Which pH is more acidic, 6.82 or 6.91? Which pH is closer to neutral, 8.41 or 5.59?

TABLE 2.3 pH Values of Selected Substances

Substance	pH Value
Gastric juice (found in the stomach)*	1.2–3.0
Lemon juice	2.3
Vinegar	3.0
Carbonated soft drink	3.0–3.5
Orange juice	3.5
Vaginal fluid*	3.5–4.5
Tomato juice	4.2
Coffee	5.0
Urine*	4.6–8.0
Saliva*	6.35–6.85
Milk	6.8
Distilled (pure) water	7.0
Blood*	7.35–7.45
Semen (fluid containing sperm)*	7.20–7.60
Cerebrospinal fluid (fluid associated with nervous system)*	7.4
Pancreatic juice (digestive juice of the pancreas)*	7.1–8.2
Bile (liver secretion that aids fat digestion)*	7.6–8.6
Milk of magnesia	10.5
Lye (sodium hydroxide)	14.0

* Denotes substances in the human body.

(or bases) ionize easily and contribute many H^+ (or OH^-) to a solution. Therefore, they can change pH drastically, which can disrupt the body's metabolism. Weak acids (or bases) do not ionize as much and contribute fewer H^+ (or OH^-). Hence, they have less effect on the pH. The chemical compounds that can convert strong acids or bases into weak ones are called **buffers**. They do so by removing or adding protons (H^+).

One important buffer system in the body is the **carbonic acid–bicarbonate buffer system**. Carbonic acid (H_2CO_3) can act as a weak acid, and the bicarbonate ion (HCO_3^-) can act as a weak base. Hence, this buffer system can compensate for either an excess or a shortage of H^+. For example, if there is an excess of H^+ (an acidic condition), HCO_3^- can function as a weak base and remove the excess H^+, as follows:

$$\underset{\text{Hydrogen ion}}{H^+} + \underset{\text{Bicarbonate ion (weak base)}}{HCO_3^-} \longrightarrow \underset{\text{Carbonic acid}}{H_2CO_3}$$

By contrast, if there is a shortage of H^+ (an alkaline condition), H_2CO_3 can function as a weak acid and provide needed H^+ as follows:

$$\underset{\text{Carbonic acid (weak acid)}}{H_2CO_3} \longrightarrow \underset{\text{Hydrogen ion}}{H^+} + \underset{\text{Bicarbonate ion}}{HCO_3^-}$$

Chapter 20 describes buffers and their roles in maintaining acid–base balance in more detail.

CHECKPOINT

13. How do inorganic compounds differ from organic compounds?

14. What functions does water perform in the body?

15. What are three ways to express the concentration of a solution?

16. How do bicarbonate ions prevent buildup of excess H^+?

2.5 Organic Compounds

OBJECTIVES

- Describe the functional groups of organic molecules.
- Identify the building blocks and functions of carbohydrates, lipids, proteins, and nucleic acids.
- Explain the importance of adenosine triphosphate (ATP).

Inorganic compounds are relatively simple. Their molecules have only a few atoms and cannot be used by cells to perform complicated biological functions. Many organic molecules, by contrast, are relatively large and have unique characteristics that allow them to carry out complex functions. Important categories of organic compounds include carbohydrates, lipids, proteins, nucleic acids, and adenosine triphosphate (ATP).

Carbon and Its Functional Groups Provide Useful Properties to Living Organisms

Carbon has several properties that make it particularly useful to living organisms. For one thing, it can form bonds with one to thousands of other carbon atoms to produce large molecules that can have many different shapes. Due to this property of carbon, the body can build many different organic compounds, each of which has a unique structure and function. Moreover, the large size of most carbon-containing molecules and the fact that some do not dissolve easily in water make them useful materials for building body structures.

Organic compounds are usually held together by covalent bonds. Carbon has four electrons in its outermost (valence) shell. It can bond covalently with a variety of atoms, including other carbon atoms, to form rings and straight or branched chains. Other elements that most often bond with carbon in organic compounds are hydrogen, oxygen, and nitrogen. Sulfur and phosphorus are also present in organic compounds. The other elements listed in TABLE 2.1 are present in a smaller number of organic compounds.

The chain of carbon atoms in an organic molecule is called the **carbon skeleton**. Many of the carbons are bonded to hydrogen atoms, yielding a **hydrocarbon**. Also attached to the carbon skeleton are distinctive **functional groups**. Each type of functional group has a specific arrangement of atoms that confers characteristic chemical properties on the organic molecule attached to it. TABLE 2.4 lists the most common functional groups of organic molecules and describes some of their properties. Because organic molecules often are big, there are shorthand methods for representing their structural formulas. FIGURE 2.12 shows two ways to indicate the structure of the sugar glucose, a molecule with a ring-shaped carbon skeleton that has several hydroxyl groups attached.

Small organic molecules can combine into very large molecules that are called **macromolecules**. Macromolecules are usually **polymers**. A polymer is a large molecule formed by the covalent bonding of many identical or similar small building-block molecules called **monomers**. The reaction that joins two monomers is called **dehydration synthesis**. In this type of reaction, a hydrogen atom is removed from one monomer and a hydroxyl group is

TABLE 2.4 Major Functional Groups

Name and Structural Formula*	Occurrence and Significance
Hydroxyl R—O—H	*Alcohols* contain an —OH group, which is polar and hydrophilic due to its electronegative O atom. Molecules with many —OH groups dissolve easily in water.
Sulfhydryl R—S—H	*Thiols* have an —SH group, which is polar and hydrophilic due to its electronegative S atom. Certain amino acids, the building blocks of proteins, contain —SH groups, which help stabilize the shape of proteins. An example is the amino acid cysteine.
Carbonyl R—C(=O)—R or R—C(=O)—H	*Ketones* contain a carbonyl group within the carbon skeleton. The carbonyl group is polar and hydrophilic due to its electronegative O atom. *Aldehydes* have a carbonyl group at the end of the carbon skeleton.
Carboxyl R—C(=O)—OH or R—C(=O)—O^-	*Carboxylic acids* contain a carboxyl group at the end of the carbon skeleton. All amino acids have a —COOH group at one end. The negatively charged form predominates at the pH of body cells and is hydrophilic.
Ester R—C(=O)—O—R	*Esters* predominate in dietary fats and oils and also occur in our body as triglycerides. Aspirin is an ester of salicylic acid, a pain-relieving molecule found in the bark of the willow tree.
Phosphate R—O—P(=O)(—O^-)—O^-	*Phosphates* contain a phosphate group (—PO_4^{2-}), which is very hydrophilic due to the dual negative charges. An important example is adenosine triphosphate (ATP), which transfers chemical energy between organic molecules during chemical reactions.
Amino R—N(—H)—H or R—N^+(—H)(—H)—H	*Amines* have an —NH_2 group, which can act as a base and pick up a hydrogen ion, giving the amino group a positive charge. At the pH of body fluids, most amino groups have a charge of 1+. All amino acids have an amino group at one end.

* R = variable group.

FIGURE 2.12 Alternative ways to write the structural formula for glucose.

In standard shorthand, carbon atoms are understood to be at locations where two bond lines intersect, and single hydrogen atoms are not indicated.

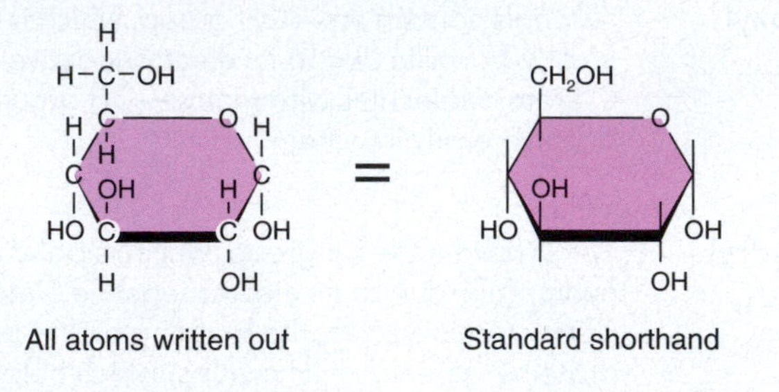

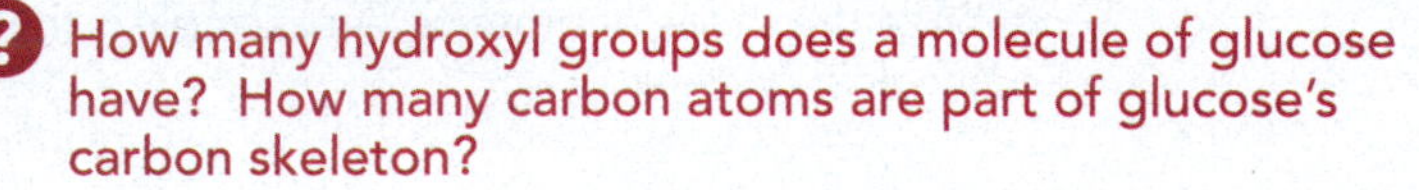

How many hydroxyl groups does a molecule of glucose have? How many carbon atoms are part of glucose's carbon skeleton?

removed from the other to form a molecule of water (FIGURE 2.13a). When two monomers are initially joined together by a dehydration synthesis reaction, a *dimer* is formed. The dimer then becomes a polymer as additional monomers are joined to it by dehydration synthesis reactions. Macromolecules such as carbohydrates, lipids, proteins, and nucleic acids are assembled in cells via dehydration synthesis reactions.

Polymers can be broken down into monomers by the addition of water molecules. This type of reaction, called **hydrolysis** (hī-DROL-i-sis; *hydro* = water; *lysis* = to break apart), is the reverse of dehydration synthesis. During hydrolysis, a water molecule is split into a hydrogen atom and a hydroxyl group, and the bond between two linked monomers is broken as the hydrogen atom is added to one monomer and the hydroxyl group is added to the other (FIGURE 2.13b). Hydrolysis reactions occur, for example, during digestion. Most of the food that we eat consists of polymers that are too large to be used by the cells of the body. The process of digestion involves hydrolysis reactions that break down the large polymers in food into monomers that can be taken up by body cells to be used for a variety of cellular activities.

Carbohydrates Are Important Sources of Chemical Energy

Carbohydrates include sugars, glycogen, starches, and cellulose. Even though they are a large and diverse group of organic compounds and have several functions, carbohydrates represent only 2–3% of your total body mass. In humans and animals, carbohydrates function mainly as a source of chemical energy for generating ATP, which is needed to drive various chemical reactions. Only a few carbohydrates are used for building structural units. One example is deoxyribose, a type of sugar that is a building block of deoxyribonucleic acid (DNA), the molecule that carries inherited genetic information.

Carbon, hydrogen, and oxygen are the elements found in carbohydrates. The ratio of hydrogen to oxygen atoms is usually 2:1, the same as in water. Although there are exceptions, carbohydrates generally contain one water molecule for each carbon atom. This is the reason they are called carbohydrates, which means "watered carbon." The three major groups of carbohydrates, based on their sizes, are monosaccharides, disaccharides, and polysaccharides (TABLE 2.5).

Monosaccharides

The monomers of carbohydrates, **monosaccharides** (mon′-ō-SAK-a-rīds), contain from three to seven carbon atoms. They are designated by names ending in "*-ose*" with a prefix that indicates the number of carbon atoms. For example, monosaccharides with three carbons are called *trioses*. There are also *tetroses* (four-carbon sugars), *pentoses* (five-carbon

FIGURE 2.13 Dehydration synthesis and hydrolysis reactions.

In a dehydration synthesis reaction, a water molecule is released as two monomers are joined; in a hydrolysis reaction, two linked monomers are split apart by the addition of a water molecule.

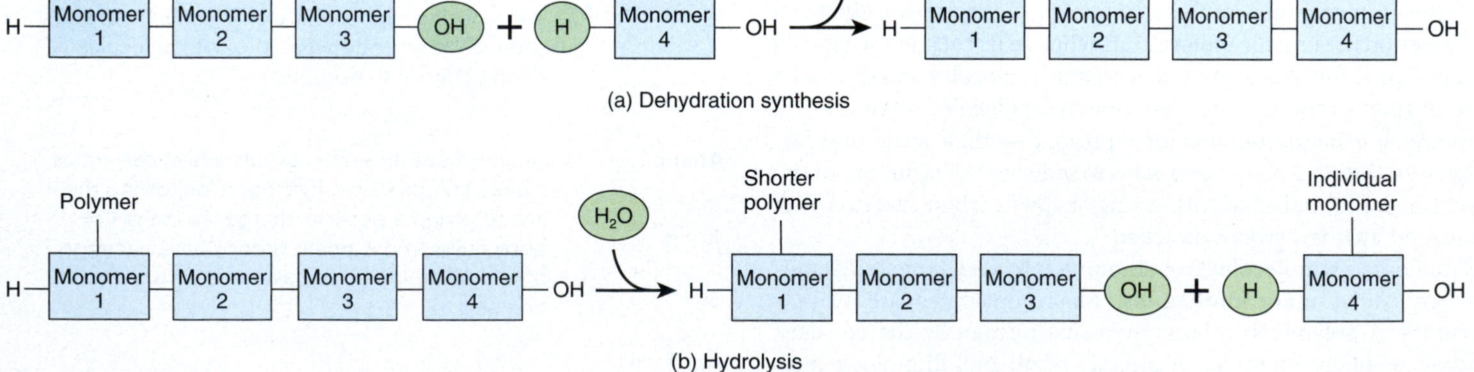

Does digestion of food involve dehydration synthesis reactions or hydrolysis reactions?

TABLE 2.5 Major Carbohydrate Groups

Type of Carbohydrate	Examples
Monosaccharides	
Carbohydrates that contain from 3 to 7 carbon atoms.	Glucose (the main blood sugar). Fructose (found in fruits). Galactose (in milk sugar). Deoxyribose (in DNA). Ribose (in RNA).
Disaccharides	
Carbohydrates formed from the combination of two monosaccharides by dehydration synthesis.	Sucrose (table sugar) = glucose + fructose. Lactose (milk sugar) = glucose + galactose. Maltose (malt sugar) = glucose + glucose.
Polysaccharides	
Carbohydrates that form from hundreds to thousands of monosaccharides joined by dehydration synthesis.	Glycogen (the stored form of carbohydrates in animals). Starch (the stored form of carbohydrate in plants and main carbohydrate in food). Cellulose (part of cell walls in plants that cannot be digested by humans but aids movement of food through intestines).

sugars), *hexoses* (six-carbon sugars), and *heptoses* (seven-carbon sugars). Examples of pentoses and hexoses are illustrated in **FIGURE 2.14**. Cells throughout the body break down the hexose glucose to produce ATP.

FIGURE 2.14 Monosaccharides. The structural formulas of selected monosaccharides are shown.

Monosaccharides are the monomers used to build larger carbohydrates.

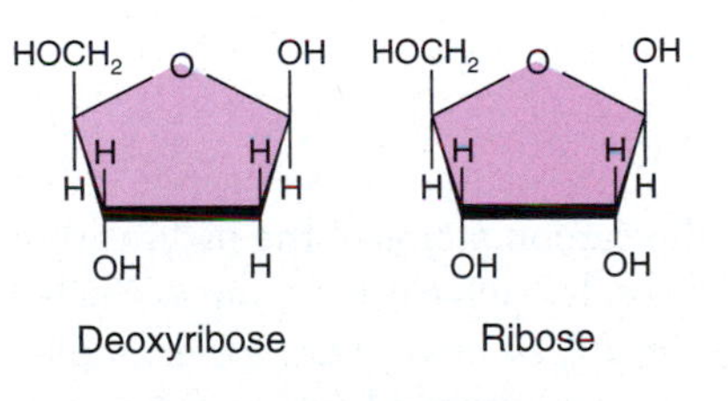

(a) Pentoses

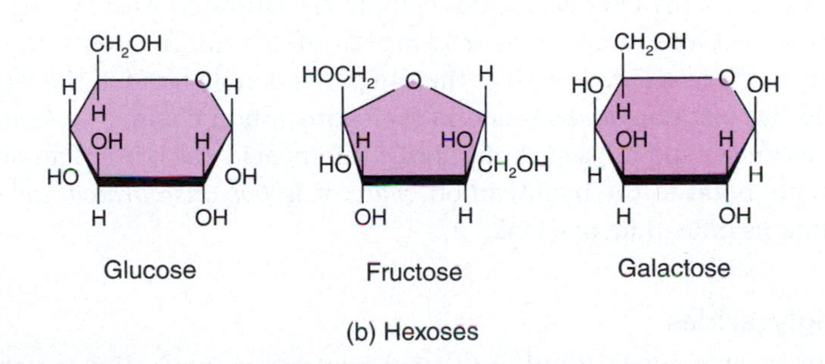

(b) Hexoses

Which of these monosaccharides are hexoses?

Disaccharides

A **disaccharide** (dī-SAK-a-rīd) is a molecule formed from the combination of two monosaccharides by dehydration synthesis. The reverse reaction, hydrolysis, splits disaccharides into monosaccharides. Examples of disaccharides are sucrose, lactose, and maltose (**FIGURE 2.15**). *Sucrose* (table sugar) consists of a glucose molecule joined to a fructose molecule; *lactose* (milk sugar) consists of a glucose molecule joined to a galactose molecule; and *maltose* (malt sugar) consists of two glucose molecules that are joined together.

Polysaccharides

The third major group of carbohydrates is the **polysaccharides** (pol′-ē-SAK-a-rīds). Each polysaccharide molecule contains hundreds to thousands of monosaccharides joined through dehydration synthesis reactions. The main polysaccharide in the human body is **glycogen**, which is made entirely of glucose monomers linked to one another in branching chains (**FIGURE 2.16**). A limited amount of carbohydrates is stored as glycogen in the liver and skeletal muscles. **Starches** are polysaccharides formed from glucose by plants. They are found in foods such as potatoes and grains. Like disaccharides, polysaccharides such as glycogen and starches can be broken down into monosaccharides through hydrolysis reactions. For example, when the blood glucose level falls, liver cells break down glycogen into glucose and release it into the blood, making it available to body cells, which break it down to synthesize ATP. **Cellulose** is a polysaccharide formed from glucose by plants that cannot be digested by humans but does provide bulk to help eliminate feces.

Lipids Contribute to Energy Storage, Membrane Structure, and Hormone Production

A second important group of organic compounds is **lipids**. Lipids make up 18–25% of body mass in lean adults. Like carbohydrates, lipids contain carbon, hydrogen, and oxygen. Unlike carbohydrates, they do not have a 2:1 ratio of hydrogen to oxygen. The proportion of electronegative oxygen atoms in lipids is usually smaller than in carbohydrates, so there are fewer polar covalent bonds. As a result, most lipids are insoluble in polar solvents such as water; they are

CLINICAL CONNECTION

Artificial Sweeteners

Some individuals use **artificial sweeteners** to limit their sugar consumption for medical reasons, while others do so to avoid calories that might result in weight gain. Examples of artificial sweeteners include aspartame (Nutrasweet® and Equal®), saccharin (Sweet'N Low®), and sucralose (Splenda®). Aspartame is 200 times sweeter than sucrose and it adds essentially no calories to the diet because only small amounts of it are used to produce a sweet taste. Saccharin is about 400 times sweeter than sucrose, and sucralose is 600 times sweeter than sucrose. Both saccharin and sucralose have zero calories because they pass through the body without being metabolized. Artificial sweeteners are also used as sugar substitutes because they do not cause tooth decay. In fact, studies have shown that using artificial sweeteners in the diet helps reduce the incidence of dental cavities.

FIGURE 2.15 Disaccharides. (a) The structural and molecular formulas for the monosaccharides glucose and fructose and the disaccharide sucrose. In dehydration synthesis, two smaller molecules, glucose and fructose, are joined to form a larger molecule of sucrose. Note the loss of a water molecule. In hydrolysis, the addition of a water molecule to the larger sucrose molecule breaks the disaccharide into two smaller molecules, glucose and fructose. Shown in (b) and (c) are the structural formulas of the disaccharides lactose and maltose, respectively.

A disaccharide consists of two monosaccharides that have combined by dehydration synthesis.

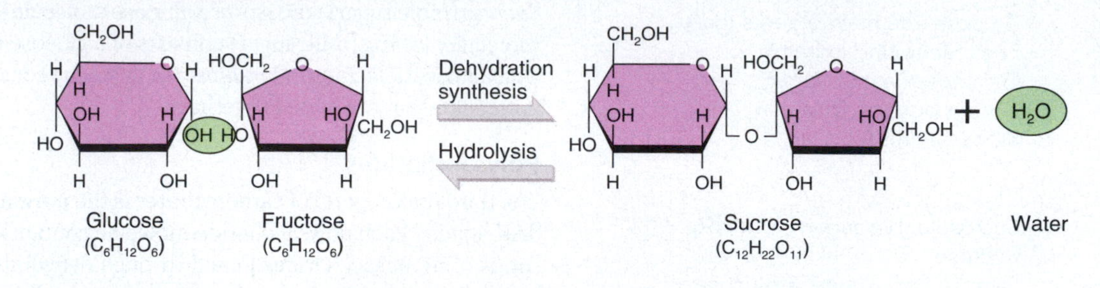

(a) Dehydration synthesis and hydrolysis of sucrose

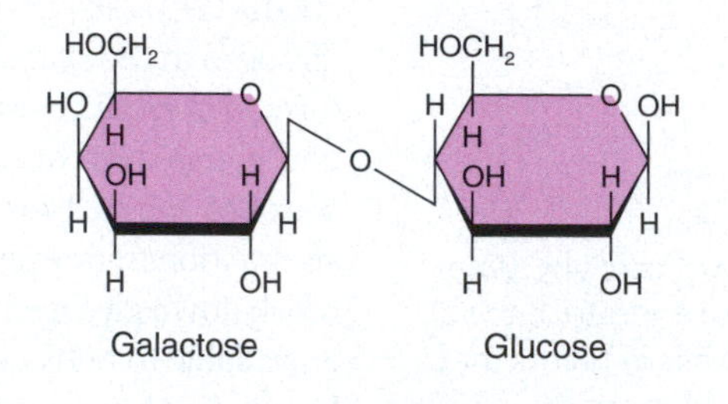

(b) Lactose

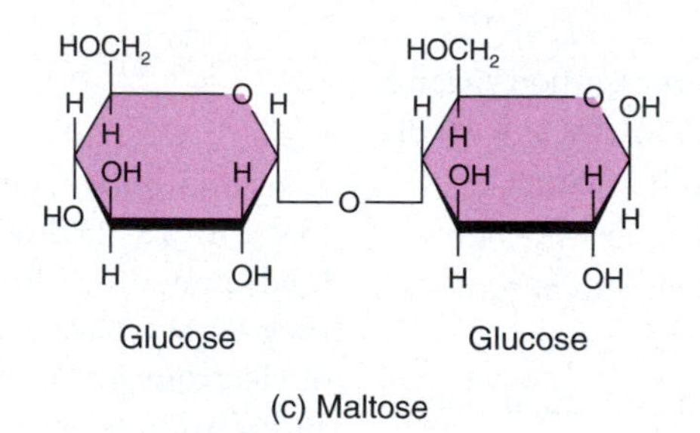

(c) Maltose

Which two types of monosaccharides are joined together to form lactose?

hydrophobic. Because they are hydrophobic, only the smallest lipids (some fatty acids) can dissolve in watery blood plasma. To become more soluble in blood plasma, most lipid molecules join with hydrophilic protein molecules. The resulting lipid/protein complexes are termed **lipoproteins** (see Section 20.9). Lipoproteins are soluble because the proteins are on the outside of the molecule and the lipids are on the inside.

The diverse lipid family includes fatty acids, triglycerides, phospholipids, steroids, and eicosanoids. TABLE 2.6 introduces the various types of lipids and highlights their roles in the human body.

Fatty Acids

Among the simplest lipids are the **fatty acids,** which are used to synthesize triglycerides and phospholipids. Fatty acids can also be broken down to generate ATP. A fatty acid consists of a carboxyl group and a hydrocarbon chain (FIGURE 2.17a). Fatty acids can be either saturated or unsaturated. A **saturated fatty acid** contains only *single covalent bonds* between the carbon atoms of the hydrocarbon chain. Because they lack double bonds, each carbon atom of the hydrocarbon chain is *saturated with hydrogen atoms* (see, for example, palmitic acid in FIGURE 2.17a). An **unsaturated fatty acid** contains one or more *double covalent bonds* between the carbon atoms of the hydrocarbon chain. Thus, the fatty acid is not completely saturated with hydrogen atoms (see, for example, oleic acid in FIGURE 2.17a). The unsaturated fatty acid has a kink (bend) at the site of the double bond. If the fatty acid has just one double bond in the hydrocarbon chain, it is *monounsaturated* and it has just one kink. If a fatty acid has more than one double bond in the hydrocarbon chain, it is *polyunsaturated* and it contains more than one kink.

Triglycerides

The most plentiful lipids in your body and in your diet are the **triglycerides** (trī-GLI-cer-īdes), also known as *triacylglycerols.* A triglyceride consists of two types of building blocks: a single glycerol

FIGURE 2.16 Part of a glycogen molecule, the main polysaccharide in the human body.

Glycogen is made up of glucose monomers and is the stored form of carbohydrate in the human body.

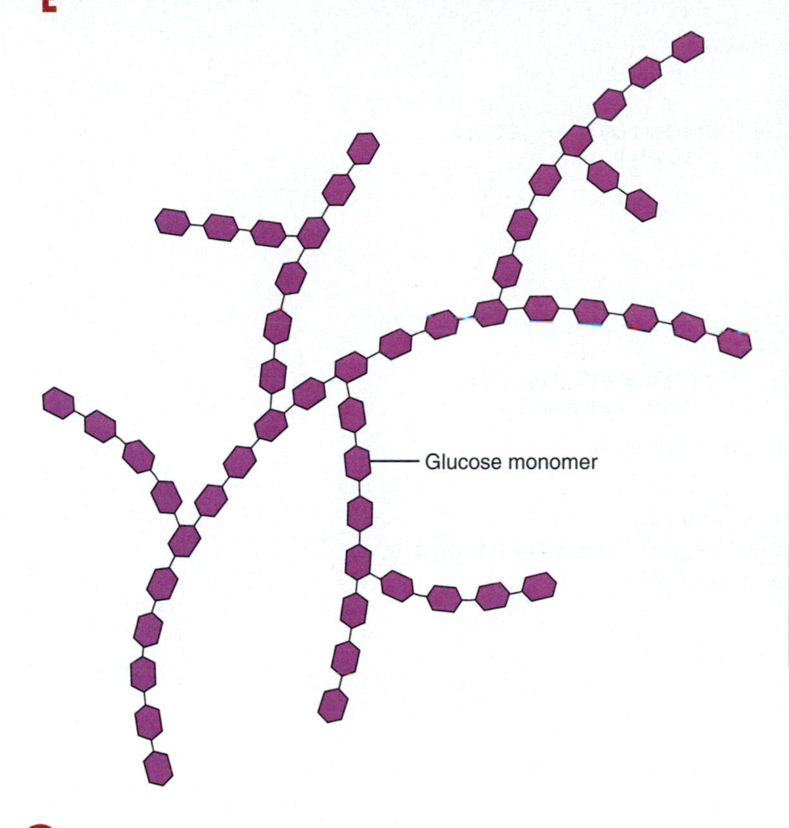

Which body cells store glycogen?

CLINICAL CONNECTION

Cis- and *Trans-*Fatty Acids

If you look closely at **FIGURE 2.17a**, you will notice that the hydrogen atoms on either side of the double bond in oleic acid are on the same side of the unsaturated fatty acid. Such an unsaturated fatty acid is called a ***cis*-fatty** acid. *Cis*-fatty acids are nutritionally beneficial, unsaturated fatty acids that are used by the body to produce hormone-like regulators and cell membranes. However, when ***cis***-fatty acids are heated, pressurized, and combined with a catalyst (usually nickel) in a process called ***hydrogenation***, they are changed to unhealthy ***trans***-fatty acids. In ***trans***-fatty acids, the hydrogen atoms are on opposite sides of the double bond of an unsaturated fatty acid. Hydrogenation is used by manufacturers to make vegetable oils solid at room temperature and less likely to turn rancid. Hydrogenated or ***trans***-fatty acids are common in commercially baked goods (crackers, cakes, and cookies), salty snack foods, some margarines, and fried foods (donuts and french fries). When oil used for frying is reused (as in fast-food french-fry machines) ***cis***-fatty acids are converted to ***trans***-fatty acids. If a product label contains the words "hydrogenated" or "partially hydrogenated," then the product contains ***trans***-fatty acids. Among the adverse effects of ***trans***-fatty acids are an increase in total cholesterol, a decrease in high-density lipoproteins (HDLs), an increase in low-density lipoproteins (LDLs), and an increase in triglycerides. These effects, which can increase the risk of heart disease and other cardiovascular diseases, are similar to those caused by saturated fats.

TABLE 2.6 Types of Lipids in the Body

Type of Lipid	Functions
Fatty acids	Used to synthesize triglycerides and phospholipids; can be broken down to generate adenosine triphosphate (ATP).
Triglycerides	Protection, insulation, energy storage.
Phospholipids	Major lipid component of cell membranes.
Steroids	
Cholesterol	Minor component of all animal cell membranes; precursor of bile salts, vitamin D, and steroid hormones.
Bile salts	Needed for digestion and absorption of dietary lipids.
Vitamin D	Helps regulate calcium level in the body; needed for bone growth and repair.
Adrenocortical hormones	Help regulate metabolism, resistance to stress, and salt and water balance.
Sex hormones	Stimulate reproductive functions and sexual characteristics.
Eicosanoids	Have diverse effects: modify responses to hormones; contribute to inflammatory and allergic responses; dilate airways to the lungs; constrict blood vessels; promote platelet activation.

FIGURE 2.17 Fatty acid structure and triglyceride synthesis. Shown in (a) are the structures of a saturated fatty acid and an unsaturated fatty acid. Each time a glycerol and a fatty acid are joined in dehydration synthesis (b), a molecule of water is removed. An ester linkage joins the glycerol to each of the three molecules of fatty acids, which vary in length and in the number and location of double bonds between carbon atoms (C=C). Shown in (c) is a triglyceride molecule that contains two saturated fatty acids and a monounsaturated fatty acid. The kink (bend) in the oleic acid occurs at the double bond.

One glycerol and three fatty acids are the building blocks of triglycerides.

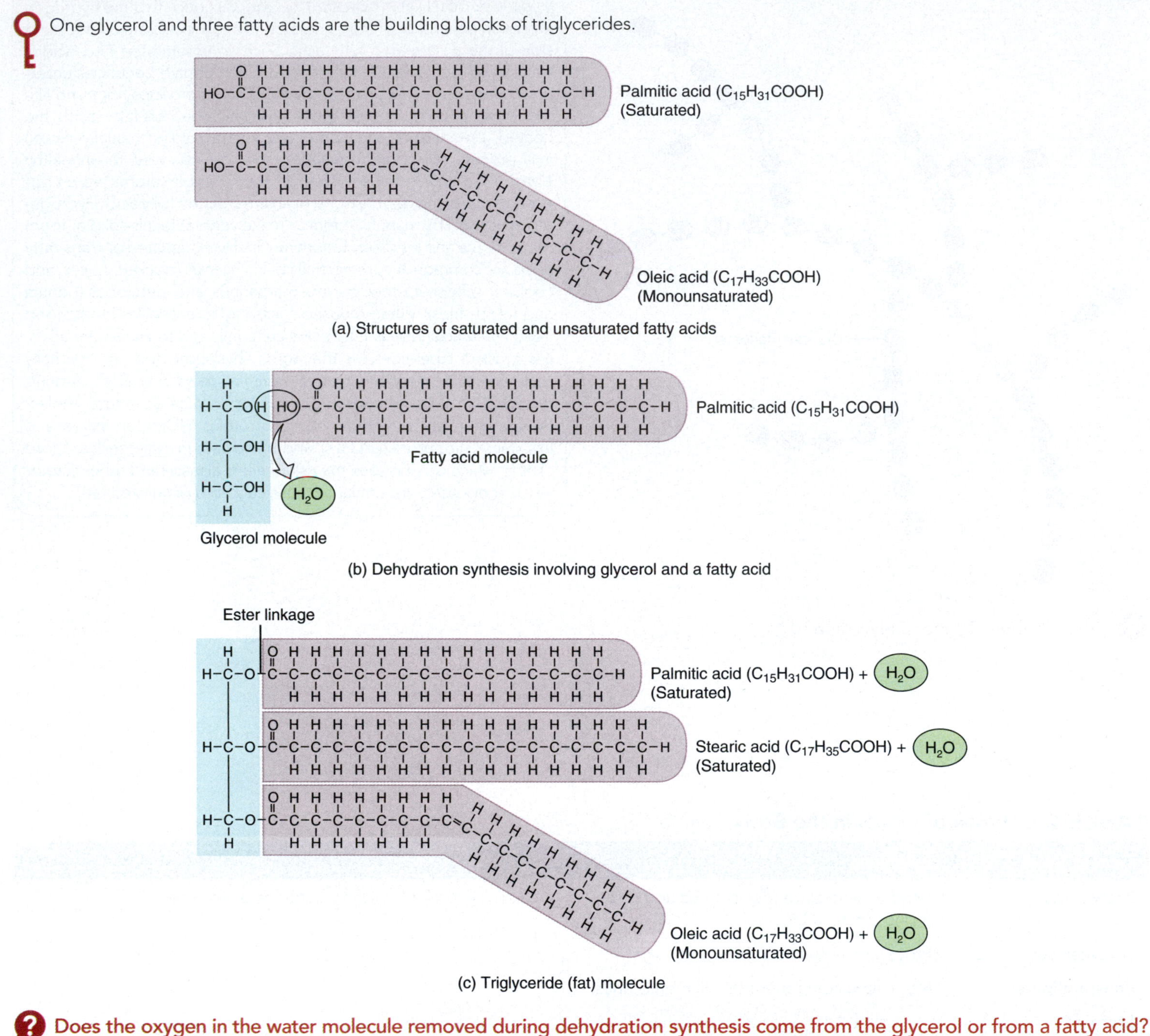

Does the oxygen in the water molecule removed during dehydration synthesis come from the glycerol or from a fatty acid?

molecule and three fatty acid molecules. A three-carbon **glycerol** molecule forms the backbone of a triglyceride (FIGURE 2.17b,c). Three fatty acids are attached by dehydration synthesis reactions, one to each carbon of the glycerol backbone. The chemical bond formed where each water molecule is removed is an *ester linkage* (see TABLE 2.4 and FIGURE 2.17c). The reverse reaction, hydrolysis, breaks down a single molecule of a triglyceride into three fatty acids and glycerol.

Triglycerides can be solids or liquids at room temperature. A **fat** is a triglyceride that is a solid at room temperature. The fatty acids of a fat are mostly saturated. Because these saturated fatty acids lack double bonds in their hydrocarbon chains, they can closely pack together and solidify at room temperature. A fat that mainly consists of saturated fatty acids is called a **saturated fat**. Saturated fats are present in lard, butter, cheese, and most animal meats (especially red meats). Diets that contain large amounts of saturated

FIGURE 2.18 Phospholipids. (a) In the synthesis of phospholipids, two fatty acids attach to the first two carbons of the glycerol backbone. A phosphate group links a small charged group to the third carbon in glycerol. In (b), the circle represents the polar head region, and the two wavy lines represent the two nonpolar tails. Double bonds in the fatty acid hydrocarbon chain often form kinks in the tail.

Phospholipids are amphipathic molecules, having both polar and nonpolar regions.

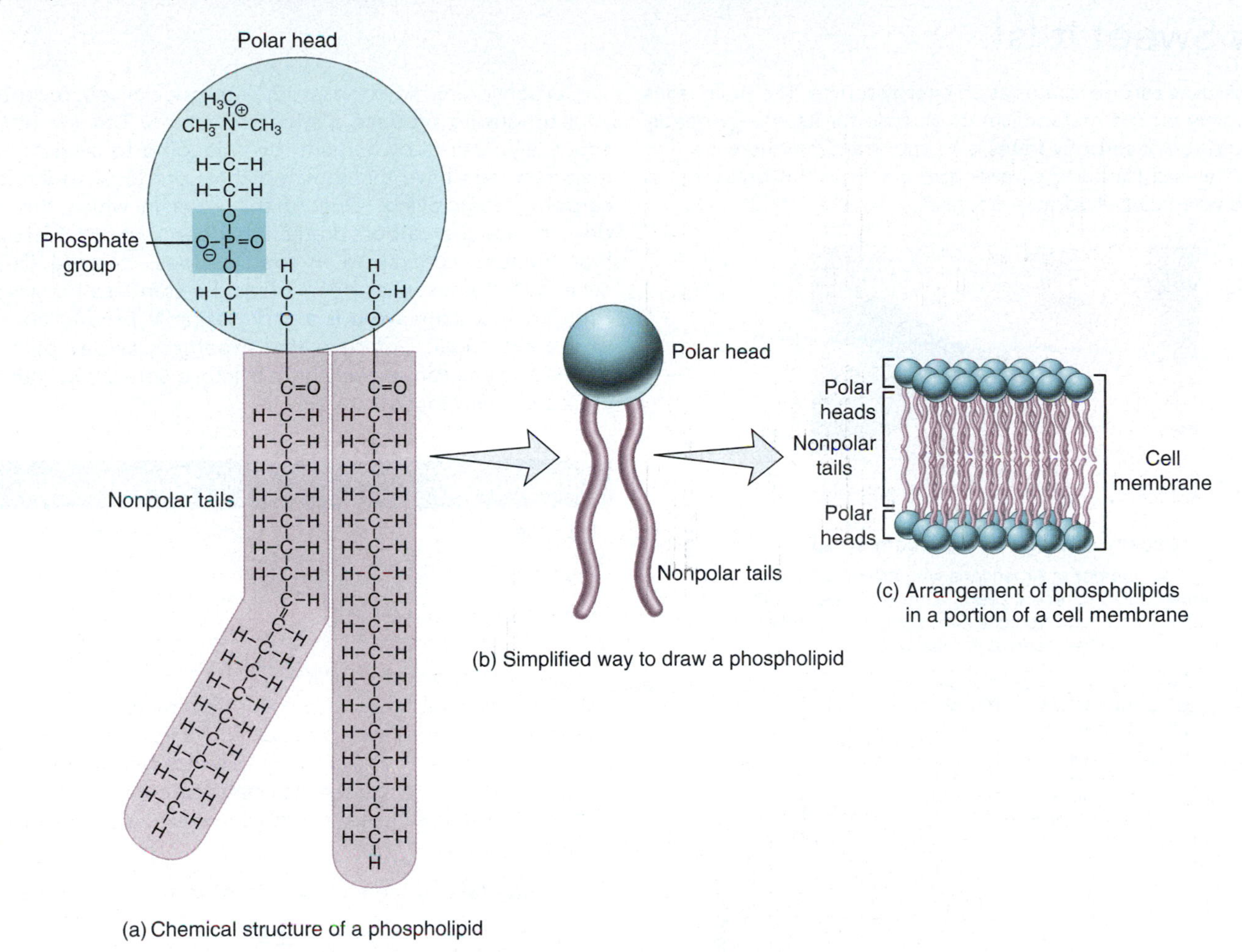

(a) Chemical structure of a phospholipid

(b) Simplified way to draw a phospholipid

(c) Arrangement of phospholipids in a portion of a cell membrane

Which portion of a phospholipid is polar, and which portion is nonpolar?

fats are associated with disorders such as heart disease and colorectal cancer.

An **oil** is a triglyceride that is a liquid at room temperature. The fatty acids of an oil are mostly unsaturated. Recall that unsaturated fatty acids contain one or more double bonds in their hydrocarbon chains. The kinks at the sites of the double bonds prevent the unsaturated fatty acids of an oil from closely packing together and solidifying. The fatty acids of an oil can be either monounsaturated or polyunsaturated. **Monounsaturated fats** contain triglycerides that consist mostly of monounsaturated fatty acids. Examples include olive oil, peanut oil, and canola oil. **Polyunsaturated fats** contain triglycerides that consist mostly of polyunsaturated fatty acids. Corn oil, sunflower oil, and fish oils are a few examples. Both monounsaturated and polyunsaturated fats are believed to decrease the risk of heart disease.

Triglycerides are the body's most highly concentrated form of chemical energy, providing more than twice as much energy per gram as carbohydrates and proteins. Our capacity to store triglycerides in adipose (fat) tissue is unlimited for all practical purposes. Excess dietary carbohydrates, proteins, and triglycerides all have the same fate: They are deposited in adipose tissue as triglycerides. Stored triglycerides not only serve as an energy source, but they also insulate and provide protection to organs and tissues of the body.

Phospholipids

Like triglycerides, **phospholipids** have a glycerol backbone and two fatty acid chains attached to the first two carbons. In the third position, however, a phosphate group (PO_4^{3-}) links a small charged group that usually contains nitrogen (N) to the backbone (FIGURE 2.18). This portion of the molecule (the "head") is polar and can form hydrogen bonds with water molecules. The two fatty acids (the "tails"), by contrast, are nonpolar and can interact only with other lipids. Molecules that have both polar and nonpolar parts are said to be **amphipathic** (am-fē-PATH-ic; *amphi-* = on both sides; *-pathic-* = feeling). Amphipathic phospholipids line up tail to tail in a double row to make up much of the membrane that surrounds each cell (FIGURE 2.18c).

Critical Thinking

How Sweet It Is!

Your body uses carbohydrates as an energy source. The brain relies almost solely on the metabolism of glucose for its energy needs. The acquisition of carbohydrates is so important that there is a biochemical reward (dopamine response) circuit in the brain that is activated when carbohydrates are eaten.

monkeybusinessimages/Getty Images

Obesity and high-fructose corn syrup

The number of Americans who are obese has quadrupled in recent years, a study shows. At the same time, high-fructose corn syrup consumption has risen at parallel rates.

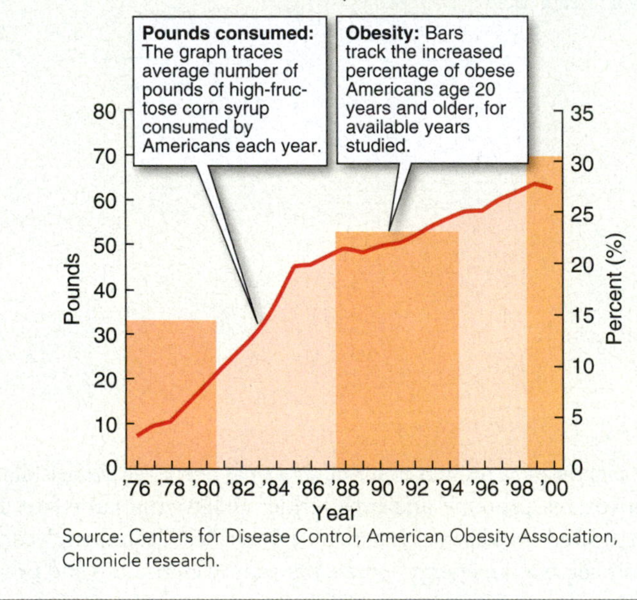

Source: Centers for Disease Control, American Obesity Association, Chronicle research.

Carbohydrates in foods stimulate specialized receptors on your tongue to produce a sweet sensation. This sweet taste is especially desired by humans, even leading to addiction. Food manufacturers have manipulated their products to exploit this carbohydrate craving. One of the ways in which this natural drive to acquire carbohydrates has been exploited is by using high-fructose corn syrup in food recipes. Sucrose (50% glucose; 50% fructose) has been set as the standard for sweetness. High-fructose corn syrup is a very desirable product for people who crave sugar. Compare the sweetness values of carbohydrates and you can see why high-fructose corn syrup is the go-to choice of many manufacturers.

Carbohydrate	Sweetness Value
Glucose	75
Fructose	140
High-fructose corn syrup	145

Soft-drink manufacturers have used high-fructose corn syrup as a sweetener in many of their products for years. In 1955, the average size of a soda was 7 ounces; today 12 ounces is considered a small drink with 16-, 26-, and 32-ounce servings being commonly available. Parallel to this increase in soft-drink sizes is an increase in U.S. adult body weight.

Why does intake of sugar cause addiction in some people?

Why does obesity follow the increase in the use of high-fructose corn syrup in foods like soft drinks?

What can be done to reverse the trend of obesity and overconsumption of sugary drinks, especially those sweetened with high-fructose corn syrup?

Steroids

Steroids are lipids that contain four interconnected hydrocarbon rings (colored gold in **FIGURE 2.19**). The various types of steroids differ structurally from one another based on the functional groups attached to the hydrocarbon rings and the location and number of the double bonds in the rings. Examples of steroids include cholesterol, estrogens, testosterone, cortisol, bile salts, and vitamin D. Steroids have important roles in the body. Cholesterol is needed for cell membrane structure; estrogens and testosterone are required for regulating sexual functions; cortisol is necessary for maintaining normal blood sugar levels; bile salts are needed for lipid digestion and absorption; and vitamin D contributes to bone growth.

Eicosanoids

Eicosanoids (ī-KŌ-sa-noids) are lipids derived from a 20-carbon fatty acid called arachidonic acid. The three principal subclasses of eicosanoids are the **prostaglandins** (pros′-ta-GLAN-dins), **thromboxanes** (throm-BOK-sānz), and **leukotrienes** (loo′-kō-TRĪ-ēnz). Prostaglandins have a wide variety of functions in the body. They modify responses to hormones, contribute to the inflammatory response, and dilate (enlarge) airways to the lungs, to name just a few. Thromboxanes constrict blood vessels and promote platelet activation. Leukotrienes participate in allergic and inflammatory responses. Because of their roles in cell signaling, eicosanoids are described in more detail in Chapter 6.

FIGURE 2.19 Steroids.

All steroids contain four interconnected hydrocarbon rings.

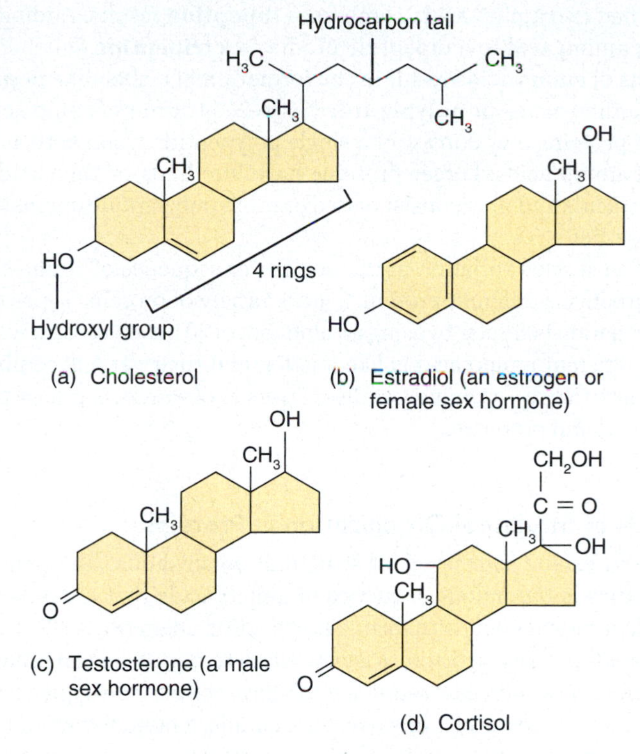

How is the structure of estradiol different from that of testosterone?

TABLE 2.7 Functions of Proteins

Type of Protein	Functions
Structural	Form structural framework of various parts of the body. *Examples:* collagen in bone and other connective tissues, and keratin in skin, hair, and fingernails.
Regulatory	Function as hormones that regulate various physiological processes; control growth and development; as neurotransmitters, mediate responses of the nervous system. *Examples:* the hormone insulin, which regulates blood glucose level, and a neurotransmitter known as substance P, which mediates sensation of pain in the nervous system.
Contractile	Allow shortening of muscle cells, which produces movement. *Examples:* myosin and actin.
Immunological	Aid responses that protect body against foreign substances and invading pathogens. *Examples:* antibodies and interleukins.
Transport	Carry vital substances throughout the body or across a membrane. *Examples:* hemoglobin, which transports most oxygen and some carbon dioxide in the blood, and the glucose transporter, which transports glucose across the plasma membrane.
Catalytic	Act as enzymes that regulate biochemical reactions. *Examples:* salivary amylase, sucrase, and ATPase.

CHECKPOINT

17. What is the difference between a dehydration synthesis reaction and a hydrolysis reaction?

18. Describe the different types of carbohydrates.

19. Distinguish among saturated, monounsaturated, and polyunsaturated fats.

20. What is the importance to the body of triglycerides, phospholipids, steroids, and eicosanoids?

Proteins Are Chains of Amino Acids That Have Diverse Roles

Proteins are large molecules that contain carbon, hydrogen, oxygen, and nitrogen. Some proteins also contain sulfur. A normal, lean adult body is 12–18% protein. Much more complex in structure than carbohydrates or lipids, proteins have many roles in the body and are largely responsible for the structure of body tissues. Some proteins are enzymes, which speed up chemical reactions. Other proteins transport substances throughout the body or across membranes. In addition, there are proteins that work as "motors" to drive muscle contraction. Antibodies are proteins that defend against invading microbes. Some hormones that regulate homeostasis are also proteins. **TABLE 2.7** describes several important functions of proteins.

Amino Acids and Polypeptides

The monomers of proteins are **amino acids** (a-MĒ-nō). Each of the 20 different amino acids has a hydrogen (H) atom and three important functional groups attached to a central carbon atom (**FIGURE 2.20a**): (1) an amino group (—NH_2), (2) a carboxyl group (—COOH), and (3) a side chain (R group). At the normal pH of body fluids, both the amino group and the carboxyl group are ionized (**FIGURE 2.20b**). The different side chains give each amino acid its distinctive chemical identity (**FIGURE 2.20c**). Amino acids are grouped into three major categories (nonpolar, polar uncharged, and polar charged) based on the properties of their side chains.

A protein is synthesized in stepwise fashion—one amino acid is joined to a second, a third is added to the first two, and so on. The covalent bond joining each pair of amino acids is a **peptide bond**. It always forms between the carbon of the carboxyl group (—COOH) of one amino acid and the nitrogen of the amino group (—NH_2) of another. As the peptide bond is formed, a molecule of water is removed

FIGURE 2.20 Amino acids.

Body proteins contain 20 different amino acids, each of which has a unique side chain.

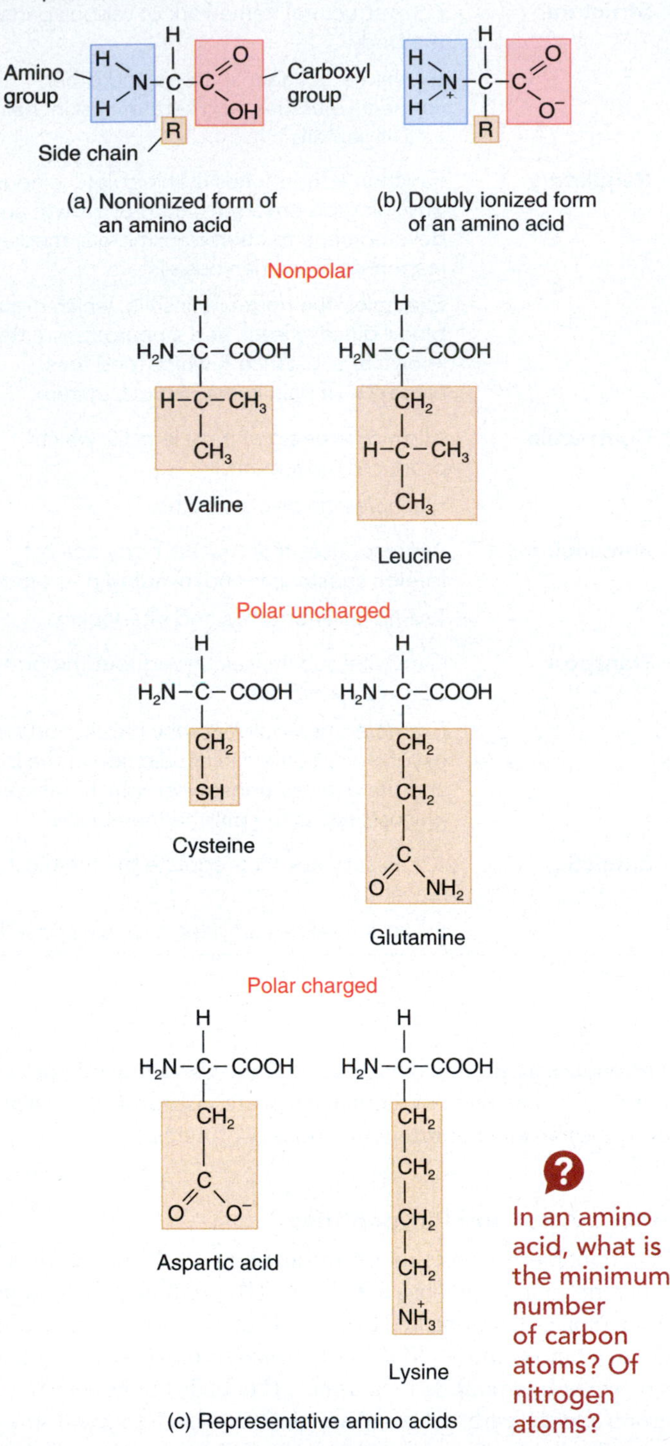

In an amino acid, what is the minimum number of carbon atoms? Of nitrogen atoms?

(FIGURE 2.21), making this a dehydration synthesis reaction. Breaking a peptide bond, as occurs during digestion of dietary proteins, is a hydrolysis reaction (FIGURE 2.21).

When two amino acids combine, a **dipeptide** results. Adding another amino acid to a dipeptide produces a **tripeptide**. Further additions of amino acids result in the formation of a chainlike **peptide** (4–9 amino acids) or **polypeptide** (10 to 2000 or more amino acids). Small proteins may consist of a single polypeptide chain with as few as 50 amino acids. Larger proteins have hundreds or thousands of amino acids and may consist of two or more polypeptide chains folded together.

Because each variation in the number or sequence of amino acids can produce a different protein, a great variety of proteins is possible. The situation is similar to using an alphabet of 20 letters to form words. Each different amino acid is like a letter, and their various combinations give rise to a seemingly endless diversity of words (peptides, polypeptides, and proteins).

Levels of Structural Organization in Proteins

Proteins exhibit four levels of structural organization. The **primary structure** is the unique sequence of amino acids that are linked by covalent peptide bonds to form a polypeptide chain (FIGURE 2.22a). A protein's primary structure is genetically determined, and any changes in a protein's amino acid sequence can have serious consequences for body cells. In **sickle-cell disease**, for example, a nonpolar amino acid (valine) replaces a polar amino acid (glutamate) through two mutations in the oxygen-carrying protein hemoglobin. This change of amino acids diminishes the water solubility of hemoglobin. As a result, the altered hemoglobin tends to form crystals inside erythrocytes (red blood cells), producing deformed, sickle-shaped cells that cannot properly squeeze through narrow blood vessels. Sickle-cell disease is described in more detail in Chapter 16 (see Section 16.2).

The **secondary structure** of a protein is the repeated twisting or folding of neighboring amino acids in the polypeptide chain (FIGURE 2.22b). Two common secondary structures are *alpha helixes* (clockwise spirals) and *beta pleated sheets*. The secondary structure of a protein is stabilized by hydrogen bonds, which form at regular intervals along the polypeptide backbone.

The **tertiary structure** (TUR-shē-er′-ē) refers to the three-dimensional shape of a polypeptide chain. Each protein has a unique tertiary structure that determines how it will function. The tertiary folding pattern may allow amino acids at opposite ends of the chain to be close neighbors (FIGURE 2.22c). Several types of bonds can contribute to a protein's tertiary structure. The strongest but least common

FIGURE 2.21 Formation of a peptide bond between two amino acids during dehydration synthesis. In this example, glycine is joined to alanine, forming a dipeptide.

Amino acids are the monomers used to build proteins.

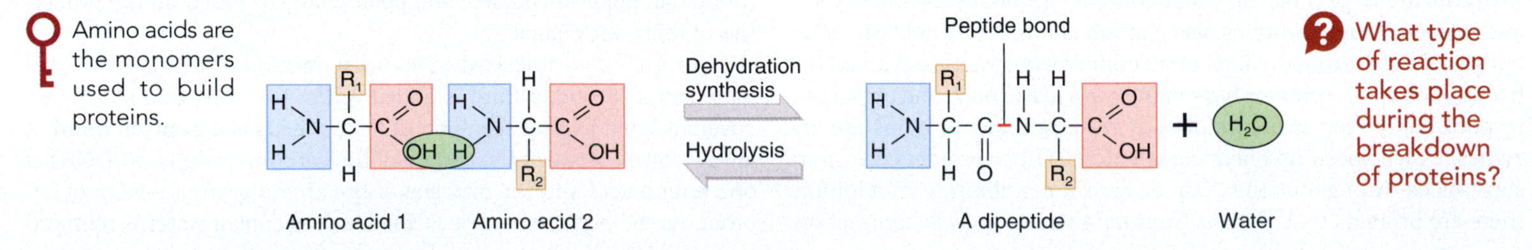

What type of reaction takes place during the breakdown of proteins?

FIGURE 2.22 Levels of structural organization in proteins. (a) The primary structure is the sequence of amino acids in the polypeptide. (b) Common secondary structures include alpha helixes and beta pleated sheets. For simplicity, the amino acid side groups are not shown here. (c) The tertiary structure is the overall folding pattern that produces a distinctive, three-dimensional shape. (d) The quaternary structure in a protein is the arrangement of two or more polypeptide chains relative to one another.

The unique shape of each protein permits it to carry out specific functions.

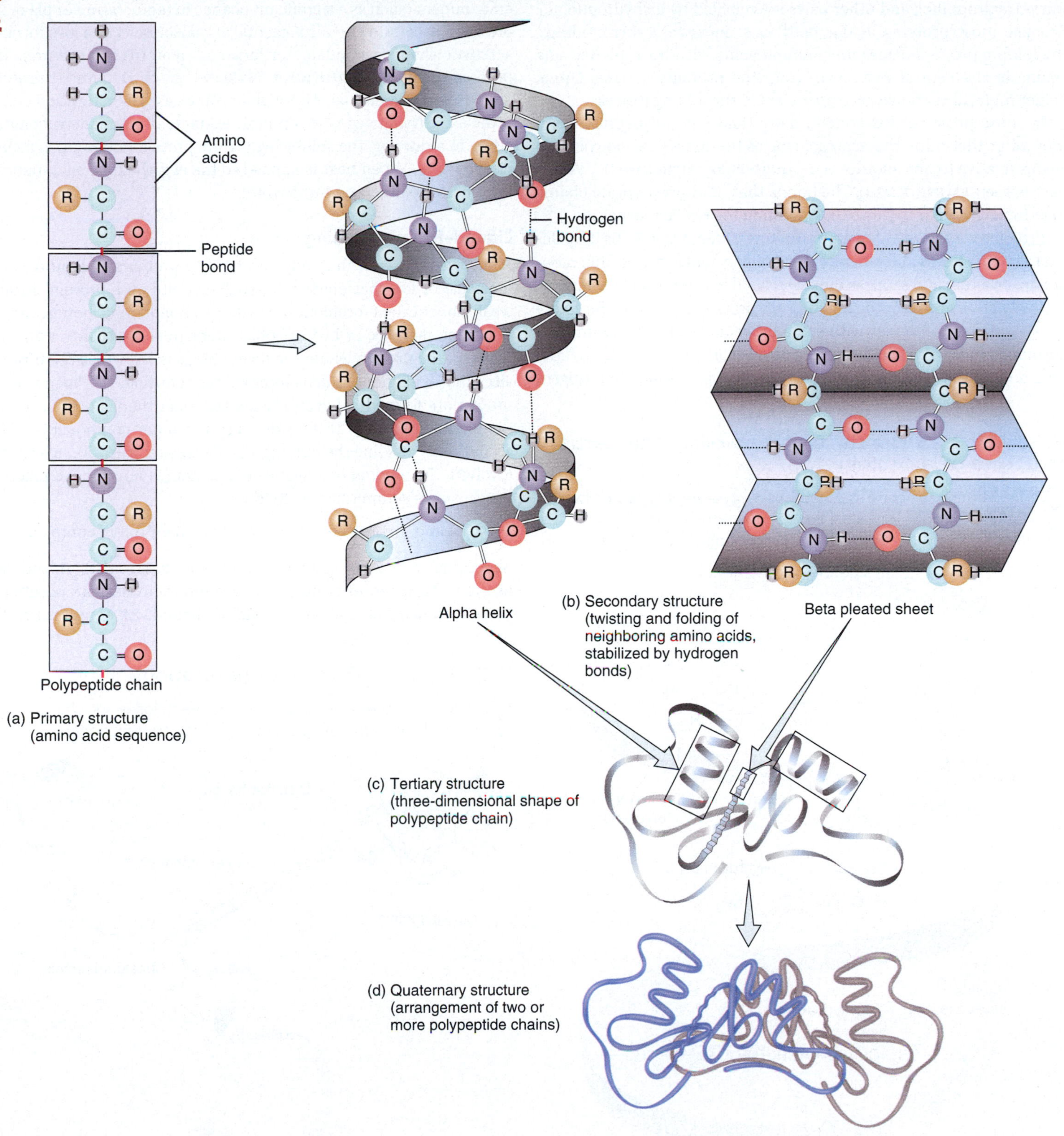

Do all proteins have a quaternary structure?

bonds, S—S covalent bonds called *disulfide bridges*, form between the sulfhydryl groups of two monomers of the amino acid cysteine. Many weak bonds—hydrogen bonds, ionic bonds, van der Waals interactions, and hydrophobic interactions—also help determine the folding pattern (**FIGURE 2.23**). Some parts of a polypeptide are attracted to water (hydrophilic), and other parts are repelled by it (hydrophobic). Because most proteins in our body exist in watery surroundings, the folding process places most amino acids with hydrophobic side chains in the central core, away from the protein's surface. Often, helper molecules known as *chaperones* aid the folding process.

In those proteins that contain more than one polypeptide chain (not all of them do), the arrangement of the individual polypeptide chains relative to one another is the **quaternary structure** (KWA-ter-ner′-ē; see **FIGURE 2.22d**). The bonds that hold polypeptide chains together are similar to those that maintain the tertiary structure.

Proteins vary tremendously in structure. Different proteins have different architectures and different three-dimensional shapes. This variation in structure and shape is directly related to their diverse functions. In practically every case, the function of a protein depends on its ability to recognize and interact with some other molecule. For example, a hormone binds to a specific protein on a cell in order to alter its function, and an antibody protein binds to a foreign substance (antigen) that has invaded the body. Thus, a protein's unique shape permits it to interact with other molecules to carry out a specific function.

Homeostatic mechanisms maintain the temperature and chemical composition of body fluids, which allow body proteins to keep their proper three-dimensional shapes. If a protein encounters an altered environment (such as a significant change in temperature or pH or the presence of certain chemical agents), it may unravel and lose its characteristic shape (secondary, tertiary, and quaternary structure). This process is called **denaturation** (**FIGURE 2.24**). Denatured proteins are no longer functional. Although in some cases denaturation can be reversed, a frying egg is a common example of permanent denaturation. In a raw egg, the soluble egg-white protein (albumin) is a clear, viscous fluid. When heat is applied to the egg, the protein denatures, becomes insoluble, and turns white.

Ligand–Protein Binding

Many substances in the body can bind to proteins. A **ligand** is any molecule or ion that binds to a particular site on a protein through weak, noncovalent interactions (hydrogen bonds, ionic bonds, van der Waals interactions, or hydrophobic interactions) (**FIGURE 2.25**). The region of the protein where the ligand binds is known as the **binding site**. The weak, noncovalent interactions that hold the ligand to the protein are formed between (1) the region of the ligand that enters the binding site and (2) the side chains or other components of the amino acids that line the binding site. Because covalent bonds are not involved, the binding of a ligand to a protein is a reversible reaction, which can be summarized as follows:

$$\text{Ligand} + \text{Protein} \rightleftharpoons \text{Ligand—Protein complex}$$

According to the *law of mass action* (see Section 2.3), an increase in the concentration of ligand or protein increases the number of protein molecules that are bound by ligand. At equilibrium, the

FIGURE 2.23 Bonds that establish the tertiary structure of a protein.

The tertiary structure is established by several types of bonds: disulfide bridges (covalent bonds), hydrogen bonds, ionic bonds, van der Waals interactions, and hydrophobic interactions.

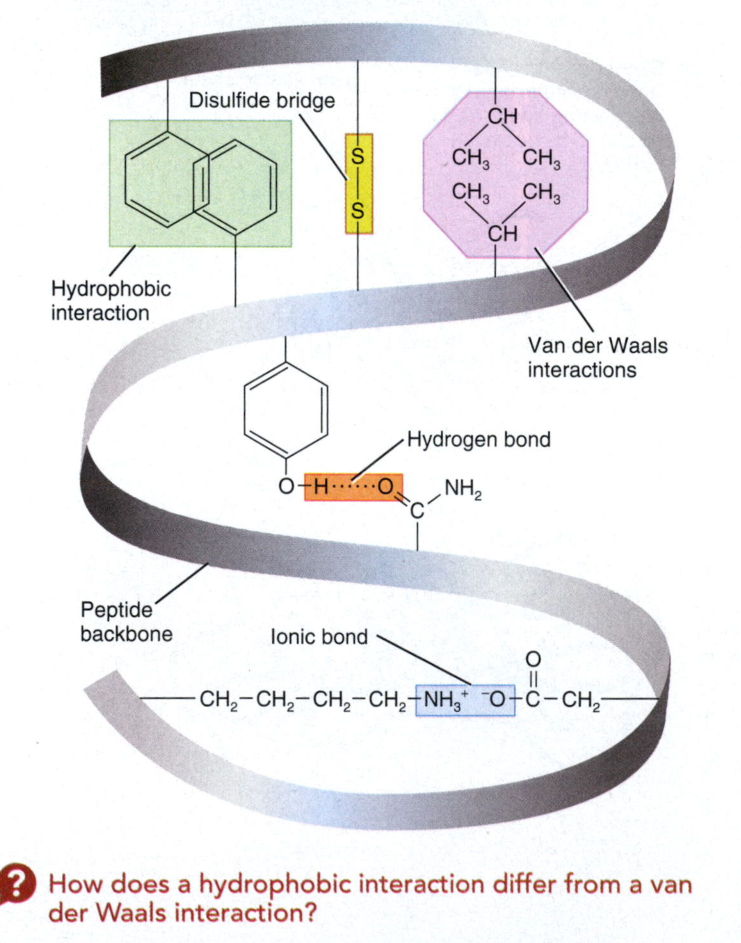

How does a hydrophobic interaction differ from a van der Waals interaction?

FIGURE 2.24 Protein denaturation.

When a protein denatures, it unravels, loses its characteristic three-dimensional shape, and becomes nonfunctional.

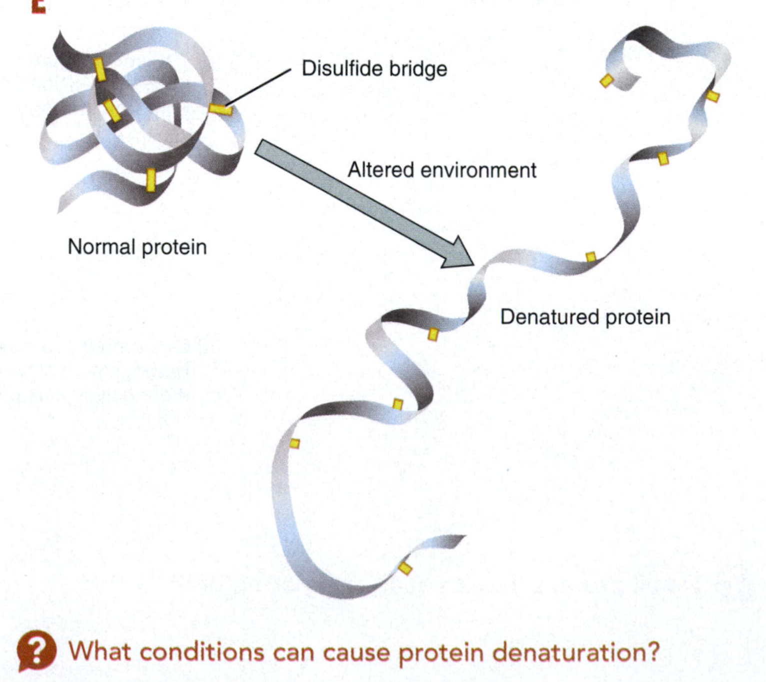

What conditions can cause protein denaturation?

FIGURE 2.25 Binding of a ligand to a protein. In this example, the ligand can bind to the protein because the region of the ligand that enters the binding site has electrical charges that form ionic bonds with charged side chains of the amino acids lining the binding site. In other cases, the ligand and protein may be held together by hydrogen bonds, van der Waals interactions, or hydrophobic interactions.

A ligand is any molecule or ion that binds to a particular site on a protein through weak, noncovalent interactions.

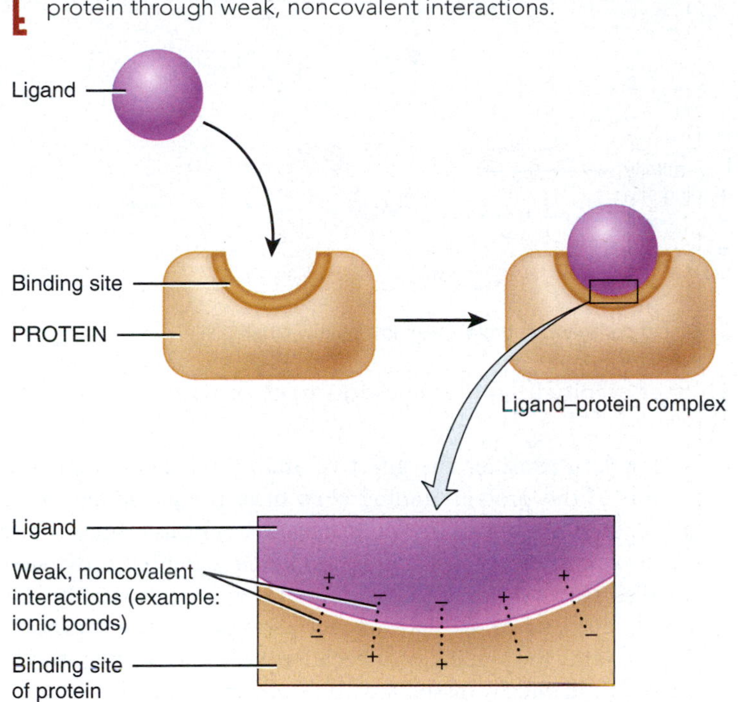

What are some examples of ligands?

rate of ligand–protein binding is equal to rate of dissociation of the ligand–protein complex.

Examples of ligands found in the body include substrates, which bind to enzymes; solutes that bind to membrane transport proteins; and hormones, which bind to receptors. Specific details about these ligands are presented in later chapters of this text (see Sections 4.3, 5.4, and 6.3, respectively). The rest of this section provides an overview of the properties and regulatory mechanisms of ligand–protein binding.

Properties of Ligand–Protein Binding The binding of a ligand to a protein exhibits four properties: specificity, affinity, saturation, and competition.

- ***Specificity.*** Only one substance or, in some cases, a small group of structurally related substances can bind to a given binding site on a protein, a property known as **specificity**. Specificity occurs because the binding site of a protein has a particular three-dimensional shape, and only a substance with a complementary shape can fit into that site. For example, FIGURE 2.26 shows a ligand trying to interact with three different proteins (proteins A, B, and C). Only protein C has a binding site that is complementary to the shape of the ligand. Therefore, the ligand will bind to protein C and not to proteins A or B.
- ***Affinity.*** The strength with which a ligand binds to a protein is a property known as **affinity**. If the binding site of a protein has a high affinity for a ligand, then the ligand binds tightly to the site and remains attached for a while before dissociating. By contrast, if the binding site of a protein has a low affinity for a ligand, then the ligand binds loosely to the site and will dissociate relatively quickly. Affinity is determined by two factors associated with the ligand and binding site: (1) the degree of shape complementarity and (2) the strength of the noncovalent interactions. The more complementary the shapes and the stronger the noncovalent interactions, the greater is the affinity. Consider the following example: Suppose that a ligand has a certain shape and a particular distribution of electrical charges. This ligand will bind tightly to a protein (protein X) that has a binding site with a perfectly complementary shape and an exact opposite distribution of electrical charges (FIGURE 2.27a). However, this same ligand binds loosely to a protein (protein Y) that has a binding site with only a moderately complementary shape and no electrical charges (FIGURE 2.27b). So the binding site of protein X has a high affinity for the ligand, whereas the binding site of protein Y has a low affinity for the ligand.
- ***Saturation.*** Another property of ligand-protein binding is **saturation**, the degree to which the binding sites of a population of protein molecules are occupied by ligand. For example, if all of the binding sites are occupied by ligand, the system is said to be fully (100%) saturated. If

FIGURE 2.26 Protein specificity for a ligand.

The binding site of a protein has a specific three-dimensional shape that will only fit a ligand with a complementary shape.

Ligand
Binding site
PROTEIN A
Binding site
PROTEIN B
Binding site
PROTEIN C

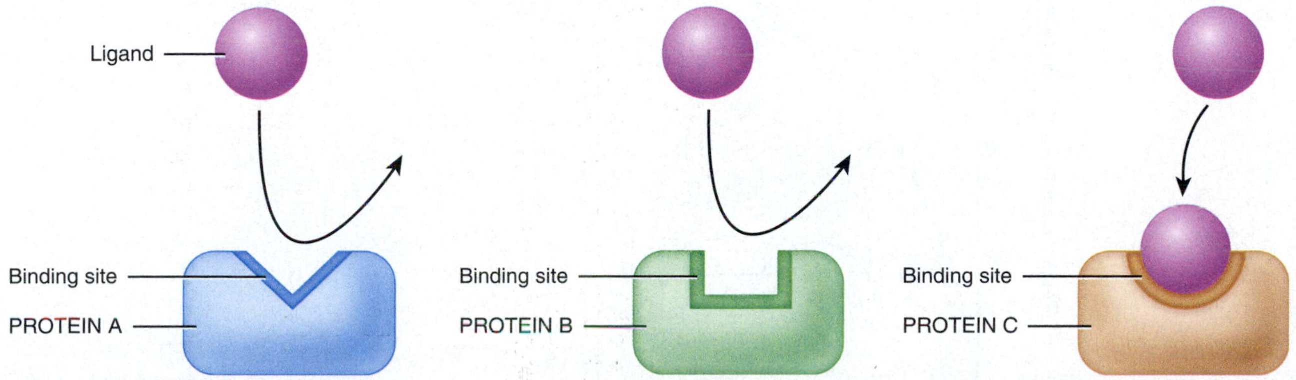

What is the shape of the ligand that most likely would bind to protein A in this figure?

FIGURE 2.27 Two proteins with different affinities for the same ligand.

Affinity refers to the strength with which a ligand binds to a protein.

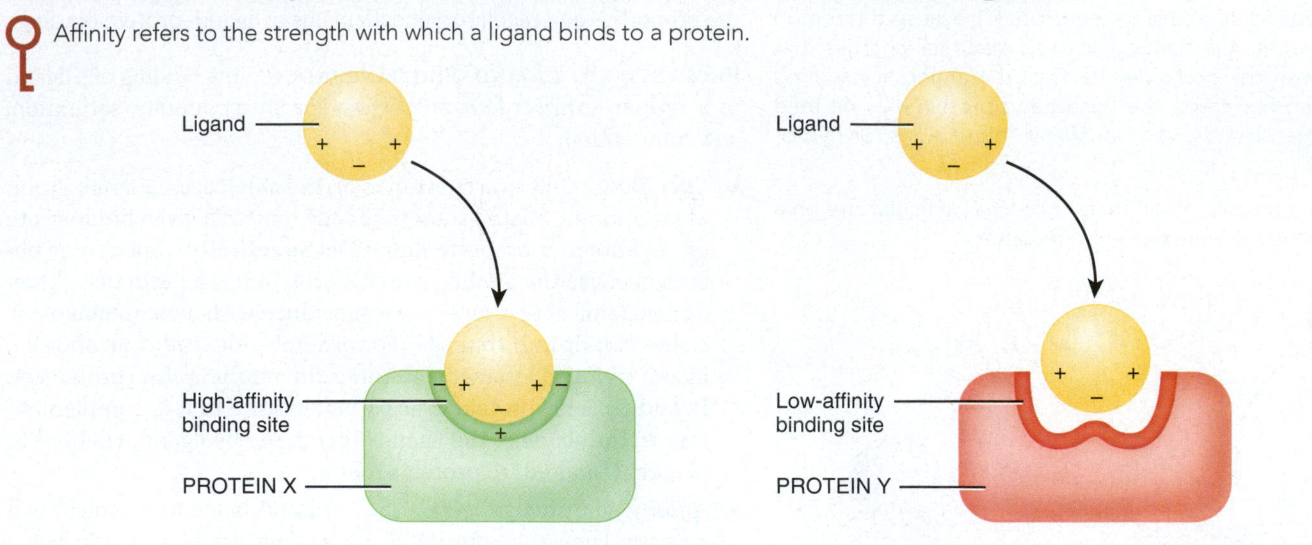

(a) Binding of ligand to protein with high-affinity binding site (b) Binding of ligand to protein with low-affinity binding site

In Figure 2.27, why does the binding site of protein X have a higher affinity than the binding site of protein Y?

only half of the available sites are occupied by ligand, the system is said to be 50% saturated, and so on. The percent saturation of binding sites depends on (1) the concentration of the ligand and (2) the affinity of the binding sites for the ligand. Let's first consider the effect of ligand concentration on the percent saturation. As the ligand concentration increases, more protein-binding sites are occupied by ligand, so the percent saturation increases (FIGURE 2.28). This relationship continues until all the binding sites are occupied by ligand and the system is 100% saturated. If the protein molecules produce a biological response after being bound by ligand, then the response also becomes greater as the percent saturation increases. However, once the system is fully saturated, the response does not become any greater despite further increases in the ligand concentration because there are no more binding sites that the ligand can occupy.

Now let's consider the effect of affinity on the percent saturation. If the protein-binding sites have a high affinity for a ligand, then a relatively low concentration of ligand can saturate the binding sites because the ligand adheres tightly to the sites and dissociates slowly. If the binding sites have a low affinity for a ligand, then a relatively high concentration of ligand will be needed to saturate the binding sites because the ligand binds loosely to the sites and quickly dissociates. As you can see in FIGURE 2.29, binding sites with a high affinity for a ligand achieve saturation at a lower ligand concentration than binding sites with a low affinity for a ligand.

- ***Competition.*** If ligands are so similar in structure that they can bind to the same site on a protein, then they may compete with one another for binding, a property known as **competition**. To

FIGURE 2.28 Effect of ligand concentration on the percent saturation of binding sites.

Saturation reflects the degree to which the binding sites of a population of protein molecules are occupied by ligand.

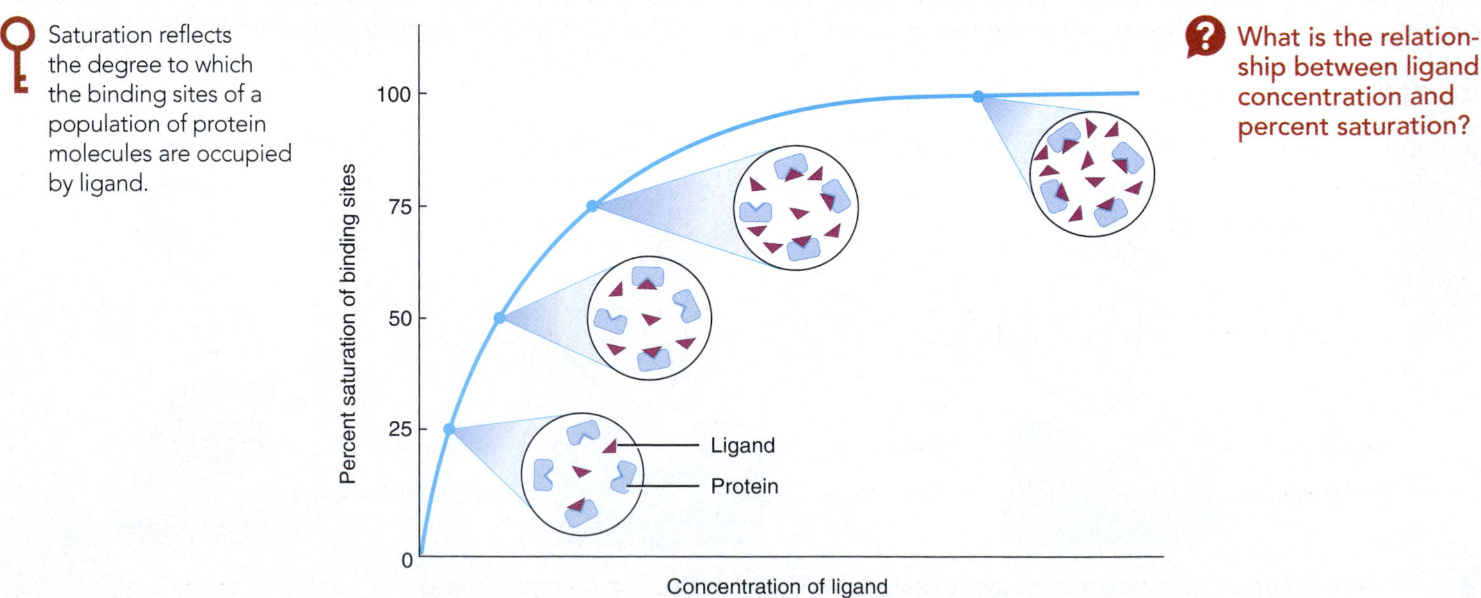

What is the relationship between ligand concentration and percent saturation?

FIGURE 2.29 Effect of affinity on the percent saturation of binding sites.

Binding sites with a high affinity for a ligand achieve saturation at a lower ligand concentration than binding sites with a low affinity for a ligand.

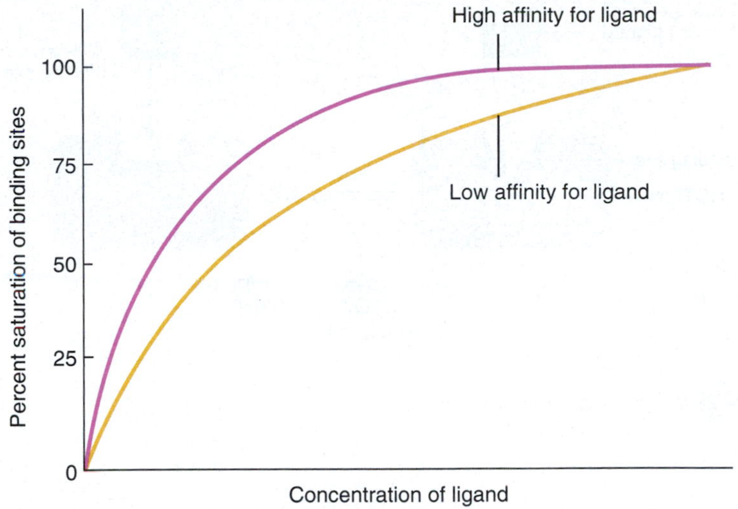

Why is a relatively high ligand concentration needed to saturate binding sites that have a low affinity for the ligand?

FIGURE 2.30 Two ligands competing for the same binding site.

Competition can occur when two ligands with similar structures have the ability to bind to the same binding site.

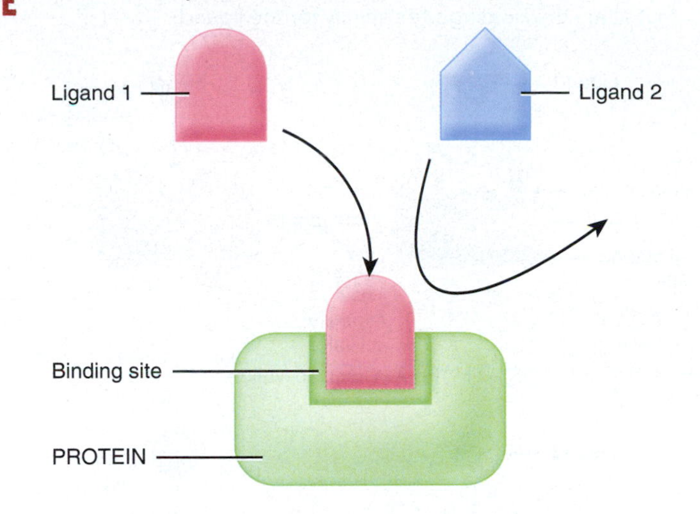

(a) Ligand 1 outcompeting ligand 2 for binding

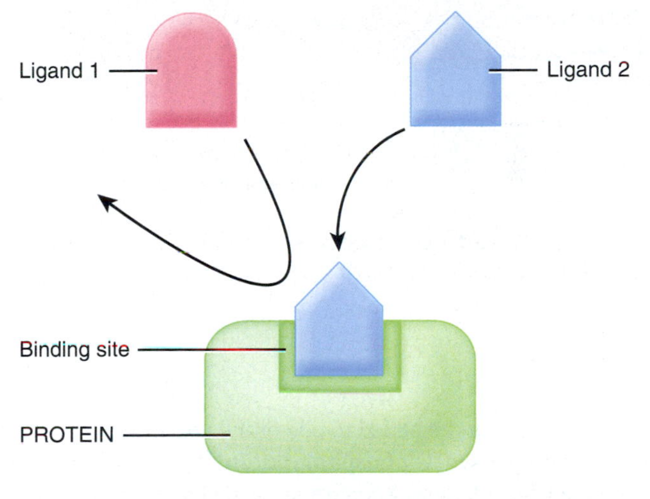

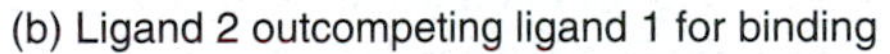

(b) Ligand 2 outcompeting ligand 1 for binding

In the figure above, what effect would increasing the concentration of ligand 1 have on its ability to compete with ligand 2 for binding?

understand how competition works, consider two structurally related ligands (ligands 1 and 2) that can occupy the same binding site on a protein. If ligand 1 makes contact with the binding site of the protein before ligand 2 does, then ligand 1 will bind to the protein (FIGURE 2.30a). By contrast, if ligand 2 interacts with the binding site before ligand 1 is able to, then ligand 2 will bind to the protein (FIGURE 2.30b). So in a solution containing many copies of the protein and both types of ligands, some of the binding sites will be bound by ligand 1 and others by ligand 2. Because both types of ligands are competing for the same binding sites, the total number of binding sites occupied by either ligand is lower than if only one type of ligand were present. Increasing the concentration of ligand 1 reduces the competitive effect of ligand 2 because more ligand 1 molecules will be available to bind to the binding sites. Alternatively, increasing the concentration of ligand 2 reduces the competitive effect of ligand 1 because a greater number of ligand 2 molecules are present for binding.

Regulation of Ligand–Protein Binding Ligand-protein binding is often subject to regulatory mechanisms. In many cases, the ability of a ligand to bind to a protein can be altered by other chemicals that interact with the protein. A **modulator** is a chemical that binds to a protein and either alters the ligand's binding ability or changes the functional activity of the protein. Two ways that ligand-protein binding can be modulated include allosteric modulation and covalent modulation.

- ***Allosteric modulation.*** In allosteric modulation, a chemical substance known as an **allosteric modulator** binds to a site called an **allosteric site** that is separate from the normal ligand binding site of the protein (FIGURE 2.31). Binding of the allosteric modulator to the allosteric site causes the protein to undergo a conformational change that alters the binding site's affinity for the ligand. An allosteric modulator may be either an activator or an inhibitor. Binding of an **allosteric activator** to an allosteric site causes a conformational change of the protein that *increases* the binding site's affinity for the ligand (FIGURE 2.31a). The increase in affinity promotes ligand binding, which enhances the protein's functional activity. By contrast, binding of an **allosteric inhibitor** to an allosteric site causes a conformational change in the protein that *decreases* the binding site's affinity for the ligand (FIGURE 2.31b). The decrease in affinity reduces ligand binding, which diminishes or inhibits the protein's activity. Because an allosteric modulator binds to the allosteric site of a protein via noncovalent bonding, the effect of allosteric modulation is reversible.
- ***Covalent modulation.*** In covalent modulation, a chemical called a **covalent modulator** binds to a protein by covalent bonding. The most common type of covalent modulator is the phosphate group

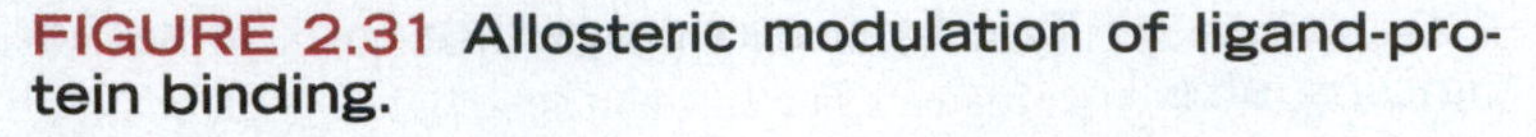

FIGURE 2.31 Allosteric modulation of ligand-protein binding.

In allosteric modulation, an allosteric activator or inhibitor binds to an allosteric site, causing a conformational change in the protein that alters the binding site's affinity for the ligand.

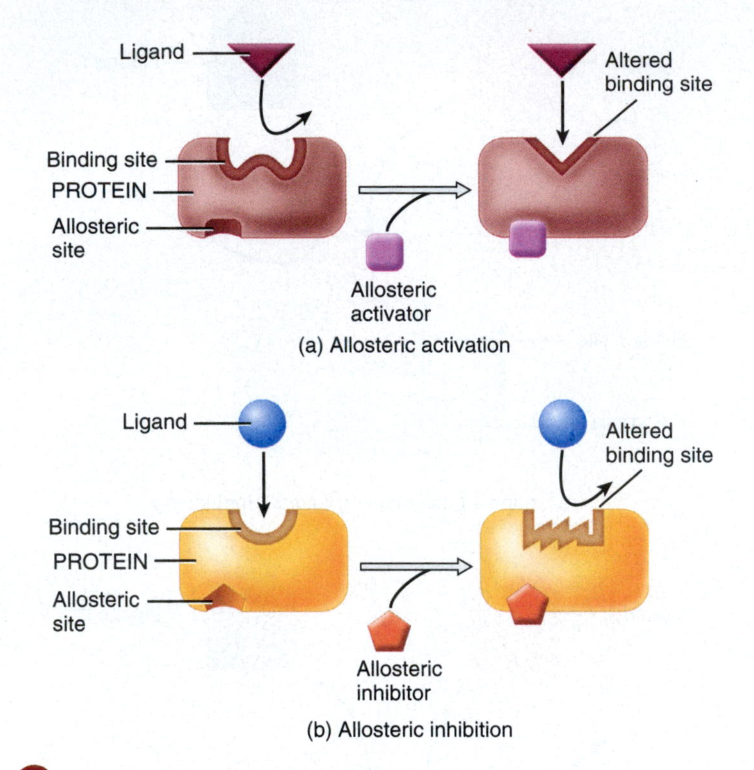

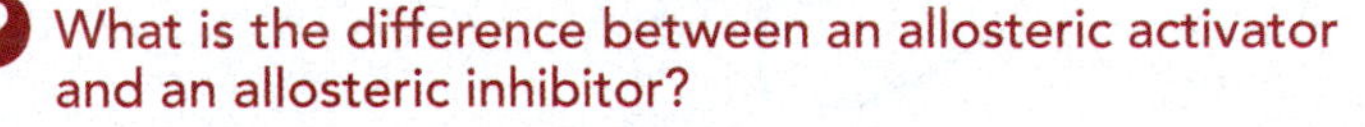

What is the difference between an allosteric activator and an allosteric inhibitor?

FIGURE 2.32 Covalent modulation of ligand-protein binding.

A covalent modulator, such as a phosphate group, can alter ligand-protein binding by attaching to the protein, causing a conformational change that alters the binding site's affinity for the ligand.

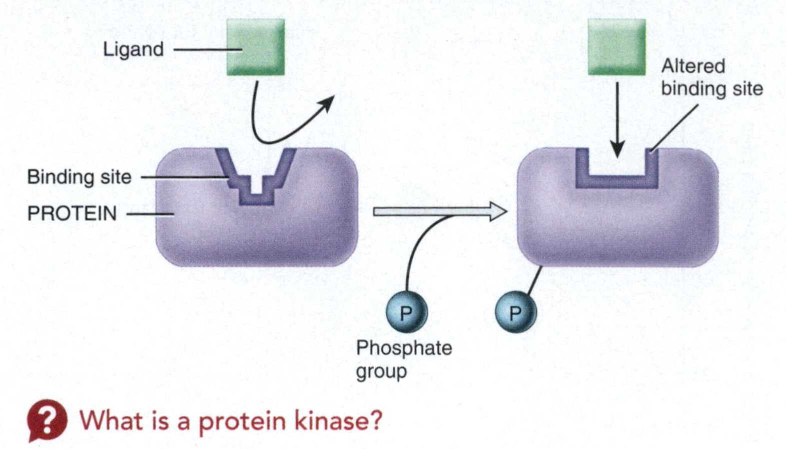

What is a protein kinase?

(PO_4^{3-}), which can be added to one of the amino acids of a protein via a process known as **phosphorylation**. When a protein is phosphorylated, it undergoes a conformational change. If the protein has a binding site for a ligand, the conformational change either increases or decreases the binding site's affinity for the ligand (**FIGURE 2.32**). The conformational change can also result in an increase or decrease in the functional activity of the protein. Phosphorylation is mediated by a type of enzyme called a *protein kinase*:

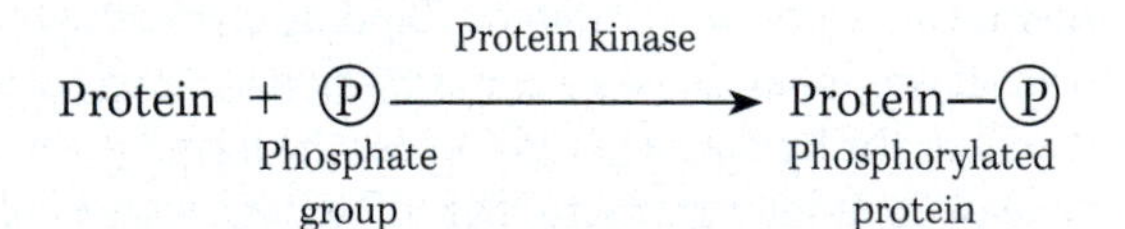

Protein + Ⓟ (Phosphate group) —Protein kinase→ Protein—Ⓟ (Phosphorylated protein)

Because a covalent bond attaches the phosphate group to the protein, the phosphate group is irreversibly bound to the protein unless the covalent bond is broken by another chemical reaction. In a process known as **dephosphorylation**, the covalent bond is broken and the phosphate group is removed from the protein. Dephosphorylation is performed by a type of enzyme known as *phosphatase*:

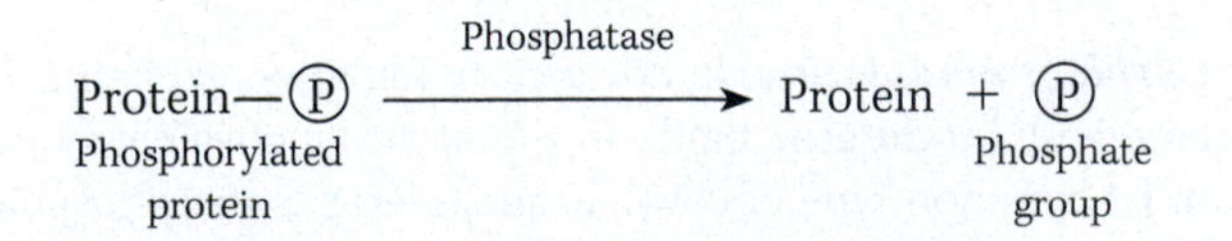

Protein—Ⓟ (Phosphorylated protein) —Phosphatase→ Protein + Ⓟ (Phosphate group)

Nucleic Acids Store and Express Genetic Information

Nucleic acids, so named because they were first discovered in the nuclei of cells, are huge organic molecules that contain carbon, hydrogen, oxygen, nitrogen, and phosphorus. Nucleic acids are of two varieties. The first, **deoxyribonucleic acid (DNA)** (dē-ok′-sē-rī-bō-nū-KLĒ-ik), forms the inherited genetic material inside each human cell. In humans, each **gene** is a segment of a DNA molecule. Our genes determine the traits we inherit and, by controlling protein synthesis, they regulate most of the activities that take place in body cells throughout our lives. When a cell divides, its hereditary information passes on to the next generation of cells. **Ribonucleic acid (RNA)**, the second type of nucleic acid, relays instructions from the genes to guide each cell's synthesis of proteins from amino acids.

CLINICAL CONNECTION

DNA Fingerprinting

A technique called **DNA fingerprinting** is used in research and in courts of law to ascertain whether a person's DNA matches the DNA obtained from samples or pieces of legal evidence such as bloodstains or hairs. In each person, certain DNA segments contain base sequences that are repeated several times. Both the number of repeat copies in one region and the number of regions subject to repeat are different from one person to another. DNA fingerprinting can be done with minute quantities of DNA—for example, from a single strand of hair, a drop of semen, or a spot of blood. It also can be used to identify a crime victim or a child's biological parents and even to determine whether two people have a common ancestor.

A nucleic acid is a chain of repeating monomers called **nucleotides**. Each nucleotide consists of three parts: a nitrogenous base, a pentose sugar, and a phosphate group (**FIGURE 2.33**).

1. ***Nitrogenous base.*** A **nitrogenous base** is a nitrogen-containing molecule arranged in a single or double ring. Nitrogenous bases that consist of a single ring are called **pyrimidines** (pī-RIM-i-dēnz), which include cytosine (C), thymine (T), and uracil (U). Nitrogenous bases that consist of a double ring are called **purines** (PŪR-ēnz), which include adenine (A) and guanine (G). A DNA nucleotide can contain cytosine, thymine, adenine, or guanine, but not uracil. An RNA nucleotide can contain cytosine, uracil, adenine, or guanine, but not thymine.
2. ***Pentose sugar.*** A five-carbon sugar attaches to each nitrogenous base of a DNA or RNA nucleotide. In DNA, the pentose sugar is **deoxyribose**; in RNA, the pentose sugar is **ribose**.
3. ***Phosphate group.*** Phosphate groups (PO_4^{3-}) alternate with pentose sugars to form the "backbone" of a DNA or RNA strand; the bases project inward from the backbone chain (see **FIGURE 2.34**).

In 1953, F. H. C. Crick of Great Britain and J. D. Watson, a young American scientist, published a brief paper describing how these three components might be arranged in DNA. Their insights into data gathered by others led them to construct a model so elegant and simple that the scientific world knew immediately that it was correct! In the Watson–Crick **double helix** model, DNA resembles a spiral ladder (**FIGURE 2.34**). Two strands of alternating phosphate groups and deoxyribose sugars form the uprights of the ladder. Paired bases, held together by hydrogen bonds, form the rungs. Because adenine always pairs with thymine, and cytosine always pairs with guanine, if you know the sequence of bases in one strand of DNA, you can predict the sequence on the complementary (second) strand. Each time DNA is copied, as when living cells divide to increase their number, the two strands unwind. Each strand serves as the template or mold on which a new second strand is constructed. Any change that occurs in the base sequence of a DNA strand is called a *mutation*. Some mutations can result in the death of a cell, cause cancer, or produce genetic defects in future generations.

Unlike DNA, RNA is not double-stranded; instead, the nucleotides of RNA are arranged in a single strand. Cells contain three different kinds of RNA: messenger RNA, ribosomal RNA, and transfer RNA. Each has a specific role to perform in carrying out the instructions encoded in DNA for protein synthesis, as will be described in Chapter 3 (see Section 3.5).

ATP Serves as the Energy Currency of Living Systems

Adenosine triphosphate (ATP) is the "energy currency" of living systems (**FIGURE 2.35**). ATP transfers energy from energy-releasing reactions to energy-requiring reactions that maintain cellular activities. Among these cellular activities are contraction of muscles, movement of chromosomes during cell division, movement of structures within cells, transport of substances across cell membranes, and synthesis of larger molecules from smaller ones.

As its name implies, ATP consists of three phosphate groups attached to adenosine, a unit composed of adenine and the five-carbon

FIGURE 2.33 Components of a nucleotide. A DNA nucleotide is shown.

Nucleotides are the repeating units of nucleic acids. Each nucleotide consists of a nitrogenous base, a pentose sugar, and a phosphate group.

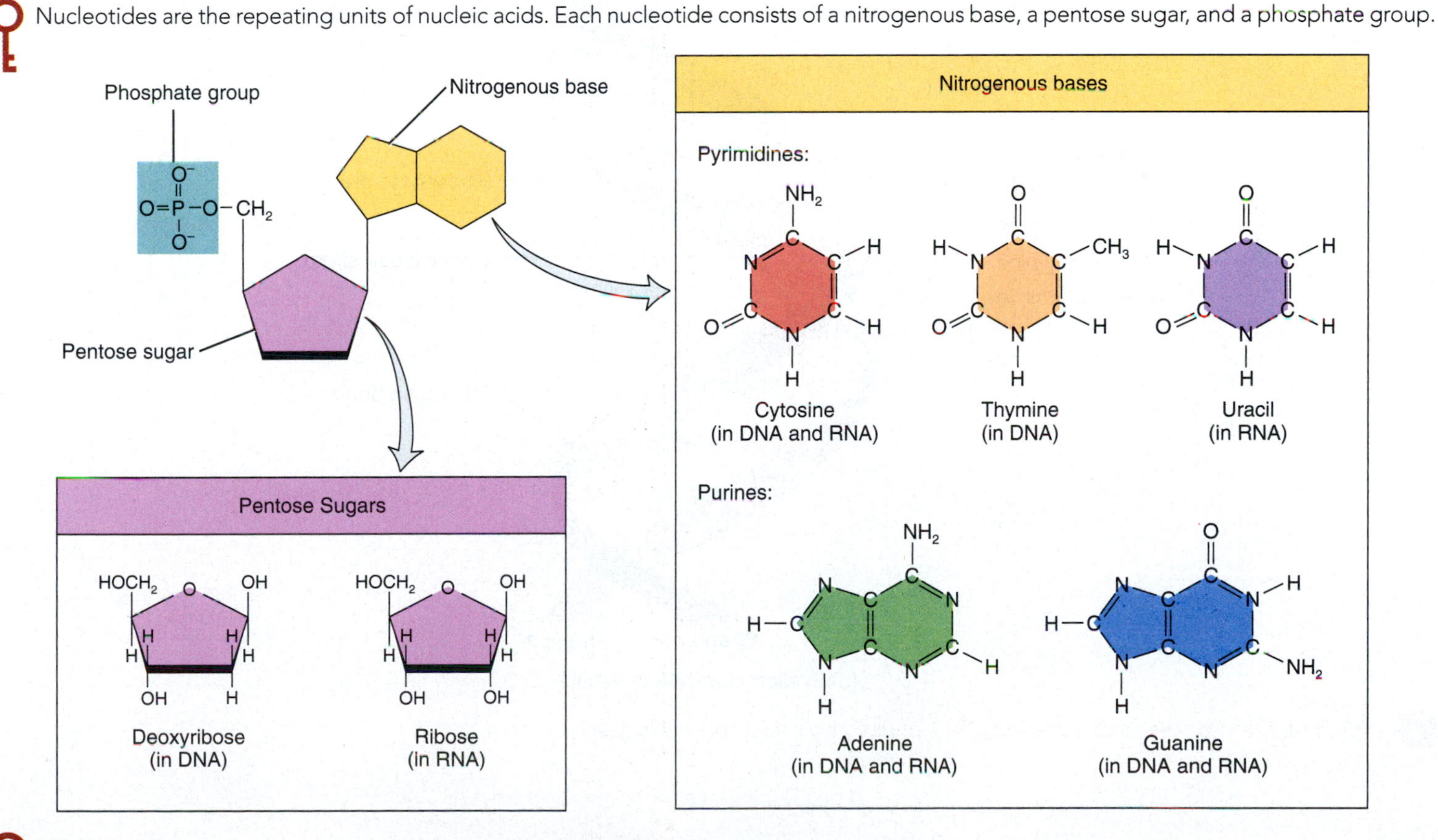

Which nitrogenous bases are present in DNA? In RNA?

FIGURE 2.34 DNA molecule. DNA is arranged in a double helix. The paired bases project toward the center of the helix. The structure is stabilized by hydrogen bonds (dotted lines) between each base pair. There are two hydrogen bonds between adenine and thymine and three between cytosine and guanine.

DNA forms the inherited genetic material inside each human cell.

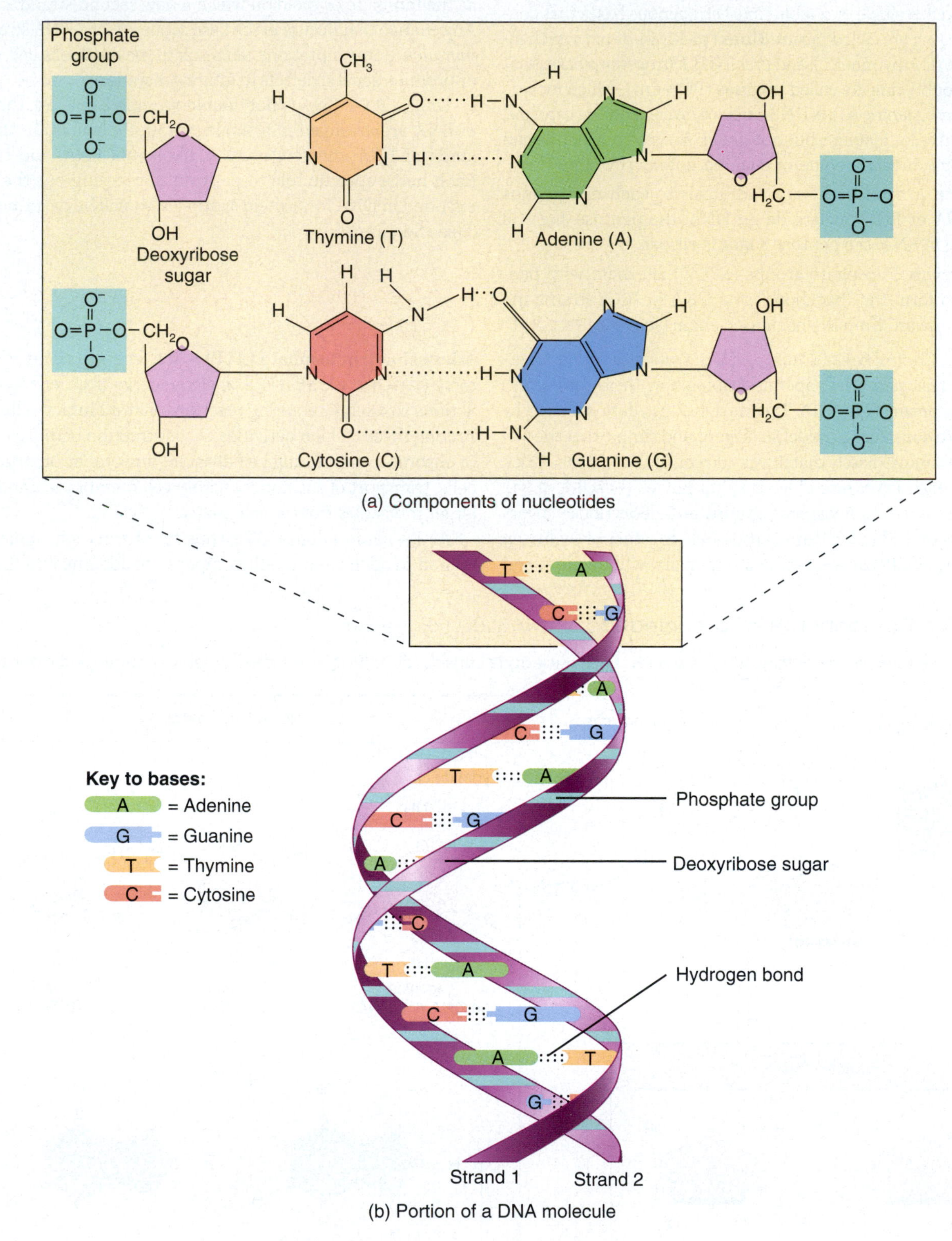

Which of the nitrogenous bases in DNA always pair with one another?

FIGURE 2.35 Structures of ATP and ADP. "Squiggles" (~) indicate the two phosphate bonds that can be used to transfer energy. Energy transfer typically involves hydrolysis of the last phosphate bond of ATP.

ATP transfers chemical energy to power cellular activities.

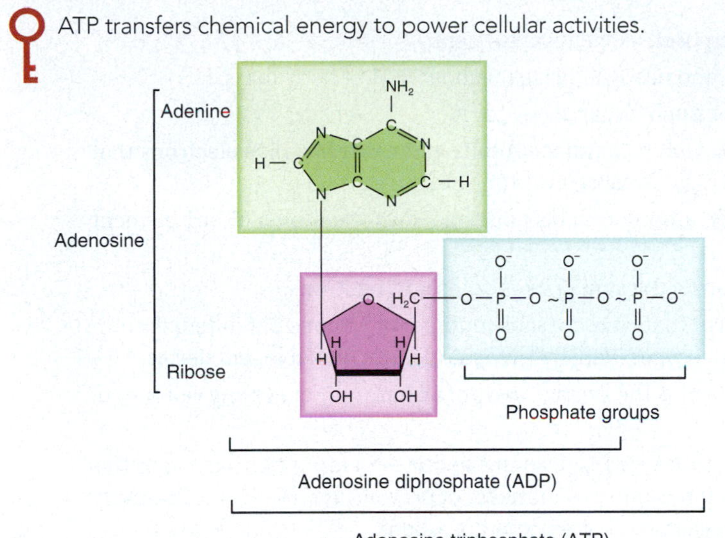

What are some cellular activities that depend on energy supplied by ATP?

sugar ribose. When a water molecule is added to ATP, the third phosphate group (PO_4^{3-}), symbolized by Ⓟ, is removed, and the overall reaction liberates energy. Removal of the third phosphate group produces a molecule called **adenosine diphosphate (ADP)**. The entire reaction is summarized as follows:

$$\underset{\text{Adenosine triphosphate}}{\text{ATP}} + \underset{\text{Water}}{H_2O} \longrightarrow \underset{\text{Adenosine diphosphate}}{\text{ADP}} + \underset{\text{Phosphate group}}{Ⓟ} + \underset{\text{Energy}}{\text{E}}$$

You will learn more about ATP's role in providing energy for chemicals reactions in Chapter 4.

CHECKPOINT

21. What are the components of an amino acid?

22. What kinds of weak interactions contribute to a protein's tertiary structure?

23. What is a ligand? What are the various properties of ligand–protein binding?

24. How do DNA and RNA differ?

25. What is the functional significance of ATP?

FROM RESEARCH TO REALITY

Antioxidants and Free Radicals

Reference

Xie, Z. et al. (2015). Functional beverage of Garcinia mangostana (mangosteen) enhances plasma antioxidant capacity in healthy adults. *Food Science Nutrition*. 3(1): 32–38.

Is mangosteen an antioxidant?

Malyshchyts Viktar/Fotalia

Many foods and beverages that we eat and drink contain important chemicals. One class of protective chemicals is antioxidants, which inactivate oxygen-derived free radicals that cause tissue damage in the body. Mangosteen is a tropical fruit that is thought to have bioactive properties, probably acting as an antioxidant in the body.

Article description:

Researchers in this study wanted to investigate a mangosteen-containing beverage to determine whether any of its functional components are absorbed into the bloodstream and to see if the beverage acts as an antioxidant in the body. The beverage that the researchers chose to study is a commercial product called *Verve*, which contains α-mangostin (derived from mangosteen fruit), green tea, aloe vera, and a variety of vitamins.

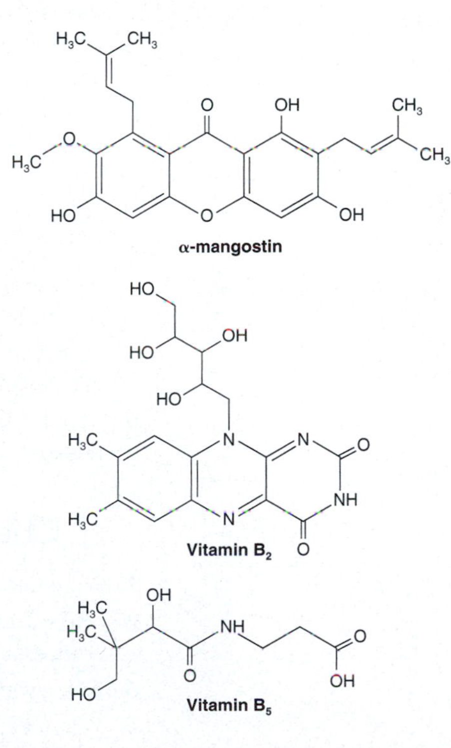

Go to WileyPLUS Learning Space and use the data from this article to answer the questions posed there and to discover more about antioxidants.

CHAPTER REVIEW

2.1 How Matter Is Organized

1. All forms of matter are composed of chemical elements.
2. Oxygen, carbon, hydrogen, and nitrogen make up about 96% of body mass.
3. Each element is made up of small units called atoms.
4. Atoms consist of a nucleus, which contains protons and neutrons, plus electrons that move about the nucleus in regions called electron shells.
5. The number of protons (the atomic number) distinguishes the atoms of one element from another.
6. The mass number of an atom is the sum of its protons and neutrons.
7. Different atoms of an element that have the same number of protons but different numbers of neutrons are called isotopes. Radioactive isotopes are unstable and decay.
8. The atomic mass of an element is the average mass of all naturally occurring isotopes of that element.
9. An atom that *gives up* or *gains* electrons becomes an ion—an atom that has a positive or negative charge because it has unequal numbers of protons and electrons. Positively charged ions are cations; negatively charged ions are anions.
10. If two atoms share electrons, a molecule is formed. Compounds contain atoms of two or more different elements.
11. A free radical is an electrically charged atom or group of atoms with an unpaired electron in its outermost shell. A common example is superoxide, which is formed by the addition of an electron to an oxygen molecule.

2.2 Chemical Bonds

1. Forces of attraction called chemical bonds hold atoms together. These bonds result from gaining, losing, or sharing electrons in the valence (outermost) electron shell.
2. Most atoms become stable when they have an octet of eight electrons in their valence shell.
3. An ionic bond forms when the force of attraction between ions of opposite charge holds them together.
4. In a covalent bond, atoms share pairs of valence electrons. Covalent bonds may be single, double, or triple and either nonpolar or polar.
5. An atom of hydrogen that forms a polar covalent bond with an oxygen atom or a nitrogen atom may also form a weaker bond, called a hydrogen bond, with an electronegative atom. The polar covalent bond causes the hydrogen atom to have a partial positive charge (δ^+) that attracts the partial negative charge (δ^-) of neighboring electronegative atoms, often oxygen or nitrogen.
6. Van der Waals interactions are weak attractions that occur between nearby molecules because of transient unequal charge distributions in these molecules.

2.3 Chemical Reactions

1. A chemical reaction occurs when atoms combine with or break apart from other atoms.
2. The starting substances of a chemical reaction are the reactants, and the ending substances are the products.
3. In a reversible reaction, the products can revert to the original reactants.
4. When a reaction is at chemical equilibrium, the rate of the forward reaction is equal to the rate of the reverse reaction.
5. The law of mass action describes the relationship between the net direction of a reversible reaction and the concentrations of the reactants and products.

2.4 Inorganic Compounds and Solutions

1. Inorganic compounds usually are small and usually lack carbon. Organic substances always contain carbon, usually contain hydrogen, and always have covalent bonds.
2. Water is the most abundant substance in the body. It is an excellent solvent and suspending medium and serves as a lubricant. Because of its many hydrogen bonds, water molecules are cohesive, which causes a high surface tension. Water also has a high capacity for absorbing heat and a high heat of vaporization.
3. Inorganic acids, bases, and salts dissociate into ions in water. An acid ionizes into hydrogen ions (H^+) and anions and is a proton donor; many bases ionize into cations and hydroxide ions (OH^-), and all are proton acceptors. A salt ionizes into cations and anions, neither of which is H^+ or OH^-.
4. Mixtures are combinations of elements or compounds that are physically blended together but are not bound by chemical bonds. Solutions, colloids, and suspensions are mixtures with different properties.
5. Three ways to express the concentration of a solution are molarity, equivalents, and mass per volume.
6. The pH of body fluids must remain fairly constant for the body to maintain homeostasis. On the pH scale, 7 represents neutrality. Values below 7 indicate acidic solutions, and values above 7 indicate alkaline solutions. Normal blood pH is 7.35–7.45.
7. Buffer systems remove or add hydrogen ions (H^+) to help maintain pH homeostasis.
8. One important buffer system is the carbonic acid–bicarbonate buffer system. The bicarbonate ion (HCO_3^-) acts as a weak base and removes excess H^+, and carbonic acid (H_2CO_3) acts as a weak acid and adds H^+.

2.5 Organic Compounds

1. Carbon, with its four valence electrons, bonds covalently with other carbon atoms to form large molecules of many different shapes. Attached to the carbon skeletons of organic molecules are functional groups that confer distinctive chemical properties.
2. Small organic molecules are joined together to form larger molecules by dehydration synthesis reactions in which a molecule of water is removed. In the reverse process, called hydrolysis, large molecules are broken down into smaller ones by the addition of water.
3. Carbohydrates provide most of the chemical energy needed to generate ATP. They may be monosaccharides, disaccharides, or polysaccharides.
4. Lipids are a diverse group of compounds that include fatty acids, triglycerides (fats and oils), phospholipids, steroids, and eicosanoids. Fatty acids are used to synthesize triglycerides and phospholipids; they can also be broken down for energy. Triglycerides protect, insulate, provide energy, and are stored. Phospholipids are important cell membrane components. Steroids are important in cell membrane structure, regulating sexual functions, maintaining normal blood sugar level, aiding lipid digestion and absorption, and helping bone growth. Eicosanoids (prostaglandins and leukotrienes) modify hormone responses, contribute to inflammation, dilate airways, and regulate body temperature.
5. Proteins are constructed from amino acids. They give structure to the body, regulate processes, provide protection, help muscles contract, transport substances, and serve as enzymes. Levels of structural organization in proteins include primary, secondary, tertiary, and (sometimes) quaternary. Variations in protein structure and shape are related to their diverse functions.
6. A variety of substances in the body can bind to proteins. A ligand is a molecule or ion that binds to a particular site on a protein through weak, noncovalent interactions. Ligand–protein binding exhibits four properties: specificity, affinity, saturation, and competition.

CHAPTER REVIEW

7. Deoxyribonucleic acid (DNA) and ribonucleic acid (RNA) are nucleic acids consisting of nitrogenous bases, five-carbon (pentose) sugars, and phosphate groups. DNA is a double helix and contains genes. RNA takes part in protein synthesis.
8. Adenosine triphosphate (ATP) is the "energy currency" of living systems; it transfers energy from energy-releasing reactions to energy-requiring reactions that maintain cellular activities.

PONDER THIS

1. You are at a gas station pumping gasoline into your car on a rainy day. Some of the gasoline drips from the nozzle onto a puddle of water on the ground. You notice that the water and gasoline do not mix; the gasoline droplets remain on the surface of the water. Why does this occur?

2. Alex is in your lab group and you are testing the effects of salivary amylase on carbohydrate digestion in the mouth. Part of your experiment is to create conditions that are optimal for the functionality of salivary amylase. One of these parameters that you must be watchful of is the pH of the solution. Salivary amylase functions best at a pH of about 7. Unfortunately, Alex is not paying attention when he mixes up the solution containing the enzyme and makes the pH too acidic (pH of 2). Explain in detail what would happen to the salivary amylase as a result of this mistake and discuss what aspects of protein–ligand functionality would be impaired.

3. Our blood plasma has a pH that is very closely regulated around 7.4. Diarrhea can often cause a condition called metabolic acidosis which results in a pH that is too acidic for the body. Explain how this could be lethal if untreated when considering metabolic pathways and enzymes.

ANSWERS TO FIGURE QUESTIONS

2.1 In carbon, the first shell contains two electrons and the second shell contains four electrons.

2.2 The four most plentiful elements in living organisms are oxygen, carbon, hydrogen, and nitrogen.

2.3 Antioxidants such as selenium, zinc, beta-carotene, vitamin C, and vitamin E can inactivate free radicals derived from oxygen.

2.4 A cation is a positively charged ion; an anion is a negatively charged ion.

2.5 An ionic bond involves the *loss* and *gain* of electrons; a covalent bond involves the *sharing* of pairs of electrons.

2.6 The N atom in ammonia is electronegative. Because it attracts electrons more strongly than do the H atoms, the nitrogen end of ammonia acquires a slight negative charge, allowing H atoms in water molecules (or in other ammonia molecules) to form hydrogen bonds with it. Likewise, O atoms in water molecules can form hydrogen bonds with H atoms in ammonia molecules.

2.7 Although a nonpolar molecule has no overall polarity, at any given instant there are transient unequal charge distributions because of transient fluctuations in electron distribution as electrons constantly move around the molecule.

2.8 The number of hydrogen atoms in the reactants must equal the number in the products—in this case, four hydrogen atoms total. Put another way, two molecules of H_2 are needed to react with each molecule of O_2 so that the number of H atoms and O atoms in the reactants is the same as the number of H atoms and O atoms in the products.

2.9 No. Because sugar dissolves easily in a polar solvent (water), you can correctly predict that it has several polar covalent bonds.

2.10 $CaCO_3$ is a salt, and H_2SO_4 is an acid.

2.11 At $pH = 6$, $[H^+] = 10^{-6}$ mol/liter and $[OH^-] = 10^{-8}$ mol/liter. A pH of 6.82 is more acidic than a pH of 6.91. Both $pH = 8.41$ and $pH = 5.59$ are 1.41 pH units from neutral ($pH = 7$).

2.12 Glucose has five —OH groups and six carbon atoms.

2.13 Digestion of food involves hydrolysis reactions.

2.14 Hexoses are six-carbon sugars; examples include glucose, fructose, and galactose.

2.15 Glucose and galactose are joined together to form lactose.

2.16 Cells in the liver and in skeletal muscle store glycogen.

2.17 The oxygen in the water molecule comes from a fatty acid.

2.18 The head of a phospholipid is polar and the tails are nonpolar.

2.19 The main structural differences between estradiol and testosterone are the number of double bonds and the types of functional groups associated with the first ring.

2.20 An amino acid has a minimum of two carbon atoms and one nitrogen atom.

2.21 Hydrolysis occurs during the breakdown of proteins.

2.22 Proteins consisting of a single polypeptide chain do not have a quaternary structure.

2.23 A hydrophobic interaction refers to the aggregation of nonpolar molecules in order to minimize exposure to water; a van der Waals interaction can occur between all types of nearby molecules (nonpolar and polar) because of unequal charge distributions in these molecules.

2.24 A change in pH or temperature or the presence of certain chemical agents can cause protein denaturation.

2.25 Examples of ligands include substrates, which bind to enzymes; solutes, which bind to membrane transport proteins; and hormones, which bind to receptors.

2.26 The ligand that binds to protein A would mostly likely have a triangular shape.

2.27 The binding site of protein X has a higher affinity because its shape is more complementary to the ligand and because it is electrically attracted to the ligand.

2.28 As the ligand concentration increases, the percent saturation of binding sites increases until all binding sites are occupied by ligand.

2.29 A relatively high concentration of ligand is needed to saturate low-affinity binding sites because the ligand binds loosely to the sites and dissociates quickly.

2.30 Increasing the concentration of ligand 1 would reduce the competitive effect of ligand 2 because more molecules of ligand 1 would be available to bind to the binding sites.

2.31 An allosteric activator binds to an allosteric site, causing a conformational change of the protein that *increases* the binding site's affinity for the ligand; an allosteric inhibitor binds to an allosteric site, resulting in a conformational change that *decreases* the binding site's affinity for the ligand.

2.32 A protein kinase is an enzyme that phosphorylates (adds a phosphate group) to a protein.

2.33 Cytosine, thymine, adenine, and guanine are the nitrogenous bases present in DNA; cytosine, uracil, adenine, and guanine are the nitrogenous bases present in RNA.

2.34 Thymine always pairs with adenine, and cytosine always pairs with guanine.

2.35 Cellular activities that depend on energy supplied by ATP include muscle contractions, movement of chromosomes, and transport of substances across cell membranes.

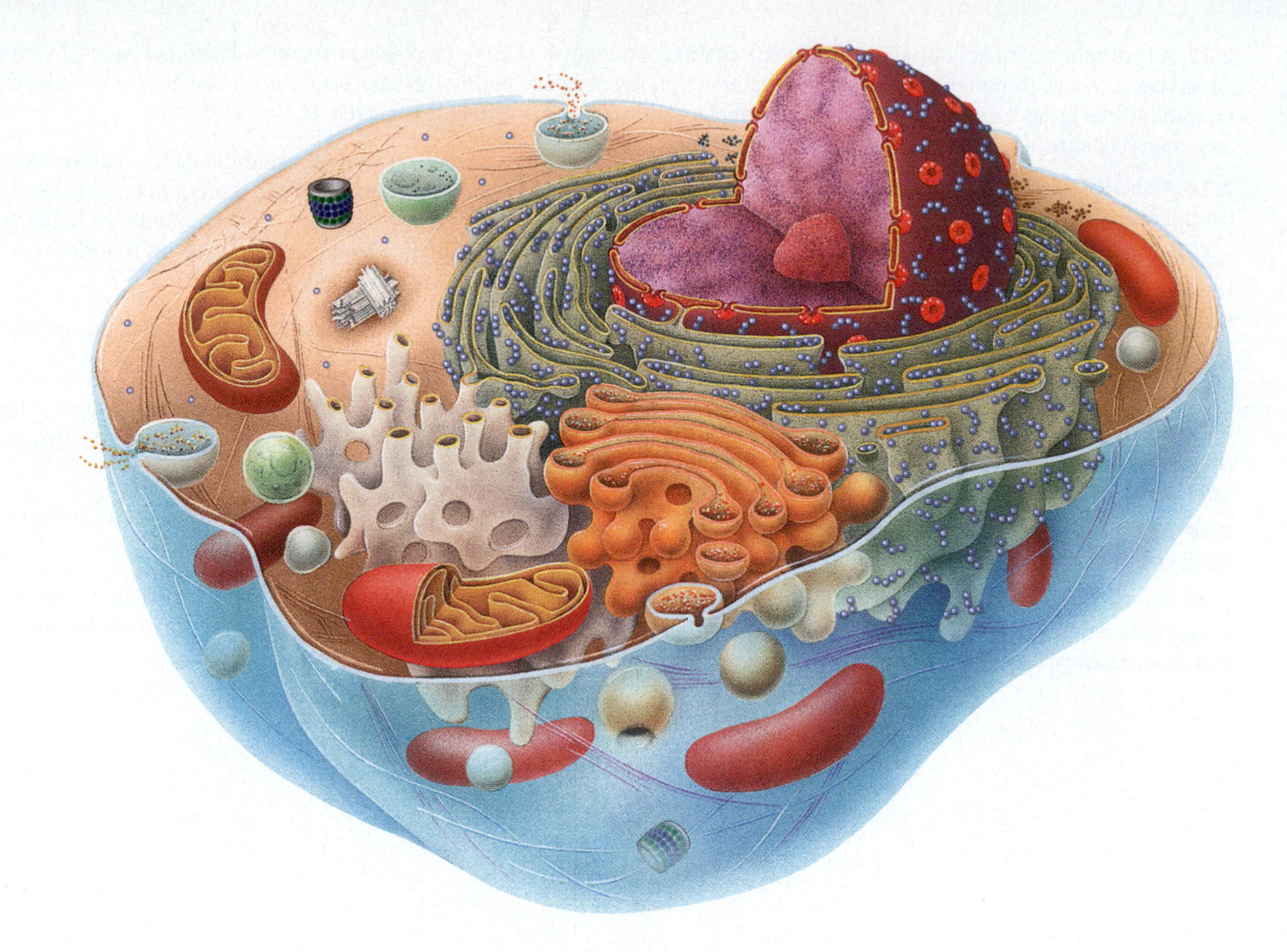

3 Cells

Cells and Homeostasis

Cells contribute to homeostasis by helping the tissues and organs that they comprise perform various functions in the body.

LOOKING BACK TO MOVE AHEAD...

- From smallest to largest, the six levels of body organization are the chemical, cellular, tissue, organ, system, and organismal levels (Section 1.2).
- Certain basic life processes distinguish living things from nonliving things: metabolism, responsiveness, movement, growth, differentiation, and reproduction (Section 1.3).
- Steroids are lipids that contain four interconnected hydrocarbon rings; examples include cholesterol, estrogens, and testosterone (Section 2.5).
- There are two major types of nucleic acids: (1) deoxyribonucleic acid (DNA), which forms the inherited genetic material inside each cell and (2) ribonucleic acid (RNA), which carries out the instructions encoded in DNA to guide each cell's synthesis of proteins from amino acids (Section 2.5).

Cells are the basic structural and functional units of an organism. They are the smallest units capable of performing all life processes. There are about 200 different types of cells in the human body. These cell types fulfill unique roles that support homeostasis and contribute to the many functional capabilities of the human organism. In this chapter, you will learn about the various parts of the cell, the expression of cellular genes into proteins, the process by which existing cells divide to produce new cells, and the organization of cells into tissues.

3.1 Components of a Cell

OBJECTIVE

- Describe the three main parts of a cell.

FIGURE 3.1 provides an overview of the typical structures found in most body cells. For ease of study, the cell is divided into three main parts: plasma membrane, cytoplasm, and nucleus.

1. The **plasma membrane** forms the cell's flexible outer surface, separating the cell's internal environment (inside the cell) from the external environment (outside the cell). It is a selective barrier that regulates the flow of materials into and out of a cell. This selectivity helps establish and maintain the appropriate environment for normal cellular activities. The plasma membrane also plays a key role in communication among cells and between cells and their external environment.
2. The **cytoplasm** consists of all the cellular contents between the plasma membrane and the nucleus. This compartment has two components: cytosol and organelles. **Cytosol**, the fluid portion of cytoplasm, contains water, dissolved solutes, and suspended particles. Surrounded by cytosol are several different types of **organelles** (*organelles* = little organs). Each type of organelle has a characteristic shape and specific functions. Examples include ribosomes, endoplasmic reticulum, Golgi complex, mitochondria, lysosomes, peroxisomes, proteasomes, and the cytoskeleton.

FIGURE 3.1 Typical structures found in body cells.

The cell is the basic structural and functional unit of the body.

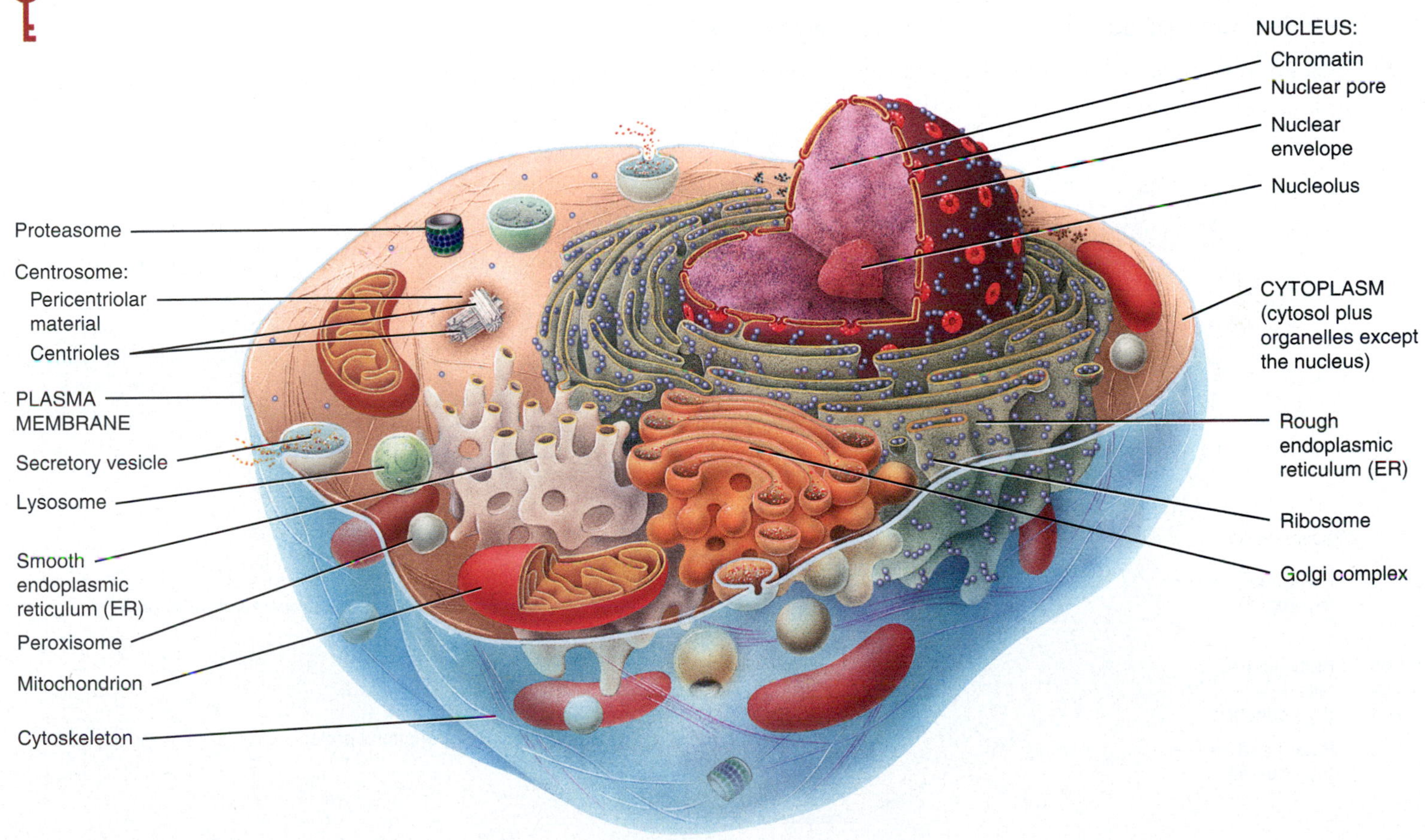

What are the three principal parts of a cell?

3. The **nucleus** is a large organelle that houses most of the cell's DNA (some DNA is also present in mitochondria). DNA is the genetic material of the cell: It contains hereditary units called **genes** that control most aspects of cellular structure and function.

CHECKPOINT

1. Explain the functions of the three main parts of a cell.

The membrane lipids allow passage of several types of lipid-soluble molecules but act as a barrier to the entry or exit of charged or polar substances. Some of the proteins in the plasma membrane allow movement of polar molecules and ions into and out of the cell. Other proteins can act as signal receptors or adhesion molecules. Transport of substances across plasma membranes is described in detail in Chapter 5; cell signaling is thoroughly discussed in Chapter 6.

3.2 The Plasma Membrane

OBJECTIVES

- Describe the fluid mosaic model of the plasma membrane.
- Identify the different types of lipids that comprise the plasma membrane.
- Discuss the functions of the various types of membrane proteins.

The **plasma membrane**, a flexible yet sturdy barrier that surrounds the cytoplasm of a cell, is best described by using a structural model called the **fluid mosaic model**. According to this model, the molecular arrangement of the plasma membrane resembles an ever-moving sea of fluid lipids that contains a mosaic of many different proteins (FIGURE 3.2). Some proteins float freely like icebergs in the lipid sea, whereas others are anchored at specific locations like boats at a dock.

The Plasma Membrane Consists of a Lipid Bilayer and a Variety of Proteins

The Lipid Bilayer

The basic structural framework of the plasma membrane is the **lipid bilayer**, two back-to-back layers made up of three types of lipid molecules—phospholipids, cholesterol, and glycolipids (FIGURE 3.2). About 75% of the membrane lipids are **phospholipids**, lipids that contain phosphorus. Present in smaller amounts are **cholesterol** (about 20%), a steroid with an attached —OH (hydroxyl) group, and various **glycolipids** (about 5%), lipids with attached carbohydrate groups.

The bilayer arrangement occurs because the lipids are **amphipathic** (am-fē-PATH-ik) molecules, which means that they have both polar and nonpolar parts. In phospholipids (see FIGURE 2.18), the polar part is the phosphate-containing "head," which is *hydrophilic* (*hydro-* = water; *-philic* = loving). The nonpolar parts are the two long fatty acid "tails," which are *hydrophobic* (*-phobic* = fearing) hydrocarbon chains.

FIGURE 3.2 Components of the plasma membrane.

The plasma membrane consists mainly of phospholipids; other components include glycolipids, cholesterol, and a variety of proteins.

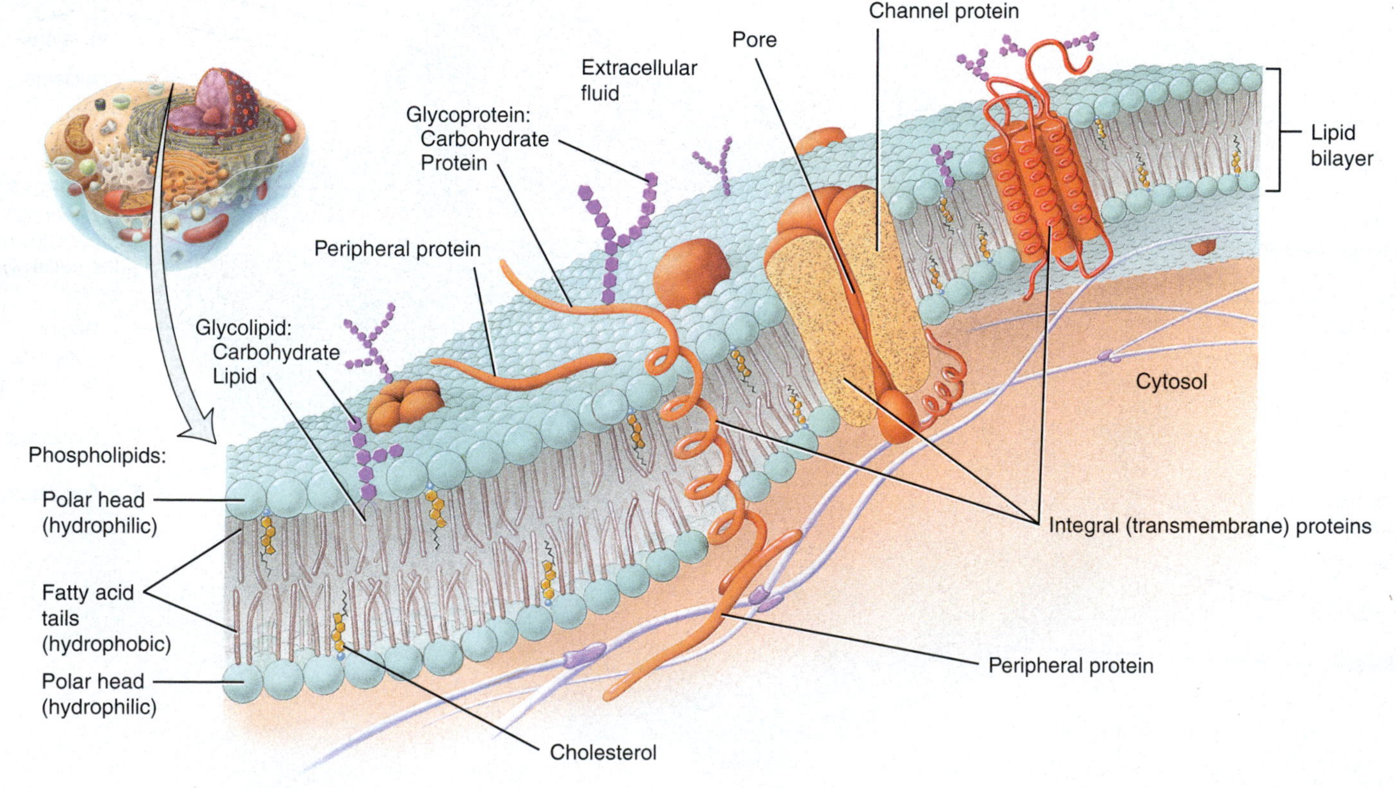

What is the glycocalyx?

Because "like seeks like," the phospholipid molecules orient themselves in the bilayer with their hydrophilic heads facing outward. In this way, the heads face a watery fluid on either side—cytosol on the inside and extracellular fluid on the outside. The hydrophobic fatty acid tails in each half of the bilayer point toward one another, forming a nonpolar, hydrophobic region in the membrane's interior.

Cholesterol molecules are weakly amphipathic (see **FIGURE 2.19a**) and are interspersed among the other lipids in both layers of the membrane. The tiny —OH group is the only polar region of cholesterol, and it forms hydrogen bonds with the polar heads of phospholipids and glycolipids. The stiff steroid rings and hydrocarbon tail of cholesterol are nonpolar; they fit among the fatty acid tails of the phospholipids and glycolipids. The carbohydrate groups of glycolipids form a polar "head"; their fatty acid "tails" are nonpolar. Glycolipids appear only in the membrane layer that faces the extracellular fluid, which is one reason the two sides of the bilayer are asymmetric, or different.

Arrangement of Membrane Proteins

Membrane proteins are classified as integral or peripheral according to whether they are firmly embedded in the membrane (**FIGURE 3.2**). **Integral proteins** extend into or through the lipid bilayer among the fatty acid tails and are firmly embedded in it. Most integral proteins are **transmembrane proteins**, which means that they span the entire lipid bilayer and protrude into both the cytosol and extracellular fluid. A few integral proteins are tightly attached to one side of the bilayer by covalent bonding to fatty acids. Like membrane lipids, integral membrane proteins are amphipathic. Their hydrophilic regions protrude into either the watery extracellular fluid or the cytosol, and their hydrophobic regions extend among the fatty acid tails.

As their name implies, **peripheral proteins** are not as firmly embedded in the membrane. They associate more loosely with the polar heads of membrane lipids or with integral proteins at the inner or outer surface of the membrane.

Many membrane proteins are **glycoproteins**, proteins with carbohydrate groups attached to the ends that protrude into the extracellular fluid. The carbohydrates are *oligosaccharides*, small chains of 2 to 60 monosaccharides that may be straight or branched. The carbohydrate portions of glycolipids and glycoproteins form an extensive sugary coat called the **glycocalyx** (glī-kō-KĀL-iks). The pattern of carbohydrates in the glycocalyx varies from one cell to another. Therefore, the glycocalyx acts like a molecular "signature" that enables cells to recognize one another. For example, a leukocyte's ability to detect a "foreign" glycocalyx is one basis of the immune response that helps destroy invading organisms.

Membrane Proteins Have Many Functions

Generally, the types of lipids in cellular membranes vary only slightly. In contrast, the membranes of different cells and various intracellular organelles have remarkably different assortments of proteins that determine many of the membrane's functions (**FIGURE 3.3**).

- Some integral membrane proteins are **ion channels**, which form *pores* or holes through which specific ions, such as potassium ions (K^+), can flow to get across the membrane. Ion channels are *selective;* they allow only a single ion or certain group of ions to pass through.

FIGURE 3.3 Functions of membrane proteins.

Membrane proteins largely reflect the functions that a cell can perform.

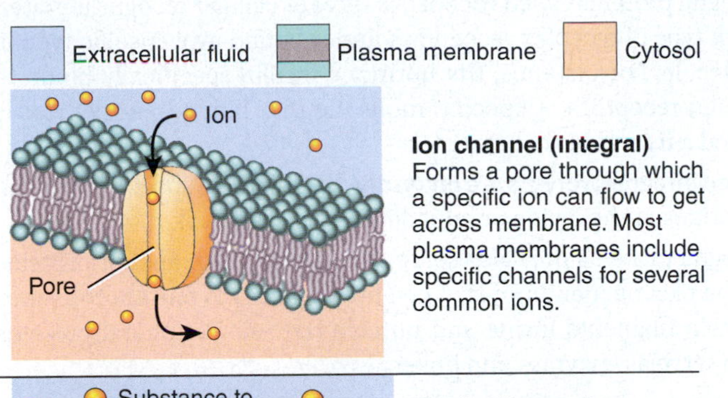

Ion channel (integral)
Forms a pore through which a specific ion can flow to get across membrane. Most plasma membranes include specific channels for several common ions.

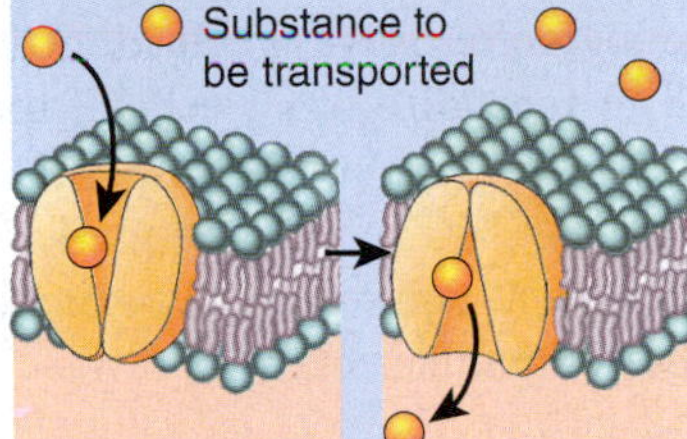

Carrier (integral)
Transports a specific substance across membrane by undergoing a change in shape. For example, amino acids, needed to synthesize new proteins, enter body cells via carriers. Carrier proteins are also known as *transporters*.

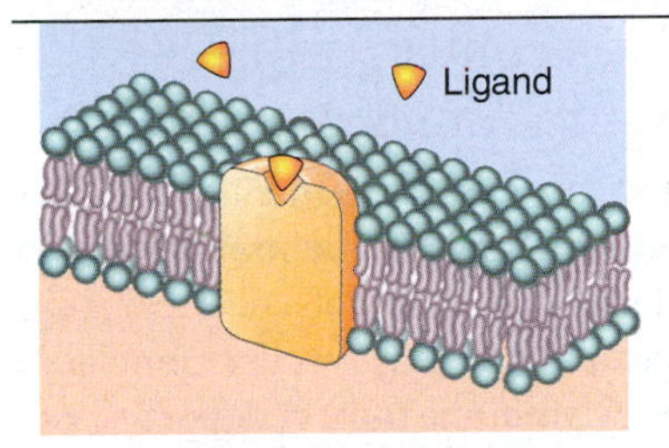

Receptor (integral)
Recognizes specific ligand and alters cell's function in some way. For example, antidiuretic hormone binds to receptors in the kidneys and changes the water permeability of certain plasma membranes.

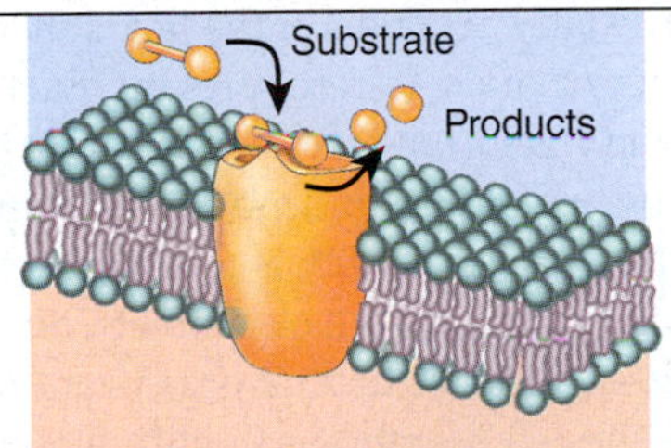

Enzyme (integral and peripheral)
Catalyzes reaction inside or outside cell (depending on which direction the active site faces). For example, lactase protruding from epithelial cells lining your small intestine splits the disaccharide lactose in the milk you drink.

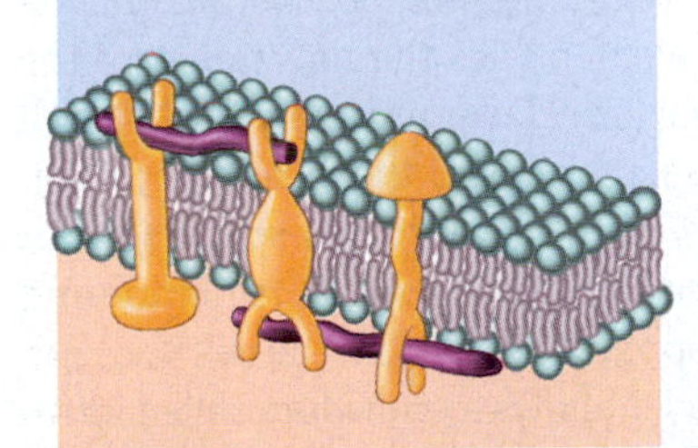

Linker (integral and peripheral)
Anchors filaments inside and outside the plasma membrane, providing structural stability and shape for the cell. May also participate in movement of the cell or link two cells together.

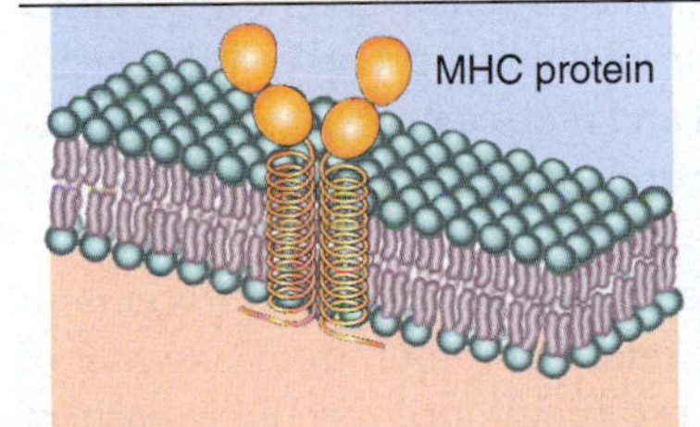

Cell identity marker (glycoprotein)
Distinguishes your cells from anyone else's (unless you are an identical twin). An important class of such markers are the major histocompatibility (MHC) proteins.

What does a carrier protein do?

- Other integral proteins act as **carriers**, selectively moving a polar substance or ion across the membrane by undergoing a conformational change (a change in shape). Carriers are also known as *transporters*.
- Integral proteins called **receptors** serve as cellular recognition sites. Each type of receptor recognizes and is bound by a specific type of molecule. For instance, the hormone insulin specifically binds to insulin receptors. A specific molecule that binds to a receptor is called a **ligand** of that receptor.
- Some integral proteins are **enzymes** that catalyze specific chemical reactions at the inside or outside surface of the cell.
- Integral proteins may also serve as **linkers**, which anchor proteins in the plasma membranes of neighboring cells to one another or to protein filaments inside and outside the cell. Peripheral proteins also serve as enzymes and linkers.
- Membrane glycoproteins and glycolipids often serve as **cell-identity markers**. They may enable a cell to recognize other cells of the same kind during tissue formation or to recognize and respond to potentially dangerous foreign cells. The ABO blood type markers are one example of cell-identity markers. When you receive a blood transfusion, the blood type must be compatible with your own.

Membrane Fluidity Allows Membrane Components to Interact and Move Around

Membranes are fluid structures; that is, most of the membrane lipids and many of the membrane proteins easily rotate and move sideways in their own half of the bilayer. Neighboring lipid molecules exchange places about 10 million times per second and may wander completely around a cell in only a few minutes! Membrane fluidity depends both on the number of double bonds in the fatty acid tails of the lipids that make up the bilayer and on the amount of cholesterol present. Each double bond puts a "kink" in the fatty acid tail (see FIGURE 2.18), which increases membrane fluidity by preventing lipid molecules from packing tightly in the membrane. Membrane fluidity is an excellent compromise for the cell; a rigid membrane would lack mobility, and a completely fluid membrane would lack the structural organization and mechanical support required by the cell. Membrane fluidity allows interactions to occur within the plasma membrane, such as the assembly of membrane proteins. It also enables the movement of the membrane components responsible for cellular processes such as cell movement, growth, division, and secretion, and the formation of cellular junctions. Fluidity allows the lipid bilayer to self-seal if torn or punctured. When a needle is pushed through a plasma membrane and pulled out, the puncture site seals spontaneously, and the cell does not burst. This property of the lipid bilayer allows a procedure called intracytoplasmic sperm injection to help infertile couples conceive a child; scientists can fertilize an oocyte by injecting a sperm cell through a tiny syringe. It also permits removal and replacement of a cell's nucleus in cloning experiments, such as the one that created Dolly, the famous cloned sheep.

Despite the great mobility of membrane lipids and proteins in their own half of the bilayer, they seldom flip-flop from one half of the bilayer to the other because it is difficult for hydrophilic parts of membrane molecules to pass through the hydrophobic core of the membrane. This difficulty contributes to the asymmetry of the membrane bilayer.

Because of the way it forms hydrogen bonds with neighboring phospholipid and glycolipid heads and fills the space between bent fatty acid tails, cholesterol makes the lipid bilayer stronger but less fluid at normal body temperature. At low temperatures, cholesterol has the opposite effect—it increases membrane fluidity.

CHECKPOINT

2. How do hydrophobic and hydrophilic regions govern the arrangement of membrane lipids in a bilayer?
3. "The proteins present in a plasma membrane determine the functions that a membrane can perform." Is this statement true or false? Explain your answer.
4. How does cholesterol affect membrane fluidity?

3.3 Cytoplasm

OBJECTIVE

- Describe the functions of the components of cytoplasm.

Cytoplasm consists of all the cellular contents between the plasma membrane and the nucleus, and has two components: (1) cytosol and (2) organelles.

The Cytosol Is the Site of Many Chemical Reactions

The **cytosol (intracellular fluid)** is the fluid portion of the cytoplasm that surrounds organelles (see FIGURE 3.1) and constitutes about 55% of total cell volume. Although it varies in composition and consistency from one part of a cell to another, cytosol is 75–90% water plus various dissolved and suspended components. Among these are different types of ions, glucose, amino acids, fatty acids, proteins, lipids, ATP, and waste products.

The cytosol is the site of many chemical reactions required for a cell's existence. For example, enzymes in cytosol catalyze *glycolysis*, a series of chemical reactions that produce a net gain of two molecules of ATP from the breakdown of one molecule of glucose (see FIGURE 4.14). Other types of cytosolic reactions provide the building blocks for maintenance of cell structures and for cell growth.

Organelles Function in Cellular Growth, Maintenance, and Reproduction

Organelles are tiny specialized structures within the cell that have characteristic shapes; they perform specific functions in cellular growth, maintenance, and reproduction. Despite the many chemical reactions going on in a cell at any given time, there is little interference among reactions because they are confined to different organelles. Each type of organelle has its own set of enzymes that carry out specific reactions, and serves as a functional compartment for specific biochemical processes. The numbers and types of organelles vary in different cells, depending on the cell's function. Although they have different functions, organelles often cooperate to maintain homeostasis. Even though the nucleus is a large organelle, it is discussed in a separate section because of its special importance in directing the life of a cell.

Ribosomes

Ribosomes (RĪ-bō-sōms) are the sites of protein synthesis. The name of these tiny organelles reflects their high content of one type of ribonucleic acid, **ribosomal RNA (rRNA)**, but each one also includes about 80 proteins. Structurally, a ribosome consists of two subunits, one about half the size of the other (FIGURE 3.4). The large and small subunits are made separately in the nucleolus, a spherical body inside the nucleus. Once produced, the large and small subunits exit the nucleus separately, then come together in the cytoplasm.

Some ribosomes, called *free ribosomes*, are unattached to any structure in the cytoplasm. Free ribosomes synthesize proteins used in the cytosol. Other ribosomes, called *bound ribosomes*, are attached to the outer surface of the endoplasmic reticulum and nucleus. These ribosomes synthesize proteins destined for export from the cell (secretion), for insertion into the plasma membrane, or for specific organelles within the cell. Ribosomes are also located within mitochondria, where they synthesize mitochondrial proteins.

Endoplasmic Reticulum

The **endoplasmic reticulum (ER)** (en′-dō-PLAS-mik re-TIK-ū-lum) is a network of membranes in the form of flattened sacs or tubules (FIGURE 3.5). The ER extends through widespread regions of the cytoplasm; it is so extensive that it constitutes more than half of the membranous surfaces within the cytoplasm of most cells.

Cells contain two distinct forms of ER that differ in structure and function. **Rough ER** extends from the nuclear envelope (membrane around the nucleus) and is folded into a series of flattened sacs (FIGURE 3.5). Rough ER appears "rough" because its outer surface is studded with ribosomes, the sites of protein synthesis. Proteins synthesized by ribosomes attached to rough ER enter spaces within the ER, where they are chemically modified. For example, carbohydrates are added to most proteins to form glycoproteins. After additional processing in the Golgi complex (described next), the proteins from the ER may be secreted from the cell, inserted into the plasma membrane of the cell, or transferred to other organelles in the cell. Thus, rough ER can be viewed as a factory that produces secretory proteins, membrane proteins, and many organellar proteins.

FIGURE 3.4 Ribosomes.

Ribosomes are the sites of protein synthesis.

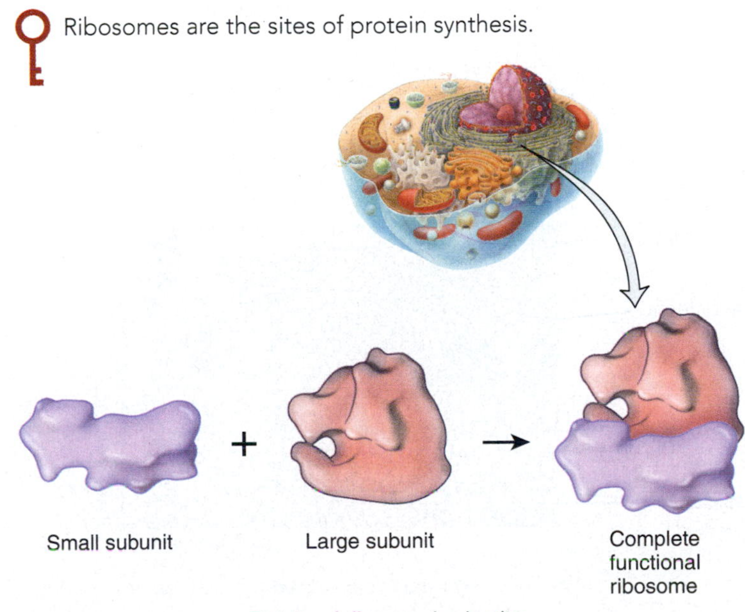

How are ribosomes produced?

FIGURE 3.5 Endoplasmic reticulum.

The endoplasmic reticulum (ER) is a network of membrane-enclosed sacs that extends through widespread regions of the cytoplasm.

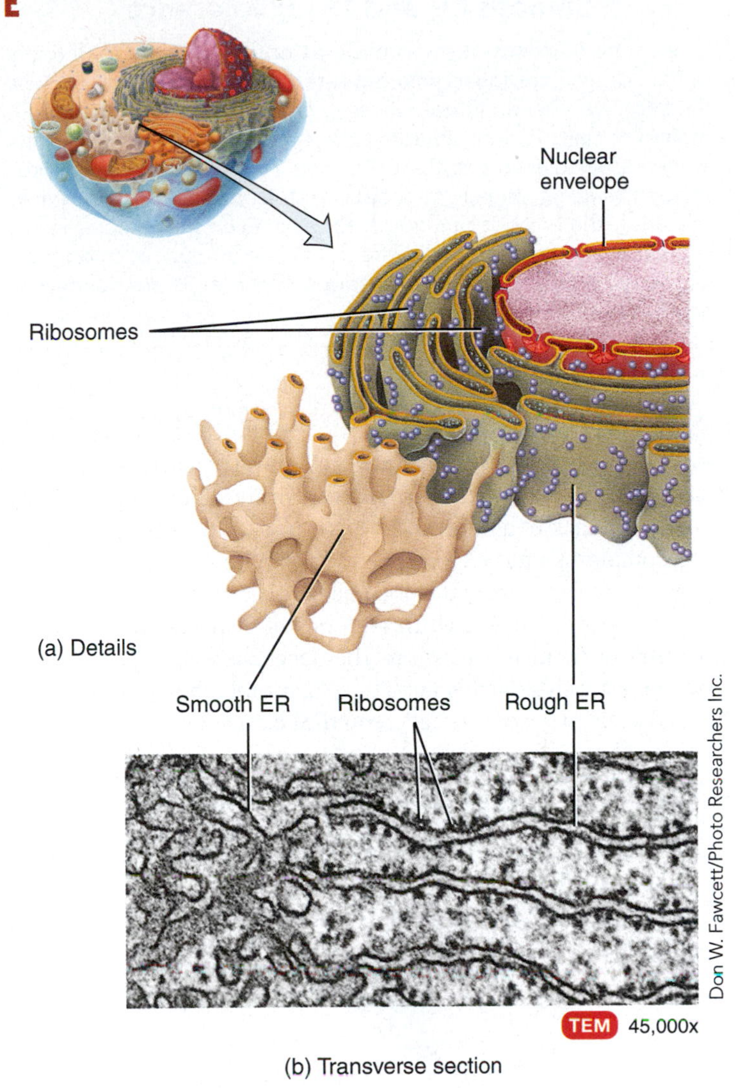

How does rough ER differ from smooth ER?

Smooth ER extends from the rough ER to form a network of membrane tubules (FIGURE 3.5). As you may already have guessed, smooth ER appears "smooth" because it lacks ribosomes. Smooth ER synthesizes fatty acids and steroids like estrogens and testosterone. In liver cells, enzymes of the smooth ER also inactivate or detoxify a variety of drugs and potentially harmful substances, such as alcohol, pesticides, and *carcinogens* (cancer-causing agents). In muscle cells, the calcium ions (Ca^{2+}) that trigger contraction are released from the sarcoplasmic reticulum, a form of smooth ER.

Golgi Complex

After proteins are processed in the rough ER, they enter an organelle called the **Golgi complex** (GOL-jē), named after the Italian physician

CLINICAL CONNECTION

Smooth ER and Drug Tolerance

One of the functions of smooth ER, as noted earlier, is to detoxify certain drugs. Individuals who repeatedly take such drugs, such as the sedative phenobarbital, develop changes in the smooth ER in their liver cells. Prolonged administration of phenobarbital results in increased tolerance to the drug: The same dose no longer produces the same degree of sedation. With repeated exposure to the drug, the amount of smooth ER and its enzymes increases to protect the cell from its toxic effects. As the amount of smooth ER increases, higher and higher dosages of the drug are needed to achieve the original effect.

Camillo Golgi, who first described it in the late 1800s. The Golgi complex consists of 3 to 20 **cisternae** (sis-TER-nē = cavities; singular is *cisterna*), small, flattened membranous sacs with bulging edges that resemble a stack of pita bread (**FIGURE 3.6**). Most cells have several Golgi complexes, and Golgi complexes are more extensive in cells that secrete proteins, a clue to the organelle's role in the cell.

Each Golgi complex has three types of cisternae (**FIGURE 3.6**). The **entry** or ***cis* face** is a cisterna that faces the rough ER. The **exit** or ***trans* face** is a cisterna that faces the plasma membrane. Sacs between the entry and exit faces are called **medial cisternae**.

The Golgi complex further modifies proteins received from the rough ER. Once the proteins are modified, the Golgi complex sorts and packages them into membrane-enclosed vesicles for transport to another destination. These functions of the Golgi complex involve the following steps (**FIGURE 3.7**):

1. Proteins synthesized by ribosomes on the rough ER are surrounded by a piece of the ER membrane, which eventually buds from the membrane surface to form **transport vesicles**.
2. Transport vesicles move toward the entry face of the Golgi complex.
3. Fusion of several transport vesicles creates the entry face of the Golgi complex and releases proteins into its lumen (space).
4. The proteins then move from the entry face into the lumen of the medial cisternae and then into the lumen of the exit face. As the proteins move through the cisternae, they are chemically modified. For example, enzymes in the cisternae alter the carbohydrate portions of glycoproteins by removing some carbohydrate groups and adding new ones.
5. After the proteins are modified, they are sorted and packaged.
6. Some of the processed proteins leave the exit face and are stored in **secretory vesicles**. These vesicles deliver the proteins to the plasma membrane, where they are discharged by exocytosis into the extracellular fluid. **Exocytosis** is the process by which materials move out of a cell by fusion with the plasma membrane of vesicles formed inside the cell. Exocytosis will be described in more detail in Chapter 5 (see **FIGURE 5.19b**).

FIGURE 3.6 Golgi complex.

The Golgi complex consists of membrane-enclosed sacs called cisternae that receive proteins synthesized by the rough ER.

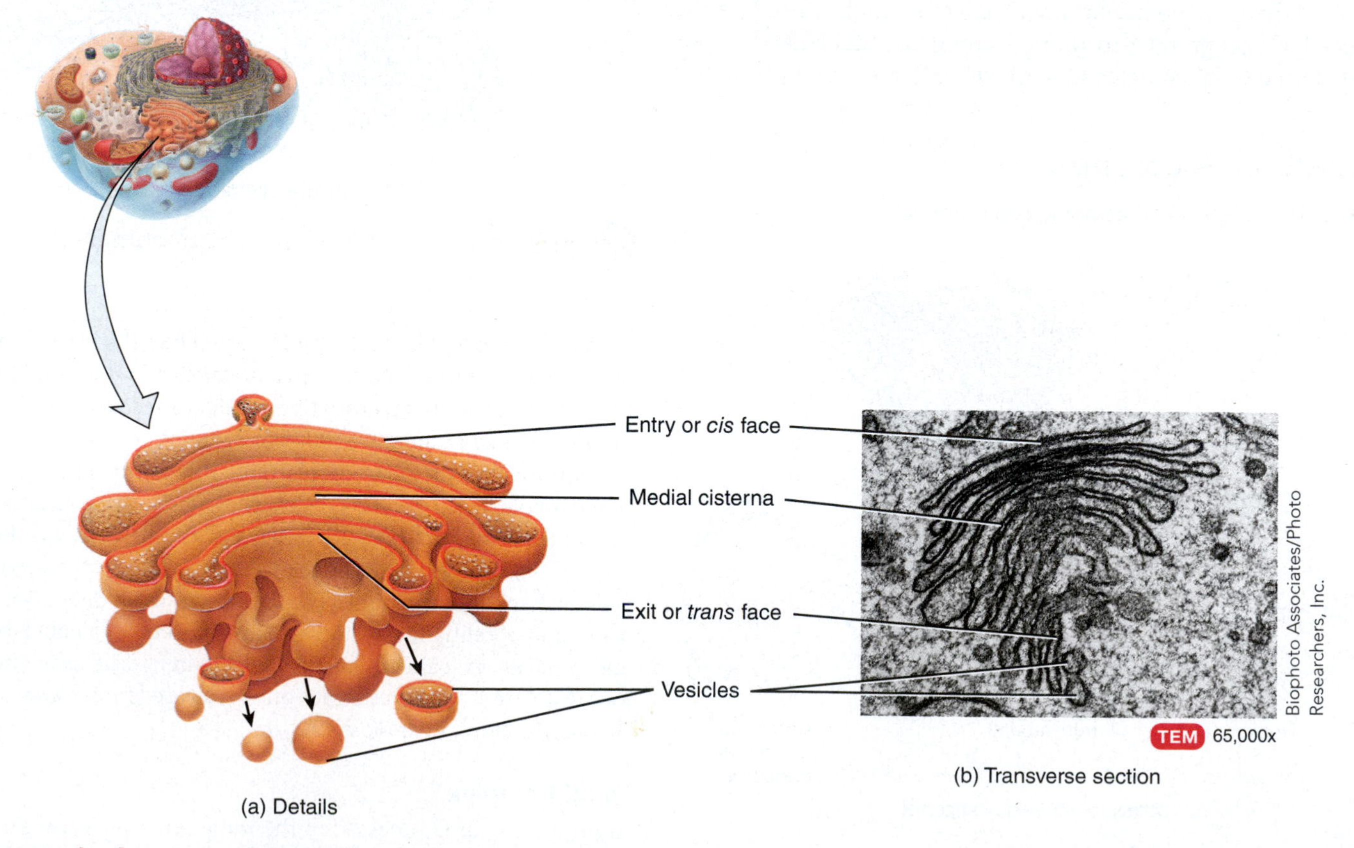

What are the functions of the Golgi complex?

FIGURE 3.7 Processing and packaging of proteins by the Golgi complex.

All proteins exported from the cell are processed in the Golgi complex.

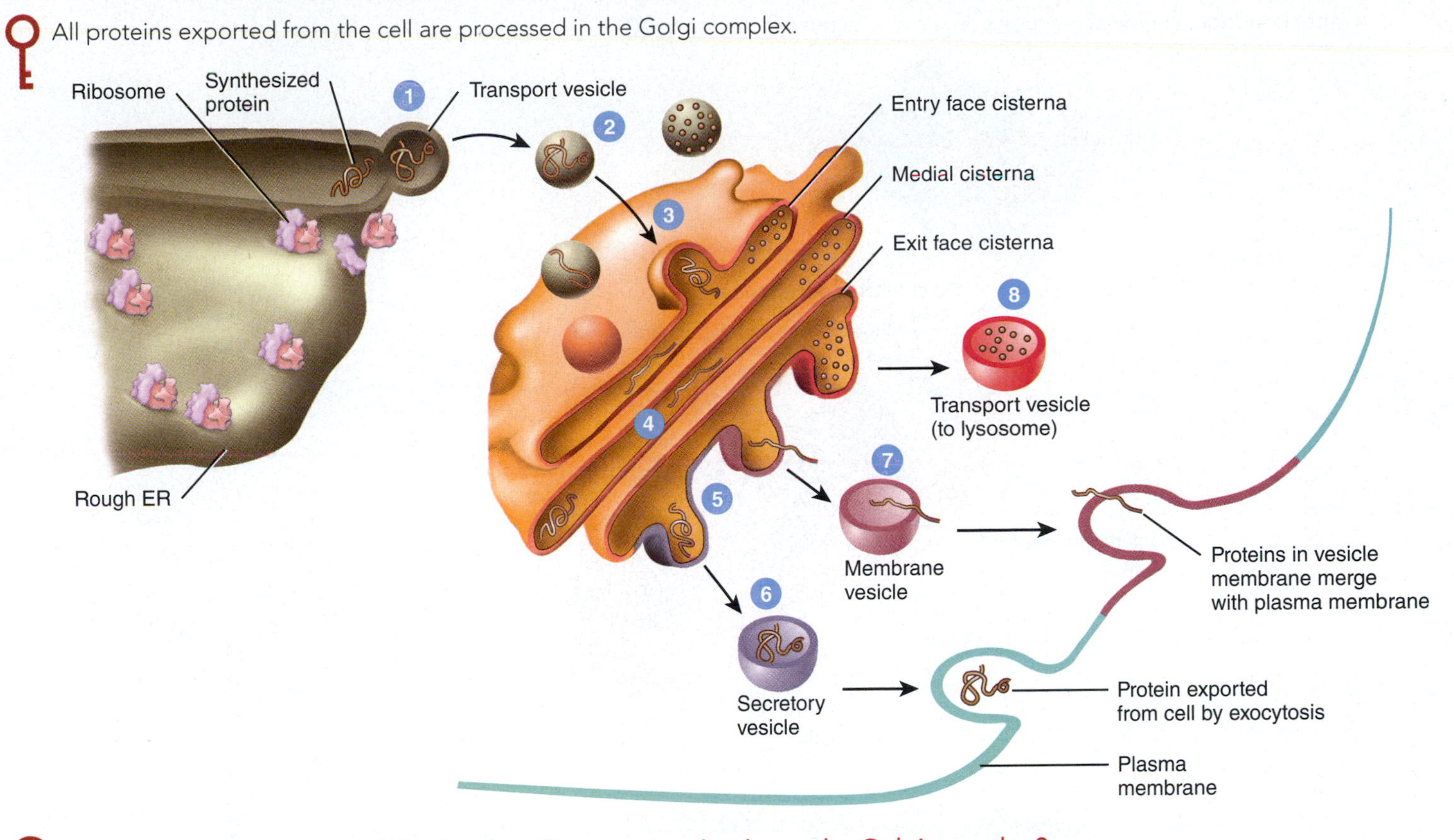

What are the three general destinations for proteins that leave the Golgi complex?

7. Other processed proteins leave the exit face in **membrane vesicles** that deliver their contents to the plasma membrane for incorporation into the membrane.
8. Finally, some processed proteins leave the exit face in transport vesicles that will carry the proteins to other organelles such as lysosomes.

Mitochondria

Because they generate most of the ATP through aerobic (oxygen-requiring) respiration, **mitochondria** (mī-tō-KON-drē-a; singular is *mitochondrion*) are referred to as the "powerhouses" of the cell. A cell may have as few as a hundred or as many as several thousand mitochondria, depending on how active it is. For example, very active cells such as those found in muscles, the liver, and kidneys use ATP at a high rate and have large numbers of mitochondria.

A mitochondrion consists of an **outer mitochondrial membrane** and an **inner mitochondrial membrane**, with a small space, known as the **intermembrane space**, located between them (FIGURE 3.8). Both membranes are similar in structure to the plasma membrane. The inner mitochondrial membrane contains a series of folds called **cristae** (KRIS-tē; singular is *crista*). The central fluid-filled cavity of a mitochondrion, enclosed by the inner mitochondrial membrane, is the **matrix**. As you will learn in Chapter 4, aerobic respiration consists of the chemical reactions of the Krebs cycle and the electron transport chain. The Krebs cycle occurs in the mitochondrial matrix; the electron transport chain is located along the cristae of the inner mitochondrial membrane.

Mitochondria self-replicate, a process that occurs during times of increased cellular energy demand or before cell division. Synthesis of some of the proteins needed for mitochondrial functions occurs on the ribosomes that are present in the mitochondria matrix. Mitochondria even have their own DNA, in the form of multiple copies of a circular DNA molecule that contains 37 genes. These mitochondrial genes control the synthesis of 2 ribosomal RNAs, 22 transfer RNAs, and 13 proteins that build mitochondrial components.

CLINICAL CONNECTION

Inheritance of Mitochondrial DNA

Although the nucleus of each somatic cell contains genes from both your mother and your father, mitochondrial genes are inherited only from your mother because all mitochondria in a cell are descendants of those that were present in the oocyte (egg) during the fertilization process. The head of a sperm (the part that penetrates and fertilizes an oocyte) normally lacks most organelles, such as mitochondria, ribosomes, endoplasmic reticulum, and the Golgi complex, and any sperm mitochondria that do enter the oocyte are soon destroyed. Because all mitochondrial genes are inherited from the maternal parent, mitochondrial DNA can be used to trace maternal lineage (in other words, to determine whether two or more individuals are related through their mother's side of the family).

FIGURE 3.8 Mitochondrion.

Within a mitochondrion, chemical reactions of aerobic respiration generate ATP.

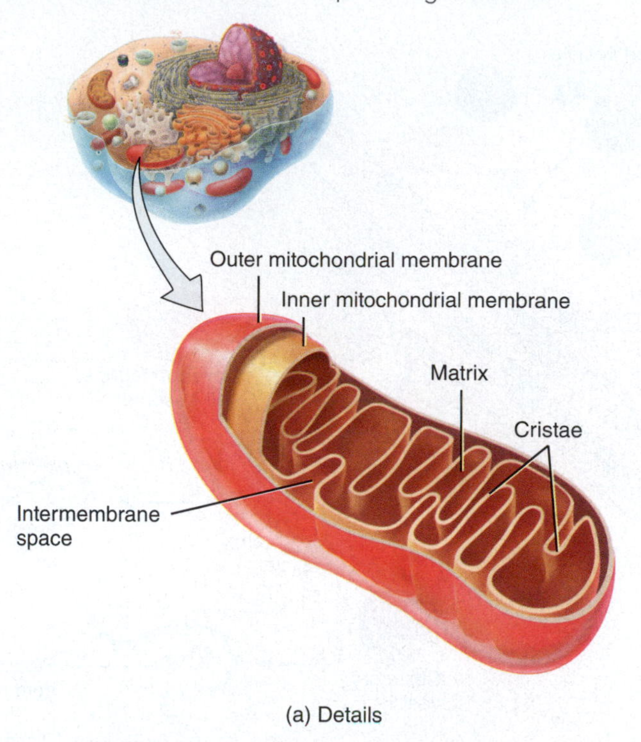

(a) Details

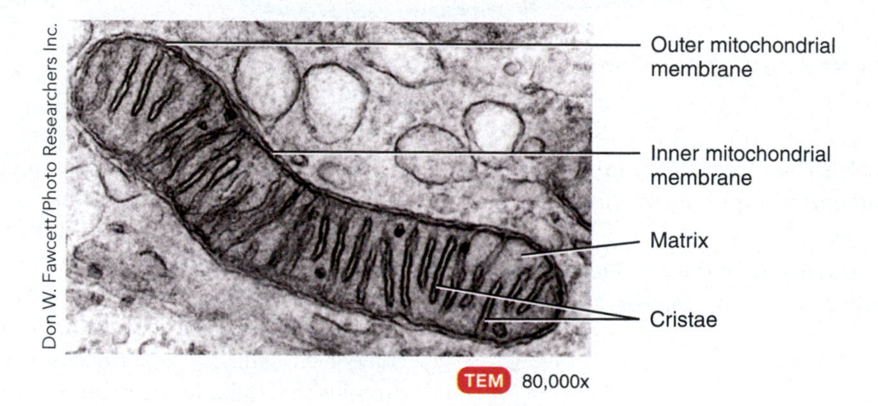

(b) Transverse section

How do the cristae of a mitochondrion contribute to its ATP-producing function?

Lysosomes

Lysosomes (LĪ-sō-sōms) are membrane-enclosed vesicles that contain digestive enzymes (**FIGURE 3.9**). These enzymes can break down a wide variety of molecules once lysosomes fuse with vesicles formed during endocytosis. **Endocytosis** is the process by which materials move into a cell in a vesicle formed from the plasma membrane; you will learn more about endocytosis in Chapter 5 (see **FIGURE 5.19a**). The lysosomal membrane also contains transporters that move the final products of digestion—such as glucose, fatty acids, and amino acids—into the cytosol.

Lysosomal enzymes also help recycle worn-out cell structures. A lysosome can engulf another organelle, digest it, and return the digested components to the cytosol for reuse. In this way, old organelles are continually replaced. The process by which entire worn-out organelles are digested is called **autophagy** (aw-TOF-a-jē;). During autophagy, the organelle to be digested is enclosed by a membrane

CLINICAL CONNECTION

Tay-Sachs Disease

Some disorders are caused by faulty or absent lysosomal enzymes. For instance, **Tay-Sachs disease**, which most often affects children of Ashkenazi (eastern European Jewish) descent, is an inherited condition characterized by the absence of a single lysosomal enzyme called Hex A. This enzyme normally breaks down a membrane glycolipid called ganglioside G_{M2} that is especially prevalent in neurons. As the excess ganglioside G_{M2} accumulates, the neurons function less efficiently. Children with Tay-Sachs disease typically experience seizures and muscle rigidity. They gradually become blind, demented, and uncoordinated and usually die before the age of 5. Tests can now reveal whether an adult is a carrier of the defective gene.

FIGURE 3.9 Lysosomes.

Lysosomes contain several types of powerful digestive enzymes.

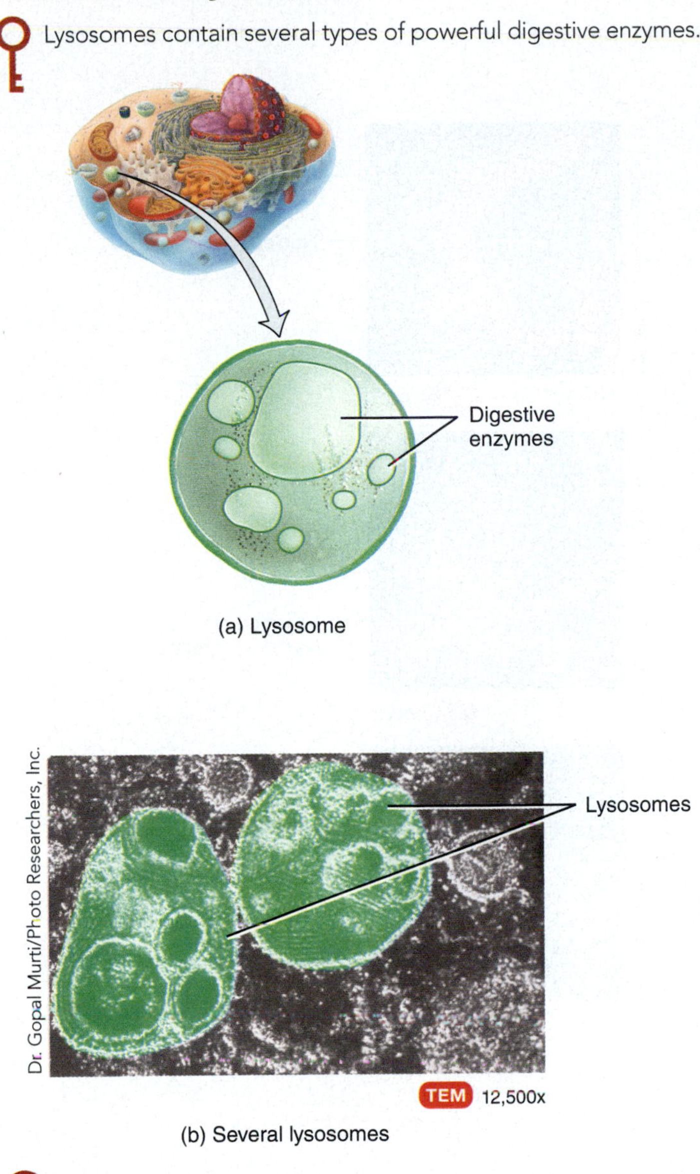

(a) Lysosome

(b) Several lysosomes

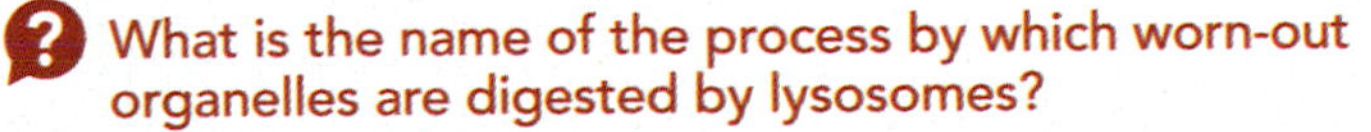

What is the name of the process by which worn-out organelles are digested by lysosomes?

derived from the ER to create a vesicle called an **autophagosome** (aw-tō-FĀ-gō-sōm); the vesicle then fuses with a lysosome. In this way, a human liver cell, for example, recycles about half of its cytoplasmic contents every week. Lysosomal enzymes may also destroy the entire cell that contains them, a process known as **autolysis** (aw-TOL-i-sis). Autolysis occurs in some pathological conditions and is also responsible for the tissue deterioration that occurs immediately after death.

Peroxisomes

Another group of organelles similar in structure to lysosomes, but smaller, are the **peroxisomes** (pe-ROKS-i-sōms; see **FIGURE 3.1**). Peroxisomes, also called *microbodies*, contain several *oxidases*, enzymes that can oxidize (remove hydrogen atoms from) various organic substances. For instance, amino acids and fatty acids are oxidized in peroxisomes as part of normal metabolism. In addition, enzymes in peroxisomes oxidize toxic substances, such as alcohol. Thus, peroxisomes are very abundant in the liver, where detoxification of alcohol and other damaging substances occurs. A by-product of the oxidation reactions is hydrogen peroxide (H_2O_2), a potentially toxic compound. However, peroxisomes also contain the enzyme *catalase*, which decomposes H_2O_2. Because production and degradation of H_2O_2 occur within the same organelle, peroxisomes protect other parts of the cell from the toxic effects of H_2O_2.

Proteasomes

Although lysosomes degrade proteins delivered to them in vesicles, proteins in the cytosol also require disposal at certain times in the life of a cell. Continuous destruction of unneeded, damaged, or faulty proteins is the function of **proteasomes** (PRŌ-tē-a-sōmes), which are tiny barrel-shaped structures consisting of four stacked rings of proteins around a central core (see **FIGURE 3.1**). For example, proteins that are part of metabolic pathways are degraded after they have accomplished their function. Such protein destruction plays a part in negative feedback by halting a pathway once the appropriate response has been achieved. A typical body cell contains many thousands of proteasomes, in both the cytosol and the nucleus. Discovered only recently because they are far too small to discern under the light microscope and do not show up well in electron micrographs, proteasomes were so named because they contain myriad *proteases*, enzymes that cut proteins into small peptides. Once the enzymes of a proteasome have chopped up a protein into smaller chunks, other enzymes then break down the peptides into amino acids, which can be recycled into new proteins.

Some diseases could result from failure of proteasomes to degrade abnormal proteins. For example, clumps of misfolded proteins accumulate in brain cells of people with Parkinson's disease and Alzheimer's disease. Discovering why the proteasomes fail to clear these abnormal proteins is a goal of ongoing research.

The Cytoskeleton

The **cytoskeleton** is a network of protein filaments that extends throughout the cytosol (see **FIGURE 3.1**). The cytoskeleton provides a structural framework for the cell, serving as a scaffold that helps to determine a cell's shape and to organize cellular contents. It also aids movement of whole cells and of components within cells. In the order of their increasing diameter, the components of the cytoskeleton are microfilaments, intermediate filaments, and microtubules.

MICROFILAMENTS **Microfilaments** are the thinnest elements of the cytoskeleton. They are composed of the protein *actin*, and are most prevalent at the edge of a cell (**FIGURE 3.10a**). Microfilaments have two general functions: They help generate movement and provide mechanical support.

With respect to movement, microfilaments are involved in muscle contraction, cell division, and cell locomotion, such as occurs during the migration of embryonic cells during development, the invasion of tissues by leukocytes to fight infection, or the migration of skin cells during wound healing.

Microfilaments provide much of the mechanical support that is responsible for the basic strength and shapes of cells. They anchor the cytoskeleton to integral proteins in the plasma membrane. Microfilaments also provide mechanical support for cell extensions called

FIGURE 3.10 Cytoskeleton.

The cytoskeleton is a network of three types of protein filaments that extend throughout the cytoplasm: microfilaments, intermediate filaments, and microtubules.

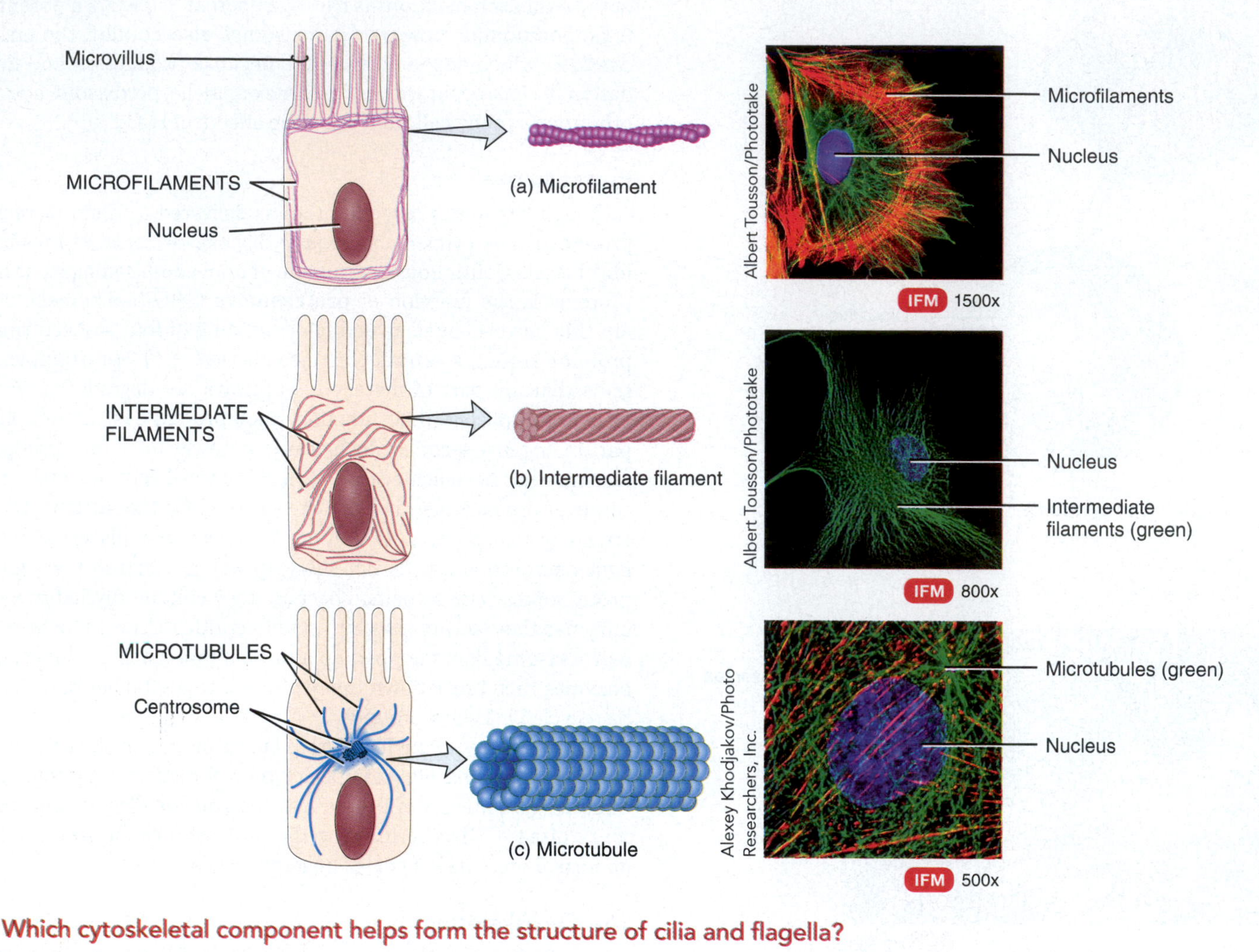

Which cytoskeletal component helps form the structure of cilia and flagella?

microvilli, nonmotile, microscopic fingerlike projections of the plasma membrane. Within each microvillus is a core of parallel microfilaments that supports it. Because they greatly increase the surface area of the cell, microvilli are abundant on cells involved in absorption, such as the epithelial cells that line the small intestine.

Intermediate Filaments As their name suggests, **intermediate filaments** are intermediate in size: They are thicker than microfilaments but thinner than microtubules (FIGURE 3.10b). Several different proteins can compose intermediate filaments, which are exceptionally strong. They are found in parts of cells subject to mechanical stress, and they help stabilize the position of organelles such as the nucleus and attach cells to one another.

Microtubules **Microtubules**, the largest of the cytoskeletal components, are long, unbranched hollow tubes composed mainly of the protein *tubulin*. The assembly of microtubules begins in an organelle called the centrosome (discussed shortly). The microtubules grow outward from the centrosome toward the periphery of the cell (FIGURE 3.10c). Microtubules help determine cell shape. They also function in the movement of organelles, secretory vesicles, chromosomes, and specialized cell projections such as cilia and flagella.

Centrosome

The **centrosome**, located near the nucleus, consists of two components: a pair of centrioles and pericentriolar material (FIGURE 3.11a). The two **centrioles** are cylindrical structures, each composed of nine clusters of three microtubules (triplets) arranged in a circular pattern (FIGURE 3.11b). The long axis of one centriole is at a right angle to the long axis of the other (FIGURE 3.11c). Surrounding the centrioles is **pericentriolar material**, which contains hundreds of ring-shaped complexes composed of the protein *tubulin*. These tubulin complexes are the organizing centers for the mitotic

FIGURE 3.11 Centrosome.

Located near the nucleus, the centrosome consists of a pair of centrioles and pericentriolar material.

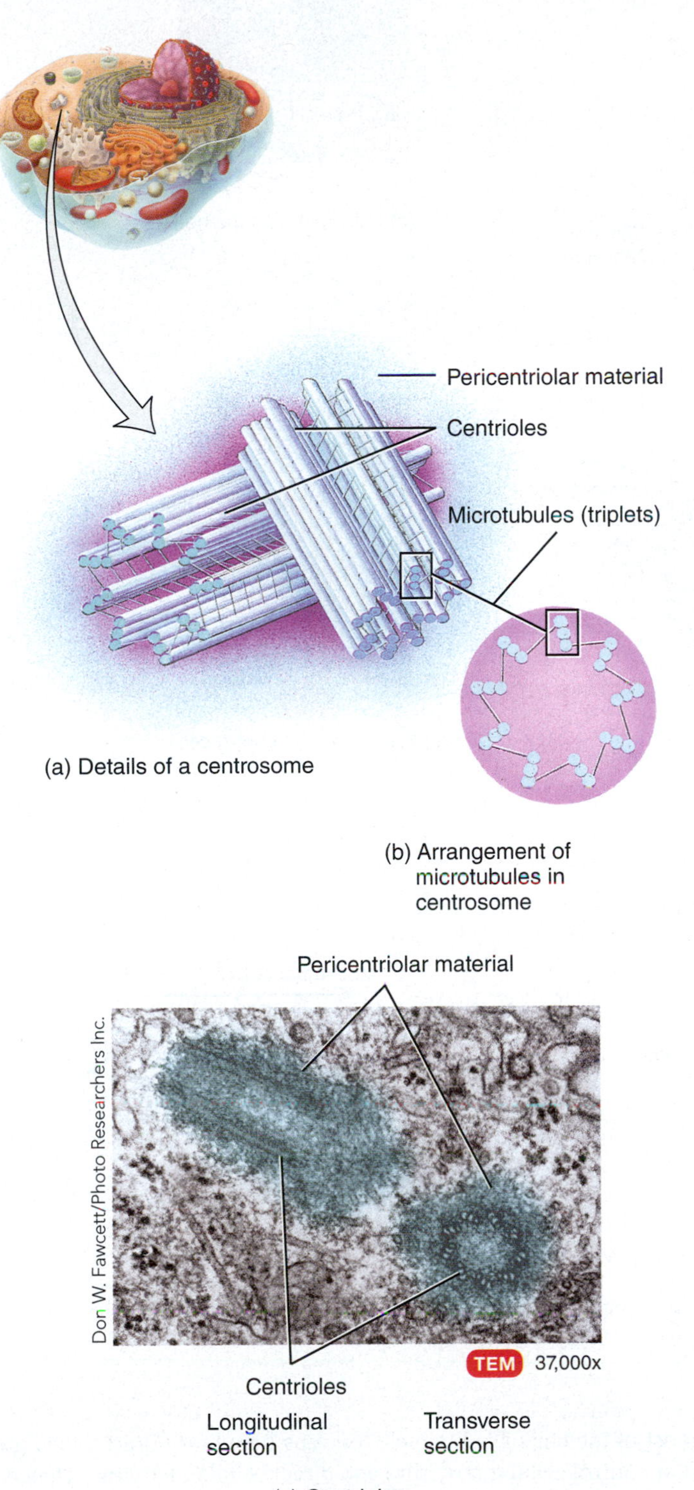

(a) Details of a centrosome

(b) Arrangement of microtubules in centrosome

(c) Centrioles

If you observed a cell that did not have a centrosome, what could you predict about its capacity for cell division?

spindle, which plays a critical role in cell division, and for microtubule formation in nondividing cells. During cell division, centrosomes replicate so that succeeding generations of cells have the capacity for cell division.

Cilia and Flagella

Microtubules are the dominant components of cilia and flagella, which are motile projections of the cell surface (FIGURE 3.12). **Cilia** (singular is *cilium*) are numerous, short, hairlike projections that extend from the surface of certain types of cells (FIGURE 3.12b). Each cilium contains a core of 20 microtubules surrounded by plasma membrane (FIGURE 3.12a). The microtubules are arranged so that one pair in the center is surrounded by nine clusters of two fused microtubules (doublets). Each cilium is anchored to a *basal body* just below the surface of the plasma membrane. A basal body is similar in structure to a centriole and functions in initiating the assembly of cilia and flagella.

Cilia propel fluids across the surfaces of cells that are firmly anchored in place. Each cilium displays an oarlike pattern of beating; it is relatively stiff during the power stroke (oar digging into the water), but more flexible during the recovery stroke (oar moving above the water preparing for a new stroke) (FIGURE 3.12d). The coordinated movement of many cilia on the surface of a cell causes the steady movement of fluid along the cell's surface. Many cells of the respiratory tract, for example, have hundreds of cilia that help sweep foreign particles trapped in mucus away from the lungs. In cystic fibrosis, extremely thick mucus secretions interfere with ciliary action and the normal functions of the respiratory tract. The movement of cilia is also paralyzed by nicotine in cigarette smoke. For this reason, smokers cough often to remove foreign particles from their airways. Cells that line the fallopian (uterine) tubes also have cilia that sweep oocytes (egg cells) toward the uterus, and females who smoke have an increased risk of ectopic (outside the uterus) pregnancy.

Flagella (singular is *flagellum*) are similar in structure to cilia but are typically much longer. Flagella usually move an entire cell. A flagellum generates forward motion along its axis by rapidly wiggling in a wavelike pattern (FIGURE 3.12e). The only example of a flagellum in the human body is a sperm cell's tail, which propels the sperm toward the oocyte in the fallopian tube (FIGURE 3.12c).

CHECKPOINT

5. What chemicals are present in cytosol?
6. What are the functions of the endoplasmic reticulum and Golgi complex?
7. What happens in the matrix and on the cristae of mitochondria?
8. How are the functions of lysosomes and peroxisomes similar? How are they different?
9. Why are proteasomes important?
10. What are the functions of the three types of proteins that comprise the cytoskeleton?

FIGURE 3.12 Cilia and flagella.

A cilium or flagellum consists of a core of microtubules with one pair in the center surrounded by nine clusters of doublet microtubules.

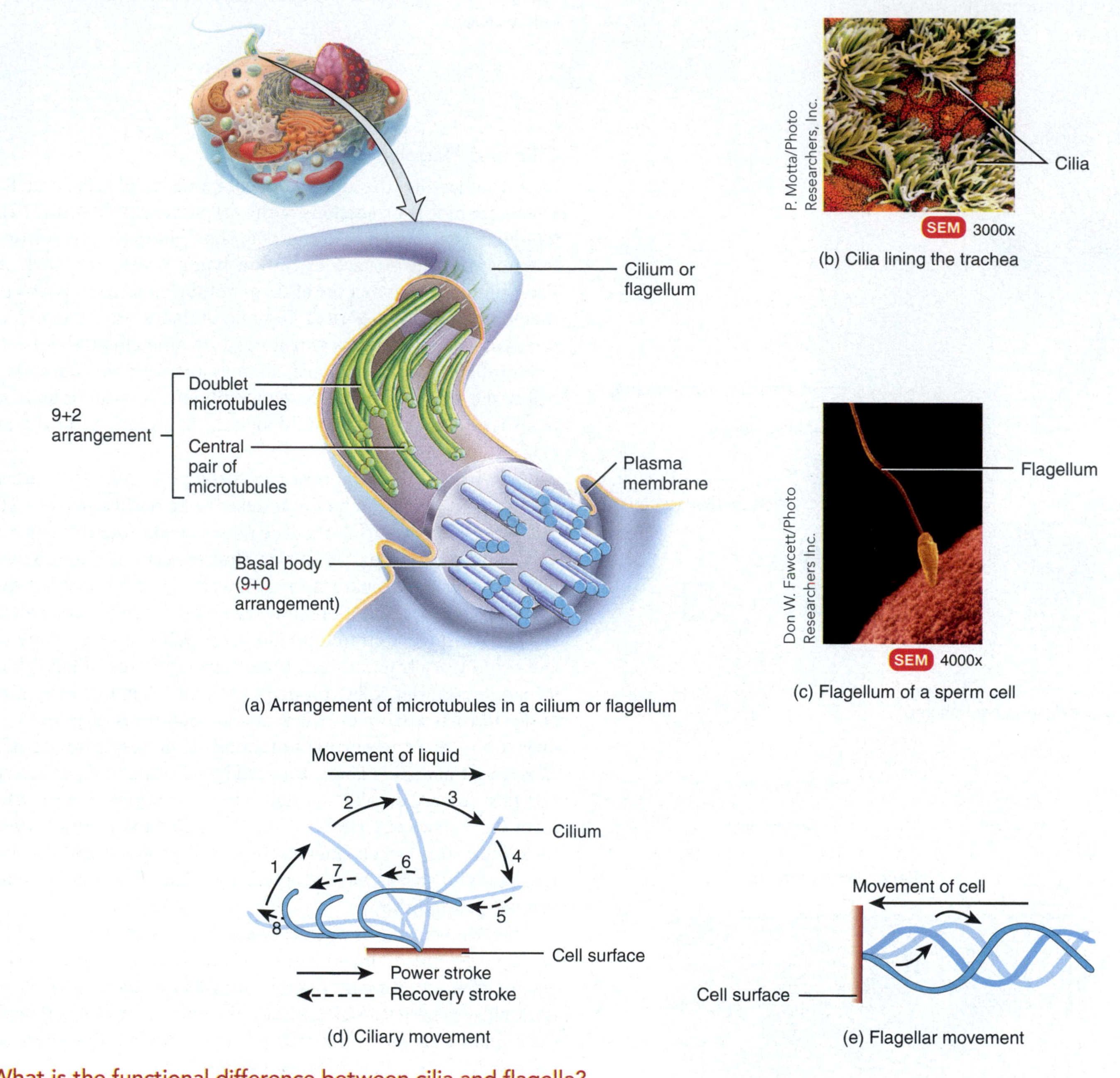

(a) Arrangement of microtubules in a cilium or flagellum

(b) Cilia lining the trachea

(c) Flagellum of a sperm cell

(d) Ciliary movement

(e) Flagellar movement

What is the functional difference between cilia and flagella?

3.4 Nucleus

OBJECTIVE

- Describe the functions of the nucleus.

The **nucleus** is a spherical or oval-shaped organelle that is usually the most prominent feature of a cell (FIGURE 3.13). Within the nucleus is most of the cell's DNA, which contains hereditary units called **genes** that control cellular structure and direct cellular activities. Hence, the nucleus functions as the control center of the cell.

Most cells have a single nucleus, although some, such as mature erythrocytes, have none. By contrast, skeletal muscle cells and a few other types of cells have multiple nuclei. A double membrane called the **nuclear envelope** separates the nucleus from the cytoplasm (FIGURE 3.13a,b). Both layers of the nuclear envelope are lipid bilayers similar to the plasma

FIGURE 3.13 Nucleus.

The nucleus is the control center of the cell.

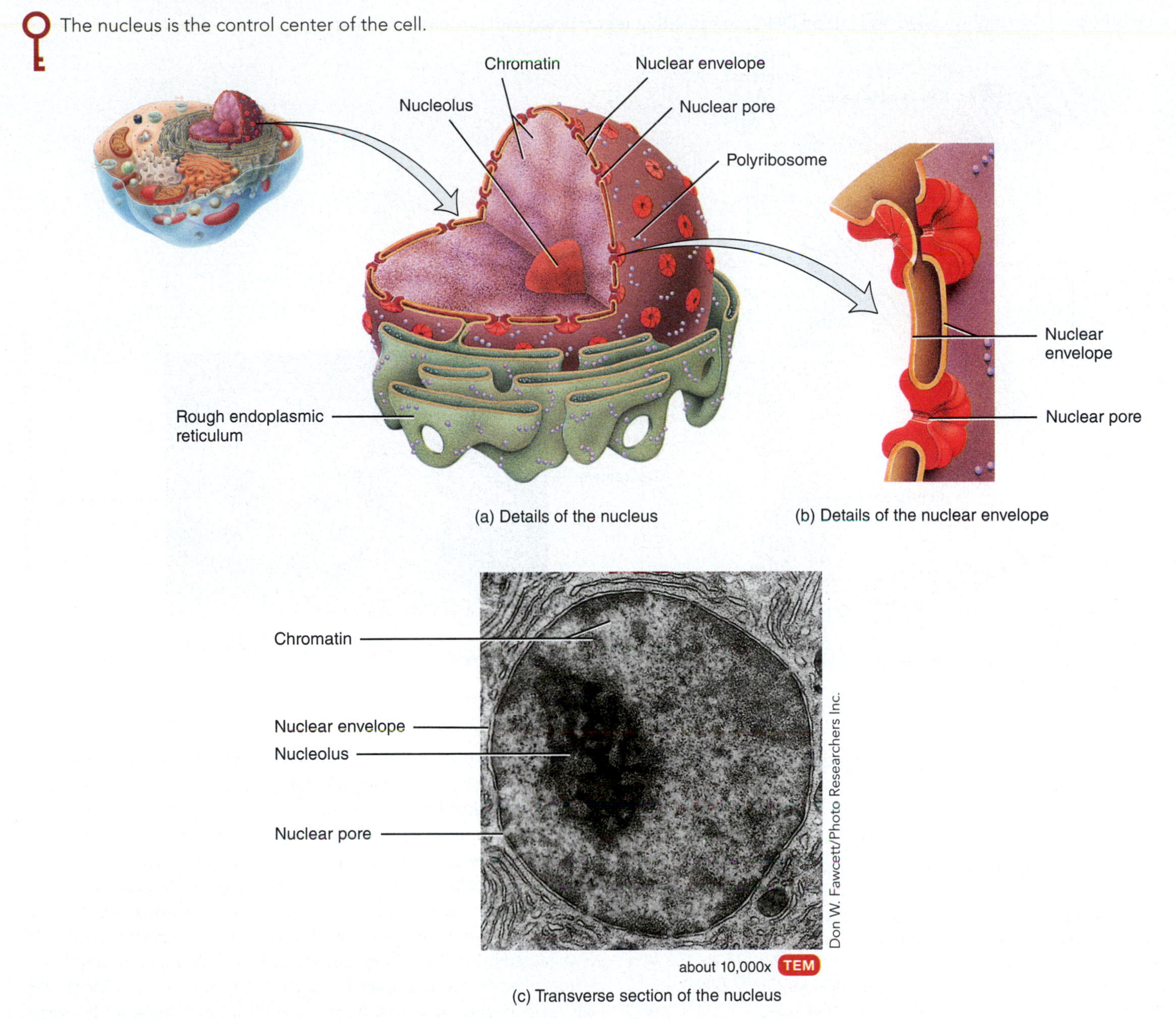

(a) Details of the nucleus

(b) Details of the nuclear envelope

(c) Transverse section of the nucleus

What types of substances pass through the nuclear pores of the nuclear envelope?

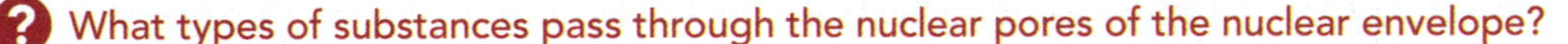

membrane. The outer membrane of the nuclear envelope is continuous with rough ER and resembles it in structure. Many openings called **nuclear pores** extend through the nuclear envelope. Each nuclear pore consists of a circular arrangement of proteins surrounding a large central opening that is about 10 times wider than the pore of a channel protein in the plasma membrane (FIGURE 3.13b). Nuclear pores allow certain substances to move between the nucleus and the cytoplasm. For example, proteins needed for nuclear functions move from the cytoplasm into the nucleus, and newly formed RNA molecules move from the nucleus into the cytoplasm.

One or more dark-staining areas called **nucleoli** (noo′-KLĒ-ō-lī; singular is *nucleolus*) are present inside the nucleus (FIGURE 3.13a,c). Each nucleolus is a cluster of DNA, RNA, and protein; it is not enclosed by a membrane. Nucleoli are the sites of synthesis of rRNA and assembly of rRNA and proteins into ribosomal subunits. Once the ribosomal subunits are assembled, they exit the nucleus through nuclear pores and enter the cytoplasm, where they form functional ribosomes. Nucleoli are quite prominent in cells that synthesize large amounts of protein, such as muscle and liver cells. Nucleoli disperse and disappear during cell division and reorganize once new cells are formed.

In cells that are not dividing, the DNA in the nucleus is associated with proteins to form a diffuse mass of fine threads called **chromatin** (FIGURE 3.14). Electron micrographs reveal that chromatin has a beads-on-a-string structure. Each bead is a **nucleosome** and consists

FIGURE 3.14 Packing of DNA into a chromosome in a dividing cell.

A chromosome is a highly coiled and folded DNA molecule that is combined with protein molecules.

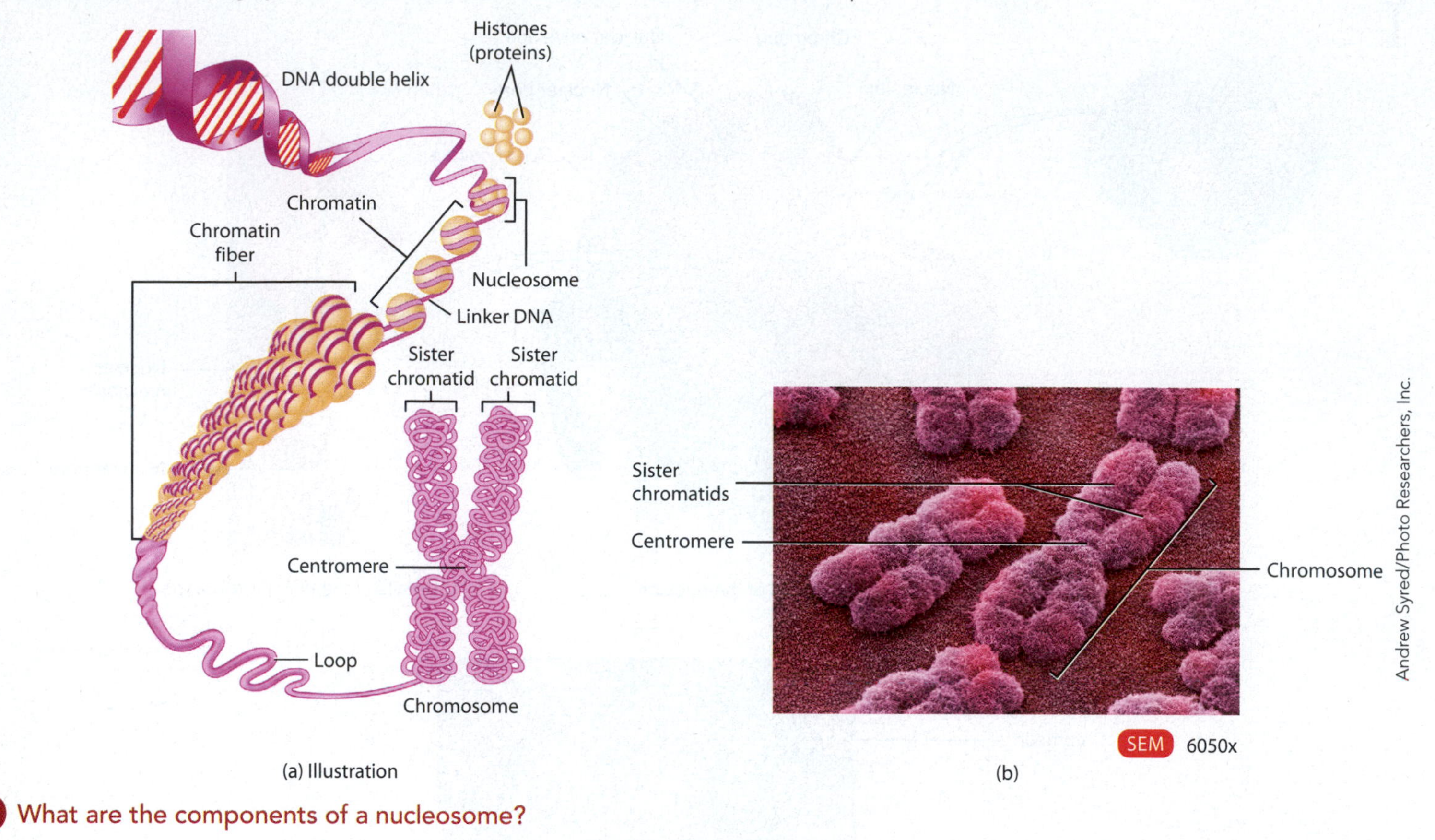

(a) Illustration

(b)

What are the components of a nucleosome?

CLINICAL CONNECTION

Genomics

In the last decade of the twentieth century, the genomes of humans, mice, fruit flies, and more than 50 microbes were sequenced. As a result, research in the field of **genomics**, the study of the relationships between the genome and the biological functions of an organism, has flourished. The Human Genome Project began in June 1990 as an effort to sequence all of the nearly 3.2 billion nucleotides of our genome, and was completed in April 2003. More than 99.9% of the nucleotide bases are identical in everyone. Less than 0.1% of our DNA (1 in each 1000 bases) accounts for inherited differences among humans. Surprisingly, at least half of the human genome consists of repeated sequences that do not code for proteins, so-called junk DNA. The average gene consists of 3000 nucleotides, but sizes vary greatly. The largest known human gene, with 2.4 million nucleotides, codes for the protein dystrophin. Scientists now know that the total number of genes in the human genome is about 35,000, far fewer than the 100,000 previously predicted to exist. Research involving the human genome and how it is affected by the environment seeks to identify and discover the functions of the specific genes that play a role in genetic diseases. Genomic medicine also aims to design new drugs and to provide screening tests to enable physicians to provide more effective counseling and treatment for disorders with significant genetic components such as hypertension (high blood pressure), obesity, diabetes, and cancer.

of DNA wrapped twice around a core of eight proteins called **histones**, which help organize the coiling and folding of DNA. The string between the beads is **linker DNA**, which holds adjacent nucleosomes together. Another histone promotes coiling of nucleosomes into a larger-diameter **chromatin fiber**, which then folds into large loops. Just before cell division takes place, the DNA replicates (duplicates) and the loops condense even more to form short, thick structures called **chromosomes**. Because the DNA has been replicated, a single chromosome consists of a pair of identical strands called **sister chromatids**, which are connected at a region called the **centromere** (FIGURE 3.14). Human somatic (body) cells have 46 chromosomes, 23 inherited from each parent. The total genetic information carried in a cell or an organism is its **genome**.

CHECKPOINT

11. Why is the nucleus referred to as the control center of the cell?
12. What is the purpose of nuclear pores?
13. How is DNA packed into the nucleus?

3.5 Gene Expression

OBJECTIVE

- Describe how a protein is synthesized using DNA as a template.

Although cells synthesize many chemicals to maintain homeostasis, much of the cellular machinery is devoted to synthesizing large numbers of diverse proteins. The proteins in turn determine the physical and chemical characteristics of cells and therefore of the organisms formed from them. Some proteins help assemble cellular structures such as the plasma membrane, the cytoskeleton, and other organelles. Others serve as hormones, antibodies, and contractile elements in muscle tissue. Still others act as enzymes, regulating the rates of the numerous chemical reactions that occur in cells, or transporters, carrying various materials in the blood. Just as genome means all of the genes in an organism, **proteome** (PRŌ-tē-ōm) refers to all of an organism's proteins.

In a process called **gene expression**, a gene's DNA is used as a template for synthesis of a specific protein. Gene expression involves two major steps—transcription and translation (FIGURE 3.15). During *transcription*, the information encoded in a specific region of DNA is copied into a specific molecule of RNA. During *translation*, the RNA attaches to a ribosome, where the information contained in RNA is used to direct the assembly of amino acids into a protein. Transcription occurs in the nucleus, whereas translation occurs in the cytoplasm.

DNA and RNA store genetic information as sets of three nucleotides. A sequence of three such nucleotides in DNA is called a **base triplet**. Each DNA base triplet is transcribed into RNA as a complementary sequence of three nucleotides, called a **codon**. A given codon specifies a particular amino acid. The **genetic code** is the set of rules that relate the base triplet sequence of DNA to the corresponding codons of RNA and the amino acids they specify.

FIGURE 3.15 Overview of gene expression. Synthesis of a specific protein requires transcription of a gene's DNA into RNA and translation of RNA into a corresponding sequence of amino acids.

Transcription occurs in the nucleus; translation occurs in the cytoplasm.

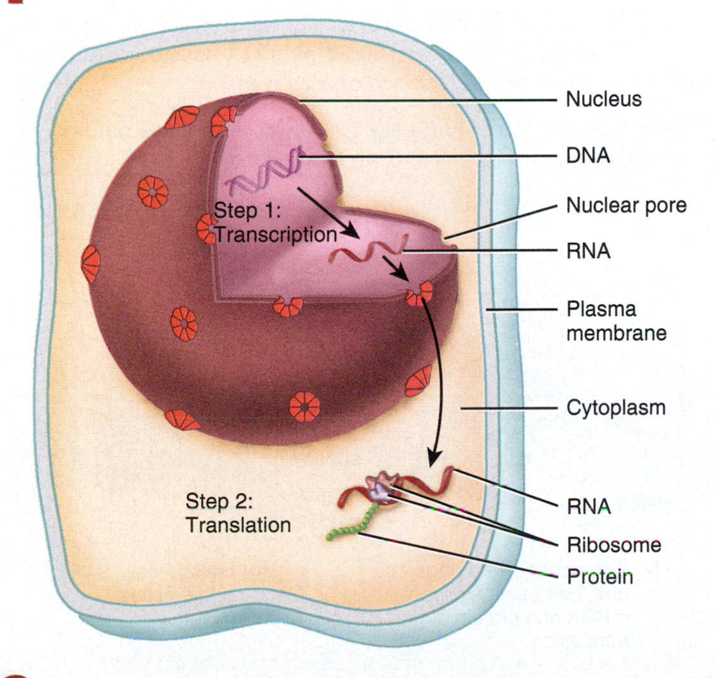

Why are proteins important in the life of a cell?

Transcription Conveys Genetic Information from DNA to RNA

In the process of **transcription**, the genetic information represented by the sequence of base triplets in DNA serves as a template for copying the information into a complementary sequence of codons (FIGURE 3.16).

FIGURE 3.16 Transcription. Transcription of DNA begins at a promoter and ends at a terminator.

During transcription, the genetic information in DNA is copied to RNA.

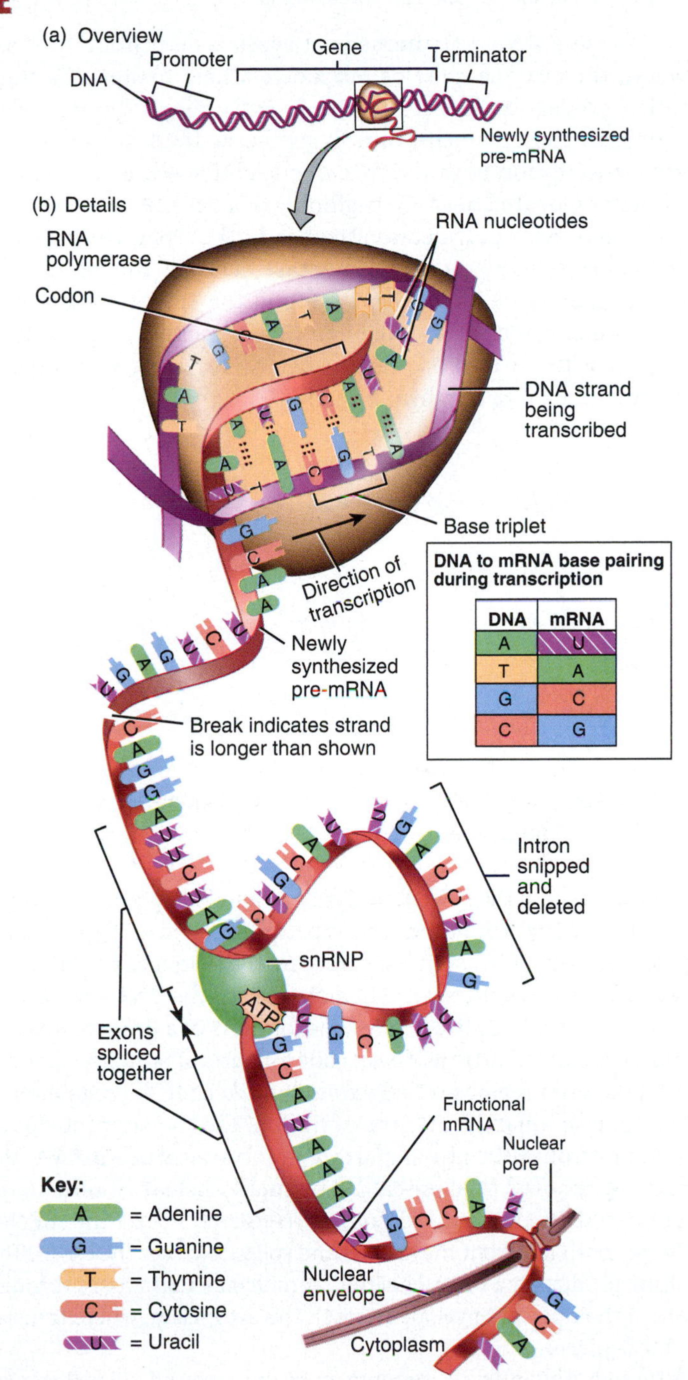

If the DNA template had the base sequence AGCT, what would be the mRNA base sequence, and what enzyme would catalyze the transcription process?

Three types of RNA are made from the DNA template:

1. **Messenger RNA (mRNA)** directs the synthesis of a protein.
2. **Ribosomal RNA (rRNA)** joins with ribosomal proteins to make ribosomes.
3. **Transfer RNA (tRNA)** binds to an amino acid and holds it in place on a ribosome until it is incorporated into a protein during translation. One end of the tRNA carries a specific amino acid, and the opposite end consists of a triplet of nucleotides called an **anticodon.** By pairing between complementary bases, the tRNA anticodon attaches to the mRNA codon. Each of the more than 20 different types of tRNA binds to only one of the 20 different amino acids.

The enzyme **RNA polymerase** catalyzes transcription of DNA. However, the enzyme must be instructed where to start the transcription process and where to end it. Only one of the two DNA strands serves as a template for RNA synthesis. The segment of DNA where transcription begins, a special nucleotide sequence called a **promoter**, is located near the beginning of a gene (FIGURE 3.16a). This is where RNA polymerase attaches to the DNA. During transcription, bases pair in a complementary manner: The bases cytosine (C), guanine (G), and thymine (T) in the DNA template pair with guanine, cytosine, and adenine (A), respectively, in the RNA strand (FIGURE 3.16b). However, adenine in the DNA template pairs with uracil (U), not thymine, in RNA:

Template DNA base sequence		Complementary RNA base sequence
A		U
T		A
G		C
	→	
C		G
A		U
T		A

Transcription of the DNA strand ends at another special nucleotide sequence called a **terminator**, which specifies the end of the gene (FIGURE 3.16a). When RNA polymerase reaches the terminator, the enzyme detaches from the transcribed RNA molecule and the DNA strand.

Not all parts of a gene actually code for parts of a protein. Regions within a gene called **introns** do *not* code for parts of proteins. They are located between regions called **exons** that *do* code for segments of a protein. Immediately after transcription, the transcript includes information from both introns and exons and is called **pre-mRNA.** The introns are removed from pre-mRNA by **small nuclear ribonucleoproteins** (snRNPs, pronounced "snurps") (FIGURE 3.16b). The snRNPs are enzymes that cut out the introns and splice together the exons. The resulting product is a functional mRNA molecule that passes through a pore in the nuclear envelope to reach the cytoplasm, where translation takes place.

Although the human genome contains around 35,000 genes, there are probably 500,000 to 1 million human proteins. How can so many proteins be coded for by so few genes? Part of the answer lies in **alternative splicing** of mRNA, a process in which the pre-mRNA transcribed from a gene is spliced in different ways to produce several different mRNAs. The different mRNAs are then translated into different proteins. In this way, one gene may code for 10 or more different proteins. In addition, chemical modifications are made to proteins after translation, for example, as proteins pass through the Golgi complex. Such chemical alterations can produce two or more different proteins from a single translation.

Translation Uses Genetic Information Carried by mRNA to Synthesize a Protein

In the process of **translation**, the nucleotide sequence in an mRNA molecule specifies the amino acid sequence of a protein. Ribosomes in the cytoplasm carry out translation. A ribosome has a binding site for mRNA and three binding sites for tRNA molecules: a P site, an A site, and an E site (FIGURE 3.17). The **P (peptidyl) site** binds the tRNA carrying the growing polypeptide chain. The **A (aminoacyl) site** binds the tRNA carrying the next amino acid to be added to the growing polypeptide. The **E (exit) site** binds tRNA just before it is released from the ribosome. Translation occurs in the following way (FIGURE 3.18):

1. An mRNA molecule binds to the small ribosomal subunit at the mRNA binding site. A special tRNA, called *initiator tRNA*, binds to the start codon (AUG) on mRNA, where translation begins. The tRNA anticodon (UAC) attaches to the mRNA codon (AUG) by pairing between the complementary bases. Besides being the start codon, AUG is also the codon for the amino acid methionine. Thus, methionine is always the first amino acid in a growing polypeptide.
2. Next, the large ribosomal subunit attaches to the small ribosomal subunit–mRNA complex, creating a functional ribosome. The initiator tRNA, with its amino acid (methionine), fits into the P site of the ribosome.

FIGURE 3.17 Translation. During translation, an mRNA molecule binds to a ribosome. Then the mRNA nucleotide sequence specifies the amino acid sequence of a protein.

A ribosome has a binding site for mRNA and three binding sites for tRNA molecules (a P site, an A site, and an E site).

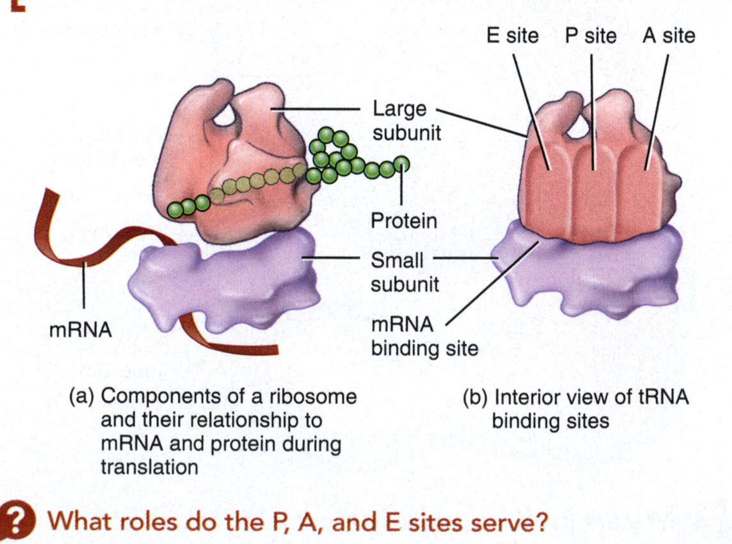

What roles do the P, A, and E sites serve?

FIGURE 3.18 Protein elongation and termination of protein synthesis during translation.

During protein synthesis, the small and large ribosomal subunits join to form a functional ribosome. When the process is complete, they separate.

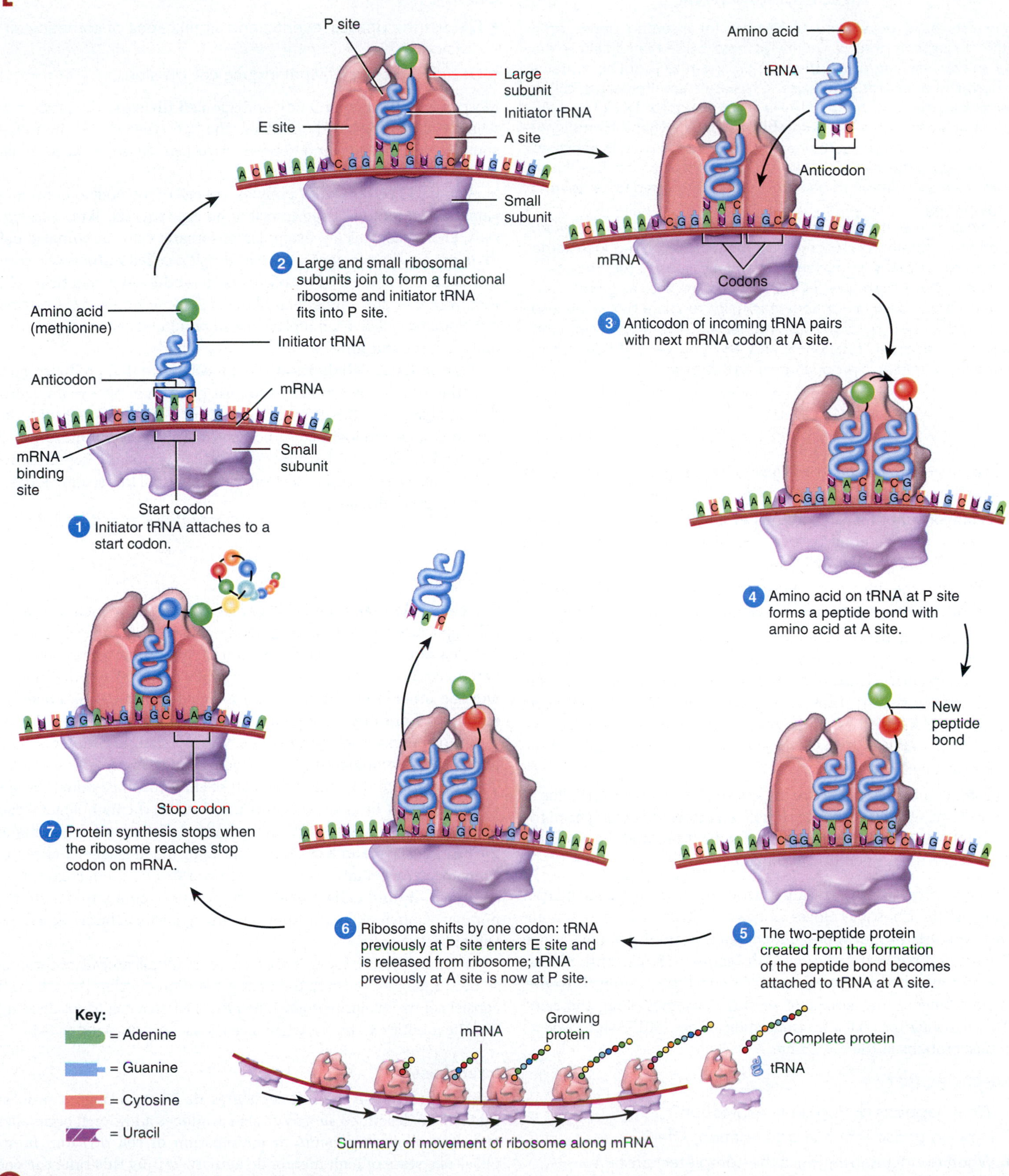

What is the function of a stop codon?

CLINICAL CONNECTION

Recombinant DNA

Scientists have developed techniques for inserting genes from other organisms into a variety of host cells. Manipulating the cell in this way can cause the host organism to produce proteins it normally does not synthesize. Organisms so altered are called **recombinants**, and their DNA—a combination of DNA from different sources—is called **recombinant DNA**. When recombinant DNA functions properly, the host synthesizes the protein specified by the new gene it has acquired. The technology that has arisen from the manipulation of genetic material is referred to as **genetic engineering**.

The practical applications of recombinant DNA technology are enormous. Strains of recombinant bacteria now produce large quantities of many important therapeutic substances, including ***human growth hormone (hGH)***, required for normal growth and metabolism; ***insulin***, a hormone that helps regulate blood glucose level and is used by diabetics; ***interferon (IFN)***, an antiviral (and possibly anticancer) substance; and ***erythropoietin (EPO)***, a hormone that stimulates production of erythrocytes.

3. The anticodon of another tRNA with its attached amino acid pairs with the second mRNA codon at the A site of the ribosome.
4. A component of the large ribosomal subunit catalyzes the formation of a peptide bond between methionine and the amino acid carried by the tRNA at the A site.
5. Following the formation of the peptide bond, the resulting two-peptide protein becomes attached to the tRNA at the A site.
6. After peptide bond formation, the ribosome shifts the mRNA strand by one codon. The tRNA in the P site enters the E site and is subsequently released from the ribosome. The tRNA in the A site bearing the two-peptide protein shifts into the P site, allowing another tRNA with its amino acid to bind to a newly exposed codon at the A site. Steps 3 through 6 occur repeatedly, and the protein lengthens progressively.
7. Protein synthesis ends when the ribosome reaches a stop codon at the A site, which causes the completed protein to detach from the final tRNA. In addition, tRNA vacates the P site, and the ribosome splits into its large and small subunits.

Protein synthesis progresses at a rate of about 15 peptide bonds per second. As the ribosome moves along the mRNA and before it completes synthesis of the whole protein, another ribosome may attach behind it and begin translation of the same mRNA strand. Several ribosomes attached to the same mRNA constitute a **polyribosome**. The simultaneous movement of several ribosomes along the same mRNA molecule permits the translation of one mRNA into several identical proteins at the same time.

CHECKPOINT

14. What happens during gene expression?
15. How do transcription and translation differ?
16. When during translation is the completed protein released from the ribosome?

3.6 Cell Division

OBJECTIVES

- Discuss the phases, events, and significance of somatic cell division.
- Describe the signals that induce cell division.

Most cells of the human body undergo **cell division**, the process by which cells reproduce themselves. The two types of cell division—somatic cell division and reproductive cell division—accomplish different goals for the organism.

A **somatic cell** (*soma* = body) is any cell of the body other than a gamete. A **gamete** is a sperm cell or an egg (oocyte). Skin cells, liver cells, and leukocytes are examples of somatic cells. In **somatic cell division**, a cell undergoes a nuclear division called **mitosis** and a cytoplasmic division called **cytokinesis** to produce two identical cells, each with the same number and kind of chromosomes as the original cell. Somatic cell division replaces dead or injured cells and adds new ones during tissue growth.

Reproductive cell division is the mechanism that produces gametes, the cells needed to form the next generation of sexually reproducing organisms. This process consists of a special type of nuclear division called **meiosis**, in which the number of chromosomes is reduced by half, and cytokinesis (cytoplasmic division). Reproductive cell division is described in Section 23.1; the rest of this section focuses on somatic cell division.

Somatic Cell Division Produces Two Identical Cells

The **cell cycle** is an orderly sequence of events by which a somatic cell duplicates its contents and divides in two. Somatic cells contain 23 pairs of chromosomes, for a total of 46 chromosomes. The two chromosomes of each pair—one originally derived from the mother and the other from the father—are called **homologous chromosomes** or **homologs**. They contain similar genes arranged in the same (or almost the same) order. When examined under a light microscope, homologous chromosomes generally look very similar. The exception to this rule is one pair of chromosomes called the **sex chromosomes**, designated X and Y. In females the homologous pair of sex chromosomes consists of two large X chromosomes; in males the pair consists of an X chromosome and a much smaller Y chromosome. Because somatic cells contain two sets of chromosomes, they are called **diploid cells**. Geneticists use the symbol n to denote the number of different chromosomes in an organism; in humans, $n = 23$. Diploid cells are $2n$.

When a cell reproduces, it must replicate (duplicate) all its chromosomes to pass its genes to the next generation of cells. The cell cycle consists of two major periods: interphase, when a cell is not dividing, and the mitotic (M) phase, when a cell is dividing (FIGURE 3.19).

Interphase

During **interphase** the cell replicates its DNA through a process that will be described shortly. It also produces additional organelles and cytosolic components in anticipation of cell division. Interphase is a state of high metabolic activity; during this time the cell does most of its growing. Interphase consists of three phases: G_1, S, and G_2 (**FIGURE 3.19**). The S stands for *synthesis* of DNA. Because the G phases are periods when there is no activity related to DNA duplication, they are thought of as *gaps* or interruptions in DNA duplication.

The **G_1 phase** is the interval between the mitotic phase and the S phase. During G_1, the cell is metabolically active; it replicates most of its organelles and cytosolic components but not its DNA. Replication of centrosomes also begins in the G_1 phase. Virtually all the cellular activities described in this chapter happen during G_1. For a cell with a total cell cycle time of 24 hours, G_1 lasts 8 to 10 hours. However, the duration of this phase is quite variable. It is very short in many embryonic cells or cancer cells. Cells that remain in G_1 for a very long time, perhaps destined never to divide again, are said to be in the **G_0 phase**. Most neurons are in the G_0 phase. Once a cell enters the S phase, however, it is committed to go through cell division.

The **S phase**, the interval between G_1 and G_2, lasts about 8 hours. During the S phase, DNA replication occurs. As a result, the two identical cells formed during cell division will have the same genetic material.

The **G_2 phase** is the interval between the S phase and the mitotic phase. It lasts 4 to 6 hours. During G_2, cell growth continues, enzymes and other proteins are synthesized in preparation for cell division, and replication of centrosomes is completed.

When DNA replicates during the S phase, its helical structure partially uncoils, and the two strands separate at the points where hydrogen bonds connect base pairs (**FIGURE 3.20**). Each exposed base of the old DNA strand then pairs with the complementary base of a newly synthesized nucleotide. A new DNA strand takes shape as chemical bonds form between neighboring nucleotides. The uncoiling and complementary base pairing continues until each of the two original DNA strands is joined with a newly formed complementary DNA strand. The original DNA molecule has become two identical DNA molecules.

FIGURE 3.19 The cell cycle. Not illustrated is cytokinesis, division of the cytoplasm, which usually begins during late anaphase of the mitotic phase.

In a complete cell cycle, a starting cell duplicates its contents and divides into two identical cells.

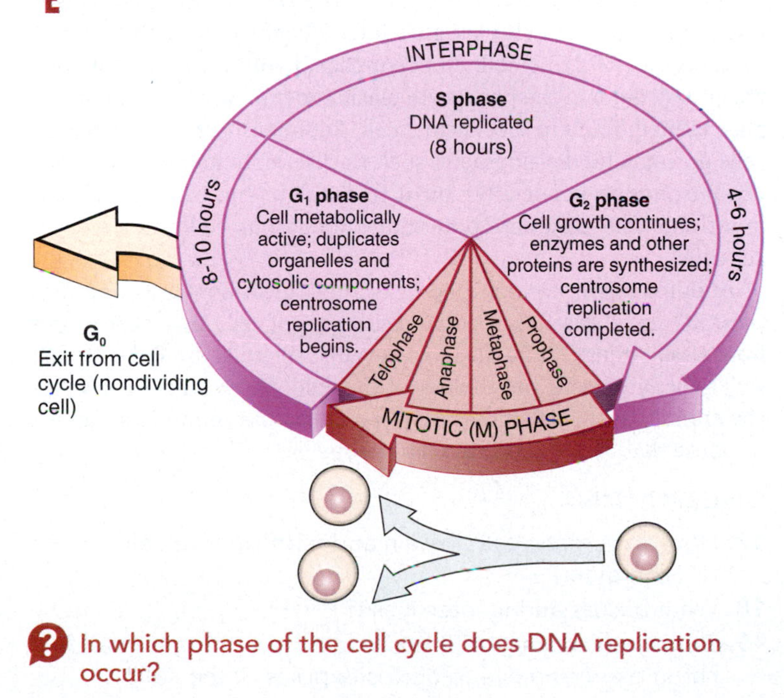

In which phase of the cell cycle does DNA replication occur?

FIGURE 3.20 Replication of DNA. The two strands of the double helix separate by breaking the hydrogen bonds (shown as dotted lines) between nucleotides. New, complementary nucleotides attach at the proper sites, and a new strand of DNA is synthesized alongside each of the original strands. Arrows indicate hydrogen bonds forming again between pairs of bases.

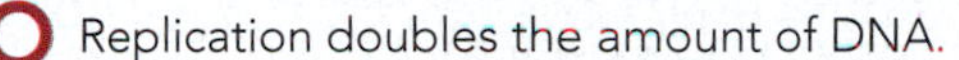

Replication doubles the amount of DNA.

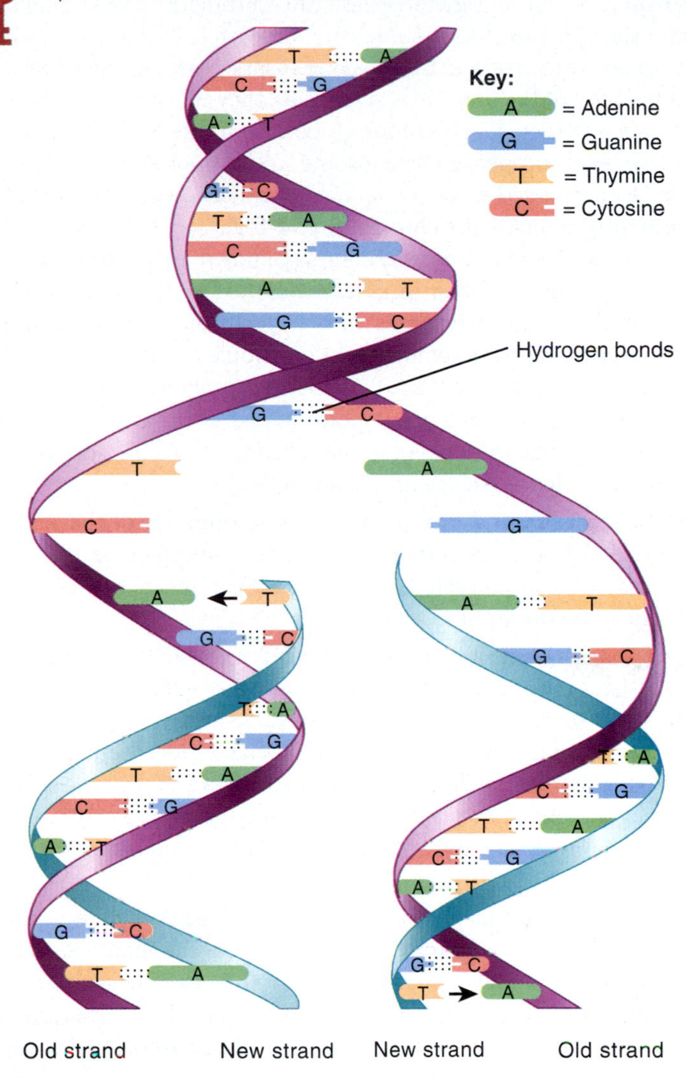

Why is it crucial that DNA replication occurs before cytokinesis in somatic cell division?

A microscopic view of a cell during interphase shows a clearly defined nuclear envelope, a nucleolus, and a tangled mass of chromatin (**FIGURE 3.21a**). Once a cell completes its activities during the G_1, S, and G_2 phases of interphase, the mitotic phase begins.

Mitotic Phase

The **mitotic (M) phase** of the cell cycle consists of a nuclear division (mitosis) and a cytoplasmic division (cytokinesis) to form two identical cells. The events that occur during mitosis and cytokinesis are plainly visible under a microscope because chromatin condenses into discrete chromosomes.

Nuclear Division: Mitosis **Mitosis**, as noted earlier, is the distribution of two sets of chromosomes into two separate nuclei. The process results in the *exact* partitioning of genetic information. For convenience, biologists divide mitosis into four phases: prophase, metaphase, anaphase, and telophase. However, mitosis is a continuous process; one phase merges directly into the next.

1. **Prophase**. During early prophase, the chromatin fibers condense and shorten into chromosomes that are visible under the light microscope (**FIGURE 3.21b**). The condensation process may prevent entangling of the long DNA strands as they move during mitosis. Because DNA replication took place during the S phase of interphase, each prophase chromosome consists of a pair of identical strands called *sister chromatids*. A constricted region called a **centromere** holds the chromatid pair together. At the outside of each centromere is a protein complex known as the **kinetochore** (ki-NET-ō-kor). Later in prophase, tubulins in the pericentriolar material of the centrosomes start to form the **mitotic spindle**, a football-shaped assembly of microtubules that attach to the kinetochore (**FIGURE 3.21b**). As the microtubules lengthen, they push the centrosomes to the poles (ends) of the cell so that the spindle extends from pole to pole. Then the nucleolus disappears and the nuclear envelope breaks down.
2. **Metaphase**. During metaphase, the microtubules of the mitotic spindle align the centromeres of the chromatid pairs at the exact center of the mitotic spindle (**FIGURE 3.21c**). This midpoint region is called the **metaphase plate**.
3. **Anaphase**. During anaphase, the centromeres split, separating the two sister chromatids of each chromosome, which move toward opposite poles of the cell (**FIGURE 3.21d**). Once separated, the sister chromatids are considered individual chromosomes. As the chromosomes are pulled by the microtubules of the mitotic spindle during anaphase, they appear V-shaped because the centromeres lead the way, dragging the trailing arms of the chromosomes toward the pole.
4. **Telophase**. The final phase of mitosis, telophase, begins after chromosomal movement stops (**FIGURE 3.21e**). The identical sets of chromosomes, now at opposite poles of the cell, uncoil and revert to the threadlike chromatin form. A nuclear envelope forms around each chromatin mass, nucleoli reappear in the identical nuclei, and the mitotic spindle breaks up.

Cytoplasmic Division: Cytokinesis As noted earlier, division of a cell's cytoplasm and organelles into two identical cells is called **cytokinesis**. This process usually begins in late anaphase with the formation of a **cleavage furrow**, a slight indentation of the plasma membrane, and is completed after telophase. The cleavage furrow usually appears midway between the centrosomes and extends around the periphery of the cell (**FIGURE 3.21d,e**). Actin microfilaments that lie just inside the plasma membrane form a *contractile ring* that pulls the plasma membrane progressively inward. The ring constricts the center of the cell, like tightening a belt around the waist, and ultimately pinches it in two. Because the plane of the cleavage furrow is always perpendicular to the mitotic spindle, the two sets of chromosomes end up in separate cells. When cytokinesis is complete, interphase begins (**FIGURE 3.21f**).

The sequence of events can be summarized as follows:

$$G_1 \rightarrow \text{S phase} \rightarrow G_2 \text{ phase} \rightarrow \text{mitosis} \rightarrow \text{cytokinesis}$$

Cell Destiny Is Controlled by Many Factors

A cell has three possible destinies—to remain alive and functioning without dividing, to grow and divide, or to die. Homeostasis is maintained when there is a balance between cell proliferation and cell death. The signals that tell a cell when to exist in the G_0 phase, when to divide, and when to die have been the subjects of intense and fruitful research in recent years.

Within a cell, enzymes called **cyclin-dependent protein kinases (Cdks)** transfer a phosphate group from ATP to a protein to activate the protein; other enzymes can remove the phosphate group from the protein to deactivate it. The activation and deactivation of Cdks at the appropriate time is crucial in the initiation and regulation of DNA replication, mitosis, and cytokinesis.

Switching the Cdks on and off is the responsibility of cellular proteins called **cyclins** (SĪK-lins), so named because their levels rise and fall during the cell cycle. The joining of a specific cyclin and Cdk molecule triggers various events that control cell division.

The activation of specific cyclin–Cdk complexes is responsible for progression of a cell from G_1 to S to G_2 to mitosis in a specific order. If any step in the sequence is delayed, all subsequent steps are delayed in order to maintain the normal sequence. The levels of cyclins in the cell are very important in determining the timing and sequence of events in cell division. For example, the level of the cyclin that helps drive a cell from G_2 to mitosis rises throughout the G_1, S, and G_2 phases and into mitosis. The high level triggers mitosis, but toward the end of mitosis, the level declines rapidly and mitosis ends. Destruction of this cyclin, as well as others in the cell, is by proteasomes.

Cellular death is also regulated. Throughout the lifetime of an organism, certain cells undergo **apoptosis** (ap-op-TŌ-sis), an orderly, genetically programmed death. In apoptosis, a triggering agent from either outside or inside the cell causes "cell-suicide" genes to produce enzymes that damage the cell in several ways, including disruption of its cytoskeleton and nucleus. As a result, the cell shrinks and pulls away from neighboring cells. Although the plasma membrane remains intact, the DNA within the nucleus fragments and the cytoplasm shrinks. Phagocytes in the vicinity then ingest the dying cell, an event that involves the binding of a receptor protein in the phagocyte plasma membrane to a lipid in the plasma membrane of the suicidal cell. Apoptosis removes unneeded cells during fetal development, such as the webbing between digits. It continues to occur after birth to regulate the number of cells in a tissue and eliminate potentially dangerous cells such as cancer cells.

Apoptosis is a normal type of cell death; in contrast, **necrosis** (ne-KRŌ-sis = death) is a pathological type of cell death that results from tissue injury. In necrosis, many adjacent cells swell, burst, and spill their cytoplasm into the interstitial fluid. The cellular debris usually stimulates an inflammatory response by the immune system, a response that does not occur in apoptosis.

CHECKPOINT

17. How do somatic cell division and reproductive cell division differ?
18. What occurs during interphase?
19. During which stage of mitosis do centromeres split and chromosomes move to opposite poles of the cell?

FIGURE 3.21 Cell division: mitosis and cytokinesis.

In somatic cell division, a single starting cell divides to produce two identical diploid cells.

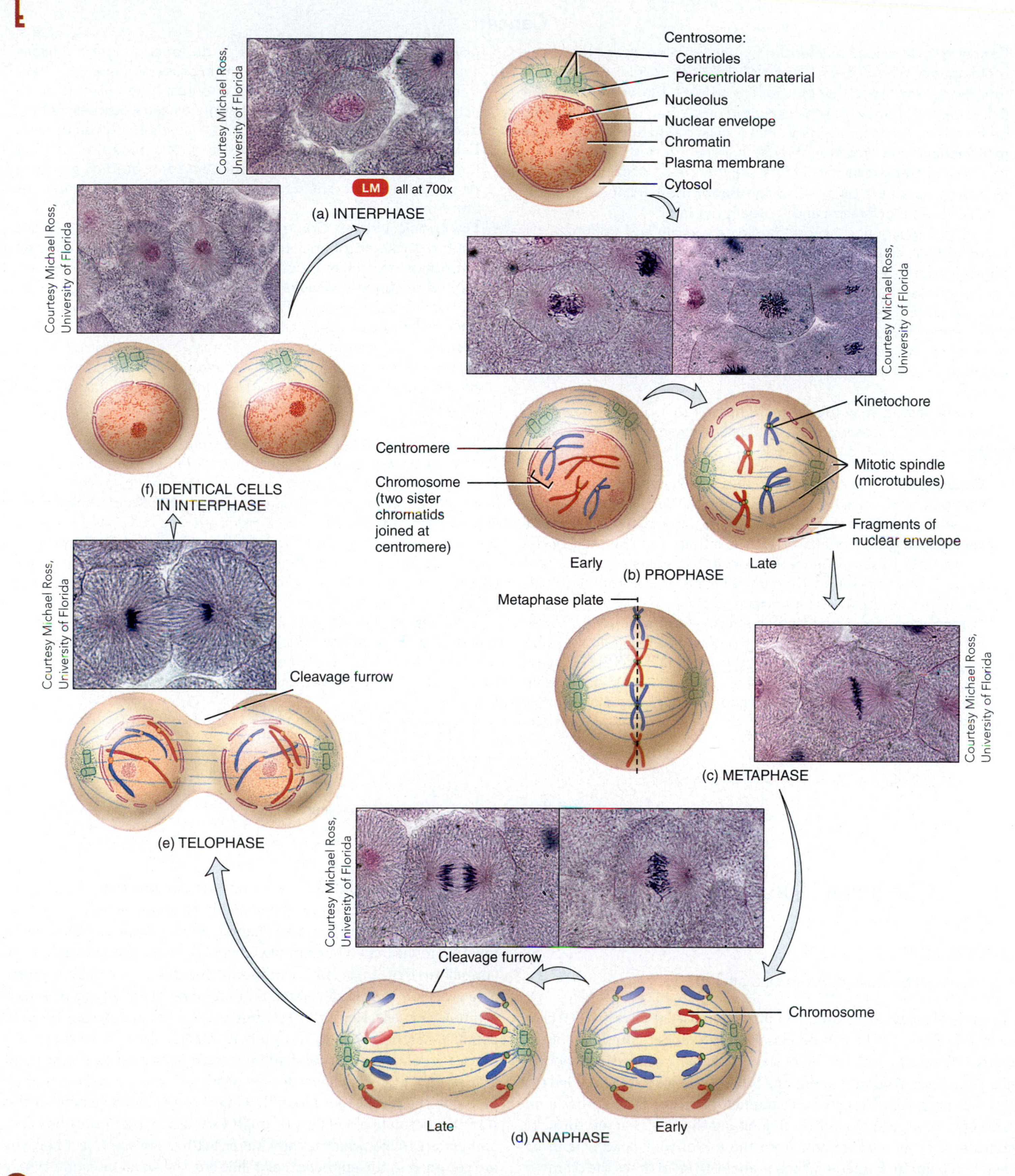

When does cytokinesis begin?

CLINICAL CONNECTION

Cancer

Cancer is a disease characterized by uncontrolled or abnormal cell proliferation. When cells in a part of the body divide without control, the excess tissue that develops is called a **tumor** or **neoplasm** (NĒ-ō-plazm). Tumors may be cancerous and often fatal, or they may be harmless. A cancerous neoplasm is called a **malignant tumor** or **malignancy**. One property of most malignant tumors is their ability to undergo **metastasis** (me-TAS-ta-sis), the spread of cancerous cells to other parts of the body. A **benign tumor** (bē-NĪN) is a neoplasm that does not metastasize and is usually not fatal.

Cells of malignant tumors duplicate rapidly and continuously. As malignant cells invade surrounding tissues, they often trigger **angiogenesis**, the growth of new networks of blood vessels. As the cancer grows, it begins to compete with normal tissues for space and nutrients. Eventually, the normal tissue decreases in size and dies. Some malignant cells may detach from the initial (primary) tumor and invade a body cavity or enter the blood or lymph, then circulate to and invade other body tissues, establishing secondary tumors.

Several factors may trigger a normal cell to lose control and become cancerous. Among these factors are carcinogens, oncogenes, and oncogenic viruses.

- ***Carcinogens.*** One cause of cancer is environmental agents: substances in the air we breathe, the water we drink, and the food we eat. A chemical agent or radiation that produces cancer is called a **carcinogen**. Carcinogens induce **mutations**, permanent changes in the DNA base sequence of a gene. Examples of carcinogens are hydrocarbons found in cigarette tar, radon gas from the earth, and ultraviolet (UV) radiation in sunlight.
- ***Oncogenes.*** Intensive research efforts are now directed toward studying cancer-causing genes, or **oncogenes** (ON-kō-jēnz). When inappropriately activated, these genes have the ability to transform a normal cell into a cancerous cell. Most oncogenes derive from normal genes called **proto-oncogenes** that regulate growth and development. The proto-oncogene undergoes some change that causes it to be expressed inappropriately, to make its products in excessive amounts, or to make its products at the wrong time. Some oncogenes cause excessive production of growth factors, chemicals that stimulate cell growth. Others may trigger changes in a cell-surface receptor, causing it to send signals as though it were being activated by a growth factor. As a result, the growth pattern of the cell becomes abnormal.
- ***Oncogenic viruses.*** Some cancers have a viral origin. Viruses are tiny packages of nucleic acids, either RNA or DNA, that can reproduce only while inside the cells they infect. Some viruses, termed **oncogenic viruses**, cause cancer by stimulating abnormal proliferation of cells. For instance, the ***human papillomavirus (HPV)*** causes virtually all cervical cancers in women. This virus produces a protein that causes proteasomes to destroy p53, a protein that normally suppresses unregulated cell division. In the absence of this suppressor protein, cells proliferate without control.

Treatment for cancer usually involves surgery to remove the tumor. Cancer treatment may also include chemotherapy and radiation therapy. Chemotherapy involves administering drugs that cause death of cancerous cells. Radiation therapy breaks chromosomes, thus blocking cell division. Because cancerous cells divide rapidly, they are more vulnerable to the destructive effects of chemotherapy and radiation therapy than are normal cells. Unfortunately for the patients, hair follicle cells, red bone marrow cells, and cells lining the gastrointestinal tract also are rapidly dividing. Hence, the side effects of chemotherapy and radiation therapy include hair loss due to death of hair follicle cells, vomiting and nausea due to death of cells lining the stomach and intestines, and susceptibility to infection due to slowed production of leukocytes in red bone marrow.

3.7 Cellular Diversity

OBJECTIVE

- Describe how cells differ in size and shape.

The body of an average human adult is composed of nearly 100 trillion cells. All of these cells can be classified into about 200 different cell types. Cells vary considerably in size. The largest cell, a single oocyte, has a diameter of about 140 μm and is barely visible to the unaided eye. By contrast, an erythrocyte has a diameter of 8 μm and requires a microscope to be seen. To better understand the size of an erythrocyte, consider that an average hair from the top of your head is approximately 100 μm in diameter, which is about 12 to 13 times the diameter of an erythrocyte.

The shapes of cells also vary considerably (**FIGURE 3.22**). They may be round, oval, flat, cube-shaped, column-shaped, elongated, star-shaped, cylindrical, or disc-shaped. A cell's shape is related to its function in the body. For example, a sperm cell has a long whiplike tail (flagellum) that it uses for locomotion. The disc shape of an erythrocyte gives it a large surface area that enhances its ability to pass oxygen to other cells. Neurons have long extensions that permit them to conduct electrical signals over great distances. The spindle shape of a relaxed smooth muscle cell becomes more spherical as it contracts. This change in shape allows groups of smooth muscle cells to narrow or widen the passage for blood flowing through blood vessels. In this way, they regulate blood flow through various tissues. Some body cells contain microvilli, which greatly increase their surface area. Microvilli are common in the epithelial cells that line the small intestine, where the large surface area speeds the absorption of digested food.

FIGURE 3.22 Diverse sizes and shapes of human cells. The relative difference in size between the smallest and largest cells is actually much greater than shown here.

The nearly 100 trillion cells in an average adult can be classified into 200 different cell types.

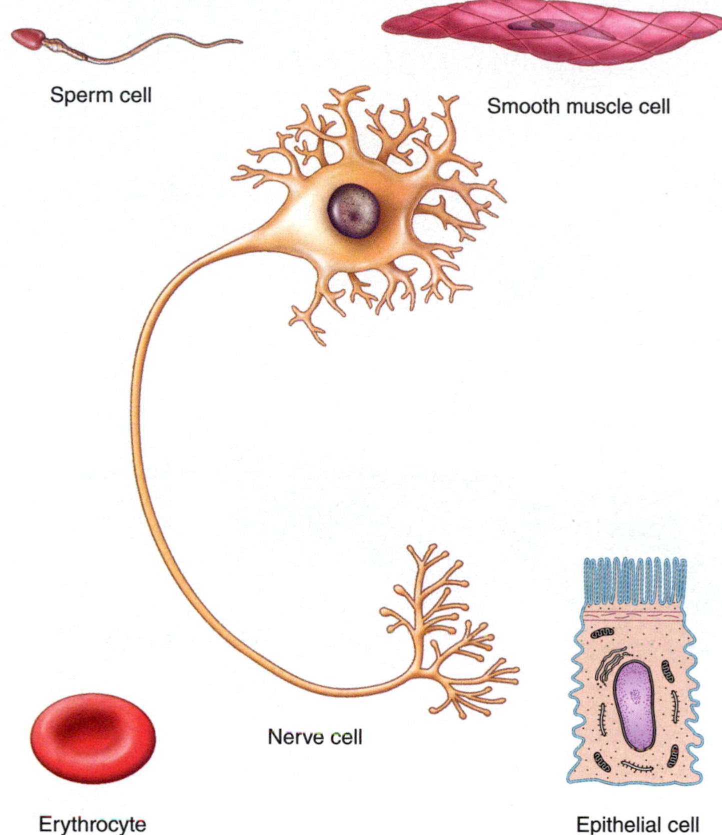

Why are sperm cells the only body cells that need to have a flagellum?

CHECKPOINT

20. How is the shape of an erythrocyte related to its function?

3.8 Organization of Cells into Tissues

OBJECTIVES

- Describe the four basic types of tissue that make up the human body and state the functions of each.
- Discuss the functions of the five main types of cell junctions.

Cells are highly organized living units, but they typically do not function alone. Instead, cells work together in groups called tissues. A **tissue** is a group of similar cells that function together to carry out specialized activities. Body tissues are classified into four basic types: epithelial tissue, connective tissue, muscle tissue, and nervous tissue.

Epithelial Tissue Serves as a Barrier, Secretes Substances, and Absorbs Materials

Epithelial tissue, or **epithelium**, covers body surfaces, lines hollow organs and ducts, and forms glands. An epithelium consists of closely packed cells with little extracellular material between them, and the cells are arranged in continuous sheets, in either single or multiple layers (FIGURE 3.23). Just beneath an epithelium is a **basement membrane**, an extracellular structure composed mostly of protein fibers. The basement membrane connects epithelial tissue to underlying connective tissue and provides support to epithelial cells (FIGURE 3.23).

Epithelial tissue has three major functions: (1) It serves as a physical barrier that protects against harmful agents in the environment; (2) it secretes substances onto the body surface or into hollow organs or the blood; and (3) it absorbs materials into the bloodstream or lymph. There are two major types of epithelial tissue: (1) covering and lining epithelium and (2) glandular epithelium.

Covering and Lining Epithelium

Covering and lining epithelium forms the outer covering of the skin (the epidermis) and some internal organs. It also lines the lumen (interior space) of hollow structures such as blood vessels, ducts, and organs of the respiratory, digestive, urinary, and reproductive tracts. Examples of covering and lining epithelium are shown in FIGURE 3.24.

Glandular Epithelium

Glandular epithelium makes up the secreting portion of glands. A **gland** may consist of a single epithelial cell or a group of epithelial cells that secrete substances into ducts or into the blood. All glands

FIGURE 3.23 Overview of epithelium.

Epithelium consists of closely packed cells with little extracellular material between them.

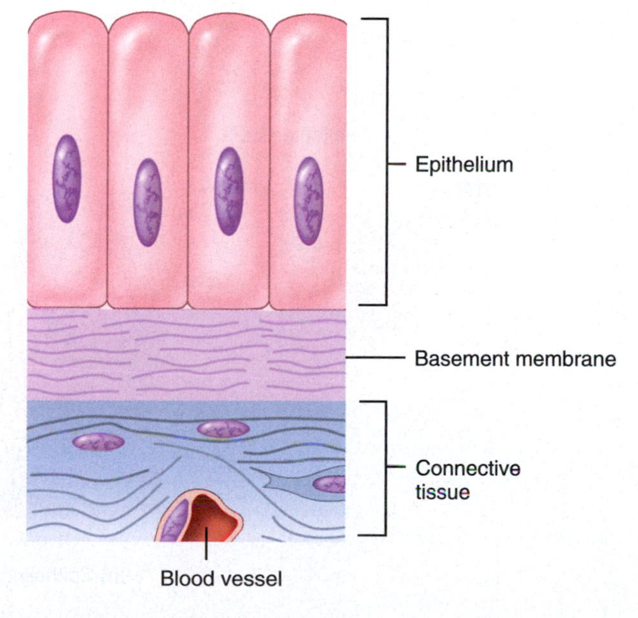

What is the functional significance of the basement membrane?

FIGURE 3.24 Covering and lining epithelium.

Covering and lining epithelium forms the outer covering of the skin (epidermis) and some internal organs; it also lines the lumen of hollow structures such as blood vessels, ducts, and organs of the respiratory, digestive, urinary, and reproductive tracts.

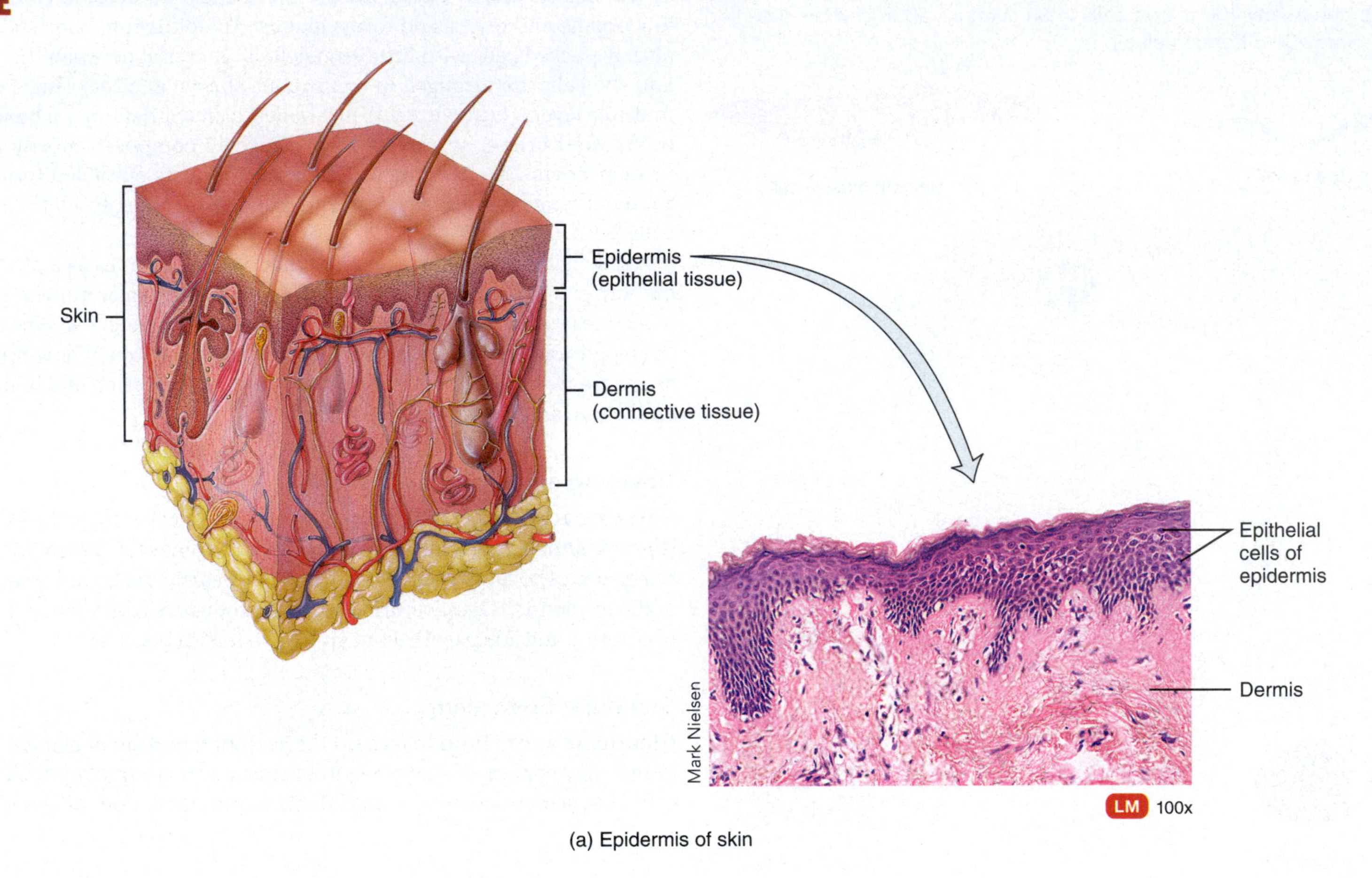

(a) Epidermis of skin

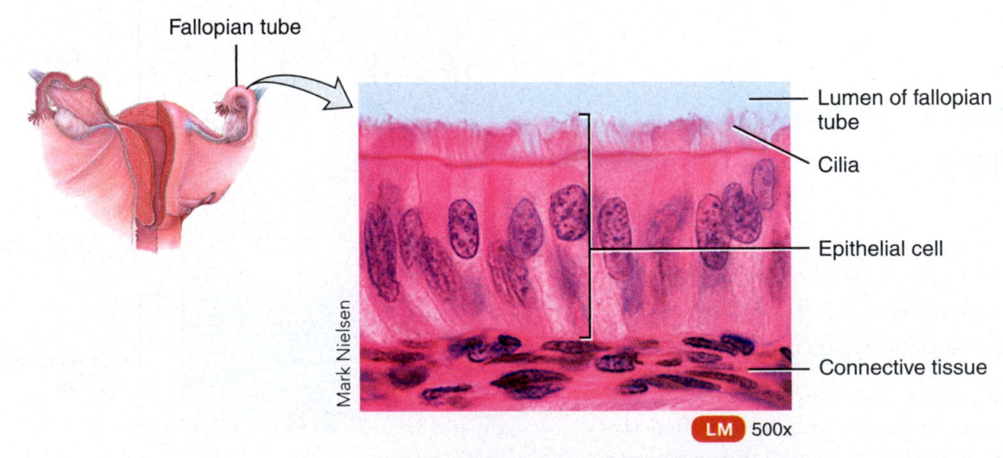

(b) Epithelium lining the fallopian tube

Of the various functions that epithelia perform, which is the most likely function of the epidermis?

of the body are classified as either endocrine or exocrine. The secretions of **endocrine glands** enter the interstitial fluid and then diffuse directly into the bloodstream without flowing through a duct. These secretions, called *hormones*, regulate many metabolic and physiological activities to maintain homeostasis. The pituitary, thyroid (FIGURE 3.25a), and adrenal glands are examples of endocrine glands. The functions of endocrine glands will be described in detail in Chapter 13.

Exocrine glands secrete their products into ducts that empty onto the surface of the body or into the lumen of a hollow organ. The secretions of exocrine glands include sweat, oil, saliva, digestive enzymes, and milk. Examples of exocrine glands include sudoriferous (sweat) glands (FIGURE 3.25b), sebaceous (oil) glands, salivary glands, gastric glands, and mammary glands. As you will learn later in this text, some glands of the body, such as the pancreas, are mixed glands that contain both endocrine and exocrine tissue.

Connective Tissue Supports, Insulates, and Protects the Organs of the Body

Connective tissue is one of the most abundant and widely distributed tissues in the body. In its various forms, connective tissue has a variety of functions. It binds together, supports, and strengthens other body tissues; protects and insulates internal organs; serves as the major transport system within the body (blood, a fluid connective tissue); is the primary location of stored energy reserves (adipose, or fat, tissue); and is the main source of immune responses.

Connective tissue consists of two basic elements: (1) extracellular matrix and (2) cells (FIGURE 3.26).

FIGURE 3.25 Glandular epithelium.

All glands of the body are classified as either endocrine or exocrine.

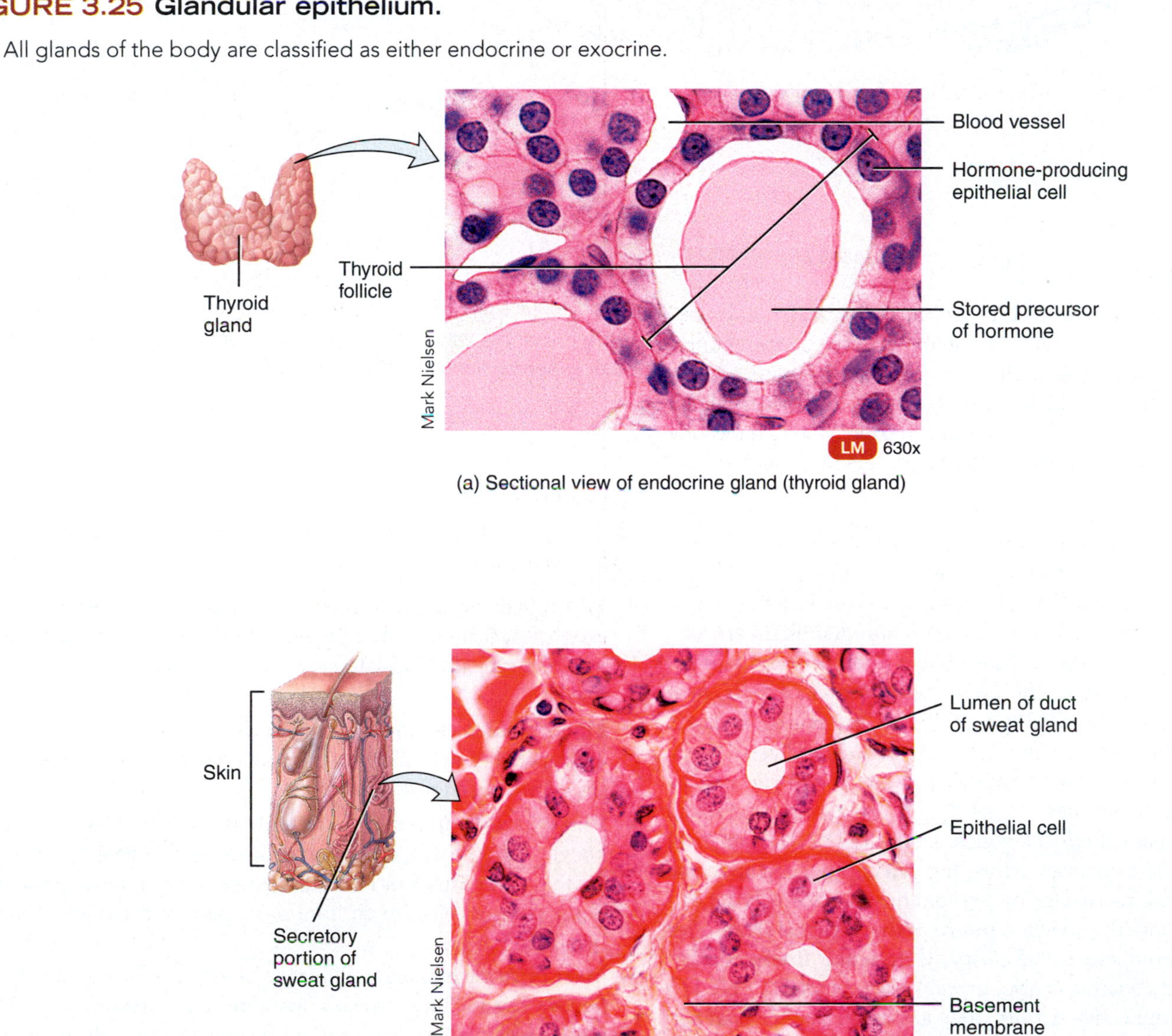

(a) Sectional view of endocrine gland (thyroid gland)

(b) Sectional view of exocrine gland (sweat gland)

What is the difference between an endocrine gland and an exocrine gland?

FIGURE 3.26 Representative cells and fibers present in connective tissues.

Fibroblasts are usually the most numerous connective tissue cells.

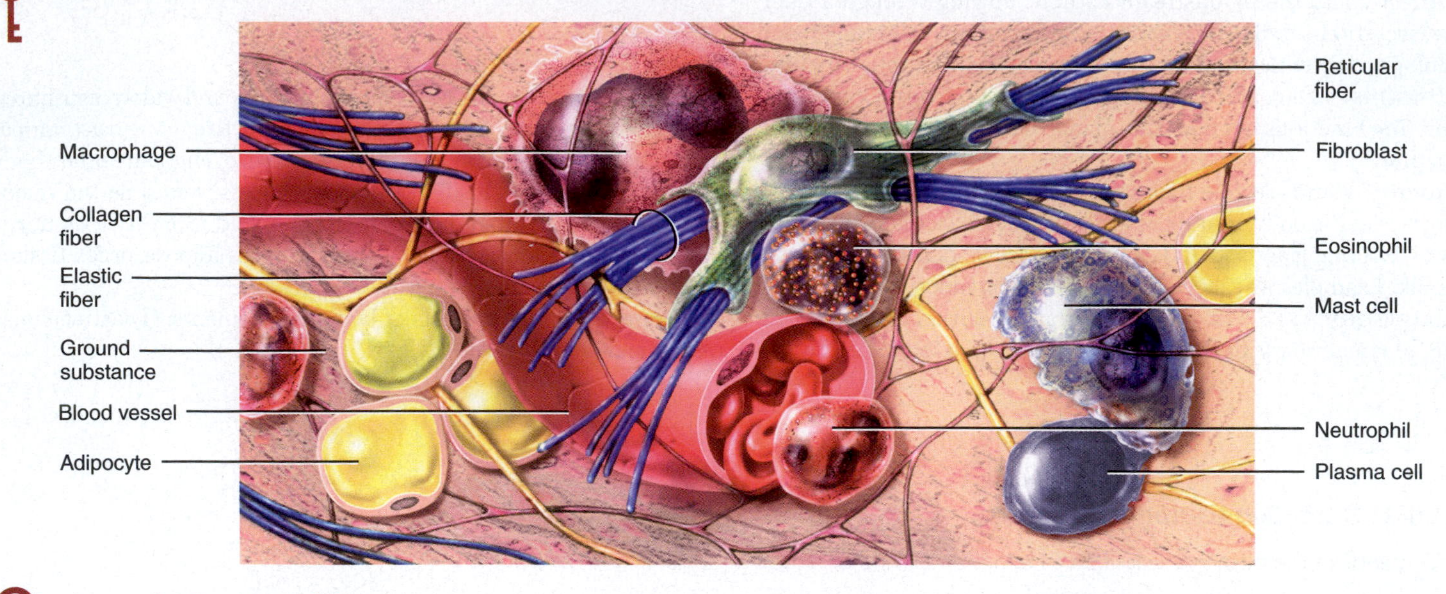

What is the function of fibroblasts?

Extracellular Matrix

A connective tissue's **extracellular matrix** is the material located between its widely spaced cells. It is usually secreted by the connective tissue's cells and determines the tissue's qualities. For instance, in cartilage, the extracellular matrix is firm but pliable. The extracellular matrix of bone, by contrast, is hard and inflexible. The extracellular matrix consists of two major components: protein fibers and ground substance.

PROTEIN FIBERS Three types of protein fibers can be embedded in the extracellular matrix between connective tissue cells: collagen fibers, elastic fibers, and reticular fibers (FIGURE 3.26). They function to strengthen and support connective tissues. **Collagen fibers** are very strong and resist pulling forces. These fibers often occur in bundles lying parallel to one another. The bundle arrangement affords great strength. Chemically, collagen fibers consist of the protein *collagen*. **Elastic fibers**, which are smaller in diameter than collagen fibers, branch and join together to form a network within the tissue. An elastic fiber consists of molecules of a protein called *elastin* surrounded by a glycoprotein named *fibrillin*, which is essential to the stability of an elastic fiber. Elastic fibers are strong but can be stretched up to one and a half times their relaxed length without breaking. Equally important, elastic fibers have the ability to return to their original shape after being stretched, a property called *elasticity*. **Reticular fibers**, consisting of *collagen* and a coating of glycoprotein, are arranged in a net-like fashion in some connective tissues. They are much thinner than collagen fibers. However, like collagen fibers, they provide support and strength.

GROUND SUBSTANCE The material between the cells and fibers of connective tissue is known as the **ground substance** (FIGURE 3.26). Depending on the connective tissue, the ground substance may be a fluid, gel, or solid. The ground substance supports cells, binds them together, stores water, and provides a medium through which substances are exchanged between the blood and cells.

Connective Tissue Cells

The types of connective tissue cells vary according to the type of tissue and include the following (FIGURE 3.26):

- **Fibroblasts** are present in several connective tissues and usually are the most numerous. They migrate through the connective tissue, secreting the fibers and ground substance of the extracellular matrix.
- **Macrophages** develop from monocytes, a type of leukocyte. They are capable of engulfing bacteria and cellular debris by phagocytosis, a form of endocytosis.
- **Plasma cells** are small cells that develop from a type of leukocyte called a B lymphocyte. Plasma cells secrete antibodies, proteins that attack or neutralize foreign substances in the body. Thus, plasma cells are an important part of the body's immune response.
- **Mast cells** are abundant alongside blood vessels that supply connective tissue. They produce histamine, a chemical that dilates blood vessels as part of the inflammatory response, the body's reaction to injury or infection.
- **Adipocytes**, also called fat cells or adipose cells, are connective tissue cells that store triglycerides (fats). They are found below the skin and around organs such as the heart and kidneys.

Leukocytes are not normally found in significant numbers in connective tissues. However, in response to certain conditions, leukocytes can leave blood and enter connective tissues. For example, *neutrophils* gather at sites of infection; *eosinophils* migrate to sites of parasitic invasion and allergic responses.

Types of Connective Tissues

Many types of connective tissues exist in the body, including areolar connective tissue, adipose tissue, dense regular connective tissue, dense irregular connective tissue, cartilage, bone, and blood. **FIGURE 3.27** illustrates these connective tissues.

FIGURE 3.27 Connective tissues.

Connective tissue binds together, supports, and strengthens other body tissues; protects and insulates internal organs; serves as the major transport system within the body; is the primary location of stored energy (fat) reserves; and is the main source of immune responses.

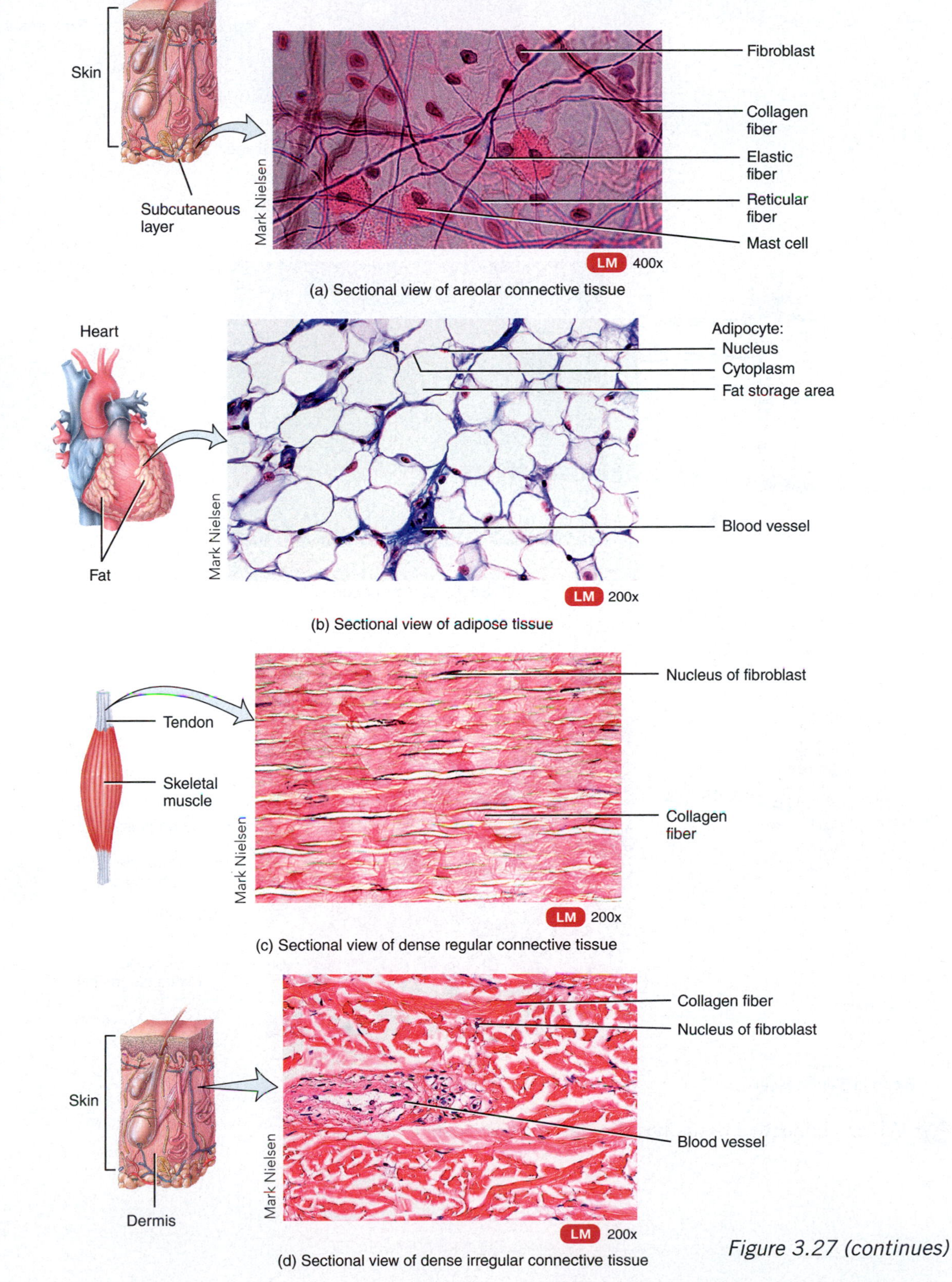

(a) Sectional view of areolar connective tissue

(b) Sectional view of adipose tissue

(c) Sectional view of dense regular connective tissue

(d) Sectional view of dense irregular connective tissue

Figure 3.27 (continues)

Figure 3.27 Continued

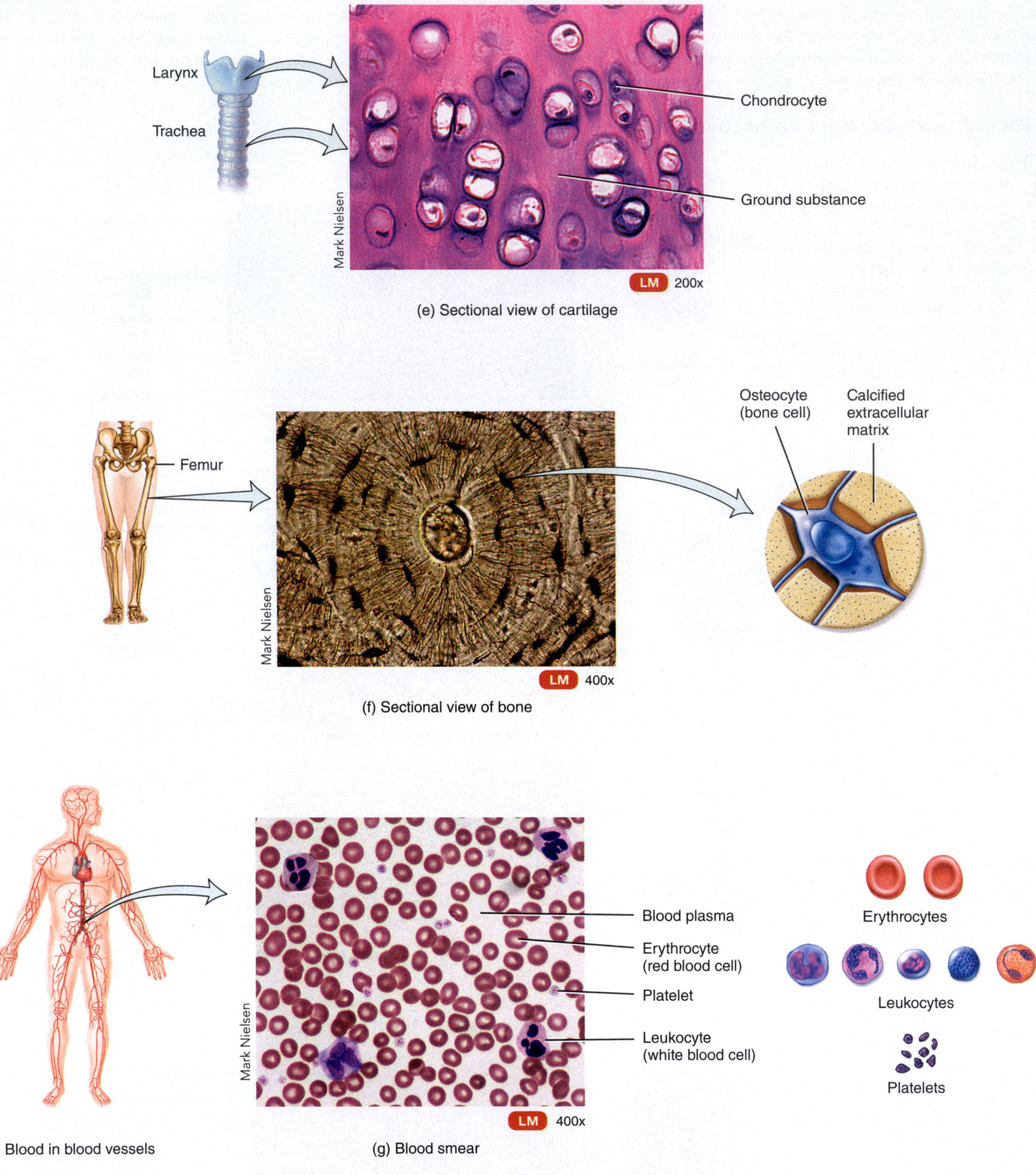

(e) Sectional view of cartilage

(f) Sectional view of bone

Blood in blood vessels

(g) Blood smear

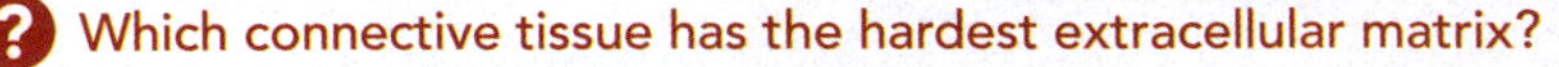

? Which connective tissue has the hardest extracellular matrix?

FIGURE 3.28 Skeletal muscle tissue.

Muscle tissue is highly specialized to contract.

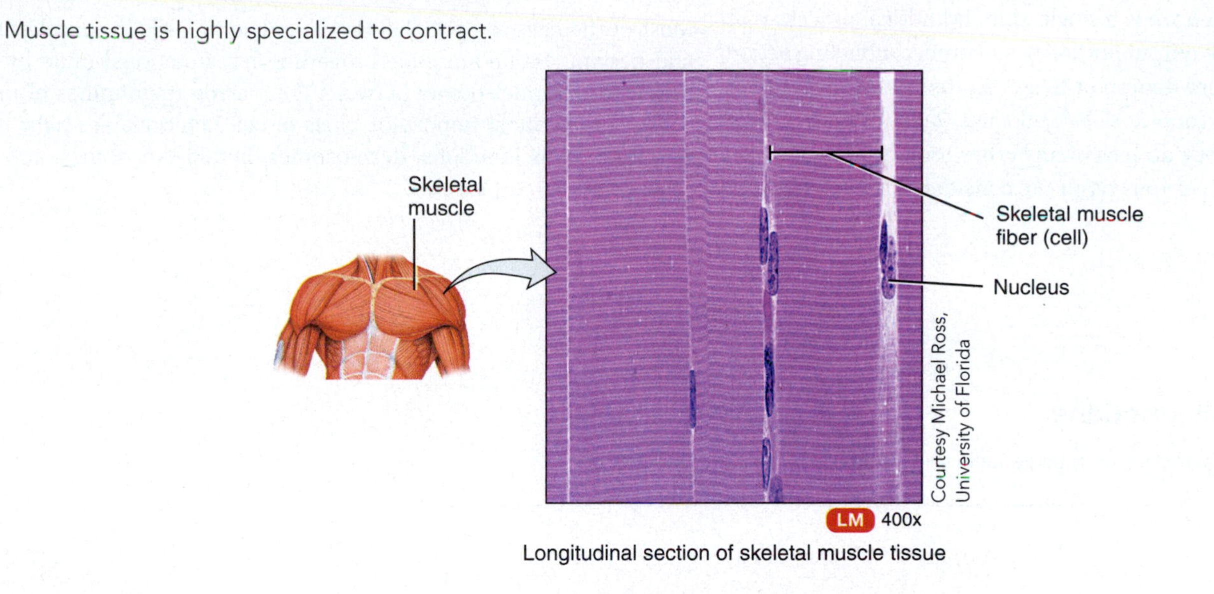

Longitudinal section of skeletal muscle tissue

What are the functions of skeletal muscle?

Muscle Tissue Specializes in Contraction

Muscle tissue consists of elongated cells called *muscle fibers* that are highly specialized to contract. Based on its location and certain structural and functional characteristics, muscle tissue is classified into three types: skeletal, cardiac, and smooth. **Skeletal muscle** (FIGURE 3.28) is usually attached to bones of the skeleton and is responsible for producing movement, maintaining posture, and generating heat. **Cardiac muscle** forms the bulk of the heart wall; its rhythmic contractions pump blood to all parts of the body. **Smooth muscle** is located in the walls of hollow internal structures such as blood vessels, airways to the lungs, the stomach, intestines, and urinary bladder. Its contractions regulate the movement of blood through blood vessels, food through the gastrointestinal tract, air through the bronchial tubes of the lungs, and urine through the urinary tract. You will learn more about the functions of the various types of muscle tissue in Chapter 11.

Nervous Tissue Detects and Responds to Changes in the Environment

Nervous tissue detects changes in the body's external or internal environment and then elicits an appropriate response. Two main types of cells comprise nervous tissue: neurons and neuroglia (FIGURE 3.29). **Neurons**, or *nerve cells*, are sensitive to a variety of stimuli. They convert stimuli into electrical signals called **action potentials**, *or nerve impulses*, and conduct these action potentials to other neurons, muscle fibers, or glands. Most neurons consist of three basic parts: a cell body, dendrites, and an axon. The *cell body* contains the nucleus and other organelles. *Dendrites* are short,

FIGURE 3.29 Nervous tissue.

Nervous tissue detects and responds to changes in the body's external or internal environment.

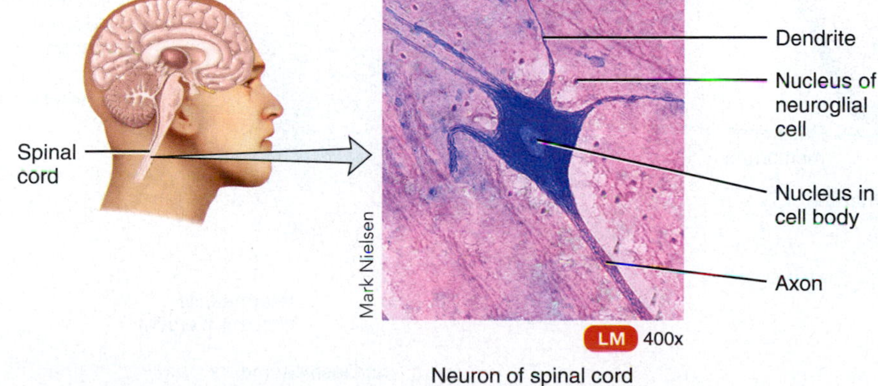

Neuron of spinal cord

Which components of nervous tissue produce action potentials?

branched processes that are the major receiving or input portion of a neuron. The *axon* of a neuron is a single, thin, cylindrical process that may be very long. It is the output portion of a neuron, conducting action potentials toward another neuron or to some other tissue. In contrast to neurons, **neuroglia** (noo-RŌG-lē-a) do not generate or conduct action potentials, but they do have many other important supportive functions. The details of nervous tissue are considered in Chapter 7.

Cell Junctions Connect Adjacent Cells

Most of the cells in epithelial tissue and some cells in muscle tissue and nervous tissue are joined together into functional units by **cell junctions**, contact points between the plasma membranes of tissue cells. The five most important types of cell junctions are tight junctions, adherens junctions, desmosomes, hemidesmosomes, and gap junctions (**FIGURE 3.30**).

FIGURE 3.30 Cell junctions.

Most epithelial cells and some muscle cells and neurons contain cell junctions.

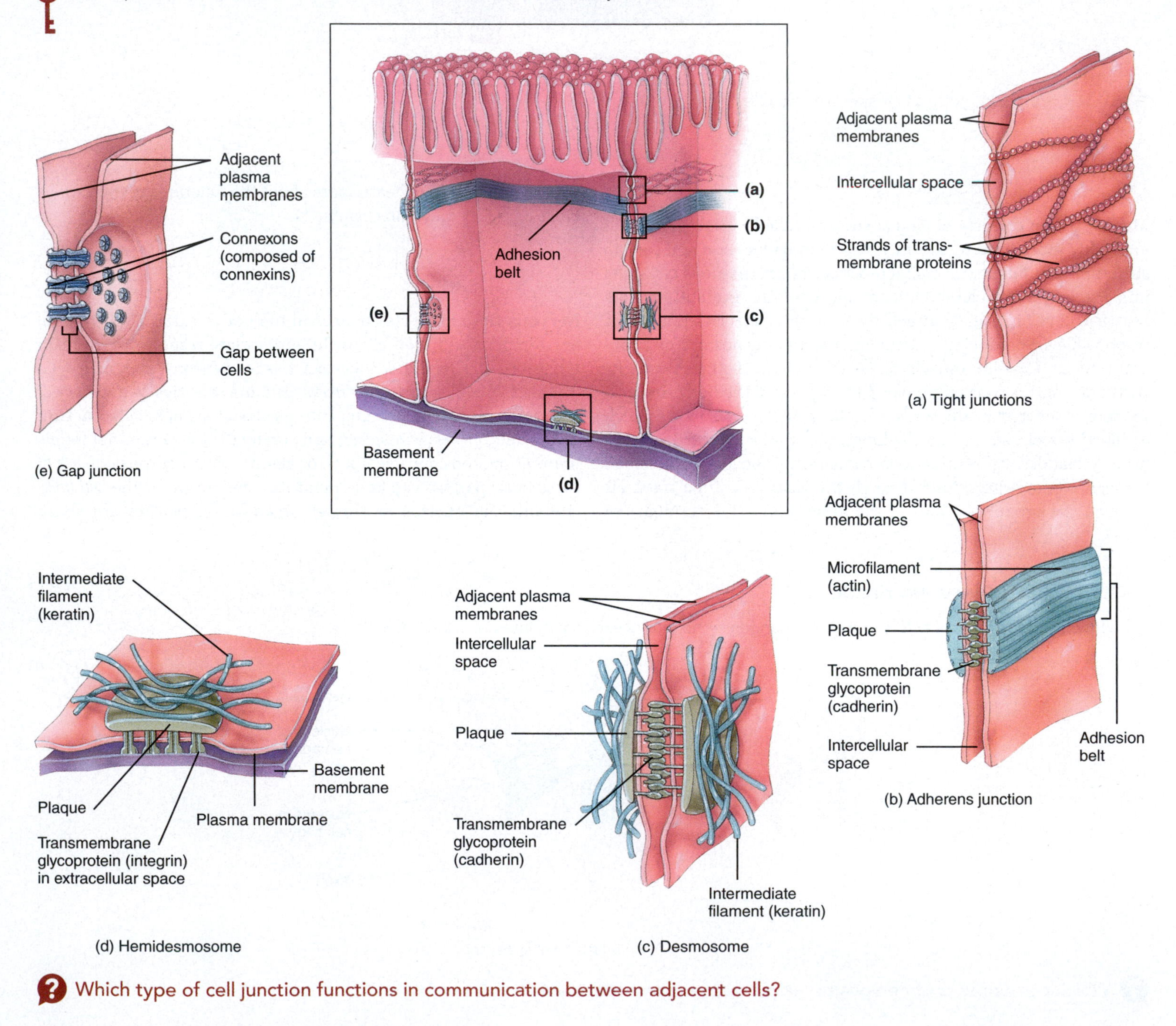

Which type of cell junction functions in communication between adjacent cells?

Tight Junctions

Tight junctions consist of weblike strands of transmembrane proteins that fuse together the outer surfaces of adjacent plasma membranes to seal off passageways between adjacent cells (FIGURE 3.30a). Cells of epithelial tissues that line the stomach, intestines, and urinary bladder have many tight junctions to retard the passage of substances between cells and prevent the contents of these organs from leaking into the blood or surrounding tissues.

Adherens Junctions

Adherens junctions (ad-HER-ens) contain **plaque**, a dense layer of proteins on the inside of the plasma membrane that attaches both to membrane proteins and to microfilaments of the cytoskeleton (FIGURE 3.30b). Transmembrane glycoproteins called **cadherins** join the cells. Each cadherin inserts into the plaque from the opposite side of the plasma membrane, partially crosses the intercellular space (the space between the cells), and connects to cadherins of an adjacent cell. In epithelial cells, adherens junctions often form extensive zones called **adhesion belts** because they encircle the cell similar to the way a belt encircles your waist. Adherens junctions help epithelial surfaces resist separation during various contractile activities, as when food moves through the intestines.

Desmosomes

Like adherens junctions, **desmosomes** (DEZ-mō-sōms) contain plaque and have transmembrane glycoproteins (cadherins) that extend into the intercellular space between adjacent cell membranes and attach cells to one another (FIGURE 3.30c). However, unlike adherens junctions, the plaque of desmosomes does not attach to microfilaments. Instead, a desmosome plaque attaches to intermediate filaments that consist of the protein keratin. The intermediate filaments extend from desmosomes on one side of the cell across the cytosol to desmosomes on the opposite side of the cell. This structural arrangement contributes to the stability of the cells and tissue. These spot-weld-like junctions are common among the cells that make up the epidermis (the outermost layer of the skin) and among cardiac muscle cells in the heart. Desmosomes prevent epidermal cells from separating under tension and cardiac muscle cells from pulling apart during contraction.

Hemidesmosomes

Hemidesmosomes resemble desmosomes but they do not link adjacent cells. The name arises from the fact that they look like half of a desmosome (FIGURE 3.30d). However, the transmembrane glycoproteins in hemidesmosomes are **integrins** rather than cadherins. On the inside of the plasma membrane, integrins attach to intermediate filaments made of the protein keratin. On the outside of the plasma membrane, the integrins attach to the protein **laminin**, which is present in the basement membrane. Thus, hemidesmosomes anchor cells not to each other but to the basement membrane.

Gap Junctions

At **gap junctions**, membrane proteins called **connexins** form tiny fluid-filled tunnels called **connexons** that connect neighboring cells (FIGURE 3.30e). The plasma membranes of gap junctions are not fused together as in tight junctions but are separated by a very narrow intercellular gap (space). Through the connexons, ions and small molecules can diffuse from the cytosol of one cell to another. Gap junctions are found in many tissues throughout the body. Their purpose is to allow the cells in a tissue to communicate with one another. For example, mature bone cells communicate through gap junctions to transfer nutrients and other substances to each other, an activity that helps bone maintain homeostasis. Because ions can pass through gap junctions, cells that communicate through gap junctions are electrically coupled. This enables electrical activity to spread rapidly among cells, a process that is crucial for the normal operation of some parts of the nervous system and for the contraction of muscle in the heart, gastrointestinal tract, and uterus.

CHECKPOINT

21. How do the functions of epithelial tissue and connective tissue differ?

22. What is the purpose of tight junctions?

Critical Thinking

Getting Inked

Greg has decided to get a tattoo. He finds a safe, reputable tattoo studio—a place where the tattoo artist wears gloves, sterilizes the equipment, and uses a fresh needle from a sealed package. Greg walks in and talks with the tattoo artist about the type of tattoo that he would like to have. After a few hours, the tattoo artist completes the job and Greg leaves the studio with his new tattoo. Greg is so excited that he shows off his tattoo to his family and friends.

BlueSkyImages/Fotali

SOME THINGS TO KEEP IN MIND:

A *tattoo* is a permanent coloration of the skin in which a foreign pigment is deposited with a needle into the dermis. It is created by injecting ink with a needle that punctures the epidermis, moves between 50 and 3000 times per minute, and deposits the ink in the dermis. Because the dermis is stable (unlike the epidermis, which is shed about every four to six weeks), tattoos are permanent. However, they can fade over time due to exposure to sunlight, improper healing, picking scabs, and flushing away of ink particles by the lymphatic system.

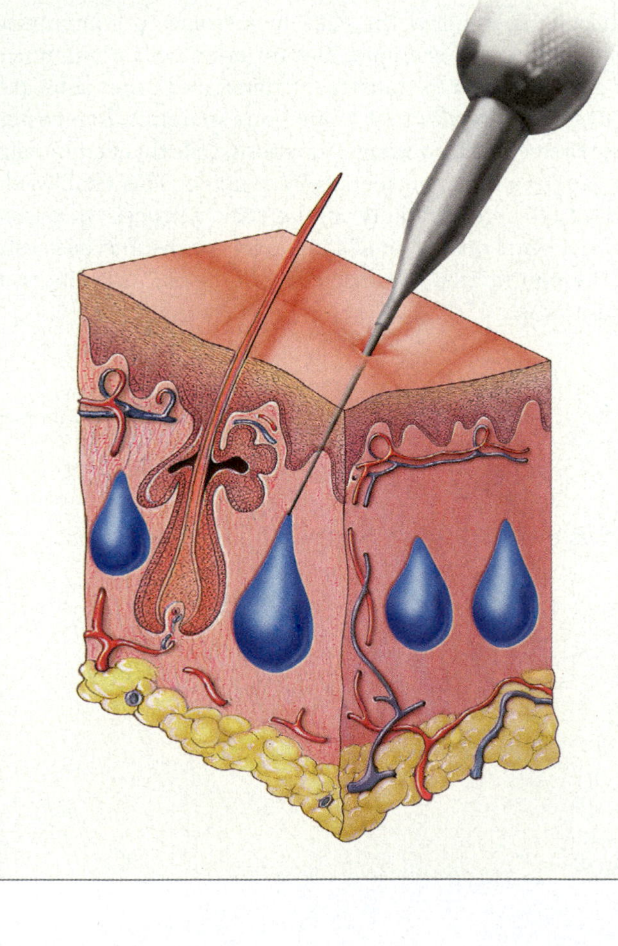

SOME INTERESTING FACTS:

The art of tattooing originated in ancient Egypt between 4000 and 2000 BC.

Today, tattooing is performed in one form or another by nearly all peoples of the world.

It is estimated that about one in five U.S. college students has one or more tattoos.

Although tattoos are popular and seemingly harmless, they have several potential complications, including allergic reactions (in response to the ink pigments), infections (such as HIV or hepatitis from the use of dirty, contaminated needles or equipment), tissue scarring, and granulomas (small bumps that form around the deposited ink).

What are the two major layers of the skin? What are their functions?

If Greg decides later that he no longer wants his tattoo, why can't he wash off the tattoo or wait for it to "disappear" from his skin as his epidermis periodically sheds?

Why was it so important that Greg find a safe, reputable tattoo studio to get his tattoo?

What are some long-term cellular effects of tattoos?

FROM RESEARCH TO REALITY

Proteasomes and Neurodegenerative Disease

Reference
Ding, Q. et al. (2003). Characterization of chronic low-level proteasome inhibition on neural homeostasis. *Journal of Neurochemistry.* 86: 489–497.

Can an understanding of low-level proteasome inhibition be the key to finding the cause of Parkinson's and Alzheimer's diseases?

Neurodegenerative diseases like Parkinson's disease (PD) and Alzheimer's disease (AD) are devastating. They have been at the center of much research that is trying to determine their cause. Both have been linked to defects in the function of proteasomes. Complete inhibition of proteasomes leads to quick and certain neuron death, but PD and AD don't generally demonstrate such quick cell death. Could chronic low-level proteasome inhibition be the key to the neurodegeneration evident in these diseases?

Article description:
The authors of this paper exposed neural clonal cell lines to 12 weeks of chronic, low-level proteasome inhibition and then ran a variety of tests on the cell lines. They wanted to see if their results were consistent with what is seen in individuals with Parkinson's disease and Alzheimer's disease, namely protein oxidation (covalent modification of proteins via interactions with reactive oxygen species) and protein aggregation (clusters of proteins).

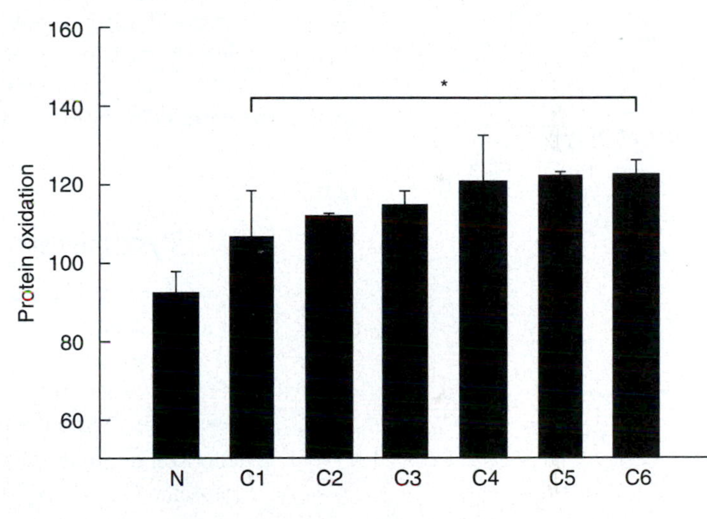

Go to WileyPLUS Learning Space and use the data from this article to answer the questions posed there and to discover more about proteasomes and neurodegenerative disease.

CHAPTER REVIEW

3.1 Components of a Cell

1. FIGURE 3.1 provides an overview of the typical structures found in most body cells.
2. A cell is divided into three main parts: plasma membrane, cytoplasm, and nucleus, each of which has different functions.

3.2 The Plasma Membrane

1. The plasma membrane surrounds the cytoplasm of a cell and is composed of lipids and proteins.
2. According to the fluid mosaic model, the plasma membrane is a mosaic of proteins floating like icebergs in a lipid bilayer sea.
3. The lipid bilayer consists of two back-to-back layers of phospholipids, cholesterol, and glycolipids. The bilayer arrangement occurs because the lipids are amphipathic, having both polar and nonpolar parts.
4. Integral proteins extend into or through the lipid bilayer; peripheral proteins associate with membrane lipids or integral proteins at the inner or outer surface of the membrane.
5. Many integral proteins are glycoproteins, with sugar groups attached to the ends that face the extracellular fluid. Together with glycolipids, the glycoproteins form a glycocalyx on the extracellular surface of cells.
6. Membrane proteins have a variety of functions. They act as channels and carriers that help specific solutes cross the membrane; receptors that serve as cellular recognition sites; enzymes that catalyze specific chemical reactions; linkers that anchor proteins in the plasma membranes to protein filaments inside and outside the cell; and cell-identity markers that distinguish your cells from foreign cells.
7. Membrane fluidity is greater when there are more double bonds in the fatty acid tails of the lipids that make up the bilayer. Cholesterol makes the lipid bilayer stronger but less fluid at normal body temperature. Membrane fluidity allows interactions to occur within the plasma membrane, enables the movement of membrane components, and permits the lipid bilayer to self-seal when torn or punctured.

3.3 Cytoplasm

1. Cytoplasm consists of all the cellular contents between the plasma membrane and the nucleus. It has two components: cytosol and organelles. Cytosol is the fluid portion of cytoplasm, containing water, ions, glucose, amino acids, fatty acids, proteins, lipids, ATP, and waste products. It is the site of many chemical reactions required for a cell's existence. Organelles are specialized structures that have specific functions.
2. Ribosomes are the sites of protein synthesis. They are composed of ribosomal RNA and proteins.
3. Endoplasmic reticulum (ER) is a network of membranes that form flattened sacs or tubules; it extends through widespread regions of the cytoplasm. Rough ER is "rough" because it is studded with ribosomes. Rough ER synthesizes and modifies proteins that are secreted by the cell, inserted into the plasma membrane of the cell, or transferred to other organelles within the cell. Smooth ER is "smooth" because it lacks ribosomes. Smooth ER synthesizes fatty acids and steroids; inactivates or detoxifies drugs and other potentially harmful substances; and releases calcium ions that trigger contraction in muscle cells.
4. The Golgi complex consists of flattened sacs called cisternae. The Golgi complex modifies and sorts proteins received from the rough ER. It also packages proteins for transport in secretory vesicles, membrane vesicles, or transport vesicles to different cellular destinations.
5. Mitochondria consist of a smooth outer membrane, an inner membrane containing cristae, and a fluid-filled cavity called the matrix. These so-called powerhouses of the cell produce most of a cell's ATP.
6. Lysosomes are membrane-enclosed vesicles that contain digestive enzymes. Lysosomes digest substances that enter the cell via endocytosis. They can also carry out autophagy (digestion of worn-out organelles) and autolysis (digestion of entire cell).

7. Peroxisomes contain oxidases that oxidize amino acids, fatty acids, and toxic substances; the hydrogen peroxide produced in the process is destroyed by catalase.
8. Proteasomes are barrel-shaped structures that contain proteases, which continually degrade unneeded, damaged, or faulty proteins by cutting them into small peptides.
9. The cytoskeleton is a network of proteins (microfilaments, intermediate filaments, and microtubules) that extends throughout the cytosol. It provides a structural framework for the cell. It also aids movement of whole cells and of components within cells.
10. The centrosome consists of pericentriolar material and a pair of centrioles. The pericentriolar material organizes microtubules in nondividing cells and the mitotic spindle in dividing cells.
11. Cilia and flagella are motile projections of the cell surface; both are composed mostly of microtubules. Cilia move fluid along the cell surface; flagella move an entire cell.

3.4 Nucleus

1. The nucleus is the control center of the cell: It contains genes that control cellular structure and direct cellular activities.
2. The nucleus is separated from the cytoplasm by a nuclear envelope. Many nuclear pores extend through the nuclear envelope; they control the movement of substances between the nucleus and cytoplasm.
3. Also present in the nucleus are nucleoli, which produce ribosomes.
4. In cells that are not dividing, the DNA in the nucleus exists as chromatin. Just before cell division takes place, the DNA replicates and condenses to form chromosomes. Human somatic cells have 46 chromosomes, 23 inherited from each parent.

3.5 Gene Expression

1. Gene expression is the process by which a gene's DNA is used as a template for synthesis of a specific protein. Gene expression involves two major steps: (1) transcription, which occurs in the nucleus, and (2) translation, which occurs in the cytoplasm.
2. The genetic code is the set of rules that relates the base triplet sequences of DNA to the corresponding codons of RNA and the amino acids they specify.
3. In transcription, the genetic information in the sequence of base triplets in DNA serves as a template for copying the information into a complementary sequence of codons in messenger RNA. Transcription begins on DNA in a region called a promoter. Regions of DNA that code for protein synthesis are called exons; those that do not are called introns. Newly synthesized pre-mRNA is modified before leaving the nucleus.
4. In the process of translation, the nucleotide sequence of mRNA specifies the amino acid sequence of a protein. The mRNA binds to a ribosome, specific amino acids attach to tRNA, and anticodons of tRNA bind to codons of mRNA, thus bringing specific amino acids into position on a growing polypeptide. Translation begins at the start codon and ends at the stop codon.

3.6 Cell Division

1. Cell division is the process by which cells reproduce themselves. There are two types of cell division: somatic cell division and reproductive cell division.
2. Somatic cell division replaces dead or injured cells and adds new ones during tissue growth. It consists of a nuclear division called mitosis and a cytoplasmic division called cytokinesis.

3. Reproductive cell division results in the production of gametes (sperm and ova). It consists of a two-step division called meiosis.
4. The cell cycle, an orderly sequence of events in which a somatic cell duplicates its contents and divides in two, consists of interphase and a mitotic phase.
5. Human somatic cells contain 23 pairs of homologous chromosomes and are thus diploid ($2n$).
6. Before the mitotic phase, the DNA molecules, or chromosomes, replicate themselves so that identical sets of chromosomes can be passed on to the next generation of cells. A cell between divisions that is carrying on every life process except division is said to be in interphase, which consists of three phases: G_1, S, and G_2. During the G_1 phase, the cell replicates its organelles and cytosolic components, and centrosome replication begins; during the S phase, DNA replication occurs; during the G_2 phase, enzymes and other proteins are synthesized and centrosome replication is completed.
7. Mitosis is the splitting of the chromosomes and the distribution of two identical sets of chromosomes into separate and equal nuclei; it consists of prophase, metaphase, anaphase, and telophase.
8. In cytokinesis, which usually begins in late anaphase and ends once mitosis is complete, a cleavage furrow forms at the cell's metaphase plate and progresses inward, pinching in through the cell to form two separate portions of cytoplasm.
9. A cell can either remain alive and function without dividing, grow and divide, or die. The control of cell division depends on specific cyclin-dependent protein kinases and cyclins.
10. Apoptosis is normal, programmed cell death. It first occurs during embryological development and continues throughout the lifetime of an organism.

3.7 Cellular Diversity

1. The almost 200 different types of cells in the body vary considerably in size and shape.
2. Cells in the body range from 8 μm to 140 μm in size.
3. A cell's shape is related to its function.

3.8 Organization of Cells into Tissues

1. A tissue is a group of similar cells that function together to carry out specialized activities. The various tissues of the body are classified into four basic types: epithelial, connective, muscle, and nervous tissue.
2. Epithelial tissue covers the body, lines hollow organs and ducts, and forms glands. An epithelium consists mostly of cells with little extracellular material between adjacent plasma membranes. The two major types of epithelial tissue are (1) covering and lining epithelium and (2) glandular epithelium.
3. Connective tissue binds together, strengthens, and supports other body tissues; protects and insulates internal organs; serves as the major transport system within the body; is the primary location of stored energy (fat) reserves; and is the main source of immune responses. Connective tissue consists of two basic elements: extracellular matrix and cells. There are several types of connective tissues, including areolar connective tissue, adipose tissue, dense regular connective tissue, dense irregular connective tissue, cartilage, bone, and blood.
4. Muscle tissue consists of cells called muscle fibers that are specialized for contraction. It provides movement, maintenance of posture, and heat production. There are three types of muscle tissue—skeletal muscle, cardiac muscle, and smooth muscle.

CHAPTER REVIEW

5. Nervous tissue detects and responds to changes in the body's external or internal environment. It is composed of neurons (nerve cells) and neuroglia (protective and supporting cells).
6. Cell junctions are points of contact between adjacent plasma membranes. Tight junctions form fluid-tight seals between cells; adherens junctions, desmosomes, and hemidesmosomes anchor cells to one another or to the basement membrane; and gap junctions permit electrical and chemical signals to pass between cells.

PONDER THIS

1. Dr. Lazarus is developing a new drug and would like to begin his clinical trials. Before doing so, Dr. Lazarus tests his drug on cells from various parts of the body. He realizes that the cells are no longer able to synthesize proteins. Perturbed by this finding, Dr. Lazarus explores further and discovers that transcription is carried out without any problems. What could the drug be interacting with? Explain your answer.

2. You have just discovered a new organism and decide to do an analysis of its different tissue types. You are particularly interested in the tissues below the skin. You come upon a very interesting layer just superficial to the muscles. The cells of this tissue are bound together to support and strengthen the surrounding tissues. In addition, the cells have the ability to generate action potentials. How would you classify this tissue type based upon what you know about human tissues? Why?

3. A toxin is introduced into the body that results in the destruction of all the transmembrane proteins embedded within the plasma membrane. What loss of function would occur as a result of this toxin?

ANSWERS TO FIGURE QUESTIONS

3.1 The three main parts of a cell are the plasma membrane, cytoplasm, and nucleus.

3.2 The glycocalyx is the sugary coat on the extracellular surface of the plasma membrane. It is composed of the carbohydrate portions of membrane glycolipids and glycoproteins.

3.3 A carrier protein is a type of membrane protein that transports a specific substance across the membrane by undergoing a change in conformation.

3.4 The subunits of ribosomes are produced separately in the nucleolus of the nucleus and then come together in the cytoplasm to form functional ribosomes.

3.5 Rough ER has attached ribosomes; smooth ER does not. Rough ER synthesizes proteins that will be exported from the cell; smooth ER is associated with lipid synthesis, detoxification of harmful substances, and storage of calcium ions.

3.6 The Golgi complex modifies, sorts, and packages proteins received from the rough ER. It also forms vesicles that transport the processed proteins to other destinations.

3.7 Some proteins are secreted from the cell by exocytosis, some are incorporated into the plasma membrane, and some are transported to other organelles such as lysosomes.

3.8 Mitochondrial cristae contain the proteins of the electron transport chain, which are some of the proteins that are needed for ATP production.

3.9 Digestion of worn-out organelles by lysosomes is called autophagy.

3.10 Microtubules help form cilia and flagella.

3.11 A cell without a centrosome probably would not be able to undergo cell division.

3.12 Cilia move fluids across cell surfaces; flagella move an entire cell.

3.13 Substances that pass through the nuclear pores include proteins needed for nuclear functions that move from the cytoplasm into the nucleus, and newly formed mRNA molecules that move from the nucleus into the cytoplasm.

3.14 A nucleosome consists of DNA wrapped twice around a core of eight histones (proteins).

3.15 Proteins determine the physical and chemical characteristics of cells.

3.16 The DNA base sequence AGCT would be transcribed into the RNA base sequence UCGA by RNA polymerase.

3.17 The P site binds the tRNA carrying the growing polypeptide chain; the A site binds the tRNA carrying the next amino acid to be added to the growing polypeptide; and the E site binds tRNA just before it is released from the ribosome.

3.18 When a ribosome encounters a stop codon at the A site, it releases the completed protein from the final tRNA.

3.19 DNA replicates during the S phase of the cell cycle.

3.20 DNA replication occurs before cytokinesis so that each of the new cells will have a complete genome.

3.21 Cytokinesis usually starts in late anaphase.

3.22 Sperm, which use the flagella for locomotion, are the only body cells required to move considerable distances.

3.23 The basement membrane connects epithelial tissue to underlying connective tissue and provides support to epithelial cells

3.24 The epidermis serves a protective function, allowing the body to resist the abrasive influences of the environment.

3.25 An endocrine gland lacks a duct and releases secretions called hormones that ultimately empty into the bloodstream; an exocrine gland contains a duct and releases its secretions through the duct onto the body surface or into a cavity.

3.26 Fibroblasts secrete the fibers and ground substance of the extracellular matrix.

3.27 Bone is the connective tissue with the hardest extracellular matrix.

3.28 Skeletal muscle produces movement, maintains posture, and generates heat.

3.29 Neurons are the components of nervous tissue that produce action potentials.

3.30 Gap junctions allow cellular communication via passage of chemical and electrical signals between adjacent cells.

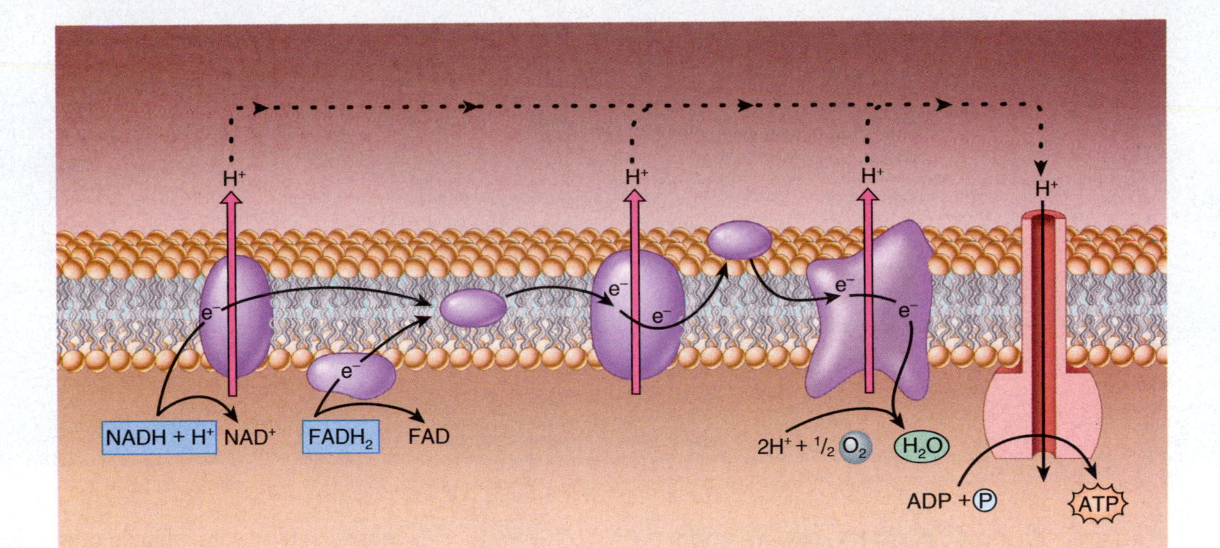

4 Metabolism

Metabolism and Homeostasis

Metabolic reactions contribute to homeostasis by harvesting chemical energy from consumed nutrients for use in the body's growth, repair, and normal functioning.

LOOKING BACK TO MOVE AHEAD...

- Chemical bonds are the forces that hold together the atoms of a molecule or compound; four types of chemical bonds are ionic bonds, covalent bonds, hydrogen bonds, and van der Waals interactions (Section 2.2).
- Valence electrons are the electrons located in the outermost shell of an atom; the likelihood that an atom will form a chemical bond with another atom depends on its number of valence electrons (Section 2.2.).
- A chemical reaction occurs when new chemical bonds form or old bonds break between atoms; the starting substances of a chemical reaction are known as reactants and the ending substances are referred to as products (Section 2.3).
- There are five categories of organic compounds that carry out important functions in the body: carbohydrates, lipids, proteins, nucleic acids, and adenosine triphosphate (ATP) (Section 2.5).

Plants use the green pigment chlorophyll to trap energy in sunlight. We don't have a similarly functioning pigment in our skin, so the food we eat is our only source of energy for activities such as running, walking, and even breathing. Carbohydrates, lipids, and proteins in our food are digested and then absorbed into the bloodstream from the gastrointestinal tract. Once food molecules have been absorbed, they have three main fates:

1. Most food molecules are used to *supply energy* for sustaining life processes, such as active transport, DNA replication, protein synthesis, muscle contraction, maintenance of body temperature, and cell division.
2. Some food molecules *serve as building blocks* for the synthesis of more complex structural or functional molecules, such as muscle proteins, hormones, and enzymes.
3. Other food molecules are *stored for future use*. For example, glycogen is stored in liver cells, and triglycerides are stored in adipose cells.

In this chapter you will explore how metabolic reactions harvest the chemical energy stored in carbohydrates, lipids, and proteins. You will also discover how each of these substances contributes to the body's growth, repair, and energy needs.

4.1 An Overview of Metabolism

OBJECTIVE

- Define metabolism.

Metabolism refers to all of the chemical reactions that occur in the body. There are two types of metabolism: catabolism and anabolism. Those chemical reactions that break down complex organic molecules into smaller molecules are collectively known as **catabolism**. Examples of catabolic reactions include the disassembly of a protein into individual amino acids (**FIGURE 4.1a**); the cleavage of a triglyceride into fatty acids and glycerol; and the breakdown of glucose, a fatty acid, or an amino acid into carbon dioxide and water.

Chemical reactions that combine smaller molecules to form larger, more complex molecules are collectively known as **anabolism**. Examples of anabolic reactions are the formation of peptide bonds between amino acids during protein synthesis; the building of fatty acids into phospholipids that form the plasma membrane bilayer; and the linkage of glucose monomers to form glycogen (**FIGURE 4.1b**).

A molecule synthesized in an anabolic reaction has a limited life span. With few exceptions, it will eventually be broken down and its component atoms recycled into other molecules or excreted from the body. Recycling of biological molecules occurs continuously in living tissues, more rapidly in some than in others. Individual cells may be refurbished molecule by molecule, or a whole tissue may be rebuilt cell by cell.

CHECKPOINT

1. What is metabolism?
2. Distinguish between catabolism and anabolism, and give examples of each.

4.2 Energy and Metabolism

OBJECTIVES

- Describe the various forms of energy.
- Compare exergonic and endergonic reactions.
- Explain the role of activation energy and catalysts in chemical reactions.

Energy Exists in Different Forms

Chemical reactions, such as those associated with metabolism, involve changes in **energy**, the capacity to do work. Two principal forms of energy are **potential energy**, energy stored by matter due to its position, and **kinetic energy**, the energy associated with matter in motion.

FIGURE 4.1 Metabolic reactions.

In a catabolic reaction, a complex organic molecule is broken down into smaller molecules; in an anabolic reaction, smaller molecules combine to form larger, more complex molecules.

Amino acids
Protein
Individual amino acids

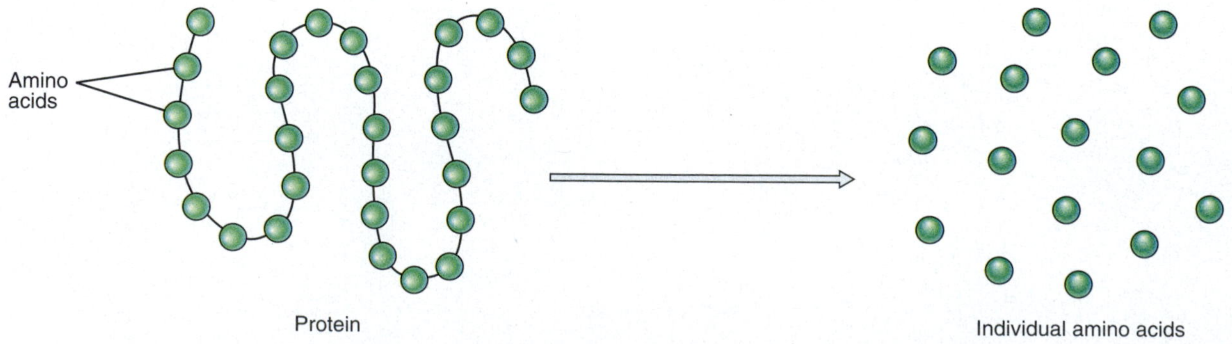

(a) Example of a catabolic reaction

Figure 4.1 (continues)

Figure 4.1 Continued

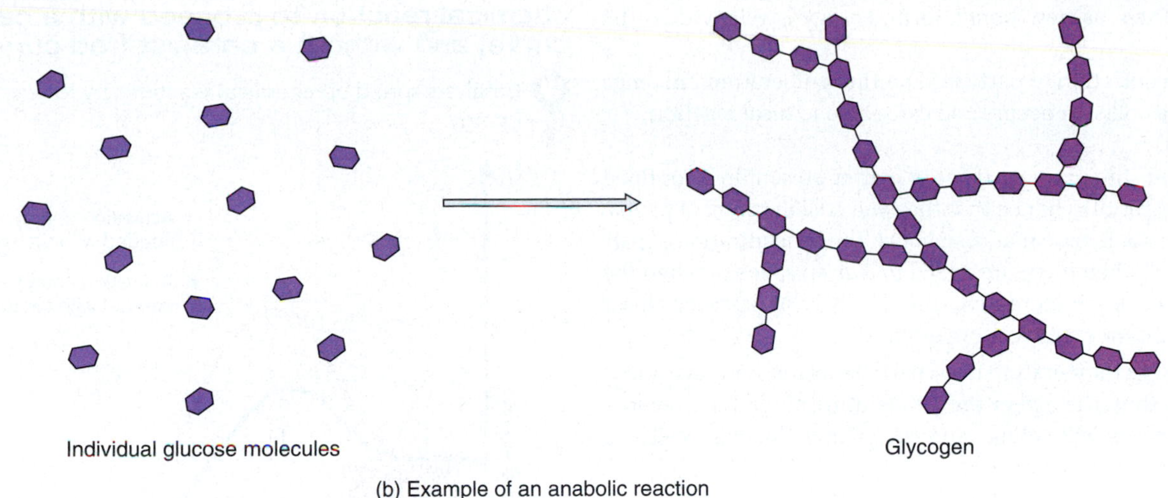

(b) Example of an anabolic reaction

Does digestion of food involve catabolic or anabolic reactions?

For example, the energy stored in water behind a dam or in a person poised to jump down some steps is potential energy. When the gates of the dam are opened or the person jumps, potential energy is converted into kinetic energy. **Chemical energy** is a form of potential energy that is stored in the bonds of compounds and molecules. The total amount of energy present at the beginning and end of a chemical reaction is the same. Although energy can be neither created nor destroyed, it may be converted from one form to another, a principle known as the **law of conservation of energy**. For example, some of the chemical energy in the foods we eat is eventually converted into various forms of kinetic energy, such as mechanical energy used to walk and talk. Conversion of energy from one form to another generally releases heat, some of which is used to maintain normal body temperature.

Chemical Reactions Release or Absorb Energy

Chemical bonds represent stored chemical energy, and chemical reactions occur when new bonds are formed or old bonds are broken between atoms. The *overall reaction* may either release energy or absorb energy. **Exergonic reactions** release more energy than they absorb. By contrast, **endergonic reactions** absorb more energy than they release. With respect to metabolism, catabolic reactions are exergonic: The chemical energy stored in the bonds of complex molecules is released as the molecules are broken down into smaller components. Anabolic reactions, however, are endergonic because they consume more energy than they produce as smaller molecules are combined to form larger molecules.

A key feature of the body's metabolism is the coupling of exergonic reactions and endergonic reactions. Energy released from an exergonic reaction is often used to drive an endergonic one. In general, exergonic reactions occur as nutrients, such as glucose, are broken down. Some of the energy released may be trapped in the covalent bonds of adenosine triphosphate (ATP), which is described more fully later in this chapter. If a molecule of glucose is broken down completely, the chemical energy in its bonds can be used to produce as many as 32 molecules of ATP. The energy transferred to the ATP molecules is then used to drive endergonic reactions needed to build body structures, such as muscles and bones. The energy in ATP is also used to do the mechanical work involved in the contraction of muscle or the movement of substances into or out of cells.

Activation Energy Is Needed to Start a Chemical Reaction

Because particles of matter such as atoms, ions, and molecules have kinetic energy, they are continuously moving and colliding with one another. A sufficiently forceful collision can disrupt the movement of valence electrons, causing an existing chemical bond to break or a new one to form. The collision energy needed to break the chemical bonds of the reactants is called the **activation energy** of the reaction (FIGURE 4.2). This initial energy "investment" is needed to start a reaction. The reactants must absorb enough energy for their chemical

FIGURE 4.2 Activation energy.

Activation energy is the energy needed to break chemical bonds in the reactant molecules so a reaction can start.

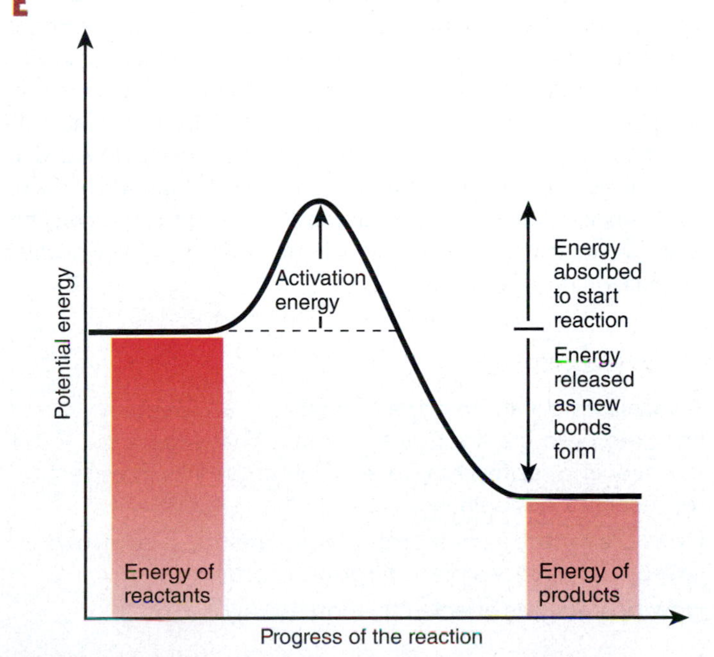

Why is the reaction illustrated here exergonic?

bonds to become unstable and their valence electrons to form new combinations. Then, as new bonds form, energy is released to the surroundings.

Both the concentration of particles and the temperature influence the chance that a collision occurs and causes a chemical reaction.

- ***Concentration.*** The more particles of matter present in a confined space, the greater the chance that they will collide (think of people crowding into a subway car at rush hour). The concentration of particles increases when more are added to a given space or when the pressure on the space increases, which forces the particles closer together so that they collide more often.
- ***Temperature.*** As temperature rises, particles of matter move about more rapidly. Thus, the higher the temperature of matter, the more forcefully particles will collide, and the greater the chance that a collision will produce a reaction.

Catalysts Lower the Activation Energy of Chemical Reactions

As you have just learned, chemical reactions occur when chemical bonds break or form after atoms, ions, or molecules collide with one another. Body temperature and the concentrations of molecules in body fluids, however, are far too low for most chemical reactions to occur rapidly enough to maintain life. Raising the temperature and the number of reacting particles of matter in the body could increase the frequency of collisions and thus increase the rate of chemical reactions, but doing so could also damage or kill the body's cells.

Substances called catalysts solve this problem. **Catalysts** are chemical compounds that speed up chemical reactions by lowering the activation energy needed for a reaction to occur (**FIGURE 4.3**). A catalyst does not alter the difference in potential energy between the reactants and the products. Rather, it lowers the amount of energy needed to start the reaction. For chemical reactions to occur, some particles of matter—especially large molecules—not only must collide with sufficient force, but they must also hit one another at precise spots. A catalyst helps to orient the colliding particles properly so that they interact at the spots that make the reaction happen. Although the action of a catalyst helps to speed up a chemical reaction, the catalyst itself is unchanged at the end of the reaction. A single catalyst molecule can assist one chemical reaction after another. The most important catalysts in the body are enzymes, which are the focus of the next section.

CHECKPOINT

3. A vase located at the edge of a table is accidentally knocked onto the floor by a toddler. What aspect of this scenario represents potential energy? Which aspect represents kinetic energy?
4. Does the synthesis of a protein from amino acids involve exergonic reactions or endergonic reactions?
5. How do catalysts affect activation energy?

FIGURE 4.3 Comparison of energy needed for a chemical reaction to proceed with a catalyst (blue curve) and without a catalyst (red curve).

Catalysts speed up chemical reactions by lowering the activation energy.

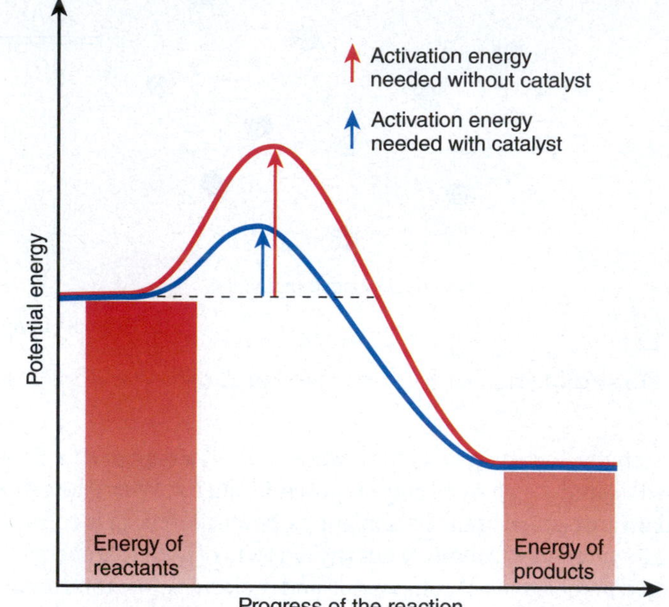

Does a catalyst change the potential energies of the products and reactants? Explain your answer.

4.3 Enzymes

OBJECTIVES

- List the properties of enzymes.
- Explain how an enzyme functions.
- Identify the various factors that influence the rate of an enzymatic reaction.
- Describe the concept of a metabolic pathway.

In living cells, most catalysts are protein molecules called **enzymes**. Enzymes catalyze chemical reactions: They accelerate the conversion of reactants into products by lowering the amount of activation energy needed for the reaction to occur. The reactant molecules on which an enzyme acts are known as **substrates**:

$$\underset{\text{Substrates}}{A + B} \xrightarrow{\text{Enzyme}} \underset{\text{Products}}{C + D}$$

The substrates of an enzyme are examples of ligands. Recall from Section 2.5 that a *ligand* is a molecule that binds to a particular site on a protein. Hence, the interaction between a substrate and an enzyme exhibits characteristics of ligand–protein binding—specificity, affinity, saturation, and competition.

The names of enzymes usually end in the suffix *-ase*. Examples include lactase, carbonic anhydrase, and elastase. Some enzymes, however, have names that do not end in *-ase*. Examples include lysozyme, trypsin, and pepsin.

Enzymes can be grouped according to the types of chemical reactions they catalyze. For example, *oxidases* add oxygen, *kinases* add phosphate, *dehydrogenases* remove hydrogen, *ATPases* split ATP, *anhydrases* remove water, *proteases* break down proteins, and *lipases* break down triglycerides (lipids).

Enzymes Have Important Properties

There are three major properties of enzymes:

1. ***Enzymes are highly specific.*** Of the more than 1000 known enzymes in your body, each has a characteristic three-dimensional shape with a specific surface configuration, which only permits binding of certain specific substrates. Not only is an enzyme matched to a particular substrate, it also catalyzes a specific reaction. From among the large number of diverse molecules in a cell, an enzyme must recognize the correct substrate and then take it apart or merge it with another substrate to form one or more specific products.
2. ***Enzymes are very efficient.*** Under optimal conditions, enzymes can catalyze reactions at rates that are from 100 million to 10 billion times more rapid than those of similar reactions occurring without enzymes. The number of substrate molecules that a single enzyme molecule can convert to product molecules in one second is generally between 1 and 10,000 and can be as high as 600,000.
3. ***Enzymes are subject to a variety of controls.*** Their rate of synthesis and their concentration at any given time are under the control of a cell's genes. Substances within the cell may either enhance or inhibit the activity of a given enzyme. For those enzymes that function outside cells, their activity can be regulated by the chemical environment of extracellular fluid.

Enzymes Catalyze Reactions by Helping Molecules Interact

Enzymes lower the activation energy of a chemical reaction by decreasing the "randomness" of the collisions between molecules. They also help bring the substrates together in the proper orientation so that the reaction can occur. **FIGURE 4.4** depicts how an enzyme works:

FIGURE 4.4 How an enzyme works.

An enzyme speeds up a chemical reaction without being altered or consumed.

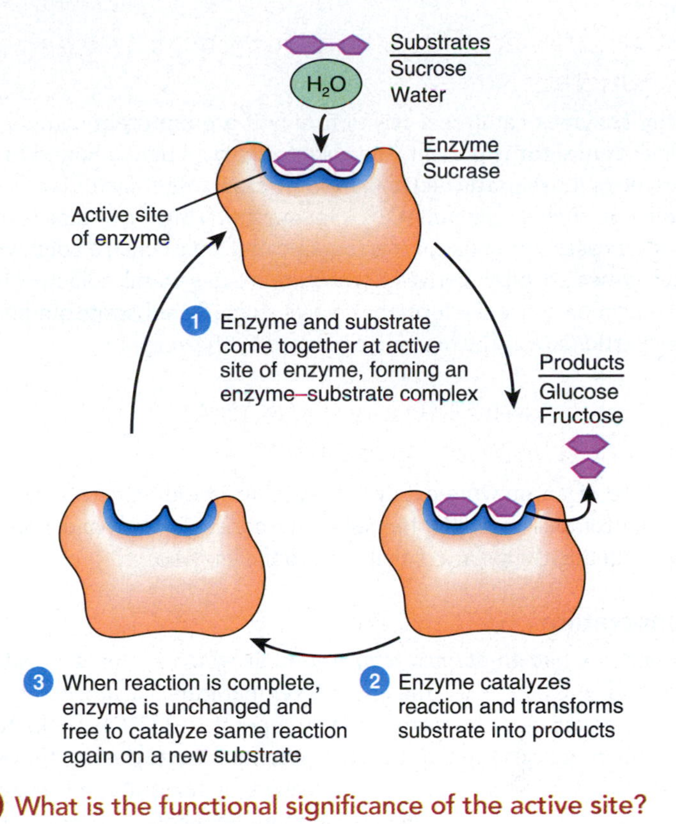

What is the functional significance of the active site?

1. The substrates (in this reaction, sucrose and water) bind to the **active site**, the part of the enzyme that catalyzes the reaction. It is a pocket or groove on the enzyme surface that is lined by certain amino acids. The three-dimensional shape of the active site accounts for the enzyme's specificity. In some cases, the active site is thought to fit the substrate like a key fits in a lock. This mode of enzyme–substrate interaction is referred to as the **lock-and-key model**. In other cases the active site slightly changes its shape to fit snugly around the substrate after the substrate enters the active site. This method of enzyme–substrate interaction is known as the **induced-fit model**. After the substrate binds to the active site, a temporary intermediate compound called the **enzyme–substrate complex** is formed. The substrate is held in the active site by weak (noncovalent) intermolecular forces, such as hydrogen bonds and ionic bonds.
2. The substrate molecules are transformed by the rearrangement of existing atoms, the breakdown of the substrate molecule, or the combination of several substrate molecules into the products of the reaction. Here the products are two monosaccharides: glucose and fructose.
3. After the reaction is completed and the reaction products move away from the enzyme, the unchanged enzyme is free to attach to other substrate molecules.

If a chemical reaction is reversible, the enzyme can catalyze the reaction in either direction, depending on the relative amounts of the substrates and products. For example, the enzyme *carbonic anhydrase* catalyzes the following reversible reaction:

$$\underset{\text{Carbon dioxide}}{CO_2} + \underset{\text{Water}}{H_2O} \xrightleftharpoons{\text{Carbonic anhydrase}} \underset{\text{Carbonic acid}}{H_2CO_3}$$

During exercise, when more CO_2 is produced and released into the blood, the reaction flows to the right, increasing the amount of carbonic acid in the blood. Then, as you exhale CO_2, its level in the blood falls and the reaction flows to the left, converting carbonic acid to CO_2 and H_2O. Recall that the relationship between the net direction of a reversible reaction and the concentrations of the reactants (in this case, the substrates) and products is known as the *law of mass action* (see Section 2.3).

Many Enzymes Require a Cofactor in Order to Function

Many enzymes catalyze a reaction only if a nonprotein component called a **cofactor** is present. The cofactor may be tightly bound to the enzyme or loosely attached to it. In some cases, the cofactor is an inorganic ion, such as calcium (Ca^{2+}), magnesium (Mg^{2+}), or zinc (Zn^{2+}). In other cases, the cofactor is an organic molecule called a **coenzyme**. Coenzymes are often derived from vitamins. In general, cofactors help the enzyme's active site form the correct shape for substrate binding or they participate in the reaction catalyzed by the enzyme.

Various Factors Influence the Rate of an Enzyme-Catalyzed Reaction

The rate of an enzyme-catalyzed reaction is influenced by several factors: temperature, pH, the substrate concentration, and nonsubstrate chemical substances that bind to the enzyme.

Temperature

An enzyme has an *optimal temperature* at which its reaction rate is maximal (**FIGURE 4.5**). The optimal temperature for most enzymes in the body is close to normal body temperature (37°C). As the temperature increases toward the optimal temperature, the reaction rate increases. This occurs because molecules of substrate and enzyme move about more rapidly and collide more frequently with each other as the temperature rises, increasing the likelihood that a reaction will occur. However, at temperatures above the optimal temperature, the reaction rate decreases drastically. This is because the higher temperatures break the weak bonds (hydrogen bonds, ionic bonds, van der Waals interactions, and hydrophobic interactions) that help determine the three-dimensional shape of the enzyme, causing the enzyme to denature.

pH

An enzyme also has an *optimal pH* at which its reaction rate is greatest (**FIGURE 4.6**). Most enzymes of the body have an optimal pH close to 7.4, which is the pH of most body fluids. At pHs that are higher or lower than the optimal pH, the reaction rate decreases. This is due to the fact that a significant change in the hydrogen ion (H^+) concentration (1) alters the charges of amino acids at the active site, causing the active site to become less active or completely nonfunctional and (2) disrupts the hydrogen bonds and ionic bonds that help establish the enzyme's three-dimensional shape, which results in enzyme denaturation. Not all enzymes have an optimal pH near 7.4. The digestive enzyme pepsin, for example, has an optimal pH of 2, allowing it to function in the highly acidic environment of the stomach.

Substrate Concentration

In addition to temperature and pH, the rate of an enzyme-catalyzed reaction is influenced by the concentration of substrate. Initially, the reaction rate increases as the substrate concentration increases (**FIGURE 4.7**). At a certain point, however, the reaction rate reaches a maximum and further increases in the substrate concentration do not increase the reaction rate. When this point is reached, the enzyme is said to be *saturated*, which means that the active site of every enzyme molecule is occupied by substrate.

FIGURE 4.5 Effect of temperature on the rate of an enzyme-catalyzed reaction.

At an enzyme's optimal temperature, the reaction rate is maximal; as the temperature increases toward the optimal temperature, the reaction rate increases; and at temperatures above the optimum temperature, the reaction rate decreases.

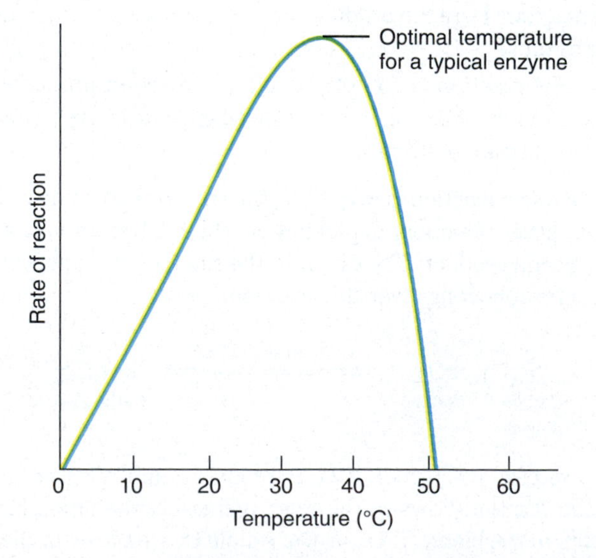

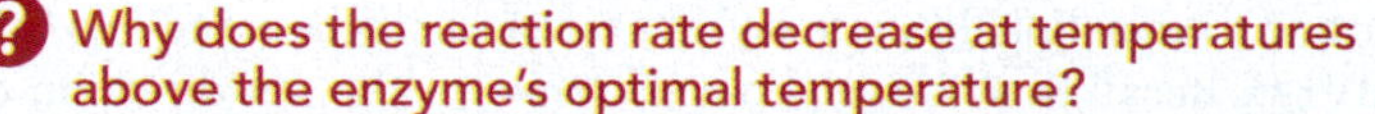

Why does the reaction rate decrease at temperatures above the enzyme's optimal temperature?

FIGURE 4.6 Effect of pH on the rate of an enzyme-catalyzed reaction.

At an enzyme's optimal pH, the reaction rate is greatest; at pHs higher or lower than the optimal pH, the reaction rate decreases.

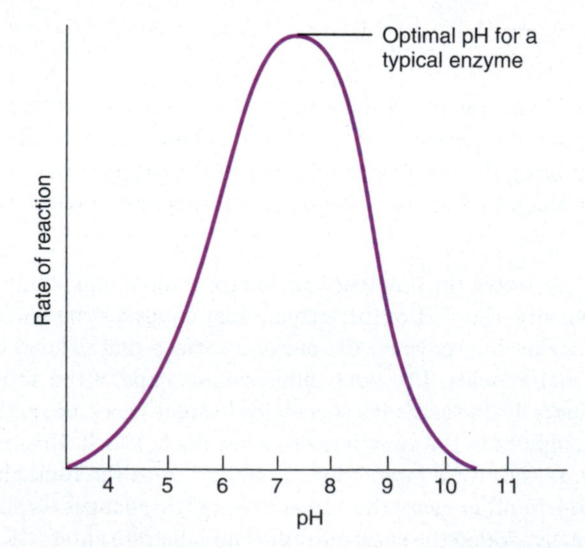

What effect does a significant change in hydrogen ion (H^+) concentration have on the active site of an enzyme?

FIGURE 4.7 Effect of substrate concentration on the rate of an enzyme-catalyzed reaction.

As the substrate concentration increases, the reaction rate increases until it reaches a maximal rate; at this point, the enzyme is saturated.

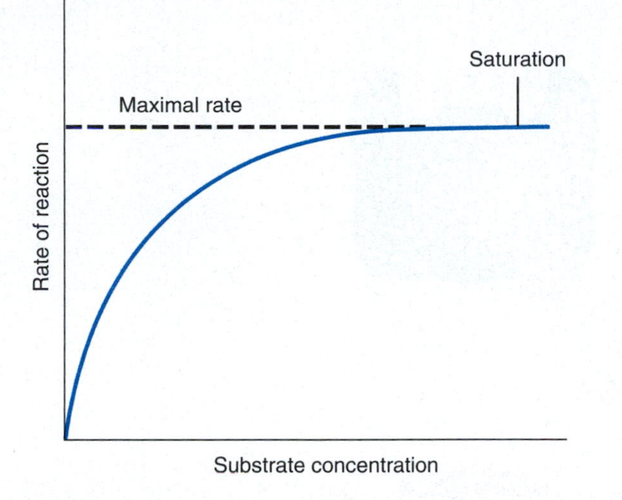

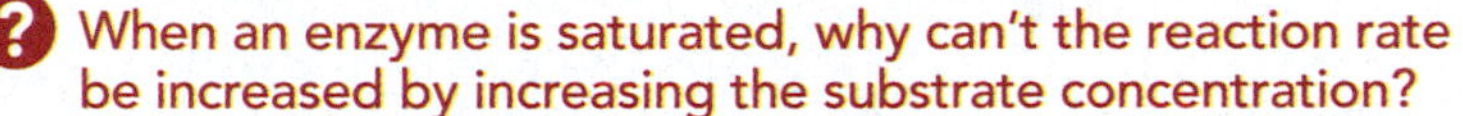

Nonsubstrate Chemical Substances That Bind to the Enzyme

The rate of an enzyme-catalyzed reaction can also be influenced by nonsubstrate chemical substances that selectively bind to the enzyme and alter its activity. A substance that increases the reaction rate is referred to as an *activator*. A substance that decreases the reaction rate or prevents the reaction from occurring altogether is termed an *inhibitor*. Some substances bind to enzymes irreversibly, in which case covalent bonds are involved. Others bind to enzymes reversibly via weak (noncovalent) bonds. Examples of nonsubstrate chemical substances that can bind to an enzyme and alter the reaction rate include competitive inhibitors and allosteric modulators.

Competitive Inhibitors A **competitive inhibitor** is a chemical substance that resembles the substrate and binds reversibly to the active site of the enzyme (**FIGURE 4.8**). It is so-named because it competes with the substrate for the active site. However, the active site does not catalyze a reaction with the competitive inhibitor as it does with the normal substrate. Thus, in the presence of both the substrate and competitive inhibitor, the reaction rate decreases. Increasing the substrate concentration increases the reaction rate because more substrate molecules are available to bind to the active site.

Allosteric Modulators An **allosteric modulator** is a chemical substance that binds reversibly to a site called an **allosteric site**, which is separate from the active site of the enzyme (**FIGURE 4.9**). Binding of the allosteric modulator to the allosteric site causes a conformational change of the enzyme that alters the affinity of the active site for the substrate. An allosteric modulator functions either as an activator or an inhibitor. Binding of an *allosteric activator* to an allosteric site causes a conformational change of the enzyme that increases the affinity of the active site for the substrate, thereby increasing the reaction rate

FIGURE 4.8 Competitive inhibition.

A competitive inhibitor is a chemical substance that competes with the substrate for the active site of an enzyme.

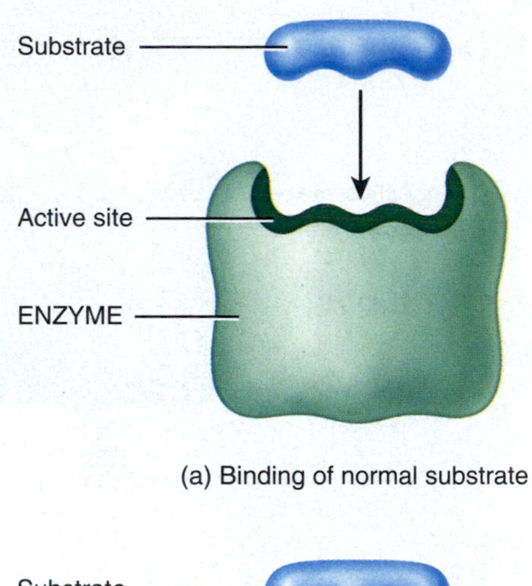

(a) Binding of normal substrate

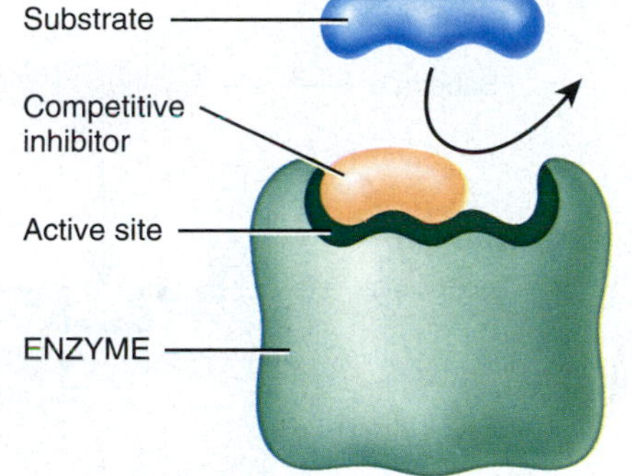

(b) Binding of a competitive inhibitor

What happens to the rate of an enzyme-catalyzed reaction if a competitive inhibitor is present along with the substrate? How can competitive inhibition be overcome?

(**FIGURE 4.9a**). Binding of an *allosteric inhibitor* to an allosteric site causes a conformational change in the enzyme that decreases the affinity of the active site for the substrate, resulting in a decrease in the reaction rate (**FIGURE 4.9b**).

A Sequence of Enzymatic Reactions Constitutes a Metabolic Pathway

A **metabolic pathway** is a sequence of reactions in which an *initial substrate* is converted into an *end product* via a series of *intermediates*. Each reaction is catalyzed by a separate enzyme and the product of one reaction serves as the substrate for the next reaction. To understand how a metabolic pathway is organized, consider the following generalized example:

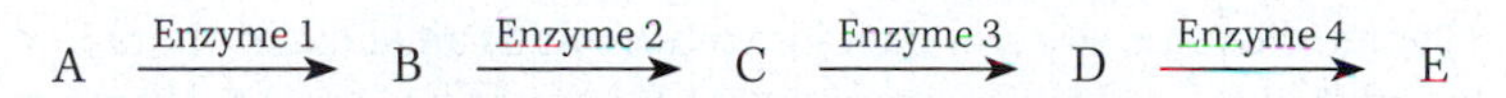

Note that A is the initial substrate; B, C, and D are the intermediates; and E is the end product, also known as the *final product*. Specific examples of metabolic pathways are presented later in this chapter.

FIGURE 4.9 Allosteric modulation of an enzyme.

In allosteric modulation of an enzyme, an allosteric activator or inhibitor binds to an allosteric site, which causes a conformational change of the enzyme that alters the affinity of the active site for the substrate.

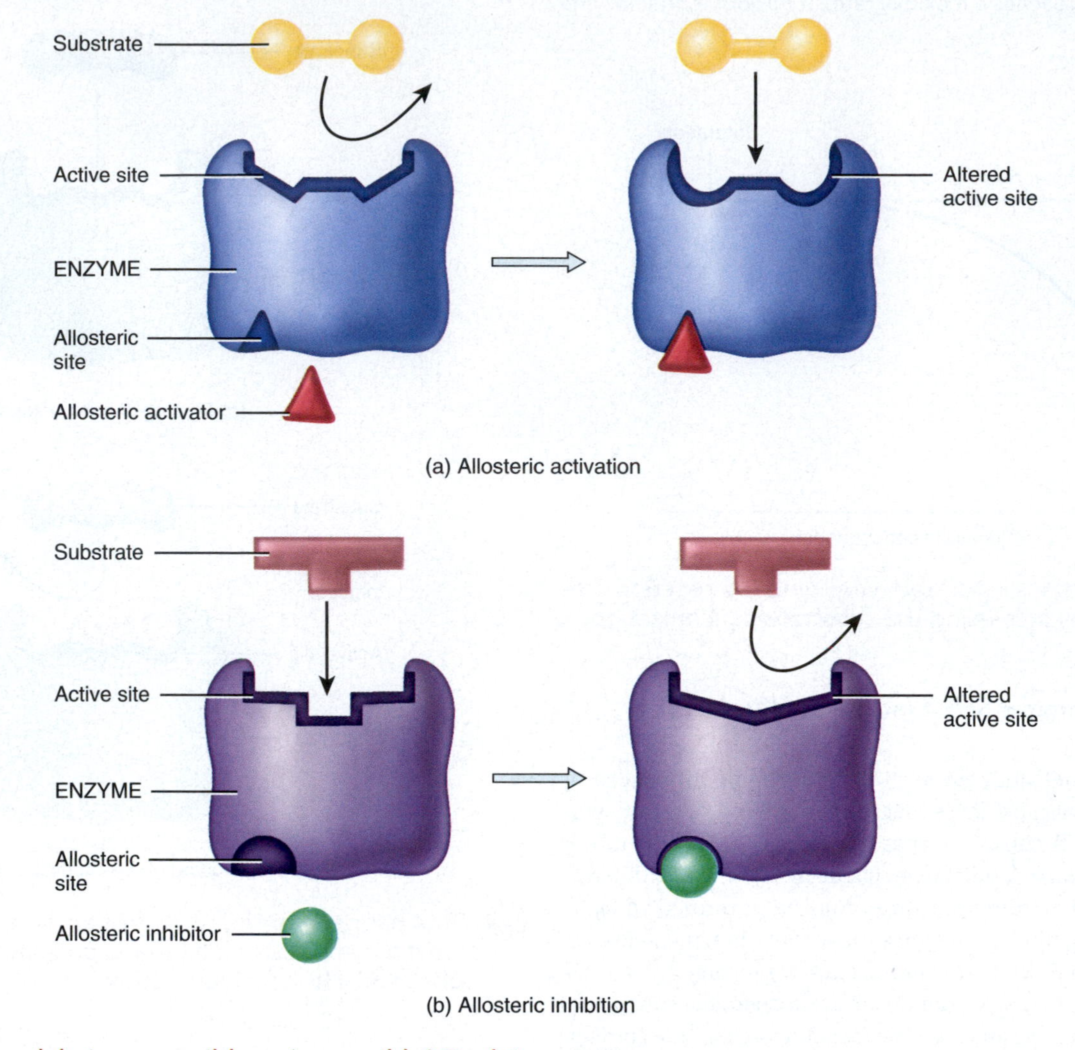

Is allosteric modulation reversible or irreversible? Explain your answer.

Metabolic Pathways Are Shut down by Feedback Inhibition

In **feedback inhibition**, or *end-product inhibition*, the end product of a metabolic pathway shuts down the pathway by binding to and inhibiting an enzyme that catalyzes one of the earlier reactions of the pathway:

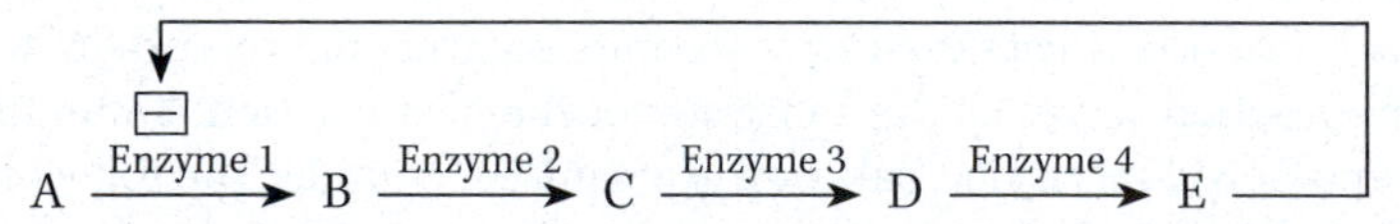

The end product (E) typically binds to the enzyme (in this case, enzyme 1) at an allosteric site and therefore functions as an allosteric inhibitor. By inhibiting enzyme 1, intermediate B cannot be formed. Likewise, intermediate C cannot form if intermediate B is not produced, and intermediate D will not be formed if intermediate C is not generated. Thus, by blocking the initial step of the metabolic pathway, the end product turns off the entire pathway.

Feedback inhibition is a type of negative feedback that maintains the concentration of end product at a level that is sufficient for the needs of the cell. When an excess of end product accumulates, feedback inhibition occurs and the reactions of the metabolic pathway come to a halt. When the concentration of the end product is low, feedback inhibition no longer occurs, and the reactions of the metabolic pathway resume.

CHECKPOINT

6. What are the three main properties of enzymes?
7. How do the lock-and-key model and induced-fit model of enzyme-substrate interaction differ?
8. Compare the effects of temperature and pH on enzyme activity.
9. What is an allosteric modulator?
10. What is a metabolic pathway?
11. Why is feedback inhibition important?

4.4 Role of ATP in Metabolism

OBJECTIVES

- Explain the role of ATP in anabolism and catabolism.
- Compare substrate-level phosphorylation and oxidative phosphorylation.
- Describe the role of NAD^+ and FAD in the generation of ATP.
- Outline the reactions that comprise cellular respiration.

Catabolism and Anabolism Are Coupled by ATP

Metabolism is an energy-balancing act between catabolic reactions and anabolic reactions. The molecule that participates most often in energy exchanges in living cells is **adenosine triphosphate (ATP)**, which couples energy-releasing catabolic reactions to energy-requiring anabolic reactions.

ATP serves as the "energy currency" of a living cell. Like money, it is readily available to "buy" cellular activities; it is spent and earned over and over. A typical cell has about a billion molecules of ATP, each of which typically lasts for less than a minute before being used. Thus, ATP is not a long-term-storage form of currency, like gold in a vault, but rather convenient cash for moment-to-moment transactions.

Recall from Chapter 2 that a molecule of ATP consists of an adenine molecule, a ribose molecule, and three phosphate groups bonded to one another (see **FIGURE 2.35**). **FIGURE 4.10** shows how ATP links anabolic and catabolic reactions. When the terminal phosphate group is split from ATP, adenosine diphosphate (ADP) and a phosphate group (symbolized as Ⓟ) are formed and energy is released:

$$\text{ATP} \rightarrow \text{ADP} + \text{Ⓟ} + \text{energy}$$

Some of the energy released is used to drive anabolic reactions such as the formation of glycogen from glucose. By contrast, when complex molecules are split apart (for example, the breakdown of a protein into its component amino acids), some of the energy released from these catabolic reactions is used to combine ADP and a phosphate group to resynthesize ATP:

$$\text{ADP} + \text{Ⓟ} + \text{energy} \rightarrow \text{ATP}$$

About 40% of the energy released in catabolism is used for cellular functions; the rest is converted to heat, some of which helps maintain normal body temperature. Excess heat is lost to the environment. Compared with machines, which typically convert only 10–20% of energy into work, the 40% efficiency of the body's metabolism is impressive. Still, the body has a continuous need to take in and process external sources of energy so that cells can synthesize enough ATP to sustain life.

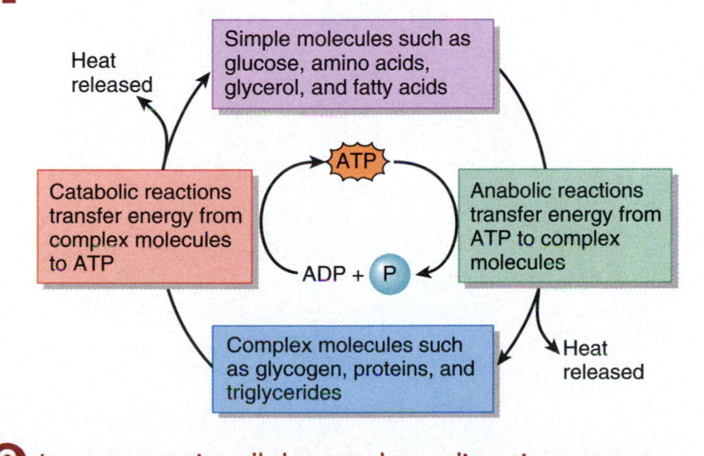

FIGURE 4.10 Role of ATP in linking anabolic and catabolic reactions. When complex molecules and polymers are split apart (catabolism, at left), some of the energy is transferred to form ATP and the rest is given off as heat. When simple molecules and monomers are combined to form complex molecules (anabolism, at right), ATP provides the energy for synthesis, and again some energy is given off as heat.

The coupling of energy-releasing and energy-requiring reactions is achieved through ATP.

In a pancreatic cell that produces digestive enzymes, does anabolism or catabolism predominate?

ATP Is Generated via Substrate-Level Phosphorylation and Oxidative Phosphorylation

As you have just learned, ATP is formed when a phosphate group (Ⓟ) is added to ADP along with an input of energy. The two high-energy phosphate bonds that can be used to transfer energy are indicated by "squiggles" (~):

$$\underbrace{\text{Adenosine} - \text{Ⓟ} \sim \text{Ⓟ}}_{\text{ADP}} + \text{Ⓟ} + \text{energy} \rightarrow \underbrace{\text{Adenosine} - \text{Ⓟ} \sim \text{Ⓟ} \sim \text{Ⓟ}}_{\text{ATP}}$$

The high-energy phosphate bond that attaches the third phosphate group contains the energy stored in this reaction. The addition of a phosphate group to a molecule, called **phosphorylation**, increases its potential energy. Cells of the body use two mechanisms of phosphorylation to generate ATP (**FIGURE 4.11**):

1. **Substrate-level phosphorylation** generates ATP by transferring a phosphate group from a phosphorylated metabolic intermediate—a substrate—directly to ADP (**FIGURE 4.11a**). This process occurs in the cytosol and the mitochondrial matrix of cells.
2. **Oxidative phosphorylation** produces ATP by adding a phosphate group to ADP using energy derived from a series of electron carriers, with oxygen serving as the final electron acceptor (**FIGURE 4.11b**). The electron carriers are collectively known as the *electron transport chain*, which is located in the inner mitochondrial membrane. During oxidative phosphorylation, hydrogen atoms are removed from certain coenzymes and split into hydrogen ions (H^+) and electrons. The H^+ ions are used to establish a hydrogen ion gradient, whereas the electrons are passed from one component of the electron transport chain to another. As H^+ ions move down their gradient, energy is released and the enzyme *ATP synthase* uses this energy to add a phosphate group to ADP to form ATP. The process by which an H^+ gradient is used to produce ATP is known as *chemiosmosis*. Together, chemiosmosis and the electron transport chain constitute oxidative phosphorylation.

FIGURE 4.11 Mechanisms of ATP production. In the reaction shown in part (a), the substrates are the phosphorylated intermediate and ADP; the products are the dephosphorylated intermediate and ATP.

The two mechanisms of ATP generation are substrate-level phosphorylation and oxidative phosphorylation.

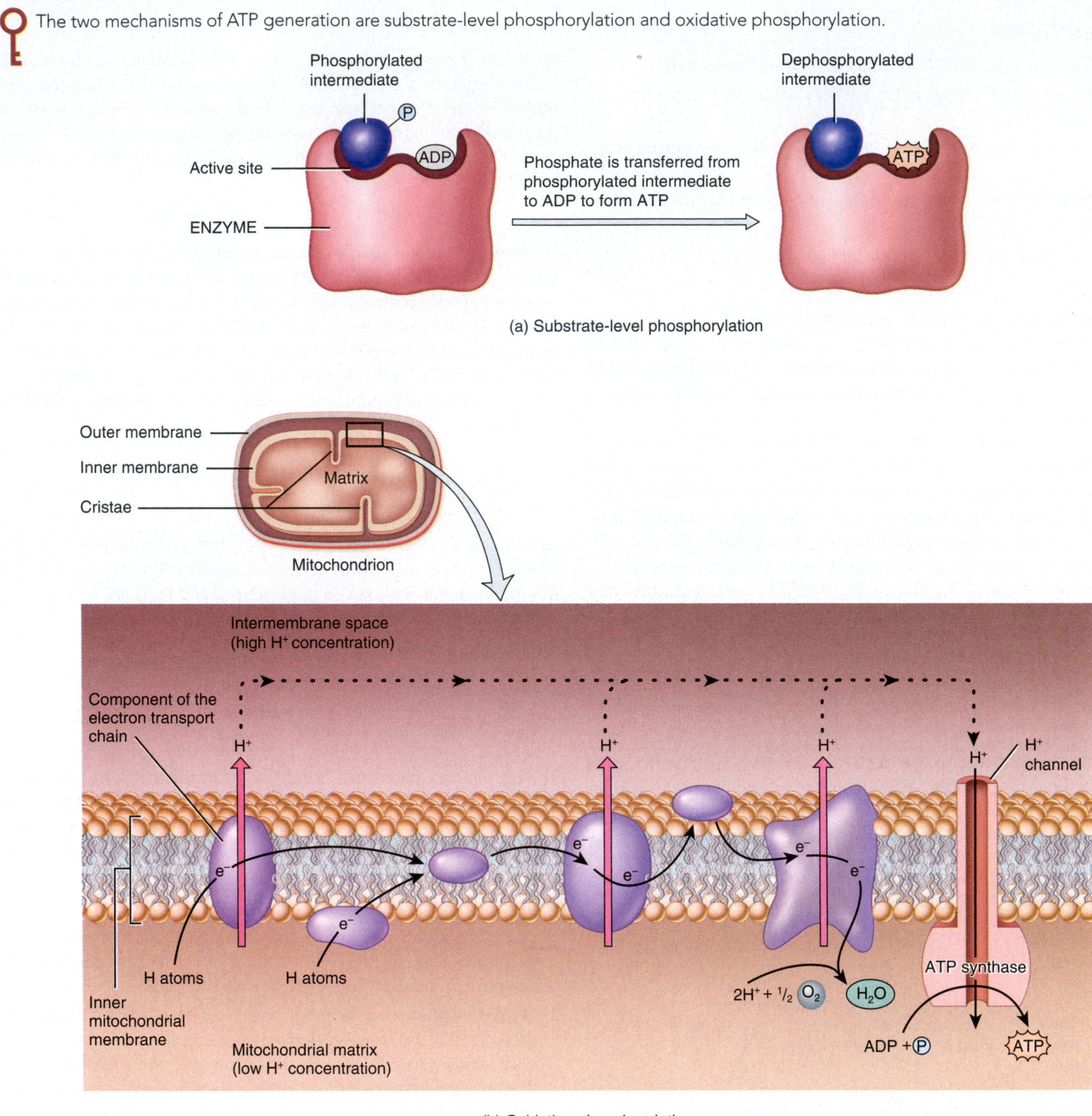

(a) Substrate-level phosphorylation

(b) Oxidative phosphorylation

What is chemiosmosis?

NAD^+ and FAD Help Generate ATP by Carrying Hydrogen Atoms to the Electron Transport Chain

Earlier in this chapter you learned that coenzymes are organic cofactors. Two examples of coenzymes important to metabolism are **nicotinamide adenine dinucleotide (NAD^+)** (FIGURE 4.12a), a derivative of the vitamin niacin, and **flavin adenine dinucleotide (FAD)** (FIGURE 4.12b), a derivative of the vitamin riboflavin. Two hydrogen atoms can be added to a molecule of NAD^+ or FAD to form NADH + H^+ or $FADH_2$, respectively:

FIGURE 4.12 Nicotinamide adenine dinucleotide (NAD^+) and flavin adenine dinucleotide (FAD).

NAD^+ and FAD are coenzymes that pick up electrons in the form of hydrogen atoms and carry them to the electron transport chain.

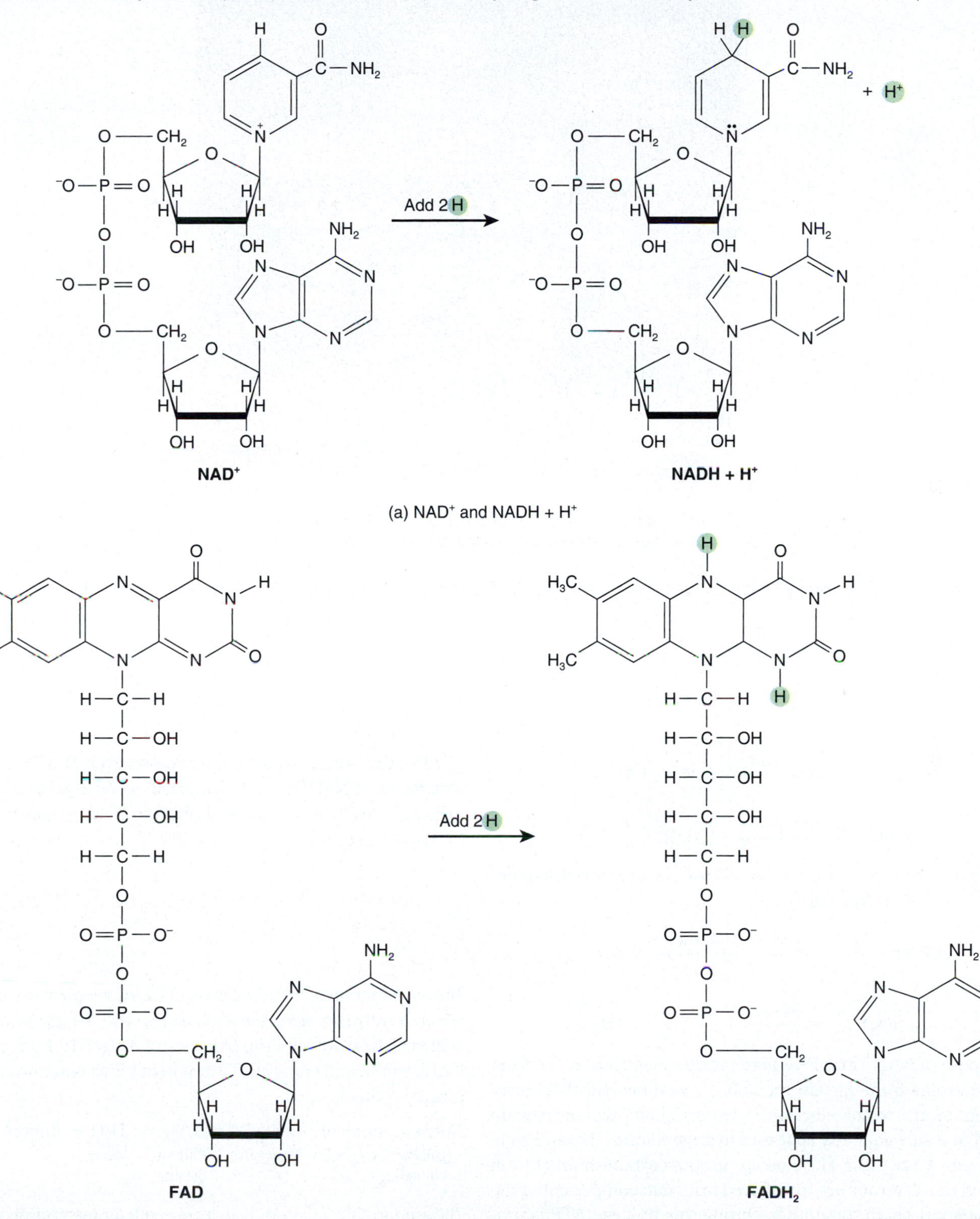

Figure 4.12 (continues)

Figure 4.12 Continued

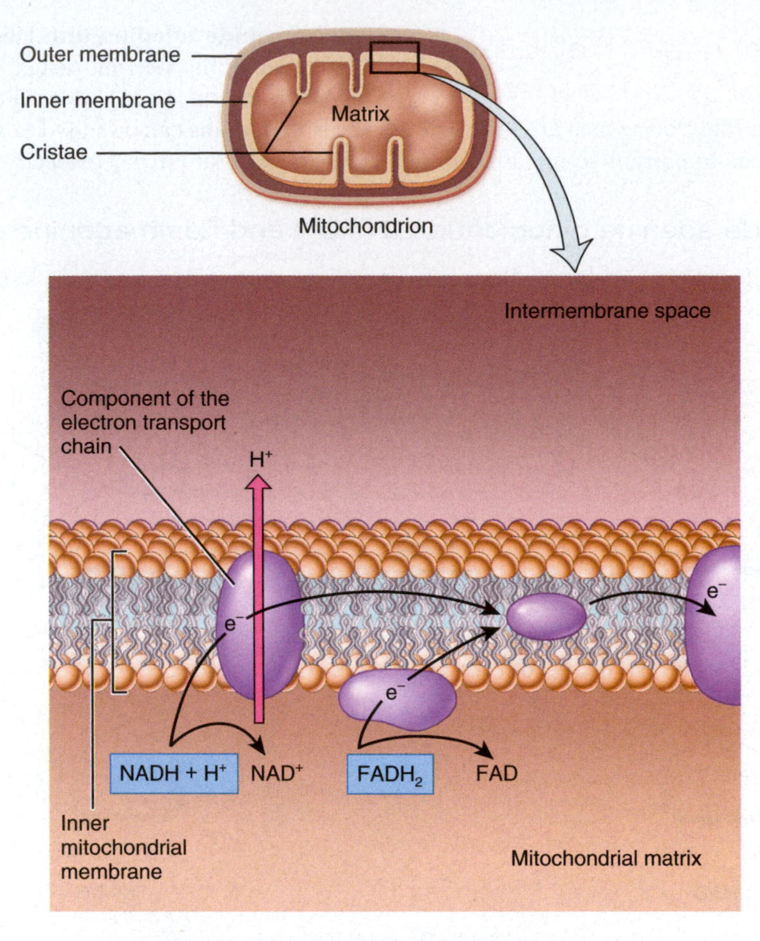

(c) Role of NAD^+ and FAD in carrying electrons in the form of hydrogen atoms to the electron transport chain

When NAD^+ or FAD picks up two hydrogen atoms, what is the name of the molecule that is formed?

$$NAD^+ \xrightarrow{\text{Add 2H atoms}} NADH + H^+$$

$$FAD \xrightarrow{\text{Add 2H atoms}} FADH_2$$

This reaction can be reversed: If the two H atoms are removed, a molecule of NAD^+ or FAD is formed again:

$$NADH + H^+ \xrightarrow{\text{Remove 2H atoms}} NAD^+$$

$$FADH_2 \xrightarrow{\text{Remove 2H atoms}} FAD$$

The purpose of NAD^+ and FAD is to pick up electrons in the form of hydrogen atoms from certain metabolic reactions and then carry the hydrogen atoms to the electron transport chain, where they are dropped off and subsequently split into hydrogen ions (H^+) and electrons (**FIGURE 4.12c**). The H^+ ions are used to establish an H^+ ion gradient, and the electrons are transferred from one component of the electron transport chain to another. During this process, ATP is produced via oxidative phosphorylation.

After the hydrogen atoms are released from $NADH + H^+$ or $FADH_2$, a molecule of NAD^+ or FAD is formed again: The NAD^+ or FAD is free to go back to a reaction of metabolism to pick up more hydrogen atoms to repeat the cycle.

Cellular Respiration Produces ATP by Breaking Down a Nutrient Molecule in the Presence of Oxygen

The cornerstone of metabolism is **cellular respiration**, the process by which a nutrient molecule such as glucose, a fatty acid, or an amino acid is broken down in the presence of oxygen to form carbon dioxide (CO_2), water, and energy (ATP and heat). This reaction can be summarized as follows:

$$\underset{\text{(glucose, fatty acid or amino acid)}}{\text{Nutrient molecule}} + \underset{\text{Oxygen}}{O_2} \longrightarrow \underset{\text{Carbon dioxide}}{CO_2} + \underset{\text{Water}}{H_2O} + \text{Energy (ATP + heat)}$$

The amount of ATP produced during cellular respiration varies depending on the type of nutrient molecule that is catabolized. For example,

Critical Thinking

A Cut Above the Rest

Marcus is a competitive body builder. He has been competing for a couple of years but has had limited success. The critiques he receives are all very similar: good mass, good performance, but not enough muscle definition. Muscle definition is often referred to as being "cut" or "ripped". In competitions, the more cut and defined each muscle is, the higher the score from the judges.

2,4-dinitrophenol

Marcus has tried removing carbohydrates and fats from his diet as he prepares for a competition but still can't seem to get that cut appearance he desires. He looked for weight loss substances on the Internet and found a reference to a chemical called 2,4-dinitrophenol (DNP). Marcus learned that DNP causes very rapid weight loss without significant loss of muscle mass. He also discovered that the drug was sold as an over-the-counter weight loss supplement during the early 1900s, but it is no longer sold this way because it has serious side effects: acute toxicity and risk of death. Despite these risks, Marcus was determined to try DNP and found a site on the Internet where he could purchase it.

SOME THINGS TO KEEP IN MIND:

DNP works in the body by uncoupling the process of oxidative phosphorylation; that is, it decreases the formation of high-energy phosphate bonds in ATP at the level of the mitochondria. DNP acts as an *ionophore*, a lipid-soluble molecule that increases the permeability of the cell membrane and facilitates the transport of specific ions (in this case H^+ ions) across the plasma membrane. This interrupts the normal flow of proton processing in the mitochondria, leading to a loss of energy as heat rather than in forming ATP molecules.

If Marcus uses this chemical for weight loss, he will most likely experience potentially lethal increases in body temperature. Why?

Why do you think that DNP causes rapid weight loss but does not interfere with muscle mass?

DNP can still be found for sale on the Internet. What advice would you give an athlete considering its use?

catabolism of fatty acids generates more ATP than the catabolism of glucose or amino acids.

Because glucose is the most common nutrient molecule catabolized for energy, the equation for cellular respiration is typically written for glucose:

$$\underset{\text{Glucose}}{C_6H_{12}O_6} + \underset{\text{Oxygen}}{6\,O_2} \longrightarrow \underset{\text{Carbon dioxide}}{6\,CO_2} + \underset{\text{Water}}{6\,H_2O} + \text{Energy (30 or 32 ATP + heat)}$$

Cellular respiration of glucose involves four sets of reactions: glycolysis, the formation of acetyl coenzyme A, the Krebs cycle, and the electron transport chain (**FIGURE 4.13a**).

1. ***Glycolysis*** is a set of reactions occurring in the cytosol in which one glucose molecule is converted into two molecules of pyruvic acid. The reactions also produce ATP molecules via substrate level phosphorylation and NADH + H^+.
2. ***Formation of acetyl coenzyme A*** is a transition step that prepares pyruvic acid for entrance into the Krebs cycle. First, pyruvic acid enters the mitochondrial matrix and is converted to a two-carbon fragment called an *acetyl group*. The acetyl group is then attached to coenzyme A, forming a substance known as *acetyl coenzyme A* (*acetyl CoA*). Two molecules of acetyl CoA are produced per molecule of glucose catabolized. As each molecule of acetyl CoA is formed, CO_2 and NADH + H^+ are produced.

FIGURE 4.13 Cellular respiration.

Cellular respiration is the process by which a nutrient molecule such as glucose, a fatty acid, or an amino acid is broken down in the presence of oxygen into carbon dioxide (CO_2), water, and energy (ATP and heat).

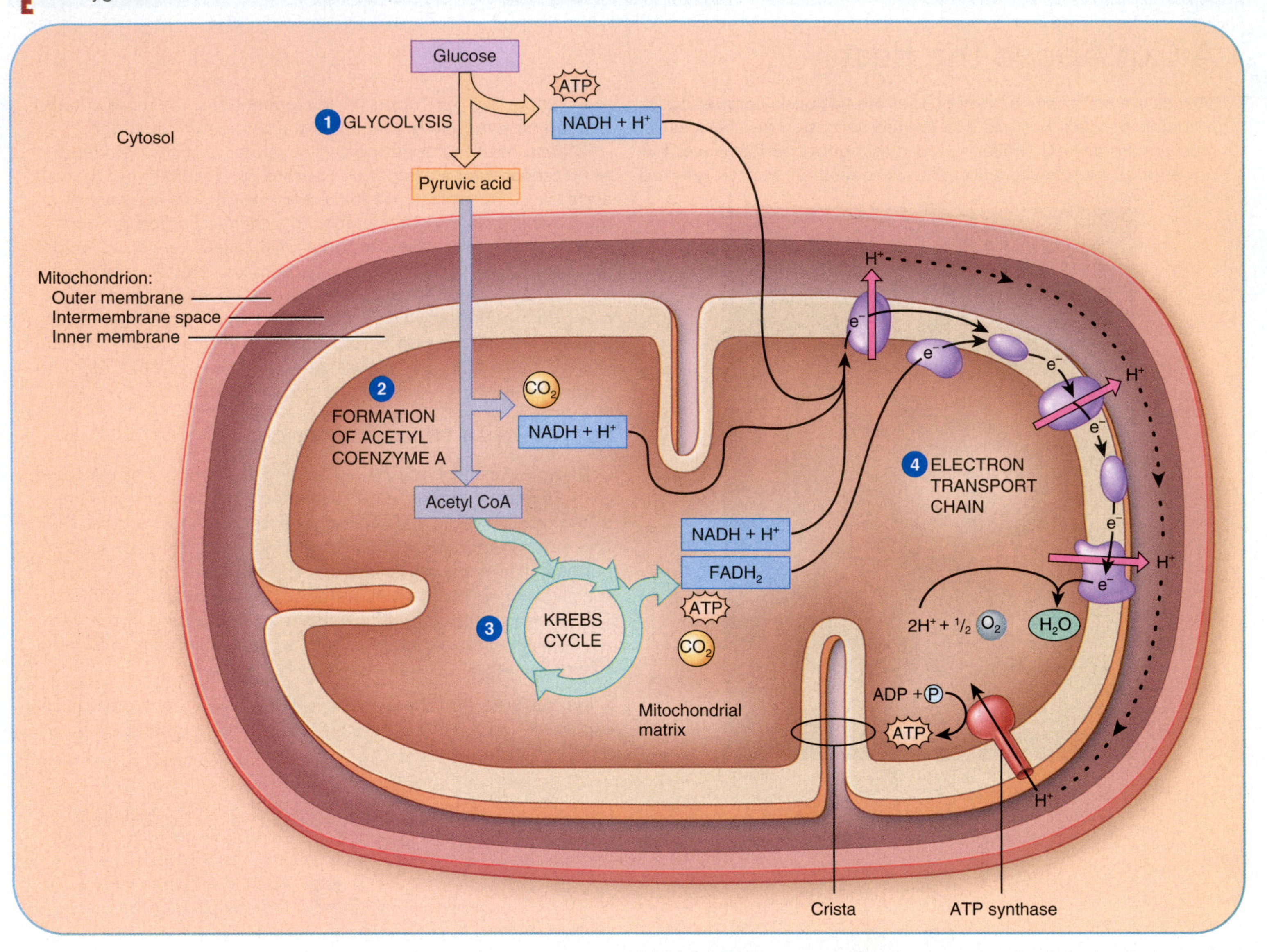

(a) Cellular respiration of glucose

3 The ***Krebs cycle*** is a series of reactions occurring in the mitochondrial matrix that involves the sequential conversion of one acid to another. The first reaction begins with the entrance of acetyl CoA into the cycle. The two molecules of acetyl CoA that are formed per molecule of glucose catabolized cause the Krebs cycle to occur twice. Each "turn" of the Krebs cycle results in the production of NADH + H^+, $FADH_2$, ATP (via substrate-level phosphorylation), and CO_2.

4 Through the reactions of the ***electron transport chain***, ATP is produced via oxidative phosphorylation. The electron transport chain is located in the inner mitochondrial membrane. Folds of this membrane called *cristae* (singular is *crista*) increase its surface area, accommodating thousands of copies of the transport chain in each mitochondrion. During oxidative phosphorylation, the hydrogen atoms of the coenzymes NADH + H^+ and $FADH_2$ that were formed during glycolysis, the formation of acetyl coenzyme A, and the Krebs cycle are removed and split into H^+ ions and electrons. The H^+ ions are used to establish an H^+ gradient, and the electrons are passed from one component of the electron transport chain to another, with oxygen serving as the final electron acceptor. As H^+ ions move down their concentration gradient, ATP is produced (chemiosmosis). Recall that chemiosmosis and the electron transport chain constitute oxidative phosphorylation. Most of the ATP produced in the catabolism of

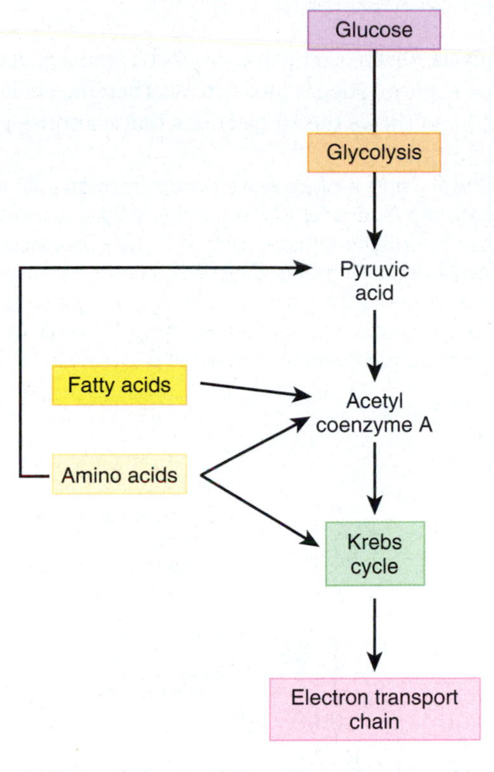

(b) Entry of glucose, fatty acids, and amino acids into cellular respiration

What is the most common fuel catabolized for energy?

glucose occurs during this step. In the process of generating this ATP, water is also produced.

It is important to note that the catabolism of glucose into CO_2, water, and energy (ATP and heat) occurs in these four stages and not in a single reaction as a way to control how much heat is produced. If the complete catabolism of glucose occurred in one step, so much heat would be released that the reaction would be explosive.

Because glycolysis does not require oxygen, it can occur under **aerobic** ("with oxygen") or **anaerobic** ("without oxygen") conditions. By contrast, the reactions of the Krebs cycle and electron transport chain require oxygen and are collectively referred to as **aerobic respiration**. Thus, when oxygen is present, all four phases occur: glycolysis, formation of acetyl CoA, the Krebs cycle, and the electron transport chain. However, if oxygen is not available or is present at a low concentration, the pyruvic acid generated during glycolysis is converted to a substance called *lactic acid* (see FIGURE 4.15) and the remaining steps of cellular respiration do not occur. When glycolysis occurs by itself under anaerobic conditions, it is referred to as **anaerobic glycolysis**.

Fatty acids and amino acids enter the reactions of cellular respiration at certain steps without entering glycolysis (FIGURE 4.13b). Fatty acids are converted into acetyl CoA, which in turn enters the Krebs cycle. Amino acids are deaminated (the amino group is removed) and the remainder of the amino acid enters cellular respiration as either pyruvic acid, acetyl CoA, or an intermediate of the Krebs cycle.

CHECKPOINT

12. How does ATP link catabolism and anabolism?
13. In what two ways can ATP be generated?
14. What do NADH + H^+ and $FADH_2$ carry to the electron transport chain?
15. What is cellular respiration?
16. Which reactions of cellular respiration occur under aerobic conditions? Which occur under anaerobic conditions?

4.5 Carbohydrate Metabolism

OBJECTIVE

- Describe the fate, metabolism, and functions of carbohydrates.

During digestion, polysaccharide and disaccharide carbohydrates are catabolized to monosaccharides—glucose, fructose, and galactose—which are absorbed into the bloodstream from the small intestine (see Section 21.6). Shortly after their absorption, however, hepatocytes (liver cells) convert fructose and galactose to glucose. So the story of carbohydrate metabolism is really the story of glucose metabolism.

Because glucose is the body's preferred source for synthesizing ATP, its use depends on the needs of body cells. If cells require immediate energy, glucose is catabolized to produce ATP. Glucose not needed for immediate ATP production may be converted to glycogen for storage in hepatocytes and skeletal muscle fibers. If these glycogen stores are full, the hepatocytes can transform the glucose to triglycerides for storage in adipose tissue. At a later time, when the cells need more ATP, glycogen and the glycerol component of triglycerides can be converted back to glucose. Cells throughout the body can also use glucose to make certain amino acids, the building blocks of proteins.

Glucose Catabolism Can Generate 30 or 32 ATP

During cellular respiration, glucose is catabolized in the presence of oxygen to produce carbon dioxide (CO_2), water, and ATP:

$$\underset{\text{Glucose}}{C_6H_{12}O_6} + \underset{\text{Oxygen}}{6\,O_2} \longrightarrow \underset{\text{Carbon dioxide}}{6\,CO_2} + \underset{\text{Water}}{6\,H_2O} + \text{Energy (30 or 32 ATP + heat)}$$

Recall that four interconnecting sets of reactions contribute to the cellular respiration of glucose: glycolysis, the formation of acetyl coenzyme A, the Krebs cycle, and the electron transport chain.

Glycolysis

During **glycolysis**, which occurs in the cytosol, chemical reactions split a 6-carbon molecule of glucose into two 3-carbon molecules of pyruvic acid. FIGURE 4.14 shows the 10 reactions that comprise glycolysis. In

FIGURE 4.14 The reactions of glycolysis. (1) Glucose is phosphorylated, using a phosphate group from an ATP molecule to form glucose 6-phosphate. (2) Glucose 6-phosphate is converted to fructose 6-phosphate. (3) A second ATP is used to add a second phosphate group to fructose 6-phosphate to form fructose 1,6-bisphosphate. (4) and (5) Fructose 1,6-bisphosphate splits into two 3-carbon molecules, glyceraldehyde 3-phosphate (G 3-P), and dihydroxyacetone phosphate, each having one phosphate group. Dihydroxyacetone phosphate is then converted into G 3-P. Therefore, the net result of step (5) is that two molecules of G 3-P are formed. (6) Each molecule of G 3-P gives up a hydrogen atom to NAD^+, forming NADH + H^+. In addition, a second phosphate group attaches to each molecule of G 3-P, forming 1,3-bisphosphoglyceric acid (BPG). (7) through (10) These reactions generate four molecules of ATP and produce two molecules of pyruvic acid (pyruvate).*

During glycolysis, glucose is converted to two molecules of pyruvic acid.

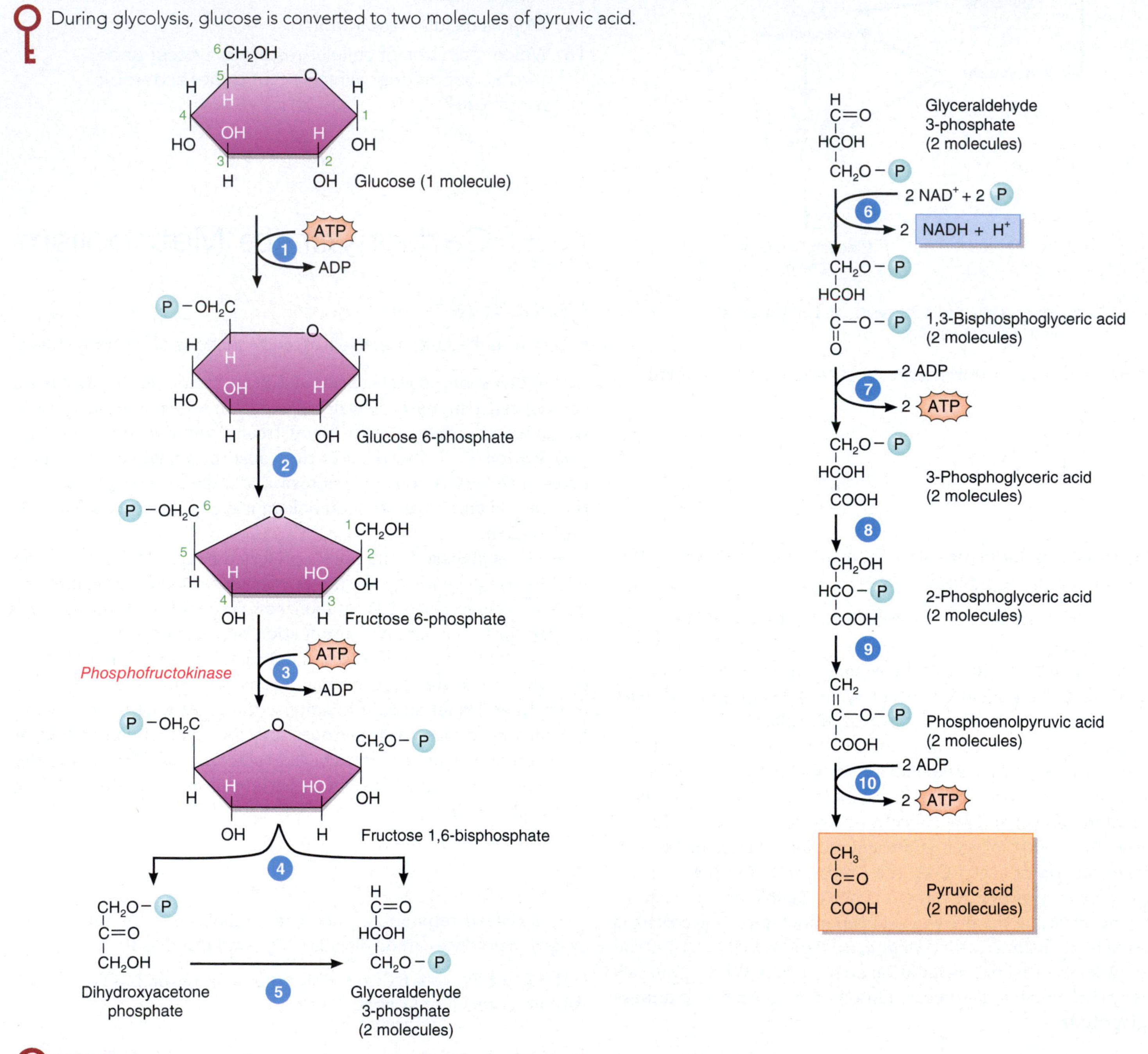

Why is the enzyme that catalyzes step 3 called a kinase?

*The carboxyl groups (—COOH) of intermediates in glycolysis and in the Krebs cycle are mostly ionized at the pH of body fluids to —COO^-. The suffix *-ic acid* indicates the nonionized form, whereas the ending *-ate* indicates the ionized form. Although the *-ate* names are more correct, the *acid* names will be used because these terms are more familiar.

the first half of the sequence (reactions 1 through 5), energy in the form of ATP is "invested" and the 6-carbon glucose molecule is split into two 3-carbon molecules of glyceraldehyde 3-phosphate. In the second half of the sequence (reactions 6 through 10), the two glyceraldehyde 3-phosphate molecules are converted to two pyruvic acid molecules, and ATP is generated. Even though glycolysis consumes two ATP molecules, it produces four ATP molecules via substrate-level phosphorylation, for a net gain of two ATP molecules for each glucose molecule that is catabolized. Glycolysis also produces two molecules of NADH + H^+ that carry electrons in the form of hydrogen atoms to the electron transport chain to generate ATP via oxidative phosphorylation.

Phosphofructokinase, the enzyme that catalyzes step 3, is the key regulator of the rate of glycolysis (FIGURE 4.14). The activity of this enzyme is high when ADP concentration is high, in which case ATP is produced rapidly. When the activity of phosphofructokinase is low, most glucose does not enter the reactions of glycolysis but instead undergoes conversion to glycogen for storage.

The fate of the pyruvic acid produced during glycolysis depends on the availability of oxygen (FIGURE 4.15). If oxygen is scarce (anaerobic conditions)—for example, in skeletal muscle fibers during strenuous exercise—then pyruvic acid is converted to **lactic acid** (lactate) via an anaerobic pathway. As lactic acid is produced, it rapidly diffuses out of the cell and enters the blood. Hepatocytes remove lactic acid from the blood and convert it back to pyruvic acid and then to glucose. A buildup of lactic acid may be one factor that contributes to muscle fatigue (see Section 11.4).

When oxygen is plentiful (aerobic conditions), most cells convert pyruvic acid to acetyl coenzyme A. This molecule links glycolysis, which occurs in the cytosol, with the Krebs cycle, which occurs in the mitochondrial matrix. Pyruvic acid enters the mitochondrial matrix with the help of a special transporter protein. Because they lack mitochondria, red blood cells can produce ATP only through glycolysis.

Formation of Acetyl Coenzyme A

Each step in the catabolism of glucose requires a different enzyme, and often a coenzyme as well. The coenzyme used at this point in cellular respiration is **coenzyme A (CoA)**, which is derived from pantothenic acid, a B vitamin. During the transitional step between glycolysis and the Krebs cycle, pyruvic acid is prepared for entrance into the cycle. The enzyme *pyruvate dehydrogenase*, which is located exclusively in the mitochondrial matrix, converts pyruvic acid to a two-carbon fragment called an **acetyl group** (FIGURE 4.15). The acetyl group then attaches to coenzyme A, producing a molecule called **acetyl coenzyme A (acetyl CoA)**. During the conversion of pyruvic acid to acetyl CoA, a molecule of CO_2 and a molecule of NADH + H^+ are produced. Recall that the catabolism of one glucose molecule produces two molecules of pyruvic acid, so for each molecule of glucose that is broken down, two molecules of carbon dioxide and two molecules of NADH + H^+ are generated during this stage of cellular respiration.

FIGURE 4.15 Fate of pyruvic acid.

When oxygen is plentiful, pyruvic acid enters mitochondria, is converted to acetyl coenzyme A, and enters the Krebs cycle (aerobic pathway). When oxygen is scarce, most pyruvic acid is converted to lactic acid via an anaerobic pathway.

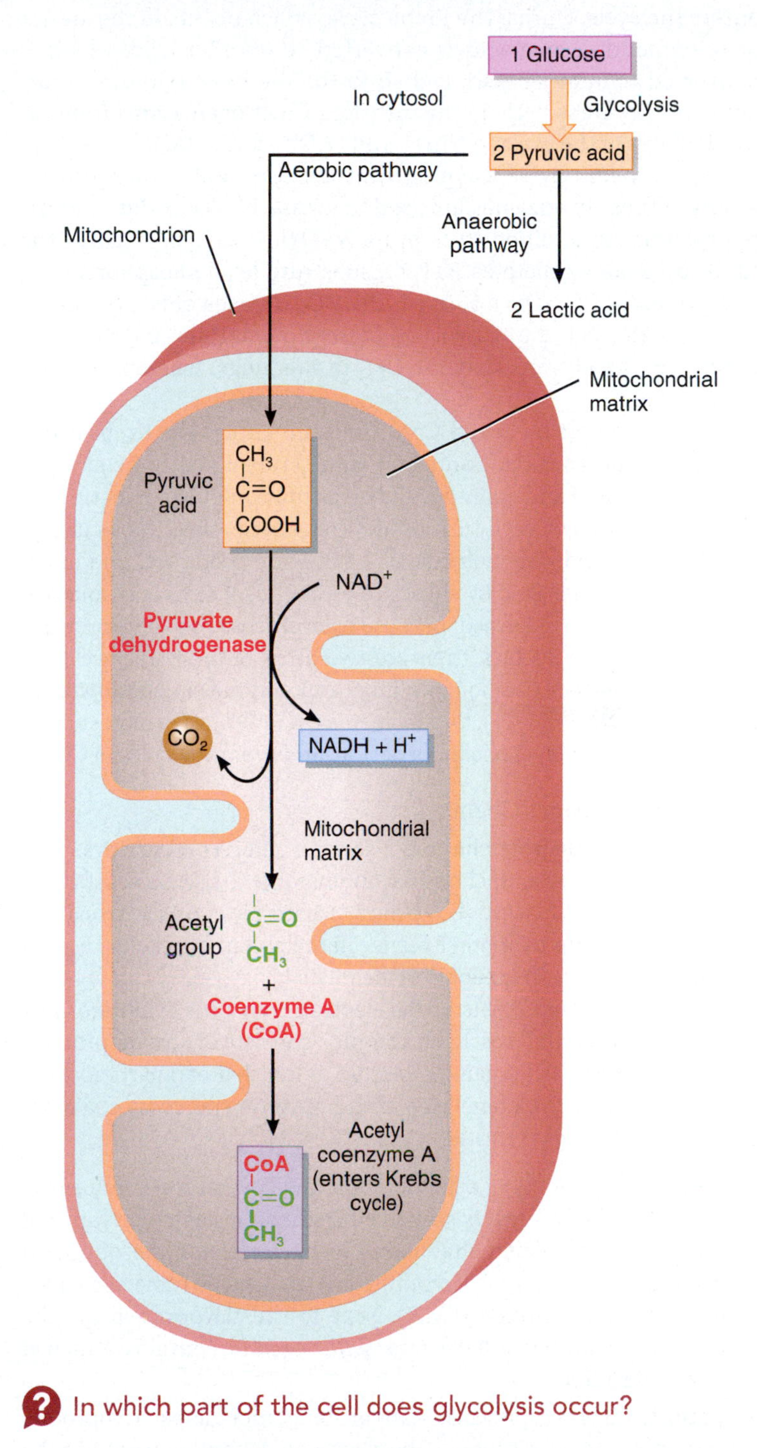

In which part of the cell does glycolysis occur?

Krebs Cycle

Once acetyl CoA is formed, it is ready to enter the Krebs cycle (**FIGURE 4.16**). The **Krebs cycle**—named for the biochemist Hans Krebs, who described these reactions in the 1930s—is also known as the **citric acid cycle** for the first molecule formed after acetyl CoA enters the cycle. During the Krebs cycle, which occurs in the matrix of mitochondria, one acid is converted to another acid, which is converted to another acid, and so forth. The most important outcome of the Krebs cycle is the transfer of hydrogen atoms from intermediates of the cycle to NAD^+ and FAD to form NADH + H^+ and $FADH_2$, respectively. In the electron transport chain, the hydrogen atoms of these coenzymes are used to form ATP via oxidative phosphorylation. In addition to forming NADH + H^+ and $FADH_2$, the Krebs cycle also generates ATP via substrate-level phosphorylation and produces CO_2. Once formed, the CO_2 diffuses out of the mitochondria, through the cytosol and plasma membrane, and then into the blood. Blood transports the CO_2 to the lungs, where it is eventually exhaled.

Each time that an acetyl CoA molecule enters the Krebs cycle, the cycle undergoes one complete "turn", starting with the production of citric acid and ending with the formation of oxaloacetic acid (**FIGURE 4.16**). For each turn of the Krebs cycle, three molecules of NADH + H^+ and one molecule of $FADH_2$ are produced, one molecule of ATP is generated by substrate-level phosphorylation, and two molecules of CO_2 are formed. Because each glucose molecule provides two acetyl CoA molecules, there are two turns of the Krebs cycle per molecule of glucose catabolized. This results in the production of six molecules of NADH + H^+, two molecules of $FADH_2$, two molecules of ATP via substrate-level phosphorylation, and four molecules of CO_2.

Electron Transport Chain

The **electron transport chain** is a series of **electron carriers**, molecules in the inner mitochondrial membrane that can accept and donate electrons (**FIGURE 4.17**). Each carrier picks up electrons and then gives them up to another carrier in the chain. In cellular respiration, the final electron acceptor of the chain is oxygen.

Most of the components of the electron transport chain are clustered into four complexes. Each complex consists of several proteins that have an attached organic molecule or ion that can participate in electron transfer. The complexes of the electron transport chain are organized as follows (**FIGURE 4.17**):

- **Complex I** consists of a flavoprotein and several iron–sulfur proteins. The *flavoprotein* is so-named because it has a derivative of the vitamin riboflavin attached to it. In complex I, the riboflavin derivative bound to the flavoprotein is an organic molecule called flavin mononucleotide (FMN). Next to the flavoprotein are the iron–sulfur proteins. Each *iron–sulfur protein* contains attached iron and sulfide ions.
- **Complex II** is comprised of many proteins, including a flavoprotein and an iron–sulfur protein. The riboflavin derivative bound to the flavoprotein in complex II is the organic molecule flavin adenine dinucleotide (FAD).
- **Complex III** contains cytochrome b, cytochrome c_1, and an iron–sulfur protein. A *cytochrome* is a brightly colored, pigmented protein with an iron-containing heme group. The heme group of a cytochrome is similar to the heme group of hemoglobin, a protein present in erythrocytes (see Section 16.2). However, in a cytochrome, the iron ion binds to electrons, whereas in hemoglobin, the iron ion binds to an oxygen molecule. Several types of cytochromes exist, each with a slightly different heme group.
- **Complex IV**, also known as the *cytochrome oxidase complex*, is composed of cytochrome a, cytochrome a_3, and proteins with attached copper ions.

In addition to the four complexes just described, the electron transport chain has two other components (**FIGURE 4.17**):

- **Coenzyme Q**, symbolized as **Q** and also referred to as **ubiquinone**, is a low-molecular-weight, hydrophobic carrier that is mobile in the lipid bilayer of the inner mitochondrial membrane. It is the only constituent of the electron transport chain that is not a protein. Its role is to transfer electrons from complexes I and II to complex III.
- **Cytochrome c** is a mobile electron carrier located between complexes III and IV. Its function is to transfer electrons between these two complexes.

Complexes I, III, and IV not only transport electrons but also act as pumps that expel H^+ ions from the mitochondrial matrix into the intermembrane space between the inner and outer mitochondrial membranes (**FIGURE 4.17**). These complexes therefore function as **proton pumps**; the term *proton pump* is used because H^+ ions consist of a single proton. As electrons pass through the electron transport chain, a series of exergonic reactions release small amounts of energy; this energy is used to drive the proton pumps. The H^+ ions that are pumped into the intermembrane space are derived from the hydrogen atoms carried to the electron transport chain via NADH + H^+ and $FADH_2$. NADH + H^+ delivers hydrogen atoms to complex I; $FADH_2$ delivers hydrogen atoms to complex II. These complexes in turn split the H atoms into H^+ ions and electrons. The electrons are transferred along the electron transport chain, whereas the H^+ ions are released into the mitochondrial matrix and then quickly picked back up and pumped into the intermembrane space.

The pumping of H^+ ions produces a concentration gradient: A high concentration of H^+ accumulates in the intermembrane space, leaving a low concentration of H^+ in the mitochondrial matrix. This concentration gradient has potential energy, called the *proton motive force*. Proton channels in the inner mitochondrial membrane allow H^+ to flow back across the membrane, driven by the proton motive force. As H^+ ions flow back, they generate ATP because the H^+ channels also include an enzyme called **ATP synthase** (**FIGURE 4.17**). The enzyme uses the energy released as H^+ moves down its concentration gradient to synthesize ATP from ADP and Ⓟ. Because stored energy in the form of a H^+ gradient is used to form ATP, this process is referred to as **chemiosmosis** (*chemi-* = chemical; *-osmosis* = pushing). Recall that chemiosmosis and the electron transport chain together constitute oxidative phosphorylation. As chemiosmosis occurs, complex IV (the cytochrome oxidase complex) passes two electrons to one-half of a molecule of oxygen (O_2), which becomes negatively charged and then picks up two H^+ ions from the surrounding medium to form H_2O. This is the only point in cellular respiration where O_2 is consumed. **Cyanide** is a deadly poison because it binds to complex IV and blocks this last step in electron transport.

For every molecule of NADH + H^+ that drops off hydrogen atoms to the electron transport chain, two or three molecules of ATP

FIGURE 4.16 The reactions of the Krebs cycle. (1) Acetyl coenzyme A (acetyl CoA) enters the cycle. The chemical bond that attaches the acetyl group to coenzyme A (CoA) breaks, and the 2-carbon acetyl group attaches to a 4-carbon molecule of oxaloacetic acid to form a 6-carbon molecule called citric acid. CoA is free to combine with another acetyl group from pyruvic acid and repeat the process. (2) Citric acid undergoes isomerization to isocitric acid, which has the same molecular formula as citrate. Notice, however, that the hydroxyl group (—OH) is attached to a different carbon. (3) Isocitric acid loses two hydrogen atoms and a molecule of CO_2, forming alpha-ketoglutaric acid. The H atoms lost from isocitric acid are passed on to NAD^+ to form NADH + H^+. (4) Alpha-ketoglutaric acid loses two hydrogen atoms and a molecule of CO_2 and picks up CoA to form succinyl CoA. The lost H atoms are picked up by another molecule of NAD^+, forming NADH + H^+. (5) CoA is displaced by a phosphate group, which is then transferred to guanosine diphosphate (GDP) to form guanosine triphosphate (GTP). GTP can donate a phosphate group to ADP to form ATP, an example of substrate-level phosphorylation. (6) Succinic acid loses two of its hydrogen atoms to form fumaric acid. The two H atoms are transferred to FAD to form $FADH_2$. (7) Fumaric acid is converted to malic acid by the addition of a molecule of water. (8) In the final step in the cycle, malic acid loses two hydrogen atoms to re-form oxaloacetic acid. The lost hydrogen atoms are transferred to NAD^+, forming NADH + H^+. The regenerated oxaloacetic acid can combine with another molecule of acetyl CoA, beginning a new cycle.

The main outcomes of the Krebs cycle are the production of NADH + H^+ and $FADH_2$, the generation of ATP via substrate-level phosphorylation, and the formation of CO_2.

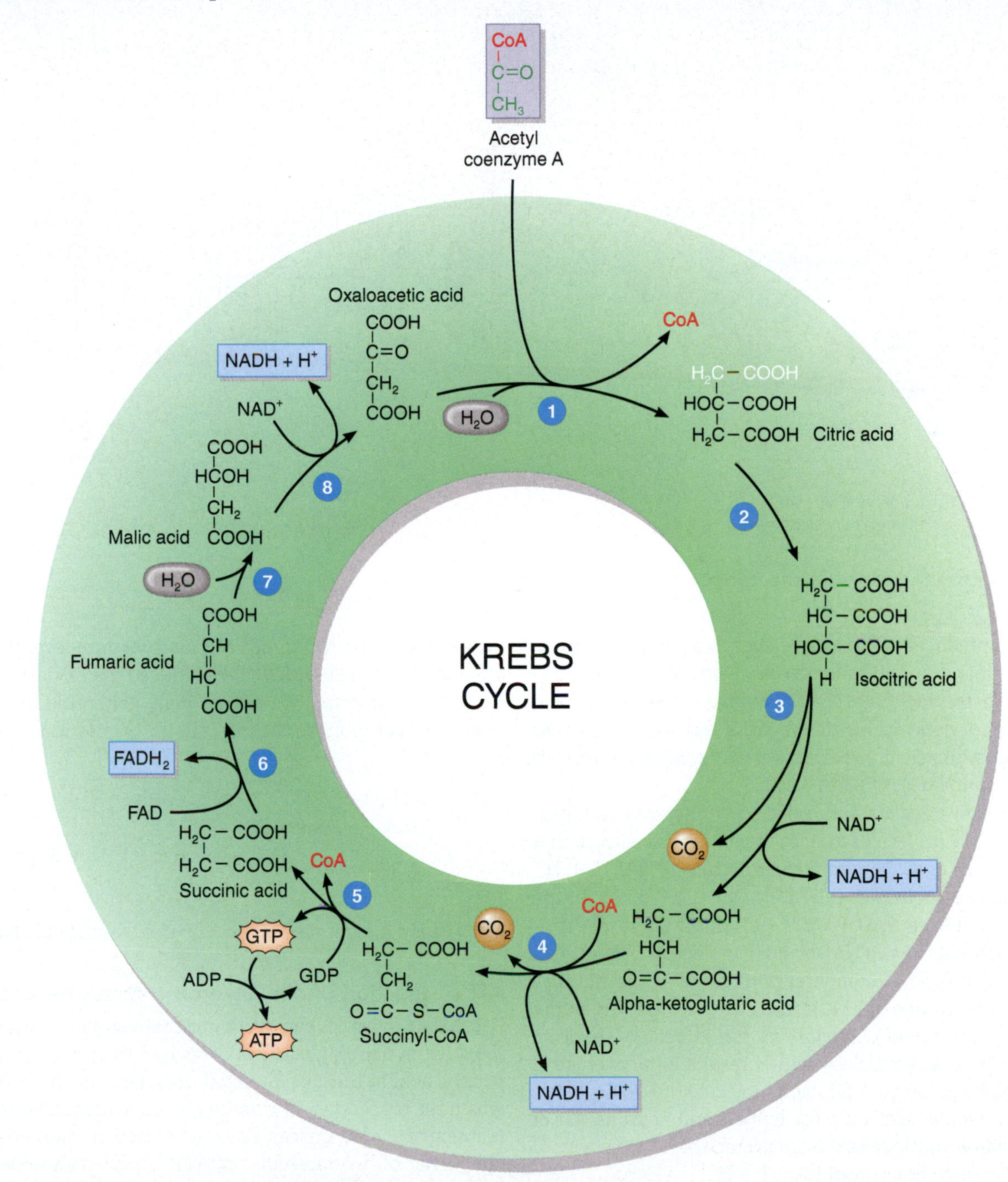

Why is the production of NADH + H^+ and $FADH_2$ important in the Krebs cycle?

FIGURE 4.17 Electron transport chain.

As the components of the electron transport chain pass electrons to one another, they also move protons (H^+) from the matrix into the intermembrane space of the mitochondrion. As protons flow back into the mitochondrial matrix through the H^+ channel in ATP synthase, ATP is synthesized.

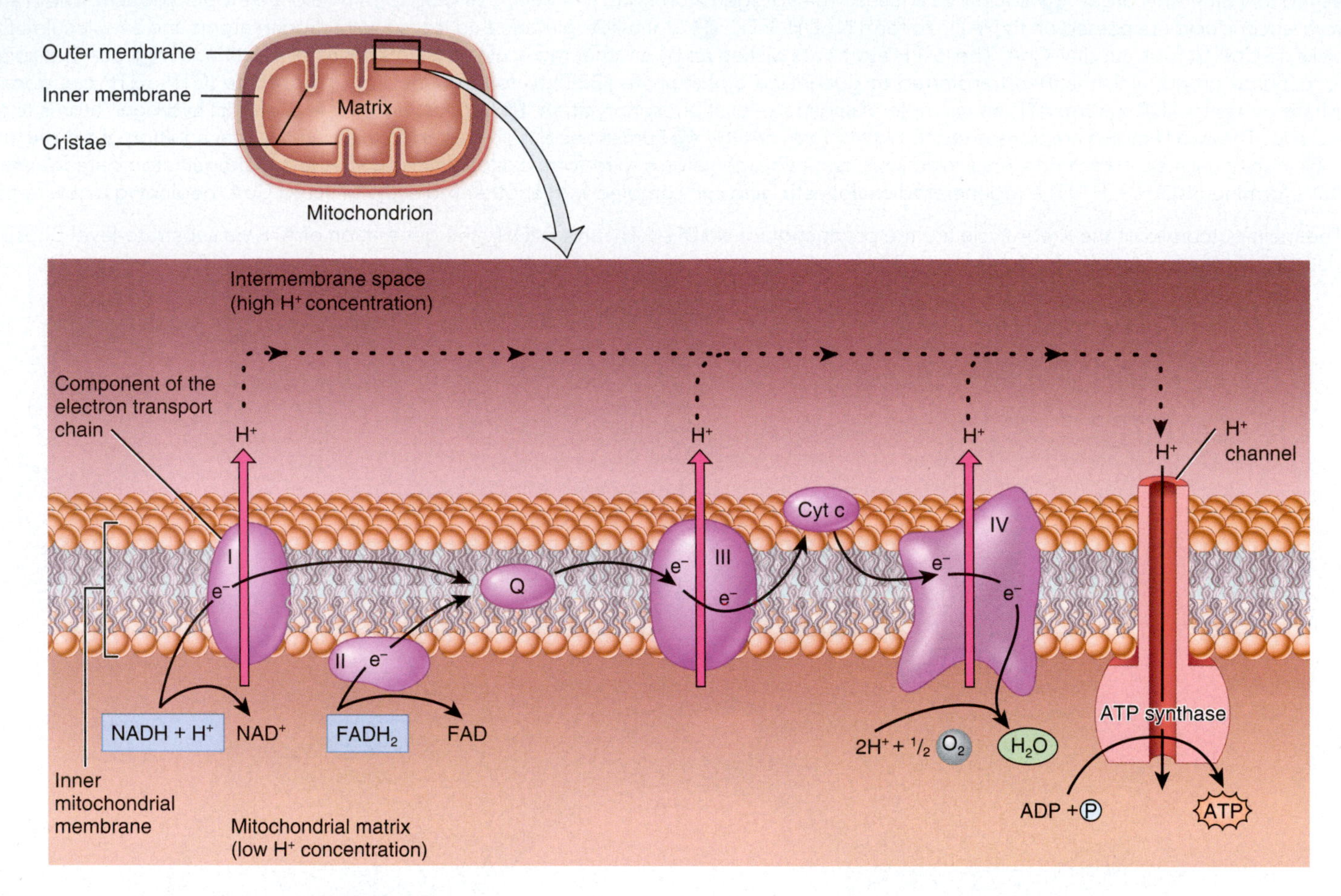

Where in the mitochondrion is the concentration of H^+ highest?

(average = 2.5) are produced via oxidative phosphorylation. For every molecule of $FADH_2$ that drops off hydrogen atoms to the electron transport chain, only one or two molecules of ATP (average = 1.5) are produced via oxidative phosphorylation. This difference is due to the fact that $FADH_2$ drops off its hydrogen atoms at a lower energy level (complex II) than NADH + H^+ (complex I) (**FIGURE 4.17**).

The various electron transfers in the electron transport chain generate either 26 or 28 ATP molecules from each molecule of glucose that is catabolized: either 23 or 25 from the ten molecules of NADH + H^+ and three from the two molecules of $FADH_2$. The discrepancy in the number of ATP formed from NADH + H^+ via oxidative phosphorylation is due to the fact that the two NADH + H^+ molecules produced in the cytosol during glycolysis cannot enter mitochondria. Instead they donate their electrons to one of two transfer shuttles known as the *malate shuttle* and the *glycerol phosphate shuttle*. In cells of the liver, kidneys, and heart, use of the malate shuttle results in an average of 2.5 molecules of ATP synthesized for each molecule of NADH + H^+. In other body cells, such as skeletal muscle fibers and neurons, use of the glycerol phosphate shuttle results in an average of 1.5 molecules of ATP synthesized for each molecule of NADH + H^+.

Recall that four ATP molecules are produced via substrate-level phosphorylation (two from glycolysis and two from the Krebs cycle). If the four ATP produced via substrate-level phosphorylation are added to the 26 or 28 ATP produced via oxidative phosphorylation, a total of either 30 or 32 ATP are generated for each molecule of glucose catabolized during cellular respiration. The overall reaction is:

$$\underset{\text{Glucose}}{C_6H_{12}O_6} + \underset{\text{Oxygen}}{6\,O_2} + 30 \text{ or } 32 \text{ ADPs} + 30 \text{ or } 32 \text{ Ⓟ} \longrightarrow \underset{\text{Carbon dioxide}}{6\,CO_2} + \underset{\text{Water}}{6\,H_2O} + 30 \text{ or } 32 \text{ ATPs}$$

TABLE 4.1 summarizes the ATP yield during cellular respiration.

ATP Production Under Anaerobic Versus Aerobic Conditions

Recall that glycolysis can occur under anaerobic or aerobic conditions. If oxygen is scarce, glucose is converted to pyruvic acid and then the pyruvic acid in turn is converted to lactic acid. The entire process by which the breakdown of glucose gives rise to lactic acid when oxygen is absent or at a low concentration is referred to as **anaerobic glycolysis**. Only a net of two molecules of ATP is produced under these conditions. The overall reaction is:

$$\text{glucose} + 2\,\text{ADP} + 2\,\text{Ⓟ} \longrightarrow 2\,\text{lactic acid} + 2\,H_2O + 2\,\text{ATP}$$

TABLE 4.1 Summary of ATP Produced per Molecule of Glucose Catabolized in Cellular Respiration

Source	ATP Produced via Substrate-level Phosphorylation	Coenzymes Formed	ATP Produced via Oxidative Phosphorylation
Glycolysis	2 ATP (net yield)	2 NADH + H^+	3 or 5 ATP*
Formation of acetyl coenzyme A	None	2 NADH + H^+	5 ATP†
Krebs cycle	2 ATP	6 NADH + H^+ 2 $FADH_2$	15 ATP† 3 ATP‡
	4 ATP		26 or 28 ATP
	Grand Total:	30 or 32 ATP	

* One molecule of NADH + H^+ produced in the cytosol yields an average of 1.5 or 2.5 molecules of ATP via oxidative phosphorylation, depending on which shuttle system transports the coenzyme's electrons into the mitochondrion.

† One molecule of NADH + H^+ produced in the mitochondrion yields an average of 2.5 molecules of ATP via oxidative phosphorylation.

‡ One molecule of $FADH_2$ yields an average of 1.5 molecules of ATP via oxidative phosphorylation.

When oxygen is available, however, pyruvic acid is converted to acetyl coenzyme A, and the reactions of the Krebs cycle and the electron transport chain occur. Recall that the Krebs cycle and the electron transport chain are collectively referred to as **aerobic respiration**. Thus, when oxygen is present, all four phases of cellular respiration occur: glycolysis, formation of acetyl CoA, the Krebs cycle, and the electron transport chain. This results in the production of 30 or 32 ATP, which is considerably much more ATP than is produced via anaerobic glycolysis.

Glucose Anabolism Includes Glycogenesis and Gluconeogenesis

Even though most of the glucose in the body is catabolized to generate ATP, glucose may take part in or be formed via several anabolic reactions. One is the synthesis of glycogen; another is the synthesis of new glucose molecules from some of the products of protein and lipid breakdown.

If glucose is not needed immediately for ATP production, it combines with many other molecules of glucose to form **glycogen**, a polysaccharide that is the only stored form of carbohydrate in our bodies. The synthesis of glycogen is referred to as **glycogenesis**. The body can store about 500 g (about 1.1 lb) of glycogen, roughly 75% in skeletal muscle and the rest in the liver. During glycogenesis, glucose is first phosphorylated to glucose 6-phosphate by the enzyme *hexokinase* (FIGURE 4.18). Glucose 6-phosphate is converted to glucose 1-phosphate, then to uridine diphosphate glucose, and finally to glycogen.

FIGURE 4.18 Glycogenesis and glycogenolysis.

The glycogenesis pathway converts glucose into glycogen; the glycogenolysis pathway breaks down glycogen into glucose.

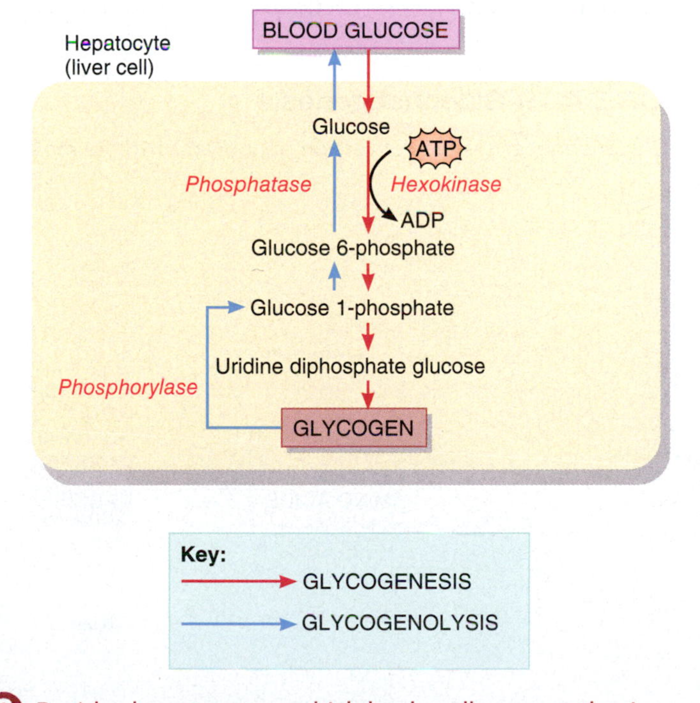

Besides hepatocytes, which body cells can synthesize glycogen? Why can't they release glucose into the blood?

When body activities require ATP, glycogen stored in hepatocytes is broken down into glucose and released into the blood to be transported to cells, where it will be catabolized by the processes of cellular respiration already described. The process of splitting glycogen into its glucose subunits is called **glycogenolysis** (glī′-kō-je-NOL-e-sis).

Glycogenolysis is not a simple reversal of the steps of glycogenesis (FIGURE 4.18). It begins by splitting glucose molecules off the branched glycogen molecule via phosphorylation to form glucose 1-phosphate. This reaction is catalyzed by the enzyme *phosphorylase*. Glucose 1-phosphate is then converted to glucose 6-phosphate. The enzyme *phosphatase*, in turn, removes phosphate from glucose 6-phosphate to form glucose, which leaves the cell via transporters in the plasma membrane. The action of phosphatase is necessary because phosphorylated glucose molecules cannot ride aboard the transporters for glucose. Phosphatase is present in liver cells but not in skeletal muscle fibers. Thus, hepatocytes can release glucose derived from glycogen to the bloodstream, but skeletal muscle cells cannot. In skeletal muscle cells, glycogen is broken down into glucose 6-phosphate, which is then catabolized for ATP production via glycolysis and the Krebs cycle. However, the lactic acid produced by glycolysis in muscle cells can be converted to glucose in the liver. In this way, muscle glycogen can be an indirect source of blood glucose.

When your liver runs low on glycogen, it is time to eat. If you don't, your body starts catabolizing triglycerides (fats) and proteins. Actually, the body normally catabolizes some of its triglycerides and proteins, but large-scale triglyceride and protein catabolism does not happen unless you are starving, eating very few carbohydrates, or suffering from an endocrine disorder.

Lactic acid, certain amino acids, and the glycerol part of triglycerides can be converted in the liver to glucose (FIGURE 4.19). The process by which glucose is formed from these noncarbohydrate sources is called **gluconeogenesis** (gloo′-kō-nē′-ō-JEN-e-sis). An easy way to distinguish this term from glycogenesis or glycogenolysis is to remember that, in this case, glucose is not converted back from glycogen but is instead *newly formed*. Lactic acid is converted into glucose via the following pathway: lactic acid → pyruvic acid → glucose. Amino acids such as alanine, cysteine, glycine, serine, and threonine are converted to glucose once they have been deaminated (their amino groups removed) to form substances called *keto acids*. These reactions are summarized as follows: amino acids → keto acids → pyruvic acid → glucose. Glycerol is converted into glucose after being turned into glyceraldehyde-3-phosphate. This pathway is summarized in the following way: glycerol → glyceradehyde-3-phosphate → glucose.

CLINICAL CONNECTION

Carbohydrate Loading

The amount of glycogen stored in the liver and skeletal muscles varies and can be completely exhausted during long-term athletic endeavors. Thus, many marathon runners and other endurance athletes follow a precise regimen of diet and exercise that includes eating large amounts of complex carbohydrates, such as pasta and potatoes, in the three days before an event. This practice, called **carbohydrate loading**, helps maximize the amount of glycogen available for ATP production in muscles. For athletic events lasting more than an hour, carbohydrate loading has been shown to increase an athlete's endurance. The increased endurance is due to increased glycogenolysis, which results in more glucose that can be catabolized for energy.

CHECKPOINT

17. What happens during glycolysis?

18. How is acetyl coenzyme A formed?

19. Outline the principal events and outcomes of the Krebs cycle.

20. What happens in the electron transport chain?

21. Under what circumstances do glycogenesis and glycogenolysis occur?

22. What is gluconeogenesis, and why is it important?

FIGURE 4.19 Gluconeogenesis.

Gluconeogenesis is the conversion of noncarbohydrate molecules (lactic acid, amino acids, or glycerol) into glucose.

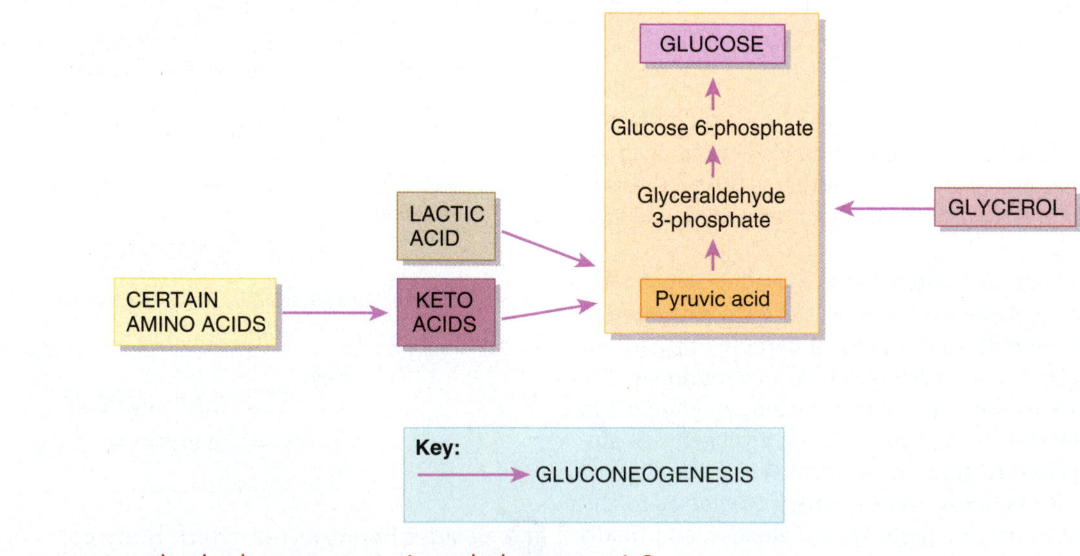

What cells can carry out both gluconeogenesis and glycogenesis?

4.6 Lipid Metabolism

OBJECTIVE

- Describe the fate, metabolism, and functions of lipids.

Lipids, like carbohydrates, may be catabolized to produce ATP. If the body has no immediate need to use lipids in this way, they are stored as triglycerides in adipose tissue throughout the body and in the liver. A few lipids are used as structural molecules or to synthesize other essential substances. Examples include phospholipids, which are constituents of plasma membranes, and lipoproteins, which are used to transport cholesterol and other lipids throughout the body.

Lipid Catabolism Involves Lipolysis and Beta Oxidation

Before cells can produce ATP from triglycerides, the triglycerides must first be split into glycerol and fatty acids, a process called **lipolysis** (FIGURE 4.20). The glycerol and fatty acids in turn can be further catabolized to CO_2, water, and ATP.

The glycerol and fatty acids that result from lipolysis are catabolized via different pathways (FIGURE 4.20). Glycerol is converted by many cells of the body to glyceraldehyde 3-phosphate, one of the compounds also formed during the catabolism of glucose. If the ATP supply in a cell is high, glyceraldehyde 3-phosphate is converted into glucose, an example of gluconeogenesis. If ATP supply in a cell is low, glyceraldehyde 3-phosphate enters the catabolic pathway to form pyruvic acid.

Fatty acid catabolism begins with a series of reactions, collectively called **beta oxidation**, that occurs in the mitochondrial matrix. Enzymes remove two carbon atoms at a time from the long chain of carbon atoms composing a fatty acid and attach the resulting two-carbon fragment to coenzyme A, forming acetyl CoA. Then acetyl CoA enters the Krebs cycle (FIGURE 4.20). A 16-carbon fatty acid such as palmitic acid can yield as many as 129 ATPs upon its complete catabolism via beta oxidation, the Krebs cycle, and the electron transport chain.

As part of normal fatty acid catabolism, the liver converts some acetyl CoA molecules into substances known as **ketone bodies**, which include acetoacetic acid, beta-hydroxybutyric acid, and acetone. The formation of these three substances is called **ketogenesis**. Once they are produced, the ketone bodies leave the liver to enter body

FIGURE 4.20 Pathways of lipid metabolism. Glycerol may be converted to glyceraldehyde 3-phosphate, which can then be converted to glucose or enter the Krebs cycle. Fatty acids undergo beta oxidation and enter the Krebs cycle via acetyl coenzyme A. The synthesis of lipids from glucose or amino acids is called lipogenesis.

Glycerol and fatty acids are catabolized in separate pathways.

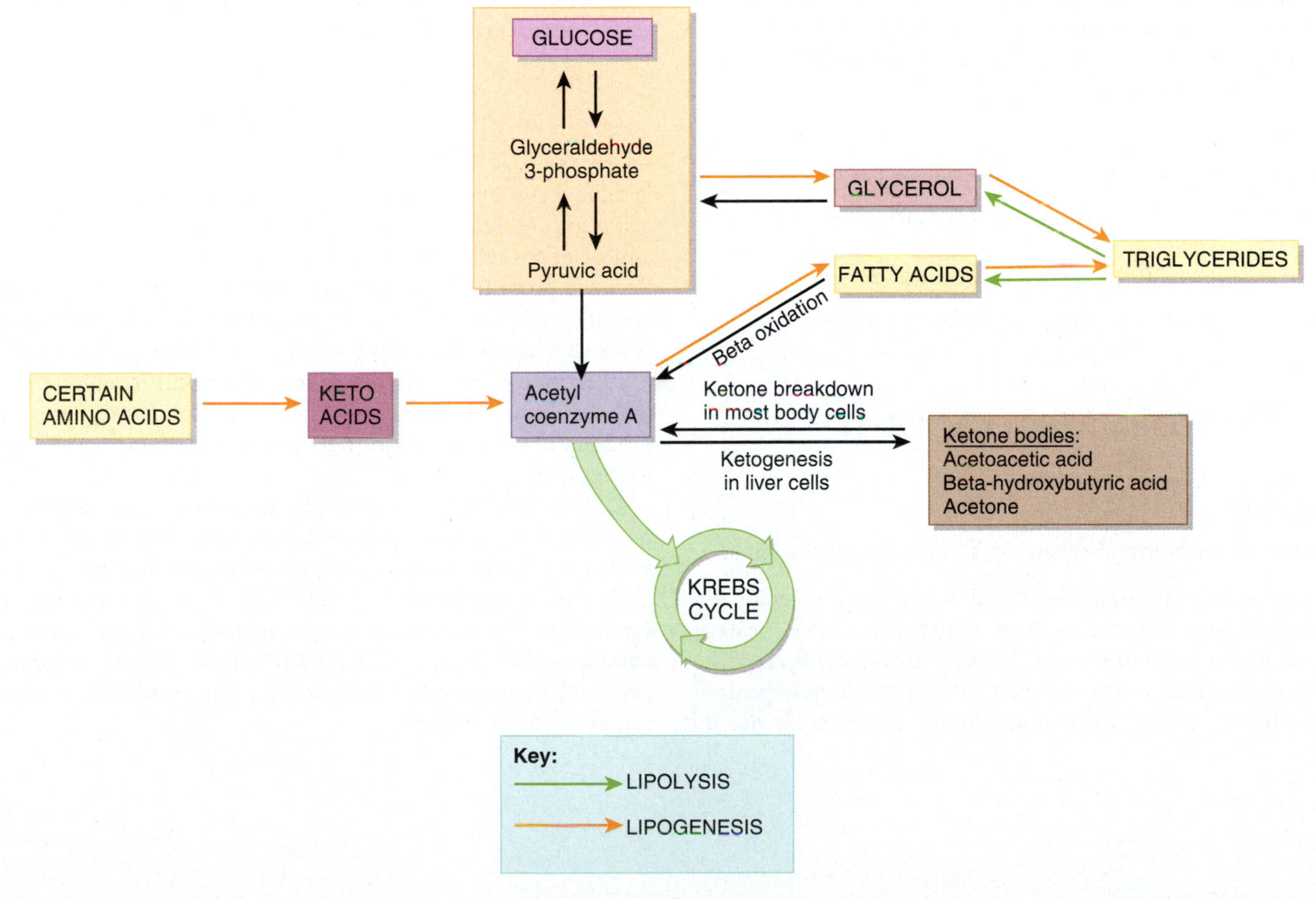

What is ketogenesis?

cells, where they are broken down into acetyl CoA, which then enters the Krebs cycle.

The level of ketone bodies in the blood normally is very low because other tissues use them for ATP production as fast as they are formed. During periods of excessive beta oxidation, however, the production of ketone bodies exceeds their uptake and use by body cells. This can occur after a meal rich in triglycerides, or during fasting or starvation, because few carbohydrates are available for catabolism. Excessive beta oxidation may also occur in poorly controlled or untreated diabetes mellitus. When the concentration of ketone bodies in the blood rises above normal, the ketone bodies, most of which are acids, must be buffered. If too many ketone bodies accumulate, blood pH falls, a condition known as **ketoacidosis**. The decreased blood pH in turn causes depression of the central nervous system, which can result in disorientation, coma, and even death if the condition is not treated.

Lipid Anabolism Occurs via Lipogenesis

Liver cells and adipose cells can synthesize lipids, a process known as **lipogenesis** (FIGURE 4.20). Lipogenesis occurs when individuals consume more calories than are needed to satisfy their ATP needs. Excess dietary carbohydrates, proteins, and fats all have the same fate—they are converted into triglycerides. Certain amino acids can undergo the following reactions: amino acids → keto acids → acetyl CoA → fatty acids → triglycerides. The use of glucose to form lipids takes place via two pathways: (1) glucose → glyceraldehyde 3-phosphate → glycerol and (2) glucose → glyceraldehyde 3-phosphate → pyruvic acid → acetyl CoA → fatty acids. The resulting glycerol and fatty acids can undergo anabolic reactions to become stored triglycerides, or they can go through a series of anabolic reactions to produce other lipids such as lipoproteins, phospholipids, and cholesterol.

CHECKPOINT

23. Where are triglycerides stored in the body?
24. Explain the principal events of the catabolism of glycerol and fatty acids.
25. What are ketone bodies? What is ketosis?
26. Define lipogenesis and explain its importance.

4.7 Protein Metabolism

OBJECTIVE

- Describe the fate, metabolism, and functions of proteins.

During digestion, proteins are broken down into amino acids. Unlike carbohydrates and triglycerides, which are stored, proteins are not warehoused for future use. Instead, amino acids are either catabolized to produce ATP or used to synthesize new proteins for body growth and repair. Excess dietary amino acids are not excreted in the urine or feces but instead are converted into glucose or triglycerides.

Protein Catabolism Leads to Deamination and Keto Acid Breakdown

Protein catabolism begins with the breakdown of proteins into amino acids. The amino acids in turn can be broken down even further into CO_2, water, and ATP. Before an amino acid can be catabolized, its amino group (NH_2) must first be removed—a process called **deamination**. Deamination occurs in the liver and produces ammonia (NH_3) and a keto acid (FIGURE 4.21a). **Ammonia** is a highly toxic substance that is converted to **urea** (a relatively harmless substance) by hepatocytes (liver cells), released into the blood, filtered by the kidneys, and then partially excreted into urine (FIGURE 4.21b). A **keto acid** is a general term that refers to the remaining portion of an amino acid after it has been deaminated. Depending on which amino acid is deaminated, the keto acid formed may be pyruvic acid, an intermediate of the Krebs cycle, or a molecule that is converted to acetyl CoA (FIGURE 4.21c). These substances enter cellular respiration at the appropriate point, and ATP is subsequently generated (FIGURE 4.22).

As an alternative to catabolizing amino acids for ATP, hepatocytes can convert some amino acids to glucose or fatty acids. The conversion of amino acids to glucose (gluconeogenesis) occurs in the following way: amino acids → keto acids → pyruvic acid → glucose (see FIGURE 4.19). The conversion of amino acids to fatty acids for lipid synthesis (lipogenesis) occurs as follows: amino acids → keto acids → acetyl CoA → fatty acids (see FIGURE 4.20).

Protein Anabolism Results in the Synthesis of New Proteins

Protein anabolism, the formation of peptide bonds between amino acids to produce new proteins, is carried out on the ribosomes of almost every cell in the body, directed by the cells' DNA and RNA. Because proteins are a main component of most cell structures, adequate dietary protein is especially essential during the growth years, during pregnancy, and when tissue has been damaged by disease or injury. Once dietary intake of protein is adequate, eating more protein will not increase bone or muscle mass; only a regular program of forceful, weight-bearing muscular activity accomplishes that goal.

Of the 20 amino acids in the human body, 10 are essential amino acids: They must be present in the diet because they cannot be synthesized in the body in adequate amounts. Nonessential amino acids can be synthesized by body cells. They are formed by **transamination**, the transfer of an amino group from an amino acid to pyruvic acid or to an acid in the Krebs cycle. Once the appropriate essential and nonessential amino acids are present in cells, protein synthesis occurs rapidly.

FIGURE 4.21 Amino acid catabolism.

Deamination of an amino acid produces ammonia and a keto acid.

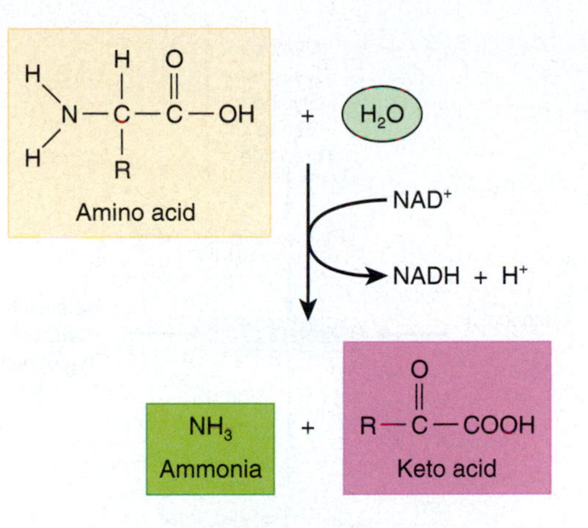

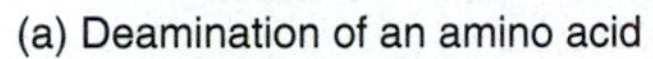

(a) Deamination of an amino acid

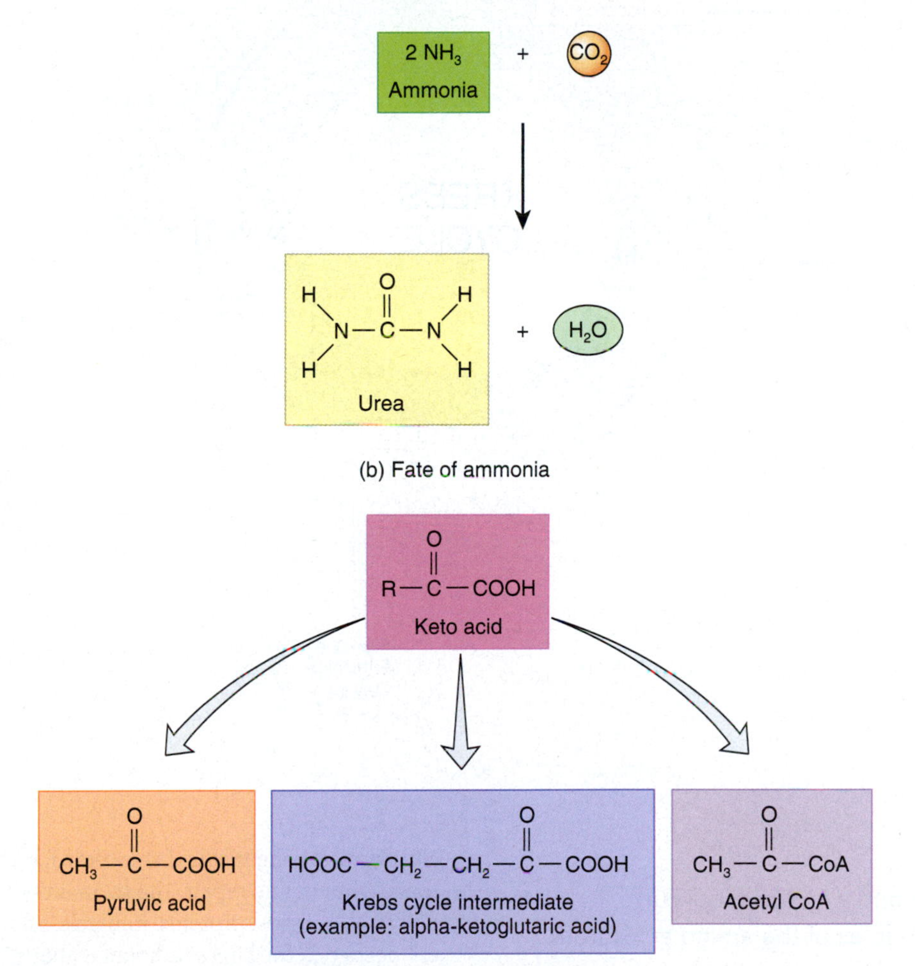

(b) Fate of ammonia

(c) Possible fates of a keto acid

In which organ does deamination occur?

FIGURE 4.22 Various points at which amino acids enter cellular respiration for catabolism. Amino acids are shown in yellow boxes.

The keto acid formed from amino acid degradation enters cellular respiration as pyruvic acid, a Krebs cycle intermediate, or acetyl CoA.

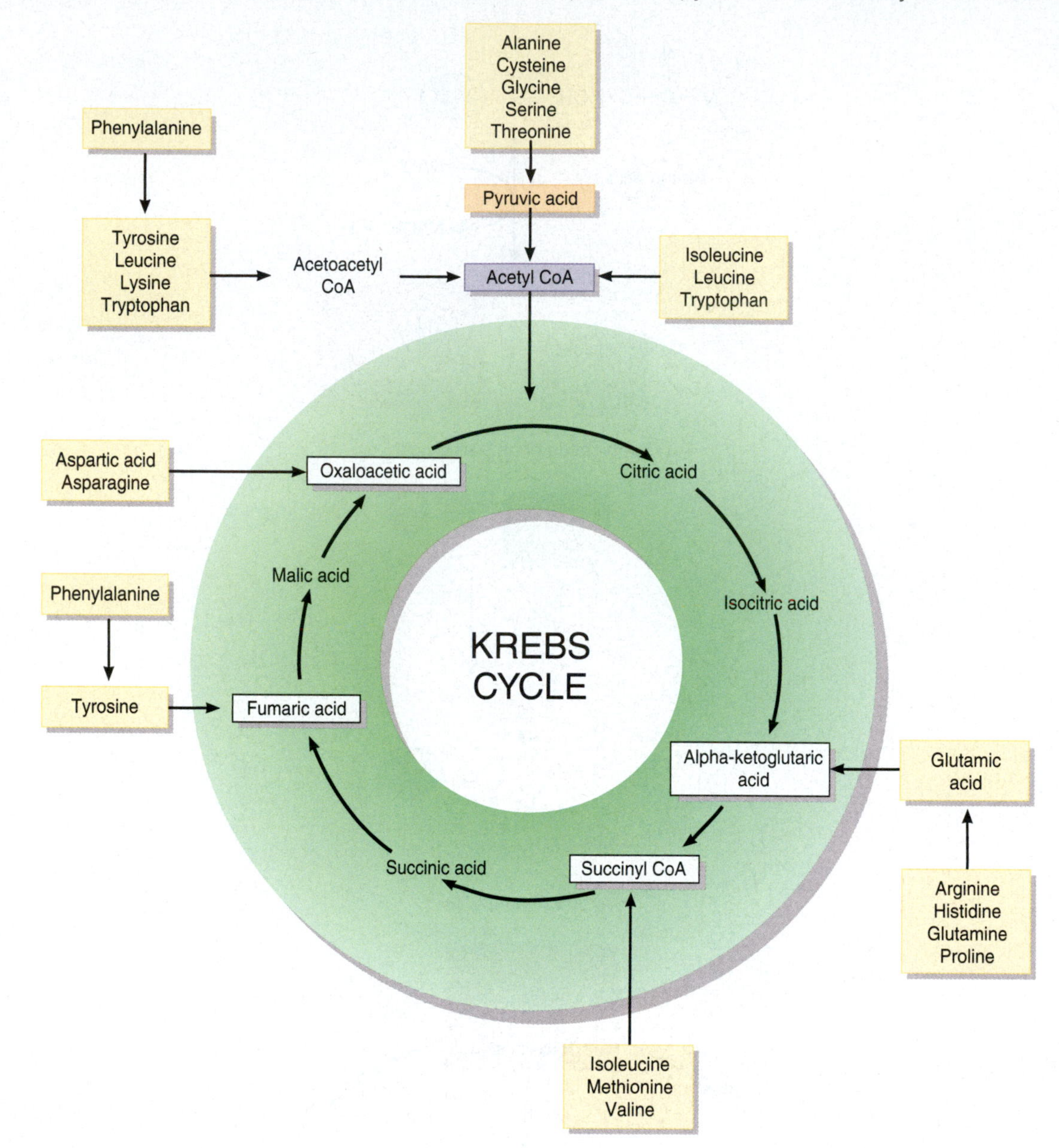

What is a keto acid?

CHECKPOINT

27. What is deamination, and why does it occur?

28. What are the possible fates of the amino acids from protein catabolism?

4.8 Nutrition and Metabolism

OBJECTIVE

- Explain the importance of minerals and vitamins in metabolism.

Nutrients are chemical substances in food that body cells use for growth, maintenance, and repair. The six main types of nutrients are water, carbohydrates, lipids, proteins, minerals, and vitamins. Water is the nutrient needed in the largest amount—about 2–3 liters per day. As the most abundant compound in the body, water provides the medium in which most metabolic reactions occur, and it participates in some reactions (for example, hydrolysis reactions). The important roles of water in the body can be reviewed in Section 2.4. Three organic nutrients—carbohydrates, lipids, and proteins—provide the energy needed for metabolic reactions and serve as building blocks to make body structures. Some minerals and many vitamins are components of the enzyme systems that catalyze metabolic reactions. *Essential nutrients* are specific nutrient molecules that the body cannot make in sufficient quantity to meet its needs and thus must be obtained from the diet. Some amino acids, fatty acids, minerals, and vitamins are essential nutrients.

Many Minerals Have Known Functions in the Body

Minerals are inorganic elements that constitute about 4% of total body mass and are concentrated most heavily in the skeleton. Minerals with known functions in the body include calcium, phosphorus, potassium, sulfur, sodium, chloride, magnesium, iron, iodide, manganese, copper, cobalt, zinc, fluoride, selenium, and chromium. Other minerals—aluminum, boron, silicon, and molybdenum—are also present, but their functions are unclear. Note that the body generally uses the ions of the minerals rather than the nonionized form. Typical diets supply adequate amounts of potassium, sodium, chloride, and magnesium. Some attention must be paid to eating foods that provide enough calcium, phosphorus, iron, and iodide. Excess amounts of most minerals are excreted in the urine and feces.

Calcium and phosphorus form part of the matrix of bone. Because minerals do not form long-chain compounds, they are otherwise poor building materials. A major role of minerals is to help regulate enzymatic reactions. Calcium, iron, magnesium, and manganese are constituents of some coenzymes. Magnesium also serves as a catalyst for the conversion of ADP to ATP. Minerals such as sodium and phosphorus work in buffer systems, which help control the pH of body fluids. Sodium also helps regulate the osmosis of water and, along with other ions, is involved in the generation of action potentials. **TABLE 4.2** describes the functions of minerals that are vital to the body.

Vitamins Help Maintain Growth and Normal Metabolism

Organic nutrients required in small amounts to maintain growth and normal metabolism are called **vitamins**. Unlike carbohydrates, lipids, or proteins, vitamins do not provide energy or serve as the

TABLE 4.2 Minerals Vital to the Body

Mineral	Functions
Calcium	Formation of bones and teeth, blood clotting, normal muscle and nerve activity, endocytosis and exocytosis, cellular motility, chromosome movement during cell division, glycogen metabolism, and release of neurotransmitters and hormones.
Phosphorus	Formation of bones and teeth. Phosphates ($H_2PO_4^-$, HPO_4^-, and PO_4^{3-}) constitute a major buffer of blood. Plays important role in muscle contraction and nerve activity. Component of many enzymes. Involved in energy transfer (ATP). Component of DNA and RNA.
Potassium	Needed for generation and conduction of action potentials in neurons and muscle fibers.
Sulfur	Component of many proteins, electron carriers in electron transport chain, and some vitamins.
Sodium	Strongly affects distribution of water through osmosis. Part of bicarbonate buffer system. Functions in nerve and muscle action potential conduction.
Chloride	Plays a role in acid–base balance of blood, water balance, and formation of HCl in stomach.
Magnesium	Required for normal functioning of muscle and nervous tissue. Participates in bone formation. Constituent of many coenzymes.
Iron	Component of hemoglobin, which reversibly binds O_2. Component of cytochromes involved in electron transport chain.
Iodide	Component of thyroid hormones, which regulate metabolic rate.
Manganese	Activates several enzymes. Needed for hemoglobin synthesis, urea formation, growth, reproduction, lactation, bone formation, and possibly production and release of insulin, and inhibition of cell damage.
Copper	As part of vitamin B_{12}, required for erythropoiesis.
Zinc	Component of certain enzymes, including carbonic anhydrase, which is important in carbon dioxide metabolism, and peptidase, which is involved in protein digestion. Necessary for normal growth and wound healing, normal taste sensations and appetite, and normal sperm counts in males.
Fluoride	Appears to improve tooth structure and inhibit tooth decay.
Selenium	Component of certain enzymes. Needed for synthesis of thyroid hormones, sperm motility, and proper functioning of the immune system. Also functions as an antioxidant. Prevents chromosome breakage and may play a role in preventing certain birth defects, miscarriage, prostate cancer, and coronary artery disease.
Chromium	Needed for normal activity of insulin in carbohydrate and lipid metabolism.

body's building materials. Most vitamins with known functions are coenzymes.

Most vitamins cannot be synthesized by the body and must be ingested in food. Other vitamins, such as vitamin K, are produced by bacteria in the gastrointestinal (GI) tract and then absorbed. The body can assemble some vitamins if the raw materials, called **provitamins**, are provided. For example, vitamin A is produced by the body from the provitamin beta-carotene, a chemical present in yellow vegetables such as carrots and in dark green vegetables such as spinach. No single food contains all of the required vitamins—one of the best reasons to eat a varied diet.

Vitamins are divided into two main groups: fat-soluble and water-soluble. The **fat-soluble vitamins**, vitamins A, D, E, and K, are absorbed into the bloodstream along with other dietary lipids in the small intestine. They cannot be absorbed in adequate quantities unless they are ingested with other lipids. Fat-soluble vitamins may be stored in cells, particularly hepatocytes. The **water-soluble vitamins**, including several B vitamins and vitamin C, are dissolved in body fluids. Excess quantities of these vitamins are not stored but instead are excreted in the urine.

Besides their other functions, three vitamins—C, E, and beta-carotene (a provitamin)—are termed **antioxidant vitamins** because they inactivate oxygen free radicals. Recall that free radicals are highly reactive ions or molecules that carry an unpaired electron in their outermost electron shell (see FIGURE 2.3b). Free radicals damage cell membranes, DNA, and other cellular structures and contribute to the formation of artery-narrowing atherosclerotic plaques. Some free radicals arise naturally in the body, and others come from environmental hazards such as tobacco smoke and radiation. Antioxidant vitamins are thought to play a role in protecting against some kinds of cancer, reducing the buildup of atherosclerotic plaque, delaying some effects of aging, and decreasing the chance of cataract formation in the lenses of the eyes. TABLE 4.3 lists the major vitamins and their functions.

TABLE 4.3 The Principal Vitamins

Vitamin	Functions
Fat-soluble	
A	Maintains general health and vigor of epithelial cells. Beta-carotene acts as an antioxidant to inactivate free radicals. Essential for formation of light-sensitive pigments in photoreceptors of retina. Aids growth of bone and teeth.
D	Essential for absorption of calcium and phosphorus from the gastrointestinal tract. Works with parathyroid hormone (PTH) to maintain calcium (Ca^{2+}) homeostasis.
E (tocopherols)	Inhibits catabolism of certain fatty acids that help form cell structures, especially membranes. Involved in formation of DNA, RNA, and red blood cells. May promote wound healing, contributes to the normal structure and functioning of the nervous system, and prevents scarring. Acts as an antioxidant to inactivate free radicals.
K	Coenzyme essential for synthesis of several clotting factors, including prothrombin, by the liver.
Water-soluble	
B_1 (thiamine)	Acts as a coenzyme for many different enzymes that break carbon-to-carbon bonds and are involved in carbohydrate metabolism of pyruvic acid to CO_2 and H_2O. Essential for synthesis of the neurotransmitter acetylcholine.
B_2 (riboflavin)	Component of certain coenzymes (for example, FAD and FMN) in carbohydrate and protein metabolism, especially in cells of the eye, integument, mucosa of intestine, and blood.
Niacin	Essential component of NAD and NADP, coenzymes in oxidation–reduction reactions. In lipid metabolism, inhibits production of cholesterol and assists in triglyceride breakdown.
B_6 (pyridoxine)	Essential coenzyme for normal amino acid metabolism. Assists production of circulating antibodies. Many function as coenzyme in triglyceride metabolism.
B_{12} (cyanocobalamin)	Coenzyme necessary for red blood cell formation, formation of the amino acid methionine, entrance of some amino acids into the Krebs cycle, and manufacture of choline (used to synthesize acetylcholine).
Pantothenic acid	Constituent of coenzyme A, which is essential for transfer of acetyl group from pyruvic acid into the Krebs cycle, conversion of lipids and amino acids into glucose, and synthesis of cholesterol and steroid hormones.
Folic acid (folate, folacin)	Component of enzyme systems synthesizing nitrogenous bases of DNA and RNA. Essential for normal production of red and white blood cells.
Biotin	Essential coenzyme for conversion of pyruvic acid to oxaloacetic acid and synthesis of fatty acids and purines.
C (ascorbic acid)	Promotes protein synthesis, including laying down of collagen in formation of connective tissue. As coenzyme, may combine with poisons, rendering them harmless until excreted. Works with antibodies, promotes wound healing, and functions as an antioxidant.

CLINICAL CONNECTION

Vitamin and Mineral Supplements

Most nutritionists recommend eating a balanced diet that includes a variety of foods rather than taking vitamin or mineral supplements, except in special circumstances. Common examples of necessary supplementations include iron for women who have excessive menstrual bleeding; folic acid (folate) for all women who may become pregnant to reduce the risk of fetal neural tube defects; calcium for most adults because they do not receive the recommended amount in their diets; and vitamin B_{12} for strict vegetarians, who eat no meat. Because most North Americans do not ingest in their food the high levels of antioxidant vitamins thought to have beneficial effects, some experts recommend supplementing vitamins C and E. More is not always better, however; larger doses of vitamins or minerals can be very harmful.

CHECKPOINT

29. What is a mineral? Briefly describe the functions of the following minerals: calcium, potassium, sodium, and iron.

30. Define vitamin. Distinguish between a fat-soluble vitamin and a water-soluble vitamin.

FROM RESEARCH TO REALITY

How Diet Can Affect Metabolism

Reference

Beisswenger, B. et al. (2005). Ketosis leads to increased methylglyoxal production on the Atkins diet. *Annals of the New York Academy of Sciences.*1043: 201–210.

How does the Atkins diet change your metabolism?

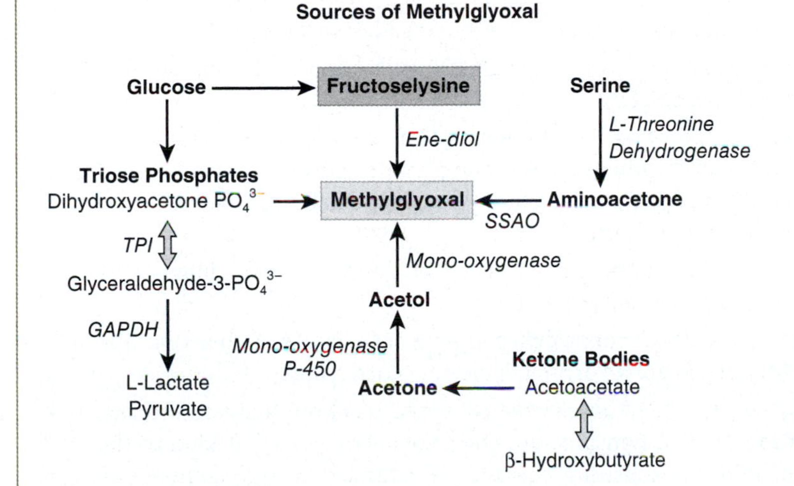

Changing your diet can change your body's metabolism. Could some of the changes to your metabolism be less than beneficial? This activity explores the production of less favorable metabolic by-products resulting from a ketogenic diet.

Article description:

The investigators in this study monitored subjects, both male and female, who chose to start the Atkins diet in an attempt to lose weight. They wanted to see if following the Atkins diet would cause a significant increase in the production of methylglyoxal, a cytotoxic chemical normally produced by the body in small quantities.

Go to WileyPLUS Learning Space and use the data from this article to answer the questions posed there and to discover more about how diet can change metabolism.

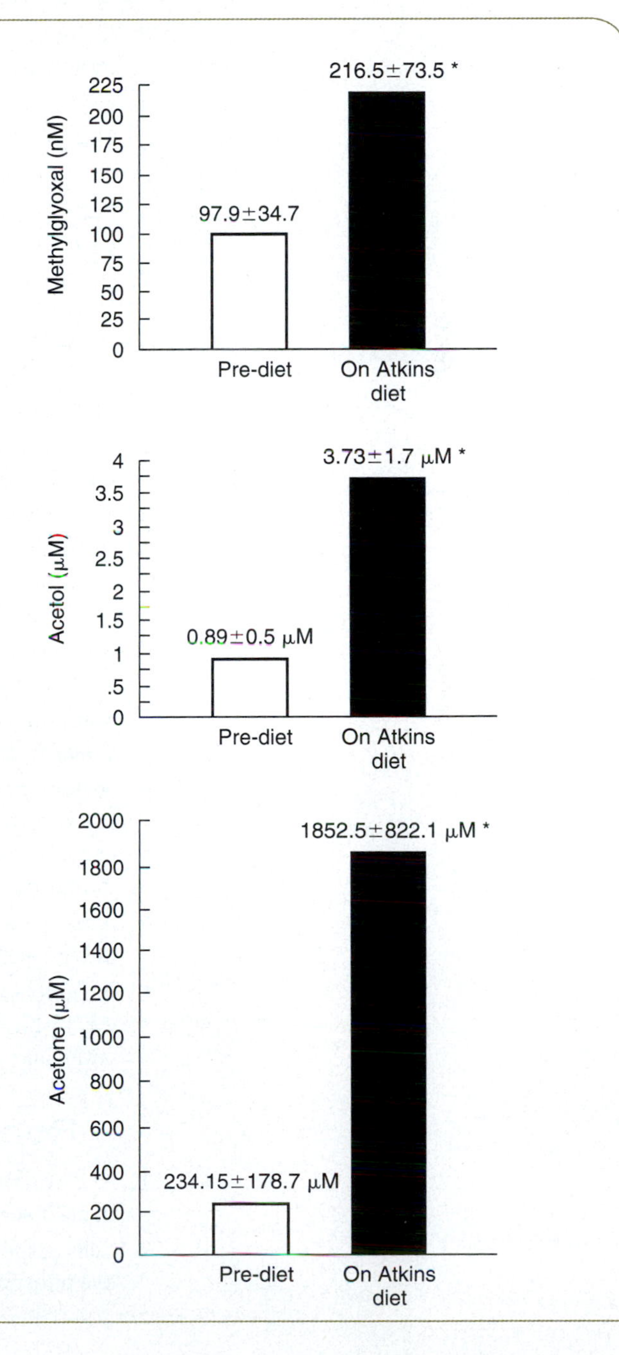

CHAPTER REVIEW

4.1 An Overview of Metabolism

1. Metabolism refers to all chemical reactions of the body and is of two types: catabolism and anabolism.
2. Catabolism is the term for reactions that break down complex organic molecules into smaller molecules. Anabolism refers to reactions that combine smaller molecules into more complex ones that form the body's structural and functional components.

4.2 Energy and Metabolism

1. Energy, the capacity to do work, is of two principal kinds: potential (stored) energy and kinetic energy (energy of motion).
2. Chemical energy is a form of potential energy that is stored in the bonds of compounds and molecules.
3. Exergonic reactions release more energy than they absorb; endergonic reactions absorb more energy than they release.
4. Overall, catabolic reactions are exergonic and anabolic reactions are endergonic.
5. The coupling of anabolism and catabolism occurs via ATP.
6. The initial energy investment needed to start a reaction is the activation energy. Reactions are more likely when the concentrations and the temperatures of the reacting particles are higher.
7. Catalysts accelerate chemical reactions by lowering the activation energy.

4.3 Enzymes

1. Most catalysts in living organisms are protein molecules called enzymes.
2. Enzymes have three important properties: They are (1) highly specific; (2) very efficient; and (3) subject to a variety of controls.
3. An enzyme functions in the following way: the substrate makes contact with the active site of the enzyme, forming an enzyme–substrate complex; the substrate is transformed into a product; and then the product is released. The enzyme remains unchanged and is free to attach to other substrate molecules.
4. An enzyme may catalyze a reversible reaction in either direction, depending on the relative amounts of the substrates and products.
5. Many enzymes require a cofactor (a nonprotein component) to catalyze a reaction. The cofactor may be an inorganic ion or an organic molecule (coenzyme).
6. Several factors influence the rate of an enzyme-catalyzed reaction: temperature, pH, the substrate concentration, and nonsubstrate chemical substances that bind to the enzyme. Examples of such nonsubstrate chemical substances include competitive inhibitors and allosteric modulators.
7. A metabolic pathway is a sequence of reactions in which an initial substrate is converted into an end product via a series of intermediates.
8. In feedback inhibition, the end product of a metabolic pathway shuts down the pathway by binding to and inhibiting an enzyme that catalyzes one of the earlier steps of the pathway.

4.4 Role of ATP in Metabolism

1. ATP couples energy-releasing catabolic reactions to energy-requiring anabolic reactions. It serves as the "energy currency" of a living cell.
2. Cells of the body generate ATP in two ways: substrate-level phosphorylation and oxidative phosphorylation.

3. Substrate-level phosphorylation generates ATP by transferring a phosphate group from a phosphorylated intermediate directly to ADP.
4. Oxidative phosphorylation produces ATP by adding a phosphate group to ADP using energy derived from a series of electron carriers, with oxygen serving as the final electron acceptor. Oxidative phosphorylation collectively refers to the electron transport chain and the process of chemiosmosis.
5. Two coenyzmes important to metabolism are NAD^+ and FAD. They pick up electrons in the form of hydrogen atoms, forming $NADH + H^+$ and $FADH_2$, respectively, and then carry the hydrogen atoms to the electron transport chain so that ATP can be produced via oxidative phosphorylation.
6. Cellular respiration is the process by which a nutrient molecule (glucose, fatty acid, or an amino acid) is broken down in the presence of oxygen into carbon dioxide, water, and energy (ATP and heat).
7. Cellular respiration of glucose involves four sets of reactions: glycolysis, the formation of acetyl coenzyme A, the Krebs cycle, and the electron transport chain.
8. Fatty acids enter cellular respiration after being converted into acetyl CoA.
9. Before an amino acid can enter cellular respiration, it must be deaminated, with the remainder of the amino acid entering cellular respiration as either pyruvic acid, acetyl CoA, or an acid of the Krebs cycle.

4.5 Carbohydrate Metabolism

1. During digestion, polysaccharides and disaccharides are catabolized into the monosaccharides glucose, fructose, and galactose; the latter two are then converted to glucose. Some glucose is catabolized by cells to provide ATP. Glucose can also be used to synthesize amino acids, glycogen, and triglycerides.
2. During cellular respiration, glucose is catabolized in the presence of oxygen to produce carbon dioxide (CO_2), water, and ATP. Four reactions contribute to the cellular respiration of glucose: glycolysis, formation of acetyl CoA, the Krebs cycle, and the electron transport chain.
3. Glycolysis is the breakdown of glucose into two molecules of pyruvic acid; there is a net production of two molecules of ATP via substrate-level phosphorylation and two molecules of $NADH + H^+$. When oxygen is in short supply, pyruvic acid is converted to lactic acid; under aerobic conditions, pyruvic acid ultimately enters the Krebs cycle.
4. Pyruvic acid is prepared for entrance into the Krebs cycle by conversion to a two-carbon acetyl group followed by the addition of coenzyme A to form acetyl coenzyme A. Two molecules of acetyl CoA are produced per molecule of glucose catabolized. As these two molecules of acetyl CoA are formed, two molecules of CO_2 and two molecules of $NADH + H^+$ are produced.
5. The Krebs cycle is a series of reactions that involves the sequential conversion of one acid to another. The two molecules of acetyl CoA that are formed per molecule of glucose catabolized cause the Krebs cycle to occur twice. The two turns of the Krebs cycle result in the production of six molecules of $NADH + H^+$, two molecules of $FADH_2$, two molecules of ATP via substrate-level phosphorylation, and four molecules of CO_2.
6. Through the reactions of the electron transport chain, ATP is produced via oxidative phosphorylation. During oxidative phosphorylation, hydrogen atoms are removed from $NADH + H^+$ and $FADH_2$ and then split into hydrogen ions (H^+) and electrons. The electrons are passed from one component of the electron transport chain to another, whereas the H^+ ions are used to establish an H^+ gradient. As H^+ ions move back down their gradient (chemiosmosis), energy is released and the enzyme ATP synthase uses this energy to add a phosphate group to ADP to form ATP. Per molecule of glucose catabolized, about 26 or 28 ATP are generated via oxidative phosphorylation.

7. If the four ATP produced via substrate-level phosphorylation during glycolysis are added to the 26 or 28 ATP produced via oxidative phosphorylation, a total of either 30 or 32 ATP are generated per one molecule of glucose catabolized during cellular respiration. The complete catabolism of glucose is represented as follows:

$$\underset{\text{Glucose}}{C_6H_{12}O_6} + \underset{\text{Oxygen}}{6\,O_2} + 30 \text{ or } 32 \text{ ADPs} + 30 \text{ or } 32 \text{ Ⓟ} \longrightarrow$$

$$\underset{\text{Carbon dioxide}}{6\,CO_2} + \underset{\text{Water}}{6\,H_2O} + 30 \text{ or } 32 \text{ ATPs}$$

8. The conversion of glucose to glycogen for storage in the liver and skeletal muscle is called glycogenesis.
9. The conversion of glycogen to glucose is called glycogenolysis.
10. Gluconeogenesis is the conversion of noncarbohydrate molecules into glucose.

4.6 Lipid Metabolism

1. Lipids may be catabolized to produce ATP or stored as triglycerides in adipose tissue.
2. In lipolysis, triglycerides are split into fatty acids and glycerol, and released from adipose tissue.
3. Glycerol can be converted into glucose by conversion into glyceraldehyde 3-phosphate.
4. In beta-oxidation of fatty acids, carbon atoms are removed in pairs from fatty acid chains; the resulting molecules of acetyl coenzyme A enter the Krebs cycle.
5. The conversion of glucose or amino acids into lipids is called lipogenesis.

4.7 Protein Metabolism

1. During digestion, proteins are catabolized into amino acids.
2. Before amino acids can be catabolized, they must be deaminated and converted to substances that can enter the Krebs cycle.
3. Amino acids may also be converted into glucose and fatty acids.
4. During protein anabolism, peptide bonds form between amino acids to produce new proteins.

4.8 Nutrition and Metabolism

1. Nutrients include water, carbohydrates, lipids, proteins, minerals, and vitamins.
2. Minerals known to perform essential functions in the body include calcium, phosphorus, potassium, sulfur, sodium, chloride, magnesium, iron, iodide, manganese, copper, cobalt, zinc, fluoride, selenium, and chromium.
3. Vitamins are organic nutrients that maintain growth and normal metabolism. Many function in enzyme systems.
4. Fat-soluble vitamins are absorbed with fats and include vitamins A, D, E, and K; water-soluble vitamins include the B vitamins and vitamin C.

PONDER THIS

1. You have discovered a new strain of cells. You notice that these cells do not have any mitochondria. Compare and contrast these cells with the cells of the human body with regard to cellular metabolism. How could this new strain produce the same amount of ATP as the human cells?

2. Sally goes into the doctor's office because she has not been feeling well. After several analyses, the doctor discovers that Sally has very low levels of vitamins A, D, E, and K. Explain what Sally could potentially do to improve her condition without taking any supplements, and why this would work.

3. A researcher wants to develop a cell that can produce as much ATP anaerobically as cells are capable of producing under aerobic conditions. Describe one thing that this researcher could do in order to accomplish this task.

ANSWERS TO FIGURE QUESTIONS

4.1 Digestion of food involves catabolic reactions that break down food into smaller components.

4.2 This reaction is exergonic because the reactants have more potential energy than the products.

4.3 No. A catalyst does not change the potential energies of the products and reactants; it only lowers the activation energy needed to get the reaction going.

4.4 The active site is the part of the enzyme that catalyzes the reaction.

4.5 The higher temperatures break the weak bonds that help determine the three-dimensional shape of the enzyme, resulting in denaturation of the enzyme.

4.6 A major change in the H^+ concentration alters the charges of amino acids at the active site, which causes the active site to become less active or nonfunctional.

4.7 When an enzyme is saturated, increasing the substrate concentration cannot increase the reaction rate because every enzyme molecule is already occupied by substrate.

4.8 If a competitive inhibitor is present along with the substrate, the reaction rate sharply decreases. Competitive inhibition can be overcome by increasing the substrate concentration, so more substrate will be available to bind to the active site.

4.9 Allosteric modulation is a reversible reaction because the allosteric modulator binds to the allosteric site via noncovalent bonds.

4.10 In pancreatic acinar cells, anabolism predominates because the primary activity is synthesis of complex molecules (digestive enzymes).

4.11 Chemiosmosis is the process by which an H^+ gradient is used to produce ATP; together, chemiosmosis and the electron transport chain constitute oxidative phosphorylation.

4.12 When a molecule of NAD^+ or FAD picks up two hydrogen atoms, a molecule or NADH + H^+ or $FADH_2$ is formed, respectively.

4.13 Glucose is the most common fuel catabolized for energy.

4.14 Kinases are enzymes that phosphorylate (add a phosphate to) their substrate.

4.15 Glycolysis occurs in the cytosol.

4.16 The production of NADH + H^+ and $FADH_2$ is important in the Krebs cycle because they will subsequently yield ATP in the electron transport chain via oxidative phosphorylation.

4.17 The concentration of H^+ is highest in the intermembrane space of the mitochondrion.

4.18 Skeletal muscle fibers can synthesize glycogen, but they cannot release glucose into the blood because they lack the enzyme phosphatase required to remove the phosphate group from glucose 6-phosphate.

4.19 Hepatocytes can carry out gluconeogenesis and glycogenesis.

4.20 Ketogenesis refers to the formation of ketone bodies (acetoacetic acid, beta-hydroxybutyric acid, and acetone) in liver cells.

4.21 Deamination occurs in the liver.

4.22 A keto acid is a general term that refers to the remaining portion of an amino acid after it has been deaminated. The structure of a keto acid varies, depending on the amino acid that is deaminated.

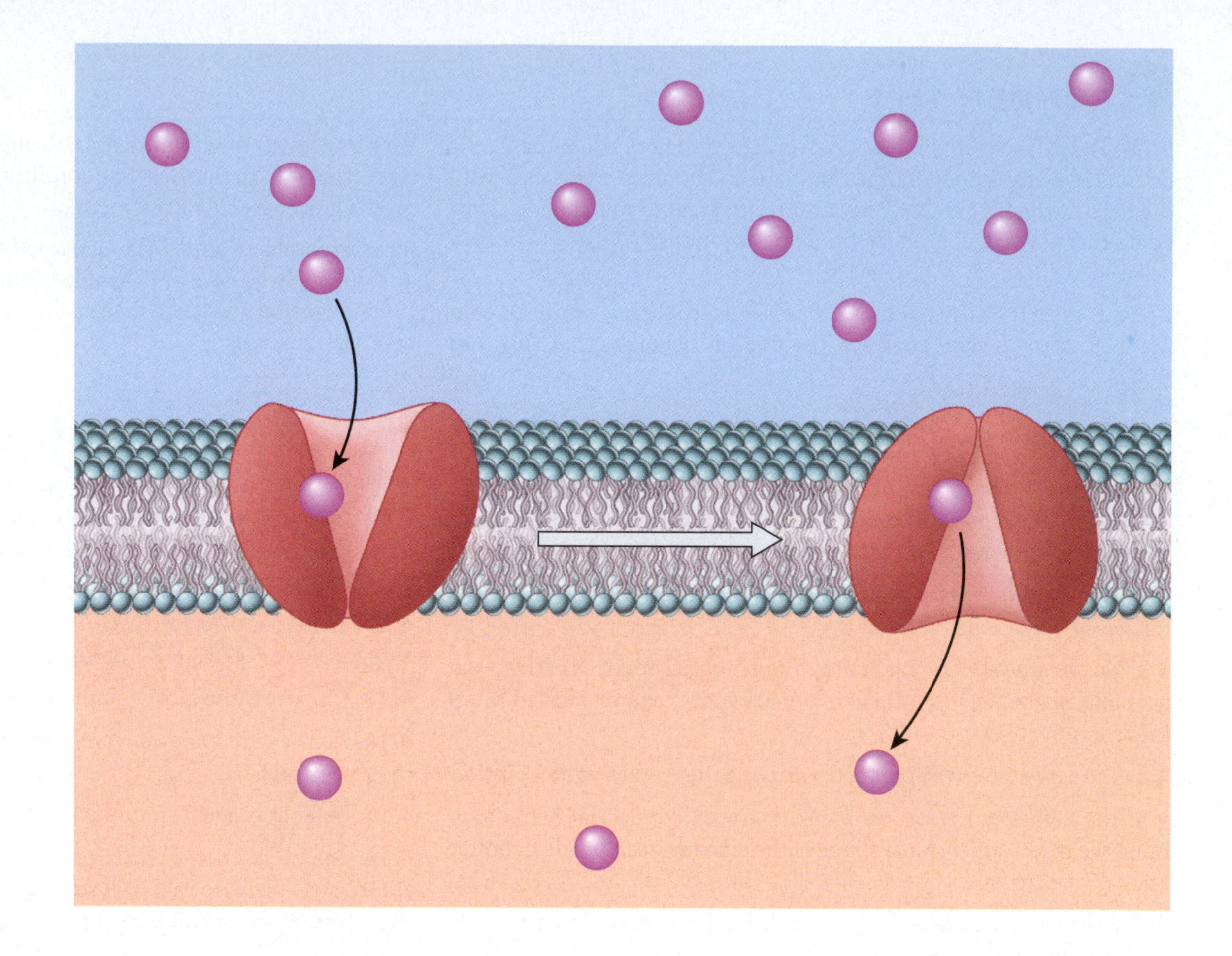

5 Transport Across the Plasma Membrane

Transport Across the Plasma Membrane and Homeostasis

Transport across the plasma membrane contributes to homeostasis by regulating which substances move into or out of the cell.

LOOKING BACK TO MOVE AHEAD...

- Hydrophilic substances, such as ions and molecules containing polar covalent bonds, dissolve easily in water; by contrast, hydrophobic substances, which include molecules that contain mainly nonpolar covalent bonds, are not water-soluble (Section 2.4).
- The basic structural framework of the plasma membrane is the lipid bilayer, which consists of two layers made up of three types of lipid molecules—phospholipids, cholesterol, and glycolipids. Because of the way the phospholipids are arranged, the interior of the plasma membrane is hydrophobic (Section 3.2).
- Integral proteins are proteins that are firmly embedded in the plasma membrane; most integral proteins are transmembrane proteins that extend through the entire lipid bilayer and protrude into both the cytosol and extracellular fluid (Section 3.2).
- Membrane proteins perform a variety of functions: they can serve as channels, carriers, receptors, enzymes, linkers, or cell-identity markers (Section 3.2).

Transport of materials across the plasma membrane is essential to the life of a cell. Certain substances must move into the cell to support metabolic reactions. Other substances that have been produced by the cell for export or as cellular waste products must move out of the cell. In this chapter, you will examine the selective permeability of the plasma membrane to different substances. Then you will learn about two types of gradients that form across the plasma membrane: the concentration gradient and the electrical gradient. The rest of this chapter focuses on the various types of membrane transport processes used by the cell to move substances across the plasma membrane.

5.1 Selective Permeability of the Plasma Membrane

OBJECTIVE

- Explain the concept of selective permeability.

A membrane is said to be *permeable* to substances that can pass through it and *impermeable* to those that cannot. The permeability of the plasma membrane to different substances varies, so some substances pass through the plasma membrane more readily than others. This property of membranes is known as **selective permeability**.

The lipid bilayer portion of the plasma membrane is highly permeable to nonpolar molecules, such as oxygen (O_2), carbon dioxide (CO_2), and steroids; moderately permeable to small, uncharged polar molecules, such as water and urea (a waste product from the breakdown of amino acids); and impermeable to ions and large, uncharged polar molecules, such as glucose (FIGURE 5.1). The permeability characteristics of the plasma membrane are due to the nonpolar, hydrophobic interior of the lipid bilayer (see FIGURE 2.18c). The more hydrophobic or lipid-soluble a substance is, the greater the membrane's permeability to that substance. Thus, the hydrophobic interior of the plasma membrane allows nonpolar molecules to rapidly pass through, but prevents passage to ions and large, uncharged polar molecules. The permeability of the lipid bilayer to water and urea is an unexpected property given that they are polar molecules. These two molecules are thought to pass through the lipid bilayer in the following way. As the fatty acid tails of membrane phospholipids and glycolipids randomly move about, small gaps briefly appear in the hydrophobic environment of the membrane's interior. Because water and urea are small polar molecules that have no overall charge, they can move from one gap to another until they have crossed the membrane.

Although in most body cells the lipid bilayer is moderately permeable to water, there are some body cells in which the bilayer is essentially impermeable to water. The degree of water permeability is influenced by the cholesterol content of the bilayer, which can vary from cell to cell. The higher the bilayer's cholesterol content, the less permeable the bilayer is to water: When large numbers of cholesterol molecules are present in a bilayer, they fill in the gaps between the fatty acid tails of the phospholipids, preventing water from passing through. This is thought to be the reason why portions of the plasma membranes of

FIGURE 5.1 Selective permeability of the plasma membrane. For simplicity, nonpolar molecules and small, uncharged polar molecules are shown moving across the plasma membrane only from the extracellular fluid (ECF) into the cytosol, but the actual direction of movement of a substance depends on its concentration in the ECF and cytosol. If the substance has a higher concentration in ECF than in cytosol, it will move from the ECF, pass through the plasma membrane, and then enter the cytosol. If the substance has a higher concentration in cytosol than in ECF, it will move from the cytosol, pass through the plasma membrane, and enter the ECF.

The lipid bilayer portion of the plasma membrane is highly permeable to nonpolar molecules; moderately permeable to small, uncharged polar molecules; and impermeable to ions and large, uncharged polar molecules.

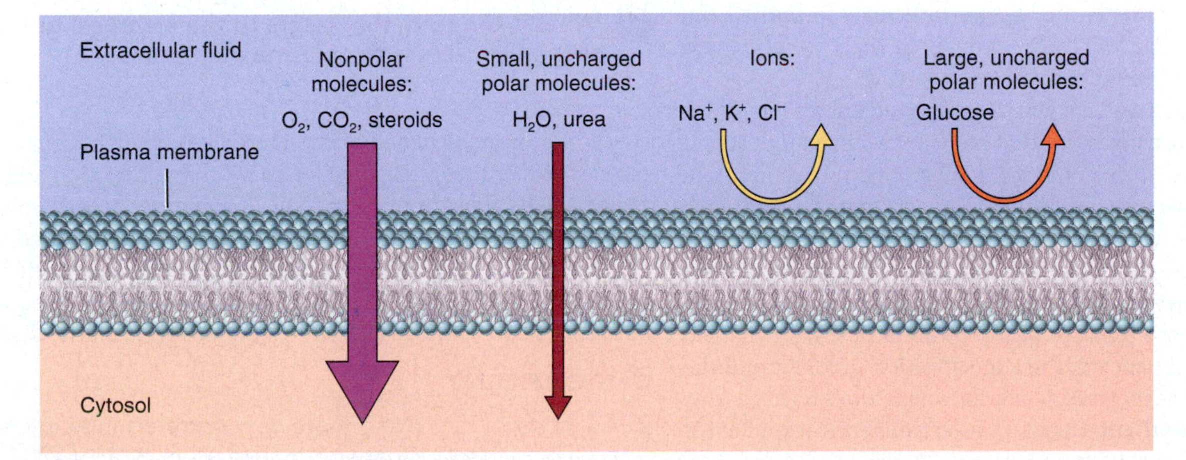

Why is the plasma membrane permeable to small, uncharged polar molecules even though the interior of the membrane is hydrophobic?

certain kidney cells are virtually impermeable to water unless another mechanism that provides passage to water is present.

Transmembrane proteins that act as channels and carriers increase the plasma membrane's permeability to a variety of ions and uncharged polar molecules that, unlike water and urea molecules, cannot cross the lipid bilayer unassisted. Channels and carriers are very selective. Each one helps a specific molecule or ion to cross the membrane. Macromolecules, such as proteins, are also unable to cross the lipid bilayer unassisted. The only way that they can move across the membrane is through the use of vesicles (membranous sacs). The movement of substances across the membrane via ion channels, carriers, and vesicles is discussed later in this chapter.

CHECKPOINT

1. What are some substances to which the plasma membrane is permeable? To which it is impermeable?

5.2 Gradients Across the Plasma Membrane

OBJECTIVE

- Distinguish between a concentration gradient and an electrical gradient.

The selective permeability of the plasma membrane allows a living cell to maintain different concentrations of certain substances on either side of the plasma membrane. A **concentration gradient** is a difference in the concentration of a chemical from one place to another, such as from the inside to the outside of the plasma membrane. Many ions and molecules are more concentrated in either the cytosol or the extracellular fluid. For instance, oxygen molecules and sodium ions (Na^+) are more concentrated in the extracellular fluid than in the cytosol; the opposite is true of carbon dioxide molecules and potassium ions (K^+) (**FIGURE 5.2a**).

The plasma membrane also creates a difference in the distribution of positively and negatively charged ions between the two sides of the plasma membrane. Typically, the inner surface of the plasma membrane is more negatively charged and the outer surface is more positively charged. A difference in electrical charges between two regions constitutes an **electrical gradient** (**FIGURE 5.2b**). Because it occurs across the plasma membrane, this charge difference is termed the **membrane potential**. You will learn more about the generation and significance of the membrane potential in Chapter 7.

Both the concentration gradient and electrical gradient are important because they serve as driving forces that help move substances across the plasma membrane. In many cases a substance moves across a plasma membrane *down its concentration gradient*. That is, a substance moves "downhill," from where it is more concentrated to where it is less concentrated, to reach equilibrium. Similarly, a positively charged substance tends to move toward a negatively charged area, and a negatively charged substance tends to move toward a positively charged area. The combined influence of the concentration gradient and the electrical gradient on the movement of a particular ion is referred to as its **electrochemical gradient**. Hence, the electrochemical gradient is the net driving force that acts on an ion.

The two components of the electrochemical gradient may occur in the same direction or in opposite directions (**FIGURE 5.3**). If both the concentration gradient and electrical gradient are in the same direction, the electrochemical gradient will also be in that direction and its magnitude will be equal to the sum of the magnitudes of the two gradients (**FIGURE 5.3a**). By contrast, if the concentration gradient and electrical gradient oppose each other, the electrochemical gradient will be in the direction of the larger gradient and its magnitude will be equal to the difference between the magnitudes of the two gradients (**FIGURE 5.3b**).

FIGURE 5.2 Gradients across the plasma membrane. (a) Sodium ions and oxygen molecules are more concentrated in extracellular fluid, whereas potassium ions and carbon dioxide are more concentrated in cytosol. (b) The inner surface of the plasma membrane of most cells is negative relative to the outer surface.

A concentration gradient is a difference in the concentration of a chemical from one place to another; an electrical gradient is a difference in electrical charges between two regions.

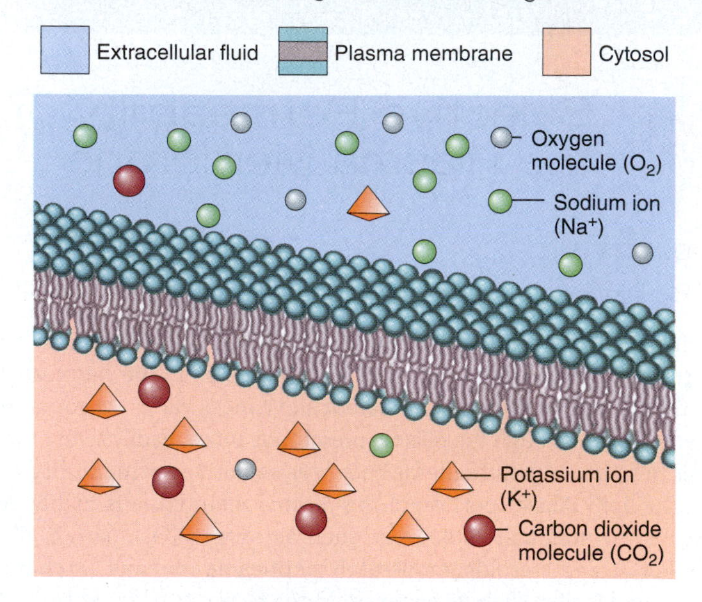

(a) Concentration gradients

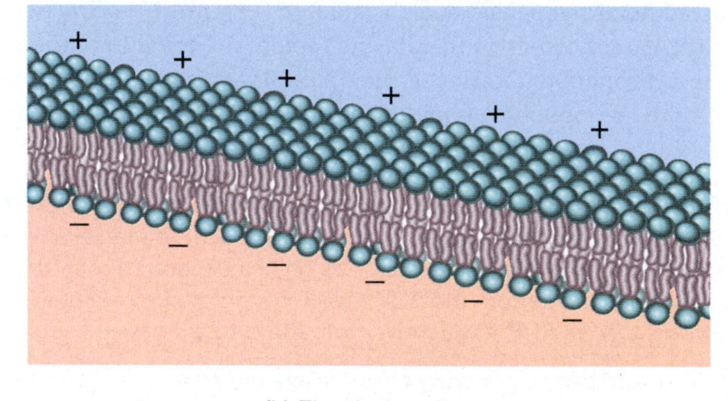

(b) Electrical gradient

What is the term that refers to the electrical gradient that exists across the plasma membrane?

CHECKPOINT

2. Are sodium ions (Na^+) more concentrated in extracellular fluid or cytosol? What about potassium (K^+) ions?
3. What is the functional significance of the electrochemical gradient?

FIGURE 5.3 Examples of electrochemical gradients.

The electrochemical gradient refers to the combined influence of the concentration gradient and the electrical gradient on movement of a particular ion across a membrane.

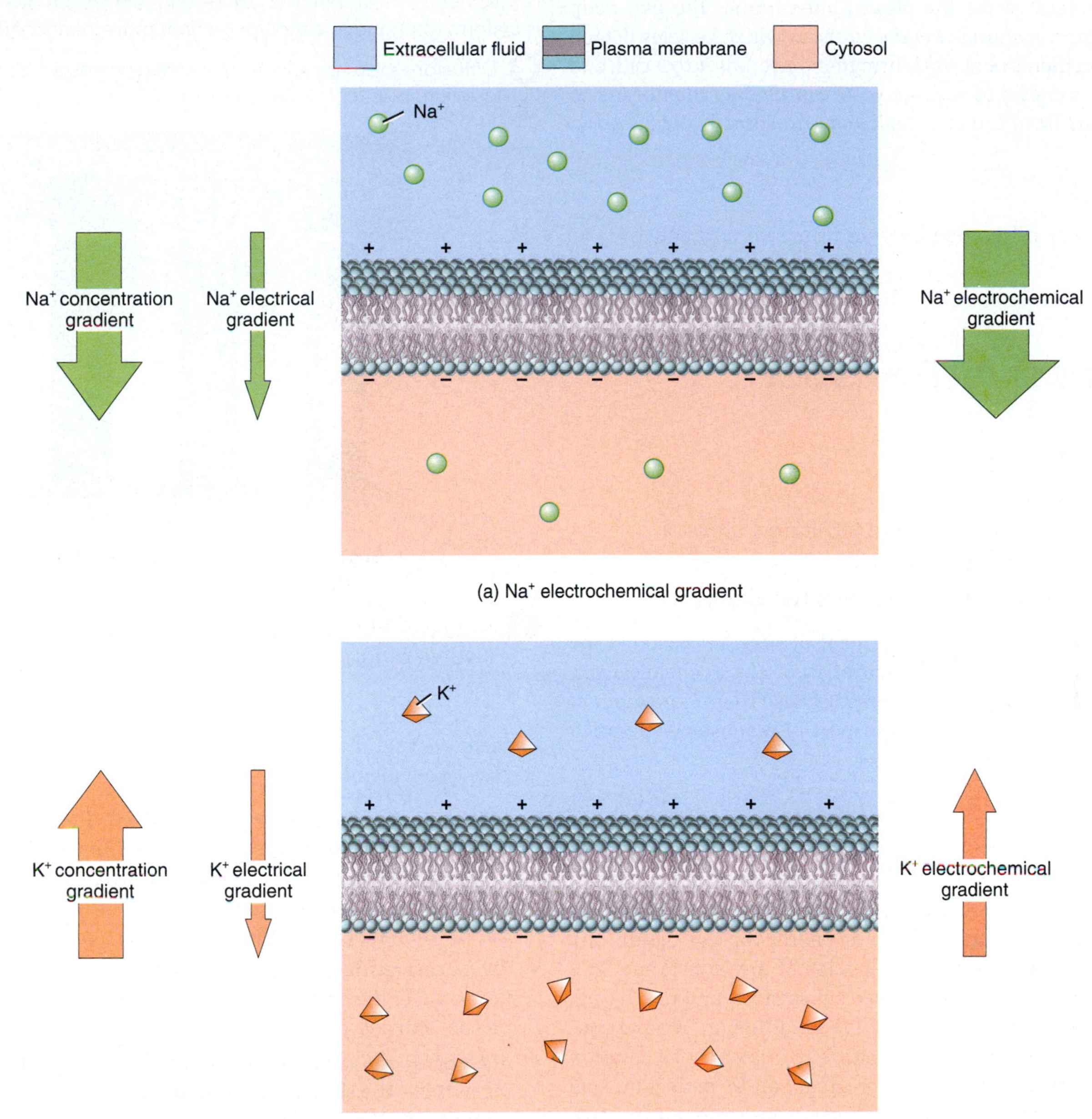

(a) Na^+ electrochemical gradient

(b) K^+ electrochemical gradient

Why does the K^+ electrochemical gradient occur in the same direction as the K^+ concentration gradient and not in the direction of the K^+ electrical gradient?

5.3 Classification of Membrane Transport Processes as Passive or Active

OBJECTIVE

- Distinguish between passive processes and active processes.

Substances generally move across cellular membranes via transport processes that can be classified as passive or active, depending on whether they require cellular energy. In **passive processes**, a substance moves across the plasma membrane without any energy input from the cell. An example is *passive transport*, in which a substance moves "downhill" along its concentration or electrochemical gradient to cross the membrane using only its own kinetic energy (energy of motion). You will learn more about passive transport in Section 5.4. In **active processes**, cellular energy is used to move a substance across the plasma membrane. The cellular energy used is usually in the form of ATP. An example of an active process is *active transport*, in which cellular energy is used to drive a substance "uphill" against its concentration or electrochemical gradient. Active transport is examined in Section 5.5. Another example of an active process is *vesicular transport*, in which tiny vesicles (membranous sacs) are used to move substances across the plasma membrane. The two main types of vesicular transport are *endocytosis*, in which vesicles detach from the plasma membrane while bringing materials into a cell, and *exocytosis*, the merging of vesicles with the plasma membrane to release materials from the cell. Vesicular transport is discussed in Section 5.6.

CHECKPOINT

4. What is the key difference between passive and active processes?

5.4 Passive Transport

OBJECTIVES

- Define diffusion.
- Distinguish simple diffusion from facilitated diffusion.
- Explain the significance of osmosis.
- Compare a solution's osmolarity to its tonicity.

Passive transport refers to the movement of substances across the plasma membrane down their concentration or electrochemical gradients. It is considered a passive process because there is no input of energy from the cell. The passive transport of a substance across a membrane occurs by diffusion.

Diffusion Relies on the Kinetic Energy That Is Intrinsic to All Particles

Diffusion is the random mixing of particles from one location to another because of the particles' kinetic energy (energy of motion). Kinetic energy is intrinsic to the particles of any substance. In a solution, both the *solutes*, the dissolved substances, and the *solvent*, the liquid that does the dissolving, undergo diffusion. If a particular solute is present in high concentration in one area of a solution and in low concentration in another area, solute molecules will diffuse toward the area of lower concentration—they move *down their concentration gradient*. After some time, the particles become evenly distributed throughout the solution and the solution is said to be at equilibrium. The particles continue to move about randomly due to their kinetic energy, but their concentrations do not change.

For example, when you place a crystal of dye in a water-filled container (**FIGURE 5.4a**), the color is most intense in the area closest to the dye because its concentration is higher there. At increasing distances, the color is lighter and lighter because the dye concentration is lower (**FIGURE 5.4b**). Sometime later, the solution of water and dye will have a uniform color because the dye molecules and water molecules have diffused down their concentration gradients until they are evenly mixed in solution—they are at equilibrium (**FIGURE 5.4c**).

In this simple example, no membrane was involved. Substances may also diffuse through a membrane if the membrane is permeable

FIGURE 5.4 Principle of diffusion. At the beginning of the experiment, a crystal of dye placed in a cylinder of water dissolves (a) and then diffuses from the region of higher dye concentration to regions of lower dye concentration (b). At equilibrium (c), the dye concentration is uniform throughout, although random movement continues.

Diffusion is the movement of a substance down its concentration gradient due to its kinetic energy.

Andy Washnik

Beginning (a) | Intermediate (b) | Equilibrium (c)

How would having a fever affect body processes that involve diffusion?

to them. Several factors influence the diffusion rate of substances across plasma membranes:

- ***Steepness of the concentration gradient.*** In general, the greater the difference in concentration between the two sides of the membrane, the higher the rate of diffusion. When charged particles are diffusing, the steepness of the electrochemical gradient determines the diffusion rate across the membrane.
- ***Temperature.*** The higher the temperature, the faster the rate of diffusion. All of the body's diffusion processes occur more rapidly in a person with a fever.
- ***Mass of the diffusing substance.*** The larger the mass of the diffusing particle, the slower its diffusion rate. Smaller molecules diffuse more rapidly than larger ones.
- ***Surface area.*** The larger the membrane surface area available for diffusion, the faster the diffusion rate. For example, the air sacs of the lungs have a large surface area available for diffusion of oxygen from the air into the blood. Some lung diseases, such as emphysema, reduce the surface area. This slows the rate of oxygen diffusion and makes breathing more difficult.
- ***Diffusion distance.*** The greater the distance over which diffusion must occur, the longer it takes. Diffusion across a plasma membrane takes only a fraction of a second because the membrane is so thin. In pneumonia, fluid collects in the lungs; the additional fluid increases the diffusion distance because oxygen must move through both the built-up fluid and the membrane to reach the bloodstream.

The factors that affect the diffusion rate of a particle across a membrane are incorporated into an equation known as Fick's law of diffusion (see the Physiological Equation box).

PHYSIOLOGICAL EQUATION

FICK'S LAW OF DIFFUSION

Fick's law of diffusion mathematically expresses the various factors that determine how quickly a particle diffuses across a membrane from one region to another. This law is expressed as follows:

$$J = -DA\frac{\Delta C}{\Delta X}$$

where J is the net diffusion rate of the particle across the membrane,
D is the diffusion coefficient of the particle,
A is the membrane surface area,
ΔC is the concentration gradient of the particle, and
ΔX is the membrane thickness.

The minus sign in this equation indicates the direction of diffusion. When a particle moves from higher to lower concentration, $\Delta C/\Delta X$ is negative; therefore, multiplying by $-DA$ provides a positive value.

It is important to point out that the diffusion coefficient takes into account the temperature, size of the particle, and viscosity of the medium through which the particle passes. For spherical particles, the diffusion coefficient is approximated by the **Stokes-Einstein equation:**

$$D = \frac{kT}{6\pi r\eta}$$

where D is the diffusion coefficient of the particle,
k is Boltzmann's constant,
T is temperature,
r is the radius of the particle, and
η is the viscosity of the medium.

If you take the Stokes-Einstein equation and substitute it for the diffusion coefficient (D) in the Fick's law equation, then Fick's law can be expressed in the following way:

$$J = -\left(\frac{kT}{6\pi r\eta}\right) \times A \times \left(\frac{\Delta C}{\Delta X}\right)$$

Based on this expanded version of Fick's law of diffusion, you can make the following conclusions:

- ***The net diffusion rate (J) is directly proportional to temperature (T), membrane surface area (A), and the particle's concentration gradient (ΔC).*** This means that if temperature, membrane surface area, or the steepness of the concentration gradient increases, the net diffusion rate increases (assuming all other parameters in the equation remain constant).
- ***The net diffusion rate (J) is inversely proportional to the particle's size (r) and the membrane thickness (ΔX).*** This means that as particle size or membrane thickness increases, the rate of net diffusion decreases (assuming that the other parameters in the equation stay the same).

Now that you have a basic understanding of the nature of diffusion, it is time to consider three types of diffusion: simple diffusion, facilitated diffusion, and osmosis.

Simple Diffusion Occurs When a Solute Moves down Its Gradient Without Any Help

Simple diffusion is a passive process in which solutes move freely through the lipid bilayer of the plasma membranes of cells without the help of membrane transport proteins (**FIGURE 5.5a**). Nonpolar, hydrophobic molecules move across the lipid bilayer through the process of simple diffusion. Such molecules include oxygen, carbon dioxide, and nitrogen gases; fatty acids; steroids; and fat-soluble vitamins (A, D, E, and K). Small, uncharged polar molecules such as water, urea, and small alcohols also pass through the lipid bilayer by simple diffusion. Simple diffusion through the lipid bilayer is important in the movement of oxygen and carbon dioxide between blood and body

FIGURE 5.5 Simple diffusion, channel-mediated facilitated diffusion, and carrier-mediated facilitated diffusion.

In simple diffusion, a solute moves across the lipid bilayer of the plasma membrane without the help of membrane transport proteins; in facilitated diffusion, a solute moves across the lipid bilayer aided by a channel protein or a carrier protein.

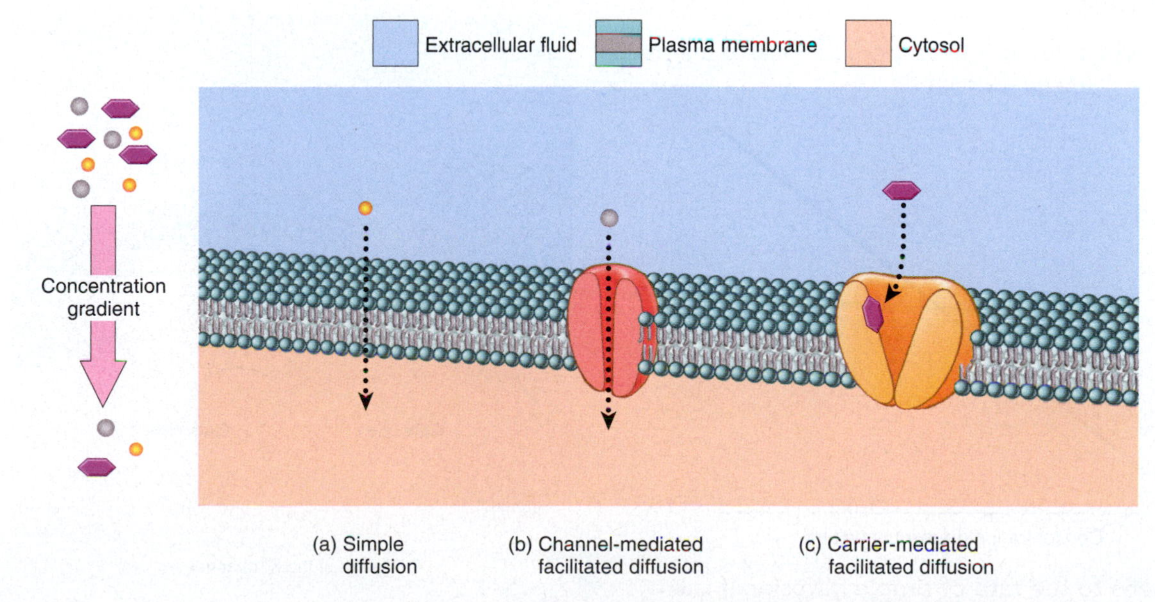

What types of molecules move across the lipid bilayer of the plasma membrane via simple diffusion?

cells, and between blood and air within the lungs during breathing. It is also the route for absorption of some nutrients and excretion of some wastes by body cells. The rate of simple diffusion is always directly proportional to the magnitude of the concentration gradient (**FIGURE 5.6**). Therefore, the steeper the concentration gradient, the higher the diffusion rate.

Facilitated Diffusion Uses a Protein to Move a Solute down Its Gradient

Solutes that are too polar or highly charged to move through the lipid bilayer by simple diffusion can cross the plasma membrane by a passive process called **facilitated diffusion**. In this process, an integral membrane protein assists a specific substance across the membrane. The integral membrane protein can be either a membrane channel or a carrier.

Channel-Mediated Facilitated Diffusion

In **channel-mediated facilitated diffusion**, a solute moves down its concentration or electrochemical gradient across the lipid bilayer through a membrane channel (see **FIGURE 5.5b**). Most membrane channels are **ion channels**, integral transmembrane proteins containing pores that allow passage of small, inorganic ions that are too hydrophilic to penetrate the nonpolar interior of the lipid bilayer. Most ion channels are selective for one type of ion; some channels, however, allow a select group of ions to pass through. The selectivity of an ion channel depends on the size of the pore within the channel and the types of amino acids that line the pore. An ion is able to pass through a given channel only if it is small enough to fit through the pore and can form attractive electrical interactions with the particular set of amino acids that line the pore.

In typical plasma membranes, the most numerous ion channels are selective for K^+ (potassium ions) or Cl^- (chloride ions); fewer channels are available for Na^+ (sodium ions) or Ca^{2+} (calcium ions). Some plasma membranes also have cation channels that allow cations (mainly Na^+, K^+, and Ca^{2+} ions) to pass through. Diffusion of ions through channels is generally slower than free diffusion through the lipid bilayer because channels occupy a smaller fraction of the membrane's total surface area than lipids. Still, facilitated diffusion through channels is a very fast process: More than a million potassium ions can flow through a K^+ channel in one second!

A channel is said to be *gated* when part of the channel protein acts as a "plug" or "gate," changing shape in one way to open the pore and in another way to close it (**FIGURE 5.7**). Some gated channels randomly alternate between the open and closed positions; others are regulated by chemical or electrical changes inside and outside the cell. Ion channels can vary in the number of gates that they have; most channels have just one gate. When all existing gates of a channel are open, ions diffuse into or out of cells, down their electrochemical gradients. The plasma membranes of different types of cells may have different numbers of ion channels and thus display different permeabilities to various ions.

Carrier-Mediated Facilitated Diffusion

In **carrier-mediated facilitated diffusion**, a **carrier** (also called a **transporter**) is used to move a solute down its concentration or electrochemical gradient across the plasma membrane (see **FIGURE 5.5c**). The solute binds to a specific carrier on one side of the membrane and is released on the other side after the carrier undergoes a conformational change (a change in shape) (**FIGURE 5.8**). The solute binds more often

FIGURE 5.6 Relationship between the rate of diffusion and the concentration gradient of a solute that is transported across the plasma membrane by simple diffusion.

The rate of simple diffusion is directly proportional to the steepness of the concentration gradient.

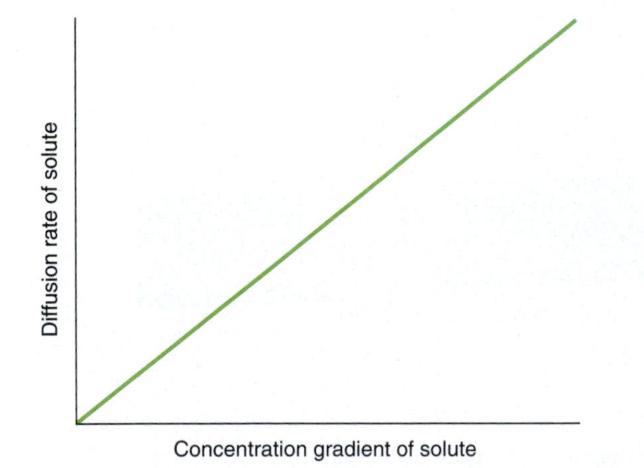

What happens to the rate of simple diffusion if the magnitude of the concentration gradient decreases?

FIGURE 5.7 Channel-mediated facilitated diffusion of potassium ions (K^+) through a gated K^+ channel. In this figure and several upcoming figures, the use of brackets around a solute's name indicates the concentration of that solute. For example, [K^+] indicates K^+ concentration. So high [K^+] means high K^+ concentration and low [K^+] means low K^+ concentration.

A gated channel is one in which a portion of the channel protein acts as a gate to open or close the channel's pore to the passage of ions.

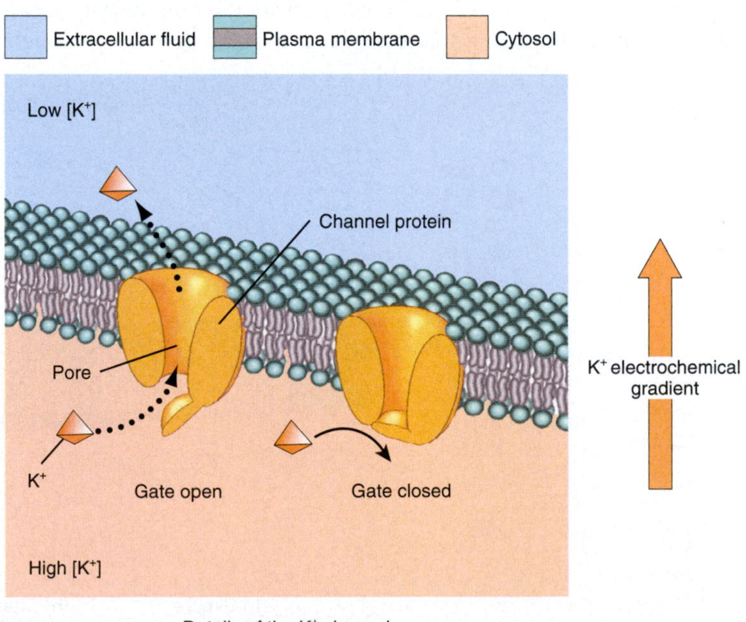

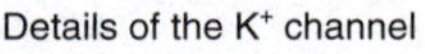

What determines the selectivity of a given ion channel?

FIGURE 5.8 Operation of a carrier protein.

A carrier protein binds to a substance on one side of the membrane, undergoes a conformational change, and then releases the substance on the opposite side of the membrane.

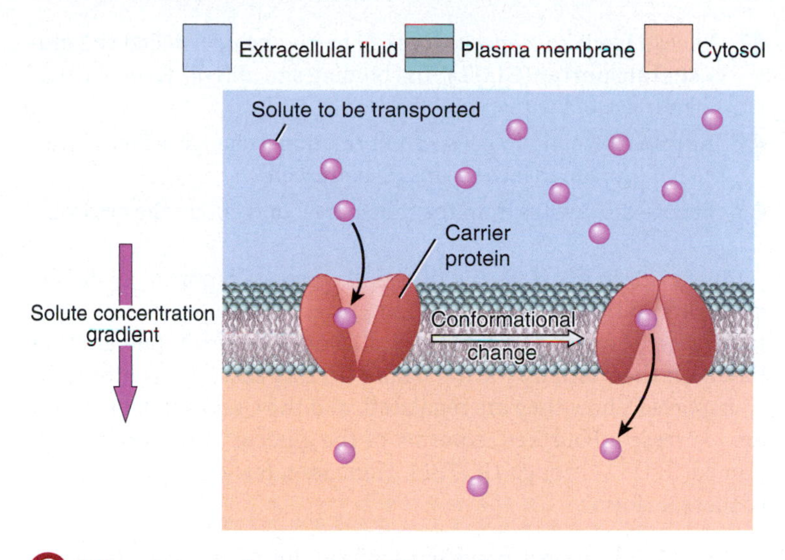

What is a conformational change?

FIGURE 5.9 Specificity of a carrier protein for solute.

A given carrier protein transports only one solute or a group of solutes that are structurally related.

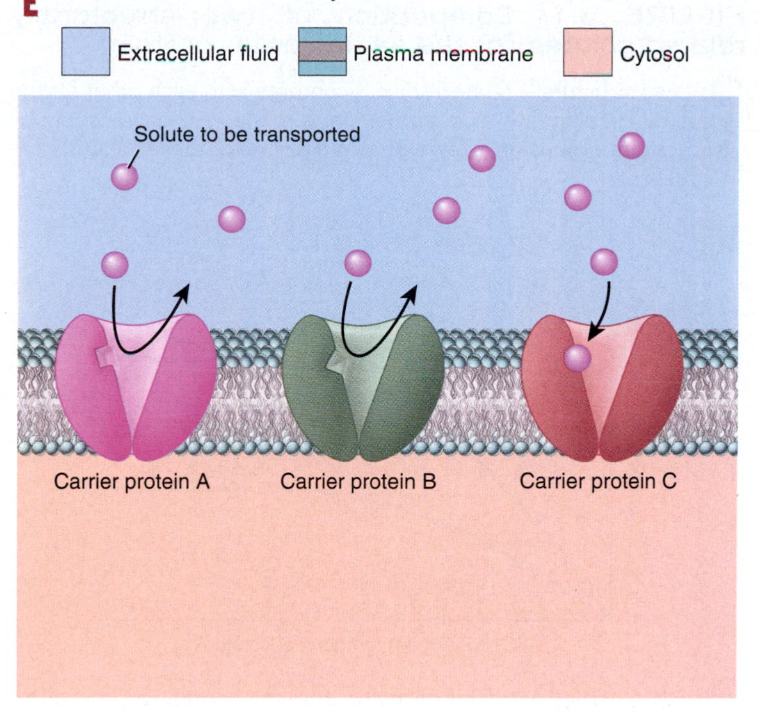

In the figure above, why can carrier protein C bind to the solute but carrier proteins A and B cannot?

to the carrier on the side of the membrane with a higher concentration of solute. Once the concentration is the same on both sides of the membrane, solute molecules bind to the carrier on the extracellular side and move into the cytosol as rapidly as they bind to the carrier on the cytosolic side and move out to the extracellular fluid.

The solutes that bind to carrier proteins are examples of ligands. Thus, carrier-mediated facilitated diffusion exhibits the four characteristics of ligand-protein binding that you learned about in Chapter 2—specificity, affinity, saturation, and competition (see Section 2.5).

1. ***Specificity.*** Each carrier protein transports only one solute or a group of structurally related solutes. The binding site of a given carrier protein has a particular three-dimensional shape, and only a solute with a complementary shape can bind to it (FIGURE 5.9).
2. ***Affinity.*** The solute binds to the binding site of the carrier protein with a certain strength, or *affinity*. In carrier-mediated facilitated diffusion, the affinity of the binding site for the solute is the same regardless of which side of the membrane the binding site faces. So the net direction of solute transport across the membrane is determined by the concentration gradient.
3. ***Saturation.*** There is a limited number of carrier proteins available in the plasma membrane of a cell. This places an upper limit, called the **transport maximum (T_m)**, on the rate at which a solute is transported by facilitated diffusion. As long as the T_m has not been reached, an increase in the concentration gradient increases the rate of transport. However, once all of the carriers are occupied, the T_m is reached and a further increase in the concentration gradient does not increase the transport rate, which means that some solutes will not be moved across the membrane (FIGURE 5.10). Thus, much like a sponge that can absorb no more water, the process of carrier-mediated facilitated diffusion displays *saturation*.
4. ***Competition.*** If a carrier protein can transport a group of solutes that are similar in structure, the solutes may compete with one

FIGURE 5.10 Saturation of a carrier protein by solute.

When carrier proteins are saturated by solute, the rate of transport by facilitated diffusion reaches an upper limit or transport maximum (T_m).

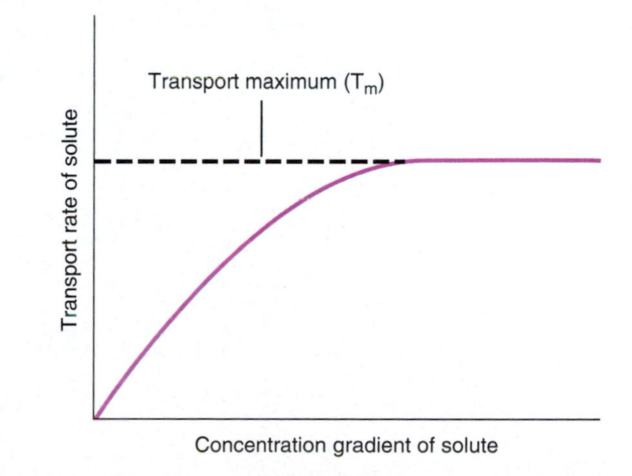

In carrier-mediated facilitated diffusion, what is the relationship between the rate of transport and the steepness of the concentration gradient?

another for the same binding site. When competition occurs, the rate of transport by facilitated diffusion is lower than if only one of these solutes were present (FIGURE 5.11).

FIGURE 5.11 Competition of two structurally related solutes for the same carrier protein.

If two structurally related solutes compete with each other for the same binding site on a carrier protein, the rate of transport by facilitated diffusion is lower than if only one of these solutes were present.

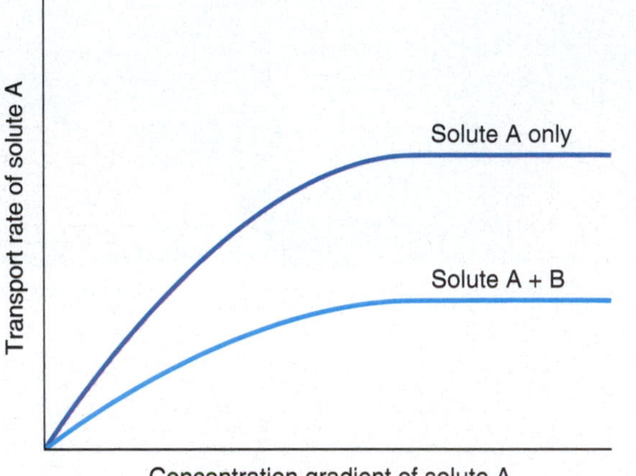

What effect would increasing the concentration of solute B have on transport rate of solute A?

Solutes that move across the plasma membrane by carrier-mediated facilitated diffusion include glucose, fructose, galactose, and some vitamins. Glucose enters many body cells by carrier-mediated facilitated diffusion as follows (FIGURE 5.12):

1. Glucose binds to a specific type of carrier protein called the **glucose transporter (GLUT)**. The binding site initially faces the outside surface of the membrane.
2. The transporter undergoes a conformational change, which causes the binding site to face the inside of the cell.
3. Glucose dissociates from the transporter and enters the cytosol.

There are many isoforms of the glucose transporter. **Isoforms** are structural variants of the same protein that have similar functions. The isoforms of a given protein can differ from one another based on their affinity for the ligand (in this case, the solute to be transported), how they are regulated, and the tissues in which they are expressed. Fourteen isoforms of the glucose transporter exist, named GLUT1 through GLUT14. Here are a few examples of some of these isoforms:

- GLUT1 is the most prevalent isoform, found in most tissues of the body.
- GLUT2 is found in the liver and transport epithelia of the kidneys and intestine. It has a much higher affinity for glucose than GLUT1, meaning that GLUT 2 will take up more glucose than GLUT1 at any given glucose concentration.

FIGURE 5.12 Carrier-mediated facilitated diffusion of glucose across a plasma membrane.

The glucose transporter (GLUT) is a carrier protein that binds to glucose on one side of the plasma membrane and releases it on the other side.

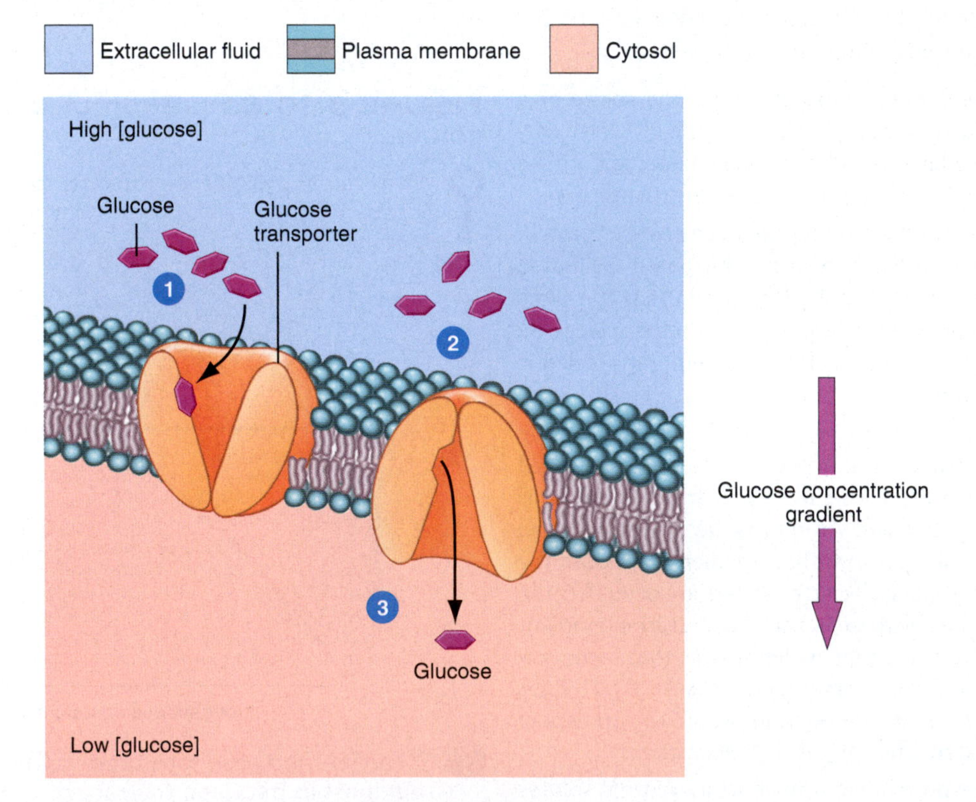

How does insulin alter glucose transport by facilitated diffusion?

- GLUT3 is expressed primarily in neurons and also has a high affinity for glucose.
- GLUT4 is found adipocytes, muscle cells, and other tissues. Unlike the first three isoforms, GLUT4 is regulated by the hormone insulin: The binding of insulin to the insulin receptor promotes the insertion of many copies of GLUT4 into the plasma membranes of fat and muscle cells, thereby increasing glucose uptake in these cells.

As you can see, the GLUT isoforms do not have the exact same properties. The existence of different isoforms of the glucose transporter allows glucose transport to be regulated in different ways in the tissues throughout the body. The glucose transporter is not the only membrane transport protein that has isoforms. In fact, most membrane transport proteins have several isoforms.

Osmosis Results in Net Movement of Water Across a Selectively Permeable Membrane

Osmosis is a type of diffusion in which there is net movement of a solvent through a selectively permeable membrane. Like the other types of diffusion, osmosis is a passive process. In living systems, the solvent is water, which moves by osmosis across plasma membranes from an area of *higher water concentration* to an area of *lower water concentration.* Another way to understand this idea is to consider the solute concentration: In osmosis, water moves through a selectively permeable membrane from an area of *lower solute concentration* to an area of *higher solute concentration.* During osmosis, water molecules pass through a plasma membrane in two ways: (1) by moving through the lipid bilayer via simple diffusion, as previously described, and (2) by moving through **aquaporins**, integral membrane proteins that function as water channels.

Osmosis occurs only when a membrane is permeable to water but is not permeable to certain solutes. A simple experiment can demonstrate osmosis. Consider a U-shaped tube in which a selectively permeable membrane separates the left and right arms of the tube. A volume of pure water is poured into the left arm, and the same volume of a solution containing a solute that cannot pass through the membrane is poured into the right arm (**FIGURE 5.13a**). Because the *water* concentration is higher on the left and lower on the right, net movement of water molecules—osmosis—occurs from left to right, so that the water is moving down its concentration gradient. At the same time, the membrane prevents diffusion of the solute from the right arm into the left arm. As a result, the volume of water in the left arm decreases, and the volume of solution in the right arm increases (**FIGURE 5.13b**).

You might think that osmosis would continue until no water remained on the left side, but this is *not* what happens. In this experiment, the higher the column of solution in the right arm becomes, the more pressure it exerts on its side of the membrane. Pressure exerted in this way by a liquid, known as **hydrostatic pressure**, forces water molecules to move back into the left arm. Equilibrium is reached when just as many water molecules move from right to left due to the hydrostatic pressure as move from left to right due to osmosis (**FIGURE 5.13b**).

Now consider what would happen if a piston were used to apply more pressure to the fluid in the right arm of the tube in **FIGURE 5.13**. With enough pressure, the volume of fluid in each arm could be restored to the starting volume, and the concentration of solute in the right arm would be the same as it was at the beginning of the experiment

FIGURE 5.13 Principle of osmosis. Water molecules move through the selectively permeable membrane; the solute molecules in the right arm cannot pass through the membrane. (a) As the experiment starts, water molecules move from the left arm into the right arm, down the water concentration gradient. (b) After some time, the volume of water in the left arm has decreased and the volume of solution in the right arm has increased. At equilibrium, there is no net osmosis: Hydrostatic pressure forces just as many water molecules to move from right to left as osmosis forces water molecules to move from left to right. (c) If pressure is applied to the solution in the right arm, the starting conditions can be restored. This pressure, which stops osmosis from occurring, is equal to the osmotic pressure.

Osmosis is the movement of water molecules through a selectively permeable membrane.

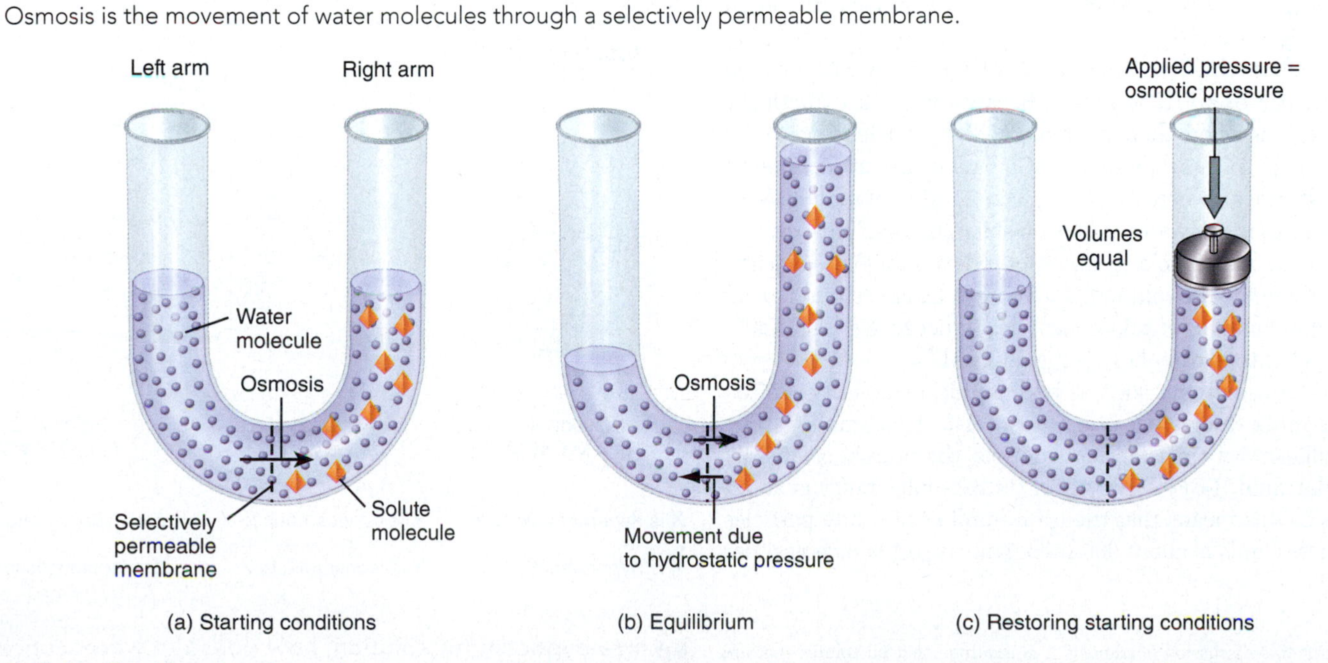

Will the fluid level in the right arm rise until the water concentrations are the same in both arms? Explain your answer.

(**FIGURE 5.13c**). The amount of pressure needed to completely stop the osmotic movement of water from the left tube into the right tube and restore the starting conditions is known as the **osmotic pressure**. The osmotic pressure of a solution is proportional to the concentration of solutes that cannot cross the membrane—the higher the concentration, the higher the solution's osmotic pressure. This is due to the fact that as the concentration of impermeable solutes in a solution increases, osmotic movement of water into that solution increases and a greater osmotic pressure will be needed to oppose osmosis. So the osmotic pressure of a solution is a measure of the tendency of that solution to gain water by osmosis. However, osmotic pressure does not produce the movement of water that occurs during osmosis. Rather, it is the pressure that would *prevent* such water movement.

The osmotic movement of water to an area of high solute concentration depends on the number of solute particles in that solution and not on the chemical nature of the solute. For example, if one glucose molecule is dissolved in water, it yields only one particle because glucose does not dissociate. However, if one NaCl completely dissolves in water, it yields two particles: one Na^+ ion and one Cl^- ion.* Each of these solute particles promotes the movement of water by osmosis. So one NaCl has a greater osmotic effect than one glucose molecule. To express the concentration of solute particles in a solution, the concept of osmolarity is used. The **osmolarity** of a solution is a measure of the total number of dissolved particles per liter of solution. The particles may be molecules, ions, or a mixture of both. The unit used to report osmolarity is osmoles per liter (osmol/L) or osmolar (OsM).

Recall that molarity refers to the number of moles of solute per liter of solution (see Section 2.4), but it does not tell you anything about the number of solute particles in that solution. To calculate osmolarity, multiply molarity by the number of particles per molecule once the molecule dissolves in solution:

$$\text{Osmolarity} = \text{Molarity} \times \text{Number of particles per molecule dissolved}$$

Here are a few examples of how to calculate osmolarity. In a 1 molar solution of glucose, each glucose molecule produces 1 particle, so the osmolarity is 1 osmol/L or 1 OsM. In a 1 molar solution of NaCl, each NaCl molecule produces two particles: one Na^+ ion and one Cl^- ion, so the osmolarity is 2 osmol/L or 2 OsM. In a 1 molar solution of $CaCl_2$, each $CaCl_2$ produces three particles: one Ca^{2+} ion and two Cl^- ions, so the osmolarity is 3 osmol/L or 3 OsM. Because water is osmotically attracted to regions where there are more solute particles (and therefore less water), the 3 OsM solution of $CaCl_2$ osmotically attracts about one and a half times as much water as the 2 OsM solution of NaCl and three times as much water as the 1 OsM solution of glucose. It is important to note that osmolarity does not indicate the types of particles that are present in the solution. Therefore, a 1 OsM solution could contain any combination of solute particles (glucose, Na^+, Cl^-, Ca^{2+}, etc.) as long as the total osmolarity equals 1 OsM.

For very dilute solutions, such as body fluids, osmolarity is often expressed as milliosmoles per liter (mosmol/L; 1 mosmol = 0.001 osmol) or milliosmolar (mOsM). For example, the normal osmolarity of extracellular fluid (ECF) and cytosol (intracellular fluid) is about 300 mosmol/L. This means that the total number of solute particles in each of these fluids is about 300 milliosmoles (0.300 osmoles) per liter of solution. Note that the actual osmolarity of the ECF or cytosol ranges from 280 to 295 mosmol/L, but this number is usually rounded up to about 300 mosmol/L for simplicity.

Another term related to osmolarity is **osmolality**, the number of solute particles per *kilogram* of water. Because it is easier to measure volumes of solutions than to determine the mass of water they contain, osmolarity is used more commonly than osmolality. Most body fluids and solutions used clinically are dilute, in which case there is less than a 1% difference between the two measures.

The osmolarities of different solutions can be compared to one another. If two solutions have the same osmolarity (same number of solute particles per unit volume), then the solutions are said to be **isoosmotic** (*iso-* = same). Because the solute concentrations are equal, the water concentrations in both solutions are also equal. A solution with a higher osmolarity (more solute particles per unit volume) than another solution is said to be **hyperosmotic** (*hyper-* = greater than). Because the solute concentration is higher, the water concentration in such a solution is lower. A solution with a lower osmolarity (fewer solute particles per unit volume) than another solution is said to be **hypoosmotic** (*hypo-* = less than). Because the solute concentration is lower, the water concentration in this type of solution is higher. **FIGURE 5.14** illustrates how the terms isoosmotic, hyperosmotic, and hypoosmotic are used to compare the osmolarities of three different solutions.

When two solutions are separated by a membrane, some of the solutes on one side of the membrane may be able to pass through the membrane, whereas others might not be able to do so because the membrane is impermeable to them. Because the osmolarity of a solution refers to the total number of solute particles in that solution, whenever the term osmolarity or any of its comparative forms (isoosmotic,

FIGURE 5.14 Comparing the osmolarities of three different solutions. For simplicity, not all solute molecules are shown in each solution, but their relative proportions are.

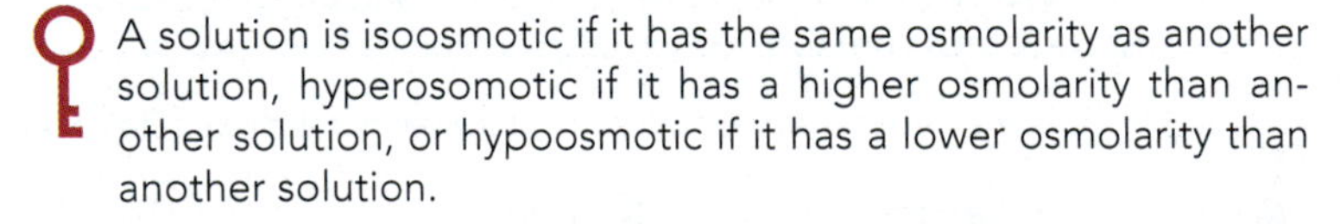

A solution is isoosmotic if it has the same osmolarity as another solution, hyperosomotic if it has a higher osmolarity than another solution, or hypoosmotic if it has a lower osmolarity than another solution.

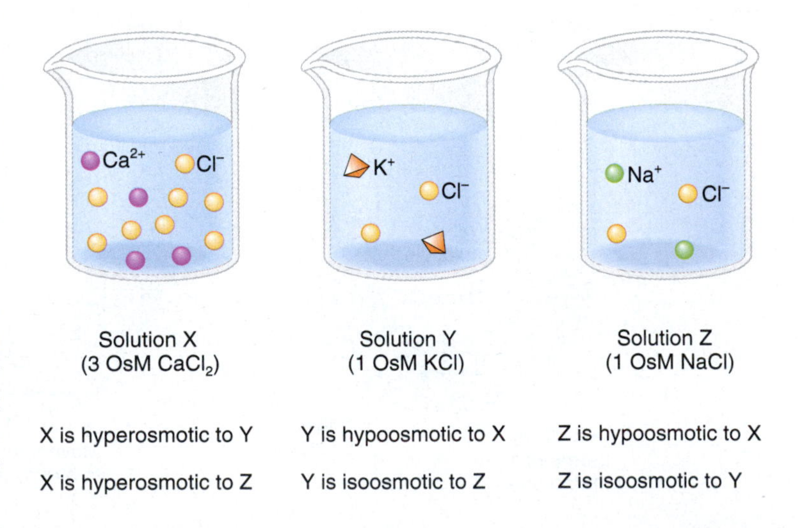

In a hypoosmotic solution, how does the water concentration compare to the concentration of solute particles?

*For simplicity, in this discussion of osmosis it is assumed that all solutes that are capable of dissociating undergo complete dissociation.

hyperosmotic, or hypoosmotic) are used, they are referring to the concentration of both penetrating solutes and nonpenetrating solutes in that solution. **Penetrating solutes** are able to pass through a membrane. An example is urea, which can pass through the plasma membranes of most cells. **Nonpenetrating solutes** are unable to pass through a membrane. Examples of nonpenetrating solutes found in extracellular fluid include Na^+ ions and Cl^- ions. Na^+ ions move into cells through ion channels or transporters, but then are immediately moved back out of the cell by the Na^+/K^+ ATPase, a carrier protein that mediates primary active transport (discussed in Section 5.5 of this chapter). Likewise, Cl^- ions that enter the cell through ion channels in the plasma membrane are also quickly transported back out of the cell. So Na^+ and Cl^- ions behave as if they cannot penetrate the membrane. Cytosol (intracellular fluid) also has nonpenetrating solutes—K^+ ions and protein molecules. K^+ ions can diffuse out of the cell through ion channels. However, the Na^+/K^+ ATPase quickly brings them back into the cell. Many protein molecules that are present within a cell lack transport mechanisms to move across the plasma membrane. Hence, K^+ ions and intracellular protein molecules behave as if they are unable to penetrate the membrane.

Nonpenetrating solutes such as Na^+, Cl^-, K^+ and intracellular proteins play an important role in determining a solution's tonicity. **Tonicity** (*tonic* = tension) is a measure of a solution's ability to change the volume of cells by altering their water content. *Tonicity is determined by the concentration of nonpenetrating solutes in the solution.* Because nonpenetrating solutes are unable to move across a membrane, unequal concentrations of nonpenetrating solutes in extracellular fluid or cytosol cause net movement of water across the membrane, changing the volume of the cell. Penetrating solutes do not affect tonicity because they move across the plasma membrane and equally distribute between extracellular fluid and the cytosol. Hence, there is no net movement of water across the membrane and no net change in cell volume.

Based on tonicity, a solution may be isotonic, hypotonic, or hypertonic. By convention, tonicity refers to the solution surrounding the cell. Any solution in which a cell—for example, an erythrocyte (red blood cell)—maintains its normal shape and volume is an **isotonic solution** (FIGURE 5.15). In an isotonic solution, the concentrations of nonpenetrating solutes are the same on both sides of the membrane. For instance, a 0.9% NaCl solution (0.9 grams of sodium chloride in 100 mL of solution), called a *normal (physiological) saline solution*, is isotonic for erythrocytes. When erythrocytes are bathed in 0.9% NaCl, water molecules enter and exit at the same rate, allowing the erythrocytes to keep their normal shape and volume. A different situation results if erythrocytes are placed in a **hypotonic solution**, a solution that has a lower concentration of nonpenetrating solutes than the cytosol inside the erythrocytes (FIGURE 5.15). In this case, water molecules enter the cells faster than they leave, causing the erythrocytes to swell and eventually to burst. The rupture of erythrocytes in this manner is called **hemolysis** (hē-MOL-i-sis; *hemo-* = blood; *-lysis* = to loosen or split apart); the rupture of other types of cells due to placement in a hypotonic solution is referred to simply as **lysis**. Pure water is very hypotonic and causes rapid hemolysis. A **hypertonic solution** has a higher concentration of nonpenetrating solutes than the cytosol inside erythrocytes (FIGURE 5.15). One example of a hypertonic solution is a 2% NaCl solution. In such a solution, water molecules move out of the cells faster than they enter, causing the cells to shrink. Such shrinkage of cells is called **crenation** (kre-NĀ-shun).

FIGURE 5.15 Tonicity and its effects on erythrocytes. The arrows indicate the direction and degree of water movement into and out of the cells.

Cells placed in an isotonic solution maintain their shape because there is no net water movement into or out of the cell.

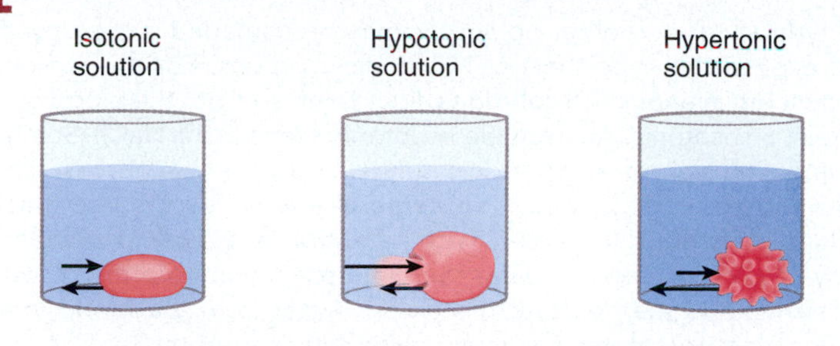

(a) Illustrations showing direction of water movement

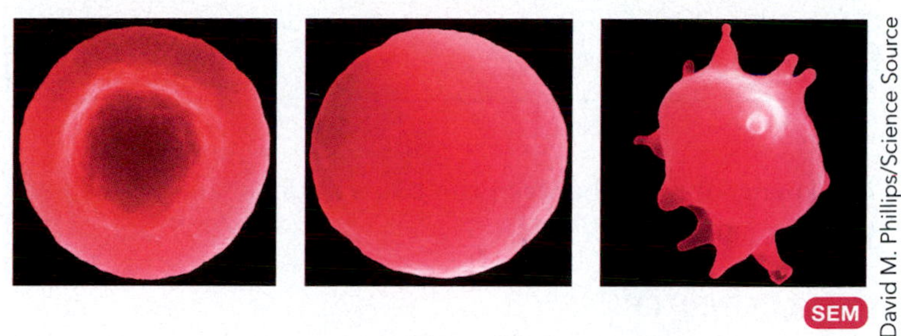

Normal erythrocyte shape | Erythrocyte undergoes hemolysis | Erythrocyte undergoes crenation

(b) Scanning electron micrographs (all 15,000x)

Will a 2% solution of NaCl cause hemolysis or crenation of erythrocytes? Explain your answer.

It is important to understand that osmolarity and tonicity are not synonymous. Furthermore, just simply knowing the osmolarity of a solution does not predict what will happen to the volume of a cell if the cell is surrounded by that solution. To understand the difference between osmolarity and tonicity, consider the following example. Suppose that a 340 mOsM solution containing 300 mosmol/L of NaCl (a nonpenetrating solute) and 40 mosmol/L of urea (a penetrating solute) surrounds a cell. Also suppose that the cytosol has a normal osmolarity of 300 mOsM—all derived from nonpenetrating solutes present inside the cell (K^+ and intracellular proteins). So the extracellular solution is hyperosmotic to the cytosol, and the cytosol is hypoosmotic to the extracellular solution. However, because urea is a penetrating solute, urea will move across the membrane until its concentration reaches equilibrium. There will be no net movement of water across the membrane because of the equal distribution of urea. So the cell's volume does not change. Hence, the extracellular solution is both hyperosmotic (referring to total osmolarity) and isotonic (referring to the tonicity).

Recall that extracellular fluid (ECF) and cytosol normally have the same osmolarity of 300 mosmol/L. So ECF and cytosol are isoosmotic with respect to one another. ECF must also be isotonic to cytosol to keep the volume of the cell from changing. When body cells are at osmotic equilibrium, as they usually are, the number of nonpenetrating solutes in the extracellular fluid equals the number of nonpenetrating solutes in cytosol, so there is no net

CLINICAL CONNECTION

Medical Uses of Isotonic, Hypertonic, and Hypotonic Solutions

Erythrocytes and other body cells may be damaged or destroyed if exposed to hypertonic or hypotonic solutions. For this reason, most **intravenous (IV) solutions**, liquids infused into the blood of a vein, are isotonic. An example is isotonic saline (0.9% NaCl). Sometimes infusion of a hypertonic solution such as mannitol is useful to treat patients who have ***cerebral edema***, excess interstitial fluid in the brain. Infusion of such a solution relieves fluid overload by causing osmosis of water from interstitial fluid into the blood. The kidneys then excrete the excess water from the blood into the urine. Hypotonic solutions, given either orally or through an IV, can be used to treat people who are dehydrated. The water in the hypotonic solution moves from the blood into interstitial fluid and then into body cells to rehydrate them. Water and most sports drinks that you consume to rehydrate after a workout are hypotonic relative to your body cells.

movement of water between the two solutions, and the cell volume stays the same.

CHECKPOINT

5. What factors can increase the rate of diffusion?
6. How does simple diffusion compare with facilitated diffusion?
7. What is osmotic pressure?
8. What is the difference between osmolarity and molarity?
9. Why are nonpenetrating solutes important?
10. What is the difference between the terms *hyperosmotic* and *hypertonic*?

5.5 Active Transport

OBJECTIVES

- Compare primary active transport and secondary active transport.
- Describe the functional significance of the Na^+/K^+ ATPase.
- Provide specific examples of carrier proteins that mediate secondary active transport.

Some polar or charged solutes that must enter or leave body cells cannot cross the plasma membrane through any form of passive transport because they would need to move "uphill," *against* their concentration or electrochemical gradients. Such solutes may be able to cross the membrane by a process called **active transport**. Active transport is considered an active process because energy is required to move solutes across the membrane. Two sources of cellular energy can be used to drive active transport: (1) Energy obtained from hydrolysis of adenosine triphosphate (ATP) is the source in *primary active transport*; (2) energy stored in an ionic electrochemical gradient is the source in *secondary active transport*.

Active transport of solutes is mediated by carrier proteins. Like carrier-mediated facilitated diffusion, active transport exhibits specificity, affinity, saturation, and competition. For a carrier protein that mediates active transport, the binding site has a high affinity for solute when the binding site faces the low-concentration side of the membrane and has a low affinity for solute when the binding site faces the high-concentration side of the membrane. This is in contrast to a carrier protein that mediates facilitated diffusion, in which the affinity of the binding site for solute is the same on either side of the membrane. Solutes actively transported across the plasma membrane include several ions, such as Na^+, K^+, H^+, Ca^{2+}, I^- (iodide), and Cl^-; amino acids; and monosaccharides. (Note that some of these substances also cross the membrane via facilitated diffusion when the proper channel proteins or carriers are present.)

Primary Active Transport Uses Energy from ATP to Move a Solute Against Its Gradient

In **primary active transport**, energy derived from hydrolysis of ATP changes the shape of a carrier protein, which pumps a solute across a plasma membrane against its concentration or electrochemical gradient. Indeed, carrier proteins that mediate primary active transport are often called **pumps**. A typical body cell expends about 40% of the ATP it generates on primary active transport. Chemicals that turn off ATP production—for example, the poison cyanide—are lethal because they shut down active transport in cells throughout the body.

The most prevalent primary active transport mechanism expels sodium ions (Na^+) from cells and brings potassium ions (K^+) into cells. Because of the specific ions it moves, this carrier is called the **sodium–potassium pump**. Because part of the sodium–potassium pump acts as an *ATPase*, an enzyme that hydrolyzes ATP, another name for this pump is **Na^+/K^+ ATPase**. All cells have thousands of sodium–potassium pumps in their plasma membranes. These sodium–potassium pumps maintain a low concentration of Na^+ in the cytosol by pumping these ions into the extracellular fluid against the Na^+ electrochemical gradient. At the same time, the pumps move K^+ into cells against the K^+ electrochemical gradient. Because K^+ and Na^+ slowly leak back across the plasma membrane down their electrochemical gradients—through passive transport or secondary active transport—the sodium–potassium pumps must work nonstop to maintain a low concentration of Na^+ and a high concentration of K^+ in the cytosol.

FIGURE 5.16 depicts the operation of the sodium–potassium pump:

1. Three Na^+ ions bind to high-affinity binding sites on the cytosolic side of the pump protein. The pump also has two binding sites for K^+ ions, but the sites have a low affinity for K^+ at this time, making it unlikely that K^+ will bind.
2. Binding of Na^+ triggers the hydrolysis of ATP into ADP, a reaction that also attaches a phosphate group (P) to the pump protein. This chemical reaction causes a conformational change in the pump protein that exposes the Na^+ binding sites to the extracellular fluid and reduces the affinity of the Na^+ binding sites. As a result, the three Na^+ are expelled into the extracellular fluid. The conformational change also increases the affinity of the K^+ binding sites, favoring the binding of two K^+ in the extracellular fluid to the pump protein.

FIGURE 5.16 The sodium–potassium pump (Na^+/K^+ ATPase). The sodium–potassium pump expels sodium ions (Na^+) and brings potassium ions (K^+) into the cell. For simplicity, the Na^+ electrochemical gradient and the K^+ electrochemical gradient are depicted as having the same magnitudes. However, the magnitude of the Na^+ electrochemical gradient is actually larger than that of the K^+ electrochemical gradient because both the Na^+ electrical and concentration gradients occur in the same direction, whereas the K^+ electrical and concentration gradients occur in opposite directions (see **FIGURE 5.3**).

Sodium–potassium pumps maintain a low intracellular concentration of sodium ions.

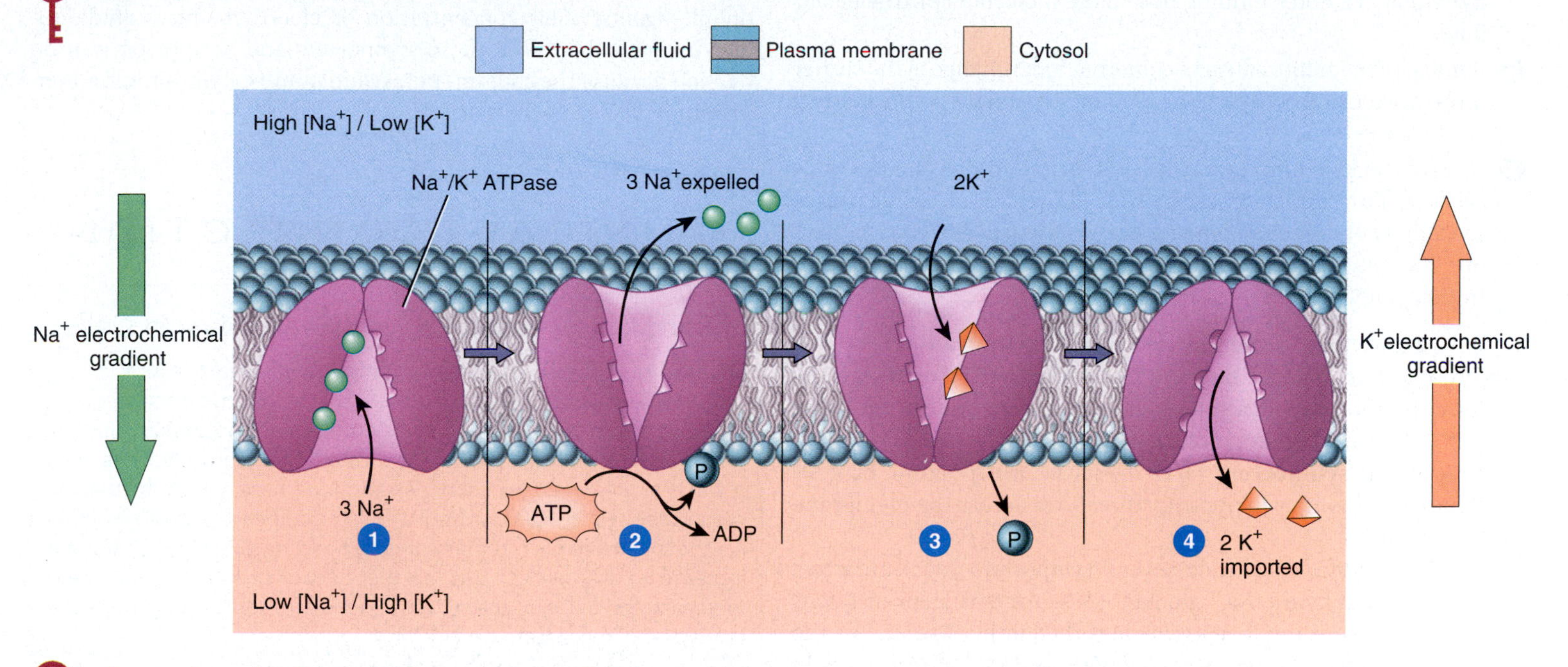

What is the role of ATP in the operation of this pump?

3. The binding of K^+ triggers release of the phosphate group from the pump protein. This reaction again causes a conformational change of the pump protein.
4. As the pump protein reverts to its original shape, the K^+ binding sites are exposed to the cytosol, the affinity of the sites decreases, and K^+ ions are released into the cytosol. In addition, the conformational changes increase the affinity of the Na^+ binding sites. At this point, the pump is again ready to bind three Na^+, and the cycle repeats.

The different concentrations of Na^+ and K^+ in cytosol and extracellular fluid are crucial for maintaining normal cell volume and for the ability of some cells to generate electrical signals such as action potentials. Recall that the tonicity of a solution is proportional to the concentration of its solute particles that cannot penetrate the membrane. Because sodium ions that diffuse into a cell or enter through secondary active transport are immediately pumped out, it is as if they never entered. In effect, sodium ions behave as if they cannot penetrate the membrane. Thus, sodium ions are an important contributor to the tonicity of the extracellular fluid. A similar condition holds for K^+ in the cytosol. By helping to maintain normal tonicity on each side of the plasma membrane, the sodium–potassium pumps ensure that cells neither shrink nor swell due to the movement of water by osmosis out of or into cells.

Secondary Active Transport Uses Energy from an Ionic Gradient to Move a Solute Against Its Gradient

In **secondary active transport**, the energy stored in an ionic electrochemical gradient is used to drive other solutes across the membrane against their own concentration or electrochemical gradients. Most secondary active transport systems are driven by the Na^+ electrochemical gradient. Because a Na^+ gradient is established by primary active transport, secondary active transport *indirectly* uses energy obtained from the hydrolysis of ATP.

The sodium–potassium pump maintains a steep concentration gradient of Na^+ across the plasma membrane. As a result, the sodium ions have stored or potential energy, just like water behind a dam. Accordingly, if there is a route for Na^+ to leak back in, some of the stored energy can be converted to kinetic energy (energy of motion) and used to transport other substances against their concentration or electrochemical gradients. In essence, secondary active transport proteins harness the energy in the Na^+ electrochemical gradient by providing routes for Na^+ to leak into cells.

As an example of secondary active transport, consider a carrier protein (transporter) that uses the energy from the Na^+ electrochemical gradient to move a solute against its concentration gradient from extracellular fluid to the cytosol (FIGURE 5.17):

1. A Na^+ ion binds to a Na^+ binding site on the extracellular side of the carrier protein.
2. Binding of Na^+ causes the solute binding site to increase its affinity, which promotes binding of a solute molecule in extracellular fluid.
3. Binding of the solute causes a conformational change in the carrier protein, exposing both the Na^+ binding site and the solute binding site to the cytosol.
4. Na^+ dissociates from the carrier protein and moves into the cytosol down its electrochemical gradient. The release of Na^+ decreases the affinity of the solute binding site, resulting in release of the solute. The solute then enters the cytosol, moving against its concentration gradient.

As you have just learned, in secondary active transport, a carrier protein binds to Na^+ and another solute and then changes its conformation so that both solutes cross the membrane at the same time. If these transporters move two solutes in the same direction, they are called **symporters** (*sym-* = same); **antiporters**, by contrast, move two solutes in opposite directions across the membrane (*anti-* = against).

Plasma membranes contain several symporters and antiporters that are powered by the Na^+ gradient (FIGURE 5.18). For instance, dietary glucose and amino acids are absorbed into cells that line the small intestine by Na^+/glucose symporters and Na^+/amino acid symporters (FIGURE 5.18a, b). By contrast, the concentration of calcium ions (Ca^{2+}) is low in the cytosol because Na^+/Ca^{2+} antiporters eject calcium ions (FIGURE 5.18c). Likewise, Na^+/H^+ antiporters help regulate the cytosol's pH (H^+ concentration) by expelling excess H^+ (FIGURE 5.18d). In each case, sodium ions are moving down their electrochemical gradient, while the other solutes move "uphill," against their concentration or electrochemical gradients. Keep in mind that all of these symporters and antiporters can do their job because the sodium–potassium pumps maintain a low concentration of Na^+ in the cytosol.

CLINICAL CONNECTION

Digitalis Increases Ca^{2+} in Heart Muscle Cells

Digitalis is a drug often given to patients with ***heart failure***, a condition of weakened pumping action by the heart. Digitalis exerts its effect by slowing the action of the sodium–potassium pumps, which lets more Na^+ accumulate inside heart muscle cells. The result is a decreased Na^+ concentration gradient across the plasma membrane, which causes the Na^+/Ca^{2+} antiporters to slow down. As a result, more Ca^{2+} remains inside heart muscle cells. The slight increase in the level of Ca^{2+} in the cytosol of heart muscle cells increases the force of their contractions and thus strengthens the force of the heartbeat.

FIGURE 5.17 Operation of a carrier protein that mediates secondary active transport.

Secondary active transport mechanisms use the energy stored in an ionic electrochemical gradient (usually Na^+); because primary active transport pumps that hydrolyze ATP maintain the gradient, secondary active transport mechanisms consume ATP indirectly.

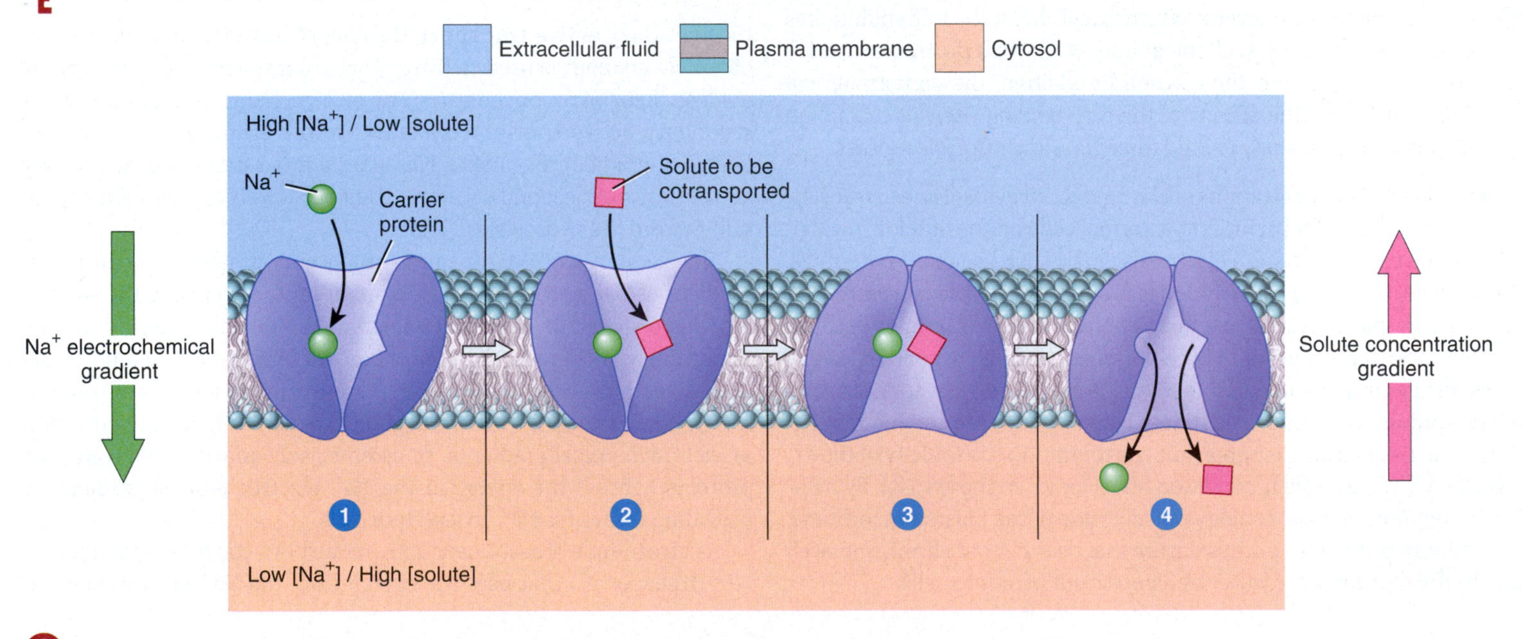

What is the main difference between primary and secondary active transport mechanisms?

FIGURE 5.18 Examples of secondary active transport. $[Na^+]$ = Na^+ concentration; [glucose] = glucose concentration; [amino acids] = amino acid concentration; $[Ca^{2+}]$ = Ca^{2+} concentration; and $[H^+]$ = H^+ concentration.

Symporters carry two substances across the membrane in the same direction; antiporters carry two substances across the membrane in opposite directions.

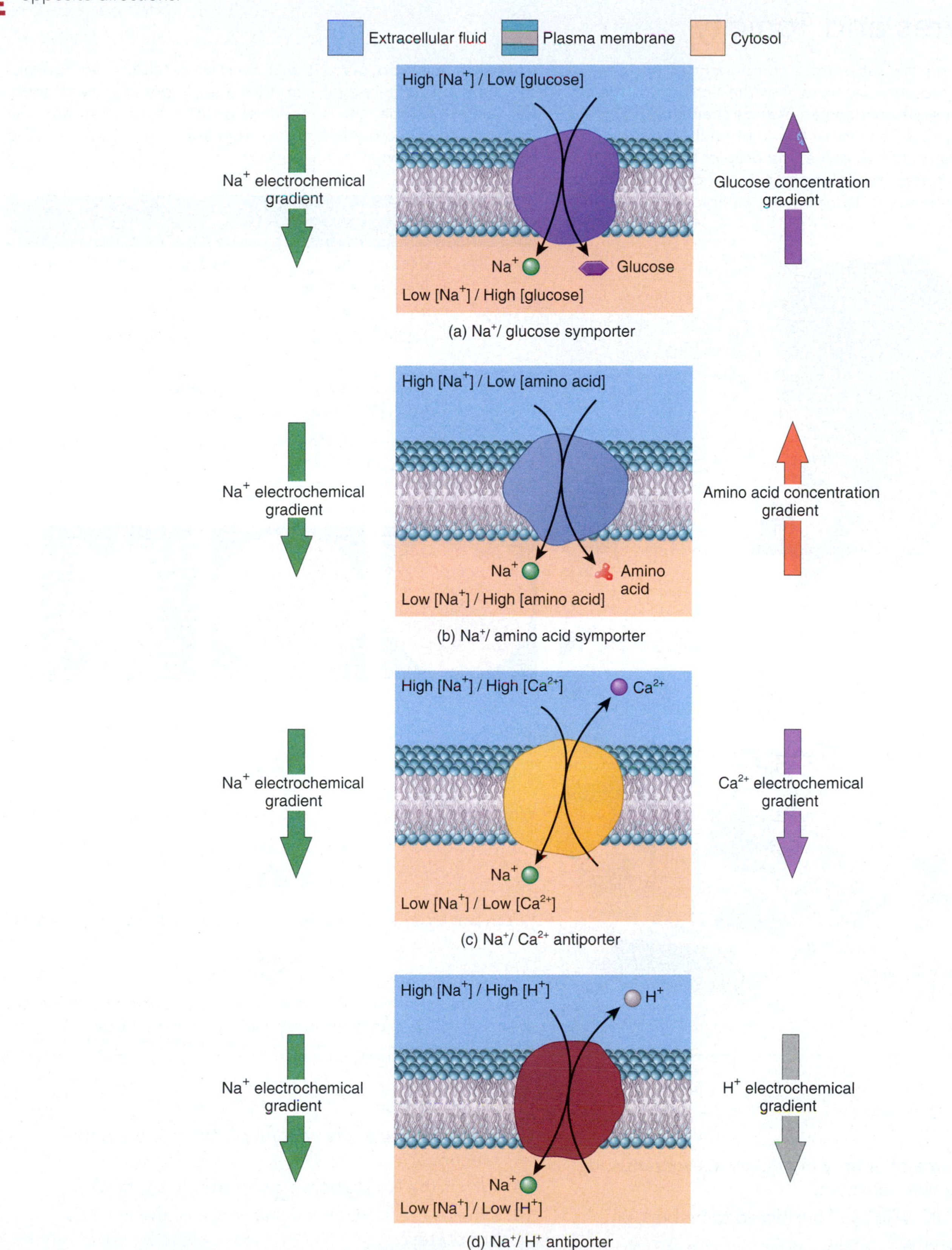

(a) Na^+/ glucose symporter

(b) Na^+/ amino acid symporter

(c) Na^+/ Ca^{2+} antiporter

(d) Na^+/ H^+ antiporter

What is the functional significance of the Na^+/H^+ antiporter?

Critical Thinking

Erythrocytes and Tonicity

Angela is attending the laboratory component of her physiology class. Her professor mentions that the activity for the day is to observe how erythrocytes can change their shape and volume depending on the tonicity of the solution that bathes these cells. Angela is instructed to place one drop of blood in each of three test tubes containing solutions of varying solute concentrations. She is then asked to remove a sample of the solution from each test tube, place it on a microscope slide, and then observe the appearance of the erythrocytes under the microscope. For each test tube, the concentration of the solution and the appearance of the erythrocytes in that solution are summarized in the table below:

JGI/Daniel Grill/Getty Images, Inc.

Test Tube Number	Concentration of Solution	Appearance of Erythrocytes
1	2 mL of 1.5% NaCl (256.8 mM)	Crenated: shriveled with spikes protruding.
2	2 mL of 0.9% NaCl (154 mM)	Normal appearance (biconcave disc); no change in shape or volume.
3	2 mL of a 0.5% NaCl (85.5 mM)	Swollen; resemble little red beach balls.
4	2 mL of a 0.25% NaCl (42.8 mM)	Swollen to the point of lysis (bursting).

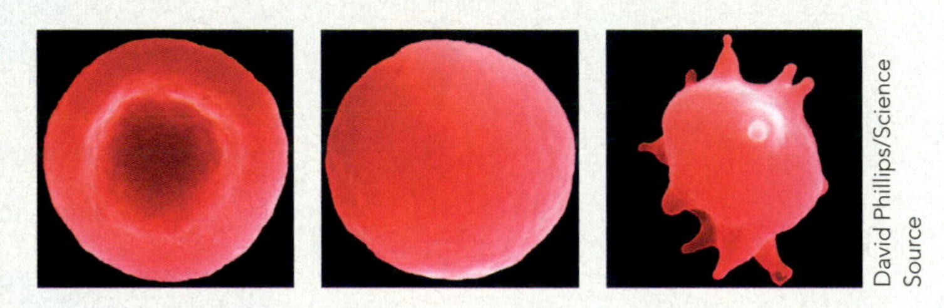

David Phillips/Science Source

What conditions promote osmosis across erythrocyte membranes?

Why don't the erythrocytes in test tube 2 change their volume or shape?

Which test tube contains a hypertonic solution? How do you know?

Why do the erythrocytes in test tube 3 only become swollen but the erythrocytes in test tube 4 become swollen to the point of lysis?

CHECKPOINT

11. What is the source of energy in primary active transport? In secondary active transport?

12. Why is the Na^+/K^+ ATPase considered to be both an ATPase and a pump?

13. How do symporters and antiporters carry out their functions?

14. Why is the Na^+/glucose symporter important?

15. What is the functional significance of the Na^+/Ca^{2+} antiporter?

5.6 Vesicular Transport

OBJECTIVES

- Describe the different types of endocytosis.
- Explain how exocytosis occurs.

A **vesicle** is a small, membranous sac. In **vesicular transport**, vesicles are used to move substances across the plasma membrane. The two main categories of vesicular transport are endocytosis and exocytosis. During **endocytosis** (*endo-* = within), materials move into a cell in a vesicle formed from the plasma membrane (FIGURE 5.19a). In **exocytosis** (*exo-* = out), materials move out of a cell by the fusion with the plasma membrane of vesicles formed inside the cell (FIGURE 5.19b). Both endocytosis and exocytosis require energy supplied by ATP. Thus, vesicular transport is an active process.

Endocytosis Allows Ligands, Large Solid Particles, and Droplets of Extracellular Fluid to Enter Cells

There are three types of endocytosis: receptor-mediated endocytosis, phagocytosis, and bulk-phase endocytosis.

Receptor-Mediated Endocytosis

Receptor-mediated endocytosis is a highly selective type of endocytosis by which cells take up specific ligands. During this process, a vesicle forms after receptor proteins in the plasma membrane are bound by particular ligands in the extracellular fluid. For instance, cells take up cholesterol-containing low-density lipoproteins (LDLs), transferrin (an iron-transporting protein in the blood), some vitamins, antibodies, and certain hormones by receptor-mediated endocytosis. Receptor-mediated endocytosis of LDLs (and other ligands) occurs as follows (FIGURE 5.20):

CLINICAL CONNECTION

Viruses and Receptor-Mediated Endocytosis

Although receptor-mediated endocytosis normally imports needed materials, some viruses are able to use this mechanism to enter and infect body cells. For example, the human immunodeficiency virus (HIV), which causes acquired immunodeficiency syndrome (AIDS), can attach to a receptor called CD4. This receptor is present in the plasma membrane of leukocytes called helper T cells. After binding to CD4, HIV enters the helper T cell via receptor-mediated endocytosis.

1. ***Binding.*** On the extracellular side of the plasma membrane, an LDL particle that contains cholesterol binds to a specific receptor in the plasma membrane to form a receptor–LDL complex. The receptors are integral membrane proteins concentrated in regions of the plasma membrane called *clathrin-coated pits*. Here, a protein called *clathrin* attaches to the membrane on its cytoplasmic side. Many clathrin molecules come together, forming a basketlike structure around the receptor–LDL complexes that causes the membrane to invaginate (fold inward).
2. ***Vesicle formation.*** The invaginated edges of the membrane around the clathrin-coated pit fuse, and a small piece of the membrane pinches off. The resulting vesicle, known as a *clathrin-coated vesicle*, contains the receptor–LDL complexes.
3. ***Uncoating.*** Almost immediately after it is formed, the clathrin-coated vesicle loses its clathrin coat to become an *uncoated vesicle*. Clathrin molecules either return to the inner surface of the plasma membrane or help form coats on other vesicles inside the cell.

FIGURE 5.19 Vesicular transport: endocytosis and exocytosis.

During endocytosis, material moves into the cell in a vesicle formed from the plasma membrane; in exocytosis, a vesicle merges with the plasma membrane to release materials from the cell.

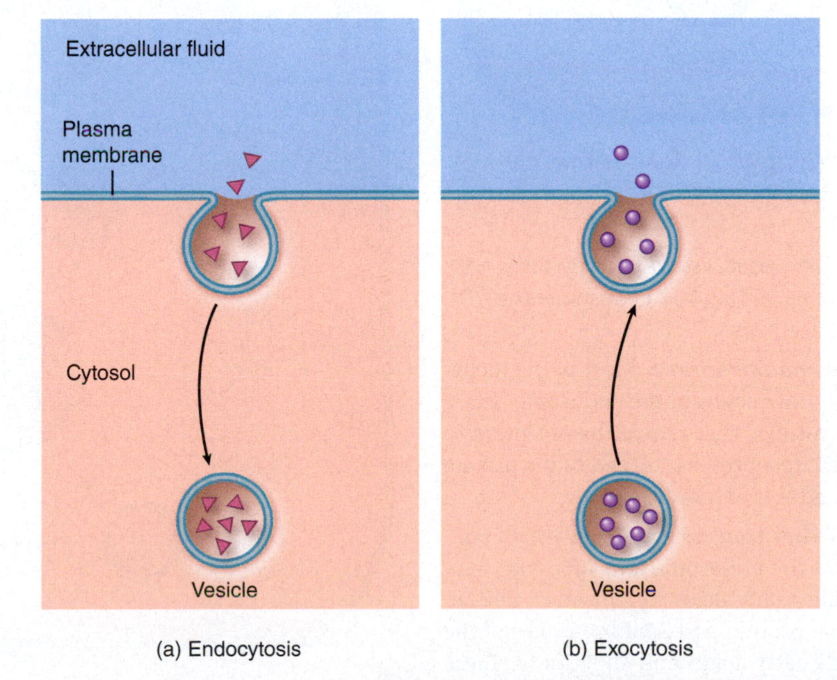

Are endocytosis and exocytosis passive processes or active processes? Explain your answer.

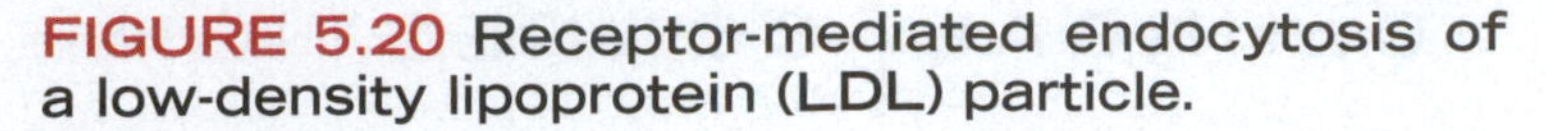

FIGURE 5.20 Receptor-mediated endocytosis of a low-density lipoprotein (LDL) particle.

Receptor-mediated endocytosis imports materials that are needed by cells.

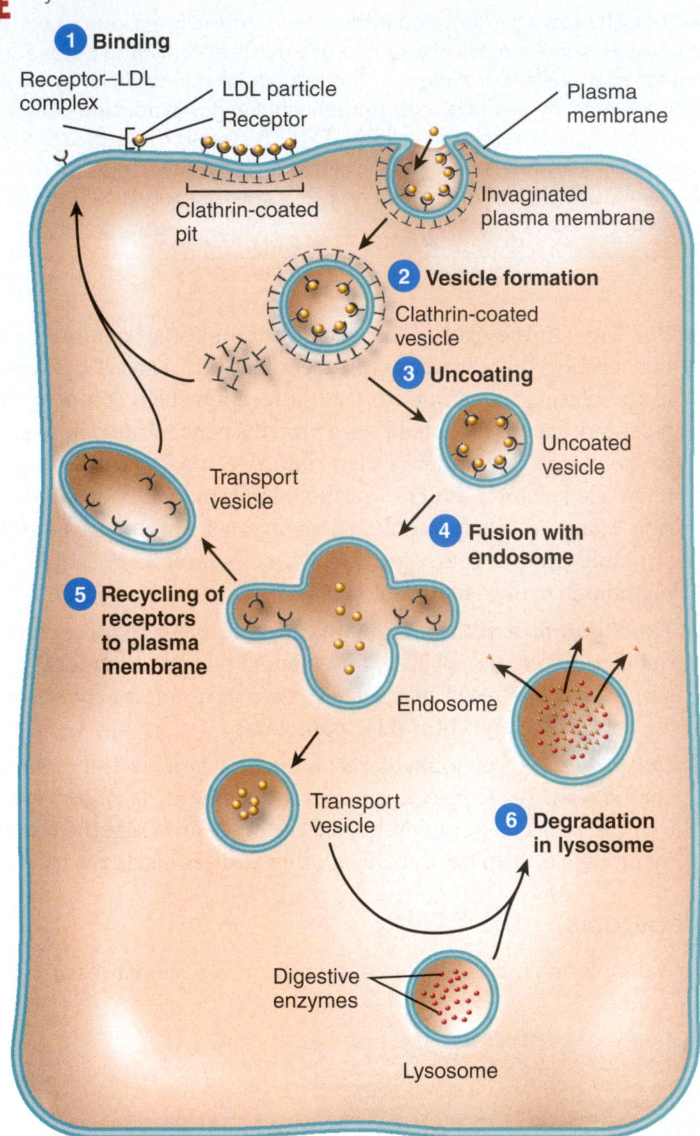

What are several other examples of ligands that can undergo receptor-mediated endocytosis?

4. ***Fusion with endosome.*** The uncoated vesicle quickly fuses with a vesicle known as an *endosome*. Within the endosome, the LDL particles separate from their receptors.
5. ***Recycling of receptors to plasma membrane.*** Most of the receptors accumulate in elongated protrusions of the endosome. These pinch off, forming transport vesicles that return the receptors to the plasma membrane. An LDL receptor is returned to the plasma membrane about 10 minutes after it enters a cell.
6. ***Degradation in lysosomes.*** Other transport vesicles, which contain the LDL particles, bud off the endosome and soon fuse with a *lysosome*. Lysosomes contain many digestive enzymes. Certain enzymes break down the large protein and lipid molecules of the LDL particle into amino acids, fatty acids, and cholesterol. These smaller molecules then leave the lysosome. The cell uses cholesterol for rebuilding its membranes and for synthesis of steroids, such as estrogens. Fatty acids and amino acids can be used for ATP production or to build other molecules needed by the cell.

Phagocytosis

Phagocytosis (*phago-* = to eat) is a form of endocytosis in which the cell engulfs large solid particles, such as worn-out cells, whole bacteria, or viruses (FIGURE 5.21). Only a few body cells, termed **phagocytes**, are able to carry out phagocytosis. Phagocytes include certain types of leukocytes—mainly *macrophages* and *neutrophils*. Phagocytosis begins when the particle binds to a plasma membrane receptor on the phagocyte, causing it to extend **pseudopods** (SOO-dō-pods; *pseudo-* = false; *-pods* = feet), projections of its plasma membrane and cytoplasm. Pseudopods surround the particle outside the cell, and the membranes fuse to form a vesicle called a *phagosome*, which enters the cytoplasm. The phagosome fuses with one or more lysosomes, and lysosomal enzymes break down the ingested material. In most cases, any undigested materials in the phagosome remain indefinitely in a vesicle called a *residual body*. The process of phagocytosis is a vital defense mechanism that helps protect the body from disease. Through phagocytosis, macrophages dispose of invading microbes and billions of aged, worn-out erythrocytes every day; neutrophils also help rid the

FIGURE 5.21 Phagocytosis. Pseudopods surround a particle, and the membranes fuse to form a phagosome.

Phagocytosis is a vital defense mechanism that helps protect the body from disease.

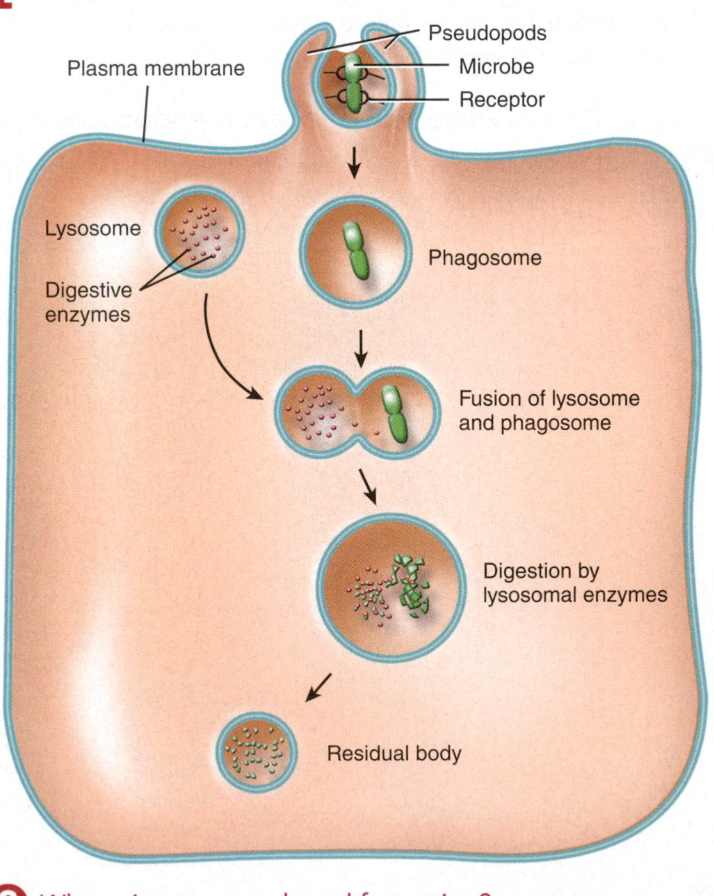

What triggers pseudopod formation?

body of invading microbes. Pus is a mixture of dead neutrophils, macrophages, and tissue cells and fluid in an infected wound.

Bulk-Phase Endocytosis

Most body cells carry out **bulk-phase endocytosis**, also called **pinocytosis** (pi-nō-sī-TŌ-sis; *pino-* = to drink), a form of endocytosis in which tiny droplets of extracellular fluid are taken up (**FIGURE 5.22**). No receptor proteins are involved; all solutes dissolved in the extracellular fluid are brought into the cell. During bulk-phase endocytosis, the plasma membrane folds inward and forms a vesicle containing a droplet of extracellular fluid. The vesicle detaches or "pinches off" from the plasma membrane and enters the cytosol. Within the cell, the vesicle fuses with a lysosome, where enzymes degrade the engulfed solutes. The resulting smaller molecules, such as amino acids and fatty acids, leave the lysosome to be used elsewhere in the cell. Bulk-phase endocytosis occurs in most cells, especially absorptive cells in the intestines and kidneys.

Exocytosis Is Used to Release Digestive Enzymes, Hormones, and Neurotransmitters from Certain Cells

In contrast with endocytosis, which brings materials into a cell, **exocytosis** releases materials from a cell. All cells carry out exocytosis, but it is especially important in two types of cells: (1) secretory cells that liberate digestive enzymes, hormones, mucus, or other secretions and (2) neurons that release substances called *neurotransmitters*. In some cases, wastes are also released by exocytosis. During exocytosis, a membrane-enclosed vesicle called a *secretory vesicle* forms inside the cell, fuses with the plasma membrane, and releases its contents into the extracellular fluid. This process involves a group of proteins known as **SNAREs** (*S*oluble *N*-ethylmaleimide-sensitive factor *A*ttachment protein *RE*ceptors) (see **FIGURE 5.23**, step 1). There are two categories

FIGURE 5.22 Bulk-phase endocytosis. The plasma membrane folds inward, forming a vesicle.

Most body cells carry out bulk-phase endocytosis, the nonselective uptake of tiny droplets of extracellular fluid.

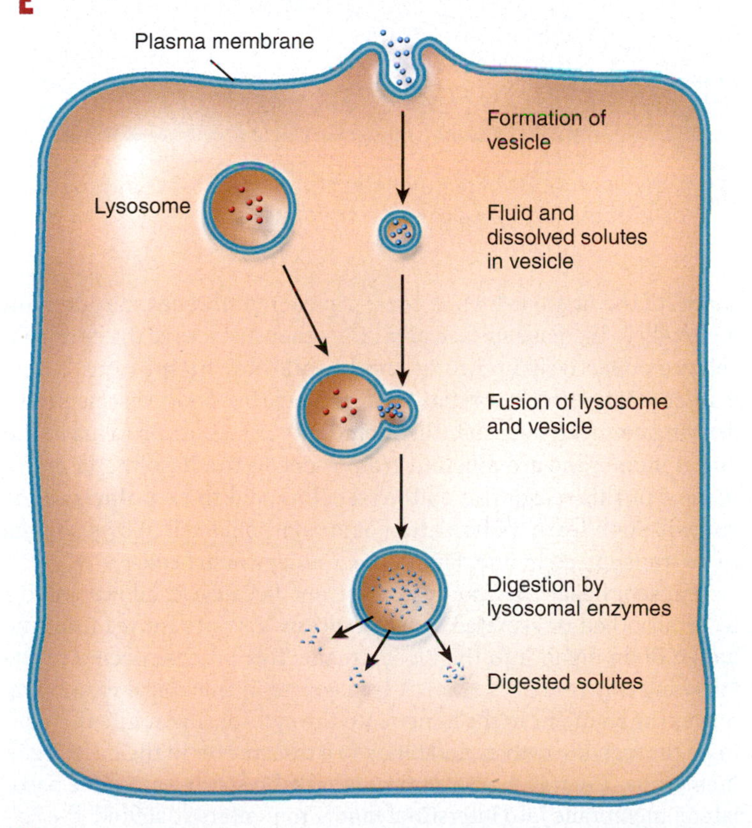

How do receptor-mediated endocytosis and phagocytosis differ from bulk-phase endocytosis?

FIGURE 5.23 Mechanism of exocytosis.

During exocytosis, proteins called SNAREs fuse the vesicle membrane to the plasma membrane, resulting in the release of vesicle contents into extracellular fluid.

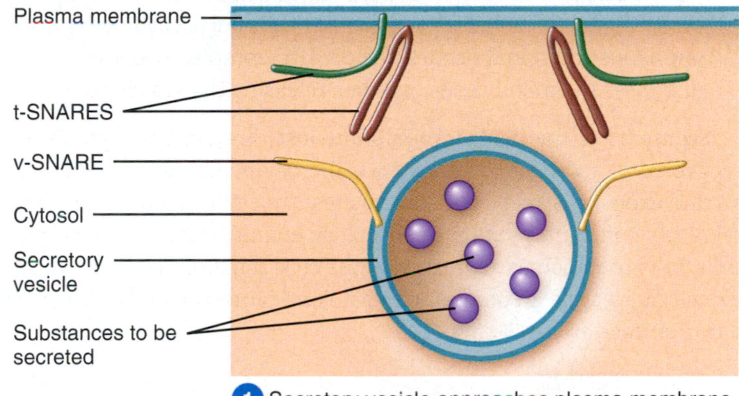

1 Secretory vesicle approaches plasma membrane.

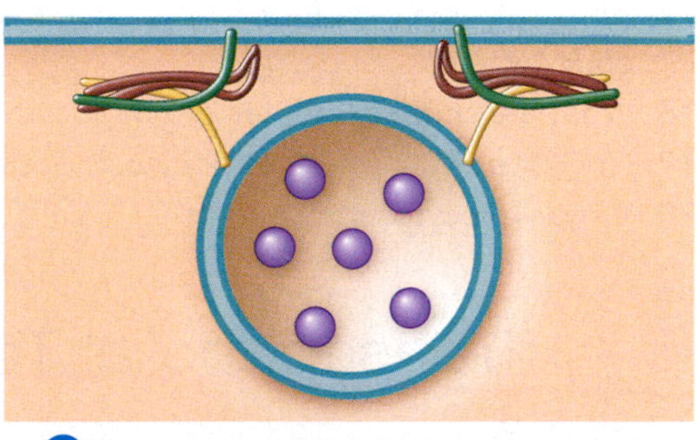

2 The v- and t-SNAREs bind, forming a tightly interwoven complex.

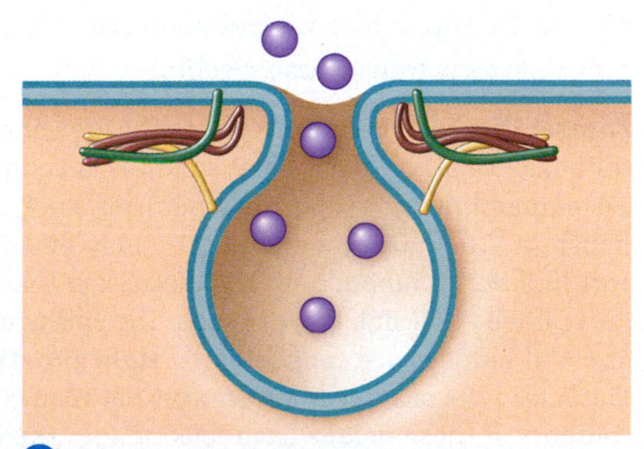

3 Membrane of secretory vesicle fuses with plasma membrane, causing release of vesicle contents.

What is the difference between a v-SNARE and a t-SNARE?

of SNAREs: a single **v-SNARE** (*v* for vesicle), which is located in the secretory vesicle membrane, and two **t-SNAREs** (*t* for target), which are located in the target membrane (the plasma membrane). SNAREs promote fusion of the secretory vesicle membrane with plasma membrane in the following way (**FIGURE 5.23**):

1. The secretory vesicle approaches the plasma membrane.
2. The single v-SNARE and two t-SNAREs bind together to form a tightly interwoven complex of four proteins (one of the t-SNAREs has two subunits).
3. In unregulated exocytosis, the complex of v- and -t-SNAREs immediately pulls the two membranes together, resulting in the release of vesicle contents into the extracellular fluid. In regulated exocytosis, the SNARE complex enters an inactive state and the vesicle remains docked at the plasma membrane until the appropriate signal (usually an increase in intracellular calcium) is present. Once the signal occurs, the SNARE complex becomes active and fuses the membranes together, releasing the vesicle contents.

Segments of the plasma membrane lost through endocytosis are recovered or recycled by exocytosis. The balance between endocytosis and exocytosis keeps the surface area of a cell's plasma membrane relatively constant. Membrane exchange is quite extensive in certain cells. In your pancreas, for example, the cells that secrete digestive enzymes can recycle an amount of plasma membrane equal to the cell's entire surface area in 90 minutes.

CHECKPOINT

16. How are endocytosis and exocytosis similar? How are they different?
17. What role does clathrin play in receptor-mediated endocytosis?
18. What are some examples of phagocytosis?
19. How does bulk-phase endocytosis occur?
20. Why is exocytosis important?

5.7 Transepithelial Transport

OBJECTIVE

- Define transepithelial transport.
- Describe the difference between absorption and secretion.
- Explain how transcytosis occurs.

Transepithelial transport refers to the movement of solutes across epithelial cells. This solute movement is accomplished through the use of membrane transport mechanisms, such as passive or active transport. The intestine and kidneys rely on transepithelial transport in order to function, so the discussion that follows focuses on the epithelia that line these organs. The epithelial cells of the intestine and kidneys are connected by **tight junctions**, cell junctions that retard the passage of substances between cells and prevent the contents of these organs from leaking into the blood or surrounding tissues (**FIGURE 5.24**). Tight junctions also serve as the boundary between the two major surfaces of the epithelial cell: the apical membrane and the basolateral membrane. The **apical membrane** is the surface of the epithelial cell that faces the lumen (interior

FIGURE 5.24 Overview of transepithelial transport.

Transepithelial transport refers to the movement of solutes across epithelial cells.

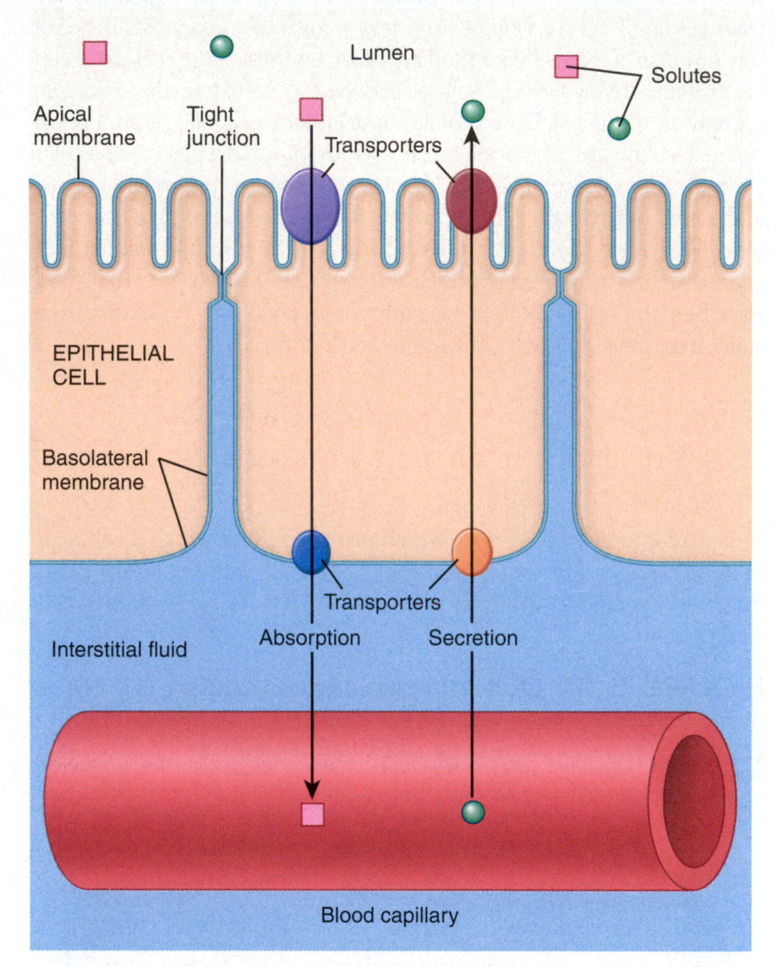

Why is an epithelial cell lining the lumen of the intestine or kidney tubule said to be polarized?

space) of the organ (**FIGURE 5.24**). Apical membranes may contain microvilli to increase surface area. The other surfaces of the epithelial cell are collectively known as the **basolateral membrane**, which is in contact with interstitial fluid and nearby blood vessels in underlying connective tissue. The membrane transport proteins in the apical membrane are different from those in the basolateral membrane. For this reason, an epithelial cell is said to be **polarized**. As you will soon learn, polarization of an epithelial cell allows solutes to be transported in one direction across epithelial cells.

Transepithelial transport is important because it allows solutes to be absorbed or secreted. In **absorption**, a solute moves from the lumen of an organ into the bloodstream. This process occurs in the following way (**FIGURE 5.24**): A transporter in the apical membrane moves the solute from the lumen into the cytosol. The solute then diffuses through the cytosol and binds to a transporter in the basolateral membrane. The transporter in turn moves the solute across the basolateral membrane into interstitial fluid. From interstitial fluid, the solute diffuses into the bloodstream. Through transepithelial transport in the intestine, the nutrients in our food are absorbed. Once absorbed, these substances circulate to cells throughout the body. The epithelial cells lining the tubules of the kidneys can also transport solutes from the lumen into the bloodstream. However, the term **reabsorption** is used instead of absorption to refer to this aspect of kidney function because a solute present in the lumen of a kidney tubule has already been absorbed into the bloodstream by the intestine and then lost from the blood by the filtering mechanism of the kidneys.

The opposite process of absorption is secretion. In this context, **secretion** refers to the movement of a solute from the bloodstream into the lumen of an organ. This process occurs as follows (FIGURE 5.24): The solute diffuses from the blood into interstitial fluid. From interstitial fluid, a transporter moves the solute across the basolateral membrane into the cytosol. The solute then diffuses through the cytosol and binds to a transporter in the apical membrane. The transporter subsequently moves the solute into the lumen. If this is the lumen of a kidney tubule, then the solute becomes part of the fluid that eventually is excreted as urine.

The absorption of glucose across an epithelial cell in the small intestine provides a great example of transepithelial transport (FIGURE 5.25):

FIGURE 5.25 Transepithelial transport of glucose.

During transepithelial transport, glucose moves across the apical membrane of an epithelial cell via a Na^+/glucose symporter and then moves across the basolateral membrane via a glucose transporter.

High [Na^+] / Low [glucose]
Lumen
Apical membrane
Na^+
Glucose
Na^+/ glucose symporter
1
Na^+
Glucose
EPITHELIAL CELL
Low [Na^+] / High [glucose]
Basolateral membrane
K^+
2
3
Na^+/ K^+ ATPase
Glucose transporter
Interstitial fluid
Na^+
K^+
Glucose
High [Na^+] / Low [glucose]
4
Blood capillary

Why is transepithelial transport of glucose important?

1. Glucose moves across the apical membrane of an intestinal epithelial cell via a Na^+/glucose symporter, which mediates secondary active transport.
2. The low intracellular Na^+ concentration needed to drive the symporter is maintained by the Na^+/K^+ ATPase (primary active transport) in the basolateral membrane.
3. Glucose diffuses through the cytosol and then moves across the basolateral membrane via a glucose transporter (GLUT2), which mediates facilitated diffusion.
4. Glucose enters interstitial fluid and then diffuses into the blood.

Transport in vesicles may also be used to move a substance successively into, across, and out of a cell. In this active process, called **transcytosis**, vesicles undergo endocytosis on one side of a cell, move across the cell, and then undergo exocytosis on the opposite side (FIGURE 5.26). As the vesicles fuse with the plasma membrane, their contents are released into the extracellular fluid. Transcytosis occurs most often across the endothelial cells that line blood vessels and is a means for materials to move between blood plasma and interstitial fluid (see FIGURE 15.8). For instance, when a woman is pregnant, some of her antibodies cross the placenta into the fetal circulation via transcytosis.

TABLE 5.1 summarizes the processes by which materials move into and out of cells.

FIGURE 5.26 Transcytosis.

In transcytosis, vesicles undergo endocytosis on one side of a cell, move across the cell, and then undergo exocytosis on the opposite side.

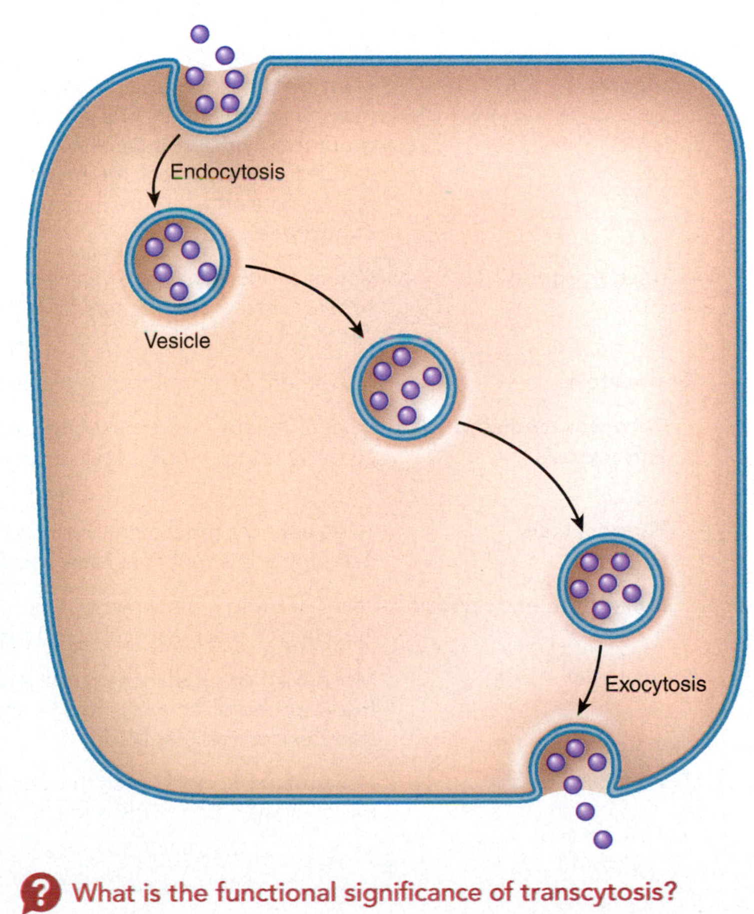

What is the functional significance of transcytosis?

TABLE 5.1 Transport of Materials into and out of Cells

Transport Process	Description	Substances Transported
Passive Processes	Movement of solutes across a membrane without any energy input from the cell.	
Passive transport	Passive process in which a solute moves down a concentration or electrochemical gradient due to its own kinetic energy until equilibrium is reached.	
Simple diffusion	Passive movement of a solute down its concentration gradient through the lipid bilayer of the plasma membrane without the help of membrane transport proteins.	Nonpolar, hydrophobic substances: oxygen, carbon dioxide, and nitrogen gases; fatty acids; steroids; and fat-soluble vitamins. Polar molecules such as water, urea, and small alcohols.
Facilitated diffusion	Passive movement of a solute down its concentration or electrochemical gradient through the lipid bilayer by transmembrane proteins that function as channels or carriers.	Polar or charged solutes: glucose; fructose; galactose; some vitamins; and ions such as K^+, Cl^-, Na^+, and Ca^{2+}.
Osmosis	Passive movement of water molecules across a selectively permeable membrane from an area of higher water concentration to an area of lower water concentration.	Solvent: water in living systems.
Active Processes	Movement of solutes across a membrane using cellular energy in the form of ATP.	
Active transport	Active process in which a cell expends energy to move a solute across the membrane against its concentration or electrochemical gradient by transmembrane proteins that function as carriers.	Polar or charged solutes.
Primary active transport	Active process in which a solute moves across the membrane against its concentration or electrochemical gradient by pumps (carriers) that use energy supplied by hydrolysis of ATP.	Na^+, K^+, Ca^{2+}, H^+, I^-, Cl^-, and other ions.
Secondary active transport	Active process in which the energy supplied by an ionic electrochemical gradient (usually from Na^+) is used to move a solute against its concentration or electrochemical gradient. Symporters move Na^+ and another solute in the same direction across the membrane; antiporters move Na^+ and another solute in opposite directions across the membrane.	Antiport: Ca^{2+}, H^+ out of cells. Symport: glucose, amino acids into cells.
Vesicular Transport	Active process in which vesicles are used to move substances across the plasma membrane; requires energy supplied by ATP.	
Endocytosis	Movement of substances into a cell in vesicles.	
Receptor-mediated endocytosis	Ligand–receptor complexes trigger infolding of a clathrin-coated pit that forms a vesicle containing ligands.	Ligands: transferrin, low-density lipoproteins (LDLs), some vitamins, certain hormones, and antibodies.
Phagocytosis	"Cell eating"; movement of a solid particle into a cell after pseudopods engulf it to form a phagosome.	Bacteria, viruses, and aged or dead cells.
Bulk-phase endocytosis	"Cell drinking"; movement of extracellular fluid into a cell by infolding of plasma membrane to form a vesicle.	Solutes in extracellular fluid.
Exocytosis	Movement of substances out of a cell in secretory vesicles that fuse with the plasma membrane and release their contents into the extracellular fluid.	Neurotransmitters, hormones, and digestive enzymes.
Transcytosis	Movement of a substance through a cell as a result of endocytosis on one side and exocytosis on the opposite side.	Substances, such as antibodies, across endothelial cells. This is a common route for substances to pass between blood plasma and interstitial fluid.

CHECKPOINT

21. What is the role of tight junctions in transepithelial transport?
22. What are the different transporters and epithelial surfaces involved in the absorption of glucose from the intestinal epithelium?
23. Where in the body does transcytosis occur most often?

FROM RESEARCH TO REALITY

Tea Catechins and Glucose Transport

Reference
Shimizu, M. et al. (2000). Regulation of intestinal glucose transport by tea catechins. *BioFactors*. 13: 61–65.

Can substances in tea inhibit glucose transport in the small intestine?

Kasiam/Getty Images, Inc.

Dietary carbohydrates are an important source of energy for the body. However, overconsumption of carbohydrates can lead to obesity, diabetes mellitus, and other diseases. What if there were some type of chemical in our food or beverages that could help reduce the amount of glucose normally absorbed by the small intestine? This chemical could potentially help regulate the level of glucose in the bloodstream and therefore contribute to blood glucose homeostasis.

Article description:
Researchers in this study investigated the ability of chemicals in tea to regulate glucose transport in the small intestine. These chemicals included catechins, theaflavins, and thearubigin. Catechins are present in green tea; theaflavins and thearubigin are present in black tea.

Go to WileyPLUS Learning Space and use the data from this article to answer the questions posed there and to discover more about tea catechins and glucose transport.

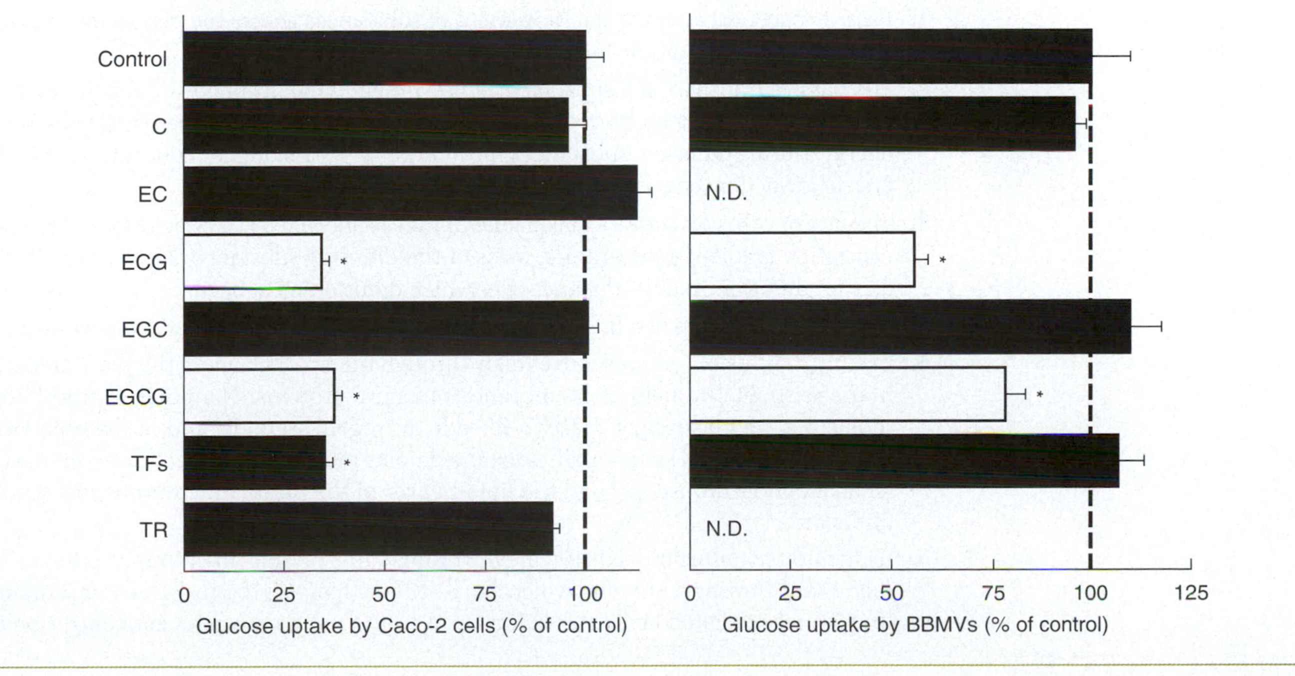

CHAPTER REVIEW

5.1 Selective Permeability of the Plasma Membrane

1. The selective permeability of the plasma membrane permits some substances to pass more readily than others.
2. The lipid bilayer is permeable to nonpolar molecules and to small, uncharged polar molecules. It is impermeable to ions and large, uncharged polar molecules.
3. Channels and carriers increase the plasma membrane's permeability to a variety of ions and uncharged polar molecules that cannot cross the lipid bilayer on their own.

5.2 Gradients Across the Plasma Membrane

1. The selective permeability of the plasma membrane supports the existence of concentration gradients, differences in the concentrations of chemicals between one side of the membrane and the other.
2. An electrical gradient (a difference in electrical charges) also exists across the plasma membrane.
3. An electrochemical gradient is the combined influence of the concentration gradient and the electrical gradient on movement of a particular ion across the membrane.

5.3 Classification of Membrane Transport Processes as Passive or Active

1. In passive processes, a substance moves across the plasma membrane without any energy input from the cell. An example is passive transport.
2. In active processes, cellular energy (usually in the form of ATP) is used to move substances across the plasma membrane. Examples include active transport and vesicular transport (endocytosis and exocytosis).

5.4 Passive Transport

1. Passive transport refers to the movement of substances across the plasma membrane down their concentration or electrochemical gradients.
2. The passive transport of a substance across a membrane occurs by diffusion, the random mixing of particles from one location to another because of the particles' kinetic energy. During diffusion, substances move from an area of higher concentration to an area of lower concentration until equilibrium is reached.
3. The rate of diffusion across a plasma membrane is affected by the steepness of the concentration gradient, temperature, mass of the diffusing substance, surface area available for diffusion, and the distance over which diffusion must occur.
4. There are three types of diffusion: simple diffusion, facilitated diffusion, and osmosis.
5. In simple diffusion, solutes move freely through the lipid bilayer of the plasma membrane without the help of membrane transport proteins. Nonpolar, hydrophobic molecules such as oxygen, carbon dioxide, nitrogen, steroids, and fat-soluble vitamins (A, E, D, and K) plus small, uncharged polar molecules such as water, urea, and small alcohols diffuse through the lipid bilayer of the plasma membrane via simple diffusion.
6. In facilitated diffusion, solutes move through the plasma membrane assisted by membrane transport proteins, which can be either channel proteins or carrier proteins. In channel-mediated facilitated diffusion, a solute moves down its concentration or

electrochemical gradient across the lipid bilayer through a membrane channel. Examples include ion channels that allow specific ions such as K^+, Cl^-, Na^+, or Ca^{2+} to move across the plasma membrane. In carrier-mediated facilitated diffusion, a solute such as glucose binds to a specific carrier protein on one side of the membrane and is released on the other side after the carrier undergoes a conformational change. Carrier-mediated facilitated diffusion exhibits specificity, affinity, saturation, and competition.

7. Osmosis is a type of diffusion in which there is net movement of water through a selectively permeable membrane from an area of higher water concentration to an area of lower water concentration.
8. The osmotic movement of water to an area of high solute concentration depends on the osmolarity of the solution. Osmolarity refers to the total number of dissolved solute particles per liter of solution.
9. The osmolarities of different solutions can be compared to one another. A solution is isoosmotic if it has the same number of solute particles as another solution, hyperosmotic if it has a greater number of solute particles than another solution, or hypoosmotic if it has fewer solute particles than another solution.
10. When two solutions are separated by a membrane, some of the solute particles in either solution may be penetrating or nonpenetrating. Penetrating solutes are able to pass through the membrane; nonpenetrating solutes cannot. Nonpenetrating solutes determine a solution's tonicity. Tonicity is a measure of a solution's ability to change the volume of cells by altering their water content. In an isotonic solution, cells maintain their normal shape; in a hypotonic solution, they undergo lysis; in a hypertonic solution, they undergo crenation.

5.5 Active Transport

1. Active transport refers to the movement of substances across the plasma membrane against their concentration or electrochemical gradients. It is mediated by transport proteins that function as carriers. Actively transported substances include ions such as Na^+, K^+, H^+, Ca^{2+}, I^-, and Cl^-; amino acids; and monosaccharides.
2. Two sources of energy are used to drive active transport: Energy obtained from hydrolysis of ATP is the source in primary active transport, and energy stored in an ionic electrochemical gradient (usually from Na^+) is the source in secondary active transport.
3. The most prevalent primary active transport pump is the sodium–potassium pump, also known as Na^+/K^+ ATPase, which actively transports three Na^+ ions out of the cell and two K^+ ions into the cell using energy derived from ATP hydrolysis. By maintaining the different concentrations of Na^+ and K^+ ions in extracellular fluid and the cytosol, the sodium–potassium pump plays a crucial role in maintaining normal cell volume and providing the ability of some cells to generate electrical signals such as action potentials.
4. In secondary active transport, a carrier protein (transporter) binds to a Na^+ ion and another solute and uses the energy stored in the Na^+ electrochemical gradient to move that other solute against its own concentration or electrochemical gradient. If the transporter moves the two solutes in the same direction, it is called a symporter; if it moves the two solutes in opposite directions, it is referred to as an antiporter. Examples of carrier proteins that mediate secondary active transport include the Na^+/glucose symporter, Na^+/amino acid symporter, Na^+/Ca^{2+} antiporter, and Na^+/H^+ antiporter.

5.6 Vesicular Transport

1. Vesicular transport is the process by which vesicles (membranous sacs) are used to transport substances across the plasma membrane.
2. There are two types of vesicular transport: endocytosis and exocytosis. In endocytosis, tiny vesicles detach from the plasma membrane to move materials across the membrane into a cell. Three forms of endocytosis exist: receptor-mediated endocytosis, phagocytosis, and bulk-phase endocytosis. Receptor-mediated endocytosis is the selective uptake of large molecules and particles (ligands) that bind to specific receptors in membrane areas called clathrin-coated pits. Phagocytosis is the ingestion of solid particles. Some leukocytes destroy microbes that enter the body in this way. In bulk-phase endocytosis (pinocytosis), the ingestion of extracellular fluid, a vesicle surrounds the fluid to take it into the cell.
3. In exocytosis, vesicles merge with the plasma membrane to move materials out of a cell. This process involves a group of proteins called SNAREs. As a vesicle approaches the plasma membrane, a v-SNARE in the vesicle membrane binds to t-SNAREs in the plasma membrane to form a tightly interwoven complex. The complex in turn fuses the two membranes together, resulting in the release of the vesicular contents into extracellular fluid.

5.7 Transepithelial Transport

1. Transepithelial transport refers to the movement of solutes across epithelial cells. The intestine and kidneys rely on transepithelial transport in order to function.
2. The epithelial cells of the intestine and kidneys are connected by tight junctions. The tight junctions also serve as the boundary between the apical and basolateral membranes of these cells. Different membrane transport proteins are found in the apical and basolateral membranes of the epithelial cells, so the epithelial cells are polarized. Polarization of the epithelium allows solutes to be transported in only one direction.
3. Transepithelial transport is important because it allows solutes to be absorbed or secreted. Absorption refers to movement of solutes from the lumen of an organ into the bloodstream. Secretion refers to the movement of solutes from the bloodstream into the lumen of an organ.
4. Transport in vesicles can also be used to move a substance successively into, across, and out of a cell, a process known as transcytosis.

PONDER THIS

1. Some recreational drugs can lead to swelling of the brain and, in some severe cases, death as a result of the swelling. Use your knowledge of tonicity and osmolarity to discuss what this drug might be doing to the body that leads to this result.

2. Jennifer is rushed to the ER after running a marathon in 98 degree weather without replenishing with water during the race. Upon arrival, the paramedics tell the ER doctor that Jennifer had been sweating profusely until her skin became very dry and warm, at which point she lost consciousness. The paramedics had determined in the field that Jennifer had sweat out a very hypotonic solution. When determining an IV regimen, what should the doctor choose and why?

3. Ouabain is a poison found in a specific type of African plant. To learn more about the effects of this poison, you decide to expose it to epithelial cells of the digestive tract in vitro. One of the observations that you make is that after an hour, the cells stop transporting glucose across the apical surface. Upon further analysis, you realize that the sodium glucose transporter on the apical surface is fine and are baffled by the fact that the glucose is no longer being transported. How might ouabain be acting in these cells? Explain your answer.

ANSWERS TO FIGURE QUESTIONS

5.1 Small, uncharged polar molecules, such as water and urea, are able to pass through the hydrophobic interior of the plasma membrane by moving between the gaps that form as the fatty acid tails of the membrane lipids move about randomly.

5.2 The term *membrane potential* refers to the electrical gradient that exists across the plasma membrane.

5.3 The K^+ electrochemical gradient occurs in the same direction as the K^+ concentration gradient and not in the direction of the K^+ electrical gradient because the magnitude of the K^+ concentration gradient is greater than that of the K^+ electrical gradient.

5.4 Because fever involves an increase in body temperature, and an increase in temperature causes particles to move more quickly, the rates of all diffusion processes would increase.

5.5 Nonpolar, hydrophobic molecules (oxygen, carbon dioxide, and nitrogen gases; fatty acids; steroids; and fat-soluble vitamins) and small, uncharged polar molecules (water, urea, and small alcohols) move across the lipid bilayer of the plasma membrane through the process of simple diffusion.

5.6 If the magnitude of the concentration gradient decreases, the rate of simple diffusion also decreases.

5.7 The selectivity of an ion channel is due to the size of the pore within the channel and the types of amino acids that line the pore. An ion can pass through a given channel only if it is small enough to fit through the pore and can form attractive electrical interactions with the amino acids lining the pore.

5.8 A conformational change is a change in shape.

5.9 The binding site of carrier protein C has the correct three-dimensional shape to fit the solute, whereas the binding sites of carrier proteins A and B do not.

5.10 Before the transport maximum (T_m) is reached, the rate of transport is directly proportional to the steepness of the concentration gradient. Once the T_m has been reached, the transport rate no longer increases with an increase in the concentration gradient because the carrier proteins are saturated.

5.11 Increasing the concentration of solute B would decrease the transport rate of solute A because more solute B molecules would be available to bind to the carrier proteins.

5.12 Insulin promotes the insertion of glucose transporters (GLUT4) in the plasma membranes of fat and muscle cells, which increases cellular uptake of glucose by carrier-mediated facilitated diffusion.

5.13 The water concentrations can never be the same in the two arms because the left arm contains pure water and the right arm contains a solution that is less than 100% water.

5.14 In a hypoosmotic solution, the water concentration is higher than the concentration of solute particles.

5.15 A 2% solution of NaCl will cause crenation of erythrocytes because it is hypertonic.

5.16 ATP adds a phosphate group to the pump protein, which changes the pump's conformation. ATP transfers energy to power the pump.

5.17 In secondary active transport, hydrolysis of ATP is used indirectly to drive the activity of symporter or antiporter proteins; this reaction directly powers the pump protein in primary active transport.

5.18 Na^+/H^+ antiporters help regulate the cytosol's pH (H^+ concentration) by expelling excess H^+.

5.19 Endocytosis and exocytosis are active processes because they require energy (ATP).

5.20 Transferrin, vitamins, and hormones are other examples of ligands that can undergo receptor-mediated endocytosis.

5.21 The binding of particles to a plasma membrane receptor triggers pseudopod formation.

5.22 Receptor-mediated endocytosis and phagocytosis involve receptor proteins; bulk-phase endocytosis does not.

5.23 A v-SNARE is found in the membrane of the secretory vesicle, whereas a t-SNARE is found in the target membrane (in this case, the plasma membrane).

5.24 An epithelial cell lining the lumen of the intestine or kidney tubule is said to be polarized because the membrane transport proteins present in the apical membrane of this cell are different than the membrane transport proteins found in the cell's basolateral membrane.

5.25 Transepithelial transport of glucose is the way that glucose is absorbed in the small intestine and reabsorbed in the kidneys.

5.26 Transcytosis is an important method of moving materials across blood capillary walls.

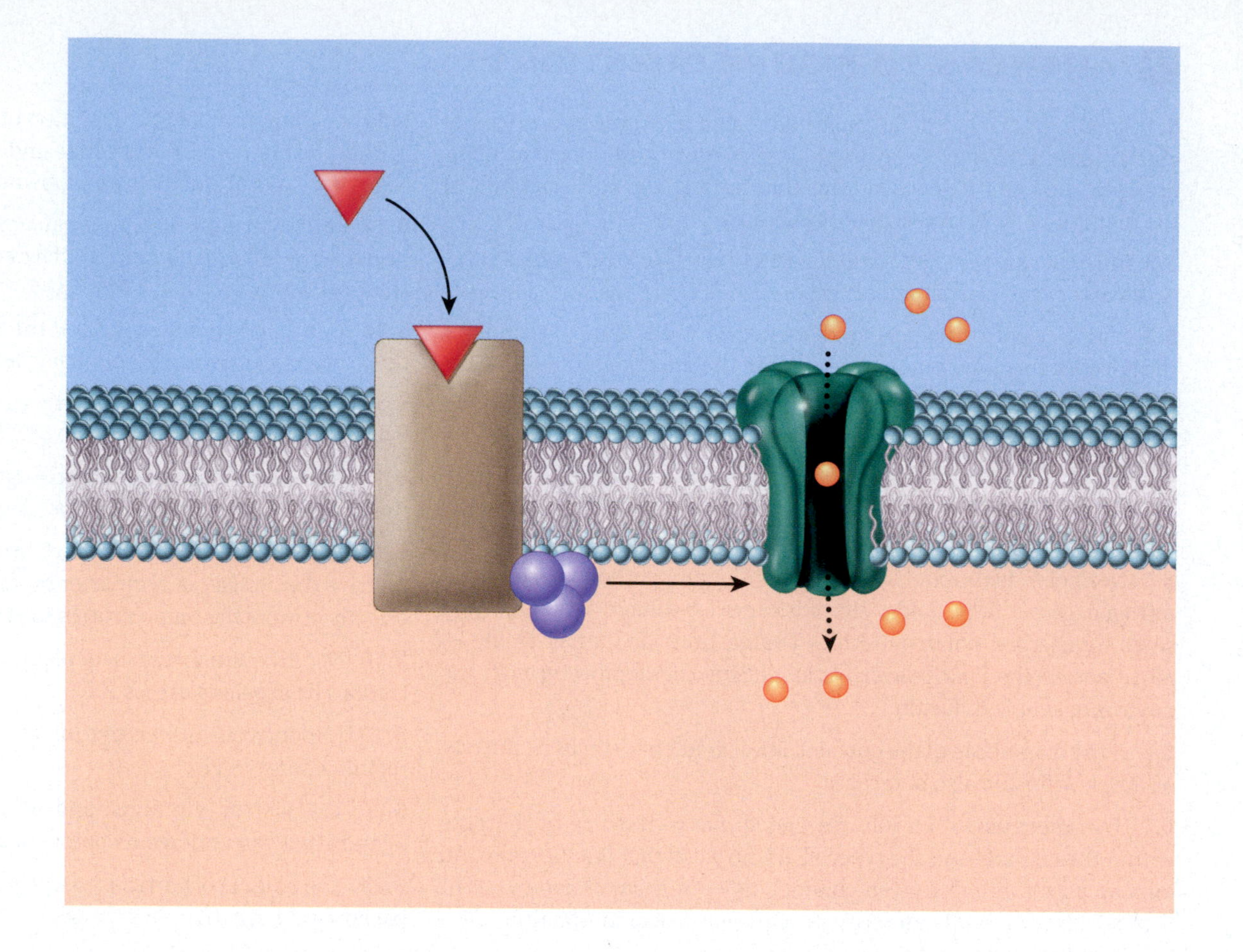

Cell Signaling

Cell Signaling and Homeostasis

Cell signaling contributes to homeostasis by allowing cells to communicate with one another in order to coordinate body activities.

LOOKING BACK TO MOVE AHEAD...

- Extracellular fluid, the fluid outside cells, consists of two components: (1) interstitial fluid, the fluid that fills the narrow spaces between cells and (2) plasma, the fluid portion of blood (Section 1.4).
- Receptors are proteins that serve as cellular recognition sites; each type of receptor recognizes and is bound by a specific ligand (Section 3.2).
- A kinase is an enzyme that adds a phosphate group to a substrate (Section 4.3).
- The plasma membrane is highly permeable to nonpolar molecules, such as oxygen (O_2) and steroids; moderately permeable to small, uncharged polar molecules, such as water; and impermeable to ions and large, uncharged polar molecules, such as glucose (Section 5.1).

The cells that comprise the body must communicate with each other to coordinate body activities. For example, if you step on a tack, neurons from your foot communicate this painful sensory input to neurons of the spinal cord, which in turn communicate output information to muscle cells of the lower limb that contract to move your foot away from the tack. Consider another example: If the oxygen level in your tissues decreases below normal, kidney cells communicate with cells in red bone marrow to produce more erythrocytes, which results in an increase in oxygen delivery to your tissues. As you can see from these two examples, communication between cells is vital to achieving and maintaining homeostasis. The communication that exists between the cells of the body is referred to as **cell signaling**, which is the subject of this chapter.

6.1 Methods of Cell-to-Cell Communication

OBJECTIVE

- Describe the different ways that cells can communicate with one another.

There are three general methods of communication between cells: (1) gap junctions, (2) cell-to-cell binding, and (3) extracellular chemical messengers.

Gap Junctions Electrically Couple Cells Together

One way that a cell can communicate with another cell is through gap junctions. At **gap junctions**, membrane proteins called **connexins** form tunnels called **connexons** that connect neighboring cells (FIGURE 6.1a). Through the connexons, ions and small molecules can diffuse from the cytosol of one cell to the cytosol of another cell. Gap junctions are found in a variety of tissues, including nervous tissue between some neurons, cardiac muscle, and some types of smooth muscle. When gap junctions are present between neurons, the movement of ions through the connexons electrically couples the cells, which facilitates and synchronizes neural communication. Likewise, gap junctions present in muscle tissue electrically couple the muscle cells together. This allows the muscle cells to be excited simultaneously and to contract together as a single unit. You will learn more about the role of gap junctions in neural communication when you read Chapter 7 (see Section 7.4); the role of gap junctions in muscle excitation is described in detail in Chapter 11 (see Sections 11.7 and 11.8).

Cell-to-Cell Binding Is Important for Development and Defense

Another method of communication between cells is through cell-to-cell binding (FIGURE 6.1b). Cells typically have a variety of surface molecules that extend from their plasma membranes. In cell-to-cell binding, a surface molecule on one cell binds to a surface molecule on another cell. This interaction is important during development and for leukocytes to recognize and respond to potentially dangerous foreign cells. Specific examples of cell-to-cell binding are presented in the discussion of immunity in Chapter 17.

FIGURE 6.1 Methods of cell-to-cell communication.

Cells can communicate with one another through gap junctions, cell-to-cell binding, or extracellular chemical messengers.

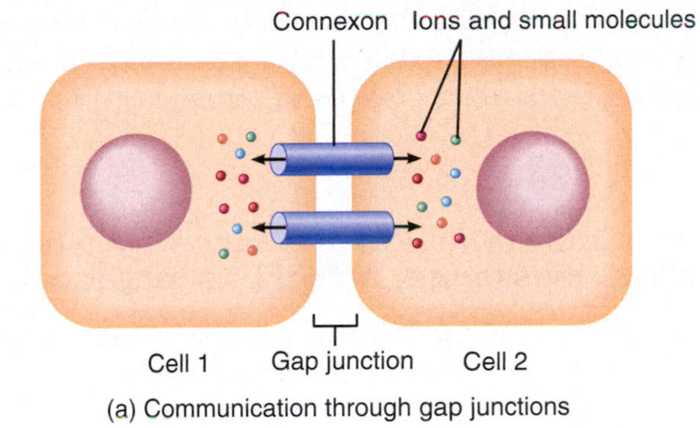

(a) Communication through gap junctions

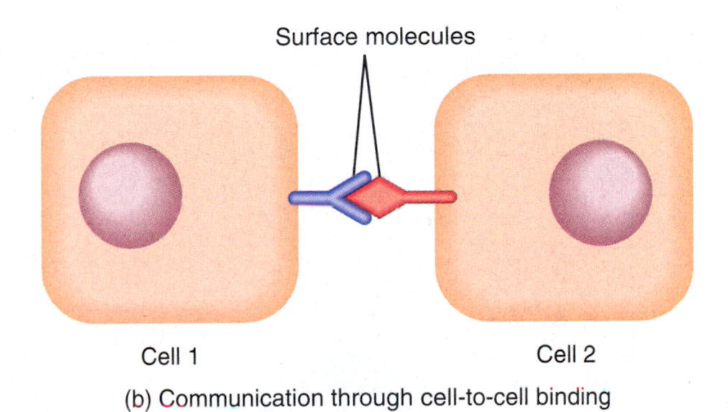

(b) Communication through cell-to-cell binding

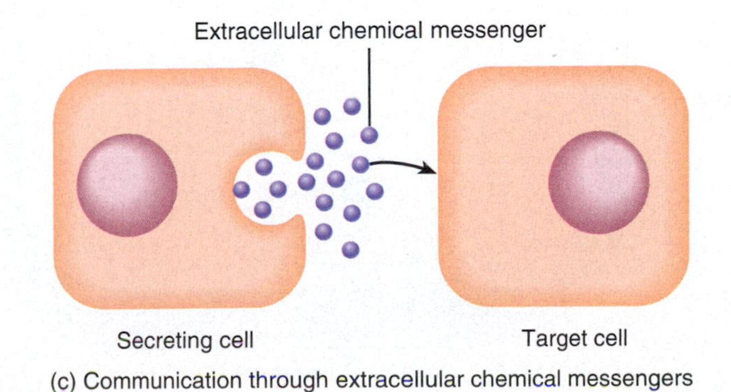

(c) Communication through extracellular chemical messengers

What is the most common method of cell-to-cell communication?

Communication Through Extracellular Chemical Messengers Permits a Wide Variety of Responses

Although cells can communicate with one another through gap junctions or cell-to-cell binding, in most cases, communication between cells occurs through the use of an *extracellular chemical messenger* (**FIGURE 6.1c**). This process begins when a cell secretes a chemical messenger into extracellular fluid (ECF). The chemical messenger then diffuses through the ECF and may randomly come in contact with many different types of cells. However, the extracellular messenger has an effect only on specific target cells. Hence, a **target cell** is a cell that can respond to the extracellular messenger.

An extracellular chemical messenger causes an effect on its target cell via the following steps (**FIGURE 6.2**):

1. ***Binding of the extracellular chemical messenger to a receptor.*** The extracellular messenger binds to a specific protein called a *receptor*. A given extracellular messenger therefore functions as a *ligand*, a molecule that binds to a particular site on a protein. Only the target cell has the correct receptor for the extracellular messenger (ligand). The receptor may be present in the target cell's plasma membrane or inside the target cell.
2. ***Signal transduction.*** Binding of the extracellular messenger to the receptor causes *signal transduction*, the process by which the signal molecule (the extracellular messenger) is transduced (converted) into a response by the target cell. The sequence of events that occurs during signal transduction is referred to as a *signal transduction pathway*. Signal transduction pathways often involve several molecules (mainly proteins) in series, with each molecule causing a change in the next molecule in line.
3. ***Cellular response.*** The signal transduction pathway culminates in a *cellular response* by the target cell. The cellular response varies, depending on the target cell and the signal transduction pathway activated. Examples of cellular responses include cell growth, protein synthesis, secretion, muscle contraction, and transport of substances across the plasma membrane.

Because cell signaling through extracellular chemical messengers is the most common method of communication between cells, it will be the focus of the rest of this chapter.

FIGURE 6.2 Overview of the steps by which an extracellular messenger affects its target cell.

An extracellular messenger binds to a specific protein receptor in order to initiate a response by its target cell.

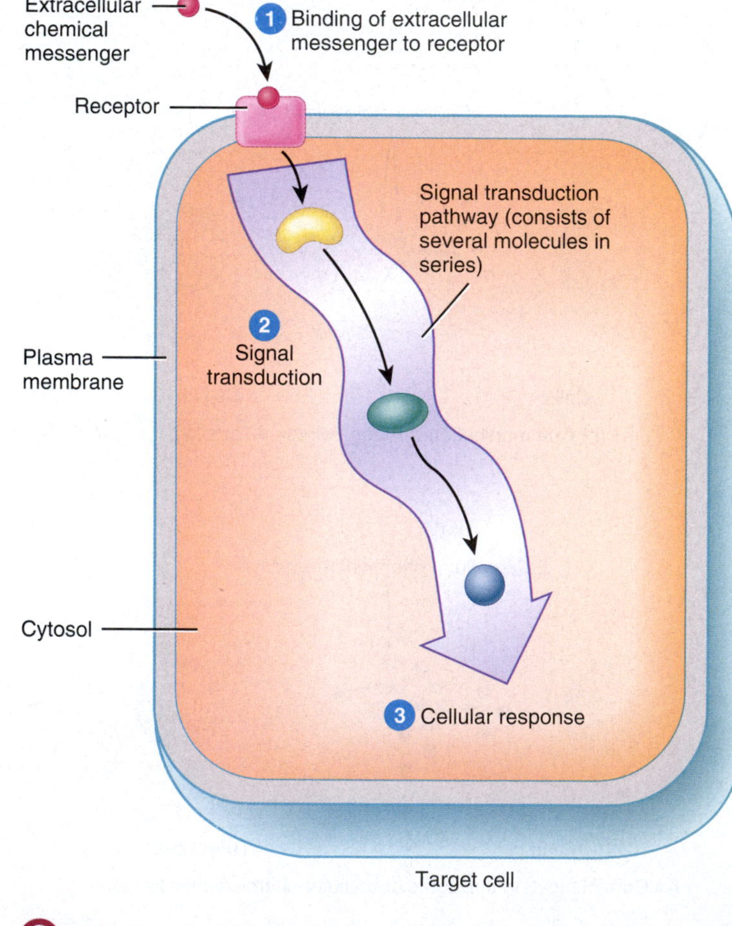

What is an example of a cellular response that can occur after an extracellular messenger binds to its receptor?

CHECKPOINT

1. What is the functional significance of gap junctions?
2. Why is cell-to-cell binding important?
3. What steps are involved when a cell communicates with another cell through an extracellular chemical messenger?

6.2 Extracellular Chemical Messengers

OBJECTIVES

- Compare the three main types of extracellular chemical messengers.
- Explain the difference between water-soluble extracellular messengers and lipid-soluble extracellular messengers.
- Describe how extracellular messengers reach their target cells.

An **extracellular chemical messenger** is a molecule that is released by a cell, enters extracellular fluid (interstitial fluid and, in many cases, blood), and then binds to a receptor on or in its target cell to cause a response. In this section, you will learn about the different types of extracellular messengers, the chemical classes of extracellular messengers, and the manner by which extracellular messengers reach their target cells.

Different Types of Extracellular Chemical Messengers Exist

There are three major types of extracellular chemical messengers: hormones, neurotransmitters, and local mediators.

Hormones

Hormones are extracellular chemical messengers that are carried by the blood to distant target cells (**FIGURE 6.3a**). Initially, the hormone

FIGURE 6.3 Types of extracellular chemical messengers.

There are three major types of extracellular chemical messengers: (1) hormones, which mediate endocrine signaling; (2) neurotransmitters, which mediate synaptic signaling; and (3) local mediators (paracrines and autocrines), which mediate local signaling.

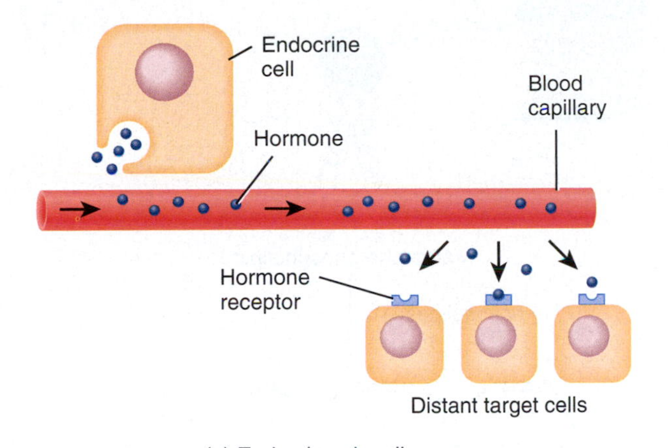

(a) Endocrine signaling

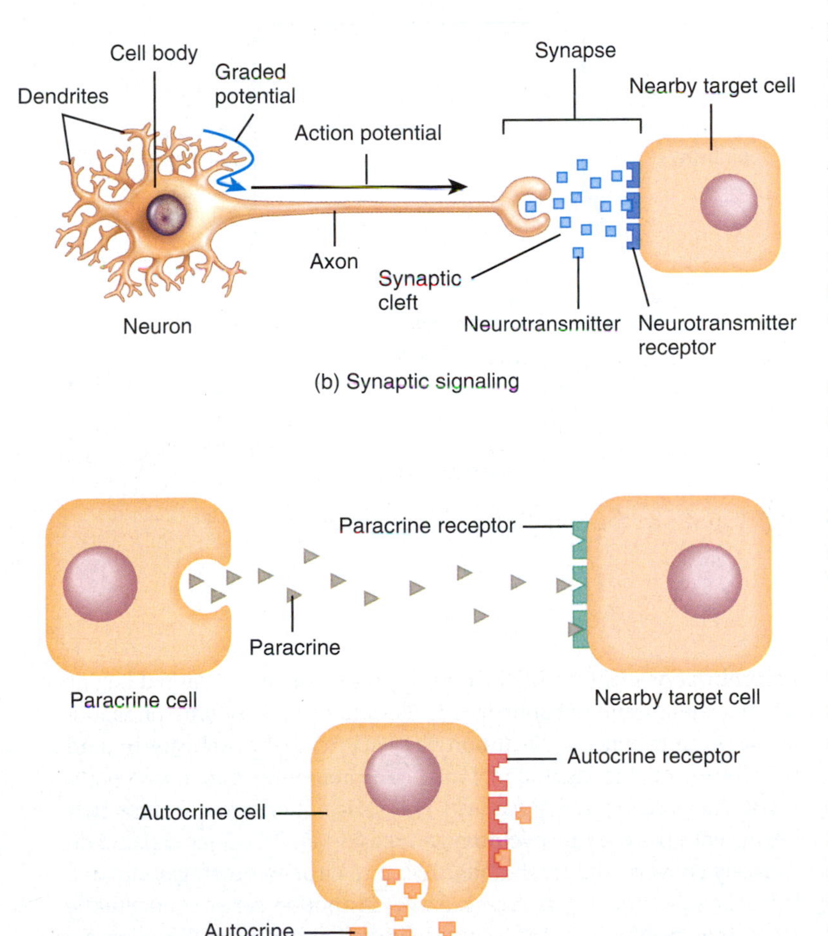

(b) Synaptic signaling

(c) Local signaling

What types of cells can be the target cell for a neuron at a synapse?

is secreted by a cell into interstitial fluid. From there, the hormone diffuses into the bloodstream. Although blood carries the hormone to virtually all cells of the body, only target cells have the correct receptors to respond to the hormone.

An example of a hormone is *glucagon*, which is secreted by certain cells of the pancreas into interstitial fluid. Glucagon then diffuses from interstitial fluid into the bloodstream, which distributes the hormone to its target cells. One of glucagon's target cells is the hepatocyte (liver cell). In response to glucagon, a hepatocyte breaks down stored glycogen into glucose, which in turn is released into the bloodstream. The glucose can then be taken up by cells that need it for ATP production.

An epithelial cell that secretes a hormone is known as an *endocrine cell*. Endocrine cells are often clustered together to form *endocrine glands* such as the pituitary, thyroid, and adrenal glands. Endocrine cells are also found in organs and tissues that are not exclusively classified as endocrine glands; examples include the kidneys, stomach, and pancreas. The endocrine glands, organs, and tissues of the body collectively form the *endocrine system*, which is the focus of Chapter 13. Cell signaling that is mediated through hormones is referred to as **endocrine signaling**.

Neurotransmitters

Neurotransmitters are extracellular chemical messengers that are released from a neuron into a synapse in order to reach a nearby target cell (FIGURE 6.3b). A *synapse* is the junction between a neuron and its target cell, which can be another neuron, a muscle cell, or a gland cell. At a synapse, the neuron that releases the neurotransmitter is called the *presynaptic neuron*, and the cell that receives the neurotransmitter is called the *postsynaptic cell*. Although the plasma membranes of a presynaptic neuron and a postsynaptic cell are close, they do not actually touch; instead, they are separated by a narrow *synaptic cleft* filled with interstitial fluid.

The nervous system functions using electrical and chemical signals. For example, when a neuron is excited, an electrical signal known as a *graded potential* is produced in the dendrites or cell body of the neuron (FIGURE 6.3b). If the graded potential is strong enough, it spreads to the beginning of the neuron's axon, where it triggers the formation of another type of electrical signal called an *action potential*. After it is generated, the action potential conducts along the axon in the direction of the synapse, ultimately causing the presynaptic neuron to release neurotransmitters (a chemical signal) into the interstitial fluid of the synaptic cleft (FIGURE 6.3b). The neurotransmitters then diffuse across the cleft through the fluid and bind to receptors in the plasma membrane of the postsynaptic cell (the target cell) to cause a response.

An example of a neurotransmitter is *dopamine*, which is released at certain synapses in the brain. Depending on the synapse, dopamine can cause either excitation or inhibition of the postsynaptic cell. Dopamine has several functions in the brain: It is involved is generating emotional responses; it plays a role in the formation of addictive behaviors and pleasurable experiences; and it helps regulate skeletal muscle tone and some aspects of movement. You will learn more about dopamine and other neurotransmitters in Chapter 7 (see Section 7.4).

Cell signaling that is mediated through neurotransmitters is called **synaptic signaling**. Because the presynaptic neuron and postsynaptic cell are close to each other, synaptic signaling is a form of short-distance communication.

It is important to note that graded potentials and action potentials are also forms of cell-to-cell communication because a neuron would not be able to release neurotransmitter to communicate with a target cell if these electrical signals did not occur. A graded potential is a type of short-distance communication because it spreads only along small areas of membrane in the dendrites or cell body of the neuron. An action potential is a form of long-distance communication because it spreads along the membrane of the entire length of the axon, which can be very long (a meter or more) in many neurons. Graded potentials and action potentials are examined further in Chapter 7 (see Section 7.3).

Local Mediators

Local mediators, also known as *local regulators* or *local agents*, are extracellular chemical messengers that act on nearby target cells without entering the bloodstream (**FIGURE 6.3c**). They are released from a cell into interstitial fluid and then diffuse through the fluid to act locally on neighboring cells or on the same cell that secreted them. Local mediators that act on neighboring cells are called **paracrines** (*para-* = beside or near), and those that act on the same cell that secreted them are called **autocrines** (*auto-* = self). Cell signaling that occurs through local mediators is called **local signaling**. Because of the close proximity of the local mediator and the target cell, local signaling is a type of short-distance communication. Synaptic signaling is actually a form of local signaling, with neurotransmitters functioning as paracrines. However, synaptic signaling is considered to be in its own category because it is a form of communication that specifically occurs between a neuron and a postsynaptic cell. Examples of local mediators include cytokines, nitric oxide, eicosanoids, and growth factors.

Cytokines An important group of local mediators in the body is the **cytokines**, molecules that regulate many cell functions, including cell growth and differentiation. An example of a cytokine is **interleukin-2 (IL-2)**, which is released by helper T cells (a type of white blood cell) during immune responses (see Chapter 17). IL-2 helps activate other nearby immune cells, a paracrine effect. It also acts as an autocrine by stimulating the same cell that released it to proliferate. This action generates more helper T cells that can secrete even more IL-2 and thus strengthen the immune response.

Nitric Oxide Another example of a local mediator is the gas **nitric oxide (NO)**, which is released by endothelial cells lining blood vessels. NO causes relaxation of nearby smooth muscle fibers in blood vessels, which in turn causes vasodilation (increase in blood vessel diameter). The effects of such vasodilation range from a lowering of blood pressure to erection of the penis in males.

Eicosanoids The **eicosanoids** (ī-KŌ-sa-noydz) are molecules released by many cells of the body in response to chemical or mechanical stimuli. They act as local mediators (paracrines or autocrines) on nearby target cells. The three major types of eicosanoids are the **prostaglandins** (pros′-ta-GLAN-dins), **leukotrienes** (loo-kō-TRĪ-ēnz), and **thromboxanes** (throm-BOK-sānz). Synthesis of an eicosanoid begins when a membrane phospholipid is cleaved by the enzyme phospholipase A_2 to produce a 20-carbon fatty acid called **arachidonic acid** (**FIGURE 6.4**). The arachidonic acid can then enter one of two pathways. The *cyclooxygenase pathway* is a series of enzymatic steps that begins with the enzyme **cyclooxygenase** (sī-klō-OKS-ē-jen-ās) and ultimately converts the arachidonic acid into either a prostaglandin or a thromboxane. The *lipoxygenase pathway* is another series of enzymatic steps that begins with the enzyme **lipoxygenase** (lip-OKS-ē-jen-ās) and eventually converts the arachidonic acid into a leukotriene. Different eicosanoids are produced by different cells, depending on which set of enzymes is present in those cells.

FIGURE 6.4 Synthesis of eicosanoids.

Eicosanoid synthesis begins when a phospholipid is cleaved by the enzyme phospholipase A_2 to form arachidonic acid, which in turn enters either the cyclooxygenase pathway to form a prostaglandin or a thromboxane, or the lipoxygenase pathway to form a leukotriene.

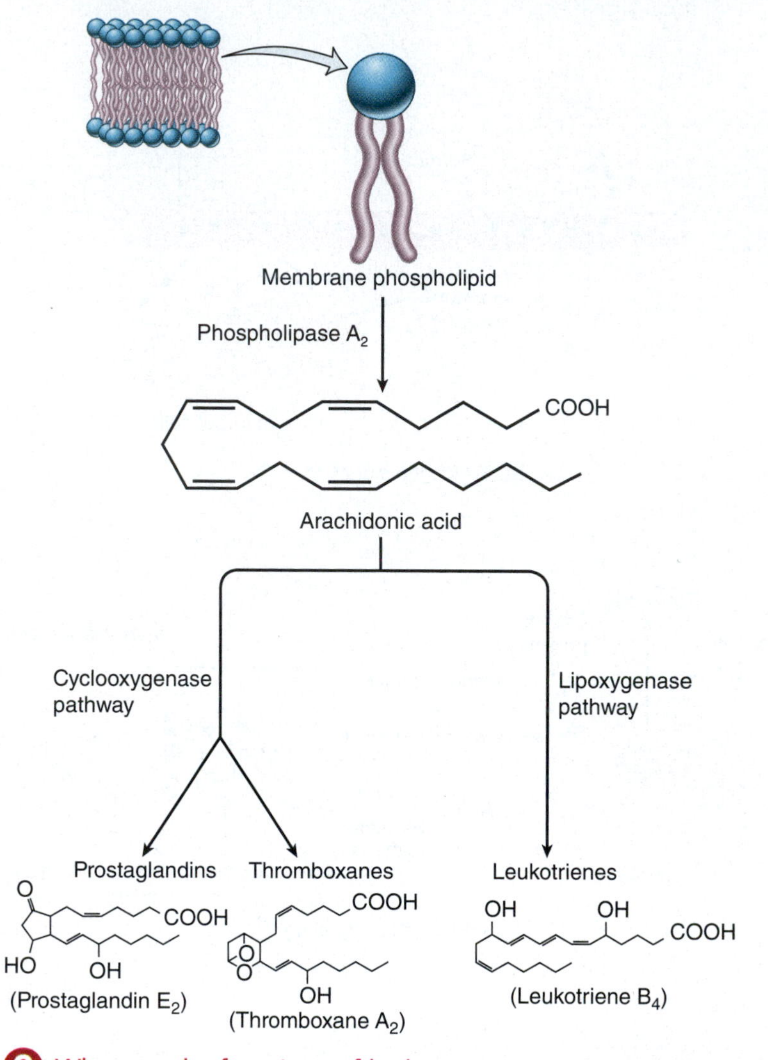

What are the functions of leukotrienes?

CLINICAL CONNECTION

Nonsteroidal Anti-Inflammatory Drugs

In 1971, scientists solved the long-standing puzzle of how aspirin works. Aspirin and related **nonsteroidal anti-inflammatory drugs (NSAIDs)**, such as ibuprofen (Motrin®), inhibit cyclooxygenase, a key enzyme involved in prostaglandin synthesis. Because NSAIDs do not inhibit lipoxygenase, synthesis of leukotrienes is not affected. NSAIDs are used to treat a wide variety of inflammatory disorders, from rheumatoid arthritis to tennis elbow. The success of NSAIDs in reducing fever, pain, and inflammation shows how prostaglandins contribute to these woes.

Leukotrienes stimulate chemotaxis (attraction to a chemical stimulus) of leukocytes and mediate inflammation. Prostaglandins alter smooth muscle contraction, glandular secretions, blood flow, reproductive processes, platelet function, respiration, action potential transmission, lipid metabolism, and immune responses. They also have roles in promoting inflammation and fever, and in intensifying pain. Thromboxanes constrict blood vessels and promote platelet activation.

GROWTH FACTORS **Growth factors** are substances that play important roles in tissue development, growth, and repair. Growth factors are *mitogenic* substances—they cause growth by stimulating cell division. Many growth factors act locally, as autocrines or paracrines. Examples of growth factors include epidermal growth factor (EGF), platelet-derived growth factor (PDGF), fibroblast growth factor (FGF), nerve growth factor (NGF), and transforming growth factor (TGF).

Different Functional Capacities for the Same Extracellular Messenger

A given extracellular chemical messenger can function as a hormone, neurotransmitter, or local mediator, depending on where it is secreted. For example, a neuron typically releases an extracellular messenger into the synapse between the neuron and the postsynaptic cell (the target cell). When this occurs, the messenger functions as a neurotransmitter. However, some neurons release messengers into the bloodstream, which in turn carries the messenger to its target cell. In such cases, the extracellular messenger functions as a hormone. Norepinephrine is a good example of an extracellular chemical messenger that can function both as a neurotransmitter and as a hormone. In the brain, norepinephrine is released at some synapses, functioning as a neurotransmitter. However, neurons in the inner part of the adrenal gland, which is just above the kidney, also release norepinephrine, but the norepinephrine is released into the bloodstream and therefore functions as a hormone.

Extracellular Messengers Are Chemically Classified as Water-Soluble or Lipid-Soluble

Based on their chemical structure, extracellular messengers can be divided into two broad classes: those that are soluble in water and those that are soluble in lipids (**FIGURE 6.5**). The majority of messengers are **water-soluble extracellular messengers**, which include peptide or protein hormones (such as oxytocin); amine hormones (such as norepinephrine); nearly all of the neurotransmitters; and a large number of local mediators, including eicosanoids and growth factors. **Lipid-soluble extracellular messengers** include steroid hormones (such as testosterone and estrogens), thyroid hormones, and nitric oxide. This chemical classification of extracellular messengers is useful functionally because, as you will soon see, the two classes exert their effects differently.

Extracellular Messengers Travel Through Interstitial Fluid and/or Blood to Reach Their Target Cells

After an extracellular messenger is secreted by a cell, it enters interstitial fluid and begins the journey to its target cell. Neurotransmitters and local mediators simply diffuse through interstitial fluid to reach their target cells. Hormones initially diffuse through interstitial fluid and then enter the bloodstream. Most water-soluble hormones circulate in the watery blood plasma in a "free" form (not attached to other molecules), but most lipid-soluble hormones are bound to **transport**

FIGURE 6.5 Chemical classes of extracellular messengers.

Extracellular messengers can be divided into two broad chemical classes: those that are soluble in water and those that are soluble in lipids.

Water-soluble extracellular messengers

- Include peptide or protein hormones, amine hormones, nearly all neurotransmitters, and most local mediators
- Example:

Amino acids

Oxytocin, a peptide hormone

Lipid-soluble extracellular messengers

- Include steroid hormones, thyroid hormones, and nitric oxide
- Example:

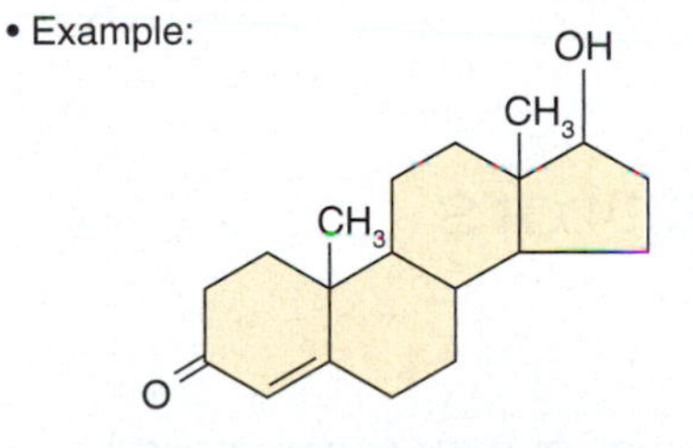

Testosterone, a steroid hormone

Is norepinephrine water-soluble or lipid-soluble?

FIGURE 6.6 Circulation of water-soluble and lipid-soluble hormones through the bloodstream.

Most water-soluble hormones circulate through the blood in "free" form, but most lipid-soluble hormones are bound to transport proteins.

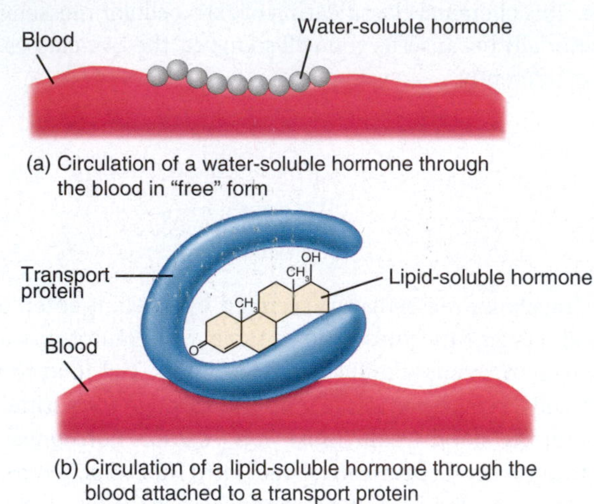

? Why do lipid-soluble hormones require a transport protein to circulate through the bloodstream?

proteins (FIGURE 6.6). The transport proteins, which are synthesized by cells in the liver, have three functions:

1. They make lipid-soluble hormones temporarily water-soluble, increasing their solubility in blood.
2. They retard passage of small hormone molecules through the filtering mechanism in the kidneys, slowing the rate of hormone loss in the urine.
3. They provide a ready reserve of hormone, already present in the bloodstream.

In general, 0.1–10% of the molecules of a lipid-soluble hormone are not bound to a transport protein. This **free fraction** diffuses out of capillaries, binds to receptors, and triggers responses. As free hormone molecules leave the blood and bind to their receptors, transport proteins abide by the *law of mass action* (see Section 2.3) and release new ones to replenish the free fraction.

CHECKPOINT

4. What is a hormone?
5. At a synapse, how does a presynaptic neuron communicate with a postsynaptic cell?
6. How do paracrines differ from autocrines?
7. Why are eicosanoids important?
8. What are some examples of water-soluble extracellular messengers?

6.3 Receptors

OBJECTIVES

- Discuss the properties of messenger–receptor binding.
- Describe where receptors may be located in a target cell.
- Explain how receptors are regulated.

Extracellular chemical messengers influence their target cells by binding to specific protein **receptors**. When a given extracellular messenger binds to its specific receptor, the receptor undergoes a conformational change and is activated. The activated receptor in turn triggers a signal transduction pathway that causes a cellular response by the target cell. Hence, the role of a receptor in cell signaling is to detect a signal molecule (the extracellular messenger) and then initiate the signal transduction process. The binding of the extracellular messenger to the receptor is reversible, so when the chemical messenger leaves the receptor, the cellular response ultimately comes to an end. In this section you will learn about the properties of messenger–receptor binding, where receptors are located in target cells, and how receptors are regulated.

Messenger–Receptor Binding Exhibits Several Properties

Because an extracellular messenger is a ligand and a receptor is a protein, messenger–receptor binding displays properties that are characteristic of ligand-protein binding—specificity, affinity, saturation, and competition.

Specificity

A given receptor permits binding of only one extracellular messenger or, in some cases, a small group of structurally related extracellular messengers. This property of messenger–receptor binding is known as *specificity*. Specificity occurs because the binding site of a receptor protein has a particular three-dimensional shape, and only a substance with a complementary shape can bind to that site (FIGURE 6.7). Although an extracellular

FIGURE 6.7 Receptor specificity for messenger.

Specificity is a property of messenger–receptor binding in which a given receptor permits binding of only one extracellular messenger or a small number of structurally related messengers.

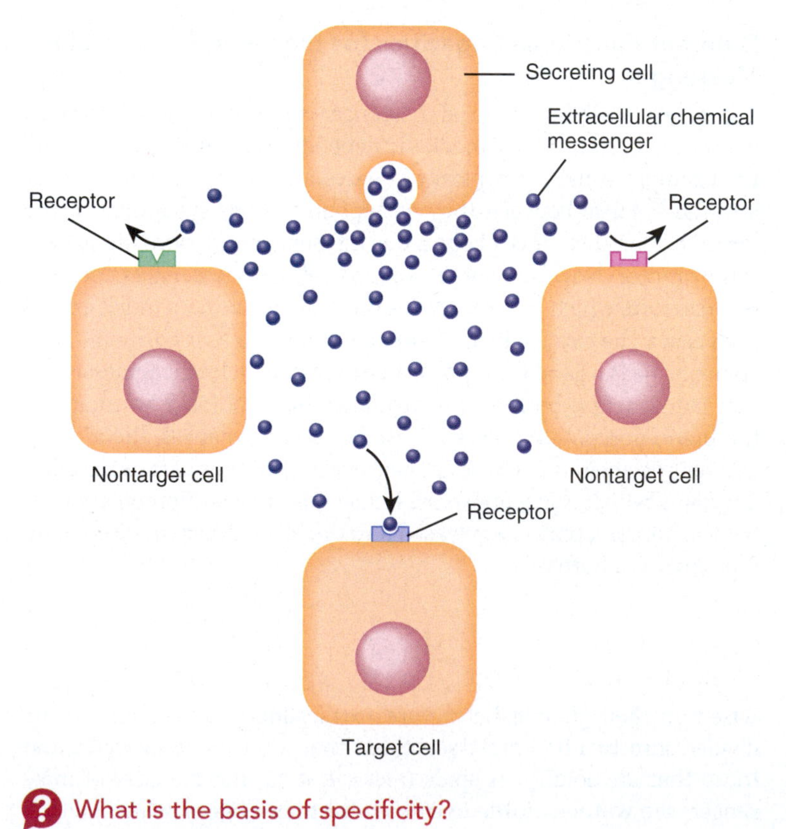

? What is the basis of specificity?

messenger may come in contact with many cells in the body, only its target cells have the correct receptors to which the messenger can bind.

A given cell can serve as the target cell for several different extracellular messengers. When this is the case, the target cell has several types of receptors, each responding only to its specific extracellular messenger. Furthermore, each messenger may cause a different response by that target cell. Generally, a target cell has 2000 to 100,000 receptors for a particular messenger. If the cell is a target cell for many different messengers, it may have millions of receptors.

Affinity

The binding of an extracellular messenger to a receptor involves reversible, noncovalent intermolecular forces such as hydrogen bonds, ionic bonds, van der Waals interactions, and hydrophobic interactions. The strength with which a messenger binds to a receptor is called **affinity**. If a receptor has a high affinity for a messenger, then the messenger binds tightly to the receptor, and a relatively low concentration of the messenger causes a response by the target cell. By contrast, if a receptor has a low affinity for the messenger, then the messenger binds loosely to the receptor, and a relatively high concentration of the messenger is needed to cause a response by the target cell.

If extracellular messengers with similar structures are able to bind to the same receptor, the receptor may have different affinities for each messenger. As an example, consider a group of receptors called *adrenergic receptors* that respond to the binding of either norepinephrine or epinephrine—two extracellular messengers that have almost identical chemical structures (see FIGURE 7.32). The two main types of adrenergic receptors are *alpha* (α) *receptors* and *beta* (β) *receptors*. These receptors are further classified into subtypes—α_1, α_2, β_1, β_2, and β_3. The α receptors and β_3 receptors have a higher affinity for norepinephrine than epinephrine. β_2 receptors, however, have a higher affinity for epinephrine than norepinephrine. β_1 receptors have nearly equal affinity for norepinephrine and epinephrine.

Saturation

As mentioned earlier, a target cell usually has thousands of copies of a receptor for a given extracellular messenger. Most of these receptors are not initially occupied by messenger because the concentration of messenger is usually very low. Therefore, as the concentration of extracellular messenger increases, more receptors are bound by the messenger, resulting in a greater cellular response (FIGURE 6.8). At a certain point, however, the cellular response does not become any greater, despite further increases in the messenger concentration, because there is a limited number of receptors in the target cell and they are all bound by messenger. When this point is reached, the receptors are said to be fully saturated. *Saturation* is a property of messenger–receptor binding that reflects the degree to which receptors are bound by messenger. So the receptors are fully saturated if 100% of the receptors are bound by messenger (FIGURE 6.8). If only half of the receptors are bound by messenger, the receptors are said to be 50% saturated.

Competition

If extracellular messengers are so similar in structure that they can bind to the same receptor, the messengers may compete with one another for that receptor. This property of messenger–receptor binding is known as *competition* (FIGURE 6.9). For example, norepinephrine and epinephrine

FIGURE 6.8 Saturation of receptors with messenger.

Saturation is a property of messenger–receptor binding that reflects the degree to which receptors are bound by messenger.

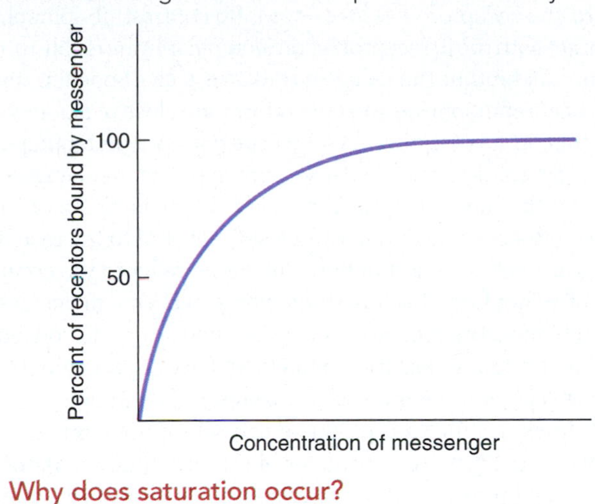

Why does saturation occur?

FIGURE 6.9 Competition between structurally related messengers for the same receptor. In this figure, norepinephrine and epinephrine compete with each other to bind to the beta-1 (β_1) adrenergic receptor.

Competition is a property of messenger–receptor binding in which two or more structurally similar messengers compete with one another to bind to the same receptor.

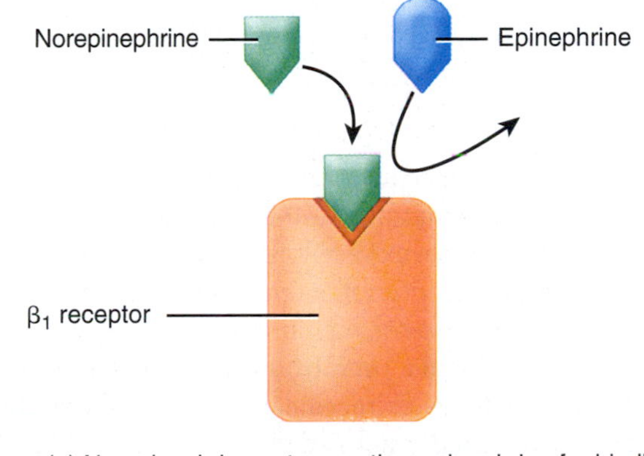

(a) Norepinephrine outcompeting epinephrine for binding

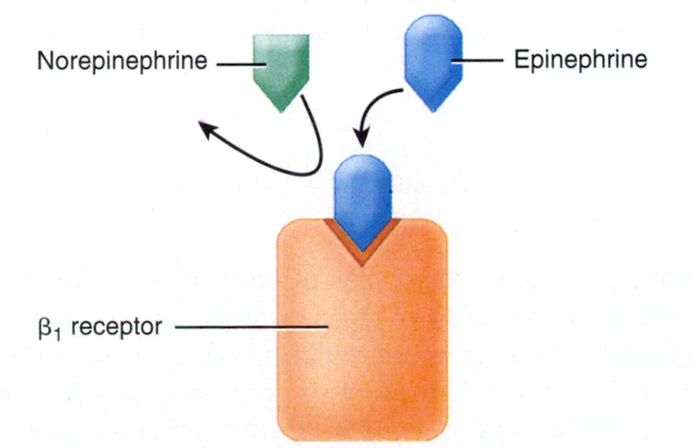

(b) Epinephrine outcompeting norepinephrine for binding

What is an agonist? An antagonist?

compete with each other to bind to adrenergic receptors. Recall that the beta-1 (β_1) receptor is a type of adrenergic receptor that has an equal affinity for norepinephrine and epinephrine. If norepinephrine interacts with the β_1 receptor before epinephrine is able to, then norepinephrine will bind to the receptor (**FIGURE 6.9a**). By contrast, if epinephrine makes contact with the β_1 receptor before norepinephrine is able to, then epinephrine will bind to the receptor (**FIGURE 6.9b**). So when several molecules of norepinephrine and epinephrine are close to a population of β_1 receptors, some of the receptors will be bound by norepinephrine and others by epinephrine. Because both types of messengers are competing for the same receptors, the relative concentrations of these messengers influence which one will most likely bind to the receptors. Increasing the concentration of norepinephrine reduces the competitive effect of epinephrine because more norepinephrine molecules are available to bind to the receptors. Likewise, increasing the concentration of epinephrine reduces the competitive effect of norepinephrine because more epinephrine molecules are present for binding.

Several drugs and natural products can selectively compete with extracellular messengers to activate or block receptors. An **agonist** is a substance that binds to and activates a receptor, in the process mimicking the effect of the endogenous messenger. Phenylephrine, a common ingredient in cold and sinus medications, is an agonist to alpha-1 (α_1) receptors that bind norepinephrine or epinephrine. Because it constricts blood vessels in the nasal mucosa, phenylephrine reduces production of mucus, thus relieving nasal congestion. An **antagonist** is a substance that binds to and blocks a receptor, thereby preventing the endogenous messenger from exerting its effect. An example is propranolol (Inderal®), a beta blocker that binds to beta receptors and prevents their activation by norepinephrine or epinephrine. The effect of beta blockers is to reduce blood pressure.

Receptors Are Located in the Target Cell's Plasma Membrane or Inside the Target Cell

The receptor for an extracellular chemical messenger may be present either in the plasma membrane of the target cell or inside the target cell. Because water-soluble extracellular messengers are unable to pass through the hydrophobic interior of the plasma membrane, their receptors are located in the plasma membrane. These receptors are appropriately known as **plasma membrane receptors** (**FIGURE 6.10a**). By contrast, lipid-soluble extracellular messengers are able to pass through the plasma membranes of their target cells. Therefore, their receptors are present inside the target cell, either in the cytosol or in the nucleus. Such receptors are known as **intracellular receptors** (**FIGURE 6.10b**).

Receptors Are Subject to Down-Regulation and Up-Regulation

Receptors, like many other types of proteins, are subject to regulation. Two important forms of receptor regulation are down-regulation and up-regulation. If an extracellular messenger is present in excess, the number of target-cell receptors may decrease—an effect called **down-regulation**. The decrease in the number of receptors occurs in the following way: Some receptors are removed from the plasma membrane via endocytosis and then degraded in lysosomes (**FIGURE 6.11a**). Another method of down-regulation involves chemically modifying the

FIGURE 6.10 Receptor location.

Water-soluble extracellular messengers bind to plasma membrane receptors, whereas lipid-soluble extracellular messengers bind to intracellular receptors located in either the cytosol or nucleus.

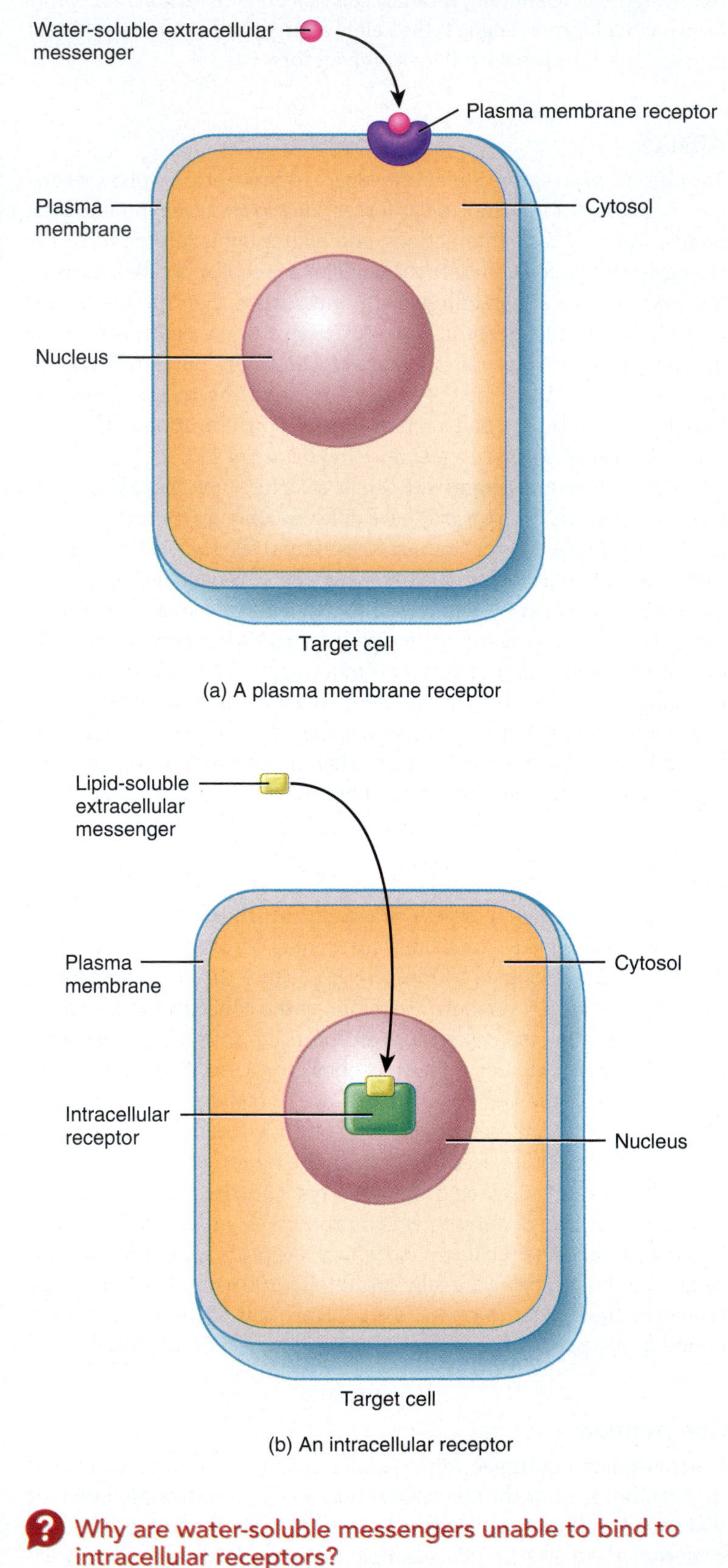

Why are water-soluble messengers unable to bind to intracellular receptors?

FIGURE 6.11 Receptor regulation.

Down-regulation involves a decrease in the number of receptors in response to an excess of messenger; up-regulation is an increase in the number of receptors in response to a deficiency of messenger.

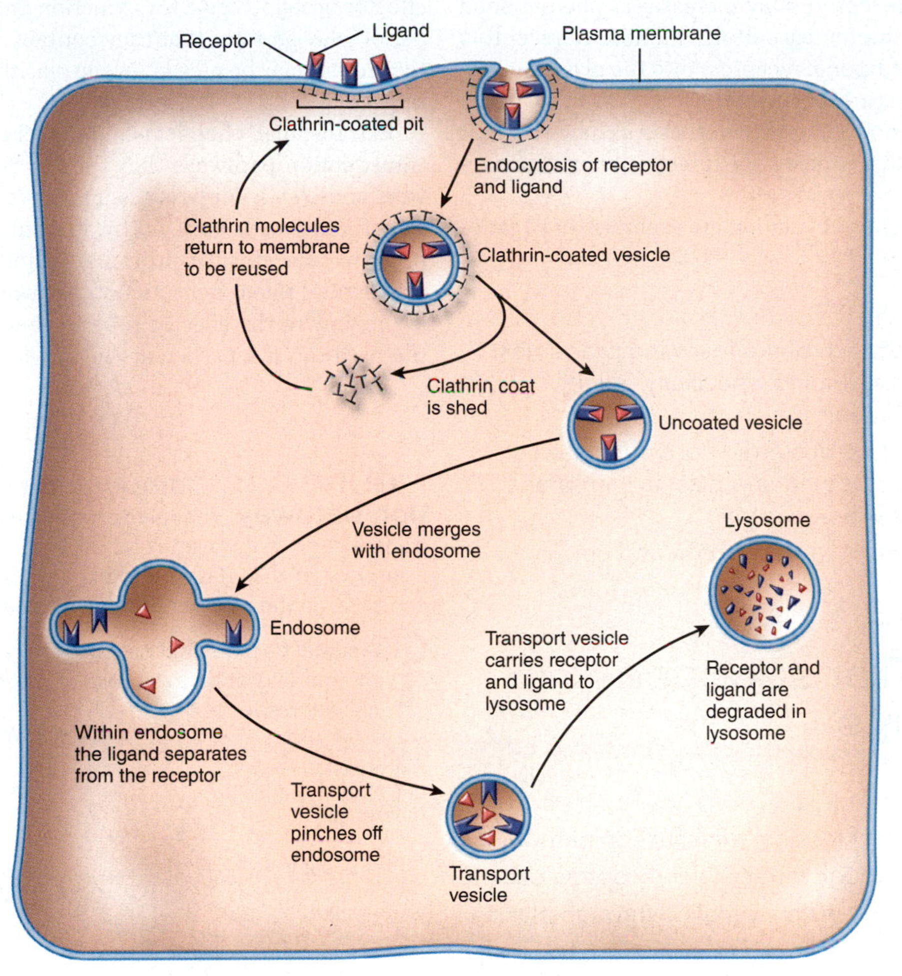

(a) Down-regulation

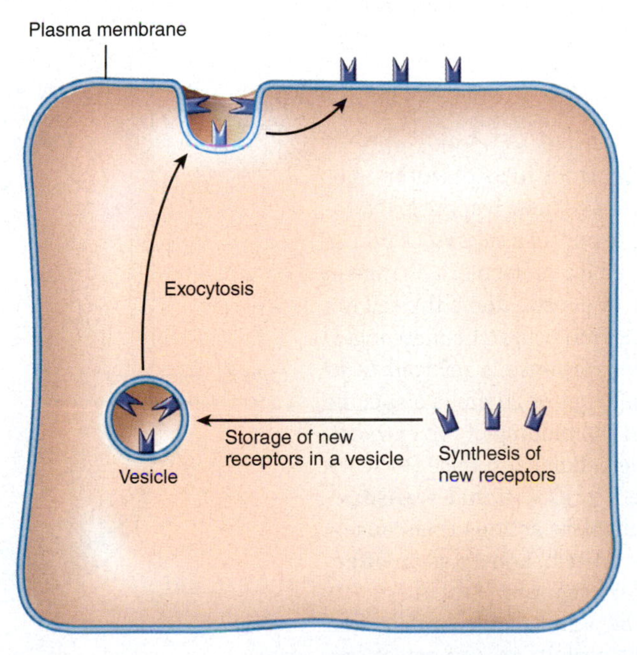

(b) Up-regulation

Is down-regulation an example of negative or positive feedback?

receptor to disrupt its ability to function. So down-regulation ultimately makes the target cell *less sensitive* to the messenger.

In contrast to down-regulation, when a messenger is deficient, the number of target-cell receptors may increase—a phenomenon known as **up-regulation**. The increase in the number of receptors occurs by the insertion of additional receptors into the plasma membrane from internal stores via exocytosis (FIGURE 6.11b). Up-regulation can also occur by increasing the synthesis of new receptor molecules. Hence, up-regulation makes the target cell *more sensitive* to the messenger.

Both down-regulation and up-regulation are examples of negative feedback.

CHECKPOINT

9. Provide an example of each of the following properties of messenger–receptor binding: specificity, affinity, saturation, and competition.
10. When aldosterone, a type of steroid hormone, interacts with its target cell, does it bind to a plasma membrane receptor or an intracellular receptor?
11. Compare and contrast down-regulation and up-regulation.

6.4 Signal Transduction Pathways

OBJECTIVES

- Identify the components of a signal transduction pathway.
- Explain how signaling through intracellular receptors occurs.
- Discuss the various ways by which signaling through plasma membrane receptors can occur.
- Describe how signal transduction pathways are amplified.
- Explain how signal transduction pathways are terminated.

Signal Transduction Begins with Binding of Messenger and Ends with a Cellular Response

After an extracellular chemical messenger binds to its receptor on or in its target cell, the target cell produces a cellular response. Examples of cellular responses include the transport of a substance across the plasma membrane, the synthesis of a new molecule, a change in the rate of a specific metabolic reaction, or contraction if the cell is a muscle cell. The process by which a signal molecule (the messenger) is transduced (converted) into a cellular response is referred to as **signal transduction**. During signal transduction, there is a specific sequence of events that occurs between the binding of the extracellular messenger to the receptor and the cellular response (FIGURE 6.12). This sequence of events, referred to as a **signal transduction pathway**, or **signaling pathway**, involves actions that cause a change in a key protein of the cell. This protein, known as an **effector protein**, in turn causes the cellular response (FIGURE 6.12). For example, the effector protein may be a contractile protein that causes contraction, an ion channel that permits movement of certain ions across the plasma membrane, or an enzyme that promotes a specific metabolic reaction. In many signal transduction pathways, a series of **relay proteins** conveys the signal between the receptor and the effector protein. When this occurs, each relay protein causes a change in the next protein in line until there is a change in the effector protein. Signal transduction pathways can differ in the number of relay proteins that they contain. Relay proteins and the effector protein may be present in the plasma membrane or cytosol of the target cell.

Extracellular chemical messengers vary in the types of signal transduction pathways that they activate. Some messengers activate completely different signal transduction pathways in their target cells, with each pathway containing a unique set of relay proteins and effector proteins. Other messengers, however, activate similar signal transduction pathways in their target cells, differing mainly in the effector protein that is altered in the last step of the pathway. If a particular cell serves as the target cell for several

FIGURE 6.12 Components of a signal transduction pathway. The relay proteins and effector protein of a signal transduction pathway may be present in the cytosol (as shown in this figure) or in the plasma membrane. Signal transduction pathways can differ in the number of relay proteins that they contain.

A signal transduction pathway is the sequence of events between the binding of an extracellular messenger to a receptor and the cellular response.

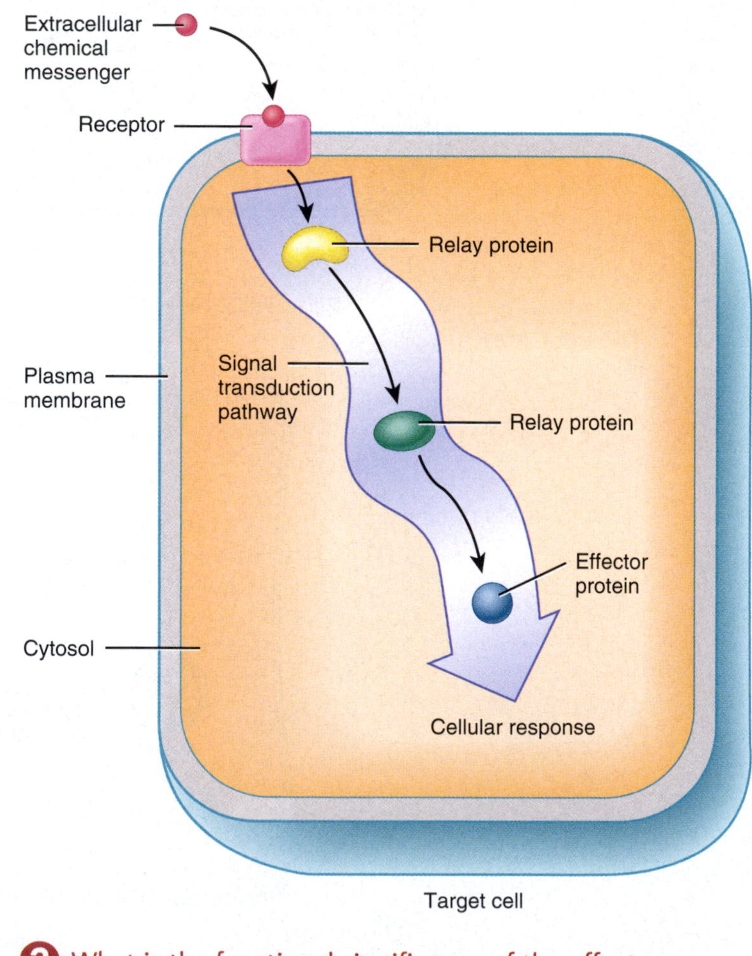

different messengers, the messengers usually activate different signal transduction pathways in the target cell and cause different responses in that cell.

In many, but not all, signaling pathways, after the extracellular messenger binds to its receptor, an intracellular messenger is generated in the cytosol or plasma membrane of the target cell to help mediate the transduction process. In such cases, the extracellular messenger is referred to as the **first messenger**, and the intracellular messenger is known as a **second messenger** (FIGURE 6.13). Second messengers are usually generated by special types of receptors that have enzymatic activity or by certain enzymes that are relay proteins. Once formed, the second messenger typically binds to and activates the next relay protein of the signaling pathway—usually a protein kinase, which will be described shortly. Examples of second messengers include cyclic AMP (cAMP), cyclic GMP (cGMP), diacylglycerol (DAG), inositol trisphosphate (IP_3), and calcium ions (Ca^{2+}). You will learn more about second messengers later in this section as specific signal transduction pathways are discussed.

Signal transduction pathways may also involve one or more protein kinases. A **protein kinase** is an enzyme that phosphorylates (adds a phosphate group to) a target protein. Phosphorylation either activates or inhibits the target protein. In a signal transduction pathway, a protein kinase is often one of the relay proteins, and its target protein may be another relay protein or the effector protein. If a protein kinase phosphorylates an effector protein, the change in the activity of the effector protein that results from the phosphorylation causes the cellular response (FIGURE 6.14).

There are different types of protein kinases. Some are named based on what activates them. For example, *protein kinase A* is activated by the second messenger cAMP, and *protein kinase G* is activated by the second messenger cGMP. Other protein kinases are named based on which amino acids of a protein they phosphorylate. For example, *tyrosine*

FIGURE 6.13 Role of a second messenger in signal transduction. The flow of the signal transduction pathway is represented by purple arrows. Breaks in the pathway indicate that specific steps have been omitted for simplicity.

A second messenger is a molecule that is generated inside the target cell to help mediate the signal transduction process for an extracellular messenger (the primary messenger).

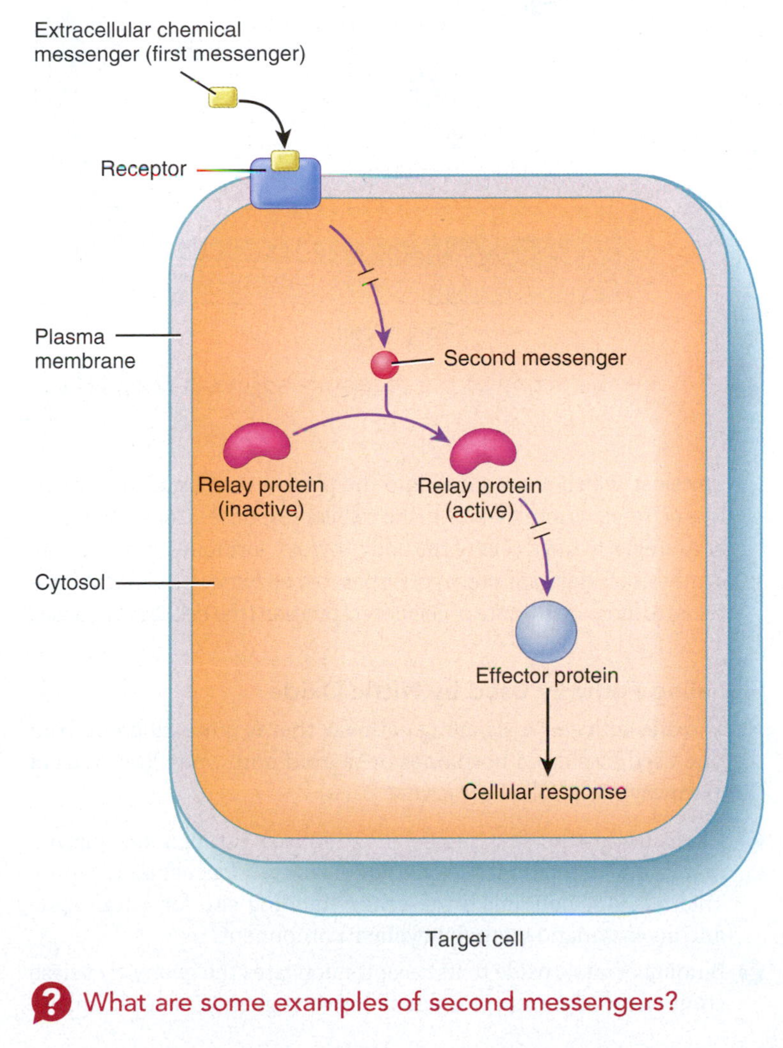

What are some examples of second messengers?

FIGURE 6.14 Role of a protein kinase in signal transduction. The flow of the signal transduction pathway is represented by purple arrows. Breaks in the pathway indicate that specific steps have been omitted for simplicity.

A protein kinase is an enzyme that phosphorylates a target protein.

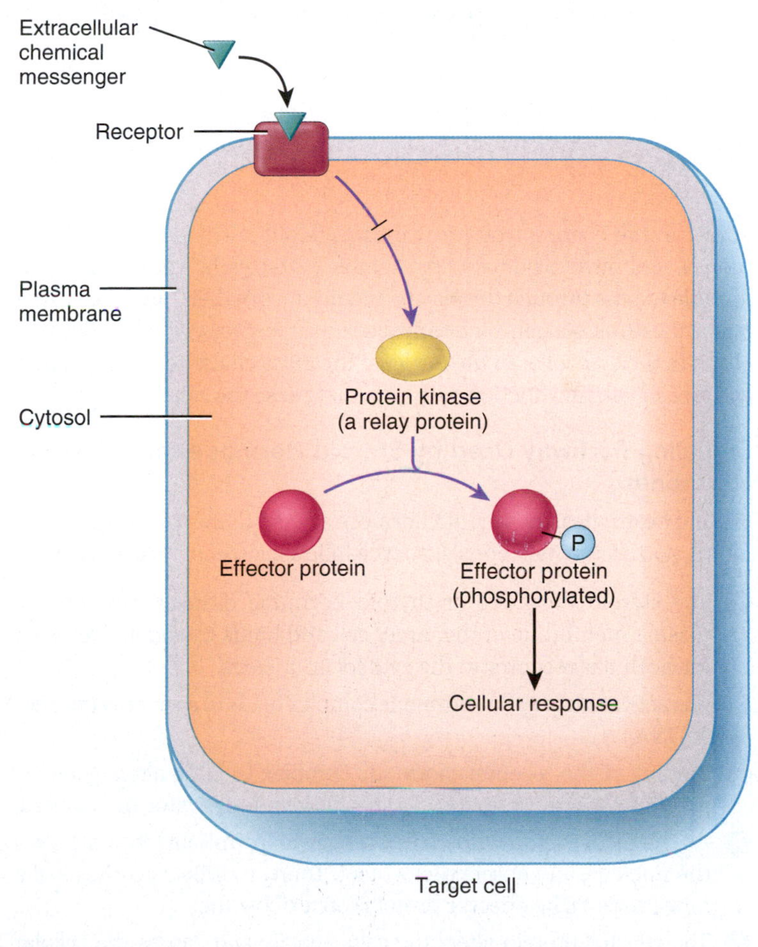

What happens to the target protein after it has been phosphorylated?

kinases phosphorylate tyrosine amino acids on their target proteins, and *serine/threonine kinases* phosphorylate serine and/or threonine amino acids on their target proteins. Different target cells may have different types of protein kinases. Although protein kinases are very common in signal transduction pathways, they are not present in every pathway.

The time required for the cellular response to occur after a messenger interacts with its target cell can vary, depending on whether the cellular response requires the synthesis of new molecules of the effector protein or a change in the activity of preexisting molecules of the effector protein. For example, in several signaling pathways, the concentration of the effector protein must increase in order for the cellular response to occur. When this is the case, the signaling pathway stimulates the transcription of the gene for the effector protein. After transcription is completed, translation occurs, resulting in the synthesis of new molecules of the effector protein. In such signaling pathways that produce new molecules of the effector protein, the cellular response is relatively slow (often several minutes to hours) because it takes time for transcription and translation to occur. By contrast, there are many signaling pathways in which the concentration of preexisting molecules of the effector protein is sufficient, but the activity of the effector proteins must be altered to achieve the cellular response. As you have just learned, a common method of altering the activity of an effector protein is phosphorylation, which may either activate or inhibit the protein. In signaling pathways that involve phosphorylation of effector proteins, the cellular response is relatively quick (seconds to minutes) because a protein kinase can rapidly add a phosphate group to an effector protein.

FIGURE 6.15 Signaling pathway used by steroid hormones and thyroid hormones.

A steroid hormone or thyroid hormone ultimately causes the synthesis of effector proteins in its target cell.

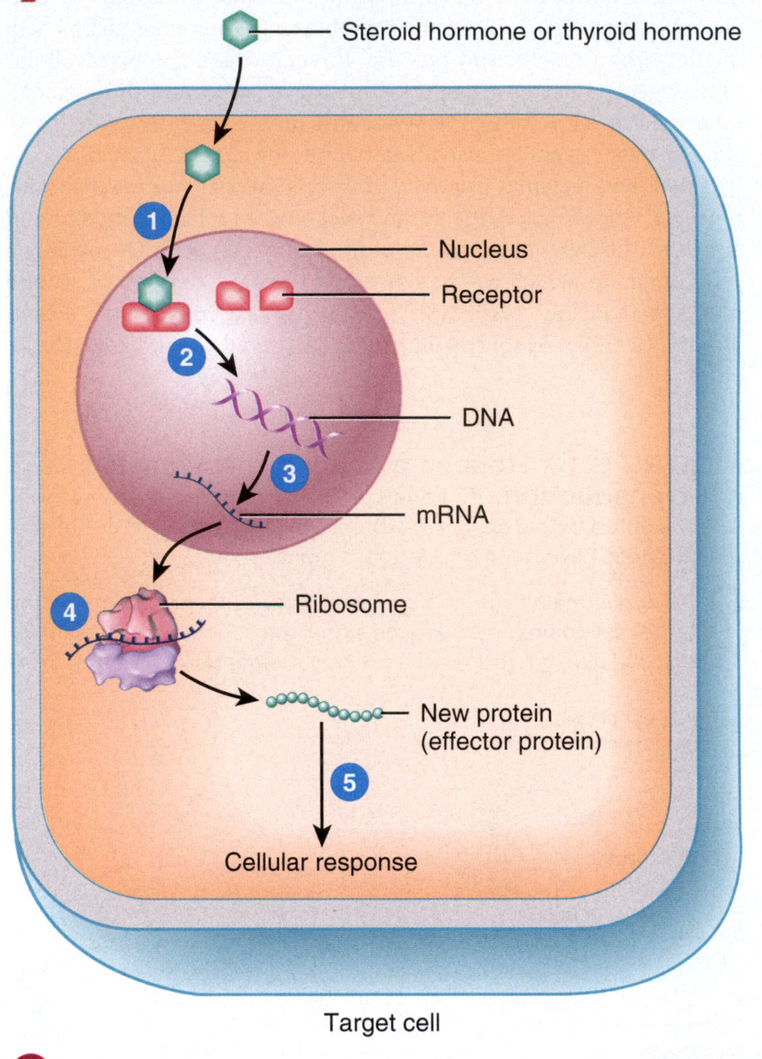

What is the action of the receptor–hormone complex?

Lipid-Soluble Extracellular Messengers Activate Signaling Pathways by Binding to Intracellular Receptors

Earlier in the chapter, you learned that steroid hormones, thyroid hormones, and nitric oxide are lipid-soluble extracellular messengers that are able to pass through the plasma membranes of their target cells. As a result, these messengers bind to intracellular receptors, which are located either in the cytosol or in the nucleus. The intracellular receptors in turn activate signal transduction pathways that cause the cellular response.

Signaling Pathway Used by Steroid Hormones and Thyroid Hormones

Steroid hormones and thyroid hormones activate the same type of signaling pathway, which consists of the following steps (FIGURE 6.15):

1. The steroid hormone or thyroid hormone diffuses through the plasma membrane of the target cell and binds to and activates an intracellular receptor in the cytosol or nucleus.
2. The activated receptor–hormone complex binds to specific sequences in DNA.
3. Binding of the receptor–hormone complex to DNA alters gene expression: The transcription of a specific gene is activated or inhibited.
4. As the DNA is transcribed, messenger RNA (mRNA) forms, leaves the nucleus, and enters the cytosol. There, it directs synthesis of a new protein (the effector protein) on a ribosome.
5. The effector protein alters the cell's activity and causes the cellular response. For example, in certain kidney cells, the steroid hormone aldosterone promotes the synthesis of Na^+ channels (the effector proteins), which are inserted into the plasma membrane to prevent loss of Na^+ ions into the urine (the cellular response). Consider another example: In most cells of the body, thyroid hormones stimulate the synthesis of additional copies of respiratory enzymes (the effector proteins) to increase the rate of cellular respiration (the cellular response).

Signaling Pathway Used by Nitric Oxide

Nitric oxide activates a signaling pathway that is quite different from the one used by steroid hormones or thyroid hormones. The steps of this pathway are as follows (FIGURE 6.16):

1. Nitric oxide (the first messenger) diffuses through the plasma membrane of the target cell and binds to an intracellular receptor. This receptor contains a messenger-binding site for nitric oxide and an enzymatic **guanylyl cyclase** component.
2. Binding of nitric oxide to its receptor activates the guanylyl cyclase component of the receptor, which converts guanosine triphosphate

FIGURE 6.16 Signaling pathway used by nitric oxide. MLCK = myosin light chain kinase.

Nitric oxide stimulates the production of cyclic GMP in a smooth muscle cell.

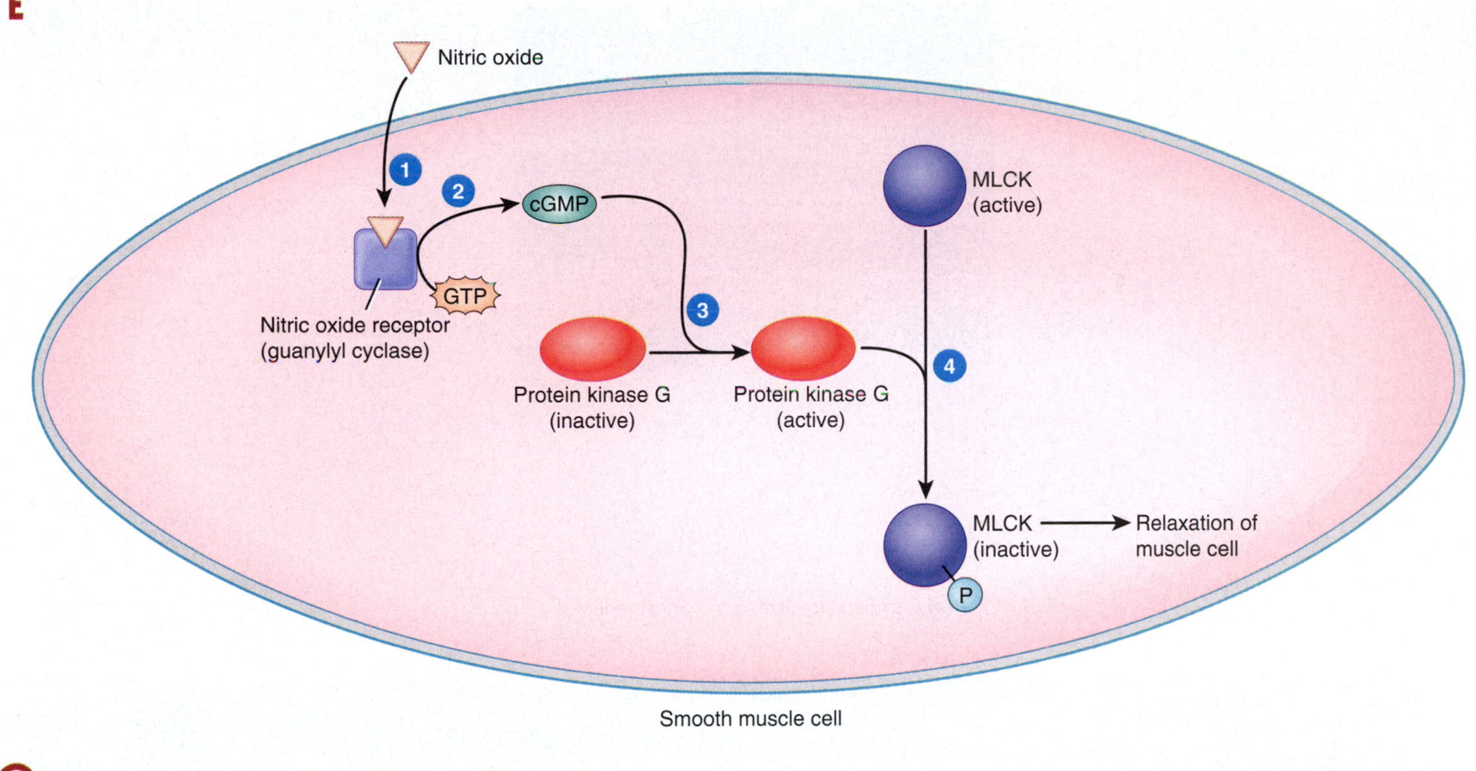

What is the functional significance of MLCK?

(GTP) into the second messenger **cyclic guanosine monophosphate**, also referred to as **cyclic GMP** or **cGMP**.

3. Cyclic GMP binds to and activates a protein known as **protein kinase G (PKG)**, which is so-named because it is a cGMP-dependent protein kinase. PKG is a type of serine/threonine kinase, which means that it phosphorylates serine and/or threonine amino acids on its target protein.

4. Activated protein kinase G phosphorylates the effector protein to cause the cellular response. When nitric oxide's target cell is a smooth muscle cell in the wall of a blood vessel, the effector protein is *myosin light chain kinase (MLCK)*, a protein that normally promotes the interaction of the contractile proteins myosin and actin to cause contraction of the cell (see **FIGURE 11.26**). When MLCK is phosphorylated by PKG, it is inhibited. This prevents myosin and actin from interacting, which causes the smooth muscle cell to relax (the cellular response). Because this smooth muscle cell is part of the wall of a blood vessel, when the smooth muscle relaxes, the blood vessel dilates.

As you will soon learn, a signaling pathway must be terminated at some point to prevent overstimulation of the target cell. The signaling pathway triggered by nitric oxide in a smooth muscle cell is terminated in part by an enzyme called *cGMP phosphodiesterase*, which breaks down cGMP. Without cGMP, the rest of the nitric oxide pathway will not occur. The drug *Viagra*® (sildenafil) enhances the effects of nitric oxide in the penis by inhibiting cGMP phosphodiesterase. As a result, cGMP levels remain elevated and the smooth muscle within the walls of blood vessels in the penis stays relaxed for a longer period of time. This allows more blood to remain in the penis, which prolongs erection.

Water-Soluble Extracellular Messengers Activate Signaling Pathways by Binding to Plasma Membrane Receptors

Because peptide or protein hormones, amine hormones, nearly all neurotransmitters, and most local mediators are water-soluble extracellular messengers, they cannot diffuse through the lipid bilayer of the plasma membrane. Consequently, they bind to plasma membrane receptors that protrude from the target cell surface. There are different types of plasma membrane receptors, which in turn give rise to different types of signal transduction pathways. The receptors for water-soluble extracellular messengers are categorized into four types: ion channel receptors, enzyme receptors, enzyme-coupled receptors, and G protein-coupled receptors.

Signaling Through Ion Channel Receptors

An **ion channel receptor** is a type of plasma membrane receptor that contains both a messenger-binding site and an ion channel (**FIGURE 6.17a**). The messenger-binding site is on the extracellular side of the plasma membrane, and the ion channel component extends through the plasma membrane. In the absence of extracellular messenger, the ion channel component of the receptor is usually closed. When the messenger binds to the receptor, the ion channel

FIGURE 6.17 An ion channel receptor.

An ion channel receptor contains a messenger-binding site and an ion channel.

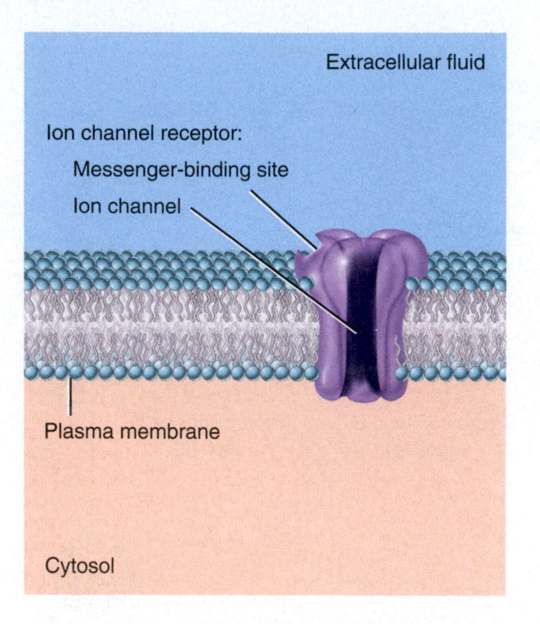

(a) Components of an ion channel receptor

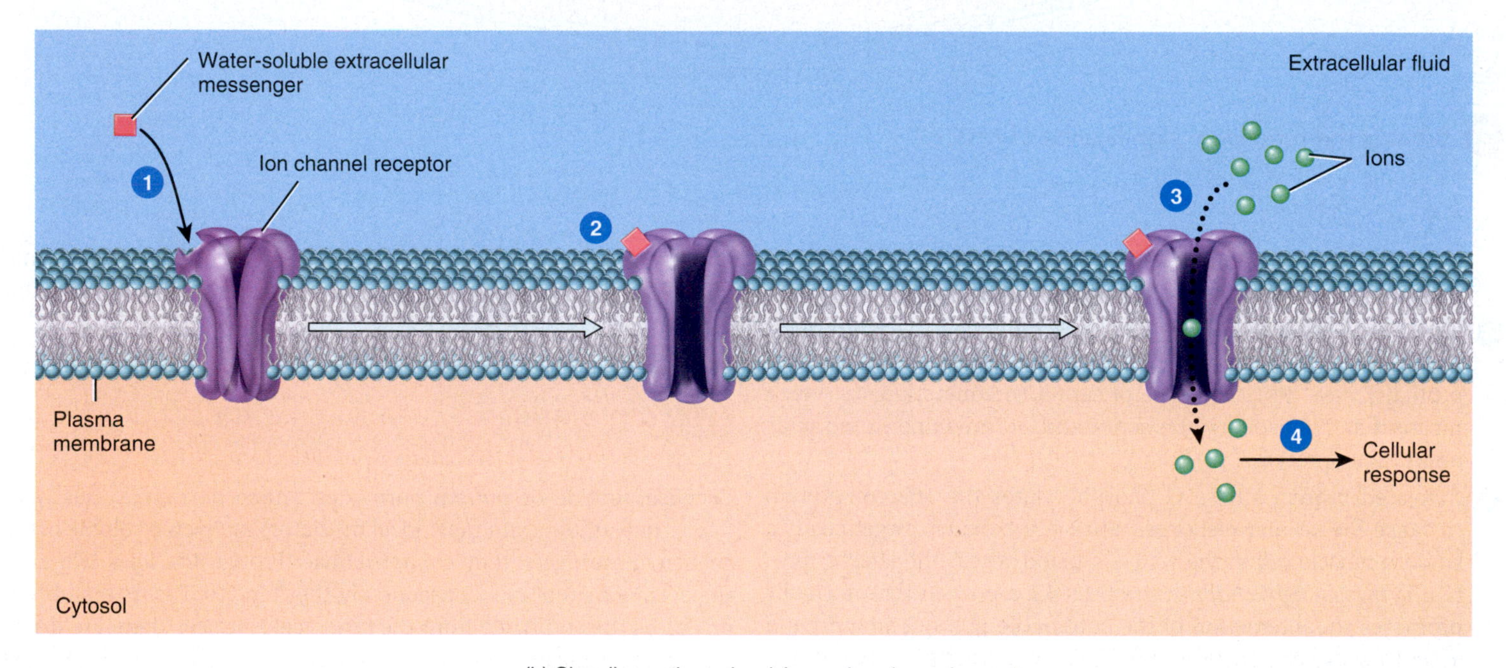

(b) Signaling pathway involving an ion channel receptor

When a messenger binds to an ion channel receptor, and ions subsequently enter the target cell, how do the ions cause a cellular response?

opens. The steps involved in this type of signaling pathway are as follows (FIGURE 6.17b):

1. A water-soluble extracellular messenger binds to an ion channel receptor at the exterior surface of the target cell's plasma membrane.
2. Binding of the messenger opens the ion channel component of the receptor.
3. Opening the ion channel allows specific ions to move across the plasma membrane into the cell, thereby increasing the ion permeability of the membrane.
4. The influx of ions causes the cellular response. In most cases, the cellular response is a change in the electrical properties of the cell, which can lead to either cell excitation or cell inhibition.

Many neurotransmitters bind to ion channel receptors. An example is acetylcholine, which is released at the synapse between a neuron

Critical Thinking

Are You Getting Enough Vitamin D?

Since Margaret retired from her job, she has become a homebody. She rarely goes outside and buys essentially everything she needs (including her groceries) online. She has also altered her diet, taking out milk and other dairy products because she can't digest lactose (lactose intolerance). Lately, Margaret has been having bone pain and muscle weakness. She goes to the doctor, who runs a battery of tests. A blood test reveals that Margaret has a low vitamin D level. In addition, a bone X-ray and a bone biopsy indicate that small cracks have developed in her bones. The doctor tells Margaret that she has a vitamin D deficiency that is causing her bones to soften—a condition known as osteomalacia (adult rickets).

SOME THINGS TO KEEP IN MIND:

Vitamin D is important to homeostasis because it promotes absorption of calcium (Ca^{2+}) and phosphate (HPO_4^{2-}) ions from food in the lumen of the small intestine into the bloodstream. Although vitamin D can be obtained from the diet (fortified milk, oily fish, and egg yolks), the main source of vitamin D is sunlight. When the body is exposed to the ultraviolet (UV) rays of the sun, a biochemical pathway occurs that ultimately causes the kidneys to secrete calcitriol, the active form of vitamin D. Once released from the kidneys, calcitriol travels through the bloodstream to the small intestine to act on absorptive epithelial cells to cause Ca^{2+} and HPO_4^{2-} absorption. Because it passes through the bloodstream to act on its target cells, calcitriol (vitamin D) is considered to be a hormone.

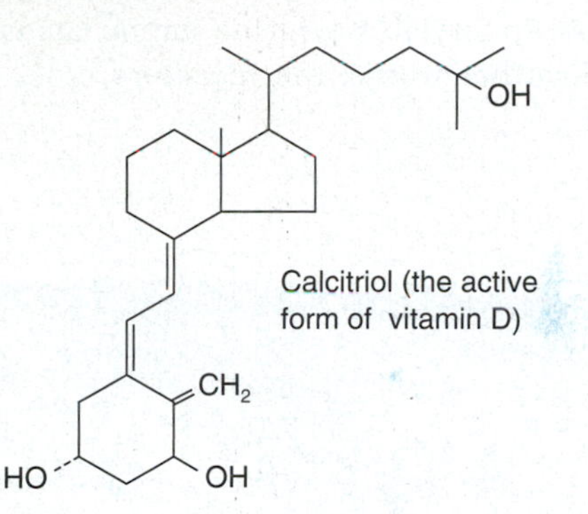

Calcitriol (the active form of vitamin D)

SOME INTERESTING FACTS:

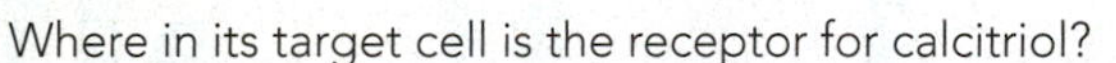

Calcitriol is categorized as a steroid hormone because it has a structure very similar to that of a steroid (contains hydrocarbon rings).

Only a small amount of exposure to UV rays of the sun (about 10 to 15 minutes, three times a week) is required for normal calcitriol synthesis.

Calcitriol promotes Ca^{2+} and HPO_4^{2-} absorption by causing the synthesis of transport proteins for these ions in absorptive cells of the small intestine.

Because osteomalacia is due to a vitamin D (calcitriol) deficiency, treatment of this condition involves replenishing low vitamin D levels. Thus, Margaret must take vitamin D supplements until her vitamin D level is normal and her bones are stronger.

Is calcitriol a water-soluble hormone or a lipid-soluble hormone?

Where in its target cell is the receptor for calcitriol?

Does the calcitriol signaling pathway involve a second messenger?

How does calcitriol cause the synthesis of transport proteins for Ca^{2+} and HPO_4^{2-} in its target cell?

and a skeletal muscle cell. When acetylcholine binds to an acetylcholine receptor, the ion channel component of the receptor opens, allowing cations—mainly sodium ions (Na^+)—to enter the muscle cell. This causes excitation of the muscle cell and ultimately results in muscle contraction.

In some signaling pathways involving ion channel receptors, the ion channel component of the receptor is open in the absence of the messenger and closes when the messenger binds to the receptor. When this occurs, specific ions are not able to pass through the ion channel. This causes a decrease in the ion permeability of the membrane, which also results in a change in the electrical properties of the cell.

Ion channel receptors are also referred to as **ligand-gated channels** because a ligand (the extracellular messenger) regulates the opening or closure of the channels. You will learn more about ligand-gated channels in the discussion of the nervous system in Chapter 7 (see Section 7.3).

Signaling Through Enzyme Receptors

An **enzyme receptor** is a type of plasma membrane receptor that contains a messenger-binding site, a transmembrane segment, and a site that functions as an enzyme. There are two major categories of enzyme receptors: receptor tyrosine kinases and receptor guanylyl cyclases.

Receptor Tyrosine Kinases A **receptor tyrosine kinase** is a receptor that contains a messenger-binding site; a transmembrane segment; and an enzymatic tyrosine kinase component, which phosphorylates tyrosine amino acids on target proteins. The messenger-binding site of the receptor is on the extracellular side of the plasma membrane, the transmembrane segment extends through the plasma membrane, and the tyrosine kinase component is on the cytosolic side of the plasma membrane (FIGURE 6.18a). A signaling pathway that uses a receptor tyrosine kinase involves the following steps (FIGURE 6.18b):

1. A water-soluble extracellular messenger binds to a receptor tyrosine kinase at the exterior surface of the target cell's plasma membrane.
2. The receptor tyrosine kinase dimerizes with a nearby receptor tyrosine kinase that has also become bound with another copy of the same messenger.
3. Dimerization of the receptors activates the tyrosine kinase components of the receptors. The tyrosine kinase of each receptor phosphorylates tyrosine amino acids on the other receptor. In other words, the receptors cross-phosphorylate each other.

FIGURE 6.18 A receptor tyrosine kinase.

A receptor tyrosine kinase contains a messenger-binding site, a transmembrane segment, and an enzymatic tyrosine kinase component.

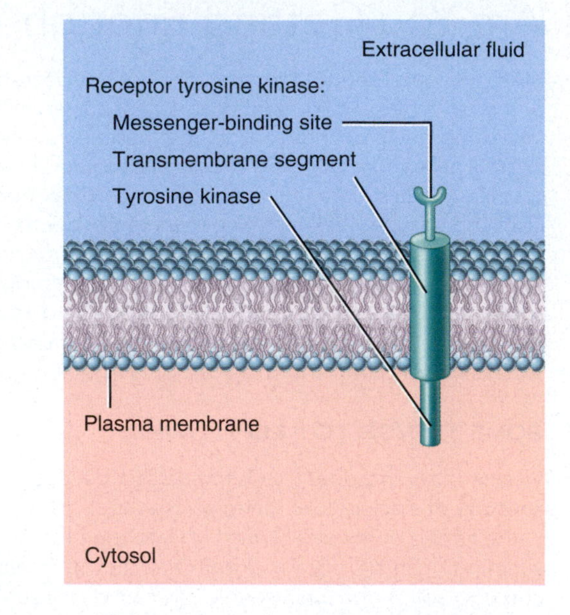

(a) Components of a receptor tyrosine kinase

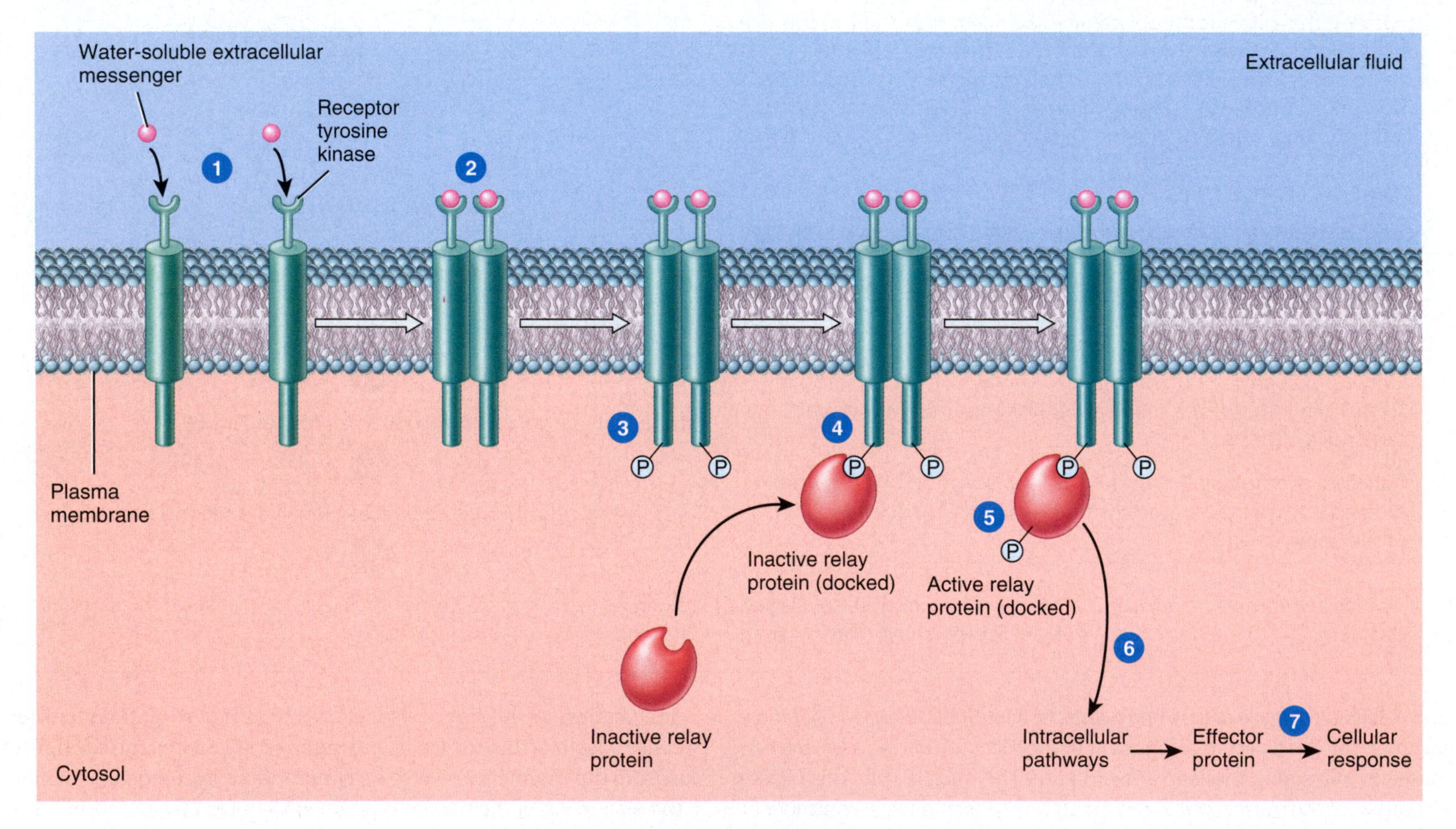

(b) Signaling pathway involving a receptor tyrosine kinase

What are some examples of water-soluble extracellular messengers that bind to receptor tyrosine kinases?

4. The phosphorylated sites on the receptor tyrosine kinases serve as docking sites where relay proteins in the cell can bind.
5. When relay proteins bind to the docking sites on the receptors, the tyrosine kinase components of the receptors phosphorylate tyrosine amino acids on these proteins. Phosphorylation activates the relay proteins.
6. The activated relay proteins trigger intracellular pathways that ultimately cause the synthesis or activation of an effector protein.
7. The effector protein causes the cellular response.

Messengers that activate signal transduction pathways involving receptor tyrosine kinases include most growth factors: epidermal growth factor (EGF), platelet-derived growth factor (PDGF), fibroblast growth factor (FGF), and nerve growth factor (NGF). Recall that growth factors are local mediators that play important roles in tissue development, growth, and repair. The hormone insulin also binds to a receptor that has tyrosine kinase activity. However, unlike the receptor tyrosine kinases activated by growth factors, the insulin receptor is a tetramer that consists of two extracellular subunits and two transmembrane subunits (see **FIGURE 13.5d**). The hormone-binding site is on the extracellular side of the plasma membrane, and the tyrosine kinase components are on the cytosolic side of the membrane. Binding of insulin to its receptor activates the receptor tyrosine kinase components of the receptor, causing the components to cross-phosphorylate each other. The rest of the signaling pathway is the same as that of the other receptor tyrosine kinases: The phosphorylated sites on the receptor serve as docking sites for relay proteins, which bind and become phosphorylated. The phosphorylated relay proteins trigger pathways that ultimately cause the synthesis or activation of the effector protein, which causes to the cellular response. You will learn more about insulin and the insulin receptor in Chapter 13.

RECEPTOR GUANYLYL CYCLASES A **receptor guanylyl cyclase** is a receptor that contains a messenger-binding site; a transmembrane segment; and an enzymatic guanylyl cyclase component, which produces the second messenger cyclic GMP (cGMP). The messenger-binding site is on the extracellular side of the plasma membrane, the transmembrane segment passes through the plasma membrane, and the guanylyl cyclase component is on the cytosolic side of the plasma membrane (**FIGURE 6.19a**). A signaling pathway that involves a receptor guanylyl cyclase typically occurs as follows (**FIGURE 6.19b**):

1. A water-soluble extracellular messenger (the first messenger) binds to a receptor guanylyl cyclase at the exterior surface of the target cell's plasma membrane.
2. The receptor guanylyl cyclase dimerizes with a nearby receptor guanylyl cyclase that has also become bound with another copy of the same messenger.
3. Dimerization of the receptors activates the guanylyl cyclase components of the receptors to convert guanosine triphosphate (GTP) into the second messenger cGMP.
4. Cyclic GMP binds to and activates protein kinase G (PKG). Recall that PKG is a cGMP-dependent protein kinase that adds phosphate groups to serine and/or threonine amino acids on its target protein.
5. Activated PKG phosphorylates the effector protein. Phosphorylation either activates or inhibits the effector protein.
6. The phosphorylated effector protein causes the cellular response.

An example of a messenger that activates a signal transduction pathway involving a receptor guanylyl cyclase is atrial natriuretic peptide (ANP). ANP is released by atrial cells of the heart and functions in the kidneys to promote Na^+ and water excretion into the urine (natriuresis).

Recall that in smooth muscle cells, the intracellular receptor for nitric oxide has an enzymatic component that functions as guanylyl cyclase (see **FIGURE 6.16**). Hence, this intracellular receptor is a soluble, cytosolic form of a receptor guanylyl cyclase.

Signaling Through Enzyme-Coupled Receptors

An **enzyme-coupled receptor** is a type of plasma membrane receptor that contains a messenger-binding site, a transmembrane segment, and a site coupled to a separate enzyme. The messenger-binding site is on the extracellular side of the plasma membrane, the transmembrane segment extends through the plasma membrane, and the site coupled to the enzyme is on the cytosolic side of the plasma membrane (**FIGURE 6.20a**). Note that in an enzyme-coupled receptor, the receptor and the enzyme are *different* proteins. This is in contrast to an enzyme receptor, where the receptor and the enzyme are parts of the *same* protein.

For many enzyme-coupled receptors, the enzyme that is linked to the receptor is a **janus kinase (JAK)**, a type of tyrosine kinase (**FIGURE 6.20a**). Because a JAK contains two structural components that are almost identical, it is named after Janus, the Roman god with two faces. There are several types of JAKs, and different receptors are coupled to different JAKs. A signaling pathway that

FIGURE 6.19 A receptor guanylyl cyclase.

A receptor guanylyl cyclase contains a messenger-binding site, a transmembrane segment, and an enzymatic guanylyl cyclase component.

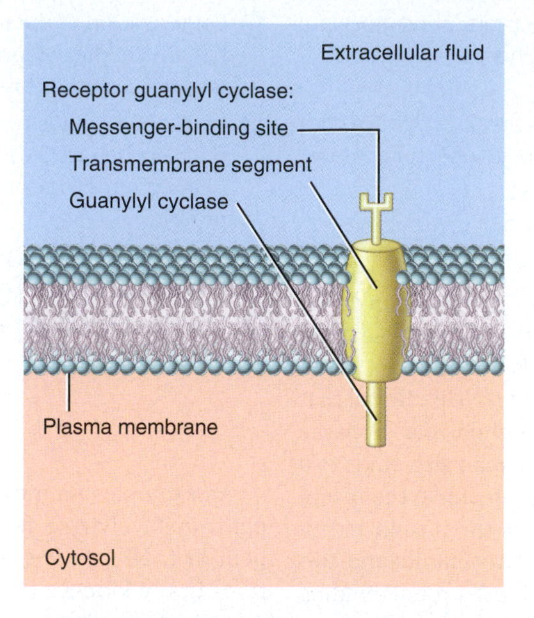

(a) Components of a receptor guanylyl cyclase

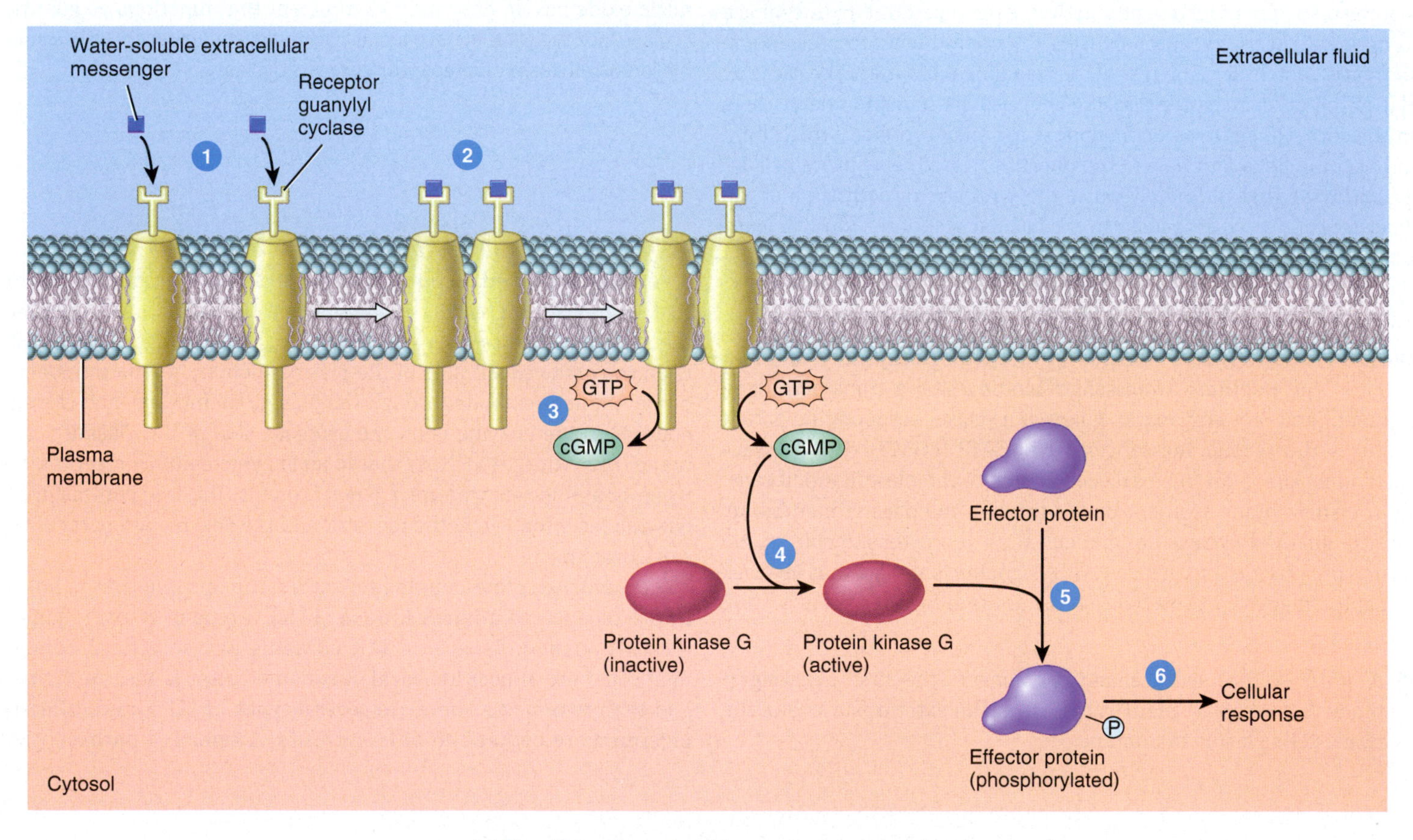

(b) Signaling pathway involving a receptor guanylyl cyclase

What is an example of a water-soluble extracellular messenger that binds to a receptor guanylyl cyclase?

FIGURE 6.20 An enzyme-coupled receptor. In part (b), STAT = signal transducer and activator of transcription.

An enzyme-coupled receptor contains a messenger-binding site, a transmembrane segment, and a site coupled to a separate enzyme.

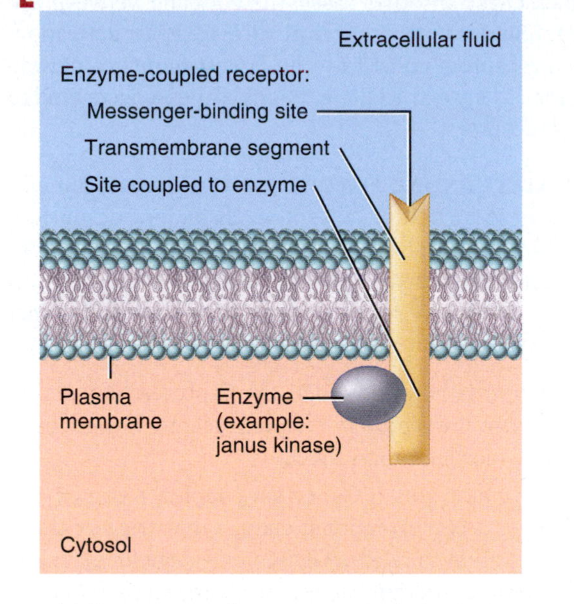

(a) Components of an enzyme-coupled receptor

contains a JAK-coupled receptor typically involves the following steps (**FIGURE 6.20b**):

1. A water-soluble extracellular messenger binds to a JAK-coupled receptor at the exterior surface of a target cell's plasma membrane.
2. The receptor dimerizes with a nearby receptor that has also become bound with another copy of the same messenger.
3. Dimerization of the receptors brings the two JAKs close together so that each JAK can phosphorylate tyrosine amino acids on the other JAK (cross-phosphorylation). The JAKs become activated and then phosphorylate tyrosine amino acids on the cytosolic portions of the receptors.
4. The phosphorylated sites on the receptors serve as docking sites where certain relay proteins called **signal transducers and activators of transcription (STATs)** can bind.
5. When STATs bind to the docking sites on the receptors, the JAKs phosphorylate the tyrosine amino acids on these proteins. Phosphorylation activates the STATs.
6. The activated STATs leave their docking sites, enter the nucleus of the cell, and bind to DNA to stimulate transcription of specific genes.

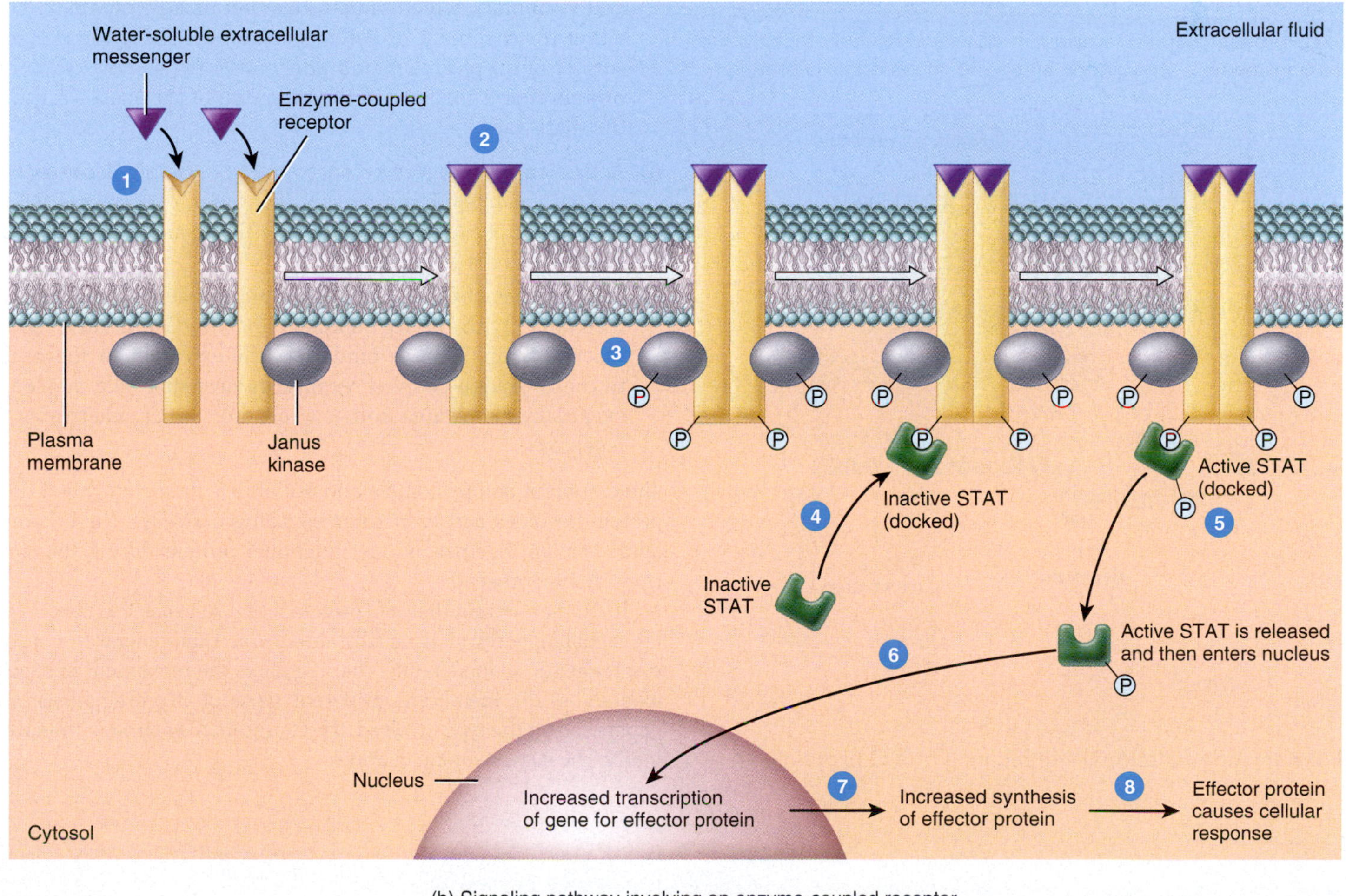

(b) Signaling pathway involving an enzyme-coupled receptor

What is the functional significance of a janus kinase (JAK)?

7. Activation of gene transcription leads to the synthesis of effector proteins.
8. The effector proteins cause the cellular response.

Among the messengers that activate signal transduction pathways involving JAK-coupled receptors are cytokines and several hormones, including growth hormone, prolactin, and erythropoietin. Cytokines play a role in mediating immune responses. Growth hormone causes growth and development of tissues such as muscle and bone. Prolactin causes milk production in the mammary glands of the breasts. Erythropoietin stimulates production of erythrocytes.

Signaling Through G Protein-Coupled Receptors

A **G protein-coupled receptor** is a type of plasma membrane receptor that contains a messenger-binding site, a transmembrane segment, and a site coupled to a G protein (FIGURE 6.21). The messenger-binding site is on the extracellular side of the plasma membrane, the transmembrane segment extends through the plasma membrane, and the site coupled to the G protein is on the cytosolic side of the plasma membrane. When a messenger binds to a G protein-coupled receptor, the receptor undergoes a conformational change that activates the G protein. The G protein in turn triggers the next step of the signal transduction pathway. Thus, a G protein serves as a relay protein that helps mediate the signal transduction process. In many cases, the G protein is already coupled to the receptor before the receptor is activated by messenger. In other cases, the coupling between the receptor and G protein does not occur until after receptor activation. G protein-coupled receptors constitute the largest family of plasma membrane receptors. Many extracellular chemical messengers bind to G protein-coupled receptors.

FIGURE 6.21 A G protein-coupled receptor.

A G protein-coupled receptor contains a messenger-binding site, a transmembrane segment, and a site coupled to a G protein.

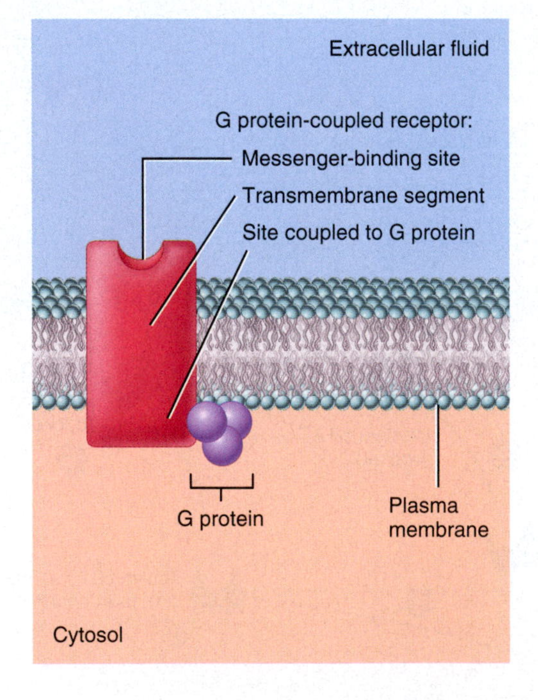

What is the functional significance of a G protein?

G Protein Organization and Function A **G protein** is a membrane protein that is so-named because it binds guanosine nucleotides, either guanosine diphosphate (GDP) or guanosine triphosphate (GTP). Three subunits comprise a G protein: alpha (α), beta (β), and gamma (γ) (FIGURE 6.22a). These components of the G protein function in the following way (FIGURE 6.22b):

1. When the α subunit is bound to GDP, the G protein is inactive. Such is the case when the G protein-coupled receptor has not been activated by an extracellular messenger.
2. Once a messenger activates a G protein-coupled receptor, the receptor undergoes a conformational change that triggers the α subunit of the G protein to release GDP in exchange for GTP. The binding of GTP to the α subunit causes the G protein to undergo a conformational change and become activated.
3. Upon activation, the α subunit typically separates from the β and γ subunits, which remain together as a $\beta\gamma$ complex. Then either the α subunit or $\beta\gamma$ complex moves along the cytosolic surface of the plasma membrane to alter the activity of a target protein that is involved in the next step of the signal transduction pathway.
4. Activation of the G protein does not last very long because the α subunit exhibits GTPase activity that quickly hydrolyzes GTP to GDP and a phosphate group. The GDP remains bound to the α subunit, and the phosphate group is released.
5. The presence of GDP on the α subunit causes the α subunit to recombine with the $\beta\gamma$ complex, and the G protein becomes inactive again. Because binding of the extracellular messenger to the G protein-coupled receptor is reversible, the messenger eventually dissociates from the receptor, which inactivates the receptor.

The α subunit and $\beta\gamma$ complex do not always dissociate when the G protein becomes active. In some signaling pathways, these components remain together as one functional protein during the signal transduction process.

The target protein that a G protein alters to cause the next step of the signaling pathway is usually an enzyme that generates a second messenger or an ion channel that regulates the movement of specific ions across the membrane. Different types of G proteins exist and, depending on the type involved, the G protein may either stimulate or inhibit the enzyme or ion channel.

FIGURE 6.22 The organization and function of a G protein.

A G protein contains three subunits: alpha (α), beta (β), and gamma (γ).

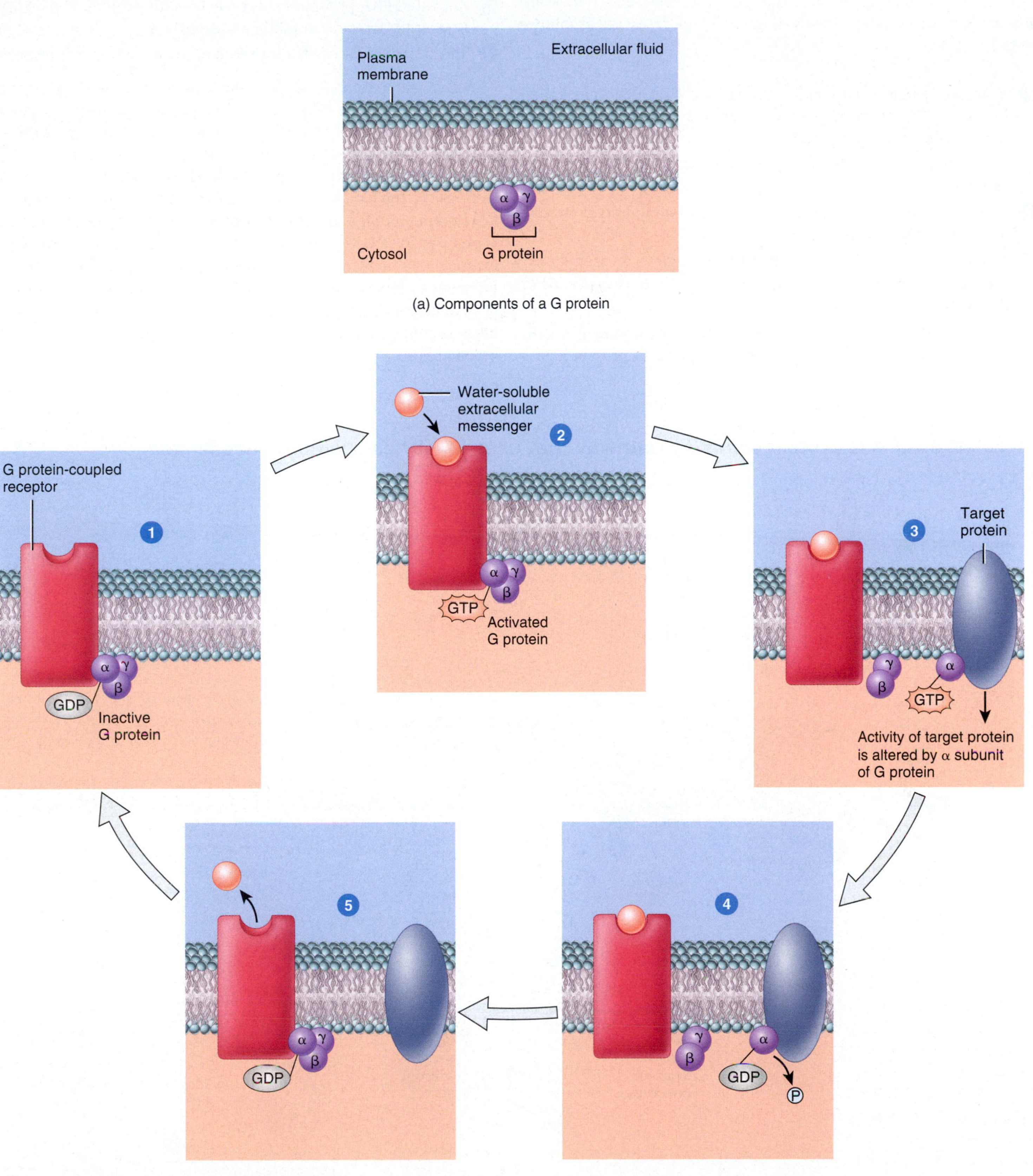

(a) Components of a G protein

(b) G protein function

Which subunit of a G protein binds to either GDP or GTP?

G Protein Signaling Pathways That Use Cyclic AMP as a Second Messenger In many G protein signaling pathways, the G protein stimulates an enzyme that generates the second messenger cyclic AMP. The steps involved in such a pathway are as follows (FIGURE 6.23):

1. A water-soluble extracellular messenger (the first messenger) binds to a G protein-coupled receptor at the exterior surface of a target cell's plasma membrane.
2. The messenger–receptor complex activates a type of G protein called $\mathbf{G_s}$ (s = stimulatory).
3. The activated G_s protein stimulates a membrane-bound enzyme called **adenylyl cyclase**.
4. Adenylyl cyclase converts ATP into the second messenger **cyclic adenosine monophosphate**, also known as **cyclic AMP** or **cAMP**.
5. Cyclic AMP binds to and activates a protein kinase known as **protein kinase A (PKA)**. PKA is a cAMP-dependent protein kinase that phosphorylates serine and/or threonine amino acids on its target protein.
6. Activated PKA phosphorylates the effector protein. Phosphorylation either activates or inhibits this protein.
7. The phosphorylated effector protein causes the cellular response.

Many water-soluble extracellular messengers, such as thyroid-stimulating hormone (TSH), exert their physiological effects through *increased* synthesis of cAMP via activation of G_s. TSH causes the synthesis and secretion of thyroid hormones by the thyroid gland.

In some signaling pathways, the level of cyclic AMP *decreases* in response to the binding of a water-soluble extracellular messenger to a G protein-coupled receptor. This process involves the activation of a G protein called $\mathbf{G_i}$ (i = inhibitory), which inhibits adenylyl cyclase. An example of a messenger that activates this type of pathway is growth hormone–inhibiting hormone (GHIH). GHIH inhibits the release of growth hormone from the pituitary gland, a function that is important because it helps regulate the concentration of growth hormone in the bloodstream.

FIGURE 6.23 A G protein signaling pathway that uses cAMP as a second messenger.

Cyclic AMP helps mediate signal transduction by activating protein kinase A.

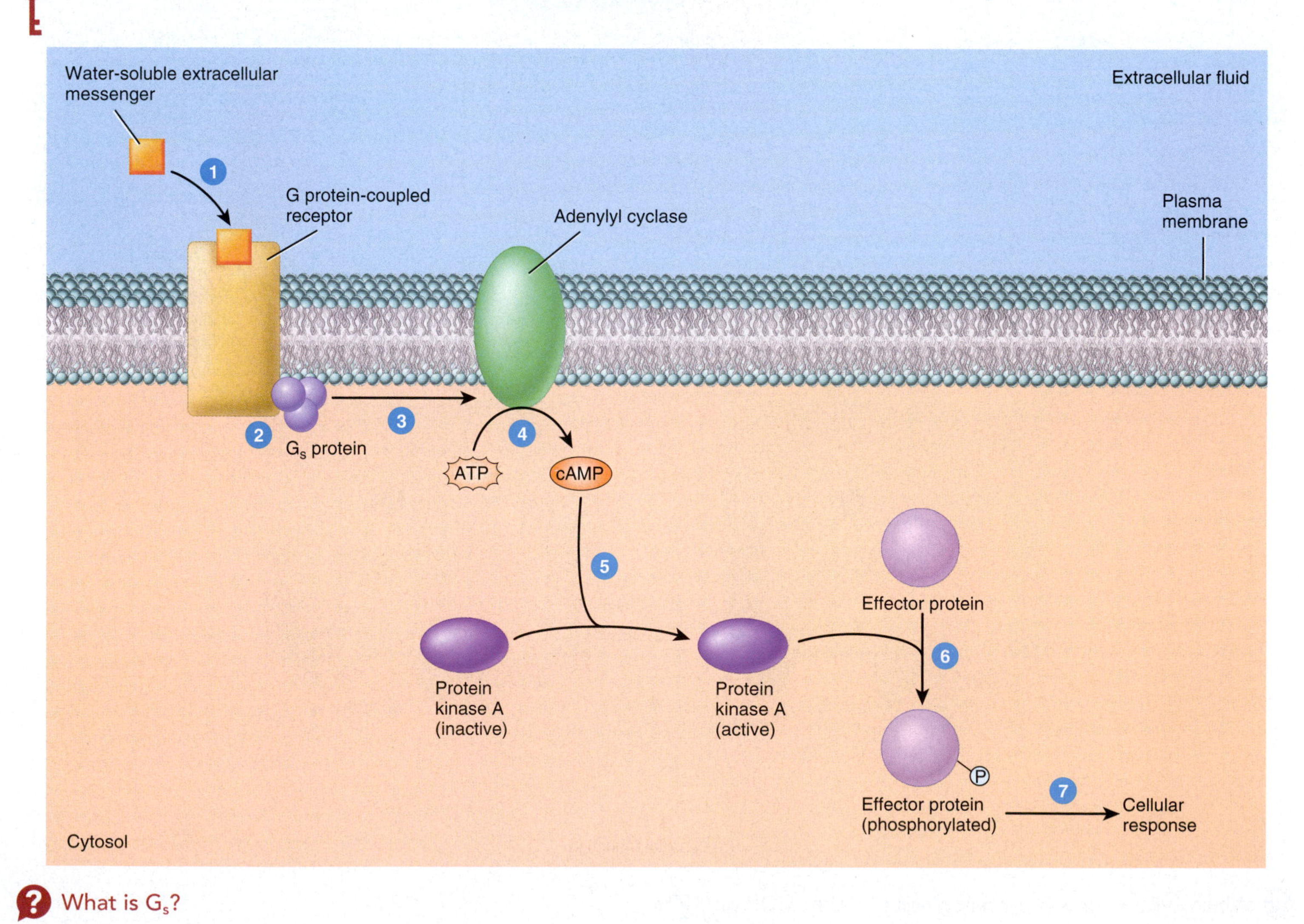

What is G_s?

G Protein Signaling Pathways That Use IP_3, DAG, and Calcium as Second Messengers Other substances besides cAMP can be used as second messengers in G protein signaling pathways. Examples include inositol trisphosphate (IP_3), diacylglycerol (DAG), and calcium ions (Ca^{2+}). When these substances function as second messengers, they are often released as part of the same signal transduction pathway. The steps of this pathway are listed below (FIGURE 6.24):

1. A water-soluble extracellular messenger (the first messenger) binds to a G protein-coupled receptor at the exterior surface of a target cell's plasma membrane.
2. The messenger–receptor complex activates a type of G protein called $\mathbf{G_q}$.
3. The activated G_q protein activates a membrane-bound enzyme called **phospholipase C**.
4. Phospholipase C cleaves a membrane phospholipid called phosphatidylinositol bisphosphate (PIP_2) to form the second messengers **inositol trisphosphate (IP_3)** and **diacylglycerol (DAG)**.
5. IP_3 is released from the plasma membrane, diffuses through the cytosol, and then binds to and opens IP_3-gated channels in the membrane of the endoplasmic reticulum (ER). The ER is a membranous organelle that stores calcium ions (Ca^{2+}).
6. After the IP_3-gated channels open, Ca^{2+} is released from the lumen of the ER into the cytosol to act as another second messenger.
7. Both DAG, which remains in the plasma membrane, and Ca^{2+} bind to and activate **protein kinase C (PKC)**. PKC is a calcium-sensitive protein kinase that phosphorylates serine and/or threonine amino acids on its target protein.
8. Activated PKC phosphorylates the effector protein. Phosphorylation either activates or inhibits this protein.
9. The phosphorylated effector protein causes the cellular response.

An example of a water-soluble extracellular messenger that activates a G protein signaling pathway that uses IP_3, DAG, and Ca^{2+} as second messengers is thrombin. Thrombin stimulates blood clotting

FIGURE 6.24 A G protein signaling pathway that uses IP_3, DAG, and Ca^{2+} as second messengers. IP_3 = inositol trisphosphate; DAG = diacylglycerol; and Ca^{2+} = calcium ion.

In some G protein signaling pathways, IP_3, DAG, and Ca^{2+} function as second messengers.

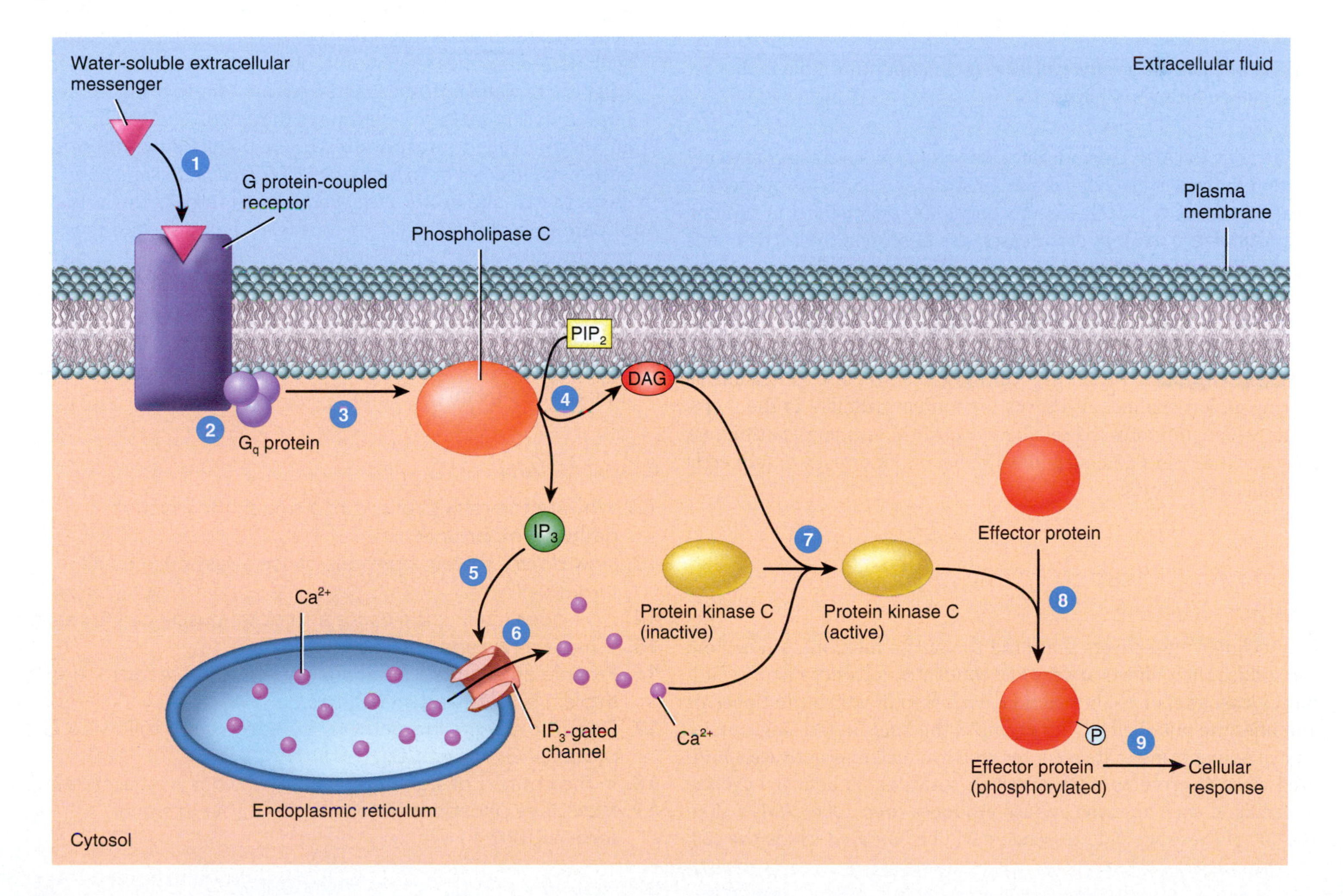

What is the purpose of IP_3?

CLINICAL CONNECTION

Cholera Toxin, Pertussis Toxin, and G Proteins

Two G proteins, G_s and G_i, can be altered by bacterial toxins. **Cholera toxin** is produced by the bacterium ***Vibrio cholerae***, which causes cholera, an intestinal disorder characterized by severe diarrhea. Cholera toxin chemically modifies G_s in intestinal cells in such a way that G_s becomes activated and cannot not be shut off. As a result, there is sustained activation of adenylyl cyclase, and elevated levels of cAMP are produced. The increased cAMP concentration leads to the secretion of large amounts of salt and water by intestinal cells, causing severe diarrhea. **Pertussis toxin** is produced by the bacterium ***Bordetella pertussis***, which causes whooping cough (pertussis). Pertussis toxin chemically modifies G_i, causing it to become permanently inactive. As a result, G_i is unable to inhibit adenylyl cyclase. This inability to inhibit adenylyl cyclase ultimately leads to whooping cough, but the mechanism by which this occurs is not well understood. Cholera toxin and pertussis toxin are important experimentally because they can be used to determine if a G_s protein or G_i protein is involved in a particular signaling pathway. For example, if a G protein in a given pathway is sensitive to cholera toxin, then G_s is involved; if a G protein in another pathway is sensitive to pertussis toxin, then that pathway involves G_i.

and platelet activation. You will learn more about thrombin in the discussion on clotting in Chapter 16.

G PROTEIN REGULATION OF ION CHANNELS In several signal transduction pathways involving G proteins, an ion channel is opened. In some cases, the G protein directly opens the ion channel by binding to it (**FIGURE 6.25a**). In other cases, the G protein indirectly opens the ion channel by activating a second messenger pathway (**FIGURE 6.25b**). Regardless of which of these mechanisms is used, the net effect is that an ion channel is opened. This alters the permeability of the plasma membrane to certain ions, which in turn causes the cellular response. G protein signaling pathways can also close ion channels. Closing an ion channel also alters the ion permeability of the membrane, which promotes the cellular response. Many neurotransmitters activate G protein signaling pathways that either directly or indirectly regulate ion channels.

Signal Amplification Promotes a Greater Cellular Response

Extracellular messengers that bind to plasma membrane receptors can induce their effects at very low concentrations because they initiate a **cascade**, or chain reaction, each step of which multiplies or amplifies the initial effect. For example, the binding of a single molecule of epinephrine to its receptor on a liver cell may activate a hundred or so G proteins, each of which activates an adenylyl cyclase molecule. If each adenylyl cyclase produces even 1000 cAMPs, then 100,000 of these second messengers will be liberated inside the cell. Each cAMP may activate a protein kinase, which in turn can act on hundreds or thousands of target proteins. Some of the protein kinases phosphorylate and activate the effector protein glycogen phosphorylase, an enzyme needed for glycogen breakdown. The end result of the binding of a single molecule of epinephrine to its receptor is the breakdown of millions of glycogen molecules into glucose monomers.

Signal Termination Prevents Overstimulation of the Target Cell

A signal transduction pathway must be terminated to prevent the target cell from being overstimulated. For signal termination to occur, the extracellular messenger must dissociate from the receptor, and the components of the signal transduction pathway must be inactivated.

- ***Dissociation of the extracellular messenger from the receptor.*** Recall that the binding of an extracellular messenger to a receptor involves noncovalent intermolecular forces. The binding of the chemical messenger to the receptor is reversible, so when the chemical messenger leaves the receptor, the receptor is unable to activate the rest of the signaling pathway.
- ***Inactivation of components of the signal transduction pathway.*** Although the receptor is no longer activated in the absence of extracellular messenger, components of the signaling pathway such as second messengers and phosphorylated proteins may still be active and must be inactivated in order to bring the cellular response to an end. In pathways that involve cAMP or cGMP, an enzyme called **phosphodiesterase** breaks down these second messengers. For example, *cAMP phosphodiesterase* breaks down cAMP, and *cGMP phosphodiesterase* breaks down cGMP. In pathways that use calcium ions (Ca^{2+}) as a second messenger, reuptake of Ca^{2+} back into the endoplasmic reticulum decreases the concentration of this ion in the cytosol. In pathways that involve phosphorylation, the phosphorylated proteins are dephosphorylated by enzymes called **phosphatases**. Dephosphorylation alters the activity of these proteins—in most cases, it inactivates them.

CHECKPOINT

12. Why are the relay proteins of a signal transduction pathway important?
13. How does a steroid hormone cause a response by its target cell?
14. How does nitric oxide cause a smooth muscle cell to relax?
15. What is a tyrosine kinase?
16. Provide an example of an enzyme-coupled receptor and explain how it exerts its effects.
17. Describe a signaling pathway for each of the following G proteins: G_s, G_i, and G_q.
18. In the context of signal transduction, what is a cascade?
19. How does phosphodiesterase play a role in signal termination?

FIGURE 6.25 G protein regulation of ion channels. The break in the pathway in part (b) indicates that specific steps have been omitted for simplicity.

A G protein can regulate the opening (or closing) of an ion channel either directly (by binding to it) or indirectly (by activating a second messenger pathway).

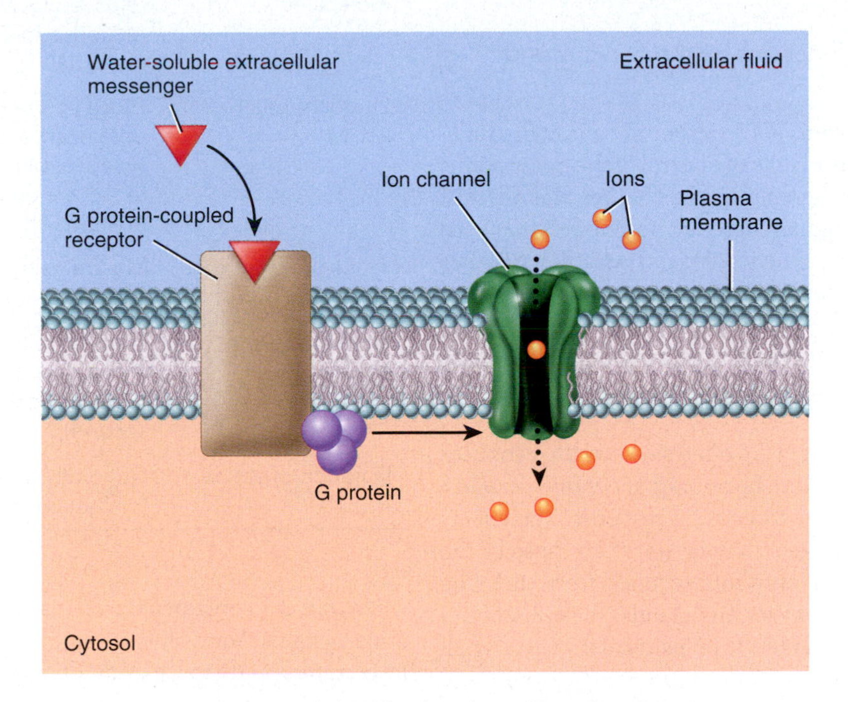

(a) Direct regulation of an ion channel by a G protein

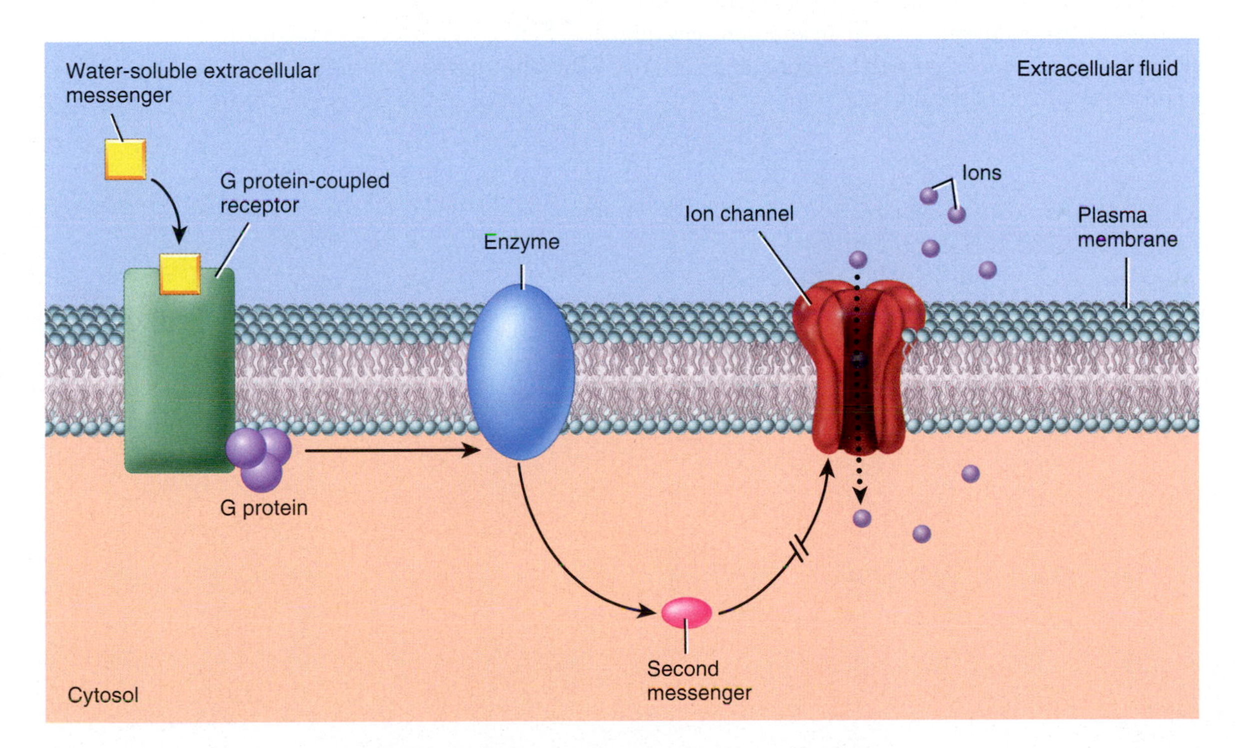

(b) Indirect regulation of an ion channel by a G protein

Which group of extracellular messengers most often activates G protein signaling pathways that regulate ion channels?

6.5 Comparison of the Nervous and Endocrine Systems

OBJECTIVE

- Compare control of body functions by the nervous and endocrine systems.

The nervous and endocrine systems act together to coordinate functions of all body systems, but the means of control of the two systems are very different. As mentioned earlier in the chapter, the nervous system acts through electrical signals—graded potentials (conveyed along dendrites or cell bodies of neurons) and action potentials (conducted along axons of neurons). At synapses, action potentials trigger the release of neurotransmitters, which alter the activity of nearby target cells. The endocrine system controls body activities by releasing hormones, which are carried by the bloodstream to distant target cells.

Responses of the endocrine system are often slower than responses of the nervous system; although some hormones act within seconds, most take several minutes or more to cause a response. The effects of nervous system activation are generally briefer than those of the endocrine system (usually taking only milliseconds to occur). The nervous system acts on specific muscles and glands. The influence of the endocrine system is much broader; it helps regulate virtually all types of body cells.

The nervous and endocrine systems can also influence each other. For example, certain parts of the nervous system stimulate or inhibit the release of hormones by the endocrine system. In addition, several hormones, especially thyroid hormones, growth hormone, and insulin influence growth and development of the nervous system.

TABLE 6.1 Comparison of Control by the Nervous and Endocrine Systems

Characteristic	Nervous System	Endocrine System
Messenger molecules	Neurotransmitters released locally in response to action potentials.	Hormones delivered to tissues throughout the body by the blood.
Site of mediator action	Close to site of release, at a synapse; binds to receptors on or in postsynaptic cell.	Far from site of release (usually); binds to receptors on or in target cells.
Types of target cells	Muscle cells, gland cells, other neurons.	Cells throughout the body.
Time to onset of action	Typically within milliseconds (thousandths of a second).	Seconds to hours or days.
Duration of action	Generally briefer (milliseconds).	Generally longer (seconds to days).

TABLE 6.1 compares the characteristics of the nervous and endocrine systems. You will learn much more about the nervous system in Chapters 7–10. In Chapter 13, you will learn about the endocrine system.

CHECKPOINT

20. What types of molecules are involved in nervous system control of the body? In control by the endocrine system?

FROM RESEARCH TO REALITY

Cell Receptors and Disease

Reference

Chen, W. et al. (2013). Deficiency of G protein-coupled bile acid receptor Gpbar1 (TGR5) enhances chemically induced liver carcinogenesis. *Hepatology*. 57: 656–666.

Could cell receptors play a role in certain diseases?

The receptors of our cells can make us more or less susceptible to diseases. The authors of this study hope to learn more about a G protein-coupled receptor found in the liver and what role it might play in susceptibility to liver damage and cancer. Could the presence, or absence, of this cell receptor make someone more likely to develop liver disease?

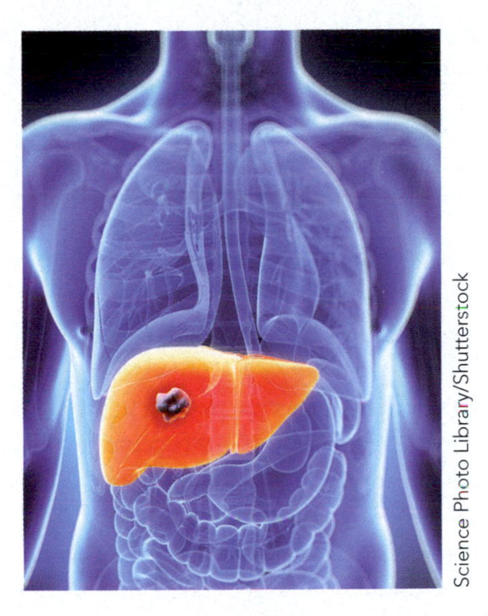

Science Photo Library/Shutterstock

Article description:

The authors of this study were interested in learning more about liver cancer. Specifically they wanted to investigate the effect of a G protein-coupled receptor (TGR5, the bile acid receptor) on liver damage and cancer.

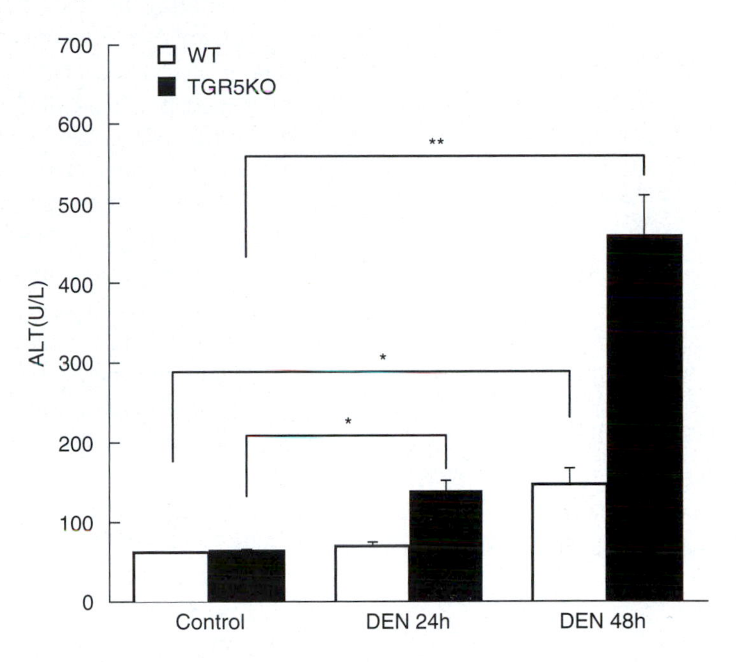

Go to WileyPLUS Learning Space and use the data from this article to answer the questions posed there and to learn more about cell receptors and liver damage.

CHAPTER REVIEW

6.1 Methods of Cell-to-Cell Communication

1. Cell signaling refers to the communication between the cells of the body.
2. There are three general methods that cells can use to communicate with one another: (1) gap junctions, (2) cell-to-cell binding, and (3) extracellular chemical messengers.
3. Signaling through extracellular messengers is the most common method of communication between cells.

6.2 Extracellular Chemical Messengers

1. An extracellular chemical messenger is a molecule that is released by a cell, enters extracellular fluid, and then binds to a receptor on or inside its target cell to cause a response.

2. The three main types of extracellular messengers are hormones, neurotransmitters, and local mediators.
3. Hormones are extracellular messengers that are carried by the blood to distant target cells.
4. Neurotransmitters are extracellular messengers that are released from a neuron into a synapse in order to reach a nearby target cell.
5. Local mediators are extracellular messengers that act on nearby target cells without entering the bloodstream. Paracrines are local mediators that act on neighboring cells. Autocrines are local mediators that act on the same cell that secreted them.
6. A given extracellular messenger can function as a hormone, neurotransmitter, or local mediator, depending on where it is secreted. For example, when certain neurons of the brain release norepinephrine at a synapse, norepinephrine functions as a neurotransmitter; when neurons in the inner adrenal gland release norepinephrine into the bloodstream, norepinephrine functions as a hormone.
7. Extracellular messengers can be divided into two broad chemical classes: water-soluble messengers and lipid-soluble messengers.
8. To get to their target cells after being secreted, neurotransmitters and local mediators simply diffuse through interstitial fluid. Hormones initially diffuse through interstitial fluid and then enter the bloodstream to reach their target cells. While in the blood, water-soluble hormones are in free form, whereas lipid-soluble hormones are bound to transport proteins.

6.3 Receptors

1. Extracellular chemical messengers influence their target cells by binding to protein receptors.
2. The binding of messenger to a receptor exhibits four properties: specificity, affinity, saturation, and competition.
3. The receptors for water-soluble extracellular messengers are present in the plasma membrane of the target cell (plasma membrane receptors); the receptors for lipid-soluble extracellular messengers are present inside the target cell, either in the cytosol or nucleus (intracellular receptors).
4. Receptors are subject to regulation. In down-regulation, the number of target-cell receptors decreases; in up-regulation, the number of target-cell receptors increases.

6.4 Signal Transduction Pathways

1. A signal transduction pathway, or signaling pathway, refers to the sequence of events that occurs between the binding of an extracellular messenger to its receptor and the cellular response by the target cell. It usually involves a series of relay proteins and an effector protein. The effector protein is a key protein that causes the cellular response.
2. In many signaling pathways, binding of an extracellular messenger (the primary messenger) to its receptor leads to the generation of a second messenger inside the cell to help mediate the signal transduction process.
3. Signal transduction pathways may also involve a protein kinase, an enzyme that phosphorylates a target protein. The target protein may be activated or inhibited, depending on which target protein is phosphorylated.
4. Steroid hormones, thyroid hormones, and nitric oxide initiate signal transduction pathways by binding to intracellular receptors. Steroid hormones and thyroid hormones cause cellular responses in their target cells by altering gene expression. Nitric oxide causes a cellular response in its target cell by causing the generation of cGMP.

5. Peptide or protein hormones, amine hormones, nearly all neurotransmitters, and most local mediators initiate signal transduction pathways by binding to plasma membrane receptors. There are four types of plasma membrane receptors: ion channel receptors, enzyme receptors, enzyme-coupled receptors, and G protein-coupled receptors.
6. In signaling through an ion channel receptor, binding of an extracellular messenger to the receptor opens or closes an ion channel component of the receptor, which alters the ion permeability of the membrane.
7. In signaling through an enzyme receptor, binding of an extracellular messenger to the receptor activates an enzyme component of the receptor (either a tyrosine kinase or guanylyl cyclase), which, in turn, triggers the rest of the signaling pathway.
8. In signaling through an enzyme-coupled receptor, binding of an extracellular messenger to the receptor activates a separate enzyme (often a janus kinase), which promotes the next step of the signaling pathway.
9. In signaling through a G protein-coupled receptor, binding of an extracellular messenger to the receptor activates a G protein, which in turn causes the next part of the signaling pathway. A G protein consists of three subunits: alpha (α), beta (β), and gamma (γ). In many signaling pathways, when the G protein is activated, the α subunit separates from the $\beta\gamma$ complex and then either the α subunit or $\beta\gamma$ complex alters the activity of an enzyme or ion channel.
10. In many G protein signaling pathways, cAMP is used as a second messenger. In other G protein signaling pathways, inositol trisphosphate (IP_3), diacylglycerol (DAG), and calcium ions (Ca^{2+}) are used as second messengers. A G protein may also directly or indirectly regulate the opening or closing of an ion channel.
11. During signal amplification, there is a cascade (chain reaction), in which each step multiplies or amplifies the initial effect.
12. For signal termination to occur, the extracellular messenger must dissociate from the receptor, and the components of the signal transduction pathway must be inactivated.

6.5 Comparison of the Nervous and Endocrine Systems

1. The nervous system controls homeostasis through action potentials and neurotransmitters, which act locally and quickly. The endocrine system uses hormones, which act more slowly in distant parts of the body.
2. The nervous system controls neurons, muscle cells, and gland cells; the endocrine system regulates virtually all body cells.

PONDER THIS

1. Mike is interested in developing a new drug that will target the cells and induce intracellular changes by means of gene transcription. After several months of observing his drug and its mechanism of action in cells *in vitro*, Mike is pleased with his results and is ready to enter into human trials. Much to his dismay, his new drug does not have significant effects in his subjects. Predict which factor associated with the drug itself might be causing this observation and explain what Mike could do to make the effects of his drug more significant.

2. You have discovered a new organism that displays a very strange type of smooth muscle contraction. These muscle fibers all contract individually rather than as a unit. What method of cell-to-cell communication would you predict that this animal does not have? Explain your answer.

3. Upon stimulation of a G protein-coupled receptor that uses cAMP as a second messenger, you notice that the cell does not produce an intracellular response. In order to figure out what might be going wrong with this cell, you run several tests and realize that while cAMP is activated, it is quickly degraded. This prevents cAMP levels from getting high enough to act effectively as the second messenger. What chemical could be elevated in this cell, and how could you fix the problem?

ANSWERS TO FIGURE QUESTIONS

6.1 The most common method of cell-to-cell communication is communication through extracellular chemical messengers.

6.2 An example of a cellular response that can occur after an extracellular messenger binds to its receptor is cell growth. Other possible examples include protein synthesis, muscle contraction, and transport of substances across the plasma membrane.

6.3 The target cell for a neuron at a synapse can be another neuron, a muscle cell, or a gland cell.

6.4 Leukotrienes stimulate chemotaxis of white blood cells and mediate inflammation.

6.5 Norepinephrine, which is an amine hormone and a neurotransmitter, is water-soluble.

6.6 Lipid-soluble hormones are overall hydrophobic (nonpolar), whereas blood plasma is hydrophilic (polar) because it is mainly composed of water. Therefore, lipid-soluble hormones require a transport protein to move through the bloodstream.

6.7 Specificity occurs because the binding site of a receptor has a particular three-dimensional shape, and only a messenger with a compatible shape can bind to that site.

6.8 Saturation occurs because of a limited number of receptors for a given messenger in a target cell; as the messenger concentration increases, more receptors are bound by messenger until they are all occupied.

6.9 An agonist binds to and activates a receptor, mimicking the effect of the endogenous messenger. An antagonist binds to and blocks a receptor, which prevents the endogenous messenger from exerting its effect.

6.10 Because water-soluble extracellular messengers are unable to pass through the hydrophobic interior of the plasma membrane, they must bind to plasma membrane receptors rather than intracellular receptors.

6.11 Down-regulation is an example of negative feedback.

6.12 The effector protein is the component of the signal transduction pathway that causes the cellular response. Relay proteins convey the signal between the receptor and the effector protein.

6.13 Examples of second messengers include cyclic AMP (cAMP), cyclic GMP (cGMP), diacylglycerol (DAG), inositol trisphosphate (IP_3), and calcium ions (Ca^{2+}).

6.14 After a protein kinase phosphorylates its target protein, the target protein is either activated or inhibited.

6.15 The receptor–hormone complex alters gene expression by activating or inhibiting the transcription of a specific gene.

6.16 Myosin light chain kinase (MLCK) is a protein that promotes the interaction of myosin and actin to cause contraction of a smooth muscle cell. When MLCK is inactive due to phosphorylation, myosin and actin do not interact, which causes the smooth muscle cell to relax.

6.17 The entry of ions into the target cell causes a cellular response by changing the electrical properties of the cell. For example, the cell may become excited or inhibited.

6.18 Growth factors and insulin are examples of water-soluble extracellular messengers that bind to receptor tyrosine kinases.

6.19 Atrial natriuretic peptide (ANP) is an example of a water-soluble extracellular messenger that binds to a receptor guanylyl cyclase.

6.20 A janus kinase (JAK) is an enzyme that can be coupled to a receptor. When activated, the JAK phosphorylates the cytosolic portion of the receptor, creating docking sites for STATs.

6.21 A G protein serves as a relay protein that helps mediate signal transduction.

6.22 The alpha (α) subunit of a G protein can bind to either GDP or GTP.

6.23 G_s is a type of G protein that stimulates adenylyl cyclase (s = stimulatory).

6.24 IP_3 binds to and opens IP_3-gated channels in the endoplasmic reticulum (ER), causing Ca^{2+} to be released from the lumen of the ER into the cytosol.

6.25 Many neurotransmitters activate G protein signaling pathways that regulate ion channels.

7 The Nervous System and Neuronal Excitability

The Nervous System, Neuronal Excitability, and Homeostasis

Neuronal excitability contributes to homeostasis by allowing the nervous system to communicate with and regulate most body organs.

LOOKING BACK TO MOVE AHEAD...

- Ion channels are membrane proteins that contain pores through which specific ions can pass (Section 3.2).
- The Na^+/K^+ ATPase is a carrier protein that actively transports three Na^+ ions out of a cell and two K^+ ions into a cell (Section 5.5).
- Synaptic signaling refers to communication between a neuron and a target cell by the release of neurotransmitter molecules at a synapse (Section 6.2).
- A G protein is a membrane protein that regulates the activity of membrane enzymes or ion channels during signal transduction (Section 6.4).

The nervous system plays a major role in maintaining homeostasis: It keeps most controlled variables within limits that sustain life. When stimuli cause controlled variables to deviate from their set points, neurons of the nervous system respond rapidly by forming electrical signals (graded potentials and action potentials). The ability to produce action potentials in response to stimuli is referred to as **electrical excitability**. Once generated, neurons use action potentials to communicate with other neurons, muscle fibers, or gland cells in one or more organs of the body. This, in turn, regulates the organ's activity and ultimately restores homeostatic conditions.

Besides helping to maintain homeostasis, the nervous system is also responsible for your sensations, behaviors, and memories, and initiates all voluntary movements. Because the nervous system is quite complex, different aspects of its function will be considered in several related chapters. This chapter focuses on the organization of the nervous system and the electrical excitability of neurons. Chapter 8 examines the functions of the central nervous system (brain and spinal cord), and Chapter 9 discusses how the sensory systems detect, conduct, and process incoming sensory information. Exploration of the nervous system concludes with a discussion of how the autonomic and somatic nervous systems convey motor output to muscles and glands (Chapter 10).

7.1 Overview of the Nervous System

OBJECTIVES

- Describe the subdivisions of the nervous system.
- Explain the functions of the nervous system.

The Nervous System Is Organized into the Central Nervous System and the Peripheral Nervous System

Collectively, the nervous tissues of the body constitute the **nervous system**. With a mass of only 2 kg (4.5 lb), about 3% of the total body weight, the nervous system is one of the smallest and yet the most complex of the 12 body systems. There are two main subdivisions of the nervous system: the central nervous system and the peripheral nervous system.

Central Nervous System

The **central nervous system (CNS)** consists of the brain and spinal cord (FIGURE 7.1a). The *brain* is the part of the CNS that is located in the skull. The *spinal cord* is connected to the brain and is enclosed by the bones of the vertebral column. The CNS processes many different kinds of incoming sensory information. It is also the source of thoughts, emotions, and memories. Most signals that stimulate muscles to contract and glands to secrete originate in the CNS.

Peripheral Nervous System

The **peripheral nervous system (PNS)** consists of all nervous tissue outside the CNS (FIGURE 7.1a). Components of the PNS include nerves and sensory receptors. A *nerve* is a bundle of axons (described shortly) that lies outside the brain and spinal cord. Twelve pairs of *cranial nerves* emerge from the brain and thirty-one pairs of *spinal nerves* emerge from the spinal cord. A *sensory receptor* is a structure that monitors changes in the external or internal environment. Examples of sensory receptors include touch receptors in the skin, olfactory (smell) receptors in the nose, and stretch receptors in the stomach wall.

The PNS is divided into afferent and efferent divisions. The **afferent division** of the PNS conveys input into the CNS from sensory receptors in the body (FIGURE 7.1b). This division provides the CNS with sensory information about the *somatic senses* (tactile, thermal, pain, and proprioceptive sensations) and *special senses* (smell, taste, vision, hearing, and equilibrium).

The **efferent division** of the PNS conveys output from the CNS to **effectors** (muscles and glands). This division is further subdivided into a somatic nervous system and an autonomic nervous system (FIGURE 7.1b). The **somatic nervous system** conveys output from the CNS to *skeletal muscles* only. Because its motor responses can be consciously controlled, the action of this part of the PNS is *voluntary*. The **autonomic nervous system (ANS)** conveys output from the CNS to *smooth muscle*, *cardiac muscle*, and *glands*. Because its motor responses are not normally under conscious control, the action of the ANS is *involuntary*. The ANS is comprised of two main branches, the **parasympathetic nervous system** and the **sympathetic nervous system**. With a few exceptions, effectors receive innervation from both of these branches, and usually the two branches have opposing actions. For example, neurons of the sympathetic nervous system increase heart rate, and neurons of the parasympathetic nervous system slow it down. In general, the parasympathetic nervous system takes care of "rest-and-digest" activities, and the sympathetic nervous system helps support exercise or emergency actions—the so-called "fight-or-flight" responses. A third branch of the autonomic nervous system is the **enteric nervous system (ENS)**, an extensive network of neurons confined to the wall of the gastrointestinal (GI) tract. The ENS helps regulate the activity of the smooth muscle and glands of the GI tract. Although the ENS can function independently, it communicates with and is regulated by the other branches of the ANS.

The Nervous System Performs Sensory, Integrative, and Motor Functions

The diverse activities of the nervous system can be grouped into three basic functions: sensory, integrative, and motor.

- ***Sensory function.*** Sensory receptors *detect* external or internal stimuli, such as a raindrop landing on your arm or an increase in blood acidity. This sensory information is then conveyed through cranial and spinal nerves of the PNS into the brain and spinal cord of the CNS.
- ***Integrative function.*** The CNS processes sensory information by analyzing it and making decisions for appropriate responses—an activity known as **integration**.

FIGURE 7.1 Organization of the nervous system. (a) Subdivisions of the nervous system. (b) Nervous system organizational chart.

The two main subdivisions of the nervous system are (1) the central nervous system (CNS), which consists of the brain and spinal cord, and (2) the peripheral nervous system, which consists of all nervous tissue outside the CNS.

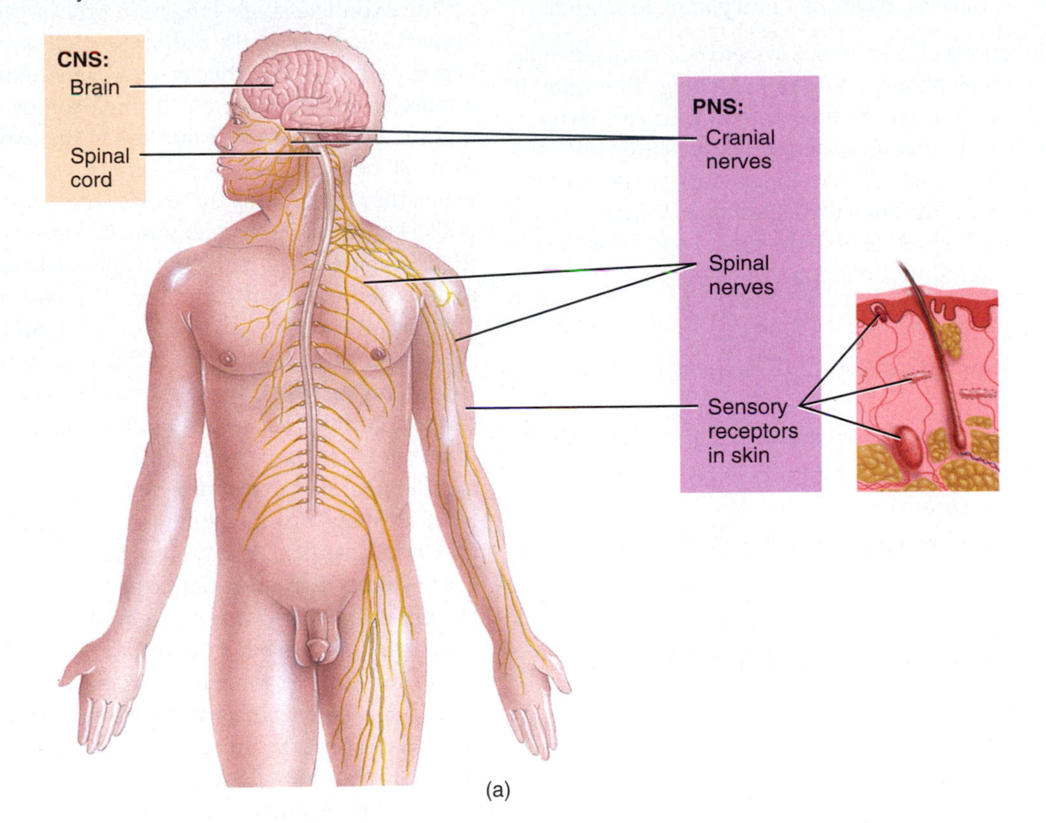

(a)

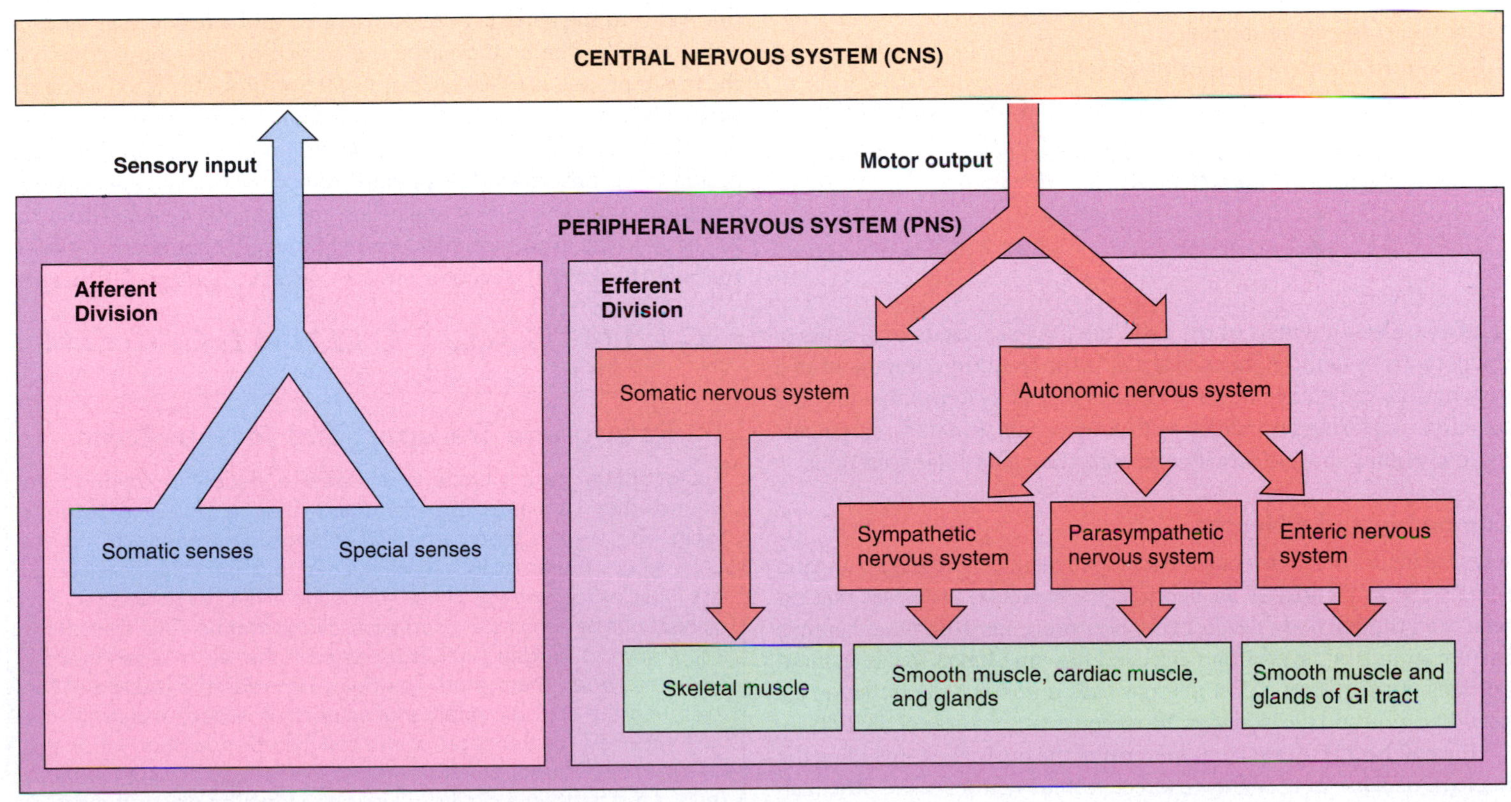

(b)

What is the function of the efferent division of the PNS?

- ***Motor function.*** Once sensory information is integrated, the CNS may elicit an appropriate motor response. For this to occur, motor information is conveyed from the CNS through cranial and spinal nerves of the PNS to effectors (muscles and glands). Stimulation of the effectors causes muscles to contract and glands to secrete.

The three basic functions of the nervous system occur, for example, when you answer your cell phone after hearing it ring. The sound of the ringing phone stimulates sensory receptors in your ears (sensory function). This auditory information is subsequently relayed into your brain, where it is processed and the decision to answer the phone is made (integrative function). The brain then stimulates the contraction of specific muscles that allow you to grab the phone and press the appropriate button to answer it (motor function).

CHECKPOINT

1. What are the components of the CNS? Why is the CNS important?
2. What purpose does the afferent division of the PNS serve?
3. How does the somatic nervous system differ from the autonomic nervous system?
4. Explain the concept of integration and provide an example.

7.2 Cells of the Nervous System

OBJECTIVES

- Discuss the functions of neurons.
- List the roles of neuroglia.
- Explain the importance of myelination.
- Describe the ability to repair neurons in the CNS and PNS.

The nervous system consists of two principal types of cells: neurons and neuroglia (supporting cells).

Neurons Are Responsible for the Main Functions of the Nervous System

Neurons, also known as **nerve cells**, are the basic functional units of the nervous system. (A *functional unit* is the smallest component of a system that can carry out the functions of that system.) Examples of functions performed by neurons include sensing, thinking, remembering, controlling muscle activity, and regulating glandular secretions.

Components of a Neuron

A typical neuron has three major parts: a cell body, dendrites, and an axon (FIGURE 7.2). **Dendrites** are short, highly branched processes that extend from the cell body. Because they receive signals from other neurons or from stimuli in the environment, dendrites function as the main input portions of the neuron. Most neurons have numerous dendrites, an aspect that substantially increases the receptive surface area of the cell.

The **cell body (soma)** contains most of the organelles, including the nucleus. Because of its ability to direct protein synthesis and other cellular activities, the cell body functions as the control center of the neuron. Like dendrites, the cell body also serves as an input portion of the neuron because it can receive signals from other neurons. Throughout the nervous system, the cell bodies of adjacent neurons are often clustered together. A cluster of neuronal cell bodies in the PNS is called a **ganglion** (plural is *ganglia*); a similar arrangement of neuronal cell bodies in the CNS is known as a **nucleus** (plural is *nuclei*).

The **axon** is a single long, thin process that extends from the cell body. It functions as the output portion of the neuron by generating action potentials and then propagating them toward another neuron, a muscle fiber, or a gland cell. The axon usually connects to the cell body at a cone-shaped region called the **axon hillock** (FIGURE 7.2). In most neurons, action potentials arise at the axon hillock, from which they travel along the axon to their destination. The axon hillock is also known as the **trigger zone** because of its role in the generation of action potentials. Along the length of an axon, side branches called **axon collaterals** may extend off. The axon and its collaterals end by dividing into smaller processes called **axon terminals**.

In most neurons, the tips of the axon terminals swell into **synaptic end bulbs**, which are so-named because these bulb-shaped structures can form synapses with other cells. A *synapse* (SIN-aps) is a site of communication between a neuron and a target cell, which can be another neuron, a muscle fiber, or a gland cell. Within the synaptic end bulbs are many tiny membrane-enclosed sacs called **synaptic vesicles** that store chemical *neurotransmitters* (FIGURE 7.2). The arrival of an action potential at the synaptic end bulb ultimately causes the release of neurotransmitters from the synaptic vesicles. The released neurotransmitter molecules, in turn, excite or inhibit the target cell.

For an axon to function, materials must move between the cell body and axon terminals, a process known as **axonal transport**. Axonal transport uses proteins called **kinesins** and **dyneins** as "motors" to transport materials along the surfaces of microtubules of the neuron's cytoskeleton (FIGURE 7.2). Each of these motor proteins has a region that binds to the particle to be transported and a region that binds to a microtubule. The bound particle is carried by the motor protein as the motor protein uses energy from ATP hydrolysis to "walk" along the surface of the microtubule. Axonal transport moves materials in both directions—away from and toward the cell body. Axonal transport that occurs in an **anterograde** (forward) direction involves kinesins. Anterograde transport moves organelles and synaptic vesicles from the cell body to the axon terminals. Axonal transport that occurs in a **retrograde** (backward) direction involves dyneins. Retrograde transport

CLINICAL CONNECTION

Retrograde Transport and Tetanus Toxin

If *Clostridium tetani* bacteria are near a deep cut or puncture wound, they can enter the damaged area and release **tetanus toxin**. This toxin is then carried by **retrograde transport** into the CNS, where it ultimately causes activation of neurons that stimulate muscles to contract. These contractions lead to prolonged and painful muscle spasms, a condition called **tetanus**. This disorder is also known as *lockjaw* because spasms in the jaw muscles make it difficult to open the mouth. The delay between the release of the toxin and the first appearance of symptoms is due, in part, to the time required for transport of the toxin to the cell body. For this reason, a deep cut or puncture wound in the head or neck is a more serious matter than a similar injury in the leg. The closer the site of injury is to the brain, the shorter the transit time, so treatment must begin quickly.

FIGURE 7.2 **Components of a neuron.**

The basic parts of a neuron are the cell body, dendrites, and axon.

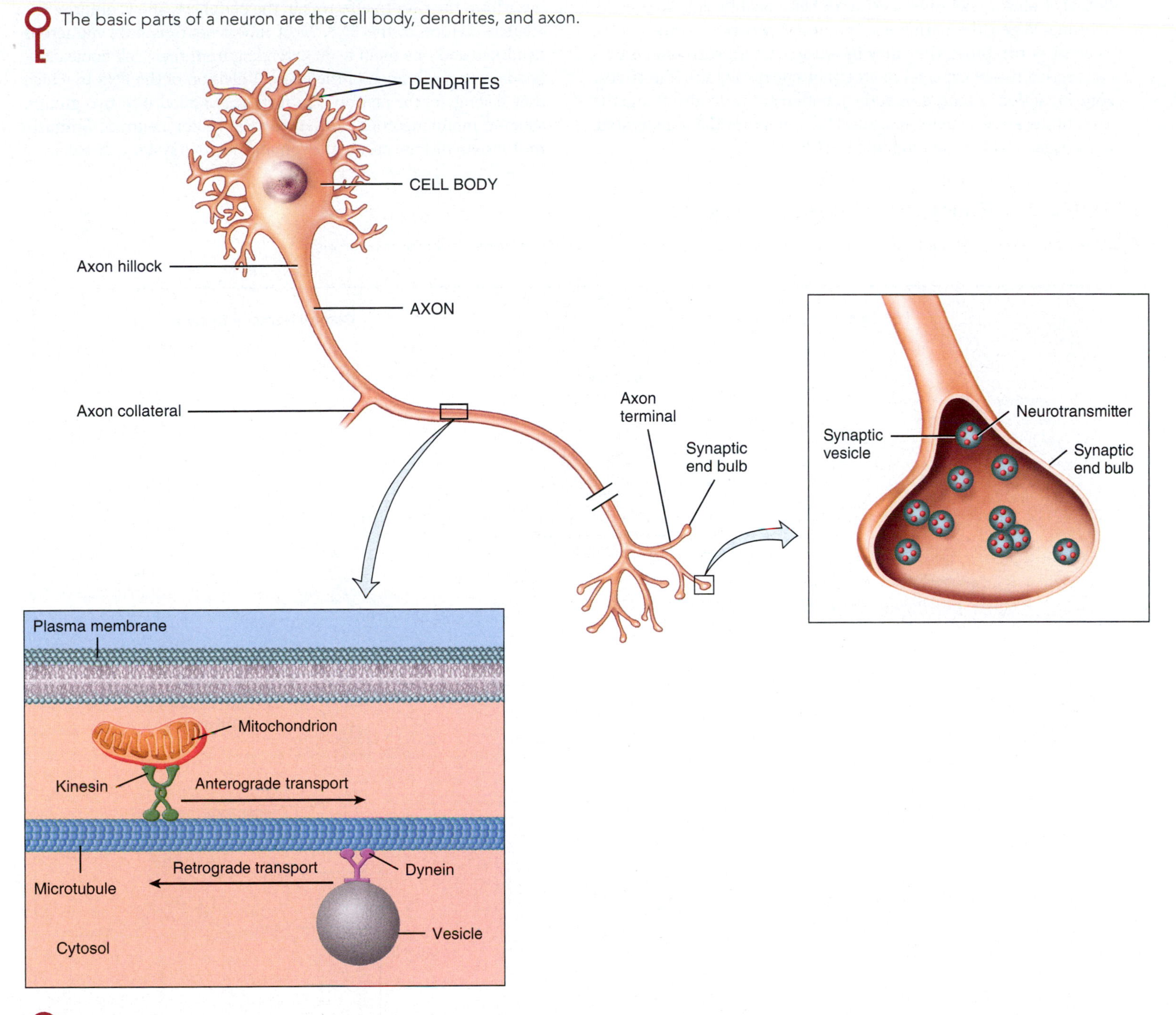

What is the function of each of the three main parts of the neuron?

moves membrane vesicles and other cellular materials from the axon terminals to the cell body to be degraded or recycled. Substances that enter the neuron at the axon terminals are also moved to the cell body by retrograde transport. These substances include (1) **trophic chemicals** such as nerve growth factor and (2) harmful agents such as tetanus toxin and the viruses that cause rabies, herpes simplex, and polio.

The length of an axon can vary from one neuron to another. Some neurons have axons as short as just a few microns. Other neurons have axons as long as a meter or more. Adjacent axons with similar lengths are often bundled together with connective tissue. A **nerve** is a bundle of axons in the PNS, whereas a **tract** is a bundle of axons in the CNS.

Functional Classes of Neurons

Neurons are divided into three functional classes based on the direction in which the action potential is conveyed relative to the CNS (FIGURE 7.3):

1. **Sensory** or **afferent neurons** (AF-er-ent) convey action potentials *into* the CNS. They constitute the afferent division of the PNS. Most sensory neurons have only one process that extends from their cell bodies. This single process is an axon that has dendrites at its peripheral end. Sensory neurons are associated with **sensory receptors** that detect a sensory stimulus such as touch, pressure, light,

or sound. Sensory receptors are either the peripheral endings (dendrites) of sensory neurons or separate cells located close to sensory neurons. When the peripheral endings of sensory neurons serve as sensory receptors, they may be encapsulated (surrounded by a connective tissue capsule) or free (not encapsulated). The trigger zone for action potentials is at the junction of the dendrites and the axon of the sensory neuron. Once an action potential is generated, it propagates along the axon into the CNS.

2. **Motor** or **efferent neurons** (EF-e-rent) convey action potentials *away* from the CNS to effectors in the periphery. They comprise the efferent division of the PNS. Most motor neurons have numerous dendrites and one main axon extending from their cell bodies. Depending on the branch of the efferent division of the PNS to which they belong, motor neurons are further classified into two groups: somatic motor neurons and autonomic motor neurons. **Somatic motor neurons** are part of the somatic nervous system; they convey

FIGURE 7.3 Functional classes of neurons.

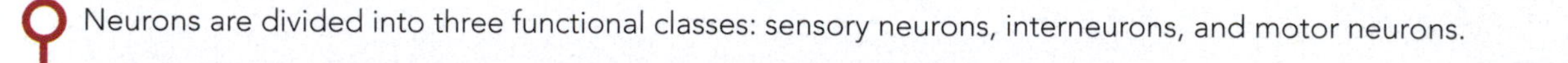

Neurons are divided into three functional classes: sensory neurons, interneurons, and motor neurons.

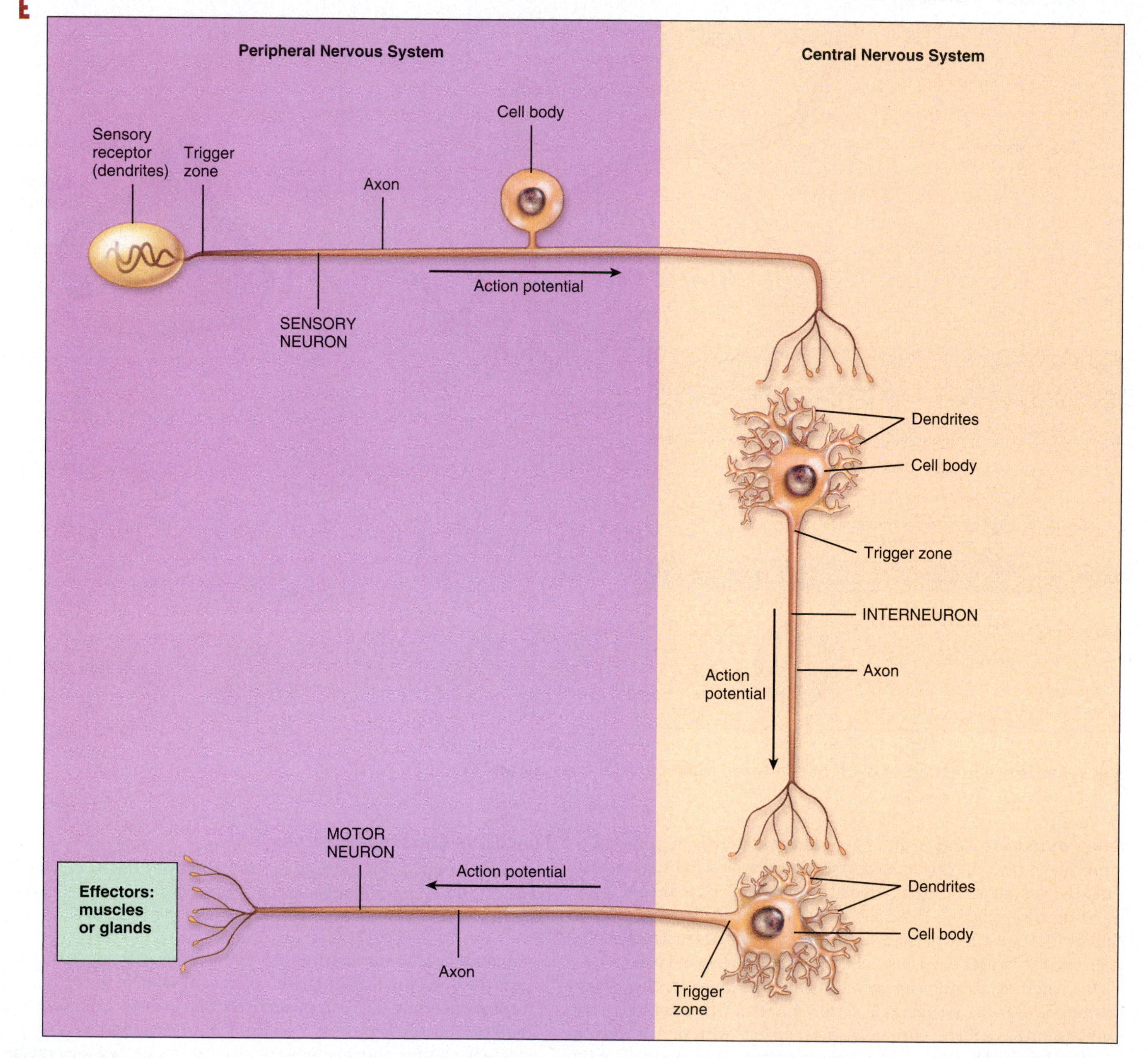

Which functional class of neurons is responsible for integration?

action potentials to skeletal muscles. **Autonomic motor neurons**, which are components of the autonomic nervous system, convey action potentials to cardiac muscle, smooth muscle, or glands.

3. **Interneurons** or **association neurons** are located entirely within the CNS between sensory and motor neurons. Interneurons are responsible for integration—they process incoming sensory information from sensory neurons and then may elicit a motor response by activating the appropriate motor neurons. Like motor neurons, interneurons usually have numerous dendrites and one main axon extending from their cell bodies. About 99% of all neurons in the body are interneurons.

Neuroglia Provide Physical, Nutritional, and Metabolic Support to Neurons

Neuroglia (noo-RŌG-lē-a; *-glia* = glue) or **glia** make up about half the volume of the CNS. Their name derives from the idea of early histologists that they were simply the "glue" that held nervous tissue together, providing physical support to neurons. It is now known that neuroglia support neurons in a variety of other ways: They nourish and protect neurons and maintain homeostasis in the interstitial fluid that bathes them. Generally, neuroglia are smaller than neurons, and they are 5 to 50 times more numerous. In contrast to neurons, glia do not generate or propagate action potentials, and they can multiply and divide in the mature nervous system. (Mature neurons are unable to divide because they remain in the G_0 phase of interphase; see Section 3.6.) In cases of injury or disease, neuroglia multiply to fill in the spaces formerly occupied by neurons. Brain tumors derived from glia, called **gliomas**, tend to be highly malignant and to grow rapidly. Different types of neuroglia are present in the CNS and PNS.

Neuroglia of the CNS

There are four types of neuroglia in the CNS: astrocytes, oligodendrocytes, microglia, and ependymal cells (**FIGURE 7.4**).

FIGURE 7.4 Neuroglia of the central nervous system (CNS).

The four types of CNS neuroglia are astrocytes, oligodendrocytes, microglia, and ependymal cells.

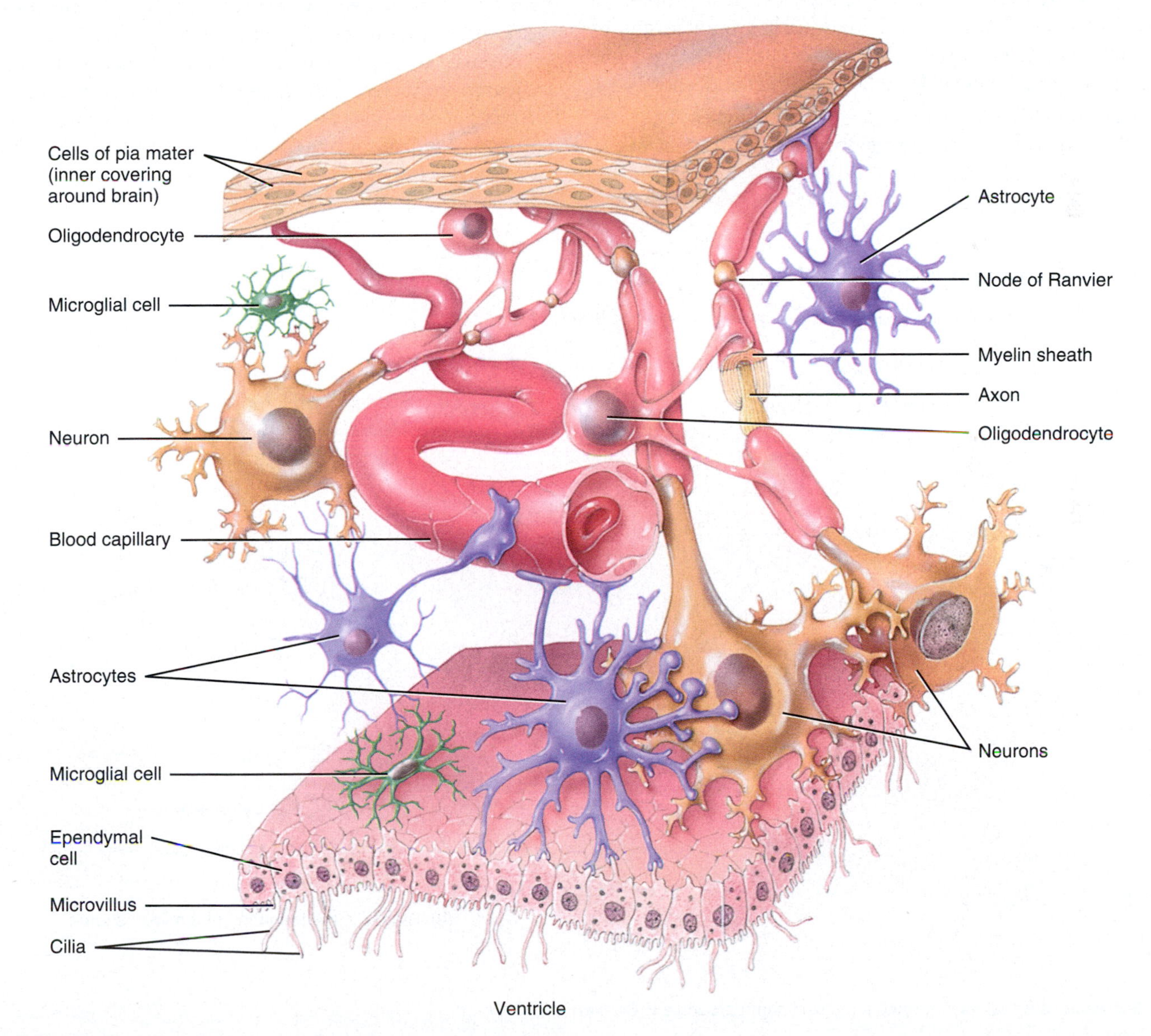

Which CNS neuroglia function as phagocytes?

1. **Astrocytes** (AS-trō-sīts) are the most numerous of the neuroglia. They have processes that wrap around capillaries (the smallest blood vessels) in the CNS. The walls of brain capillaries consist of endothelial cells (see **FIGURE 8.5b**) that are joined together by tight junctions. In effect, the tight junctions between the endothelial cells create a *blood–brain barrier*, which isolates neurons of the CNS from harmful agents and other substances in the blood. Astrocyte processes surrounding brain capillaries secrete chemicals that maintain the "tightness" of these tight junctions. Details of the blood–brain barrier are discussed in Chapter 8. Astrocytes also help to maintain the appropriate chemical environment for the generation of action potentials. For example, they regulate the concentration of important ions such as K^+, take up excess neurotransmitters, and serve as a conduit for the passage of nutrients and other substances between capillaries and neurons. In the embryo, astrocytes secrete chemicals that appear to regulate the growth, migration, and interconnections among neurons in the brain. Astrocytes may also play a role in the formation of neural synapses.
2. **Oligodendrocytes** (OL-i-gō-den′-drō-sīts) are responsible for forming and maintaining the myelin sheath around axons of neurons in the CNS. The *myelin sheath* is a multilayered lipid and protein covering that will be described in more detail shortly.
3. **Microglia** (mī-KROG-lē-a) function as phagocytes. They remove cellular debris formed during normal development of the nervous system and phagocytize microbes and damaged nervous tissue.
4. **Ependymal cells** (ep-EN-de-mal) line the ventricles of the brain and central canal of the spinal cord. (Ventricles and the central canal are spaces filled with cerebrospinal fluid, which protects and nourishes the brain and spinal cord.) Functionally, ependymal cells produce and assist in the circulation of cerebrospinal fluid.

Neuroglia of the PNS

Schwann cells (SCHVON or SCHWON) (**FIGURE 7.5a**) are a type of neuroglia found only in the PNS. They form the myelin sheath around axons of PNS neurons. Schwann cells also participate in axon regeneration, which is more easily accomplished in the PNS than in the CNS.

Myelination Increases the Speed of Action Potential Conduction

The axons of many neurons are surrounded by a **myelin sheath**, a multilayered covering composed of lipids and proteins. Like insulation covering an electrical wire, the myelin sheath insulates the axon of a neuron and increases the speed of conduction of action potentials. Two types of neuroglia produce myelin sheaths: Schwann cells (in the PNS) and oligodendrocytes (in the CNS). In the PNS, each Schwann

FIGURE 7.5 Schwann cells.

Schwann cells form the myelin sheath around axons of neurons in the PNS.

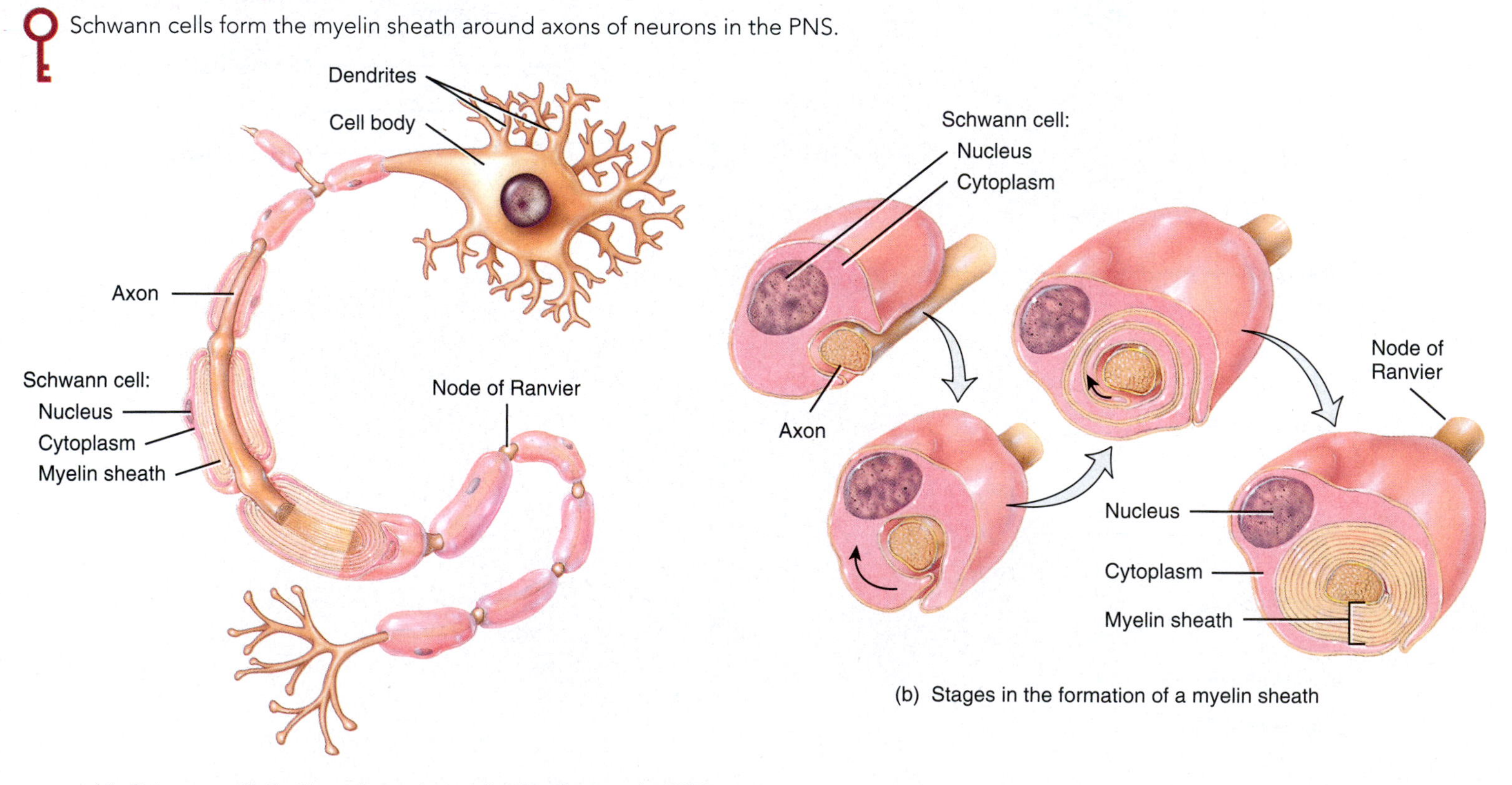

(b) Stages in the formation of a myelin sheath

(a) Schwann cells that have formed a myelin sheath around a PNS axon

What is the functional significance of the myelin sheath?

CLINICAL CONNECTION

Multiple Sclerosis

Multiple sclerosis (MS) is a disease that causes a progressive destruction of myelin sheaths of neurons in the CNS. The condition's name describes its pathology: In ***multiple*** regions the myelin sheaths deteriorate to ***scleroses***, which are hardened scars or plaques. The destruction of myelin sheaths slows and then short-circuits conduction of action potentials. MS is an autoimmune disease—the body's own immune system spearheads the attack. This disease usually appears between the ages of 20 and 40. The first symptoms may include a feeling of heaviness or weakness in the muscles, abnormal sensations, or double vision. An attack is followed by a period of remission during which the symptoms temporarily disappear. One attack follows another over the years, usually every year or two. The result is a progressive loss of function interspersed with remission periods, during which symptoms abate.

cell wraps about 1 millimeter (mm) of a single axon's length by spiraling many times around the axon (FIGURE 7.5b). Eventually, as many as 100 layers of Schwann cell membrane surround the axon to form the myelin sheath. Gaps in the myelin sheath, called **nodes of Ranvier** (RON-vē-ā), appear at intervals along the axon. Each Schwann cell wraps one axon segment between two nodes. In the CNS, an oligodendrocyte myelinates parts of several axons (see FIGURE 7.4). Each oligodendrocyte puts forth about 15 broad, flat processes that spiral around CNS axons to form the myelin sheath. Nodes of Ranvier are present, but they are fewer in number. Axons in the CNS or PNS that have a myelin sheath are said to be **myelinated**, and those without it are said to be **unmyelinated**.

Within the brain and spinal cord are regions that look white, known as **white matter**, and regions that appear gray, called **gray matter** (FIGURE 7.6). White matter is composed primarily of myelinated axons. The whitish color of myelin gives white matter its name. The gray matter of the nervous system contains neuronal cell bodies, dendrites, unmyelinated axons, axon terminals, and neuroglia. It appears grayish rather than white because of the absence of myelin in these areas. The functional roles of gray matter and white matter are described in the discussion of the central nervous system in Chapter 8.

Damaged Neurons Have a Limited Ability to Repair Themselves

Throughout your life, your nervous system exhibits **plasticity**, the capability to change based on experience. At the level of individual neurons, the changes that can occur include the sprouting of new dendrites, synthesis of new proteins, and changes in synaptic contacts with other neurons. Undoubtedly, both chemical and electrical signals drive the changes that occur. Despite this plasticity, mammalian neurons have very limited powers of **regeneration**, the capability to replicate or repair themselves. In the PNS, an axon may undergo repair if the cell body is intact and if the Schwann cells that produce myelination remain active. Schwann cells aid the repair process by forming a **regeneration tube** that guides and stimulates regrowth of the axon. Therefore, a person who injures axons of a nerve in an upper limb, for example, has a good chance of regaining nerve function.

In the CNS, there is little or no repair of an axon after injury. This seems to result from two factors: (1) inhibitory proteins secreted by neuroglia, particularly oligodendrocytes, and (2) absence of growth-stimulating cues that were present during fetal development. Axons in the CNS are myelinated by oligodendrocytes rather than Schwann cells, and this CNS myelin is one of the factors inhibiting regeneration of neurons. Also, after axonal damage, nearby astrocytes proliferate rapidly, forming a type of scar tissue that acts as a physical barrier to regeneration. Thus, injury of the brain or spinal cord usually is permanent.

FIGURE 7.6 Distribution of gray and white matter in the spinal cord and brain.

White matter consists primarily of myelinated axons of neurons; gray matter consists of neuronal cell bodies, dendrites, unmyelinated axons, axon terminals, and neuroglia.

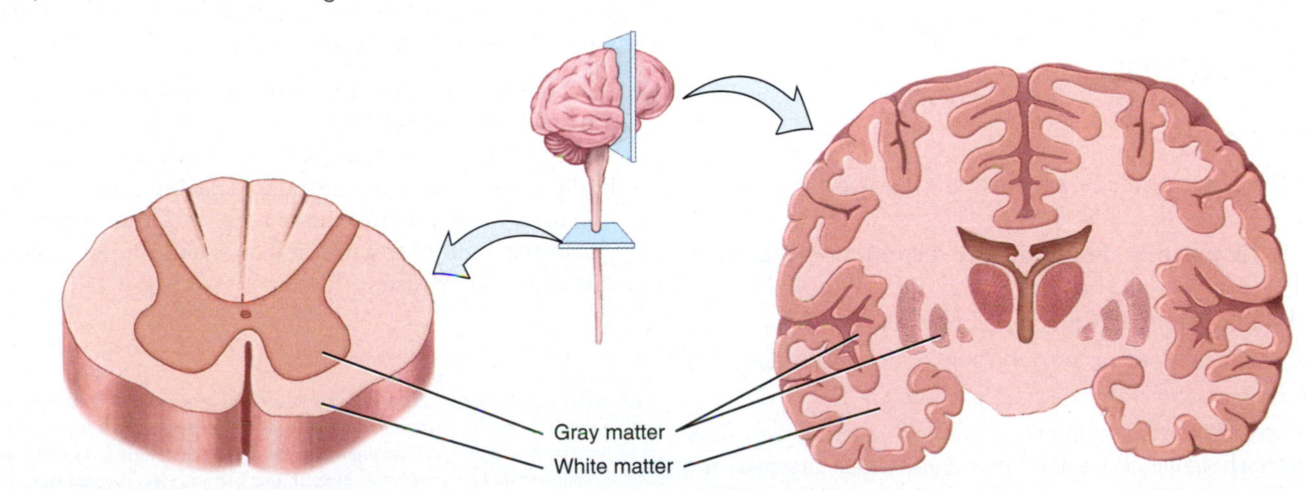

What is responsible for the white appearance of white matter?

CLINICAL CONNECTION

Neurogenesis

Neurogenesis—the birth of new neurons from undifferentiated stem cells—occurs regularly in some animals. For example, new neurons appear and disappear every year in the adult brains of some songbirds. The functional significance of these neurons is not well understood, but it is speculated that they contribute to the songbird's ability to acquire and delete new song patterns. Since the 1970s, it has been known that human olfactory receptor cells (which are neurons) are periodically generated from basal stem cells in the olfactory epithelium of the nose. This allows us to produce new olfactory receptor cells, which live for only about two months before being replaced. Until recently, however, the dogma in humans and other primates was "no new neurons" in the adult brain. Then, in 1998, scientists discovered **neural stem cells** that give rise to new neurons in the adult human ***hippocampus***, an area of the brain that is crucial for learning. Recent evidence also indicates that the adult human ***subventricular zone*** near the lateral ventricles of the brain is another source of neural stem cells. These cells migrate to the olfactory bulbs to develop into neurons that play a role in olfactory processing. Neural stem cells in the adult human brain have important clinical implications. Researchers are currently trying to find ways to stimulate neural stem cells to replace neurons lost through damage or disease and to develop tissue-cultured neurons that can be used for transplantation purposes.

CHECKPOINT

5. What is the importance of the axon hillock?

6. Which type of axonal transport involves kinesins?

7. Which type of neuron conveys action potentials from the CNS to effectors in the periphery?

8. Which type of neuroglial cell forms the myelin sheath around axons of neurons in the PNS?

9. Why is there little or no repair of a CNS axon after an injury?

7.3 Electrical Signals in Neurons

OBJECTIVES

- Describe the basic types of ion channels involved in neuron function.
- Explain the factors that determine the resting membrane potential.
- Discuss the importance of graded potentials.
- Outline the events that occur during an action potential.
- Describe how action potentials are propagated.

Neurons and muscle fibers communicate with one another using two types of electrical signals: (1) *graded potentials*, which are used for short-distance communication only, and (2) *action potentials*, which allow communication over long distances in the body. Graded potentials and action potentials are deviations in *membrane potential* (*voltage*), the difference in electrical charges that exists just across the plasma membrane.

Neurons and muscle fibers are considered to be **excitable cells** because they exhibit **electrical excitability**, the ability to respond to a stimulus and convert it into an action potential. An action potential that occurs in a neuron is called a *nerve action potential* or *nerve impulse*. In most neurons, an action potential causes the release of neurotransmitters, which allow the neuron to communicate with another neuron, a muscle fiber, or a gland cell. An action potential that occurs in a muscle fiber is called a *muscle action potential* or *muscle impulse*. When an action potential occurs in a muscle fiber, the muscle fiber contracts. The details of the nerve action potential are described in this section, and the details of the muscle action potential are discussed in Chapter 11.

To understand the functions of graded potentials and action potentials, consider how the nervous system allows you to feel the smooth surface of a pen that you have picked up off a table (**FIGURE 7.7**):

1. As you touch the pen, a graded potential develops in a sensory receptor in the skin of the fingers.
2. The graded potential triggers the axon of the sensory neuron to form a nerve action potential, which travels along the axon into the CNS and ultimately causes the release of neurotransmitter at a synapse with an interneuron.
3. The neurotransmitter stimulates the interneuron to form a graded potential in its dendrites and cell body.
4. In response to the graded potential, the axon of the interneuron forms a nerve action potential. The nerve action potential travels along the axon, which results in neurotransmitter release at the next synapse with another interneuron.
5. This process of neurotransmitter release at a synapse followed by the formation of a graded potential and then a nerve action potential occurs over and over as interneurons in higher parts of the brain (such as the thalamus and cerebral cortex) are activated. Once interneurons in the *cerebral cortex*, the outer part of the brain, are activated, perception occurs and you are able to feel the smooth surface of the pen touch your fingers. As you will learn in Chapter 8, *perception*, the conscious awareness of a sensation, is primarily a function of the cerebral cortex.

Suppose that you want to use the pen to write a letter. The nervous system would respond in the following way (**FIGURE 7.7**):

6. A stimulus in the brain causes a graded potential to form in the dendrites and cell body of an *upper motor neuron*,* a type of neuron that synapses with a lower motor neuron farther down in the CNS in order to contract a skeletal muscle. The graded potential subsequently causes a nerve action potential to occur in the axon of the upper motor neuron, followed by neurotransmitter release.

---

* An upper motor neuron is actually an interneuron because it is completely located within the CNS. So the term *upper motor neuron* is a misnomer, but it continues to be used because this type of cell originates in the upper part of the CNS and regulates the activity of lower motor neurons. A lower motor neuron, which originates in the lower part of the CNS, is a true motor neuron: It conveys action potentials from the CNS to skeletal muscle fibers in the periphery.

FIGURE 7.7 Overview of nervous system functions.

Graded potentials, nerve action potentials, and muscle action potentials are involved in the relay of sensory stimuli, integrative functions, and motor activities.

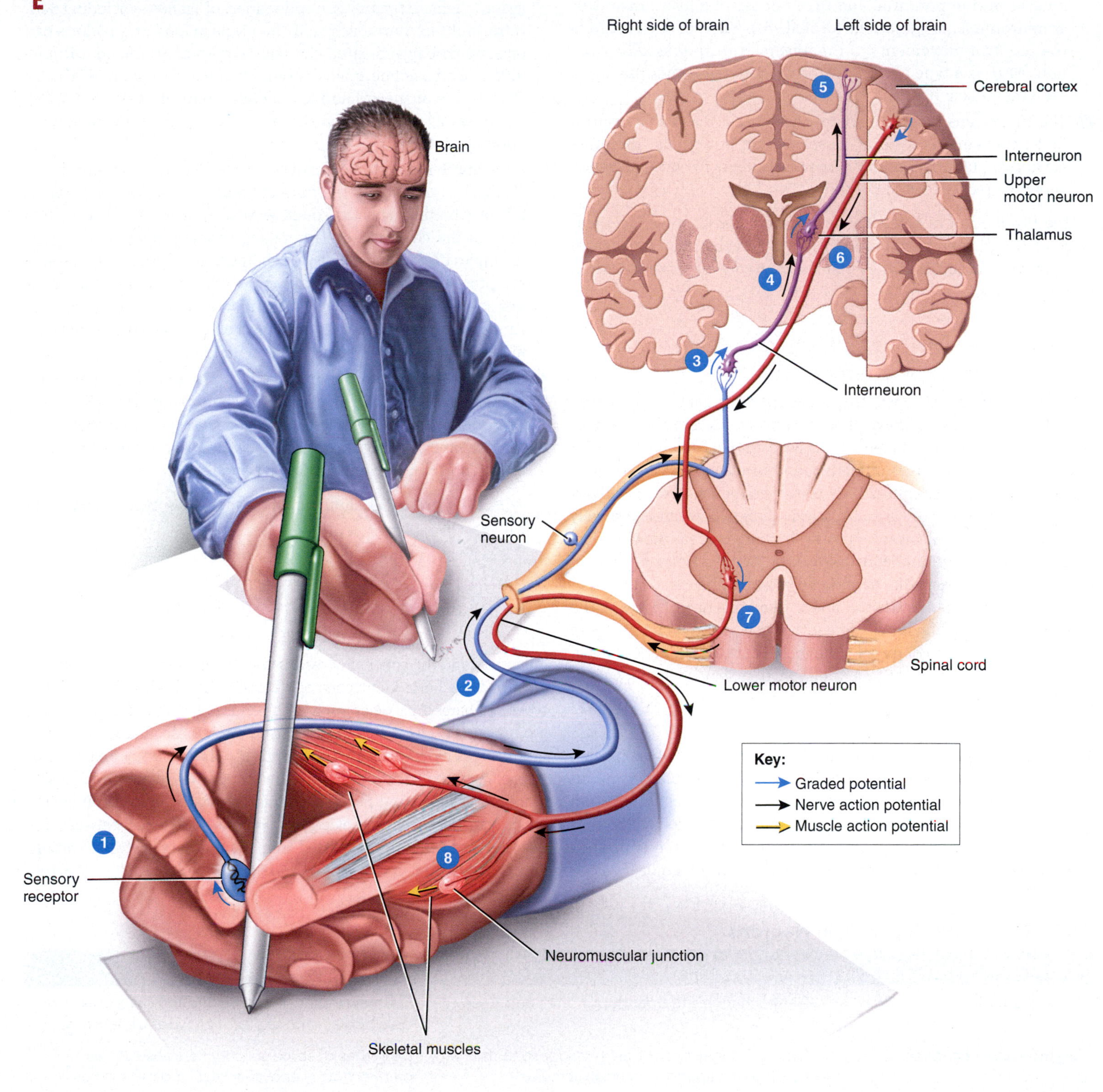

In which region of the brain does perception primarily occur?

7. The neurotransmitter generates a graded potential in a *lower motor neuron*, a type of neuron that directly supplies skeletal muscle fibers. The graded potential triggers the formation of a nerve action potential and then release of neurotransmitter at neuromuscular junctions formed with skeletal muscle fibers that control movements of the fingers. A *neuromuscular junction (NMJ)* is a type of synapse formed between a neuron and a skeletal muscle fiber.
8. The neurotransmitter stimulates the muscle fibers that control finger movements to form muscle action potentials. The muscle action potentials cause these muscle fibers to contract, which allows you to write with the pen.

The production of electrical signals depends on two basic features of the plasma membrane of excitable cells: the presence of specific types of ion channels and the existence of a resting membrane potential.

Ion Channels Permit Certain Ions to Move Across the Plasma Membrane

When ion channels are open, they allow specific ions to move across the plasma membrane, down their *electrochemical gradient*—a concentration (chemical) difference plus an electrical difference. Recall that ions move from areas of higher concentration to areas of lower concentration (the chemical part of the gradient). Also, positively charged cations move toward a negatively charged area, and negatively charged anions move toward a positively charged area (the electrical aspect of the gradient). As ions move, they create a flow of electrical current that can change the membrane potential.

Ion channels open and close due to the presence of "gates." The gate is a part of the channel protein that can seal the channel pore shut or move aside to open the pore (see **FIGURE 5.7**). Four types of ion channels are important to neuron function: leak channels, ligand-gated channels, mechanically-gated channels, and voltage-gated channels (**FIGURE 7.8**).

1. The gates of **leak channels** randomly alternate between open and closed positions (**FIGURE 7.8a**). Typically, plasma membranes have many more potassium ion (K^+) leak channels than sodium ion (Na^+) leak channels, and the K^+ leak channels are leakier than the Na^+ leak channels. Thus, the membrane's permeability to K^+ is much higher than its permeability to Na^+. Leak channels are important for establishing the resting membrane potential.
2. A **ligand-gated channel** opens or closes in response to a specific ligand (chemical) stimulus. A wide variety of ligands—including neurotransmitters, hormones, and chemicals in food or an odor—can open or close ligand-gated channels. For example, the neurotransmitter acetylcholine opens cation channels that allow Na^+ and Ca^{2+} to diffuse inward and K^+ to diffuse outward (**FIGURE 7.8b**). Ligand-gated channels participate in the generation of graded potentials.
3. A **mechanically-gated channel** opens or closes in response to mechanical stimulation in the form of touch, pressure, tissue stretching, or vibration (such as sound waves) (**FIGURE 7.8c**). The force distorts the channel from its resting position, opening the gate. Examples of mechanically-gated channels are those found in touch receptors and pressure receptors in the skin, in receptors that monitor stretching of internal organs, and in auditory receptors in the ears. Like ligand-gated channels, mechanically-gated channels are involved in the formation of graded potentials.
4. A **voltage-gated channel** opens in response to a change in membrane potential (voltage). Examples include voltage-gated K^+ channels (**FIGURE 7.8d**), voltage-gated Na^+ channels, and voltage-gated Ca^{2+} channels. Voltage-gated channels are responsible for the generation and conduction of action potentials.

TABLE 7.1 presents a summary of the four major types of ion channels in neurons.

Resting Membrane Potential Is the Voltage That Exists Across the Plasma Membrane in an Excitable Cell at Rest

To understand the concept of resting membrane potential, it is important to have a basic knowledge of electricity. The body contains many types of charged particles such as ions, proteins, and the phosphate groups of ATP. Electrical forces exist between these charged particles. Like charges repel each other, and opposite charges attract each other. In some cases, a partition may separate opposite charges. Such a separation of positive and negative charges is a form of potential energy, which can do work. The electrical potential difference between opposite charges that are separated from each other is termed **voltage**, which is measured in units called **volts** or **millivolts** (1 mV = 0.001 V).

TABLE 7.1 Ion Channels in Neurons

Type of Ion Channel	Description	Location
Leak channels	Gated channels that randomly open and close.	Found in nearly all cells, including dendrites, cells bodies, and axons of all types of neurons.
Ligand-gated channels	Gated channels that open in response to the binding of a ligand (chemical) stimulus.	Dendrites of some sensory neurons such as pain receptors and dendrites and cell bodies of interneurons and motor neurons.
Mechanically-gated channels	Gated channels that open in response to a mechanical stimulus (such as touch, pressure, tissue stretching, or vibration).	Dendrites of some sensory neurons such as touch receptors, pressure receptors, and some pain receptors.
Voltage-gated channels	Gated channels that open in response to a voltage stimulus (change in membrane potential).	Axons of all types of neurons.

FIGURE 7.8 Ion channels in the plasma membrane.

Neuron function relies on four types of ion channels: leak channels, ligand-gated channels, mechanically-gated channels, and voltage-gated channels.

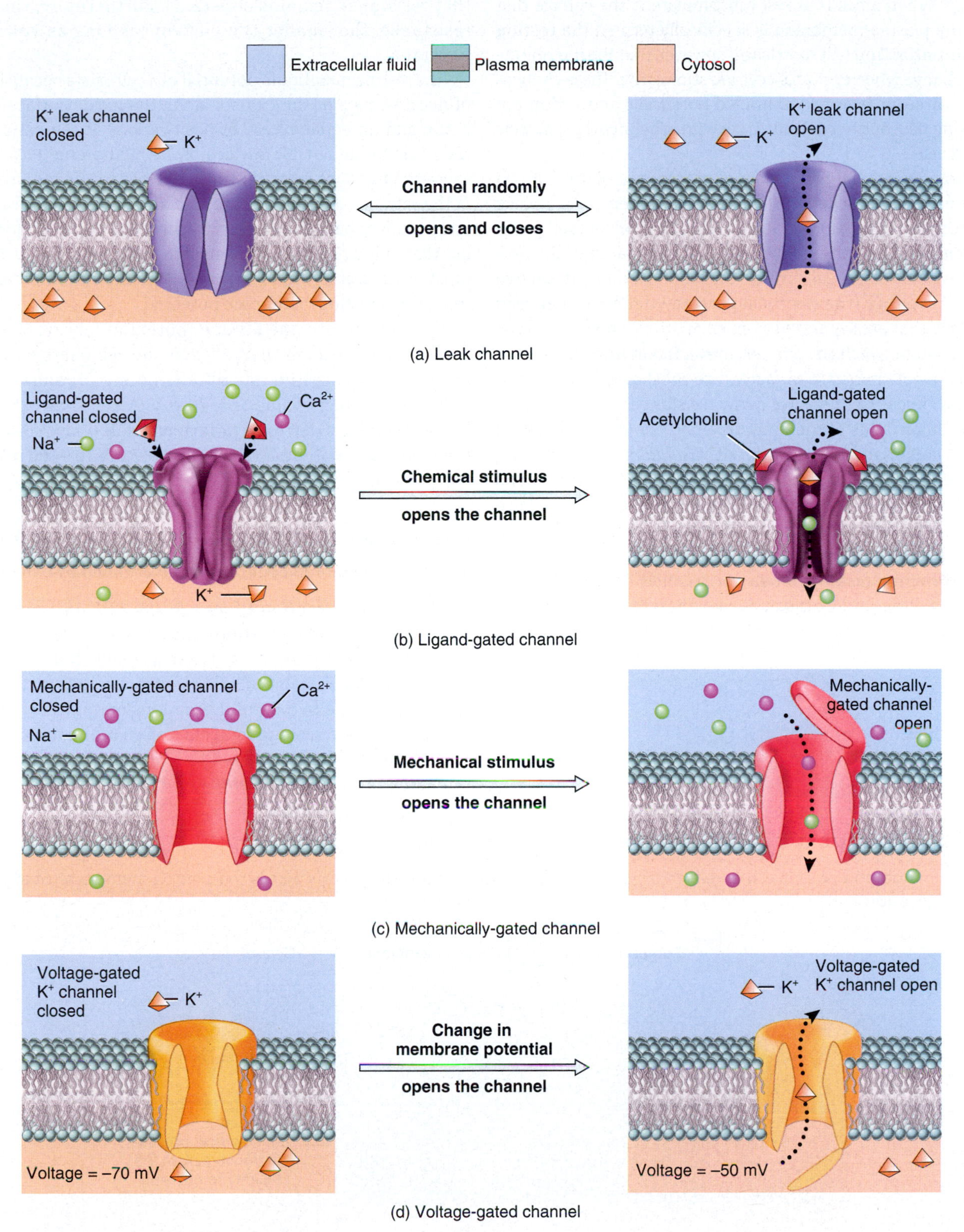

What type of gated channel is activated by a touch on the arm?

Most cells, including excitable cells, have a separation of positive and negative charges just across their plasma membranes. The voltage that exists across the plasma membrane of a cell is called **membrane potential (V_m)**. When a cell is at rest (unstimulated), the voltage that exists across the plasma membrane is specifically termed the **resting membrane potential**. You will soon learn, however, that the membrane potential can change when excitable cells are stimulated. These changes in membrane potential give rise to graded potentials and action potentials, allowing neurons to communicate with other neurons, muscle fibers, or gland cells.

The membrane potential is like voltage stored in a battery. If you connect the positive and negative terminals of a battery with a piece of wire, electrons will flow along the wire. This flow of charged particles is called **current**. In living cells, the flow of ions (rather than electrons) constitutes the electrical current. Current flow depends on two main factors: (1) voltage (the electrical potential difference between opposite charges that are separated from each other) and (2) the type of substance through which the charges move. **Resistance** is the hindrance to the flow of charges. *Conductors* are substances that permit fast current flow because they have a low resistance. The extracellular and intracellular fluids of the body are good conductors because they consist of ions that carry the current. *Insulators* are substances that decrease current flow because they have a high resistance. The plasma membrane is a good insulator since membrane lipids have few charged groups and cannot carry current. Because the plasma membrane is a good electrical insulator, the main paths for current to flow across the membrane are through ion channels. The relationship among current, voltage, and resistance is expressed by an equation called **Ohm's law**:

$$I = \frac{V}{R}$$

where I is current,
V is voltage, and
R is resistance.

As you can see from the equation, current (I) is directly proportional to voltage (V) and inversely proportional to resistance (R). So the greater the voltage, the greater the current (assuming that resistance remains constant). Furthermore, the greater the resistance, the smaller the current (as long as voltage remains constant).

The resting membrane potential of a cell exists because of an excess of negative ions in the cytosol along the inside surface of the membrane and an equal excess of positive ions in the extracellular fluid (ECF) along the outside surface of the membrane (**FIGURE 7.9**). The excess charges are located only very close to the membrane; the rest of the extracellular fluid or cytosol contains equal numbers of positive and negative charges and is electrically neutral. It is important to point out that only a *tiny* fraction of all of the charges in the ECF and cytosol must be separated across the plasma membrane in order to establish the normal resting membrane potential.

By convention, membrane potential always compares the amount of net excess charge *inside* the cell relative to the outside. In neurons, the resting membrane potential ranges from −40 to −90 mV. A typical value is −70 mV. A cell that exhibits a membrane potential is said to be **polarized**. Most body cells are polarized; the membrane potential varies from +5 mV to −100 mV in different types of cells.

Determinants of the Resting Membrane Potential

The resting membrane potential is determined by two major factors:

1. ***Unequal distribution of ions in the ECF and cytosol.*** The concentrations of major cations and anions are different outside and inside cells (**FIGURE 7.10**). Extracellular fluid (ECF) is rich in Na^+ and chloride ions (Cl^-). In cytosol, however, the main cation is K^+, and the two dominant anions are proteins and the phosphate ions attached to molecules such as the three phosphates in ATP. The symbol A^- is used to refer collectively to the negatively charged (anionic) proteins and phosphate ions of the cytosol.

FIGURE 7.9 Concept of the resting membrane potential.

The resting membrane potential is an electrical potential difference (voltage) that exists across the plasma membrane of an excitable cell under resting conditions.

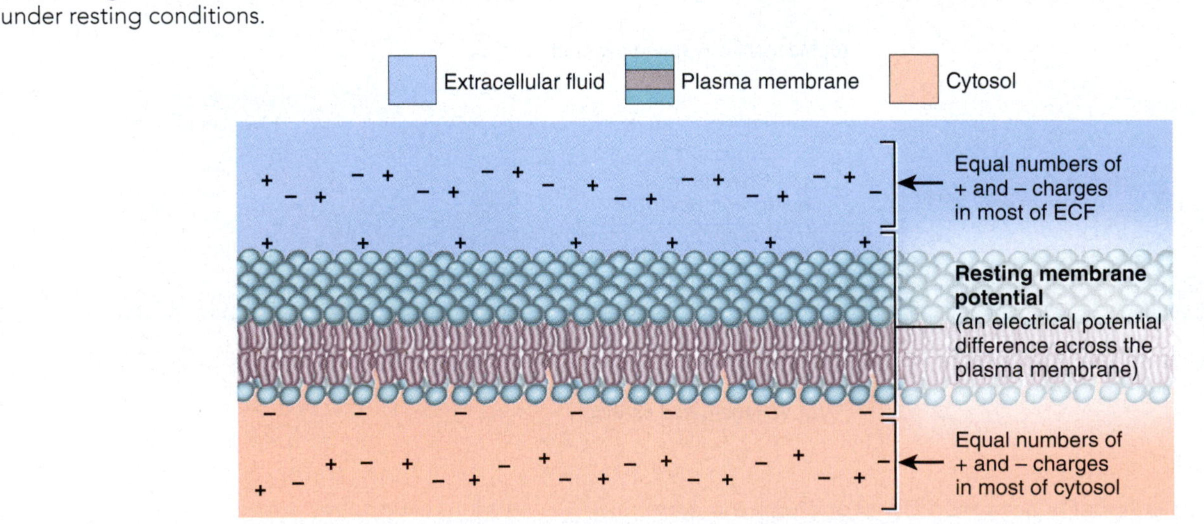

Does it take a large number of charges to be separated across the plasma membrane in order to establish the normal resting membrane potential? Explain your answer.

FIGURE 7.10 Factors that contribute to the resting membrane potential. The ECF is rich in Na^+ and Cl^- ions and the cytosol is rich in K^+ ions and A^- ions (anionic proteins and phosphates). The plasma membrane has a large number of K^+ leak channels and a smaller number of Na^+ leak channels.

The resting membrane potential is determined by two major factors: (1) unequal distribution of ions in the ECF and cytosol and (2) differences in membrane permeability to various ions.

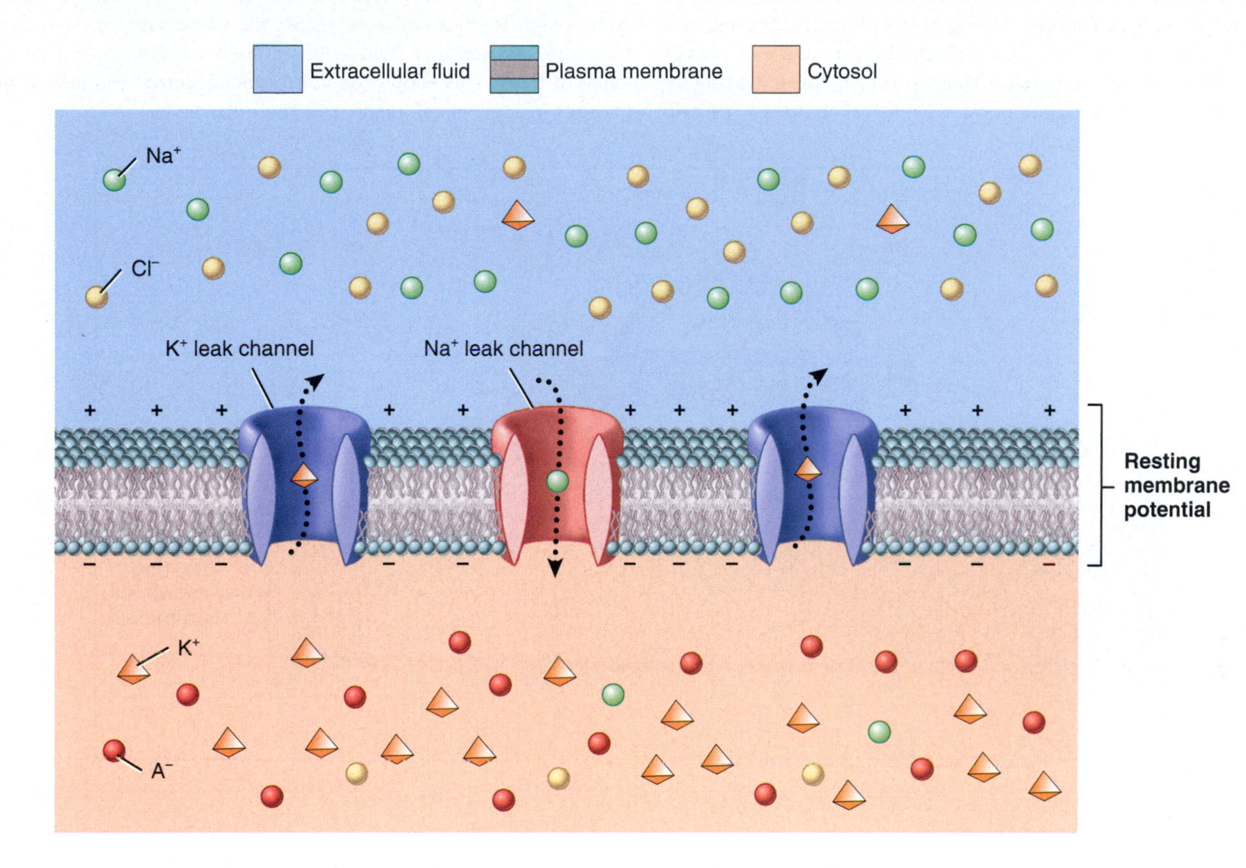

Which ion has the greatest influence on the resting membrane potential of a neuron? Why?

The concentrations of selected solutes in extracellular fluid and cytosol (intracellular fluid) are listed in **TABLE 7.2**. As you will soon learn, the unequal distribution of ions in the ECF and cytosol establishes the concentration gradients that certain ions use to help generate the resting membrane potential.

TABLE 7.2 Concentrations of Selected Solutes in Extracellular and Intracellular Fluids

Ion	Extracellular Fluid (mM*)	Intracellular Fluid (mM)
K^+	5	150
Na^+	145	15
Cl^-	115	5
Anions (A^-)	10	100
Ca^{2+}	2	0.0001

* mM = millimolar, which represents the number of millimoles per liter (mmol/L).

2. ***Differences in membrane permeability to various ions.*** In neurons at rest, there are differences in membrane permeability to various ions because some ions are able to pass through the plasma membrane via specific leak channels, whereas other ions do not have transport mechanisms allowing them passage through the membrane. In general, the more permeable the plasma membrane is to a particular ion, the greater the influence that ion has on the resting membrane potential. This is because ion movement across the membrane can alter the number of negative or positive charges that are located along the inside and outside surfaces of the membrane. Of all of the solutes present in the ECF and cytosol, neurons at rest are most permeable to K^+ ions because K^+ leak channels are the most abundant type of leak channel in their plasma membranes (**FIGURE 7.10**). This high K^+ permeability allows K^+ ions to influence the resting membrane potential to a much greater extent than any other type of ion. Resting neurons are also permeable to Na^+ ions because their plasma membranes contain Na^+ leak channels (**FIGURE 7.10**). However, Na^+ permeability is lower than K^+ permeability because there are fewer Na^+ leak channels than K^+ leak channels. This means that Na^+ ions do influence the resting membrane potential of a neuron but

TOOL OF THE TRADE

THE VOLTMETER

The resting membrane potential of a neuron can be measured using an instrument known as a **voltmeter**. During this procedure, the tip of a recording microelectrode is inserted inside the neuron and a reference (ground) electrode is placed outside the neuron in the extracellular fluid. (*Electrodes* are devices that conduct electrical charges.) The two electrodes are connected to the voltmeter, which detects the electrical potential difference (voltage) across the plasma membrane. By convention, the extracellular fluid is designated as the ground and given a value of 0 mV. This allows the voltmeter to detect the excess electrical charges inside the cell relative to the outside.

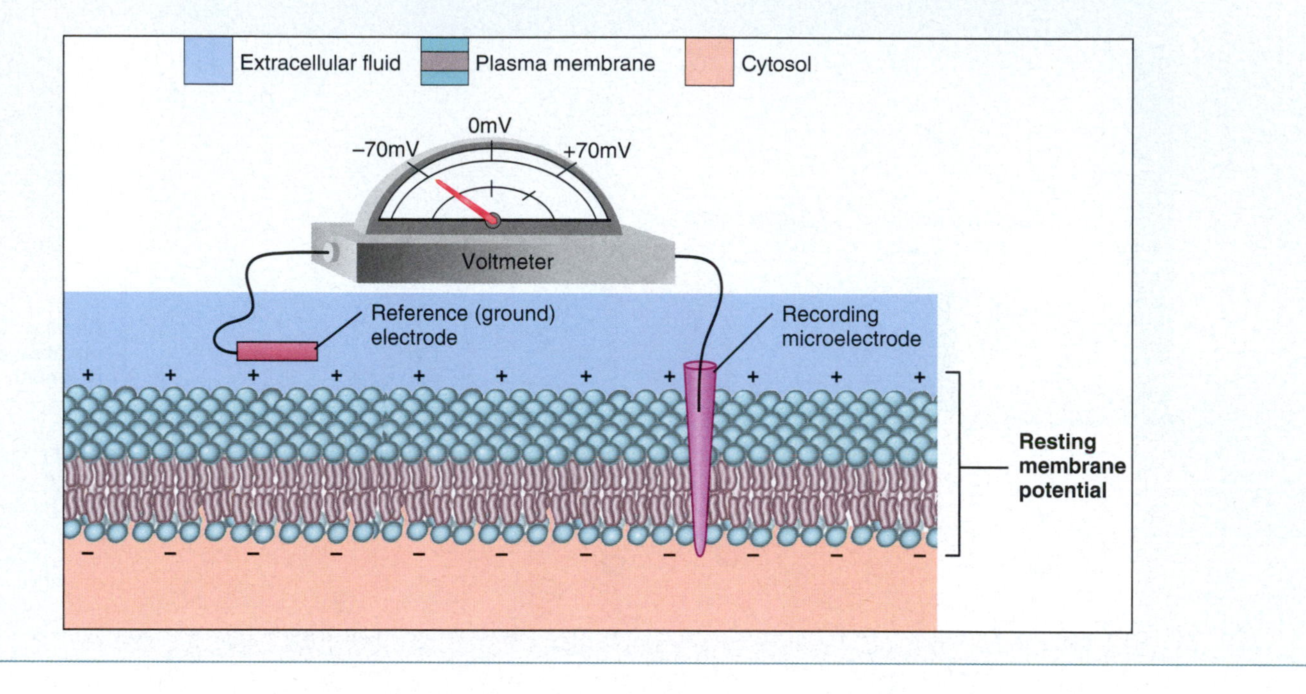

to a lesser extent than K^+ ions. Neurons at rest are also permeable to Cl^- ions due to the presence of Cl^- leak channels in their plasma membranes. However, for reasons that will be discussed later in this section, chloride ions do not make a significant contribution to the resting membrane potential in most neurons. Although resting neurons are permeable to K^+, Na^+, and Cl^- ions, they are essentially impermeable to Ca^{2+} ions and the anionic proteins and phosphates (A^- ions) of the cytosol because plasma membranes of most resting neurons do not have leak channels or other transport mechanisms for these ions. Therefore, Ca^{2+} and A^- ions do not have a direct impact on the resting membrane potential. In summary, the resting membrane potential of most neurons is influenced primarily by K^+ ions and to a lesser extent by Na^+ ions, and is not significantly influenced by Cl^-, Ca^{2+}, or A^- ions.

Membrane Potential Generation

To understand how K^+ and Na^+ ions influence the resting membrane potential, first consider what would happen if a neuron were permeable only to K^+. The passage of K^+ through K^+ leak channels would allow the K^+ ions to move down their concentration gradient from the cytosol into the ECF, generating an inside-negative membrane potential (**FIGURE 7.11a**). As K^+ ions continue to move out of the neuron, the interior of the membrane becomes more negative, creating an electrical gradient that favors movement of K^+ ions from the ECF back into the cytosol (**FIGURE 7.11b**). Initially, however, there is net movement of K^+ ions *out* of the neuron because the magnitude of the K^+ concentration gradient is greater than the magnitude of the K^+ electrical gradient. As the membrane potential becomes even more negative, the magnitude of the K^+ electrical gradient increases. Eventually, the K^+ electrical gradient becomes equal in magnitude to the opposing K^+ concentration gradient and there is no net movement of K^+ ions into or out of the neuron (**FIGURE 7.11c**). The membrane potential that exists at this equilibrium is called the **K^+ equilibrium potential (E_K)** and it is equal to −90 mV. In general, an **equilibrium potential** is the membrane potential at which the concentration gradient and electrical gradient for a particular ion are equal in magnitude but opposite in direction and there is no net movement of that ion across the plasma membrane. The K^+ equilibrium potential is close to, but not exactly equal to, the resting membrane potential (−70 mV), which means that K^+ ions are not the only ions that contribute to the resting membrane potential.

Now suppose that the neuron is permeable only to Na^+ ions. The passage of Na^+ through Na^+ leak channels allows Na^+ ions to move down their concentration gradient from the ECF into the cytosol, generating an inside-positive membrane potential

FIGURE 7.11 Membrane potential generation in a neuron that is permeable only to K^+ ions.

In a hypothetical neuron that is permeable only to K^+ ions, an inside-negative membrane potential is generated if there is net movement of K^+ ions from the cytosol into the ECF.

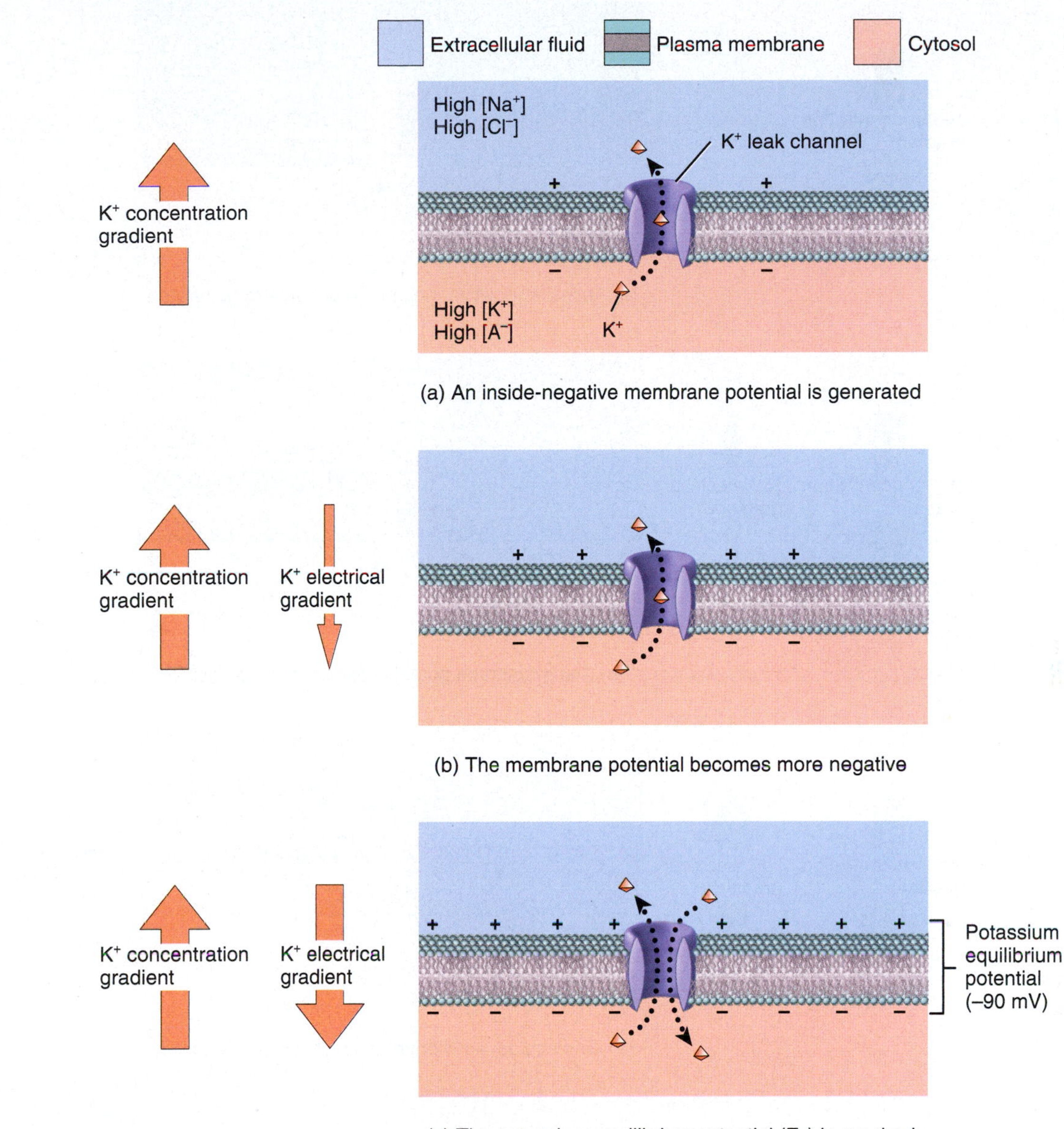

(a) An inside-negative membrane potential is generated

(b) The membrane potential becomes more negative

(c) The potassium equilibrium potential (E_K) is reached

Why does the K^+ electrical gradient favor movement of K^+ ions from the ECF into the cytosol when an inside-negative membrane potential exists?

(FIGURE 7.12a). As Na^+ ions continue to move into the neuron, the membrane potential becomes more positive, producing an electrical gradient that favors movement of Na^+ ions from the cytosol back into the ECF (FIGURE 7.12b). At this point, however, there is net movement of Na^+ ions *into* the neuron because the magnitude of the Na^+ concentration gradient is greater than the magnitude of the Na^+ electrical gradient. As the membrane potential becomes even more positive, the magnitude of the Na^+ electrical gradient increases. Eventually, the Na^+ electrical gradient becomes equal in magnitude to the opposing Na^+ concentration gradient and there is no net movement of Na^+ ions into or out of the neuron (FIGURE 7.12c). The membrane potential that exists at this equilibrium is called the **Na^+ equilibrium potential (E_{Na})** and it is equal to +60 mV. The fact that the Na^+ equilibrium potential is relatively far away from the resting membrane potential (−70 mV) lets you know that Na^+ ions contribute to but are not as important as K^+ ions in establishing the resting membrane potential. The equilibrium potential for Na^+, K^+, or any other ion can be calculated by using the Nernst equation (see the Physiological Equation box).

FIGURE 7.12 Membrane potential generation in a neuron that is permeable only to Na^+ ions.

In a hypothetical neuron that is permeable only to Na^+ ions, an inside-positive membrane potential is generated if there is net movement of Na^+ ions from the ECF into the cytosol.

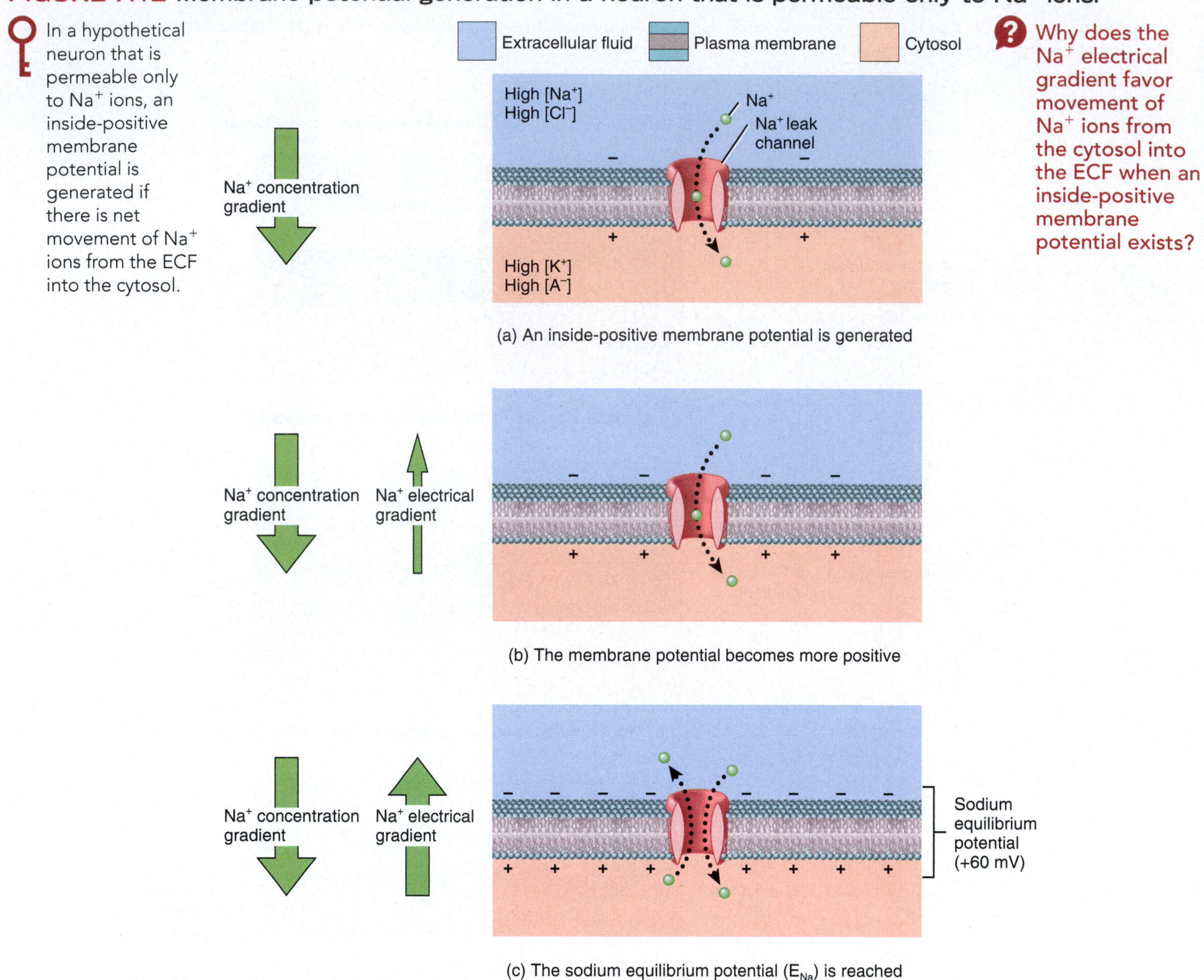

(a) An inside-positive membrane potential is generated

(b) The membrane potential becomes more positive

(c) The sodium equilibrium potential (E_{Na}) is reached

Why does the Na^+ electrical gradient favor movement of Na^+ ions from the cytosol into the ECF when an inside-positive membrane potential exists?

PHYSIOLOGICAL EQUATION

NERNST EQUATION

The equilibrium potential is the membrane potential at which the concentration gradient and the electrical gradient for a particular ion are equal in magnitude and opposite in direction and there is no net movement of that ion across the plasma membrane. If the extracellular and intracellular concentrations of an ion are known, the equilibrium potential for that ion can be calculated by using the **Nernst equation:**

$$E_x = \frac{61}{z} \log \frac{[X]_{out}}{[X]_{in}}$$

where E_x is the equilibrium potential of ion X in millivolts at 37°C,
$[X]_{out}$ is the extracellular concentration of ion X in millimoles/liter (mM),
$[X]_{in}$ is the intracellular concentration of ion X in millimoles/liter (mM), and
z is the valence of ion X (for K^+ and Na^+ ions, the valence is +1).

Using the K^+ and Na^+ concentrations listed in Table 7.2, the K^+ and Na^+ equilibrium potentials can be calculated using the Nernst equation in the following way:

$$E_K = \frac{61}{+1} \log \frac{5}{150}$$
$$= -90 \text{ mV}$$
$$E_{Na} = \frac{61}{+1} \log \frac{145}{15}$$
$$= +60 \text{ mV}$$

As you know, neurons are permeable to *both* K^+ and Na^+ ions. Because the plasma membrane of a neuron is more permeable to K^+ ions than to Na^+ ions, the number of K^+ ions that move out of the neuron through K^+ leak channels is greater than the number of Na^+ ions that move into the neuron through Na^+ leak channels. In other words, there is net movement of positive charge *out* of the neuron, which generates an inside-negative membrane potential (**FIGURE 7.13a**). The membrane potential becomes increasingly negative with the continuous movement of more K^+ out of the neuron and less Na^+ into the neuron. Eventually, K^+ movement out of

FIGURE 7.13 Membrane potential generation in a neuron that is more permeable to K^+ ions than to Na^+ ions.

In a real neuron, which is more permeable to K^+ ions than to Na^+ ions, an inside-negative membrane potential is generated because the number of K^+ ions that move out of the neuron is greater than the number of Na^+ ions that move into the neuron.

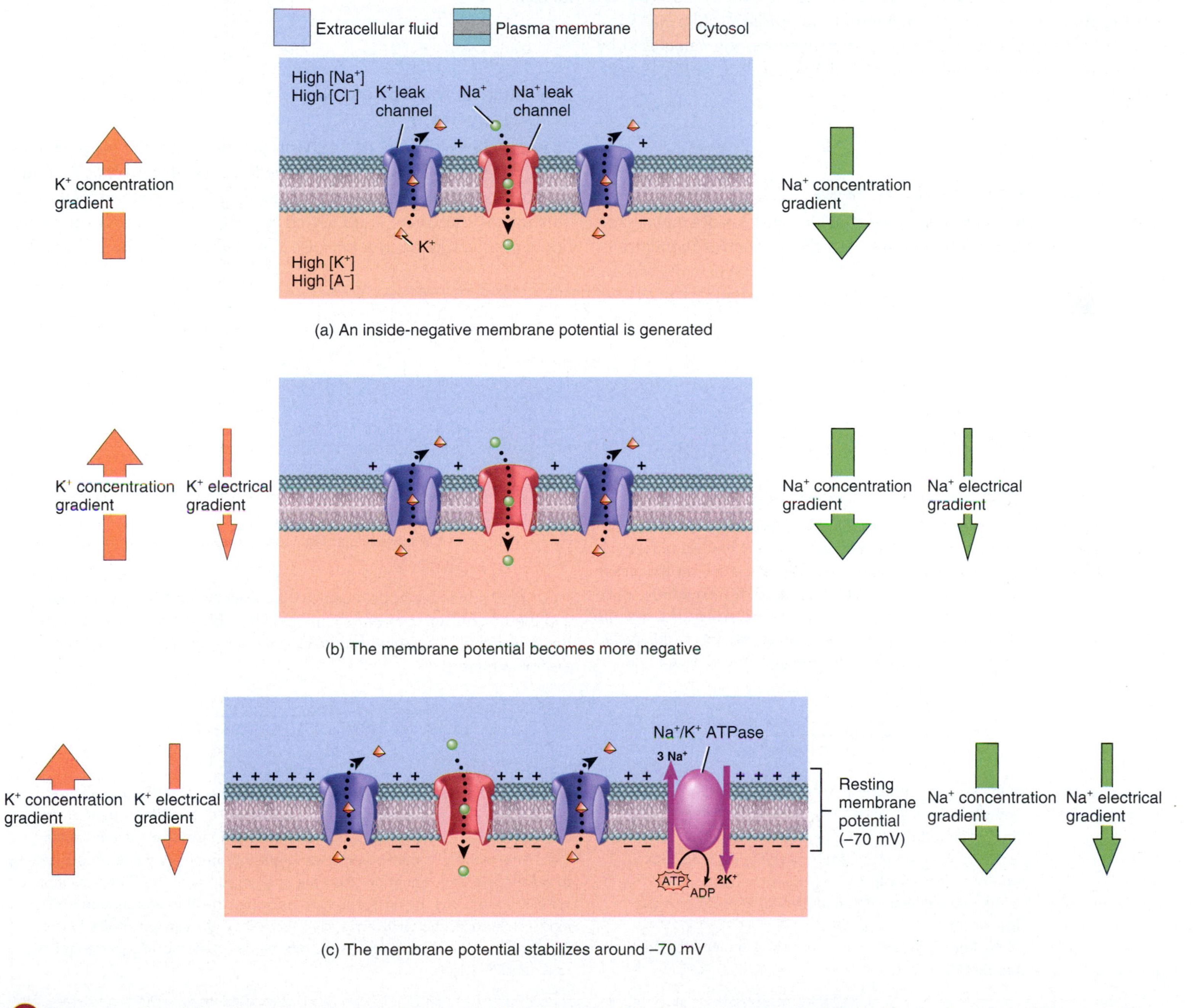

(a) An inside-negative membrane potential is generated

(b) The membrane potential becomes more negative

(c) The membrane potential stabilizes around –70 mV

Suppose that the plasma membrane of a neuron has more Na^+ leak channels than K^+ leak channels. What effect would this have on the resting membrane potential?

the neuron slows down and Na^+ entry into the neuron speeds up because the negative membrane potential creates an electrical gradient that favors movement of both Na^+ and K^+ ions into the neuron (FIGURE 7.13b). There is still net movement of K^+ ions *out* of the neuron, however, because the magnitude of the K^+ concentration gradient is greater than the magnitude of the K^+ electrical gradient. With the help of Na^+/K^+ ATPases (described shortly), the resting membrane potential of the neuron stabilizes when it reaches about −70 mV (FIGURE 7.13c). At this point, K^+ movement out of the neuron is exactly balanced by Na^+ movement into the neuron and there is no net movement of charge across the membrane. As you have already discovered, the resting membrane potential (−70 mV) is *closer* to the K^+ equilibrium potential (−90 mV) than to the Na^+ equilibrium potential (+60 mV). This is because a resting neuron is more permeable to K^+ ions than it is to Na^+ ions, which allows K^+ ions to have a greater influence on the resting membrane potential than Na^+ ions. The resting membrane potential of a neuron can be calculated by using the Goldman-Hodgkin-Katz (GHK) equation (see the Physiological Equation box).

Contribution of the Na^+/K^+ ATPases

Because the resting membrane potential (−70 mV) is not equal to either the K^+ equilibrium potential (−90 mV) or the Na^+ equilibrium potential (+60 mV), some K^+ continuously leaks out of the neuron and some Na^+ continuously leaks into the neuron. Left unchecked, the small outward K^+ leak and the small inward Na^+ leak would eventually dissipate the K^+ and Na^+ concentration gradients. This does not happen because the inward Na^+ leak and outward K^+ leak are offset by the Na^+/K^+ ATPases (sodium-potassium pumps) (see FIGURE 7.13c). Recall from Chapter 5 that the Na^+/K^+ ATPases maintain the K^+ and Na^+ concentration gradients by expelling three Na^+ ions for each two K^+ ions imported using energy derived from the hydrolysis of ATP (see Section 5.5). Because these pumps remove more positive charges from the cell than they bring into the cell, they are **electrogenic**, which means they contribute to the negativity of the resting membrane potential. However, their total contribution is very small, only a few millivolts of the total −70 mV resting membrane potential in a typical neuron.

Existence of the Resting Neuron in a Steady State

As noted above, the Na^+/K^+ ATPases help maintain the resting membrane potential by using energy derived from ATP hydrolysis to pump out Na^+ as fast as it leaks in and to bring in K^+ as fast as it leaks out. This means that there is no net charge movement across the membrane of a resting neuron because Na^+ leakage into the cell and K^+

PHYSIOLOGICAL EQUATION

GOLDMAN-HODGKIN-KATZ EQUATION

Resting membrane potential is determined by two major factors: (1) an unequal distribution of ions in the ECF and cytosol and (2) differences in membrane permeability to various ions. These factors are expressed mathematically by the **Goldman-Hodgkin-Katz (GHK) equation**, which can be used to calculate the resting membrane potential of a neuron. Because the resting membrane potential of most neurons is influenced mainly by K^+ and Na^+ ions, a simplified version of the GHK equation is as follows:

$$Vm = 61 \log \frac{P_K[K^+]_{out} + P_{Na}[Na^+]_{out}}{P_K[K^+]_{in} + P_{Na}[Na^+]_{in}}$$

where
- Vm is the resting membrane potential in mV at 37°C,
- P_K is the potassium membrane permeability value (in most neurons, $P_K = 1$),
- P_{Na} is the sodium membrane permeability value (in most neurons, $P_{Na} = 0.04$),
- $[K^+]_{out}$ is the extracellular K^+ concentration in millimoles per liter (mM),
- $[K^+]_{in}$ is the intracellular K^+ concentration in millimoles per liter (mM),
- $[Na^+]_{out}$ is the extracellular Na^+ concentration in millimoles per liter (mM), and
- $[Na^+]_{in}$ is the intracellular Na^+ concentration in millimoles per liter (mM).

Using the membrane permeability values for K^+ and Na^+ listed above and the K^+ and Na^+ concentrations listed in Table 7.2, the resting membrane potential of a neuron can be calculated using the GHK equation in the following way:

$$Vm = 61 \log \frac{1\,(5) + 0.04\,(145)}{1\,(150) + 0.04\,(15)}$$

$$= -70 \text{ mV}$$

Note that if the potassium membrane permeability value (P_K) is set to zero in the GHK equation, then the potassium terms in the GHK equation are eliminated, and the GHK equation becomes the Nernst equation for Na^+ ions. Alternatively, if the sodium membrane permeability value (P_{Na}) is set to zero in the GHK equation, then the sodium terms in the GHK equation are eliminated, and the GHK equation becomes the Nernst equation for K^+ ions. In essence, the GHK equation is an extended version of the Nernst equation that takes into account the differences in membrane permeability to K^+ and Na^+ ions.

leakage out of the cell are exactly balanced by the constant activity of the Na^+/K^+ ATPases. Because there is no net charge movement across the membrane and energy is required to maintain this constant condition, the resting neuron exists in a steady state. Recall that when a *steady state* exists, energy is required to keep a particular condition constant (see Section 1.4).

Chloride Ions and the Resting Membrane Potential

Neurons at rest are permeable to Cl^- ions because their plasma membranes contain Cl^- leak channels. In most neurons, however, Cl^- ions do not significantly influence the resting membrane potential because there is no net movement of these ions across the plasma membrane through the Cl^- leak channels. This is because the plasma membranes of these cells lack active transport mechanisms for maintaining the Cl^- concentration gradient. Instead, Cl^- ions passively distribute across the membrane until the chloride equilibrium potential (E_{Cl}) becomes the same as the resting membrane potential. In other words, at −70 mV, the Cl^- concentration gradient (from ECF to cytosol) is equal in magnitude to the opposing Cl^- electrical gradient (from cytosol to ECF) and there is no net Cl^- movement across the membrane.

Some neurons, however, do have active transport mechanisms that move Cl^- ions against their concentration gradient from the cytosol into the ECF. This generates a stronger Cl^- concentration gradient compared to what would be the case without active Cl^- transport. Consequently, the magnitude of the Cl^- concentration gradient becomes greater than the magnitude of the Cl^- electrical gradient, causing net movement of Cl^- ions from the ECF into the cytosol of the neuron (via Cl^- leak channels in the membrane) under resting conditions. This means that E_{Cl} is not equal to the resting membrane potential; in fact, E_{Cl} is more negative than the resting membrane potential in neurons that actively transport Cl^- ions. Because there is net movement of Cl^- ions across the membrane, Cl^- ions make a significant contribution to the resting membrane potential in these cells.

Graded Potentials Can Be Depolarizing or Hyperpolarizing

A **graded potential** is a small deviation from the membrane potential that makes the membrane either less polarized (inside less negative) or more polarized (inside more negative). When the response makes the membrane less polarized (inside less negative), it is termed a **depolarizing graded potential** (FIGURE 7.14a). When the response makes the membrane more polarized (inside more negative), it is termed a **hyperpolarizing graded potential** (FIGURE 7.14b). The duration of a depolarizing or hyperpolarizing graded potential can last from several milliseconds (msec) to several minutes.

A graded potential occurs when a stimulus causes mechanically-gated channels or ligand-gated channels to open or close in an excitable cell's plasma membrane (FIGURE 7.15). Mechanically-gated channels and ligand-gated channels can be present in the dendrites of sensory neurons, and ligand-gated channels are numerous in the dendrites and cell bodies of interneurons and motor neurons (see TABLE 7.1). Hence, graded potentials occur mainly in the dendrites and cell body of a neuron.

FIGURE 7.14 Graded potentials. Most graded potentials occur in the dendrites and cell body (areas colored blue in the inset).

During a depolarizing graded potential, the membrane potential is inside less negative than the resting level; during a hyperpolarizing graded potential, the membrane potential is inside more negative than the resting level.

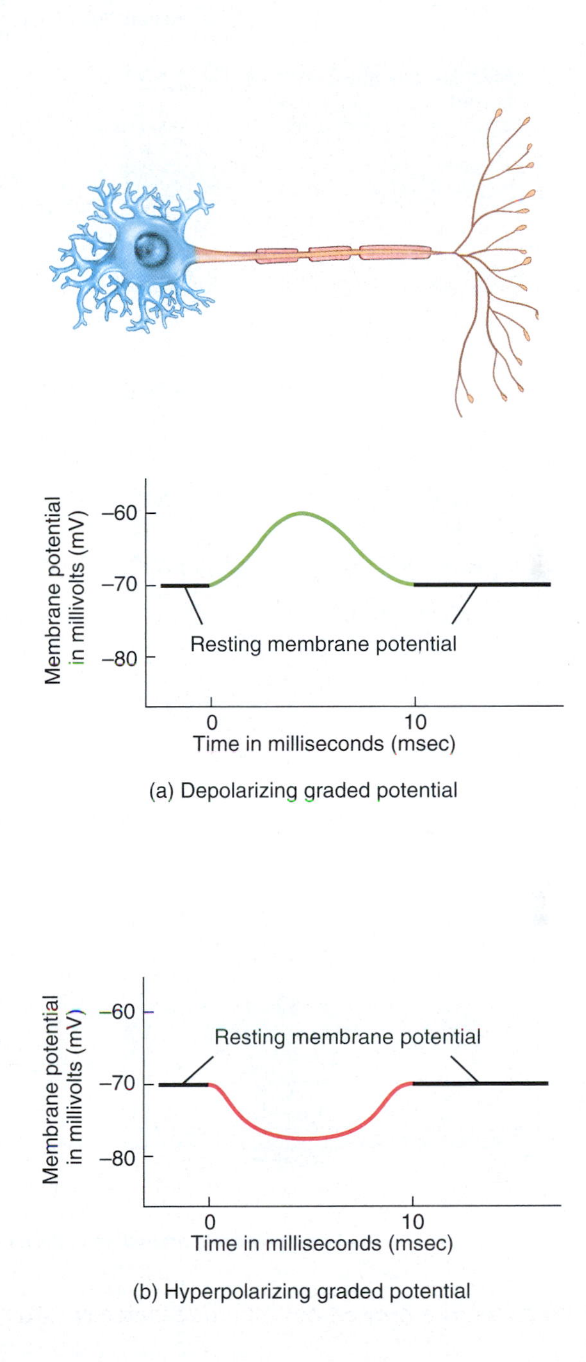

Which type of graded potential describes a change in membrane potential from −70 to −60 mV? From −70 to −80 mV?

FIGURE 7.15 Generation of graded potentials in response to the opening of mechanically-gated channels or ligand-gated channels. (a) A mechanical stimulus (pressure) opens a mechanically-gated channel that allows passage of cations (mainly Na^+ and Ca^{2+}) into the cell, causing a depolarizing graded potential. (b) The neurotransmitter acetylcholine (a ligand stimulus) opens a cation channel that allows passage of Na^+, K^+, and Ca^{2+}; Na^+ inflow is greater than either Ca^{2+} inflow or K^+ outflow, causing a depolarizing graded potential. (c) The neurotransmitter glycine (a ligand stimulus) opens a Cl^- channel that allows passage of Cl^- ions into the cell, causing a hyperpolarizing graded potential.

A graded potential forms in response to the opening of mechanically-gated channels or ligand-gated channels.

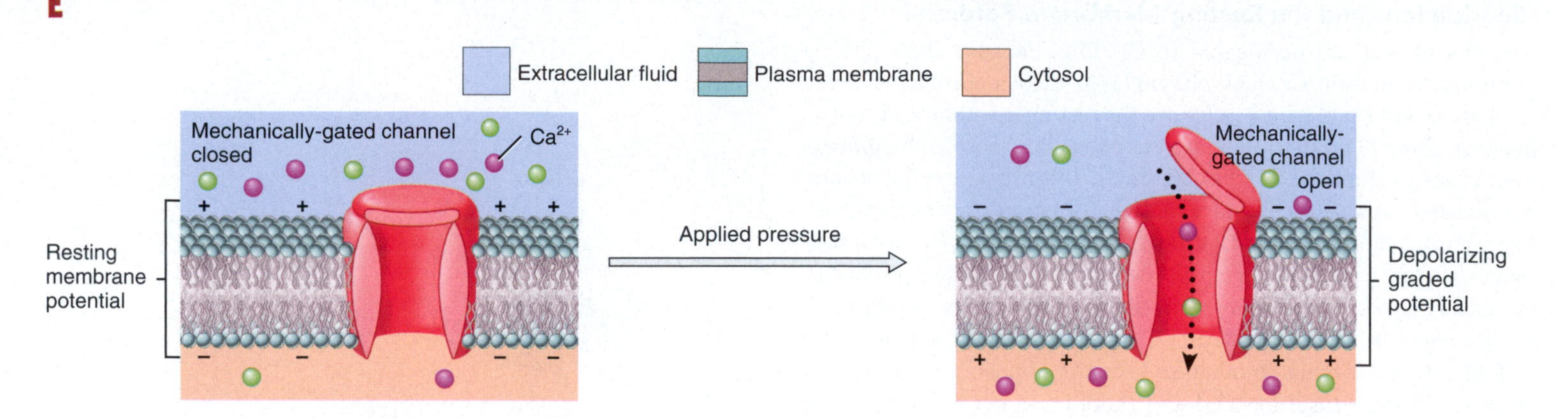

(a) Depolarizing graded potential caused by pressure, a mechanical stimulus

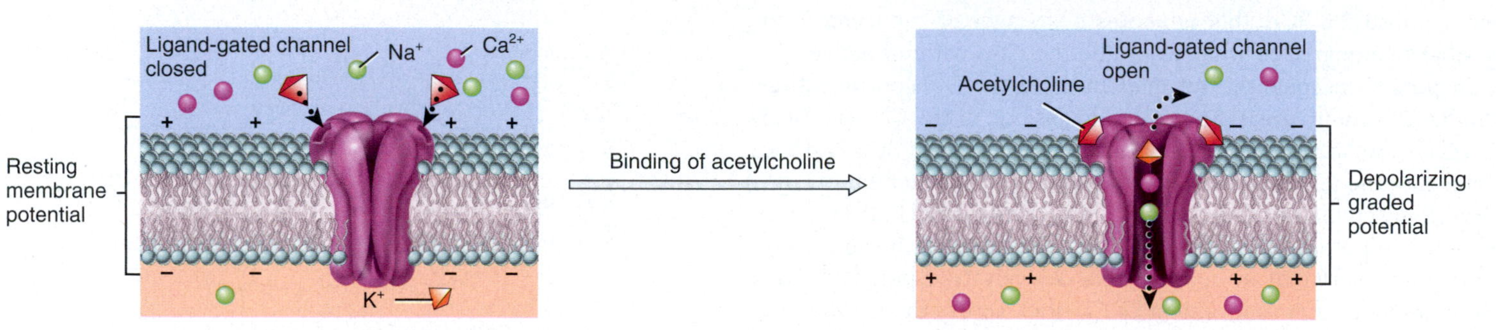

(b) Depolarizing graded potential caused by the neurotransmitter acetylcholine, a ligand stimulus

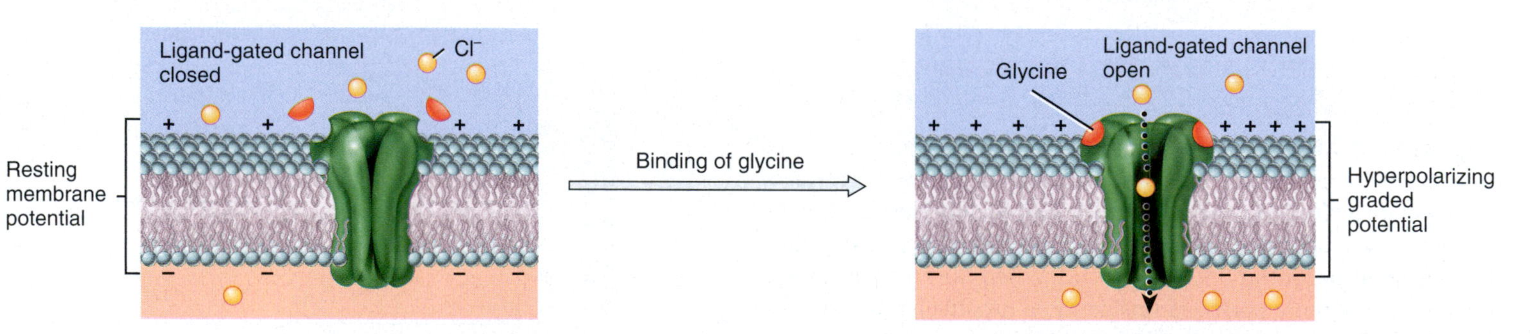

(c) Hyperpolarizing graded potential caused by the neurotransmitter glycine, a ligand stimulus

Which parts of a neuron contain mechanically-gated channels? Ligand-gated channels?

To say that these electrical signals are *graded* means that they vary in amplitude (size), depending on the strength of the stimulus (FIGURE 7.16). Graded potentials are larger or smaller depending on how many ligand-gated or mechanically-gated channels have opened (or closed) and how long each remains open. The opening or closing of these ion channels alters the number of specific ions that move across the plasma membrane to cause the graded potential. The amplitude of a graded potential can vary from less than 1 mV to more than 50 mV.

After it is generated, a graded potential spreads along the membrane in both directions away from the stimulus source. The spread of a graded

FIGURE 7.16 The graded nature of graded potentials. As the stimulus strength increases (stimuli 1, 2, and 3), the amplitude (size) of each resulting depolarizing graded potential increases. Although not shown in the figure, a similar relationship exists between stimulus strength and the amplitude of a hyperpolarizing graded potential.

The amplitude of a graded potential depends on the stimulus strength. The greater the stimulus strength, the larger the amplitude of the graded potential.

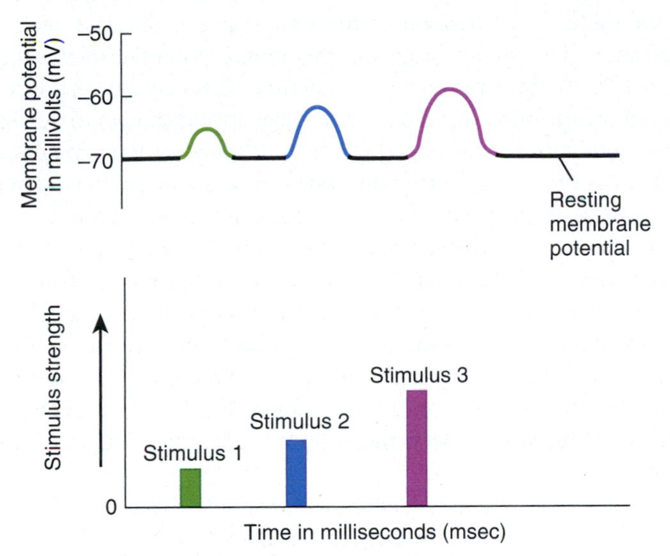

Why does a stronger stimulus cause a larger graded potential than a weaker stimulus?

FIGURE 7.17 Local current flow. The black arrows represent local current flow.

Local current flow is the passive movement of charges from one membrane region to adjacent membrane regions because of differences in membrane potential in these areas.

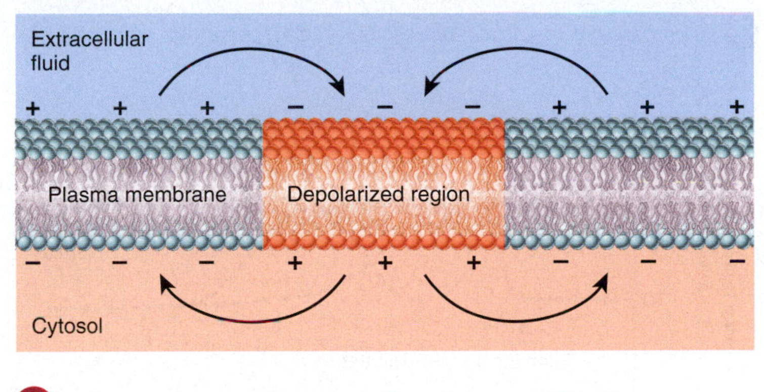

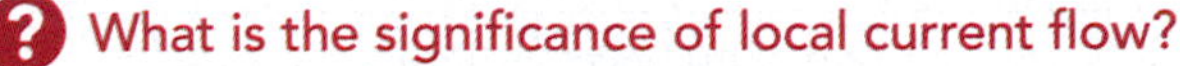
What is the significance of local current flow?

potential is accomplished by local current flow. **Local current flow** refers to the passive movement of charges from one region of membrane to adjacent regions of membrane due to differences in membrane potential in these areas. To understand how local current flow occurs, consider a region of membrane that is depolarized (FIGURE 7.17). Local current flow occurs in the *cytosol* as positive charges move from the depolarized membrane region to the more negative adjacent membrane regions that are still at resting membrane potential. At the same time, local current flow occurs in the *ECF* as positive charges move from adjacent membrane regions to the more negative depolarized membrane region. As a result of local current flow, adjacent membrane regions become depolarized. In other words, the graded potential spreads along the membrane in both directions away from the stimulus source.

As a graded potential spreads to adjacent regions of membrane by local current flow, it gradually dies out because its charges are lost across the membrane through leak channels. This mode of travel by which graded potentials die out as they spread along the membrane is known as **decremental conduction**. FIGURE 7.18 shows that the amplitude of a graded potential decreases as the distance from the graded potential's point of origin increases. Because they die out within a few millimeters of where they originate, graded potentials are useful for short-distance communication only.

Although an individual graded potential undergoes decremental conduction, it can become stronger and last longer by summating with other graded potentials. **Summation** is the process by which graded potentials add together. If two depolarizing graded potentials summate, the net result is a larger depolarizing graded potential (FIGURE 7.19). If two hyperpolarizing graded potentials summate, the net result is a larger hyperpolarizing graded potential. If two equal but opposite graded potentials summate (one depolarizing and the other hyperpolarizing), they cancel each other out and the overall graded potential disappears. You will learn more about the process of summation later in this chapter.

Graded potentials have different names depending on which type of stimulus causes them and where they occur. For example, when a graded potential occurs in the dendrites or cell body of a neuron in

FIGURE 7.18 The decremental nature of a graded potential.

The amplitude of a graded potential decreases as the distance from the graded potential's point of origin increases.

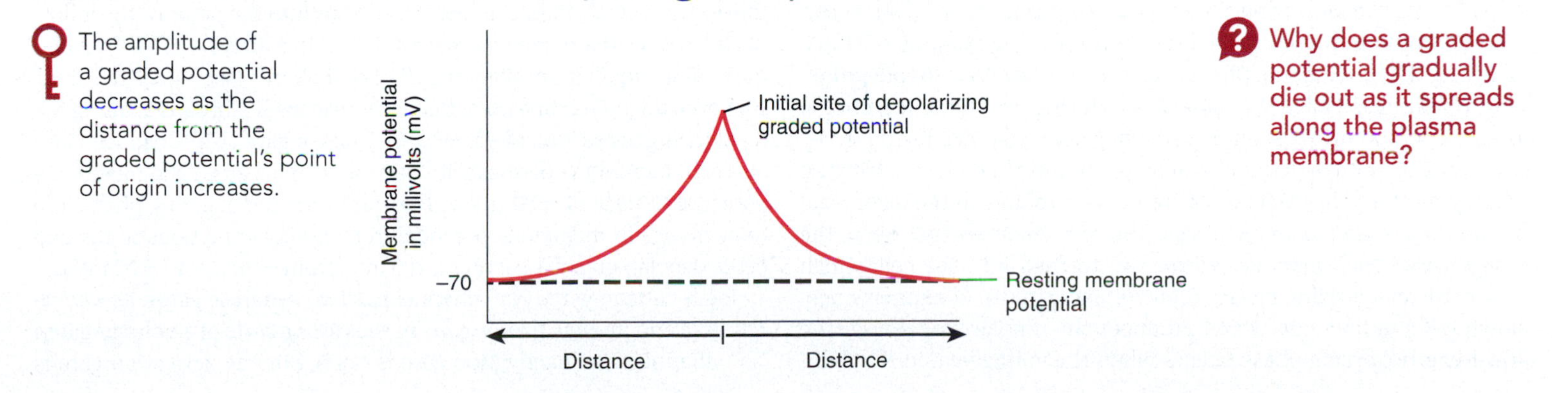

Why does a graded potential gradually die out as it spreads along the plasma membrane?

FIGURE 7.19 Summation of graded potentials. Summation of two depolarizing graded potentials in response to two stimuli of the same strength that occur very close together in time is shown. The dotted lines represent the individual depolarizing graded potentials that would form if summation did not occur.

Summation is the process by which graded potentials add together.

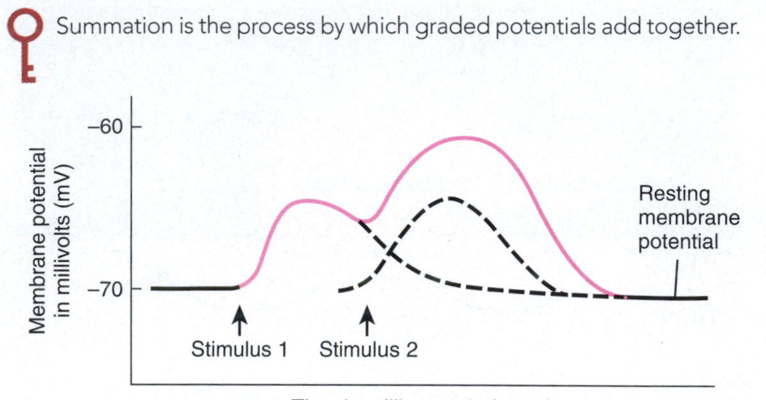

What would happen if summation of graded potentials in a neuron did not occur?

response to a neurotransmitter, it is called a *postsynaptic potential* (explained in Section 7.4). The graded potentials that occur in sensory receptors are termed *receptor potentials* (explained in Chapter 9). When a graded potential occurs in the plasma membrane of a skeletal muscle fiber at the neuromuscular junction (NMJ), it is called an *end plate potential (EPP)* (explained in Chapter 10).

Action Potentials Are Generated When the Axon Membrane Reaches Threshold

An **action potential (AP)** or **impulse** is a sequence of rapidly occurring events that decrease and reverse the membrane potential and then eventually restore it to the resting state. An action potential has two main phases: a depolarizing phase and a repolarizing phase (FIGURE 7.20a). During the **depolarizing phase**, or *rising phase*, the negative membrane potential becomes less negative, reaches zero, and then becomes positive. The depolarizing phase reaches its peak at +30 mV. The part of the depolarizing phase between 0 mV and +30 mV is called the *overshoot*. During the **repolarizing phase**, or *falling phase*, the membrane potential is restored to the resting state of −70 mV. Following the repolarizing phase, there may be an **after-hyperpolarizing phase**, also called the *undershoot*, during which the membrane potential temporarily becomes more negative than the resting level. Two types of voltage-gated channels open and then close during an action potential. These channels are present mainly in the axon plasma membrane and axon terminals. The first channels that open, the voltage-gated Na^+ channels, allow Na^+ to rush into the cell, which causes the depolarizing phase. Then voltage-gated K^+ channels open, allowing K^+ to flow out, which produces the repolarizing phase. The after-hyperpolarizing phase occurs when the voltage-gated K^+ channels remain open after the repolarizing phase ends. The duration of the action potential in most neurons is about 1–2 msec.

An action potential occurs in the membrane of the axon when depolarization reaches a certain level termed the **threshold** (about −55 mV in many neurons) (FIGURE 7.21). Different neurons may have different thresholds for action potential generation, but the threshold in a particular neuron is usually constant. The generation of an action potential depends on whether a particular stimulus is able to bring the membrane potential to threshold. An action potential does not occur in response to a **subthreshold stimulus**, a stimulus that is a weak depolarization that cannot bring the membrane potential to threshold (FIGURE 7.21). However, an action potential does occur in response to a **threshold stimulus**, a stimulus that is just strong enough to depolarize the membrane to threshold (FIGURE 7.21). Several action potentials form in response to a **suprathreshold stimulus**, a stimulus that is strong enough to depolarize the membrane *above* threshold (FIGURE 7.21). Each of the action potentials caused by a suprathreshold stimulus has the same amplitude (size) as an action potential caused by a threshold stimulus. Therefore, once an action potential is generated, the amplitude of an action potential is always the same and does not depend on stimulus intensity. Instead, the greater the stimulus strength above threshold, the greater the frequency of the action potentials until a maximum frequency is reached as determined by the absolute refractory period (described shortly).

So an action potential is generated in response to a threshold stimulus, but it does not form when there is a subthreshold stimulus. In other words, an action potential either completely occurs or it does not occur at all. This characteristic of an action potential is known as the **all-or-none principle**. The all-or-none principle of the action potential is similar to pushing the first domino in a long row of standing dominoes. When the push on the first domino is strong enough (when depolarization reaches threshold), that domino falls against the second domino, and the *entire* row topples (an action potential occurs). Stronger pushes on the first domino produce the identical effect—toppling of the entire row. Thus, pushing on the first domino produces an all-or-none event: The dominoes all fall or none fall.

Depolarizing Phase

During the **depolarizing phase** of the action potential, membrane permeability to Na^+ ions increases (see FIGURE 7.20b). The depolarizing phase begins when a depolarizing graded potential or some other stimulus causes the membrane of the axon to depolarize to threshold. Once threshold is reached, voltage-gated Na^+ channels open rapidly. Because both the Na^+ concentration and electrical gradients favor inward movement of Na^+, there is a rush of Na^+ ions into the neuron. The inflow of Na^+ causes the membrane potential to move above −55 mV toward the sodium equilibrium potential (E_{Na}) of +60 mV (see FIGURE 7.20a). However, the membrane potential never reaches E_{Na} because, during the repolarizing phase (described next), the voltage-gated Na^+ channels close and Na^+ membrane permeability decreases. This causes the membrane potential to peak at +30 mV at the end of the depolarizing phase. The total change in membrane potential from resting conditions to the end of the depolarizing phase is about 100 mV (from −70 mV to +30 mV).

Each voltage-gated Na^+ channel has two separate gates, an *activation gate* and an *inactivation gate*. In the *resting state* of a voltage-gated Na^+ channel, the inactivation gate is open, but the activation gate is

FIGURE 7.20 The action potential (AP). The action potential arises at the trigger zone and then propagates along the axon to the axon terminals. The green-colored regions of the neuron in the inset indicate the parts that typically have voltage-gated Na^+ and K^+ channels (axon plasma membrane and axon terminals).

An action potential consists of a depolarizing phase and a repolarizing phase, which may be followed by an after-hyperpolarizing phase.

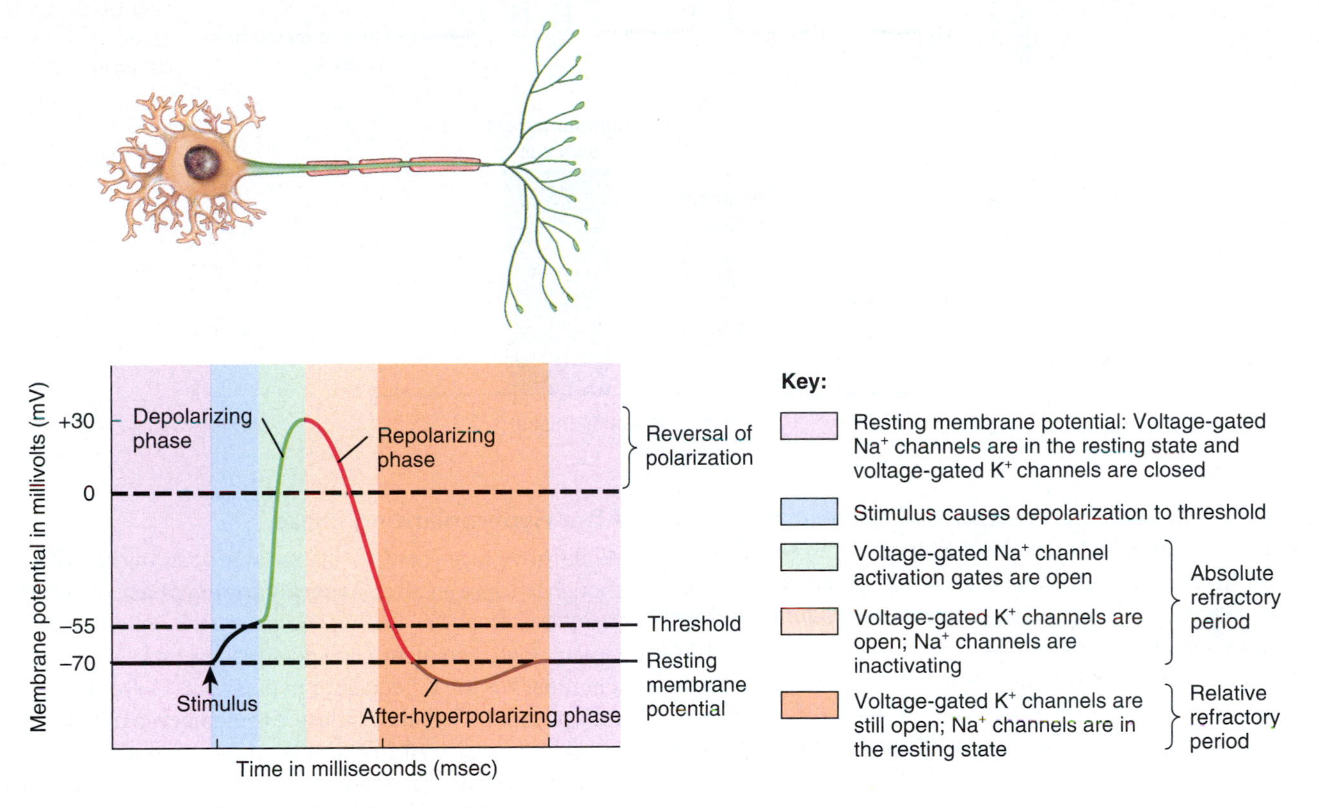

(a) Phases of the action potential

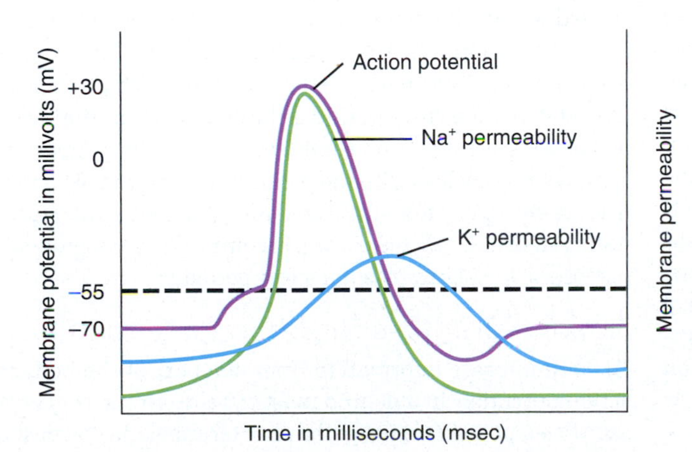

(b) Na^+ and K^+ permeability changes during the action potential

Which channels are open during the depolarizing phase? During the repolarizing phase?

FIGURE 7.21 Stimulus strength and action potential generation. For simplicity, the after-hyperpolarizing phase of the action potential is not shown.

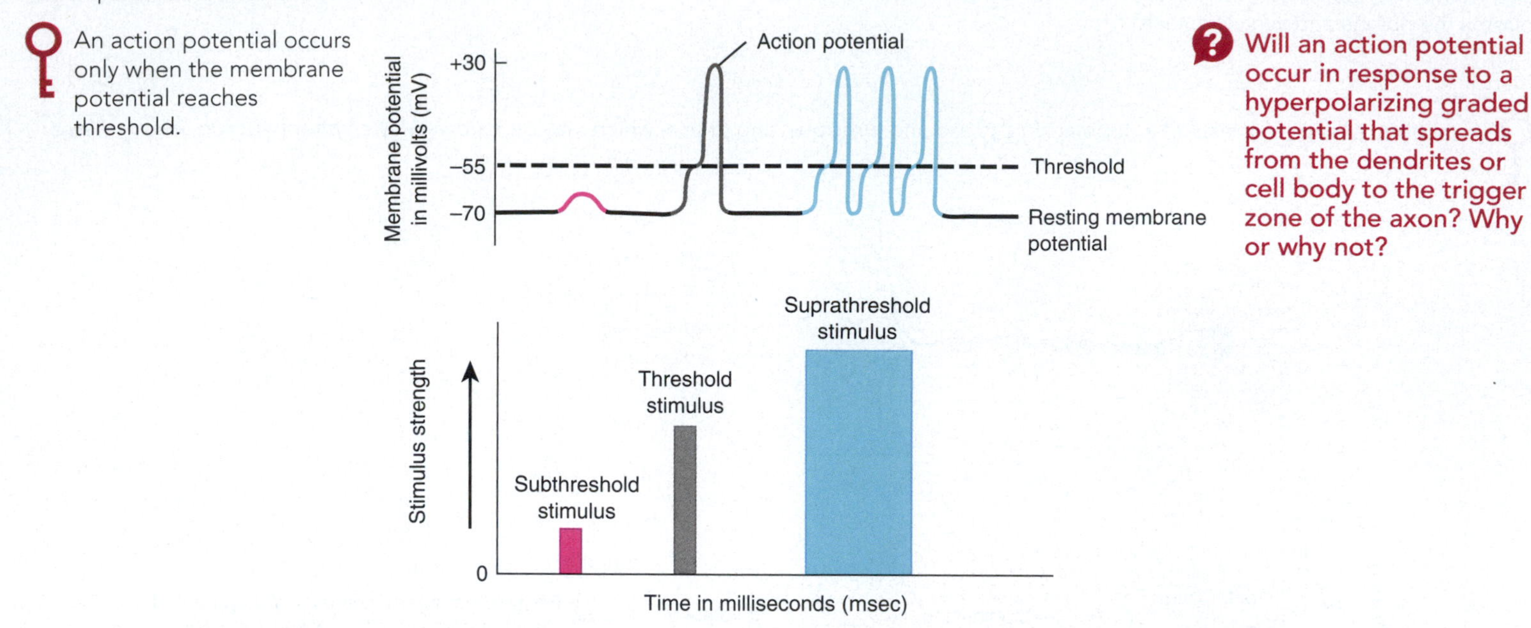

closed (step 1 in FIGURE 7.22). As a result, Na^+ cannot move into the cell through these channels. At threshold, voltage-gated Na^+ channels are activated. In the *activated state* of a voltage-gated Na^+ channel, both the activation and inactivation gates in the channel are open and Na^+ inflow begins (step 2 in FIGURE 7.22). As more channels open, Na^+ inflow increases, the membrane depolarizes further, and more Na^+ channels open. This is an example of a positive feedback mechanism. During the few ten-thousandths of a second that the voltage-gated Na^+ channel is open, about 10,000 Na^+ ions flow across the membrane and change the membrane potential considerably, but the concentration of Na^+ hardly changes because of the millions of Na^+ present in the ECF. The sodium-potassium pumps easily bail out the 10,000 or so Na^+ ions that enter the cell during a single action potential and maintain the low concentration of Na^+ inside the cell.

Repolarizing Phase

During the **repolarizing phase** of the action potential, membrane permeability to Na^+ decreases and membrane permeability to K^+ increases (see FIGURE 7.20b). The repolarizing phase begins when the inactivation gates of the voltage-gated Na^+ channels close (step 3 in FIGURE 7.22). Now the voltage-gated Na^+ channel is in an *inactivated state.* The threshold-level depolarization that opened the voltage-gated Na^+ channels also opens voltage-gated K^+ channels (steps 3 and 4 in FIGURE 7.22). Because the voltage-gated K^+ channels open more slowly, their opening occurs at about the same time that the voltage-gated Na^+ channels are closing. The slower opening of voltage-gated K^+ channels and the closing of previously open voltage-gated Na^+ channels produce the repolarizing phase of the action potential. As the Na^+ channels are inactivated, Na^+ inflow slows. At the same time, the K^+ channels are opening, accelerating K^+ outflow. Slowing of Na^+ inflow and acceleration of K^+ outflow causes the membrane potential to change from +30 mV to −70 mV. As the membrane potential approaches −70 mV, the inactivated Na^+ channels revert to their resting state.

After-Hyperpolarizing Phase

While the voltage-gated K^+ channels are open, outflow of K^+ may be large enough to cause an **after-hyperpolarizing phase** of the action potential (see FIGURE 7.20a). During this phase, the voltage-gated K^+ channels remain open. The membrane potential becomes even more negative as it approaches the K^+ equilibrium potential (E_K) of about −90 mV. Once voltage-gated K^+ channels close, the membrane potential returns to the resting level of −70 mV. Unlike voltage-gated Na^+ channels, most voltage-gated K^+ channels do not enter an inactivated state. Instead, they alternate between closed (resting) and open (activated) states.

Refractory Period

The period of time after an action potential begins during which an excitable cell cannot generate another action potential in response to a *normal* threshold stimulus is called the **refractory period** (see key in FIGURE 7.20a). During the **absolute refractory period**, even a very strong stimulus cannot initiate a second action potential. This period coincides with the period of Na^+ channel activation and inactivation (steps 2 and 3 in FIGURE 7.22). Inactivated Na^+ channels cannot reopen; they first must return to the resting state (step 1 in FIGURE 7.22). The **relative refractory period** is the period of time during which a second action potential can be initiated, but only by a larger-than-normal stimulus. It coincides with the period when the voltage-gated K^+ channels are still open after inactivated Na^+ channels have returned to their resting state (see FIGURE 7.20a). In contrast to action potentials, graded potentials do not exhibit a refractory period.

Action Potentials Undergo Propagation

To communicate information from one part of the body to another, action potentials in a neuron must travel from the trigger zone of the axon (their point of origin) to the axon terminals. In contrast to a graded potential, an action potential is not decremental (it does not die out). Instead, an action potential maintains its strength as its spreads along the membrane. This mode of conduction, called **propagation**, depends on positive feedback. When sodium ions flow in, they cause

FIGURE 7.22 Changes in ion flow through voltage-gated channels during the depolarizing and repolarizing phases of an action potential. Leak channels and sodium-potassium pumps are not shown.

Inflow of sodium ions (Na^+) causes the depolarizing phase of the action potential, and outflow of potassium ions (K^+) causes the repolarizing phase.

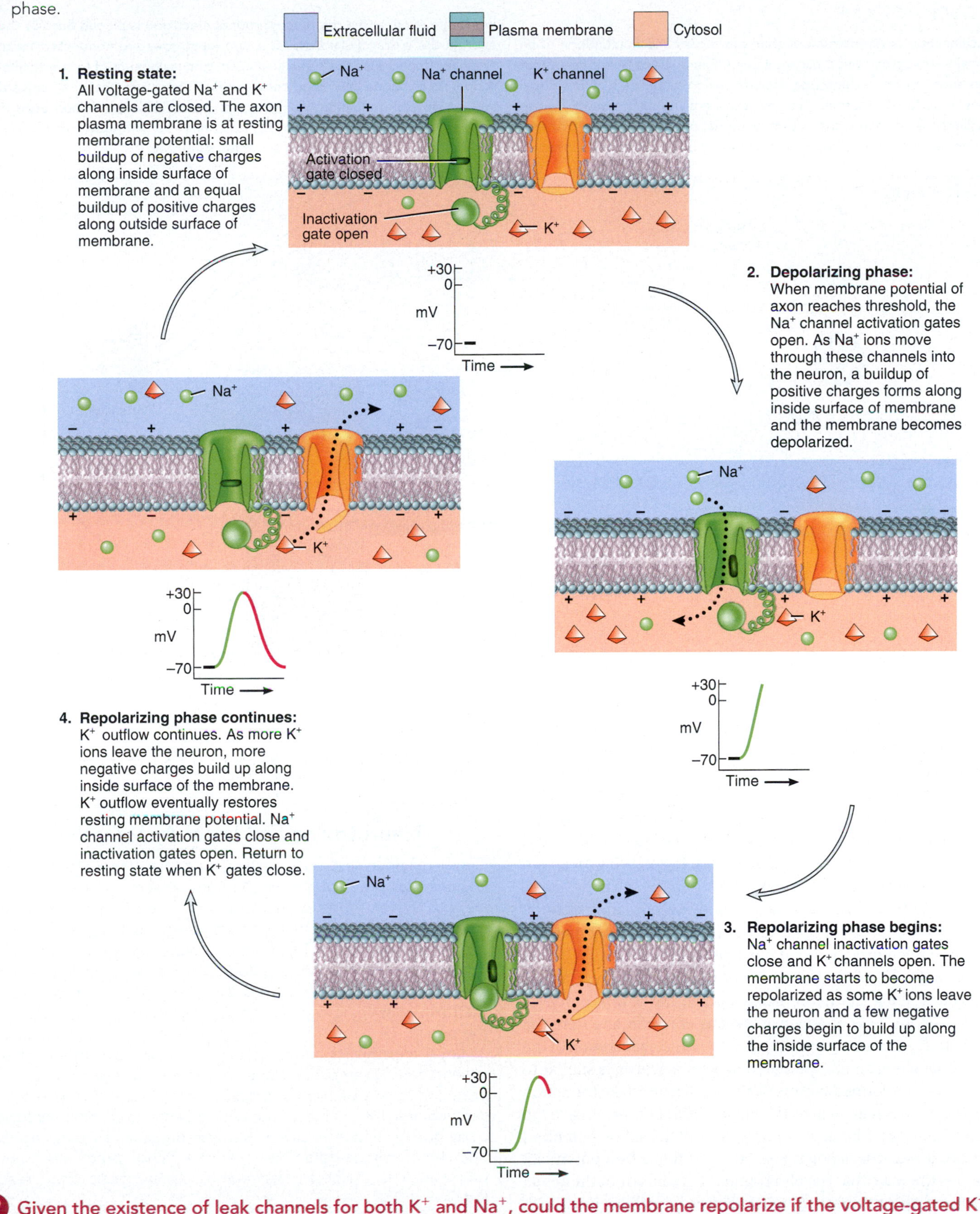

Given the existence of leak channels for both K^+ and Na^+, could the membrane repolarize if the voltage-gated K^+ channels did not exist? Explain your answer.

TOOL OF THE TRADE

THE OSCILLOSCOPE

An **oscilloscope** is an instrument that can record the fluctuations of an electrical event across a fluorescent screen. Examples of phenomena that can be seen on an oscilloscope include action potentials of individual neurons, muscle contractions, and electrical activity of the heart. To use an oscilloscope to record the action potential of a neuron, the following procedure is performed. The tip of a recording microelectrode is inserted inside the axon, and a reference (ground) electrode is placed outside the axon in the extracellular fluid. The two electrodes are connected to an amplifier (to increase the signal), which in turn is connected to the oscilloscope. As a change in voltage occurs during the action potential, upward and downward deflections are seen on the oscilloscope. When used in this way, the oscilloscope essentially functions as a voltmeter that records rapid changes occurring during an action potential.

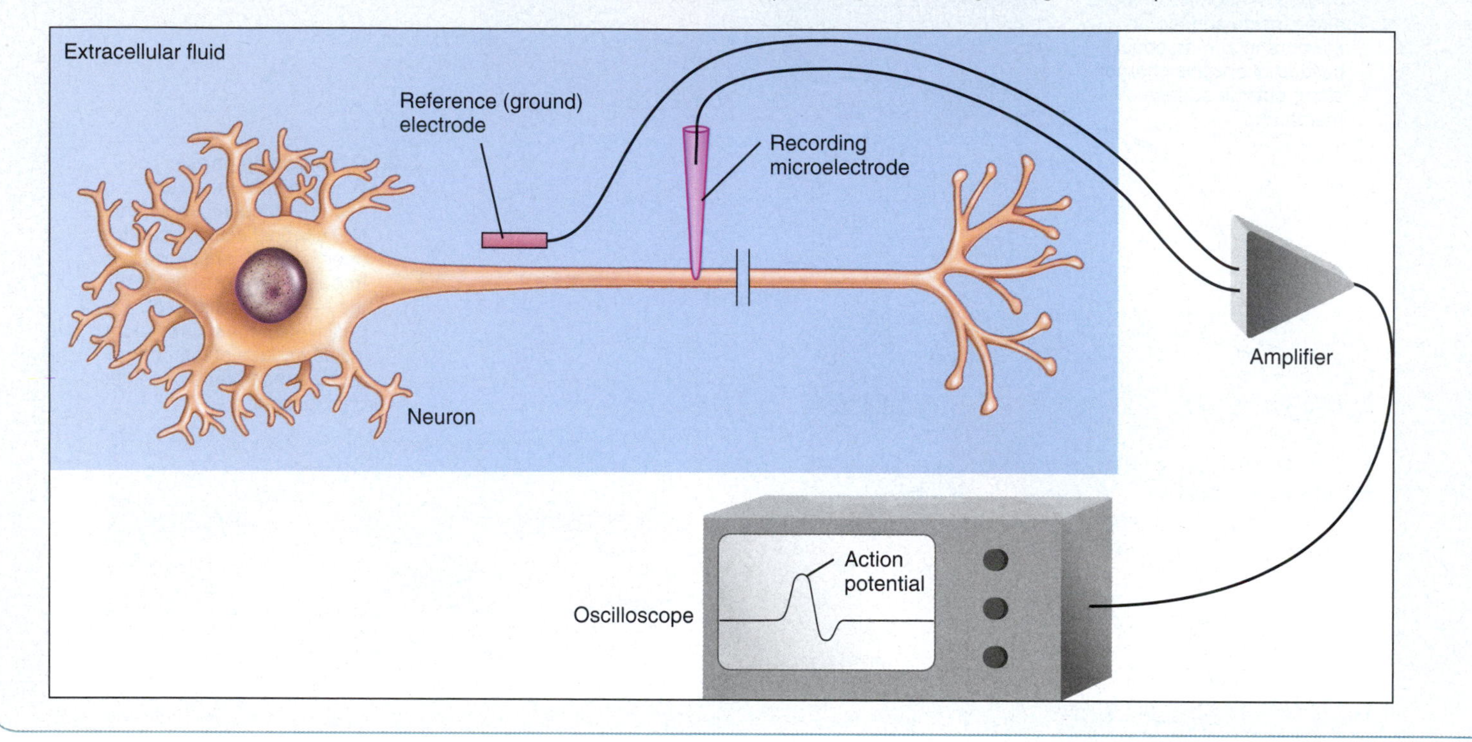

voltage-gated Na^+ channels in adjacent segments of the membrane to open. Thus, the action potential travels along the membrane rather like the activity of that long row of dominoes. The propagation of an action potential is accomplished by local current flow (see **FIGURE 7.17**). Recall that local current flow refers to the passive movement of charges from one membrane region to adjacent membrane regions due to differences in membrane potential in these areas. In a neuron, an action potential can propagate along the axon away from the cell body only—it cannot propagate back toward the cell body because any region of membrane that has just undergone an action potential is temporarily in the absolute refractory period and cannot generate another action potential. You should realize that it is not the same action potential that propagates along the entire axon. Rather, the action potential *regenerates* over and over at adjacent regions of membrane from the trigger zone to the axon terminals (**FIGURE 7.23**). Regeneration of the action potential along an unmyelinated axon of a neuron is similar to the "wave" that is performed by fans at a football game. Like the action potential, the wave is regenerated throughout the stadium as each fan sequentially stands up (the depolarizing phase of the action potential) and then sits down (the repolarizing phase of the action potential). Therefore, it is the wave that travels around the stadium, not the actual fans. Because they can travel along a membrane without dying out, action potentials function in communication over long distances.

CLINICAL CONNECTION

Neurotoxins and Local Anesthetics

Certain shellfish and other organisms contain **neurotoxins**, substances that produce their poisonous effects by acting on the nervous system. One particularly lethal neurotoxin is tetrodotoxin (TTX), present in the gonads, liver, intestine, and skin of Japanese puffer fish. TTX effectively blocks action potentials by inserting itself into voltage-gated Na^+ channels, which prevents them from opening. Chefs must be specially trained to remove the toxin-containing organs of puffer fish before serving them to the public.

Local anesthetics are drugs that block pain and other somatic sensations. Examples include procaine (Novocaine®) and lidocaine, which may be used to produce anesthesia in the skin during suturing of a gash, in the mouth during dental work, or in the lower body during childbirth. Like TTX, these drugs act by blocking the opening of voltage-gated Na^+ channels. Action potentials cannot propagate past the obstructed region, so pain signals do not reach the CNS.

FIGURE 7.23 Regeneration of the three phases of the action potential along an unmyelinated axon. Parts (a) to (d) of the figure represent the same section of axon membrane at different points in time. The black arrows represent either Na^+ movement into the neuron or K^+ movement out of the neuron. For simplicity, voltage-gated channels and local current flow are not shown.

Each phase of the action potential sequentially regenerates along the axon of a neuron from the trigger zone to the axon terminals.

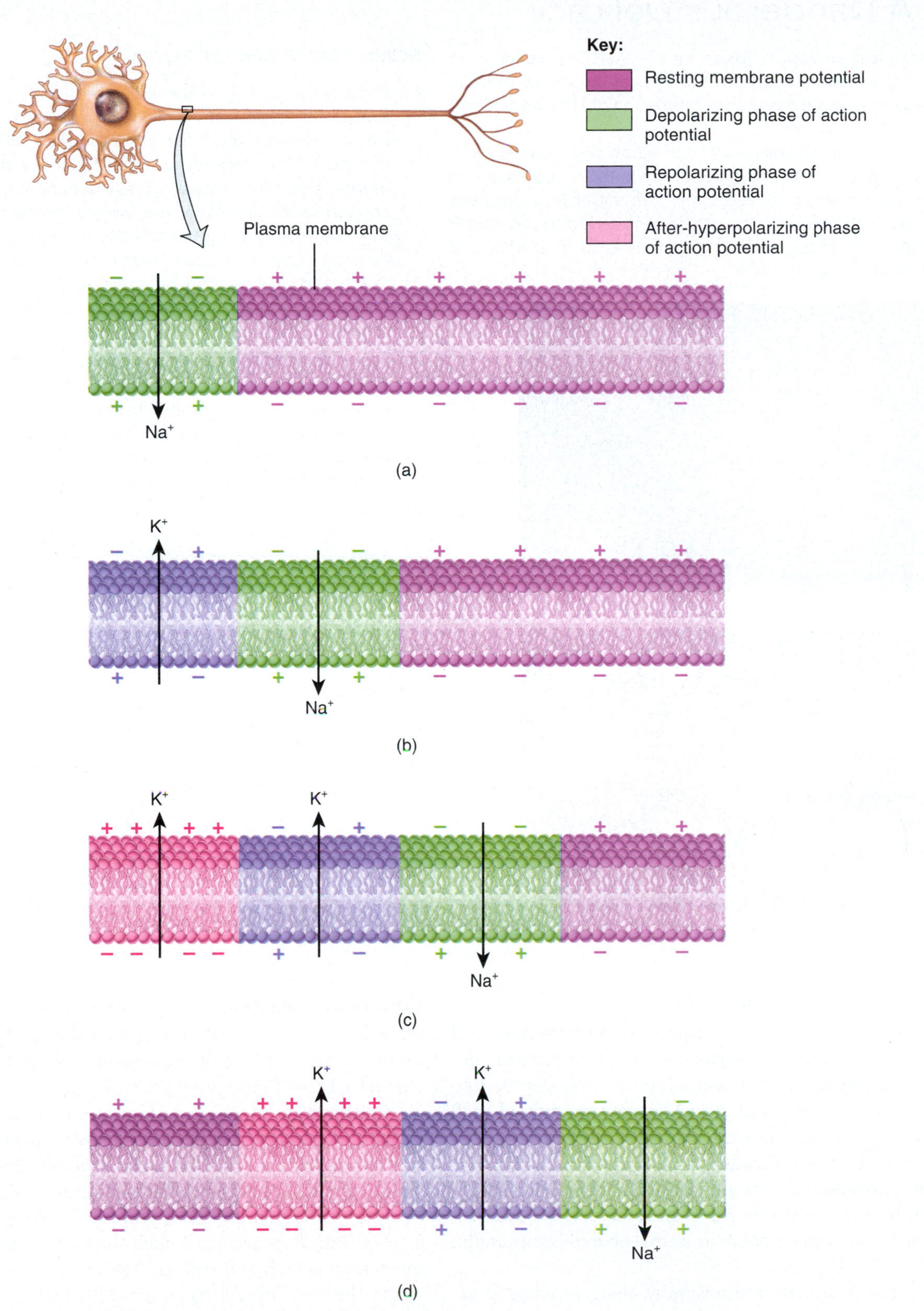

Why doesn't an action potential undergo decremental conduction?

Critical Thinking

Fugu: A Dangerous Delicacy

Max and Sarah just arrived in Tokyo for the start of a vacation in Japan. The couple wants to experience all that Tokyo has to offer, and they decide to go out for a night on the town. They begin their evening in a sushi restaurant.

After eating several pieces each, both Sarah and Max begin to experience a tingling and burning sensation in their mouths, and a numbness of their faces. The restaurant staff members have seen only one or two instances where patrons responded in this way. If these are symptoms of fugu poisoning, they know that Max and Sarah could be in serious trouble.

SOME FACTS ABOUT FUGU:

1. Puffer fish—or fugu—are considered to be among the most poisonous vertebrates in the world, second only to the golden poison frog, but these fish are also long considered a delicacy in Japan. More recently, in other locations such as the United States, they are consumed as sashimi (raw) and in various cooked forms. In Tokyo, the long-standing practice has been that chefs serving fugu needed to be highly trained; they must possess a city certificate indicating their expertise in the preparation and presentation for consumption of this fish. Today, in Tokyo, some want to relax these rules and allow less accomplished individuals to serve fugu products.
2. Tetrodotoxin (TTX) is a neurotoxin that inhibits sodium ion movement through the Na^+ channels, thereby inhibiting the production of action potentials. It is synthesized by certain aquatic bacteria and when consumed by puffer fish, it accumulates in the gonads, liver, intestine, and skin. These repository sites contain enough TTX to kill adult men and women if they consume them.
3. Consumption of TTX can lead to a series of symptoms, such as burning of the mouth and tongue, numbness of the face, drowsiness, and incoherent speech within 30 minutes. In severe cases people have displayed ataxia, hypotension, cardiac arrhythmias, muscle paralysis, and even death. There is no known antidote for TTX poisoning, so supportive measures are recommended.

Why would TTX exposure cause the symptoms experienced by Max and Sarah?

Some people have died within 17 minutes after eating fugu. What conditions could lead to the transition from intoxication to death?

IF TTX is so lethal (10 to 100 times more lethal than black widow spider venom, 10,000 times more lethal than cyanide), why don't the fugu die when they eat the toxic bacteria?

Continuous and Saltatory Conduction

There are two types of propagation: continuous conduction and saltatory conduction. The type of action potential propagation described so far is **continuous conduction**, which involves step-by-step depolarization, repolarization, and after-hyperpolarization of each adjacent segment of the plasma membrane (**FIGURE 7.24a**). In continuous conduction, ions flow through their voltage-gated channels in each adjacent segment of the membrane. Note that the action potential propagates only a relatively short distance in a few milliseconds. Continuous conduction occurs in unmyelinated axons and in muscle fibers.

Action potentials propagate more rapidly along myelinated axons than along unmyelinated axons. If you compare parts (a) and (b) in **FIGURE 7.24**, you will see that the action potential propagates much farther along the myelinated axon in the same period of time. **Saltatory conduction** (SAL-ta-tō-rē; *saltat* = *leaping*), the special mode of action potential propagation that occurs along myelinated axons, occurs because of the uneven distribution of voltage-gated channels. Few voltage-gated channels are present in regions where a myelin sheath covers the axon plasma membrane. By contrast, at the nodes of Ranvier (where there is no myelin sheath), the axon plasma membrane has many voltage-gated channels. Hence, current carried by Na^+ and K^+ flows across the membrane mainly at the nodes.

When an action potential propagates along a myelinated axon, local current flow occurs through the extracellular fluid surrounding the myelin sheath and through the cytosol from one node of Ranvier to the next. The action potential at the first node generates local current flow in the cytosol and extracellular fluid that depolarizes the membrane to threshold, opening voltage-gated Na^+ channels at the second node. The resulting ionic flow through the opened channels

FIGURE 7.24 Continuous conduction and saltatory conduction. Dotted lines indicate local current flow. The insets show the path of current flow.

Unmyelinated axons exhibit continuous conduction; myelinated axons exhibit saltatory conduction.

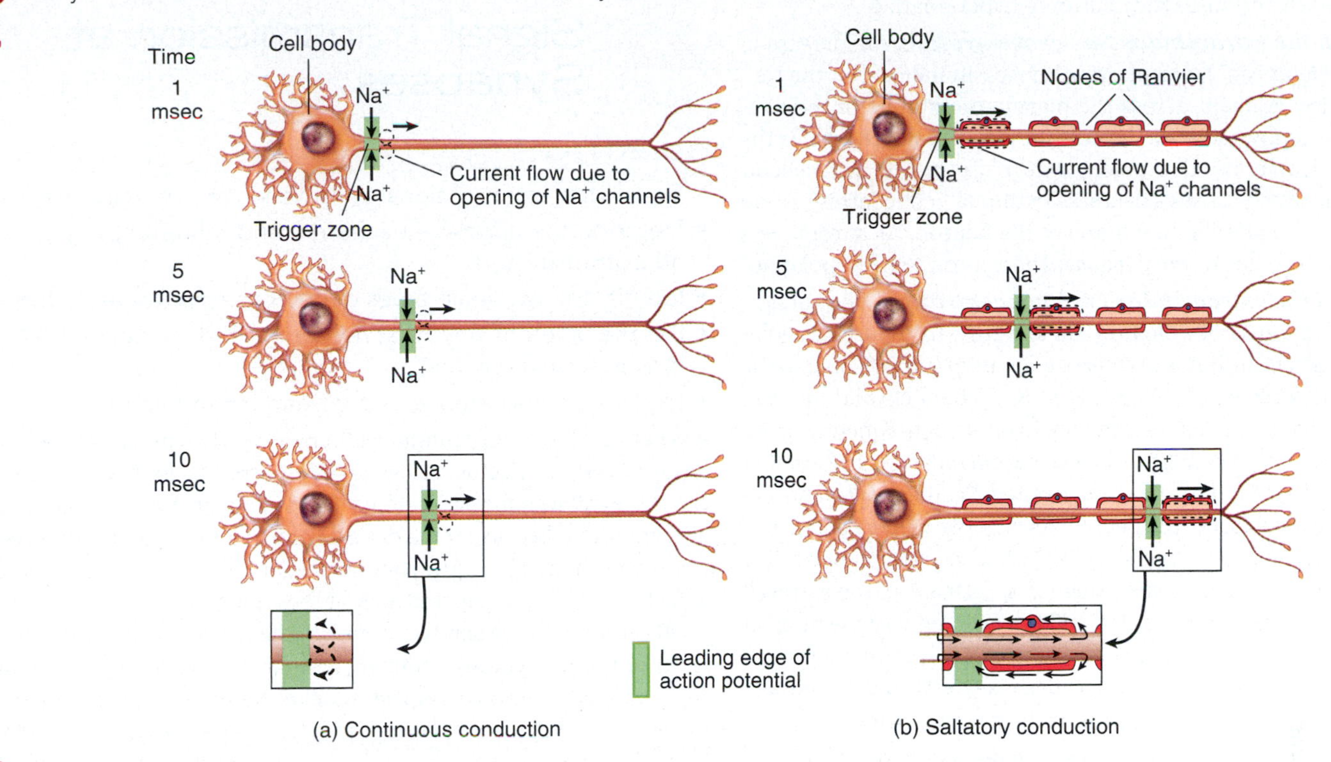

What factors determine the speed of propagation of an action potential?

constitutes an action potential at the second node. Then the action potential at the second node generates local current flow that opens voltage-gated Na^+ channels at the third node, and so on. After it depolarizes, each node repolarizes, hyperpolarizes, and then is restored back to resting membrane potential.

The flow of current across the membrane only at the nodes of Ranvier has two consequences:

1. The action potential appears to "leap" from node to node as each nodal area depolarizes to threshold, thus the name *saltatory*. Because an action potential leaps across long segments of myelinated axon plasma membrane as current flows from one node to the next, it travels much faster than it would in an unmyelinated axon of the same diameter.
2. Opening a smaller number of channels only at the nodes, rather than many channels in each adjacent segment of membrane, is a more energy-efficient mode of conduction. Because only small regions of the membrane depolarize and repolarize, minimal inflow of Na^+ and outflow of K^+ occurs each time an action potential passes by. Thus, less ATP is used by sodium-potassium pumps to maintain the low intracellular concentration of Na^+ and the low extracellular concentration of K^+.

Factors That Affect Conduction Velocity

The velocity of action potential conduction is affected by two major factors:

1. ***Axon diameter.*** The larger the diameter of the axon, the faster the action potential is conducted. An axon with a large diameter offers less resistance to local current flow, which allows adjacent regions of membrane to be brought to threshold more quickly.
2. ***Presence or absence of myelin.*** Conduction of action potentials is more rapid along myelinated axons than along unmyelinated axons. Recall that this is because an action potential leaps across long segments of membrane of a myelinated axon, whereas it must travel through each adjacent segment of membrane of an unmyelinated axon.

Large-diameter, myelinated axons called **A fibers** conduct action potentials at velocities ranging from 12–130 m/sec (27–280 mi/hr). They carry urgent information such as sensory signals associated with touch, pressure, and position of joints, and motor signals that cause contraction of skeletal muscles. By contrast, small-diameter, unmyelinated axons known as **C fibers** conduct action potentials at velocities ranging from 0.5–2 m/sec (1–4 mi/hr). They carry information that is less critical, such as motor signals that cause contraction of smooth muscle in digestive organs.

The Extracellular Concentrations of Several Ions Influence Neuronal Excitability

The excitability of neurons is influenced by the extracellular concentrations of K^+, Na^+, and Ca^{2+} ions.

- ***Changes in the extracellular K^+ concentration.*** An increase in the extracellular K^+ concentration causes a decrease in the K^+ concentration gradient across the plasma membrane of a neuron. Consequently, less K^+ leaves the neuron, which causes the neuron to depolarize. By contrast, a decrease in the extracellular K^+ concentration causes an increase in the K^+ concentration gradient across the plasma membrane of the neuron. As a result, more K^+ leaves the neuron, which causes the neuron to hyperpolarize.

- ***Changes in the extracellular Na^+ concentration.*** An increase in the extracellular Na^+ concentration causes an increase in the Na^+ concentration gradient across the plasma membrane of the neuron. Consequently, more Na^+ enters the neuron, which causes the neuron to depolarize. Conversely, a decrease in the extracellular Na^+ concentration causes a decrease in the Na^+ concentration gradient across the plasma membrane of the neuron. As a result, less Na^+ enters the neuron, which causes the neuron to hyperpolarize.
- ***Changes in the extracellular Ca^{2+} concentration.*** The voltage-gated Na^+ channels of a neuron are sensitive to the extracellular Ca^{2+} concentration. This is because Ca^{2+} ions in the ECF bind to the extracellular surfaces of voltage-gated Na^+ channels and increase the voltage that these channels require to open. Consequently, an increase in the extracellular Ca^{2+} concentration increases the number of Ca^{2+} ions that bind to the voltage-gated Na^+ channels. This decreases the excitability of the neuron because the voltage-gated Na^+ channels now require a higher voltage than normal to open. The opposite situation occurs when there is a decrease in the extracellular Ca^{2+} concentration. Less Ca^{2+} in the ECF reduces the number of Ca^{2+} ions that bind to the voltage-gated Na^+ channels. This increases the excitability of the neuron because the voltage-gated Na^+ channels are able to open at a lower voltage than normal.

TABLE 7.3 presents a summary of the differences between graded potentials and action potentials in neurons.

CHECKPOINT

10. Of the different types of ion channels present in neurons, which type opens in response to a chemical stimulus?
11. Explain how the resting membrane potential of a neuron is generated.
12. Compare and contrast graded potentials and action potentials.
13. How is saltatory conduction different from continuous conduction?
14. What effect does increasing the extracellular K^+ concentration have on neuronal excitability?

7.4 Signal Transmission at Synapses

OBJECTIVES

- Explain the events of signal transmission at a chemical synapse.
- Describe the action of excitatory and inhibitory neurotransmitters.
- Identify the two main types of neurotransmitter receptors.
- List the different ways that neurotransmitters can be removed from a synapse.
- Explain the importance of presynaptic modulation.

A **synapse** is the site of communication between two neurons or between a neuron and an effector cell. Synapses are essential for homeostasis because they allow information to be filtered and integrated. Synapses are also important because some diseases and neurological disorders result from disruptions of synaptic communication, and many therapeutic and addictive chemicals affect the body at these junctions.

At a synapse between two neurons, the neuron sending the signal is called the **presynaptic neuron**, and the neuron receiving the message is called the **postsynaptic neuron**. Neural synapses are named based on the parts of the neurons that form the synapse and the direction of information flow. Examples include **axodendritic** (from axon to dendrite), **axosomatic** (from axon to cell body), and **axoaxonic** (from axon to axon) synapses (FIGURE 7.25). Most neural synapses are either axodendritic or axosomatic.

A synapse may be electrical or chemical. These two types of synapses differ both structurally and functionally.

Electrical Synapses Involve Gap Junctions

At an **electrical synapse**, action potentials conduct directly between adjacent cells through **gap junctions**. Each gap junction contains tubular *connexons*, which function as tunnels that connect the cytosol of the two cells directly (see FIGURE 6.1a). As ions flow from one cell to

TABLE 7.3 Comparison of Graded Potentials and Action Potentials in Neurons

Characteristic	Graded Potentials	Action Potentials
Origin	Arise mainly in dendrites and cell body.	Arise at trigger zone and propagate along axon.
Types of channels	Ligand-gated or mechanically-gated ion channels.	Voltage-gated channels for Na^+ and K^+.
Conduction	Decremental (not propagated); permit communication over short distances.	Propagate and thus permit communication over longer distances.
Amplitude (size)	Depending on stimulus strength, varies from less than 1 mV to more than 50 mV.	All or none; typically about 100 mV.
Duration	Typically longer, ranging from several msec to several min.	Shorter, ranging from 1–2 msec.
Polarity	May be hyperpolarizing (inhibitory to generation of an action potential) or depolarizing (excitatory to generation of an action potential).	Always consists of depolarizing phase followed by repolarizing phase and return to resting membrane potential.
Refractory period	Not present; thus summation can occur.	Present; thus summation cannot occur.

FIGURE 7.25 Synapses between neurons.

Neural synapses can be axodendritic (from axon to dendrite), axosomatic (from axon to cell body), or axoaxonic (from axon to axon).

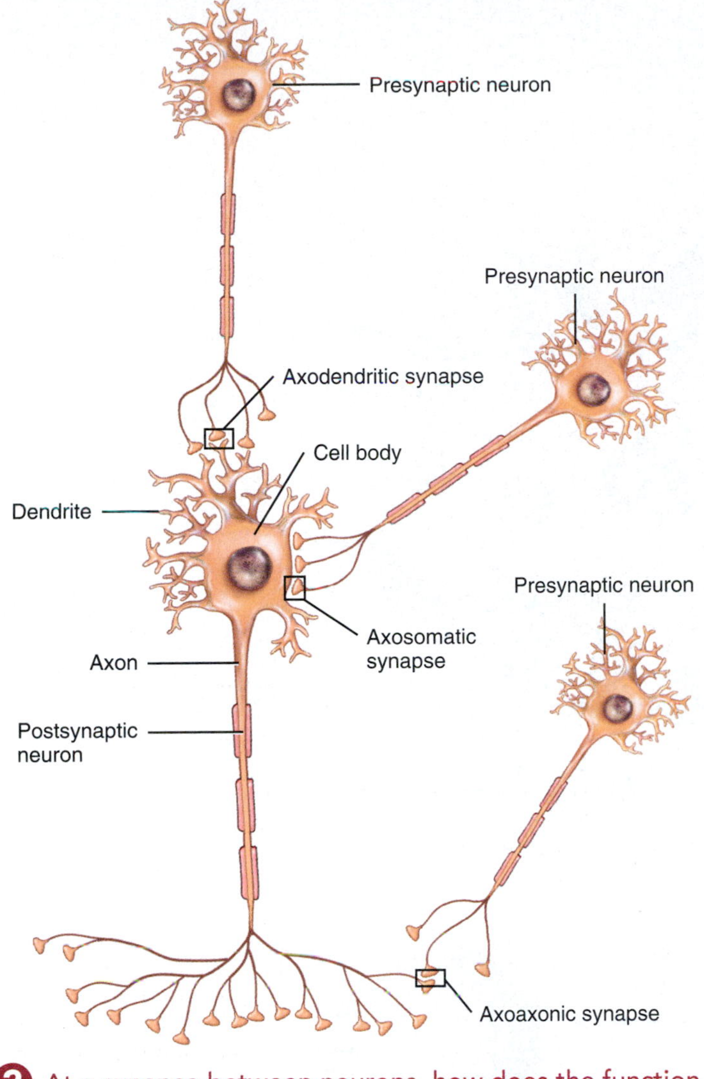

? At a synapse between neurons, how does the function of the presynaptic neuron differ from the postsynaptic neuron?

the next through the connexons, the action potential spreads from cell to cell. At many electrical synapses, the flow of ions through gap junctions is bidirectional; at other electrical synapses, ions flow through gap junctions in one direction only. Gap junctions are common in cardiac muscle and visceral smooth muscle. They also occur in the CNS.

Electrical synapses have two main advantages:

1. ***Faster communication.*** Because action potentials conduct directly through gap junctions, electrical synapses are faster than chemical synapses. At an electrical synapse, the action potential passes directly from the presynaptic cell to the postsynaptic cell. The events that occur at a chemical synapse take some time and delay communication slightly.
2. ***Synchronization.*** Electrical synapses can synchronize the activity of a group of neurons or muscle fibers. In other words, a large number of neurons or muscle fibers can produce action potentials in unison if they are connected by gap junctions. The value of synchronized action potentials in the heart or in visceral smooth muscle is coordinated contraction of these fibers to produce a heartbeat or move food through the gastrointestinal tract.

Chemical Synapses Involve the Release of Neurotransmitter into a Synaptic Cleft

Although the plasma membranes of presynaptic and postsynaptic neurons in a **chemical synapse** are close, they do not touch. They are separated by the **synaptic cleft**, a space of 20 - 50 nm that is filled with interstitial fluid. Action potentials cannot conduct across the synaptic cleft, so an alternate, indirect form of communication occurs. In response to an action potential, the presynaptic neuron releases a neurotransmitter that diffuses through the fluid in the synaptic cleft and binds to receptors in the plasma membrane of the postsynaptic neuron. The postsynaptic neuron receives the chemical signal and, in turn, produces a *postsynaptic potential*, a type of graded potential. Thus, the presynaptic neuron converts an electrical signal (action potential) into a chemical signal (released neurotransmitter). The postsynaptic neuron receives the chemical signal and, in turn, generates an electrical signal (postsynaptic potential). The time required for these processes at a chemical synapse, a **synaptic delay** of about 0.5 msec, is the reason that chemical synapses relay signals more slowly than electrical synapses.

A typical chemical synapse transmits a signal as follows (FIGURE 7.26a):

1. An action potential arrives at a synaptic end bulb of a presynaptic axon.
2. The membrane of the synaptic end bulb contains **voltage-gated Ca^{2+} channels** in addition to the voltage-gated Na^+ and K^+ channels found in other parts of the axon. The depolarizing phase of the action potential opens not only the voltage-gated Na^+ channels but also the voltage-gated Ca^{2+} channels. Because calcium ions are more concentrated in the extracellular fluid, Ca^{2+} flows inward through the open voltage-gated Ca^{2+} channels.
3. An increase in the Ca^{2+} concentration inside the synaptic end bulb serves as a signal that triggers exocytosis of the synaptic vesicles. This process involves several proteins, including **synaptotagmin** and the **SNAREs** (*S*oluble *N*-ethylmaleimide-sensitive factor *A*ttachment protein *RE*ceptors) (FIGURE 7.26b). Synaptotagmin is found in the membrane of the synaptic vesicle. One of the SNAREs, *synaptobrevin*, is also present in the synaptic vesicle membrane. Two other SNAREs, *syntaxin* and *SNAP-25*, are in the presynaptic membrane. Recall from Chapter 5 that SNAREs are classified into two categories: v-SNAREs and t-SNAREs. Synaptobrevin is an example of a *v-SNARE* (*v* for vesicle) because it is located in the vesicle membrane. Syntaxin and SNAP-25 are examples of *t-SNAREs* (*t* for target) because they are present in the target cell membrane, which, in this case, is the presynaptic membrane. As the SNAREs interact with one another, they form a tightly interwoven complex that docks the synaptic vesicle on the presynaptic membrane. After Ca^{2+} enters the synaptic end bulb during the depolarizing phase of the action potential, it binds to synaptotagmin. This, in turn, causes synaptotagmin to bind to the SNARE proteins, resulting in fusion of the synaptic vesicle with the presynaptic membrane. As fusion occurs, neurotransmitter molecules within the synaptic vesicles are released into the synaptic cleft. Each synaptic vesicle contains several thousand molecules of neurotransmitter.

FIGURE 7.26 Signal transmission at a chemical synapse. Through exocytosis of synaptic vesicles, a presynaptic neuron releases neurotransmitter molecules. After diffusing across the synaptic cleft, the neurotransmitter binds to receptors in the plasma membrane of the postsynaptic neuron and produces a postsynaptic potential.

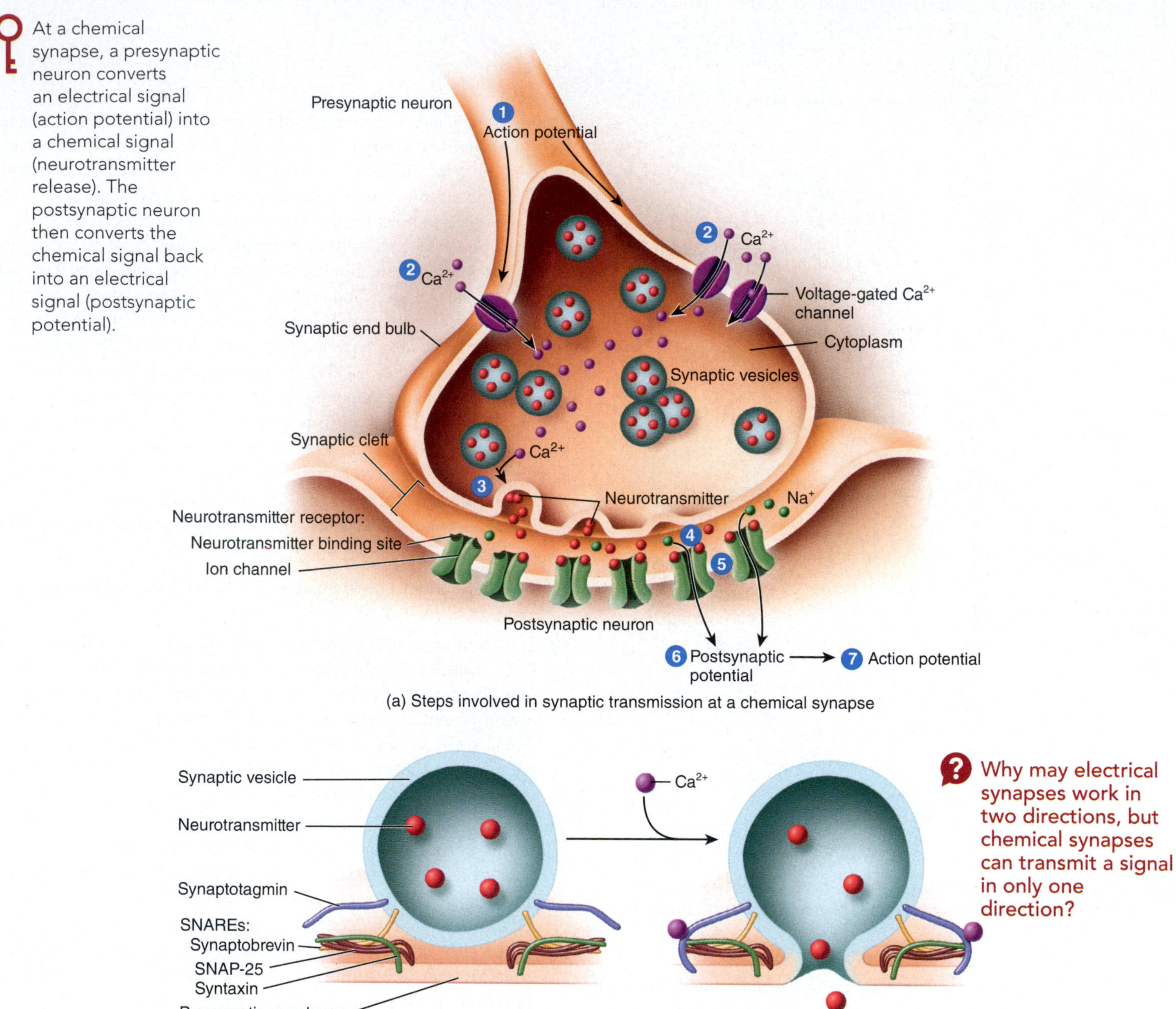

(b) Role of Ca^{2+}, SNAREs, and synaptotagmin in neurotransmitter release

4. The neurotransmitter molecules diffuse across the synaptic cleft and bind to **neurotransmitter receptors** in the postsynaptic neuron's plasma membrane. The receptor shown in FIGURE 7.26a is part of a ligand-gated channel; in other cases, the receptor may be coupled to a G protein that regulates the opening of an ion channel or causes another effect in the cell.
5. Binding of neurotransmitter molecules to the receptor sites on ligand-gated channels opens the channels and allows particular ions to flow across the membrane.
6. As ions flow through the opened channels, the voltage across the membrane changes. This change in membrane voltage is a **postsynaptic potential**. Depending on which ions the channels admit, the postsynaptic potential may be a depolarization or hyperpolarization. For example, opening of Na^+ channels allows inflow of Na^+, which causes depolarization. However, opening of Cl^- or K^+ channels causes hyperpolarization. Opening Cl^- channels permits Cl^- to move into the cell, while opening the K^+ channels allows K^+ to move out—in either event, the inside of the cell becomes more negative.

7 When a depolarizing postsynaptic potential reaches threshold, it triggers an action potential in the axon of the postsynaptic neuron.

At most chemical synapses, only *one-way information transfer* can occur—from a presynaptic neuron to a postsynaptic neuron or an effector, such as a muscle fiber or a gland cell. For example, synaptic transmission at an NMJ proceeds from a somatic motor neuron to a skeletal muscle fiber (but not in the opposite direction). Only synaptic end bulbs of presynaptic neurons can release neurotransmitter, and only the postsynaptic neuron's membrane has the receptor proteins that can recognize and bind that neurotransmitter. As a result, action potentials move in one direction.

Neurotransmitters Excite or Inhibit the Postsynaptic Cell

A neurotransmitter has either an excitatory or inhibitory effect on the membrane of the postsynaptic cell. If the binding of a neurotransmitter to a neurotransmitter receptor opens or closes ion channels in the postsynaptic membrane, the neurotransmitter either depolarizes or hyperpolarizes the postsynaptic cell. A neurotransmitter that *depolarizes* the postsynaptic membrane is excitatory because it brings the membrane closer to threshold. A depolarizing postsynaptic potential is called an **excitatory postsynaptic potential (EPSP)**. Although a single EPSP normally does not initiate an action potential, the postsynaptic cell does become more excitable. Because it is partially depolarized, it is more likely to reach threshold when the next EPSP occurs.

A neurotransmitter that causes *hyperpolarization* of the postsynaptic membrane is inhibitory. During hyperpolarization, generation of an action potential is more difficult than usual because the membrane potential becomes inside more negative and thus even farther from threshold than in its resting state. A hyperpolarizing postsynaptic potential is termed an **inhibitory postsynaptic potential (IPSP)**.

Neurotransmitters do not always change the membrane potential of the postsynaptic cell because the binding of some neurotransmitters to their receptors does not result in the opening of ion channels in the postsynaptic membrane. Instead, there may be another response, such as synthesis of new proteins or changes in the intracellular Ca^{2+} levels. This is often the case for those neurotransmitters that bind to receptors in effector cells (muscle fibers and gland cells). In these situations, the neurotransmitter has an excitatory effect if it causes contraction of the muscle fiber or stimulates gland cell secretion. The neurotransmitter has an inhibitory effect if it causes relaxation of the muscle fiber or inhibits gland cell secretion.

There Are Two Main Types of Neurotransmitter Receptors: Ionotropic and Metabotropic

As you have just learned, neurotransmitters released from a presynaptic neuron bind to neurotransmitter receptors in the plasma membrane of a postsynaptic cell. Neurotransmitter receptors are classified into two main categories: ionotropic receptors and metabotropic receptors.

Ionotropic Receptors

An **ionotropic receptor** is a type of neurotransmitter receptor that contains both a neurotransmitter binding site and an ion channel as part of its structure. Ionotropic receptors are examples of ligand-gated channels (see FIGURE 7.8b). In the absence of neurotransmitter (the ligand), the ion channel component of the ionotropic receptor is closed. When the correct neurotransmitter binds to the ionotropic receptor, the ion channel opens, and an EPSP or IPSP occurs in the postsynaptic cell.

Many excitatory neurotransmitters bind to ionotropic receptors that contain cation channels (FIGURE 7.27a). EPSPs result from opening these cation channels. When cation channels open, they allow passage of the three most plentiful cations (Na^+, K^+, and Ca^{2+}) through the postsynaptic cell membrane. However, the amount of Na^+ that enters the postsynaptic cell is greater than the amount of K^+ that leaves it because the resting membrane potential of the postsynaptic cell is closer to the potassium equilibrium potential (E_K) than to the sodium equilibrium potential (E_{Na}). In addition, more Na^+ ions than Ca^{2+} ions enter the postsynaptic cell because the concentration of Na^+ ions in the ECF is higher than the concentration of Ca^{2+} ions in the ECF. Therefore, the net effect of opening cation channels in the postsynaptic cell is that Na^+ inflow is greater than either K^+ outflow or Ca^{2+} inflow, and the inside of the postsynaptic cell becomes less negative (depolarized).

Many inhibitory neurotransmitters bind to ionotropic receptors that contain chloride channels (FIGURE 7.27b). In cells that actively transport Cl^- ions into the ECF, IPSPs result from opening these Cl^- channels. When Cl^- channels open, there is net movement of chloride ions into the cell because the chloride equilibrium potential (E_{Cl}) is not equal to the resting membrane potential. The inward flow of Cl^- ions causes the inside of the postsynaptic cell to become more negative (hyperpolarized). In cells that do not actively transport Cl^- ions, opening Cl^- channels does not change the membrane potential because E_{Cl} is equal to the resting membrane potential and there is no net movement of Cl^- ions into or out of the cell. However, opening these Cl^- channels does stabilize the membrane potential and makes it more difficult for excitatory neurotransmitters to depolarize the cell.

Metabotropic Receptors

A **metabotropic receptor** is a type of neurotransmitter receptor that contains a neurotransmitter binding site and a site that is coupled to a G protein. The G protein, in turn, either directly opens (or closes) an ion channel or it activates a second messenger pathway that opens (or closes) an ion channel or causes another response in the cell, such as increasing the synthesis of new proteins, modifying the activity of existing proteins, or increasing the intracellular Ca^{2+} levels.

Some inhibitory neurotransmitters bind to metabotropic receptors that cause K^+ channels to open (FIGURE 7.27c). When the K^+ channels open, a larger number of potassium ions diffuses outward. The outward flow of K^+ ions causes the inside of the postsynaptic cell to become more negative (hyperpolarized), resulting in the formation of an IPSP.

Different Postsynaptic Effects for the Same Neurotransmitter

The same neurotransmitter can be excitatory at some synapses and inhibitory at others, depending on the structure of the neurotransmitter receptor to which it binds. For example, at some excitatory synapses, acetylcholine (ACh) binds to ionotropic receptors containing cation

FIGURE 7.27 Ionotropic and metabotropic neurotransmitter receptors. (a) The nicotinic acetylcholine receptor is a type of ionotropic receptor. It contains two binding sites for the neurotransmitter acetylcholine (ACh) and a cation channel. Binding of ACh to this receptor causes the cation channel to open, allowing passage of the three most plentiful cations (Na^+, K^+, and Ca^{2+}) through the postsynaptic cell membrane. Because Na^+ inflow is greater than either Ca^{2+} inflow or K^+ outflow, an excitatory postsynaptic potential (EPSP) is generated. (b) The $GABA_A$ receptor is a type of ionotropic receptor. It contains two binding sites for the neurotransmitter gamma-aminobutyric acid (GABA) and a Cl^- channel. Binding of GABA to this receptor causes the Cl^- channel to open, allowing a larger number of chloride ions to diffuse inward and an inhibitory postsynaptic potential (IPSP) to be generated. (c) The muscarinic acetylcholine receptor is a type of metabotropic receptor. It contains a binding site for the neurotransmitter ACh. Binding of ACh to this receptor activates a G protein, which in turn opens a K^+ channel, allowing a larger number of potassium ions to diffuse out of the cell and an IPSP to form.

An ionotropic receptor is a type of neurotransmitter receptor that contains a neurotransmitter binding site and an ion channel; a metabotropic receptor is a type of neurotransmitter receptor that contains a neurotransmitter binding site and is coupled to a separate ion channel by a G protein.

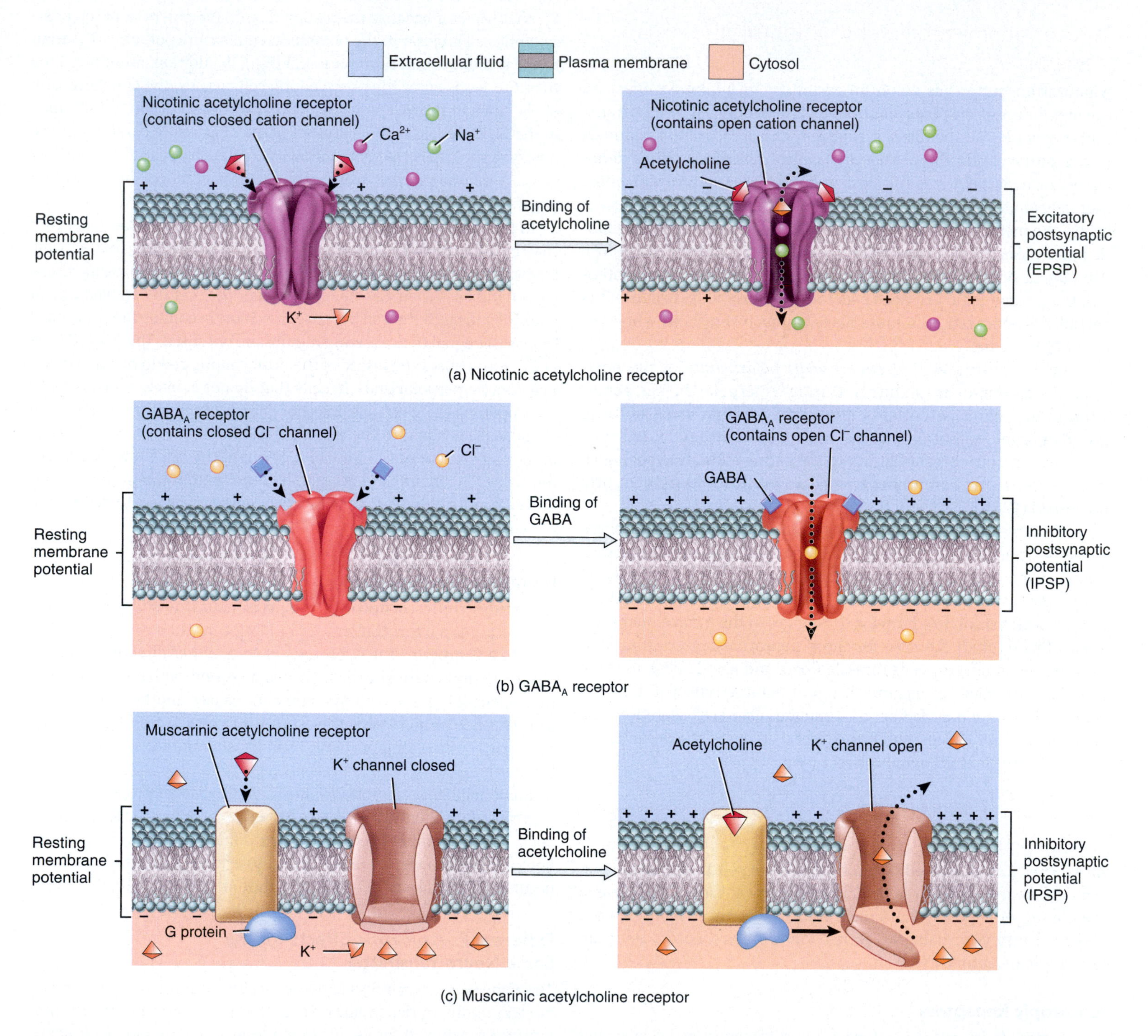

How can the neurotransmitter acetylcholine (ACh) be excitatory at some synapses and inhibitory at other synapses?

channels that open and subsequently generate EPSPs in the postsynaptic cell (FIGURE 7.27a). By contrast, at some inhibitory synapses ACh binds to metabotropic receptors coupled to G proteins that open K^+ channels, resulting in the formation of IPSPs in the postsynaptic cell (FIGURE 7.27c).

Fast and Slow Responses in Postsynaptic Cells

The time required for a neurotransmitter to induce a postsynaptic potential (either an EPSP or an IPSP) and the duration of that postsynaptic potential are both influenced by whether the neurotransmitter binds to an ionotropic receptor or to a metabotropic receptor. In general, binding of a neurotransmitter to an ionotropic receptor causes a *fast response* in the postsynaptic cell because activation of an ionotropic receptor quickly opens or closes ion channels in the postsynaptic membrane, and the subsequent postsynaptic potential that is generated usually lasts for only a few milliseconds or less. By contrast, binding of neurotransmitter to a metabotropic receptor causes a *slow response* in the postsynaptic cell because activation of a metabotropic receptor opens or closes ion channels more slowly (because G protein and second messengers are involved) and the subsequent postsynaptic potential that is generated typically lasts for hundreds of milliseconds to several minutes or even longer.

Neurotransmitter Can Be Removed in Different Ways

Removal of the neurotransmitter from the synaptic cleft is essential for normal synaptic function. If a neurotransmitter could linger in the synaptic cleft, it would influence the postsynaptic neuron, muscle fiber, or gland cell indefinitely. Neurotransmitter is removed in three possible ways:

- ***Diffusion.*** Some of the released neurotransmitter molecules diffuse away from the synaptic cleft (FIGURE 7.28a). Once a

FIGURE 7.28 Removal of neurotransmitter.

A neurotransmitter can be removed in three ways: (1) diffusion, (2) enzymatic degradation, and (3) uptake by cells.

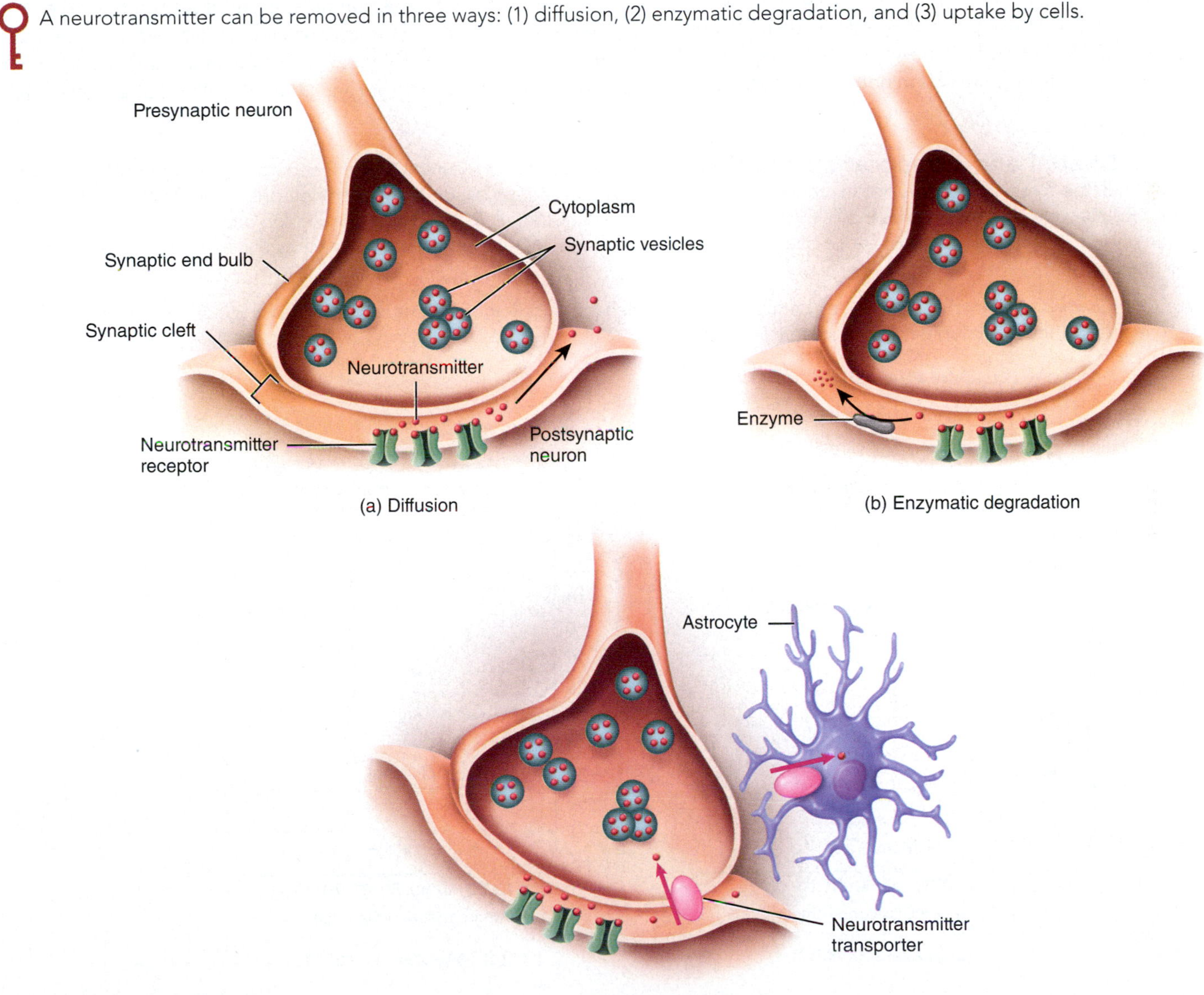

What is the difference between reuptake and uptake of a neurotransmitter?

neurotransmitter molecule is out of reach of its receptors, it can no longer exert an effect.

- ***Enzymatic degradation.*** Certain neurotransmitters are inactivated through enzymatic degradation (**FIGURE 7.28b**). For example, the enzyme acetylcholinesterase, which is located on the postsynaptic membrane, breaks down acetylcholine in the synaptic cleft.
- ***Uptake by cells.*** Many neurotransmitters are actively transported back into the neuron that released them (reuptake) (**FIGURE 7.28c**). Others are transported into neighboring neuroglia (uptake). The neurons that release norepinephrine, for example, rapidly take up the norepinephrine and recycle it into new synaptic vesicles. The membrane proteins that accomplish such uptake are called *neurotransmitter transporters.*

Postsynaptic Potentials Are Summated

A typical neuron in the CNS receives input from 1,000 to 10,000 synapses. Integration of these inputs involves summation of the postsynaptic potentials that form in the postsynaptic neuron. Recall that summation is the process by which graded potentials add together. The greater the summation of EPSPs, the greater the chance that threshold will be reached. At threshold, one or more action potentials arise.

There are two types of summation: spatial summation and temporal summation. **Spatial summation** is summation of postsynaptic potentials in response to stimuli that occur at different *locations* in the membrane of a postsynaptic cell at the same time. For example, spatial summation results from the buildup of neurotransmitter released simultaneously by *several* presynaptic end bulbs (**FIGURE 7.29a**). **Temporal summation** is summation of postsynaptic potentials in response to stimuli that occur at the same location in the membrane of the postsynaptic cell but at different *times*. For example, temporal summation results from buildup of neurotransmitter released by a *single* presynaptic end bulb two or more times in rapid succession (**FIGURE 7.29b**). Because a typical EPSP lasts about 15 msec, the second (and subsequent) release of neurotransmitter must occur soon after the first one if temporal summation is to occur. Summation is rather like voting on the Internet. Many people voting yes or no on an issue at the same time can be compared to spatial summation. One person voting repeatedly and rapidly is like temporal summation. Most of the time, spatial and temporal summations are acting together to influence the chance that a neuron fires an action potential.

A single postsynaptic neuron receives input from many presynaptic neurons, some of which release excitatory neurotransmitters and some of which release inhibitory neurotransmitters (**FIGURE 7.30**). The sum of all of the excitatory and inhibitory effects at any given time

FIGURE 7.29 Spatial and temporal summation. (a) When presynaptic neurons 1 and 2 separately cause EPSPs (arrows) in postsynaptic neuron 3, the threshold level is not reached in neuron 3. Spatial summation occurs only when neurons 1 and 2 act simultaneously on neuron 3; their EPSPs sum to reach the threshold level and trigger an action potential. (b) Temporal summation occurs when stimuli applied to the same axon in rapid succession (arrows) cause overlapping EPSPs that sum. When depolarization reaches the threshold level, an action potential is triggered.

Spatial summation results from the buildup of neurotransmitter released simultaneously by several presynaptic end bulbs; temporal summation results from the buildup of neurotransmitter released by a single presynaptic end bulb two or more times in rapid succession.

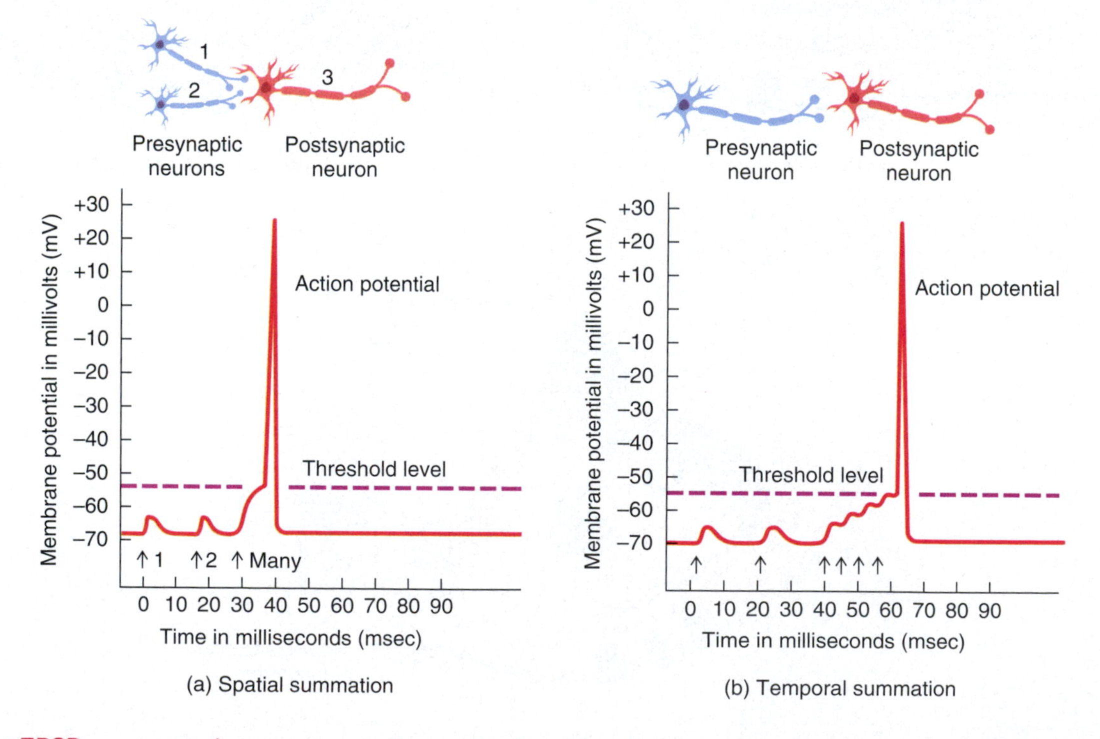

Suppose that EPSPs summate in a postsynaptic neuron in response to simultaneous stimulation by the neurotransmitters glutamate, serotonin, and acetylcholine released by three separate presynaptic neurons. Is this an example of spatial or temporal summation? Explain your answer.

FIGURE 7.30 Summation of postsynaptic potentials at the trigger zone of a postsynaptic neuron. Presynaptic neurons 1, 3, and 5 release excitatory neurotransmitters (purple dots) that generate excitatory postsynaptic potentials (EPSPs) (purple arrows) in the membrane of a postsynaptic neuron. Presynaptic neurons 2 and 4 release inhibitory neurotransmitters (red dots) that generate inhibitory postsynaptic potentials (IPSPs) (red arrows) in the membrane of the postsynaptic neuron. The net summation of these EPSPs and IPSPs determines whether an action potential is generated at the trigger zone of the postsynaptic neuron.

If the net summation of EPSPs and IPSPs is a depolarization that reaches threshold, then an action potential occurs at the trigger zone of a postsynaptic neuron.

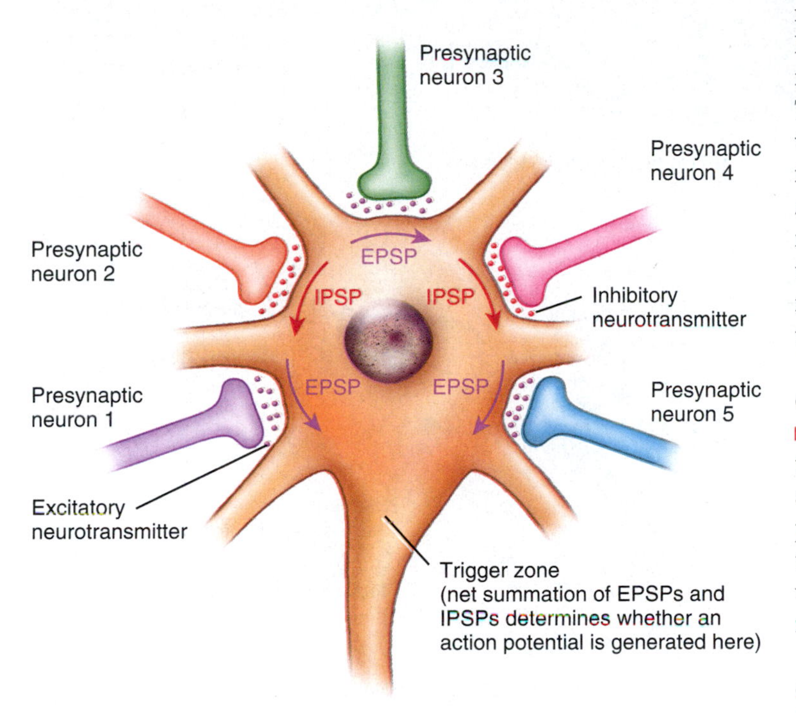

Suppose that the net summation of the EPSPs and IPSPs shown in this figure is a depolarization that brings the membrane potential of the trigger zone of the postsynaptic neuron to −60 mV. Will an action potential occur in the postsynaptic neuron? Why or why not?

determines the effect on the postsynaptic neuron, which may respond in the following ways:

- ***EPSP.*** If the total excitatory effects are greater than the total inhibitory effects but less than the threshold level of stimulation, the result is an EPSP that does not reach threshold. Following an EPSP, subsequent stimuli can more easily generate an action potential through summation because the neuron is partially depolarized.
- ***Action potential(s).*** If the total excitatory effects are greater than the total inhibitory effects and threshold is reached, one or more action potentials will be triggered. Action potentials continue to be generated as long as the EPSP is at or above the threshold level.
- ***IPSP.*** If the total inhibitory effects are greater than the excitatory effects, the membrane hyperpolarizes (IPSP). The result is inhibition of the postsynaptic neuron and an inability to generate an action potential.

Presynaptic Modulation Regulates Neurotransmitter Release

Earlier in this chapter you learned that an axoaxonic synapse is a type of neural synapse in which the axon of one neuron communicates with the axon of another neuron (see **FIGURE 7.25**). An axoaxonic synapse is often used by the nervous system to modulate the amount of neurotransmitter that is released at another synapse. For example, in **FIGURE 7.31**, a presynaptic neuron (neuron A) forms an axodendritic synapse with a postsynaptic neuron (neuron C) and is at the receiving end of an axoaxonic synapse with another neuron (neuron B). When neuron B is inactive, neuron A releases a normal amount of neurotransmitter in response to an action potential (**FIGURE 7.31a**). The neurotransmitter in turn binds to neurotransmitter receptors in the membrane of neuron C to generate a particular postsynaptic effect. By contrast, when neuron B is active, it functions as a *modulating neuron* that can either increase or decrease the amount of neurotransmitter released from neuron A (**FIGURE 7.31b, c**). The alteration in the amount of neurotransmitter released by a presynaptic neuron is referred to as **presynaptic modulation**. Types of presynaptic modulation include presynaptic inhibition and presynaptic facilitation.

During **presynaptic inhibition**, there is a *decrease* in the amount of neurotransmitter released from the presynaptic neuron (neuron A in **FIGURE 7.31b**). This occurs because neuron B releases neurotransmitter that binds to **presynaptic receptors** in the membrane of the synaptic end bulb of neuron A. The binding of the neurotransmitter to the presynaptic receptors in turn decreases the amount of Ca^{2+} that enters the synaptic end bulb of neuron A through voltage-gated Ca^{2+} channels when there is an action potential. As a result, neuron A releases less neurotransmitter, and the postsynaptic effect on neuron C is reduced.

During **presynaptic facilitation**, there is an *increase* in the amount of neurotransmitter released from the presynaptic neuron (neuron A in **FIGURE 7.31c**). This also occurs because neuron B releases neurotransmitter that binds to presynaptic receptors in the membrane of the synaptic end bulb of neuron A. In this case, however, binding of the neurotransmitter to the presynaptic receptors increases the amount of Ca^{2+} that enters the synaptic end bulb of neuron A through voltage-gated Ca^{2+} channels when there is an action potential. As a result, neuron A releases more neurotransmitter and the postsynaptic effect on neuron C is enhanced.

Note that neuron B typically causes *either* presynaptic inhibition of neuron A or presynaptic facilitation of neuron A but not both types of modulation. The factors that determine whether the modulating neuron causes presynaptic inhibition or presynaptic facilitation of neuron A are (1) the type of neurotransmitter released by the modulating neuron and (2) the type of presynaptic receptors in the membrane of the synaptic end bulb of neuron A. It is also important to note that during either presynaptic inhibition or presynaptic facilitation, neuron B has no direct effect on neuron C. However, neuron B does *indirectly* affect the activity of neuron C by modulating the amount of neurotransmitter released from neuron A. Presynaptic modulation is an important mechanism utilized by the nervous system because it allows selective regulation of one specific input to a particular neuron. In the case of the synapses shown in **FIGURE 7.31**, there is selective regulation of input from neuron A to neuron C.

FIGURE 7.31 Presynaptic modulation.

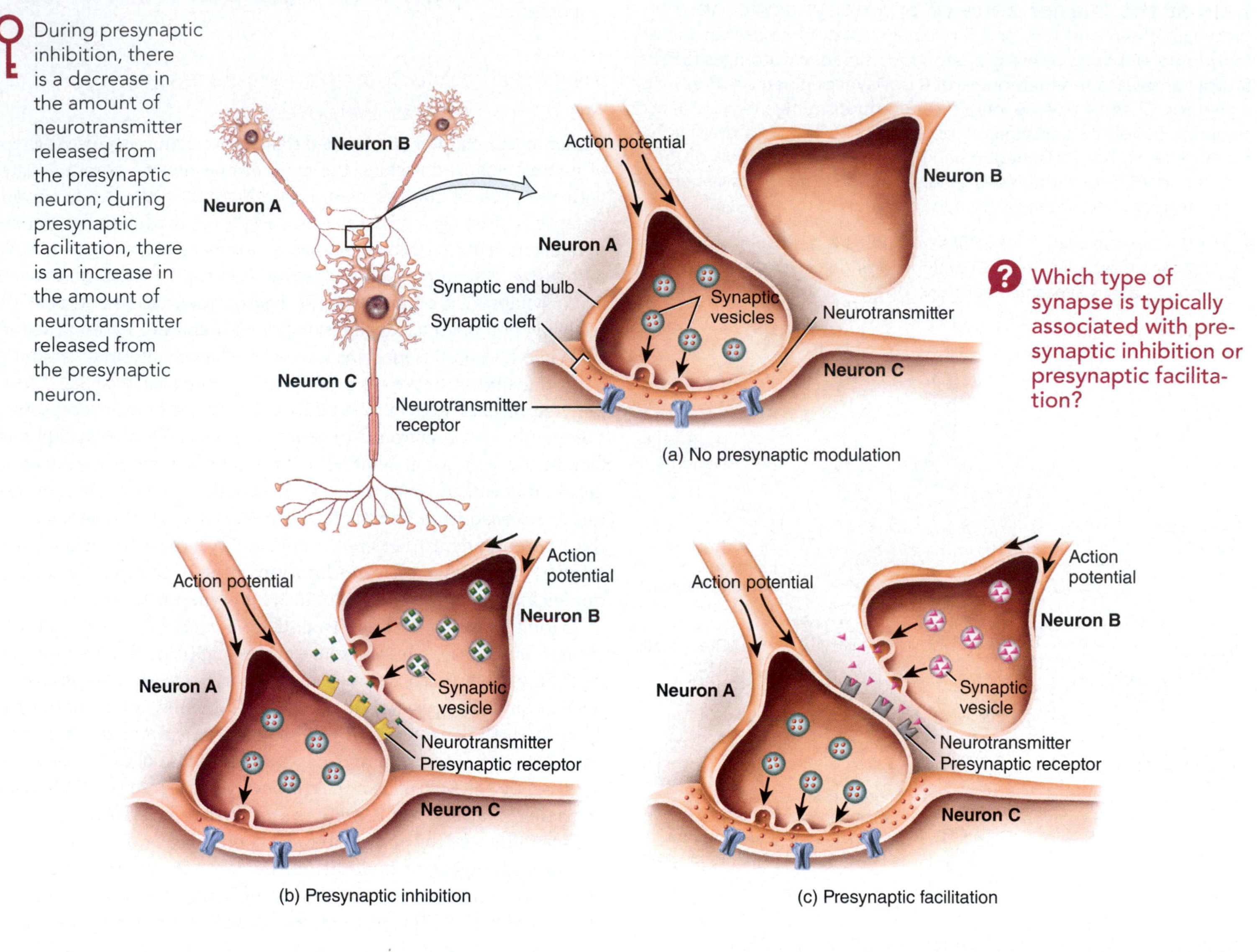

(a) No presynaptic modulation

(b) Presynaptic inhibition

(c) Presynaptic facilitation

During presynaptic inhibition, there is a decrease in the amount of neurotransmitter released from the presynaptic neuron; during presynaptic facilitation, there is an increase in the amount of neurotransmitter released from the presynaptic neuron.

Which type of synapse is typically associated with presynaptic inhibition or presynaptic facilitation?

CHECKPOINT

15. Why is Ca^{2+} important to the release of neurotransmitter at a chemical synapse?

16. How are excitatory and inhibitory postsynaptic potentials similar and different?

17. Why are action potentials said to be "all or none"? Why are EPSPs and IPSPs described as "graded"?

18. How does an ionotropic receptor differ from a metabotropic receptor?

19. How is neurotransmitter removed from the synaptic cleft?

20. Which type of summation is due to summation of stimuli that occur at the same location but at different times?

21. What is the difference between presynaptic inhibition and presynaptic facilitation?

7.5 Neurotransmitters

OBJECTIVES

- Describe the functions of the different types of small-molecule neurotransmitters.
- Provide specific examples of neuropeptides.

Neurotransmitters are the chemical substances that neurons use to communicate with other neurons, muscle fibers, and gland cells. They are divided into two classes based on size: small-molecule neurotransmitters and neuropeptides (FIGURE 7.32). The **small-molecule neurotransmitters** include acetylcholine, amino acids, biogenic amines, purines, gases, and endocannabinoids. The **neuropeptides** are larger in size; they consist of many amino acids linked together by peptide

FIGURE 7.32 Neurotransmitters.

Neurotransmitters are divided into two classes based on size: small-molecule neurotransmitters and neuropeptides.

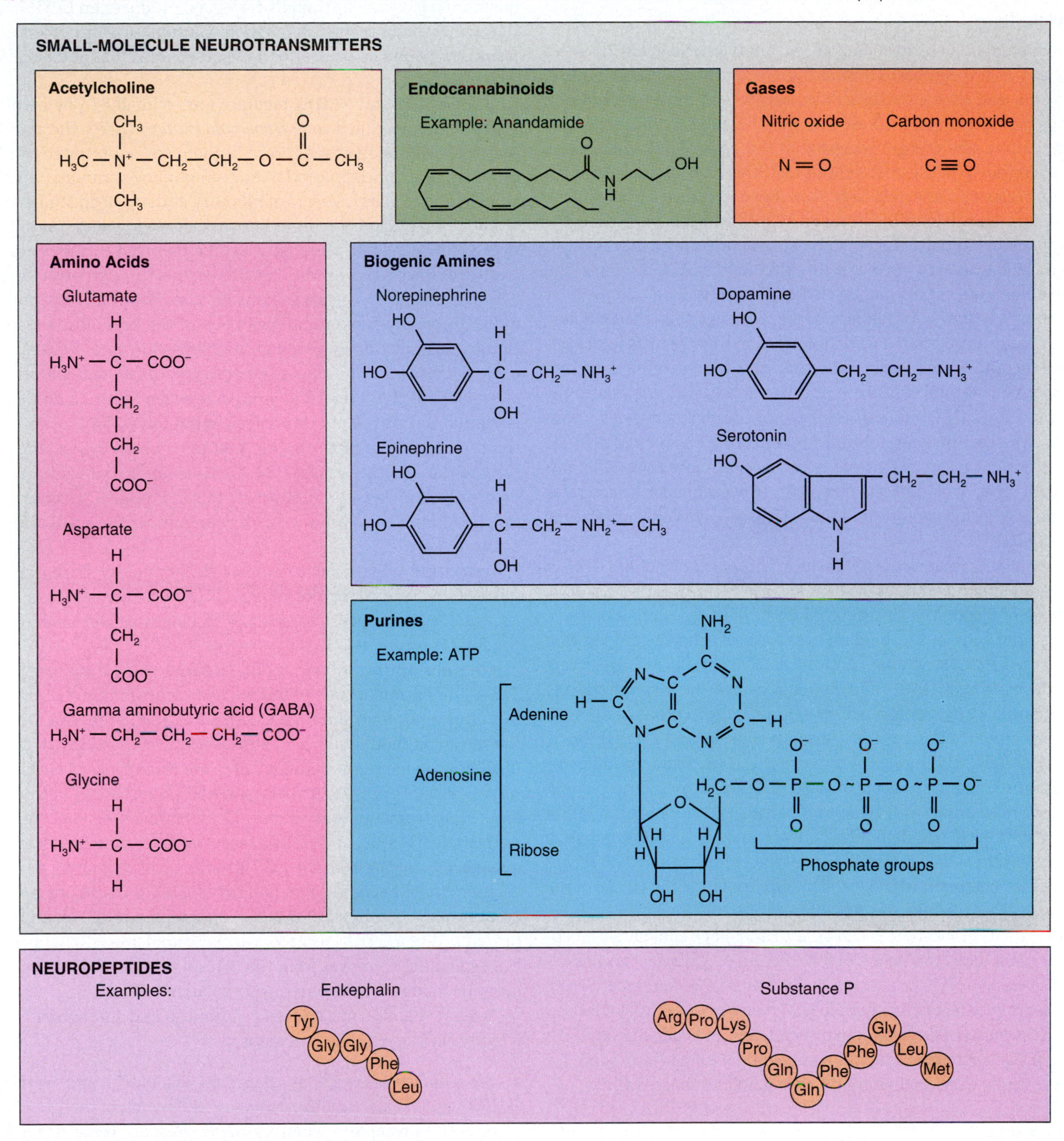

Why are norepinephrine, epinephrine, dopamine, and serotonin classified as biogenic amines?

bonds. Most small-molecule neurotransmitters cause EPSPs or IPSPs by opening or closing ion channels in the postsynaptic membrane. By contrast, some small-molecule neurotransmitters and most neuropeptides do not change the membrane potential of the postsynaptic cell. Instead, these neurotransmitters function as **neuromodulators**, substances that do not generate EPSPs or IPSPs but alter the strength of a particular synaptic response. For example, a neuromodulator may act on the postsynaptic cell to alter the cell's response to a specific neurotransmitter. Alternatively, the neuromodulator may act on the presynaptic cell to alter the synthesis, release, or reuptake of a specific neurotransmitter. The synaptic effects of neuromodulators usually are long-lasting, having a duration of days, months, or even years.

There Are Several Types of Small-Molecule Neurotransmitters

Acetylcholine

The best-studied neurotransmitter is **acetylcholine (ACh)** (FIGURE 7.32). ACh is synthesized from the precursors acetyl coenzyme A (acetyl CoA) and choline. Neurons that release ACh, called **cholinergic neurons** (kō′-lin-ER-jik), are present in the CNS and PNS. Once released, ACh binds to a **cholinergic receptor** in the postsynaptic membrane, causing a response in the postsynaptic cell. This response is short-lived, however, because ACh is quickly broken down into acetate and choline by the enzyme **acetylcholinesterase (AChE)**, which is located on the postsynaptic membrane. The choline is transported back into the synaptic end bulb where it is used to synthesize another ACh molecule. The acetate diffuses out of the synaptic cleft and into the blood.

There are two types of cholinergic receptors: nicotinic acetylcholine receptors and muscarinic acetylcholine receptors. **Nicotinic acetylcholine receptors** are so named because the drug nicotine is an agonist. *Nicotine*, a natural substance in tobacco leaves, is not a naturally occurring substance in humans and is not normally present in nonsmokers. **Muscarinic acetylcholine receptors** are so named because the mushroom poison *muscarine* is an agonist. Note that ACh activates both types of cholinergic receptors. However, nicotine activates only nicotinic ACh receptors, and muscarine activates only muscarinic ACh receptors.

Nicotinic ACh receptors are present in some neurons of the CNS, in certain autonomic neurons of the PNS, and in skeletal muscle at the NMJ. Two types of nicotinic ACh receptors exist. Both types are ionotropic receptors that contain two binding sites for acetylcholine and a cation channel (see FIGURE 7.27a). Activation of nicotinic ACh receptors causes depolarization (EPSP) and thus excitation of the postsynaptic cell.

Muscarinic ACh receptors are present in some neurons of the CNS and in effectors (cardiac muscle, smooth muscle, and glands) innervated by certain autonomic neurons of the PNS. There are different types of muscarinic ACh receptors and they are all metabotropic. Activation of muscarinic ACh receptors causes excitation or inhibition of the postsynaptic cell, depending on which type of muscarinic ACh receptor is activated. For example, one type of muscarinic ACh receptor opens K^+ channels, which leads to hyperpolarization (IPSP) of the postsynaptic cell (see FIGURE 7.27c).

Some chemical substances can bind to and block cholinergic receptors. The plant derivative *curare* blocks nicotinic ACh receptors in skeletal muscle at the NMJ. As a result, skeletal muscle becomes paralyzed. The drug *atropine* blocks muscarinic ACh receptors. Clinically, it is used to dilate the pupils, reduce glandular secretions, and relax smooth muscle in the gastrointestinal tract. It can also be used as an antidote for chemical warfare agents that inhibit acetylcholinesterase (AChE).

Amino Acids

Several amino acids are neurotransmitters in the CNS: glutamate, aspartate, gamma-aminobutyric acid (GABA), and glycine (see FIGURE 7.32).

Glutamate (glutamic acid) has powerful excitatory effects. Most excitatory neurons in the CNS and perhaps half of the synapses in the brain communicate via glutamate. There are many types of receptors for glutamate. Examples include the **AMPA receptor** and the **NMDA receptor**, which are named after the agonists that activate them: AMPA (α-amino-3-hydroxy-5-methyl-4-isoxazole-propionate) and NMDA (N-methyl-D-aspartate). The AMPA and NMDA receptors are ionotropic receptors that contain cation channels. Binding of glutamate to these receptors opens the cation channels, and the consequent inflow of cations (mainly Na^+ ions) produces an EPSP. Inactivation of glutamate occurs via reuptake. Glutamate transporters actively transport glutamate back into the synaptic end bulbs and neighboring neuroglia.

Both AMPA and NMDA receptors are thought to be involved in a phenomenon called *long-term potentiation (LTP)*, the process in which transmission at synapses is enhanced (potentiated) for hours or weeks after a brief period of high-frequency stimulation. LTP has been linked to some aspects of memory and has been studied extensively in a region of the brain known as the hippocampus (see Section 8.3). NMDA receptors may also be involved in a process called **excitotoxicity**—the destruction of neurons through prolonged activation of excitatory synaptic transmission. The most common cause of excitotoxicity is oxygen deprivation of the brain due to ischemia (inadequate blood flow), which happens during a stroke. Lack of oxygen causes the glutamate transporters to fail, and glutamate accumulates in the interstitial spaces between neurons and glia. The accumulated glutamate excessively stimulates NMDA receptors in the membranes of postsynaptic neurons, which causes these neurons to die. In other words, the neurons are literally stimulated to death. Clinical trials are underway to see if NMDA receptor antagonists administered after a stroke can offer some protection from excitotoxicity.

Aspartate (aspartic acid) is an excitatory neurotransmitter that is released by certain neurons of the CNS. Aspartate activates some of the same types of neurotransmitter receptors as glutamate (namely, the NMDA receptors).

Gamma-aminobutyric acid (GABA) (GAM-ma am-i-nō-bū-TIR-ik) is an important inhibitory neurotransmitter that is derived from the amino acid glutamate. GABA is found in the CNS, where it is the most common inhibitory neurotransmitter. As many as one-third of all brain synapses use GABA. There are three types of GABA receptors: $GABA_A$, $GABA_B$, and $GABA_C$. $GABA_A$ and $GABA_C$ receptors are ionotropic receptors that contain Cl^- channels (see FIGURE 7.27b). Activation of $GABA_A$ and $GABA_C$ receptors opens the Cl^- channels. As a result, Cl^- moves from the ECF into the cytosol, and the postsynaptic membrane becomes hyperpolarized (IPSP). $GABA_B$ receptors are metabotropic receptors that are often coupled to K^+ channels by G proteins. When $GABA_B$ receptors are activated, the K^+ channels open. Consequently, K^+ moves from the cytosol into the ECF, and the postsynaptic membrane becomes hyperpolarized (IPSP).

Several chemical substances can bind to and subsequently modulate the activity of $GABA_A$ receptors:

- *Benzodiazepines* such as diazepam (Valium®) and clonazepam (Klonopin®) increase the *frequency* of opening of the Cl^- channels of $GABA_A$ receptors when GABA is present. These drugs have a tranquilizing effect and are used to reduce anxiety, promote sleep, and treat epilepsy.
- *Barbiturates* such as phenobarbital and pentobarbital increase the *duration* of opening of the Cl^- channels of the $GABA_A$ receptors when GABA is present. These drugs are used for their sedative and anesthetic effects and to control epilepsy.
- *Ethanol*, an alcohol, enhances the activity of $GABA_A$ receptors. Consequently, there is an overall inhibition of the nervous system when a person consumes significant amounts of alcohol.

CLINICAL CONNECTION

Strychnine Poisoning

The importance of inhibitory neurons can be appreciated by observing what happens when their activity is blocked. Normally, inhibitory neurons in the spinal cord called ***Renshaw cells*** release the neurotransmitter glycine at inhibitory synapses with somatic motor neurons. This inhibitory input to their motor neurons prevents excessive contraction of skeletal muscles. **Strychnine** is a lethal poison that binds to and blocks glycine receptors. The normal delicate balance between excitation and inhibition of motor neurons is disturbed, which causes the motor neurons to generate action potentials without restraint. All skeletal muscles, including the diaphragm, contract fully and remain contracted. Because the diaphragm cannot relax, the victim cannot inhale, resulting in suffocation.

Like GABA, the amino acid **glycine** is an inhibitory neurotransmitter. About half of the inhibitory synapses in the spinal cord use glycine; the rest use GABA. Like $GABA_A$ and $GABA_C$ receptors, glycine receptors are ionotropic receptors that contain Cl^- channels. Activation of glycine receptors causes the Cl^- channels to open. As a result, Cl^- moves from the ECF into the cytosol, and the postsynaptic membrane becomes hyperpolarized (IPSP).

Biogenic Amines

Certain amino acids are modified and decarboxylated (carboxyl group removed) to produce the **biogenic amines**, which include norepinephrine, epinephrine, dopamine, and serotonin (see **FIGURE 7.32**). Norepinephrine, epinephrine, and dopamine are chemically classified as **catecholamines** (cat-e-KŌL-a-mēns) because they all contain an *amino group* ($-NH_2$) and a *catechol ring*, which is composed of six carbons and two adjacent hydroxyl (-OH) groups. The catecholamines are synthesized from the amino acid tyrosine. Removal of catecholamines from synapses occurs via reuptake into synaptic end bulbs or enzymatic degradation. Two enzymes that break down catecholamines are **catechol-*O*-methyltransferase (COMT)** (kat′-e-kōl-ō-meth-il-TRANS-fer-ās) and **monoamine oxidase (MAO)** (mon-ō-AM-īn OK-si-dās).

Norepinephrine (NE), also known as **noradrenaline**, is involved in arousal (awakening from sleep), attention, and regulating mood. Norepinephrine is released by certain neurons in the brain stem and by some autonomic neurons of the PNS. *Amphetamines* promote the release of norepinephrine from synaptic end bulbs. The stimulant effects that a person experiences after taking amphetamines reflects norepinephrine's role in arousal. **Epinephrine**, also known as **adrenaline**, is released by only a small number of neurons in the brain and has the lowest concentration in the brain of all of the catecholamines. Both norepinephrine and epinephrine also serve as hormones. Cells of the adrenal medulla, the inner portion of the adrenal gland, release them into the blood.

Norepinephrine and epinephrine bind to **adrenergic receptors** (ad′-ren-ER-jik) in the postsynaptic membrane. Adrenergic receptors are present in some neurons of the CNS and in effectors (cardiac muscle, smooth muscle, and glands) innervated by certain autonomic neurons of the PNS. There are two main groups of adrenergic receptors: **alpha (α) receptors** and **beta (β) receptors**. These receptors are further classified into subtypes—α_1, α_2, β_1, β_2, and β_3—based on the specific responses they elicit and by their selective binding of drugs that activate or block them. All adrenergic receptors are metabotropic. Norepinephrine stimulates alpha receptors more strongly than beta receptors; epinephrine is a potent stimulator of both alpha and beta receptors. Activation of adrenergic receptors can cause either excitation or inhibition of the postsynaptic cell, depending on which type of adrenergic receptor is activated. A large variety of drugs can activate or block specific adrenergic receptors. Examples include *phenylephrine*, which is an agonist at α_1 receptors, and *propranolol*, which is a nonselective antagonist at β receptors.

Neurons that release the neurotransmitter **dopamine (DA)** are present in the brain, especially in the substantia nigra and ventral tegmental area of the midbrain. There are several types of dopamine receptors, all of which are metabotropic. Activation of dopamine receptors can either cause excitation or inhibition of the postsynaptic cell, depending on which type of dopamine receptor is activated. Dopamine is involved is generating emotional responses. In fact, the behavioral disorder **schizophrenia** has been linked to the accumulation of excess dopamine. People with schizophrenia may have inappropriate or absent emotions, delusions, distortions of reality, paranoia, and hallucinations. Dopamine also plays a role in the formation of addictive behaviors and pleasurable experiences. The drug *cocaine* produces euphoria—intensely pleasurable feelings—by blocking transporters for dopamine reuptake. This action allows dopamine to linger in synaptic clefts, producing excessive stimulation of certain brain regions. Dopamine is also involved in the regulation of skeletal muscle tone and some aspects of movement due to contraction of

CLINICAL CONNECTION

Depression

Depression is a disorder that affects over 18 million people each year in the United States. People who are depressed feel sad and helpless, lack interest in activities that they once enjoyed, and experience suicidal thoughts. There are several types of depression. A person with **major depression** experiences symptoms of depression that last for more than two weeks. A person with **bipolar disorder**, formerly referred to as ***manic-depressive illness***, experiences recurrent episodes of depression and extreme elation (mania). A person with **seasonal affective disorder (SAD)** experiences depression during the winter months, when day length is short. Although the exact cause of depression is unknown, research suggests that depression is linked to an imbalance of the neurotransmitters serotonin, norepinephrine, and dopamine in the brain. Factors that may contribute to depression include heredity, stress, chronic illnesses, certain personality traits (such as low self-esteem), and hormonal changes. Medication is the most common treatment for depression. For example, **selective serotonin reuptake inhibitors (SSRIs)** are drugs that provide relief from some forms of depression. By inhibiting reuptake of serotonin by serotonin transporters, SSRIs increase the duration of serotonin function at brain synapses without affecting the activities of the other biogenic amines. SSRIs include fluoxetine (Prozac®), paroxetine (Paxil®), and sertraline (Zoloft®).

skeletal muscles. The tremors (shaking), rigidity (muscular stiffness), and slow movements that occur in **Parkinson's disease** are due to degeneration of neurons that release dopamine. Many patients with Parkinson's disease benefit from taking the drug L-dopa because it is a precursor of dopamine. For a limited time, taking L-dopa boosts dopamine production in affected brain areas.

Serotonin, also known as **5-hydroxytryptamine (5-HT)**, is released by neurons in the brain stem. Unlike the catecholamines, which are synthesized from the amino acid tyrosine, serotonin is synthesized from the amino acid tryptophan. There are several types of serotonin receptors, most of which are metabotropic. Activation of serotonin receptors can cause either excitation or inhibition of the postsynaptic cell, depending on the type of serotonin receptor that is activated. Removal of serotonin from the synapse occurs by reuptake into synaptic end bulbs. Then the serotonin is degraded by MAO. Serotonin is thought to be involved in sensory perception, temperature regulation, control of mood, appetite, and the induction of sleep. The drug *lysergic acid diethylamine (LSD)* is an agonist of a specific type of serotonin receptor called the $5\text{-HT}2_A$ receptor. Activation of these receptors by LSD results in powerful hallucinations.

Purines

The *purines*, which include adenosine and its triphosphate, diphosphate, and monophosphate derivatives (ATP, ADP, and AMP), are named for the purine ring that comprises adenine (see **FIGURE 7.32**). The purines function as neurotransmitters in both the CNS and the PNS. Once they are released, purines bind to **purinergic receptors** in the postsynaptic membrane. Some purinergic receptors are ionotropic; others are metabotropic. Adenosine plays a role in inducing sleep by binding to purinergic receptors and inhibiting certain neurons of the reticular activating system of the brain that participate in arousal (see Section 8.3).

Gases

Two gases can function as neurotransmitters: nitric oxide and carbon monoxide (see **FIGURE 7.32**). The simple gas **nitric oxide (NO)** is an important neurotransmitter that has widespread effects throughout the body. The enzyme **nitric oxide synthase (NOS)** catalyzes the formation of NO from the amino acid arginine. Unlike all previously identified neurotransmitters, NO is not synthesized in advance and packaged into synaptic vesicles. Instead, it is formed on demand and acts immediately. Its action is brief because NO is a highly reactive free radical that exists for less than 10 seconds before it combines with oxygen and water to form inactive nitrates and nitrites.

Studies have shown that blockage of NO signaling pathways in the hippocampus of the brain prevents long-term potentiation. This suggests that NO may play a role in learning and memory. NO is also released by certain autonomic neurons of the PNS. For example, parasympathetic neurons that innervate the erectile tissue of the penis release NO as a neurotransmitter. The NO in turn causes relaxation of smooth muscle in the walls of the arterioles supplying the penis, resulting in penile erection. The drug Sildenafil (Viagra®) alleviates erectile dysfunction (impotence) by enhancing the effect of NO.

Some neurons of the brain release extremely small quantities of the gas **carbon monoxide (CO)** as a neurotransmitter. Like nitric oxide, CO is not synthesized in advance or packaged into synaptic vesicles. Instead, it is formed on demand and acts very quickly.

Endocannabinoids

Lipids known as the **endocannabinoids** can function as neurotransmitters. Examples of endocannabinoids are the fatty acids anandamide (see **FIGURE 7.32**) and 2-arachidonylglycerol. These neurotransmitters are formed from the breakdown of lipids in the plasma membrane and then released from the neuron that produces them. The endocannabinoids bind to **cannabinoid receptors**, which are present throughout the brain. Research suggests that endocannaboinoids play roles in learning and memory, regulation of motor activity, pain processing, and appetite stimulation. The active ingredient in marijuana, *Δ^9-tetrahydrocannabinol (THC)*, binds to and activates specific cannabinoid receptors called CB1 receptors. This causes the euphoria, relaxation, analgesic effects, altered perception, and increased appetite associated with smoking marijuana.

Neuropeptides Are Composed of Amino Acids Linked by Peptide Bonds

Neurotransmitters consisting of 3 to 40 amino acids linked by peptide bonds are called **neuropeptides** (noor-ō-PEP-tīds) (see **FIGURE 7.32**). They are numerous and widespread in both the CNS and the PNS. Neuropeptides are formed in the neuron cell body, packaged into vesicles, and transported to axon terminals. Virtually all receptors for neuropeptides are metabotropic. Besides their role as neurotransmitters, many neuropeptides serve as hormones that regulate physiological responses elsewhere in the body.

Scientists discovered that certain brain neurons have plasma membrane receptors for opiate drugs such as morphine and heroin. The quest to find the naturally occurring substances that use these receptors brought to light the first neuropeptides: two molecules, each a chain of five amino acids, named **enkephalins** (en-KEF-a-lins). Their potent analgesic (pain-relieving) effect is 200 times stronger than morphine. Other so-called *opioid peptides* include the **endorphins** (en-DOR-fins) and **dynorphins** (dī-NOR-fins). It is thought that opioid peptides are the body's natural painkillers. Acupuncture may produce analgesia (loss of pain sensation) by increasing the release of opioids. These neuropeptides have also been linked to improved memory and learning; feelings of pleasure or euphoria; control of body temperature; regulation of hormones that affect the onset of puberty, sexual drive, and reproduction; and mental illnesses such as depression and schizophrenia.

Another neuropeptide, **substance P**, is released by neurons that transmit pain-related input from peripheral pain receptors into the central nervous system, enhancing the perception of pain. Enkephalin and endorphin suppress the release of substance P, thus decreasing the number of action potentials relayed to the brain for pain sensations. Substance P has also been shown to counter the effects of certain nerve-damaging chemicals, prompting speculation that it might prove useful as a treatment for nerve degeneration.

TABLE 7.4 provides brief descriptions of these neuropeptides, as well as others that will be discussed in later chapters.

TABLE 7.4 Neuropeptides

Substance	Description
Substance P	Found in sensory neurons, spinal cord pathways, and parts of brain associated with pain; enhances perception of pain.
Enkephalins	Inhibit pain impulses by suppressing release of substance P; may have a role in memory and learning, control of body temperature, sexual activity, and mental illness.
Endorphins	Inhibit pain by blocking release of substance P; may have a role in memory and learning, sexual activity, control of body temperature, and mental illness.
Dynorphins	May be related to controlling pain and registering emotions.
Hypothalamic releasing and inhibiting hormones	Produced by the hypothalamus; regulate the release of hormones by the anterior pituitary.
Angiotensin II	Stimulates thirst; may regulate blood pressure in the brain. As a hormone causes vasoconstriction and promotes release of aldosterone, which increases the rate of salt and water reabsorption by the kidneys.
Cholecystokinin (CCK)	Found in the brain and small intestine; may regulate feeding as a "stop eating" signal. As a hormone, regulates pancreatic enzyme secretion during digestion, and contraction of smooth muscle in the gastrointestinal tract.
Neuropeptide Y	Stimulates food intake; may play a role in the stress response.

CHECKPOINT

22. For each of the following neurotransmitter receptors, name a neurotransmitter that binds to that receptor: (a) purinergic receptor, (b) AMPA receptor, (c) β_1 adrenergic receptor, (d) muscarinic cholinergic receptor, and (e) 5-HT2_A receptor.

23. In what ways is nitric oxide different from all previously identified neurotransmitters?

24. What are some examples of neuropeptides?

7.6 Neural Circuits

OBJECTIVE

- Identify the various types of neural circuits in the nervous system.

The CNS contains billions of neurons organized into complicated networks called **neural circuits**, functional groups of neurons that process specific types of information. In a **simple series circuit**, a presynaptic neuron stimulates a single postsynaptic neuron. The second neuron then stimulates another, and so on. However, most neural circuits are far more complex.

A single presynaptic neuron may synapse with several postsynaptic neurons. Such an arrangement, called **divergence**, permits one presynaptic neuron to influence several postsynaptic neurons (or several muscle fibers or gland cells) at the same time. In a **diverging circuit**, the action potential from a single presynaptic neuron causes the stimulation of increasing numbers of cells along the circuit (FIGURE 7.33a). For example, a small group of neurons in the brain that governs a particular body movement stimulates a much larger number of neurons in the spinal cord. Sensory signals are also arranged in diverging circuits, allowing sensory input to be relayed to several regions of the brain. This arrangement amplifies the signal.

In another arrangement, called **convergence**, several presynaptic neurons synapse with a single postsynaptic neuron. This arrangement permits more effective stimulation or inhibition of the postsynaptic neuron. In a **converging circuit** (FIGURE 7.33b), the postsynaptic neuron receives action potentials from several different sources. For example, a single motor neuron that synapses with skeletal muscle fibers at neuromuscular junctions receives input from several pathways that originate in different brain regions.

Some circuits are constructed so that, once the presynaptic cell is stimulated, it causes the postsynaptic cell to transmit a series of action potentials. One such circuit is called a **reverberating circuit**

CLINICAL CONNECTION

Epilepsy

Epilepsy is characterized by short, recurrent attacks of motor, sensory, or psychological malfunction, although it almost never affects intelligence. The attacks, called ***epileptic seizures***, afflict about 1% of the world's population. They are initiated by abnormal, synchronous electrical discharges from millions of neurons in the brain, perhaps resulting from abnormal reverberating circuits. The discharges stimulate many of the neurons to send action potentials over their conduction pathways. As a result, lights, noise, or smells may be sensed when the eyes, ears, and nose have not been stimulated. In addition, the skeletal muscles of a person having a seizure may contract involuntarily. ***Partial seizures*** begin in a small focus on one side of the brain and produce milder symptoms; ***generalized seizures*** involve larger areas on both sides of the brain and result in loss of consciousness. Epileptic seizures can often be eliminated or alleviated by anti-epileptic drugs, such as benzodiazepines (see Section 7.5).

FIGURE 7.33 Neural circuits.

A neural circuit is a functional group of neurons that processes a specific kind of information.

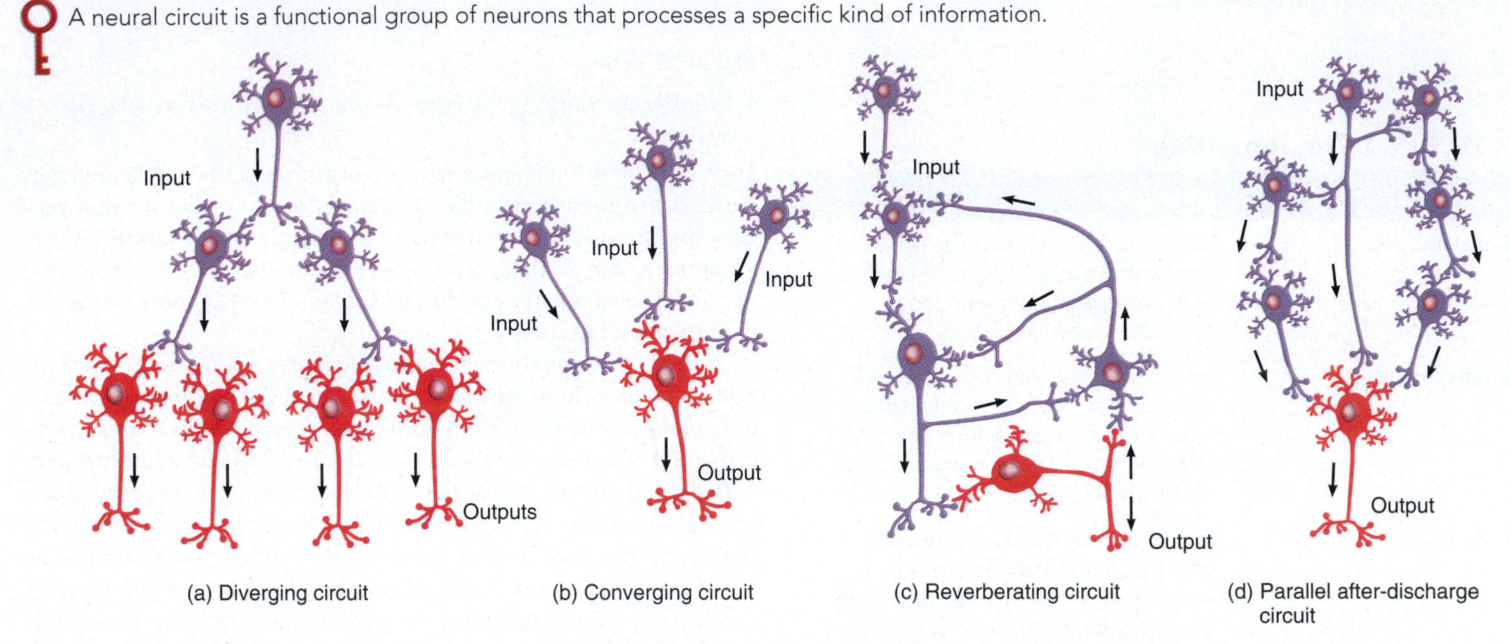

(a) Diverging circuit (b) Converging circuit (c) Reverberating circuit (d) Parallel after-discharge circuit

A motor neuron in the spinal cord typically receives input from neurons that originate in several different regions of the brain. Is this an example of convergence or divergence? Explain your answer.

(FIGURE 7.33c). In this pattern, the incoming action potential stimulates the first neuron, which stimulates the second, which stimulates the third, and so on. Branches from later neurons synapse with earlier ones. This arrangement sends action potentials back through the circuit again and again. The output signal may last from a few seconds to many hours, depending on the number of synapses and the arrangement of neurons in the circuit. Inhibitory neurons may turn off a reverberating circuit after a period of time. Among the body responses thought to be the result of output signals from reverberating circuits are breathing, coordinated muscular activities, waking up, and short-term memory.

A fourth type of circuit is the **parallel after-discharge circuit** (FIGURE 7.33d). In this circuit, a single presynaptic cell stimulates a group of neurons, each of which synapses with a common postsynaptic cell. A differing number of synapses between the first and last neurons imposes varying synaptic delays so that the last neuron exhibits multiple EPSPs or IPSPs. If the input is excitatory, the postsynaptic neuron then can send out a stream of action potentials in quick succession. Parallel after-discharge circuits may be involved in precise activities such as mathematical calculations.

CHECKPOINT

25. What are the functions of diverging, converging, reverberating, and parallel after-discharge circuits?

FROM RESEARCH TO REALITY

Using Neurons to Beat Binge Eating

Reference

Cao, X. et al. (2014). Estrogens stimulate serotonin neurons to inhibit binge-like eating in mice. *The Journal of Clinical Investigation.* 124 (10): 4351–4362.

Could understanding serotonin's effect on the brain help researchers find a cure for binge eating?

Donna Day/Getty Images, Inc.

Binge eating is a disorder characterized by eating large quantities of food in a short period of time, seemingly without control. It is typically associated with other undesirable side effects, including weight gain and depression. What effect does estrogen and serotonin-releasing neurons have on binge-eating behavior? What can be done to help prevent this destructive behavior?

Article description:
The following study investigates the effect of estrogen on binge-eating behavior in female mice. Investigators hoped to determine whether that effect was mediated by estrogen receptors on the serotonin neurons found in a specific region of the brain stem, the dorsal raphe nuclei, which has a high population of serotonin neurons. Ideally the researchers hope to develop a treatment for binge eating that does not require the use of estrogen supplementation.

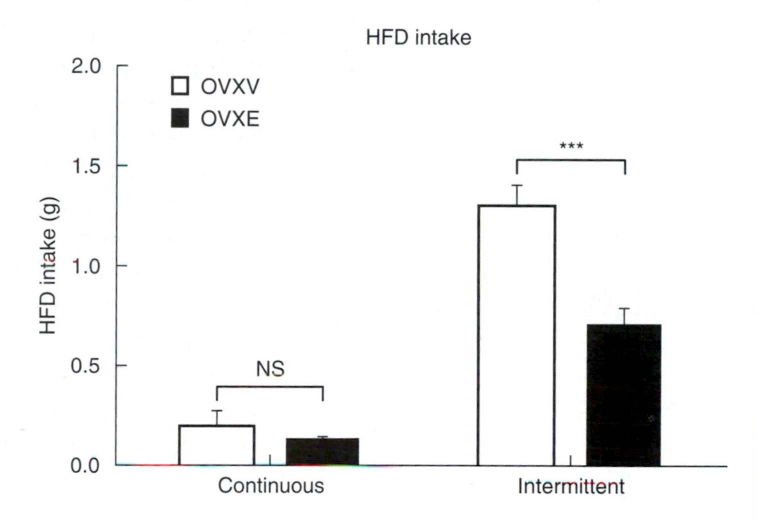

Go to WileyPLUS Learning Space and use the data from this article to answer the questions posed there and to learn more about estrogen's effect on serotonin neurons.

CHAPTER REVIEW

7.1 Overview of the Nervous System

1. The nervous system is organized into two main subdivisions: the central nervous system (CNS) and the peripheral nervous system (PNS). The CNS consists of the brain and spinal cord. The PNS consists of all nervous tissue outside the CNS; it includes nerves and sensory receptors.
2. The PNS is divided into an afferent division and an efferent division. The afferent division conveys sensory input into the CNS from sensory receptors, and the efferent division conveys motor output from the CNS to effectors (muscles and glands). The efferent division of the PNS is further subdivided into a somatic nervous system (conveys motor output from the CNS to skeletal muscles only) and an autonomic nervous system (conveys motor output from the CNS to smooth muscle, cardiac muscle, and glands).

3. The nervous system helps maintain homeostasis and regulates all body activities by sensing changes (sensory function), by analyzing them and making decisions for appropriate responses (integrative function), and by reacting to them (motor function).

7.2 Cells of the Nervous System

1. The nervous system consists of two principal types of cells: neurons (nerve cells) and neuroglia (supporting cells).
2. Most neurons have three parts. The dendrites are the main receiving or input region. The cell body functions as the control center and can also serve as an input region. The output part typically is a single axon, which propagates action potentials toward another neuron, a muscle fiber, or a gland cell.
3. Axonal transport allows materials to be conveyed to and from the cell body and axon terminals.
4. Neurons are functionally classified as sensory (afferent) neurons, motor (efferent) neurons, and interneurons. Sensory neurons carry sensory information into the CNS. Motor neurons carry information out of the CNS to effectors (muscles and glands). Interneurons are located within the CNS between sensory and motor neurons.
5. Neuroglia provide physical, nutritional, and metabolic support to neurons. Neuroglia in the CNS include astrocytes, oligodendrocytes, microglia, and ependymal cells. Neuroglia in the PNS include Schwann cells.
6. Two types of neuroglia produce myelin sheaths: Oligodendrocytes myelinate axons in the CNS, and Schwann cells myelinate axons in the PNS.
7. A damaged axon in the PNS may undergo repair if the cell body is intact and if Schwann cells are functional.
8. White matter consists of aggregates of myelinated axons; gray matter contains cell bodies, dendrites, unmyelinated axons, and neuroglia.

7.3 Electrical Signals in Neurons

1. Neurons communicate with one another using graded potentials, which are used for short-distance communication only, and action potentials, which allow communication over long distances.
2. Neuron function relies on four kinds of ion channels: leak channels, ligand-gated channels, mechanically-gated channels, and voltage-gated channels.
3. A resting membrane potential exists across the plasma membrane of excitable cells that are unstimulated (at rest). The resting membrane potential exists because of a small buildup of negative ions in the cytosol along the inside surface of the membrane, and an equal buildup of positive ions in the extracellular fluid along the outside surface of the membrane. A typical value for the resting membrane potential of a neuron is -70 mV. A cell that exhibits a membrane potential is polarized.
4. The resting membrane potential is determined by two major factors: (1) unequal distribution of ions in the ECF and cytosol and (2) differences in membrane permeability to various ions.
5. A graded potential is a small deviation from the resting membrane potential that occurs because ligand-gated or mechanically-gated channels open or close. A depolarizing graded potential makes the membrane less polarized (inside less negative); a hyperpolarizing graded potential makes the membrane more polarized (inside more negative). The amplitude of a graded potential varies depending on the strength of the stimulus.

6. An action potential is a sequence of rapidly occurring events that decrease and reverse the membrane potential and then eventually restore it to the resting state. According to the all-or-none principle, if a stimulus is strong enough to generate an action potential, the action potential generated is of a constant size. A stronger stimulus does not generate a larger action potential. Instead, the greater the stimulus strength above threshold, the greater the frequency of the action potentials.
7. During an action potential, voltage-gated Na^+ and K^+ channels open and close in sequence. This results first in depolarization, the reversal of membrane polarization (from -70 mV to $+30$ mV). Then repolarization, the recovery of the resting membrane potential (from $+30$ mV to -70 mV), occurs.
8. During the first part of the refractory period, another action potential cannot be generated at all (absolute refractory period); a little later, it can be triggered only by a larger-than-normal stimulus (relative refractory period).
9. Because an action potential travels from point to point along the membrane without getting smaller, it is useful for long-distance communication.
10. Action potential propagation in which the action potential leaps from one node of Ranvier to the next along a myelinated axon is saltatory conduction. Saltatory conduction is faster than continuous conduction. Large-diameter, myelinated axons conduct action potentials at higher speeds than do small-diameter, unmyelinated axons.

7.4 Signal Transmission at Synapses

1. A synapse is the functional junction between one neuron and another, or between a neuron and an effector such as a muscle or a gland. The two types of synapses are electrical and chemical.
2. A chemical synapse produces one-way information transfer—from a presynaptic neuron to a postsynaptic neuron.
3. When a neurotransmitter is released by a presynaptic neuron, it binds to neurotransmitter receptors in the postsynaptic membrane and causes a postsynaptic potential (a type of graded potential).
4. An excitatory neurotransmitter depolarizes the postsynaptic neuron's membrane, bringing the membrane potential closer to threshold. A depolarizing postsynaptic potential is called an excitatory postsynaptic potential (EPSP).
5. An inhibitory neurotransmitter hyperpolarizes the membrane of the postsynaptic neuron, moving it farther from threshold. A hyperpolarizing postsynaptic potential is called an inhibitory postsynaptic potential (IPSP).
6. There are two major types of neurotransmitter receptors: ionotropic receptors and metabotropic receptors. An ionotropic receptor contains a neurotransmitter binding site and an ion channel. A metabotropic receptor contains a neurotransmitter binding site and a site coupled to a G protein.
7. Neurotransmitter is removed from the synaptic cleft in three ways: diffusion, enzymatic degradation, and uptake by cells (neurons and neuroglia).
8. Postsynaptic potentials undergo summation. Summation may be spatial or temporal.
9. The postsynaptic neuron is an integrator. It receives excitatory and inhibitory signals, integrates them, and then responds accordingly.
10. During presynaptic modulation, the amount of neurotransmitter released by a presynaptic neuron is altered by another neuron. Types of presynaptic modulation include presynaptic facilitation and presynaptic inhibition.

CHAPTER REVIEW

7.5 Neurotransmitters

1. Both excitatory and inhibitory neurotransmitters are present in the CNS and the PNS. A given neurotransmitter may be excitatory in some locations and inhibitory in others.
2. Neurotransmitters can be divided into two classes based on size: (1) small-molecule neurotransmitters (acetylcholine, amino acids, biogenic amines, purines, gases, and endocannabinoids), and (2) neuropeptides, which are composed of 3 to 40 amino acids linked by peptide bonds.

7.6 Neural Circuits

1. Neurons in the central nervous system are organized into networks called neural circuits.
2. Neural circuits include simple series, diverging, converging, reverberating, and parallel after-discharge circuits.

PONDER THIS

1. Birth control pills often cause a problem with K^+ concentration in the extracellular fluid. These pills can result in a condition known as hyperkalemia, an increased concentration of K^+ in the extracellular fluid. Why should this condition be of concern?

2. You are walking around your house and neglect to notice the thumbtack located in front of you. As you walk toward your room, you step on the thumbtack and immediately withdraw your foot. Discuss the pathways involved in the perception of your initial sensation as well as your response. What parts of the nervous system were responsible for facilitating the transmission of this information?

3. Dr. Salazar, a prestigious taxonomist, has just discovered a new species of frog. In order to learn more about the nervous system of this species, Dr. Salazar sacrifices one organism (luckily he discovered a large population) and extracts a few neurons from the sciatic nerve of its leg. Dr. Salazar runs several tests on this neuron and compares his results to the well documented data on the unmyelinated sciatic nerve of the bullfrog. Dr. Salazar concludes that the conduction velocity of the neurons in the new species is ten times faster than the conduction velocity of the bullfrog neurons. What must be different in the neurons of the new species compared to the neurons of the bullfrog? Explain your answer.

ANSWERS TO FIGURE QUESTIONS

7.1 The efferent division of the PNS conveys motor output from the CNS to effectors (muscles and glands).

7.2 Dendrites receive input; the cell body functions as the control center, and it can receive input; the axon sends output in the form of action potentials to another neuron or effector cell by releasing a neurotransmitter at its synaptic end bulbs.

7.3 Interneurons are responsible for integration.

7.4 Microglia function as phagocytes in the central nervous system.

7.5 The myelin sheath electrically insulates the axon of a neuron and increases the speed of conduction of action potentials.

7.6 Myelin makes white matter appear white.

7.7 Perception occurs primarily in the cerebral cortex.

7.8 A touch on the arm activates mechanically-gated channels.

7.9 No. Only a tiny fraction of all of the charges in the ECF and cytosol must be separated across the plasma membrane in order to establish the normal resting membrane potential.

7.10 K^+ ions have the greatest influence on the resting membrane potential of a neuron because neurons at rest are most permeable to K^+ ions.

7.11 When an inside-negative membrane potential exists, the K^+ electrical gradient favors movement of K^+ ions from the ECF into the cytosol because the K^+ ions are attracted to the negative charges along the inside surface of the plasma membrane and are repelled by the positive charges along the outside surface of the plasma membrane.

7.12 When an inside-positive membrane potential exists, the Na^+ electrical gradient favors movement of Na^+ ions from the cytosol into the ECF because the Na^+ ions are attracted to the negative charges along the outside surface of the plasma membrane and are repelled by the positive charges along the inside surface of the plasma membrane.

7.13 More Na^+ would leak into the cell and less K^+ ions would leak out of the cell, which would make the resting membrane more inside-positive.

7.14 A change in membrane potential from −70 to −60 mV is a depolarizing graded potential because the membrane potential is inside less negative than at rest. A change in membrane potential from −70 to −80 mV is a hyperpolarizing graded potential because the membrane potential is inside more negative than at rest.

7.15 Ligand-gated channels and mechanically-gated channels can be present in the dendrites of sensory neurons, and ligand-gated channels are numerous in the dendrites and cell bodies of interneurons and motor neurons.

7.16 A stronger stimulus opens more mechanically-gated channels or ligand-gated channels than a weaker stimulus.

7.17 Local current flow allows a graded potential to spread along the plasma membrane in both directions away from the stimulus source.

7.18 As a graded potential spreads along the plasma membrane, it gradually dies out because its charges are lost across the membrane through leak channels.

7.19 Because individual graded potentials undergo decremental conduction, they would die out as they spread through the dendrites and cell body if summation did not occur and an action potential would not be generated at the trigger zone of the axon.

7.20 Voltage-gated Na^+ channels are open during the depolarizing phase, and voltage-gated K^+ channels are open during the repolarizing phase.

7.21 An action potential will not occur in response to a hyperpolarizing graded potential because a hyperpolarizing graded potential causes the membrane potential to become inside more negative and therefore farther away from threshold (−55 mV).

7.22 Yes. The leak channels would still allow K^+ to exit more rapidly than Na^+ could enter the axon.

7.23 An action potential does not undergo decremental conduction because it propagates (keeps its strength as it spreads along the membrane).

7.24 The diameter of an axon and the presence or absence of a myelin sheath determine the speed of propagation of an action potential.

7.25 At a synapse between neurons, the presynaptic neuron sends the signal and the postsynaptic neuron receives the message.

7.26 In many electrical synapses (gap junctions), ions may flow equally well in either direction, so either neuron may function as the presynaptic neuron. At a chemical synapse, one neuron releases neurotransmitter and the other neuron has receptors that bind this chemical. Thus, the signal can proceed in only one direction.

7.27 At some excitatory synapses, ACh binds to ionotropic receptors that contain cation channels that open and subsequently generate EPSPs in the postsynaptic cell. At some inhibitory synapses, ACh binds to metabotropic receptors coupled to G proteins that open K^+ channels, resulting in the formation of IPSPs in the postsynaptic cell.

7.28 Reuptake is transport of the neurotransmitter back into the neuron that originally released it; uptake is transport of the neurotransmitter into another cell and not into the neuron that originally released it.

7.29 This is an example of spatial summation because the summation results from the buildup of neurotransmitter released simultaneously by several presynaptic end bulbs.

7.30 Because −60 mV is below threshold (−55 mV), an action potential will not occur in the postsynaptic neuron.

7.31 An axoaxonic synapse is typically associated with presynaptic inhibition or presynaptic facilitation.

7.32 Norepinephrine, epinephrine, dopamine, and serotonin are classified as biogenic amines because they are derived from amino acids that have been chemically modified.

7.33 A motor neuron receiving input from several other neurons is an example of convergence.

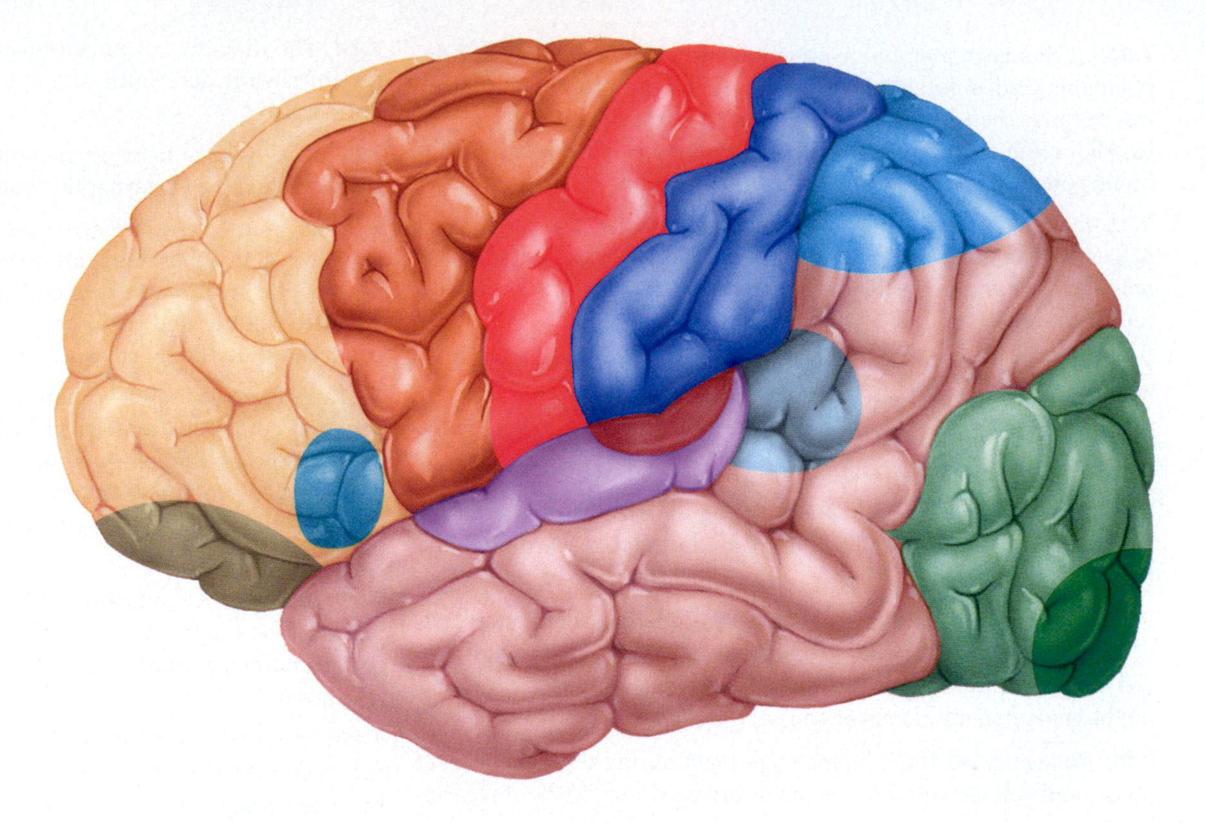

8 The Central Nervous System

The Central Nervous System and Homeostasis

The central nervous system contributes to homeostasis by receiving sensory input, integrating new and stored information, making decisions, and causing motor activities.

LOOKING BACK TO MOVE AHEAD...

- The plasma membrane is highly permeable to nonpolar molecules, such as oxygen (O_2) and steroids; moderately permeable to small, uncharged polar molecules, such as water; and impermeable to ions and large, uncharged polar molecules, such as glucose (Section 5.1).
- Somatic motor neurons convey action potentials to skeletal muscles; autonomic motor neurons convey action potentials to cardiac muscle, smooth muscle, and glands (Section 7.2).
- A postsynaptic potential is a graded potential that occurs in a postsynaptic cell; in a postsynaptic neuron, the postsynaptic potential is called an excitatory postsynaptic potential (EPSP) if it is depolarizing or an inhibitory postsynaptic potential (IPSP) if it is hyperpolarizing (Section 7.4).
- An ionotropic receptor is a type of neurotransmitter receptor that contains both a neurotransmitter binding site and an ion channel (Section 7.4).

The **central nervous system (CNS)**, which consists of the brain and spinal cord, has many functions: It processes sensory information; generates thoughts, emotions, and memories; and stimulates muscles to contract and glands to secrete. Our study of the CNS will begin with the relatively simple functions of the spinal cord and then progress to the more complex functions of the brain.

8.1 Spinal Cord

OBJECTIVES

- Identify the protective coverings of the spinal cord.
- Discuss the importance of spinal nerves.
- Explain how the spinal cord processes sensory input and motor output.
- Describe the functions of the major sensory and motor tracts of the spinal cord.
- List the functional components of a spinal reflex arc.

The **spinal cord** is a cylinder of nervous tissue that extends from the brain. In adults it is about 45 cm (18 in.) in length and 2 cm (0.75 in.) in diameter.

The Spinal Cord Is Protected by Vertebrae and Meninges

Two types of coverings—vertebrae and meninges—surround and protect the delicate nervous tissue of the spinal cord (FIGURE 8.1a). The spinal cord is located within the vertebral canal of the vertebral column. The bony vertebrae provide a sturdy shelter for the enclosed spinal cord. The **meninges** (me-NIN-jēz) are tough, connective tissue coverings that encircle the spinal cord and brain. The three spinal meninges are the outer *dura mater* (DOO-ra MĀ-ter), the middle *arachnoid mater* (a-RAK-noyd MĀ-ter), and the inner *pia mater* (PĒ-a MĀ-ter). Of the three meningeal layers, only the pia mater adheres to the surface of the spinal cord and brain. Between the arachnoid mater and the pia mater is the *subarachnoid space*, which contains cerebrospinal fluid (CSF) that serves as a shock absorber and suspension system for the nervous tissue of the CNS. Inflammation of the meninges, called **meningitis**, is usually caused by a bacterial or viral infection. Symptoms include fever, headache, stiff neck, vomiting, confusion, lethargy, and drowsiness. Bacterial meningitis is much more serious than viral meningitis and may be fatal if not treated promptly. Viral meningitis requires no specific treatment and usually resolves on its own within a few weeks. A vaccine is available to help protect against some types of bacterial meningitis.

Spinal Nerves Link the Spinal Cord to Sensory Receptors and Effectors

Thirty-one pairs of **spinal nerves** emerge from the spinal cord at regular intervals through the spaces between adjacent vertebrae (FIGURE 8.1a, b). Spinal nerves are part of the peripheral nervous system (PNS); they connect the spinal cord to sensory receptors and effectors (muscles and glands) in most parts of the body. Spinal nerves are named according to the region of the vertebral column from which they emerge. There are 8 pairs of *cervical nerves*, 12 pairs of *thoracic nerves*, 5 pairs of *lumbar nerves*, 5 pairs of *sacral nerves*, and 1 pair of *coccygeal nerves*.

A spinal nerve is connected to the spinal cord by two bundles of axons, called **roots** (FIGURE 8.1c). The **dorsal root** contains only sensory axons, which conduct action potentials from sensory receptors in the skin, muscles, and internal organs into the CNS. Each dorsal root has a swelling, the **dorsal root ganglion**, which contains the cell bodies of sensory neurons. The **ventral root*** contains axons of motor neurons, which conduct action potentials from the CNS to effectors. Because the dorsal root contains sensory axons and the ventral root contains motor axons, a spinal nerve is classified as a *mixed nerve*.

The Internal Organization of the Spinal Cord Allows Processing of Sensory Input and Motor Output

The spinal cord is organized into regions of white matter that surround an inner core of gray matter (FIGURE 8.1c). The white matter of the spinal cord consists primarily of bundles of myelinated axons of neurons. The gray matter of the spinal cord consists of dendrites and cell bodies of neurons, unmyelinated axons, and neuroglia. In the center of the gray matter is a small space called the *central canal*; it extends the entire length of the spinal cord and is filled with cerebrospinal fluid.

The gray matter on each side of the spinal cord is subdivided into regions called **horns** (FIGURE 8.1c). The **dorsal gray horns** contain axons of incoming sensory neurons as well as cell bodies and axons of interneurons. The **ventral gray horns** contain cell bodies of somatic motor neurons that convey action potentials to skeletal muscles. Between the dorsal and ventral gray horns are the **lateral gray horns**, which are present only in thoracic and upper lumbar segments of the spinal cord. The lateral gray horns contain cell bodies of autonomic motor neurons that convey action potentials to cardiac muscle, smooth muscle, and glands.

The white matter of the spinal cord is subdivided into three areas called **columns:** (1) **dorsal white columns**, (2) **ventral white columns**, and (3) **lateral white columns** (FIGURE 8.1c). Each column contains tracts (bundles of axons) that may extend long distances up or down the spinal cord (see FIGURE 8.3). **Sensory (ascending) tracts** consist of axons that conduct action potentials *toward* the brain. **Motor (descending) tracts** consist of axons that carry action potentials *away* from the brain. Sensory and motor

* In humans, the terms *ventral* and *anterior* both mean nearer to or at the front of the body, and the terms *dorsal* and *posterior* both mean nearer to or at the back of the body. So *ventral* and *anterior* can be used interchangeably and so can *dorsal* and *posterior*. For example, the ventral root is also known as the anterior root. Likewise, the dorsal root ganglion can also be called the posterior root ganglion.

FIGURE 8.1 The spinal cord and associated structures. For simplicity, the dendrites of neurons in part (c) are not shown in this and several other illustrations of cross sections of the spinal cord. Blue, red, and green arrows indicate the direction of action potential propagation.

Within the gray matter of the spinal cord are (1) dorsal gray horns, which contain axons of sensory neurons and cell bodies of interneurons; (2) lateral gray horns, which contain cell bodies of autonomic motor neurons; and (3) ventral gray horns, which contain cell bodies of somatic motor neurons.

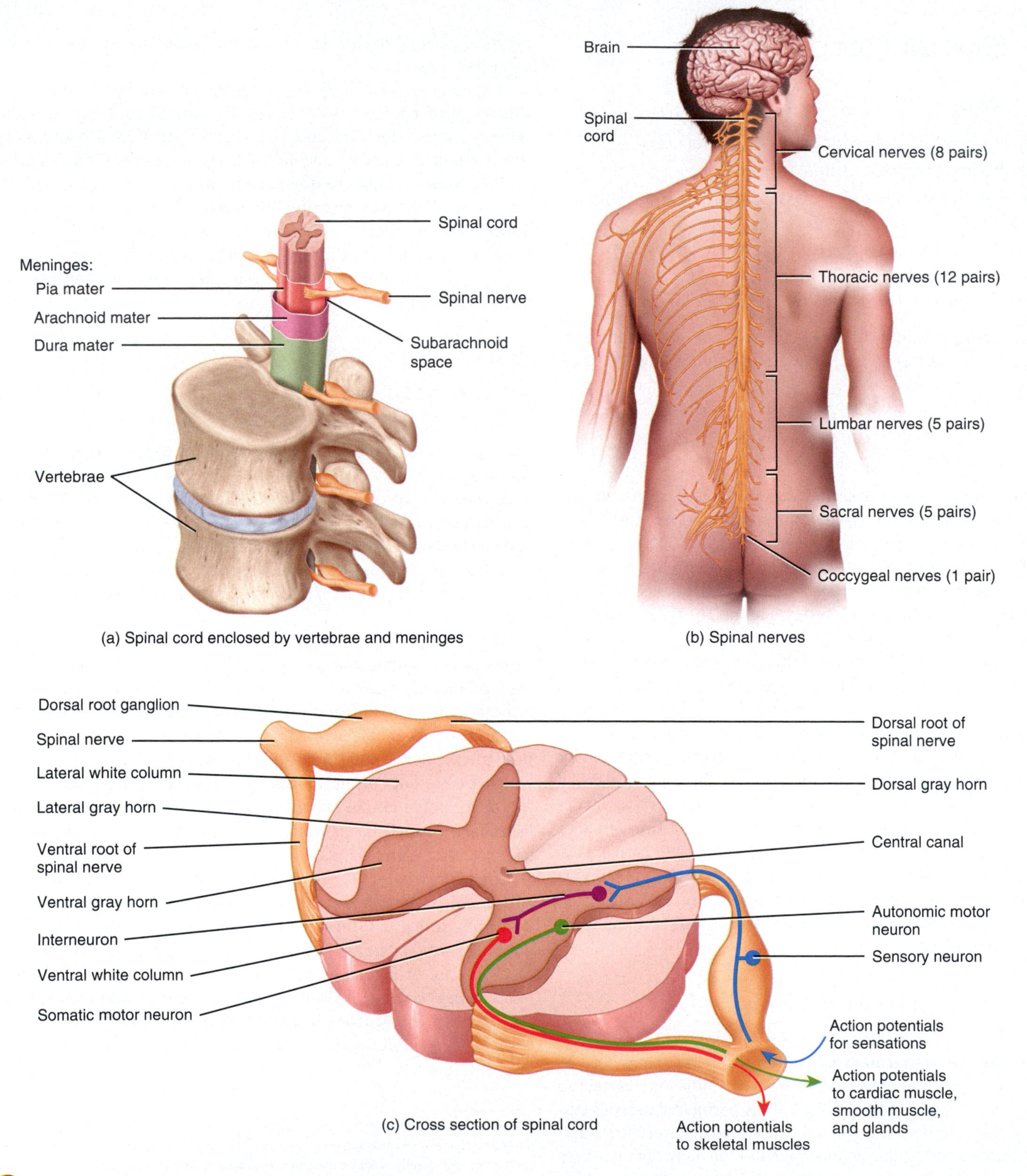

(a) Spinal cord enclosed by vertebrae and meninges

(b) Spinal nerves

(c) Cross section of spinal cord

What is the significance of the dorsal root ganglion?

tracts of the spinal cord are continuous with sensory and motor tracts of the brain.

Because of its internal organization, the spinal cord processes sensory input and motor output in the following way (**FIGURE 8.2**):

1. Sensory receptors detect a sensory stimulus.
2. Sensory neurons convey this sensory input in the form of action potentials along their axons, which extend from sensory receptors into the spinal nerve and then into the dorsal root. From the dorsal root, axons of sensory neurons may proceed along three possible paths (see steps 3, 4, and 5).
3. Axons of sensory neurons may extend into the white matter of the spinal cord and ascend to the brain as part of a sensory tract.
4. Axons of sensory neurons may enter the dorsal gray horn and synapse with interneurons whose axons extend into the white matter of the spinal cord and then ascend to the brain as part of a sensory tract.

FIGURE 8.2 Processing of sensory input and motor output by the spinal cord.

Sensory input is conveyed from sensory receptors to the dorsal gray horns of the spinal cord, whereas motor output is conveyed from the ventral and lateral gray horns of the spinal cord to effectors (muscles and glands).

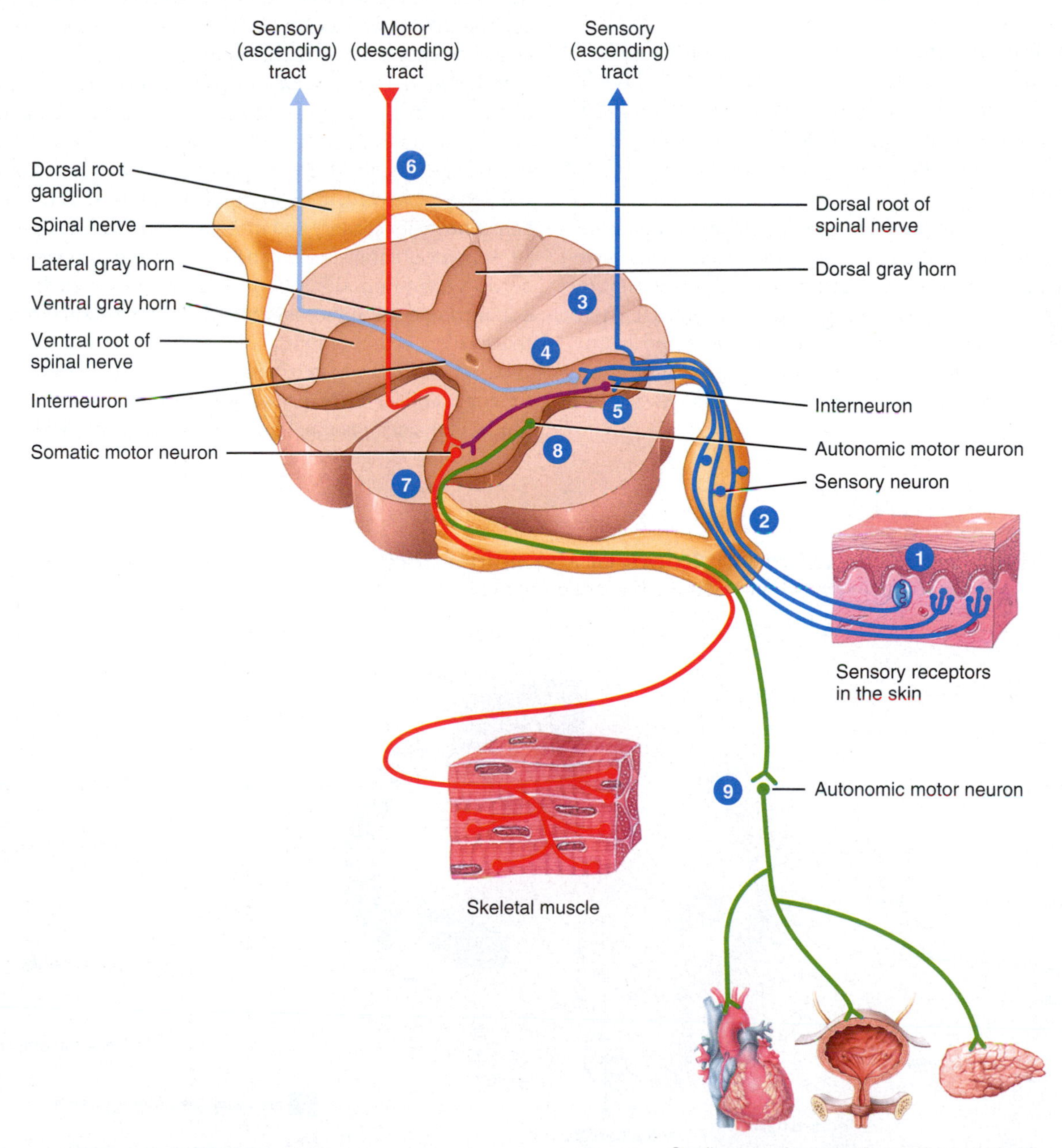

What happens to sensory information during step 3 of this figure?

5. Axons of sensory neurons may enter the dorsal gray horn and synapse with interneurons that in turn synapse with somatic motor neurons that are involved in spinal reflex pathways. Spinal reflex pathways are described in more detail later in this section and in Chapter 12.
6. Motor output from the spinal cord to skeletal muscles involves somatic motor neurons of the ventral gray horn. Many somatic motor neurons are regulated by the brain. Axons from higher brain centers form motor tracts that descend from the brain into the white matter of the spinal cord and then synapse with somatic motor neurons.
7. When activated, somatic motor neurons convey action potentials along their axons, which sequentially pass through the ventral gray horn and ventral root to enter the spinal nerve. From the spinal nerve, axons of somatic motor neurons extend to skeletal muscles of the body.
8. Motor output from the spinal cord to cardiac muscle, smooth muscle, and glands involves autonomic motor neurons of the lateral gray horn. When activated, autonomic motor neurons convey action potentials along their axons, which sequentially pass through the lateral gray horn, ventral gray horn, and ventral root to enter the spinal nerve.
9. While passing through the spinal nerve, axons of the autonomic motor neurons from the spinal cord synapse with another group of autonomic motor neurons located in the peripheral nervous system. The axons of this second group of autonomic motor neurons in turn synapse with cardiac muscle, smooth muscle, and glands. You will learn more about autonomic motor neurons when the autonomic nervous system is described in Chapter 10.

The Spinal Cord Propagates Signals Along Sensory and Motor Tracts and Coordinates Reflexes

The spinal cord has two principal functions in maintaining homeostasis: action potential propagation and integration of information. The *white matter tracts* in the spinal cord serve as the "highways" for action potential propagation. Sensory input travels along these tracts toward the brain, and motor output travels from the brain along these tracts toward skeletal muscles and other effector tissues. The *gray matter* of the spinal cord receives and integrates incoming and outgoing information.

Sensory and Motor Tracts

One way the spinal cord promotes homeostasis is by propagating action potentials along tracts. The name of a tract often indicates its position in the white matter and where it begins and ends. For example, the ventral corticospinal tract is located in the *ventral* white column; it begins in the *cerebral cortex* (a region of the brain) and ends in the *spinal cord.* Notice that the location of the axon terminals comes last in the name. This regularity in naming allows you to determine the direction of information flow along any tract named according to this convention. Because the ventral corticospinal tract conveys action potentials from the brain toward the spinal cord, it is a motor (descending) tract. **FIGURE 8.3** highlights the major sensory and motor tracts in the spinal cord.

Action potentials from sensory receptors propagate up the spinal cord to the brain along two main routes on each side: the dorsal columns and the spinothalamic tract. The **dorsal columns** are tracts that convey action potentials for touch, pressure, vibration, and proprioception (awareness of the positions and movements of muscles, tendons, and joints). The **spinothalamic tract** conveys action potentials for sensing pain, temperature, itch, and tickle. Other sensory

FIGURE 8.3 Major sensory and motor tracts of the spinal cord. Sensory tracts are indicated on one half and motor tracts on the other half of the cord, but actually all tracts are present on both sides.

The name of the tract often indicates its location in the white matter and where it begins and ends.

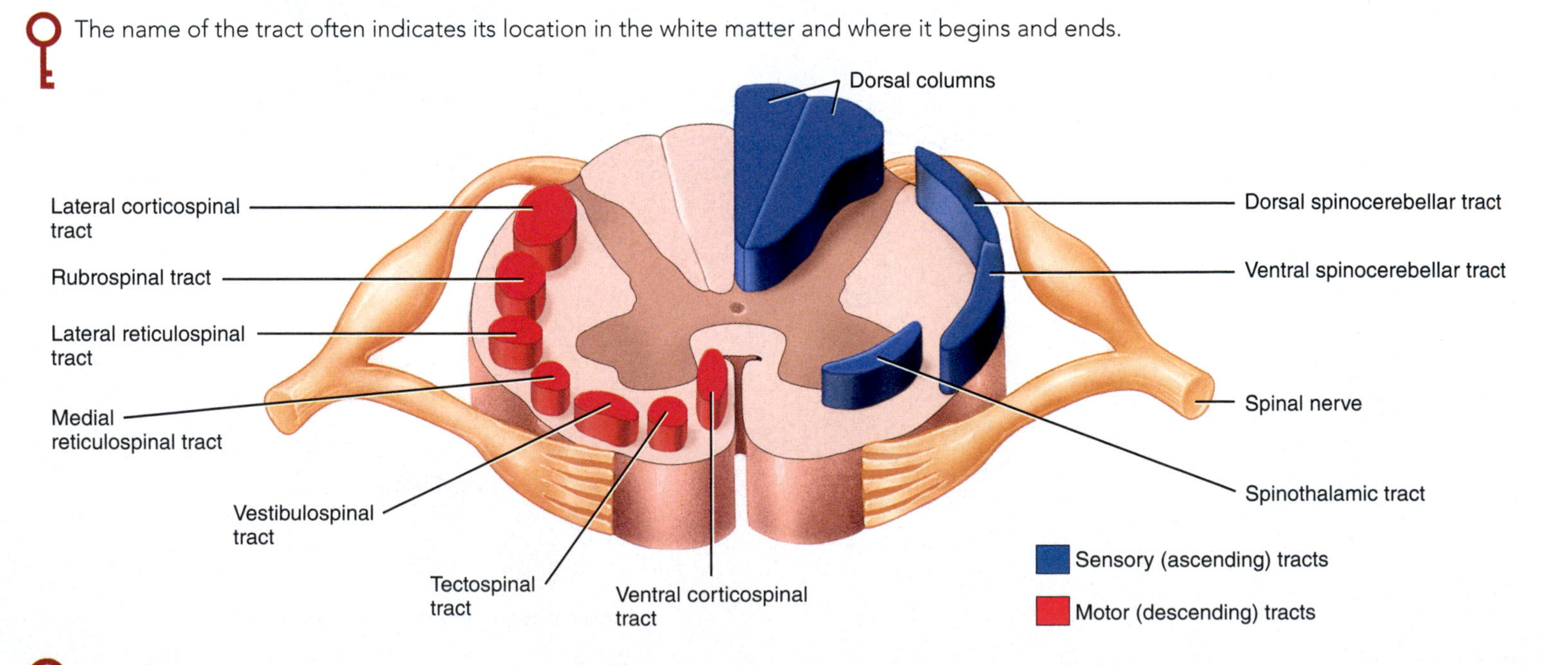

Based on its name, indicate the origin and destination of the spinothalamic tract. Is this a sensory or a motor tract?

tracts in the spinal cord include the **dorsal** and **ventral spinocerebellar** tracts, which also convey action potentials for proprioception.

The sensory systems keep the CNS informed of changes in the external and internal environments. The sensory information is integrated by interneurons in the spinal cord and brain. Responses to the integrative decisions are brought about by motor activities (muscular contractions and glandular secretions). The cerebral cortex, the outer part of the brain, plays a major role in controlling precise voluntary muscular movements. Other brain regions regulate involuntary movements. Motor output to skeletal muscles travels down the spinal cord in two types of descending pathways: direct and indirect. The **direct motor pathways**, also known as the *pyramidal pathways*, include the **lateral corticospinal**, **ventral (anterior) corticospinal**, and **corticobulbar tracts**. They convey action potentials that originate in the cerebral cortex and are destined to cause *voluntary* movements of skeletal muscles. The **indirect motor pathways**, also referred to as the *extrapyramidal pathways*, include the **tectospinal** (TEK-tō-spī-nal), **vestibulospinal** (ves-TIB-ū-lō-spī-nal), **rubrospinal** (ROO-brō-spī-nal), **lateral reticulospinal** (re-TIK-ū-lō-spī-nal), and **medial reticulospinal tracts**. They convey action potentials from the brain stem to cause *involuntary movements* that regulate posture, balance, and muscle tone. These sensory and motor tracts of the spinal cord are described in detail in Chapters 9 and 12, respectively.

Reflexes

The second way the spinal cord promotes homeostasis is by serving as an integrating center for some reflexes. A **reflex** is a fast, involuntary, unplanned sequence of actions that occurs in response to a particular stimulus. Some reflexes are inborn, such as pulling your hand away from a hot surface before you even feel that it is hot. Other reflexes are learned or acquired. For instance, you learn many reflexes while acquiring driving expertise. Slamming on the brakes in an emergency is one example. When integration takes place in the spinal cord gray matter, the reflex is a **spinal reflex.** Familiar examples of spinal reflexes include the *patellar reflex (knee jerk)*, which occurs in response to a tap on the knee at the patellar tendon, and the *flexor (withdrawal) reflex*, which moves a limb away from a painful stimulus. If integration occurs in the brain stem rather than the spinal cord, the reflex is called a **cranial reflex.** An example is the tracking movements of your eyes as you read this sentence. You are probably most aware of **somatic reflexes**, which involve contraction of skeletal muscles. Equally important, however, are the **autonomic (visceral) reflexes**, which generally are not consciously perceived. They involve responses of smooth muscle, cardiac muscle, and glands. Body functions such as heart rate, digestion, urination, and defecation are controlled by the autonomic nervous system through autonomic reflexes.

Action potentials propagating into, through, and out of the CNS follow specific pathways, depending on the type of information, its origin, and its destination. The pathway followed by action potentials that produce a reflex is called a **reflex arc (reflex circuit)**. A reflex arc includes five functional components (**FIGURE 8.4**):

1. **Sensory receptor**. The distal end of a sensory neuron or an associated sensory structure serves as a sensory receptor. It responds to a specific **stimulus**—a change in the external or internal

FIGURE 8.4 General components of a reflex arc. Arrows show the direction of action potential propagation.

A reflex is a fast, predictable sequence of involuntary actions that occurs in response to certain changes in the environment.

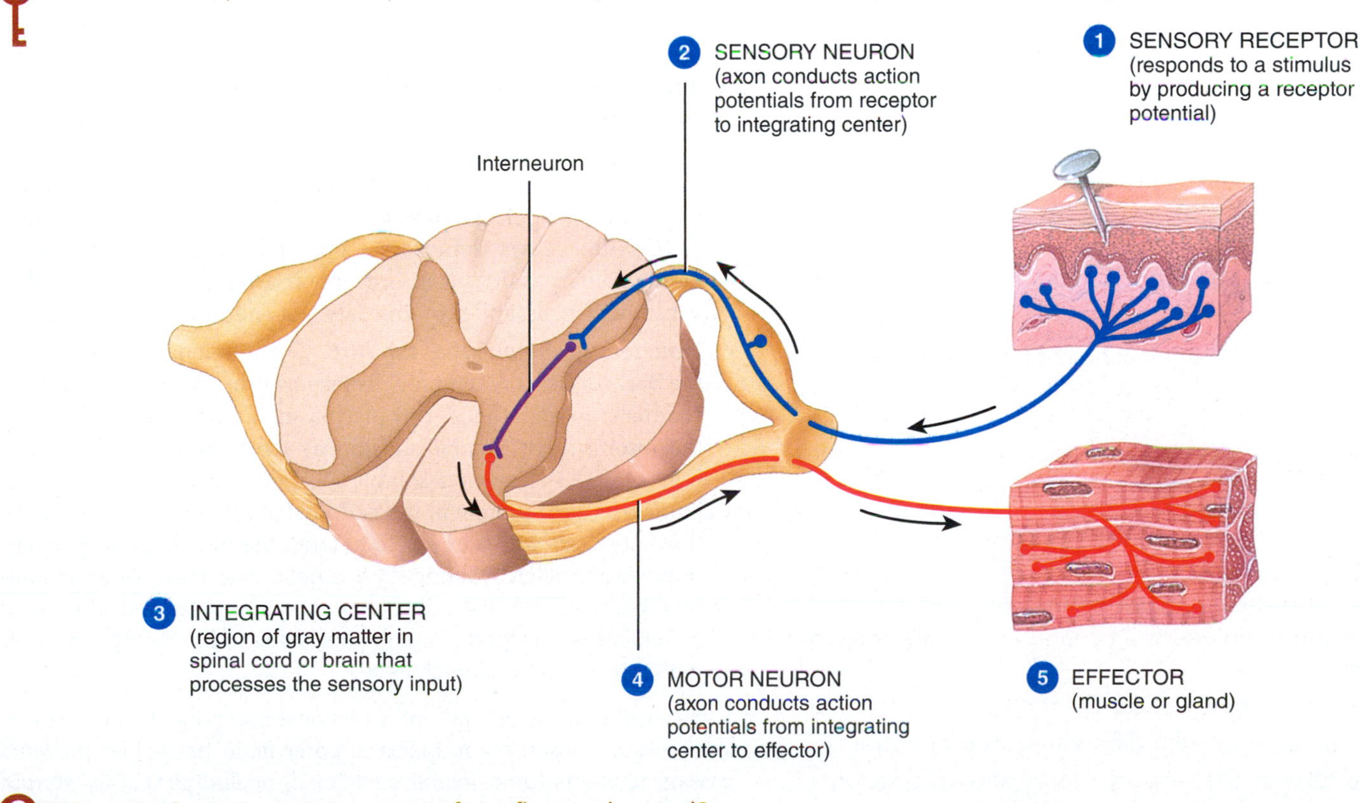

Where is the integrating center of a reflex arc located?

environment—by producing a graded potential known as a *receptor potential.*

2. **Sensory neuron.** If the receptor potential reaches the threshold level of depolarization, it will trigger one or more action potentials in the axon of the sensory neuron. The action potentials propagate along the axon into the CNS.
3. **Integrating center.** The integrating center is a region of gray matter within the brain stem or spinal cord that processes the incoming sensory information. The integrating center may involve just one synapse between the sensory neuron and a motor neuron, or it may involve one or more interneurons that relay signals from the sensory neuron to the motor neuron.
4. **Motor neuron.** Action potentials triggered by the integrating center propagate out of the CNS along the axon of a motor neuron to the effector.
5. **Effector.** The effector is the part of the body that responds to the motor neuron. It is usually a muscle or a gland. If the effector is skeletal muscle, the reflex is a somatic reflex. If the effector is smooth muscle, cardiac muscle, or a gland, the reflex is an autonomic reflex. You will learn more about autonomic reflexes in Chapter 10 and somatic reflexes in Chapter 12.

It is important to point out that while a somatic reflex is occurring, axon collaterals of the sensory neuron also relay the sensory input to the appropriate area of the brain that allows conscious awareness of the sensory information. However, because it takes longer for the sensory input to ascend to higher brain centers than to pass through the reflex arc, by the time you perceive the sensory stimulus, the effector has already been activated and the reflex is over. This is advantageous because if the sensory stimulus is harmful (for example, the hot burner on a stove), there would be more tissue damage if you had to wait to perceive the stimulus before you could activate the effector to move your limb away from the stimulus.

CHECKPOINT

1. Why are the vertebrae and meninges important to the spinal cord?
2. What is the purpose of a spinal nerve?
3. How does the spinal cord processes sensory input and motor output?
4. Which spinal cord tracts are ascending tracts? Which are descending tracts?
5. What is the functional significance of a spinal reflex?

8.2 Brain

OBJECTIVES

- Explain the various ways that the brain is protected.
- Discuss why the brain needs a continuous supply of oxygen and glucose.
- Describe the importance of cranial nerves.
- Discuss the functions of the different parts of the brain.
- Identify the roles of the sensory, motor, and association areas of the cerebral cortex.

The **brain** is the portion of the central nervous system contained within the cranium (skull). About 100 billion neurons and 10–50 trillion neuroglia make up the brain, which has a mass of about 1300 g (almost 3 lb) in adults.

The Brain Is Protected in Many Ways

Several features protect the brain: the cranium, meninges, blood–brain barrier, and cerebrospinal fluid.

Cranium and Meninges

The cranium and the meninges surround and protect the brain (FIGURE 8.5a). The brain is located in the cranial cavity, which is the space within the cranium. The bony structure of the cranium serves as a protective armor for the brain. The cranial meninges are continuous with and have the same names as the spinal meninges: an outer dura mater, a middle arachnoid mater, and an inner pia mater. However, the cranial dura mater has two layers; the spinal dura mater has only one. The two dural layers around the brain are fused together except where they separate to enclose the dural venous sinuses, which are cavities that drain venous blood from the brain.

The Blood–Brain Barrier

The existence of a **blood–brain barrier (BBB)** protects brain cells from harmful agents by preventing passage of many substances from blood into brain tissue. The BBB has both structural and functional components:

- ***Structural component of the BBB.*** The structural component of the BBB consists of tight junctions that seal together the endothelial cells of brain capillaries (FIGURE 8.5b). The processes of many astrocytes, which you learned in Chapter 7 are a type of neuroglia, press up against the capillaries and secrete chemicals that maintain the "tightness" of the tight junctions. Because an endothelial cell plasma membrane, like any other plasma membrane, has a hydrophobic interior, lipid-soluble substances are able to diffuse across brain capillary walls, whereas most water-soluble substances are unable to do so. Furthermore, since there are no pores (spaces) between the tightly-sealed endothelial cells, large substances such as cells and proteins are also not able to move across brain capillary walls. Therefore, the structural component of the BBB allows diffusion of lipid-soluble substances across brain capillary walls, but it prevents movement of water-soluble substances, cells, and proteins between endothelial cells.
- ***Functional component of the BBB.*** The functional component of the BBB consists of membrane transport proteins (carriers and ion channels) in endothelial cell plasma membranes that selectively move some water-soluble substances across brain capillary walls (FIGURE 8.5b). If a water-soluble substance does not have a membrane transport protein in the endothelial cells of brain capillaries, that substance will not be able to cross the BBB. The main exception is water itself. Although it is a polar molecule, water can diffuse across endothelial cell plasma membranes without the use of a membrane transport protein. This is due to the fact that a water molecule is small and uncharged.

Because of its structural and functional components, the BBB allows certain substances in blood to enter brain tissue and prevents passage to others. Lipid-soluble substances (including O_2, CO_2, steroid hormones, alcohol, barbiturates, nicotine, and caffeine) and water

FIGURE 8.5 Protective features of the brain.

The brain is protected by the cranium, meninges, blood–brain barrier, and cerebrospinal fluid.

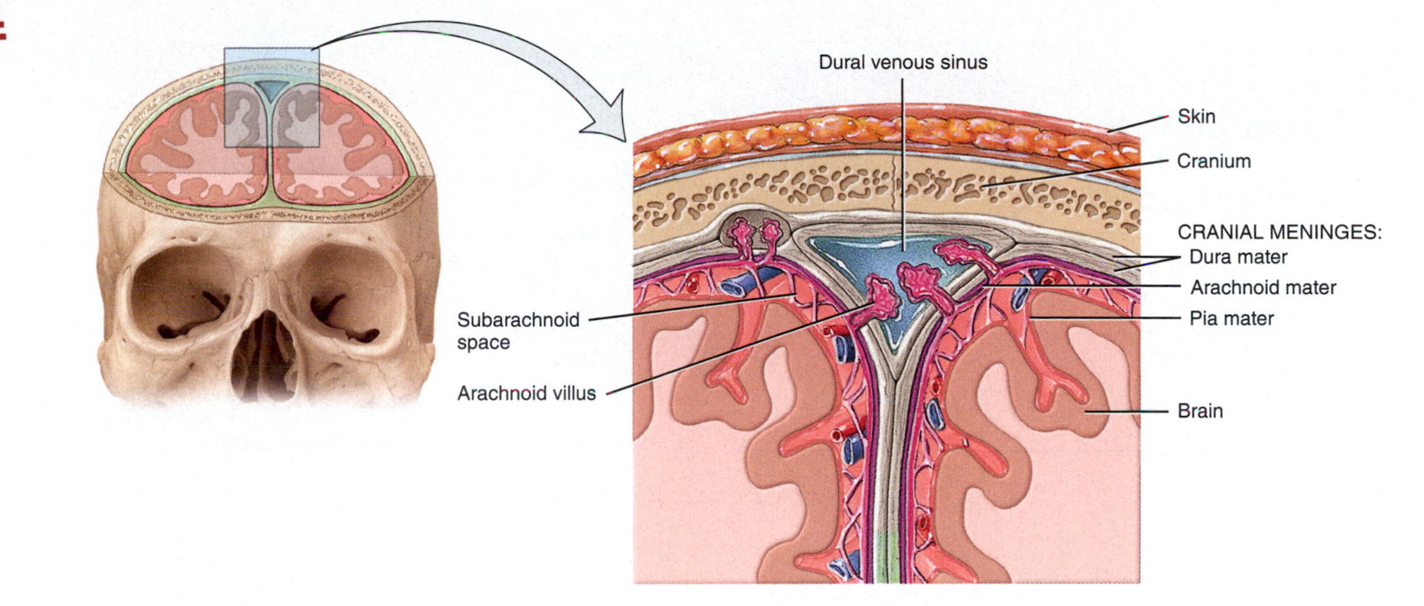

(a) The cranium and cranial meninges

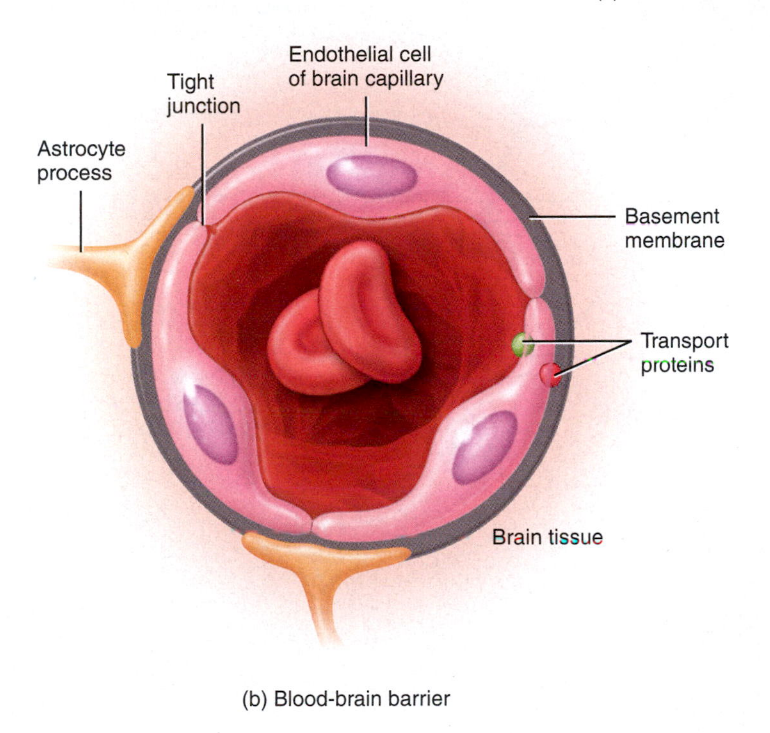

(b) Blood-brain barrier

Figure 8.5 (continues)

CLINICAL CONNECTION

Breaching the Blood–Brain Barrier

Although the blood–brain barrier (BBB) protects the brain from potentially harmful substances, a consequence of the BBB's efficient protection is that it prevents the passage of certain drugs that could be therapeutic for brain cancer or other CNS disorders. Researchers are exploring ways to move drugs past the BBB. In one method, the drug is injected in a concentrated sugar solution. The high osmotic pressure of the sugar solution causes the endothelial cells of the capillaries to shrink, which opens gaps between their tight junctions and makes the BBB more leaky. As a result, the drug can enter the brain tissue.

molecules easily cross the BBB by diffusing across the lipid bilayer of endothelial cell plasma membranes. A few water-soluble substances, such as glucose, quickly cross the BBB by facilitated transport. Other water-soluble substances, such as most ions, are transported across the BBB very slowly. Still other substances—proteins and most antibiotic drugs—do not pass at all from the blood into brain tissue. Trauma, certain toxins, and inflammation can cause a breakdown of the BBB.

Cerebrospinal Fluid

Cerebrospinal fluid (CSF) is a clear, colorless liquid that protects the brain and spinal cord from chemical and physical injuries. It also carries oxygen, glucose, and other needed chemicals from the blood to neurons and neuroglia. CSF continuously circulates around the brain and spinal cord in the subarachnoid space, through the central canal of the spinal cord, and through cavities

Figure 8.5 Continued

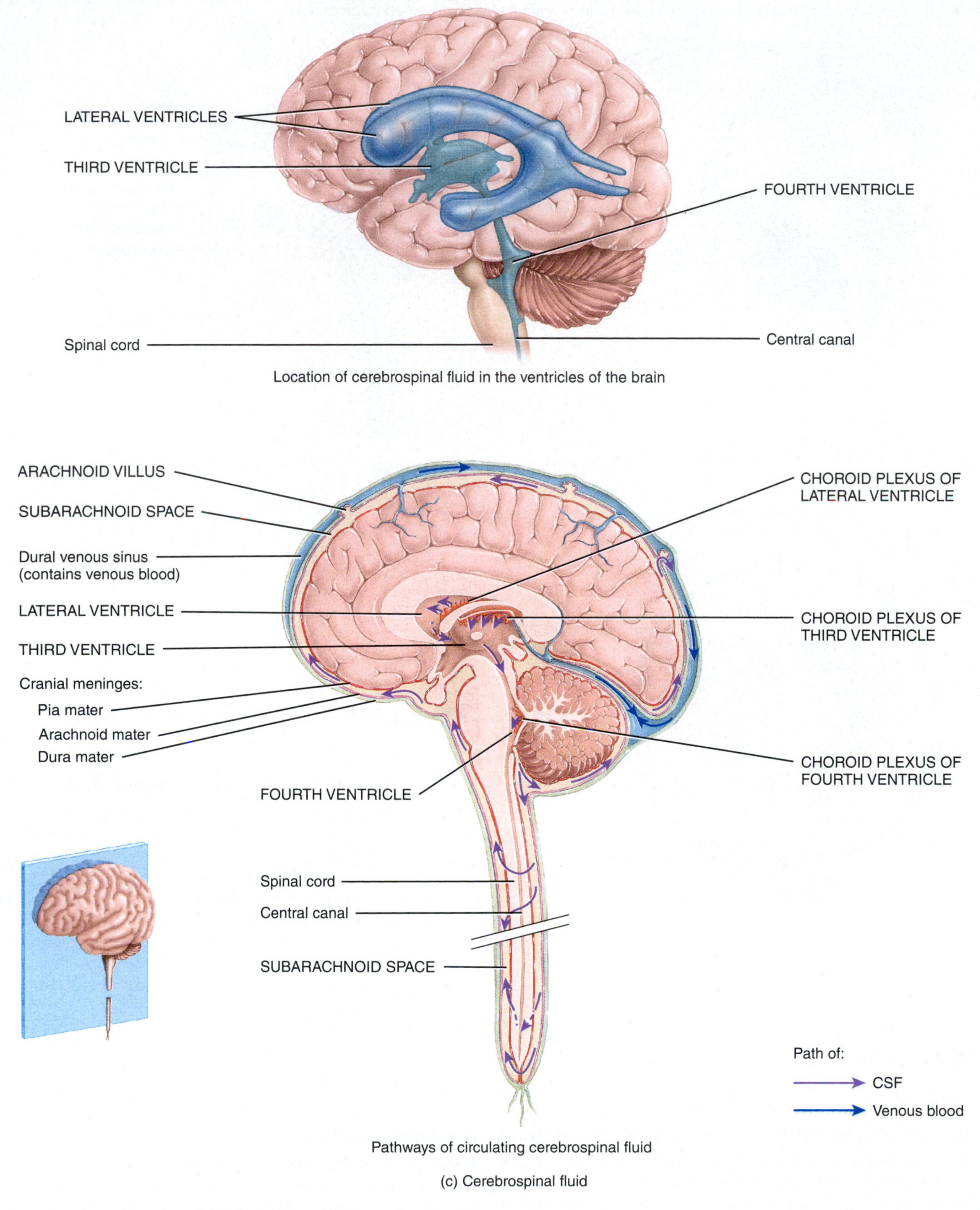

Location of cerebrospinal fluid in the ventricles of the brain

Pathways of circulating cerebrospinal fluid

(c) Cerebrospinal fluid

How does cerebrospinal fluid protect the brain?

in the brain known as **ventricles**. There are four ventricles: two **lateral ventricles**, one **third ventricle**, and one **fourth ventricle** (FIGURE 8.5c). Openings connect the ventricles with one another, with the central canal of the spinal cord, and with the subarachnoid space.

The total volume of CSF is 80 to 150 mL (3 to 5 oz) in an adult. CSF contains glucose, proteins, lactic acid, urea, cations (Na^+, K^+, Ca^{2+}, Mg^{2+}), and anions (Cl^- and HCO_3^-); it also contains some leukocytes. CSF contributes to homeostasis in three main ways:

1. ***Mechanical protection***. CSF serves as a shock-absorbing medium that protects the delicate tissues of the brain and spinal cord from jolts that would otherwise cause them to hit the bony walls of the cranial cavity and vertebral canal. The fluid also buoys the brain so that it "floats" in the cranial cavity.
2. ***Chemical protection***. CSF provides an optimal chemical environment for accurate neuronal signaling. Even slight changes in the ionic composition of CSF within the brain can seriously disrupt production of action potentials and postsynaptic potentials.
3. ***Circulation***. CSF allows exchange of nutrients and waste products between the blood and nervous tissue.

The sites of CSF production are the **choroid plexuses** (KŌ-royd), networks of blood capillaries in the walls of the ventricles (FIGURE 8.5c). The capillaries are covered by **ependymal cells** that form cerebrospinal fluid from blood plasma by filtration and secretion. Because the ependymal cells are joined by tight junctions, materials entering CSF from choroid capillaries cannot leak between these cells; instead, they must pass through the ependymal cells. This **blood–cerebrospinal fluid barrier** permits certain substances to enter the CSF but excludes others, protecting the brain and spinal cord from potentially harmful blood-borne substances. From the ventricles, CSF flows into the central canal of the spinal cord and into the subarachnoid space. CSF is gradually reabsorbed into the blood through **arachnoid villi** (singular is **villus**), fingerlike extensions of the arachnoid that project into the dural venous sinuses (FIGURE 8.5a, c). Normally, CSF is reabsorbed as rapidly as it is formed by the choroid plexuses, at a rate of about 20 mL/hour (about 480 mL/day). This means that the entire volume of CSF is replaced about three times per day. Because the rates of formation and reabsorption are the same, the pressure of CSF (about 10 mm Hg) is usually constant.

CLINICAL CONNECTION

Hydrocephalus

Abnormalities in the brain—tumors, inflammation, or developmental malformations—can interfere with the drainage of CSF from the ventricles into the subarachnoid space. When excess CSF accumulates in the ventricles, the CSF pressure rises. Elevated CSF pressure causes a condition called **hydrocephalus** (hī′-drō-SEF-a-lus). In a baby whose fontanels (the soft spots between the cranial bones of an infant's skull) have not yet closed, the head bulges due to the increased pressure. If the condition persists, the fluid buildup compresses and damages the delicate nervous tissue. Hydrocephalus is relieved by draining the excess CSF. A neurosurgeon may implant a drain line, called a shunt, into the lateral ventricle to divert CSF into a vein or the abdominal cavity, where it can be absorbed by the blood. In adults, hydrocephalus may occur after head injury, meningitis, obstruction by cysts or tumors, or subarachnoid hemorrhage. This condition can quickly become life-threatening and requires immediate intervention because an adult's skull bones have already fused.

The Brain Depends on a Continuous Supply of Oxygen and Glucose by the Blood

In an adult, the brain represents only 2% of total body weight, but it consumes about 20% of the oxygen and glucose used even at rest. Neurons synthesize ATP almost exclusively from glucose via reactions that use oxygen. When activity of neurons and neuroglia increases in a region of the brain, blood flow to that area also increases. Even a brief slowing of blood flow to the brain may cause unconsciousness. Typically, an interruption in blood flow for 1 or 2 minutes impairs neuronal function, and total deprivation of oxygen for about 4 minutes causes permanent injury. Because virtually no glucose is stored in the brain, the supply of glucose also must be continuous. If blood entering the brain has a low level of glucose, mental confusion, dizziness, convulsions, and loss of consciousness may occur.

CLINICAL CONNECTION

Stroke

An interruption of the blood supply to a part of the brain is known as a **stroke** or **cerebrovascular accident (CVA)**. When a part of the brain does not receive an adequate blood supply, brain cells in the affected region begin to die within several minutes. A stroke is characterized by the sudden onset of persisting neurological signs and symptoms, such as numbness and muscle paralysis of the face, arm, or leg (typically on one side of the body); severe headache; mental confusion; difficulty speaking or understanding speech; trouble seeing in one or both eyes; and trouble walking due to dizziness and a loss of balance. A stroke usually occurs when there is a blockage or rupture of a blood vessel that supplies the brain. Common causes of strokes include atherosclerosis (formation of cholesterol-containing plaques that block blood flow) of cerebral arteries, emboli (blood clots or other particles that break away from the wall of an artery of an organ, travel in the bloodstream, and then lodge in smaller arteries of another organ), and intracerebral hemorrhage (bleeding from a blood vessel in the pia mater or brain). A clot-dissolving drug called *tissue plasminogen activator (t-PA)* is currently being used to open up blocked blood vessels in the brain. The drug is most effective when administered ***within three hours*** of the onset of the stroke, however, and is helpful only for strokes due to a blood clot. Use of t-PA can decrease the permanent disability associated with these types of strokes by 50%. Studies show that "cold therapy" may be successful in limiting the amount of residual damage from a stroke. These "cooling" therapies developed from knowledge obtained following examination of cold-water drowning victims. States of hypothermia seem to trigger a survival response in which the body requires less oxygen. Some commercial companies now provide "stroke survival kits," which include cooling blankets that can be kept in the home.

Cranial Nerves Link the Brain to Sensory Receptors and Effectors

Extending from the brain are 12 pairs of **cranial nerves**, which are part of the peripheral nervous system. They connect the brain to sensory receptors and effectors in the head, neck, and many organs in the thoracic and abdominal cavities. Cranial nerves are classified as *sensory nerves* if they contain only axons of sensory neurons, *motor nerves* if they contain only axons of motor neurons, or *mixed nerves* if they contain axons of both sensory and motor neurons. TABLE 8.1 presents a summary of the cranial nerves.

CHECKPOINT

6. How does the blood–brain barrier function?
7. What structures produce CSF and where they are located?
8. What happens to neuronal function if the neurons of the brain are totally deprived of oxygen for 4 minutes or longer?
9. What are the functions of the facial nerve? Vagus nerve? Oculomotor nerve? Hypoglossal nerve? (*Hint*: Refer to TABLE 8.1).

TABLE 8.1 Summary of Cranial Nerves

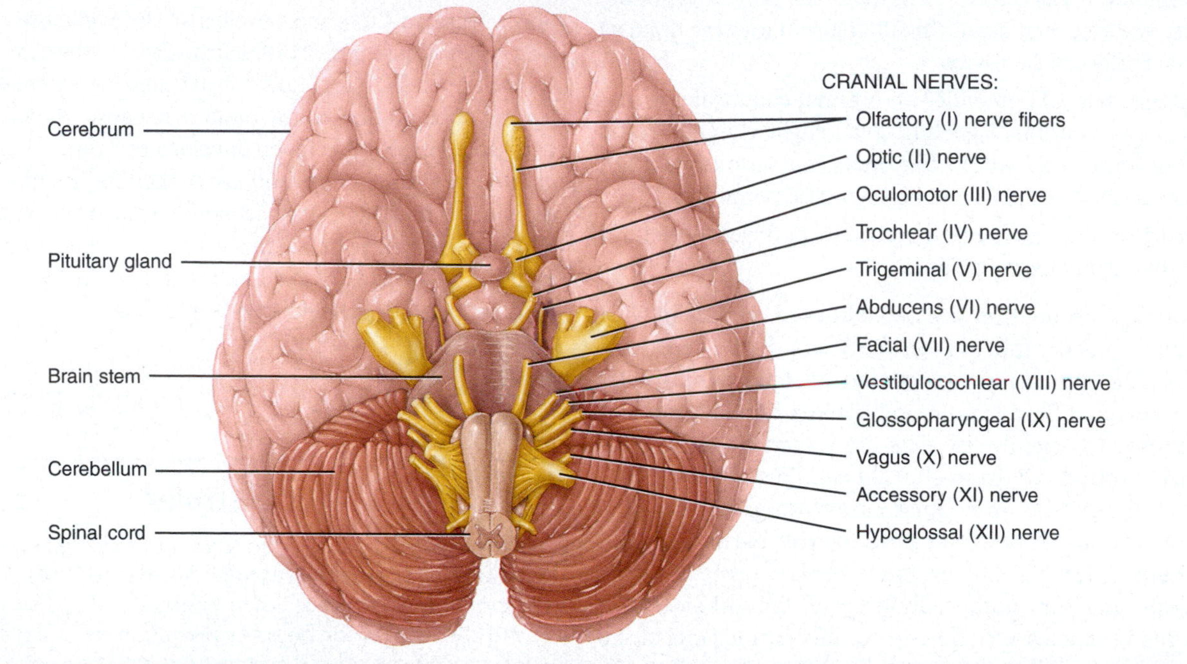

Number	Name	Type	Function
I	Olfactory	Sensory	Conveys olfactory (smell) input from the nose.
II	Optic	Sensory	Conveys visual input from the eye.
III	Oculomotor	Motor	Controls eye movements, pupillary constriction, and accommodation of the lens for near vision.
IV	Trochlear	Motor	Controls eye movements.
V	Trigeminal	Mixed	Conveys touch, pressure, vibration, pain, and temperature input from the face, nose, and mouth; controls chewing muscles.
VI	Abducens	Motor	Controls eye movements.
VII	Facial	Mixed	Conveys taste input from anterior two-thirds of tongue; controls muscles of facial expression; stimulates secretion of saliva and tears.
VIII	Vestibulocochlear	Sensory	Conveys audition (hearing) and equilibrium (balance) input from the inner ear.
IX	Glossopharyngeal	Mixed	Conveys taste input from posterior one-third of tongue; blood pressure input from baroreceptors; input about the levels of O_2, CO_2, and H^+ ions from chemoreceptors; controls muscles of pharynx involved in swallowing; stimulates secretion of saliva.
X	Vagus	Mixed	Conveys visceral sensory input from thoracic and abdominal organs; blood pressure input from baroreceptors; input about the levels of O_2, CO_2, and H^+ ions from chemoreceptors; and taste input from pharynx and epiglottis of larynx; controls muscles in thoracic and abdominal organs and muscles of pharynx and larynx involved in swallowing and vocalization.
XI	Accessory	Motor	Controls muscles that move the head and neck.
XII	Hypoglossal	Motor	Controls tongue movements.

The Different Parts of the Brain Perform a Variety of Functions

The brain is comprised of four major parts: brain stem, cerebellum, diencephalon, and cerebrum (**FIGURE 8.6a**). The *brain stem* is continuous with the spinal cord and consists of the medulla oblongata, pons, and midbrain. Located behind the brain stem is the *cerebellum* (ser′-e-BEL-um = little brain). Above the brain stem is the *diencephalon* (dī′-en-SEF-a-lon), which consists of the thalamus, hypothalamus, and pineal gland. Supported on the diencephalon and brain stem is the *cerebrum* (se-RĒ-brum = brain), the largest part of the brain. Together, the cerebrum and diencephalon are referred to as the *forebrain*.

Brain Stem

As you have just learned, the three parts of the brain stem are the medulla oblongata, pons, and midbrain. The **medulla oblongata** (me-DOOL-la ob′-long-GA-ta), or more simply the **medulla**, is the lower portion of the brain stem (**FIGURE 8.6a**). The medulla contains several important nuclei (clusters of neuronal cell bodies). Some of these nuclei control vital body functions, including the **cardiovascular center**, which regulates the rate and force of the heartbeat and the diameter of blood vessels, and the **medullary respiratory center**, which regulates breathing. Other nuclei in the medulla control reflexes for vomiting, swallowing, sneezing, coughing, and hiccupping. Within the white matter of the medulla are sensory (ascending) tracts and motor (descending) tracts extending between the spinal cord and other parts of the brain.

The **pons** is the middle portion of the brain stem (**FIGURE 8.6a**). Like the medulla, the pons consists of nuclei and tracts. Signals for voluntary movements from motor areas of the cerebral cortex are relayed through several **pontine nuclei** (PON-tīn) into the cerebellum. Also within the pons is the **pontine respiratory center**, which functions together with the medullary respiratory center to control breathing.

The **midbrain** or **mesencephalon** (mes′-en-SEF-a-lon) is the upper portion of the brain stem (**FIGURE 8.6a**). Like the medulla and the pons, the midbrain is comprised of nuclei and tracts. Nuclei of the midbrain include the **substantia nigra** (sub-STAN-shē-a = substance; NĪ-gra = black), which is darkly pigmented, and the **red nucleus**, which looks reddish due to an iron-containing pigment and a rich blood supply (**FIGURE 8.6c**). The substantia nigra and red nucleus both play a role in controlling body movements. On the back surface of the midbrain are rounded elevations that contain nuclei called the superior colliculi and the inferior colliculi (**FIGURE 8.6b**). The **superior colliculi** (ko-LIK-ū-lī = little hills; singular is *colliculus*) serve as reflex centers for certain visual activities. Through neural circuits that extend from the retina of the eye to the superior colliculi to the extrinsic eye muscles, visual stimuli elicit eye movements for tracking moving images (such as a moving car) and scanning stationary images (as you are doing to read this sentence). The superior colliculi are also responsible for reflexes that turn the head and trunk in the direction of a sudden visual stimulus, such as an insect that rapidly crawls along a surface in your field of view. Below the superior colliculi are the **inferior colliculi**, which are part of the auditory pathway that relays input from the receptors for hearing to the brain. The inferior colliculi also send auditory input to the superior colliculi to promote the *startle reflex*, which turns the head and trunk in the direction of a sudden auditory stimulus, such as a gunshot.

Another important component of the brain stem is the **reticular formation**, a netlike region of interspersed gray and white matter that extends from the upper part of the spinal cord, throughout the brain stem, and into the lower part of the diencephalon (see **FIGURE 8.11**). Neurons within the reticular formation have both ascending (sensory) and descending (motor) functions. Part of the reticular formation, called the *reticular activating system (RAS)*, consists of sensory axons that project to the cerebral cortex. The RAS helps maintain consciousness and is active during awakening from sleep. For example, you awaken to the sound of

FIGURE 8.6 Parts of the brain. The pituitary gland is discussed with the endocrine system in Chapter 13.

The four principal parts of the brain are the brain stem, cerebellum, diencephalon, and cerebrum.

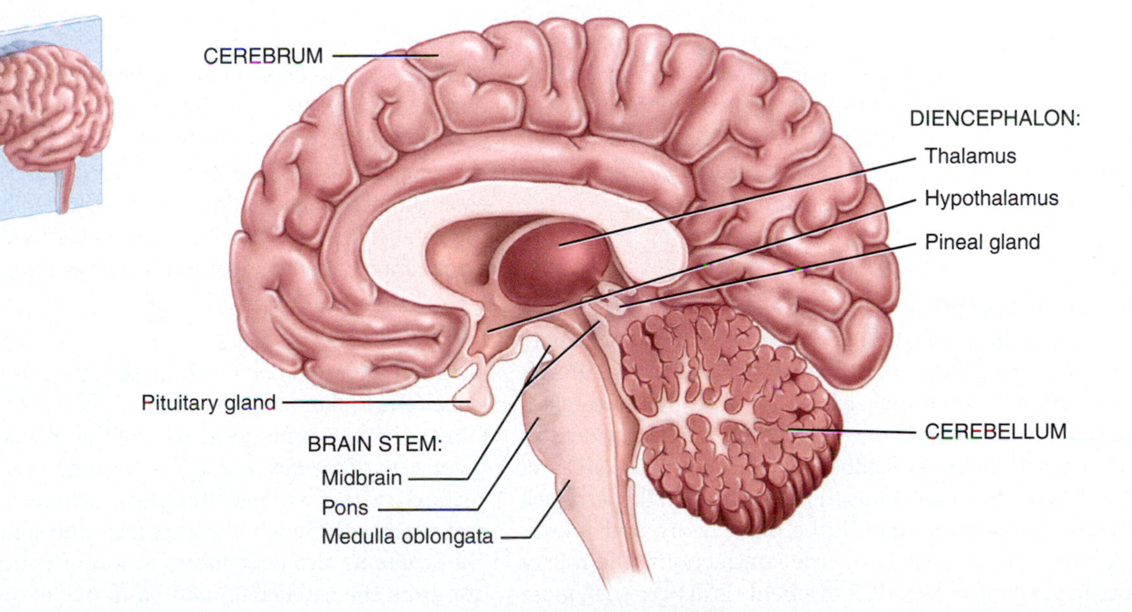

(a) Sagittal section of the brain

Figure 8.6 (continues)

Figure 8.6 Continued

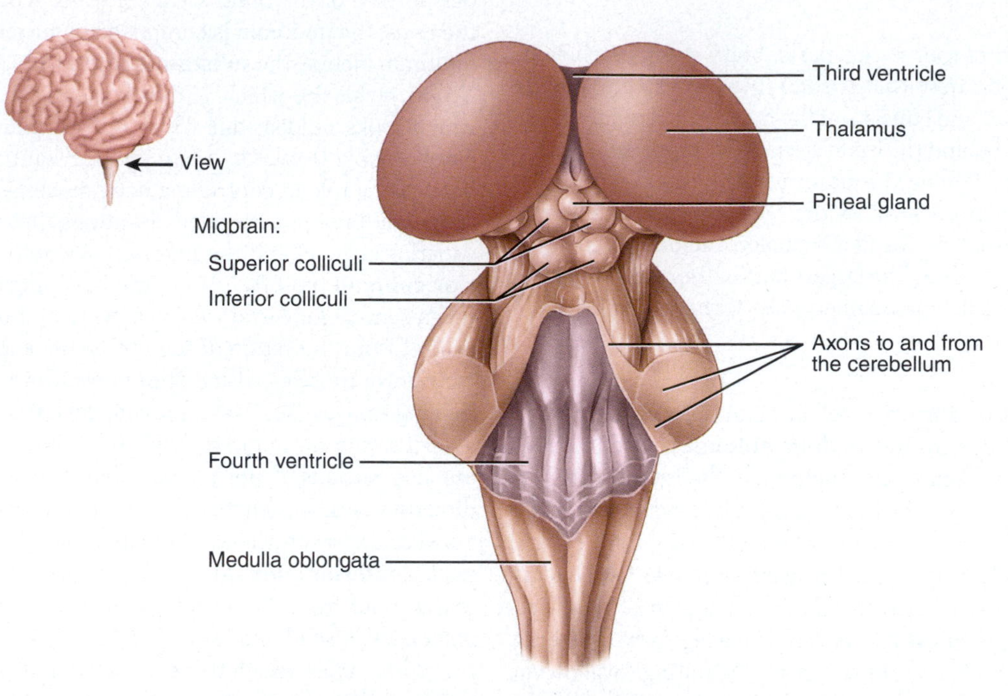

(b) Posterior view of brain stem and diencephalon with cerebellum removed

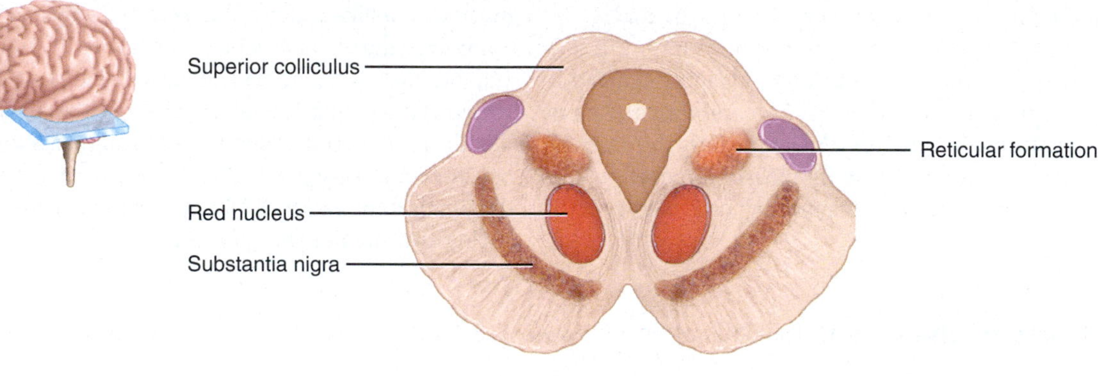

(c) Cross section of midbrain

an alarm clock, a flash of lightning, or a painful pinch because of RAS activity that arouses the cerebral cortex. You will learn more about the RAS later in this chapter (see Section 8.3). The descending functions of the reticular formation are to help regulate posture and *muscle tone*, the slight degree of contraction in normal resting muscles.

Cerebellum

The **cerebellum** is the part of the brain located behind the brain stem (FIGURE 8.6a). The primary function of the cerebellum is to evaluate how well movements initiated by motor areas in the cerebrum are actually being carried out. When movements initiated by the cerebral motor areas are not being carried out correctly, the cerebellum detects the discrepancies. It then sends feedback signals to motor areas of the cerebral cortex via its connections to the thalamus. The feedback signals help correct the errors, smooth the movements, and coordinate complex sequences of skeletal muscle contractions. The role of the cerebellum in the control of body movements is discussed in more detail in Chapter 12. Aside from its role in the coordination of skilled movements, the cerebellum is the main brain region that regulates posture and balance. These aspects of cerebellar function make possible all skilled muscular activities, from catching a baseball to dancing to speaking. The presence of reciprocal connections between the cerebellum and association areas of the cerebral cortex suggests that the cerebellum may also have nonmotor functions such as cognition (acquisition of knowledge) and language processing.

Diencephalon

The **diencephalon** extends from the brain stem to the cerebrum; it includes the thalamus, hypothalamus, and pineal gland. The **thalamus** (THAL-a-mus) is the upper part of the diencephalon (FIGURE 8.6a). It consists of masses of gray matter organized into nuclei with interspersed tracts of white matter. The thalamus relays most sensory input that reaches the cerebral cortex from the spinal cord and brain stem. The thalamus also contributes to motor control by relaying information from the cerebellum and basal nuclei to the motor areas of the cerebral cortex. As sensory or motor information is relayed through

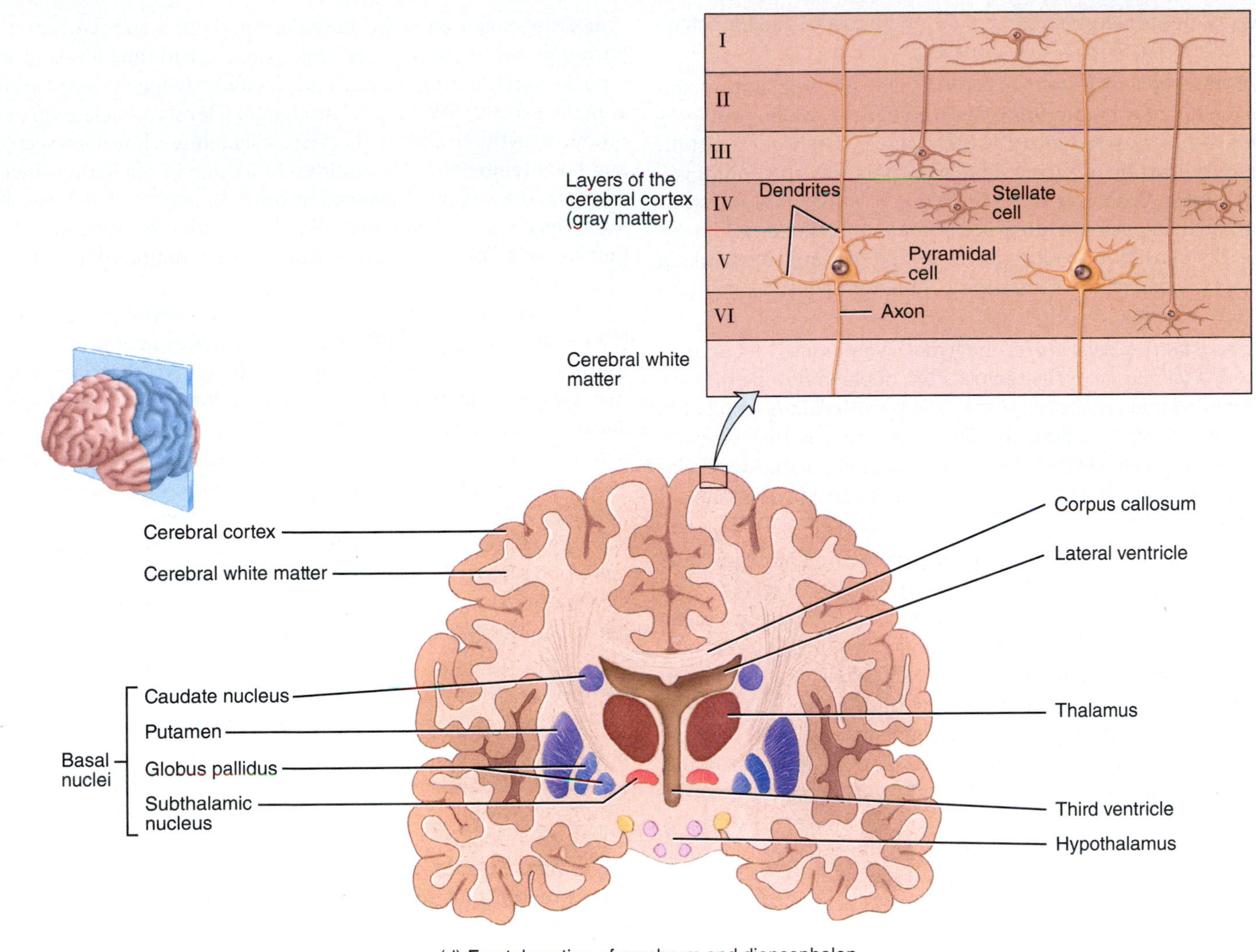

(d) Frontal section of cerebrum and diencephalon

Which component of the diencephalon plays a role in regulating feeding behavior?

the thalamus, neurons in the thalamus can selectively modify or filter the information.

The **hypothalamus** is a small part of the diencephalon located below the thalamus (FIGURE 8.6a). It is composed of a dozen or so nuclei that control many body activities, most of which are related to homeostasis. The hypothalamus receives sensory input from receptors for vision, taste, and smell. Other receptors within the hypothalamus itself continually monitor osmotic pressure, glucose level, certain hormone concentrations, and the temperature of blood. The hypothalamus has several very important connections with the pituitary gland and produces a variety of hormones, which are described in detail in Chapter 13. Some functions of the hypothalamus can be attributed to specific nuclei, but others are not so precisely localized. Important functions of the hypothalamus include the following:

- ***Control of the autonomic nervous system (ANS).*** The hypothalamus controls and integrates activities of the ANS, which regulates contraction of smooth muscle and cardiac muscle and the secretions of many glands. Axons extend from the hypothalamus to parasympathetic and sympathetic nuclei in the brain stem and spinal cord. Through the ANS, the hypothalamus is a major regulator of visceral activities, including regulation of heart rate, movement of food through the gastrointestinal tract, and contraction of the urinary bladder.
- ***Production of hormones.*** The hypothalamus produces several hormones and has two types of important connections with the pituitary gland, an endocrine gland located below the hypothalamus. First, hypothalamic hormones known as *releasing hormones* and *inhibiting hormones* are released into the bloodstream. The bloodstream subsequently carries these hormones directly to the anterior lobe of the pituitary, where they stimulate or inhibit secretion of anterior pituitary hormones (see FIGURE 13.9b). Second, axons extend from the **paraventricular nucleus** and **supraoptic nucleus** of the hypothalamus (see FIGURE 13.14) into the posterior lobe of the pituitary. The cell bodies of both of these nuclei synthesize the hormones *oxytocin* and *antidiuretic hormone (ADH).* Once oxytocin and ADH are synthesized, they are transported down the axons of these nuclei to the posterior pituitary, where they are released.
- ***Regulation of emotional and behavioral patterns.*** Together with the limbic system (described shortly), the hypothalamus participates

in expressions of rage, aggression, pain, and pleasure, and the behavioral patterns related to sexual arousal.

- ***Regulation of eating and drinking.*** The hypothalamus regulates food intake. It contains a **feeding center**, which promotes eating, and a **satiety center**, which causes a sensation of fullness and cessation of eating. The hypothalamus also contains a **thirst center**. When certain cells in the hypothalamus are stimulated by rising osmotic pressure of the extracellular fluid, they cause the sensation of thirst. The intake of water by drinking restores the osmotic pressure to normal, removing the stimulation and relieving the thirst.
- ***Control of body temperature.*** The hypothalamus also functions as the body's *thermostat*. If the temperature of blood flowing through the hypothalamus is above normal, the hypothalamus directs the autonomic nervous system to stimulate activities that promote heat loss. When blood temperature is below normal, the hypothalamus generates action potentials that promote heat production and retention.
- ***Regulation of circadian rhythms.*** The **suprachiasmatic nucleus (SCN)** (soo′-pra-kī-as-MA-tik) of the hypothalamus serves as the body's internal biological clock because it establishes **circadian rhythms**, patterns of biological activity (such as the sleep–wake cycle, secretion of certain hormones, and slight fluctuations in body temperature) that occur on a 24-hour cycle. This nucleus receives input from the eyes (retina) and sends output to other hypothalamic nuclei, the reticular formation, and the pineal gland. The visual input to the SCN entrains (synchronizes) the neurons of the SCN to the light–dark cycle associated with day and night. Without this input, the SCN still promotes biological rhythms, but the rhythms become progressively out of sync with the normal light–dark cycle because the inherent activity of the SCN creates cycles that last about 25 hours instead of 24. Therefore, the SCN must receive light–dark cues from the external environment in order to create rhythms that occur on a 24-hour cycle. The molecular mechanism responsible for the internal clock in an SCN neuron is due to the rhythmic turning on and off of **clock genes** in the cell's nucleus, resulting in alternating levels of **clock proteins** in the cell's cytosol. The clock genes are self-starting: They are turned on automatically and then are transcribed and translated. The resulting clock proteins accumulate in the cytosol and then enter the nucleus to turn off the clock genes. Gradually, the clock proteins degrade, and without these proteins present, the clock genes are activated again and the cycle repeats, with each cycle corresponding to a 24-hour period. The alternating levels of clock proteins cause rhythmic changes in the output of SCN neurons, which in turn causes rhythmic changes in other parts of the body, especially the pineal gland (described next).

The **pineal gland** (PĪN-ē-al) is a pea-sized gland located behind the thalamus (FIGURE 8.6a). The pineal gland is considered part of the endocrine system because it secretes the hormone **melatonin**. Melatonin helps regulate circadian rhythms, which, as you have just learned, are established by the suprachiasmatic nucleus (SCN) of the hypothalamus. In response to visual input from the eyes (retina), the SCN stimulates the pineal gland (via neural connections with sympathetic neurons of the autonomic nervous system) to secrete the hormone melatonin in a rhythmic pattern, with low levels of melatonin secreted during the day and significantly higher levels secreted at night (see FIGURE 13.23). The changing levels of melatonin in turn promote rhythmic changes in sleep, wakefulness, hormone secretion, and body temperature. In addition to its role in regulating circadian rhythms, melatonin is involved in other functions. It induces sleep, serves as an antioxidant, and inhibits reproductive functions in certain animals. You will learn more about the functions of melatonin in Chapter 13.

Parts of the diencephalon, called **circumventricular organs (CVOs)** (ser′-kum-ven-TRIK-ū-lar) because they lie in the wall of the third ventricle, can monitor chemical changes in the blood because they lack a blood–brain barrier. CVOs include part of the hypothalamus, the pineal gland, and the pituitary gland. These regions coordinate homeostatic activities of the nervous and endocrine systems, such as the regulation of blood pressure, fluid balance, hunger, and thirst. CVOs are also thought to be the sites of entry into the brain of the human immunodeficiency virus (HIV), the virus that causes acquired immunodeficiency syndrome (AIDS). Once in the brain, HIV may cause dementia (irreversible deterioration of mental state) and other neurological disorders.

Cerebrum

The cerebrum is the seat of intelligence. It provides us with the ability to read, write, and speak; to make calculations and compose music; and to remember the past, plan for the future, and imagine things that have never existed before. The cerebrum consists of an outer rim of cortex, an internal region of white matter, and nuclei deep within the white matter (FIGURE 8.6d). The cerebral nuclei are regions of gray matter that include some of the components of the basal nuclei and limbic system (described shortly).

Although only 2–4 mm (0.08–0.16 in.) thick, the cerebral cortex contains billions of neurons arranged into six distinct layers (FIGURE 8.6d). Types of cortical neurons include stellate cells and pyramidal cells. **Stellate cells**, also known as *granule cells*, are the main input cells of the cerebral cortex. They receive and process incoming sensory information. Stellate cells are present in cortical layer IV. **Pyramidal cells**, named for their pyramid-shaped cell bodies, are the main output cells of the cerebral cortex. Although the cell bodies of pyramidal cells are present in cortical layer V, their axons extend down the rest of the cerebral cortex, pass through the cerebral white matter, and then enter the brain stem and spinal cord, where they terminate on motor neurons that supply skeletal muscle fibers. As you will soon learn, different regions of the cerebral cortex have different functions. These differences are attributed to variations in the arrangement of the cortical layers and the types of neural connections formed with cortical neurons. For example, regions of cerebral cortex that perceive sensations have a prominent layer IV (stellate cell layer) and a thin layer V (pyramidal cell layer). By contrast, regions of cerebral cortex involved in executing motor activities have a thin layer IV and a prominent layer V.

During embryonic development, when brain size increases rapidly, the gray matter of the cerebral cortex enlarges much faster than the underlying white matter. As a result, the cerebral cortex rolls and folds upon itself so that it can fit into the cranial cavity. The folds are called *gyri* (JĪ-rī; singular is *gyrus*) (**FIGURE 8.7a**). The shallow grooves between the folds are termed *sulci* (SUL-sī; singular is *sulcus*); the deep grooves between the folds are known as *fissures*. The main fissure, the *longitudinal fissure*, separates the cerebrum into right and left halves called **cerebral hemispheres**. The cerebral hemispheres are connected internally by the **corpus callosum** (kal-LŌ-sum), a band of white matter containing axons that extend between the hemispheres (see **FIGURES 8.6d** and **8.10**). The corpus callosum allows the two cerebral hemispheres to communicate with each other.

Each cerebral hemisphere has four major lobes named after the cranial bones that cover them: **frontal lobe**, **parietal lobe**, **temporal lobe**, and **occipital lobe** (**FIGURE 8.7a, b**). The *central sulcus* (SUL-kus) separates the frontal lobe from the parietal lobe, the *lateral cerebral sulcus* separates the frontal lobe from the temporal lobe, and the *parieto-occipital sulcus* separates the parietal lobe from the occipital lobe. A fifth part of the cerebrum, the **insula**, cannot be seen at the surface of the brain because it is located deep in the parietal, frontal, and temporal lobes (**FIGURE 8.7b**).

FIGURE 8.7 Lobes of the cerebral hemispheres. Because the insula cannot be seen externally, it has been projected to the surface in part (b).

Each cerebral hemisphere consists of four major lobes (frontal, parietal, temporal, and occipital) and a deeper area called the insula.

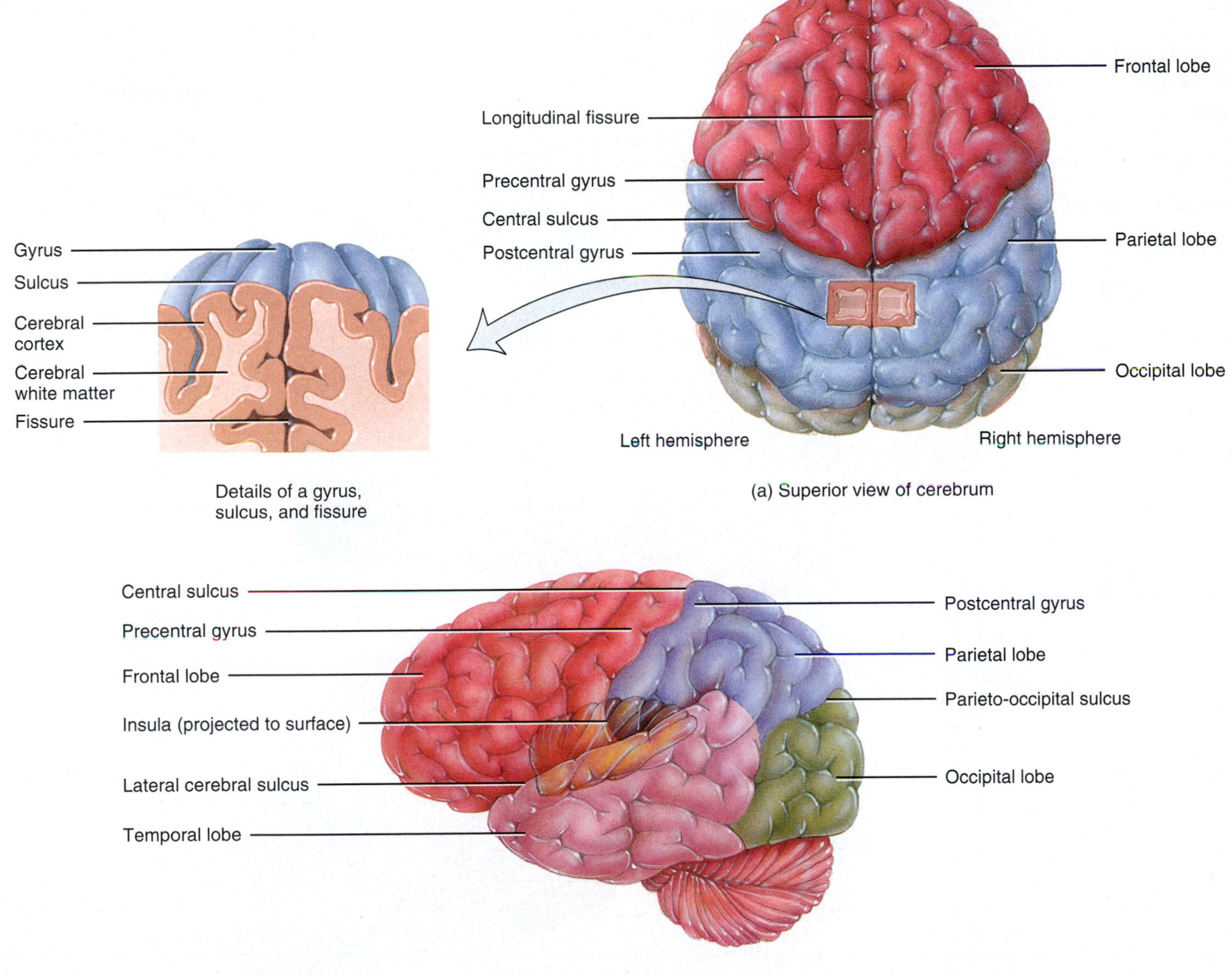

Details of a gyrus, sulcus, and fissure

(a) Superior view of cerebrum

(b) Lateral view of cerebrum

What is the purpose of the corpus callosum?

Specific types of sensory, motor, and integrative signals are processed in certain regions of the cerebral cortex known as **functional areas** (FIGURE 8.8). Generally, **sensory areas** receive sensory information and are involved in **perception**, the conscious awareness of a sensation; **motor areas** control the execution of voluntary movements; and **association areas** deal with more complex integrative functions such as memory, emotions, reasoning, will, judgment, personality traits, and intelligence.

Sensory Areas Sensory areas of the cerebral cortex receive sensory information that has been relayed from peripheral sensory receptors through lower regions of the brain. The following are some important sensory areas (FIGURE 8.8):

- The **primary somatosensory cortex** is located in the postcentral gyrus of the parietal lobe. The *postcentral gyrus* refers to the fold of parietal cortex that is just behind the central sulcus. The primary somatosensory cortex receives sensory information for touch, pressure, vibration, temperature (coldness and warmth), pain, and proprioception (muscle and joint position) and is involved in the perception of these somatic sensations. Somatic sensory input that occurs on one side of the body is conveyed to

FIGURE 8.8 Functional areas of the cerebral cortex.

Particular areas of the cerebral cortex process sensory, motor, or integrative signals.

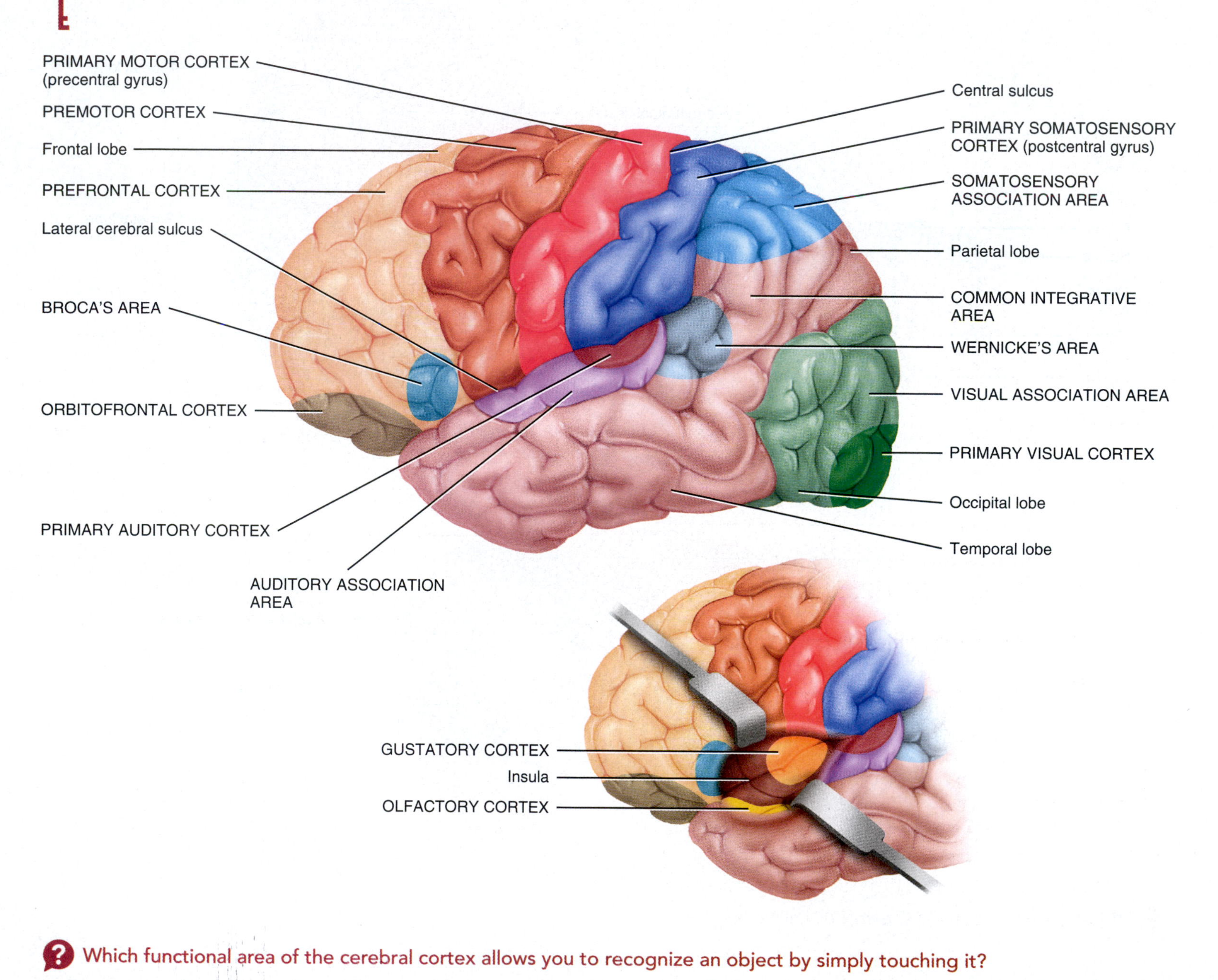

Which functional area of the cerebral cortex allows you to recognize an object by simply touching it?

the primary somatosensory cortex of the cerebral hemisphere on the opposite side. This occurs because ascending somatic sensory pathways cross over to the opposite side of the body before they reach the primary somatosensory cortex. The primary somatosensory cortex allows you to pinpoint where somatic sensations originate so that you know exactly where on your body to swat that mosquito. The primary somatosensory cortex also permits you to determine the size, shape, texture, and weight of an object by feeling it and to sense the relationship of one body part to another.

- The **primary visual cortex**, located in the occipital lobe, receives visual information and is involved in visual perception.
- The **primary auditory cortex**, located in the temporal lobe, receives information for sound and is involved in auditory perception.
- The **gustatory cortex**, located in the insula, receives information for taste and is involved in gustatory perception and taste discrimination.
- The **olfactory cortex**, located on the inner surface of the temporal lobe, receives information for smell and is involved in olfactory perception.

MOTOR AREAS Motor areas of the cerebral cortex provide output that ultimately results in contraction of specific skeletal muscles of the body. Among the most important motor areas are the following (**FIGURE 8.8**):

- The **primary motor cortex** is located in the precentral gyrus of the frontal lobe. The *precentral gyrus* refers to the fold of frontal cortex that is just in front of the central sulcus. Each region in the primary motor cortex controls voluntary contractions of specific muscles or groups of muscles. Electrical stimulation of any point of the primary motor cortex causes contraction of specific skeletal muscle fibers on the opposite side of the body. This occurs because somatic motor tracts originating in the primary motor cortex cross over to the opposite side of the body as they descend through the brain stem and spinal cord.
- **Broca's area** (BRŌ-kaz), located in the frontal lobe, is involved in the articulation of speech. It is active as you translate thoughts into spoken words. In most people, Broca's area is present only in the *left* cerebral hemisphere.

ASSOCIATION AREAS Association areas are often adjacent to primary sensory areas. They usually receive input from both primary sensory areas and other brain regions. Association areas integrate sensory experiences to generate meaningful patterns of recognition and awareness. For example, a person with damage in the *primary* visual cortex would be blind in at least part of his visual field, but a person with damage to the visual *association* area might see normally yet be unable to recognize ordinary objects such as a lamp or a toothbrush just by looking at them.Important association areas include the following (**FIGURE 8.8**):

- The **somatosensory association area**, located in the parietal lobe, receives input from the primary somatosensory cortex. This area stores memories of past somatic sensory experiences, enabling you to compare current sensations with previous experiences. For example, the somatosensory association area allows you to recognize objects such as a pencil and a paperclip simply by touching them.
- The **visual association area**, located in the occipital lobe, receives input from the primary visual cortex. It relates present and past visual experiences and is essential for recognizing and evaluating what is seen. For example, the visual association area allows you to recognize an object such as a spoon simply by looking at it.
- The **facial recognition area**, located in the inferior temporal lobe (not visible in **FIGURE 8.8**), receives input from the visual association area. This area stores information about faces, and it allows you to recognize people by their faces. The facial recognition area in the *right* hemisphere is usually more dominant than the corresponding region in the left hemisphere.
- The **auditory association area**, located in the temporal lobe, allows you to recognize a particular sound as speech, music, or noise.
- The **orbitofrontal cortex**, located in the frontal lobe, receives input from the olfactory cortex. This area allows you to identify odors and to discriminate among different odors. During olfactory processing, the orbitofrontal cortex of the *right* hemisphere exhibits greater activity than the corresponding region in the left hemisphere.
- **Wernicke's area** (VER-ni-kē-z), located in the temporal lobe, interprets the meaning of words. It is active as you translate words into thoughts. In most people, Wernicke's area is present only in the *left* cerebral hemisphere. Recall that this is also the case for Broca's area. The regions in the *right* hemisphere that correspond to Wernicke's and Broca's areas in the left hemisphere contribute to verbal communication by adding emotional content, such as anger or joy, to spoken words.
- The **common integrative area**, located in the parietal lobe, receives input from the somatosensory, visual, and auditory association areas of the cerebral cortex. This area integrates sensory interpretations from the association areas, allowing the formation of thoughts based on a variety of sensory inputs. It then transmits signals to other parts of the brain for the appropriate response to the sensory signals it has interpreted.
- The **prefrontal cortex**, also known as the *frontal association area*, is an extensive area located in the frontal lobe. It is concerned with the makeup of a person's personality, intellect, complex learning abilities, recall of information, initiative, judgment, foresight, reasoning, conscience, intuition, mood, planning for the future, and development of abstract ideas. A person with bilateral damage to the prefrontal cortices typically becomes rude, inconsiderate, incapable of accepting advice, moody, inattentive, less creative, unable to plan for the future, and incapable of anticipating the consequences of rash or reckless words or behavior.
- The **premotor cortex** is a motor association area located in the frontal lobe. It deals with learned motor activities of a complex and sequential nature. By activating the appropriate neurons of the primary motor cortex, the premotor cortex causes specific groups of muscles to contract in a specific sequence, as when you write your name. The premotor cortex also serves as a memory bank for such movements.

Although the two cerebral hemispheres share performance of many functions, each hemisphere also dominates or specializes in performing certain unique functions. This functional asymmetry is termed **hemispheric lateralization**. In the most obvious example of hemispheric lateralization, the left hemisphere receives somatic sensory signals from and controls muscles on the right side of the body, whereas the right hemisphere receives sensory signals from and controls muscles on the left side of the body. In most people, the left hemisphere is more important for reasoning, numerical and scientific skills, spoken and written language, and the ability to use and understand sign language. Conversely, the right hemisphere is more specialized for musical and artistic awareness; spatial and pattern perception; recognition of faces and emotional content of language; discrimination of different smells; and generating mental images of sight, sound, touch, taste, and smell to compare relationships among them. **TABLE 8.2** summarizes some of the functional differences between the two cerebral hemispheres.

Much of what is known about hemispheric lateralization comes from studies of *split-brain patients* who have had the corpus callosum surgically severed to treat severe epilepsy. The corpus callosum is a broad band of white matter consisting of about 300 million axons that interconnect the cerebral hemispheres. When the corpus callosum and other interconnecting pathways are completely cut, the cerebral hemispheres become separated from each other, and the unique functions of the hemispheres are revealed.

TABLE 8.2 Functional Differences Between the Two Cerebral Hemispheres

Right Hemisphere Functions	Left Hemisphere Functions
Receives somatic sensory signals from and controls muscles on the left side of the body.	Receives somatic sensory signals from and controls muscles on the right side of the body.
Musical and artistic awareness.	Reasoning.
Space and pattern perception.	Numerical and scientific skills.
Recognition of faces and emotional content of facial expressions.	Ability to use and understand sign language.
Generating emotional content of language.	Spoken and written language.
Generating mental images to compare spatial relationships.	
Identifying and discriminating among odors.	

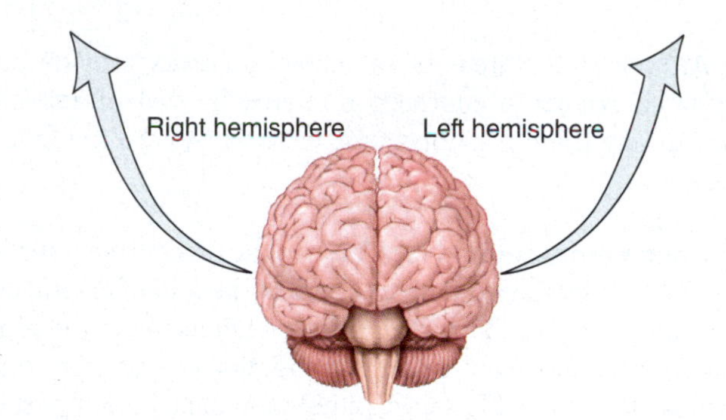

Basal Nuclei

The **basal nuclei**, also referred to as the *basal ganglia*, are several masses of gray matter found in different parts of the brain: (1) the **globus pallidus** (GLŌ-bus PAL-i-dus), **putamen** (pū-TĀ-men), and **caudate nucleus** (KAW-dāt) located deep within the cerebral hemispheres; (2) the **subthalamic nucleus** found just below the thalamus; and (3) the substantia nigra located in the midbrain (**FIGURE 8.9**). The basal nuclei help initiate body movements, suppress unwanted movements,

FIGURE 8.9 The basal nuclei.

The basal nuclei consist of (1) the caudate nucleus, putamen, and globus pallidus in the cerebral hemispheres; (2) the subthalamic nucleus found just below the thalamus; and (3) the substantia nigra located in the midbrain.

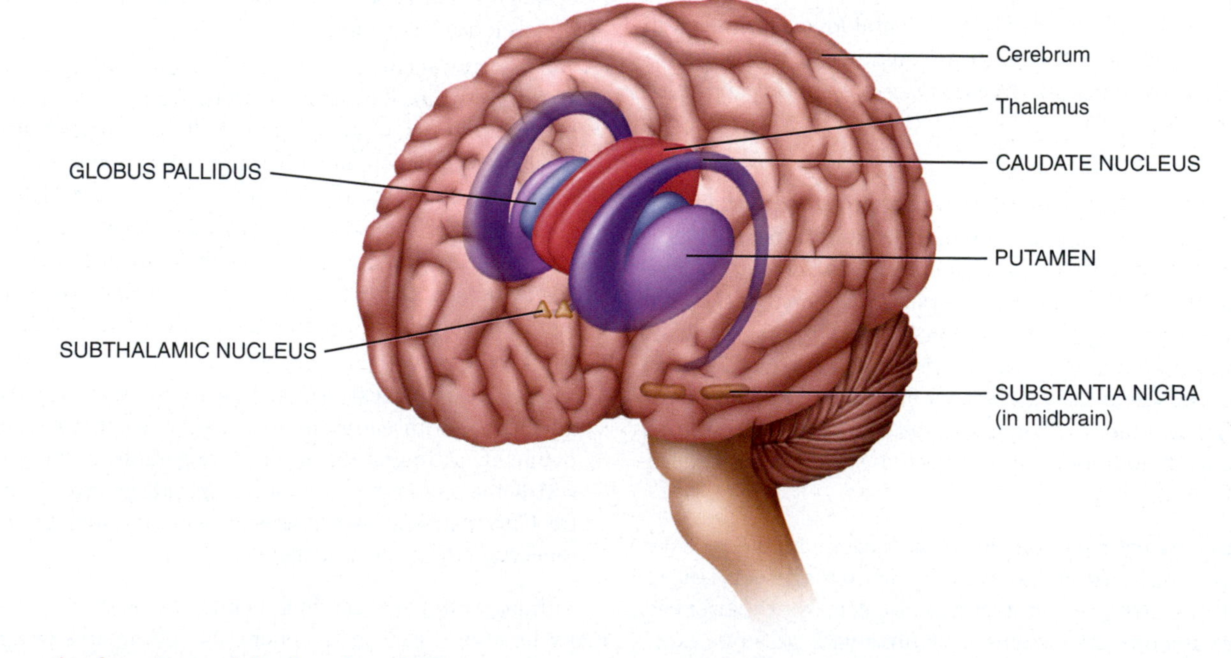

What are the functions of the basal nuclei?

and regulate muscle tone. In addition, they influence several nonmotor aspects of cortical function, including sensory, limbic, cognitive, and linguistic functions. Basal nuclei functions and disorders such as Parkinson's disease and schizophrenia are examined more fully in Chapter 12.

The Limbic System

Encircling the upper part of the brain stem and the corpus callosum is a ring of structures that constitutes the **limbic system**. It includes the **cingulate gyrus** (SIN-gyu-lat), **amygdala** (a-MIG-da-la), **hippocampus**, **dentate gyrus**, and **parahippocampal gyrus** of the cerebrum; portions of the thalamus and hypothalamus; and the olfactory bulbs (FIGURE 8.10). The limbic system plays a primary role in a range of emotions, including pleasure, pain, aggression, docility, affection, fear, and anger. It is also involved in olfaction (smell) and memory. You will learn more about these functions of the limbic system later in this chapter (see Section 8.3).

FIGURE 8.10 Components of the limbic system and surrounding structures. The limbic system components are shaded green.

The limbic system governs emotional aspects of behavior.

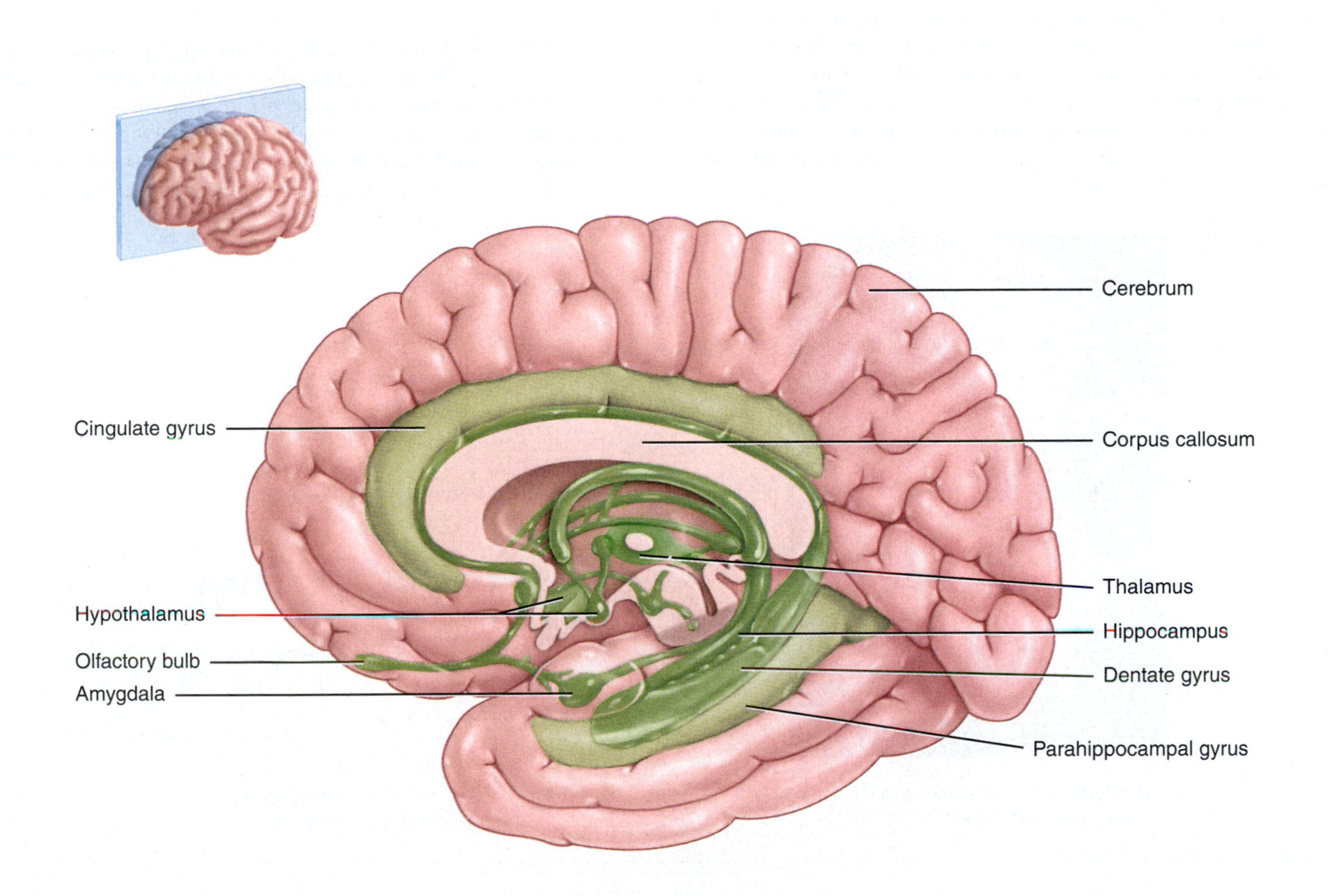

What emotions are regulated by the limbic system?

TOOL OF THE TRADE

ELECTROENCEPHALOGRAPHY

At any instant, the dendrites and cell bodies of brain neurons are generating millions of postsynaptic potentials (EPSPs and IPSPs). Taken together, these electrical signals are called **brain waves**. Brain waves generated by neurons close to the brain surface, mainly neurons in the cerebral cortex, can be detected by sensors called electrodes placed on the forehead and scalp (**FIGURE A**). The process of recording brain waves is called **electroencephalography** (e-lek'-trō-en-sef-a-LOG-ra-fē). A record of such waves is called an **electroencephalogram (EEG)** (e-lek'-trō-en-SEF-a-lō-gram).

Brain wave frequency and amplitude vary with the degree of neural activity in the cerebral cortex. Brain wave patterns have two important features: (1) In general, the more alert a person is, the higher the frequency of the brain wave. For example, the frequency of brain waves is high when a person is awake, but brain wave frequency is low during most stages of sleep. (2) The more synchronized cortical neurons are when generating electrical signals, the larger the amplitude (size) of the brain wave. For example, brain waves have a large amplitude during most stages of sleep because cortical neurons generate electrical activity in a synchronous pattern. However, brain waves have a small amplitude when a person is awake and paying attention to sensory input. This means that the electrical activity generated by cortical neurons is desynchronized because sensory input selectively activates different parts of the cerebral cortex.

Four types of brain waves are typically associated with the EEG: alpha, beta, theta, and delta waves (**FIGURE B**).

- **Alpha waves.** These waves occur at a frequency of 8–13 Hz (1 hertz [Hz] = 1 cycle per second) and have a small amplitude. Alpha waves are present in the EEGs of nearly all normal individuals when they are awake and resting with their eyes closed. These waves disappear entirely during sleep.
- **Beta waves.** These waves occur at a frequency of 14–30 Hz and have the smallest amplitude compared to alpha, theta, and delta waves. Beta waves generally appear when the nervous system is active—that is, during periods of sensory input and mental activity.
- **Theta waves.** These waves occur at a frequency of 4–7 Hz and have an amplitude that is larger than alpha or beta waves. Theta waves normally occur in children. They also occur in adults experiencing emotional stress.
- **Delta waves.** These waves occur at a frequency of 1–5 Hz and have the largest amplitude compared to alpha, beta, and theta waves. Delta waves occur during deep sleep in adults, but they are normal in awake infants. When produced by an awake adult, they indicate brain damage.

Electroencephalograms are useful both in studying normal brain functions, such as changes that occur during sleep, and in diagnosing a variety of brain disorders, such as epilepsy, tumors, trauma, metabolic abnormalities, and degenerative diseases. The EEG is also utilized to determine if brain activity is present. Clinically, the absence of brain waves (a flat EEG) is one of the criteria used to confirm that **brain death** has occurred.

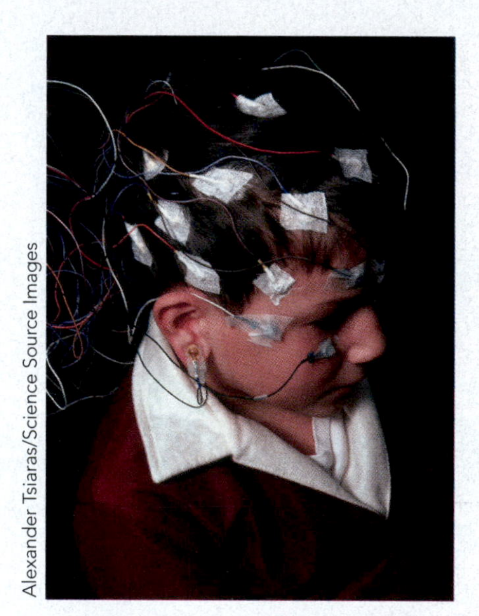

Alexander Tsiaras/Science Source Images

(A) Attachment of electrodes to the scalp during electroencephalography

Alpha
Beta
Theta
Delta
1 sec

(B) Types of brain waves recorded in an electroencephalogram

CHECKPOINT

10. What are the functions of the medulla oblongata? Pons? Midbrain?
11. What is the main function of the cerebellum?
12. Why is the thalamus important?
13. What are the functions of the hypothalamus?
14. Compare the functions of the sensory, motor, and association areas of the cerebral cortex.
15. What is hemispheric lateralization?
16. What are the functions of the basal nuclei?
17. Why is the limbic system functionally significant?

Critical Thinking

Do I Know You?

Recently Paul was in a car accident and experienced severe head trauma. His family was by his side in the hospital every day as he recovered. Each day that his family came to visit, they were shocked to find out that Paul did not recognize any of them. Even after he was released from the hospital, Paul could not recognize his other family members or friends. He could not even recognize his own face when he looked into a mirror. However, his ability to recognize objects such as a cell phone, a fork, or a lamp was still intact. Evidently, the car accident caused damage to the facial recognition area of Paul's brain.

Pixland/Getty Images, Inc.

SOME THINGS TO KEEP IN MIND:

Prosopagnosia refers to the inability to recognize faces even though other aspects of visual recognition remain functional. There are two major forms of prosopagnosia: acquired and congenital. *Acquired prosopagnosia* results from damage to the inferior temporal lobe (usually in the right cerebral hemisphere but sometimes also in the left). This damage is often due to stroke, brain trauma, or certain neurodegenerative diseases. *Congenital prosopagnosia* is present at birth and is usually present during childhood and thereafter. It is not caused by brain damage but rather by genetic factors and can be inherited.

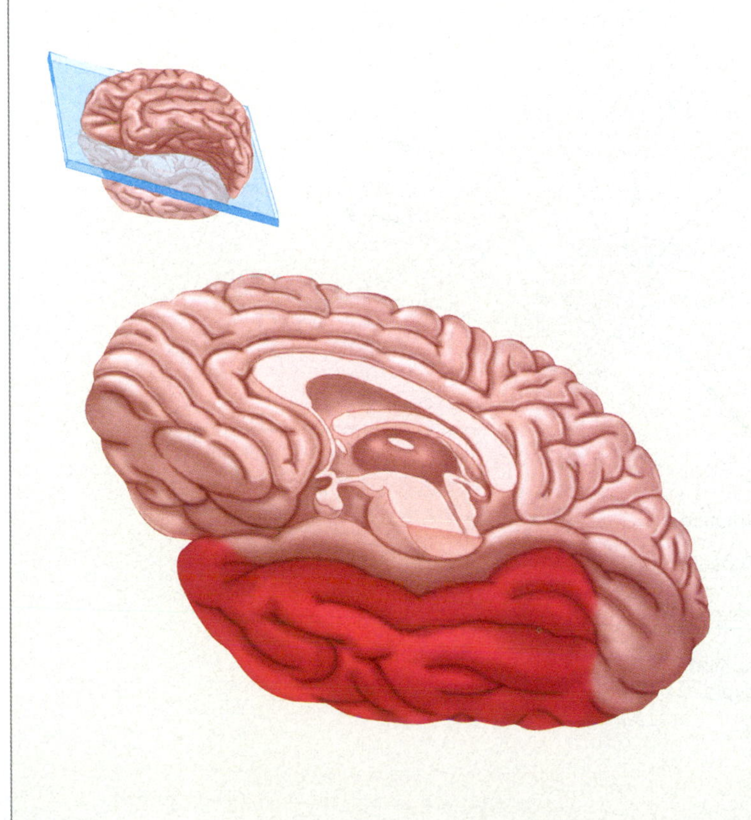

SOME INTERESTING FACTS:

Prosopagnosia dates back to antiquity; the term derives from the Greek words "prosopon" (face) and "agnosia" (without knowledge).

Although once thought to be a rare disorder, recent studies indicate that prosopagnosia may affect about 2% of the general population.

People with prosopagnosia often rely on nonfacial clues such as hair, body shape, voice, gait, jewelry, or clothing to identify people.

Which form of prosopagnosia does Paul have?

Why can Paul still recognize objects such as a cell phone or a lamp but not the faces of the people he knows?

If the brain trauma occurred more in the left temporal lobe than in the right, how would that affect the extent of Paul's prosopagnosia?

What effect would prosopagnosia have on Paul's ability to visually recognize the actors on his favorite television shows? To listen to his favorite musical artists on the radio?

8.3 Integrative Functions of the Cerebrum

OBJECTIVES

- Identify the physiological changes that occur during NREM and REM sleep.
- Outline the neural pathways involved in comprehension and expression of language.
- Describe the generation of emotional behaviors.
- Explain the significance of the mesolimbic dopamine pathway.
- Compare the formation of short-term and long-term memories.

Let's now turn our attention to a fascinating function of the cerebrum: **integration**, the processing of sensory information by analyzing and storing it and making decisions for various responses. The **integrative functions** of the cerebrum include wakefulness and sleep, language, emotions, motivation, and learning and memory.

Wakefulness and Sleep Involve Multiple Areas of the Brain

Humans sleep and awaken in a circadian rhythm (24-hour cycle). A person who is awake is in a state of readiness and is able to react consciously to various stimuli. EEG recordings show that the cerebral cortex is very active during wakefulness and less active during most stages of sleep.

The Role of the Reticular Activating System in Awakening

How does your nervous system make the transition between wakefulness and sleep? Because stimulation of some of its parts increases activity of the cerebral cortex, a portion of the reticular formation is known as the **reticular activating system (RAS)** (FIGURE 8.11). When this area is active, many action potentials are transmitted to widespread areas of the cerebral cortex, both directly and via the thalamus. The effect is a generalized increase in cortical activity.

Arousal, or awakening from sleep, involves increased activity in the RAS. For arousal to occur, the RAS must be stimulated. Many sensory stimuli can activate the RAS: painful stimuli detected by nociceptors, touch and pressure on the skin, movement of the limbs, bright light, the buzz of an alarm clock, or the smell of coffee brewing. This occurs because the RAS receives input from somatic sensory receptors in the skin, muscles, and joints and from sensory receptors in the eyes, ears, and nose (FIGURE 8.11). Once the RAS is activated, the cerebral cortex is also activated, and arousal occurs. The result is a state of wakefulness called **consciousness**.

FIGURE 8.11 Reticular activating system.

The reticular activating system (RAS) consists of neurons whose axons project from the reticular formation through the thalamus to the cerebral cortex.

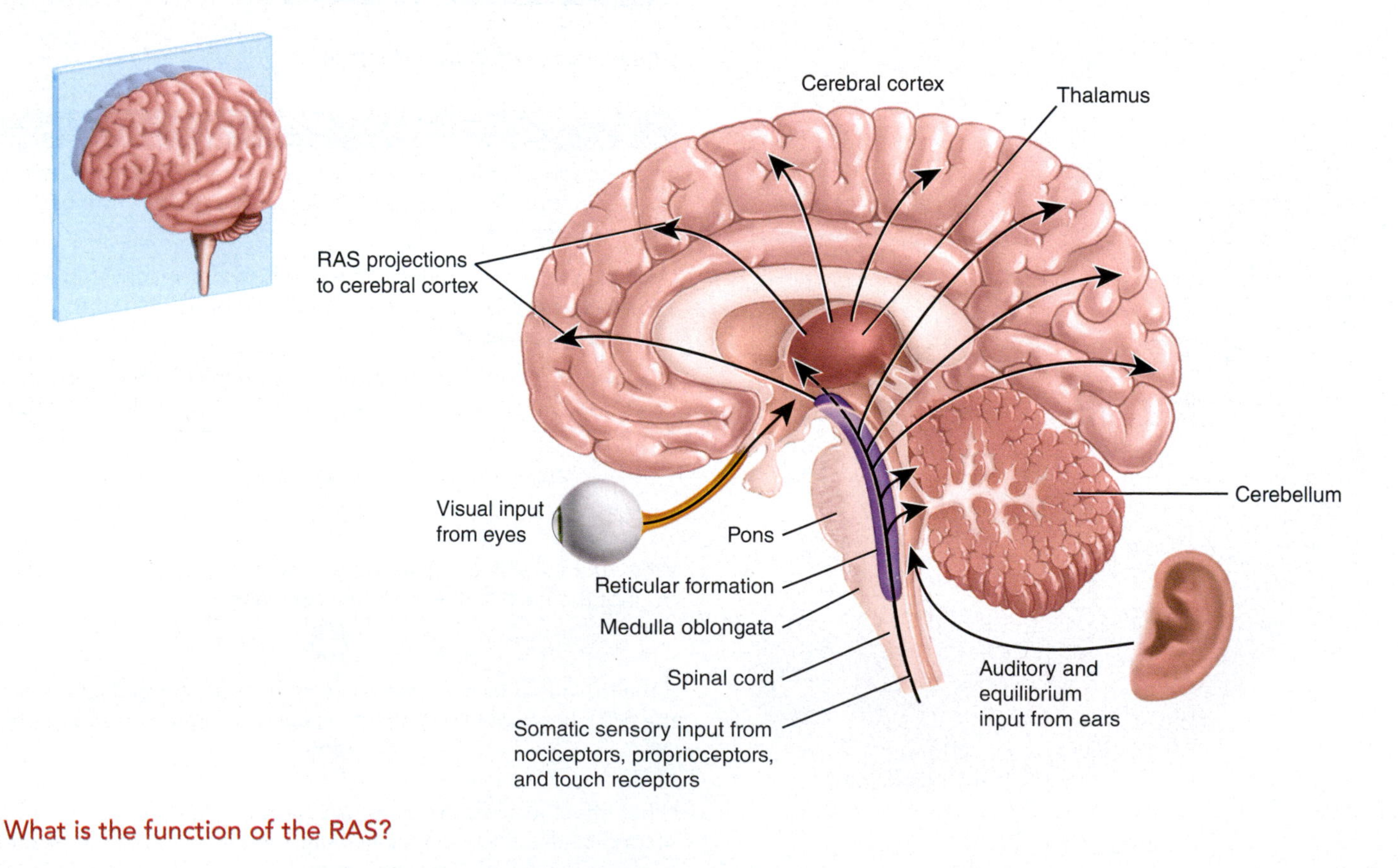

What is the function of the RAS?

Sleep

Sleep is a state of unconsciousness from which an individual can be aroused by stimuli. An adult typically spends about 8 hours a day sleeping. This means that sleep consumes about one-third of our lives. Normal sleep consists of two phases: nonrapid eye movement (NREM) sleep and rapid eye movement (REM) sleep.

NREM sleep consists of four gradually merging stages, each of which is characterized by a different kind of EEG activity (FIGURE 8.12a):

1. *Stage 1* is a transition stage between wakefulness and sleep. The person is relaxed with eyes closed and has fleeting thoughts. On the EEG, alpha waves are replaced by brain waves that have a lower frequency and a slightly larger amplitude. People awakened during this stage often say that they have not been sleeping.
2. *Stage 2* or *light sleep* is the first stage of true sleep. The EEG shows *sleep spindles*—bursts of sharply pointed waves that occur at a frequency of 12–14 Hz and last for periods of 1–2 seconds. It is easy to awaken a person at this time.
3. *Stage 3* is a period of moderately deep sleep. The EEG shows a mixture of sleep spindles and larger, lower-frequency waves. It is a little more difficult to awaken a person during this stage.
4. *Stage 4* or *slow-wave sleep* is the deepest level of sleep. Slow, large-amplitude delta waves dominate the EEG. During this stage of sleep, it is very difficult to awaken a person.

Several physiological changes occur during NREM sleep. There are decreases in heart rate, respiratory rate, and blood pressure. Muscle tone also decreases but only slightly. As a result, there is a moderate amount of muscle tone during NREM sleep, which allows the sleeping person to shift body positions while in bed. Dreaming sometimes takes place during NREM sleep but only occasionally. You will soon learn that most dreaming occurs during REM sleep. When dreaming does occur during NREM sleep, the dreams are usually less vivid, less emotional, and more logical than REM dreams.

During **REM sleep**, the eyes move rapidly back and forth under closed eyelids. REM sleep is also known as *paradoxical sleep* because EEG readings taken during this time show high-frequency, small-amplitude waves, which are similar to those of a person who is awake (FIGURE 8.12a). Surprisingly, neuronal activity is high during REM sleep—brain blood flow and oxygen use are actually higher during REM sleep than during intense mental or physical activity while awake! In spite of this high amount of neuronal activity, it is even more difficult to awaken a person during REM sleep than during any of the stages of NREM sleep.

REM sleep is associated with several physiological changes. For example, heart rate, respiratory rate, and blood pressure increase during REM sleep. In addition, most somatic motor neurons are inhibited during REM sleep, which causes a significant decrease in muscle tone and even paralyzes the skeletal muscles. The main exceptions to this inhibition are those somatic motor neurons that govern breathing and eye movements. REM sleep is also the period when most dreaming occurs. Brain imaging studies on people going through REM sleep reveal that there is increased activity in both the visual association area (which is involved in recognition of visual images) and limbic system (which plays a major role in generation of emotions) and decreased activity in the prefrontal cortex (which is concerned with reasoning). These studies help to explain why dreams during REM sleep are often full of vivid imagery, emotional responses, and situations that may be illogical or even bizarre. Erection of the penis and enlargement of the clitoris may also occur during REM sleep, even when dream content is not sexual. The presence of penile erections during REM sleep in a man with erectile dysfunction (inability to attain an erection while awake) indicates that his problem has a psychological, rather than

FIGURE 8.12 Stages of sleep. (a) EEG recordings during sleep stages. In an awake person whose eyes are closed, alpha waves predominate; slow, high-amplitude delta waves characterize stage 4 sleep. (b) Alternating intervals of NREM and REM sleep during a typical sleep period.

As nonrapid eye movement (NREM) sleep progresses from stage 1 to stage 4, the EEG waves become slower (lower frequency) and higher (larger amplitude).

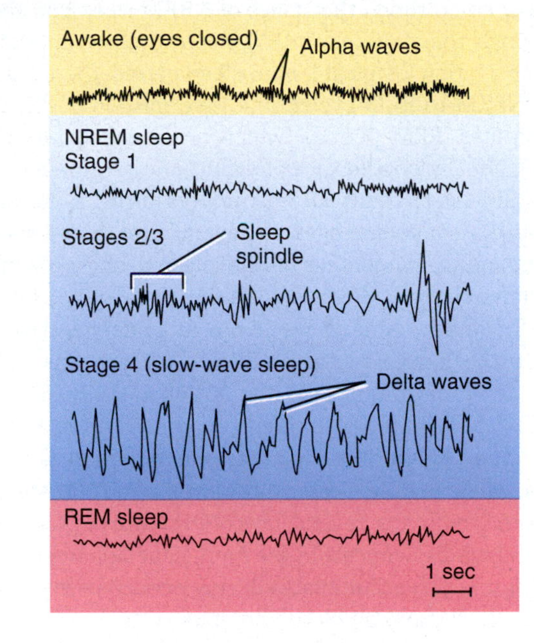

(a) EEG waves during sleep stages

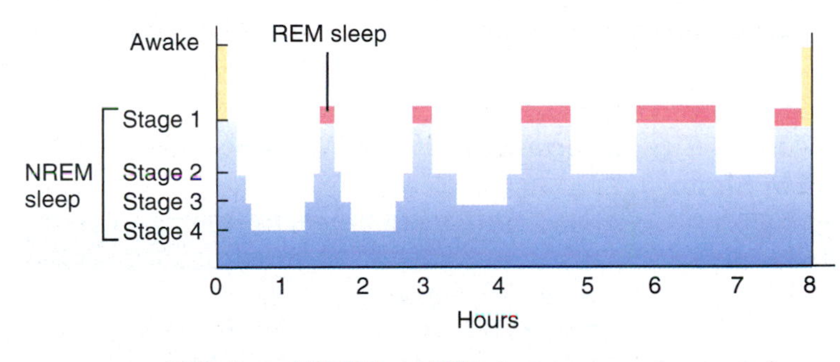

(b) Pattern of NREM and REM sleep over one sleep period

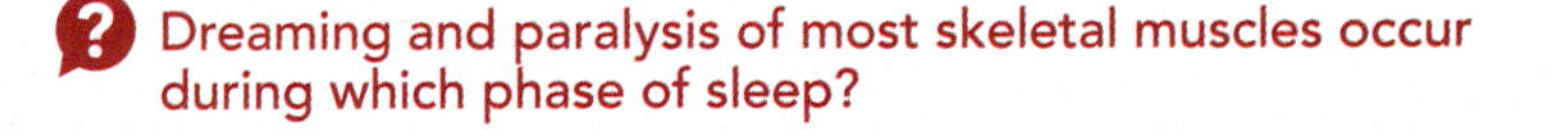
Dreaming and paralysis of most skeletal muscles occur during which phase of sleep?

a physical, cause. TABLE 8.3 presents a summary of the differences between NREM sleep and REM sleep.

Intervals of NREM and REM sleep alternate throughout the night (FIGURE 8.12b). Initially, a person falls asleep by sequentially going through the stages of NREM sleep (from stage 1 to stage 4) in about 45 minutes. Then the person goes through the stages of NREM sleep in reverse order (from stage 4 to stage 1) in about the same amount of time before entering a period of REM sleep. Afterward, the person again descends through the stages of NREM sleep and then ascends back through the stages of NREM sleep to enter another period of REM sleep. During a typical 8-hour sleep period, there are four or five of these NREM-to-REM cycles. The first episode of REM sleep lasts 10–20 minutes. REM periods, which occur approximately every 90 minutes, gradually lengthen, with the final one lasting about 50 minutes. In adults, REM sleep totals 90–120 minutes during a typical eight-hour sleep period. As a person ages, the average total time spent sleeping decreases, and the percentage of REM sleep declines. As much as 50% of an infant's sleep is REM sleep, as opposed to 35% for 2-year-olds and 25% for adults. Although we do not yet understand the function of REM sleep, the high percentage of REM sleep in infants and children is thought to be important for the maturation of the brain.

Different parts of the brain mediate NREM and REM sleep. NREM sleep is induced by **NREM sleep centers** in the hypothalamus and basal forebrain, whereas REM sleep is promoted by a **REM sleep center** in the pons and midbrain. Several lines of evidence suggest the existence of sleep-inducing chemicals in the brain. One apparent sleep-inducer is adenosine, which accumulates during periods of high usage of ATP (adenosine triphosphate) by the nervous system. Adenosine inhibits neurons of the RAS that participate in arousal. Adenosine achieves this function by binding to specific purinergic receptors, called A1 receptors in the membranes of these neurons (see Section 7.5 for a discussion of purinergic receptors). Thus, activity in the RAS during sleep is low due to the inhibitory effect of adenosine. Caffeine (in coffee) and theophylline (in tea)—substances known for their ability to maintain wakefulness—bind to and block the A1 receptors, preventing adenosine from binding and inducing sleep.

Sleep is essential to the normal functioning of the body. Studies have shown that sleep deprivation impairs attention, memory, performance, and immunity; if the lack of sleep lasts long enough, it can lead to mood swings, hallucinations, and even death. Although it is essential, the exact functions of sleep are still unclear. There has been considerable debate in the scientific community about the importance of sleep, but some proposed functions of sleep are widely accepted: (1) restoration, providing time for the body to repair itself; (2) consolidation of memories; (3) enhancement of immune system function; and (4) maturation of the brain.

TABLE 8.3 Comparison of NREM Sleep and REM Sleep

Characteristic	NREM Sleep	REM Sleep
EEG	Shows low-frequency, high-amplitude waves.	Shows high-frequency, small-amplitude waves that are similar to those of an awake person.
Muscle tone	Moderate amount, which allows shifts in body position during sleep.	Trivial amount, which results in no movement (skeletal muscle paralysis). Exceptions include muscles that control breathing and eye movements.
Heart Rate, Respiratory Rate, and Blood Pressure	Decreased.	Increased.
Dreaming	Occurs occasionally.	Occurs frequently.

Coma

Recall that sleep is state of unconsciousness from which an individual can be aroused by stimuli. By contrast, a **coma** is a state of unconsciousness in which an individual has little or no response to stimuli. Causes of coma include head injuries, damage to the reticular activating system (RAS), brain infections, alcohol intoxication, and drug overdoses. If brain damage is minor or reversible, a person may come out of a coma and recover fully; if brain damage is severe and irreversible, recovery is unlikely.

After a few weeks of being in a coma, some patients enter into a **persistent vegetative state** in which the patient has normal sleep–wake cycles but does not have an awareness of the surroundings. Individuals in this state are unable to speak or to respond to commands. They may smile, laugh, or cry, but do not understand the meaning of these actions.

It is important to point out that people who are in a coma or a persistent vegetative state are not brain dead because their EEGs still exhibit waveform activity. As you learned earlier in this chapter, one of the criteria used to confirm that brain death has occurred is the absence of brain waves (flat EEG).

CLINICAL CONNECTION

Sleep Disorders

Sleep disorders affect over 70 million Americans each year. Common sleep disorders include insomnia, sleep apnea, and narcolepsy. A person with **insomnia** (in-SOM-nē-a) has difficulty in falling asleep or staying asleep. Possible causes of insomnia include stress, excessive caffeine intake, disruption of circadian rhythms (for example, working the night shift instead of the day shift at a job), and depression. **Sleep apnea** (AP-nē-a) is a disorder in which a person repeatedly stops breathing for 10 or more seconds while sleeping. Most often, it occurs because a loss of muscle tone in respiratory muscles allows the airway to collapse. **Narcolepsy** (NAR-kō-lep-sē) is a condition in which REM sleep cannot be inhibited during waking periods. As a result, involuntary periods of sleep that last about 15 minutes occur throughout the day. Recent studies have revealed that people with narcolepsy have a deficiency of the neuropeptide ***orexin***, also known as ***hypocretin***. Orexin is released from certain neurons of the hypothalamus and has a role in promoting wakefulness.

Language Is an Important Form of Communication

Animals as diverse as ants, birds, whales, and humans have developed ways to communicate with members of their own species. Humans use language to communicate with one another. **Language** is a system of vocal sounds and symbols that conveys information. Most commonly it is spoken and/or written.

The cerebral cortex contains two **language areas**—Wernicke's area and Broca's area, which are usually present only in the *left* cerebral hemisphere (see FIGURE 8.8). *Wernicke's area*, an association area found in the temporal lobe, interprets the meaning of written or spoken words. It essentially translates words into thoughts. Wernicke's area receives input from the primary visual cortex (for written words) and from the primary auditory cortex (for spoken words). *Broca's area*, a motor area located in the frontal lobe, is active as you translate thoughts into speech. To accomplish this function, Broca's area receives input from Wernicke's area and then generates a motor pattern for activation of muscles needed for the words that you want to say. The motor pattern is transmitted from Broca's area to the primary motor cortex, which in turn activates the appropriate speech muscles. The contractions of your speech muscles enable you to speak your thoughts.

To further understand how the language areas function, consider the neural pathways that are used when you see or hear a particular word and then say that word (FIGURE 8.13):

1. Information about the word is conveyed to Wernicke's area. If the word is written, Wernicke's area receives input about the word from the primary visual cortex. If the word is spoken, Wernicke's area receives input about the word from the primary auditory cortex.
2. Once Wernicke's area receives this information, it translates the written or spoken word into the appropriate thought.
3. For a person to say this word, Wernicke's area transmits information about the word to Broca's area.
4. Broca's area receives this input and then develops a motor pattern for activation of the muscles needed to say the word.
5. The motor pattern is conveyed from Broca's area to the primary motor cortex, which subsequently activates the appropriate muscles of speech. Contraction of the speech muscles allows the word to be spoken.

FIGURE 8.13 Neural pathways used for saying a written or spoken word.

The cerebral cortex contains two language areas: (1) Wernicke's area, which translates words into thoughts, and (2) Broca's area, which translates thoughts into speech.

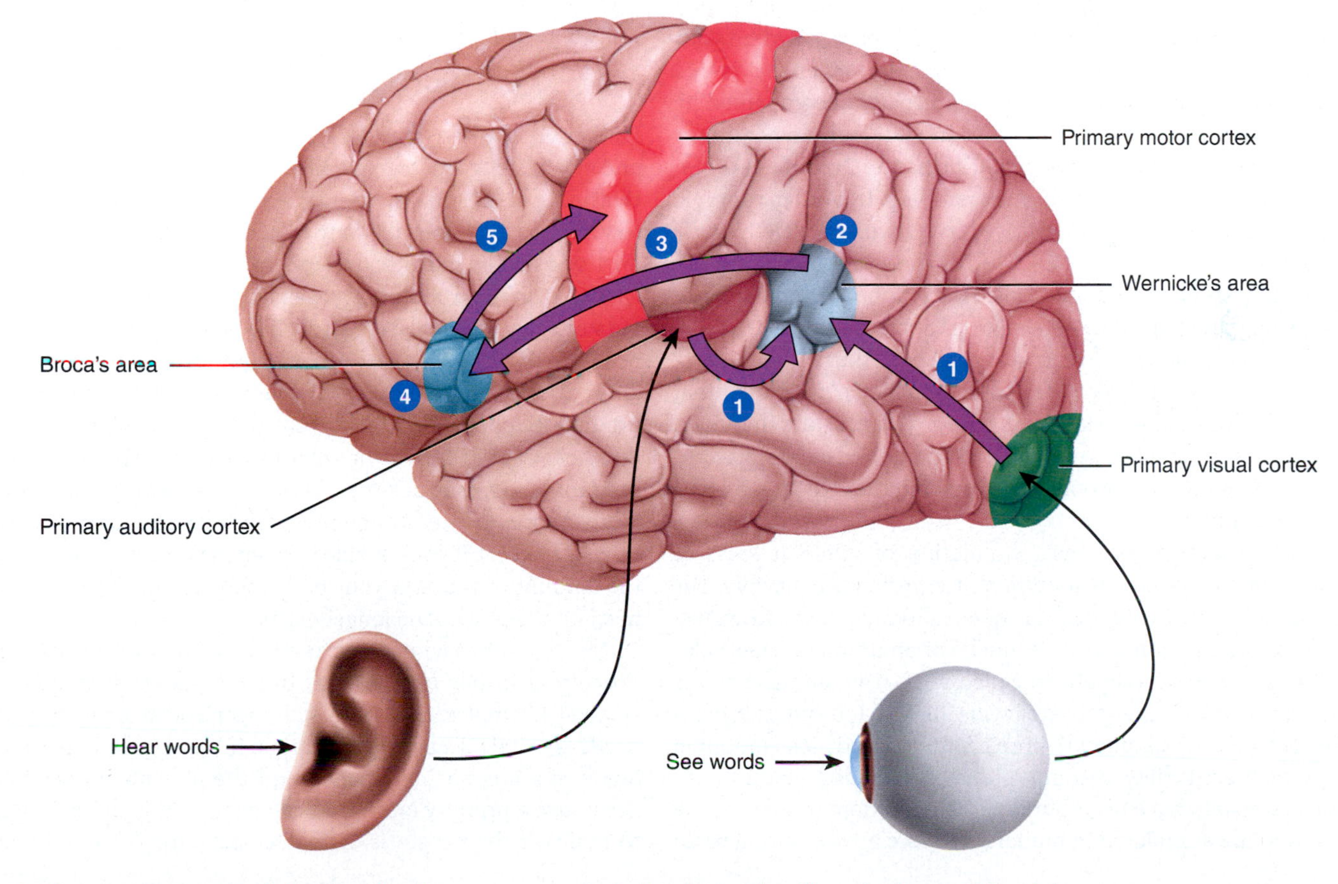

Are the language areas of the cerebral cortex present in both cerebral hemispheres? Explain your answer.

CLINICAL CONNECTION

Aphasia

Much of what we know about language areas comes from studies of patients with language or speech disturbances that have resulted from brain damage. Injury to language areas of the cerebral cortex results in **aphasia** (a-FĀ-zē-a), an inability to use or comprehend words. Damage to Broca's area results in **expressive aphasia**, an inability to properly articulate or form words. People with this type of aphasia know what they wish to say but have difficulty speaking. Damage to Wernicke's area results in **receptive aphasia**, characterized by faulty understanding of spoken or written words. A person experiencing this type of aphasia may fluently produce strings of words that have no meaning ("word salad"). For example, someone with receptive aphasia might say, "I car river dinner light rang pencil jog." The underlying deficit may be **word deafness** (an inability to understand spoken words), **word blindness** (an inability to understand written words), or both.

Emotions Can Trigger Autonomic and Somatic Motor Responses

Emotions are subjective feelings that you have about your own sense of well-being when a particular situation arises. For example, you might experience fear if you have to stand up in front of a large audience and give a speech. Or you might feel angry when your computer crashes as you are trying to complete a paper that is due later that same day in class. Emotions are usually accompanied by a variety of physical responses. Examples include autonomic motor responses (changes in heart rate, blood pressure, breathing, digestion, etc.) and somatic motor responses (smiling, frowning, running, jumping, or any other type of movement that involves skeletal muscle contraction).

The **limbic system** is sometimes called the emotional brain because it plays a primary role in a range of emotions, including pleasure, pain, aggression, docility, affection, fear, and anger. Recall that the limbic system is a ring of structures that encircles the upper part of the brain stem and the corpus callosum (see **FIGURE 8.10**). It includes the amygdala, hippocampus, cingulate gyrus, dentate gyrus, and parahippocampal gyrus of the cerebrum; portions of the thalamus and hypothalamus; and the olfactory bulbs.

Experiments have shown that stimulation of different areas of an animal's limbic system elicits different emotional behaviors. For example, stimulation of an animal's amygdala produces fear and aggression, whereas stimulation of certain nuclei of an animal's hypothalamus produces a behavioral pattern called rage—the animal extends its claws, raises its tail, opens its eyes wide, hisses, and spits. Stimulation of other areas of an animal's limbic system produces emotional behaviors such as docility, extreme pleasure, affection, and intense pain. Similar results have been obtained when different areas of the limbic system are stimulated in humans who are awake during brain surgery. For example, patients under these conditions report experiencing fear when the amygdala is stimulated and pleasure when other areas are stimulated.

The components of the limbic system have neural connections with other parts of the brain. These connections are used to generate emotional behavior in the following way (**FIGURE 8.14**):

1. Sensory stimuli are conveyed to the cerebral cortex. This sensory input provides the cerebral cortex with information about a particular situation that may affect your sense of well-being.
2. The cerebral cortex conveys the information to the limbic system.
3. The limbic system then forms an emotion about the situation.
4. Information about the emotion is sent from the limbic system to the cerebral cortex, which allows the emotion to be perceived.
5. Information about the emotion is also sent to the hypothalamus.
6. The hypothalamus produces the involuntary autonomic and somatic motor responses that accompany the emotion. It accomplishes this function by activating specific autonomic motor neurons and somatic motor neurons in the brain stem and spinal cord through relays in the reticular formation.
7. Any voluntary somatic motor responses that occur during the emotional behavior involve relatively direct connections between the cerebral cortex and somatic motor neurons in the brain stem and spinal cord.

Motivation Is Promoted by the Mesolimbic Dopamine Pathway

Finding food when you are hungry, studying for a physiology exam, and establishing a career are all examples of behaviors related to **motivation**, the internal processes that direct an individual's behavior toward a particular goal. Many motivated behaviors start with a *need*, anything that an individual deems necessary to have. The need in turn causes a *drive*—a motivational state that involves the creation of a plan of action to achieve a goal. The goal, of course, is to satisfy the need. Once the goal is reached, the motivated behavior usually ends. To understand how motivated behaviors work, consider the following example. Suppose that you become dehydrated while running outside on a hot summer day. As the volume of water in your body decreases, a need for fluid intake develops. This need causes a thirst drive, meaning that you become thirsty and then develop a plan to find something to drink (the goal). If you are able to quench your thirst (reach your goal), the amount of water in your body increases back to normal and the need for fluid intake no longer exists.

Motivated behaviors are classified as either primary or secondary. *Primary motivated behaviors* are based on needs that are required for survival. Examples include thirst, hunger, and avoidance of pain. These needs are vital because they maintain homeostasis. If they are not met, homeostasis is not maintained and the individual does not survive. Sex is also a primary motivated behavior. Although sex is not required to maintain homeostasis, it is necessary for the continuation, and

FIGURE 8.14 **Pathways involved in emotional behavior.**

The limbic system generates emotional behavior through neural connections with the cerebral cortex, hypothalamus, brain stem, and spinal cord.

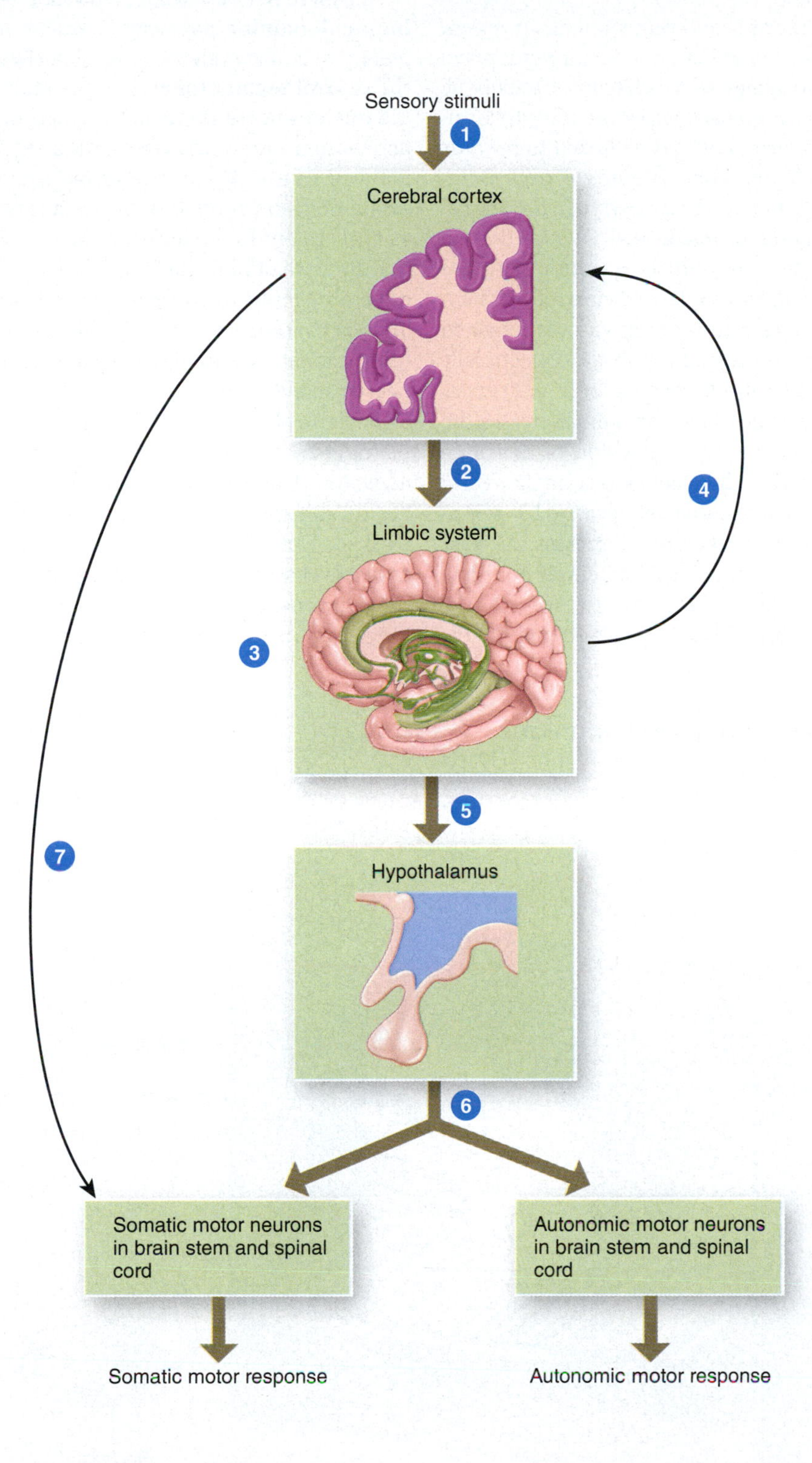

Which part of the brain is responsible for the involuntary somatic and autonomic responses that accompany an emotion?

therefore survival, of the human species. Primary motivated behaviors are innate and are found in many other animals besides humans. *Secondary motivated behaviors* are based on needs that are not required for survival. Examples include achievement, power, and success. Many secondary motivated behaviors are learned and are often influenced by emotions, experience, and habit.

Motivation is associated with rewards and punishments. A *reward* is an event that increases the frequency of a behavior; a *punishment* is an event that decreases the frequency of a behavior. A technique known as *brain self-stimulation* has been used experimentally to localize areas of the brain that are involved in behaviors linked to rewards and punishments (**FIGURE 8.15**). In this technique, an electrode is implanted in a particular region of an animal's brain and then the animal is placed into a box that has a lever on the inside wall. Outside the box is an electronic stimulator that has an input connection with the lever and an output connection with the implanted electrode. Whenever the animal presses the lever, the electronic stimulator is activated and the area of the brain with the implanted electrode is electrically stimulated. At first the animal randomly presses the lever as it moves around in the box. However, the rate at which the animal presses the lever increases or decreases, depending on the region of the brain in which the electrode is implanted. If the electrode is implanted in certain areas of the brain, the animal experiences pleasurable sensations after pressing the lever. As a result, the animal presses the lever over and over again—in some cases as much as 5000 times per hour. In fact, the animal may even go without food and water for the sake of pressing the lever! Areas of the brain that give rise to pleasurable sensations when stimulated are referred to as **reward centers**. If the electrode is implanted in other regions of the brain, the animal experiences unpleasant sensations after pressing the lever and will avoid pressing it at all costs. Areas of the brain that give rise to unpleasant sensations when stimulated are known as **punishment centers**.

The main neural system involved in reward is known as the **mesolimbic dopamine pathway** (*meso-* = mesencephalon, or midbrain; *limbic-* = limbic system), which consists of dopaminergic neurons in the **ventral tegmental area** of the midbrain that send their axons to the **nucleus accumbens** in the cerebrum, parts of the limbic system, and the prefrontal cortex (**FIGURE 8.16**). Stimulation of neurons in this pathway causes repeated lever pressing during brain self-stimulation experiments on animals or in humans that have implanted electrodes as a treatment for epilepsy or chronic pain. Studies have also shown that the rate of brain self-stimulation decreases in response to dopamine antagonists and increases in response to dopamine agonists. Thus, activation of the mesolimbic dopamine pathway in humans and other animals causes pleasurable sensations that are mediated by the neurotransmitter dopamine.

Because of its role in inducing pleasure, the mesolimbic dopamine pathway is frequently activated when an individual develops a drug addiction. Examples of addictive drugs include cocaine, morphine, heroin, and amphetamines. Although these drugs have different mechanisms of action, they all ultimately increase the amount of dopamine available at synapses in the mesolimbic dopamine pathway. As a result, the individual experiences extreme pleasure (euphoria), which causes the person to want more of any of these drugs.

FIGURE 8.15 Brain self-stimulation technique.

The brain self-stimulation technique is used to localize reward or punishment centers in the brain.

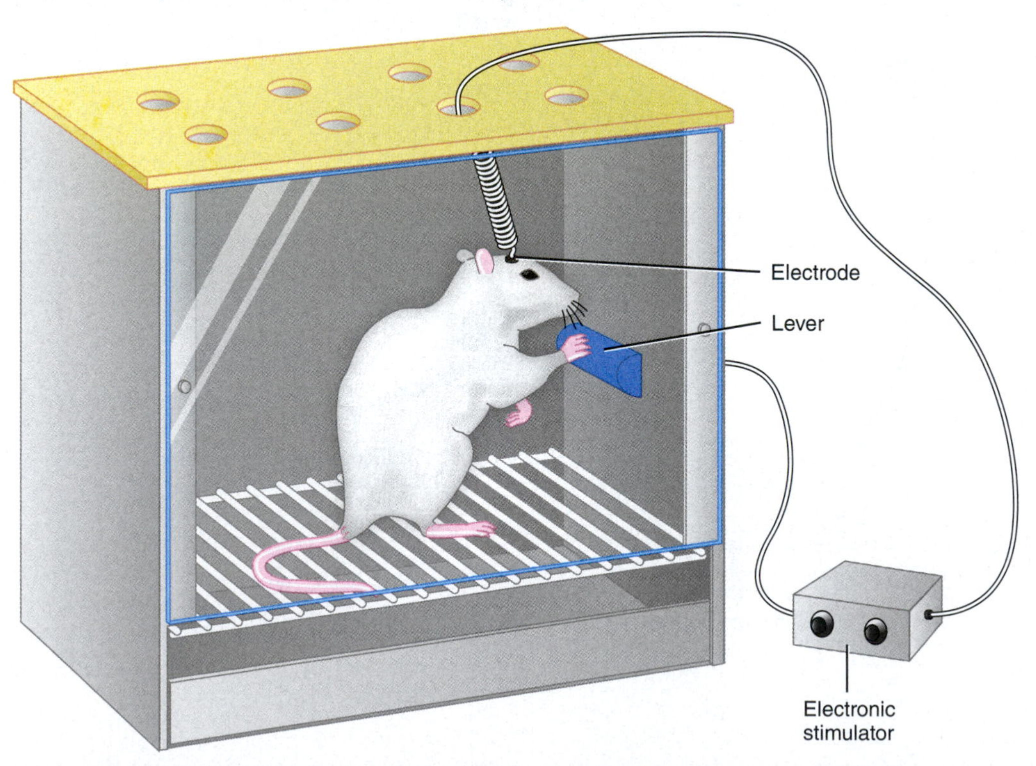

What happens to the frequency of lever pressing by the rat shown in this figure if the electrode is implanted into a punishment center?

FIGURE 8.16 Mesolimbic dopamine pathway.

The mesolimbic dopamine pathway is the main neural system involved in reward.

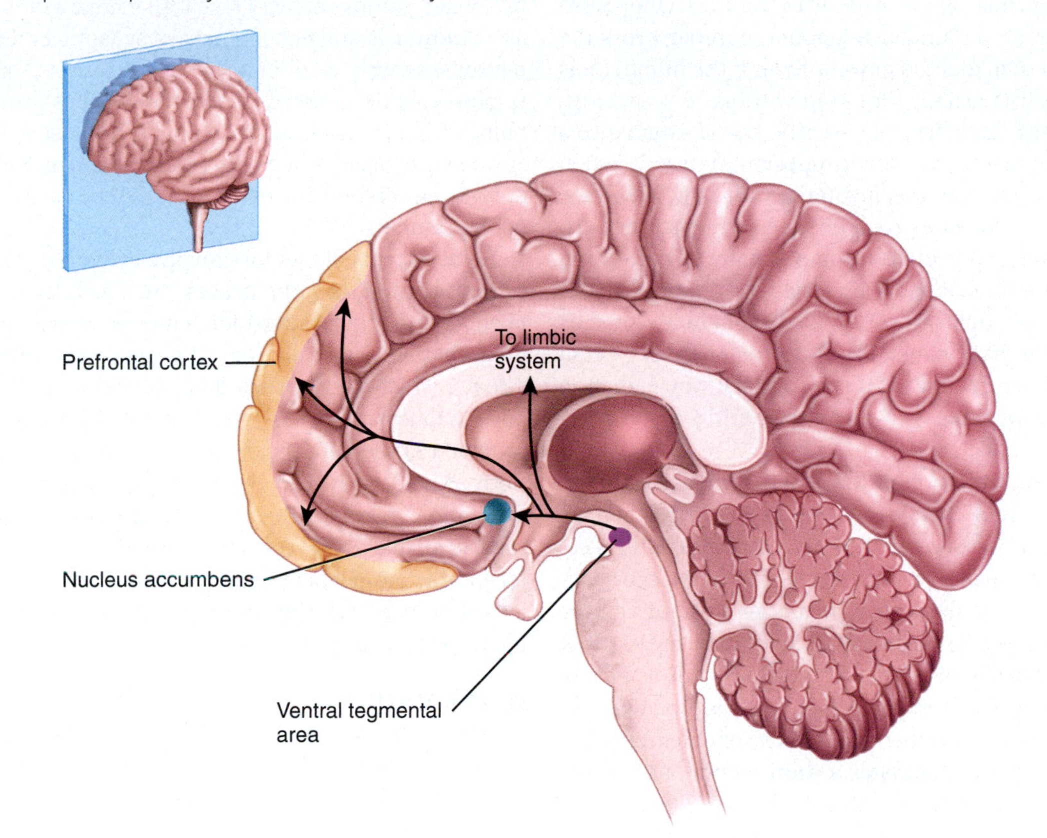

How do addictive drugs such as cocaine, morphine, heroin, and amphetamines affect the mesolimbic dopamine pathway?

Learning and Memory Allow Us to Acquire, Store, and Recall Information

Without learning and memory, you would not be able to discover new things in your surroundings nor would you be able to repeat your successes or accomplishments except by chance. Although both learning and memory have been studied extensively, a completely satisfactory explanation for how we recall information or remember events still does not exist. However, scientists do know something about how information is acquired and stored, and it is clear that there are different categories of memory.

Learning is the ability to acquire new information or skills through instruction or experience. There are two main categories of learning: associative learning and nonassociative learning. **Associative learning** occurs when a connection is made between two stimuli. The Russian physiologist Ivan Pavlov provided a classic example of associative learning when he observed that ringing a bell stimulated the salivation reflex in dogs. When he first began this experiment, Pavlov rang the bell and then provided food for the dogs. The presence of the food caused the dogs to salivate. After repeating this activity several times, Pavlov observed that the dogs would still salivate even if he did not provide them with any food, which indicated that the dogs learned to associate food with the bell ringing. **Nonassociative learning** occurs when repeated exposure to a single stimulus causes a change in behavior. There are two types of nonassociative learning: habituation and sensitization. In **habituation**, repeated exposure to an irrelevant stimulus causes a *decreased* behavioral response. For example, when you first hear a loud sound, it may make you jump. However, if this loud sound occurs over and over again, you may eventually stop paying attention to it. Habituation demonstrates that an animal has learned to ignore an unimportant stimulus. In **sensitization**, repeated exposure to a noxious stimulus causes an *increased* behavioral response. For example, if a limb is damaged repeatedly by a painful stimulus, the flexor (withdrawal) reflex for the affected limb becomes more vigorous. Sensitization demonstrates that an animal has learned to respond more quickly to a harmful stimulus.

Memory is the process by which information acquired through learning is stored and retrieved. There are two main types of memory: declarative memory and procedural memory. **Declarative (explicit) memory** is the memory of experiences that can be verbalized (declared) such as facts, events, objects, names, and places. This type of memory requires conscious recall and is stored in the association areas of the cerebral cortex. For example, visual memories are stored in the visual association area, and auditory memories are stored in the auditory association area. **Procedural (implicit) memory** is the memory of motor skills, procedures, and rules. Examples include riding a bike, serving a tennis ball, and performing the steps of your favorite dance. This type of memory does not require conscious recall, and it is stored in the basal nuclei, cerebellum, and premotor cortex.

Memory, whether declarative or procedural, occurs in stages over a period of time. **Short-term memory** is the temporary ability to recall a few pieces of information for seconds to minutes. One example is when you look up an unfamiliar telephone number, cross the room to the phone, and then dial the new number. If the number has no special significance, it is usually forgotten within a few seconds. Information in short-term memory may later be transformed into a more permanent type of memory, called **long-term memory**, which lasts from days to years. For example, if you use that new telephone number often enough, it becomes part of long-term memory. Although the brain receives many stimuli, you normally pay attention to only a few of them at a time. It has been estimated that only 1% of all of the information that comes to your consciousness is stored as long-term memory. Note that memory does not record every detail as if it were a DVR recorder. Even when details are lost, you can often explain the idea or concept using your own words and ways of viewing things.

Some evidence supports the notion that short-term memory depends more on electrical and chemical events in the brain than on structural changes at synapses. Several conditions that inhibit the electrical activity of the brain, such as anesthesia, coma, and electroconvulsive therapy (ECT), disrupt short-term memories without altering previously established long-term memories. Studies also suggest that short-term memory may involve a temporary increase in the activity of preexisting synapses, especially those that are components of reverberating circuits. Recall that, in a reverberating circuit, one neuron stimulates a second neuron, which stimulates a third neuron, and so on. Branches from later neurons synapse with earlier ones. This arrangement sends action potentials back through the circuit again and again (see Section 7.6).

The process by which a short-term memory is transformed into a long-term memory is called **memory consolidation**. The hippocampus plays a major role in the consolidation of declarative memories. It serves as a temporary storage facility for new long-term declarative memories and then transfers these memories to the appropriate areas of the cerebral cortex for permanent storage. A key factor that contributes to memory consolidation is repetition. Therefore, you remember more information if you review every day for an upcoming physiology exam instead of cramming for the exam the night before!

For an experience to become part of long-term memory, it must produce persistent structural and functional changes that represent the experience in the brain. This capability for change associated with learning is termed **plasticity**. It involves changes in individual neurons as well as changes in the strengths of synaptic connections among neurons. For example, electron micrographs of neurons subjected to prolonged, intense activity reveal an increase in the number of presynaptic terminals and enlargement of synaptic end bulbs in presynaptic neurons, as well as an increase in the number of dendritic branches in postsynaptic neurons. Moreover, neurons grow new synaptic end bulbs with increasing age, presumably because of increased use. Opposite changes occur when neurons are inactive. For example, the visual area of the cerebral cortex of animals that have lost their eyesight becomes thinner.

Some aspects of memory involve a phenomenon called **long-term potentiation (LTP)**, the process by which transmission at synapses is enhanced (potentiated) for hours or weeks after a brief period of high-frequency stimulation. The neurotransmitter released is glutamate, which acts on two different types of glutamate receptors in postsynaptic neurons: AMPA (α-amino-3-hydroxy-5-methyl-4-isoxazole-propionate) receptors and NMDA (N-methyl-D-aspartate) receptors. Both AMPA and NMDA receptors are ionotropic receptors that contain cation channels. The cation channel of the AMPA receptor is permeable mainly to Na^+ ions. However, the cation channel of the NMDA receptor is permeable to both Na^+ and Ca^{2+} ions. LTP has been studied extensively in the hippocampus. A possible mechanism of LTP is as follows (**FIGURE 8.17**):

1. ***Initial activation of AMPA receptors.*** An action potential arriving in the synaptic end bulb causes exocytosis of synaptic vesicles and release of glutamate into the synaptic cleft. Glutamate diffuses across the synaptic cleft and binds to both AMPA and NMDA receptors in the postsynaptic membrane. Binding of glutamate to AMPA receptors opens channels that allow Na^+ ions to enter the cell. This causes a depolarization (EPSP) that raises the membrane potential to just a few millivolts above the resting membrane potential of -70 mV. Unlike the AMPA receptor, the NMDA receptor is not activated at this point. NMDA receptor activation requires glutamate binding *and* a depolarization that raises the membrane potential to about -60 mV. The *slight* depolarization that occurs when the AMPA receptors initially open is not enough to activate NMDA receptors. As long as the membrane potential is below -60 mV, a magnesium (Mg^{2+}) ion blocks the channel of the NMDA receptor, which prevents ions from moving through the channel.
2. ***Subsequent activation of NMDA receptors.*** As Na^+ ions continue to pass through the channels of the AMPA receptors, the postsynaptic membrane becomes more depolarized. Once the membrane potential reaches -60 mV, the Mg^{2+} ion moves out of the channels of the NMDA receptors. The channels of the NMDA receptors are now open, which allows Na^+ and Ca^{2+} ions to enter the cell. As a result, the postsynaptic membrane becomes even more depolarized and the intracellular Na^+ and Ca^{2+} concentrations increase.
3. ***Long-lasting changes in synaptic transmission.*** The increase in intracellular Ca^{2+} concentration in the postsynaptic cell activates second messenger pathways that ultimately cause long-lasting changes in synaptic transmission. These changes include an increase in sensitivity of AMPA receptors to glutamate and the release of a paracrine agent (possibly nitric oxide) that acts on the synaptic end bulb of the presynaptic neuron to release more glutamate.

CLINICAL CONNECTION

Amnesia

Amnesia (am-NĒ-zē-a) refers to the lack or loss of memory. It is a total or partial inability to remember past experiences. In ***anterograde amnesia***, there is memory loss for events that occur *after* the trauma or disease that caused the condition. In other words, it is an inability to form new memories. In ***retrograde amnesia***, there is a memory loss for events that occurred ***before*** the trauma or disease that caused the condition. In other words, it is an inability to recall past events.

FIGURE 8.17. Mechanism of long-term potentiation.

Long-term potentiation (LTP) is the process by which transmission at synapses is enhanced (potentiated) for hours or weeks after a brief period of high-frequency stimulation.

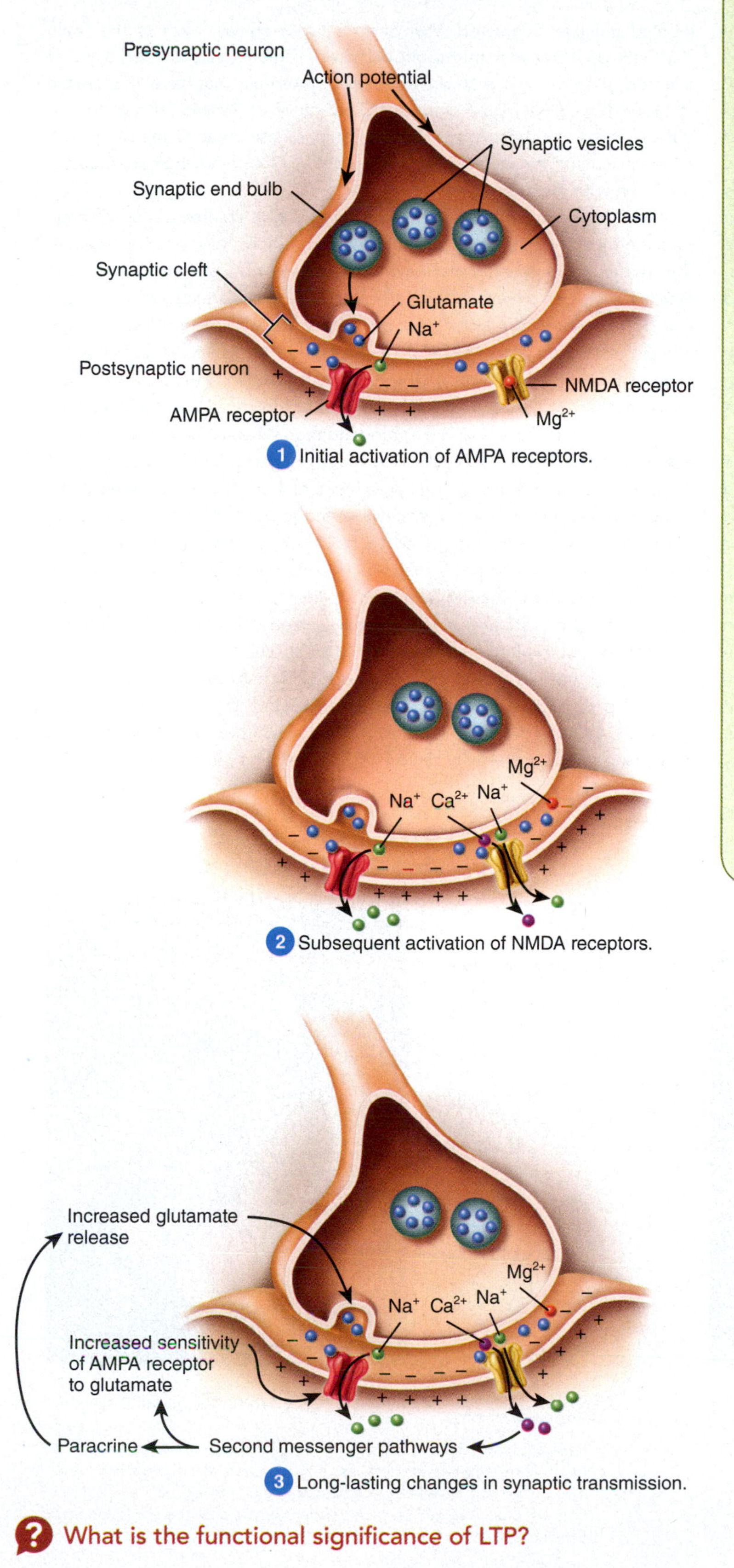

What is the functional significance of LTP?

CLINICAL CONNECTION

Alzheimer's Disease

Severe memory deficits occur in individuals who have **Alzheimer's disease (AD)**. AD is the most common form of ***senile dementia***, the age-related loss of intellectual capabilities (including impairment of memory, judgment, abstract thinking, and changes in personality). The cause of most AD cases is still unknown, but evidence suggests that it is due to a combination of genetic factors, environmental or lifestyle factors, and the aging process. Individuals with AD initially have trouble remembering recent events. They then become confused and forgetful, often repeating questions or getting lost while traveling to familiar places. Disorientation grows; memories of past events disappear; and episodes of paranoia, hallucination, or violent changes in mood may occur. As their minds continue to deteriorate, people with AD lose their ability to read, write, talk, eat, or walk. The disease culminates in dementia. A person with AD usually dies of some complication that afflicts bedridden patients, such as pneumonia. At autopsy, brains of AD victims show four distinct structural abnormalities:

1. ***Loss of neurons that liberate acetylcholine.*** A major center of neurons that liberate acetylcholine is the nucleus basalis, which is below the globus pallidus. Axons of these neurons project widely throughout the cerebral cortex and limbic system. Their destruction is a hallmark of Alzheimer's disease.
2. ***Deterioration of the hippocampus.*** Recall that the hippocampus plays a major role in memory formation.
3. ***Beta-amyloid plaques.*** These are clusters of abnormal proteins deposited outside neurons.
4. ***Neurofibrillary tangles.*** These are abnormal bundles of filaments inside neurons in affected brain regions.

Drugs that inhibit acetylcholinesterase (AChE), the enzyme that inactivates acetylcholine, improve alertness and behavior in some AD patients.

TOOL OF THE TRADE

FUNCTIONAL BRAIN IMAGING

When a person performs a particular sensory, motor, or mental task, any area of the brain involved in the task increases its metabolic activity. Consequently, blood flow to this metabolically active area of the brain increases, whereas blood flow to less metabolically active areas of the brain decreases. **Functional brain imaging** refers to techniques used to create a map of brain function by detecting changes in metabolism or cerebral blood flow that are associated with performing a particular task. These techniques allow the active area to "light up" on an image of the brain shown on a video monitor. Whatever task the patient performs when the area on the brain image lights up is considered to be the function of that area. Examples of functional brain imaging include positron emission tomography and functional magnetic resonance imaging.

Positron emission tomography (PET) creates a map of brain function by detecting radioactively labeled substances that accumulate in brain areas that have an increase in metabolic activity or blood flow when a particular task is performed. In this technique, a radioisotope that emits positrons (positively charged particles) is incorporated into either glucose or a water molecule and then injected into the bloodstream. When an area of the brain is activated by a particular task, glucose consumption and blood flow increase in that area. Consequently, the radioactively labeled glucose or water (the main component of blood) accumulates in the active brain area. As the radioisotope decays, the emitted positrons collide with negatively charged electrons in the surrounding tissue. This collision produces gamma rays (similar to X-rays) that are detected by gamma cameras positioned around the subject. A computer receives signals from the gamma cameras and constructs a *PET scan* image displayed in color on a video monitor. Exactly what the PET scan shows depends on which radioactively labeled substance is injected into the subject. If radioactively labeled glucose is injected, then the PET scan shows areas of the brain that have an increase in metabolic activity. If radioactively labeled water is injected, then the PET scan shows areas of the brain that have an increase in blood flow. **FIGURE A** shows PET scans of an individual performing different tasks. In these scans, dark colors indicate areas of the brain with minimal activity; the yellow, orange, red, and white colors indicate areas of the brain with increasingly greater activity.

Functional magnetic resonance imaging (fMRI) creates a map of brain function by detecting the high ratio of oxyhemoglobin to deoxyhemoglobin that occurs in brain areas that have an increase in blood flow when a particular task is performed. In this technique, the brain is exposed to a high-energy magnetic field, which causes nuclei of atoms within brain tissue to align themselves in relation to the field. Then a pulse of radio waves displaces the aligned nuclei out of position. As the atomic nuclei realign themselves with the magnetic field, they emit a signal that is used to form a color-coded *magnetic resonance image* of the brain on a video monitor. When an area of the brain is activated by a particular task, blood flow and oxygen delivery increase in that area. As a result, the active brain area has a higher ratio of oxyhemoglobin to deoxyhemoglobin. These local changes in the concentration of oxyhemoglobin and deoxyhemoglobin lead to *blood oxygenation level dependent (BOLD)* signals that affect the magnetic resonance image displayed on the monitor. In the fMRI image shown in **FIGURE B**, the bright orange, yellow, and red colors indicate brain areas that have a high ratio of oxyhemoglobin to deoxyhemoglobin when the subject performs a visual task.

An important advantage of fMRI over PET is that fMRI does not require the injection of radioactive substances because it detects endogenous hemoglobin. In addition, fMRI has better temporal resolution than PET, meaning that changes in neural activity can be measured over a shorter period of time.

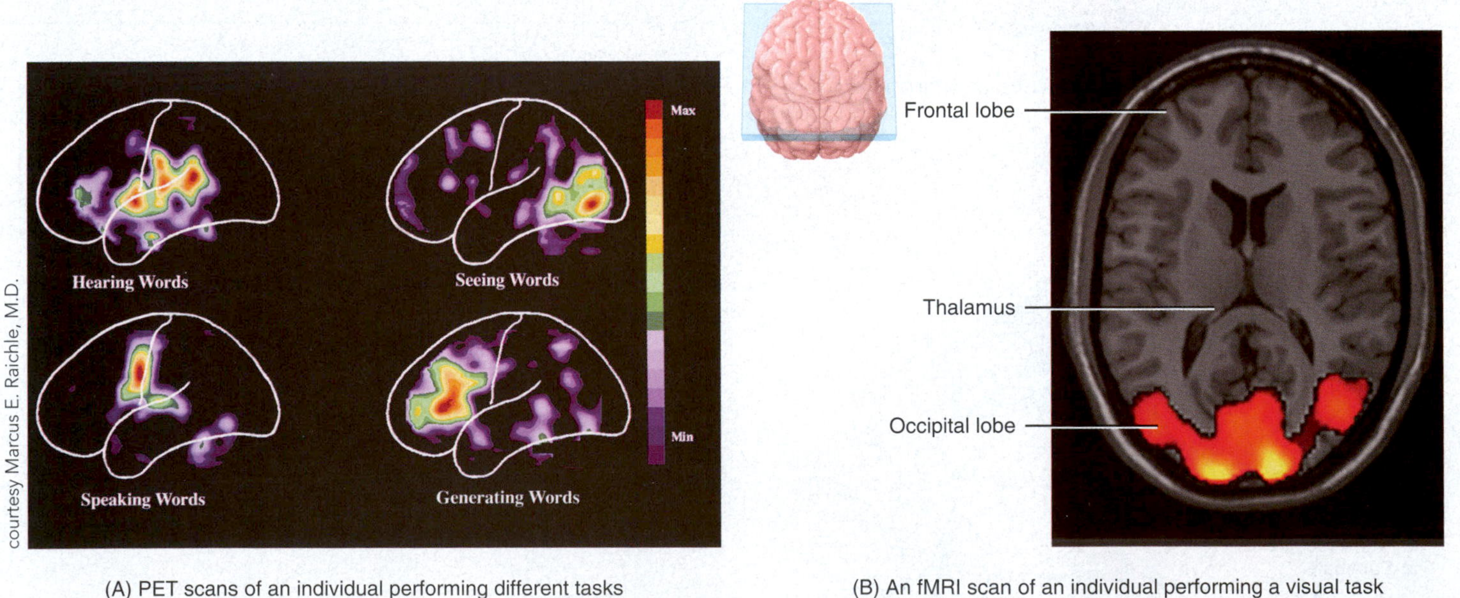

(A) PET scans of an individual performing different tasks

(B) An fMRI scan of an individual performing a visual task

Reprinted from Biol Psychiatry. 2004 Aug 15;56(4):284-91, Willson M.C. et al. *Dextroamphetamine causes a change in regional brain activity in vivo during cognitive tasks: a functional magnetic resonance imaging study of blood oxygen level-dependent response* with permission from Elsevier

CHECKPOINT

18. What is the function of the reticular activating system (RAS)?

19. What are the four stages of nonrapid eye movement (NREM) sleep? How is NREM sleep distinguished from rapid eye movement (REM) sleep?

20. List two functional areas of the cerebral cortex that provide input to Wernicke's area.

21. Describe the emotional behavior that occurs when an animal's amygdala is stimulated.

22. What are the two main categories of learning?

23. What is the difference between a declarative memory and a procedural memory?

24. What is long-term potentiation?

FROM RESEARCH TO REALITY

Gene Transfer and Narcolepsy

Reference

Liu, M. et al. (2008). Orexin (hypocretin) gene transfer diminishes narcoleptic sleep behavior in mice. *European Journal of Neuroscience*. 28: 1382–1393.

Can gene transfer be used to treat narcolepsy?

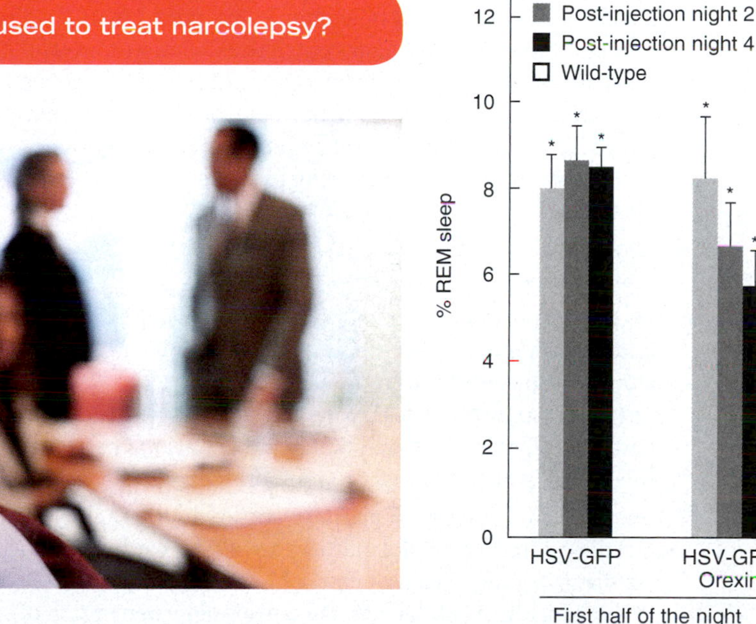

© Radius Images/Alamy Inc

We're always told how important it is to get enough sleep. Some people, however, get too much sleep because they have narcolepsy, a condition characterized by involuntary periods of rapid eye movement (REM) sleep that occur throughout the day. Narcolepsy results from the loss of hypothalamic neurons that produce the neuropeptide orexin, also called hypocretin.

Article description:

The authors of this study wanted to determine the effect of orexin gene transfer on narcoleptic rats. They used a replication-defective herpes simplex virus (type 1) as a vector to transfer the orexin gene to knockout mice lacking that gene and then monitored the sleep behavior in these mice.

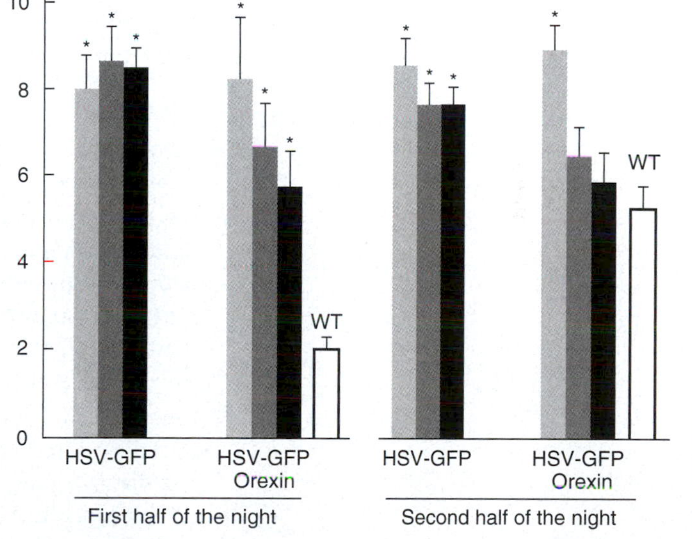

Go to WileyPLUS Learning Space and use the data from this article to answer the questions posed there and to discover more about gene transfer and narcolepsy.

CHAPTER REVIEW

8.1 Spinal Cord

1. The spinal cord is the part of the central nervous system that extends from the brain. The cord is protected by vertebrae and spinal meninges.
2. Thirty-one pairs of spinal nerves emerge from the spinal cord; they connect the spinal cord to sensory receptors and effectors in most parts of the body.
3. The spinal cord conveys sensory information by way of ascending tracts and motor information by way of descending tracts.
4. The spinal cord also serves as an integrating center for spinal reflexes. This integration occurs in the gray matter of the spinal cord.

8.2 Brain

1. Several features protect the brain: the cranium, meninges, blood–brain barrier, and cerebrospinal fluid (CSF).
2. Twelve pairs of cranial nerves emerge from the brain; they connect the brain to sensory receptors and effectors in the head, neck, and many organs in the thoracic and abdominal cavities.
3. The four major parts of the brain are the brain stem, cerebellum, diencephalon, and cerebrum.
4. The brain stem consists of the medulla oblongata, pons, and the midbrain. The medulla oblongata regulates cardiovascular activities, breathing, vomiting, swallowing, sneezing, coughing, and hiccupping. The pons functions along with the medulla to regulate breathing and also contributes to motor control. The midbrain regulates body movements and coordinates movements of the head and trunk in response to visual or auditory stimuli.The core of the brain stem consists of a netlike region of gray and white matter called the reticular formation, which helps maintain consciousness, causes awakening from sleep, and contributes to regulation of muscle tone.
5. The cerebellum smoothes and coordinates the contractions of skeletal muscles. It also maintains posture and balance.
6. The diencephalon consists of the thalamus, hypothalamus, and pineal gland. The thalamus relays most sensory input to the cerebral cortex from the spinal cord and brain stem. It also contributes to motor control. The hypothalamus controls the autonomic nervous system; produces hormones; and regulates eating, drinking, body temperature, circadian rhythms, and emotional and behavioral patterns. The pineal gland secretes melatonin, which is thought to promote sleep.
7. The cerebrum consists of an outer cortex, an internal region of white matter, and nuclei deep within the white matter. The neurons of the cerebral cortex are arranged into six distinct layers. Types of cortical neurons include stellate cells and pyramidal cells. The sensory areas of the cerebral cortex allow perception of sensory information. The motor areas control the execution of voluntary movements. The association areas are concerned with more complex integrative functions such as memory, personality traits, and intelligence. Hemispheric lateralization refers to the functional asymmetry that exists between the two cerebral hemispheres.
8. The basal nuclei are a group of nuclei in the cerebral hemispheres, diencephalon, and midbrain. They help initiate movements, suppress unwanted movements, and regulate muscle tone.
9. The limbic system encircles the upper part of the brain stem and the corpus callosum. It functions in emotional aspects of behavior.

CHAPTER REVIEW

8.3 Integrative Functions of the Cerebrum

1. Integrative functions of the cerebrum include wakefulness and sleep, language, emotions, motivation, and learning and memory.
2. Wakefulness involves increased activity of the reticular activating system (RAS).
3. Normal sleep consists of two phases: nonrapid eye movement (NREM) sleep and rapid eye movement (REM) sleep.
4. Comprehension and expression of language involves neural connections that extend from the primary visual cortex or primary auditory cortex to Wernicke's area and then to Broca's area, which in turn activates the primary motor cortex.
5. The emotions generated by the limbic system can give rise to a variety of autonomic and somatic motor responses. This involves connections among the limbic system, cerebral cortex, hypothalamus, and motor neurons in the brain stem and spinal cord.
6. Motivation is associated with rewards and punishments. The main neural system involved in reward is the mesolimbic dopamine pathway.
7. Associative learning and nonassociative learning are the two main categories of learning. Associative learning occurs when a connection is made between two stimuli; nonassociative learning occurs when repeated exposure to a single stimulus causes a change in behavior.
8. There are two types of memory: declarative memory and procedural memory. Memory occurs in stages over time. Short-term memory lasts for seconds to minutes. Long-term memory lasts for days to years. Some aspects of memory involve long-term potentiation (LTP), the process by which transmission at synapses is enhanced (potentiated) for hours or weeks after a period of high-frequency stimulation.

1. Sarah has suffered an accident, which seems to have caused some neurological damage. In order to figure out what has gone wrong as a result of her accident, Sarah is put through some standard neurological tests, one of which includes testing her reflexes. Sarah's doctor does a simple patellar tendon reflex and he finds that Sarah's leg does not respond. Doing further analysis, he discovers that both the motor neuron and sensory neuron have normal function. What has been damaged in Sarah's accident? Explain your answer.

2. A patient comes into the doctor's office and is able to interpret words that are heard or read, but is not able to form a proper sentence that makes sense. Functional MRI shows that the occipital lobe, precentral gyrus, and postcentral gyrus are functioning normally. Given this information, what aspect of the association areas dealing with language of the brain isn't working properly?

3. Steven has suffered a stroke. As a result, his body is having a hard time maintaining his sleep/wake cycles, regulating body temperature, regulating certain cardiovascular functions, and regulating water balance. What part of the brain was damaged as a result of the stroke? Explain your answer.

ANSWERS TO FIGURE QUESTIONS

8.1 The dorsal root ganglion contains cell bodies of sensory neurons.

8.2 During step 3 of this figure, axons of sensory neurons form an ascending tract that carries sensory information from the spinal cord to the brain.

8.3 The spinothalamic tract originates in the spinal cord and ends in the thalamus (a region of the brain). Because *spino-* comes first in the name, you know it contains ascending axons and thus is a sensory tract.

8.4 The integrating center of a reflex arc is located in the gray matter of the brain stem or spinal cord.

8.5 Cerebrospinal fluid (CSF) protects the brain by serving as a shock-absorbing medium, providing an optimal chemical environment for neuronal signaling, and allowing the exchange of nutrients and waste products between the blood and nervous tissue.

8.6 The hypothalamus is the part of the diencephalon that regulates feeding behavior: It contains both a feeding center and a satiety center.

8.7 The corpus callosum is a band of white matter (axons) that extends between the two cerebral hemispheres, allowing the hemispheres to communicate with each other.

8.8 The somatosensory association area allows you to recognize an object by simply touching it.

8.9 The basal nuclei help initiate body movements; suppress unwanted movements; regulate muscle tone; and influence several nonmotor aspects of cortical function, including sensory, limbic, cognitive, and linguistic functions.

8.10 The limbic system regulates pleasure, pain, aggression, docility, affection, fear, and anger.

8.11 The RAS promotes arousal (awakening from sleep).

8.12 Dreaming and paralysis of most skeletal muscles occur during REM sleep.

8.13 No. Wernicke's area and Broca's area are usually present only in the left cerebral hemisphere.

8.14 The hypothalamus is responsible for the involuntary somatic and autonomic responses that accompany an emotion.

8.15 If the electrode is implanted into a punishment center, the frequency of lever pressing by the rat will decrease because electrical stimulation of this area will give rise to unpleasant sensations.

8.16 Addictive drugs such as cocaine, morphine, heroin, and amphetamines ultimately increase the amount of dopamine available at synapses in the mesolimbic dopamine pathway.

8.17 LTP is believed to underlie some aspects of memory.

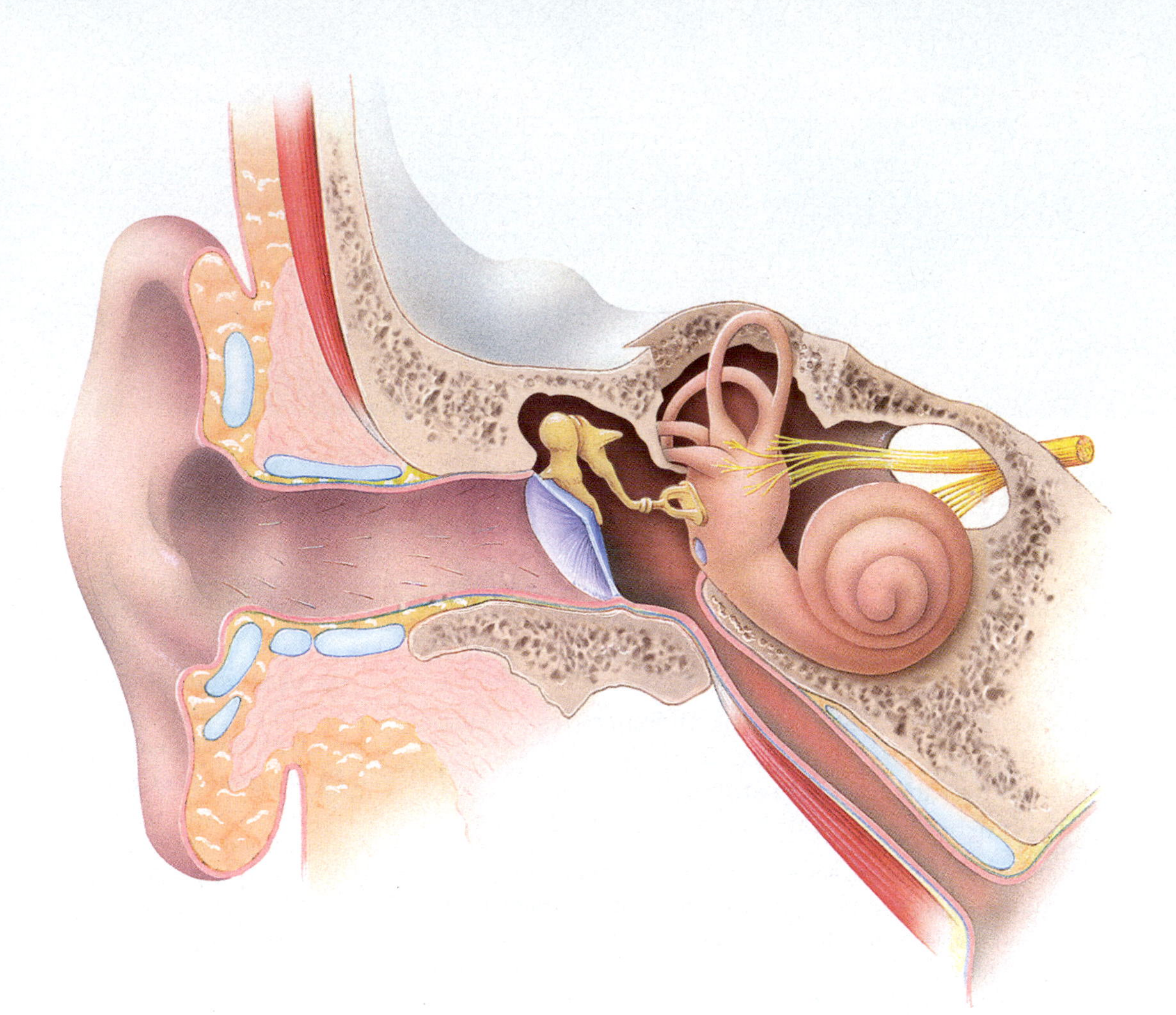

9 Sensory Systems

Sensory Systems and Homeostasis

Sensory systems contribute to homeostasis by detecting, conducting, and processing information about the external or internal environment.

LOOKING BACK TO MOVE AHEAD...

- The external environment is the space that surrounds the entire body; the internal environment refers to extracellular fluid—the fluid that surrounds and bathes the cells within the body (Section 1.4).
- Sensory (afferent) neurons convey information into the central nervous system (CNS) from sensory receptors in the body (Section 7.2).
- Neurons communicate with one another using two types of electrical signals: (1) graded potentials, which are used for short-distance communication only and (2) action potentials, which are used for long-distance communication (Section 7.3).
- A ligand-gated channel opens in response to a ligand (chemical) stimulus such as a neurotransmitter, a chemical in food, or an odor; a mechanically-gated channel opens in response to a mechanical stimulus such as touch, pressure, tissue stretching, or vibration (Section 7.3).

A **sensory system** is a component of the nervous system that consists of (1) sensory receptors that detect information in the external or internal environment, (2) neural pathways that convey the sensory information from the receptors to the CNS (brain and spinal cord), and (3) the parts of the CNS that process the information. Sensory receptors and the neural pathways that convey sensory input into the CNS constitute the afferent division of the peripheral nervous system (PNS). Recall that this division provides the CNS with information about the **somatic senses** (tactile, thermal, pain, and proprioceptive sensations) and **special senses** (smell, taste, vision, hearing, and equilibrium) (see Figure 7.1b). Once this sensory information enters the CNS, the CNS integrates (processes) the input by analyzing it and making decisions for appropriate responses. The major sensory systems include the **somatic sensory, visual, olfactory** (smell), **gustatory** (taste), **auditory** (hearing), and **vestibular** (equilibrium) **systems.** In this chapter you will explore the functions of each of these sensory systems.

9.1 Overview of Sensation

OBJECTIVES

- Explain the events that take place in order for a sensation to occur.
- Compare the receptive fields of different types of sensory neurons.
- Discuss how the following attributes of a stimulus are encoded: modality, location, intensity, and duration.

In its broadest definition, **sensation** is the conscious or subconscious awareness of changes in the external or internal environment. The nature of the sensation and the type of reaction generated vary according to the ultimate destination of the action potentials (nerve impulses) that convey sensory information to the CNS. Sensory information that reaches the spinal cord may serve as input for spinal reflexes, such as the flexor (withdrawal) reflex that moves a limb away from a painful stimulus. Sensory input that reaches the brain stem elicits more complex reflexes, such as changes in heart rate or breathing rate. When sensory information reaches the cerebral cortex, you become consciously aware of the sensory stimuli and can precisely locate and identify specific sensations such as touch, smell, hearing, or taste. **Perception** is the conscious awareness and interpretation of sensations and is primarily a function of the cerebral cortex. You have no perception of some sensory information because it never reaches the cerebral cortex. For example, certain sensory receptors constantly monitor the pressure of blood in blood vessels. Because the action potentials conveying blood pressure information propagate to the cardiovascular center in the medulla oblongata rather than to the cerebral cortex, blood pressure is not consciously perceived.

The Process of Sensation Involves Four Events

For a sensation to arise, sensory information must be transformed into the "language" of the nervous system—namely graded potentials and action potentials. The process of sensation typically involves the following events (**FIGURE 9.1**):

1. ***Stimulation of the sensory receptor.*** The process of sensation begins in a **sensory receptor**, a structure of the nervous system that is associated with a sensory (afferent) neuron. A given sensory receptor responds mainly to one particular type of stimulus. A **stimulus** is a change in the external or internal environment that can activate the sensory receptor.
2. ***Transduction of the stimulus.*** The sensory receptor converts the energy in the stimulus into a graded potential, a process known as **transduction**. Recall that a graded potential is a change in membrane potential that causes the membrane to become either depolarized or hyperpolarized (see Section 7.3).
3. ***Generation of action potentials.*** If a graded potential in a sensory neuron reaches threshold, it triggers one or more action potentials, which then propagate into the CNS.
4. ***Integration of sensory input.*** As sensory input is relayed from one synapse to another in the CNS, the information is integrated (processed). This means that the information is either modified, allowed to continue on as is, or filtered out, depending on how important the information is. Sensory input that reaches the level of consciousness is integrated in the cerebral cortex.

There Are Different Types of Sensory Receptors

A sensory receptor is activated by a particular stimulus. Most stimuli are in the form of mechanical energy, such as pressure changes or sound waves; electromagnetic energy, such as light or heat; or chemical energy, such as in a molecule of glucose. The type of stimulus to which a sensory receptor responds best is known as its **adequate stimulus**. Sensory receptors are classified into five major groups according to their adequate stimuli:

- **Mechanoreceptors** are sensitive to mechanical stimuli such as the deformation, stretching, or bending of cells. Mechanoreceptors provide sensations of touch, pressure, vibration, proprioception (muscle and joint position), and hearing and equilibrium. They also monitor the stretching of blood vessels and internal organs.
- **Thermoreceptors** detect changes in temperature.
- **Photoreceptors** detect light that strikes the retina of the eye.
- **Chemoreceptors** detect chemicals in the mouth (taste), nose (smell), and body fluids.
- **Nociceptors** (nō′-sē-SEP-tors; *noci-* = harmful) respond to painful stimuli resulting from physical or chemical damage to tissues.

Although a sensory receptor is most responsive to its adequate stimulus, it can also respond to other stimuli if the intensity is high enough. For example, photoreceptors of the eye are most responsive to light. However, an intense mechanical stimulus such as a blow to the eye activates photoreceptors and causes you to "see stars."

Sensory receptors are either (1) the peripheral endings (dendrites) of sensory neurons or (2) separate cells that synapse with sensory neurons.

FIGURE 9.1 The major events of sensation.

During the process of sensation, sensory information is transformed into electrical signals (graded potentials and action potentials), conveyed into the CNS, and then integrated.

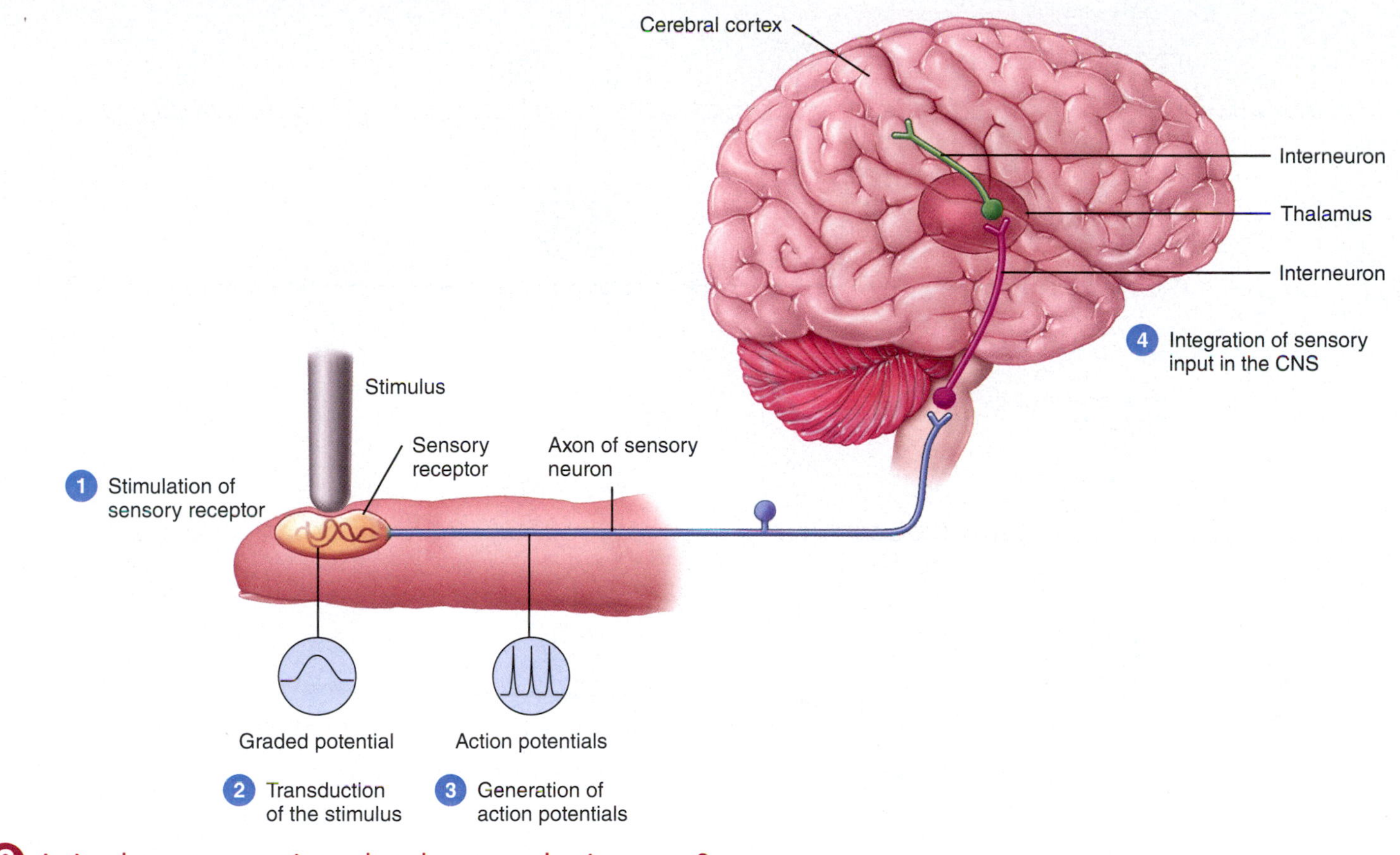

As it relates to sensation, what does transduction mean?

When the **peripheral endings** of sensory neurons serve as sensory receptors, they may be encapsulated (surrounded by a connective tissue capsule) or bare (not encapsulated); these two types of peripheral endings are referred to as *encapsulated nerve endings* and *free nerve endings*, respectively (FIGURE 9.2a). Encapsulated nerve endings include the receptors for pressure, vibration, and some touch sensations. The presence of the connective tissue capsule enhances the sensitivity or specificity of the receptor. Examples of free nerve endings include the receptors for pain, thermal, itch, tickle, and some touch sensations.

The receptors for most special senses are specialized, **separate cells** that synapse with sensory neurons (FIGURE 9.2b). These include gustatory receptor cells in taste buds, photoreceptors in the retina of the eye for vision, and hair cells for hearing and equilibrium in the inner ear. The olfactory receptors for the special sense of smell are not separate cells; instead they are located in olfactory cilia, which are hairlike structures that project from the dendrite of an olfactory receptor cell (a type of neuron) (see FIGURE 9.20).

During transduction, a sensory receptor responds to a stimulus by generating a graded potential. The graded potential that forms in a sensory receptor is referred to as a **receptor potential** (FIGURE 9.2a,b). A stimulus causes a receptor potential by opening or closing ion channels in the membrane of the sensory receptor (either directly or indirectly by activating a second messenger pathway). In many sensory systems, the stimulus opens cation channels that allow Na^+ and Ca^{2+} ions to enter the sensory receptor, resulting in depolarization of the sensory receptor's membrane. In the visual system, however, cation channels close in response to the stimulus (light) and the sensory receptors (photoreceptors) are hyperpolarized.

In sensory receptors that are peripheral endings of sensory neurons, if the receptor potential is large enough to reach threshold, it triggers one or more action potentials in the axon of the sensory neuron (FIGURE 9.2a). The action potentials then propagate along the axon into the CNS. In sensory receptors that are separate cells, the receptor potential triggers release of neurotransmitter through exocytosis of synaptic vesicles (FIGURE 9.2b). The neurotransmitter molecules liberated from the synaptic vesicles diffuse across the synaptic cleft and produce a postsynaptic potential (PSP), a type of graded potential, in the sensory neuron. If threshold is reached, the PSP will trigger one or more action potentials, which propagate along the axon into the CNS.

Sensory Neurons Have Receptive Fields

An important aspect of a sensory neuron is its receptive field. What constitutes the receptive field varies depending on the type of sensory neuron involved. For example, the receptive field of a somatic sensory neuron is the area of the body where stimulation causes a response in

FIGURE 9.2 Sensory receptors.

Sensory receptors are either (1) the peripheral endings of sensory neurons or (2) separate cells that synapse with sensory neurons.

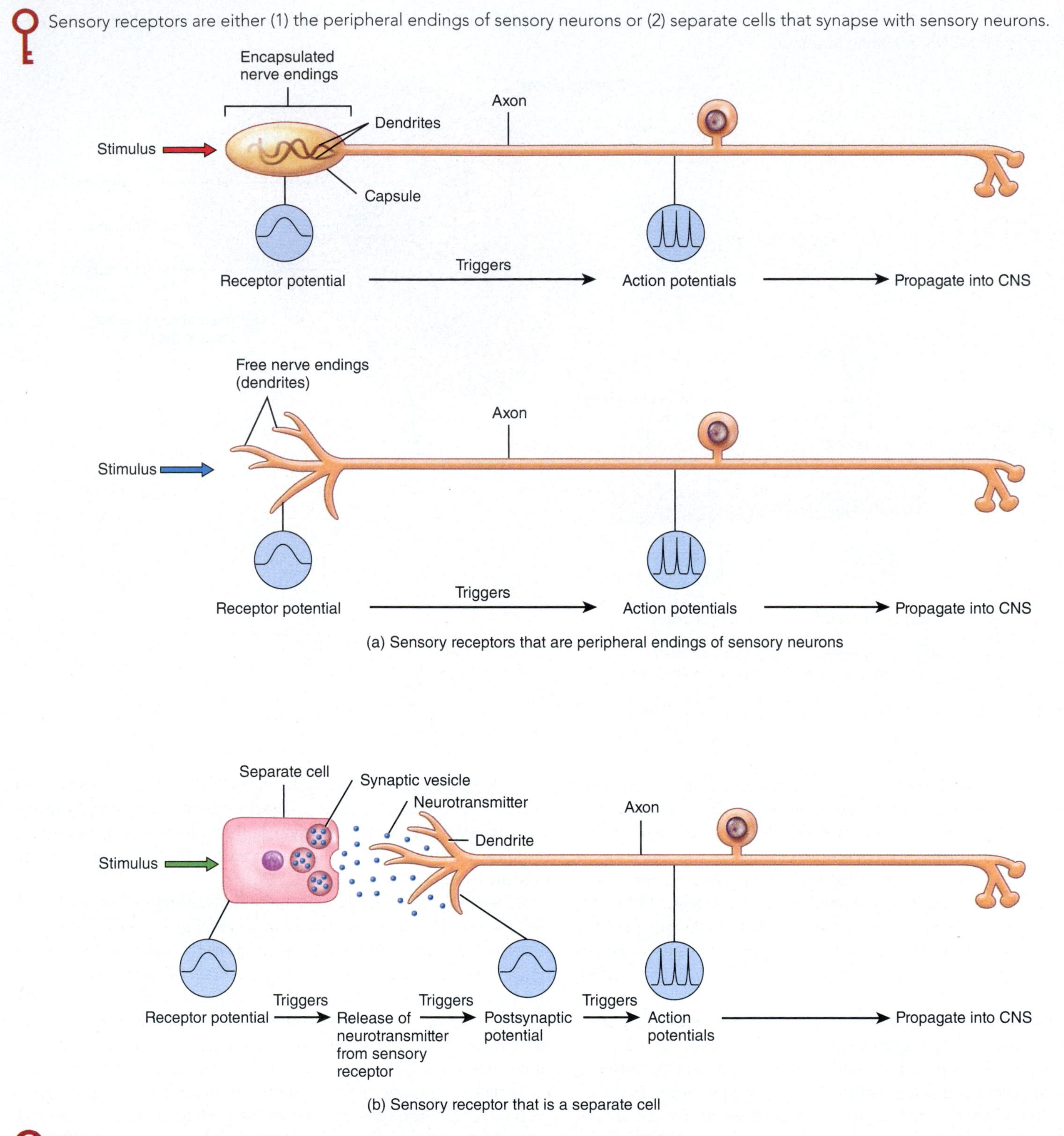

(a) Sensory receptors that are peripheral endings of sensory neurons

(b) Sensory receptor that is a separate cell

Which senses are served by sensory receptors that are separate cells?

that neuron. In the case of a somatic sensory neuron that supplies the skin, the receptive field corresponds to the region of skin where the neuron's sensory receptors are located (FIGURE 9.3a). An adequate stimulus that occurs anywhere within this skin region will cause a response in the somatic sensory neuron. For sensations that are not somatic, the receptive field has a different meaning. In the visual system, the receptive field is the area of visual space where stimulation by light causes a response in a visual neuron of the eye (FIGURE 9.3b). In the olfactory system, the receptive field is the select group of odorants (chemicals in odors) to which an olfactory receptor cell of the nose can

FIGURE 9.3 Receptive fields of different types of sensory neurons.

The receptive field of a neuron is the stimulated physical area, specific group of chemicals, or particular set of sound frequencies that causes a response in that neuron.

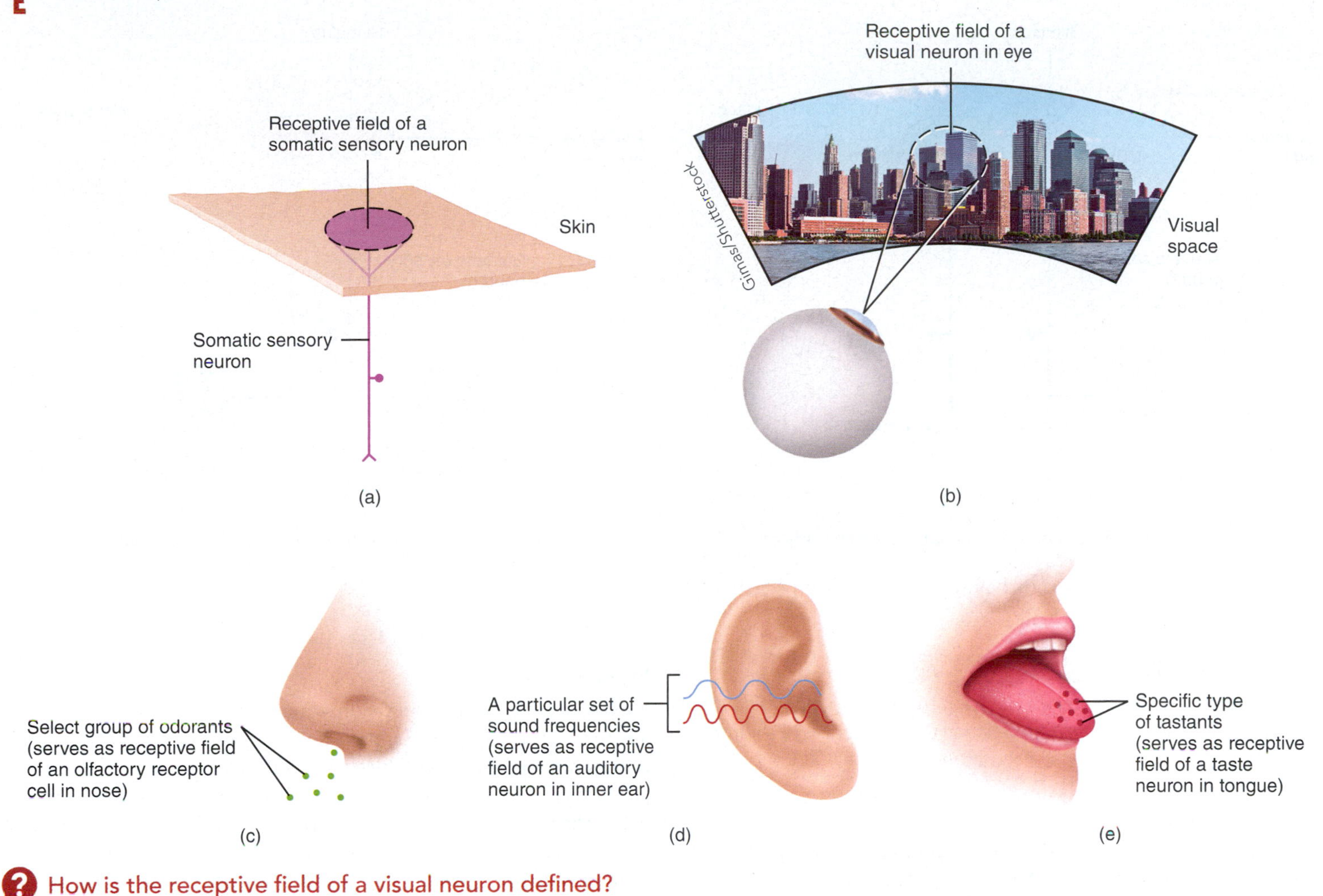

How is the receptive field of a visual neuron defined?

respond (FIGURE 9.3c). In the auditory system, the receptive field is the particular set of sound frequencies that elicits a response in an auditory neuron innervating a hair cell of the inner ear (FIGURE 9.3d). In the gustatory system, the receptive field is the specific type of tastants (chemicals in food) that causes a response in a taste neuron supplying a gustatory receptor cell in the tongue (FIGURE 9.3e). Considering all of the different types of sensations that the body experiences, the **receptive field** of a sensory neuron can be defined as the stimulated physical area, specific group of chemicals, or particular set of sound frequencies that causes a response in that neuron.

In the somatic sensory and visual systems, receptive fields are physical areas, and the receptive fields of neighboring sensory neurons may be in separate locations or they may overlap. If sensory neurons have separate receptive fields, then a given neuron will respond only if there is a stimulus present in the receptive field associated with that neuron (FIGURE 9.4a). If sensory neurons have overlapping receptive fields, then all participating neurons will respond to a stimulus that extends into the region of overlap, but the response of each neuron is proportional to the relative position of the stimulus (FIGURE 9.4b). So if the stimulus falls more in one receptive field than another, then the neuron most affected by the stimulus will have the greatest response. Because a stimulus usually elicits differential responses in sensory neurons with overlapping receptive fields, the brain can compare the difference in activation of these neurons to determine where the stimulus is located. In other words, the presence of overlapping receptive fields improves stimulus localization. By contrast, when a sensory neuron has a separate receptive field, the brain has no way of knowing exactly where in that receptive field the stimulus is located and there are no signals from other sensory neurons that can be used for comparison (FIGURE 9.4a). Receptive fields frequently overlap in the somatic sensory and visual systems, which enhances the brain's ability to localize somatic sensory and visual stimuli.

Sensory Coding Distinguishes the Attributes of a Stimulus

As you have already learned, sensory information is ultimately converted into action potentials before it is conveyed into the CNS for integration. If all action potentials are the same, how does the nervous system tell the difference between one stimulus and another? This

FIGURE 9.4 Separate and overlapping receptive fields.

In the somatic sensory and visual systems, the receptive fields of neighboring sensory neurons may occupy separate physical areas or they may overlap.

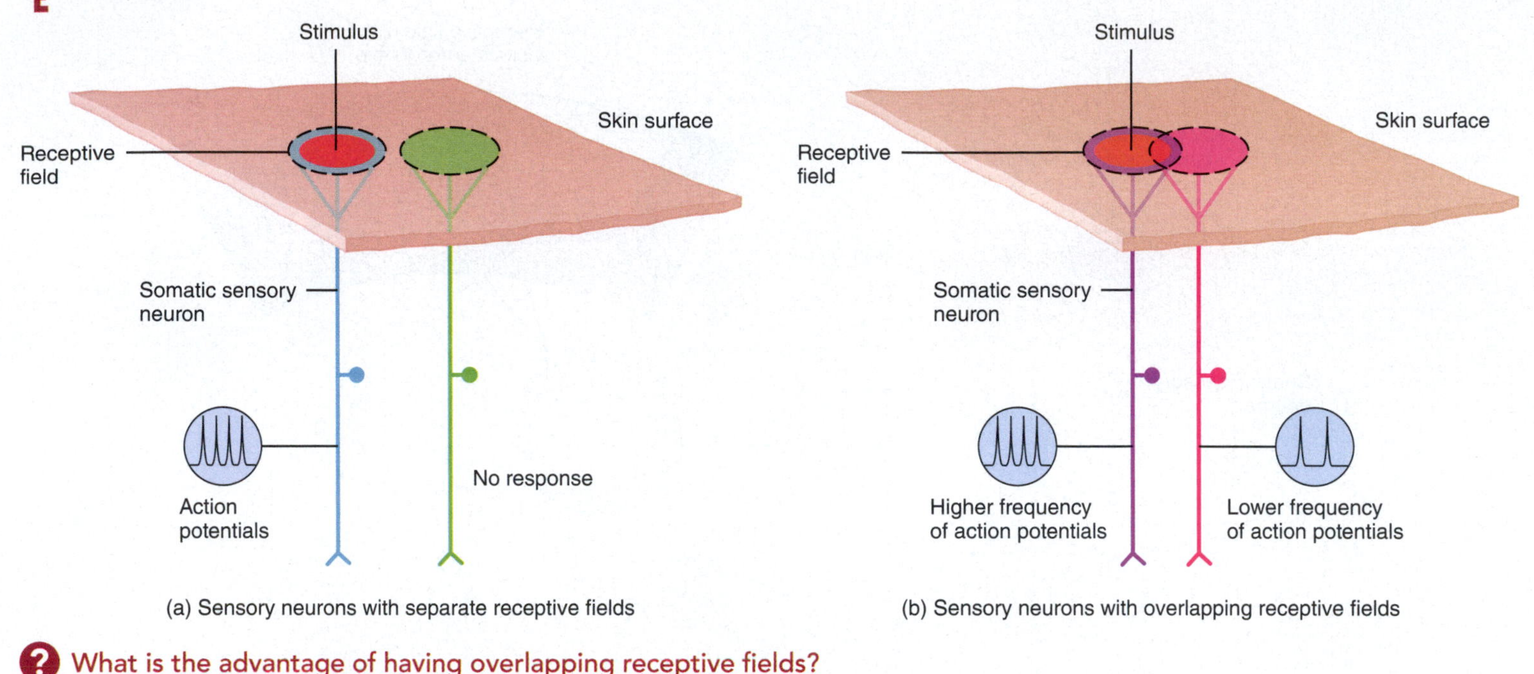

(a) Sensory neurons with separate receptive fields

(b) Sensory neurons with overlapping receptive fields

What is the advantage of having overlapping receptive fields?

is accomplished by **sensory coding**, the use of organizational and functional features of the nervous system to represent specific details about a stimulus. Sensory systems encode four attributes of a stimulus: modality, location, intensity, and duration.

Stimulus Modality

Each unique type of sensation—such as touch, pain, vision, taste, or hearing—is called a **modality**. Modality is encoded by which sensory receptor and neural pathway are activated when a stimulus is applied. For example, the modality of touch is perceived whenever a stimulus activates a touch receptor and its neural pathway to the cerebral cortex. Likewise, the modality of taste is perceived whenever a stimulus activates a gustatory receptor and its pathway to the cortex. Different types of sensory information are conveyed to different parts of the cerebral cortex for perception. The neural pathways that convey information about modality from peripheral receptors to specific regions of the cerebral cortex are called **labeled lines** (FIGURE 9.5). Each labeled line consists of a set of neurons that transmits information for only one modality. Therefore, you seem to see with your eyes, hear with your ears, or feel pressure on your arm because stimulation of receptors in each of these parts of the body activates a separate labeled line to a specific region of the cerebral cortex, which perceives the modality associated with that labeled line. The association of a modality with the activation of a particular labeled line is called **labeled line coding**. Activation of a given labeled line always causes perception of its modality, regardless of how the pathway is stimulated. For example, you normally perceive light when light waves activate the visual pathway to the cerebral cortex. Recall, however, that you also perceive light in the form of "stars" when there is a blow to the eye. Although the blow to the eye is not a visual stimulus, it still causes you to see light because the visual pathway to the cortex is activated.

Stimulus Location

For somatic sensations and vision, stimulus location is encoded by the location of the activated receptive field. Sensory receptors at the body surface and in the retina of the eye are organized in such a way that adjacent receptors give rise to pathways that project to adjacent regions of the cerebral cortex. This orderly arrangement in which the relationship between adjacent sensory receptors is maintained as information is processed in the CNS is referred to as a *topographic pattern*. So when a stimulus activates one or more sensory receptors in the receptive field of a somatic sensory or visual neuron, the brain uses that input to identify a specific topographic location for the stimulus.

In the auditory and olfactory systems, receptive fields do not provide information about stimulus location because the receptive fields of auditory sensory neurons and olfactory sensory neurons are not physical areas. Recall that the receptive field of an auditory sensory neuron is a particular set of sound frequencies, and the receptive field of an olfactory receptor cell is a select group of odorants. So how is stimulus location coded in the auditory and olfactory systems? The answer is based on slight timing differences in the arrival of sound waves at the two ears or odorants at the two nostrils. The brain compares the timing of receptor activation in the two ears or two sides of the nose to determine the location of sound or smell, respectively. For example, if receptors in the right ear are activated before receptors in the left ear, input from the right ear will reach the brain before input from the left ear. The brain uses the time difference of this input to determine that the sound is coming from the right.

Now that you have an understanding of how stimulus location is encoded, you are ready to learn about acuity, an important property related to stimulus location. **Acuity** is sharpness of perception—in other words, the ability to precisely locate and distinguish one stimulus from another. In the somatic sensory and visual systems, two major

FIGURE 9.5 Labeled lines. Labeled lines can vary in the number of neurons that they contain.

A labeled line is a set of neurons that conveys sensory information about a particular modality from a peripheral receptor to a specific region of the cerebral cortex.

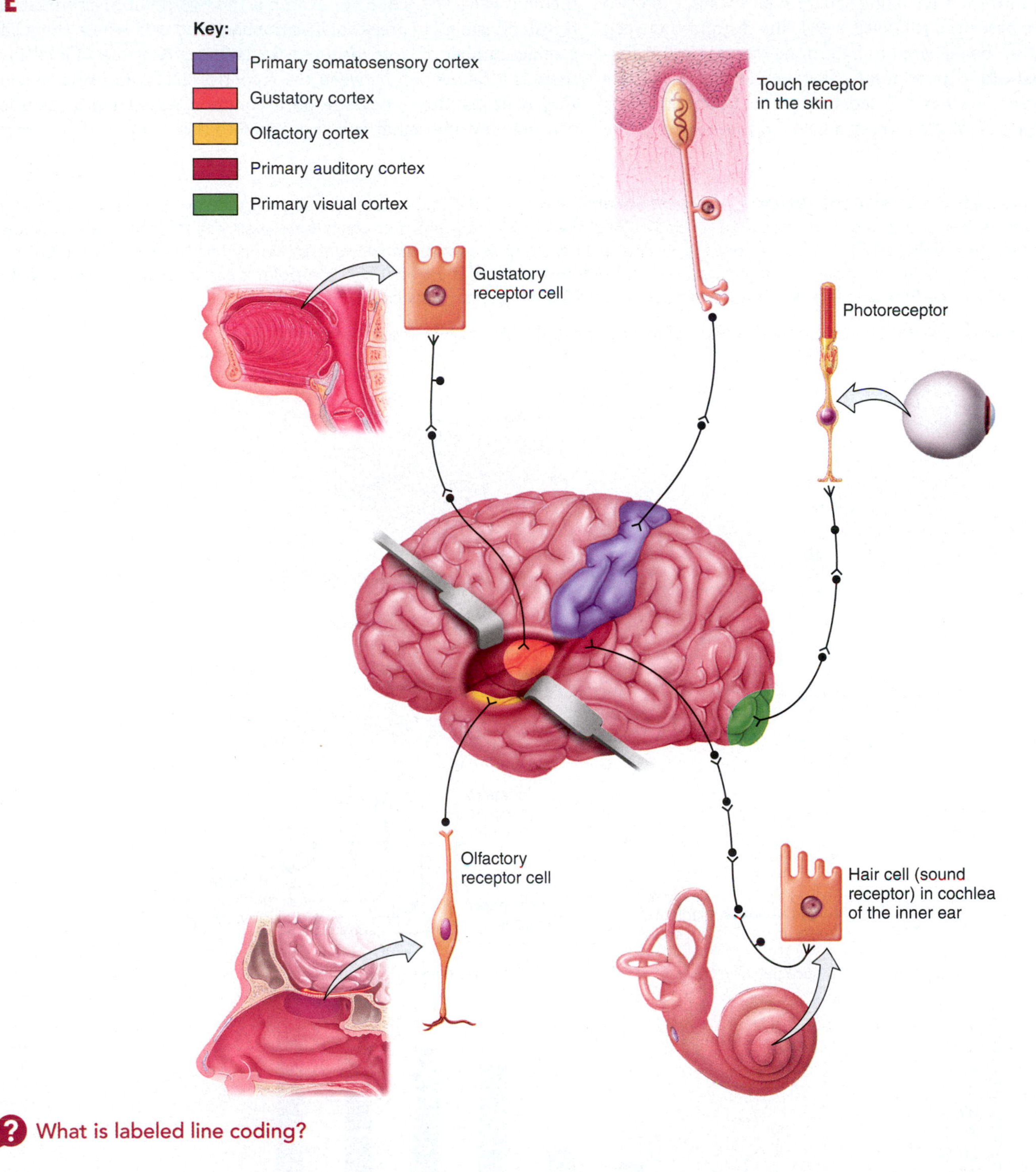

What is labeled line coding?

factors affect acuity: (1) the size of the receptive field and (2) lateral inhibition.

Size of the Receptive Field The size of the receptive field varies inversely with the number of sensory receptors that it contains. The smaller the receptive field, the more densely packed it is with sensory receptors and the greater its acuity. Conversely, the larger the receptive field, the less densely packed it is with sensory receptors and the lower its acuity. A measure of *tactile acuity*, or sharpness of touch perception, is **two-point discrimination**—the ability to perceive two points applied to the skin as two separate points. Two-point discrimination can be demonstrated by applying the two points of a caliper to the skin. If the two caliper points stimulate the same receptive field, then only one point of touch is perceived (FIGURE 9.6a). If the two caliper points stimulate different receptive fields and this input is conveyed into the CNS along separate pathways, then two points of touch are perceived (FIGURE 9.6b). The distance between the caliper points can be adjusted to determine the **two-point discrimination threshold**—the minimum distance at which the two caliper points are perceived as two separate points of touch, which is a reflection of the size of the receptive fields in that region of the body. If the distance between the two caliper points is less than the two-point discrimination threshold, then only one point of touch will be felt. The two-point discrimination threshold varies throughout the body (FIGURE 9.6c). For example, the two-point discrimination threshold is high in areas such as the back and calf, where there are a small number of large receptive fields (FIGURE 9.6a). In addition, there is little overlap between the receptive fields, and several sensory neurons often converge on a common postsynaptic neuron, causing only one signal to be conveyed to the brain. In these areas,

FIGURE 9.6 Two-point discrimination. Two-point discrimination can be demonstrated by applying the two points of a caliper to the skin. (a) If the two caliper points stimulate the same receptive field, then only one point of touch is perceived. (b) If the two caliper points stimulate different receptive fields and this input is conveyed into the CNS along separate pathways, then two points of touch are perceived. (c) A bar graph showing the two-point discrimination thresholds for different parts of the body. The two-point discrimination threshold is the minimum distance at which the two caliper points are perceived as two separate points of touch.

Two-point discrimination is the ability to perceive two points applied to the skin as two separate points.

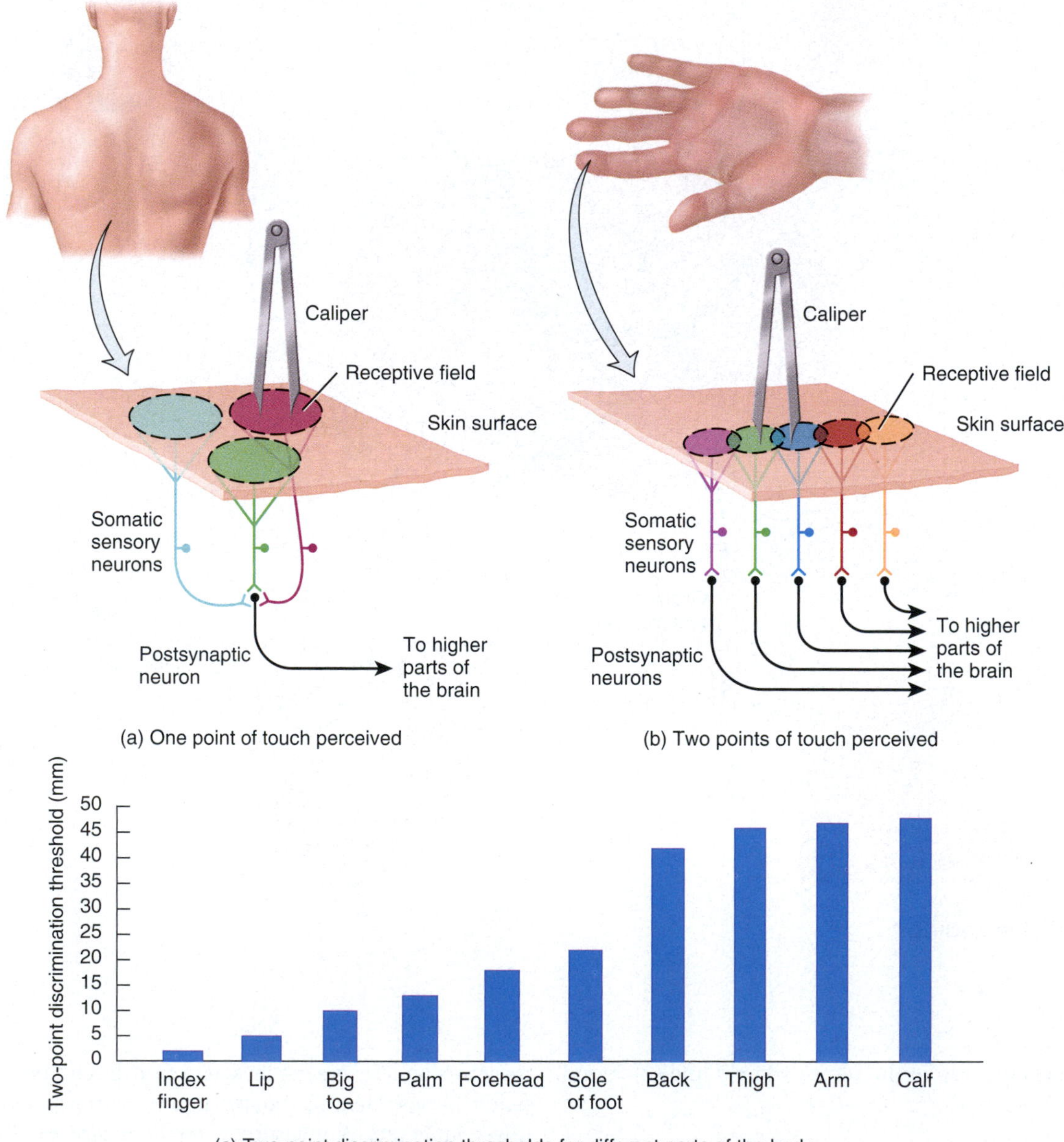

(c) Two-point discrimination thresholds for different parts of the body

The Skin Senses by S. Weinstein and D.R. Kenshalo (editors), Charles C. Thomas Publisher, 1968.

The two-point discrimination threshold of the sole of the foot is 22 mm. What does this mean?

two points as far apart as 40 mm are perceived as just one point. Therefore, tactile acuity is poor in regions of the body that have a high two-point discrimination threshold. By contrast, the two-point discrimination threshold is low in areas such as the fingertips, where there are a large number of small receptive fields (FIGURE 9.6b). The receptive fields often overlap, which increases the ability to localize tactile stimuli. In addition, there is little convergence of sensory neurons, allowing input to be conveyed into the CNS along separate pathways. Two points placed as close together as 2 mm on the fingertips are perceived as two separate points. Thus, tactile acuity is high in regions of the body that have a low two-point discrimination threshold. The high degree of tactile acuity in the fingertips allows a person to read Braille, which consists of raised dots that are separated from each other by 2.5 mm.

Lateral Inhibition Acuity is also influenced by a phenomenon known as lateral inhibition. In **lateral inhibition**, input from sensory receptors along the border of a stimulus is substantially inhibited compared to input from sensory receptors at the center of the stimulus. An example of lateral inhibition occurs when the point of a pencil is pressed against the skin (FIGURE 9.7). The sensory neuron whose receptors are in the center area where the pencil point strongly deforms the skin is activated more than the sensory neurons whose receptors are in the periphery of the stimulated area where the pencil point causes less skin deformation. As a result, the sensory neuron from the center area generates action potentials at a higher frequency than the sensory neurons from the peripheral area. To enhance this contrast, the sensory neuron from the center area inhibits the pathways from the peripheral area via inhibitory interneurons (FIGURE 9.7). Such laterally interconnecting inhibitory interneurons are common between sensory neurons that serve neighboring receptive fields. The peripheral sensory neurons also inhibit the sensory neuron in the center area via inhibitory interneurons, but this inhibition has less of an effect because the peripheral sensory neurons are only weakly activated compared to the sensory neuron of the center area. Because of lateral inhibition, (1) there is a further decrease in the number of action potentials transmitted by the pathways from the peripheral area and (2) the pathway from the center area is only slightly inhibited and continues to transmit a higher frequency of action potentials than the peripheral area. Both of these effects allow the brain to determine exactly where the pencil point is located. If lateral inhibition did not occur, a significant number of action potentials would be transmitted along all the pathways from the stimulated region (both the center and periphery), which would make it more difficult for the brain to localize the pencil point.

FIGURE 9.7 Lateral inhibition.

Lateral inhibition is the phenomenon by which input from sensory receptors along the border of a stimulus is substantially inhibited compared to input from sensory receptors at the center of the stimulus.

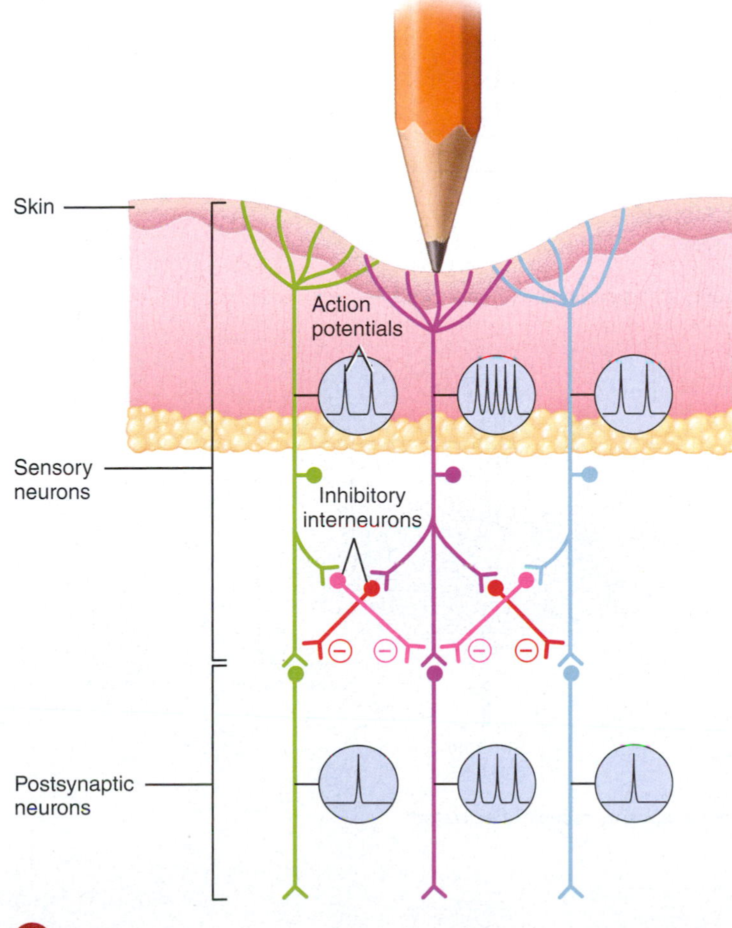

What is the purpose of lateral inhibition?

Stimulus Intensity

Stimulus intensity is encoded by two main factors: (1) the frequency of action potentials generated in response to the stimulus and (2) the number of sensory receptors activated by the stimulus. As stimulus intensity increases, the amplitude of the receptor potentials formed in the sensory receptor increases. As long as each of these receptor potentials reaches threshold, the sensory neuron generates a higher frequency of action potentials (FIGURE 9.8a) until a maximum level is reached, as determined by the absolute refractory period. The increased action potential frequency of the sensory neuron is interpreted by the nervous system as an increase in stimulus intensity. The use of action potential frequency to determine stimulus intensity is known as **frequency coding**.

An increase in stimulus intensity also causes activation of more sensory receptors because a stronger stimulus usually affects a larger region of the body than a weaker stimulus (FIGURE 9.8b). The additional sensory receptors that are activated may be associated with the same sensory neuron or with different sensory neurons that are located close to each other. In either case, the nervous system interprets the increase in sensory receptor activation as an increase in stimulus intensity.

Stimulus Duration

Stimulus duration is encoded by the duration of action potentials in the sensory neuron. In general, the longer the stimulus lasts, the longer the sensory neuron produces action potentials. If the stimulus persists, however, most sensory neurons do not continue to generate the same frequency of action potentials because their sensory receptors adapt. **Adaptation** is a decrease in the response of a sensory receptor during a maintained, constant stimulus. This means

FIGURE 9.8 Stimulus intensity.

Stimulus intensity is encoded by (1) the frequency of action potentials generated in response to the stimulus and (2) the number of sensory receptors activated by the stimulus.

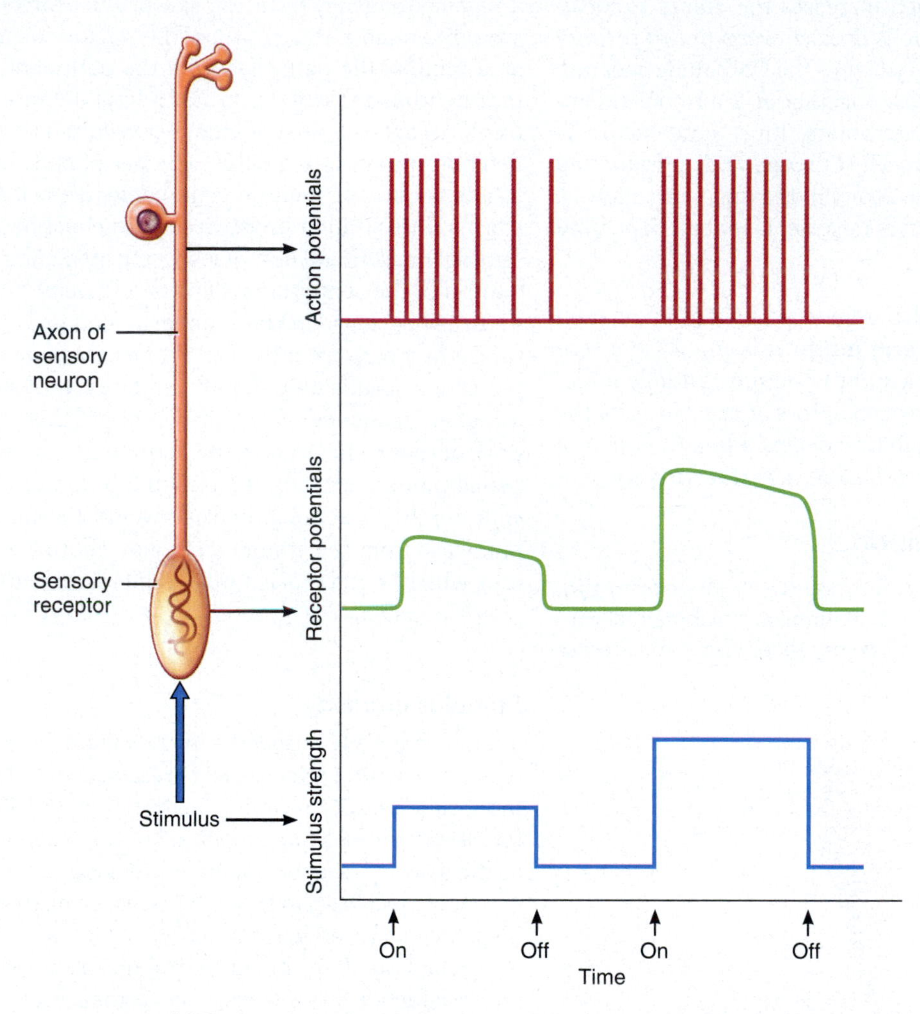

(a) Coding of stimulus intensity by the frequency of action potentials generated

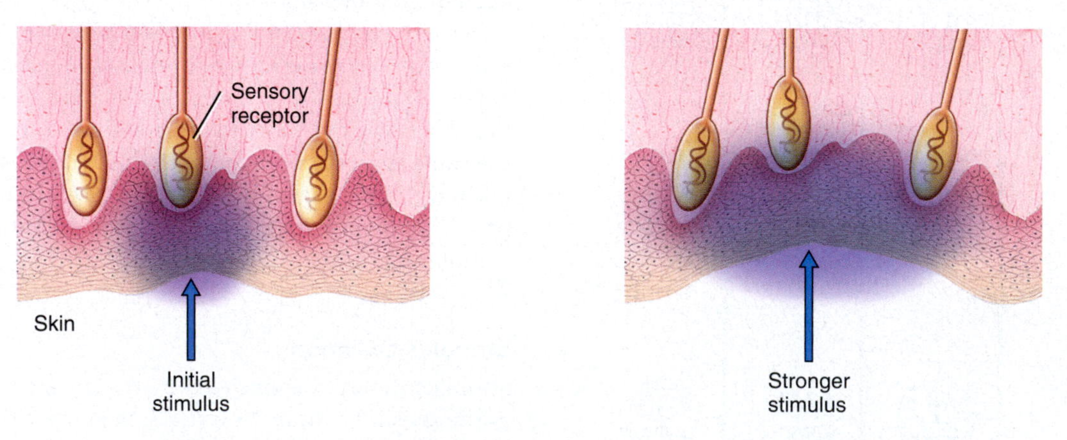

(b) Coding of stimulus intensity by the number of sensory receptors activated

What is frequency coding?

that the receptor potential diminishes while the prolonged stimulus is present, causing the frequency of action potentials in the sensory neuron to decrease. Because of adaptation, the perception of a sensation may fade or disappear even though the stimulus persists. For example, when you first step into a hot shower, the water may feel very hot, but soon the sensation decreases to one of comfortable warmth even though the stimulus (the high temperature of the water) does not change.

Receptors vary in how quickly they adapt. **Slowly adapting receptors**, also called *tonic receptors*, adapt slowly and continue to produce a significant response as long as the stimulus persists (FIGURE 9.9a). Slowly adapting receptors are associated with stimuli that require constant monitoring. For example, the brain must be continuously informed about the level of arterial blood pressure in order to maintain the driving force for movement of blood through body tissues. Receptors known as *baroreceptors* (*baro-* = pressure), which provide this input to the brain, are slowly adapting. **Rapidly adapting receptors**, also called *phasic receptors*, adapt very quickly (FIGURE 9.9b). They respond when the stimulus is first applied—the *on response*—and then cease to respond while the stimulus is maintained. Some rapidly adapting receptors also produce a second response called the *off response* when the stimulus is removed. Rapidly adapting receptors are specialized for signaling *changes* in a stimulus. For example, you normally notice your clothes touching your body when you first put them on and then are unaware of them after that because of rapidly adapting touch receptors present in the skin.

The mechanism of adaptation can vary, depending on the sensory receptor. In many cases, adaptation occurs because cation channels in the membrane of the sensory receptor inactivate after being open for a period of time. Recall that the depolarizing receptor potential that forms in a sensory receptor is often due to the opening of cation channels that allow Na^+ and Ca^{2+} ions to enter the cell. If these cation channels inactivate, inflow of Na^+ and Ca^{2+} is reduced, the depolarization is diminished, and generation of action potentials by the sensory neuron decreases or completely stops. In sensory receptors that are encapsulated, the connective tissue capsule may influence how long the cation channels stay open.

FIGURE 9.9 Slowly adapting and rapidly adapting receptors.

Slowly adapting receptors produce a significant response as long as the stimulus persists. Rapidly adapting receptors respond when the stimulus is first applied (the on response) and then cease to respond while the stimulus is maintained; they may also produce a second response (the off response) when the stimulus is removed.

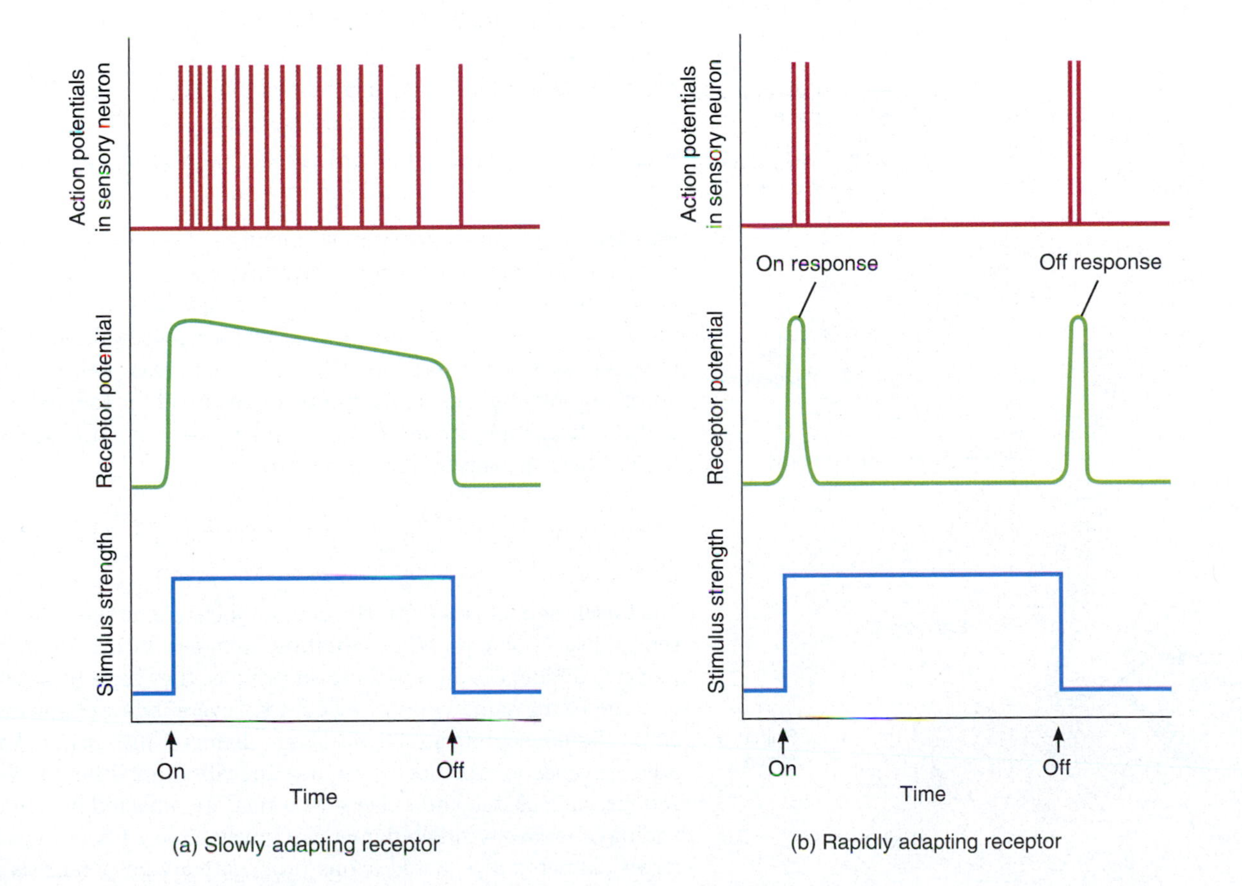

(a) Slowly adapting receptor

(b) Rapidly adapting receptor

Which type of receptor (slowly adapting or rapidly adapting) is important for signaling changes in a stimulus?

A Sensory Pathway Conveys Sensory Information

A **sensory pathway** is a group of parallel chains of neurons that conveys sensory information from sensory receptors in the periphery to the cerebral cortex (FIGURE 9.10). Each chain of the sensory pathway is a *labeled line* that conveys sensory information about one particular modality. The neurons of a sensory pathway are referred to as first-, second-, third-, fourth-, or higher-order neurons based on the order in which they occur in the chain (FIGURE 9.10). Integration (processing) of information occurs at each synapse along the sensory pathway.

First-order neurons are sensory neurons, which convey information into the CNS from sensory receptors in the body. All remaining neurons of a sensory pathway are interneurons, which are located completely within the CNS. Axons of first-order neurons synapse with **second-order neurons**, which are usually located in either the brain stem or spinal cord. The axons of most second-order neurons ascend to the thalamus, where they synapse with **third-order neurons**. The third-order neurons in turn project their axons to **fourth-order neurons** in a primary sensory area of the cerebral cortex, where perception of the sensation occurs. From the primary sensory area, axons of fourth-order neurons extend to **higher-order neurons** in association areas of the cerebral cortex, where complex integration of sensory information takes place. Different sensory pathways extend to and activate different areas of the cerebral cortex. Although most sensory pathways are organized in the manner just described, sensory pathways can vary with respect to the number of neurons in their chains and the parts of the brain or spinal cord where they form synapses. The details of the various sensory pathways will be presented in later sections of this chapter.

One other point worth mentioning is that most sensory pathways **decussate** (cross over to the opposite side) as they course through the spinal cord or brain stem (FIGURE 9.10). When this occurs, sensory information from one side of the body is perceived by a specific region of cerebral cortex on the *opposite* side of the brain. For those sensory pathways that do not decussate, sensory information on one side of the body is perceived by a specific region of cerebral cortex on the *same* side of the brain.

FIGURE 9.10 Components of a sensory pathway. For simplicity, fourth- and higher-order neurons are not shown.

A sensory pathway is a group of parallel chains of neurons that conveys sensory information from sensory receptors in the periphery to the cerebral cortex.

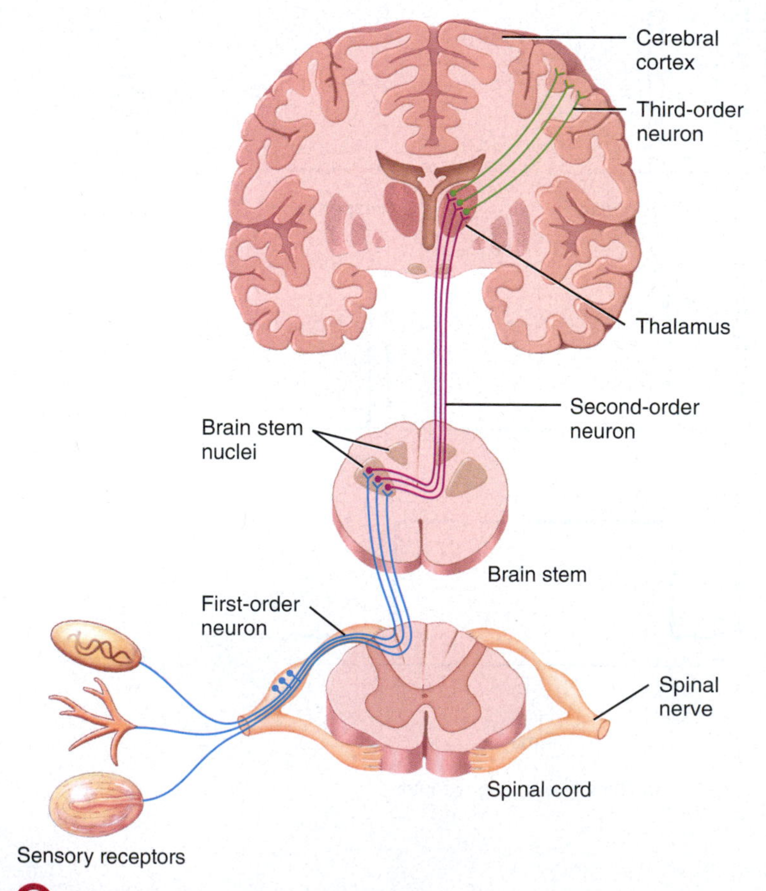

What does it mean if a sensory pathway decussates?

CHECKPOINT

1. What is a receptor potential and how does it give rise to an action potential?
2. What is the receptive field of an auditory neuron?
3. What is a labeled line?
4. How is stimulus intensity encoded?
5. Why are slowly adapting receptors important?

9.2 The Somatic Sensory System

OBJECTIVES

- Discuss the functions of the different types of somatic sensory receptors.
- Describe the two major pathways that convey somatic sensory information to the cerebral cortex.
- Explain the roles of the primary somatosensory cortex and the somatosensory association area.

Somatic sensations arise from stimulation of sensory receptors embedded in the skin, subcutaneous layer, skeletal muscles, tendons, and joints. The sensory receptors for somatic sensations are distributed unevenly—some parts of the body surface are densely populated with receptors and others contain only a few. The areas with the highest density of somatic sensory receptors are the tip of the tongue, the lips, and the fingertips. There are four modalities of somatic sensation: tactile, thermal, pain, and proprioceptive.

Tactile Sensations Allow Us to Feel Touch, Pressure, Vibration, Itch, and Tickle

The **tactile sensations** (TAK-tīl; *tact-* = touch) encompass a variety of sensations—touch, pressure, vibration, itch, and tickle. Although we perceive differences among these sensations, they arise by activation of some of the same types of receptors. Several types of encapsulated mechanoreceptors attached to large-diameter myelinated A fibers mediate sensations of touch, pressure, and vibration. Other tactile sensations, such as itch and tickle sensations, are detected by free nerve endings attached to small-diameter, unmyelinated C fibers. Recall that larger diameter, myelinated axons propagate action potentials more rapidly than do smaller diameter, unmyelinated axons.

Tactile receptors in the skin or subcutaneous layer include Meissner corpuscles, hair root plexuses, Merkel discs, Ruffini corpuscles, pacinian corpuscles, and free nerve endings (FIGURE 9.11). For those tactile receptors that are in the skin, some are located in the *epidermis* (outer layer of the skin) and others are found in the *dermis* (inner layer of the skin). Although not shown in FIGURE 9.11, the body actually contains two major types of skin: glabrous and hairy. *Glabrous skin* (GLĀY-brus) is smooth and hairless; it is found on the lips, fingertips, palms of the hands, soles of the feet, and some parts of the external genitalia. The rest of the skin of the body is *hairy skin*, and the amount of hair that forms can vary from person to person. Most tactile receptors are present in both types of skin; the exceptions are Meissner corpuscles, which are found only in glabrous skin, and hair root plexuses, which are found only in hairy skin.

Touch

Sensations of **touch** generally result from stimulation of tactile receptors in the skin. There are two types of rapidly adapting touch receptors: Meissner corpuscles and hair root plexuses. **Meissner corpuscles** (MĪS-ner), or *corpuscles of touch*, are touch receptors located in the upper part of the dermis of glabrous skin. Each corpuscle is an egg-shaped mass of nerve endings enclosed by a capsule of connective tissue. Because Meissner corpuscles are rapidly adapting receptors, they generate action potentials mainly at the onset of a touch. **Hair root plexuses** are rapidly adapting touch receptors found in hairy skin; they consist of free nerve endings wrapped around hair follicles. Hair root plexuses detect movements on the skin surface that disturb hairs. For example, an insect landing on a hair causes movement of the hair shaft that stimulates the free nerve endings.

There are also two types of slowly adapting touch receptors: Merkel discs and Ruffini corpuscles. **Merkel discs**, also called *type I cutaneous mechanoreceptors*, are saucer-shaped free nerve endings located at the border of the epidermis and dermis. They respond to continuous touch, such as holding an object in your hand for an extended period of time. **Ruffini corpuscles**, also known as *type II cutaneous mechanoreceptors*, are elongated, encapsulated receptors located in the dermis, subcutaneous layer, and other tissues of the body. They are highly sensitive to skin stretching, such as when a masseuse stretches your skin during a massage.

Pressure

Pressure, a sustained sensation that is felt over a larger area than touch, occurs with deeper deformation of the skin and subcutaneous layer. The receptors that contribute to sensations of pressure are Merkel discs and Ruffini corpuscles. These receptors are able to respond to a steady pressure stimulus because they are slowly adapting.

Vibration

Sensations of **vibration** result from rapid, repetitive sensory signals from tactile receptors. The receptors for vibration sensations are pacinian corpuscles and Meissner corpuscles. A **pacinian corpuscle** (pa-SIN-ē-an), or *lamellated corpuscle*, consists of a nerve ending surrounded by a multilayered connective tissue capsule that resembles a sliced onion. Like Meissner corpuscles, pacinian corpuscles adapt rapidly. They are found in the dermis, subcutaneous layer, and other body tissues. Pacinian corpuscles respond to high-frequency vibrations, such as the vibrations you feel when you use a power drill or other electric tools. Meissner corpuscles also detect vibrations, but they respond to low-frequency vibrations. An example is the vibrations you feel when your hand moves across a textured object such as a basket or paneled door.

FIGURE 9.11 Sensory receptors in the skin and subcutaneous layer.

The somatic sensations of touch, pressure, vibration, itch, tickle, warmth, cold, and pain arise from sensory receptors that are present in the skin and subcutaneous layer.

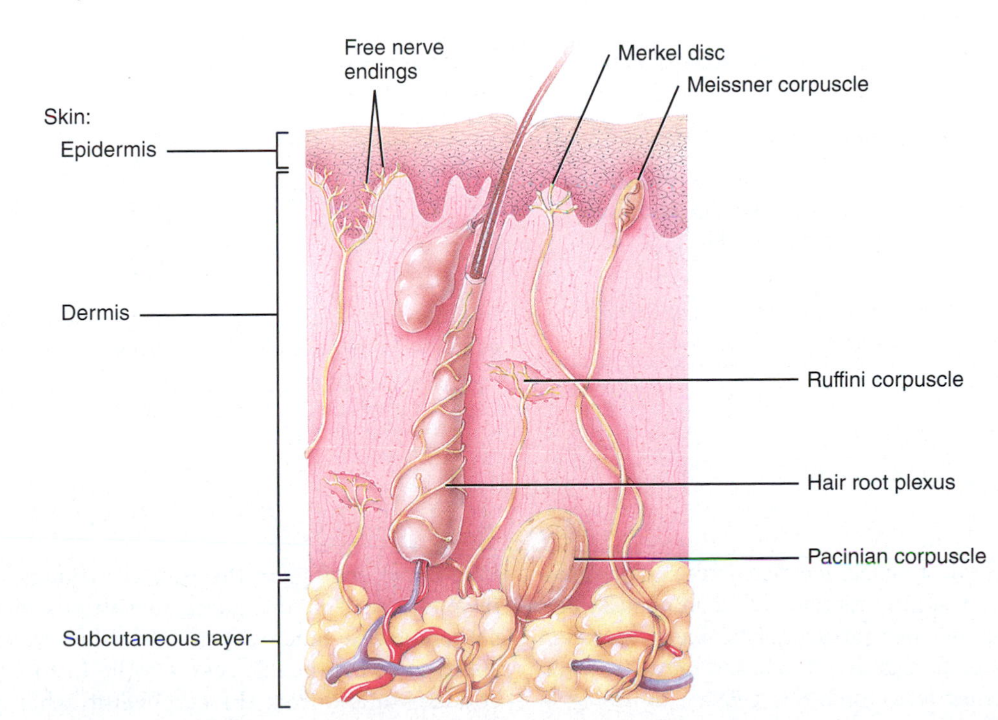

Which somatic sensory receptor generates action potentials mainly at the onset of a touch?

Critical Thinking

Mixed Signals

When Philip hears the word *baseball,* he thinks he is tasting chocolate. When Jessica hears a dog barking, she perceives the color purple. When Adam thinks about an ordered sequence of numbers, each number has a particular personality type (outgoing, shy, mischievous, etc.). What do Philip, Jessica, and Adam have in common? They all have synesthesia.

SOME THINGS TO KEEP IN MIND:

Synesthesia is a neurological phenomenon in which some senses overlap or mix. Stimulation of one sensory pathway results in an automatic stimulation of a different sensory pathway. Synesthesia is a rare condition: About 1 in 2000 people are synesthetes (people who have synesthesia). Synesthesia is a gene-related phenomenon and can result from a single nucleotide exchange.

SOME INTERESTING FACTS:

There are over 60 types of synesthesia; the following table lists several of the different types:

Type of Synesthesia	Description
Grapheme–color	Letters and/or numbers (collectively known as graphemes) are perceived to have a specific color.
Sound-to-color	Sounds or tones stimulate the perception of color.
Number–form	Numbers take on a three-dimensional spacial map.
Order–linguistic personification	Ordered sequences have particular personality types.
Lexical–gustatory	Taste is perceived when words are heard.
Mirror–Touch	Seeing someone else being touched (a pinch, tickle, pat on the back, etc.) evokes perception of the same touch sensation.
Auditory–tactile	Sounds stimulate the perception of tactile sensations in the body.

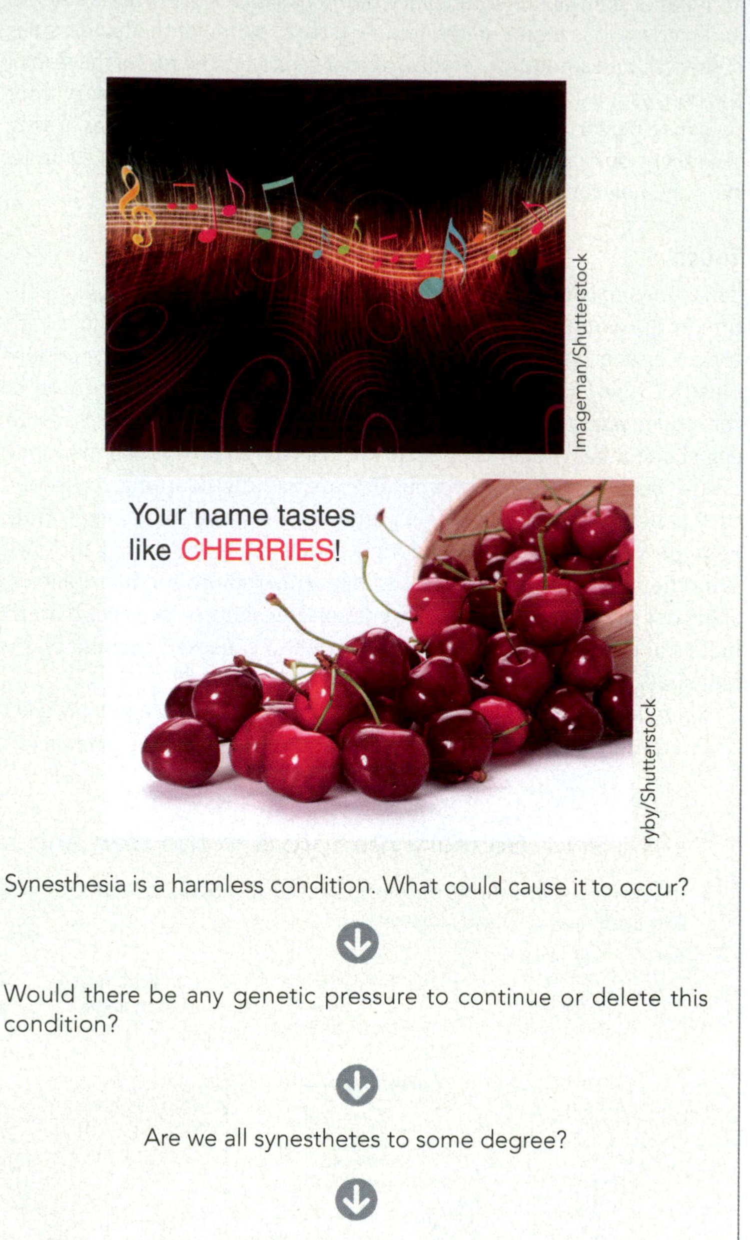

Synesthesia is a harmless condition. What could cause it to occur?

Would there be any genetic pressure to continue or delete this condition?

Are we all synesthetes to some degree?

People using LSD often report experiencing the same type of sensory overlap as synesthetes. Why?

Itch

The **itch** sensation results from stimulation of free nerve endings in the skin by certain chemicals, such as histamine or bradykinin, often because of a local inflammatory response. Receptors for histamine or bradykinin are present in free nerve endings that detect itch, and the activation of these receptors causes a *pruritogenic* (proo-ri-tō-GEN-ic), or *itching, response.* Scratching usually alleviates itching by activating a pathway that blocks transmission of the itch signal through the spinal cord.

Tickle

Free nerve endings in the skin are thought to mediate the **tickle** sensation. This intriguing sensation typically arises only when someone else touches you, not when you touch yourself. The solution to this puzzle seems to lie in the action potentials that conduct to and from the cerebellum when you are moving your fingers and touching yourself that don't occur when someone else is tickling you.

CLINICAL CONNECTION

Phantom Limb Sensation

Patients who have had a limb amputated may still experience sensations such as itching, pressure, tingling, or pain as if the limb were still there. This phenomenon is called **phantom limb sensation**. Although the limb has been removed, severed endings of sensory axons are still present in the remaining stump. If these severed endings are activated, the cerebral cortex interprets the sensation as coming from the sensory receptors in the missing limb. Another explanation for phantom limb sensation is that the area of the cerebral cortex that previously received sensory input from the missing limb undergoes extensive functional reorganization that allows it to respond to stimuli from another body part.The remodeling of this cortical area is thought to give rise to false sensory perceptions from the missing limb. Phantom limb pain can be very distressing to an amputee. Many report that the pain is severe or extremely intense, and that it often does not respond to traditional pain medication therapy. In such cases, alternative treatments may include electrical nerve stimulation, acupuncture, and biofeedback.

Transduction of Tactile Stimuli

Much of what is known about transduction in tactile receptors has been obtained from studies on pacinian corpuscles. Recall that a pacinian corpuscle consists of a nerve ending (dendrite) surrounded by a multilayered connective tissue capsule. Within the membrane of the nerve ending are cation channels that are permeable mainly to Na^+ and Ca^{2+} ions. When the pacinian corpuscle is at rest, these cation channels are closed (**FIGURE 9.12a**). However, when the pacinian corpuscle is stimulated by high-frequency vibration, the vibration stimulus deforms the capsule, which in turn deforms the nerve ending membrane, causing the cation channels to open (**FIGURE 9.12b**). Opening the cation channels allows Na^+ and Ca^{2+} ions to enter the cytosol of the nerve ending, resulting in the formation of a depolarizing receptor potential. If the depolarization brings the membrane to threshold, an action potential is generated in the axon of the sensory neuron.

The capsule of the pacinian corpuscle plays an important role in how the corpuscle adapts. When the stimulus is initially present, the capsule is deformed, the nerve fiber is deformed, cation channels open, and a depolarization is generated (the on response). In the presence of a maintained stimulus, however, the different layers of the capsule, which have fluid between them, slip past one another and relieve the deformation of the nerve ending, causing the cation channels to close and the depolarization to dissipate. When the stimulus is removed, the capsule elastically rebounds to its original shape, temporarily deforming the nerve ending membrane in the process and causing the cation channels to open and another depolarization to form (the off response). Thus, the capsule allows the pacinian corpuscle to respond to a rapid, repetitive stimulus (vibration) but not to a sustained stimulus such as steady pressure.

Thermal Sensations Provide Information About How Hot or Cold the Skin Is

Thermal sensations—coldness and warmth—are detected by **thermoreceptors**, which are free nerve endings that are present in the

FIGURE 9.12 Transduction in a pacinian corpuscle.

Transduction in a pacinian corpuscle involves the opening of cation channels in response to a mechanical stimulus (vibration).

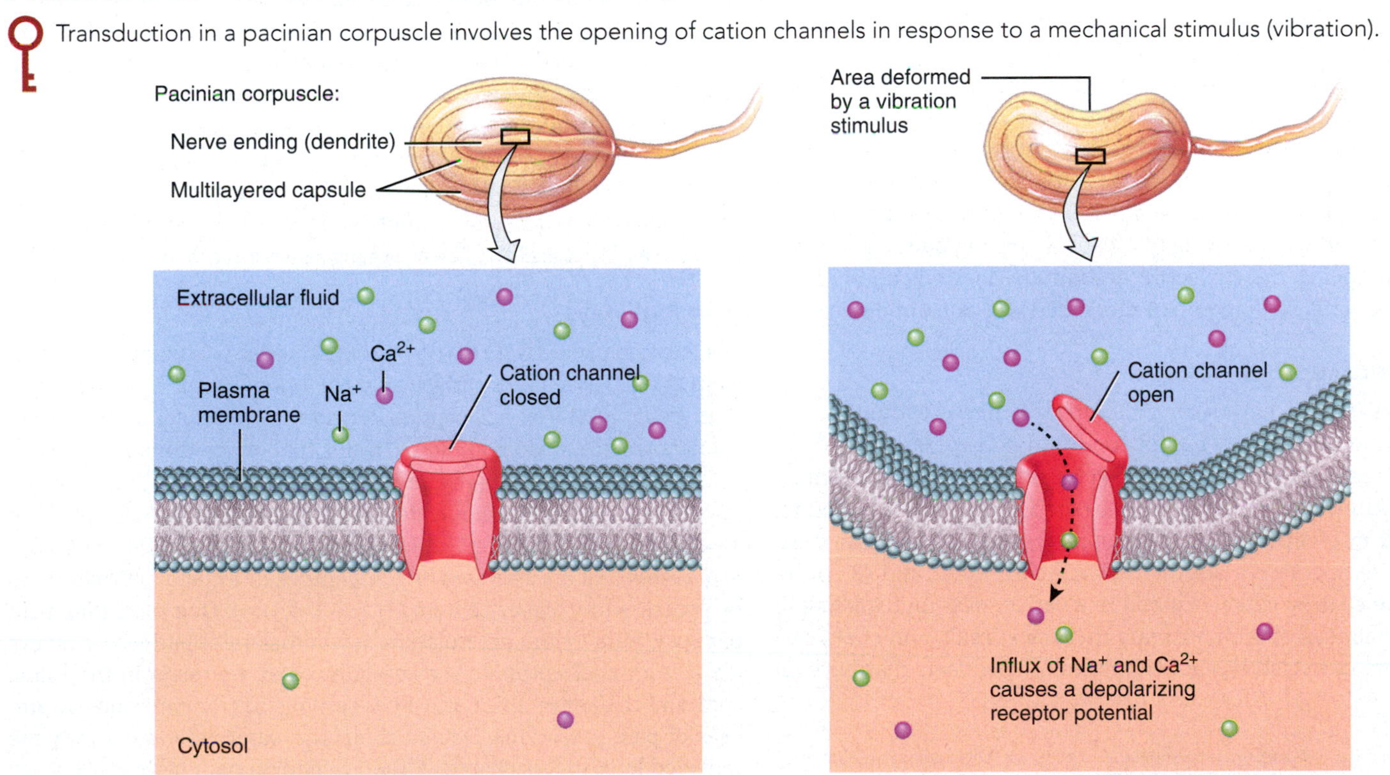

(a) A pacinian corpuscle at rest

(b) Transduction in a pacinian corpuscle

How does the capsule of the pacinian corpuscle contribute to adaptation?

skin. There are two types of thermoreceptors: cold receptors and warm receptors. **Cold receptors** are activated by temperatures between 10° and 35°C (50–95°F). **Warm receptors**, which are not as abundant as cold receptors, are activated by temperatures between 30° and 45°C (86–113°F). Cold and warm receptors both adapt rapidly at the onset of a stimulus, but then they continue to generate action potentials at a lower frequency throughout a prolonged stimulus. Temperatures below 10°C and above 45°C primarily stimulate pain receptors rather than thermoreceptors, producing painful sensations.

Transduction in thermoreceptors involves a class of proteins known as **transient receptor potential (TRP) channels** that are permeable to certain cations. As you will soon learn, TRP channels also play a role in the transduction mechanisms of other types of sensory stimuli, including pain and gustation. Different types of TRP channels are responsible for the transduction mechanisms of different sensory stimuli. The TRP channels present in thermoreceptors are temperature-sensitive, which means that they open in response to certain temperatures. Examples of these channels include *TRPV3*, which is present in warm receptors, and *TRPM8*, which is present in cold receptors.* Warm temperatures stimulate the opening of TRPV3, and cool temperatures stimulate the opening of TRPM8. In either case, opening these channels allows Na^+ ions to enter the cell, resulting in a depolarizing receptor potential and, if threshold is reached, the formation of an action potential in the sensory neuron. TRPV3 and TRPM8 are also sensitive to chemical stimuli. The chemical *camphor* (a topical agent) opens TRPV3, causing a warming sensation; the chemical *menthol* (often present in cough drops) opens TRPM8, which induces a cooling sensation.

Pain Sensations Protect the Body from Stimuli That Can Cause Tissue Damage

Pain is indispensable for survival. It serves a protective role by signaling the presence of noxious stimuli that are causing or are about to cause tissue damage. The amount of pain perceived by a person is highly subjective and can vary among individuals based on their past or present experiences. For example, pain perception may be elevated when a person who is afraid of needles goes to the doctor and has to have blood withdrawn. By contrast, pain perception may be suppressed in an injured person who is trying to escape a burning building. From a medical standpoint, the subjective description and indication of the location of pain may help pinpoint the underlying cause of disease.

Types of Nociceptors

Nociceptors, the receptors for pain, are free nerve endings found in every tissue of the body except the brain and spinal cord. There are three types of nociceptors: (1) **mechanical nociceptors**, which respond to intense mechanical stimuli, such as a pinch or puncture; (2) **thermal nociceptors**, which respond to extreme thermal stimuli such as temperatures above 45°C or below 10°C; and (3) **polymodal nociceptors**, which respond to a variety of stimuli, including intense mechanical stimuli, extreme thermal stimuli, and chemicals released from damaged tissues. Examples of chemicals that have an effect on polymodal nociceptors include bradykinin, potassium ions (K^+), histamine, and prostaglandins. Bradykinin, K^+ ions, and histamine activate nociceptors, whereas prostaglandins *sensitize* nociceptors, which means that they cause nociceptors to become more responsive to noxious stimuli by reducing the nociceptor's threshold for activation. An increased sensitivity to painful stimuli is known as **hyperalgesia**. Pain may persist even after a pain-producing stimulus is removed because pain-mediating chemicals linger and because nociceptors exhibit very little adaptation.

Transduction in Nociceptors

Transduction of noxious stimuli into electrical signals involves ion channels that are present in the membrane of the nociceptor. One example is *TRPV1*†, a member of the transient receptor potential (TRP) family. TRPV1 channels are found in the membrane of a polymodal nociceptor. They are cation channels that open in response to extreme heat or to *capsaicin*—the ingredient that makes chilli peppers painfully hot. Opening these cation channels allows Na^+ and Ca^{2+} ions to enter the nociceptor, which causes a depolarizing receptor potential to form in the nociceptor's membrane. If threshold is reached, an action potential is generated in the axon of the sensory neuron.

Fast and Slow Pain

There are two types of pain: fast and slow. **Fast pain** is a sharp, pricking sensation that is well localized. For example, the pain felt from a needle puncture or a knife cut to the skin is fast pain. Signals for fast pain are transmitted along **A-delta (Aδ) fibers**, small, myelinated axons with conduction velocities ranging from 12 to 30 m/sec. Sensory neurons associated with mechanical or thermal nociceptors have Aδ fibers and are the source of the signals for fast pain. **Slow pain**, by contrast, is a dull, aching sensation that is poorly localized. An example is the pain associated with a minor toothache. Signals for slow pain are transmitted along **C fibers**, small, unmyelinated axons that have conduction velocities ranging from 0.5 to 2 m/sec. Sensory neurons associated with polymodal nociceptors have C fibers and are the source of the signals for slow pain. You can perceive the difference in onset of fast and slow pain when you injure a body part that is far from the brain because the conduction distance is long. When you stub your toe, for example, you first feel the sharp sensation of fast pain and then feel the slower, aching sensation of slow pain.

Pain Pathways

Nociceptors can activate two types of pathways: (1) spinal reflex pathways and (2) ascending pathways to the brain. *Spinal reflex pathways* that are activated by nociceptors provide unconscious protective responses when a noxious stimulus begins to damage the body (**FIGURE 9.13a**). For example, stepping on a tack or touching a hot burner on the stove elicits the flexor reflex, which quickly withdraws the affected limb away from the painful stimulus (see **FIGURE 12.5**). *Ascending pathways to the brain* that are activated by nociceptors allow pain information to be processed by higher centers (**FIGURE 9.13b**). One ascending pain pathway, namely the anterolateral (spinothalamic) pathway, conveys input from nociceptors to the cerebral cortex via relays in the spinal cord and thalamus. This pathway is responsible for conscious awareness of pain sensations. Details about the anterolateral pathway are provided later in this chapter. Other ascending pain pathways convey input from nociceptors to the reticular formation, limbic system, and

* TRP channels are categorized into subfamilies such as the vanilloid subfamily or melastatin subfamily. In addition, each TRP channel is given a number according to its order in the subfamily (member 1, member 3, etc.). Hence, TRPV3 = Transient Receptor Potential, Vanilloid subfamily, member 3; TRPM8 = Transient Receptor Potential, Melastatin subfamily, member 8.

† TRPV1 = Transient Receptor Potential, Vanilloid subfamily, member 1.

FIGURE 9.13 Pain pathways.

Nociceptors can activate two types of pathways: (1) spinal reflex pathways and (2) ascending pathways to the brain.

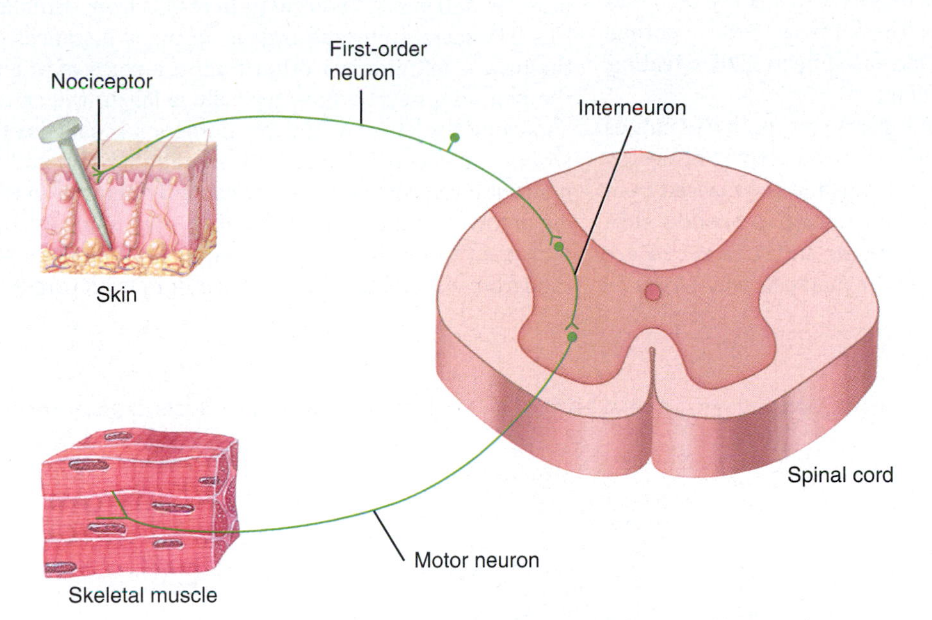

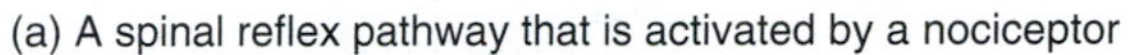

(a) A spinal reflex pathway that is activated by a nociceptor

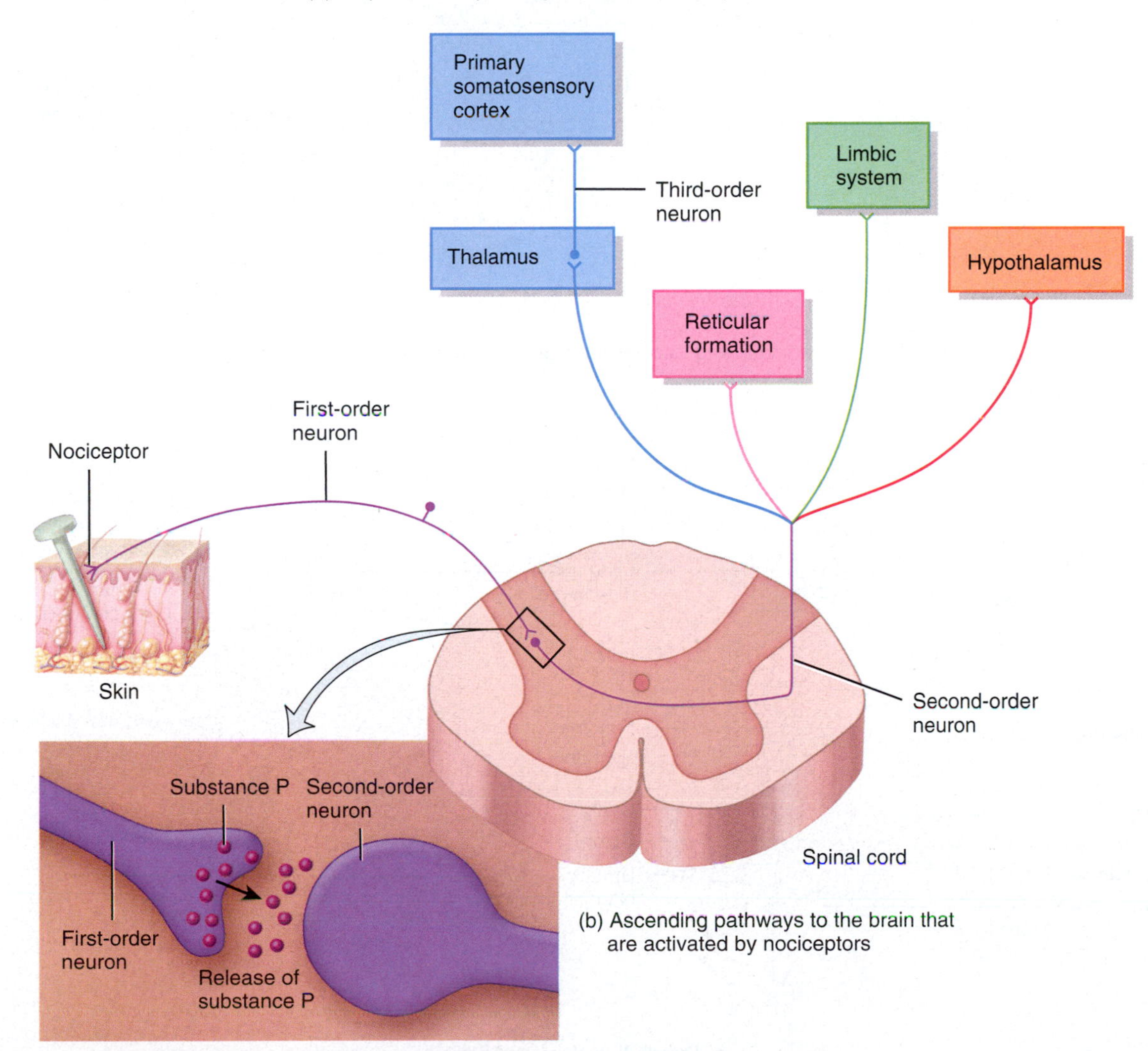

(b) Ascending pathways to the brain that are activated by nociceptors

Which parts of the brain, other than the thalamus and cerebral cortex, receive input from ascending pain pathways?

hypothalamus. The pathway to the reticular formation increases your level of arousal in response to a painful stimulus; the pathway to the limbic system causes the emotional responses (fear, anxiety, etc.) that may occur after a painful experience; and the pathway to the hypothalamus elicits the autonomic responses (increased heart rate, sweating, etc.) that may accompany a painful incident.

Activation of first-order neurons of a given pain pathway causes the release of neurotransmitters that influence the activity of nearby postsynaptic neurons. The two main neurotransmitters released from first-order pain neurons are glutamate and the neuropeptide **substance P** (FIGURE 9.13b, inset). Once glutamate and substance P are released, they activate the next neurons of the pain pathway.

Somatic, Visceral, and Referred Pain

Somatic pain arises from stimulation of nociceptors in skin, skeletal muscles, and joints. **Visceral pain** results from stimulation of nociceptors in visceral (internal) organs. In many instances of visceral pain, the pain is felt at a site other than the place of origin. For example, the pain of a heart attack typically is felt in the skin over the heart and along the left arm. This phenomenon is called **referred pain**. Referred pain occurs because both somatic sensory and visceral sensory neurons often converge on second-order neurons of the same ascending pathway to the brain (FIGURE 9.14a). Since the brain is more accustomed to receiving sensory input from somatic sensory neurons than from visceral sensory neurons, it may incorrectly interpret pain

FIGURE 9.14 Referred pain.

Referred pain occurs because both somatic sensory and visceral sensory neurons often converge on second-order neurons of the same ascending pathway to the brain.

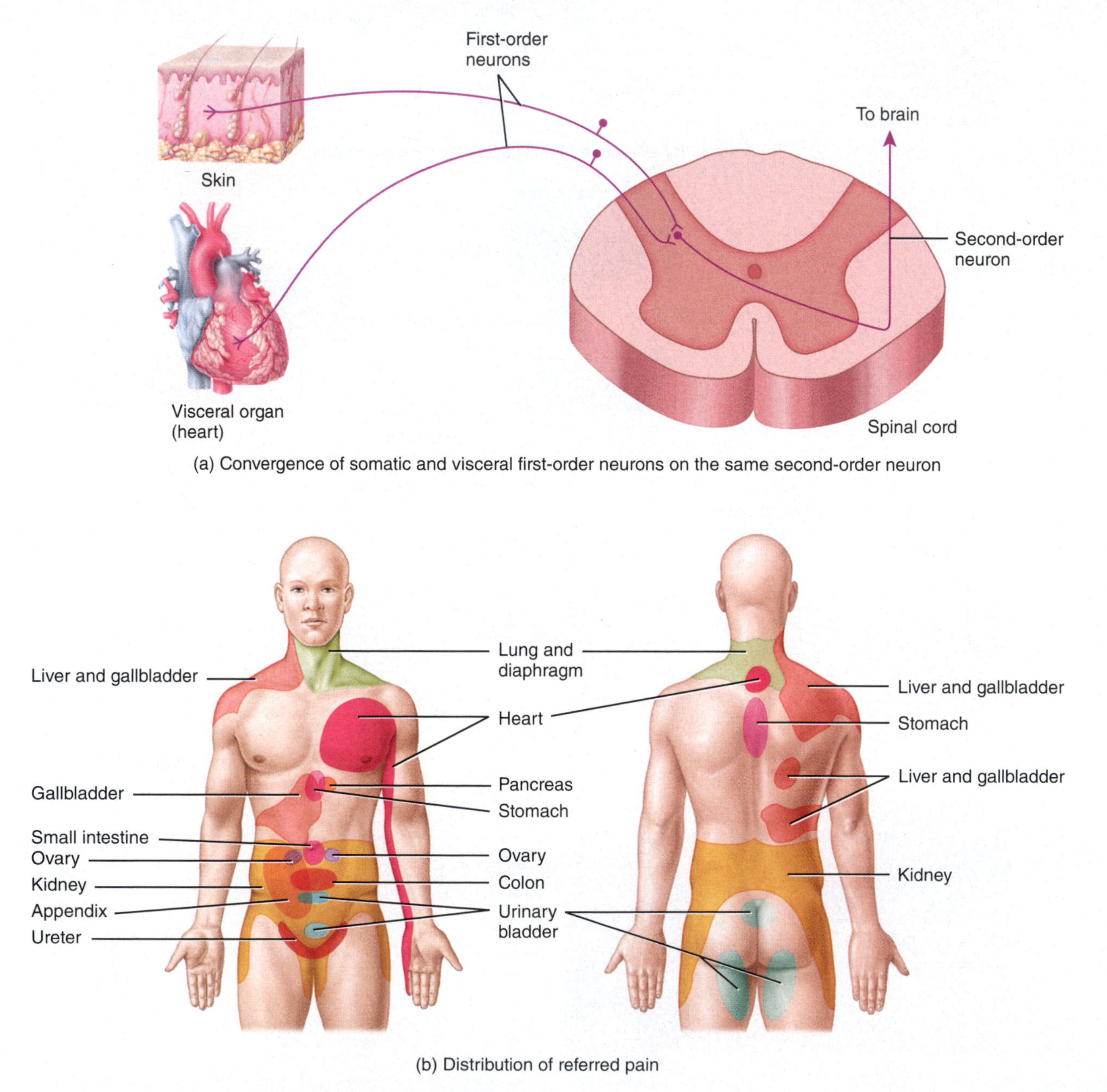

(a) Convergence of somatic and visceral first-order neurons on the same second-order neuron

(b) Distribution of referred pain

Which visceral organ has the broadest area for referred pain?

from a visceral organ as having a somatic origin. **FIGURE 9.14b** shows the skin regions to which visceral pain may be referred.

Suppression of Pain Sensations

Although pain serves a protective role, there may be instances when pain sensations are so severe or occur so frequently that an individual needs pain relief in order to function normally. The individual may seek medical treatment that suppresses pain or may rely on the body's natural mechanisms of pain suppression. There are three main ways that pain sensations can be suppressed: drug therapy, application of a mechanical stimulus to the painful area, and the endogenous analgesia system.

DRUG THERAPY Pain sensations can be suppressed by two types of drugs: analgesics and anesthetics. An *analgesic* is a drug that causes **analgesia**, pain relief that occurs without affecting other modalities of sensation or consciousness. Aspirin and ibuprofen are analgesics that block the formation of prostaglandins, which sensitize nociceptors. Another example of an analgesic is the opiate drug morphine, which suppresses the transmission of pain signals at the synapse between first- and second-order neurons of a pain pathway. **Anesthesia** refers to the absence of all modalities of sensations, including pain sensations. A *general anesthetic* is a drug that affects the entire body and results in loss of consciousness, whereas a *local anesthetic* is a drug that affects only a small area of the body and does not result in loss of consciousness. Local anesthetics, such as procaine (Novocaine®) and lidocaine, provide short-term pain relief by blocking the opening of voltage-gated Na^+ channels in the membranes of axons of all types of sensory neurons, including those that are sensitive to pain. Action potentials cannot propagate past the obstructed region, so pain signals do not reach the CNS.

APPLICATION OF A MECHANICAL STIMULUS TO THE PAINFUL AREA A mechanical stimulus such as touch, pressure, or vibration can suppress pain sensations, a concept known as the **gate control theory of pain**. Within the spinal cord are interneurons that normally inhibit second-order neurons of ascending pain pathways—an action that prevents transmission of pain signals to the brain. When part of the body experiences a painful stimulus without an accompanying mechanical stimulus, C fibers carrying information about the painful stimulus from nociceptors excite second-order pain neurons and completely inhibit the activity of the inhibitory interneurons (**FIGURE 9.15a**). As a result, second-order pain neurons are fully activated and transmit pain signals to the brain. If part of the body experiences a painful stimulus along with a mechanical stimulus, C fibers still activate second-order pain neurons and inhibit the activity of the inhibitory interneurons; however, the inhibitory interneurons are also excited by Aβ fibers (large diameter axons) carrying information about the mechanical stimulus from mechanoreceptors (**FIGURE 9.15b**). Because the inhibitory interneurons receive excitatory input in addition to inhibitory input, they become partially activated. Consequently, second-order pain neurons are partially inhibited and transmit fewer pain signals to the brain. The gate control theory of pain explains why rubbing a damaged part of the body helps to reduce pain in that area: rubbing (a mechanical stimulus) activates Aβ fibers that ultimately decrease the activity of second-order pain neurons. The gate control theory of pain is also the rationale for using **transcutaneous electrical nerve stimulation (TENS)** for pain relief. In TENS, current is applied to a painful area through electrodes placed on the skin. This method provides pain relief by activating nearby Aβ fibers.

FIGURE 9.15 Gate control theory of pain.

The gate control theory of pain states that a mechanical stimulus such as touch, pressure, or vibration can suppress pain sensations.

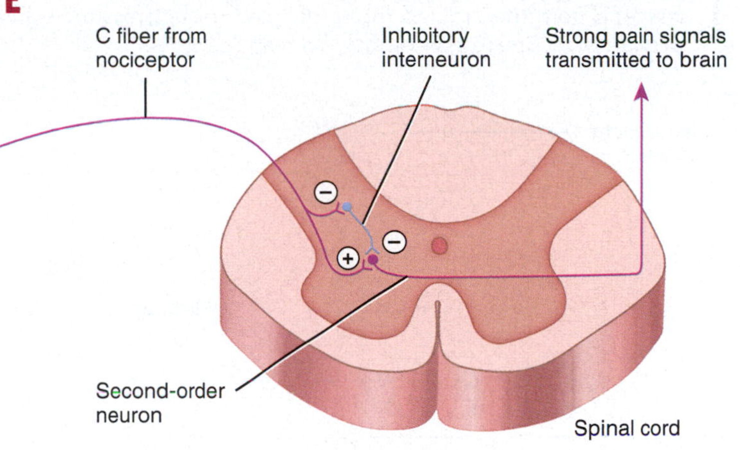

(a) Transmission of strong pain signals when a mechanical stimulus is absent

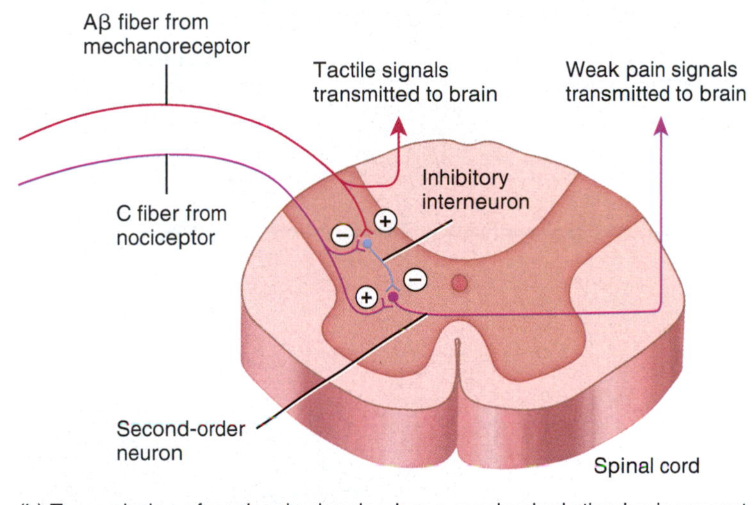

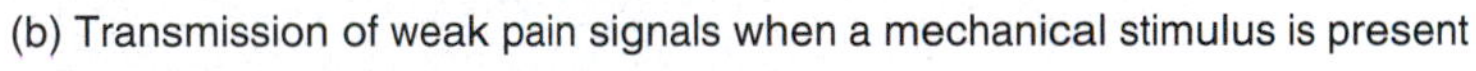

(b) Transmission of weak pain signals when a mechanical stimulus is present

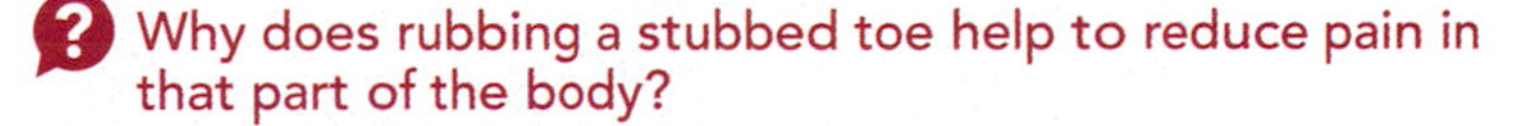

Why does rubbing a stubbed toe help to reduce pain in that part of the body?

ENDOGENOUS ANALGESIA SYSTEM Another way that pain sensations can be suppressed is by the **endogenous analgesia system** of the brain. This system consists of descending pathways that suppress the transmission of incoming pain signals from nociceptors. There are two main components of the brain involved in the endogenous analgesia system: the *periaqueductal gray matter* (a region of gray matter located in the midbrain) and the *nucleus raphe magnus* (a region of gray matter located in the reticular formation of the medulla) (**FIGURE 9.16**). Electrical stimulation of either of these areas produces profound pain relief, a phenomenon known as *stimulation-produced analgesia*. Stress can also activate the components of the endogenous analgesia system, which helps to explain why a wounded soldier may not feel any pain while engaged in battle. An example of one of the pathways of the endogenous analgesia system

FIGURE 9.16 An example of a pathway of the endogenous analgesia system.

The endogenous analgesia system consists of pathways that descend from the brain to the spinal cord to suppress the transmission of incoming pain signals from nociceptors.

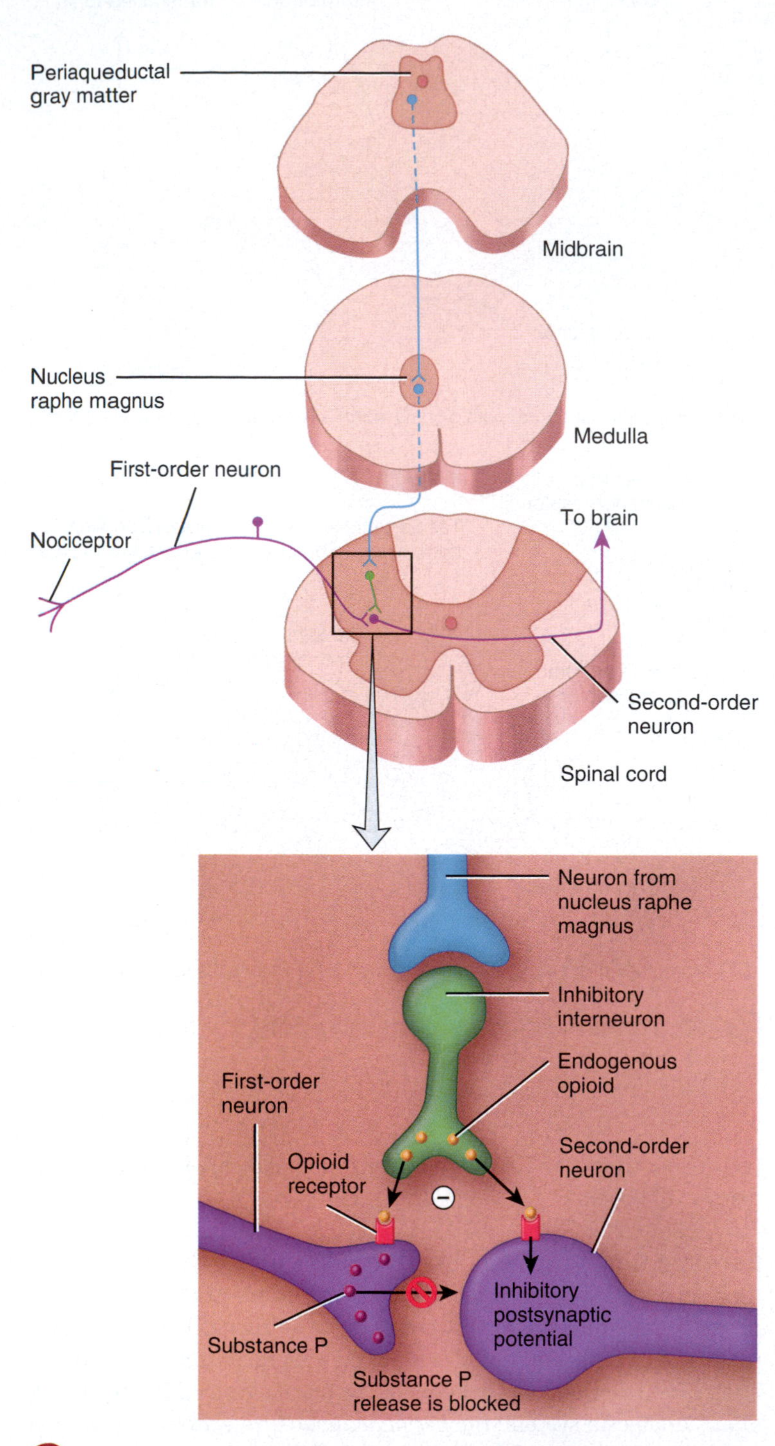

What are the three types of endogenous opioids?

is as follows (FIGURE 9.16): Neurons from the periaqueductal gray matter synapse with neurons in the nucleus raphe magnus. Axons of neurons from the nucleus raphe magnus in turn descend to the spinal cord, where they synapse with inhibitory interneurons. The axon terminals of the inhibitory interneurons are located close to the synapses between first- and second-order pain neurons. When activated, the inhibitory interneurons suppress transmission of pain signals at these synapses. A key feature of the endogenous analgesia system is that it uses three types of neuropeptides that have morphine-like actions—*enkephalins*, *endorphins*, and *dynorphins*, which are collectively known as the **endogenous opioids**. Endogenous opioids are released by neurons of the endogenous analgesia system and by the inhibitory interneurons in the spinal cord that are associated with this system. Endogenous opioids bind to and activate **opioid receptors**, which are present in the axon terminals of first-order pain neurons and in the dendrites and cell bodies of second-order pain neurons. Activation of opioid receptors has two effects (FIGURE 9.16): (1) It blocks the release of substance P from first-order pain neurons via presynaptic inhibition and (2) it induces postsynaptic inhibition of second-order pain neurons by causing these cells to form inhibitory postsynaptic potentials (IPSPs). The net result of these effects is suppression of pain signal transmission to the brain. Like the endogenous opioids, the exogenous drug morphine produces analgesia by binding to opioid receptors in first- and second-order pain neurons.

Proprioceptive Sensations Provide Information About Muscle and Joint Position

Proprioceptive sensations allow us to know where our limbs are located and how they are moving even if we are not looking at them, so that we can walk, type, or dress without using our eyes. **Kinesthesia** (kin′-es-THĒ-zē-a; *kin-* = motion; *-esthesia* = perception) is the perception of body movements. Proprioceptive sensations arise in receptors termed **proprioceptors** (PRŌ-prē-ō-sep′-tors). Those proprioceptors embedded in muscles (especially postural muscles) and tendons inform us of the degree to which muscles are contracted, the amount of tension on tendons, and the positions of joints. Because most proprioceptors adapt slowly and only slightly, the brain continually receives action potentials related to the position of different body parts and makes adjustments to ensure coordination.

CLINICAL CONNECTION

Acupuncture

Acupuncture is a type of therapy that originated in China over 2000 years ago. It is based on the idea that vital energy called ***qi*** (pronounced chee) flows through the body along pathways called ***meridians***. Practitioners of acupuncture believe that illness results when the flow of qi along one or more meridians is blocked or out of balance. Acupuncture is performed by inserting fine needles into the skin at specific locations in order to unblock and rebalance the flow of qi. A main purpose for using acupuncture is to provide pain relief. It is thought that acupuncture relieves pain by activating sensory neurons that ultimately trigger the release of endogenous opioids from the endogenous analgesia system of the brain. Studies have shown that acupuncture is a safe procedure as long as it is administered by a trained professional who uses a sterile needle for each application site. Therefore, many members of the medical community consider acupuncture to be a viable alternative to traditional methods for relieving pain.

Proprioceptors also allow **weight discrimination**, the ability to assess the weight of an object. This type of information helps you to determine the muscular effort necessary to perform a task. For example, as you pick up a shopping bag, you quickly realize whether it contains books or clothes, and you then exert the correct amount of effort needed to lift it.

Three types of proprioceptors will now be discussed: muscle spindles, tendon organs, and joint kinesthetic receptors.

Muscle Spindles

Muscle spindles are proprioceptors located within skeletal muscles. Each **muscle spindle** consists of a connective tissue capsule that surrounds several sensory nerve endings wrapped around 3 to 10 **intrafusal muscle fibers** (*intrafusal* = inside a spindle) (**FIGURE 9.17**). Intrafusal fibers are specialized muscle fibers that contain contractile filaments at each end but not in the center. There are two types of intrafusal fibers: (1) **nuclear bag fibers**, which have a dilated central region containing a large number of nuclei and (2) **nuclear chain fibers**, which have a thin central region with nuclei arranged in a row. Muscle spindles are interspersed among most skeletal muscle fibers and aligned parallel to them.

Muscle spindles perform two functions: They detect *static* muscle length as well as *changes* in muscle length. To accomplish these functions, two types of sensory endings terminate on the intrafusal fibers. **Primary (annulospiral) endings** wrap around the central regions of the nuclear bag and nuclear chain fibers. These endings are rapidly adapting and respond to changes in muscle length. **Secondary (flower-spray) endings** terminate along the contractile portions of the nuclear chain fibers. They are slowly adapting and respond to static muscle length. Either sudden or prolonged stretching of the intrafusal muscle fibers stimulates the sensory nerve endings. This occurs by opening mechanically-gated cation channels in the sensory nerve ending membrane. Opening these channels allows mainly Na^+ and Ca^{2+} ions to enter the cytosol of the nerve ending, forming a depolarizing receptor potential. The resulting action potentials propagate into the CNS along the axon of the sensory neuron. Information from muscle spindles is used to participate in stretch reflexes (see **FIGURE 12.3**). In addition, information from muscle spindles arrives at the somatosensory cortex, which allows conscious awareness of limb positions and movements. At the same time, action potentials from muscle spindles pass to the cerebellum, where the input is used to coordinate muscle contractions.

In addition to their sensory nerve endings near the middle of intrafusal fibers, muscle spindles receive innervation from motor neurons called **gamma motor neurons**. These motor neurons terminate near both ends of the intrafusal fibers and adjust the tension in a muscle spindle to variations in the length of the muscle (**FIGURE 9.17**). For example, when a muscle shortens, gamma motor neurons stimulate the ends of the intrafusal fibers to contract slightly. This keeps the intrafusal fibers taut and maintains the sensitivity of the muscle spindle to stretching of the muscle. As the frequency of action potentials in its gamma motor neuron increases, a muscle spindle becomes more sensitive to stretching of its midregion.

Surrounding muscle spindles are ordinary skeletal muscle fibers, called **extrafusal muscle fibers** (*extrafusal* = outside a spindle), which

FIGURE 9.17 Two types of proprioceptors: a muscle spindle and a tendon organ.

Muscle spindles detect static muscle length and changes in muscle length; tendon organs detect changes in muscle tension.

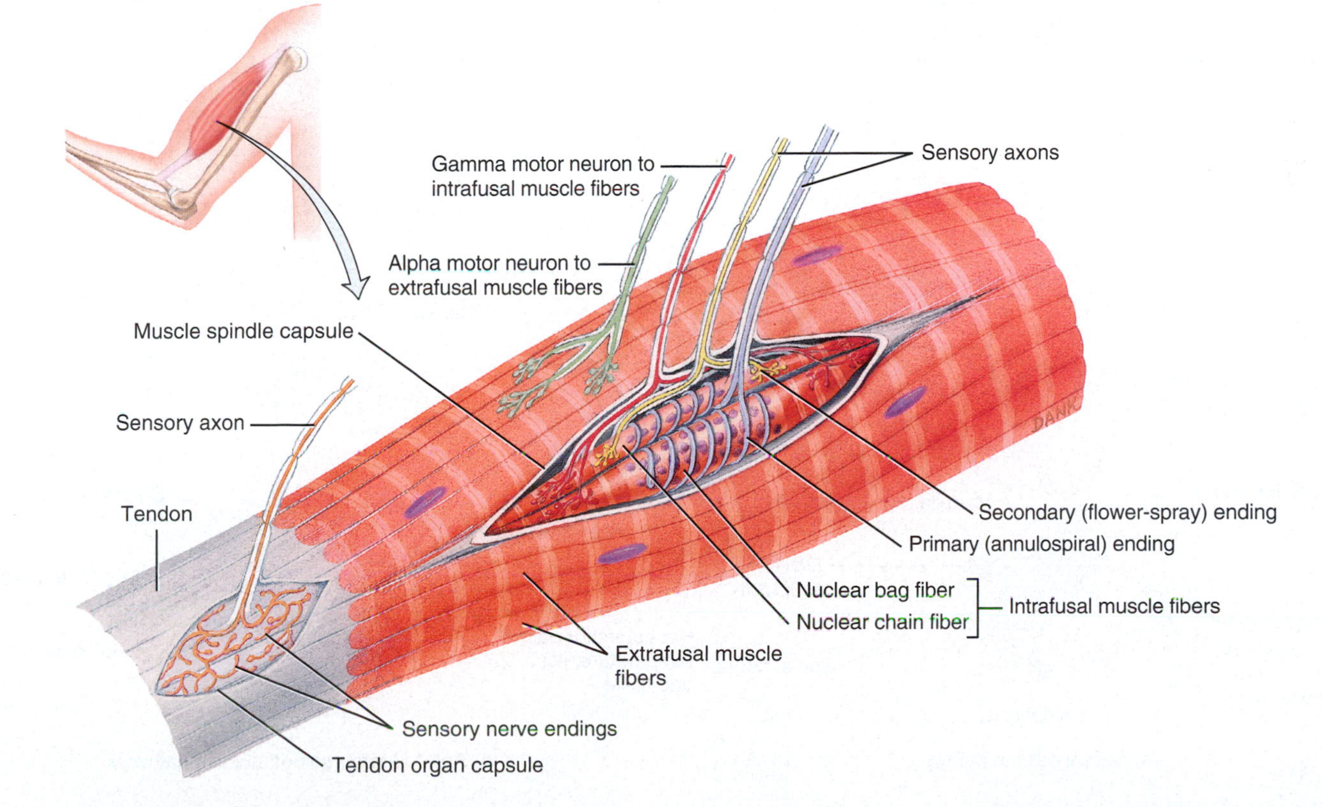

How is a muscle spindle activated?

are supplied by **alpha motor neurons** (FIGURE 9.17). The cell bodies of both gamma and alpha motor neurons are located in the ventral gray horn of the spinal cord (or in the brain stem for muscles in the head). During the stretch reflex, action potentials in muscle spindle sensory axons propagate into the spinal cord and brain stem and activate alpha motor neurons that connect to extrafusal muscle fibers in the same muscle. In this way, activation of its muscle spindles causes contraction of a skeletal muscle, which relieves the stretching.

Tendon Organs

Tendon organs are slowly adapting receptors located at the junction of a tendon and a muscle. By initiating tendon reflexes (see FIGURE 12.4), tendon organs protect tendons and their associated muscles from damage due to excessive tension. (When a muscle contracts, it exerts a force that pulls the points of attachment of the muscle at either end toward each other. This force is the muscle tension.) Each **tendon organ** consists of a connective tissue capsule that surrounds one or more sensory nerve endings entwined around bundles of collagen fibers within the tendon (FIGURE 9.17). When tension is applied to a muscle, the tendon organs generate action potentials that propagate into the CNS, providing information about changes in muscle tension. Tendon reflexes decrease muscle tension by causing muscle relaxation.

Joint Kinesthetic Receptors

Several types of **joint kinesthetic receptors** are present within and around the capsules of joints. Free nerve endings and Ruffini corpuscles in joint capsules respond to pressure. Ruffini corpuscles also respond to joint rotation. Small pacinian corpuscles around joint capsules respond to acceleration and deceleration of joints during movement. Articular ligaments contain receptors similar to tendon organs that adjust reflex inhibition of the adjacent muscles when excessive strain is placed on the joint.

Two Major Pathways Convey Somatic Sensory Input to the Primary Somatosensory Cortex

Somatic sensory pathways relay information from somatic sensory receptors to the primary somatosensory cortex in the parietal lobe (FIGURE 9.18). The two major somatic sensory pathways are the dorsal column pathway and the anterolateral (spinothalamic) pathway. These pathways decussate either in the brain stem or spinal cord. Therefore,

FIGURE 9.18 Somatic sensory pathways.

Somatic sensory pathways convey information from somatic sensory receptors to the primary somatosensory cortex.

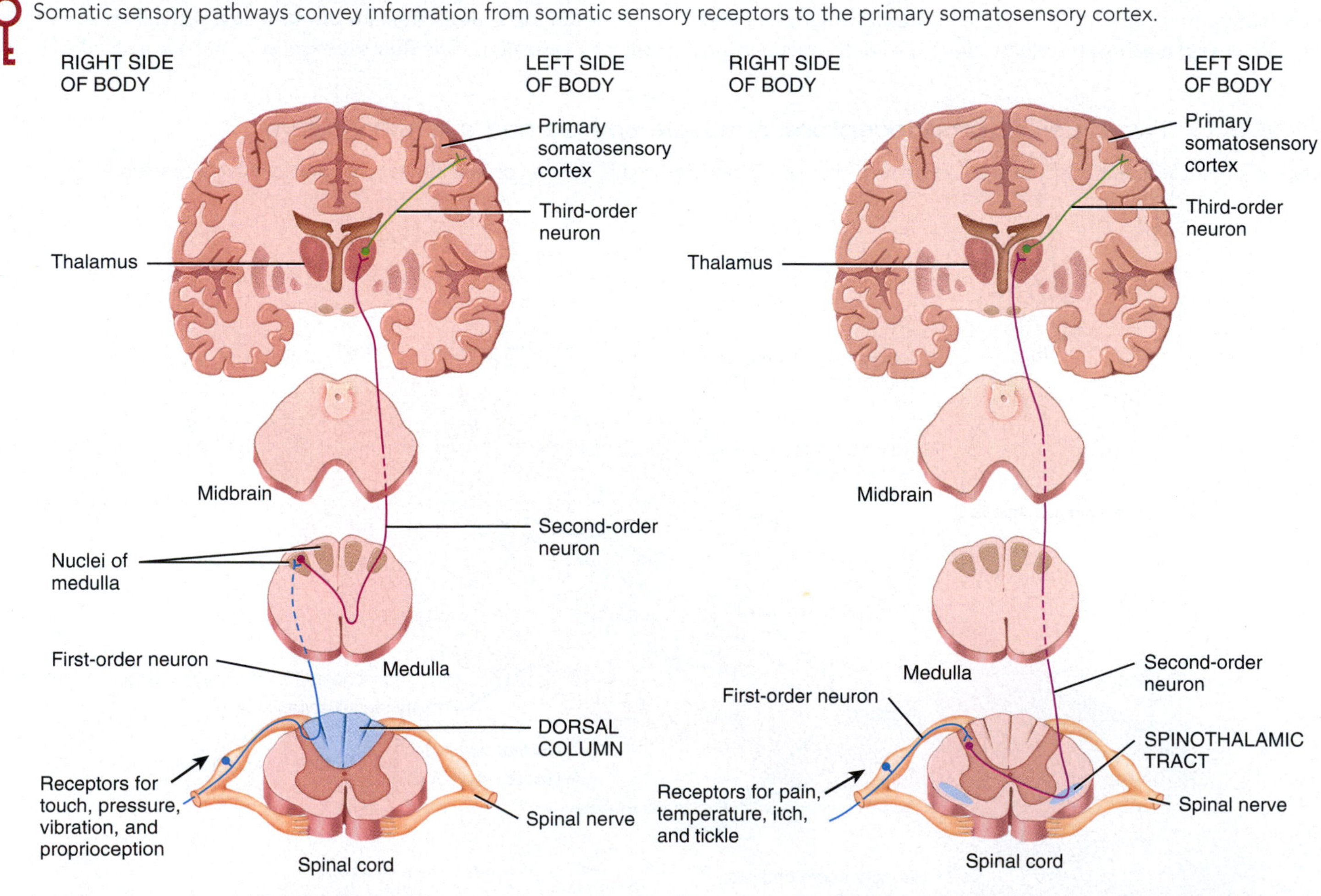

(a) Dorsal column pathway

(b) Anterolateral (spinothalamic) pathway

What types of sensory deficits could be produced by damage to the right spinothalamic tract?

somatic sensations that occur on one side of the body are perceived by the primary somatosensory cortex on the opposite side of the brain.

Dorsal Column Pathway

The **dorsal column pathway** conveys action potentials for touch, pressure, vibration, and proprioception to the cerebral cortex (**FIGURE 9.18a**). This pathway begins with first-order neurons that extend from peripheral sensory receptors into the spinal cord. After entering the spinal cord, axons of these first-order neurons ascend to the medulla via tracts known as the **dorsal columns**. Once the axons of the first-order neurons are in the medulla, they synapse with second-order neurons. Axons of the second-order neurons decussate in the medulla and then ascend to the thalamus, where they synapse with third-order neurons. Axons of the third-order neurons in turn project to the primary somatosensory cortex.

Anterolateral Pathway

The **anterolateral (spinothalamic) pathway** conveys action potentials for pain, temperature, itch, and tickle to the cerebral cortex (**FIGURE 9.18b**). In this pathway, first-order neurons extend from peripheral sensory receptors into the spinal cord, where they synapse with second-order neurons. The axons of the second-order neurons decussate in the spinal cord and then ascend to the thalamus as the **spinothalamic tract**. In the thalamus, the axons of the second-order neurons synapse with third-order neurons, which project their axons to the primary somatosensory cortex.

The Primary Somatosensory Cortex Allows Precise Localization of Somatic Sensory Stimuli

Precise localization of somatic sensations occurs when action potentials arrive at the **primary somatosensory cortex**. This enables you to identify the exact parts of the body where somatic sensations originate. The primary somatosensory cortex also permits you to determine the size, shape, texture, and weight of an object by feeling it and to sense the relationship between one part of the body and another.

A "map" of the entire body exists in the primary somatosensory cortex. This means that each region within this area receives somatic sensory input from a different part of the body (**FIGURE 9.19**). Note that some parts of the body—mainly the face, lips, tongue, and hand—provide input to large regions in the primary somatosensory cortex. Other parts of the body, such as the trunk and lower limbs, project to much smaller cortical regions. The relative sizes of these regions in the primary somatosensory cortex are proportional to the number of sensory receptors within the corresponding part of the body. For example, there are many somatic sensory receptors in the skin of the lips but few in the skin of the trunk. This distorted somatic sensory map of the body is known as the **sensory homunculus** (*homunculus* = little man). The size of the cortical region that represents a body part may expand or shrink somewhat, depending on the quantity of sensory input received from the body part. For example, people who learn to read Braille eventually have a larger cortical region in the primary somatosensory cortex to represent the fingertips.

The Somatosensory Association Area Permits Recognition of Somatic Sensory Stimuli

From the primary somatosensory cortex, neurons project axons to the **somatosensory association area** in the parietal lobe of the cerebral cortex (see **FIGURE 8.8**). Its role is to interpret the meaning of somatic sensory information. The somatosensory association area stores memories of somatic sensory experiences and compares current sensations with previous ones. This allows you to recognize an object by using somatic sensory cues (for example, touching or feeling the object). So when you reach into your pocket, you are able to recognize items such as coins or keys simply by touching them because of the somatosensory association area.

FIGURE 9.19 The sensory homunculus.

The sensory homunculus is the distorted map of the body that exists in the primary somatosensory cortex.

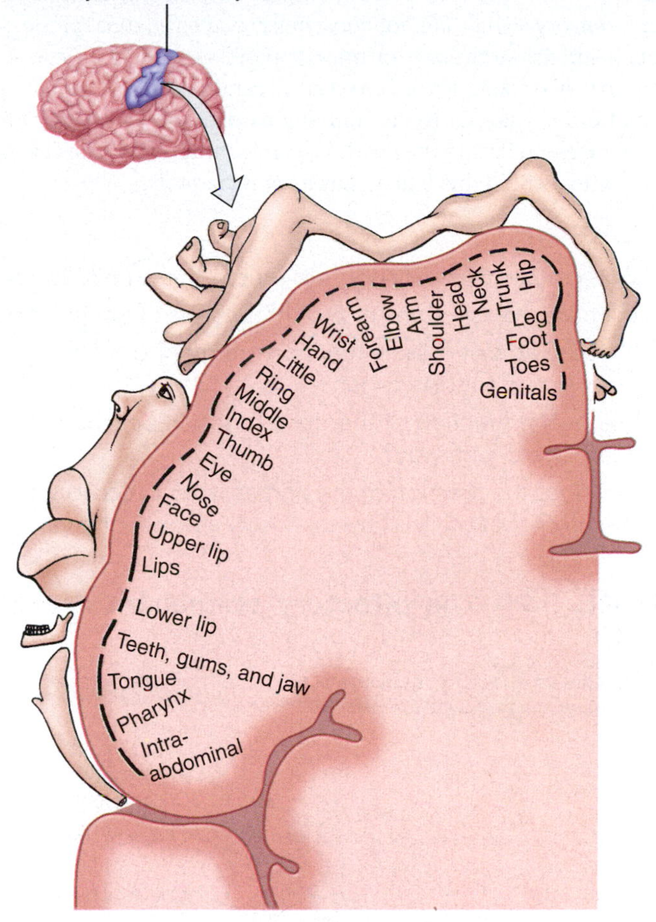

Which areas of the body have the greatest representation in the primary somatosensory cortex? Why?

Visceral Sensations Provide Input About Internal Conditions

Although somatic sensations deal with sensory input from the skin, subcutaneous layer, muscles, and joints, by convention they do not include sensations from the viscera (internal organs). **Visceral sensations** provide information about conditions in the visceral organs of the body. The sensory neurons that convey visceral sensations into the CNS are associated with **interoceptors**, sensory receptors that respond to *internal* stimuli. Interoceptors are located in visceral organs, blood vessels, and specific regions of the nervous system. Examples of interoceptors include mechanoreceptors that detect the degree of stretch in the walls of organs or blood vessels, chemoreceptors that monitor the blood CO_2 level, and nociceptors that detect visceral pain. By contrast, sensory receptors such as touch receptors that detect stimuli from the external environment are referred to as **exteroceptors**. Unlike the sensory signals triggered by a flower's perfume, a beautiful painting, or a delicious meal, visceral sensory signals are not consciously perceived most of the time, although intense activation of interoceptors may produce conscious sensations. Two examples of perceived visceral sensations are pain sensations from damaged viscera and angina pectoris (chest pain) from inadequate blood flow to the heart. Visceral sensations will be discussed in association with individual organs in later chapters.

CHECKPOINT

6. Which tactile receptors detect touch? Pressure? Vibration?
7. What are the functions of the three types of nociceptors?
8. What aspects of muscle function are monitored by muscle spindles and tendon organs?
9. What is the function of the dorsal column pathway? The anterolateral pathway?
10. How does the function of the primary somatosensory cortex differ from that of the somatosensory association area?

9.3 The Olfactory System

OBJECTIVES

- Describe the components of the olfactory epithelium.
- Explain the process of olfactory transduction.
- Discuss the olfactory pathway to the brain.

Both smell and taste are chemical senses because they arise from the interaction of chemicals with smell or taste receptors. Since action potentials for smell and taste propagate to the limbic system (and to higher cortical areas as well), certain odors and tastes can evoke strong emotional responses or a flood of memories.

The Olfactory Epithelium Contains the Receptors for Smell

The receptors for **olfaction**, the sense of smell, are located in a 5-cm^2 patch called the *olfactory epithelium* in the upper part of the nasal cavity. The olfactory epithelium consists of three types of cells: olfactory receptor cells, supporting cells, and basal cells (FIGURE 9.20).

FIGURE 9.20 The olfactory epithelium. The olfactory epithelium consists of olfactory receptor cells, supporting cells, and basal cells.

The cilia of olfactory receptor cells contain olfactory receptor proteins that detect odorants.

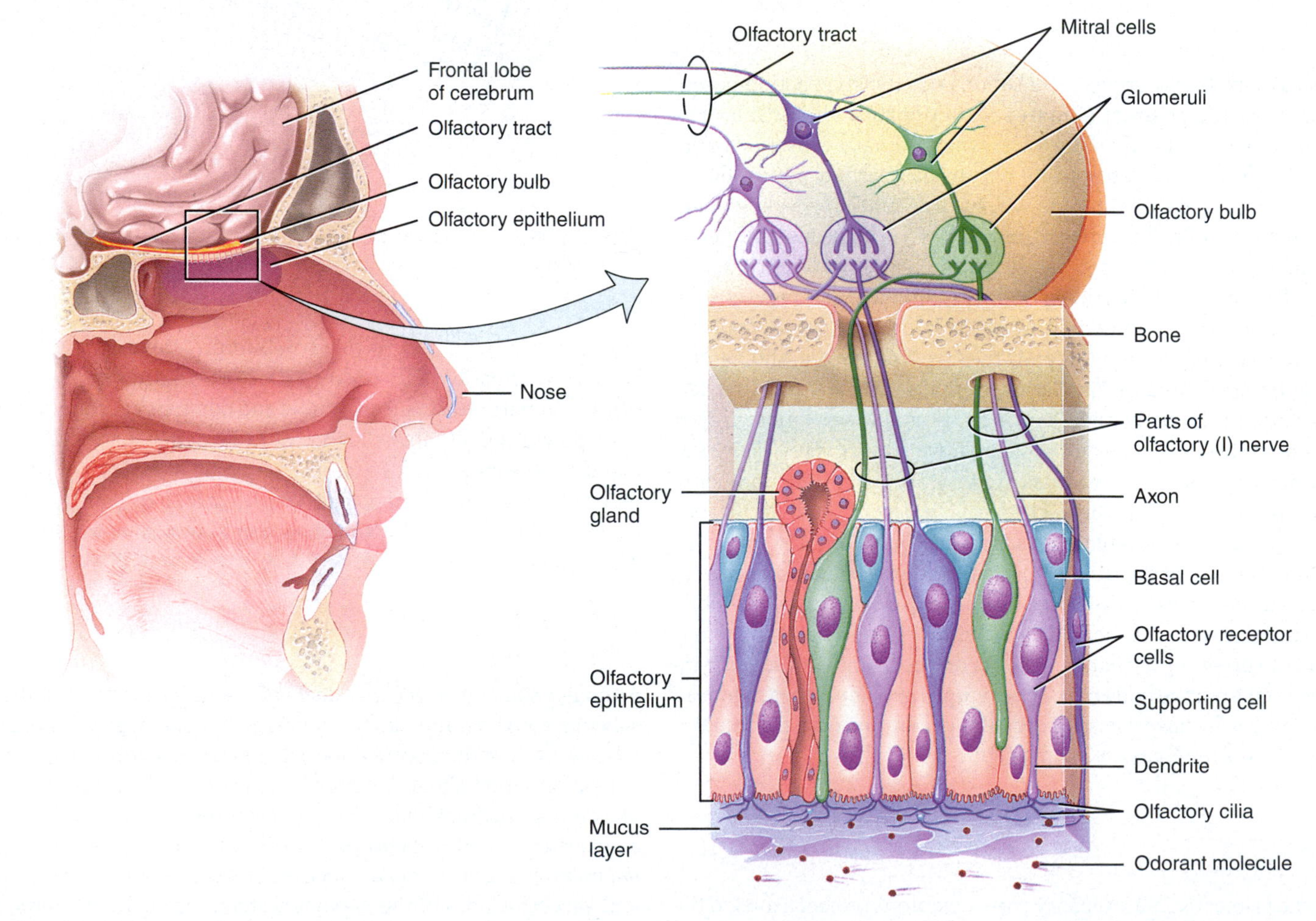

What are odorants?

Olfactory receptor cells are sensory neurons that respond to olfactory stimuli. Extending from the dendrite of an olfactory receptor cell are several nonmotile cilia, which are the sites of olfactory transduction. Within the plasma membranes of the olfactory cilia are *olfactory receptor* proteins that detect inhaled chemicals. Chemicals that bind to and stimulate the receptors in the olfactory cilia are called **odorants**. Stimulation of olfactory receptors by odorants initiates the olfactory response. **Supporting cells** provide physical support to the olfactory receptor cells and help detoxify chemicals that come in contact with the olfactory epithelium. **Basal cells** are stem cells that continually undergo cell division to produce new olfactory receptor cells, which live for only about two months before being replaced. This process is remarkable considering that olfactory receptor cells are neurons and, as you have already learned, mature neurons are generally not replaced. The olfactory epithelium also contains **olfactory glands**, which produce mucus that moistens the surface of the olfactory epithelium and dissolves odorants so that transduction can occur. When you have a cold, this layer of mucus becomes thicker and the ability to smell decreases because a thicker layer of mucus makes it more difficult for odorants to reach the olfactory receptors.

Olfactory Transduction Converts an Olfactory Stimulus into a Receptor Potential

Olfactory receptors react to odorants in the same way that most sensory receptors react to their stimuli: A depolarizing receptor potential develops and triggers one or more action potentials. The olfactory transduction process involves the following steps (**FIGURE 9.21**):

1. An odorant binds to an olfactory receptor protein in the plasma membrane of an olfactory cilium.

FIGURE 9.21 Olfactory transduction. Binding of an odorant molecule to an olfactory receptor protein activates a G protein and adenylyl cyclase, resulting in the production of cAMP. Cyclic AMP opens a cation channel that allows Na^+ and Ca^{2+} ions to enter the olfactory receptor. The resulting depolarization may generate an action potential, which propagates along the axon of the olfactory receptor cell.

Odorants can produce depolarizing receptor potentials, which can lead to action potentials.

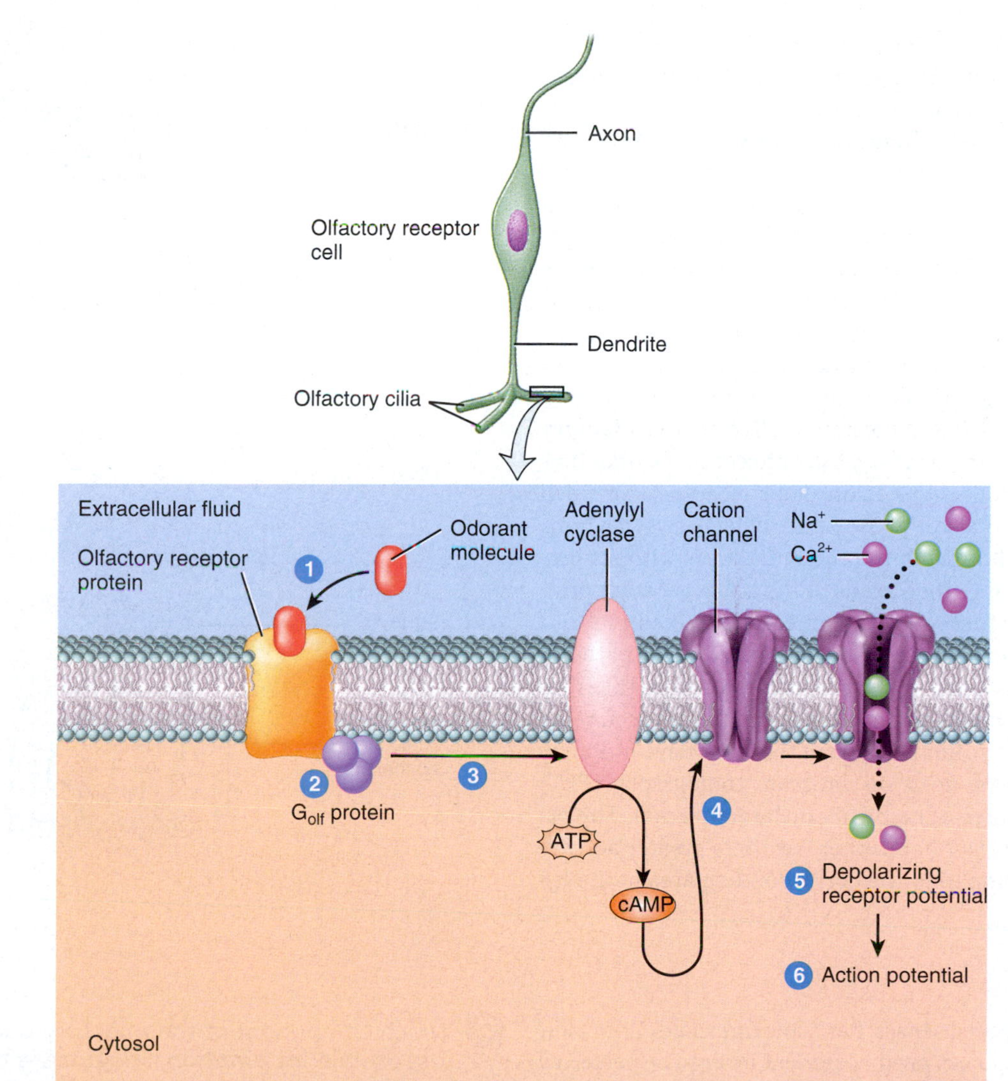

How many different types of olfactory receptor proteins exist in the human olfactory system?

2. Binding of the odorant to the olfactory receptor protein stimulates a G protein known as $\mathbf{G_{olf}}$ (*olf* = olfactory).
3. G_{olf} activates adenylyl cyclase to produce the second messenger cAMP.
4. cAMP opens a cation channel that allows mainly Na^+ and Ca^{2+} ions to enter the cytosol.
5. The influx of Na^+ and Ca^{2+} causes a depolarizing receptor potential to form in the membrane of the olfactory receptor cell.
6. If the depolarization reaches threshold, an action potential is generated along the axon of the olfactory receptor cell.

The human nose contains about 10 million olfactory receptors, of which there are about 400 different functional types. Each type of olfactory receptor can react to only a select group of odorants. Only one type of receptor is found in any given olfactory receptor cell. Therefore, 400 different types of olfactory receptor cells are present in the olfactory epithelium. Our ability to recognize about 10,000 different odors depends on patterns of activity in the brain that arise from activation of many different combinations of the olfactory receptor cells.

The Olfactory Pathway Extends from Olfactory Receptors to the Olfactory Areas of the Brain

The olfactory receptor cells are the first-order neurons of the olfactory pathway. On each side of the nose, bundles of axons of olfactory receptor cells form the right and left **olfactory (I) nerves** (FIGURE 9.22). The olfactory nerves extend to parts of the brain known as the **olfactory bulbs**, which contain ball-like arrangements called **glomeruli** (glō-MER-ū-lī = little balls; singular is *glomerulus*). Within each glomerulus, axons of olfactory receptor cells converge onto **mitral cells**—the second-order neurons of the olfactory pathway. Each glomerulus receives input from only one type of olfactory receptor (see FIGURE 9.20). This allows the mitral cells of a particular glomerulus to convey information about a select group of odorants to the remaining parts of the olfactory pathway. The axons of the mitral cells form the **olfactory tract**. Some of the axons of the olfactory tract project to the **olfactory cortex** in the temporal lobe, where conscious awareness of smell occurs (FIGURE 9.22; see also FIGURE 8.8). Olfactory sensations are the only sensations that reach the cerebral cortex without first synapsing in the thalamus. Other axons of the olfactory tract project to the limbic system; these neural connections account for our emotional responses to odors. From the olfactory cortex, a pathway extends via the thalamus to the **orbitofrontal cortex** in the frontal lobe, where odor identification and discrimination occur (see FIGURE 8.8). People who suffer damage in this area have difficulty identifying different odors. Positron emission tomography (PET) studies suggest some degree of hemispheric lateralization: The orbitofrontal cortex of the *right* hemisphere exhibits greater activity during olfactory processing than the corresponding area in the *left* hemisphere.

FIGURE 9.22 Olfactory pathway.

The olfactory pathway conveys olfactory information from olfactory receptors in the nose to processing centers in the cerebral cortex and limbic system.

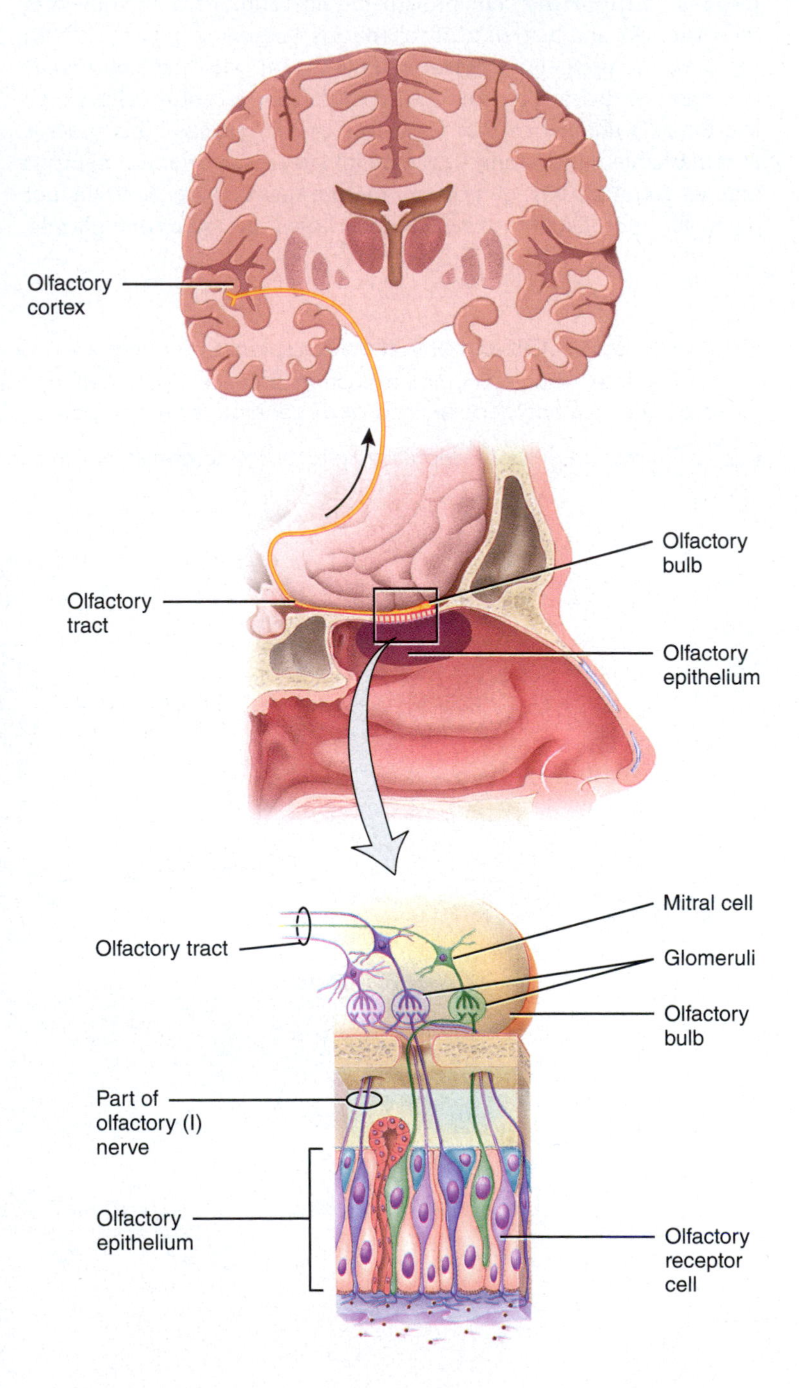

Which component of the olfactory pathway is responsible for emotional responses to odors?

The Threshold for Detecting Odors Is Low

Olfaction, like all of the special senses, has a low threshold. Only a few molecules of certain substances need be present in air to be perceived

CLINICAL CONNECTION

Hyposmia

Hyposmia (hī-POZ-mē-a; *-osmi-* = smell, odor) is a reduced ability to smell. A major factor that contributes to hyposmia is aging. This is due to a gradual loss of olfactory receptor cells coupled with their slower rate of replacement as we age. Hyposmia affects about half of those over age 65 and 75% of those over age 80. Other causes of hyposmia include neurological changes, such as a head injury, Alzheimer's disease, or Parkinson's disease; certain drugs, such as antihistamines, analgesics, or steroids; and the damaging effects of smoking.

as an odor. A good example is the chemical methyl mercaptan, which smells like rotten cabbage and can be detected in concentrations as low as 1/25 billionth of a milligram per milliliter of air. Because the natural gas used for cooking and heating is odorless but lethal and potentially explosive if it accumulates, a small amount of methyl mercaptan is added to natural gas to provide olfactory warning of gas leaks.

The Olfactory System Rapidly Adapts

Adaptation (decreasing sensitivity) to odors occurs rapidly. Olfactory receptors adapt by about 50% in the first second or so after stimulation but adapt very slowly thereafter. At the receptor level, adaptation seems to involve the phosphorylation of odorant receptor proteins by a protein kinase. This reduces the sensitivity of the olfactory receptor to odorant binding. Another mechanism of adaptation occurs by decreasing the opening of cAMP-gated cation channels, causing the olfactory transduction process to be less responsive to odorant. Still, complete insensitivity to certain strong odors occurs about a minute after exposure. This involves an adaptation process in the central nervous system: Olfactory areas of the brain activate pathways that inhibit transmission of olfactory signals through the olfactory bulb.

CHECKPOINT

11. Outline the sequence of events from the activation of an olfactory receptor by the smell of fresh flowers to the arrival of an action potential in the orbitofrontal cortex.
12. What is the significance of the glomeruli in the olfactory bulb?
13. How does your sense of smell adapt to a bad odor?

9.4 The Gustatory System

OBJECTIVES

- Identify the five primary tastes.
- Explain the process of taste transduction.
- Describe the gustatory pathway to the brain.

Like olfaction, **gustation**, or taste, is a chemical sense. However, gustation is much simpler than olfaction in that only five primary tastes can be distinguished: *salty, sour, sweet, bitter*, and *umami* (oo-MAH-mē).Salty taste is caused by the presence of sodium ions (Na^+) in food. A common dietary source of Na^+ is NaCl (table salt). Sour taste is produced by hydrogen ions (H^+) released from acids. Lemons have a sour taste because they contain citric acid. Sweet taste is elicited by sugars such as glucose, fructose, and sucrose and by artificial sweeteners such as saccharin, aspartame, and sucralose. Bitter taste is caused by a wide variety of substances, including caffeine, morphine, and quinine. In addition, many poisonous substances like strychnine have a bitter taste. When something tastes bitter, a natural response is to spit it out, a reaction that serves to protect you from ingesting potentially harmful substances. The umami taste, first reported by Japanese scientists, is described as "meaty" or "savory." It is elicited by amino acids (especially glutamate) that are present in food. This is the reason why the additive monosodium glutamate (MSG) is used as a flavor enhancer in many foods. All other flavors, such as chocolate, pepper, and coffee, are combinations of the five primary tastes, plus any accompanying olfactory, tactile, and thermal sensations. Odors from food can pass upward from the mouth into the nasal cavity, where they stimulate olfactory receptors. Because olfaction is much more sensitive than taste, a given concentration of a food substance may stimulate the olfactory system thousands of times more strongly than it stimulates the gustatory system. When you have a cold or are suffering from allergies and cannot taste your food, it is actually olfaction that is blocked, not taste.

Taste Buds Contain the Receptors for Taste

The receptors for sensations of taste are located in the taste buds. Most of the nearly 10,000 taste buds of a young adult are on the tongue, but some are also on the pharynx (throat) and epiglottis (cartilage lid over the voice box). Taste buds are found in elevations on the tongue called **papillae** (pa-PIL-ē; singular is *papilla*), which provide a rough texture to the upper surface of the tongue (**FIGURE 9.23a,b**). Each **taste bud** consists of three types of cells: supporting cells, gustatory receptor cells, and basal cells (**FIGURE 9.23c**). The **supporting cells** surround about 50 **gustatory receptor cells** in each taste bud. Several microvilli project from each gustatory receptor cell to the tongue's surface through the **taste pore**, an opening in the taste bud. **Basal cells** are stem cells that produce supporting cells, which then develop into gustatory receptor cells. Each gustatory receptor cell has a life span of about 10 days, which is why it doesn't take the tongue too long to recover from being burned by that too hot cup of coffee or cocoa. Unlike olfactory receptor cells, gustatory receptors cells are not neurons. Instead, they are modified epithelial cells that synapse with first-order taste neurons of the gustatory pathway.

Taste Transduction Converts a Gustatory Stimulus into a Receptor Potential

Chemicals that stimulate gustatory receptor cells are known as **tastants**. Once a tastant is dissolved in saliva, it can enter a taste pore and interact with the gustatory microvilli, which are the sites of taste transduction. The result is a depolarizing receptor potential that causes the release of excitatory neurotransmitter from the gustatory receptor cell. The liberated neurotransmitter molecules in turn trigger graded potentials that elicit action potentials in the first-order taste neuron associated with the gustatory receptor cell.

FIGURE 9.23 The relationship of gustatory receptor cells in taste buds to tongue papillae.

Gustatory receptor cells are located in taste buds, along with supporting cells and basal cells.

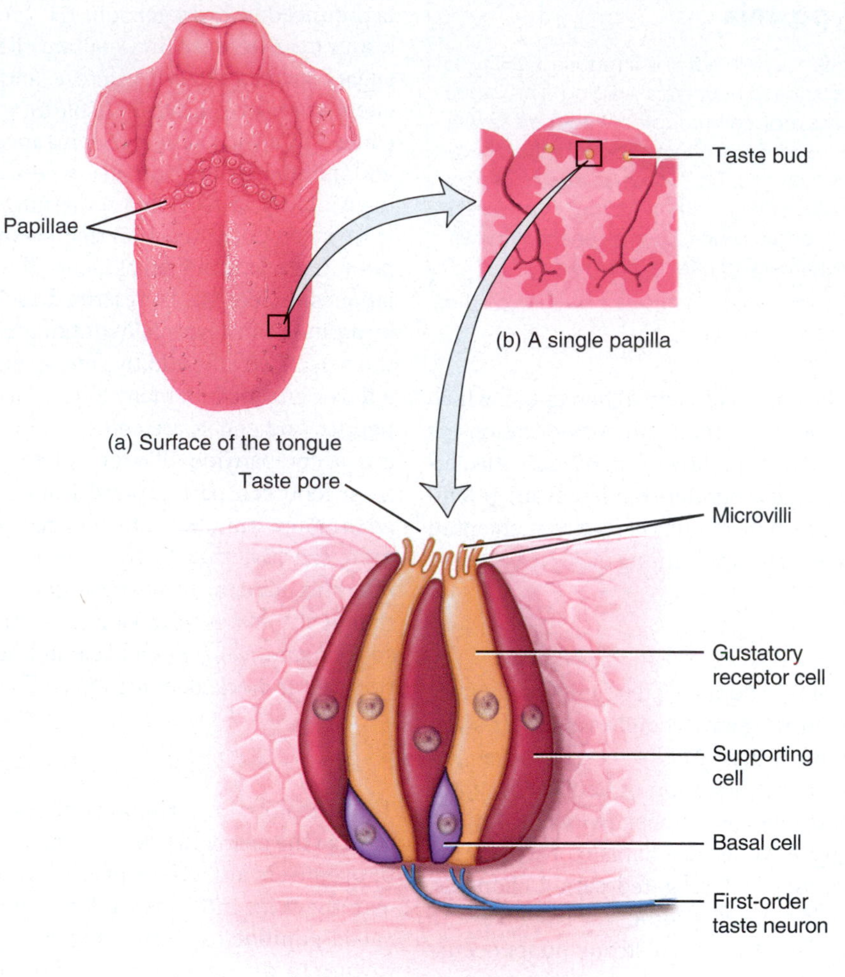

(c) Components of a taste bud

What role do basal cells play in taste buds?

The depolarizing receptor potential arises differently for different tastants. Transduction of salty or sour tastants involves the passage of these stimuli directly into the gustatory receptor cell through ion channels present in the plasma membrane. This occurs in the following way (**FIGURE 9.24a**):

1. The Na^+ ions in a salty tastant or the H^+ ions in a sour tastant enter a gustatory receptor cell through Na^+ channels or H^+ channels, respectively.
2. The movement of Na^+ or H^+ ions into the cell causes a depolarizing receptor potential to form.
3. The depolarization in turn causes voltage-gated Ca^{2+} channels in the plasma membrane to open, allowing Ca^{2+} ions to flow into the cell.
4. The increase in intracellular Ca^{2+} stimulates the release of neurotransmitter. Once released, the neurotransmitter molecules excite the first-order taste neuron that synapses with the gustatory receptor cell.

Unlike salty or sour tastants, sweet, bitter, and umami tastants do not themselves enter gustatory receptor cells. Instead, they bind to G protein-coupled receptors located in the plasma membrane. Stimulation of these receptors in turn activates an inositol trisphosphate (IP_3) second messenger pathway. Transduction of sweet, bitter, and umami tastants occurs via the following steps (**FIGURE 9.24b**):

1. A sweet, bitter, or umami tastant binds to a specific receptor that is coupled to a G protein known as **gustducin**. Gustducin then activates the enzyme phospholipase C to produce the second messenger inositol trisphosphate (IP_3).
2. IP_3 binds to and opens **transient receptor potential (TRP) channels** (TRPM5*) that are present in the plasma membrane.
3. Opening the TRP channels mainly allows Na^+ ions to enter the cell, resulting in the formation of a depolarizing receptor potential.
4. The depolarization in turn causes voltage-gated Ca^{2+} channels in the plasma membrane to open, allowing Ca^{2+} ions to enter the cell.
5. IP_3 also binds to and opens Ca^{2+} channels in the membrane of the endoplasmic reticulum (ER). The ER is a membranous organelle that stores calcium ions (Ca^{2+}). So, opening these **IP_3-gated Ca^{2+} channels** causes the release of Ca^{2+} from the lumen of the ER into the cytosol.

---

* TRPM5 = Transient Receptor Potential, Melastatin subfamily, member 5.

FIGURE 9.24 Taste transduction. Despite its appearance, a given gustatory receptor cell can respond to only one type of tastant.

Transduction of salty or sour tastants involves movement of these stimuli through ion channels; transduction of sweet, bitter, or umami tastants involves binding of these tastants to G protein-coupled receptors.

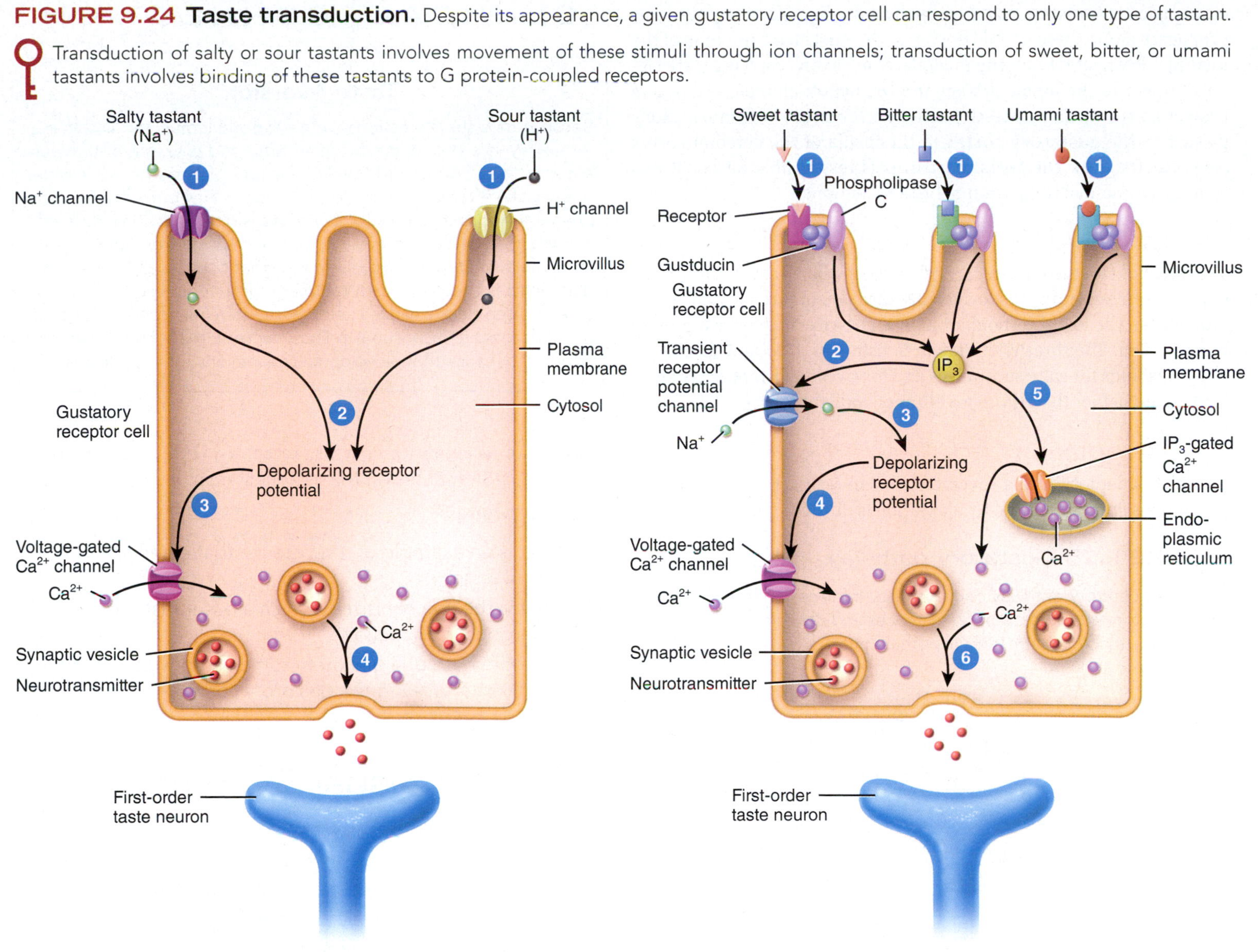

(a) Transduction of salty and sour tastants

(b) Transduction of sweet, bitter, and umami tastants

Which second messenger is common to the transduction mechanisms for sweet, bitter, and umami tastants?

6 The increase in intracellular Ca^{2+} due to the opening of voltage-gated Ca^{2+} channels and IP_3-gated Ca^{2+} channels triggers the release of neurotransmitter from the gustatory receptor cell. Then the liberated neurotransmitter molecules excite the first-order taste neuron that synapses with the gustatory receptor cell.

Contrary to what is shown in **FIGURE 9.24**, an individual gustatory receptor cell responds to only one type of tastant. This occurs because the membrane of a gustatory receptor cell has either ion channels or receptors for only one of the primary tastes. For example, a gustatory receptor cell that detects bitter tastants only has receptors for these tastants and cannot respond to salty, sour, sweet, or umami tastants. Thus, each gustatory receptor cell is "tuned" to detect a specific primary taste, and this segregation is maintained as the specific taste information is relayed along its own labeled line into the brain. It is also important to mention that a given taste bud contains gustatory receptor cells for each type of tastant, allowing all of the primary tastes to be detected in all parts of the tongue.

If all tastants cause release of neurotransmitter from gustatory receptor cells, why do foods taste different? The answer to this question is thought to lie in the patterns of activity in the brain that arise when gustatory receptor cells are activated. Different tastes arise from activation of different combinations of gustatory receptor cells. For example, the tastants in chocolate activate a certain combination of gustatory receptor cells, and the resultant pattern of activity in the brain is interpreted as the flavor chocolate. By contrast, the tastants in vanilla activate a different combination of gustatory receptor cells, and the resultant pattern of activity in the brain is interpreted as the flavor vanilla.

The Gustatory Pathway Extends from Taste Receptors to the Gustatory Cortex

Three cranial nerves contain axons of first-order taste neurons that innervate the taste buds: The facial (VII) and glossopharyngeal (IX) nerves serve the tongue, and the vagus (X) nerve serves the pharynx and epiglottis (**FIGURE 9.25**). From the taste buds, action potentials propagate along these cranial nerves to the **gustatory nucleus** of the medulla oblongata. From the medulla, some axons carrying taste signals project to the limbic system and the hypothalamus, and others project to the thalamus (**FIGURE 9.25**). From the thalamus, axons project to the **gustatory cortex** in the insula of the cerebral cortex (see **FIGURE 8.8**). The gustatory cortex is responsible for conscious awareness and discrimination of taste sensations.

The Threshold for Detecting Taste Can Vary

The threshold for taste varies for each of the primary tastes. The threshold for bitter substances is lowest. Because poisonous substances are often bitter, the low threshold (or high sensitivity) may have a protective function. The threshold for sour substances is somewhat higher. The thresholds for salty substances and for sweet substances are similar, and are higher than those for bitter or sour substances.

Taste Adaptation Occurs at Many Levels

Complete adaptation to a specific taste can occur in 1–5 minutes of continuous stimulation. Taste adaptation is due to changes that occur in the taste receptors, in olfactory receptors, and in neurons of the gustatory pathway in the central nervous system.

CHECKPOINT

14. What is an example of a substance that gives rise to the umami taste?

15. What is the mechanism by which a bitter tastant undergoes transduction?

16. Trace the path of a gustatory stimulus from contact of a tastant with saliva to the gustatory cortex.

CLINICAL CONNECTION

Taste Aversion

Because of taste projections to the hypothalamus and limbic system, there is a strong link between taste and pleasant or unpleasant emotions. Sweet foods evoke reactions of pleasure, while bitter ones cause expressions of disgust even in newborn babies. This phenomenon is the basis for **taste aversion**, in which people and animals quickly learn to avoid a food if it upsets the digestive system. The advantage of avoiding foods that cause such illness is longer survival. The drugs and radiation treatments used to combat cancer often cause nausea and gastrointestinal upset regardless of which foods are consumed. Thus, cancer patients may lose their appetite because they develop taste aversions for most foods.

FIGURE 9.25 Gustatory pathway.

The gustatory pathway conveys taste information from gustatory receptor cells in taste buds to processing centers in the cerebral cortex, limbic system, and hypothalamus.

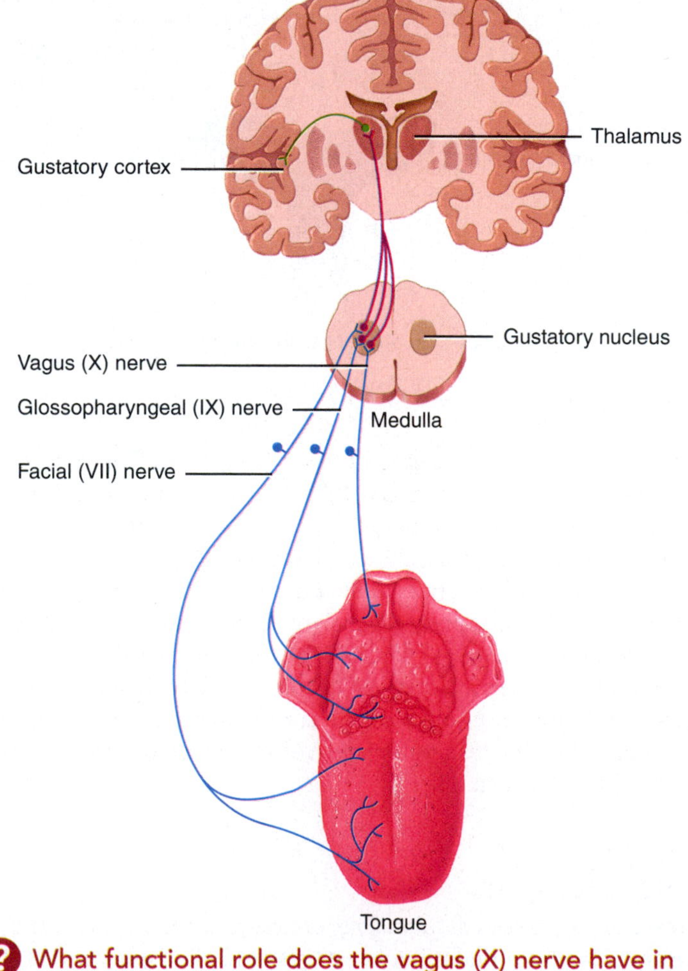

What functional role does the vagus (X) nerve have in the gustatory pathway?

9.5 The Visual System

OBJECTIVES

- Describe the functions of the components of the eye.
- Outline the steps involved in image formation.
- Explain the events of phototransduction.
- Discuss how visual processing begins in the retina.
- Describe the visual pathway to the brain.
- Compare the functions of the primary visual cortex and the visual association area.

Vision, the act of seeing, is extremely important to human survival. More than half of the sensory receptors in the human body are located in the eyes, and a large part of the cerebral cortex is devoted to processing visual information. In this section of the chapter, you will learn about visible light, the accessory structures of the eye, the components of the eye itself, and the functions of these components.

Visible Light Is the Part of the Electromagnetic Spectrum That the Eye Can Detect

Electromagnetic radiation is energy in the form of waves that radiates from the sun. There are many types of electromagnetic radiation, including gamma rays, X-rays, ultraviolet (UV) rays, visible light, infrared radiation, microwaves, and radio waves. This range of electromagnetic radiation is known as the **electromagnetic spectrum** (**FIGURE 9.26**). The distance between two consecutive peaks of an electromagnetic

FIGURE 9.26 The electromagnetic spectrum.

Visible light is the part of the electromagnetic spectrum with wavelengths ranging from about 400 to 700 nm.

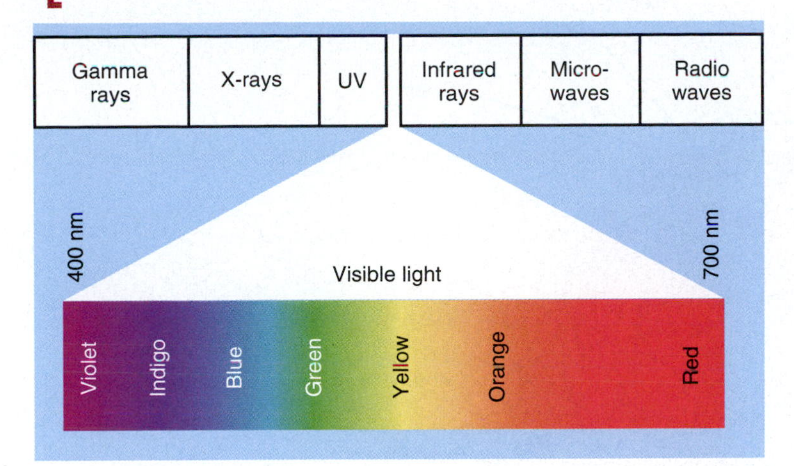

(a) Electromagnetic spectrum

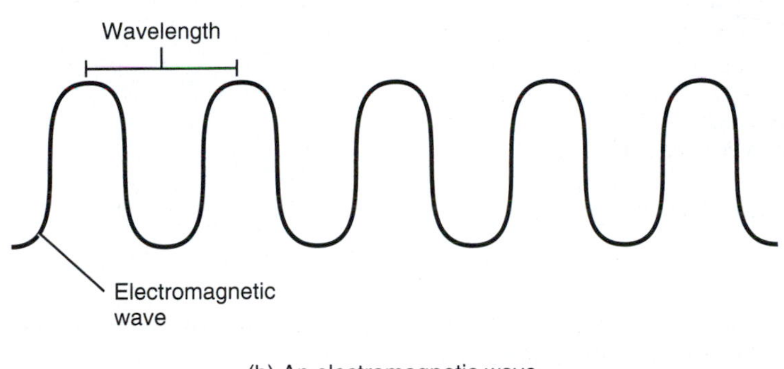

(b) An electromagnetic wave

Visible light that has a wavelength of 700 nm is what color?

wave is the *wavelength*. Wavelengths range from short to long; for example, gamma rays have wavelengths smaller than a nanometer (nm), and most radio waves have wavelengths greater than a meter.

The eyes are responsible for the detection of **visible light**, the part of the electromagnetic spectrum with wavelengths ranging from about 400 to 700 nm. Visible light exhibits colors: The color of visible light depends on its wavelength. For example, light that has a wavelength of 400 nm is violet, and light that has a wavelength of 700 nm is red. An object can absorb certain wavelengths of visible light and reflect others; the object will appear the color of the wavelength that is reflected. For example, grass appears green because it reflects mostly green light and absorbs most other wavelengths of visible light. An object appears white because it reflects all wavelengths of visible light. An object appears black because it absorbs all wavelengths of visible light.

Accessory Structures Protect, Lubricate, and Move the Eye

The **accessory structures** of the eye include the eyebrows, eyelashes, eyelids, lacrimal apparatus, and extrinsic eye muscles (**FIGURE 9.27**). The *eyebrows* and *eyelashes* help protect the eyes from foreign objects, perspiration, and direct rays of the sun. The upper and lower *eyelids* shade the eyes during sleep and protect the eyes from excessive light and foreign objects. The *lacrimal apparatus* is a group of glands, canals, and ducts that produce and drain **lacrimal fluid** or **tears**. The *lacrimal glands* are the components of the lacrimal apparatus that secrete tears. After being secreted, tears pass over the surface of the eyeball toward the nose into two lacrimal canals and a nasolacrimal duct, which allow the tears to drain into the nasal cavity. Tears lubricate and cleanse the eye. They also contain a bacterial-killing enzyme called **lysozyme** that helps protect the eye from infection. Other important accessory structures of the eye are the *extrinsic eye muscles*, which attach to the outer surface of the eyeball. These skeletal muscles are responsible for moving the eye in various directions.

FIGURE 9.27 Accessory structures of the eye. The components of the lacrimal apparatus are shaded green.

Accessory structures of the eye include the eyebrows, eyelashes, eyelids, lacrimal apparatus, and extrinsic eye muscles.

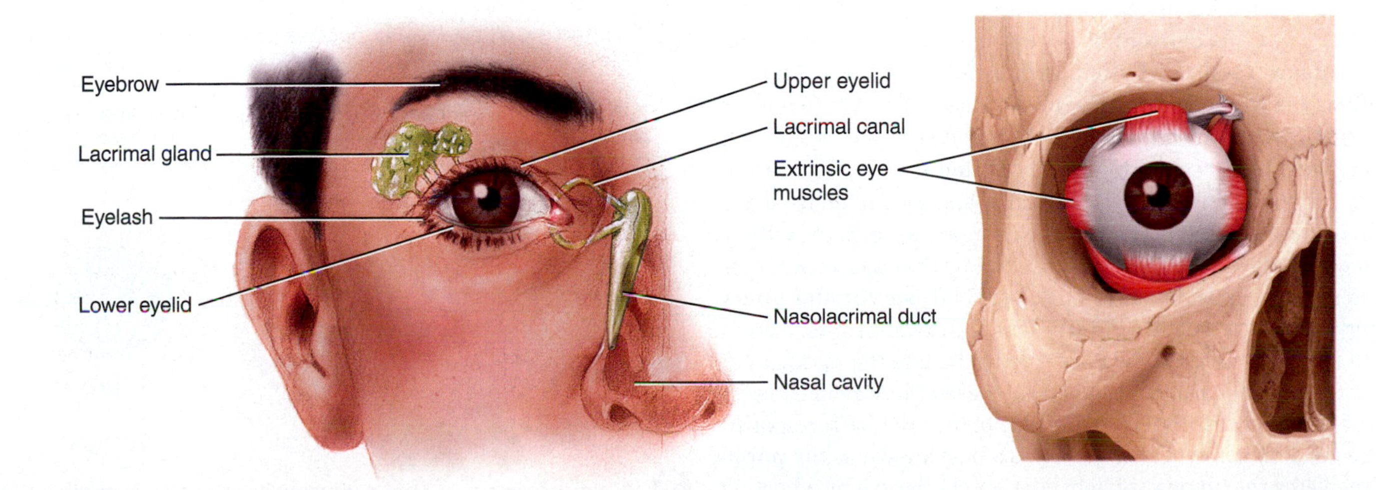

What are the functions of tears?

FIGURE 9.28 Components of the eye.

The eye is a sphere that consists of three layers, a lens, and two cavities.

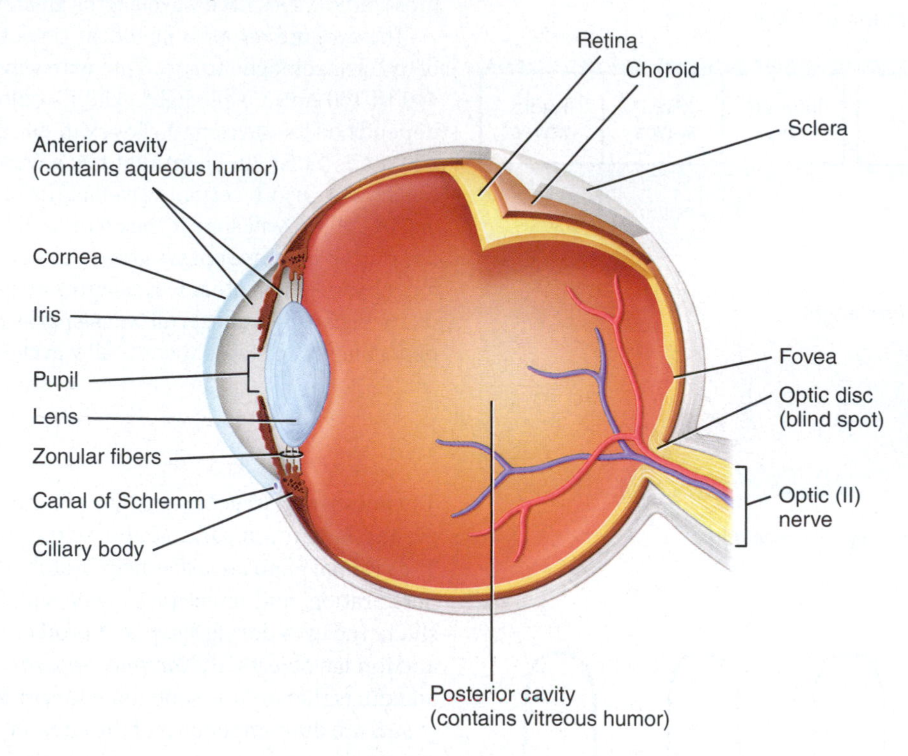

What is the purpose of the choroid?

The Eye Consists of Several Functional Components

The eye is a sphere composed of three layers, a lens, and two cavities (FIGURE 9.28). The outer layer of the eye consists of an anterior cornea and a posterior sclera. The **cornea** is a transparent structure that admits light into the eye. It also refracts (bends) the incoming light rays. As you will soon learn, refraction of light by the cornea and lens helps focus light onto the retina, the part of the eye that detects light. The **sclera**, the "white" of the eye, is a tough coat of connective tissue that covers the entire eyeball except the cornea. The sclera gives shape to the eye and protects its internal parts.

The middle layer of the eye consists of the choroid, ciliary body, and iris (FIGURE 9.28). The **choroid** lines most of the inner surface of the sclera. It contains blood vessels that provide nutrients to the retina. It also contains the pigment melanin, which causes this layer to appear dark in color. Melanin in the choroid absorbs stray light rays, which prevents reflection and scattering of light within the eye. As a result, the image cast on the retina by the cornea and lens remains sharp and clear. At the front end of the eye, the choroid becomes the **ciliary body**, which is responsible for secreting a fluid called aqueous humor (described shortly). Extending from the ciliary body are **zonular fibers** (*suspensory ligaments*) that attach to the lens. Contraction or relaxation of smooth muscle present in the ciliary body changes the tightness of the zonular fibers, which alters the shape of the lens for viewing objects up close or at a distance. The **iris** is the part of the eye that is responsible for eye color. In the center of the iris is a hole known as the **pupil**. The iris regulates the amount of light that enters the eye by adjusting the diameter of the pupil. Changes in pupil diameter involve two types of smooth muscle present in the iris: **circular muscles** (*sphincter pupillae*) and **radial muscles** (*dilator pupillae*) (FIGURE 9.29). When the eye is stimulated by bright light, the circular muscles contract, which decreases the size of the pupil (constriction). When the eye must adjust to dim light, the radial muscles contract, which increases the size of the pupil (dilation). The smooth muscles of the iris are under the control of the autonomic nervous system. Parasympathetic fibers of the oculomotor (III)

FIGURE 9.29 Responses of the pupil to light of varying brightness.

Contraction of the circular muscles causes constriction of the pupil; contraction of the radial muscles causes dilation of the pupil.

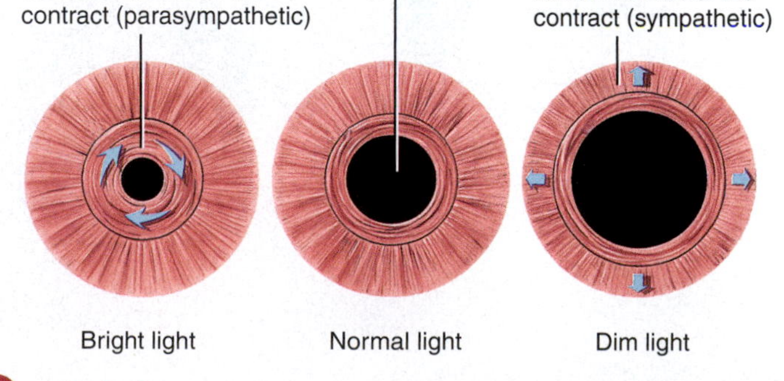

Which division of the autonomic nervous system causes pupillary constriction? Which causes pupillary dilation?

nerve cause contraction of the circular muscles, whereas sympathetic nerve fibers cause contraction of the radial muscles.

The inner layer of the eye is the **retina**, which is responsible for converting light into action potentials (see **FIGURE 9.28**). The retina is further subdivided into a pigmented layer and a neural layer (**FIGURE 9.30a**). The *pigmented layer* consists of epithelial cells that contain melanin. The melanin in the pigmented layer of the retina, like in the choroid, helps to absorb stray light rays. The *neural layer* of the retina is a multilayered outgrowth of the brain. It consists of three distinct layers of cells that sequentially process visual signals—the photoreceptor layer, bipolar cell layer, and ganglion cell layer (**FIGURE 9.30a**). **Photoreceptors**, which include *rods* and *cones*, are sensory receptors

FIGURE 9.30 Organization of the retina. The downward blue arrow at the left in part (a) indicates the direction of the signals passing through the neural layer of the retina. Eventually, action potentials arise in ganglion cells and propagate along their axons, which make up the optic (II) nerve.

In the retina, visual signals pass from photoreceptors to bipolar cells to ganglion cells.

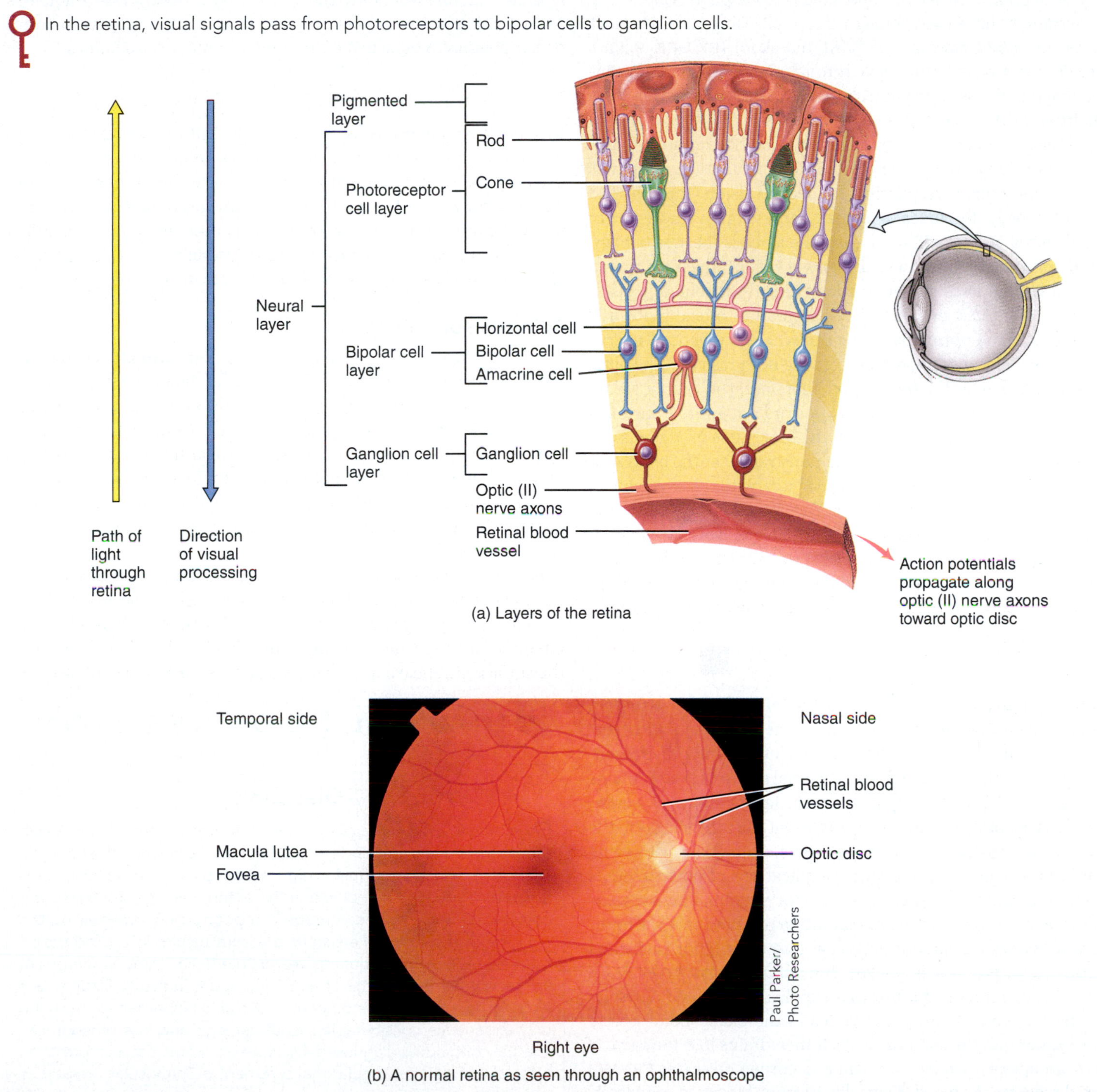

(a) Layers of the retina

(b) A normal retina as seen through an ophthalmoscope

What is the main function of the retina?

that detect light and convert it into receptor potentials. **Bipolar cells** are neurons that convey signals from photoreceptors to ganglion cells. **Ganglion cells** are neurons that generate action potentials in response to signals from bipolar cells. The axons of the ganglion cells give rise to the optic (II) nerve, which carries sensory information about light in the form of action potentials from the eye to the brain. Note that when light enters the eye, it passes through the ganglion and bipolar cell layers before it enters the photoreceptor layer. Two other types of cells present in the bipolar cell layer of the retina are **horizontal cells** and **amacrine cells** (FIGURE 9.30a). These cells form laterally directed neural circuits that modify the signals being transmitted along the pathway from photoreceptors to bipolar cells to ganglion cells.

The retina can be viewed through the pupil using an instrument known as an *ophthalmoscope* (of-THAL-mō-skōp) (FIGURE 9.30b). Features of the retina that are visible through the ophthalmoscope include retinal blood vessels, the macula lutea, fovea, and optic disc. The **macula lutea** (MAK-ū-la LOO-tē-a; *macula* = spot; *lute-* = yellowish) is an oval area located in the center of the posterior retina. It is yellowish in color due to the presence of yellow pigment (xanthophyll). This part of the retina is responsible for central vision (the ability to see straight ahead). The **fovea** is a small depression in the center of the macula lutea. As you will soon learn, the fovea is the area of highest **visual acuity** or **resolution** (sharpness of vision). The main reason that you move your head and eyes to look directly at something is to place images of interest on your fovea—as you do to read the words in this sentence! The **optic disc** is the site where the optic (II) nerve exits the eyeball. Because photoreceptors are not present in the optic disc, it is also known as the **blind spot**. Therefore, you cannot see an image that strikes this region. Normally, you are not aware of having a blind spot because visual processing in the brain "fills in" the missing information. However, you can easily demonstrate that the blind spot exists. Hold this page about 20 in. from your face, with the cross shown below directly in front of your right eye. You should be able to see the cross and the square when you close your left eye. Now, keeping the left eye closed, slowly bring the page closer to your face while keeping the right eye on the cross. At a certain distance the square will disappear from your field of vision because its image falls on the blind spot.

+ ■

Behind the pupil and iris of the eye is the **lens**, an elastic structure that refracts light rays (see FIGURE 9.28). The lens normally is perfectly transparent because its cells lose their nuclei and other organelles and gradually become filled with a special group of clear proteins called *crystallins*. A loss of transparency of the lens is known as a **cataract**. The lens becomes cloudy (less transparent) due to changes in the structure of the crystallin proteins. Cataracts often occur with aging but may also be caused by injury or complications of other diseases (for example, diabetes mellitus). Treatment for a cataract involves surgical removal of the old lens and implantation of a new artificial one.

The lens divides the interior of the eye into an anterior cavity and a posterior cavity (see FIGURE 9.28). The **anterior cavity** contains a clear fluid called **aqueous humor** that supplies oxygen and nutrients to the lens and cornea. Blood capillaries in the ciliary body secrete aqueous humor into the anterior cavity. It then drains into the *canal of Schlemm*, an opening where the sclera and cornea meet (see FIGURE 9.28), and reenters the blood. Normally, aqueous humor is completely replaced about every 90 minutes. The **posterior cavity** contains a clear, jelly-like substance called **vitreous humor** that maintains the shape of the eye and keeps the retina attached to the choroid. Vitreous humor forms during embryonic life and is not replaced thereafter. Occasionally, collections of debris within the vitreous humor may cast a shadow on the retina and create the appearance of specks that dart in and out of the field of vision. These *vitreous floaters*, which are more common in older individuals, are usually harmless and do not require treatment.

The pressure in the eye, called **intraocular pressure**, is produced mainly by the aqueous humor and partly by the vitreous humor; normally it is about 16 mmHg (millimeters of mercury). The intraocular pressure maintains the shape of the eyeball and prevents it from collapsing. Puncture wounds to the eyeball may cause the loss of aqueous humor and the vitreous humor. This in turn causes a decrease in intraocular pressure, a detached retina, and in some cases blindness.

The Eye Forms Images of Objects on the Retina

In some ways the eye is like a camera: Its optical elements focus an image of some object on a light-sensitive "film"—the retina—while ensuring the correct amount of light to make the proper "exposure." To understand how the eye forms clear images of objects on the retina, three processes must be examined: (1) the refraction or bending of light by the lens and cornea; (2) accommodation, the change in the shape of the lens; and (3) constriction or narrowing of the pupil.

Refraction of Light Rays

When light rays traveling through a transparent medium (such as air) pass into a second transparent medium with a different density (such as water), they bend at the interface between the two media. This bending is called **refraction**. Light rays are refracted as long as they strike the interface at any angle other than perpendicular (FIGURE 9.31a). Light rays that strike the interface perpendicularly are not refracted and pass straight through (FIGURE 9.31b). When light rays are refracted, the amount of refraction depends on two factors: (1) the difference in densities between the two media and (2) the angle at which the light rays strike the interface. As light rays enter the eye, they are refracted at the outer and inner surfaces of the cornea and at both surfaces of the lens.

When parallel light rays strike a curved surface such as a lens, the direction in which the light rays are refracted depends on whether the surface is concave or convex. If a surface curves inward, it is said

CLINICAL CONNECTION

Glaucoma

Glaucoma (glaw-KŌ-ma) is the most common cause of blindness in the United States, afflicting about 2% of the population over age 40. Glaucoma is an abnormally high intraocular pressure due to a buildup of aqueous humor within the anterior cavity. The fluid compresses the lens into the vitreous humor and puts pressure on the neurons of the retina. Persistent pressure results in a progression from mild visual impairment to irreversible destruction of neurons of the retina, damage to the optic nerve, and blindness. Glaucoma is typically painless, and the other eye compensates largely, so a person may experience considerable retinal damage and loss of vision before the condition is diagnosed. Because glaucoma occurs more often with advancing age, regular measurement of intraocular pressure is an increasingly important part of an eye exam as people grow older.

to be *concave* and causes parallel light rays to diverge (spread apart) (**FIGURE 9.31c**). If a surface curves outward, it is said to be *convex* and causes parallel light rays to converge (come together) at a single point known as the *focal point* (**FIGURE 9.31d**). The distance between the center of a lens and its focal point is called the *focal length*. The cornea and lens of the eye have convex surfaces that converge the light rays reflected from an object being viewed. As a result, a miniature image of the object is focused on the retina.

Images focused on the retina are inverted (upside down) (**FIGURE 9.31e**). They also undergo right-to-left reversal; that is, light from the right side of an object strikes the left side of the retina, and vice versa. The reason the world does not look inverted and reversed is that the brain "learns" early in life to coordinate visual images with the orientations of objects. The brain stores the inverted and reversed images we acquired when we first reached for and touched objects and interprets those visual images as being correctly oriented in space.

FIGURE 9.31 Refraction of light rays.

Images focused on the retina are inverted and right-to-left reversed.

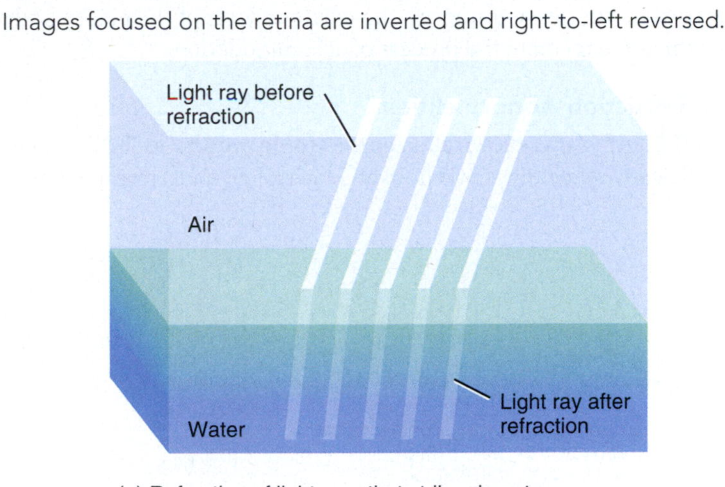

(a) Refraction of light rays that strike air-water interface at angle other than perpendicular

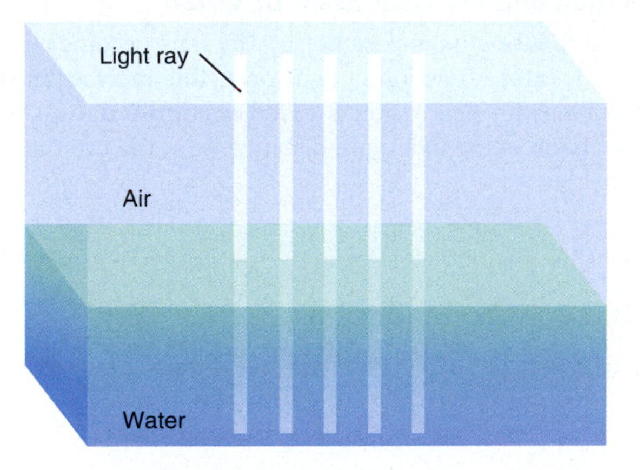

(b) No refraction of light rays that strike air-water interface perpendicularly

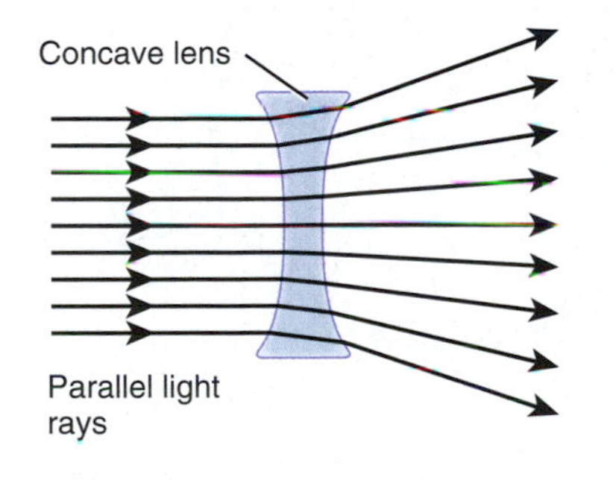

(c) Divergence of light rays by a concave lens

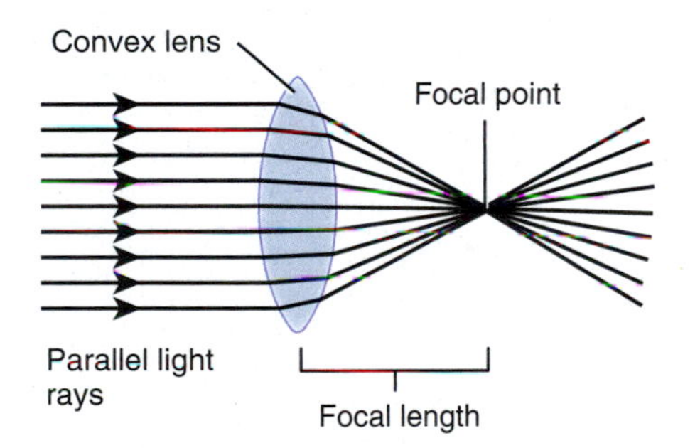

(d) Convergence of light rays by a convex lens

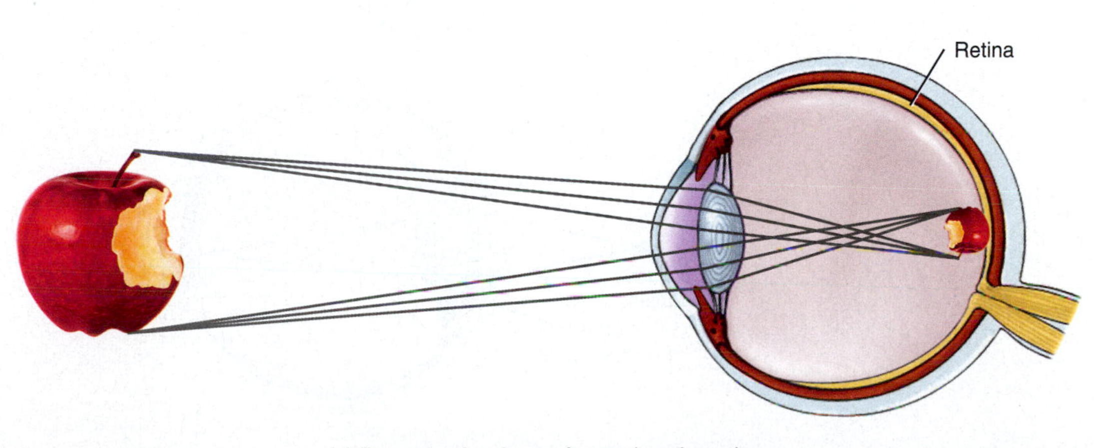

(e) Example of an image focused on the retina

Is the lens of the eye concave or convex?

About two-thirds of the total refraction of light occurs at the cornea. The lens provides the remaining one-third of focusing power, and it can change its curvature to view near or distant objects. When an object is 6 m (20 ft) or more away from the viewer, the light rays reflected from the object are nearly parallel to one another (FIGURE 9.32a). The lens must bend these parallel rays just enough so that they fall exactly focused on the retina. Because light rays that are reflected from objects closer than 6 m (20 ft) are divergent rather than parallel (FIGURE 9.32b), the rays must be refracted more if they are to be focused on the retina. This additional refraction is accomplished through a process called accommodation.

Accommodation and the Near Point of Vision

When the eye is focusing on a close object, the lens becomes more curved, causing greater refraction of light rays. This increase in the curvature of the lens for near vision is called **accommodation**. The **near point of vision** is the minimum distance from the eye that an object can be clearly focused with maximum accommodation. This distance is about 10 cm (4 in.) in a young adult.

How does accommodation occur? When you are viewing distant objects, the ciliary muscle of the ciliary body is relaxed and the lens is flatter because it is stretched in all directions by taut zonular fibers (FIGURE 9.32a). When you view a close object, the ciliary muscle contracts, which pulls the ciliary body and choroid forward toward the lens. This action releases tension on the lens and zonular fibers (FIGURE 9.32b). Because it is elastic, the lens becomes more spherical (more convex), which increases its focusing power and causes greater convergence of the light rays. Parasympathetic fibers of the oculomotor (III) nerve innervate the ciliary muscle of the ciliary body and therefore mediate the process of accommodation.

Refraction Abnormalities

The normal eye, known as an **emmetropic eye** (em′-e-TROP-ik), can sufficiently refract light rays from an object 6 m (20 ft) away so that a clear

FIGURE 9.32 Accommodation.

During accommodation, the lens becomes more spherical (more convex), which causes greater refraction of light rays.

(a) Viewing a distant object

(b) Viewing a close object via accommodation

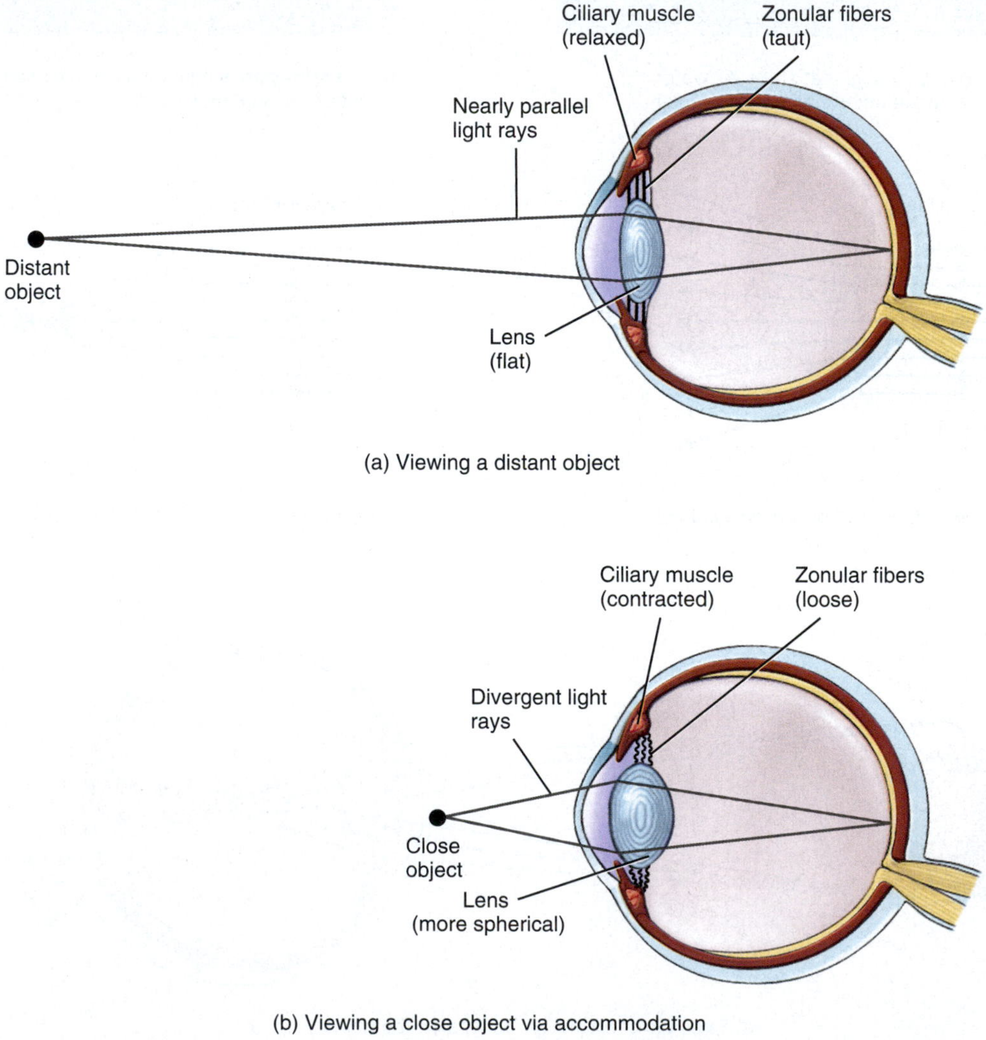

What is the shape of the lens in a person who is viewing a distant object?

CLINICAL CONNECTION

Presbyopia

With aging, the lens loses elasticity and thus its ability to curve to focus on objects that are close. Therefore, older people cannot read print at the same close range as can younger people. This condition is called **presbyopia** (prez-bē-Ō-pē-a). By age 40, the near point of vision may have increased to 20 cm (8 in.), and at age 60 it may be as much as 80 cm (31 in.). Presbyopia usually begins in the mid-forties. At about that age, people who have not previously worn glasses begin to need them for reading. Those who already wear glasses typically start to need bifocals, lenses that can focus for both distant and close vision.

image is focused on the retina (FIGURE 9.33a). However, many people lack this ability because of refraction abnormalities. Among these abnormalities are **myopia** (mī-Ō-pē-a), or *nearsightedness*, which occurs when the eyeball is too long relative to the focusing power of the cornea and lens, or when the lens is thicker than normal, so an image is focused in front of the retina (FIGURE 9.33b). Myopic individuals can see close objects clearly but not distant objects. Myopia can be corrected by a concave lens, which diverges entering light rays so that they come into focus directly on the retina (FIGURE 9.33c). In **hyperopia** (hī-per-Ō-pē-a) or *farsightedness*, also known as **hypermetropia** (hī′-per-me-TRŌ-pē-a), the eyeball length is short relative to the focusing power of the cornea and lens, or the lens is thinner than normal, so an image is focused behind the retina (FIGURE 9.33d). Hyperopic individuals can see distant objects clearly but not close ones. Hyperopia can be corrected by a convex lens, which converges entering light rays so that they focus directly on the retina (FIGURE 9.33e). Another refraction abnormality is **astigmatism** (a-STIG-ma-tizm), in which either the cornea or the lens has an irregular curvature. As a result, parts of the image are out of focus, and thus vision is blurred or distorted. Astigmatism can be corrected by a special cylindrical lens that compensates for the irregular curvature of the cornea or lens of the eye.

Constriction of the Pupil

The circular muscle fibers of the iris also have a role in the formation of clear retinal images. As you have already learned, **constriction of the pupil** is a narrowing of the diameter of the hole through which light enters the eye due to the contraction of the circular muscles of the iris.

LASIK

An increasingly popular alternative to wearing glasses or contact lenses is refractive surgery to correct the curvature of the cornea for conditions such as farsightedness, nearsightedness, and astigmatism. The most common type of refractive surgery is laser-assisted in-situ keratomileusis (**LASIK**). After anesthetic drops are placed in the eye, a circular flap of tissue from the center of the cornea is cut. The flap is folded out of the way, and the underlying layer of cornea is reshaped with a laser, one microscopic layer at a time. A computer assists the physician in removing very precise layers of the cornea. After the sculpting is complete, the corneal flap is repositioned over the treated area. A patch is placed over the eye overnight and the flap quickly reattaches to the rest of the cornea.

FIGURE 9.33 Refraction abnormalities in the eyeball and their correction. (a) Normal (emmetropic) eye. (b) In the nearsighted or myopic eye, the image is focused in front of the retina. The condition may result from an elongated eyeball or thickened lens. (c) Correction of myopia is by use of a concave lens that diverges entering light rays so that they come into focus directly on the retina. (d) In the farsighted or hyperopic eye, the image is focused behind the retina. The condition results from a shortened eyeball or a thin lens. (e) Correction of hyperopia is by use of a convex lens that converges entering light rays so that they focus directly on the retina.

In myopia (nearsightedness), only close objects can be seen clearly; in hyperopia (farsightedness), only distant objects can be seen clearly.

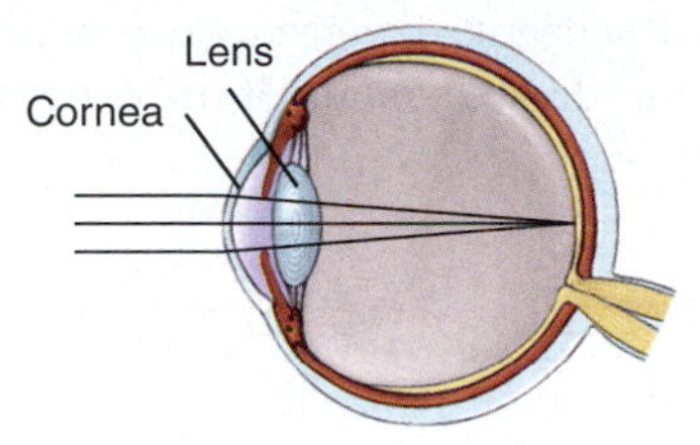

(a) Normal (emmetropic) eye

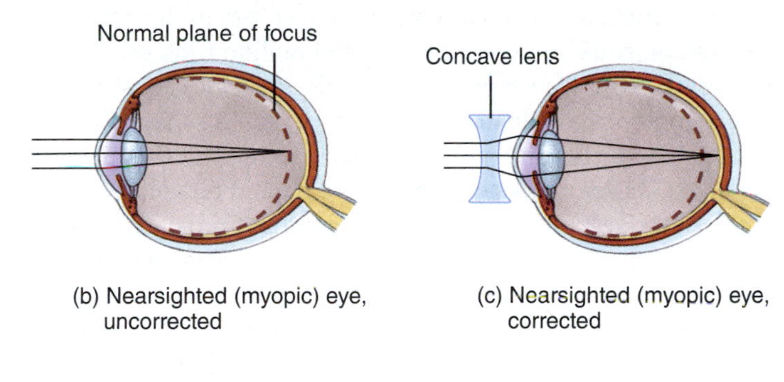

(b) Nearsighted (myopic) eye, uncorrected

(c) Nearsighted (myopic) eye, corrected

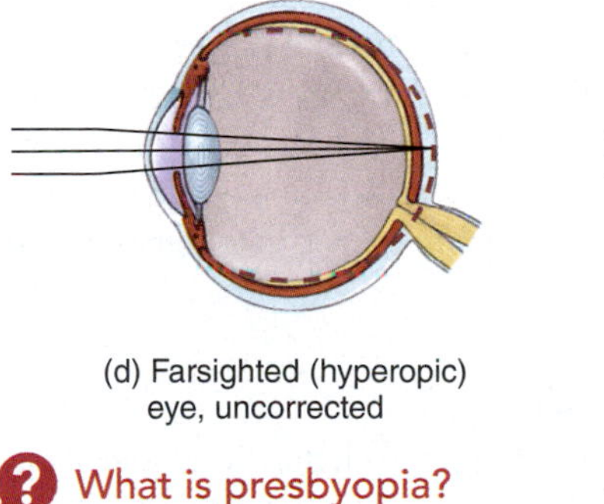

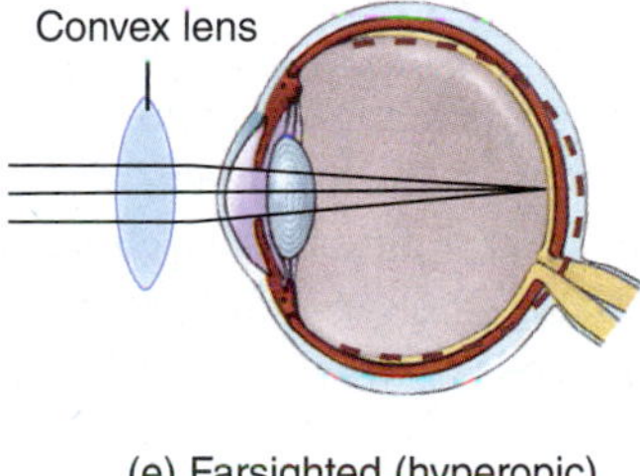

(d) Farsighted (hyperopic) eye, uncorrected

(e) Farsighted (hyperopic) eye, corrected

What is presbyopia?

This autonomic reflex occurs simultaneously with accommodation and prevents light rays from entering the eye through the periphery of the lens. Light rays entering at the periphery would not be brought to focus on the retina and would result in blurred vision. The pupil, as noted earlier, also constricts in bright light.

Convergence of the Eyes Maintains Binocular Vision

Because of the position of their eyes in their heads, many animals, such as horses and goats, see one set of objects off to the left through one eye, and an entirely different set of objects off to the right through the other. In humans, both eyes focus on only one set of objects—a characteristic called **binocular vision**. This feature of our visual system allows the perception of depth and an appreciation of the three-dimensional nature of objects.

Binocular vision occurs when light rays from an object strike corresponding points on the two retinas. When we stare straight ahead at a distant object, the incoming light rays are aimed directly at both pupils and are refracted to comparable spots on the retinas of both eyes. As we move closer to an object, however, the eyes must rotate toward the nose if the light rays from the object are to strike the same points on both retinas. The term **convergence** refers to this movement of the two eyes so that both are directed toward the object being viewed, for example, tracking a pencil moving toward your eyes. The nearer the object, the greater the degree of convergence needed to maintain binocular vision. The coordinated action of the extrinsic eye muscles brings about convergence.

Two Types of Photoreceptors Are Required for Normal Vision

Organization of Photoreceptors

Recall that there are two types of photoreceptors in the neural layer of the retina: **rods** and **cones**. About 120 million rods and 6 million cones are present in each retina. Both rods and cones consist of three parts: an outer segment, an inner segment, and a synaptic terminal (**FIGURE 9.34**). The *outer segment* is located next to the pigmented layer of the retina. Rods and cones were named for the appearance of their outer segments. The outer segments of rods are cylindrical or rod-shaped; those of cones are tapered or cone-shaped (**FIGURE 9.34**). The outer segments of both types of photoreceptors are folded into membranous discs. In cones, the discs are continuous with the plasma membrane. In rods, the discs pinch off from the plasma membrane and become stacked like coins inside a wrapper. As you will soon learn, the conversion of light energy into a receptor potential occurs in the outer segments of both rods and cones. The *inner segment* of a photoreceptor contains several organelles, including the nucleus, Golgi complex, and many mitochondria. Extending from the inner segment is a *synaptic terminal*, which is filled with synaptic vesicles. The synaptic terminal forms synapses with bipolar cells, the next cells of the visual pathway.

FIGURE 9.34 Organization of rods and cones.

A photoreceptor consists of three parts: an outer segment, an inner segment, and a synaptic terminal.

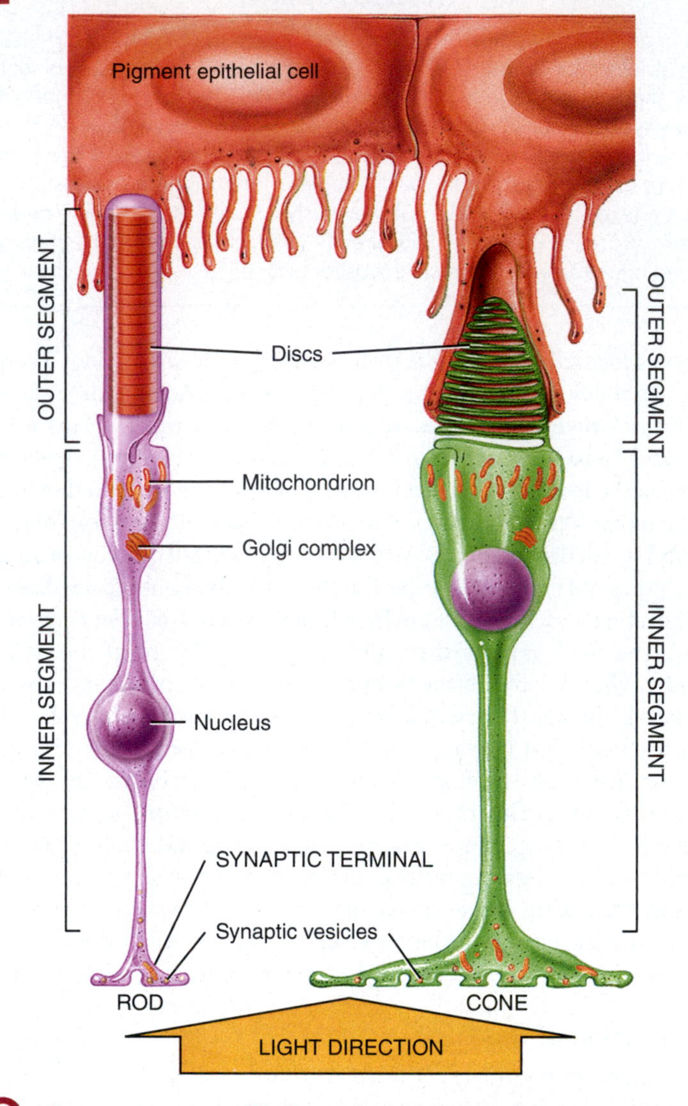

The conversion of light energy into a receptor potential occurs in which part of a photoreceptor?

Photoreceptor Sensitivity to Light

Rods and cones have different sensitivities to light. Rods are highly sensitive to light: As little as a single photon (the smallest unit of light) can elicit a response in a rod. Thus, rods allow you to see in the dim light present at night (night vision). Cones do not contribute to night vision because they have a low sensitivity to light. Hundreds of photons are required to evoke a response in a cone. Therefore, cones allow you to see in the bright light present during the day (day vision). Rods contribute very little to day vision because they become saturated (completely bleached) in bright light.

Photoreceptors and Photopigments

Rods and cones are able to respond to light because they contain photopigments in the membranes of the discs of their outer segments. A **photopigment** is a substance that absorbs light and then undergoes structural changes that lead to the development of a receptor potential. The photopigment in rods is called **rhodopsin**. Rhodopsin absorbs most wavelengths of light, with absorption being the greatest at a wavelength of 500 nm (**FIGURE 9.35**). Because all rods have the same photopigment, their response to a given wavelength of light is always the same. Consequently, rods cannot distinguish between different colors, providing vision in only black and white. Cones, however, do provide color vision. This is because there are three types of cones (blue cones, green cones, and red cones), each containing a different photopigment that maximally absorbs light in a different part of the visible spectrum (**FIGURE 9.35**). As a result, the three types of cones have different responses to a given wavelength of light. **Blue cones** contain *blue-sensitive pigment*, which maximally absorbs light at a wavelength of 420 nm; **green cones** contains *green-sensitive pigment*, which maximally absorbs light at a wavelength of 530 nm; and **red cones** contain *red-sensitive pigment*, which maximally absorbs light at a wavelength of 560 nm. Although each type of cone photopigment best absorbs a particular wavelength of light, it can also respond with reduced sensitivity to a range of other wavelengths. In addition, the absorption spectra of the different cone photopigments overlap considerably, which is why we are able to see more than just three colors. The brain determines color by comparing the responses of the three types of cones. For example, when green light with a wavelength of 490 nm strikes the retina, it activates red cones at 25% of their maximum level of activity, green cones at 63%, and blue cones at 33%. This pattern of cone activation is represented by the ratio 25:63:33 and is interpreted by the brain as the color green. When orange light with a wavelength of 590 nm strikes the retina, it activates red cones at 85% of their maximum level of activity and green cones at 33%, but has no effect at all on blue cones. This pattern of cone activation, represented by the ratio 85:33:0, is interpreted as the color orange. The ratio for blue light with a wavelength of 430 nm is 0:0:95, and the ratio for yellow light at a wavelength of 540 nm is 84:95:0. Most of our experiences are mediated by the cone system, the loss of which produces legal blindness. A person who loses rod vision mainly has difficulty seeing in dim light and thus should not drive at night.

The photopigments present in rods and cones contain two parts: a derivative of vitamin A called **retinal** and a protein known as **opsin** (FIGURE 9.36). Retinal is the light-absorbing part of all visual photopigments. The human retina contains four different opsins: one present in rods and one in each of the three different types of cones. Small variations in the amino acid sequences of the different opsins permit the rods and cones to maximally absorb different wavelengths of light.

FIGURE 9.35 Absorption spectra of the four different photopigments.

Each photopigment maximally absorbs a particular wavelength of light but also responds with reduced sensitivity to a range of other wavelengths.

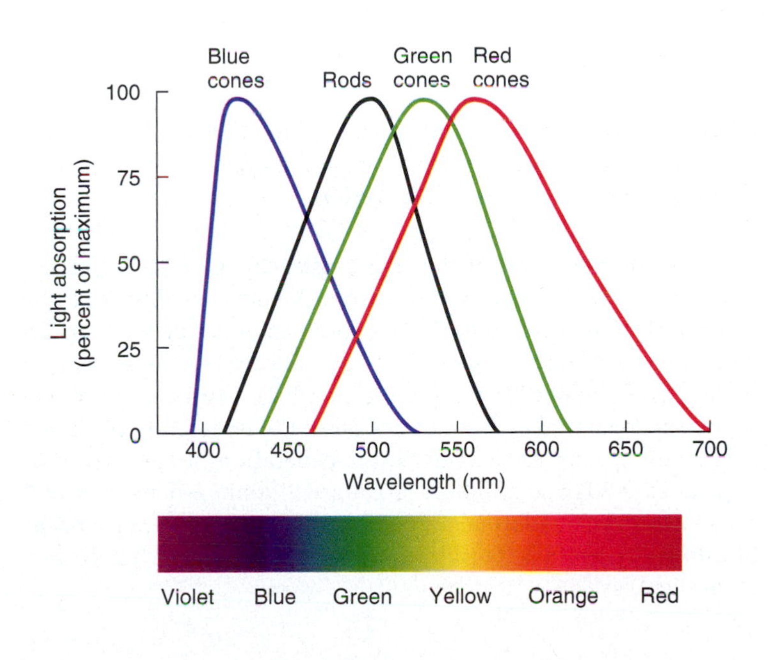

Why are cones able to provide color vision? Why are rods able to provide vision only in black and white?

FIGURE 9.36 Components of a photopigment. The photopigment shown is rhodopsin.

A photopigment consists of two main components: a vitamin A derivative called retinal and a protein known as opsin.

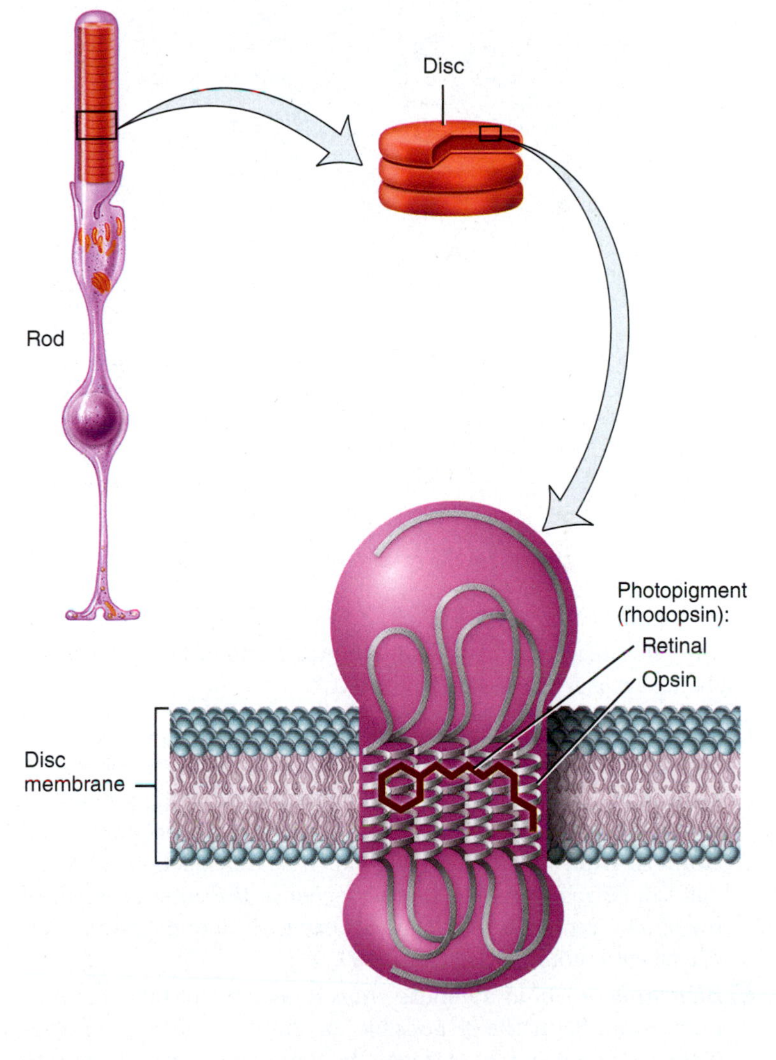

Which photopigment component absorbs light?

CLINICAL CONNECTION

Color Blindness and Night Blindness

Most forms of **color blindness**, an inherited inability to distinguish between certain colors, result from the absence or deficiency of one of the three types of cones. The most common type is ***red–green color blindness***, in which red cones or green cones are missing. As a result, the person cannot distinguish between red and green. For example, an individual with red–green color blindness is unable to see the number 74 in the Ishihara color chart below:

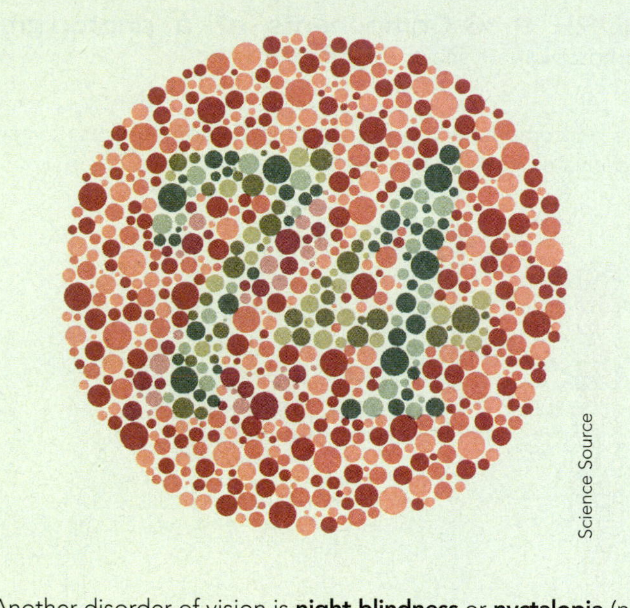

Another disorder of vision is **night blindness** or **nyctalopia** (nik′-ta-LŌ-pē-a), an inability to see well at low light levels. Night blindness is caused by a below-normal amount of rhodopsin, usually due to prolonged vitamin A deficiency.

FIGURE 9.37 The cyclical bleaching and regeneration of photopigment. Blue arrows indicate bleaching steps; black arrows indicate regeneration steps.

When *cis*-retinal absorbs a photon of light, it is converted to *trans*-retinal.

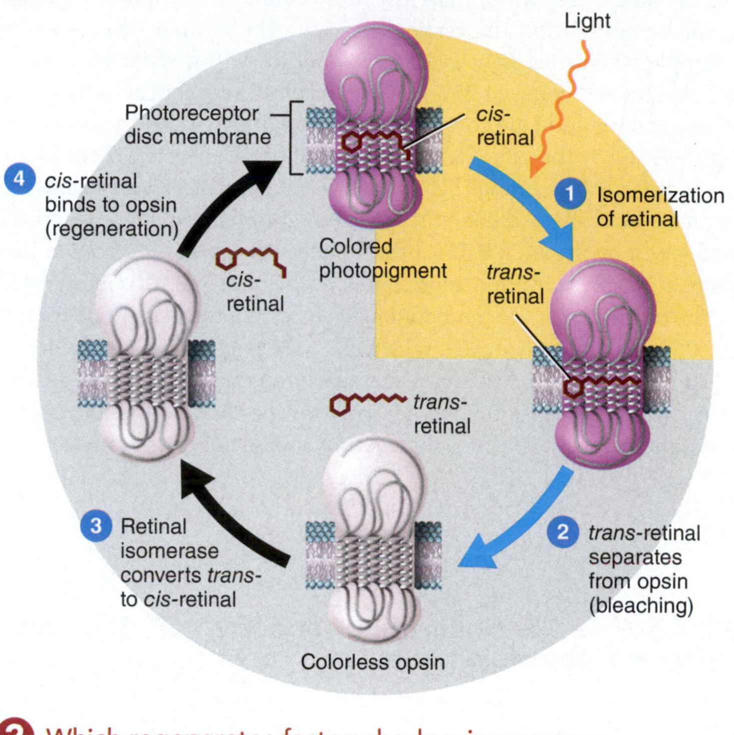

Which regenerates faster: rhodopsin or a cone photopigment?

The photopigments of rods and cones respond to light via the following cyclical process (FIGURE 9.37):

1. ***Isomerization.*** In darkness, retinal has a bent shape, called *cis*-retinal, which fits snugly into the opsin portion of the photopigment. When *cis*-retinal absorbs a photon of light, it straightens out to a shape called *trans*-retinal. This *cis*-to-*trans* conversion is called **isomerization** and is the first step in visual transduction. After retinal isomerizes, chemical changes occur in the outer segment of the photoreceptor. These chemical changes lead to the production of a receptor potential (see FIGURE 9.41).
2. ***Bleaching.*** In about a minute, *trans*-retinal completely separates from opsin. Retinal is responsible for the color of the photopigment, so the separation of trans-retinal from opsin causes opsin to look colorless. Because of the color change, this part of the cycle is termed **bleaching** of photopigment.
3. ***Conversion.*** An enzyme called **retinal isomerase** converts *trans*-retinal back to *cis*-retinal.
4. ***Regeneration.*** The *cis*-retinal then can bind to opsin, reforming a functional photopigment. This part of the cycle—resynthesis of a photopigment—is called **regeneration**.

The pigmented layer of the retina adjacent to the photoreceptors stores a large quantity of vitamin A and contributes to the regeneration process in rods. Cone photopigments regenerate much more quickly than the rhodopsin in rods and are less dependent on the pigmented layer. After complete bleaching, regeneration of half of the rhodopsin takes 5 minutes; half of the cone photopigments regenerate in only 90 seconds. Full regeneration of bleached rhodopsin takes 30 to 40 minutes. Prolonged vitamin A deficiency and the resulting below-normal amount of rhodopsin may cause **night blindness** or **nyctalopia** (nik′-ta-LŌ-pē-a), an inability to see well at low light levels.

Distribution of Photoreceptors in the Retina

Rods and cones have different patterns of distribution in the retina. The *fovea*, a small depression in the center of the macula lutea, contains only cones (FIGURE 9.38). The fovea is the site of highest visual acuity (sharpness of vision). Recall that you move your head and eyes while looking at something in order to place images of

FIGURE 9.38 The fovea.

The fovea is the site of highest visual acuity.

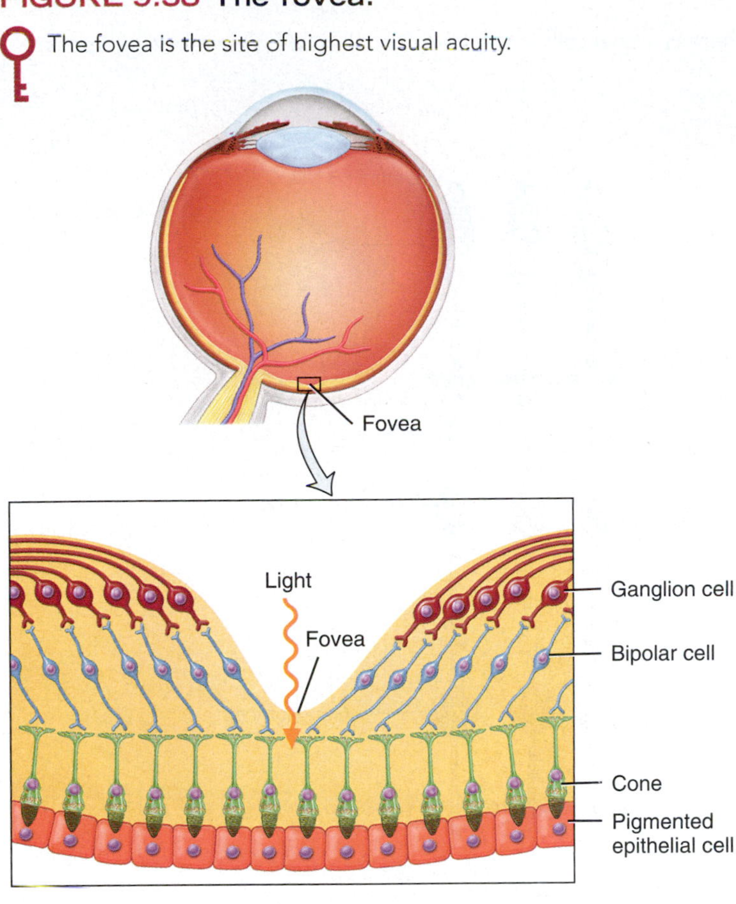

Why is vision sharpest at the fovea?

interest on your fovea. Vision is sharpest at the fovea for the following reasons:

1. The fovea has the highest density of photoreceptors. There are more photoreceptors (all cones) per square millimeter (mm^2) of the fovea compared to rest of the retina.
2. The layers of bipolar and ganglion cells, which scatter light to some extent, do not cover the cones in the fovea; instead, these layers are displaced to the side, allowing light to strike the cones directly.
3. Most of the retinal cells in the fovea have one-to-one synapses; in other words, one cone synapses with one bipolar cell, which in turn synapses with one ganglion cell. This provides a direct pathway for visual information to enter the brain.

Cone density sharply declines with distance from the fovea to the peripheral retina; rods are absent from the fovea, but the density of rods sharply increases in the surrounding area and then gradually declines toward the peripheral retina (FIGURE 9.39). Because rod vision is more sensitive than cone vision, you can see a faint object (such as a dim star) better if you gaze slightly to one side rather than looking directly at it.

Photoreceptors and Convergence of Visual Input

Within the retina, certain features of visual input are enhanced while other features may be discarded. Input from several cells may either converge upon a smaller number of postsynaptic neurons or diverge to a large number. Overall, convergence predominates: There are only 1 million ganglion cells but 126 million photoreceptors in the human eye. In the peripheral retina, where there are mainly rods, many pho-

FIGURE 9.39 Distribution of rods and cones in the retina. Note that cones in the fovea are more densely packed and have smaller diameters than the cones in the peripheral retina.

Cone density is highest in the fovea and then sharply declines toward the peripheral retina; rod density sharply increases in the area surrounding the fovea and then gradually declines toward the peripheral retina.

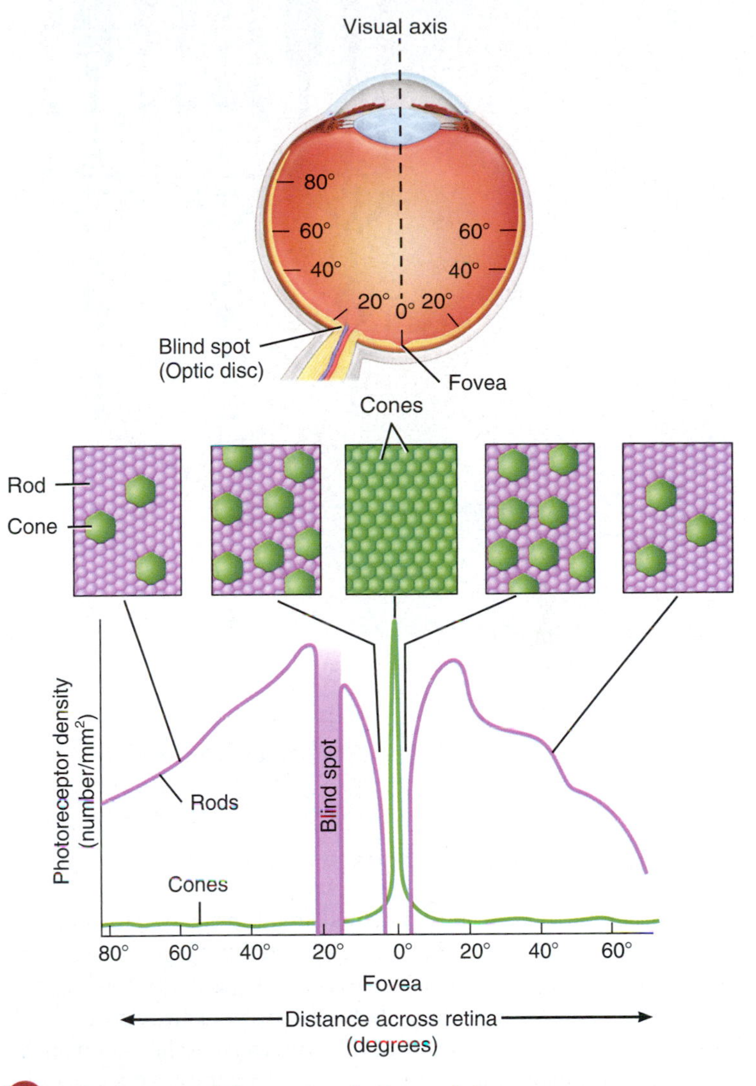

Which part of the retina lacks rods?

toreceptors synapse with a single bipolar cell, and several bipolar cells in turn synapse with a single ganglion cell (FIGURE 9.40). The high degree of convergence in rod pathways increases the light sensitivity of rod vision but slightly blurs the image that is perceived. As you have just learned, most cone pathways exhibit little convergence. In the fovea, which has the highest concentration of cones, a single cone usually synapses with a single bipolar cell, which synapses with a single ganglion cell (FIGURE 9.40). Cone vision, although less sensitive, is sharper because of these one-to-one synapses.

Phototransduction Converts Light into a Receptor Potential

Phototransduction is the process by which light energy is converted into a receptor potential in the outer segment of a photoreceptor. In most sensory systems, activation of a sensory receptor by its adequate stimulus

FIGURE 9.40 Convergence in the retina.

Convergence of retinal cells occurs mainly in the peripheral retina, whereas retinal cells in the fovea usually have one-to-one synapses.

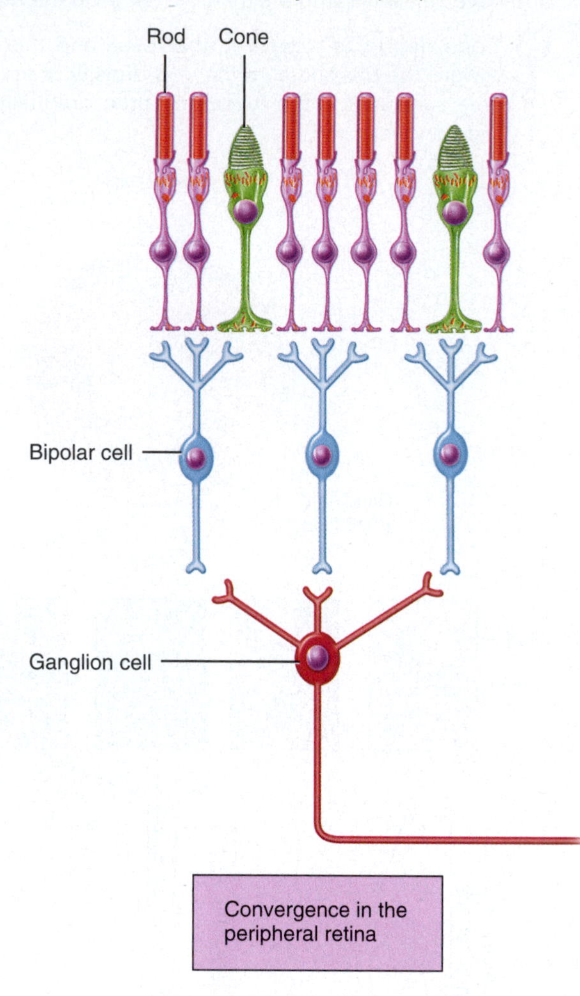

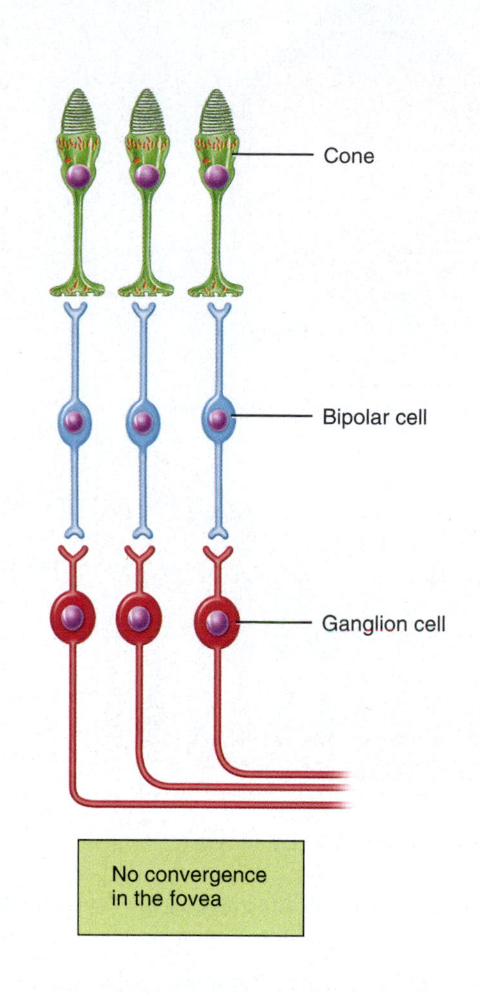

What effect does convergence have on visual acuity?

triggers a depolarizing receptor potential. In the visual system, however, activation of a photoreceptor by its adequate stimulus (light) causes a hyperpolarizing receptor potential. Just as surprising is that, when the photoreceptor is at rest—that is, in the dark—the cell is relatively depolarized. To understand how phototransduction occurs, you must first examine the operation of a photoreceptor in the absence of light (FIGURE 9.41a):

1. In darkness, *cis*-retinal is the form of retinal associated with the photopigment of the photoreceptor. Photopigment molecules are present in the disc membranes of the photoreceptor outer segment.
2. Another important occurrence during darkness is that there is a high concentration of the second messenger **cyclic GMP (cGMP)** in the cytosol of the photoreceptor outer segment. This is due to the continuous production of cGMP by the enzyme **guanylyl cyclase** in the disc membrane.
3. After it is produced, cGMP binds to and opens nonselective cation channels in the outer segment membrane. These **cGMP-gated channels** mainly allow Na^+ ions to enter the cell.
4. The inflow of Na^+, called the **dark current**, depolarizes the photoreceptor. As a result, in darkness, the membrane potential of a photoreceptor is about −40 mV. This is much closer to zero than a typical neuron's resting membrane potential of −70 mV.
5. The depolarization during darkness spreads from the outer segment to the synaptic terminal, which contains **voltage-gated Ca^{2+} channels** in its membrane. The depolarization keeps these channels open, allowing Ca^{2+} to enter the cell. The entry of Ca^{2+} in turn triggers exocytosis of synaptic vesicles, resulting in tonic release of large amounts of neurotransmitter from the synaptic terminal.

The absorption of light and isomerization of retinal initiates chemical changes in the photoreceptor outer segment that allow phototransduction to occur (FIGURE 9.41b):

1. When light strikes the retina, *cis*-retinal undergoes isomerization to *trans*-retinal.
2. Isomerization of retinal causes activation of a G protein known as **transducin** that is located in the disc membrane.
3. Transducin in turn activates an enzyme called **cGMP phosphodiesterase**, which is also present in the disc membrane.
4. Once activated, cGMP phosphodiesterase breaks down cGMP. The breakdown of cGMP lowers the concentration of cGMP in the cytosol of the outer segment.
5. As a result, the number of open cGMP-gated channels in the outer segment membrane is reduced and Na^+ inflow decreases.

FIGURE 9.41 Phototransduction.

Phototransduction is the process by which light energy is converted into a receptor potential.

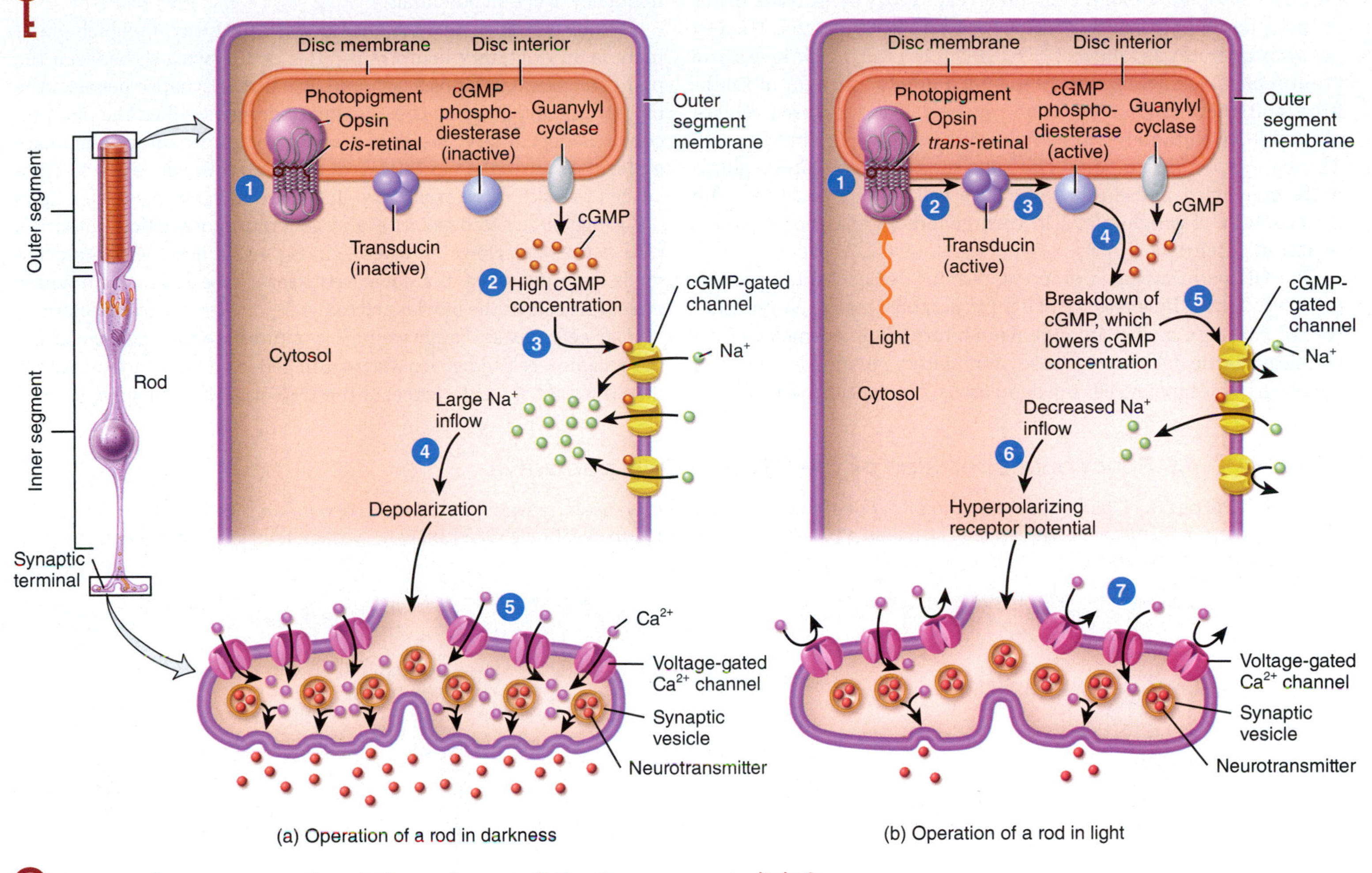

(a) Operation of a rod in darkness

(b) Operation of a rod in light

Does a photoreceptor depolarize or hyperpolarize in response to light?

6 The decreased Na^+ inflow causes the membrane potential to drop to about −65 mV, thereby producing a hyperpolarizing receptor potential.

7 The hyperpolarization spreads from the outer segment to the synaptic terminal, causing a decrease in the number of open voltage-gated Ca^{2+} channels. Ca^{2+} entry into the cell is reduced, which decreases the release of neurotransmitter from the synaptic terminal. Dim lights cause small and brief receptor potentials that partially turn off neurotransmitter release; brighter lights elicit larger and longer receptor potentials that more completely shut down neurotransmitter release.

Recall that the discs of rods form by pinching off from the plasma membrane of the outer segment; in cones, the discs are continuous with the outer segment membrane. Therefore, in rods, molecules of photopigment, transducin, cGMP phosphodiesterase, and guanylyl cyclase are located in a different membrane than the cGMP-gated channels; in cones, all of these proteins are located in the same membrane.

Visual Processing Begins in the Retina

Visual signals in the retina undergo considerable processing at the synapses among its various types of cells. Once receptor potentials arise in the outer segments of photoreceptors (rods and cones), they spread through the inner segments to the synaptic terminals, where they affect the release of neurotransmitter molecules. The neurotransmitter released by photoreceptors is the amino acid glutamate. Glutamate molecules induce local graded potentials in bipolar cells. Some bipolar cells are excited by glutamate; other bipolar cells are inhibited. The effect of glutamate on bipolar cells depends on the type of glutamate receptor present in the bipolar cell membrane. When glutamate has an inhibitory effect, it binds to metabotropic glutamate receptors that close cation channels, resulting in hyperpolarization of the bipolar cell. When glutamate has an excitatory effect, it binds to ionotropic glutamate receptors that open cation channels, causing depolarization of the bipolar cell. Once bipolar cells are excited, they release their own neurotransmitter, which is also glutamate, at synapses with ganglion cells. The effect of glutamate on ganglion cells is usually excitatory because most ganglion cells contain ionotropic glutamate receptors that open cation channels. Ganglion cells at rest spontaneously produce action potentials at a steady, low baseline rate. The frequency of action potentials increases above the baseline rate when the ganglion cell is excited and decreases below this rate when the ganglion cell is inhibited. Ganglion cells are the only cells of the retina that can generate action potentials; the other retinal cells produce only graded potentials.

In one type of retinal pathway called the **ON pathway**, glutamate is an inhibitory neurotransmitter at the synapse between the photoreceptor and bipolar cell and an excitatory neurotransmitter at the synapse between the bipolar cell and ganglion cell. The ON pathway operates as follows (**FIGURE 9.42a**): During darkness, the photoreceptor is depolarized, causing a greater amount of inhibitory glutamate molecules to be released. The increased release of inhibitory glutamate molecules hyperpolarizes the bipolar cell. The hyperpolarized bipolar cell releases fewer excitatory glutamate molecules, causing the ganglion cell to hyperpolarize. The hyperpolarized ganglion cell in turn generates a lower frequency of action potentials.

The ON pathway has a different response when light is present (**FIGURE 9.42b**): The presence of light hyperpolarizes the photoreceptor cell, causing a smaller amount of inhibitory glutamate molecules to be released. The decreased release of inhibitory glutamate molecules depolarizes the bipolar cell. The depolarized bipolar cell releases more excitatory glutamate molecules, causing the ganglion cell to depolarize. In response to depolarization, the ganglion cell generates a higher frequency of action potentials.

Within the retina there is also an **OFF pathway**, in which glutamate is an excitatory neurotransmitter at the synapse between the photoreceptor and bipolar cell, and also at the synapse between the bipolar cell and ganglion cell. So, during darkness when the photoreceptor is depolarized, the increased release of excitatory glutamate molecules causes the bipolar cell to depolarize; the depolarized bipolar cell releases more excitatory glutamate molecules, which causes the ganglion cell to depolarize and generate more action potentials (**FIGURE 9.42c**). However, when light is present and the photoreceptor is hyperpolarized, the decreased release of excitatory glutamate molecules causes the bipolar cell to hyperpolarize; the hyperpolarized bipolar cell releases fewer excitatory glutamate molecules, causing the ganglion cell to hyperpolarize and generate fewer action potentials (**FIGURE 9.42d**). Hence, in the OFF pathway, the ganglion cell

FIGURE 9.42 Functional operation of the ON and OFF pathways.

In the ON pathway, glutamate is an inhibitory neurotransmitter at the synapse between the photoreceptor and bipolar cell, and an excitatory neurotransmitter at the synapse between a bipolar cell and ganglion cell. In the OFF pathway, glutamate is an excitatory neurotransmitter at both synapses.

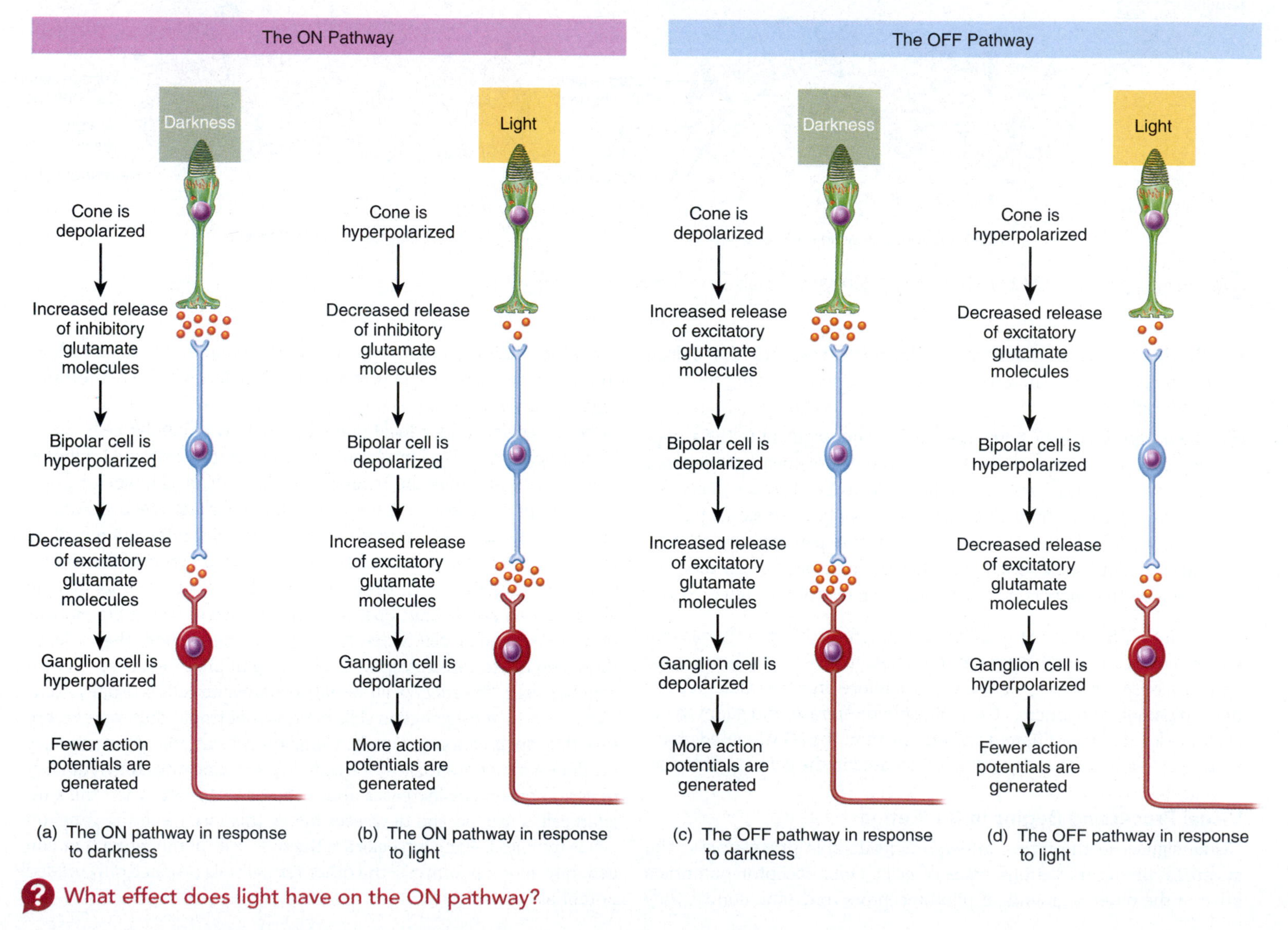

(a) The ON pathway in response to darkness (b) The ON pathway in response to light (c) The OFF pathway in response to darkness (d) The OFF pathway in response to light

What effect does light have on the ON pathway?

depolarizes during darkness and hyperpolarizes when light is present. This contrasts with the ON pathway, in which the ganglion cell hyperpolarizes during darkness and depolarizes when light is present. ON and OFF pathways are usually side by side in the retina. The existence of both ON and OFF pathways allows retinal cells to respond differently to light or darkness. As a result, the visual system is sensitive to changes in illumination rather than to absolute darkness or brightness.

Synaptic activity between photoreceptors and bipolar cells is influenced by horizontal cells (see **FIGURE 9.30a**). Horizontal cells form synapses with photoreceptors and have only indirect effects on bipolar cells. In adjacent areas of the retina, one photoreceptor usually forms an excitatory synapse (via glutamate) with a horizontal cell, and the horizontal cell in turn forms an inhibitory synapse (via the neurotransmitter GABA) with the presynaptic terminals of another photoreceptor (see **FIGURE 9.43c,d**). In this way, one photoreceptor can excite the horizontal cell, which can then inhibit the other photoreceptor, decreasing the amount of neurotransmitter that is released onto a bipolar cell. Hence, horizontal cells can transmit laterally directed inhibitory signals to photoreceptors. This lateral inhibition helps to improve visual contrast between adjacent areas of the retina.

Synaptic activity between bipolar cells and ganglion cells is influenced by amacrine cells (see **FIGURE 9.30a**). Amacrine cells transmit laterally directed inhibitory signals (lateral inhibition) at synapses formed with bipolar cells and ganglion cells. The inhibitory neurotransmitters released

FIGURE 9.43 Ganglion cell receptive fields.

In an ON-center/OFF-surround field, the ganglion cell is excited when light is present in the center and inhibited when light is present in the surround; in an OFF-center/ON-surround field, the ganglion cell is inhibited when light is present in the center and excited when light is present in the surround.

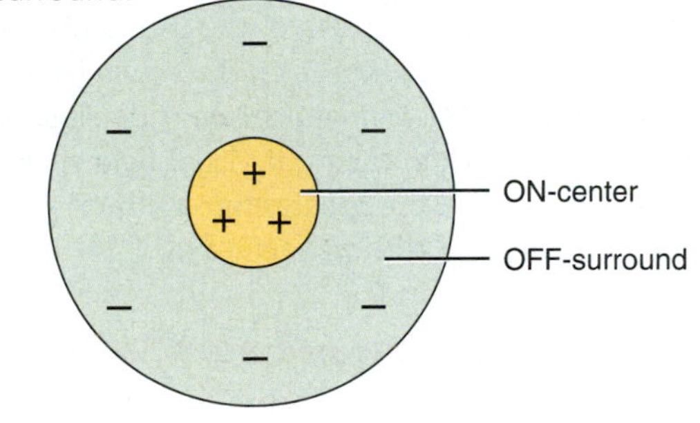

(a) ON-center/OFF-surround field

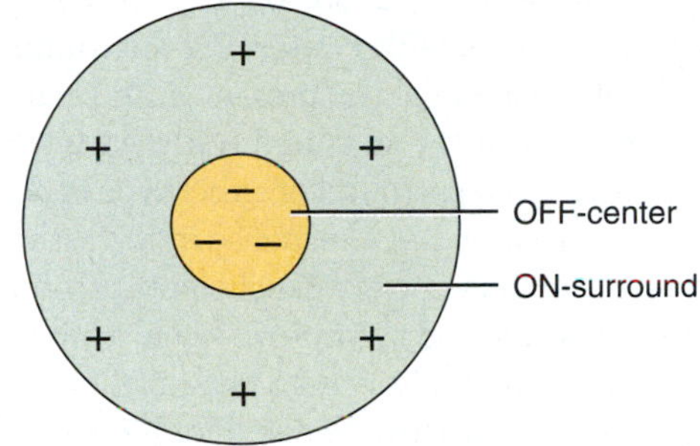

(b) OFF-center/ON-surround field

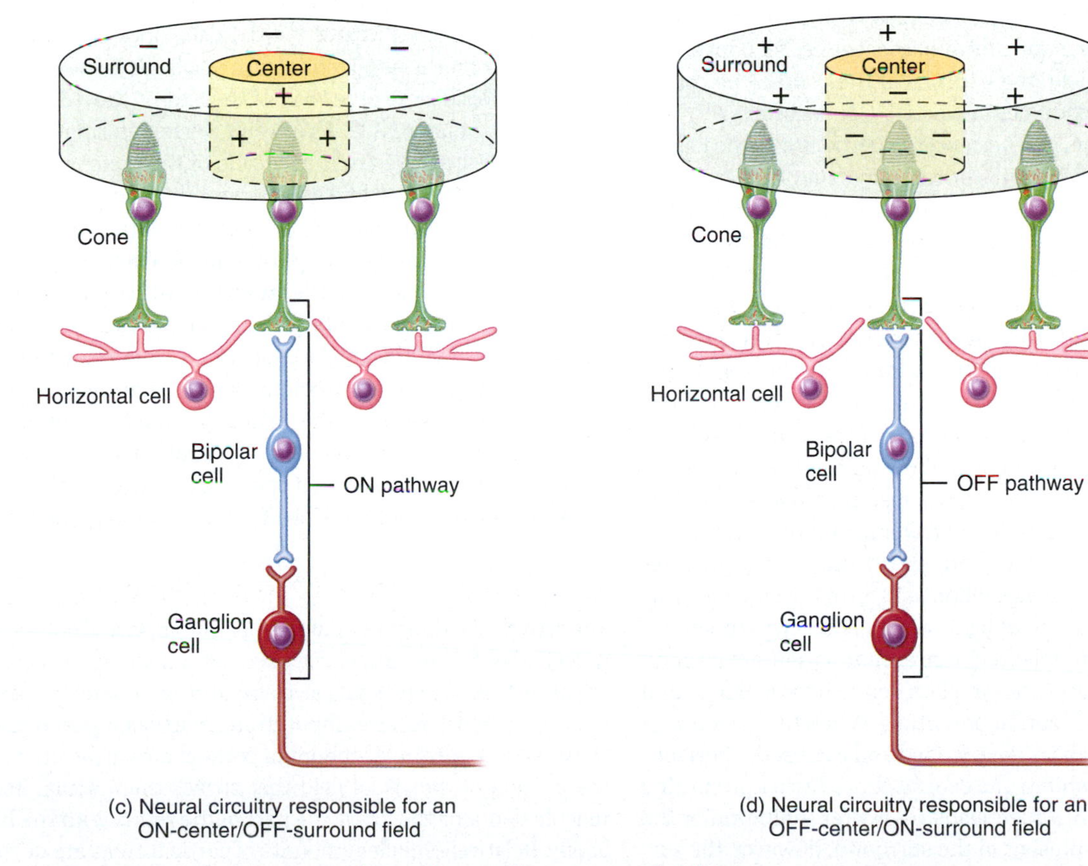

(c) Neural circuitry responsible for an ON-center/OFF-surround field

(d) Neural circuitry responsible for an OFF-center/ON-surround field

What is the functional significance of ganglion cell receptive fields?

by amacrine cells at these synapses include GABA and glycine. There are many different types of amacrine cells, and they have a variety of functions. Depending on which amacrine cells are involved, they can respond to a change in the level of illumination in the retina, the onset or offset of a visual signal, or movement of a visual signal in a particular direction.

In the presence of light, a ganglion cell can be either excited (generates more action potentials) or inhibited (generates fewer action potentials), depending on where in its receptive field the light falls. The receptive field of a ganglion cell refers to the area of visual space where light elicits a response in the ganglion cell. It also includes all of the retinal cells (photoreceptors, bipolar cells, etc.) that provide input to that ganglion cell. Ganglion cell receptive fields are shaped like a doughnut and consist of two parts: an inner *center* (the doughnut hole) and an outer *surround* (the doughnut itself). There are two types of ganglion cell receptive fields: ON-center/OFF-surround and OFF-center/ON-surround. In an **ON-center/OFF-surround field**, the ganglion cell is excited when light is present in the center and inhibited when light is present in the surround (**FIGURE 9.43a**). The opposite response occurs in an **OFF-center/ON-surround field**: The ganglion cell is inhibited when light is present in the center and excited when light is present in the surround (**FIGURE 9.43b**). Diffuse light over both the center and surround of either type of ganglion cell receptive field causes only a minimal response in the associated ganglion cell. Therefore, the organization of ganglion cell receptive fields is designed to enhance contrast and to improve visual acuity.

In a simplistic case, the center/surround organization of a ganglion receptive field is due to the presence of one photoreceptor (such as a cone) in the center and two cones in the surround (**FIGURE 9.43c, d**). The cone in the center synapses with a bipolar cell, which in turn synapses with a ganglion cell. The cones in the surround do not synapse with the bipolar cell but instead synapse with horizontal cells, which synapse with the presynaptic terminals of the center cone. In an ON-center/OFF-surround field, the neurons in the center (the center cone, bipolar cell, and ganglion cell) form the ON pathway (**FIGURE 9.43c**), so the presence of light in the center hyperpolarizes the center cone, causing it to release fewer inhibitory glutamate molecules, which ultimately causes the ganglion cell to depolarize and generate a higher frequency of action potentials. When the surround is in darkness, the surround cone is depolarized and excites the horizontal cell. The horizontal cell then inhibits the central cone, further enhancing the cone's hyperpolarization, which leads to increased excitation of the ganglion cell. However, when light is present in the surround, the surround cone is inhibited and cannot excite the horizontal cell. The horizontal cell hyperpolarizes and can no longer inhibit the central cone. This has a depolarizing effect on the central cone, allowing the cone to release more of its inhibitory glutamate molecules, resulting in the inhibition of the ganglion cell and the generation of fewer action potentials.

In an OFF-center/ON-surround field, the neurons in the center (the center cone, bipolar cell, and ganglion cell) form the OFF pathway (**FIGURE 9.43d**), so the presence of light in the center hyperpolarizes the center cone, causing it to release fewer excitatory glutamate molecules, which ultimately causes the ganglion cell to hyperpolarize and generate a lower frequency of action potentials. When the surround is in darkness, the surround cone is depolarized and excites the horizontal cell. The horizontal cell inhibits the central cone, further promoting the cone's hyperpolarization, which leads to greater inhibition of the ganglion cell. When light is present in the surround, however, the surround cones hyperpolarize and are not able to excite the horizontal cell. The horizontal cell hyperpolarizes and cannot inhibit the center cone. The lack of inhibition by the horizontal cell has a depolarizing effect on the cone, permitting the cone to release more of its excitatory glutamate molecules. As a result, the ganglion cell is excited and generates a higher frequency of action potentials.

Both Eyes Receive Input from the Left and Right Visual Fields

Everything that can be seen by one eye is that eye's **visual field**. As noted earlier, because the eyes are located in the front of the head, the visual fields overlap considerably (**FIGURE 9.44**). Humans have binocular vision due to the large region where the visual fields of the two eyes overlap—the **binocular visual field**. The part of the visual field that can be seen only with one eye is called the **monocular portion**.

If the visual fields of both eyes are merged to form one large visual field, then that visual field can be divided into two halves: the left visual field and the right visual field. Input from the left or right visual fields strikes different parts of the retina (**FIGURE 9.44**). Input from the left visual field strikes the *nasal retina* (side closest to the nose) of the left eye and the *temporal retina* (side closest to the temple) of the right eye. Input from the right visual field strikes the temporal retina of the left eye and the nasal retina of the right eye. Hence, both eyes receive input from both visual fields.

The Visual Pathway Extends from Photoreceptors to the Visual Areas of the Brain

Axons of retinal ganglion cells provide output from the retina to the brain, exiting the eyeball as the **optic (II) nerve**. The two optic nerves meet at the base of the brain near the hypothalamus to form the **optic chiasm** (kī-AZ-m) (**FIGURE 9.44**). In the optic chiasm, half of the axons (those from the nasal retina) from each eye decussate (cross over) to the opposite side; the other half of the axons (those from the temporal retina) remain uncrossed. As a result, visual information from the *right* visual field is conveyed to the *left* side of the brain, and visual information from the *left* visual field is conveyed to the *right* side of the brain (**FIGURE 9.44**). After passing through the optic chiasm, the axons enter the brain and are now called the **optic tract**. Some of the fibers of the optic tract terminate in the **superior colliculus**, which controls eye movements and movements of the head and trunk in response to sudden visual stimuli. The great majority of the optic tract fibers terminate in the **lateral geniculate nucleus** of the thalamus.From the thalamus, neurons project axons to the **primary visual cortex** in the occipital lobe, where conscious awareness of visual sensations occurs (**FIGURE 9.44**). From the primary visual cortex, axons extend to the **visual association area** in the occipital lobe for complex integration of visual input.

The Primary Visual Cortex Perceives Light

The arrival of action potentials in the primary visual cortex allows you to perceive light. The primary visual cortex has a map of visual space: Each region within the cortex receives input from a different part of the retina, which in turn receives input from a particular part of the visual field (**FIGURE 9.44**). A large amount of cortical area is devoted to input from the portion of the visual field that strikes the macula. Recall that the macula contains the fovea, the part of the retina with the highest visual acuity. Relatively smaller amounts of cortical areas are devoted to those portions of the visual field that strike the peripheral parts of the retina.

FIGURE 9.44 The visual pathway.

At the optic chiasm, half of the axons from each eye cross over to the opposite side of the brain.

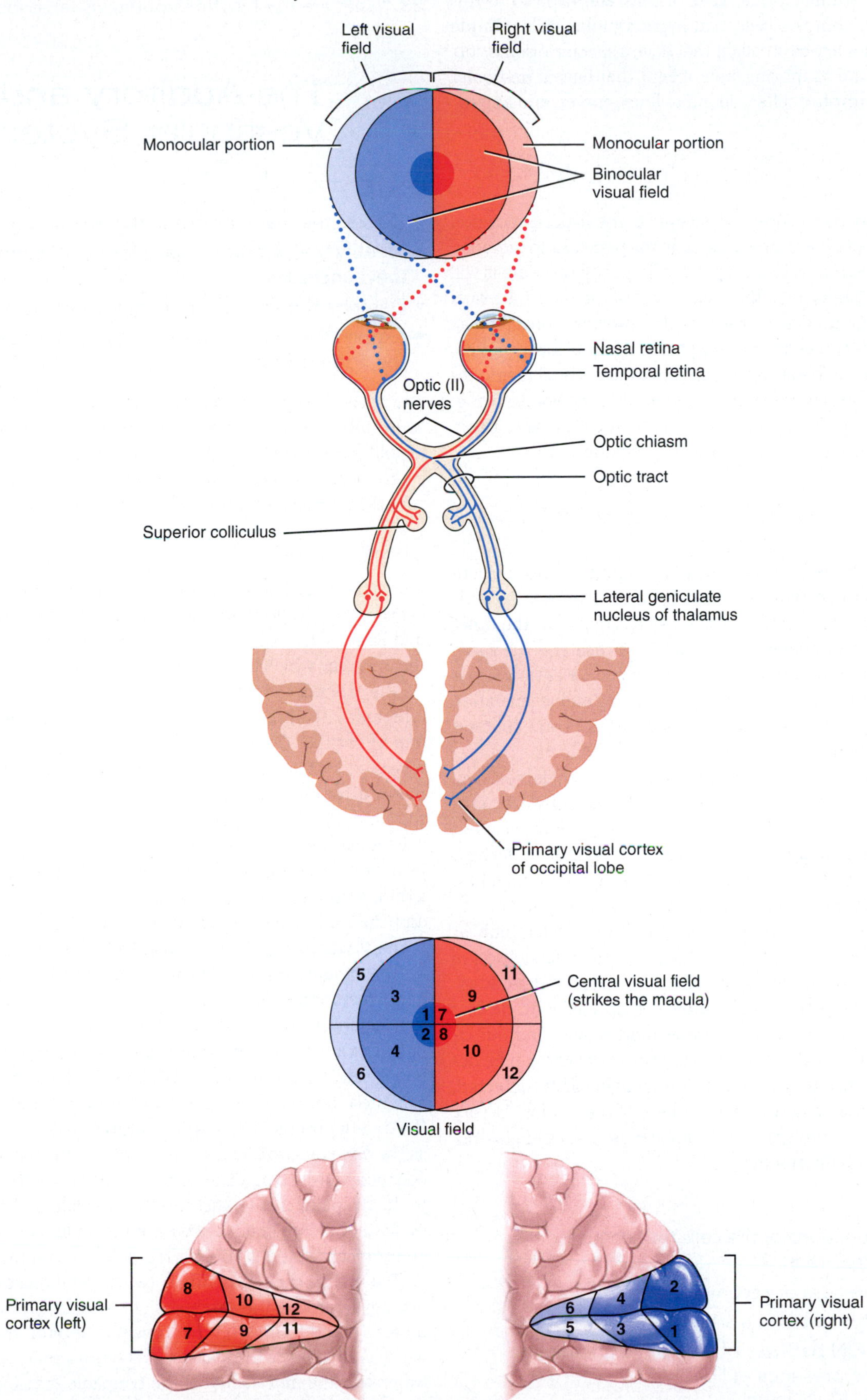

What is the correct order of structures that carry action potentials from the retina to the primary visual cortex?

Cells in the primary visual cortex begin to process the shape and orientation of visual stimuli. Three types of cells are involved in this process: simple cells, complex cells, and hypercomplex cells. **Simple cells** respond to stationary bars of light that have a specific orientation. **Complex cells** respond to moving bars of light that have a particular orientation. **Hypercomplex cells** respond to lines, curves, and angles.

The Visual Association Area Performs Several Functions

Input from the primary visual cortex is conveyed to the visual association area in the occipital lobe. There are also areas in the parietal and temporal lobes that receive and process visual input; for simplicity, these areas will be considered as an extension of the visual association area. The visual association area further processes visual input to provide more complex visual patterns, such as three-dimensional position, overall form, motion, and color. In addition, the visual association area stores visual memories and relates past and present visual experiences, allowing you to recognize what you are seeing. For example, the visual association area allows you to recognize an object such as pencil just by looking at it.

The Visual System Can Undergo Light and Dark Adaptation

When you emerge from dark surroundings (say, a tunnel) into the sunshine, **light adaptation** occurs—your visual system adjusts in seconds to the brighter environment by decreasing its sensitivity. On the other hand, when you enter a darkened room such as a theater, your visual system undergoes **dark adaptation**—its sensitivity increases slowly over many minutes. The difference in the rates of bleaching and regeneration of the photopigments in the rods and cones accounts for some (but not all) of the sensitivity changes during light and dark adaptation.

As the light level increases, more and more photopigment is bleached. While light is bleaching some photopigment molecules, however, others are being regenerated. In daylight, regeneration of rhodopsin cannot keep up with the bleaching process, so rods contribute little to daylight vision. In contrast, cone photopigments regenerate rapidly enough that some of the *cis* form is always present, even in very bright light.

If the light level decreases abruptly, sensitivity increases rapidly at first and then more slowly. In complete darkness, full regeneration of cone photopigments occurs during the first 8 minutes of dark adaptation. During this time, a threshold (barely perceptible) light flash is seen as having color. Rhodopsin regenerates more slowly, and our visual sensitivity increases until even a single photon of light can be detected. In that situation, although much dimmer light can be detected, threshold flashes appear gray-white, regardless of their color. At very low light levels, such as starlight, objects appear as shades of gray because only the rods are functioning.

CHECKPOINT

17. What are the functions of the cells that comprise the neural layer of the retina?

18. Describe the process of accommodation.

19. How do receptor potentials arise in photoreceptors?

20. How does the ON pathway of the retina function in darkness? In the presence of light?

21. What is an ON-center/OFF-surround ganglion receptive field?

22. What is the binocular visual field?

23. What are the functions of the visual association area?

9.6 The Auditory and Vestibular Systems

OBJECTIVES

- Discuss the major events in the physiology of hearing.
- Identify the receptor organs for equilibrium and explain how they function.
- Describe the auditory and equilibrium pathways to the brain.
- Explain the importance of the primary auditory cortex and the auditory association area.

The ear is an engineering marvel; its sensory receptors can transduce sound vibrations with amplitudes as small as the diameter of an atom of gold (0.3 nm) into electrical signals 1000 times faster than photoreceptors can respond to light. Besides receptors for sound waves, the ear also contains receptors for equilibrium.

The Ear Has Different Functional Components

The ear is divided into three main regions (FIGURE 9.45a): (1) the external ear, which collects sound waves and channels them inward; (2) the middle ear, which conveys sound vibrations to the oval window; and (3) the inner ear, which houses the receptors for hearing and equilibrium.

The *external ear* consists of the pinna, external auditory canal, and tympanic membrane (FIGURE 9.45a). The **pinna** is a skin-covered flap of cartilage located on each side of the head. It collects sound waves and directs them into the external auditory canal. The **external auditory canal** is a curved tube that conveys sound waves from the pinna to the tympanic membrane. The canal contains a sticky substance called **cerumen** (earwax), which helps prevent foreign objects from entering the ear. The **tympanic membrane**, or *eardrum*, is a thin, semitransparent structure between the external auditory canal and the middle ear. When sound waves strike the tympanic membrane, it vibrates and then transmits the vibrations to the middle ear.

The *middle ear* is a small, air-filled cavity located between the tympanic membrane and inner ear (FIGURE 9.45a, b). Within the partition separating the middle and inner ears are two membrane-covered openings: an upper **oval window** and a lower **round window**. Three tiny bones called **auditory ossicles** (OS-si-kuls) extend across the middle ear. These bones, which are named for their shapes, are the **malleus** (MAL-ē-us), **incus** (ING-kus), and **stapes** (STĀ-pēz), commonly known as the *hammer*, *anvil*, and *stirrup*, respectively. The auditory ossicles are connected to each other. In addition, one end of the malleus is attached to the tympanic membrane, and the footplate of the stapes fits into the oval window. The main function of the auditory ossicles is to transmit and amplify vibrations from the tympanic membrane to the oval window.

The middle ear also contains two skeletal muscles that contract reflexively to protect the structures of the inner ear from damage by loud noises (FIGURE 9.45b). The **tensor tympani muscle** (TIM-pan-ē), which is attached to the malleus, protects the inner ear from loud sounds by limiting the movements of the tympanic membrane. The **stapedius muscle** (sta-PĒ-de-us), which is attached to the stapes, protects the inner ear from loud noises by dampening the movements of the stapes in

FIGURE 9.45 Components of the ear. In part (c), the outer, tan-colored area is part of the bony labyrinth; the inner, pink-colored area is the membranous labyrinth.

The ear consists of three main parts: the external ear, middle ear, and inner ear.

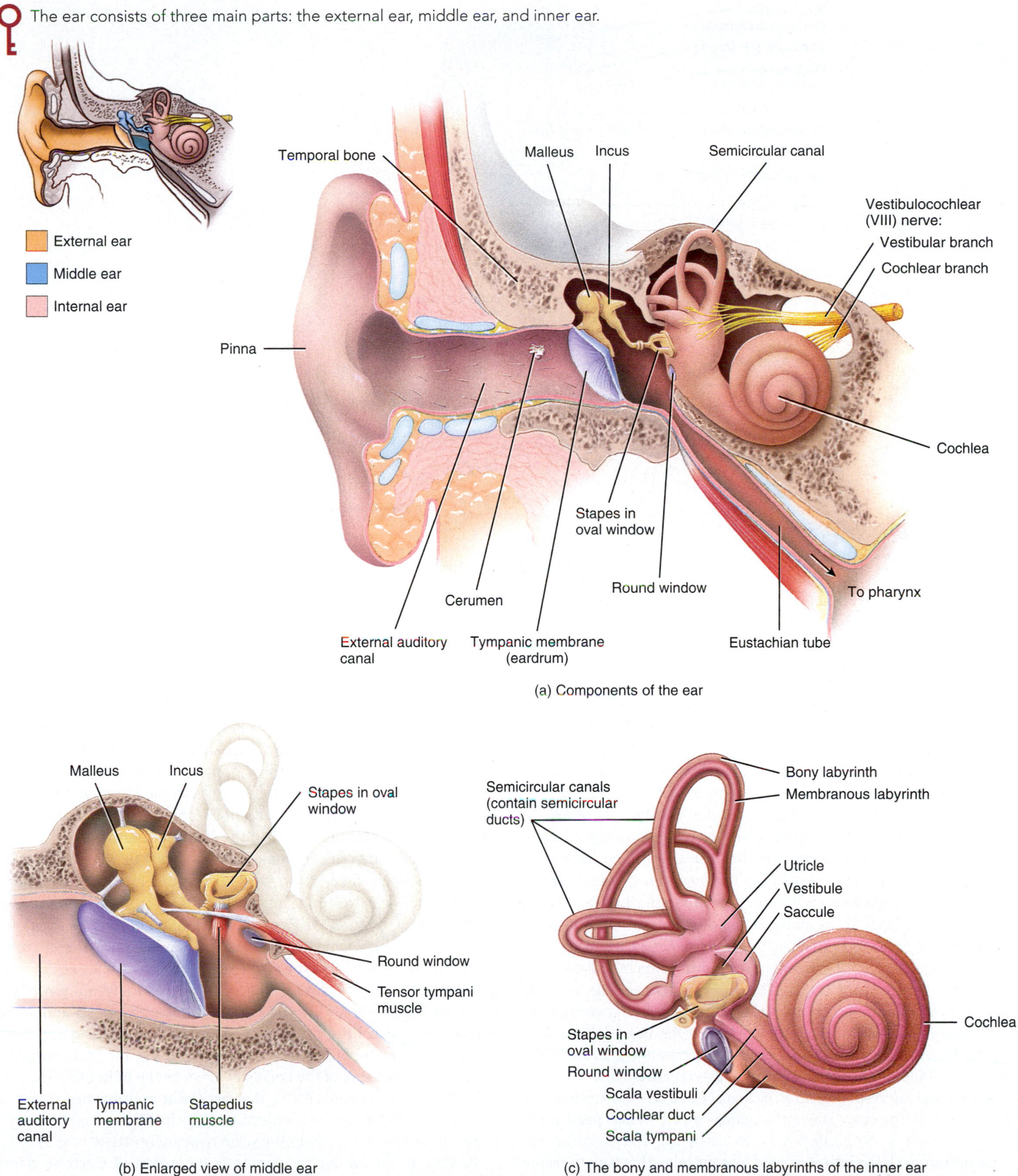

(a) Components of the ear

(b) Enlarged view of middle ear

(c) The bony and membranous labyrinths of the inner ear

Figure 9.45 (continues)

Figure 9.45 Continued

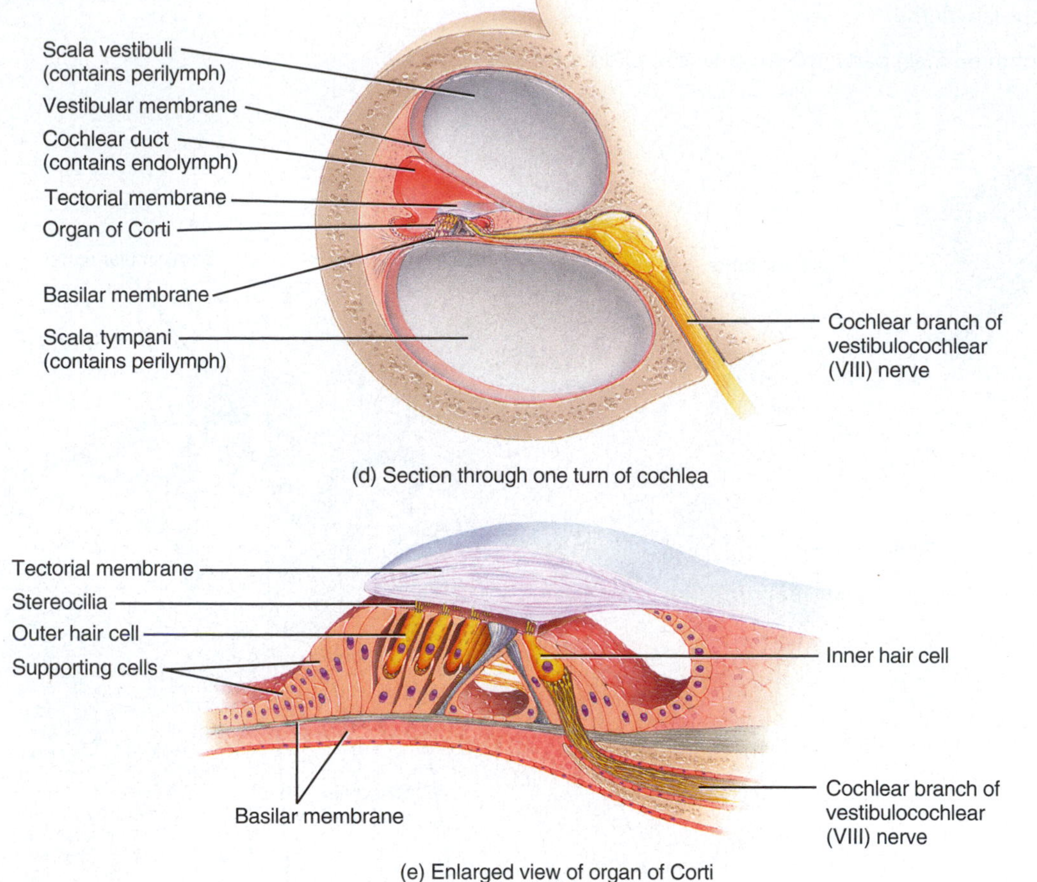

(d) Section through one turn of cochlea

(e) Enlarged view of organ of Corti

? Which part of the ear contains the receptors for hearing and equilibrium?

the oval window. Because it takes a fraction of a second for the tensor tympani and stapedius muscles to contract, they can protect the inner ear from prolonged loud noises but not from brief ones such as a gunshot.

The middle ear is connected to the pharynx (throat) by the **eustachian tube**, also known as the *auditory tube* (FIGURE 9.45a). The eustachian tube is normally closed at the end closest to the pharynx. During swallowing and yawning, it opens, allowing air to enter or leave the middle ear until the pressure in the middle ear equals the atmospheric pressure. This explains why yawning can help equalize the pressure changes that occur while flying in an airplane. Most of us have experienced our ears popping as the pressures equalize. When the pressures are balanced, the tympanic membrane vibrates freely as sound waves strike it. If the pressure is not equalized, intense pain, hearing impairment, and ringing in the ears could develop.

The final component of the ear is the *inner ear*, which is also known as the *labyrinth* because of its complicated series of canals. The inner ear consists of two main divisions: an outer bony labyrinth that encloses an inner membranous labyrinth (FIGURE 9.45c). The **bony labyrinth** is a series of cavities in the temporal bone of the cranium; it includes the cochlea, vestibule, and semicircular canals. The cochlea is the sense organ for hearing, and the vestibule and semicircular canals are the sense organs for equilibrium. The bony labyrinth contains a fluid called **perilymph**. This fluid, which is chemically similar to cerebrospinal fluid, surrounds the **membranous labyrinth**, a series of sacs and tubes that have the same general shape as the bony labyrinth. The membranous labyrinth contains a fluid called **endolymph**. The level of potassium ions (K^+) in endolymph is unusually high for an extracellular fluid, and potassium ions play a role in the generation of auditory signals (described shortly).

The **vestibule** (VES-ti-būl) is the middle part of the bony labyrinth (FIGURE 9.45c). The membranous labyrinth in the vestibule consists of two sacs called the **utricle** (Ū-tri-kl) and the **saccule** (SAK-ūl). Behind the vestibule are the three bony **semicircular canals**, each of which lies at approximately right angles to the other two (FIGURE 9.45c). The portions of the membranous labyrinth that lie inside the bony semicircular canals are called the **semicircular ducts**.

A section through the **cochlea** (KOK-lē-a), a bony spiral canal that resembles a snail's shell, shows that it is divided into three channels: cochlear duct, scala vestibuli, and scala tympani (FIGURE 9.45d). The **cochlear duct**, or *scala media*, is a continuation of the membranous labyrinth into the cochlea; it is filled with endolymph. The channel above the cochlear duct is the **scala vestibuli**, which ends at the oval window. The channel below the cochlear duct is the **scala tympani**, which ends at the round window. Both the scala vestibuli and scala tympani are part of the bony labyrinth of the cochlea and are filled with perilymph. The scala vestibuli is connected to the scala tympani at a region at the apex of the cochlea known as the **helicotrema** (hel-i-kō-TRĒ-ma; see FIGURE 9.47). The **vestibular membrane** separates the scala vestibuli from the cochlear duct; the **basilar membrane** separates the cochlear duct from the scala tympani (FIGURE 9.45d).

Resting on the basilar membrane is the **organ of Corti** or *spiral organ* (FIGURE 9.45d, e). It consists of supporting cells and hair cells. There are two groups of hair cells: a single row of **inner hair cells** and three rows of **outer hair cells**. At the apical tip of each hair cell are stereocilia, which are actually microvilli arranged in several rows of graded height. The stereocilia are embedded in a flexible, gelatinous covering called the **tectorial membrane**. At their basal ends, hair cells receive innervation from sensory and motor neurons of the cochlear branch of the vestibulocochlear (VIII) nerve. Most of the sensory neurons, which are first-order auditory neurons, synapse with inner hair cells; the motor neurons synapse mainly with outer hair cells.* Inner and outer hair cells have different functional roles. Inner hair cells are the receptors for hearing: They convert the mechanical vibrations of sound into electrical signals. Outer hair cells do not serve as hearing receptors; instead, they increase the sensitivity of the inner hair cells.

Sound Waves Are Generated from a Vibrating Object

To understand the physiology of hearing, it is necessary to learn about its input, which occurs in the form of sound waves. **Sound waves** are alternating high- and low-pressure areas that originate from a vibrating object and travel through some medium (such as air). For example, when a tuning fork is struck, it vibrates, producing sound waves in the surrounding air. The sound waves consist of areas of *compression*, where the air molecules are close together and the pressure is high, alternating with areas of *rarefaction*, where the air molecules are farther apart and the pressure is low (**FIGURE 9.46a**). As the air molecules vibrate back and forth, they bump other air molecules in adjacent regions, creating new areas of compression and rarefaction. As a result, the sound waves spread in all directions from the tuning fork (**FIGURE 9.46b**) in much the same way that ripples travel over the surface of a pond when you toss a stone into it. With increasing distance from the tuning fork, more energy is lost and the sound waves eventually die out.

Two important physical properties of a sound are pitch and intensity.

- ***Pitch.*** The **pitch**, or **tone**, of a sound is determined by the *frequency* of the sound waves. The higher the wave frequency, the higher is the pitch (**FIGURE 9.46c**). Pitch is measured in units called **hertz (Hz)** (1 Hz = 1 cycle per second). The sounds heard most acutely by our ears have frequencies between 500 and 5000 Hz. The entire audible range of human hearing extends from 20 to 20,000 Hz. Sounds of speech primarily contain frequencies

* Although the vestibulocochlear (VIII) nerve contains axons of both sensory and motor neurons, it is classified as a sensory nerve and not a mixed nerve because the motor axons do not innervate muscle or gland tissue.

FIGURE 9.46 Properties of sound waves.

Sound waves are alternating high- and low-pressure areas that originate from a vibrating object and travel through the air or some other medium.

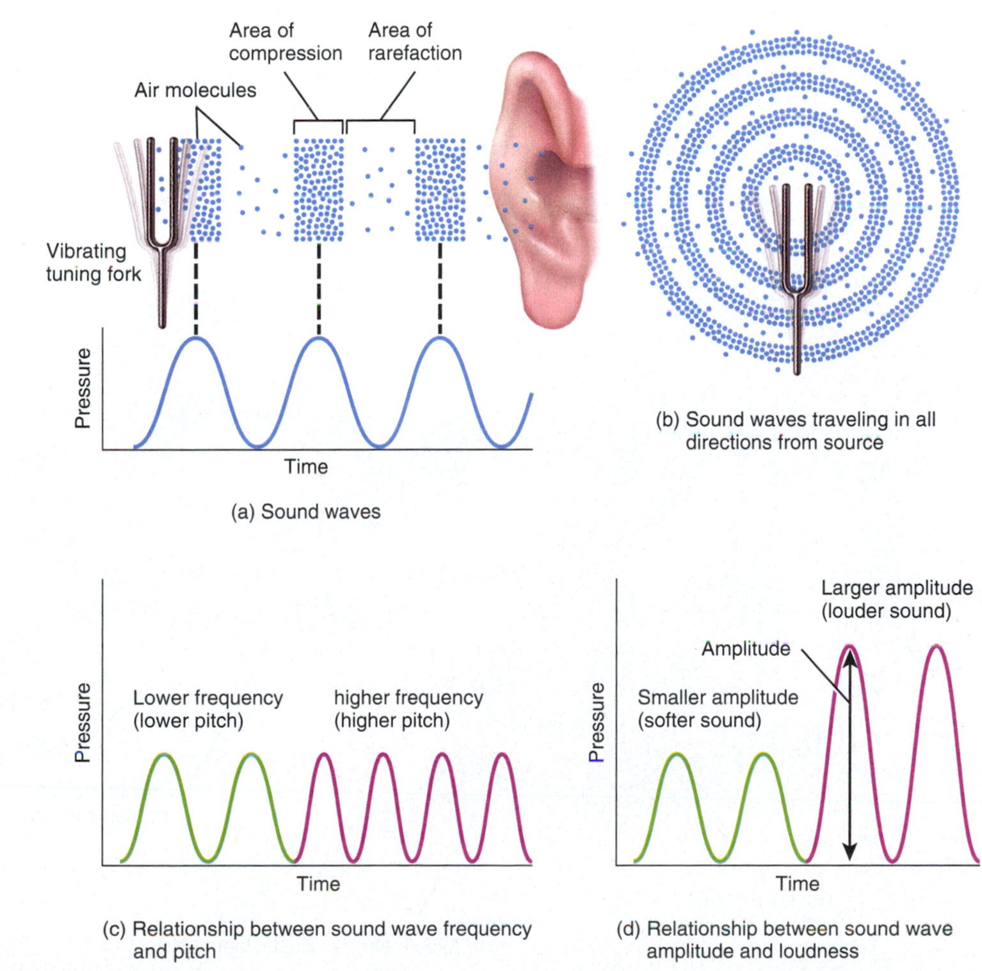

What range of sound wave frequencies can the human ear detect?

between 100 and 3000 Hz, and the "high C" sung by a soprano has a dominant frequency at 1048 Hz. The sounds from a jet plane several miles away range from 20 to 100 Hz. Unlike a tuning fork, which produces a pure tone with a single frequency, most sounds that we hear are mixtures of tones of various frequencies that are superimposed on a main, fundamental frequency. These superimposed tones, called *overtones*, are responsible for a sound's **quality** or **timber** (TAM-ber; also TIM-ber). Timber allows you to tell the difference between a violin and a flute that are playing the same note; it also lets you distinguish among the different voices of your family and friends.

- ***Intensity.*** The **intensity** of a sound is determined by the *amplitude* (size) of the sound waves, which is the difference in pressures between the areas of compression and rarefaction. We interpret intensity as **loudness**. The larger the amplitude of the sound waves, the louder is the sound (**FIGURE 9.46d**). Sound intensity is measured in logarithmic units called **decibels (dB)**. Every 10-dB increase represents a tenfold increase in sound intensity. The hearing threshold—the point at which an average young adult can just distinguish sound from silence—is defined as 0 dB at 1000 Hz. Rustling leaves have a decibel level of 15; whispered speech, 30; normal conversation, 60; a vacuum cleaner, 75; shouting, 80; and a nearby motorcycle or jackhammer, 90. Sound becomes uncomfortable to a normal ear at about 120 dB, and it is painful above 140 dB.

Transmission of Sound Waves Through the Ear Involves Several Steps

The following events are involved in hearing (**FIGURE 9.47**):

1. The pinna directs sound waves into the external auditory canal.
2. When sound waves strike the tympanic membrane, the alternating high and low pressure of the air causes the tympanic membrane to vibrate back and forth. The distance it moves, which is very small, depends on the intensity and frequency of the sound waves. The tympanic membrane vibrates slowly in response to low-frequency (low-pitched) sounds and rapidly in response to high-frequency (high-pitched) sounds.
3. The central area of the tympanic membrane connects to the malleus, which also starts to vibrate. The vibration is transmitted from the malleus to the incus and then to the stapes.
4. As the stapes moves back and forth, it pushes the membrane of the oval window in and out. The oval window vibrates about 20 times more vigorously than the tympanic membrane because the auditory ossicles efficiently transmit small vibrations spread over a large surface area (tympanic membrane) into larger vibrations of a smaller surface (oval window).
5. The movement of the stapes at the oval window sets up fluid pressure waves in the perilymph of the cochlea. As the oval window bulges inward, it pushes on the perilymph of the scala vestibuli.

FIGURE 9.47 Transmission of sound waves through the ear. The cochlea has been uncoiled to more easily visualize sound wave transmission and the resulting distortion of the vestibular and basilar membranes of the cochlear duct.

Vibration of the basilar membrane activates hair cells of the organ of Corti.

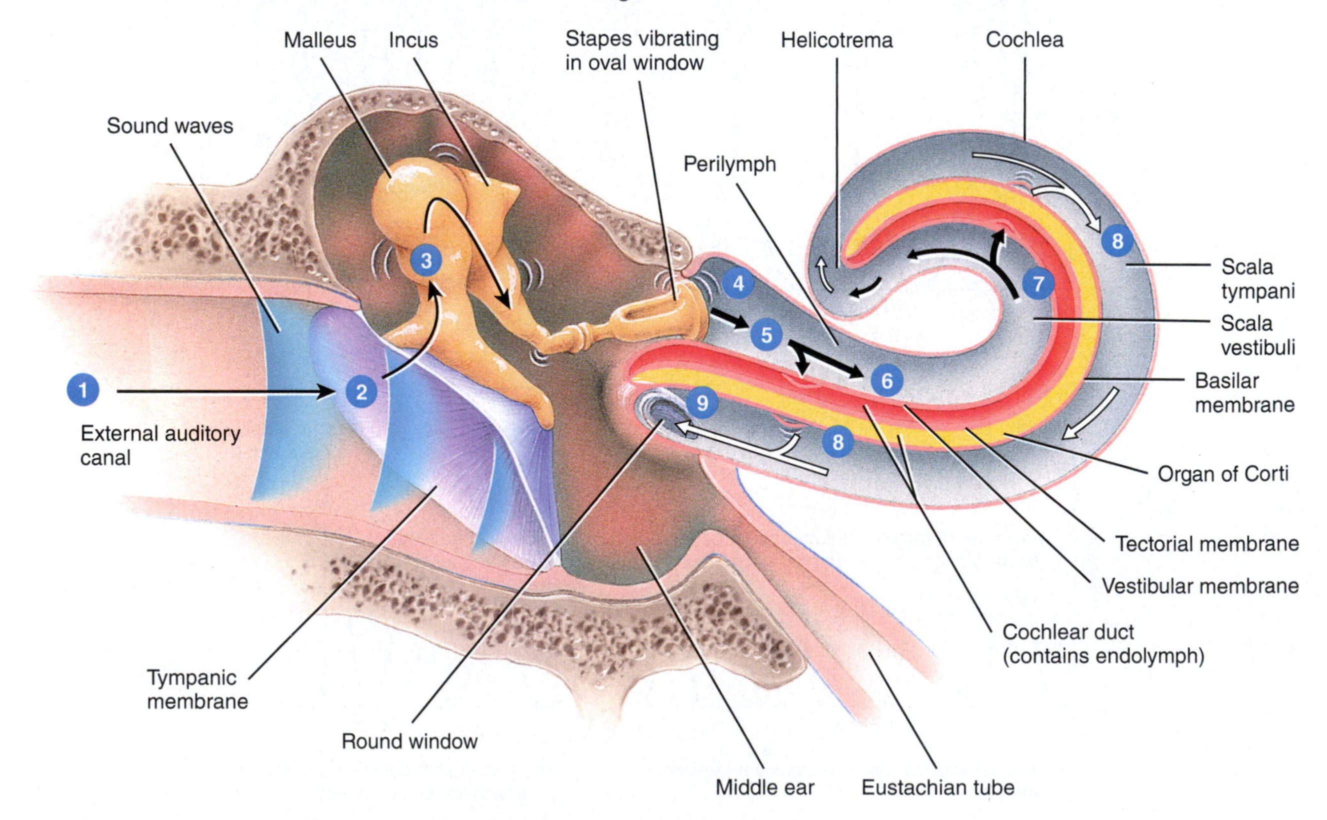

What causes the basilar membrane to vibrate?

6. Pressure waves are transmitted from the scala vestibuli to the scala tympani and eventually to the round window, causing it to bulge outward into the middle ear. (See step 9 in the figure.)
7. As the pressure waves deform the walls of the scala vestibuli and scala tympani, they also push the vestibular membrane back and forth, creating pressure waves in the endolymph inside the cochlear duct.
8. The pressure waves in the endolymph cause the basilar membrane to vibrate, which moves the hair cells of the organ of Corti against the tectorial membrane. This leads to bending of the hair cell stereocilia, resulting in the production of receptor potentials that ultimately lead to the generation of action potentials.

Inner Hair Cells Are Responsible for Sound Transduction

Inner hair cells transduce mechanical vibrations into electrical signals (**FIGURE 9.48**). As the basilar membrane vibrates, the stereocilia at the apex of the hair cell bend back and forth and slide against one another. Mechanically-gated cation channels are located in the membrane of the

FIGURE 9.48 Sound transduction.

Hair cells of the organ of Corti convert a mechanical vibration into a receptor potential.

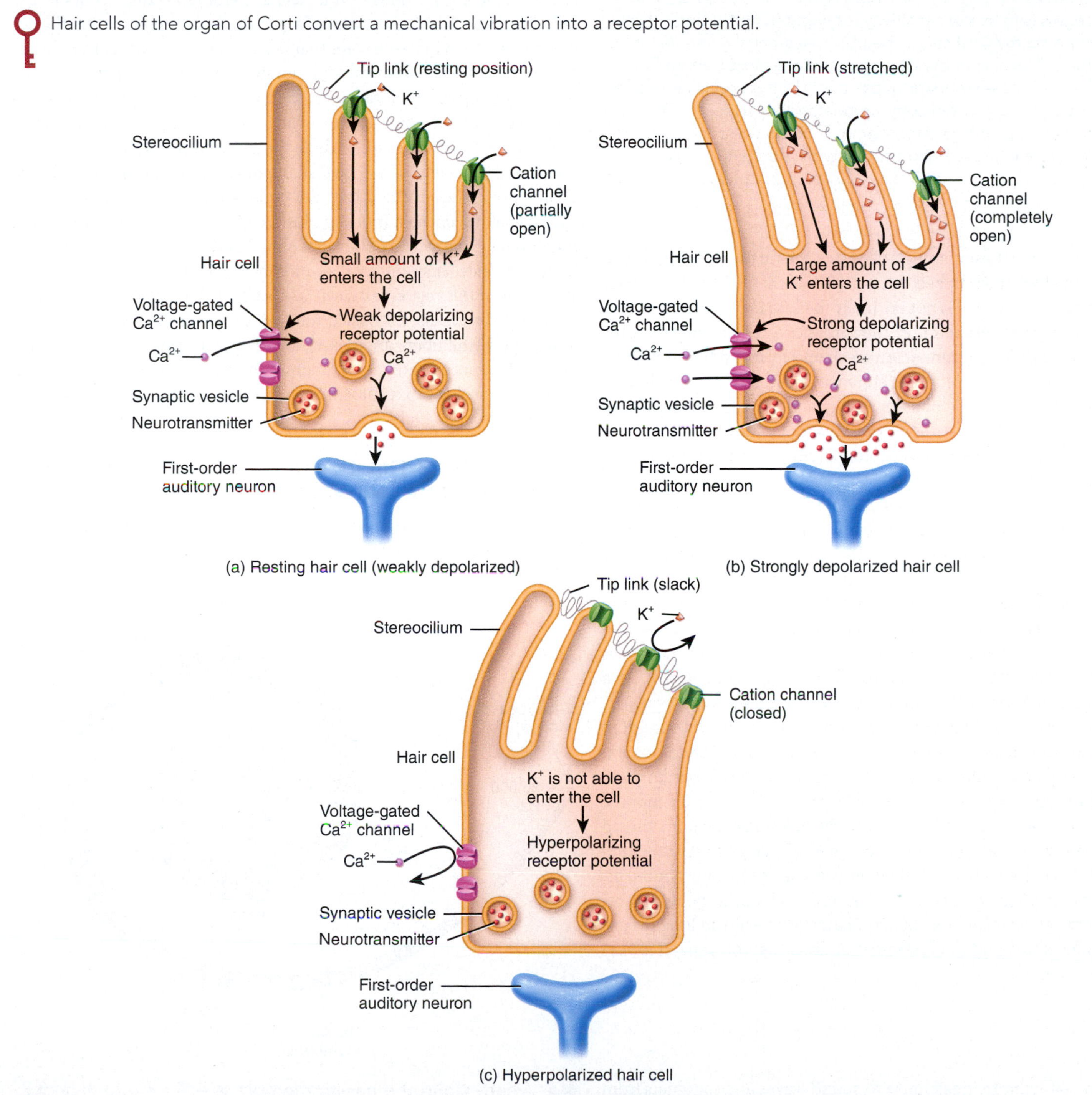

What is the purpose of the tip link proteins associated with the hair cells of the organ of Corti?

CLINICAL CONNECTION

Loud Sounds and Hair Cell Damage

Exposure to loud music and the engine roar of jet planes, revved-up motorcycles, lawn mowers, and vacuum cleaners damages hair cells of the cochlea. Because prolonged noise exposure causes hearing loss, employers in the United States must require workers to use hearing protectors when occupational noise levels exceed 90 dB. Rock concerts and even inexpensive headphones can easily produce sounds over 110 dB. Continued exposure to high-intensity sounds is one cause of ***deafness,*** a significant or total hearing loss. The louder the sounds, the more rapid is the hearing loss. Deafness usually begins with loss of sensitivity for high-pitched sounds. If you are listening to music through head-phones and bystanders can hear it, the dB level is in the damaging range. Most people fail to notice their progressive hearing loss until destruction is extensive and they begin having difficulty understanding speech. Wearing earplugs with a noise-reduction rating of 30 dB while engaging in noisy activities can protect the sensitivity of your ears.

stereocilia. Opening these channels allows cations in the endolymph, primarily K^+, to enter the hair cell cytosol. (Recall that K^+ levels in endolymph are very high, which is not normally the case in other extracellular fluids of the body.) As cations enter, they produce a depolarizing receptor potential. A *tip link* protein connects a mechanically-gated cation channel in a stereocilium to the tip of its taller stereocilium neighbor. When the hair cell is at rest, the stereocilia point straight up and the cation channels are in a partially open state (**FIGURE 9.48a**). This allows a few K^+ to enter the cell, causing a weak depolarizing receptor potential. The weak depolarization spreads along the plasma membrane and opens a few voltage-gated Ca^{2+} channels in the base of the cell. As a result, a small amount of Ca^{2+} enters the cell and triggers exocytosis of a small number of synaptic vesicles containing neurotransmitter. The low level of neurotransmitter release generates a low frequency of action potentials in the first-order auditory neuron that synapses with the hair cell. When vibration of the basilar membrane causes the stereocilia to bend toward the tallest stereocilium, the tip links are stretched and tug on the cation channels, causing the cation channels to completely open (**FIGURE 9.48b**). As a result, a larger amount of K^+ enters the cell, causing a strong depolarizing receptor potential. This leads to the opening of more voltage-gated Ca^{2+} channels and release of more neurotransmitter. The increase in neurotransmitter release generates a higher frequency of action potentials in the first-order auditory neuron. When vibration of the basilar membrane causes the stereocilia to bend away from the tallest stereocilium, the tip links become slack and all of the cation channels close (**FIGURE 9.48c**) Because K^+ is not able to enter the hair cell, the cell becomes more inside-negative (compared to when it is at rest), and a hyperpolarizing receptor potential develops. This hyperpolarization results in little release of neurotransmitter, and the first-order auditory neuron generates very few action potentials.

Outer Hair Cells Enhance the Sensitivity of Inner Hair Cells

Outer hair cells vibrate in response to sound waves and to signals from motor neurons. As they depolarize and hyperpolarize, outer hair cells rapidly shorten and lengthen, a phenomenon known as **electromotility**. Outer hair cell electromotility amplifies the motion of the basilar membrane, making it easier for inner hair cells to detect sounds. This is similar to the way that jumping more vigorously on a trampoline makes it easier to reach a greater height. Hence, the outer hair cells act as *cochlear amplifiers* that enhance the sensitivity of inner hair cells.

Pitch Discrimination Depends on Which Region of the Basilar Membrane Vibrates

The entire basilar membrane does not respond the same way to a given sound frequency. Different sound frequencies cause maximum vibrations of different portions of the basilar membrane. This occurs in the following way: As a sound wave of a particular frequency enters the cochlea, it creates a pressure wave that travels along the basilar membrane to the portion that can maximally vibrate at the same frequency. As the membrane vigorously vibrates, the energy in the wave dissipates, causing the wave to die out and not travel any further along the basilar membrane. Two factors determine where the maximum vibration occurs: the width and flexibility of the basilar membrane. The basilar membrane consists of collagen fibers that are embedded in a gel-like matrix. At the base of the cochlea (portion closer to the oval window), the basilar membrane is narrow (its fibers are short) and stiff (the fibers are taut); high-frequency (high-pitched) sounds near 20,000 Hz induce maximal vibrations in this region (**FIGURE 9.49**). Toward the apex of the cochlea near the helicotrema, the basilar membrane is wide (its fibers are longer) and flexible (the fibers are looser); low-frequency (low-pitched) sounds near 20 Hz cause maximal vibration of the basilar membrane there. All other frequencies cause maximum vibrations at specific regions of the

FIGURE 9.49 Relationship between the basilar membrane and sound frequency.

Each region of the basilar membrane is "tuned" to a particular pitch (frequency).

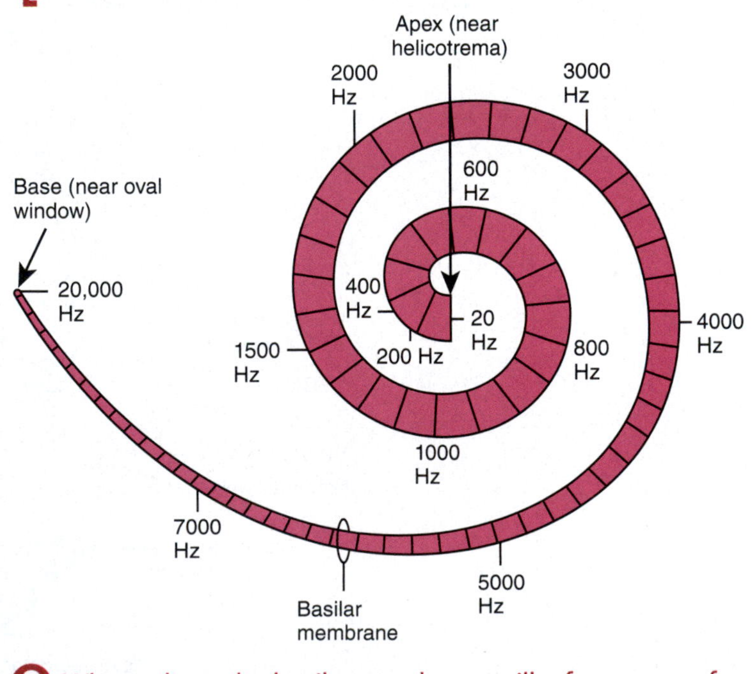

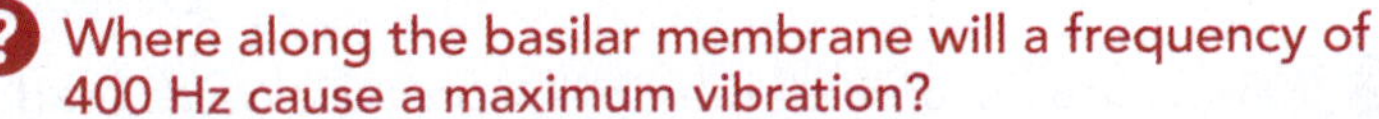

Where along the basilar membrane will a frequency of 400 Hz cause a maximum vibration?

basilar membrane between the base and apex. Hence, each portion of the basilar membrane and its associated hair cells and auditory neurons are "tuned" to a particular pitch (frequency). This does not mean that these structures cannot respond to other frequencies; in fact, they actually respond to a broad range of frequencies. However, there is one specific frequency that causes a maximum response in these structures.

Loudness Discrimination Depends on How Much the Basilar Membrane Vibrates

As noted previously, loudness is determined by the intensity of sound waves. High-intensity sound waves cause larger vibrations of the basilar membrane. This in turn causes stereocilia of hair cells to bend more, causing greater opening of ion channels. As a result, there is more flow of cations (mainly K^+) into the cells and greater release of neurotransmitter, leading to a higher frequency of action potentials reaching the brain. The brain interprets the increased frequency of action potentials as a louder sound. Louder sounds may also stimulate a larger number of hair cells.

The Auditory Pathway Conveys Sound Input to the Auditory Cortex

The release of neurotransmitter from hair cells of the organ of Corti ultimately generates action potentials in the first-order auditory neurons that innervate the hair cells. The axons of these neurons form the cochlear branch of the vestibulocochlear (VIII) nerve. These axons synapse with neurons in the **cochlear nuclei** in the medulla oblongata (**FIGURE 9.50**). Some of the axons from the cochlear nuclei terminate

FIGURE 9.50 The auditory pathway.

From hair cells of the cochlea, auditory information is conveyed along the cochlear branch of the vestibulocochlear (VIII) nerve and then to the brain stem, thalamus, and cerebral cortex.

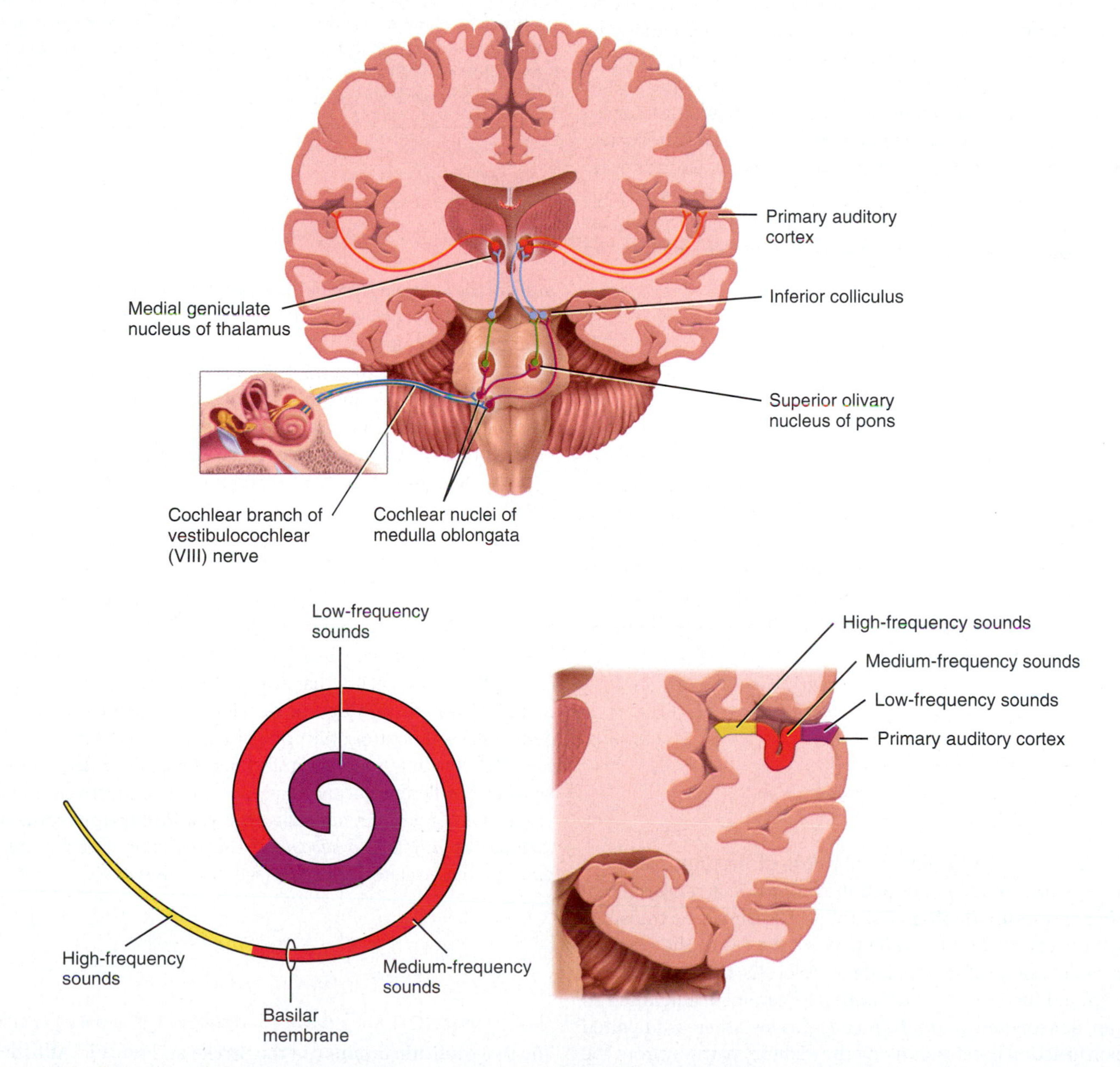

What is the function of the superior olivary nucleus of the pons?

in the **inferior colliculus** of the midbrain. Other axons from the cochlear nuclei end in the **superior olivary nucleus** of the pons. Slight differences in the timing of action potentials arriving from the two ears at the superior olivary nuclei allow us to locate the source of a sound. Axons from the superior olivary nuclei ascend to the midbrain, where they terminate in the inferior colliculi. From each inferior colliculus, axons extend to the **medial geniculate nucleus** of the thalamus. Neurons in the thalamus in turn project axons to the **primary auditory cortex** in the temporal lobe, where conscious awareness of sound occurs (FIGURE 9.50; see also FIGURE 8.8). From the primary auditory cortex, axons extend to the **auditory association area** in the temporal lobe (see FIGURE 8.8) for more complex integration of sound input.

The Primary Auditory Cortex Perceives Sound

The arrival of action potentials in the primary auditory cortex allows you to perceive sound. One aspect of sound that is perceived by this area is pitch (frequency). The primary auditory cortex is mapped according to pitch: Input about pitch from each portion of the basilar membrane is conveyed to a different part of the primary auditory cortex. High-frequency sounds activate one part of the cortex, low-frequency sounds activate another part, and medium-frequency sounds activate the region in between (FIGURE 9.50). Hence, different cortical neurons respond to different pitches. Neurons in the primary auditory cortex also allow you to perceive other aspects of sound such as loudness and duration.

The Auditory Association Area Allows You to Recognize a Sound

From the primary auditory cortex, auditory information is conveyed to the auditory association area in the temporal lobe. This area stores auditory memories and compares present and past auditory experiences, allowing you to recognize a particular sound as speech, music, or noise. If the sound is speech, input in the auditory association is relayed to Wernicke's area in the adjacent part of the temporal lobe, which interprets the meaning of words, translating them into thoughts (see FIGURES 8.8 and 8.13).

Deafness Results from Defects in the Conductive or Neural Pathways Associated with the Ear

Deafness refers to a significant or total loss of hearing. There are two main types of deafness: conduction deafness and sensorineural deafness. In **conduction deafness**, sound waves cannot be transmitted through the external and middle parts of the ear to the cochlea. Causes of conduction deafness include otosclerosis, the deposition of new bone around the oval window; impacted cerumen; and injury to the eardrum. **Sensorineural deafness** occurs when there is impaired transmission of action potentials along the auditory pathway from the cochlea to the cerebral cortex. It may be caused by damage to one of more of the following: the hair cells of the cochlea, the cochlear branch of the vestibulocochlear (VIII) nerve, or the areas in the brain that process auditory input.

By about age 60, around 25% of individuals experience a noticeable hearing loss, especially for higher-pitched sounds. The age-associated progressive loss of hearing in both ears is called **presbycusis** (pres′-bī-KŪ-sis). It may be related to damaged and lost hair cells in the organ of Corti or degeneration of the nerve pathway for hearing.

CLINICAL CONNECTION

Cochlear Implants

A **cochlear implant** is a device that translates sounds into electrical signals that can be interpreted by the brain. Such a device is useful for people with deafness caused by damage to hair cells in the cochlea. The external parts of a cochlear implant consist of (1) a ***microphone,*** which is worn around the ear and picks up sound waves; (2) a ***sound processor,*** which may be placed in a shirt pocket and converts sound waves into electrical signals; and (3) a ***transmitter***, which is worn behind the ear and receives signals from the sound processor and passes them to an internal receiver. The internal parts of a cochlear implant are the (1) ***internal receiver***, which relays signals to (2) ***electrodes*** implanted in the cochlea, where they trigger action potentials in sensory neurons in the cochlear branch of the vestibulocochlear (VIII) nerve. These artificially induced action potentials propagate over their normal pathways to the brain. The perceived sounds are crude compared to normal hearing, but they provide a sense of rhythm and loudness; information about certain noises, such as those made by telephones and automobiles; and the pitch and cadence of speech. Some patients hear well enough with a cochlear implant to use a phone.

Equilibrium Is the Sense of Balance

The ear not only detects hearing but also changes in equilibrium (balance). Body movements that stimulate the receptors for equilibrium include linear acceleration or deceleration, such as when a car suddenly takes off or stops; tilting the head forward or backward, as if to say "yes"; and rotational (angular) acceleration or deceleration, such as when a rollercoaster takes a quick curve. Collectively, the receptor organs for equilibrium are called the **vestibular apparatus** and they include the *utricle* and *saccule* of the vestibule and the *semicircular ducts* of the semicircular canals (see FIGURE 9.45c).

The Otolithic Organs Detect Linear Acceleration or Deceleration and Head Tilt

The two **otolithic organs** are the utricle and saccule. Attached to the inner walls of both the utricle and the saccule is a small, thickened region called the **macula** (**FIGURE 9.51a**). The two *maculae* (plural) contain the receptors for linear acceleration or deceleration and the position of the head (head tilt). The maculae consist of two types of cells: **hair cells**, which are the sensory receptors, and **supporting cells**. Hair cells have on their surface stereocilia (which are actually microvilli) of graduated height, plus one *kinocilium*, a conventional cilium that extends beyond the longest stereocilium (**FIGURE 9.51b**). As in the cochlea, the stereocilia are connected by tip links. The hairs (stereocilia and kinocilium) of the hair cells project into a thick, gelatinous, glycoprotein layer called the **otolithic membrane**. A layer of dense calcium carbonate crystals, called **otoliths**, extends over the entire surface of the otolithic membrane.

The maculae of the utricle and saccule are perpendicular to one another. When the head is in an upright position, the macula of the utricle is oriented horizontally and the macula of the saccule is oriented vertically. Because of these orientations, the utricle and saccule have

FIGURE 9.51 Utricle and saccule.

The utricle and saccule contain receptors for linear acceleration or deceleration and head tilt.

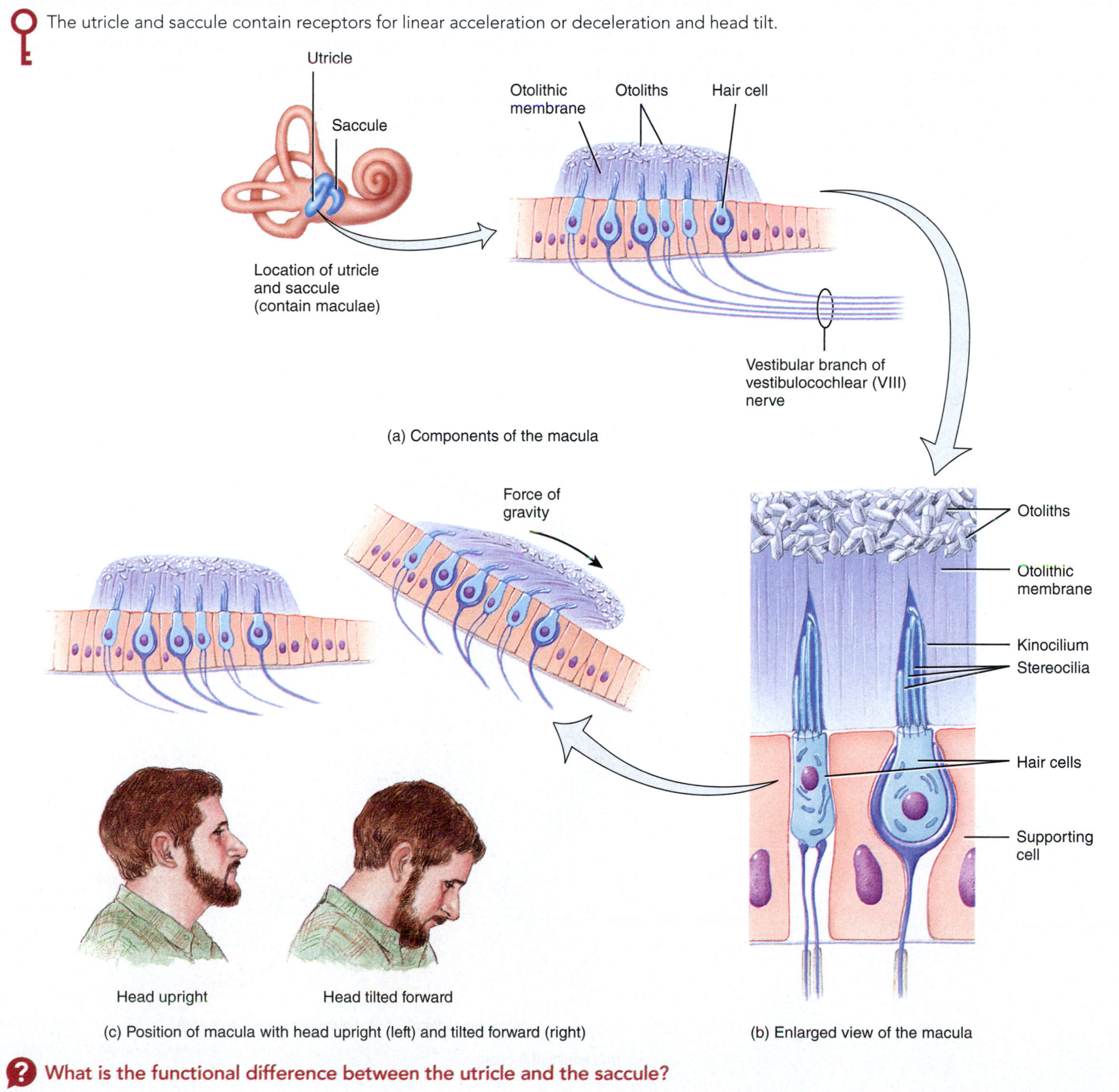

What is the functional difference between the utricle and the saccule?

different functional roles. The utricle responds to linear acceleration or deceleration that occurs in a horizontal direction, such as when the body is being moved in a car that is speeding up or slowing down. The utricle also responds when the head tilts forward or backward. The saccule responds to linear acceleration or deceleration that occurs in a vertical direction, such as when the body is being moved up or down in an elevator.

Because the otolithic membrane sits on top of the macula, the otolithic membrane (along with the otoliths) is pulled by gravity if you tilt your head forward, (FIGURE 9.51c). It slides down over the hair cells in the direction of the tilt, which causes the hairs of the hair cells to bend. However, if you are sitting upright in a car that suddenly jerks forward, the otolithic membrane lags behind the head movement due to inertia, pulls on the hairs, and makes them bend in the other direction. Bending of the hairs in one direction stretches the tip links, which pulls open cation channels, producing depolarizing receptor potentials; bending in the opposite direction closes the cation channels and produces hyperpolarization.

As the hair cells depolarize and hyperpolarize, they release neurotransmitter at a faster or slower rate. The hair cells synapse with first-order vestibular neurons of the vestibular branch of the vestibulocochlear (VIII) nerve. These neurons generate action potentials at a slow or rapid pace depending on the amount of neurotransmitter present.

The Semicircular Ducts Detect Rotational Acceleration or Deceleration

The three membranous semicircular ducts lie at right angles to one another (FIGURE 9.52). This positioning permits detection of rotational acceleration or deceleration in all possible directions. The dilated portion of each duct, the **ampulla**, contains a small elevation called the **crista** (FIGURE 9.52a). Each crista consists of a group of **hair cells** and **supporting cells**. The hair cells contain a kinocilium and stereocilia, and the stereocilia are interconnected via tip links. The hairs (stereocilia and kinocilium) of the hair cells project into an overlying mass of gelatinous material called the **cupula** (KŪ-pū-la).

When the head rotates, the attached semicircular ducts and hair cells move with it (FIGURE 9.52b). However, the endolymph within the ampulla is not attached and lags behind due to inertia. The drag of the endolymph causes the cupula and the hairs that project into it to bend in the direction opposite to that of the head movement. If the

FIGURE 9.52 Semicircular ducts.

The semicircular ducts contain receptors for rotational acceleration or deceleration.

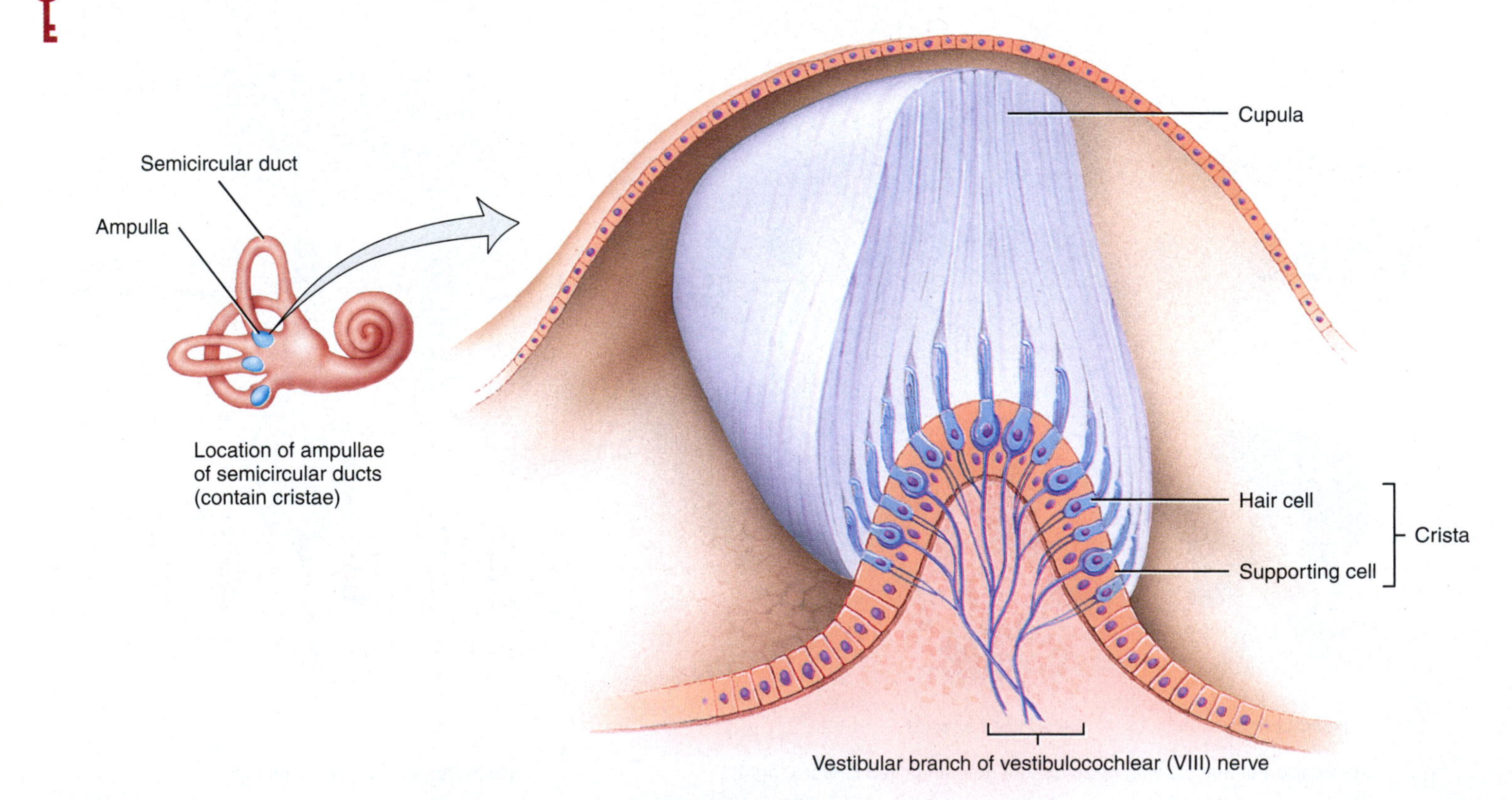

(a) Details of a crista

CLINICAL CONNECTION

Disorders of Equilibrium

There are several **disorders of equilibrium**, which affect the body's ability to maintain proper orientation with respect to gravity. **Vertigo** (VER-ti-gō = dizziness) is a sensation of spinning or movement in which the world seems to revolve or the person seems to revolve in space; it is often associated with nausea and, in some cases, vomiting. It may be caused by an infection of the vestibular apparatus, strokes, tumors, certain drugs, or other inner ear disorders. **Motion sickness** is a condition that results when there is a conflict among the senses with regard to motion. For example, if you are in the cabin of a moving ship, your vestibular apparatus informs the brain that there is movement, but your eyes don't see any movement, resulting in conflicting information from the eyes and inner ear. Motion sickness can also be experienced in a car, airplane, train, or amusement park ride. Symptoms of motion sickness include nausea, dizziness, weakness, and malaise. Usually, when motion stops, symptoms improve. Over-the-counter or prescription medications for motion sickness are available to help alleviate the symptoms. **Ménière's disease** (men′-ē-ĀRZ) results from an abnormal buildup of endolymph that enlarges the membranous labyrinth. The cause of the endolymph buildup is unknown, but it is thought that allergies, abnormal immune responses, or viral infections may play a role. Among the symptoms of Ménière's disease are fluctuating hearing loss (caused by distortion of the basilar membrane of the cochlea), roaring tinnitus (ringing) in the ears, and spinning or whirling vertigo. Almost total destruction of hearing may occur over a period of years.

head continues to move at a steady pace, the endolymph begins to move at the same rate as the rest of the head. This causes the cupula and its embedded hairs to stop bending and to return to their resting positions. Once the head stops moving, the endolymph temporarily keeps moving due to inertia, which causes the cupula and its hairs to bend in the same direction as the preceding head movement. At some point the endolymph stops moving, and the cupula and its hairs return to their resting, unbent positions. Note that bending the hairs in one direction depolarizes the hair cells; bending in the opposite direction hyperpolarizes the cells. The hair cells synapse with first-order vestibular neurons of the vestibular branch of the vestibulocochlear (VIII) nerve. When hair cells are depolarized, there is a greater frequency of action potentials generated in the vestibulocochlear (VIII) nerve than when hair cells are hyperpolarized.

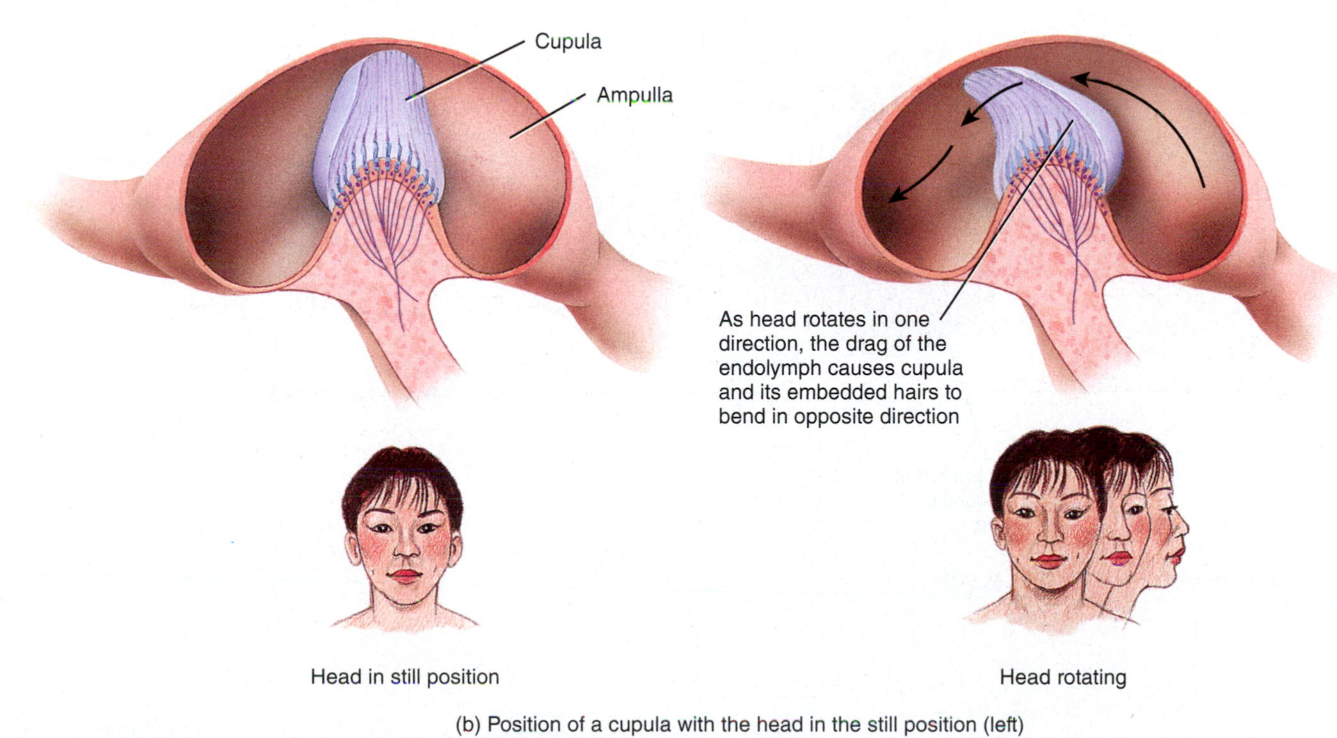

(b) Position of a cupula with the head in the still position (left) and when the head rotates (right)

? Why are the three semicircular ducts positioned at right angles to one another?

Equilibrium Pathways Convey Vestibular Input to Different Parts of the Brain

The release of neurotransmitter from hair cells of the utricle, saccule, or semicircular ducts ultimately generates action potentials in the first-order vestibular neurons that innervate the hair cells. The axons of these neurons form the vestibular branch of the vestibulocochlear (VIII) nerve. Most of these axons synapse with neurons in the **vestibular nuclei** of the medulla and pons; the remaining axons enter the cerebellum (**FIGURE 9.53**). Bidirectional pathways interconnect the cerebellum and vestibular nuclei. The vestibular nuclei also receive input from the eyes and somatic receptors, especially proprioceptors in the neck muscles that indicate the position of the head. The vestibular nuclei integrate information from vestibular, visual, and somatic receptors and then send commands to: (1) cranial nerves that control eye movements with those of the head to help maintain focus on the visual field; (2) the vestibulospinal tract, a descending tract that conveys action potentials down the spinal cord to maintain muscle tone in skeletal muscles to help maintain

FIGURE 9.53 The equilibrium pathway.

From hair cells of the utricle, saccule, and semicircular ducts, vestibular information is conveyed along the vestibular branch of the vestibulocochlear (VIII) nerve and then to the brain stem, cerebellum, thalamus, and cerebral cortex.

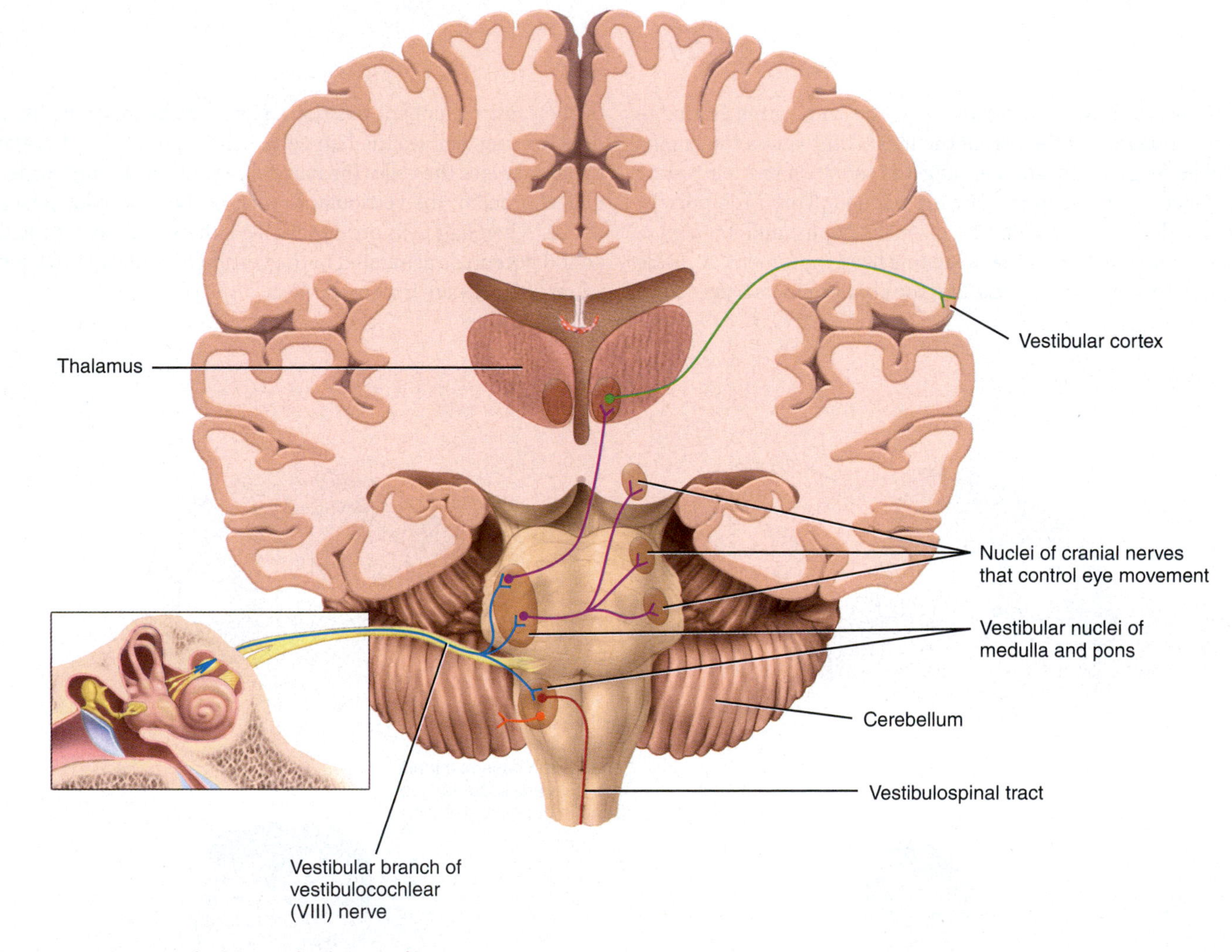

Where are the vestibular nuclei located?

equilibrium (see FIGURE 12.11); and (3) the thalamus and then to the vestibular cortex in the parietal lobe to provide us with the conscious awareness of the position and movement of the head.

CHECKPOINT

24. How are sound waves transmitted from the pinna to the organ of Corti?

25. How do hair cells in the cochlea and vestibular apparatus transduce mechanical vibrations into electrical signals?

26. What is the pathway for action potentials from the cochlea to the cerebral cortex?

27. What are otoliths?

28. Which parts of the brain receive vestibular input?

FROM RESEARCH TO REALITY

Can Your Tongue "See"?

Reference
Sampaioa, E. et al. (2001). Brain plasticity: 'visual' acuity of blind persons via the tongue. *Brain Research*. 908(2): 204–207.

Is technology making it possible to "see" with your tongue?

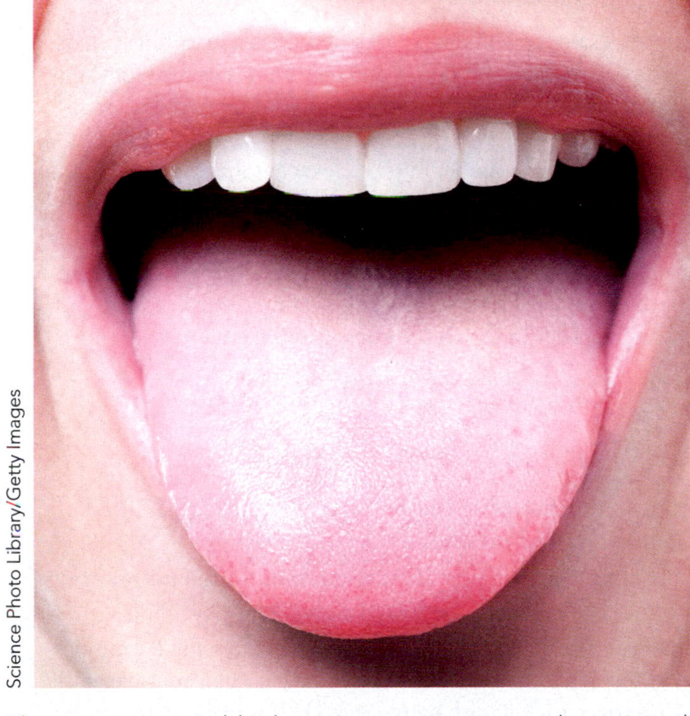

The eyes are arguably the most important special sense in the body. Vision gives us a great deal of information about the world around us. Losing that sense or never having had it changes the way that someone is able to relate to the world. Researchers in the present study used a vision substitution system that allows subjects, both blind and sighted, to see the world with their tongue. Were subjects actually able to "see" the world with their tongue?

Article description:
Blind and sighted individuals were fitted with a vision substitution system that uses a TV camera, a tongue electrode array, a human–machine interface, and a computer. The camera was pointed at whatever the subject was trying to "see." The Snellen falling E was used in its various orientations, some large and some quite small, to determine visual acuity of the subjects using the vision substitution system.

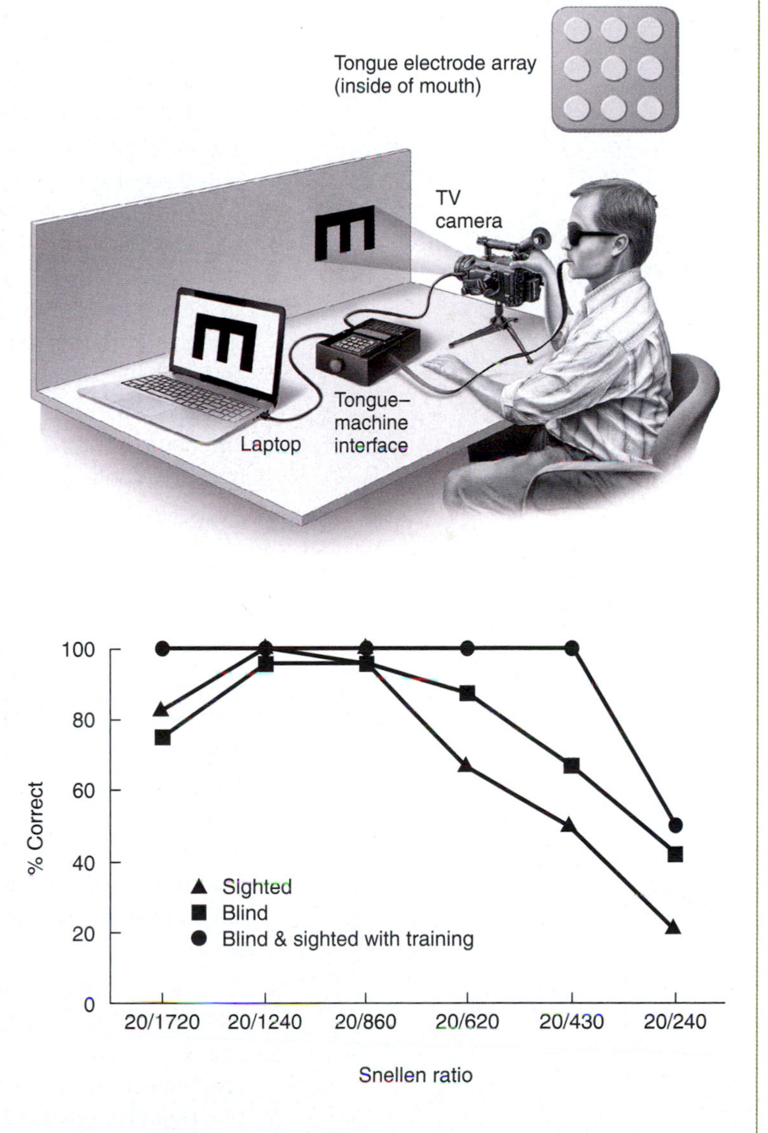

Go to WileyPLUS Learning Space and use the data from this article to answer the questions posed there and to learn more about the vision substitution system.

CHAPTER REVIEW

9.1 Overview of Sensation

1. Sensation is the conscious or subconscious awareness of changes in the external or internal environment. Perception is the conscious awareness and interpretation of sensations and is primarily a function of the cerebral cortex.
2. During the process of sensation, four events typically occur: stimulation of a sensory receptor, transduction of the stimulus, generation of action potentials, and integration of sensory input.
3. Sensory receptors are either the peripheral endings of sensory neurons or separate cells that synapse with sensory neurons.
4. Sensory receptors respond to stimuli by producing receptor potentials. In sensory receptors that are peripheral endings of sensory neurons, if the receptor potential reaches threshold, it will trigger one or more action potentials in the axons of the sensory neuron. In sensory receptors that are separate cells, the receptor potential triggers the release of neurotransmitter. The neurotransmitter in turn causes a postsynaptic potential to form in the adjacent sensory neuron, and if the postsynaptic potential reaches threshold, an action potential forms in the sensory neuron.
5. Sensory neurons have receptive fields. The receptive field of a sensory neuron is the stimulated physical area, specific group of chemicals, or particular set of sound frequencies that causes a response in that neuron.
6. Sensory systems encode four attributes of a stimulus: modality, location, intensity, and duration. Modality is encoded by which sensory receptor and neural pathway are activated when a stimulus is applied. Stimulus location is encoded by the location of the activated receptive field. Stimulus intensity is encoded by the frequency of action potentials generated in response to the stimulus and by the number of sensory receptors activated by the stimulus. Stimulus duration is encoded by the duration of action potentials in the sensory neuron.
7. A sensory pathway is a group of neurons in series that conveys sensory information. The neurons of a sensory pathway are referred to as first-, second-, third-, fourth-, or higher-order neurons based on the order in which they occur in the chain.

9.2 The Somatic Sensory System

1. Somatic sensations arise from stimulation of sensory receptors embedded in the skin, subcutaneous layer, skeletal muscles, tendons, and joints.
2. There are four modalities of somatic sensations: tactile, thermal, pain, and proprioceptive.
3. Tactile sensations include touch, pressure, vibration, itch, and tickle. Meissner corpuscles respond mainly to the onset of touch and to low-frequency vibrations; hair root plexuses detect movements on the skin surface that disturb hairs; Merkel discs respond to continuous touch and to pressure; Ruffini corpuscles detect skin stretching and pressure; pacinian corpuscles respond to high-frequency vibrations; free nerve endings detect itch and tickle.
4. Thermal sensations—coldness and warmth—are detected by thermoreceptors. The two types of thermoreceptors are cold receptors and warm receptors.
5. Pain sensations are detected by nociceptors. There are two types of pain: fast and slow. Fast pain is sharp, pricking pain that is well localized. Slow pain is a dull, aching sensation that is poorly localized.
6. Referred pain is pain that is felt at a site other than the place of origin.
7. Pain sensations may be suppressed by drug therapy, application of a mechanical stimulus to a painful area, or the endogenous analgesia system.

8. Receptors for proprioceptive sensations are located in muscles, tendons, and joints. Muscle spindles are proprioceptors that detect static muscle length and changes in muscle length; tendon organs are proprioceptors that detect changes in muscle tension.
9. Somatic sensory pathways relay information from somatic sensory receptors to the primary somatosensory cortex of the cerebral cortex. The two main somatic sensory pathways are the dorsal column pathway and the anterolateral (spinothalamic) pathway.
10. The primary somatosensory cortex is responsible for precise localization of somatic sensory stimuli. This cortical region has a distorted map of the body. Areas with the largest representation include the face, lips, tongue, and hands. The adjacent somatosensory association area allows recognition of somatic sensory stimuli.

9.3 The Olfactory System

1. Olfactory receptor cells are sensory neurons that respond to olfactory stimuli; they are located in the olfactory epithelium of the nose.
2. Extending from the dendrite of an olfactory receptor cell are olfactory cilia, the sites of olfactory transduction.
3. Chemicals called odorants stimulate olfactory receptor cells by binding to receptor proteins in olfactory cilia.
4. A receptor potential develops in response to the binding of an odorant to an olfactory receptor protein. The receptor potential in turn triggers one or more action potentials in the axon of the olfactory receptor cell.
5. Axons of olfactory receptor cells form the olfactory (I) nerves, which convey olfactory signals to the olfactory bulbs. From the olfactory bulbs, olfactory signals are conveyed to the olfactory cortex of the cerebral cortex and limbic system.
6. The threshold of smell is low, and adaptation to odors occurs quickly.

9.4 The Gustatory System

1. There are five primary tastes: salty, sour, sweet, bitter, and umami. All other flavors are combinations of the five primary tastes, plus any accompanying olfactory, tactile, and thermal sensations.
2. Gustatory receptor cells respond to taste stimuli; they are located in taste buds.
3. Several microvilli project from each gustatory receptor cell to the tongue's surface via a taste pore.
4. Dissolved chemicals, called tastants, stimulate a gustatory receptor cell by interacting with a membrane protein (either an ion channel or receptor) in the plasma membrane of the microvilli.
5. Receptor potentials developed in gustatory receptor cells cause the release of neurotransmitter, which can generate action potentials in first-order taste neurons.
6. The threshold varies with the taste involved, and adaptation to taste occurs quickly.
7. Gustatory receptor cells trigger action potentials in cranial nerves VII, IX, and X. Taste signals then pass to the medulla oblongata, thalamus, primary gustatory cortex of the cerebral cortex, and limbic system.

9.5 The Visual System

1. Visible light is the part of the electromagnetic spectrum with wavelengths ranging from about 400 to 700 nm.
2. Accessory structures of the eyes include the eyebrows, eyelids, eyelashes, lacrimal apparatus, and extrinsic eye muscles.

3. The eye is a sphere that is composed of three layers, a lens, and two cavities.
4. The retina consists of a pigmented layer and a neural layer that includes a photoreceptor layer, bipolar cell layer, ganglion cell layer, horizontal cells, and amacrine cells.
5. Image formation on the retina involves refraction of light rays by the cornea and lens, which focus an inverted and reversed image on the retina.
6. For viewing close objects, the lens increases its curvature (accommodation) and the pupil constricts to prevent light rays from entering the eye through the periphery of the lens.
7. Photoreceptors are sensory receptors that respond to light. There are two types of photoreceptors: rods and cones.
8. In darkness, photoreceptors are depolarized. In light, photoreceptors are hyperpolarized. The process by which light energy is converted into a receptor potential is called phototransduction.
9. Visual signals from photoreceptors are sequentially processed by bipolar cells and ganglion cells. Ganglion cells are the only cells of the retina that can form action potentials. Axons of ganglion cells form the optic (II) nerve.
10. Action potentials from ganglion cells are conveyed into the optic (II) nerve, through the optic chiasm and optic tract, and then to the thalamus. From the thalamus, action potentials are conveyed to the primary visual cortex.

9.6 The Auditory and Vestibular Systems

1. The ear is divided into three main regions: the external ear, middle ear, and inner ear.
2. Sound waves enter the external auditory canal, strike the tympanic membrane, pass through the ossicles, strike the oval window, set up waves in the perilymph, strike the vestibular membrane, increase pressure in the endolymph, vibrate the basilar membrane, and stimulate the hair cells of the organ of Corti.
3. Hair cells convert mechanical vibrations into a receptor potential, which causes the release of neurotransmitter and the generation of action potentials in first-order auditory neurons of the cochlear branch of the vestibulocochlear (VIII) nerve.
4. Axons of the cochlear branch of the vestibulocochlear (VIII) nerve terminate in the medulla oblongata. Auditory signals then pass to the pons, inferior colliculus, thalamus, and primary auditory cortex of the cerebral cortex.
5. The vestibular apparatus detects changes in equilibrium. It consists of the utricle, saccule, and semicircular ducts. The utricle and saccule detect linear acceleration or deceleration and head tilt. The semicircular ducts detect rotational acceleration or deceleration.
6. The release of neurotransmitter from hair cells in the utricle, saccule, or semicircular ducts ultimately causes the generation of action potentials in first-order vestibular neurons that give rise to the vestibular branch of the vestibulocochlear (VIII) nerve.
7. Most axons of the vestibular branch of the vestibulocochlear (VIII) nerve enter the brain stem and terminate in the medulla and pons; other axons enter the cerebellum. Vestibular input is also conveyed to the vestibular area in the parietal lobe of the cerebral cortex.

PONDER THIS

1. Marie is out playing soccer with her friends. She is not paying attention to the game at one point and is instead texting her sister. At that moment, Marie is hit hard in the back of the head with a soccer ball and is knocked unconscious. When she comes to, she explains to everyone around her that she saw flashes of light right when she was hit with the ball. How can you explain this phenomenon? You should use the terms modality and labeled line coding to explain your answer.

2. Steven has been listening to loud music most of his life. Now, at the age of 50, it is very difficult for him to hear any sound that is low pitch. What area/portion of his basilar membrane has been damaged as a result of this continuous stress? Explain your answer.

3. You have been tasked with designing a drug that interferes with the muscle spindle reflex. You do not want to alter neuron function, nor do you want to alter the function of the brain or spinal cord. What type of drug would you make?

ANSWERS TO FIGURE QUESTIONS

9.1 As it relates to sensation, transduction is the conversion of the energy in a stimulus into a graded potential.

9.2 The receptors for gustation, vision, audition, and equilibrium are separate cells.

9.3 The receptive field of a visual neuron is the area of visual space where light causes a response in that neuron.

9.4 The presence of overlapping receptive fields improves stimulus localization because a stimulus usually elicits differential responses in sensory neurons with overlapping receptive fields, and the brain can compare the differential activation of these neurons to determine where the stimulus is located.

9.5 Labeled line coding is the association of a modality with the activation of a particular labeled line.

9.6 The two-point discrimination threshold of the sole of the foot is 22 mm, which means that 22 mm is the minimum distance that two caliper points applied to the skin of the foot can be perceived as two separate points of touch. If the distance between the two caliper points is less than 22 mm, then only one point of touch can be perceived.

9.7 The purpose of lateral inhibition is to increase stimulus localization by enhancing contrast between the center and periphery of the stimulated area.

9.8 Frequency coding is the use of action potential frequency to determine stimulus intensity.

9.9 Rapidly adapting receptors are important for signaling changes in a receptor.

9.10 If a sensory pathway decussates, that means that, at some point along its course through the CNS, the sensory pathway crosses over to the side of the body that is opposite where the stimulus occurred. The result of decussation is that sensory input that occurs on one side of the body is perceived by a specific region of the cortex on the opposite side of the brain.

9.11 Meissner corpuscles are rapidly adapting receptors that generate action potentials mainly at the onset of a touch.

9.12 The capsule of the pacinian corpuscle contributes to adaptation in the following way: When a stimulus is initially present, the capsule is deformed, the nerve ending is deformed, and a depolarization is generated (the on response); when the stimulus is maintained, the layers of the capsule slip past one another, the nerve ending is no longer deformed, and the depolarization dissipates; when the stimulus is removed, the capsule rebounds, the nerve ending is temporarily deformed, and another depolarization is generated (the off response).

9.13 In addition to the thalamus and cerebral cortex, ascending pain pathways provide input to the reticular formation, limbic system, and hypothalamus.

9.14 The kidneys have the broadest area for referred pain.

9.15 Rubbing a stubbed toe activates Aβ fibers, which ultimately decrease the activity of second-order pain neurons that transmit pain signals from the injured area.

9.16 The three types of endogenous opioids are enkephalins, endorphins, and dynorphins.

9.17 Muscle spindles are activated when the intrafusal fibers are stretched.

9.18 Damage to the right spinothalamic tract could result in loss of pain, thermal, itch, and tickle sensations on the left side of the body.

9.19 The face, lips, tongue, and hand have the greatest representation in the primary somatosensory cortex because these parts of the body have the largest number of sensory receptors.

9.20 Odorants are chemicals that bind to and stimulate olfactory receptors.

9.21 The human olfactory system contains about 400 different types of olfactory receptor proteins, each of which can react to only a select group of odorants.

9.22 Neural connections between the olfactory pathway and the limbic system are responsible for the emotional responses to odors.

9.23 Basal cells are stem cells that develop into supporting cells and then gustatory receptor cells of the taste buds.

9.24 The transduction mechanisms for sweet, bitter, and umami tastants all involve the second messenger inositol trisphosphate (IP_3).

9.25 The vagus (X) nerve carries taste signals from taste buds in the throat and epiglottis to the medulla of the brain.

9.26 Visible light that has a wavelength of 700 nm is red.

9.27 Tears clean, lubricate, and moisten the eyeball.

9.28 The choroid provides nutrients to the retina and absorbs scattered light rays.

9.29 The parasympathetic division of the ANS causes pupillary constriction; the sympathetic division of the ANS causes pupillary dilation.

9.30 The retina converts light into action potentials.

9.31 The lens of the eye is biconvex (convex on both surfaces).

9.32 The lens is flat (less convex) when a person is viewing a distant object.

9.33 Presbyopia is the loss of lens elasticity that occurs with aging.

9.34 The conversion of light energy into a receptor potential occurs in the outer segment of a photoreceptor.

9.35 Cones are able to provide color vision because each of the three types (blue cones, green cones, and red cones) contains a different photopigment that maximally absorbs light in a different part of the visible spectrum, giving them different responses to a given wavelength of light. The brain determines color vision by comparing the responses of the three cone types. There is only one type of rod and all rods contain the same photopigment (rhodopsin), so all rods respond in the same way to a given wavelength of light, allowing you to see in only black and white.

9.36 Retinal is the component of a photopigment that absorbs light.

9.37 All cone photopigments regenerate faster than rhodopsin.

9.38 Vision is sharpest (greatest visual acuity) at the fovea because light has direct access to cones.

9.39 The fovea lacks rods; it contains only cones.

9.40 Convergence increases the light sensitivity but slightly blurs the image that is perceived, so visual acuity is not as sharp compared to when retinal cells have one-to-one synapses.

9.41 A photoreceptor hyperpolarizes in response to light.

9.42 In the ON pathway, light ultimately activates the ganglion cell, causing the generation of action potentials.

9.43 Ganglion cell receptive fields enhance contrast and improve visual acuity.

9.44 Structures carrying action potentials from the retina: axons of ganglion cells → optic (II) nerve → optic chiasm → optic tract → thalamus → primary visual cortex of cerebral cortex.

9.45 The inner ear contains the receptors for hearing and equilibrium.

9.46 The audible range of sound wave frequencies for the human ear is from 20 to 20,000 Hz.

9.47 The basilar membrane vibrates in response to the following events: Pressure waves in the scala vestibuli distort the vestibular membrane, creating pressure waves in the endolymph in the cochlear duct. The pressure waves in the endolymph, in turn, cause the basilar membrane to vibrate.

9.48 A tip link protein connects a cation channel in a stereocilium to the tip of its taller stereocilium neighbor. When the stereocilia bend toward the tallest stereocilium, the tip link is stretched and tugs on the cation channel, causing it to completely open. This allows large amounts of K^+ to enter the hair cell, resulting in the formation of a strong depolarizing receptor potential.

9.49 A frequency of 400 Hz causes a maximum vibration of the part of the basilar membrane close to the apex.

9.50 The superior olivary nucleus is the part of the auditory pathway that allows a person to locate the source of a sound.

9.51 The utricle detects head tilt and linear acceleration or deceleration that occurs in a horizontal direction; the saccule detects linear acceleration or deceleration that occurs in a vertical direction.

9.52 The positions of the semicircular ducts allows for the detection of rotational acceleration or deceleration in all possible directions.

9.53 The vestibular nuclei are located in the medulla and pons.

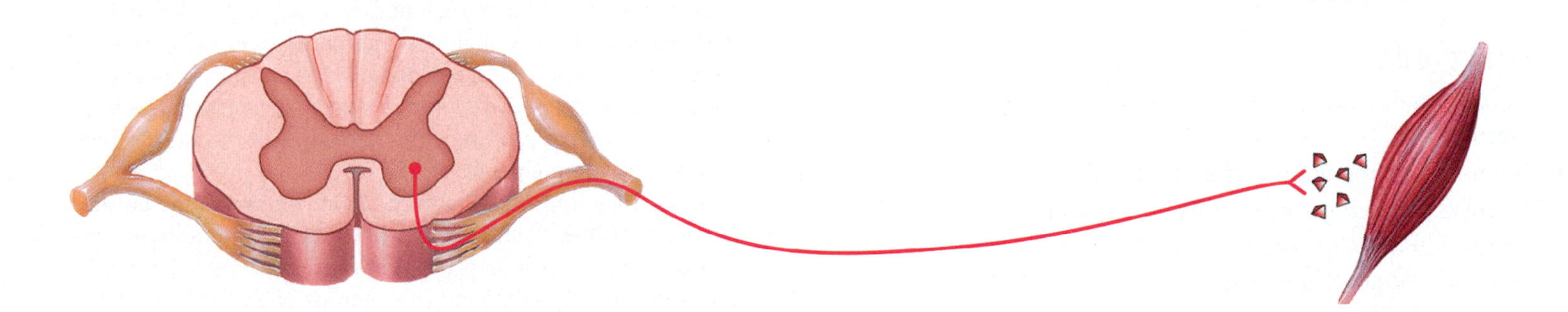

10 Autonomic and Somatic Nervous Systems

Autonomic and Somatic Nervous Systems and Homeostasis

The autonomic and somatic nervous systems contribute to homeostasis by conveying motor output from the central nervous system to effectors (muscles and glands) for appropriate responses to integrated sensory information.

LOOKING BACK TO MOVE AHEAD...

- In signaling pathways that begin with G protein-coupled receptors, different G proteins have different effects on second messenger pathways: G_s stimulates adenylyl cyclase to increase cyclic AMP (cAMP) levels; G_i inhibits adenylyl cyclase to decrease cAMP levels; and G_q stimulates phospholipase C to trigger the inositol trisphosphate (IP_3)/diacylglycerol (DAG) pathway (Section 6.4).
- A synapse is a site of communication between a neuron and a target cell, which can be another neuron, a muscle fiber, or a gland cell (Section 7.4).
- There are two major categories of neurotransmitter receptors: (1) ionotropic receptors, which contain an ion channel as part of their structure and (2) metabotropic receptors, which are coupled to G proteins (Section 7.4).
- The hypothalamus is the part of the brain located below the thalamus; it plays a key role in several body functions, including regulating autonomic activities (Section 8.2).

In Chapter 9 you learned how the sensory systems detect, convey, and process information from the external or internal environment. Once the central nervous system (CNS) integrates this sensory information, it elicits an appropriate response by activating the efferent division of the peripheral nervous system (PNS), which conveys motor output from the CNS to effectors (muscles and glands) in the body. Recall that the efferent division of the PNS is further subdivided into a somatic nervous system and an autonomic nervous system (see Figure 7.1b). The **somatic nervous system** conveys output from the CNS to skeletal muscles. Because its motor responses can be consciously controlled, the action of the somatic nervous system is voluntary. The **autonomic nervous system** conveys output from the CNS to smooth muscle, cardiac muscle, and glands. Because its motor responses are not normally under conscious control, the action of the autonomic nervous system is involuntary. In this chapter, you will learn about the functions of both the autonomic and somatic nervous systems.

10.1 Autonomic Nervous System

OBJECTIVES

- Explain the functional significance of the autonomic nervous system.
- Outline the components of an autonomic motor pathway.
- Describe the neurotransmitters and receptors of the ANS.
- Compare the major responses of the body to stimulation of the parasympathetic and sympathetic branches of the ANS.

The Autonomic Nervous System Regulates the Activity of Smooth Muscle, Cardiac Muscle, and Glands

The autonomic nervous system (ANS) innervates smooth muscle, cardiac muscle, and glands. These tissues are often referred to as **visceral effectors** because they are usually associated with the viscera (internal organs) of the body. Autonomic motor neurons regulate visceral activities by either increasing (exciting) or decreasing (inhibiting) ongoing activities in their effectors. Adjustment of the rate and force of the heartbeat, dilation or constriction of the bronchial tubes of the lungs, and secretion of digestive glands are examples of autonomic responses. Unlike skeletal muscle, tissues innervated by the ANS often function to some extent even if their nerve supply is damaged or cut. The heart continues to beat when it is removed for transplantation into another person, smooth muscle in the lining of the gastrointestinal tract contracts rhythmically on its own, and glands produce some secretions in the absence of ANS control.

The ANS usually operates without conscious control. For example, you probably cannot voluntarily slow down your heart rate; instead, your heart rate is subconsciously regulated. For this reason, some autonomic responses are the basis for *polygraph* ("lie detector") tests. The ANS was so-named because it was thought to function autonomously or in a self-governing manner (*autonomic* = autonomous), without control by the CNS. However, as you will soon learn, centers in the brain and spinal cord do regulate autonomic activities.

The ANS consists of two main branches: the **parasympathetic nervous system** and the **sympathetic nervous system.** Most organs receive nerves from both of these branches, an arrangement known as **dual innervation**. In general, one branch stimulates the organ to increase its activity (excitation), and the other branch decreases the organ's activity (inhibition). For example, neurons of the sympathetic nervous system increase heart rate, and neurons of the parasympathetic nervous system slow it down. The parasympathetic nervous system enhances *rest-and-digest* activities, which conserve and restore body energy during times of rest and recovery. By contrast, the sympathetic nervous system promotes the *fight-or-flight* response, which prepares the body for emergency situations.

The ANS is also comprised of a third branch known as the **enteric nervous system (ENS)**. The ENS consists of millions of neurons in plexuses that extend most of the length of the gastrointestinal (GI) tract. Its operation is involuntary. Although the neurons of the ENS can function autonomously, they can also be regulated by the other branches of the ANS. The ENS contains sensory neurons, interneurons, and motor neurons. Enteric sensory neurons monitor chemical changes within the GI tract as well as the stretching of its walls. Enteric interneurons integrate information from the sensory neurons and provide input to motor neurons. Enteric motor neurons govern contraction of GI tract smooth muscle and secretion of GI tract glands. Despite the presence of sensory neurons and interneurons, the ENS is considered to be part of the efferent (motor) division of the PNS because of its association with the ANS. The ENS is described in greater detail in the discussion of the digestive system in Chapter 21. The rest of this section is devoted to the parasympathetic and sympathetic branches of the ANS.

An Autonomic Motor Pathway Is Comprised of Two Autonomic Motor Neurons and a Visceral Effector

Autonomic motor pathways (both parasympathetic and sympathetic) consist of two autonomic motor neurons in series and a visceral effector (smooth muscle, cardiac muscle, or a gland) (**FIGURE 10.1**). The first neuron, called the **preganglionic neuron,** has its cell body in the brain or spinal cord. Its axon exits the CNS via a cranial or spinal nerve and then extends to an **autonomic ganglion**, where it synapses with the second neuron. (Recall that a *ganglion* is a cluster of neuronal cell bodies in the PNS). The second neuron, called the **postganglionic neuron**, lies entirely in the PNS. Its cell body is located in the autonomic ganglion, and its axon extends from the ganglion to the visceral effector. Thus, preganglionic neurons convey action potentials from the CNS to autonomic ganglia, and postganglionic neurons relay the action potentials from autonomic ganglia to visceral effectors.

FIGURE 10.1 Autonomic motor pathways. Note that most autonomic motor neurons release either acetylcholine (ACh) or norepinephrine (NE).

An autonomic motor pathway consists of two autonomic motor neurons in series: (1) a preganglionic neuron, which extends from the CNS to an autonomic ganglion and (2) a postganglionic neuron, which extends from the autonomic ganglion to the effector (smooth muscle, cardiac muscle or a gland).

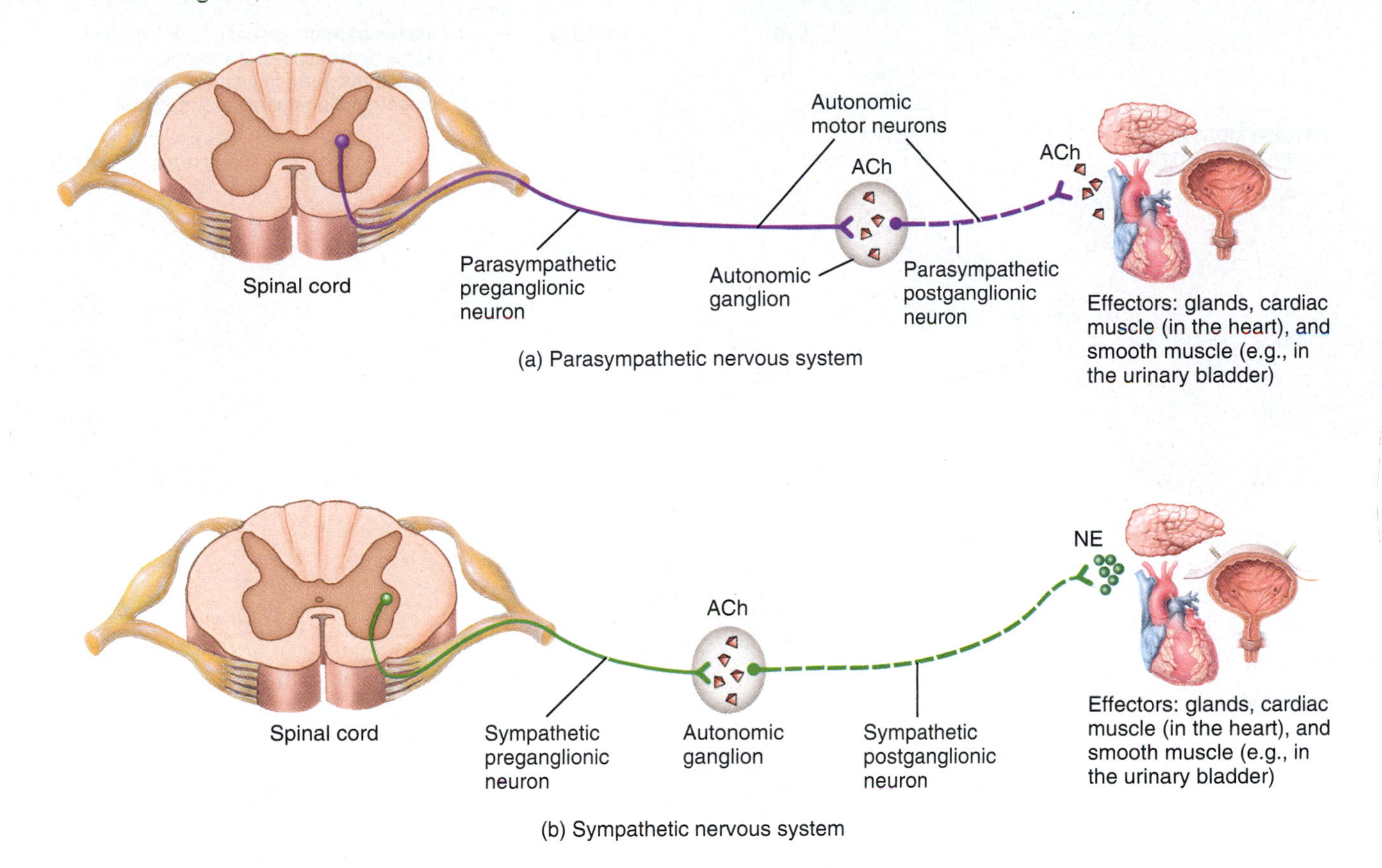

Does activation of autonomic motor pathways always result in excitation of their effectors? Explain your answer.

In the parasympathetic nervous system, preganglionic neurons have their cell bodies in the brain stem and sacral regions of the spinal cord (FIGURE 10.2). Hence, the parasympathetic nervous system is also known as the *craniosacral division* of the ANS. Parasympathetic preganglionic axons exit the CNS through four cranial nerves (III, VII, IX, and X) and several sacral spinal nerves (S2-S4). The axons then extend to parasympathetic postganglionic neurons in **terminal ganglia**, which are located close to or within the wall of the visceral effector (FIGURE 10.2). From the terminal ganglia, parasympathetic postganglionic axons extend to cells in the visceral organ. Because terminal ganglia are located either close to or in the wall of the visceral effector, parasympathetic preganglionic axons are long, and parasympathetic postganglionic axons are short.

In the sympathetic nervous system, preganglionic neurons have their cell bodies in the thoracic and upper lumbar regions of the spinal cord (FIGURE 10.2). For this reason, the sympathetic nervous system is also called the *thoracolumbar division* of the ANS. Sympathetic preganglionic axons exit the CNS through thoracic and lumbar spinal nerves. After leaving the CNS, most sympathetic preganglionic axons extend to sympathetic postganglionic neurons in the **sympathetic trunk**, a chain of ganglia located on either side of the spinal cord (FIGURE 10.2). Other sympathetic preganglionic axons extend to sympathetic postganglionic neurons in **collateral ganglia**, individual ganglia that are not associated with the sympathetic trunk (FIGURE 10.2). From the sympathetic trunk or collateral ganglia, sympathetic postganglionic axons extend to the visceral effector. Because sympathetic trunk ganglia are located near the spinal cord, most sympathetic preganglionic axons are short and most sympathetic postganglionic axons are long.

Some sympathetic preganglionic axons extend to specialized cells called **chromaffin cells** in the adrenal medulla (inner portion of the adrenal gland) without synapsing in either the sympathetic trunk or collateral ganglia (FIGURE 10.3; see also FIGURE 10.2). The adrenal medulla develops from the same embryonic tissue as the sympathetic ganglia. Therefore, chromaffin cells are modified sympathetic postganglionic neurons that lack dendrites and axons. Rather than extending to another organ, however, these cells release hormones into the blood. Upon stimulation by sympathetic preganglionic neurons, the chromaffin cells of the adrenal medulla release a mixture of catecholamine hormones—about 80% **epinephrine**, 20% **norepinephrine**, and a trace amount of **dopamine** (FIGURE 10.3). These hormones circulate through the body and intensify responses elicited by the sympathetic nervous system.

FIGURE 10.2 Organization of the parasympathetic and sympathetic nervous systems. Note that the spinal cord consists of cervical (C1–C8), thoracic (T1–T12), lumbar (L1–L5), sacral (S1–S5), and coccygeal (Co) regions.

The parasympathetic nervous system originates in the brain stem and sacral region of the spinal cord; the sympathetic nervous system originates in the thoracic and lumbar regions of the spinal cord.

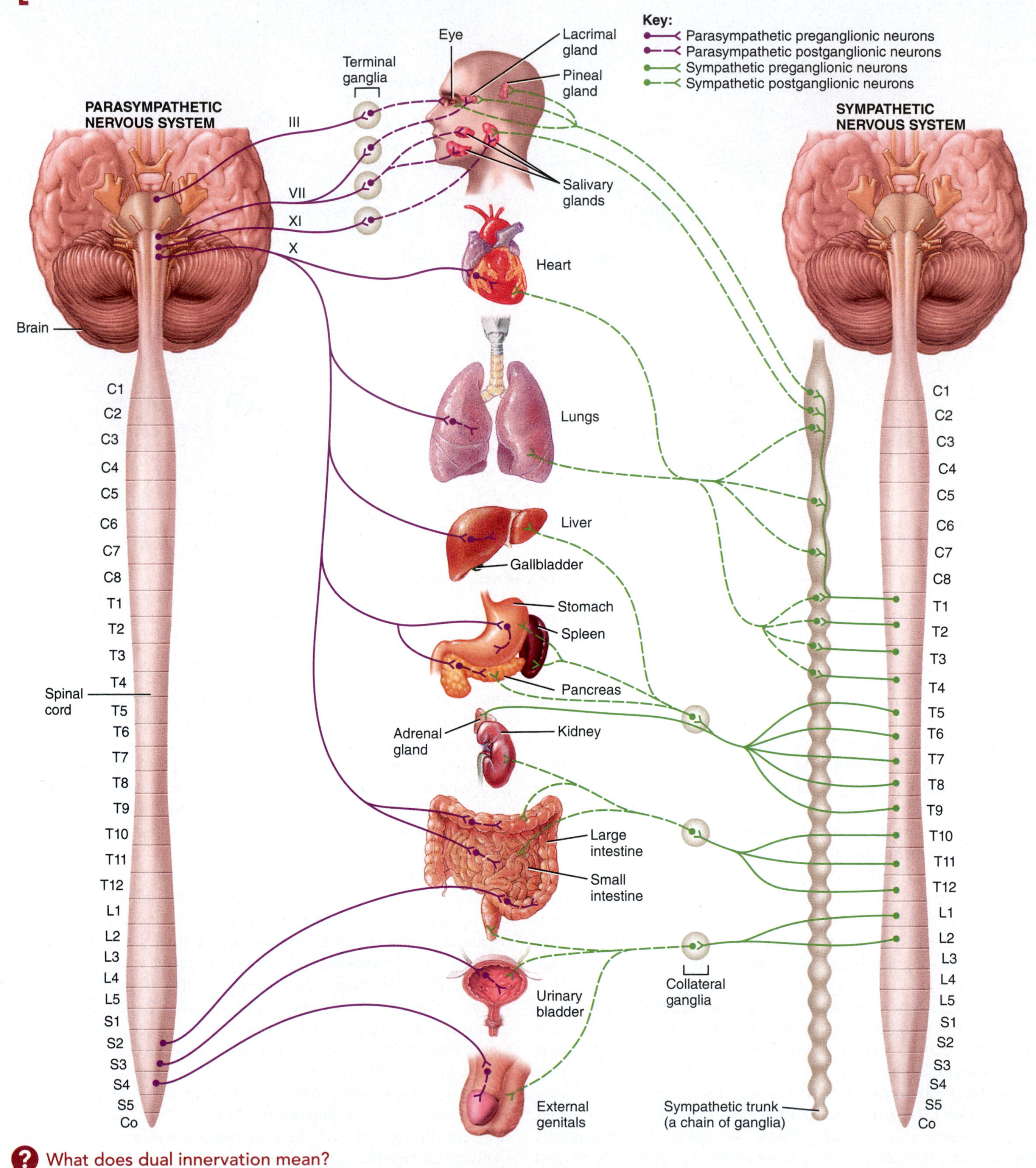

What does dual innervation mean?

FIGURE 10.3 Autonomic motor pathway to the adrenal medulla.

Chromaffin cells are modified sympathetic postganglionic neurons that release the hormones epinephrine and norepinephrine (NE) into the blood.

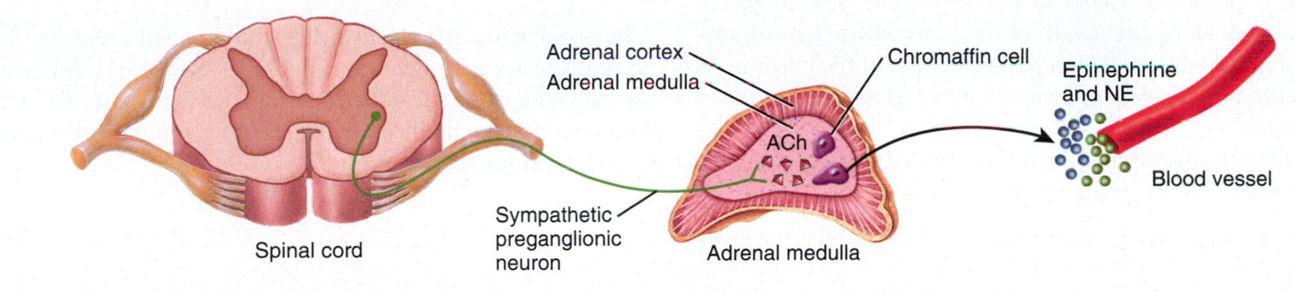

How do chromaffin cells differ from other sympathetic postganglionic neurons?

The Neuroeffector Junction Is the Site Where an Autonomic Postganglionic Neuron Communicates with a Visceral Effector

The synapse between an autonomic postganglionic neuron and a visceral effector is called the **neuroeffector junction (NEJ)** (FIGURE 10.4).

The organization of the NEJ differs from a typical neuron-to-neuron synapse in two major ways:

1. The axon terminals of the postganglionic neuron lack synaptic end bulbs; instead, they exhibit swollen regions called **varicosities**, which contain synaptic vesicles with neurotransmitter.

FIGURE 10.4 Autonomic varicosities.

Autonomic postganglionic neurons release neurotransmitters from varicosities, which are swollen regions found at the ends of axon terminals.

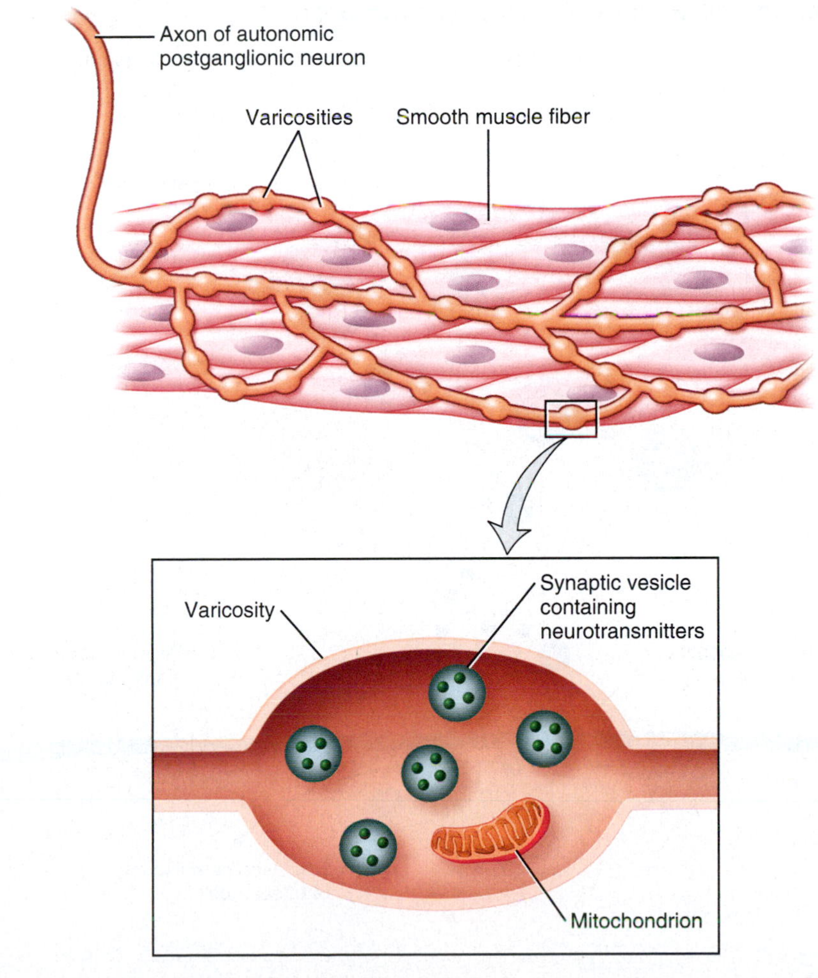

How does a neuroeffector junction (NEJ) differ from a typical neuron-to-neuron synapse?

2. In the effector, the receptors for the neurotransmitters are not confined to a specific receptor region; rather, they are located along the entire surface of the cell.

When an action potential occurs in the autonomic postganglionic neuron, varicosities along the length of the axon release neurotransmitters, which in turn diffuse to receptors throughout the effector. As a result, the autonomic postganglionic neuron affects a large area of the effector tissue.

Release of neurotransmitter at the NEJ occurs in the same way as at a typical neuron-to-neuron synapse (**FIGURE 10.5**):

1. An action potential arrives at a varicosity of the autonomic postganglionic axon.
2. The depolarizing phase of the action potential opens **voltage-gated Ca^{2+} channels**, which are present in the varicosity membrane. Because calcium ions are more concentrated in extracellular fluid, Ca^{2+} flows inward through the opened channels.
3. An increase in the Ca^{2+} concentration inside the varicosity serves as a signal that triggers exocytosis of the synaptic vesicles, causing the release of neurotransmitter into the synapse.
4. The neurotransmitter molecules diffuse across the synapse and bind to neurotransmitter receptors in the effector cell's plasma membrane.
5. Binding of neurotransmitter to its receptor activates a G protein, ultimately leading to a response that either excites or inhibits the effector cell, depending on the type of receptor and G protein pathway that are activated.

Removal of the neurotransmitter from the NEJ occurs by the same mechanisms as in a neuron-to-neuron synapse: (1) diffusion away from the synapse, (2) degradation by enzymes in extracellular fluid or plasma membrane of the effector cell, or (3) uptake into a nearby cell via active transport.

The Autonomic Nervous System Uses Different Types of Neurotransmitters and Receptors

Based on the neurotransmitters they produce and release, most autonomic neurons are classified as either cholinergic or adrenergic. The receptors for the neurotransmitters are integral membrane proteins located in the plasma membrane of the postsynaptic neuron or effector cell.

Cholinergic Neurons and Receptors

Cholinergic neurons release the neurotransmitter **acetylcholine (ACh)**. In the ANS, cholinergic neurons include (1) all parasympathetic and sympathetic preganglionic neurons, (2) most parasympathetic

FIGURE 10.5 Signal transmission at a neuroeffector junction.

Binding of neurotransmitter to receptor molecules in the effector causes a response that leads to excitation or inhibition of the effector cell.

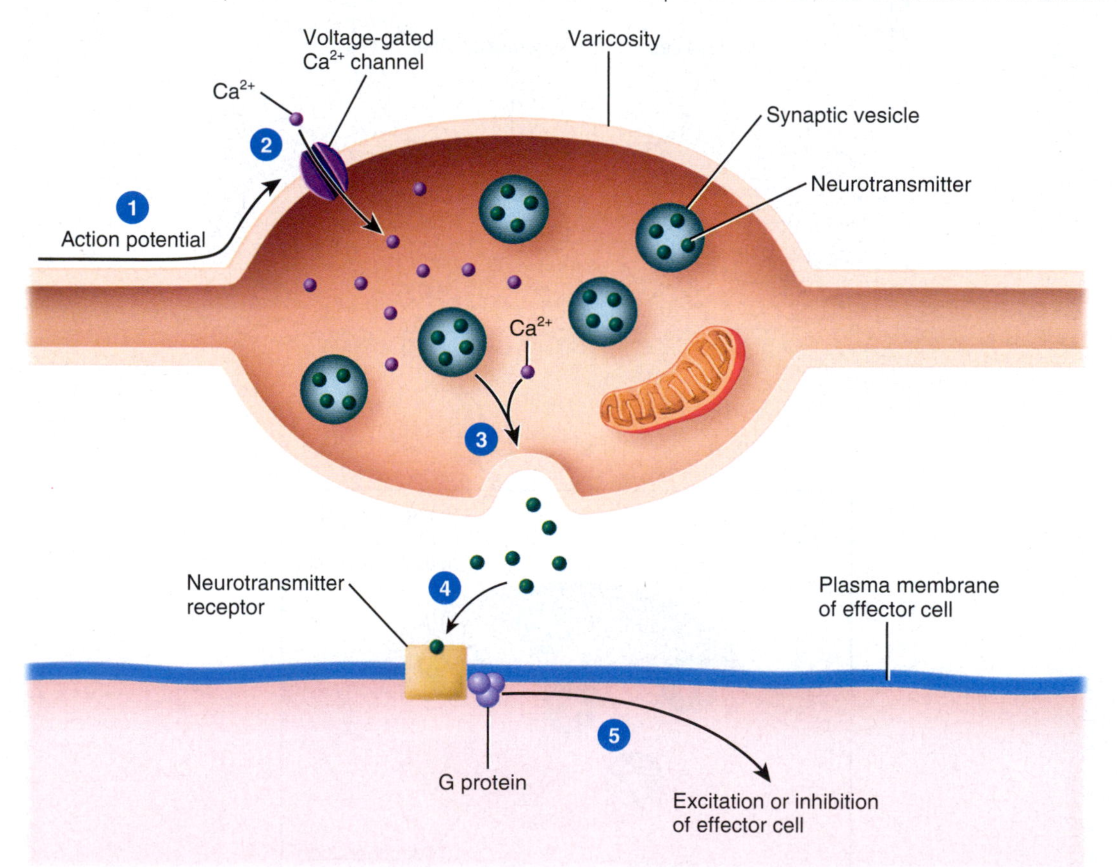

What determines whether the effector cell is excited or inhibited by the binding of neurotransmitter to its receptor?

postganglionic neurons, and (3) sympathetic postganglionic neurons that innervate most sweat glands (FIGURE 10.6).

Once ACh is released from a cholinergic neuron, it diffuses across the synaptic cleft and binds with specific **cholinergic receptors**, integral membrane proteins in the *postsynaptic* plasma membrane. Recall that there are two types of cholinergic receptors: nicotinic acetylcholine receptors and muscarinic acetylcholine receptors (see Section 7.5). **Nicotinic acetylcholine receptors** are present in the plasma membranes of the dendrites and cell bodies of both parasympathetic and sympathetic postganglionic neurons (FIGURE 10.6a–c), in the plasma

FIGURE 10.6 Cholinergic neurons and adrenergic neurons in the parasympathetic and sympathetic nervous systems.

Cholinergic neurons release acetylcholine (ACh); adrenergic neurons release norepinephrine (NE).

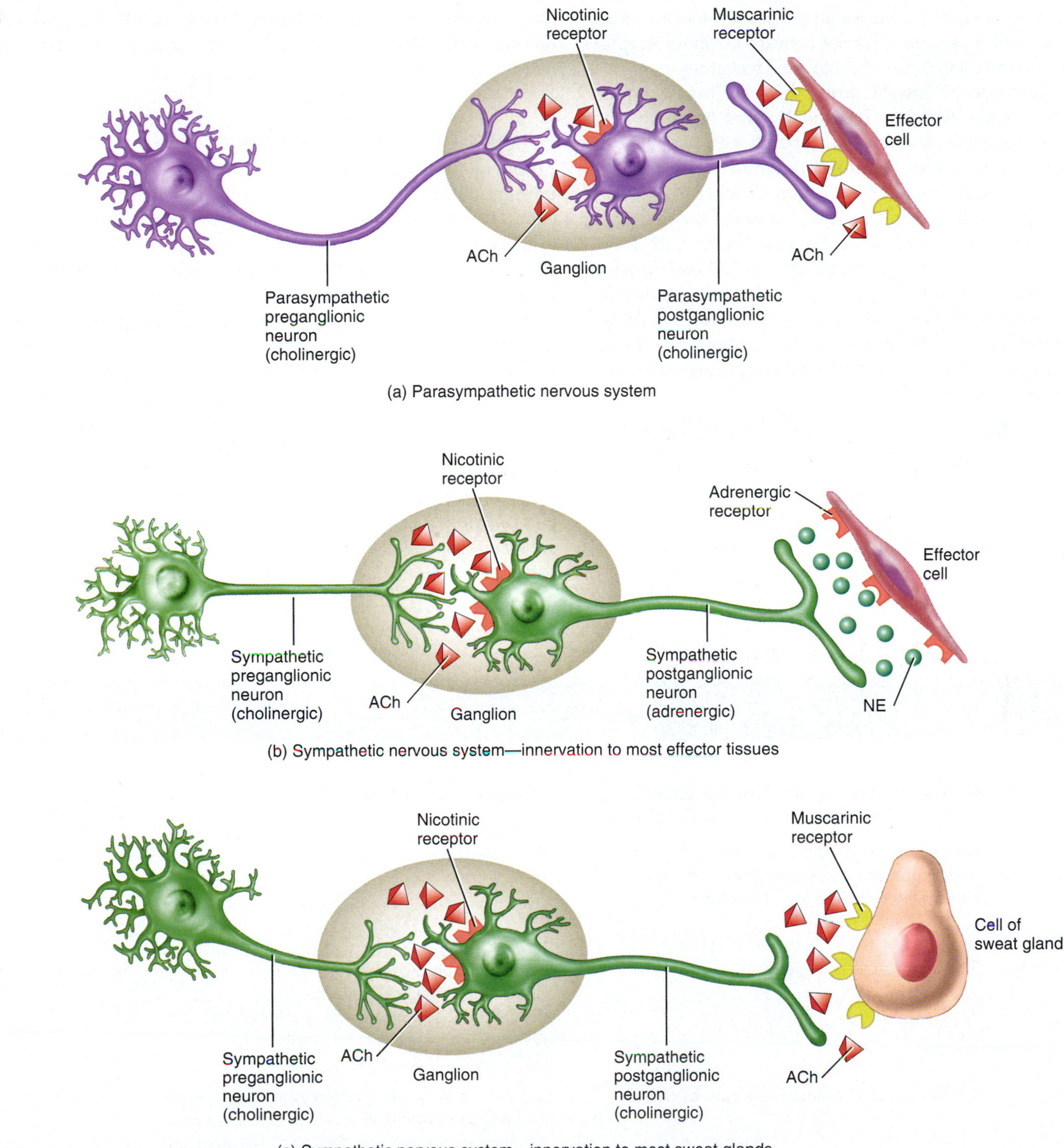

Which ANS neurons are adrenergic?

membranes of chromaffin cells of the adrenal medulla, and in skeletal muscle at the neuromuscular junction (NMJ). They are so-named because the drug *nicotine* mimics the action of ACh by binding to these receptors. **Muscarinic acetylcholine receptors** are present in the plasma membranes of the effectors (smooth muscle, cardiac muscle, and glands) innervated by parasympathetic postganglionic axons (FIGURE 10.6a). In addition, most sweat glands receive their innervation from *cholinergic* sympathetic postganglionic neurons and possess muscarinic ACh receptors (FIGURE 10.6c). These receptors are so-named because a mushroom poison called *muscarine* mimics the actions of ACh by binding to them. Nicotine does not activate muscarinic receptors, and muscarine does not activate nicotinic receptors, but ACh does activate both types of cholinergic receptors.

Based on responses to specific drugs, there are two subtypes of nicotinic acetylcholine receptors (N_1 and N_2) and five subtypes of muscarinic acetylcholine receptors (M_1 through M_5). Both subtypes of nicotinic ACh receptors are ionotropic receptors that contain two binding sites for acetylcholine and a cation channel (see FIGURE 7.27a). Opening the cation channel allows passage of cations (mainly Na^+ and K^+) through the membrane. Because Na^+ is farther away from equilibrium than K^+, Na^+ inflow is greater than K^+ outflow, resulting in depolarization. Activation of nicotinic ACh receptors therefore excites the postsynaptic cell. The nicotinic ACh receptor subtype found in skeletal muscle at the neuromuscular junction is N_1. The subtype present in parasympathetic and sympathetic postganglionic neurons and in chromaffin cells is N_2.

All five subtypes of muscarinic acetylcholine receptors are G protein-coupled (metabotropic) receptors. Activation of muscarinic ACh receptors either excites or inhibits the effector cell, depending on which subtype of muscarinic ACh receptor is activated. M_1, M_3, and M_5 receptors activate a G_q protein and phospholipase C to trigger the inositol trisphosphate (IP_3)/diacylglycerol (DAG) second messenger pathway that increases intracellular Ca^{2+} levels, resulting in excitation of the effector cell. M_2 and M_4 receptors activate a G_i protein, which inhibits adenylyl cyclase and decreases cAMP levels, resulting in the opening of K^+ channels and inhibition of the effector cell. In the heart, which has M_2 receptors, the G_i protein can directly open K^+ channels without altering adenylyl cyclase activity or the cAMP level.

The different types of nicotinic and muscarinic ACh receptors are summarized in TABLE 10.1.

The activity of acetylcholine at a synapse is terminated by enzymatic degradation via **acetylcholinesterase (AChE)**. Because AChE quickly breaks down acetylcholine, effects triggered by cholinergic neurons are brief.

Adrenergic Neurons and Receptors

Adrenergic neurons release **norepinephrine (NE)**, also known as **noradrenaline**. Most sympathetic postganglionic neurons are adrenergic (FIGURE 10.6b). Once NE is released from an adrenergic neuron, it diffuses across the synaptic cleft and binds to specific adrenergic receptors on the postsynaptic membrane, causing either excitation or inhibition of the effector cell.

As you learned in Chapter 7, **adrenergic receptors** respond to the binding of either norepinephrine or epinephrine (see Section 7.5). In the ANS, norepinephrine can be released as a neurotransmitter (by sympathetic postganglionic neurons) or as a hormone (by chromaffin cells of the adrenal medulla); epinephrine is released as a hormone. The two main types of adrenergic receptors are **alpha (α) receptors** and **beta (β) receptors**. In the ANS, these receptors are found on visceral effectors innervated by most sympathetic postganglionic axons (FIGURE 10.6B). Alpha and beta receptors are further classified into subtypes—α_1, α_2, β_1, β_2, and β_3—based on

TABLE 10.1 Types of Cholinergic Receptors

Receptor Type	Location	Mechanism of Action	Effect on Effector Organ
Nicotinic			
N_1	Skeletal muscle at neuromuscular junction.	Opens a cation channel.	Excitation
N_2	Dendrites and cell bodies of parasympathetic and sympathetic postganglionic neurons; chromaffin cells in adrenal medulla.	Opens a cation channel.	Excitation
Muscarinic			
M_1, M_3, and M_5	Smooth muscle and glands.	Activates a G_q protein, which stimulates phospholipase C to generate the IP_3/DAG second messenger pathway that results in an increase in intracellular Ca^{2+}.	Excitation
M_2 and M_4	Smooth muscle, glands, and cardiac muscle.	Activates a G_i protein, which inhibits adenylyl cyclase, causing cAMP levels to decrease and K^+ channels to open. In the heart, M_2 receptors activate a G_i protein that directly opens K^+ channels.	Inhibition

the specific responses they elicit and by their selective binding of drugs that activate or block them. The α receptors have a higher affinity for norepinephrine than epinephrine. β_1 receptors have nearly equal affinity for norepinephrine and epinephrine; β_2 receptors have a higher affinity for epinephrine than norepinephrine; and β_3 receptors have a higher affinity for norepinephrine than epinephrine.

Adrenergic receptors are metabotropic receptors that trigger second messenger pathways by coupling to G proteins. Different types of adrenergic receptors couple to different types of G proteins. The α_1 receptors activate G_q proteins that trigger the IP_3/DAG second messenger pathway, which increases intracellular Ca^{2+} levels. The α_2 receptors activate G_i proteins, which inhibit adenylyl cyclase and decrease cAMP levels. β_1, β_2, and β_3 receptors all activate G_s proteins, which activate adenylyl cyclase and increase cAMP levels. Although there are some exceptions, activation of α_1 and β_1 receptors generally produces excitation, and activation of α_2 and β_2 receptors causes inhibition of effector tissues. β_3 receptors are present mainly on the cells of brown adipose tissue, where their activation causes thermogenesis (heat production). Cells of most effectors contain either alpha or beta receptors; some visceral effector cells contain both.

The different types of adrenergic receptors are summarized in TABLE 10.2.

The activity of norepinephrine at a synapse is terminated either when the NE is taken up by the axon that released it or when the NE is enzymatically inactivated by **catechol-O-methyltransferase (COMT)** or **monoamine oxidase (MAO)**. Compared to ACh, norepinephrine lingers in the synaptic cleft for a longer time. Thus, effects triggered by adrenergic neurons typically are longer lasting than those triggered by cholinergic neurons.

CLINICAL CONNECTION

Receptor Agonists and Antagonists of the ANS

A large variety of drugs and natural products can selectively activate or block specific cholinergic or adrenergic receptors. An **agonist** is a substance that binds to and activates a receptor, in the process mimicking the effect of a natural neurotransmitter or hormone. Phenylephrine, an adrenergic agonist at α_1 receptors, is a common ingredient in cold and sinus medications. Because it constricts blood vessels in the nasal mucosa, phenylephrine reduces production of mucus, thus relieving nasal congestion. An **antagonist** is a substance that binds to and blocks a receptor, thereby preventing a natural neurotransmitter or hormone from exerting its effect. For example, atropine dilates the pupils, reduces glandular secretions, and relaxes smooth muscle in the gastrointestinal tract by blocking muscarinic ACh receptors. As a result, it is used to dilate the pupils during eye examinations and to treat smooth muscle disorders such as intestinal hypermotility, and as an antidote for chemical warfare agents that inactivate acetylcholinesterase.

Propranolol (Inderal®) is often prescribed for patients with hypertension (high blood pressure). It is a nonselective beta blocker, meaning it binds to all types of beta receptors and prevents their activation by epinephrine and norepinephrine. The desired effects of propranolol are due to its ***blockade*** of β_1 receptors—namely, decreased heart rate and force of contraction and a consequent decrease in blood pressure. Undesirable effects due to blockade of β_2 receptors may include hypoglycemia (low blood glucose), resulting from decreased glycogen breakdown and decreased gluconeogenesis (conversion of a noncarbohydrate into glucose in the liver), and mild bronchoconstriction (narrowing of the airways). If these side effects pose a threat to the patient, a selective β_1 blocker such as metoprolol (Lopressor®) can be prescribed instead of propranolol.

TABLE 10.2 Types of Adrenergic Receptors

Receptor Type	Location	Relative Affinities	Mechanism of Action	Effect on Effector Organ
Alpha				
α_1	Most sympathetic target tissues.	NE > E*	Activates a G_q protein, which stimulates phospholipase C to generate the IP_3/DAG second messenger pathway that results in an increase in intracellular Ca^{2+}.	Excitation
α_2	Digestive glands and smooth muscle in certain parts of digestive tract.	NE > E	Activates a G_i protein, which inhibits adenylyl cyclase, causing cAMP levels to decrease.	Inhibition
Beta				
β_1	Cardiac muscle and kidneys.	NE = E	Activates a G_s protein, which stimulates adenylyl cyclase, causing cAMP levels to increase.	Excitation
β_2	Smooth muscle in walls of airways, some blood vessels, and certain visceral organs, such as the urinary bladder.	E > NE	Activates a G_s protein, which stimulates adenylyl cyclase, causing cAMP levels to increase.	Inhibition
β_3	Mainly adipose tissue	NE > E	Activates a G_s protein, which stimulates adenylyl cyclase, causing cAMP levels to increase.	Thermogenesis (heat production)

* NE = norepinephrine; E = epinephrine.

Critical Thinking

The Nightshade, Beautiful but Deadly

John was a typical young teenager. He had lots of energy and a fairly rebellious streak that often expressed itself as a conflict with his parents. Today was especially chaotic, and John decided that he would run away from home. John lived in a fairly rural area and there were lots of places to go, but few restaurants or stores where he could buy a snack or a drink. John was beginning to feel hungry and thirsty, but had neglected to pack any food for this "trip." As he walked along a cool shady area, he noticed a few tall, deeply green plants with nice fat purple berries. He couldn't resist and tried a few! He swallowed two or three, but even though they were somewhat sweet, he didn't have any more. He decided to just go home where he could get something to eat. As he walked, he noticed that he was feeling very hot. His vision was blurry, and he felt as though his heart was going to jump out of his chest. Then he passed out. Luckily for John, his family had started to look for him as he had been gone for a few hours. They found him collapsed in a field.

vainillaychile/123RF

SOME THINGS TO KEEP IN MIND:

The reason why John became ill was because he had eaten the berries of the deadly nightshade, *Atropa belladonna.* This plant is especially dangerous because it contains several alkaloids, including atropine, scopolamine, and hyoscyamine. What makes these products so dangerous is their anticholinergic activities. These are antagonists of the muscarinic receptors, effectively blocking the endogenous actions of acetylcholine (ACh).

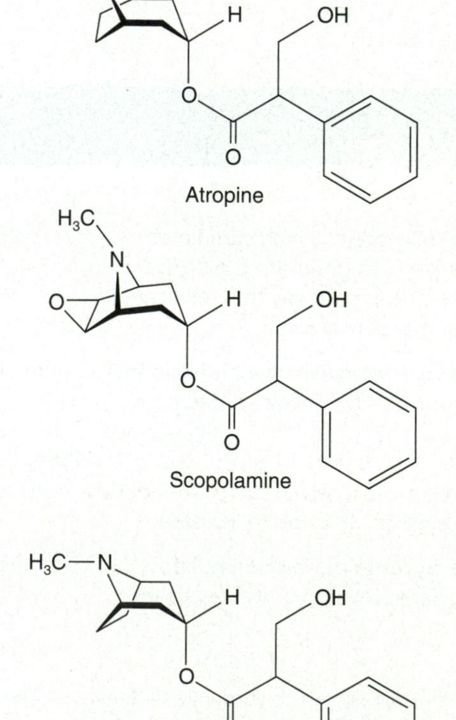

SOME INTERESTING FACTS:

Atropa belladonna has bell-shaped, light purple flowers with green tinges and large dark purple berries.

The term *belladonna* is derived from Italian and means "beautiful woman" because oil from the plant was used by women to dilate their pupils to make the eyes appear more seductive.

Atropa belladonna commonly grows in the wild in several regions of the world (including parts of the United States) and is sometimes cultivated because of its pretty appearance.

Once John was found by his parents, he was taken to the hospital where he was treated and fully recovered. He learned his lesson about running away from home and eating foreign plants.

Using your knowledge of the autonomic nervous system, what other symptoms should John have displayed from being poisoned with anticholinergic substances?

Would John have any effects on his skeletal muscles, which are also controlled by ACh?

The motor pathway of the autonomic nervous system consists of a preganglionic neuron, a postganglionic neuron, and a visceral effector. Which of these components is affected by the alkaloids in the nightshade berries?

Which branch of the autonomic nervous system is affected by this poisoning?

What treatment should John receive?

Nonadrenergic, Noncholinergic Neurons

Some autonomic neurons release neither norepinephrine nor acetylcholine. These neurons are called **nonadrenergic, noncholinergic neurons**. Neurotransmitters that can be released by these neurons include adenosine, ATP, nitric oxide, somatostatin, substance P, and vasoactive intestinal polypeptide (VIP). Nonadrenergic, noncholinergic neurons are classified as part of the parasympathetic nervous system or sympathetic nervous system based on where the preganglionic neurons exit the CNS.

The Autonomic Nervous System Performs a Variety of Functions

Autonomic Tone

As noted earlier, most body organs receive innervation from both parasympathetic and sympathetic branches of the ANS, which typically work in opposition to one another. The balance between parasympathetic and sympathetic activity, called **autonomic tone**, is regulated by the hypothalamus. Typically, the hypothalamus turns down parasympathetic tone at the same time it turns up sympathetic tone, and vice versa. The two branches can affect body organs differently because their postganglionic neurons release different neurotransmitters and because the effector organs possess different cholinergic and adrenergic receptors. A few structures receive only sympathetic innervation—sweat glands, arrector pili muscles attached to hair follicles in the skin, the kidneys, the spleen, most blood vessels, and the adrenal medullae of the adrenal glands. In these structures, there is no opposition from the parasympathetic nervous system. Still, an increase in sympathetic tone has one effect, and a decrease in sympathetic tone produces the opposite effect.

Parasympathetic Responses

The parasympathetic nervous system enhances **rest-and-digest** activities. Parasympathetic responses support body functions that conserve and restore body energy during times of rest and recovery. When the body is relaxed, parasympathetic input to the digestive glands and the smooth muscle of the gastrointestinal tract predominate over sympathetic input. This allows energy-supplying food to be digested and absorbed. At the same time, parasympathetic responses reduce body functions that support physical activity.

The acronym *SLUDD* can be helpful in remembering five parasympathetic responses. It stands for salivation (S), lacrimation (L), urination (U), digestion (D), and defecation (D). All of these activities are stimulated by the parasympathetic nervous system. Other important parasympathetic responses are "three decreases": decreased heart rate, decreased diameter of the bronchial tubes of the lungs (bronchoconstriction), and decreased diameter of the pupils (pupillary constriction).

Sympathetic Responses

During physical or emotional stress, the sympathetic nervous system dominates the parasympathetic nervous system. High sympathetic tone favors body functions that can support vigorous physical activity and rapid production of ATP. At the same time, the sympathetic nervous system reduces body functions that favor the storage of energy. In addition to physical exertion, various emotions—such as fear, embarrassment, or rage—stimulate the sympathetic nervous system. Visualizing body changes that occur during "E situations" such as exercise, emergency, excitement, and embarrassment will help you remember most of the sympathetic responses. Activation of the sympathetic nervous system and release of hormones by the adrenal medulla set in motion a series of physiological responses collectively called the **fight-or-flight response**, which includes the following effects:

- The diameter of the pupils increases (pupillary dilation).
- Heart rate, force of heart contraction, and blood pressure increase.
- The diameter of the bronchial tubes increases (bronchodilation), allowing faster movement of air into and out of the lungs.
- The blood vessels that supply the kidneys and gastrointestinal tract constrict, which decreases blood flow through these tissues. The result is a slowing of urine formation and digestive activities, which are not essential during emergency situations.
- Blood vessels that supply organs involved in exercise or fighting off danger—skeletal muscles, cardiac muscle, liver, and adipose tissue—dilate, allowing greater blood flow through these tissues.
- Liver cells break down glycogen into glucose, and adipose tissue cells break down triglycerides into fatty acids and glycerol, providing molecules that can be used by body cells for ATP production.
- Release of glucose by the liver increases blood glucose level.
- Processes that are not essential for dealing with the stressful situation are inhibited. For example, muscular movements of the gastrointestinal tract and digestive secretions slow down or even stop.

The effects of sympathetic stimulation are longer lasting and more widespread than the effects of parasympathetic stimulation for three reasons:

1. Sympathetic postganglionic axons diverge more extensively; as a result, many tissues are activated simultaneously.
2. Acetylcholinesterase quickly inactivates acetylcholine, but norepinephrine lingers in the synaptic cleft for a longer period.
3. Epinephrine and norepinephrine secreted into the blood from the adrenal medulla intensify and prolong the responses caused by norepinephrine liberated from sympathetic postganglionic axons. These blood-borne hormones circulate throughout the body, affecting all tissues that have alpha and beta receptors. In time, blood-borne norepinephrine and epinephrine are inactivated by enzymatic destruction in the liver.

TABLE 10.3 compares the main features of the parasympathetic and sympathetic branches of the ANS. TABLE 10.4 lists the responses of glands, cardiac muscle, and smooth muscle to stimulation by the parasympathetic and sympathetic branches of the ANS.

TABLE 10.3 Comparison of Parasympathetic and Sympathetic Branches of the ANS

	Parasympathetic Nervous System	Sympathetic Nervous System
Origin of preganglionic neuron	Brain stem and sacral spinal cord (S2–S4).	Thoracic and upper lumbar spinal cord (T1–L2).
Origin of postganglionic neuron	Terminal ganglion.	Sympathetic trunk ganglion, collateral ganglion, or adrenal medulla.
Axon length	Long preganglionic axons; short postganglionic axons.	Short preganglionic axons; long postganglionic axons.
Neurotransmitters	Preganglionic and postganglionic neurons release acetylcholine (ACh).	Preganglionic neurons release ACh; most postganglionic neurons release norepinephrine (NE). Postganglionic neurons that innervate most sweat glands and some blood vessels in skeletal muscle release ACh.
Physiological effects	Rest-and-digest activities.	Fight-or-flight responses.

TABLE 10.4 Effects of Sympathetic and Parasympathetic Branches of the ANS

Visceral Effector	Effect of Parasympathetic Stimulation (Muscarinic ACh Receptors)	Effect of Sympathetic Stimulation (Adrenergic Receptors, Except as Noted)*
GLANDS		
Sweat	No known effect.	Increases sweating in most body regions (muscarinic ACh receptors); sweating on palms and soles (α_1).
Lacrimal (tear)	Secretion of tears.	No known effect.
Salivary	Secretion of large volume of watery saliva.	Secretion of small volume of thick, viscous saliva (α_1).
Pancreas	Secretion of digestive enzymes and the hormone insulin.	Inhibits secretion of digestive enzymes and the hormone insulin (α_2); promotes secretion of the hormone glucagon (β_2).
Gastric	Secretion of gastric juice.	Inhibits secretion of gastric juice (α_2).
Intestinal	Secretion of intestinal juice.	Inhibits secretion of intestinal juice (α_2).
Adrenal medulla	No known effect.	Secretion of epinephrine and norepinephrine (nicotinic ACh receptors).
Liver†	Glycogen synthesis; increases bile secretion.	Glycogenolysis (breakdown of glycogen into glucose); gluconeogenesis (conversion of noncarbohydrates into glucose); and decreased bile secretion (α and β_2).
Adipose tissue†	No known effect.	Lipolysis (breakdown of triglycerides into fatty acids and glycerol) (β_1); release of fatty acids into blood (β_1 and β_3).
Kidneys, juxtaglomerular cells†	No known effect.	Secretion of renin (β_1).
CARDIAC (HEART) MUSCLE		
	Decreased heart rate and decreased rate of action potential conduction from atria to ventricles.	Increased heart rate, increased rate of action potential conduction from atria to ventricles, and increased ventricular contraction (β_1).
SMOOTH MUSCLE		
Iris, circular smooth muscle	Contraction → constriction of the pupil.	No known effect.
Iris, radial smooth muscle	No known effect.	Contraction → dilation of the pupil (α_1).
Ciliary muscle of eye	Contraction for near vision.	Relaxation for far vision (β_2).
Lungs, bronchial muscle	Contraction → airway constriction.	Relaxation → airway dilation (β_2)
Stomach and intestines	Increased motility and tone; relaxation of sphincters.	Decreased motility and tone (α_1, α_2, β_2); contraction of sphincters (α_1).

TABLE 10.4 (continues)

TABLE 10.4 Continued

Visceral Effector	Effect of Parasympathetic Stimulation (Muscarinic ACh Receptors)	Effect of Sympathetic Stimulation (Adrenergic Receptors, Except as Noted)*
SMOOTH MUSCLE (*continued*)		
Gallbladder and ducts	Contraction → release of bile into small intestine	Relaxation to facilitate storage of bile in the gallbladder (β_2).
Spleen	No known effect.	Contraction and discharge of stored blood into general circulation (α_1).
Urinary bladder	Contraction of muscular wall; relaxation of internal urethral sphincter.	Relaxation of muscular wall (β_2); contraction of internal urethral sphincter (α_1).
Sex organs	Vasodilation; erection of penis (males) and clitoris (females).	In males: contraction of smooth muscle of vas deferens, prostate, and seminal vesicle, resulting in ejaculation (α_1).
Hair follicles, arrector pili muscle	No known effect.	Contraction → erection of hairs resulting in goosebumps (α_1).
VASCULAR SMOOTH MUSCLE		
Coronary (heart) arterioles	Contraction → vasoconstriction.	Relaxation → vasodilation (β_2).
Skin arterioles	Vasodilation, which may not be physiologically significant.	Contraction → vasoconstriction (α_1).
Skeletal muscle arterioles	No known effect.	Relaxation → vasodilation (β_2).
Abdominal viscera arterioles	No known effect.	Contraction → vasoconstriction (α_1).
Kidney (arterioles)	No known effect.	Constriction of blood vessels → decreased urine volume (α_1).
Systemic veins	No known effect.	Contraction → constriction (α_1); relaxation → dilation (β_2).

* Subcategories of α and β receptors are listed if known.

† Listed with glands because they release substances into the blood.

Autonomic Reflexes Help Maintain Homeostasis

Autonomic (visceral) reflexes are fast, involuntary responses that occur when action potentials pass through an autonomic reflex arc. These reflexes play a key role in homeostasis by regulating cardiovascular activities, digestion, defecation, and micturition (urination). An autonomic reflex arc consists of the following components (FIGURE 10.7):

- **Sensory receptor.** The sensory receptor is the distal end of a sensory neuron; it responds to a stimulus by producing a receptor potential.

FIGURE 10.7 Components of an autonomic reflex arc. Arrows show the direction of action potential propagation.

An autonomic reflex arc consists of a sensory receptor, sensory neuron, integrating center, autonomic motor neurons, and a visceral effector.

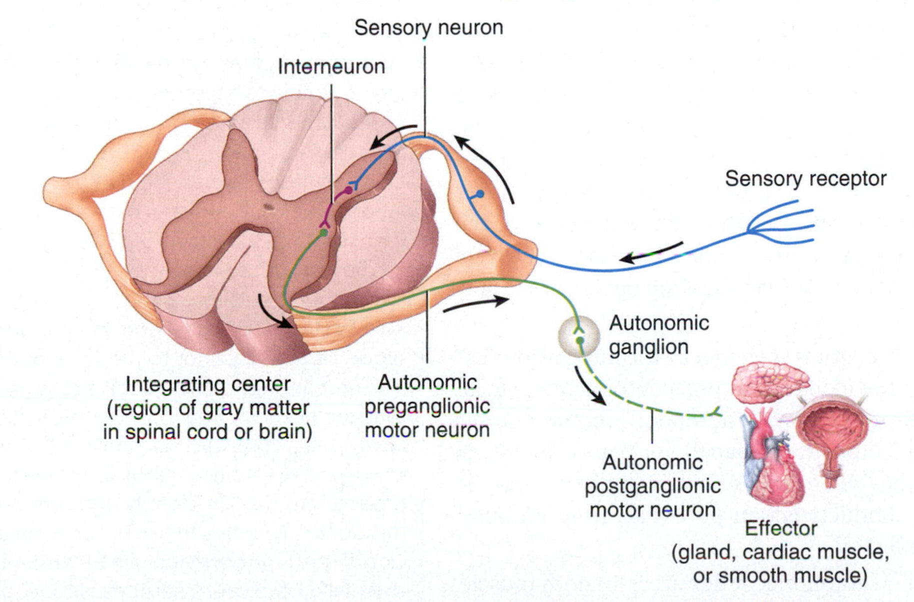

What are some examples of activities in the body that are regulated by autonomic reflexes?

CLINICAL CONNECTION

Raynaud Phenomenon

In **Raynaud phenomenon** (rā-NŌ), the digits (fingers and toes) become ischemic (lack blood) after exposure to cold or with emotional stress. The condition is due to excessive sympathetic stimulation of smooth muscle in the arterioles of the digits and a heightened response to stimuli that cause vasoconstriction. When arterioles in the digits vasoconstrict in response to sympathetic stimulation, blood flow is greatly diminished. As a result, the digits may blanch (look white due to blockage of blood flow) or become cyanotic (look blue due to deoxygenated blood in capillaries). In extreme cases, the digits may become necrotic from lack of oxygen and nutrients. With rewarming after cold exposure, the arterioles may dilate, causing the fingers and toes to look red. Many patients with Raynaud phenomenon have low blood pressure. Some have increased numbers of alpha adrenergic receptors. Raynaud phenomenon is most common in young women and occurs more often in cold climates. Patients with Raynaud phenomenon should avoid exposure to cold, wear warm clothing, and keep the hands and feet warm. Drugs used to treat Raynaud phenomenon include nifedipine, a calcium channel blocker that relaxes vascular smooth muscle, and prazosin, which relaxes smooth muscle by blocking alpha receptors. Smoking and the use of alcohol or illicit drugs can exacerbate the symptoms of this condition.

The sensory receptors of an autonomic reflex are usually interoceptors, which respond to internal stimuli, such as the degree of stretch of an organ wall or the chemical composition of a body fluid.

- **Sensory neuron.** If the receptor potential in the sensory receptor reaches threshold, it will generate one or more action potentials in the axon of the sensory neuron. The action potentials are conveyed along the axon into the CNS.
- **Integrating center.** The integrating center is a region of gray matter within the CNS that processes the incoming sensory input. The integrating center may involve just one synapse between the sensory neuron and a motor neuron, or it may involve one or more interneurons that relay signals from the sensory neuron to the motor neuron. The integrating centers for most autonomic reflexes are located in the hypothalamus and brain stem. However, some autonomic reflexes have integrating centers in the spinal cord. The reflex is a *cranial reflex* if integration occurs in the gray matter of the brain; the reflex is a *spinal reflex* if integration occurs in the gray matter of the spinal cord.
- **Motor neurons.** Action potentials triggered by the integrating center propagate out of the CNS along autonomic motor neurons to an effector. In an autonomic reflex arc, two autonomic motor neurons connect the CNS to an effector: The preganglionic neuron conducts action potentials from the CNS to an autonomic ganglion, and the postganglionic neuron conducts action potentials from an autonomic ganglion to the effector.
- **Effector.** The effector of an autonomic reflex arc is smooth muscle, cardiac muscle, or a gland.

Autonomic Control Centers Are Present in the Brain and Spinal Cord

Normally, we are not aware of muscular contractions of our digestive organs, our heartbeat, changes in the diameter of our blood vessels, and pupil dilation and constriction because the integrating centers for these autonomic responses are in the spinal cord or lower regions of the brain. Examples of autonomic control centers include the cardiovascular, respiratory, deglutition (swallowing), salivation, vomiting, and pupillary reflex centers in the brain stem, and the erection, ejaculation, defecation, and micturition (urination) centers in the spinal cord (**FIGURE 10.8**). Sensory neurons deliver input to these centers, and autonomic motor neurons provide output that adjusts activity in the visceral effectors, usually without our conscious awareness.

The hypothalamus is the major control and integration center of the ANS (**FIGURE 10.8**). The hypothalamus receives sensory input related to visceral functions, olfaction (smell), and gustation (taste), as well as changes in temperature, osmolarity, and levels of various substances in blood. It also receives input relating to emotions from the limbic system. Output from the hypothalamus influences autonomic centers in both the brain stem and spinal cord.

The hypothalamus is connected to both parasympathetic and sympathetic branches of the ANS by axons of neurons with dendrites and cell bodies in various hypothalamic nuclei. The axons form tracts from the hypothalamus to parasympathetic and sympathetic nuclei in the brain stem and spinal cord through relays in the reticular

CLINICAL CONNECTION

Autonomic Dysreflexia

Autonomic dysreflexia is an exaggerated response of the sympathetic division of the ANS that occurs in about 85% of individuals with spinal cord injury at or above the level of T6. The condition is seen after recovery from spinal shock and occurs due to interruption of the control of ANS neurons by higher centers. When certain sensory input, such as that resulting from stretching of a full urinary bladder, is unable to ascend the spinal cord, mass stimulation of the sympathetic nerves inferior to the level of injury occurs. Other triggers include stimulation of pain receptors and the visceral contractions resulting from sexual stimulation, labor/delivery, and bowel stimulation. One of the effects of increased sympathetic activity is severe vasoconstriction, which elevates blood pressure. In response, the cardiovascular center in the medulla oblongata (1) increases parasympathetic output via the vagus (X) nerve, which decreases heart rate, and (2) decreases sympathetic output, which causes dilation of blood vessels superior to the level of the injury.

The condition is characterized by a pounding headache; hypertension; flushed, warm skin with profuse sweating above the injury level; pale, cold, dry skin below the injury level; and anxiety. This emergency condition requires immediate intervention. The first approach is to quickly identify and remove the problematic stimulus. If this does not relieve the symptoms, an antihypertensive drug such as clonidine or nitroglycerin can be administered. Untreated autonomic dysreflexia can cause seizures, stroke, or heart attack.

FIGURE 10.8 Autonomic control centers.

The hypothalamus, brain stem, and spinal cord contain centers that control autonomic activities.

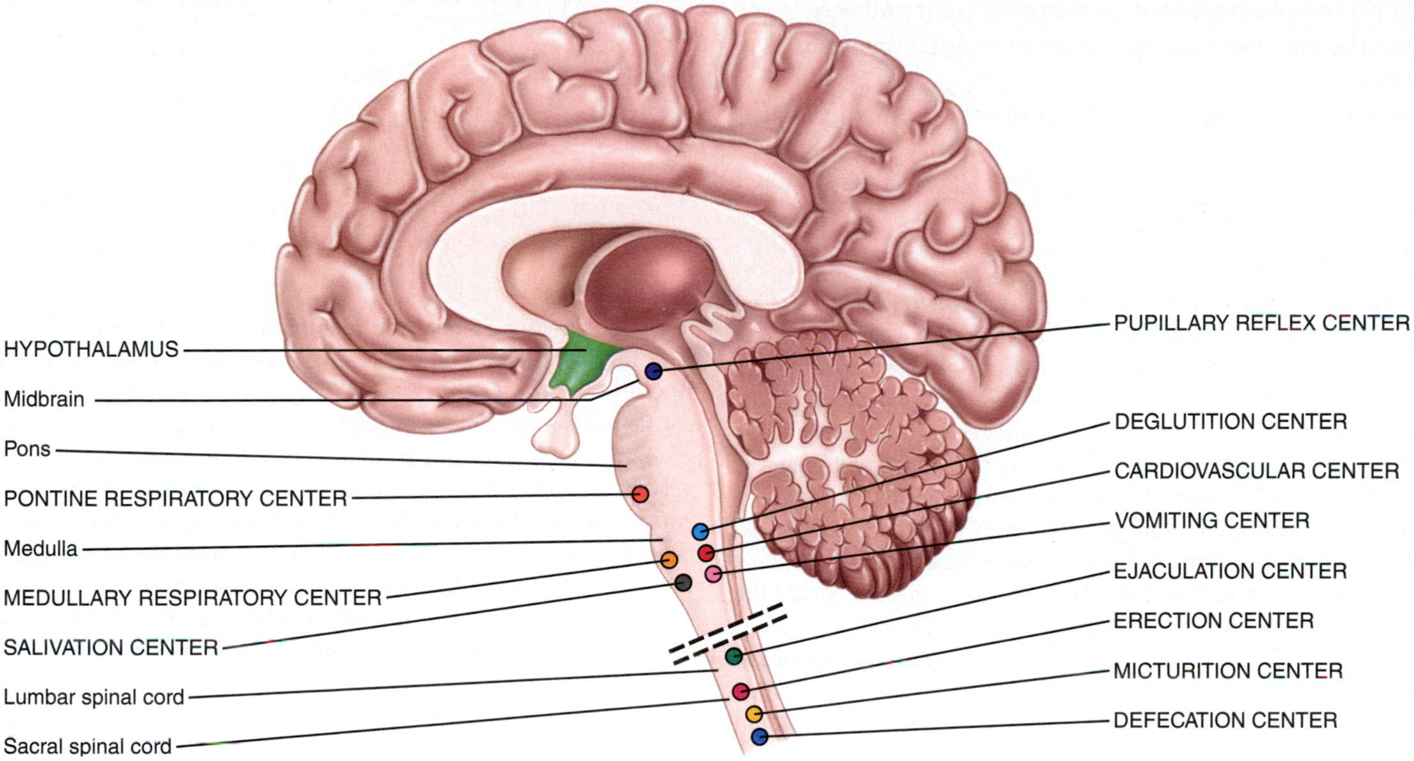

Which part of the brain is considered to be the major control and integrating center of the ANS?

formation. The anterior and medial parts of the hypothalamus control the parasympathetic nervous system. Stimulation of these areas results in a decrease in heart rate, lowering of blood pressure, constriction of the pupils, and increased secretion and motility of the gastrointestinal tract. By contrast, the posterior and lateral parts of the hypothalamus control the sympathetic nervous system. Stimulation of these areas produces an increase in heart rate and force of contraction, a rise in blood pressure due to constriction of blood vessels, an increase in body temperature, dilation of the pupils, and inhibition of the gastrointestinal tract.

Some Autonomic Responses Can Be Voluntarily Controlled via Biofeedback

Although most people are unable to consciously alter or suppress autonomic responses to any great degree, practitioners of yoga or other techniques of meditation may learn how to regulate at least some of their autonomic activities through long practice. **Biofeedback**, in which electronic-monitoring devices display information about a body function such as heart rate or blood pressure, enhances the ability to learn such conscious control. The monitoring devices provide visual or auditory signals about the autonomic responses. By concentrating on positive thoughts, individuals learn to alter autonomic responses. For example, biofeedback has been used to decrease heart rate and blood pressure in order to decrease the severity of migraine headaches.

CHECKPOINT

1. Describe the functional significance of each of the following terms: terminal ganglia, sympathetic trunk, collateral ganglia, and chromaffin cells.
2. Which ANS neurons are cholinergic? Which are adrenergic?
3. How do the various types of adrenergic receptors differ from one another?
4. What are some examples of the antagonistic effects of the parasympathetic and sympathetic branches of the ANS?
5. How does an autonomic reflex arc differ from a somatic reflex arc?

10.2 Somatic Nervous System

OBJECTIVES

- Discuss the importance of the somatic nervous system.
- Identify the components of a somatic motor pathway.
- Describe the functional operation of the neuromuscular junction.
- Explain how chemical agents alter events at the neuromuscular junction.

The Somatic Nervous System Regulates the Activity of Skeletal Muscle

The somatic nervous system innervates the skeletal muscles of the body. When a somatic motor neuron stimulates a skeletal muscle, it contracts. If somatic motor neurons cease to stimulate a skeletal muscle, the result is a paralyzed, limp muscle that has no muscle tone.

The somatic nervous system usually operates under voluntary (conscious) control. Voluntary control of movement involves motor areas of the cerebral cortex that activate somatic motor neurons whenever you have a desire to move. For example, if you want to perform a particular movement (kick a ball, turn a screwdriver, smile for a picture, etc.), neural pathways from the motor cortex activate somatic motor neurons that cause the appropriate skeletal muscles to contract. The somatic nervous system is not always under voluntary control, however. The somatic motor neurons that innervate skeletal muscles involved in posture, balance, breathing, and somatic reflexes (such as the flexor reflex) are involuntarily controlled by integrating centers in the brain stem and spinal cord. You will learn more about the control of body movement in Chapter 12.

A Somatic Motor Pathway Is Comprised of a Somatic Motor Neuron and Skeletal Muscle

Somatic motor pathways consist of a single somatic motor neuron that originates in the CNS and extends its axon to one or more fibers (cells) of a skeletal muscle (**FIGURE 10.9**). The cell bodies of somatic motor neurons are located in either the brain stem or ventral gray horn of the spinal cord. From the brain stem, axons of somatic motor neurons extend through *cranial nerves* to innervate skeletal muscles of the face and head. From the spinal cord, axons of somatic motor neurons extend through *spinal nerves* to innervate skeletal muscles of the limbs and trunk.

The Neuromuscular Junction Is the Site Where a Somatic Motor Neuron Communicates with a Skeletal Muscle Fiber

The synapse formed between a somatic motor neuron and a skeletal muscle fiber is called the **neuromuscular junction (NMJ)** (**FIGURE 10.10**). At the NMJ, a terminal branch of the somatic motor neuron's axon divides into a cluster of synaptic end bulbs, which contain synaptic vesicles (**FIGURE 10.10a, b**). Inside each synaptic vesicle are thousands of molecules of acetylcholine (ACh), the neurotransmitter released at the NMJ. ACh has an excitatory effect on the NMJ, ultimately causing the skeletal muscle fiber to contract. The region of the muscle fiber plasma membrane opposite the synaptic end bulbs is called the **motor end plate** (**FIGURE 10.10b, c**). Within the motor end plate are 30 to 40 million **acetylcholine receptors** (of the nicotinic type) that bind specifically to ACh. These receptors are abundant in *junctional folds*, deep grooves in the motor end plate that provide a large surface area for ACh. Although the synaptic end bulbs and motor end plate are close to each other, they do not actually touch; instead, they are separated by a small space called the **synaptic cleft**. A neuromuscular junction thus includes all of the synaptic end bulbs on one side of the synaptic cleft, the synaptic cleft itself, plus the motor end plate of the muscle fiber on the other side. A skeletal muscle fiber has only one NMJ and it is usually located near the midpoint of the fiber.

FIGURE 10.9 Somatic motor pathway. Note that the somatic motor neuron releases the neurotransmitter acetylcholine (ACh).

In a somatic motor pathway, a single somatic motor neuron extends to a skeletal muscle.

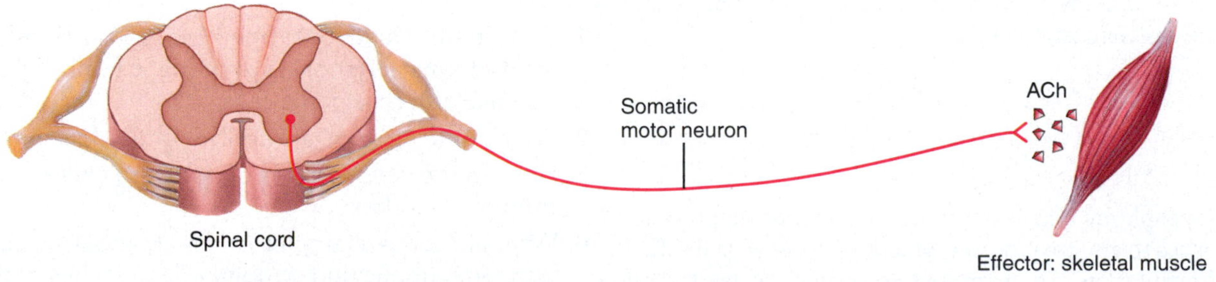

What effect does the ACh released by the somatic motor neuron have on skeletal muscle?

FIGURE 10.10 Organization of the neuromuscular junction.

The neuromuscular junction refers to the synapse between a somatic motor neuron and a skeletal muscle fiber.

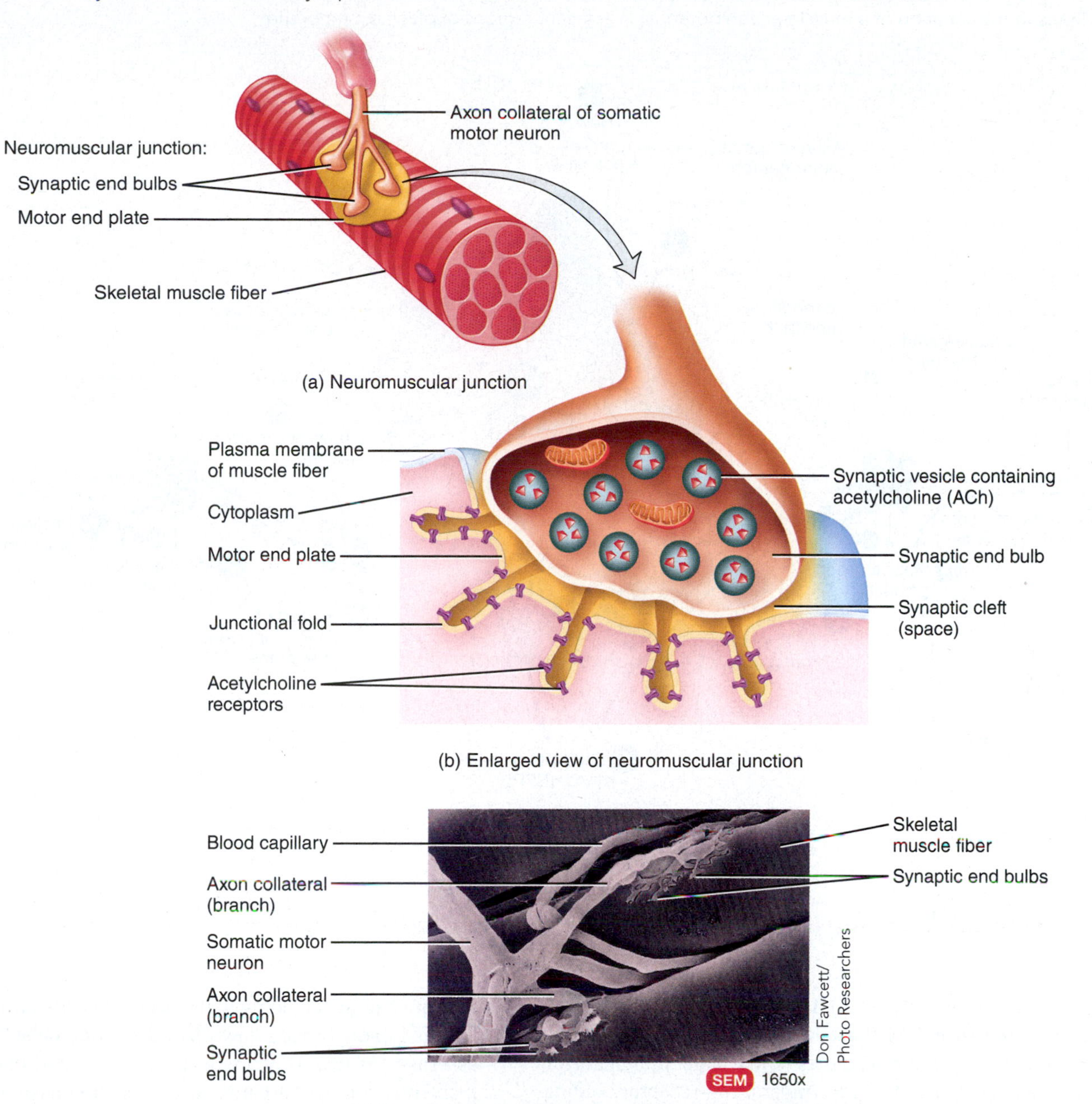

(a) Neuromuscular junction

(b) Enlarged view of neuromuscular junction

(c) Scanning electron micrograph of two neuromuscular junctions

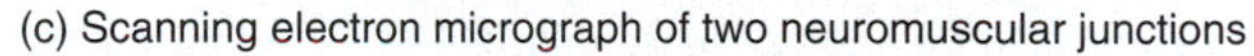

Which part of the muscle plasma membrane contains acetylcholine receptors?

A nerve action potential in a somatic motor neuron elicits a muscle action potential in a skeletal muscle fiber in the following way (**FIGURE 10.11**):

1. A nerve action potential arrives at a synaptic end bulb of a somatic motor neuron.
2. **Voltage-gated Ca^{2+} channels** present in the membrane of the synaptic end bulb open in response to the nerve action potential. Because calcium ions are more concentrated in the extracellular fluid, Ca^{2+} flows inward through the opened channels.
3. An increase in the Ca^{2+} concentration inside the synaptic end bulb serves as a signal that triggers exocytosis of the synaptic vesicles, liberating ACh into the synaptic cleft.
4. ACh diffuses across the synaptic cleft and binds to nicotinic ACh receptors on the motor end plate. Recall that a nicotinic ACh receptor is a type of ionotropic receptor that contains two binding sites for ACh and a cation channel. Binding of two ACh molecules to the nicotinic ACh receptor opens the cation channel. Opening the cation channel allows passage of cations (mainly Na^+ and K^+) through the end plate membrane, but Na^+ inflow is greater than K^+ outflow.
5. The net influx of Na^+ ions into the muscle fiber through the open nicotinic ACh receptors causes the motor end plate to depolarize. This change in membrane potential is called an **end plate potential (EPP)**. An EPP is a type of graded potential that is similar to an excitatory postsynaptic potential (EPSP), which forms at synapses

FIGURE 10.11 Signal transmission at a neuromuscular junction.

Binding of acetylcholine (ACh) to nicotinic ACh receptors on the motor end plate causes the formation of an end plate potential (EPP), which in turn leads to the initiation of a muscle action potential in adjacent regions of plasma membrane.

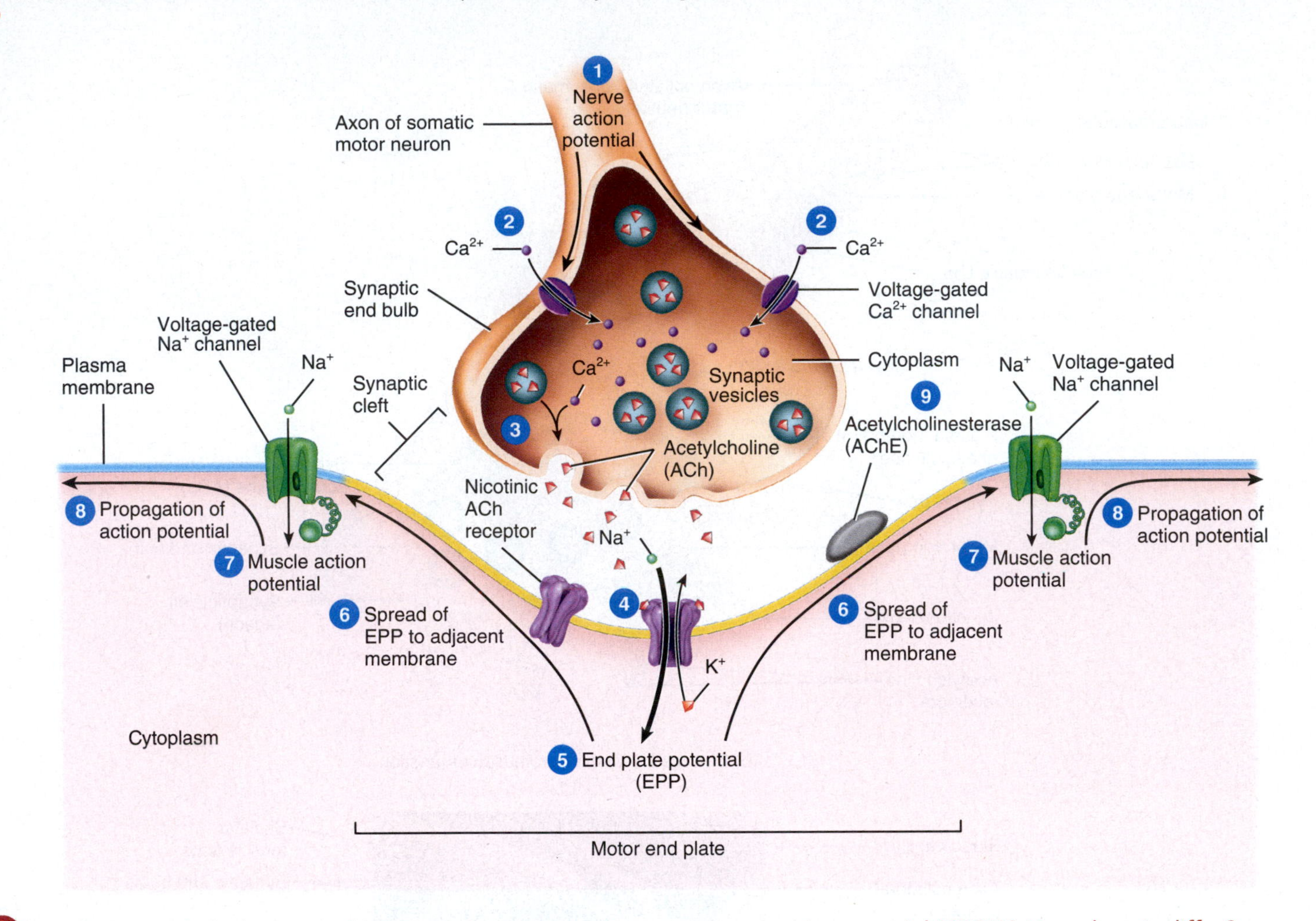

How is an end plate potential (EPP) similar to an excitatory postsynaptic potential (EPSP)? How does it differ?

between neurons (see Section 7.4). However, an EPP has a larger amplitude (size) than an EPSP because there are more neurotransmitter receptors in the motor end plate and a larger number of ion channels open in response to neurotransmitter–receptor binding. Consequently, a single EPP typically is large enough to depolarize a muscle fiber to threshold (see step 6), whereas a single EPSP normally is too small to depolarize a neuron to threshold.

6. The EPP spreads by local current flow to an adjacent region of plasma membrane on each side of the motor end plate and depolarizes these areas to threshold.
7. The adjacent membrane areas contain voltage-gated Na^+ channels, which open in response to the threshold-level depolarization. The resultant inflow of Na^+ into the muscle fiber through the open voltage-gated Na^+ channels initiates a **muscle action potential**.
8. Since the NMJ is usually near the midpoint of the muscle fiber, once the muscle action potential arises, it propagates throughout the muscle fiber membrane in both directions away from the NMJ toward the ends of the fiber. The action potential triggers a chain of events that ultimately leads to contraction of the muscle fiber.
9. The effect of ACh binding to its receptor lasts only briefly because ACh is rapidly broken down by an enzyme called **acetylcholinesterase (AChE),** which is located on the end plate membrane. AChE breaks down ACh into acetyl and choline, products that cannot individually activate the nicotinic ACh receptor.

If another nerve action potential releases more acetylcholine, steps 4 through 8 repeat. When action potentials in the motor neuron cease, ACh is no longer released, and AChE rapidly breaks down the ACh already present in the synaptic cleft. This ends the production of action potentials in the skeletal muscle fiber.

The Events at the NMJ Can Be Altered by Chemicals

Several chemical agents selectively alter certain events at the neuromuscular junction (NMJ). Among these substances are botulinum toxin, α-latrotoxin, curare, and organophosphates. *Botulinum toxin*, produced by the bacterium *Clostridium botulinum*, blocks exocytosis of synaptic vesicles at the NMJ. As a result, acetylcholine (ACh) is not released, and muscle contraction does not occur. The bacteria proliferate

CLINICAL CONNECTION

Myasthenia Gravis

Myasthenia gravis (mī-as-THĒ-nē-a GRAV-is) is an autoimmune disease that causes chronic, progressive damage of the neuromuscular junction. The immune system inappropriately produces antibodies that bind to and block some acetylcholine (ACh) receptors, thereby decreasing the number of functional ACh receptors at the motor end plates of skeletal muscles (see **FIGURE 10.10**). As the disease progresses, more ACh receptors are lost. Thus, muscles become increasingly weaker, fatigue more easily, and may eventually cease to function. Treatment may involve the use of anticholinesterase drugs that act as inhibitors of acetylcholinesterase (AChE), the enzyme that breaks down ACh. These inhibitors raise the level of ACh that is available to bind with still-functional receptors. More recently, steroid drugs such as prednisone have been used with success to reduce antibody levels. Another treatment is plasmapheresis, a procedure that removes the antibodies from the blood.

in improperly canned foods, and their toxin is one of the most lethal chemicals known. A tiny amount can cause death by paralyzing skeletal muscles. Breathing stops because respiratory muscles (which are skeletal muscles) are paralyzed. Despite these lethal effects, botulinum toxin can be used in small, diluted amounts as a medicine (Botox®). Injections of Botox into the affected muscles can help patients who have strabismus (crossed eyes), uncontrollable blinking, or spasms of the vocal cords that interfere with speech. Botox is also used as a cosmetic treatment to relax muscles that cause facial wrinkles and to alleviate chronic back pain due to muscle spasms.

The venom of the black widow spider contains α-*latrotoxin,* which induces massive exocytosis of synaptic vesicles at the NMJ. This causes the release of excessive amounts of ACh, which leads to overstimulation of skeletal muscles. Prolonged stimulation of respiratory muscles can lead to respiratory failure and death.

The plant derivative *curare,* a poison used by South American Indians on arrows and blowgun darts, causes muscle paralysis by binding to and blocking nicotinic ACh receptors on the motor end plate. In the presence of curare, the cation channel of the nicotinic ACh receptor does not open. Curare-like drugs are often used during surgery to relax skeletal muscles.

Organophosphates are chemicals that inhibit acetylcholinesterase (AChE). Inhibition of AChE at the NMJ initially causes uncontrolled contractions of skeletal muscles and then causes muscle paralysis as muscle fibers become refractory to further stimulation. Organophosphates include the nerve gas sarin, which is used in chemical warfare, and malathion, which is an ingredient used in certain insecticides.

This concludes our discussion of the somatic nervous system. **TABLE 10.5** provides a comparison of the autonomic and somatic nervous systems.

CHECKPOINT

6. What is the effector of a somatic motor neuron pathway? Is this effector usually controlled voluntarily or involuntarily?
7. What is the motor end plate? What happens there?
8. What effect does curare have on transmission at the neuromuscular junction (NMJ)?
9. Organophosphates inhibit which important enzyme at the NMJ?

Now that the various components of the nervous system have been explored, you can appreciate the many ways that this system contributes to homeostasis of other body systems by examining *Focus on Homeostasis: Contributions of the Nervous System.*

TABLE 10.5 Comparison of the Autonomic and Somatic Nervous Systems

	Autonomic Nervous System	Somatic Nervous System
Effectors	Smooth muscle, cardiac muscle, and glands.	Skeletal muscle.
Motor neuron pathway	Usually a two-neuron pathway: (1) a preganglionic neuron that extends from CNS to an autonomic ganglion and (2) a postganglionic neuron that extends from the autonomic ganglion to the effector. Alternatively, a preganglionic neuron may extend from CNS to synapse with chromaffin cells of the adrenal medulla.	One-neuron pathway: a single somatic motor neuron extends from the CNS to synapse directly with the effector.
Neurotransmitters	Most autonomic motor neurons release either acetylcholine (ACh) or norepinephrine (NE). Chromaffin of the adrenal medulla releases epinephrine and NE into bloodstream as hormones.	All somatic motor neurons release ACh.
Receptor type on effector organ	Usually cholinergic or adrenergic.	Cholinergic.
Action of neurotransmitter on effector	May be excitatory (causes contraction of smooth muscle, increases rate and force of contraction of cardiac muscle, or increases secretions of glands) or inhibitory (causes relaxation of smooth muscle, decreases rate and force of contraction of cardiac muscle, or decreases secretions of glands).	Always excitatory (causes contraction of skeletal muscle).
Control of motor output	Involuntary control from hypothalamus, brain stem, and spinal cord.	Voluntary control from cerebral cortex, with contributions from the basal nuclei, cerebellum, brain stem, and spinal cord.

FOCUS on HOMEOSTASIS

INTEGUMENTARY SYSTEM

- Sympathetic nerves of the autonomic nervous system (ANS) control contraction of smooth muscles attached to hair follicles and secretion of perspiration from sweat glands

SKELETAL SYSTEM

- Pain receptors in bone tissue warn of bone trauma or damage

MUSCULAR SYSTEM

- Somatic motor neurons receive instructions from motor areas of the brain and stimulate contraction of skeletal muscles to bring about body movements
- Basal nuclei and reticular formation set level of muscle tone
- Cerebellum coordinates skilled movements

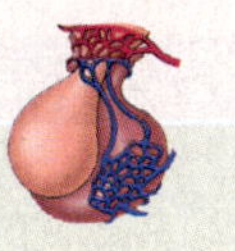

ENDOCRINE SYSTEM

- Hypothalamus regulates secretion of hormones from anterior pituitary
- ANS regulates secretion of hormones from adrenal medulla and pancreas

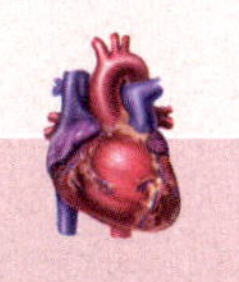

CARDIOVASCULAR SYSTEM

- Cardiovascular center in the medulla oblongata provides action potentials to ANS that govern heart rate, forcefulness of the heartbeat, blood pressure, and blood flow through blood vessels

CONTRIBUTIONS OF THE NERVOUS SYSTEM

FOR ALL BODY SYSTEMS

- Action potentials of the nervous system provide communication with and regulation of most body organs

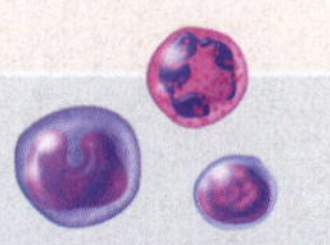

IMMUNE SYSTEM

- Certain neurotransmitters help regulate immune responses
- Activity in nervous system may increase or decrease immune responses

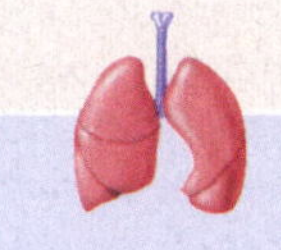

RESPIRATORY SYSTEM

- Respiratory areas in brain stem control breathing rate and depth
- ANS helps regulate diameter of airways

DIGESTIVE SYSTEM

- Enteric division of the ANS helps regulate digestion
- Parasympathetic division of ANS stimulates many digestive processes

URINARY SYSTEM

- ANS helps regulate blood flow to kidneys, thereby influencing the rate of urine formation
- Brain and spinal cord centers govern emptying of the urinary bladder

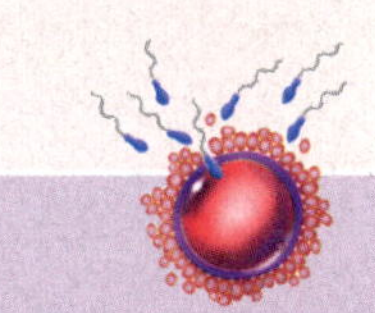

REPRODUCTIVE SYSTEMS

- Hypothalamus and limbic system govern a variety of sexual behaviors
- ANS brings about erection of penis in males and clitoris in females and ejaculation of semen in males
- Hypothalamus regulates release of anterior pituitary hormones that control gonads (ovaries and testes)
- Action potentials elicited by touch stimuli from suckling infant cause release of oxytocin and milk ejection in nursing mothers

FROM RESEARCH TO REALITY

Biofeedback and Blood Pressure

Reference
Cejnar, M. et al. (1988). Voluntary blood pressure control using continuous systolic blood pressure biofeedback. *Clinical and Experimental Pharmacology & Physiology*. 15: 265–269.

Is biofeedback a viable method of controlling blood pressure?

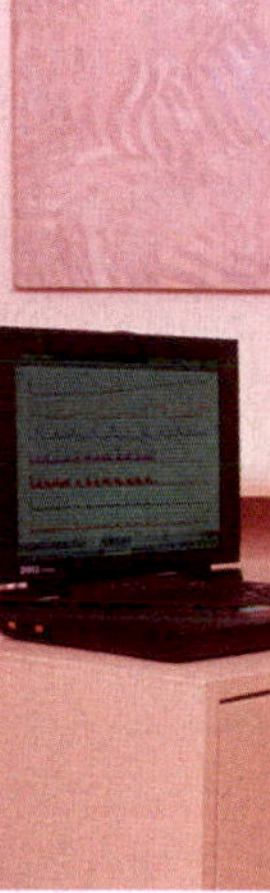

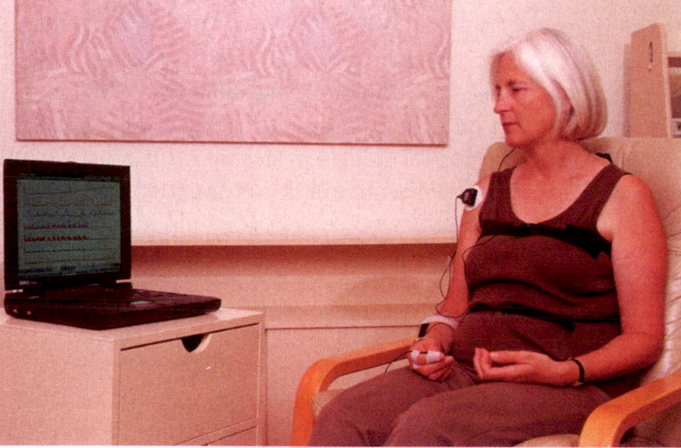

Biofeedback can be used to regulate some autonomic responses such as blood pressure. The ability to alter blood pressure can help treat cardiovascular disorders such as hypertension or hypotension. Can biofeedback significantly decrease or even increase blood pressure?

Article description:
The authors of this study wanted to determine the effectiveness of biofeedback on altering a person's systolic or diastolic blood pressure. Thirteen subjects with an average blood pressure of 127/82 mmHg were included in this study. A noninvasive biofeedback apparatus was attached to each subject's finger, and it displayed the subject's systolic blood pressure. Instructions were then given for subjects to raise or lower their blood pressures. Subjects participated in 30 trials of blood pressure biofeedback in six sessions over three weeks. Each trial consisted of blood pressure–raising and blood pressure–lowering periods of 45 seconds and 90 seconds, respectively, with intervening 45-second baseline periods.

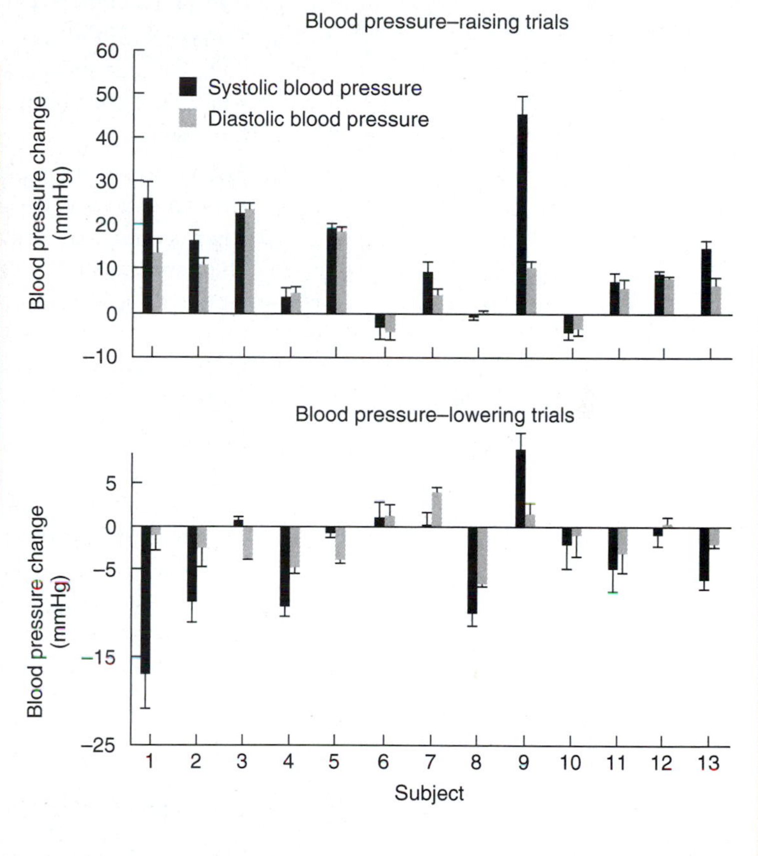

Go to WileyPLUS Learning Space and use the data from this article to answer the questions posed there and to discover more about biofeedback and blood pressure.

CHAPTER REVIEW

10.1 Autonomic Nervous System

1. The autonomic nervous system (ANS) innervates visceral effectors—smooth muscle, cardiac muscle, and glands. The ANS usually operates without conscious control.

2. The ANS consists of two main branches: the parasympathetic nervous system and the sympathetic nervous system. A third branch of the ANS, the enteric nervous system, is confined to the wall of the gastrointestinal tract.
3. Autonomic motor pathways (both parasympathetic and sympathetic) consist of two autonomic motor neurons in series and a visceral effector. The first neuron is the preganglionic neuron, which extends from the CNS to an autonomic ganglion. The second neuron is the postganglionic neuron, which extends from the autonomic ganglion to the visceral effector.
4. Autonomic ganglia in the parasympathetic nervous system are known as terminal ganglia. Autonomic ganglia in the sympathetic nervous system include the sympathetic trunk and collateral ganglia.
5. Some sympathetic preganglionic neurons extend to chromaffin cells in the adrenal medulla without synapsing in either the sympathetic trunk or collateral ganglia. When stimulated, the chromaffin cells release mainly epinephrine and norepinephrine into the blood.
6. Autonomic postganglionic neurons release neurotransmitters along their axons at swollen regions called varicosities. When an action potential reaches a varicosity, it triggers exocytosis of synaptic vesicles, causing the release of stored neurotransmitter. The neurotransmitter molecules diffuse across the neuroeffector junction (NEJ) and bind to specific receptors, which leads to a response that excites or inhibits the effector cell.
7. Most autonomic motor neurons release either acetylcholine (ACh) or norepinephrine (NE).
8. In the ANS, ACh and NE can cause excitatory or inhibitory responses in the visceral effectors. Excitatory responses include contraction of smooth muscle, increased rate and force of contraction of cardiac muscle, and increased secretions of glands. Inhibitory responses include relaxation of smooth muscle, decreased rate and force of contraction of cardiac muscle, and decreased secretions of glands.
9. The parasympathetic nervous system regulates activities that conserve and restore body energy; the sympathetic nervous system favors body functions that can support vigorous physical activity and rapid production of ATP (fight-or-flight response).

10.2 Somatic Nervous System

1. The somatic nervous system innervates the skeletal muscles of the body.
2. The somatic nervous system usually operates under voluntary control.
3. Somatic motor pathways consist of a single somatic motor neuron that extends from the CNS all the way to the skeletal muscle.
4. The neurotransmitter released by a somatic motor neuron is acetylcholine (ACh) and the effect of ACh at the neuromuscular junction is always excitatory, which causes contraction of the skeletal muscle.
5. The neuromuscular junction (NMJ) is the synapse between a somatic motor neuron and a skeletal muscle fiber. The NMJ includes the axon terminals and synaptic end bulbs of a motor neuron, plus the adjacent motor end plate of the muscle fiber plasma membrane.
6. When an action potential reaches the synaptic end bulbs of a somatic motor neuron, it triggers exocytosis of the synaptic vesicles, which releases ACh. ACh diffuses across the synaptic cleft and binds to ACh receptors, initiating a muscle action potential.
7. Several chemicals can alter events at the NMJ, including botulinum toxin, α-latrotoxin, curare, and organophosphates.

PONDER THIS

1. If you were to administer a drug that blocked all of the nicotinic receptors found in the peripheral nervous system, would it affect the autonomic nervous system or the somatic nervous system? Explain your answer.

2. Compare and contrast the autonomic nervous system with the somatic nervous system. You need to discuss the receptors, organs, neurotransmitters, and whether or not the system is under voluntary control.

3. You and your friend are walking through the woods and suddenly your friend gets bitten by an insect that neither of you has ever seen. All of a sudden, you notice interesting symptoms in your friend. Your friend is: Breathing with ease, pupils are dilated, heart rate has increased, and blood pressure has increased. You recognize these symptoms and realize that this insect must have a toxin that has activated the autonomic nervous system. Which branch of the autonomic nervous system was activated and why?

ANSWERS TO FIGURE QUESTIONS

10.1 No. Activation of autonomic motor pathways results in excitation of some effectors and inhibition of others.

10.2 Dual innervation means that an organ receives neural innervation from both the parasympathetic and sympathetic branches of the ANS.

10.3 Unlike most sympathetic postganglionic neurons, chromaffin cells lack dendrites and axons and they release hormones into the blood.

10.4 A neuroeffector junction (NEJ) differs from a typical neuron-to-neuron synapse in two main ways: (1) The axon terminals exhibit varicosities instead of synaptic end bulbs and (2) the neurotransmitter receptors are located along the entire surface of the effector cell instead of being confined to a specific receptor region.

10.5 In response to neurotransmitter binding to its receptor, the effector cell is either excited or inhibited, depending on the type of receptor and G protein pathway that are activated.

10.6 Most (but not all) sympathetic postganglionic neurons are adrenergic.

10.7 Autonomic reflexes help regulate cardiovascular activities, digestion, defecation, and micturition (urination).

10.8 The hypothalamus is the major control and integrating center of the ANS.

10.9 ACh released by a somatic motor neuron always causes excitation (contraction) of skeletal muscle.

10.10 The part of the muscle plasma membrane that contains acetylcholine receptors is the motor end plate.

10.11 An EPP and an EPSP are both depolarizing graded potentials. However, an EPP occurs at a neuromuscular junction, whereas an EPSP occurs at a synapse between neurons. In addition, an EPP has a larger amplitude (size) than an EPSP, so a single EPP typically is large enough to depolarize a muscle fiber to threshold, but a single EPSP normally is too small to depolarize a neuron to threshold.

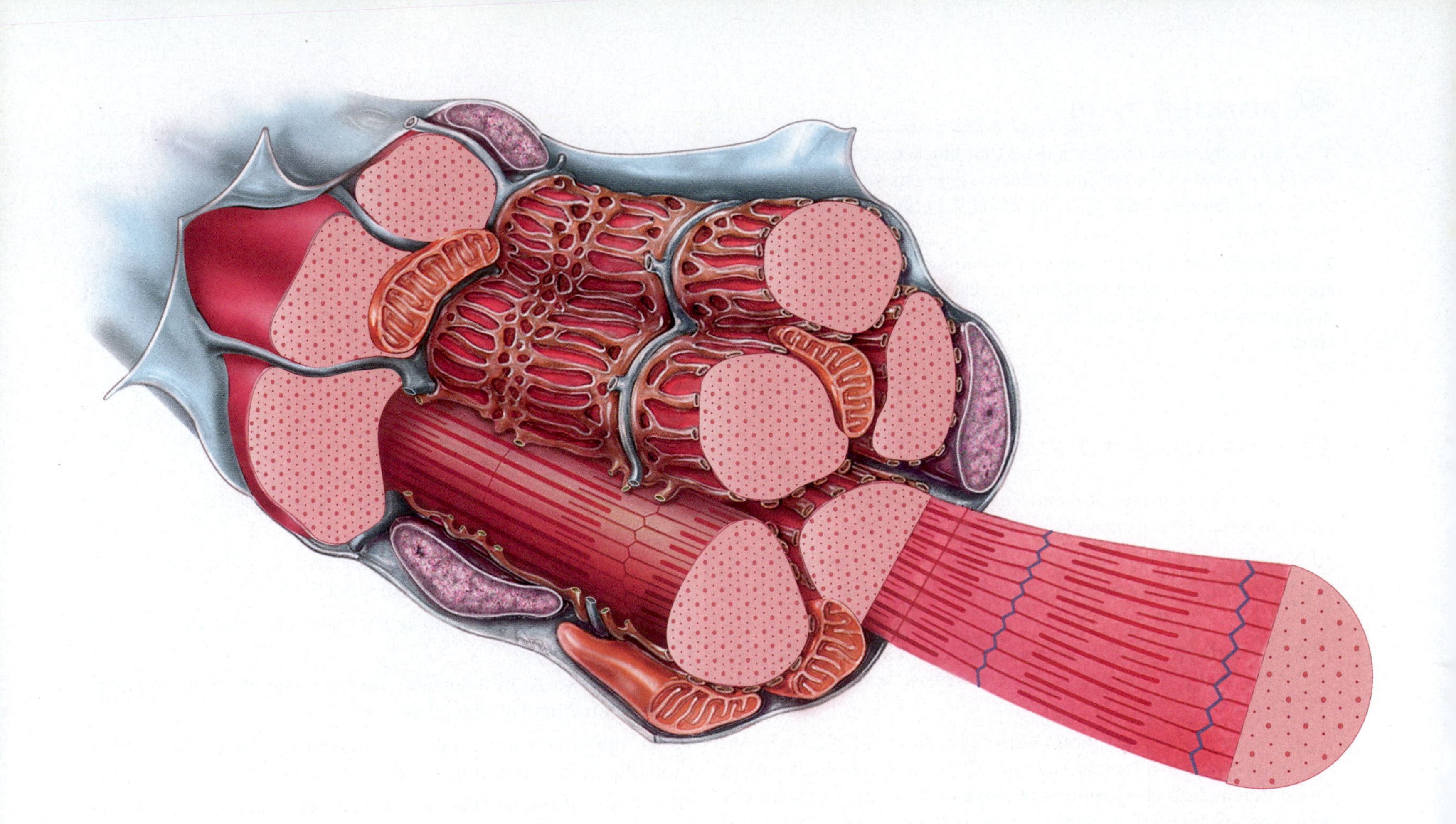

11 Muscle

Muscle and Homeostasis

Muscle contributes to homeostasis by producing body movements, moving substances through the body, and producing heat to maintain normal body temperature.

LOOKING BACK TO MOVE AHEAD...

- The Krebs cycle and the electron transport chain both require oxygen to produce ATP and are collectively known as aerobic respiration (Section 4.4).
- Somatic motor neurons convey action potentials to skeletal muscles, whereas autonomic motor neurons convey action potentials to cardiac muscle, smooth muscle, and glands (Section 7.2).
- The nerve action potential consists of two main phases: a depolarizing phase and a repolarizing phase (Section 7.3).
- The refractory period is the period of time after an action potential begins when an excitable cell cannot generate another action potential in response to a normal threshold stimulus (Section 7.3).

Movements such as throwing a ball, riding a bike, and walking result from the contraction of muscle, which makes up 40–50% of total adult body weight. Your muscular strength reflects the primary function of muscle—the transformation of chemical energy into mechanical energy to generate force, perform work, and produce movement. In addition, muscle stabilizes body position, regulates organ volume, generates heat, and propels fluids and food matter through various body systems.

11.1 Overview of Muscle

OBJECTIVES

- Compare the three types of muscle.
- Explain the functions of muscle.
- Describe the properties of muscle.

Three Types of Muscle Exist in the Body

There are three types of muscle tissue—skeletal, cardiac, and smooth. The different types of muscle tissue differ from one another in their microscopic structure, their location, and how they are controlled by the nervous system.

As its name suggests, most **skeletal muscle** is attached to bones and moves parts of the skeleton (FIGURE 11.1a). It is *striated*; that is, **striations**, or alternating light and dark bands, are visible under a light microscope. Skeletal muscle works mainly in a *voluntary* manner: Its activity can be consciously controlled by motor neurons that are part of the somatic nervous system. Many skeletal muscles are also controlled subconsciously to some extent. For example, your diaphragm continues to alternately contract and relax without conscious control so that you don't stop breathing. In addition, the skeletal muscles that maintain posture or stabilize body position contract without conscious control. Together, all of the skeletal muscles of the body comprise the **muscular system**.

Only the heart contains **cardiac muscle**, which forms most of the heart wall (FIGURE 11.1b). Like skeletal muscle, cardiac muscle is *striated*.

FIGURE 11.1 Types of muscle.

There are three types of muscle—skeletal, cardiac, and smooth.

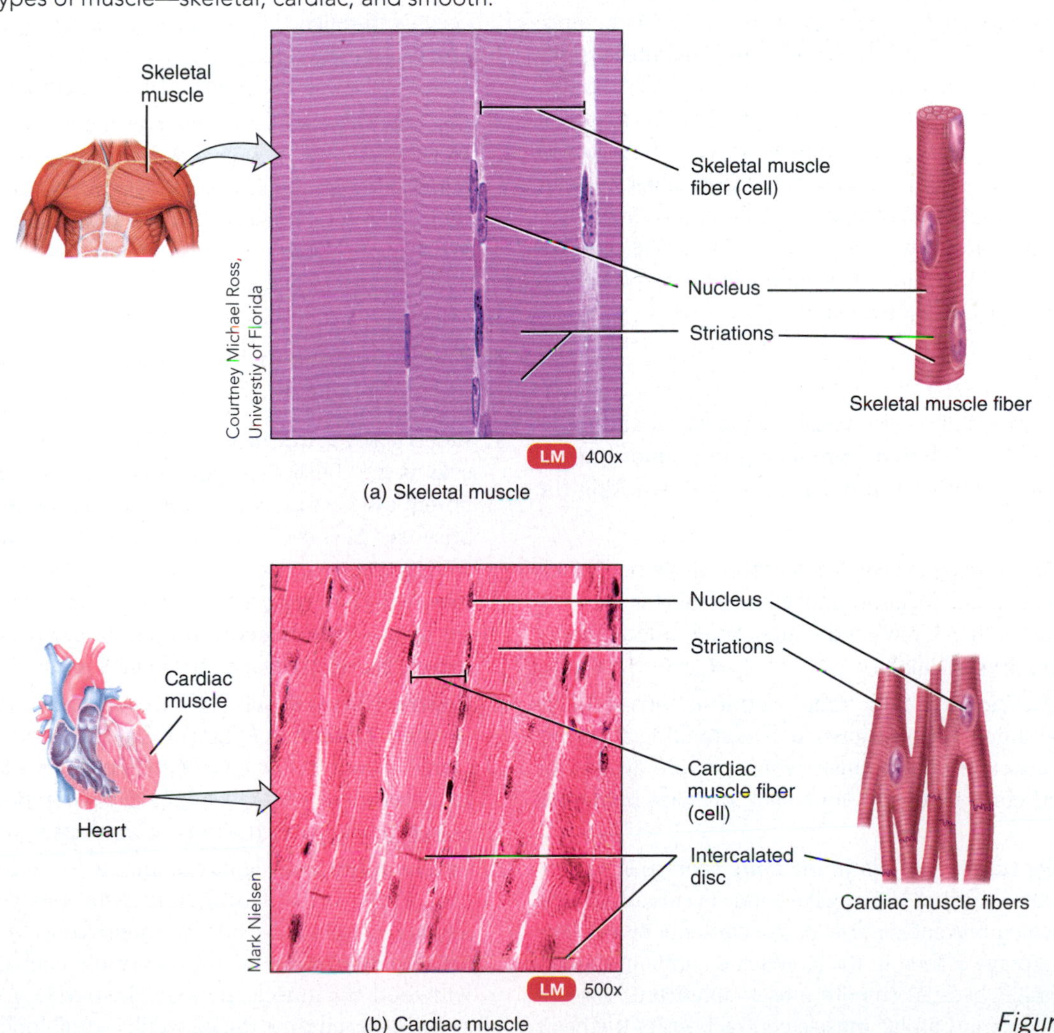

Figure 11.1 (continues)

Figure 11.1 Continued

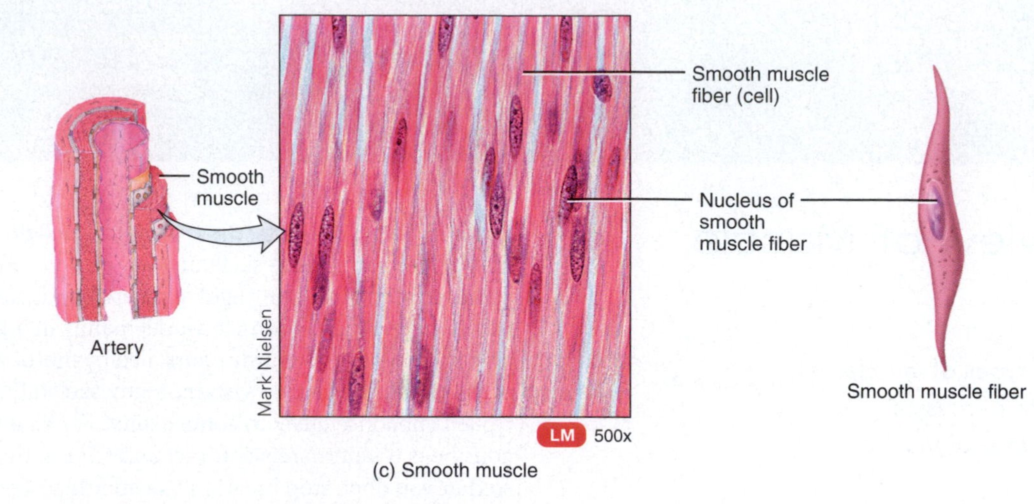

(c) Smooth muscle

Which type of muscle is striated and involuntarily controlled?

However, unlike skeletal muscle, cardiac muscle is *involuntary:* Its contractions are not under conscious control. Instead, the heart beats because it has a pacemaker that initiates each contraction. This built-in rhythm is termed *autorhythmicity*. Several neurotransmitters and hormones can adjust heart rate by speeding up or slowing down the pacemaker.

Smooth muscle is located in the walls of hollow internal structures, such as blood vessels, the airways, stomach, intestines, and uterus (FIGURE 11.1c). Under the light microscope, smooth muscle lacks the striations that are present in skeletal and cardiac muscle. For this reason, it looks *nonstriated*, which is why it is referred to as *smooth*. The action of smooth muscle is usually *involuntary*, and some smooth muscle tissue, such as the muscles that propel food through your gastrointestinal tract, has autorhythmicity. Both cardiac muscle and smooth muscle are regulated by motor neurons that are part of the autonomic nervous system and by hormones released by endocrine glands.

Muscle Performs a Variety of Functions

Through sustained contraction or alternating contraction and relaxation, muscle has four key functions: producing body movements, stabilizing body positions, storing and moving substances within the body, and generating heat.

1. ***Producing body movements.*** Movements of the whole body such as walking and running, and localized movements such as grasping a pencil or nodding the head, rely on the integrated functioning of skeletal muscles, bones, and joints.
2. ***Stabilizing body positions.*** Skeletal muscle contractions stabilize joints and help maintain body positions, such as standing or sitting. Postural muscles contract continuously when you are awake; for example, sustained contractions of your neck muscles hold your head upright.
3. ***Storing and moving substances within the body.*** Storage is accomplished by sustained contractions of ringlike bands of smooth muscle called *sphincters*, which prevent outflow of the contents of a hollow organ. Temporary storage of food in the stomach or urine in the urinary bladder is possible because smooth muscle sphincters close off the outlets of these organs. Cardiac muscle contractions of the heart pump blood through the blood vessels of the body. Contraction and relaxation of smooth muscle in the walls of blood vessels help adjust blood vessel diameter and thus regulate the rate of blood flow. Smooth muscle contractions also move food and substances such as bile and enzymes through the gastrointestinal tract, push gametes (sperm and oocytes) through the passageways of the reproductive systems, and propel urine through the urinary system. Skeletal muscle contractions promote the flow of lymph and aid the return of blood to the heart.
4. ***Generating heat.*** As muscle contracts, it produces heat, a process known as **thermogenesis**. Much of the heat generated by muscle is used to maintain normal body temperature. Involuntary contractions of skeletal muscles, known as *shivering*, can increase the rate of heat production.

Muscle Has Several Important Properties

Muscle has four special properties that enable it to function and contribute to homeostasis:

1. **Electrical excitability**, a property of both neurons and muscle cells, is the ability to respond to certain stimuli by producing action potentials. Chapter 7 provided details about how action potentials arise (see Section 7.3). For muscle cells, two main types of stimuli trigger action potentials. One is chemical stimuli, such as neurotransmitters released by neurons and hormones distributed by the blood. The other is autorhythmic electrical signals arising in the muscle tissue itself, as in the heart's pacemaker.
2. **Contractility** is the ability of muscle to contract forcefully when adequately stimulated. When a muscle contracts, it generates **tension** (force of contraction) while pulling on its attachment points. If the tension generated is great enough to overcome the resistance of the object to be moved, the muscle shortens and movement occurs.
3. **Extensibility** is the ability of muscle to stretch without being damaged. Extensibility allows a muscle to contract forcefully even if it is already stretched. Normally, smooth muscle is subject to the greatest amount of stretching. For example, each time your stomach fills with food, the muscle in its wall is stretched. Cardiac muscle is also stretched each time the heart fills with blood.

4. **Elasticity** is the ability of muscle to return to its original length and shape after contraction or extension.

CHECKPOINT

1. What features distinguish the three types of muscle?
2. What are the general functions of muscle?
3. Which property of muscle allows tension to be generated?

11.2 Organization of Skeletal Muscle

OBJECTIVES

- Describe the components of a skeletal muscle fiber.
- Explain the functions of the different types of muscle proteins.

Each of your skeletal muscles is composed of hundreds to thousands of muscle cells, called **muscle fibers**, arranged parallel to one another (FIGURE 11.2). An individual muscle fiber is surrounded by a sheath of connective tissue. In addition, connective tissue surrounds groups of 10 to 100 or more muscle fibers, separating them into bundles called **fascicles** (FAS-i-kuls = little bundles) (FIGURE 11.2). Connective tissue also surrounds the entire muscle itself. All of the connective tissues of a muscle are continuous with its tendons. A **tendon** is a cord of connective tissue that attaches the muscle to a bone (FIGURE 11.2).

The Components of a Skeletal Muscle Fiber Allow Contraction to Occur

A mature skeletal muscle fiber is a long, cylindrical structure with a diameter that ranges from 10 to 100 μm and a length that is typically about 10 cm (4 in.), although some are as long as 30 cm (12 in.). Because each skeletal muscle fiber arises during embryonic development from the fusion of many small, undifferentiated cells called

FIGURE 11.2 Organization of skeletal muscle and associated connective tissues.

A skeletal muscle consists of individual muscle fibers (cells) bundled into fascicles.

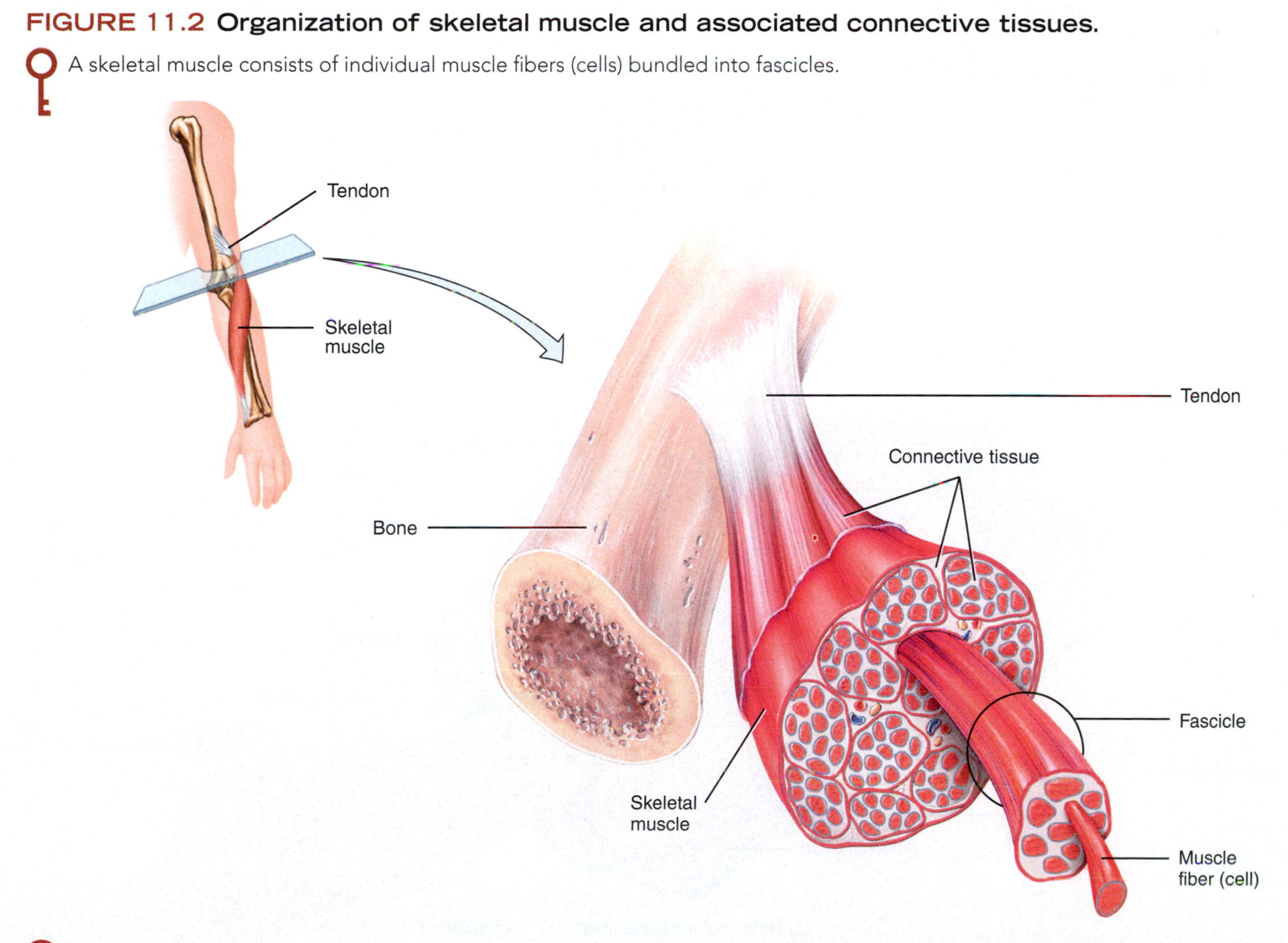

What is a tendon?

myoblasts (**FIGURE 11.3a**), a mature skeletal muscle fiber has multiple nuclei. Once fusion has occurred, the muscle fiber loses its ability to undergo cell division. Thus, the number of skeletal muscle fibers is set before you are born, and most of these cells will last you a lifetime. A few myoblasts do persist in mature skeletal muscle as *satellite cells* (**FIGURE 11.3a, b**). These cells retain the capacity to fuse with one another or with damaged muscle fibers to regenerate functional muscle fibers. However, the number of new skeletal muscle fibers formed is not enough to compensate for significant skeletal muscle damage or degeneration.

FIGURE 11.3 Microscopic organization of skeletal muscle. (a) During embryonic development, many myoblasts fuse to form one skeletal muscle fiber. Once fusion has occurred, a skeletal muscle fiber loses the ability to undergo cell division, but satellite cells retain this ability. (b–d) The sarcolemma of the fiber encloses sarcoplasm and myofibrils, which are striated. Sarcoplasmic reticulum wraps around each myofibril. Thousands of transverse tubules, filled with extracellular fluid, invaginate from the sarcolemma toward the center of the muscle fiber. A triad is a transverse tubule and the two terminal cisternae of the sarcoplasmic reticulum on either side of it.

The contractile elements of muscle fibers, the myofibrils, contain overlapping thick and thin filaments.

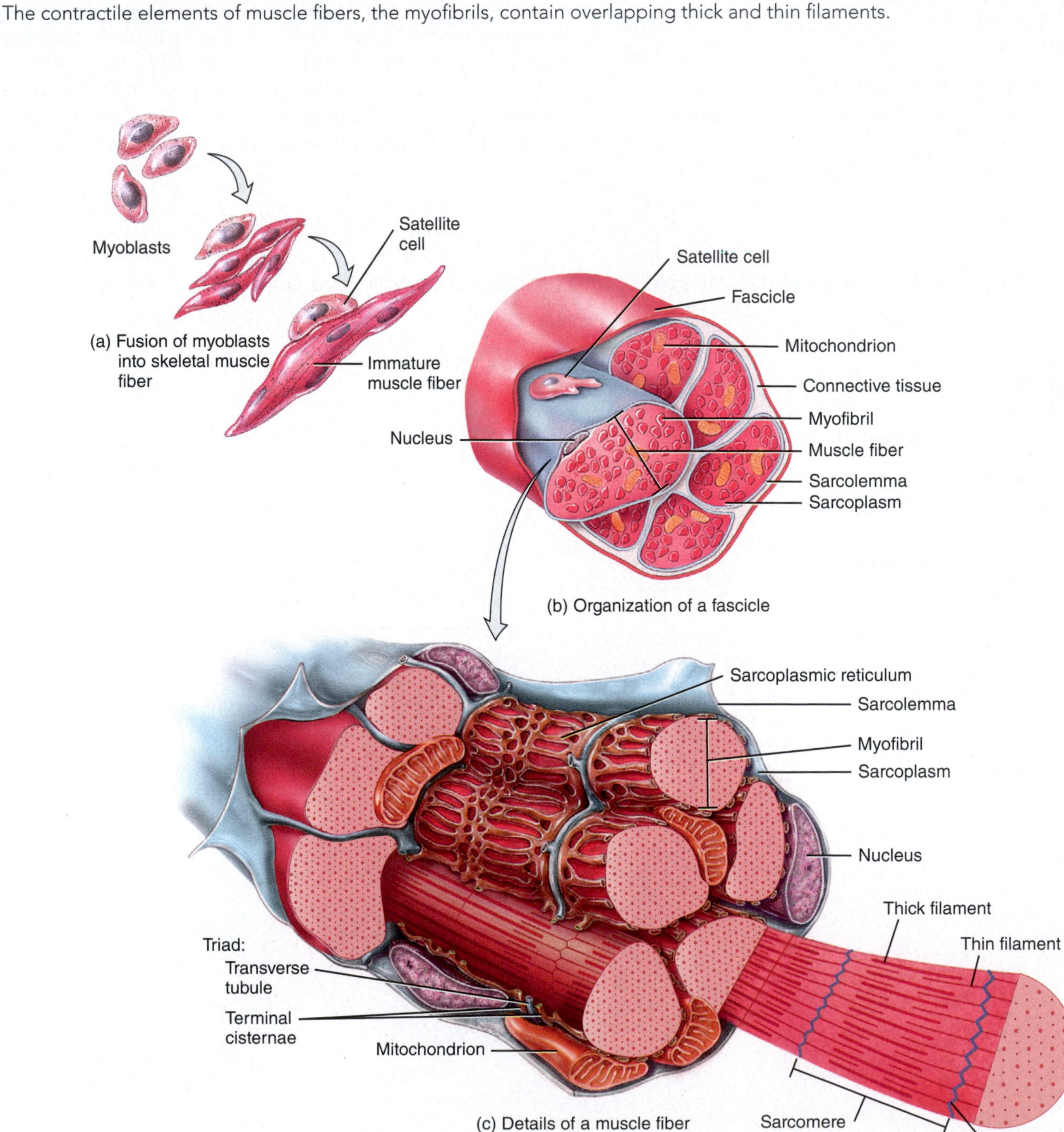

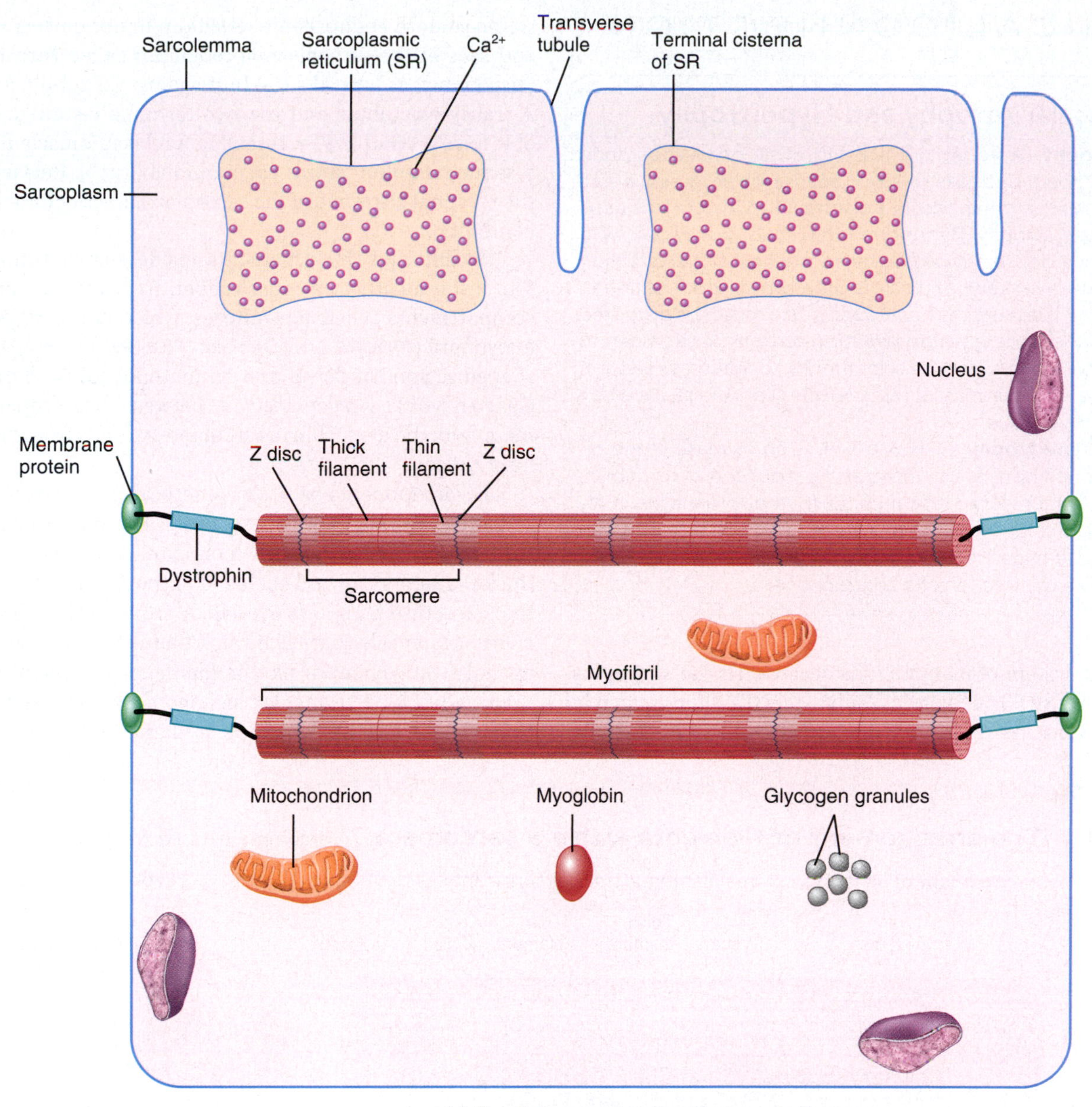

(d) Simplistic representation of a muscle fiber

Which structure shown here releases calcium ions to trigger muscle contraction?

The multiple nuclei of a skeletal muscle fiber are located just beneath the **sarcolemma**, the plasma membrane of a muscle fiber (FIGURE 11.3b–d). Thousands of tiny invaginations of the sarcolemma, called **transverse (T) tubules**, tunnel in from the surface toward the center of each muscle fiber. Because T tubules are open to the outside of the fiber, they are filled with extracellular fluid. Muscle action potentials travel along the sarcolemma and through the T tubules, quickly spreading throughout the muscle fiber. This arrangement ensures that an action potential excites all parts of the muscle fiber at essentially the same instant.

The sarcolemma surrounds the **sarcoplasm**, the cytoplasm of a muscle fiber (FIGURE 11.3b–d). Within the sarcoplasm are mitochondria, which produce large amounts of ATP for muscle contraction. The sarcoplasm also contains **glycogen**, a large polysaccharide consisting of thousands of glucose molecules covalently linked together. Glycogen serves as a storage form of glucose. It can be broken down into individual glucose molecules that can be used to synthesize ATP. Also present in the sarcoplasm are molecules of **myoglobin** (mī-ō-GLŌB-in), a red-colored, oxygen-binding protein that is found only in muscle. Myoglobin stores oxygen until it is needed by mitochondria to generate ATP.

Extending throughout the sarcoplasm are **myofibrils** (mī-ō-FĪ-brils), the contractile elements of the skeletal muscle fiber (FIGURE 11.3c, d). Within myofibrils are smaller structures called **filaments**, which can have either a thin or thick diameter. **Thin filaments** are 8 nm in diameter and 1–2 μm long, while **thick filaments** are 16 nm in diameter and 1–2 μm long. Both thin and thick filaments are directly involved in the contraction process.

CLINICAL CONNECTION

Muscular Atrophy and Hypertrophy

Muscular atrophy (A-trō-fē) is a wasting away of muscles. Individual muscle fibers decrease in size because of progressive loss of myofibrils. Atrophy that occurs because muscles are not used is termed ***disuse atrophy***. Bedridden individuals and people with casts experience disuse atrophy because the flow of nerve action potentials to inactive skeletal muscle is greatly reduced. The condition is reversible. If instead the nerve supply to a muscle is disrupted or cut, the muscle undergoes ***denervation atrophy***. Over a period of 6 months to 2 years, the muscle shrinks to about one-fourth its original size, and the muscle fibers are irreversibly replaced by fibrous connective tissue.

Muscular hypertrophy (hī-PER-trō-fē) is an increase in the diameter of muscle fibers due to increased production of myofibrils, mitochondria, sarcoplasmic reticulum, and other organelles. It results from very forceful, repetitive muscular activity, such as strength training. Because hypertrophied muscles contain more myofibrils, they are capable of more forceful contractions.

A fluid-filled system of membranous sacs called the **sarcoplasmic reticulum (SR)** (sar′-kō-PLAZ-mik re-TIK-ū-lum) encircles each myofibril (FIGURE 11.3c, d). This elaborate system is similar to the smooth endoplasmic reticulum in nonmuscular cells. Dilated end sacs of the sarcoplasmic reticulum called **terminal cisternae** (also known as **lateral sacs**) butt against a T tubule from both sides. A transverse tubule and the two terminal cisternae on either side of it form a **triad** (*tri-* = three). In a relaxed muscle fiber, the sarcoplasmic reticulum stores calcium ions (Ca^{2+}). Release of Ca^{2+} from the terminal cisternae of the sarcoplasmic reticulum triggers muscle contraction.

The thin and thick filaments inside a myofibril do not extend the entire length of a muscle fiber. Instead, they are arranged in compartments called **sarcomeres**, which are the repeating units of a myofibril (FIGURE 11.4; see also FIGURE 11.3c, d). Narrow, plate-shaped regions of dense protein material called **Z discs** (*Z* for the German word *zwischenscheibe* = between discs) separate one sarcomere from the next. Thus, a sarcomere extends from one Z disc to the next Z disc.

The components of a sarcomere are organized into a variety of bands and zones (FIGURE 11.4). The dark middle part of the sarcomere is the **A band**, which extends the entire length of the thick filaments. Toward each end of the A band is a zone where the thick and thin filaments overlap. A cross section through this *zone of overlap* reveals that each thick filament is surrounded by a hexagonal arrangement of six thin filaments, and each thin filament is surrounded by a triangular arrangement of three thick filaments (FIGURE 11.4). The A band is so-named because it is *anisotropic*

FIGURE 11.4 The arrangement of filaments within a sarcomere. A sarcomere extends from one Z disc to the next.

Because of the arrangement of its thick and thin filaments, a sarcomere contains an A band, I bands, an H zone, and an M line.

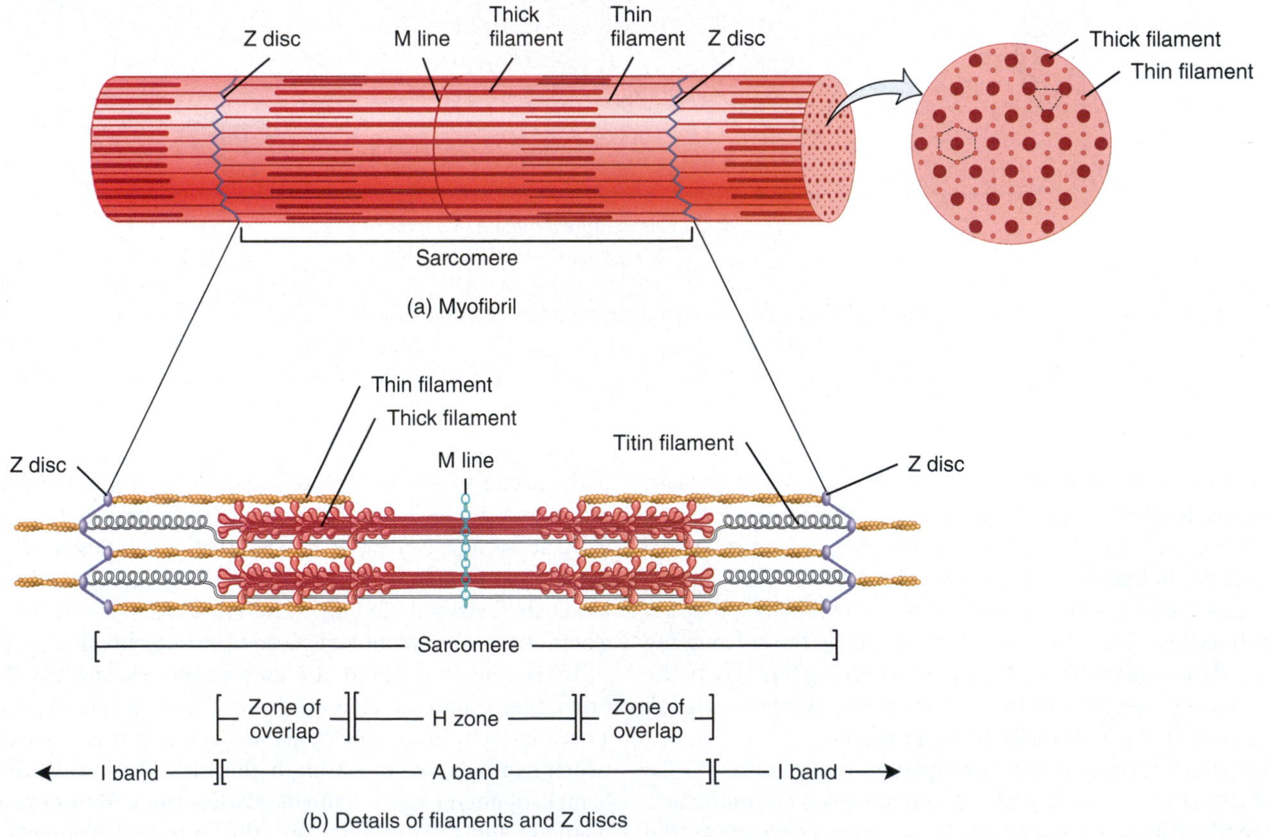

Which components of a sarcomere are responsible for the striations in skeletal and cardiac muscle fibers?

CLINICAL CONNECTION

Exercise-Induced Muscle Damage

Comparison of electron micrographs of muscle tissue taken from athletes before and after intense exercise reveals considerable **exercise-induced muscle damage**, including torn sarcolemmas in some muscle fibers, damaged myofibrils, and disrupted Z discs. Microscopic muscle damage after exercise is also indicated by increases in blood levels of proteins, such as myoglobin and the enzyme creatine kinase, that are normally confined within muscle fibers. From 12 to 48 hours after a period of strenuous exercise, skeletal muscles often become sore. Such **delayed onset muscle soreness (DOMS)** is accompanied by stiffness, tenderness, and swelling. Although the causes of DOMS are not completely understood, microscopic muscle damage and the inflammation that accompanies this damage appear to be major factors. In response to exercise-induced muscle damage, muscle fibers undergo repair: New regions of sarcolemma are formed to replace torn sarcolemmas, and more muscle proteins (including those of the myofibrils) are synthesized in the sarcoplasm of the muscle fibers.

to polarized light, meaning that it refracts polarized light unevenly in different directions. On either side of the A band is a light area called the **I band**, which contains the rest of the thin filaments but no thick filaments. A Z disc passes through the center of each I band. The name of the I band is derived from the fact that it is *isotropic* to polarized light, which means that it refracts polarized light uniformly in all directions. The alternating dark A bands and light I bands create the striations that can be seen in both myofibrils and in whole skeletal and cardiac muscle fibers. A narrow **H zone** (*H* for the German word *helles* = clear) in the center of each A band contains thick filaments but no thin filaments. Supporting proteins that hold the thick filaments together at the center of the H zone form the **M line** (*M* for the German word *mittel* = middle), named for its position in the middle of the sarcomere.

TABLE 11.1 summarizes the components of the sarcomere.

TABLE 11.1 Components of the Sarcomere

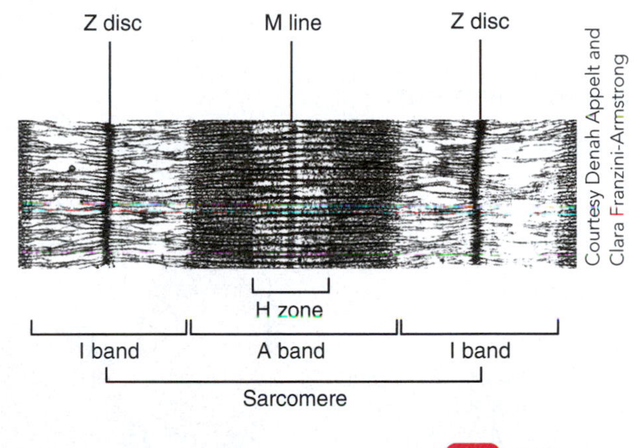

TEM 21,600x

Component	Description
Z discs	Narrow, plate-shaped regions of dense material that separate one sarcomere from the next.
A band	The dark, middle part of the sarcomere that extends the entire length of the thick filaments and also includes those parts of the thin filaments that overlap with the thick filaments.
I band	The lighter, less dense area of the sarcomere that contains the rest of the thin filaments but no thick filaments. A Z disc passes through the center of each I band.
H zone	A narrow region in the center of each A band that contains thick filaments but no thin filaments.
M line	A region in the center of the H zone that contains proteins that hold the thick filaments together at the center of the sarcomere.

There Are Three Types of Muscle Proteins

Myofibrils are built from three kinds of proteins: (1) contractile proteins, which generate force during contraction; (2) regulatory proteins, which help switch the contraction process on and off; and (3) structural proteins, which keep the thick and thin filaments in the proper alignment, give the myofibril extensibility and elasticity, and link the myofibrils to the sarcolemma and extracellular matrix.

The two *contractile proteins* in muscle are myosin and actin, which are the main components of thick and thin filaments, respectively. **Myosin** functions as a *motor protein* in all three types of muscle tissue. Motor proteins push or pull various cellular structures to achieve movement by converting the chemical energy in ATP to the mechanical energy of motion or the production of force. A myosin molecule consists of six polypeptide chains: two large **heavy chains** and four small **light chains**. These polypeptide chains are arranged in such a way that the myosin molecule has a rodlike tail and two globular heads (**FIGURE 11.5a**). The *myosin tail* is composed of portions of the heavy chains that wrap around each other to form a double-stranded helix. The *myosin heads* contain the light chains and the remaining portions of the heavy chains. Each myosin head has two binding sites (**FIGURE 11.5a**): (1) an *actin-binding site* and (2) an *ATP-binding site*. The ATP-binding site also functions as an *ATPase*—an enzyme that hydrolyzes ATP to generate energy for muscle contraction. A flexible *hinge* is located where the myosin heads join the myosin tail (**FIGURE 11.5a**). The hinge region allows the myosin head to pivot during the contraction process. Overall, a myosin molecule is shaped like two golf clubs twisted together: the myosin tail resembles the twisted golf club handles and the myosin heads look like the golf club heads.

About 300 molecules of myosin form a single thick filament (**FIGURE 11.5b**). The myosin tail of each myosin molecule points toward the M line in the center of the sarcomere. Tails of neighboring myosin molecules lie parallel to one another, forming the shaft of the thick filament. The myosin heads project outward from the shaft in a spiraling fashion, each extending toward one of the six thin filaments that surround each thick filament.

The main component of the thin filament is the protein **actin** (**FIGURE 11.6**). Individual actin molecules are known as **G actin** because they are *globular* proteins. On each G actin molecule is a *myosin-binding site*, where a myosin head can attach. G actin molecules are linked together to form a long polymer called **F actin**, which is so-named because it has a *fibrous* structure. A single thin filament contains two F actin strands arranged in a double helix like two strands of pearls that are twisted together.

Smaller amounts of two *regulatory proteins*—tropomyosin and troponin—are also part of the thin filament (**FIGURE 11.6**). **Tropomyosin**

FIGURE 11.5 Components of a thick filament.

A thick filament contains about 300 myosin molecules.

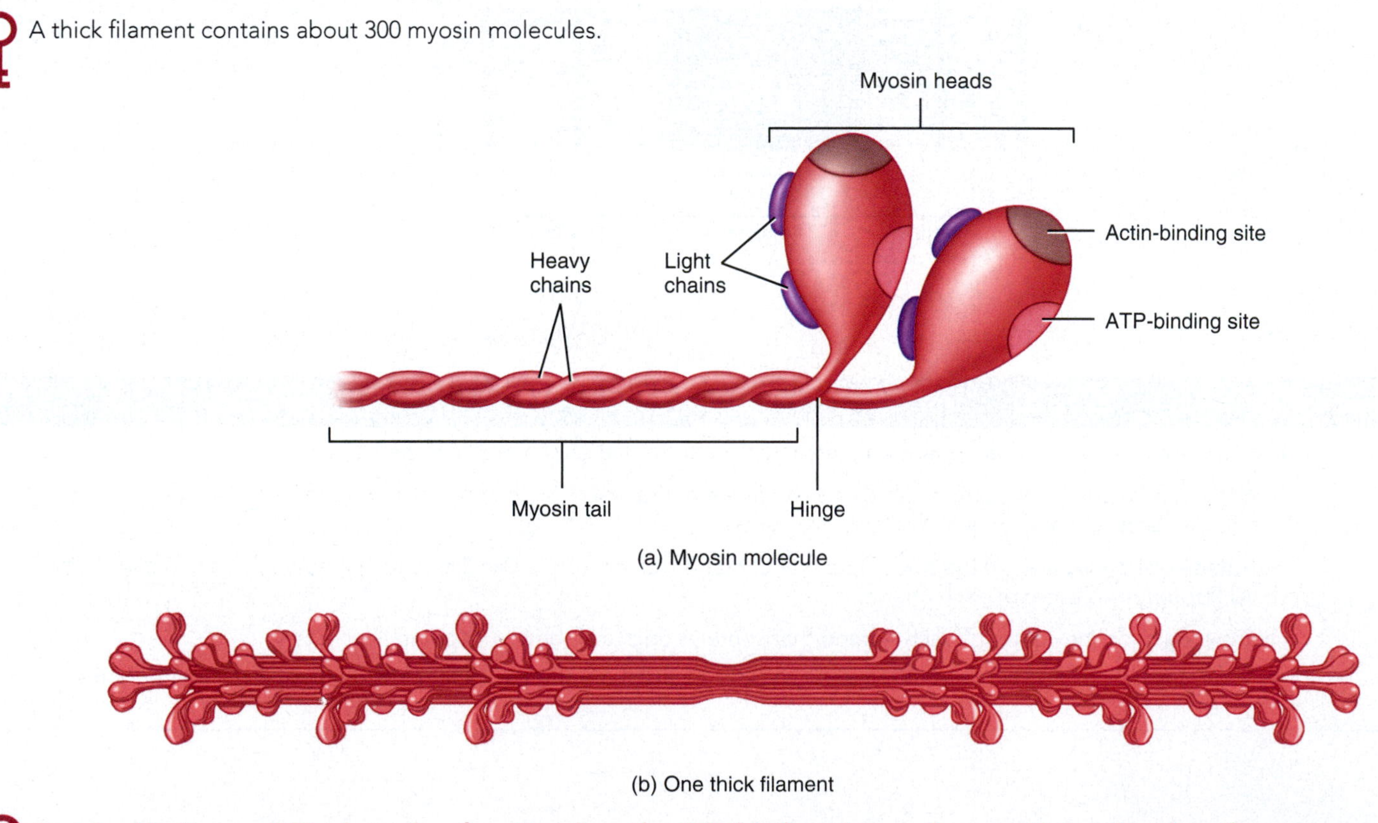

Besides binding to ATP, what other function does the ATP-binding site on the myosin head perform?

FIGURE 11.6 Components of a thin filament.

A thin filament contains actin, troponin, and tropomyosin.

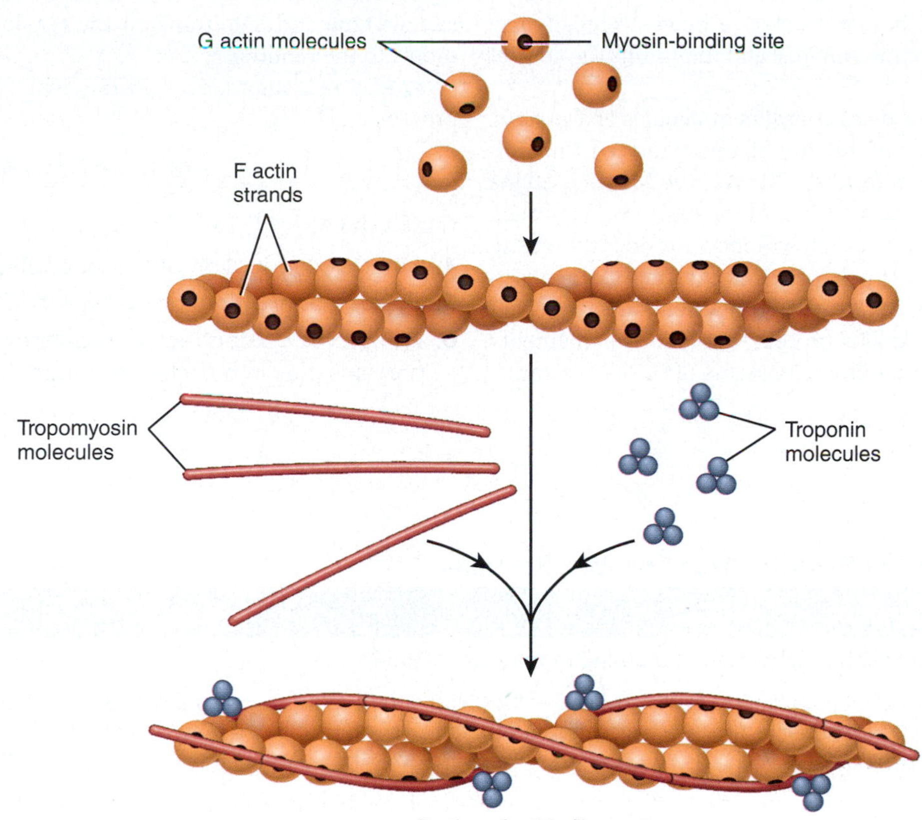

Which component of the thin filament prevents myosin from binding to actin when a muscle fiber is relaxed?

is a rod-shaped protein that joins with other tropomyosin molecules to form two long strands that wrap around the F actin double helix. Each of the tropomyosin molecules within a given strand extends along approximately seven G actin molecules. In a relaxed muscle, myosin is blocked from binding to actin because the strands of tropomyosin cover the myosin-binding sites on actin. The tropomyosin strands in turn are held in place by molecules of troponin. **Troponin** is a protein that consists of three globular subunits—one that binds to tropomyosin, one that binds to actin, and one that has binding sites for calcium ions (Ca^{2+}). When Ca^{2+} binds to troponin, troponin undergoes a change in shape; this conformational change moves tropomyosin away from the myosin-binding sites on actin, and muscle contraction subsequently begins as myosin attaches to actin.

In addition to contractile and regulatory proteins, muscle contains about a dozen *structural proteins*, which contribute to the alignment, stability, extensibility, and elasticity of myofibrils. Several key structural proteins are titin, α-actinin, myomesin, nebulin, and dystrophin. *Titin* (*titan* = gigantic) is the third most plentiful protein in skeletal muscle (after actin and myosin). This molecule's name reflects its huge size. With a molecular weight of about 3 million daltons, titin is 50 times larger than an average-sized protein. Each titin molecule spans half a sarcomere, from a Z disc to an M line (see **FIGURE 11.4b**), a distance of 1 to 1.2 μm in relaxed muscle. Each titin molecule connects a Z disc to the M line of the sarcomere, thereby helping stabilize the position of the thick filament. The part of the titin molecule that extends from the Z disc is very elastic. Because

CLINICAL CONNECTION

Muscular Dystrophy

The term **muscular dystrophy** (DIS-trō-fē) refers to a group of inherited muscle-destroying diseases that cause progressive degeneration of skeletal muscle fibers. The most common form of muscular dystrophy is ***Duchenne muscular dystrophy*** (doo-SHĀN) or ***DMD***. In DMD, the gene that codes for the protein dystrophin is mutated, so little or no dystrophin is present in the muscle fiber. Without the reinforcing effect of dystrophin, the sarcolemma tears easily during muscle contraction, causing muscle fibers to rupture and die. Because the mutated dystrophin gene is on the X chromosome, and males have only one, DMD is a sex-linked disorder that strikes boys almost exclusively. DMD usually becomes apparent between the ages of 2 and 5, when parents notice that the child falls often and has difficulty running, jumping, and hopping. By age 12 most boys with DMD are unable to walk. Respiratory or cardiac failure usually causes death by age 20. The dystrophin gene was discovered in 1987, and by 1990 the first attempts were made to treat DMD patients with gene therapy. The muscles of three boys with DMD were injected with myoblasts bearing functional dystrophin genes, but only a few muscle fibers gained the ability to produce dystrophin. Similar clinical trials with additional patients have also failed. An alternative approach to the problem is to find a way to induce muscle fibers to produce the protein ***utrophin***, which is similar to dystrophin. Experiments with dystrophin-deficient mice suggest that this approach may work.

it can stretch to at least four times its resting length and then spring back unharmed, titin accounts for much of the extensibility and elasticity of myofibrils. Titin probably helps the sarcomere return to its resting length after a muscle has contracted or been stretched, may help prevent overextension of sarcomeres, and maintains the central location of the A bands.

The dense material of the Z discs contains molecules of *α-actinin*, which bind to actin molecules of the thin filament and to titin. Molecules of the protein *myomesin* form the M line. The M line proteins bind to titin and connect adjacent thick filaments to one another. *Nebulin* is a long, nonelastic protein wrapped around the entire length of each thin filament. It helps anchor the thin filaments to the Z discs and regulates the length of thin filaments during development. *Dystrophin* is a cytoskeletal protein that links thin filaments of the sarcomere to integral membrane proteins of the sarcolemma, which are attached in turn to proteins in the connective tissue extracellular matrix that surrounds muscle fibers (see **FIGURE 11.3d**). Dystrophin and its associated proteins are thought to reinforce the sarcolemma and help transmit the tension generated by the sarcomeres to the tendons.

TABLE 11.2 summarizes the different types of skeletal muscle fiber proteins.

CHECKPOINT

4. What is the function of the sarcoplasmic reticulum?
5. Define the following terms: A band, I band, H zone, and M line.
6. What roles do contractile, regulatory, and structural proteins play in muscle contraction and relaxation?

TABLE 11.2 Summary of Skeletal Muscle Fiber Proteins

Type of Protein	Description
Contractile proteins	Proteins that generate force during muscle contractions.
Myosin	A contractile protein that makes up the thick filament. A myosin molecule consists of a tail and two myosin heads, which bind to myosin-binding sites on actin molecules of a thin filament during muscle contraction.
Actin	A contractile protein that is the main component of the thin filament. Each actin molecule has a myosin-binding site to which a myosin head of a thick filament binds during muscle contraction.
Regulatory proteins	Proteins that help switch the muscle contraction process on and off.
Tropomyosin	A regulatory protein that is a component of the thin filament. In a relaxed skeletal muscle fiber, tropomyosin covers the myosin-binding sites on actin molecules, preventing myosin from binding to actin.
Troponin	A regulatory protein that is a component of the thin filament. When calcium ions (Ca^{2+}) bind to troponin, troponin undergoes a conformational change that moves tropomyosin away from myosin-binding sites on actin molecules, and muscle contraction subsequently begins as myosin binds to actin.
Structural proteins	Proteins that keep the thick and thin filaments of the myofibrils in proper alignment, give the myofibrils elasticity and extensibility, and link the myofibrils to the sarcolemma and extracellular matrix.
Titin	A structural protein that connects a Z disc to the M line of the sarcomere, helping to stabilize the position of the thick filament. Because it can stretch and then spring back unharmed, titin accounts for much of the elasticity and extensibility of myofibrils.
α-actinin	A structural protein of the Z discs that attaches to actin molecules of thin filaments and to titin molecules.
Myomesin	A structural protein that forms the M line of the sarcomere; it binds to titin molecules and connects adjacent thick filaments to one another.
Nebulin	A structural protein that wraps around the entire length of each thin filament; it helps anchor the thin filaments to the Z discs and regulates the length of the thin filaments during development.
Dystrophin	A structural protein that links the thin filaments of the sarcomere to integral membrane proteins in the sarcolemma, which are attached in turn to proteins in the connective tissue matrix that surrounds muscle fibers. It is thought that dystrophin helps reinforce the sarcolemma and helps transmit tension generated by sarcomeres to tendons.

11.3 Contraction and Relaxation of Skeletal Muscle Fibers

OBJECTIVES

- Explain the significance of the sliding filament mechanism.
- Outline the steps of the contraction cycle.
- Summarize the events by which a somatic motor neuron causes a skeletal muscle fiber to generate an action potential.
- Describe excitation–contraction coupling in a skeletal muscle fiber.
- Discuss how a skeletal muscle fiber relaxes after a period of contraction.

Muscle Contraction Occurs by the Sliding Filament Mechanism

When scientists examined the first electron micrographs of skeletal muscle in the mid-1950s, they were surprised to see that the lengths of the thick and thin filaments were the same in both relaxed and contracted muscle. It had been thought that muscle contraction must be a folding process, somewhat like closing an accordion. Instead, researchers discovered that skeletal muscle shortens during contraction because the thick and thin filaments slide past one another. The model describing this process is known as the **sliding filament mechanism** of muscle contraction.

Muscle contraction occurs because myosin heads attach to and "walk" along the thin filaments at both ends of a sarcomere, progressively pulling the thin filaments toward the M line (FIGURE 11.7). As a result, the thin filaments slide inward and meet at the center of a sarcomere. They may even move so far inward that their ends overlap (FIGURE 11.7c). As the thin filaments slide inward, the I band and H zone narrow and eventually disappear altogether when the muscle is maximally contracted. However, the width of the A band and the individual lengths of the thick and thin filaments remain unchanged. Since the thin filaments on each side of the sarcomere are attached to Z discs, when the thin filaments slide inward, the Z discs come closer together and the sarcomere shortens. Shortening of the sarcomeres causes shortening of the whole muscle fiber, which in turn leads to shortening of the entire muscle.

The Contraction Cycle Involves Four Major Steps

At the onset of contraction, the sarcoplasmic reticulum releases calcium ions (Ca^{2+}) into the sarcoplasm. There, they bind to troponin. Troponin then moves tropomyosin away from the myosin-binding sites on actin. Once the binding sites are "free," the **contraction cycle**—the

FIGURE 11.7 Sliding filament mechanism of muscle contraction, as it occurs in two adjacent sarcomeres.

During muscle contraction, thin filaments move toward the M line of each sarcomere.

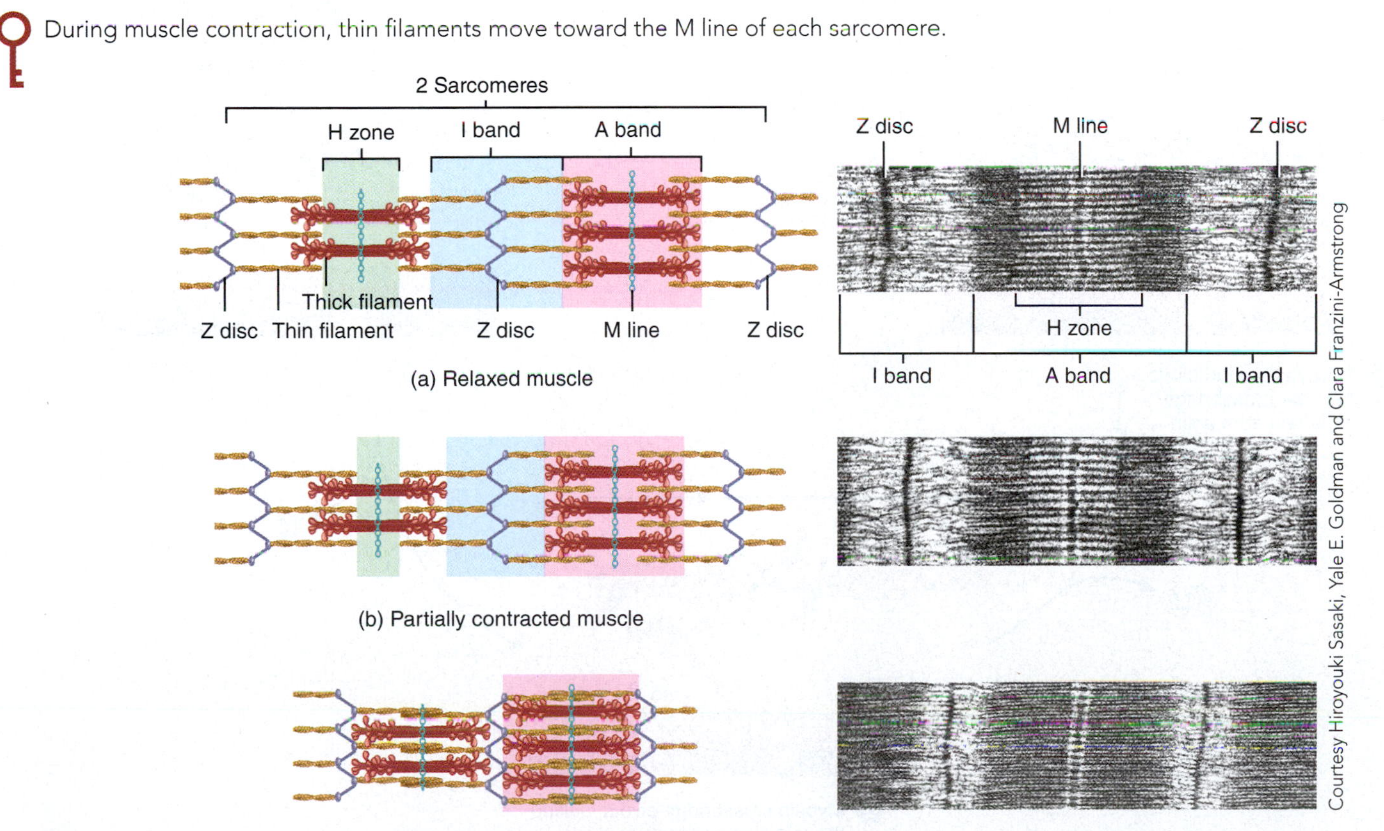

What happens to the I band and H zone as muscle contracts? Do the lengths of the thick and thin filaments change?

repeating sequence of events that causes the filaments to slide—begins. The contraction cycle consists of four steps (**FIGURE 11.8**):

1. ***ATP hydrolysis.*** As mentioned earlier, a myosin head includes an ATP-binding site that functions as an ATPase—an enzyme that hydrolyzes ATP into ADP (adenosine diphosphate) and a phosphate group. The energy generated from this hydrolysis reaction is stored in the myosin head for later use during the contraction cycle. The myosin head is said to be *energized* when it contains stored energy. The energized myosin head assumes a "cocked" position, like a stretched spring. In this position, the myosin head is perpendicular (at a 90° angle) relative to the thick and thin filaments and has the proper orientation to bind to an actin molecule. Notice that the products of ATP hydrolysis—ADP and a phosphate group—are still attached to the myosin head.
2. ***Attachment of myosin to actin.*** The energized myosin head attaches to the myosin-binding site on actin and releases the previously hydrolyzed phosphate group. When a myosin head attaches to actin during the contraction cycle, the myosin head is referred to as a **crossbridge**. Although a single myosin molecule has a double head, only one head binds to actin at a time.
3. ***Power stroke.*** After a crossbridge forms, the myosin head pivots, changing its position from a 90° angle to a 45° angle relative to the thick and thin filaments. As the myosin head changes to its new position, it pulls the thin filament past the thick filament toward the center of the sarcomere, generating tension (force) in the process. This event is known as the **power stroke**. The energy required for the power stroke is derived from the energy stored in the myosin

FIGURE 11.8 The contraction cycle. Sarcomeres exert force and shorten through repeated cycles, during which the myosin heads attach to actin (forming crossbridges), pivot, and detach.

During the power stroke of the contraction cycle, crossbridges pivot, pulling the thin filaments past the thick filaments toward the center of the sarcomere.

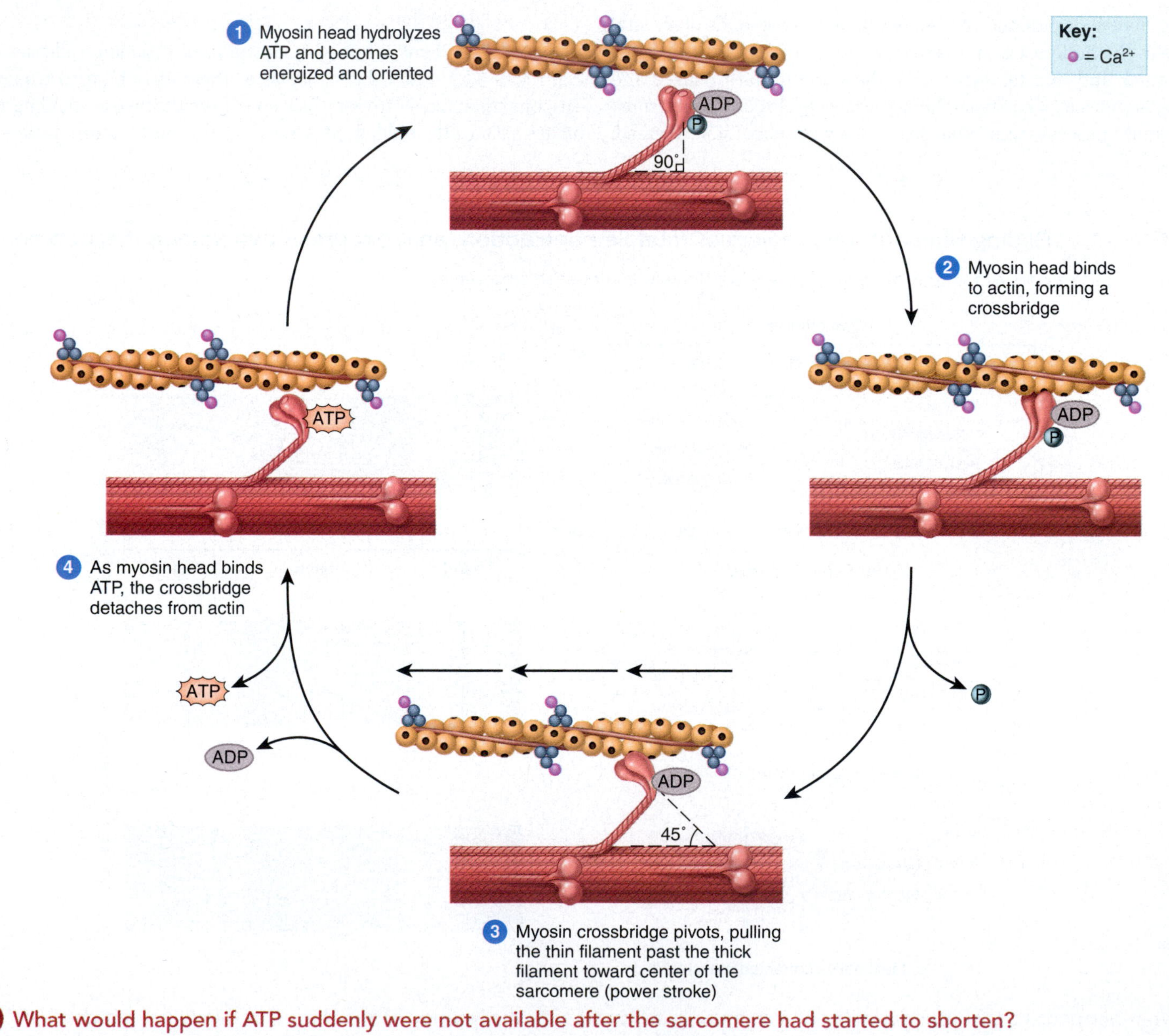

What would happen if ATP suddenly were not available after the sarcomere had started to shorten?

head from the hydrolysis of ATP (see step 1). Once the power stroke occurs, ADP is released from the myosin head.

4. ***Detachment of myosin from actin.*** At the end of the power stroke, the crossbridge remains firmly attached to actin until it binds another molecule of ATP. As ATP binds to the ATP-binding site on the myosin head, the myosin head detaches from actin.

Hydrolysis of the newly bound ATP molecule by the myosin ATPase causes the myosin head to pivot back to its cocked position (perpendicular relative to the thick and thin filaments), and a new contraction cycle begins. The contraction cycle repeats as long as ATP is available and the Ca^{2+} level near the thin filament is sufficiently high.

All of the myosin heads of a given thick filament do not proceed through the steps of the contraction cycle at the same time. At any one instant, some myosin heads are attaching to actin, others are engaging in power strokes and generating tension (force), and still others are detaching from actin and getting ready to bind to another actin molecule farther along the thin filament. With repeated contraction cycles, sequential power strokes from different myosin heads of a thick filament continuously pull the thin filament closer and closer toward the center of the sarcomere. The asynchronous activity of the myosin heads during repeated contraction cycles is responsible for the continuous inward movement of the thin filament: If all of the myosin heads attached to and then detached from actin at the same time, the thin filament would slide back to its resting position between power strokes. Because the myosin heads at opposite ends of a thick filament are arranged in opposite directions, as power strokes from each end pull a thin filament toward the center of the sarcomere, the Z discs are drawn toward each other, and the sarcomere shortens. During a maximal muscle contraction, the distance between two Z discs can decrease to half the resting length. The Z discs in turn pull on neighboring sarcomeres, and the whole muscle fiber shortens.

The Neuromuscular Junction Is the Synapse Between a Somatic Motor Neuron and a Skeletal Muscle Fiber

Before a skeletal muscle fiber can contract, it must be stimulated by a somatic motor neuron. The synapse between a somatic motor neuron and a skeletal muscle fiber is known as the **neuromuscular junction (NMJ)**. In Chapter 10 you learned about the organization and operation of the NMJ. Recall that the NMJ has three components: (1) the synaptic end bulbs of a terminal branch of the somatic motor neuron; (2) a synaptic cleft; and (3) the **motor end plate**, the region of muscle fiber membrane opposite the synaptic end bulbs (see FIGURE 10.10). The neurotransmitter released at the NMJ, acetylcholine (ACh), has an excitatory effect on neuromuscular transmission (see FIGURE 10.11): When an action potential occurs in a somatic motor neuron, it releases ACh molecules stored in the synaptic vesicles of the synaptic end bulbs. ACh then diffuses across the synaptic cleft and binds to nicotinic ACh receptors on the motor end plate, generating a depolarizing graded potential called an **end plate potential (EPP)**. A single EPP is typically large enough to depolarize adjacent regions of sarcolemma to threshold, resulting in the generation of a muscle action potential. Because the NMJ is usually near the midpoint of the muscle fiber, once the muscle action potential arises, it propagates through the muscle fiber membrane in both directions away from the NMJ toward the ends of the fiber. As the muscle action potential passes through the membrane (sarcolemma and T tubules), it triggers a chain of events that ultimately leads to contraction of the muscle fiber. Thus, a single action potential in a somatic motor neuron elicits a single action potential in a skeletal muscle fiber, which in turn causes the skeletal muscle fiber to contract. The events just described repeat as long as nerve action potentials continue to occur in the somatic motor neuron, causing the release of ACh. Once nerve action potentials in the somatic motor neuron cease, ACh is no longer released and the enzyme **acetylcholinesterase (AChE)**, located on the end plate membrane, breaks down any ACh that is present in the synaptic cleft. Without ACh, end plate potentials are not generated, and the production of muscle action potentials ends.

The Skeletal Muscle Action Potential Has Two Main Phases: A Depolarizing Phase and a Repolarizing Phase

The resting membrane potential of a skeletal muscle fiber is about −90 mV. As you have already learned, when a region of sarcolemma adjacent to the motor end plate is depolarized to threshold by an end plate potential, an action potential is generated in the skeletal muscle fiber. As with a nerve action potential, a skeletal muscle action potential consists of two main phases: a *depolarizing (rising) phase*, caused by the opening of voltage-gated Na^+ channels, and a *repolarizing (falling) phase*, caused by closure of the voltage-gated Na^+ channels and the opening of voltage-gated K^+ channels (FIGURE 11.9). The

FIGURE 11.9 The skeletal muscle action potential.

The skeletal muscle action potential consists of two main phases: a depolarizing phase and a repolarizing phase.

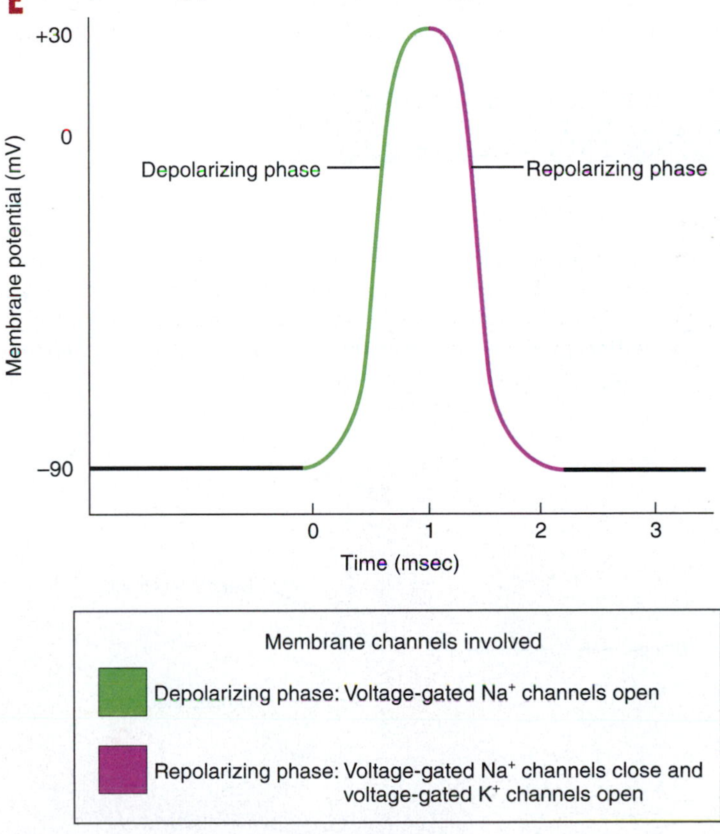

Are the voltage-gated channels involved in the skeletal muscle action potential the same as or different from those that are involved in the nerve action potential?

CLINICAL CONNECTION

Rigor Mortis

After death, cellular membranes become leaky. Calcium ions leak out of the sarcoplasmic reticulum into the sarcoplasm and allow myosin heads to bind to actin. ATP synthesis ceases shortly after breathing stops, however, so the myosin crossbridges cannot detach from actin. The resulting condition, in which muscles are in a state of rigidity (cannot contract or stretch), is called **rigor mortis** (rigidity of death). Rigor mortis begins 3–4 hours after death and lasts about 24 hours; then it disappears as proteolytic enzymes from lysosomes digest the myosin crossbridges.

duration of the skeletal muscle action potential is very brief, lasting only about 1–2 msec. Once an action potential is generated in a skeletal muscle fiber, it propagates along the sarcolemma and T tubules via continuous conduction, the same mechanism used for propagation of nerve action potentials along unmyelinated axons (see Section 7.3).

Excitation–Contraction Coupling in Skeletal Muscle Links the Muscle Action Potential to Muscle Contraction

An increase in Ca^{2+} concentration in the sarcoplasm starts muscle contraction, and a decrease stops it. When a muscle fiber is relaxed, the concentration of Ca^{2+} in its sarcoplasm is very low, only about 0.1 micromole per liter (0.1 μm/L). However, a huge amount of Ca^{2+} is stored inside the sarcoplasmic reticulum (SR) (FIGURE 11.10a). As a muscle action potential propagates along the sarcolemma and into the T tubules, it causes the release of Ca^{2+} from the SR into the sarcoplasm, and muscle contraction begins. The sequence of events that links the muscle action potential to muscle contraction is known as **excitation–contraction (EC) coupling**. EC coupling occurs at the triads of the skeletal muscle fiber. Recall that a *triad* consists of a transverse (T) tubule and two opposing terminal cisternae (lateral sacs) of the SR. At a given triad, the T tubule and terminal cisternae are mechanically linked together by two groups of integral membrane proteins: dihydropyridine receptors and Ca^{2+} release channels (FIGURE 11.10a). **Dihydropyridine (DHP) receptors**, which are so-named because the drug dihydropyridine binds to them, are located in the T-tubule membrane. They are **L-type voltage-gated Ca^{2+} channels** (L for *long-lasting* because they remain open for a relatively long period of time in response to activation) that are arranged in clusters of four known as *tetrads*. Although these channels allow extracellular Ca^{2+} to enter the sarcoplasm when activated, their main role in EC coupling is to serve as voltage sensors that trigger the opening of the Ca^{2+} release channels. **Ca^{2+} release channels** are present in the terminal cisternal membrane of the SR. They are also known as **ryanodine receptors** because the plant alkaloid ryanodine binds to them. Each Ca^{2+} release channel has four projections called *foot processes* or *junctional feet*. Each foot process comes in contact with one of the four DHP receptors in a tetrad. When the T tubule is at resting membrane potential, the part of the Ca^{2+} release channel that extends into the sarcoplasm is blocked by a given cluster of DHP receptors, which prevents Ca^{2+} from leaving the SR (FIGURE 11.10a). When an action potential travels along the T tubule, the DHP receptors detect the change in voltage and undergo a conformational change that ultimately causes the Ca^{2+} release channels to open (FIGURE 11.10b). Once these channels open,

TOOL OF THE TRADE

ELECTROMYOGRAPHY

Electromyography (EMG) (e-lek′-trō-mī-OG-ra-fē) is a test that measures the electrical activity (muscle action potentials) in resting and contracting muscles. Normally, resting muscle produces no electrical activity, a slight contraction produces some electrical activity, and a more forceful contraction produces increased electrical activity. In the procedure, a ground electrode is placed over the muscle to be tested to eliminate background electrical activity (FIGURE A). Then a fine needle electrode attached by wires to a recording instrument is inserted into the muscle. The electrical activity of the muscle is displayed as waves on an oscilloscope (FIGURE B) and heard through a loudspeaker.

EMG helps to determine if muscle weakness or paralysis is due to a malfunction of the muscle itself or the nerves supplying the muscle. EMG is also used to diagnose certain muscle disorders, such as muscular dystrophy.

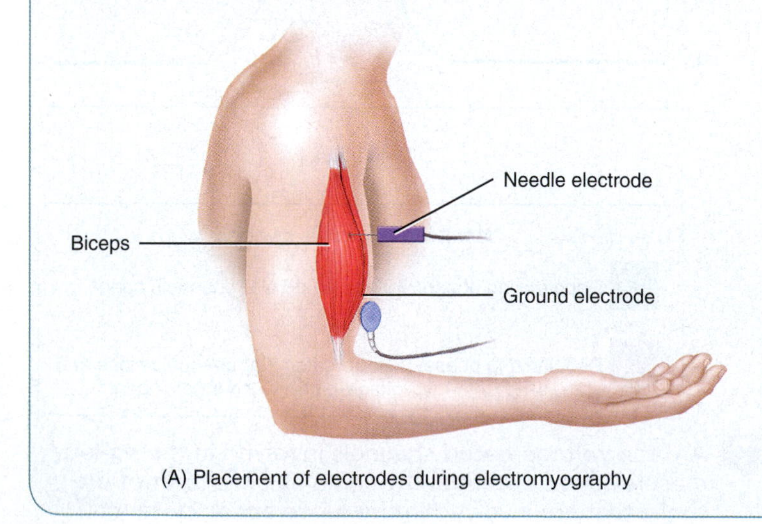

(A) Placement of electrodes during electromyography

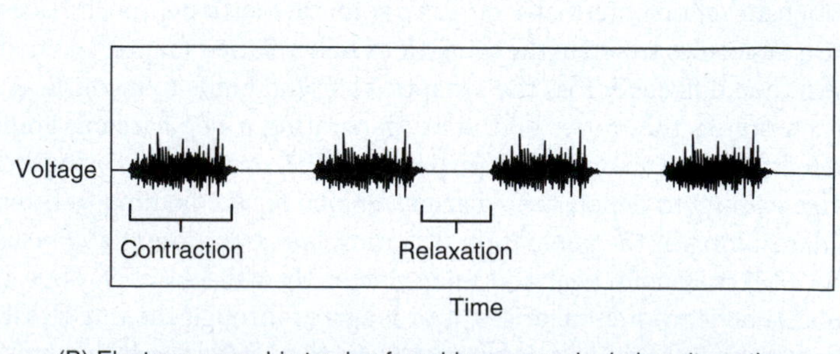

(B) Electromyographic tracing from biceps muscle during alternating periods of contraction and relaxation

large amounts of Ca^{2+} flow out of the SR into the sarcoplasm around the thick and thin filaments. As a result, the Ca^{2+} concentration in the sarcoplasm rises tenfold or more. The released calcium ions combine with troponin, which in turn undergoes a conformational change that causes tropomyosin to move away from the myosin-binding sites on actin. Once these binding sites are free, myosin heads bind to them to form crossbridges, and the contraction cycle begins.

Skeletal Muscle Relaxes in Response to a Decrease in the Sarcoplasmic Ca^{2+} Concentration

The membrane of the sarcoplasmic reticulum (SR) also contains active transport proteins called **Ca^{2+}-ATPase pumps** that constantly transport Ca^{2+} from the sarcoplasm into the SR (**FIGURE 11.10**). While

FIGURE 11.10 Mechanism of excitation–contraction coupling in a skeletal muscle fiber. (a) During relaxation, the level of Ca^{2+} in the sarcoplasm is low because dihydropyridine (DHP) receptors in the transverse (T) tubule membrane block Ca^{2+} release channels in the terminal cisternal membrane of the sarcoplasmic reticulum (SR). When the calcium concentration in the sarcoplasm is low, tropomyosin covers the myosin-binding sites on actin, preventing myosin and actin from interacting. (b) A muscle potential propagating along a T tubule causes the DHP receptors to undergo a conformational change that ultimately leads to the opening of the Ca^{2+} release channels. As a result, Ca^{2+} is released from the SR into the sarcoplasm. The released calcium ions bind to troponin, which in turn undergoes a conformational change that causes tropomyosin to move away from the myosin-binding sites on actin. Myosin molecules subsequently bind to actin and the contraction cycle begins.

Excitation–contraction coupling refers to the sequence of events that connects the muscle action potential to muscle contraction.

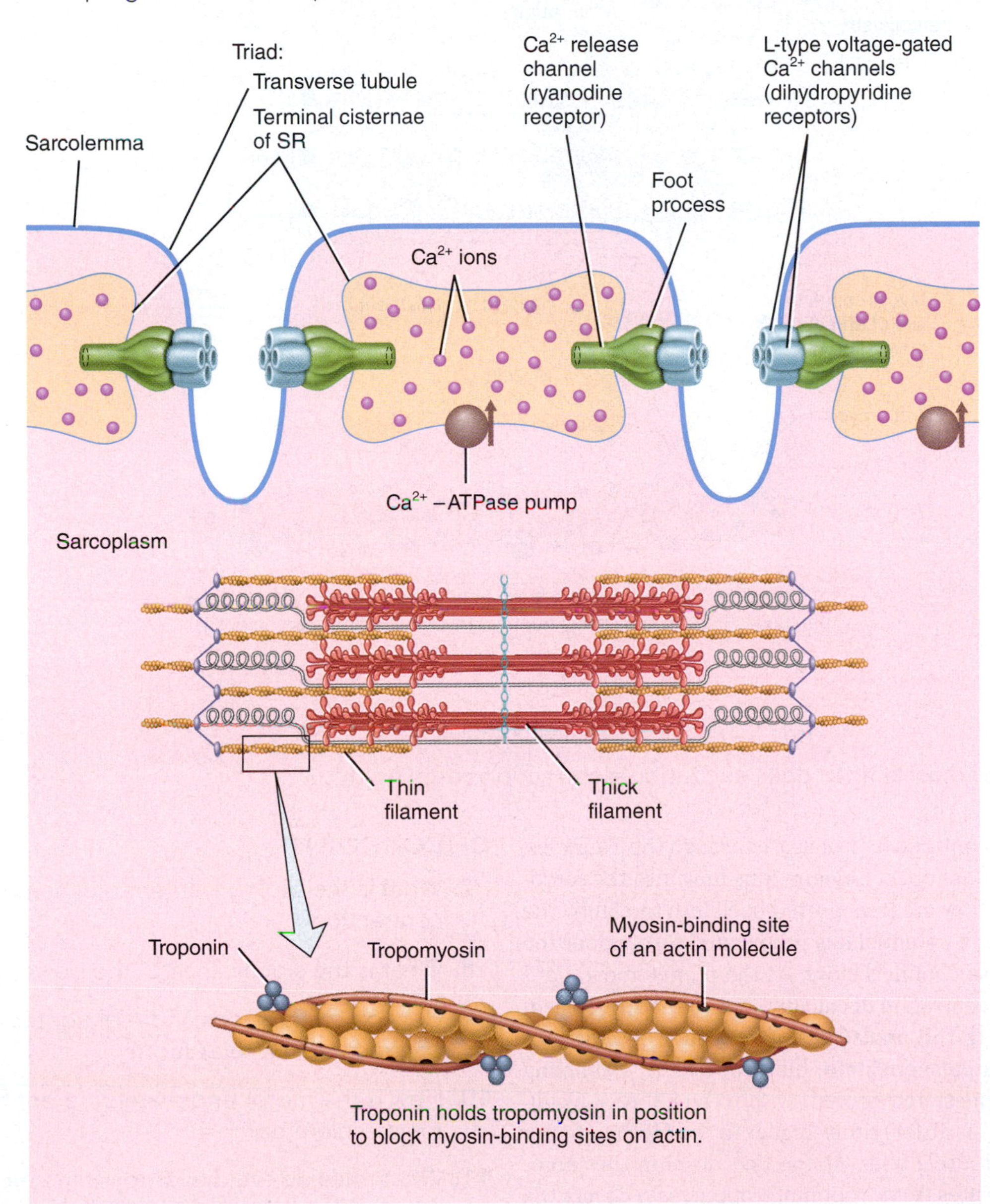

(a) Relaxation

Figure 11.10 (continues)

Figure 11.10 Continued

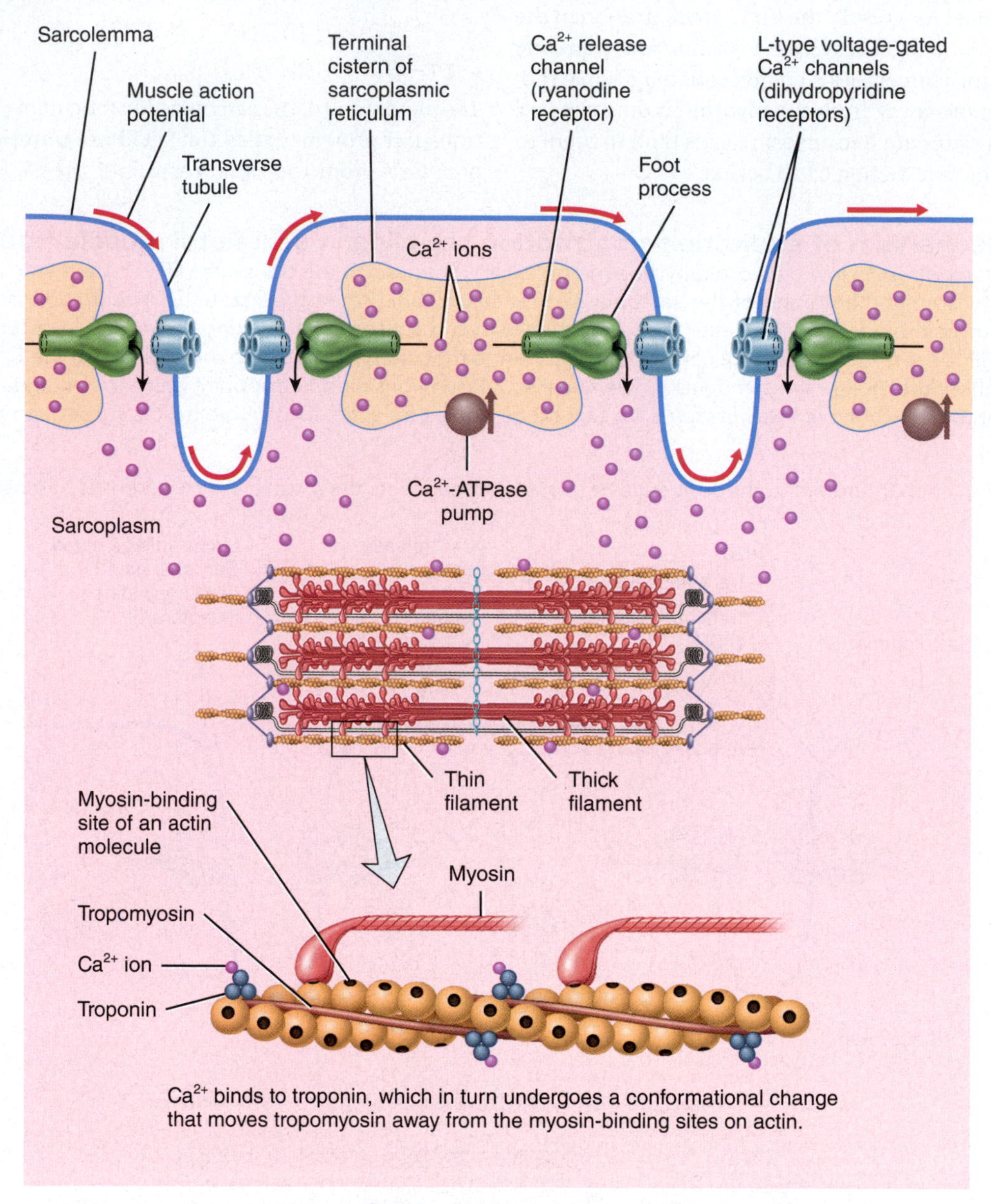

(b) Contraction

? Where in a skeletal muscle fiber does excitation–contraction coupling occur?

muscle action potentials continue to propagate through the T tubules, the Ca^{2+} release channels are open. Calcium ions flow into the sarcoplasm more rapidly than they are transported back into the SR by the pumps. After the last action potential has propagated throughout the T tubules, the Ca^{2+} release channels close. As the pumps move Ca^{2+} back into the SR, the concentration of calcium ions in the sarcoplasm quickly decreases. Inside the SR, molecules of a calcium-binding protein, appropriately called **calsequestrin**, bind to the Ca^{2+}, enabling even more Ca^{2+} to be sequestered (stored) within the SR. As a result, the concentration of Ca^{2+} is 10,000 times higher in the SR than in the sarcoplasm in a relaxed muscle fiber. As the Ca^{2+} level in the sarcoplasm drops, Ca^{2+} dissociates from troponin, tropomyosin covers the myosin-binding sites on actin, and the muscle fiber relaxes.

FIGURE 11.11 summarizes the events that occur during contraction and relaxation of a skeletal muscle fiber.

CHECKPOINT

7. What is the sliding filament mechanism of muscle contraction?
8. What is the significance of the power stroke?
9. How do calcium ions and ATP contribute to muscle contraction and relaxation?
10. How is the motor end plate different from other parts of the sarcolemma?
11. What roles do the dihydropyridine receptors and Ca^{2+} release channels play in excitation–contraction coupling in a skeletal muscle fiber?
12. What is the purpose of calsequestrin?

FIGURE 11.11 Summary of the events of contraction and relaxation in a skeletal muscle fiber.

Acetylcholine released at the neuromuscular junction triggers a muscle action potential, which leads to muscle contraction.

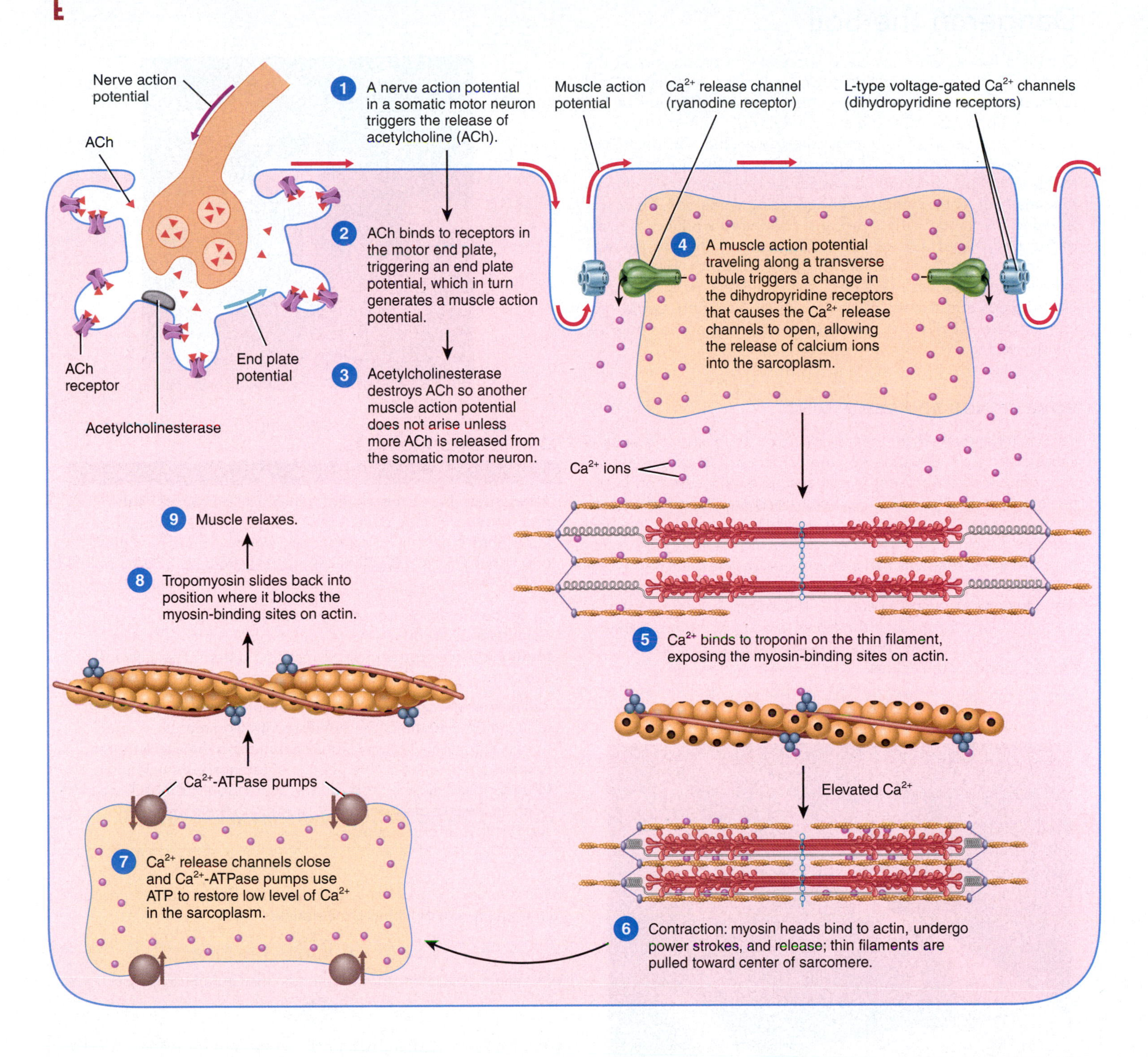

What are three function of ATP in muscle contraction?

Critical Thinking

Danger in the Soil

Cecil is a farmer who has been successful in producing all the food his family needs. To cut down on farming expenses, Cecil routinely recycles the waste from his horses and chickens as fertilizer for his crops. After the soil is fertilized, Cecil usually grabs some of it to see how it clumps when squeezed. This lets Cecil know when the soil is moist enough to start putting seeds into the ground. This spring as he was studying his soil, Cecil ignored the small cut in his hand. After squeezing some soil, he merely wiped his hands with a rag and continued working the rest of the day. About a week later, Cecil noticed that his jaw was stiff, making it difficult for him to open his mouth. He was also experiencing stiffness in his neck and shoulders, muscle spasms in his biceps and triceps muscles, a fever, sweating, and a rapid heart rate. Cecil's wife called the ambulance so he could be transported to the hospital.

Graham Taylor/Getty Images

SOME THINGS TO KEEP IN MIND:

Cecil has contracted *tetanus*, also known as *lockjaw*. Tetanus is caused by a toxin produced by the bacterium *Clostridium tetani*, an organism often present in soil and especially common in the feces of farm animals such as horses and chickens. This organism was present in the soil that Cecil handled; he was infected through the cut in his hand. When the *C. tetani* are in anaerobic conditions, the cells can take on their vegetative form. These cells produce endospores, which are nonmetabolizing structures that allow the *C. tetani* to survive. The endospores produce toxins that, if introduced into the body, can affect the mechanisms of muscle contractions, producing the stiffness and spasms like those experienced by Cecil.

sot/Getty Images

SOME INTERESTING FACTS:

Tetanospasmin is a very potent neurotoxin produced by the endospores of *Clostridium tetani*. The estimated minimum lethal human dose is 2.5 ng/kg body weight. This means that only about 175 ng (0.0000000006 ounces) could kill a 70-kg (150-lb) man.

Once the endospores enter the body through a cut, they germinate and produce the tetanospasmin, which is moved through the bloodstream and lymphatic system. The closer the site of the injury is to the brain, the shorter the transit time.

Tetanospasmin can act at motor end plates, in the spinal cord, or in the brain to affect the signals controlling muscle contractions. Often, the inhibitory signals controlling the alpha motor neurons are blocked by the actions of tetanospasmin, leading to unopposed muscle contractions of antagonistic muscle pairs (such as the biceps and triceps), producing the commonly seen body contorting spasms.

What role does axonal transport play in the delivery of tetanospasmin from the site of injury to the nervous system?

What effect does tetanospasmin have on the neuromuscular junction?

What kind of treatment will Cecil receive in the hospital for his infection?

How could Cecil have avoided coming in contact with *Clostridium tetani*?

11.4 ATP Production in Skeletal Muscle

OBJECTIVES

- Describe the reactions by which skeletal muscle fibers produce ATP.
- Explain the factors that contribute to muscle fatigue.

Unlike most cells of the body, skeletal muscle fibers often switch between a low level of activity, when they are relaxed and using only a modest amount of ATP, and a high level of activity, when they are contracting and using ATP at a rapid pace. A huge amount of ATP is needed to power the contraction cycle, to pump Ca^{2+} into the sarcoplasmic reticulum, and for other metabolic reactions involved in muscle contraction. However, the ATP present inside muscle fibers is enough to power contraction for only a few seconds. If muscle contractions continue past that time, the muscle fibers must make more ATP. Skeletal muscle fibers have three ways to produce ATP: (1) from creatine phosphate, (2) by anaerobic glycolysis, and (3) by aerobic respiration (**FIGURE 11.12**).

Creatine Phosphate Is the First Source of ATP During Muscle Contraction

While muscle fibers are relaxed, they produce more ATP than they need for resting metabolism. Most of the excess ATP is used to synthesize **creatine phosphate**, an energy-rich molecule that is found in muscle fibers. The enzyme *creatine kinase (CK)* catalyzes the transfer of one of the high-energy phosphate groups from ATP to creatine, forming creatine phosphate and ADP. This reversible reaction is summarized as follows:

$$\text{Creatine} + \text{ATP} \underset{}{\overset{\text{CK}}{\rightleftharpoons}} \text{Creatine phosphate} + \text{ADP}$$

Creatine is a small, amino acid–like molecule that is synthesized in the liver, kidneys, and pancreas and then transported to muscle fibers. Creatine phosphate is three to six times more plentiful than ATP in the sarcoplasm of a relaxed muscle fiber. Based on the law of mass action, when contraction begins and the ADP level starts to rise, CK catalyzes the transfer of a high-energy phosphate group from creatine phosphate back to ADP. This direct phosphorylation reaction quickly regenerates new ATP molecules (**FIGURE 11.12**). Because the formation of ATP

FIGURE 11.12 ATP production in skeletal muscle.

Skeletal muscle fibers have three ways to produce ATP: (1) from creatine phosphate, (2) by anaerobic glycolysis, and (3) by aerobic respiration.

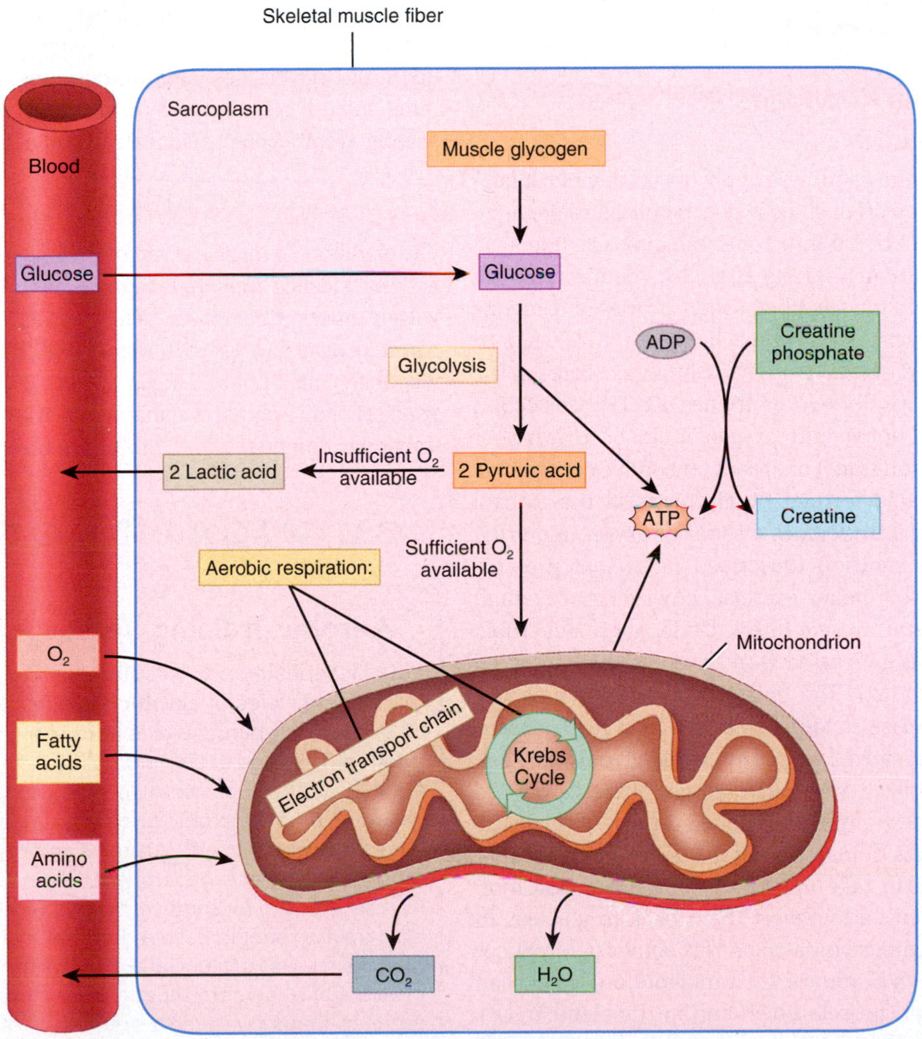

How is most ATP produced in a skeletal muscle fiber during a long-term event such as a marathon race?

CLINICAL CONNECTION

Creatine Supplementation

Creatine is both synthesized in the body (in the liver, kidneys, and pancreas) and derived from foods such as milk, red meat, and some fish. Adults need to synthesize and ingest a total of about 2 grams of creatine daily to make up for the urinary loss of creatinine, the breakdown product of creatine. Some studies have demonstrated improved performance during explosive movements, such as sprinting, in people who have ingested creatine supplements. Other studies, however, have failed to find any performance-enhancing effect from creatine supplementation. Ingesting extra creatine decreases the body's own synthesis of creatine, and it is not known whether natural synthesis recovers after long-term creatine supplementation. In addition, creatine supplementation can cause dehydration and may cause kidney dysfunction. Further research is needed to determine both the long-term safety and the value of creatine supplementation.

from creatine phosphate occurs very rapidly, creatine phosphate is the first source of energy when muscle contraction begins. The other energy-generating mechanisms in a muscle fiber (anaerobic glycolysis and aerobic respiration) take more time to produce ATP. Together, stores of creatine phosphate and ATP provide enough energy for muscles to contract maximally for about 15 seconds.

Anaerobic Glycolysis Produces ATP When Oxygen Levels Are Low

When muscle activity continues and the supply of creatine phosphate within the muscle fiber is depleted, glucose is catabolized to generate ATP. Glucose passes from the blood into contracting muscle fibers via facilitated diffusion (see Section 5.4), and it is also produced by the breakdown of glycogen within muscle fibers. Then a series of reactions known as **glycolysis** quickly breaks down each glucose molecule into two molecules of pyruvic acid (FIGURE 11.12). Glycolysis occurs in the cytosol and produces a net gain of two molecules of ATP (see Section 4.5). Because glycolysis does not require oxygen, it can occur whether oxygen is present (aerobic conditions) or absent (anaerobic conditions).

Ordinarily, the pyruvic acid formed by glycolysis in the cytosol enters mitochondria, where it undergoes a series of oxygen-requiring reactions called aerobic respiration (described next) that produce a large amount of ATP. During heavy exercise, however, not enough oxygen is available to skeletal muscle fibers. Under these anaerobic conditions, the pyruvic acid generated from glycolysis is converted to **lactic acid** (FIGURE 11.12). The entire process by which the breakdown of glucose gives rise to lactic acid when oxygen is absent or at a low concentration is referred to as **anaerobic glycolysis.** Each molecule of glucose catabolized via anaerobic glycolysis yields two molecules of lactic acid and two molecules of ATP. Most of the lactic acid produced by this process diffuses out of the skeletal muscle fiber into the blood. Liver cells can take up some of the lactic acid molecules from the bloodstream and convert them back to glucose. In addition to providing new glucose molecules, this conversion reduces the acidity of the blood. When produced at a rapid rate, lactic acid can accumulate in active skeletal muscle fibers and in the bloodstream. This buildup of lactic acid is thought to be responsible for the muscle soreness that is felt during strenuous exercise. Compared to aerobic respiration, anaerobic glycolysis produces fewer ATPs, but it is faster and can occur when oxygen levels are low. Anaerobic glycolysis provides enough energy for about 2 minutes of maximal muscle activity.

Aerobic Respiration Generates ATP When Sufficient Oxygen Is Available

If sufficient oxygen is present, the pyruvic acid formed by glycolysis enters the mitochondria, where it undergoes **aerobic respiration**, a series of oxygen-requiring reactions that produce ATP, carbon dioxide, water, and heat (FIGURE 11.12). Recall from Chapter 4 that aerobic respiration includes the reactions of the **Krebs cycle** and the **electron transport chain** (see Section 4.4). Thus, when oxygen is present, glycolysis, the Krebs cycle, and the electron transport chain occur. Although aerobic respiration is slower than anaerobic glycolysis, it yields much more ATP. Each molecule of glucose catabolized under aerobic conditions yields about 30 or 32 molecules of ATP.

Muscle tissue has two sources of oxygen: (1) oxygen that diffuses into muscle fibers from the blood and (2) oxygen released by myoglobin within muscle fibers. Both myoglobin (found only in muscle cells) and hemoglobin (found only in erythrocytes) are oxygen-binding proteins. They bind oxygen when it is plentiful and release oxygen when it is scarce.

Aerobic respiration supplies enough ATP for muscles during periods of rest or light to moderate exercise, provided sufficient oxygen and nutrients are available. These nutrients include the pyruvic acid obtained from the glycolysis of glucose, fatty acids from the breakdown of triglycerides, and amino acids from the breakdown of proteins. In activities that last from several minutes to an hour or more, aerobic respiration provides nearly all of the needed ATP.

Several Factors Contribute to Muscle Fatigue

The inability of a muscle to maintain force of contraction after prolonged activity is called **muscle fatigue.** Fatigue results mainly from changes within muscle fibers. Even before actual muscle fatigue occurs, a person may have feelings of tiredness and the desire to cease activity; this response, called *central fatigue*, is caused by changes in the central nervous system (brain and spinal cord). Although its exact mechanism is unknown, fatigue may be a protective mechanism to stop a person from

CLINICAL CONNECTION

Aerobic Training Versus Anaerobic Training

Regular, repeated activities such as jogging or walking on the treadmill are examples of **aerobic training**, which increases the supply of oxygen-rich blood available to skeletal muscles for aerobic respiration. This type of training builds endurance for prolonged activities. By contrast, activities such as weight lifting rely more on ATP production via anaerobic glycolysis. Such **anaerobic training** stimulates synthesis of muscle proteins and result, over time, in increased muscle size (muscle hypertrophy). Hence, anaerobic training builds muscle strength for short-term feats. Athletes who engage in anaerobic training should have a diet that includes an adequate amount of proteins. This protein intake allows the body to synthesize muscle proteins and increase muscle mass. **Interval training** is a workout regimen that incorporates both types of training—for example, alternating sprints (anaerobic) with jogging (aerobic).

exercising before muscles become damaged. As you will learn shortly, certain types of skeletal muscle fibers fatigue more quickly than others.

Although the precise mechanisms that cause muscle fatigue are still not clear, several factors are thought to contribute. One is inadequate release of calcium ions from the SR, resulting in a decline of Ca^{2+} concentration in the sarcoplasm. Depletion of creatine phosphate is also associated with fatigue, but surprisingly the ATP levels in fatigued muscle often are not much lower than those in resting muscle. Other factors that contribute to muscle fatigue include insufficient oxygen, depletion of glycogen and other nutrients, buildup of lactic acid and ADP, and failure of action potentials in the motor neuron to release enough acetylcholine.

Oxygen Consumption Increases for a While After Exercise

During prolonged periods of muscle contraction, increases in breathing rate and blood flow enhance oxygen delivery to muscle tissue. After muscle contraction has stopped, heavy breathing continues for a while, and oxygen consumption remains above the resting level. Depending on the intensity of the exercise, the recovery period may be just a few minutes, or it may last as long as several hours. The term **oxygen debt** refers to the added oxygen, over and above the resting oxygen consumption, that is taken into the body after exercise. This extra oxygen is used to "pay back" or restore metabolic conditions to the resting level in three ways: (1) to convert lactic acid back into glycogen stores in the liver, (2) to resynthesize creatine phosphate and ATP in muscle fibers, and (3) to replace the oxygen removed from myoglobin.

The metabolic changes that occur *during exercise* can account for only some of the extra oxygen used *after exercise*. Only a small amount of glycogen resynthesis occurs from lactic acid. Instead, most glycogen is made much later from dietary carbohydrates. Much of the lactic acid that remains after exercise is converted back to pyruvic acid and used for ATP production via aerobic respiration in the heart, liver, kidneys, and skeletal muscle. Oxygen use after exercise is also boosted by ongoing changes. First, the elevated body temperature after strenuous exercise increases the rate of chemical reactions throughout the body. Faster reactions use ATP more rapidly, and more oxygen is needed to produce the ATP. Second, the heart and the muscles used in breathing are still working harder than they were at rest, and thus they consume more ATP. Third, tissue repair processes are occurring at an increased pace. For these reasons, **recovery oxygen uptake** is a better term than oxygen debt for the elevated use of oxygen after exercise.

CHECKPOINT

13. What is the functional significance of creatine phosphate?

14. What is skeletal muscle's main source of ATP during rest? During periods of light to moderate exercise? During heavy exercise?

15. What factors contribute to muscle fatigue?

11.5 Skeletal Muscle Mechanics

OBJECTIVES

- Describe the importance of a motor unit.
- Outline the phases of a twitch.
- Discuss the factors that determine muscle tension.
- Use the terms lever, fulcrum, and effort to explain how skeletal muscles produce body movements by pulling on bones.
- Distinguish between isotonic and isometric contractions.

A Motor Unit Is Comprised of a Somatic Motor Neuron and Its Muscle Fibers

A skeletal muscle receives neural input from many somatic motor neurons. As the axon of a somatic motor neuron enters a skeletal muscle, it divides into a number of branches, each supplying a different muscle fiber. A single somatic motor neuron innervates several muscle fibers, but each muscle fiber is innervated by only one somatic motor neuron. A **motor unit** consists of a somatic motor neuron plus all the muscle fibers it innervates (**FIGURE 11.13**). When a somatic motor neuron is activated, all of the muscle fibers in its motor unit contract in unison. Thus, the motor unit serves as the functional contractile unit of skeletal muscle. Typically, the muscle fibers of a motor unit are dispersed throughout a muscle rather than clustered together.

A Muscle Twitch Consists of Latent, Contraction, and Relaxation Periods

A **twitch** is the brief contraction of a group of muscle fibers within a muscle in response to a single action potential. In the laboratory, a twitch can be produced by surgically removing a muscle from an animal and then electrically stimulating it. The record of a muscle contraction, called a **myogram**, is shown in **FIGURE 11.14**. Twitches of skeletal muscle fibers last anywhere from 20 to 200 msec. This is very long compared to the brief 1–2 msec that a muscle action potential lasts.

A twitch consists of three sequential phases: the latent period, the contraction period, and the relaxation period (**FIGURE 11.14**). The **latent period**, which lasts about 2 msec, is a brief delay that occurs between application of the stimulus (time zero on the graph) and the beginning of contraction. During this time, the events of excitation–contraction coupling occur: The muscle action potential sweeps along the sarcolemma and into the T tubules, causing the release of calcium ions

FIGURE 11.13 Motor units. Two somatic motor neurons (one purple and one green) are shown, each supplying the muscle fibers of its motor unit.

A motor unit consists of a somatic motor neuron plus all the muscle fibers it innervates.

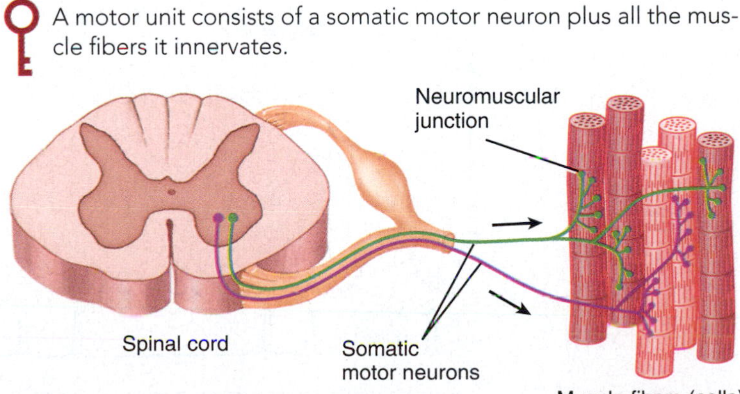

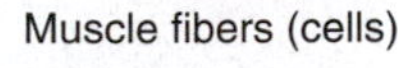

What is the effect of the size of a motor unit on its strength of contraction? (Assume that each muscle fiber can generate about the same amount of tension.)

FIGURE 11.14 Myogram of a muscle twitch. The arrow indicates the time at which the stimulus occurred.

A myogram is a record of a muscle contraction.

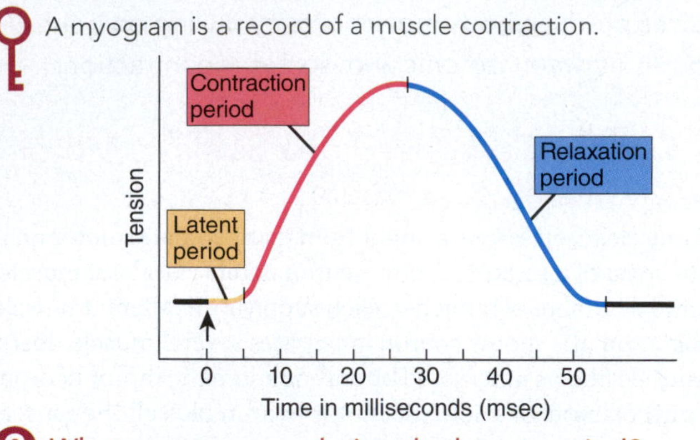

What events occur during the latent period?

from the sarcoplasmic reticulum. During the **contraction period**, which lasts 10–100 msec, Ca^{2+} binds to troponin, myosin-binding sites on actin are exposed, and myosin crossbridges form. As a result, peak tension develops in the muscle fiber. During the **relaxation period**, also lasting 10–100 msec, Ca^{2+} is actively transported back into the sarcoplasmic reticulum, myosin-binding sites are covered by tropomyosin, myosin heads detach from actin, and tension in the muscle fiber decreases. The actual duration of these periods depends on the type of skeletal muscle fiber. Some fibers, such as the fast-twitch fibers that move the eyes (described shortly), have contraction periods as brief as 10 msec and equally brief relaxation periods. Others, such as the slow-twitch fibers that move the legs, have contraction and relaxation periods of about 100 msec each.

Graded Contractions Can Occur in Skeletal Muscle

A single nerve action potential in a somatic motor neuron elicits a single muscle action potential in all the skeletal muscle fibers with which it forms synapses. Action potentials always have the same size in a given neuron or muscle fiber. By contrast, the force of muscle fiber contraction can vary; a muscle fiber is capable of generating a much greater force than the one that results from a single twitch. Hence, skeletal muscles can produce **graded contractions**, contractions that vary in strength depending on how much force is needed by the muscle to support a particular object.

Numerous Factors Determine Muscle Tension

The total force or tension that a *single* muscle fiber can produce depends on the frequency of stimulation (the rate at which the muscle fiber is stimulated by a motor neuron), the length of the muscle fiber before contraction begins, and the diameter of the muscle fiber. The total tension a *whole* muscle can produce depends on the number of muscle fibers that are contracting in unison, which is determined by the size and number of motor units that are activated. The factors that determine muscle tension will now be described in more detail.

Frequency of Stimulation

If two stimuli are applied to a skeletal muscle fiber, one immediately after the other, the muscle fiber will respond to the first stimulus but not to the second. When a muscle fiber receives enough stimulation to contract, it temporarily loses its excitability and cannot respond for a time. The period of lost excitability, called the **refractory period**, is a characteristic of all muscle and nerve cells. The duration of the refractory period varies with the muscle involved. Skeletal muscle has a short refractory period of about 1 msec; cardiac muscle has a longer refractory period of about 250 msec.

If a second stimulus occurs in a skeletal muscle fiber after the refractory period of the first stimulus is over but before the muscle fiber has relaxed, the second contraction will actually be stronger than the first (FIGURE 11.15a, b). This phenomenon, in which stimuli

FIGURE 11.15 Myograms showing the effects of different frequencies of stimulation. (a) Single twitch. (b) When a second stimulus occurs before the muscle fiber has relaxed, the second contraction is stronger than the first, a phenomenon called wave summation. (The dashed line indicates the force of contraction expected in a single twitch.) (c) Unfused tetanus produces a jagged curve due to the slight relaxation of the muscle fiber between stimuli. (d) In fused tetanus, which occurs when the muscle fibers do not relax at all between stimuli, the myogram line, like the contraction force, is steady and sustained.

Due to wave summation, the tension produced during a sustained contraction is greater than that produced by a single twitch.

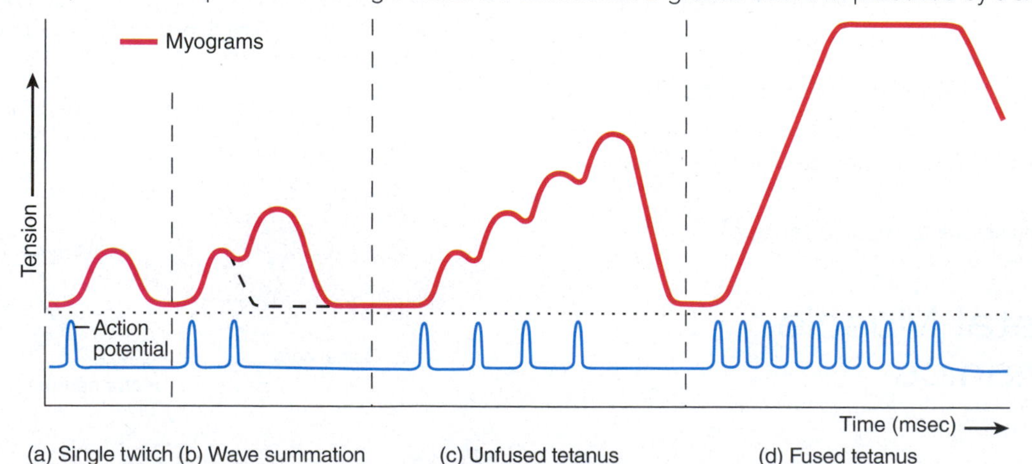

Would the peak tension of the second contraction in (b) be larger or smaller if the second stimulus were applied a few milliseconds later?

arriving at different times cause contractions with greater tension, is called **wave summation**. If a skeletal muscle fiber is stimulated at a higher rate, it can relax only slightly between stimuli (FIGURE 11.15c). The result is a sustained but wavering contraction called **unfused (incomplete) tetanus**. If a skeletal muscle fiber is stimulated at an even higher rate, it does not relax at all (FIGURE 11.15d). The result is **fused (complete) tetanus**, a smooth, sustained contraction in which individual twitches cannot be detected and maximum tension is reached. (Do not mistake unfused or fused tetanus for the disease called tetanus, which is characterized by prolonged and painful muscle spasms; see the Critical Thinking box *Danger in the Soil* at the end of Section 11.3).

Wave summation and both kinds of tetanus (unfused and fused) occur when additional Ca^{2+} is released from the sarcoplasmic reticulum by subsequent stimuli while the levels of Ca^{2+} in the sarcoplasm are still elevated from the first stimulus. Because of the buildup in the Ca^{2+} level, the peak tension generated during fused tetanus is 5 to 10 times larger than the peak tension produced during a single twitch. Even so, fused tetanus rarely occurs in muscles of the body, except for those occasions when tremendous strength is necessary (for example, attempting to lift part of a car away from someone who is trapped beneath one of its tires). Most smooth, sustained voluntary muscle contractions in the body are achieved by out-of-synchrony unfused tetanus in different motor units.

Muscle Fiber Length

FIGURE 11.16 shows the **length–tension relationship** for a skeletal muscle fiber, which indicates how the forcefulness of muscle contraction (tension) depends on the length of the sarcomeres within a muscle fiber *before contraction begins*. At a sarcomere length of about 2.0–2.4 μm (which is very close to the resting length in most muscles), the zone of overlap in each sarcomere is optimal, and the muscle fiber can develop maximum tension (100%). As the sarcomeres of a muscle fiber are stretched to a longer length (overstretched), the zone of overlap shortens, and fewer myosin heads can make contact with thin filaments. Therefore, the tension the fiber can produce decreases. When a skeletal muscle fiber is stretched to 170% of its optimal length, there is no overlap between the thick and thin filaments. Because none of the myosin heads can bind to thin filaments, the muscle fiber cannot contract, and tension is zero. As sarcomere lengths become increasingly shorter than the optimum (understretched), the tension that can develop again decreases. This is because the thick filaments crumple as they are compressed by the Z discs, resulting in fewer myosin heads making contact with thin filaments. Normally, resting muscle fiber length is held very close to the optimum by firm attachments of skeletal muscle to bones (via their tendons) and to other inelastic tissues.

Muscle Fiber Diameter

Muscle fibers that have a thicker diameter have more myofibrils and can generate more tension compared to muscle fibers that have a thinner diameter. This relationship between muscle fiber diameter and muscle strength is evident in a bodybuilder, whose huge, thick muscles can exert more force than the smaller, thinner muscles of an average person.

Motor Unit Size

The size of the motor units that are activated in a muscle affects the amount of tension the muscle can generate. Muscles that control precise movements, which involve only small amounts of force, contain small motor units. For instance, muscles of the larynx (voice box) that control voice production have as few as two or three muscle fibers per motor unit, and muscles controlling eye movements may have 10 to 20 muscle fibers per motor unit. By contrast, muscles responsible for large-scale and powerful movements have large motor units. For example, the biceps muscle in the arm and the gastrocnemius muscle in the calf of the leg have as many as 2000 to 3000 muscle fibers in some motor units.

Motor Unit Recruitment

When a muscle needs to generate more force during a contraction, more of its motor units are activated. The process of increasing the number of active motor units is called **motor unit recruitment**. Recruitment is one factor responsible for producing smooth movements rather than a series of jerks. Typically, the different motor units of an entire muscle are not stimulated to contract in unison. While some motor units are contracting, others are relaxed, a phenomenon known as **asynchronous recruitment** of motor units. This pattern of motor unit activity delays muscle fatigue and allows contraction of a whole muscle to be sustained for long periods. The

FIGURE 11.16 Length–tension relationship in a skeletal muscle fiber. Maximum tension during contraction occurs when the resting sarcomere length is 2.0–2.4 μm.

A muscle fiber develops its greatest tension when there is an optimal zone of overlap between thick and thin filaments.

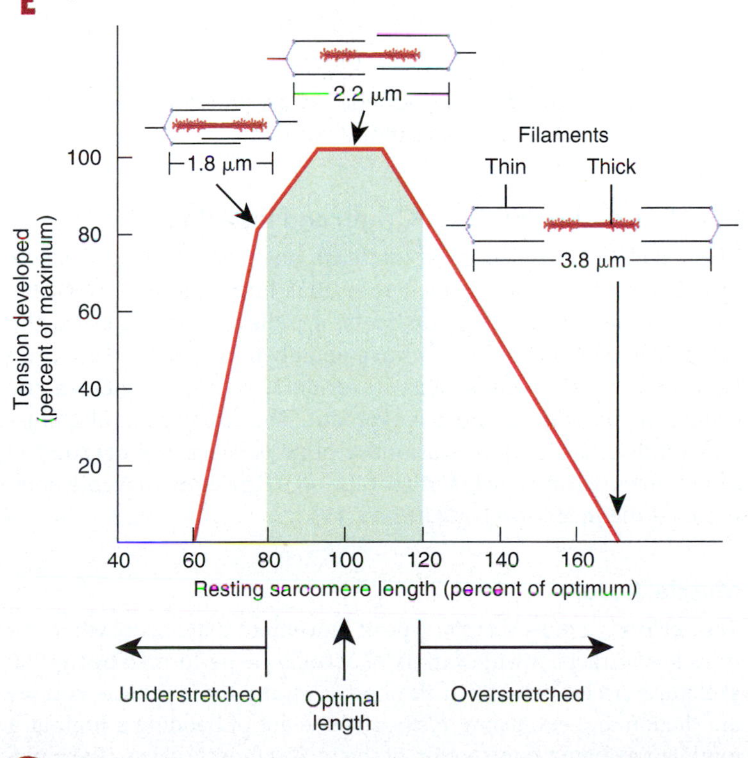

Why is tension maximal at a sarcomere length of 2.2 μm?

smallest motor units are recruited first, with progressively larger motor units added if the task requires more force. For this reason, when a motor unit is recruited or turned off, only slight changes occur in muscle tension.

Muscle Tone Is Established by Different Motor Units That Are Alternately Active and Inactive

Even at rest, a skeletal muscle exhibits **muscle tone**, a small amount of tautness or tension in the muscle due to weak, involuntary contractions of its motor units. Recall that skeletal muscle contracts only after it is activated by acetylcholine released by action potentials in its motor neurons. Hence, muscle tone is established by neurons in the brain and spinal cord that excite the muscle's motor neurons. When the motor neurons serving a skeletal muscle are damaged or cut, the muscle becomes **flaccid** (FLAS-sid = flabby), a state of limpness in which muscle tone is lost. To sustain muscle tone, small groups of motor units are alternately active and inactive in a constantly shifting pattern. Muscle tone keeps skeletal muscles firm, but it does not result in a force strong enough to produce movement. For example, when the muscles in the back of the neck are in normal tonic contraction, they keep the head upright and prevent it from slumping forward on the chest. Muscle tone is also important in smooth muscle tissues, such as those found in the gastrointestinal tract, where the walls of the digestive organs maintain a steady pressure on their contents. The tone of smooth muscle fibers in the walls of blood vessels plays a crucial role in maintaining blood pressure.

Movement Involves the Interaction of Skeletal Muscles, Tendons, and Bones

Skeletal muscles produce movements by exerting force on tendons, which in turn pull on the bones of the skeleton.

Muscle Attachment Sites: Origin and Insertion

Most skeletal muscles cross at least one joint and are usually attached to the articulating bones that form the joint (FIGURE 11.17). When the muscle contracts, it pulls one bone toward the other. The two bones do not move equally in response to contraction. One bone remains near its original position, whereas the other bone undergoes more movement. The attachment of a muscle's tendon to the more stationary bone is called the **origin**; the attachment of the muscle's other tendon to the more movable bone is called the **insertion** (FIGURE 11.17).

Muscle Actions

The **action** of a muscle is the type of movement that occurs when the muscle contracts. A wide variety of actions are performed by the 700 skeletal muscles present in the body. Among these muscle actions are flexion and extension. *Flexion* is the act of bending a limb at a joint. For example, contraction of the biceps muscle causes flexion of the forearm at the elbow joint, which means that the forearm moves toward the arm (FIGURE 11.18a). Muscles such as the biceps are referred to as *flexors* because they promote flexion of a limb. *Extension* is the act of straightening out a limb at a joint. For example, contraction of the triceps muscle causes extension of the forearm at the elbow joint, meaning that the forearm moves away from the arm (FIGURE 11.18b). Muscles such as the triceps that cause extension of a limb are known as *extensors.*

CLINICAL CONNECTION

Hypotonia and Hypertonia

Hypotonia refers to decreased or lost muscle tone. Such muscles are said to be flaccid. Flaccid muscles are loose and appear flattened rather than rounded. Certain disorders of the nervous system and disruptions in the balance of electrolytes (especially sodium, calcium, and—to a lesser extent—magnesium) may result in **flaccid paralysis**, which is characterized by loss of muscle tone, loss or reduction of tendon reflexes, and atrophy (wasting away) and degeneration of muscles.

Hypertonia refers to increased muscle tone and is expressed in two ways: spasticity or rigidity. **Spasticity** (spas-TIS-i-tē) is characterized by increased muscle tone (stiffness) associated with an increase in tendon reflexes and pathological reflexes (such as the Babinski sign, in which the great toe extends with or without fanning of the other toes in response to stroking the outer margin of the sole). Certain disorders of the nervous system and electrolyte imbalances such as those previously noted may result in **spastic paralysis**, partial paralysis in which the muscles exhibit spasticity. **Rigidity** refers to increased muscle tone in which reflexes are not affected, as occurs in the disease tetanus.

FIGURE 11.17 Muscle attachment sites to bone.

The origin and the insertion are the sites where the tendons of a muscle attach to bone.

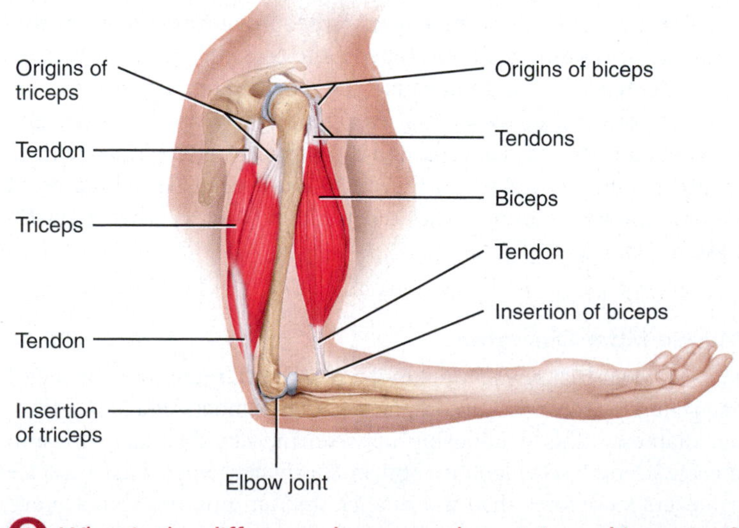

What is the difference between the origin and insertion?

FIGURE 11.18 Flexion and extension at the elbow joint.

Flexion is the act of bending a limb at a joint, whereas extension is the act of straightening a limb at a joint.

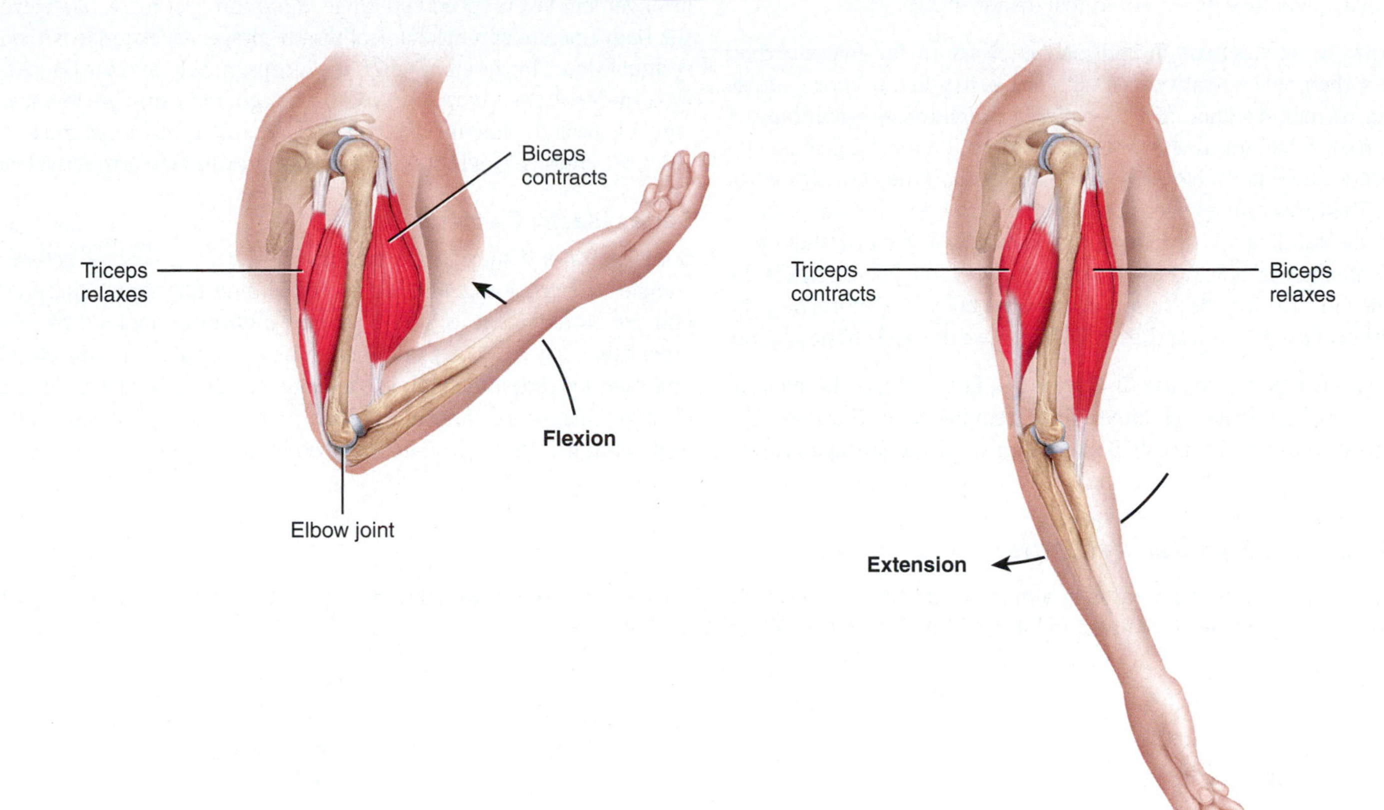

Why are the biceps and triceps referred to as antagonistic muscles?

The biceps and triceps muscles are examples of **antagonistic muscles** because they promote opposite actions at the same joint. With a given pair of antagonistic muscles, when one muscle contracts, the other muscle relaxes. For example, during flexion of the forearm, the biceps contracts and the triceps relaxes (FIGURE 11.18a). However, during extension of the forearm, the triceps contracts and the biceps relaxes (FIGURE 11.18b). If both muscles of an antagonistic pair contracted at the same time with equal force, there would be no net movement.

Lever Systems and Leverage

In producing movement, bones act as levers, and joints function as the fulcrums of these levers. A **lever** is a rigid structure that can move around a fixed point called a **fulcrum**, symbolized by F. A lever is acted on at two different points by two different forces: the **effort** (E), which causes movement, and the **load** L or **resistance**, which opposes movement. The effort is the force (tension) exerted by muscular contraction; the load is typically the weight of the body part that is moved. Motion occurs when the effort applied to the bone at the insertion exceeds the load. Consider the biceps muscle flexing the forearm as an object is lifted (FIGURE 11.19). When the forearm is raised, the elbow joint is the fulcrum. The weight of the forearm plus the weight of the object in the hand is the load. The force of contraction of the biceps muscle pulling the forearm up is the effort.

FIGURE 11.19 Movement of the forearm as illustrated by the lever–fulcrum principle.

Skeletal muscles produce movements by pulling on bones. Bones serve as levers, and joints act as fulcrums for the levers.

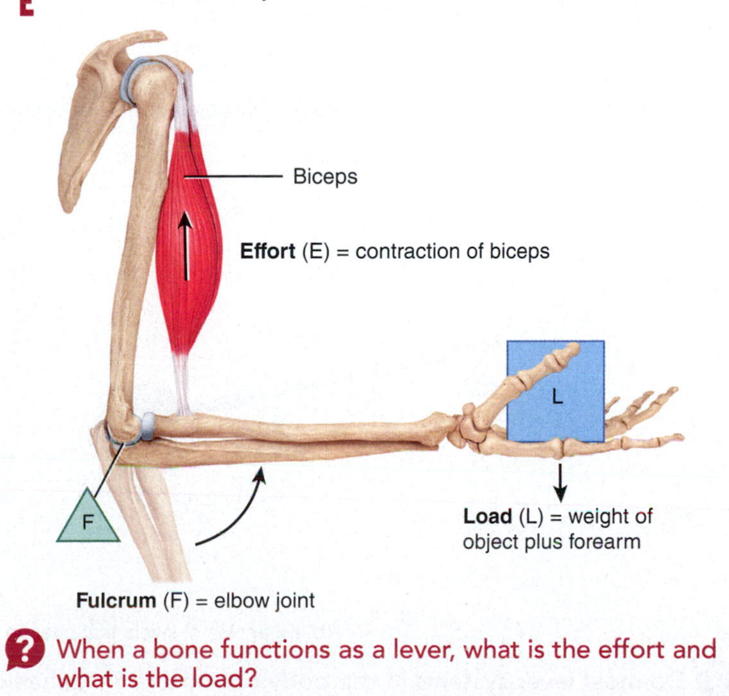

When a bone functions as a lever, what is the effort and what is the load?

The relative distance between the fulcrum and load and the point at which the effort is applied determine if a given lever operates at a mechanical advantage or a mechanical disadvantage.

Mechanical Advantage If the load is closer to the fulcrum than the effort, then only a relatively small effort is required to move a large load over a small distance. This is called a **mechanical advantage**. A wheelbarrow functions as a lever that operates at a mechanical advantage because the load is closer to the fulcrum than the effort (FIGURE 11.20a). This arrangement sacrifices speed and range of motion for force. In the body, standing up on your toes is an example of a lever that operates at a mechanical advantage (FIGURE 11.20a). The fulcrum (F) is the ball of the foot. The load (L) is the weight of the body. The effort (E) is the contraction of the muscles of the calf, which raise the heel off the ground.

Mechanical Disadvantage If the load is farther from the fulcrum than the effort, then a relatively large effort is required to move a small load (but at greater speed). This is called a **mechanical disadvantage**. A shovel functions as lever that operates at a mechanical disadvantage because the load is farther from the fulcrum than the effort (FIGURE 11.20b). This arrangement favors speed and range of motion over force. Most levers in the body operate at a mechanical disadvantage. An example is the lever system formed by the elbow joint, the biceps muscle, and the bones of the forearm (FIGURE 11.20b). As you have seen, in flexing the forearm, the elbow joint is the fulcrum (F), the contraction of the biceps muscle provides the effort (E), and the weight of the hand and forearm is the load (L).

Series Elastic Elements

Some parts of a muscle are elastic: They stretch slightly before they transfer the tension generated by the sliding filaments. These elastic components, known as **series elastic elements**, include titin molecules (see FIGURE 11.4), connective tissue around the muscle fibers, and tendons that attach muscle to bone. As the cells of a skeletal muscle start to shorten, they first pull on their connective tissue coverings and tendons. The coverings and tendons stretch and then become

FIGURE 11.20 Lever systems.

In a lever system that operates at a mechanical advantage, the load is closer to the fulcrum than the effort; in a lever system that operates at a mechanical disadvantage, the load is farther from the fulcrum than the effort.

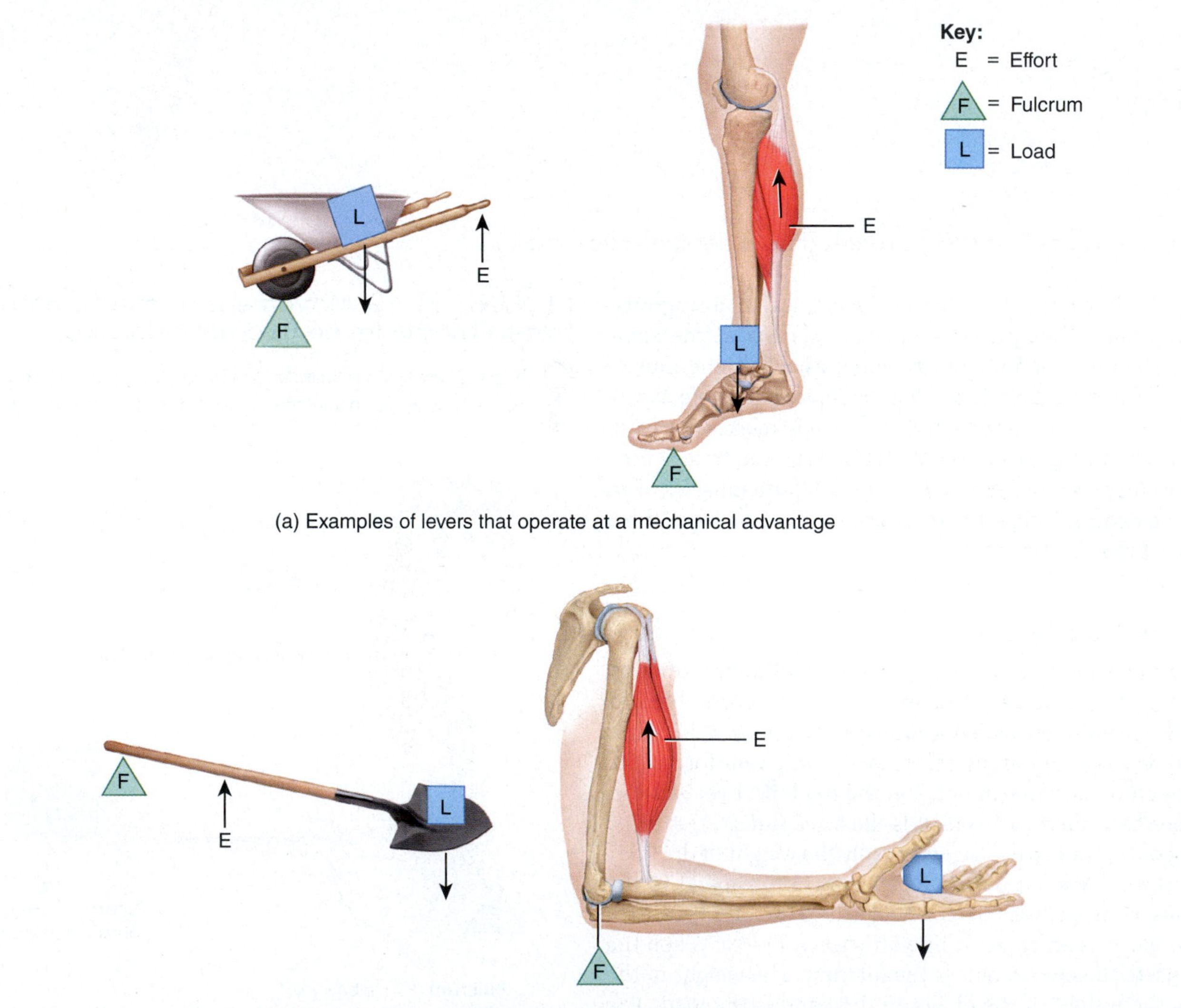

(a) Examples of levers that operate at a mechanical advantage

(b) Examples of levers that operate at a mechanical disadvantage

Do most lever systems in the body operate at a mechanical advantage or a mechanical disadvantage?

FIGURE 11.21 Comparison between isotonic (concentric and eccentric) and isometric contractions. Parts (a) and (b) show isotonic contractions of the biceps muscle in the arm; part (c) shows isometric contractions of shoulder and arm muscles.

In an isotonic contraction, tension remains constant as muscle length decreases or increases; in an isometric contraction, tension increases greatly without a change in muscle length.

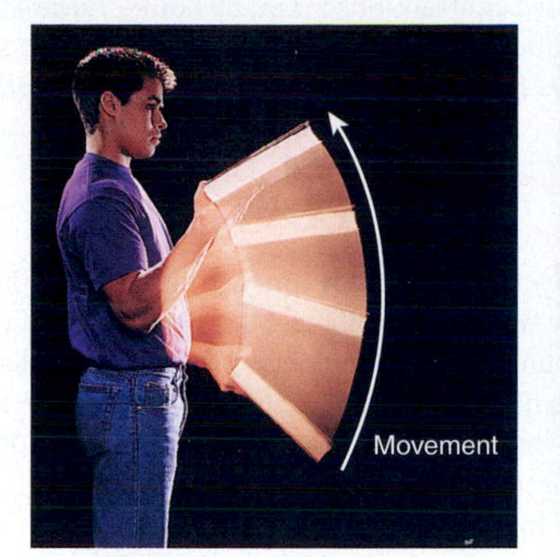

(a) Concentric contraction while picking up a book

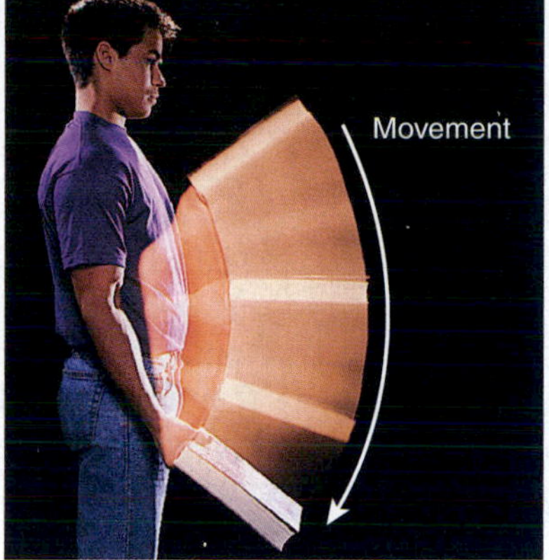

(b) Eccentric contraction while lowering a book

(c) Isometric contraction while holding a book steady

What type of contraction occurs in your neck muscles to keep your head upright while you are walking?

taut, and the tension passed through the tendons pulls on the bones to which they are attached. The result is movement of a part of the body. You are just about to learn, however, that the contraction cycle does not always result in shortening of the muscle fibers and the whole muscle. In some contractions, the myosin crossbridges pivot and generate tension, but the thin filaments cannot slide inward because the tension they generate is not large enough to move the load on the muscle.

There Are Two Major Categories of Muscle Contractions: Isotonic and Isometric

Muscle contractions are classified as either isotonic or isometric. In an **isotonic contraction** (*isotonic* = constant tension), the tension developed by the muscle remains almost constant while the muscle changes its length. Isotonic contractions are used for body movements and for moving objects. The two types of isotonic contractions are concentric and eccentric. In a **concentric isotonic contraction**, the tension generated is great enough to exceed the load and the muscle shortens, pulling on another structure (such as a tendon) to produce movement. Picking a book up off a table involves concentric isotonic contractions of the biceps muscle in the arm (FIGURE 11.21a). By contrast, as you lower the book to place it back on the table, the previously shortened biceps lengthens in a controlled manner while it continues to contract. When the length of a muscle increases during a contraction, the contraction is an **eccentric isotonic contraction** (FIGURE 11.21b). During an eccentric contraction, the tension exerted by the myosin crossbridges resists movement of a load (the book, in this case) and slows the lengthening process. For reasons that are not well understood, repeated eccentric isotonic contractions (for example, walking downhill) produce more muscle damage and more delayed-onset muscle soreness than concentric isotonic contractions.

In an **isometric contraction** (*isometric* = constant length), the tension generated is not enough to exceed the load, and the muscle does not change its length. Isometric contractions occur when you try to lift an object that is too heavy for you to move. Muscles also contract isometrically in order to maintain posture and for supporting objects in a fixed position. An example of the latter is holding a book steady using an outstretched arm (FIGURE 11.21c). The book pulls the arm downward, stretching the shoulder and arm muscles. The isometric contraction of the shoulder and arm muscles counteracts the stretch. Although isometric contractions do not result in body movement, energy is still expended. Most activities that occur throughout the day include both isotonic and isometric contractions.

A major factor that determines the velocity of muscle shortening is the load. For concentric isotonic contractions, the load and velocity of shortening are *inversely* related. This **load–velocity relationship** works as follows (FIGURE 11.22): (1) When the load is zero, the

FIGURE 11.22 Load–velocity relationship.

For a concentric isotonic contraction, the load and the velocity of muscle shortening are inversely related.

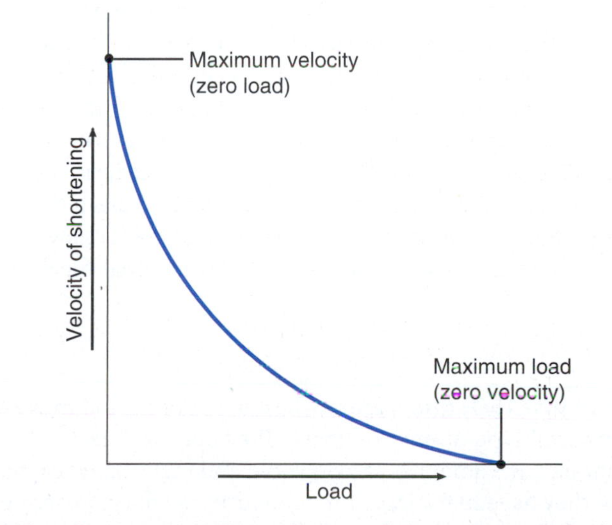

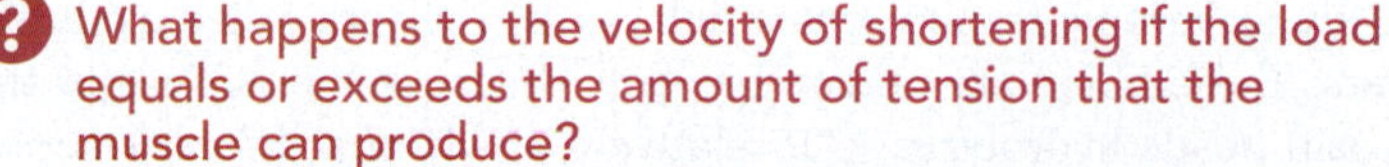

What happens to the velocity of shortening if the load equals or exceeds the amount of tension that the muscle can produce?

velocity of shortening is maximal; (2) as the load increases, the velocity of shortening decreases; and (3) when the load is equal to or exceeds the maximum tetanic tension that the muscle can produce, the velocity of shortening is zero and the contraction becomes isometric. You may not know it, but you are already familiar with the load–velocity relationship from common experience: You can lift a light object (a magazine) very quickly, a moderately weighted object (a textbook) less quickly, and a heavy object (a box of textbooks) only very slowly.

CHECKPOINT

16. What is the difference among twitch, wave summation, unfused tetanus, and fused tetanus?
17. What is the significance of the length–tension relationship for a skeletal muscle fiber?
18. What is motor unit recruitment? Why is it important?
19. Describe a lever system in the body that operates at a mechanical advantage and one that operates at a mechanical disadvantage.
20. Define each of the following terms: concentric isotonic contraction, eccentric isotonic contraction, and isometric contraction.
21. Explain the load–velocity relationship for concentric isotonic contractions.

11.6 Types of Skeletal Muscle Fibers

OBJECTIVE

- Describe the differences among the three types of skeletal muscle fibers.

Skeletal muscle fibers are not all alike in composition and function. For example, muscle fibers vary in their content of myoglobin, the red-colored protein that binds oxygen in muscle fibers. Skeletal muscle fibers that have a high myoglobin content are termed *red muscle fibers* and appear darker (the dark meat in chicken legs and thighs); those that have a low content of myoglobin are called *white muscle fibers* and appear lighter (the white meat in chicken breasts). Red muscle fibers also contain more mitochondria and are supplied by more blood capillaries.

Skeletal muscle fibers also contract and relax at different speeds, vary in which metabolic reactions they use to generate ATP, and differ in how quickly they fatigue. For example, a fiber is categorized as either slow or fast depending on how rapidly the ATPase in its myosin heads hydrolyzes ATP. Based on all of these structural and functional characteristics, skeletal muscle fibers are classified into three main types: (1) slow oxidative fibers, (2) fast oxidative–glycolytic fibers, and (3) fast glycolytic fibers.

Slow Oxidative Fibers Have a High Resistance to Fatigue

Slow oxidative (SO) fibers are smallest in diameter and thus are the least powerful type of muscle fibers. They appear dark red because they contain large amounts of myoglobin and many blood capillaries. Because they have many large mitochondria, SO fibers generate ATP mainly by aerobic respiration, which is why they are called oxidative fibers. These fibers are said to be "slow" because the ATPase in the myosin heads hydrolyzes ATP relatively slowly and the contraction cycle proceeds at a slower pace than in "fast" fibers. As a result, SO fibers have a slow speed of contraction. Their twitch contractions last from 100–200 msec, and they take longer to reach peak tension. However, slow fibers are very resistant to fatigue and are capable of prolonged, sustained contractions for many hours. These slow-twitch, fatigue-resistant fibers are adapted for maintaining posture and for aerobic, endurance-type activities such as running a marathon.

Fast Oxidative–Glycolytic Fibers Have a Moderate Resistance to Fatigue

Fast oxidative–glycolytic (FOG) fibers are intermediate in diameter between the other two types of fibers. Like slow oxidative fibers, they contain large amounts of myoglobin and many blood capillaries. Thus, they also have a dark red appearance. FOG fibers can generate considerable ATP by aerobic respiration, which gives them a moderate resistance to fatigue. Because their intracellular glycogen level is high, they also generate ATP by anaerobic glycolysis. FOG fibers are fast because the ATPase in their myosin heads hydrolyzes ATP three to five times faster than the myosin ATPase in SO fibers, which makes their speed of contraction faster. Thus, twitches of FOG fibers reach peak tension more quickly than those of SO fibers but are briefer in duration—less than 100 msec. FOG fibers contribute to activities such as walking and sprinting.

Fast Glycolytic Fibers Have a Low Resistance to Fatigue

Fast glycolytic (FG) fibers are largest in diameter and contain the most myofibrils. Hence, they can generate the most powerful contractions. FG fibers have low myoglobin content, relatively few blood capillaries, few mitochondria, and appear white in color. They contain large amounts of glycogen and generate ATP mainly by anaerobic glycolysis. Due to their large size and their ability to hydrolyze ATP rapidly, FG fibers contract strongly and quickly. These fast-twitch fibers are adapted for intense anaerobic movements of short duration, such as weight lifting or throwing a ball, but they fatigue quickly. Strength training programs that engage a person in activities requiring great strength for short times increase the size, strength, and glycogen content of fast glycolytic fibers. The FG fibers of a weight lifter may be 50% larger than those of a sedentary person or an endurance athlete. The increase in size is due to increased synthesis of muscle proteins. The overall result is muscle enlargement due to hypertrophy of the FG fibers.

The Three Types of Skeletal Muscle Fibers Are Distributed Differently and Recruited in a Certain Order

Most skeletal muscles are a mixture of all three types of skeletal muscle fibers; about half the fibers in a typical skeletal muscle are SO fibers. However, the proportions vary somewhat, depending on the action of the muscle, the person's training regimen, and genetic factors. For example, the continually active postural muscles of the neck, back, and legs have a high proportion of SO fibers. Muscles of the shoulders and arms, in contrast, are not constantly active but are used briefly now and then to produce large amounts of tension, such as in lifting and throwing. These muscles have a high proportion of FG fibers. Leg muscles, which not only support the body but are also used for walking and running, have large numbers of both SO and FOG fibers.

Within a particular motor unit, all of the skeletal muscle fibers are of the same type. The different motor units in a muscle are recruited in a specific order, depending on need. For example, if weak contractions suffice to perform a task, only SO motor units are activated. If more force is needed, the motor units of FOG fibers are also recruited. Finally, if maximal force is required, motor units of FG fibers are also called into action. Activation of various motor units is controlled by the brain and spinal cord.

Exercise Can Induce Changes in the Different Types of Skeletal Muscle Fibers

The relative ratio of FG and SO fibers in each muscle is genetically determined and helps account for individual differences in physical performance. For example, people with a higher proportion of FG fibers often excel in activities that require periods of intense activity, such as weight lifting or sprinting. People with higher percentages of SO fibers are better at activities that require endurance, such as long-distance running.

Although the total number of skeletal muscle fibers usually does not increase, the characteristics of those present can change to some extent. Various types of exercises can induce changes in the fibers in a skeletal muscle. Endurance-type (aerobic) exercises, such as running or swimming, cause a gradual transformation of some FG fibers into FOG fibers. The transformed muscle fibers show slight increases in diameter, number of mitochondria, blood supply, and strength. Endurance exercises also result in cardiovascular and respiratory changes that cause skeletal muscles to receive better supplies of oxygen and nutrients, but they do not increase muscle mass. By contrast, exercises that require great strength for short periods produce an increase in the size and strength of FG fibers. The increase in size is due to increased synthesis of thick and thin filaments. The overall result is muscle enlargement (hypertrophy), as evidenced by the bulging muscles of bodybuilders.

TABLE 11.3 summarizes the characteristics of the three types of skeletal muscle fibers.

TABLE 11.3 Characteristics of the Three Types of Skeletal Muscle Fibers

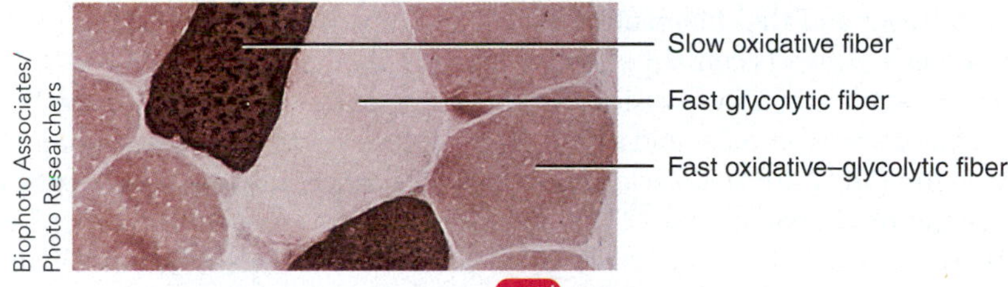

Transverse section of three types of skeletal muscle fibers

Characteristic	Slow Oxidative (SO) Fibers	Fast Oxidative–Glycolytic (FOG) Fibers	Fast Glycolytic (FG) Fibers
Fiber diameter	Smallest.	Intermediate.	Largest.
Myoglobin content	Large amount.	Large amount.	Small amount.
Mitochondria	Many.	Many.	Few.
Capillaries	Many.	Many.	Few.
Color	Red.	Red-pink.	White (pale).
Capacity for generating ATP and method used	High capacity, by aerobic respiration.	Intermediate capacity, by both aerobic respiration and anaerobic glycolysis.	Low capacity, by anaerobic glycolysis.
Rate of ATP hydrolysis by myosin ATPase	Slow.	Fast.	Fast.
Contraction velocity	Slow.	Fast.	Fast.
Fatigue resistance	High.	Intermediate.	Low.
Creatine kinase	Lowest amount.	Intermediate amount.	Highest amount.
Glycogen stores	Low.	Intermediate.	High.
Order of recruitment	First.	Second.	Third.
Location where fibers are abundant	Postural muscles such as those of the neck.	Lower limb muscles.	Upper limb muscles.
Primary functions of fibers	Maintaining posture and aerobic endurance activities.	Walking, sprinting.	Rapid, intense movements of short duration.

CHECKPOINT

22. Why are some skeletal muscle fibers classified as "fast" and others are said to be "slow"?
23. In what order are the various types of skeletal muscle fibers recruited when you sprint to make it to the bus stop?
24. On a cellular level, what causes muscle hypertrophy?

11.7 Cardiac Muscle

OBJECTIVE

- Describe the properties of cardiac muscle.

Cardiac muscle is found only in the heart, where it forms the bulk of the heart wall (see **FIGURE 11.1b**). Individual cardiac muscle fibers are about 50–100 μm long and 14 μm in diameter. They are branched and usually have only one centrally located nucleus. Like skeletal muscle fibers, cardiac muscle fibers are striated due to the presence of repeating sarcomeres consisting of thick and thin filaments that have a regular pattern of overlap. The thick filaments contain myosin, and the thin filaments contain actin, troponin, and tropomyosin. Also present in cardiac muscle fibers are transverse (T) tubules and a moderately developed sarcoplasmic reticulum (SR). Cardiac muscle fibers rely almost exclusively on aerobic respiration to generate the ATP they need for muscle contraction. As a result, these fibers contain large numbers of mitochondria.

Unique to cardiac muscle fibers are *intercalated discs* (in-TER-ka-lāt-ed) (see **FIGURE 11.1b**). These microscopic structures are irregular transverse thickenings of the sarcolemma that connect the ends of cardiac muscle fibers to one another. The discs contain *desmosomes*, which hold the fibers together, and *gap junctions*, which allow muscle action potentials to spread from one cardiac muscle fiber to another. Because cardiac muscle fibers are electrically coupled by gap junctions, when an action potential is generated in a mass of cardiac muscle fibers, it quickly spreads to all of the muscle fibers in that mass and then the muscle fibers contract together. Such a mass of interconnected muscle fibers acts as a single, coordinated unit or **functional syncytium** (sin-SISH-ē-um; plural is *syncytiums* or *syncytia*). The atria (upper chambers) and ventricles (lower chambers) of the heart behave as two distinct functional syncytiums. Thus, the atria and ventricles contract independently of each other, with the atria contracting before the ventricles. This allows the ventricles to fill with blood from the atria before the ventricles eject blood out of the heart to the rest of the body.

You have already learned that skeletal muscle contracts only when stimulated by acetylcholine released by a nerve action potential in a somatic motor neuron. By contrast, cardiac muscle does not require any external stimulation to contract. Contractions occur because action potentials within cardiac muscle itself are spontaneously generated on a periodic basis. This built-in rhythm, termed **autorhythmicity**, is a major physiological difference between cardiac muscle and skeletal muscle. Cardiac muscle, as a functional syncytium, consists of two types of cells: a small number of *autorhythmic fibers* (also known as *pacemaker cells*) and a large number of *contractile fibers*. Autorhythmic fibers spontaneously generate action potentials. They are unable to contract, however, because they contain essentially no myofibrils. Contractile fibers have the necessary myofibrils to contract, but do not have the ability to initiate action potentials. Instead, they become excited and then contract together in response to action potentials conducted to them from autorhythmic fibers via gap junctions.

Once an action potential occurs in a contractile muscle fiber, it ultimately causes the muscle fiber to contract by increasing the Ca^{2+} concentration in the sarcoplasm. Recall that the sequence of events that connects the muscle action potential to muscle contraction is called *excitation–contraction (EC) coupling*. Like skeletal muscle, EC coupling in cardiac muscle involves L-type voltage-gated Ca^{2+} channels (dihydropyridine receptors) in the membrane of transverse (T) tubules and Ca^{2+} release channels (ryanodine receptors) in the terminal cisternal membrane of the sarcoplasmic reticulum (SR). However, in cardiac muscle fibers, the L-type voltage-gated channels are not arranged in tetrads nor are they mechanically coupled to the Ca^{2+} release channels. Instead, each L-type voltage-gated Ca^{2+} channel in the T-tubule membrane is close to, and has an effect on, only one Ca^{2+} release channel in the SR membrane (**FIGURE 11.23**). Excitation–contraction coupling in cardiac muscle fibers begins as the cardiac muscle action potential travels along the sarcolemma and into the T tubules, where it causes the L-type voltage-gated Ca^{2+} channels to open. The entering Ca^{2+} functions as *trigger* Ca^{2+} that binds to and opens the Ca^{2+} release channels in the SR membrane. As a result, additional Ca^{2+} enters the sarcoplasm from the SR. The process by which extracellular Ca^{2+} triggers the release of additional Ca^{2+} from the SR is known as **Ca^{2+}-induced Ca^{2+} release (CICR)**. About 90% of the Ca^{2+} needed for contraction of a cardiac muscle fiber comes from the SR via CICR; the remaining 10% comes from the extracellular fluid. So, in cardiac muscle, the Ca^{2+} required for contraction is derived from two sources: the SR and extracellular fluid. This is contrast to skeletal muscle, in which all of the requisite Ca^{2+} comes from the SR. Although extracellular Ca^{2+} does enter the sarcoplasm of a skeletal muscle fiber through L-type voltage-gated Ca^{2+} channels when the cell is excited, studies have shown that EC coupling in a skeletal muscle fiber can occur whether extracellular Ca^{2+} is present or absent.

After the Ca^{2+} concentration in the sarcoplasm increases, cardiac muscle fibers contract in the same way as skeletal muscle fibers: Ca^{2+} binds to troponin, moving tropomyosin away from myosin-binding sites on actin, and the thick and thin filaments begin sliding past one another. Cardiac muscle fibers can produce graded contractions (contractions that vary in strength) by increasing the amount of Ca^{2+} that enters the sarcoplasm, which in turn increases myosin crossbridge formation.

Contraction of cardiac muscle fibers ends when the Ca^{2+} concentration in the sarcoplasm decreases to resting levels. This occurs via (1) Ca^{2+}-ATPase pumps in the SR membrane that actively transport Ca^{2+} from the sarcoplasm into the SR and (2) Na^+–Ca^{2+} exchangers in the sarcolemma that actively transport Ca^{2+} out of the cell in exchange for Na^+ movement into the cell.

Like neurons and skeletal muscle fibers, cardiac muscle fibers have a *refractory period*, the period of time after an action potential

FIGURE 11.23 Excitation–contraction coupling in cardiac muscle.

In a contractile cardiac muscle fiber, an action potential opens L-type voltage-gated Ca^{2+} channels; extracellular Ca^{2+} moves into the cell through these channels and then causes Ca^{2+}-induced Ca^{2+} release (CICR) from the sarcoplasmic reticulum.

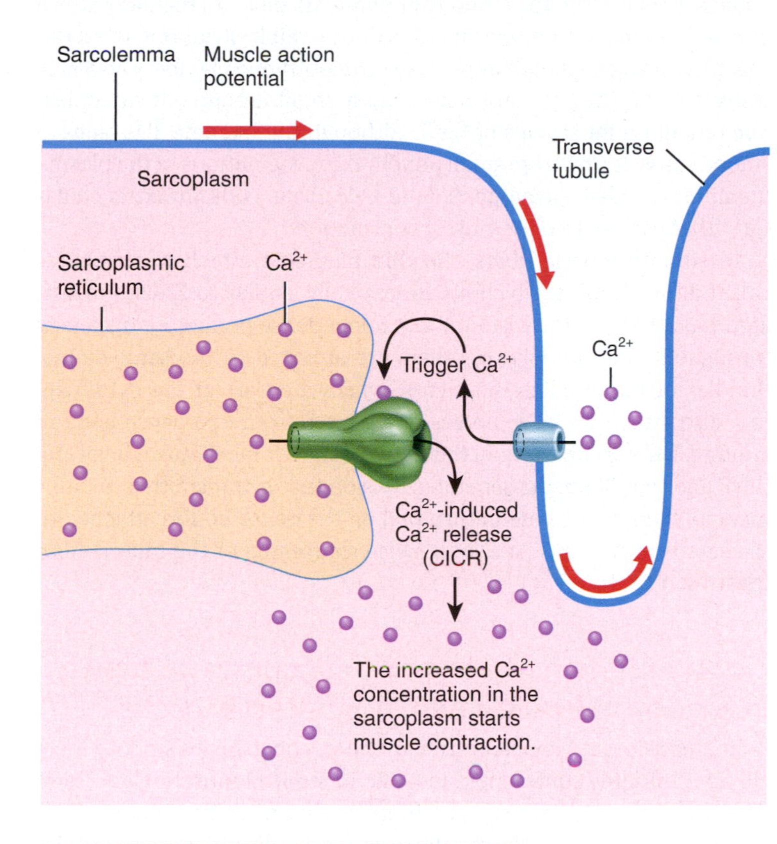

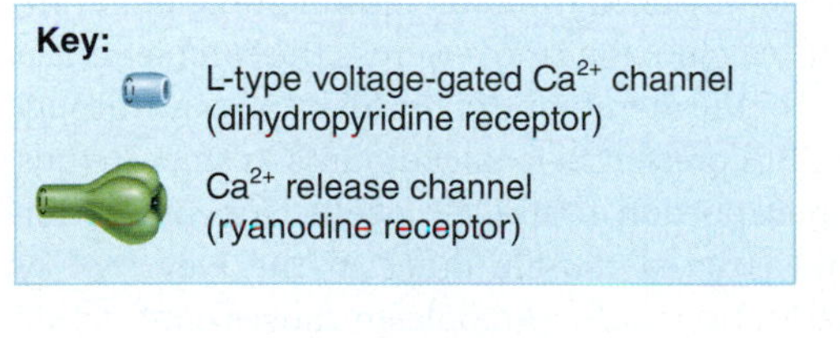

How does extracellular Ca^{2+} cause CICR?

begins when an excitable cell temporarily loses its excitability. The refractory period in contractile cardiac muscle fibers is long: It lasts about 250 msec, which is almost as long as the duration of contraction (300 msec) (**FIGURE 11.24**). Consequently, a cardiac muscle fiber cannot be reexcited until its previous contraction is almost over. For this reason, summation of contractions and tetanus do not occur in cardiac muscle. If cardiac muscle could undergo tetanus, the heart would not be able to function as a pump because it would not have a chance to relax and fill up with blood, a situation that would be lethal.

Under normal resting conditions, cardiac muscle tissue contracts and relaxes about 75 times a minute. Although cardiac muscle can contract on its own in the absence of external stimulation, it is innervated by the autonomic nervous system (ANS), which alters the *rate* and *strength* of cardiac muscle contractions. These effects of the ANS on cardiac muscle activity usually occur without conscious control. Other factors that influence cardiac muscle activity include hormones (such as epinephrine and norepinephrine from the adrenal glands); the concentrations of K^+, Ca^{2+}, and Na^+ in extracellular fluid; and body temperature.

You will learn more details about cardiac muscle in the discussion of the heart in Chapter 14.

CHECKPOINT

25. Why is cardiac muscle said to behave as a functional syncytium?

26. Compare and contrast excitation–contraction coupling in cardiac muscle and skeletal muscle.

27. What is the significance of the long refractory period in cardiac muscle?

FIGURE 11.24 Relationship among the action potential, refractory period, and tension developed in a contractile cardiac muscle fiber.

In cardiac muscle, the refractory period (about 250 msec) lasts almost as long as the duration of contraction (300 msec), which prevents a contractile fiber from being reexcited until its previous contraction is almost over; consequently, summation of contractions and tetanus do not occur.

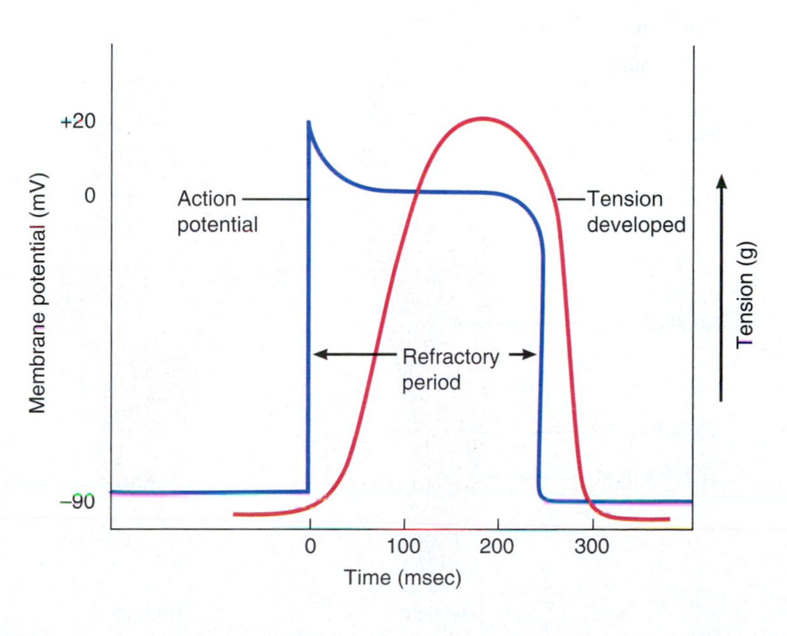

What would happen if cardiac muscle could undergo tetanus?

11.8 Smooth Muscle

OBJECTIVES

- Describe how smooth muscle contracts.
- Distinguish between single-unit smooth muscle and multi-unit smooth muscle.
- Explain the mechanism of excitation–contraction coupling in smooth muscle.

Smooth muscle is found mainly in the walls of hollow organs and tubes, where it contracts to move substances through the interior spaces of these structures. Of the three types of muscle, smooth muscle is the most variable because it has such a wide range of properties.

The Components of a Smooth Muscle Fiber Allow Considerable Tension to Develop During Contraction

A relaxed smooth muscle fiber is 30–200 μm long and 3–8 μm in diameter. It is spindle-shaped (thickest in the middle and tapered at each end) and has a single, centrally located nucleus (FIGURE 11.25). The sarcoplasm of smooth muscle fibers contains both thick filaments and thin filaments, but they are not arranged in orderly sarcomeres, as in striated muscle. The thick filaments contain myosin, and the entire length of each thick filament has myosin heads projecting from it—an arrangement that helps smooth muscle exert a considerable amount of tension during contraction. (Recall that in striated muscle, myosin heads are not present in the center of the thick filament.) The thin filaments contain actin and tropomyosin, but troponin is absent. Thin filaments are more abundant in smooth muscle fibers than in striated muscle fibers, with 10 to 15 thin filaments per thick filament in smooth muscle compared to two thin filaments per thick filament in striated muscle. Because the thick and thin filaments have no regular pattern of overlap, smooth muscle fibers do not exhibit striations (FIGURE 11.25), causing a smooth appearance. Smooth muscle fibers also lack transverse (T) tubules and have only a small amount of sarcoplasmic reticulum for storage of Ca^{2+}. Although there are no T tubules in smooth muscle, there are small pouch-like invaginations of the plasma membrane called **caveolae** (kav′-ē-Ō-lē) that contain extracellular Ca^{2+} that can be used for muscle contraction.

In smooth muscle fibers, the thin filaments attach to structures called **dense bodies**, which are functionally similar to Z discs in striated muscle fibers (FIGURE 11.25). Some dense bodies are dispersed throughout the sarcoplasm; others are attached to the sarcolemma. Bundles of intermediate filaments, which are part of the cytoskeleton, also attach to dense bodies and stretch from one dense body to another. During contraction, the sliding filament mechanism involving thick and thin filaments generates tension that is transmitted to intermediate filaments. These in turn pull on the dense bodies attached to the sarcolemma, causing a lengthwise shortening of the muscle fiber (FIGURE 11.25).

FIGURE 11.25 Microscopic organization of a smooth muscle fiber. A photomicrograph of smooth muscle is shown in FIGURE 11.1c.

Smooth muscle fibers have thick and thin filaments but no transverse tubules and only a small amount of sarcoplasmic reticulum.

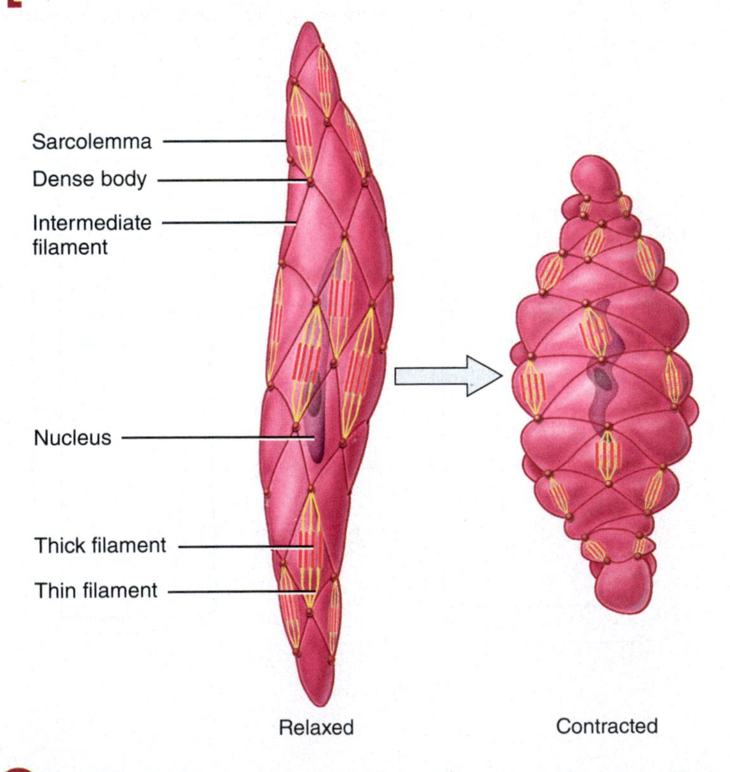

Do the thick and thin filaments of a smooth muscle fiber have a regular pattern of overlap? Explain your answer.

Contraction and Relaxation Occur More Slowly in Smooth Muscle Than in Striated Muscle

An increase in Ca^{2+} concentration in the sarcoplasm of a smooth muscle fiber initiates contraction, just like in striated muscle. Ca^{2+} flows into smooth muscle sarcoplasm from two sources: Most comes from extracellular fluid and the rest from the sarcoplasmic reticulum (SR). To enter the sarcoplasm, calcium ions move across the sarcolemma of the smooth muscle fiber or the membrane of the SR by passing through ion channels. Because SR is present in small amounts in smooth muscle, it provides only a small portion of the Ca^{2+} needed for contraction.

Smooth muscle and striated muscle differ in the way that an increase in Ca^{2+} concentration in the sarcoplasm causes contraction. As you already know, in striated muscle fibers, Ca^{2+} binds to troponin, causing tropomyosin to move away from myosin-binding sites on actin. Once the myosin-binding sites are exposed, myosin attaches to actin, and muscle contraction begins. In smooth muscle fibers, the thin filaments lack troponin and, although they contain tropomyosin, tropomyosin does not cover the myosin-binding sites on actin. In addition, myosin molecules in smooth muscle can bind to actin only after phosphate groups are added to light chains in the myosin heads. An increase in Ca^{2+} concentration in the sarcoplasm of a smooth muscle fiber causes contraction in the following way (FIGURE 11.26):

1. Ca^{2+} binds to **calmodulin**, a regulatory protein in the sarcoplasm that is similar in structure to troponin.
2. The Ca^{2+}–calmodulin complex activates an enzyme called **myosin light chain kinase (MLCK)**, which is also present in the sarcoplasm.
3. Activated MLCK in turn phosphorylates (adds a phosphate group to) light chains in the myosin heads.

FIGURE 11.26 Contraction of smooth muscle. The abbreviation MLCK stands for myosin light chain kinase.

In smooth muscle, contraction occurs once the myosin heads of the thick filaments are phosphorylated.

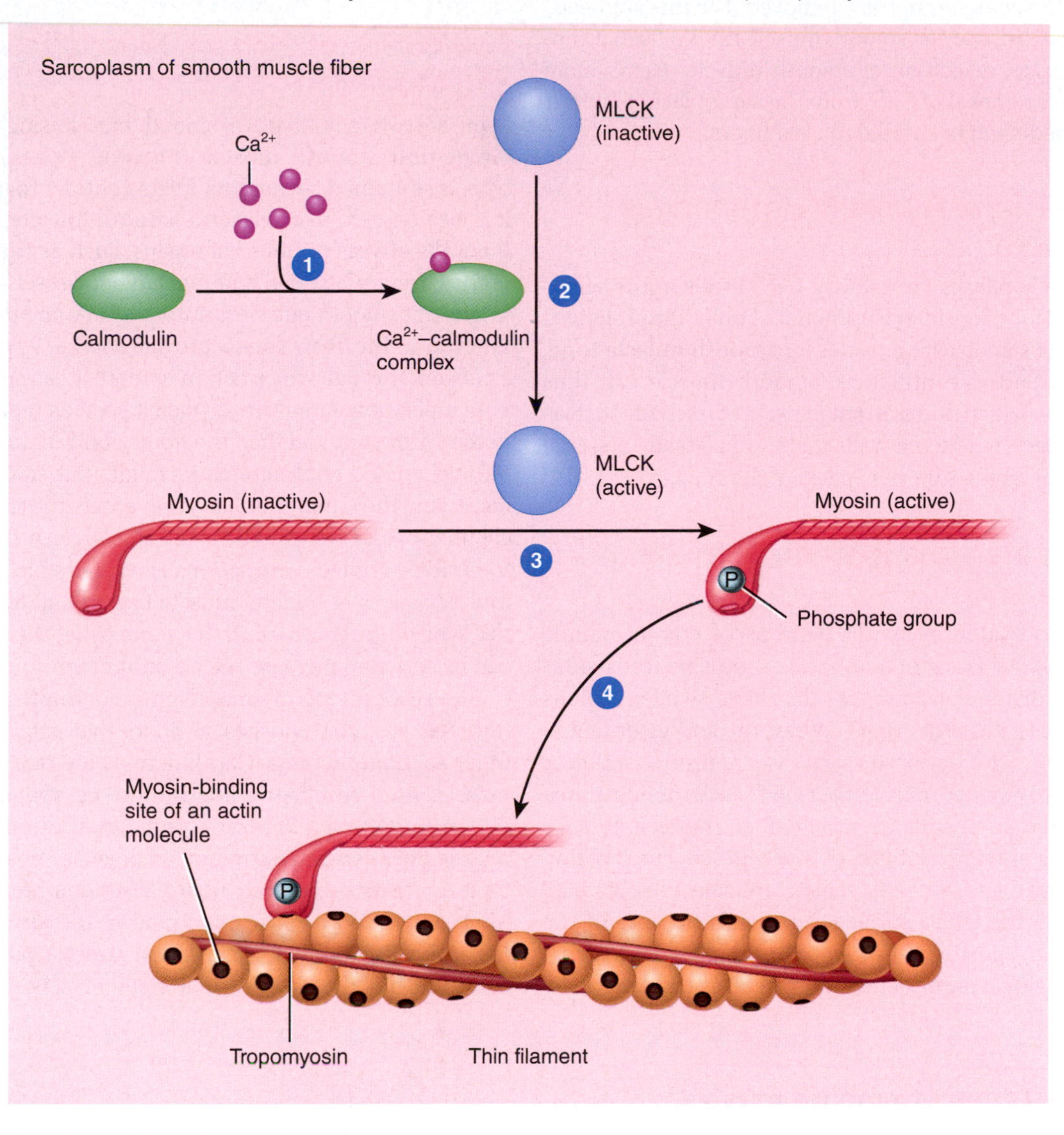

What does MLCK do and how is it activated?

4 The phosphorylated myosin heads bind to actin, and muscle contraction begins.

Therefore, contraction in smooth muscle is triggered by calcium-induced changes in the *thick* filaments, whereas contraction in striated muscle is triggered by calcium-induced changes in the *thin* filaments.

Contraction of smooth muscle starts more slowly and lasts much longer than contraction of striated muscle. A smooth muscle fiber contraction can last as long as 3 seconds. By contrast, a skeletal muscle fiber contraction lasts about 20–200 msec, and a cardiac muscle fiber contraction lasts about 300 msec. The slow start and longer duration of contraction in smooth muscle occur because of several factors:

1. There are no transverse tubules in smooth muscle fibers, so it takes longer for Ca^{2+} to reach the filaments in the center of the fiber and trigger the contraction process.
2. Myosin light chain kinase works rather slowly.
3. The ATPase activity of myosin heads in smooth muscle is much slower than in striated muscle.
4. The myosin heads in smooth muscle can enter a **latch state**, in which they stay attached to actin for a longer portion of the cross-bridge cycle compared to myosin heads in striated muscle. Although the latch state is poorly understood, it is important because it allows smooth muscle to maintain tension for long periods of time with minimal ATP consumption.

Relaxation of smooth muscle involves two main steps: (1) decreasing the Ca^{2+} concentration in the sarcoplasm to resting levels and (2) dephosphorylating (removing the phosphate group from) light chains in the myosin heads. As in cardiac muscle fibers, Ca^{2+} is removed from the sarcoplasm of a smooth muscle fiber by Ca^{2+}-ATPase pumps in the SR membrane and by Na^+–Ca^{2+} exchangers in the sarcolemma. As the Ca^{2+} concentration in the sarcoplasm decreases, Ca^{2+} dissociates from calmodulin, and myosin light chain kinase becomes inactive.

Dephosphorylation of myosin heads occurs via the enzyme **myosin phosphatase**, which is present in the sarcoplasm of the smooth muscle fiber. Once the phosphate groups are removed, the myosin heads are unable to bind to actin, and the smooth muscle fiber relaxes. Compared to striated muscle, relaxation of smooth muscle occurs more slowly, mainly because removal of Ca^{2+} from the sarcoplasm is slower in smooth muscle fibers than in striated muscle fibers.

Smooth Muscle Tone Allows Maintenance of Steady Pressure

Because smooth muscle relaxes very slowly, Ca^{2+} levels in the sarcoplasm remain elevated for a considerable amount of time. The prolonged presence of Ca^{2+} in the sarcoplasm provides for **smooth muscle tone**, a state of continued partial contraction. Smooth muscle can thus sustain long-term tone, which is important in parts of the body such as the gastrointestinal tract, where the walls maintain a steady pressure on food material in the lumen (interior space) of the tract.

The Autonomic Nervous System Regulates Smooth Muscle

Smooth muscle is innervated by motor neurons of the autonomic nervous system (ANS). As axons of autonomic motor neurons enter smooth muscle, they divide into branches that have swollen regions called **varicosities** (see FIGURE 10.4). When an action potential occurs along the axon, the varicosities release neurotransmitters, which in turn diffuse to smooth muscle fibers in the surrounding area (see FIGURE 10.5). In a given smooth muscle fiber, the receptors for the neurotransmitters are not confined to a specific receptor region like they are at the motor end plate of a skeletal muscle fiber; instead, they are located along the entire surface of the cell. Because of the way the varicosities are positioned in smooth muscle, neurotransmitters released by a single autonomic motor neuron regulate many smooth muscle fibers, and each smooth muscle fiber may be regulated by neurotransmitters from different autonomic motor neurons.

Two Forms of Smooth Muscle Exist: Single-Unit and Multi-Unit

There are two major types of smooth muscle: single-unit and multi-unit. **Single-unit smooth muscle** (FIGURE 11.27a), the most common type, is so-named because its fibers contract together as a single unit. It is also referred to as **visceral smooth muscle** because it is found in the walls of viscera (internal organs) such as the stomach, intestines, uterus, urinary bladder, and many blood vessels. Like cardiac muscle, single-unit smooth muscle is autorhythmic and behaves as a functional syncytium. The fibers connect to one another by gap junctions, forming a network through which action potentials can spread. When an action potential is generated in one fiber, it spreads rapidly to all of the fibers in the syncytium and then the fibers contract in unison. As in cardiac muscle, graded contractions in single-unit smooth muscle occur by increasing the amount of Ca^{2+} that enters the sarcoplasm of the muscle fibers. Note that recruitment of a larger number of muscle fibers to produce graded contractions is not possible in single-unit smooth muscle as it is in skeletal muscle because all of the existing fibers in the syncytium contract at the same time; therefore, no more fibers can be added to increase the amount of tension generated.

The second type of smooth muscle, **multi-unit smooth muscle** (FIGURE 11.27b), consists of fibers that act independently of each other as multiple units. Gap junctions are rare in multi-unit smooth muscle. As a result, the fibers must be stimulated individually by nerves to contract. Whereas stimulation of one single-unit smooth muscle fiber causes contraction of adjacent fibers, stimulation of one multi-unit smooth muscle fiber causes contraction of that fiber only. Multi-unit smooth muscle is found in the airways to the lungs, the iris and ciliary body of the eye, the arrector pili muscles of the skin, and some blood vessels. As in skeletal muscle, graded contractions in

FIGURE 11.27 Types of smooth muscle.

Single-unit (visceral) smooth muscle fibers connect to one another by gap junctions and contract in unison; multi-unit smooth muscle fibers lack gap junctions and contract independently.

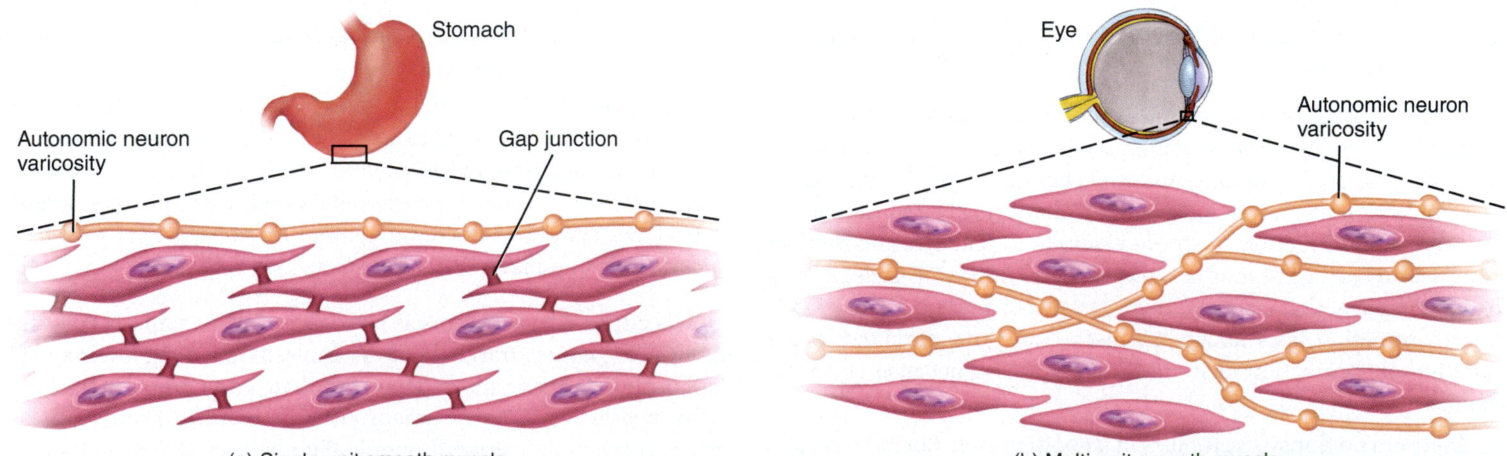

(a) Single-unit smooth muscle

(b) Multi-unit smooth muscle

Which type of smooth muscle is more like cardiac muscle than skeletal muscle, with respect to both its structure and function?

multi-unit smooth muscle occur by recruitment of additional muscle fibers. Compared to single-unit smooth muscle, multi-unit smooth muscle has a richer supply of ANS nerve endings (**FIGURE 11.27b**).

Smooth Muscle Can Exhibit Autorhythmicity

Like cardiac muscle, single-unit smooth muscle does not require external stimulation to contract because action potentials within the muscle itself are spontaneously generated. Single-unit smooth muscle, as a functional syncytium, consists of two types of cells: a small number of *autorhythmic fibers (pacemaker cells)*, which are usually grouped together, and a large number of *contractile fibers*. The autorhythmic fibers cannot contract but instead generate action potentials spontaneously on a periodic basis; this built-in rhythm is called *autorhythmicity*. The contractile fibers are able to contract, but they cannot initiate action potentials. They become excited and then contract together in response to action potentials conducted to them from autorhythmic fibers. Recall that multi-unit smooth muscle is not autorhythmic nor does it behave as a functional syncytium. All of the cells in multi-unit smooth muscle are contractile fibers that act as independent units.

Autorhythmic fibers of single-unit smooth muscle are able to initiate their own action potentials because they have unstable resting membrane potentials. The membrane potential begins at about −50 mV and then spontaneously depolarizes. If threshold is reached, an action potential is generated. Two types of spontaneous depolarizations can occur in autorhythmic fibers of single-unit smooth muscle: pacemaker potentials and slow-wave potentials. A **pacemaker potential** is a spontaneous depolarization that always reaches threshold and therefore triggers the production of an action potential (**FIGURE 11.28a**). After repolarization, the pacemaker potential starts to develop again and the cycle repeats. The pacemaker potential in an autorhythmic smooth muscle fiber is caused by either an increase in Ca^{2+} movement into the cell or a decrease in K^+ movement out of the cell. **Slow-wave potentials** are cycles of alternating depolarization and repolarization that do not necessarily reach threshold (**FIGURE 11.28b**). Sometimes, threshold is reached and an action potential is generated; on other occasions, threshold is not reached and an action potential does not occur. The mechanism underlying slow-wave potentials is thought to involve fluctuations in Na^+ movement out of the cell caused by periodic changes in the activity of Na^+/K^+ pumps.

The action potential generated in an autorhythmic single-unit smooth muscle fiber consists of two phases: a depolarizing phase and a repolarizing phase (**FIGURE 11.28a, b**). The depolarizing phase is caused by the opening of L-type voltage-gated Ca^{2+} channels, whereas the repolarizing phase is caused by the closure of the L-type voltage-gated Ca^{2+} channels and the opening of voltage-gated K^+ channels.

Contractile Smooth Muscle Fibers Can Produce Action Potentials When Excited by Autorhythmic Signals or Other Stimuli

Unlike autorhythmic fibers of smooth muscle, contractile fibers of smooth muscle have a stable resting membrane potential of about −50 mV. Contractile fibers of single-unit smooth muscle produce action potentials when they are depolarized to threshold by an autorhythmic signal or certain other stimuli. Contractile fibers of multi-unit smooth muscle, by contrast, do not usually produce action potentials,

FIGURE 11.28 Spontaneous depolarizations in autorhythmic smooth muscle fibers.

Two types of spontaneous depolarizations can occur in autorhythmic smooth muscle fibers: pacemaker potentials and slow-wave potentials.

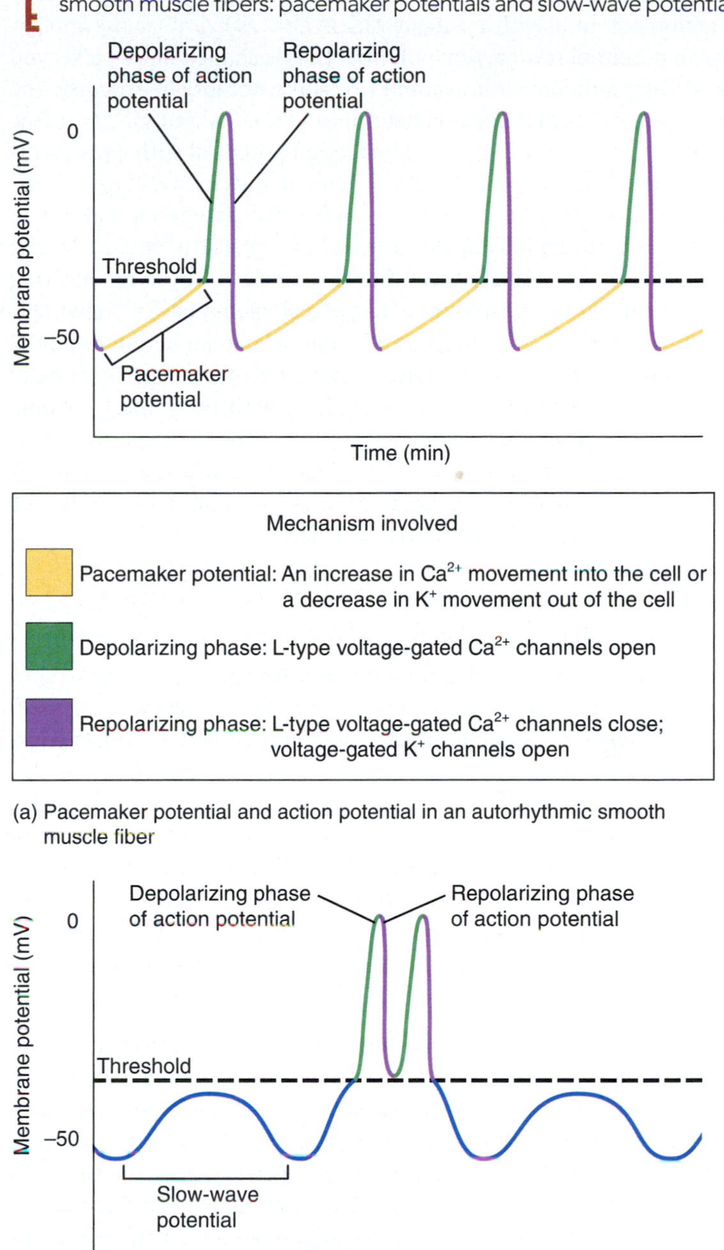

(a) Pacemaker potential and action potential in an autorhythmic smooth muscle fiber

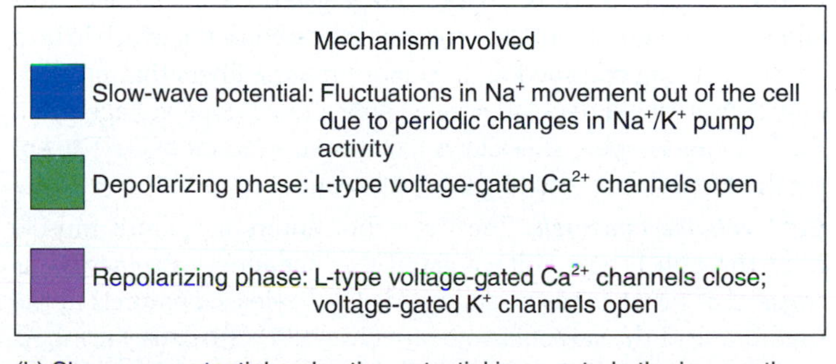

(b) Slow-wave potential and action potential in an autorhythmic smooth muscle fiber

What is the difference between a pacemaker potential and a slow-wave potential?

but they can respond to some stimuli by forming graded potentials (depolarizations or hyperpolarizations).

When a contractile single-unit smooth muscle fiber produces an action potential, the action potential can be either a spike potential or an action potential with a plateau (**FIGURE 11.29**). As its name implies, a spike potential is an action potential that is shaped like a spike; you are already familiar with examples of spike potentials—namely, the action potentials that occur in neurons and skeletal muscle fibers (see **FIGURE 7.21** and **FIGURE 11.9**). An action potential with a plateau is similar to the action potential that occurs in contractile cardiac muscle fibers (see **FIGURE 11.24**). For both spike potentials and action potentials with a plateau, the depolarizing phase is caused by the opening of L-type voltage-gated Ca^{2+} channels, and the repolarizing phase is caused by the closure of L-type voltage-gated Ca^{2+} channels and the opening of voltage-gated K^+ channels. In an action potential with a plateau, the plateau phase is caused by prolonged opening of L-type voltage-gated Ca^{2+} channels, along with the partial opening of voltage-gated K^+ channels. The duration of the two types of action potentials in contractile single-unit smooth muscle fibers varies: A spike potential usually lasts about 50 msec, and an action potential with a plateau typically lasts about 200 msec.

Excitation–Contraction Coupling in Smooth Muscle Involves Several Mechanisms

In striated muscle, an action potential must be generated before contraction can begin. In smooth muscle, however, contraction can occur over a relatively broad range of membrane potentials. A smooth muscle fiber can contract in response to an action potential; it can also contract in response to a subthreshold depolarization that never gives rise to an action potential; and it can even contract when there is no change in membrane potential at all. In all three cases, contraction occurs because the Ca^{2+} concentration in the sarcoplasm increases. The increase in sarcoplasmic Ca^{2+} concentration that triggers contraction of a smooth muscle fiber can be caused by the opening of one or more of the following types of channels (**FIGURE 11.30**):

- ***Voltage-gated channels.*** The sarcolemma of a smooth muscle fiber contains L-type voltage-gated Ca^{2+} channels that open in response to membrane depolarization, allowing Ca^{2+} to move from extracellular fluid into the sarcoplasm (**FIGURE 11.30**). The L-type voltage-gated Ca^{2+} channels in smooth muscle open in a graded fashion: As the strength of the depolarization increases, more channels open. In smooth muscle fibers that produce action potentials, the strong depolarization associated with the initial phase of the action potential opens a large number of L-type voltage-gated Ca^{2+} channels. This allows a large amount of Ca^{2+} to enter the sarcoplasm, which in turn causes a strong contraction. In smooth muscle fibers that produce only subthreshold depolarizations, only a few L-type voltage-gated Ca^{2+} channels open. This allows just a small amount of Ca^{2+} to enter the sarcoplasm, resulting in a weak contraction.
- ***Ca^{2+} release channels.*** The Ca^{2+} that enters a smooth muscle fiber through L-type voltage-gated Ca^{2+} channels also serves as *trigger Ca^{2+}* that binds to and opens Ca^{2+} release channels in the membrane of the sarcoplasmic reticulum (SR) (**FIGURE 11.30**). As a result, more Ca^{2+} is released into the sarcoplasm from the SR. Recall that the process by which extracellular Ca^{2+} triggers the release of additional Ca^{2+} from the SR is called *Ca^{2+}-induced Ca^{2+} release (CICR).* You first learned about CICR in cardiac muscle, where it

FIGURE 11.29 Action potentials in contractile smooth muscle fibers.

In contractile smooth muscle fibers that are capable of producing action potentials, the action potential can be either a spike potential or an action potential with a plateau.

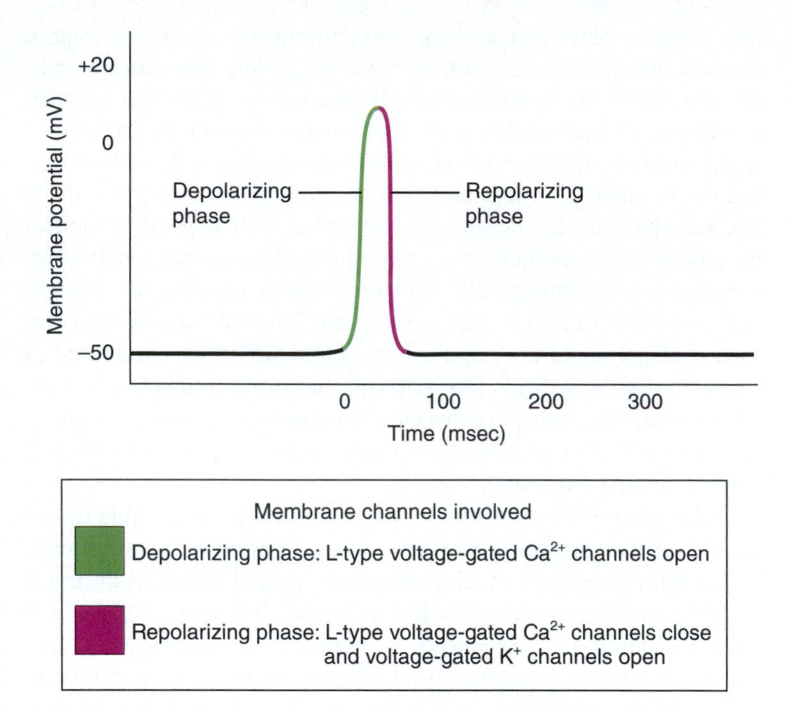

(a) Spike potential in a contractile smooth muscle fiber

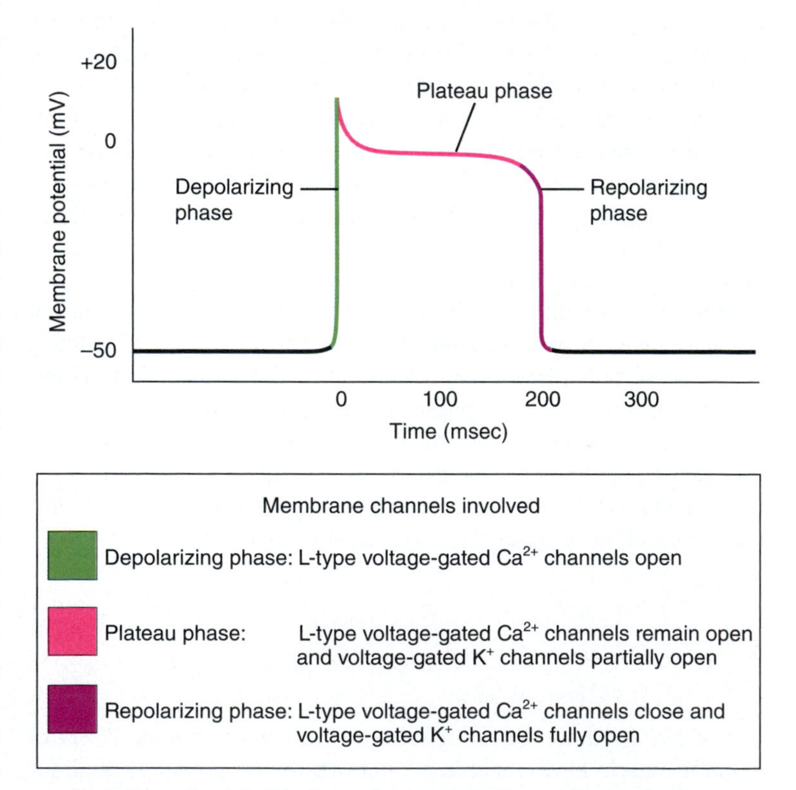

(b) Action potential with plateau in a contractile smooth muscle fiber

What other type of excitable cell has an action potential with a plateau?

FIGURE 11.30 Excitation–contraction coupling in smooth muscle.

The increase in sarcoplasmic Ca^{2+} concentration that triggers contraction of a smooth muscle fiber can be caused by the opening of one or more of the following types of channels: voltage-gated, Ca^{2+} release, receptor-activated, IP_3-gated, store-operated, and mechanically-gated.

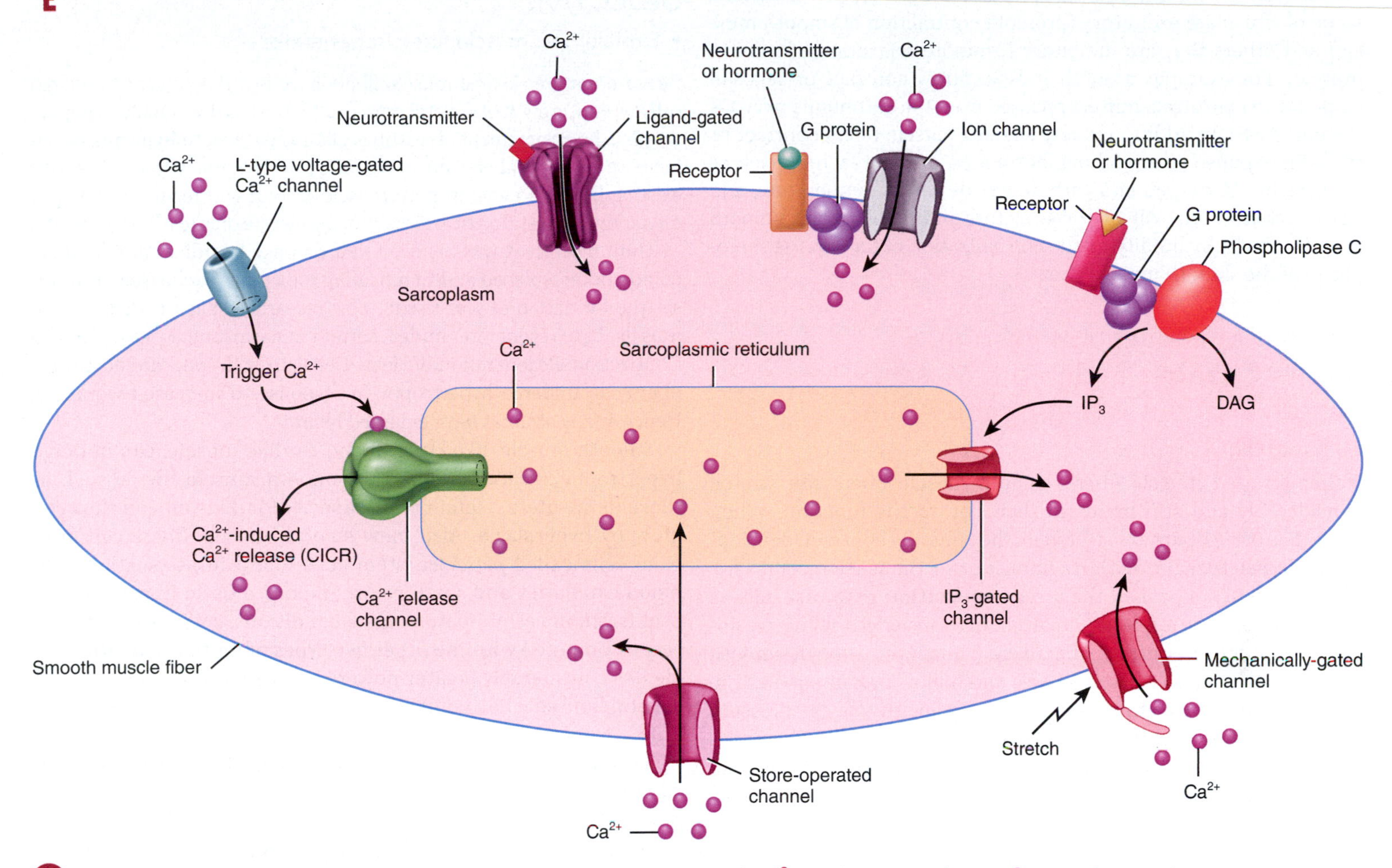

How do store-operated channels increase the concentration of Ca^{2+} in the sarcoplasm of smooth muscle?

provides the majority of the Ca^{2+} needed for contraction since cardiac muscle fibers have a moderately extensive SR with a large intracellular reserve of Ca^{2+}. In smooth muscle, however, CICR provides only a small amount of the Ca^{2+} required for contraction because the SR is present in small amounts and therefore has only a small intracellular Ca^{2+} reserve. Most of the Ca^{2+} needed for contraction in smooth muscle comes from extracellular fluid.

- ***Receptor-activated channels.*** Some neurotransmitters and hormones open receptor-activated channels in the sarcolemma of a smooth muscle fiber. Examples include ligand-gated channels and channels associated with G protein-coupled receptors (FIGURE 11.30). When these channels open, Ca^{2+} moves into the sarcoplasm from extracellular fluid.
- ***Inositol trisphosphate (IP_3)-gated channels.*** Although some neurotransmitters and hormones increase the sarcoplasmic Ca^{2+} concentration in smooth muscle by opening receptor-activated channels, others can increase the sarcoplasmic Ca^{2+} concentration by activating a second messenger pathway that opens inositol trisphosphate (IP_3)-gated channels (FIGURE 11.30). During this process, the neurotransmitter or hormone binds to a G protein-coupled receptor that activates the enzyme phospholipase C, which in turn causes the production of the second messengers, inositol trisphosphate (IP_3) and diacylglycerol (DAG). Once IP_3 is generated, it binds to and opens IP_3-gated Ca^{2+} channels in the SR membrane, causing the release of Ca^{2+} ions from the SR into the sarcoplasm.
- ***Store-operated channels.*** When the intracellular reserves of Ca^{2+} in the SR are depleted, a signal is relayed from the SR to the sarcolemma, where it causes **store-operated channels** to open (FIGURE 11.30). Opening these channels allows Ca^{2+} to enter the sarcoplasm from extracellular fluid. The entering Ca^{2+} can be used for contraction or to replenish the depleted Ca^{2+} stores in the SR.
- ***Mechanically-gated channels.*** The sarcolemma of a smooth muscle fiber contains mechanically-gated channels that are sensitive to stretch (FIGURE 11.30). Therefore, when a smooth muscle fiber is stretched, the mechanically-gated channels open, allowing extracellular Ca^{2+} to move into the sarcoplasm.

Several Factors Can Regulate Smooth Muscle Activity

The activity of smooth muscle is regulated by a variety of factors, some of which are excitatory (promote contraction of smooth muscle) and others that are inhibitory (promote relaxation of smooth muscle). For example, smooth muscle fibers contract or relax in response to neurotransmitters released from the autonomic nervous system (ANS). In addition, many smooth muscle fibers contract or relax in response to stretching, hormones, or local factors such as changes in pH, oxygen and carbon dioxide levels, temperature, and ion concentrations. All of these factors ultimately affect smooth muscle activity by modifying the concentration of Ca^{2+} in the sarcoplasm of the smooth muscle fiber.

The Stress–Relaxation Response Allows Changes in Smooth Muscle Length Without Affecting the Ability to Contract

Unlike striated muscle fibers, smooth muscle fibers can stretch considerably and still maintain their contractile function. When smooth muscle fibers are stretched, they initially contract, developing increased tension. Within a minute or so, the tension decreases. This phenomenon, called the **stress–relaxation response**, allows smooth muscle to undergo great changes in length while retaining the ability to contract effectively. Thus, even though smooth muscle in the walls of blood vessels and hollow organs such as the stomach, intestines, and urinary bladder can stretch, the pressure on the contents within them changes very little. After the organ empties, the smooth muscle in the wall rebounds, and the wall retains its firmness.

Smooth Muscle Produces ATP by Aerobic Respiration and Anaerobic Glycolysis

Because its energy demands are low, smooth muscle does not consume as much ATP as striated muscle. The principal method of ATP production in smooth muscle fibers is aerobic respiration. ATP can also be produced in these cells by anaerobic glycolysis.

CHECKPOINT

28. What role does myosin light chain kinase have in smooth muscle contraction?
29. What is the functional significance of the latch state?
30. Which type of smooth muscle contains fibers that function independently of one another?
31. What is the difference between a pacemaker potential and a slow-wave potential?
32. List examples of channels that can open in order to increase the sarcoplasmic Ca^{2+} concentration in a smooth muscle fiber.

11.9 Regeneration of Muscle

OBJECTIVE

- Explain how muscle fibers regenerate.

Because mature skeletal muscle fibers have lost the ability to undergo cell division, growth of skeletal muscle after birth is due mainly to **hypertrophy**, the enlargement of existing cells, rather than to **hyperplasia**, an increase in the number of fibers. Satellite cells divide slowly and fuse with existing fibers to assist in both muscle growth and repair of damaged fibers. Thus, skeletal muscle tissue can regenerate only to a limited extent.

Until recently it was believed that damaged cardiac muscle fibers could not be replaced and that healing took place exclusively by fibrosis, the formation of scar tissue. New research indicates that cardiac muscle can regenerate under certain circumstances (see Clinical Connection: Regeneration of Heart Cells). In addition, cardiac muscle fibers can undergo hypertrophy in response to increased workload. Hence, many athletes have enlarged hearts.

Smooth muscle, like skeletal and cardiac muscle, can undergo hypertrophy. In addition, certain smooth muscle fibers, such as those in the uterus, retain their capacity for division and thus can grow by hyperplasia. Also, new smooth muscle fibers can arise from cells called *pericytes*, stem cells found in association with blood capillaries and small veins. Smooth muscle fibers can also proliferate in certain pathological conditions, such as atherosclerosis. Compared with the other two types of muscle, smooth muscle has considerably greater powers of regeneration. Such powers are still limited when compared with other tissues, such as epithelium.

TABLE 11.4 summarizes the major characteristics of the three types of muscle.

CLINICAL CONNECTION

Regeneration of Heart Cells

A recent study of heart transplant recipients by American and Italian scientists provides evidence for significant replacement of heart cells. The researchers studied men who had received a heart from a female, and then looked for the presence of a Y chromosome in heart cells. (All female cells except gametes have two X chromosomes and lack the Y chromosome.) Several years after the transplant surgery, between 7% and 16% of the heart cells in the transplanted tissue, including cardiac muscle fibers and endothelial cells in coronary arterioles and capillaries, had been replaced by the recipient's own cells, as evidenced by the presence of a Y chromosome. The study also revealed cells with some of the characteristics of stem cells in both transplanted hearts and control hearts. Evidently, stem cells can migrate from the blood into the heart and differentiate into functional muscle and endothelial cells. Researchers hope to learn how to "turn on" such regeneration of heart cells to treat people with heart failure or cardiomyopathy (diseased heart).

TABLE 11.4 Summary of the Major Characteristics of the Three Types of Muscle

Characteristic	Skeletal Muscle	Cardiac Muscle	Smooth Muscle
Microscopic appearance and features	Striated.	Striated.	Not striated (smooth).
Location	Most commonly attached by tendons to bones.	Heart	Walls of hollow viscera, airways, iris and ciliary body of eye, arrector pili muscles of skin.
Fiber diameter	Very large (10–100 μm).	Large (10–20 μm).	Small (3–8 μm).
Fiber length	100 μm–30 cm.	50–100 μm.	30–200 μm.
Contractile proteins organized into sarcomeres	Yes.	Yes.	No.
Sarcoplasmic reticulum	Abundant	Some.	Small amount.
Transverse tubules present	Yes.	Yes.	No.
Autorhythmicity	No.	Yes.	Yes, in single-unit smooth muscle.
Source of Ca^{2+} for contraction	Sarcoplasmic reticulum.	Sarcoplasmic reticulum and extracellular fluid.	Sarcoplasmic reticulum and extracellular fluid.
Regulator proteins for contraction	Troponin and tropomyosin.	Troponin and tropomyosin.	Calmodulin and myosin light chain kinase.
Speed of contraction	Fast.	Moderate.	Slow.
Nervous control	Voluntary (somatic nervous system).	Involuntary (autonomic nervous system).	Involuntary (autonomic nervous system).
Contraction regulated by . . .	Acetylcholine released by somatic motor neurons.	Acetylcholine and norepinephrine released by autonomic motor neurons.	Acetylcholine and norepinephrine released by autonomic motor neurons; several hormones; local chemical changes; stretching.
Capacity for regeneration	Limited, via satellite cells.	Limited, under certain conditions.	Considerable, via pericytes (compared with other muscle tissues, but limited compared with epithelium).

CHECKPOINT

33. Which type of muscle has the greatest ability to regenerate?

To appreciate the many ways that the muscular system contributes to homeostasis of other body systems, examine *Focus on Homeostasis: Contributions of the Muscular System.*

FOCUS on HOMEOSTASIS

INTEGUMENTARY SYSTEM

- Pull of skeletal muscles on attachments to skin of face causes facial expressions
- Exercise increases skin blood flow

SKELETAL SYSTEM

- Skeletal muscle causes movement of body parts by pulling on attachments to bones
- Skeletal muscle provides stability for bones and joints

NERVOUS SYSTEM

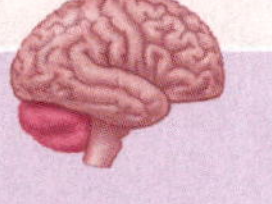

- Skeletal, cardiac, and smooth muscles carry out commands for the nervous system
- Shivering—involuntary contraction of skeletal muscles that is regulated by the brain—generates heat to raise body temperature

ENDOCRINE SYSTEM

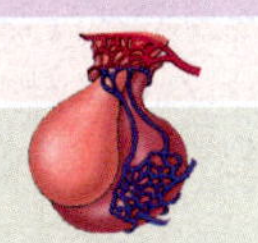

- Regular exercise improves action and signaling mechanisms of some hormones, such as insulin
- Muscles protect some endocrine glands

CARDIOVASCULAR SYSTEM

- Cardiac muscle powers pumping action of heart
- Contraction and relaxation of smooth muscle in blood vessel walls help adjust the amount of blood flowing through various body tissues
- Contraction of skeletal muscles in the legs assists return of blood to the heart
- Regular exercise causes cardiac hypertrophy (enlargement) and increases heart's pumping efficiency
- Lactic acid produced by active skeletal muscles may be used for ATP production by the heart

CONTRIBUTIONS OF THE MUSCULAR SYSTEM

FOR ALL BODY SYSTEMS

- Produces body movements
- Stabilizes body positions
- Moves substances within the body
- Produces heat that helps maintain normal body temperature

IMMUNE SYSTEM

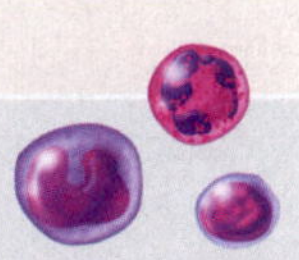

- Exercise may increase or decrease some immune responses

RESPIRATORY SYSTEM

- Skeletal muscles involved with breathing cause air to flow into and out of the lungs
- Smooth muscle fibers adjust size of airways
- Vibrations in skeletal muscles of larynx control air flowing past vocal cords, regulating voice production
- Coughing and sneezing, due to skeletal muscle contractions, help clear airways
- Regular exercise improves efficiency of breathing

DIGESTIVE SYSTEM

- Skeletal muscles protect and support organs in the abdominal cavity
- Alternating contraction and relaxation of skeletal muscles power chewing and initiate swallowing
- Smooth muscle sphincters control volume of organs of the gastrointestinal (GI) tract
- Smooth muscles in walls of GI tract mix and move its contents through the tract

URINARY SYSTEM

- Skeletal and smooth muscle sphincters and smooth muscle in wall of urinary bladder control whether urine is stored in the urinary bladder or voided (urination)

REPRODUCTIVE SYSTEMS

- Skeletal and smooth muscle contractions eject semen from male
- Smooth muscle contractions propel oocyte along fallopian tube, help regulate flow of menstrual blood from uterus, and force baby from uterus during childbirth
- During intercourse, skeletal muscle contractions are associated with orgasm and pleasurable sensations in both sexes

FROM RESEARCH TO REALITY

Exercise and Duchenne Muscular Dystrophy

Reference

Gordon, B. et al. (2014). Exercise increases utrophin protein expression in the *mdx* mouse model of Duchenne muscular dystrophy. *Muscle Nerve*. 49: 915–918.

Can exercise benefit people who have Duchenne muscular dystrophy?

Odua Images/Shutterstock

Duchenne muscular dystrophy (DMD) is an X-linked disorder that affects boys almost exclusively. It is caused by a mutation of the gene for dystrophin. Without this protein, the sarcolemma tears easily, leading to severe muscle damage and muscle dysfunction. The protein utrophin, which is similar to dystrophin, can help compensate for the lack of dystrophin in individuals with DMD. Although it is feared that exercise will cause additional muscle damage in boys with DMD, studies have suggested that submaximal aerobic exercise may actually help reduce the pathology of the disease, and it may do so by increasing the utrophin content in exercised muscle. Does exercise increase the utrophin content in muscle? Do different muscles that are being exercised have the same increase in utrophin content?

Article description:

Investigators in this study wanted to determine if exercise increased the utrophin content of muscles in *mdx* mice. The *mdx* mouse is an animal model for Duchenne muscular dystrophy because it has a mutation in the dystrophin gene. In this study, *mdx* mice were divided into two groups: those that exercised and those that did not (sedentary). Exercised mice had access to running wheels, whereas sedentary (control) mice had access to locked running wheels. Mice were subjected to these conditions for 12 weeks. Afterward, the utrophin content was measured in the quadriceps and soleus muscles—two muscles that are activated by voluntary wheel running.

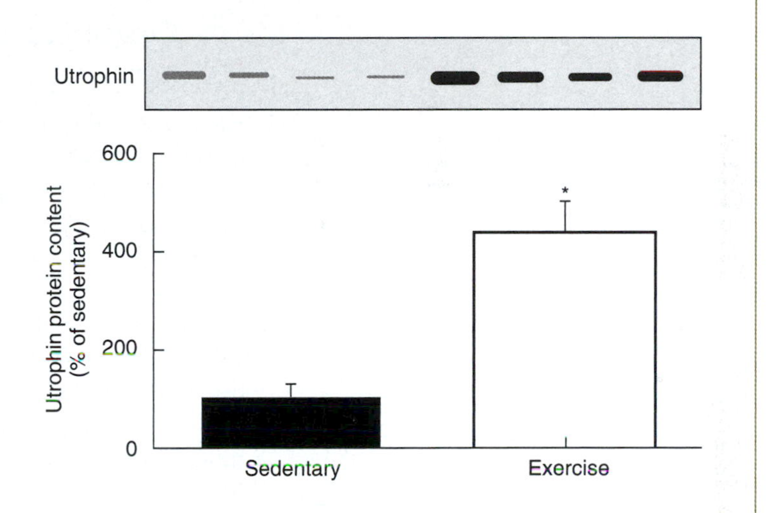

Go to WileyPLUS Learning Space and use the data from this article to answer the questions posed there and to discover more about exercise and Duchenne muscular dystrophy.

CHAPTER REVIEW

11.1 Overview of Muscle

1. Movement results from the contraction of muscle, which constitutes 40–50% of total body weight.
2. The primary function of muscle is changing chemical energy into mechanical energy to perform work.
3. The three types of muscle are skeletal, cardiac, and smooth. Skeletal muscle is primarily attached to bones; it is striated and voluntary. Cardiac muscle forms the wall of the heart; it is striated and involuntary. Smooth muscle is located primarily in internal organs; it is nonstriated (smooth) and involuntary.
4. Through contraction and relaxation, muscle performs four important functions: producing body movements, stabilizing body positions, storing and moving substances within the body, and generating heat.
5. Four special properties of muscle are (1) electrical excitability, the property of responding to stimuli by producing action potentials; (2) contractility, the ability to generate tension to do work; (3) extensibility, the ability to be stretched without being damaged; and (4) elasticity, the ability to return to the original shape after contraction or extension.

11.2 Organization of Skeletal Muscle

1. A skeletal muscle is composed of hundreds to thousands of muscle fibers (cells).
2. Each muscle fiber has multiple nuclei because it arises from the fusion of many myoblasts. The sarcolemma is a muscle fiber's plasma membrane; it surrounds the sarcoplasm. Transverse tubules are invaginations of the sarcolemma.
3. Extending throughout the sarcoplasm are myofibrils, the contractile elements of skeletal muscle. Sarcoplasmic reticulum surrounds each myofibril. Within a myofibril are thin and thick filaments, arranged in compartments called sarcomeres.
4. The components of a sarcomere are organized into bands and zones. Dark A bands alternate with light I bands, creating the striations that can be seen in skeletal and cardiac muscle fibers.
5. Myofibrils are composed of three types of proteins: contractile, regulatory, and structural. The contractile proteins are myosin (thick filament) and actin (thin filament). Regulatory proteins are tropomyosin and troponin, both of which are part of the thin filament. Structural proteins include titin, myomesin, nebulin, and dystrophin.

11.3 Contraction and Relaxation of Skeletal Muscle Fibers

1. Muscle contraction occurs because myosin heads attach to and "walk" along the thin filaments at both ends of a sarcomere, progressively pulling the thin filaments toward the center of a sarcomere. As the thin filaments slide inward, the Z discs come closer together, and the sarcomere shortens.
2. The contraction cycle is the repeating sequence of events that causes sliding of the filaments: (1) Myosin ATPase hydrolyzes ATP and becomes energized; (2) the myosin head attaches to actin, forming a crossbridge; (3) the crossbridge pulls the thin filament toward the center of the sarcomere (power stroke); and (4) binding of ATP to the myosin head detaches it from actin. The myosin head again hydrolyzes the ATP, returns to its original position, and binds to a new site on actin as the cycle continues.
3. The neuromuscular junction (NMJ) is the synapse between a somatic motor neuron and a skeletal muscle fiber. The NMJ includes the synaptic end bulbs of a terminal branch of the somatic motor neuron's axon, the synaptic cleft, plus the motor end plate of the skeletal muscle fiber.

4. When a nerve action potential reaches the synaptic end bulbs of a somatic motor neuron, it triggers exocytosis of the synaptic vesicles, which causes the release of acetylcholine (ACh). ACh diffuses across the synaptic cleft and binds to ACh receptors, initiating a muscle action potential. Acetylcholinesterase then quickly breaks down ACh into its component parts.
5. An increase in Ca^{2+} concentration in the sarcoplasm starts filament sliding; a decrease turns off the sliding process.
6. When an action potential travels along the T-tubule system, dihydropyridine receptors detect the change in voltage and undergo a conformational change that causes Ca^{2+} release channels in the SR membrane to open. Calcium ions diffuse from the SR into the sarcoplasm and combine with troponin. This binding causes tropomyosin to move away from the myosin-binding sites on actin.
7. Ca^{2+}-ATPase pumps continually remove Ca^{2+} from the sarcoplasm into the SR. When the concentration of calcium ions in the sarcoplasm decreases, tropomyosin slides back over and blocks the myosin-binding sites, and the muscle fiber relaxes.

11.4 ATP Production in Skeletal Muscle

1. Skeletal muscle fibers have three sources of ATP production: creatine phosphate, anaerobic glycolysis, and aerobic respiration.
2. Creatine kinase catalyzes the transfer of a high-energy phosphate group from creatine phosphate to ADP to form new ATP molecules. Together, stores of creatine phosphate and ATP provide enough energy for muscles to contract maximally for about 15 seconds.
3. During glycolysis, glucose is broken down into two molecules of pyruvic acid, with a net gain of two molecules of ATP produced. When oxygen is absent or at a low concentration (anaerobic conditions), pyruvic acid is converted to lactic acid. The process by which the breakdown of glucose gives rise to lactic acid is called anaerobic glycolysis. Anaerobic glycolysis provides enough energy for about 2 minutes of maximal muscle activity.
4. If sufficient oxygen is present, pyruvic acid formed by glycolysis undergoes aerobic respiration, which includes the reactions of the Krebs cycle and the electron transport chain. Each molecule of glucose catabolized via aerobic respiration yields about 30 to 32 molecules of ATP. Aerobic respiration provides energy for those activities that last from several minutes to an hour or more.
5. The inability of a muscle to contract forcefully after prolonged activity is muscle fatigue.
6. Elevated oxygen use after exercise is called recovery oxygen uptake.

11.5 Skeletal Muscle Mechanics

1. A somatic motor neuron and the muscle fibers it innervates form a motor unit.
2. A twitch is the brief contraction of a group of muscle fibers within a muscle in response to a single action potential.
3. A record of a muscle contraction is called a myogram. It consists of a latent period, a contraction period, and a relaxation period.
4. Skeletal muscles can produce graded contractions, which are contractions that vary in strength depending on how much force is needed by the muscle to support a particular object.
5. Several factors determine muscle tension: frequency of stimulation, muscle fiber length, muscle fiber diameter, motor unit size, and motor unit recruitment.
6. Skeletal muscles produce movements by exerting force on tendons, which in turn pull on the bones of the skeleton. The attachment of a muscle's tendon to the more

stationary bone is the origin; the attachment of the muscle's tendon to the more movable bone is the insertion.

7. The action of a muscle is the type of movement that occurs when the muscle contracts. Examples of muscle actions include flexion and extension.
8. Bones serve as levers, and joints serve as fulcrums. Two different forces act on the lever: load (resistance) and effort.
9. In an isotonic contraction, the tension developed by the muscle remains almost constant while the muscle changes its length. The two types of isotonic contractions are concentric and eccentric.
10. In an isometric contraction, tension is generated without a change in muscle length.

11.6 Types of Skeletal Muscle Fibers

1. On the basis of their structure and function, skeletal muscle fibers are classified as slow oxidative (SO), fast oxidative–glycolytic (FOG), and fast glycolytic (FG) fibers.
2. Most skeletal muscles contain a mixture of all three fiber types. Their proportions vary with the typical action of the muscle.
3. The motor units of a muscle are recruited in the following order: first SO fibers, then FOG fibers, and finally FG fibers.
4. Various types of exercises can induce changes in skeletal muscle fibers. Endurance-type (aerobic) exercises cause a gradual transformation of some fast glycolytic (FG) fibers into fast oxidative–glycolytic (FOG) fibers.
5. Exercises that require great strength for short periods produce an increase in the size and strength of fast glycolytic (FG) fibers. The increase in size is due to increased synthesis of thick and thin filaments.

11.7 Cardiac Muscle

1. Cardiac muscle is found only in the heart. Cardiac muscle fibers have the same arrangement of actin and myosin and the same bands, zones, and Z discs as skeletal muscle fibers. The fibers connect to one another through intercalated discs, which contain both desmosomes and gap junctions.
2. Cardiac muscle fibers contract in the same way as skeletal muscle fibers: Ca^{2+} binds to troponin, moving tropomyosin away from myosin-binding sites on actin, and the thick and thin filaments begin sliding past one another.
3. Cardiac muscle, as a functional syncytium, consists of two types of cells: (1) autorhythmic fibers, which spontaneously generate action potentials, and (2) contractile fibers, which become excited and then contract in response to action potentials conducted to them from autorhythmic fibers via gap junctions.
4. During excitation–contraction coupling in cardiac muscle, an action potential opens L-type voltage-gated Ca^{2+} channels in the sarcolemma; extracellular Ca^{2+} moves into the cell through these channels and then causes Ca^{2+}-induced Ca^{2+} release (CICR) by binding to and opening Ca^{2+} release channels in the membrane of the sarcoplasmic reticulum.
5. Cardiac muscle has a long refractory period, which prevents tetanus.

11.8 Smooth Muscle

1. Smooth muscle is nonstriated because its thick and thin filaments do not have a regular pattern of overlap.
2. Smooth muscle fibers contain intermediate filaments and dense bodies; the function of dense bodies is similar to that of the Z discs in striated muscle.

3. Smooth muscle and striated muscle differ in the way that an increase in the sarcoplasmic Ca^{2+} concentration triggers contraction. In smooth muscle, the following events take place: (1) Ca^{2+} binds to calmodulin; (2) the Ca^{2+}–calmodulin complex activates myosin light chain kinase (MLCK); (3) MLCK phosphorylates light chains in myosin heads; and (4) the phosphorylated myosin binds to actin.
4. There are two major types of smooth muscle: single-unit (visceral) and multi-unit. Single-unit smooth muscle is autorhythmic and behaves as a functional syncytium. Multi-unit smooth muscle fibers act as independent units.
5. Single-unit smooth muscle, as a function syncytium, consists of two types of fibers: (1) autorhythmic fibers, which spontaneously generate action potentials, and (2) contractile fibers, which become excited and contract in response to action potentials conducted to them from autorhythmic fibers via gap junctions.
6. Multi-unit smooth muscle is not autorhythmic nor does it behave as a functional syncytium. All cells of multi-unit smooth muscle are contractile fibers.
7. Two types of spontaneous depolarizations can occur in autorhythmic smooth muscle fibers: pacemaker potentials and slow-wave potentials.
8. Action potentials in contractile smooth muscle fibers can be either spike potentials or action potentials with a plateau.
9. During excitation–contraction coupling in smooth muscle, the increase in sarcoplasmic Ca^{2+} concentration that triggers contraction can occur by opening one or more of the following channels: voltage-gated, Ca^{2+} release, receptor-activated, inositol trisphosphate (IP_3)–gated, store-operated, and mechanically-gated.

11.9 Regeneration of Muscle

1. Skeletal muscle fibers cannot divide and have limited powers of regeneration; cardiac muscle fibers can regenerate under limited circumstances; and smooth muscle fibers have the best capacity for division and regeneration.

PONDER THIS

1. You are working in a lab developing a new pharmaceutical. Upon testing this new drug, you notice that it interferes with skeletal muscle function such that skeletal muscles do not contract. Investigating this further, you note that the motor neuron works well, acetylcholine is released, and it binds normally to a nicotinic receptor. Additionally, you note that the muscle membrane depolarizes normally. What is one thing that the drug could possibly be interfering with?

2. If the extracellular levels of calcium in the body were drastically reduced, which type of muscle (smooth, skeletal, and/or cardiac) would this interfere with? Explain your answer. Assume that all nervous function remains normal.

3. You genetically engineer a mouse, which is a knockout for gap junctions. This would lead to this species being unable to make gap junction proteins. Which muscle type (smooth, skeletal, and/or cardiac) would this interfere with and how? Explain your answer.

ANSWERS TO FIGURE QUESTIONS

11.1 Cardiac muscle is striated and involuntarily controlled.

11.2 A tendon is a cord of connective tissue that attaches muscle to a bone.

11.3 The sarcoplasmic reticulum releases calcium ions to trigger muscle contraction.

11.4 The striations that can be seen in skeletal and cardiac muscle fibers are created by alternating dark A bands and light I bands.

11.5 The ATP-binding site also functions as an ATPase that hydrolyzes ATP to generate energy for muscle contraction.

11.6 In a relaxed muscle, tropomyosin prevents myosin from binding to actin.

11.7 The I bands and H zones narrow and eventually disappear when muscle is maximally contracted; the lengths of the thin and thick filaments do not change.

11.8 If ATP were not available, the myosin crossbridges would not be able to detach from actin. The muscles would remain in a state of rigidity, as occurs in rigor mortis.

11.9 The voltage-gated channels involved in the skeletal muscle action potential are the same as those involved in the nerve action potential.

11.10 Excitation–contraction coupling occurs at the triads of a skeletal muscle fiber. A triad consists of a transverse (T) tubule and two opposing terminal cisternae of the sarcoplasmic reticulum (SR).

11.11 Three functions of ATP in muscle contraction include the following: (1) hydrolysis of ATP by myosin ATPase activates the myosin head so it can bind to actin and undergo the power stroke, (2) binding of ATP to myosin causes detachment from actin after the power stroke, and (3) ATP powers the pumps that transport Ca^{2+} from the sarcoplasm back into the sarcoplasmic reticulum.

11.12 During a long-term event such as a marathon race, most ATP is produce by aerobic respiration.

11.13 Motor units having many muscle fibers are capable of more forceful contractions than those having only a few fibers.

11.14 During the latent period, the events of excitation–contraction coupling occur: The muscle action potential sweeps along the sarcolemma and into the T tubules, causing the release of calcium ions from the sarcoplasmic reticulum.

11.15 If the second stimulus were applied a little later, the second contraction would be smaller than the one illustrated in part (b) of FIGURE 11.15.

11.16 A sarcomere length of 2.2 μm gives a generous zone of overlap between the parts of the thick filaments that have myosin heads and the thin filaments, without the overlap being so extensive that sarcomere shortening is limited.

11.17 The origin is the attachment of a muscle tendon to a relatively stationary bone; the insertion is the attachment of a muscle tendon to a movable bone.

11.18 The biceps and triceps are called antagonistic muscles because they promote opposite actions (flexion versus extension) at the same joint.

11.19 The effort is the force (tension) exerted by muscle contraction; the load is the weight of the bone plus the weight of any object carried by the bone.

11.20 Most lever systems in the body operate at a mechanical disadvantage.

11.21 Holding your head upright without movement involves mainly isometric contractions of neck muscles.

11.22 If the load becomes equal to or exceeds the amount of tension that a muscle can produce, the velocity of shortening is zero and the contraction becomes isometric.

11.23 Extracellular Ca^{2+} causes CICR by entering the cell and then binding to Ca^{2+} release channels in the SR. Binding of Ca^{2+} in turn causes the Ca^{2+} release channels to open, allowing an even larger amount of Ca^{2+} to enter the sarcoplasm.

11.24 If cardiac muscle could undergo tetanus, the heart would no longer be able to function as a pump because it would not have a chance to relax and refill with blood.

11.25 No. The thick and thin filaments of a smooth muscle fiber do not have a regular pattern of overlap; that is why they have a smooth (nonstriated) appearance under a microscope.

11.26 MLCK phosphorylates light chains in myosin heads. MLCK is activated by the Ca^{2+}–calmodulin complex.

11.27 Single-unit smooth muscle is more like cardiac muscle; both contain gap junctions, which allow action potentials to spread from each cell to its neighbors.

11.28 Pacemaker potentials are spontaneous depolarizations that always reach threshold; slow-wave potentials are cycles of alternating depolarization and repolarization that do not necessarily reach threshold.

11.29 A contractile cardiac muscle fiber has an action potential with a plateau.

11.30 When intracellular Ca^{2+} reserves in the SR are depleted, store-operated channels in the sarcolemma of a smooth muscle fiber open, allowing Ca^{2+} to enter the sarcoplasm. The entering Ca^{2+} can be used for contraction or to replenish the depleted Ca^{2+}stores in the SR.

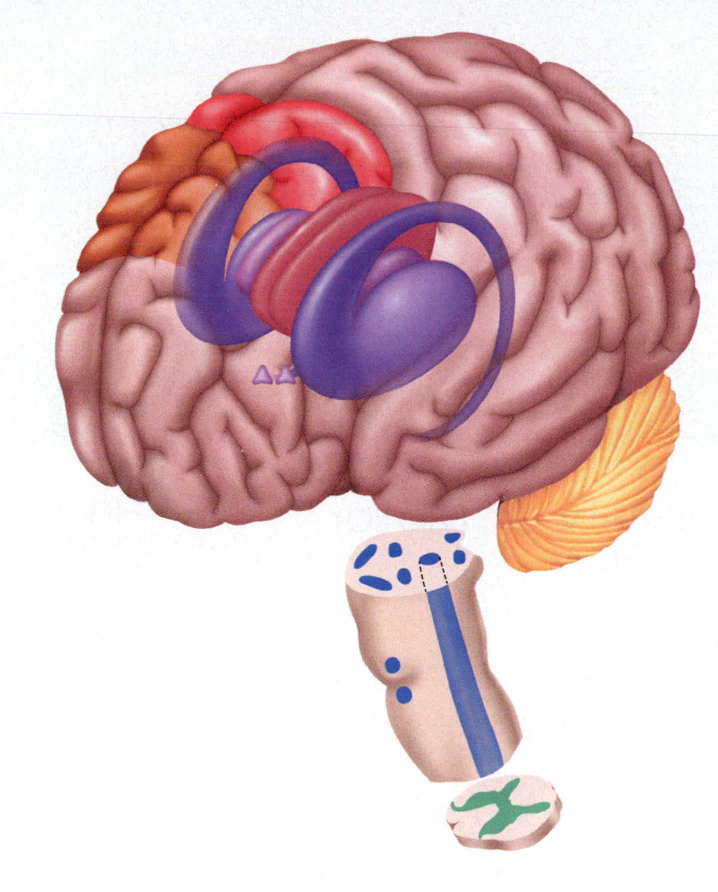

12 Control of Body Movement

Control of Body Movement and Homeostasis

Control of body movement contributes to homeostasis by providing the appropriate motor responses that allow the body to accomplish a particular task.

LOOKING BACK TO MOVE AHEAD...

- The reticular formation is a region of interspersed gray and white matter that extends through the core of the brain stem; in addition to playing a role in wakefulness, it contributes to motor control by helping to regulate posture and muscle tone (Section 8.2).
- Muscle spindles and tendon organs are types of proprioceptors: Muscle spindles detect static muscle length and changes in muscle length; tendon organs detect changes in muscle tension (Section 9.2).
- The somatic nervous system is the division of the peripheral nervous system (PNS) that conveys output from the central nervous system (CNS) to skeletal muscle fibers; it is usually under voluntary control but can be regulated involuntarily (Section 10.2).
- In an isotonic contraction, the tension developed by the muscle remains almost constant while the muscle changes its length; in an isometric contraction, tension is generated without the muscle changing its length (Section 11.5).

Throwing a ball, hanging a picture, and lifting your foot off a tack are all examples of **body movements**, which occur because of isotonic contractions of skeletal muscles. In each case, the contracting muscle generates enough tension to cause movement of the appropriate part of the body. By contrast, isometric contractions of skeletal muscles produce no movement. This occurs, for example, when muscles of the body contract to maintain posture. Although isometric contractions do not result in movement of a body part, for convenience, they are still included along with isotonic contractions when the term *body movement* is used from this point forward.

Body movements are subject to voluntary or involuntary control. In the examples just presented, throwing the ball and hanging the picture are voluntary acts, whereas lifting your foot off the tack and maintaining posture are involuntary responses. Body movements, whether voluntary or involuntary, involve the interaction between multiple parts of the brain and the motor neurons of the somatic nervous system. In this chapter, you will learn about the various neural mechanisms that participate in the control of body movement.

12.1 Overview of Motor Control

OBJECTIVES

- Explain the importance of lower motor neurons.
- Describe the sources of input to lower motor neurons.

Lower Motor Neurons Provide Output from the CNS to Skeletal Muscle Fibers

Neural circuits in the brain and spinal cord orchestrate all movements of the body. Ultimately, the excitatory and inhibitory signals that control all voluntary movements and some involuntary movements converge on somatic motor neurons. Because somatic motor neurons have their cell bodies in the lower parts of the CNS (in either the brain stem or spinal cord), they are also known as **lower motor neurons**. From the brain stem, axons of lower motor neurons extend through *cranial nerves* to innervate skeletal muscles of the face and head. From the spinal cord, axons of lower motor neurons extend through *spinal nerves* to innervate skeletal muscles of the limbs and trunk. Only lower motor neurons provide output from the CNS to skeletal muscle fibers. For this reason, they are also called the **final common pathway**.

There Are Four Sources of Input to Lower Motor Neurons

Neurons in four distinct but highly interactive neural circuits participate in control of movement by providing input to lower motor neurons (FIGURE 12.1):

1. ***Local circuit neurons.*** Input arrives at lower motor neurons from nearby interneurons called **local circuit neurons**. These neurons are located close to the lower motor neuron cell bodies in the brain stem and spinal cord. Local circuit neurons coordinate many types of somatic reflexes and play a major role in locomotion (walking and running).
2. ***Upper motor neurons.*** Both local circuit neurons and lower motor neurons receive input from **upper motor neurons**,* neurons that have their cell bodies in motor processing centers in the upper parts of the CNS. Most upper motor neurons synapse with local circuit neurons, which in turn synapse with lower motor neurons. A few upper motor neurons synapse directly with lower motor

FIGURE 12.1 Neural circuits that regulate lower motor neurons. Lower motor neurons receive input directly from (1) local circuit neurons (purple arrow) and (2) upper motor neurons in the cerebral cortex and brain stem (green arrows). Neurons in the basal nuclei (3) and cerebellum (4) regulate activity of upper motor neurons (red arrows).

Because lower motor neurons provide all output to skeletal muscles, they are called the final common pathway.

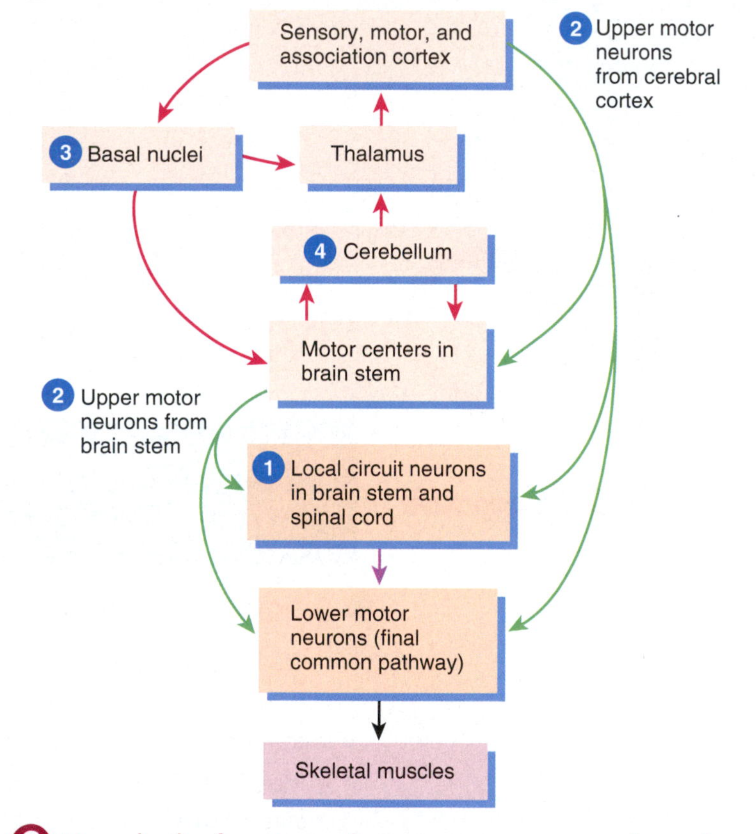

How do the functions of upper motor neurons from the cerebral cortex and brain stem differ?

* Recall from Chapter 7 that an upper motor neuron is an interneuron and not a true motor neuron; it is so-named because the cell originates in the upper part of the CNS and regulates the activity of lower motor neurons. Only a lower motor neuron is a true motor neuron because it conveys action potentials from the CNS to skeletal muscles in the periphery.

neurons. Upper motor neurons from the cerebral cortex are essential for the planning and execution of voluntary movements of the body. Other upper motor neurons originate in motor centers of the brain stem: the vestibular nuclei, reticular formation, superior colliculus, and red nucleus. Upper motor neurons from the brain stem help regulate posture, balance, muscle tone, and reflexive movements of the head and trunk.

3. ***Basal nuclei.*** Neurons of the basal nuclei assist movement by providing input to upper motor neurons. Neural circuits interconnect the basal nuclei with motor areas of the cerebral cortex (via the thalamus) and the brain stem. These circuits help initiate movements, suppress unwanted movements, and establish a normal level of muscle tone.
4. ***Cerebellum.*** Neurons of the cerebellum also aid movement by controlling the activity of upper motor neurons. Neural circuits interconnect the cerebellum with motor areas of the cerebral cortex (via the thalamus) and the brain stem. A prime function of the cerebellum is to monitor differences between intended movements and movements actually performed. Then, it issues commands to upper motor neurons to reduce errors in movement. The cerebellum thus coordinates body movements and helps maintain normal posture and balance.

Now that you have a general understanding of how body movement is controlled, the remaining sections of this chapter will describe the different components of motor control in more detail.

CHECKPOINT

1. Why are lower motor neurons called the final common pathway?
2. What are the four sources that provide input to lower motor neurons in order to control movement?

12.2 Local Level of Motor Control

OBJECTIVES

- Describe the importance of a somatic reflex arc.
- Provide specific examples of somatic reflexes.
- Discuss the functional significance of central pattern generators.

The local level of motor control is coordinated by local circuit neurons (interneurons) present in the brain stem or spinal cord close to the cell bodies of lower motor neurons. Local circuit neurons receive input from somatic sensory receptors, such as nociceptors and muscle spindles, as well as from higher centers in the brain. In response to this input, the local circuit neurons promote somatic reflexes and the rhythmic movements of locomotion.

Somatic Reflexes Allow Fast, Involuntary Contractions of Skeletal Muscle

Somatic reflexes are fast, involuntary responses that occur when action potentials pass through a somatic reflex arc (pathway). These reflexes are important because they allow skeletal muscle to contract quickly in response to certain stimuli, such as pain, overstretching of muscle, and excessive muscle tension. A somatic reflex arc consists of the following components (FIGURE 12.2):

- **Sensory receptor.** The sensory receptor is the distal end of a sensory neuron; it responds to a stimulus and produces a receptor potential. The type of sensory receptor associated with a somatic reflex

FIGURE 12.2 General components of a somatic reflex arc. The arrows show the direction of action potential propagation.

A somatic reflex arc consists of a sensory receptor, sensory neuron, integrating center, somatic motor neuron, and effector.

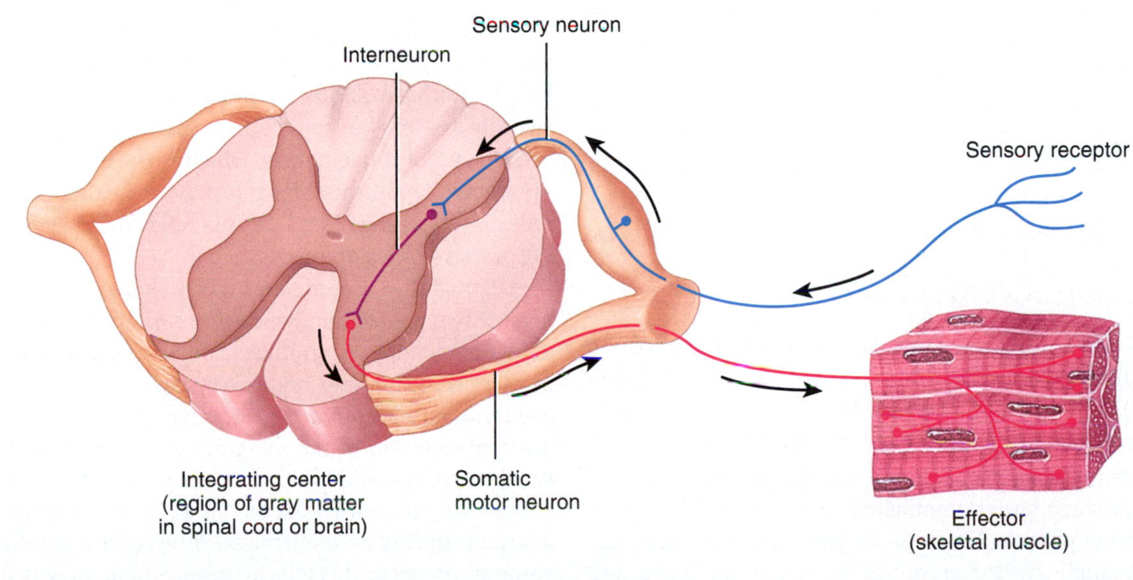

Why are somatic reflexes important?

arc can vary; examples include nociceptors, muscle spindles, and tendon organs.

- **Sensory neuron**. If the receptor potential in the sensory receptor reaches threshold, it triggers one or more action potentials in the axon of the sensory neuron. The action potentials are conveyed along the axon into the CNS.
- **Integrating center**. The integrating center is a region of gray matter within the brain stem or spinal cord that processes the incoming sensory information. If integration takes place in the spinal cord gray matter, the reflex is a *spinal reflex*; if integration occurs in the brain stem gray matter, the reflex is a *cranial reflex*. A reflex pathway having only one synapse in the CNS is termed a **monosynaptic reflex arc**. More often, the integrating center consists of one or more interneurons, which may relay action potentials to other interneurons as well as to a motor neuron. A **polysynaptic reflex arc** involves more than two types of neurons and more than one CNS synapse.
- **Motor neuron**. Action potentials triggered by the integrating center propagate out of the CNS along a somatic motor neuron to an effector.
- **Effector**. In a somatic reflex arc, the effector is skeletal muscle.

Because reflexes are normally so predictable, they provide useful information about the health of the nervous system and can greatly aid diagnosis of disease. Damage or disease anywhere along its reflex arc can cause a reflex to be absent or abnormal. If a reflex ceases to function or functions abnormally, the physician may suspect that the damage lies somewhere along a particular conduction pathway. Somatic reflexes generally can be tested simply by tapping or stroking the body surface.

Let's now examine four important somatic spinal reflexes: the stretch reflex, the tendon reflex, the flexor (withdrawal) reflex, and the crossed extensor reflex.

The Stretch Reflex

A **stretch reflex** causes contraction of a skeletal muscle (the effector) in response to stretching of the muscle. This type of reflex occurs via a monosynaptic reflex arc. The reflex can occur by activation of a single sensory neuron that forms one synapse in the CNS with a single motor neuron. Stretch reflexes can be elicited by tapping on tendons attached to muscles at the elbow, wrist, knee, and ankle joints. An example of a stretch reflex is the **patellar (knee jerk) reflex**. This reflex involves extension of the leg at the knee joint by contraction of the quadriceps femoris muscle of the thigh in response to tapping the patellar ligament. Recall that *extension* is the act of straightening out a limb at a joint. Muscles such as the quadriceps femoris that cause extension of a limb are known as *extensors*.

A stretch reflex operates as follows (FIGURE 12.3):

1. Slight stretching of a muscle stimulates sensory receptors in the muscle called **muscle spindles** (shown in more detail in FIGURE 9.17). The spindles monitor the *length* of the muscle (both static muscle length and changes in muscle length).
2. In response to being stretched, a muscle spindle generates a receptor potential. If the receptor potential reaches threshold, it triggers one or more action potentials that propagate along a somatic sensory neuron through the dorsal root of the spinal nerve and into the spinal cord.

CLINICAL CONNECTION

Abnormal Patellar Reflex

A normal patellar (knee jerk) reflex involves reflex extension of the leg at the knee joint in response to tapping the patellar ligament. However, the patellar reflex may be either absent or exaggerated, resulting in an **abnormal patellar reflex.** Absence of the patellar reflex could indicate damage to sensory or motor nerves supplying the quadriceps femoris muscle of the thigh or to the integrating centers in the lumbar spinal cord. The patellar reflex is exaggerated in disease or injury involving certain motor tracts descending from the higher centers of the brain to the spinal cord.

3. In the spinal cord (integrating center), the sensory neuron makes an excitatory synapse with and thereby activates a motor neuron in the ventral gray horn.
4. If the excitation is strong enough, one or more action potentials arise in the motor neuron and propagate along its axon, which extends from the spinal cord into the ventral root and through peripheral nerves to the stimulated muscle. The axon terminals of the motor neuron form neuromuscular junctions (NMJs) with skeletal muscle fibers of the stretched muscle.
5. Acetylcholine released by action potentials at the NMJs triggers one or more action potentials in the stretched muscle (effector), and the muscle contracts. Thus, muscle stretch is followed by muscle contraction, which relieves the stretching.

In the reflex arc just described, sensory input enters the spinal cord on the same side from which motor output leaves it. This arrangement is called an **ipsilateral reflex** (ip′-si-LAT-er-al). All monosynaptic reflexes are ipsilateral.

In addition to the large-diameter motor neurons (called **alpha motor neurons**) that innervate typical skeletal muscle fibers, smaller-diameter motor neurons (known as **gamma motor neurons**) innervate the intrafusal muscle fibers associated with the muscle spindles themselves (see FIGURE 9.17). The brain regulates muscle spindle sensitivity through pathways to these smaller motor neurons. This regulation ensures proper muscle spindle signaling over a wide range of muscle lengths during voluntary and reflex contractions. By adjusting how vigorously a muscle spindle responds to stretching, the brain sets an overall level of **muscle tone**, which is the small degree of contraction present while the muscle is at rest. Because the stimulus for the stretch reflex is stretching of muscle, this reflex helps avert injury by preventing overstretching of muscles.

Although the stretch reflex pathway itself is monosynaptic (involving just two neurons and one synapse), a polysynaptic reflex arc to the antagonistic (opposing) muscles operates at the same time. This arc involves three neurons and two synapses. An axon collateral (branch) from the muscle spindle sensory neuron also synapses with an inhibitory interneuron in the integrating center. In turn, the interneuron synapses with and inhibits a motor neuron that normally excites the antagonistic muscles (FIGURE 12.3). Thus, when the stretched muscle contracts during a stretch reflex, antagonistic muscles that oppose the contraction relax. This type of arrangement, in which the components of a neural circuit simultaneously cause contraction of one muscle and

FIGURE 12.3 Stretch reflex. This monosynaptic reflex arc has only one synapse in the CNS—between a single sensory neuron and a single motor neuron. A polysynaptic reflex arc to antagonistic muscles that includes two synapses in the CNS and one interneuron is also illustrated. Plus signs (+) indicate excitatory synapses; the minus sign (−) indicates an inhibitory synapse.

The stretch reflex causes contraction of a muscle that has been stretched.

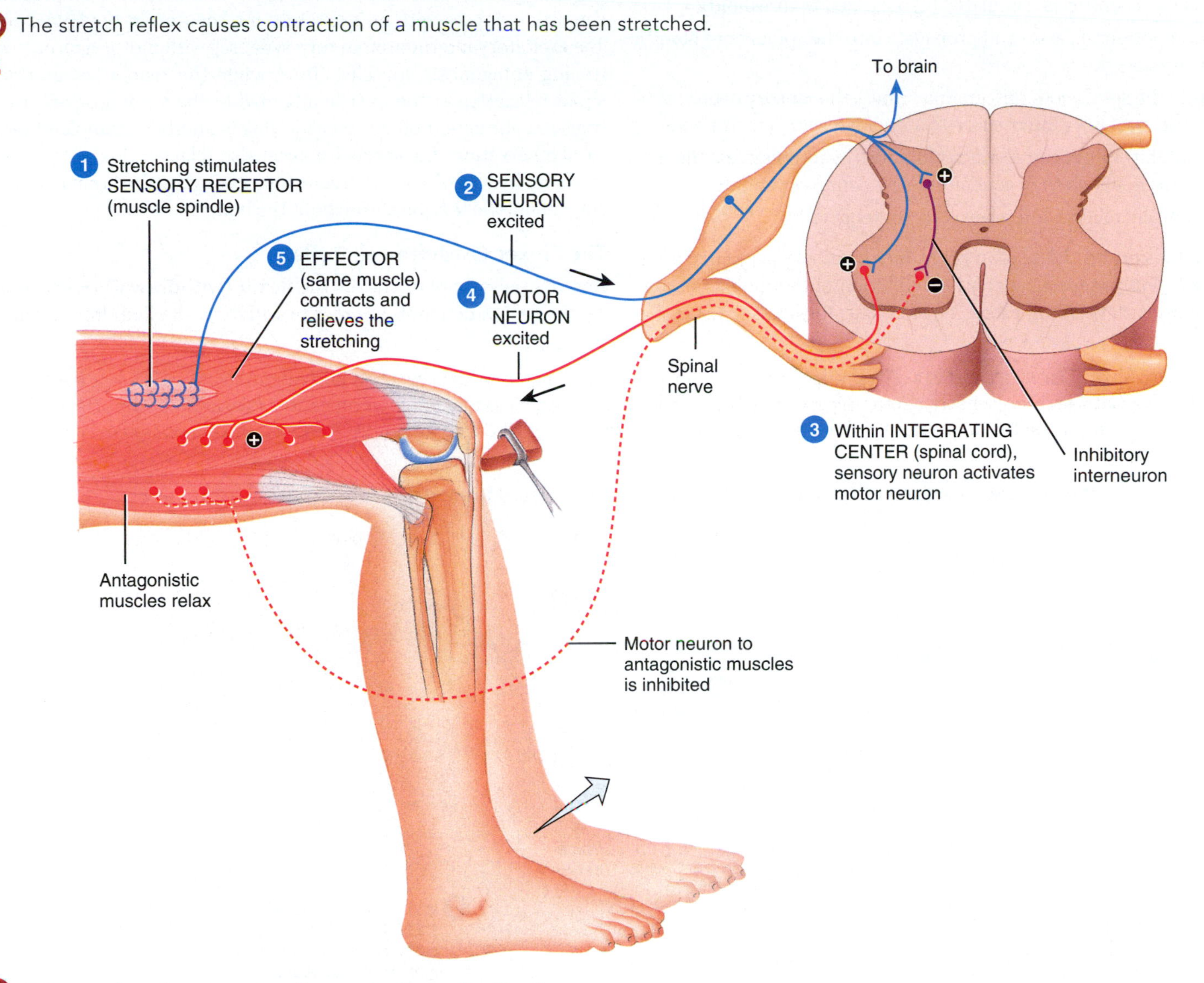

What makes the stretch reflex an ipsilateral reflex?

relaxation of its antagonists, is termed **reciprocal innervation**. Reciprocal innervation prevents conflict between opposing muscles and is vital in coordinating body movements.

Axon collaterals of the muscle spindle sensory neuron also relay action potentials to the brain over specific ascending pathways. In this way, the brain receives input about the state of stretch or contraction of skeletal muscles, enabling it to coordinate muscular movements. The action potentials that pass to the brain also allow conscious awareness that the reflex has occurred.

The stretch reflex can also help maintain posture. For example, if a standing person begins to lean forward, the muscles of the calf (posterior leg) are stretched. Consequently, stretch reflexes are initiated in these muscles, which cause them to contract and reestablish the body's upright posture. Similar types of stretch reflexes occur in the muscles of the shin (anterior leg) when a standing person begins to lean backward.

The Tendon Reflex

The stretch reflex operates as a feedback mechanism to control muscle *length* by causing muscle contraction. In contrast, the **tendon reflex** operates as a feedback mechanism to control muscle *tension* by causing muscle relaxation before muscle force becomes so great that tendons might be torn. Although the tendon reflex is less sensitive than the stretch reflex, it can override the stretch reflex when tension is great, making you drop a very heavy weight, for example. Like the stretch reflex, the tendon reflex is ipsilateral. The sensory receptors for this reflex are called **tendon organs**, or *Golgi tendon organs*, which lie within a tendon near its junction with a muscle. In contrast to muscle spindles, which are sensitive to static muscle length and changes in muscle length, tendon organs detect and respond to changes in muscle tension that are caused by passive stretch or muscular contraction.

A tendon reflex operates as follows (**FIGURE 12.4**):

1. As the tension applied to a tendon increases, the tendon organ (sensory receptor) is stimulated (depolarized to threshold).
2. Action potentials arise and propagate into the spinal cord along a sensory neuron.
3. Within the spinal cord (integrating center), the sensory neuron activates an inhibitory interneuron that synapses with a motor neuron.
4. The inhibitory neurotransmitter inhibits (hyperpolarizes) the motor neuron, which then generates fewer action potentials.
5. The muscle relaxes and relieves excess tension.

Thus, as tension on the tendon organ increases, the frequency of inhibitory action potentials increases; inhibition of the motor neurons to the muscle developing excess tension (effector) causes relaxation of the muscle. In this way, the tendon reflex protects the tendon and muscle from damage due to excessive tension.

Note in **FIGURE 12.4** that the sensory neuron from the tendon organ also synapses with an excitatory interneuron in the spinal cord. The excitatory interneuron in turn synapses with motor neurons controlling antagonistic muscles. Thus, while the tendon reflex brings about relaxation of the muscle attached to the tendon organ, it also triggers contraction of antagonists. This is another example of reciprocal innervation. The sensory neuron also relays action potentials to the brain by way of sensory tracts, thus informing the brain about the state of muscle tension throughout the body.

The Flexor (Withdrawal) Reflex

Another polysynaptic reflex is the **flexor (withdrawal) reflex**, which causes flexion of a limb in order to withdraw the limb from a painful

FIGURE 12.4 Tendon reflex. This reflex arc is polysynaptic—more than one CNS synapse and more than two different neurons are involved in the pathway. The sensory neuron synapses with two interneurons. An inhibitory interneuron causes relaxation of the effector, and a stimulatory interneuron causes contraction of the antagonistic muscle. Plus signs (+) indicate excitatory synapses; the minus sign (−) indicates an inhibitory synapse.

The tendon reflex causes relaxation of the muscle attached to the stimulated tendon organ.

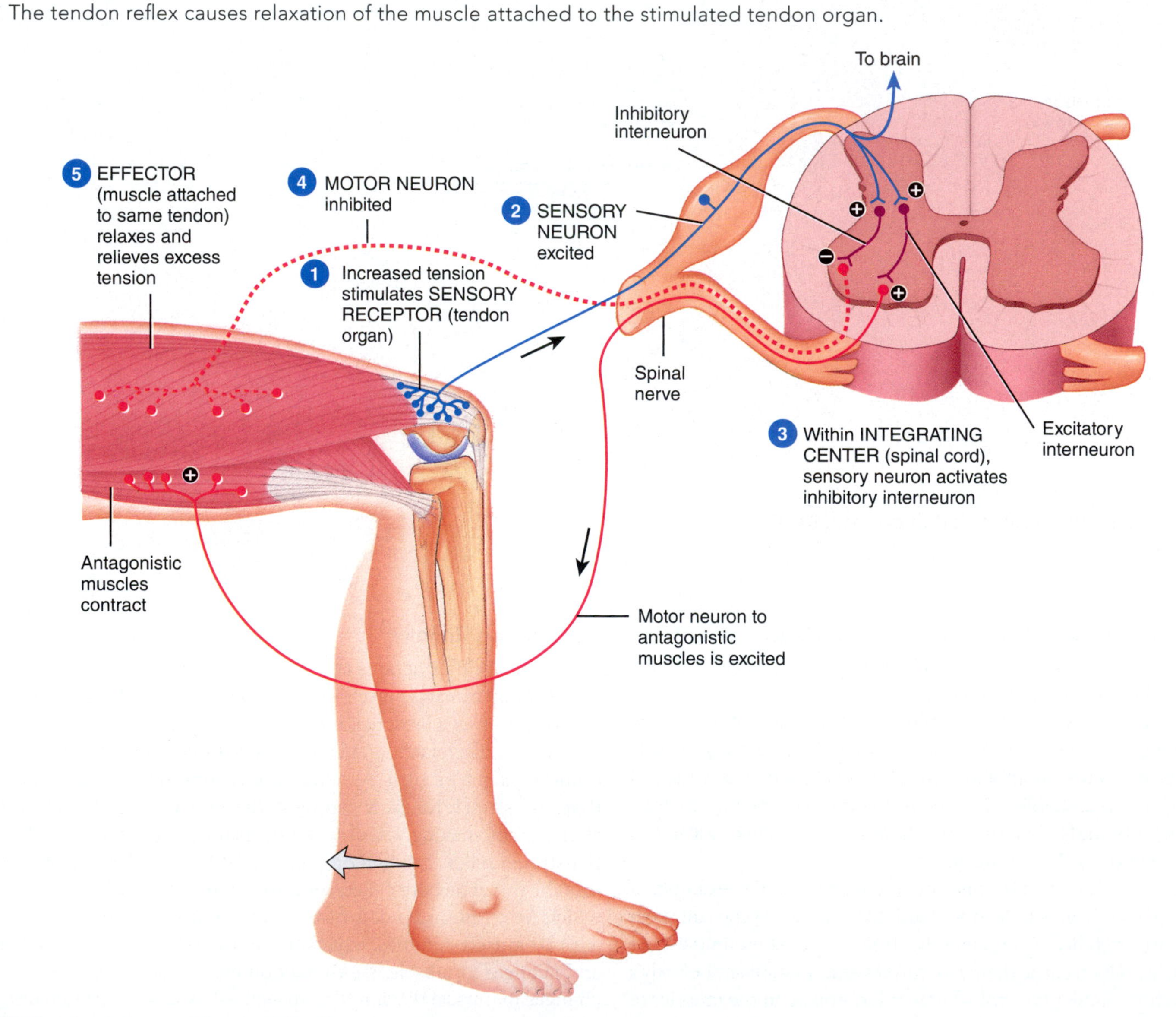

What is reciprocal innervation?

stimulus. For example, if you step on a tack, you immediately flex (bend) your leg in order to withdraw your leg from the tack. Recall that *flexion* is the act of bending a limb at a joint. Muscles that cause flexion of a limb are known as *flexors.* The flexor (withdrawal) reflex operates as follows (**FIGURE 12.5**):

1. Stepping on a tack stimulates the dendrites (sensory receptor) of a pain-sensitive neuron.
2. This sensory neuron then generates action potentials, which propagate into the spinal cord.
3. Within the spinal cord (integrating center), the sensory neuron activates interneurons that extend to several spinal cord segments.
4. The interneurons activate motor neurons in several spinal cord segments. As a result, the motor neurons generate action potentials, which propagate toward the axon terminals.
5. Acetylcholine released by the motor neurons causes the flexor muscles in the thigh (effectors) to contract, producing withdrawal of the leg. This reflex is protective because contraction of flexor muscles moves a limb away from the source of a possibly damaging

FIGURE 12.5 Flexor (withdrawal) reflex. This reflex arc is polysynaptic and ipsilateral. Plus signs (+) indicate excitatory synapses.

The flexor reflex causes withdrawal of a part of the body in response to a painful stimulus.

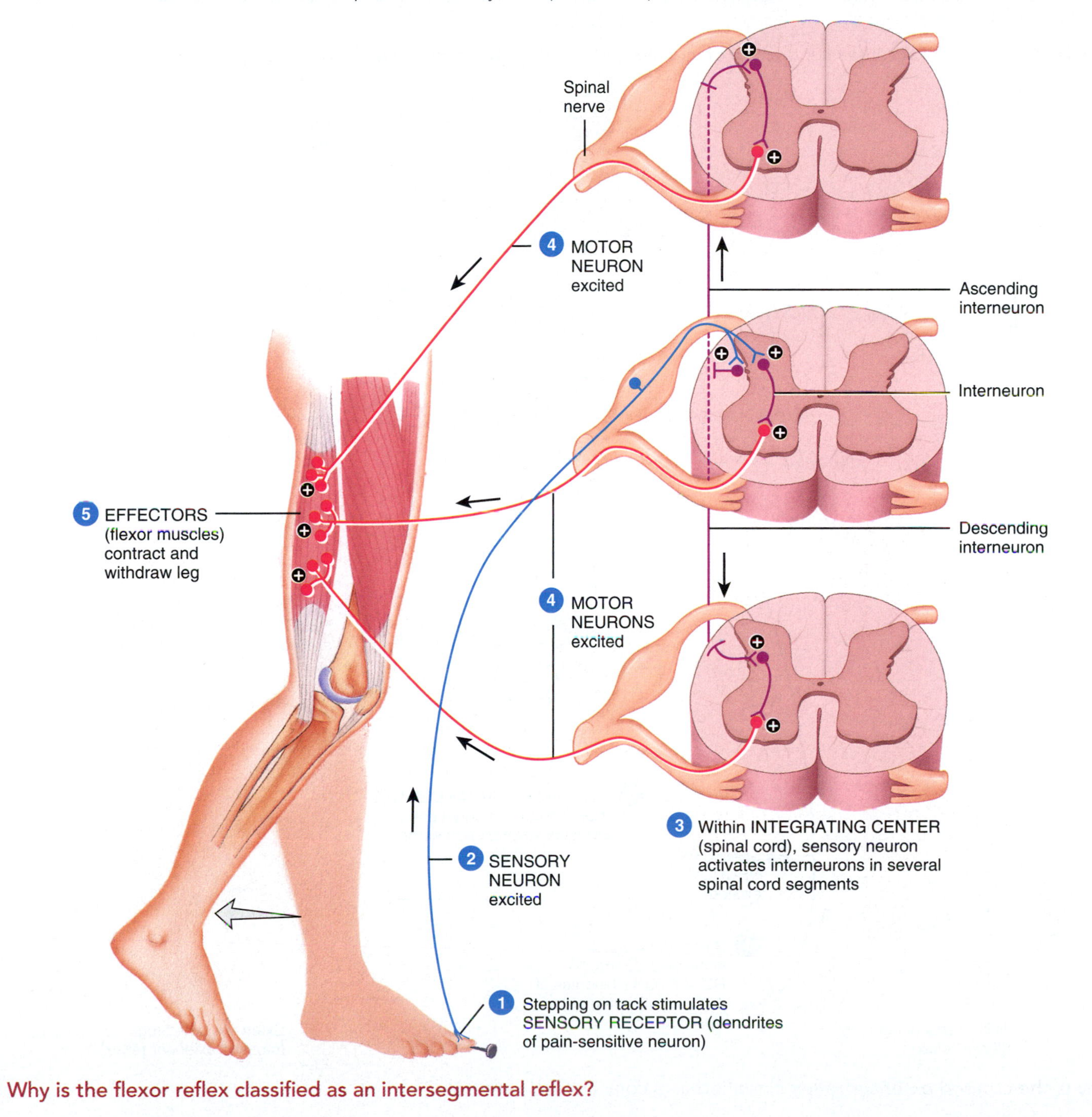

Why is the flexor reflex classified as an intersegmental reflex?

stimulus. A comparable reflex occurs with painful stimulation of either upper limb.

The flexor reflex, like the stretch reflex, is ipsilateral—the incoming and outgoing action potentials propagate into and out of the same side of the spinal cord. The flexor reflex also illustrates another feature of polysynaptic reflex arcs. Moving your entire lower or upper limb away from a painful stimulus involves contraction of more than one muscle group. Hence, several motor neurons must simultaneously convey action potentials to several limb muscles. Because action potentials from one sensory neuron ascend and descend in the spinal cord and activate interneurons in several segments of the spinal cord, this type of reflex is called an **intersegmental reflex arc**. Through intersegmental reflex arcs, a single sensory neuron can activate several motor neurons, thereby stimulating more than one effector. The monosynaptic stretch reflex, in contrast, involves muscles receiving action potentials from one spinal cord segment only.

The Crossed Extensor Reflex

Something else may happen when you step on a tack: You may start to lose your balance as your body weight shifts to the other foot. In addition to initiating the flexor reflex that causes you to withdraw the limb, the pain impulses from stepping on the tack also initiate a **crossed extensor reflex**, which causes extension of the opposite limb to help you maintain balance. The crossed extensor reflex operates as follows (**FIGURE 12.6**):

FIGURE 12.6 Crossed extensor reflex. The flexor reflex arc is shown (at left) for comparison with the crossed extensor reflex arc. Plus signs (+) indicate excitatory synapses.

A crossed extensor reflex causes contraction of muscles that extend joints in the limb opposite to a painful stimulus.

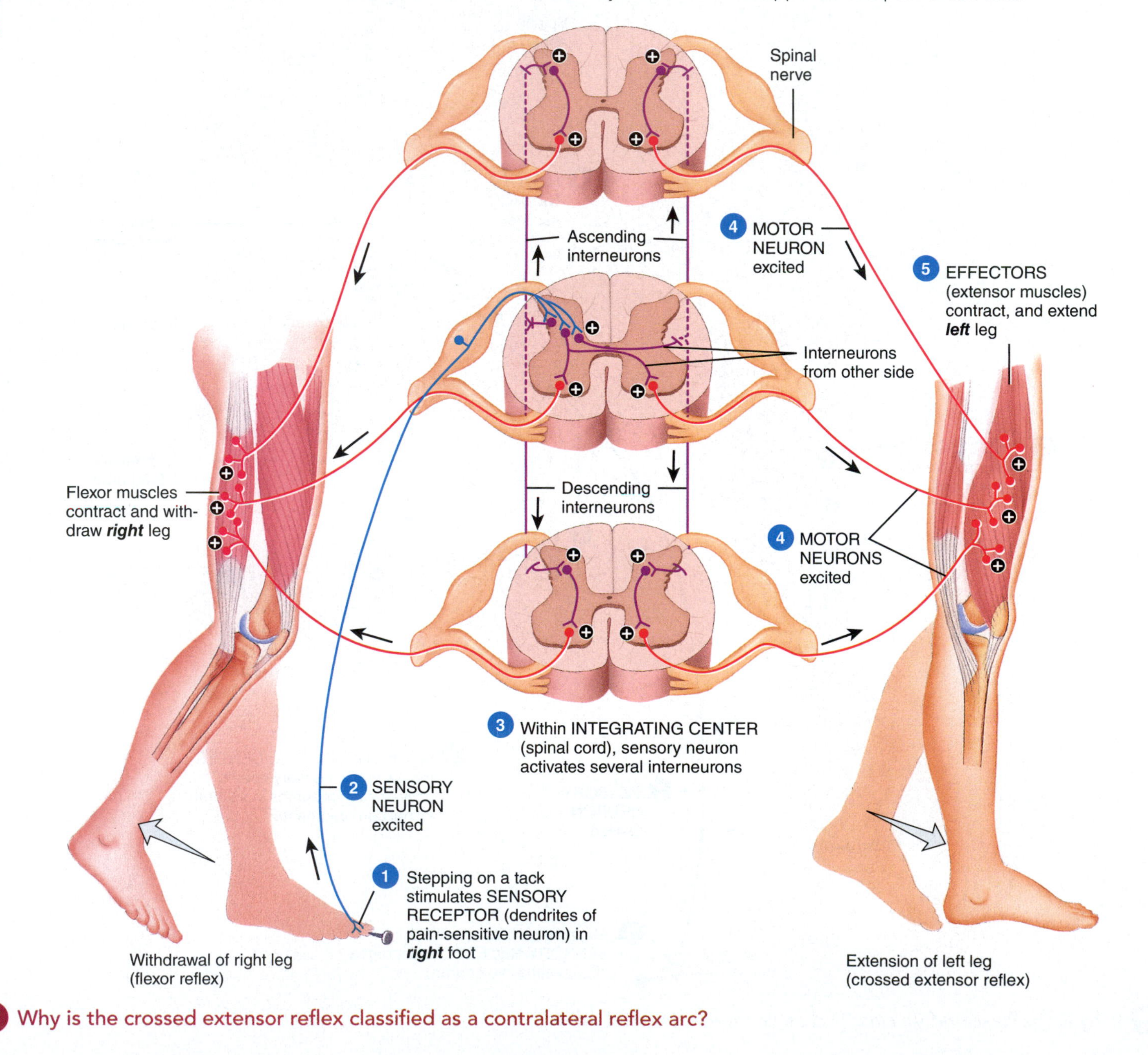

Why is the crossed extensor reflex classified as a contralateral reflex arc?

1. Stepping on a tack stimulates the sensory receptor of a pain-sensitive neuron in the right foot.
2. This sensory neuron then generates action potentials, which propagate into the spinal cord.
3. Within the spinal cord (integrating center), the sensory neuron activates several interneurons that synapse with motor neurons on the left side of the spinal cord in several spinal cord segments. Thus, incoming pain signals cross to the opposite side through interneurons at that level and at several levels above and below the point of entry into the spinal cord.
4. The interneurons excite motor neurons in several spinal cord segments that innervate extensor muscles. The motor neurons in turn generate more action potentials, which propagate toward the axon terminals.
5. Acetylcholine released by the motor neurons causes extensor muscles in the thigh (effectors) of the unstimulated left limb to contract, producing extension of the left leg. In this way, weight can be placed on the foot that must now support the entire body. A comparable reflex occurs with painful stimulation of the left lower limb or either upper limb.

Unlike the flexor reflex, which is an ipsilateral reflex, the crossed extensor reflex involves a **contralateral reflex arc** (kon′-tra-LAT-er-al): Sensory input enters one side of the spinal cord and motor output exits on the opposite side. Thus, a crossed extensor reflex synchronizes the extension of the contralateral limb with the withdrawal (flexion) of the stimulated limb. Reciprocal innervation also occurs in both the flexor reflex and the crossed extensor reflex. In the flexor reflex, when the flexor muscles of a painfully stimulated lower limb are contracting, the extensor muscles of the same limb are relaxing to some degree. If both sets of muscles contracted at the same time, the two sets of muscles would pull on the bones in opposite directions, which might immobilize the limb. Because of reciprocal innervation, one set of muscles contracts while the other relaxes.

Central Pattern Generators Are Responsible for Locomotion

Local circuit neurons also play a major role in locomotion (walking and running). The lumbar spinal cord contains networks of local circuit neurons called **central pattern generators (CPGs)**, which are responsible for the rhythmic movements (alternating flexion and extension) of the limbs that occur during locomotion. CPGs coordinate the output of the lower motor neurons that control the muscles of the limbs while the body is walking or running. You can voluntarily start or stop locomotor movements because of input from the cerebral cortex. Once locomotion begins, however, the CPGs sustain limb movements on their own without any additional input from higher centers in the brain. This has been shown experimentally in cats whose spinal cords have been transected (cut) at the thoracic level: The cat's limbs will still make walking movements if the cat is placed and supported on a moving treadmill.

CHECKPOINT

3. What are the components of a somatic reflex arc?
4. Describe the mechanism and function of the stretch reflex, tendon reflex, flexor (withdrawal) reflex, and crossed extensor reflex.
5. What does each of the following terms mean in relation to reflex arcs? Monosynaptic, ipsilateral, polysynaptic, intersegmental, contralateral, and reciprocal innervation.
6. What is the difference between alpha and gamma motor neurons?
7. How does locomotion occur?

12.3 Control of Movement by the Cerebral Cortex

OBJECTIVES

- Discuss the functions of the motor areas of the cerebral cortex.
- Describe the roles of the corticospinal and corticobulbar tracts.

Control of body movement involves neural pathways that begin in the cerebral cortex. Two areas of the cerebral cortex that play important roles in motor control are the **primary motor cortex** and its adjacent association area, the **premotor cortex** (FIGURE 12.7).

The Premotor Cortex Creates a Motor Plan

The idea or desire to move is generated in one or more cortical association areas, such as the prefrontal cortex, somatosensory association area, auditory association area, or visual association area. This information is sent to the basal nuclei, which process the information and then send it to the thalamus and then to the premotor cortex, where a motor plan is developed. This plan identifies which muscles should contract, how much they need to contract, and in what order. From the premotor cortex, the plan is transmitted to the primary motor cortex for execution. The premotor cortex also stores information about learned motor activities. By activating the appropriate neurons of the

FIGURE 12.7 Motor areas of the cerebral cortex.

The two major motor areas of the cerebral cortex are the primary motor cortex and the premotor cortex.

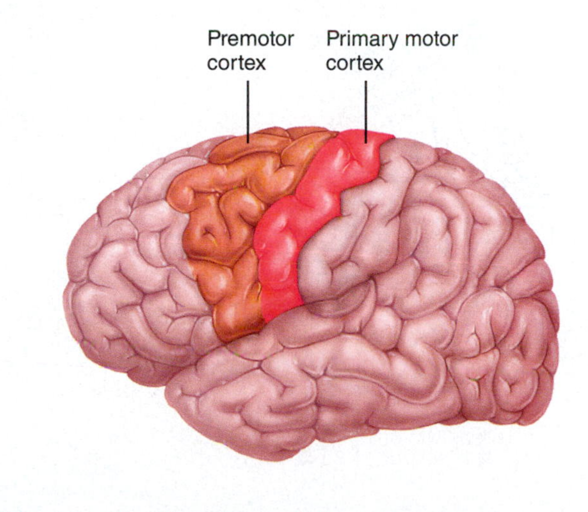

What are the functions of the primary motor cortex and the premotor cortex?

primary motor cortex, the premotor cortex causes specific groups of muscles to contract in a specific sequence.

The Primary Motor Cortex Controls the Execution of Voluntary Movements

The primary motor cortex is the major control region for the execution of voluntary movements. Electrical stimulation of any point in the primary motor cortex causes contraction of specific muscles on the opposite side of the body. The primary motor cortex controls muscles by forming descending pathways that extend to the spinal cord and brain stem (described shortly). As is true for somatic sensory representation in the primary somatosensory cortex, a "map" of the body is present in the primary motor cortex: Each point within the area controls muscle fibers in a different part of the body. Different muscles are represented unequally in the primary motor cortex (**FIGURE 12.8**). More cortical area is devoted to those muscles involved in skilled, complex, or delicate movement. Muscles in the thumb, fingers, lips, tongue, and vocal cords have large representations; the trunk has a much smaller representation. This distorted muscle map of the body is called the **motor homunculus**.

FIGURE 12.8 The motor homunculus.

Each region within the primary motor cortex controls muscle fibers in a different part of the body.

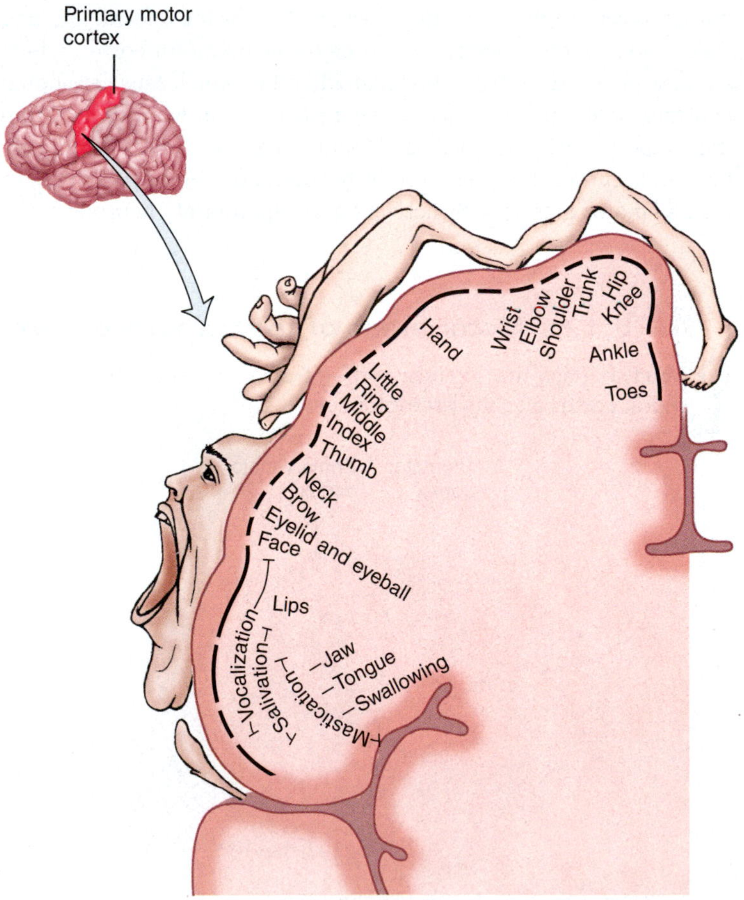

Muscles in which parts of the body have a large representation in the primary motor cortex? Why?

The Primary Motor Cortex Gives Rise to the Direct Motor Pathways

The axons of upper motor neurons extend from the brain to lower motor neurons via two types of pathways—direct and indirect. Direct motor pathways provide input to lower motor neurons via axons that extend directly from the cerebral cortex. Indirect motor pathways provide input to lower motor neurons from motor centers in the brain stem. Direct and indirect pathways both govern the generation of action potentials in the lower motor neurons, the neurons that stimulate contraction of skeletal muscles. The rest of this section focuses on the direct motor pathways; indirect pathways are described in Section 12.4.

Action potentials for voluntary movements propagate from the cerebral cortex to lower motor neurons via the **direct motor pathways**. Also known as the *pyramidal pathways*, the direct motor pathways consist of axons that descend from pyramidal cells of the primary motor cortex and premotor cortex. *Pyramidal cells* are upper motor neurons that have pyramid-shaped cell bodies (see **FIGURE 8.6d**). They are the main output cells of the cerebral cortex. The direct motor pathways consist of the corticospinal pathways and the corticobulbar pathway.

The Corticospinal Pathways

The **corticospinal pathways** (kor′-ti-kō-SPĪ-nal) conduct action potentials for voluntary control of skeletal muscles of the limbs and trunk. Axons of upper motor neurons in the cerebral cortex form the **corticospinal tracts**, which descend from the cerebral cortex to the spinal cord (**FIGURE 12.9a**). As the corticospinal tracts pass through the medulla oblongata, they form ventral bulges known as the *pyramids*. There are two types of corticospinal tracts: the lateral corticospinal tract and the ventral corticospinal tract. About 90% of the corticospinal axons decussate (cross over) in the medulla and then descend into the spinal cord, where they form the **lateral corticospinal tract** (**FIGURE 12.9a**). Axons of this tract synapse with lower motor neurons that supply skeletal muscles in the distal parts of the limbs. The distal muscles are responsible for precise, agile, and highly skilled movements of the hands and feet. Examples include the movements needed to button a shirt or play the piano. The 10% of the corticospinal axons that do not decussate in the medulla descend into the spinal cord to form the **ventral corticospinal tract** (**FIGURE 12.9a**). Many axons of this tract decussate in the spinal cord and then synapse with lower motor neurons that supply skeletal muscles in the trunk and proximal parts of the limbs. Muscles of the trunk are those located in the chest, back, and abdomen; proximal muscles are present in the upper parts of the limbs. Because most axons of the corticospinal tracts decussate, the right cerebral cortex controls muscles on the left side of the body, and the left cerebral cortex controls muscles on the right side of the body.

The Corticobulbar Pathway

The **corticobulbar pathway** (kor′-ti-kō-BUL-bar) conducts action potentials for voluntary control of skeletal muscles in the head. Axons of upper motor neurons from the cerebral cortex form the **corticobulbar tract**, which descends from the cerebral cortex to the brain stem (**FIGURE 12.9b**). Some of the axons of the corticobulbar tract decussate; others do not. The axons of this tract synapse with lower motor neurons associated with nine pairs of cranial nerves: oculomotor (III), trochlear (IV), trigeminal (V), abducens (VI), facial (VII),

FIGURE 12.9 The direct motor pathways. For simplicity, only two cranial nerves are illustrated for the corticobulbar pathway.

The direct motor pathways conduct action potentials for voluntary control of skeletal muscles of the body.

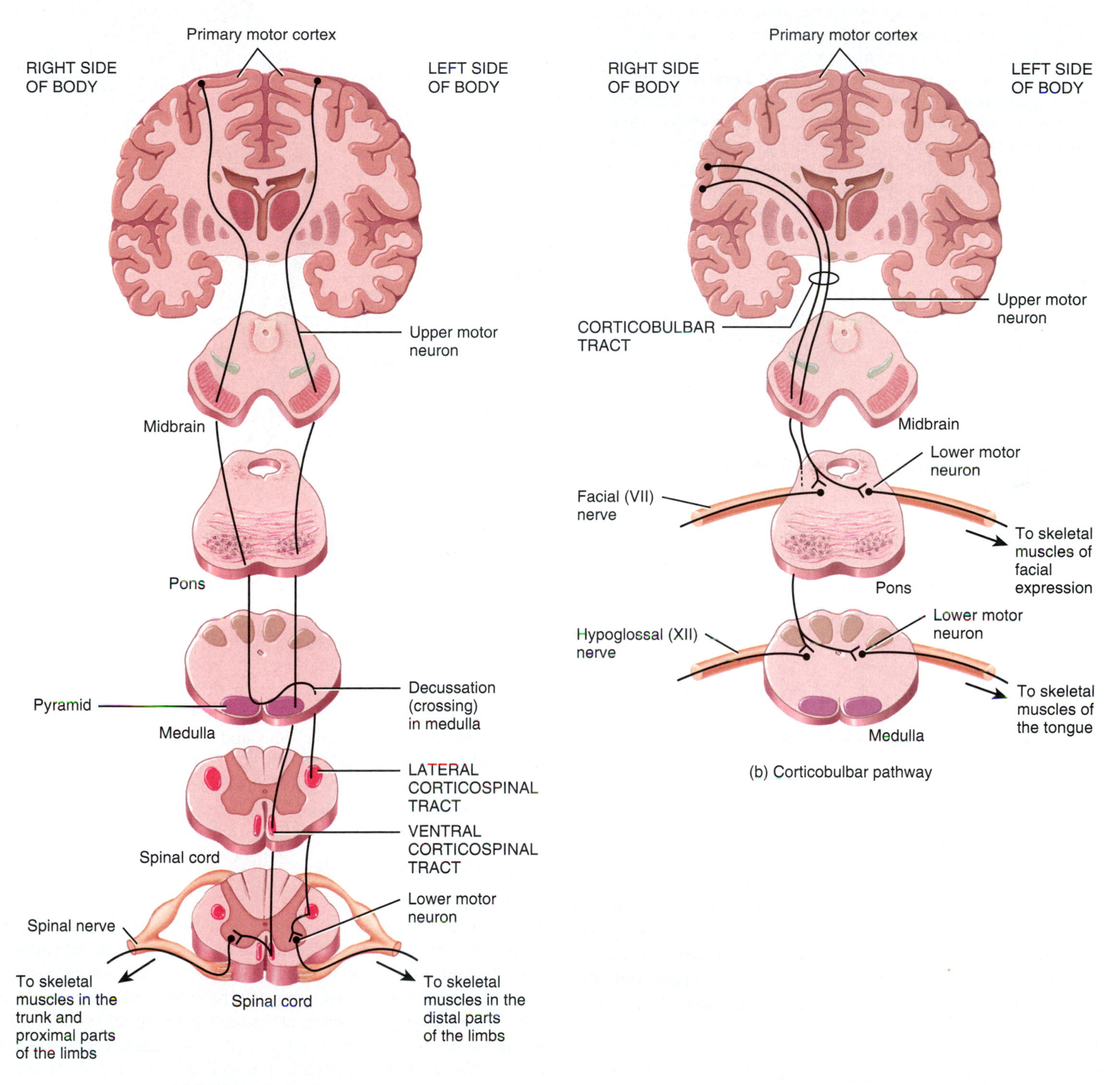

Which tract conveys action potentials that result in voluntary contractions of the muscles in the hands and feet?

CLINICAL CONNECTION

Paralysis

Damage or disease of *lower* motor neurons produces **flaccid paralysis** of muscles on the same side of the body. There is neither voluntary nor reflex action of the innervated muscle fibers, muscle tone is decreased or lost, and the muscle remains limp or flaccid. Injury or disease of ***upper*** motor neurons in the cerebral cortex removes inhibitory influences that some of these neurons have on lower motor neurons, which causes **spastic paralysis** of muscles on the opposite side of the body. In this condition, muscle tone is increased, reflexes are exaggerated, and pathological reflexes such as the Babinski sign appear. The Babinski sign occurs when the great toe extends with or without fanning of the other toes in response to stroking the outer margin of the sole.

glossopharyngeal (IX), vagus (X), accessory (XI), and hypoglossal (XII). These lower motor neurons supply skeletal muscles that control facial expression, chewing, speech, and movements of the eyes, tongue, and neck.

CHECKPOINT

8. Where in the cerebral cortex is the idea or thought to move generated?
9. Which motor area of the cerebral cortex is responsible for developing a motor plan and storing learned muscle information?
10. Which tract from the cerebral cortex allows you to tie you shoes? To contract your abdominal muscles? To smile?

FIGURE 12.10 Motor centers of the brain stem.

The four major motor centers of the brain stem are the vestibular nuclei, reticular formation, superior colliculus, and red nucleus.

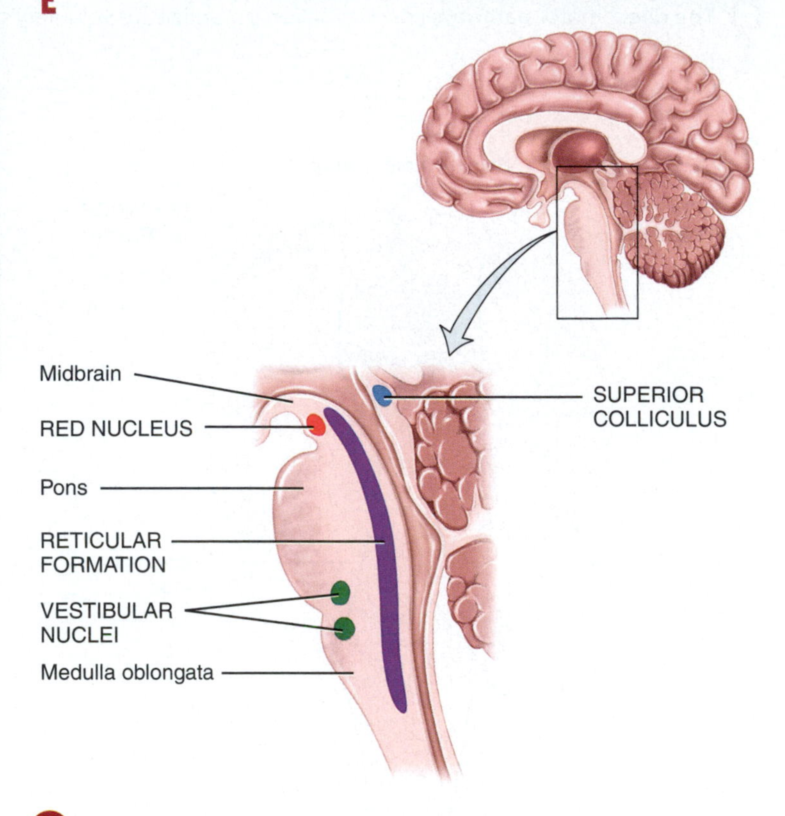

How does the reticular formation contribute to motor control?

12.4 Control of Movement by the Brain Stem

OBJECTIVES

- Identify the motor centers of the brain stem.
- Explain the functions of the various tracts of the indirect motor pathways.

The brain stem is another region important to motor control. It contains four major motor centers that help regulate body movements—(1) the **vestibular nuclei** in the medulla and pons; (2) the **reticular formation** located throughout the brain stem; (3) the **superior colliculus** in the midbrain, and (4) the **red nucleus**, also present in the midbrain (FIGURE 12.10).

Brain Stem Motor Centers Give Rise to the Indirect Motor Pathways

Upper motor neurons whose cell bodies are in the brain stem motor centers extend axons that form the **indirect motor pathways**, also known as the *extrapyramidal pathways*. Several tracts comprise the indirect motor pathways: the vestibulospinal, the medial and lateral reticulospinal, the tectospinal, and the rubrospinal tracts (FIGURE 12.11). In general, the indirect motor pathways convey action potentials from the brain stem to cause involuntary movements that regulate posture, balance, muscle tone, and reflexive movements of the head and trunk. An exception is the rubrospinal tract, which plays an ancillary role to the lateral corticospinal tract in the regulation of voluntary movements of the upper limbs.

The Vestibular Nuclei Help Control Posture in Response to Changes in Balance

Many postural muscles of the trunk and limbs are reflexively controlled by upper motor neurons in the brain stem. **Postural reflexes** keep the body in an upright and balanced position. Input for postural reflexes comes from three sources: (1) the eyes, which provide visual information about the position of the body in space; (2) the vestibular apparatus of the inner ear, which provides information about the position of the head, and (3) proprioceptors in muscles and joints, which provide information about the position of the limbs. In response to this sensory input, upper motor neurons in the brain stem activate lower motor neurons, which in turn cause the appropriate postural muscles to contract in order keep the body properly oriented in space. Posture is also maintained by the stretch and crossed extensor reflexes, which are regulated by reflex centers in the spinal cord.

FIGURE 12.11 The indirect motor pathways. For simplicity, the vestibular nucleus is shown only in the pons, the reticular formation is shown only in the medulla, and only one reticulospinal tract is shown in the spinal cord.

In general, the indirect motor pathways conduct action potentials to cause involuntary movements that regulate posture, balance, muscle tone, and reflexive movements of the head and trunk.

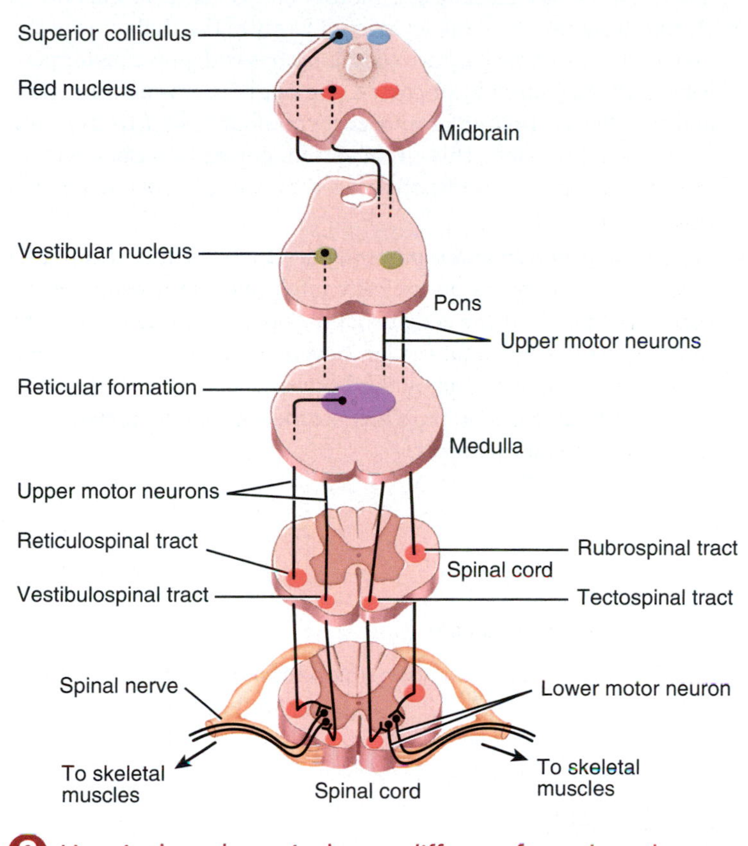

How is the rubrospinal tract different from the other tracts of the indirect motor pathways?

The vestibular nuclei play an important role in the regulation of posture. They receive neural input from the vestibulocochlear (VIII) nerve regarding the state of equilibrium (balance) of the body (mainly the head) and neural input from the cerebellum. In response to this input, the vestibular nuclei generate action potentials along the axons of the **vestibulospinal tract**, which conveys signals to skeletal muscles of the trunk and proximal parts of the limbs (FIGURE 12.11). The vestibulospinal tract causes contraction of these muscles in order to maintain posture in response to changes in equilibrium.

The Reticular Formation Helps Regulate Posture and Muscle Tone During Ongoing Movements

The reticular formation also helps control posture. In addition, it can alter muscle tone. The reticular formation receives input from several sources, including the eyes, ears, cerebellum, and basal nuclei. In response to this input, discrete nuclei in the reticular formation generate action potentials along the **medial reticulospinal tract** and **lateral reticulospinal tract**, both of which convey signals to skeletal muscles of the trunk and proximal limbs (FIGURE 12.11). Although the pathways are similar, the medial reticulospinal tract *excites* the skeletal muscles of the trunk and extensor muscles of the proximal limbs, whereas the lateral reticulospinal tract *inhibits* the skeletal muscles of the trunk and extensor muscles of the proximal limbs. The medial and lateral reticulospinal tracts work together to maintain posture and regulate muscle tone *during ongoing movements*. For example, as you use the biceps muscle in your arm to pick up a heavy weight when working out at the gym, other muscles of the trunk and limbs must contract (or relax) to maintain your posture. Those muscles that need to contract will be activated by the medial reticulospinal tract, whereas those muscles that need to relax will be inhibited by the lateral reticulospinal tract.

The Superior Colliculus Promotes Reflexive Movements of the Head and Trunk and Saccadic Eye Movements

The superior colliculus, which is located in the *tectum* (roof) of the midbrain, receives visual input from the eyes and auditory input from the ears (via connections with the inferior colliculus). When this input occurs in a sudden, unexpected manner, the superior colliculus produces action potentials along the **tectospinal tract**, which conveys neural signals that activate skeletal muscles in the head and trunk (FIGURE 12.11). This allows the body to turn in the direction of the sudden visual stimulus (such as a bug darting across the floor) or the sudden auditory stimulus (such as a bolt of thunder). These responses serve to protect you from potentially dangerous stimuli.

The superior colliculus is also an integrating center for **saccades** (sa-kādz′), small, rapid jerking movements of the eyes that occur as a person looks at different points in the visual field. Although you typically do not realize it, your eyes are constantly making saccades as you read the sentences on the pages of this book or as you look at different parts of a picture or statue. In addition to upper motor neurons that give rise to the tectospinal tract, the superior colliculus also contain upper motor neurons that synapse with local circuit neurons in the **gaze centers** in the reticular formation of the midbrain and pons. The local circuit neurons in the gaze centers in turn synapse with lower motor neurons in the nuclei of the three cranial nerves that regulate the extrinsic eye muscles: oculomotor (III), trochlear (IV), and abducens (VI). Contractions of different combinations of these eye muscles cause horizontal and/or vertical saccades.

The Red Nucleus Helps Control Voluntary Movements of the Upper Limbs

The red nucleus receives input from the cerebral cortex and the cerebellum. In response to this input, the red nucleus generates action potentials along the axons of the **rubrospinal tract**, which conveys neural signals that activate skeletal muscles that cause fine, precise, voluntary movements of the distal parts of the upper limbs (FIGURE 12.11). Note that skeletal muscles in the distal parts of the lower limbs are not activated by the rubrospinal tract. Recall that the lateral corticospinal tract from the cerebral cortex also causes fine, precise movements of the distal parts of the *upper* and *lower* limbs. Compared to the lateral corticospinal tract, the rubrospinal tract plays only a minor role in contracting muscles of the distal parts of the upper limbs. However, the rubrospinal tract becomes functionally significant if the lateral corticospinal tract is damaged.

CHECKPOINT

11. Why are postural reflexes important? Which tracts from the brain stem help regulate posture?
12. What is the purpose of the tectospinal tract?
13. How are the rubrospinal tract and lateral corticospinal tract functionally similar? How are they functionally different?

12.5 The Basal Nuclei and Motor Control

OBJECTIVE

- Describe the functions of the basal nuclei.

The **basal nuclei**, also known as the *basal ganglia*, consist of (1) several masses of gray matter deep within the cerebral hemispheres: the **globus pallidus** (GLŌ-bus PAL-i-dus), **putamen** (pū-TĀ-men), and **caudate nucleus** (KAW-dāt); (2) the **subthalamic nucleus** found just below the thalamus; and (3) the **substantia nigra** (sub-STAN-shē-a NĪ-gra) located in the midbrain (FIGURE 12.12). The basal nuclei influence movement in several ways. This influence is mediated through upper motor neurons rather than lower motor neurons or local circuit neurons. The functions of the basal nuclei include the following:

- ***Initiation of movements.*** The basal nuclei play a major role in initiating movements. Neurons of the basal nuclei receive input from sensory, association, and motor areas of the cerebral cortex. Output from the basal nuclei is sent by way of the thalamus to the premotor cortex, which in turn communicates with upper motor neurons in the primary motor cortex. The upper motor neurons then activate the corticospinal and corticobulbar tracts to promote movement. Therefore, this circuit—from cortex to basal nuclei to thalamus to cortex—is responsible for the initiation of movements (FIGURE 12.13).
- ***Suppression of unwanted movements.*** The basal nuclei suppress unwanted movements by tonically inhibiting the neurons of the thalamus that affect the activity of the upper motor neurons in the motor cortex. When a particular movement is desired, the inhibition of thalamic neurons by the basal nuclei is removed, which allows the thalamic neurons to activate the appropriate upper motor neurons in the motor cortex.

FIGURE 12.12 The basal nuclei. The components of the basal nuclei are shown in purple.

The basal nuclei include the (1) caudate nucleus, putamen, and globus pallidus in the cerebral hemispheres; (2) subthalamic nuclei located below the thalamus; and (3) substantia nigra of the midbrain.

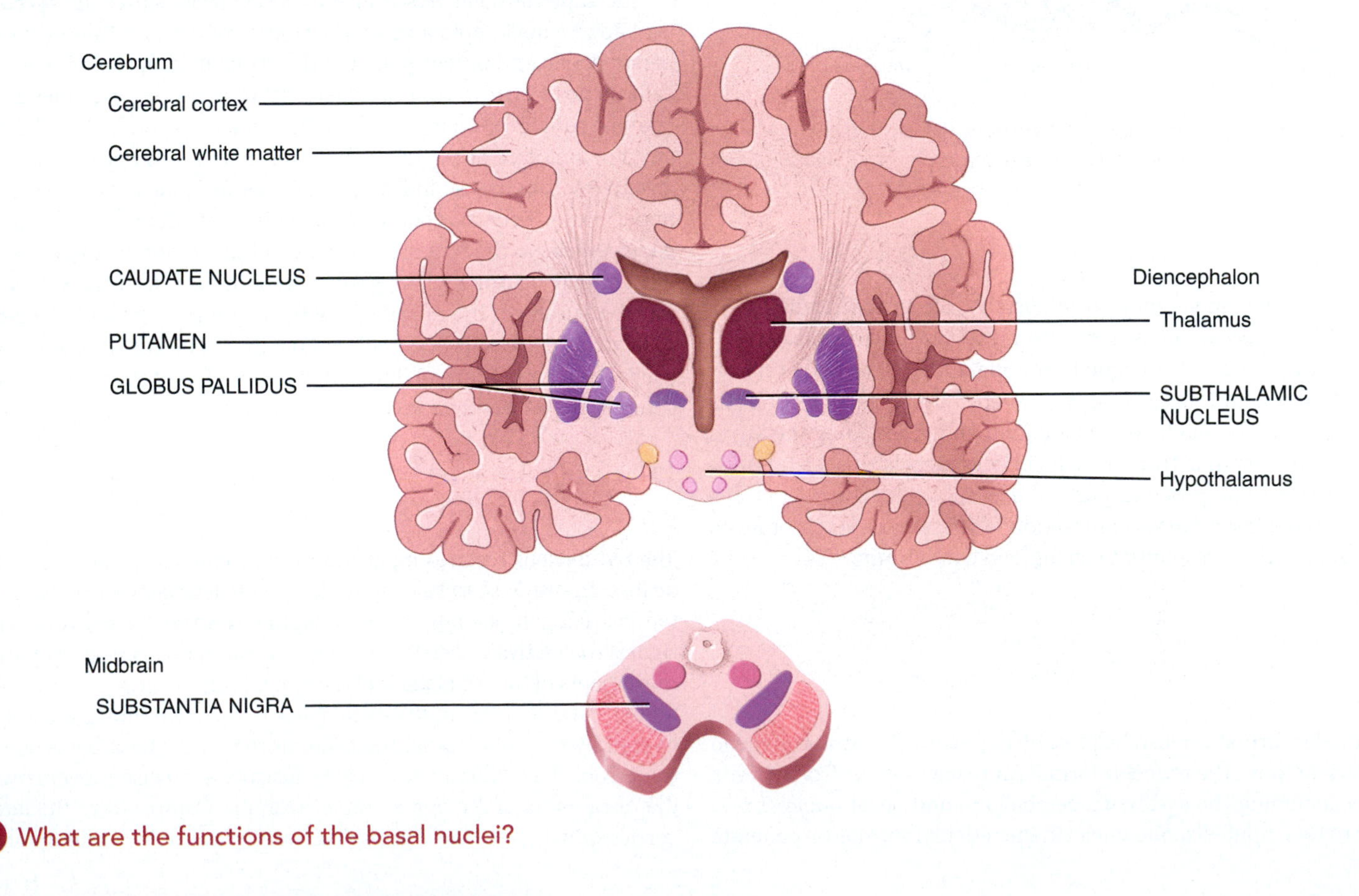

What are the functions of the basal nuclei?

FIGURE 12.13 Pathway by which the basal nuclei initiate movement.

The circuit from the cortex to basal nuclei to thalamus to cortex is responsible for the initiation of movement.

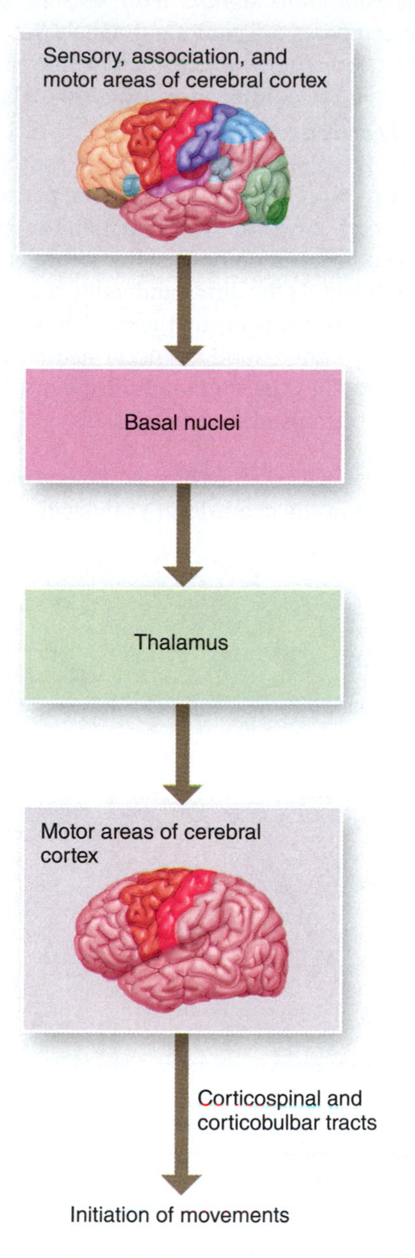

How do the basal nuclei suppress unwanted movements?

- ***Regulation of muscle tone.*** The basal nuclei influence muscle tone. Neurons of the basal nuclei send action potentials into the reticular formation that reduce muscle tone via the medial and lateral reticulospinal tracts. Damage or destruction of some basal nuclei connections causes a generalized increase in muscle tone.
- ***Regulation of nonmotor processes.*** The basal nuclei influence several nonmotor aspects of cortical function, including sensory, limbic, cognitive, and linguistic functions. For example, the basal nuclei help initiate and terminate some cognitive processes, such as attention, memory, and planning. In addition, the basal nuclei may act with the limbic system to regulate emotional behaviors.

CLINICAL CONNECTION

Disorders of the Basal Nuclei

Disorders of the basal nuclei can affect body movements, cognition, and behavior. **Parkinson's disease (PD)** is a progressive disorder of the CNS that typically begins to affect its victims around age 60. In PD, dopamine-releasing neurons that extend from the substantia nigra to the putamen and caudate nucleus degenerate. The loss of dopamine causes ***tremor*** (uncontrollable shaking); ***bradykinesia*** (slowness of movements); and rigidity (stiffness) of the face, arms, and legs. Rigidity of the facial muscles gives the face a masklike appearance. The expression is characterized by a wide-eyed, unblinking stare and a slightly opened mouth with uncontrolled drooling. Treatment of PD is directed toward increasing the levels of dopamine in the brain. Many patients with PD benefit from the use of the drug ***levodopa (L-dopa)*** because it is a precursor of dopamine. However, this drug does not slow the progression of the disease. As more and more affected brain cells die, the drug becomes less and less effective. PD is also treated with drugs that inhibit monoamine oxidase, the enzyme that degrades catecholamine neurotransmitters such as dopamine. Other therapies for PD include transplantation of dopamine-rich fetal nervous tissue into the basal nuclei; lesioning (destroying) areas of the basal nuclei that generate tremors or produce muscle rigidity; and ***deep-brain stimulation (DBS),*** which involves the implantation of an electrode into the brain. The electrical currents released by the implanted electrode reduce many of the symptoms of PD.

Huntington's disease (HD) is an inherited disorder in which the caudate nucleus and putamen degenerate, with loss of neurons that normally release GABA or acetylcholine. A key sign of HD is **chorea** (KŌ-rē-a = a dance), in which rapid, jerky movements occur involuntarily and without purpose. Progressive mental deterioration also occurs. Symptoms of HD often do not appear until age 30 or 40. Death occurs 10 to 20 years after symptoms first appear.

Tourette's syndrome is a disorder that is characterized by involuntary body movements (motor tics) and the use of inappropriate or unnecessary sounds or words (vocal tics). Although the cause is unknown, research suggests that this disorder involves a dysfunction of the cognitive neural circuits between the basal nuclei and the prefrontal cortex.

Some psychiatric disorders, such as schizophrenia and obsessive–compulsive disorder, are thought to involve dysfunction of the behavioral neural circuits between the basal nuclei and the limbic system. In **schizophrenia**, excess dopamine activity in the brain causes a person to experience delusions, distortions of reality, paranoia, and hallucinations. People who have **obsessive–compulsive disorder (OCD)** experience repetitive thoughts (obsessions) that cause repetitive behaviors (compulsions) that they feel obligated to perform. For example, a person with OCD might have repetitive thoughts about someone breaking into the house; these thoughts might drive that person to check the doors of the house over and over again (for minutes or hours at a time) to make sure that they are locked.

CHECKPOINT

14. How do the basal nuclei initiate movements? Suppress unwanted movements? Reduce muscle tone?
15. What are some of the nonmotor functions of the basal nuclei?

Modulation of Movement by the Cerebellum

OBJECTIVE

- Describe how the cerebellum controls body movements.

The final contributor to motor control is the cerebellum. Like the basal nuclei, the cerebellum exerts its influence on body movements by acting on upper motor neurons. In addition to maintaining proper posture and balance, the cerebellum is active in both learning and performing rapid, coordinated, highly skilled movements such as hitting a golf ball, speaking, and swimming. Cerebellar function involves four activities (FIGURE 12.14):

1. ***Monitoring intentions for movement.*** The cerebellum monitors intentions for movement by receiving input from the motor cortex and basal nuclei via **pontine nuclei** in the pons regarding what movements are planned (purple lines).
2. ***Monitoring actual movement.*** The cerebellum monitors actual movement by receiving input from proprioceptors in joints and muscles that reveals what is actually happening (blue line). These action potentials travel along the **ventral** and **dorsal spinocerebellar tracts**. Action potentials from the vestibular (equilibrium-sensing) apparatus in the inner ear and from the eyes also enter the cerebellum.
3. ***Comparing command signals with sensory information.*** The cerebellum compares the command signals (intentions for movement) *with sensory information* (actual movement performed). This is known as the **comparator function** of the cerebellum.
4. ***Sending out corrective feedback.*** If there is a discrepancy between intended and actual movement, nuclei deep within the cerebellum send out corrective feedback signals to upper motor neurons in the cerebral cortex (via the thalamus) and in brain stem motor centers (green lines). The upper motor neurons then alter movements accordingly via the direct and indirect motor pathways (red lines). As the movements occur, the cerebellum continuously provides error corrections to upper motor neurons, which decreases errors and smoothes the motion. It also contributes over longer periods to the learning of new motor skills.

Skilled activities such as tennis or volleyball provide good examples of the contribution of the cerebellum to movement. To make a good

FIGURE 12.14 Role of the cerebellum in the control of movement.

The cerebellum coordinates and smoothes contractions of skeletal muscles during skilled movements and helps maintain posture and balance.

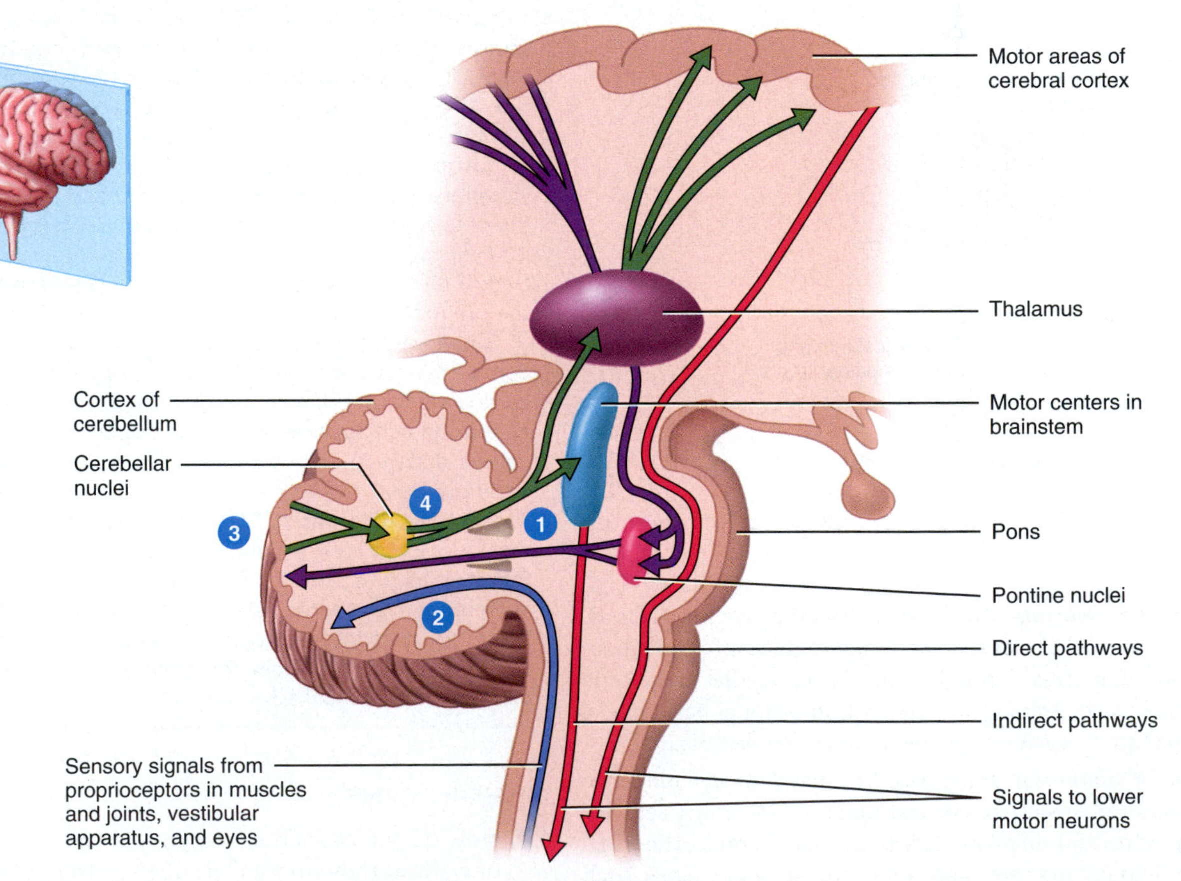

How does the cerebellum send out corrective feedback to upper motor neurons in the motor cortex when there is a discrepancy between intended movement and actual movement?

Critical Thinking

A Stellar Performance

Elizabeth is an acrobat for a circus that travels around the country. She performs spectacular feats such as walking on a tightrope and swinging from a trapeze. She is also known for another skill that she performs along a rope: She hangs by her arms or legs from a rope suspended above the stage and then performs dance-like maneuvers as she ascends and descends the rope, all of which is choreographed to music.

Byelikova Oksana/Shutterstock

SOME THINGS TO KEEP IN MIND:

Elizabeth's artistic performance and more importantly, her safety depend on her brain's ability to control movements precisely. She must be able to determine where she is in space, how her interaction with the rope is changing with each maneuver, and the relationship of her movements to the music. Several parts of Elizabeth's central nervous system (CNS) work together to control her body movements; these areas include the premotor cortex, primary motor cortex, basal nuclei, cerebellum, motor centers in the brain stem, and the spinal cord. Through neural connections with each other and with skeletal muscles, these areas of the CNS ultimately control which muscles contract, when they contract, and how much they contract.

Key:

- Premotor cortex
- Primary motor cortex
- Basal nuclei
- Cerebellum
- Brain stem motor centers
- Spinal cord

Which part of Elizabeth's brain develops a motor plan for the movements that she performs along the rope?

What role does the primary motor cortex play in Elizabeth's ability to move?

How do the basal nuclei contribute to Elizabeth's movements?

What effect would dysfunction of the cerebellum have on Elizabeth's ability to synchronize her movements to music?

Which components of Elizabeth's brain allow her to maintain the proper posture as she ascends or descends the rope?

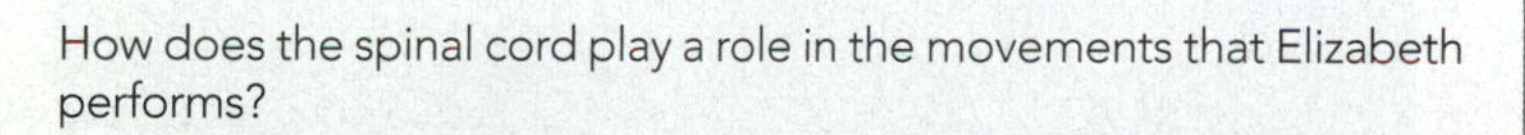

How does the spinal cord play a role in the movements that Elizabeth performs?

CLINICAL CONNECTION

Ataxia

Damage to the cerebellum through trauma or disease disrupts muscle coordination, a condition called **ataxia** (a-TAK-sē-a). Blindfolded people with ataxia cannot touch the tip of their nose with a finger because they cannot coordinate movement with their sense of where a body part is located. Another sign of ataxia is changed speech pattern due to uncoordinated speech muscles. Cerebellar damage may also result in staggering or abnormal walking movements. People who consume too much alcohol show signs of ataxia because alcohol inhibits activity of the cerebellum. Such individuals have difficulty passing sobriety tests. Ataxia can also occur as a result of degenerative diseases (multiple sclerosis and Parkinson's disease), trauma, brain tumors, genetic factors, and as a side effect of medication prescribed for bipolar disorder.

serve or to block a spike, you must bring your racket or arms forward just far enough to make solid contact. How do you stop at exactly the right point? Before you even hit the ball, the cerebellum has sent action potentials to the cerebral cortex and basal nuclei informing them where your swing must stop. In response to action potentials from the cerebellum, the cortex and basal nuclei transmit action potentials to opposing body muscles to stop the swing.

CHECKPOINT

16. How does the cerebellum know which movements are being planned?
17. What is the purpose of the ventral and dorsal spinocerebellar tracts?
18. Why are the cerebellar nuclei important?

FROM RESEARCH TO REALITY

Nicotine and Plasticity of the Motor Cortex

Reference

Gonzalez, C. et al. (2005). Nicotine stimulates dendritic arborization in motor cortex and improves concurrent motor skill but impairs subsequent motor learning. *Synapse*. 55: 183–191.

Can nicotine cause changes in the motor cortex?

Neurons in the brain can exhibit plasticity: They may undergo structural and functional changes based on the presence or absence of certain stimuli. Could a certain drug cause plasticity in the motor cortex? If so, how would this plasticity affect the ability of the motor cortex to control voluntary movements?

Article description:

Researchers in this study investigated the effect of nicotine on neural plasticity in the rat motor cortex and on motor learning. They injected rats with nicotine or saline (control) and then determined how successfully the rats performed a tray reaching task that was previously learned and a single pellet reaching task that was subsequently learned. In addition, pyramidal cells in the rat motor cortex were examined for dendritic length and arborization (branching).

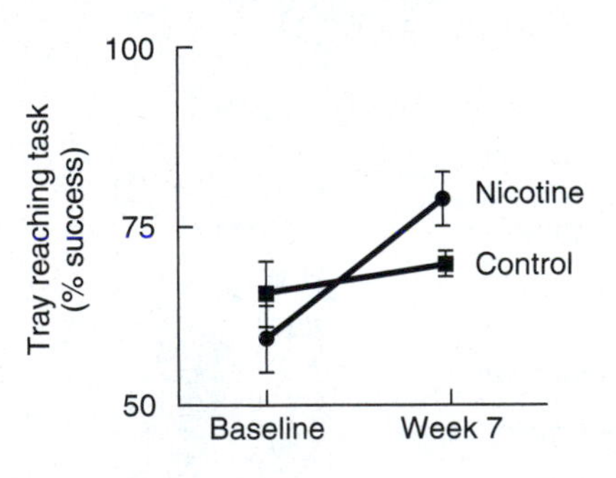

Go to WileyPLUS Learning Space and use the data from this article to answer the questions posed there and to discover more about nicotine and plasticity of the motor cortex.

CHAPTER REVIEW

12.1 Overview of Motor Control

1. All excitatory and inhibitory signals that control movement converge on somatic motor neurons, also known as lower motor neurons or the final common pathway.
2. Neurons in four neural circuits participate in control of movement by providing input to lower motor neurons: local circuit neurons, upper motor neurons, basal nuclei, and the cerebellum.

12.2 Local Level of Motor Control

1. The local level of motor control is coordinated by local circuit neurons, which promote somatic reflexes and locomotion.
2. A somatic reflex is a fast, involuntary response that occurs when action potentials pass through a somatic reflex arc.
3. The components of a somatic reflex arc include a sensory receptor, sensory neuron, integrating center, somatic motor neuron, and effector (skeletal muscle).
4. Somatic spinal reflexes include the stretch reflex, the tendon reflex, the flexor (withdrawal) reflex, and the crossed extensor reflex; all exhibit reciprocal innervation.
5. A two-neuron or monosynaptic reflex arc consists of one sensory neuron and one motor neuron. A stretch reflex, such as the patellar reflex, is one example.
6. The stretch reflex is ipsilateral and is important in maintaining muscle tone.
7. A polysynaptic reflex arc contains sensory neurons, interneurons, and motor neurons. The tendon reflex, flexor (withdrawal) reflex, and crossed extensor reflexes are examples.
8. The tendon reflex is ipsilateral and prevents damage to muscles and tendons when muscle force becomes too extreme. The flexor reflex is ipsilateral and moves a limb away from the source of a painful stimulus. The crossed extensor reflex extends the limb contralateral to a painfully stimulated limb, allowing the weight of the body to shift when a supporting limb is withdrawn.
9. Locomotion occurs because of rhythmic movements (alternating flexion and extension) of the limbs. Central pattern generators (networks of local circuit neurons) in the spinal cord play a key role in locomotion.

12.3 Control of Movement by the Cerebral Cortex

1. The idea or desire to move is generated in one or more association areas of the cerebral cortex.
2. Once the idea is formed, the information is sent to the basal nuclei, which process it and send it to the thalamus and then to the premotor cortex, where a motor plan is developed.
3. From the premotor cortex, the plan is sent to the primary motor cortex, the major control region for the execution of voluntary movements.
4. A somatotopic map is present in the primary motor cortex: Each point within the area controls muscles in a different part of the body. Different muscles are represented unequally; more cortical area is devoted to those muscles involved in skilled, complex or delicate movement.
5. The primary motor cortex controls skeletal muscles by forming descending pathways. Axons of upper motor neurons extend from the cerebral cortex via direct motor pathways, which include the corticospinal and corticobulbar pathways. The corticospinal pathways convey action potentials from the motor cortex to skeletal muscles in the limbs and trunk. The corticobulbar pathway conveys action potentials from the motor cortex to skeletal muscles in the head.

CHAPTER REVIEW

12.4 Control of Movement by the Brain Stem

1. The brain stem contains four major motor centers that help regulate body movements—the vestibular nuclei, reticular formation, superior colliculus, and red nucleus.
2. Axons of upper motor neurons whose cell bodies are in these brain stem motor centers extend axons to form the indirect motor pathways, which include the tectospinal, vestibulospinal, rubrospinal, and medial and lateral reticuospinal tracts. In general, the indirect motor pathways regulate posture, balance, muscle tone, and reflexive movements of the head and trunk.
3. Through the vestibulospinal tract, the vestibular nuclei help control posture in response to changes in equilibrium (balance).
4. Through the medial and lateral reticulospinal tracts, the reticular formation helps regulate posture and muscle tone during ongoing movements.
5. Through the tectospinal tract, the superior colliculus promotes reflexive movements of the head and trunk in response to sudden visual or auditory stimuli; the superior colliculus also promotes saccadic eye movements in a separate pathway that involves gaze centers in the reticular formation.
6. Through the rubrospinal tract, the red nucleus helps control voluntary movements of the upper limbs.

12.5 The Basal Nuclei and Motor Control

1. The basal nuclei include (1) the globus pallidus, putamen, and caudate nucleus in the cerebral hemispheres; (2) the subthalamic nuclei below the thalamus; and (3) the substantia nigra of the midbrain.
2. Neurons of the basal nuclei assist movement by providing input to the upper motor neurons. They help initiate movements, suppress unwanted movements, establish a normal level of muscle tone, and regulate several nonmotor processes.

12.6 Modulation of Movement by the Cerebellum

1. The cerebellum is active in learning and performing rapid, coordinated, highly skilled movements. It also contributes to maintaining balance and posture.
2. Cerebellar function involves four major activities: (1) monitoring intentions for movements; (2) monitoring actual movement; (3) comparing command signals with sensory information; and (4) sending out corrective feedback.

1. An athlete has suffered an injury to his triceps muscle. He undergoes many tests to see what the level of damage is. It turns out that when he tries to get his triceps muscle to contract in response to excessive force, the muscle reflexively relaxes but when the triceps tendon is tapped, nothing happens. What sensory aspect of muscle control is damaged? Explain your answer.

2. A common roadside test is to ask individuals thought to be under the influence of alcohol to close their eyes and touch their noses. Failure to do so effectively indicates intoxication. With which part of the brain is alcohol interfering that would allow for this roadside test to be used?

3. Compare and contrast the four different sources of input to lower motor neurons and explain how these are important in the control of motor movement.

ANSWERS TO FIGURE QUESTIONS

12.1 Upper motor neurons from the cerebral cortex are essential for the planning and execution of voluntary movements of the body; upper motor neurons from the brain stem regulate involuntary movements such as posture, balance, muscle tone, and reflexive movement of the head and trunk.

12.2 Somatic reflexes allow fast, involuntary contractions of skeletal muscles in response to stimuli such as pain, overstretching of muscle, and excessive muscle tension.

12.3 In an ipsilateral reflex, the sensory and motor neurons are on the same side of the spinal cord.

12.4 Reciprocal innervation is a type of arrangement of a neural circuit involving simultaneous contraction of one muscle and relaxation of its antagonist.

12.5 The flexor reflex is intersegmental because action potentials go out over motor neurons located in several spinal nerves, each arising from a different segment of the spinal cord.

12.6 The crossed extensor reflex is a contralateral reflex arc because the motor output leaves the spinal cord on the side opposite the entry of sensory input.

12.7 The premotor cortex is responsible for planning voluntary movements and storing information about learned movements; the primary motor cortex controls the execution of voluntary movements.

12.8 Muscles in the thumb, fingers, lips, tongue, and vocal cords have large representations because more cortical area is devoted to those muscles involved in skilled, complex, or delicate movement.

12.9 The lateral corticospinal tract conducts action potentials that result in contractions of the muscles in the hands and feet.

12.10 The reticular formation helps regulate posture and muscle tone during ongoing movements.

12.11 The rubrospinal tract helps promote voluntary contractions of the upper limbs, whereas the rest of the indirect motor pathways cause involuntary contractions of muscles in the body.

12.12 The basal nuclei initiate movements, suppress unwanted movements, regulate muscle tone, and regulate several nonmotor processes.

12.13 The basal nuclei suppress unwanted movements by inhibiting the neurons of the thalamus that affect the activity of the upper motor neurons in the motor cortex.

12.14 When there is a discrepancy between intended and actual movement, the cerebellum sends corrective feedback signals to upper motor neurons in the motor cortex (via the thalamus) and in the brain stem. The upper motor neurons then modify the movements via the direct and indirect motor pathways.

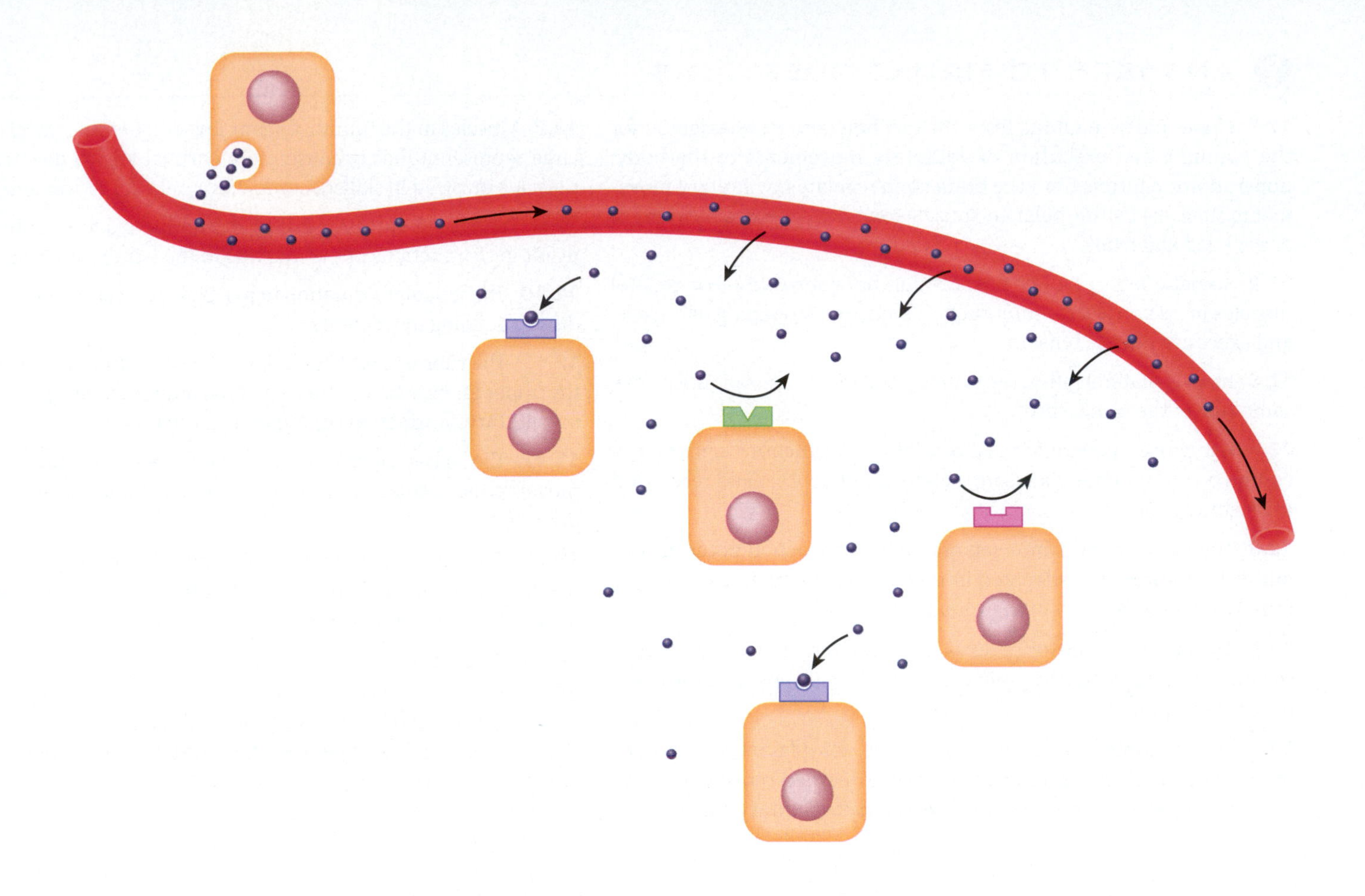

13 The Endocrine System

The Endocrine System and Homeostasis

The hormones of the endocrine system contribute to homeostasis by regulating the activity and growth of target cells in the body.

LOOKING BACK TO MOVE AHEAD...

- The body regulates controlled variables, such as blood pressure, blood glucose level, and stretch of the uterine cervix, through feedback systems; a negative feedback system reverses a change in a controlled variable, whereas a positive feedback system reinforces a change in a controlled variable (Section 1.4).
- There are two types of glands in the body: (1) exocrine glands, which secrete substances into ducts that empty onto the body surface or into the lumen of an organ and (2) endocrine glands, which secrete hormones that enter interstitial fluid and then into the bloodstream without flowing through a duct (Section 3.8).
- An extracellular chemical messenger is a molecule that is released by a cell, enters extracellular fluid, and then binds to a receptor on or in its target cell to cause a response; the three types of extracellular chemical messengers are hormones, neurotransmitters, and local mediators (paracrines and autocrines) (Section 6.2).
- The hypothalamus is a region of the brain that performs a variety of functions, including production of hormones and regulation of circadian rhythms (Section 8.2).

As males and females enter puberty, they start to develop striking differences in physical appearance and behavior. Perhaps no other period in life so dramatically shows the impact of the endocrine system in directing development and regulating body functions. In females, estrogens promote accumulation of adipose tissue in the breasts and hips, sculpting a feminine shape. At the same time or a little later, increasing levels of testosterone in males begin to help build muscle mass and enlarge the vocal cords, producing a lower-pitched voice. These changes are just a few examples of the powerful influence of endocrine secretions. Less dramatically, perhaps, multitudes of hormones help maintain homeostasis on a daily basis. They regulate the activity of smooth muscle, cardiac muscle, and some glands; alter metabolism; spur growth and development; influence reproductive processes; and participate in circadian (daily) rhythms established by the suprachiasmatic nucleus of the hypothalamus.

In this chapter, you will learn about the major hormone-producing glands and tissues that comprise the endocrine system. You will also examine how hormones govern body activities. In addition, you will have several opportunities to see how the nervous and endocrine systems function together as an interlocking "supersystem." For example, certain parts of the nervous system stimulate or inhibit the release of hormones by the endocrine system.

13.1 Overview of the Endocrine System

OBJECTIVES

- Identify the components of the endocrine system.
- Compare the two chemical classes of hormones.
- Discuss the mechanisms of hormone action.
- Explain the different types of hormone interactions.
- Describe the ways by which hormone secretion can be controlled.

The Endocrine System Consists of All Glands, Organs, and Tissues That Contain Hormone-Secreting Cells

Endocrine glands are ductless glands that secrete **hormones**, chemical messengers that are carried by the blood to distant target cells. Initially, the hormone is secreted by a cell into interstitial fluid. From there, the hormone diffuses into the bloodstream. Because most hormones are required in very small amounts, circulating levels typically are low, usually in the picomolar (10^{-12} M) to nanomolar (10^{-9} M) range. The endocrine glands of the body include the pituitary, thyroid, parathyroid, adrenal, and pineal glands (FIGURE 13.1). In addition, several organs and tissues are not exclusively classified as endocrine glands but contain cells that secrete hormones. These include the hypothalamus, skin, thymus, heart, liver, stomach, pancreas, kidneys, small intestine, ovaries, testes, adipose tissue, and placenta. Taken together, all endocrine glands and hormone-secreting cells constitute the **endocrine system**.

FIGURE 13.1 Components of the endocrine system.

The endocrine system consists of all glands, organs, and tissues that contain hormone-secreting cells.

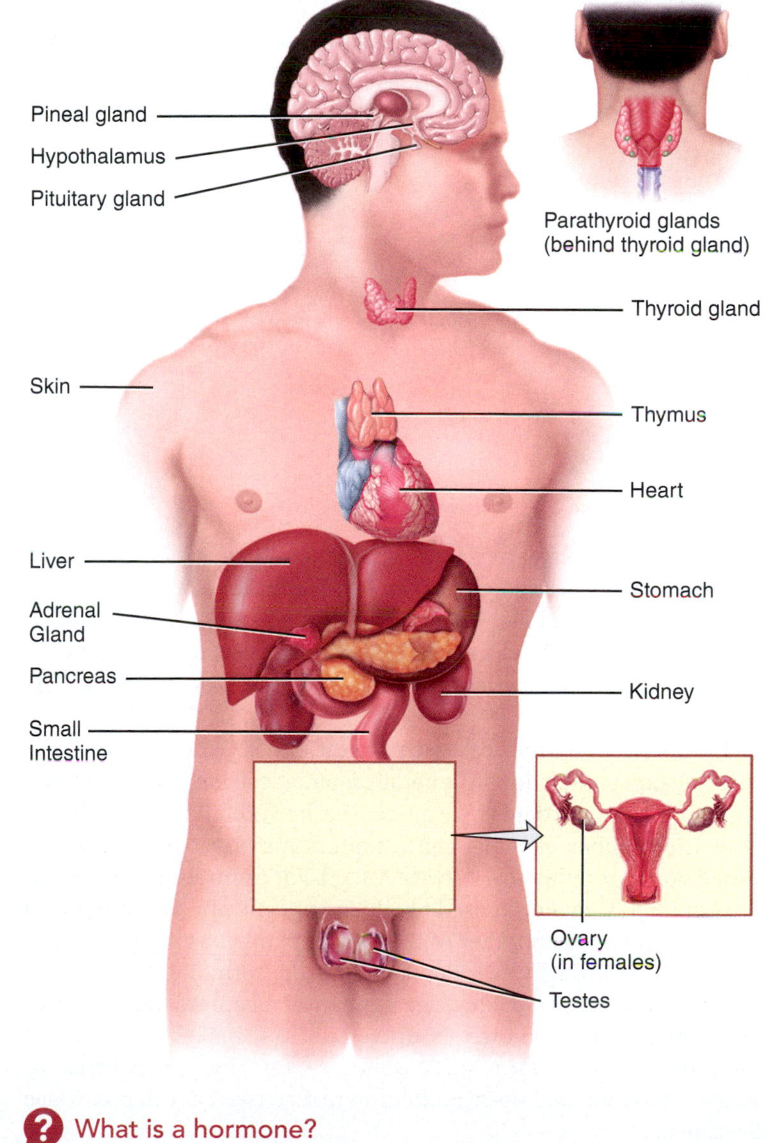

What is a hormone?

Hormones Influence Target Cells by Binding to Receptors

Although a given hormone travels throughout the body in the blood, it affects only specific target cells. Hormones influence their target cells by binding to specific protein **receptors** (FIGURE 13.2). The binding of hormones to their receptors causes the target cells to produce a cellular response, such as cell growth, protein synthesis, secretion, muscle

FIGURE 13.2 Circulation of a hormone through the blood from an endocrine cell to its target cells.

The binding of a hormone to the appropriate receptor on its target cell triggers a cellular response.

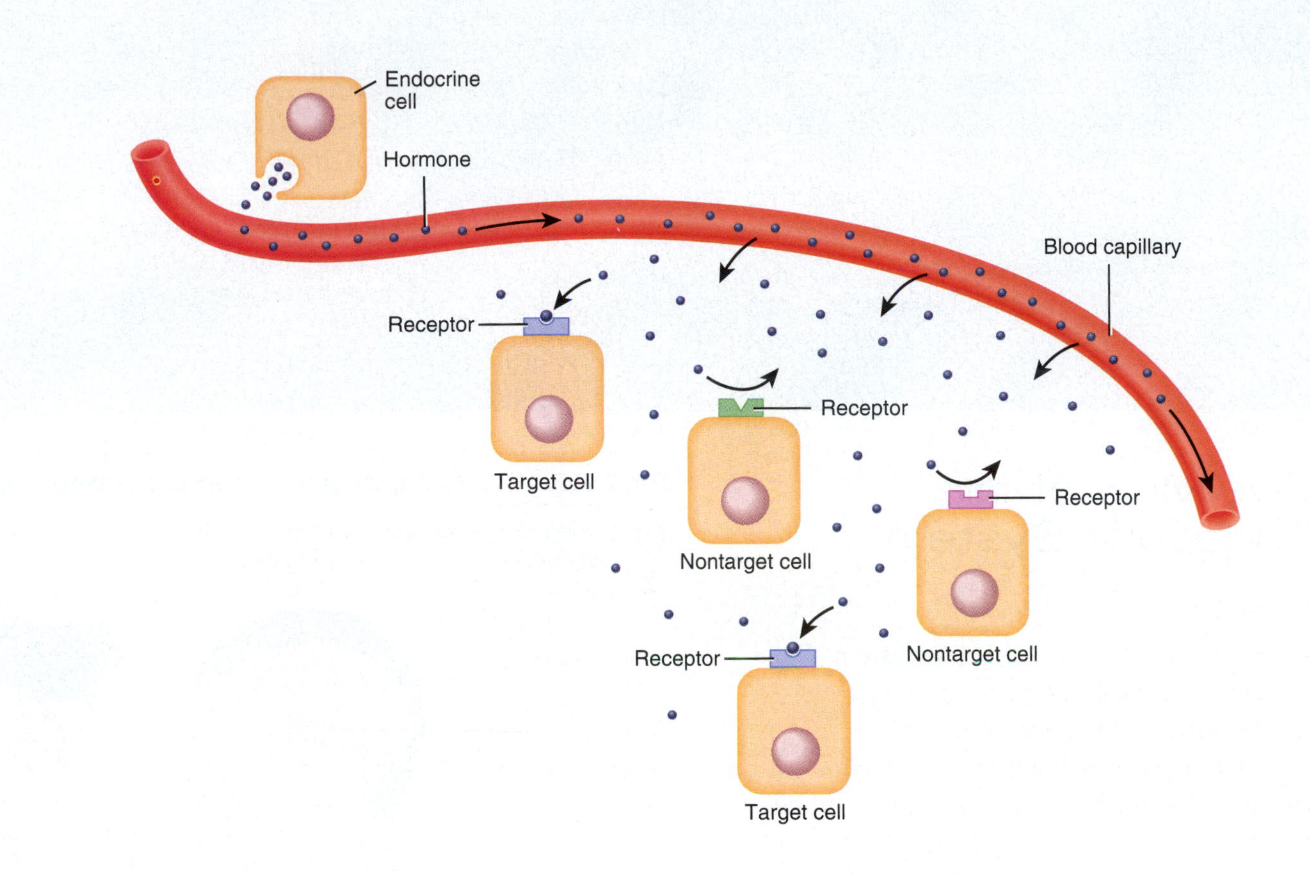

What are some examples of cellular responses that can occur in the target cells of a hormone?

contraction, or transport of substances across the plasma membrane. Only the target cells for a given hormone have receptors that bind and recognize that hormone. For example, thyroid-stimulating hormone (TSH) binds to receptors on cells of the thyroid gland, but it does not bind to cells of the ovaries because ovarian cells do not have TSH receptors.

Receptors, like other cellular proteins, are constantly being synthesized and broken down. Generally, a target cell has 2000 to 100,000 receptors for a particular hormone. If a hormone is present in excess, the number of target-cell receptors may decrease—an effect called **down-regulation** (FIGURE 13.3a). For example, when certain cells of the testes are exposed to a high concentration of luteinizing hormone (LH), the number of LH receptors decreases. Down-regulation makes a target cell *less sensitive* to a hormone. By contrast, when a hormone is deficient, the number of receptors may increase. This phenomenon, known as **up-regulation**, makes a target cell *more sensitive* to a hormone (FIGURE 13.3b). The mechanisms for down-regulation and up-regulation were discussed in Chapter 6 (see Section 6.3).

Hormones Are Chemically Classified as Lipid-Soluble or Water-Soluble

Hormones can be divided into two broad chemical classes: those that are lipid-soluble (hydrophobic) and those that are water-soluble (hydrophilic). This chemical classification is also useful functionally because the two classes exert their effects differently.

Lipid-Soluble Hormones

The lipid-soluble hormones include steroid hormones and thyroid hormones.

- **Steroid hormones** are derived from cholesterol. They are lipid-soluble because they contain four interconnected hydrocarbon rings. Each steroid hormone is unique due to the presence of different chemical groups attached at various sites on the four rings. These small differences allow for a large diversity of functions. Steroid hormones include aldosterone, cortisol, dehydroepiandrosterone (adrenal androgen), testosterone, estrogens, and progesterone.

FIGURE 13.3 Down-regulation and up-regulation of hormone receptors.

Down-regulation is a decrease in the number of receptors in response to excess hormone levels; up-regulation is an increase in the number of receptors in response to a deficiency of hormone.

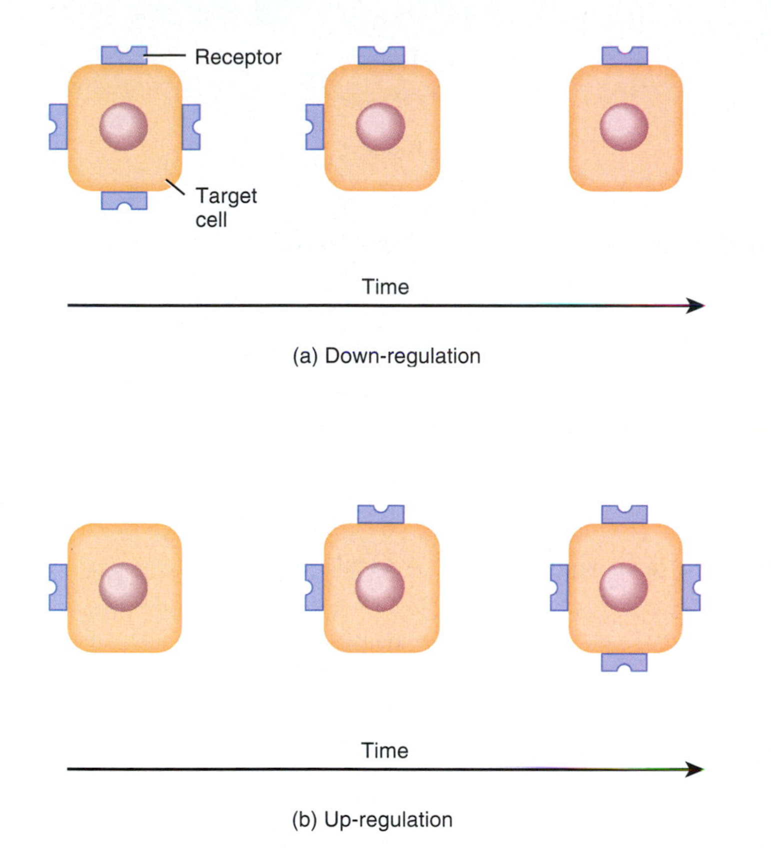

What is the purpose of down-regulation? Of up-regulation?

Although calcitriol (the active form of vitamin D) is not a steroid, it is included in this category because its structure is somewhat similar to that of a steroid.

- The two **thyroid hormones** (T_3 and T_4) are synthesized by attaching iodine to the amino acid tyrosine. T_3 and T_4 are lipid-soluble because they contain two hydrocarbon rings.

Water-Soluble Hormones

The water-soluble hormones include amine hormones and peptide/protein hormones.

- **Amine hormones** are synthesized by modifying certain amino acids. The catecholamines—epinephrine, norepinephrine, and dopamine—are amine hormones derived from the amino acid tyrosine. Melatonin is an amine hormone derived from the amino acid tryptophan.
- **Peptide/protein hormones** are amino acid polymers. Most hormones belong to this category. The smaller peptide hormones consist of chains of 3 to 49 amino acids; the larger protein hormones include 50 to 200 amino acids. Examples of peptide hormones are antidiuretic hormone and oxytocin; protein hormones include insulin and growth hormone. Several protein hormones, such as thyroid-stimulating hormone, have attached carbohydrate groups and thus are **glycoprotein hormones**.

TABLE 13.1 summarizes the classes of lipid-soluble and water-soluble hormones and provides an overview of the major hormones and their sites of secretion.

Hormones Circulate Through the Blood in Free Form or Bound to Transport Proteins

Most water-soluble hormones circulate through the watery blood plasma in a "free" form (not attached to other molecules), but most lipid-soluble hormones are bound to *transport proteins* because they are not soluble in blood (FIGURE 13.4). About 0.1–10% of the molecules of a lipid-soluble hormone are not bound to a transport protein. This *free fraction* diffuses out of capillaries, binds to receptors, and triggers responses. As free hormone molecules leave the blood and bind to their receptors, transport proteins release new ones to replenish the free fraction.

A Hormone's Mechanism of Action Depends on the Signaling Pathway That It Activates

Binding of a hormone to its specific receptor activates a signaling pathway in the target cell. The signaling pathway is a sequence of events that ultimately causes a change in a key protein of the cell—the effector protein—which in turn causes the cellular response. For example, the effector protein may be a contractile protein that causes contraction, an ion channel that permits movement of certain ions across the plasma membrane, or an enzyme that promotes a specific metabolic reaction. In many signaling pathways, a series of relay proteins (such as a G protein, adenylyl cyclase, and a protein kinase) conveys the signal between the receptor and the effector protein. When this occurs, each relay protein causes a change in the next protein in line until there is a change in the effector protein.

The receptors for a hormone may be present either in the plasma membrane of the target cell (plasma membrane receptors) or inside the target cell (intracellular receptors). Because water-soluble hormones are unable to pass through the hydrophobic interior of the plasma membrane, they bind to plasma membrane receptors. By contrast, lipid-soluble hormones are able to pass through the plasma membranes of target cells; hence, they bind to intracellular receptors, which are located either in the cytosol or in the nucleus.

Mechanisms of Action of Lipid-Soluble Hormones

When a lipid-soluble hormone, such as a steroid hormone or thyroid hormone, binds to an intracellular receptor, the signaling pathway

TABLE 13.1 Summary of Hormones by Chemical Class

Chemical Class	Hormones	Site of Secretion
LIPID-SOLUBLE		
Steroid hormones (Aldosterone structure)	Aldosterone, cortisol, and dehydroepiandrosterone (adrenal androgen).	Adrenal cortex.
	Testosterone.	Testes.
	Estrogens and progesterone.	Ovaries.
	Calcitriol.	Kidneys.
Thyroid hormones (Triiodothyronine (T_3) structure)	T_3 (triiodothyronine) and T_4 (thyroxine).	Thyroid gland.
WATER-SOLUBLE		
Amines (Norepinephrine structure)	Epinephrine, norepinephrine, and dopamine (catecholamines).	Adrenal medulla.
	Melatonin.	Pineal gland.
Peptides and proteins (Oxytocin structure: Glutamine—Isoleucine, Asparagine, Tyrosine, Cysteine—S—S—Cysteine, Proline, Leucine, Glycine, NH_2)	All hypothalamic releasing and inhibiting hormones.	Hypothalamus.
	Oxytocin, antidiuretic hormone.	Posterior pituitary.
	Growth hormone, thyroid-stimulating hormone, adrenocorticotropic hormone, follicle-stimulating hormone, luteinizing hormone, prolactin, melanocyte-stimulating hormone.	Anterior pituitary.
	Insulin, glucagon, somatostatin, pancreatic polypeptide.	Pancreas.
	Parathyroid hormone.	Parathyroid glands.
	Calcitonin.	Thyroid gland.
	Gastrin and ghrelin.	Stomach.
	Secretin, cholecystokinin, GIP (glucose-dependent insulinotropic peptide), and GLP (glucagon-like peptide).	Small intestine.
	Erythropoietin.	Kidneys.
	Leptin.	Adipose tissue.

that is activated typically alters gene expression (FIGURE 13.5a). The receptor–hormone complex binds to DNA, either stimulating or inhibiting the transcription of a specific gene. As DNA is transcribed, mRNA is formed and then translated, resulting in the formation of a new protein (the effector protein) that alters the cell's activity and causes the cellular response. Lipid-soluble hormones can also have *nongenomic effects*, meaning that they can activate signaling pathways that change the activity of proteins in the target cell without altering gene expression.

Mechanisms of Action of Water-Soluble Hormones

When a water-soluble hormone binds to a plasma membrane receptor, the signaling pathway that is activated depends on the type of receptor involved. The receptors for water-soluble hormones include G protein-coupled receptors, receptor tyrosine kinases, receptor guanylyl cyclases, and janus kinase-coupled receptors.

- ***G protein-coupled receptors.*** Most water-soluble hormones cause a cellular response in their target cells by binding to G protein-coupled

FIGURE 13.4 Circulation of water-soluble and lipid-soluble hormones through the bloodstream.

Most water-soluble hormones circulate through the blood in "free" form, but most lipid-soluble hormones are bound to transport proteins.

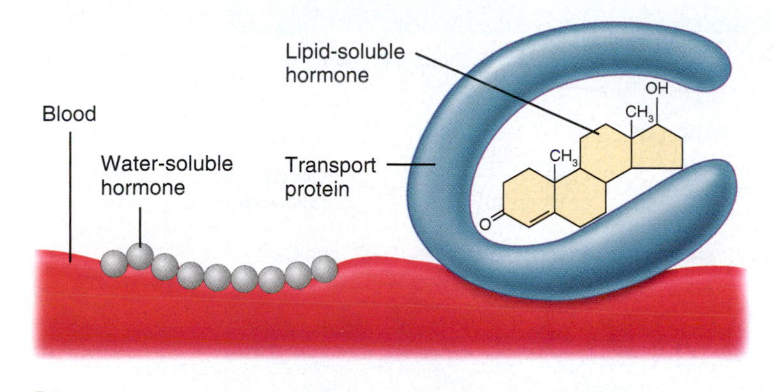

Why do lipid-soluble hormones require a transport protein to circulate through the bloodstream?

receptors. When a hormone binds to a G protein-coupled receptor, the receptor undergoes a conformational change that activates the G protein. The G protein in turn alters the activity of an enzyme that generates a second messenger. In many G protein-signaling pathways, a G_s protein stimulates the enzyme adenylyl cyclase to produce the second messenger cyclic AMP (cAMP) (**FIGURE 13.5b**). Cyclic AMP then activates protein kinase A, which phosphorylates an effector protein. The phosphorylated effector protein then promotes the cellular response. In other G protein pathways, a G_q protein stimulates the enzyme phospholipase C to cleave the membrane phospholipid phosphatidylinositol bisphosphate (PIP_2) to produce the second messengers inositol trisphosphate (IP_3) and diacylglycerol (DAG) (**FIGURE 13.5c**). IP_3 in turn causes the release of Ca^{2+}, another second messenger, from the endoplasmic reticulum. Once released, Ca^{2+}, along with DAG, binds to and activates protein kinase C, which phosphorylates an effector protein. The phosphorylated effector protein then causes the cellular response. Not all G protein pathways activated by water-soluble hormones result in an increased production of second messengers. Some G protein pathways contain a G_i protein, which inhibits adenylyl cyclase, reducing the concentration of cAMP. The decreased level of cAMP ultimately causes a change in the activity of the effector protein in the target cell. Examples of water-soluble hormones that activate G protein-coupled receptors include antidiuretic hormone (ADH), which uses G_s; oxytocin, which uses G_q; and growth-hormone-inhibiting hormone, which uses G_i.

- ***Receptor tyrosine kinases.*** The water-soluble hormones insulin and insulin-like growth factors (IGFs) bind to receptor tyrosine kinases (**FIGURE 13.5d**). In this type of signaling pathway, binding of the hormone to the receptor activates the tyrosine kinase components of the receptor, causing them to phosphorylate each other at tyrosine amino acids (cross phosphorylation). The phosphorylated sites on the receptor serve as docking sites where relay proteins bind and are

FIGURE 13.5 Mechanisms of hormone action.

Hormones can have different mechanisms of action depending on the types of signaling pathways that they activate in their target cells.

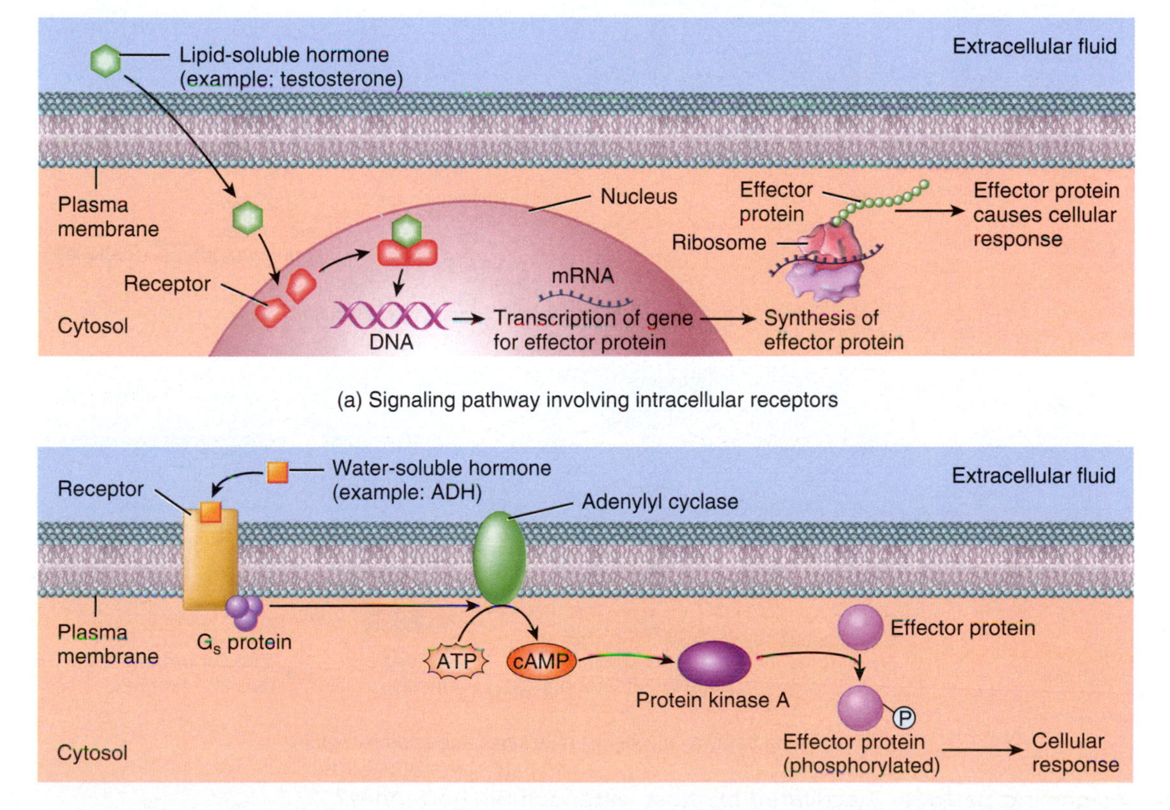

(a) Signaling pathway involving intracellular receptors

(b) Signaling pathway involving a G protein-coupled receptor and the second messenger cAMP

Figure 13.5 (continues)

Figure 13.5 Continued

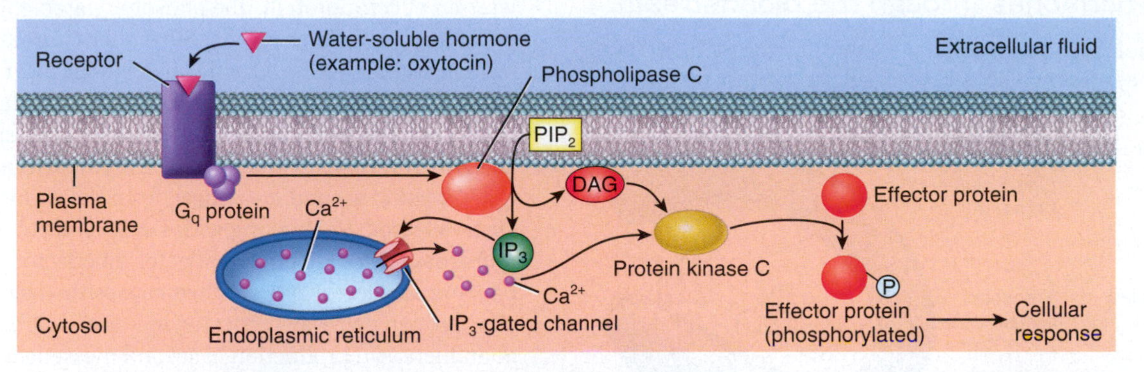

(c) Signaling pathway involving a G protein-coupled receptor and the second messengers IP_3, DAG, and Ca^{2+}

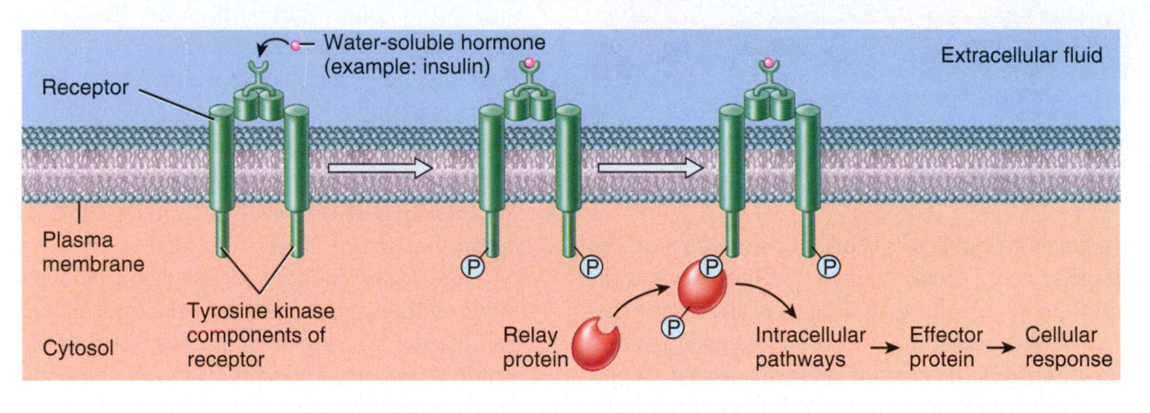

(d) Signaling pathway involving receptor tyrosine kinases

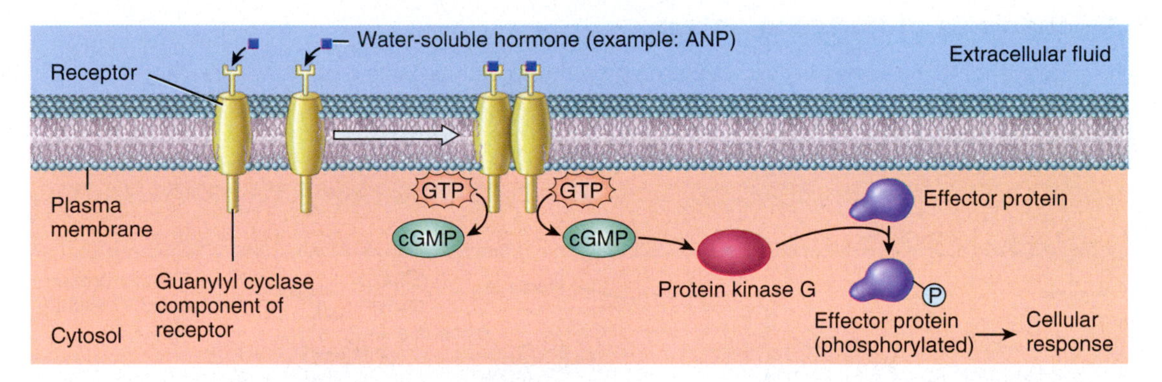

(e) Signaling pathway involving receptor guanylyl cyclases

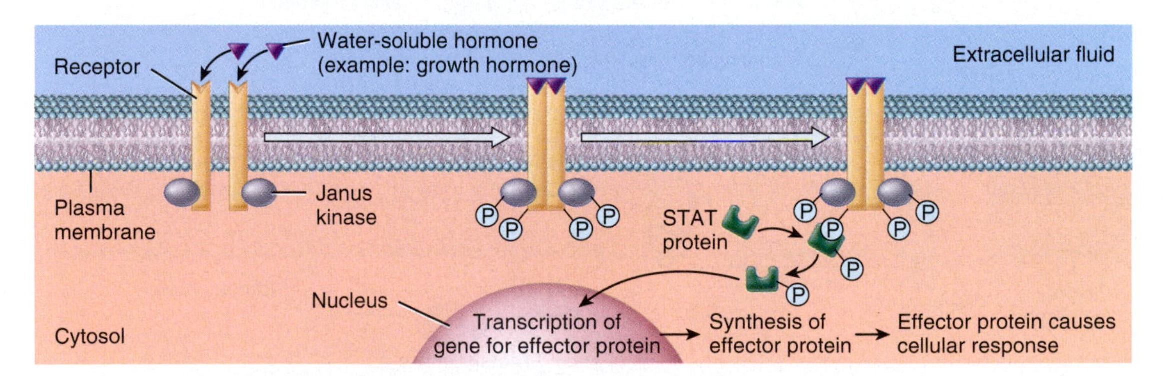

(f) Signaling pathway involving janus kinase-coupled receptors

? Which type of signaling pathway is activated by most water-soluble hormones?

phosphorylated on their tyrosine amino acids. Once phosphorylated, the relay proteins are activated and trigger intracellular pathways that ultimately cause the synthesis or activation of an effector protein. The effector protein in turn causes the cellular response.

- ***Receptor guanylyl cyclases.*** Some water-soluble hormones, such as atrial natriuretic peptide (ANP), bind to receptor guanylyl cyclases (FIGURE 13.5e). In this type of pathway, two receptor guanylyl cyclases dimerize after the water-soluble hormones bind to their receptor sites. Dimerization of the receptors activates the guanylyl cyclase components of the receptors, causing them to produce cGMP. The cGMP activates protein kinase G, which phosphorylates the effector protein. The phosphorylated effector protein causes the cellular response.
- ***Janus kinase-coupled receptors.*** Several water-soluble hormones, including growth hormone, prolactin, and erythropoietin, bind to receptors coupled to the enzyme janus kinase (JAK) (FIGURE 13.5f). In this type of signaling pathway, two JAK-coupled receptors dimerize after the water-soluble hormones have become bound. Dimerization allows the two JAKs to cross-phosphorylate each other at tyrosine amino acids. The JAKs become activated and then phosphorylate tyrosine amino acids on the cytosolic portions of the receptors. The phosphorylated receptor sites serve as docking sites where relay proteins called *signal transducers and activators of transcription (STATs)* can bind. The STATs are phosphorylated on their tyrosine amino acids. The phosphorylated STATs then leave their docking sites, enter the nucleus, and bind to DNA, stimulating the transcription of specific genes. The transcribed genes are translated, resulting in the synthesis of effector proteins. The effector proteins in turn cause the cellular response.

It is important to understand that a given hormone can have several different target cells. In addition, a given cell can serve as the target cell for several different hormones. When this is the case, the target cell has several types of receptors, each responding only to its specific hormone. Furthermore, each hormone may cause a different response in that target cell or it may cause the same response, depending on the signaling pathway and effector proteins involved.

Hormones Can Have Permissive, Synergistic, or Antagonistic Effects

The responsiveness of a target cell to a hormone depends on three factors: (1) the hormone's concentration, (2) the number of the target cell's hormone receptors, and (3) influences exerted by other hormones. A target cell responds more vigorously when the level of a hormone rises or when it has more receptors. In addition, some hormones require a simultaneous or recent exposure to a second hormone to cause a greater response in their target cells; in such cases, the second hormone is said to have a **permissive effect** (FIGURE 13.6a). For example, epinephrine alone only weakly stimulates lipolysis (the breakdown of triglycerides) in adipose tissue cells (adipocytes), but when small amounts of thyroid hormones (T_3 and T_4) are present, the same amount of epinephrine stimulates lipolysis much more powerfully. Permissive effects occur because the permissive hormone either increases the number of receptors (up-regulation) for the other hormone or it promotes the synthesis of an enzyme required for the expression of the other hormone's effects.

When the effect of two hormones acting together is greater than the sum of their individual effects, the two hormones are said to have a **synergistic effect** (FIGURE 13.6b). For example, both glucagon and

FIGURE 13.6 Permissive, synergistic, and antagonistic effects of hormones.

A hormone has a permissive effect if it is required for another hormone to elicit a greater response in its target cell; hormones have a synergistic effect if the effect of the hormones acting together is greater than the sum of their individual effects; and hormones have antagonistic effects if the hormones have opposing actions.

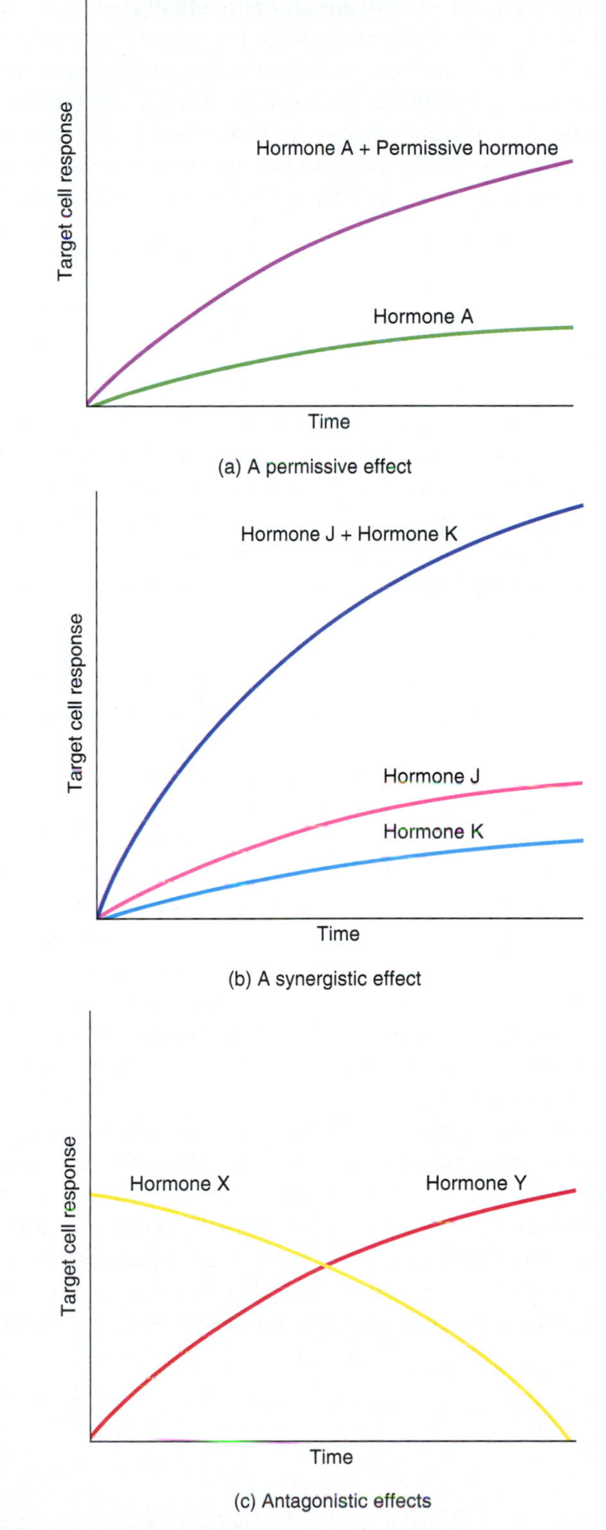

(a) A permissive effect

(b) A synergistic effect

(c) Antagonistic effects

What are the possible cellular mechanisms responsible for permissive effects?

epinephrine increase the blood glucose concentration by stimulating the breakdown of glycogen in liver cells. When both hormones are present, the increase in blood glucose concentration is greater than the sum of the individual hormone responses. Synergistic effects are thought to occur because the hormones activate pathways that lead to formation of the same types of second messengers, thereby amplifying the cellular response.

When one hormone opposes the actions of another hormone, the two hormones are said to have **antagonistic effects** (FIGURE 13.6c). An example of a pair of hormones with antagonistic effects is insulin and glucagon: Insulin promotes the synthesis of glycogen by liver cells, and glucagon stimulates the breakdown of glycogen in the liver. Antagonistic effects occur because the hormones activate pathways that cause opposite cellular responses or one hormone decreases the number of receptors (down-regulation) for the other hormone.

Hormone Secretion Is Controlled in a Variety of Ways

The release of most hormones occurs in short bursts, with little or no secretion between bursts. When stimulated, an endocrine gland releases its hormone in more frequent bursts, increasing the concentration of the hormone in the blood. In the absence of stimulation, the blood level of the hormone decreases. Regulation of secretion normally prevents overproduction or underproduction of any given hormone.

Hormone secretion can be regulated by (1) signals from the nervous system, (2) chemical changes in the blood, (3) distension (stretch) of an organ, or (4) other hormones. For example, activation of the sympathetic nervous system causes the release of epinephrine, norepinephrine, and dopamine from the adrenal medulla (FIGURE 13.7a); a lower-than-normal level of Ca^{2+} in the blood stimulates secretion of parathyroid hormone (PTH) from the parathyroid glands (FIGURE 13.7b); stretching the uterine cervix causes secretion of oxytocin from the posterior pituitary gland (FIGURE 13.7c); and adrenocorticotropic hormone (ACTH) secreted from the anterior pituitary gland stimulates the release of cortisol from the adrenal cortex (FIGURE 13.7d). ACTH is an example of a tropic hormone. **Tropic hormones** (TRŌ-pik), or **tropins**, are hormones that act on other endocrine glands or tissues to regulate the secretion of another hormone.

Most hormonal regulatory systems work via negative feedback. For example, after the parathyroid glands release PTH in response to low blood Ca^{2+}, PTH acts on its target cells to increase the blood Ca^{2+} level. Once the blood Ca^{2+} level is back to normal, PTH secretion is inhibited (FIGURE 13.7b). Consider another example of negative feedback regulation of hormonal secretion: The ACTH released from the anterior pituitary causes the adrenal cortex to secrete cortisol. However, if the cortisol level in the blood becomes too high, the elevated cortisol level inhibits ACTH secretion (FIGURE 13.7d).

Although most hormone regulatory systems work via negative feedback, a few operate via positive feedback. For example, during childbirth, stretching of the uterine cervix by the fetus stimulates oxytocin release. Oxytocin then stimulates contractions of the uterus, and the contractions push the fetus farther down the uterus, which stretches the cervix even more. This stretching of the cervix in turn stimulates more oxytocin release, a positive feedback effect (FIGURE 13.7c).

Now that you have a general understanding of the endocrine system, it is time to discuss the various endocrine glands and the hormones they secrete.

CHECKPOINT

1. What are the components of the endocrine system?
2. What is the difference between down-regulation and up-regulation?
3. What are the two main chemical classes of hormones? List an example of each.
4. How are hormones transported in the blood?
5. What is the mechanism of action of each of the following hormones: aldosterone, oxytocin, prolactin, ADH, and insulin?
6. How does a permissive effect differ from a synergistic effect?
7. What are the four ways that hormone secretion can be controlled?

13.2 Pituitary Gland

OBJECTIVES

- Describe the functional relationship between the pituitary gland and the hypothalamus.
- Discuss the functions of the hormones of the pituitary gland.

The **pituitary gland** or *hypophysis* (hī-POF-i-sis) is a pea-sized structure that extends from the brain. It consists of two lobes: a larger

FIGURE 13.7 Control of hormone secretion. The dashed line and negative sign indicate negative feedback; the dashed line and positive sign indicate positive feedback.

Hormone secretion can be regulated by signals from the nervous system, chemical changes in the blood, distension (stretch) of an organ, or other hormones.

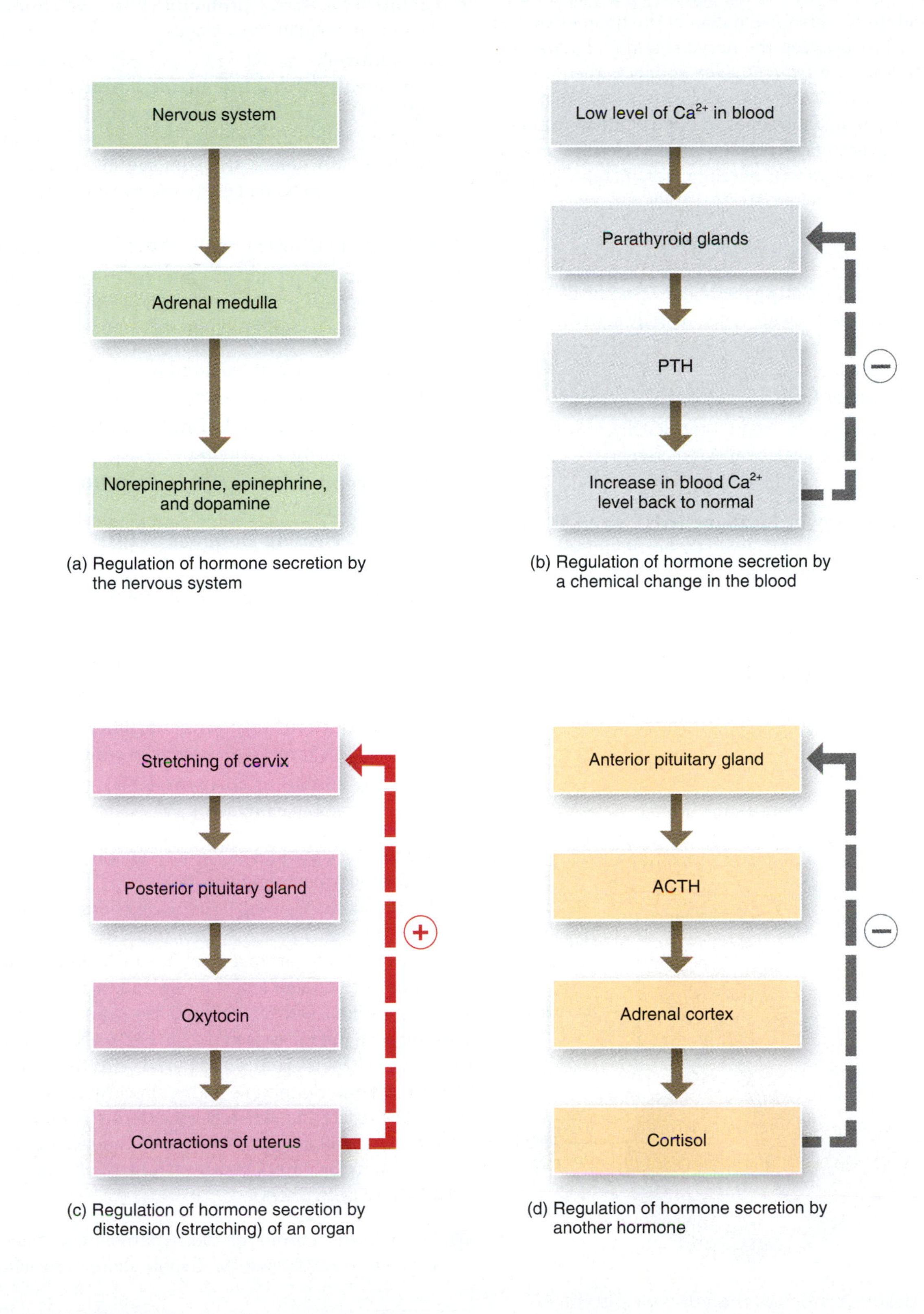

(a) Regulation of hormone secretion by the nervous system

(b) Regulation of hormone secretion by a chemical change in the blood

(c) Regulation of hormone secretion by distension (stretching) of an organ

(d) Regulation of hormone secretion by another hormone

What is a tropic hormone?

anterior pituitary (anterior lobe) and a smaller **posterior pituitary (posterior lobe)** (FIGURE 13.8). The anterior pituitary consists of glandular epithelial tissue, whereas the posterior pituitary consists of nervous tissue.

For many years, the pituitary gland was called the master endocrine gland because it secretes several hormones that control other endocrine glands. It is now known that the pituitary gland itself has a master—the **hypothalamus**. This small region of the brain below the thalamus is the major link between the nervous and endocrine systems. A stalk-like structure, the *infundibulum*, connects the pituitary gland to the hypothalamus (FIGURE 13.8). Cells in the hypothalamus synthesize at least nine different hormones, and the pituitary gland secretes six. Together, these hormones play important roles in the regulation of virtually all aspects of growth, development, metabolism, and homeostasis.

The Anterior Pituitary Secretes Many Hormones

The **anterior pituitary** or *adenohypophysis* (ad′-e-nō-hī-POF-i-sis) secretes hormones that regulate a wide range of body activities, from growth to reproduction.

Types of Anterior Pituitary Cells

Five types of anterior pituitary cells—somatotrophs, thyrotrophs, corticotrophs, lactotrophs, and gonadotrophs—secrete six hormones (FIGURE 13.9a):

- **Somatotrophs** secrete **growth hormone (GH)**, also known as *somatotropin* (sō′-ma-tō-TRŌ-pin). Growth hormone stimulates general body growth and regulates aspects of metabolism.

FIGURE 13.8 Pituitary gland.

The pituitary gland consists of two parts: the anterior pituitary and posterior pituitary.

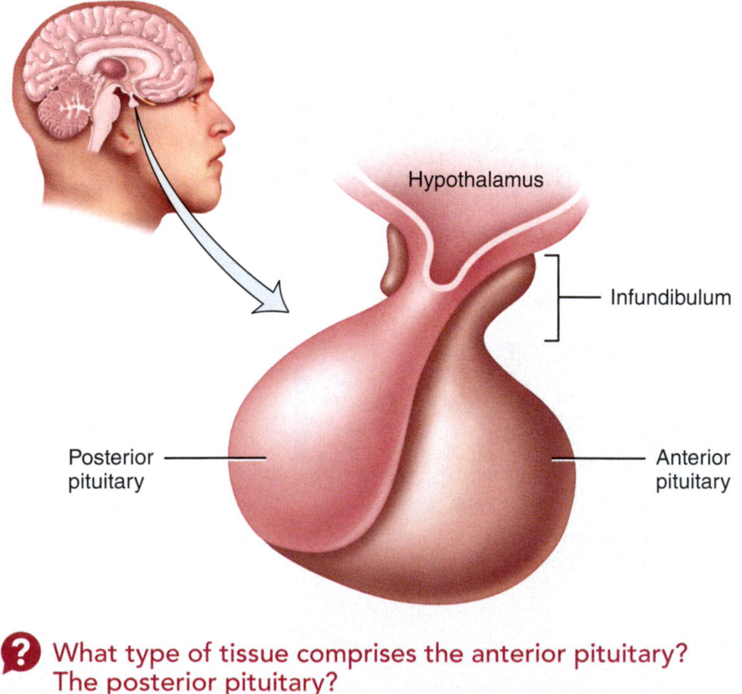

What type of tissue comprises the anterior pituitary? The posterior pituitary?

- **Thyrotrophs** secrete **thyroid-stimulating hormone (TSH)**, also called *thyrotropin* (thī-rō-TRŌ-pin). TSH controls the secretions and other activities of the thyroid gland.
- **Corticotrophs** secrete **adrenocorticotropic hormone (ACTH)**, also referred to as *corticotropin* (kor′-ti-kō-TRŌ-pin), which stimulates the adrenal cortex to secrete glucocorticoids such as cortisol.
- **Lactotrophs** secrete **prolactin (PRL)**, which initiates milk production in the mammary glands.
- **Gonadotrophs** secrete two *gonadotropins*: **follicle-stimulating hormone (FSH)** and **luteinizing hormone (LH)** (LOO-tē-in′-īz-ing). FSH and LH both act on the gonads (testes and ovaries). In men, they stimulate the testes to produce sperm and to secrete testosterone. In women, they stimulate the ovaries to produce eggs (oocytes) and to secrete estrogens and progesterone.

Hypothalamic Control of the Anterior Pituitary

Release of anterior pituitary hormones is regulated in part by the hypothalamus. The hypothalamus secretes five **releasing hormones**, which stimulate secretion of anterior pituitary hormones:

- **Growth hormone–releasing hormone (GHRH)**, also known as *somatocrinin*, stimulates secretion of growth hormone.
- **Thyrotropin-releasing hormone (TRH)** stimulates secretion of thyroid-stimulating hormone.
- **Corticotropin-releasing hormone (CRH)** stimulates secretion of adrenocorticotropic hormone.
- **Prolactin-releasing hormone (PRH)*** stimulates secretion of prolactin.
- **Gonadotropin-releasing hormone (GnRH)** stimulates secretion of FSH and LH.

The hypothalamus also produces two **inhibiting hormones**, which suppress secretion of anterior pituitary hormones:

- **Growth hormone–inhibiting hormone (GHIH)**, also known as *somatostatin*, suppresses secretion of growth hormone.
- **Prolactin-inhibiting hormone (PIH)**, which is dopamine, suppresses secretion of prolactin.

TABLE 13.2 summarizes the anterior pituitary hormones and their control by the hypothalamus.

Hypothalamic releasing or inhibiting hormones reach the anterior pituitary through a portal system (FIGURE 13.9b). A **portal system** is a type of vascular arrangement in which blood flows from one capillary network through a portal vein and then into a second capillary network without first returning to the heart. In the **hypothalamic–hypophyseal portal system** (hī′-pō-FIZ-ē-al), blood flows from capillaries in the hypothalamus into portal veins that carry blood to capillaries in the anterior pituitary. This direct route permits hypothalamic hormones to act immediately on anterior pituitary cells before the hormones are diluted or destroyed in the general circulation. Regulation of anterior pituitary secretion by the hypothalamus occurs as follows (FIGURE 13.9b):

1. Specialized neurons of the hypothalamus called **neurosecretory cells** synthesize the hypothalamic releasing or inhibiting

* Although PRH is thought to exist, its exact nature is uncertain.

FIGURE 13.9 Anterior pituitary cells and hypothalamic control of anterior pituitary hormone secretion. For simplicity, cells are not drawn to scale, and only one of each type of anterior pituitary cell is shown.

Hypothalamic hormones are an important link between the nervous and endocrine systems.

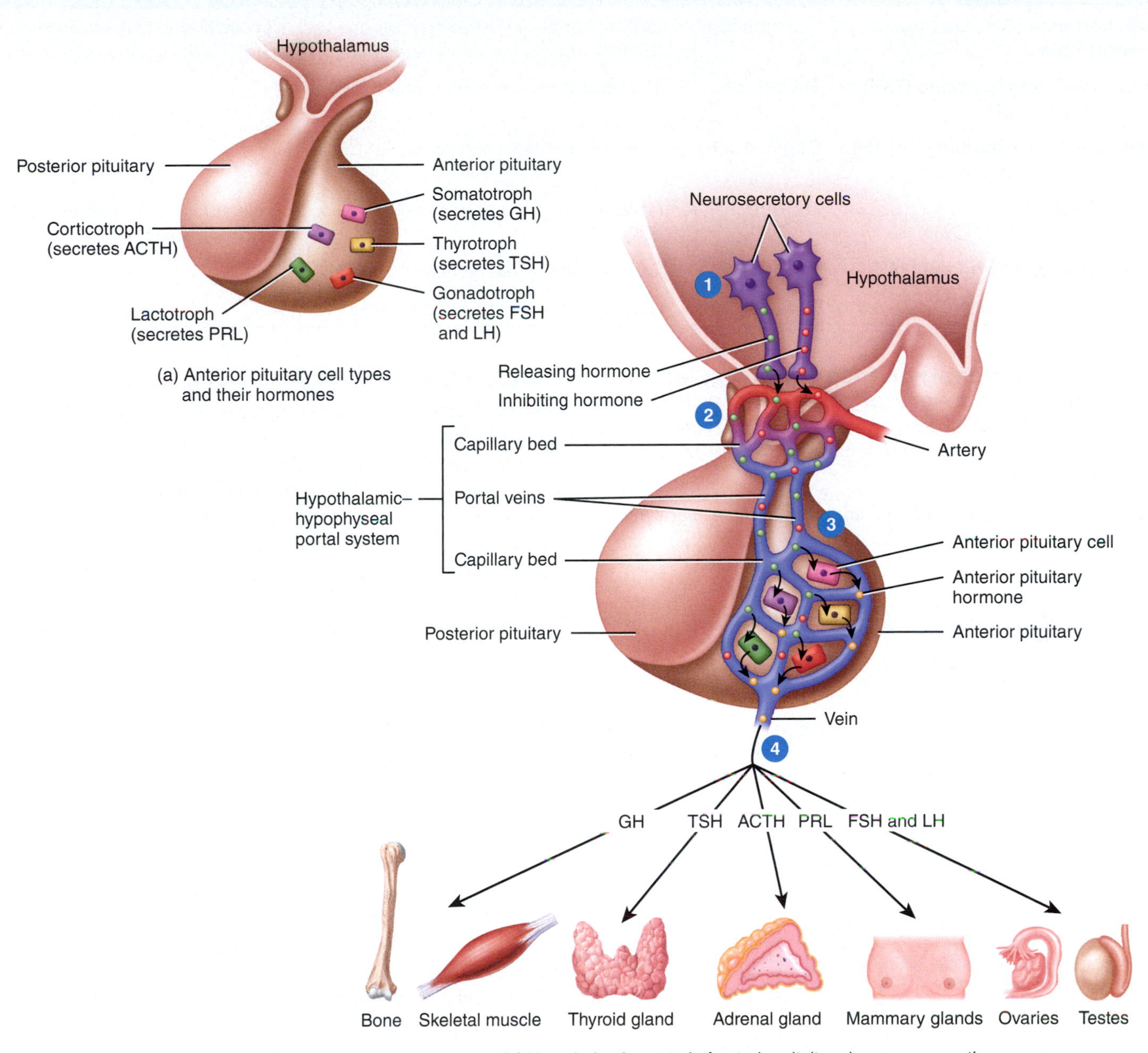

(a) Anterior pituitary cell types and their hormones

(b) Hypothalamic control of anterior pituitary hormone secretion

What is the functional importance of the hypothalamic–hypophyseal portal system?

hormones in their cell bodies and package the hormones inside vesicles. The vesicles move by axonal transport to the axon terminals for storage.

2 When the neurosecretory cells of the hypothalamus are excited, action potentials trigger exocytosis of the vesicles, causing the hormones to be released. The hormones then diffuse into the hypothalamic–hypophyseal portal system.

3 After the hypothalamic hormones are conveyed to the anterior pituitary, they diffuse out of the bloodstream and interact with anterior pituitary cells. When stimulated by the appropriate hypothalamic releasing hormones, the anterior pituitary cells secrete hormones into the capillaries of the anterior pituitary, which drain into venous blood.

4 The anterior pituitary hormones then travel through the bloodstream to their target organs.

TABLE 13.2 Hormones of the Anterior Pituitary and Their Control by the Hypothalamus

Hormone	Secreted by	Releasing Hormone (Stimulates Secretion)	Inhibiting Hormone (Suppresses Secretion)
Growth hormone (GH), also known as **somatotropin**	Somatotrophs.	Growth hormone–releasing hormone (GHRH), also known as somatocrinin.	Growth hormone–inhibiting hormone (GHIH), also known as somatostatin.
Thyroid-stimulating hormone (TSH), also called **thyrotropin**	Thyrotrophs.	Thyrotropin-releasing hormone (TRH).	—
Adrenocorticotropic hormone (ACTH), also referred to as **corticotropin**	Corticotrophs.	Corticotropin-releasing hormone (CRH).	—
Prolactin (PRL)	Lactotrophs.	Prolactin-releasing hormone (PRH)	Prolactin-inhibiting hormone (PIH), which is dopamine.
Follicle-stimulating hormone (FSH)	Gonadotrophs.	Gonadotropin-releasing hormone (GnRH)	—
Luteinizing hormone (LH)	Gonadotrophs.	Gonadotropin-releasing hormone (GnRH).	—

FIGURE 13.10 Negative feedback regulation of hypothalamic neurosecretory cells and anterior pituitary corticotrophs. Solid green arrows show stimulation of secretions; dashed gray arrows and negative sign show inhibition of secretion via negative feedback.

Cortisol secreted by the adrenal cortex suppresses secretion of CRH and ACTH.

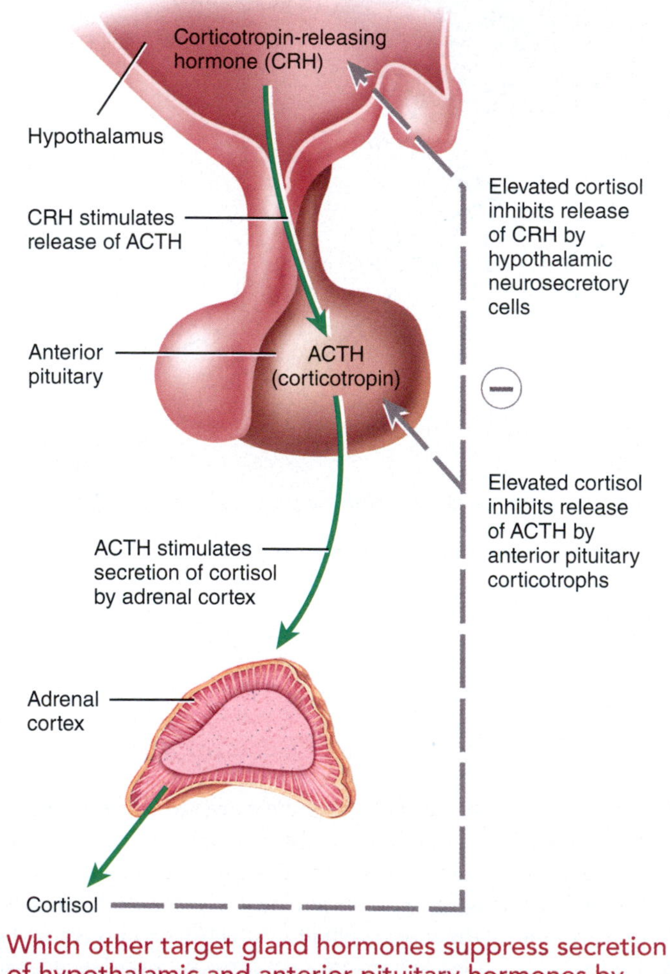

Which other target gland hormones suppress secretion of hypothalamic and anterior pituitary hormones by negative feedback?

Feedback Control of the Anterior Pituitary

Release of anterior pituitary hormones is regulated not only by the hypothalamus but also by negative feedback. The secretory activity of three types of anterior pituitary cells (thyrotrophs, corticotrophs, and gonadotrophs) decreases when blood levels of their target gland hormones rise. For example, adrenocorticotropic hormone (ACTH) stimulates the cortex of the adrenal gland to secrete glucocorticoids, mainly cortisol. In turn, an elevated blood level of cortisol decreases secretion of both ACTH (corticotropin) and corticotropin-releasing hormone (CRH) by suppressing the activity of the anterior pituitary corticotrophs and hypothalamic neurosecretory cells (FIGURE 13.10).

Growth Hormone

Somatotrophs are the most numerous cells in the anterior pituitary, and growth hormone (GH) is the most plentiful anterior pituitary hormone. GH promotes growth of body tissues, including bones and skeletal muscles, and it regulates certain aspects of metabolism. GH exerts its growth-promoting effects indirectly through small protein hormones called **insulin-like growth factors (IGFs)** or *somatomedins* (sō′-ma-tō-MĒ-dins). In response to growth hormone, cells in the liver, skeletal muscle, cartilage, and bone secrete IGFs. IGFs synthesized in the liver enter the bloodstream as hormones that circulate to target cells throughout the body to cause growth. IGFs produced in skeletal muscle, cartilage, and bone act locally as autocrines or paracrines to cause growth of those tissues. Unlike the effects of GH on body growth, the effects of GH on metabolism are direct, meaning that GH interacts directly with target cells to cause specific metabolic reactions.

To understand how IGFs promote bone growth, you must first understand how bone is organized (FIGURE 13.11):

- A long bone, such as the humerus (arm bone), consists of two **epiphyses** (the circular ends of the bone; singular is *epiphysis*); a **diaphysis** (the shaft or main portion of the bone); and a **marrow cavity** (the space within the diaphysis that stores fatty yellow bone marrow). Between the diaphysis and each epiphysis is the **epiphyseal (growth) plate** (ep′-i-FIZ-ē-al), a layer of cartilage that is responsible for lengthwise growth of bone. When bone ceases to grow in length at about ages 18–21, the cartilage in the epiphyseal plate is replaced by bone; the resulting bony structure is known as the *epiphyseal line*.

FIGURE 13.11 Organization of bone.

A long bone contains two epiphyses, a diaphysis, and a marrow cavity.

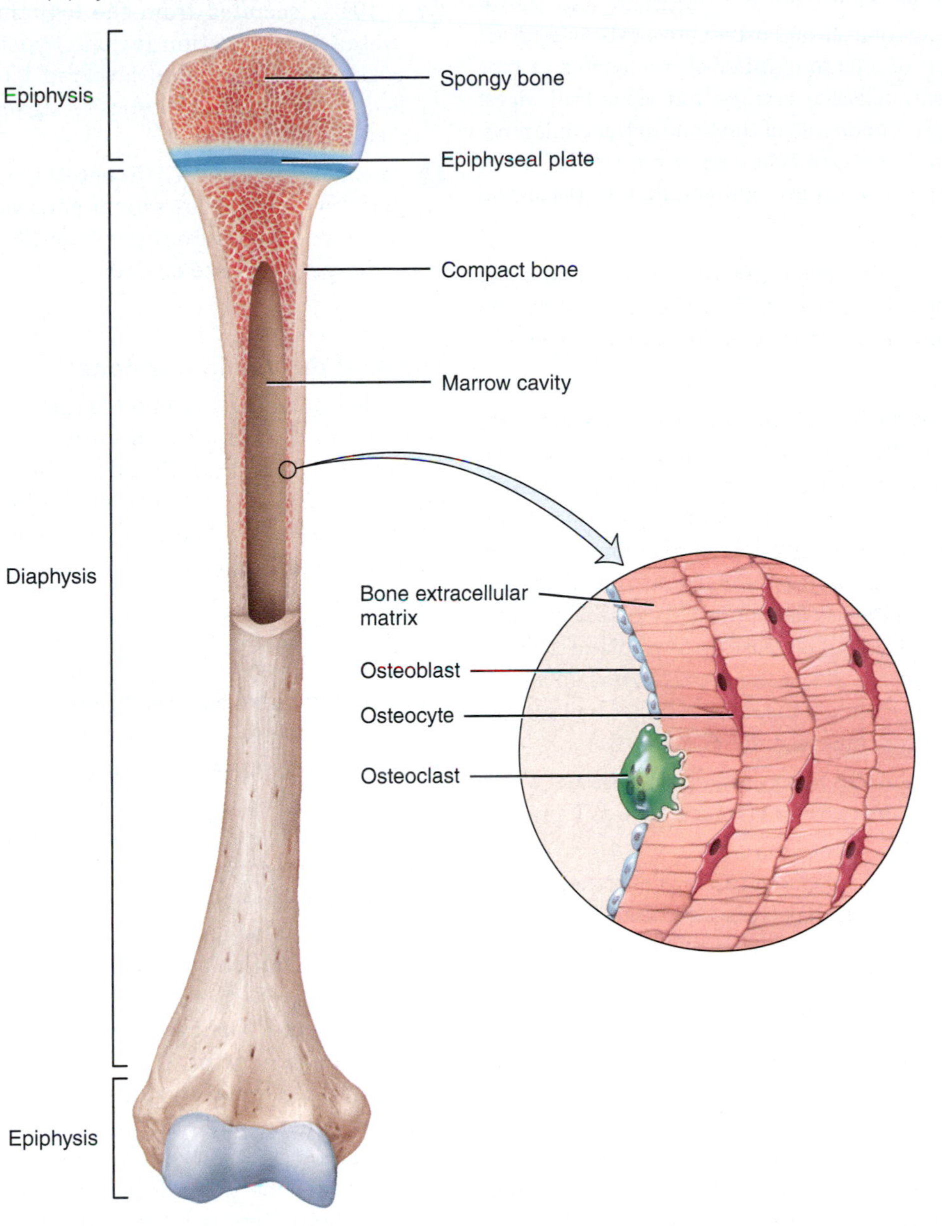

Which type of bone cell is responsible for bone resorption?

- Contrary to how it looks, bone is not completely solid but has many small spaces. Some of these spaces serve as channels for blood vessels that supply bone cells with nutrients; other spaces function as storage areas for bone marrow. Depending on the size and distribution of the spaces, the regions of a bone may be categorized as compact or spongy. **Compact bone** contains relatively few spaces and is the strongest form of bone tissue. It makes up the bulk of the diaphysis of long bones. Compact bone tissue provides protection and support and resists the stresses produced by weight and movement. **Spongy bone** contains a large number of spaces. The spaces within spongy bone make bones lighter and can sometimes be filled with red bone marrow. Spongy bone makes up most of the epiphyses of long bones and forms a narrow rim around the marrow cavity of the diaphysis of long bones. Spongy bone is always covered by a layer of compact bone for protection.
- Like other connective tissues, bone tissue contains an abundant extracellular matrix that surrounds widely separated cells. The extracellular matrix is responsible for the hardness of bone; it is comprised of about 65% crystallized mineral salts (mainly calcium phosphate) called **hydroxyapatite** [$Ca_{10}(PO_4)_6(OH)_2$], 25% collagen fibers, and 10% water. Three important types of cells are present within bone tissue: osteoblasts, osteocytes, and osteoclasts. **Osteoblasts** are bone-building cells. They synthesize and secrete collagen fibers and promote the deposition of mineral salts into bone extracellular matrix, a process known as **bone deposition**. As osteoblasts surround themselves with extracellular matrix, they become

trapped in their secretions and turn into **osteocytes**, mature bone cells. Osteocytes are the main cells in bone tissue and maintain its daily metabolism, such as the exchange of nutrients and wastes with the blood. Unlike osteoblasts and osteocytes, **osteoclasts** are derived from the fusion of a large number of monocytes (a type of leukocyte). Osteoclasts release enzymes and acids that digest the protein and mineral components of the bone extracellular matrix. This breakdown of bone extracellular matrix, termed **bone resorption**, is part of the normal development, maintenance, and repair of bone.

Using IGFs as mediators, GH causes growth of bones and other tissues of the body. Through direct effects, GH helps regulate certain metabolic reactions in body cells. The specific functions of IGFs and GH include the following:

- ***Increase growth of bones and soft tissues.*** In bones, IGFs stimulate osteoblasts, promote cell division at the epiphyseal plate, and enhance synthesis of the proteins needed to build more bone matrix. In soft tissues such as skeletal muscle, the kidneys, and intestines, IGFs cause cells to grow by increasing uptake of amino acids into cells and accelerating protein synthesis. IGFs also decrease the breakdown of proteins and the use of amino acids for ATP production. Due to the effects of IGFs, GH increases growth of the skeleton and soft tissues during childhood and the teenage years. In adults, GH (acting via IGFs) helps maintain the mass of bones and soft tissues and promotes healing of injuries and tissue repair.
- ***Enhance lipolysis.*** GH enhances lipolysis in adipose tissue, which results in increased use of the released fatty acids for ATP production by body cells.
- ***Decrease glucose uptake.*** GH influences carbohydrate metabolism by decreasing glucose uptake, which decreases the use of glucose for ATP production by most body cells. This action spares glucose so that it is available to neurons for ATP production in times of glucose scarcity. GH also stimulates liver cells to release glucose into the blood.

Somatotrophs in the anterior pituitary release bursts of growth hormone every few hours, especially during sleep. Their secretory activity is controlled mainly by two hypothalamic hormones: (1) growth hormone–releasing hormone (GHRH) promotes secretion of growth hormone and (2) growth hormone–inhibiting hormone (GHIH) suppresses it. Regulation of growth hormone secretion by GHRH and GHIH occurs in the following way: (**FIGURE 13.12**):

1. GHRH is secreted from the hypothalamus. Factors that promote GHRH secretion include hypoglycemia (low blood glucose concentration); decreased blood levels of fatty acids; increased blood levels of amino acids; deep sleep (stages 3 and 4 of nonrapid eye movement sleep); increased activity of the sympathetic nervous system, such as might occur with stress or vigorous physical exercise; and other hormones, including testosterone, estrogens, thyroid hormones, and ghrelin.
2. Once secreted, GHRH enters the hypothalamic–hypophyseal portal system and flows to the anterior pituitary, where it stimulates somatotrophs to secrete GH.
3. GH acts directly on various cells to promote certain metabolic reactions. In liver, bone, skeletal muscle, and cartilage, GH is converted to IGFs, which in turn promote growth of bones, skeletal muscle, and other tissues.
4. Elevated levels of GH and IGFs inhibit release of GHRH and GH (negative feedback inhibition).
5. GHIH is secreted from the hypothalamus. Factors that promote GHIH secretion include hyperglycemia (high blood glucose); increased blood levels of fatty acids; decreased blood levels of amino acids; obesity; aging; and high blood levels of GH and IGFs.
6. After being secreted, GHIH enters the hypothalamic–hypophyseal portal system and flows to the anterior pituitary, where it prevents the somatotrophs from secreting GH by interfering with the signaling pathway used by GHRH.

Thyroid-Stimulating Hormone

Thyroid-stimulating hormone (TSH) stimulates the synthesis and secretion of the two thyroid hormones, triiodothyronine (T_3) and thyroxine (T_4), both produced by the thyroid gland. Thyrotropin-releasing hormone (TRH) from the hypothalamus controls TSH secretion. Release of TRH depends on blood levels of T_3 and T_4; high levels of T_3 and T_4 inhibit secretion of both TRH and TSH via negative feedback.

Adrenocorticotropic Hormone

Corticotrophs secrete adrenocorticotropic hormone (ACTH). ACTH controls the production and secretion of cortisol and other glucocorticoids by the cortex (outer portion) of the adrenal glands. Corticotropin-releasing hormone (CRH) from the hypothalamus stimulates secretion of ACTH by corticotrophs. CRH is released in response to stress-related stimuli, such as low blood glucose or physical trauma, and interleukin-1, a substance produced by macrophages. Glucocorticoids inhibit CRH and ACTH release via negative feedback.

ACTH is synthesized as part of a larger precursor molecule called **pro-opiomelanocortin (POMC)** (prō-ō′-pē-ō-MEL-a-nō-kor′-tin) (**FIGURE 13.13**). In addition to ACTH, POMC can serve as the precursor for several other peptides, including α- and γ-melanocyte-stimulating hormone (MSH), β- and γ-lipotropin, β-endorphin, corticotropin-like intermediate-lobe peptide (CLIP), and a few peptide fragments. α- and γ-MSH stimulate the dispersion of the pigment melanin in the skin, thereby influencing skin color; α-MSH also functions in the brain to suppress appetite. β- and γ-lipotropin contribute to lipid metabolism by mobilizing lipids from adipose tissue. β-endorphin is an endogenous opioid that suppresses pain transmission in the central nervous system (CNS). CLIP and the peptide fragments have no known functions. Several tissues (the pituitary gland, skin, and brain) produce POMC and, depending on the types of processing enzymes present in these tissues, POMC is cleaved (spliced) in different ways to form various combinations of the peptides just described. In corticotrophs of the anterior pituitary, POMC is cleaved to form peptide fragments, ACTH, and β-lipotropin. In the intermediate lobe of the pituitary, which is found only in the fetal brain, POMC is cleaved to form peptide fragments, γ-MSH, α-MSH, CLIP, γ-lipotropin, and β-endorphin. In the skin, POMC is cleaved to form α-MSH, which helps promote skin color. In certain neurons of the hypothalamus, POMC is cleaved to form α-MSH, which reduces food intake. In other neurons of the CNS, POMC is cleaved to produce

FIGURE 13.12 Regulation of growth hormone (GH) secretion. Each dashed arrow and negative sign indicates negative feedback.

Secretion of GH is stimulated by growth hormone–releasing hormone (GHRH) and inhibited by growth hormone–inhibiting hormone (GHIH).

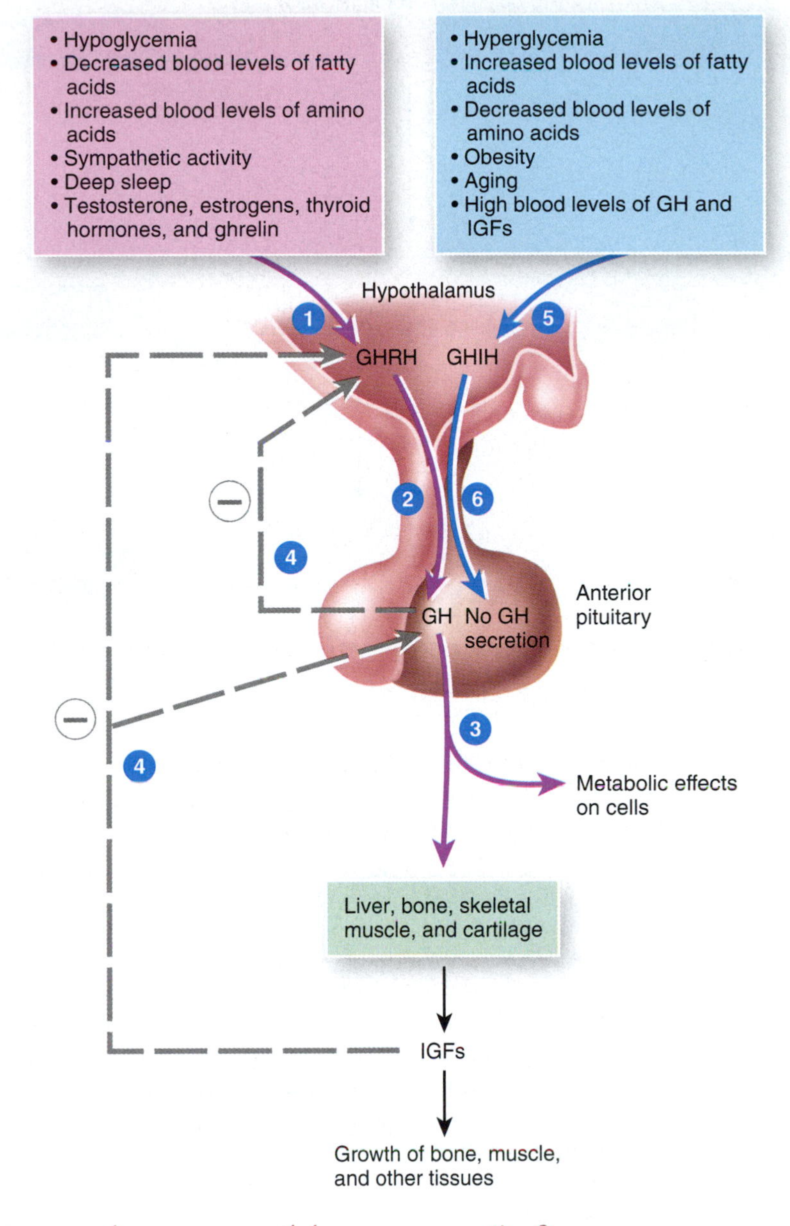

Does hyperglycemia increase or decrease growth hormone secretion?

β-endorphin, which suppresses pain signals. Some of the peptides cleaved from POMC have more than one biological effect. For example, ACTH not only has ACTH activity, it also has MSH activity because part of the ACTH peptide contains the amino acid sequence for MSH (**FIGURE 13.13**). The MSH activity of ACTH is evident when the ACTH level in the blood is high. For example, in Addison's disease, the blood ACTH level is elevated and the MSH activity of ACTH causes the skin to darken (see Clinical Connection: Adrenal Gland Disorders in Section 13.5).

Prolactin

Prolactin (PRL), together with other hormones, initiates and maintains milk production by the mammary glands. By itself, prolactin has only a weak effect. Only after the mammary glands have been primed by estrogens, progesterone, glucocorticoids, growth hormone, thyroxine, and insulin, which exert permissive effects, does PRL bring about milk production. Ejection of milk from the mammary glands depends on the hormone oxytocin, which is released from the posterior pituitary. Together, milk production and ejection constitute **lactation**.

FIGURE 13.13 Pro-opiomelanocortin and the peptides that can form from it. ACTH = adrenocorticotropic hormone; MSH = melanocyte-stimulating hormone; and CLIP = corticotropin-like intermediate-lobe peptide.

Pro-opiomelanocortin (POMC) can be cleaved (spliced) in different ways to give rise to several peptides, including ACTH, α- and γ-MSH, β- and γ-lipotropin, β-endorphin, CLIP, and a few peptide fragments.

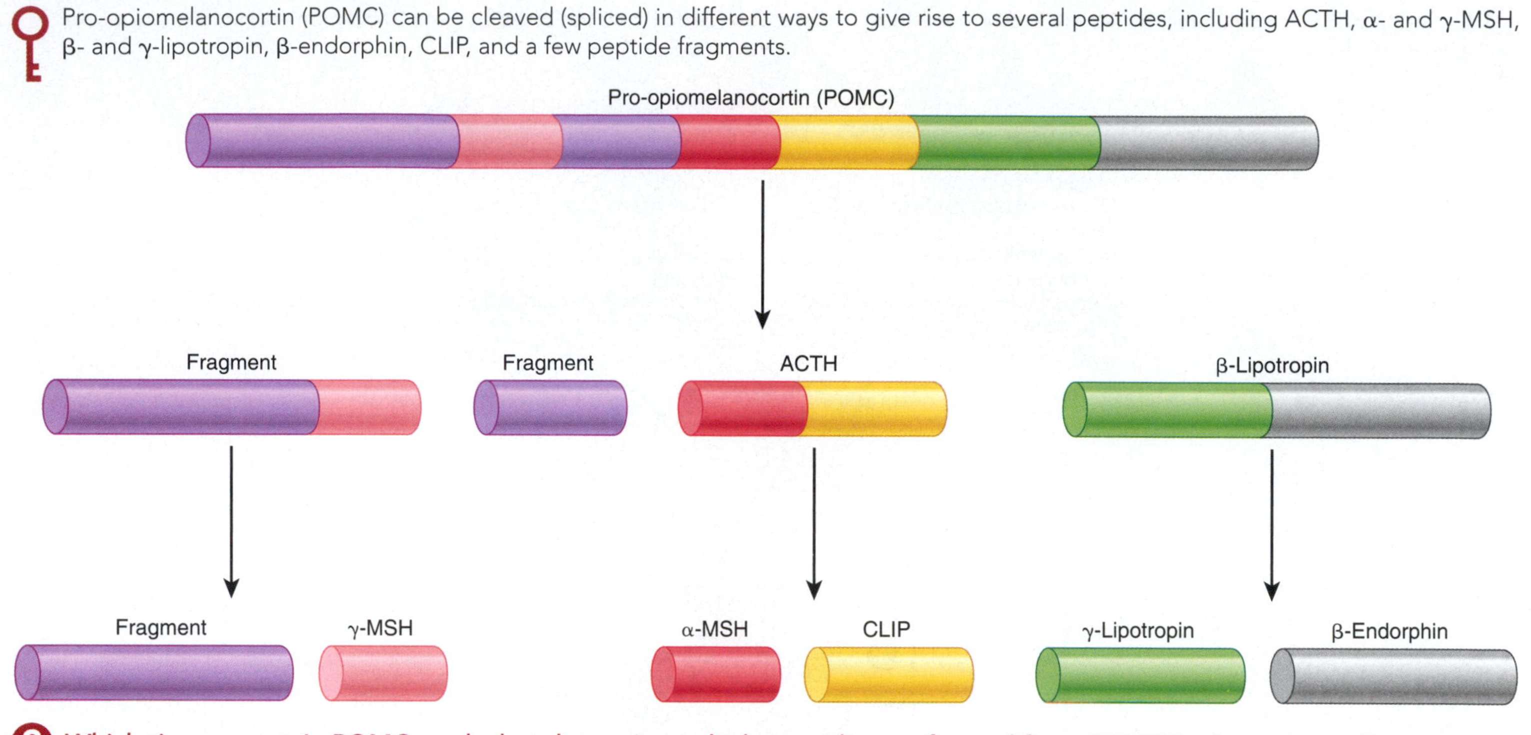

Which tissues contain POMC, and what determines which peptides are formed from POMC in these tissues?

The hypothalamus secretes both inhibitory and excitatory hormones that regulate prolactin secretion. In females, prolactin-inhibiting hormone (PIH), which is dopamine, inhibits the release of prolactin from the anterior pituitary most of the time. Each month, just before menstruation begins, the secretion of PIH diminishes and the blood level of prolactin rises, but not enough to stimulate milk production. Breast tenderness just before menstruation may be caused by elevated prolactin. As the menstrual cycle begins anew, PIH is again secreted and the prolactin level drops. During pregnancy, the prolactin level rises, stimulated by prolactin-releasing hormone (PRH) from the hypothalamus. The suckling action of a nursing infant decreases hypothalamic secretion of PIH and increases secretion of PRH.

The function of prolactin in males is not known, but its hypersecretion causes erectile dysfunction (impotence, the inability to achieve or maintain an erection of the penis). In females, hypersecretion of prolactin causes galactorrhea (inappropriate lactation) and amenorrhea (absence of menstrual cycles).

Follicle-Stimulating Hormone

In females, the ovaries are the targets for follicle-stimulating hormone (FSH). Each month FSH initiates the development of several ovarian follicles, saclike arrangements of secretory cells that surround a developing egg (oocyte). FSH also stimulates follicular cells to secrete estrogens (female sex hormones). In males, FSH stimulates sperm production in the testes. Gonadotropin-releasing hormone (GnRH) from the hypothalamus stimulates FSH release. Release of GnRH and FSH is suppressed by estrogens in females and by testosterone (the principal male sex hormone) in males through negative feedback systems.

Luteinizing Hormone

In females, luteinizing hormone (LH) triggers **ovulation**, the release of an egg by an ovary. LH also stimulates formation of the corpus luteum (structure formed after ovulation) in the ovary and the secretion of estrogens and progesterone (another female sex hormone) by the corpus luteum. Estrogens and progesterone prepare the uterus for implantation of a fertilized ovum. In males, LH stimulates cells in the testes to secrete testosterone. Secretion of LH, like that of FSH, is controlled by gonadotropin-releasing hormone (GnRH) and suppressed by estrogen in females and testosterone in males.

TABLE 13.3 summarizes the principal actions of the anterior pituitary hormones.

The Posterior Pituitary Releases Oxytocin and Antidiuretic Hormone

Although the **posterior pituitary** or *neurohypophysis* does not *synthesize* hormones, it does *store* and *release* two hormones. It consists of axons and axon terminals of more than 10,000 hypothalamic neurosecretory cells. The cell bodies of the neurosecretory cells are organized into two clusters in the hypothalamus: the **paraventricular nucleus** and the **supraoptic nucleus** (FIGURE 13.14). The axons of these neurosecretory cells form the **hypothalamic–hypophyseal tract**, which begins in the hypothalamus and ends near blood capillaries in the posterior pituitary. Release of hormones from the posterior pituitary occurs as follows (FIGURE 13.14):

1. Neurosecretory cells in the paraventricular and supraoptic nuclei of the hypothalamus synthesize **oxytocin** and **antidiuretic hormone (ADH)**, also called **vasopressin** (vā-sō-PRES-in). Each

TABLE 13.3 Summary of the Principal Actions of Anterior Pituitary Hormones

Hormone	Principal Actions
Growth hormone (GH)	GH stimulates liver, muscle, cartilage, and bone to synthesize and secrete insulin-like growth factors (IGFs), which in turn promote growth of body tissues. GH acts directly on target cells to enhance lipolysis and decreases glucose uptake.
Thyroid-stimulating hormone (TSH)	Stimulates the synthesis and secretion of thyroid hormones by the thyroid gland.
Adrenocorticotropic hormone (ACTH)	Stimulates secretion of glucocorticoids (mainly cortisol) by the adrenal cortex.
Prolactin (PRL)	Promotes milk production by the mammary glands.
Follicle-stimulating hormone (FSH)	In females, initiates development of oocytes and induces ovarian secretion of estrogen. In males, stimulates testes to produce sperm.
Luteinizing hormone (LH)	In females, stimulates secretion of estrogen and progesterone, ovulation, and formation of corpus luteum. In males, stimulates testes to produce testosterone.

FIGURE 13.14 Steps involved in the release of hormones from the posterior pituitary gland.

The hypothalamic–hypophyseal tract consists of axons that extend from the paraventricular and supraoptic nuclei of the hypothalamus to the posterior pituitary.

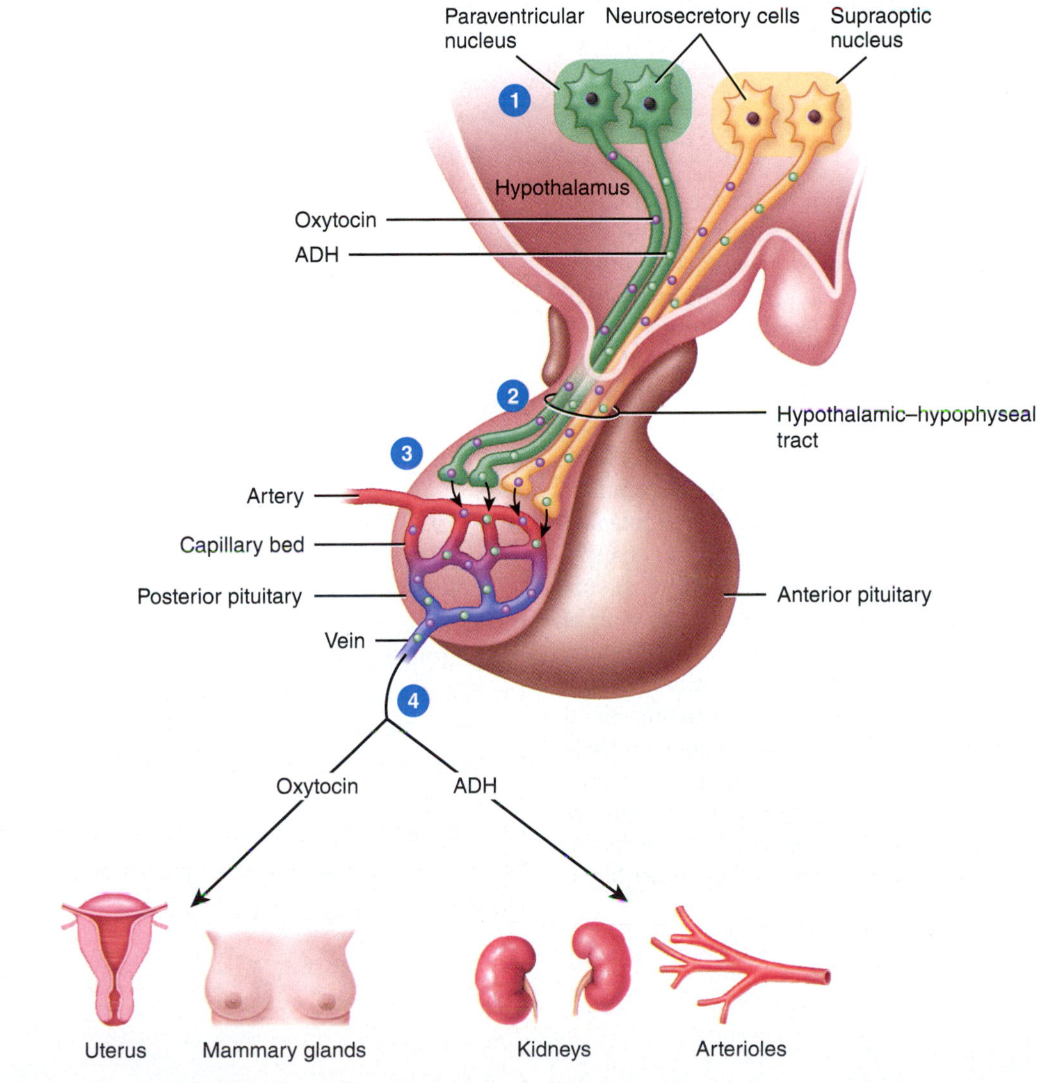

Which two hormones are released from the posterior pituitary?

hormone is made by a separate cell type and both cell types are present in the paraventricular nucleus and in the supraoptic nucleus. These hormones are produced in the cell bodies of the neurosecretory cells and then packaged into vesicles.

2. The vesicles move by axonal transport along the hypothalamic–hypophyseal tract to the axon terminals in the posterior pituitary, where they are stored.
3. When the appropriate stimulus excites the hypothalamus, action potentials trigger exocytosis and release of oxytocin or ADH into the bloodstream.
4. The released oxytocin or ADH then travels to its target tissues in the body.

Oxytocin

During and after delivery of a baby, oxytocin affects two target tissues: the mother's uterus and breasts. During delivery, oxytocin enhances contraction of smooth muscle cells in the wall of the uterus; after delivery, it stimulates milk ejection ("letdown") from the mammary glands in response to the mechanical stimulus provided by a suckling infant. The function of oxytocin in males and in nonpregnant females is not clear. Evidence suggests that it has actions within the brain that promote bonding (feelings of attachment) between couples and between parents and children. It may also be responsible in part for the feelings of sexual pleasure during and after intercourse.

Antidiuretic Hormone

As its name implies, an *antidiuretic* (*anti-* = against; *dia* = throughout; *ouresis* = urination) is a substance that decreases urine production. Antidiuretic hormone (ADH) causes the kidneys to return more water to the blood, thus decreasing urine volume. In the absence of ADH, urine output increases tremendously, from the normal 1 to 2 liters to about 36 liters a day. ADH also causes constriction of arterioles, which increases blood pressure. This hormone's other name, **vasopressin**, reflects this effect on blood pressure (*vaso-* = vessel; *pressum* = pressure).

Two major stimuli promote ADH secretion: a rise in blood osmolarity and a decrease in blood volume. High blood osmolarity is detected by **osmoreceptors**, neurons in the hypothalamus that monitor changes in blood osmolarity. Decreased blood volume is detected by volume receptors in the atria of the heart and by baroreceptors in the walls of certain blood vessels. Once stimulated, osmoreceptors, atrial volume receptors, and baroreceptors activate the hypothalamic neurosecretory cells that synthesize and release ADH into the bloodstream. Blood carries ADH to two target tissues: the kidneys and smooth muscle in blood vessel walls. The kidneys respond by retaining more water, which decreases urine output. Smooth muscle in the walls of arterioles (small arteries) contracts in response to high levels of ADH, which constricts (narrows) the lumen of these blood vessels and increases blood pressure.

TABLE 13.4 Summary of Posterior Pituitary Hormones

Hormone	Control of Secretion	Principal Actions
Oxytocin (OT)	Neurosecretory cells of hypothalamus secrete OT in response to uterine distension and stimulation of nipples.	Stimulates contraction of smooth muscle cells of the uterus during childbirth; stimulates contraction of myoepithelial cells in the mammary glands to cause milk ejection.
Antidiuretic hormone (ADH), also called **vasopressin**	Neurosecretory cells of hypothalamus secrete ADH in response to elevated blood osmolarity, loss of blood volume, pain, or stress. Low blood osmolarity, high blood volume, and alcohol inhibit ADH secretion.	Conserves body water by decreasing urine volume; decreases water loss through perspiration; raises blood pressure by constricting arterioles.

Secretion of ADH can be altered in other ways as well. Pain, stress, trauma, anxiety, acetylcholine, nicotine, and drugs such as morphine, tranquilizers, and some anesthetics stimulate ADH secretion. Drinking alcohol often causes frequent and copious urination because alcohol inhibits secretion of ADH. This dehydrating effect of alcohol may cause both the thirst and the headache typical of a hangover.

TABLE 13.4 lists the posterior pituitary hormones, control of their secretion, and their principal actions.

CHECKPOINT

8. In what respect is the pituitary gland actually two glands?
9. How do hypothalamic releasing or inhibiting hormones influence secretions of the anterior pituitary?
10. What is the functional significance of insulin-like growth factors (IGFs)?
11. Which peptides can be cleaved from pro-opiomelanocortin (POMC)?
12. What are the functional roles of FSH and LH in women? In men?
13. Why is the hypothalamic–hypophyseal tract important?
14. How do prolactin and oxytocin promote lactation?
15. What is the purpose of ADH?

Pituitary Gland Disorders

Disorders of the endocrine system often involve either **hyposecretion**, inadequate release of a hormone, or **hypersecretion**, excessive release of a hormone. In other cases, the problem is faulty hormone receptors, an inadequate number of receptors, or defects in second-messenger systems. Because hormones are distributed in the blood to target tissues throughout the body, problems associated with endocrine dysfunction may also be widespread.

Several disorders of the anterior pituitary involve growth hormone (GH). Hyposecretion of GH during the growth years slows bone growth, and the epiphyseal plates close before normal height is reached. This condition is called **pituitary dwarfism**. Other organs of the body also fail to grow, and the body proportions are childlike. The GH deficiency may be due to pituitary dysfunction or lack of GHRH from the hypothalamus. Treatment requires administration of GH during childhood, before the epiphyseal plates close. Hypersecretion of GH during childhood causes **giantism**, an abnormal increase in the length of long bones. The person grows to be very tall, but body proportions are about normal (**FIGURE A**). Hypersecretion of GH during adulthood is called **acromegaly** (ak′-rō-MEG-a-lē). Although GH cannot produce further lengthening of the long bones because the epiphyseal plates are already closed, the bones of the hands, feet, and jaws thicken and other tissues enlarge (**FIGURE B**). The excess level of GH associated with giantism or acromegaly is usually caused by a tumor of the pituitary gland. Treatment involves surgical removal of the tumor or use of drugs that block the effects of GH.

The most common abnormality associated with dysfunction of the posterior pituitary is **diabetes insipidus (DI)**. This disorder is due to defects in antidiuretic hormone (ADH) receptors or an inability to secrete ADH. ***Neurogenic diabetes insipidus*** results from hyposecretion of ADH, usually caused by a brain tumor, head trauma, or brain surgery that damages the posterior pituitary or the hypothalamus. In ***nephrogenic diabetes insipidus***, the kidneys do not respond to ADH. The ADH receptors may be nonfunctional, or the kidneys may be damaged. A common symptom of both forms of DI is excretion of large volumes of urine, with resulting dehydration and thirst. Bedwetting is common in afflicted children. Because so much water is lost in the urine, a person with DI may die of dehydration if deprived of water for only a day or so. Treatment of neurogenic diabetes insipidus involves hormone replacement, usually for life. Either subcutaneous injection or nasal spray application of ADH analogs is effective. Treatment of nephrogenic diabetes insipidus is more complex and depends on the nature of the kidney dysfunction. Restriction of salt in the diet and, paradoxically, the use of certain diuretic drugs are helpful.

From New England Journal of Medicine, Massachusettes Medical Society, February 18, 1999, vol. 340, No. 7, page 524

(A) A 22-year-old man with pituitary giantism shown beside his identical twin

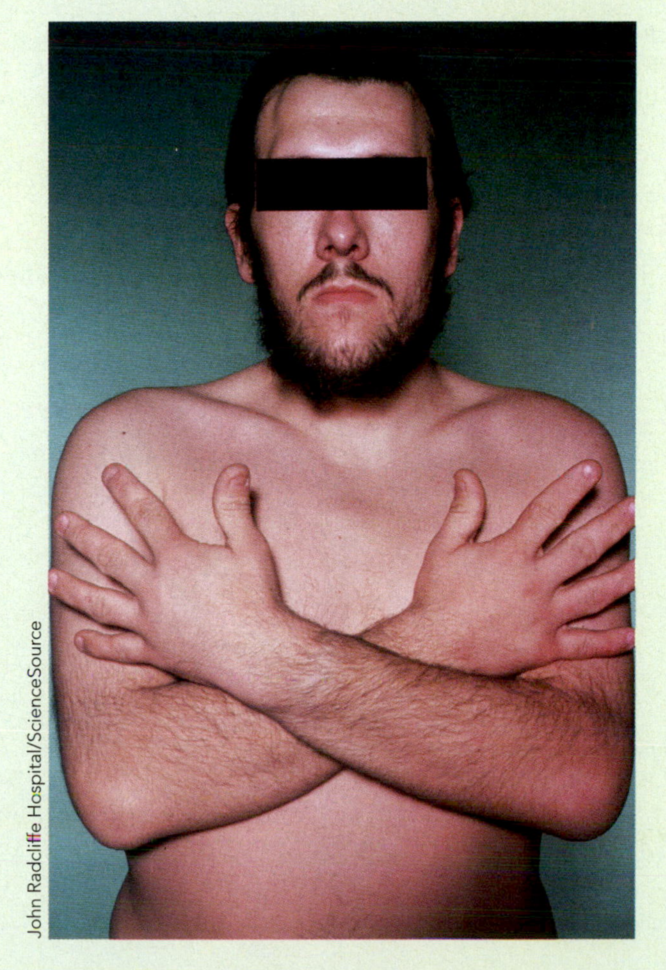

John Radcliffe Hospital/ScienceSource

(B) Acromegaly

13.3 Thyroid Gland

OBJECTIVE

- Describe the functions of the hormones produced by the thyroid gland.

The butterfly-shaped **thyroid gland** is located just below the larynx (voice box). Microscopic spherical sacs called **thyroid follicles** make up most of the thyroid gland (FIGURE 13.15). The wall of each follicle consists primarily of cells called **follicular cells**, which produce two hormones: **thyroxine** (thī-ROK-sēn), which is also called **tetraiodothyronine (T_4)** (tet-ra-ī-ō-dō-THĪ-rō-nēn) because it contains four atoms of iodine, and **triiodothyronine (T_3)** (trī-ī′-ō-dō-THĪ-rō-nēn), which contains three atoms of iodine. T_3 and T_4 together are also known as **thyroid hormones**. A few cells called **parafollicular cells** or **C cells** lie between the follicles of the thyroid gland. They produce the hormone **calcitonin** (kal-si-TŌ-nin), which helps regulate calcium homeostasis.

Thyroid Follicles Form, Store, and Release Thyroid Hormones

Thyroid hormones are produced by follicular cells by adding iodine to tyrosine amino acids. This process involves several steps (FIGURE 13.16):

1. ***Iodide trapping.*** Thyroid follicular cells trap iodide ions (I^-) by actively transporting them from the blood into the cytosol. As a result, the thyroid gland normally contains most of the iodide in the body.
2. ***Synthesis of thyroglobulin.*** While the follicular cells are trapping I^-, they are also synthesizing **thyroglobulin (TGB)**, a large glycoprotein that is produced in the rough endoplasmic reticulum, modified in the Golgi complex, and packaged into secretory vesicles. The vesicles then undergo exocytosis, which releases TGB into the lumen of the follicle.
3. ***Oxidation of iodide.*** Some of the amino acids in TGB are tyrosines that will become iodinated. However, negatively charged iodide ions cannot bind to tyrosine until they undergo oxidation (removal of electrons) to iodine: $I^- \rightarrow I^0$. The enzyme that catalyzes this reaction is known as **thyroid peroxidase**, located at the luminal membrane of the follicular cell. As the iodide ions are being oxidized, they pass through the membrane into the lumen of the follicle.
4. ***Iodination of tyrosine.*** As iodine atoms (I^0) form, they react with tyrosines that are part of thyroglobulin molecules. Binding of one iodine atom yields **monoiodotyrosine (MIT)**, and a second iodination produces **diiodotyrosine (DIT)**. The TGB with attached iodine atoms is termed **colloid**, a sticky material that accumulates in the lumen of the thyroid follicle.
5. ***Coupling of MIT and DIT.*** During the last step in the synthesis of thyroid hormone, one MIT and one DIT join to form T_3 or two DIT molecules join to form T_4. Large amounts of T_3 and T_4 are stored in this form, attached to the colloid. Thyroid follicles normally contain enough colloid to provide about a 100-day supply of thyroid hormones.
6. ***Pinocytosis and digestion of colloid.*** Droplets of colloid reenter follicular cells by pinocytosis and merge with lysosomes. Digestive enzymes in the lysosomes break down TGB, cleaving off molecules of T_3 and T_4.

FIGURE 13.15 Thyroid gland.

Thyroid hormones increase basal metabolic rate, enhance the actions of catecholamines, and regulate development and growth of nervous tissue and bones.

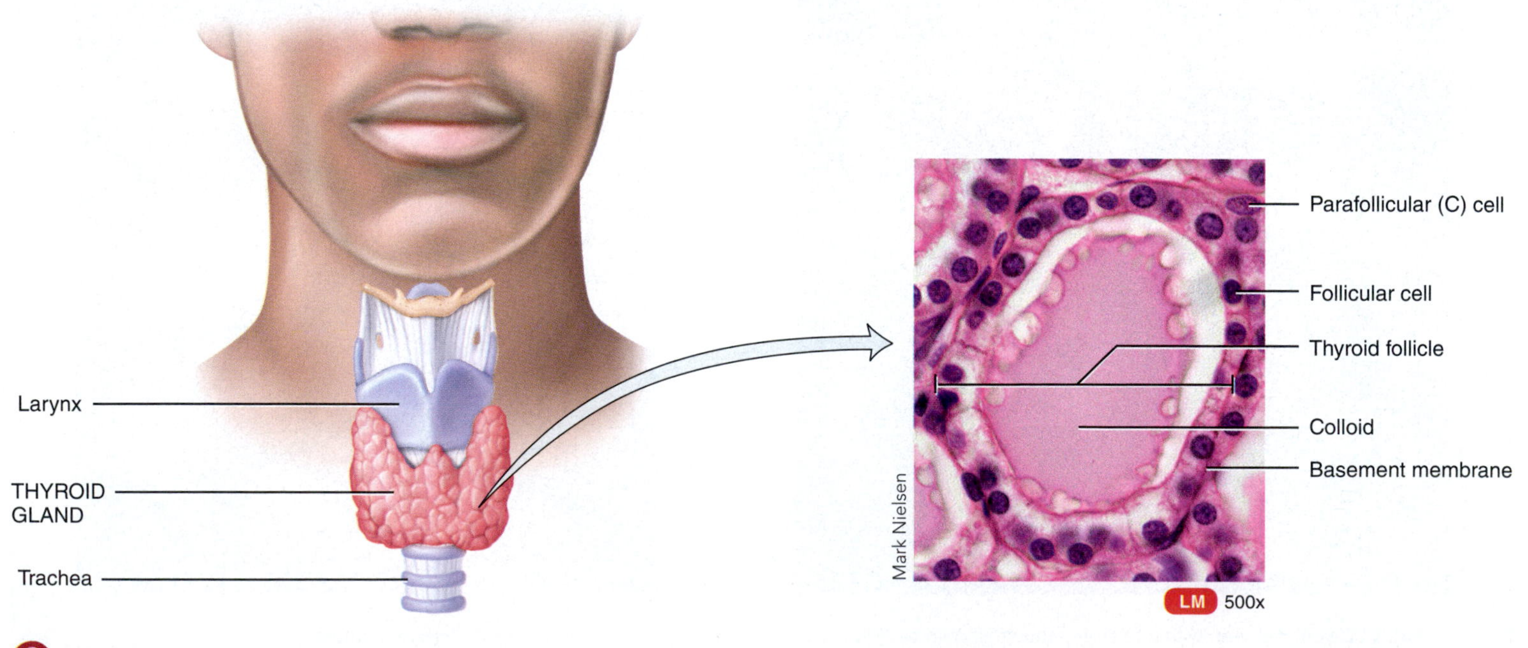

Which cells of the thyroid gland secrete T_3 and T_4? Which secrete calcitonin? Which of these hormones are also called thyroid hormones?

FIGURE 13.16 Steps in the synthesis and secretion of thyroid hormones.

Thyroid hormones are synthesized by attaching iodine atoms to the amino acid tyrosine.

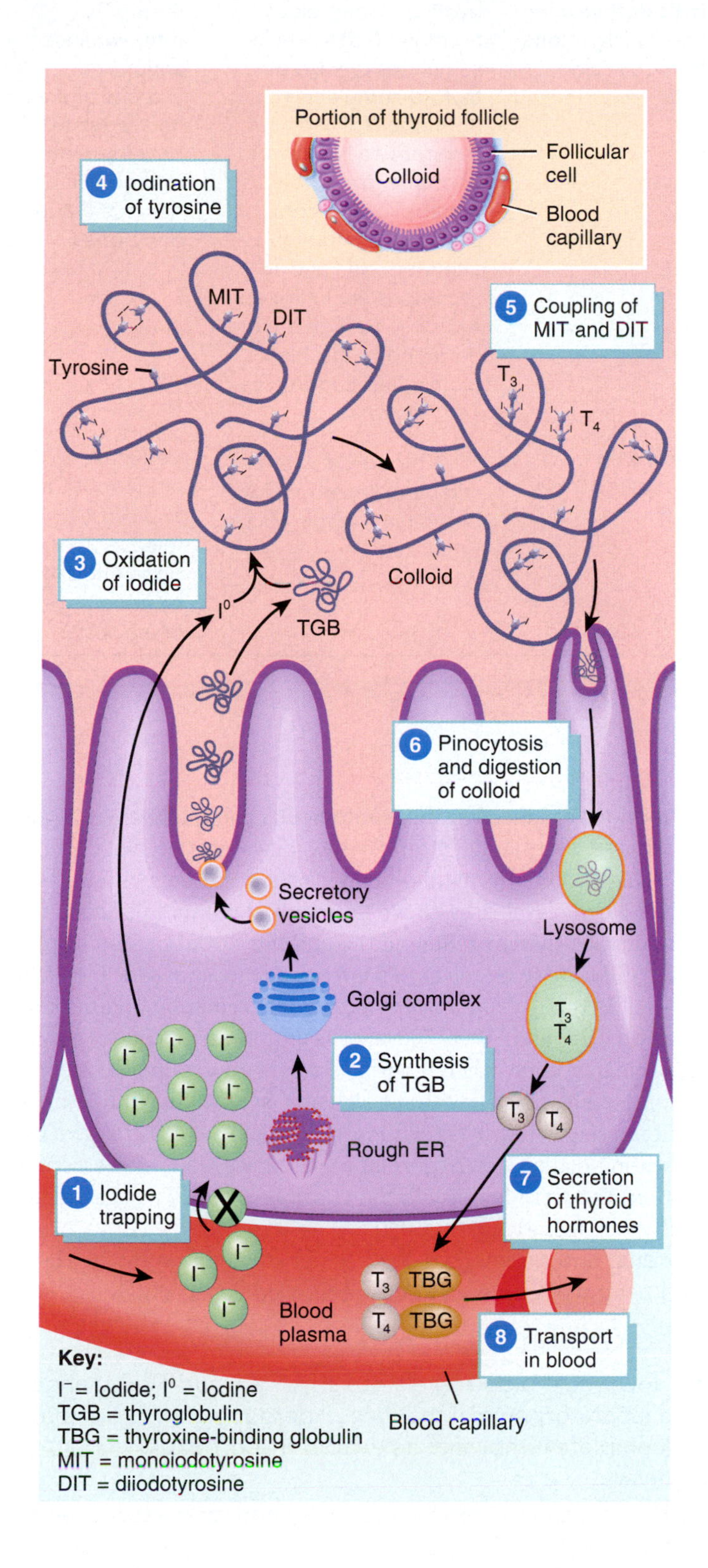

What is the significance of the colloid that accumulates in the lumen of the thyroid follicle?

7. ***Secretion of thyroid hormones.*** Because T_3 and T_4 are lipid-soluble, they diffuse through the plasma membrane into interstitial fluid and then into the blood. T_4 normally is secreted in greater quantity than T_3, but T_3 is several times more potent. Moreover, after T_4 enters a body cell, most of it is converted to T_3 by removal of one iodine via enzymes called **deiodinases**.
8. ***Transport in the blood.*** More than 99% of both the T_3 and the T_4 combine with transport proteins in the blood, mainly **thyroxine-binding globulin (TBG).**

Thyroid Hormones Increase Basal Metabolic Rate and Have Other Effects

Because most body cells have receptors for thyroid hormones, T_3 and T_4 affect tissues throughout the body. Thyroid hormones act on their target cells mainly by inducing gene transcription and protein synthesis. The newly formed proteins in turn carry out the cellular response. Functions of thyroid hormones include the following:

- ***Increase basal metabolic rate.*** Thyroid hormones raise the **basal metabolic rate (BMR)**, the rate of energy expenditure under standard or basal conditions (awake, at rest, and fasting). When BMR increases, cellular metabolism of carbohydrates, lipids, and proteins increases. Thyroid hormones increase BMR in several ways: (1) They stimulate synthesis of additional Na^+/K^+ ATPases, which use large amounts of ATP to continually eject sodium ions (Na^+) from cytosol into extracellular fluid and potassium ions (K^+) from extracellular fluid into cytosol; (2) they increase the concentrations of enzymes involved in cellular respiration, which increases the breakdown of organic fuels and ATP production; and (3) they increase the number and activity of mitochondria in cells, which also increases ATP production. As cells produce and use more ATP, BMR increases, more heat is given off, and body temperature rises, a phenomenon called the **calorigenic effect**. In this way, thyroid hormones play an important role in the maintenance of normal body temperature. Normal mammals can survive in freezing temperatures, but those whose thyroid glands have been removed cannot.
- ***Enhance actions of catecholamines.*** Thyroid hormones have permissive effects on the catecholamines (epinephrine and norepinephrine) because they up-regulate β-adrenergic receptors. Recall that catecholamines bind to beta-adrenergic receptors, promoting sympathetic responses. Therefore, symptoms of excess levels of thyroid hormone include increased heart rate, more forceful heartbeats, and increased blood pressure.
- ***Regulate development and growth of nervous tissue and bones.*** Thyroid hormones are necessary for the development of the nervous system: They promote synapse formation, myelin production, and growth of dendrites. Thyroid hormones are also required for growth of the skeletal system: They promote formation of ossification centers in developing bones, synthesis of many bone proteins, and secretion of growth hormone (GH) and insulin-like growth factors (IGFs). Deficiency of thyroid hormones during fetal development, infancy, or childhood causes severe mental retardation and stunted bone growth.

Secretion of Thyroid Hormones Is Regulated by the Hypothalamus and Anterior Pituitary

Thyrotropin-releasing hormone (TRH) from the hypothalamus and thyroid-stimulating hormone (TSH) from the anterior pituitary stimulate secretion of thyroid hormones, as shown in **FIGURE 13.17**:

1. Low blood levels of T_3 and T_4 or low metabolic rate stimulate the hypothalamus to secrete TRH.
2. TRH enters the hypothalamic–hypophyseal portal system and flows to the anterior pituitary, where it stimulates thyrotrophs to secrete TSH.
3. TSH stimulates virtually all aspects of thyroid follicular cell activity, including iodide trapping, hormone synthesis and secretion, and growth of the follicular cells.
4. The thyroid follicular cells release T_3 and T_4 into the blood until the metabolic rate returns to normal.
5. An elevated level of T_3 inhibits release of TRH and TSH (negative feedback inhibition).

Conditions that increase ATP demand—a cold environment, hypoglycemia, high altitude, and pregnancy—increase the secretion of the thyroid hormones.

FIGURE 13.17 Regulation of secretion of thyroid hormones. TRH = thyrotropin-releasing hormone, TSH = thyroid-stimulating hormone, T_3 = triiodothyronine, and T_4 = thyroxine (tetraiodothyronine). The dashed arrows and negative sign indicate negative feedback.

TSH promotes release of thyroid hormones (T_3 and T_4) by the thyroid gland.

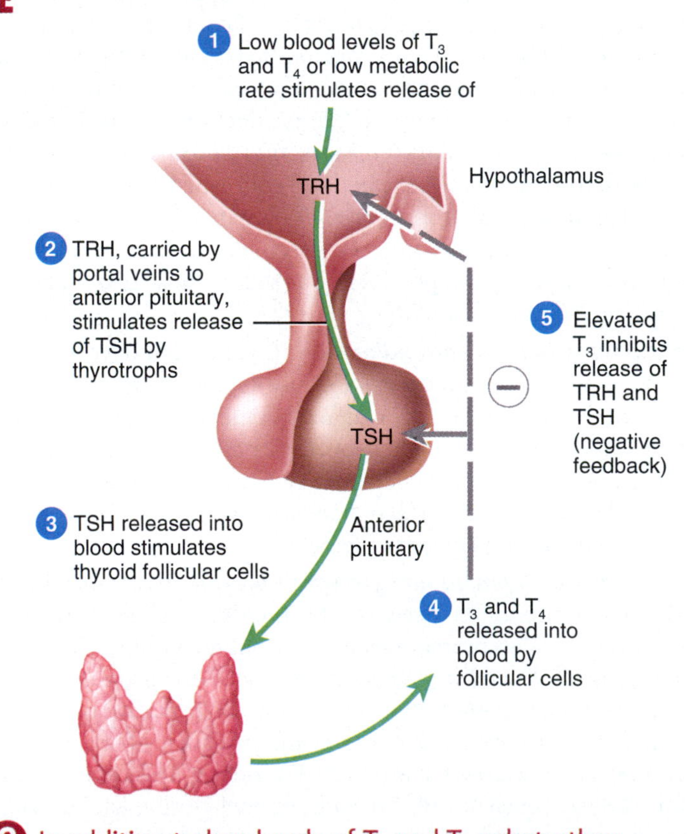

In addition to low levels of T_3 and T_4, what other conditions increase secretion of thyroid hormones?

TABLE 13.5 Summary of Thyroid Gland Hormones

Hormone and Source	Control of Secretion	Principal Actions
T_3 (triiodothyronine) and **T_4 (thyroxine)** from follicular cells	Secretion is increased by thyrotropin-releasing hormone (TRH), which stimulates release of thyroid-stimulating hormone (TSH) in response to low thyroid hormone levels, low metabolic rate, cold, pregnancy, and high altitudes. TRH and TSH secretions are inhibited in response to high thyroid hormone levels; high iodine level suppresses T_3/T_4 secretion.	Increase basal metabolic rate, enhance actions of catecholamines, and regulate development and growth of nervous tissue and bones.
Calcitonin from parafollicular cells	High blood Ca^{2+} levels stimulate secretion; low blood Ca^{2+} levels inhibit secretion.	Reduces amount of Ca^{2+} and HPO_4^{2-} released from bone into blood by inhibiting osteoclasts.

Calcitonin Lowers the Blood Calcium Level

The hormone produced by the parafollicular cells of the thyroid gland (see **FIGURE 13.15**) is calcitonin, which plays a role in blood calcium (Ca^{2+}) regulation. The parafollicular cells secrete calcitonin when the blood Ca^{2+} level rises above normal. Calcitonin inhibits activity of osteoclasts, thereby reducing the amount of Ca^{2+} and phosphate (HPO_4^{2-}) ions that are released from bone matrix into the blood. The net result is that calcitonin promotes bone formation and decreases the blood Ca^{2+} and HPO_4^{2-} levels. Despite these effects, the role of calcitonin in normal calcium homeostasis is uncertain because it can be completely absent without causing symptoms. Nevertheless, calcitonin harvested from salmon (Miacalcin®) is an effective drug for treating osteoporosis (porous bones) because it slows bone resorption. The secretion of calcitonin is controlled by a negative feedback system (see **FIGURE 13.19**).

TABLE 13.5 summarizes the hormones produced by the thyroid gland, control of their secretion, and their principal actions.

CHECKPOINT

16. How would blood levels of T_3/T_4, TSH, and TRH change in a laboratory animal that has undergone a thyroidectomy (complete removal of its thyroid gland)? Explain your answer.
17. How are the thyroid hormones synthesized, stored, and secreted?
18. What are the physiological effects of the thyroid hormones?
19. How is the secretion of T_3 and T_4 regulated?

CLINICAL CONNECTION

Thyroid Gland Disorders

Thyroid gland disorders affect all major body systems and are among the most common endocrine disorders. The two main categories of thyroid disorders are hyperthyroidism and hypothyroidism. **Hyperthyroidism** is the excess production of thyroid hormones. The most common form of hyperthyroidism is **Graves' disease**, an autoimmune disorder in which the person produces antibodies called **thyroid-stimulating immunoglobulins (TSIs)** that mimic the action of thyroid-stimulating hormone (TSH) by binding to TSH receptors on the thyroid gland. The antibodies continually stimulate the thyroid gland to grow and produce thyroid hormones. A primary sign is a **goiter**, an enlarged thyroid gland (FIGURE A). The goiter is often two or three times the size of the thyroid gland, but in severe cases, it can become even larger. The elevated levels of thyroid hormone also cause an abnormally high basal metabolic rate (BMR). This leads to excessive heat production, profuse sweating, and heat intolerance. In addition, there is net breakdown of carbohydrates, lipids, and proteins in body tissues, resulting in weight loss and muscle weakness. Because thyroid hormones have permissive effects on catecholamines, the high levels of thyroid hormones result in rapid heart rate and heart palpitations. Graves' disease patients often have a peculiar edema behind the eyes, called **exophthalmos** (ek'-sof-THAL-mos), which causes the eyes to protrude (FIGURE B). The edema is caused by an immune reaction between the thyroid-stimulating immunoglobulins and the skeletal muscles attached to the outside surface of the eyeball. Treatment of Graves' disease may include surgical removal of part or all of the thyroid gland (thyroidectomy), the use of radioactive iodine (^{131}I) to selectively destroy thyroid tissue, and the use of antithyroid drugs to block synthesis of thyroid hormones.

Hypothyroidism is the deficiency of thyroid hormone production. It may be caused by (1) inadequate secretion of thyroid hormones due to thyroid gland dysfunction; (2) a lack of TRH from the hypothalamus, TSH from the anterior pituitary, or both; or (3) insufficient dietary intake of iodine. The deficiency in thyroid hormone secretion causes an abnormally low basal metabolic rate, which in turn causes low body temperature, sensitivity to cold temperatures, general lethargy, and a tendency to gain weight easily. Because protein synthesis is reduced, there are fewer proteins in the skin and hair, causing these structures to become dry and thin. Another hallmark of this disorder is **myxedema** (mix-e-DĒ-ma), a condition in which the face, hands, and feet swell and look puffy. Myxedema is caused by the buildup of glycosaminoglycans (long, unbranched polysaccharide chains) in connective tissue beneath the skin. Normally, thyroid hormones prevent overproduction of these polysaccharides. However, when thyroid hormone levels are low, glycosaminoglycans are able to accumulate and attract water, resulting in tissue swelling. Hypothyroidism can also lead to the formation of a goiter. In some places in the world, dietary iodine intake is inadequate; the resultant low level of thyroid hormones in the blood stimulates secretion of TRH and TSH (due to release of negative feedback inhibition of the hypothalamus and anterior pituitary), which causes thyroid gland enlargement. A goiter can also form if hypothyroidism is caused by inadequate secretion of thyroid hormones because of thyroid gland dysfunction. Just like with iodine deficiency, the low levels of thyroid hormones stimulate TRH and TSH secretion, resulting in the enlargement of the thyroid gland. Other effects of hypothyroidism are specific to the age at which the disorder occurs. If a person is born with hypothyroidism, a condition known as **congenital hypothyroidism** (previously termed **cretinism**), the deficiency of thyroid hormones causes mental retardation and stunted bone growth. If hypothyroidism does not occur until adulthood, then mental retardation does not occur because the brain has already reached maturity, and height is not affected because the skeletal system has developed normally. The low levels of thyroid hormones will, however, slow nervous system activity, causing the individual to become less alert. The treatment for most forms of hypothyroidism is oral thyroid hormones. If congenital hypothyroidism exists, oral thyroid hormone treatment must be started soon after birth and continued for life. For those cases where hypothyroidism is caused by iodine deficiency, iodine supplements are taken. Most people in the United States never develop iodine-deficient hypothyroidism because the table salt that we add to our food is iodized (has iodine added to it) to make sure that we have enough iodine in our diets.

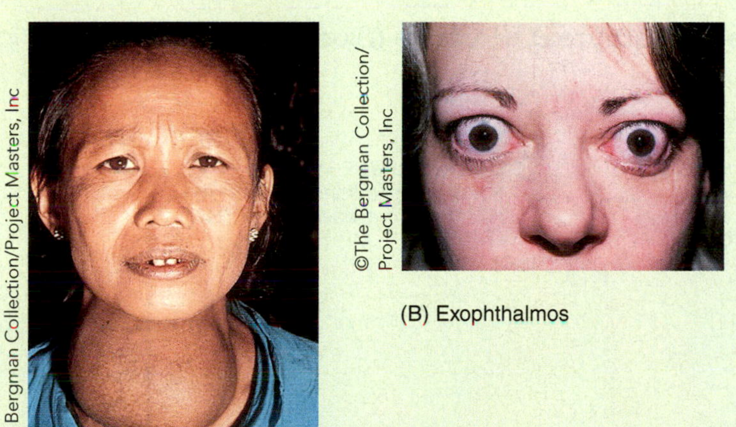

(A) Goiter

(B) Exophthalmos

13.4 Parathyroid Glands

OBJECTIVE

- Identify the hormone produced by the parathyroid glands and describe its functions.

The **parathyroid glands** are small, round masses of glandular tissue that are partially embedded in the back surface of the thyroid gland (FIGURE 13.18). Within the parathyroid glands are secretory cells called **chief cells** that release **parathyroid hormone (PTH)**. PTH is the major regulator of the levels of calcium (Ca^{2+}), magnesium (Mg^{2+}), and phosphate (HPO_4^{2-}) ions in the blood. One specific action of PTH is to increase the number and activity of osteoclasts. The result is elevated bone resorption, which releases ionic calcium (Ca^{2+}) and phosphates (HPO_4^{2-}) into the blood. PTH also acts on the kidneys. First, it slows the rate at which Ca^{2+} and Mg^{2+} are lost from blood into the urine. Second, it increases

FIGURE 13.18 Parathyroid glands.

The parathyroid glands produce parathyroid hormone (PTH).

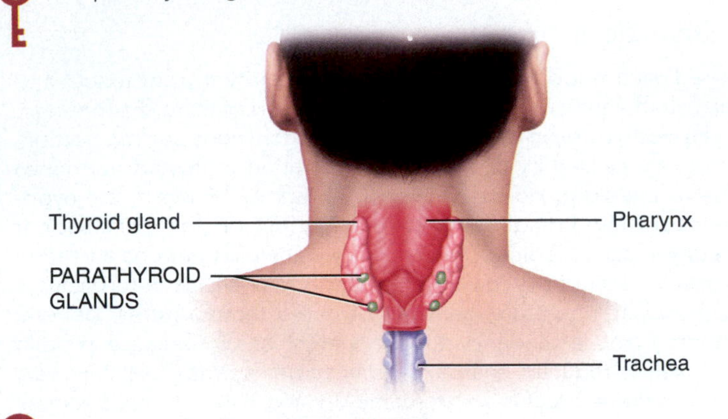

What are the functions of PTH?

loss of HPO_4^{2-} from blood into the urine. Because more HPO_4^{2-} is lost in the urine than is gained from the bones, PTH decreases blood HPO_4^{2-} level and increases blood Ca^{2+} and Mg^{2+} levels. A third effect of PTH on the kidneys is to promote formation of the hormone *calcitriol*, the active form of vitamin D. Calcitriol increases the rate of Ca^{2+} and HPO_4^{2-} absorption from the gastrointestinal tract into the blood.

The blood calcium level directly controls the secretion of both calcitonin and parathyroid hormone via negative feedback loops that do not involve the pituitary gland (FIGURE 13.19):

1. A higher-than-normal level of calcium ions (Ca^{2+}) in the blood stimulates parafollicular cells of the thyroid gland to release more calcitonin.
2. Calcitonin inhibits the activity of osteoclasts, thereby decreasing the blood Ca^{2+} level.
3. A lower-than-normal level of Ca^{2+} in the blood stimulates chief cells of the parathyroid gland to release more PTH.
4. PTH promotes resorption of bone extracellular matrix, which releases Ca^{2+} into the blood and slows loss of Ca^{2+} in the urine, raising the blood level of Ca^{2+}.
5. PTH also stimulates the kidneys to synthesize calcitriol, the active form of vitamin D.

CLINICAL CONNECTION

Parathyroid Gland Disorders

Hypoparathyroidism—too little parathyroid hormone—leads to a deficiency of blood Ca^{2+}, which causes neurons and muscle fibers to depolarize and produce action potentials spontaneously. This in turn leads to twitches, spasms, and **tetany** (maintained contraction) of skeletal muscle. The leading cause of hypoparathyroidism is accidental damage to the parathyroid glands or to their blood supply during thyroidectomy surgery.

Hyperparathyroidism, an elevated level of parathyroid hormone, most often is due to a tumor of one of the parathyroid glands. An elevated level of PTH causes excessive resorption of bone matrix, raising the blood levels of calcium and phosphate ions, and causing bones to become soft and easily fractured. High blood calcium level also promotes formation of kidney stones.

FIGURE 13.19 The roles of calcitonin (green arrows), parathyroid hormone (blue arrows), and calcitriol (orange arrows) in calcium homeostasis.

With respect to regulation of blood Ca^{2+} level, PTH and calcitonin have antagonistic effects.

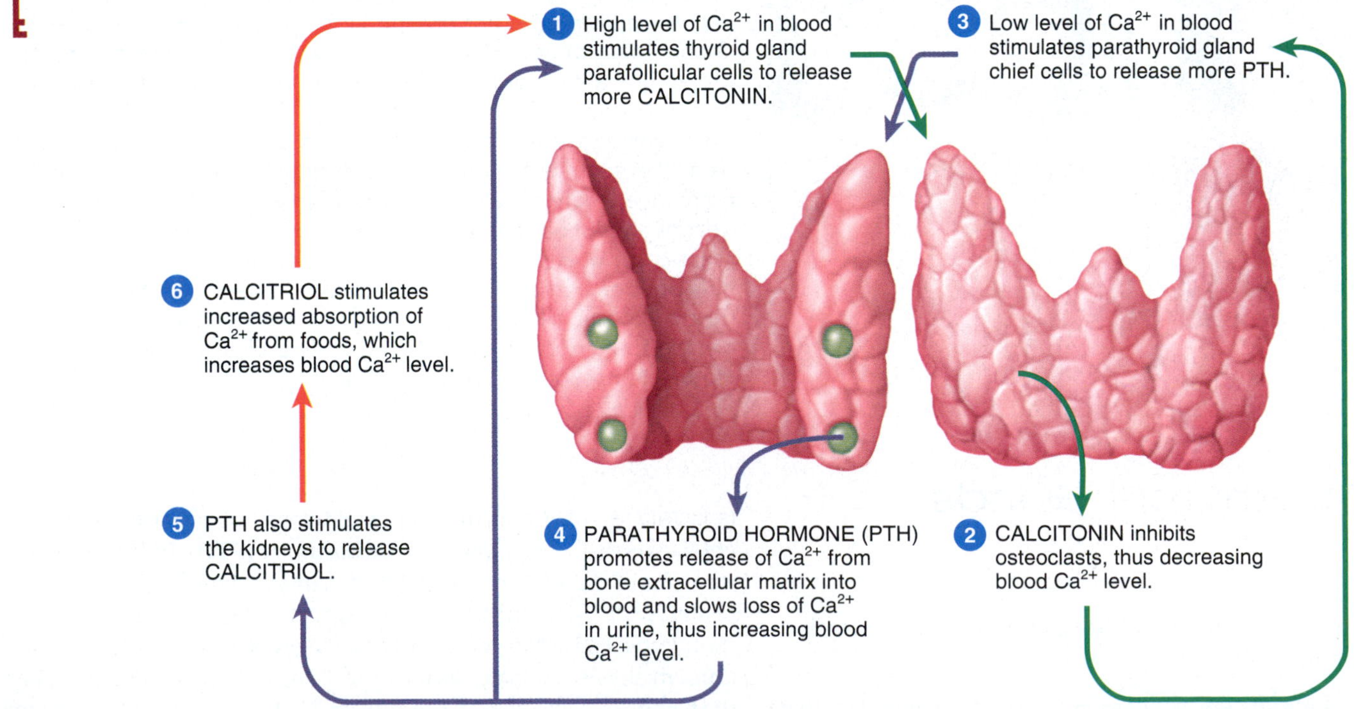

1 High level of Ca^{2+} in blood stimulates thyroid gland parafollicular cells to release more CALCITONIN.

3 Low level of Ca^{2+} in blood stimulates parathyroid gland chief cells to release more PTH.

6 CALCITRIOL stimulates increased absorption of Ca^{2+} from foods, which increases blood Ca^{2+} level.

5 PTH also stimulates the kidneys to release CALCITRIOL.

4 PARATHYROID HORMONE (PTH) promotes release of Ca^{2+} from bone extracellular matrix into blood and slows loss of Ca^{2+} in urine, thus increasing blood Ca^{2+} level.

2 CALCITONIN inhibits osteoclasts, thus decreasing blood Ca^{2+} level.

What are the primary target tissues of PTH, calcitonin, and calcitriol?

TABLE 13.6 Summary of Parathyroid Gland Hormone

Hormone and Source	Control of Secretion	Principal Actions
Parathyroid hormone (PTH) from chief cells	Low blood Ca^{2+} levels stimulate secretion; high blood Ca^{2+} levels inhibit secretion.	Increases blood Ca^{2+} and Mg^{2+} levels and decreases blood HPO_4^{2-} level; increases bone resorption by osteoclasts; increases Ca^{2+} reabsorption and HPO_4^{2-} excretion by kidneys; and promotes formation of calcitriol (active form of vitamin D), which increases rate of dietary Ca^{2+} and HPO_4^{2-} absorption from gastrointestinal tract.

6 Calcitriol stimulates increased absorption of Ca^{2+} from foods in the gastrointestinal tract, which helps increase the blood level of Ca^{2+}.

TABLE 13.6 summarizes control of secretion and the principal actions of parathyroid hormone.

CHECKPOINT

20. How is secretion of parathyroid hormone regulated?

21. In what ways are the actions of PTH and calcitriol similar? How are they different?

13.5 Adrenal Glands

OBJECTIVE

- Describe the functions of the hormones of the adrenal glands.

There are two **adrenal glands**, one lying atop each kidney (FIGURE 13.20). Each adrenal gland consists of two regions: an outer **adrenal cortex**, which makes up about 85% of the gland, and an inner **adrenal medulla**. The adrenal cortex produces a variety of steroid hormones. The adrenal medulla produces three catecholamine hormones—epinephrine, norepinephrine, and a trace amount of dopamine.

The Adrenal Cortex Consists of Three Zones That Secrete Hormones

The adrenal cortex is subdivided into three zones, each of which secretes different hormones (FIGURE 13.20). The outer zone is the **zona glomerulosa** (glo-mer′-ū-LŌ-sa). Its cells secrete hormones called **mineralocorticoids** (min′-er-al-ō-KOR-ti-koyds) because they affect mineral homeostasis. The middle zone, or **zona fasciculata** (fa-sik′-ū-LA-ta), is the widest of the three zones. The cells of the zona fasciculata secrete mainly **glucocorticoids** (gloo′-kō-KOR-ti-koyds), so named because they affect glucose homeostasis. The cells of the inner zone, the **zona reticularis** (re-tik′-ū-LAR-is), synthesize small amounts of weak **androgens**, steroid hormones that have masculinizing effects.

FIGURE 13.20 Adrenal glands.

Each adrenal gland consists of two main parts: an outer adrenal cortex and an inner adrenal medulla.

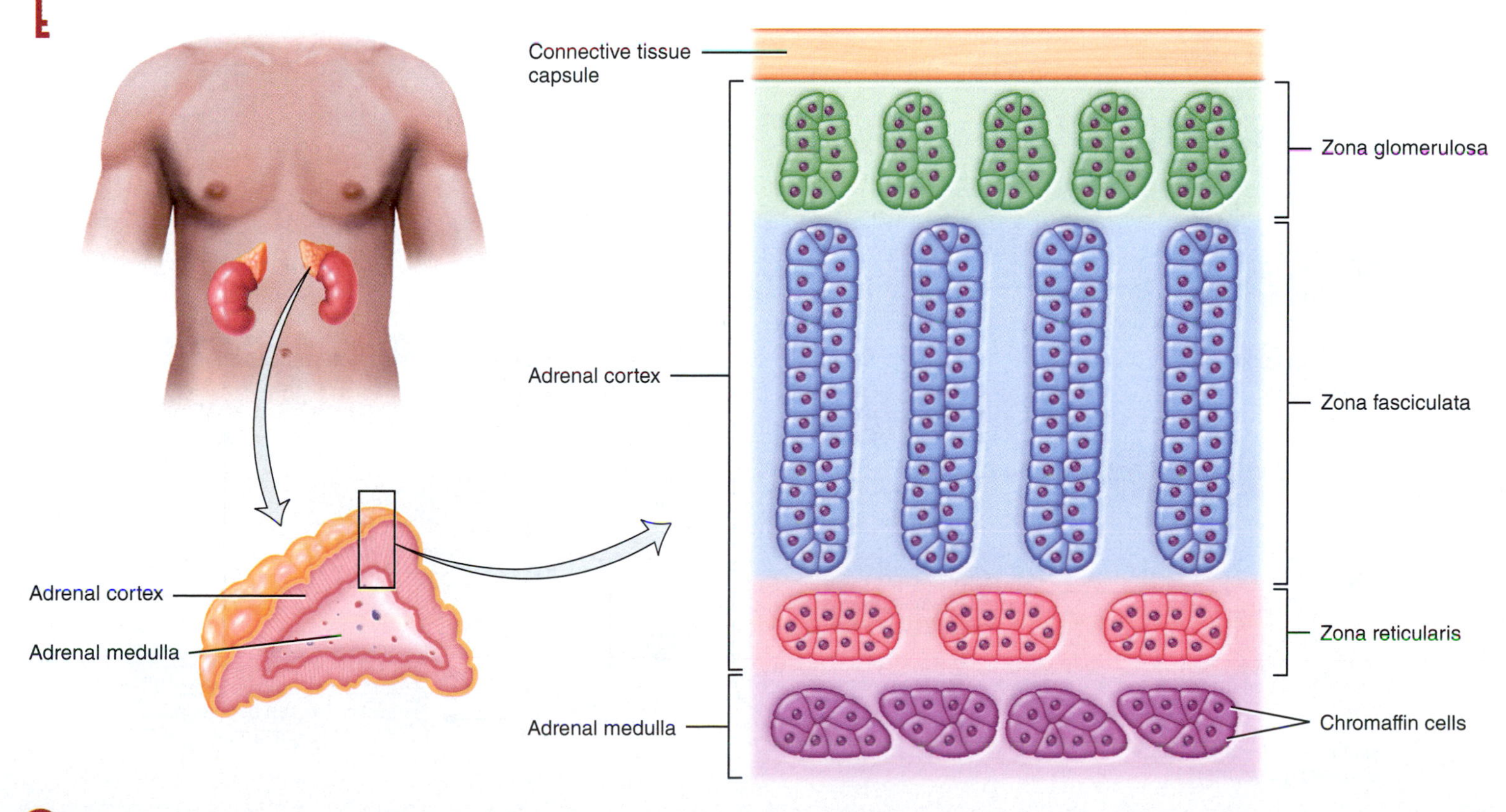

What hormones are secreted by the adrenal cortex? By the adrenal medulla?

Mineralocorticoids

Aldosterone is the major mineralocorticoid. It regulates homeostasis of two mineral ions, namely sodium ions (Na^+) and potassium ions (K^+), and helps adjust blood pressure and blood volume. Aldosterone also promotes excretion of H^+ in the urine; this removal of acids from the body can help prevent acidosis (blood pH below 7.35), which is discussed in Chapter 20.

The **renin–angiotensin–aldosterone (RAA) pathway** controls secretion of aldosterone (FIGURE 13.21):

1. Stimuli that initiate the renin–angiotensin–aldosterone pathway include dehydration, Na^+ deficiency, or hemorrhage.
2. These conditions cause a decrease in blood volume.
3. Decreased blood volume leads to decreased blood pressure.
4. Lowered blood pressure stimulates certain cells of the kidneys, called juxtaglomerular cells, to secrete the enzyme **renin** (RĒ-nin).
5. The level of renin in the blood increases.
6. Renin converts **angiotensinogen** (an′-jē-ō-ten-SIN-ō-jen), a plasma protein produced by the liver, into **angiotensin I**.
7. Blood containing increased levels of angiotensin I circulates in the body.
8. As blood flows through capillaries, particularly those of the lungs, the enzyme **angiotensin-converting enzyme (ACE)** converts angiotensin I into the hormone **angiotensin II**.
9. Blood level of angiotensin II increases.
10. Angiotensin II stimulates the adrenal cortex to secrete aldosterone.
11. Blood containing increased levels of aldosterone circulates to the kidneys.
12. In the kidneys, aldosterone increases reabsorption of Na^+ so that less is lost in the urine. The osmotic consequence of reabsorbing more Na^+ is that more water is reabsorbed (as long as antidiuretic hormone is present). Aldosterone also stimulates the kidneys to increase secretion of K^+ and H^+ into the urine.
13. With increased water reabsorption by the kidneys, blood volume increases.
14. As blood volume increases, blood pressure increases to normal.
15. Angiotensin II also stimulates contraction of smooth muscle in the walls of arterioles. The resulting vasoconstriction of the arterioles increases blood pressure and thus helps raise blood pressure to normal.
16. Besides angiotensin II, a second stimulator of aldosterone secretion is an increase in the K^+ concentration of blood (or interstitial fluid). A decrease in the blood K^+ level has the opposite effect.

Glucocorticoids

The glucocorticoids, which regulate metabolism and resistance to stress, include **cortisol (hydrocortisone)**, **corticosterone**, and

FIGURE 13.21 Regulation of aldosterone secretion by the renin–angiotensin–aldosterone (RAA) pathway.

Aldosterone helps regulate blood volume; blood pressure; and levels of Na^+, K^+, and H^+ in the blood.

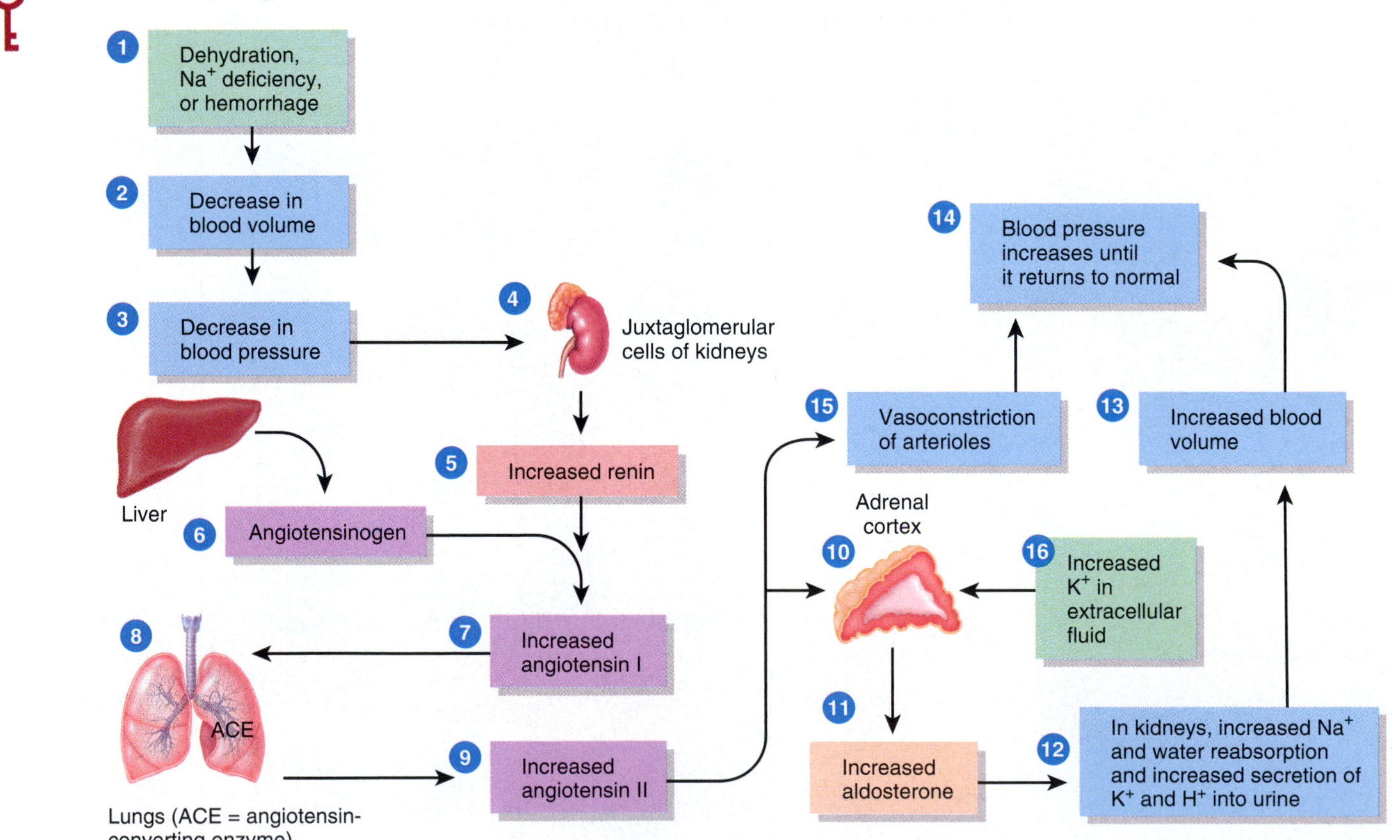

In what two ways can angiotensin II increase blood pressure, and what are its target tissues in each case?

cortisone. Of these three hormones secreted by the zona fasciculata, cortisol is the most abundant, accounting for about 95% of glucocorticoid activity.

Control of glucocorticoid secretion occurs via a typical negative feedback system (FIGURE 13.22). Low blood levels of glucocorticoids, mainly cortisol, stimulate neurosecretory cells in the hypothalamus to secrete corticotropin-releasing hormone (CRH). CRH (along with a low level of cortisol) promotes the release of ACTH from the anterior pituitary. ACTH flows in the blood to the adrenal cortex, where it stimulates glucocorticoid secretion. (To a much smaller extent, ACTH also stimulates secretion of aldosterone.) The discussion of stress at the end of the chapter describes how the hypothalamus also increases CRH release in response to a variety of physical and emotional stresses (see Section 13.11).

Glucocorticoids have the following effects:

- ***Protein breakdown.*** Glucocorticoids increase the rate of protein breakdown, mainly in muscle fibers, and thus increase the liberation of amino acids into the bloodstream. The amino acids may be used by body cells for synthesis of new proteins or for ATP production.
- ***Glucose formation.*** Upon stimulation by glucocorticoids, liver cells may convert certain amino acids, lactic acid, or glycerol to glucose, which neurons and other cells can use for ATP production. The formation of glucose from noncarbohydrate sources is called **gluconeogenesis**.
- ***Lipolysis.*** Glucocorticoids stimulate **lipolysis**, the breakdown of triglycerides and release of fatty acids and glycerol from adipose tissue into the blood.
- ***Resistance to stress.*** Glucocorticoids work in many ways to provide resistance to stress. The additional glucose supplied by liver cells provides tissues with a ready source of ATP to combat a range of stresses, including exercise, fasting, fright, temperature extremes, high altitude, bleeding, infection, surgery, trauma, and disease. Because glucocorticoids make blood vessels more sensitive to other hormones that cause vasoconstriction, they raise blood pressure. This effect would be an advantage in cases of severe blood loss, which causes blood pressure to drop.
- ***Anti-inflammatory effects.*** Glucocorticoids inhibit leukocytes that participate in inflammatory responses. Unfortunately, glucocorticoids also retard tissue repair and, as a result, they slow wound healing. Although high doses can cause severe mental disturbances, glucocorticoids are very useful in the treatment of chronic inflammatory disorders such as rheumatoid arthritis.
- ***Depression of immune responses.*** High doses of glucocorticoids depress immune responses. For this reason, glucocorticoids are prescribed for organ transplant recipients to retard tissue rejection by the immune system.

Adrenal Androgens

In both males and females, the adrenal cortex secretes small amounts of weak androgens. The major androgen secreted by the adrenal gland is **dehydroepiandrosterone (DHEA)** (dē-hī-drō-ep′-ē-an-DROS-ter-ōn). After puberty in males, the androgen testosterone is also released in much greater quantity by the testes, and in some tissues, testosterone is converted into another androgen called dihydrotestosterone (DHT). Thus, the amount of androgens secreted by the adrenal

FIGURE 13.22 Negative feedback regulation of glucocorticoid secretion.

A high level of CRH and a low level of glucocorticoids promote the release of ACTH, which stimulates glucocorticoid secretion by the adrenal cortex.

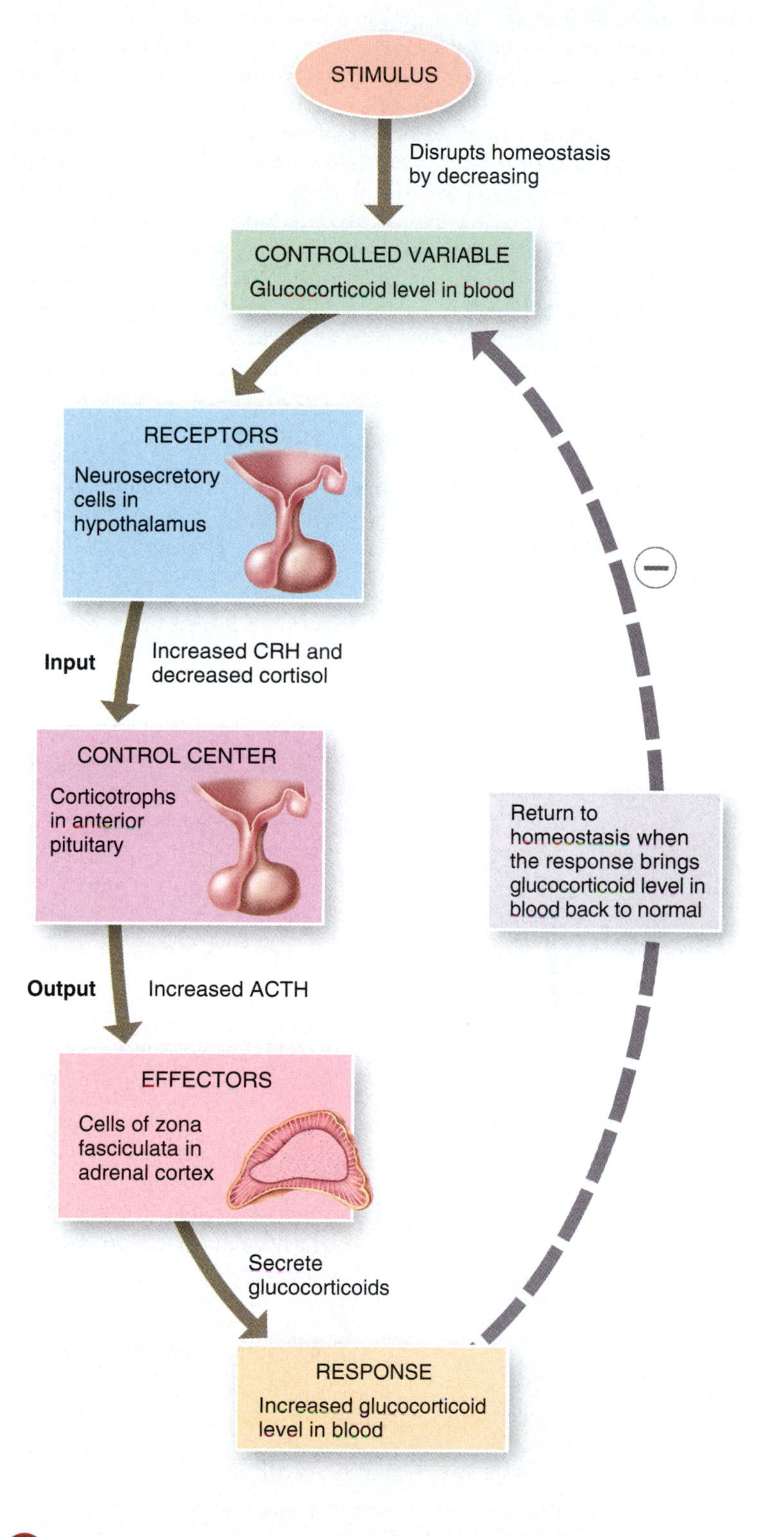

If a heart transplant patient receives prednisone (a glucocorticoid) to help prevent rejection of the transplanted tissue, will blood levels of ACTH and CRH be high or low? Explain your answer.

CLINICAL CONNECTION

Adrenal Gland Disorders

Hypersecretion of cortisol by the adrenal cortex produces **Cushing's syndrome**. Causes include a tumor of the adrenal gland that secretes cortisol, or a tumor elsewhere that secretes adrenocorticotropic hormone (ACTH), which in turn stimulates excessive secretion of cortisol. Cushing's syndrome is characterized by breakdown of muscle proteins and redistribution of body fat, resulting in spindly (thin and weak) arms and legs accompanied by a rounded moon face, a buffalo hump of fat on the upper back, and a pendulous (hanging) abdomen (**FIGURE A**). Facial skin is flushed, and the skin covering the abdomen develops stretch marks. The person also bruises easily, and wound healing is poor. The elevated level of cortisol causes hyperglycemia, osteoporosis, weakness, hypertension, increased susceptibility to infection, decreased resistance to stress, and mood swings. People who need long-term glucocorticoid therapy—for instance, to prevent rejection of a transplanted organ—may develop a cushinoid appearance.

Hyposecretion of glucocorticoids and mineralocorticoids causes **Addison's disease (chronic adrenocortical insufficiency)**. The majority of cases are autoimmune disorders in which antibodies cause adrenal cortex destruction or block binding of ACTH to its receptors. Pathogens, such as the bacterium that causes tuberculosis, also may trigger adrenal cortex destruction. Loss of cortisol results in hypoglycemia, mental lethargy, weight loss, and muscular weakness. The cortisol deficiency also leads to an elevated ACTH level in the blood due to reduced negative feedback at the anterior pituitary. Because ACTH has melanocyte-stimulating hormone (MSH) activity (see **FIGURE 13.13**), the high blood level of ACTH causes the skin to darken, giving the skin a "bronzed" appearance that is often mistaken for a suntan (**FIGURE B**). Loss of aldosterone in Addison's disease also causes a number of effects: elevated potassium and decreased sodium in the blood, low blood pressure, dehydration, decreased cardiac output, arrhythmias, and even cardiac arrest. Treatment of Addison's disease consists of replacing glucocorticoids and mineralocorticoids and increasing sodium in the diet.

A **pheochromocytoma** (fē-ō-krō′-mō-si-TŌ-ma) is a rare, usually benign tumor that develops from chromaffin cells of the adrenal medulla. This tumor causes hypersecretion of epinephrine and norepinephrine. The result is a prolonged version of the fight-or-flight response: rapid heart rate, high blood pressure, high levels of glucose in blood and urine, nervousness, sweating, and decreased gastrointestinal motility. Treatment involves surgical removal of the tumor.

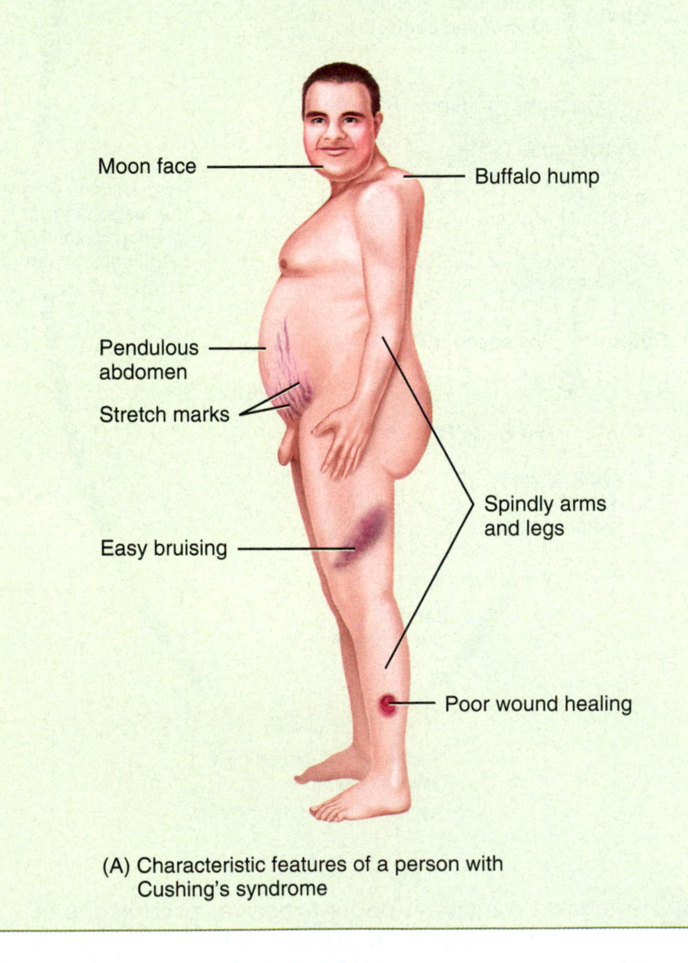

(A) Characteristic features of a person with Cushing's syndrome

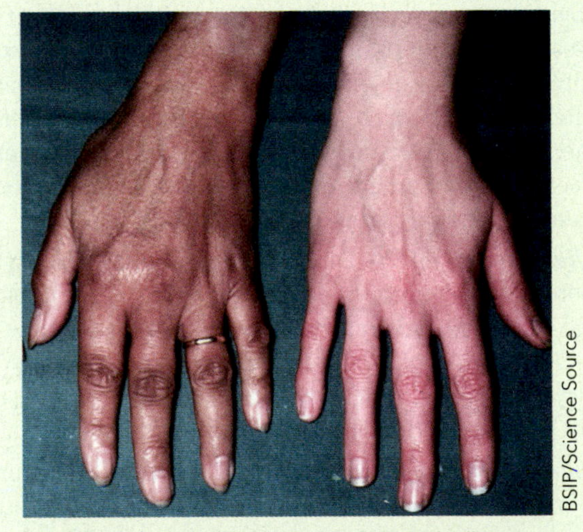

(B) Bronzed hand of person with Addison's disease (left) compared with normal hand (right) of a person with comparable ethnicity

TABLE 13.7 Summary of Adrenal Gland Hormones

Hormones and Source	Control of Secretion	Principal Actions
ADRENAL CORTEX HORMONES		
Mineralocorticoids (mainly **aldosterone**) from zona glomerulosa cells	Angiotensin II and increased blood K^+ level stimulate secretion.	Increase blood level of Na^+ and decrease blood level of K^+.
Glucocorticoids (mainly **cortisol**) from zona fasciculata cells	ACTH stimulates release; corticotropin-releasing hormone (CRH) promotes ACTH secretion in response to stress and low blood levels of glucocorticoids.	Increase protein breakdown (except in liver), stimulate gluconeogenesis and lipolysis, provide resistance to stress, dampen inflammation, and depress immune responses.
Androgens (mainly **dehydroepiandrosterone [DHEA]**) from zona reticularis cells	ACTH stimulates secretion.	Assist in early growth of axillary and pubic hair in both sexes; in females, contribute to libido and are source of estrogens after menopause.
ADRENAL MEDULLA HORMONES		
Epinephrine and **norepinephrine** from chromaffin cells	Sympathetic preganglionic neurons stimulate secretion.	Produce effects that enhance those of the sympathetic division of the autonomic nervous system (ANS) during stress.

gland in males is usually so low that their effects are insignificant. In females, however, adrenal androgens play important roles. They promote libido (sex drive) and are converted into estrogens (feminizing sex steroids) by other body tissues. After menopause, when ovarian secretion of estrogens ceases, all female estrogens come from conversion of adrenal androgens. Adrenal androgens also stimulate growth of axillary and pubic hair in males and females and contribute to the prepubertal growth spurt. Although control of adrenal androgen secretion is not fully understood, the main hormone that stimulates its secretion is ACTH.

The Adrenal Medulla Is an Extension of the Sympathetic Nervous System That Secretes Hormones

The inner region of the adrenal gland, the **adrenal medulla**, is a modified sympathetic ganglion of the autonomic nervous system (ANS). It develops from the same embryonic tissue as all other sympathetic ganglia, but its cells, which lack dendrites and axons, form clusters around large blood vessels. Rather than releasing a neurotransmitter, the cells of the adrenal medulla secrete hormones. The hormone-producing cells, called **chromaffin cells** (see FIGURE 13.20), are innervated by sympathetic preganglionic neurons of the ANS. Because the ANS exerts direct control over the chromaffin cells, hormone release can occur very quickly.

The two major hormones synthesized by the adrenal medulla are **epinephrine** and **norepinephrine (NE)**, also called *adrenaline* and *noradrenaline*, respectively. The chromaffin cells of the adrenal medulla secrete an unequal amount of these hormones—about 80% epinephrine and 20% norepinephrine. In stressful situations and during exercise, action potentials from the hypothalamus stimulate sympathetic preganglionic neurons, which in turn stimulate the chromaffin cells to secrete epinephrine and norepinephrine. These two hormones greatly augment the fight-or-flight response of the sympathetic nervous system that you learned about in Chapter 10. By increasing heart rate and force of contraction, epinephrine and norepinephrine increase the output of the heart, which increases blood pressure. They also increase blood flow to the heart, liver, skeletal muscles, and adipose tissue; dilate airways to the lungs; and increase blood levels of glucose and fatty acids.

TABLE 13.7 summarizes the hormones produced by the adrenal glands, control of their secretion, and their principal actions.

CHECKPOINT

22. What is the function of each hormone secreted by the adrenal cortex?
23. How is secretion of adrenal cortex hormones regulated?
24. How is the adrenal medulla related to the autonomic nervous system?

13.6 Pineal Gland

OBJECTIVE

- Describe the functions of the hormone produced by the pineal gland.

The **pineal gland** (PĪN-ē-al) is a small endocrine gland associated with the brain (FIGURE 13.23). It secretes **melatonin**, an amine hormone derived from serotonin. Melatonin has several functions, which include the following:

- ***Influences circadian rhythms.*** Melatonin helps regulate **circadian rhythms** (ser-KĀ-dē-an; *circa-* = about; *-dia* = day), patterns of biological activity (such as the sleep–wake cycle, secretion of certain hormones, and slight fluctuations in body temperature) that occur on a 24-hour cycle. Recall that the **suprachiasmatic nucleus (SCN)** of the hypothalamus serves as the body's internal biological clock and is ultimately responsible for establishing circadian rhythms (see Section 8.2). This nucleus receives input from the eyes (retina) and sends output to other hypothalamic nuclei, the reticular formation, and the pineal gland. The visual input to the SCN

FIGURE 13.23 Control of melatonin secretion by the pineal gland.

The suprachiasmatic nucleus of the hypothalamus alters secretion of melatonin by the pineal gland in response to visual input from the eyes.

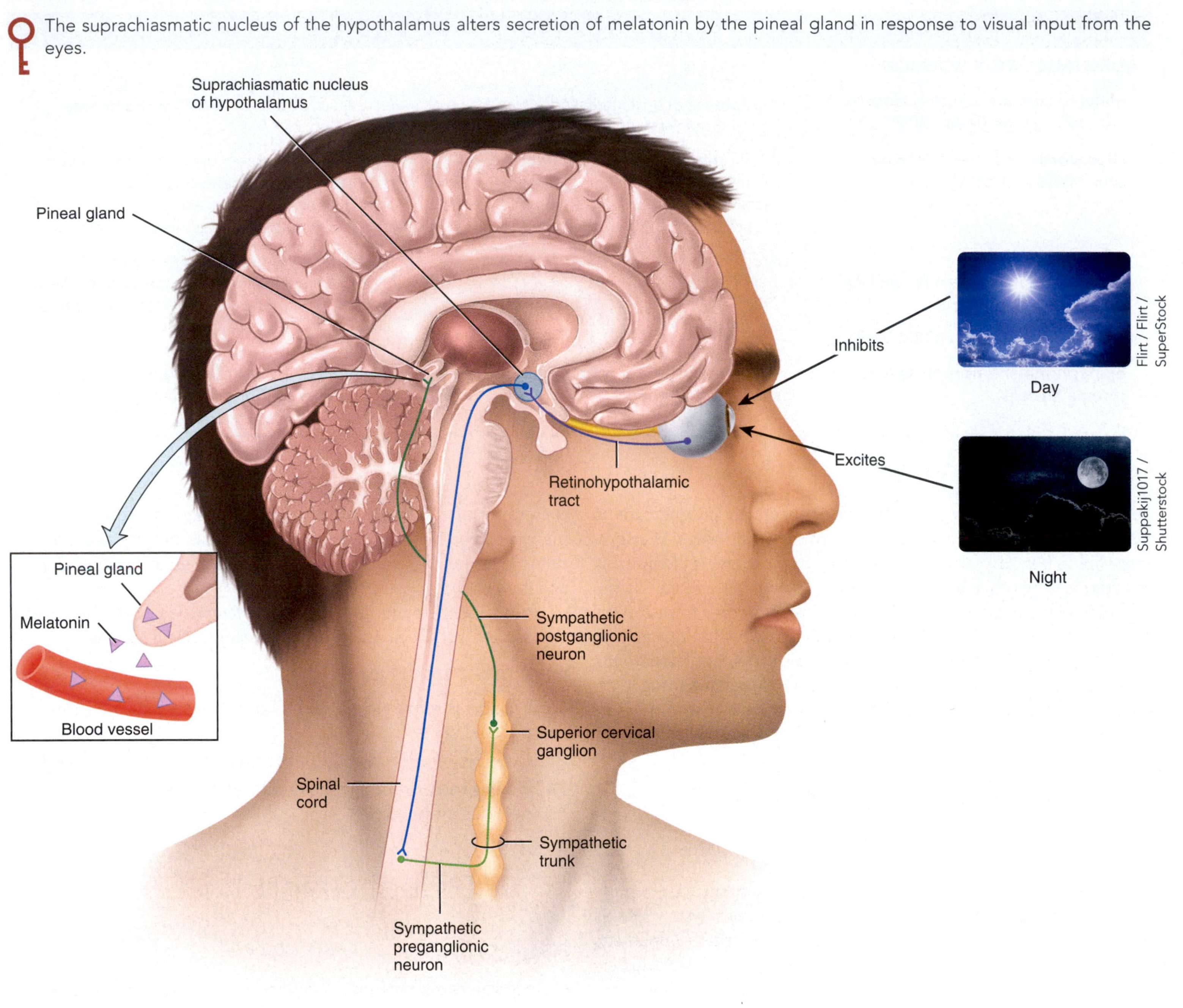

What is the functional significance of the sympathetic preganglionic and postganglionic neurons that are part of the neural pathway from the hypothalamus to the pineal gland?

is important because it entrains (synchronizes) the neurons of the SCN to the light–dark cycle associated with day and night. Without this input, the SCN still promotes biological rhythms, but the rhythms become progressively out of phase with the normal light–dark cycle because the inherent activity of the SCN creates cycles that last about 25 hours instead of 24. Hence, the SCN must receive light–dark cues from the external environment in order to create rhythms that follow a 24-hour cycle. One way that the SCN establishes circadian rhythms is by altering the amount of melatonin secreted by the pineal gland over a 24-hour period. The changing blood levels of melatonin in turn promote rhythmic variations in biological processes such as sleep, wakefulness, body temperature, and hormone secretion. The SCN does not have direct connections to the pineal gland; instead it activates the gland via connections with sympathetic neurons of the autonomic nervous system. The entire neural pathway involved in melatonin secretion occurs as follows (FIGURE 13.23): Visual input from the eyes (retina) is conveyed to the SCN via the **retinohypothalamic tract**. From the SCN, neurons extend axons to sympathetic preganglionic neurons in the thoracic spinal cord. The sympathetic preganglionic neurons in turn project axons to sympathetic postganglionic neurons in the superior cervical ganglion. From the superior cervical ganglion, axons of the sympathetic postganglionic neurons extend to the pineal gland to stimulate secretion of melatonin. Melatonin secretion occurs in a rhythmic pattern, with low levels of melatonin secreted during the day and significantly higher levels secreted at night. This is because the neural pathway responsible for melatonin secretion is inhibited by light and excited by darkness (FIGURE 13.23). The photopigment that detects the amount of environmental illumination in order to entrain the SCN to a 24-hour cycle is called **melanopsin**. Melanopsin is found in a small subset of retinal ganglion cells, and it is the axons of these neurons that project to the SCN via the retinohypothalamic tract.

- ***Induces sleep.*** As more melatonin is liberated during darkness than in light, this hormone is thought to promote sleepiness. During sleep, plasma levels of melatonin increase tenfold and then decline to a low level again before a person awakens. Small doses of melatonin given orally can induce sleep and reset daily rhythms, which might benefit workers whose shifts alternate between daylight and nighttime hours.
- ***Protects against free radicals.*** Melatonin is a potent antioxidant that protects against oxygen-derived free radicals, such as the hydroxyl (•OH) and superoxide ($O_2•^-$) radicals. Recall that a free radical is an atom or molecule with an unpaired electron in the outermost shell (see FIGURE 2.3b). Free radicals are highly reactive and disrupt homeostasis by breaking apart important body molecules, resulting in cell and tissue damage. Melatonin protects against free radicals by converting the radicals to stable products that do not react with other molecules.
- ***Inhibits reproductive functions.*** In animals that breed during specific seasons, melatonin inhibits reproductive functions, but it is unclear whether melatonin influences human reproductive function. Melatonin levels are higher in children and decline with age into adulthood, but there is no evidence that changes in melatonin secretion correlate with the onset of puberty and sexual maturation. Nevertheless, because melatonin causes atrophy of the gonads in several animal species, the possibility of adverse effects on human reproduction must be studied before its use to reset daily rhythms can be recommended.

CLINICAL CONNECTION

Seasonal Affective Disorder and Jet Lag

Seasonal affective disorder (SAD) is a type of depression that afflicts some people during the winter months, when day length is short. It is thought to be caused in part by overproduction of melatonin. Full-spectrum bright-light therapy—repeated doses of several hours of exposure to artificial light as bright as sunlight—provides relief for some people by inhibiting melatonin secretion. Three to six hours of exposure to bright light also appears to speed recovery from jet lag, the fatigue suffered by travelers who quickly cross several time zones.

CHECKPOINT

25. What is the relationship between melatonin and sleep?

13.7 Pancreas

OBJECTIVE

- Describe the functions of the hormones produced by the pancreas.

The **pancreas** is an elongated, tapered gland located behind the stomach (FIGURE 13.24). It contains both an exocrine portion and an endocrine portion. The *exocrine portion*, which forms about 99% of the pancreas, secretes fluid containing digestive enzymes into ducts. The *endocrine portion*, which comprises the remaining 1% of the pancreas, consists of clusters of cells called **pancreatic islets (islets of Langerhans)** that secrete hormones. The endocrine functions of the pancreas are discussed here, and the exocrine functions of the pancreas are described in the discussion of the digestive system in Chapter 21.

The Pancreas Secretes Several Hormones, Including Insulin and Glucagon

Each pancreatic islet contains four types of hormone-secreting cells (FIGURE 13.24):

1. **Alpha (α) cells** constitute about 17% of pancreatic islet cells and secrete **glucagon**.
2. **Beta (β) cells** constitute about 70% of pancreatic islet cells and secrete **insulin**.
3. **Delta (δ) cells** constitute about 7% of pancreatic islet cells and secrete **somatostatin**.
4. **F cells** constitute the remainder of pancreatic islet cells and secrete **pancreatic polypeptide**.

The interactions of the four pancreatic hormones are complex and not completely understood. What is known is that glucagon raises the blood glucose level, and insulin lowers it. Somatostatin acts in a paracrine manner to inhibit both insulin and glucagon release from neighboring beta and alpha cells. It may also act as a circulating hormone to slow absorption of nutrients from the gastrointestinal tract. In addition, somatostatin inhibits the secretion of growth hormone. Pancreatic polypeptide inhibits somatostatin secretion, gallbladder contraction, and secretion of digestive enzymes by the pancreas.

Insulin Stimulates Uptake of Glucose, Fatty Acids, and Amino Acids and Synthesis of Glycogen, Triglycerides, and Proteins

Insulin performs a variety of functions in the body. Most of these functions promote anabolism, the synthesis of larger, more complex molecules from smaller molecules.

FIGURE 13.24 Pancreas.

The endocrine portion of the pancreas produces glucagon, insulin, somatostatin, and pancreatic polypeptide.

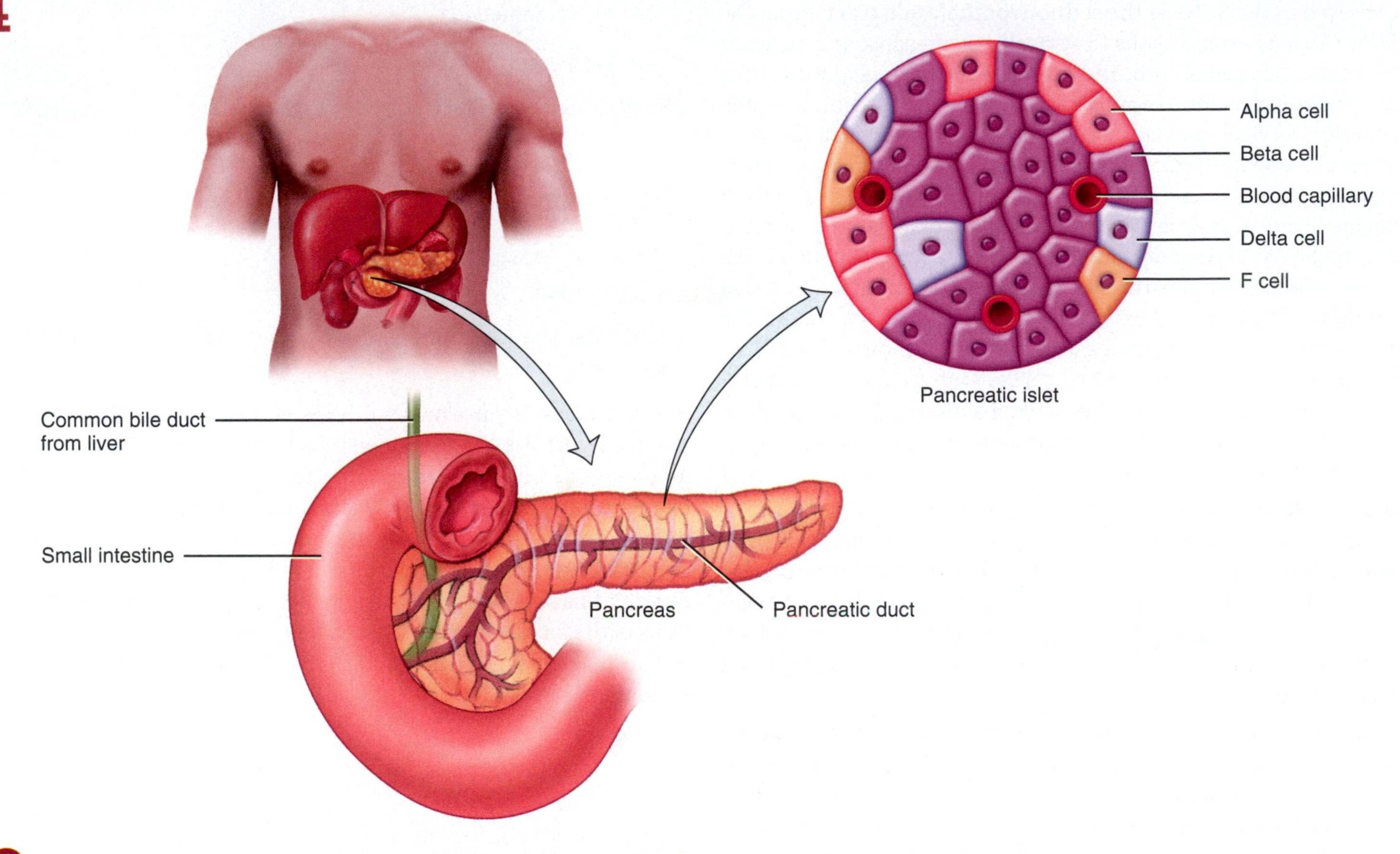

What effect does insulin have on the blood glucose level?

- ***Uptake of glucose.*** Insulin accelerates the facilitated diffusion of glucose into body cells, most notably muscle fibers and adipocytes. The mechanism involved in glucose uptake occurs as follows (**FIGURE 13.25**):

  1. Insulin binds to an insulin receptor in the plasma membrane of its target cell. Recall that the insulin receptor is a receptor tyrosine kinase.
  2. Binding of insulin to its receptor activates the tyrosine kinase components of the receptor, causing them to phosphorylate each other at tyrosine amino acids (cross phosphorylation).
  3. The phosphorylated sites on the receptor serve as docking sites where relay proteins known as **insulin-receptor substrates (IRSs)** can bind and are phosphorylated on their tyrosine amino acids.
  4. Once phosphorylated, the IRS proteins are activated and trigger intracellular pathways.
  5. One of the intracellular pathways causes the insertion of **glucose transporters (GLUTs)** from internal vesicles into the plasma membrane of the cell via exocytosis. The movement of transporters from internal vesicles to the plasma membrane is known as **translocation**. Recall that there are 14 isoforms of the glucose transporter, named GLUT1 through GLUT14 (see Section 5.4). The isoform that is insulin-dependent is **GLUT4**, which is present in muscle fibers, adipocytes, and other tissues. Note that GLUT4 is not found in neurons, hepatocytes (liver cells), or transport epithelia of the kidneys and intestine; instead, these cells have other isoforms of the glucose transporter that are already present in the plasma membrane and do not require insulin to function.
  6. Insertion of GLUT4 molecules into the plasma membrane enhances the facilitated diffusion of glucose into the cell. Once inside the cell, glucose is phosphorylated to glucose-6-phosphate, which prevents it from going back across the plasma membrane because the glucose transporter cannot transport phosphorylated glucose. The glucose can then be used in metabolic reactions occurring inside the cell. As long as insulin remains bound to its receptor, GLUT4 molecules remain in the plasma membrane. Once insulin dissociates from its receptor, however, the GLUT4 molecules are moved back to intracellular vesicles via endocytosis, and glucose uptake by the cell markedly decreases.

- ***Formation of glycogen.*** Insulin stimulates the conversion of glucose into glycogen (glycogenesis) in liver and skeletal muscle. Insulin accomplishes this function by activating **glycogen synthase**, the key enzyme involved in glycogen synthesis.
- ***Inhibition of processes that release glucose.*** Insulin inhibits catabolic processes that result in the release of glucose: In liver and

FIGURE 13.25 Mechanism of glucose uptake by a target cell in response to insulin. GLUT4 = glucose transporter isoform 4.

Insulin promotes glucose uptake into target cells by causing the translocation of GLUT4 molecules from internal vesicles to the plasma membrane.

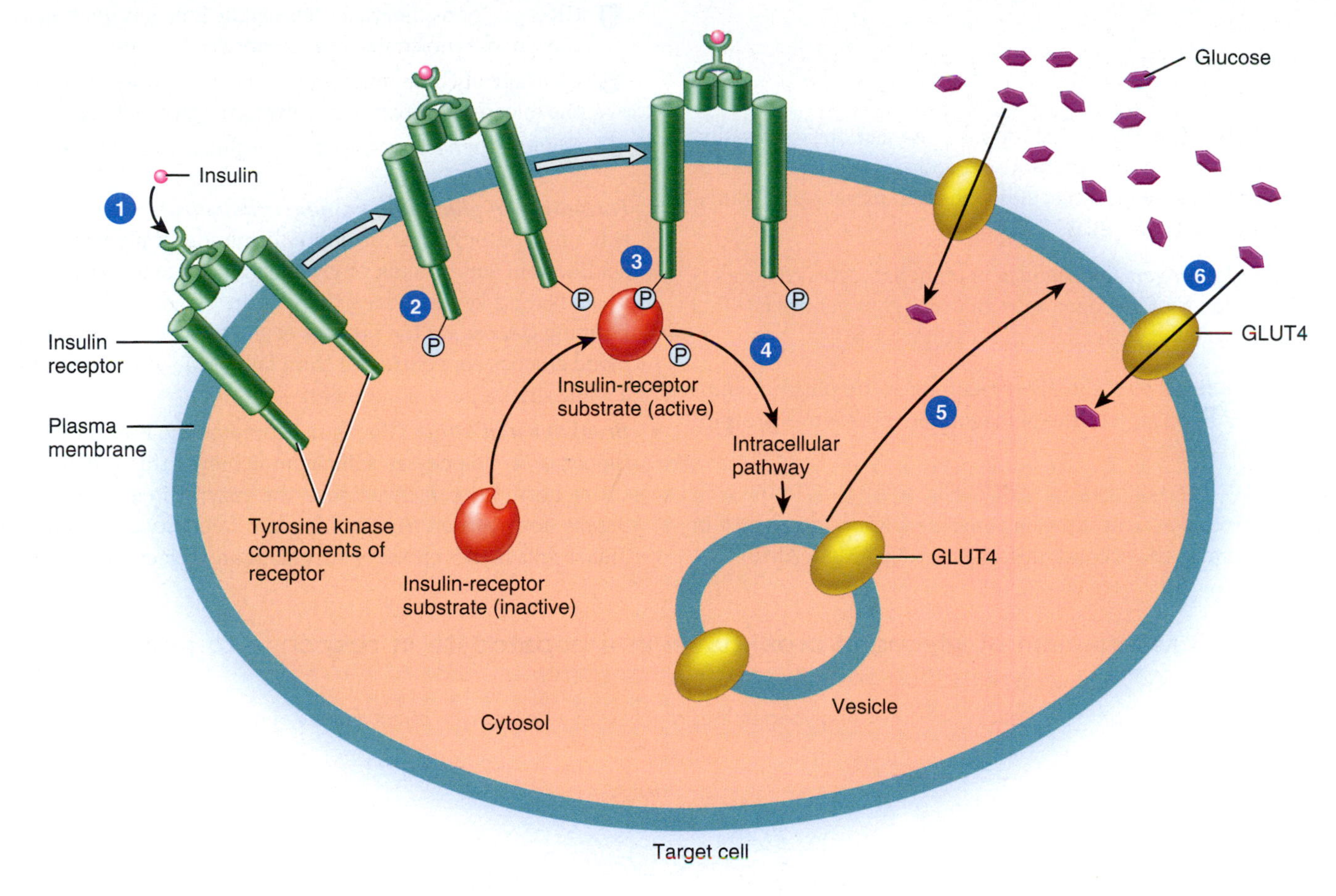

What will happen to the GLUT4 molecules when insulin dissociates from its receptor?

skeletal muscle, insulin prevents the breakdown of glycogen into glucose (glycogenolysis), and in the liver it suppresses the formation of glucose from noncarbohydrate sources such as lactic acid, certain amino acids, or glycerol (gluconeogenesis). Insulin achieves these functions by inhibiting enzymes required for glycogenolysis and gluconeogenesis.

- ***Uptake of fatty acids and formation of triglycerides.*** Insulin promotes uptake of fatty acids into adipocytes. Although fatty acids can diffuse freely across the plasma membrane of an adipocyte, transporters help mediate this process, and insulin stimulates the translocation of fatty acid transporters to the plasma membrane. Once inside the cell, the fatty acids are combined with glycerol to form triglycerides. Insulin also promotes the uptake of glucose into adipocytes by stimulating the translocation of GLUT4 molecules to the plasma membrane. After entering the cell, the glucose molecules can be used to synthesize triglycerides via two pathways: (1) glucose is converted to glycerol and then combined with fatty acids to form triglycerides or (2) glucose is converted into fatty acids and then combined with glycerol to form triglycerides. The formation of triglycerides (lipogenesis) from fatty acids and glycerol or from glucose involves many steps; insulin promotes lipogenesis by activating several of the enzymes that catalyze these steps. Once the triglycerides are formed, they are stored as fat within the cytosol of the adipocyte. Given that insulin promotes the synthesis of triglycerides (lipogenesis), it should not be a surprise that insulin also suppresses the breakdown of triglycerides (lipolysis). Insulin does this by inhibiting **hormone-sensitive lipase**, an enzyme in adipose tissue that breaks down stored triglycerides into free fatty acids and glycerol.
- ***Uptake of amino acids and formation of proteins.*** Insulin increases the uptake of amino acids into skeletal muscle fibers as well as other body cells. One way that insulin accomplishes this function is by stimulating the translocation of amino acid transporters to the plasma membrane. Another way is by activating genes, resulting in the formation of new amino acid transporters that are then inserted into the plasma membrane. In addition to promoting uptake of amino acids, insulin promotes the synthesis of proteins from amino acids that have entered the cell. Insulin achieves this goal by enhancing the activity of initiation factors and enzymes that are involved in protein synthesis. Another function of insulin is to

inhibit protein catabolism: Insulin suppresses activity in lysosomes and proteasomes, organelles of the cell that degrade proteins. Because of all of its effects on protein metabolism, insulin contributes to overall body growth.

Glucagon Has Functions That Are Antagonistic to Those of Insulin

Glucagon performs functions that oppose the actions of insulin. Hence, most of these functions promote catabolism, the breakdown of larger molecules into smaller molecules. The main target cell of glucagon is the hepatocyte (liver cell).

- ***Breakdown of glycogen.*** Glucagon acts on hepatocytes to accelerate the breakdown of glycogen into glucose (glycogenolysis). This occurs via the following mechanism (**FIGURE 13.26**):

  1. Glucagon binds to a glucagon receptor in the plasma membrane of a hepatocyte. The glucagon receptor is a G protein-coupled receptor.
  2. Binding of glucagon to its receptor activates a G_s protein.
  3. The G_s protein activates the enzyme adenylyl cyclase, which in turn generates the second messenger cyclic AMP (cAMP).
  4. Cyclic AMP binds to and activates the enzyme protein kinase A (PKA).
  5. PKA phosphorylates the cytosolic enzyme **glycogen phosphorylase**, which becomes activated.
  6. Glycogen phosphorylase cleaves the linkages holding together the glucose molecules that comprise glycogen.
  7. Cleavage of these linkages by glycogen phosphorylase causes the release of glucose from glycogen (glycogenolysis).
  8. Glucose then leaves the cell via a glucose transporter (GLUT2).

- ***Formation of glucose from noncarbohydrate sources.*** Glucagon acts on hepatocytes to promote formation of glucose from lactic acid, certain amino acids, or glycerol (gluconeogenesis). This occurs by activating key enzymes that are involved in gluconeogenesis. Forming glucose via the process of gluconeogenesis helps keep the blood glucose concentration from falling too low when it has been a while since a person has eaten.
- ***Breakdown of lipids.*** Glucagon promotes the breakdown of lipids (lipolysis) in adipocytes. Glucagon achieves this function by activating hormone-sensitive lipase, the enzyme in adipose tissue that breaks down stored triglycerides into free fatty acids and glycerol. This results in the release of fatty acids and glycerol from adipocytes

FIGURE 13.26 Mechanism of glycogen breakdown in a hepatocyte in response to glucagon. GLUT2 = glucose transporter isoform 2.

Glucagon promotes glycogenolysis (breakdown of glycogen into glucose) in a hepatocyte by ultimately stimulating the enzyme glycogen phosphorylase.

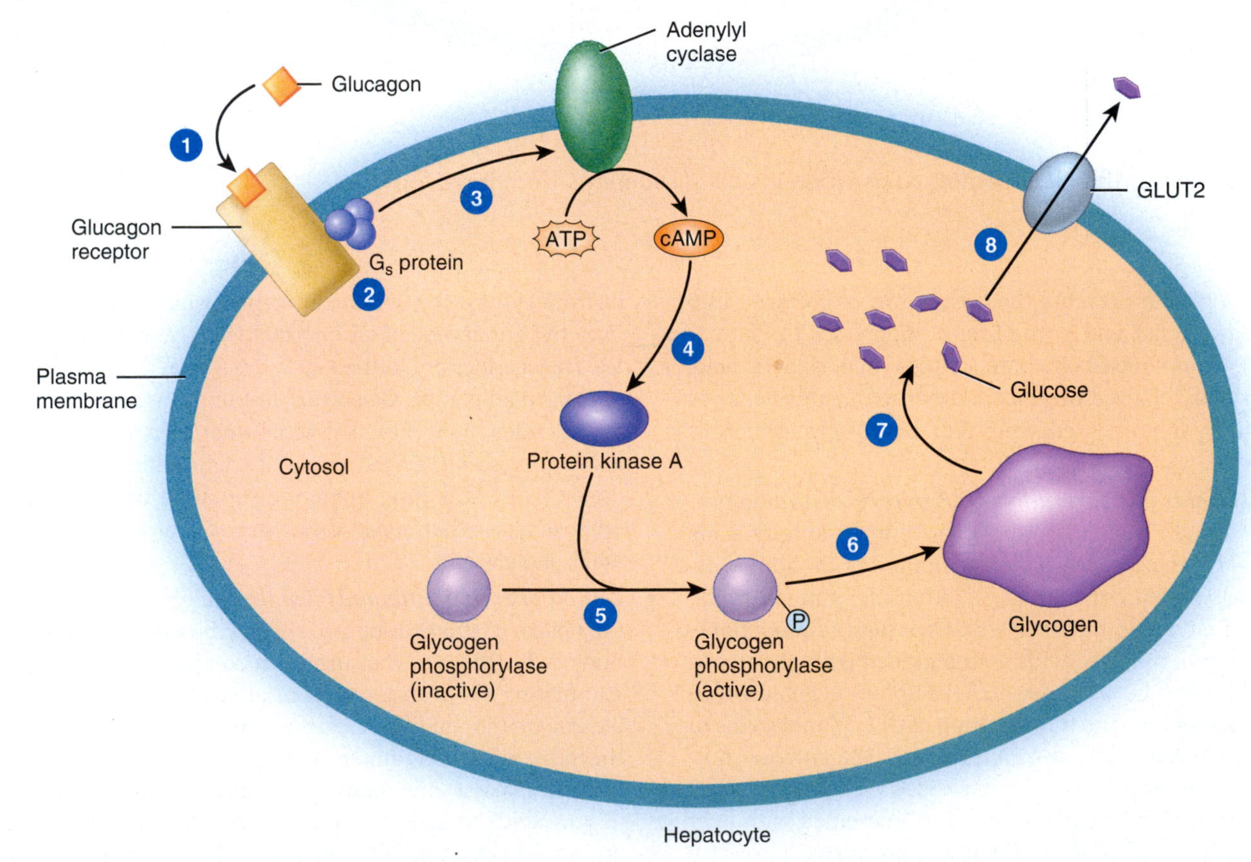

Why is the glucose transporter (GLUT2) in step 8 transporting glucose out of the cell instead of into the cell?

into the blood. The fatty acids and glycerol in turn can be catabolized by cells for energy.

- ***Inhibition of protein synthesis.*** Glucagon inhibits protein synthesis in hepatocytes. Glucagon does this by suppressing enzymes that carry out protein synthesis. As a result, glucagon provides a free pool of amino acids that can be used by liver cells for energy or converted to glucose (gluconeogenesis).

Insulin and Glucagon Secretion Are Regulated by the Blood Glucose Concentration and Other Factors

The principal action of insulin is to lower the blood glucose level when it is too high (rises above normal). Glucagon, on the other hand, increases the blood glucose level when it is too low (falls below normal). The level of blood glucose controls secretion of insulin and glucagon via negative feedback (**FIGURE 13.27**):

1. Hyperglycemia (high blood glucose) stimulates secretion of insulin by beta cells of the pancreatic islets.
2. Insulin promotes uptake of glucose by body cells and synthesis of glycogen (glycogenesis) in liver and skeletal muscle.
3. As a result, blood glucose level falls.
4. If blood glucose level drops below normal, low blood glucose inhibits release of insulin (negative feedback) and stimulates release of glucagon.
5. Hypoglycemia (low blood glucose) stimulates secretion of glucagon from alpha cells of the pancreatic islets.
6. Glucagon acts on hepatocytes to accelerate breakdown of glycogen into glucose (glycogenolysis) and to promote formation of glucose from lactic acid, certain amino acids, or glycerol (gluconeogenesis).
7. As a result, hepatocytes release glucose into the blood more rapidly, and blood glucose level rises.
8. If blood glucose continues to rise, high blood glucose level inhibits release of glucagon (negative feedback) and stimulates release of insulin.

Through this negative feedback system, insulin and glucagon function antagonistically to maintain the normal blood glucose level at about 70 to 110 milligrams per deciliter (mg/dL).

Although blood glucose level is the most important regulator of insulin and glucagon, several hormones and neurotransmitters also regulate the release of these two hormones. In addition to the

FIGURE 13.27 Negative feedback regulation of the secretion of insulin (blue arrows) and glucagon (orange arrows).

Hyperglycemia stimulates secrete of insulin; hypoglycemia stimulates secretion of glucagon.

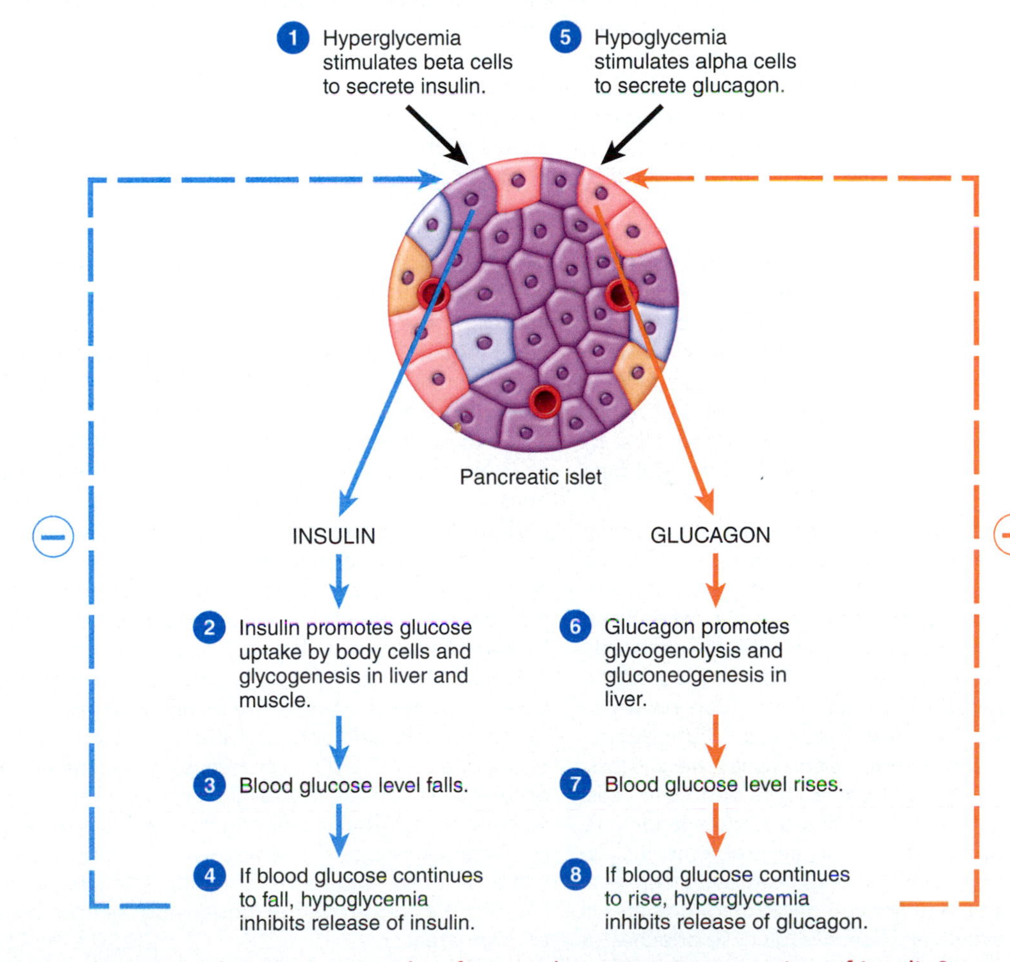

Other than hyperglycemia, what are some other factors that promote secretion of insulin?

responses to blood glucose level just described, glucagon stimulates insulin release directly; insulin has the opposite effect, suppressing glucagon secretion. As blood glucose level declines and less insulin is secreted, the alpha cells of the pancreas are released from the inhibitory effect of insulin so they can secrete more glucagon. Indirectly, growth hormone (GH) and adrenocorticotropic hormone (ACTH) stimulate secretion of insulin because they act to elevate blood glucose.

Insulin secretion is also stimulated by:

- Increased activity of the parasympathetic nerves that supply the pancreatic islets.
- An elevated blood amino acid level, which occurs after a protein-containing meal.
- The hormones glucose-dependent insulinotropic peptide (GIP) and glucagon-like peptide (GLP), which are released by the small intestine in response to the presence of food; these hormones, collectively called **incretins**, provide a type of feedforward control that anticipates the increase in blood glucose that occurs after a typical meal.

Thus, digestion and absorption of food containing both carbohydrates and proteins provide strong stimulation for insulin release.

Glucagon secretion is also stimulated by:

- Increased activity of the sympathetic nerves that supply the pancreatic islets, as occurs during exercise.
- A rise in blood amino acids if blood glucose level is low, which could occur after a meal that contained mainly protein.

CLINICAL CONNECTION

Diabetes Mellitus

The most common endocrine disorder is **diabetes mellitus** (MEL-i-tus), caused by an inability to produce or use insulin. Diabetes mellitus is the fourth leading cause of death by disease in the United States, primarily because of its damage to the cardiovascular system. Because insulin is unavailable to aid transport of glucose into body cells, blood glucose level is high and glucose "spills" into the urine, a condition known as **glucosuria**. Hallmarks of diabetes mellitus are the three "polys": polyuria, polydipsia, and polyphagia. **Polyuria**, excessive urine production, occurs because the presence of glucose in urine osmotically attracts water, causing large volumes of urine to form. **Polydipsia**, excessive thirst, occurs due to the large water loss from the body via urine, resulting in dehydration. **Polyphagia**, excessive eating, occurs because of the inability of cells to use glucose, resulting in an increased appetite.

Both genetic and environmental factors contribute to onset of the two types of diabetes mellitus—type 1 and type 2—but the exact mechanisms are still unknown. In **type 1 diabetes**, the insulin level is low because the person's immune system destroys the pancreatic beta cells. This type of diabetes was previously known as ***insulin-dependent diabetes mellitus (IDDM)*** because insulin injections are required to prevent death. Most commonly, type 1 diabetes develops in people younger than age 20, though it persists throughout life. By the time symptoms of type 1 diabetes arise, 80–90% of the islet beta cells have been destroyed.

The cellular metabolism of an untreated type 1 diabetic is similar to that of a starving person. Because insulin is not present to aid the entry of glucose into body cells, most cells use fatty acids to produce ATP. Stores of triglycerides in adipose tissue are catabolized to yield fatty acids and glycerol. The by-products of fatty acid breakdown—organic acids called ketones or ketone bodies—accumulate. Buildup of ketones causes blood pH to fall, a condition known as **ketoacidosis**. Unless treated quickly, ketoacidosis can cause death.

The breakdown of stored triglycerides also causes weight loss. As lipids are transported by the blood from storage depots to cells, lipid particles are deposited on the walls of blood vessels, leading to atherosclerosis and a multitude of cardiovascular problems, including cerebrovascular insufficiency, ischemic heart disease, peripheral vascular disease, and gangrene. A major complication of diabetes is loss of vision due either to cataracts (excessive glucose attaches to lens proteins, causing cloudiness) or to damage to blood vessels of the retina. Severe kidney problems also may result from damage to renal blood vessels.

Type 1 diabetes is treated through self-monitoring of blood glucose level, regular meals containing 45–50% carbohydrates and less than 30% fats, exercise, and periodic insulin injections. Several implantable pumps are available to provide insulin without the need for repeated injections.

Type 2 diabetes, formerly called ***non-insulin-dependent diabetes mellitus (NIDDM)***, is much more common than type 1, representing more than 90% of all cases. Type 2 diabetes most often occurs in obese people who are over age 35. However, the number of obese children and teenagers with type 2 diabetes is increasing. Clinical symptoms are mild, and the high glucose levels in the blood can often be controlled by diet, exercise, and weight loss. Exercise helps control type 2 diabetes because skeletal muscle contractions promote the translocation of GLUT4 molecules to the plasma membranes of skeletal muscle fibers. Although some type 2 diabetics need insulin, many have a sufficient amount (or even a surplus) of insulin in the blood. For these people, diabetes arises not from a shortage of insulin but because target cells become less sensitive to it due to down-regulation of insulin receptors.

Hyperinsulinism most often results when a diabetic injects too much insulin. The main symptom is hypoglycemia (decreased blood glucose level), which occurs because the excess insulin stimulates too much uptake of glucose by body cells. The resulting hypoglycemia stimulates the secretion of epinephrine, glucagon, and human growth hormone. As a consequence, anxiety, sweating, tremor, increased heart rate, hunger, and weakness occur. When blood glucose falls, brain cells are deprived of the steady supply of glucose they need to function effectively. Severe hypoglycemia leads to mental disorientation, convulsions, unconsciousness, and shock. Shock due to an insulin overdose is termed **insulin shock**. Death can occur quickly unless blood glucose level is raised. From a clinical standpoint, a diabetic suffering from either a hyperglycemia or a hypoglycemia crisis can have very similar symptoms—mental changes, coma, seizures, and so on. It is important to quickly and correctly identify the cause of the underlying symptoms and treat them appropriately.

TABLE 13.8 Summary of Pancreatic Islet Hormones

Hormone and Source	Control of Secretion	Principal Actions
Glucagon from alpha cells of pancreatic islets	Decreased blood level of glucose, exercise, and meals containing mainly protein; somatostatin and insulin inhibit secretion.	Raises blood glucose level by accelerating breakdown of glycogen into glucose in liver (glycogenolysis), converting other nutrients into glucose in liver (gluconeogenesis), and releasing glucose into the blood.
Insulin from beta cells of pancreatic islets	Increased blood level of glucose, increased parasympathetic activity; elevated blood amino acid level, glucagon, GIP, GLP, GH, and ACTH stimulate secretion; somatostatin inhibits secretion.	Lowers blood glucose level by accelerating transport of glucose into cells, converting glucose into glycogen (glycogenesis), and decreasing glycogenolysis and gluconeogenesis; also increases lipogenesis and stimulates protein synthesis.
Somatostatin from delta cells of pancreatic islets	Pancreatic polypeptide inhibits secretion.	Inhibits secretion of insulin and glucagon and slows absorption of nutrients from the gastrointestinal tract.
Pancreatic polypeptide from F cells of pancreatic islets	Meals containing protein, fasting, exercise, and acute hypoglycemia stimulate secretion; somatostatin and elevated blood glucose level inhibit secretion.	Inhibits somatostatin secretion, gallbladder contraction, and secretion of pancreatic digestive enzymes.

TABLE 13.8 summarizes the hormones produced by the pancreas, control of their secretion, and their principal actions.

CHECKPOINT

26. How are the functions of insulin different from those of glucagon?
27. What are some specific examples of enzymes that are regulated by insulin? By glucagon?
28. How are the blood levels of insulin and glucagon controlled?
29. What are incretins?

13.8 Ovaries and Testes

OBJECTIVE

- Describe the functions of female and male reproductive hormones.

Gonads are the organs that produce gametes—eggs (oocytes) in females and sperm in males. In addition to their reproductive function, the gonads secrete hormones. The female gonads, the **ovaries**, are paired oval structures located in the female pelvic cavity. They produce several steroid hormones, including two **estrogens** (estradiol and estrone) and **progesterone**. These female sex hormones, along with FSH and LH from the anterior pituitary, regulate the menstrual cycle and maintain pregnancy. Estrogens also promote enlargement of the breasts and widening of the hips at puberty, and help maintain these female secondary sex characteristics. In addition to estrogens and progesterone, the ovaries produce the hormones inhibin and relaxin. **Inhibin** inhibits secretion of FSH. **Relaxin** inhibits contractions of the uterus, making it easier for a fertilized egg to implant in the uterine wall.

The male gonads, the **testes**, are oval glands located in the scrotum. The main hormone produced and secreted by the testes is the androgen **testosterone** (the principal male sex hormone). Testosterone stimulates descent of the testes before birth, regulates production of sperm, and stimulates the development and maintenance of male secondary sex characteristics, such as beard growth and deepening of the voice. The testes also produce inhibin, which inhibits secretion of FSH. The detailed functions of the ovaries and testes and the specific roles of the sex hormones are discussed in Chapter 23.

CHECKPOINT

30. What are the functions of estrogens? Of testosterone?

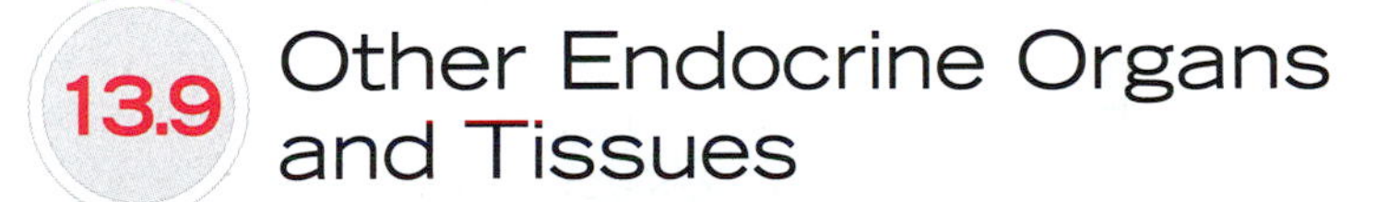

13.9 Other Endocrine Organs and Tissues

OBJECTIVE

- Describe the functions of the hormones secreted by cells in organs and tissues other than endocrine glands.

Cells in organs and tissues other than those usually classified as endocrine glands have an endocrine function and secrete hormones. You have already learned about several of these components: the hypothalamus, pancreas, ovaries, and testes. In this section, you will learn about the remainder of these endocrine organs and tissues: the skin, thymus, heart, liver, stomach, small intestine, kidneys, adipose tissue, and placenta.

The Skin Secretes Cholecalciferol

The skin produces **cholecalciferol**, or *vitamin* D_3, a substance that plays a role in the synthesis of calcitriol—the active form of vitamin D. Calcitriol synthesis involves several steps; the first step occurs as 7-dehydrocholesterol, a substance present in the skin, is converted

Critical Thinking

Endocrine Disruptors in the Environment

Endocrine disruptors are chemicals that interfere with the endocrine system. These substances alter endocrine signals, causing changes in normal body functions. Endocrine disruptors are associated with adverse health effects such as developmental disorders, organ dysfunction, cancer, and autoimmune diseases. These effects have occurred in laboratory animals, wildlife, and humans after exposure to endocrine disruptors, even at very low concentrations.

SOME THINGS TO KEEP IN MIND:

Endocrine disruptors can affect the endocrine system in a variety of ways: (1) They can mimic the natural hormone; (2) they can block the natural hormone from binding to its receptor; or (3) they can inhibit or stimulate the gland that produces the natural hormone. Examples of endocrine disruptors include atrazine, bisphenol A (BPA), and di(2-ethylhexyl) phthalate (DEHP).

Fotokostic / Shutterstock

R. Mackay Photography llc /Shutterstock

SOME INTERESTING FACTS:

Atrazine is an agricultural herbicide. It is used in the United States to control the growth of unwanted weeds and grass. In animals and humans, atrazine has estrogenic activity: It stimulates aromatase, an enzyme involved in estrogen synthesis. Studies have shown that atrazine causes hermaphroditism in frogs. In addition, in one Illinois town, relatively high levels of atrazine in drinking water have been linked to irregular menstrual cycles in women.

Bisphenol A (BPA) is used in the production of polycarbonate plastics (found in certain plastic bottles) and epoxy resins (such as the white lining in some canned foods). BPA has been shown to mimic estrogen: It binds to the same estrogen receptor as the endogenous hormone. Studies have indicated that BPA can have effects on the reproductive system, the brain, and behavior. It has also been linked to cancer, obesity, thyroid dysfunction, and heart disease.

Di(2-ethylhexyl) phthalate (DEHP) is found in some food packaging, certain toys, some medical devices with polyvinylchloride (PVC), some building materials, and certain fragrances. Studies indicate that, in rats, DEHP reduces testosterone synthesis during sexual differentiation, leading to impaired development of testes and production of sperm. DEHP may pose a potential risk to human male development.

How could atrazine cause changes in the menstrual cycle in women?

Could atrazine, BPA, or DEHP affect the development of a baby during gestation?

Why are endocrine disruptors able to cause adverse effects at low concentrations?

to cholecalciferol when the skin is exposed to ultraviolet (UV) rays in sunlight (**FIGURE 13.28**). After it is formed, cholecalciferol circulates to the liver, where it is converted to **25-hydroxycholecalciferol**, also known as *25-hydroxy vitamin* D_3. The 25-hydroxycholecalciferol in turn circulates to the kidneys, where it is converted to **calcitriol**, also referred to as *1,25-dihydroxycholecalciferol* or *1,25-dihydroxy vitamin* D_3. Calcitriol then acts on the small intestine to increase the absorption of calcium (Ca^{2+}) and phosphate (HPO_4^{2-}) ions from food in the intestinal tract into the blood. Because cholecalciferol, 25-hydroxycholecalciferol, and calcitriol travel through the bloodstream to act on their target organs, they are all considered to be hormones. Note that only a small amount of exposure to UV light (about 10 to 15 minutes at least three times a week) is required for vitamin D synthesis.

FIGURE 13.28 Steps involved in the synthesis of calcitriol.

Synthesis of calcitriol, the active form of vitamin D, begins in the skin, continues in the liver, and is completed in the kidneys.

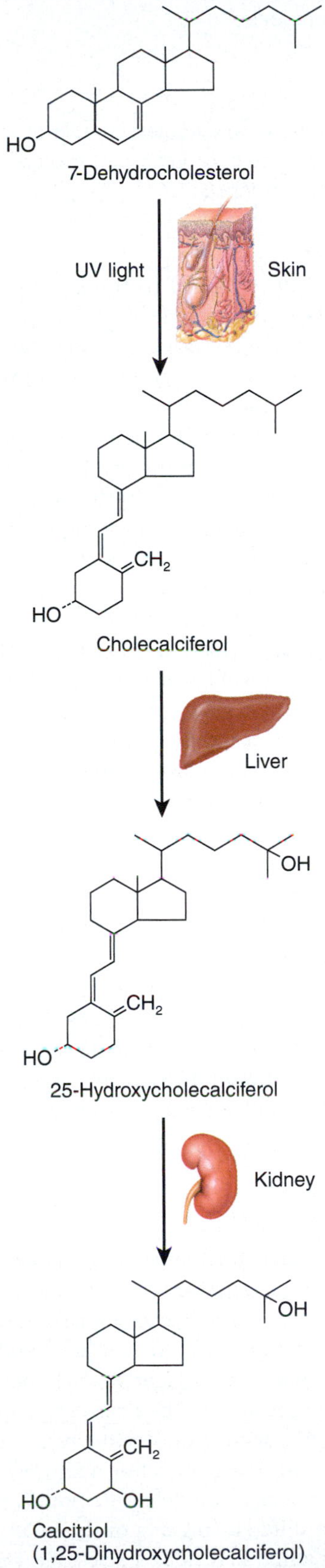

What is the functional significance of calcitriol?

The Thymus Secretes Hormones That Promote Immune Function

The **thymus** is located just above the heart (see FIGURE 13.1). Two major hormones are produced by the thymus: **thymosin** and **thymopoietin**. These thymic hormones promote the maturation of T cells (a type of leukocyte that destroys microbes and other foreign substances). Thymic hormones may also play a role in slowing down the aging process. You will learn more about thymic hormones in the discussion of the immune system in Chapter 17.

The Heart Secretes Atrial Natriuretic Peptide

The heart produces the hormone **atrial natriuretic peptide (ANP)**, which inhibits reabsorption of sodium ions (Na^+) and water by the kidneys so more is lost into the urine. These actions increase excretion of Na^+ in urine and increase urine output, which decreases blood volume and blood pressure. The functions of ANP are described in greater detail in the discussion of the urinary system in Chapter 19.

The Liver Produces Two Hormones

The liver produces the hormone **25-hydroxycholecalciferol**, which is involved in the synthesis of calcitriol (see FIGURE 13.28). Another hormone produced by the liver is **thrombopoietin**, which stimulates the production of platelets. The role of thrombopoietin in platelet formation is explained more thoroughly in the discussion of the blood in Chapter 16.

The Stomach and Small Intestine Secrete Hormones That Regulate Digestive Activities

The stomach produces the hormone **gastrin**, which stimulates secretion of gastric juice and increases motility (contractions) of the stomach. Gastrin also promotes reflexes that trigger movement of digested materials through the intestines. Another hormone produced by the stomach, **ghrelin**, plays a role in stimulating appetite.

The small intestine produces four hormones: secretin, cholecystokinin, glucose-dependent insulinotropic peptide, and glucagon-like peptide. **Secretin** stimulates secretion of pancreatic juice that is rich in bicarbonate ions (HCO_3^-). **Cholecystokinin (CCK)** stimulates secretion of pancreatic juice that is rich in digestive enzymes, promotes contraction of the gallbladder to release stored bile, and brings about a feeling of fullness after eating. **Glucose-dependent insulinotropic peptide (GIP)** and **glucagon-like peptide (GLP)** stimulate the release of insulin from the pancreas. The hormones secreted by the stomach and small intestine are fully explored in the discussion of the digestive system in Chapter 21.

The Kidneys Secrete Calcitriol and Erythropoietin

The kidneys produce the hormone **calcitriol**, which is the active form of vitamin D. Calcitriol increases the absorption of calcium (Ca^{2+}) and phosphate (HPO_4^{2-}) ions from food in the small intestine into the bloodstream. Recall that synthesis of calcitriol begins in the skin, continues in the liver, and ends in the kidneys (see FIGURE 13.28). Another hormone produced by the kidneys is **erythropoietin (EPO)**, which increases the rate of erythrocyte production. The kidneys secrete EPO in response to low levels of oxygen in the tissues. The function of EPO is described in more detail in the discussion of the blood in Chapter 16.

Adipose Tissue Secretes Leptin

Adipose tissue produces **leptin**, a hormone that helps decrease total body fat. As fat stores increase, more leptin is secreted by adipose tissue. Leptin in turn acts on the brain to suppress appetite, thereby reducing food intake. Leptin also increases energy expenditure, which helps break down body fat reserves. You will learn more about the functions of leptin in the discussion of food intake in Chapter 22.

The Placenta Secretes Several Hormones That Help Maintain Pregnancy and Prepare the Maternal Body for Birth

The placenta forms during pregnancy and serves as the site of exchange of nutrients and wastes between the mother and fetus. Several hormones are produced by the placenta: estrogens, progesterone, relaxin, human chorionic somatomammotropin, and corticotropin-releasing hormone. Estrogens and progesterone maintain the lining of the uterus during pregnancy. Relaxin increases the flexibility of the pubic symphysis and helps dilate the uterine cervix during labor and delivery. These actions help ease the baby's passage by enlarging the birth canal. **Human chorionic somatomammotropin** helps prepare the mammary glands of the breasts for lactation. Corticotropin-releasing hormone (CRH) is thought to play a role in establishing the timing of birth. The functions of the placenta and the hormones it produces are described in greater detail in the discussion of the female reproductive system in Chapter 23.

CHECKPOINT

31. What steps are involved in the synthesis of calcitriol?
32. Which hormones are produced by the small intestine?
33. Why is leptin important?
34. How does relaxin contribute to labor and delivery?

13.10 Endocrine Control of Growth

OBJECTIVES

- Identify the cellular activities that occur during the growth process.
- Explain how bones grow in length and in width.
- Describe the hormones that help control growth of bone.
- Discuss factors other than hormones that contribute to bone growth.

Growth is a process that involves net protein synthesis and cell division. As a result, cells of the body increase in size **(hypertrophy)** and increase in number **(hyperplasia)**. Growth is particularly evident in the skeletal system, as the bones of the body grow in length and width. Multiple hormones participate in the control of growth. Other factors that help control growth include genetics and dietary intake of minerals and vitamins. Although growth occurs in all tissues of the body, this section focuses on the factors that control growth of bone, with an emphasis on the contributions provided by hormones. To appreciate how bone growth is regulated, you must first understand how bone grows and how it remodels.

Bones Grow in Length and in Thickness

During infancy, childhood, and adolescence, long bones grow in length and bones throughout the body grow in thickness. During adulthood, bones may continue to thicken.

Growth in Length

Bone growth in length is related to the activity of the **epiphyseal (growth) plate**, a layer of cartilage located at the junction between the diaphysis and each epiphysis of a growing bone (**FIGURE 13.29a**). Within the epiphyseal plate is a group of **chondrocytes** (cartilage-producing cells) that are constantly dividing. As a bone grows in length, new chondrocytes are formed on the epiphyseal side of the plate, while old chondrocytes on the diaphyseal side of the plate are replaced by bone (**FIGURE 13.29a**). In this way the thickness of the epiphyseal plate remains relatively constant, but the bone on the diaphyseal side increases in length. At about age 18 in females and 21 in males, the epiphyseal plates close; the epiphyseal cartilage cells stop dividing, and bone replaces all of the cartilage. The epiphyseal plate fades, leaving a bony structure called the **epiphyseal line**. The appearance of the epiphyseal line signifies that the bone has stopped growing in length. If a bone fracture damages the epiphyseal plate, the fractured bone may be shorter than normal once adult stature is reached because damage to cartilage, which is avascular (lacks blood vessels), accelerates closure of the epiphyseal plate, thus inhibiting lengthwise growth of the bone.

Growth in Thickness

As long bones lengthen, they also grow in thickness (width). Growth in thickness occurs in the following way (**FIGURE 13.29b**): At the bone's outer surface, osteoblasts secrete bone extracellular matrix. At the same time, osteoclasts deep within the bone destroy bone tissue lining the marrow cavity. Bone destruction on the inside of the bone by osteoclasts occurs at a slower rate than bone formation on the outside of the bone. Thus, the marrow cavity enlarges as the bone increases in thickness.

Bones Undergo Remodeling

Bone forms before birth but continually renews itself thereafter. **Bone remodeling** is the ongoing replacement of old bone tissue by new bone tissue. It involves bone resorption (the removal of minerals and collagen fibers from bone by osteoclasts) and bone deposition (the addition of minerals and collagen fibers to bone by osteoblasts). Thus, bone resorption results in the destruction of bone extracellular matrix, while bone deposition results in the formation of bone extracellular matrix. At any given time, about 5% of the total bone mass in the body is being remodeled. Remodeling also takes place at different rates in different regions of the body. Even after bones have reached their adult shapes and sizes, old bone is continually destroyed and new bone is formed in its place. Remodeling also

FIGURE 13.29 **Bone growth.**

Bones can grow in length and in thickness (width).

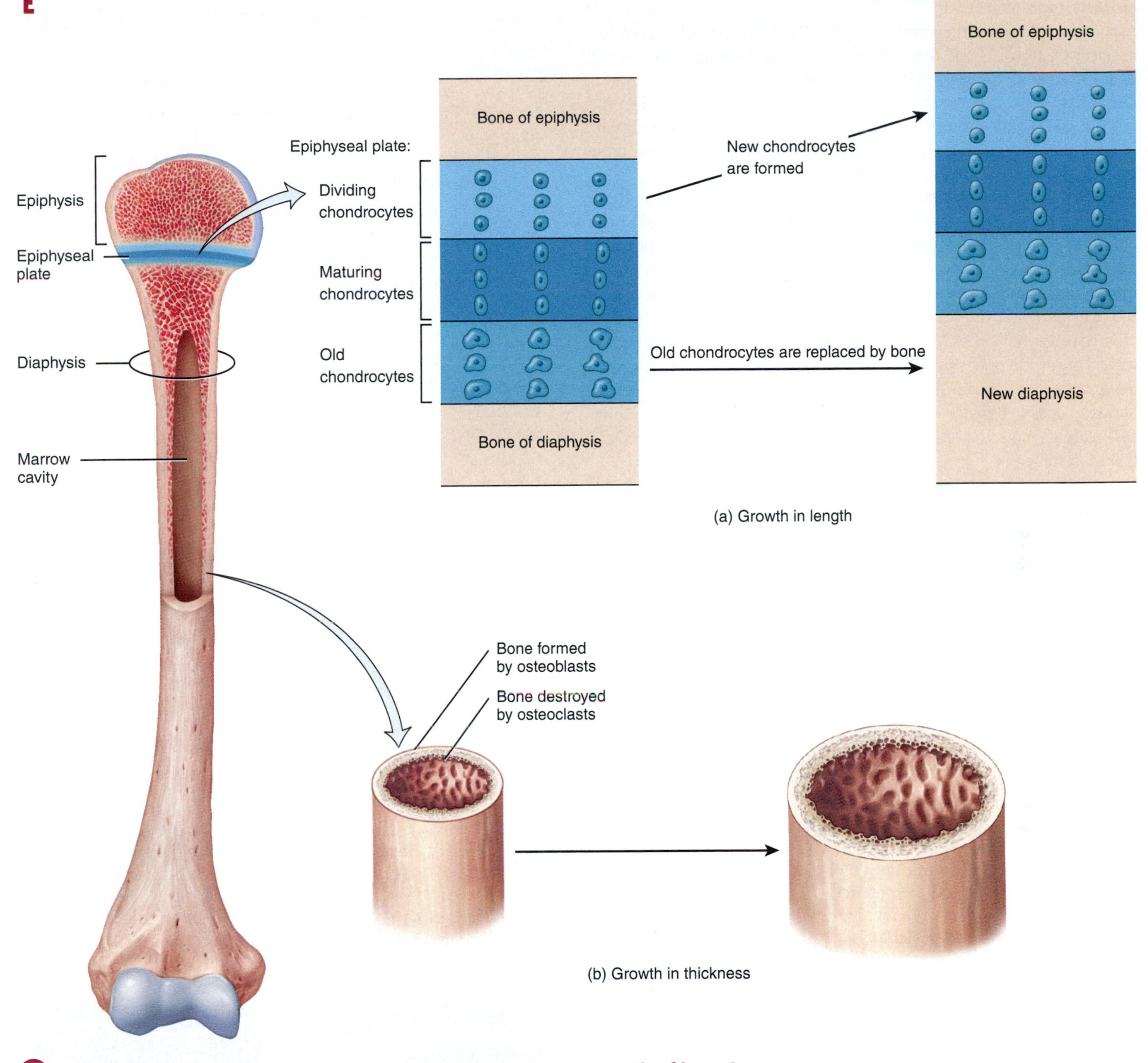

(a) Growth in length

(b) Growth in thickness

How does the epiphyseal plate account for the lengthwise growth of bone?

removes injured bone, replacing it with new bone tissue. Remodeling may be triggered by factors such as exercise, sedentary lifestyle, and changes in diet.

A delicate balance exists between the actions of osteoclasts and osteoblasts. Should too much new tissue be formed, the bones become abnormally thick and heavy. If too much mineral material is deposited in the bone, the surplus may form thick bumps, called *spurs*, on the bone that interfere with movement at joints. Excessive loss of calcium or tissue weakens the bones, and they may break, as occurs in osteoporosis, or they may become too flexible, as in rickets and osteomalacia (see Clinical Connection: Bone Disorders at the end of this section).

Several Hormones Affect Bone Growth

Bone growth is affected by a variety of hormones. During childhood, the hormones most important to bone growth are the insulin-like growth factors (IGFs), which are produced by the liver, bone, and other tissues in response to secretion of growth hormone (GH) from the anterior pituitary. Recall that IGFs stimulate osteoblasts, promote cell division at the epiphyseal plate, and enhance synthesis of the proteins needed to build new bone. Thyroid hormones (T_3 and T_4) from the thyroid gland also promote bone growth by stimulating osteoblasts. In addition, the hormone insulin from the pancreas promotes bone growth by increasing the synthesis of bone proteins.

At puberty, the secretion of sex hormones causes a dramatic effect on bone growth. The sex hormones include estrogens (produced by the ovaries) and testosterone (produced by the testes). The sex hormones are responsible for increased osteoblast activity and synthesis of bone extracellular matrix and the sudden "growth spurt" that occurs during the teenage years. The sex hormones also promote changes in the skeleton that are typical of males or females, such as a narrow pelvis (in males) or a wide pelvis (in females). Ultimately the sex hormones shut down growth at epiphyseal plates, causing lengthwise growth of the bones to cease. Because estrogens cause faster closure of the epiphyseal plate than testosterone, females typically do not reach the same height as males.

During adulthood, the sex hormones contribute to bone remodeling by slowing resorption of old bone and promoting deposition of new bone. One way that estrogens slow resorption is by promoting apoptosis (programmed death) of osteoclasts.

Two other hormones contribute to bone growth and remodeling: PTH and calcitonin. PTH promotes bone resorption by osteoclasts, enhances recovery of calcium from urine, and promotes formation of the active form of vitamin D (calcitriol). Calcitonin inhibits bone resorption by osteoclasts.

Factors Other Than Hormones Also Influence Bone Growth

In addition to hormones, several other factors influence bone growth:

- ***Genetics.*** The genes that you inherit from your parents help determine your adult stature. Although you have the *potential* to grow to a certain height based on genetics, you may not actually achieve that height if the other factors that influence growth are not present.
- ***Minerals.*** Large amounts of calcium and phosphorus are needed while bones are growing, as are smaller amounts of magnesium, fluoride, and manganese. These minerals are also necessary during bone remodeling.
- ***Vitamins.*** Vitamin A stimulates activity of osteoblasts. Vitamin C is needed for synthesis of collagen, the main bone protein. Vitamin D helps build bone by increasing the absorption of calcium and phosphate from food in the small intestine into the blood. Vitamins K and B_{12} are also needed for synthesis of bone proteins.
- ***Aging.*** As the level of sex hormones diminishes during middle age to older adulthood, especially in women after menopause, bone resorption by osteoclasts outpaces bone deposition by osteoblasts, which leads to a decrease in bone mass and an increased risk of osteoporosis.
- ***Exercise.*** Weight-bearing activities stimulate osteoblasts and consequently help build thicker, stronger bones and retard loss of bone mass that occurs as people age.

CLINICAL CONNECTION

Bone Disorders

Several disorders affect the growth and remodeling of bones. One such disorder is **osteoporosis** (os′-tē-ō-pō-RŌ-sis), a condition of porous bones (**FIGURE A, B**). The basic problem is that bone resorption outpaces bone deposition. In large part this is due to depletion of calcium from the body—more calcium is lost in urine, feces, and sweat than is absorbed from the diet. Bone mass becomes so depleted that bones fracture, often spontaneously, under the mechanical stresses of everyday living. For example, a hip fracture might result from simply sitting down too quickly. Osteoporosis primarily affects middle-aged and elderly people, 80% of them women. Older women suffer from osteoporosis more often than men for two reasons: (1) Women's bones are less massive than men's bones, and (2) production of estrogens in women declines dramatically at menopause, but production of the main androgen, testosterone, wanes gradually and only slightly in older men. (Recall that estrogens and testosterone stimulate osteoblast activity and synthesis of bone extracellular matrix.) Treatment options for osteoporosis include a diet high in calcium to reduce the risk of fractures, regularly performing weight-bearing exercises to maintain and build bone mass, and taking medications that slow down progression of bone loss and/or promote increasing bone mass.

Rickets and **osteomalacia** (oste-ō-ma-LĀ-shē-a) are two forms of the same disease that result from inadequate calcification of the extracellular bone matrix, usually caused by a vitamin D deficiency. Rickets is a disease of children in which the growing bones become "soft" or rubbery and are easily deformed. Osteomalacia is the adult counterpart of rickets, sometimes called ***adult rickets***. New bone formed during remodeling fails to calcify, and the person experiences varying degrees of pain and tenderness in bones. Prevention and treatment for rickets and osteomalacia consists of the administration of adequate vitamin D and exposure to moderate amounts of sunlight.

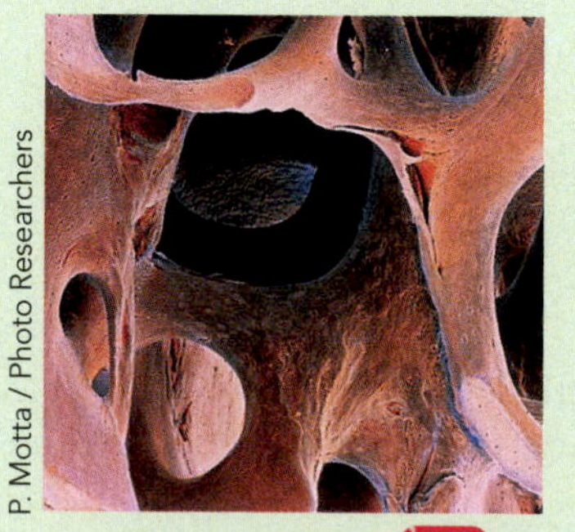

P. Motta / Photo Researchers

SEM 30x

(A) Normal bone

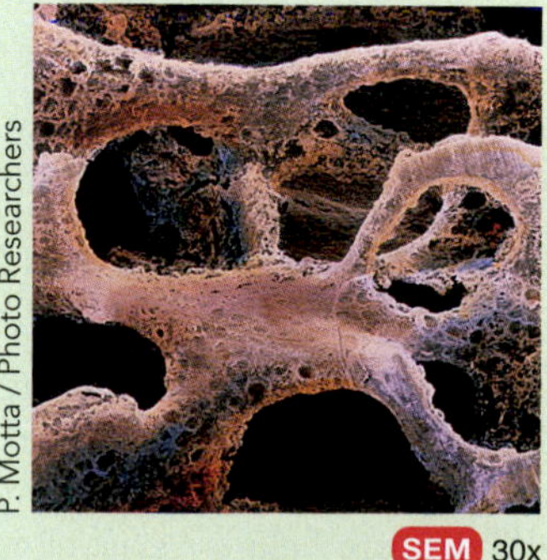

P. Motta / Photo Researchers

SEM 30x

(B) Osteoporotic bone

CHECKPOINT

35. What is the difference between hyperplasia and hypertrophy?

36. What are the roles of osteoblasts and osteoclasts in the process of bone remodeling?

37. What effects do the sex hormones have on bone growth?

13.11 The Stress Response

OBJECTIVE

- Describe how the body responds to stress.

It is impossible to remove all stress from our everyday lives. Some stress, called **eustress**, prepares us to meet certain challenges and thus is helpful. Other stress, called **distress**, is harmful. Any stimulus that produces a stress response is called a **stressor**. A stressor may be almost any disturbance of the human body—heat or cold, environmental poisons, toxins given off by bacteria, heavy bleeding from a wound or surgery, a strong emotional reaction, or hypoglycemia. The responses to stressors may be pleasant or unpleasant, and they vary among people and even within the same person at different times.

Your body's homeostatic mechanisms attempt to counteract stress. When they are successful, the internal environment remains within normal physiological limits. If stress is extreme, unusual, or long-lasting, the normal mechanisms may not be enough. In 1936, Hans Selye, a pioneer in stress research, showed that a variety of stressful conditions or noxious agents elicit a similar sequence of bodily changes. These changes, called the **stress response** or **general adaptation syndrome (GAS)**, are controlled mainly by the hypothalamus. The stress response occurs in three stages: (1) an initial fight-or-flight response, (2) a slower resistance reaction, and eventually (3) exhaustion.

The Fight-or-Flight Response Allows the Body to Handle Stress Quickly

The **fight-or-flight response** is initiated by action potentials from the hypothalamus to the sympathetic division of the autonomic nervous system (ANS), including the adrenal medulla. It quickly mobilizes the body's resources for immediate physical activity (FIGURE 13.30a). The fight-or-flight response brings huge amounts of glucose and oxygen to the organs that are most active in warding off danger: the brain, which must become highly alert; the skeletal muscles, which may have to fight off an attacker or flee; and the heart, which must work vigorously to pump enough blood to the brain and muscles. During the fight-or-flight response, nonessential body functions such as digestive, urinary, and reproductive activities are inhibited. Reduction of blood flow to the kidneys promotes release of renin, which sets into motion the renin–angiotensin–aldosterone pathway (see FIGURE 13.21). Aldosterone causes the kidneys to retain Na^+, which leads to water retention and elevated blood pressure. Water retention also helps preserve body fluid volume in the case of severe bleeding.

The Resistance Reaction Provides a Longer-lasting Response to Stress

The second stage in the stress response is the **resistance reaction** (FIGURE 13.30b). Unlike the short-lived fight-or-flight response, which is initiated by action potentials from the hypothalamus, the resistance reaction is a longer-lasting response initiated in large part by hypothalamic releasing hormones. The hormones involved are corticotropin-releasing hormone (CRH), growth hormone–releasing hormone (GHRH), and thyrotropin-releasing hormone (TRH).

CRH stimulates the anterior pituitary to secrete ACTH, which in turn stimulates the adrenal cortex to increase release of cortisol. Cortisol then stimulates gluconeogenesis by liver cells, breakdown of triglycerides into fatty acids (lipolysis), and catabolism of proteins into amino acids. Tissues throughout the body can use the resulting glucose, fatty acids, and amino acids to produce ATP or to repair damaged cells. Cortisol also reduces inflammation.

A second hypothalamic releasing hormone, GHRH, causes the anterior pituitary to secrete growth hormone (GH). GH stimulates lipolysis and glycogenolysis, the breakdown of glycogen to glucose, in the liver.

A third hypothalamic releasing hormone, TRH, stimulates the anterior pituitary to secrete thyroid-stimulating hormone (TSH). TSH promotes secretion of thyroid hormones, which stimulate the increased use of glucose for ATP production. The combined actions of GH and TSH supply additional ATP for metabolically active cells throughout the body.

The resistance stage helps the body continue fighting a stressor long after the fight-or-flight response dissipates. This is why your heart continues to pound for several minutes even after the stressor is removed. Generally, it is successful in seeing us through a stressful episode, and our bodies then return to normal. Occasionally, however, the resistance stage fails to combat the stressor, and the body moves into the state of exhaustion.

After the Resistance Reaction Is Exhaustion

The resources of the body may eventually become so depleted that they cannot sustain the resistance stage, and **exhaustion** ensues. Prolonged exposure to high levels of cortisol and other hormones involved in the resistance reaction causes wasting of muscle, suppression of the immune system, ulceration of the gastrointestinal tract, and failure of pancreatic beta cells. In addition, pathological changes may occur because resistance reactions persist after the stressor has been removed.

Stress Can Lead to Disease

Although the exact role of stress in human diseases is unknown, it is clear that stress can lead to particular diseases by temporarily inhibiting certain components of the immune system. Stress-related disorders include hypertension, irritable bowel syndrome, asthma, rheumatoid arthritis, migraine headaches, anxiety, and depression. People under stress are at a greater risk of developing chronic disease or dying prematurely.

Interleukin-1, a substance secreted by macrophages of the immune system, is an important link between stress and immunity. One action of interleukin-1 is to stimulate secretion of ACTH, which in turn stimulates the production of cortisol. Not only does cortisol provide resistance to stress and inflammation, but it also suppresses further production of interleukin-1. Thus, the immune system turns on the stress response, and the resulting cortisol then turns off one immune system mediator. This negative feedback system keeps the immune response in check once it has accomplished its goal. Because of this activity, cortisol and other glucocorticoids are used as immunosuppressive drugs for organ transplant recipients.

FIGURE 13.30 Responses to stressors during the stress response. Red arrows (hormonal responses) and green arrows (neural responses) in (a) indicate immediate fight-or-flight reactions; black arrows in (b) indicate long-term resistance reactions.

Stressors stimulate the hypothalamus to initiate the stress response through the fight-or-flight response and the resistance reaction.

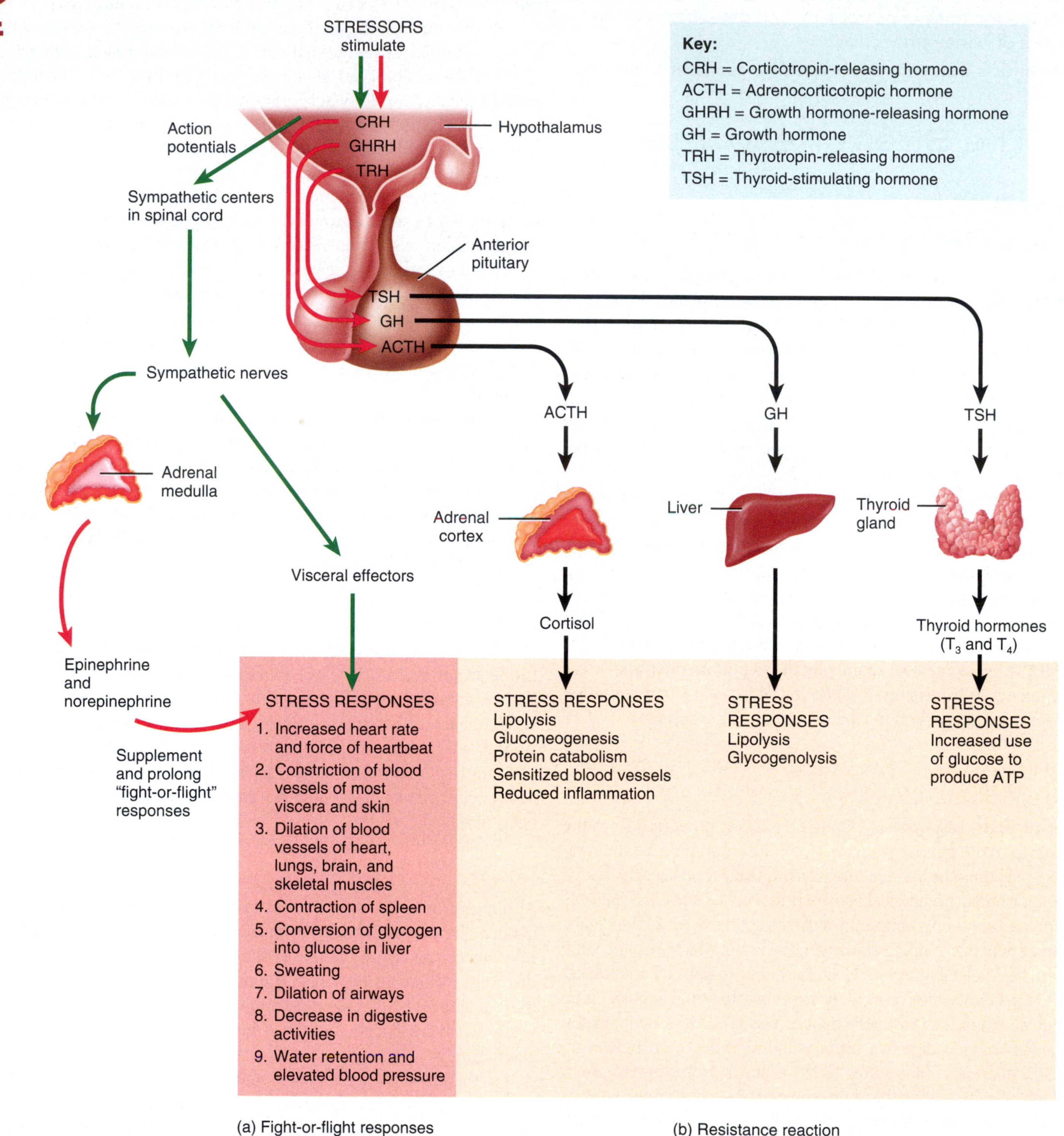

What is the basic difference between the stress response and homeostasis?

CHECKPOINT

38. What is the central role of the hypothalamus during stress?

39. What body reactions occur during the fight-or-flight response, the resistance reaction, and exhaustion?

40. What is the relationship between stress and immunity?

To appreciate the many ways the endocrine system contributes to homeostasis of other body systems, examine *Focus on Homeostasis: Contributions of the Endocrine System.*

FOCUS on HOMEOSTASIS

INTEGUMENTARY SYSTEM

- Androgens stimulate growth of axillary and pubic hair and activation of sebaceous (oil) glands

SKELETAL SYSTEM

- Growth hormone (GH) and insulin-like growth factors (IGFs) stimulate bone growth
- Estrogens and testosterone cause closure of the epiphyseal (growth) plates at the end of puberty and help maintain bone mass in adults
- Parathyroid hormone (PTH) and calcitonin regulate levels of calcium and other minerals in bone matrix and blood
- Thyroid hormones are needed for normal development and growth of the skeleton

MUSCULAR SYSTEM

- Epinephrine and norepinephrine help increase blood flow to exercising muscle
- PTH maintains proper level of Ca^{2+}, needed for muscle contraction
- Glucagon, insulin, and other hormones regulate metabolism in muscle fibers
- GH, IGFs, and thyroid hormones help maintain muscle mass

NERVOUS SYSTEM

- Several hormones, especially thyroid hormones, insulin, and growth hormone, influence growth and development of the nervous system
- PTH maintains proper level of Ca^{2+}, needed for generation and conduction of action potentials

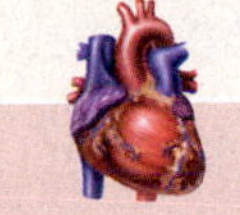

CARDIOVASCULAR SYSTEM

- Erythropoietin (EPO) promotes formation of erythrocytes (red blood cells)
- Aldosterone and antidiuretic hormone (ADH) increase blood volume
- Epinephrine and norepinephrine increase heart rate and force of contraction
- Several hormones elevate blood pressure during exercise and other stresses

CONTRIBUTIONS OF THE ENDOCRINE SYSTEM

FOR ALL BODY SYSTEMS

- Hormones of the endocrine system regulate activity and growth of target cells throughout the body
- Several hormones regulate metabolism, uptake of glucose, and molecules used for ATP production by body cells

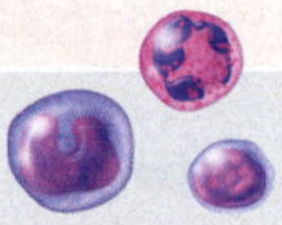

IMMUNE SYSTEM

- Glucocorticoids such as cortisol depress inflammation and immune responses
- Thymic hormones promote maturation of T cells (a type of leukocyte)

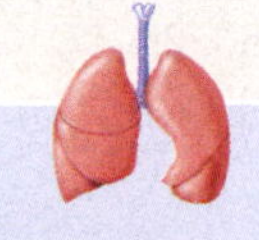

RESPIRATORY SYSTEM

- Epinephrine and norepinephrine dilate (widen) airways during exercise and other stresses
- Erythropoietin regulates amount of oxygen carried in blood by adjusting number of erythrocytes

DIGESTIVE SYSTEM

- Epinephrine and norepinephrine depress activity of the digestive system
- Gastrin, cholecystokinin, secretin, and glucose-dependent insulinotropic peptide (GIP) help regulate digestion
- Calcitriol promotes absorption of dietary calcium
- Leptin suppresses appetite

URINARY SYSTEM

- ADH, aldosterone, and atrial natriuretic peptide (ANP) adjust the rate of loss of water and ions in the urine, thereby regulating blood volume and ion content of the blood

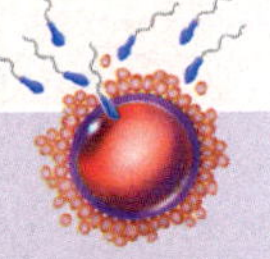

REPRODUCTIVE SYSTEMS

- Hypothalamic releasing and inhibiting hormones, follicle-stimulating hormone (FSH), and luteinizing hormone (LH) regulate development, growth, and secretions of the gonads (ovaries and testes)
- Estrogens and testosterone contribute to development of oocytes and sperm and stimulate development of secondary sex characteristics
- Prolactin promotes milk secretion in mammary glands
- Oxytocin causes contraction of the uterus and ejection of milk from the mammary glands

FROM RESEARCH TO REALITY

Melatonin and Adult Neurogenesis

Reference

Ramirez-Rodriguez, G. et al. (2011). Chronic treatment with melatonin stimulates dendrite maturation and complexity in adult hippocampal neurogenesis of mice. *Journal of Pineal Research.* 50: 29–37.

Does melatonin promote neurogenesis in the adult hippocampus?

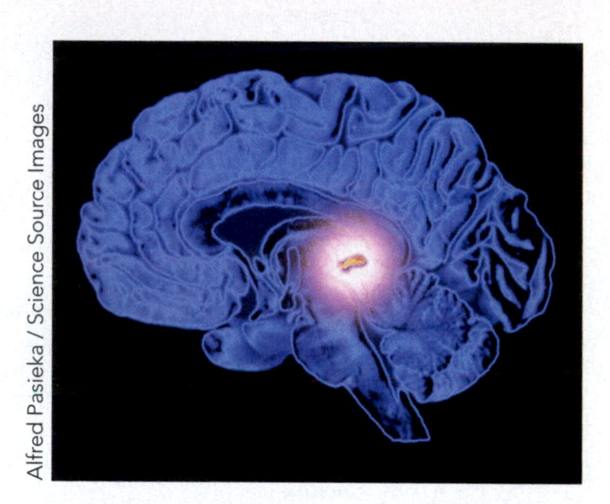

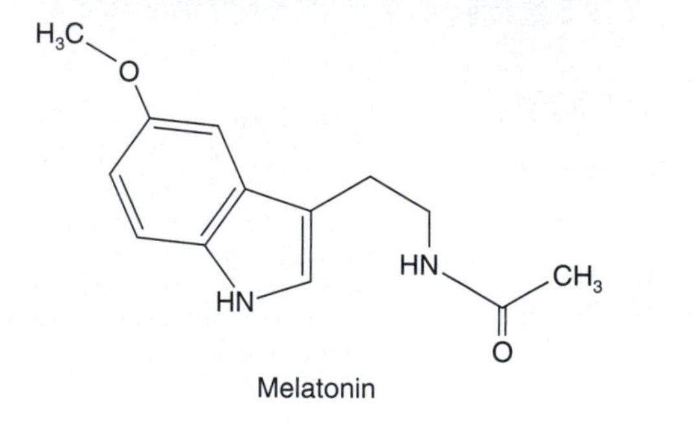

Neurogenesis (formation of new neurons from undifferentiated stem cells) in adults is limited, occurring in only a few parts of the nervous system such as the hippocampus of the brain. Finding ways to promote neurogenesis in adults is important because of neurodegenerative diseases such as Alzheimer's disease and the decline in neurons with the aging process. Is there a hormone that can promote adult neurogenesis?

Article description:

Researchers in this study investigated whether the hormone melatonin stimulated neurogenesis in the hippocampus of the brain in adult mice. Indicators of neurogenesis such as the production of the protein doublecortin, branching of dendrites, and growth of the granular cell layer of the hippocampus were examined.

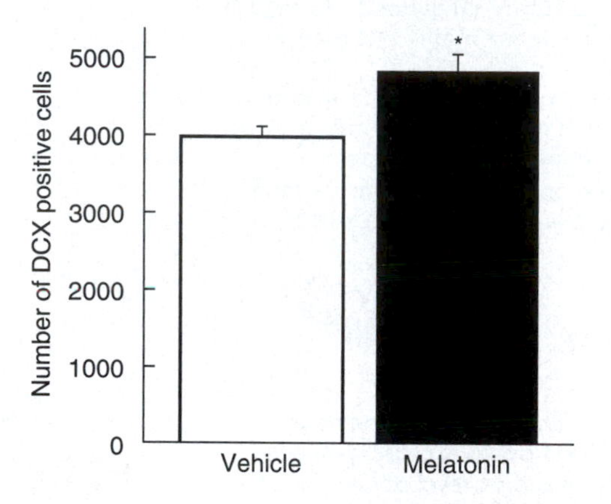

Go to WileyPLUS Learning Space and use the data from this article to answer the questions posed there and to discover more about melatonin and adult neurogenesis.

CHAPTER REVIEW

13.1 Overview of the Endocrine System

1. The endocrine system consists of endocrine glands (pituitary, thyroid, parathyroid, adrenal, and pineal) and other hormone-secreting tissues (hypothalamus, skin, thymus, heart, liver, stomach, pancreas, kidneys, small intestine, ovaries, testes, adipose tissue, and placenta).
2. Hormones affect only specific target cells that have receptors to recognize (bind) a given hormone. The number of hormone receptors may decrease (down-regulation) or increase (up-regulation).
3. Chemically, hormones are either lipid-soluble (steroids and thyroid hormones) or water-soluble (amines, peptides, and proteins).

4. Most water-soluble hormone molecules circulate through the watery blood plasma in a "free" form (not attached to plasma proteins); most lipid-soluble hormones are bound to transport proteins because they are not soluble in blood.
5. Hormonal interactions can have three types of effects: permissive, synergistic, or antagonistic.
6. Hormone secretion can be controlled by signals from the nervous system, chemical changes in blood, distension (stretch) of an organ, or other hormones.
7. Most hormonal regulatory systems operate via negative feedback.

13.2 Pituitary Gland

1. The pituitary gland consists of two lobes: an anterior pituitary and a posterior pituitary.
2. The hypothalamus is the major integrating link between the nervous and endocrine systems.
3. Secretion of anterior pituitary hormones is stimulated by releasing hormones and suppressed by inhibiting hormones from the hypothalamus.
4. Hypothalamic releasing or inhibiting hormones are carried from the hypothalamus to the anterior pituitary via the hypothalamic–hypophyseal portal system.
5. The anterior pituitary consists of somatotrophs that produce growth hormone (GH); lactotrophs that produce prolactin (PRL); corticotrophs that secrete adrenocorticotropic hormone (ACTH); thyrotrophs that secrete thyroid-stimulating hormone (TSH); and gonadotrophs that synthesize follicle-stimulating hormone (FSH) and luteinizing hormone (LH).
6. Growth hormone (GH) stimulates body growth through insulin-like growth factors (IGFs). Secretion of GH is inhibited by GHIH (growth hormone–inhibiting hormone, or somatostatin) and promoted by GHRH (growth hormone–releasing hormone).
7. TSH regulates thyroid gland activities. Its secretion is stimulated by TRH (thyrotropin-releasing hormone).
8. FSH and LH regulate the activities of the gonads—ovaries and testes. Their secretion is controlled by GnRH (gonadotropin-releasing hormone).
9. Prolactin (PRL) helps initiate milk secretion. Prolactin-inhibiting hormone (PIH) suppresses secretion of PRL; prolactin-releasing hormone (PRH) stimulates PRL secretion.
10. ACTH regulates the activities of the adrenal cortex and is controlled by CRH (corticotropin-releasing hormone).
11. The posterior pituitary contains axon terminals of neurosecretory cells whose cell bodies are in the hypothalamus. The axons of these neurosecretory cells form the hypothalamic–hypophyseal tract.
12. Hormones made by the hypothalamus and stored in the posterior pituitary include oxytocin (OT), which stimulates contraction of the uterus and ejection of milk from the breasts, and antidiuretic hormone (ADH), which stimulates water reabsorption by the kidneys and constriction of arterioles.
13. Oxytocin secretion is stimulated by uterine stretching and suckling during nursing; ADH secretion is controlled by blood osmolarity and blood volume.

13.3 Thyroid Gland

1. The thyroid gland consists of thyroid follicles composed of follicular cells, which secrete the thyroid hormones triiodothyronine (T_3) and thyroxine (T_4), and parafollicular cells, which secrete calcitonin (CT).

2. Thyroid hormones are synthesized from iodine and tyrosine within thyroglobulin (TGB). They are transported in the blood bound to plasma proteins, mostly thyroxine-binding globulin (TBG).
3. Thyroid hormones increase basal metabolic rate, enhance the actions of catecholamines, and regulate development and growth of nervous tissue and bones.
4. Secretion of thyroid hormones is controlled by TRH from the hypothalamus and thyroid-stimulating hormone (TSH) from the anterior pituitary.

13.4 Parathyroid Glands

1. The parathyroid glands are embedded in the back surface of the thyroid gland; they consist of chief cells that secrete parathyroid hormone (PTH).
2. Parathyroid hormone (PTH) regulates the homeostasis of calcium, phosphate, and magnesium ions.

13.5 Adrenal Glands

1. The adrenal glands are located above the kidneys; each consists of an outer adrenal cortex and inner adrenal medulla.
2. The adrenal cortex is divided into a zona glomerulosa, a zona fasciculata, and a zona reticularis; the adrenal medulla consists of chromaffin cells.
3. Cortical secretions include mineralocorticoids, glucocorticoids, and androgens.
4. Mineralocorticoids (mainly aldosterone) increase sodium and decrease potassium reabsorption. Secretion is controlled by the renin–angiotensin–aldosterone (RAA) pathway and by K^+ level in the blood.
5. Glucocorticoids (mainly cortisol) promote protein breakdown, gluconeogenesis, and lipolysis; help resist stress; and serve as anti-inflammatory substances. Their secretion is controlled by ACTH.
6. Androgens secreted by the adrenal cortex stimulate growth of axillary and pubic hair, aid the prepubertal growth spurt, and contribute to libido.
7. The adrenal medulla secretes epinephrine and norepinephrine, which are released during stress and produce effects similar to sympathetic responses.

13.6 Pineal Gland

1. The pineal gland is an endocrine gland associated with the brain.
2. It secretes melatonin, which influences circadian rhythms, induces sleep, protects against free radicals, and inhibits reproductive functions in some animals.

13.7 Pancreas

1. The endocrine portion of the pancreas consists of pancreatic islets (islets of Langerhans), which secrete several hormones, including insulin and glucagon.
2. Insulin performs many functions: It accelerates uptake of glucose; stimulates glycogenesis, lipogenesis, and protein synthesis; and inhibits glycogenolysis, gluconeogenesis, and protein catabolism.
3. Glucagon performs functions that are antagonistic to those of insulin: It promotes glycogenolysis, gluconeogenesis, and lipolysis, and inhibits protein synthesis.

13.8 Ovaries and Testes

1. The ovaries produce estrogens, progesterone, and inhibin. These sex hormones govern the development and maintenance of female secondary sex characteristics, reproductive cycles, pregnancy, lactation, and normal female reproductive functions.
2. The testes produce testosterone and inhibin. These sex hormones govern the development and maintenance of male secondary sex characteristics and normal male reproductive functions.

13.9 Other Endocrine Organs and Tissues

1. Cells in organs and tissues other than those normally classified as endocrine glands secrete hormones.
2. In addition to the hypothalamus, pancreas, ovaries, and testes, these components include the skin, which secretes cholecalciferol; the thymus, which secretes thymosin and thymopoietin; the heart, which secretes atrial natriuretic peptide; the liver, which secretes 25-hydroxycholecalciferol and thrombopoietin; the kidneys, which secrete calcitriol and erythropoietin; the stomach, which secretes gastrin and ghrelin; the small intestine, which secretes secretin, cholecystokinin, glucose-dependent insulinotropic peptide, and glucagon-like peptide; adipose tissue, which secretes leptin; and the placenta, which secretes estrogens, progesterone, relaxin, human chorionic somatomammotropin, and corticotropin-releasing hormone.

13.10 Endocrine Control of Growth

1. Growth involves net protein synthesis and cell division. It results in an increase in size and number of body cells and is particularly evident in the bones of the skeleton.
2. Bone growth in length involves cell division in the epiphyseal (growth) plate.
3. Bone growth in thickness (width) is due to the addition of bone tissue at the outer bone surface and removal of bone tissue along the marrow cavity.
4. Bone remodeling is an ongoing process in which osteoclasts carve out small tunnels in old bone tissue and then osteoblasts rebuild it.
5. Hormones, along with other factors, play important roles in the control of growth. Growth hormone (via insulin-like growth factors), thyroid hormones, and insulin stimulate bone growth. Sex hormones cause the growth spurt during puberty and the teenage years and then stop growth in length around age 18–21.
6. Genetics plays a role in determining adult stature.
7. Dietary minerals (especially calcium and phosphorus) and vitamins (A, C, D, K, and B_{12}) are needed for bone growth and maintenance.

13.11 The Stress Response

1. Productive stress is termed eustress, and harmful stress is termed distress.
2. If stress is extreme, it triggers the stress response (general adaptation syndrome), which occurs in three stages: the fight-or-flight response, resistance reaction, and exhaustion.
3. The stimuli that produce the stress response are called stressors. Stressors include surgery, poisons, infections, fever, and strong emotional responses.
4. The fight-or-flight response is initiated by action potentials from the hypothalamus to the sympathetic division of the autonomic nervous system and the adrenal medulla. This response rapidly increases circulation, promotes ATP production, and decreases nonessential activities.

5. The resistance reaction is initiated by releasing hormones secreted by the hypothalamus, most importantly CRH, TRH, and GHRH. Resistance reactions are longer lasting and accelerate breakdown reactions to provide ATP for counteracting stress.
6. Exhaustion results from depletion of body resources during the resistance stage.
7. Stress may trigger certain diseases by inhibiting the immune system. An important link between stress and immunity is interleukin-l, produced by macrophages; it stimulates secretion of ACTH.

PONDER THIS

1. Hormone A is released and it causes an increase in steroid hormone production. Hormone B is released and it also causes an increase in steroid hormone production. When these two hormones are released together, the production of steroid hormone increases exponentially such that the effect of the two hormones together is more than additive. What type of hormone interaction does this scenario represent?

2. Two patients come in with hypothyroidism. You draw blood from both patients and find that patient A has high levels of thyrotropin-releasing hormone, low levels of thyroid-stimulating hormone, and low levels of thyroid hormone. Patient B has high levels of thyrotropin-releasing hormone, high levels of thyroid-stimulating hormone, and low levels of thyroid hormone. Which of these two patients would benefit from a medicine that functions as a thyroid-stimulating hormone agonist? Explain your answer.

3. If you were to surgically remove the parathyroid glands, what would happen to the extracellular concentrations of calcium? Why?

ANSWERS TO FIGURE QUESTIONS

13.1 A hormone is a chemical messenger carried by the blood to distant target cells.

13.2 Examples of cellular responses that can occur in target cells include cell growth, protein synthesis, secretion, muscle contraction, or transport of substances across the plasma membrane

13.3 Down-regulation makes the target cell less sensitive to the hormone; up-regulation makes the target cell more sensitive to the hormone.

13.4 Lipid-soluble hormones are overall hydrophobic (nonpolar), whereas blood plasma is hydrophilic (polar) because it is mainly composed of water. Therefore, lipid-soluble hormones require a transport protein to move through the bloodstream.

13.5 Most water-soluble hormones activate a signaling pathway that involves a G protein-coupled receptor and the second messenger cAMP.

13.6 Permissive effects occur because the permissive hormone either increases the number of receptors for the other hormone or it promotes the synthesis of an enzyme required for the expression of the other hormone's effects.

13.7 A tropic hormone (tropin) is a hormone that acts on another endocrine gland or tissue to regulate the secretion of another hormone.

13.8 The anterior pituitary consists of glandular epithelial tissue; the posterior pituitary consists of nervous tissue.

13.9 The hypothalamic–hypophyseal portal system carries blood from the hypothalamus, where hypothalamic releasing or inhibiting hormones are secreted, to the anterior pituitary, where these hormones act.

13.10 Thyroid hormones suppress secretion of TSH by thyrotrophs and of TRH by hypothalamic neurosecretory cells; gonadal hormones suppress secretion of FSH and LH by gonadotrophs and of GnRH by hypothalamic neurosecretory cells.

13.11 Osteoclasts are responsible for bone resorption.

13.12 Hyperglycemia decreases growth hormone secretion.

13.13 The pituitary gland, skin, and brain contain POMC, and different combinations of peptides can be cleaved from POMC depending on the types of processing enzymes that are present in these tissues.

13.14 Oxytocin and antidiuretic hormone are released from the posterior pituitary gland.

13.15 Follicular cells secrete T_3 and T_4, also known as thyroid hormones; parafollicular cells secrete calcitonin.

13.16 The colloid serves as a storage form of T_3 and T_4.

13.17 Low metabolic rate, a cold environment, hypoglycemia, high altitude, and pregnancy also increase the secretion of thyroid hormones.

13.18 PTH increases the blood calcium (Ca^{2+}) and magnesium (Mg^{2+}) levels and decreases the blood phosphate (HPO_4^{2-}) level.

13.19 Target tissues for PTH are bones and the kidneys; target tissue for calcitonin is bone; target tissue for calcitriol is the gastrointestinal tract.

13.20 The adrenal cortex secretes steroid hormones (aldosterone, cortisol and dehydroepiandrosterone); the adrenal medulla secretes norepinephrine, epinephrine, and a trace amount of dopamine.

13.21 Angiotensin II acts to constrict blood vessels by causing contraction of vascular smooth muscle, and it stimulates secretion of aldosterone (by zona glomerulosa cells of the adrenal cortex), which in turn causes the kidneys to increase reabsorption of Na^+. As long as the hormone ADH is present, water follows the reabsorbed Na^+, resulting in an increase in blood volume and blood pressure.

13.22 A transplant recipient who takes prednisone will have low blood levels of ACTH and CRH due to negative feedback suppression of the anterior pituitary and hypothalamus by the prednisone.

13.23 The sympathetic preganglionic and postganglionic neurons connect the hypothalamus to the pineal gland and ultimately cause the pineal gland to secrete melatonin.

13.24 Insulin lowers the blood glucose concentration.

13.25 When insulin dissociates from its receptor, the GLUT4 molecules will move back to internal vesicles via endocytosis.

13.26 The direction of glucose transport via a glucose transporter, such as GLUT2, depends on the glucose gradient. Because there is a higher level of glucose inside the cell than outside the cell due to glycogenolysis occurring in the cytosol, glucose moves across the membrane via GLUT2 from the cytosol to the extracellular fluid.

13.27 Other than hyperglycemia, insulin secretion is stimulated by increased parasympathetic activity, an elevated blood amino acid level, and the incretin hormones glucose-dependent insulinotropic peptide (GIP) and glucagon-like peptide (GLP).

13.28 Calcitriol acts on the small intestine to increase absorption of calcium (Ca^{2+}) and phosphate (HPO_4^{2-}) ions from food in the intestinal tract into the bloodstream.

13.29 The lengthwise growth of bone is caused by cell divisions of chondrocytes on the epiphyseal side and replacement of old chondrocytes with bone on the diaphyseal side.

13.30 Homeostasis maintains controlled conditions typical of a normal internal environment; the stress response resets controlled conditions at a different level to cope with various stressors.

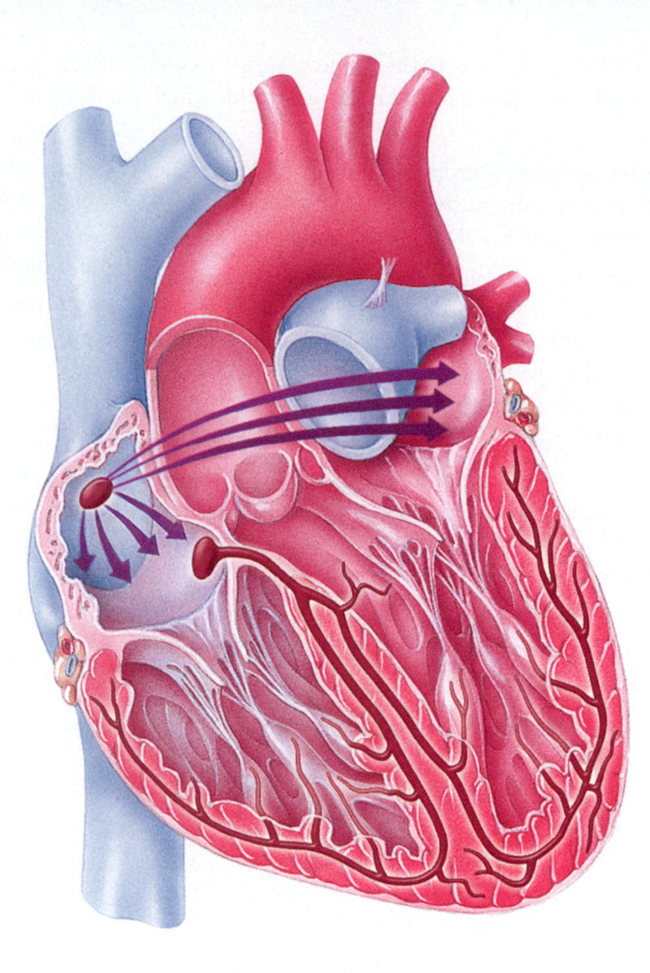

14 The Cardiovascular System: The Heart

The Heart and Homeostasis

The heart contributes to homeostasis by pumping blood through blood vessels to the tissues of the body.

LOOKING BACK TO MOVE AHEAD...

- Equilibrium potential is the membrane potential at which the concentration gradient and electrical gradient for a particular ion are equal in magnitude but opposite in direction and there is no net movement of that ion across the plasma membrane (Section 7.3).
- Acetylcholine binds to cholinergic receptors, which include muscarinic receptors and nicotinic receptors; norepinephrine and epinephrine bind to adrenergic receptors, which include alpha receptors (α_1 and α_2) and beta receptors (β_1, β_2, and β_3) (Section 7.5).
- During muscle contraction, myosin heads of thick filaments bind to and pull on actin molecules of thin filaments, causing the thick and thin filaments to slide past one another and tension (force) to be generated (Section 11.3).
- In an isotonic contraction, the tension developed by the muscle remains almost constant while the muscle changes its length; in an isometric contraction, tension is generated without the muscle changing its length (Section 11.5).

The **cardiovascular system** consists of three interrelated components: the heart, blood vessels, and blood (FIGURE 14.1). The *heart* serves as a pump that generates the pressure needed to circulate blood to the tissues of the body. To accomplish this, the heart beats about 100,000 times every day, which adds up to 35 million beats in a year and about 3 billion beats in an average lifetime. *Blood vessels* are tubular structures through which blood flows from the heart to body tissues and then back to the heart. *Blood* is a fluid that delivers oxygen and nutrients to cells and removes carbon dioxide and other wastes from cells. It also regulates pH and body temperature, and provides protection against disease. This chapter explores the unique properties of the heart that allow it to function as a pump for a lifetime without rest. The next two chapters will examine blood vessels and the blood, respectively.

FIGURE 14.1 Components of the cardiovascular system.

The cardiovascular system consists of the heart, blood vessels, and blood.

Heart
Blood vessels:
Artery
Vein
Leukocyte (white blood cell)
Erythrocyte (red blood cell)
Platelet
Blood plasma
Mark Nielsen
LM 400x
Blood

What is the purpose of the heart?

Basic Design of the Cardiovascular System

OBJECTIVES

- Distinguish between the pulmonary and systemic circulations.
- Describe parallel flow through the systemic circulation.

To appreciate how the heart functions, you first need to understand the basic design of the cardiovascular system.

The Heart Pumps Blood Through the Pulmonary and Systemic Circulations

With each beat, the heart pumps blood into two closed circuits—the pulmonary circulation and the systemic circulation (FIGURE 14.2). The **pulmonary circulation** consists of blood vessels that carry blood from the right side of the heart to the alveoli (air sacs) of the lungs and then back to the left side of the heart. The **systemic circulation** consists of blood vessels that carry blood from the left side of the heart to all organs and tissues of the body except the alveoli and then back to

FIGURE 14.2 Pulmonary and systemic circulations. Throughout this book, blood vessels that carry oxygenated blood are colored red, whereas those that carry deoxygenated blood are colored blue.

The pulmonary circulation carries blood from the right side of the heart to the alveoli of the lungs and then back to the left side of the heart; the systemic circulation carries blood from the left side of the heart to all organs and body tissues except the alveoli and then back to the right side of the heart.

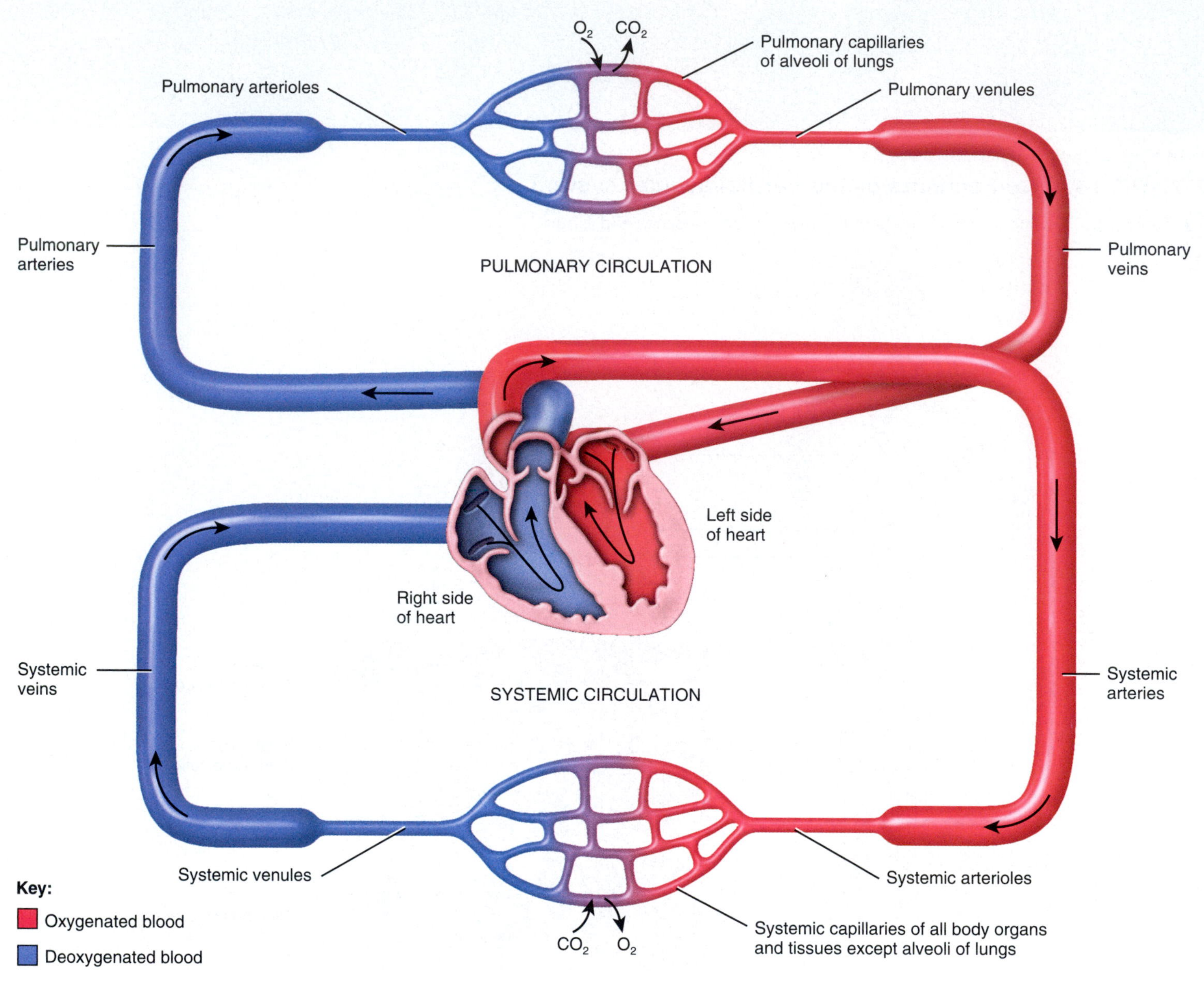

Where in the systemic circulation does blood become deoxygenated?

the right side of the heart. In both circuits, blood is carried away from and then returned to the heart in the following way (FIGURE 14.2):

- Large blood vessels called *arteries* carry blood away from the heart.
- Arteries branch to form smaller vessels called *arterioles*.
- Arterioles give rise to even smaller vessels called *capillaries*.
- From capillaries, blood enters larger vessels called *venules*.
- Venules give rise to even larger vessels called *veins*, which carry blood back to the heart.

Capillaries are the smallest blood vessels of the body; they serve as the sites of gas, nutrient, and waste exchange between blood and surrounding tissues. In pulmonary capillaries, blood becomes *oxygenated* as it picks up oxygen (O_2) from inhaled air in the alveoli of the lungs and drops off some molecules of carbon dioxide (CO_2), which are exhaled from the body. Oxygenated blood is bright red in color. In systemic capillaries, blood becomes *deoxygenated* as it drops off some of its O_2 to cells and picks up CO_2, a waste product of cellular metabolism. Deoxygenated blood is dark red. By convention, blood vessels that contain oxygenated blood are colored red and blood vessels that contain deoxygenated blood are colored blue (FIGURE 14.2). Because of the way gas exchange takes place in the pulmonary and systemic capillaries, blood in pulmonary veins, the left side of the heart, and systemic arteries is oxygenated, whereas blood in systemic veins, the right side of the heart, and pulmonary arteries is deoxygenated.

Blood Is Distributed in the Systemic Circulation Mainly via Parallel Flow

In the systemic circulation, blood flows through pathways that are *parallel* to each other, a design feature known as **parallel flow** (FIGURE 14.3). In most of these pathways, a given portion of blood flows through an artery to only one organ and enters only one set of capillaries before returning to the heart through a vein. Because of this arrangement, the same portion of blood does not flow from one organ to the next. The parallel flow of blood through the systemic circulation is important for two reasons: (1) It allows each organ to receive its own supply of freshly oxygenated blood, and (2) it allows blood flow to different organs to be regulated independently.

FIGURE 14.3 Parallel flow through the systemic circulation. Arrows represent the direction of blood flow.

In the systemic circulation, blood flows through parallel pathways, each usually carrying a portion of blood through an artery to only one organ and entering only one set of capillaries before returning to the heart through a vein.

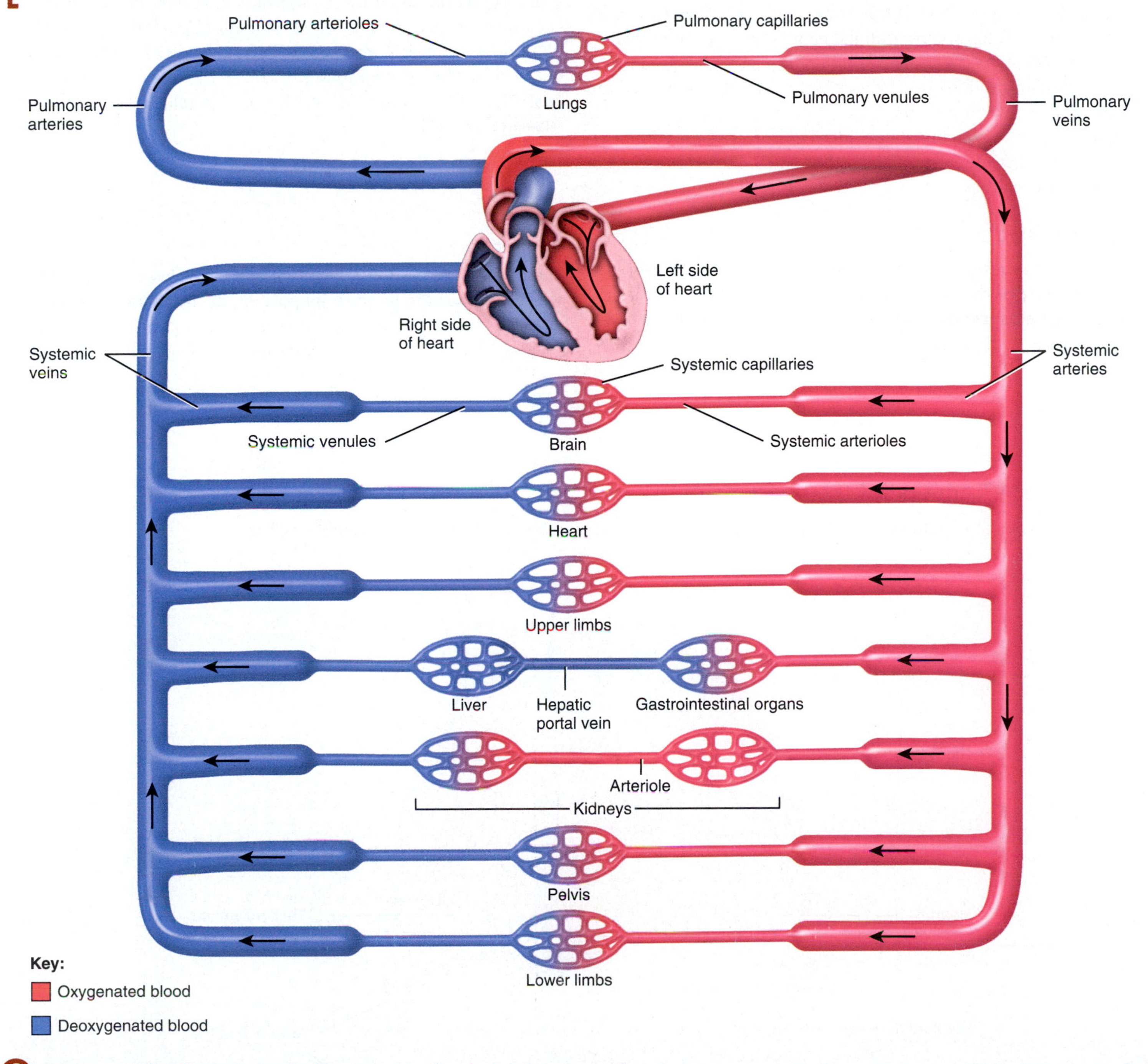

Why is parallel flow through the systemic circulation important?

There are a few exceptions to parallel flow. In some organs, blood flows between two sets of capillaries that are arranged in *series* (one right after the other). You are already familiar with the hypothalamic–hypophyseal portal system that delivers hormones in blood from capillaries in the hypothalamus to capillaries in the anterior pituitary gland (see FIGURE 13.9b). Recall that in a **portal system**, blood flows from one capillary network into a portal vein and then into a second capillary network before returning to the heart. Another example of a portal system is the hepatic portal circulation, which carries absorbed nutrients in blood from capillaries in gastrointestinal organs to capillaries in the liver via the hepatic portal vein (FIGURE 14.3). This arrangement allows the liver to store or modify some of the absorbed nutrients before they pass into the general circulation. A final exception to parallel flow in the systemic circulation occurs in the kidneys, where two sets of capillaries (the glomerulus and peritubular capillaries) are connected by an arteriole (FIGURE 14.3). The functional roles of the glomerulus and peritubular capillaries are described in the discussion of the urinary system in Chapter 19.

Now that you have an understanding of the layout of the cardiovascular system, it is time to learn about the heart, which is the focus of the rest of this chapter.

CHECKPOINT

1. Determine whether the following blood vessels contain oxygenated or deoxygenated blood: pulmonary arteries, systemic veins, pulmonary veins, and systemic arteries.
2. What is a portal system?

14.2 Organization of the Heart

OBJECTIVES

- Identify the layers of the heart wall.
- Distinguish the four chambers of the heart.
- Explain how the valves of the heart function.
- Describe the fibrous skeleton of the heart.
- Discuss the blood supply of the heart.

The **heart** is a hollow, muscular organ that is about the size of a closed fist. It is located in the thoracic (chest) cavity, with most of its mass lying to the left of the body's midline (FIGURE 14.4). The heart is bordered by the sternum (breastbone) anteriorly, by the lungs laterally, by the vertebral column posteriorly, and by the diaphragm inferiorly. You can visualize the heart as a cone lying on its side: The upper, broad portion of the cone is known as the *base*; the lower, pointed tip of the cone is referred to as the *apex* (FIGURE 14.4).

The Pericardium Protects and Anchors the Three-Layered Heart

The heart is enclosed in a membranous sac called the **pericardium**, which confines the heart to its position in the thoracic cavity while allowing sufficient freedom of movement for vigorous and rapid contraction. The pericardium consists of an outer parietal layer and an

FIGURE 14.4 Location of the heart.

The heart is located in the thoracic (chest) cavity, with most of its mass lying to the left of the body's midline.

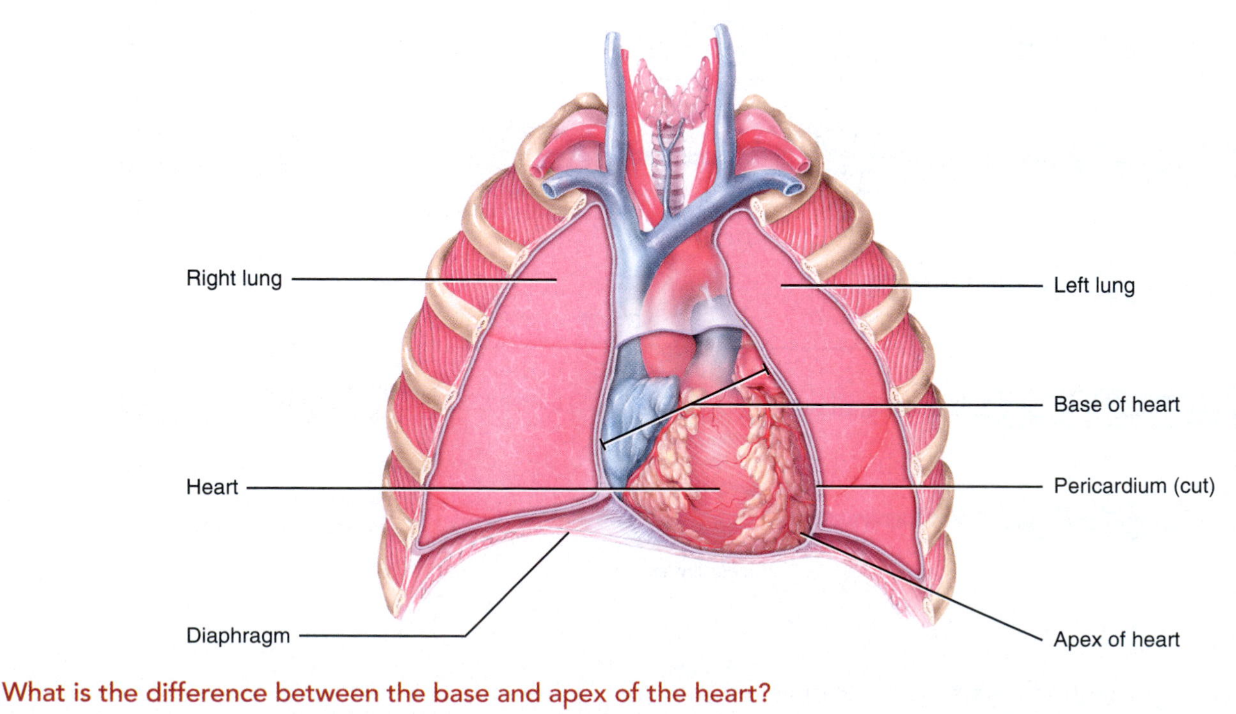

What is the difference between the base and apex of the heart?

CLINICAL CONNECTION

Cardiopulmonary Resuscitation

Because the heart lies between two rigid structures—the sternum and the vertebral column—external pressure on the chest (compression) can be used to force blood out of the heart and into the circulation. In cases in which the heart suddenly stops beating, **cardiopulmonary resuscitation (CPR)**—properly applied cardiac compressions, performed with artificial ventilation of the lungs via mouth-to-mouth respiration—saves lives. CPR keeps oxygenated blood circulating until the heart can be restarted.

Researchers have found that chest compressions alone are equally as effective as, if not better than, traditional CPR with lung ventilation. This is good news because it is easier for an emergency dispatcher to give instructions limited to chest compressions to frightened, nonmedical bystanders. As public fear of contracting contagious diseases such as hepatitis, HIV, and tuberculosis continues to rise, bystanders are much more likely to perform chest compressions alone than treatment involving mouth-to-mouth rescue breathing.

CLINICAL CONNECTION

Pericarditis and Cardiac Tamponade

Inflammation of the pericardium is known as **pericarditis**. In this condition, the visceral and parietal layers of the pericardium rub against each other, resulting in sharp chest pain. Pericarditis can also lead to the buildup of pericardial fluid within the pericardial cavity, and the fluid buildup compresses the heart. This compression, known as **cardiac tamponade** (tam′-pon-ĀD), impairs the heart's ability to pump blood. Treatment of cardiac tamponade involves draining the excess fluid through a needle passed into the pericardial cavity.

inner visceral layer (FIGURE 14.5). The **parietal layer** is surrounded by a tough fibrous coat of connective tissue that anchors the heart in place by attaching to nearby structures in the thoracic cavity. The **visceral layer**, also called the **epicardium**, adheres to the surface of the heart. Between the parietal and visceral layers of the pericardium is a space called the **pericardial cavity** that is filled with a thin film of lubricating fluid. This fluid, known as **pericardial fluid**, reduces friction within the pericardium as the heart moves.

The wall of the heart consists of three layers (FIGURE 14.5): the epicardium (outer layer), the myocardium (middle layer), and the endocardium (inner layer). The **epicardium**, also known as the visceral layer of the pericardium, consists of epithelium and connective tissue. The **myocardium** forms the bulk of the heart wall. It consists of cardiac muscle and is responsible for the pumping action of the heart. The **endocardium** is a thin layer of epithelium that lines the chambers of the heart and covers heart valves. It is also continuous with the epithelial cells lining the blood vessels that are attached to the heart. Epithelial cells that line the heart, blood vessels, and lymphatic vessels of the body are referred to as **endothelial cells** or simply as **endothelium**.

FIGURE 14.5 Pericardium and heart wall.

The pericardium is a membranous sac that surrounds the heart; the heart wall is composed of three layers: epicardium, myocardium, and endocardium.

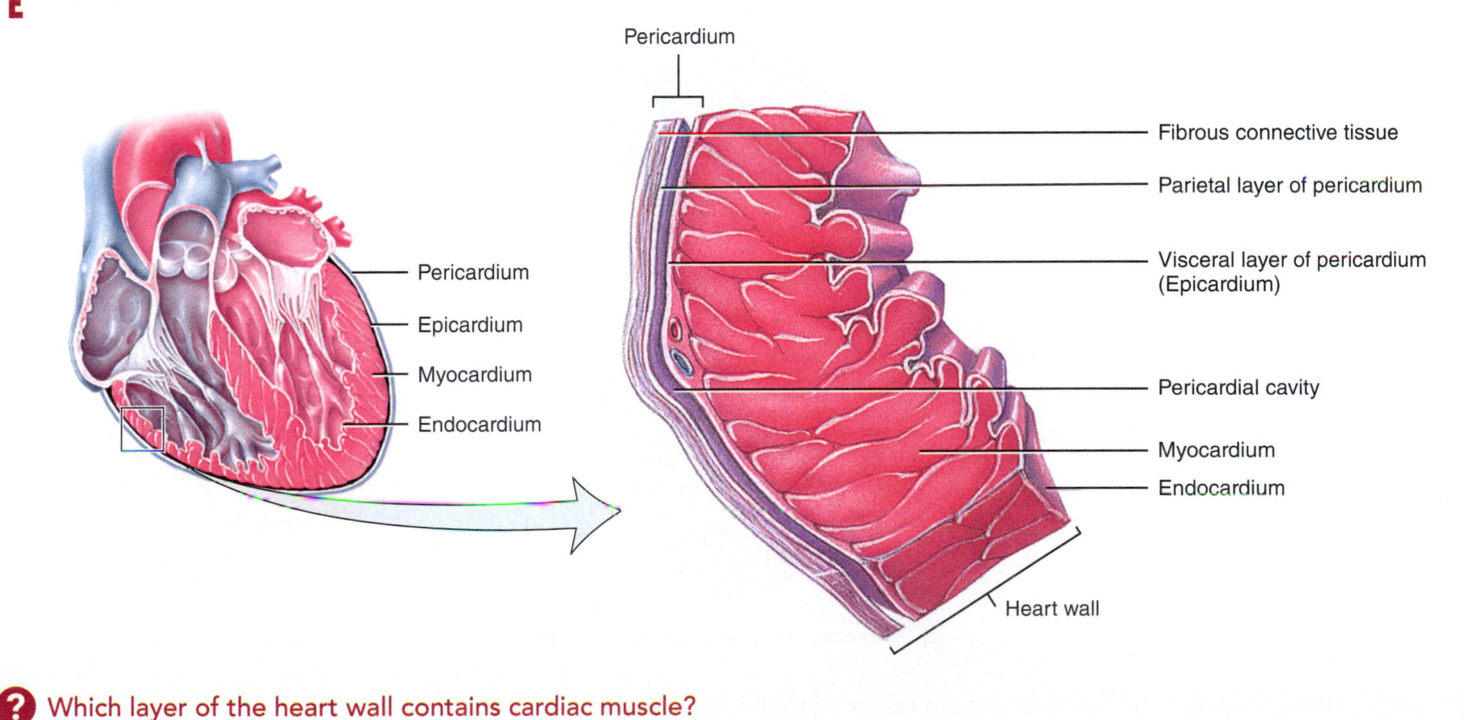

Which layer of the heart wall contains cardiac muscle?

The Heart Contains Four Chambers and Is Associated with Several Great Vessels

The heart consists of four chambers: two upper **atria** (singular is *atrium*) and two lower **ventricles** (FIGURE 14.6a, b). Functionally, the heart can be divided into right and left sides, with each side consisting of an atrium and a ventricle. The right side of the heart serves as the pump for the pulmonary circulation; the left side of the heart serves as the pump for the systemic circulation. Therefore, the heart is actually two separate pumps located within the same organ. A muscular partition,

FIGURE 14.6 Organization of the heart.

The heart consists of four chambers: two upper atria and two lower ventricles.

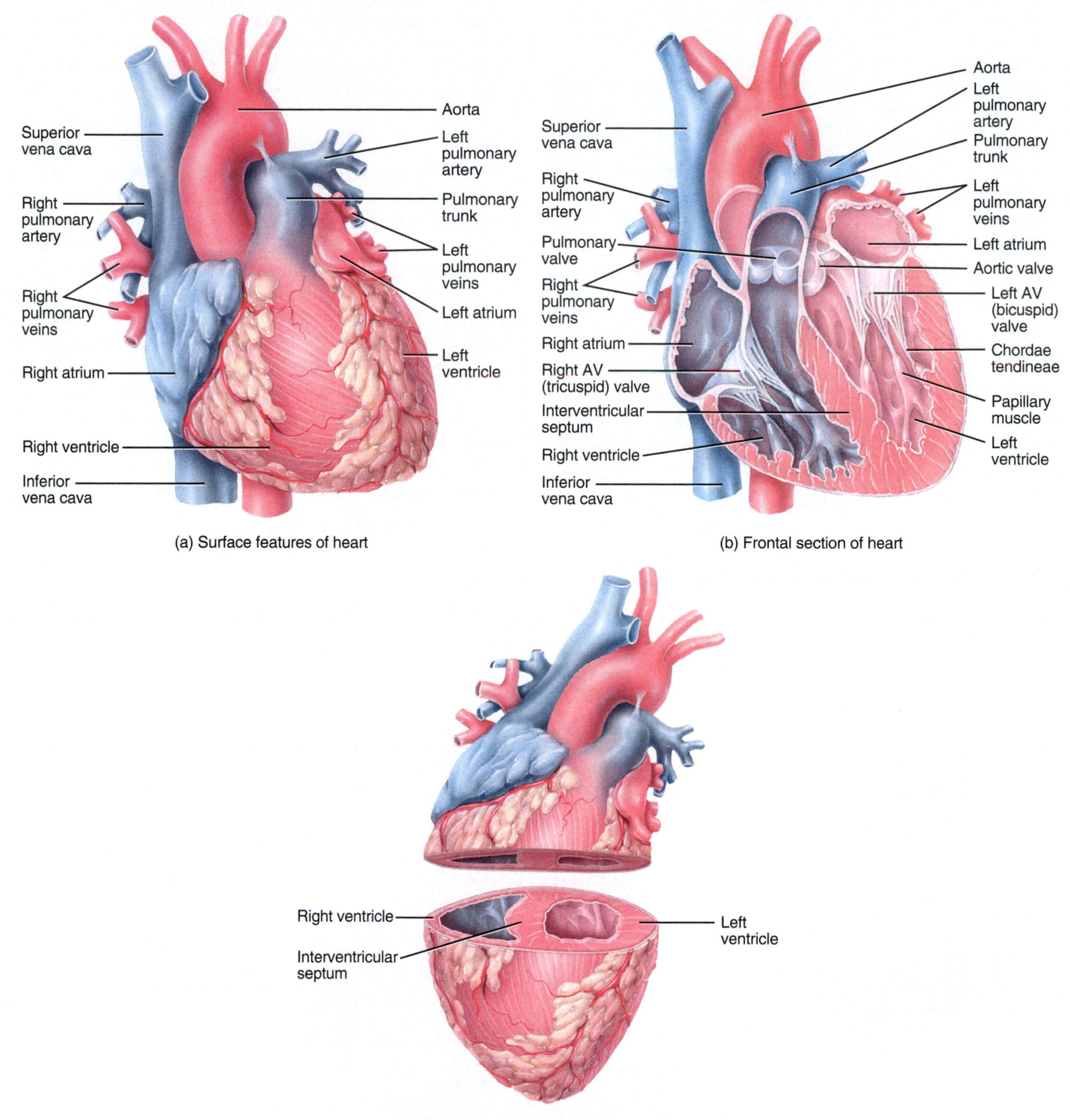

How does the thickness of the myocardium relate to the workload of a heart chamber?

or *septum*, separates the right and left sides of the heart. The septum prevents blood from mixing between the two sides of the heart. The part of the septum located between the two atria is called the *interatrial septum*; the part of the septum between the two ventricles is known as the *interventricular septum* (FIGURE 14.6b).

The thickness of the myocardium of the chambers varies according to the amount of work each chamber has to perform (FIGURE 14.6b). The atria have thin walls because they deliver blood under less pressure into the ventricles. The ventricles have thicker walls because they pump blood out of the heart under higher pressure and over greater distances. Although the right and left ventricles act as two separate pumps that simultaneously eject equal volumes of blood, the right side has a much smaller workload. It pumps blood a short distance to the alveoli of the lungs at lower pressure, and the resistance to blood flow is small. The left ventricle pumps blood great distances to all other parts of the body at higher pressure, and the resistance to blood flow is larger. Therefore, the left ventricle works much harder than the right ventricle to maintain the same rate of blood flow. The structure of the two ventricles confirms this functional difference—the muscular wall of the left ventricle is considerably thicker than the wall of the right ventricle (FIGURE 14.6c).

The chambers of the heart are associated with several large or *great* blood vessels. The right atrium receives deoxygenated blood through two major veins: the **superior vena cava**, which brings blood mainly from parts of the body above the heart, and the **inferior vena cava**, which brings blood mostly from parts of the body below the heart (FIGURE 14.6b). The right atrium then delivers the deoxygenated blood into the right ventricle, which pumps it into an artery called the **pulmonary trunk**. The pulmonary trunk divides into the **pulmonary arteries**, which carry the deoxygenated blood to the lungs. While in the alveoli of the lungs, blood becomes oxygenated as it picks up oxygen (O_2) and unloads some of its carbon dioxide (CO_2). The oxygenated blood is then carried to the left atrium via **pulmonary veins**. From the left atrium the oxygenated blood passes into the left ventricle, which pumps the blood into a large artery called the **aorta**. The aorta branches into several smaller arteries that carry the oxygenated blood to all parts of the body except the alveoli. While in the tissues of the body, blood becomes deoxygenated as it drops off some of its O_2 and picks up CO_2.

Heart Valves Ensure One-Way Blood Flow

As each chamber of the heart contracts, it pushes a volume of blood into a ventricle or out of the heart into an artery. To prevent the blood from flowing backward, the heart has four valves: two atrioventricular valves and two semilunar valves. The valves of the heart are composed of connective tissue covered by endocardium, and they open and close in response to *pressure changes* as the heart contracts and relaxes. Each of the four valves helps ensure the one-way flow of blood by opening to let blood through and then closing to prevent its backflow.

Atrioventricular Valves

As the name implies, **atrioventricular (AV) valves** lie between the atria and ventricles (FIGURE 14.6b). The **right AV valve** is located between the right atrium and right ventricle. It is also called the **tricuspid valve** because it consists of three *cusps* or flaps. The **left AV valve** is located between the left atrium and left ventricle. Because it consists of two cusps, it is also known as the **bicuspid valve**. Another name for the left AV valve is the **mitral valve** because it resembles a bishop's miter (hat), which is two-sided.

The cusps of the AV valves are connected to tendonlike cords called **chordae tendineae** (KOR-dē ten-DIN-ē-ē), which in turn are connected to **papillary muscles**, cone-shaped muscle projections located on the inner surface of the ventricles (FIGURE 14.6b). The chordae tendineae prevent the valve cusps from everting (opening into the atria) when the ventricles contract and are aligned to allow the valve cusps to tightly close the valve.

The AV valves open when pressure in the atria exceeds pressure in the ventricles. When the ventricles are relaxed, the papillary muscles are relaxed, the chordae tendineae are slack, and blood pushes the AV valves open, moving from a higher pressure in the atria to a lower pressure in the ventricles (FIGURE 14.7a). When the

FIGURE 14.7 Operation of the heart valves. The pulmonary valve in parts (c) and (d) actually has three cusps, but only two are shown for simplicity.

Heart valves prevent the backflow of blood.

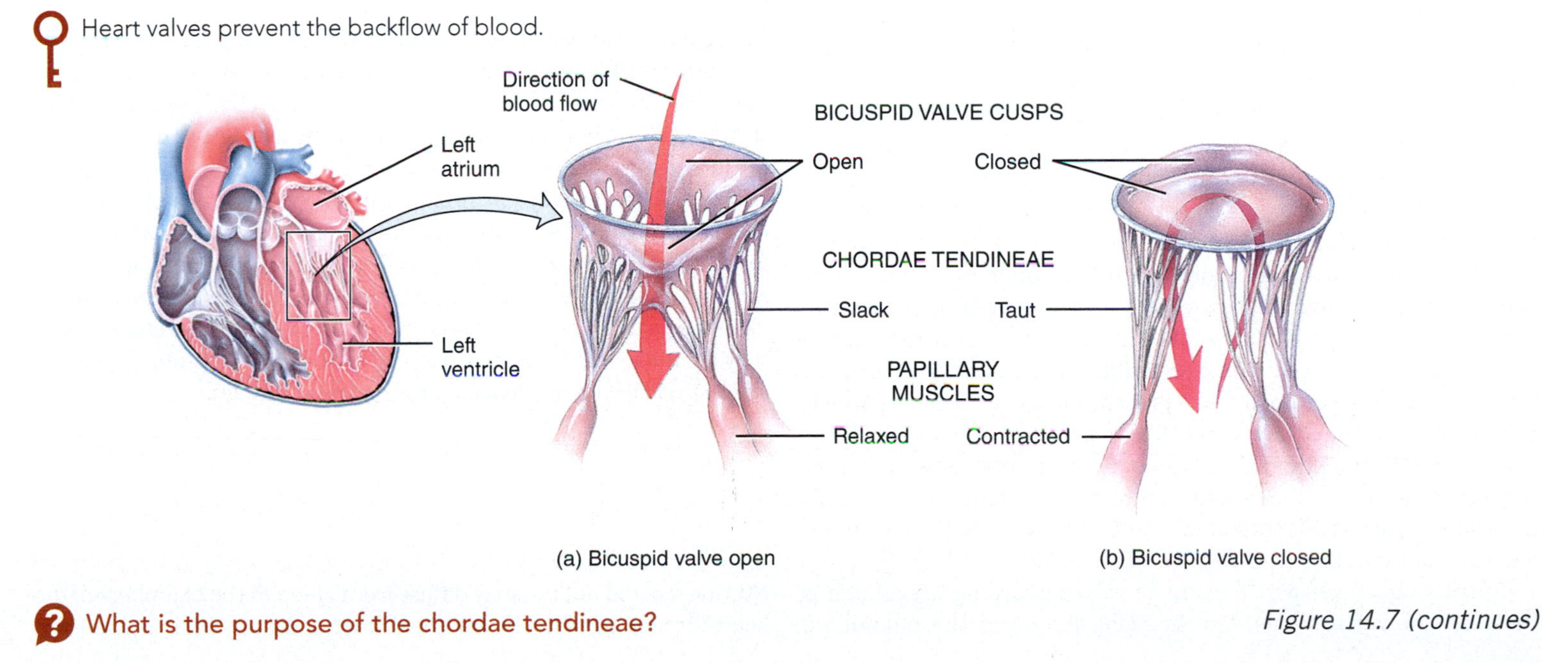

What is the purpose of the chordae tendineae?

Figure 14.7 (continues)

Figure 14.7 Continued

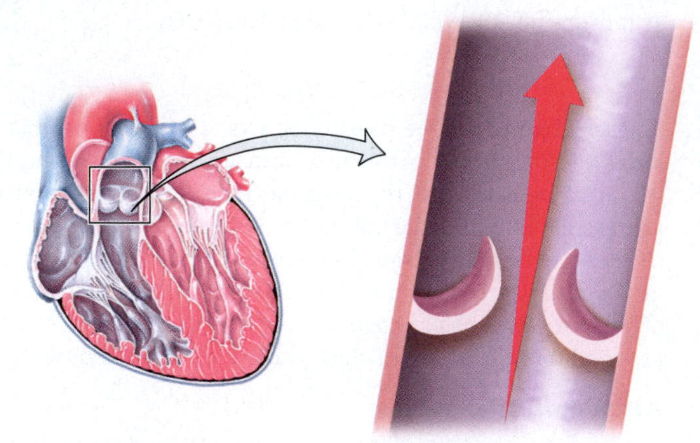

(c) Pulmonary valve open

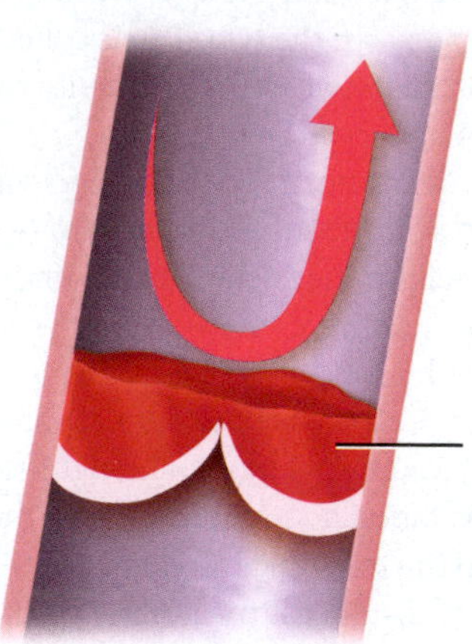

(d) Pulmonary valve closed

What is the purpose of the chordae tendineae?

ventricles contract, the pressure of the blood drives the cusps upward until their edges meet and close the opening (**FIGURE 14.7b**). At the same time, the papillary muscles contract, which pulls on and tightens the chordae tendineae. This prevents the valve cusps from swinging upward and opening into the atria in response to the high ventricular pressure. If the AV valves or chordae tendineae are damaged, blood may regurgitate (flow back) into the atria when the ventricles contract.

Semilunar Valves

The two **semilunar (SL) valves** of the heart are the **pulmonary valve,** which is located between the right ventricle and the pulmonary trunk, and the **aortic valve,** which is located between the left ventricle and the aorta (see **FIGURE 14.6b**). These valves are referred to as *semilunar* because they are both made up of three cusps that are shaped like half moons (*semi-* = half; *-lunar* = moon-shaped). The cusps are attached to the walls of the pulmonary trunk and aorta, and project into the lumen of each of these arteries. The SL valves allow ejection of blood from the heart into arteries but prevent backflow of blood into the ventricles. When the ventricles contract, pressure builds up within the chambers. The semilunar valves open when pressure in the ventricles exceeds the pressure in the arteries, permitting ejection of blood from the ventricles into the pulmonary trunk and aorta (**FIGURE 14.7c**). As the ventricles relax, blood starts to flow back toward the heart. This backflowing blood fills the valve cusps, which causes the semilunar valves to close tightly (**FIGURE 14.7d**).

Surprisingly perhaps, there are no valves guarding the junctions between the venae cavae and the right atrium or the pulmonary veins and the left atrium. As the atria contract, a small amount of blood does flow backward from the atria into these vessels. However, backflow is minimized by a different mechanism: As the atrial muscle contracts, it compresses and nearly collapses the venous entry points.

A summary of blood flow through the chambers and valves of the heart is provided in **FIGURE 14.8**.

The Fibrous Skeleton of the Heart Prevents Overstretching of Heart Valves

In addition to cardiac muscle, the heart wall contains connective tissue that forms the **fibrous skeleton of the heart** (**FIGURE 14.9**). Essentially, the fibrous skeleton consists of four dense connective tissue rings that surround the valves of the heart, fuse with one another, and merge with the interventricular septum. In addition to forming a structural foundation for the heart valves, the fibrous skeleton prevents overstretching of the valves as blood passes through them. It also serves as an attachment for bundles of cardiac muscle fibers. During atrial contraction, the two atria are pulled downward toward the fibrous skeleton; during ventricular contraction, the fibrous skeleton stabilizes the lower chambers as they contract. The connective tissue of the fibrous skeleton also acts as an electrical insulator between the atria and ventricles.

The Coronary Circulation Supplies Blood to the Heart Wall

Nutrients could not possibly diffuse from blood in the chambers of the heart through all of the layers of cells that make up the heart wall. For

FIGURE 14.8 Blood flow through the heart.

The right side of the heart receives deoxygenated blood from the systemic circulation and pumps it into the pulmonary circulation; the left side of the heart receives oxygenated blood from the pulmonary circulation and pumps it into the systemic circulation.

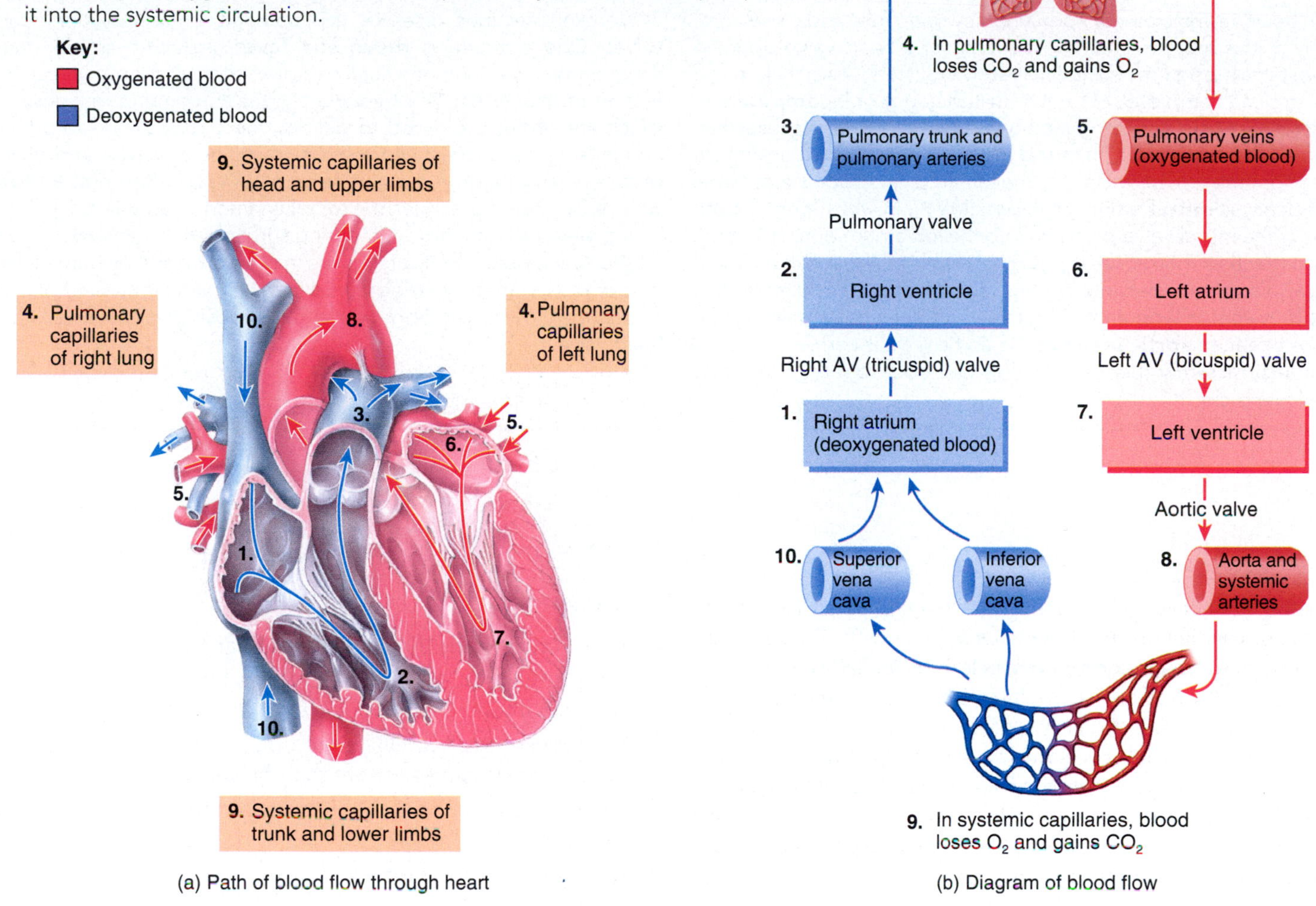

(a) Path of blood flow through heart

(b) Diagram of blood flow

Which major blood vessel carries oxygenated blood away from the left ventricle of the heart?

FIGURE 14.9 Fibrous skeleton of the heart. Elements of the fibrous skeleton are shown in capital letters.

Fibrous rings support the four valves of the heart and are fused to one another.

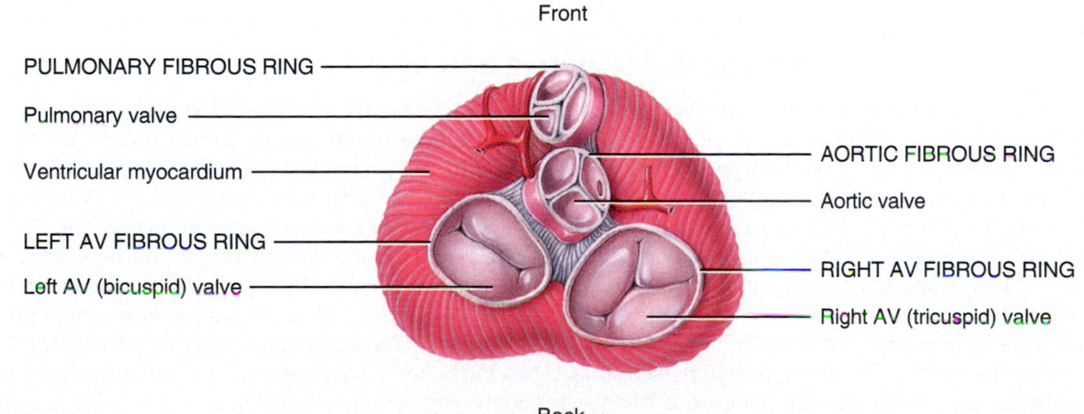

Superior view of the heart with the atria removed

In what two ways does the fibrous skeleton contribute to the functioning of heart valves?

CLINICAL CONNECTION

Heart Valve Disorders

When heart valves operate normally, they open fully and close completely at the proper times. A narrowing of a heart valve opening that restricts blood flow is known as **stenosis** (ste-NŌ-sis); failure of a valve to close completely is termed **insufficiency** or **incompetence**. In **mitral stenosis**, scar formation or a congenital defect causes narrowing of the mitral valve. In **mitral insufficiency** there is backflow of blood from the left ventricle into the left atrium. One cause of mitral insufficiency is **mitral valve prolapse (MVP)**, in which one or both cusps of the mitral valve protrude into the left atrium during ventricular contraction. Mitral valve prolapse is one of the most common valvular disorders, affecting as much as 30% of the population. It is more prevalent in women than in men, and does not always pose a serious threat. In **aortic stenosis** the aortic valve is narrowed, and in **aortic insufficiency** there is backflow of blood from the aorta into the left ventricle.

Certain infectious diseases can damage or destroy the heart valves. One example is **rheumatic fever**, an acute systemic inflammatory disease that usually occurs after a streptococcal infection of the throat. The bacteria trigger an immune response in which antibodies produced to destroy the bacteria instead attack and inflame the connective tissues in joints, heart valves, and other organs. Even though rheumatic fever may weaken the entire heart wall, most often it damages the mitral and aortic valves.

If a heart valve cannot be repaired surgically, then the valve must be replaced. Tissue (biologic) valves may be provided by human donors or pigs; sometimes mechanical (artificial) valves made of plastic or metal are used. The aortic valve is the most commonly replaced heart valve.

this reason, the heart wall has its own network of blood vessels, the **coronary** or **cardiac circulation** (*coronary* = crown). The **coronary arteries** branch from the aorta and encircle the heart like a crown encircles the head (**FIGURE 14.10a**). While the heart is contracting, little blood flows through the coronary arteries because they are squeezed shut. When the heart relaxes, however, the high pressure of blood in the aorta propels blood through the coronary arteries, into capillaries, and then into **coronary (cardiac) veins** (**FIGURE 14.10b**). The blood from the coronary veins drains into a large vein called the **coronary sinus**, which empties into the right atrium.

CHECKPOINT

3. Which layer forms the bulk of the heart wall?
4. What is the relationship between wall thickness and function among the various chambers of the heart?
5. What causes the heart valves to open and to close? What supporting structures ensure that the valves operate properly?
6. What is the functional significance of the fibrous skeleton of the heart?
7. Why does the heart have its own blood supply?

CLINICAL CONNECTION

Myocardial Ischemia and Infarction

Partial obstruction of blood flow in the coronary arteries may cause **myocardial ischemia** (is-KĒ-mē-a), a condition of reduced blood flow to the myocardium. Usually, ischemia causes **hypoxia** (reduced oxygen supply), which may weaken cells without killing them. **Angina pectoris** (an-JĪ-na, or AN-ji-na, PEK-to-ris), which literally means "strangled chest," is a severe pain that usually accompanies myocardial ischemia. Typically, sufferers describe it as a tightness or squeezing sensation, as though the chest were in a vise. The pain associated with angina pectoris is often referred to the neck, chin, or down the left arm to the elbow. **Silent myocardial ischemia**, ischemic episodes without pain, is particularly dangerous because the person has no forewarning of an impending heart attack.

A complete obstruction to blood flow in a coronary artery may result in a **myocardial infarction (MI)** (in-FARK-shun), commonly called a ***heart attack***. ***Infarction*** means the death of an area of tissue because of interrupted blood supply. Because the heart tissue distal to the obstruction dies and is replaced by noncontractile scar tissue, the heart muscle loses some of its strength. Depending on the size and location of the infarcted (dead) area, an infarction may disrupt the conduction system of the heart (described shortly) and cause sudden death by triggering ventricular fibrillation. Treatment for a myocardial infarction may involve injection of a thrombolytic (clot-dissolving) agent such as streptokinase or tissue plasminogen activator (tPA), plus heparin (an anticoagulant), or performing coronary angioplasty or coronary artery bypass grafting. Fortunately, heart muscle can remain alive in a resting person if it receives as little as 10–15% of its normal blood supply.

FIGURE 14.10 The coronary circulation.

The coronary arteries deliver blood to the heart wall; the coronary veins drain blood from the heart wall.

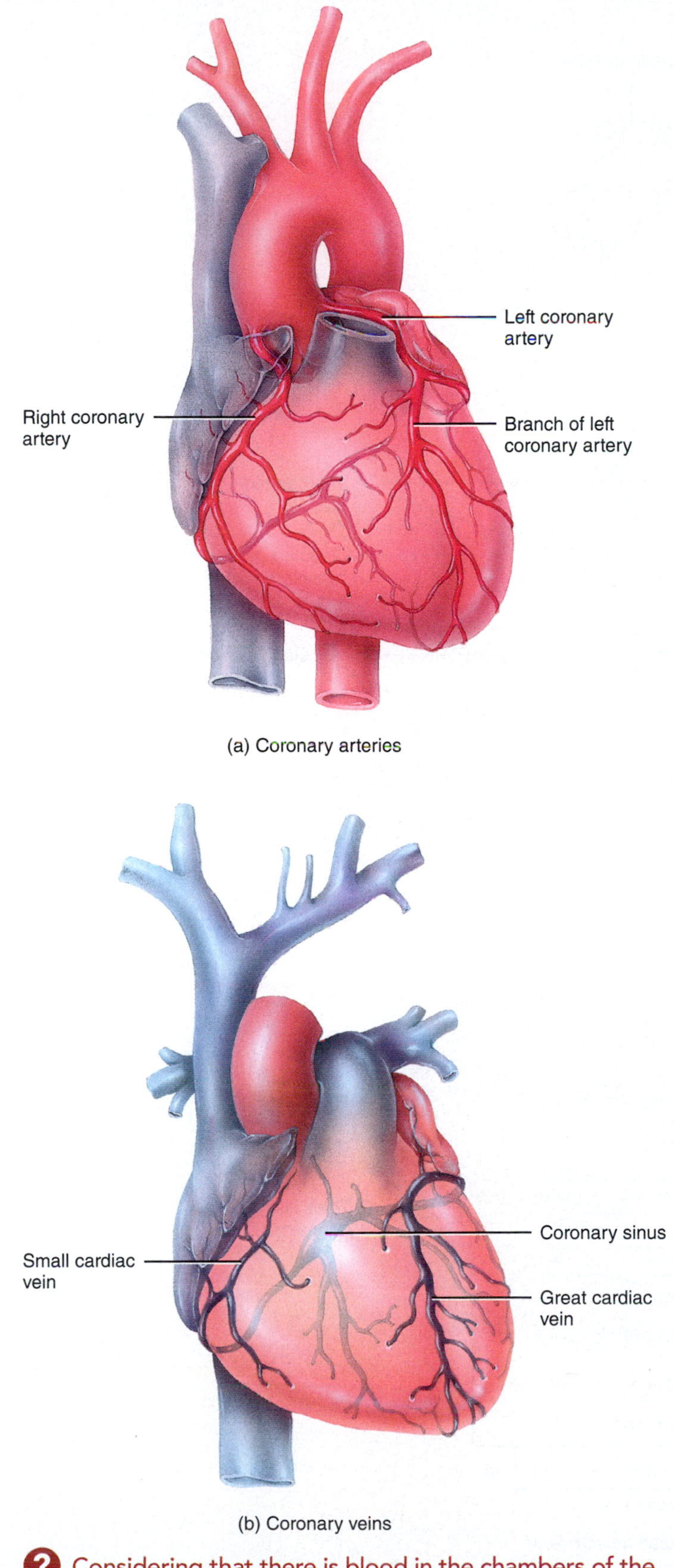

Considering that there is blood in the chambers of the heart, why does the heart wall need its own blood supply?

14.3 Cardiac Muscle Tissue and the Cardiac Conduction System

OBJECTIVES

- Describe the functional characteristics of cardiac muscle tissue and the cardiac conduction system.
- Outline the phases of an action potential in a contractile cardiac muscle fiber.
- Discuss how excitation–contraction coupling occurs in a cardiac muscle fiber.
- Explain the significance of the long refractory period in cardiac muscle.
- Discuss the electrical events of a normal electrocardiogram (ECG).

Interconnected Cardiac Muscle Fibers Act as a Functional Syncytium

A typical cardiac muscle fiber is 50–100 μm long and has a diameter of about 14 μm. Compared to skeletal muscle fibers, most cardiac muscle fibers are shorter in length and smaller in diameter. They also exhibit branching, which gives individual cardiac muscle fibers a stair-step appearance (**FIGURE 14.11**). A cardiac muscle fiber usually has one nucleus, although an occasional cell may contain two nuclei. Like skeletal muscle fibers, cardiac muscle fibers are striated due to repeating sarcomeres consisting of thick filaments and thin filaments that have a regular pattern of overlap. In addition, the sarcomeres of cardiac muscle fibers have the same zones and bands as those of skeletal muscle fibers due to the arrangement of the thick and thin filaments. The thick filaments contain myosin, and the thin filaments contain actin, troponin, and tropomyosin. The proteins of the thick and thin filaments function in the same way as in skeletal muscle fibers. Transverse (T) tubules and sarcoplasmic reticulum (SR) are also present in cardiac muscle fibers; however, compared to skeletal muscle, the T tubules are less abundant and the SR is somewhat smaller.

Unlike skeletal muscle fibers, the ends of cardiac muscle fibers are connected to one another by irregular transverse thickenings of the sarcolemma called **intercalated discs** (in-TER-kā-lāt-ed). An intercalated disc contains two types of cell junctions: desmosomes and gap junctions (**FIGURE 14.11**). **Desmosomes** mechanically bind cardiac muscle fibers together. They are resistant to mechanical stress, a property that prevents cardiac muscle fibers from pulling apart during contraction. **Gap junctions** electrically couple cardiac muscle fibers to each other. They allow action potentials to conduct from one cardiac muscle fiber to its neighbors. Because cardiac muscle fibers are interconnected by gap junctions, when an action potential is generated in a mass of cardiac muscle fibers, the action potential quickly spreads to all of the muscle fibers in that mass and then the muscle fibers contract together. Such a mass of interconnected muscle fibers acts as a single, coordinated unit or **functional syncytium** (sin-SISH-ē-um; plural is *syncytiums* or *syncytia*). Because the fibrous skeleton of the heart electrically insulates the atria from the ventricles, the atria and ventricles behave as two distinct functional syncytiums and therefore contract independently of each other. You will soon learn that the atria contract before the ventricles.

FIGURE 14.11 Microscopic organization of cardiac muscle fibers. A photomicrograph of cardiac muscle fibers is shown in Figure 11.1b.

Cardiac muscle fibers connect to neighboring fibers by intercalated discs, which contain desmosomes and gap junctions.

(a) Location of cardiac muscle in the heart wall

(b) Cardiac muscle fibers

(c) Arrangement of components in a cardiac muscle fiber

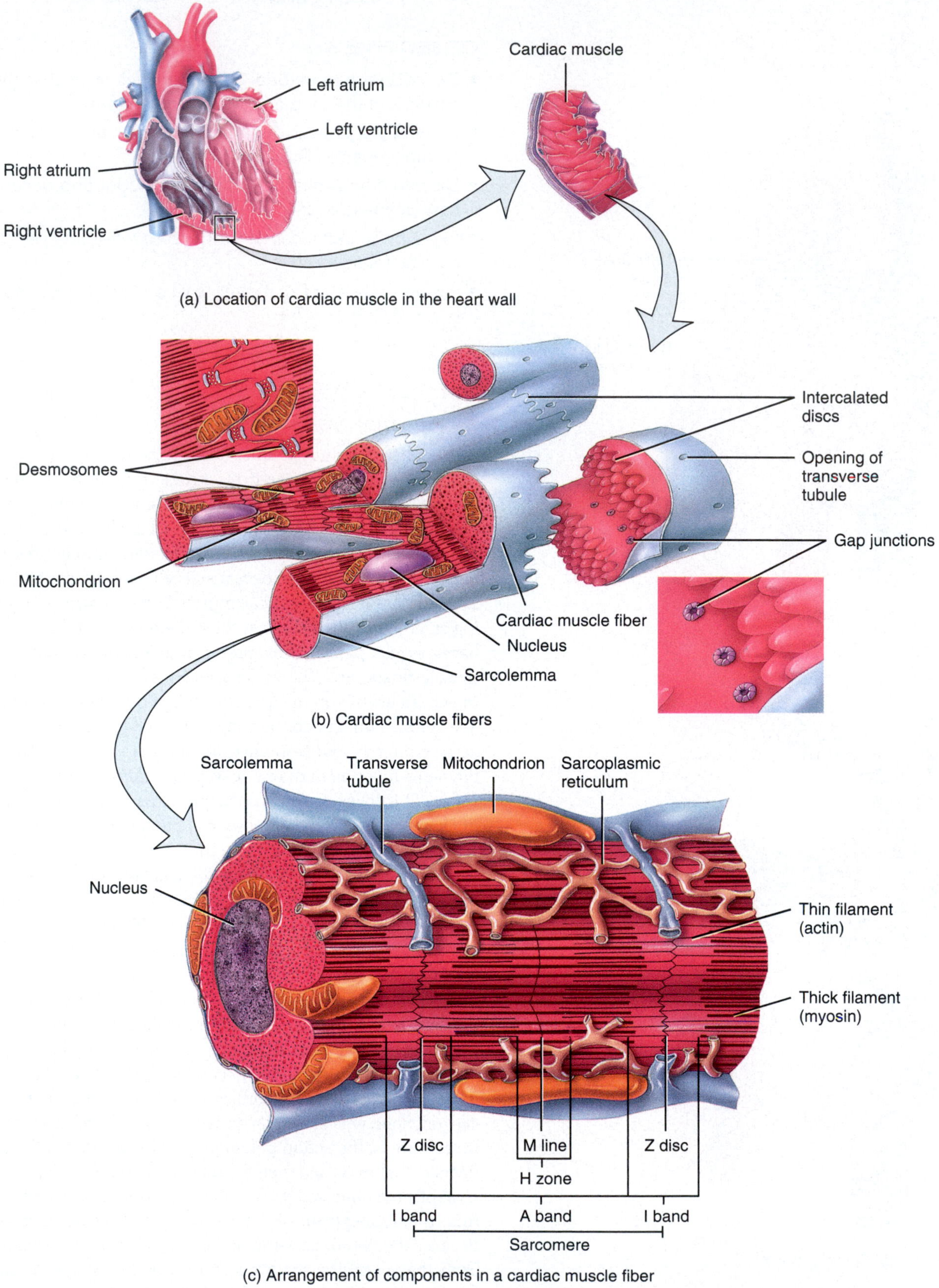

What are the functions of the desmosomes and gap junctions in the intercalated discs?

This allows the ventricles to fill with blood from the atria before the ventricles eject blood out of the heart to the rest of the body.

Mitochondria are larger and more numerous in cardiac muscle fibers than in skeletal muscle fibers. In a cardiac muscle fiber, they take up 25% of the volume of the sarcoplasm; in a skeletal muscle fiber, only 2% of the volume of the sarcoplasm is occupied by mitochondria. This structural feature means that cardiac muscle depends largely on aerobic respiration to generate ATP and consequently requires a constant supply of oxygen.

The Conduction System of the Heart Ensures Coordinated Contraction

Cardiac muscle does not require any external stimulation to contract. Contractions occur because action potentials within cardiac muscle itself are spontaneously generated on a periodic basis. This built-in rhythm is termed **autorhythmicity**.

Cardiac muscle, as a functional syncytium, consists of two types of muscle fibers: autorhythmic fibers and contractile fibers. *Autorhythmic fibers*, also known as *pacemaker cells*, spontaneously generate action potentials. They account for only a very small number of cells in the functional syncytium and are usually grouped together. Because autorhythmic fibers contain essentially no myofibrils, they are unable to contract. *Contractile fibers* constitute the great majority of cells in the functional syncytium. They have the necessary myofibrils to contract but do not have the ability to initiate action potentials. Instead, they become excited and then contract together in response to action potentials conducted to them from autorhythmic fibers via gap junctions.

The following components of the heart contain autorhythmic fibers (**FIGURE 14.12**):

- **Sinoatrial (SA) node** in the wall of the right atrium close to the opening of the superior vena cava.
- **Atrioventricular (AV) node** in the interatrial septum.
- **Atrioventricular (AV) bundle**, also known as the **bundle of His**, in the upper part of the interventricular septum.
- **Right** and **left bundle branches** in the interventricular septum.
- **Purkinje fibers** in the ventricular wall.

The autorhythmic fibers of the heart have two important functions:

1. They act as a **pacemaker**, setting the rhythm of electrical excitation that causes contraction of the heart.
2. They form the **conduction system**, the pathway that rapidly delivers action potentials throughout the heart muscle. Action potentials are able to be conducted throughout the heart by the conduction system because gap junctions connect the components of the conduction system to each other and to the contractile fibers of the heart. The conduction system ensures that cardiac chambers become stimulated to contract in a coordinated manner, which makes the heart an effective pump. As you will see later in the chapter, problems with the conduction system can result in arrhythmias (abnormal rhythms) in which the heart beats irregularly, too fast, or too slowly.

Cardiac action potentials propagate through the conduction system in the following sequence (**FIGURE 14.12**):

1. Cardiac excitation normally begins in the SA node, when SA node cells spontaneously depolarize to threshold by producing a pacemaker potential (described shortly). Once threshold is reached,

FIGURE 14.12 The conduction system of the heart.

The conduction system ensures that the chambers of the heart contract in a coordinated manner.

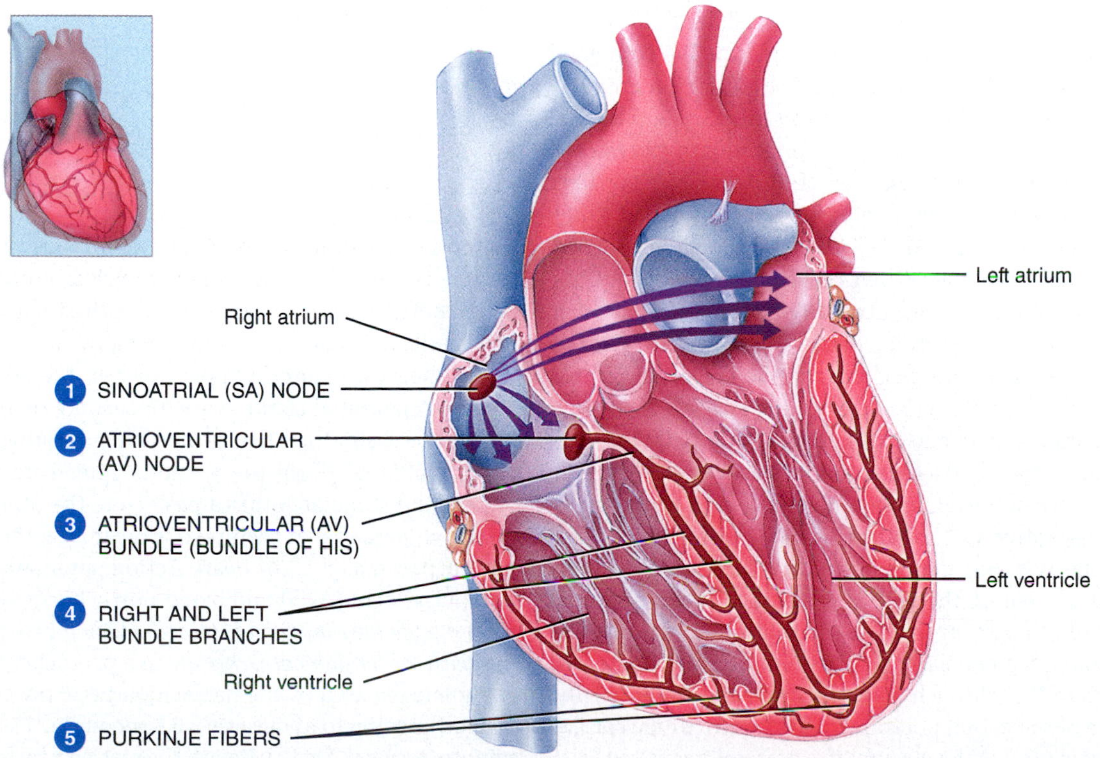

Which component of the conduction system provides the only electrical connection between the atria and the ventricles?

an action potential is generated and then propagates throughout both atria. Following the action potential, the atria contract.

2. The action potential propagates from the atria to the atrioventricular (AV) node.
3. From the AV node, the action potential enters the AV bundle (bundle of His). This bundle is the only site where action potentials can conduct from the atria to the ventricles. Recall that elsewhere the fibrous skeleton of the heart electrically insulates the atria from the ventricles.
4. After propagating along the AV bundle, the action potential enters both the right and left bundle branches.
5. From the bundle branches, the action potential propagates to the Purkinje fibers, which in turn conduct the action potential beginning at the apex of the heart to the remainder of the ventricular myocardium. Then the ventricles contract, pushing the blood upward toward the semilunar valves.

Autorhythmic fibers can initiate their own action potentials because they have unstable resting membrane potentials. The membrane potential starts at about −60 mV and then spontaneously depolarizes to threshold (−40 mV), at which point an action potential is generated. After repolarization, the membrane potential again starts to depolarize and the cycle repeats. The spontaneous depolarization to threshold that occurs in an autorhythmic fiber of cardiac muscle is known as a **pacemaker potential** (FIGURE 14.13). The first half of the pacemaker potential is caused by (1) the closure of voltage-gated K^+ channels that were open during the repolarizing phase of the previous action potential and (2) the opening of **F-type channels** (so-named because they have *funny* or unusual properties), which are mainly permeable to Na^+ ions. Closure of the voltage-gated K^+ channels decreases movement of K^+ out of the cell (K^+ has a higher concentration in the sarcoplasm than in extracellular fluid); opening of the F-type channels increases movement of Na^+ from extracellular fluid (which has a higher Na^+ concentration) into the sarcoplasm. The combined effects of these channel activities cause the membrane potential to start drifting slowly above −60 mV. Before the membrane potential reaches threshold, however, the F-type channels close and a new set of channels open: **T-type voltage-gated Ca^{2+} channels** (T for *transient* because they remain open for only a relatively short period of time). Opening of the T-type voltage-gated Ca^{2+} channels causes the second half of the pacemaker potential. When these channels open, Ca^{2+} enters the cell because the Ca^{2+} concentration is higher in extracellular fluid than in the sarcoplasm. The influx of Ca^{2+} depolarizes the membrane even further, eventually bringing it to threshold. Once threshold is reached, an action potential occurs.

In an autorhythmic cardiac muscle fiber, an action potential consists of a depolarizing phase and a repolarizing phase (FIGURE 14.13). The depolarizing phase of the action potential is caused by the opening of **L-type voltage-gated Ca^{2+} channels** (L for *long-lasting* because they open for a relatively long period of time). When these channels open, additional Ca^{2+} enters the cell, and the membrane potential rises above threshold to a positive value. (Recall that in neurons and skeletal muscle fibers, the depolarizing phase of the action potential is due to the influx of Na^+ through voltage-gated Na^+ channels.) The repolarizing phase of the action potential in an autorhythmic cardiac muscle fiber is caused by (1) the closure of L-type voltage-gated Ca^{2+} channels and (2) the opening of voltage-gated K^+ channels. Opening the voltage-gated K^+ channels allows K^+ ions to leave the cell, which decreases the membrane potential back to around −60 mV. Once an action potential is generated in an autorhythmic fiber, it spreads to the contractile fibers of cardiac muscle via gap junctions.

On their own, autorhythmic fibers in the SA node would initiate an action potential about every 0.6 second, or 100 times per minute. This rate is faster than that of any other autorhythmic fibers. Because action potentials from the SA node spread through the conduction system and stimulate other areas before the other areas can generate an action potential at their own, slower rate, the SA node acts as the natural pacemaker of the heart. Action potentials from the autonomic nervous system (ANS) and blood-borne hormones (such as epinephrine) *modify the timing and strength* of each heartbeat, but they *do not establish the fundamental rhythm.* In a person at rest, for example, acetylcholine released by the parasympathetic division of the ANS slows SA node pacing to about every 0.8 second or 75 action potentials per minute (FIGURE 14.13). Hence, the resting heart rate is about 75 beats per minute.

FIGURE 14.13 Pacemaker potentials and action potentials in an autorhythmic cardiac muscle fiber.

Autorhythmic cardiac muscle fibers can initiate their own action potentials because they have unstable resting membrane potentials.

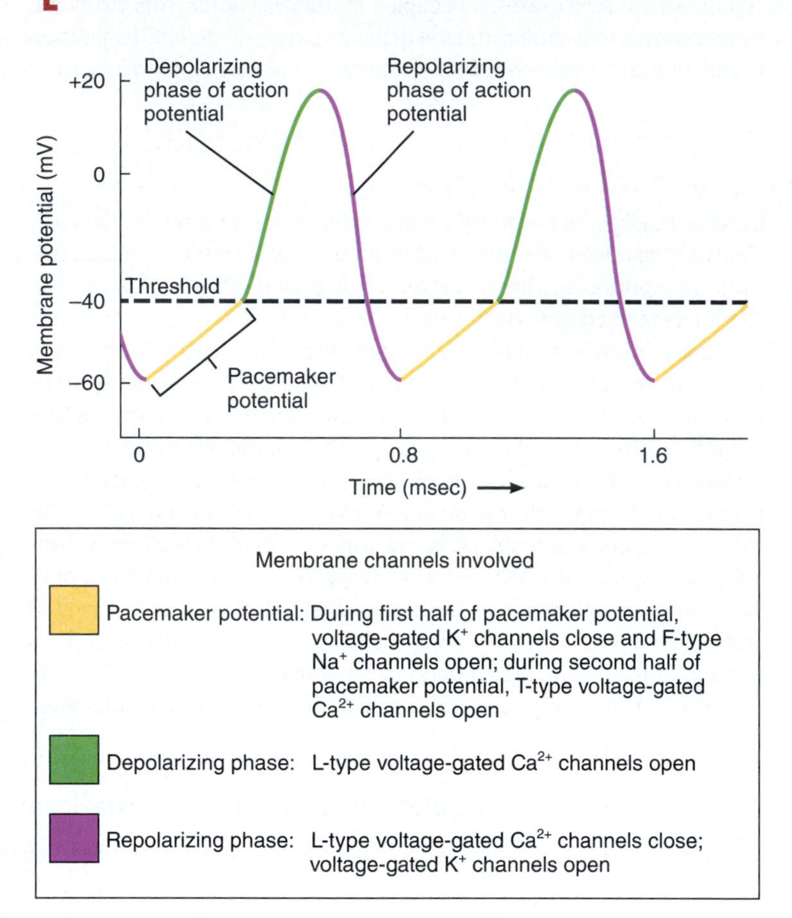

What is the pacemaker potential?

If the SA node becomes damaged or diseased, the AV node can pick up the pacemaking task. With pacing by the AV node, however, heart rate is slower, only 40 to 60 beats per minute. If activity of both nodes is suppressed, the heartbeat may still be maintained by the AV bundle, a bundle branch, or Purkinje fibers. These fibers generate action potentials very slowly, about 20 to 35 times per minute. At such a low heart rate, blood flow to the brain is inadequate.

Contractile Fibers Produce Action Potentials in Response to Autorhythmic Fibers

Unlike autorhythmic fibers, contractile cardiac muscle fibers have a stable resting membrane potential of about −90 mV. This value is a result of the fact that resting contractile fibers are highly permeable to K^+ ions and not very permeable to other ions. So the resting membrane potential stabilizes around the K^+ equilibrium potential of −90 mV. When a contractile cardiac muscle fiber is depolarized to threshold by an action potential initiated by an autorhythmic fiber, it produces its own action potential. The action potentials that occur in contractile cardiac muscle fibers consist of four phases: a depolarizing phase, initial repolarizing phase, plateau phase, and final repolarizing phase (**FIGURE 14.14**).

Depolarizing Phase

During the depolarizing phase, **fast voltage-gated Na^+ channels** open. These channels are referred to as *fast* because they open very rapidly in response to a threshold-level depolarization. They are the same type of voltage-gated Na^+ channels that are present in neurons and skeletal muscle fibers. Opening of these channels increases the membrane permeability to Na^+ ions, allowing Na^+ to flow into the cell. This produces a rapid depolarization that increases the membrane potential to about +20 mV.

Initial Repolarizing Phase

Within a few milliseconds, the fast Na^+ channels automatically inactivate, reducing the membrane permeability to Na^+. As a result, Na^+ inflow decreases. Of the several different types of voltage-gated K^+ channels present in a contractile cardiac muscle fiber, a subset known as **fast voltage-gated K^+ channels*** opens at this time, allowing K^+ ions to leave the cell. The closure of fast voltage-gated Na^+ channels and the opening of fast voltage-gated K^+ channels cause the initial repolarizing phase of the action potential. During this phase, the membrane potential begins to decrease.

Plateau Phase

The next phase of the action potential is the **plateau**, a period of sustained depolarization. It is due in part to the opening of L-type voltage-gated Ca^{2+} channels. When these channels open, calcium ions move from extracellular fluid into the cell. While this is occurring, fast voltage-gated K^+ channels close and slow voltage-gated K^+ channels begin to open. The **slow voltage-gated K^+ channels**† are so-named because they are activated when the membrane initially depolarizes but are slow to open. They are the same type of voltage-gated K^+ channels found in neurons and skeletal muscle fibers. Because the fast voltage-gated K^+ channels are completely closed and the slow voltage-gated K^+ channels are only partially open, the membrane permeability to K^+ is relatively low at this time. However, there is just enough efflux of K^+ through the slow voltage-gated K^+ channels to balance the Ca^{2+} influx through the L-type voltage-gated Ca^{2+} channels, causing the action potential curve to flatten out like a plateau. The plateau phase lasts for about 0.2 sec, and the membrane potential of the contractile fiber is close to 0 mV. By comparison, depolarization in a neuron or skeletal muscle fiber is much briefer, about 1 msec, because it lacks a plateau phase.

FIGURE 14.14 An action potential in a contractile cardiac muscle fiber.

An action potential in a contractile cardiac muscle fiber has four phases: a depolarizing phase, initial repolarizing phase, plateau phase, and final repolarizing phase.

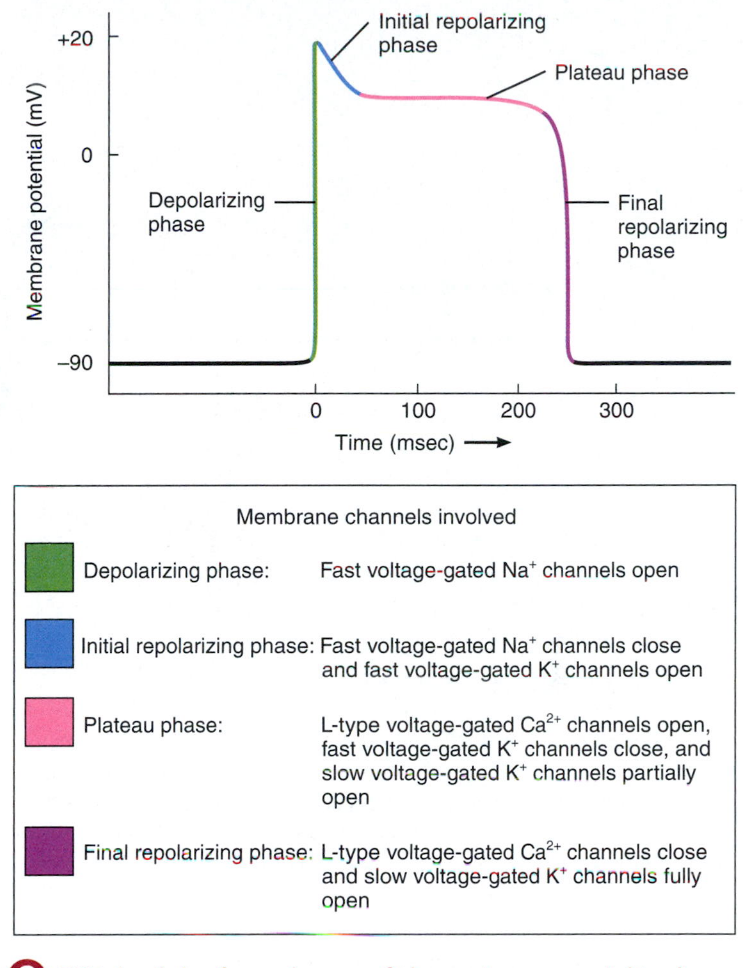

Which of the four phases of the action potential is the longest?

Final Repolarizing Phase

During the final repolarizing phase of the action potential, the slow voltage-gated K^+ channels fully open. This increases the membrane

* Fast voltage-gated K^+ channels are officially known as transient outward current K^+ channels.

† Slow voltage-gated K^+ channels are officially known as delayed rectifier K^+ channels.

CLINICAL CONNECTION

Artificial Pacemakers

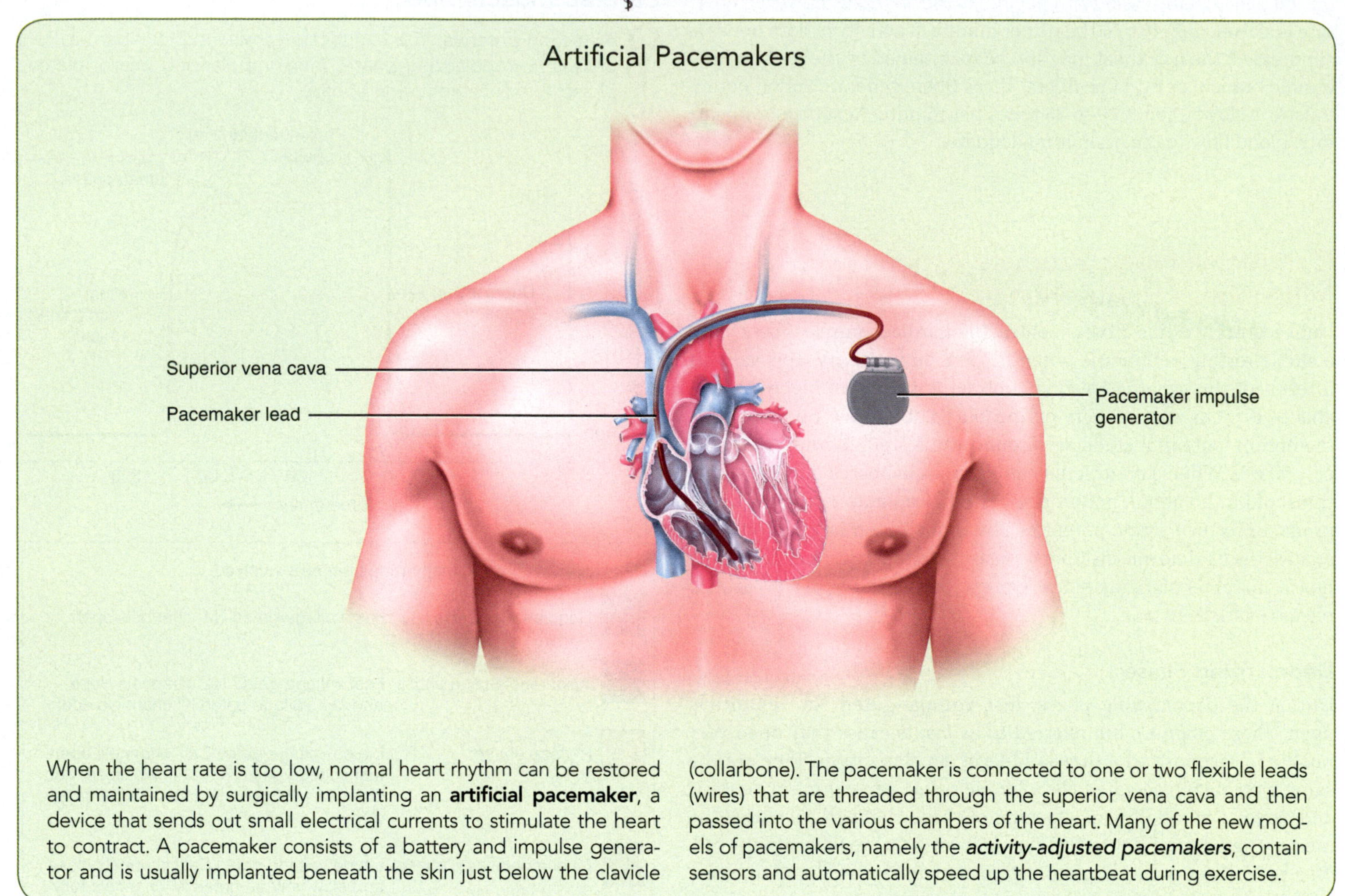

When the heart rate is too low, normal heart rhythm can be restored and maintained by surgically implanting an **artificial pacemaker**, a device that sends out small electrical currents to stimulate the heart to contract. A pacemaker consists of a battery and impulse generator and is usually implanted beneath the skin just below the clavicle (collarbone). The pacemaker is connected to one or two flexible leads (wires) that are threaded through the superior vena cava and then passed into the various chambers of the heart. Many of the new models of pacemakers, namely the ***activity-adjusted pacemakers***, contain sensors and automatically speed up the heartbeat during exercise.

permeability to K^+, accelerating K^+ outflow. At the same time, the L-type voltage-gated Ca^{2+} channels close. This decreases the membrane permeability to Ca^{2+}, reducing Ca^{2+} inflow. The increase in K^+ outflow and decrease in Ca^{2+} inflow rapidly restore the membrane potential to -90 mV. Once the membrane potential reaches the resting level, the slow voltage-gated K^+ channels close.

Excitation–Contraction Coupling Links Cardiac Action Potentials to Cardiac Contraction

Once an action potential is generated in a contractile cardiac muscle fiber, it ultimately causes the muscle fiber to contract by increasing the Ca^{2+} concentration in the sarcoplasm. The sequence of events that connects the muscle action potential to muscle contraction is known as **excitation–contraction (EC) coupling.** EC coupling in cardiac muscle involves L-type voltage-gated Ca^{2+} channels in the membrane of transverse (T) tubules and nearby Ca^{2+} release channels in the terminal cisternal membrane of the sarcoplasmic reticulum (SR) (FIGURE 14.15). As the cardiac muscle action potential travels along the sarcolemma and into the T tubules, L-type voltage-gated Ca^{2+} channels open, allowing Ca^{2+} to move from extracellular fluid into the sarcoplasm. The entering Ca^{2+} functions as *trigger* Ca^{2+} that binds to Ca^{2+} release channels in the sarcoplasmic reticulum (SR), causing the channels to open and release an even larger amount of Ca^{2+} into the sarcoplasm. The process by which extracellular Ca^{2+} triggers the release of additional Ca^{2+} from the SR is called **Ca^{2+}-induced Ca^{2+} release (CICR)**. About 90% of the calcium needed for contraction of a cardiac muscle fiber comes from the sarcoplasmic reticulum via CICR, and the remaining 10% of the requisite Ca^{2+} comes from extracellular fluid.

After the Ca^{2+} concentration in the sarcoplasm increases, cardiac muscle fibers contract in the same way as skeletal muscle fibers: Ca^{2+} binds to troponin, which in turn undergoes a conformational change that causes tropomyosin to move away from myosin-binding sites on actin (FIGURE 14.15). Once the binding sites are exposed, myosin binds to actin, and the thick and thin filaments begin sliding past one another (sliding filament mechanism).

Cardiac muscle is capable of producing graded contractions (contractions that vary in strength). These graded contractions do not involve recruitment of a larger number of muscle fibers because cardiac muscle is a functional syncytium: All muscle fibers in the syncytium contract at the same time, so no other muscle fibers can be added to increase the amount of tension generated. Instead, cardiac muscle produces graded contractions by increasing the strength of contraction of the existing muscle fibers in the syncytium. This occurs by the addition of more Ca^{2+} to the sarcoplasm, which in turn increases crossbridge formation.

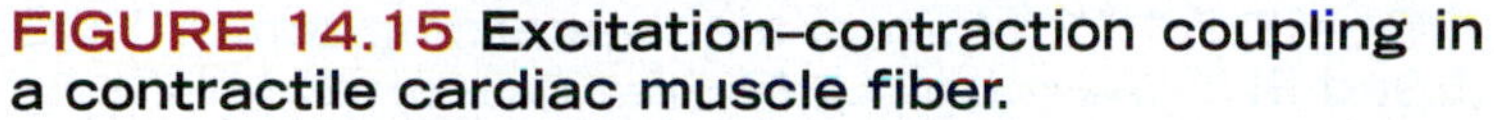

FIGURE 14.15 Excitation–contraction coupling in a contractile cardiac muscle fiber.

Excitation–contraction coupling refers to the sequence of events that connects the muscle action potential to muscle contraction.

What is Ca^{2+}-induced Ca^{2+} release (CICR)?

Relaxation of cardiac muscle fibers involves decreasing the Ca^{2+} concentration in the sarcoplasm to resting levels. The sarcoplasmic reticulum (SR) of a cardiac muscle fiber contains Ca^{2+}–ATPase pumps that actively transport Ca^{2+} from the sarcoplasm into the SR. In addition, the sarcolemma of a cardiac muscle fiber contains Na^{+}–Ca^{2+} exchangers that actively transport Ca^{2+} out of the cell in exchange for Na^{+} movement into the cell. As the Ca^{2+} concentration in the sarcoplasm drops, Ca^{2+} dissociates from troponin, tropomyosin covers the myosin-binding sites on actin, and the cardiac muscle fiber relaxes.

Cardiac Muscle Fibers Have a Long Refractory Period

Cardiac muscle fibers, like neurons and skeletal muscle fibers, have a *refractory period*, the period of time after an action potential begins when an excitable cell temporarily loses its excitability. The refractory period occurs because voltage-gated Na^{+} channels that are initially activated during the depolarizing phase of the action potential quickly become inactivated and must wait until the membrane repolarizes and returns to the resting state before they are capable of being activated again. Recall that in skeletal muscle fibers, the refractory period is about 1 msec, which is much shorter than the duration of contraction (20–200 msec). Because the refractory period is so short, a skeletal muscle fiber can be re-excited before its previous contraction is over. Consequently, skeletal muscle contractions can summate (add together in strength) and produce *tetanus*, a sustained contraction in which the muscle relaxes only slightly between stimuli (incomplete tetanus) or does not relax at all between stimuli (complete tetanus). In contractile cardiac muscle fibers, however, the refractory period is long (about 250 msec) due to the prolonged plateau phase of the action potential (FIGURE 14.16). The refractory period of these fibers lasts almost as long as the duration of contraction (300 msec). As a result, a contractile cardiac muscle fiber cannot be re-excited until its previous contraction is almost over. For this reason, summation of contractions and tetanus do not occur in cardiac muscle. The advantage is apparent if you consider that the pumping action of the heart depends on alternating contraction (when the heart ejects blood) and relaxation (when the heart refills). If cardiac muscle could undergo tetanus, the heart would no longer be able to function as a pump because it would not have a chance to relax and refill with blood—a situation that would be fatal.

FIGURE 14.16 Relationship among the action potential, refractory period, and tension developed in a contractile cardiac muscle fiber.

In cardiac muscle, the refractory period (about 250 msec) lasts almost as long as the duration of contraction (300 msec).

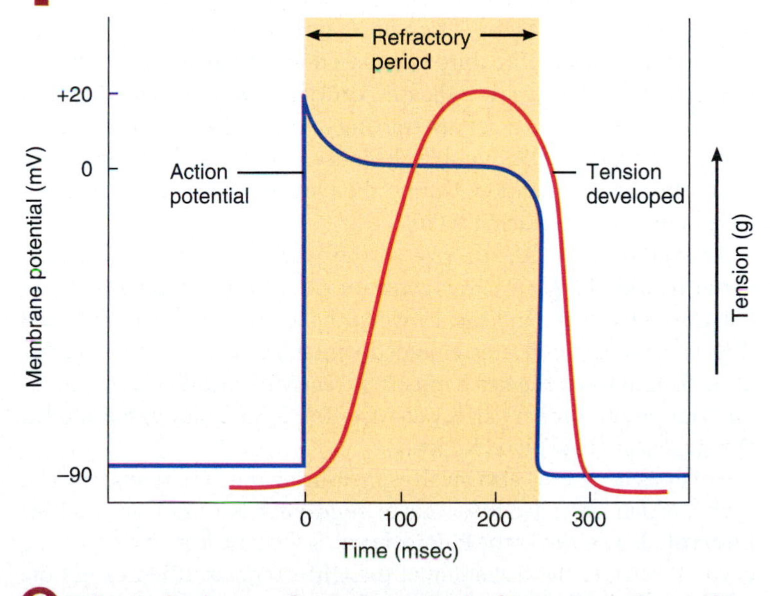

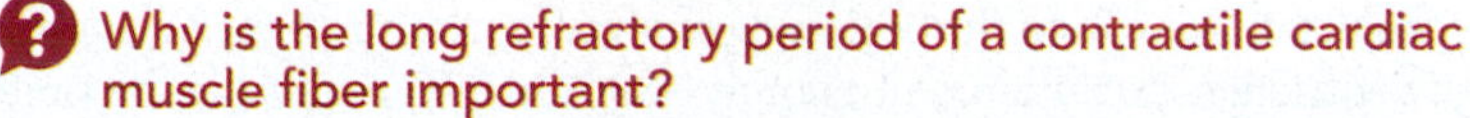

Why is the long refractory period of a contractile cardiac muscle fiber important?

Cardiac Muscle Produces ATP via Aerobic Respiration

In contrast to skeletal muscle, cardiac muscle produces little of the ATP it needs by anaerobic glycolysis (see **FIGURE 11.12**). Instead, it relies almost exclusively on aerobic respiration in its numerous mitochondria. The needed oxygen diffuses from blood in the coronary circulation and is released from myoglobin inside cardiac muscle fibers. Cardiac muscle fibers use several fuels to power mitochondrial ATP production. In a person at rest, cardiac muscle's ATP comes mainly from catabolism of fatty acids (60%) and glucose (35%), with smaller contributions from lactic acid, amino acids, and ketone bodies. During exercise, cardiac muscle's use of lactic acid, produced by actively contracting skeletal muscles, rises.

Like skeletal muscle, cardiac muscle also produces some ATP from creatine phosphate. One sign that a myocardial infarction (heart attack; see Clinical Connection: Myocardial Ischemia and Infarction in Section 14.2) has occurred is the presence in blood of creatine kinase (CK), the enzyme that catalyzes transfer of a phosphate group from creatine phosphate to ADP to make ATP. Normally, CK and other enzymes are confined within cells. Injured or dying cardiac or skeletal muscle fibers release CK into the blood.

The Electrocardiogram Records Electrical Signals Generated by the Heart

As action potentials propagate through the heart, they generate electrical currents that can be detected at the surface of the body. An **electrocardiogram** (e-lek′-trō-KAR-dē-ō-gram), abbreviated either **ECG** or **EKG** (from the German word *elektrokardiogram*), is a recording of these electrical signals. The ECG is a composite record of action potentials produced by all of the heart muscle fibers during each heartbeat.

FIGURE 14.17 Normal electrocardiogram or ECG (Lead II). P wave = atrial depolarization; QRS complex = ventricular depolarization; T wave = ventricular repolarization.

An ECG is a recording of the electrical activity that occurs with each heartbeat.

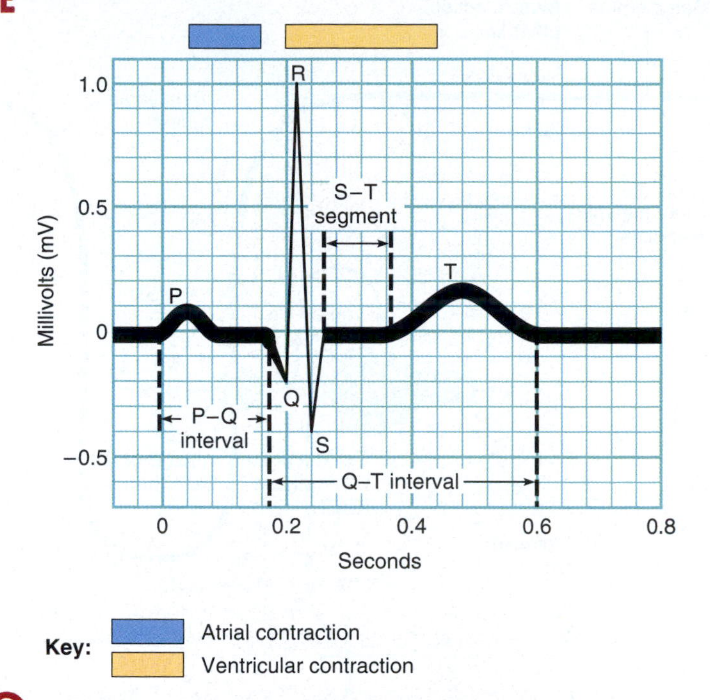

What is the significance of an enlarged Q wave?

In a typical ECG, three clearly recognizable waves appear with each heartbeat (**FIGURE 14.17**). The first, called the **P wave**, is a small upward deflection on the ECG. The P wave represents **atrial depolarization**, which spreads from the SA node through contractile fibers in both atria. The second wave, called the **QRS complex**, begins as a downward deflection; continues as a large, upright, triangular wave; and ends as a downward wave. The QRS complex represents **ventricular depolarization**, as the action potential spreads through ventricular contractile fibers. The third wave is a dome-shaped upward deflection called the **T wave.** It indicates **ventricular repolarization** and occurs just as the ventricles are starting to relax. The T wave is smaller and wider than the QRS complex because repolarization occurs more slowly than depolarization. During the plateau phase of steady depolarization, the ECG tracing is flat.

In reading an ECG, the size of the waves can provide clues to abnormalities. Larger P waves indicate enlargement of an atrium; an enlarged Q wave may indicate a myocardial infarction; and an enlarged R wave generally indicates enlarged ventricles. The T wave is flatter than normal when the heart muscle is receiving insufficient oxygen—for example, in coronary artery disease. The T wave may be elevated in hyperkalemia (high blood K^+ level).

Analysis of an ECG also involves measuring the time spans between waves, which are called **intervals** or **segments**. For example, the **P–Q interval**, also known as the **P–R interval**, is the time from the beginning of the P wave to the beginning of the QRS complex. It represents the conduction time from the beginning of atrial excitation to the beginning of ventricular excitation. Put another way, the P–Q interval is the time required for the action potential to travel through the atria, AV node, and the remaining fibers of the conduction system. When the action potential is forced to detour around scar tissue caused by disorders such as coronary artery disease and rheumatic fever, the P–Q interval lengthens.

The **S–T segment**, which begins at the end of the S wave and ends at the beginning of the T wave, represents the time when the ventricular contractile fibers are depolarized during the plateau phase of the action potential. The S–T segment is elevated (above the baseline) in acute myocardial infarction and depressed (below the baseline) when the heart muscle receives insufficient oxygen. The **Q–T interval** extends from the start of the QRS complex to the end of the T wave. It is the time from the beginning of ventricular depolarization to the end of ventricular repolarization. The Q–T interval may be lengthened by myocardial damage, myocardial ischemia (decreased blood flow), or conduction abnormalities.

ECG Waves Predict the Timing of Atrial and Ventricular Systole and Diastole

As you have learned, the atria and ventricles depolarize and then contract at different times because the conduction system routes cardiac action potentials along a specific pathway. The term **systole** (SIS-tō-lē = contraction) refers to the phase of contraction; the phase of relaxation is **diastole** (dī-AS-tō-lē = dilation or expansion). The ECG waves predict the timing of atrial and ventricular systole and diastole. At a heart rate of 75 beats per minute, the timing is as follows (**FIGURE 14.18**):

1. A cardiac action potential arises in the SA node. It propagates throughout the atrial muscle and down to the AV node in about

FIGURE 14.18 Timing and route of action potential depolarization and repolarization through the conduction system and myocardium. Green indicates depolarization, and red indicates repolarization.

Depolarization causes contraction and repolarization causes relaxation of cardiac muscle fibers.

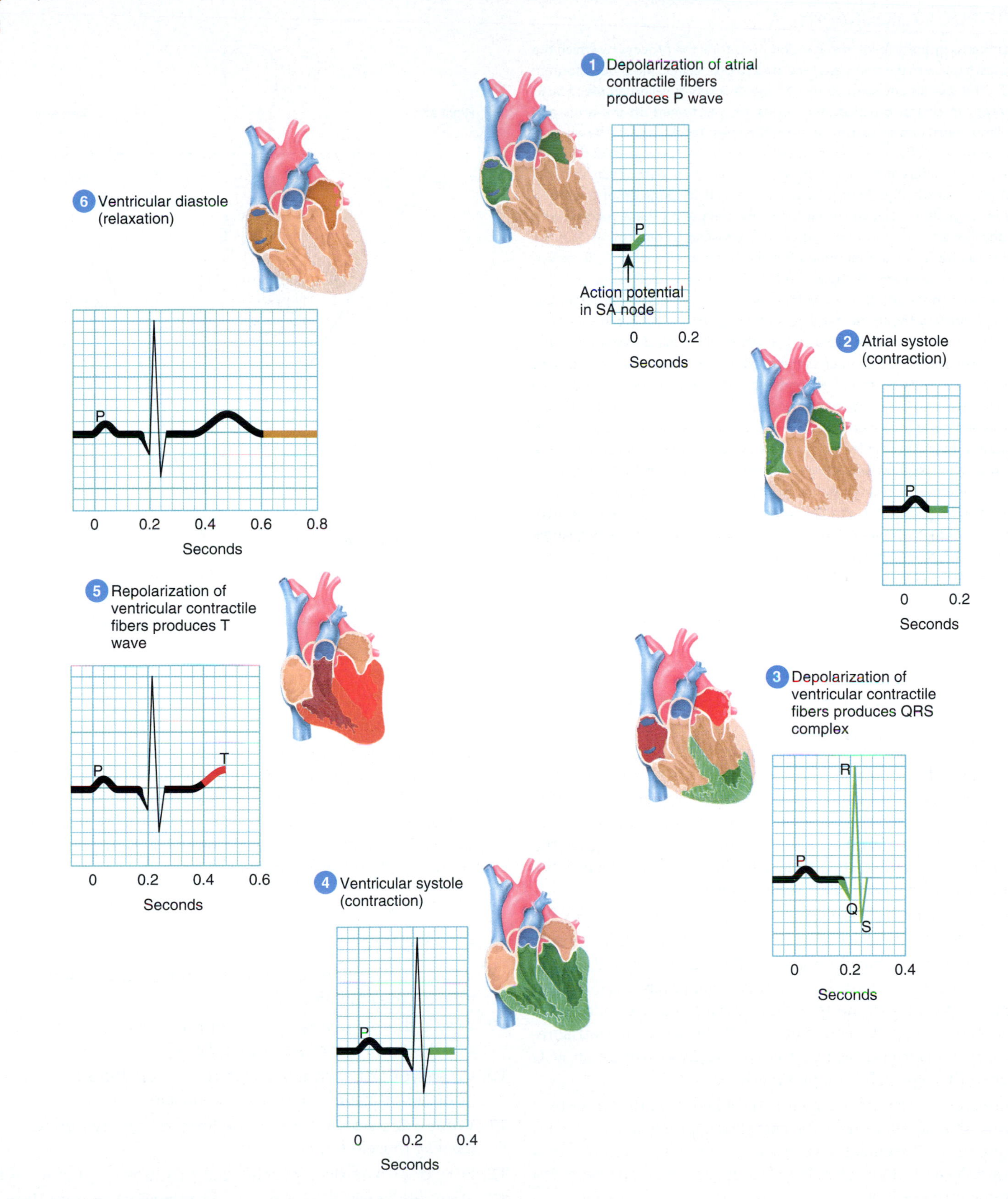

Where in the conduction system do action potentials propagate most slowly?

TOOL OF THE TRADE

ELECTROCARDIOGRAPHY

Electrocardiography (e-lek′-trō-kar-dē-OG-ra-fē) is the process by which the electrical signals of the heart are recorded to produce an electrocardiogram (ECG). The instrument used to record the changes is called an **electrocardiograph.** In clinical practice, electrodes are positioned on the arms and legs (limb leads) and at six positions on the chest (chest leads). The electrocardiograph amplifies the heart's electrical signals and produces 12 different tracings from different combinations of limb and chest leads. Each limb and chest lead records slightly different electrical activity because of the difference in its position relative to the heart. By comparing these records with one another and with normal records, it is possible to determine (1) if the conduction pathway is abnormal, (2) if the heart is enlarged, (3) if certain regions of the heart are damaged, and (4) the cause of chest pain.

Although there are 12 conventional leads used in electrocardiography, for simplicity only the three standard limb leads are shown in the FIGURE. Each standard limb lead consists of two electrodes (one designated as the positive electrode and the other as the negative electrode) that are attached to the surface of the body. In limb lead I, the positive electrode is connected to the left arm and the negative electrode to the right arm; in limb lead II, the positive electrode is connected to the left leg and the negative electrode to the right arm; and in limb lead III, the positive electrode is connected to the left leg and the negative electrode to the left arm. Each limb lead detects the electrical potential difference between the positive and negative electrodes and provides a separate electrocardiogram (ECG) recording. The three standard limb leads form an imaginary triangle known as ***Einthoven's triangle*** that extends between the arms and left leg to surround the heart. Note that the right leg is not part of the triangle but instead serves as the ground.

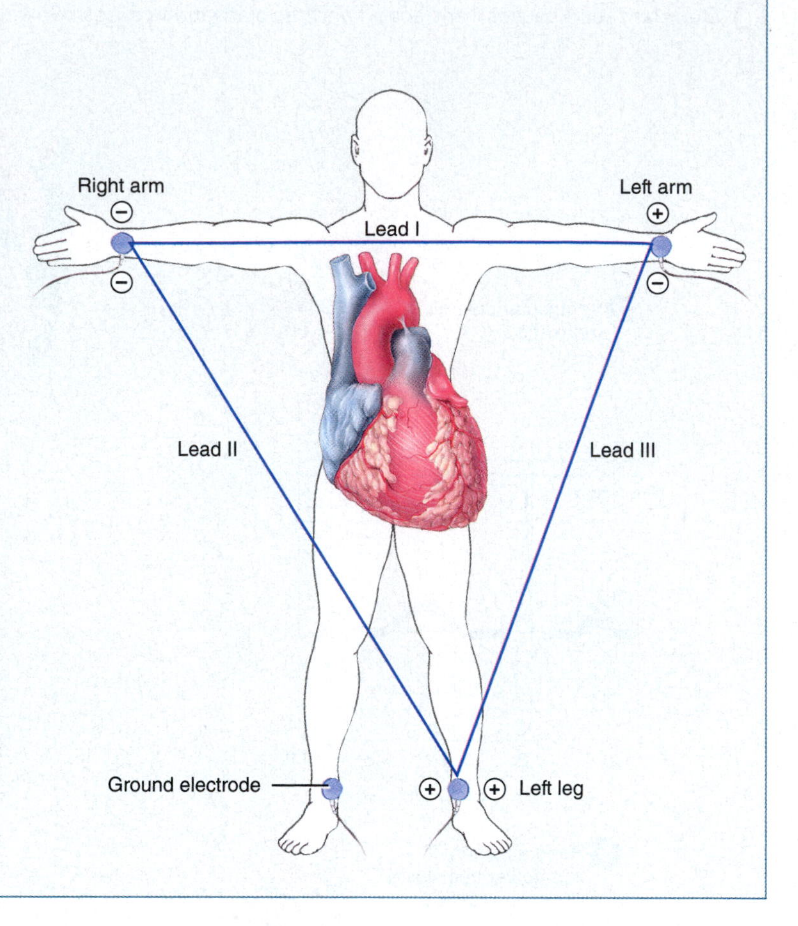

0.03 sec. As the atrial contractile fibers depolarize, the P wave appears on the ECG.

2. After the P wave begins, the atria contract (atrial systole). Conduction of the action potential slows at the AV node because the fibers there have much smaller diameters and fewer gap junctions. (Traffic slows in a similar way where a four-lane highway narrows to one lane in a construction zone!) The resulting 0.1-sec delay gives the atria time to contract, thus adding to the volume of blood in the ventricles, before ventricular systole begins.
3. The action potential propagates rapidly again after entering the AV bundle. About 0.2 sec after onset of the P wave, it has propagated through the bundle branches, Purkinje fibers, and the entire ventricular myocardium. Depolarization progresses down the septum, upward from the apex, and outward from the endocardial surface, producing the QRS complex. At the same time, atrial repolarization is occurring, but it is not usually evident in an ECG because the larger QRS complex masks it.
4. Contraction of ventricular contractile fibers (ventricular systole) begins shortly after the QRS complex appears and continues during the S–T segment. As contraction proceeds from the apex toward the base of the heart, blood is squeezed upward toward the semilunar valves.
5. Repolarization of ventricular contractile fibers begins at the apex and spreads throughout the ventricular myocardium. This produces the T wave in the ECG about 0.4 sec after the onset of the P wave.
6. Shortly after the T wave begins, the ventricles start to relax (ventricular diastole). By 0.6 sec, ventricular repolarization is complete and ventricular contractile fibers are relaxed. During the next 0.2 sec, contractile fibers in both the atria and ventricles are relaxed. At 0.8 sec, the P wave appears again on the ECG, the atria begin to contract, and the cycle repeats.

CHECKPOINT

8. What are the components of the conduction system and how do they function?
9. Why is there a plateau phase during the action potential of a contractile cardiac muscle fiber?
10. What types of channels play a role in excitation–contraction coupling in cardiac muscle?
11. What is the significance of the long refractory period in cardiac muscle fiber?
12. How does a cardiac muscle fiber produce most of its ATP?
13. How does each ECG wave, interval, and segment relate to contraction (systole) and relaxation (diastole) of the atria and ventricles?

Arrhythmias

The usual rhythm of heartbeats, established by the SA node, is called **normal sinus rhythm** (**FIGURE A**). The term **arrhythmia** (a-RITH-mē-a) or ***dysrhythmia*** refers to an abnormal rhythm as a result of a defect in the conduction system of the heart. The heart may beat irregularly, too quickly, or too slowly.

Arrhythmias are categorized by their speed, rhythm, and origination of the problem. **Tachycardia** (tak-i-KAR-dē-a; *tachy-* = quick) refers to a rapid heart rate (over 100 beats per minute); **bradycardia** (brād-e-KAR-dē-a; *brady-* = slow) refers to a slow heart rate (below 60 beats per minute); and **fibrillation** (fi-bri-LĀ-shun) refers to rapid, uncoordinated heartbeats.

- **Heart block** is an arrhythmia that occurs when the electrical pathways between the atria and ventricles are blocked, slowing the transmission of action potentials. The most common site of blockage is the atrioventricular node, a condition called ***atrioventricular (AV) block.*** In ***first-degree AV block***, the P–Q interval is prolonged, usually because conduction through the AV node is slower than normal. In ***second-degree AV block***, some of the action potentials from the SA node are not conducted through the AV node. The result is "dropped" beats because excitation doesn't always reach the ventricles. Consequently, there are fewer QRS complexes than P waves on the ECG (**FIGURE B**). In ***third-degree (complete) AV block***, no SA node action potentials get through the AV node. Autorhythmic fibers in the atria and ventricles pace the upper and lower chambers separately. With complete AV block, the ventricular contraction rate is less than 40 beats/min.
- **Atrial fibrillation (AF)** is a common arrhythmia, affecting mostly older adults, in which contraction of the atrial fibers is asynchronous (not in unison) so that atrial pumping ceases altogether. The atria may beat 300–600 beats/min. The ventricles may also speed up, resulting in a rapid heartbeat (up to 160 beats/min). The ECG of an individual with atrial fibrillation typically has no detectable P waves and irregularly spaced QRS complexes (**FIGURE C**). Because the atria and ventricles do not beat in rhythm, the heartbeat is irregular in timing and strength. In an otherwise strong heart, atrial fibrillation reduces the pumping effectiveness of the heart by 20–30%. The most dangerous complication of atrial fibrillation is stroke because blood may stagnate in the atria and form blood clots. A stroke occurs when part of a blood clot occludes an artery supplying the brain.
- **Ventricular fibrillation (VF** or **V-fib)** is the most deadly arrhythmia, in which contractions of the ventricular fibers are completely asynchronous so that the ventricles quiver rather than contract in a coordinated way. As a result, ventricular pumping stops, blood ejection ceases, and circulatory failure and death occur unless there is immediate medical intervention. During ventricular fibrillation, the ECG has no detectable P waves, QRS complexes, or T waves (**FIGURE D**). The most common cause of ventricular fibrillation is inadequate blood flow to the heart due to coronary artery disease, as occurs during a myocardial infarction. Other causes are cardiovascular shock, electrical shock, and very low potassium levels. Ventricular fibrillation causes unconsciousness in seconds and, if untreated, seizures occur and irreversible brain damage may occur after five minutes. Death soon follows. Treatment involves cardiopulmonary resuscitation (CPR) and defibrillation. In **defibrillation** (dē-fib-re-LĀ-shun), also called **cardioversion** (kar′-dē-ō-VER-shun), a strong, brief electrical current is passed to the heart and often can stop the ventricular fibrillation. The electrical shock is generated by a device called a **defibrillator** (de-FIB-ri-LĀ-tor) and applied via two large paddle-shaped electrodes pressed against the skin of the chest. Patients who face a high risk of dying from heart rhythm disorders now can receive an **automatic implantable cardioverter defibrillator (AICD)**, an implanted device that monitors their heart rhythm and delivers a small shock directly to the heart when a life-threatening rhythm disturbance occurs. Also available are **automated external defibrillators (AEDs)** that function like AICDs, except that they are external devices. About the size of a laptop computer, AEDs are used by emergency response teams and are found increasingly in public places such as stadiums, casinos, airports, hotels, and shopping malls. Defibrillation may also be used as an emergency treatment for cardiac arrest.

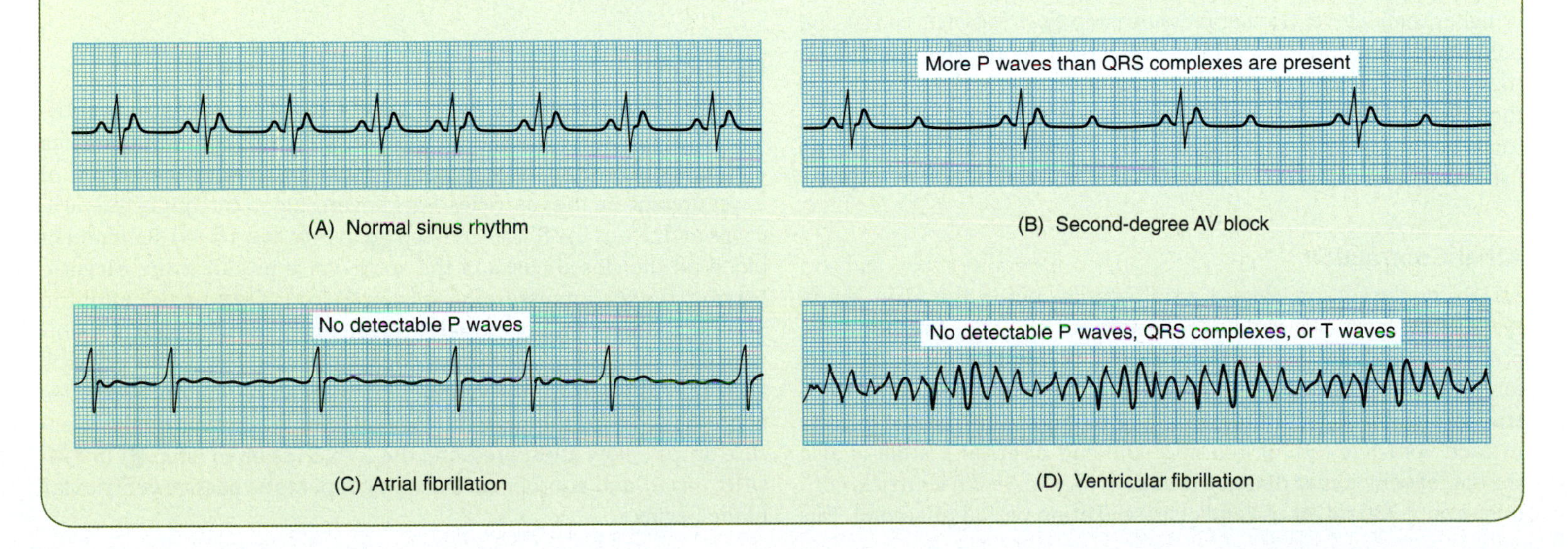

(A) Normal sinus rhythm

(B) Second-degree AV block

(C) Atrial fibrillation

(D) Ventricular fibrillation

14.4 The Cardiac Cycle

OBJECTIVES

- Describe the phases of the cardiac cycle.
- Discuss the duration of the cardiac cycle.
- Relate the timing of heart sounds to the ECG waves and pressure changes during systole and diastole.

A single **cardiac cycle** includes all of the events associated with one heartbeat. Thus, a cardiac cycle consists of diastole (relaxation) and systole (contraction) of the atria plus diastole and systole of the ventricles.

The Cardiac Cycle Has Five Phases

The cardiac cycle is divided into five phases: (1) passive ventricular filling, (2) atrial contraction, (3) isovolumetric ventricular contraction, (4) ventricular ejection, and (5) isovolumetric ventricular relaxation. FIGURE 14.19 shows the various phases of the cycle as well as the relationship between the heart's electrical signals and changes in pressure and volume during each phase. The pressures given in FIGURE 14.19 apply to the left side of the heart; pressures on the right side are considerably lower. Each ventricle, however, expels the same volume of blood per beat, and the same pattern exists for both pumping chambers.

Passive Ventricular Filling

Our discussion of the cardiac cycle begins during the period when both the atria and ventricles are in diastole. Atrial pressure is higher than ventricular pressure (step 1 in FIGURE 14.19) because the atria are filling with blood returning to the heart by veins. As a result of the pressure difference, the atrioventricular (AV) valves open, and blood flows from the atria into the ventricles. This phase of the cardiac cycle is known as **passive ventricular filling**. The term *passive* is used because no muscle contractions are involved. About 80% of ventricular filling occurs during this phase; the remaining 20% of ventricular filling occurs during atrial contraction. It is worth mentioning that the semilunar (SL) valves are closed at this time because aortic pressure is higher than left ventricular pressure (step 2 in FIGURE 14.19), and pulmonary trunk pressure is higher than right ventricular pressure. At the end of atrial diastole, an action potential arises in the SA node and then propagates throughout the atria, causing the atria to depolarize. Atrial depolarization is indicated by the P wave on the ECG (step 3 in FIGURE 14.19).

Atrial Contraction

Atrial depolarization causes atrial systole. While the atria are in systole, the ventricles remain in diastole. As the atria contract, atrial pressure increases (step 4 in FIGURE 14.19) and more blood is forced through the open AV valves into the ventricles. **Atrial contraction** contributes a final 25 mL of blood to the volume already in each ventricle (about 105 mL). The end of atrial systole is also the end of ventricular diastole (relaxation). Thus, each ventricle contains about 130 mL at the end of its relaxation period (diastole). This blood volume is called the **end-diastolic volume (EDV)** (step 5 in FIGURE 14.19). Toward the end of atria systole, the QRS complex appears on the ECG, marking the onset of ventricular depolarization (step 6 in FIGURE 14.19).

Isovolumetric Ventricular Contraction

Ventricular depolarization causes ventricular systole. While the ventricles are in systole, the atria are in diastole. As ventricular systole begins, pressure rises inside the ventricles and pushes blood up against the AV valves, forcing them shut (step 7 in FIGURE 14.19). For a brief moment, both the AV and SL valves are closed. This phase of the cardiac cycle is referred to as **isovolumetric ventricular contraction** (*iso-* = same). During this interval, cardiac muscle fibers are contracting and exerting force but are not yet shortening. Thus, the muscle contraction is isometric (same length). Moreover, because all four valves are closed, ventricular volume remains the same (isovolumic).

Ventricular Ejection

Continued contraction of the ventricles causes pressure inside the chambers to rise sharply. When left ventricular pressure surpasses aortic pressure at about 80 mmHg and right ventricular pressure rises above pulmonary trunk pressure (about 20 mmHg), both SL valves open (step 8 in FIGURE 14.19). At this point, the **ventricular ejection** phase of the cardiac cycle begins. During this phase, blood is pumped out of the heart. The left ventricle ejects about 70 mL of blood into the aorta, and the right ventricle ejects the same volume of blood into the pulmonary trunk. The volume remaining in each ventricle at the end of systole, about 60 mL, is the **end-systolic volume (ESV)** (step 9 in FIGURE 14.19). **Stroke volume**, the volume ejected per beat from each ventricle, equals end-diastolic volume minus end-systolic volume: SV = EDV − ESV. At rest, the stroke volume is about 130 mL − 60 mL = 70 mL (a little more than 2 oz). The percentage of the end-diastolic volume that is ejected with each stroke volume is called the **ejection fraction (EF)**: EF = SV/EDV × 100. Under normal resting conditions, EF is about 54% (70 mL/130 mL × 100). Changes in stroke volume alter the ejection fraction. Near the end of ventricular systole, the T wave appears on the ECG, marking the onset of ventricular repolarization (step 10 in FIGURE 14.19).

Isovolumetric Ventricular Relaxation

Ventricular repolarization causes ventricular diastole. As the ventricles relax, pressure within the chambers falls, and blood in the aorta and pulmonary trunk begins to flow backward toward the regions of lower pressure in the ventricles. Backflowing blood catches in the valve cusps and closes the SL valves (step 11 in FIGURE 14.19). Rebound of blood off the closed cusps of the aortic valve produces the **dicrotic wave** on the aortic pressure curve (step 12 in FIGURE 14.19). After the SL valves close, there is a brief interval when ventricular blood volume does not change because all four valves are closed. This phase is known as **isovolumetric ventricular relaxation**. As the ventricles continue to relax, the pressure falls quickly. When ventricular pressure drops below atrial pressure, the AV valves open (step 13 in FIGURE 14.19), and another cardiac cycle repeats as passive ventricular filling begins.

FIGURE 14.19 Cardiac cycle. (a) ECG. (b) Changes in left atrial pressure (green line), left ventricular pressure (blue line), and aortic pressure (red line) as they relate to the opening and closing of heart valves. (c) Heart sounds. (d) Changes in left ventricular volume. (e) Phases of the cardiac cycle.

A cardiac cycle includes all of the events associated with a single heartbeat.

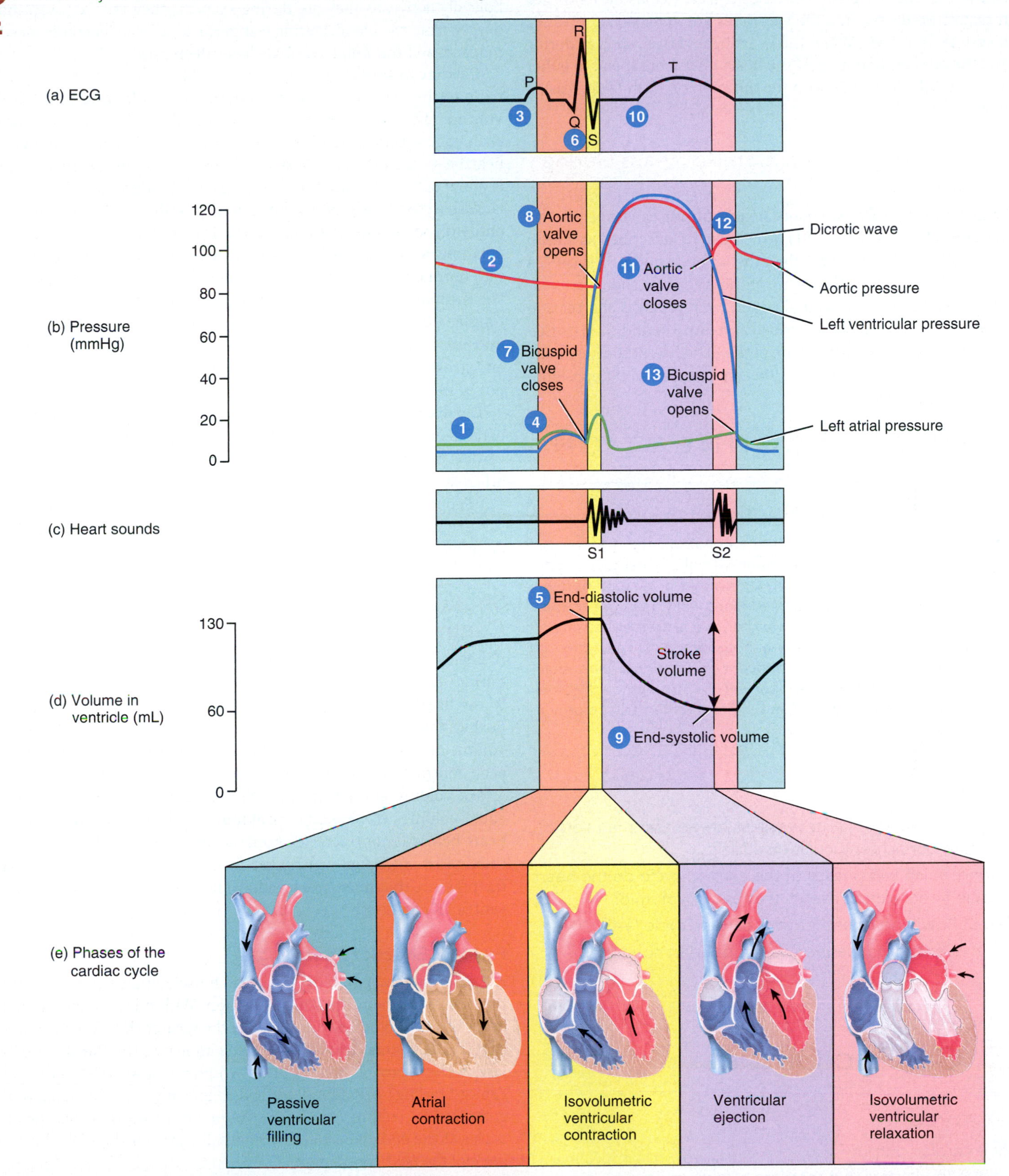

How much blood remains in each ventricle at the end of ventricular diastole in a resting person? What is this volume called?

...st a Cardiac Cycle Lasts About ...Seconds

...rest, when the heart rate is 75 beats per minute, a cardiac cycle lasts about 0.8 sec. In one complete cycle, the atria are in diastole for 0.7 sec and in systole for 0.1 sec, and the ventricles are in diastole for 0.5 sec and in systole for 0.3 sec. When the heart beats faster, such as during exercise, the amount of time that the heart chambers spend in either diastole or systole decreases, with the duration of diastole decreasing the most.

Two Major Heart Sounds Can Be Heard During Each Cardiac Cycle

During each cardiac cycle, two major **heart sounds** can be heard with a stethoscope. The first sound (S1), which can be described as a **lubb** sound, is louder and a bit longer than the second sound. S1 is caused by vibrations associated with closure of the AV valves soon after ventricular systole begins. The second sound (S2), which is shorter and not as loud as the first, can be described as a **dupp** sound. S2 is caused by vibrations associated with closure of the SL valves at the beginning of ventricular diastole. Thus, the heartbeat is heard as lubb-dupp, lubb-dupp, lubb-dupp, and so on.

CLINICAL CONNECTION

Heart Murmurs

Heart sounds provide valuable information about the mechanical operation of the heart. A **heart murmur** is an abnormal sound consisting of a clicking, rushing, or gurgling noise that is heard before, between, or after the normal heart sounds, or may mask the normal heart sounds. Heart murmurs in children are extremely common and usually do not represent a health condition. These types of heart murmurs are referred to as ***innocent*** or ***functional heart murmurs***; they often subside or disappear with growth. Although some heart murmurs in adults are innocent, most often an adult murmur indicates a valve disorder.

CHECKPOINT

14. Why must left ventricular pressure be greater than aortic pressure during ventricular ejection?

15. During which two phases of the cardiac cycle do the heart muscle fibers exhibit events that are isovolumic?

16. About how long does a normal cardiac cycle last?

17. What events produce the two major heart sounds?

14.5 Cardiac Output

OBJECTIVES

- Define cardiac output.
- Describe the factors that regulate stroke volume.
- Discuss the factors that regulate heart rate.

Although the heart has autorhythmic fibers that enable it to beat independently, its operation is governed by events occurring throughout the body. All body cells must receive a certain amount of oxygenated blood each minute to maintain health and life. When cells are metabolically active, as they are during exercise, they take up even more oxygen from the blood. During rest periods, cellular metabolic need is reduced, and the workload of the heart decreases.

Cardiac output (CO) is the volume of blood ejected from *each* ventricle of the heart per minute (**FIGURE 14.20**). It equals the **stroke volume (SV)**, the volume of blood ejected by the ventricle during each contraction, multiplied by the **heart rate (HR)**, the number of heartbeats per minute. In a typical adult male at rest, cardiac output is about 5.25 L/min (see the Physiological Equation box). This volume is close to the total blood volume, which is about 5 liters. Thus, your entire blood volume flows through your pulmonary and systemic circulations each minute. Factors that increase stroke volume or heart rate normally increase CO. During mild exercise, for example, stroke volume may increase to 100 mL/beat, and heart rate to 100 beats/min. Cardiac output then would be 10 L/min. During intense (but still not maximal) exercise, the heart rate may accelerate to 150 beats/min, and stroke volume may rise to 130 mL/beat, resulting in a cardiac output of 19.5 L/min.

Cardiac reserve is the difference between a person's maximum cardiac output and cardiac output at rest. The average person has a cardiac reserve of four or five times the resting value. Top endurance athletes may have a cardiac reserve seven or eight times their resting CO. People with severe heart disease may have little or no cardiac reserve, which limits their ability to carry out even the simple tasks of daily living.

Stroke Volume Is Regulated by Preload, Contractility, and Afterload

A healthy heart pumps out the blood that entered its chambers during the previous diastole. In other words, if more blood returns to the heart during diastole, then more blood is ejected during the next systole. At rest, the stroke volume is 50–60% of the end-diastolic volume because 40–50% of the blood remains in the ventricles after each contraction (end-systolic volume). Three major factors regulate stroke volume and ensure that the left and right ventricles pump equal volumes of blood: (1) **preload**, the degree of stretch on the heart before it contracts; (2) **contractility**, the forcefulness of contraction of individual ventricular muscle fibers; and (3) **afterload**, the pressure that must be exceeded before ejection of blood from the ventricles can occur.

Preload: Effect of Stretching

A greater preload (stretch) on cardiac muscle fibers prior to contraction increases their force of contraction. Within limits, the more the heart fills with blood during diastole, the greater the force of contraction during systole. This relationship is known as the **Frank–Starling law of the heart**, in honor of the two physiologists (Otto Frank and Ernest Starling) who first described it. Preload is proportional to the end-diastolic volume (EDV) (the volume of blood that fills the ventricles at the end of diastole). Normally, the greater the EDV, the more forceful the next contraction.

The Frank–Starling law is a manifestation of the length–tension relationship for cardiac muscle because EDV influences the

FIGURE 14.20 Cardiac output.

Cardiac output, the volume of blood ejected from each ventricle of the heart per minute, is equal to the product of the stroke volume and the heart rate.

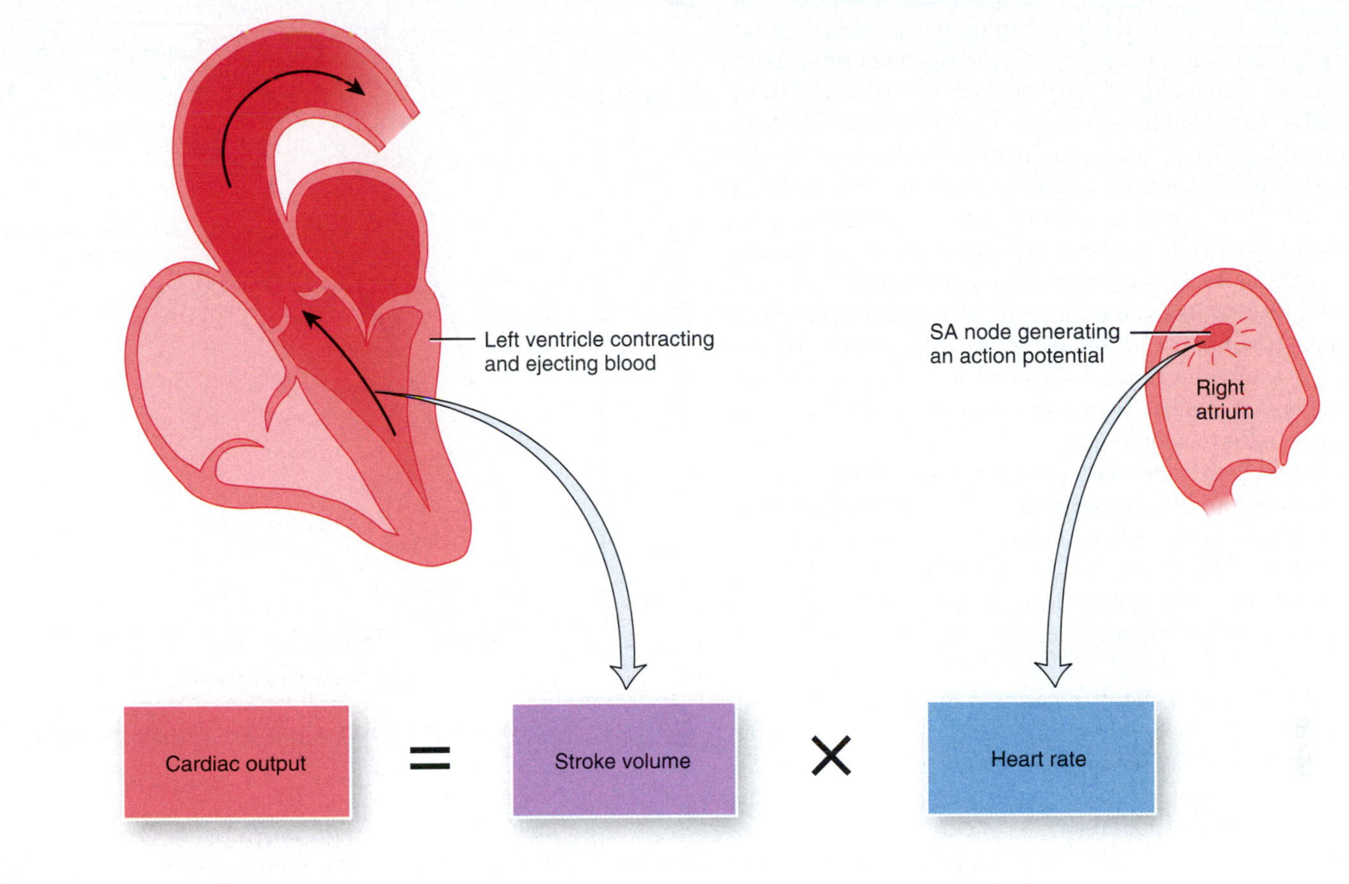

What is stroke volume?

length of the sarcomeres just before contraction begins. Recall that in the length–tension relationship for skeletal muscle, resting sarcomeres are held close to their optimal lengths (see FIGURE 11.16). The zone of overlap between the thick and thin filaments of each sarcomere is ideal, and the muscle fibers can develop maximum tension. When skeletal muscle fibers are stretched beyond their optimal lengths, there is less overlap between the thick and thin filaments. As a result, the tension that the fibers can produce

PHYSIOLOGICAL EQUATION

CALCULATION OF CARDIAC OUTPUT

Cardiac output, the volume of blood ejected from each ventricle of the heart per minute, is calculated using the following equation:

$$CO = SV \times HR$$

where CO is the cardiac output in mL/min,
SV is the stroke volume in mL/beat, and
HR is the heart rate in beats/min.

Recall that stroke volume is equal to the difference between the end diastolic volume (EDV) and the end systolic volume (ESV):

$$SV = EDV - ESV$$

Under resting conditions, EDV is 130 mL/beat and ESV is 60 mL/beat, so stroke volume is equal to 70 mL/beat:

$$SV = 130\ \text{mL/beat} - 60\ \text{mL/beat} = 70\ \text{mL/beat}$$

In a young adult male at rest, the normal heart rate is about 75 beats/min. Thus, cardiac output is

$$CO = 70\ \text{mL/beat} \times 75\ \text{beats/min} = 5250\ \text{mL/min} = 5.25\ \text{L/min}$$

decreases. In cardiac muscle, resting sarcomeres are held at a length that is shorter than the optimum (FIGURE 14.21). At such a length, the thin filaments from each side of the sarcomere overlap, reducing the amount of interaction between the thick and thin filaments. This results in a relatively low amount of tension development during contraction that is responsible for the normal stroke volume. However, when cardiac muscle fibers are stretched by an increase in EDV, the stretch causes the sarcomeres to get closer to their optimal lengths, resulting in greater tension development when the muscle fibers contract and an increase in stroke volume. If cardiac muscle fibers are stretched beyond their optimal lengths, there are fewer interactions between the thick and thin filaments, the tension that the fibers can produce decreases, and stroke volume falls. In a healthy heart, cardiac muscle is usually prevented from stretching beyond its optimal length by connective tissues in the wall of the heart, the fibrous skeleton, and the pericardium. Thus, cardiac muscle almost always operates along the ascending portion of the length–tension curve.

Two key factors determine EDV: (1) **filling time**, the duration of ventricular diastole, and (2) **venous return**, the volume of blood returning to the right ventricle. When heart rate increases, filling time is shorter. Less filling time means a smaller EDV, and the ventricles may contract before they are adequately filled. By contrast, when venous return increases, a greater volume of blood flows into the ventricles, and the EDV is increased.

When heart rate exceeds about 160 beats/min, stroke volume usually declines due to the short filling time. At such rapid heart rates, EDV is less, and the preload is lower. People who have slow resting heart rates usually have large resting stroke volumes because filling time is prolonged and preload is larger.

The Frank–Starling law of the heart equalizes the output of the right and left ventricles and keeps the same volume of blood flowing to both the systemic and pulmonary circulations. If the left side of the heart pumps a little more blood than the right side, the volume of blood returning to the right ventricle (venous return) increases. The increased EDV causes the right ventricle to contract more forcefully on the next beat, bringing the two sides back into balance.

FIGURE 14.21 Relationship between end-diastolic volume and stroke volume (Frank–Starling law of the heart).

Cardiac muscle usually functions along the ascending limb of the length–tension curve; so, as end-diastolic volume increases, stroke volume increases.

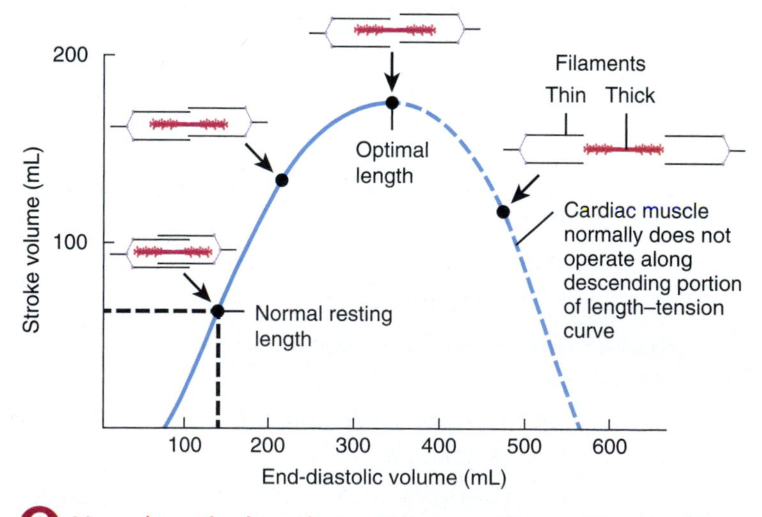

How does the length–tension curve for cardiac muscle differ from that for skeletal muscle?

FIGURE 14.22 Inotropic effects.

A positive inotropic effect is an increase in contractility; a negative ionotropic effect is a decrease in contractility.

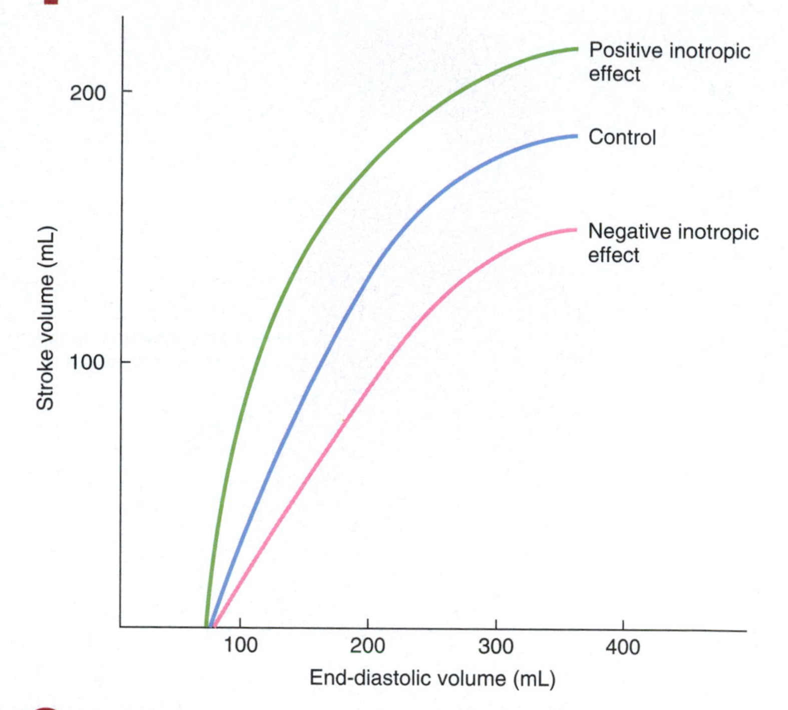

How do positive inotropic agents increase contractility?

Contractility

The second factor that influences stroke volume is myocardial **contractility**, the strength of contraction at any given preload. Agents that alter contractility are said to have an **inotropic effect**: Those that increase contractility have a **positive inotropic effect**, and those that decrease contractility have a **negative inotropic effect**. Thus, for a constant preload, a positive inotropic effect causes an increase in stroke volume, and a negative inotropic effect causes a decrease in stroke volume (FIGURE 14.22).

Stimulation of the sympathetic nervous system, hormones such as epinephrine and norepinephrine, increased extracellular Ca^{2+} levels, and the drug digitalis all have positive inotropic effects. Positive inotropic agents enhance contractility by increasing the amount of Ca^{2+} available in the sarcoplasm during cardiac action potentials. As an example, consider how an increase in contractility is caused by sympathetic stimulation (FIGURE 14.23):

1. Norepinephrine released from sympathetic nerve endings binds to β_1-adrenergic receptors in the sarcolemma of a contractile muscle fiber in the ventricular myocardium.

FIGURE 14.23 Mechanism of the sympathetic positive inotropic effect.

In response to sympathetic stimulation of the heart, norepinephrine activates a G-protein-signaling pathway that increases contractility by phosphorylating several target proteins.

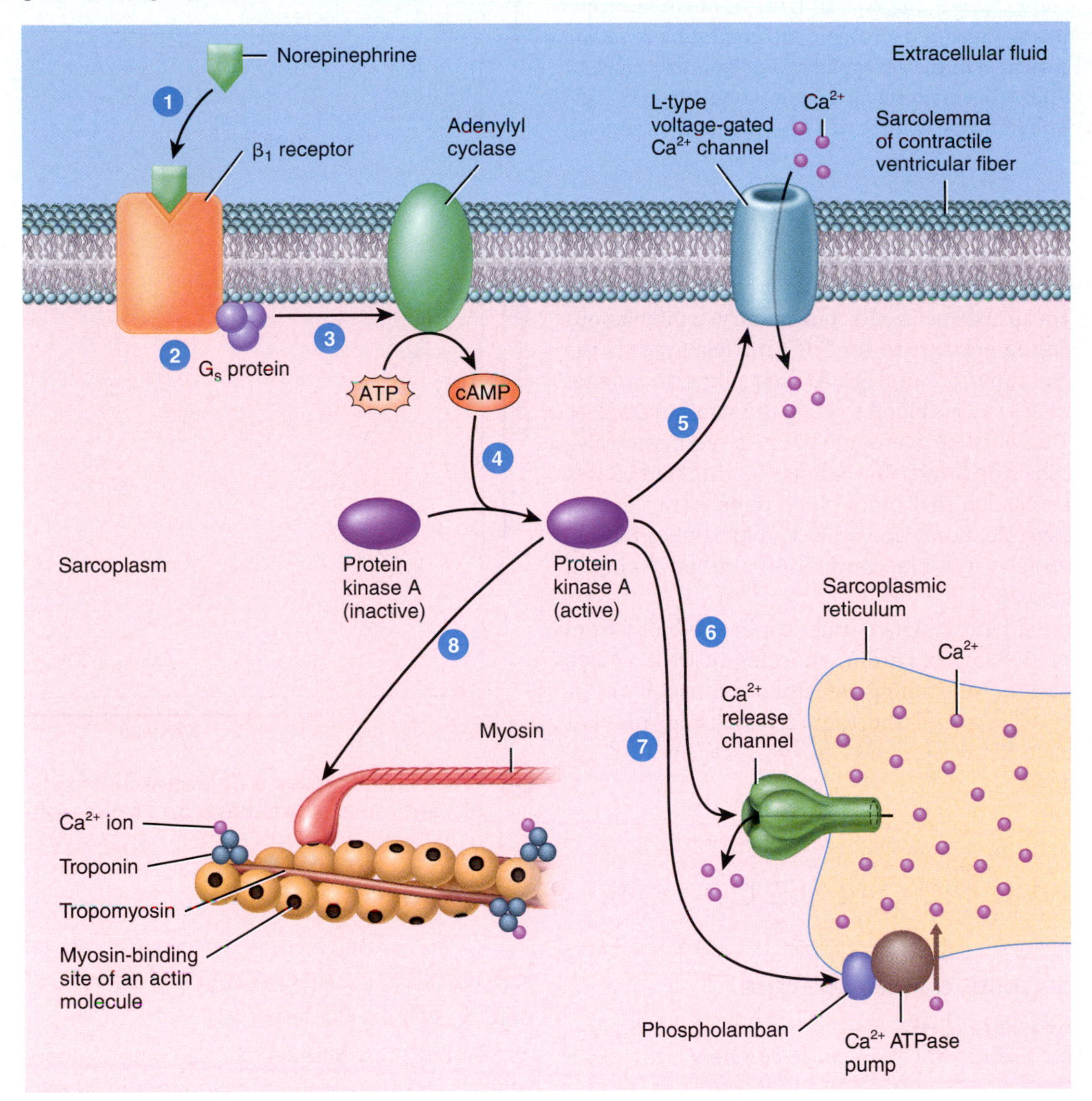

What are the target proteins in this pathway that are phosphorylated to promote an increase in contractility?

2 Binding of norepinephrine to the receptor activates a stimulatory G protein (G_s).

3 Activated G_s stimulates adenylyl cyclase to produce the second messenger cyclic AMP (cAMP).

4 cAMP binds to and activates a protein kinase (protein kinase A). The activated protein kinase in turn phosphorylates a variety of proteins with effects that ultimately cause an increase in contractility (see steps 5 through 8).

5 Phosphorylation of L-type voltage-gated Ca^{2+} channels in the sarcolemma increases the duration of the open state of these channels, allowing more Ca^{2+} to move from extracellular fluid into the sarcoplasm for contraction.

6 Phosphorylation of Ca^{2+} release channels in the membrane of the sarcoplasmic reticulum (SR) enhances Ca^{2+} release from the SR lumen into the sarcoplasm, providing additional Ca^{2+} for contraction.

7 Phosphorylation of phospholamban in the SR membrane results in an increase in Ca^{2+} uptake into the SR lumen by Ca^{2+}-ATPase pumps. This speeds relaxation and makes more Ca^{2+} available to be released for the next contraction. **Phospholamban** is a regulatory protein that normally inhibits the Ca^{2+}-ATPase. When phospholamban is phosphorylated, it is inactivated, and the inhibition of the Ca^{2+}-ATPase is removed.

8 Phosphorylation of myosin heads of the thick filaments enhances myosin ATPase activity, which increases the rate of crossbridge cycling.

In contrast to positive inotropic effects, negative inotropic effects decrease contractility. Inhibition of the sympathetic nervous system, excess H^+ ions (from acidosis), increased extracellular K^+ levels, and drugs known as *calcium channel blockers* all have negative inotropic effects. Negative inotropic agents decrease contractility by reducing the amount of Ca^{2+} available in the sarcoplasm. For example, calcium channel blockers inhibit the opening of L-type voltage-gated Ca^{2+} channels in the sarcolemma of contractile fibers, thereby reducing Ca^{2+} inflow.

Afterload

Ejection of blood from the heart begins when pressure in the right ventricle exceeds the pressure in the pulmonary trunk (about 20 mmHg) and when the pressure in the left ventricle exceeds the pressure in the aorta (about 80 mmHg). At that point, the higher pressure in the ventricles causes blood to push the semilunar valves open. The pressure that must be overcome before a semilunar valve can open is termed the **afterload**. An increase in afterload causes stroke volume to decrease so that more blood remains in the ventricles at the end of systole. Conditions that can increase afterload include hypertension (elevated blood pressure) and narrowing of arteries by atherosclerosis.

The effect of afterload on stroke volume can be predicted from the load–velocity relationship of cardiac muscle (**FIGURE 14.24**). As afterload increases, the shortening velocity of ventricular muscle fibers decreases. A decreased shortening velocity means that the ventricular muscle fibers eject less blood, resulting in a decrease in stroke volume.

FIGURE 14.24 Effect of afterload on velocity of shortening in ventricular muscle fibers.

As afterload increases, the velocity of shortening of ventricular muscle fibers decreases.

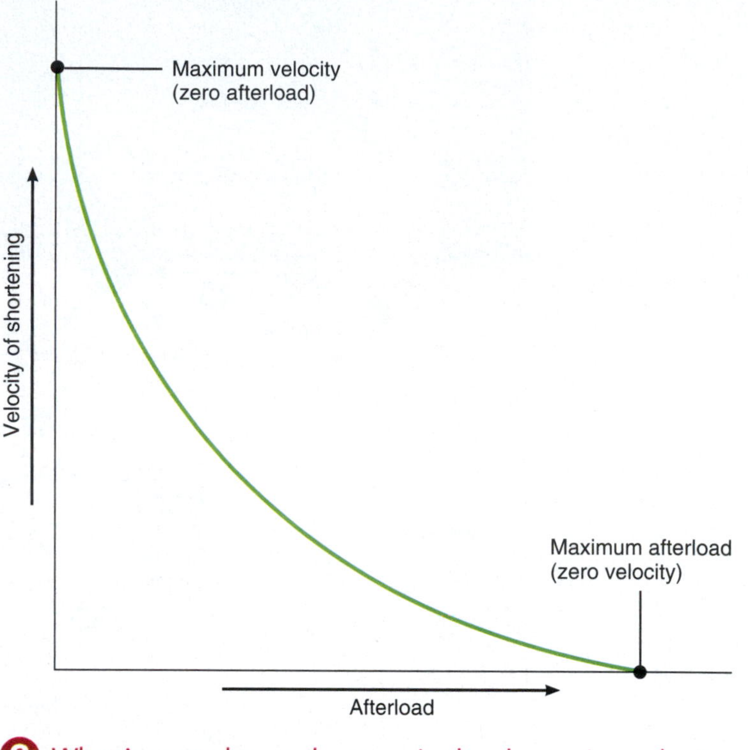

What impact does a decrease in the shortening velocity of ventricular fibers have on stroke volume?

CLINICAL CONNECTION

Congestive Heart Failure

In **congestive heart failure (CHF)**, there is a loss of pumping efficiency by the heart. Causes of CHF include coronary artery disease, congenital defects, long-term high blood pressure (which increases the afterload), myocardial infarctions (regions of dead heart tissue due to a previous heart attack), and valve disorders. As the pump becomes less effective, more blood remains in the ventricles at the end of each cycle, and gradually the end-diastolic volume (preload) increases. Initially, increased preload may promote increased force of contraction (the Frank–Starling law of the heart), but as the preload increases further, the ventricles become overstretched (enlarged) and contract less forcefully. The result is a potentially lethal positive feedback loop: Less-effective pumping leads to even lower pumping capability.

Often, one side of the heart starts to fail before the other. If the left ventricle fails first, it can't pump out all of the blood it receives. As a result, blood backs up in the lungs and causes ***pulmonary edema***, fluid accumulation in the lungs that can cause suffocation if left untreated. If the right ventricle fails first, blood backs up in the systemic veins. The resulting ***peripheral edema*** usually is most noticeable in the feet and ankles.

Heart Rate Is Regulated Mainly by the ANS and Certain Chemicals

As you have just learned, cardiac output depends on both stroke volume and heart rate. Adjustments in heart rate are important in the short-term control of cardiac output and blood pressure. The sinoatrial (SA) node initiates contraction and, if left to itself, would set a constant heart rate of about 100 beats/min. However, tissues require different volumes of blood flow under different conditions. During exercise, for example, cardiac output rises to supply working tissues with increased amounts of oxygen and nutrients. Stroke volume may fall if the ventricular myocardium is damaged or if blood volume is reduced by bleeding. In these cases, homeostatic mechanisms maintain adequate cardiac output by increasing the heart rate and contractility. The two main factors that regulate heart rate are the autonomic nervous system and certain chemicals (including several hormones and ions). Other factors that contribute to regulation of heart rate include age, gender, physical fitness, and body temperature. Agents that alter the heart rate are said to have a **chronotropic effect**. Those that increase heart rate have a **positive chronotropic effect**; those that decrease heart rate have a **negative chronotropic effect**.

Autonomic Regulation of Heart Rate

The heart receives innervation from both the sympathetic and parasympathetic divisions of the autonomic nervous system (ANS) (FIGURE 14.25). Sympathetic innervation of the heart begins with sympathetic preganglionic neurons, which extend from the thoracic spinal cord to sympathetic trunk ganglia, where they synapse with sympathetic postganglionic neurons. Axons of the sympathetic postganglionic neurons in turn form sympathetic nerves known as **cardiac accelerator nerves** that extend to the SA node, AV node, and ventricular myocardium. Action potentials in cardiac accelerator nerves trigger the release of norepinephrine, which binds to β_1-adrenergic receptors on cardiac cells.

Parasympathetic innervation of the heart begins with parasympathetic preganglionic neurons, which extend from the brain stem via the vagus (X) nerves to synapse with parasympathetic postganglionic neurons in the heart wall. Axons of the parasympathetic postganglionic neurons in turn terminate mainly in the SA node and AV node, with essentially no inputs to the ventricles. Action potentials in parasympathetic postganglionic axons trigger the release of acetylcholine, which binds to muscarinic cholinergic receptors on cardiac cells.

Sympathetic Regulation Stimulation of the sympathetic nervous system has three main effects on the heart: (1) it increases heart rate; (2) it increases the rate of action potential conduction from the atria to the ventricles; and (3) it increases contractility.

- ***Increase in heart rate.*** Sympathetic stimulation of the SA node causes an increase in heart rate. This occurs by increasing the rate of spontaneous depolarization in SA node cells so that threshold for an action

FIGURE 14.25 Autonomic innervation of the heart.

The heart is innervated by both the sympathetic and parasympathetic divisions of the autonomic nervous system.

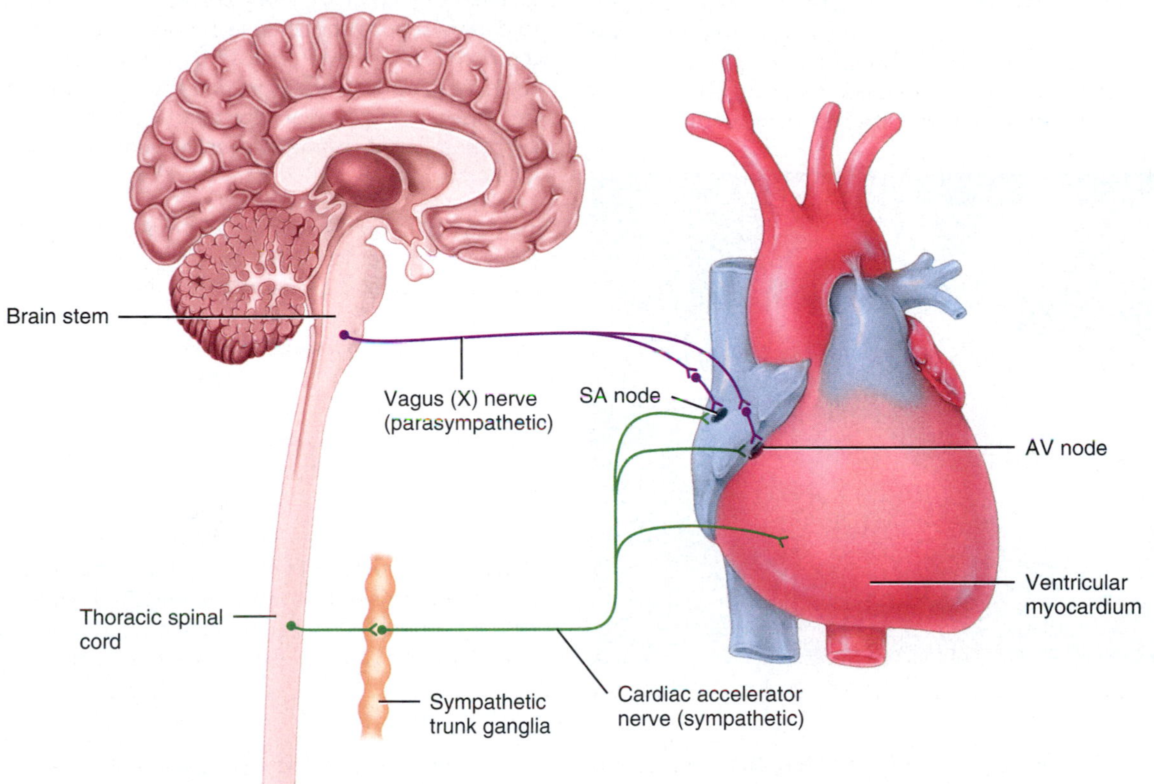

Which part of the heart has sympathetic innervation but essentially no parasympathetic innervation?

Critical Thinking

A Failing Heart

Alex is a 70-year-old male who has been diagnosed with congestive heart failure. Congestive heart failure is characterized by an inability of the heart to pump blood efficiently throughout the body. In Alex's case, his heart failure is severe enough that his only option is a heart transplant.

SOME THINGS TO KEEP IN MIND:

Congestive heart failure is a condition that usually results from other cardiovascular diseases. To understand why Alex has congestive heart failure, let's examine his medical history. When Alex was younger, he began to develop hypercholesterolemia (high blood cholesterol). Alex tried to lower his high blood cholesterol but unfortunately was not able to and began to suffer some long-term consequences. For example, Alex would feel minor chest pains that would come and go on a regular basis. On one particular occasion, Alex experienced severe chest pains and was rushed to the emergency room, where he was told that he had suffered a myocardial infarction (heart attack). After consulting his cardiologist, Alex discovered that he had coronary artery disease—a condition that can result from increased cholesterol and triglycerides in the blood. Alex's cardiologist

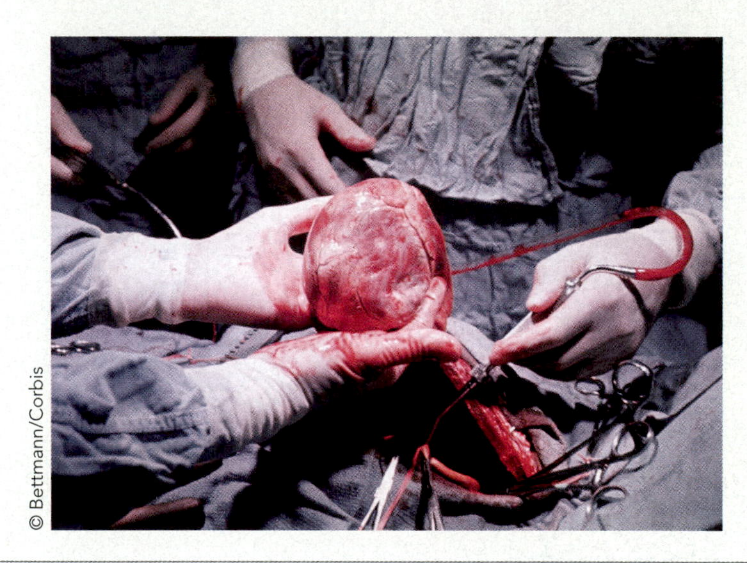
© Bettmann/Corbis

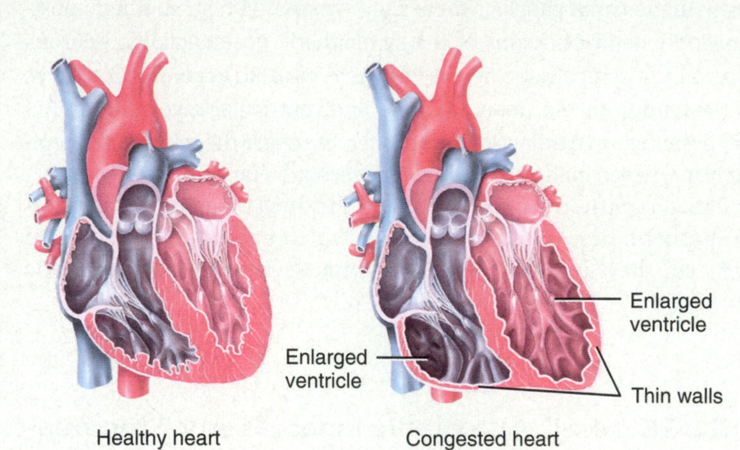

explained that coronary artery disease often leads to heart attacks and a more serious condition known as ischemic heart disease. The ischemic heart disease is caused by a low oxygenation of his heart muscle, leading to a weakening in the muscle itself. The coronary artery disease, myocardial infarction, and ischemic heart disease all took their toll on Alex's heart over the years, resulting in his congestive heart failure.

After Alex suffered his heart attack, many of the cardiac contractile cells in his left ventricle were damaged, which in turn led to overall muscle damage and a weakening of the myocardium. How would you expect force of contraction to change (increase/decrease/no change)? Explain your reasoning.

What effect does the ventricular muscle damage have on stroke volume and cardiac output?

As more blood remains in Alex's ventricles due to congestive heart failure, the ventricles become overstretched (enlarged) and do not function properly. What will happen to both end diastolic volume and end systolic volume? Explain your reasoning.

potential is reached more rapidly (**FIGURE 14.26a**). The specific steps involved in this process are as follows (**FIGURE 14.26b**):

1. Norepinephrine from sympathetic postganglionic neurons binds to β_1-adrenergic receptors in the sarcolemma of an SA node cell.
2. Binding of norepinephrine to the receptor activates a stimulatory G protein (G_s).
3. Activated G_s stimulates the enzyme adenylyl cyclase to produce cyclic AMP (cAMP).
4. cAMP directly binds to F-type Na^+ channels in the sarcolemma. Recall that F-type channels are involved in producing the *pacemaker potential*, the spontaneous depolarization to threshold that occurs in pacemaker cells (see **FIGURE 14.13**). Binding of cAMP to F-type channels keeps the channels open for a longer period of time, allowing more Na^+ ions to enter the cell.
5. The increased influx of Na^+ through F-type channels increases the rate of spontaneous depolarization, allowing threshold to be reached more quickly. As a result, the SA node generates action potentials at a higher frequency and heart rate increases.

- ***Increase in rate of action potential conduction from atria to ventricles.*** Sympathetic stimulation of the AV node speeds up conduction velocity through the AV node, which increases the rate at which action potentials are conducted from the atria to the ventricles. The effect of sympathetic stimulation of the AV node occurs

FIGURE 14.26 Sympathetic regulation of SA node activity.

Sympathetic stimulation of the SA node increases the rate of spontaneous depolarization so that the nodal cells generate action potentials more rapidly and heart rate increases.

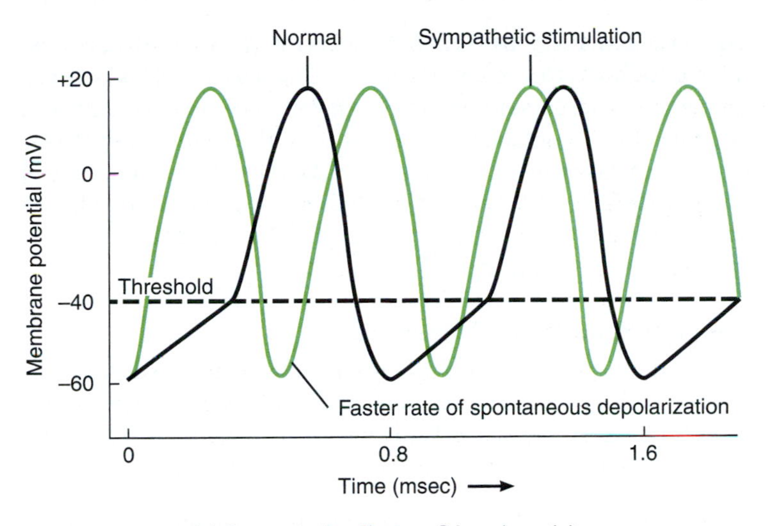

(a) Sympathetic effect on SA node activity

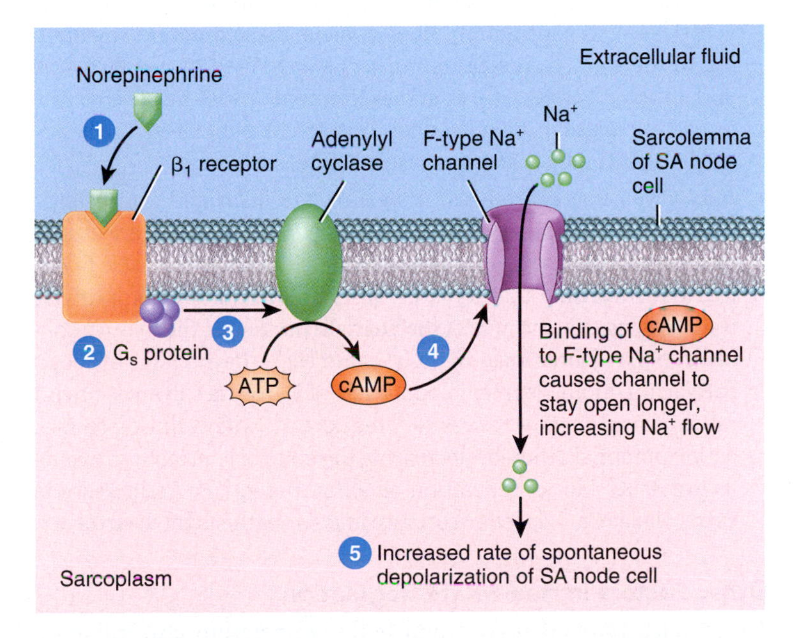

(b) Mechanism of sympathetic effect on SA node activity

What effect does cAMP have on F-type Na^+ channels?

because of enhanced opening of F-type Na^+ channels, increasing Na^+ inflow. This depolarizes the AV node membrane, making it easier for AV node cells to be excited by incoming action potentials from the atria. As a result, conduction velocity through the AV node increases, and action potentials travel from the atria to the ventricles at a faster rate.

- ***Increase in contractility.*** Sympathetic stimulation of the ventricular myocardium causes an increase in contractility. Recall that binding of norepinephrine to β_1-adrenergic receptors in contractile ventricular fibers activates a G protein signaling pathway that ultimately increases the Ca^{2+} levels in the sarcoplasm, thereby increasing contractility (see FIGURE 14.23). As a result, a greater volume of blood is ejected during systole. With a moderate increase in heart rate, stroke volume does not decline because the increased contractility offsets the decreased preload. With maximal sympathetic stimulation, however, heart rate may reach 200 beats/min in a 20-year-old person. At such a high heart rate, stroke volume is lower than at rest due to the very short filling time.

PARASYMPATHETIC REGULATION Stimulation of the parasympathetic nervous system has two main effects on the heart: (1) it decreases heart rate and (2) it decreases the rate of action potential conduction from the atria to the ventricles. Because only a few parasympathetic fibers innervate ventricular muscle, parasympathetic stimulation has little or no effect on contractility of the ventricles.

- ***Decrease in heart rate.*** Parasympathetic stimulation of the SA node decreases heart rate by decreasing the rate of spontaneous depolarization in SA node cells so that threshold for an action potential is reached more slowly (FIGURE 14.27a). The specific steps involved occur as follows (FIGURE 14.27b):

1. Acetylcholine (ACh) released from parasympathetic postganglionic neurons binds to muscarinic receptors in the sarcolemma of an SA node cell.
2. Binding of ACh to its receptor activates an inhibitory G protein (G_i).
3. Activated G_i directly opens a subset of K^+ channels known as **acetylcholine-regulated K^+ channels** or **K^+_{ACh} channels**. Opening the K^+ channels allows more K^+ ions than normal to leave the cell. As a result, the cell membrane hyperpolarizes, causing the pacemaker potential of the SA node cell to start at a more negative value.
4. The activated G_i protein also inhibits adenylyl cyclase, causing the production of cyclic AMP (cAMP) to decrease. The decreased concentration of cAMP accelerates the closure of F-type Na^+ channels. (Recall that binding of cAMP to F-type channels increases their duration of opening; when fewer cAMP molecules are available, the channels have a greater probability of closing.) Closure of the F-type Na^+ channels reduces entry of Na^+ ions into the SA node cell during the pacemaker potential.

FIGURE 14.27 Parasympathetic regulation of SA node activity.

Parasympathetic stimulation of the SA node decreases the rate of spontaneous depolarization so that the nodal cells generate action potentials more slowly and heart rate decreases.

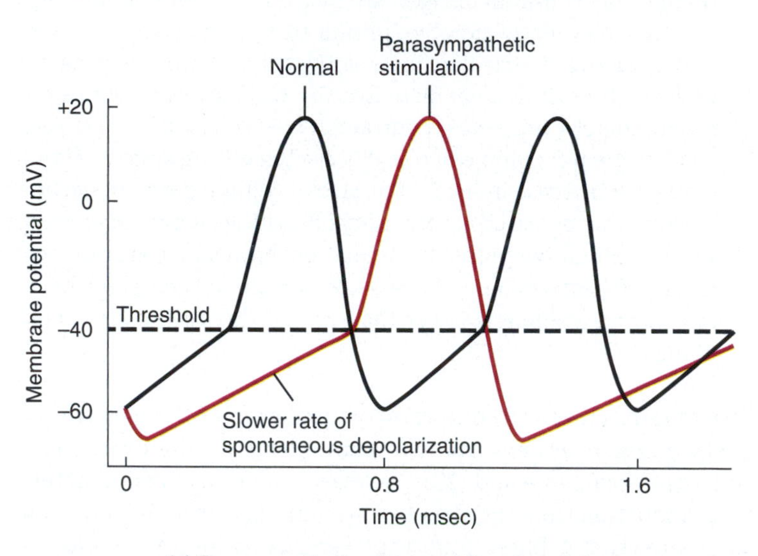

(a) Parasympathetic effect on SA node activity

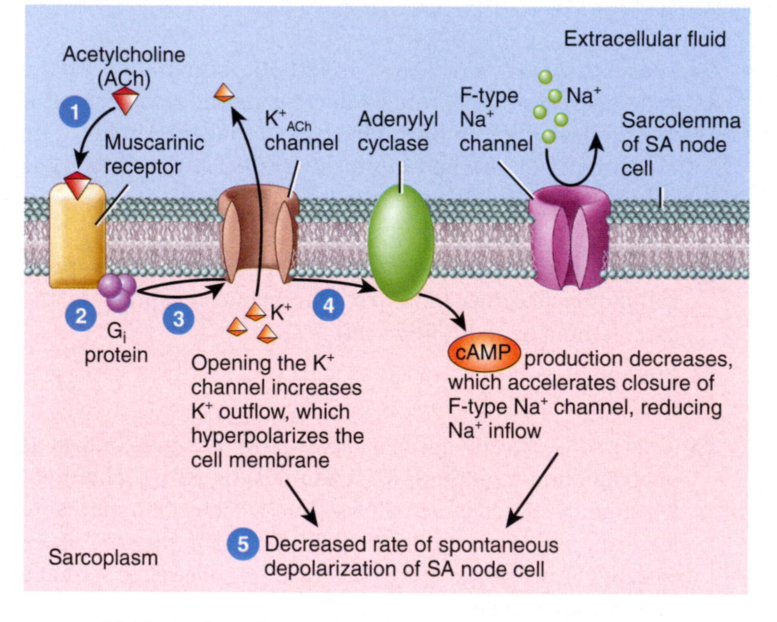

(b) Mechanism of parasympathetic effect on SA node activity

Why is the K^+_{ACh} channel so-named?

5 Hyperpolarization of the SA node membrane (step 3) and decreased Na^+ entry during the pacemaker potential (step 4) both have the same effect on the SA node cell: They decrease the rate of spontaneous depolarization. As a result, the SA node generates action potentials at a lower frequency, and heart rate decreases.

- ***Decrease in rate of action potential conduction from atria to ventricles.*** Parasympathetic stimulation of the AV node slows conduction velocity through the AV node, which decreases the rate at which action potentials are conducted from the atria to the ventricles. The effect of parasympathetic stimulation of the AV node is due to augmented opening of K^+_{ACh} channels, increasing K^+ outflow. This hyperpolarizes the AV node membrane, making it more difficult for the AV node to be excited by action potentials from the atria. So, conduction velocity through the AV node decreases and action potentials travel from the atria to the ventricles at a slower rate.

SHIFTING BALANCE BETWEEN SYMPATHETIC AND PARASYMPATHETIC REGULATION A continually shifting balance exists between sympathetic and parasympathetic stimulation of the heart. At rest, parasympathetic stimulation predominates. The resting heart rate—about 75 beats/min—is usually lower than the autorhythmic rate of the SA node (about 100 beats/min). With maximal stimulation by the parasympathetic division, the heart can slow to 20 or 30 beats/min or can even stop momentarily.

Chemical Regulation of Heart Rate

Certain chemicals influence both the basic physiology of cardiac muscle and the heart rate. For example, hypoxia (lowered oxygen level), acidosis (low pH), and alkalosis (high pH) all depress cardiac activity. Several hormones and ions have major effects on the heart:

- ***Hormones.*** Epinephrine and norepinephrine (from the adrenal medullae) enhance the heart's pumping effectiveness. These hormones affect cardiac muscle fibers in the same way as norepinephrine released by cardiac accelerator nerves—they increase heart rate, the rate of action potential conduction from the atria to the ventricles, and contractility. They achieve these functions by binding to the same β_1 receptors in cardiac cells (see **FIGURES 14.23** and **14.26**). Exercise, stress, and excitement cause the adrenal medullae to release more hormones. Thyroid hormones also enhance cardiac contractility and increase heart rate.
- ***Ions.*** Given that differences between intracellular and extracellular concentrations of several ions (for example, Na^+ and K^+) are crucial for the production of action potentials in all nerve and muscle fibers, it is not surprising that ionic imbalances can quickly compromise the pumping effectiveness of the heart. In particular, the relative concentrations of three ions—K^+, Ca^{2+}, and Na^+—have a large effect on cardiac function. Elevated blood levels of K^+ or Na^+ decrease heart rate and contractility. Excess Na^+ blocks Ca^{2+} inflow during cardiac action potentials, thereby decreasing the force of contraction, whereas excess K^+ blocks generation of action potentials. An increase in extracellular Ca^{2+} speeds heart rate and strengthens the heartbeat.

Other Factors in Heart Rate Regulation

Age, gender, physical fitness, and body temperature also influence resting heart rate. A newborn baby is likely to have a resting heart rate over 120 beats/min; the rate then gradually declines throughout life. Adult females often have slightly higher resting heart rates than adult males, although regular exercise tends to bring resting heart rate down in both sexes. A physically fit person may even exhibit bradycardia (a resting heart rate under 60 beats/min). This is a beneficial effect of endurance-type training because a slowly beating heart is more energy efficient than one that beats more rapidly.

FIGURE 14.28 Factors that influence cardiac output.

A variety of factors influence cardiac output by affecting stroke volume and heart rate.

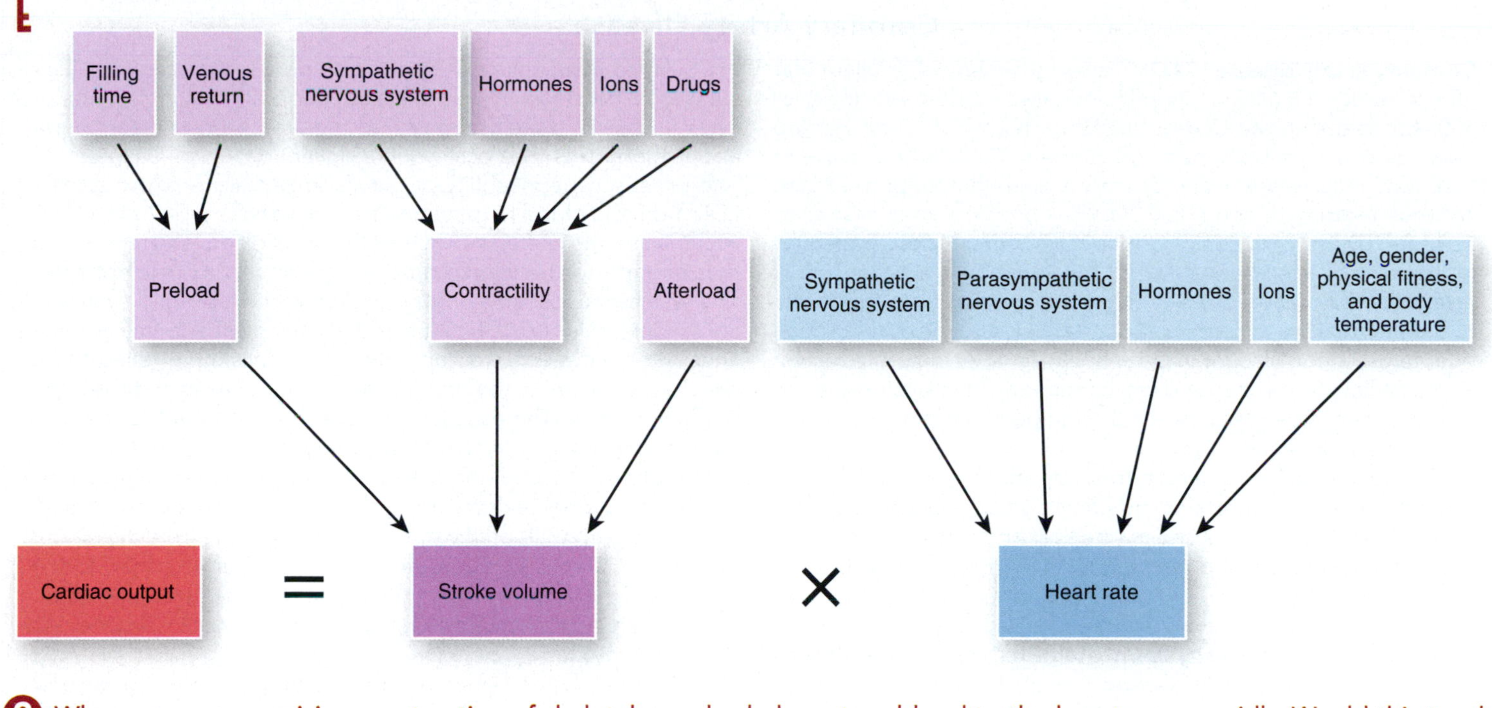

When you are exercising, contraction of skeletal muscles helps return blood to the heart more rapidly. Would this tend to increase or decrease stroke volume? What about cardiac output?

Increased body temperature, as occurs during a fever or strenuous exercise, causes the SA node to discharge action potentials more quickly, thereby increasing heart rate. Decreased body temperature decreases heart rate and strength of contraction.

FIGURE 14.28 summarizes the various factors that influence cardiac output.

CHECKPOINT

18. How is cardiac output calculated?
19. Define stroke volume (SV) and explain the factors that regulate it.
20. What is the significance of the Frank–Starling law of the heart?
21. How do the sympathetic and parasympathetic divisions of the autonomic nervous system affect heart rate?

14.6 Exercise and the Heart

OBJECTIVE

- Explain the relationship between exercise and the heart.

Regardless of current fitness level, a person's cardiovascular fitness can be improved at any age with regular exercise. Some types of exercise are more effective than others for improving the health of the cardiovascular system. **Aerobics**, any activity that works large body muscles for at least 20 minutes, elevates cardiac output and accelerates metabolic rate. Three to five such sessions a week are usually recommended for improving the health of the cardiovascular system. Brisk walking, running, bicycling, cross-country skiing, and swimming are examples of aerobic activities.

Sustained exercise increases the oxygen demand of the muscles. Whether the demand is met depends mainly on the adequacy of cardiac output and proper functioning of the respiratory system. After several weeks of training, a healthy person increases maximal cardiac output, thereby increasing the maximal rate of oxygen delivery to the tissues. Oxygen delivery also rises because skeletal muscles develop more capillary networks in response to long-term training.

During strenuous activity, a well-trained athlete can achieve a cardiac output double that of a sedentary person, in part because training causes hypertrophy (enlargement) of the heart. Even though the heart of a well-trained athlete is larger, *resting* cardiac output is about the same as in a healthy untrained person because stroke volume is increased while heart rate is decreased. The resting heart rate of a trained athlete is often only 40–60 beats per minute (*resting bradycardia*). Regular exercise also helps to reduce blood pressure, anxiety, and depression; control weight; and increase the body's ability to dissolve blood clots by increasing fibrinolytic activity.

CHECKPOINT

22. What are some of the cardiovascular benefits of regular exercise?

CLINICAL CONNECTION

Coronary Artery Disease

Coronary artery disease (CAD) is a serious medical problem that affects about 13 million people annually. Responsible for over 500,000 deaths in the United States each year, it is the leading cause of death for both men and women. The principal cause of CAD is atherosclerosis of the coronary arteries that supply the heart. **Atherosclerosis** (ath-er-ō-skle-RŌ-sis) is a progressive disease characterized by the formation of lesions called **atherosclerotic plaques** in the walls of arteries (FIGURE A). It is the most common form of **arteriosclerosis** (ar-tē-rē-ō-skle-RŌ-sis), a general term that refers to the thickening and hardening of arterial walls. The accumulation of atherosclerotic plaques in coronary arteries leads to a reduction in blood flow to the myocardium. Some individuals have no signs or symptoms; others experience angina pectoris (chest pain), and still others suffer heart attacks.

To understand how atherosclerosis develops, you need to learn about the role of **lipoproteins**, spherical particles consisting of lipid and protein that are produced by the liver and small intestine. Like most lipids, cholesterol does not dissolve in water and must be made water-soluble in order to be transported in the blood. This is accomplished by combining it with lipoproteins. Two major lipoproteins are **low-density lipoproteins (LDLs)** and **high-density lipoproteins (HDLs)**. LDLs transport cholesterol from the liver to body cells for use in the synthesis of cell membranes and to cells in the testes, ovaries, and adrenal glands for use in the production of steroid hormones. Although LDLs perform important functions in the body, excessive amounts of LDLs promote atherosclerosis, so the cholesterol in these particles is commonly known as "bad cholesterol." HDLs, on the other hand, remove excess cholesterol from body cells and the blood, and transport it to the liver for elimination. Because HDLs decrease blood cholesterol level, the cholesterol in HDLs is commonly referred to as "good cholesterol." Basically, you want your LDL concentration to be low and your HDL concentration to be high.

Studies suggest that atherosclerosis results from a chronic inflammatory response initiated by agents that repeatedly injure the arterial wall. Examples of these agents include hypertension (high blood pressure), chemicals in cigarette smoke, free radicals, and certain viruses and bacteria. Atherosclerosis starts as the damaged arterial wall undergoes ***inflammation***, a defensive response of the body to tissue injury. During this response, the arterial wall increases its permeability and releases chemicals that attract monocytes (a type of white blood cell) to the area. (You will learn more about the events that occur during inflammation in Chapter 17.) The increased permeability of the damaged artery allows excess LDLs from the blood to accumulate in the vessel's wall. As the LDLs accumulate, the cholesterol in the LDLs is oxidized. The monocytes that migrate to the injured area are converted to macrophages. The macrophages then ingest and become so filled with the lipid-laden, oxidized LDL particles that they have a foamy appearance when viewed microscopically. These cells, now called **foam cells**, accumulate in the arterial wall to form a **fatty streak**, the beginning of an atherosclerotic plaque. Following fatty streak formation, smooth muscle cells of the artery migrate to the top of the atherosclerotic plaque (the side closest to the lumen), where they proliferate (divide) and deposit a fibrous cap of connective tissue. This fibrous cap walls off the atherosclerotic plaque from the blood. In time, the plaque becomes larger as more lipid is added to the core. As the plaque becomes larger, it protrudes into the lumen of the artery. The plaque also becomes harder because calcium is deposited into it.

Although atherosclerotic plaques, as they grow, can reduce or completely block blood flow in a coronary artery, their expansion into the vessel lumen is responsible for relatively few heart attacks. Instead, most heart attacks occur when the cap over the plaque breaks open, triggering the formation of a blood clot that is large enough to significantly decrease or stop the flow of blood in a coronary artery.

People who possess combinations of certain risk factors are more likely to develop CAD. ***Risk factors*** (characteristics, symptoms, or signs present in a disease-free person that are statistically associated with a greater chance of developing a disease) include smoking, high blood pressure, diabetes, high cholesterol levels, obesity, and sedentary lifestyle. Most of these can be modified by changing diet and other habits or can be controlled by taking medications. However, other risk factors are unmodifiable (beyond our control), including genetic predisposition (family history of CAD), age, and gender. For example, adult males are more likely than adult females to develop CAD; however, after age 70 the risks are

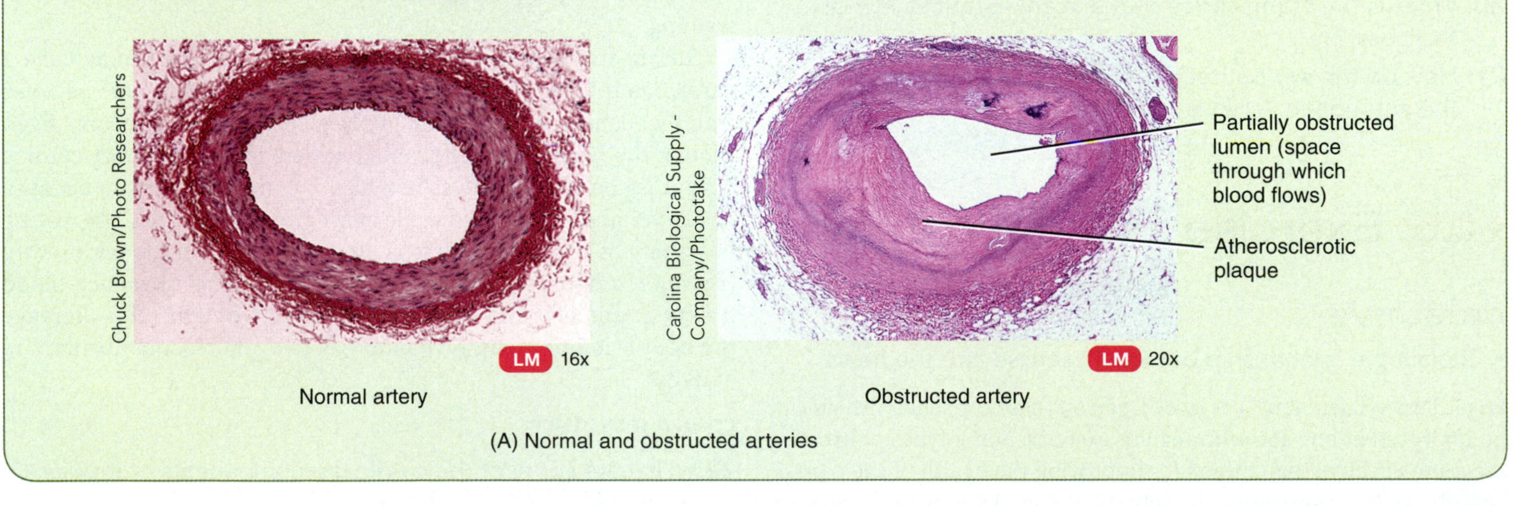

(A) Normal and obstructed arteries

roughly equal. Smoking is undoubtedly the number-one risk factor in all CAD-associated diseases, roughly doubling the risk of morbidity and mortality.

In recent years, a number of new risk factors (all modifiable) have been identified as significant predictors of CAD. **C-reactive proteins (CRPs)** are proteins produced by the liver or present in blood in an inactive form and are converted to an active form during inflammation. CRPs may play a direct role in the development of atherosclerosis by promoting the uptake of LDLs by macrophages. **Lipoprotein (a)** is an LDL-like particle that binds to endothelial cells, macrophages, and blood platelets; may promote the proliferation of smooth muscle fibers; and inhibits the breakdown of blood clots. **Homocysteine** is an amino acid that may induce blood vessel damage by promoting platelet aggregation and smooth muscle fiber proliferation.

Treatment options for CAD include **drugs** (clot-dissolving agents, cholesterol-lowering drugs, and vasodilator drugs such as **nitroglycerin**) and various surgical and nonsurgical procedures designed to increase the blood supply to the heart. **Coronary artery bypass grafting (CABG)** is a surgical procedure in which a blood vessel from another part of the body is attached ("grafted") to a coronary artery to bypass an area of blockage. A piece of the grafted blood vessel is sutured between the aorta and the unblocked portion of the coronary artery (FIGURE B). A nonsurgical procedure used to treat CAD is **balloon angioplasty** (*angio-* = blood vessel; *-plasty* = to mold or to shape). In this procedure, a catheter (plastic tube) with a balloon at the tip is inserted into an artery of an arm or leg and gently guided into a coronary artery to the point of obstruction (FIGURE C). The balloon is then inflated with air to squash the plaque against the blood vessel wall. To prevent restenosis (renarrowing) of the artery, a stent may be inserted via a catheter. A **stent** is a metallic, fine wire tube that is placed in an artery to keep the artery *patent* (open), permitting blood to circulate (FIGURE D).

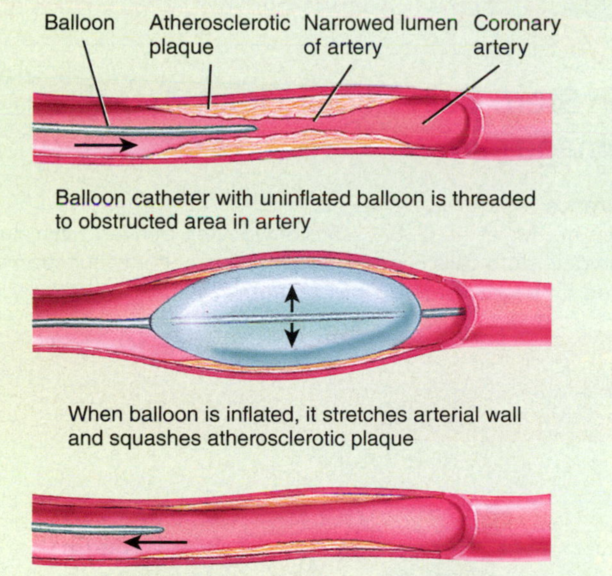

(C) Balloon angioplasty

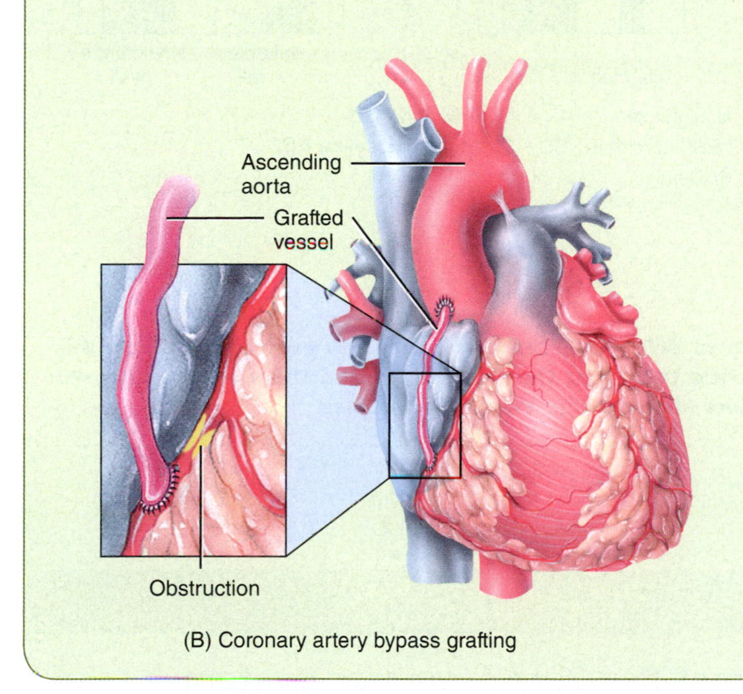

(B) Coronary artery bypass grafting

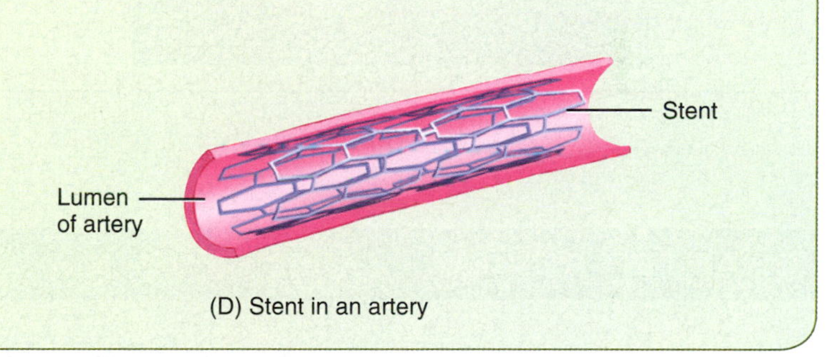

(D) Stent in an artery

FROM RESEARCH TO REALITY

Can Stem Cells Repair Heart Attack Damage?

Reference

Laflamme, M. et al. (2007). Cardiomyocytes derived from human embryonic stem cells in pro-survival factors enhance function of infarcted rat hearts. *Nature Biotechnology*. 25(9): 1015–1024.

Can stem cells repair a heart damaged by a heart attack?

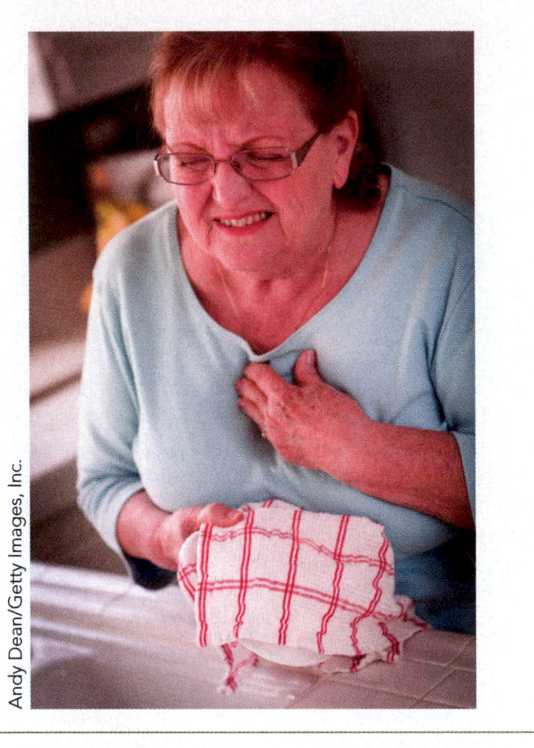

Heart attacks can be quite devastating; quick action is required to sustain life. Serious lifestyle changes are often required after someone suffers from a heart attack in order to minimize the probability of suffering from a second one. Treating patients after a heart attack and helping them to recover, in the absence of lifestyle changes, is the topic of this article. Can the use of embryonic stem cells help to repair the damage to the heart caused by a heart attack?

Article description:

The purpose of this study was to investigate how well human embryonic stem cell grafts improved the function of rat hearts that had undergone a myocardial infarction.

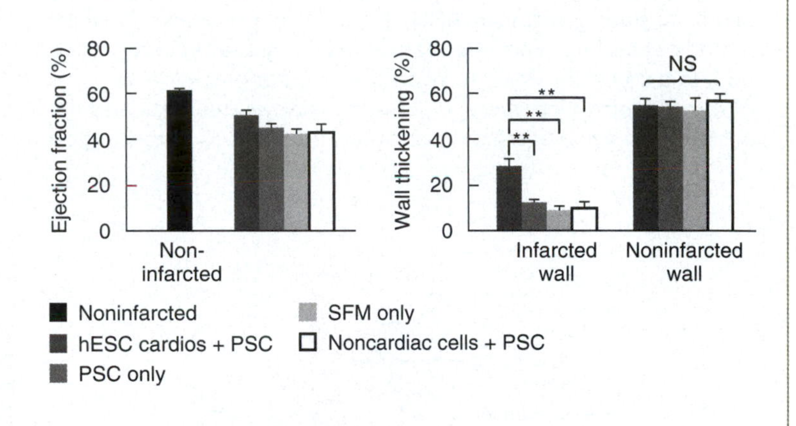

Go to WileyPLUS Learning Space and use the data from this article to answer the questions posed there and to discover more about cardiac stem cell therapies.

CHAPTER REVIEW

14.1 Basic Design of the Cardiovascular System

1. The cardiovascular system consists of the heart, blood vessels, and blood.
2. The heart pumps blood into two closed circuits—the pulmonary circulation and the systemic circulation. In both circuits, blood is carried away from and then back to the heart through the following sequence of blood vessels: arteries, arterioles, capillaries, venules, and veins.
3. The pulmonary circulation consists of blood vessels that carry blood from the right side of the heart to the alveoli of the lungs and then back to the left side of the heart.
4. The systemic circulation consists of blood vessels that carry blood from the left side of the heart to all organs and tissues of the body except the alveoli of the lungs and then back to the right side of the heart.

5. In the systemic circulation, blood flow usually is parallel, which means that a given portion of blood flows through an artery to only one organ and enters only one set of capillaries before returning to the heart through a vein.

14.2 Organization of the Heart

1. The heart is located in the thoracic (chest) cavity, mostly to the left of the body's midline.
2. The pericardium is a membranous sac that surrounds the heart. It confines the heart to its position in the thoracic cavity while allowing sufficient movement for contraction.
3. Three layers make up the wall of the heart: epicardium (outer layer), myocardium (middle and thickest layer), and endocardium (inner layer).
4. The heart consists of four chambers: two upper atria and two lower ventricles.
5. The thickness of the myocardium of the four chambers varies according to the chamber's function. The left ventricle, with the highest workload, has the thickest wall.
6. Heart valves prevent backflow of blood within the heart. The atrioventricular (AV) valves, which lie between atria and ventricles, are the tricuspid valve on the right side of the heart and the bicuspid (mitral) valve on the left. The semilunar (SL) valves are the aortic valve, at the entrance to the aorta, and the pulmonary valve, at the entrance to the pulmonary trunk.
7. The fibrous skeleton of the heart is connective tissue that surrounds and supports the valves of the heart.
8. The coronary circulation supplies blood to the myocardium. The main arteries of the coronary circulation are the left and right coronary arteries; the main veins are the cardiac veins and the coronary sinus.

14.3 Cardiac Muscle Tissue and the Cardiac Conduction System

1. Cardiac muscle fibers are striated, are organized into sarcomeres, and have the same zones and bands as skeletal muscle.
2. Cardiac muscle fibers are connected end-to-end via intercalated discs, which contain desmosomes and gap junctions.
3. Cardiac muscle is a functional syncytium that consists of two types of fibers: autorhythmic fibers and contractile fibers.
4. The action potential in a contractile cardiac muscle fiber consists of four phases: depolarizing phase, initial repolarizing phase, plateau phase, and final repolarizing phase.
5. Cardiac muscle tissue has a long refractory period, which prevents tetanus.
6. Autorhythmic fibers form the conduction system, the pathway that rapidly delivers action potentials throughout the heart.
7. Components of the conduction system include the sinoatrial (SA) node (pacemaker), atrioventricular (AV) node, atrioventricular (AV) bundle (bundle of His), bundle branches, and Purkinje fibers.
8. The record of electrical changes during each cardiac cycle is called an electrocardiogram (ECG). A normal ECG consists of a P wave (atrial depolarization), a QRS complex (ventricular depolarization), and a T wave (ventricular repolarization).
9. The P–Q interval represents the conduction time from the beginning of atrial excitation to the beginning of ventricular excitation. The S–T segment represents the time when ventricular contractile fibers are fully depolarized.

14.4 The Cardiac Cycle

1. A cardiac cycle includes of all the events associated with a single heartbeat.
2. The phases of the cardiac cycle are (1) passive ventricular filling, (2) atrial contraction, (3) isovolumetric ventricular contraction, (4) ventricular ejection, and (5) isovolumetric ventricular relaxation.
3. S1, the first heart sound (lubb), is caused by vibrations associated with closure of the atrioventricular valves. S2, the second heart sound (dupp), is caused by vibrations associated with closure of the semilunar valves.

14.5 Cardiac Output

1. Cardiac output (CO) is the amount of blood ejected per minute by the left ventricle into the aorta (or by the right ventricle into the pulmonary trunk). It equals the stroke volume (SV) multiplied by the heart rate (HR).
2. Stroke volume (SV) is the amount of blood ejected by a ventricle during each systole.
3. Cardiac reserve is the difference between a person's maximum cardiac output and his or her cardiac output at rest.
4. Stroke volume is related to preload (stretch on the heart before it contracts), contractility (forcefulness of contraction), and afterload (pressure that must be exceeded before ventricular ejection can begin).
5. According to the Frank–Starling law of the heart, a greater preload (end-diastolic volume) stretching cardiac muscle fibers just before they contract increases their force of contraction until the stretching becomes excessive.
6. The heart receives innervation from both the sympathetic and parasympathetic divisions of the heart. The sympathetic nerves to the heart supply the SA node, AV node, and ventricular myocardium. The parasympathetic nerves to the heart supply the SA node and AV node but not the ventricles.
7. Sympathetic stimulation has three main effects on the heart: (1) it increases heart rate; (2) it increases action potential conduction from the atria to the ventricles; and (3) it increases contractility.
8. Parasympathetic stimulation has two main effects on the heart: (1) it decreases heart rate and (2) it decreases action potential conduction from the atria to the ventricles.
9. Heart rate is also affected by hormones (epinephrine, norepinephrine, thyroid hormones), ions (Na^+, K^+, Ca^{2+}), age, gender, physical fitness, and body temperature.

14.6 Exercise and the Heart

1. Sustained exercise increases oxygen demand on muscles.
2. Among the benefits of aerobic exercise are increased cardiac output, decreased blood pressure, weight control, and increased fibrinolytic activity.

PONDER THIS

1. John goes in to see his cardiologist for a routine follow-up after his heart attack a few months ago. One of the tests was an electrocardiogram (ECG). The analysis shows that there are no P waves but there are abnormal QRS complexes and abnormal T waves. What does his ECG tell you about the conduction system in his heart? How would you predict his heart rate to differ from that of a person with a normal ECG? Why?

2. Compare and contrast the different channels and ions involved with an action potential in an autorhythmic cell and in a contractile cell. Include the similarities among and differences between these cells with respect to the action potential phases, the channels and ions involved, and the overall functions of each cell type.

3. Nick suffers from myocarditis, an inflammatory condition of the cardiac muscle. Unfortunately, Nick does not seek prompt treatment for his condition and some of the muscle tissue in his left ventricle dies. How would this dead muscle tissue affect the heart's pumping abilities and his overall circulation? Explain your answer.

4. If potassium chloride is injected into the heart, it will first cause arrhythmias and eventually cause the heart to stop. Using your knowledge of the different phases of the action potential of an autorhythmic cell as well as your knowledge of membrane potentials, explain why an injection of KCl would result in cardiac arrest.

5. Sarah comes into the emergency room (ER) with a very rapid heart rate. The ER nurse needs to inject a drug that will act as a negative chronotropic agent to reduce the heart rate. What type of drug should the nurse choose? Explain your answer.

ANSWERS TO FIGURE QUESTIONS

14.1 The heart is a muscular pump that creates the pressure needed to circulate blood through blood vessels to body tissues.

14.2 Blood becomes deoxygenated in the systemic capillaries.

14.3 Parallel flow through the systemic circulation allows each organ to receive its own supply of freshly oxygenated blood, and it allows blood flow to different organs to be regulated independently.

14.4 The heart's base is its upper, broad portion; the heart's apex is its lower, pointed tip.

14.5 The myocardium is the layer of the heart wall that contains cardiac muscle.

14.6 The greater the workload of a heart chamber, the thicker its myocardium.

14.7 The chordae tendineae prevent the cusps of the AV valves from everting into the atria when the ventricles contract.

14.8 The aorta is the large artery that carries oxygenated blood away from the left ventricle of the heart.

14.9 The fibrous skeleton (1) attaches to the heart valves and (2) prevents overstretching of the valves as blood passes through them.

14.10 The heart wall has its own blood supply because nutrients are unable to diffuse from blood in the chambers through all of the layers of cells that comprise the wall.

14.11 The desmosomes mechanically bind cardiac muscle fibers together, which prevents the fibers from pulling apart during contraction; the gap junctions electrically couple cardiac muscle fibers to each other, which allows cardiac muscle to behave as a functional syncytium.

14.12 The only electrical connection between the atria and the ventricles is the atrioventricular bundle (bundle of His).

14.13 The pacemaker potential is the spontaneous depolarization to threshold that occurs in an autorhythmic fiber.

14.14 The plateau phase is the longest phase of an action potential in a contractile cardiac muscle fiber.

14.15 Ca^{2+}-induced Ca^{2+}release (CICR) refers to the phenomenon by which extracellular Ca^{2+} triggers the release of additional Ca^{2+} from the SR.

14.16 The long refractory period of a contractile cardiac muscle fiber prevents the cell from being re-excited until its previous contraction is almost over; consequently, summation of contractions and tetanus do not occur.

14.17 An enlarged Q wave may indicate a myocardial infarction (heart attack).

14.18 Action potentials propagate most slowly through the AV node.

14.19 The amount of blood in each ventricle at the end of ventricular diastole—called the end-diastolic volume—is about 130 mL in a resting person.

14.20 Stroke volume is the amount of blood ejected from the ventricle during each contraction.

14.21 In the length–tension relationship for skeletal muscle, resting sarcomeres are held close to their optimal lengths; in the length–tension relationship for cardiac muscle, resting sarcomeres are shorter than the optimum.

14.22 Positive inotropic agents increase contractility by increasing the amount of Ca^{2+} available in the sarcoplasm for contraction.

14.23 The phosphorylated proteins include L-type voltage-gated Ca^{2+} channels, Ca^{2+} release channels, phospholamban, and myosin.

14.24 A decreasing in shortening velocity of ventricular muscle fibers causes a decrease in stroke volume.

14.25 The ventricles have sympathetic innervation but lack parasympathetic innervation.

14.26 cAMP binds to F-type Na^{+} channels, keeping them open for a longer period of time; this allows more Na^{+} ions to enter the cell.

14.27 The name of the $K^{+}{}_{ACh}$ channel is derived from the fact that this channel is regulated by acetylcholine via activation of a G protein (G_i).

14.28 The skeletal muscle pump increases stroke volume by increasing preload (end-diastolic volume). It also increases cardiac output, assuming that there is no change in heart rate.

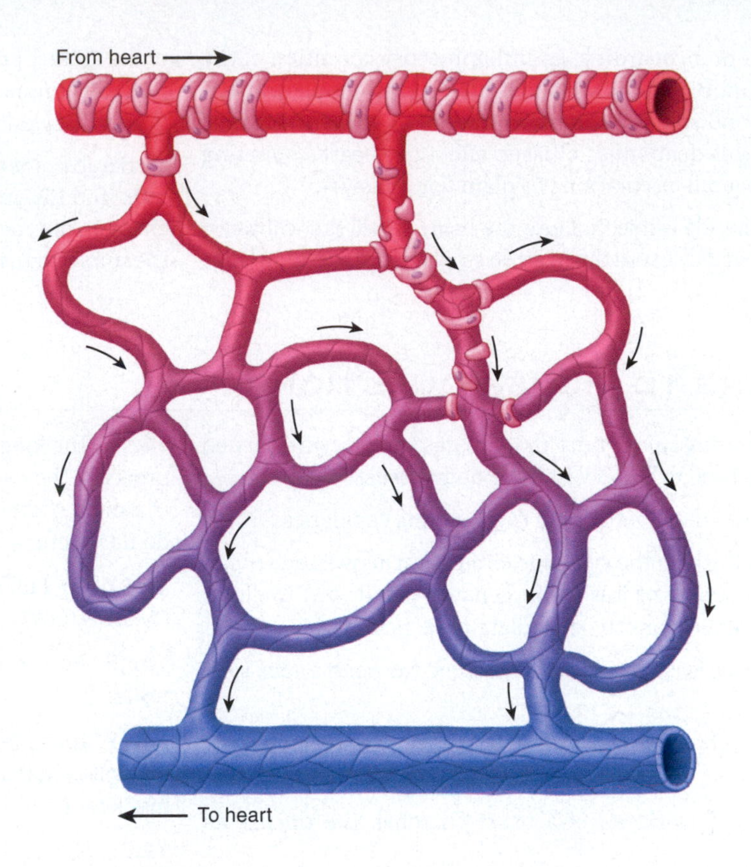

15 The Cardiovascular System: Blood Vessels and Hemodynamics

Blood Vessels, Hemodynamics, and Homeostasis

Blood vessels contribute to homeostasis by serving as the conduits through which blood flows from the heart to body tissues and then back to the heart. They also play an important role in adjusting the velocity and volume of blood flow.

LOOKING BACK TO MOVE AHEAD...

- Extracellular fluid (ECF) is the fluid outside body cells; it consists of two components: (1) interstitial fluid, the fluid that fills the narrow spaces between cells and (2) plasma, the fluid portion of blood (Section 1.4).
- Tight junctions consist of weblike strands of transmembrane proteins that fuse together the outer surfaces of adjacent plasma membranes to seal off passageways between adjacent cells (Section 3.8).
- Transcytosis is a membrane transport process that successively moves a substance into, across, and out of a cell (Section 5.7).
- Cardiac output (CO), the volume of blood ejected from each ventricle of the heart per minute, is equal to stroke volume (SV) multiplied by heart rate (HR) (Section 14.5).

The **blood vessels** of the body form a closed system of tubes that carries blood away from the heart, transports it to the tissues of the body, and then returns it to the heart. The heart pumps blood through an estimated 100,000 km (60,000 mi) of blood vessels. Collectively, the blood vessels of the body are referred to as the **vasculature**. In this chapter, you will explore the functions of the various types of blood vessels and you will learn about *hemodynamics*, the forces involved in circulating blood throughout the body.

15.1 Overview of the Vasculature

OBJECTIVES

- Describe the functions of arteries, arterioles, capillaries, venules, and veins.
- Identify the different types of capillaries.
- Discuss the distribution of blood in the cardiovascular system.

The five main types of blood vessels are arteries, arterioles, capillaries, venules, and veins (FIGURE 15.1). *Arteries* carry blood away from the heart. Large, elastic arteries leave the heart and divide into medium-sized, muscular arteries that distribute blood to the various organs of the body. As a muscular artery enters an organ, it divides into small arteries called *arterioles*. These in turn branch into numerous *capillaries*, which are the smallest blood vessels of the body. The thin, porous walls of capillaries allow the exchange of substances between the blood and body tissues. Groups of capillaries within a tissue unite to form small veins called *venules*, which in turn merge to form progressively larger veins. *Veins* are blood vessels that carry blood from tissues back to the heart.

FIGURE 15.1 Organization of the blood vessels of the body.

All blood vessels contain an inner layer of endothelium surrounded by a basement membrane; except for capillaries, blood vessels also contain layers of smooth muscle and connective tissue that surround the basement membrane.

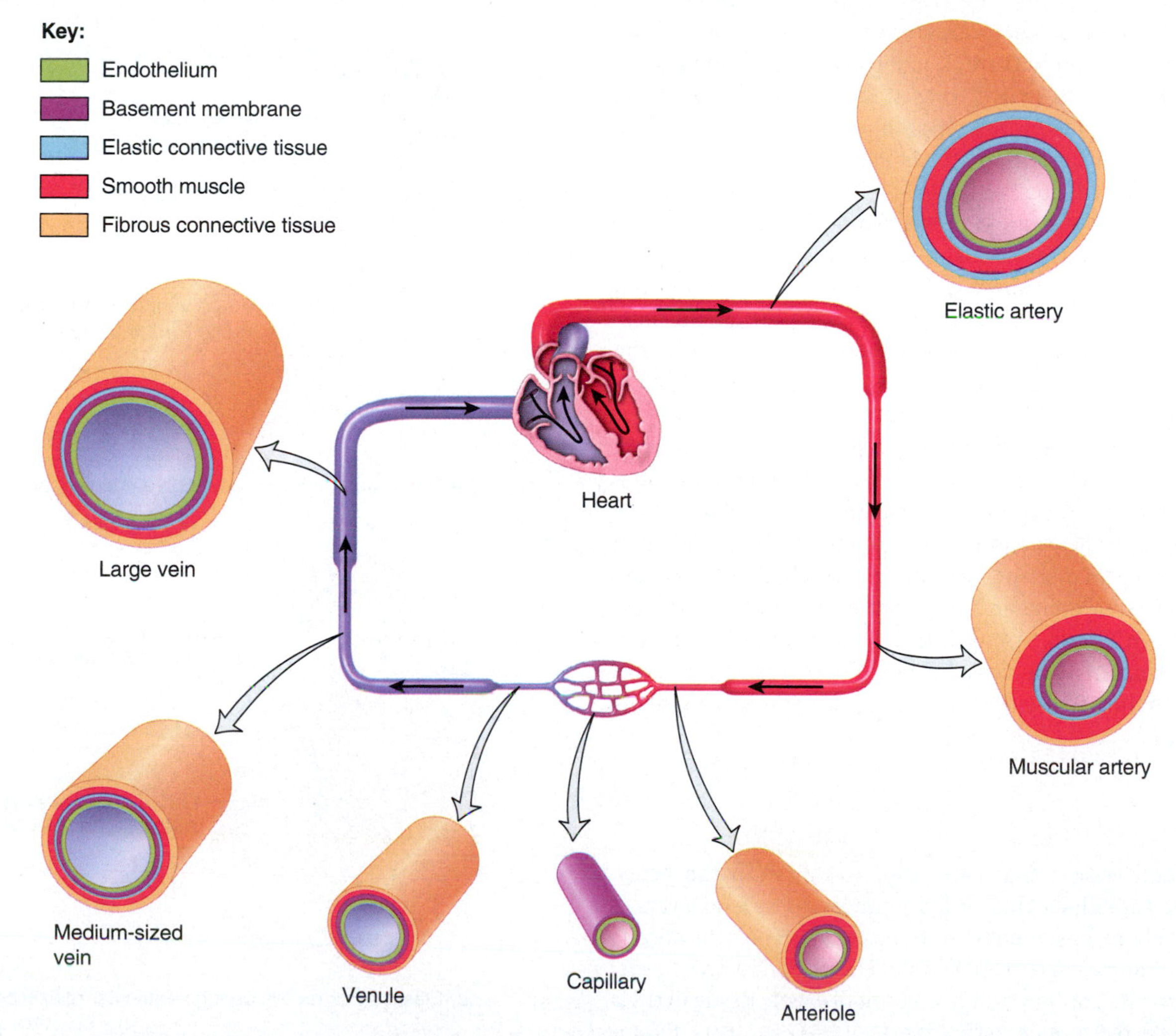

What are some of the functions of the endothelium of a blood vessel?

The Layers of a Blood Vessel Contribute to Vessel Function

All blood vessels contain an inner layer of epithelial cells called the **endothelium** (FIGURE 15.1), which is continuous with the endocardial lining of the heart. The endothelium is in direct contact with the blood that flows through the lumen (interior opening) of the vessel. Until recently, endothelial cells were regarded as little more than a passive barrier between the blood and the remainder of the vessel wall. It is now known that endothelial cells are active participants in a variety of vessel-related activities, including physically influencing blood flow, secreting locally acting chemical mediators that influence the contractile state of the vessel's overlying smooth muscle, and assisting with capillary permeability. Surrounding the endothelium of a blood vessel is a thin layer of extracellular material called the *basement membrane*, which provides support to the endothelial cells. With the exception of capillaries, all blood vessels contain layers of smooth muscle and connective tissue that surround the basement membrane, and these layers vary in thickness in different types of blood vessels (FIGURE 15.1).

The primary role of the smooth muscle of a blood vessel is to regulate the diameter of the lumen. Sympathetic fibers of the autonomic nervous system innervate vascular smooth muscle. An increase in sympathetic stimulation typically causes the smooth muscle to contract, squeezing the vessel wall and narrowing the lumen. Such a decrease in the diameter of the lumen of a blood vessel is called **vasoconstriction**. By contrast, when sympathetic stimulation decreases, or in the presence of certain chemicals (such as nitric oxide and H^+ ions), smooth muscle fibers relax. The resulting increase in lumen diameter is called **vasodilation**. The smooth muscle of most blood vessels is at least partially contracted at all times. The ability of a blood vessel's smooth muscle to maintain a state of partial contraction is referred to as *vascular tone*. As you will learn in more detail shortly, the rate of blood flow through different parts of the body is regulated by the extent of smooth muscle contraction in the walls of particular vessels. Furthermore, the extent of smooth muscle contraction in particular vessel types is crucial to the regulation of blood pressure.

Two types of connective tissue may be present in the walls of blood vessels: elastic connective tissue and fibrous connective tissue. *Elastic connective tissue* contains elastic fibers, which allow blood vessels to stretch in response to incoming blood and then to return to their original shape after being stretched. *Fibrous connective tissue* contains collagen fibers, which provide blood vessels with significant tensile strength that can sustain the pressure that blood exerts against the vessel walls. Blood vessels vary with respect to the amounts of elastic and fibrous connective tissues that they contain.

Arteries Carry Blood Away from the Heart

Arteries are blood vessels that carry blood away from the heart. Large-diameter arteries have thick walls composed of several layers of tissue: an endothelium, basement membrane, smooth muscle, fibrous connective tissue, and a high proportion of elastic connective tissue. Because these arteries contain a large amount of elastic tissue in their walls, they are referred to as **elastic arteries** (FIGURE 15.1). Examples include the two major trunks that exit the heart (the aorta and the pulmonary trunk) and the major branches of the aorta. The aorta and pulmonary trunk are the largest elastic arteries, with diameters of about 2.5 cm each; the other elastic arteries have diameters of about 1 cm. Elastic arteries perform an important function: They serve as **pressure reservoirs** that maintain the driving force for blood flow while the ventricles are relaxing. As blood is ejected from the heart into elastic arteries during systole, their highly elastic walls stretch due to the increased blood volume. As the walls stretch, they momentarily store some of the pressure generated by the contraction of the ventricles (FIGURE 15.2a). Then, while the ventricles are relaxing during diastole, the walls of the elastic arteries recoil as the stored pressure is released. This *elastic recoil* propels blood onward, ensuring that blood continues to move through the remaining arteries of the circulation even though the ventricles are relaxing and not ejecting blood (FIGURE 15.2b).

FIGURE 15.2 Pressure reservoir function of elastic arteries.

Elastic arteries function as pressure reservoirs that stretch during ventricular contraction, momentarily storing some of the pressure generated by the contraction, and then recoil during ventricular relaxation, releasing the stored pressure.

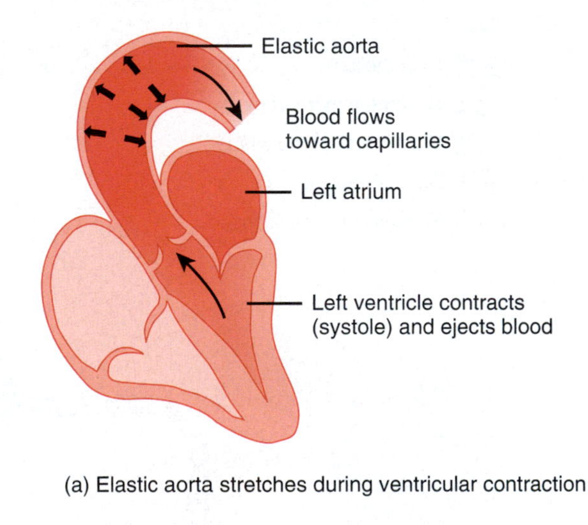

(a) Elastic aorta stretches during ventricular contraction

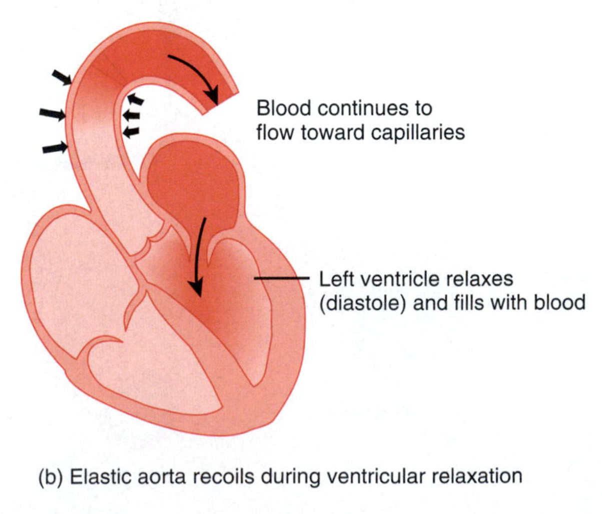

(b) Elastic aorta recoils during ventricular relaxation

Why is recoil of elastic arteries important?

Elastic arteries conduct blood from the heart to medium-sized arteries. Medium-sized arteries contain more smooth muscle and less elastic connective tissue in their walls than elastic arteries. Such arteries are called **muscular arteries** (see FIGURE 15.1). Because of the reduced amount of elastic tissue in the walls of muscular arteries, these vessels have a lower ability to recoil and propel blood compared to elastic arteries. Instead, muscular arteries are capable of greater vasoconstriction and vasodilation. Most of the arteries in the arterial circuit are muscular arteries. Their diameters span a wide range of sizes, from as little as 0.5 mm to as much as 1 cm. Muscular arteries distribute blood to the organs of the body. As a muscular artery enters an organ, it branches into many arterioles.

Arterioles Deliver Blood to Capillaries

Literally meaning "small arteries," **arterioles** are abundant microscopic vessels that deliver blood to capillaries. The approximately 400 million arterioles have diameters that range in size from 15 μm to 100 μm. The walls of large arterioles consist of an endothelium, basement membrane, smooth muscle, and a small amount of fibrous connective tissue (see FIGURE 15.1). Smaller arterioles consist of little more than an endothelium and basement membrane surrounded by an incomplete layer of smooth muscle fibers (see FIGURE 15.3).

Arterioles are known as the **resistance vessels** of the circulation because their small diameters provide the greatest resistance to blood flow. Although the diameter of a capillary is smaller than that of an arteriole, the number of capillaries in the body is so large that they collectively offer less resistance to blood flow than arterioles. In a blood vessel, resistance is due mainly to friction between blood and the inner walls of blood vessels. When the blood vessel diameter is smaller, the friction is greater, so there is more resistance. Contraction of the smooth muscle of an arteriole causes vasoconstriction, which increases resistance even more and decreases blood flow into capillaries supplied by that arteriole. By contrast, relaxation of the smooth muscle of an arteriole causes vasodilation, which decreases resistance and increases blood flow into capillaries. A change in arteriole diameter can also affect blood pressure: Vasoconstriction of arterioles increases blood pressure, and vasodilation of arterioles decreases blood pressure.

Capillaries Are the Sites of Nutrient and Waste Exchange

Capillaries are microscopic vessels that connect arterioles to venules (FIGURE 15.3). Because arterioles, capillaries, and venules require a microscope to be seen, they are collectively known as the **microcirculation**. Capillaries are the smallest blood vessels of the body, each having a diameter of only 5–10 μm and a length of only 1 mm. Because erythrocytes have a diameter of 8 μm, they must often fold upon themselves in order to pass single file through the lumens of these vessels. The body contains an estimated 10 to 40 billion capillaries. This large number of capillaries provides an enormous total surface area (about 600 m^2) to make contact with the body's cells.

Capillaries are known as **exchange vessels** because their primary function is the exchange of nutrients and wastes between the blood

FIGURE 15.3 The microcirculation. For simplicity, the basement membrane is not shown.

Arterioles, capillaries, and venules are collectively known as the microcirculation because they require a microscope to be seen.

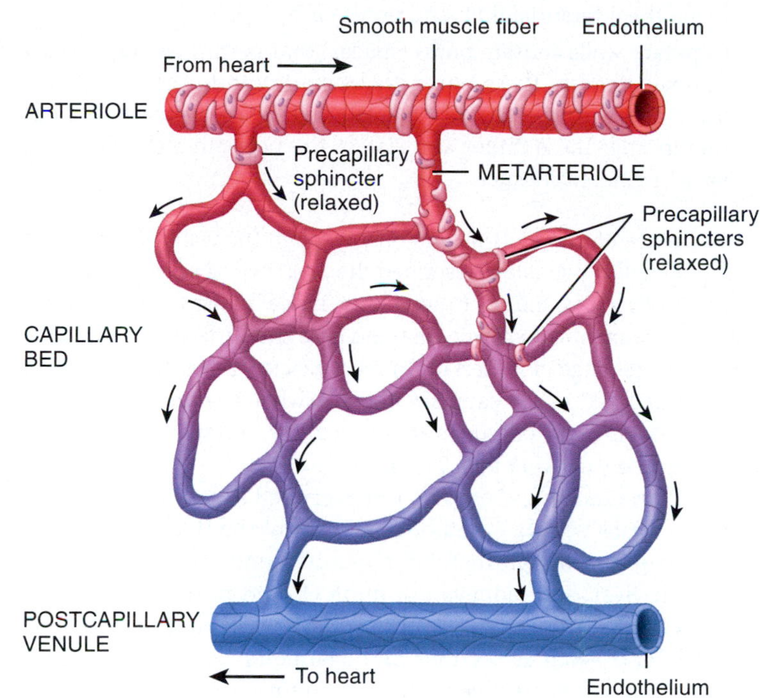

(a) Sphincters relaxed: blood flowing through capillaries

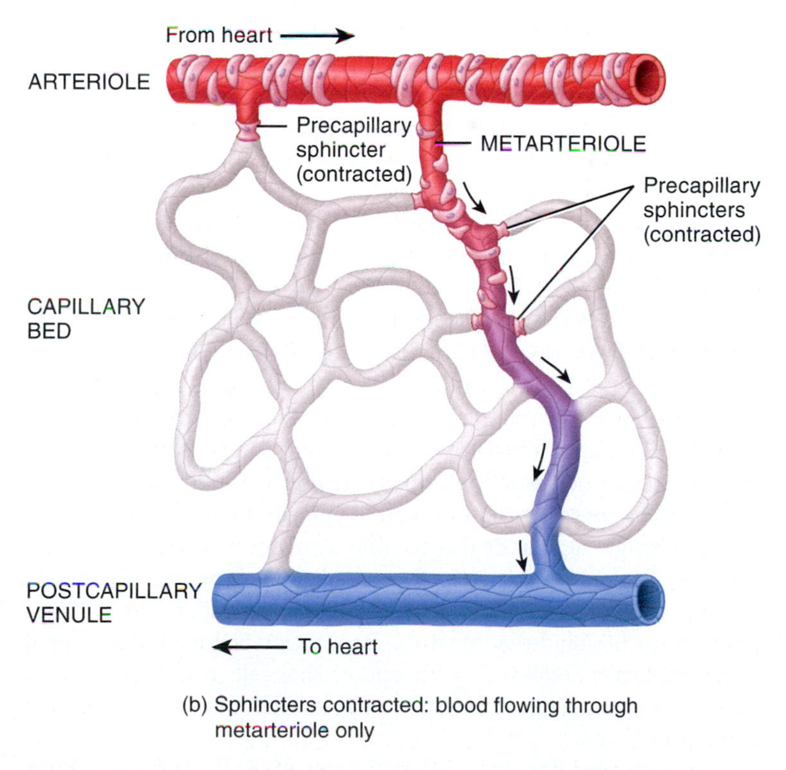

(b) Sphincters contracted: blood flowing through metarteriole only

What is the purpose of precapillary sphincters?

and tissue cells. The structure of capillaries is well suited to this function for two reasons:

1. Capillary walls are thin, composed of only a single layer of endothelial cells surrounded by a basement membrane; no smooth muscle or connective tissue is present (see **FIGURE 15.1**). Thus, a substance in the blood must pass through just one cell layer to reach the interstitial fluid and tissue cells.
2. Capillary walls contain pores (spaces) that permit passage to certain substances. These pores are located both in the endothelial layer and throughout the extracellular matrix of the basement membrane. The number and size of the pores in a capillary can vary in different tissues.

Exchange of materials occurs only through the walls of capillaries and postcapillary venules (described shortly); the walls of arteries, arterioles, most venules, and veins present too thick a barrier. Capillaries form extensive branching networks that increase the surface area available for rapid exchange of materials. In most tissues, blood flows through only a small part of the capillary network when metabolic needs are low. However, when a tissue is active, such as contracting muscle, the entire capillary network fills with blood.

Capillaries are found near almost every cell in the body, but their number varies with the metabolic activity of the tissue they serve. Body tissues with high metabolic requirements, such as muscles, the brain, liver, and kidneys use more O_2 and nutrients and thus have extensive capillary networks. Tissues with lower metabolic requirements, such as tendons and ligaments, contain fewer capillaries. Capillaries are absent in a few tissues, such as all covering and lining epithelia (see Section 3.8), the cornea and lens of the eye, and cartilage.

Throughout the body, capillaries function as part of a **capillary bed** (**FIGURE 15.3**), a network of 10 to 100 capillaries. In many cases, capillaries branch directly from an arteriole and then rejoin to form a venule. Alternatively, capillaries branch from and then reconnect with a blood vessel called a **metarteriole** that extends from an arteriole to a venule. The metarteriole serves as a shunt that allows blood to bypass the capillaries. The structure of a metarteriole is between that of an arteriole and a capillary: Its wall contains an endothelium, basement membrane, and scattered smooth muscle fibers.

At the junctions where a capillary branches from an arteriole or a metarteriole are rings of smooth muscle fibers called **precapillary sphincters** that control the flow of blood through the capillaries. When the precapillary sphincters are relaxed (open), blood flows into the capillaries (**FIGURE 15.3a**); when precapillary sphincters contract (close or partially close), blood flow through the capillaries decreases or ceases (**FIGURE 15.3b**). Typically, blood flows intermittently through capillaries due to alternating contraction and relaxation of the smooth muscle of precapillary sphincters. This intermittent contraction and relaxation, which may occur 5 to 10 times per minute, is called **vasomotion**. Vasomotion is due in part to local chemical factors released by the endothelial cells (see Section 15.4). At any given time, blood flows through only about 25% of the body's capillaries.

The body contains three different types of capillaries: continuous capillaries, fenestrated capillaries, and sinusoids (**FIGURE 15.4**). Most capillaries are **continuous capillaries**, in which the plasma membranes of endothelial cells form a continuous tube that is interrupted only by **intercellular clefts**, slit-shaped pores about 10 nm in diameter that are located between neighboring endothelial cells (**FIGURE 15.4a**). These capillaries are permeable to water and small solutes such as sodium ions (Na^+) and glucose. Continuous capillaries are found in muscle, connective tissue, and the lungs. They are also present in the brain. However, in most brain areas, only a few substances move across capillary walls because the endothelial cells of most brain capillaries are sealed together by tight junctions. The resulting blockade to movement of materials into and out of brain capillaries is known as the *blood–brain barrier* (see Section 8.2).

Other capillaries of the body are **fenestrated capillaries**. These capillaries are so-named because they have **fenestrations**, cylindrical pores ranging from 20 to 100 nm in diameter that extend through the endothelial cells (**FIGURE 15.4b**) Intercellular clefts are also present between the endothelial cells. Compared to continuous capillaries, fenestrated capillaries have a higher permeability to water and small solutes. Fenestrated capillaries are found in the kidneys, small intestine, and endocrine glands.

Sinusoids are wider and more winding than other capillaries. Their endothelial cells may have unusually large fenestrations (**FIGURE 15.4c**). In addition, sinusoids have very large intercellular clefts ranging from 100 to 1000 nm in diameter and an incomplete or absent basement membrane. As a result, sinusoids are not only permeable to water and small solutes but also to relatively large substances in blood such as proteins and blood cells. Sinusoids are present in bone marrow, the spleen, and the liver.

Venules Drain Blood from Capillaries

Venules are microscopic veins that drain capillary blood and begin the return of blood back toward the heart. As noted earlier, venules that initially receive blood from capillaries are called *postcapillary venules*. They are the smallest venules, measuring 10 μm to 50 μm in diameter. The walls of postcapillary venules consist of little more than an endothelium surrounded by a basement membrane (see **FIGURE 15.3**). The endothelial cells have loosely organized intercellular junctions (the weakest endothelial contacts encountered along the entire vascular tree) and thus are very porous. Postcapillary venules function as significant sites of exchange of nutrients and wastes and leukocyte emigration, and for this reason form part of the microcirculatory exchange unit along with the capillaries. Postcapillary venules merge to form larger venules, which acquire one or two layers of smooth muscle and a thin layer of fibrous connective tissue (see **FIGURE 15.1**). Larger venules, which measure about 50 μm to 100 μm in diameter, have thicker walls across which exchanges with the interstitial fluid can no longer occur.

Veins Carry Blood back to the Heart

Veins are blood vessels that convey blood from tissues back to the heart. They range in size from 0.5 mm in diameter for small veins to 3 cm for the large veins that enter the right atrium of the heart (the superior vena cava and inferior vena cava). Veins are similar to arteries in that their walls contain several tissue layers: an endothelium, basement membrane, smooth muscle, and elastic and fibrous connective tissues (see **FIGURE 15.1**). However, veins have thinner walls and less smooth muscle and elastic tissue compared to arteries.

FIGURE 15.4 Types of capillaries.

The three types of capillaries in the body are continuous capillaries, fenestrated capillaries, and sinusoids.

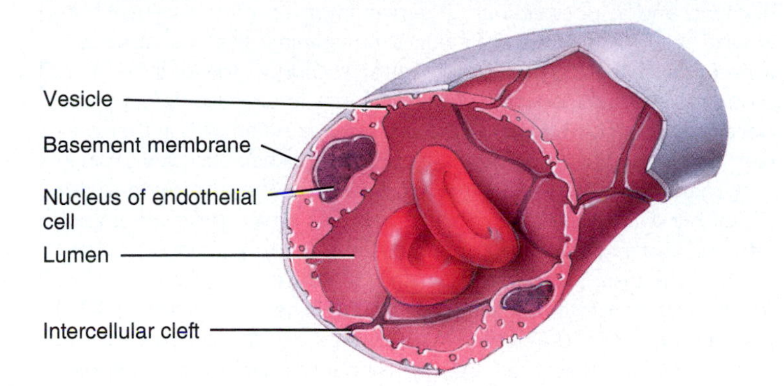

(a) Continuous capillary

Fenestration
Basement membrane
Lumen
Nucleus of endothelial cell
Intercellular cleft
Vesicle

(b) Fenestrated capillary

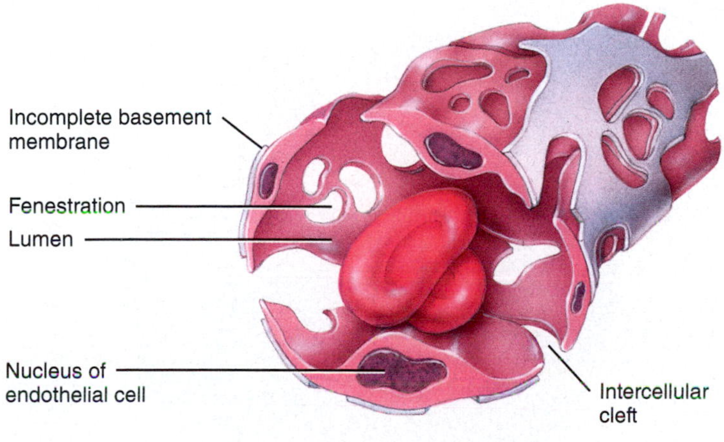

(c) Sinusoid

What is the difference between intercellular clefts and fenestrations?

Many veins, especially those in the limbs, contain **valves**, thin folds of endothelium and connective tissue that form flaplike cusps (FIGURE 15.5). The valve cusps project into the lumen, pointing toward the heart. The valves aid movement of venous blood to the heart by preventing backflow of blood. Leaky venous valves can cause veins to become dilated and twisted in appearance, a condition called **varicose veins** or **varices** (VAR-i-sēz). The valvular defect may be congenital or may result from mechanical stress (prolonged standing or pregnancy) or aging.

The contraction of skeletal muscles in the lower limbs also helps boost movement of venous blood back to the heart. This is known as the **skeletal muscle pump**, which operates as follows (FIGURE 15.5):

1. While you are standing at rest, both the venous valve closer to the heart (proximal valve) and the one farther from the heart (distal valve) in this part of the leg are open, and blood flows upward toward the heart.
2. Contraction of leg muscles, such as when you stand on your tiptoes or take a step, compresses the vein. The compression pushes blood through the proximal valve, an action called *milking*. At the same time, the distal valve in the uncompressed segment of the vein closes as some blood is pushed against it. People who are immobilized through injury or disease lack these contractions of

FIGURE 15.5 Venous valves and the skeletal muscle pump.

The skeletal muscle pump refers to skeletal muscle contractions that compress veins of the limbs, increasing movement of venous blood back to the heart.

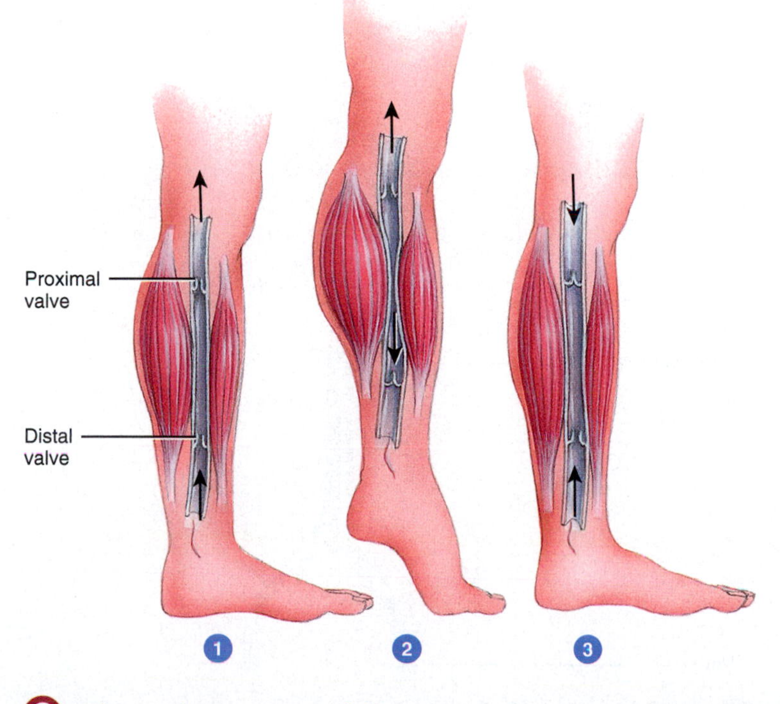

What is the meaning of the term *milking* as it relates to the skeletal muscle pump?

leg muscles. As a result, the return of venous blood to the heart is slower and circulation problems may develop.

3. Just after muscle relaxation, pressure falls in the previously compressed section of vein, which causes the proximal valve to close. The distal valve now opens because blood pressure in the foot is higher than in the leg. When the vein fills with blood from the foot, the proximal valve reopens.

Pressure changes in the thoracic and abdominal cavities during inspiration also promote movement of venous blood back to the heart, a phenomenon known as the **respiratory pump** (FIGURE 15.6). During inspiration (breathing in), the diaphragm moves downward, which causes a decrease in pressure in the thoracic cavity and an increase in pressure in the abdominal cavity. This pressure gradient enhances movement of blood from abdominal veins into thoracic veins and then into the right atrium, thereby increasing the volume of venous blood that returns to the heart. When the pressures reverse during expiration (breathing out), venous valves of the limbs prevent backflow of blood.

Systemic Veins and Venules Function as Blood Reservoirs

The largest portion of your blood volume at rest—about 64%—is in systemic veins and venules (FIGURE 15.7). Systemic arteries and arterioles hold about 13% of the blood volume, systemic capillaries hold about 7%, pulmonary blood vessels hold about 9%, and the heart holds about 7%. Because systemic veins and venules contain a large percentage of the blood volume, they function as **blood reservoirs** from which blood can be diverted quickly if the need arises. For example, during increased muscular activity, the cardiovascular center in the brain stem increases the activity of the sympathetic nerves that innervate veins. The result is *venoconstriction* (constriction of veins), which reduces the volume of blood in reservoirs and allows a greater blood volume to flow to skeletal muscles, where it is needed most. A similar mechanism operates in cases of hemorrhage, when blood volume and pressure decrease; in this case, venoconstriction helps counteract the drop in blood pressure.

CLINICAL CONNECTION

Angiogenesis and Disease

Angiogenesis (an′-jē-ō-JEN-e-sis) refers to the growth of new blood vessels. It is an important process in embryonic and fetal development, and in postnatal life it serves important functions such as wound healing, formation of a new uterine lining after menstruation, formation of the corpus luteum after ovulation, development of blood vessels in skeletal muscles in response to exercise training, and development of blood vessels around obstructed arteries in the coronary circulation. Several proteins (peptides) are known to play a role in angiogenesis. Examples include ***vascular endothelial growth factor (VEGF)*** and ***fibroblast growth factor (FGF)***, which promote angiogenesis, and ***angiostatin*** and ***endostatin***, which inhibit angiogenesis.

Angiogenesis is important clinically because cells of a malignant tumor secrete proteins called ***tumor angiogenesis factors (TAFs)*** that stimulate blood vessel growth to provide nourishment for the tumor cells. Scientists are currently using certain inhibitors of angiogenesis to help stop the growth of tumors. In diabetic retinopathy, angiogenesis may be important in the development of fragile retinal blood vessels that easily break open and leak blood. The bleeding causes scar tissue to form, which can lead to retinal detachment, severe vision loss, and even blindness. So the use of angiogenesis inhibitors could prevent the loss of vision associated with diabetes.

FIGURE 15.6 The respiratory pump.

The respiratory pump refers to pressure changes in the thoracic and abdominal cavities during inspiration that promote increased movement of venous blood back to the heart.

How does the diaphragm contribute to the respiratory pump?

FIGURE 15.7 Blood distribution in the cardiovascular system at rest.

Because systemic veins and venules contain more than half of the total blood volume, they are called blood reservoirs.

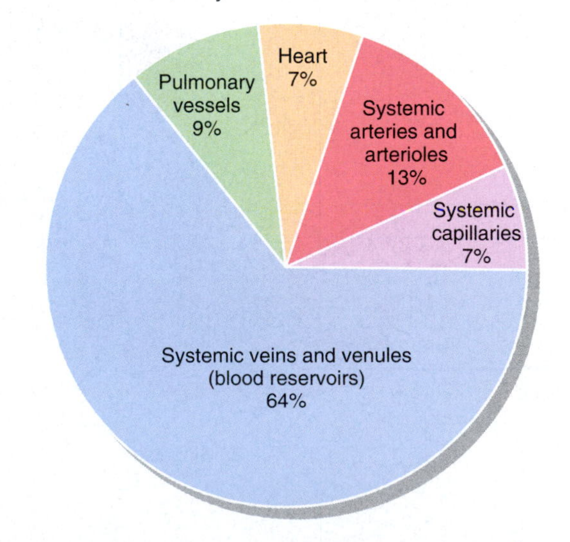

If your total blood volume is 5 liters, what volume is in your veins and venules right now? In your capillaries?

CHECKPOINT

1. Which blood vessels are resistance vessels? Exchange vessels? Blood reservoirs? Pressure reservoirs?
2. What is the purpose of a metarteriole?
3. Which group of blood vessels collectively has the largest percentage of blood?

15.2 Capillary Exchange and Lymphatics

OBJECTIVES

- Explain the mechanisms by which materials move across capillary walls.
- Describe the pressures that cause movement of fluids between capillaries and interstitial spaces.
- Discuss the formation and flow of lymph through the lymphatic system.

The mission of the entire cardiovascular system is to keep blood flowing through capillaries to allow **capillary exchange**, the movement of substances between blood and interstitial fluid. The 7% of the blood in systemic capillaries at any given time is continually exchanging materials with interstitial fluid. Substances enter and leave capillaries by three basic mechanisms: diffusion, transcytosis, and bulk flow.

Diffusion Is the Most Important Method of Capillary Exchange

Many substances, such as oxygen (O_2), carbon dioxide (CO_2), glucose, amino acids, and hormones, enter and leave capillaries by simple diffusion. Because O_2 and nutrients normally are present in higher concentrations in blood, they diffuse down their concentration gradients into interstitial fluid and then into body cells. CO_2 and other wastes released by body cells are present in higher concentrations in interstitial fluid, so they diffuse into blood.

Substances in blood or interstitial fluid can cross the walls of a capillary by diffusing through the endothelial cells or by diffusing through pores (intercellular clefts and fenestrations) (**FIGURE 15.8**).

FIGURE 15.8 Diffusion of substances across the capillary wall. The double-headed arrows indicate that a substance can diffuse in either direction depending on its concentration in blood and interstitial fluid. If there is a higher concentration of the substance in blood, the substance diffuses from blood, crosses the capillary wall, and enters interstitial fluid. If there is a higher concentration of the substance in interstitial fluid, the substance diffuses from interstitial fluid, crosses the capillary wall, and enters blood.

In a capillary, nutrients, gases, and wastes are exchanged between blood and interstitial fluid.

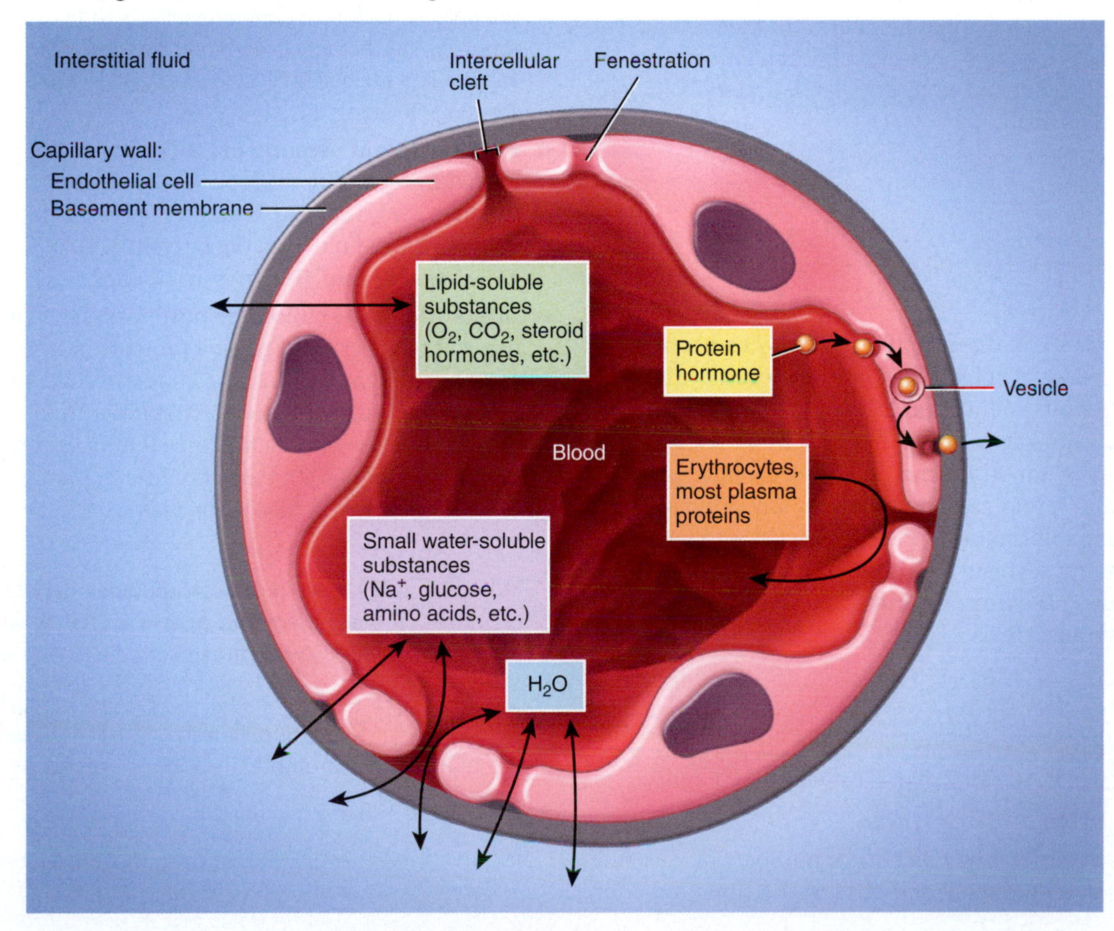

How do substances move across a capillary wall?

Lipid-soluble (nonpolar) substances, such as O_2, CO_2, and steroid hormones, may pass across capillary walls directly through the lipid bilayer of endothelial cell plasma membranes. Small water-soluble substances, which include ions and polar molecules such as glucose and amino acids, pass across capillary walls through intercellular clefts or fenestrations. Water itself can diffuse across capillary walls in all possible ways: by moving through the intercellular clefts or fenestrations or by moving through endothelial cell plasma membranes. (Recall from Chapter 5 that the plasma membrane of a cell is typically permeable not only to nonpolar molecules but also to small, uncharged polar molecules such as water.) Erythocytes cannot pass through the walls of continuous or fenestrated capillaries because they are too large to fit through the intercellular clefts and fenestrations. Plasma proteins are also too large to fit through the intercellular clefts of continuous or fenestrated capillaries. In tissues of some organs, such as the kidneys, plasma proteins are able to pass through the fenestrations. However, negative charges in the basement membrane repel the plasma proteins, most of which are anionic, preventing the proteins from passing across the capillary wall.

In sinusoids, the intercellular clefts are so large that they allow even plasma proteins and blood cells to pass through their walls. For example, hepatocytes (liver cells) synthesize and release many plasma proteins, such as fibrinogen (the main clotting protein) and albumin, which then diffuse into the bloodstream through sinusoids. In red bone marrow, blood cells are formed (hematopoiesis) and then enter the bloodstream through sinusoids.

In contrast to sinusoids, most capillaries of the brain allow only a few substances to move across their walls because the capillaries are sealed together by tight junctions that help form the blood–brain barrier. In brain areas that lack the blood–brain barrier, for example, the hypothalamus, pineal gland, and pituitary gland, materials undergo capillary exchange more freely.

Transcytosis Allows Transport of Large Molecules Across Capillary Walls

A small quantity of material crosses capillary walls by **transcytosis**. In this process, substances in blood become enclosed within tiny vesicles that first enter endothelial cells by endocytosis, then move across the cell and exit on the other side by exocytosis. This method of transport is important mainly for large, lipid-insoluble molecules that cannot cross capillary walls in any other way. For example, many protein hormones leave the bloodstream and enter interstitial fluid by transcytosis (**FIGURE 15.8**), and certain antibodies (also proteins) pass from the maternal circulation into the fetal circulation by transcytosis.

Bulk Flow Regulates the Relative Volumes of Blood and Interstitial Fluid

Bulk flow is a passive process in which *large* numbers of ions, molecules, or particles in a fluid move together in the same direction. The substances move at rates far greater than can be accounted for by diffusion alone. Bulk flow occurs from an area of higher pressure to an area of lower pressure, and it continues as long as a pressure difference exists. Diffusion is more important for *solute exchange* between blood and interstitial fluid, but bulk flow is more important for regulation of the *relative volumes* of blood and interstitial fluid. Pressure-driven movement of fluid and solutes *from* blood capillaries *into* interstitial fluid is called **filtration**. Pressure-driven movement *from* interstitial fluid *into* blood capillaries is called **reabsorption**.

Bulk flow across blood capillary walls is determined by four pressures, which are collectively referred to as **Starling forces** in honor of the physiologist Ernest Starling who first described them. Of the four Starling forces, two promote filtration and two promote reabsorption (**FIGURE 15.9**):

1. **Capillary hydrostatic pressure (P_C)** is the pressure that water in blood exerts against the inner surface of capillary walls. It promotes filtration by forcing fluid out of capillaries into interstitial fluid. P_C is about 37 mmHg at the arterial end of a capillary and about 17 mmHg at the capillary's venous end. The drop in hydrostatic pressure along the length of a capillary occurs because of energy loss due to friction as blood interacts with the capillary wall.
2. **Interstitial fluid hydrostatic pressure (P_{IF})** is the pressure that water in interstitial fluid exerts against the outer surface of capillary walls. It promotes reabsorption by forcing fluid from interstitial spaces back into capillaries. P_{IF} is difficult to measure, and its reported values vary from small positive values to small negative values. For purposes of discussion, it is assumed that P_{IF} equals 1 mmHg all along the capillaries.
3. **Plasma colloid osmotic pressure (π_P)** is the pressure due to the colloidal suspension in blood of plasma proteins, which are unable to move across capillary walls. π_P promotes reabsorption by causing osmosis of fluid from the interstitial spaces into the capillaries. π_P averages 25 mmHg in most capillaries.
4. **Interstitial fluid colloid osmotic pressure (π_{IF})** is the pressure due to the presence of plasma proteins in interstitial fluid. It promotes filtration by causing osmosis of fluid from blood into the interstitial spaces. Typically, interstitial fluid contains very little protein, so π_{IF} is considered to be 0 mmHg. If the protein content of interstitial fluid increases because of damage or disease of the blood capillary wall, then π_{IF} increases.

Whether fluids leave or enter capillaries depends on the balance of pressures. The **net filtration pressure (NFP)**, which indicates the direction of fluid movement, is calculated as follows:

$$\text{NFP} = \underbrace{(P_C + \pi_{IF})}_{\text{Pressures that promote filtration}} - \underbrace{(\pi_P + P_{IF})}_{\text{Pressures that promote reabsorption}}$$

At the arterial end of a capillary,

$$\begin{aligned}\text{NFP} &= (37 + 0)\text{ mmHg} - (25 + 1)\text{ mmHg} \\ &= 37 - 26\text{ mmHg} = 11\text{ mmHg}\end{aligned}$$

FIGURE 15.9 Dynamics of capillary exchange.

Bulk flow across capillary walls is determined by four pressures collectively known as Starling forces.

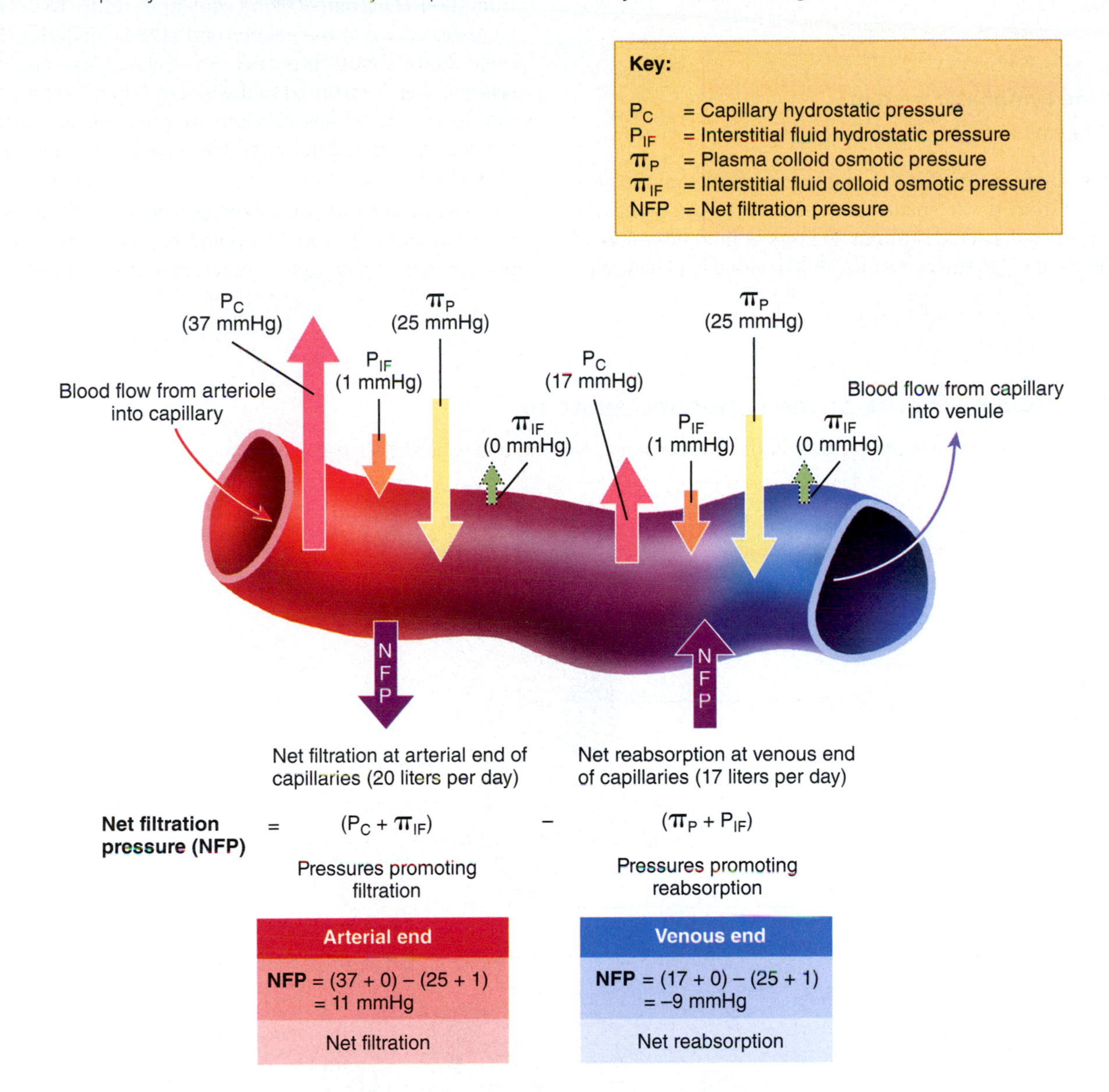

How does a deficit of plasma proteins affect plasma colloid osmotic pressure, and what is the effect on capillary filtration and reabsorption?

Thus, at the arterial end of a capillary, there is a *net outward pressure* of 11 mmHg, and fluid moves out of the capillary into interstitial spaces (filtration).

At the venous end of a capillary,

$$\begin{aligned} \text{NFP} &= (17 + 0)\ \text{mmHg} - (25 + 1)\ \text{mmHg} \\ &= 17 - 26\ \text{mmHg} = -9\ \text{mmHg} \end{aligned}$$

At the venous end of a capillary, the negative value (−9 mmHg) represents a *net inward pressure*, and fluid moves into the capillary from tissue spaces (reabsorption).

Every day about 20 liters of fluid filter out of capillaries in tissues throughout the body. On average, 85% of this fluid (about 17 L) is reabsorbed; the remaining 15% (about 3 L) enters the lymphatic system (described next) and is eventually returned to the bloodstream.

The Lymphatic System Begins in the Tissues Close to Blood Capillaries

The **lymphatic system** consists of a fluid called *lymph*; tubes called *lymphatic vessels* that transport the lymph; and *lymphoid organs* *and tissues*, including lymph nodes, bone marrow, thymus, spleen, tonsils, Peyer's patches of the small intestine, and the appendix (FIGURE 15.10).

Functions of the Lymphatic System

The lymphatic system has four primary functions:

1. ***Drains excess interstitial fluid.*** Lymphatic vessels drain excess interstitial fluid from tissue spaces and return it to the blood. As you have just learned, each day about 20 liters of fluid filter across blood capillary walls into the interstitial fluid located between cells. This fluid must be returned to the cardiovascular system to maintain normal blood volume. About 17 liters of the fluid filtered daily from the arterial end of blood capillaries return to the blood directly by reabsorption at the venous end of the capillaries. The excess filtered fluid—about 3 liters per day—passes first into lymphatic vessels and then is returned to the blood. When interstitial fluid enters into lymphatic vessels, it is known as **lymph**. The major difference between interstitial fluid and lymph is location: Interstitial fluid is found between cells, and lymph is located within lymphatic vessels.
2. ***Returns filtered plasma proteins back to the blood.*** Interstitial fluid contains only a small amount of protein because most plasma proteins are unable to be filtered across blood capillary walls. The

FIGURE 15.10 Components of the lymphatic system.

The lymphatic system consists of lymph, lymphatic vessels, and lymphoid organs and tissues.

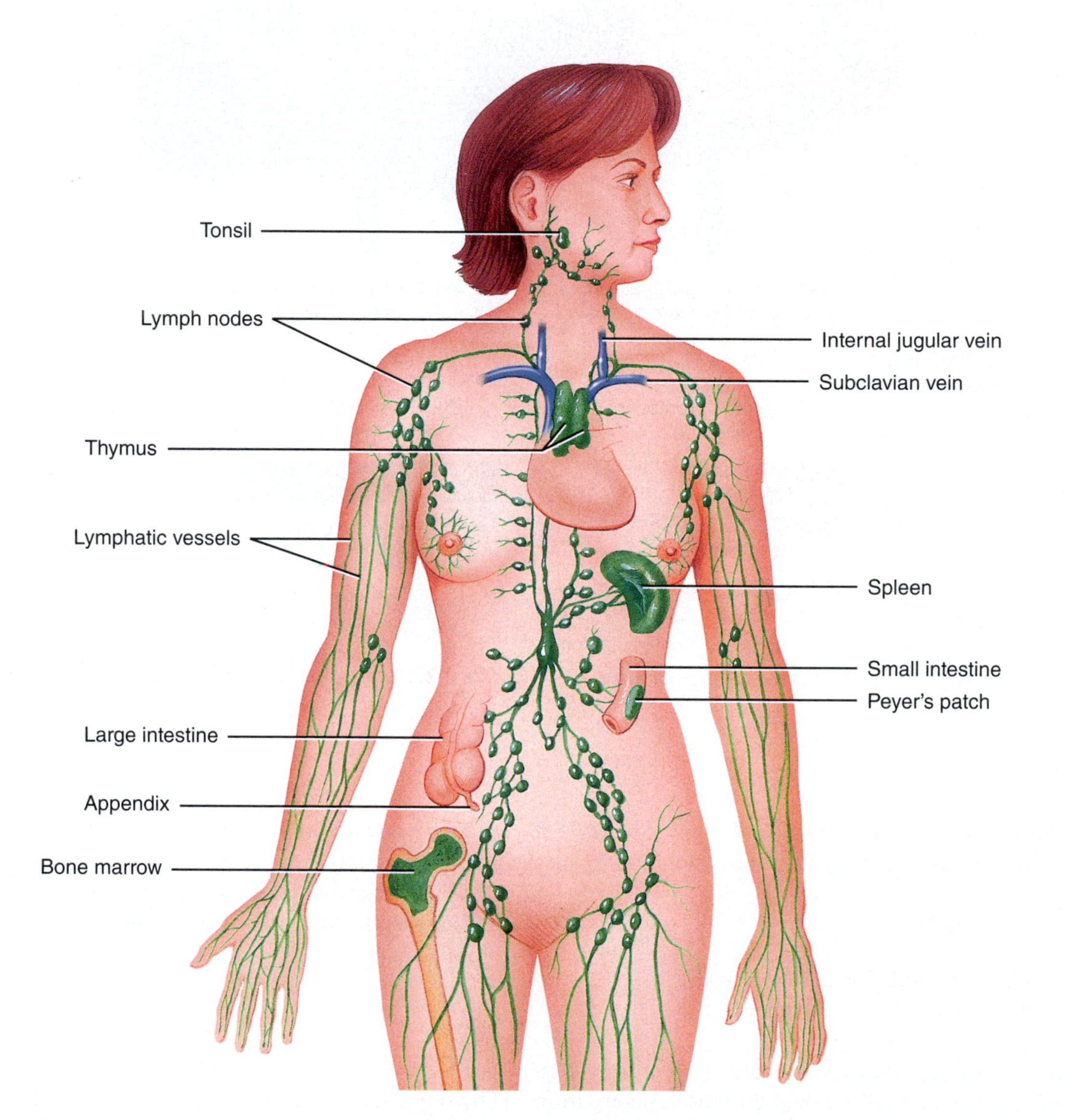

What are the functions of the lymphatic system?

relatively few proteins that are filtered across blood capillary walls cannot return to the blood by diffusion because the concentration gradient (high level of proteins inside blood capillaries, low level outside) opposes such movement. These proteins can, however, move readily into lymph by entering highly permeable lymphatic vessels called lymphatic capillaries (described shortly). From lymph, the plasma proteins eventually move back into the bloodstream.

3. ***Carries out immune responses.*** Lymphoid organs and tissues initiate immune responses directed against microbes or abnormal cells. The roles of these structures in immunity are described in conjunction with the immune system in Chapter 17.
4. ***Transports dietary lipids.*** The lymphatic system transports absorbed lipids from the gastrointestinal tract to the blood. The lipids first enter the lymphatic system by passing into specialized lymphatic vessels called *lacteals*. From lacteals the lipids enter larger lymphatic vessels and then move into the bloodstream. The lipid transport function of the lymphatic system is described in more detail in conjunction with the digestive system in Chapter 21.

Lymphatic Vessels and Lymph Circulation

Lymphatic vessels begin as **lymphatic capillaries**. These tiny vessels, which are located in the spaces between cells, are closed at one end (FIGURE 15.11). Lymphatic capillaries are slightly larger than blood capillaries and have a unique structure that permits interstitial fluid to flow into them but not out. The endothelial cells that make up the wall of a lymphatic capillary are not attached end to end; rather, the ends overlap. When pressure is greater in interstitial fluid than in lymph, the endothelial cells separate, forming spaces between them, and interstitial fluid enters the lymphatic capillary. When pressure is greater inside the lymphatic capillary, the endothelial cells adhere more closely and lymph cannot escape back into interstitial fluid. The pressure is relieved as lymph moves further down the lymphatic capillary. A lymphatic capillary is more permeable than a blood capillary. This is because the spaces between the endothelial cells of a lymphatic capillary are larger than the pores of a blood capillary. As a result, large substances that may be present in interstitial fluid such as proteins, viruses, bacteria, leukocytes, and cancer cells can enter a lymphatic capillary.

Unlike blood capillaries, which link two larger blood vessels that form part of a circuit, lymphatic capillaries begin in the tissues and carry the lymph that forms there toward a larger lymphatic vessel. Just as blood capillaries unite to form venules and veins, lymphatic capillaries unite to form larger and larger **lymphatic vessels**

FIGURE 15.11 Lymphatic capillaries.

Lymphatic vessels begin as lymphatic capillaries, which are larger in size and are more permeable than blood capillaries.

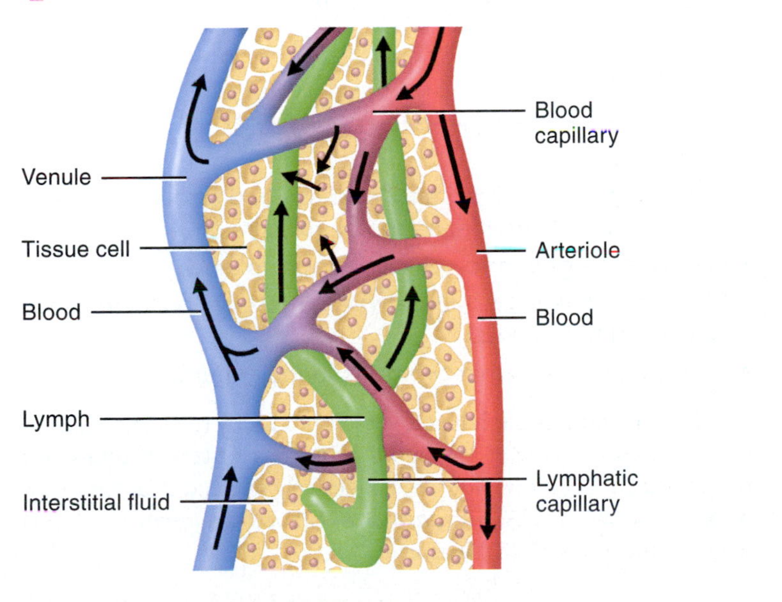

(a) Relationship of lymphatic capillaries to tissue cells and blood capillaries

Lymph
Endothelium of lymphatic capillary
Tissue cell
Interstitial fluid
Opening

(b) Details of a lymphatic capillary

Is lymph more similar to plasma or interstitial fluid? Explain your answer.

FIGURE 15.12 Schematic diagram showing the relationship of the lymphatic system to the cardiovascular system. The arrows indicate the direction of flow of lymph and blood.

The sequence of fluid flow is blood capillaries (blood) → interstitial spaces (interstitial fluid) → lymphatic capillaries (lymph) → larger lymphatic vessels (lymph) → venous circulation.

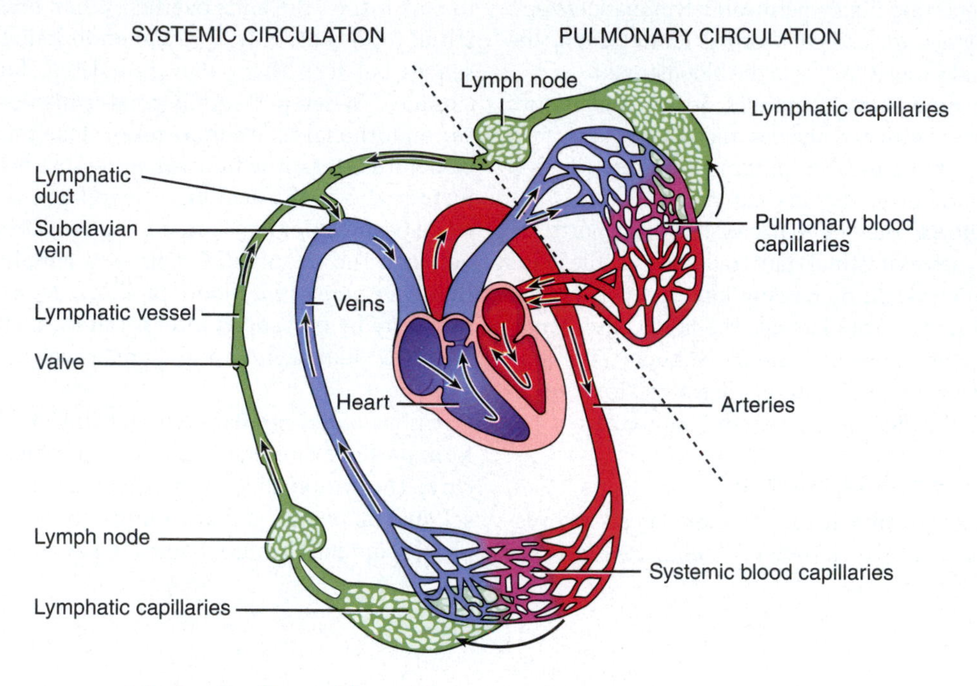

Does inhalation promote or hinder the flow of lymph?

(**FIGURE 15.12**), which resemble veins in structure but have thinner walls. Located at intervals along lymphatic vessels are lymph nodes, where lymph is filtered of foreign substances. Lymphatic vessels ultimately empty into the venous system. Thus, lymph drains back into the blood. Like veins, lymphatic vessels contain valves, which ensure the one-way movement of lymph.

Several mechanisms maintain the flow of lymph:

1. ***Smooth muscle contractions.*** When a large lymphatic vessel distends because of the presence of lymph, the smooth muscle in its wall contracts, which helps move lymph from one segment of the vessel to the next.
2. ***Skeletal muscle pump.*** The milking action of skeletal muscle contractions compresses lymphatic vessels (as well as veins) and forces lymph toward the venous system.
3. ***Respiratory pump.*** Lymph flow is also maintained by pressure changes that occur during inspiration. Lymph flows from the abdominal region, where the pressure is higher, toward the thoracic region, where it is lower. When the pressures reverse during expiration, the valves of the lymphatic vessels prevent backflow of lymph.

CHECKPOINT

4. How can substances enter and leave blood?
5. How do hydrostatic and osmotic pressures determine fluid movement across the walls of capillaries?
6. Which functions of the lymphatic system are linked to the cardiovascular system?

15.3 Hemodynamics

OBJECTIVES

- Explain how blood flow is affected by pressure and resistance.
- Define blood pressure, systolic pressure, diastolic pressure, pulse pressure, and mean arterial pressure.
- Explain the significance of vascular compliance.
- Describe the relationship between cross-sectional area and velocity of blood flow.
- Discuss the factors that promote venous return.

The flow of blood through blood vessels abides by the same physical principles as the flow of any liquid through a system of tubes. The factors that affect blood flow are collectively referred to as **hemodynamics** (hē-mō-dī-NAM-iks; *hemo-* = blood; *-dynamics* = power), which is the focus of this section.

Blood Flow Is Affected by Pressure and Resistance

Blood flow is the volume of blood that flows through any tissue in a given time period (often expressed in mL/min or cm^3/sec). It depends on two major factors: (1) the *pressure gradient* (difference in pressure) that drives blood flow through a tissue and (2) the *resistance* to blood flow in specific blood vessels. The relationship between blood flow, the pressure gradient, and resistance is given by the **flow equation**:

$$F = \frac{\Delta P}{R}$$

where F is blood flow,
ΔP is the pressure gradient, and
R is resistance to blood flow.

As you can see from this equation, blood flow (F) is directly proportional to the pressure gradient (ΔP) and inversely proportional to resistance (R). Blood flows from regions of higher pressure to regions of lower pressure; the greater the pressure gradient, the greater the blood flow. However, the higher the resistance, the smaller the blood flow.

To understand how a pressure gradient affects blood flow, consider the flow of blood through a given blood vessel of the body. Flow through the vessel is directly proportional to the pressure gradient between the two ends of the vessel ($\Delta P = P_1 - P_2$, where P_1 and P_2 are the pressures of the vessel inlet and outlet, respectively) (**FIGURE 15.13a**). The greater the pressure gradient across a blood vessel, the greater the flow rate. For example, suppose that pressure in a vessel is 60 mmHg at one end and 20 mmHg at the other end ($\Delta P = 40$ mmHg), whereas pressure in another vessel of the same size is 90 mmHg at one end and 10 mmHg at the other end ($\Delta P = 80$ mmHg). Because the second vessel has a greater pressure gradient, it has a greater flow (**FIGURE 15.13b**). If a pressure gradient does not exist across a blood vessel, then there will be no flow through that vessel. For example, suppose that pressure at both ends of a vessel is 60 mmHg; because $\Delta P = 0$, no flow occurs (**FIGURE 15.13c**). So, it is the *difference* in pressure between the two ends of the blood vessel, and not the absolute pressures within the vessel, that determines the flow rate. It is also important to note that two vessels of the same size can have different absolute pressures but the same flow. For example, suppose that pressure in a vessel is 80 mmHg at one end and 60 mmHg at the other end ($\Delta P = 20$ mmHg), whereas pressure in another vessel is 40 mmHg at one end and 20 mmHg at other end ($\Delta P = 20$ mmHg). Because both vessels have the same pressure gradients, they have equal flow rates (**FIGURE 15.13d**).

As noted earlier in the chapter, **resistance** is the opposition to blood flow due to friction between blood and the walls of blood vessels. Resistance depends on three factors: (1) blood viscosity, (2) blood vessel length, and (3) blood vessel radius. The factors that determine resistance can be expressed by the following relationship, known as the **resistance equation**:

$$R = \frac{\eta L}{r^4}$$

where
R is resistance,
η is blood viscosity,
L is blood vessel length, and
r is blood vessel radius.

1. ***Blood viscosity.*** Resistance to blood flow is directly proportional to the viscosity (thickness) of blood. The higher the blood's viscosity, the higher the resistance. The viscosity of blood depends mostly on the ratio of erythrocytes to plasma (fluid) volume, and to a smaller extent on the concentration of proteins in plasma. Normally, these factors are essentially constant, so

FIGURE 15.13 Relationship between the pressure gradient across a blood vessel and blood flow.

The greater the pressure gradient across a blood vessel, the greater the blood flow through that vessel.

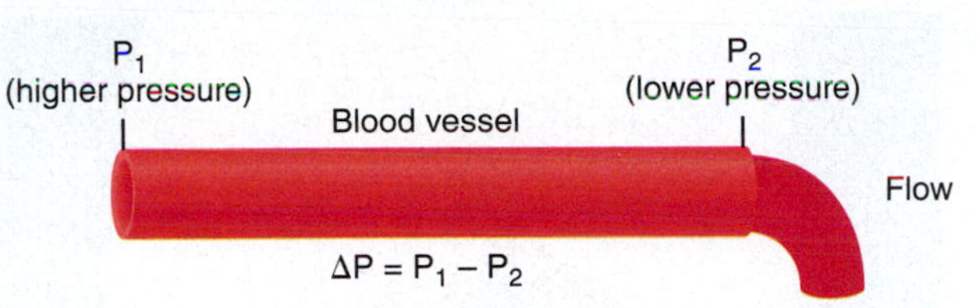

(a) Flow through blood vessel when a pressure gradient (ΔP) exists

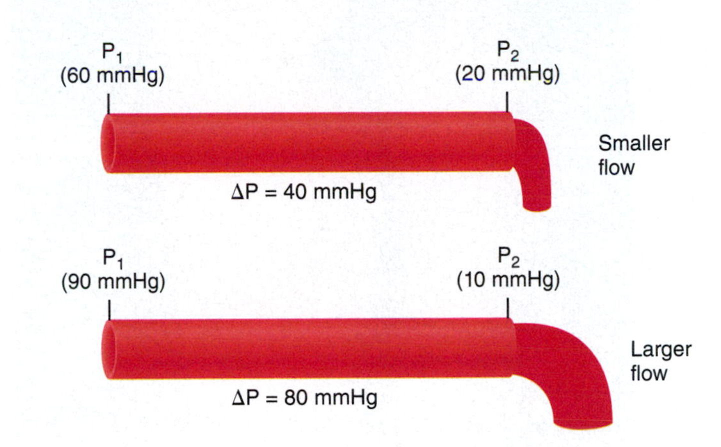

(b) Comparison of flow rates in blood vessels that have a different ΔP

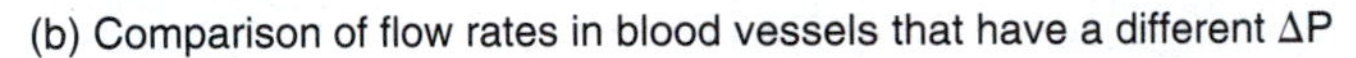

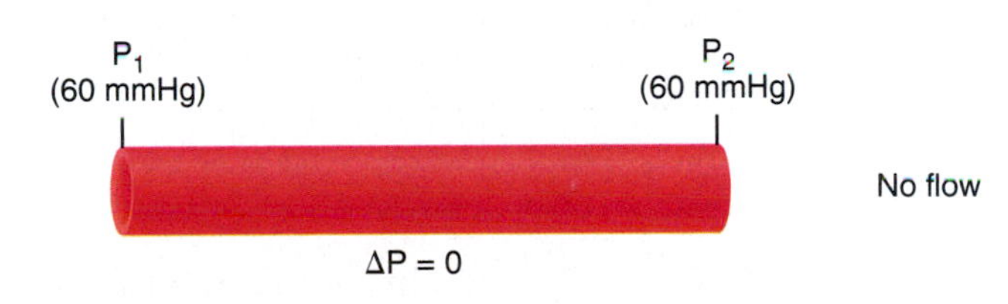

(c) No flow through blood vessel when ΔP is zero

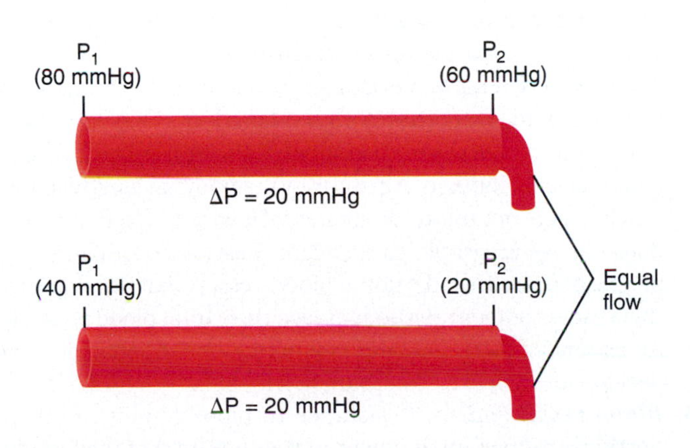

(d) Comparison of flow rates in blood vessels that have the same ΔP

What happens to flow if the pressure gradient across a blood vessel is zero?

CLINICAL CONNECTION

Edema

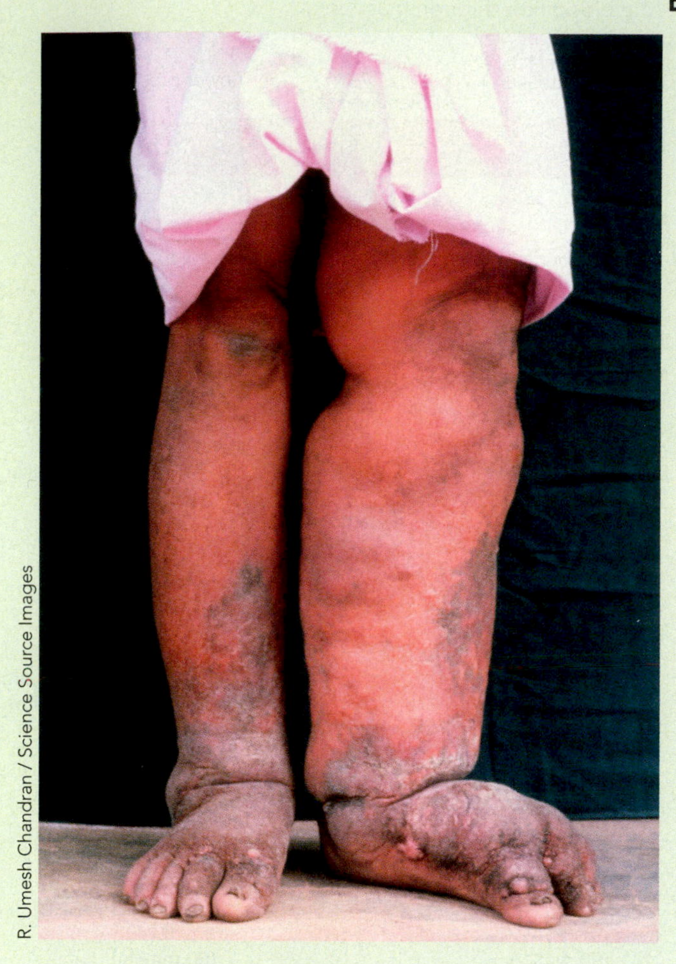
R. Umesh Chandran / Science Source Images

If filtration greatly exceeds reabsorption, the result is **edema**, an abnormal increase in interstitial fluid volume. Edema can result from excess filtration or inadequate reabsorption. Two situations may cause excess filtration:

- ***Increased capillary hydrostatic pressure*** causes more fluid to be filtered from capillaries.
- ***Increased permeability of capillaries*** raises interstitial fluid colloid osmotic pressure by allowing some plasma proteins to escape. Such leakiness may be caused by the destructive effects of chemical, bacterial, thermal, or mechanical agents on capillary walls.

One situation commonly causes inadequate reabsorption:

- ***Decreased concentration of plasma proteins*** lowers the plasma colloid osmotic pressure. Inadequate synthesis or dietary intake or loss of plasma proteins is associated with liver disease, burns, malnutrition, and kidney disease.

Edema can also be caused by inadequate lymphatic drainage, in which case the term **lymphedema** is used. It may occur when lymph nodes and lymphatic vessels near a tumor are removed during surgery for cancer. Another potential cause of lymphedema is radiation therapy, which can lead to a buildup of fibrous scar tissue that blocks lymphatic vessels. Lymphedema may also result from parasitic infections. For example, in the disease known as **filariasis** (fil-a-RĪ-a-sis), threadlike parasitic worms infect lymphatic vessels, which prevents the vessels from draining properly. This leads to a type of lymphedema called **elephantiasis** (el-e-fan-TĪ-a-sis) that produces extreme swelling of the limbs and external genitalia (see FIGURE in this box).

viscosity does not usually play a major role in resistance. However, any condition that increases the viscosity of blood, such as dehydration or polycythemia (an unusually high number of erythrocytes), increases resistance. A depletion of plasma proteins or erythrocytes, due to anemia or hemorrhage, decreases viscosity and thus decreases resistance.

2. ***Blood vessel length.*** Resistance to blood flow through a vessel is also directly proportional to the length of the blood vessel. The longer the blood vessel, the greater the resistance. Blood vessel length usually remains constant in the body, so length is not a significant contributor to resistance. However, blood vessel length does change as people gain weight. Resistance is higher in obese people because the additional blood vessels that develop to supply their excess adipose tissue increase their total blood vessel length. An estimated 650 km (about 400 miles) of additional blood vessels develop for each extra kilogram (2.2 lb) of fat.
3. ***Blood vessel radius.*** Resistance to blood flow is inversely proportional to the fourth power of the radius (r) of the blood vessel ($R \propto 1/r^4$). The smaller the radius of the blood vessel, the greater the resistance it offers to blood flow. This is because a given volume of blood has more contact with the vessel wall in a blood vessel that has a smaller radius, resulting in more frictional resistance. Because blood vessel radius is raised to the fourth power, it is the main determinant of resistance. So, small changes in radius size cause large changes in resistance. As an example, consider three blood vessels (vessels A, B, and C) that have different radii but the same pressure gradient (FIGURE 15.14). The radius, resistance, and flow of vessel A are arbitrarily equal to 1. The radius of vessel B is one-half that of vessel A; as a result, the resistance of vessel B is 16 times greater ($R \propto 1/(\frac{1}{2})^4 = 16$) and flow is 16 times lower ($F \propto 1/R = 1/16$) than vessel A. Vessel C has a radius that is one-fourth that of vessel A; consequently the resistance of vessel C is 256 times greater ($R \propto 1/(\frac{1}{4})^4 = 256$) and flow is 256 times lower ($F \propto 1/R = 1/256$) than vessel A. In the circulation, changes in blood vessel radius occur by vasoconstriction or vasodilation. Vasoconstriction decreases blood vessel radius and vasodilation increases it. Normally, moment-to-moment fluctuations in blood flow through a given tissue are due to vasoconstriction and vasodilation of the tissue's arterioles. As arterioles dilate, resistance decreases, blood flow increases, and blood pressure falls. As arterioles constrict, resistance increases, blood flow decreases, and blood pressure rises.

FIGURE 15.14 Relationship between resistance and blood flow.

The greater the resistance, the smaller the blood flow.

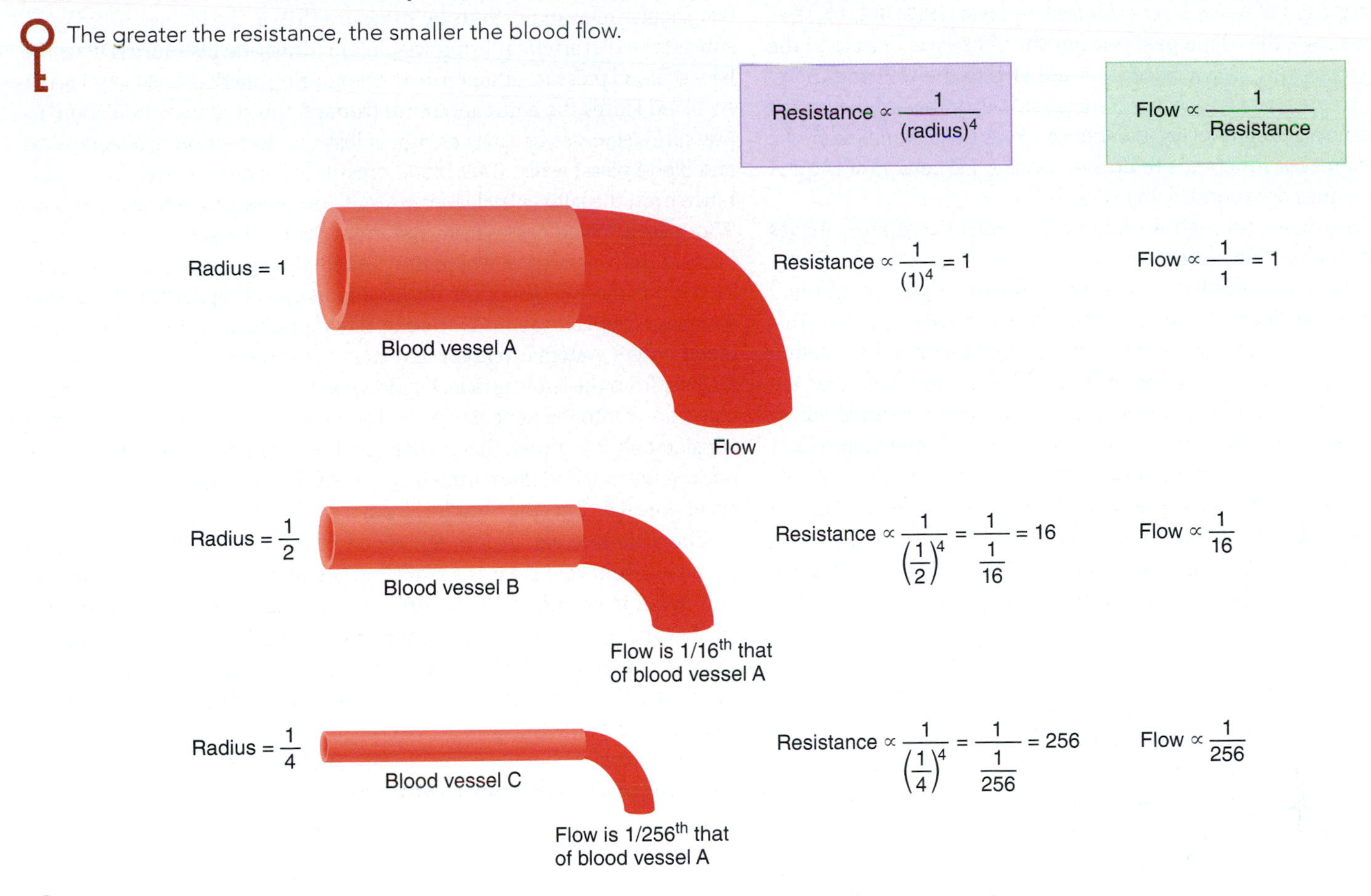

What is the resistance and flow in a blood vessel that has a radius twice that of vessel A?

Total peripheral resistance (TPR), also known as *systemic vascular resistance*, refers to all of the vascular resistances offered by systemic blood vessels. The radii of arteries and veins are large, so their resistances are very small because most of the blood does not come into physical contact with the walls of these blood vessels. Capillaries and venules contribute more resistance, and arterioles contribute the most resistance of all the blood vessel types. A major function of arterioles is to control TPR—and therefore blood pressure and blood flow to particular tissues—by changing their radii. Arterioles need to vasodilate or vasoconstrict only slightly to have a large effect on TPR. The main center for regulation of TPR is the vasomotor center in the brain stem (described in Section 15.5).

Blood Flow Through Blood Vessels Can Be Laminar or Turbulent

Blood normally flows through a blood vessel in a smooth, streamlined manner that is parallel to the vessel axis, a phenomenon known as

PHYSIOLOGICAL EQUATION

POISEUILLE'S LAW

You have already learned that blood flow (F) is directly proportional to the pressure gradient (ΔP) and inversely proportional to resistance (R):

$$F = \frac{\Delta P}{R}$$

In addition, you have just learned that resistance (R) is directly proportional to blood viscosity (η) and blood vessel length (L) and inversely proportional to the fourth power of the radius of the blood vessel (r^4):

$$R = \frac{\eta L}{r^4}$$

If you take this second equation (the resistance equation), substitute it for R in the first equation (the flow equation), and then add $\pi/8$ as a constant of proportionality, you will derive **Poiseuille's law** (pwah-ZŪ-ēz):

$$F = \frac{\pi \Delta P r^4}{8\eta L}$$

This equation tells you that blood flow (F) is directly proportional to the pressure gradient (ΔP) and to the fourth power of the radius of the blood vessel (r^4) and inversely proportional to blood viscosity (η) and blood vessel length (L).

laminar flow. Laminar flow is so-named because the fluid behaves as if it were comprised of many layers (*lamina-* = layer) (**FIGURE 15.15a**). As the layers move, they slide past one another. The layer closest to the vessel wall moves very slowly because it adheres to the wall and therefore has the greatest resistance. Each successive layer toward the center of the vessel moves progressively faster because, as the distance from the vessel wall increases, resistance to flow decreases. Laminar flow is quiet and therefore does not produce any sounds.

When blood flows through an abnormally constricted area, moves over a rough surface, makes a sharp turn, or exceeds a critical velocity, laminar flow becomes turbulent. In **turbulent flow**, components of blood move at various angles to the axis of the vessel (**FIGURE 15.15b**). This causes blood to mix, forming vortices (whorls) that increase the interaction between blood and the vessel wall, resulting in vibrations that are heard as sounds. For example, turbulent flow through abnormal valves of the heart produces heart murmurs (see Clinical Connection: Heart Murmurs in Section 14.4). In addition, turbulent flow is responsible for the sounds heard during blood pressure measurements (see Tool of the Trade: Measuring Blood Pressure in this section). Because turbulent flow promotes more interaction between blood and the vessel wall, there is greater resistance during turbulent flow than during laminar flow.

Blood Pressure Is Hydrostatic Pressure Exerted by Blood on Blood Vessel Walls

Earlier it was mentioned that blood flows from regions of higher pressure to regions of lower pressure; the greater the pressure gradient, the greater the blood flow. Contraction of the ventricles generates **blood pressure (BP)**, the hydrostatic pressure exerted by blood on the walls of blood vessels. BP is determined by cardiac output, blood volume, and vascular resistance. BP is highest in the aorta and other large systemic arteries; in a resting, healthy young adult, BP rises to about 110 mmHg during systole (ventricular contraction) and drops to about 70 mmHg during diastole (ventricular relaxation). **Systolic pressure (SP)** is the highest blood pressure attained in arteries during systole, and **diastolic pressure (DP)** is the lowest blood pressure attained in arteries during diastole (**FIGURE 15.16**).

As blood leaves the aorta and flows through the systemic circulation, its pressure decreases because energy is lost due to friction between blood and blood vessel walls. Thus, blood pressure falls progressively as the distance from the left ventricle increases. Blood pressure decreases to about 37 mmHg as blood passes from systemic arteries through systemic arterioles and into capillaries, where the pressure fluctuations associated with systole and diastole disappear. At the venous end of capillaries, blood pressure has dropped to about 17 mmHg. Blood pressure continues to drop as blood enters systemic venules and then veins because these vessels are farthest from the left ventricle. Finally, blood pressure reaches 0 mmHg as blood flows into the right atrium. So the average blood pressure in veins is considerably lower than that in arteries. The difference in pressure can be noticed when blood flows from a cut vessel. Blood leaves a cut vein in an even, slow flow but spurts rapidly from a cut artery.

The alternate expansion and recoil of an artery after each systole of the left ventricle creates a traveling pressure wave called the **pulse**. The pulse is strongest in the arteries closest to the heart, becomes weaker in the arterioles, and disappears altogether in the capillaries. The radial artery at the wrist is most commonly used to palpate (feel) the pulse. The pulse rate normally is the same as the heart rate, about 75 beats per minute at rest.

The difference between the systolic pressure (SP) and the diastolic pressure (DP) is called **pulse pressure (PP)**:

$$PP = SP - DP$$

FIGURE 15.15 Laminar and turbulent flow.

In laminar flow, blood moves in a smooth, streamlined manner that is parallel to the axis of the blood vessel; in turbulent flow, blood moves at various angles to the vessel axis.

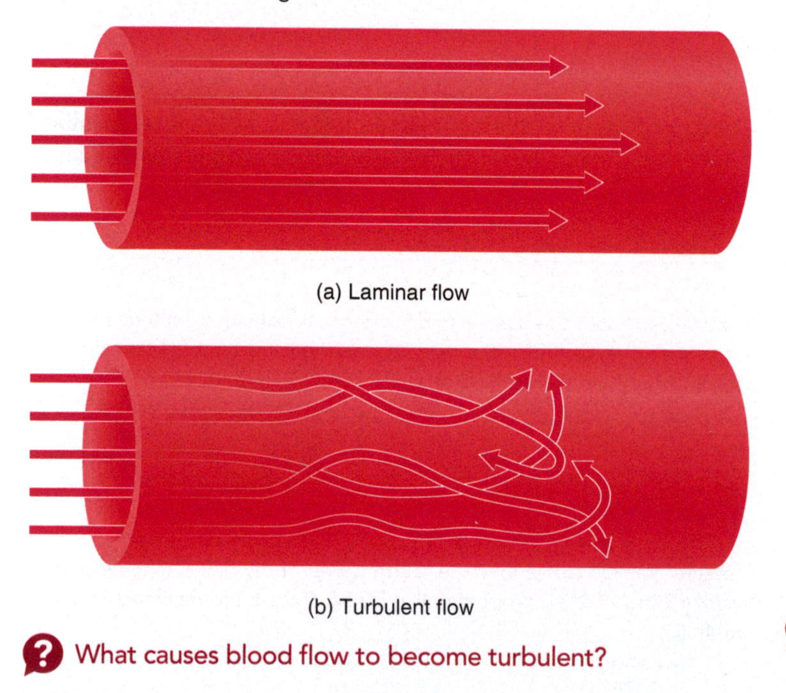

(a) Laminar flow

(b) Turbulent flow

What causes blood flow to become turbulent?

FIGURE 15.16 Blood pressures in various parts of the cardiovascular system. The dashed line is the mean (average) blood pressure in the aorta, arteries, and arterioles.

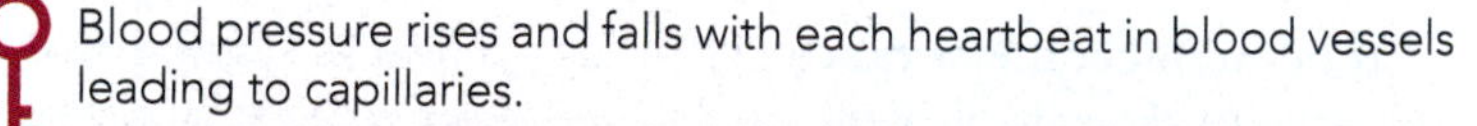

Blood pressure rises and falls with each heartbeat in blood vessels leading to capillaries.

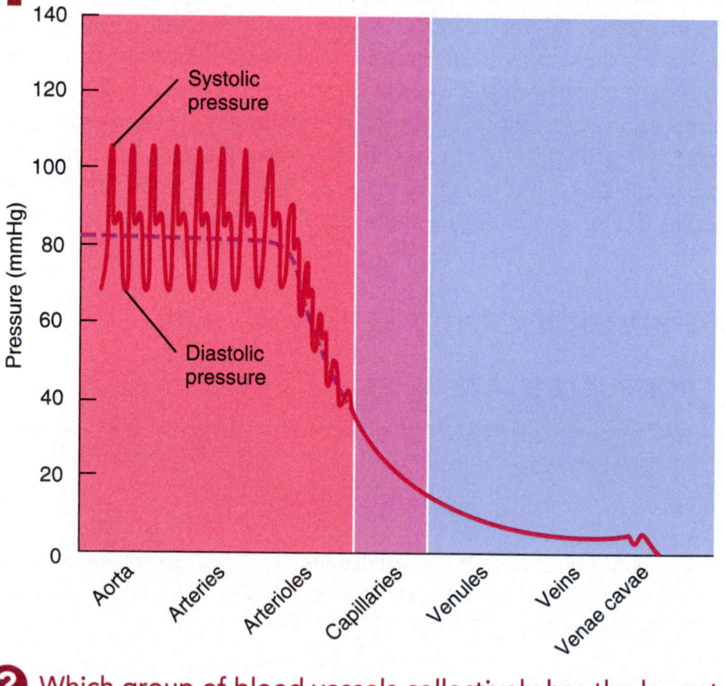

Which group of blood vessels collectively has the lowest pressure?

TOOL OF THE TRADE

MEASURING BLOOD PRESSURE

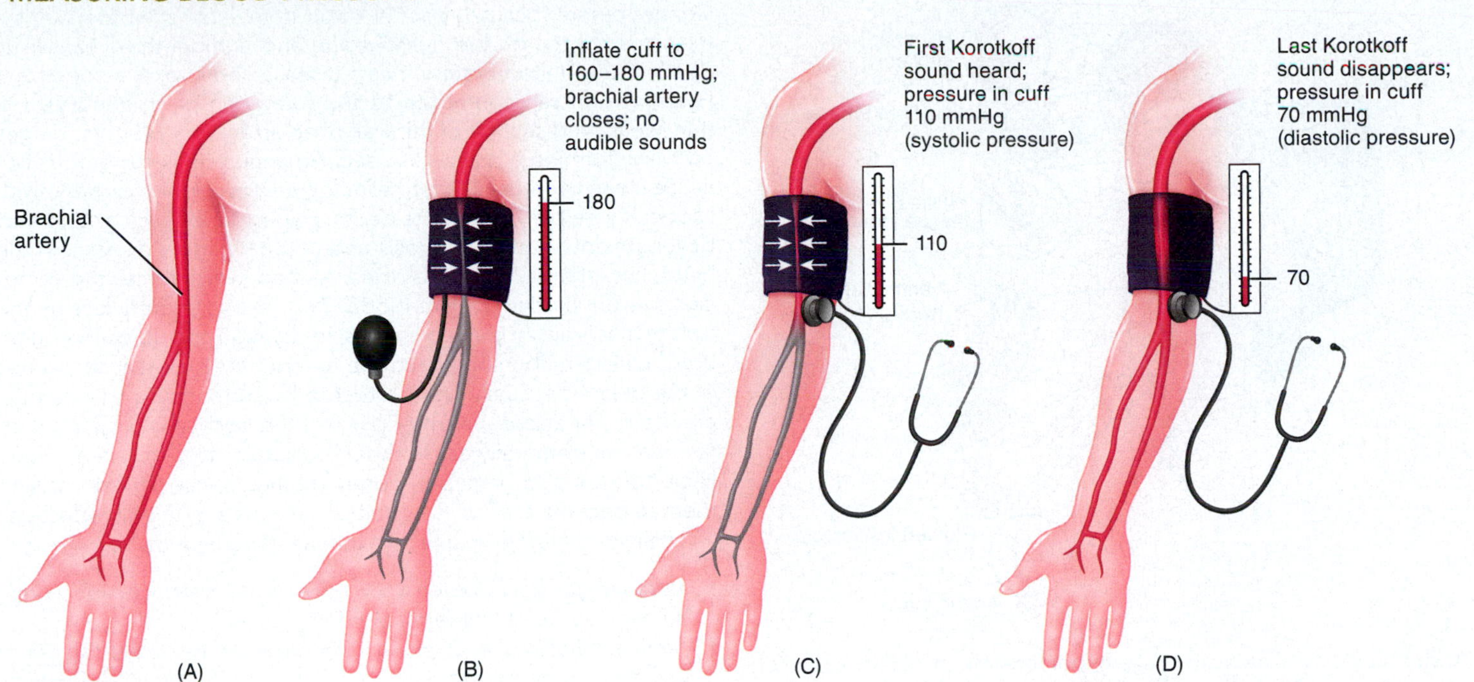

In clinical use, the term **blood pressure** usually refers to the pressure in arteries generated by the left ventricle during systole and the pressure remaining in the arteries when the ventricle is in diastole. Blood pressure is typically measured in the brachial artery in the left arm (**FIGURE A**). The device used to measure blood pressure is called a **sphygmomanometer** (sfig′-mō-ma-NOM-e-ter). It consists of a rubber cuff connected to a rubber bulb that is used to inflate the cuff, and a meter that registers the pressure in the cuff. With the arm resting on a table so that it is about the same level as the heart, the cuff of the sphygmomanometer is wrapped around a bare arm and a stethoscope is placed below the cuff on the brachial artery. At this point, blood flow through the brachial artery is laminar, so no sounds can be heard. The cuff is inflated by squeezing the bulb until the brachial artery is completely compressed and blood flow stops during all stages of the cardiac cycle, about 30 mmHg higher than the person's usual systolic pressure (**FIGURE B**). If the normal systolic pressure is unknown, the cuff can be inflated to about 160 to 180 mmHg. Because the brachial artery is closed and no blood is flowing through it, no sounds can be heard through the stethoscope. The technician then slowly deflates the cuff. When the cuff is deflated enough to allow the brachial artery to partially open during systole, a spurt of blood passes through. The spurt of blood is turbulent, resulting in the first sound heard through the stethoscope (**FIGURE C**). This sound corresponds to ***systolic pressure***, the force of blood pressure on arterial walls just after ventricular contraction. Note that after blood spurts through the constricted brachial artery during systole, the artery closes again during diastole. For the next few heartbeats, this pattern repeats: The brachial artery partially opens during systole and then collapses during diastole. As the cuff is deflated further, the brachial artery opens up even more and for a longer portion of the cardiac cycle. The sounds become louder and more pronounced, then softer and muffled. Once the brachial artery fully opens for the entire duration of systole and diastole, laminar flow resumes and the sounds disappear (**FIGURE D**). This level, called the ***diastolic pressure***, represents the force exerted by the blood remaining in arteries during ventricular relaxation. The sounds that are heard while taking blood pressure are called **Korotkoff sounds** (kō-ROT-kof).

The normal blood pressure of a young adult male is less than 120 mmHg systolic and less than 80 mmHg diastolic. For example, "110 over 70" (written as 110/70) is a normal blood pressure. In young adult females, the pressures are 8 to 10 mmHg less. People who exercise regularly and are in good physical condition may have even lower blood pressures. Thus, blood pressure slightly lower than 120/80 may be a sign of good health and fitness.

Pulse pressure, which is normally about 40 mmHg (110 − 70), provides information about the condition of the cardiovascular system. For example, conditions such as atherosclerosis greatly increase pulse pressure.

Mean arterial pressure (MAP) is the average blood pressure in arteries. Because diastole lasts longer than systole during a cardiac cycle, MAP is approximately equal to the diastolic pressure plus one-third of the pulse pressure:

$$\text{MAP} = \text{DP} + \frac{1}{3}\,\text{PP}$$

Thus, in a person whose BP is 110/70 mmHg, MAP is about 83 mmHg [70 + ⅓(40)].

Recall that blood flow (F) is directly proportional to the pressure gradient (ΔP) and inversely proportional to resistance (R):

$$F = \frac{\Delta P}{R}$$

Total blood flow is cardiac output (CO), the volume of blood that circulates through systemic (or pulmonary) blood vessels each minute.

Hypertension

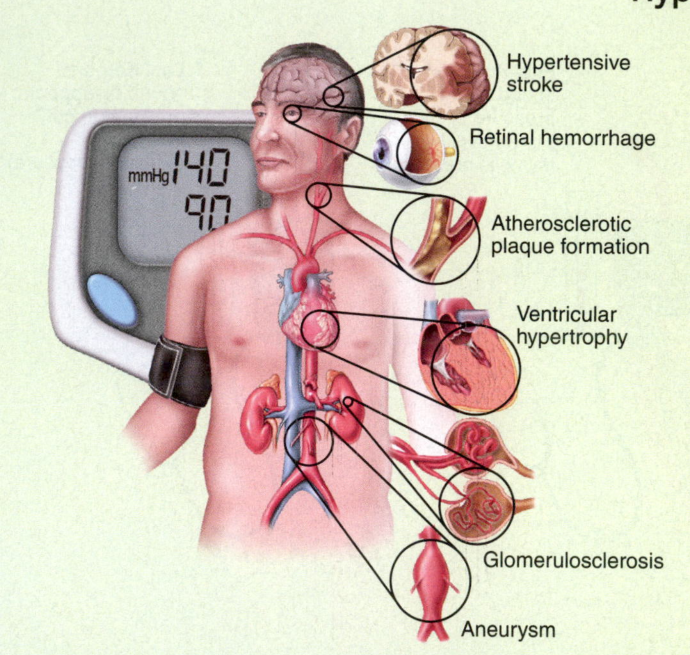

About 50 million Americans have **hypertension**, or persistently high blood pressure (140/90 mmHg or higher). Hypertension is the most common disorder affecting the heart and blood vessels and is the major cause of heart failure, kidney disease, and stroke. The current guidelines for blood pressure are as follows:

Category	Systolic (mmHg)		Diastolic (mmHg)
Normal	Less than 120	*and*	Less than 80
Prehypertension	120–139	*or*	80–89
Stage 1 hypertension	140–159	*or*	90–99
Stage 2 hypertension	Greater than 160	*or*	Greater than 100

Between 90 and 95% of all cases of hypertension are **primary hypertension**, a persistently elevated blood pressure that cannot be attributed to any identifiable cause. The remaining 5–10% of cases are **secondary hypertension**, which has an identifiable underlying cause. Several disorders cause secondary hypertension:

- ***Obstruction of renal blood flow*** or disorders that damage renal tissue may cause the kidneys to release excessive amounts of renin into the blood. The resulting high level of angiotensin II causes vasoconstriction, thus increasing total peripheral resistance.
- ***Hypersecretion of aldosterone***—resulting, for instance, from a tumor of the adrenal cortex—stimulates excess reabsorption of salt and water by the kidneys, which increases body fluid volume.
- ***Hypersecretion of epinephrine and norepinephrine*** by a **pheochromocytoma** (fē-ō-krō′-mō-sī-TŌ-ma), a tumor of the adrenal medulla, leads to an increase in heart rate and contractility and an increase in total peripheral resistance.

High blood pressure is known as the "silent killer" because it can cause considerable damage to the blood vessels, heart, brain, and kidneys before it causes pain or other noticeable symptoms. It is a major risk factor for the number-one and number-three causes of death in the United States—heart disease and stroke, respectively. Hypertension can contribute to the formation of an **aneurysm**, a thin, weakened section of the wall of an artery or a vein that bulges outward, forming a balloonlike sac. An aneurysm is dangerous because it can eventually burst, resulting in massive hemorrhage with shock, severe pain, stroke, or death. Hypertension also accelerates development of atherosclerosis and coronary artery disease. In the heart, hypertension increases the afterload, which forces the ventricles to work harder to eject blood. This leads to ventricular hypertrophy that is accompanied by muscle damage and fibrosis (a buildup of collagen fibers between the muscle fibers). Because arteries in the brain are usually less protected by surrounding tissues than are the major arteries in other parts of the body, prolonged hypertension can eventually cause them to rupture, resulting in a stroke. Hypertension also damages kidney glomeruli (capillaries), causing them to become scarred (a condition known as glomerulosclerosis).

Lifestyle changes are effective in managing hypertension:

- ***Lose weight***. Loss of even a few pounds helps reduce blood pressure in overweight hypertensive individuals.
- ***Limit alcohol intake***. Drinking in moderation may lower the risk of coronary heart disease.
- ***Exercise***. Becoming more physically fit by engaging in moderate activity (such as brisk walking) several times a week for 30 to 45 minutes can lower systolic blood pressure by about 10 mmHg.
- ***Reduce intake of sodium (salt)***. Roughly half the people with hypertension are "salt-sensitive." For them, a high-salt diet appears to promote hypertension, and a low-salt diet can lower their blood pressure.
- ***Maintain recommended dietary intake of potassium, calcium, and magnesium***. Higher levels of potassium, calcium, and magnesium in the diet are associated with a lower risk of hypertension.
- ***Don't smoke***. Smoking has devastating effects on the heart and can augment the damaging effects of high blood pressure by promoting vasoconstriction.
- ***Manage stress***. Various meditation and biofeedback techniques help some people reduce high blood pressure. These methods may work by decreasing the daily release of epinephrine and norepinephrine by the adrenal medulla.

Drugs with several different mechanisms of action are effective in lowering blood pressure. Many people are successfully treated with ***diuretics***, agents that decrease blood pressure by decreasing blood volume because they increase elimination of water and salt in the urine. ***ACE*** (***angiotensin converting enzyme***) ***inhibitors*** block formation of angiotensin II and thereby promote vasodilation and decrease the secretion of aldosterone. ***Beta blockers*** reduce blood pressure by inhibiting the secretion of renin and by decreasing heart rate and contractility. ***Vasodilators*** relax the smooth muscle in arterial walls, causing vasodilation and lowering blood pressure by lowering total peripheral resistance. An important category of vasodilators are the ***calcium channel blockers***, which slow the inflow of Ca^{2+} into vascular smooth muscle cells. They reduce the heart's workload by slowing Ca^{2+} entry into pacemaker cells and regular myocardial fibers, thereby decreasing heart rate and the force of myocardial contraction.

If the entire systemic circulation is considered to be a single blood vessel that carries blood away from and back to the heart, then ΔP represents the difference between the pressure at the beginning of the vessel (the aorta) and the end of the vessel (the vena cavae at the entrance to the right atrium). Because MAP is the average pressure in the aorta and other arteries and the venous pressure at the right atrium is essentially 0 mmHg, ΔP is equal to MAP. The collective resistance offered by all systemic blood vessels is equal to the total peripheral resistance (TPR). Hence, the flow equation can be rewritten as follows:

$$CO = \frac{MAP}{TPR}$$

By rearranging the terms of this equation, you can see that mean arterial pressure is equal to cardiac output multiplied by total peripheral resistance:

$$MAP = CO \times TPR$$

From this equation, you can conclude that if cardiac output rises, the mean arterial pressure rises as long as total peripheral resistance remains steady. Likewise, a decrease in cardiac output causes a decrease in mean arterial pressure if total peripheral resistance does not change.

Blood pressure also depends on the total volume of blood in the cardiovascular system. The normal volume of blood in an adult is about 5 liters (5.3 qt). Any decrease in this volume, as from hemorrhage, decreases the amount of blood that is circulated through the arteries each minute. A modest decrease can be compensated for by homeostatic mechanisms that help maintain blood pressure (described in Section 15.5), but if the decrease in blood volume is greater than 10% of the total, blood pressure drops. Conversely, anything that increases blood volume, such as water retention in the body, tends to increase blood pressure.

Vascular Compliance Is High in Veins and Low in Arteries

Compliance is the ability of a hollow object to stretch. Strictly speaking, compliance is defined as the change in volume per unit change of transmural pressure. **Transmural pressure**, also known as *distending pressure*, is the pressure difference between the inside and outside walls of the hollow object. The relationship among compliance, volume, and pressure is given by the following equation:

$$C = \frac{\Delta V}{\Delta P}$$

where C is compliance,
ΔV is the change in volume, and
ΔP is the change in transmural pressure.

According to this equation, compliance is high when a large volume change causes only a small change in pressure, and compliance is low when a small volume change results in a large change in pressure.

Blood vessels exhibit varying degrees of compliance. If a blood vessel has a high compliance, a large increase in blood volume causes only a small increase in blood pressure because the increase in volume stretches the vessel wall rather than causing a significant increase in pressure. If a blood vessel has a low compliance, a small increase in blood volume causes a large increase in blood pressure because the vessel wall cannot stretch as much to accommodate the increase in volume.

Veins have a high compliance because they contain thin walls that are easily stretched. The high compliance of veins means that an increase in the blood volume in veins simply stretches their walls, allowing blood to pool in veins; this is why veins contain the majority of the blood volume of the body. The high compliance of veins also means that an increase in venous volume does not cause a significant increase in venous pressure (**FIGURE 15.17**). Even though veins contain the majority of blood in the body, venous pressure is very low, averaging about 10 mmHg.

Arteries have a lower compliance than veins because they contain thick walls that are not as easy to stretch. Recall that most arteries are muscular arteries that have only a low ability to stretch, and those arteries that have a greater ability to stretch, namely, the elastic arteries, do not remain in a stretched state for very long because they quickly recoil. Because arteries have a relatively low compliance, an increase in blood volume in arteries causes a significant increase in arterial pressure (**FIGURE 15.17**). This is why

FIGURE 15.17 Pressure–volume curves for the arterial compartment and the venous compartment.

Vascular compliance refers to the ability of a blood vessel to stretch; veins have a high compliance, whereas arteries have a low compliance.

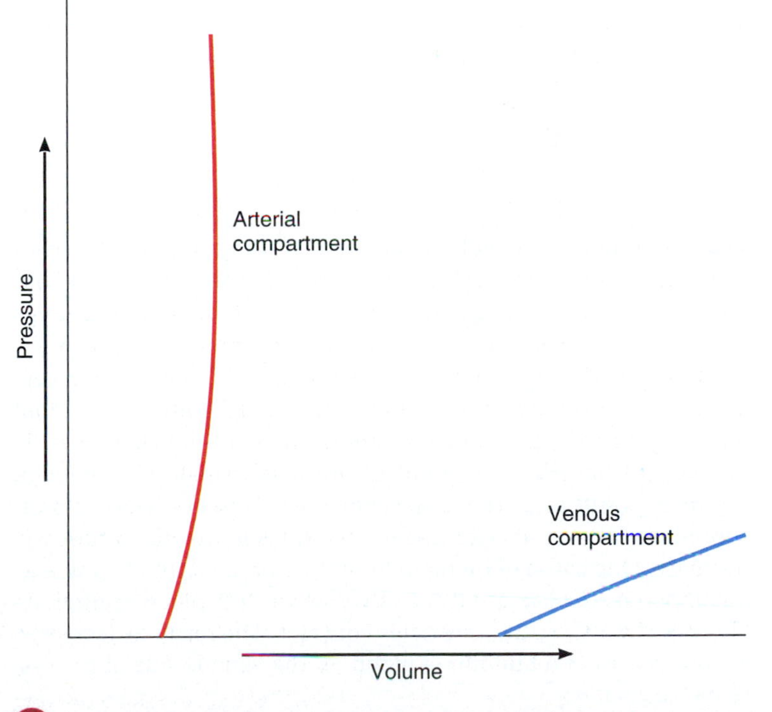

What effect does the high compliance of veins have on the amount of pressure that normally exists in veins?

arterial pressures are typically high, fluctuating between 70 mmHg and 110 mmHg.

Velocity of Blood Flow Is Inversely Proportional to the Cross-Sectional Area

Earlier you learned that blood flow is the *volume* of blood that flows through any tissue in a given time period (usually expressed in mL/min or cm^3/sec). It is also important to understand the speed or *velocity* of blood flow, which is typically expressed in cm/sec. The velocity of blood flow is inversely proportional to the cross-sectional area of the vessel or group of vessels through which the blood is flowing. Velocity is slowest where the total cross-sectional area is greatest (**FIGURE 15.18**). Each time an artery branches, the total cross-sectional area of all of its branches is greater than the cross-sectional area of the original vessel, so blood flow becomes slower and slower as blood moves further away from the heart, and is slowest in the capillaries. Conversely, when venules unite to form veins, the total cross-sectional area becomes smaller and flow becomes faster. In an adult, the cross-sectional area of the aorta is only 4 cm^2, and the average velocity of the blood there is 22 cm/sec. In capillaries, the total cross-sectional area is 4500 cm^2, and the velocity of blood flow is about 0.02 cm/sec. The total cross-sectional area of the vena cava is 6 cm^2, and the velocity is 15 cm/sec. Thus, the velocity of blood flow decreases as blood flows from the aorta to arteries to arterioles to capillaries, and increases as it leaves capillaries and returns to the heart. The relatively slow rate of flow through capillaries aids the exchange of materials between blood and interstitial fluid. The Physiological Equation box describes how to calculate the velocity of blood flow.

Circulation time is the time required for a drop of blood to pass from the right atrium, through the pulmonary circulation, to the left atrium, through the systemic circulation down to the foot, and back again to the right atrium. In a resting person, circulation time normally is about 1 minute.

FIGURE 15.18 Relationship between velocity (speed) of blood flow and total cross-sectional area in different types of blood vessels.

Velocity of blood flow is slowest in the capillaries because they have the largest total cross-sectional area.

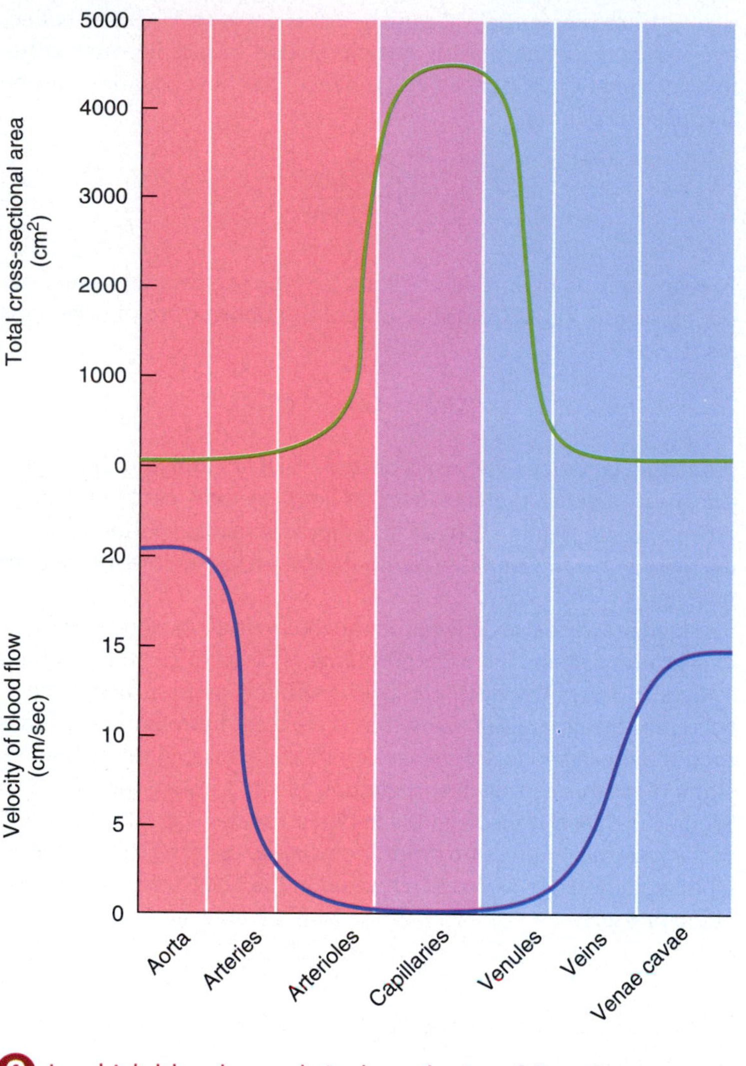

Venous Return Brings Blood Back to the Heart

Venous return is the volume of blood flowing back to the heart through systemic veins. A major factor that determines venous return is the venous pressure gradient—the driving force through systemic veins, equal to the difference in pressure between the venules and the right atrium. The pressure in venules averages about 17 mmHg, and the pressure in the right atrium is about 0 mmHg. So the venous pressure gradient is 17 mmHg ($\Delta P = 17$ mmHg $- 0$ mmHg $= 17$ mmHg). Although small, this gradient normally is sufficient to cause venous return to the heart. If pressure increases in the right atrium or ventricle, venous return will decrease. One cause of increased pressure in the right atrium is an incompetent (leaky) tricuspid valve, which lets blood regurgitate (flow backward) as the ventricles contract. The result is decreased venous return and buildup of blood on the venous side of the systemic circulation.

When you stand up, the pressure pushing blood up the veins in your lower limbs is barely enough to overcome the force of gravity pushing it back down. Besides the venous pressure gradient, additional mechanisms help return venous blood back to the heart. For example, valves in the veins of the limbs help promote venous return by preventing backflow of blood. Venous return is also enhanced by skeletal muscle contractions of the limbs (the skeletal muscle pump) and by pressure changes in the thoracic and abdominal cavities during inspiration (the respiratory pump). The details of how these pumps operate were discussed in Section 15.1 (see **FIGURES 15.5** and **15.6**). Another mechanism that increases venous return is venoconstriction, which reduces the volume of blood in reservoirs, allowing a greater blood volume to flow back to the heart and then to those organs of the body where it is needed.

The factors that promote venous return are summarized in **FIGURE 15.19**.

PHYSIOLOGICAL EQUATION

CALCULATION OF VELOCITY OF BLOOD FLOW

The velocity of blood flow is directly proportional to blood flow and inversely proportional to the cross-sectional area of the blood vessel, as expressed by the following equation:

$$V = \frac{F}{A}$$

where V is velocity of blood flow in cm/sec,
F is blood flow in cm^3/sec, and
A is the cross-sectional area of the blood vessel or a group of blood vessels in cm^2

To calculate the velocity of blood flow, you must first know the values for flow and the cross-sectional area. Flow (F) is equal to cardiac output. Recall that cardiac output is 5250 mL/min under resting conditions. Thus, F = 5250 mL/min or about 88 mL/sec. Because 1 cm^3 = 1 mL, flow can also be written as 88 cm^3/sec.

The total cross-sectional area (A) for each of the various types of blood vessels of the circulation is as follows:

Vessels	Total Cross-Sectional Area (cm^2)
Aorta	4
Arteries	20
Arterioles	400
Capillaries	4500
Venules	2000
Veins	40
Vena cava	6

As an example of how to use the velocity of blood flow equation, let's calculate the velocity of blood flow through the capillaries. The total cross-sectional area (A) of the capillaries is 4500 cm^2. Because blood flow (F) is 88 cm^3/sec, the velocity of blood flow (V) through the capillaries is equal to:

$$V = \frac{88\ cm^3/sec}{4500\ cm^2}$$

$$= 0.02\ cm/sec$$

CHECKPOINT

7. Explain how blood pressure and resistance affect blood flow.
8. What is total peripheral resistance and what factors contribute to it?
9. How does laminar flow occur?
10. In which parts of the circulation are there normal fluctuations in blood pressure and why do they occur?
11. Why do arteries have a low compliance?

FIGURE 15.19 Factors that promote venous return

Venous return is the volume of blood flowing back to the heart through systemic veins.

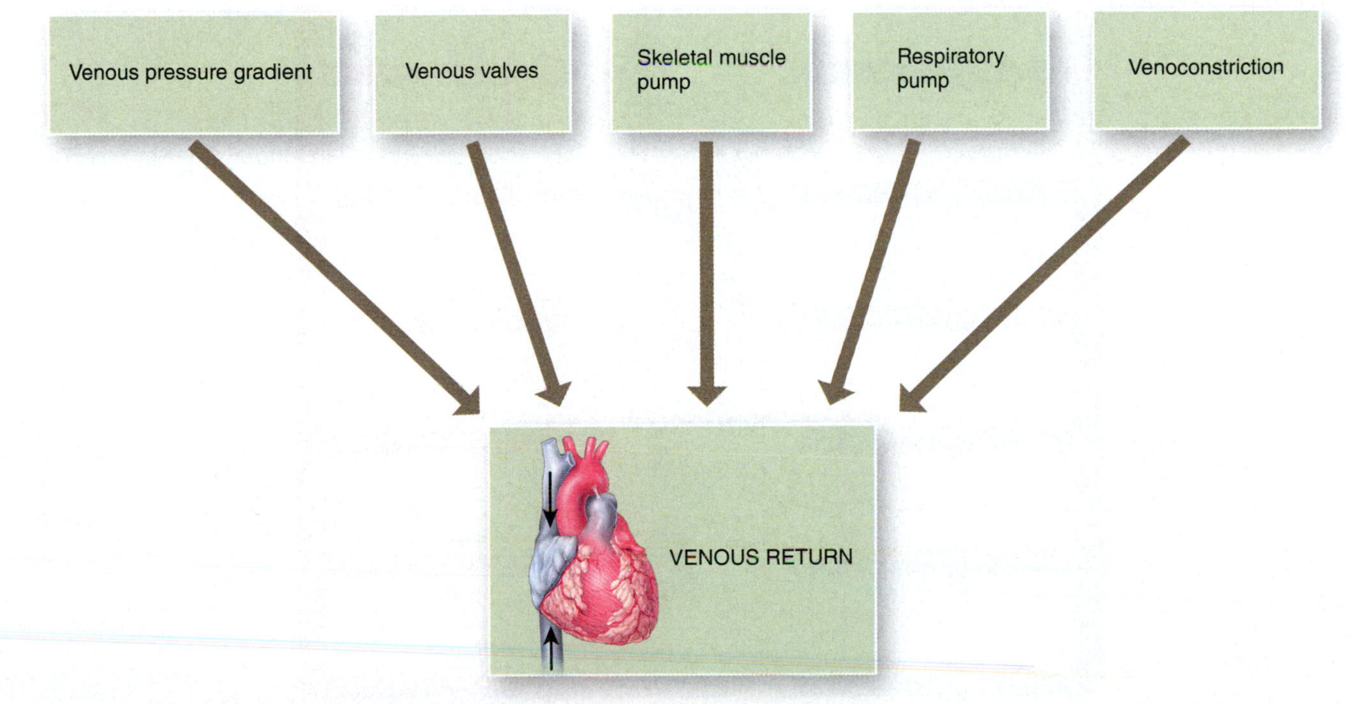

Why does venous blood often need help to get back to the heart?

12. Why is the velocity of blood flow faster in arteries and veins than in capillaries?

13. How is the return of venous blood to the heart accomplished?

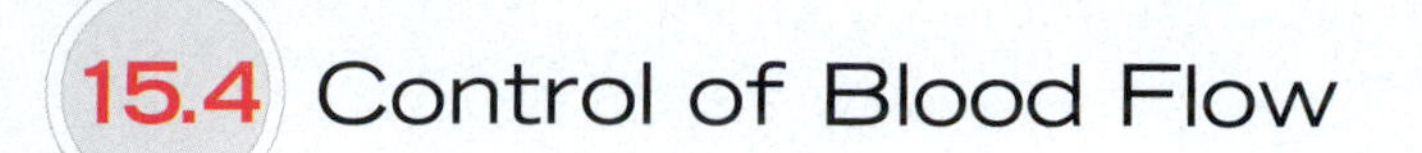

15.4 Control of Blood Flow

OBJECTIVE

- Describe the regulation of blood flow through body tissues.

Blood flow varies from one organ to another, depending on the particular needs of each organ. For example, the lungs must receive all the blood pumped out of the right ventricle (100% of the cardiac output) so that O_2 can be added to blood and CO_2 removed from it. Because of the parallel arrangement of blood vessels in the systemic circulation, the remaining organs of the body receive only a fraction of the cardiac output ejected from the left ventricle. The amount of the cardiac output distributed to these organs depends on how active they are. At rest, the digestive tract and liver receive about 27% of the cardiac output; the kidneys receive 20%, skeletal muscles 17%, brain 13%, skin 9%, and heart 4% (**FIGURE 15.20**). With the exception of the brain, blood flow to these organs increases or decreases in response to changes in metabolic demand. During heavy exercise, for

FIGURE 15.20 Distribution of blood flow at rest.

The lungs receive 100% of the cardiac output from the right ventricle; the other organs of the body receive only a fraction of the cardiac output from the left ventricle.

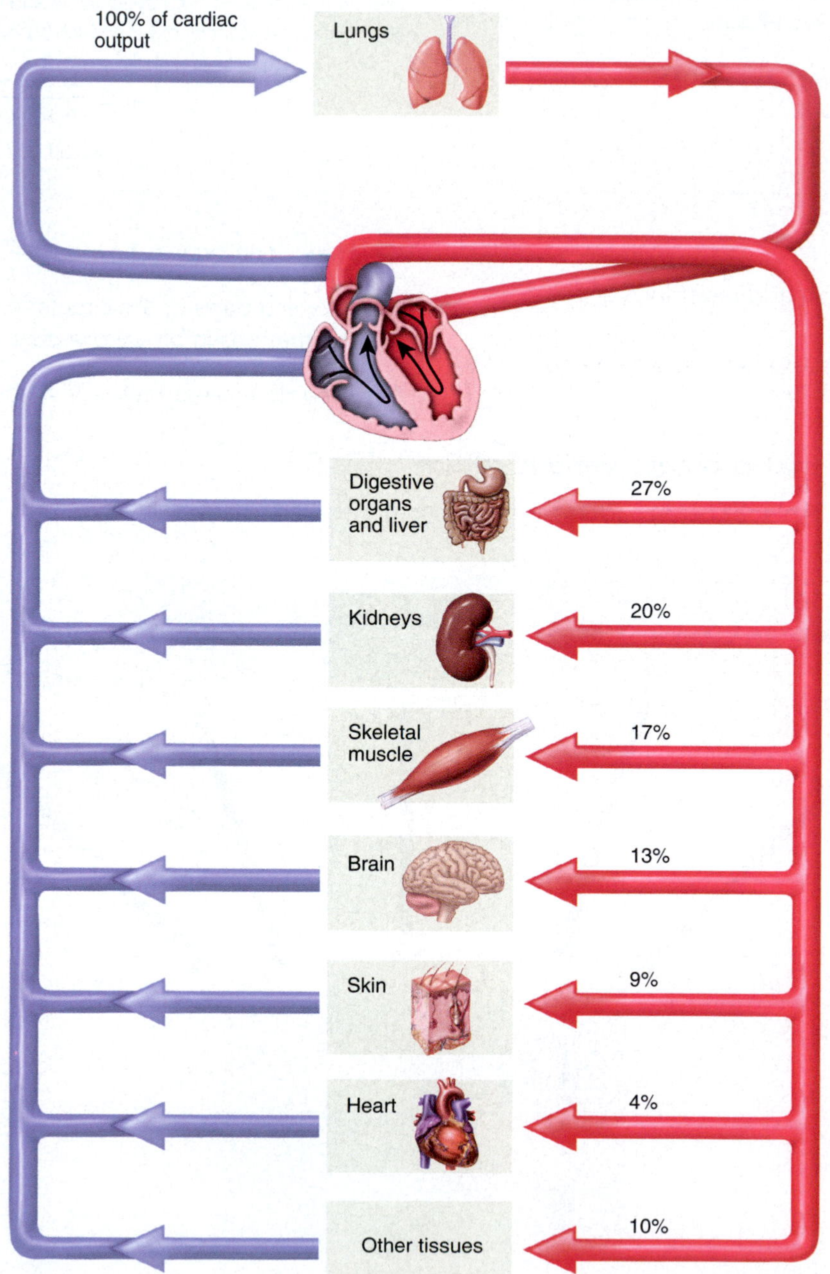

Which organ has a total constant blood flow, regardless of whether a person is resting or exercising?

example, blood flow to skeletal muscles increases substantially, to about 85%, because of the greater metabolic demand for O_2 and nutrients by contracting muscle fibers. While this is occurring, blood flow to the digestive organs and kidneys is vastly reduced because these organs are less active during exercise and therefore have fewer metabolic demands. TABLE 15.1 compares the distribution of blood flow to organs during rest and heavy exercise. Note that the brain is the only organ that has a constant total blood flow, regardless of whether a person is resting or exercising. However, changes in regional blood flow within the brain can occur at any given time, depending on which part of the brain is being used. During a conversation, for example, blood flow increases to your motor speech areas when you are talking and increases to the auditory areas when you are listening.

As you learned in the last section, flow is directly proportional to the pressure gradient and inversely proportional to resistance ($F = \Delta P/R$). Because mean arterial pressure is essentially constant from one organ to another, the driving pressure for blood flow is the same for each organ. This means that differences in blood flow are solely determined by changes in resistance. Recall that the principal determinant of resistance is blood vessel radius and arterioles are the main blood vessels that contribute to resistance. Therefore, the main way to regulate blood flow to different organs in the body is to alter the radii of the arterioles that supply these organs.

The radius of an arteriole is altered by contracting or relaxing the smooth muscle in its wall. Arteriolar smooth muscle exhibits a state of partial contraction called **vascular tone**, which establishes a baseline level from which contraction can be increased or decreased. So, increasing the contractile state of arteriolar smooth muscle causes vasoconstriction, which decreases the vessel radius, increases resistance, and decreases blood flow. By contrast, decreasing the contractile state of arteriolar smooth muscle causes vasodilation, which increases the vessel radius, decreases resistance, and increases blood flow. The radii of arterioles supplying different organs can be independently adjusted to regulate blood flow to these organs in response to changes in metabolic demands. For example, if the cells of skeletal muscle, the stomach, and bone have different levels of metabolic activity, the radii of the arterioles supplying these organs can be adjusted accordingly to make sure that each organ has the appropriate amount of blood flow (FIGURE 15.21).

TABLE 15.1 Comparison of blood flow distribution during rest and heavy exercise.

	REST		HEAVY EXERCISE	
Organ	mL/min	% Cardiac Output	mL/min	% Cardiac Output
Digestive organs and liver	1400	27	250	1
Kidneys	1050	20	250	1
Skeletal muscle	900	17	21,250	85
Brain	700	13	700	3
Skin	450	9	1300	5
Heart	200	4	1000	4
Other	550	10	250	1
Total cardiac output	5250	100	25,000	100

FIGURE 15.21 Changes in blood flow to different organs due to independent adjustments of arteriolar radii.

Adjusting the radii of arterioles is the principal method of regulating blood flow to different organs of the body.

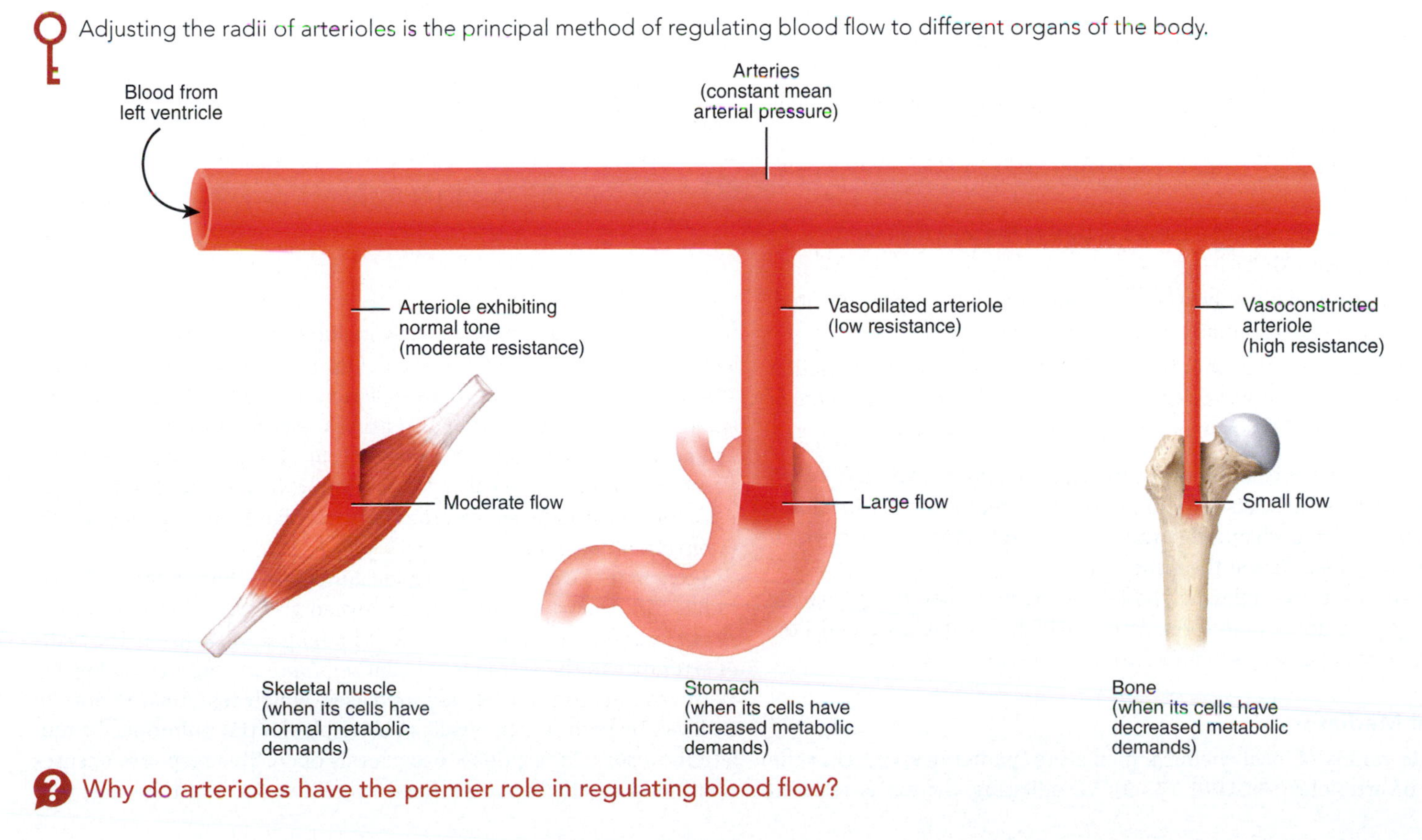

Why do arterioles have the premier role in regulating blood flow?

There are two methods of controlling the radii, and therefore the resistances, of arterioles: intrinsic control and extrinsic control. The relative importance of each method depends on which tissue is involved.

Intrinsic Control of Blood Flow Includes Physical Changes and Local Mediators

Intrinsic (local) controls are mechanisms *within* an organ that regulate blood flow by altering the radii of the arterioles that supply the organ. Intrinsic controls allow a tissue to adjust blood flow to match its metabolic demands. For example, in tissues such as skeletal muscle and the heart, where the demand for O_2 and nutrients and for the removal of wastes can increase tremendously during physical activity, intrinsic control is an important contributor to increased blood flow through the tissue. In other organs, such as the brain and kidneys, intrinsic controls help to keep blood flow relatively constant despite fluctuations in mean arterial pressure. The ability of a tissue to maintain a relatively constant blood flow in the presence of changing arterial pressure is known as **autoregulation**. There are two types of intrinsic control: physical changes and local mediators.

Physical Changes

Warming promotes vasodilation, and cooling causes vasoconstriction. In addition, smooth muscle in arteriole walls exhibits a **myogenic response**—it contracts more forcefully when it is stretched and relaxes when stretching lessens. If, for example, arterial pressure increases, the elevated blood pressure stretches the walls of the arterioles. This causes the arteriolar smooth muscle to contract and produce vasoconstriction, which keeps blood flow constant. The molecular mechanism underlying this myogenic response occurs in the following way (**FIGURE 15.22**):

1. The stretch of smooth muscle cells in the arteriolar wall by the elevated arterial pressure causes mechanically-gated channels in the muscle membrane to open, allowing extracellular calcium ions (Ca^{2+}) to enter the cells.
2. The entering Ca^{2+} binds to calmodulin to form a Ca^{2+}–calmodulin complex.
3. The Ca^{2+}–calmodulin complex activates the enzyme myosin-light chain kinase (MLCK).
4. MLCK in turn phosphorylates myosin, which becomes activated. Activated myosin binds to actin to cause contraction.
5. Contraction of the smooth muscle cells causes vasoconstriction of the arteriole. This vasoconstriction negates the initial stretch of the arteriole, thereby maintaining constant blood flow.

Here is another example of the myogenic response: If arterial blood pressure drops, the walls of the arterioles are stretched less. This causes the arteriolar smooth muscle to relax and produce vasodilation, which helps keep blood flow at the same level.

From these examples, it should be evident that the myogenic response is a form of autoregulation, which normalizes blood flow when there are changes in mean arterial pressure.

FIGURE 15.22 Mechanism of the myogenic response when there is an increase in arterial pressure.

In the myogenic response, smooth muscle of an arteriole contracts more forcefully when it is stretched, and it relaxes when stretching lessens.

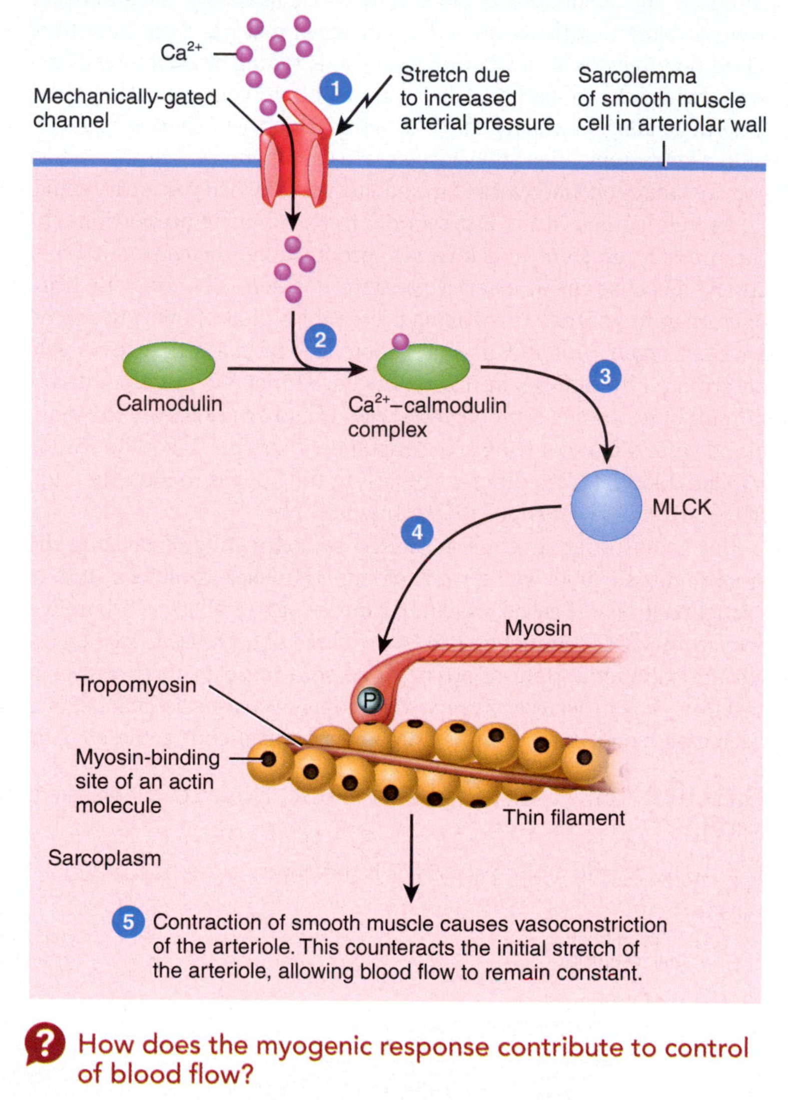

How does the myogenic response contribute to control of blood flow?

Local Mediators

A wide variety of local chemical mediators (paracrines) can alter the radii of arterioles (**FIGURE 15.23**). Vasodilating chemicals released by metabolically active tissue cells include CO_2, K^+, H^+ (from acids such as lactic acid), and adenosine (from ATP). Another important vasodilator is nitric oxide, a gas released from endothelial cells. Tissue trauma or inflammation also causes release of vasodilators such as bradykinin, histamine, and prostacyclin (a type of prostaglandin). Vasoconstricting chemicals released from cells include thromboxane A_2, superoxide radicals, serotonin (from platelets), and endothelin (from endothelial cells).

Oxygen (O_2) can also act as a local mediator that alters arteriolar radius. An important difference between the systemic and pulmonary circulations is their response to changes in O_2 level. The walls of arterioles in the systemic circulation *dilate* in response to low O_2. With vasodilation, O_2 delivery increases, which restores the normal O_2 level. By contrast, the walls of arterioles in the pulmonary circulation *constrict* in response to low levels of O_2. This response ensures that blood mostly bypasses those alveoli (air sacs) in the lungs that

FIGURE 15.23 The effect of local mediators on arteriolar radius.

Local mediators can alter arteriolar radius; some mediators increase arteriolar radius (vasodilation) and others decrease it (vasoconstriction).

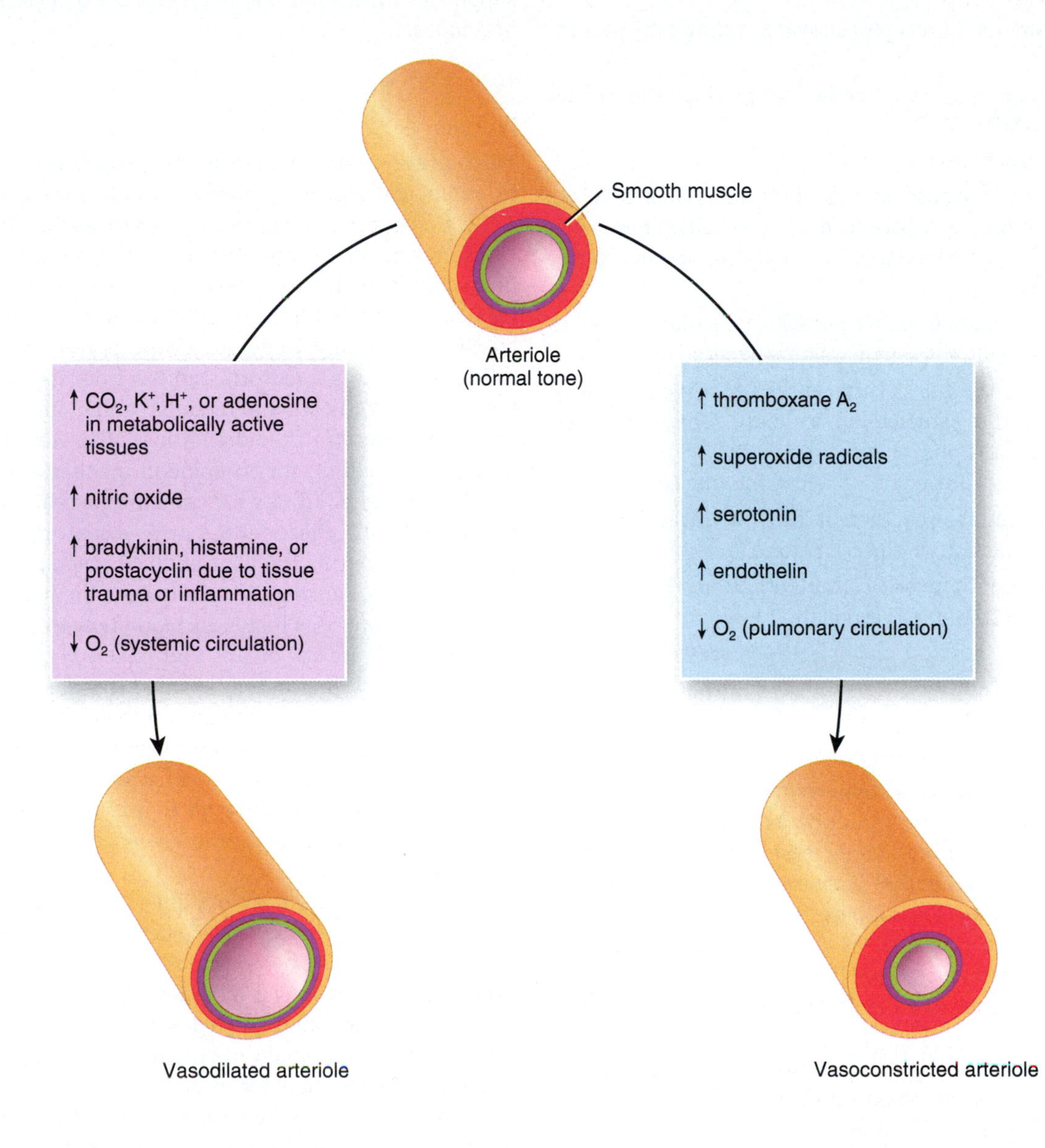

What effect does a low level of O_2 have on arteriolar radius in the systemic circulation? In the pulmonary circulation?

are poorly ventilated by fresh air. Thus, most blood flows to better-ventilated areas of the lung.

In addition to acting on arterioles to alter vessel radius, local mediators can act on precapillary sphincters to regulate blood flow through capillary beds. Vasodilators relax precapillary smooth muscle, increasing flow through capillary beds, whereas vasoconstrictors contract precapillary smooth muscle, reducing blood flow through capillaries.

As an example of how a local mediator functions, consider how adenosine promotes vasodilation of a coronary arteriole (**FIGURE 15.24**):

1. Adenosine binds to an **A_{2A} purinergic receptor** in the sarcolemma of a smooth muscle cell located in the wall of a coronary arteriole.
2. Binding of adenosine to the receptor activates a stimulatory G protein (G_s).
3. Activated G_s stimulates adenylyl cyclase to produce the second messenger cyclic AMP (cAMP).
4. cAMP binds to and activates a protein kinase.
5. The activated protein kinase in turn phosphorylates a **K^+_{ATP} channel** in the sarcolemma. This channel is so-named because it is sensitive to ATP. Phosphorylation of the K^+_{ATP} channel causes the channel to open.
6. Opening the K^+_{ATP} channels allows more K^+ ions than normal to leave the cell. As a result, the cell membrane hyperpolarizes.
7. Hyperpolarization of the cell in turn causes L-type voltage-gated Ca^{2+} channels in the sarcolemma to close. These channels are normally open as long as the membrane is depolarized. Closure of the channels prevents Ca^{2+} entry into the cell.
8. Without Ca^{2+} entering the cell, there is little Ca^{2+} available in the sarcoplasm for contraction, so the cell relaxes.
9. Relaxation of this smooth muscle cell and others in the wall of the arteriole causes the arteriole to vasodilate.

Hyperemia

Intrinsic control of blood flow is evident when blood flow to a tissue increases, a phenomenon known as **hyperemia**. There are two types of hyperemia: active and reactive. In **active hyperemia**, blood flow to a tissue increases in response to an increase in metabolic activity. When tissues are active, metabolism increases. Cells consume more O_2 and release large amounts of metabolites such as CO_2, K^+, and adenosine. These local mediators trigger vasodilation of nearby arterioles and precapillary sphincters, increasing blood flow to the tissue. The increase in blood flow brings in more O_2 and nutrients and removes the metabolites.

In **reactive hyperemia**, blood flow to a tissue increases in response to a temporary blockage of the blood supply to that area. When the blood supply to a tissue is blocked, cells run out of oxygen and an oxygen debt accrues. In addition, metabolites released from the cells accumulate because there is no blood flow to remove them. The decreased O_2 and built-up metabolites cause vasodilation, but the occlusion prevents blood from entering. However, once the blockage is removed, there is an increase in blood flow to the area. The increase in blood flow continues until the O_2 debt is paid back and the metabolites are flushed out.

Active hyperemia and reactive hyperemia operate via the same mechanisms—release of local mediators that cause vasodilation of arterioles and precapillary sphincters. The difference between the two types of hyperemia is based on the *cause*. In active hyperemia, the cause is an increase in metabolic activity; in reactive hyperemia, the cause is a blocked blood supply.

Extrinsic Control of Blood Flow Includes Nerves and Hormones

Extrinsic controls are mechanisms originating *outside* an organ that regulate blood flow by altering the radii of the arterioles that supply the organ. There are two types of extrinsic control: nerves and hormones.

Nerves

Most arterioles are innervated by sympathetic nerves and contain α_1-adrenergic receptors in the smooth muscle within their walls. Activation of the α_1 receptors causes the arterioles to vasoconstrict. Unlike arterioles in most tissues of the body, the arterioles in the heart and skeletal muscle have an abundance of β_2-adrenergic receptors. Activation of the β_2 receptors causes the arterioles to vasodilate. Both types of adrenergic receptors respond to norepinephrine (released either as a neurotransmitter from sympathetic nerves or as hormone from the adrenal medulla) and epinephrine (released only as a hormone from the adrenal medulla). Recall that α_1 receptors have a higher affinity for norepinephrine than epinephrine; β_2 receptors have a higher affinity for epinephrine than norepinephrine (see Section 6.3). The effects of α_1- or β_2-receptor stimulation are especially evident during the fight-or-flight response. Epinephrine binds to β_2 receptors, causing the arterioles of the heart and skeletal muscle to vasodilate. This increases blood flow to the heart and skeletal muscle, which are key organs that enable you to deal with the dangerous or stressful situation at hand. Because the arterioles of the kidneys and digestive tract contain mainly α_1 receptors, binding of norepinephrine or epinephrine to these receptors during the fight-or-flight response causes vasoconstriction, reducing blood flow to these organs. This should come as no surprise because the kidneys and digestive organs are essentially inactive when the body is in the sympathetic mode.

There is little parasympathetic innervation of arterioles, except for the arterioles of the external genitalia. Acetylcholine and nitric oxide released from parasympathetic neurons cause vasodilation of the arterioles of the penis or clitoris, resulting in increased blood flow and erection.

It is worth mentioning that sympathetic nerves also innervate arteries and veins, which mainly have α_1-adrenergic receptors that cause vasoconstriction when activated. However, because arteries and veins contribute very little resistance, they are not significant in regulating blood flow. One final point to note is that the precapillary sphincters associated with capillary beds are not innervated, so their regulation occurs in response to local mediators.

Hormones

As you have just learned, the hormones norepinephrine and epinephrine released from the adrenal medulla in response to sympathetic stimulation can affect blood flow. Norepinephrine preferentially activates α_1-adrenergic receptors to cause vasoconstriction of arterioles in most tissues. Epinephrine preferentially binds to β_2-adrenergic receptors to cause vasodilation of arterioles in the heart and skeletal muscle.

FIGURE 15.24 Mechanism of smooth muscle cell relaxation by adenosine.

Adenosine promotes vasodilation of arterioles by causing the smooth muscle cells in the arteriolar wall to relax.

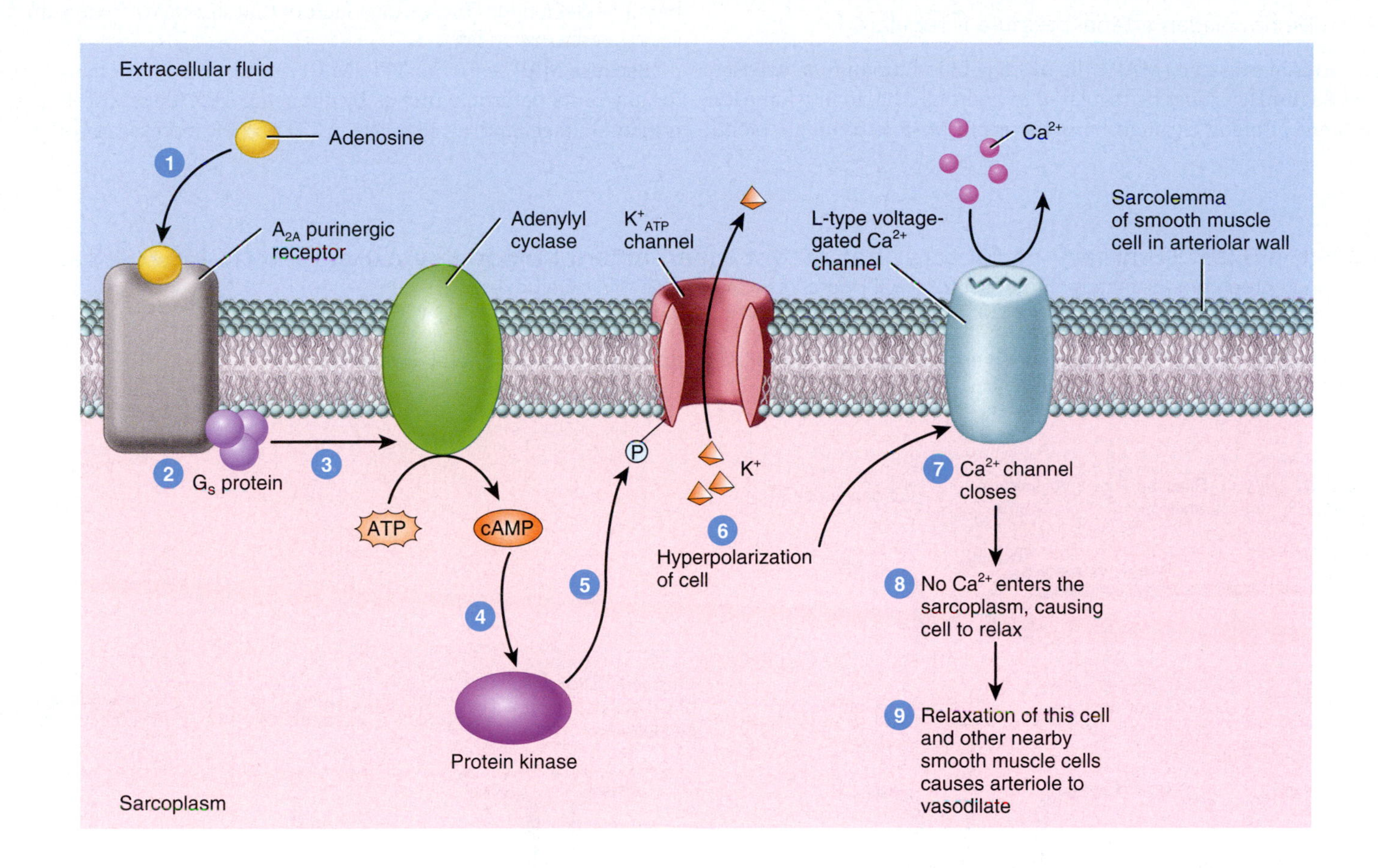

What role does hyperpolarization play in adenosine's ability to relax a smooth muscle cell?

CHECKPOINT

14. What is the significance of the myogenic response?

15. What are some examples of local mediators that alter the radii of arterioles?

16. How does active hyperemia differ from reactive hyperemia?

17. What are some examples of tissues that have arterioles with α_1-adrenergic receptors? With β_2-adrenergic receptors?

15.5 Regulation of Mean Arterial Pressure

OBJECTIVES

- Discuss the factors that affect mean arterial pressure.
- Describe how mean arterial pressure is regulated.

Mean arterial pressure (MAP), the average blood pressure in arteries (about 83 mmHg), must be regulated to maintain the driving force for movement of blood through body tissues. If MAP is too high, blood flows through blood vessels too quickly; if MAP is too low, blood moves through vessels too slowly. Recall that MAP is equal to the product of cardiac output and total peripheral resistance (MAP = CO × TPR). Cardiac output in turn is determined by heart rate and stroke volume. TPR is determined mainly by changes in blood vessel radius (vasoconstriction or vasodilation) and to a lesser extent by blood viscosity and blood vessel length. The various factors that affect MAP are summarized in **FIGURE 15.25**.

Because MAP = CO × TPR, MAP can be altered by changing CO (or any of its determinants) or by changing TPR (or any of its determinants). In general, an increase in CO or TPR increases MAP, and a

FIGURE 15.25 Summary of factors that affect mean arterial blood pressure. Factors that affect cardiac output are shown in green boxes; factors that affect total peripheral resistance appear in blue boxes.

The two major determinants of mean arterial pressure are cardiac output and total peripheral resistance.

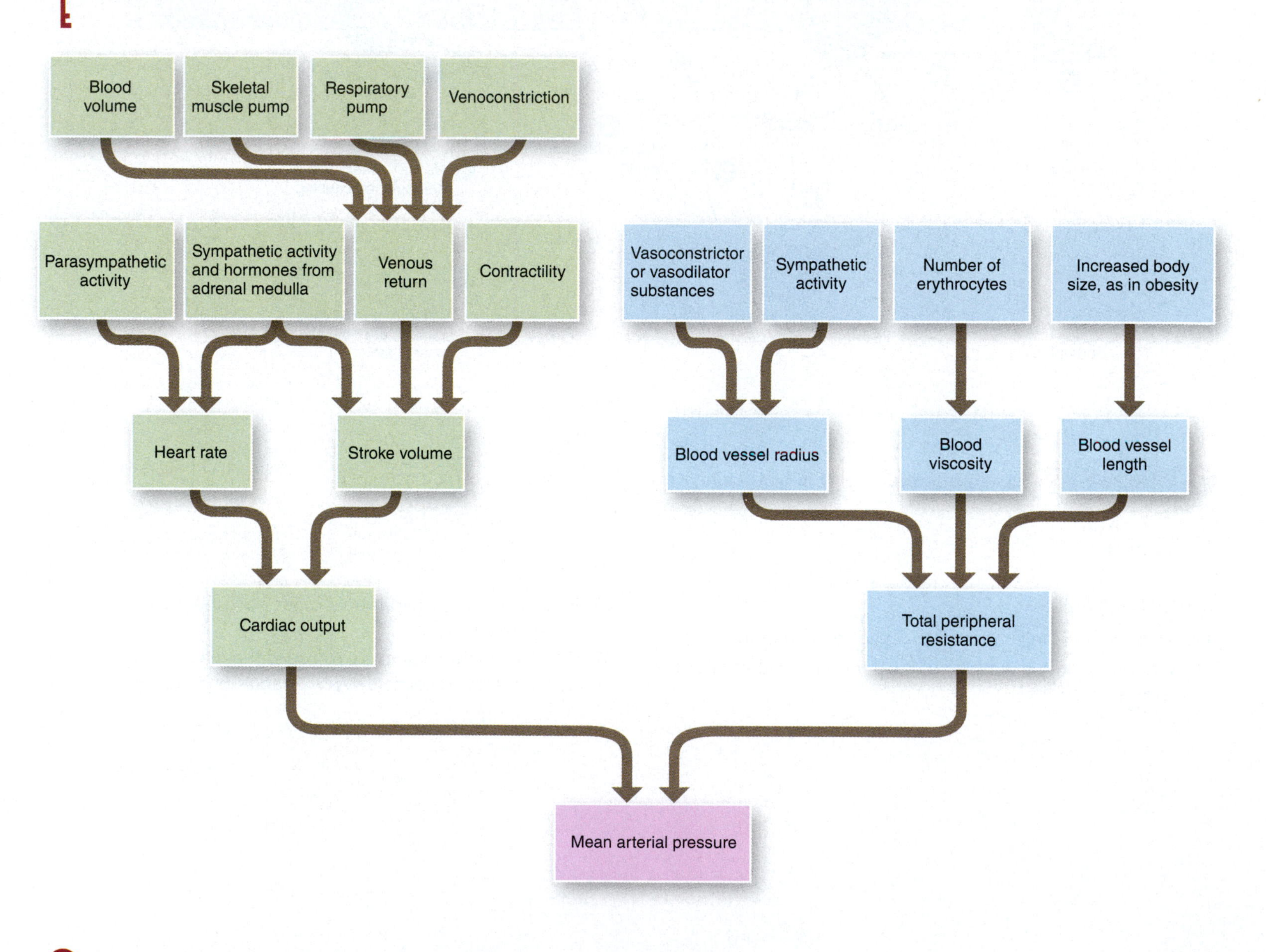

Why must mean arterial pressure be regulated?

decrease in CO or TPR decreases MAP. In this section, you will learn that several interconnected negative feedback systems control mean arterial pressure by adjusting CO, TPR, or one or more of their determinants. Some systems allow rapid adjustments to cope with sudden changes, such as the drop in blood pressure in the brain that occurs when you get out of bed; others act more slowly to provide long-term regulation of blood pressure.

The Nervous System Is an Important Regulator of Mean Arterial Pressure

The nervous system regulates mean arterial pressure via negative feedback loops that involve the cardiovascular center and two types of reflexes: baroreceptor reflexes and chemoreceptor reflexes.

Role of the Cardiovascular Center

The **cardiovascular (CV) center** in the medulla oblongata (**FIGURE 15.26**) helps regulate mean arterial blood pressure by altering heart rate, contractility, and blood vessel radius. Scattered within the CV center are several groups of neurons: Some neurons stimulate the heart (cardiostimulatory center), others inhibit the heart (cardioinhibitory center), and still others control blood vessel radius by causing vasoconstriction or vasodilation (vasomotor center). Because the CV center neurons communicate with one another, function together, and are not clearly separated anatomically, they are discussed here as a group.

The CV center receives input both from higher brain regions and from sensory receptors (**FIGURE 15.26**). Action potentials descend from the cerebral cortex, limbic system, and hypothalamus to affect the CV center. For example, even before you start to run a race, your heart rate may increase due to action potentials conveyed from the limbic system to the CV center. If your body temperature rises during a race, the hypothalamus sends action potentials to the CV center. The resulting vasodilation of skin blood vessels allows heat to dissipate more rapidly from the surface of the skin. The three main types of sensory receptors that provide input to the CV center are proprioceptors, baroreceptors, and chemoreceptors. *Proprioceptors* monitor the positions of limbs and muscles and provide input to the CV center during physical activity. Their activity accounts for the rapid increase in heart rate at the beginning of exercise. *Baroreceptors* monitor changes in pressure and stretch in the walls of blood vessels, and *chemoreceptors* monitor the concentration of various chemicals in the blood.

Output from the CV center to the heart and blood vessels flows along parasympathetic and sympathetic neurons of the ANS (**FIGURE 15.26**). The parasympathetic neurons that supply the heart originate in the CV center, and their axons extend to the heart via the vagus (X) nerves (**FIGURE 15.27**). An increase in parasympathetic stimulation decreases heart rate. The CV center controls the sympathetic neurons that supply the heart in the following way: Neurons from the CV center extend down to the spinal cord, where they synapse with sympathetic neurons that give rise to the cardiac accelerator nerves that supply the heart (**FIGURE 15.27**). An increase in sympathetic stimulation increases heart rate and contractility.

The cardiovascular center also sends output to smooth muscle in blood vessel walls through sympathetic neurons. Neurons from the CV center extend down to the spinal cord, where they synapse with sympathetic neurons that give rise to **vasomotor nerves** that

FIGURE 15.26 The cardiovascular center.

The cardiovascular center is the main region for nervous system regulation of the heart and blood vessels.

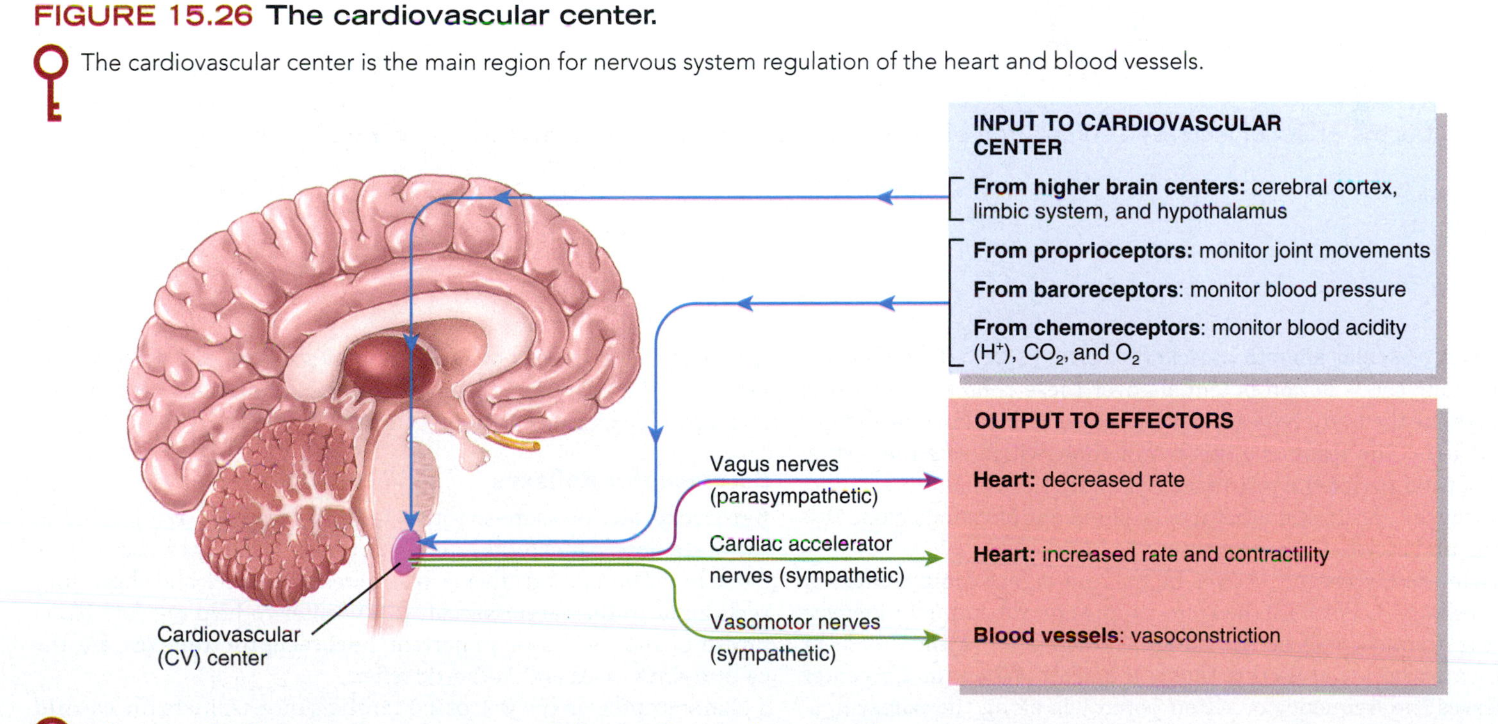

What are the functional components of the cardiovascular center?

FIGURE 15.27 Detailed input and output connections of the cardiovascular (CV) center.

The cardiovascular center has output connections with the heart and blood vessels, allowing it to regulate heart rate, contractility, and blood vessel diameter.

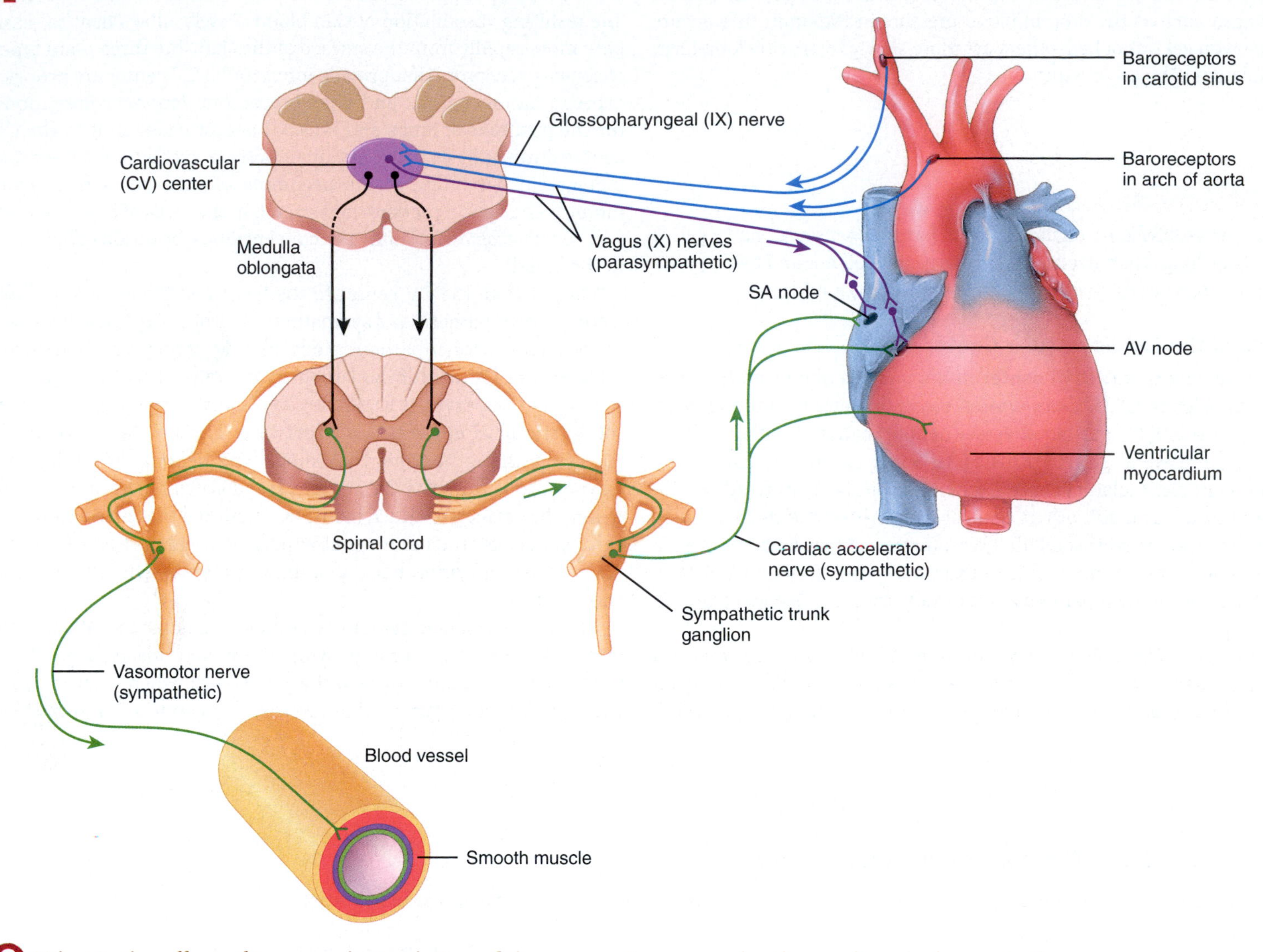

What is the effect of increased stimulation of the vasomotor nerves by the cardiovascular center?

innervate vascular smooth muscle (FIGURE 15.27). The CV center continually sends action potentials over these routes to arterioles throughout the body, but especially to those in the skin and abdominal viscera. The result is a vasomotor tone, which sets the resting level of total peripheral resistance. Sympathetic stimulation of most veins causes constriction that moves blood out of venous blood reservoirs and increases blood pressure.

Once the CV center receives input from higher brain centers or from sensory receptors, it then directs appropriate output by increasing the frequency of action potentials along the sympathetic or parasympathetic nerves that supply the heart. When the CV center increases the frequency of action potentials along the parasympathetic nerves to the heart, heart rate decreases. When the CV center increases the frequency of action potentials along the sympathetic nerves to the heart and blood vessels, heart rate and contractility increase and vasoconstriction occurs.

Baroreceptor Reflexes

Baroreceptors, pressure-sensitive sensory receptors, are located in the aorta, internal carotid arteries (arteries in the neck that supply blood to the brain), and other large arteries in the neck and chest. They send input to the cardiovascular (CV) center to help regulate blood pressure. The two most important **baroreceptor reflexes** are the carotid sinus reflex and the aortic reflex.

Baroreceptors in the wall of the carotid sinuses initiate the **carotid sinus reflex**, which helps regulate blood pressure in the brain. A

carotid sinus is a small widening of each internal carotid artery just above the point where it branches from the common carotid artery (FIGURE 15.27). Baroreceptors in the wall of the arch of the aorta initiate the **aortic reflex**, which regulates systemic blood pressure. The CV center receives input from carotid sinus baroreceptors via the glossopharyngeal (IX) nerves and from aortic arch baroreceptors via the vagus (X) nerves.

Blood pressure stretches the walls of the carotid sinus or aortic arch baroreceptors. When blood pressure falls, the baroreceptors are stretched less, and the sensory (afferent) neurons associated with the baroreceptors send action potentials at a slower rate to the CV center (FIGURE 15.28). In response, the CV center decreases parasympathetic stimulation of the heart by way of motor axons of the vagus nerves and increases sympathetic stimulation of the heart via cardiac accelerator nerves. As a result, heart rate and contractility increase, which causes cardiac output to increase. Another consequence of increased sympathetic stimulation is increased activity of the vasomotor nerves, resulting in greater vasoconstriction and total peripheral resistance. Increased cardiac output and increased total peripheral resistance cause blood pressure to increase to the normal level.

Conversely, when an increase in blood pressure is detected, the baroreceptors are stretched more, and the associated sensory neurons send action potentials at a faster rate to the CV center. The CV center responds by increasing parasympathetic stimulation and decreasing sympathetic stimulation. The resulting decreases in heart rate and contractility reduce cardiac output. The CV center also slows the rate at which it sends sympathetic output along vasomotor nerves that cause vasoconstriction. The resulting vasodilation lowers total peripheral resistance. Decreased cardiac output and decreased total peripheral resistance both lower systemic arterial blood pressure to the normal level.

Moving from a prone (lying down) to an erect position decreases blood pressure and blood flow in the head and upper part of the body. The baroreceptor reflexes, however, quickly counteract the drop in pressure. Sometimes these reflexes operate more slowly than normal, especially in the elderly or in certain disease conditions; in these cases, the person can faint from reduced brain blood flow upon standing up too quickly. The excessive lowering of systemic blood pressure when a person assumes an erect or semi-erect posture is referred to as **orthostatic (postural) hypotension**. It may be caused by excessive fluid loss, certain drugs, and cardiovascular or neurogenic factors.

Chemoreceptor Reflexes

Chemoreceptors, sensory receptors that monitor the chemical composition of blood, are located close to the baroreceptors of the carotid sinus and arch of the aorta in small structures called **carotid bodies** and **aortic bodies**, respectively (see FIGURE 18.17). These chemoreceptors detect changes in blood levels of O_2, CO_2, and H^+. *Hypoxia* (lowered O_2 availability), *acidosis* (an increase in H^+ concentration), or *hypercapnia* (excess CO_2) stimulates the chemoreceptors to send input to the cardiovascular center. In response, the CV center increases sympathetic stimulation to arterioles and veins, producing vasoconstriction and an increase in blood pressure. These chemoreceptors also provide input to the respiratory center in the brain stem to adjust the rate of breathing.

FIGURE 15.28 Negative feedback regulation of blood pressure via baroreceptor reflexes.

Baroreceptors detect changes in blood pressure.

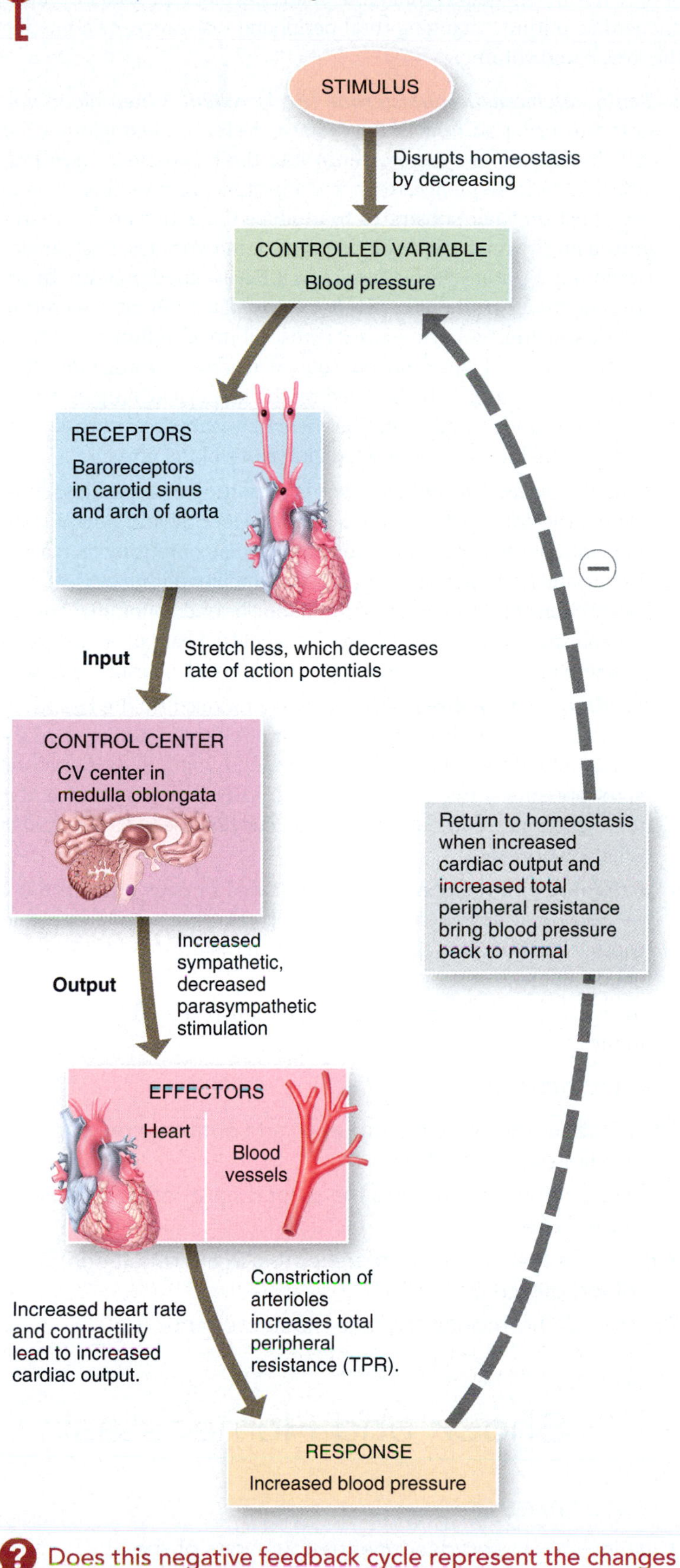

Does this negative feedback cycle represent the changes that occur when you lie down or when you stand up? Explain your answer.

Hormones Also Regulate Mean Arterial Pressure

Several hormones help regulate mean arterial blood pressure by altering cardiac output, changing total peripheral resistance, or adjusting the total blood volume:

1. ***Renin–angiotensin–aldosterone (RAA) system.*** When blood volume falls or blood flow to the kidneys decreases, juxtaglomerular cells in the kidneys secrete **renin** into the bloodstream (see FIGURE 13.21). In sequence, renin and angiotensin converting enzyme (ACE) act on their substrates to produce the active hormone **angiotensin II**, which raises blood pressure in two ways. First, angiotensin II is a potent vasoconstrictor; it raises blood pressure by increasing total peripheral resistance. Second, it stimulates secretion of **aldosterone**, which increases reabsorption of sodium ions (Na^+) by the kidneys. The osmotic consequence of reabsorbing more Na^+ is that more water is reabsorbed by the kidneys as long as antidiuretic hormone is present. The increased water reabsorption increases total blood volume, which increases blood pressure.
2. ***Epinephrine and norepinephrine.*** In response to sympathetic stimulation, the adrenal medulla releases epinephrine and norepinephrine. Both of these hormones increase cardiac output by increasing the rate and force of heart contractions. Norepinephrine causes vasoconstriction of arterioles and veins in the skin and abdominal organs; epinephrine causes vasodilation of arterioles in cardiac and skeletal muscle, which helps increase blood flow to muscle during exercise.
3. ***Antidiuretic hormone (ADH).*** ADH is produced by the hypothalamus and released from the posterior pituitary in response to increased blood osmolarity, decreased blood volume, or decreased blood pressure. Among other actions, ADH causes vasoconstriction, which increases blood pressure. For this reason ADH is also called **vasopressin**.
4. ***Atrial natriuretic peptide (ANP).*** Released by cells in the atria of the heart, ANP lowers blood pressure by causing vasodilation and by promoting the loss of salt and water in the urine, which reduces blood volume.

TABLE 15.2 summarizes the regulation of blood pressure by hormones.

CHECKPOINT

18. What are the principal inputs to and outputs from the cardiovascular center?
19. Explain the operation of the carotid sinus reflex and the aortic reflex.
20. What is the role of chemoreceptors in the regulation of blood pressure?
21. How do hormones regulate blood pressure?

15.6 Shock and Homeostasis

OBJECTIVES

- Define shock, and describe the four types of shock.
- Explain how the body's response to shock is regulated by negative feedback.

TABLE 15.2 Blood Pressure Regulation by Hormones

Factor Influencing Blood Pressure	Hormone	Effect on Blood Pressure
Cardiac Output		
Increased heart rate and contractility	Norepinephrine Epinephrine	Increase
Total Peripheral Resistance		
Vasoconstriction	Angiotensin II Antidiuretic hormone (vasopressin) Norepinephrine* Epinephrine†	Increase
Vasodilation	Atrial natriuretic peptide Epinephrine† Nitric oxide	Decrease
Blood Volume		
Blood volume increase	Aldosterone Antidiuretic hormone	Increase
Blood volume decrease	Atrial natriuretic peptide	Decrease

*Acts at α_1 receptors in arterioles of abdomen and skin.

†Acts at β_2 receptors in arterioles of cardiac and skeletal muscle; norepinephrine has a much smaller vasodilating effect.

Shock is a failure of the cardiovascular system to deliver enough O_2 and nutrients to meet cellular metabolic needs. The causes of shock are many and varied, but all are characterized by inadequate blood flow to body tissues. With inadequate oxygen delivery, cells switch from aerobic to anaerobic production of ATP, and lactic acid accumulates in body fluids. If shock persists, cells and organs become damaged, and cells may die unless proper treatment begins quickly.

There Are Four Types of Shock

Shock can be of four different types: (1) **hypovolemic shock** due to decreased blood volume, (2) **cardiogenic shock** due to poor heart function, (3) **vascular shock** due to inappropriate vasodilation, and (4) **obstructive shock** due to obstruction of blood flow.

A common cause of hypovolemic shock is acute (sudden) hemorrhage. The blood loss may be external, as occurs in trauma, or internal, as in rupture of an aortic aneurysm. Loss of body fluids through excessive sweating, diarrhea, or vomiting also can cause hypovolemic shock. Other conditions—for instance, diabetes mellitus—may cause excessive loss of fluid in the urine. Sometimes, hypovolemic shock is due to inadequate intake of fluid. Whatever the cause, when the volume of body fluids falls, venous return to the heart declines, filling of the heart lessens, stroke volume decreases, and cardiac output decreases.

Replacing fluid volume as quickly as possible is essential in managing hypovolemic shock. An immediate response by the body to hypovolemic shock is a fluid shift from interstitial fluid to blood. This minimizes the decrease in blood pressure and volume until other compensatory mechanisms become involved.

In cardiogenic shock, the heart fails to pump adequately, most often because of a myocardial infarction (heart attack). Other causes of cardiogenic shock include poor perfusion of the heart (ischemia), heart valve problems, excessive preload or afterload, impaired contractility of heart muscle fibers, and arrhythmias.

Even with normal blood volume and cardiac output, vascular shock may occur if blood pressure drops due to a decrease in total peripheral resistance. A variety of conditions can cause inappropriate dilation of arterioles or venules. In *anaphylactic shock*, a severe allergic reaction—for example, to a bee sting—releases histamine and other mediators that cause vasodilation. In *neurogenic shock*, vasodilation may occur following trauma to the head that causes malfunction of the cardiovascular center in the medulla. Shock stemming from certain bacterial toxins that produce vasodilation is termed *septic shock*. In the United States, septic shock causes more than 100,000 deaths per year and is the most common cause of death in hospital critical care units.

Obstructive shock occurs when blood flow through a portion of the circulation is blocked. The most common cause is *pulmonary embolism*, a blood clot lodged in a blood vessel of the lungs.

The Body's Response to Shock Involves Several Compensatory Mechanisms

The major mechanisms of compensation in shock are *negative feedback systems* that work to return cardiac output and arterial blood pressure to normal. When shock is mild, compensation by homeostatic mechanisms prevents serious damage. In an otherwise healthy person, compensatory mechanisms can maintain adequate blood flow and blood pressure despite an acute blood loss of as much as 10% of total volume. **FIGURE 15.29** shows several of the negative feedback systems that respond to hypovolemic shock.

1. ***Activation of the renin–angiotensin–aldosterone system.*** Decreased blood flow to the kidneys causes the kidneys to secrete renin and initiates the renin–angiotensin–aldosterone (RAA) system (see **FIGURE 13.21**). Recall that angiotensin II causes vasoconstriction and stimulates the adrenal cortex to secrete aldosterone, a hormone that increases Na^+ reabsorption by the kidneys. The Na^+ reabsorption in turn causes water reabsorption via osmosis (as long as antidiuretic hormone is present), resulting in an increase in blood volume. The increases in total peripheral resistance and blood volume by the RAA system help raise blood pressure.
2. ***Secretion of antidiuretic hormone.*** In response to decreased blood pressure, the posterior pituitary releases antidiuretic hormone (ADH). ADH enhances water reabsorption by the kidneys, which conserves remaining blood volume. It also causes vasoconstriction, which increases total peripheral resistance (see **FIGURE 13.14**.)
3. ***Activation of the sympathetic division of the ANS.*** As blood pressure decreases, the aortic and carotid baroreceptors initiate powerful sympathetic responses throughout the body. One result is marked vasoconstriction of arterioles and veins of the skin, kidneys, and other abdominal viscera. (Vasoconstriction does not occur in the brain or heart.) The constriction of arterioles increases total peripheral resistance, and the constriction of veins increases venous return. Both effects help maintain an adequate blood pressure. Sympathetic stimulation also increases heart rate and contractility, and increases secretion of epinephrine and norepinephrine by the adrenal medulla. These hormones intensify vasoconstriction and increase heart rate and contractility, all of which help raise blood pressure.
4. ***Release of local vasodilators.*** In response to hypoxia, cells liberate vasodilators—including K^+, H^+, adenosine, and nitric oxide—that dilate arterioles and relax precapillary sphincters. Such vasodilation increases local blood flow and may restore the O_2 level to normal in part of the body. However, vasodilation also has the potentially harmful effect of decreasing total peripheral resistance and thus lowering the blood pressure.

If blood volume drops more than 10–20%, or if the heart cannot bring blood pressure up sufficiently, compensatory mechanisms may fail to maintain adequate blood flow to tissues. At this point, shock becomes life threatening as damaged cells start to die.

Shock Has Many Signs and Symptoms

Even though the signs and symptoms of shock vary with the severity of the condition, most can be predicted in light of the responses generated by the negative feedback systems that attempt to correct the problem. Among the signs and symptoms of shock are the following:

- Systolic blood pressure is lower than 90 mmHg.
- Resting heart rate is rapid due to sympathetic stimulation and increased blood levels of epinephrine and norepinephrine.
- Pulse is weak and rapid due to reduced cardiac output and fast heart rate.
- Skin is cool, pale, and clammy due to sympathetic constriction of skin blood vessels and sympathetic stimulation of sweating.
- Mental state is altered due to reduced oxygen supply to the brain.
- Urine formation is reduced due to increased levels of aldosterone and antidiuretic hormone (ADH).
- The person is thirsty due to loss of extracellular fluid.
- The pH of blood is low (acidosis) due to buildup of lactic acid.
- The person may have nausea because of impaired blood flow to the digestive organs from sympathetic vasoconstriction.

FIGURE 15.29 Negative feedback systems that can restore normal blood pressure during hypovolemic shock.

Homeostatic mechanisms can compensate for an acute blood loss of as much as 10% of total blood volume.

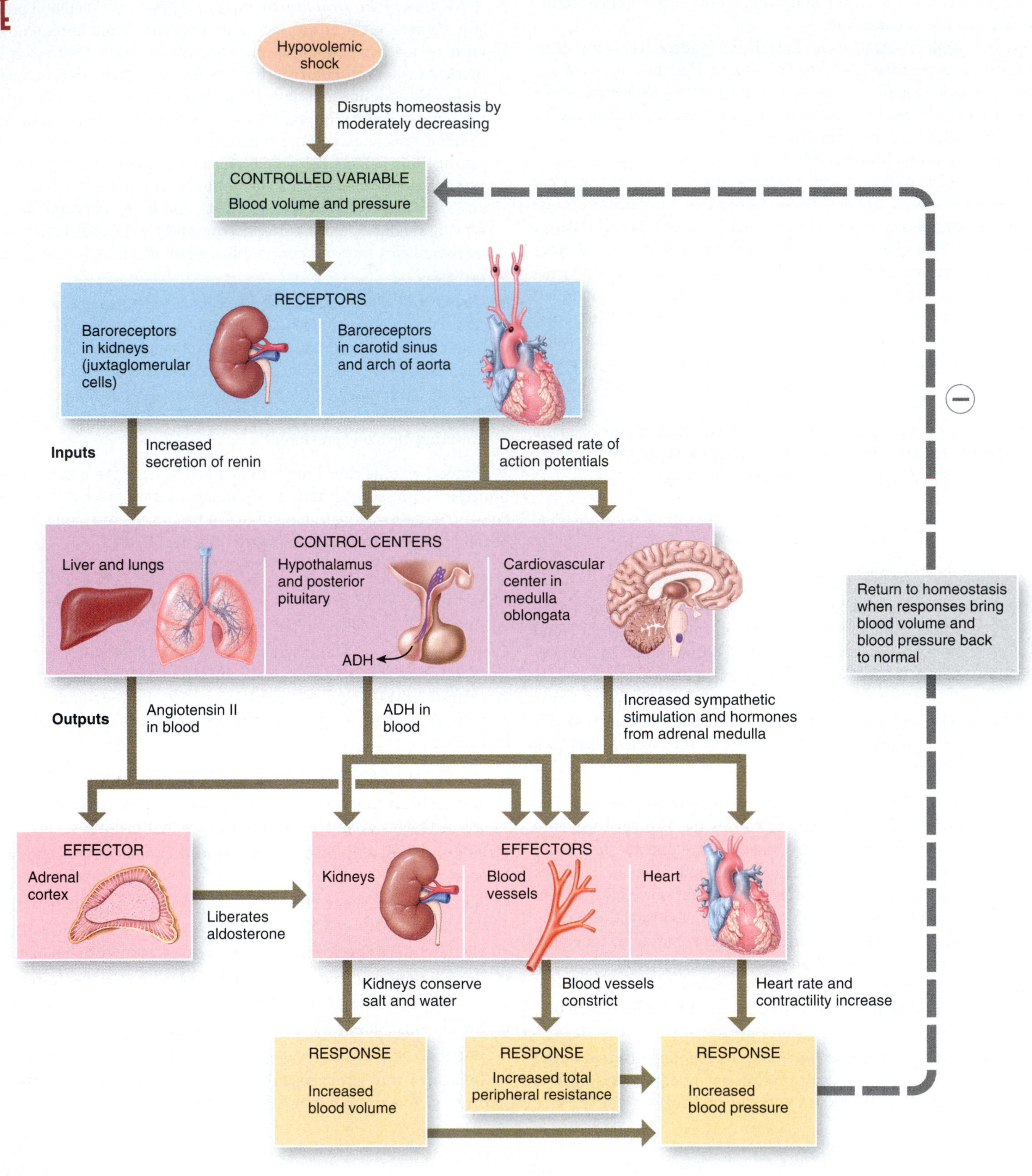

Does almost-normal blood pressure in a person who has lost blood indicate that the patient's tissues are receiving adequate perfusion (blood flow)? Explain your answer.

Critical Thinking

An Emergency Situation

Adam is running late for work. He has only 10 minutes to be on time but lives 20 minutes away from his job. He gets into his car and speeds along his way. As he is driving frantically, he does not notice the sharp curve in the road ahead. He drives too fast for the curve and ends up losing control of his vehicle. The car becomes airborne and then crashes upside down on the side of the road. The windows are broken, the frame is badly damaged, and Adam is trapped within the wreckage. He is barely conscious and severely bleeding.

SOME THINGS TO KEEP IN MIND:

The broken shards of glass from the car crash have caused Adam to hemorrhage. He has numerous lacerations on his body. Large amounts of blood are flowing from his left arm and the side of his neck. In fact, about two liters of blood have pooled around Adam's body. In addition, Adam's blood pressure has fallen drastically, and his heart rate and force of contraction have significantly increased. A passing motorist happens upon the scene of the crash, calls 911, and attempts to apply pressure to Adam's wounds to stop the bleeding. Within a few minutes, the paramedics arrive on the scene. They check Adam's vital signs, including blood pressure and heart rate, and stabilize him by immediately administering plasma and applying pressure bandages to his wounds. They then transport Adam in the ambulance to the hospital, where he eventually recovers.

Soundsnaps / Shutterstock

Paolo Cipriani / Getty Images

Did the accident cause Adam to go into circulatory shock? If so, which type?

Adam was hemorrhaging a large amount of blood because of the lacerations. Which homeostatic mechanisms usually minimize blood loss and why were they not succeeding in the locations on Adam's body where blood loss was the greatest?

In the accident scenario described above, Adam's blood pressure was dangerously low, but his heart rate and force of contraction had increased. This represents an example of which type of homeostatic feedback mechanism?

Which homeostatic parameters is Adam's body attempting to maintain in the face of the major loss of blood?

How is this homeostatic response actually exacerbating the impact of the wounds and posing a more serious threat to Adam's survival?

If you were one of the paramedics on the scene and you had measured Adam's blood pressure before any additional treatment was given, what blood pressure values would you be looking for as an indicator of acute hypotension?

Why did the paramedics administer blood plasma to Adam immediately?

CHECKPOINT

22. What are some possible causes of cardiogenic shock?

23. What are four negative feedback systems that respond to hypovolemic shock?

FROM RESEARCH TO REALITY

Chocolate and Blood Pressure

Reference

Litterio, M. et al. (2012). Blood pressure-lowering effect of dietary epicatechin administration in L-NAME-treated rats is associated with restored nitric oxide levels. *Free Radical Biology and Medicine*. 53: 1894–1902.

Does eating chocolate reduce your risk of developing high blood pressure?

Dorian Gray / Getty Images

High blood pressure is often referred to as the "silent killer" because it is frequently asymptomatic but very damaging to your organs and vascular system. Fortunately, healthy lifestyle choices can go a long way toward preventing hypertension as well as combating it. This study explores the effect of a chemical found in chocolate on blood pressure. Could that chocolate bar you ate the other day help prevent high blood pressure?

Article description:

Researchers in this study explored the effect of epicatechin on high blood pressure in rats. Epicatechin is a chemical present in foods such as chocolate, green tea, and certain fruits. Rats were first treated with L-NAME, a chemical that induces high blood pressure by preventing the production of the vasodilator nitric oxide. Rats were then given epicatechin to see if this chemical would lower the high blood pressure induced by L-NAME.

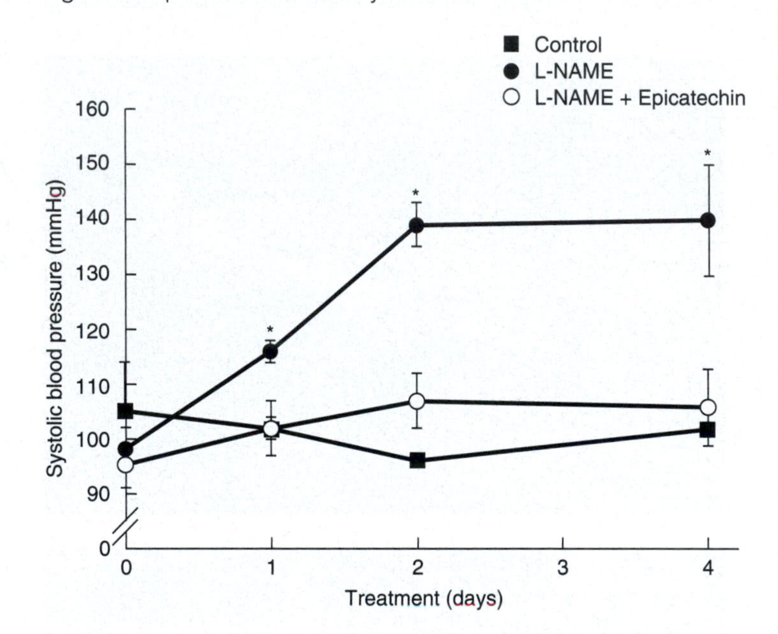

Go to WileyPLUS Learning Space and use the data from this article to answer the questions posed there and to discover more about chocolate and its effect on blood pressure.

CHAPTER REVIEW

15.1 Overview of the Vasculature

1. The five main types of blood vessels are arteries, arterioles, capillaries, venules, and veins.
2. Arteries carry blood away from the heart. Large-diameter, elastic arteries serve as pressure reservoirs that maintain the driving force for blood while the ventricles are relaxing (diastole). Medium-sized, muscular arteries distribute blood to the organs of the body.
3. Arterioles are small arteries that deliver blood to capillaries. Through vasoconstriction and vasodilation, arterioles assume a key role in regulating blood flow and in altering blood pressure.
4. Capillaries are microscopic blood vessels through which materials are exchanged between blood and tissue cells.
5. Capillaries branch to form an extensive network throughout a tissue. This network increases surface area, allowing a rapid exchange of large quantities of materials. Precapillary sphincters regulate blood flow through capillaries.
6. The body contains three types of capillaries: continuous capillaries, fenestrated capillaries, and sinusoids.
7. Venules are small vessels that continue from capillaries and merge to form veins.
8. Veins carry blood back to the heart. Veins of the limbs contain valves to prevent backflow of blood.
9. Systemic veins and venules are collectively called blood reservoirs because they hold the largest portion of the blood volume at rest. If the need arises, this blood can be shifted into other blood vessels through venoconstriction.

15.2 Capillary Exchange and Lymphatics

1. Substances enter and leave capillaries by diffusion, transcytosis, or bulk flow.
2. The movement of water and solutes (except proteins) through capillary walls depends on four pressures collectively known as Starling forces.
3. The four Starling forces are capillary hydrostatic pressure (P_C), interstitial fluid hydrostatic pressure (P_{IF}), plasma colloid osmotic pressure (π_P), and interstitial fluid colloid osmotic pressure (π_{IF}).
4. The lymphatic system consists of lymph, lymphatic vessels, and lymphoid organs and tissues.
5. The lymphatic system drains excess interstitial fluid, returns filtered plasma proteins back to the blood, and carries out immune responses.

15.3 Hemodynamics

1. Blood flow is directly proportional to the pressure difference across a blood vessel and inversely proportional to resistance to flow.
2. Resistance is directly proportional to blood viscosity and blood vessel length and inversely proportional to the fourth power of the radius of the blood vessel. Total peripheral resistance (TPR) refers to all of the resistances offered by systemic blood vessels.
3. Blood flow through blood vessels normally is laminar (streamlined); however, blood flow can become turbulent if blood moves through an abnormally constricted area, passes over a rough surface, makes a sharp turn, or exceeds a critical velocity.
4. Blood pressure is the hydrostatic pressure exerted by blood on the walls of a blood vessel.
5. Systolic blood pressure is the highest pressure attained in arteries during systole; diastolic pressure is the lowest arterial pressure during diastole. A normal blood pressure is 110 mmHg (systolic) and 70 mmHg (diastolic), written as 110/70 mmHg.

6. As blood leaves the aorta and flows through the systemic circulation, its pressure progressively falls to 0 mmHg by the time it reaches the right atrium.
7. Mean arterial pressure is the average blood pressure in arteries. In a person whose blood pressure is 110/70 mmHg, mean arterial pressure is 83 mmHg.
8. Compliance is the ability of a hollow object to stretch. It is equal to the change in volume per unit change of transmural pressure. Compliance is high when a large volume change causes only a small change in pressure; compliance is low when a small volume change results in a large change in pressure. Veins have a high compliance because they can accommodate a large volume with only a small pressure change. Arteries have a lower compliance than veins because a small change in arterial volume causes a relatively large change in pressure.
9. The velocity of blood flow is inversely related to the cross-sectional area of blood vessels; blood flows slowest where cross-sectional area is greatest. The velocity of blood flow decreases from the aorta to arteries to capillaries and increases in venules and veins.
10. Venous return is the volume of blood flowing back to the heart through systemic veins. Several factors help return venous blood to the heart, including the venous pressure gradient, venous valves, skeletal muscle pump, respiratory pump, and venoconstriction.

15.4 Control of Blood Flow

1. Blood flow varies from one organ to another, depending on the particular needs of each organ. The lungs receive 100% of the cardiac output pumped out of the right ventricle; the remaining organs receive only a fraction of the cardiac output pumped out of the left ventricle.
2. The main way to regulate blood flow to the different organs of the body is to alter the radii of the arterioles that supply the organs.
3. There are two methods of controlling arteriolar radii: intrinsic control and extrinsic control.
4. Intrinsic controls include physical changes (temperature changes and the myogenic response) and local mediators (paracrines that promote either vasoconstriction or vasodilation). Extrinsic controls include nerves and hormones.

15.5 Regulation of Mean Arterial Pressure

1. Mean arterial pressure (MAP) must be regulated to maintain the driving force for movement of blood through body tissues. MAP can be altered by changing cardiac output (or any of its determinants) or total peripheral resistance (or any of its determinants).
2. The nervous system regulates MAP via negative feedback loops that involve the cardiovascular center and two types of reflexes: baroreceptor reflexes and chemoreceptor reflexes.
3. The cardiovascular (CV) center is a group of neurons in the medulla oblongata that regulates heart rate, contractility, and blood vessel radius. The CV center sends output to the heart through parasympathetic and sympathetic neurons of the ANS. It also sends output to vascular smooth muscle through sympathetic neurons. An increase in parasympathetic stimulation decreases heart rate. An increase in sympathetic stimulation increases heart rate and contractility, and increases the degree of vasoconstriction. The CV center receives input from higher brain regions and sensory receptors (baroreceptors and chemoreceptors).
4. Baroreceptors monitor blood pressure, and chemoreceptors monitor blood levels of O_2, CO_2, and hydrogen ions. The carotid sinus reflex helps regulate blood pressure in the brain. The aortic reflex regulates general systemic blood pressure.

5. Several hormones contribute to regulation of MAP, including epinephrine, norepinephrine, ADH (vasopressin), angiotensin II, and ANP.

15.6 Shock and Homeostasis

1. Shock is a failure of the cardiovascular system to deliver enough O_2 and nutrients to meet the metabolic needs of cells.
2. Types of shock include hypovolemic, cardiogenic, vascular, and obstructive.
3. Homeostatic responses to shock include activation of the renin–angiotensin–aldosterone system, secretion of antidiuretic hormone, activation of the sympathetic division of the ANS, and release of local vasodilators.

PONDER THIS

1. You want to design a new organ and have a particular series of functions in mind. One thing that is very important to the organ's function is that it be highly permeable. Substances produced by this organ will be large in size and these products must be able to leave the cells, enter the bloodstream, and be distributed throughout the body. What type of capillaries would you have to use in this new organ to ensure that it is able to accomplish these functions? Explain your answer.

2. Sam has entered into an exercise physiology study. As part of the study, the researchers want to measure cardiac output at different workloads. One of these activities is a treadmill run. Sam has been running on the treadmill for several minutes and one of the researchers tells him that his end diastolic volume is high, making his stroke volume high. When Sam asks how this happens, the researcher tells him that it has a lot to do with venous return. Describe two aspects of the venous system that directly contribute to an increase in venous return.

3. While working with the Peace Corps, you come across a patient who is very malnourished. One thing you notice is that this person has a great deal of edema (swelling in the hands, feet, and abdomen). You also note that the patient is very dehydrated. You need to administer IV fluids and must choose among the following options: (1) saline solution containing only sodium chloride (salt), (2) a solution of distilled water, or (3) a saline solution infused with albumin. Which would be the best choice for this patient? Explain your answer.

4. Sally is very hypotensive due to extensive systemic vasodilation. Upon further analysis, doctors notice that Sally's heart rate and stroke volume are within normal range. What could the doctors give her to help her regain a normal blood pressure without directly affecting her cardiac function? Explain your answer.

5. Mrs. Salazar goes in to see her cardiologist to get an angiogram (a type of imaging that examines blood flow through blood vessels) of her right coronary artery. The results of Mrs. Salazar's angiogram show that her right coronary artery has a substantial amount of plaque buildup, reducing the radius of the artery by half. Describe what has happened to resistance and blood flow through this artery as a result of the plaque buildup. Why is this potentially dangerous?

ANSWERS TO FIGURE QUESTIONS

15.1 The endothelium of a blood vessel can physically influence blood flow, secrete locally acting chemical mediators that influence the contractile state of the vessel's smooth muscle, and assist with capillary permeability.

15.2 Recoil of elastic arteries keeps blood flowing during ventricular relaxation (diastole).

15.3 Precapillary sphincters regulate the flow of blood through capillary beds.

15.4 Intercellular clefts are slit-shaped pores between neighboring endothelial cells; fenestrations are cylindrical pores that extend through endothelial cells.

15.5 Milking refers to skeletal muscle contractions that drive blood through a venous valve toward the heart.

15.6 The diaphragm contributes to the respiratory pump in the following way: When it contracts during inspiration, it moves downward, causing the pressure in the thoracic cavity to decrease and the pressure in the abdominal cavity to increase. This pressure gradient in turn promotes greater movement of blood from abdominal veins to thoracic veins and then into the right atrium.

15.7 Blood volume in veins is about 64% of 5 liters, or 3.2 liters; blood volume in capillaries is about 7% of 5 liters, or 350 mL.

15.8 Materials cross capillary walls through the plasma membranes of endothelial cells, through intercellular clefts and fenestrations, and via transcytosis.

15.9 Plasma colloid osmotic pressure is lower than normal in a person with a low level of plasma proteins, and therefore capillary reabsorption is low; the result is edema.

15.10 The lymphatic system drains excess interstitial fluid, returns filtered plasma proteins back to the blood, carries out immune responses, and transports dietary lipids.

15.11 Lymph is more similar to interstitial fluid than to plasma because the protein content of lymph is low.

15.12 Inhalation promotes movement of lymph from abdominal lymphatic vessels toward the thoracic region because the pressure in the vessels of the thoracic region is lower than the pressure in the abdominal region when you inhale.

15.13 If the pressure gradient across a blood vessel is zero, no flow occurs through the vessel.

15.14 If a blood vessel has a radius twice that of vessel A (i.e., $r = 2$), then resistance would be 1/16 ($R \propto 1/(2)^4 = 1/16$) and flow would be 16 ($F \propto 1/R = 16$).

15.15 Blood flow becomes turbulent when blood flows through an abnormally constricted area, moves over a rough area, makes a sharp turn, or exceeds a critical velocity.

15.16 Veins are the blood vessels with the lowest pressure in the circulation.

15.17 Because veins are highly compliant vessels, the pressure in veins is normally low, averaging about 10 mmHg.

15.18 Velocity of blood flow is fastest in the aorta and arteries.

15.19 Venous blood needs help returning to the heart because the pressure pushing up the veins in the lower limbs is barely enough to overcome the force of gravity pushing it back down, especially when you are standing up.

15.20 The brain has a total constant blood flow, regardless of the activity level of the individual.

15.21 Arterioles have the premier role in regulating blood flow to different organs because they are the main resistance vessels of the circulation.

15.22 The myogenic response maintains constant blood flow through an arteriole: If arterial pressure increases, the arteriolar walls are stretched, causing the arteriole to vasoconstrict to keep blood flow constant; if arterial pressure decreases, arteriolar walls are stretched less, causing the arteriole to vasodilate in order to maintain constant flow.

15.23 In the systemic circulation, a low level of O_2 causes arterioles to vasodilate; in the pulmonary circulation, a low level of O_2 causes arterioles to vasoconstrict.

15.24 Adenosine activates a signaling pathway that hyperpolarizes the smooth muscle cell, causing L-type-voltage-gated Ca^{2+} channels to close. As a result, Ca^{2+} does not enter the cell and the cell relaxes.

15.25 Mean arterial pressure (MAP) must be regulated because it is the driving force for movement of blood through body tissues. If MAP is too high, blood flows too quickly; if it is too low, blood flows too slowly.

15.26 The cardiovascular center consists of (1) a cardiostimulatory center, (2) a cardioinhibitory center, and (3) a vasomotor center.

15.27 Increased stimulation of the vasomotor nerves by the cardiovascular center causes vasoconstriction.

15.28 The negative feedback cycle represents a change that occurs when you stand up because gravity causes pooling of blood in leg veins once you are upright, which in turn decreases the blood pressure in your upper body.

15.29 Almost-normal blood pressure in a person who has lost blood does not necessarily indicate that the patient's tissues are receiving adequate blood flow; if total peripheral resistance has increased greatly, tissue perfusion may still be inadequate.

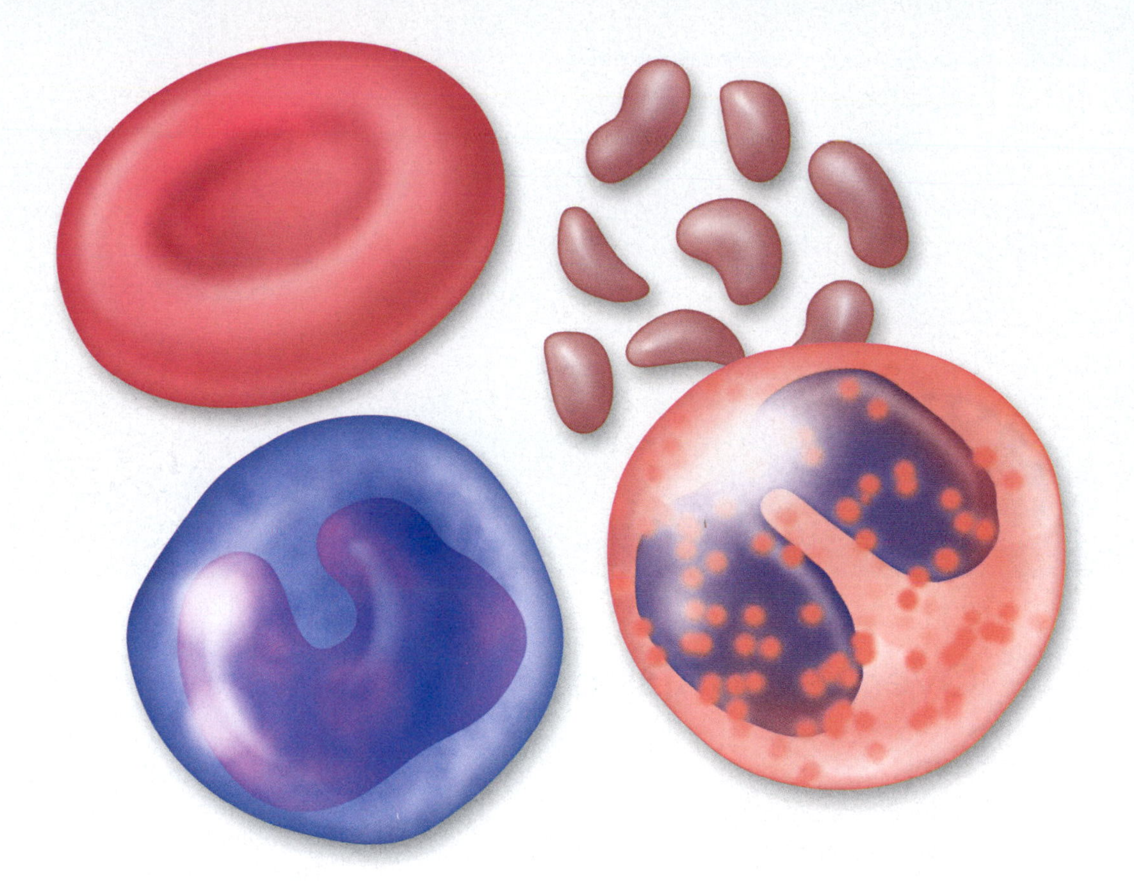

16 The Cardiovascular System: The Blood

Blood and Homeostasis

Blood contributes to homeostasis by transporting substances to and from your body's cells. It also helps regulate pH and body temperature, and provides protection against disease through a variety of defensive responses.

LOOKING BACK TO MOVE AHEAD...

- Stem cells are unspecialized cells that have the ability to divide for indefinite periods and give rise to specialized cells (Section 1.3).
- Phagocytosis is a form of endocytosis in which the cell engulfs large solid particles such as worn-out cells, bacteria, or viruses (Section 5.6).
- Bulk flow of fluid across the wall of a blood capillary is determined by four Starling forces: capillary hydrostatic pressure, interstitial fluid hydrostatic pressure, plasma colloid osmotic pressure, and interstitial fluid colloid osmotic pressure (Section 15.2).
- The viscosity (thickness) of blood is one factor that affects the resistance to blood flow; the higher the viscosity of blood, the higher the resistance to flow (Section 15.3).

Most cells of a multicellular organism cannot move around to obtain oxygen and nutrients or eliminate carbon dioxide and other wastes. Instead, these needs are met by the circulation of blood. **Blood** is a connective tissue that consists of a liquid portion called plasma and a cellular portion composed of various cells and cell fragments. The constant movement of blood keeps the cells and cell fragments dispersed throughout the plasma. Blood transports oxygen from the lungs and nutrients from the gastrointestinal tract to the cells of the body. Carbon dioxide and other wastes move in the reverse direction, from body cells to the blood. Blood then transports the wastes to the lungs and kidneys for elimination from the body. In addition to its transportation function, blood helps regulate pH and body temperature, and provides protection from disease.

Blood has several physical characteristics that are noteworthy. It is denser and more viscous than water and has a slightly alkaline pH, ranging from 7.35 to 7.45 (average = 7.4). The color of blood varies with its oxygen content: It is bright red when saturated with oxygen and dark red when the oxygen content is low. Blood constitutes about 8% of the total body mass and has a volume of about 5 liters in an average-sized adult.

In this chapter you will learn about the functions of the different components of blood. You will also discover how new blood cells are formed and how the body prevents the loss of blood through the clotting process. Your exploration of blood concludes with a discussion of the ABO and Rh blood groups.

16.1 Overview of Blood

OBJECTIVES

- Discuss the functions of blood.
- Describe the principal components of blood.
- Explain the origin of blood cells.

Blood Performs Important Functions in the Body

Blood, a liquid connective tissue, has three general functions:

1. ***Transportation***. Blood transports oxygen from the lungs to the cells of the body and carbon dioxide from body cells to the lungs for exhalation. It carries nutrients from the gastrointestinal tract to body cells and hormones from endocrine glands to target cells. Blood also transports nitrogenous wastes such as urea, creatinine, and uric acid from body cells to the kidneys, where they are removed from the body via urine.
2. ***Regulation***. Circulating blood helps maintain homeostasis of all body fluids. Blood helps regulate pH through the use of buffers. It also helps adjust body temperature through the heat-absorbing and coolant properties of the water in plasma and its variable rate of flow through the skin, where excess heat can be lost from the blood to the environment. In addition, plasma colloid osmotic pressure influences the movement of fluid across blood capillary walls.
3. ***Protection***. Blood can clot, which protects against its excessive loss from the cardiovascular system after an injury. In addition, its leukocytes protect against disease by carrying on phagocytosis and other defensive responses. Several types of blood proteins, including antibodies, interferons, and complement, help protect against disease in a variety of ways.

Plasma and Cellular Elements Are the Two Main Components of Blood

Blood has two main components: (1) plasma, the fluid portion, and (2) cellular elements, which are cells and cell fragments. If a sample of blood is centrifuged (spun) in a small glass tube, the cells sink to the bottom of the tube while the lighter-weight plasma forms a layer on top (**FIGURE 16.1a**). Blood is about 55% plasma and 45% cellular elements. Normally, more than 99% of the cellular elements are erythrocytes, which have a red color. Pale, colorless leukocytes and platelets occupy less than 1% of the cellular elements. Because they are less dense than erythrocytes but more dense than plasma, they form a very thin **buffy coat** layer between the packed erythrocytes and plasma in centrifuged blood.

The percentage of total blood volume occupied by erythrocytes is called the **hematocrit** (he-MAT-ō-krit). For example, a hematocrit of 40 indicates that 40% of the volume of blood is composed of erythrocytes. The normal range of hematocrit for women is 38–46% (average = 42); for men, it is 40–54% (average = 47). The hormone testosterone, present in much higher concentration in men than in women, stimulates synthesis of erythropoietin, the hormone that in turn stimulates production of erythrocytes. Thus, testosterone contributes to higher hematocrits in men. Lower values in women during their reproductive years also may be due to excessive loss of blood during menstruation.

Plasma

When the cellular elements are removed from blood, a straw-colored liquid called **plasma** remains. Plasma is about 92% water, 7% proteins, and 1% solutes other than proteins (**FIGURE 16.1b**). Some of the proteins in plasma are also found elsewhere in the body, but those confined to blood are called **plasma proteins**. Among other functions, plasma proteins play a role in maintaining proper plasma colloid osmotic pressure. Recall that *plasma colloid osmotic pressure* is a force caused by the colloidal suspension of nondiffusible proteins in plasma. The effect of this pressure is to "pull" fluid from interstitial spaces into capillaries (see Section 15.2).

Hepatocytes (liver cells) synthesize most of the plasma proteins, which include albumins, globulins, and fibrinogen. **Albumins** account for about 60% of the plasma proteins. Because they are the most abundant plasma proteins, albumins are the main contributors to plasma colloid osmotic pressure. In addition, albumins serve as transport proteins for several steroid hormones and for fatty acids. **Globulins** constitute about 35% of the plasma proteins. There are

FIGURE 16.1 Components of blood in a normal adult.

Blood consists of plasma (liquid) and cellular elements (cells and cell fragments).

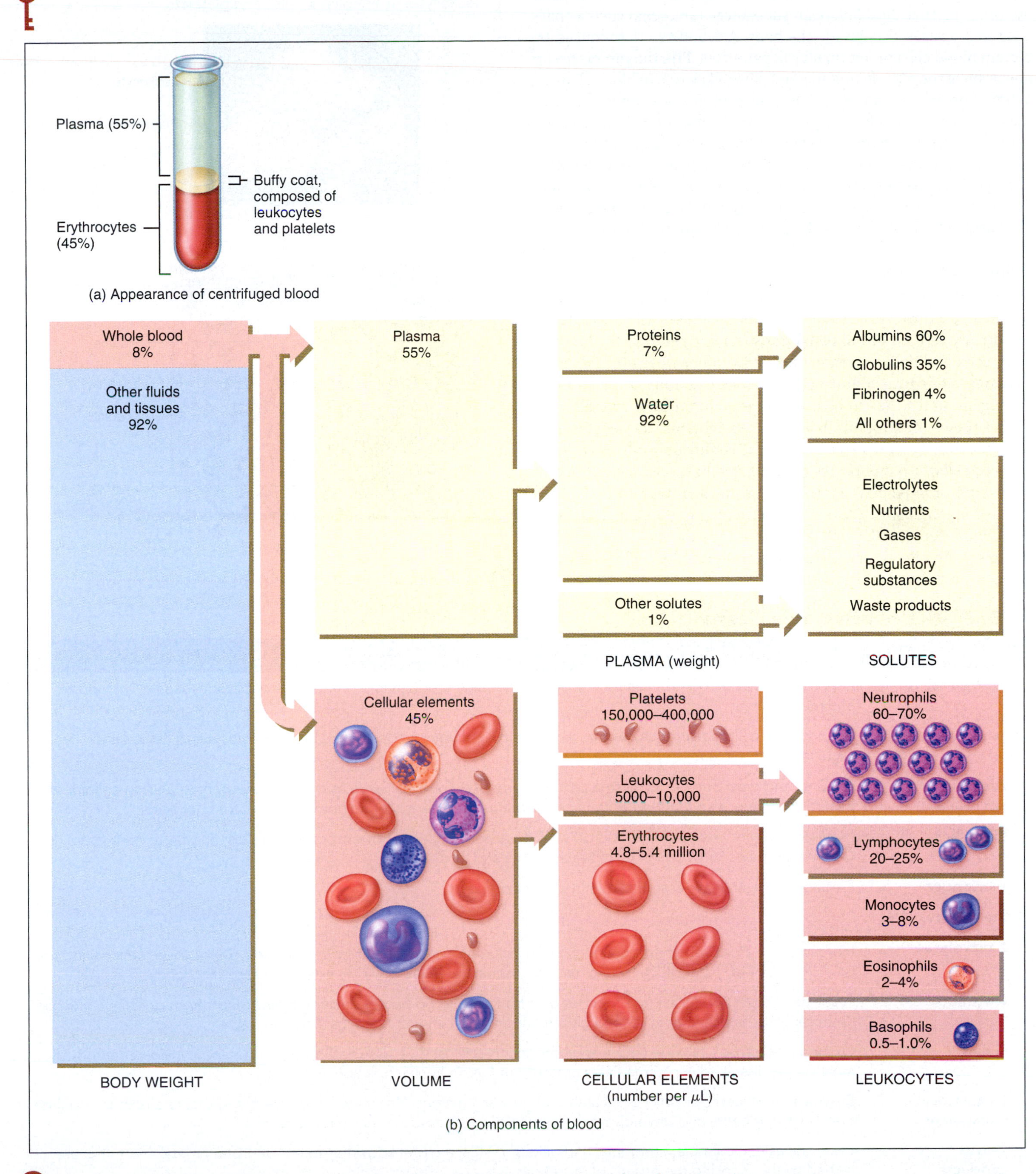

Which type of plasma protein is most abundant, and what are some of its functions?

three types of globulins: alpha, beta, and gamma globulins. **Alpha** and **beta globulins** transport iron, lipids, and fat-soluble vitamins. **Gamma globulins**, also known as **immunoglobulins** or **antibodies**, are proteins that disable foreign substances (antigens) such as bacteria and viruses that invade the body. Antibodies are produced by certain blood cells rather than by hepatocytes. **Fibrinogen** makes up about 4% of the plasma proteins and plays a key role in blood clotting. When fibrinogen and other clotting proteins are removed from plasma during the clotting process, the remaining fluid is called serum. Therefore, **serum** is plasma without its clotting proteins.

Besides proteins, other solutes in plasma include electrolytes, nutrients, gases, regulatory substances such as hormones and enzymes, and waste products such as urea, uric acid, creatinine, ammonia, and bilirubin. TABLE 16.1 describes the chemical composition of plasma.

Cellular Elements

The **cellular elements** of the blood include three principal components: erythrocytes, leukocytes, and platelets (FIGURE 16.2). **Erythrocytes (red blood cells)** transport oxygen from the lungs to body cells and deliver carbon dioxide from body cells to the lungs. **Leukocytes (white blood cells)** protect the body from invading pathogens and other foreign substances. There are several types of leukocytes: neutrophils, basophils, eosinophils, monocytes, and lymphocytes. Lymphocytes are further subdivided into B lymphocytes (B cells), T lymphocytes (T cells), and natural killer (NK) cells. Each type of leukocyte contributes in its own way to the body's defense mechanisms. **Platelets**, the final type of cellular element, are fragments of cells that do not have a nucleus. Among other actions, they release chemicals that promote blood clotting when blood vessels are damaged. Platelets are the functional equivalent

FIGURE 16.2 The cellular elements of blood.

The cellular elements of blood are erythrocytes (red blood cells), leukocytes (white blood cells), and platelets.

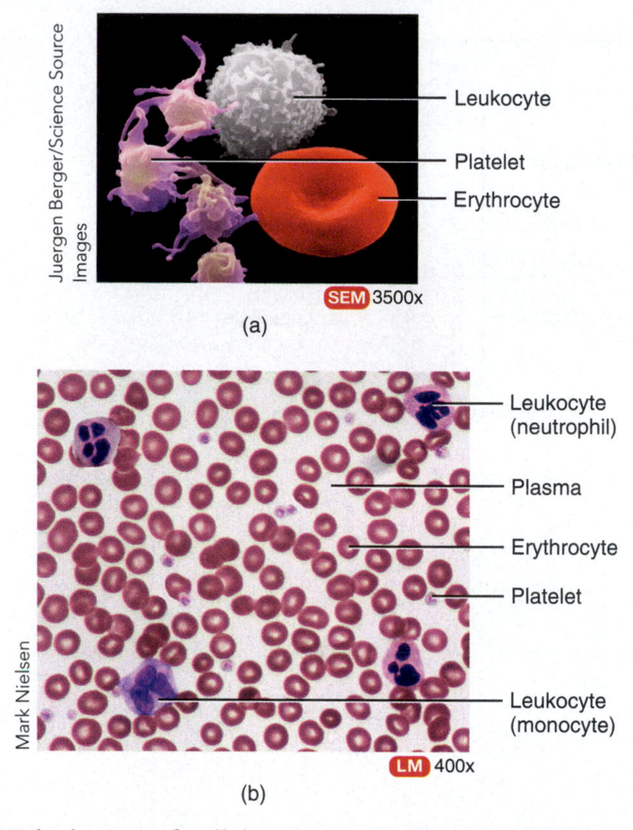

Which type of cellular element helps clot blood?

TABLE 16.1 Substances in Plasma

Constituent	Description
Water (92%)	Liquid portion of blood. Acts as solvent and suspending medium for components of blood; absorbs, transports, and releases heat.
Plasma proteins (7%)	Exert plasma colloid osmotic pressure, which helps maintain water balance between blood and tissues and regulates blood volume.
Albumins	Smallest and most numerous plasma proteins; produced by liver. Serve as main contributors to plasma colloid osmotic pressure; also function as transport proteins for several steroid hormones and for fatty acids.
Globulins	Produced by liver and certain blood cells. Alpha and beta globulins transport iron, lipids, and fat-soluble vitamins. Gamma globulins (antibodies) help attack viruses and bacteria.
Fibrinogen	Produced by liver. Plays essential role in blood clotting.
Other solutes (1%)	
Electrolytes	Inorganic salts. Positively charged ions (cations) include Na^+, K^+, Ca^{2+}, Mg^{2+}; negatively charged ions (anions) include Cl^-, HPO_4^{2-}, SO_4^{2-}, and HCO_3^-. Help maintain osmotic pressure and play essential roles in the function of cells.
Nutrients	Products of digestion pass into blood for distribution to all body cells. Include amino acids (from proteins), glucose (from carbohydrates), fatty acids and glycerol (from triglycerides), vitamins, and minerals.
Gases	Oxygen (O_2), carbon dioxide (CO_2), and nitrogen (N_2). Most O_2 is associated with hemoglobin inside erythrocytes; most CO_2 is dissolved in plasma. N_2 is present but has no known function in the body.
Regulatory substances	Enzymes, produced by body cells, catalyze chemical reactions. Hormones, produced by endocrine glands, regulate metabolism, growth, and development. Vitamins are cofactors for enzymatic reactions.
Waste products	Urea, uric acid, creatinine, ammonia, and bilirubin. Most are breakdown products of protein metabolism that are carried by the blood to the organs of excretion.

of *thrombocytes*, nucleated cells found in lower vertebrates that prevent blood loss by clotting blood.

Blood Cells Are Formed in Bone Marrow

Although some lymphocytes have a lifetime measured in years, most cellular elements of the blood last only hours, days, or weeks and must be replaced continually. Negative feedback systems regulate the total number of erythrocytes and platelets in circulation, and their numbers normally remain steady. The abundance of the different types of leukocytes, however, varies in response to challenges by invading pathogens and other foreign substances.

The process by which blood cells are formed is called **hematopoiesis** (hē′-ma-tō-poy-Ē-sis) or **hemopoiesis**. Before birth, hematopoiesis first occurs in the yolk sac of an embryo and later in the liver and spleen of a fetus. **Bone marrow**, a soft tissue that fills the internal spaces of bones, becomes the primary site of hematopoiesis in the last three months before birth and continues as the source of blood cells after birth and throughout life. There are two types of bone marrow: (1) *red bone marrow*, which is a highly vascularized tissue that contains stem cells capable of developing into blood cells and (2) *yellow bone marrow*, which consists largely of adipose (fat) cells that store triglycerides. Only red bone marrow is capable of hematopoiesis. In a newborn, all bone marrow is red and thus active in hematopoiesis. With increasing age, much of the bone marrow changes from red to yellow. Consequently, red bone marrow is present only in a limited number of bones of the adult skeleton, such as the ribs, sternum (breastbone), vertebrae, pelvis, and the upper ends of the humerus and femur.

About 0.05–0.1% of red bone marrow cells are **pluripotent hematopoietic stem cells (PHSCs)**, which have the capacity to develop into all types of blood cells (FIGURE 16.3). PHSCs first develop into

FIGURE 16.3 Origin and development of blood cells.

Hematopoiesis is the process by which blood cells are formed.

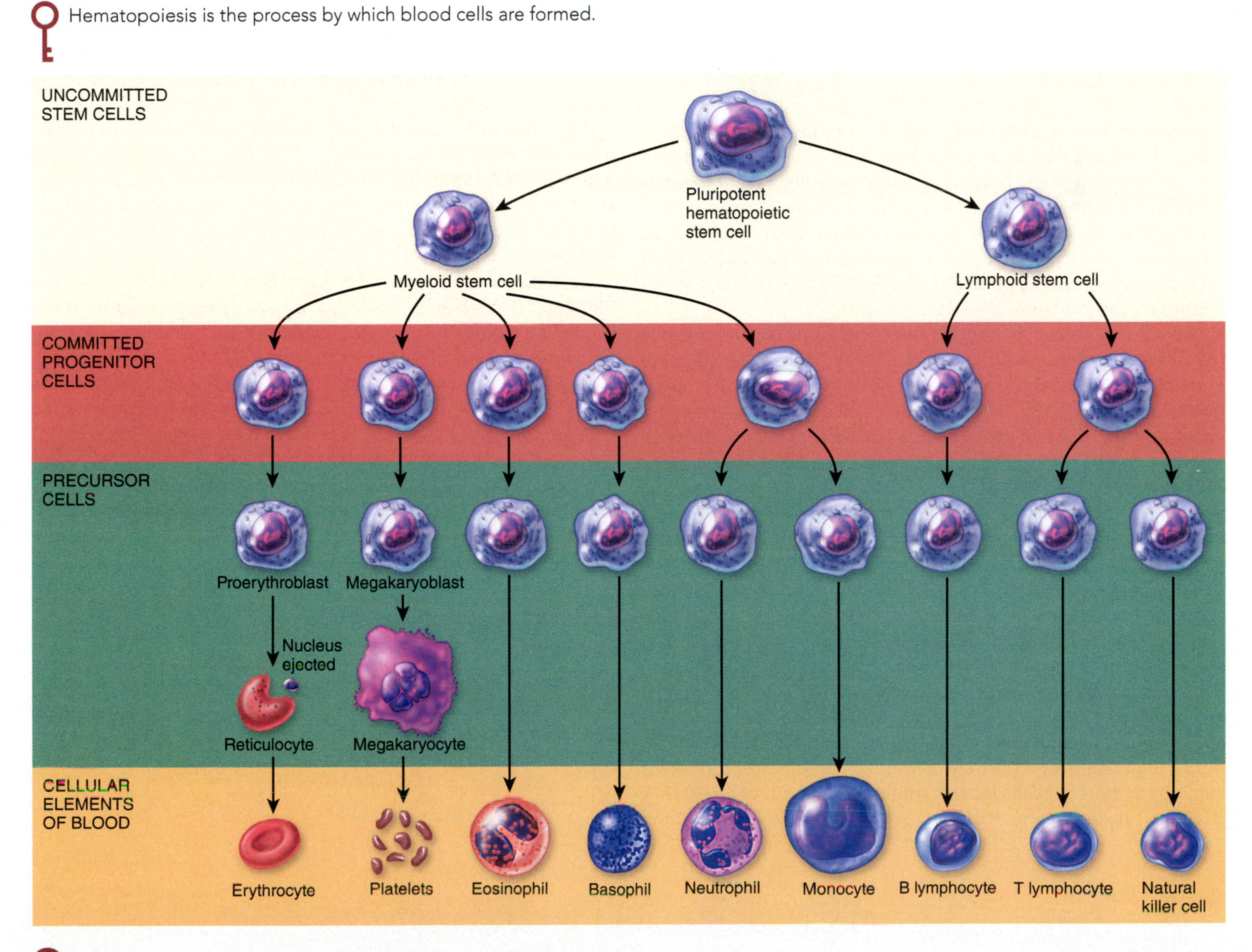

What are three examples of hematopoietic growth factors?

CLINICAL CONNECTION

Medical Uses of Hematopoietic Growth Factors

Hematopoietic growth factors made available through recombinant DNA technology hold tremendous potential for medical uses when a person's natural ability to form new blood cells is diminished or defective. The artificial form of erythropoietin (Epoetin alfa) is very effective in treating the diminished erythrocyte production that accompanies end-stage kidney disease. Colony stimulating factors (CSFs) are given to stimulate leukocyte formation in cancer patients who are undergoing chemotherapy, which kills red bone marrow cells as well as cancer cells because both cell types are undergoing mitosis. Thrombopoietin can be used to prevent the depletion of platelets during chemotherapy. In addition, CSFs and thrombopoietin improve the outcome of patients who receive bone marrow transplants. Hematopoietic growth factors are also used to treat certain clotting disorders and various types of anemia. Research on these medications is ongoing and shows a great deal of promise.

myeloid stem cells and *lymphoid stem cells*. The myeloid or lymphoid stem cells in turn give rise to **progenitor cells** that are committed to developing into just one or two types of blood cells. **Precursor cells** arise from progenitor cells and eventually develop into the specific cellular elements of the blood. Several terms are used to refer to the formation of specific blood cells: *erythropoiesis* refers to the formation of erythrocytes, *leukopoiesis* is the formation of leukocytes, and *thrombopoiesis* is the formation of platelets.

Substances known as **hematopoietic growth factors** regulate the differentiation and proliferation of blood cells. The hormone **erythropoietin (EPO)** (e-rith′-rō-POY-ē-tin) stimulates the formation of erythrocytes by increasing the number of erythrocyte precursors in red bone marrow. EPO is produced by the kidneys in response to low levels of oxygen in the tissues (see FIGURE 16.5). **Thrombopoietin (TPO)** (throm′-bō-POY-ē-tin) is a hormone that stimulates the formation of platelets from megakaryocytes in red bone marrow. TPO is produced by the liver when the platelet concentration in the blood decreases below normal levels. Several different cytokines influence the development of different blood cell types. **Cytokines** are local hormones (paracrines or autocrines) that regulate many normal cell functions, such as cell growth and differentiation. They are secreted by a variety of cells, including red bone marrow cells, leukocytes, and endothelial cells. Two important families of cytokines are **colony-stimulating factors (CSFs)** and **interleukins**. CSFs and interleukins affect blood cell formation in the following ways: (1) they convert uncommitted pluripotent hematopoietic stem cells into committed progenitor cells; (2) they stimulate the development of leukocytes from progenitor cells; and (3) they regulate the functional activities of mature leukocytes (see Chapter 17).

CHECKPOINT

1. What substances does blood transport?
2. What is the functional significance of the hematocrit?
3. What roles do progenitor cells and precursor cells have during hematopoiesis?

16.2 Erythrocytes

OBJECTIVE

- Describe the functions, production, and life cycle of erythrocytes.

Erythrocytes, or **red blood cells (RBCs)**, contain the oxygen-carrying protein **hemoglobin**, which is a pigment that gives whole blood its red color. A healthy adult male has about 5.4 million erythrocytes per microliter (μL) of blood,* and a healthy adult female has about 4.8 million. (One drop of blood is about 50 μL.) To maintain normal numbers of erythrocytes, new mature cells must enter the circulation at the astonishing rate of at least 2 million per second, a pace that balances the equally high rate of erythrocyte destruction.

Erythrocytes are biconcave discs with a diameter of 7–8 μm (FIGURE 16.4a). Mature erythrocytes have a simple structure. Their plasma membrane is both strong and flexible, which allows them to deform without rupturing as they squeeze through narrow capillaries. As you will see later, certain glycolipids and proteins in the plasma membrane of erythrocytes are antigens that account for the various blood groups such as the ABO and Rh groups. Erythrocytes lack a nucleus and other organelles and can neither reproduce nor carry on extensive metabolic activities. The cytosol of erythrocytes contains hemoglobin molecules; these important molecules are synthesized before loss of the nucleus during erythrocyte production and constitute about 33% of the cell's weight.

Erythrocytes Transport Oxygen and Carbon Dioxide

Erythrocytes are highly specialized for their oxygen transport function. Because mature erythrocytes have no nucleus, all of their internal space is available for oxygen transport. In addition, because erythrocytes lack mitochondria and generate ATP anaerobically, they do not use up any

*1 μL = 1 mm^3 = 10^{-6} liter.

FIGURE 16.4 The shapes of an erythrocyte and a hemoglobin molecule. In (b), note that each of the four polypeptide chains of a hemoglobin molecule (blue) has one heme group (gold), which contains an iron ion (Fe^{2+}), shown in red.

The iron portion of a heme group binds oxygen for transport by hemoglobin.

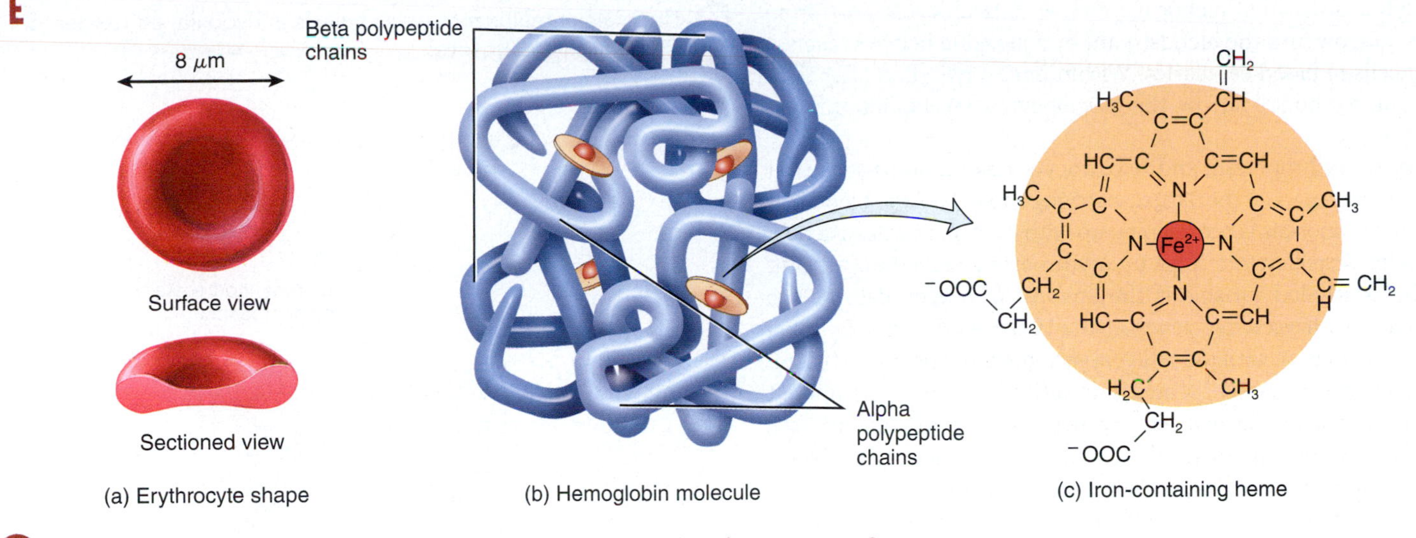

How many molecules of O_2 can one hemoglobin molecule transport?

of the oxygen they transport. Even the shape of an erythrocyte facilitates its function. A biconcave disc has a much greater surface area for the diffusion of gas molecules into and out of the erythrocyte than would, say, a sphere or a cube.

Each erythrocyte contains about 280 million hemoglobin molecules. Hemoglobin transports nearly all of the oxygen (about 98.5%) in blood; the rest of the oxygen (1.5%) is dissolved in plasma. A hemoglobin molecule has two main parts: (1) **globin**, a protein composed of four polypeptide chains (two alpha and two beta chains) and (2) **heme**, a ringlike nonprotein pigment that is bound to each of the four chains (FIGURE 16.4b). At the center of each heme ring is an iron ion (Fe^{2+}) that can combine reversibly with one oxygen molecule (FIGURE 16.4c), allowing each hemoglobin molecule to bind four oxygen molecules. Each oxygen molecule picked up from the lungs is bound to an iron ion. Hemoglobin that has combined with oxygen is called **oxyhemoglobin** and is bright red. As blood flows through tissue capillaries, the iron–oxygen reaction reverses. Hemoglobin releases some of its oxygen, which diffuses first into the interstitial fluid and then into cells. Hemoglobin that has released oxygen is called **deoxyhemoglobin** and is dark red. Arterial blood is bright red because nearly all of its hemoglobin molecules exist in the oxyhemoglobin form. Venous blood is dark red because it contains a significant number of deoxyhemoglobin molecules.

Hemoglobin also transports about 23% of the total carbon dioxide in blood; the remaining carbon dioxide is dissolved in plasma (7%) or carried as bicarbonate ions (70%). Hemoglobin's role in transporting carbon dioxide occurs in the following way: Blood flowing through tissue capillaries picks up carbon dioxide, some of which combines with amino acids in the globin part of hemoglobin, forming a complex called **carbaminohemoglobin**. As blood flows through the lungs, the carbon dioxide is released from hemoglobin and then exhaled.

In addition to its key role in transporting oxygen and carbon dioxide, hemoglobin plays a role in the regulation of blood flow and blood pressure. The gaseous hormone **nitric oxide (NO)**, produced by the endothelial cells that line blood vessels, binds to hemoglobin. Under some circumstances, hemoglobin releases NO. The released NO causes *vasodilation*, an increase in blood vessel diameter that occurs when the smooth muscle in the vessel wall relaxes. Vasodilation improves blood flow and enhances oxygen delivery to cells near the site of NO release.

Besides hemoglobin, erythrocytes also contain the enzyme **carbonic anhydrase (CA)**, which catalyzes the conversion of carbon dioxide and water to carbonic acid, which in turn dissociates into a hydrogen ion and a bicarbonate ion. The entire reaction is reversible and is summarized as follows:

This reaction is significant because it allows most of the carbon dioxide (about 70%) in blood to be transported in the form of bicarbonate ions (HCO_3^-). You will learn more about the transport of carbon dioxide and oxygen in the discussion of the respiratory system in Chapter 18 (see Section 18.5).

Erythrocytes Are Produced via Erythropoiesis

The formation of erythrocytes is referred to as **erythropoiesis** (e-rith′-rō-poy-Ē-sis). During this process, a pluripotent hematopoietic stem cell develops into a myeloid stem cell, which gives rise to a progenitor cell. The progenitor cell in turn develops into a precursor cell called a **proerythroblast** (see FIGURE 16.3). The proerythroblast divides several times, producing cells that begin to synthesize hemoglobin. Ultimately, a cell near the end of the development sequence ejects its nucleus and becomes a **reticulocyte** (re-TIK-ū-lō-sīt). Loss of the nucleus causes the center of the cell to indent, producing the erythrocyte's distinctive biconcave shape. Reticulocytes pass from red bone marrow into the bloodstream by squeezing between the endothelial cells of blood capillaries. Within one to two days after their release from red bone marrow, the reticulocytes develop into mature erythrocytes.

Normally, erythropoiesis and erythrocyte destruction proceed at roughly the same pace. If the oxygen-carrying capacity of the blood falls because erythropoiesis is not keeping up with erythrocyte destruction, a negative feedback system steps up erythrocyte production (FIGURE 16.5). The controlled variable is the amount of oxygen delivered to body tissues. An oxygen deficiency at the tissue level, called **hypoxia** (hī-POKS-ē-a), may occur if too little oxygen enters the blood. For example, the lower oxygen content of air at high altitudes reduces the amount of oxygen in the blood. Oxygen delivery may also fall due to anemia (see Clinical Connection: Anemia) or as a result of circulatory problems that reduce blood flow to tissues. Whatever the cause, hypoxia stimulates the kidneys to increase the release of the hormone *erythropoietin*, which speeds the development of proerythroblasts into reticulocytes in the red bone marrow. As the number of circulating erythrocytes increases, more oxygen can be delivered to body tissues.

The Erythrocyte Life Cycle Is About 120 Days

Erythrocytes live only about 120 days because of the wear and tear their plasma membranes undergo as they squeeze through blood capillaries. Without a nucleus and other organelles, erythrocytes cannot synthesize new components to replace damaged ones. The plasma membrane becomes more fragile with age, and the cells are more likely to burst, especially as they squeeze through narrow channels in the spleen. Ruptured erythrocytes are removed from circulation and destroyed by fixed phagocytic macrophages in the spleen and liver, and the breakdown products are recycled as follows (FIGURE 16.6):

1. Macrophages in the spleen, liver, or red bone marrow phagocytize ruptured and worn-out erythrocytes.
2. The globin and heme portions of hemoglobin are split apart.
3. Globin is broken down into amino acids, which can be reused to synthesize other proteins.
4. Iron is removed from the heme portion in the form of Fe^{3+}, which associates with the plasma protein **transferrin**, a transporter for Fe^{3+} in the bloodstream.
5. In the liver, spleen, and skeletal muscles, Fe^{3+} detaches from transferrin and attaches to an iron-storage protein called **ferritin**.
6. Upon release from a storage site, Fe^{3+} reattaches to transferrin.
7. The Fe^{3+}–transferrin complex is then carried to red bone marrow, where erythrocyte precursor cells take it up through receptor-mediated endocytosis (see Section 5.6) for use in hemoglobin synthesis. Iron is needed for the heme portion of the hemoglobin molecule, and amino acids are needed for the globin portion. Vitamin B_{12} and folic acid are also needed for erythrocyte precursor cells to develop.
8. Erythropoiesis in red bone marrow results in the production of erythrocytes, which enter the circulation.

FIGURE 16.5 Negative feedback regulation of erythropoiesis. Lower oxygen content of air at high altitudes, anemia, and circulatory problems may reduce oxygen delivery to body tissues.

The main stimulus for erythropoiesis is hypoxia, an oxygen deficiency at the tissue level.

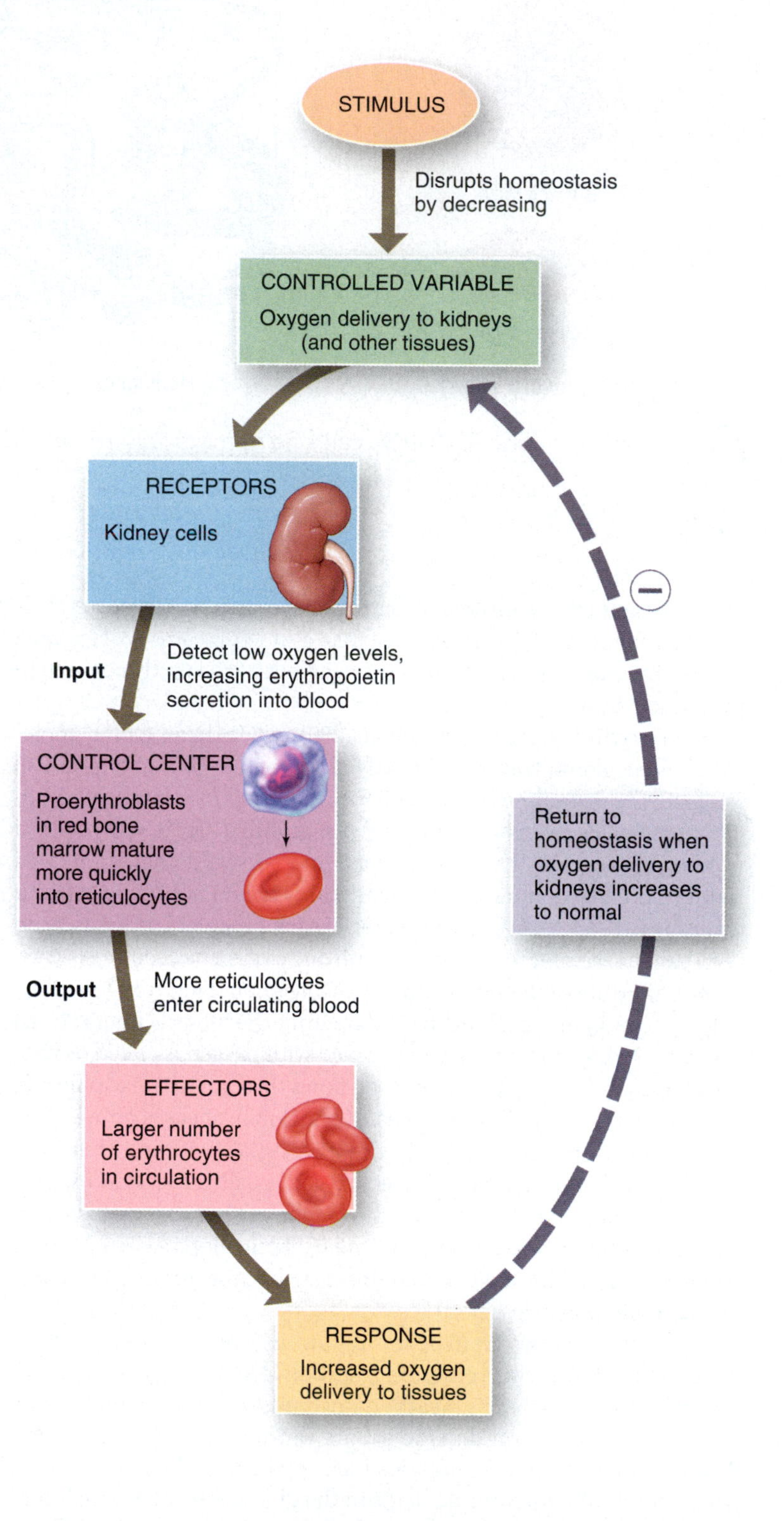

How might your hematocrit change if you moved from a town at sea level to a high mountain village?

FIGURE 16.6 Formation and destruction of erythrocytes, and the recycling of hemoglobin components. Erythrocytes circulate for about 120 days after leaving red bone marrow before they are phagocytized by macrophages.

Normally the rate of erythrocyte formation by red bone marrow equals the rate of erythrocyte destruction by macrophages.

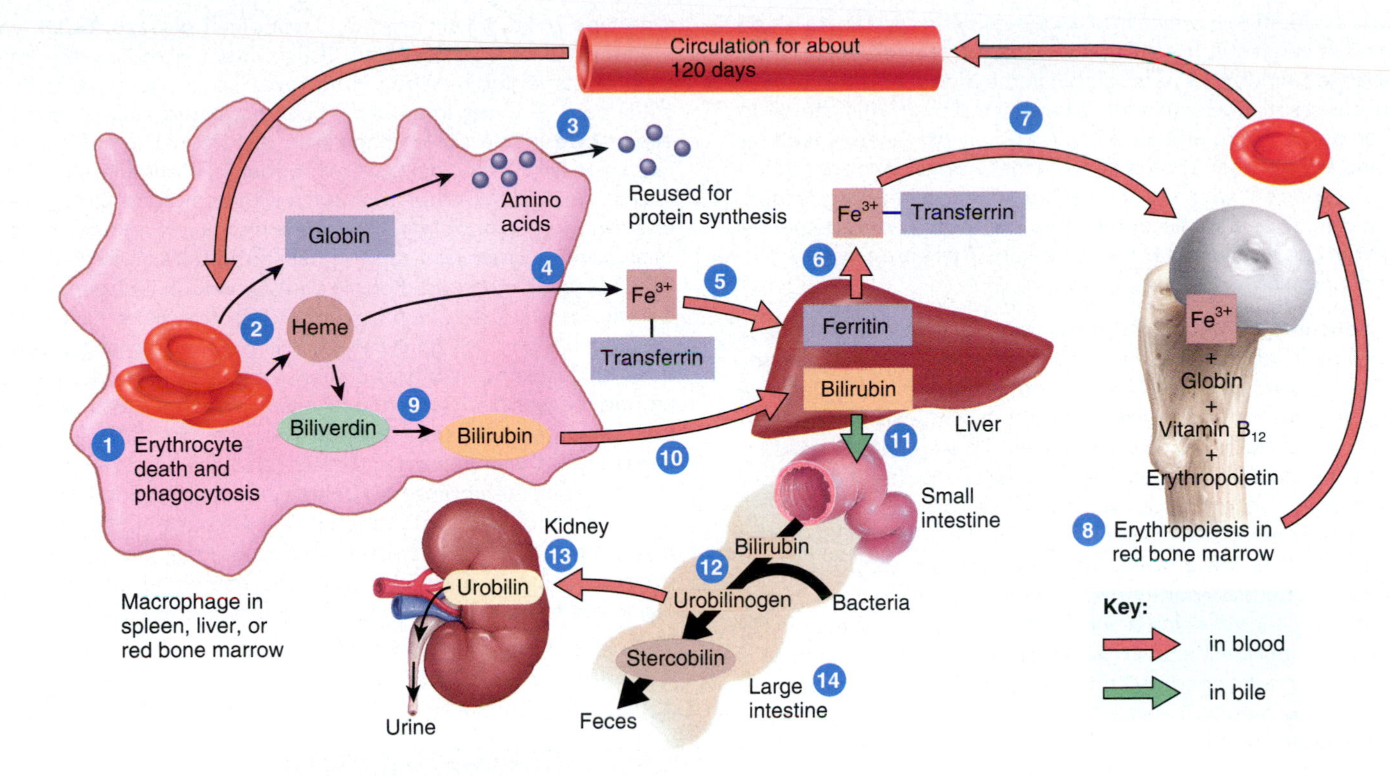

What is the function of transferrin?

CLINICAL CONNECTION

Polycythemia

Polycythemia (pol'-ē-sī-THĒ-mē-a) is a condition in which the concentration of erythrocytes in the blood is higher than normal. It is associated with an increase in the hematocrit, which may be 55% or higher. There are different types of polycythemia. **Primary polycythemia**, also known as **polycythemia vera**, occurs when abnormal stem cells in red bone marrow produce too many erythrocytes. In some cases, the hematocrit may be as high as 80%. The increase in hematocrit increases the viscosity of blood, which in turn increases the resistance to flow and makes blood more difficult for the heart to pump.

Secondary polycythemia occurs when an excess of erythrocytes is produced in response to extended periods of hypoxia. The prolonged hypoxia causes the kidneys to increase secretion of erythropoietin (EPO), which in turn acts on red bone marrow to stimulate production of more erythrocytes. The erythrocyte count in secondary polycythemia typically is lower than that in primary polycythemia. In people who have low tissue oxygen levels due to smoking, heart or lung disease, or living at high altitudes, secondary polycythemia is a normal, adaptive response that increases the amount of oxygen delivered to the tissues.

Because erythrocytes are the main transport vehicle for oxygen, over the years many athletes have tried several means of increasing their hematocrit, causing **induced polycythemia**, to gain a competitive edge. Training at higher altitudes produces a natural increase in hematocrit because the lower amount of oxygen in inhaled air triggers release of more EPO. Some athletes perform **blood doping** to elevate their hematocrit. In this procedure, blood cells are removed from the body and stored for a month or so, during which time the hematocrit returns to normal. Then, a few days before an athletic event, the erythrocytes are reinjected to increase the hematocrit. Another way that some athletes enhance their erythrocyte production is by injecting Epoetin alfa (Procrit® or Epogen®), a synthetic form of EPO that is used to treat anemia. Practices that increase hematocrit are dangerous, however, because they increase the viscosity of the blood, which can lead to cardiovascular disorders. Since blood doping and the use of Epoetin alfa give athletes an unfair advantage and cause adverse side effects, these practices are banned by the International Olympic Committee and other competitive sports organizations.

CLINICAL CONNECTION

Anemia

Anemia is a condition in which the oxygen-carrying capacity of blood is reduced. It can result from reduced numbers of erythrocytes and/or a decreased amount of hemoglobin in the blood. The decrease in erythrocytes associated with anemia is reflected by a significant drop in the hematocrit (normal is 38–54%). A person with anemia feels fatigued and is intolerant of cold. Both of these symptoms are related to a lack of oxygen, which is needed for ATP and heat production. Also, the skin appears pale due to the low content of red-colored hemoglobin circulating in skin blood vessels. There are six major categories of anemia:

1. **Iron-deficiency anemia**, the most common type of anemia, is caused by insufficient intake or absorption of iron or by excessive loss of iron. Because iron is a component of hemoglobin, a shortage of iron slows down hemoglobin synthesis and eventually impairs production of erythrocytes. Those erythrocytes that can be produced are small and pale because they have a low hemoglobin concentration. Women are at greater risk for iron-deficiency anemia due to menstrual blood losses and increased iron demands of the growing fetus during pregnancy.
2. **Pernicious anemia** results from an inability of the stomach to produce intrinsic factor, which is needed for absorption of vitamin B_{12} in the small intestine. Because vitamin B_{12} is required for erythrocytes to develop, a deficiency in vitamin B_{12} leads to decreased production of erythrocytes.
3. **Hemorrhagic anemia** is caused by excessive loss of erythrocytes from the body due to bleeding. It can occur as a result of large wounds, stomach ulcers, or especially heavy menstruation.
4. **Aplastic anemia** results from the failure of red bone marrow to produce enough erythrocytes. It occurs in response to destruction of red bone marrow by radiation, toxic chemicals, or some pathogens.
5. **Renal anemia** is caused by insufficient secretion of erythropoietin (EPO) due to kidney disease. Because EPO stimulates erythropoiesis, insufficient EPO secretion results in decreased production of erythrocytes.
6. **Hemolytic anemia** is caused by the rupture (hemolysis) of a large number of erythrocytes. This condition may result from inherited defects in the erythrocyte or from outside agents such as parasites, toxins, or antibodies from incompatible transfused blood.

One type of hemolytic anemia is **sickle-cell disease (SCD)**. The erythrocytes of a person with SCD contain Hb-S, an abnormal type of hemoglobin. When Hb-S gives up oxygen to the interstitial fluid, it forms long, stiff, rodlike structures that bend the erythrocyte into a sickle shape (see FIGURE). The sickled cells rupture easily. Even though erythropoiesis is stimulated by the loss of the cells, it cannot keep pace with hemolysis. Sickled cells also tend to get stuck in blood capillaries, depriving body organs of sufficient oxygen and causing pain and tissue damage. Sickle-cell disease is inherited. People with two sickle-cell genes have severe anemia; those with only one defective gene have minor problems. Sickle-cell genes are found primarily among populations or descendants of populations that live in the malaria belt around the world, including parts of sub-Saharan Africa, Mediterranean Europe, and tropical Asia. The gene responsible for the tendency of the erythrocytes to sickle also alters the permeability of the plasma membranes of sickled cells, causing potassium ions to leak out. Low levels of potassium kill the malaria parasites that may infect sickled cells. Because of this effect, a person with one normal gene and one sickle-cell gene has higher-than-average resistance to malaria. The possession of a single sickle-cell gene thus confers a survival advantage.

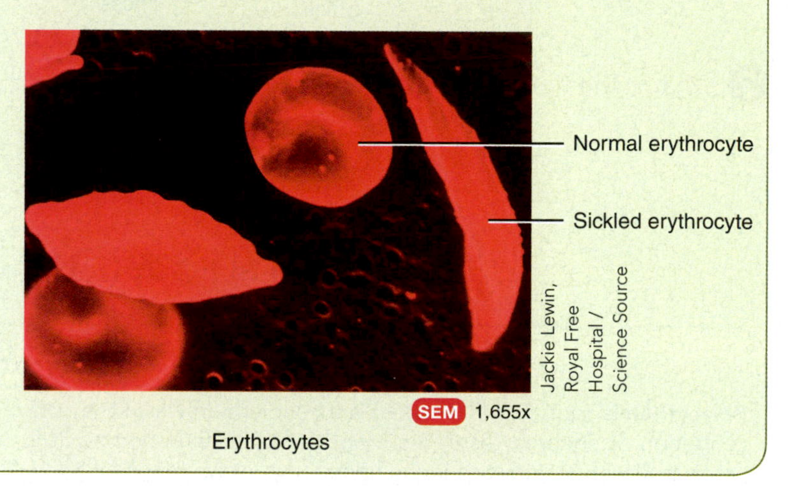

Erythrocytes

9. When iron is removed from heme, the non-iron portion of heme is converted to **biliverdin** (bil′-i-VER-din), a green pigment, and then into **bilirubin** (bil′-i-ROO-bin), a yellow-orange pigment.
10. Bilirubin enters the blood and is transported to the liver.
11. Within the liver, bilirubin is released by liver cells into bile, which passes into the small intestine and then into the large intestine.
12. In the large intestine, bacteria convert bilirubin into **urobilinogen** (ūr-ō-bī-LIN-ō-jen).
13. Some urobilinogen is absorbed back into the blood, converted to a yellow pigment called **urobilin** (ūr-ō-BĪ-lin), and excreted in urine.
14. Most urobilinogen is eliminated in feces in the form of a brown pigment called **stercobilin** (ster′-kō-BĪ-lin), which gives feces its characteristic color.

CHECKPOINT

4. Describe the functions of erythrocytes.
5. What is erythropoiesis? How does erythropoiesis affect hematocrit? What factors speed up and slow down erythropoiesis?
6. How is hemoglobin recycled?

Critical Thinking

Moving to a Higher State

John competed in a marathon running event (1 marathon = 26.2 miles) in Miami, Florida, and had an impressive run time of 2 hours 45 minutes and 8 seconds (2:45:08). This week he moved to Denver, Colorado, to be closer to his family and friends. A few days after he arrived, he decided to go for a run in effort to prepare for an upcoming marathon in Denver. Within a few minutes, he was breathing harder than normal and soon felt quite fatigued. He also noticed that it took longer than usual to return to his normal breathing rate. After about a week, John was much less fatigued during his runs and by 1 month, he was back to his normal routine.

© RosaIreneBetancourt 7 / Alamy Stock Photo

Cheryl Senter/ AP Photos

SOME THINGS TO KEEP IN MIND:

John was fatigued and breathing hard during his initial runs in Denver because the altitude is higher and the partial pressure of oxygen in the air is lower compared to Miami. Partial pressure is an indication of the oxygen concentration in the air; the higher the elevation above sea level, the lower the partial pressure of oxygen in the atmosphere.

SOME INTERESTING FACTS:

City	Elevation Above Sea Level	Partial Pressure of O_2 In the Air
Miami, Florida	5 feet	159 mmHg
Denver, Colorado	5280 feet	132 mmHg

Erythrocytes carry O_2 in the blood and deliver it to the tissues. What is the mechanism employed by the erythrocytes to carry O_2?

Running in higher altitudes presented John's body with a new stress—a decreased O_2 level in the air, which in turn caused a decrease O_2 level in his tissues. How would John's kidneys respond to this disruption of homeostasis?

After living in Denver for a while, how would John's hematocrit change? Would this be an example of polycythemia? If so, which type?

Blood doping is the removal and storage of one's own erythrocytes. Once the body replaces the erythrocytes that were removed and hematocrit is restored to normal, these erythrocytes can be re-injected into the blood, providing an enhanced hematocrit. If this is performed at the appropriate time before an athletic event, that athlete may have an advantage over his or her competitors. What could be the problem with this type of manipulation? Is it legal?

Suppose that a marathon runner is accused of using a performance-enhancing drug, such as Epoetin alfa. Theoretically, how would this drug affect that person's athletic performance?

16.3 Leukocytes

OBJECTIVES

- Distinguish between granulocytes and agranulocytes.
- Discuss the functions of the different types of leukocytes.
- Describe the production and life span of leukocytes.

Unlike erythrocytes, **leukocytes**, or **white blood cells (WBCs)**, have nuclei and other organelles but do not contain hemoglobin. Because hemoglobin is not present, leukocytes are colorless under a light microscope unless they are stained with specific dyes. Leukocytes are far less numerous than erythrocytes; at about 5000–10,000 cells per μL of blood, they are outnumbered by erythrocytes by about 700:1.

The Five Types of Leukocytes Are Grouped into Two Categories

There are five distinct types of leukocytes—neutrophils, basophils, eosinophils, monocytes, and lymphocytes—each with a unique microscopic appearance (FIGURE 16.7). The different types of leukocytes are grouped into two major categories: granulocytes and agranulocytes. Eosinophils, basophils, and neutrophils are categorized as **granulocytes** because they contain prominent granules (chemical-filled vesicles) in their cytoplasm. The three types of granulocytes are distinguished based on the staining properties of their granules. The granules within an **eosinophil** (ē-ō-SIN-ō-fil) are *eosinophilic* (= eosin-loving)—they stain red-orange with the acidic dye *eosin* (FIGURE 16.7a). The granules usually do not cover the nucleus, which has two distinct lobes. The granules of a **basophil** (BĀ-sō-fil) are *basophilic* (= basic loving)—they stain blue-purple with a basic (alkaline) dye (FIGURE 16.7b). The granules commonly obscure the nucleus, which is shaped like the letter S. The granules of a **neutrophil** (NOO-trō-fil) are neutral: They have little affinity for either acidic or basic dyes. As a result, the granules stain poorly, having only a pale lilac color (FIGURE 16.7c). Young neutrophils have a rod-shaped nucleus and are known as *band cells.* As a neutrophil matures, parts of the nucleus constrict until the nucleus has two to five lobes. Because older neutrophils have several, differently shaped nuclear lobes, they are often called *polymorphonuclear leukocytes (PMNs)*, polymorphs, or "polys." Neutrophils are the most prevalent type of leukocyte, accounting for about 60–70% of the circulating leukocyte population.

Monocytes and lymphocytes are known as **agranulocytes** because they do not have prominent cytoplasmic granules. A **monocyte** is larger than a lymphocyte and has a nucleus that is shaped liked a horseshoe or a kidney (FIGURE 16.7d). A **lymphocyte** (LIM-fō-sīt) has a large, round nucleus surrounded by a thin ring of cytoplasm (FIGURE 16.7e). Lymphocytes are the second most abundant type of leukocyte, constituting about 20–25% of the circulating leukocyte population.

Leukocytes Defend the Body Against Invading Pathogens and Abnormal Cells

The body is continuously exposed to microbes and other harmful foreign substances. **Immunity** is the body's ability to ward off damage

FIGURE 16.7 Types of leukocytes.

The shapes of their nuclei and the staining properties of their cytoplasmic granules distinguish leukocytes from one another.

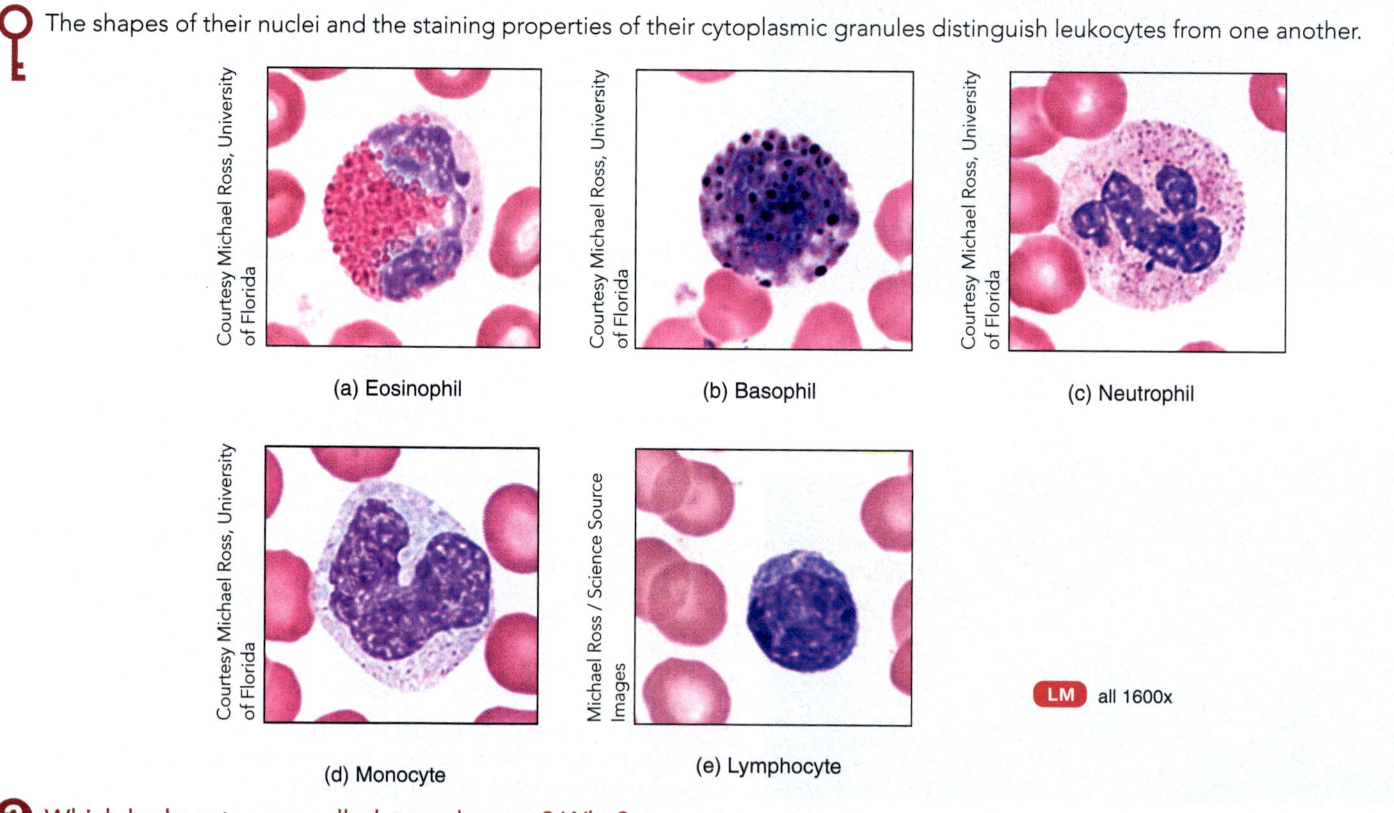

(a) Eosinophil (b) Basophil (c) Neutrophil (d) Monocyte (e) Lymphocyte

Which leukocytes are called granulocytes? Why?

or disease caused by these destructive agents. The body system that provides immunity is known as the **immune system**. Leukocytes are key components of the immune system: They combat invading pathogens and abnormal cells with a variety of defensive responses. To engage in battle, many leukocytes leave the bloodstream and gather at sites of pathogen invasion or tissue damage. Thus, leukocytes use the bloodstream primarily for rapid transportation to areas of the body that need protection.

Each type of leukocyte has a particular role in immunity. Neutrophils serve as phagocytes that engulf and destroy microbes and their toxins. Among leukocytes, neutrophils respond most quickly to tissue destruction by bacteria. Monocytes migrate to a site of infection and then enlarge and differentiate into huge phagocytes called *macrophages*. Because macrophages phagocytize more bacteria than neutrophils, they are more potent phagocytes than neutrophils. Eosinophils destroy parasites such as worms. In addition, they play a role in allergic reactions and can act as weak phagocytes. Basophils release chemicals such as histamine that promote inflammation when a tissue is damaged. *Inflammation*, a response characterized by redness, pain, heat, and swelling, serves as a signal that attracts phagocytes to the region of tissue damage; the phagocytes can then destroy any invading microbes in that area. Last but certainly not least, lymphocytes provide defensive responses that are critical to the normal functioning of the immune system. There are three types of lymphocytes: B lymphocytes (B cells), T lymphocytes (T cells), and natural killer (NK) cells. B cells differentiate into specialized cells that produce *antibodies*, proteins that disable foreign substances that are in body fluids. T cells and NK cells attack infected body cells and destroy cancer cells. You will learn more about the functions of the different types of leukocytes in Chapter 17.

Leukocytes Are Produced via Leukopoiesis

The formation of leukocytes is known as **leukopoiesis** (loo′-kō-poy-Ē-sis). During this process, pluripotent hematopoietic stem cells develop into myeloid stem cells and lymphoid stem cells (see **FIGURE 16.3**). The myeloid stem cells give rise to cells that develop into granulocytes and monocytes, and the lymphoid stem cells give rise to cells that develop into lymphocytes. Once they are formed, leukocytes enter the bloodstream and then are transported by the blood to areas of infection or injury.

In a healthy body, some leukocytes, especially lymphocytes, can live for several months or years. However, most leukocytes live only a few days because a phagocyte can ingest only a certain amount of material before the cell's own metabolic activities are disrupted. During a period of infection, phagocytic leukocytes may live only a few hours.

CHECKPOINT

7. Is a lymphocyte classified as a granulocyte or an agranulocyte? Why?
8. What functions do eosinophils, basophils, B cells, and natural killer cells perform?
9. Why do phagocytic leukocytes typically have a short life span?

CLINICAL CONNECTION

Leukemia

The term **leukemia** (loo-KĒ-mē-a) refers to a group of red bone marrow cancers in which abnormal leukocytes multiply uncontrollably. The accumulation of cancerous leukocytes in red bone marrow interferes with the production of normal erythrocytes, leukocytes, and platelets. As a result, anemia develops, the individual is more susceptible to infection, and blood clotting is abnormal. Treatment options include chemotherapy, radiation, stem cell transplantation, and blood transfusion.

16.4 Platelets

OBJECTIVE

- Describe the function of platelets.

Besides the immature cell types that develop into erythrocytes and leukocytes, pluripotent hematopoietic stem cells also differentiate into cells that produce platelets. During this process, myeloid stem cells develop into progenitor cells that give rise to precursor cells called *megakaryoblasts* (see **FIGURE 16.3**). Megakaryoblasts transform into megakaryocytes, huge cells that splinter into 2000 to 3000 fragments. Each fragment, enclosed by a piece of the plasma membrane, is a **platelet**. The process by which platelets are formed is known as **thrombopoiesis** (throm′-bō-poy-Ē-sis). Platelets break off from the megakaryocytes in red bone marrow and then enter the blood circulation. Between 150,000 and 400,000 platelets are present in each µL of blood. Each is disc-shaped, 2–4 µm in diameter, and has many vesicles but no nucleus.

Platelets help stop blood loss from damaged blood vessels by forming a platelet plug. Their granules also contain chemicals that, once released, promote blood clotting. Platelets have a short life span, normally just 5 to 9 days. Aged and dead platelets are removed by fixed macrophages in the spleen and liver.

TABLE 16.2 summarizes the cellular elements of blood.

CHECKPOINT

10. What is the main role of platelets?

16.5 Hemostasis

OBJECTIVES

- Describe the three mechanisms that contribute to hemostasis.
- Identify the stages of blood clotting.
- Explain the various factors that promote and inhibit blood clotting.

Hemostasis (hē-mō-STĀ-sis) is a sequence of responses that stops bleeding. When blood vessels are damaged or ruptured, the hemostatic response must be quick, localized to the region of damage, and

TABLE 16.2 Summary of the Cellular Elements of Blood

Name and Appearance	Characteristics*	Functions
Erythrocytes	Biconcave discs; lack nuclei; live for about 120 days.	Hemoglobin within erythrocytes transports most of the oxygen and part of the carbon dioxide in the blood.
Leukocytes	Most live for a few hours to a few days.†	Combat pathogens and other foreign substances that enter the body.
Granular leukocytes		
Neutrophils	Nucleus has 2–5 lobes; cytoplasm has very fine, pale lilac granules.	Phagocytosis.
Eosinophils	Nucleus has 2 lobes; large, red-orange granules fill the cytoplasm.	Destroy certain parasitic worms; play a role in allergic reactions; can serve as weak phagocytes.
Basophils	Nucleus is S-shaped; large cytoplasmic granules appear deep blue-purple.	Liberate chemicals that promote inflammation; also participate in allergic reactions.
Agranular leukocytes		
Lymphocytes (B cells, T cells, and natural killer cells)	Nucleus is round; cytoplasm forms a ring around the nucleus.	B cells differentiate into specialized cells that produce antibodies; T cells and natural killer cells attack infected body cells and destroy cancer cells.
Monocytes	Nucleus is shaped like a kidney or a horseshoe.	Phagocytosis (after transforming into macrophages).
Platelets	Cell fragments that live for 5–9 days; contain many vesicles but no nucleus.	Form platelet plug in hemostasis; release chemicals that promote vascular spasm and blood clotting.

* Colors are those seen when using Wright's stain.

† Some lymphocytes, called B and T memory cells, can live for many years once they are established.

carefully controlled in order to be effective. Three mechanisms reduce blood loss: (1) vascular spasm, (2) platelet plug formation, and (3) blood clotting (coagulation). When successful, hemostasis prevents **hemorrhage**, the loss of a large amount of blood from blood vessels. Hemostatic mechanisms can prevent hemorrhage from smaller blood vessels, but extensive hemorrhage from larger vessels usually requires medical intervention.

Vascular Spasm Occurs Immediately After Vessel Injury

When arteries or arterioles are damaged, the circularly arranged smooth muscle in their walls contracts immediately, a reaction called **vascular spasm**. This reduces blood loss for several minutes to several hours, during which time the other hemostatic mechanisms go into operation. The spasm is caused by damage to the smooth muscle, by substances released from activated platelets, and by reflexes initiated by pain receptors.

A Platelet Plug Is a Mass of Platelets That Accumulates at the Site of Injury

Considering their small size, platelets store an impressive array of chemicals in cytoplasmic granules (vesicles). *Dense granules* contain ADP, Ca^{2+}, and serotonin. *Alpha* (α) *granules* contain clotting factors and **platelet-derived growth factor (PDGF)**, a protein that causes proliferation of vascular smooth muscle fibers and stimulates fibroblasts to help repair damaged blood vessel walls. Platelets also synthesize the eicosanoid **thromboxane A_2**, which is not stored because it is released immediately upon production.

Platelet plug formation involves the following steps (FIGURE 16.8):

1. ***Platelet adhesion.*** Initially, platelets contact and stick to parts of a damaged blood vessel, such as collagen fibers in the blood vessel wall that become exposed when the endothelial lining is disrupted. Platelets adhere to the exposed collagen fibers with the help of **von Willebrand factor (VWF)**, a protein secreted into plasma by endothelial cells. Von Willebrand factor binds both to collagen and to platelet surface receptors, and therefore serves as a bridge between the injured area of the blood vessel wall and the platelets.
2. ***Platelet activation.*** Due to adhesion and the presence of thrombin (a substance that also plays a role in blood clotting), platelets become activated. During activation, platelets undergo a change in shape: They extend many projections that enable them to contact and interact with one another. In addition, they begin to liberate the contents of their vesicles and other substances that have been synthesized, an event known as the **platelet release reaction**. Liberated ADP and thromboxane A_2 play a major role by activating nearby platelets. Serotonin and thromboxane A_2 function as vasoconstrictors, causing and sustaining contraction of vascular smooth muscle, which decreases blood flow through the injured vessel.
3. ***Platelet aggregation.*** The release of ADP makes other platelets in the area sticky, and the stickiness of the newly recruited and activated platelets causes them to adhere to the originally activated platelets. This gathering of platelets is called **platelet aggregation**. Eventually, the accumulation and attachment of large numbers of platelets form a mass called a **platelet plug**.

A platelet plug is very effective in preventing blood loss in a small vessel. Although initially the platelet plug is loose, it becomes quite tight when reinforced by fibrin threads formed during clotting (see FIGURE 16.10). A platelet plug can stop blood loss completely if the hole in a blood vessel is not too large.

As you have just learned, platelet activation is vital to the formation of a platelet plug. Substances that promote platelet activation include VWF (as part of platelet adhesion) and thrombin: Binding of either VWF or thrombin to platelet receptors triggers signal transduction pathways that result in changes in platelet shape and release of platelet chemicals. To understand the molecular steps involved in platelet

FIGURE 16.8 Platelet plug formation.

A platelet plug can stop blood loss completely if the hole in a blood vessel is small enough.

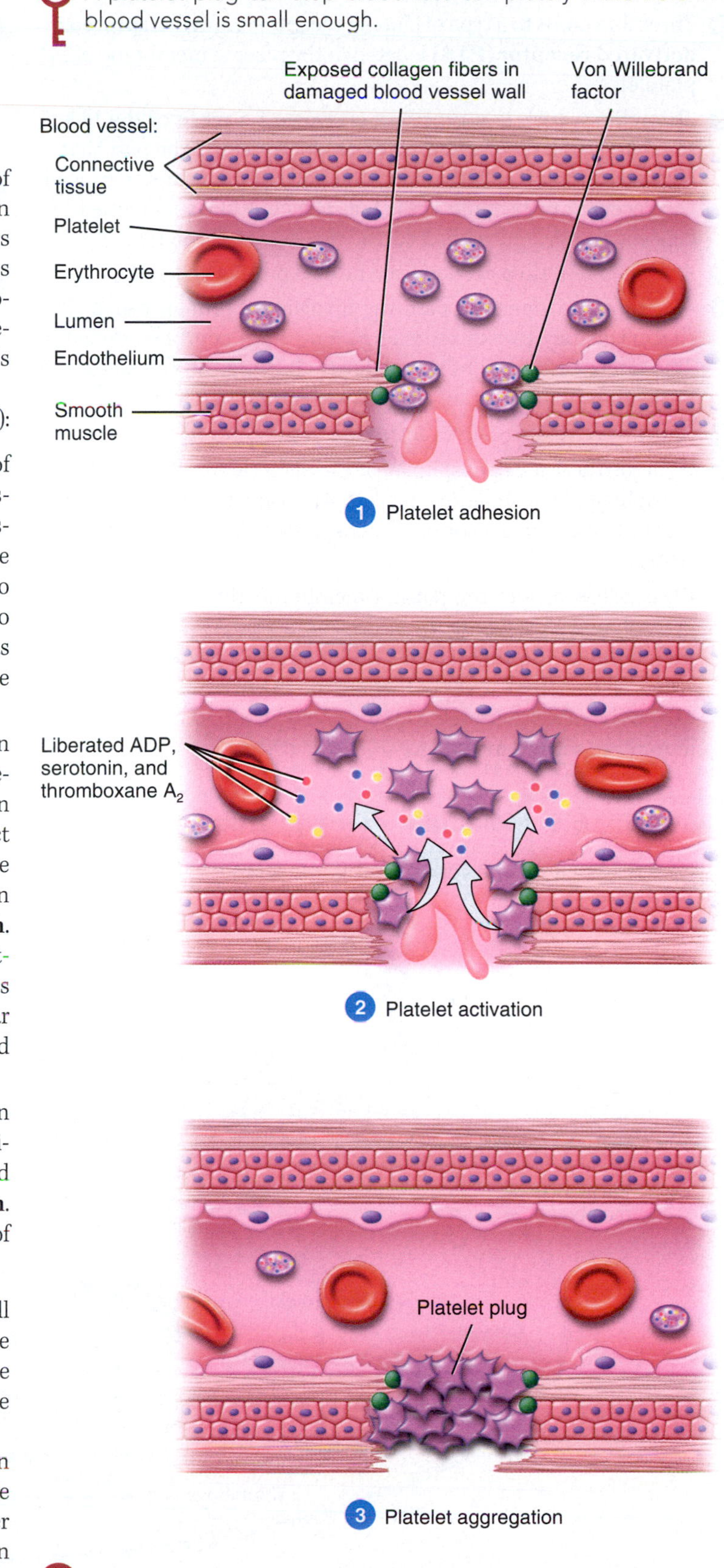

Along with platelet plug formation, which two mechanisms contribute to hemostasis?

activation, let's look at the signaling pathways that thrombin induces (**FIGURE 16.9**):

1. Thrombin binds to a type of platelet receptor known as a **protease-activated receptor (PAR)** located in the plasma membrane of the platelet.
2. Thrombin, which is a protease, cleaves off a piece of the receptor. As a result, the portion of the receptor next to the part that is cleaved off becomes tethered to the body of the receptor, causing the receptor to become activated. Hence, PAR is so-named because it is activated via proteolytic cleavage by thrombin.
3. Activated PAR then activates a type of G protein called $\mathbf{G_{12}}$.
4. G_{12} triggers a signaling pathway involving a protein called **Rho** that ultimately causes reorganization of the actin cytoskeleton. This alters the shape of the platelet, allowing the cell to extend projections that facilitate contact with other platelets.
5. Activated PAR also causes activation of the G protein G_q.
6. G_q in turn activates phospholipase C, which cleaves the membrane phospholipid phosphatidylinositol bisphosphate (PIP_2) to form the second messengers inositol trisphosphate (IP_3) and diacylglycerol (DAG).
7. IP_3 is released from the plasma membrane, diffuses through the cytosol, and then binds to and opens IP_3-gated channels in the membrane of the endoplasmic reticulum (ER). Recall that the ER stores calcium ions (Ca^{2+}). Opening the IP_3-gated channels allows Ca^{2+} to be released from the lumen of the ER into the cytosol to act as another second messenger.
8. Both DAG, which remains in the plasma membrane, and Ca^{2+} bind to and activate protein kinase C (PKC).
9. Activated protein kinase C phosphorylates intracellular proteins, including certain SNARE proteins. Recall that SNAREs (*S*oluble *N*-ethylmaleimide-sensitive factor *A*ttachment protein *RE*ceptors) play a key role in fusion of vesicle membranes to the plasma membrane (see Section 5.6).
10. Phosphorylation of SNARE proteins triggers exocytosis of platelet granules (vesicles), causing the chemicals stored inside to be released from the cell.
11. The increase in intracellular Ca^{2+} that occurs due to the opening of IP_3-gated channels in the ER also increases the activity of phospholipase A_2, which converts a membrane phospholipid to arachidonic acid. This in turn increases the activity of the cyclooxygenase pathway responsible for producing thromboxane A_2. Once formed, thromboxane A_2 is released from the cell.

Blood Clotting Results in the Formation of Fibrin Threads at the Injured Area

Normally, blood remains in its liquid form as long as it stays within its vessels. If it is drawn from the body, however, it thickens and forms a gel. Eventually, the gel separates from the liquid. The straw-colored liquid, called **serum**, is simply plasma minus the clotting proteins. The gel is

FIGURE 16.9 Molecular mechanisms of platelet cell activation by thrombin.

Thrombin activates platelets by triggering signal transduction pathways that involve Rho, phospholipase C, and phospholipase A_2.

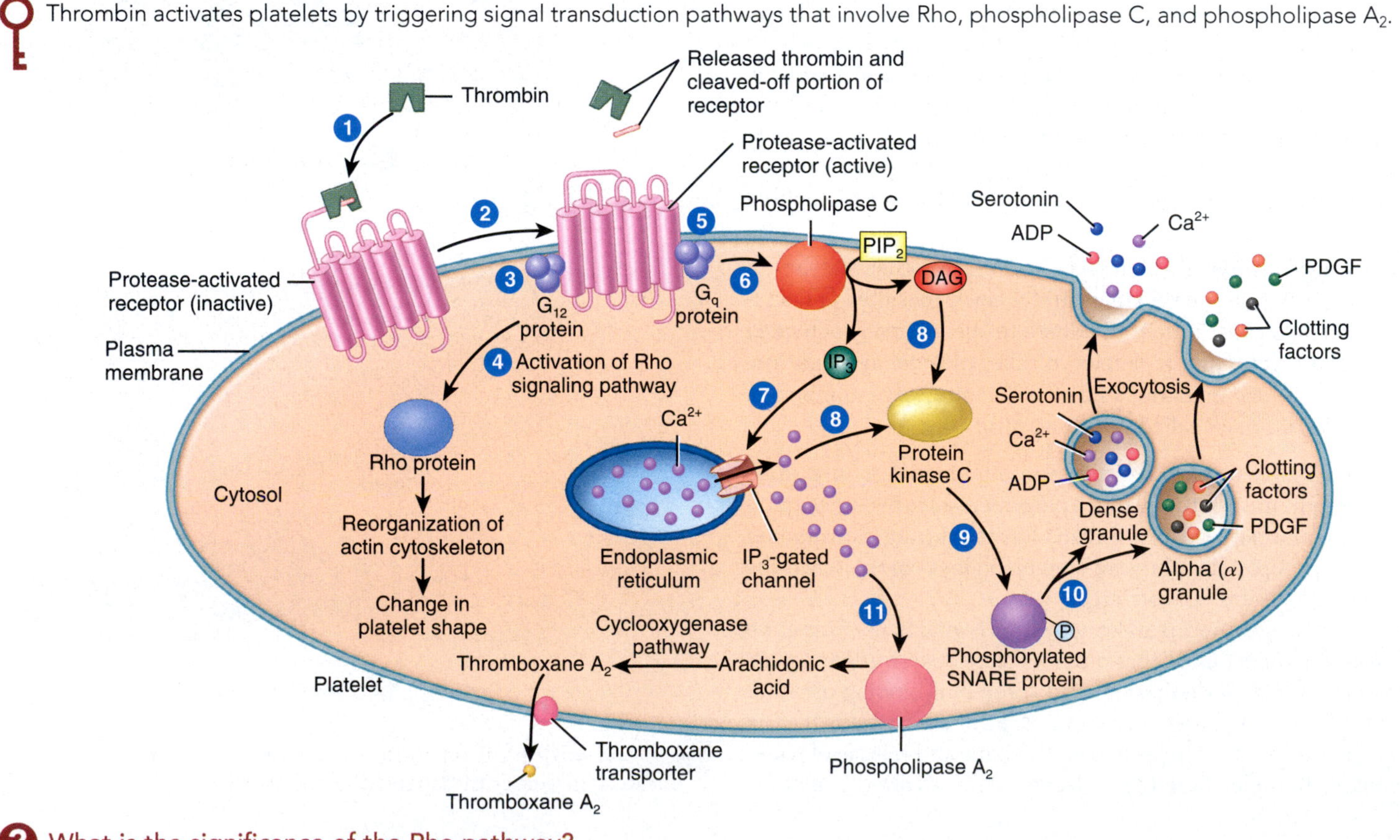

What is the significance of the Rho pathway?

called a **clot**. It consists of a network of insoluble protein fibers called fibrin in which the cellular elements of blood are trapped (FIGURE 16.10).

The process of gel formation, called **clotting** or **coagulation**, is a series of chemical reactions that culminates in formation of fibrin threads. If blood clots too easily, the result can be **thrombosis**—clotting in an undamaged blood vessel. If the blood takes too long to clot, hemorrhage can occur.

Clotting involves several substances known as **clotting (coagulation) factors**. These factors include calcium ions (Ca^{2+}), several inactive enzymes that are synthesized by hepatocytes (liver cells) and released into the bloodstream, and various molecules associated with platelets or released by damaged tissues. Most clotting factors are identified by Roman numerals that indicate the order of their discovery (not necessarily the order of their participation in the clotting process).

Clotting is a complex cascade of enzymatic reactions in which each clotting factor activates many molecules of the next one in a fixed sequence. Finally, a large quantity of product (the insoluble protein fibrin) is formed. Clotting can be divided into three stages (FIGURE 16.11):

1. Two pathways, called the extrinsic pathway (FIGURE 16.11a) and the intrinsic pathway (FIGURE 16.11b), which will be described shortly, lead to the formation of prothrombinase. Once prothrombinase is formed, the steps involved in the next two stages of clotting are the same for both the extrinsic and intrinsic pathways, and together these two stages are referred to as the common pathway.

FIGURE 16.10 Blood clot formation. Notice the platelet and erythrocytes entrapped in fibrin threads.

A blood clot is a gel that contains cellular elements of the blood entangled in fibrin threads.

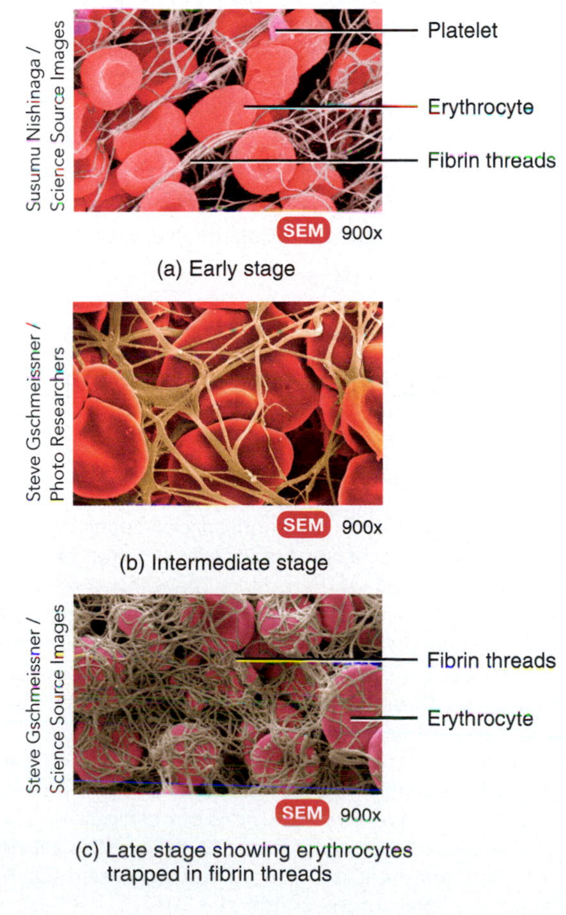

What is serum?

FIGURE 16.11 The blood-clotting cascade.

In blood clotting, coagulation factors are activated in a particular sequence, resulting in a cascade of reactions that includes positive feedback cycles.

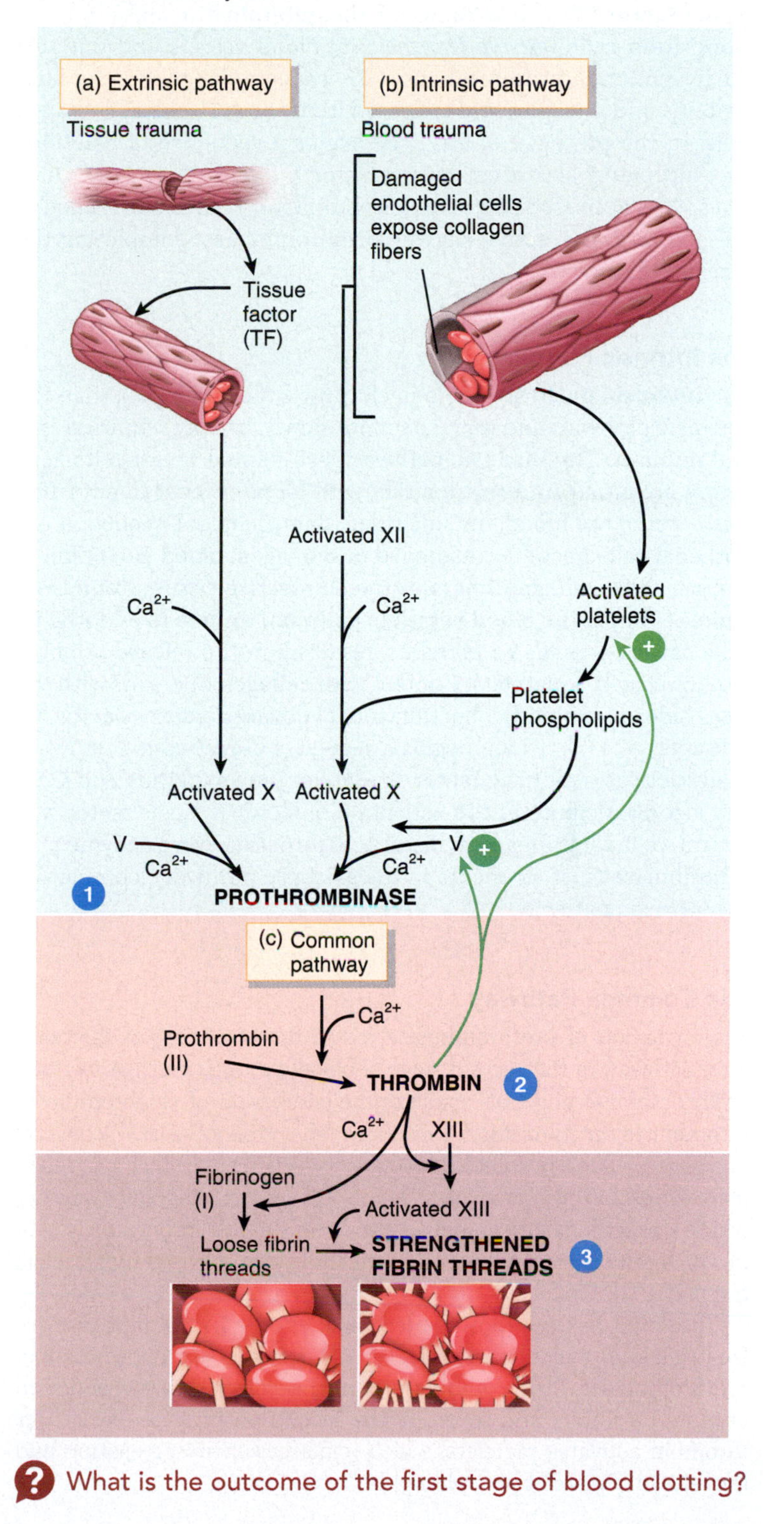

What is the outcome of the first stage of blood clotting?

2. Prothrombinase converts prothrombin (a plasma protein formed by the liver) into the enzyme thrombin.
3. Thrombin converts soluble fibrinogen (another plasma protein formed by the liver) into insoluble fibrin. Fibrin forms the threads of the clot.

The Extrinsic Pathway

The **extrinsic pathway** of blood clotting has fewer steps than the intrinsic pathway and occurs rapidly—within a matter of seconds if trauma is severe. It is so-named because a tissue protein called **tissue factor (TF)**, also known as **thromboplastin**, leaks into the blood from cells *outside (extrinsic to)* blood vessels and initiates the formation of prothrombinase. TF is a complex mixture of lipoproteins and phospholipids released from the surfaces of damaged cells. In the presence of Ca^{2+}, TF begins a sequence of reactions that ultimately activates clotting factor X (FIGURE 16.11a). Once factor X is activated, it combines with factor V in the presence of Ca^{2+} to form the active enzyme prothrombinase, completing the extrinsic pathway.

The Intrinsic Pathway

The **intrinsic pathway** of blood clotting is more complex than the extrinsic pathway, and it occurs more slowly, usually requiring several minutes. The intrinsic pathway is so-named because its activators are either in direct contact with blood or contained *within (intrinsic to)* the blood; outside tissue damage is not needed. If endothelial cells become roughened or damaged, blood can come in contact with collagen fibers in the connective tissue around the endothelium of the blood vessel. In addition, trauma to endothelial cells causes damage to platelets, resulting in the release of phospholipids by the platelets. Contact with collagen fibers (or with the glass sides of a blood collection tube) activates clotting factor XII (FIGURE 16.11b), which begins a sequence of reactions that eventually activates clotting factor X. Platelet phospholipids and Ca^{2+} can also participate in the activation of factor X. Once factor X is activated, it combines with factor V to form the active enzyme prothrombinase (just as occurs in the extrinsic pathway), completing the intrinsic pathway.

The Common Pathway

The formation of prothrombinase marks the beginning of the common pathway. In the second stage of blood clotting (FIGURE 16.11c), prothrombinase and Ca^{2+} catalyze the conversion of prothrombin to thrombin. In the third stage, thrombin, in the presence of Ca^{2+}, converts fibrinogen, which is soluble, to loose fibrin threads, which are insoluble. Thrombin also activates factor XIII (fibrin stabilizing factor), which strengthens and stabilizes the fibrin threads into a sturdy clot. Plasma contains some factor XIII, which is also released by platelets trapped in the clot.

Thrombin has two positive feedback effects. In the first positive feedback loop, which involves factor V, it accelerates the formation of prothrombinase. Prothrombinase in turn accelerates the production of more thrombin, and so on. In the second positive feedback loop, thrombin activates platelets, which reinforces their aggregation and the release of platelet phospholipids.

Clot Retraction

Once a clot is formed, it plugs the ruptured area of the blood vessel and thus stops blood loss. **Clot retraction** is the consolidation or tightening of the fibrin clot. The fibrin threads attached to the damaged surfaces of the blood vessel gradually contract as platelets pull on them. As the clot retracts, it pulls the edges of the damaged vessel closer together, decreasing the risk of further damage. During retraction, some serum can escape between the fibrin threads, but the cellular elements in blood cannot. Normal retraction depends on an adequate number of platelets in the clot, which release factor XIII and other factors, thereby strengthening and stabilizing the clot. Permanent repair of the blood vessel can then take place. In time, fibroblasts form connective tissue in the ruptured area, and new endothelial cells repair the vessel lining.

Role of Vitamin K in Clotting

Normal clotting depends on adequate levels of vitamin K in the body. Although vitamin K is not involved in actual clot formation, it is required for the synthesis of four clotting factors (II, VII, IX, and X). Normally produced by bacteria that inhabit the large intestine, vitamin K is a fat-soluble vitamin that can be absorbed through the lining of the intestine and into the blood if absorption of lipids is normal. People suffering from disorders that slow absorption of lipids (for example, inadequate release of bile into the small intestine) often experience uncontrolled bleeding as a consequence of vitamin K deficiency.

The various clotting factors, their sources, and the pathways of activation are summarized in TABLE 16.3.

Hemostatic Control Mechanisms

Many times each day, little clots start to form in the body, often at a site of minor roughness or at a developing atherosclerotic plaque inside a blood vessel. Because blood clotting involves amplification and

CLINICAL CONNECTION

Hemophilia

Hemophilia (hē-mō-FIL-ē-a) is an inherited deficiency of clotting in which bleeding may occur spontaneously or after only minor trauma. It is the oldest known hereditary bleeding disorder; descriptions of the disease are found as early as the second century A.D. Hemophilia usually affects males and is sometimes referred to as "the royal disease" because many descendants of Queen Victoria, beginning with one of her sons, were affected by the disease. Different types of hemophilia are due to deficiencies of different blood clotting factors (such as factor VIII or IX) and exhibit varying degrees of severity, ranging from mild to severe bleeding tendencies. Treatment involves transfusions of fresh plasma or concentrates of the deficient clotting factor to relieve the tendency to bleed. Another treatment is the drug desmopressin (DDAVP), which can boost the levels of the clotting factors.

TABLE 16.3 Clotting (Coagulation) Factors

Number*	Name(s)	Source	Pathway(s) of Activation
I	Fibrinogen.	Liver.	Common.
II	Prothrombin.	Liver.	Common.
III	Tissue factor (thromboplastin).	Damaged tissues and activated platelets.	Extrinsic.
IV	Calcium ions (Ca^{2+}).	Diet, bones, and platelets.	All.
V	Proaccelerin, labile factor, or accelerator globulin (AcG).	Liver and platelets.	Extrinsic and intrinsic.
VII	Serum prothrombin conversion accelerator (SPCA), stable factor, or proconvertin.	Liver.	Extrinsic.
VIII	Antihemophilic factor (AHF), antihemophilic factor A, or antihemophilic globulin (AHG).	Liver.	Intrinsic.
IX	Christmas factor, plasma thromboplastin component (PTC), or antihemophilic factor B.	Liver.	Intrinsic.
X	Stuart-Prower factor or thrombokinase.	Liver.	Extrinsic and intrinsic.
XI	Plasma thromboplastin antecedent (PTA) or antihemophilic factor C.	Liver.	Intrinsic.
XII	Hageman factor, glass factor, contact factor, or antihemophilic factor D.	Liver.	Intrinsic.
XIII	Fibrin-stabilizing factor (FSF).	Liver and platelets.	Common.

* There is no factor VI. Prothrombinase (prothrombin activator) is a combination of activated factors V and X.

positive feedback cycles, a clot has a tendency to enlarge, creating the potential for impairment of blood flow through undamaged vessels. The **fibrinolytic system** (fī-bri-nō-LIT-ik) dissolves small, inappropriate clots; it also dissolves clots at a site of damage once the damage is repaired. Dissolution of a clot is called **fibrinolysis** (fī-bri-NOL-i-sis). When a clot is formed, an inactive plasma enzyme called **plasminogen** is incorporated into the clot. Both body tissues and blood contain substances that can activate plasminogen to **plasmin (fibrinolysin)**, an active plasma enzyme. Among these substances are thrombin, activated factor XII, and tissue plasminogen activator (tPA), which is synthesized in endothelial cells of most tissues and released into the blood. Once plasmin is formed, it can dissolve the clot by digesting fibrin threads and inactivating substances such as fibrinogen, prothrombin, and factors V and XII.

Anticoagulants

People who are at increased risk of forming blood clots may receive anticoagulants. Examples are heparin or warfarin. Heparin is often administered during hemodialysis and open-heart surgery. **Warfarin (Coumadin®)** acts as an antagonist to vitamin K and thus blocks synthesis of four clotting factors (II, VII, IX, and X). Warfarin is slower acting than heparin. To prevent clotting in donated blood, blood banks and laboratories often add substances that remove Ca^{2+}; examples are EDTA (ethylene diamine tetraacetic acid) and CPD (citrate phosphate dextrose).

Even though thrombin has a positive feedback effect on blood clotting, clot formation normally remains localized at the site of damage. A clot does not extend beyond a wound site into the general circulation, in part because fibrin absorbs thrombin into the clot. Another reason for localized clot formation is that, because of the dispersal of some of the clotting factors by the blood, their concentrations are not high enough to bring about widespread clotting.

Several other mechanisms also control blood clotting. For example, endothelial cells and leukocytes produce a prostaglandin called **prostacyclin** (pros-ta-SĪ-klin) that opposes the actions of thromboxane A_2. Prostacyclin is a powerful inhibitor of platelet adhesion and release.

In addition, substances that delay, suppress, or prevent blood clotting, called **anticoagulants**, are present in blood. These include **antithrombin**, which blocks the action of several factors, including XII, X, and II (prothrombin). **Heparin**, an anticoagulant that is produced by mast cells and basophils, combines with antithrombin and increases its effectiveness in blocking thrombin. Another anticoagulant, **activated protein C (APC)**, inactivates the two major clotting factors not blocked by antithrombin and enhances activity of plasminogen activators. Babies that lack the ability to produce APC due to a genetic mutation usually die of blood clots in infancy.

Intravascular Clotting

Despite the anticoagulating and fibrinolytic mechanisms, blood clots sometimes form within the cardiovascular system. Such clots may be initiated by roughened endothelial surfaces of a blood vessel resulting from atherosclerosis, trauma, or infection. These conditions induce adhesion of platelets. Intravascular clots may also form when blood flows too slowly (stasis), allowing clotting factors to accumulate locally

CLINICAL CONNECTION

Aspirin and Thrombolytic Agents

In people with heart and blood vessel disease, the events of hemostasis may occur even without external injury to a blood vessel. At low doses, **aspirin** inhibits vasoconstriction and platelet aggregation by blocking synthesis of thromboxane A_2. It also reduces the chance of thrombus formation. Due to these effects, aspirin reduces the risk of strokes, myocardial infarction, and blockage of peripheral arteries.

Thrombolytic agents are chemical substances that are injected into the body to dissolve blood clots that have already formed to restore circulation. They either directly or indirectly activate plasminogen. The first thrombolytic agent, approved in 1982 for dissolving clots in the coronary arteries of the heart, was **streptokinase**, which is produced by streptococcal bacteria. A genetically engineered version of human **tissue plasminogen activator (tPA)** is now used to treat victims of both heart attacks and brain attacks (strokes) that are caused by blood clots.

in high enough concentrations to initiate coagulation. Clotting in an unbroken blood vessel (usually a vein) is called **thrombosis** (throm-BŌ-sis). The clot itself, called a **thrombus**, may dissolve spontaneously. If it remains intact, however, the thrombus may become dislodged and be swept away in the blood. A blood clot, bubble of air, fat from broken bones, or a piece of debris transported by the bloodstream is called an **embolus**. An embolus that breaks away from an arterial wall may lodge in a smaller-diameter artery downstream and block blood flow to a vital organ. When an embolus lodges in the lungs, the condition is called **pulmonary embolism**.

CHECKPOINT

11. How do vascular spasm and platelet plug formation occur?
12. How do the extrinsic and intrinsic pathways of blood clotting differ?
13. What is fibrinolysis? Why does blood rarely remain clotted inside blood vessels?
14. Define each of the following terms: anticoagulant, thrombus, embolus, and thrombolytic agent.

16.6 Blood Groups and Blood Types

OBJECTIVES

- Distinguish between the ABO and Rh blood groups.
- Explain why it is so important to match donor and recipient blood types before administering a transfusion.

The surfaces of erythrocytes contain a genetically determined assortment of **antigens** that includes carbohydrates and proteins. These antigens, called **agglutinogens** (a-gloo-TIN-ō-jens), occur in characteristic combinations. Based on the presence or absence of various antigens, blood is categorized into different **blood groups**. Within a given blood group, there may be two or more different **blood types**. There are at least 24 blood groups and more than 100 antigens that can be detected on the surface of erythrocytes. Here, two major blood groups—ABO and Rh—are discussed. Other blood groups include the Lewis, Kell, Kidd, and Duffy systems. The incidence of ABO and Rh blood types varies among different population groups, as indicated in TABLE 16.4.

The ABO Blood Group Is Determined by the Presence or Absence of A and B Antigens

The **ABO blood group** is based on two glycolipid antigens called A and B (FIGURE 16.12). People whose erythrocytes display *only antigen A* have **type A** blood. Those who have *only antigen B* are **type B**. Individuals who have *both A and B antigens* are **type AB**; those who have *neither antigen A nor B* are **type O**.

Plasma usually contains **antibodies** called **agglutinins** (a-GLOO-ti-nins) that react with the A or B antigens if the two are mixed. These are the **anti-A antibody**, which reacts with antigen A, and the **anti-B antibody**, which reacts with antigen B. The antibodies present in each of the four blood types are shown in FIGURE 16.12. You do not have antibodies that react with the antigens of your own erythrocytes, but you do have antibodies for any antigens that your erythrocytes lack. For example, if your blood type is B, you have B antigens on your erythrocytes, and you have anti-A antibodies in your plasma. Although agglutinins start to appear in the blood within a few months after birth, the reason for their presence is not clear. Perhaps they are formed in response to bacteria that normally inhabit the gastrointestinal tract. Because the antibodies are large IgM-type antibodies (see TABLE 17.3 in Section 17.3) that do not cross the placenta, ABO incompatibility between a mother and her fetus rarely causes problems.

An Incompatible Transfusion Causes Agglutination

Despite the differences in erythrocyte antigens reflected in the blood group systems, blood is the most easily shared of human tissues,

TABLE 16.4 Blood Types in the United States

	Blood Type (percentage)				
Population Group	O	A	B	AB	Rh^+
European-American	45	40	11	4	85
African-American	49	27	20	4	95
Korean-American	32	28	30	10	100
Japanese-American	31	38	21	10	100
Chinese-American	42	27	25	6	100
Native American	79	16	4	1	100

FIGURE 16.12 Antigens and antibodies of the ABO blood types.

The antibodies in your plasma do not react with the antigens on your erythrocytes.

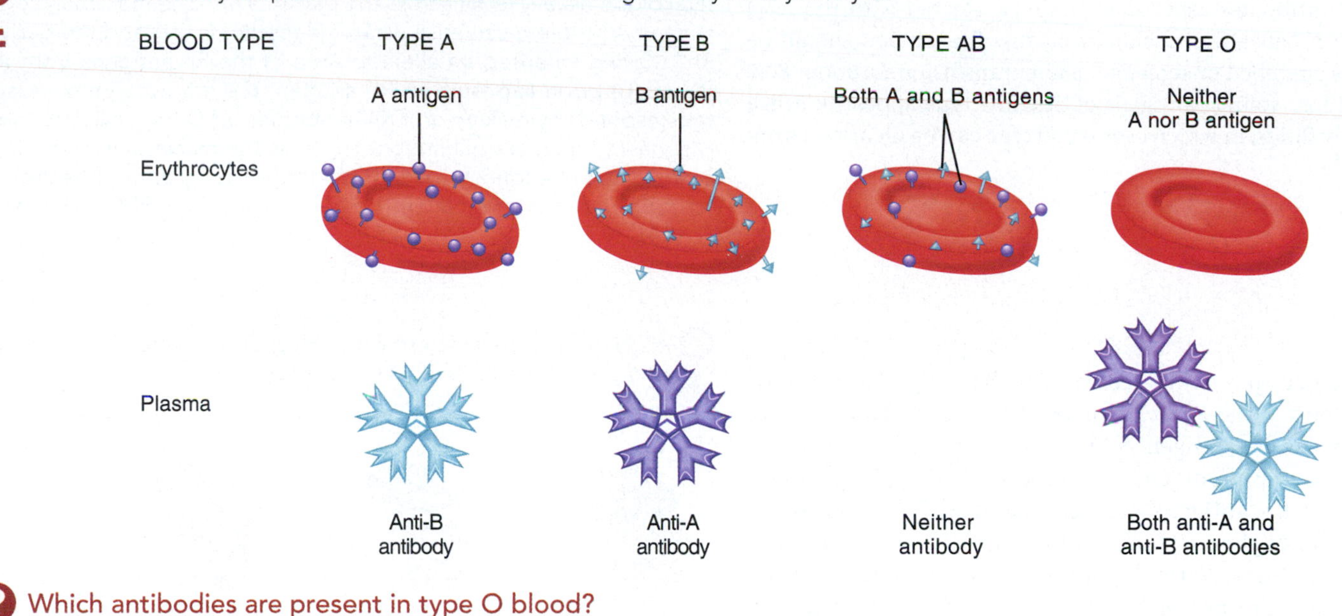

Which antibodies are present in type O blood?

saving many thousands of lives every year through transfusions. A **transfusion** is the transfer of whole blood or blood components (erythrocytes only or plasma only) into the bloodstream or directly into the red bone marrow. A transfusion is most often given to alleviate anemia, to increase blood volume (for example, after a severe hemorrhage), or to improve immunity. However, the normal components of one person's erythrocyte plasma membrane can trigger damaging antigen–antibody responses in a transfusion recipient. In an incompatible blood transfusion, antibodies in the recipient's plasma bind to the antigens on the donated erythrocytes, which causes **agglutination** (a-gloo-ti-NĀ-shun), or clumping, of the erythrocytes. Agglutination is an antigen–antibody response in which erythrocytes become cross-linked to one another. (Note that agglutination is not the same as blood clotting.) When these antigen–antibody complexes form, they activate plasma proteins of the complement family (described in Section 17.2). In essence, complement molecules make the plasma membrane of the donated erythrocytes leaky, causing **hemolysis** (rupture) of the erythrocytes and the release of hemoglobin into the plasma. The liberated hemoglobin may cause kidney damage by clogging the filtration membranes.

Consider what happens if a person with type A blood receives a transfusion of type B blood. The recipient's blood (type A) contains A antigens on the erythrocytes and anti-B antibodies in the plasma. The donor's blood (type B) contains B antigens and anti-A antibodies. In this situation, two things can happen. First, the anti-B antibodies in the recipient's plasma can bind to the B antigens on the donor's erythrocytes, causing agglutination and hemolysis of the erythrocytes. Second, the anti-A antibodies in the donor's plasma can bind to the A antigens on the recipient's erythrocytes, a less serious reaction because the donor's anti-A antibodies become so diluted in the recipient's plasma that they do not cause significant agglutination and hemolysis of the recipient's erythrocytes.

TABLE 16.5 summarizes the interactions of the four blood types of the ABO system.

People with type AB blood do not have anti-A or anti-B antibodies in their blood plasma. They are sometimes called *universal recipients* because theoretically they can receive blood from donors of all four blood types. They have no antibodies to attack antigens on donated erythrocytes (TABLE 16.5). People with type O blood have neither A nor B antigens on their erythrocytes and are sometimes called *universal donors* because theoretically they can donate blood to all four ABO blood types. Type O persons requiring blood may receive only type O

TABLE 16.5 Summary of ABO Blood Group Interactions

	Blood Type			
Characteristic	**A**	**B**	**AB**	**O**
Agglutinogen (antigen) on erythrocytes	A	B	Both A and B	Neither A nor B
Agglutinin (antibody) in plasma	Anti-B	Anti-A	Neither anti-A nor anti-B	Both anti-A and anti-B
Compatible donor blood types (no hemolysis)	A, O	B, O	A, B, AB, O	O
Incompatible donor blood types (hemolysis)	B, AB	A, AB	—	A, B, AB

blood (TABLE 16.5). In practice, use of the terms *universal recipient* and *universal donor* is misleading and dangerous. Blood contains antigens and antibodies other than those associated with the ABO system that can cause transfusion problems. Thus, blood should be carefully cross-matched or screened before transfusion. In about 80% of the population, soluble antigens of the ABO type appear in saliva and other body fluids, in which case blood type can be identified from a saliva sample.

The Rh Blood Group Is Based on the Presence or Absence of Rh Antigens

The **Rh blood group** is so-named because the antigen was discovered in the blood of the *Rhesus* monkey. The alleles of three genes may code for the Rh antigen, which is a protein. People whose erythrocytes have Rh antigens are designated Rh^+ (Rh positive); those who lack Rh antigens are designated Rh^- (Rh negative). TABLE 16.4 shows the incidence of Rh^+ (and, by omission, Rh^-) in various populations. Normally, plasma does not contain anti-Rh antibodies. If an Rh^- person receives an Rh^+ blood transfusion, however, the immune system starts to make anti-Rh antibodies that remain in the blood. If a second transfusion of Rh^+ blood is given later, the previously formed anti-Rh antibodies would cause agglutination and hemolysis of the erythrocytes in the donated blood, and a severe reaction may occur.

The most common problem with Rh incompatibility, **hemolytic disease of the newborn (HDN)**, may arise during pregnancy (FIGURE 16.13). Normally, no direct contact occurs between maternal and fetal blood while a woman is pregnant. However, if a small amount of Rh^+ blood leaks from the fetus through the placenta into the bloodstream of an Rh^- mother, the mother starts to make anti-Rh antibodies. Because the greatest possibility of fetal blood leakage into the maternal circulation occurs at delivery, the firstborn baby usually is not affected. If the mother becomes pregnant again, however, her anti-Rh antibodies can cross the placenta and enter the bloodstream of the fetus. If the fetus is Rh^-, there is no problem because Rh^- blood does not have the Rh antigen. If the fetus is Rh^+, however, agglutination and hemolysis brought on by fetal–maternal incompatibility may occur in the fetal blood. An injection of anti-Rh antibodies called anti-Rh gamma globulin (RhoGAM®) can be given to prevent HDN. Rh^- women should receive RhoGAM® before delivery and soon after every delivery, miscarriage, or abortion. These antibodies bind to and inactivate the fetal Rh antigens before the mother's immune system can respond to the foreign antigens by producing her own anti-Rh antibodies.

FIGURE 16.13 Development of hemolytic disease of the newborn (HDN). (a) At birth, a small quantity of fetal blood usually leaks across the placenta into the maternal bloodstream. A problem can arise when the mother is Rh^- and the baby is Rh^+, having inherited an allele for one of the Rh antigens from the father. (b) Upon exposure to Rh antigen, the mother's immune system responds by making anti-Rh antibodies. (c) During a subsequent pregnancy, the maternal antibodies cross the placenta into the fetal blood. If the second fetus is Rh^+, the ensuing antigen–antibody reaction causes agglutination and hemolysis of fetal RBCs. The result is HDN.

HDN occurs when maternal anti-Rh antibodies cross the placenta and cause hemolysis of fetal erythrocytes.

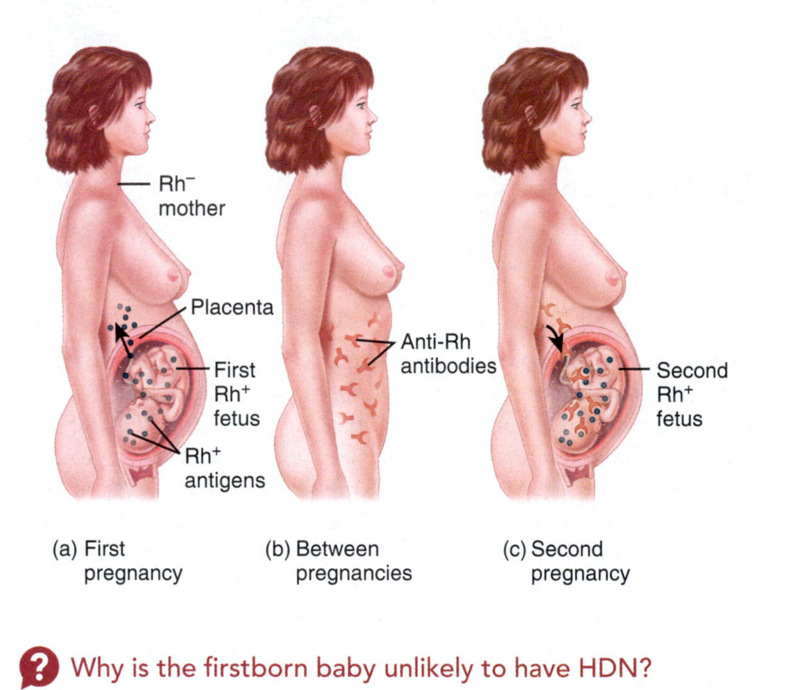

Why is the firstborn baby unlikely to have HDN?

TOOL OF THE TRADE

TYPING AND CROSS-MATCHING BLOOD FOR TRANSFUSION

To avoid blood-type mismatches, laboratory technicians type the patient's blood and then either cross-match it to potential donor blood or screen it for the presence of antibodies. In the procedure for ABO blood typing, single drops of blood are mixed with different *antisera*, solutions that contain antibodies (see FIGURE). One drop of blood is mixed with anti-A serum, which contains anti-A antibodies that agglutinate erythrocytes that possess A antigens. Another drop is mixed with anti-B serum, which contains anti-B antibodies that agglutinate erythrocytes that possess B antigens. If the erythrocytes agglutinate only when mixed with anti-A serum, the blood is type A. If the erythrocytes agglutinate only when mixed with anti-B serum, the blood is type B. The blood is type AB if both drops agglutinate; if neither drop agglutinates, the blood is type O.

In the procedure for determining Rh factor, a drop of blood is mixed with antiserum containing antibodies that agglutinate erythrocytes displaying Rh antigens. If the blood agglutinates, it is Rh^+; no agglutination indicates Rh^-.

Once the patient's blood type is known, donor blood of the same ABO and Rh type is selected. In a **cross-match**, the possible donor erythrocytes are mixed with the recipient's serum. If agglutination does not occur, the recipient does not have antibodies that will attack the donor erythrocytes. Alternatively, the recipient's serum can be **screened** against a test panel of erythrocytes having antigens known to cause blood transfusion reactions to detect any antibodies that may be present.

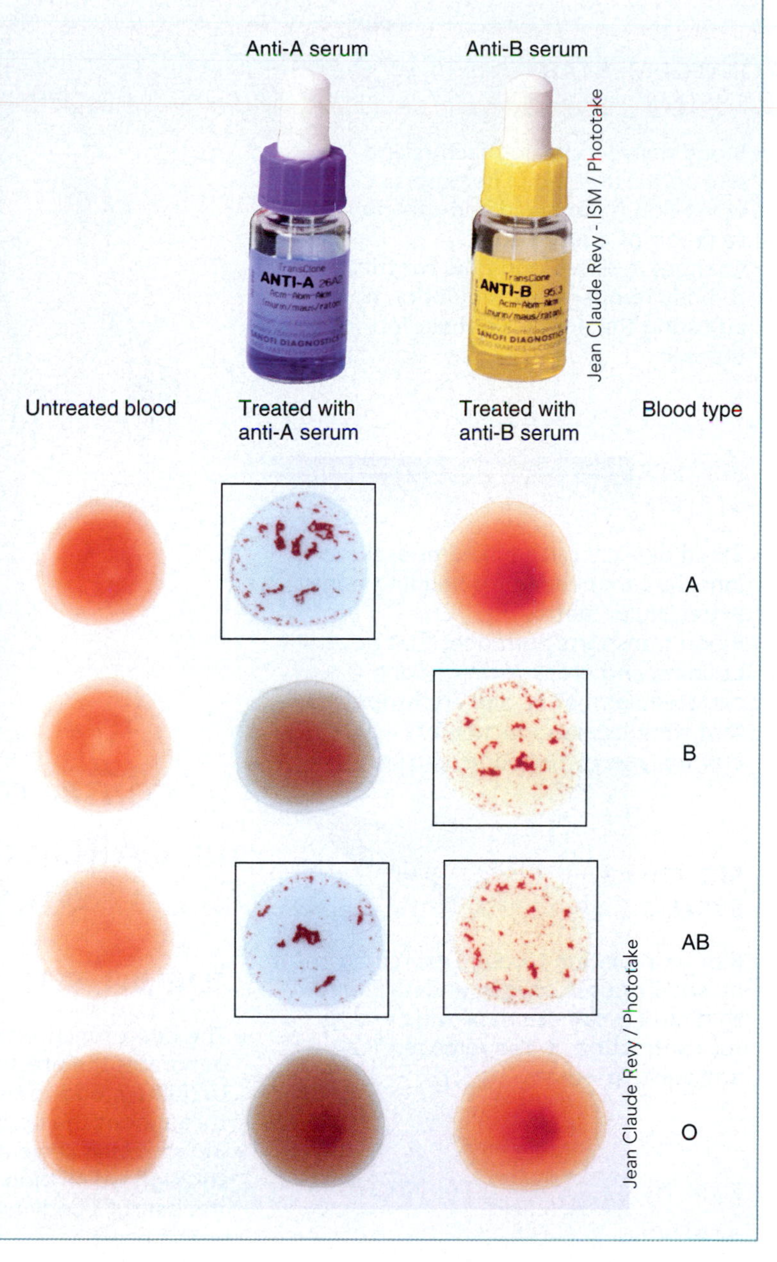

CHECKPOINT

15. What precautions must be taken before giving a blood transfusion?

16. What is hemolysis, and how can it occur after a mismatched blood transfusion?

17. Explain the conditions that may cause hemolytic disease of the newborn.

To appreciate the many ways that the heart, blood vessels, and blood contribute to homeostasis of other body systems, examine *Focus on Homeostasis: Contributions of the Cardiovascular System.*

FOCUS on HOMEOSTASIS

INTEGUMENTARY SYSTEM

- Blood delivers clotting factors and leukocytes that aid in hemostasis when skin is damaged and contribute to repair of injured skin
- Changes in skin blood flow contribute to body temperature regulation by adjusting the amount of heat loss via the skin

SKELETAL SYSTEM

- Blood delivers calcium and phosphate ions that are needed for building bone extracellular matrix
- Blood transports hormones that govern building and breakdown of bone extracellular matrix, and erythropoietin that stimulates production of erythrocytes by red bone marrow

MUSCULAR SYSTEM

- Blood circulating through exercising muscle brings in oxygen and nutrients that muscle can use to provide energy for contraction; it also removes heat and waste products

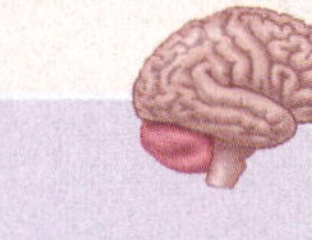

NERVOUS SYSTEM

- Endothelial cells lining choroid plexuses in brain ventricles help produce cerebrospinal fluid (CSF) and contribute to the blood–brain barrier

ENDOCRINE SYSTEM

- Circulating blood delivers most hormones to their target tissues
- Atrial cells of the heart secrete atrial natriuretic peptide

CONTRIBUTIONS OF THE CARDIOVASCULAR SYSTEM

FOR ALL BODY SYSTEMS

- The heart functions as a pump that generates the pressure needed to circulate blood to tissues throughout the body
- Blood vessels serve as the conduits through which blood travels from the heart to body tissues and back to the heart
- Blood delivers oxygen and nutrients to cells and removes carbon dioxide and other wastes from cells; it also regulates pH and body temperature and protects the body from disease

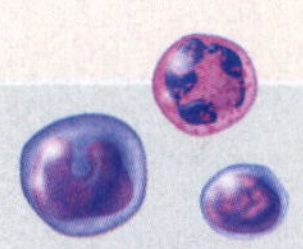

IMMUNE SYSTEM

- Circulating blood distributes lymphocytes, antibodies, and macrophages that carry out immune functions

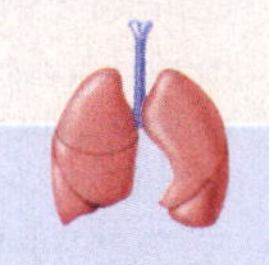

RESPIRATORY SYSTEM

- Circulating blood transports inhaled oxygen from the lungs to body cells and carbon dioxide from body cells to the lungs for exhalation

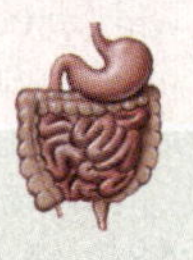

DIGESTIVE SYSTEM

- Blood carries newly absorbed nutrients and water to the liver
- Blood distributes hormones that aid digestion

URINARY SYSTEM

- Heart and blood vessels deliver 20% of the resting cardiac output to the kidneys, where blood is filtered, needed substances are reabsorbed, and unneeded substances remain as part of urine, which is excreted

REPRODUCTIVE SYSTEMS

- Vasodilation of arterioles in penis and clitoris causes erection during sexual intercourse
- Blood distributes hormones that regulate reproductive functions

FROM RESEARCH TO REALITY

Cord Blood and Megakaryocytes

Reference

Bornstein, R. et al. (2001). Cord blood megakaryocytes do not complete maturation, as indicated by impaired establishment of endomitosis and low expression of G1/S cyclins upon thrombopoietin-induced differentiation. *British Journal of Haematology*. 114: 458–465.

Do the benefits of using cord blood outweigh the risks?

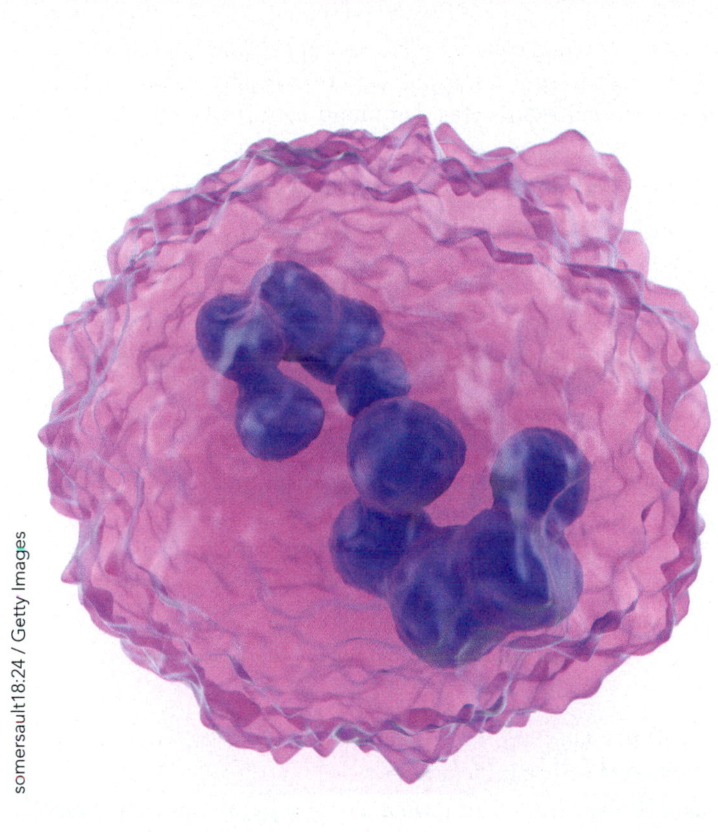

somersault18:24 / Getty Images

After a baby is born, the blood remaining in the umbilical cord is usually discarded. This cord blood is an excellent source of hematopoietic stem cells that can be used to treat a variety of diseases, including leukemia, lymphoma, and anemia. In addition, cord blood has a lower likelihood of rejection by the host and a lower transmission of infectious diseases than cells sourced from an adult. You may have seen advertisements on TV touting the benefits of banking your child's cord blood. A number of cord-blood banking services are ready to help you should you choose to save your child's cord blood. Unfortunately, cord blood doesn't always engraft as well as cells sourced from an adult. This means that, for cord blood, it takes longer for platelets to populate the recipient's blood, resulting in a greater chance of fatal bleeding. Why doesn't cord blood engraft as well as cells sourced from an adult?

Article description:

During thrombopoiesis, megakaryocytes turn into platelets under the influence of the hormone thrombopoietin. There are different developmental stages of a megakaryocyte; a mature megakaryocyte is polyploid, having a DNA content up to 128N. The high DNA content occurs because of cycles of endomitosis, in which cells repeatedly enter the synthesis (S) and mitotic (M) phases of the cell cycle but do not complete nuclear division. Several protein cyclins that play a role in the transition between the gap 1 (G1), S, gap 2 (G2), and M phases of the cell cycle are also involved in promoting endomitosis. In this study, researchers wanted to find out why cord blood megakaryocytes do not engraft as well as adult peripheral blood megakaryocytes. They hypothesized that the cord blood megakaryocytes do not reach their full maturation potential. To test their hypothesis, investigators examined the ploidy level attained in cord blood megakaryocytes and also analyzed these cells for the presence of cyclins that usually promote endomitosis.

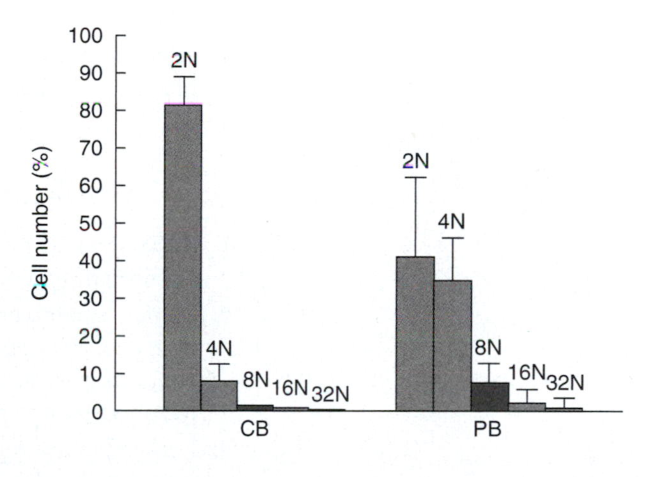

Go to WileyPLUS Learning Space and use the data from this article to answer the questions posed there and to learn more about cord blood and megakaryocytes.

CHAPTER REVIEW

16.1 Overview of Blood

1. Blood has several functions: (1) It transports oxygen, carbon dioxide, nutrients, hormones, and nitrogenous wastes; (2) it helps regulate pH, body temperature, and movement of fluid across capillary walls; and (3) it provides protection through clotting and by combating microbes and other foreign substances.
2. Blood has two main components: plasma (liquid portion) and cellular elements (cells and cell fragments. Blood is about 55% plasma and 45% cellular elements.
3. The hematocrit is the percentage of total blood volume occupied by erythrocytes.
4. Plasma consists of water, proteins, and solutes other than proteins. The plasma proteins include albumins, globulins, and fibrinogen.
5. The cellular elements of blood include erythrocytes (red blood cells), leukocytes (white blood cells), and platelets.
6. Hematopoiesis is the formation of blood cells from pluripotent hematopoietic stem cells in red bone marrow. During this process, myeloid stem cells give rise to erythrocytes, platelets, granulocytes, and monocytes; lymphoid stem cells give rise to lymphocytes.
7. Several hematopoietic growth factors stimulate differentiation and proliferation of the various blood cells.

16.2 Erythrocytes

1. Mature erythrocytes are biconcave discs that lack nuclei and contain hemoglobin.
2. The function of the hemoglobin in erythrocytes is to transport most of the oxygen and part of the carbon dioxide in blood.
3. In response to hypoxia, the kidneys release erythropoietin, which promotes erythropoiesis.
4. Erythrocytes live about 120 days. After phagocytosis of aged erythrocytes by macrophages, hemoglobin is recycled.

16.3 Leukocytes

1. Leukocytes are nucleated cells that lack hemoglobin; they are colorless under the light microscope unless they are stained with specific dyes.
2. The five major types of leukocytes are divided into two major groups (granulocytes and agranulocytes) based on the presence or absence of prominent cytoplasmic granules. Granulocytes include neutrophils, eosinophils, and basophils; agranulocytes include lymphocytes and monocytes.
3. Leukocytes are key components of the immune system. The general function of leukocytes is to combat invading pathogens and abnormal cells. Each type of leukocyte has a particular role in immunity.
4. Except for lymphocytes, which may live for years, leukocytes usually live for only a few days.

16.4 Platelets

1. Platelets are disc-shaped cell fragments that splinter from megakaryocytes.
2. Platelets help stop blood loss from damaged blood vessels by forming a platelet plug and releasing chemicals that promote blood clotting.

CHAPTER REVIEW

16.5 Hemostasis

1. Hemostasis refers to the stoppage of bleeding; it involves vascular spasm, platelet plug formation, and blood clotting (coagulation).
2. In vascular spasm, the smooth muscle of a blood vessel wall contracts, which slows blood loss.
3. Platelet plug formation involves the aggregation of platelets to stop bleeding.
4. A clot is a network of insoluble protein fibers (fibrin) in which cellular elements of blood are trapped.
5. The chemicals involved in clotting are known as clotting (coagulation) factors.
6. Blood clotting involves a cascade of reactions that may be divided into three stages: formation of prothrombinase, conversion of prothrombin into thrombin, and conversion of soluble fibrinogen into insoluble fibrin.
7. Clotting is initiated by the interplay of the extrinsic and intrinsic pathways of blood clotting.
8. Normal coagulation requires vitamin K and is followed by clot retraction (tightening of the clot) and ultimately fibrinolysis (dissolution of the clot).
9. Clotting in an unbroken blood vessel is called thrombosis. A thrombus that moves from its site of origin is called an embolus.

16.6 Blood Groups and Blood Types

1. ABO and Rh blood groups are genetically determined and based on antigen–antibody responses.
2. In the ABO blood group, the presence or absence of A and B antigens on the surface of erythrocytes determines blood type.
3. In the Rh system, individuals whose erythrocytes have Rh antigens are classified as Rh^+; those who lack the antigen are Rh^-.
4. Hemolytic disease of the newborn (HDN) can occur when an Rh^- mother is pregnant with an Rh^+ fetus.

PONDER THIS

1. Jessica wants to increase her oxygen-carrying capacity. She decides to take an exogenous form of erythropoietin. How will this affect her hematocrit? Explain your answer.

2. When plaques form in a blood vessel, they can be categorized as either stable or unstable. This is due to the fact that one type (stable) has a fibrous cap and is least likely to rupture while the other type (unstable) is weaker and more susceptible to rupture. In terms of blood and hemostasis, which type is more dangerous? Explain your answer.

3. You are designing a new medication that you hope will be a new type of treatment for high cholesterol. After a few clinical trials, you note that people on this drug have very low erythrocyte counts. Upon further analysis, you discover that they all have normal levels of erythropoietin, iron, transferrin, and ferritin. Additionally, there has been no damage to the bone marrow. What is this drug interfering with, and why is it causing low erythrocyte counts?

4. Jim and Nancy have decided to start a family. Nancy's first pregnancy went well and they were able to have a healthy baby girl. Both have decided to expand their family and Nancy is once again pregnant. This pregnancy, however, is not going as well. The doctor has told Nancy that her baby is in danger of developing hemolytic disease of the newborn (HDN). What is this condition? Why has this happened during Nancy's second pregnancy and not her first?

ANSWERS TO FIGURE QUESTIONS

16.1 Albumin is the most abundant plasma protein. It serves as the main contributor to plasma colloid osmotic pressure; it also functions as a transport protein for several steroid hormones and for fatty acids.

16.2 Platelets are the cellular elements that help clot blood.

16.3 Examples of hematopoietic growth factors include erythropoietin, thrombopoietin, and cytokines.

16.4 One hemoglobin molecule can transport a maximum of four O_2 molecules, one O_2 bound to each heme group.

16.5 Once you move to high altitude, your hematocrit increases because of increased secretion of erythropoietin.

16.6 Transferrin is a plasma protein that transports iron in the blood.

16.7 Neutrophils, eosinophils, and basophils are called granulocytes because they have prominent cytoplasmic granules that are visible through a light microscope when stained.

16.8 Along with platelet plug formation, vascular spasm and blood clotting contribute to hemostasis.

16.9 The Rho signaling pathway alters the cytoskeleton of the platelet, causing the cell to change its shape.

16.10 Serum is plasma minus the clotting proteins.

16.11 The outcome of the first stage of clotting is the formation of prothrombinase.

16.12 Type O blood contains both anti-A and anti-B antibodies.

16.13 Because the mother is most likely to start making anti-Rh antibodies after the first baby is already born, that baby suffers no damage.

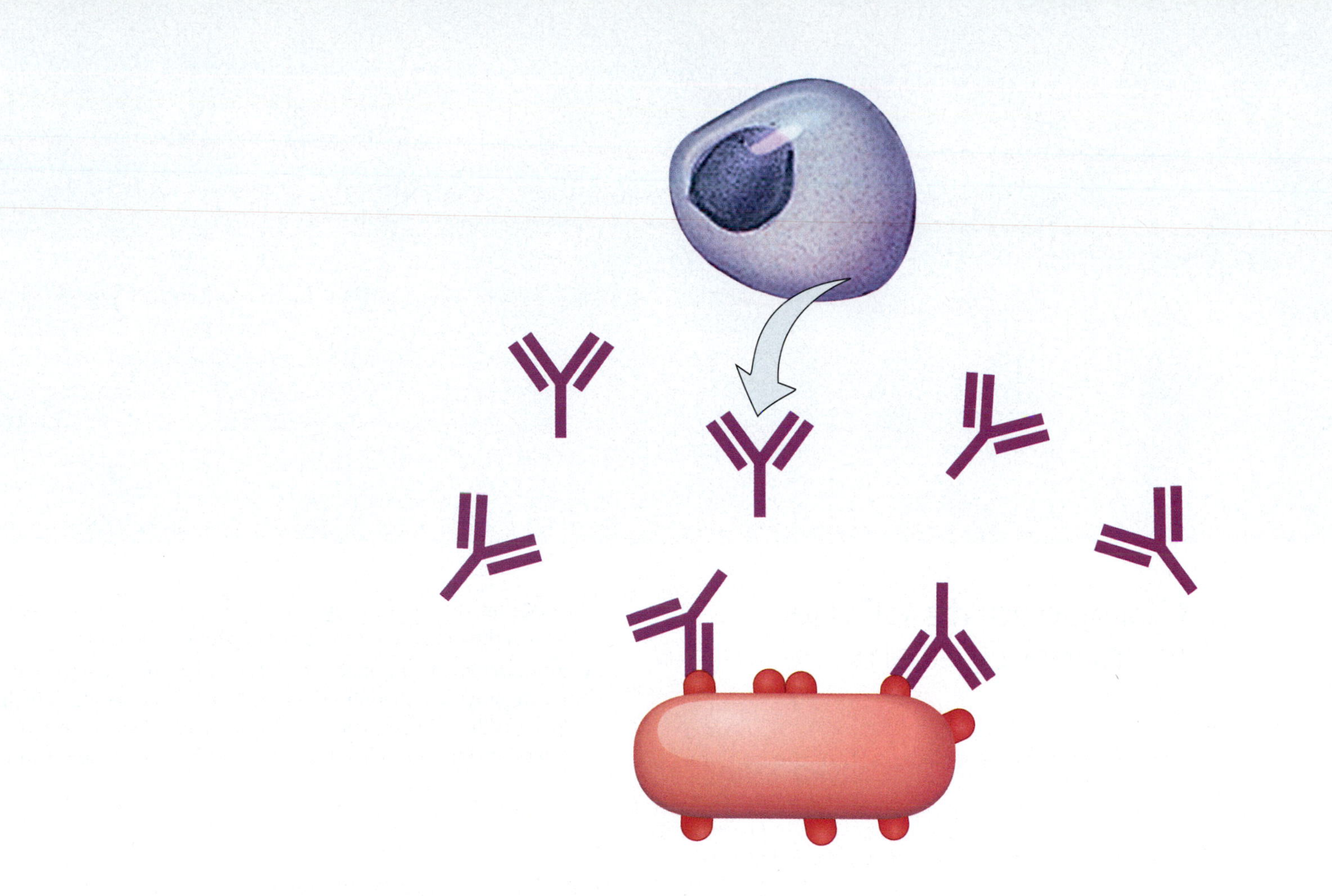

17 The Immune System

The Immune System and Homeostasis

The immune system contributes to homeostasis by providing mechanisms of defense that protect the body from microbes and other foreign substances.

LOOKING BACK TO MOVE AHEAD...

- Apoptosis refers to programmed cell death; in this process, a triggering agent from either outside or inside the cell causes "cell-suicide" genes to produce enzymes that ultimately result in the death of the cell (Section 3.6).
- Lymphatic vessels drain excess interstitial fluid from the spaces between cells and return it to the blood; after interstitial fluid enters lymphatic vessels, it is known as lymph (Section 15.2).
- Leukocytes are classified into two groups: granulocytes and agranulocytes. Granulocytes, which include neutrophils, eosinophils, and basophils, contain prominent granules (chemical-filled vesicles) that are present in the cytoplasm; agranulocytes, which include monocytes and lymphocytes, do not have prominent cytoplasmic granules (Section 16.3).
- In an incompatible blood transfusion, antibodies in the recipient's plasma bind to antigens on the donated erythrocytes, which causes agglutination (clumping) of the erythrocytes (Section 16.6).

Maintaining homeostasis in the body requires continual combat against harmful agents in our internal and external environment. Despite constant exposure to a variety of **pathogens**, disease-producing microbes such as bacteria, viruses, and fungi, most people remain healthy. The body surface also endures cuts and bumps, exposure to ultraviolet rays in sunlight, chemical toxins, and minor burns with an array of defensive ploys.

Immunity or **resistance** is the ability to ward off damage or disease through our defenses. Vulnerability or lack of resistance is termed **susceptibility**. The two general types of immunity are (1) innate and (2) adaptive. *Innate (nonspecific) immunity* refers to defenses that are present at birth. They are always present and available to provide rapid responses to protect us against disease. Innate immunity does not involve specific recognition of a microbe and acts against all microbes in the same way. In addition, innate immunity does not have a memory component; that is, it cannot recall a previous contact with a foreign molecule. *Adaptive (specific) immunity* refers to defenses that involve specific recognition of a microbe once it has breached the innate immunity defenses. Adaptive immunity is based on a specific response to a specific microbe; that is, it adapts or adjusts to handle a specific microbe. Unlike innate immunity, adaptive immunity is slower to respond, but it does have a memory component.

The body system responsible for immunity is the **immune system**. It consists of a diverse group of cells and tissues that are widely distributed throughout the body. This chapter will describe the mechanisms used by the immune system to defend the body against intruders and to promote the repair of body tissues.

17.1 Components of the Immune System

OBJECTIVE

- Describe the functions of the components of the immune system.

The immune system consists of two major components: (1) cells that provide immune responses and (2) lymphoid organs and tissues.

The Cells of the Immune System Include Leukocytes, Mast Cells, and Dendritic Cells

The main cells of the immune system are **leukocytes**, or *white blood cells*, which were first described in Chapter 16. The general function of leukocytes is to combat pathogens that have entered the tissues of the body. There are five types of leukocytes—neutrophils, eosinophils, basophils, monocytes, and lymphocytes (FIGURE 17.1a–e). Each type of leukocyte has a particular role in immunity:

1. **Neutrophils** are phagocytes, which are cells that engage in *phagocytosis*—a process by which microbes and cellular debris are engulfed and destroyed. Among leukocytes, neutrophils respond most quickly to tissue destruction by bacteria. In fact, a high neutrophil count often indicates a bacterial infection.
2. **Eosinophils** destroy parasites, such as worms, that enter body tissues. They accomplish this task by attaching to the parasite and then releasing toxic substances. Eosinophils also contribute to allergic reactions and can function as phagocytes, but their phagocytic activity is weak compared to the other types of phagocytes of the immune system.
3. **Basophils** release a variety of chemicals, including histamine, leukotrienes, and prostaglandins. These substances promote inflammation and are involved in allergic reactions.
4. **Monocytes** travel to a site of infection and then enlarge and differentiate into phagocytes called **macrophages**. Some macrophages, called **wandering macrophages**, migrate from one tissue to another. Other macrophages, called **fixed macrophages**, stand guard in specific tissues. Examples of fixed macrophages include *microglia* in the brain, *Kupffer cells* in the liver, *alveolar macrophages* in the lungs, and *histiocytes* in connective tissue.
5. **Lymphocytes** are the major soldiers in immune system battles. Three main types of lymphocytes are B cells, T cells (which include helper T cells and cytotoxic T cells), and natural killer cells. B cells differentiate into specialized cells that produce *antibodies*, proteins

FIGURE 17.1 Cells of the immune system.

The cells of the immune system include leukocytes (neutrophils, eosinophils, basophils, monocytes, and lymphocytes), mast cells, and dendritic cells.

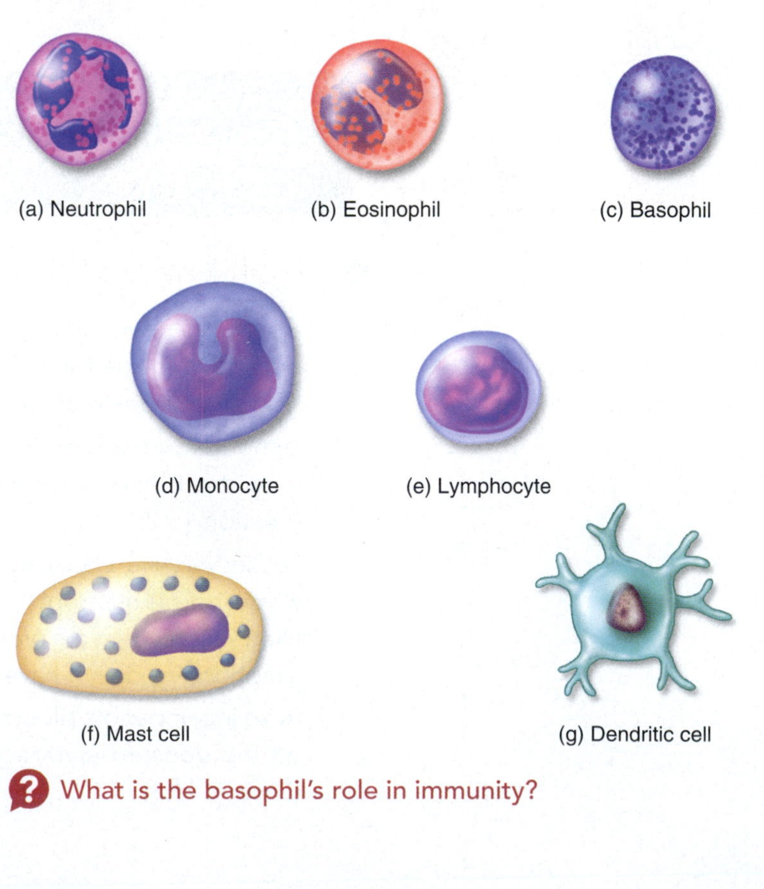

What is the basophil's role in immunity?

that disable foreign substances such as microbes. Helper T cells stimulate the proliferation of B cells and T cells, a role that helps B cells and T cells respond to foreign substances that invade the body. Cytotoxic T cells and natural killer cells destroy infected body cells and cancer cells; cytotoxic T cells can also kill foreign cells from tissue transplants. One mechanism that cytotoxic T cells and natural killer cells use to destroy a target cell is to create perforations (holes) in the target cell's membrane, which ultimately causes the target cell to lyse (burst).

To defend against pathogens that have entered body tissues, many leukocytes leave the bloodstream through blood capillary walls and collect in the interstitial spaces between cells at sites of pathogen invasion or inflammation. When neutrophils, eosinophils, basophils, and monocytes leave the bloodstream to fight injury or infection, they never return to it. Lymphocytes, on the other hand, continually recirculate—from blood to interstitial fluid to lymph and back to blood. Only 2% of the total lymphocyte population is circulating in the blood at any given time; the rest are in lymph and lymphoid organs and tissues.

In addition to leukocytes, two other types of cells are part of the immune system: mast cells and dendritic cells (FIGURE 17.1f, g). **Mast cells** are widely dispersed in the body, particularly in connective tissues of the skin and linings of the respiratory and gastrointestinal tracts. Like basophils, mast cells release histamine and other substances that are involved in inflammation and allergic reactions. **Dendritic cells**, which are named for their long, branching projections that resemble the dendrites of neurons, are located in the skin and in other organs of the body. They serve as *antigen-presenting cells*, cells that process antigens (foreign substances) and then present the processed antigens to lymphocytes to promote an immune response. As you will soon learn, macrophages and B cells can also function as antigen-presenting cells.

FIGURE 17.2 Lymphoid organs and tissues.

Lymphoid organs and tissues are structures in which lymphocytes develop, reside, or carry out immune responses.

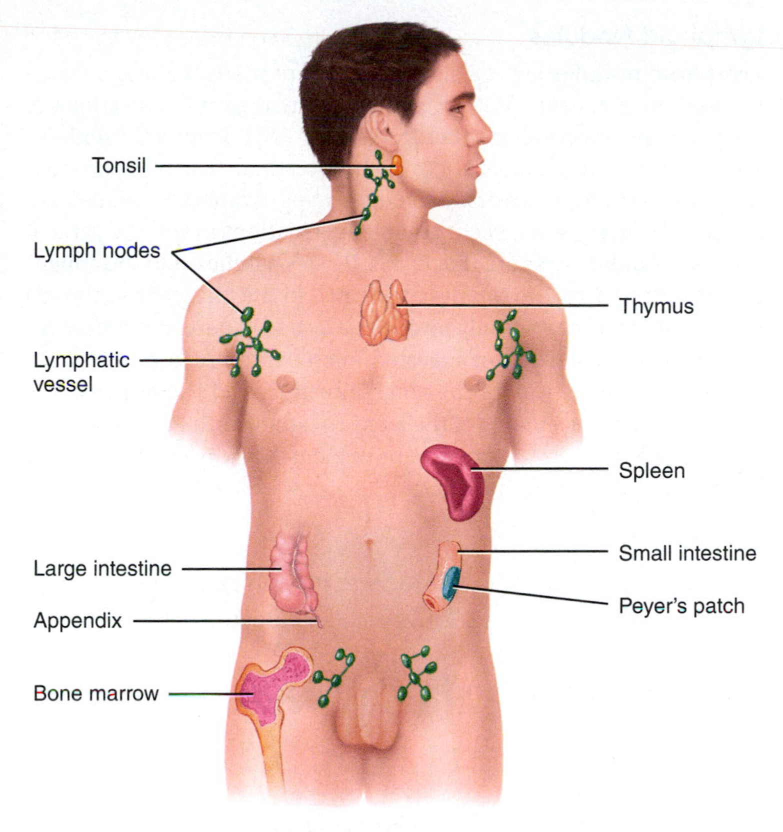

What are the functions of the spleen?

There Are Several Types of Lymphoid Organs and Tissues

Lymphoid organs and tissues are structures in which lymphocytes develop, reside, or carry out immune responses. Lymphoid organs and tissues are classified into two groups based on their functions. **Primary lymphoid organs** are the sites where stem cells divide and develop into mature B cells and T cells. The primary lymphoid organs are bone marrow and the thymus. Pluripotent stem cells in bone marrow give rise to mature B cells and immature pre-T cells. The pre-T cells in turn migrate to the thymus, where they become mature T cells. From primary lymphoid organs, mature B cells and T cells migrate to **secondary lymphoid organs and tissues**, which are the sites where most immune responses occur. Secondary lymphoid organs and tissues include the lymph nodes, spleen, and lymphoid nodules. The thymus, lymph nodes, and spleen are considered to be organs because each is surrounded by a connective tissue capsule; lymphoid nodules, by contrast, are not considered to be organs because they lack a capsule.

Thymus

The **thymus** is a bilobed organ located just above the heart (FIGURE 17.2). It consists of large numbers of T cells and scattered dendritic cells, epithelial cells, and macrophages. The thymus is the site where T cells mature. In addition, the thymus produces the hormones **thymosin** and **thymopoietin**, which promote normal T cell function in lymphoid organs and tissues. Thymic hormones may also play a role in retarding the aging process. In infants, the thymus is large, with a mass of about 70 g (2.3 oz). After puberty, fatty connective tissue begins to replace the thymic tissue. By the time a person reaches maturity, the thymus has atrophied considerably, and in old age it may weigh only 3 g (0.1 oz). Before the thymus atrophies, it populates the secondary lymphoid organs and tissues with T cells. However, some T cells continue to proliferate in the thymus throughout an individual's lifetime.

Lymph Nodes

Located along lymphatic vessels are bean-shaped **lymph nodes** (FIGURE 17.2). They are scattered throughout the body and usually occur in groups. Large groups of lymph nodes are present in the neck, armpits, and groin and near the mammary glands. Within a lymph node are lymphocytes, dendritic cells, and macrophages. Lymph nodes filter microbes and other foreign substances from lymph. As lymph flows through the node, macrophages destroy some microbes by phagocytosis, and lymphocytes destroy others by a variety of immune responses.

Spleen

The **spleen** is the largest lymphoid organ (FIGURE 17.2). It contains a variety of cells, including lymphocytes, macrophages, and dendritic cells. The spleen filters microbes and other foreign materials from blood. As blood passes through the spleen, B cells and T cells carry out immune responses, and macrophages destroy microbes by phagocytosis. The macrophages of the spleen also remove aged or defective erythrocytes from the blood.

Lymphoid Nodules

Lymphoid nodules are egg-shaped masses of tissue that are not surrounded by a capsule. Within a lymphoid nodule are collections of lymphocytes, macrophages, and dendritic cells. Lymphoid nodules are plentiful in the linings of the gastrointestinal, urinary, reproductive, and respiratory tracts. Although many lymphoid nodules are small and solitary, some occur as large aggregations in specific parts of the body. Among these are the tonsils, Peyer's patches, and the appendix (**FIGURE 17.2**). The **tonsils** are located in the pharyngeal (throat) region. They are strategically positioned to participate in immune responses against inhaled or ingested microbes. **Peyer's patches** are located within the wall of the small intestine, and the **appendix** is located near the beginning of the large intestine. Both Peyer's patches and the appendix carry out immune responses against microbes that invade the gastrointestinal tract.

CHECKPOINT

1. How do eosinophils contribute to immunity?
2. What is the difference between a monocyte and a macrophage?
3. What functions do lymph nodes, the spleen, and the tonsils serve?

17.2 Innate Immunity

OBJECTIVE

- Describe the components of innate immunity.

Innate (nonspecific) immunity is the ability of the body to defend itself against microbes and other foreign substances using mechanisms that do not involve specific recognition of the invading agents. Among the components of innate immunity are the first line of defense (the physical and chemical barriers of the body) and the second line of defense (antimicrobial substances, natural killer cells, phagocytes, inflammation, and fever).

The First Line of Defense Consists of the External Physical and Chemical Barriers of the Body

The external physical and chemical barriers of the body are the first line of defense against pathogens. They discourage microbes and other foreign substances from penetrating the body and causing disease.

The **skin** is a formidable physical barrier to the entrance of foreign substances into body tissues. The skin is composed of two parts: an outer *epidermis*, which consists of epithelial tissue and an inner *dermis*, which consists of connective tissue (**FIGURE 17.3a**). The epidermis is stratified (arranged into many rows of cells) and lacks blood vessels. Most epidermal cells are *keratinocytes*, epithelial cells that contain keratin (**FIGURE 17.3b**). *Keratin* is a tough, fibrous protein that helps protect the skin and underlying tissues from heat, microbes, and chemicals. Keratinocytes are tightly packed together, an arrangement that allows the skin to resist invasion by microbes. In addition, keratinocytes release a water-repellent sealant that decreases water entry and inhibits the entry of foreign materials. On average, the epidermis is replaced about once every four weeks. This process begins with the generation of keratinocytes from stem cells in the deepest part of the epidermis. From there, keratinocytes are pushed upward through the epidermis, eventually die, and then are shed from the surface of the

FIGURE 17.3 The skin.

The skin serves as a physical barrier to the entrance of microbes into body tissues.

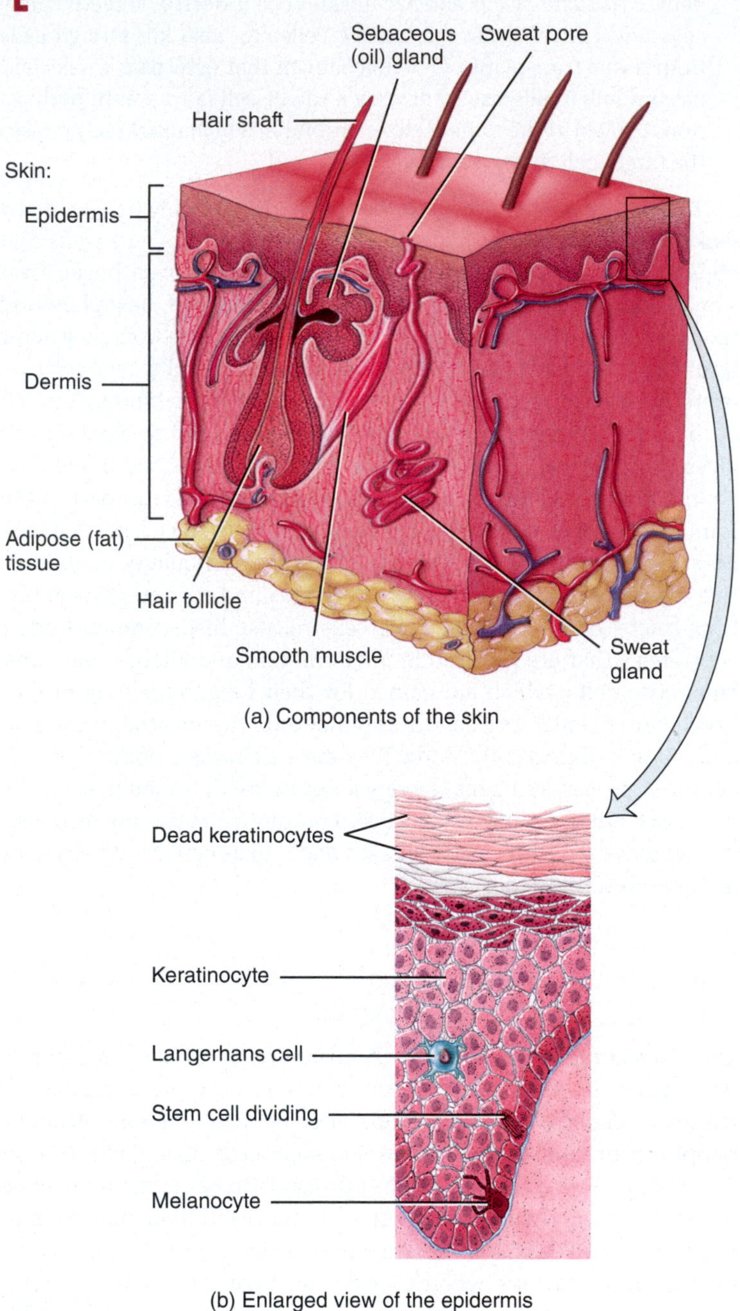

(a) Components of the skin

(b) Enlarged view of the epidermis

What are Langerhans cells?

epidermis. Therefore, the outer part of the epidermis consists of dead keratinocytes. The periodic shedding of epidermal cells helps remove microbes from the skin surface. Besides keratinocytes, two other cell types in the epidermis have protective roles: Langerhans cells and melanocytes (FIGURE 17.3b). *Langerhans cells* are a type of dendritic cell; they alert the immune system to the presence of microbes that invade the skin. *Melanocytes* produce the dark pigment *melanin*, which contributes to skin color and protects the skin from the damaging effects of ultraviolet light. Although the epidermis contributes the most to the skin's ability to serve as a physical barrier to microbes, the dermis plays a role as well. For example, macrophages in the dermis phagocytize bacteria and viruses that manage to migrate through the epidermis into the dermis.

Other physical barriers to the entrance of microbes into body tissues include mucus, hairs, and cilia. The linings of the respiratory, gastrointestinal, reproductive, and urinary tracts secrete **mucus**, a sticky substance that traps microbes and foreign substances. The nose has mucus-coated **hairs** that trap and filter microbes, dust, and pollutants from inhaled air. In addition to mucus, the lining of the upper respiratory tract contains **cilia**, microscopic hairlike projections. The waving action of cilia propels inhaled dust and microbes that have become trapped in mucus toward the throat. Coughing and sneezing accelerate movement of mucus and its entrapped pathogens out of the body. Swallowing mucus sends pathogens to the stomach where gastric juice destroys them.

Chemical barriers to the entrance of microbes into body tissues include sebum, lysozyme, gastric juice, and vaginal secretions. Sebaceous (oil) glands of the skin (FIGURE 17.3a) secrete **sebum**, an oily substance that forms a protective film over the surface of the skin. The unsaturated fatty acids in sebum inhibit the growth of certain pathogenic bacteria and fungi. The acidity of the skin (pH 3–5) is caused in part by the secretion of fatty acids and lactic acid. **Lysozyme** is an enzyme capable of breaking down the cell walls of certain bacteria. It is present in tears, saliva, perspiration, nasal secretions, and tissue fluids. **Gastric juice**, produced by the glands of the stomach, is a mixture of hydrochloric acid, enzymes, and mucus. The strong acidity of gastric juice (pH 1.2–3.0) destroys many bacteria and most bacterial toxins. **Vaginal secretions** are also slightly acidic, which discourages bacterial growth.

The Second Line of Defense Is Comprised of Various Internal Defenses

When pathogens penetrate the external physical and chemical barriers of the body, they encounter an internal second line of defense: antimicrobial substances, natural killer cells, phagocytes, inflammation, and fever.

Antimicrobial Substances

Four main types of **antimicrobial substances** discourage microbial growth: interferons, complement, iron-binding proteins, and antimicrobial proteins.

- Cells infected with viruses produce proteins called **interferons (IFNs)** (in′-ter-FĒR-ons). Once released by virus-infected cells, IFNs diffuse to uninfected neighboring cells, where they induce synthesis of antiviral proteins that interfere with viral replication (FIGURE 17.4). Although IFNs do not prevent viruses from attaching to and penetrating the neighboring cells, they do stop replication. Viruses can cause disease only if they can replicate within body cells. IFNs also activate natural killer cells and cytotoxic T cells, which kill infected body cells and cancer cells. In addition, IFNs inhibit cell division and suppress the formation of tumors. The three types of interferon are alpha-, beta-, and gamma-IFN.

FIGURE 17.4 Interferons.

A virus-infected cell produces interferons, which diffuse to uninfected neighboring cells and induce synthesis of antiviral proteins that interfere with viral replication.

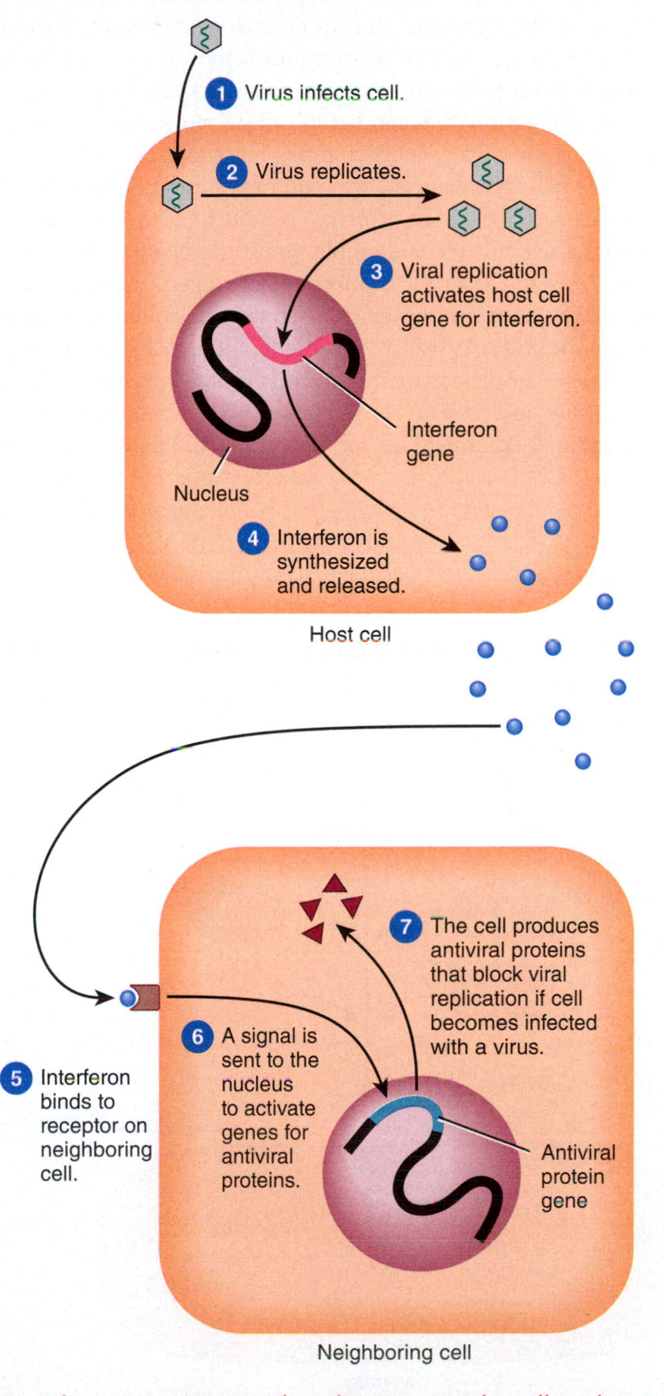

? Besides preventing viral replication inside cells, what are some other functions of interferons?

- The **complement system** consists of about 30 normally inactive proteins that are produced by the liver and circulate in blood plasma until needed to fight an infection. Most complement proteins are designated by an uppercase letter *C* and a number (for example, complement protein C1). In some cases, the name of a complement protein is also followed by a lowercase letter *a* or *b* (for example, complement protein C3a). Complement proteins act in a cascade—one reaction triggers another reaction, which in turn triggers another reaction, and so on. With each succeeding reaction, more and more product is formed so that the net effect is amplified many times. The complement system can be activated in two main ways: (1) The **classical complement pathway** starts when antibodies bind to antigens (foreign substances). The antigen–antibody complex then activates the complement system. The classical pathway is a type of adaptive immune response. (2) The **alternative complement pathway** does not involve antibodies. It is initiated by an interaction between carbohydrate molecules on the surface of a microbe and certain complement proteins. The alternative complement pathway is a type of innate immune response. Once activated, complement proteins "complement" or enhance certain immune reactions. For example, some complement proteins (C5b, C6, C7, C8, and several C9 molecules) join together and form a **membrane attack complex**, which inserts into the plasma membrane of a microbe (FIGURE 17.5). The membrane attack complex creates a channel in the plasma membrane that results in **cytolysis** (bursting) of the microbial cell due to the inflow of extracellular fluid through the channel. Another complement protein (C3b) functions as an opsonin (described shortly) that coats the surface of the microbe and enhances phagocytosis. Still, other complement proteins (C3a and C5a) bind to mast cells and cause them to release histamine, which increases blood vessel permeability during inflammation.
- **Iron-binding proteins** inhibit the growth of certain bacteria by reducing the amount of available iron. Examples include *transferrin* (found in blood and tissue fluids), *lactoferrin* (found in milk, saliva, and mucus), *ferretin* (found in the liver, spleen, and bone marrow), and *hemoglobin* (found in erythrocytes).
- **Antimicrobial proteins (AMPs)** are short peptides that have a broad spectrum of antimicrobial activity. Examples of AMPs are *dermicidin* (produced by sweat glands), *defensins* and *cathelicidins* (produced by neutrophils, macrophages, and epithelia), and *thrombocidin* (produced by platelets). Besides killing a wide range of microbes, AMPs can attract dendritic cells and mast cells, which participate in immune responses.

Natural Killer Cells

About 5–10% of lymphocytes in the blood are **natural killer (NK) cells**. They are also present in the spleen, lymph nodes, and bone marrow. NK cells nonspecifically kill infected body cells and cancer cells. The binding of NK cells to a target cell, such as a body cell infected with a virus, causes the release of toxic substances from granules present in the NK cells (FIGURE 17.6). Some granules release **perforin**, a protein that inserts into the plasma membrane of the target cell and creates channels (perforations) in the membrane. As a result, extracellular fluid flows into the target cell and cytolysis (cell bursting) occurs. Other granules of NK cells release **granzymes**, protein-digesting enzymes that enter the target cell through the perforin channels and then induce the target cell to undergo apoptosis (programmed cell death). This type of attack kills infected cells but not the microbes inside the cells; the released microbes, which may or may not be intact, can be engulfed and destroyed by phagocytes.

FIGURE 17.5 Formation of a membrane attack complex.

Complement proteins (C5b, C6, C7, C8, and several molecules of C9) join together to form a membrane attack complex that inserts into the plasma membrane of a microbe.

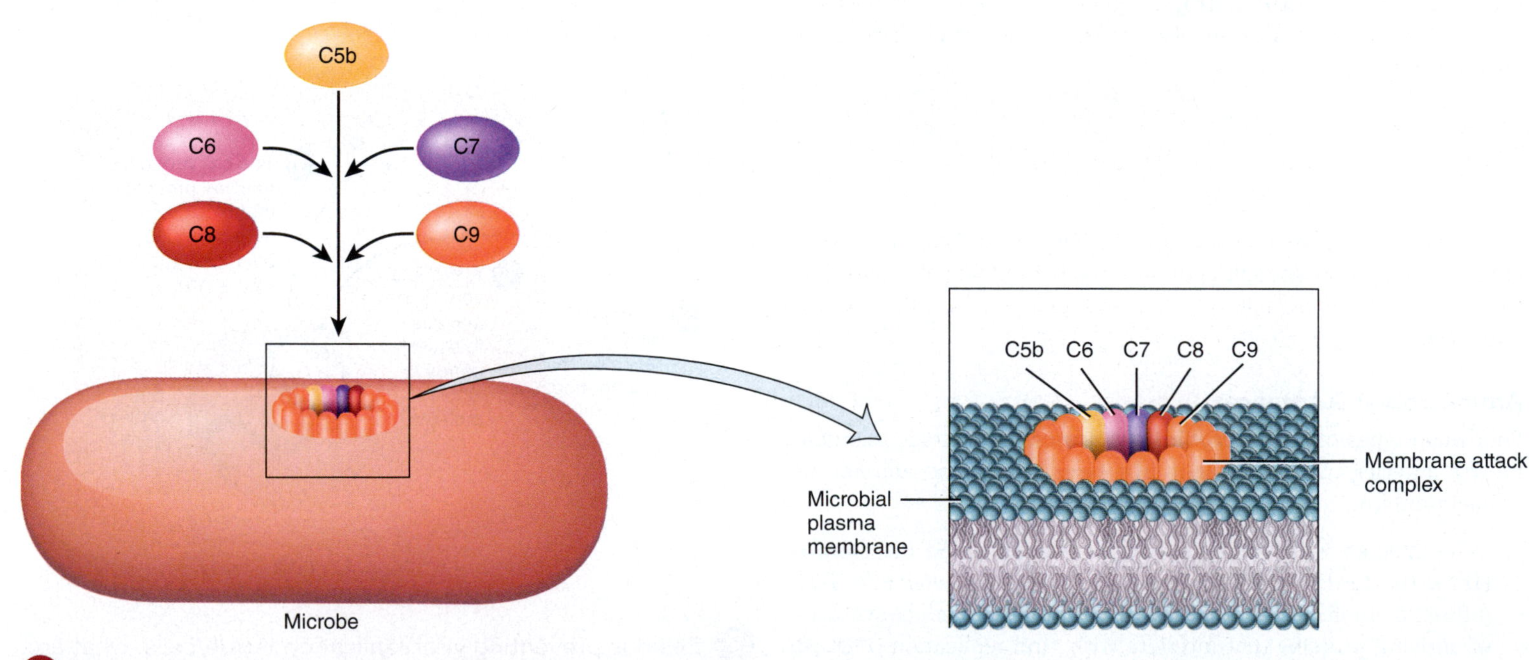

What is the purpose of the membrane attack complex?

FIGURE 17.6 Natural killer cell function.

Natural killer cells kill infected body cells and cancer cells by releasing perforin molecules and granzymes.

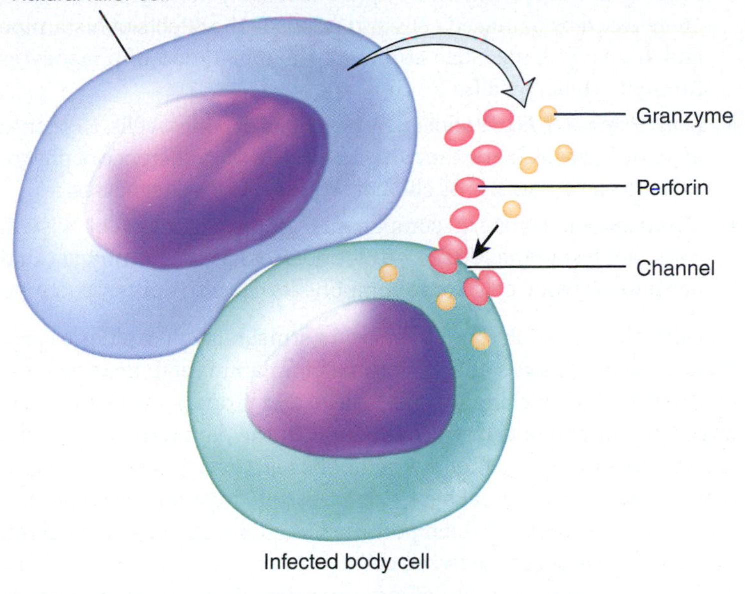

What is perforin and how does it destroy a target cell?

Phagocytes

Phagocytes are specialized cells that perform **phagocytosis**, a form of endocytosis whereby large solid particles (such as microbes and cellular debris) are engulfed and destroyed. The two major types of phagocytes are **neutrophils** and **macrophages**. When an infection occurs, neutrophils and monocytes migrate to the infected area. During this migration, the monocytes enlarge and develop into macrophages. True to their name, macrophages (*macro-* = large) are much more potent phagocytes than neutrophils. A single macrophage can engulf up to 100 bacteria during its life span, whereas a single neutrophil can engulf up to only about 20 bacteria before it dies. In addition to being an innate defense mechanism, phagocytosis plays a vital role in adaptive immunity, as discussed later in the chapter.

Phagocytosis occurs in four phases: adherence, ingestion, digestion, and killing (FIGURE 17.7):

1. ***Adherence.*** Phagocytosis begins with **adherence**, the attachment of the phagocyte to a microbe. Adherence is enhanced by **opsonins**, proteins that coat the surface of a microbe and make it easier for the phagocyte to bind to it. Examples of opsonins include antibodies and the complement protein C3b. The process by which opsonins coat the surface of a microbe to facilitate the adherence phase of phagocytosis is referred to as **opsonization**.
2. ***Ingestion.*** The plasma membrane of the phagocyte extends projections, called **pseudopods**, that engulf the microbe in a process called **ingestion**. When the pseudopods meet, they fuse, surrounding the microorganism with a sac called a **phagosome**.
3. ***Digestion.*** The phagosome enters the cytoplasm and merges with lysosomes to form a single, larger structure called a **phagolysosome**. The lysosome contributes lysozyme, which breaks down microbial cell walls, and other digestive enzymes that degrade carbohydrates, proteins, lipids, and nucleic acids. The phagocyte also forms lethal oxidants, such as superoxide anion (O_2^-), hypochlorite anion (OCl^-), and hydrogen peroxide (H_2O_2), in a process called an **oxidative burst**.
4. ***Killing.*** The chemical onslaught provided by lysozyme, digestive enzymes, and oxidants within a phagolysosome quickly kills many types of microbes. Any materials that cannot be degraded further remain in structures called **residual bodies**.

FIGURE 17.7 Phagocytosis of a microbe.

The major types of phagocytes are neutrophils and macrophages.

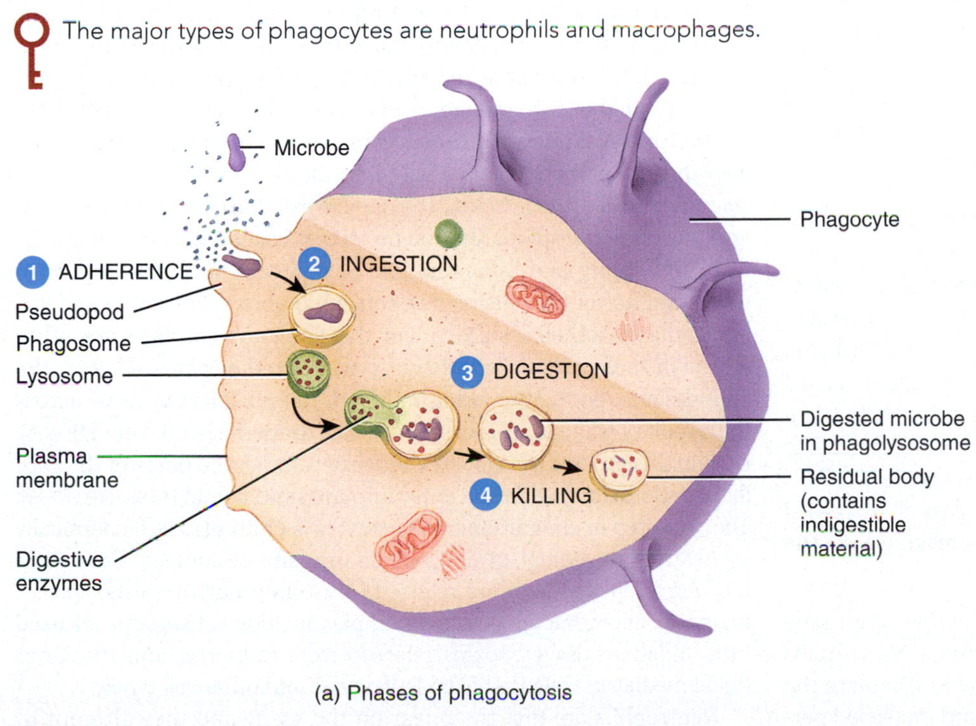

(a) Phases of phagocytosis

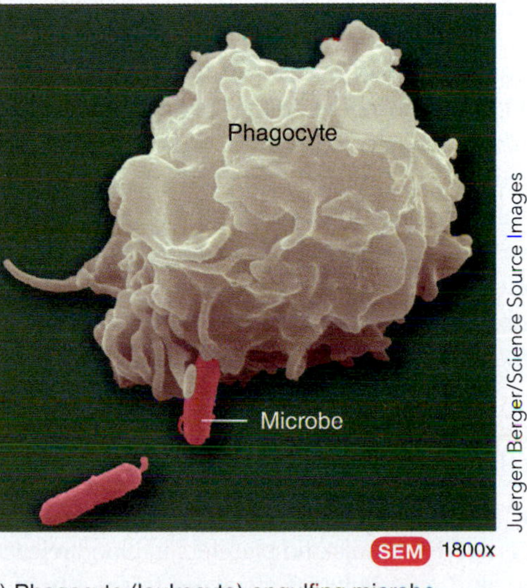

(b) Phagocyte (leukocyte) engulfing microbe

What chemicals are responsible for killing ingested microbes?

CLINICAL CONNECTION

Microbial Evasion of Phagocytosis

Some microbes, such as the bacteria that cause pneumonia, are surrounded by thick polysaccharide capsules that are difficult for phagocytes to recognize as foreign. These bacteria may avoid being captured unless their capsules are coated with opsonins to enhance the adherence phase of phagocytosis. Other microbes, such as the toxin-producing bacteria that cause one kind of food poisoning, may be ingested but not killed; instead, the toxins they produce (leukocidins) may kill the phagocytes by causing the release of the phagocyte's own lysosomal enzymes into its cytoplasm. Still other microbes—such as the bacteria that cause tuberculosis—inhibit fusion of phagosomes and lysosomes, and thus prevent exposure of the microbes to lysosomal enzymes. These bacteria apparently can also use chemicals in their cell walls to counter the effects of lethal oxidants produced by phagocytes. Subsequent multiplication of the microbes within phagosomes may eventually destroy the phagocyte.

Inflammation

Inflammation is a nonspecific, defensive response of the body to tissue damage. Among the conditions that may produce inflammation are pathogens, abrasions, chemical irritations, distortion or disturbances of cells, and extreme temperatures. The four characteristic signs and symptoms of inflammation are *redness, pain, heat,* and *swelling*. Inflammation can also cause the *loss of function* in the injured area (for example, the inability to detect sensations), depending on the site and extent of the injury. Inflammation is an attempt to dispose of microbes, toxins, or foreign material at the site of injury; to prevent their spread to other tissues; and to prepare the site for tissue repair in an attempt to restore tissue homeostasis.

Because inflammation is one of the body's nonspecific defense mechanisms, the response of a tissue to a cut is similar to the response to damage caused by burns, radiation, or bacterial or viral invasion. In each case, the inflammatory response has three basic stages: (1) vasodilation and increased permeability of blood vessels, (2) emigration (movement) of phagocytes from the blood into interstitial fluid, and ultimately (3) tissue repair.

VASODILATION AND INCREASED PERMEABILITY OF BLOOD VESSELS Two immediate changes occur in the blood vessels in a region of tissue injury: *vasodilation* (increase in the diameter) of arterioles and *increased permeability* of capillaries. Increased permeability means that substances normally retained in blood are permitted to pass from the blood vessels. Vasodilation allows more blood to flow through the damaged area, and increased permeability permits defensive proteins such as antibodies and clotting factors to enter the injured area from the blood. The increased blood flow also helps remove microbial toxins and dead cells.

Among the substances that contribute to vasodilation, increased permeability, and other aspects of the inflammatory response are the following:

- ***Histamine.*** In response to injury, mast cells in connective tissue and basophils and platelets in blood release histamine. Neutrophils and macrophages attracted to the site of injury also stimulate the release of histamine, which causes vasodilation and increased permeability of blood vessels.
- ***Kinins.*** These polypeptides, formed in blood from inactive precursors called kininogens, induce vasodilation and increased permeability and serve as chemotactic agents that attract phagocytes. An example of a kinin is bradykinin.
- ***Prostaglandins (PGs).*** These lipids, especially those of the E series, are released by damaged cells and intensify the effects of histamine and kinins. PGs may also stimulate the emigration of phagocytes through capillary walls.
- ***Leukotrienes (LTs).*** Produced by basophils and mast cells, LTs cause increased permeability; they also function in the adherence of phagocytes to pathogens and as chemotactic agents for phagocytes.
- ***Complement.*** Different components of the complement system stimulate histamine release, attract neutrophils by chemotaxis, and promote phagocytosis; some components can also destroy bacteria.

Dilation of arterioles and increased permeability of capillaries produce three of the signs and symptoms of inflammation: heat, redness (erythema), and swelling (edema). Heat and redness result from the large amount of blood that accumulates in the damaged area. As the local temperature rises slightly, metabolic reactions proceed more rapidly and release additional heat. Edema results from increased permeability of blood vessels, which permits more fluid to move from blood plasma into the spaces between cells.

Pain is a prime symptom of inflammation. It results from injury to neurons and from toxic chemicals released by microbes. Kinins affect some nerve endings, causing much of the pain associated with inflammation. Prostaglandins intensify and prolong the pain associated with inflammation. Pain may also be due to increased pressure from edema.

The increased permeability of capillaries allows leakage of blood-clotting factors into tissues. The clotting sequence is set into motion, and fibrinogen is ultimately converted to an insoluble, thick mesh of fibrin threads that localizes and traps invading microbes and blocks their spread.

EMIGRATION OF PHAGOCYTES Within an hour after the inflammatory process starts, phagocytes move from the bloodstream into interstitial fluid to gather at the site of injury. Emigration of phagocytes involves three steps: margination, diapedesis, and chemotaxis (FIGURE 17.8). During **margination**, phagocytes stick to the inner surface of the endothelium of blood capillaries. Molecules known as *adhesion molecules* help phagocytes stick to the capillary endothelium. For example, endothelial cells display adhesion molecules called *selectins* in response to nearby injury and inflammation. Selectins stick to carbohydrates on the surface of the phagocytes, causing them to slow down and roll along the endothelial surface. On the phagocyte surface are other adhesion molecules called *integrins*, which tether phagocytes to the endothelium. Soon after being tethered, the phagocytes move across the capillary wall via a process known as **diapedesis** (dī-a-pe-DĒ-sis). During this process, phagocytes squeeze through the pores of the capillary endothelium and then enter into interstitial fluid (FIGURE 17.8). The final step in emigration of phagocytes is **chemotaxis**, a chemically stimulated movement of phagocytes to a site of damage. Chemicals that attract phagocytes are referred to as **chemoattractants**, *chemotaxins*, or chemotactic agents. Examples include substances released from inflamed tissues, toxins released from microbes, and cytokines (local mediators) released from leukocytes and other cell types.

Neutrophils are first to appear on the scene and they attempt to destroy the invading microbes by phagocytosis. A steady stream of

FIGURE 17.8 Emigration of phagocytes.

Emigration of phagocytes from blood to a site of tissue injury involves three steps: margination, diapedesis, and chemotaxis.

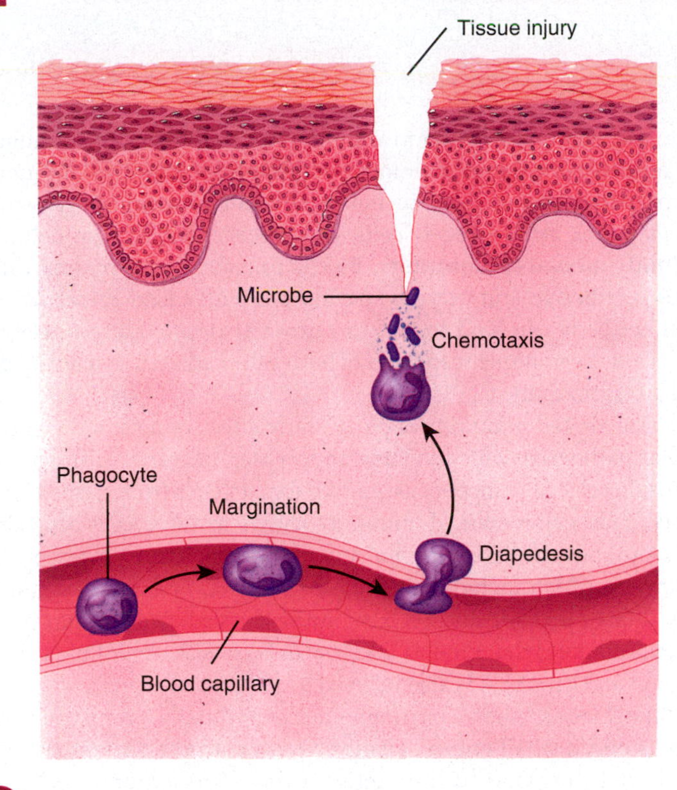

What is diapedesis?

neutrophils is ensured by the production and release of additional cells from bone marrow. Such an increase in leukocytes in the blood is termed *leukocytosis*. Although neutrophils predominate in the early stages of infection, they die off rapidly. As the inflammatory response continues, monocytes follow the neutrophils into the infected area. Once in the tissue, monocytes transform into wandering macrophages that add to the phagocytic activity of the fixed macrophages already present. They engulf invading microbes, damaged tissue, and worn-out neutrophils. Eventually, macrophages also die. Within a few days, a pocket of dead phagocytes and damaged tissue forms; this collection of dead cells and fluid is called **pus**. Pus formation occurs in most inflammatory responses and usually continues until the infection subsides. At times, pus reaches the surface of the body or drains into an internal cavity and is dispersed; on other occasions the pus remains even after the infection is terminated. In this case, the pus is gradually destroyed over a period of days and is absorbed.

Tissue Repair Once the infection has been contained and phagocytes have successfully killed the invading microbes, tissue repair of the injured area occurs. During tissue repair, worn-out, damaged, or dead cells are replaced. In the case of the skin, tissue repair may involve replacing damaged portions of the epidermis and dermis. Stem cells in the deepest part of the epidermis divide, giving rise to epithelial cells that fill in the missing tissue of the epidermis. Fibroblasts (connective tissue cells) produce scar tissue (collagen fibers and glycoproteins) that fills in the vacated tissue of the dermis. **Scar tissue** differs from normal connective tissue in that its collagen fibers are more densely arranged, it has decreased elasticity, and it has fewer blood vessels.

Fever

Fever is an abnormally high body temperature that occurs because the hypothalamic thermostat is reset. It commonly occurs during infection and inflammation. Many bacterial toxins elevate body temperature, sometimes by triggering release of fever-causing cytokines such as interleukin-1 from macrophages. Elevated body temperature intensifies the effects of interferons, inhibits the growth of some microbes, and speeds up body reactions that aid repair.

TABLE 17.1 summarizes the components of innate immunity.

TABLE 17.1 Summary of Innate Defenses

Component	Functions
FIRST LINE OF DEFENSE: PHYSICAL AND CHEMICAL BARRIERS	
Physical barriers	
Skin	Prevents microbes on body surface from entering into deeper tissues.
Mucus	Traps microbes in respiratory, gastrointestinal, reproductive, and urinary tracts.
Hairs	Filter out microbes and dust in nose.
Cilia	Together with mucus, trap and remove microbes and dust from upper respiratory tract.
Chemical barriers	
Sebum	Forms a protective acidic film over the skin surface that inhibits growth of many microbes.
Lysozyme	Antimicrobial substance in tears, perspiration, saliva, nasal secretions, and tissue fluids.
Gastric juice	Destroys bacteria and most toxins in stomach.
Vaginal secretions	Slight acidity discourages bacterial growth; flush microbes out of vagina.
SECOND LINE OF DEFENSE: INTERNAL DEFENSES	
Antimicrobial substances	
Interferons (IFNs)	Prevent viruses from replicating inside cells; activate natural killer cells and cytotoxic T cells; inhibit cell division and suppress the formation of tumors.
Complement system	Causes cytolysis of microbes, promotes phagocytosis, and contributes to inflammation.
Iron-binding proteins	Inhibit growth of certain bacteria by reducing the amount of available iron.
Antimicrobial proteins (AMPs)	Perform broad spectrum of antimicrobial activities and attract dendritic cells and mast cells.
Natural killer (NK) cells	Kill infected body cells and cancer cells.
Phagocytes	Engulf and destroy microbes and cellular debris.
Inflammation	Confines and destroys microbes and initiates tissue repair.
Fever	Intensifies effects of interferons; inhibits growth of some microbes; speeds up body reactions that aid repair.

CHECKPOINT

4. What are the external physical and chemical barriers to the entrance of microbes into body tissues?
5. What internal defenses provide protection against microbes that penetrate the physical and chemical barriers of the body?
6. What is the significance of a phagolysosome?
7. What are the main signs, symptoms, and stages of inflammation?

17.3 Adaptive Immunity

OBJECTIVES

- Explain the relationship between an antigen and an antibody.
- Compare the functions of cell-mediated immunity and antibody-mediated immunity.
- Distinguish between a primary response and a secondary response to infection.
- Describe the two major types of allergic reactions.

The ability of the body to defend itself against specific microbes and other foreign agents is called **adaptive (specific) immunity**. Substances that are recognized as foreign and provoke immune responses are called **antigens**. Two properties distinguish adaptive immunity from innate immunity: (1) *specificity* for particular foreign molecules (antigens), which also involves distinguishing self from nonself molecules, and (2) *memory* for most previously encountered antigens so that a second encounter prompts an even more rapid and vigorous response.

B Cells and T Cells Are Important Components of Adaptive Immunity

Adaptive immunity involves lymphocytes called B cells and T cells. Recall that both types of cells develop in primary lymphoid organs: B cells mature in bone marrow, whereas immature pre-T cells migrate from bone marrow to the thymus to mature. B cells and T cells are named based on the location where they become mature. In birds, B cells mature in an organ called the *bursa of Fabricius*. Although this organ is not present in humans, the term *B cell* is still used, but the letter *B* stands for *bursa equivalent*, which refers to bone marrow since that is the location in humans where maturation of B cells takes place. T cells are so named because they mature in the *thymus*.

During the maturation process, B cells and T cells develop **immunocompetence**, the ability to carry out adaptive immune responses. This means that B cells and T cells make several distinctive proteins that are inserted into their plasma membranes. Some of these proteins function as **antigen receptors**—molecules that are able to recognize specific antigens (FIGURE 17.9).

Two major types of mature T cells exit the thymus: **helper T cells** and **cytotoxic T cells** (FIGURE 17.9). Helper T cells are also known as **CD4 cells**, which means that, in addition to antigen receptors, their plasma membranes include a protein called CD4. Cytotoxic T cells are also referred to as **CD8 cells** because their plasma membranes not only contain antigen receptors but also a protein known as CD8. As you will see later in this chapter, these two types of T cells have very different functions.

Adaptive Immunity Can Be Mediated by Cells or Antibodies

There are two types of adaptive immunity: cell-mediated immunity and antibody-mediated immunity. Both types of adaptive immunity are triggered by antigens. In **cell-mediated immunity**, cytotoxic T cells attack infected body cells, cancer cells, and foreign cells. In **antibody-mediated immunity**, B cells transform into plasma cells, which synthesize and secrete specific proteins called *antibodies*. A given antibody can bind to and inactivate a specific antigen. Helper T cells aid the immune responses of both cell-mediated and antibody-mediated immunity.

Cell-mediated immunity is particularly effective against (1) intracellular pathogens, which include any viruses, bacteria, or fungi that are inside cells; (2) cancer cells; and (3) foreign cells from tissue transplants. Thus, cell-mediated immunity always involves cells attacking cells. Antibody-mediated immunity works mainly against extracellular pathogens, which include any viruses, bacteria, or fungi that are in body fluids outside cells. Because antibody-mediated immunity involves antibodies that bind to antigens in body *humors* or fluids (such as blood and lymph), it is also referred to as *humoral immunity*.

In most cases, when a particular antigen initially enters the body, there is only a small group of lymphocytes with the correct antigen receptors to respond to that antigen; this small group of cells includes a few helper T cells, cytotoxic T cells, and B cells. Depending on its location, a given antigen can provoke both types of adaptive immune responses. This is due to the fact that when a specific antigen invades the body, there are usually many copies of that antigen spread throughout the body's tissues and fluids. Some copies of the antigen may be present inside body cells (which provokes a cell-mediated immune response by cytotoxic T cells), while other copies of the antigen may be present in extracellular fluid (which provokes an antibody-mediated immune response by B cells). Thus, cell-mediated and antibody-mediated immune responses often work together to get rid of the large number of copies of a particular antigen from the body.

Clonal Selection Increases the Number of B Cells and T Cells that Can Respond to an Antigen

As you have just learned, when a specific antigen is present in the body, there are usually many copies of that antigen located throughout the body's tissues and fluids. The numerous copies of the antigen initially outnumber the small group of helper T cells, cytotoxic T cells, and B cells with the correct antigen receptors to respond to that antigen. Therefore, once each of these lymphocytes encounters a copy of the antigen and receives stimulatory cues, it subsequently undergoes clonal selection. **Clonal selection** is the process by which a lymphocyte *proliferates* (divides) and *differentiates* (forms more highly specialized cells) in response to a specific antigen. The result of clonal selection is the formation of a population of identical cells, called a **clone**, that can recognize the same specific antigen as the

FIGURE 17.9 B cells and pre-T cells arise from pluripotent stem cells in bone marrow. B cells and T cells develop in primary lymphoid organs (bone marrow and the thymus) and are activated in secondary lymphoid organs and tissues (lymph nodes, spleen, and lymphoid nodules). Once activated, each type of lymphocyte forms a clone of cells that can recognize a specific antigen. For simplicity, antigen receptors, CD4 proteins, and CD8 proteins are not shown in the plasma membranes of the cells of the lymphocyte clones.

The two types of adaptive immunity are cell-mediated immunity and antibody-mediated immunity.

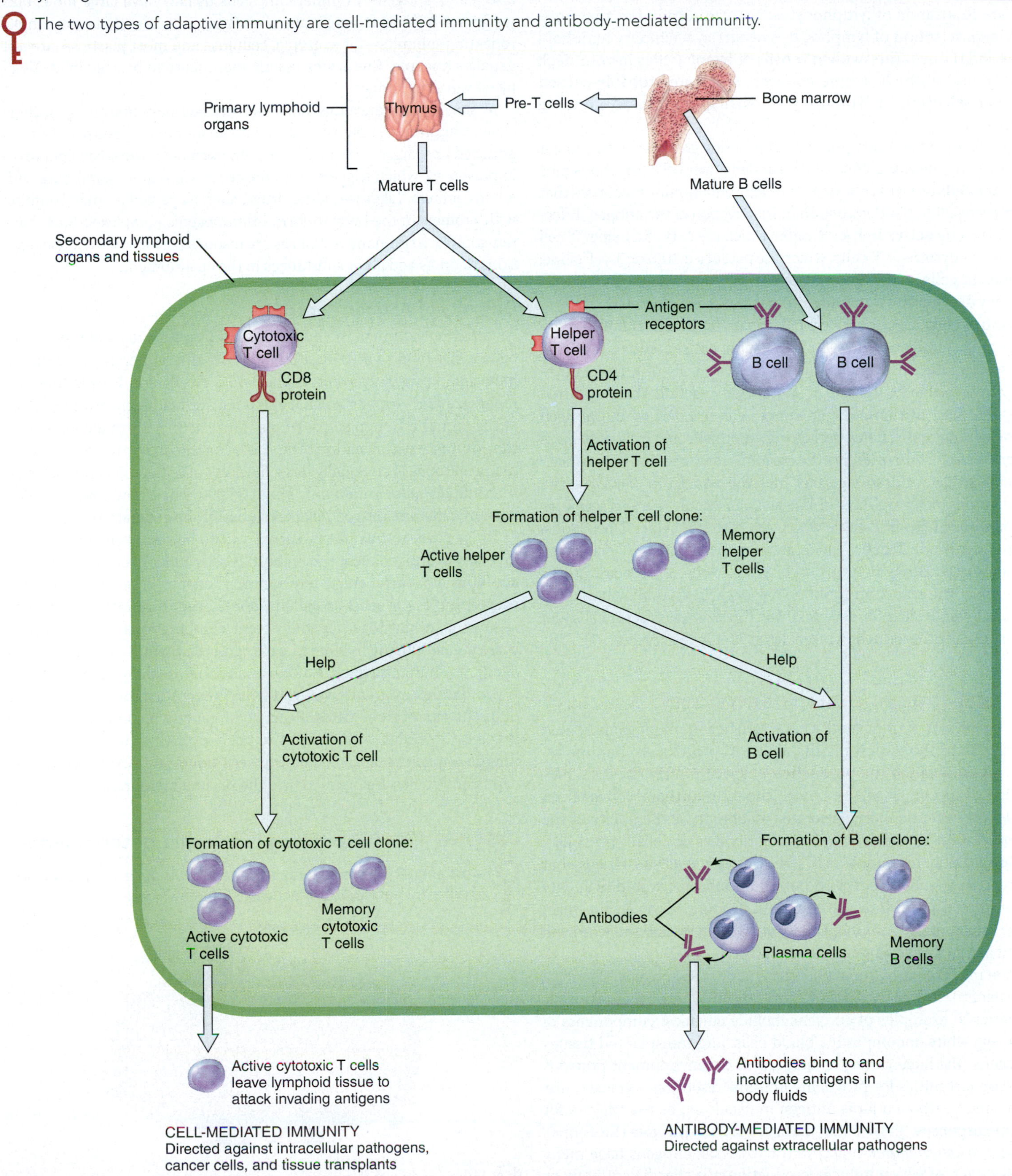

Which type of cell participates in both cell-mediated and antibody-mediated immunity?

original lymphocyte (FIGURE 17.9). This is because each cell within the clone has the same type of antigen receptor as the original lymphocyte. Before the first exposure to a given antigen, only a few lymphocytes are able to recognize it, but once clonal selection occurs, there are thousands of lymphocytes that can respond to that antigen. Clonal selection of lymphocytes occurs in secondary lymphoid organs and tissues. The swollen tonsils or lymph nodes in your neck you experienced the last time you were sick were probably caused by clonal selection of lymphocytes participating in an immune response.

A lymphocyte that undergoes clonal selection gives rise to two major types of cells in the clone: effector cells and memory cells. The thousands of **effector cells** of a lymphocyte clone carry out immune responses that ultimately result in the destruction or inactivation of the antigen. Effector cells include **active helper T cells**, which are part of a helper T cell clone; **active cytotoxic T cells**, which are part of a cytotoxic T cell clone; and **plasma cells**, which are part of a B cell clone. Most effector cells eventually die after the immune response has been completed.

Memory cells do not actively participate in the initial immune response to the antigen. However, if the same antigen enters the body again in the future, the thousands of memory cells of a lymphocyte clone are available to initiate a far swifter reaction than occurred during the first invasion. The memory cells respond to the antigen by proliferating and differentiating into more effector cells and more memory cells. Consequently, the second response to the antigen is usually so fast and so vigorous that the antigen is destroyed before any signs or symptoms of disease can occur. Memory cells include **memory helper T cells**, which are part of a helper T cell clone; **memory cytotoxic T cells**, which are part of a cytotoxic T cell clone; and **memory B cells**, which are part of a B cell clone. Most memory cells do not die at the end of an immune response. Instead, they have long life spans (often lasting for decades). The functions of effector cells and memory cells are described in more detail later in this chapter.

Antigens Trigger Immune Responses

Antigens have two important characteristics: immunogenicity and reactivity. **Immunogenicity** is the ability to provoke an immune response by stimulating the production of specific antibodies, the proliferation of specific T cells, or both. The term **antigen** derives from its function as an *anti*body *gen*erator. **Reactivity** is the ability of the antigen to react specifically with the antibodies or cells it provoked. Strictly speaking, immunologists define antigens as substances that have reactivity; substances with both immunogenicity and reactivity are considered **complete antigens**. Commonly, however, the term *antigen* implies both immunogenicity and reactivity, and the word will be used in this way throughout the text.

Entire microbes or parts of microbes may act as antigens. Chemical components of viruses and bacteria are antigenic, as are bacterial toxins. Nonmicrobial examples of antigens include chemical components of pollen, egg white, incompatible blood cells, and transplanted tissues and organs. The huge variety of antigens in the environment provides myriad opportunities for provoking immune responses. Typically, just certain small parts of a large antigen molecule act as the triggers for immune responses. These small parts are called **epitopes** (EP-i-tōps), or *antigenic determinants* (FIGURE 17.10). Most antigens have many epitopes, each of which induces production of a specific antibody or activates a specific T cell.

Chemical Nature of Antigens

Antigens are large, complex molecules. Most often, they are proteins. However, nucleic acids, lipids, and certain large polysaccharides may also act as antigens. Complete antigens usually have large molecular weights of 10,000 daltons or more, but large molecules that have simple, repeating subunits—for example, cellulose and most plastics—are not usually antigenic. This is why plastic materials can be used in artificial heart valves or joints.

A smaller substance that has reactivity but lacks immunogenicity is called a **hapten**. A hapten can stimulate an immune response only if it is attached to a larger carrier molecule. An example is the small lipid toxin in poison ivy, which triggers an immune response after combining with a body protein. Likewise, some drugs, such as penicillin, may combine with proteins in the body to form immunogenic complexes. Such hapten-stimulated immune responses are responsible for some allergic reactions to drugs and other substances in the environment.

Diversity of Antigen Receptors

An amazing feature of the human immune system is its ability to recognize and bind to at least a billion (10^9) different epitopes. Before a particular antigen ever enters the body, T cells and B cells that can recognize and respond to that intruder are ready and waiting. Cells of the immune system can even recognize artificially made molecules that do not exist in nature. The basis for the ability to recognize so many epitopes is an equally large diversity of antigen receptors. Given that human cells contain only about 35,000 genes, how could a billion or more different antigen receptors possibly be generated?

The answer to this puzzle turned out to be simple in concept. The diversity of antigen receptors in both B cells and T cells is the result of shuffling and rearranging a few hundred versions of several small gene segments. This process is called **genetic recombination**. The gene segments are put together in different combinations as the lymphocytes are developing in bone marrow and the thymus. The situation is similar to shuffling a deck of 52 cards and then dealing out three cards. If you did this over and over, you could generate many more than 52 different sets of three cards. Because of genetic recombination, each B cell or T cell has a unique set of gene segments that codes for its unique antigen receptor. After transcription and translation, the receptor molecules are inserted into the plasma membrane.

FIGURE 17.10 Epitopes (antigenic determinants).

Most antigens have several epitopes that induce the production of different antibodies or activate different T cells.

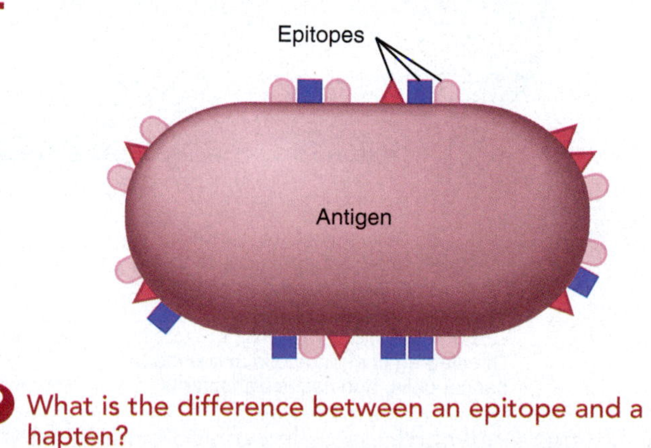

What is the difference between an epitope and a hapten?

Major Histocompatibility Complex Proteins Are Types of Self-Antigens

Located in the plasma membrane of body cells are "self-antigens", the **major histocompatibility complex (MHC)** proteins. These transmembrane glycoproteins are also called *human leukocyte antigens (HLAs)* because they were first identified on leukocytes. Unless you have an identical twin, your MHC proteins are unique. Thousands to several hundred thousand MHC molecules mark the surface of each of your body cells except erythrocytes. Although MHC proteins are the reason that tissues may be rejected when they are transplanted from one person to another, their normal function is to help T cells recognize that an antigen is foreign, not self. Such recognition is an important first step in any adaptive immune response.

The two types of major histocompatibility complex proteins are class I and class II. Class I MHC (MHC-I) molecules are built into the plasma membranes of all body cells except erythrocytes. Class II MHC (MHC-II) molecules appear on the surface of antigen-presenting cells (described in the next section).

Antigens Are Processed and Then Presented

For an immune response to occur, B cells and T cells must recognize that a foreign antigen is present. B cells can recognize and bind to antigens in lymph, interstitial fluid, or blood plasma. T cells only recognize fragments of antigenic proteins that are processed and presented in a certain way. In **antigen processing**, antigenic proteins are broken down into peptide fragments that then associate with MHC molecules. Next the antigen–MHC complex is inserted into the plasma membrane of a body cell. The insertion of the complex into the plasma membrane is called **antigen presentation**. When a peptide fragment comes from a *self-protein*, T cells ignore the antigen–MHC complex. However, if the peptide fragment comes from a *foreign protein*, T cells recognize the antigen–MHC complex as an intruder, and an immune response takes place. Antigen processing and presentation occurs in two ways, depending on whether the antigen is located outside or inside body cells.

Processing of Exogenous Antigens

Foreign antigens that are present in fluids *outside* body cells are termed *exogenous antigens*. They include intruders such as bacteria and bacterial toxins, parasitic worms, inhaled pollen and dust, and viruses that have not yet infected a body cell. A special class of cells called **antigen-presenting cells (APCs)** process and present exogenous antigens. APCs include dendritic cells, macrophages, and B cells. They are strategically located in places where antigens are likely to penetrate the innate defenses and enter the body, such as the skin (recall that Langerhans cells are a type of dendritic cell) and the linings of the respiratory, gastrointestinal, urinary, and reproductive tracts. After processing and presenting an antigen, APCs migrate from tissues via lymphatic vessels to lymph nodes.

The steps in the processing and presenting of an exogenous antigen by an antigen-presenting cell occur as follows (**FIGURE 17.11**):

1. ***Ingestion of the antigen.*** Antigen-presenting cells ingest exogenous antigens by phagocytosis or endocytosis. Ingestion could

FIGURE 17.11 Processing and presenting of exogenous antigen by an antigen-presenting cell (APC).

Fragments of exogenous antigens are processed and then presented with MHC-II molecules on the surface of an antigen-presenting cell (APC).

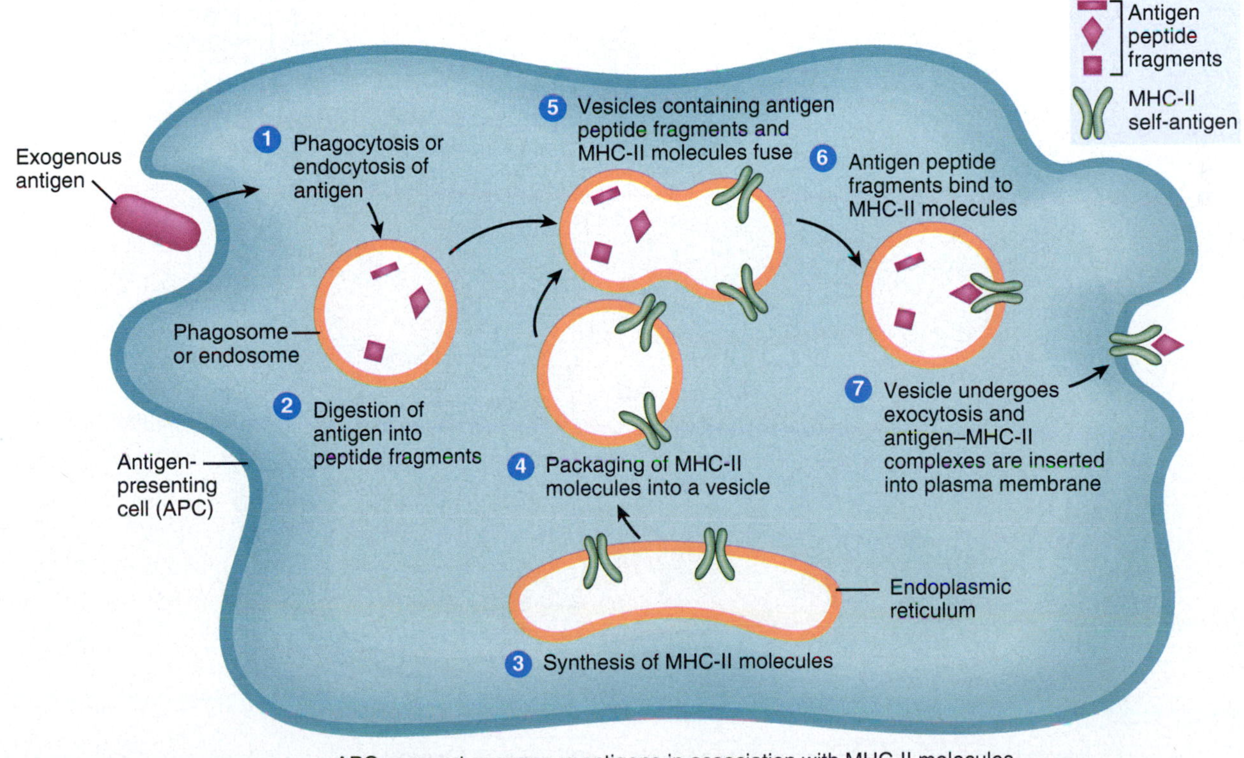

APCs present exogenous antigens in association with MHC-II molecules

What types of cells are APCs?

occur almost anywhere in the body that invaders, such as microbes, have penetrated the innate defenses.

2. ***Digestion of antigen into peptide fragments.*** Within the phagosome or endosome, protein-digesting enzymes split large antigens into short peptide fragments.
3. ***Synthesis of MHC-II molecules.*** At the same time, the APC synthesizes MHC-II molecules at the endoplasmic reticulum (ER).
4. ***Packaging of MHC-II molecules.*** Once synthesized, the MHC-II molecules are packaged into vesicles.
5. ***Fusion of vesicles.*** The vesicles containing antigen peptide fragments and MHC-II molecules merge and fuse.
6. ***Binding of peptide fragments to MHC-II molecules.*** After fusion of the two types of vesicles, antigen peptide fragments bind to MHC-II molecules.
7. ***Insertion of antigen–MHC-II complexes into the plasma membrane.*** The combined vesicle that contains antigen–MHC-II complexes undergoes exocytosis. As a result, the antigen–MHC-II complexes are inserted into the plasma membrane.

After processing an antigen, the antigen-presenting cell migrates to lymphoid tissue to present the antigen to T cells. Within lymphoid tissue, a small number of T cells that have compatibly shaped receptors recognize and bind to the antigen fragment–MHC-II complex, triggering an adaptive immune response. The presentation of exogenous antigen together with MHC-II molecules by antigen-presenting cells informs T cells that intruders are present in the body and that combative action should begin.

Processing of Endogenous Antigens

Foreign antigens that are present *inside* body cells are termed *endogenous antigens.* Such antigens may be viral proteins produced after a virus infects the cell and takes over the cell's metabolic machinery, toxins produced from intracellular bacteria, or abnormal proteins synthesized by a cancerous cell.

The steps in the processing and presenting of an endogenous antigen by an infected body cell occur as follows (FIGURE 17.12):

1. ***Digestion of antigen into peptide fragments.*** Within the infected cell, protein-digesting enzymes split the endogenous antigen into short peptide fragments.
2. ***Synthesis of MHC-I molecules.*** At the same time, the infected cell synthesizes MHC-I molecules at the endoplasmic reticulum (ER).
3. ***Binding of peptide fragments to MHC-I molecules.*** The antigen peptide fragments enter the ER and then bind to MHC-I molecules.
4. ***Packaging of antigen–MHC-I complexes.*** From the ER, antigen–MHC-I complexes are packaged into vesicles.
5. ***Insertion of antigen–MHC-I complexes into the plasma membrane.*** The vesicles that contain antigen–MHC-I complexes undergo exocytosis. As a result, the antigen–MHC-I complexes are inserted into the plasma membrane.

FIGURE 17.12 Processing and presenting of endogenous antigen by an infected body cell.

Fragments of endogenous antigens are processed and then presented with MHC-I proteins on the surface of an infected body cell.

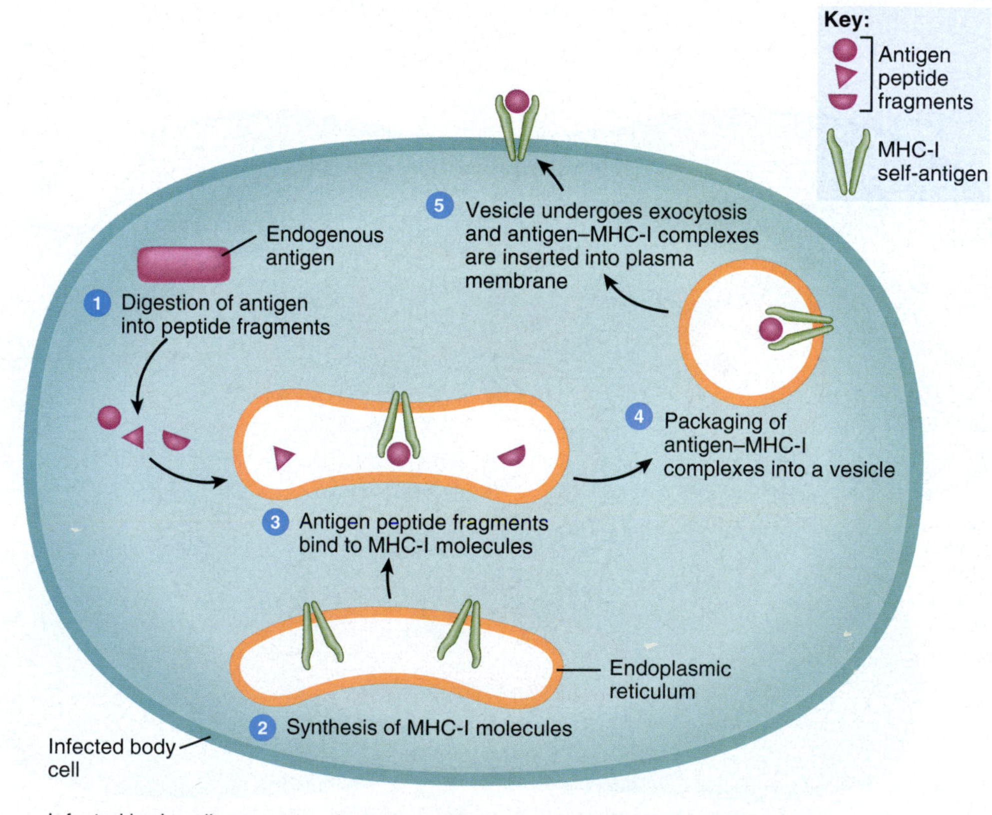

Infected body cells present endogenous antigens in association with MHC-I molecules

What are some examples of endogenous antigens?

Most cells of the body can process and present endogenous antigens. The display of an endogenous antigen bound to an MHC-I molecule signals that a cell has been infected and needs help.

Cytokines Have Important Roles in Immunity

Cytokines are local mediators (paracrines or autocrines) that regulate many normal cell functions, such as cell growth and differentiation. They are secreted by a variety of cells, including lymphocytes and antigen-presenting cells. Some cytokines stimulate proliferation of progenitor blood cells in bone marrow. Others regulate activities of cells involved in innate defenses or adaptive immune responses, as described in TABLE 17.2.

CHECKPOINT

8. What is immunocompetence, and which body cells display it?
9. How do the major histocompatibility complex class I and class II proteins function?
10. How do antigen-presenting cells process exogenous antigens?
11. What are cytokines and how do they function?

TABLE 17.2 Summary of Selected Cytokines Participating in Immune Responses

Cytokine	Origins and Functions
Interleukin-1 (IL-1)	Produced by macrophages; promotes proliferation of helper T cells; acts on hypothalamus to cause fever.
Interleukin-2 (IL-2)	Secreted by helper T cells; costimulates the proliferation of helper T cells, cytotoxic T cells, and B cells; activates NK cells.
Interleukin-4 (IL-4)	Produced by helper T cells; costimulator for B cells; causes plasma cells to secrete IgE antibodies; promotes growth of T cells.
Interleukin-5 (IL-5)	Produced by some helper T cells and mast cells; costimulator for B cells; causes plasma cells to secrete IgA antibodies.
Interleukin-6 (IL-6)	Produced by helper T cells; enhances B cell proliferation, B cell differentiation into plasma cells, and secretion of antibodies by plasma cells.
Tumor necrosis factor (TNF)	Produced mainly by macrophages; stimulates accumulation of neutrophils and macrophages at sites of inflammation and stimulates their killing of microbes.
Interferons (IFNs)	Produced by virus-infected cells to inhibit viral replication in uninfected cells; activate cytotoxic T cells and natural killer cells; inhibit cell division and suppress the formation of tumors.
Macrophage migration inhibiting factor	Produced by cytotoxic T cells; prevents macrophages from leaving site of infection.

Cell-Mediated Immunity Uses T Cells to Eliminate Specific Antigens

A cell-mediated immune response begins with *activation* of a small number of T cells by a specific antigen. Once a T cell has been activated, it undergoes clonal selection. Recall that *clonal selection* is the process by which a lymphocyte proliferates (divides several times) and differentiates (forms more highly specialized cells) in response to a specific antigen. The result of clonal selection is the formation of a *clone* of cells that can recognize the same antigen as the original lymphocyte (see FIGURE 17.9). Some of the cells of a T cell clone become effector cells, while other cells of the clone become memory cells. The effector cells of a T cell clone carry out immune responses that ultimately result in elimination of the intruder.

Activation of T Cells

At any given time, most T cells are inactive. Antigen receptors on the surface of T cells, called **T-cell receptors (TCRs)**, recognize and bind to specific foreign antigen fragments that are presented in antigen–MHC complexes. There are millions of different T cells; each has its own unique TCRs that can recognize a specific antigen–MHC complex. When an antigen enters the body, only a few T cells have TCRs that can recognize and bind to the antigen. Antigen recognition also involves other surface proteins on T cells, the CD4 or CD8 proteins. These proteins interact with the MHC proteins and help maintain the TCR–MHC coupling. For this reason, they are referred to as *coreceptors*. Antigen recognition by a TCR with CD4 or CD8 proteins is the *first signal* in activation of a T cell.

A T cell becomes activated only if it binds to the foreign antigen and at the same time receives a *second signal*, a process known as **costimulation**. Of the more than 20 known costimulators, some are cytokines, such as **interleukin-2**. Other costimulators include pairs of plasma membrane molecules, one on the surface of the T cell and a second on the surface of an antigen-presenting cell, that enable the two cells to adhere to one another for a period of time.

The need for two signals to activate a T cell is a little like starting and driving a car: When you insert the correct key (antigen) in the ignition (T-cell receptor) and turn it, the car starts (recognition of specific antigen), but it cannot move forward until you move the gearshift into drive (costimulation). The need for costimulation may prevent immune responses from occurring accidentally. Different costimulators affect the activated T cell in different ways, just as shifting a car into reverse has a different effect than shifting it into drive. Moreover, recognition (antigen binding to a receptor) without costimulation leads to a prolonged state of inactivity called *anergy* in both T cells and B cells. Anergy is rather like leaving a car in neutral gear with its engine running until it's out of gas!

Once a T cell has received the appropriate two signals (antigen recognition and costimulation), it is *activated*. An activated T cell subsequently undergoes clonal selection.

Activation and Clonal Selection of Helper T Cells

Most T cells that display CD4 develop into **helper T cells**, also known as **CD4 cells**. Inactive (resting) helper T cells recognize exogenous antigen fragments associated with major histocompatibility complex class II (MHC-II) molecules at the surface of an APC (FIGURE 17.13).

With the aid of the CD4 protein, the helper T cell and APC interact with each other (antigenic recognition), costimulation occurs, and the helper T cell becomes activated. Once activated, the helper T cell undergoes clonal selection (FIGURE 17.13). The result is the formation of a clone of helper T cells that consists of active helper T cells and memory helper T cells. Within hours after costimulation, **active helper T cells** start secreting a variety of cytokines (see TABLE 17.2). One very important cytokine produced by helper T cells is interleukin-2 (IL-2), which is needed for virtually all immune responses and is the prime trigger of T cell proliferation. IL-2 can act as a costimulator for resting helper T cells or cytotoxic T cells, and it enhances activation and proliferation of T cells, B cells, and natural killer cells. Some actions of interleukin-2 provide a good example of a beneficial positive feedback system. As noted earlier, activation of a helper T cell stimulates it to start secreting IL-2, which then acts in an autocrine manner by binding to IL-2 receptors on the plasma membrane of the cell that secreted it. One effect is stimulation of cell division. As the helper T cells proliferate, a positive feedback effect occurs because they secrete more IL-2, which causes further cell division. IL-2 may also act in a paracrine manner by binding to IL-2 receptors on neighboring helper T cells, cytotoxic T cells, or B cells. If any of these neighboring cells have already become bound to a copy of the same antigen, IL-2 serves as a costimulator.

The **memory helper T cells** of a helper T cell clone are not active cells. However, if the same antigen enters the body again in the future, memory helper T cells can quickly proliferate and differentiate into more active helper T cells and more memory helper T cells.

FIGURE 17.13 Activation and clonal selection of a helper T cell.

Once a helper T cell is activated, it forms a clone of active helper T cells and memory helper T cells.

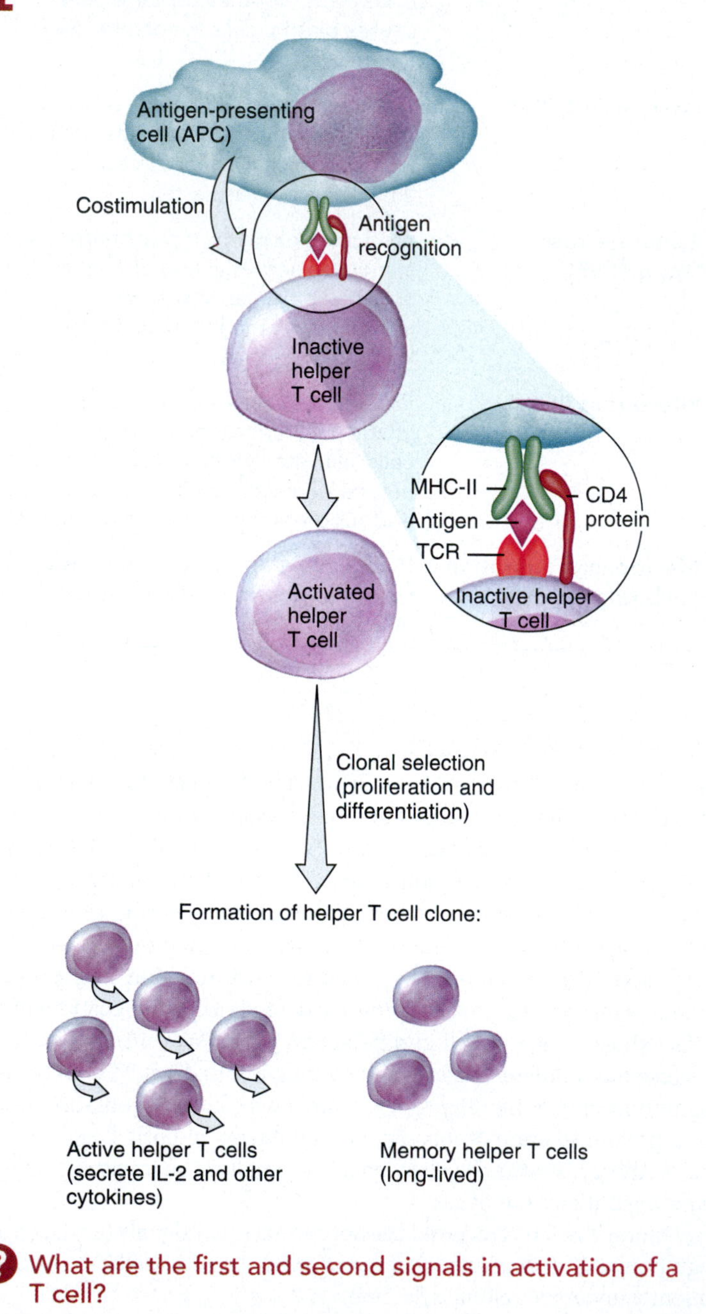

What are the first and second signals in activation of a T cell?

Activation and Clonal Selection of Cytotoxic T Cells

Most T cells that display CD8 develop into **cytotoxic T cells**, also termed **CD8 cells**. Cytotoxic T cells recognize foreign antigens combined with major histocompatibility complex class I (MHC-I) molecules on the surface of (1) body cells infected by microbes, (2) cancer cells, and (3) foreign cells of a tissue transplant (FIGURE 17.14). Recognition requires the TCR and CD8 protein to maintain the coupling with MHC-I. Following antigenic recognition, costimulation occurs. To become activated, cytotoxic T cells require costimulation by interleukin-2 or other cytokines produced by active helper T cells that have already become bound to copies of the same antigen. (Recall that helper T cells are activated by antigen associated with MHC-II molecules.) Thus, *maximal activation* of cytotoxic T cells requires presentation of antigen associated with both MHC-I and MHC-II molecules.

Once activated, the cytotoxic T cell undergoes clonal selection (FIGURE 17.14). The result is the formation of a clone of cytotoxic T cells that consists of active cytotoxic T cells and memory cytotoxic T cells. **Active cytotoxic T cells** attack other cells that display the antigen. **Memory cytotoxic T cells** do not attack cells that have the antigen. Instead, they can quickly proliferate and differentiate into more active cytotoxic T cells and more memory cytotoxic T cells if the same antigen enters the body at a future time.

Elimination of Invaders

Cytotoxic T cells are the soldiers that march forth to do battle with foreign invaders in cell-mediated immune responses. They leave secondary lymphoid organs and tissues and migrate to seek out and destroy infected body cells, cancer cells, and transplanted cells. Using receptors on their surfaces, cytotoxic T cells recognize and attach to their target cells. Then the cytotoxic T cells deliver a "lethal hit" that kills

FIGURE 17.14 Activation and clonal selection of a cytotoxic T cell.

Once a cytotoxic T cell is activated, it forms a clone of active cytotoxic T cells and memory cytotoxic T cells.

FIGURE 17.15 Cytotoxic T cell function.

Cytotoxic T cells kill infected body cells, cancer cells, and foreign cells from a tissue transplant by releasing perforin molecules and granzymes.

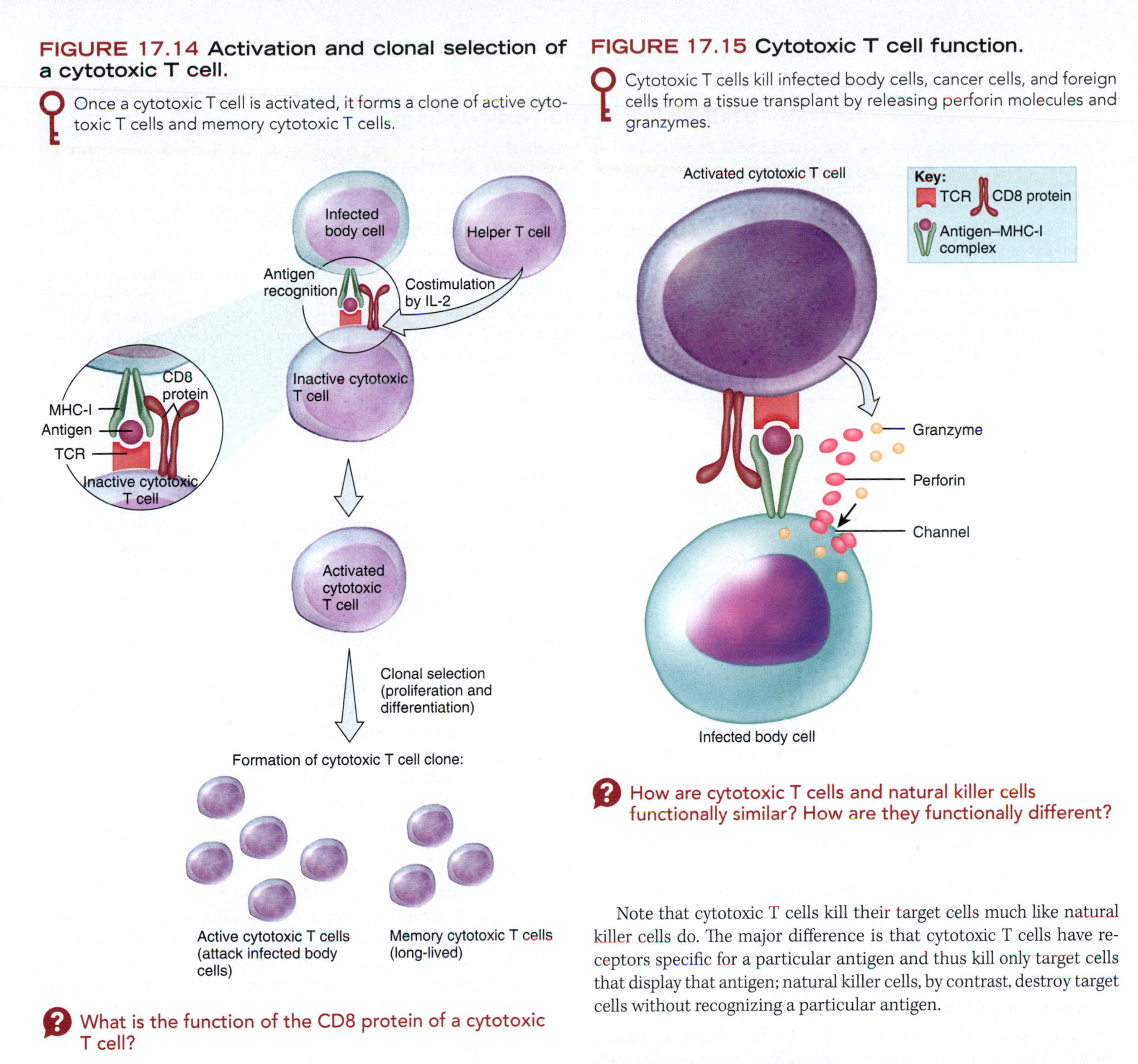

What is the function of the CD8 protein of a cytotoxic T cell?

How are cytotoxic T cells and natural killer cells functionally similar? How are they functionally different?

the target cells. Cytotoxic T cells kill target cells, such as an infected body cell, by releasing two types of toxic substances from their granules: perforin and granzymes (FIGURE 17.15). Perforin is a protein that inserts into the plasma membrane of the target cell and creates channels in the membrane. As a result, extracellular fluid flows into the target cell and cytolysis (cell bursting) occurs. Granzymes are protein-digesting enzymes that enter the target cell through the perforin channels and then induce the target cell to undergo apoptosis. Once the infected cell is destroyed, the released microbes are engulfed and destroyed by phagocytes.

Note that cytotoxic T cells kill their target cells much like natural killer cells do. The major difference is that cytotoxic T cells have receptors specific for a particular antigen and thus kill only target cells that display that antigen; natural killer cells, by contrast, destroy target cells without recognizing a particular antigen.

Immunological Surveillance

When a normal cell transforms into a cancerous cell, it often displays novel cell surface components called **tumor antigens**. These molecules are rarely, if ever, displayed on the surface of normal cells. If the immune system recognizes a tumor antigen as nonself, it can destroy any cancer cells carrying that antigen. Such immune responses, called **immunological surveillance**, are carried out by cytotoxic T cells, macrophages, and natural killer cells. Immunological surveillance is most effective in eliminating tumor cells due to cancer-causing viruses. For this reason, transplant recipients who are taking immunosuppressive drugs to prevent transplant rejection have an increased incidence of virus-associated cancers. Their risk for other types of cancer is not increased.

CLINICAL CONNECTION

Graft Rejection and Tissue Typing

Organ transplantation involves the replacement of an injured or diseased organ, such as the heart, liver, kidney, lungs, or pancreas, with an organ donated by another individual. Usually, the immune system recognizes the proteins in the transplanted organ as foreign and mounts both cell-mediated and antibody-mediated immune responses against them. This phenomenon is known as **graft rejection**.

The success of an organ or tissue transplant depends on **histocompatibility**—that is, the tissue compatibility between the donor and the recipient. The more similar the MHC proteins, the greater the histocompatibility and thus the greater the probability that the transplant will not be rejected. **Tissue typing (histocompatibility testing)** is done before any organ transplant. In the United States, a nationwide computerized registry helps physicians select the most histocompatible and needy organ transplant recipients whenever donor organs become available. The closer the match between the major histocompatibility complex proteins of the donor and recipient, the weaker is the graft rejection response.

To reduce the risk of graft rejection, organ transplant recipients receive immunosuppressive drugs. One such drug is ***cyclosporine***, derived from a fungus, which inhibits secretion of interleukin-2 by helper T cells but has only a minimal effect on B cells. Thus, the risk of rejection is diminished while resistance to some diseases is maintained.

CHECKPOINT

12. What are the functions of helper T cells, cytotoxic T cells, and memory T cells?
13. How do cytotoxic T cells kill their target cells?
14. How is immunological surveillance useful?

Antibody-Mediated Immunity Uses Antibodies to Inactivate Specific Antigens

The body contains not only millions of different T cells but also millions of different B cells, each capable of responding to a specific antigen. Cytotoxic T cells leave lymphoid organs and tissues to seek out and destroy a foreign antigen, but B cells stay put. In the presence of a foreign antigen, a specific B cell in a lymph node, the spleen, or lymphoid nodule becomes activated. Then it undergoes clonal selection, forming a clone of plasma cells and memory cells. Plasma cells are the effector cells of a B cell clone; they secrete specific antibodies, which in turn circulate in the lymph and blood to reach the site of invasion.

Activation and Clonal Selection of B Cells

During activation of a B cell, an antigen binds to **B-cell receptors (BCRs)**. These integral transmembrane proteins are structurally identical to the antibodies that eventually are secreted by plasma cells. Although B cells can respond to an unprocessed antigen present in lymph or interstitial fluid, their response is much more intense when they process the antigen. Antigen processing in a B cell occurs in the following way (FIGURE 17.16): The antigen is taken into the B cell, broken down into peptide fragments, combined with MHC-II self-antigens, and moved to the B cell plasma membrane. Helper T cells recognize the antigen–MHC-II complex and deliver the costimulation needed for B cell proliferation and differentiation. The helper T cell produces interleukin-2 and other cytokines that function as costimulators to activate B cells.

Once activated, a B cell undergoes clonal selection (FIGURE 17.16). The result is the formation of a clone of B cells that consists of plasma cells and memory B cells. **Plasma cells** secrete antibodies. A few days after exposure to an antigen, a plasma cell secretes hundreds of millions of antibodies each day for about four or five days, until the plasma cell dies. Most antibodies travel in lymph and blood to the invasion site. Interleukin-4 and interleukin-6, also produced by helper T cells, enhance B cell proliferation, B cell differentiation into plasma cells, and secretion of antibodies by plasma cells. **Memory B cells** do not secrete antibodies. Instead, they can quickly proliferate and differentiate into more plasma cells and more memory B cells should the same antigen reappear at a future time.

Different antigens stimulate different B cells to develop into plasma cells and their accompanying memory B cells. All the B cells of a particular clone are capable of secreting only one type of antibody, which is identical to the antigen receptor displayed by the B cell that first responded. Each specific antigen activates only those B cells that are predestined (by the combination of gene segments they carry) to secrete antibody specific to that antigen. Antibodies produced by a clone of plasma cells enter the circulation and form antigen–antibody complexes with the antigen that initiated their production.

Antibodies

An **antibody (Ab)** can combine specifically with the epitope on the antigen that triggered its production. The antibody's structure matches its antigen much as a lock accepts a specific key. In theory, plasma cells could secrete as many different antibodies as there are different B-cell receptors because the same recombined gene segments code for both the BCR and the antibodies eventually secreted by plasma cells.

FIGURE 17.16 Activation and clonal selection of B cells. Plasma cells are actually much larger than B cells.

Plasma cells secrete antibodies.

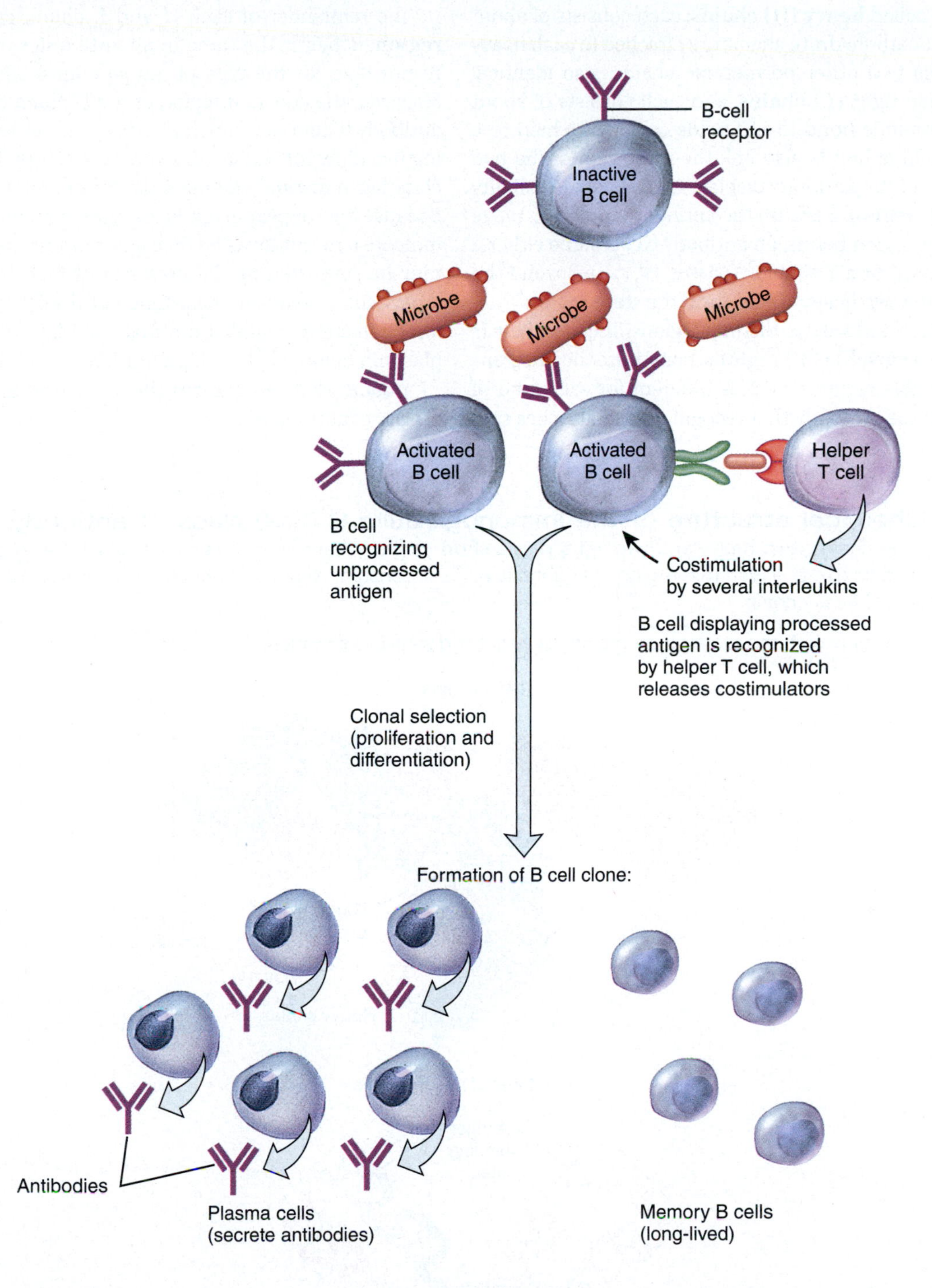

How many different kinds of antibodies will be secreted by the plasma cells in the clone shown here?

Antibody Structure Antibodies belong to a group of glycoproteins called globulins, and for this reason they are also known as **immunoglobulins (Igs)** (im′-ū-nō-GLOB-ū-lins). Most antibodies contain four polypeptide chains (FIGURE 17.17). Two of the chains are identical to each other and are called **heavy (H) chains**; each consists of about 450 amino acids. Short carbohydrate chains are attached to each heavy polypeptide chain. The two other polypeptide chains, also identical to each other, are called **light (L) chains**, and each consists of about 220 amino acids. A disulfide bond (S—S) holds each light chain to a heavy chain. Two disulfide bonds also link the midregion of the two heavy chains; this part of the antibody displays considerable flexibility and is called the **hinge region**. Because the antibody "arms" can move somewhat as the hinge region bends, an antibody can assume either a T shape (FIGURE 17.17a) or a Y shape (FIGURE 17.17b). Beyond the hinge region, parts of the two heavy chains form the **stem region**.

Within each H and L chain are two distinct regions. The tips of the H and L chains, called the **variable (V) regions**, constitute the **antigen-binding site**. The variable region, which is different for each kind of antibody, is the part of the antibody that recognizes and attaches specifically to a particular antigen. Because most antibodies have two antigen-binding sites, they are said to be *bivalent*. Flexibility at the hinge allows the antibody to bind simultaneously to two epitopes that are some distance apart—for example, on the surface of a microbe.

The remainder of each H and L chain, called the **constant (C) region**, is nearly the same in all antibodies of the same class and is responsible for the type of antigen–antibody reaction that occurs. However, the constant region of the H chain differs from one class of antibody to another, and its structure serves as a basis for distinguishing five different classes, designated IgG, IgA, IgM, IgD, and IgE. Each class has a distinct chemical structure and a specific biological role. Because they appear first and are relatively short-lived, IgM antibodies indicate a recent invasion. In a sick patient, the responsible pathogen may be suggested by the presence of high levels of IgM specific to a particular organism. Resistance of the fetus and newborn baby to infection stems mainly from maternal IgG antibodies that cross the placenta before birth and IgA antibodies in breast milk after birth.

TABLE 17.3 summarizes the structures and functions of the five classes of antibodies.

FIGURE 17.17 Chemical structure of the immunoglobulin G (IgG) class of antibody. Each molecule is composed of four polypeptide chains (two heavy and two light) plus a short carbohydrate chain attached to each heavy chain. In (a), each circle represents one amino acid. In (b), V_L = variable regions of light chain, C_L = constant region of light chain, V_H = variable region of heavy chain, and C_H = constant region of heavy chain.

An antibody combines only with the epitope on the antigen that triggered its production.

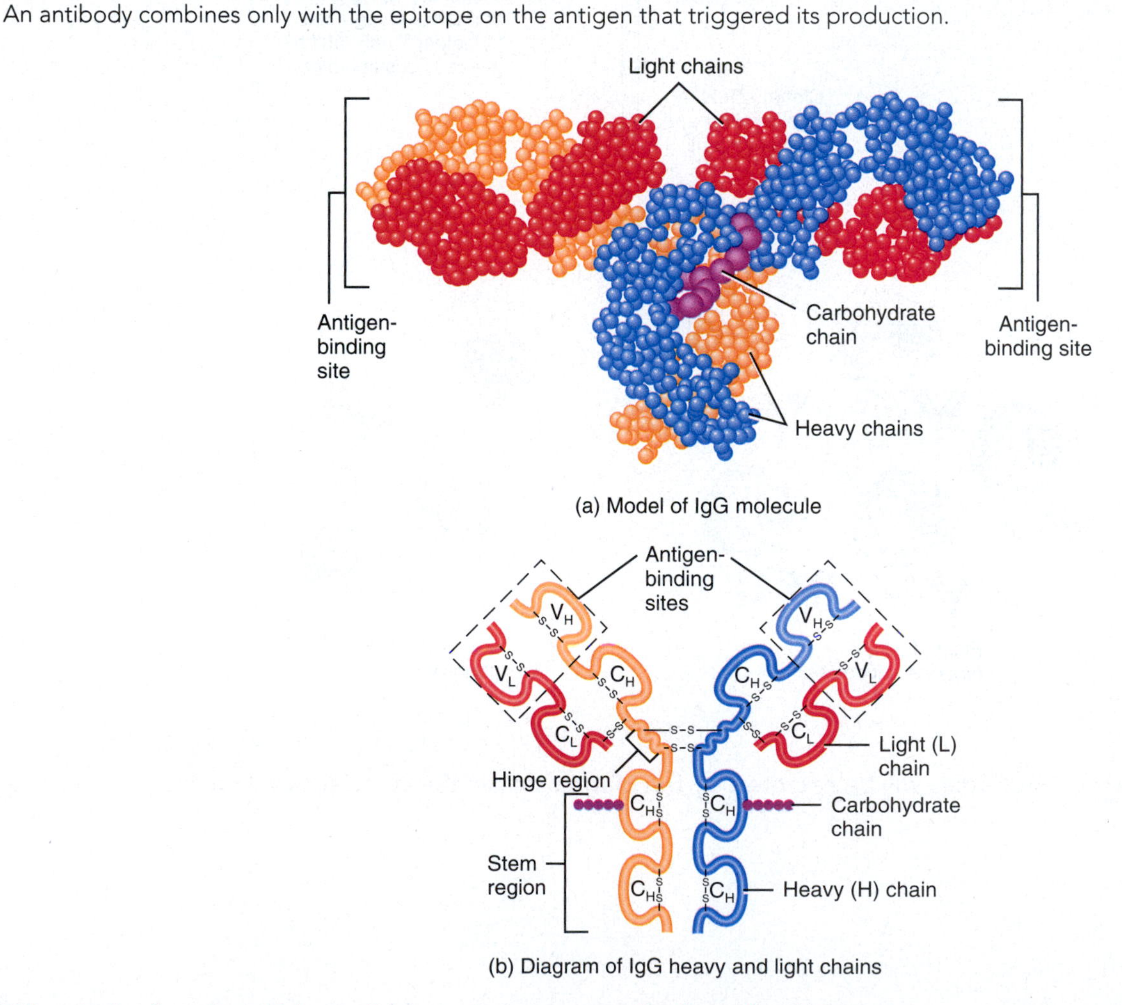

What is the function of the variable regions in an antibody molecule?

TABLE 17.3 Classes of Immunoglobulins (Igs)

Name and Structure	Characteristics and Functions
IgG	Most abundant (about 80%) of all antibodies in blood; also found in lymph and the intestines; monomer (one-unit) structure. Protects against bacteria and viruses by enhancing phagocytosis, neutralizing toxins, and triggering the complement system. It is the only class of antibody to cross the placenta from mother to fetus, conferring considerable immune protection to newborns.
IgA	Found mainly in sweat, tears, saliva, mucus, breast milk, and gastrointestinal secretions. Smaller quantities are present in blood and lymph. Makes up 10–15% of all antibodies in the blood; occurs as monomers and dimers (two units). Levels decrease during stress, lowering resistance to infection.
IgM	About 5–10% of all antibodies in the blood; also found in lymph. Occurs as pentamers (five units). First antibody class to be secreted by plasma cells after an initial exposure to any antigen. Activates complement and causes agglutination and lysis of microbes. Also present as monomers on the surfaces of B cells, where they serve as antigen receptors. In blood plasma, the anti-A and anti-B antibodies of the ABO blood group, which bind to A and B antigens during incompatible blood transfusions, are also IgM antibodies.
IgD	Mainly found on the surfaces of B cells as antigen receptors, where it occurs as monomers; involved in activation of B cells. About 0.2% of all antibodies in the blood.
IgE	Less than 0.1% of all antibodies in the blood; occurs as monomers; attach to receptors on mast cells and basophils. Involved in allergic and hypersensitivity reactions; provides protection against parasitic worms.

Antibody Actions The actions of the five classes of immunoglobulins differ somewhat, but all of them act to disable antigens in some way. Actions of antibodies include (**FIGURE 17.18**):

- ***Neutralizing antigen.*** The reaction of antibody with antigen blocks or neutralizes some bacterial toxins and prevents attachment of some viruses to body cells (**FIGURE 17.18a**).
- ***Agglutinating antigen.*** Because antibodies have two or more sites for binding to antigen, the antigen–antibody reaction may cross-link pathogens to one another, causing agglutination (clumping together) (**FIGURE 17.18b**). Phagocytic cells ingest agglutinated microbes more readily.
- ***Precipitating antigen.*** Antibodies can cross-link soluble antigens into complexes that are too large to stay in solution (**FIGURE 17.18c**). As a result, the complex precipitates out of solution, which makes it easier for the antigen to be phagocytized.
- ***Activating complement.*** Antigen–antibody complexes initiate the classical pathway of the complement system (**FIGURE 17.18d**).
- ***Opsonization.*** Antibodies can function as opsonins that coat the surface of a microbe, thereby making it easier for phagocytes to engulf and destroy the microbe (**FIGURE 17.18e**). Recall that the process by which opsonins coat the surface of a microbe to enhance phagocytosis is called *opsonization* (see Section 17.2).

Immunological Memory Allows the Immune System to Remember Previously Encountered Antigens

A hallmark of immune responses is memory for specific antigens that have triggered immune responses in the past. Immunological memory is due to the presence of long-lasting antibodies and very long-lived lymphocytes that arise during clonal selection of antigen-stimulated B cells and T cells.

Immune responses, whether cell-mediated or antibody-mediated, are much quicker and more intense after a second or subsequent exposure to an antigen than after the first exposure. Initially, only a few cells have the correct specificity to respond, and the immune response may take several days to build to maximum intensity. Because thousands of memory cells exist after an initial encounter with an antigen, the next time the same antigen appears, the memory cells can proliferate and differentiate into helper T cells, cytotoxic T cells, or plasma cells within hours.

FIGURE 17.18 Actions of antibodies.

Actions of antibodies include neutralizing antigen, agglutinating antigen, precipitating antigen, activating complement, and opsonization.

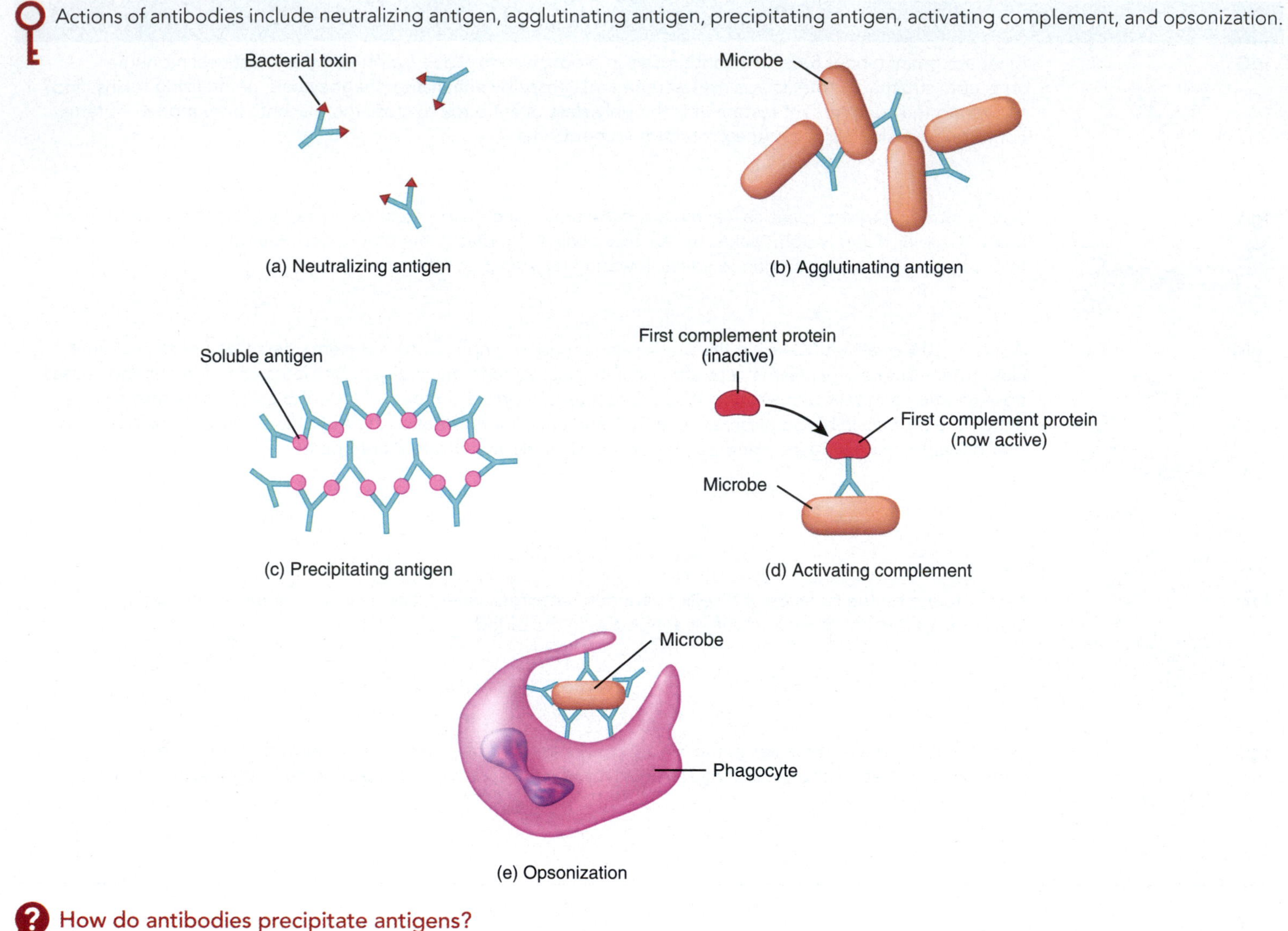

How do antibodies precipitate antigens?

CLINICAL CONNECTION

Monoclonal Antibodies

The antibodies produced against a given antigen by plasma cells can be harvested from an individual's blood. However, because an antigen typically has many epitopes, several different clones of plasma cells produce different antibodies against the antigen. If a single plasma cell could be isolated and induced to proliferate into a clone of identical plasma cells, then a large quantity of identical antibodies could be produced. Unfortunately, lymphocytes and plasma cells are difficult to grow in culture, so scientists sidestepped this difficulty by fusing B cells with tumor cells that grow easily and proliferate endlessly. The resulting hybrid cell is called a **hybridoma** (hī-bri-DŌ-ma). Hybridomas are long-term sources of large quantities of pure, identical antibodies, called **monoclonal antibodies (MAbs)** because they come from a single clone of identical cells. One clinical use of monoclonal antibodies is for measuring levels of a drug in a patient's blood. Other uses include the diagnosis of strep throat; pregnancy; allergies; and diseases such as hepatitis, rabies, and some sexually transmitted diseases. MAbs have also been used to detect cancer at an early stage and to ascertain the extent of metastasis. They may also be useful in preparing vaccines to counteract the rejection associated with transplants, to treat autoimmune diseases, and perhaps to treat AIDS.

One measure of immunological memory is *antibody titer* (TĪ-ter), the amount of antibody in serum. After an initial contact with an antigen, no antibodies are present for a period of several days. Then a slow rise in the antibody titer occurs, first IgM and then IgG, followed by a gradual decline in antibody titer (FIGURE 17.19). This is the **primary response**.

Memory cells may remain for decades. Every new encounter with the same antigen results in a rapid proliferation of memory cells. After subsequent encounters, the antibody titer is far greater than during a primary response and consists mainly of IgG antibodies (FIGURE 17.19). This accelerated, more intense response is called the **secondary response**. Antibodies produced during a secondary response have an even higher affinity for the antigen than those produced during a primary response, and thus they are more successful in disposing of it.

Primary and secondary responses occur, for example, during microbial infection. When you recover from an infection without taking antimicrobial drugs, it is usually because of the primary response. If the same microbe infects you later, the secondary response could be so swift that the microbes are destroyed before you exhibit any signs or symptoms of infection.

There Are Four Ways to Acquire Adaptive Immunity

Adaptive immunity can be acquired either actively or passively. Adaptive immunity is acquired *actively* if the person's own immune system develops antibodies and immune system cells (plasma cells, cytotoxic T cells, helper T cells, and B and T memory cells) against a specific antigen. Adaptive immunity is acquired *passively* if the person receives antibodies from another person or animal. Because there is no memory component involved, immunity that is passively acquired is only temporary and lasts until the antibodies degrade. Adaptive immunity can also be acquired naturally or artificially. Adaptive immunity is acquired *naturally* if the person develops immunity as a result of a natural event, such as exposure to an antigen by chance. Adaptive immunity is acquired *artificially* if the person develops immunity as a result of artificial means, such as deliberate exposure to an antigen through vaccination. Based on all of these factors, there are four ways to acquire adaptive immunity:

FIGURE 17.19 Production of antibodies in the primary and secondary responses to a given antigen.

Immunological memory is the basis for successful immunization by vaccination.

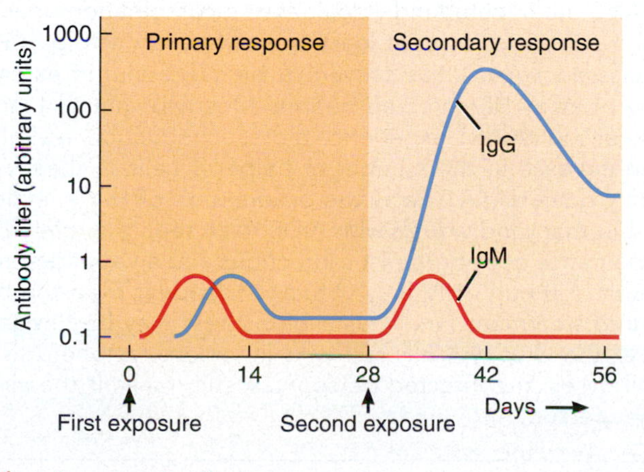

What is a vaccine?

- **Active natural immunity** develops when a person is exposed to an antigen by chance, becomes ill, and then produces antibody-secreting plasma cells, cytotoxic T cells, helper T cells, and B and T memory cells. This is the most common method of acquiring adaptive immunity.
- **Active artificial immunity** develops due to deliberate exposure to an antigen through vaccination. A **vaccine** is an attenuated (weakened) or dead antigen. The antigen, which is usually a microbe or portions of a microbe, has been pretreated to be immunogenic but not pathogenic, meaning that it will trigger an immune response but not cause significant illness. The antigens introduced through vaccination stimulate the immune system to produce antibodies and immune system cells, including memory cells.
- **Passive natural immunity** develops when antibodies are passed from mother to fetus through the placenta (IgG) or from mother to infant via breast milk (IgA). Neither the fetus nor an infant has a well-developed immune system; both are susceptible to illness. The antibodies introduced through passive natural immunity help the fetus and infant fight antigens that enter their bodies. However, these antibodies do not last forever: They eventually degrade, and the infant has to rely on his or her own developing immune system to provide protection.
- **Passive artificial immunity** occurs when a person receives serum containing antibodies from another person or an animal (such as a horse, rabbit, or goat) that has already been vaccinated against an antigen. This type of immunity provides immediate protection against an antigen and is the preferred type of adaptive immunity when there is not enough time for a person to develop his or her own antibodies and immune system cells. This protection is only temporary; the antibodies will eventually degrade and the person will have to develop his or her own immunity against the antigen.

TABLE 17.4 summarizes the various ways to acquire adaptive immunity.

TABLE 17.4 Ways to Acquire Adaptive Immunity

Method	Description
Active natural immunity	Following natural exposure to a microbe, antigen recognition by B cells and T cells and costimulation lead to the formation of antibody-secreting plasma cells, cytotoxic T cells, and B and T memory cells.
Active artificial immunity	Antigens introduced during a vaccination stimulate cell-mediated and antibody-mediated immune responses, leading to production of antibodies and immune cells. The antigens are pretreated to be immunogenic but not pathogenic.
Passive natural immunity	Transfer of IgG antibodies from mother to fetus across placenta, or of IgA antibodies from mother to baby in milk during breast feeding.
Passive artificial immunity	Intravenous injection of antibodies obtained from another human or animal donor; provides immediate protect against an antigen.

CLINICAL CONNECTION

Acquired Immunodeficiency Syndrome (AIDS)

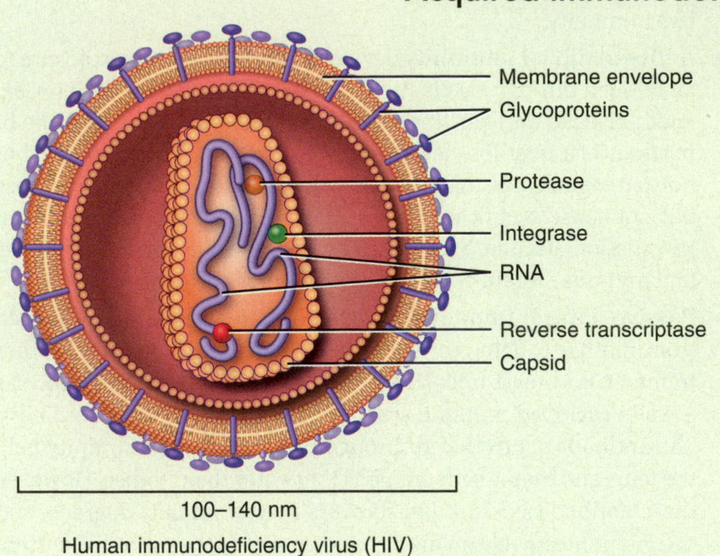

Human immunodeficiency virus (HIV)

Acquired immunodeficiency syndrome (**AIDS**) is a condition in which a person experiences a telltale assortment of infections due to the progressive destruction of immune system cells by the **human immunodeficiency virus** (**HIV**). AIDS represents the end stage of infection by HIV. A person who is infected with HIV may be symptom-free for many years, even while the virus is actively attacking the immune system.

HIV consists of several components: RNA (two strands), viral enzymes (reverse transcriptase, integrase, and protease), a capsid (protein coat), and a membrane envelope that is penetrated by glycoproteins (see FIGURE). HIV is classified as a **retrovirus** because its genetic information is carried in RNA instead of DNA. HIV infection of a host cell begins with the binding of HIV glycoproteins to receptors in the host cell's plasma membrane. This causes the host cell to transport the virus into its cytoplasm via receptor-mediated endocytosis. Once inside the host cell, HIV sheds its protein coat, and a viral enzyme called **reverse transcriptase** converts the viral RNA into DNA. The viral DNA is then integrated into the host cell's DNA via the viral enzyme **integrase**. Thus, the viral DNA is duplicated along with the host cell's DNA during normal cell division. In addition, the viral DNA can cause the infected cell to begin producing millions of copies of viral RNA and to form a capsid for each copy. Capsid formation involves the viral enzyme **protease**, which cuts proteins into pieces to assemble the capsid. Once new copies of HIV are formed, they bud off from the cell's plasma membrane and circulate in the blood to infect other cells.

HIV mainly destroys helper T cells—the very cells that orchestrate adaptive immune responses. Over 10 billion viral copies may be produced each day. The viruses bud so rapidly from the infected cell's plasma membrane that cell lysis eventually occurs. In most HIV-infected individuals, helper T cells are initially replaced as fast as they are destroyed. After a period of 2 to 10 years, however, HIV destroys enough helper T cells that most infected people begin to experience symptoms of immunodeficiency. As the immune system slowly collapses, an HIV-infected person becomes susceptible to a host of ***opportunistic infections***. These are diseases caused by microorganisms that are normally held in check but now proliferate because of the defective immune system. Examples of opportunistic infections include pneumocystis carinii pneumonia, tuberculosis, and Kaposi's sarcoma (KAP-ō-sēz) (a type of connective tissue cancer). AIDS is diagnosed when the helper T cell count drops below 200 cells per microliter of blood or when opportunistic infections arise, whichever occurs first. In time, opportunistic infections can lead to death.

Because HIV is present in the blood and some body fluids, it is transmitted most effectively by actions or practices that involve the exchange of blood or body fluids between people. HIV is transmitted in semen or vaginal fluid during unprotected (without a condom) anal, vaginal, or oral sex. HIV is also transmitted by direct blood-to-blood contact, such as occurs among intravenous drug users who share hypodermic needles. In addition, HIV can be transmitted from an HIV-infected mother to her baby at birth or during breast feeding. Within three to six months after HIV infection, plasma cells begin secreting antibodies against HIV. These antibodies are detectable in blood plasma and form the basis for some of the screening tests for HIV. When people test "HIV-positive," it usually means they have antibodies to HIV antigens in their bloodstream.

At present, infection with HIV cannot be cured. The recommended treatment for HIV-infected patients is **highly active antiretroviral therapy** (**HAART**), the use of antiretroviral drugs to inhibit HIV activity. There are three major classes of antiretroviral drugs:

1. **Reverse transcriptase inhibitors** interfere with the action of reverse transcriptase, preventing the enzyme from converting viral RNA into DNA. Among the drugs in this category are zidovudine (ZDV, previously called AZT), didanosine, and stavudine.
2. **Integrase inhibitors** interfere with the action of integrase. As a result, viral DNA is unable to insert into host cell DNA. The drug raltegravir is an example of an integrase inhibitor.
3. **Protease inhibitors** interfere with the action of protease, thereby blocking the formation of the viral capsid. Drugs in this category include nelfinavir, saquinavir, ritonavir, and indinavir.

In HAART, the infected individual takes a combination of antiretroviral medications from at least two differently acting inhibitor drug classes. HAART has proven to be successful in extending the life of many HIV-infected people. Most HIV-infected individuals receiving HAART experience a drastic reduction in viral load and an increase in the number of helper T cells in their blood. Not only does HAART delay the progression of HIV infection to AIDS, but many individuals with AIDS have seen the remission or disappearance of opportunistic infections and an apparent return to health. Although HIV may virtually disappear from the blood with drug treatment (and thus a blood test may be "negative" for HIV), the virus typically still lurks in various lymphoid tissues. In such cases, the infected person can still transmit the virus to another person.

Critical Thinking

Snake Venom—A Challenge for the Immune System

Jim is hiking in the West Virginia section of the Appalachian trial. He has been on the trail for a week and, although he is in good shape, he is feeling tired. Jim is getting ready to take a break for lunch and finds a blown-over tree to sit on. Unfortunately, Jim did not see the timber rattlesnake coiled under the tree trunk and actually stepped on the animal. While these snakes are normally not aggressive, this one responded to being stepped on by striking and biting Jim on his leg. Jim knows this is a serious situation and quickly moves away from the snake. He feels considerable pain at the bite site, but he knows he will not be able to get medical help for many hours.

SOME THINGS TO KEEP IN MIND:

The timber rattlesnake (*Crotalus horridus*) has an extensive range and can be found in the eastern United States and parts of the Midwest. These pit vipers normally feed on small mammals, frogs, and birds and may even eat other snakes. Timber rattlesnakes are dangerous because of their poisonous venom. The venom is "injected" into the victims through hollow fangs.

© Byron Jorjorian/Alamy Stock Photo

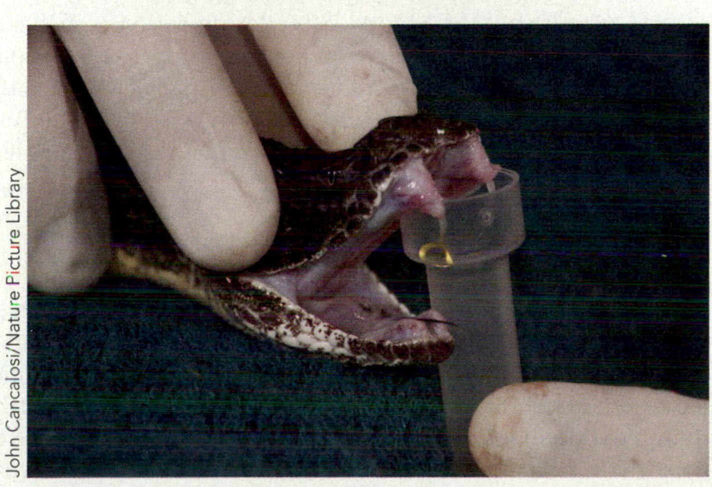

John Cancalosi/Nature Picture Library

SOME INTERESTING FACTS:

There are four variations of the type of venom that the timber rattlesnake can produce, mostly depending on the geographic range in which the snake is located.

Venom Type	Geographic Range	Venom Effects
Type A	Southern range.	Neurotoxic actions, resulting in uncontrollable muscle spasms that can lead to death because of the inability to control breathing muscles.
Type B	Northern range.	Hemotoxic actions, causing tissue destruction and massive bleeding.
Type A+B	Found where the northern and southern ranges overlap.	Characteristics of both type A venom and type B venom (neurotoxic actions and hemotoxic actions).
Type C	Varies; individual snakes in the northern range, southern range, or range of overlap can have this type of venom.	Weak actions compared to other types of venom but can be slightly neurotoxic.

How does Jim's immune system begin responding to this incident?

Because the blood vessels are dilated and more permeable, will any venom toxins move to other areas of the body?

Jim is experiencing increasing pain at the bite site that radiates through his leg. What is causing his pain?

The timber rattlesnake that bit Jim is likely to have type B venom because this is the most common form found in the northern range of the snake. What actions will one expect the toxins to cause?

What is the most successful treatment for a venomous bite?

Self-Recognition and Self-Tolerance Prevent the Immune System from Attacking the Body's Own Tissues

To function properly, your T cells must have two traits: (1) They must be able to *recognize* your own major histocompatibility complex (MHC) proteins, a process known as **self-recognition**, and (2) they must *lack reactivity* to peptide fragments from your own proteins, a condition known as **self-tolerance**. B cells also display self-tolerance. Loss of self-tolerance leads to the development of autoimmune diseases (see Clinical Connection: Autoimmune Diseases).

Pre-T cells in the thymus develop the capability for self-recognition via **positive selection**. In this process, some pre-T cells express T-cell receptors (TCRs) that interact with self-MHC proteins on epithelial cells in the thymus. Because of this interaction, the T cells can recognize the MHC part of an antigen–MHC complex. These T cells survive. Other immature T cells that fail to interact with thymic epithelial cells are not able to recognize self-MHC proteins. These cells undergo apoptosis.

The development of self-tolerance occurs by a weeding-out process called **negative selection**, in which the T cells interact with dendritic cells in the thymus. In this process, T cells with receptors that react to self-peptide fragments or other self-antigens are eliminated or inactivated. The T cells selected to survive do not respond to self-antigens, the fragments of molecules that are normally present in the body. Negative selection occurs via both deletion and anergy. In **deletion**, self-reactive T cells undergo apoptosis and die; in **anergy**, they remain alive but are unresponsive to antigenic stimulation. Only 1–5% of the immature T cells in the thymus receive the proper signals to survive apoptosis during both positive and negative selection and emerge as mature, immunocompetent T cells.

Once T cells have emerged from the thymus, they may still encounter an unfamiliar self-protein; in such cases they may also become anergic if there is no costimulator. Deletion of self-reactive T cells may also occur after they leave the thymus.

B cells also develop self-tolerance through deletion and anergy. While B cells are developing in bone marrow, those cells exhibiting antigen receptors that react to common self-antigens (such as MHC proteins or blood group antigens) are deleted. Once B cells are released into the blood, however, anergy appears to be the main mechanism for preventing responses to self-proteins. When B cells encounter an antigen not associated with an antigen-presenting cell, the necessary costimulation signal is often missing. In this case, the B cell is likely to become anergic (inactivated) rather than activated.

CLINICAL CONNECTION

Autoimmune Diseases

In an **autoimmune disease** or **autoimmunity**, the immune system fails to display self-tolerance and attacks the person's own tissues. Various mechanisms produce different autoimmune diseases. Some involve production of **autoantibodies**, antibodies that bind to and stimulate or block self-antigens. For example, autoantibodies that mimic thyroid-stimulating hormone (TSH) are present in Graves' disease and stimulate secretion of thyroid hormones (thus producing hyperthyroidism); autoantibodies that bind to and block acetylcholine receptors cause the muscle weakness characteristic of myasthenia gravis. Other autoimmune diseases involve activation of cytotoxic T cells that destroy certain body cells. Examples include type 1 diabetes mellitus, in which T cells attack the insulin-producing pancreatic beta cells, and multiple sclerosis (MS), in which T cells attack myelin sheaths around axons of neurons. Inappropriate activation of helper T cells also occurs in certain autoimmune diseases. Other examples of autoimmune disorders include rheumatoid arthritis, systemic lupus erythematosus, rheumatic fever, and Addison's disease. Therapies for various autoimmune diseases include removal of the thymus gland (thymectomy); injections of beta interferon; immunosuppressive drugs; and plasmapheresis, in which the person's blood plasma is filtered to remove antibodies and antigen–antibody complexes.

Allergic Reactions May Be Immediate or Delayed

A person who is overly reactive to a substance that is tolerated by most other people is said to be **allergic** or **hypersensitive**. Whenever an allergic reaction takes place, some tissue injury occurs. The antigens that induce an allergic reaction are called **allergens**. Common allergens include certain foods (milk, peanuts, shellfish, eggs), antibiotics (penicillin, tetracycline), venoms (bee, spider, snake), cosmetics, poison ivy, pollen, dust, and mold.

Two major types of allergic reactions are immediate hypersensitivity reactions and delayed hypersensitivity reactions. **Immediate hypersensitivity reactions** occur within a few minutes after a person who is already sensitized to an allergen is reexposed to it. In response to the first exposure to certain allergens, some people produce IgE antibodies that bind to receptors on the surface of mast cells and basophils, a process known as **sensitization** (**FIGURE 17.20**). The next time the same allergen enters the body, it attaches to the IgE antibodies already present. In response, the mast cells and basophils undergo **degranulation**, releasing histamine from preformed granules (**FIGURE 17.20**). After degranulation occurs, mast cells and basophils also release leukotrienes and prostaglandins. Collectively, histamine, leukotrienes, and prostaglandins function as inflammatory agents that dilate blood vessels (vasodilation), increase blood capillary permeability, contract smooth muscle in the airways of the lungs, increase mucus secretion, and stimulate nerve endings that cause pain and itching. As a result, a person may experience signs and symptoms of **hay fever**: nasal congestion, sneezing, runny nose, itching, watery eyes (tearing), and difficulty in breathing. Allergic reactions are usually localized to the region where the allergen enters the body. However, if large amounts of allergen or inflammatory chemicals enter the bloodstream, allergic reactions can become systemic and affect the whole body. In **anaphylactic shock**, shortness of breath and wheezing as airways constrict are usually accompanied by shock due to systemic vasodilation and fluid loss from blood. Anaphylactic shock may occur

FIGURE 17.20 Mechanism of immediate hypersensitivity reactions.

Immediate hypersensitivity reactions occur within a few minutes after a person who is already sensitized to an allergen is reexposed to it.

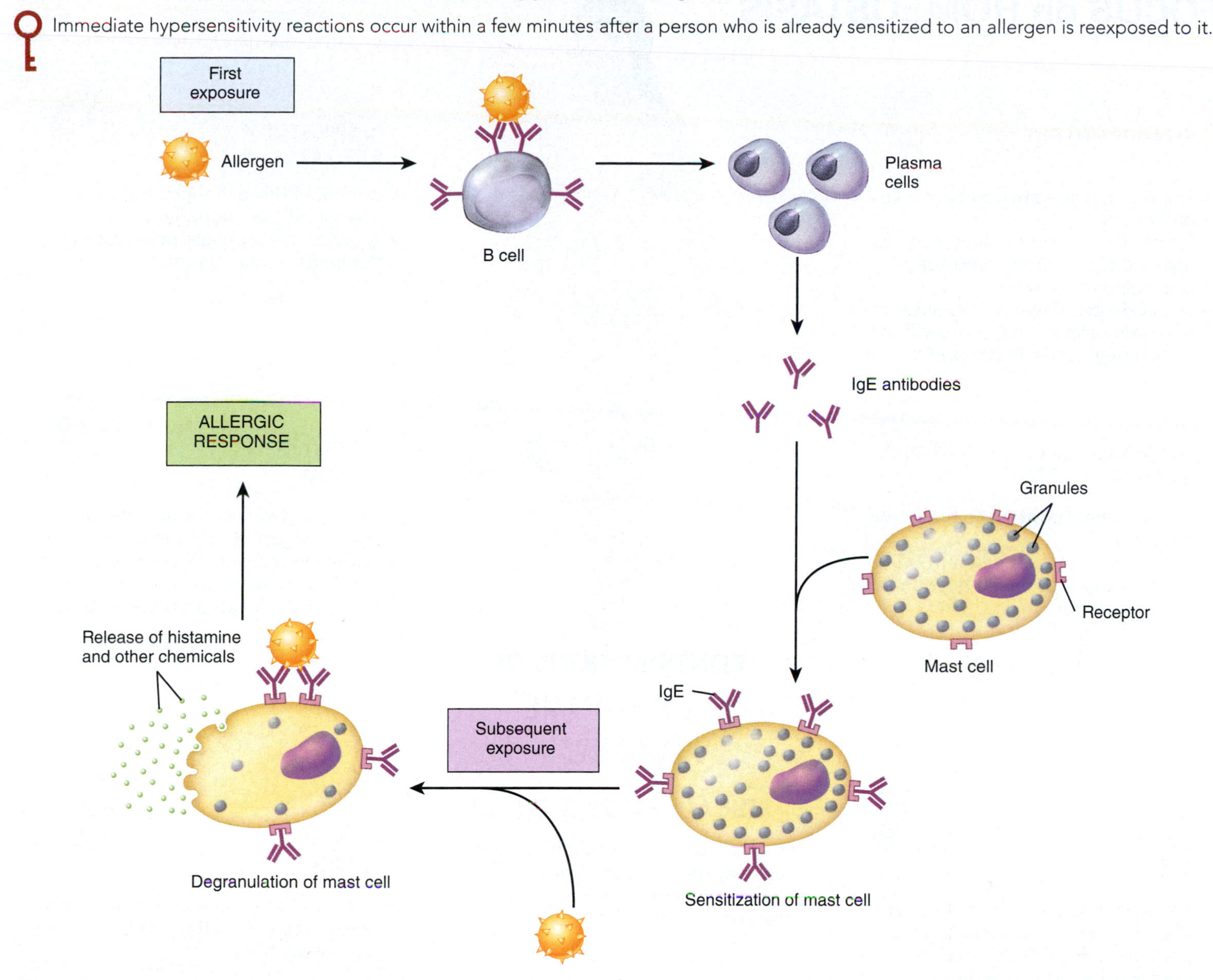

What are some examples of allergens?

in a susceptible individual who has just received a triggering drug or been stung by a bee. This life-threatening emergency is usually treated by injecting epinephrine to dilate the airways and strengthen the heartbeat.

Delayed hypersensitivity reactions usually appear 12–72 hours after exposure to an allergen. These reactions occur when allergens are taken up by antigen-presenting cells (such as Langerhans cells in the skin) that migrate to lymph nodes and present the allergen to T cells, which then proliferate. Some of the new T cells return to the site of allergen entry into the body, where they produce gamma-interferon, which activates macrophages, and tumor necrosis factor, which stimulates an inflammatory response. Intracellular bacteria such as *Mycobacterium tuberculosis* trigger this type of immune response, as do certain haptens, such as poison ivy toxin. The skin test for tuberculosis is also a delayed hypersensitivity reaction.

CHECKPOINT

15. How do the five classes of antibodies differ in structure and function?
16. How are cell-mediated and antibody-mediated immune responses similar and different?
17. How is the secondary response to an antigen different from the primary response?
18. What do positive selection, negative selection, and anergy accomplish?
19. What are some examples of allergens? How do immediate hypersensitivity reactions differ from delayed hypersensitivity reactions?

To appreciate the many ways that the immune system contributes to homeostasis of other body systems, examine *Focus on Homeostasis: Contributions of the Immune System.*

FOCUS on HOMEOSTASIS

INTEGUMENTARY SYSTEM

- The skin acts as a physical barrier to pathogens
- Langerhans cells in the skin alert the immune system to the presence of pathogens in the skin
- The acidic pH of sebum (oil) released onto skin surface inhibits growth of certain pathogenic bacteria and fungi

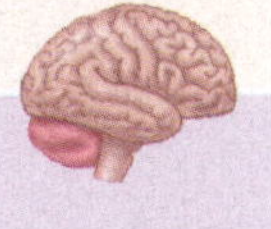

NERVOUS SYSTEM

- Microglia help protect the brain from pathogens
- Interleukin-1 (IL-1) released from macrophages acts on the hypothalamus to cause a fever

ENDOCRINE SYSTEM

- The thymus releases the hormones thymosin and thymopoietin, which promote normal T cell function in lymphoid organs and tissues

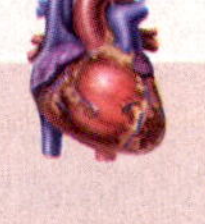

CARDIOVASCULAR SYSTEM

- Macrophages in spleen destroy pathogens that are present in the blood and remove aged erythrocytes
- Antibodies provide protection against foreign substances in the blood

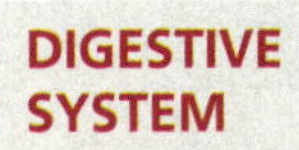

CONTRIBUTIONS OF THE IMMUNE SYSTEM

FOR ALL BODY SYSTEMS

- B cells, T cells, and antibodies protect all body systems from attack by harmful foreign invaders (pathogens), foreign cells, and cancer cells

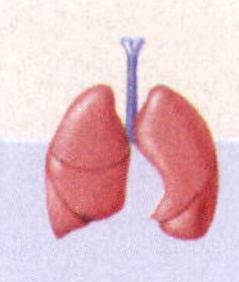

RESPIRATORY SYSTEM

- Lymphoid tissue in tonsils defends against inhaled pathogens
- Alveolar macrophages help protect the lungs from pathogens

DIGESTIVE SYSTEM

- Lymphoid tissue in tonsils defends against ingested pathogens
- Lymphoid tissue in Peyer's patches and appendix defends against pathogens that invade the gastrointestinal tract

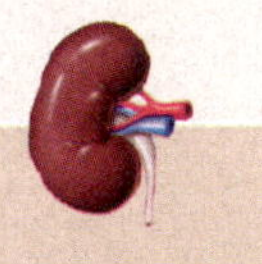

URINARY SYSTEM

- Lymphoid tissue in lining of the urethra defends against pathogens that invade the urinary tract

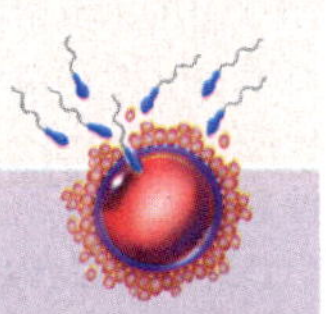

REPRODUCTIVE SYSTEMS

- Lymphoid tissue in the wall of the reproductive tract defends against pathogens that enter the body via the penis or vagina
- IgG antibodies can cross the placenta to provide protection to a developing fetus
- IgA antibodies in breast milk provide protection to a nursing infant

FROM RESEARCH TO REALITY

How Location Can Affect the Incidence of Autoimmune Disease

Reference

Kondrashova, A. et al. (2012). The "Hygiene hypothesis" and the sharp gradient in the incidence of autoimmune and allergic diseases between Russian Karelia and Finland. *Acta Pathologica Microbiologica Immunologica Scandinavica*. 121: 478–493.

Are populations of certain countries more susceptible to developing autoimmune diseases than others?

Article description:

In this study, researchers explored the incidence of type 1 diabetes in populations of children from 1–14 years of age in Finland and Russian Karelia. The two countries are close together geographically, but the populations have dramatically different socioeconomic statuses. Russian Karelia has a gross domestic product nearly 15 times lower than that of Finland. However, children in Russian Karelia experience fewer infectious diseases and parasites than children of the same age in Finland. These two neighboring populations living close to each other in different socioeconomic circumstances create an ideal setting to explore the pathogenesis of immune-mediated diseases.

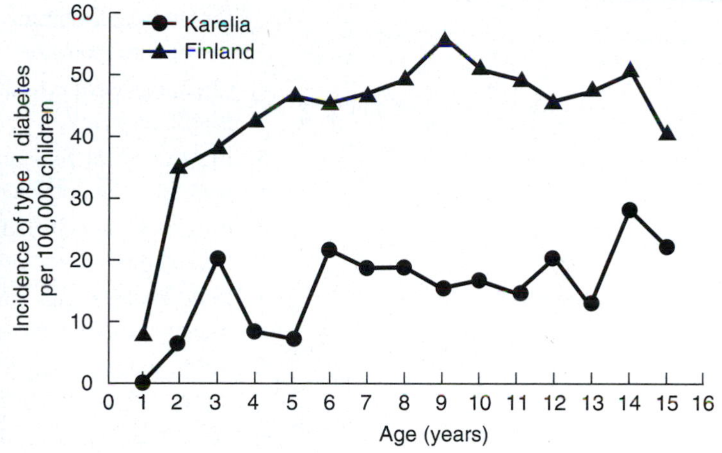

Autoimmune diseases are fairly common; they occur because the body's immune system attacks its own cells, leading to disease and sometimes death. Does where you grow up dictate your susceptibility to developing an autoimmune disease? This study explores populations in two neighboring countries and their susceptibility to developing type 1 diabetes, a type of autoimmune disorder.

Go to WileyPLUS Learning Space and use the data from this article to answer the questions posed there and to discover more about how the incidence of autoimmune disease can vary in different countries.

CHAPTER REVIEW

17.1 Components of the Immune System

1. The immune system consists of two major components: (1) cells that provide immune responses and (2) lymphoid organs and tissues.
2. The cells of the immune system include leukocytes (neutrophils, eosinophils, basophils, monocytes, and lymphocytes), mast cells, and dendritic cells. Each of these cells has a particular role in immunity.
3. Lymphoid organs and tissues are structures in which lymphocytes develop, reside, or carry out immune responses.

4. Primary lymphoid organs (bone marrow and thymus) are the sites where stem cells divide and develop into mature B cells and T cells.
5. Secondary lymphatic organs, which include lymph nodes, the spleen, and lymphoid nodules, are the sites where most immune responses occur.

17.2 Innate Immunity

1. Innate immunity is the ability of the body to defend itself against microbes and other foreign substances using mechanisms that do not involve specific recognition of the invading agents.
2. Among the components of innate immunity are the first line of defense (the physical and chemical barriers of the body) and the second line of defense (antimicrobial substances, natural killer cells, phagocytes, inflammation, and fever).
3. Physical barriers to the entrance of microbes into body tissues include the skin, mucus secreted by the linings of various hollow organs, hairs within the nose, and cilia lining the upper respiratory tract.
4. Chemical barriers to the entrance of microbes into body tissues include sebum, lysozyme, gastric juice, and vaginal secretions.
5. Antimicrobial substances include interferons, the complement system, iron-binding proteins, and antimicrobial proteins.
6. Natural killer cells kill infected body cells and cancer cells. They accomplish this function by releasing perforin molecules and granzymes.
7. Phagocytes engulf and destroy microbes and cellular debris. The two major types of phagocytes are neutrophils and macrophages.
8. Inflammation aids disposal of microbes, toxins, or foreign material at the site of an injury, and prepares the site for tissue repair.
9. Fever intensifies the antiviral effects of interferons, inhibits growth of some microbes, and speeds up body reactions that aid repair.

17.3 Adaptive Immunity

1. Adaptive immunity is the ability of the body to defend itself against specific microbes and other foreign agents. It involves lymphocytes called B cells and T cells.
2. B cells and T cells arise from stem cells in bone marrow. B cells mature in bone marrow, whereas T cells mature in the thymus. Before B cells leave bone marrow or T cells leave the thymus, they develop immunocompetence, the ability to carry out adaptive immune responses. This process involves the insertion of antigen receptors into their plasma membranes. Antigen receptors are molecules that are capable of recognizing specific antigens. Two major types of mature T cells exit the thymus: helper T cells (also known as CD4 cells) and cyotoxic T cells (also referred to as CD8 cells).
3. There are two types of adaptive immunity: cell-mediated immunity and antibody-mediated immunity. In cell-mediated immune responses, cytotoxic T cells attack infected body cells, cancer cells, and foreign cells from tissue transplants; in antibody-mediated immune responses, B cells transform into plasma cells that secrete antibodies.
4. Clonal selection is the process by which a lymphocyte proliferates and differentiates in response to a specific antigen. The result of clonal selection is the formation of a clone of cells that can recognize the same specific antigen as the original lymphocyte.
5. Antigens are chemical substances that are recognized as foreign by the immune system. Antigen receptors exhibit great diversity due to genetic recombination.

6. Self-antigens called major histocompatibility complex (MHC) proteins are unique to each person's body cells. All cells except erythrocytes display MHC-I molecules. Antigen-presenting cells (APCs) display MHC-II molecules. APCs include macrophages, B cells, and dendritic cells. Exogenous antigens (formed outside body cells) are presented with MHC-II molecules; endogenous antigens (formed inside body cells) are presented with MHC-I molecules.
7. Cytokines are local hormones (paracrines or autocrines) that may stimulate or inhibit many normal cell functions such as growth and differentiation. Other cytokines regulate immune responses.
8. A cell-mediated immune response begins with activation of a small number of T cells by a specific antigen. During the activation process, T-cell receptors (TCRs) recognize antigen fragments associated with MHC molecules on the surface of a body cell. Activation of T cells also requires costimulation, either by cytokines such as interleukin-2 or by pairs of plasma membrane molecules. Once a T cell has been activated, it undergoes clonal selection. The result of clonal selection is the formation of a clone of effector cells and memory cells.
9. Helper T cells display CD4 protein; recognize antigen fragments associated with MHC-II molecules; and secrete several cytokines, most important, interleukin-2, which acts as a costimulator for other helper T cells, cytotoxic T cells, and B cells.
10. Cytotoxic T cells display CD8 protein and recognize antigen fragments associated with MHC-I molecules. Active cytotoxic T cells eliminate infected body cells, cancer cells, and foreign cells from tissue transplants by releasing perforin molecules and granzymes.
11. An antibody-mediated immune response begins with activation of a B cell by a specific antigen. B cells can respond to unprocessed antigens, but their response is more intense when they process the antigen. Interleukin-2 and other cytokines secreted by helper T cells provide costimulation for activation of B cells. Once activated, a B cell undergoes clonal selection, forming a clone of plasma cells and memory cells. Plasma cells are the effector cells of a B cell clone; they secrete antibodies.
12. An antibody is a protein that combines specifically with the antigen that triggered its production. Antibodies consist of heavy and light chains, and variable and constant regions. Based on chemistry and structure, antibodies are grouped into five principal classes (IgG, IgA, IgM, IgD, and IgE), each with specific biological roles. Actions of antibodies include neutralization of antigen, agglutination of antigen, precipitation of antigen, activation of complement, and opsonization.
13. Immunization against certain microbes is possible because memory B cells and memory T cells remain after a primary response to an antigen. The secondary response provides protection should the same microbe enter the body again.
14. T cells undergo positive selection to ensure that they can recognize self-MHC proteins (self-recognition), and negative selection to ensure that they do not react to other self-proteins (self-tolerance). Negative selection involves both deletion and anergy. B cells develop tolerance through deletion and anergy.
15. An allergen is an antigen that induces an allergic reaction. Two major types of allergic reactions are immediate hypersensitivity reactions and delayed hypersensitivity reactions.

PONDER THIS

1. Esperanza watched as her mother got a flu shot. "Why do you need a shot if you're not sick?" she asked. "So I won't get sick," answered her mom. Explain how the influenza vaccination prevents illness and the type of adaptive immunity that it provides.

2. Due to the presence of breast cancer, Janet had a right radical mastectomy in which her right breast and underlying muscle, right axillary lymph nodes, and vessels were removed. Now she is experiencing severe swelling in her right arm. Why did the surgeon remove lymphoid tissue as well as the breast? Why is Janet's right arm swollen?

3. Tariq's little sister has the mumps. Tariq can't remember if he has had mumps or not, but he is feeling slightly feverish. How could Tariq's doctor determine if he is getting sick with mumps or if he has previously had mumps?

ANSWERS TO FIGURE QUESTIONS

17.1 Basophils promote inflammation and are involved in allergic reactions.

17.2 The spleen filters microbes and other foreign materials from the blood. It also removes aged or defective erythrocytes from the blood.

17.3 Langerhans cells are a type of dendritic cell; they alert the immune system to the presence of microbes that invade the skin.

17.4 Interferons activate natural killer cells and cytotoxic T cells (both of which kill infected body cells and cancer cells). In addition, interferons inhibit cell division and suppress the formation of tumors.

17.5 The membrane attack complex creates a channel in the microbe plasma membrane that results in cytolysis of the microbial cell due to the inflow of extracellular fluid through the channel.

17.6 Perforin is a protein that inserts into the plasma membrane of the target cell and creates channels in the membrane. As a result, extracellular fluid flows into the target cell and cytolysis occurs.

17.7 Lysozyme, digestive enzymes, and oxidants can kill microbes ingested during phagocytosis.

17.8 Diapedesis is the process by which a phagocyte squeezes through the pores in the wall of a blood capillary and then enters into interstitial fluid.

17.9 Helper T cells participate in both cell-mediated and antibody-mediated immune responses.

17.10 Epitopes are small immunogenic parts of a larger antigen; haptens are small molecules that become immunogenic only when they attach to a body protein.

17.11 APCs include dendritic cells, macrophages, and B cells.

17.12 Endogenous antigens include viral proteins, toxins from intracellular bacteria, and abnormal proteins synthesized by a cancerous cell.

17.13 The first signal in T-cell activation is antigen binding to a T cell receptor (TCR); the second signal is a costimulator, such as a cytokine or another pair of plasma membrane molecules.

17.14 The CD8 protein of a cytotoxic T cell binds to the MHC-I molecule of an infected body cell to help anchor the T cell receptor (TCR)–antigen interaction so that antigen recognition can occur.

17.15 Both cytotoxic T cells and natural killer cells kill their target cells by releasing perforin molecules and granzymes. The major difference is that cytotoxic T cells have receptors specific for a particular antigen and thus kill only target cells that display that antigen; natural killer cells destroy target cells without recognizing a particular antigen.

17.16 Because all the plasma cells in Figure 17.16 are part of the same clone, they secrete just one kind of antibody.

17.17 The variable regions recognize and bind to a specific antigen.

17.18 Antibodies can cross-link soluble antigens into complexes that are too large to stay in solution. As a result, the complex precipitates out of solution, which makes it easier for the antigen to be phagocytized.

17.19 A vaccine contains attenuated (weakened) or killed whole microbes or portions of microbes that activate B cells and T cells.

17.20 Examples of allergens include certain foods (milk, peanuts, shellfish, eggs), antibiotics (penicillin, tetracycline), venoms (bee, spider, snake), cosmetics, poison ivy, pollen, dust, and mold.

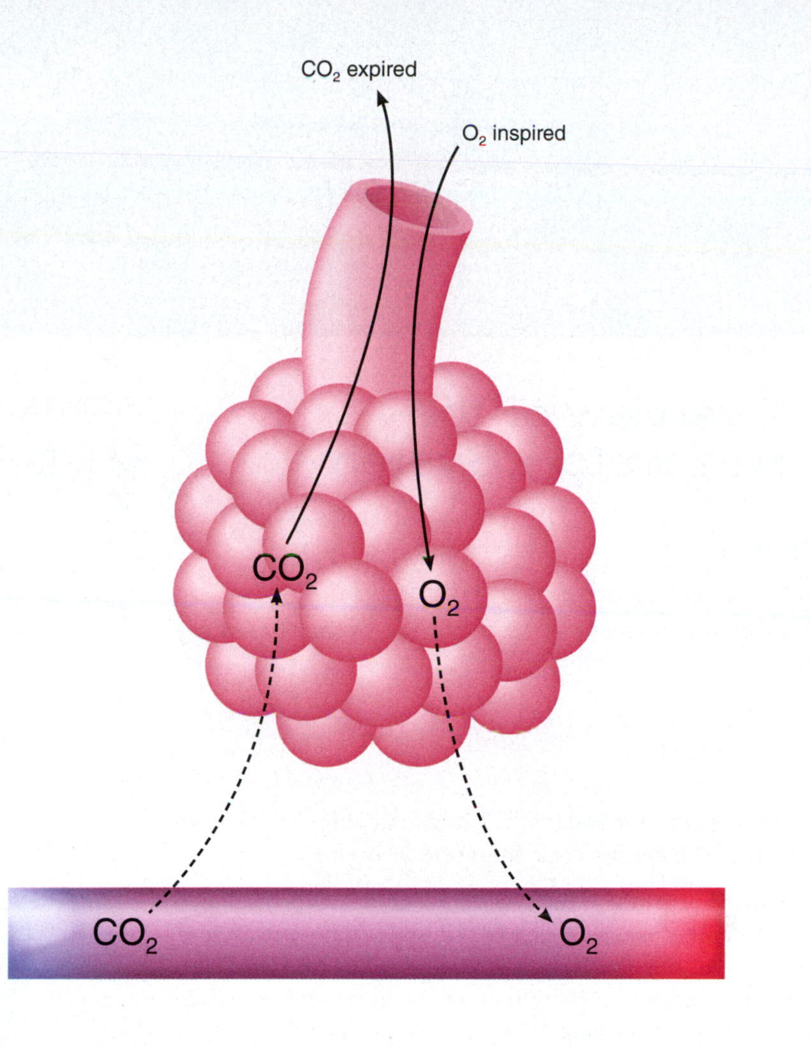

The Respiratory System

The Respiratory System and Homeostasis

The respiratory system contributes to homeostasis by providing for the exchange of gases—oxygen and carbon dioxide—between the atmospheric air, blood, and tissue cells. It also helps adjust the pH of body fluids.

LOOKING BACK TO MOVE AHEAD...

- The hydrogen bonds that link neighboring water molecules give water considerable cohesion; the cohesion of water molecules creates a very high surface tension (Section 2.2).
- Cilia are short, hairlike projections that extend from the surface of a cell; each cilium contains a core of microtubules surrounded by plasma membrane (Section 3.3).
- Cellular respiration is the process by which a nutrient molecule such as glucose, a fatty acid, or an amino acid is broken down in the presence of oxygen to form carbon dioxide, water, and energy (ATP and heat) (Section 4.4).
- Hemoglobin is a red-pigment molecule found in erythrocytes; it has binding sites for both oxygen and carbon dioxide (Section 16.2).

Your body's cells continually use oxygen (O_2) for the metabolic reactions that generate ATP from the breakdown of nutrient molecules. At the same time, these reactions release carbon dioxide (CO_2) as a waste product. Because an excessive amount of CO_2 produces acidity that can be toxic to cells, excess CO_2 must be eliminated quickly and efficiently. You inhale needed O_2 and exhale the waste product CO_2 because of the respiratory system. In addition, the respiratory system helps regulate blood pH, contains receptors for the sense of smell, filters inspired air, produces sounds, and rids the body of some water and heat in exhaled air. In this chapter, you will learn about the various functions of the respiratory system.

18.1 Overview of the Respiratory System

OBJECTIVES

- Discuss the steps that occur during respiration.
- Describe the functions of each component of the respiratory system.

Respiration Supplies the Body with O_2 and Removes CO_2

The process of supplying the body with O_2 and removing CO_2 is known as **respiration**, which has five basic steps (**FIGURE 18.1**):

1. **Ventilation (breathing).** Air flows into and out of the lungs. Movement of air into the lungs is called *inspiration (inhalation).* Movement of air out of the lungs is referred to as *expiration (exhalation).* Inspiration allows O_2 to enter the lungs and expiration permits CO_2 to leave the lungs.
2. **Pulmonary gas exchange.** Gases are exchanged between the alveoli (air sacs) of the lungs and the blood in pulmonary capillaries. In this step, pulmonary capillary blood gains O_2 and loses CO_2.
3. **Transport of O_2 and CO_2 by the blood.** The blood carries O_2 from the lungs to tissue cells and CO_2 from tissue cells to the lungs.
4. **Systemic gas exchange.** Gases are exchanged between blood in systemic capillaries and tissue cells of the body. In this step, systemic capillary blood loses O_2 and gains CO_2.
5. **Cellular respiration.** Cells consume O_2 and give off CO_2 as metabolic reactions break down nutrient molecules in order to produce ATP.

The respiratory system does not carry out all of the steps of respiration; it is responsible only for ventilation and gas exchange. Transport of O_2 and CO_2 by the blood is a function of the cardiovascular system. Cellular respiration is accomplished by metabolic reactions in the cytosol and mitochondria of any given body cell.

The Respiratory System Is Comprised of Several Organs

The **respiratory system** consists of the nose, pharynx, larynx, trachea, primary bronchi, and lungs (**FIGURE 18.2a, c**). The lungs contain smaller bronchi, bronchioles, and numerous microscopic air sacs called *alveoli*, which are the sites of gas exchange between the air and blood. All of the structures that carry air to and from the alveoli of the lungs are collectively referred to as the **airways**. As in the digestive and urinary systems, which will be covered in subsequent chapters, in the

FIGURE 18.1 The steps involved in respiration.

During the process of respiration, the body is supplied with O_2 and CO_2 is removed.

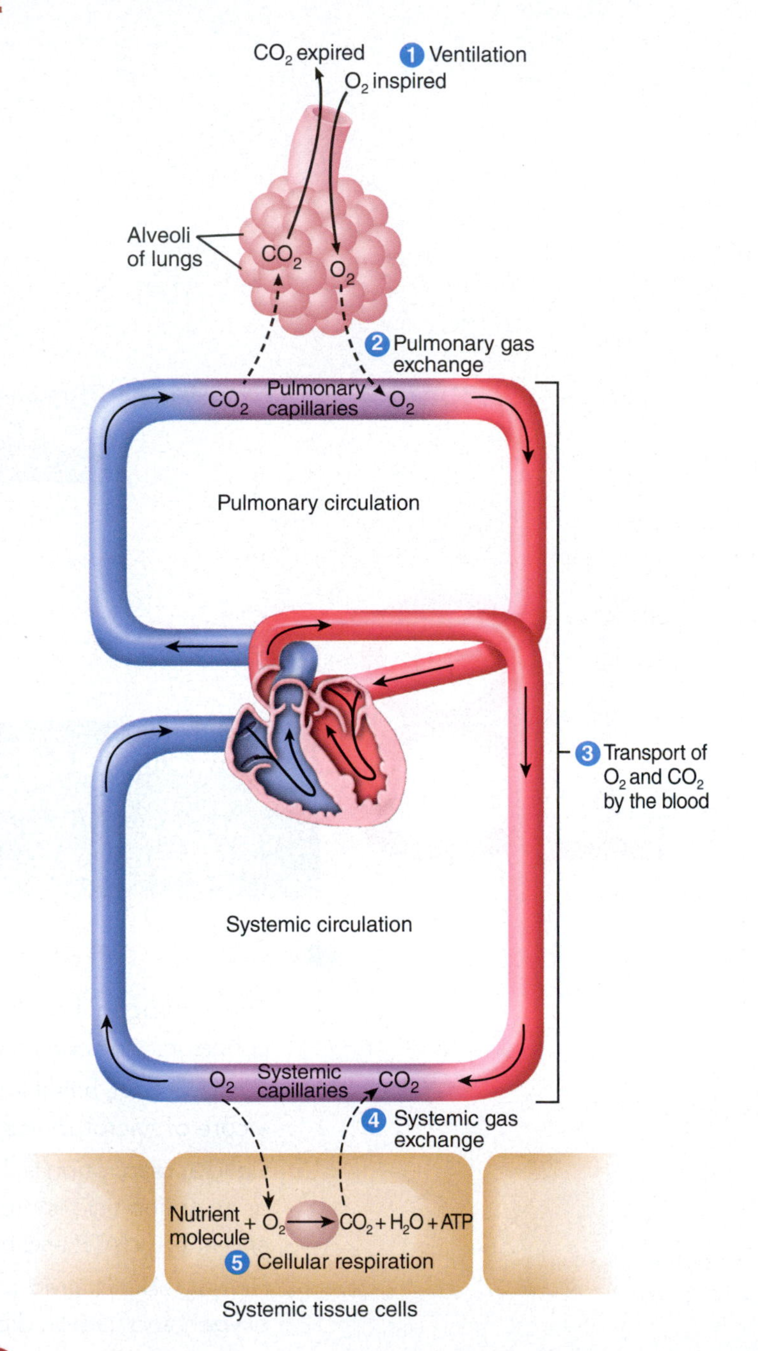

How does pulmonary gas exchange differ from systemic gas exchange?

FIGURE 18.2 Components of the respiratory system.

The respiratory system consists of the nose, pharynx, larynx, trachea, primary bronchi, and lungs.

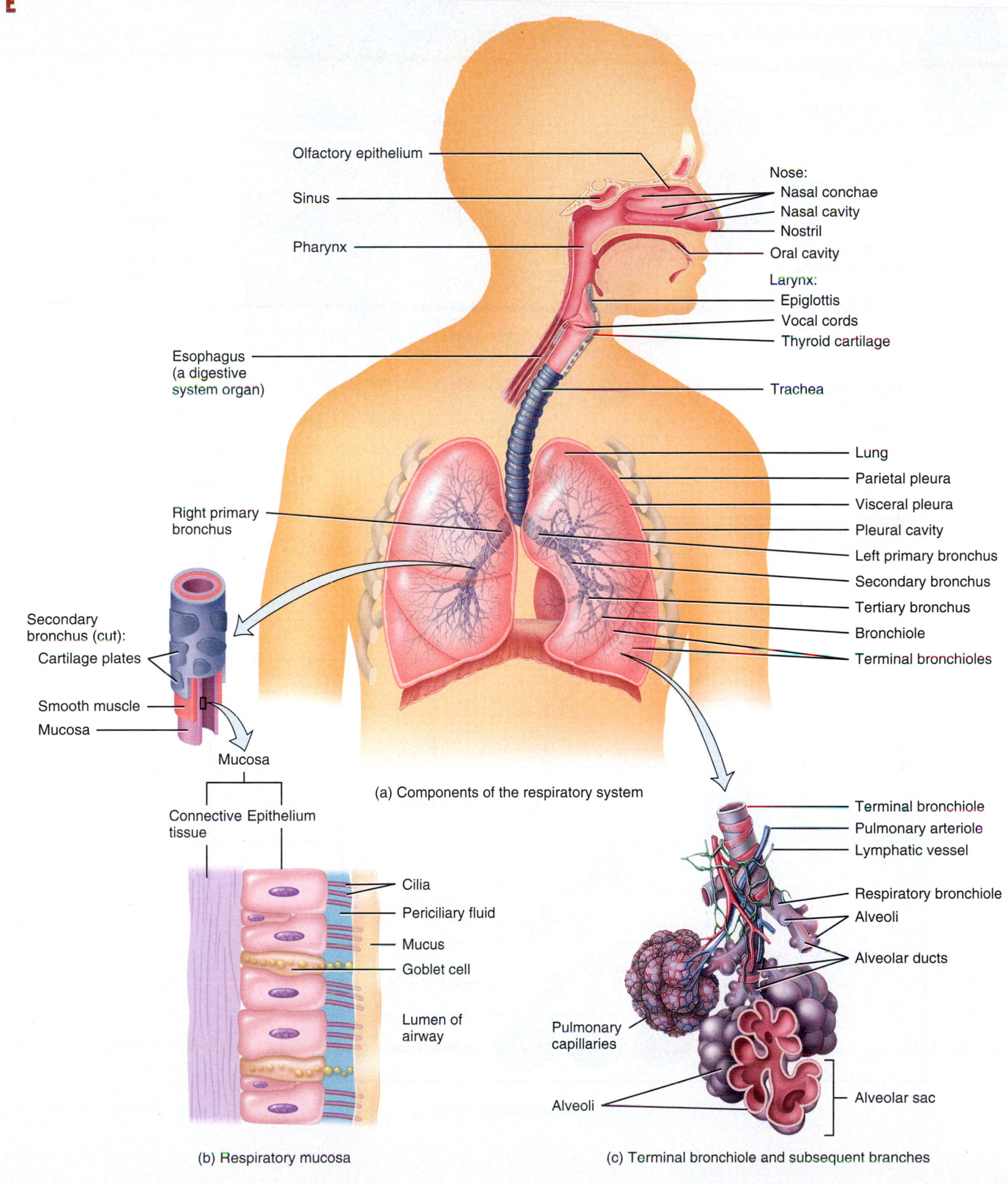

Figure 18.2 (continues)

Figure 18.2 Continued

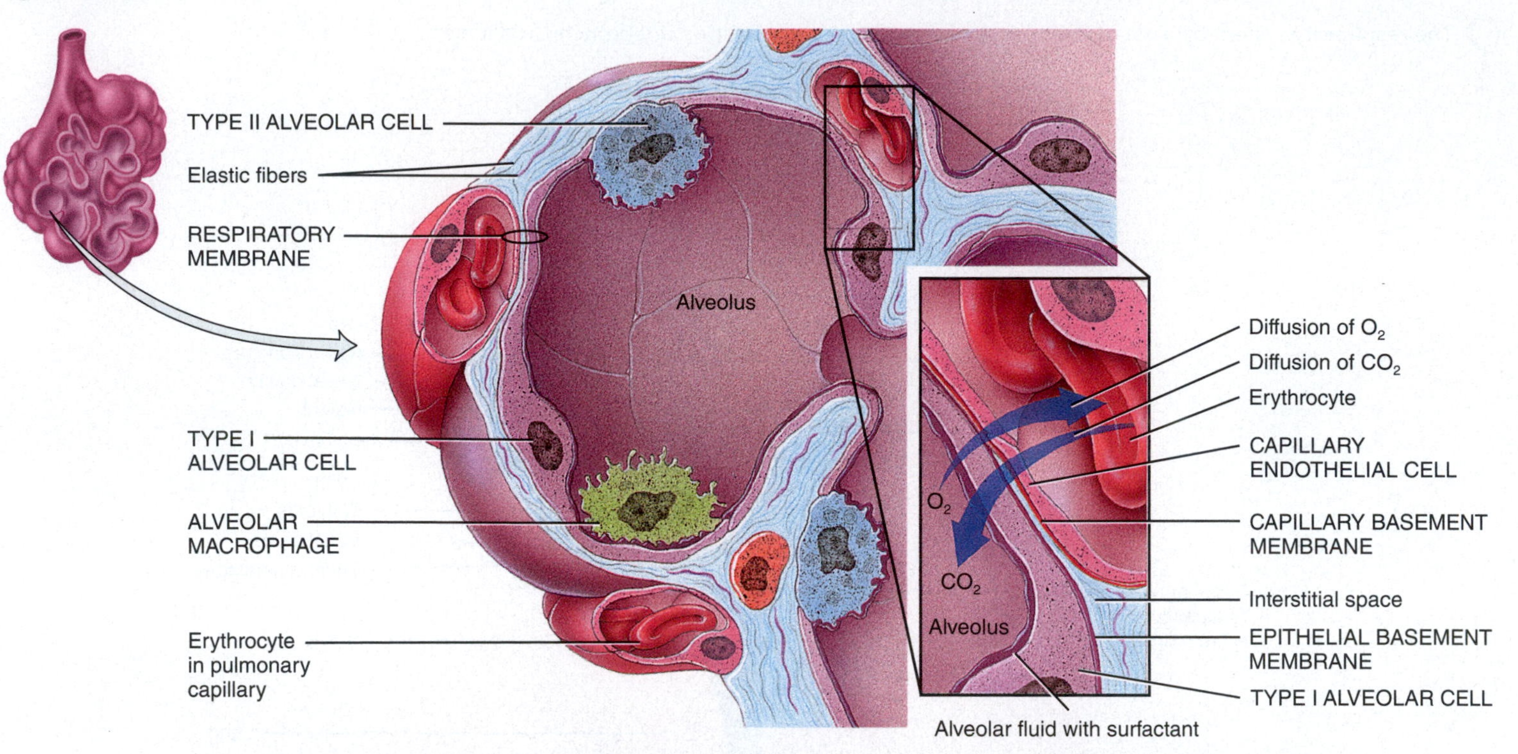

(d) Section through an alveolus showing cellular components

(e) Details of respiratory membrane

Airway branching		
	Names of branches	Generation #
Conducting zone	Trachea	0
	Primary bronchi	1
	Smaller bronchi ↓	2–10
	Bronchioles	11–16
Respiratory zone	Respiratory bronchioles	17–19
	Alveolar ducts	20–22
	Alveolar sacs	23

(f) Airway branching and functional zones

What are the two functional zones of the respiratory system and how do they differ?

respiratory system there is an extensive area of contact between the external environment and capillary blood vessels.

The Respiratory Mucosa Produces Mucus That Traps Particles and Lubricates the Airways

The airways from the nose to the bronchioles of the lungs are lined by a mucosa. A **mucosa**, or *mucous membrane*, consists of a layer of epithelial cells and an underlying layer of connective tissue (**FIGURE 18.2b**). The epithelium of many parts of the airways contains ciliated cells (cells with cilia attached to them) and scattered **goblet cells** that secrete mucus. **Mucus** is a sticky secretion that traps inhaled particles and serves as a lubricant for the lining of the respiratory tract. **Cilia** are short, hairlike projections that extend from the surface of a cell. The cilia in the nose move mucus and trapped particles down toward the pharynx; the cilia in the larynx, trachea, bronchi, and bronchioles move these substances up toward the pharynx. The term **mucociliary escalator** refers to the movement of mucus along the respiratory tract toward the pharynx. Once the mucus and trapped particles reach the pharynx, they can be swallowed or expectorated (spit out). The movement of cilia is paralyzed by nicotine in cigarette smoke. For this reason, smokers cough often to remove foreign particles from their airways.

Only the tips of the cilia of the respiratory mucosa actually make contact with mucus. Beneath the mucus and surrounding the remaining parts of the cilia is a thin, watery saline layer known as **periciliary fluid** (**FIGURE 18.2b**). Periciliary fluid facilitates movement of mucus along the respiratory tract. If the volume of periciliary fluid is reduced (see Clinical Connection: Cystic Fibrosis), the mucus thickens and entangles the cilia, the cilia are unable to move the mucus, and the mucus clogs the airways.

The Nose Brings Air into the Respiratory System

The airways begin with the **nose**, the site where air normally enters the respiratory system. Air can also enter through the mouth (oral cavity), especially when you have a cold and your nose is congested. The space within the nose is called the **nasal cavity**, which opens to the exterior as the **nostrils** (**FIGURE 18.2a**). A major function of the nose is to filter, warm, and humidify incoming air. This occurs in the following way: When air enters the nostrils, it passes coarse hairs that trap and filter out large dust particles. The air then flows over shelflike extensions of bone called **nasal conchae** that extend from the wall of the nasal cavity. The conchae cause inhaled air to become turbulent. As the air swirls, additional airborne particles are filtered out as they come in contact with the mucus that lines the nasal cavity. The turbulence also allows the incoming air to be warmed by blood circulating in abundant capillaries and to be humidified by droplets of water evaporating from the mucosal surface. Another function of the nose is to detect olfactory stimuli: Recall that olfactory receptors are located in the upper part of the nasal cavity in a region called the olfactory epithelium (see Section 9.3). The nose also modifies speech vibrations as they pass through the *sinuses*—large, hollow resonating chambers of the head. *Resonance* refers to prolonging, amplifying, or modifying a sound by vibration.

The Pharynx Is a Common Passageway for Air and Food

The **pharynx**, or throat, is a funnel-shaped tube that extends from the nasal and oral cavities to the larynx and esophagus (a structure that plays a role in digestion) (**FIGURE 18.2a**). Because of these connections, the pharynx is part of both the respiratory and digestive systems. Functionally, the pharynx serves as a passageway for air and food and acts as a resonating chamber for speech sounds.

The Larynx Routes Air and Food into the Proper Channels and Also Causes Vocalization

The **larynx**, or voice box, is a short passageway that connects the pharynx with the trachea (**FIGURE 18.2a**). You are probably familiar with the part of the larynx known as the *thyroid cartilage* or *Adam's apple*, which is a cartilage protrusion located in the anterior neck. It is present in both males and females but is usually larger in males because male sex hormones stimulate its growth during puberty. An important function of the larynx is to allow air, but not food or liquid, to flow into the rest of the airways. A component of the larynx called the **epiglottis** accomplishes this function. The epiglottis is a leaf-shaped piece of cartilage that moves up and down over the entrance to the larynx (**FIGURE 18.2a**). When only air is present in the pharynx, the epiglottis is in an upward position and the larynx is open, allowing air to pass through. However, when food or liquid is present in the pharynx, the epiglottis moves down and forms a lid over the larynx, closing it off. The closing of the larynx in this way during swallowing routes food and liquid into the esophagus and keeps them out of the larynx and the rest of the airways. When small particles of dust, smoke, food, or liquid pass into the larynx, a cough reflex occurs, usually expelling the material. Another function of the larynx is voice production. Two folds of elastic tissue known as the **vocal cords** extend from one side of the larynx to the other (**FIGURE 18.2a**). Air that passes over the vocal cords causes them to vibrate and produce sound waves. The greater the air pressure, the louder the sound. Pitch is controlled by the amount of tension on the vocal cords. If they are pulled taut (tight) by skeletal muscles of the larynx, the vocal cords vibrate more rapidly and a higher pitch results. Decreasing the muscular tension on the vocal cords causes them to vibrate more slowly and produce lower-pitched sounds. Due to the influence of male sex hormones, vocal cords are usually longer and thicker in males than in females and therefore they vibrate more slowly. This is why a man's voice generally has a lower range of pitch than a woman's voice.

The Trachea Carries Air to the Primary Bronchi

The **trachea**, or windpipe, is a tubular passageway that extends from the larynx (**FIGURE 18.2a**). The purpose of the trachea is to convey air from the larynx to the primary bronchi. The wall of the trachea contains 15–20 C-shaped rings of cartilage stacked one on top of another. These cartilage rings provide rigid support so the tracheal wall does not collapse inward and obstruct airflow.

The Primary Bronchi Transport Air to the Lungs

The end of the trachea branches into two large bronchi: a **right primary bronchus**, which goes into the right lung, and a **left primary bronchus**, which goes into the left lung (**FIGURE 18.2a**). Like the trachea, the primary bronchi contain incomplete rings of cartilage that provide support. Functionally, the primary bronchi convey air from the trachea to the smaller bronchi and bronchioles of the lungs.

CLINICAL CONNECTION

Cystic Fibrosis

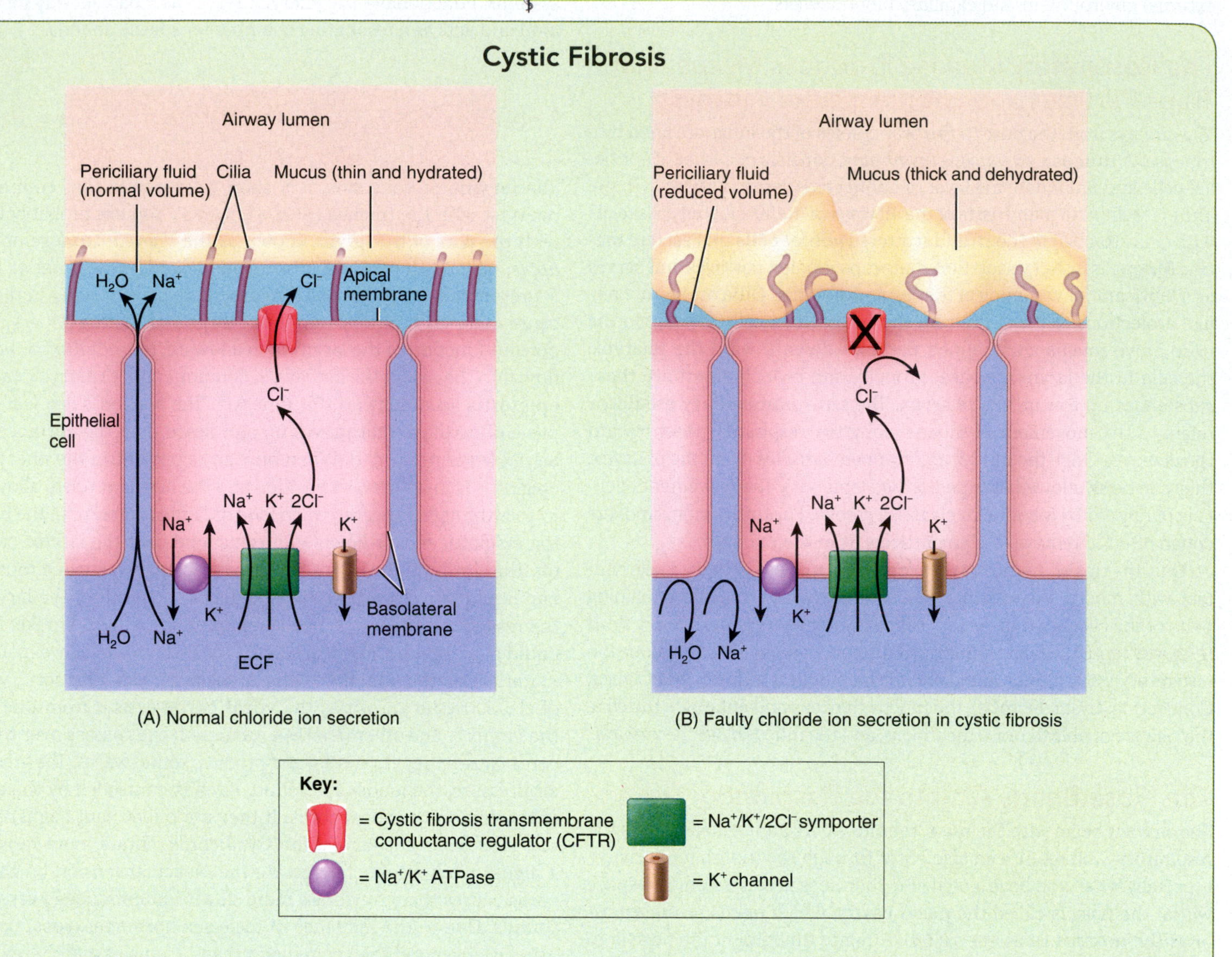

(A) Normal chloride ion secretion

(B) Faulty chloride ion secretion in cystic fibrosis

Cystic fibrosis (CF) is an inherited disease of secretory epithelia that affects the airways, liver, pancreas, small intestine, and sweat glands. It is the most common lethal genetic disease in Caucasians, occurring about once in every 2500 births. The cause of cystic fibrosis is a genetic mutation affecting a chloride (Cl^-) ion channel called the **cystic fibrosis transmembrane conductance regulator (CFTR)**, which permits movement of Cl^- ions across the plasma membranes of many epithelial cells. The mutation reduces the volume of periciliary fluid, resulting in a thick, dehydrated mucus that does not drain easily and obstructs the internal passageways of several organs. Under normal conditions, the production of periciliary fluid involves transport proteins in the basolateral and apical membranes of epithelial cells (FIGURE A). Sodium (Na^+), potassium (K^+), and Cl^- ions are actively taken up by the cell via a $Na^+/K^+/2Cl^-$ symporter located in the basolateral membrane. The gradient for these ions is established by a Na^+/K^+ ATPase that is also present in the basolateral membrane. The Cl^- ions then leave the cell through the Cl^- channel (CFTR) in the apical membrane and enter periciliary fluid. The movement of Cl^- ions creates an electrical gradient that causes Na^+ ions to flow from extracellular fluid (ECF) into periciliary fluid by passing between the cells. H_2O then flows in the same direction due to the osmotic gradient created by the Na^+ and Cl^- ions. The water movement establishes the normal volume of periciliary fluid.

In most forms of cystic fibrosis, the defective Cl^- ion channel (CFTR) is degraded by the cell and consequently is not inserted into the plasma membrane (FIGURE B). This causes Cl^- secretion by the cell to decrease, resulting in decreased movement of Na^+ and water into periciliary fluid. The volume of periciliary fluid is reduced, causing the mucus to thicken and clog the airways. The buildup of mucus also makes it easier for bacterial infections to occur because bacteria grow rapidly in the accumulated mucus. Clogging and infection of the airways leads to difficulty in breathing and eventual destruction of lung tissue. Lung disease accounts for most deaths from CF. Obstruction of small bile ducts in the liver interferes with digestion and disrupts liver function; clogging of pancreatic ducts prevents digestive enzymes from reaching the small intestine. With respect to the reproductive systems, blockage of the vas deferens leads to infertility in males; the formation of dense mucus plugs in the vagina restricts the entry of sperm into the uterus and can lead to infertility in females. Treatment of cystic fibrosis involves chest therapy (tapping on the chest to loosen mucus), taking mucus-thinning drugs, using antibiotics to treat the bacterial infections, altering the diet, and administration of exogenous pancreatic enzymes to help with digestive function. The gene for CFTR has been identified, and scientists are hoping to introduce the gene into affected epithelial cells (gene therapy) in order to restore normal Cl^- secretion.

The Lungs Contain Most of the Components of the Respiratory System

The **lungs** are paired cone-shaped organs in the thoracic cavity (FIGURE 18.2a). Each lung is enclosed by a double-layered membrane of epithelium and connective tissue called the **pleura**. The outer layer, called the **parietal pleura**, lines the wall of the thoracic cavity; the inner layer, the **visceral pleura**, covers the lungs themselves. Between the visceral and parietal pleurae is a small space, the **pleural cavity**, which contains a few milliliters of lubricating fluid secreted by the membranes. This **intrapleural fluid** reduces friction between the membranes, allowing them to slide easily over one another during breathing. Intrapleural fluid also causes the two membranes to adhere to one another just as a film of water causes two glass microscope slides to stick together. Inflammation of the pleural membranes, called **pleurisy**, may cause pain due to friction between the parietal and visceral layers of the pleura. If the inflammation persists, excess fluid accumulates in the pleural space, a condition known as **pleural effusion**.

Each lung contains all the branches of a primary bronchus (FIGURE 18.2a, c). On entering the lungs, the primary bronchi divide to form smaller bronchi—the **secondary bronchi**, one for each lobe of the lung. (The right lung has three lobes; the left lung has two.) The secondary bronchi continue to branch, forming still smaller bronchi, called **tertiary bronchi**, that divide several times, eventually giving rise to smaller **bronchioles**. Bronchioles in turn branch into even smaller tubes called **terminal bronchioles**. Terminal bronchioles subdivide into microscopic branches called **respiratory bronchioles**, which have a few alveoli that extend from their walls. As the respiratory bronchioles penetrate more deeply into the lungs, they subdivide into several **alveolar ducts**, which contain more alveoli. The alveolar ducts give rise to **alveolar sacs**, which contain large numbers of alveoli arranged in clusters. An alveolar sac is comparable to a bunch of grapes, with each grape being an alveolus. It has been estimated that the lungs contain 300 million alveoli, providing an immense total surface area of 75 m^2—about the size of a racquetball court—for gas exchange. The respiratory passages from the trachea to the alveoli contain about 23 generations of branching (FIGURE 18.2f). This extensive branching from the trachea resembles an inverted tree and is commonly referred to as the **bronchial tree**.

The larger bronchi of the lungs consist of an outer layer of cartilage plates, a middle layer of smooth muscle, and an inner layer of mucous membrane (mucosa). As the branching becomes more extensive in the bronchial tree, several structural changes occur. First, the plates of cartilage become less abundant and finally disappear in the bronchioles. Second, as the amount of cartilage decreases, the amount of smooth muscle increases; the smooth muscle encircles the lumen in spiral bands. Third, the mucosa of the bronchial tree changes from ciliated epithelium, with some goblet cells in the larger bronchi and bronchioles, to nonciliated epithelium, with no goblet cells in terminal bronchioles. In regions where cilia are absent, inhaled particles are removed by macrophages.

The lack of cartilage and the presence of smooth muscle in bronchioles allows these tubes to change their diameters, altering the flow of air to the alveoli. Bronchioles are the main sites of resistance to airflow, just as arterioles are the main sites of resistance to blood flow. Bronchiolar smooth muscle is innervated by both the sympathetic and parasympathetic divisions of the autonomic nervous system (ANS). During exercise, norepinephrine released from sympathetic neurons, and epinephrine and norepinephrine secreted from the adrenal medulla bind to β_2-adrenergic receptors, causing relaxation of bronchiolar smooth muscle, which dilates the bronchioles (bronchodilation). Because more air reaches the alveoli, lung ventilation increases. During periods of rest, acetylcholine (ACh) released from parasympathetic neurons binds to muscarinic ACh receptors, causing contraction of bronchiolar smooth muscle, which results in constriction of the bronchioles (bronchoconstriction). Because less air reaches the alveoli, lung ventilation decreases. Mediators of allergic reaction such as histamine also promote bronchoconstriction by causing contraction of bronchiolar smooth muscle. The bronchoconstriction may be so severe that very little air reaches the alveoli.

The Alveoli Are the Sites of Gas Exchange Between Air and Blood

Alveoli (singular is **alveolus**) are air-filled sacs that extend from respiratory bronchioles, alveolar ducts, and alveolar sacs (FIGURE 18.2c). An alveolus consists of little more than an epithelium supported by a basement membrane (a sheet of extracellular material). The thinness of the alveolar wall is vital to the process of gas exchange. The epithelium of an alveolus consists of two types of cells: type I alveolar cells, which are more numerous, and type II alveolar cells (FIGURE 18.2d). **Type I alveolar cells** form a nearly continuous lining of the alveolar wall; they are the main sites of gas exchange. **Type II alveolar cells**, also called *septal cells*, secrete alveolar fluid, which keeps the surface between the cells and the air moist. Included in the alveolar fluid is **surfactant** (sur-FAK-tant), a complex mixture of lipids and proteins. Surfactant lowers the surface tension of alveolar fluid, reducing the

CLINICAL CONNECTION

Asthma

Asthma is a disorder characterized by chronic airway inflammation, airway hypersensitivity to a variety of stimuli, and airway obstruction. The airway obstruction may be due to smooth muscle spasms in the walls of smaller bronchi and bronchioles, edema of the mucosa of the airways, increased mucus secretion, or damage to the epithelium of the airway. Asthmatics typically react to concentrations of agents too low to cause symptoms in people without asthma. Sometimes the trigger is an allergen such as pollen, house dust mites, molds, or a particular food. Other common triggers of asthma attacks are emotional upset, aspirin, sulfiting agents (used in wine and beer and to keep greens fresh in salad bars), exercise, and breathing cold air or cigarette smoke. Symptoms include difficult breathing, coughing, wheezing, chest tightness, tachycardia, and fatigue. An acute attack is treated by giving an inhaled β_2-adrenergic agonist (albuterol) to help relax smooth muscle in the bronchioles and open up the airways. However, long-term therapy of asthma strives to suppress the underlying inflammation. The anti-inflammatory drugs used most often are inhaled corticosteroids (glucocorticoids), cromolyn sodium (Intal®), and leukotriene blockers (Accolate®).

tendency of alveoli to collapse (described later). Associated with the alveolar epithelium are **alveolar macrophages**. These cells roam around alveolar spaces and phagocytize fine dust particles, debris, and any microbes that may be present. Beneath the alveolar cells is a thin layer of elastic fibers. This elastic tissue helps promote elastic recoil when the lungs are stretched. On the outer surface of the alveoli is an extensive network of pulmonary capillaries (FIGURE 18.2c, d). Each capillary consists of a single layer of endothelial cells and a basement membrane. The capillary supply is so dense that the alveoli are considered to be surrounded by a nearly continuous "sheet" of blood.

The exchange of O_2 and CO_2 between the air spaces in the lungs and the blood takes place by diffusion across the alveolar and capillary walls, which together form the **respiratory membrane**. Extending from the alveolar air space to the blood, the respiratory membrane consists of four layers (FIGURE 18.2e): (1) A layer of type I and type II alveolar cells that constitutes the **alveolar epithelium**, (2) an **epithelial basement membrane** underlying the alveolar epithelium, (3) a **capillary basement membrane** that is often fused to the epithelial basement membrane, and (4) the **capillary endothelium**. Despite having several layers, the respiratory membrane is very thin—only 0.5 μm thick, about one-sixteenth the diameter of an erythrocyte—to allow rapid diffusion of gases.

There Are Two Functional Zones of the Respiratory System

The respiratory system is functionally divided into two zones: the conducting zone and the respiratory zone (FIGURE 18.2f). The **conducting zone** is the part of the respiratory system from the nose to the terminal bronchioles. It filters, warms, and humidifies air and conducts it into the lungs. No gas exchange takes place in the conducting zone because alveoli are not present. The **respiratory zone** is the part of the respiratory system that contains alveoli. It includes the respiratory bronchioles, alveolar ducts, and alveolar sacs. The function of the respiratory zone is to carry out gas exchange.

The Pulmonary Circulation Has a High Rate of Flow, a Low Resistance, and a Low Pressure

As you learned in Chapter 14, the pulmonary circulation consists of blood vessels that carry blood from the heart to the alveoli of the lungs and then back to the heart. The lungs receive deoxygenated blood from the heart via pulmonary arteries. Each pulmonary artery branches into smaller pulmonary arterioles, which give rise to even smaller pulmonary capillaries that surround the alveoli. While near the alveoli, blood in pulmonary capillaries becomes oxygenated as it picks up oxygen (O_2) and unloads some of its carbon dioxide (CO_2). Oxygenated blood then enters pulmonary venules, which merge to form larger pulmonary veins. From the pulmonary veins, oxygenated blood is carried back to the heart.

The pulmonary circulation has a high rate of blood flow. In fact, blood flows through the lungs at a higher rate than in the other tissues of the body because the lungs receive the entire cardiac output (5.25 L/min) ejected from the right ventricle of the heart. Recall that the same volume of cardiac output ejected from the left ventricle of the heart supplies all remaining body tissues and that this cardiac output is divided among these tissues, causing each of them to have a smaller rate of blood flow than the lungs (see FIGURE 15.20). Hence, the lungs receive as much blood in one minute as all other tissues in the body during the same time period.

Although its rate of blood flow is high, the pulmonary circulation has a low resistance. The low resistance occurs because pulmonary blood vessels have larger diameters, thinner walls, and are more compliant compared to systemic blood vessels. Because its resistance is low, the pulmonary circulation has a low blood pressure. For example, in pulmonary arteries, the normal systolic pressure is about 25 mmHg, and the diastolic pressure is 8 mmHg (25/8), whereas in systemic arteries the normal systolic pressure is about 110 mmHg, and the diastolic pressure 70 mmHg (110/70). The lower pressure in the pulmonary circulation means that the right ventricle does not have to pump as forcefully for blood to reach the lungs. Accordingly, the peak systolic pressure in the right ventricle is only 20% of that in the left ventricle.

CHECKPOINT

1. Which steps of respiration are provided by the respiratory system?
2. What is the functional significance of the mucociliary escalator?
3. How does the larynx function in voice production?
4. Why are alveoli able to undergo gas exchange but the rest of the components of the bronchial tree cannot?
5. How do the resistance and blood pressure of the pulmonary circulation compare to those of the systemic circulation?

18.2 Ventilation

OBJECTIVES

- Describe the events that occur during ventilation.
- Discuss how ventilation is affected by factors such as surface tension of alveolar fluid, compliance of the lungs, and airway resistance.
- Explain ventilation–perfusion matching.

Ventilation, or **breathing**, is the mechanical flow of air into and out of the lungs. Three pressures are important to ventilation (FIGURE 18.3):

1. **Atmospheric pressure** is the pressure of the air in the atmosphere, which at sea level is about 760 millimeters of mercury (mmHg), or 1 atmosphere (atm).
2. **Alveolar pressure** is the pressure of air within the alveoli of the lungs. Depending on the stage of the breathing cycle, it may be equal to, lower than, or higher than atmospheric pressure. Air flows into or out of the lungs because a pressure gradient exists between the atmosphere and the alveoli. Air moves into the lungs when alveolar pressure is lower than atmospheric pressure. Air moves out of the lungs when alveolar pressure is higher than atmospheric pressure.
3. **Intrapleural pressure** is the pressure within the pleural cavity. Recall that the pleural cavity is the space between the parietal and visceral layers of the pleura (see FIGURE 18.2a). A small amount

of intrapleural fluid is present in this space. Intrapleural pressure is always a negative pressure (lower than atmospheric pressure), ranging from 754–756 mmHg during normal quiet breathing. Because the pleural cavity has a negative pressure, it essentially functions as a vacuum. The suction of this vacuum couples the lungs to the chest wall via the pleura to form the **lung–chest wall system.** Thus, if the thoracic cavity increases in size, the lungs also expand; if the thoracic cavity decreases in size, the lungs recoil (become smaller). The changes in lung volume caused by alterations in thoracic cavity size in turn cause a change in alveolar pressure.

The Breathing Cycle Has Three Phases

A single **breathing cycle**, or *respiratory cycle*, consists of three phases: rest, inspiration, and expiration (**FIGURE 18.3**).

Rest

During the brief resting phase of the breathing cycle, alveolar pressure is equal to atmospheric pressure (**FIGURE 18.3a**). Because a pressure gradient does not exist between the atmosphere and the alveoli, air does not flow into or out of the lungs.

Inspiration

Breathing in is called **inspiration (inhalation)**. For air to flow into the lungs, alveolar pressure must become lower than atmospheric pressure. This condition is achieved by increasing the volume of the lungs.

The pressure of a gas in a closed container is inversely proportional to the volume of the container. This means that if the size of a closed container is increased, the pressure of the gas inside the container decreases, and that if the size of the container is decreased, then the pressure inside it increases. This inverse relationship between volume and

FIGURE 18.3 Pressure changes during ventilation.

Air moves into the lungs when alveolar pressure is less than atmospheric pressure and out of the lungs when alveolar pressure is greater than atmospheric pressure.

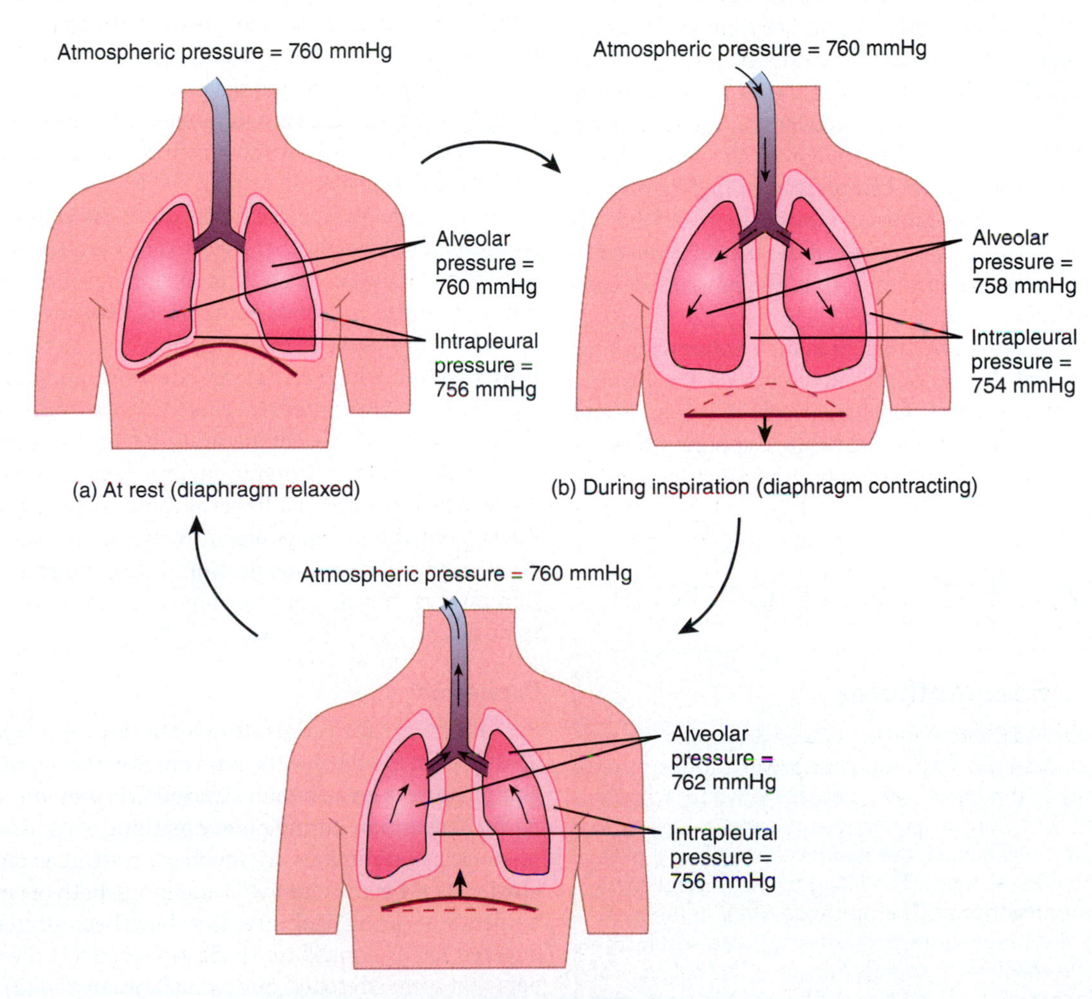

How does intrapleural pressure change during a normal, quiet breath?

FIGURE 18.4 Boyle's law.

The volume of a gas varies inversely with its pressure.

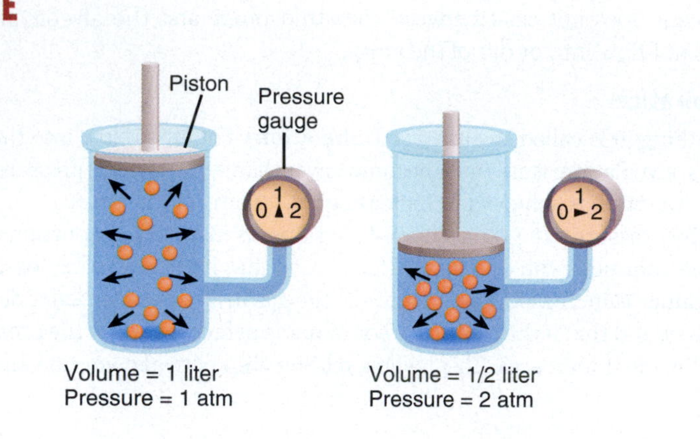

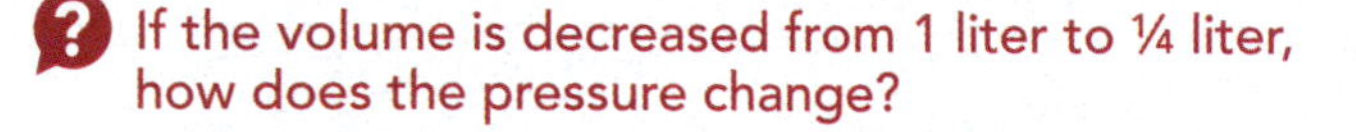

If the volume is decreased from 1 liter to ¼ liter, how does the pressure change?

pressure, called **Boyle's law**, may be demonstrated as follows (FIGURE 18.4): Suppose a gas is placed in a cylinder that has a movable piston and a pressure gauge and that the initial pressure created by the gas molecules striking the wall of the container is 1 atm. If the piston is pushed down, the gas is compressed into a smaller volume so that the same number of gas molecules strike less wall area. The gauge shows that the pressure doubles as the gas is compressed to half its original volume. In other words, the same number of molecules in half the volume produces twice the pressure. Conversely, if the piston is raised to increase the volume, the pressure decreases. Thus, the pressure of a gas varies inversely with volume.

Differences in pressure caused by changes in lung volume force air into our lungs when we inhale and out when we exhale. For inspiration to occur, the lungs must expand, which increases lung volume and thus decreases the alveolar pressure to below atmospheric pressure. The first step in expanding the lungs during normal quiet inspiration involves contraction of the main muscles of inspiration, the diaphragm and external intercostals (FIGURE 18.5).

The most important muscle of inspiration is the **diaphragm**, the dome-shaped skeletal muscle that forms the floor of the thoracic cavity. It is innervated by fibers of the **phrenic nerves**. Contraction of the diaphragm causes it to flatten, lowering its dome. This increases the volume of the thoracic cavity. During a normal, quiet inspiration, the diaphragm descends about 1 cm (0.4 in.), producing a pressure difference of 1–3 mmHg and an inhalation of about 500 mL of air. During a deep, forceful inspiration, the diaphragm may descend 10 cm (4 in.), which produces a pressure difference of 100 mmHg and the inhalation of 2–3 liters of air. Contraction of the diaphragm is responsible for about 75% of the air that enters the lungs during quiet breathing. Advanced pregnancy, excessive obesity, or confining abdominal clothing can prevent complete descent of the diaphragm.

The next most important muscles of inspiration are the **external intercostals**, which extend between the ribs. They are innervated by the **intercostal nerves**. When the external intercostals contract, they pull the ribs upward and outward, which further increases the volume of the thoracic cavity. Contraction of the external intercostals is responsible for about 25% of the air that enters the lungs during normal quiet breathing.

As the volume of the thoracic cavity increases, the volume of the lungs increases and alveolar pressure drops from 760 to 758 mmHg (see FIGURE 18.3b). A pressure difference is thus established between the atmosphere and the alveoli. Because air always flows from a region of higher pressure to a region of lower pressure, inspiration takes place. Air continues to flow into the lungs as long as a pressure gradient exists.

During a deep, forceful inspiration, the accessory muscles of inspiration also contract. These muscles are so-named because they make little, if any, contribution during normal quiet inspiration, but during exercise or forced ventilation, they may contract vigorously. The accessory muscles of inspiration include the **sternocleidomastoid muscles**, which elevate the sternum, and the **scalene muscles**, which elevate the upper two ribs (FIGURE 18.5). The contractions of the accessory muscles of inspiration cause an even greater increase in the volume of the thoracic cavity. Consequently, the lungs expand more, which results in a lower alveolar pressure, and additional air flows from the atmosphere into the lungs. Because both normal quiet inspiration and inspiration during exercise or forced ventilation involve muscle contractions, the process of inspiration is said to be *active*.

Pneumothorax

The pleural cavity is sealed off from the outside environment, which prevents intrapleural pressure from equalizing with atmospheric pressure. An opening of the chest wall, such as from thoracic surgery or a stab or gunshot wound, can expose the pleural cavity to the atmosphere. When this occurs, the negative intrapleural pressure pulls air into the pleural cavity. The filling of the pleural cavity with air is called a **pneumothorax**. The presence of air in the pleural cavity uncouples the lung from the chest wall. As a result, the lung collapses and the chest wall expands outward. The collapse of a lung is called **atelectasis** (at'-e-LEK-ta-sis). The goal of treatment is the evacuation of air from the pleural cavity, which allows the lung to reinflate. A small pneumothorax may resolve on its own, but it is often necessary to insert a chest tube to assist in evacuation.

Expiration

Breathing out, called **expiration (exhalation)**, is also due to a pressure gradient, but in this case the gradient is in the opposite direction: Alveolar pressure is greater than atmospheric pressure. Unlike inspiration, normal expiration during quiet breathing is a *passive process* because no muscle contractions are involved. Instead, expiration results from **elastic recoil** of the chest wall and lungs, both of which have a natural tendency to spring back after they have been stretched. Two inwardly directed forces contribute to elastic recoil: (1) the recoil of elastic fibers that were stretched during inspiration and (2) the inward pull of surface tension due to the film of alveolar fluid.

Expiration starts when the inspiratory muscles relax. As the diaphragm relaxes, its dome moves upward due to its elasticity; as the external intercostals relax, the ribs are depressed. These movements

FIGURE 18.5 Muscles of inspiration and expiration and their actions. For simplicity, muscles of inspiration are indicated on one side of the body and muscles of expiration on the other side, but each group is actually present on both sides of the body.

During a normal quiet inspiration, the diaphragm and external intercostals contract, the lungs expand, and air moves into the lungs; during a normal quiet expiration, the diaphragm relaxes and the lungs recoil inward, forcing air out of the lungs.

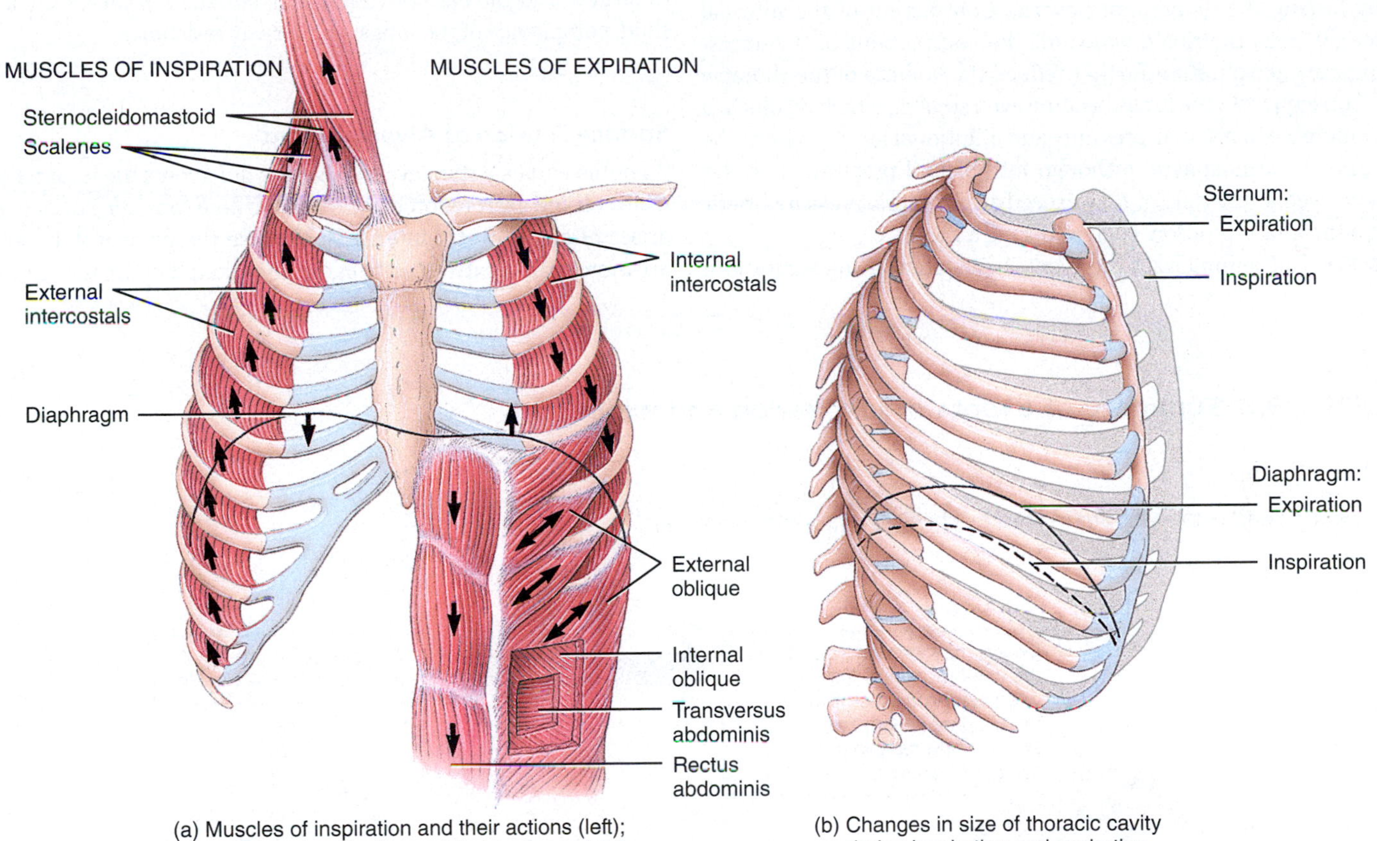

(a) Muscles of inspiration and their actions (left); muscles of expiration and their actions (right)

(b) Changes in size of thoracic cavity during inspiration and expiration

(c) During inspiration, the ribs move upward and outward like the handle on a bucket

Which muscles contract during a forceful expiration?

decrease the volume of the thoracic cavity, which decreases lung volume. In turn, the alveolar pressure increases to about 762 mmHg (see **FIGURE 18.3c**). Air then flows from the area of higher pressure in the alveoli to the area of lower pressure in the atmosphere.

Expiration becomes active only during forceful breathing, as occurs while playing a wind instrument or during exercise. During a forced expiration, the accessory muscles of expiration contract. These muscles include the **abdominal muscles** (rectus abdominis, external and internal obliques, and transversus abdominis) and the **internal intercostals** (**FIGURE 18.5**). Contraction of the abdominal muscles moves the lower ribs downward and compresses the abdominal viscera, thereby forcing the diaphragm upward. Contraction of the internal intercostals pulls the ribs downward. The contractions of the accessory muscles of expiration further reduce the volume of the thoracic cavity. Consequently, the lungs become even smaller, which results in a greater increase in alveolar pressure, and additional air flows from the lungs into the atmosphere. Although intrapleural pressure is always less than alveolar pressure, it may exceed atmospheric pressure briefly during a forceful expiration, such as during a cough.

FIGURE 18.6 summarizes the events of inspiration and expiration.

Respiratory Rate

The **respiratory rate** is the number of breathing cycles that occur per unit of time, usually expressed in breaths/min. In an average adult at rest, the respiratory rate is about 12 breaths/min. Each breath is the equivalent of one breathing cycle.

Several Factors Affect Ventilation

As you have just learned, air pressure differences drive airflow during inspiration and expiration. However, three other factors affect the rate of airflow and the ease of ventilation: surface tension of the alveolar fluid, compliance of the lungs, and airway resistance.

Surface Tension of Alveolar Fluid

As noted earlier, a thin layer of alveolar fluid coats the luminal surface of alveoli and exerts a force known as **surface tension**. Surface tension arises at all air–water interfaces because the polar water molecules are more strongly attracted to each other than they are to the nonpolar

FIGURE 18.6 Summary of events of inspiration and expiration.

Inspiration and expiration are caused by changes in alveolar pressure.

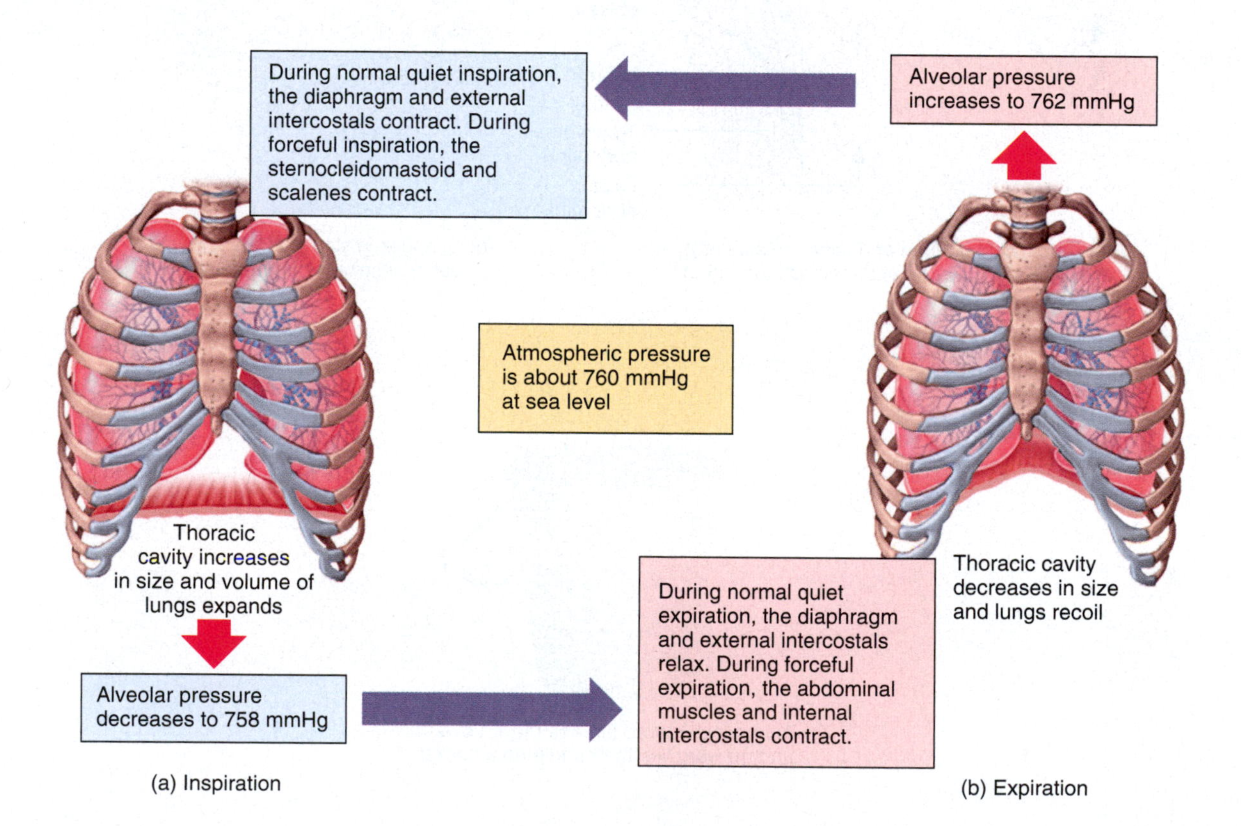

What is the normal atmospheric pressure at sea level?

gas molecules in the air. When liquid surrounds a sphere of air, as in an alveolus or a soap bubble, surface tension produces an inwardly directed force. Soap bubbles burst because they collapse inward due to surface tension. In the lungs, surface tension causes the alveoli to assume the smallest possible diameter. During breathing, surface tension must be overcome to expand the lungs during each inspiration. Surface tension also accounts for two-thirds of lung elastic recoil, which decreases the size of alveoli during expiration.

The surfactant in alveolar fluid reduces surface tension. **Surfactant**, a *surface active agent*, is a complex mixture of lipids and proteins secreted by type II alveolar cells of the lungs. It intersperses between the water molecules at the air–water interface. This disrupts the cohesive forces between water molecules, causing a marked decrease in surface tension. The presence of surfactant in alveolar fluid reduces the work of breathing and increases lung compliance (described shortly).

A deficiency of surfactant in premature infants causes **respiratory distress syndrome (RDS)**, in which the surface tension of alveolar fluid is greatly increased so that many alveoli collapse at the end of each expiration. Treatment involves the administration of surfactant directly into the lungs.

The surface tension of an alveolus and the effect of surfactant in reducing that surface tension can be calculated by using the law of Laplace (see the Physiological Equation box).

Compliance of the Lungs

Compliance refers to how much effort is required to stretch the lungs and chest wall. High compliance means that the lungs and chest wall expand easily; low compliance means that they resist expansion. By analogy, a thin balloon that is easy to inflate has high compliance, and a heavy and stiff balloon that takes a lot of effort to inflate has low compliance. In the lungs, compliance is related to two principal factors: elasticity and surface tension. The lungs normally have high compliance and expand easily because elastic fibers that are present in lung tissue are easily stretched and surfactant in alveolar fluid reduces surface tension. Decreased compliance is a common feature in pulmonary conditions that (1) scar lung tissue, (2) cause lung tissue to become filled with fluid (**pulmonary edema**), (3) produce a deficiency in surfactant, or (4) impede lung expansion in any way (for example, paralysis of the intercostal muscles). Decreased lung compliance occurs in emphysema (see Clinical Connection: Chronic Obstructive Pulmonary Disease) due to destruction of elastic fibers in alveolar walls.

Airway Resistance

Like the flow of blood through blood vessels, the rate of airflow through the airways depends on both the pressure gradient and the resistance: Airflow equals the pressure gradient between the alveoli and the atmosphere divided by the resistance. The relationship among airflow, the pressure gradient, and resistance is given by the following equation:

$$F = \frac{\Delta P}{R}$$

where F = airflow,
ΔP = the pressure gradient, and
R = the resistance to airflow.

The walls of the airways, especially the bronchioles, offer some resistance to the normal flow of air into and out of the lungs. (Larger-diameter airways have decreased resistance.) As the lungs expand during inspiration, the bronchioles enlarge because their walls are pulled outward in all directions. Airway resistance then increases during expiration as the diameter of bronchioles decreases. Airway diameter is also regulated by the degree of contraction or relaxation of smooth muscle in the walls of the airways. Signals from the sympathetic division of the autonomic nervous system (ANS) cause relaxation of bronchiolar smooth muscle, which results in **bronchodilation** and decreased resistance. Signals from the parasympathetic division of the ANS cause contraction of bronchiolar smooth muscle, resulting in **bronchoconstriction** and increased resistance.

Any condition that narrows or obstructs the airways increases resistance so that more pressure is required to maintain the same airflow. The hallmark of asthma or chronic obstructive pulmonary disease (COPD)—emphysema or chronic bronchitis—is increased airway resistance due to obstruction or collapse of airways.

CLINICAL CONNECTION

Chronic Obstructive Pulmonary Disease

Chronic obstructive pulmonary disease (COPD) is a type of respiratory disorder characterized by chronic and recurrent obstruction of airflow, which increases airway resistance. The principal types of COPD are emphysema and chronic bronchitis. In most cases, COPD is preventable because its most common cause is cigarette smoking or breathing secondhand smoke. Other causes include air pollution, pulmonary infection, occupational exposure to dusts and gases, and genetic factors.

Emphysema (em-fi-SĒ-ma) is a disorder characterized by destruction of the walls of the alveoli, producing abnormally large air spaces that remain filled with air during expiration. With less surface area for gas exchange, O_2 diffusion across the damaged respiratory membrane is reduced. Blood O_2 level is somewhat lowered, and any mild exercise that raises the O_2 requirements of the cells leaves the patient breathless. As increasing numbers of alveolar walls are damaged, lung elastic recoil decreases due to loss of elastic fibers, and an increasing amount of air becomes trapped in the lungs at the end of expiration. Over several years, added exertion during inspiration increases the size of the chest cage, resulting in a barrel chest. Emphysema is generally caused by a long-term irritation; cigarette smoke, air pollution, and occupational exposure to industrial dust are the most common irritants. Treatment consists of cessation of smoking, removal of other environmental irritants, exercise training under careful medical supervision, breathing exercises, use of bronchodilators, and oxygen therapy.

Chronic bronchitis is a disorder characterized by excessive secretion of bronchial mucus accompanied by a productive cough (sputum is raised). **Sputum** (SPŪ-tum) refers to mucus and other fluids from the air passages that are expelled by coughing. Cigarette smoking is the leading cause of chronic bronchitis. Inhaled irritants lead to chronic inflammation with an increase in the size and number of mucous glands and goblet cells in the airway epithelium. The thickened and excessive mucus produced narrows the airway and impairs the action of cilia. Thus, inhaled pathogens become embedded in airway secretions and multiply rapidly. Besides a productive cough, symptoms of chronic bronchitis are shortness of breath, wheezing, cyanosis, and pulmonary hypertension. Treatment for chronic bronchitis is similar to that for emphysema.

PHYSIOLOGICAL EQUATION

LAW OF LAPLACE

The **law of Laplace** states that the pressure inside a sphere such as an alveolus is directly proportional to the surface tension of that sphere and inversely proportional to the radius of the sphere:

$$P = \frac{2T}{r}$$

where P = pressure inside the sphere,
T = surface tension of the sphere, and
r = radius of the sphere.

From the law of Laplace, one can conclude that if two alveoli of different sizes have the same surface tension, then the pressure inside the smaller alveolus will be greater than the pressure inside the larger alveolus. To understand this concept, consider two hypothetical alveoli: alveolus A and alveolus B (**FIGURE A**). Suppose that alveolus A has a radius of 1, alveolus B has a radius of 2, and the surface tension of both alveoli is 12. According to the law of Laplace, the pressure inside alveolus A is equal to 24 and the pressure inside alveolus B is equal to 12:

Alveolus A	Alveolus B
$P = \frac{2(12)}{1}$	$P = \frac{2(12)}{2}$
$= 24$	$= 12$

Because air moves from higher pressure to lower pressure, air would flow from the smaller alveolus (alveolus A) into the larger alveolus (alveolus B) if the two alveoli are connected by a common duct, causing the smaller alveolus to collapse.

Such a situation does not normally occur in the alveoli of the lungs because surfactant is unequally distributed in alveoli of different sizes. Smaller alveoli have a greater concentration of surfactant than larger alveoli. In fact, as alveoli become smaller, the surfactant concentration increases and the surfactant molecules are closer together. Consequently, the surface tension of smaller alveoli is lower than that of larger alveoli. The lower surface tension of smaller alveoli compensates for the smaller radius, thereby equalizing the pressure among alveoli of different sizes. This stabilizes the smaller alveoli, allowing them to remain open instead of collapsing. For example, if the two alveoli mentioned earlier were present in the lungs of your body, the smaller alveolus (alveolus A) would have a greater concentration of surfactant than the larger alveolus (alveolus B) (**FIGURE B**). Suppose that the higher concentration of surfactant in alveolus A decreases the surface tension of alveolus A from 12 to 4, whereas the lower concentration of surfactant in alveolus B decreases the surface tension of alveolus B from 12 to only 8. If you insert these values into the law of Laplace, the pressure inside either alveolus A or alveolus B is calculated to be 8:

Alveolus A	Alveolus B
$P = \frac{2(4)}{1}$	$P = \frac{2(8)}{2}$
$= 8$	$= 8$

Thus, the greater concentration of surfactant in alveolus A reduces surface tension enough so that the pressure in alveolus A is equal to that in alveolus B. Because both alveoli have the same pressure, air does not flow from alveolus A to alveolus B, and alveolus A remains open.

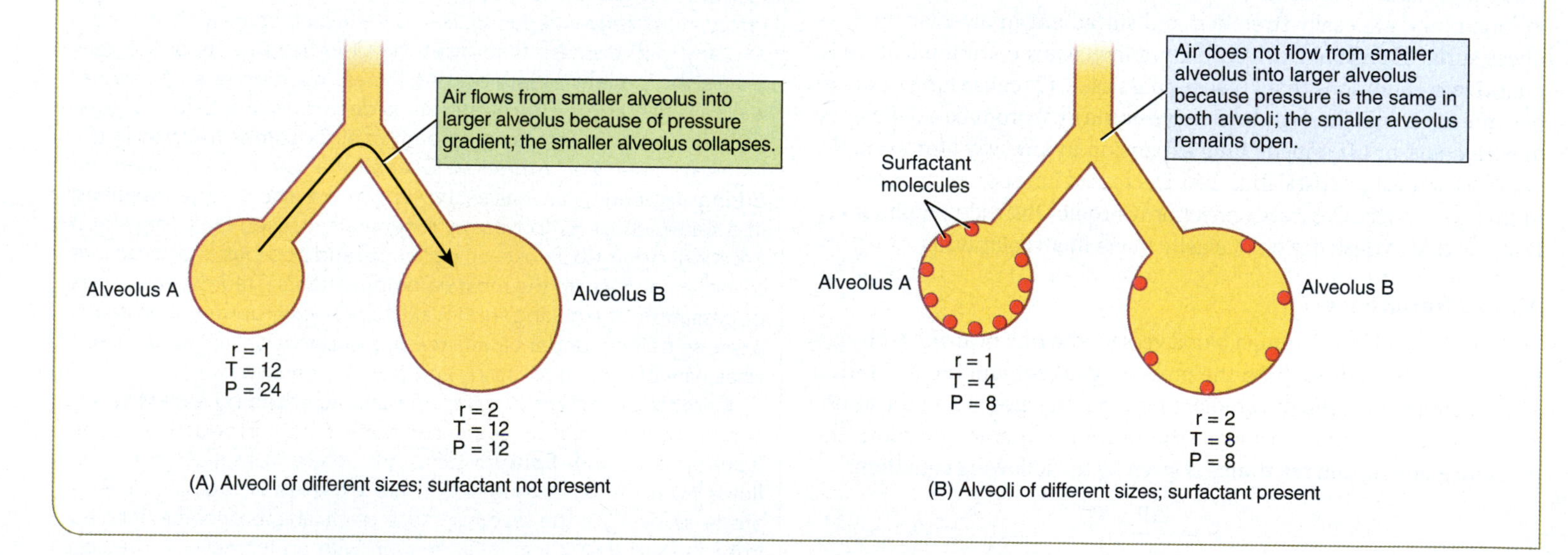

(A) Alveoli of different sizes; surfactant not present

(B) Alveoli of different sizes; surfactant present

Ventilation and Perfusion of the Alveoli Are Matched for Optimal Gas Exchange

To maximize gas exchange, ventilation of the alveoli must be matched or coupled to perfusion (blood flow), a phenomenon known as **ventilation–perfusion matching**. When there are mismatches or inequalities in ventilation and perfusion of the alveoli, gas exchange is inefficient because the amount of airflow or blood flow to the alveoli is compromised. Autoregulatory mechanisms are responsible for ventilation–perfusion matching: The two local chemical mediators that play key roles in this process are CO_2 and O_2. The CO_2 level in an alveolus controls ventilation by altering bronchiolar diameter. The O_2 level in the blood of pulmonary arterioles controls perfusion by altering arteriolar diameter.

If ventilation exceeds perfusion (**FIGURE 18.7a**), the CO_2 level in the alveolus and surrounding tissue decreases because too much CO_2 is exhaled due to the excess ventilation. In addition, the O_2 level increases because the excess ventilation brings in more O_2, which diffuses into the blood. To compensate for the ventilation–perfusion mismatch, the low level of CO_2 in the alveolus causes contraction of bronchiolar smooth muscle (bronchoconstriction), and the high level of O_2 in the blood causes relaxation of pulmonary arteriolar smooth muscle (vasodilation). The bronchoconstriction decreases airflow to match the smaller blood flow, and the vasodilation increases blood flow to the overventilated alveolus.

If perfusion exceeds ventilation (**FIGURE 18.7b**), the level of CO_2 in the alveolus and surrounding tissue increases because the excess

FIGURE 18.7 Ventilation–perfusion matching.

Ventilation–perfusion matching is the process by which ventilation of the alveoli is matched to perfusion in order to maximize gas exchange.

If ventilation exceeds perfusion, the CO_2 level ↓ and the O_2 level ↑

Bronchiole

Alveolus

Pulmonary arteriole

The low level of CO_2 causes bronchoconstriction

The high level of O_2 causes vasodilation

Ventilation and perfusion are now balanced

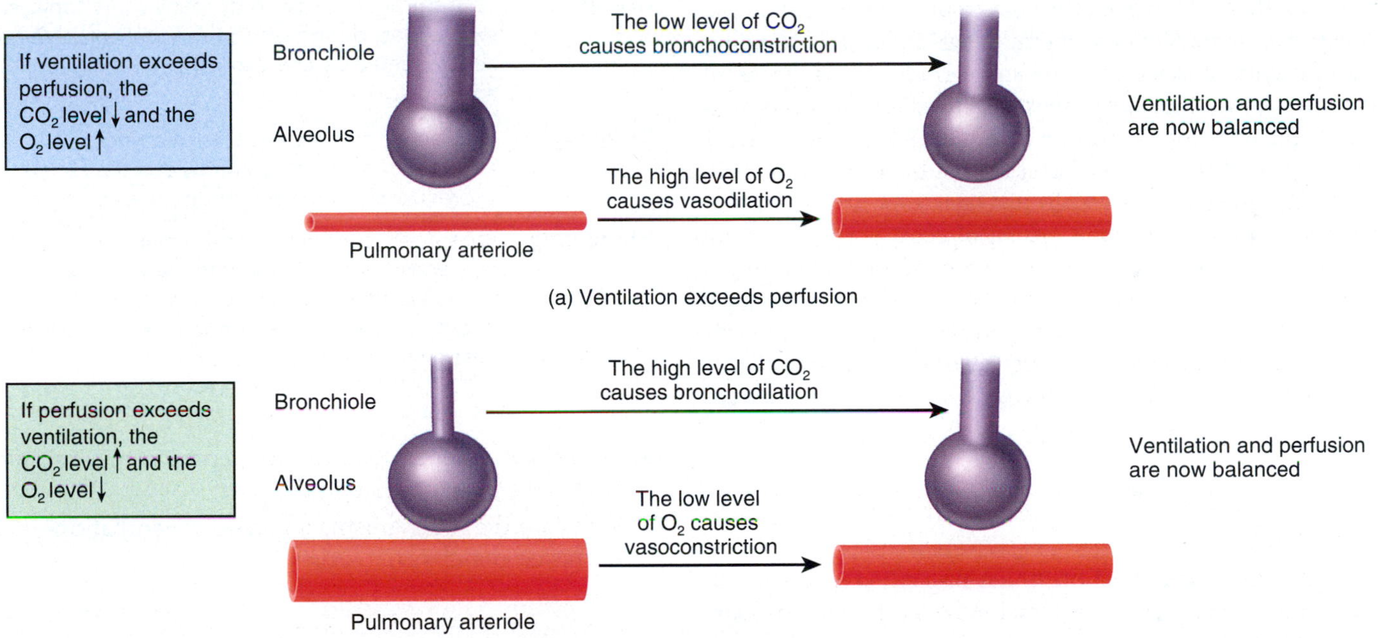

(a) Ventilation exceeds perfusion

If perfusion exceeds ventilation, the CO_2 level ↑ and the O_2 level ↓

Bronchiole

Alveolus

Pulmonary arteriole

The high level of CO_2 causes bronchodilation

The low level of O_2 causes vasoconstriction

Ventilation and perfusion are now balanced

(b) Perfusion exceeds ventilation

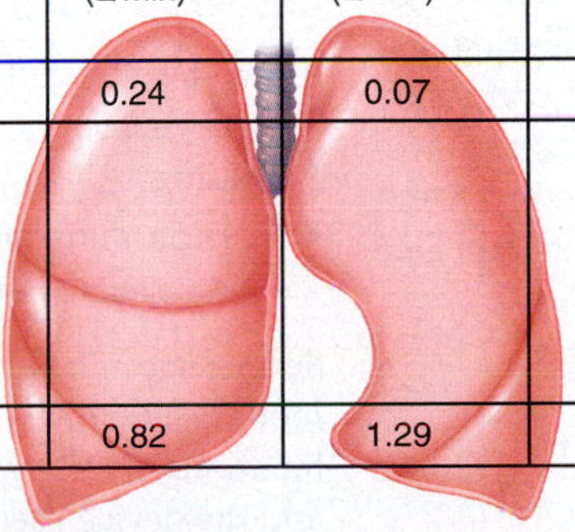

	Ventilation (L/min)	Perfusion (L/min)	Ventilation/Perfusion Ratio
Apex	0.24	0.07	3.40
Base	0.82	1.29	0.63

(c) Ventilation and perfusion rates at the apex and base of the lungs

What local chemical factors are responsible for the mechanisms involved in ventilation–perfusion matching?

blood flow drops off more CO_2 than is exhaled into the air. In addition, the level of O_2 decreases because the excess blood flow carries away more O_2 than the alveoli can bring in. To compensate for the ventilation–perfusion mismatch, the high level of CO_2 in the alveolus causes relaxation of bronchiolar smooth muscle (bronchodilation), and the low level of O_2 in the blood causes constriction of pulmonary arteriolar smooth muscle (vasoconstriction). The bronchodilation increases airflow to match the larger blood flow and the vasoconstriction reduces blood flow to the underventilated alveolus.

Note that an important difference between the pulmonary and systemic circulations is their autoregulatory response to changes in O_2 level. The walls of blood vessels in the systemic circulation *dilate* in response to low O_2. With vasodilation, O_2 delivery increases, which restores the normal O_2 level. By contrast, the walls of blood vessels in the pulmonary circulation *constrict* in response to low levels of O_2. This response ensures that blood mostly bypasses those alveoli (air sacs) in the lungs that are poorly ventilated by fresh air. Thus, most blood flows to better-ventilated areas of the lung.

Despite the autoregulatory mechanisms that promote ventilation–perfusion matching, there is still a small amount of ventilation–perfusion mismatch that occurs from the apex (top) to the base (bottom) of the lungs due to gravitational effects. When a person is standing, the force of gravity causes ventilation and perfusion to be greater at the base of the lungs than at the apex, but the effect on perfusion is greater than the effect on ventilation (FIGURE 18.7c). Consequently, the ventilation–perfusion ratios decrease from the apex to the base of the lung. This mismatch is usually not significant in a healthy individual, but it does allow some mixing of deoxygenated air with oxygenated air, resulting in a small decrease in the amount of oxygen in pulmonary veins and systemic arteries compared to alveolar air. If the lungs are diseased, the ventilation–perfusion mismatch from the apex to the base of the lungs can be exaggerated and the autoregulatory mechanisms may not be able to compensate, resulting in reduced gas exchange.

There Are Different Patterns of Respiratory Movements

Recall that during normal quiet breathing, inspiration involves contractions of the diaphragm and external intercostals, whereas expiration occurs because these muscles relax. The term for the normal pattern of quiet breathing is **eupnea** (ūp-NĒ-a).

In addition to providing air for gas exchange, respirations provide humans with methods for expressing emotions such as laughing, sighing, and crying. Moreover, respiratory air can be used to expel foreign matter from the air passages through actions such as sneezing and coughing. Respiratory movements are also modified and controlled during talking and singing. Some of the modified respiratory movements that express emotion or clear the airways are listed in TABLE 18.1. All of these movements are reflexes, but some of them can also be initiated voluntarily.

CHECKPOINT

6. What are the three pressures that are important to ventilation?
7. How does each stage of the breathing cycle occur?
8. Compare what happens during quiet versus forceful ventilation.

TABLE 18.1 Modified Respiratory Movements

Movement	Description
Coughing	A long-drawn and deep inspiration followed by a strong expiration that suddenly sends a blast of air through the upper respiratory passages. Stimulus for this reflex act may be a foreign body lodged in the larynx, trachea, or epiglottis.
Sneezing	Spasmodic contraction of muscles of expiration that forcefully expels air through the nose and mouth. Stimulus may be an irritation of the nasal mucosa.
Sighing	A long-drawn and deep inspiration immediately followed by a shorter but forceful expiration.
Yawning	A deep inspiration through the widely opened mouth producing an exaggerated depression of the mandible (lower jaw). It may be stimulated by drowsiness, or someone else's yawning, but the precise cause is unknown.
Crying	An inspiration followed by many short convulsive expirations, during which the vocal cords vibrate; accompanied by characteristic facial expressions and tears.
Laughing	The same basic movements as crying, but the rhythm of the movements and the facial expressions usually differ from those of crying.
Hiccupping	Spasmodic contraction of the diaphragm followed by a spasmodic closure of the larynx, which produces a sharp sound on inspiration. Stimulus is usually irritation of the sensory nerve endings of the gastrointestinal tract.

9. How do alveolar surface tension, compliance, and airway resistance affect ventilation?
10. What are the mechanisms involved in ventilation–perfusion matching?

18.3 Lung Volumes and Capacities

OBJECTIVES

- Explain the differences among tidal volume, inspiratory reserve volume, expiratory reserve volume, and residual volume.
- Compare inspiratory capacity, functional residual capacity, vital capacity, and total lung capacity.

Several specific **lung volumes** can be measured as a way to assess pulmonary function. A record of lung volumes and **lung capacities** (combinations of lung volumes) is known as a **spirogram** (FIGURE 18.8), and the device used to measure them is called a spirometer (see Tool of the Trade: Spirometry). In general, lung volumes are larger in males, taller individuals, and younger adults, and smaller in females, shorter individuals, and the elderly. Various disorders may be diagnosed by

FIGURE 18.8 Spirogram of lung volumes and capacities. The average values for a healthy adult male and female are indicated, with the values for a female in parentheses. Note that the spirogram is read from right (start of record) to left (end of record).

Lung capacities are combinations of lung volumes.

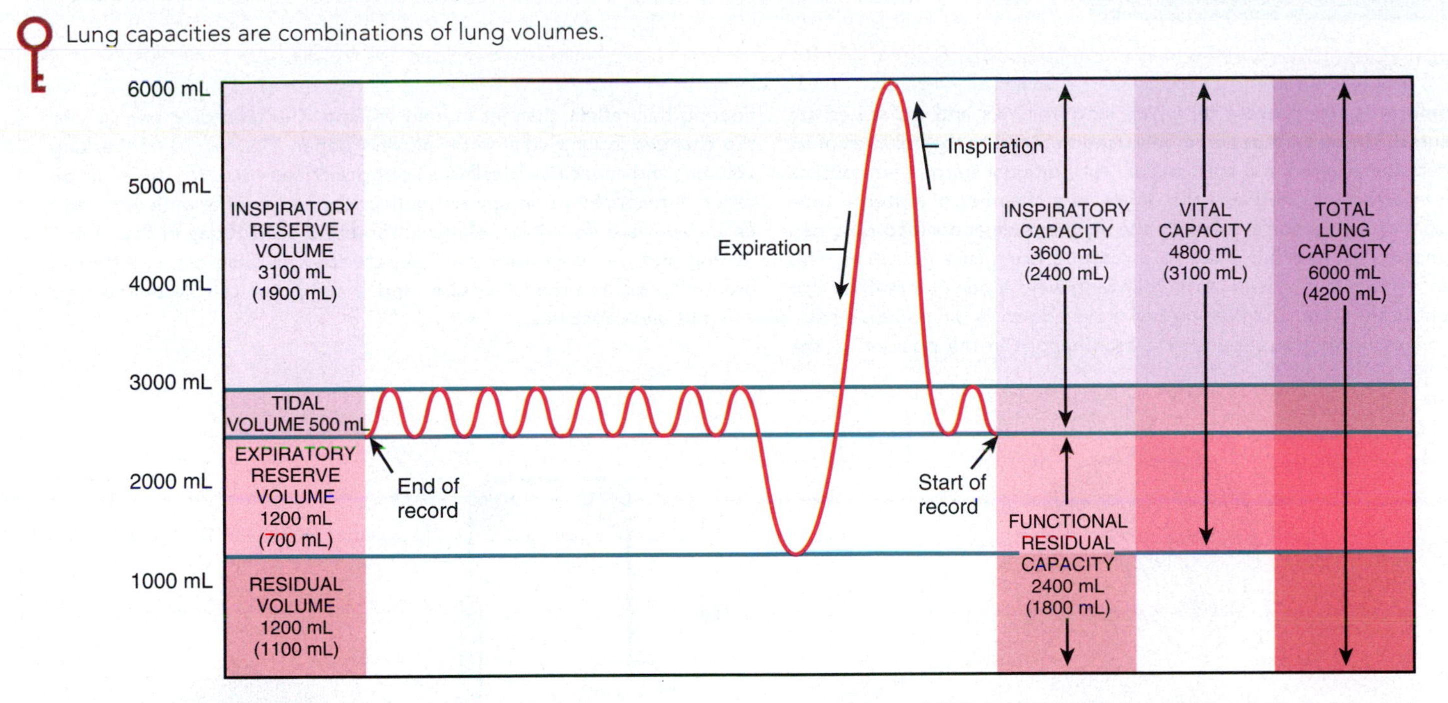

If you breathe in as deeply as possible and then exhale as much air as possible, which lung capacity have you demonstrated?

comparison of actual and predicted normal values for a patient's gender, height, and age. Examples of lung volumes are listed below. The values given are averages for young adults.

- **Tidal volume (V_T)** is the volume of air inspired or expired during a single breathing cycle under resting conditions. It equals 500 mL in an average adult male or female.
- **Inspiratory reserve volume (IRV)** is the maximum volume of air that can be inspired after a normal inspiration. It is about 3100 mL in an average adult male and 1900 mL in an average adult female.
- **Expiratory reserve volume (ERV)** is the maximum volume of air that can be expired after a normal expiration. It averages 1200 mL in males and 700 mL in females.
- **Residual volume (RV)** is the volume of air that remains in the lungs after a maximum expiration. It amounts to about 1200 mL in males and 1100 mL in females. This air remains in the lungs because the subatmospheric intrapleural pressure keeps the alveoli slightly inflated, and some air also remains in the noncollapsible airways. Because it does not leave the lungs, RV and any lung capacity that includes the RV cannot be measured with a spirometer.

Lung capacities are calculated by adding together two or more specific lung volumes (FIGURE 18.8). The following is a list of several lung capacities:

- **Functional residual capacity (FRC)** is the volume of air in the lungs at the end of a normal expiration. It is the sum of residual volume and expiratory reserve volume (1200 mL + 1200 mL = 2400 mL in males and 1100 mL + 700 mL = 1800 mL in females).
- **Inspiratory capacity (IC)** is the maximum volume of air that can be inspired after a normal expiration. It is the sum of tidal volume and inspiratory reserve volume (500 mL + 3100 mL = 3600 mL in males and 500 mL + 1900 mL = 2400 mL in females).
- **Vital capacity (VC)** is the maximum volume of air that can be expired after a maximum inspiration. It is equal to the sum of inspiratory reserve volume, tidal volume, and expiratory reserve volume (4800 mL in males and 3100 mL in females). A term related to the VC is the **forced expiratory volume in 1 second (FEV_1)**, the volume of air that can be exhaled from the lungs in 1 second with maximal effort following a maximal inspiration. Under normal conditions, the FEV_1 is about 80% of the VC. Typically, chronic obstructive pulmonary disease (COPD) greatly reduces FEV_1 because COPD increases airway resistance.
- **Total lung capacity (TLC)** is the total volume of air in the lungs after a maximum inspiration. It is the sum of vital capacity and residual volume (4800 mL + 1200 mL = 6000 mL in males and 3100 mL + 1100 mL = 4200 mL in females).

Another way to assess pulmonary function is to determine the amount of air that flows into and out of the lungs each minute. The **minute ventilation ($\dot{V}$)**—the total volume of air inspired and expired each minute—is tidal volume multiplied by respiratory rate. In a typical adult at rest, minute ventilation is about 6000 mL/min (see the Physiological Equation box). A lower-than-normal minute ventilation usually is a sign of pulmonary malfunction.

In a typical adult, about 70% of the tidal volume (350 mL) actually reaches the respiratory zone of the respiratory system—the respiratory bronchioles, alveolar ducts, and alveolar sacs—and participates

TOOL OF THE TRADE

SPIROMETRY

Spirometry is the process by which lung volumes and capacities are measured. The apparatus commonly used to measure the volumes of air inspired and expired is a **spirometer**. A traditional spirometer consists of an inverted, air-filled bell that floats in a chamber of water, a tube that connects a person's mouth to the air chamber, and a recording pen that marks on paper attached to a rotating drum (see FIGURE). This type of spirometer is used in the following way: A person breathes into and out of the tube. The floating bell moves down as the patient inhales and moves up as the patient exhales. Changes in the position of the floating bell reflect changes in lung volume. The recording pen graphs the changes in lung volume on moving paper. The record of the lung volumes and capacities is called a **spirogram** (see FIGURE 18.8). Inspiration is recorded as an upward deflection, and expiration is recorded as a downward deflection. Many spirometers used today in the clinical setting and the laboratory are computerized. In such cases, a person breathes into and out of a tube, and a computer calculates the lung volumes and capacities.

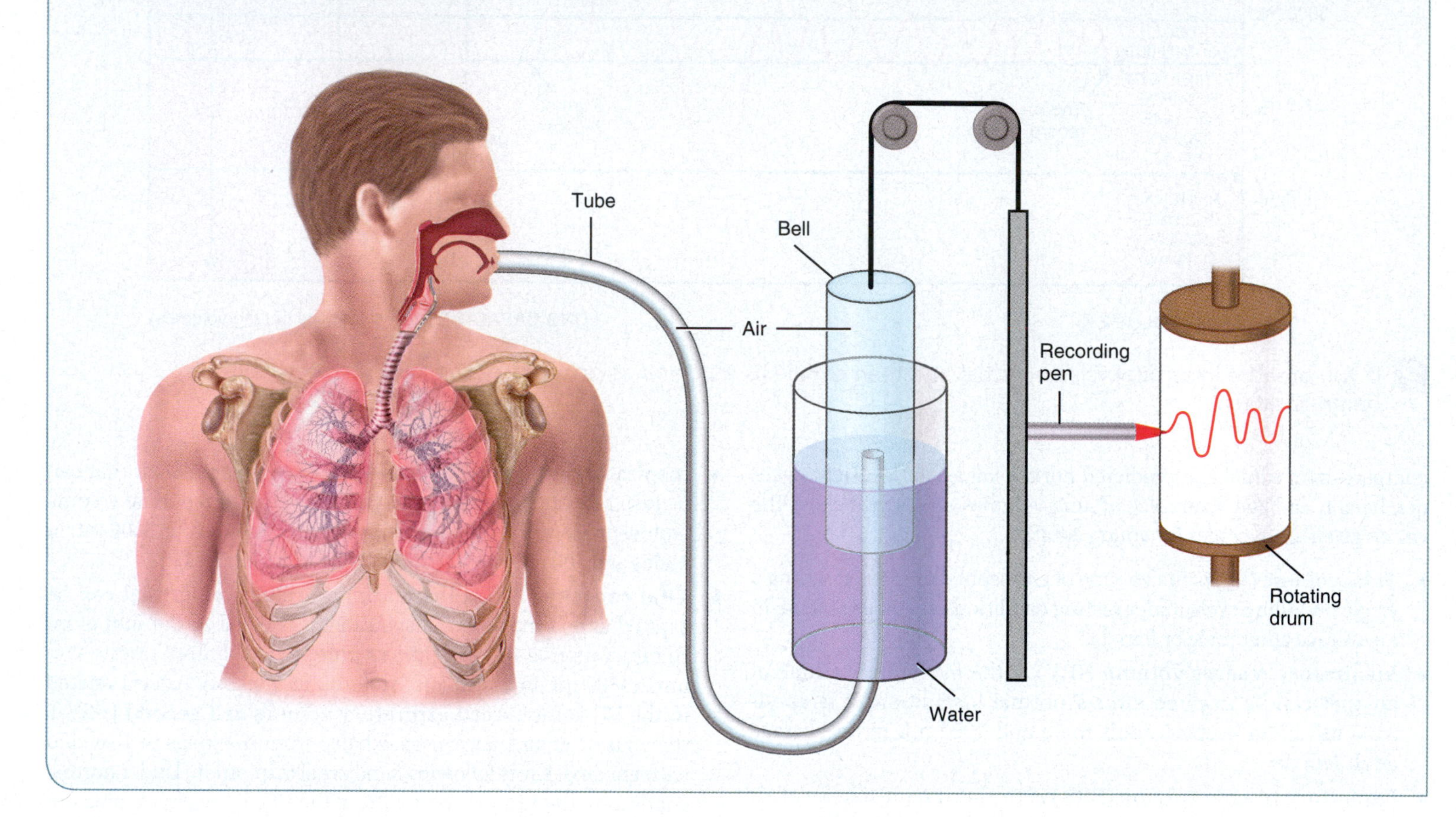

PHYSIOLOGICAL EQUATION

CALCULATION OF MINUTE VENTILATION

Minute ventilation ($\dot{V}$), the total volume of air inspired and expired per minute, is calculated using the following equation:

$$\dot{V} = V_T \times RR$$

where $\dot{V}$ is the minute ventilation in mL/min,
V_T is the tidal volume in mL/breath, and
RR is the respiratory rate in breaths/min.

In a typical resting adult, tidal volume averages 500 mL/breath, and respiratory rate is about 12 breaths/min. Thus, average minute ventilation is

$$\dot{V} = 500 \text{ mL/breath} \times 12 \text{ breaths/min}$$
$$= 6000 \text{ mL/min}$$
$$= 6.0 \text{ L/min}$$

PHYSIOLOGICAL EQUATION

CALCULATION OF ALVEOLAR VENTILATION

Alveolar ventilation ($\dot{V}_A$), the volume of air that reaches the respiratory zone per minute, is calculated as follows:

$$\dot{V}_A = (V_T - V_D) \times RR$$

where $\dot{V}_A$ is the alveolar ventilation in mL/min,
V_T is the tidal volume in mL/breath,
V_D is the anatomic dead space volume in mL/breath, and
RR is the respiratory rate in breaths/min.

In a typical resting adult, tidal volume averages 500 mL/breath, anatomic dead space is 150 mL/breath, and respiratory rate is about 12 breaths/min. Thus, alveolar ventilation is

$$\begin{aligned}\dot{V}_A &= (500\text{ mL/breath} - 150\text{ mL/breath}) \times 12\text{ breaths/min}\\ &= 350\text{ mL/breath} \times 12\text{ breaths/min}\\ &= 4200\text{ mL/min}\\ &= 4.2\text{ L/min}\end{aligned}$$

in gas exchange. The other 30% (150 mL) remains in the conducting zone (nose, pharynx, larynx, trachea, bronchi, bronchioles, and terminal bronchioles). This 150 mL volume of the conducting zone is known as the **anatomic dead space** because it contains air that does not undergo gas exchange. Hence, not all of the minute ventilation can be used in gas exchange because some of it remains in the anatomic dead space. The **alveolar ventilation ($\dot{V}_A$)** is the volume of air per minute that actually reaches the respiratory zone. Alveolar ventilation is typically about 4200 mL/min (see the Physiological Equation box).

CHECKPOINT

11. What is the difference between a lung volume and a lung capacity?
12. Define each of the following: inspiratory reserve volume, residual volume, functional residual capacity, and vital capacity.
13. How is minute ventilation calculated?
14. Define alveolar ventilation and FEV_1.

18.4 Exchange of Oxygen and Carbon Dioxide

OBJECTIVES

- Explain the significance of Dalton's law and Henry's law.
- Describe the exchange of oxygen and carbon dioxide during pulmonary and systemic gas exchange.

Two Gas Laws—Dalton's Law and Henry's Law—Are Important for Understanding How Gas Exchange Occurs

The exchange of oxygen and carbon dioxide between alveolar air and pulmonary blood occurs via passive diffusion, which is governed by the behavior of gases as described by two gas laws, Dalton's law and Henry's law. Dalton's law is important for understanding how gases move down their pressure gradients by diffusion, and Henry's law helps explain how the solubility of a gas relates to its diffusion.

Dalton's Law

According to **Dalton's law**, each gas in a mixture of gases exerts its own pressure as if no other gases were present. The pressure of a specific gas in a mixture is called its partial pressure (P_x); the subscript is the chemical formula of the gas. The total pressure of the mixture is calculated simply by adding together all of the partial pressures. Atmospheric air is a mixture of gases—nitrogen (N_2), oxygen (O_2), argon (Ar), water vapor (H_2O), and carbon dioxide (CO_2), plus other gases present in much smaller quantities. Atmospheric pressure is the sum of the pressures of all of these gases:

$$\text{Atmospheric pressure (760 mmHg)} = P_{N_2} + P_{O_2} + P_{Ar} + P_{H_2O} + P_{CO_2} + P_{\text{other gases}}$$

You can determine the partial pressure exerted by each component in the mixture by multiplying the percentage of the gas in the mixture by the total pressure of the mixture. Atmospheric air is 78.6% nitrogen, 20.9% oxygen, 0.093% argon, 0.04% carbon dioxide, and 0.06% other gases; a variable amount of water vapor is also present, about 0.3% on a cool, dry day. Thus, the partial pressures of the gases in inhaled air are as follows:

$$\begin{aligned}P_{N_2} &= 0.786 \times 760\text{ mmHg} = 597.4\text{ mmHg}\\ P_{O_2} &= 0.209 \times 760\text{ mmHg} = 158.8\text{ mmHg}\\ P_{Ar} &= 0.0009 \times 760\text{ mmHg} = 0.7\text{ mmHg}\\ P_{H_2O} &= 0.003 \times 760\text{ mmHg} = 2.3\text{ mmHg}\\ P_{CO_2} &= 0.0004 \times 760\text{ mmHg} = 0.3\text{ mmHg}\\ P_{\text{other gases}} &= 0.0006 \times 760\text{ mmHg} = 0.5\text{ mmHg}\\ \text{Total} &= 760\text{ mmHg}\end{aligned}$$

These partial pressures determine the movement of O_2 and CO_2 between the atmosphere and lungs, between the lungs and blood, and between the blood and body cells. Each gas diffuses across a permeable membrane from the area where its partial pressure is greater to the area where its partial pressure is less. The greater the difference in partial pressure, the faster the rate of diffusion.

Compared with inhaled air, alveolar air has less O_2 (13.8% versus 20.9%) and more CO_2 (5.2% versus 0.04%) for two reasons. First, gas exchange in the alveoli increases the CO_2 content and decreases the O_2 content of alveolar air. Second, when air is inhaled, it becomes humidified as it passes along the moist mucosal linings. As water vapor content of the air increases, the relative percentage that is O_2 decreases. In contrast, exhaled air contains more O_2 than alveolar air (16% versus 13.8%) and less CO_2 (4.5% versus 5.2%) because some of the exhaled air was in the anatomic dead space and did not participate in gas exchange. Exhaled air is a mixture of alveolar air and inhaled air that was in the anatomic dead space.

Henry's Law

Henry's law states that the quantity of a gas that will dissolve in a liquid is proportional to the partial pressure of the gas and its solubility. In body fluids, the ability of a gas to stay in solution is greater when its partial pressure is higher and when it has a high solubility in water. The higher the partial pressure of a gas over a liquid and the higher the solubility, the more gas will stay in solution. In comparison to oxygen, much more CO_2 is dissolved in blood plasma because the solubility of CO_2 is 24 times greater than that of O_2.

An everyday experience gives a demonstration of Henry's law. You have probably noticed that a soft drink makes a hissing sound when the top of the container is removed, and bubbles rise to the surface for some time afterward. The gas dissolved in carbonated beverages is CO_2. Because the soft drink is bottled or canned under high pressure and capped, the CO_2 remains dissolved as long as the container is unopened. Once you remove the cap, the pressure decreases and the gas begins to bubble out of solution.

Henry's law also explains an important point about the solubility of nitrogen in body fluids: Even though the air we breathe contains about 79% nitrogen, this gas has no known effect on bodily functions, and very little of it dissolves in blood plasma because its solubility is very low at sea-level pressure. As the total air pressure increases, the partial pressures of all of its gases increase. When a scuba diver breathes air under high pressure, the nitrogen in the mixture can have serious negative effects. Because the partial pressure of nitrogen is higher in a mixture of compressed air than in air at sea-level pressure, a considerable amount of nitrogen dissolves in plasma and interstitial fluid. Excessive amounts of dissolved nitrogen may produce giddiness and other symptoms similar to alcohol intoxication. The condition is called **nitrogen narcosis** or "rapture of the deep". If the diver ascends to the surface slowly, nitrogen comes out of solution slowly and is eliminated by the lungs through exhalation. However, if the ascent is too rapid, nitrogen comes out of solution too quickly and forms gas bubbles in the tissues, resulting in **decompression sickness** or **the bends**. The effects of decompression sickness typically result from bubbles in nervous tissue and can be mild or severe, depending on the number of bubbles formed. Symptoms include joint pain, especially in the arms and legs, dizziness, shortness of breath, extreme fatigue, paralysis, and unconsciousness.

CLINICAL CONNECTION

Hyperbaric Oxygenation

A major clinical application of Henry's law is **hyperbaric oxygenation**, the use of pressure to cause more O_2 to dissolve in the blood. It is an effective technique in treating patients infected by anaerobic bacteria, such as those that cause tetanus and gangrene. (Anaerobic bacteria cannot live in the presence of free O_2.) A person undergoing hyperbaric oxygenation is placed in a hyperbaric chamber, which contains O_2 at a pressure greater than one atmosphere (760 mmHg). As body tissues pick up the O_2, the bacteria are killed. Hyperbaric chambers may also be used for treating certain heart disorders, carbon monoxide poisoning, gas embolisms, crush injuries, cerebral edema, certain hard-to-treat bone infections caused by anaerobic bacteria, smoke inhalation, near-drowning, asphyxia, vascular insufficiencies, and burns.

There Are Two Types of Gas Exchange: Pulmonary and Systemic

Pulmonary gas exchange is the diffusion of O_2 from air in the alveoli of the lungs to blood in pulmonary capillaries and the diffusion of CO_2 in the opposite direction (FIGURE 18.9a). Pulmonary gas exchange converts **deoxygenated blood** (depleted of some O_2) coming from the right side of the heart into **oxygenated blood** (saturated with O_2) that returns to the left side of the heart. As blood flows through the pulmonary capillaries, it picks up O_2 from alveolar air and unloads CO_2 into alveolar air. Although this process is commonly called an exchange of gases, each gas diffuses independently from the area where its partial pressure is higher to the area where its partial pressure is lower.

As FIGURE 18.9a shows, O_2 diffuses from alveolar air, where its partial pressure is 105 mmHg, into the blood in pulmonary capillaries, where P_{O_2} is only 40 mmHg in a resting person. If you have been exercising, the P_{O_2} is even lower because contracting muscle fibers are using more O_2. Diffusion continues until the P_{O_2} of pulmonary capillary blood increases to match the P_{O_2} of alveolar air, 105 mmHg. Because there is a small amount of ventilation–perfusion mismatching from the apex to the base of the lungs due to gravitational effects (see Section 18.2), the P_{O_2} of blood in the pulmonary veins is slightly less than the P_{O_2} in pulmonary capillaries, about 100 mmHg.

While O_2 is diffusing from alveolar air into deoxygenated blood, CO_2 is diffusing in the opposite direction. The P_{CO_2} of deoxygenated blood is 45 mmHg in a resting person, and the P_{CO_2} of alveolar air is 40 mmHg. Because of this difference in P_{CO_2}, carbon dioxide diffuses from deoxygenated blood into the alveoli until the P_{CO_2} of the blood decreases to 40 mmHg. Expiration keeps alveolar P_{CO_2} at 40 mmHg. Oxygenated blood returning to the left side of the heart in the pulmonary veins thus has a P_{CO_2} of 40 mmHg.

The number of capillaries near alveoli in the lungs is very large, and blood flows slowly enough through these capillaries that it picks up a maximal amount of O_2. During vigorous exercise, when cardiac output is increased, blood flows more rapidly through both the systemic and pulmonary circulations. As a result, blood's transit time in the pulmonary capillaries is shorter. Still, the P_{O_2} of blood in the pulmonary veins normally reaches 100 mmHg. In diseases that decrease the rate of gas diffusion, however, the blood may not come into full equilibrium with alveolar air, especially during exercise. When this happens, the P_{O_2} declines and P_{CO_2} rises in systemic arterial blood.

The left ventricle pumps oxygenated blood into the aorta and through the systemic arteries to systemic capillaries. The exchange of O_2 and CO_2 between systemic capillaries and tissue cells is called **systemic gas exchange** (FIGURE 18.9b). As O_2 leaves the bloodstream, oxygenated blood is converted into deoxygenated blood. Unlike pulmonary gas exchange, which occurs only in the lungs, systemic gas exchange occurs in tissues throughout the body.

The P_{O_2} of blood pumped into systemic capillaries is higher (100 mmHg) than the P_{O_2} in tissue cells (40 mmHg at rest) because the cells constantly use O_2 to produce ATP. Due to this pressure difference, oxygen diffuses out of the capillaries into tissue cells and blood P_{O_2} drops to 40 mmHg by the time the blood exits systemic capillaries.

FIGURE 18.9 Changes in partial pressures of oxygen and carbon dioxide (in mmHg) during pulmonary and systemic gas exchange.

Gases diffuse from areas of higher partial pressure to areas of lower partial pressure.

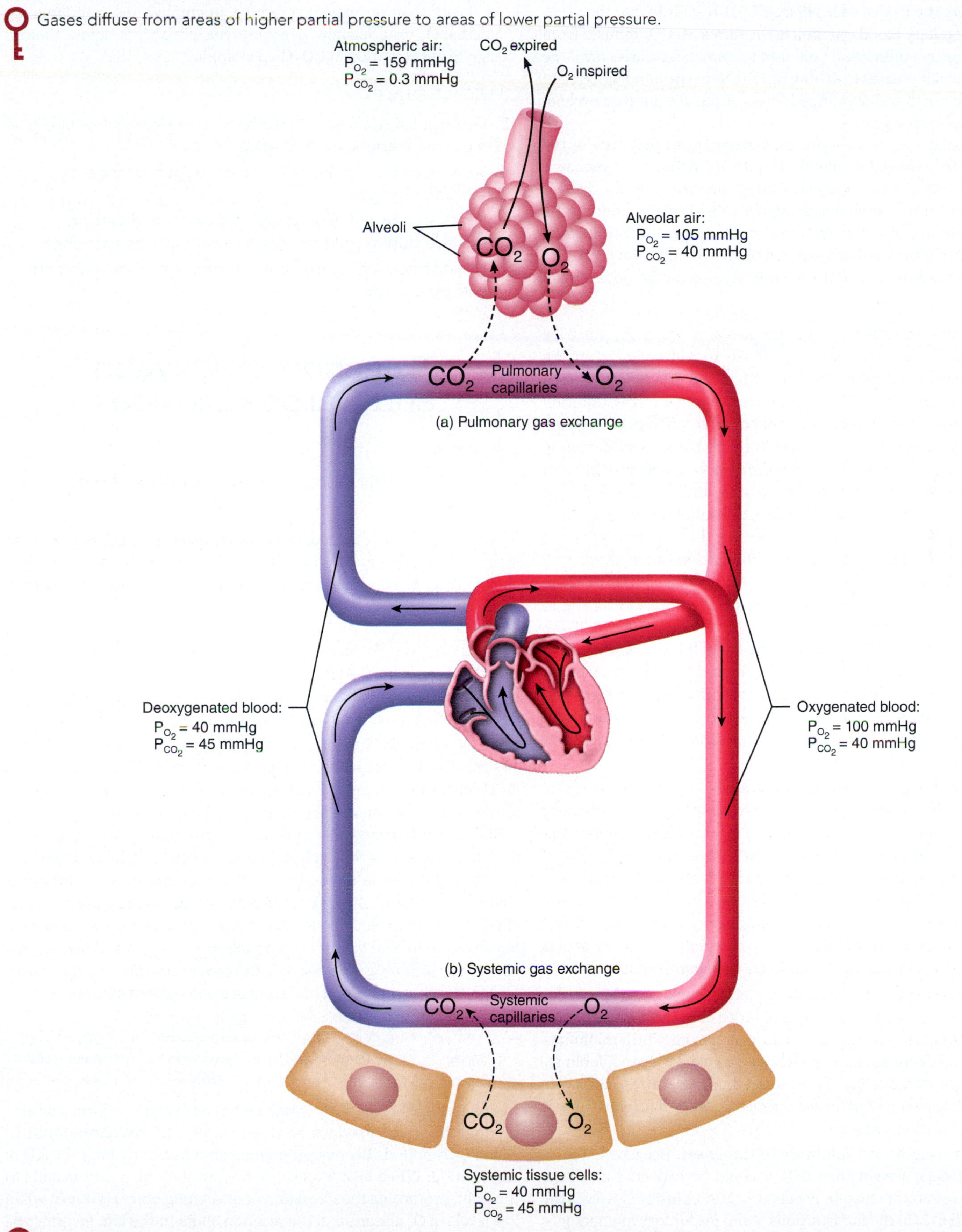

What causes oxygen to enter pulmonary capillaries from alveoli and to enter tissue cells from systemic capillaries?

While O_2 diffuses from the systemic capillaries into tissue cells, CO_2 diffuses in the opposite direction. Because tissue cells are constantly producing CO_2, the P_{CO_2} of cells (45 mmHg at rest) is higher than that of systemic capillary blood (40 mmHg). As a result, CO_2 diffuses from tissue cells through interstitial fluid into systemic capillaries until the P_{CO_2} in the blood increases to 45 mmHg. The deoxygenated blood then returns to the heart and is pumped to the lungs for another cycle of pulmonary gas exchange.

In a person at rest, tissue cells, on average, need only 25% of the available O_2 in oxygenated blood; despite its name, deoxygenated blood retains 75% of its O_2 content. During exercise, more O_2 diffuses from the blood into metabolically active cells, such as contracting skeletal muscle fibers. Active cells use more O_2 for ATP production, causing the O_2 content of deoxygenated blood to drop below 75%.

The rate of pulmonary and systemic gas exchange depends on several factors.

- **Partial pressure difference of the gases**. Alveolar P_{O_2} must be higher than blood P_{O_2} for oxygen to diffuse from alveolar air into the blood. The rate of diffusion is faster when the difference between P_{O_2} in alveolar air and pulmonary capillary blood is larger; diffusion is slower when the difference is smaller. The differences between P_{O_2} and P_{CO_2} in alveolar air versus pulmonary blood increase during exercise. The larger partial pressure differences accelerate the rates of gas diffusion. The partial pressures of O_2 and CO_2 in alveolar air also depend on the rate of airflow into and out of the lungs. Certain drugs (such as morphine) slow ventilation, thereby decreasing the amount of O_2 and CO_2 that can be exchanged between alveolar air and blood. With increasing altitude, the total atmospheric pressure decreases, as does the partial pressure of O_2—from 159 mmHg at sea level to 110 mmHg at 10,000 ft, to 73 mmHg at 20,000 ft. Although O_2 is still 20.9% of the total, the P_{O_2} of inhaled air decreases with increasing altitude. Alveolar P_{O_2} decreases correspondingly, and O_2 diffuses into the blood more slowly. The common signs and symptoms of **high altitude sickness**—shortness of breath, headache, fatigue, insomnia, nausea, and dizziness—are due to a lower level of oxygen in the blood.
- **Surface area available for gas exchange**. As you learned earlier in the chapter, the surface area of the alveoli is huge, approximately the size of a racquetball court (about 75 m^2). In addition, many capillaries surround each alveolus, so many that as much as 100 mL of blood can participate in gas exchange at any instant. Any pulmonary disorder that decreases the functional surface area of the respiratory membranes decreases the rate of pulmonary gas exchange. In emphysema, for example, alveolar walls disintegrate, so surface area is smaller than normal and pulmonary gas exchange is reduced.
- **Diffusion distance**. The respiratory membrane is very thin, so diffusion occurs quickly. Also, the capillaries are so narrow that the erythrocytes must pass through them in single file, which minimizes the diffusion distance from an alveolar air space to hemoglobin inside erythrocytes. Buildup of interstitial fluid between alveoli, as occurs in pulmonary edema, slows the rate of gas exchange because it increases diffusion distance.
- **Molecular weight and solubility of the gases**. Because O_2 has a lower molecular weight than CO_2, it could be expected to diffuse across the respiratory membrane about 1.2 times faster. However, the solubility of CO_2 in the fluid portions of the respiratory membrane is about 24 times greater than that of O_2. Taking both of these factors into account, net outward CO_2 diffusion occurs 20 times more rapidly than net inward O_2 diffusion. Consequently, when diffusion is slower than normal, for example, in emphysema or pulmonary edema, O_2 insufficiency (hypoxia) typically occurs before there is significant retention of CO_2 (hypercapnia).

CHECKPOINT

15. Distinguish between Dalton's law and Henry's law and give a practical application of each.
16. How does the partial pressure of oxygen change as altitude changes?
17. What are the diffusion paths of oxygen and carbon dioxide during pulmonary and systemic gas exchange?
18. What factors affect the rates of diffusion of oxygen and carbon dioxide?

18.5 Transport of Oxygen and Carbon Dioxide

OBJECTIVE

- Describe how the blood transports oxygen and carbon dioxide.

As you have already learned, the blood transports gases between the lungs and body tissues. When O_2 and CO_2 enter the blood, certain chemical reactions occur that aid in gas transport and gas exchange.

Oxygen Is Transported Through the Blood Mainly by Hemoglobin

Oxygen does not dissolve easily in water, so only about 1.5% of inhaled O_2 is dissolved in blood plasma, which is mostly water. About 98.5% of blood O_2 is bound to hemoglobin in erythrocytes (FIGURE 18.10). Each 100 mL of oxygenated blood contains the equivalent of 20 mL of gaseous O_2. Using the percentages just given, the amount dissolved in the plasma is 0.3 mL, and the amount bound to hemoglobin is 19.7 mL.

Recall that a hemoglobin molecule has two main parts: (1) globin, a protein composed of four polypeptide chains (two alpha and two beta chains) and (2) heme, a red pigment that is bound to each of the four chains (see FIGURE 16.4b, c). At the center of each heme group is an iron ion (Fe^{2+}) that can bind to one oxygen molecule, allowing each hemoglobin molecule to bind to four oxygen molecules. Oxygen and hemoglobin bind in an easily reversible reaction. By convention, this reaction is written for a single heme-polypeptide unit of a hemoglobin molecule:

$$\underset{\text{Oxygen}}{O_2} + \underset{\text{Deoxyhemoglobin}}{Hb} \underset{\text{Dissociation of } O_2}{\overset{\text{Binding of } O_2}{\rightleftharpoons}} \underset{\text{Oxyhemoglobin}}{Hb{-}O_2}$$

Hence, a single heme–polypeptide unit can exist in two forms: **deoxyhemoglobin (Hb)**, in which no O_2 is bound, and **oxyhemoglobin**, in which O_2 is bound. The oxygen–hemoglobin reaction obeys the law of mass action: When high levels of O_2 are present, the reaction shifts to the right, promoting the formation of oxyhemoglobin. However, when low levels of O_2 are present, the reaction shifts to the left, favoring the dissociation of O_2 and formation of deoxyhemoglobin.

FIGURE 18.10 Transport of oxygen (O_2) and carbon dioxide (CO_2) in the blood.

Most O_2 is transported by hemoglobin as oxyhemoglobin (Hb–O_2) within erythrocytes; most CO_2 is transported in blood plasma as bicarbonate ions (HCO_3^-).

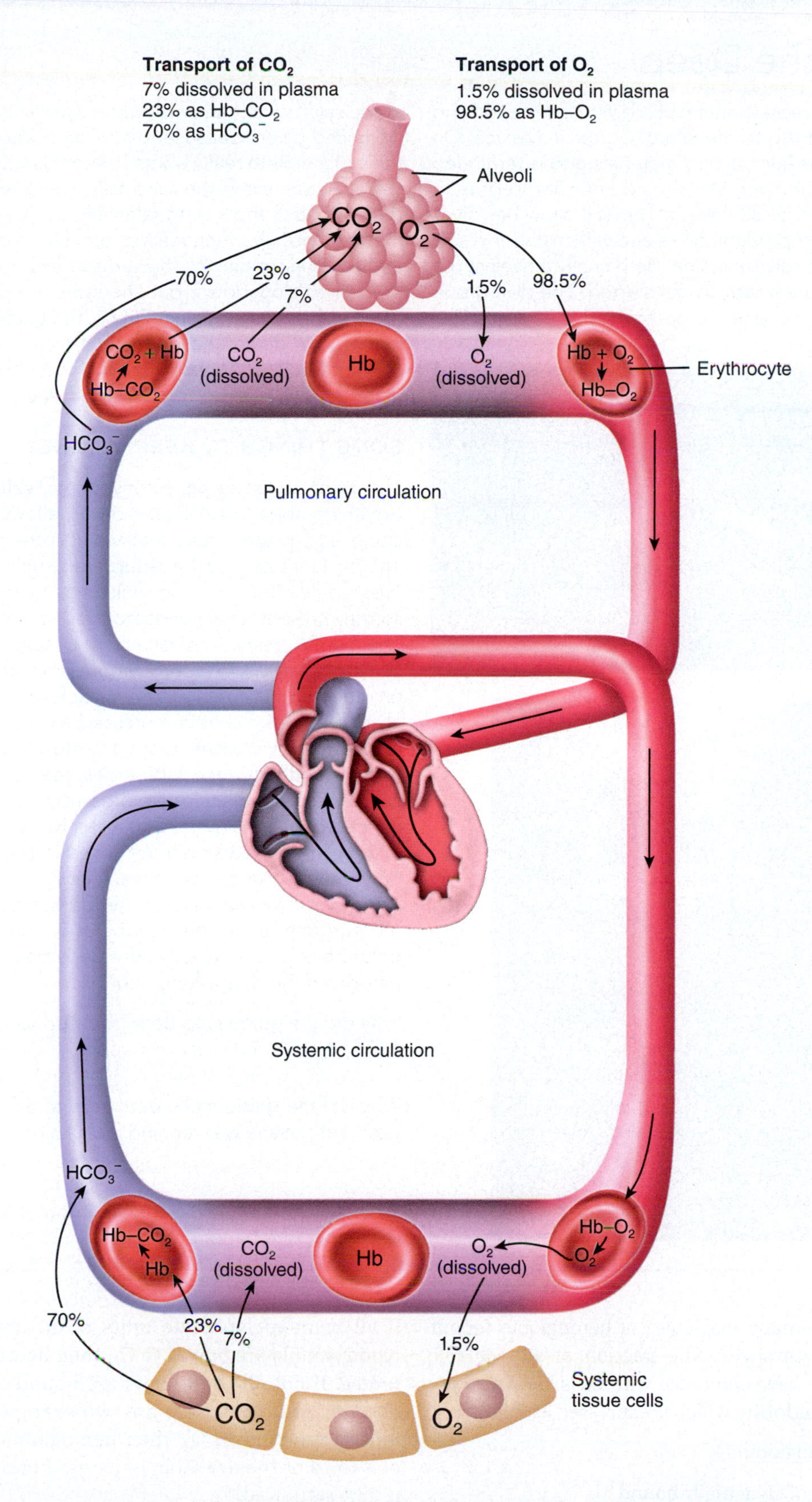

What is the most important factor that determines how much O_2 binds to hemoglobin?

Critical Thinking

Breathing in the Deep

Beth has just received her recreational scuba diver certification and has booked a guided dive trip to the Sea of Cortez in Mexico. On the plane ride, she reviews her training materials and is reminded that, as a new recreational diver, she should limit her depths to 18 meters with a maximum of 30 meters. This will allow her dives to be considered no-decompression dives and will ensure that she has a great time on her first dive vacation. Beth finally arrives at the dive shop: She is excited and ready to get started. The dive group loads the gear into the boat and heads out to the first dive site. The guide says there is an underwater sand falls, but it is limited to only advanced divers because it is located about 30–40 meters below the surface. Beth really wants to see this site and decides to go that deep. As she views the sand falls, she does not do a good job of depth control and moves relatively quickly between the 30 and 40 meter marks. She then realizes that she is using her air very quickly and starts ascending to the surface very quickly. The guide tries to get her to slow down, but she swims upward faster than her exhaled bubbles are moving. Finally, the guide catches her and forces her to slow down and does a safety stop at 5 meters for about 15 minutes. Beth actually runs out of air before the 15 minutes are up and has to use the guide's emergency "spare air."

© Dirk-Jan Mattaar / Alamy Stock Photo

Luis Javier Sandoval / Getty Images

SOME THINGS TO KEEP IN MIND:

The term *scuba* is an acronym for "self-contained underwater breathing apparatus." Scuba diving allows humans to explore exciting underwater areas that would otherwise not be attainable. The air tanks of a scuba setup are usually filled with highly compressed air. The air in the tanks may be compressed to 1800 psi, so they present a very dangerous condition if used inappropriately. The human respiratory anatomy simply cannot withstand pressures of that magnitude. Compression of the air is necessary to provide enough total air for a reasonable time underwater. As the aluminum tank full of compressed air is submerged, the pressure on that tank increases about 1 atmosphere per 10 meters of sea water. The deeper one submerges, the less time the air will last because the gas molecules are pressed more closely together by the increased environmental pressures. Moreover, the individual gases in the compressed air will be moved in and out of the bloodstream differently as the person moves through different depths and pressures. For example, nitrogen tends to accumulate in the blood as the nitrogen molecules move closer together with the increased pressure of deeper depths; the body becomes fully saturated with nitrogen if the dive is long enough.

Why did the guide stop Beth from surfacing quickly?

Why did the guide make Beth stop at 5 meters below the surface even though she was running out of air?

What could be done if Beth had continued ascending rapidly and developed the bends?

A blood sample contains many molecules of hemoglobin (about 280 million within each erythrocyte). The fraction of all available heme–polypeptide units that have combined with O_2 is known as the **percent saturation of hemoglobin**, which is expressed as follows:

$$\text{Percent saturation of hemoglobin} = \frac{\text{Amount of } O_2 \text{ actually bound}}{\text{Maximum amount of } O_2 \text{ that can potentially be bound}} \times 100$$

If all heme–polypeptide units of all hemoglobin molecules in a blood sample are bound to O_2, then hemoglobin is *fully (100%) saturated*. If not all these units are bound to O_2, then hemoglobin is *partially saturated*. Here are two examples: If half of the available units are bound to O_2, then hemoglobin is 50% saturated; if only one-third of the available units are bound to O_2, then hemoglobin is 33% saturated.

The P_{O_2} Determines How Much Oxygen Binds to Hemoglobin

The O_2 that is bound to hemoglobin is trapped inside erythrocytes, so only the dissolved O_2 can diffuse out of tissue capillaries into tissue cells. Thus, it is important to understand the factors that promote O_2 binding to and dissociation from hemoglobin.

The most important factor that determines how much O_2 binds to hemoglobin is the P_{O_2}. The higher the P_{O_2}, the more O_2 combines with hemoglobin. The relationship between the percent saturation of hemoglobin and P_{O_2} is illustrated in the **oxygen–hemoglobin dissociation curve** in FIGURE 18.11. The curve is sigmoidal, or S-shaped, which means that binding of O_2 to hemoglobin exhibits *cooperativity:* As an O_2 molecule binds to one of the subunits of hemoglobin, the hemoglobin molecule undergoes a conformational change that increases the affinity of hemoglobin for the next O_2 molecule. This means that it becomes progressively easier for additional O_2 molecules to bind to hemoglobin.

Besides cooperativity, other aspects of the oxygen–hemoglobin dissociation curve are important to understand. For example, when the P_{O_2} is high, hemoglobin binds with large amounts of O_2 and is almost 100% saturated. When P_{O_2} is low, hemoglobin is only partially saturated. In other words, the greater the P_{O_2}, the more O_2 binds to hemoglobin, until all the available hemoglobin molecules are saturated. Therefore, in pulmonary capillaries, where P_{O_2} is high, a large amount of O_2 binds to hemoglobin. In tissue capillaries, where the P_{O_2} is lower, hemoglobin does not hold as much O_2, and the dissolved O_2 is unloaded via diffusion into tissue cells. Note that hemoglobin is still 75% saturated with O_2 at a P_{O_2} of 40 mmHg, the average P_{O_2} of tissue cells in a person at rest. This is the basis for the earlier statement that only 25% of the available O_2 unloads from hemoglobin and is used by tissue cells under resting conditions.

When the P_{O_2} is between 60 and 100 mmHg (the plateau portion of the oxygen–hemoglobin dissociation curve), hemoglobin is 90% or more saturated with O_2 (FIGURE 18.11). Thus, blood picks up a nearly full load of O_2 from the lungs even when the P_{O_2} of alveolar air is as low as 60 mmHg. The oxygen–hemoglobin dissociation curve explains why people can still perform well at high altitudes or when they have certain cardiac and pulmonary diseases, even though P_{O_2} may drop as low as 60 mmHg. When the P_{O_2} is between 0 and 60 mmHg (the steep portion of the oxygen–hemoglobin dissociation curve), O_2 is unloaded from hemoglobin. At a P_{O_2} of 40 mmHg, hemoglobin is 75% saturated with O_2. Oxygen saturation of hemoglobin drops further to 35% at 20 mmHg. Between 40 and 20 mmHg, large amounts of O_2 are released from hemoglobin in response to only small decreases in P_{O_2}. In active tissues such as contracting muscles, P_{O_2} may drop well below 40 mmHg. Then a large percentage of the O_2 is released from hemoglobin, providing more O_2 to metabolically active tissues.

FIGURE 18.11 Oxygen–hemoglobin dissociation curve showing the relationship between hemoglobin saturation and P_{O_2} at normal body temperature.

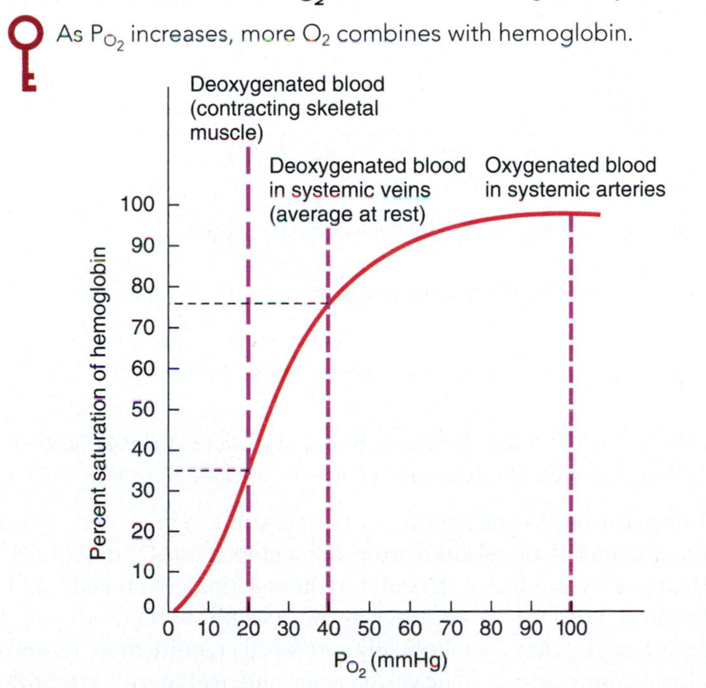

What point on the curve represents blood in your pulmonary veins right now? Which point represents blood in your pulmonary veins when you are jogging?

Several Factors Affect the Affinity of Hemoglobin for Oxygen

Although P_{O_2} is the most important factor that determines the percent saturation of hemoglobin, several other factors influence the affinity with which hemoglobin binds O_2. In effect, these factors shift the entire curve either to the left (higher affinity) or to the right (lower affinity). The changing affinity of hemoglobin for O_2 is another example of how homeostatic mechanisms adjust body activities to cellular needs. Each one makes sense if you keep in mind that metabolically active tissue cells need O_2 and produce acids, CO_2, and heat as wastes. The following four factors affect the affinity of hemoglobin for O_2:

1. **Acidity (pH).** As acidity increases (pH decreases), the affinity of hemoglobin for O_2 decreases, and O_2 dissociates more readily from hemoglobin (FIGURE 18.12a). In other words, increasing acidity enhances the unloading of oxygen from hemoglobin. The main acids produced by metabolically active tissues are lactic acid and carbonic acid. When pH decreases, the entire oxygen–hemoglobin dissociation curve shifts to the right; at any given P_{O_2}, hemoglobin is less saturated with O_2, a change termed the **Bohr effect**. The Bohr effect works both ways: An increase in H^+ in blood causes O_2 to unload from hemoglobin, and the binding of O_2 to hemoglobin causes unloading of H^+ from hemoglobin. The explanation for the Bohr effect is that hemoglobin can act as a buffer for hydrogen ions (H^+). But when H^+ ions bind to amino acids in hemoglobin, they alter its structure slightly, decreasing its oxygen-carrying capacity. Thus, lowered pH drives O_2 off hemoglobin, making more O_2 available for tissue cells. By contrast, elevated pH increases the affinity of hemoglobin for O_2 and shifts the oxygen–hemoglobin dissociation curve to the left.
2. **Partial pressure of carbon dioxide.** CO_2 can also bind to hemoglobin, and the effect is similar to that of H^+ (shifting the curve to the right). As P_{CO_2} rises, hemoglobin releases O_2 more readily (FIGURE 18.12b). P_{CO_2} and pH are related factors because low blood pH (acidity) results from high P_{CO_2}. As CO_2 enters the blood, much of it is temporarily converted to carbonic acid (H_2CO_3), a

FIGURE 18.12 Oxygen–hemoglobin dissociation curves showing the relationship of (a) pH, (b) P_{CO_2}, (c) temperature, and (d) BPG to hemoglobin saturation.

As pH decreases or the P_{CO_2}, temperature, or BPG concentration increases, the affinity of hemoglobin for O_2 declines, so less O_2 combines with hemoglobin and more is available to tissues.

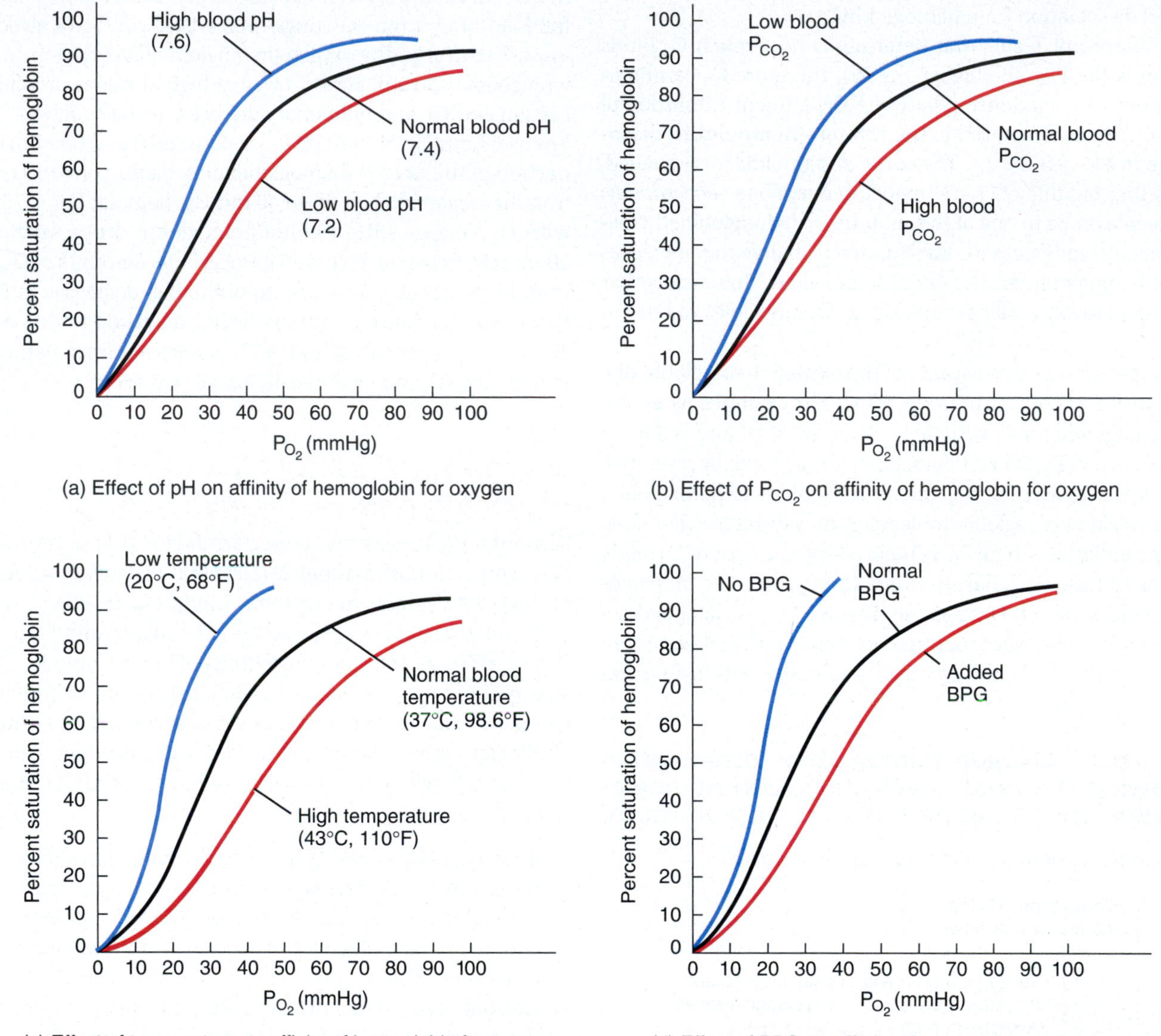

(a) Effect of pH on affinity of hemoglobin for oxygen

(b) Effect of P_{CO_2} on affinity of hemoglobin for oxygen

(c) Effect of temperature on affinity of hemoglobin for oxygen

(d) Effect of BPG on affinity of hemoglobin for oxygen

Is O_2 more available or less available to tissue cells when you have a fever? Explain your answer.

reaction catalyzed by an enzyme in erythrocytes called carbonic anhydrase (CA):

$$\underset{\text{Carbon dioxide}}{CO_2} + \underset{\text{Water}}{H_2O} \overset{CA}{\rightleftharpoons} \underset{\text{Carbonic acid}}{H_2CO_3} \rightleftharpoons \underset{\text{Hydrogen ion}}{H^+} + \underset{\text{Bicarbonate ion}}{HCO_3^-}$$

The carbonic acid thus formed in erythrocytes dissociates into hydrogen ions and bicarbonate ions. As the H^+ concentration increases, pH decreases. Thus, an increased P_{CO_2} produces a more acidic environment, which helps release O_2 from hemoglobin. During exercise, lactic acid—a by-product of anaerobic metabolism within muscles—also decreases blood pH. Decreased P_{CO_2} (and elevated pH) shifts the saturation curve to the left.

3. **Temperature**. Within limits, as temperature increases, so does the amount of O_2 released from hemoglobin (FIGURE 18.12c). Heat is a by-product of the metabolic reactions of all cells, and the heat released by contracting muscle fibers tends to raise body temperature. Metabolically active cells require more O_2 and liberate more acids and heat. The acids and heat in turn promote release of O_2 from oxyhemoglobin. Fever produces a similar result. By contrast, during hypothermia (lowered body temperature) cellular metabolism slows, the need for O_2 is reduced, and

more O_2 remains bound to hemoglobin (a shift to the left in the saturation curve).

4. **BPG.** A substance found in erythrocytes called **2,3-bisphosphoglycerate (BPG)**, previously called diphosphoglycerate (DPG), decreases the affinity of hemoglobin for O_2 and thus helps unload O_2 from hemoglobin. BPG is formed in erythrocytes when they break down glucose to produce ATP during glycolysis. When BPG combines with hemoglobin by binding to the terminal amino groups of the two beta globin chains, the hemoglobin binds O_2 less tightly at the heme group sites. The greater the level of BPG, the more O_2 is unloaded from hemoglobin (FIGURE 18.12d). Certain hormones, such as thyroxine, growth hormone, epinephrine, norepinephrine, and testosterone, increase the formation of BPG. The level of BPG is also higher in people living at higher altitudes.

FIGURE 18.13 Oxygen–hemoglobin dissociation curves comparing fetal and maternal hemoglobin.

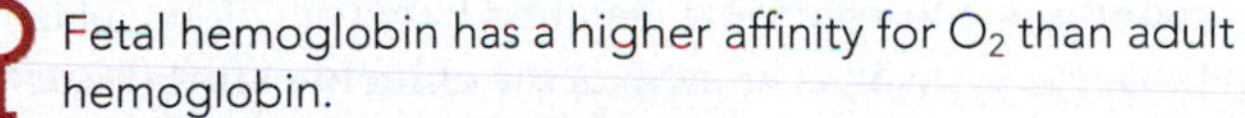

Fetal hemoglobin has a higher affinity for O_2 than adult hemoglobin.

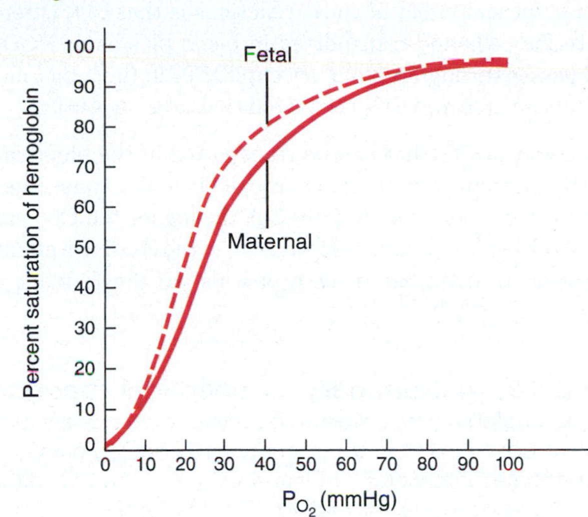

The P_{O_2} of placental blood is about 40 mmHg. What are the O_2 saturations of maternal and fetal hemoglobin at this P_{O_2}?

Fetal Hemoglobin and Adult Hemoglobin Have Different Affinities for Oxygen

Fetal hemoglobin (Hb-F) differs from **adult hemoglobin (Hb-A)** in structure and in its affinity for O_2. Hb-F has a higher affinity for O_2 because it binds BPG less strongly. Thus, when P_{O_2} is low, Hb-F can carry up to 30% more O_2 than maternal Hb-A (FIGURE 18.13). As the maternal blood enters the placenta, O_2 is readily transferred to fetal blood. This is very important because the O_2 saturation in maternal blood in the placenta is quite low, and the fetus might suffer hypoxia were it not for the greater affinity of fetal hemoglobin for O_2.

Carbon Dioxide Is Transported Through the Blood in Three Forms

Under normal resting conditions, each 100 mL of deoxygenated blood contains the equivalent of 53 mL of gaseous CO_2, which is transported in the blood in three main forms (see FIGURE 18.10):

1. **Dissolved CO_2.** The smallest percentage—about 7%—is dissolved in blood plasma. Upon reaching the lungs, it diffuses into alveolar air and is exhaled.
2. **Carbamino compounds.** A somewhat higher percentage, about 23%, combines with the amino groups of amino acids and proteins in blood to form **carbamino compounds.** Because the most prevalent protein in blood is hemoglobin (inside erythrocytes), most of the CO_2 transported in this manner is bound to hemoglobin. The main CO_2 binding sites are the terminal amino acids in the two alpha and two beta globin chains. Hemoglobin that has bound CO_2 is termed **carbaminohemoglobin (Hb–CO_2):**

$$\underset{\text{Hemoglobin}}{\text{Hb}} + \underset{\text{Carbon dioxide}}{CO_2} \rightleftharpoons \underset{\text{Carbaminohemoglobin}}{\text{Hb–}CO_2}$$

   The formation of carbaminohemoglobin is greatly influenced by P_{CO_2}. For example, in tissue capillaries P_{CO_2} is relatively high, which promotes formation of carbaminohemoglobin. But in pulmonary capillaries P_{CO_2} is relatively low, and the CO_2 readily splits apart from globin and enters the alveoli by diffusion.

3. **Bicarbonate ions.** The greatest percentage of CO_2—about 70%—is transported in blood plasma as **bicarbonate ions** (HCO_3^-). As CO_2 diffuses into systemic capillaries and enters erythrocytes, it reacts with water in the presence of the enzyme carbonic anhydrase (CA) to form carbonic acid, which dissociates into H^+ and HCO_3^-:

$$\underset{\substack{\text{Carbon}\\\text{dioxide}}}{CO_2} + \underset{\text{Water}}{H_2O} \overset{\text{CA}}{\rightleftharpoons} \underset{\substack{\text{Carbonic}\\\text{acid}}}{H_2CO_3} \rightleftharpoons \underset{\substack{\text{Hydrogen}\\\text{ion}}}{H^+} + \underset{\substack{\text{Bicarbonate}\\\text{ion}}}{HCO_3^-}$$

   Thus, as blood picks up CO_2, HCO_3^- accumulates inside erythrocytes. Some HCO_3^- moves out into the blood plasma, down its

CLINICAL CONNECTION

Carbon Monoxide Poisoning

Carbon monoxide (CO) is a colorless and odorless gas found in exhaust fumes from automobiles, gas furnaces, and space heaters and in tobacco smoke. It is a by-product of the combustion of carbon-containing materials such as coal, gas, and wood. CO binds to the heme group of hemoglobin, just as O_2 does, except that the binding of carbon monoxide to hemoglobin is over 200 times as strong as the binding of O_2 to hemoglobin. Thus, at a concentration as small as 0.1% (P_{CO} = 0.5 mmHg), CO combines with half of the available hemoglobin molecules and reduces the oxygen-carrying capacity of the blood by 50%. Elevated blood levels of CO cause **carbon monoxide poisoning**, which can cause the lips and oral mucosa to appear bright, cherry-red (the color of hemoglobin with carbon monoxide bound to it). Without prompt treatment, carbon monoxide poisoning is fatal. It is possible to rescue a victim of CO poisoning by administering pure oxygen, which speeds up the separation of carbon monoxide from hemoglobin.

concentration gradient. In exchange, chloride ions (Cl^-) move from plasma into the erythrocytes. This exchange of negative ions, which maintains the electrical balance between blood plasma and erythrocyte cytosol, is known as the **chloride shift** (FIGURE 18.14b). The net effect of these reactions is that CO_2 is removed from tissue cells and transported in blood plasma as HCO_3^-. As blood passes through pulmonary capillaries in the lungs, all these reactions reverse and CO_2 is exhaled (FIGURE 18.14a).

The amount of CO_2 that can be transported in the blood is influenced by the percent saturation of hemoglobin with oxygen. The lower the amount of oxyhemoglobin (Hb–O_2), the higher the CO_2 carrying capacity of the blood, a relationship known as the **Haldane effect**. Two characteristics of deoxyhemoglobin give rise to the Haldane effect: (1) Deoxyhemoglobin binds to and thus transports more CO_2 than does Hb–O_2 and (2) deoxyhemoglobin also buffers more H^+ than does Hb–O_2, thereby removing H^+ from solution and promoting conversion of CO_2 to HCO_3^- via the reaction catalyzed by carbonic anhydrase.

Gas Exchange and Transport Can Be Summarized

Deoxygenated blood returning to the pulmonary capillaries in the lungs (FIGURE 18.14a) contains CO_2 dissolved in blood plasma, CO_2 combined with globin as carbaminohemoglobin (Hb–CO_2), and CO_2 incorporated into HCO_3^- within erythrocytes. The erythrocytes have also picked up H^+, some of which binds to and therefore is

FIGURE 18.14 Summary of chemical reactions that occur during gas exchange. (a) As carbon dioxide (CO_2) is exhaled, hemoglobin (Hb) inside erythrocytes in pulmonary capillaries unloads CO_2 and picks up O_2 from alveolar air. Binding of O_2 to Hb–H releases hydrogen ions (H^+). Bicarbonate ions (HCO_3^-) pass into the erythrocyte and bind to released H^+, forming carbonic acid (H_2CO_3). The H_2CO_3 dissociates into water (H_2O) and CO_2, and the CO_2 diffuses from blood into alveolar air. To maintain electrical balance, a chloride ion (Cl^-) exits the erythrocyte for each HCO_3^- that enters (reverse chloride shift). (b) CO_2 diffuses out of tissue cells that produce it and enters erythrocytes, where some of it binds to hemoglobin, forming carbaminohemoglobin (Hb–CO_2). This reaction causes O_2 to dissociate from oxyhemoglobin (Hb–O_2). Other molecules of CO_2 combine with water to produce bicarbonate ions (HCO_3^-) and hydrogen ions (H^+). As Hb buffers H^+, the Hb releases O_2 (Bohr effect). To maintain electrical balance, a chloride ion (Cl^-) enters the erythrocyte for each HCO_3^- that exits (chloride shift).

Hemoglobin inside erythrocytes transports O_2, CO_2, and H^+.

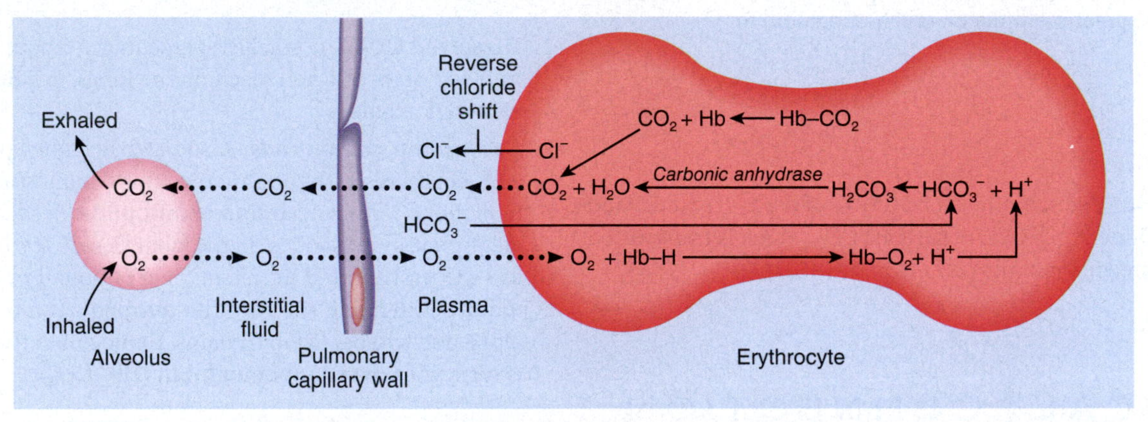

(a) Exchange of O_2 and CO_2 in pulmonary capillaries (pulmonary gas exchange)

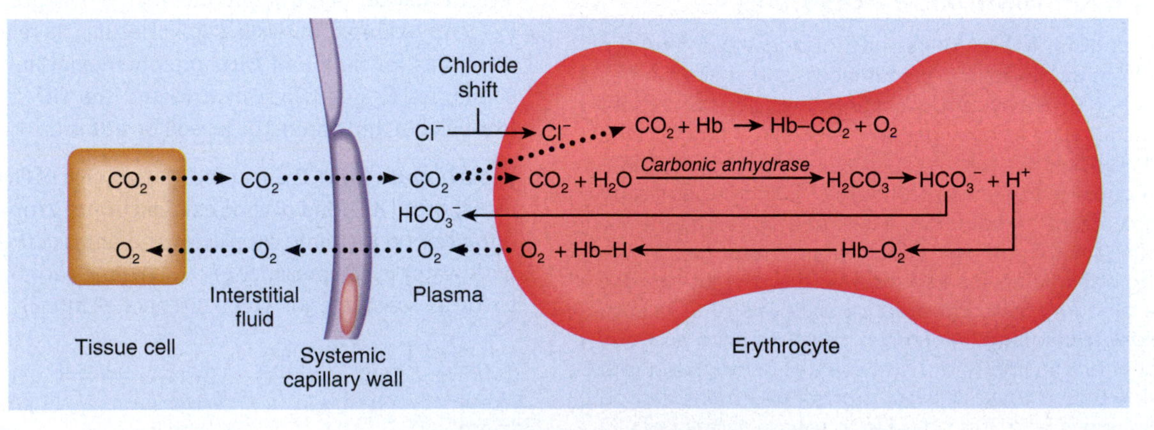

(b) Exchange of O_2 and CO_2 in systemic capillaries (systemic gas exchange)

? Would you expect the concentration of HCO_3^- to be higher in blood taken from a systemic artery or a systemic vein?

buffered by hemoglobin (Hb–H). As blood passes through the pulmonary capillaries, molecules of CO_2 dissolved in blood plasma and CO_2 that dissociates from the globin portion of hemoglobin diffuse into alveolar air and are exhaled. At the same time, inhaled O_2 diffuses from alveolar air into erythrocytes and is binding to hemoglobin to form oxyhemoglobin (Hb–O_2). Carbon dioxide is also released from HCO_3^- when H^+ combines with HCO_3^- inside erythrocytes. The H_2CO_3 formed from this reaction then splits into CO_2, which is exhaled, and H_2O. As the concentration of HCO_3^- declines inside erythrocytes in pulmonary capillaries, HCO_3^- diffuses in from the blood plasma, in exchange for Cl^-. In sum, oxygenated blood leaving the lungs has increased O_2 content and decreased amounts of CO_2 and H^+. In systemic capillaries, as cells use O_2 and produce CO_2, the chemical reactions reverse (FIGURE 18.14b).

CHECKPOINT

19. In a resting person, what is the average number of O_2 molecules attached to each hemoglobin molecule in blood in the pulmonary arteries? In blood in the pulmonary veins?
20. What is the relationship between hemoglobin and P_{O_2}? How do temperature, H^+, P_{O_2}, and BPG influence the affinity of hemoglobin for O_2?
21. Why can hemoglobin unload more oxygen as blood flows through capillaries of metabolically active tissues, such as skeletal muscle during exercise, than is unloaded at rest?

FIGURE 18.15 Components of the respiratory center.

The respiratory center is composed of neurons in the medullary respiratory center in the medulla oblongata plus the pontine respiratory center in the pons.

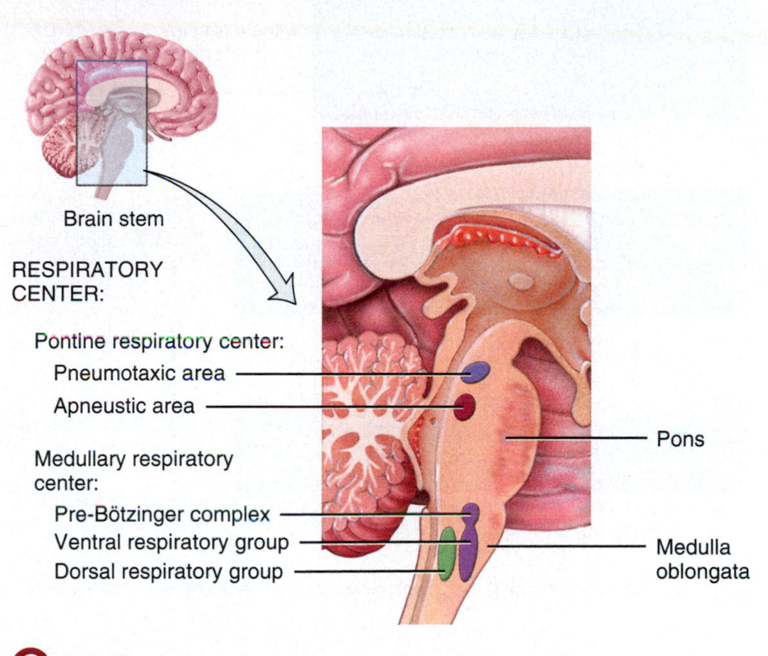

Which area contains inspiratory neurons that are active and then inactive in a repeating cycle?

18.6 Control of Ventilation

OBJECTIVES

- Explain how the nervous system controls breathing.
- List the factors that can alter the rate and depth of breathing.

At rest, about 200 mL of O_2 are used each minute by body cells. During strenuous exercise, however, O_2 use typically increases 15- to 20-fold in normal healthy adults, and as much as 30-fold in elite endurance-trained athletes. Several mechanisms help match breathing effort to metabolic demand.

The Respiratory Center Controls Breathing

The size of the thoracic cavity is altered by the action of the respiratory muscles, which contract as a result of action potentials transmitted from centers in the brain and relax in the absence of action potentials. These action potentials are sent from clusters of neurons located in the medulla oblongata and pons of the brain stem. This widely dispersed group of neurons, collectively called the **respiratory center**, can be divided into two areas on the basis of their functions: (1) the medullary respiratory center in the medulla oblongata and (2) the pontine respiratory center in the pons.

Medullary Respiratory Center

The **medullary respiratory center** is made up of two collections of neurons called the **dorsal respiratory group (DRG)** and the **ventral respiratory group (VRG)** (FIGURE 18.15). The DRG consists mainly of inspiratory neurons. During normal quiet breathing (FIGURE 18.16a), the inspiratory neurons of the DRG generate action potentials to the diaphragm via the phrenic nerves and the external intercostal muscles via the intercostal nerves. These action potentials are released in bursts, which begin weakly, increase in strength for about 2 seconds, and then stop altogether. When the action potentials reach the diaphragm and external intercostals, the muscles contract and inspiration occurs. When the DRG inspiratory neurons become inactive, the diaphragm and external intercostals relax for about 3 seconds, allowing the passive elastic recoil of the lungs and chest wall that causes expiration (FIGURE 18.16a). Then, the cycle repeats itself.

Located in the VRG is a cluster of neurons called the **pre-Bötzinger complex** (BOT-zin-ger) that is believed to be important in the generation of the rhythm of respiration (see FIGURE 18.15). This rhythm generator, analogous to the one in the heart, is composed of pacemaker cells that set the basic rhythm of breathing. The exact mechanism of these pacemaker cells is unknown and is the topic of much ongoing research. However, it is thought that the pacemaker cells provide input to the DRG, driving the rate at which DRG inspiratory neurons fire action potentials.

The VRG also contains inspiratory neurons and expiratory neurons, but these neurons do not participate in normal quiet breathing. Instead, they become activated when forceful breathing is required, such as during exercise, playing a wind instrument, or at high altitudes. During forceful inspiration (FIGURE 18.16b), action potentials from DRG inspiratory neurons not only stimulate the diaphragm and external intercostal muscles to contract, they also activate VRG inspiratory neurons to send action potentials to the accessory muscles of inspiration (sternocleidomastoid and scalene muscles). Contraction of these muscles results in forceful inspiration.

FIGURE 18.16 Mechanism by which the respiratory center controls (a) normal quiet breathing and (b) forceful breathing.

During normal quiet breathing, the ventral respiratory group is inactive; during forceful breathing, the dorsal respiratory group activates the ventral respiratory group.

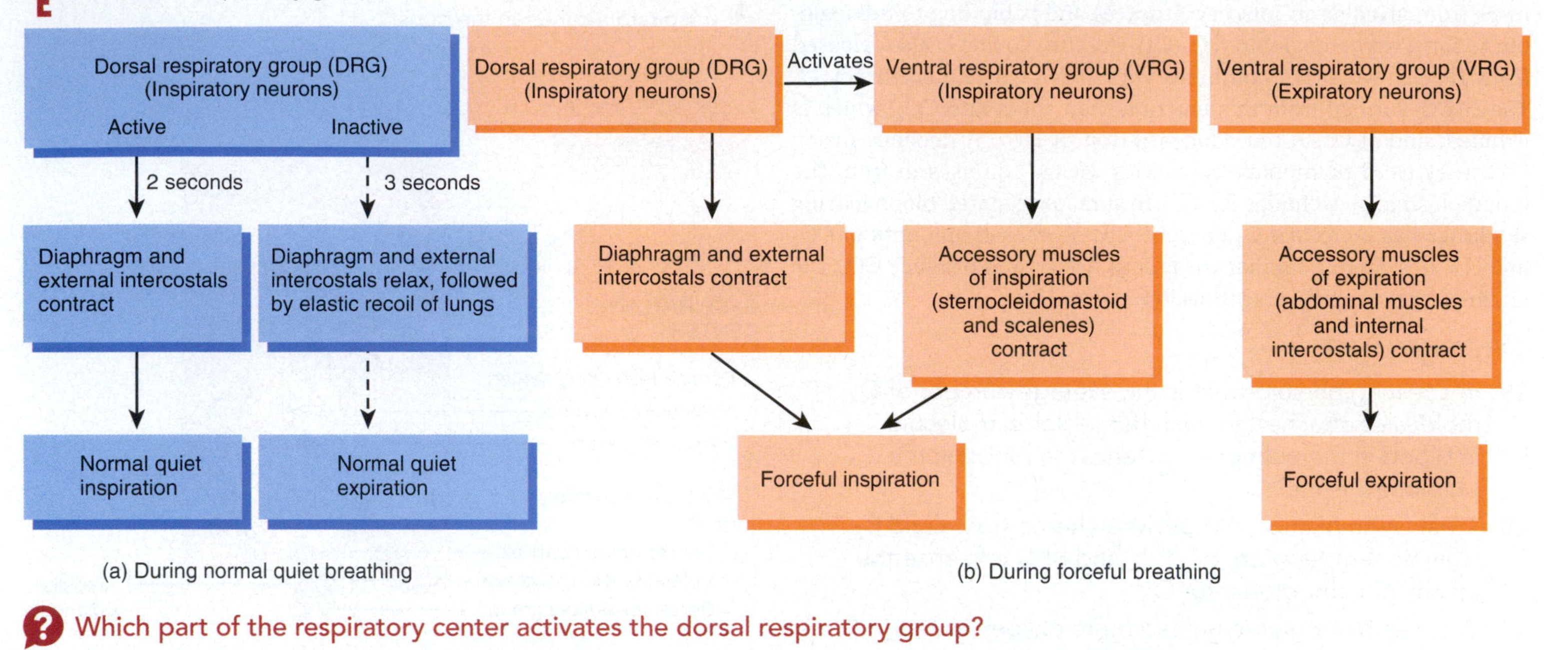

Which part of the respiratory center activates the dorsal respiratory group?

During forceful expiration (FIGURE 18.16b), DRG inspiratory neurons are inactive along with VRG inspiratory neurons, but VRG expiratory neurons become activated and send action potentials to the accessory muscles of expiration (abdominal muscles and internal intercostals). Contraction of these muscles results in forceful expiration.

Pontine Respiratory Center

The **pontine respiratory center** consists of two groups of neurons that help coordinate the transition between inspiration and expiration: the pneumotaxic area in the upper pons and the apneustic area in the lower pons (see FIGURE 18.15). The **pneumotaxic area**, also known as the *pontine respiratory group*, sends inhibitory signals to the DRG. The major effect of these signals is to help turn off the DRG before the lungs become too full of air. In other words, the signals shorten the duration of inspiration. When the pneumotaxic area is more active, breathing rate is more rapid. The **apneustic area** sends excitatory signals to the DRG that activate it and prolong inspiration. The result is a long, deep inspiration. When the pneumotaxic area is active, it overrides signals from the apneustic area.

The Respiratory Center Is Subject to Regulation

Activity of the respiratory center can be modified in response to input from higher brain regions, receptors in the peripheral nervous system, and a variety of other factors.

Cortical Influences on Breathing

Because the cerebral cortex has connections with the respiratory center, we can voluntarily alter our pattern of breathing. We can even refuse to breathe at all for a short time. Voluntary control is protective because it enables us to prevent water or irritating gases from entering the lungs. The ability not to breathe, however, is limited by the buildup of CO_2 and H^+ in the body. When P_{CO_2} and H^+ concentrations increase to a certain level, the DRG neurons of the medullary respiratory center are strongly stimulated, action potentials are sent along the phrenic and intercostal nerves to inspiratory muscles, and breathing resumes, whether the person wants it to or not. It is impossible for small children to kill themselves by voluntarily holding their breath, even though many have tried in order to get their way. If breath is held long enough to cause fainting, breathing resumes when consciousness is lost. Action potentials from the hypothalamus and limbic system also stimulate the respiratory center, allowing emotional stimuli to alter respirations as in, for example, laughing and crying.

Chemoreceptor Regulation of Breathing

Certain chemical stimuli modulate how quickly and how deeply we breathe. The respiratory system functions to maintain proper levels of CO_2 and O_2 and is very responsive to changes in the levels of these gases in body fluids. **Chemoreceptors** are sensory receptors that are responsive to chemicals. Chemoreceptors in two locations monitor levels of CO_2, H^+, and O_2 and provide input to the respiratory center (FIGURE 18.17). **Central chemoreceptors** are located in the medulla oblongata in the *central* nervous system. They respond to changes in H^+ concentration or P_{CO_2}, or both, in cerebrospinal fluid. **Peripheral chemoreceptors** are located in the **aortic bodies**, clusters of chemoreceptors located in the wall of the arch of the aorta, and in the **carotid bodies**, clusters of chemoreceptors in the wall of the left and right common carotid arteries where they divide into the internal and external carotid

FIGURE 18.17 Locations of peripheral chemoreceptors.

Chemoreceptors are sensory receptors that respond to changes in the levels of certain chemicals in the body.

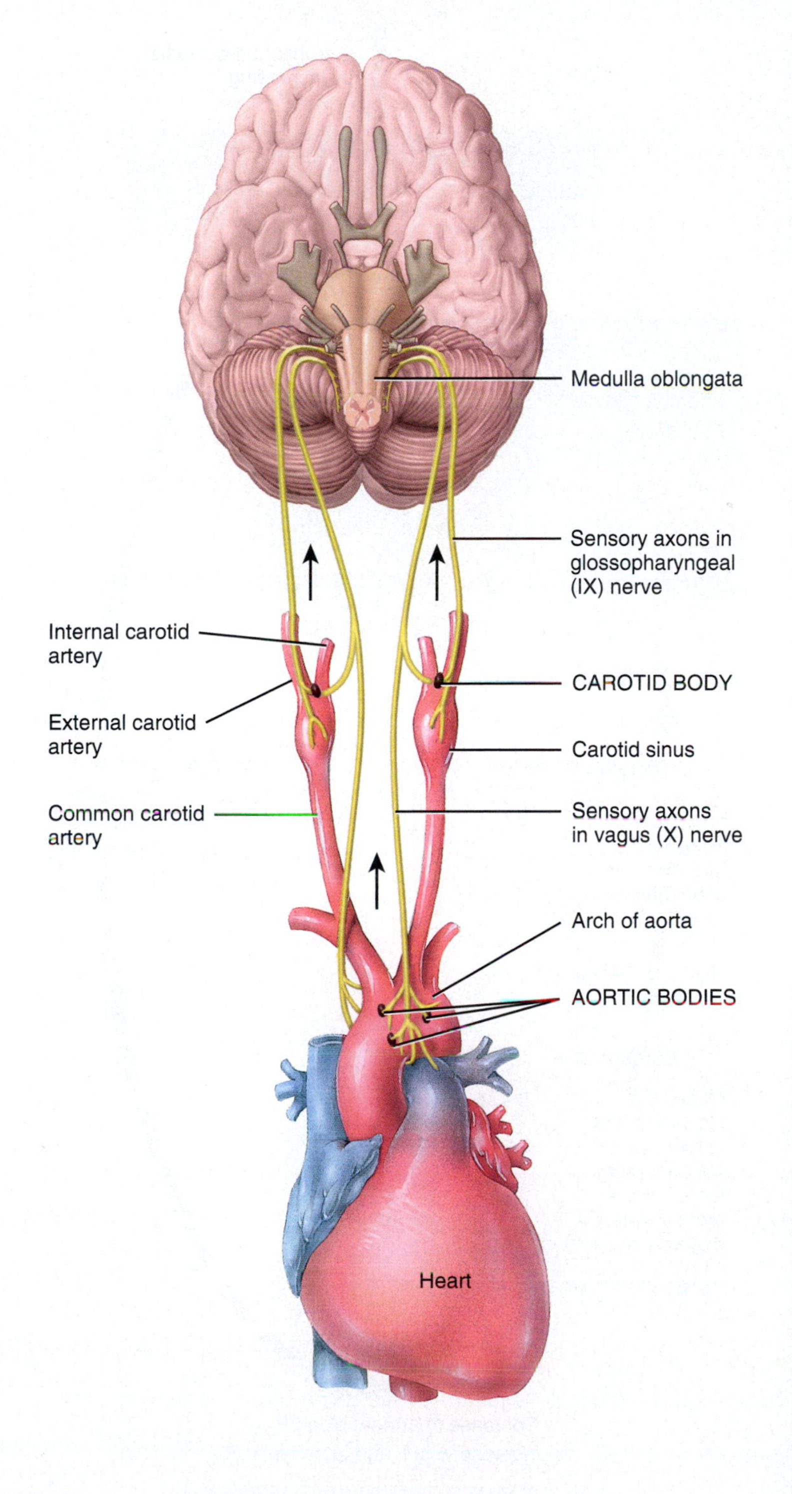

Which chemicals stimulate peripheral chemoreceptors?

arteries. (The chemoreceptors of the aortic bodies are located close to the aortic baroreceptors, and the carotid bodies are located close to the carotid sinus baroreceptors. Recall from Chapter 15 that baroreceptors are sensory receptors that monitor blood pressure.) These chemoreceptors are part of the peripheral nervous system and are sensitive to changes in P_{O_2}, H^+, and P_{CO_2} in the blood. Axons of sensory neurons from the aortic bodies are part of the vagus (X) nerves, and those from the carotid bodies are part of the glossopharyngeal (IX) nerves.

Because CO_2 is lipid-soluble, it easily diffuses into cells where, in the presence of carbonic anhydrase, it combines with water (H_2O) to form carbonic acid (H_2CO_3). Carbonic acid quickly breaks down into H^+ and HCO_3^-. Thus, an increase in CO_2 in the blood causes an increase in H^+ inside cells, and a decrease in CO_2 causes a decrease in H^+.

Normally, the P_{CO_2} in arterial blood is 40 mmHg. If even a slight increase in P_{CO_2} occurs—a condition called **hypercapnia** or *hypercarbia*—the central chemoreceptors are stimulated and respond vigorously to the resulting increase in H^+ level. The peripheral chemoreceptors are also stimulated by both the high P_{CO_2} and the rise in H^+. In addition, the peripheral chemoreceptors (but not the central chemoreceptors) respond to a deficiency of O_2. When P_{O_2} in arterial blood falls from a normal level of 100 mmHg but is still above 50 mmHg, the peripheral chemoreceptors are stimulated. Severe deficiency of O_2 depresses activity of the central chemoreceptors and DRG, which then do not respond well to any inputs and send fewer action potentials to the muscles of inspiration. As the breathing rate decreases or breathing ceases altogether, P_{O_2} falls lower and lower, establishing a positive feedback cycle with a possibly fatal result.

The central and peripheral chemoreceptors participate in a negative feedback system that regulates the levels of CO_2, O_2, and H^+ in

CLINICAL CONNECTION

Hypoxia

Hypoxia is a deficiency of O_2 at the tissue level. Based on the cause, hypoxia is classified into four types:

1. **Hypoxic hypoxia** is caused by a low P_{O_2} in arterial blood as a result of high altitude, airway obstruction, or fluid in the lungs.
2. In **anemic hypoxia**, too little functioning hemoglobin is present in the blood, which reduces O_2 transport to tissue cells. Among the causes are hemorrhage, anemia, and failure of hemoglobin to carry its normal complement of O_2, as in carbon monoxide poisoning.
3. In **ischemic hypoxia**, blood flow to a tissue is so reduced that too little O_2 is delivered to it, even though P_{O_2} and oxyhemoglobin levels are normal.
4. In **histotoxic hypoxia**, the blood delivers adequate O_2 to tissues, but the tissues are unable to use it properly because of the action of some toxic agent. One cause is cyanide poisoning, in which cyanide blocks an enzyme required for the use of O_2 during ATP synthesis.

the blood (FIGURE 18.18). As a result of increased P_{CO_2}, decreased pH (increased H^+), or decreased P_{O_2}, input from the central and peripheral chemoreceptors causes the DRG to become highly active, and the rate and depth of breathing increase. Rapid and deep breathing, called **hyperventilation**, allows the inhalation of more O_2 and expiration of more CO_2 until P_{CO_2} and H^+ are lowered to normal.

If arterial P_{CO_2} is lower than 40 mmHg—a condition called **hypocapnia** or *hypocarbia*—the central and peripheral chemoreceptors are not stimulated, and stimulatory input is not sent to the DRG. As a result, the area sets its own moderate pace until CO_2 accumulates and the P_{CO_2} rises to 40 mmHg. The DRG is stimulated more strongly when P_{CO_2} is rising above normal than when P_{O_2} is falling below normal. As a result, people who hyperventilate voluntarily and cause hypocapnia can hold their breath for an unusually long period. Swimmers were once encouraged to hyperventilate just before diving in to compete. However, this practice is risky because the O_2 level may fall dangerously low and cause fainting before the P_{CO_2} rises high enough to stimulate inspiration. If you faint on land you may suffer bumps and bruises, but if you faint in the water you could drown.

Proprioceptor Stimulation of Breathing

As soon as you start exercising, your rate and depth of breathing increase, even before changes in P_{O_2}, P_{CO_2}, or H^+ level occur. The main stimulus for these quick changes in respiratory effort is input from proprioceptors, which monitor movement of joints and muscles. Action potentials from the proprioceptors stimulate the DRG of the medulla oblongata. At the same time, axon collaterals (branches) of upper motor neurons that originate in the primary motor cortex also feed excitatory signals into the DRG.

The Inflation Reflex

Similar to those in the blood vessels, stretch-sensitive receptors are located in the walls of bronchi and bronchioles. When these receptors become stretched during overinflation of the lungs, action potentials are sent along the vagus (X) nerves to the dorsal respiratory group (DRG) in the medullary respiratory center. In response, the DRG is inhibited and the diaphragm and external intercostals relax. As a result, further inspiration is stopped and expiration begins. As air leaves the lungs during expiration, the lungs deflate and the stretch receptors are no longer stimulated. Thus, the DRG is no longer inhibited, and a new inspiration begins. This reflex is referred to as the **inflation (Hering–Breuer) reflex** (HER-ing BROY-er). In infants, the reflex appears to function in normal breathing. In adults, however, the reflex is not activated until tidal volume (normally 500 mL) reaches more than 1500 mL. Therefore, the reflex in adults is a protective mechanism that prevents excessive inflation of the lungs, for example, during severe exercise, rather than a key component in the normal control of respiration.

Other Influences on Breathing

Other factors that contribute to regulation of ventilation include the following:

- **Limbic system stimulation**. Anticipation of activity or emotional anxiety may stimulate the limbic system, which then sends excitatory input to the DRG, increasing the rate and depth of ventilation.

FIGURE 18.18 Regulation of breathing in response to changes in blood P_{CO_2}, P_{O_2}, and pH (H^+) via negative feedback control.

An increase in arterial blood P_{CO_2} stimulates the dorsal respiratory group.

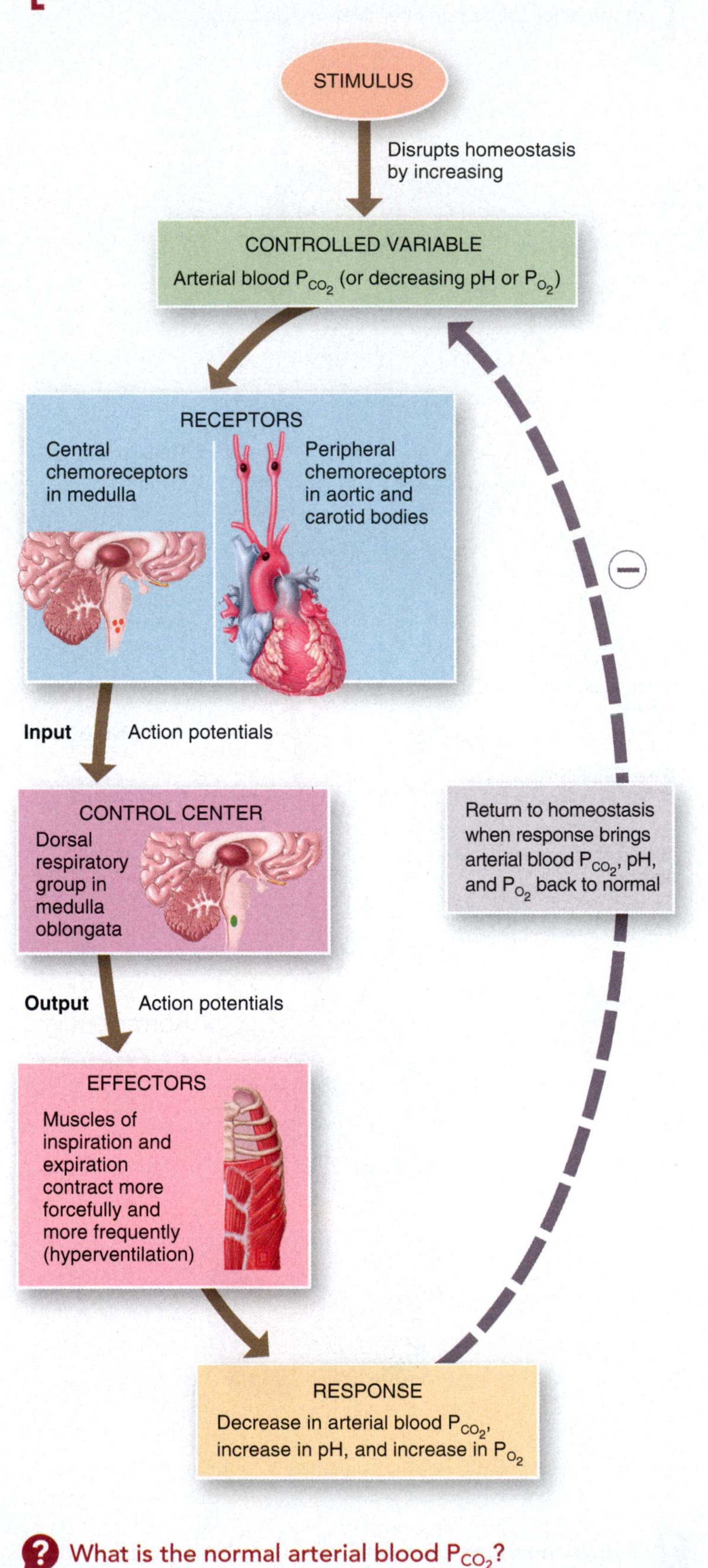

What is the normal arterial blood P_{CO_2}?

- **Temperature**. An increase in body temperature, as occurs during a fever or vigorous muscular exercise, increases the rate of ventilation. A decrease in body temperature decreases respiratory rate. A sudden cold stimulus (such as plunging into cold water) causes temporary **apnea**, an absence of breathing.
- **Pain**. A sudden, severe pain brings about brief apnea, but a prolonged somatic pain increases respiratory rate. Visceral pain may slow the rate of ventilation.
- **Stretching the anal sphincter muscle**. This action increases the respiratory rate and is sometimes used to stimulate ventilation in a newborn baby or a person who has stopped breathing.
- **Irritation of airways**. Physical or chemical irritation of the pharynx or larynx brings about an immediate cessation of breathing followed by coughing or sneezing.
- **Blood pressure**. The carotid and aortic baroreceptors that detect changes in blood pressure have a small effect on breathing. A sudden rise in blood pressure decreases the rate of respiration, and a drop in blood pressure increases the respiratory rate.

CHECKPOINT

22. How does the medullary respiratory center regulate breathing?
23. What roles do the pneumotaxic and apneustic areas of the pontine respiratory center play in the control of breathing?
24. How do the cerebral cortex, levels of CO_2 and O_2, proprioceptors, inflation reflex, temperature changes, pain, and irritation of the airways modify breathing?

18.7 Exercise and the Respiratory System

OBJECTIVE

- Describe the effects of exercise on the respiratory system.

The respiratory and cardiovascular systems make adjustments in response to both the intensity and duration of exercise. The effects of exercise on the heart are discussed in Chapter 14. This section focuses on the effects of exercise on the respiratory system.

Recall that the heart pumps the same amount of blood to the lungs as to all the rest of the body. Thus, as cardiac output rises, the blood flow to the lungs, termed **pulmonary perfusion**, increases as well. In addition, the **O_2 diffusing capacity**, a measure of the rate at which O_2 can diffuse from alveolar air into the blood, may increase threefold during maximal exercise because more pulmonary capillaries become maximally perfused. As a result, there is a greater surface area available for diffusion of O_2 into pulmonary blood capillaries.

When muscles contract during exercise, they consume large amounts of O_2 and produce large amounts of CO_2. During vigorous exercise, O_2 consumption and ventilation both increase dramatically. At the onset of exercise, an abrupt increase in ventilation is followed by a more gradual increase. With moderate exercise, the increase is due mostly to an increase in the depth of ventilation rather than to increased breathing rate. When exercise is more strenuous, the frequency of breathing also increases.

CLINICAL CONNECTION

The Effect of Smoking on Respiratory Efficiency

Smoking may cause a person to become easily "winded" during even moderate exercise because several factors decrease respiratory efficiency in smokers:

1. Nicotine constricts terminal bronchioles, which decreases airflow into and out of the lungs.
2. Carbon monoxide in smoke binds to hemoglobin and reduces its oxygen-carrying capability.
3. Irritants in smoke cause increased mucus secretion by the mucosa of the bronchial tree and swelling of the mucosal lining, both of which impede airflow into and out of the lungs.
4. Irritants in smoke also inhibit the movement of cilia and destroy cilia in the lining of the respiratory system. Thus, excess mucus and foreign debris are not easily removed, which further adds to the difficulty in breathing.
5. With time, smoking leads to destruction of elastic fibers in the lungs and is the prime cause of emphysema. These changes cause collapse of small bronchioles and trapping of air in alveoli at the end of expiration. The result is less efficient gas exchange.

The abrupt increase in ventilation at the start of exercise is due to neural changes that send excitatory signals to the dorsal respiratory group (DRG) of the medullary respiratory center. These changes include (1) anticipation of the activity, which stimulates the limbic system; (2) sensory information from proprioceptors in muscles, tendons, and joints; and (3) motor information from the primary motor cortex. The more gradual increase in ventilation during moderate exercise is due to chemical and physical changes in the bloodstream, including (1) slightly decreased P_{O_2}, due to increased O_2 consumption; (2) slightly increased P_{CO_2}, due to increased CO_2 production by contracting muscle fibers; and (3) increased temperature, due to liberation of more heat as more O_2 is utilized. During strenuous exercise, HCO_3^- buffers H^+ released by lactic acid in a reaction that liberates CO_2, which further increases P_{CO_2}.

At the end of an exercise session, an abrupt decrease in ventilation is followed by a more gradual decline to the resting level. The initial decrease is due mainly to changes in neural factors when movement stops or slows; the more gradual phase reflects the slower return of blood chemistry levels and temperature to the resting state.

CHECKPOINT

25. How does exercise affect the DRG?

To appreciate the many ways that the respiratory system contributes to homeostasis of other body systems, examine *Focus on Homeostasis: Contributions of the Respiratory System.*

FOCUS on HOMEOSTASIS

MUSCULAR SYSTEM

- Increased rate and depth of breathing support increased activity of skeletal muscles during exercise

NERVOUS SYSTEM

- Nose contains receptors for sense of smell (olfaction)
- Vibrations of air flowing across vocal cords produce sounds for speech

ENDOCRINE SYSTEM

- Angiotensin-converting enzyme (ACE) in lungs catalyzes formation of the hormone angiotensin II from angiotensin I

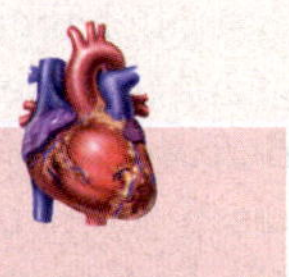

CARDIOVASCULAR SYSTEM

- During inspirations, respiratory pump aids return of venous blood to the heart

CONTRIBUTIONS OF THE RESPIRATORY SYSTEM

FOR ALL BODY SYSTEMS

- Provides oxygen and removes carbon dioxide
- Helps adjust pH of body fluids through exhalation of carbon dioxide

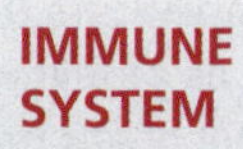

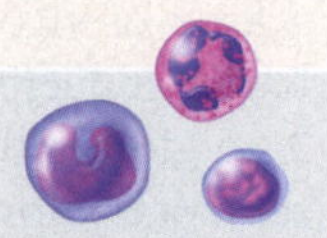

IMMUNE SYSTEM

- Hairs in nose, cilia and mucus in trachea, bronchi, and smaller airways, and alveolar macrophages contribute to nonspecific resistance to disease
- Pharynx (throat) contains tonsils

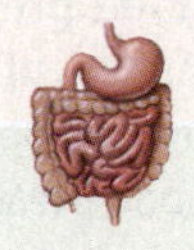

DIGESTIVE SYSTEM

- Forceful contraction of respiratory muscles can assist in defecation

URINARY SYSTEM

- Together, respiratory and urinary systems regulate pH of body fluids

REPRODUCTIVE SYSTEMS

- Increased rate and depth of breathing support activity during sexual intercourse

FROM RESEARCH TO REALITY

Pacifiers and Asthma

Reference

Morass, B. et al. (2008). The impact of early lifestyle factors on wheezing and asthma in Austrian preschool children. *Acta Pædiatrica*. 97: 337–341.

Does a pacifier's level of cleanliness influence a child's risk of developing asthma?

West Coast Surfer / Getty Images

Asthma is a very common childhood disease and its prevalence has increased in the past decades. Because of this increasing prevalence, researchers are interested in learning about the risk factors that might make a child more susceptible to developing asthma. Once risk factors are known, they can be better controlled. Could something as simple as the frequency of cleaning a child's pacifier change a child's likelihood of getting asthma?

Article description:

Researchers in this study used a questionnaire to obtain information from parents about their kindergarten-aged children. The questionnaire asked about a variety of known and possible risk factors for asthma and whether the child had asthma or demonstrated wheezing in the past 12 months. The researchers were able to find strong correlations between asthma and wheezing and certain lifestyle factors.

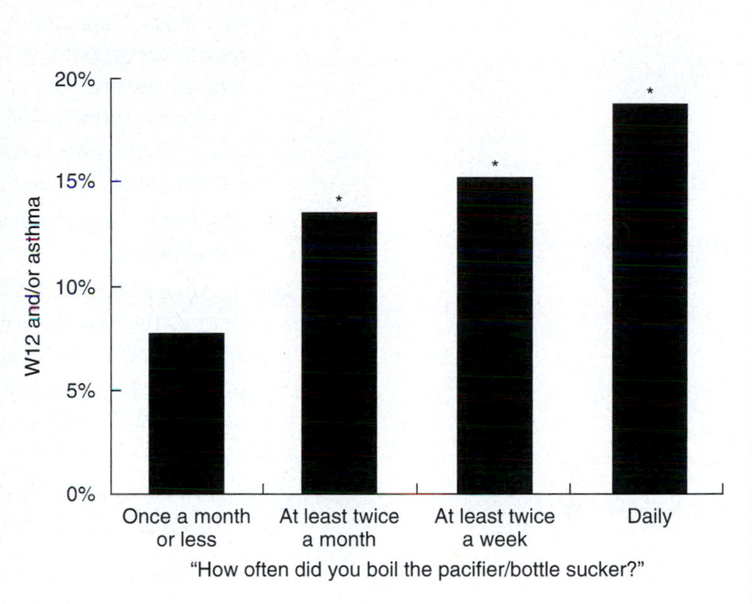

Go to WileyPLUS Learning Space and use the data from this article to answer the questions posed there and to learn more about lifestyle factors that may influence a child's susceptibility to developing asthma.

CHAPTER REVIEW

18.1 Components of the Respiratory System

1. The respiratory system consists of the nose, pharynx, larynx, trachea, primary bronchi, and lungs.
2. The airways from the nose to the bronchioles of the lungs are lined by a mucosa (mucous membrane), which consists of a layer of epithelial cells and an underlying layer of connective tissue. The mucus produced by the mucosa traps particles and lubricates the airways.
3. The nose filters, warms, and humidifies air; it also detects olfactory stimuli and helps modify speech vibrations.
4. The pharynx is a passageway for air and food and serves as a resonating chamber for speech sounds.

5. The larynx routes air into the rest of the airways and food into the esophagus; it also contains the vocal cords, which produce sound as they vibrate. Taut cords produce high pitches, and relaxed ones produce low pitches.
6. The trachea conveys air from the larynx to the primary bronchi.
7. The primary bronchi bring air from the trachea to the smaller bronchi and bronchioles of the lungs.
8. The lungs are paired organs in the thoracic cavity; each is enclosed by a membrane called the pleura. Between the parietal and visceral layers of the pleura is a pleural cavity that contains intrapleural fluid. Each lung contains all the branches of a primary bronchus—secondary bronchi, tertiary bronchi, bronchioles, terminal bronchioles, respiratory bronchioles, alveolar ducts, and alveolar sacs. Bronchioles are the main sites of resistance to airflow; bronchiolar smooth muscle is innervated by the autonomic nervous system (sympathetic and parasympathetic divisions). Alveolar walls are thin, which facilitates gas exchange. Alveolar and capillary walls together form the respiratory membrane.
9. The pulmonary circulation has a high rate of blood flow, a low resistance, and a low blood pressure.
10. There are two functional zones of the respiratory system. The conducting zone is the part of the respiratory system from the nose to the terminal bronchioles; it filters, warms, and humidifies air and conducts it into the lungs. The respiratory zone is the part of the respiratory system that contains alveoli (respiratory bronchioles, alveolar ducts, and alveolar sacs); it carries out gas exchange.

18.2 Ventilation

1. Ventilation is the process by which air flows into and out of the lungs. Three pressures that are important to ventilation include atmospheric pressure, alveolar pressure, and intrapleural pressure.
2. The movement of air into and out of the lungs depends on pressure changes governed in part by Boyle's law, which states that the volume of a gas varies inversely with pressure, assuming that temperature remains constant.
3. Inspiration occurs when alveolar pressure falls below atmospheric pressure. Contraction of the diaphragm and external intercostals increases the size of the thoracic cavity, which causes the lungs to expand. Expansion of the lungs decreases alveolar pressure so that air moves down a pressure gradient from the atmosphere into the lungs. During forceful inspiration, accessory muscles of inspiration (sternocleidomastoids and scalenes) are also used.
4. Expiration occurs when alveolar pressure is higher than atmospheric pressure. Relaxation of the diaphragm and external intercostals results in elastic recoil of the chest wall and lungs; lung volume decreases and alveolar pressure increases, so air moves from the lungs to the atmosphere. Forceful expiration involves contraction of the internal intercostal and abdominal muscles.
5. The surface tension exerted by alveolar fluid is decreased by the presence of surfactant.
6. Compliance is the ease with which the lungs and thoracic wall can expand.
7. The walls of the airways, especially the bronchioles, offer some resistance to breathing.
8. For efficient gas exchange, ventilation of the alveoli must be matched to perfusion (ventilation–perfusion matching). Autoregulatory mechanisms involving two local chemical mediators (CO_2 and O_2) play key roles in this process. CO_2 controls ventilation by altering bronchiolar diameter; O_2 control perfusion by altering arteriolar diameter.
9. Normal quiet breathing is termed eupnea. Modified respiratory movements, such as coughing, sneezing, sighing, yawning, crying, laughing, and hiccupping, are used to express emotions and to clear the airways.

CHAPTER REVIEW

18.3 Lung Volumes and Capacities

1. Specific lung volumes can be measured to assess normal pulmonary function.
2. Lung volumes include tidal volume, inspiratory reserve volume, expiratory reserve volume, and residual volume.
3. Lung capacities, the sum of two or more lung volumes, include functional residual capacity, inspiratory capacity, vital capacity, and total lung capacity.
4. Minute ventilation ($\dot{V}$) is the volume of air inspired and expired per minute; in an average adult at rest, it equals 6000 mL/min.
5. Alveolar ventilation ($\dot{V}_A$) is the volume of air per minute that actually reaches the respiratory zone; typically, it is about 4200 mL/min.

18.4 Exchange of Oxygen and Carbon Dioxide

1. The partial pressure of a gas is the pressure exerted by that gas in a mixture of gases. It is symbolized by P_x, where the subscript is the chemical formula of the gas.
2. According to Dalton's law, each gas in a mixture of gases exerts its own pressure as if all the other gases were not present.
3. Henry's law states that the quantity of a gas that will dissolve in a liquid is proportional to the partial pressure of the gas and its solubility (given that the temperature remains constant).
4. In pulmonary and systemic gas exchange, O_2 and CO_2 diffuse from areas of higher partial pressures to areas of lower partial pressures.
5. Pulmonary gas exchange is the exchange of gases between alveoli and pulmonary blood capillaries. It depends on partial pressure differences, a large surface area for gas exchange, a small diffusion distance across the respiratory membrane, and the rate of airflow into and out of the lungs.
6. Systemic gas exchange is the exchange of gases between systemic blood capillaries and tissue cells.

18.5 Transport of Oxygen and Carbon Dioxide

1. In each 100 mL of oxygenated blood, 1.5% of the O_2 is dissolved in blood plasma and 98.5% is bound to hemoglobin as oxyhemoglobin ($Hb–O_2$).
2. The binding of O_2 to hemoglobin is affected by P_{O_2}, acidity (pH), P_{CO_2}, temperature, and 2,3-bisphosphoglycerate (BPG).
3. Fetal hemoglobin differs from adult hemoglobin in structure and has a higher affinity for O_2.
4. In each 100 mL of deoxygenated blood, 7% of CO_2 is dissolved in blood plasma, 23% combines with hemoglobin as carbaminohemoglobin ($Hb–CO_2$), and 70% is converted to bicarbonate ions (HCO_3^-).
5. In an acidic environment, hemoglobin's affinity for O_2 is lower, and O_2 dissociates more readily from it (Bohr effect).
6. In the presence of O_2, less CO_2 binds to hemoglobin (Haldane effect).

18.6 Control of Ventilation

1. The respiratory center consists of a medullary respiratory center in the medulla oblongata and a pontine respiratory center in the pons.

2. The medullary respiratory center consists of the dorsal respiratory group (DRG) and the ventral respiratory group (VRG). The DRG is active during quiet inspiration and inactive during quiet expiration. The VRG contains the pre-Bötzinger complex, which establishes the basic rhythm of breathing. The VRG also contains neurons that are active during forceful breathing.
3. The pontine respiratory center consists of the pneumotaxic and apneustic areas. The pneumotaxic area shortens the duration of inspiration; the apneustic area prolongs inspiration.
4. Respirations may be modified by a number of factors, including cortical influences; the inflation reflex; chemical stimuli, such as O_2 and CO_2 and H^+ levels; proprioceptor input; blood pressure changes; limbic system stimulation; temperature; pain; and irritation to the airways.

18.7 Exercise and the Respiratory System

1. The rate and depth of ventilation change in response to both the intensity and duration of exercise.
2. An increase in pulmonary perfusion and O_2-diffusing capacity occurs during exercise.
3. The abrupt increase in ventilation at the start of exercise is due to neural changes that send excitatory input to the inspiratory area in the medulla oblongata. The more gradual increase in ventilation during moderate exercise is due to chemical and physical changes in the bloodstream.

PONDER THIS

1. Your friend gets very frightened at the sight of a spider and all of a sudden you see her start to hyperventilate. Explain what would happen to her blood pH over time if she continues to breathe this way. Use a chemical reaction to explain your answer.

2. You have set up a metabolic chamber because you wish to measure metabolic rate during an intensive bike ride. You hook up all the necessary equipment and are ready to get your subject started. You ask the subject to breathe through a spirometer, which is attached to a gas tank as a control method for oxygen delivery. As your subject gets started you note, despite being well trained, that it takes only about 5 minutes for your subject to start breathing incredibly heavily and suddenly he loses consciousness. Trying to figure out what might have gone wrong, you notice that you hooked up the wrong gas tank to the spirometer. Instead of delivering a gas mixture that contains 21% oxygen, you connected the spirometer to a tank that has only 5% oxygen. Explain why this led to the observed outcome. Discuss hemoglobin in your answer.

3. Steve suffers from chronic obstructive pulmonary disease (COPD). This disease is characterized by an increase in airway resistance and a decrease in lung compliance. Explain what would happen to Steve's minute ventilation in terms of tidal volume as a result of these issues. Additionally, explain what the levels of oxygen and carbon dioxide in his arterial system would be compared to those of a healthy person.

ANSWERS TO FIGURE QUESTIONS

18.1 Pulmonary gas exchange involves the exchange of O_2 and CO_2 between the alveoli of the lungs and the blood in pulmonary capillaries; systemic gas exchange involves the exchange of O_2 and CO_2 between the blood in systemic capillaries and tissue cells of the body.

18.2 The two zones of the respiratory system are the conducting zone and the respiratory zone. The conducting zone consists of the nose to the terminal bronchioles; it filters, warms, and humidifies air and conducts air into the lungs. The respiratory zone consists of the respiratory bronchioles, alveolar ducts, and alveolar sacs; its function is to carry out gas exchange.

18.3 At the start of inspiration, intrapleural pressure is about 756 mmHg. With contraction of the diaphragm, it decreases to about 754 mmHg as the volume of the space between the two pleural layers expands. With relaxation of the diaphragm, it increases back to 756 mmHg.

18.4 The pressure would increase fourfold, to 4 atm.

18.5 During a forceful expiration, the abdominal muscles and internal intercostals contract.

18.6 Normal atmospheric pressure at sea level is 760 mmHg.

18.7 The levels of CO_2 and O_2 in the alveolus and surrounding tissue are responsible for the mechanisms involved in ventilation–perfusion matching. CO_2 controls ventilation by altering bronchiolar diameter; O_2 controls perfusion by altering arteriolar diameter.

18.8 Breathing in and then exhaling as much air as possible demonstrates vital capacity.

18.9 A difference in P_{O_2} promotes oxygen diffusion into pulmonary capillaries from alveoli and into tissue cells from systemic capillaries.

18.10 The most important factor that determines how much O_2 binds to hemoglobin is the P_{O_2}.

18.11 Both during exercise and at rest, hemoglobin in your pulmonary veins would be fully saturated with O_2, a point that is at the upper right of the curve.

18.12 O_2 is more available to your tissue cells when you have a fever because the affinity of hemoglobin for O_2 decreases with increasing temperature.

18.13 At a P_{O_2} of 40 mmHg, fetal Hb is 80% saturated with O_2 and maternal Hb is about 75% saturated.

18.14 Blood in a systemic vein has a higher concentration of HCO_3^-.

18.15 The dorsal respiratory group contains inspiratory neurons that are active and then inactive in a repeating cycle.

18.16 The pre-Bötzinger complex provides stimulatory input to the dorsal respiratory group.

18.17 Peripheral chemoreceptors are responsive to changes in blood levels of oxygen, carbon dioxide, and H^+.

18.18 Normal arterial blood P_{CO_2} Is 40 mmHg.

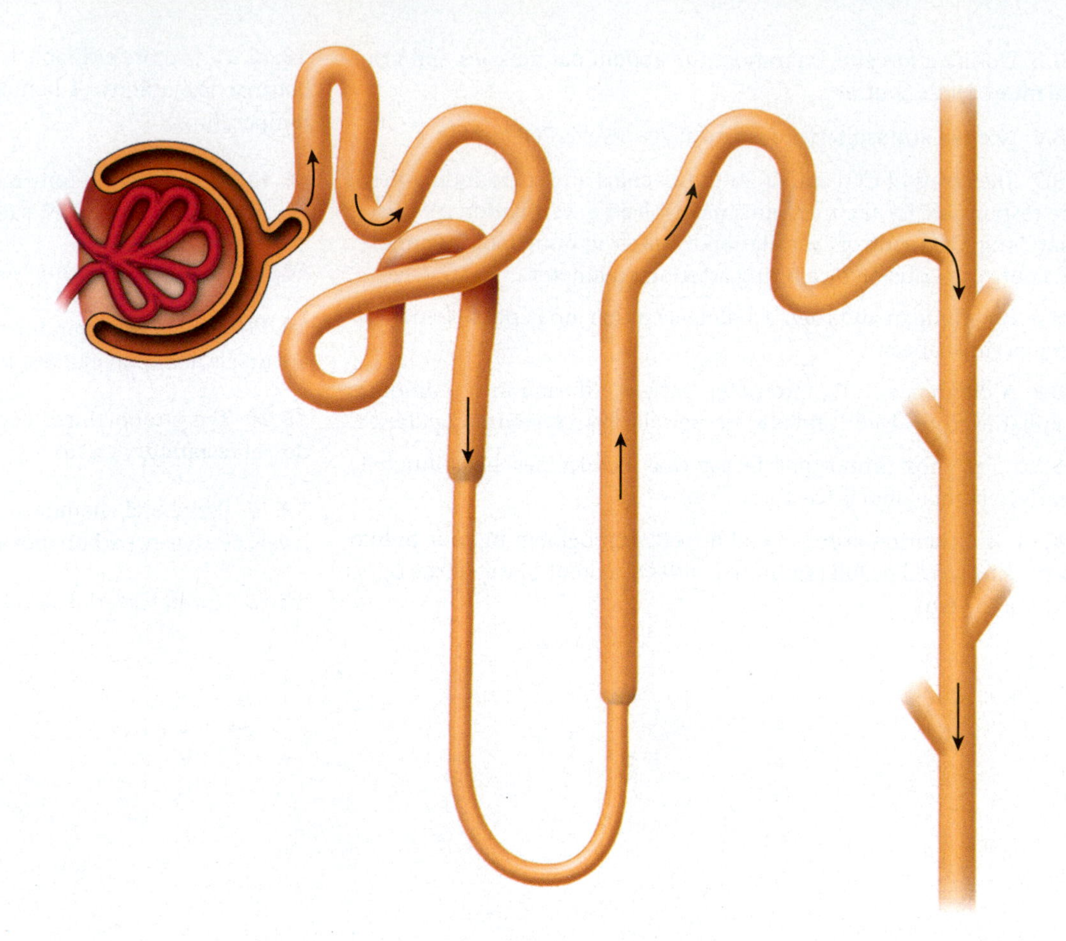

19 The Urinary System

THE URINARY SYSTEM AND HOMEOSTASIS

The urinary system contributes to homeostasis by excreting wastes; altering blood composition, pH, volume, and pressure; maintaining blood osmolarity; and producing hormones.

LOOKING BACK TO MOVE AHEAD...

- Tight junctions are proteins that fuse together the outer surfaces of adjacent plasma membranes to seal off passageways between adjacent cells (Section 3.8).
- Secondary active transport is a type of active transport in which a substance moves against its concentration or electrochemical gradient using the energy stored in an ionic electrochemical gradient (Section 5.5).
- Creatine phosphate is an energy-rich molecule in muscle fibers that transfers its phosphate group to ADP to form ATP (Section 11.4).
- Movement of fluid across blood capillary walls depends on the balance of four pressures (Starling forces): capillary hydrostatic pressure and interstitial fluid colloid osmotic pressure promote filtration, whereas plasma colloid osmotic pressure and interstitial fluid hydrostatic pressure promote reabsorption (Section 15.2).

As body cells carry out metabolic activities, they consume oxygen and nutrients and produce waste products such as carbon dioxide, urea, and uric acid. Wastes must be eliminated from the body because they can be toxic to cells if they accumulate. While the respiratory system rids the body of carbon dioxide, the urinary system disposes of most other wastes. The **urinary system** consists of two kidneys, two ureters, one urinary bladder, and one urethra. The kidneys filter blood of wastes and excrete them into a fluid called **urine**. From the kidneys, urine passes through the ureters to the urinary bladder, where it is stored until it is excreted from the body through the urethra. Disposal of wastes through the release of urine is not the only purpose of the urinary system. As you are about to discover, this system carries out a number of other important functions as well.

19.1 Overview of Kidney Functions

OBJECTIVE

- Describe the functions of the kidneys.

The kidneys do the major work of the urinary system. The other parts of this system are mainly passageways and storage areas. Functions of the kidneys include the following:

- ***Excretion of wastes.*** By forming urine, the kidneys help excrete wastes from the body. Some wastes excreted in urine result from metabolic reactions. These include urea and ammonia from the deamination of amino acids; creatinine from the breakdown of creatine phosphate; uric acid from the catabolism of nucleic acids; and urobilin from the breakdown of hemoglobin. Urea, ammonia, creatinine, uric acid, and urobilin are collectively known as **nitrogenous wastes** because they are waste products that contain nitrogen. Other wastes excreted in the urine are foreign substances that have entered the body, such as drugs and environmental toxins.
- ***Regulation of blood ionic composition.*** The kidneys help regulate the blood levels of several ions, including sodium ions (Na^+), potassium ions (K^+), calcium ions (Ca^{2+}), chloride ions (Cl^-), and phosphate ions (HPO_4^{2-}). The kidneys accomplish this task by adjusting the amounts of these ions that are excreted into the urine.
- ***Regulation of blood pH.*** The kidneys excrete a variable amount of hydrogen ions (H^+) into the urine and conserve bicarbonate ions (HCO_3^-), which are an important buffer of H^+ in the blood. Both of these activities help regulate blood pH.
- ***Regulation of blood volume.*** The kidneys adjust blood volume by returning water to the blood or eliminating it in the urine. An increase in blood volume increases blood pressure; a decrease in blood volume decreases blood pressure.
- ***Regulation of blood pressure.*** The kidneys also help regulate blood pressure by secreting the enzyme *renin*, which activates the renin–angiotensin–aldosterone pathway (see **FIGURE 13.21**). Increased renin causes an increase in blood pressure.
- ***Maintenance of blood osmolarity.*** By separately regulating loss of water and loss of solutes in the urine, the kidneys maintain a relatively constant blood osmolarity close to 300 milliosmoles per liter (mosmol/liter).*
- ***Production of hormones.*** The kidneys produce two hormones. *Calcitriol*, the active form of vitamin D, helps regulate calcium homeostasis (see **FIGURE 13.19**), and *erythropoietin* stimulates the production of erythrocytes (see **FIGURE 16.5**).

As is evident from the functions just listed, urine contains more than just waste products. It also contains water and other substances, such as ions, that have important roles in the body but are in excess of the body's needs. You will learn more about the composition of urine in Section 19.7.

CHECKPOINT

1. What are examples of wastes that may be present in urine?

19.2 Organization of the Kidneys

OBJECTIVES

- Describe the components of a nephron.
- Explain the functional significance of the blood vessels that supply the nephron.
- Define the juxtaglomerular apparatus.

The paired **kidneys** are bean-shaped organs located just above the waist, with one on each side of the vertebral column (**FIGURE 19.1a**). They are said to be *retroperitoneal* (re′-trō-per-i-tō-NĒ-al; *retro-* = behind) because they are positioned behind the peritoneum, a membrane that lines the abdominal cavity. Each kidney consists of two

* The **osmolarity** of a solution is a measure of the total number of dissolved particles per liter of solution. The particles may be molecules, ions, or a mixture of both. To calculate osmolarity, multiply molarity by the number of particles per molecule, once the molecule dissolves (see Section 5.4). A similar term, *osmolality*, is the number of particles of solute per *kilogram* of water. Because it is easier to measure volumes of solutions than to determine the mass of water they contain, osmolarity is used more commonly than osmolality. Most body fluids and solutions used clinically are dilute, in which case there is less than a 1% difference between the two measures.

FIGURE 19.1 **Components of the urinary system.**

Urine formed by the kidneys passes first into the ureters, then to the urinary bladder for storage, and finally through the urethra for elimination from the body.

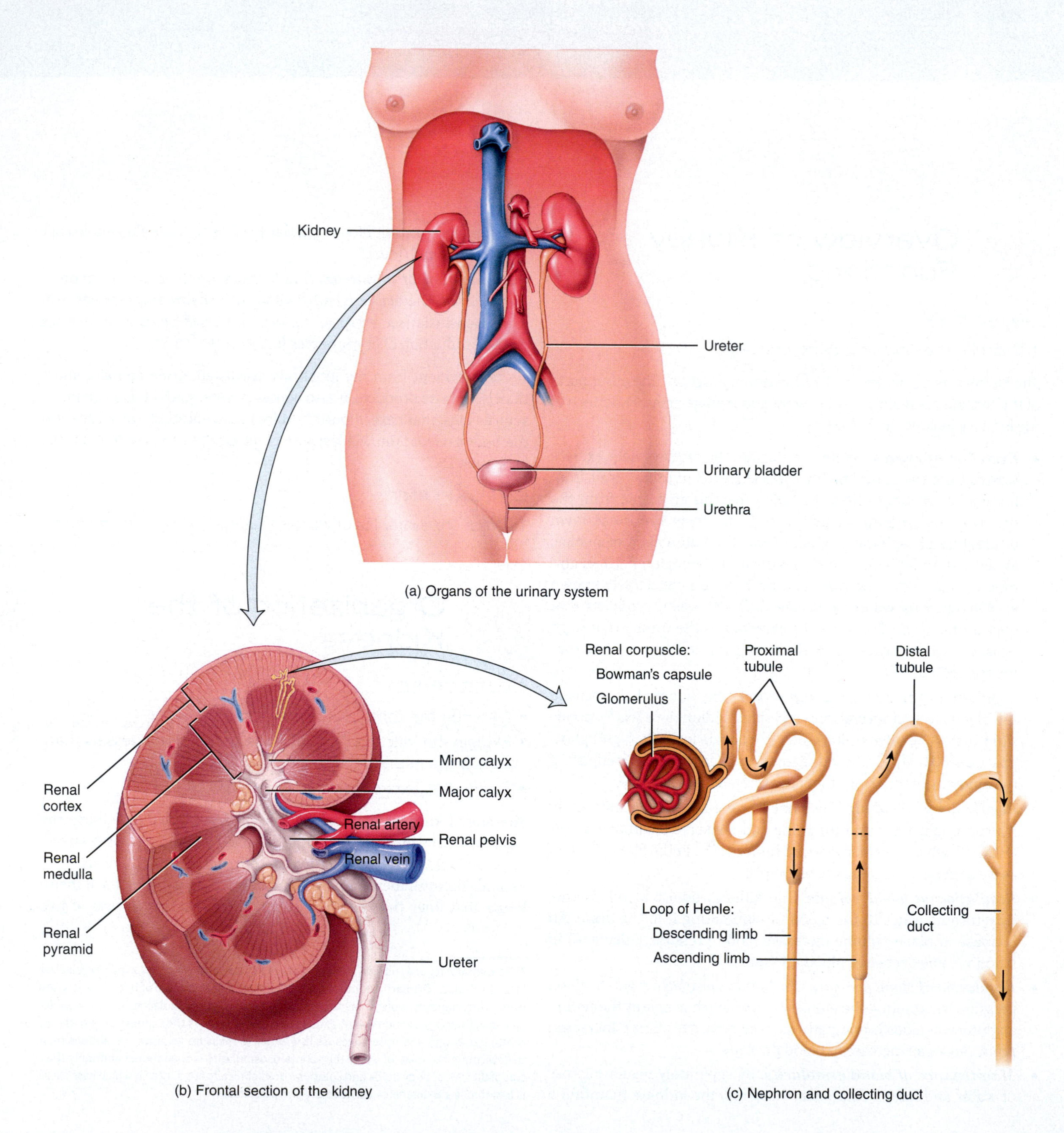

(a) Organs of the urinary system

(b) Frontal section of the kidney

(c) Nephron and collecting duct

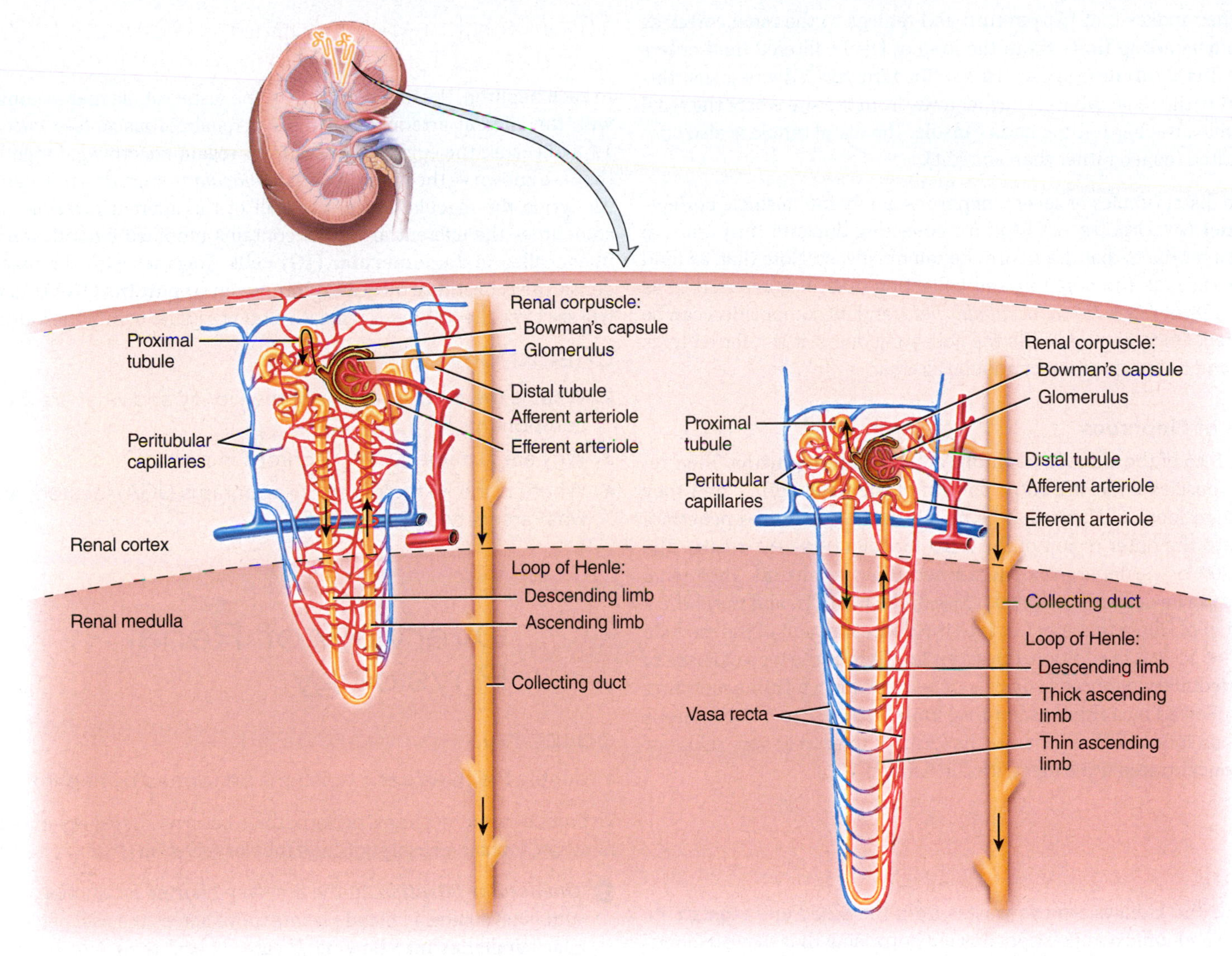

(d) Cortical nephron and blood supply

(e) Juxtamedullary nephron and blood supply

? What are the functional units of the kidneys called?

distinct regions: a superficial **renal cortex** and a deep, inner region called the **renal medulla** (FIGURE 19.1b). Within the renal medulla are several cone-shaped **renal pyramids**. From the renal pyramids, urine drains into cuplike structures called **minor** and **major calyces** (KĀ-li-sēz; singular is *calyx*). A minor calyx is smaller in size than a major calyx, and several minor calyces unite to form a major calyx. From the major calyces, urine flows into a single large cavity called the **renal pelvis** and then out through the ureter to the urinary bladder.

Nephrons Perform the Main Functions of the Kidneys

Within the renal cortex and renal medulla of each kidney are about 1 million microscopic structures called **nephrons**. Nephrons are the functional units of the kidneys.

Parts of a Nephron

Each nephron consists of two parts: (1) a renal corpuscle and (2) a renal tubule, which extends from the renal corpuscle.

1. ***Renal corpuscle.*** The **renal corpuscle** is where blood plasma is filtered. The two components of a renal corpuscle are the **glomerulus** (a capillary network) and **Bowman's capsule**, or *glomerular capsule*, a double-walled epithelial cup that surrounds the glomerular capillaries (FIGURE 19.1c)
2. ***Renal tubule.*** Once blood plasma is filtered at the renal corpuscle, the filtered fluid passes into the **renal tubule**, which consists of a single layer of epithelial cells that lines a lumen. The renal tubule has three main sections: proximal tubule, loop of Henle, and distal tubule. The renal corpuscle, proximal tubule, and distal tubule lie within the renal cortex; the loop of Henle extends into the renal medulla, makes a hairpin turn, and then returns to the renal cortex. From the renal corpuscle, filtered fluid first enters the **proximal tubule** (FIGURE 19.1c). The term *proximal* denotes that this part of the renal tubule is closest to the site where the renal tubule attaches to Bowman's capsule. Most of the proximal tubule is *convoluted*, which means that it is coiled rather than straight. From the proximal tubule, filtered fluid enters the **loop of Henle**, also known as the *nephron loop* (FIGURE 19.1c). The first part of the loop of Henle dips into the renal medulla, where it is called the **descending limb**.

It then makes that hairpin turn and returns to the renal cortex as the **ascending limb**. From the loop of Henle, filtered fluid enters the **distal tubule** (FIGURE 19.1c). The term *distal* denotes that this part of the renal tubule is farther away from the site where the renal tubule attaches to Bowman's capsule. The distal tubule is also convoluted (coiled rather than straight).

The distal tubules of several nephrons empty into a single **collecting duct** (FIGURE 19.1c). Multiple collecting ducts in turn unite to form larger ducts that drain into the minor calyces. Note that, as fluid passes through the nephron and collecting duct, it is referred to as either *filtrate, filtered fluid,* or *tubular fluid,* and its composition can be modified. Once the fluid exits the collecting ducts, it is referred to as *urine,* and its composition cannot be altered.

Types of Nephrons

About 80% of the kidney's nephrons are **cortical nephrons**. Their renal corpuscles lie in the outer portion of the renal cortex, and they have *short* loops of Henle that lie mainly in the cortex and penetrate only into the outer region of the renal medulla (FIGURE 19.1d). The other 20% of the nephrons are **juxtamedullary nephrons**. Their renal corpuscles lie deep in the cortex, close to the medulla, and they have a *long* loop of Henle that extends into the deepest region of the medulla (FIGURE 19.1e). In addition, the ascending limb of the loop of Henle of juxtamedullary nephrons consists of two portions: a **thin ascending limb** followed by a **thick ascending limb** (FIGURE 19.1e). Nephrons with long loops of Henle enable the kidneys to excrete very dilute or very concentrated urine (described in Section 19.6).

The Extensive Blood Supply of the Kidneys Contributes to Renal Function

Because the kidneys remove wastes from the blood and regulate its volume and ionic composition, it is not surprising that they are abundantly supplied with blood vessels. Although the kidneys constitute less than 0.5% of total body mass, they receive about 20% of the resting cardiac output via the right and left renal arteries. In adults, **renal blood flow**, the blood flow through both kidneys, is about 1200 mL per minute.

Within each kidney, the **renal artery** divides into smaller arteries that eventually deliver blood to the **afferent arterioles** (FIGURE 19.1d, e). Each nephron receives one afferent arteriole, which carries blood to the glomerulus. The **glomerulus** (glō-MER-ū-lus = little ball; plural is *glomeruli*) is a tangled, ball-shaped capillary network. The capillaries of the glomerulus in turn give rise to an **efferent arteriole** that carries blood away from the glomerulus. Glomerular capillaries are unique among capillaries in the body because they are positioned between two arterioles rather than between an arteriole and a venule. The efferent arteriole divides to form the **peritubular capillaries**, which surround (1) tubular parts of cortical and juxtamedullary nephrons that are located in the renal cortex (FIGURE 19.1d, e) and (2) short loops of Henle of cortical nephrons that extend from the renal cortex into the outer region of the renal medulla (FIGURE 19.1d). Extending from some efferent arterioles are long loop-shaped capillaries called **vasa recta** (VĀ-sa REK-ta), which surround long loops of Henle of juxtamedullary nephrons (FIGURE 19.1e). The peritubular capillaries eventually reunite to form small veins that give rise to a single, larger **renal vein**, which drains blood from the kidney.

The Juxtaglomerular Apparatus Consists of Part of the Distal Tubule and Afferent Arteriole

In each nephron, the initial portion of the distal tubule makes contact with the afferent arteriole serving that renal corpuscle (see FIGURE 19.3). Because the epithelial cells in this region are crowded together, they are known as the **macula densa** (*macula* = spot; *densa* = dense). Alongside the macula densa, the wall of the afferent arteriole (and sometimes the efferent arteriole) contains modified smooth muscle fibers called **juxtaglomerular (JG) cells**. Together with the macula densa, they constitute the **juxtaglomerular apparatus (JGA)**. As you will see later, the JGA helps regulate blood pressure within the kidneys.

CHECKPOINT

2. What is the functional unit of the kidney and what are its components?
3. Why are juxtamedullary nephrons important?
4. Where is the juxtaglomerular apparatus (JGA) located, and what are its two main parts?

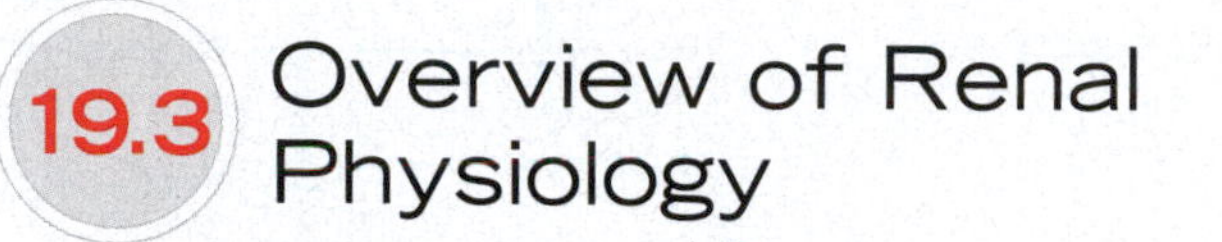

19.3 Overview of Renal Physiology

OBJECTIVE

- Identify the three basic functions performed by nephrons.

To produce urine, nephrons perform three basic processes—glomerular filtration, tubular reabsorption, and tubular secretion (FIGURE 19.2):

1. ***Glomerular filtration.*** In the first step of urine production, water and most solutes in blood plasma move across the wall of glomerular capillaries into Bowman's capsule and then into the renal tubule.
2. ***Tubular reabsorption.*** As filtered fluid flows through the renal tubule and collecting duct, tubule and duct cells reabsorb about 99% of the filtered water and many solutes. *Reabsorption* is the transfer of substances from fluid in the tubular lumen to blood in the peritubular capillaries. This process allows useful substances to be returned back to the bloodstream. The remaining 1% of filtered fluid that is not reabsorbed contains substances that the body does not need (wastes, drugs, excess ions, etc.) and will eventually be excreted into the urine.
3. ***Tubular secretion.*** As fluid flows through the renal tubule and collecting duct, tubule and duct cells also secrete wastes and other substances that are not useful to the body. *Secretion* is the transfer of substances from blood in the peritubular capillaries to fluid in the tubular lumen. This process serves as an additional mechanism for removing unneeded substances from the bloodstream.

Solutes in the fluid that drains into the renal pelvis remain in the urine and are excreted. The rate of urinary excretion of any solute is equal to its rate of glomerular filtration, plus its rate of secretion, minus its rate of reabsorption.

By filtering, reabsorbing, and secreting, nephrons help maintain homeostasis of the blood's volume and composition. The situation is somewhat analogous to a recycling center: Garbage trucks dump refuse

FIGURE 19.2 The three basic functions of the nephron: glomerular filtration, tubular reabsorption, and tubular secretion.

Glomerular filtration occurs in the renal corpuscle, whereas tubular reabsorption and tubular secretion occur all along the renal tubule and collecting duct.

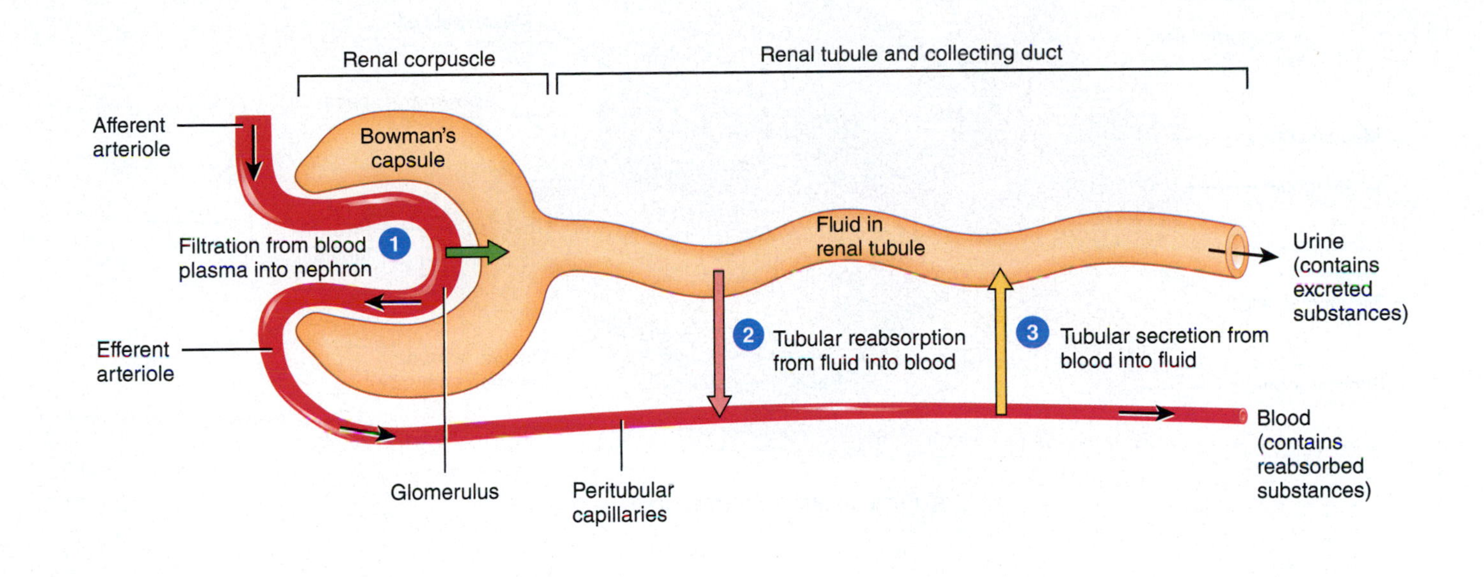

When cells of the renal tubule secrete the drug penicillin, is the drug being added to or removed from the bloodstream?

into an input hopper, where the smaller refuse passes onto a conveyor belt (glomerular filtration of plasma). As the conveyor belt carries the garbage along, workers remove useful items, such as aluminum cans, plastics, and glass containers (reabsorption). Other workers place additional garbage left at the center and larger items onto the conveyor belt (secretion). At the end of the belt, all remaining garbage falls into a truck for transport to the landfill (excretion of wastes in urine).

CHECKPOINT

5. What is the difference between tubular reabsorption and tubular secretion?

19.4 Glomerular Filtration

OBJECTIVES

- Describe the components of the filtration membrane.
- Discuss the pressures that promote and oppose glomerular filtration.
- Define the glomerular filtration rate.

As you have already learned, blood plasma is filtered at the **renal corpuscle**, which consists of the glomerulus and Bowman's capsule. Bowman's capsule is organized into two layers: an outer parietal layer and an inner visceral layer (FIGURE 19.3a). The *parietal layer* consists of a single layer of epithelial cells that forms the outer wall of Bowman's capsule. It is continuous with the epithelial cells of the renal tubule. The *visceral layer* consists of a single layer of modified epithelial cells called **podocytes** (PŌ-dō-cīts; *podo-* = foot; *-cytes* = cells) that form the inner wall of Bowman's capsule (FIGURE 19.3a, b). Extending from each podocyte are thousands of footlike processes termed **pedicels** (PED-i-sels = little feet) that wrap around the glomerular capillaries. Fluid filtered from the glomerular capillaries enters **Bowman's space**, the space between the parietal and visceral layers of Bowman's capsule. Bowman's space is continuous with the lumen of the renal tubule.

The fluid that enters Bowman's space is called the **glomerular filtrate**. It contains many of the same components found in blood, except that it lacks blood cells and contains very few plasma proteins. Because blood plasma is blood without the cellular elements (cells) (see Section 16.1), the composition of glomerular filtrate is comparable to blood plasma that is essentially protein-free. On average, the daily volume of glomerular filtrate in adults is about 180 liters. Under conditions of normal body hydration (adequate water intake), about 99% of filtered fluid (about 178–178.5 L/day) returns to the bloodstream via tubular reabsorption and the remaining 1% of the filtered fluid (about 1.5–2 L/day) is excreted as urine.

The Renal Corpuscle Contains a Filtration Membrane

Together, the glomerulus and visceral (podocyte) layer of Bowman's capsule form a leaky barrier known as the **filtration membrane**. This sandwichlike assembly permits filtration of water and small solutes but prevents filtration of blood cells and nearly all plasma proteins. Substances filtered from the blood cross three barriers—glomerular

FIGURE 19.3 The renal corpuscle and filtration membrane. The size of the endothelial fenestrations and podocyte filtration slits in part (b) have been exaggerated for emphasis.

A renal corpuscle consists of a glomerulus and Bowman's capsule; the filtration membrane is the part of the renal corpuscle through which blood is filtered.

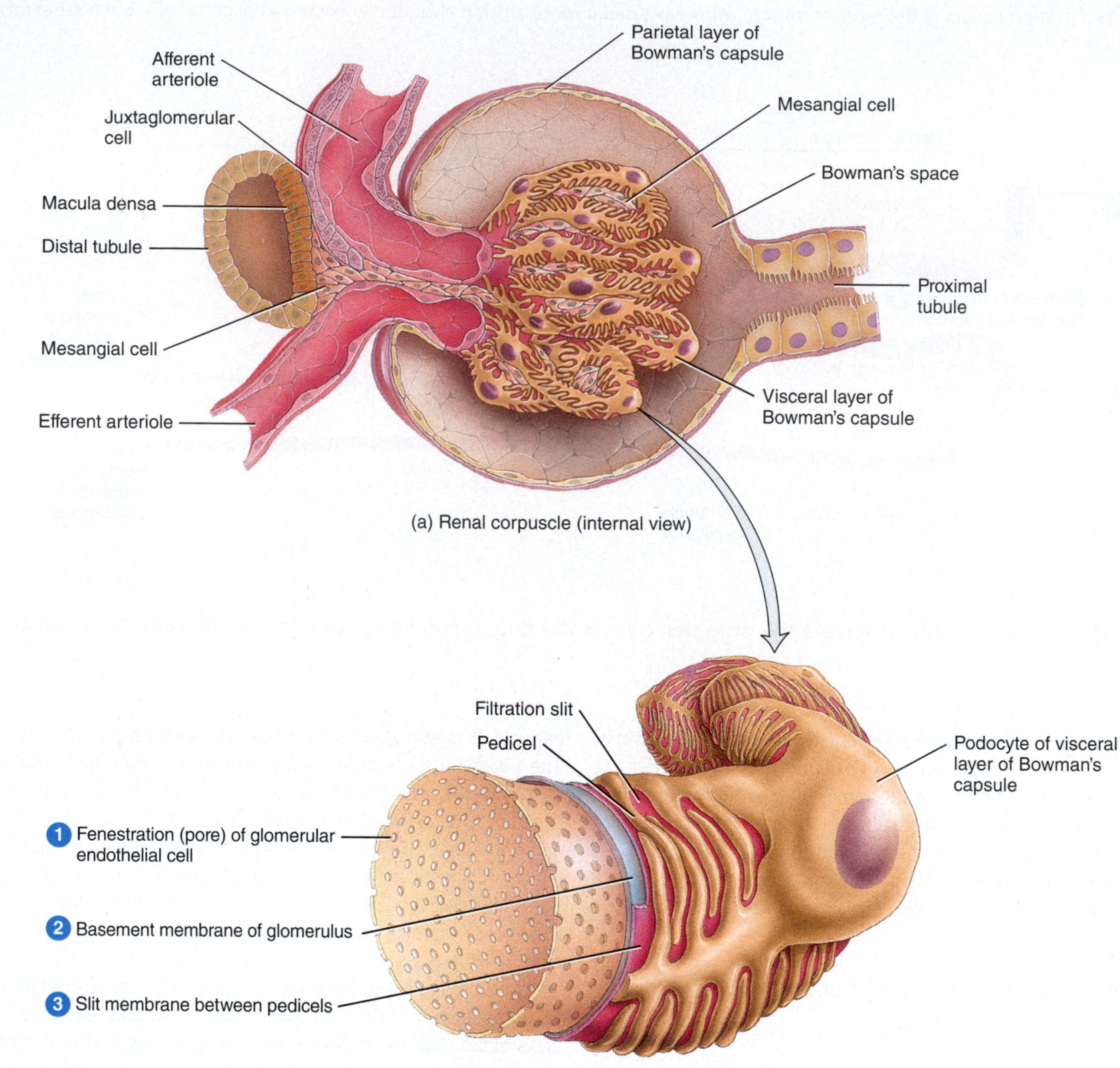

(b) Details of filtration membrane

endothelial cells, the basement membrane of the glomerulus, and filtration slits formed by podocytes (FIGURE 19.3b):

1. Glomerular endothelial cells are quite leaky because they have **fenestrations** (pores) that measure 70–100 nm in diameter. This size permits passage to water and all solutes in blood plasma. However, the fenestrations are too small to allow blood cells to filter through the endothelium. Located among the glomerular capillaries and in the cleft between afferent and efferent arterioles are **mesangial cells** (see FIGURE 19.3a). As you will soon learn, these contractile cells help regulate glomerular filtration.
2. The **basement membrane**, a porous layer of acellular material between the endothelium and the podocytes, consists of collagen fibers and negatively charged glycoproteins. The pores within the basement membrane allow water and most small solutes to pass through. However, the negative charges of the glycoproteins repel plasma proteins, most of which are anionic; the repulsion hinders filtration of these proteins.
3. Recall that each podocyte contains footlike processes called **pedicels** that wrap around glomerular capillaries. The spaces between pedicels are the **filtration slits**. A thin membrane, the **slit membrane**, extends across each filtration slit; it contains pores that permit the passage of molecules that have a diameter smaller than 6–7 nm, including water, glucose, vitamins, hormones, amino acids, urea, ammonia, and ions. Negatively charged glycoproteins

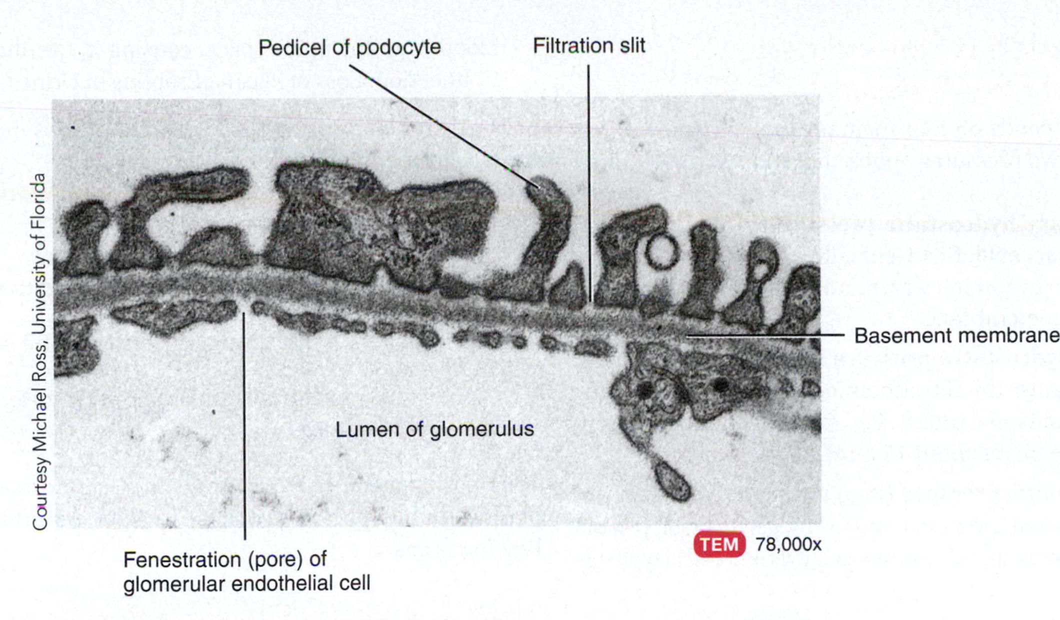

(c) Filtration membrane

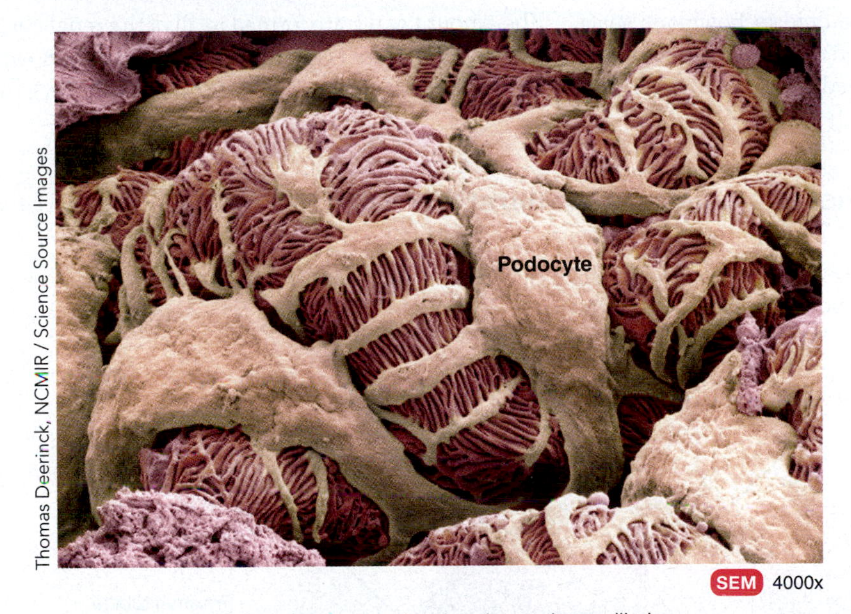

(d) A podocyte covering the glomerular capillaries

? Which component of the filtration membrane prevents erythrocytes from entering Bowman's space?

that cover the slit membrane oppose the filtration of plasma proteins. Because of the negative charges associated with the podocyte layer and basement membrane and the small size of the slit membrane pores, less than 1% of albumin, the smallest and most abundant plasma protein, is able to pass through the filtration membrane and enter the glomerular filtrate.

The principle of *filtration*—the use of pressure to force fluids and solutes through a membrane—is the same in glomerular capillaries as in capillaries elsewhere in the body (see Starling's law of the capillaries in Section 15.2). However, the volume of fluid filtered by the renal corpuscle is much larger than in other capillaries of the body for three reasons:

1. Glomerular capillaries present a large surface area for filtration because they are long and extensive. The mesangial cells regulate how much of this surface area is available for filtration. When mesangial cells are relaxed, surface area is maximal, and glomerular filtration is very high. Contraction of mesangial cells reduces the available surface area, and glomerular filtration decreases.
2. The filtration membrane is thin and porous. Despite having several layers, the thickness of the filtration membrane is only 0.1 mm (100 μm). Glomerular capillaries also are about 50 times leakier than capillaries in most other tissues, mainly because of their large fenestrations.
3. Glomerular capillary blood pressure is high. Because the efferent arteriole is smaller in diameter than the afferent arteriole, resistance to the outflow of blood from the glomerulus is high. As a result, blood pressure in glomerular capillaries is considerably higher than in capillaries elsewhere in the body.

Glomerular Filtration Is Determined by the Balance of Four Pressures

Glomerular filtration depends on four main pressures. Two pressures promote filtration and two pressures oppose filtration (**FIGURE 19.4**):

1. **Glomerular capillary hydrostatic pressure (P_{GC})** is the blood pressure in glomerular capillaries. Generally, P_{GC} is about 55 mmHg. It promotes filtration by forcing water and solutes in blood plasma through the filtration membrane.
2. **Bowman's space hydrostatic pressure (P_{BS})** is the hydrostatic pressure exerted against the filtration membrane by fluid already in Bowman's space and renal tubule. P_{BS} opposes filtration and represents a "back pressure" of about 15 mmHg.
3. **Plasma colloid osmotic pressure (π_{GC})** is due to the presence of proteins such as albumin, globulins, and fibrinogen in blood plasma of glomerular capillaries. π_{GC} also opposes filtration and is typically about 30 mmHg.
4. **Bowman's space colloid osmotic pressure (π_{BS})**, which is due to the presence of proteins in the fluid in Bowman's space, promotes filtration. Under normal conditions, the fluid in Bowman's space has very little protein, so π_{BS} is considered to be 0 mmHg. However, when the filtration membrane is damaged, protein can enter from blood into Bowman's space, causing π_{BS} to increase (see Clinical Connection: Loss of Plasma Proteins in Urine Causes Edema).

Net filtration pressure (NFP), the total pressure that promotes filtration, is determined in the following way:

$$\text{Net filtration pressure (NFP)} = \underbrace{(P_{GC} + \pi_{BS})}_{\text{Pressures that promote filtration}} - \underbrace{(\pi_{GC} + P_{BS})}_{\text{Pressures that oppose filtration}}$$

By substituting the values just given, normal NFP may be calculated:

$$\begin{aligned} \text{NFP} &= (55\text{ mmHg} + 0\text{ mmHg}) - (30\text{ mmHg} + 15\text{ mmHg}) \\ &= 10\text{ mmHg} \end{aligned}$$

Thus, a pressure of only 10 mmHg causes a normal amount of blood plasma (minus plasma proteins) to filter from the glomerulus into Bowman's space.

The Glomerular Filtration Rate Is an Important Aspect of Kidney Function

The amount of filtrate formed in all of the renal corpuscles of both kidneys each minute is the **glomerular filtration rate (GFR)**. In adults, the GFR averages about 125 mL/min (180 L/day). Homeostasis of body

FIGURE 19.4 The pressures that drive glomerular filtration. Taken together, these pressures determine net filtration pressure (NFP).

Glomerular capillary hydrostatic pressure and Bowman's space colloid osmotic pressure promote filtration, whereas plasma colloid osmotic pressure and Bowman's space hydrostatic pressure oppose filtration.

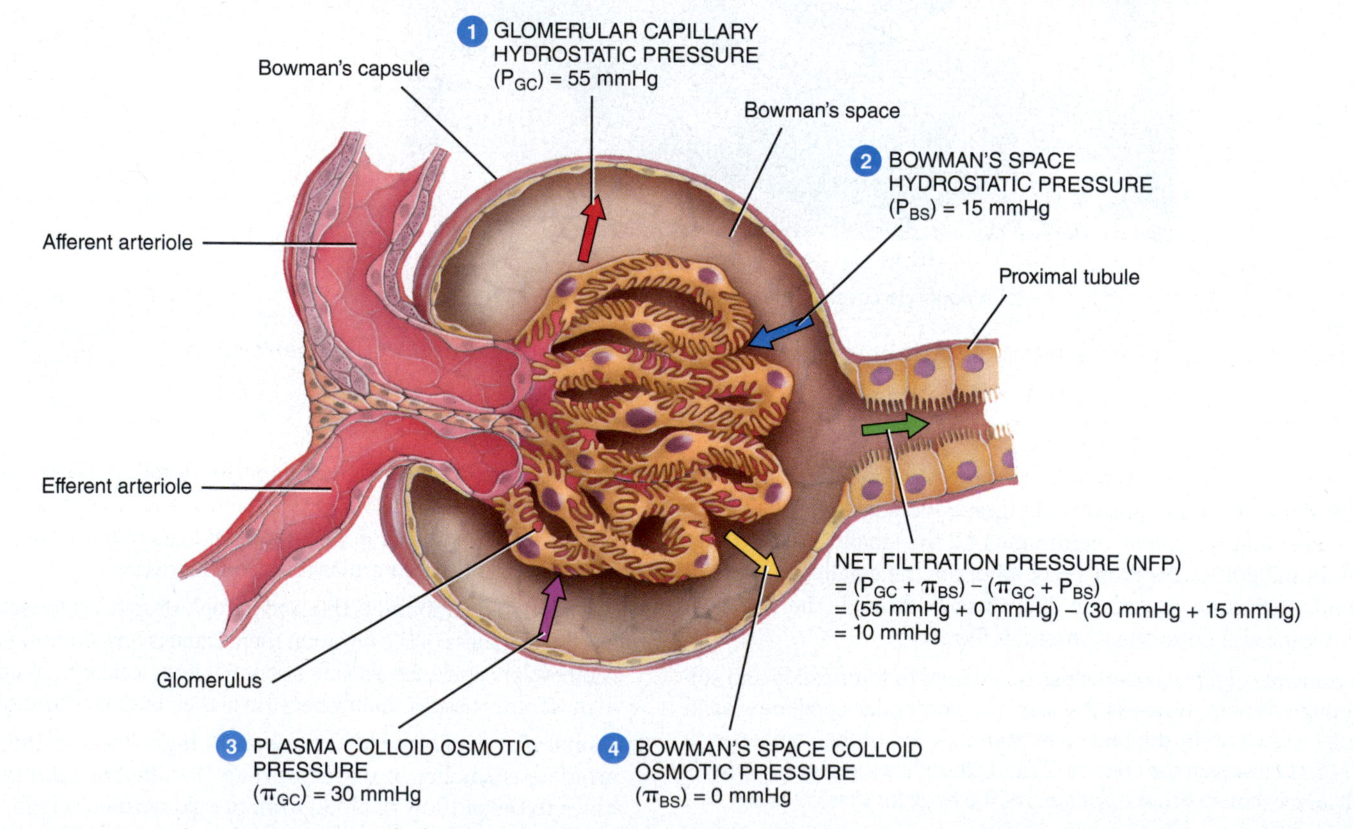

Suppose a tumor is pressing on and obstructing the right ureter. What effect might this have on Bowman's space hydrostatic pressure and thus on NFP in the right kidney? Would the left kidney also be affected?

fluids requires that the kidneys maintain a relatively constant GFR. If the GFR is too high, needed substances may pass so quickly through the renal tubules that some are not reabsorbed and are lost in the urine. If the GFR is too low, nearly all the filtrate may be reabsorbed and certain waste products may not be adequately excreted.

GFR is directly related to the pressures that determine net filtration pressure; any change in net filtration pressure affects GFR. For example, severe blood loss reduces mean arterial blood pressure and decreases the glomerular capillary hydrostatic pressure. Filtration ceases if glomerular capillary hydrostatic pressure drops to 45 mmHg because the opposing pressures add up to 45 mmHg. Amazingly, when systemic blood pressure rises above normal, net filtration pressure and GFR increase very little. GFR is nearly constant when the mean arterial blood pressure is anywhere between 80 and 180 mmHg.

The Glomerular Filtration Rate Is Regulated in Different Ways

The mechanisms that regulate glomerular filtration rate operate in two main ways: (1) by adjusting blood flow into and out of the glomerulus and (2) by altering the glomerular capillary surface area available for filtration. GFR increases when blood flow into the glomerular capillaries increases. Coordinated control of the diameter of both afferent and efferent arterioles regulates glomerular blood flow. Constriction of the afferent arteriole decreases blood flow into the glomerulus; dilation of the afferent arteriole increases it. Three mechanisms control GFR: renal autoregulation, neural regulation, and hormonal regulation.

Renal Autoregulation of GFR

The kidneys themselves help maintain a constant renal blood flow and GFR despite normal, everyday changes in blood pressure, like those that occur during exercise. This capability is called **renal autoregulation** and consists of two mechanisms—the myogenic mechanism and tubuloglomerular feedback. Working together, they can maintain nearly constant GFR over a wide range of systemic blood pressures.

The **myogenic mechanism** occurs when stretching triggers contraction of smooth muscle cells in the walls of afferent arterioles. As blood pressure rises, GFR also rises because renal blood flow increases. However, the elevated blood pressure stretches the walls of the afferent arterioles. In response, smooth muscle fibers in the wall of the afferent arteriole contract, which narrows the arteriole's lumen. As a result, renal blood flow decreases, thus reducing GFR to its previous level. Conversely, when arterial blood pressure drops, the smooth muscle cells are stretched less and thus relax. The afferent arterioles dilate, renal blood flow increases, and GFR increases. The myogenic mechanism normalizes renal blood flow and GFR within seconds after a change in blood pressure.

The second contributor to renal autoregulation, **tubuloglomerular feedback**, is so-named because part of the renal tubules—the macula densa—provides feedback to the glomerulus (**FIGURE 19.5**). When GFR is above normal due to elevated systemic blood pressure, filtered fluid flows more rapidly along the renal tubules. As a result, the proximal tubule and loop of Henle have less time to reabsorb Na^+, Cl^-, and water. Macula densa cells are thought to detect the increased delivery of Na^+, Cl^-, and water and to inhibit release of nitric oxide (NO) from cells in the juxtaglomerular apparatus (JGA). Because NO causes vasodilation, afferent arterioles constrict when the level of NO declines. As a result, less blood flows into the glomerular capillaries, and GFR decreases. When blood pressure falls, causing GFR to be lower

FIGURE 19.5 Tubuloglomerular feedback.

Macula densa cells of the juxtaglomerular apparatus provide negative feedback regulation of glomerular filtration rate.

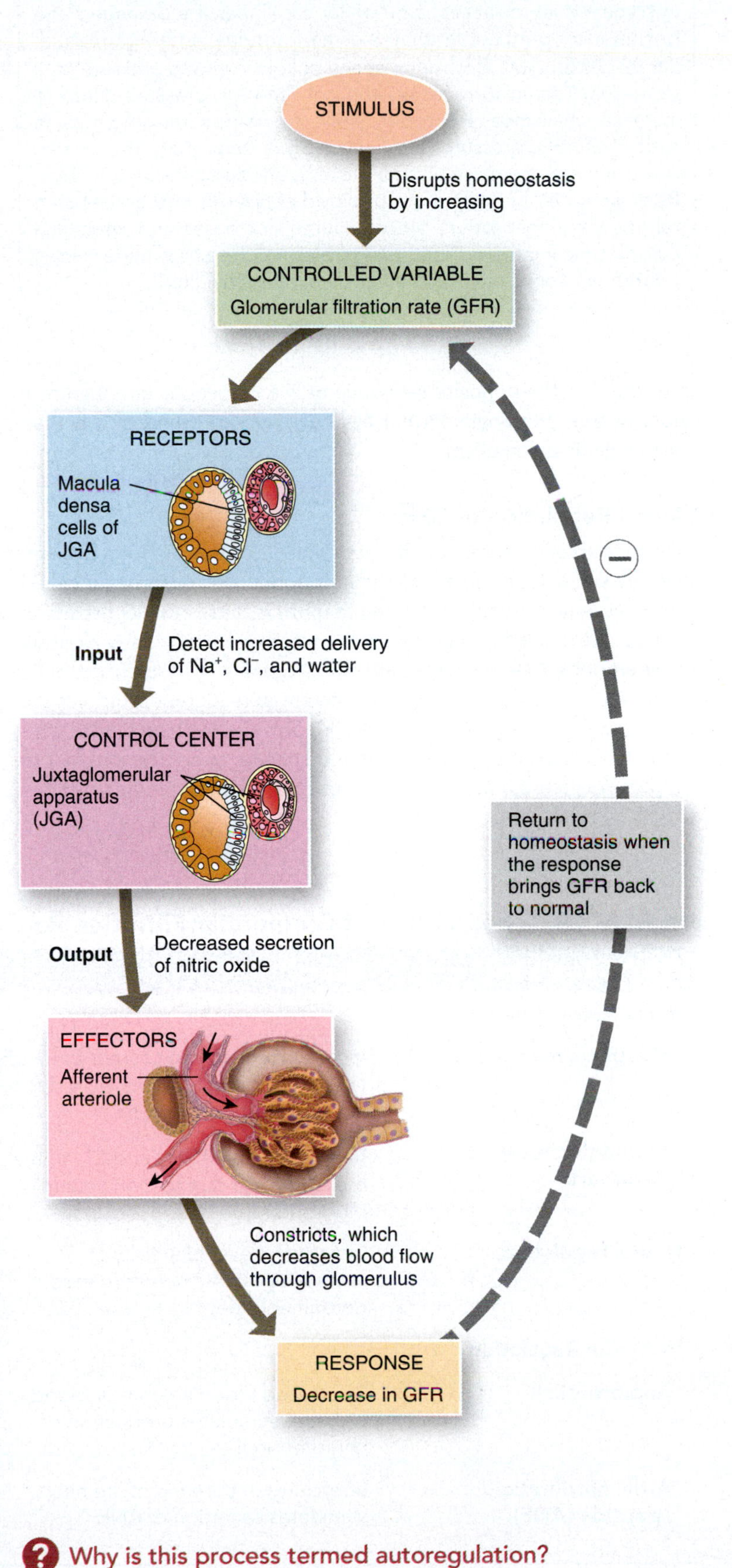

Why is this process termed autoregulation?

CLINICAL CONNECTION

Loss of Plasma Proteins in Urine Causes Edema

In some kidney diseases, glomerular capillaries are damaged and become so permeable that plasma proteins enter glomerular filtrate. As a result, Bowman's space colloid osmotic pressure (π_{BS}) increases. This in turn causes the net filtration pressure (NFP) to increase, which means more fluid is filtered. At the same time, plasma colloid osmotic pressure (see Section 15.2) throughout the circulation decreases because plasma proteins are being lost in the urine. Because more fluid filters out of blood capillaries into tissues than returns via reabsorption, blood volume decreases and interstitial fluid volume increases. Thus, loss of plasma proteins in urine causes **edema**, an abnormally high volume of interstitial fluid.

than normal, the opposite sequence of events occurs, although to a lesser degree. Tubuloglomerular feedback operates more slowly than the myogenic mechanism.

Neural Regulation of GFR

Like most blood vessels of the body, those of the kidneys are supplied by sympathetic fibers of the autonomic nervous system (ANS) that release norepinephrine. Norepinephrine causes vasoconstriction through the activation of α_1 receptors, which are particularly plentiful in the smooth muscle fibers of afferent arterioles. At rest, sympathetic stimulation is moderately low, the afferent and efferent arterioles are dilated, and renal autoregulation of GFR prevails. With moderate sympathetic stimulation, both afferent and efferent arterioles constrict to the same degree. Blood flow into and out of the glomerulus is restricted to the same extent, which decreases GFR only slightly. With greater sympathetic stimulation, however, as occurs during exercise or hemorrhage, vasoconstriction of the afferent arterioles predominates. As a result, blood flow into glomerular capillaries is greatly decreased, and GFR drops. This lowering of renal blood flow has two consequences: (1) It reduces urine output, which helps conserve blood volume. (2) It permits greater blood flow to other body tissues.

Hormonal Regulation of GFR

Two hormones contribute to regulation of GFR. Angiotensin II reduces GFR; atrial natriuretic peptide (ANP) increases GFR. **Angiotensin II** is a very potent vasoconstrictor that narrows both afferent and efferent arterioles and reduces renal blood flow, thereby decreasing GFR. Cells in the atria of the heart secrete **atrial natriuretic peptide (ANP)**. Stretching of the atria, as occurs when blood volume increases, stimulates secretion of ANP. By causing relaxation of the glomerular mesangial cells, ANP increases the capillary surface area available for filtration. Glomerular filtration rate rises as the surface area increases.

TABLE 19.1 summarizes the regulation of glomerular filtration rate.

CHECKPOINT

6. How does the composition of glomerular filtrate differ from that of blood?
7. What is the functional significance of the slit membrane that extends across the filtration slits of the podocyte layer of the filtration membrane?
8. What would happen to the net filtration pressure associated with glomerular filtration if Bowman's space hydrostatic pressure decreased?
9. How does tubuloglomerular feedback contribute to the regulation of the glomerular filtration rate?

TABLE 19.1 Regulation of Glomerular Filtration Rate (GFR)

Type of Regulation	Major Stimulus	Mechanism and Site of Action	Effect on GFR
Renal Autoregulation			
Myogenic mechanism	Increased stretching of smooth muscle fibers in afferent arteriole walls due to increased blood pressure.	Stretched smooth muscle fibers contract, thereby narrowing the lumen of the afferent arterioles.	Decrease.
Tubuloglomerular feedback	Rapid delivery of Na^+ and Cl^- to the macula densa due to high systemic blood pressure.	Decreased release of nitric oxide (NO) by the juxtaglomerular apparatus causes constriction of afferent arterioles.	Decrease.
Neural regulation	Increase in level of activity of renal sympathetic nerves releases norepinephrine.	Constriction of afferent arterioles through activation of α_1 receptors and increased release of renin.	Decrease.
Hormone Regulation			
Angiotensin II	Decreased blood volume or blood pressure stimulates production of angiotensin II.	Constriction of both afferent and efferent arterioles.	Decrease.
Atrial natriuretic peptide (ANP)	Stretching of the atria of the heart stimulates secretion of ANP.	Relaxation of mesangial cells in glomerulus increases capillary surface area available for filtration.	Increase.

19.5 Tubular Reabsorption and Tubular Secretion

OBJECTIVES

- Describe the routes and mechanisms of tubular reabsorption and secretion.
- Explain how specific segments of the renal tubule and collecting duct reabsorb water and solutes.
- Discuss how specific segments of the renal tubule and collecting duct secrete solutes into the urine.

The normal rate of glomerular filtration is so high that the volume of fluid entering the proximal tubules in half an hour is greater than the total plasma volume. Obviously some of this fluid must be returned somehow to the bloodstream. *Reabsorption*—the return of most of the filtered water and many of the filtered solutes to the bloodstream—is the second basic function of the nephron and collecting duct (see FIGURE 19.2). Normally, about 99% of the filtered water is reabsorbed. Epithelial cells all along the renal tubule and collecting duct carry out reabsorption, but proximal tubule cells make the largest contribution. Solutes that are reabsorbed by active or passive processes include glucose, amino acids, urea, and ions such as Na^+ (sodium), K^+ (potassium), Ca^{2+} (calcium), Cl^- (chloride), HCO_3^- (bicarbonate), and HPO_4^{2-} (phosphate). Once fluid passes through the proximal tubule, cells located more distally *fine-tune* the reabsorption processes to maintain homeostatic balances of water and selected ions. Most small proteins and peptides that pass through the filter are also reabsorbed, usually via pinocytosis. To appreciate the magnitude of tubular reabsorption, look at TABLE 19.2 and compare the amounts of substances that are filtered, reabsorbed, and excreted in urine.

The third function of nephrons and collecting ducts is tubular *secretion*, the transfer of materials from the blood and tubule cells into tubular fluid (see FIGURE 19.2). Secreted substances include hydrogen ions (H^+), K^+, ammonium ions (NH_4^+), creatinine, and certain drugs such as penicillin. Tubular secretion has two important outcomes: (1) The secretion of H^+ helps control blood pH, and (2) the secretion of other substances helps eliminate them from the body.

There Are Two Types of Reabsorption Routes: Paracellular and Transcellular

A substance being reabsorbed from the fluid in the tubule lumen can take one of two routes before entering a peritubular capillary: It can move *between* adjacent tubule cells or *through* an individual tubule cell

TABLE 19.2 Substances Filtered, Reabsorbed, and Excreted in Urine

Substance	Filtered* (Enters Bowman's Capsule per Day)	Reabsorbed (Returned to Blood per Day)	Urine (Excreted per Day)
Water	180 liters	178–178.5 liters	1.5–2 liters
Proteins	2.0 g	1.9 g	0.1 g
Sodium ions (Na^+)	579 g	575 g	4 g
Chloride ions (Cl^-)	640 g	633.7 g	6.3 g
Bicarbonate ions (HCO_3^-)	275 g	274.97 g	0.03 g
Glucose	162 g	162 g	0 g
Urea	54 g	24 g	30 g†
Potassium ions (K^+)	29.6 g	29.6 g	2.0 g‡
Uric acid	8.5 g	7.7 g	0.8 g
Creatinine	1.6 g	0 g	1.6 g

* Assuming GFR is 180 liters per day.

† In addition to being filtered and reabsorbed, urea is secreted.

‡ After virtually all filtered K^+ is reabsorbed in the proximal and distal tubules and loop of Henle, a variable amount of K^+ is secreted by principal cells in the late distal tubule and collecting duct.

(**FIGURE 19.6**). Along the renal tubule, tight junctions surround and join neighboring cells to one another, much like the plastic rings that hold a six-pack of soda cans together. The *apical membrane* (the tops of the soda cans) contacts the tubular fluid, and the *basolateral membrane* (the bottoms and sides of the soda cans) contacts interstitial fluid at the base and sides of the cell.

The tight junctions do not completely seal off the interstitial fluid from the fluid in the tubule lumen. Fluid can leak *between* the cells in a passive process known as **paracellular reabsorption** (*para-* = beside). In some parts of the renal tubule, the paracellular route is thought to account for up to 50% of the reabsorption of certain ions and the water that accompanies them via osmosis. In **transcellular reabsorption** (*trans-* = across), a substance passes from the fluid in the tubular lumen through the apical membrane of a tubule cell, across the cytosol, and out into interstitial fluid through the basolateral membrane.

Transport of Substances Across the Tubular Wall Often Involves the Use of Transport Proteins

When renal cells transport solutes out of or into tubular fluid, they move specific substances in one direction only. Not surprisingly, different types of transport proteins are present in the apical and basolateral membranes. The tight junctions form a barrier that prevents mixing of proteins in the apical and basolateral membrane compartments.

Reabsorption of Na^+ by the renal tubules is especially important because of the large number of sodium ions that pass through the glomerular filters. Cells lining the renal tubules, like other cells throughout the body, have a low concentration of Na^+ in their cytosol due to the activity of sodium–potassium pumps (Na^+/K^+ ATPases). These pumps are located in the basolateral membranes and eject Na^+ from the renal tubule cells (**FIGURE 19.6**). The absence of sodium–potassium pumps in the apical membrane ensures that reabsorption of Na^+ is a one-way process. Most sodium ions that cross the apical membrane are pumped into interstitial fluid at the base and sides of the cell. The amount of ATP used by sodium–potassium pumps in the renal tubules is about 6% of the total ATP consumption of the body at rest. This may not sound like much, but it is about the same amount of energy used by the diaphragm as it contracts during quiet breathing.

As you learned in Chapter 5, transport of materials across membranes may be either active or passive. Recall that, in *primary active transport*, the energy derived from hydrolysis of ATP is used to "pump" a substance across a membrane; the sodium–potassium pump is one such pump. In *secondary active transport* the energy stored in an ion's electrochemical gradient, rather than hydrolysis of ATP, drives another substance across a membrane. Secondary active transport couples the movement of an ion down its electrochemical gradient to the "uphill" movement of a second substance against its concentration or electrochemical gradient. *Symporters* are membrane proteins that move two or more substances in the same direction across a membrane. *Antiporters* move two or more substances in opposite directions across a membrane. Each type of transporter has an upper limit on how fast it can work, just as an escalator has a limit on how many people it can carry from one level to another in a given period. This limit, called the *transport maximum* (T_m), is measured in mg/min.

FIGURE 19.6 Reabsorption routes: paracellular reabsorption and transcellular reabsorption.

In paracellular reabsorption, water and solutes in tubular fluid return to the bloodstream by moving between tubule cells; in transcellular reabsorption, solutes and water in tubular fluid return to the bloodstream by passing through a tubule cell.

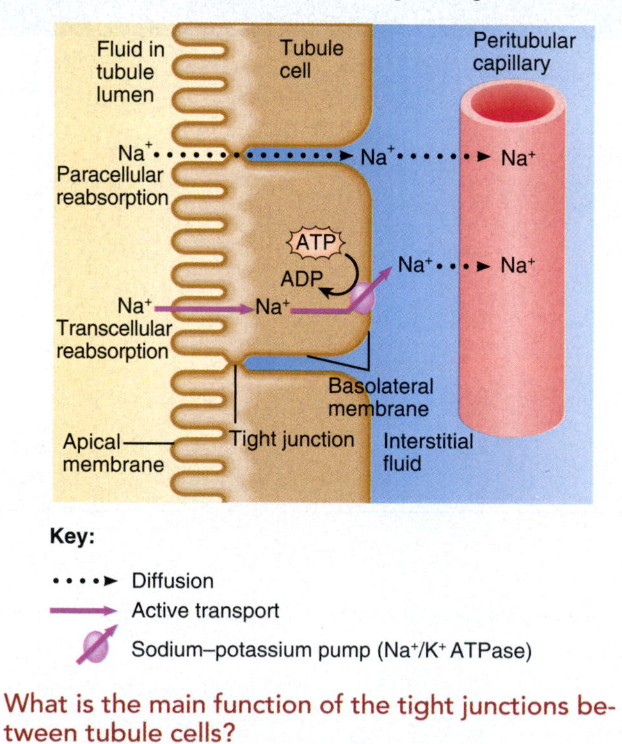

What is the main function of the tight junctions between tubule cells?

Water Reabsorption May Be Obligatory or Facultative

Solute reabsorption drives water reabsorption because all water reabsorption occurs via osmosis. About 80% of the reabsorption of water filtered by the kidneys occurs along with the reabsorption of solutes such as Na^+, Cl^-, and glucose. Water reabsorbed with solutes in tubular fluid is termed **obligatory water reabsorption** because the water is "obliged" to follow the solutes when they are reabsorbed. This type of water reabsorption occurs in the proximal tubule and the descending limb of the loop of Henle because these segments of the nephron are always permeable to water. Reabsorption of the final 20% of the water is termed **facultative water reabsorption**. The word *facultative* means "capable of adapting to a need." Facultative water reabsorption is regulated by antidiuretic hormone and occurs in the late distal tubule and throughout the collecting duct.

Different Substances Are Reabsorbed or Secreted to Varying Degrees in Different Parts of the Renal Tubule and Collecting Duct

Now that the principles of renal transport have been discussed, you will follow the filtered fluid from the proximal tubule, into the loop of Henle, on to the distal tubule, and through the collecting ducts. In each segment, you will examine where and how specific substances are reabsorbed and secreted.

Reabsorption and Secretion in the Proximal Tubule

The largest amount of solute and water reabsorption from filtered fluid occurs in the proximal tubules, which reabsorb 65% of the filtered water, Na^+, K^+, and Ca^{2+}; 100% of most filtered organic solutes such as glucose and amino acids; 50% of the filtered Cl^-; 80–90% of the filtered

HCO_3^-; 50% of the filtered urea; and a variable amount of the filtered Mg^{2+} (magnesium) and HPO_4^{2-} (phosphate). In addition, proximal tubules secrete a variable amount of H^+ ions, ammonium ions (NH_4^+), and urea. The proximal tubule consists of epithelial cells that contain a brush border of microvilli along their apical membranes. These microvilli increase the surface area for reabsorption and secretion.

Most solute reabsorption in the proximal tubule involves Na^+. Na^+ transport occurs via symport and antiport mechanisms in the proximal tubule. Normally, filtered glucose, amino acids, lactic acid, water-soluble vitamins, and other nutrients are not lost in the urine. Rather, they are completely reabsorbed in the first half of the proximal tubule by several types of **Na^+ symporters** located in the apical membrane. FIGURE 19.7 depicts the operation of one such symporter, the **Na^+–glucose symporter** in the apical membrane of a cell in the proximal tubule. Two Na^+ and a molecule of glucose attach to the symporter protein, which carries them from the tubular fluid into the tubule cell. The glucose molecules then exit the basolateral membrane via facilitated diffusion and diffuse into peritubular capillaries. Other Na^+ symporters in the proximal tubule reclaim filtered HPO_4^{2-} (phosphate) and SO_4^{2-} (sulfate) ions, all amino acids, and lactic acid in a similar way.

FIGURE 19.7 Reabsorption of glucose by Na^+–glucose symporters in cells of the proximal tubule.

Normally, all filtered glucose is reabsorbed in the proximal tubule.

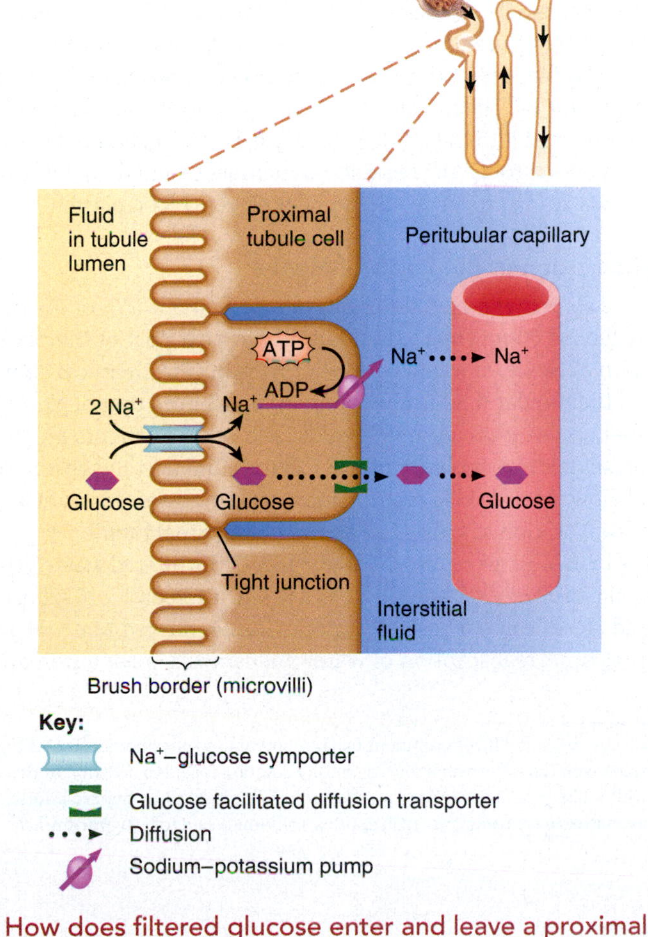

How does filtered glucose enter and leave a proximal tubule cell?

In another secondary active transport process, the **Na^+/H^+ antiporters** carry filtered Na^+ down its concentration gradient into a proximal tubule cell as H^+ is moved from the cytosol into the lumen (FIGURE 19.8a), causing Na^+ to be reabsorbed into blood and H^+ to

FIGURE 19.8 Actions of Na^+/H^+ antiporters in proximal tubule cells. (a) Reabsorption of sodium ions (Na^+) and secretion of hydrogen ions (H^+) via secondary active transport through the apical membrane. (b) Reabsorption of bicarbonate ions (HCO_3^-) via facilitated diffusion through the basolateral membrane. CO_2 = carbon dioxide; H_2CO_3 = carbonic acid; CA = carbonic anhydrase.

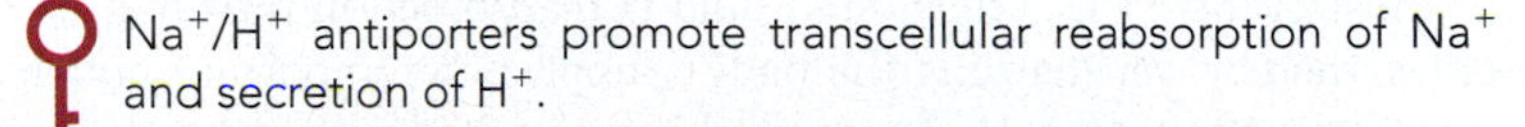
Na^+/H^+ antiporters promote transcellular reabsorption of Na^+ and secretion of H^+.

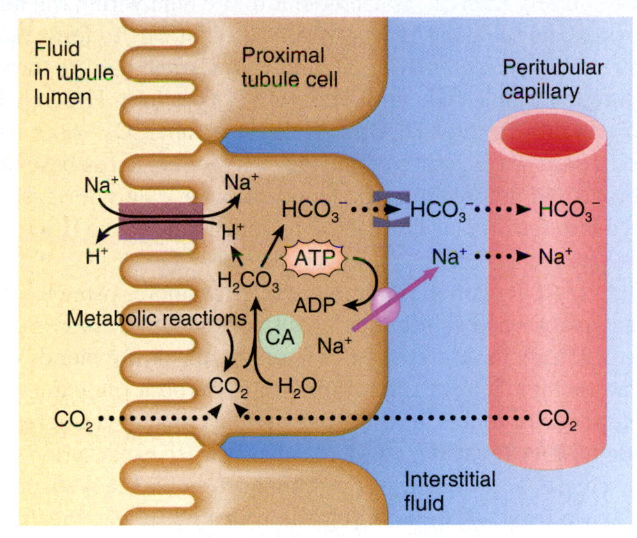

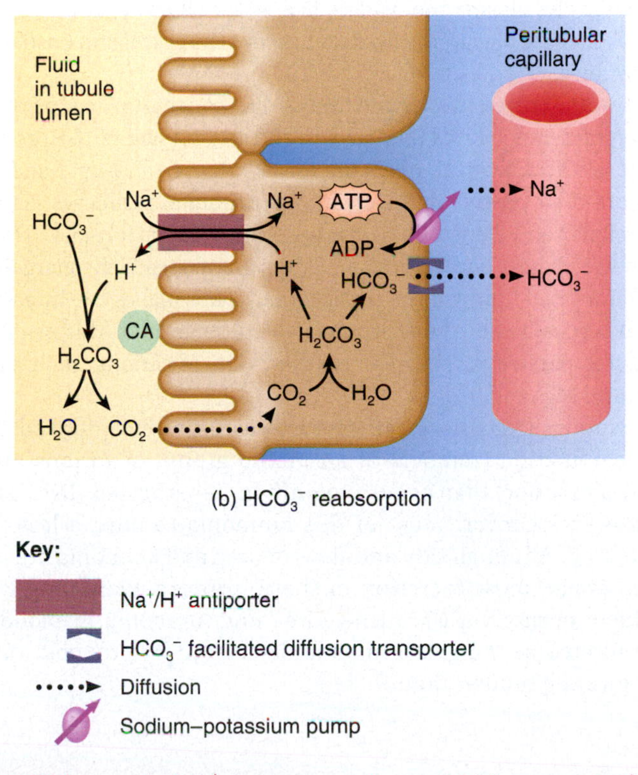

Which step in Na^+ movement in part (a) is promoted by the electrochemical gradient?

be secreted into tubular fluid. Proximal tubule cells produce the H^+ needed to keep the antiporters running in the following way. Carbon dioxide (CO_2) diffuses from peritubular blood or tubular fluid or is produced by metabolic reactions within the cells. As occurs in erythrocytes (see **FIGURE 18.14**), the enzyme *carbonic anhydrase (CA)* catalyzes the reaction of CO_2 with water (H_2O) to form carbonic acid (H_2CO_3), which then dissociates into H^+ and HCO_3^-:

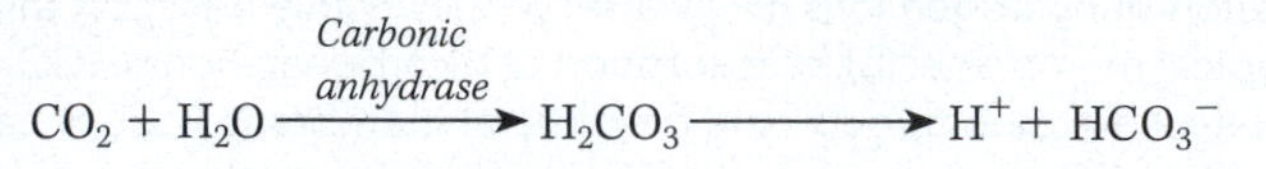

$$CO_2 + H_2O \xrightarrow{\textit{Carbonic anhydrase}} H_2CO_3 \longrightarrow H^+ + HCO_3^-$$

Most of the HCO_3^- in filtered fluid is reabsorbed in proximal tubules, thereby safeguarding the body's supply of an important buffer (**FIGURE 19.8b**). After H^+ is secreted into the fluid within the lumen of the proximal tubule, it reacts with filtered HCO_3^- to form H_2CO_3, which readily dissociates into CO_2 and H_2O. Carbon dioxide then diffuses into the tubule cells and joins with H_2O to form H_2CO_3, which dissociates into H^+ and HCO_3^-. As the level of HCO_3^- rises in the cytosol, it exits via facilitated diffusion transporters in the basolateral membrane and diffuses into the blood with Na^+. Thus, for every H^+ secreted into the tubular fluid of the proximal tubule, one HCO_3^- and one Na^+ are reabsorbed.

Solute reabsorption in proximal tubules promotes osmosis of water. Each reabsorbed solute increases the osmolarity, first inside the tubule cell, then in interstitial fluid, and finally in the blood. Water thus moves rapidly from the tubular fluid, via both the paracellular and transcellular routes, into the peritubular capillaries and restores osmotic balance (**FIGURE 19.9**). In other words, reabsorption of the solutes creates an osmotic gradient that promotes the reabsorption of water via osmosis. Cells lining the proximal tubule and the descending limb of the loop of Henle are especially permeable to water because they have many molecules of **aquaporin-1.** This integral protein in the plasma membrane is a water channel that greatly increases the rate of water movement across the apical and basolateral membranes.

As water leaves the tubular fluid, the concentrations of the remaining filtered solutes increase. In the second half of the proximal tubule, electrochemical gradients for Cl^-, K^+, Ca^{2+}, Mg^{2+}, and urea promote their passive diffusion into peritubular capillaries via both paracellular and transcellular routes. Among these ions, Cl^- is present in the highest concentration. Diffusion of negatively charged Cl^- into interstitial fluid via the paracellular route makes the interstitial fluid electrically more negative than the tubular fluid. This negativity promotes passive paracellular reabsorption of cations, such as K^+, Ca^{2+}, and Mg^{2+}.

Ammonia (NH_3) is a poisonous waste product derived from the deamination (removal of an amino group) of various amino acids, a reaction that occurs mainly in hepatocytes (liver cells). Hepatocytes convert most of this ammonia to urea, a less-toxic compound. Although tiny amounts of urea and ammonia are present in sweat, most excretion of these nitrogen-containing waste products occurs via the urine. Urea and ammonia in blood are both filtered at the glomerulus and secreted by proximal tubule cells into the tubular fluid.

FIGURE 19.9 Passive reabsorption of Cl^-, K^+, Ca^{2+}, Mg^{2+}, urea, and water in the second half of the proximal tubule.

Electrochemical gradients promote passive reabsorption of solutes via both paracellular and transcellular routes.

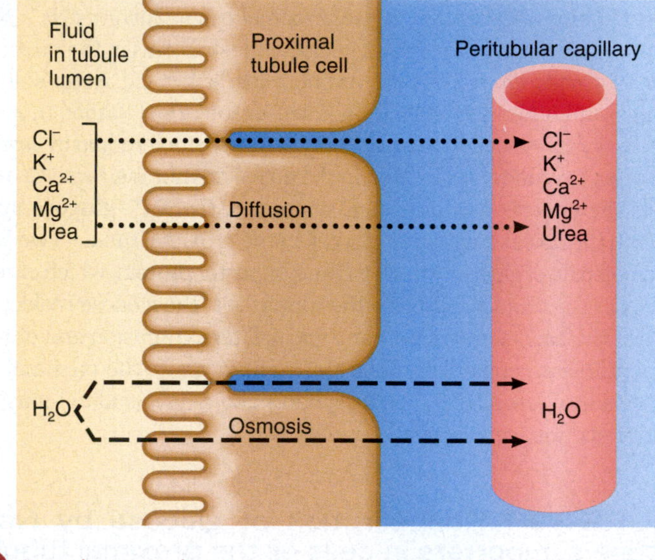

By what mechanism is water reabsorbed from tubular fluid?

Proximal tubule cells can produce additional NH_3 by deaminating the amino acid glutamine in a reaction that also generates HCO_3^-. The NH_3 quickly binds H^+ to become an ammonium ion ($NH4^+$), which can substitute for H^+ aboard Na^+/H^+ antiporters in the apical membrane and be secreted into the tubular fluid. The HCO_3^- generated in this reaction moves through the basolateral membrane and then diffuses into the bloodstream, providing additional buffers in blood plasma.

Reabsorption in the Loop of Henle

Because all of the proximal tubules reabsorb about 65% of the filtered water (about 80 mL/min), fluid enters the next part of the nephron, the loop of Henle, at a rate of 40–45 mL/min. The chemical composition of the tubular fluid is now quite different from that of glomerular filtrate because glucose, amino acids, and other nutrients are no longer present. The osmolarity of the tubular fluid is still isoosmotic* to blood, however, because reabsorption of water by osmosis keeps pace with reabsorption of solutes all along the proximal tubule.

The loop of Henle reabsorbs about 15% of the filtered water; 20–30% of the filtered Na^+, K^+, and Ca^{2+}; 35% of the filtered Cl^-; 10–20% of the filtered HCO_3^-; and a variable amount of the filtered Mg^{2+}. Here, for the first time, reabsorption of water via osmosis is *not* automatically

* The osmolarities of different solutions can be compared to one another (see Section 5.4). A solution with the same osmolarity as another solution is said to be **isoosmotic**, a solution with a higher osmolarity than another solution is said to be **hyperosmotic**, and a solution with a lower osmolarity than another solution is said to be **hypoosmotic**.

coupled to reabsorption of filtered solutes because part of the loop of Henle is relatively impermeable to water. The loop of Henle thus sets the stage for *independent* regulation of both the *volume* and *osmolarity* of body fluids.

The apical membranes of cells in the thick ascending limb of the loop of Henle have **Na^+–K^+–2Cl^- symporters** that simultaneously reclaim one Na^+, one K^+, and two Cl^- from the fluid in the tubular lumen (**FIGURE 19.10**). Na^+ that is actively transported into interstitial fluid at the base and sides of the cell diffuses into the vasa recta. Cl^- moves through leak channels in the basolateral membrane into interstitial fluid and then into the vasa recta. Because many K^+ leak channels are present in the apical membrane, most K^+ brought in by the symporters moves down its concentration gradient back into the tubular fluid. Thus, the main effect of the Na^+–K^+–2Cl^- symporters is reabsorption of Na^+ and Cl^-.

The movement of positively charged K^+ into the tubular fluid through the apical membrane channels leaves the interstitial fluid and blood with more negative charges relative to fluid in the ascending limb of the loop of Henle. This relative negativity promotes reabsorption of cations—Na^+, K^+, Ca^{2+}, and Mg^{2+}—via the paracellular route.

Although about 15% of the filtered water is reabsorbed in the *descending* limb of the loop of Henle, little or no water is reabsorbed in the *ascending* limb. In this segment of the tubule, the apical membranes are virtually impermeable to water. Because ions but not water molecules are reabsorbed, the osmolarity of the tubular fluid progressively decreases (becomes hypoosmotic compared to blood) as fluid flows toward the end of the ascending limb.

FIGURE 19.10 Na^+–K^+–2Cl^- symporter in the thick ascending limb of the loop of Henle.

The Na^+–K^+–2Cl^- symporter simultaneously moves one Na^+, one K^+, and two Cl^- from tubular fluid into a thick ascending limb cell.

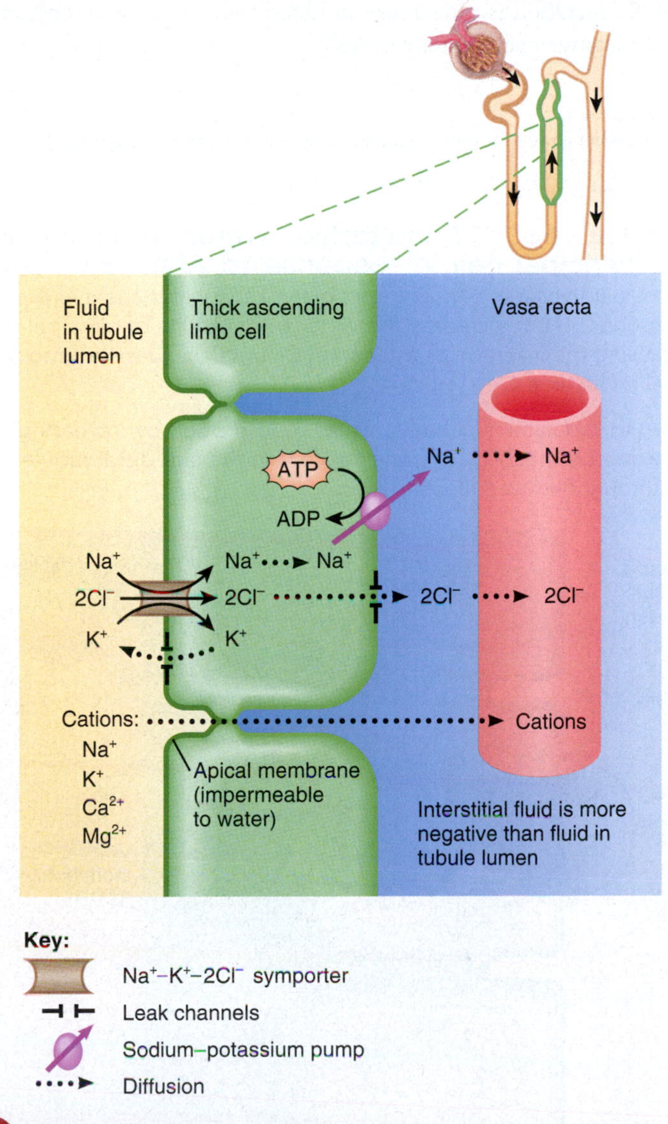

Why is this process considered secondary active transport? Does water reabsorption accompany ion reabsorption in this region of the nephron? Explain your answer.

Reabsorption in the Early Distal Tubule

Fluid enters the distal tubules at a rate of about 25 mL/min because 80% of the filtered water has now been reabsorbed. The early or initial part of the distal tubule reabsorbs about 5% of the filtered Na^+ and 5% of the filtered Cl^-. Reabsorption of Na^+ and Cl^- occurs by means of **Na^+–Cl^- symporters** in the apical membranes. Sodium–potassium pumps and Cl^- leak channels in the basolateral membranes then permit reabsorption of Na^+ and Cl^- into the peritubular capillaries. The early distal tubule is also a major site where parathyroid hormone (PTH) stimulates reabsorption of Ca^{2+}. The amount of Ca^{2+} reabsorption in this part of the nephron varies depending on the body's needs.

Reabsorption and Secretion in the Late Distal Tubule and Collecting Duct

In the late or terminal part of the distal tubule and throughout the collecting duct, two different types of cells are present—**principal cells**, the main cell type, and **intercalated cells**. The principal cells reabsorb Na^+ and secrete K^+. These cells also have receptors for aldosterone and antidiuretic hormone (ADH). The intercalated cells reabsorb HCO_3^- and secrete H^+, thereby playing a role in blood pH regulation. In addition, the intercalated cells reabsorb K^+. In the late distal tubules and collecting ducts, the amount of water and solute reabsorption and the amount of solute secretion vary depending on the body's needs.

In contrast to earlier segments of the nephron, Na^+ passes through the apical membrane of principal cells via Na^+ leak channels rather than by means of symporters or antiporters (**FIGURE 19.11**). The concentration of Na^+ in the cytosol remains low, as usual, because the sodium–potassium pumps actively transport Na^+ across the basolateral membranes. Then Na^+ passively diffuses into the peritubular capillaries from the interstitial spaces around the tubule cells.

Normally, transcellular and paracellular reabsorption in the proximal tubule and loop of Henle return most filtered K^+ to the bloodstream. To adjust for varying dietary intake of potassium and to maintain a stable level of K^+ in body fluids, principal cells secrete a variable amount of K^+ (**FIGURE 19.11**). Because the basolateral sodium–potassium pumps continually bring K^+ into principal cells, the intracellular concentration of K^+ remains high. K^+ leak channels are present in both the apical and basolateral membranes. Thus, some K^+ diffuses down its concentration gradient into the tubular fluid, where the K^+ concentration is very low. This secretion mechanism is the main source of K^+ excreted in the urine.

FIGURE 19.11 Reabsorption of Na^+ and secretion of K^+ by principal cells in the late distal tubule and the collecting duct.

In the apical membrane of principal cells, Na^+ leak channels allow entry of Na^+, while K^+ leak channels allow exit of K^+ into the tubular fluid.

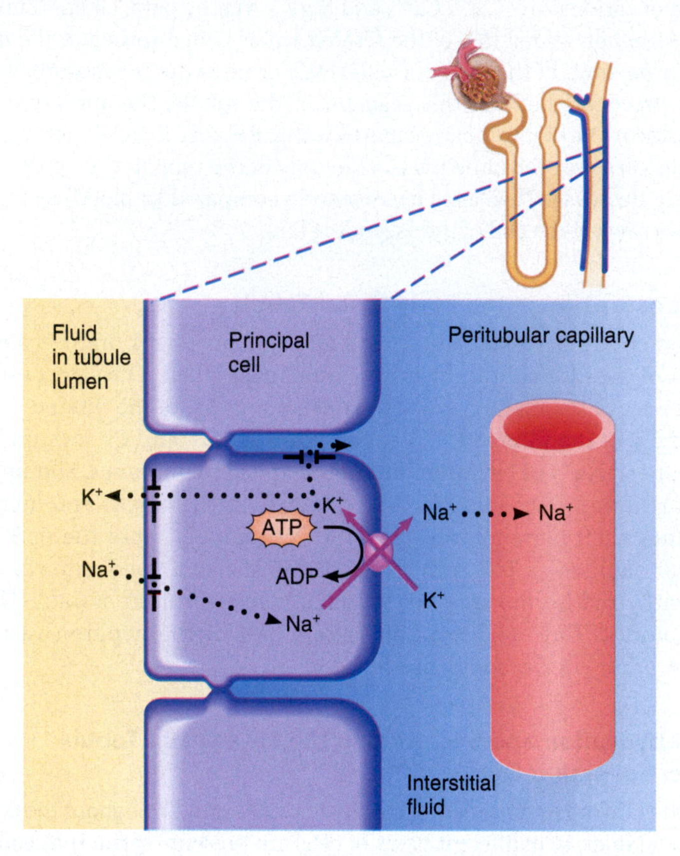

Which hormone stimulates reabsorption of Na^+ and secretion of K^+ by principal cells?

Tubular Reabsorption and Tubular Secretion Are Subject to Hormonal Regulation

Four hormones affect the extent of water, Na^+, and Ca^{2+} reabsorption as well as K^+ secretion by the renal tubules. These hormones include antidiuretic hormone, aldosterone, atrial natriuretic peptide, and parathyroid hormone.

Antidiuretic Hormone

Antidiuretic hormone (ADH), also known as *vasopressin*, is released by the posterior pituitary gland. It regulates facultative water reabsorption by increasing the water permeability of principal cells in the late distal tubule and throughout the collecting duct. In the absence of ADH, the apical membranes of principal cells have a very low permeability to water. Within principal cells are tiny vesicles containing many copies of a water channel protein known as **aquaporin-2.*** ADH stimulates insertion of the aquaporin-2–containing vesicles into the apical membranes via exocytosis (FIGURE 19.12). As a result, the water permeability of the principal cell's apical membrane increases, and water molecules move more rapidly from the tubular fluid into the cells. Because the basolateral membranes are always relatively permeable to water, water molecules then move rapidly into the blood. This results in an increase in blood volume and blood pressure. When the ADH level declines, the aquaporin-2 channels are removed from the apical membrane via endocytosis, and water permeability of the principal cells decreases.

A negative feedback system involving ADH regulates facultative water reabsorption (FIGURE 19.13). When the osmolarity or osmotic pressure of plasma and interstitial fluid increases—that is, when water concentration decreases—by as little as 1%, osmoreceptors in the hypothalamus detect the change. Their action potentials stimulate secretion of more ADH into the blood, and the principal cells become more permeable to water. As facultative water reabsorption increases, plasma osmolarity decreases to normal. A second powerful stimulus for ADH secretion is a decrease in blood volume, such as occurs in hemorrhaging or severe dehydration.

* ADH does not govern the previously mentioned water channel (aquaporin-1).

FIGURE 19.12 Facultative water reabsorption in a principal cell in response to ADH. Within principal cells are vesicles containing copies of the water channel protein aquaporin-2. ADH increases the water permeability of principal cells by causing the insertion of aquaporin-2-containing vesicles into their apical membranes.

ADH regulates facultative water reabsorption by increasing the water permeability of principal cells in the late distal tubule and throughout the collecting duct.

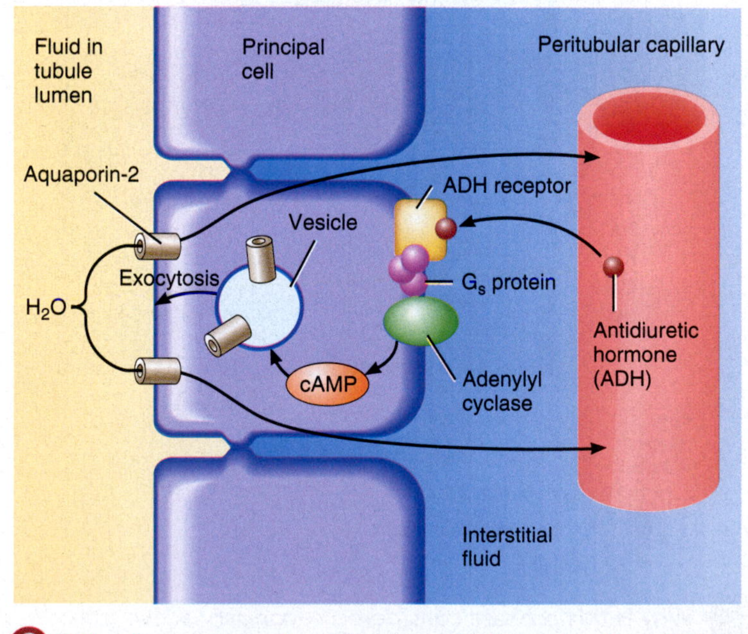

What happens to the aquaporin-2 channels when ADH levels in the blood decline?

FIGURE 19.13 Negative feedback regulation of facultative water reabsorption by ADH.

Most water reabsorption (80%) is obligatory; 20% is facultative.

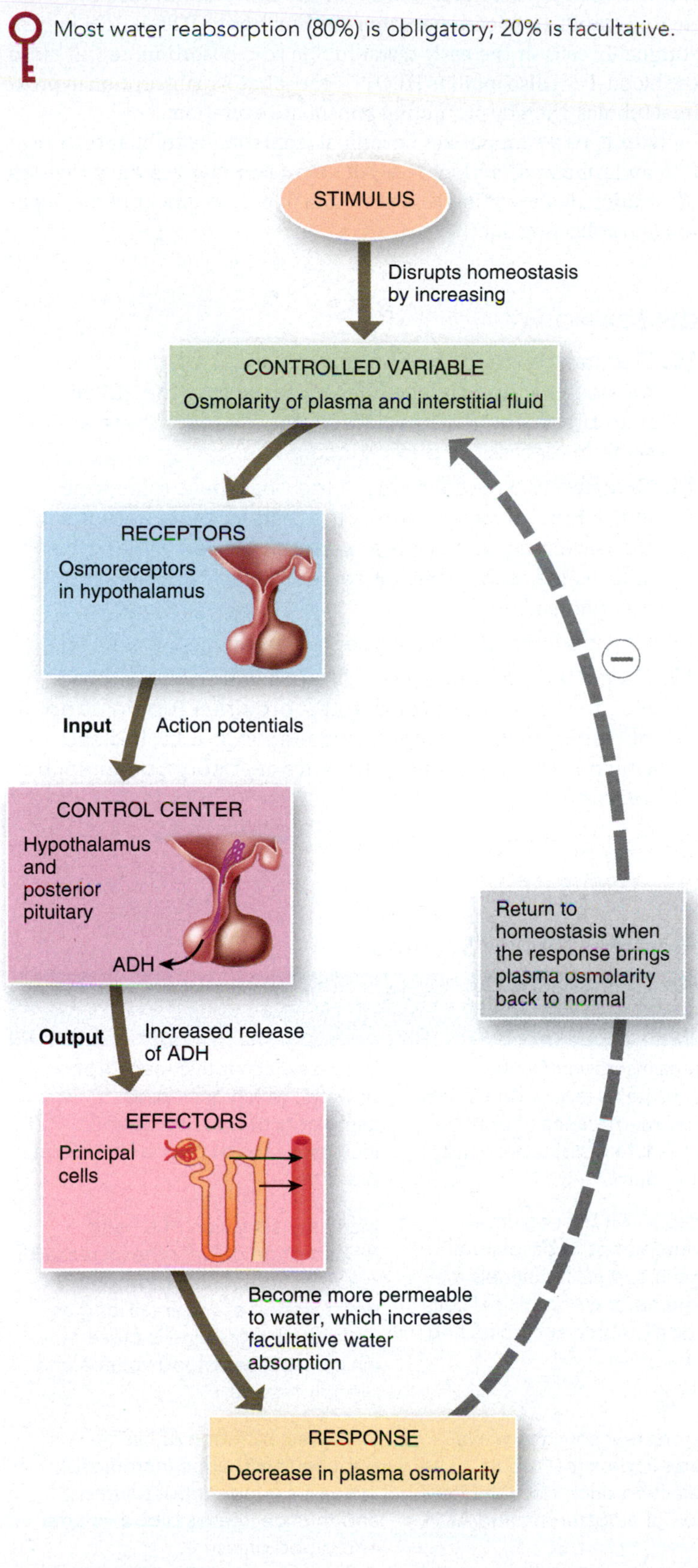

What is the difference between obligatory and facultative water reabsorption?

The degree of facultative water reabsorption caused by ADH in the late distal tubule and collecting duct depends on whether the body is normally hydrated, dehydrated, or overhydrated.

- ***Normal hydration.*** Under conditions of normal body hydration (adequate water intake), enough ADH is present in the blood to cause reabsorption of 19% of the filtered water in the late distal tubule and the collecting duct. This means that the total amount of filtered water reabsorbed in the renal tubule and collecting duct is 99%: 65% in the proximal tubule + 15% in the loop of Henle + 19% in the late distal tubule and collecting duct. The remaining 1% of the filtered water (about 1.5–2 L/day) is excreted in urine. Therefore, when the body is normally hydrated, the kidneys produce about 1.5–2 L of urine on a daily basis and the urine is slightly hyperosmotic (slightly concentrated) compared to blood.
- ***Dehydration.*** When the body is dehydrated, the concentration of ADH in the blood increases. This in turn causes an increase in the amount of filtered water that is reabsorbed in the late distal tubule and collecting duct. Depending on how much the blood ADH level increases, the amount of filtered water that is reabsorbed in the late distal tubule and collecting duct can increase from just above 19% to as high as 19.8%. As a result, less than 1% of filtered water remains unreabsorbed in the late distal tubule and collecting duct, which corresponds to a urine output *below* the normal 1.5–2 L/day. The urine produced under these circumstances is very hyperosmotic (highly concentrated) compared to blood because it contains less water than normal. In the case of severe dehydration, the amount of filtered water that is reabsorbed in the late distal tubule and collecting duct reaches a maximum limit of 19.8%. This means that the total amount of filtered water reabsorbed in the renal tubule and collecting duct is 99.8%: 65% in the proximal tubule + 15% in the loop of Henle + 19.8 % in the late distal tubule and collecting duct. The remaining 0.2% of the filtered water (about 400 mL/day) is excreted in urine. Thus, the kidneys produce a small volume of highly concentrated urine when the body is dehydrated.
- ***Overhydration.*** When the body is overhydrated (too much water intake), the concentration of ADH in the blood decreases. This in turn causes a decrease in the amount of filtered water that is reabsorbed in the late distal tubule and collecting duct. Depending on how much the blood ADH level decreases, the amount of filtered water that is reabsorbed in the late distal tubule and collecting duct can decrease from just below 19% to as low as 0%. As a result, more than 1% of filtered water remains unreabsorbed in the late distal tubule and collecting duct, which corresponds to a urine output *above* the normal 1.5–2 L/day. The urine produced under these conditions is hypoosmotic (dilute) compared to blood because it contains more water than normal. In the case of severe overhydration, no ADH is present in the blood, and the amount of water reabsorbed in the late distal tubule and collecting duct is 0%. This means that the total amount of filtered water that is reabsorbed in the renal tubule and collecting duct is 80%: 65% in the proximal tubule + 15% in the loop of Henle + 0% in the late distal tubule and collecting duct. The remaining 20% of filtered water (about 36 L/day) is excreted in urine. Hence, the kidneys produce a large volume of dilute urine when the body is overhydrated.

Renin–Angiotensin–Aldosterone System

When blood volume and blood pressure decrease, the walls of the afferent arterioles are stretched less, and the juxtaglomerular cells (see FIGURE 19.3) secrete the enzyme **renin** into the blood. Renin converts angiotensinogen, a plasma protein synthesized by the liver, into angiotensin I. *Angiotensin-converting enzyme (ACE)*, which is present on the luminal surfaces of capillaries throughout the body, converts angiotensin I to **angiotensin II**, the active form of the hormone.

Angiotensin II affects renal physiology in two main ways:

1. It decreases the glomerular filtration rate by causing vasoconstriction of the afferent and efferent arterioles.
2. It stimulates the adrenal gland to release **aldosterone**, a hormone that in turn stimulates the principal cells in the late distal tubule and the collecting duct to secrete more K^+ and to reabsorb more Na^+. The osmotic consequence of reabsorbing more Na^+ is that more water is reabsorbed (as long as antidiuretic hormone is present), which causes an increase in blood volume and blood pressure.

Atrial Natriuretic Peptide

A large increase in blood volume promotes release of **atrial natriuretic peptide (ANP)** from the heart. Although the importance of ANP in normal regulation of tubular function is unclear, it can inhibit reabsorption of Na^+ and water in the proximal tubule and collecting duct. ANP also suppresses the secretion of aldosterone and ADH. These effects increase the excretion of Na^+ in urine (natriuresis) and increase urine output (diuresis), which decreases blood volume and blood pressure.

Parathyroid Hormone

A lower-than-normal level of Ca^{2+} in the blood stimulates the parathyroid glands to release **parathyroid hormone (PTH)**. PTH in turn stimulates cells in the early distal tubule to reabsorb more Ca^{2+} into the blood. PTH also inhibits HPO_4^{2-} (phosphate) reabsorption in proximal tubules, thereby promoting phosphate excretion.

TABLE 19.3 summarizes hormonal regulation of tubular reabsorption and tubular secretion. FIGURE 19.14 summarizes the processes of filtration, reabsorption, and secretion in each segment of the nephron and collecting duct.

CHECKPOINT

10. Diagram the reabsorption of substances via the transcellular and paracellular routes across the apical membrane and the basolateral membrane. Where are the sodium–potassium pumps located?
11. Describe two mechanisms in the proximal tubule, one in the loop of Henle, one in the distal tubule, and one in the collecting duct for reabsorption of Na^+. What other solutes are reabsorbed or secreted with Na^+ in each mechanism?
12. How do intercalated cells secrete hydrogen ions?
13. Graph the percentages of filtered water and filtered Na^+ that are reabsorbed in the proximal tubule, loop of Henle, distal tubule, and collecting duct. Indicate which hormones, if any, regulate reabsorption in each segment.

TABLE 19.3 Hormonal Regulation of Tubular Reabsorption and Tubular Secretion

Hormone	Major Stimuli That Trigger Release	Mechanism and Site of Action	Effects
Antidiuretic hormone (ADH) or *vasopressin*	Increased osmolarity of extracellular fluid or decreased blood volume promote release of ADH from the posterior pituitary gland.	Stimulates insertion of water-channel proteins (aquaporin-2) into the apical membranes of principal cells in the late distal tubule and collecting duct.	Increases facultative reabsorption of water, which decreases osmolarity of body fluids and increases blood volume and blood pressure.
Aldosterone	Increased angiotensin II level and increased level of plasma K^+ promote release of aldosterone by adrenal cortex.	Enhances activity of sodium–potassium pumps in basolateral membrane and Na^+ channels in apical membrane of principal cells in late distal tubule and collecting duct.	Increases secretion of K^+ and reabsorption of Na^+; the increased reabsorption of Na^+ promotes reabsorption of water (as long as antidiuretic hormone is present), which increases blood volume and blood pressure.
Atrial natriuretic peptide (ANP)	Stretching of atria of heart stimulates secretion of ANP.	Suppresses reabsorption of Na^+ and water in proximal tubule and collecting duct; also inhibits secretion of aldosterone and ADH.	Increases excretion of Na^+ in peptide (ANP) urine (natriuresis); increases urine output (diuresis) and thus decreases blood volume and blood pressure.
Parathyroid hormone (PTH)	Decreased level of plasma Ca^{2+} promotes release of PTH from parathyroid glands.	Stimulates opening of Ca^{2+} channels in apical membranes of cells in early distal tubule.	Increases reabsorption of Ca^{2+}.

FIGURE 19.14 Summary of filtration, reabsorption, and secretion in the nephron and collecting duct.

Filtration occurs in the renal corpuscle; reabsorption occurs all along the renal tubule and collecting duct.

RENAL CORPUSCLE

Glomerular filtration rate: 125 mL/min of fluid that is isoosmotic to blood

Filtered substances: water and all solutes present in blood (except cells and proteins) including ions, glucose, amino acids, creatinine, and uric acid

PROXIMAL TUBULE

Reabsorption (into blood) of filtered:

Water	65% (osmosis)
Na^+	65% (sodium–potassium pumps, symporters, antiporters)
K^+	65% (diffusion)
Glucose	100% (symporters and facilitated diffusion)
Amino acids	100% (symporters and facilitated diffusion)
Cl^-	50% (diffusion)
HCO_3^-	80–90% (facilitated diffusion)
Urea	50% (diffusion)
Ca^{2+}	65% (diffusion)
Mg^{2+}	variable (diffusion)
HPO_4^{2-}	variable (symporters and antiporters)

Secretion (into tubular fluid) of:

H^+	variable (antiporters)
NH_4^+	variable, increases in acidosis (antiporters)
Urea	variable (diffusion)
Creatinine	small amount

At end of proximal tubule, tubular fluid is still isoosmotic to blood (300 mosmol/liter).

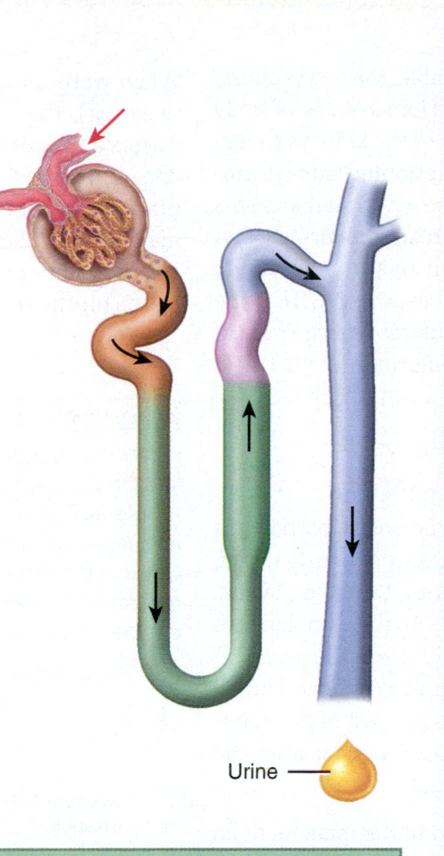

EARLY DISTAL TUBULE

Reabsorption (into blood) of filtered:

Na^+	5% (symporters)
Cl^-	5% (symporters)
Ca^{2+}	variable (stimulated by parathyroid hormone)

LATE DISTAL TUBULE AND COLLECTING DUCT

Reabsorption (into blood) of filtered:

Water	19% (insertion of water channels stimulated by ADH)
Na^+	1–4% (sodium–potassium pumps and sodium channels stimulated by aldosterone)
HCO_3^-	variable amount, depends on H^+ secretion (antiporters)
Urea	variable (recycling to loop of Henle)

Secretion (into tubular fluid) of:

K^+	variable amount to adjust for dietary intake (leak channels)
H^+	variable amount to maintain acid–base homeostasis (H^+ pumps)

Tubular fluid leaving the collecting duct is hypoosmotic when ADH level is low and hyperosmotic when ADH level is high.

LOOP OF HENLE

Reabsorption (into blood) of filtered:

Water	15% (osmosis in descending limb)
Na^+	20–30% (symporters in ascending limb)
K^+	20–30% (symporters in ascending limb)
Cl^-	35% (symporters in ascending limb)
HCO_3^-	10–20% (facilitated diffusion)
Ca^{2+}	20–30% (diffusion)
Mg^{2+}	variable (diffusion)

Secretion (into tubular fluid) of:

Urea	variable (recycling from collecting duct)

At end of loop of Henle, tubular fluid is hypoosmotic (100–150 mosmol/liter).

In which segments of the nephron and collecting duct does secretion occur?

19.6 Production of Dilute and Concentrated Urine

OBJECTIVE

- Describe how the kidneys produce dilute and concentrated urine.

Even though your fluid intake can be highly variable, the total volume of fluid in your body normally remains stable. Homeostasis of body fluid volume depends in large part on the ability of the kidneys to regulate the rate of water loss in urine. Normally functioning kidneys produce a large volume of dilute (hypoosmotic) urine when fluid intake is high, and a small volume of concentrated (hyperosmotic) urine when fluid intake is low or fluid loss is large. ADH controls whether dilute urine or concentrated urine is formed. In the absence of ADH, urine is very dilute. However, a high level of ADH stimulates reabsorption of more water into blood, producing a concentrated urine.

The Production of Dilute Urine Allows the Kidneys to Get Rid of Excess Water

Glomerular filtrate has the same ratio of water and solute particles as blood; its osmolarity is about 300 mosmol/liter. As previously noted, fluid leaving the proximal tubule is still isoosmotic to blood plasma. When *dilute* urine is being formed (**FIGURE 19.15**), the osmolarity of the fluid in the tubular lumen *increases* as it flows down the descending limb of the loop of Henle, *decreases* as it flows up the ascending limb, and *decreases* still more as it flows through the rest of the nephron and collecting duct. These changes in osmolarity result from the following conditions along the path of tubular fluid:

1. Because the osmolarity of the interstitial fluid of the renal medulla becomes progressively greater, more and more water is reabsorbed by osmosis as tubular fluid flows along the descending limb toward the tip of the loop. (The source of this medullary osmotic gradient is explained shortly.) As a result, the fluid remaining in the lumen becomes progressively more concentrated.
2. Cells lining the thick ascending limb of the loop of Henle have symporters that actively reabsorb Na^+, K^+, and Cl^- from the tubular fluid (see **FIGURE 19.10**). The ions pass from the tubular fluid into thick ascending limb cells, then into interstitial fluid, and finally some diffuse into the blood inside the vasa recta.
3. Although solutes are being reabsorbed in the thick ascending limb, the water permeability of this portion of the nephron is always quite low, so water cannot follow by osmosis. As solutes—but not water molecules—are leaving the tubular fluid, the osmolarity of the tubular fluid drops to about 150 mosmol/liter. The fluid entering the distal tubule is thus more dilute than plasma.
4. While the fluid continues flowing along the early distal tubule, more solutes, but only a small number of water molecules, are reabsorbed. The cells of the early distal tubule are not very permeable to water and are not regulated by ADH.
5. Finally, the principal cells of the late distal tubule and collecting duct are impermeable to water when the blood concentration of ADH is very low. Because additional solutes but very few water molecules are reabsorbed in these regions when there is a very low blood ADH level, tubular fluid becomes progressively more dilute as it flows onward. By the time the tubular fluid drains into the renal pelvis, its concentration can be as low as 65–70 mosmol/liter. This is four times more dilute than blood plasma or glomerular filtrate.

The Production of Concentrated Urine Allows the Kidneys to Conserve Water

When water intake is low or water loss is high (such as during heavy sweating), the kidneys must conserve water while still eliminating wastes and excess ions. Under the influence of ADH, the kidneys produce a small volume of highly concentrated urine. Urine can be four times more concentrated (up to 1200 mosmol/liter) than blood plasma or glomerular filtrate (300 mosmol/liter).

The ability of ADH to cause excretion of concentrated urine depends on the presence of an **osmotic gradient** of solutes in the in-

FIGURE 19.15 Formation of dilute urine. Numbers indicate osmolarity in milliosmoles per liter (mosmol/liter). Heavy brown lines in the ascending limb of the loop of Henle and in the early distal tubule indicate impermeability to water; heavy blue lines indicate the late distal tubule and the collecting duct, which are impermeable to water in the absence of ADH; light blue areas around the nephron represent interstitial fluid. When ADH is absent, the osmolarity of urine can be as low as 65 mosmol/liter.

When ADH level is low, urine is dilute and has an osmolarity less than the osmolarity of blood.

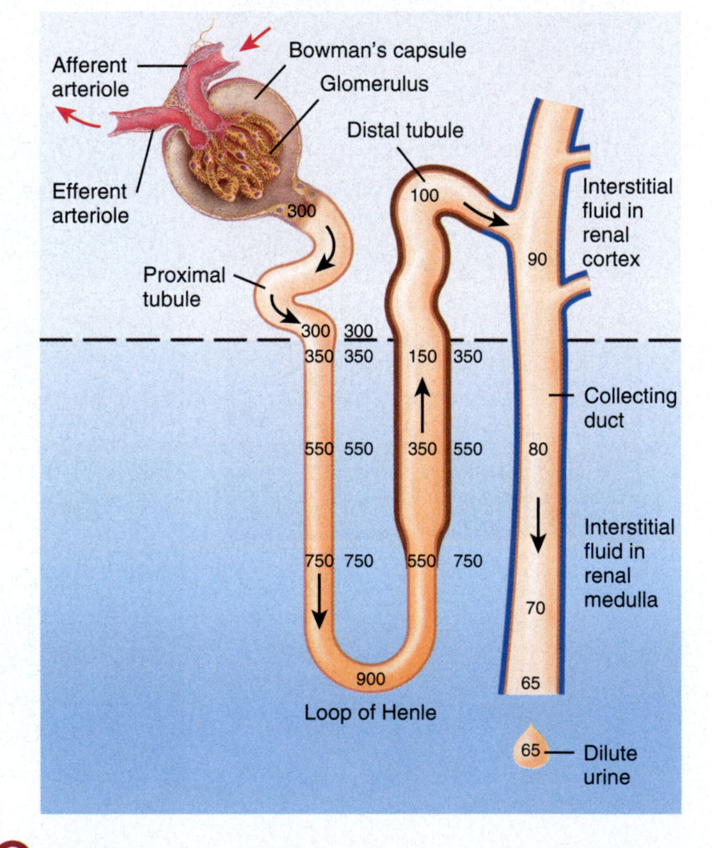

Which portions of the renal tubule and collecting duct reabsorb more solutes than water to produce dilute urine?

terstitial fluid of the renal medulla. Notice in **FIGURE 19.16** that the solute concentration of the interstitial fluid in the kidney increases from about 300 mosmol/liter in the renal cortex to about 1200 mosmol/liter deep in the renal medulla. The three major solutes that contribute to this high osmolarity are Na^+, Cl^-, and urea. Two main factors contribute to building and maintaining this osmotic

FIGURE 19.16 Mechanism of urine concentration in long-loop juxtamedullary nephrons. The green line indicates the presence of Na^+–K^+–$2Cl^-$ symporters that simultaneously reabsorb these ions into the interstitial fluid of the renal medulla; this portion of the nephron is also relatively impermeable to water and urea. All concentrations are in milliosmoles per liter (mosmol/liter).

The formation of concentrated urine depends on high concentrations of solutes in interstitial fluid in the renal medulla.

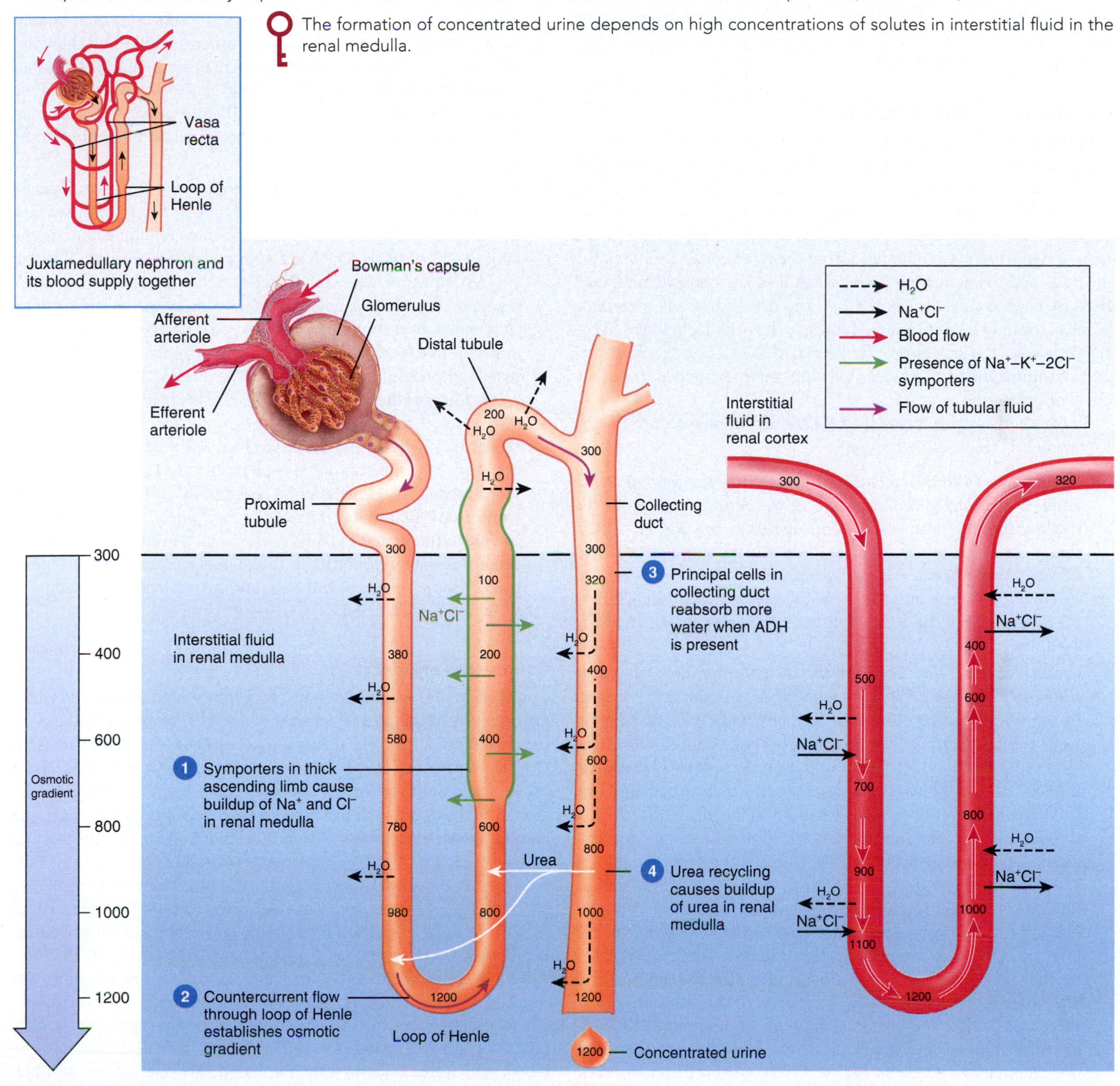

(a) Reabsorption of Na^+, Cl^-, and water in long-loop juxtamedullary nephron

(b) Recycling of salts and urea in vasa recta

Which solutes are the main contributors to the high osmolarity of interstitial fluid in the renal medulla?

gradient: (1) differences in solute and water permeability and reabsorption in different sections of the long loops of Henle and the collecting ducts, and (2) the countercurrent flow of fluid through tube-shaped structures in the renal medulla. *Countercurrent flow* refers to the flow of fluid in opposite directions. This occurs when fluid flowing in one tube runs counter (opposite) to fluid flowing in a nearby parallel tube. Examples of countercurrent flow include the flow of tubular fluid through the descending and ascending limbs of the loop of Henle and the flow of blood through the descending and ascending parts of the vasa recta. Two types of **countercurrent mechanisms** exist in the kidneys: countercurrent multiplication and countercurrent exchange.

Countercurrent Multiplication

Countercurrent multiplication is the process by which a progressively increasing osmotic gradient is formed in the interstitial fluid of the renal medulla as a result of countercurrent flow. Countercurrent multiplication involves the long loops of Henle of juxtamedullary nephrons. Note in FIGURE 19.16a that the descending limb of the loop of Henle carries tubular fluid from the renal cortex deep into the medulla, and the ascending limb carries it in the opposite direction. Because countercurrent flow through the descending and ascending limbs of the long loop of Henle establishes the osmotic gradient in the renal medulla, the long loop of Henle is said to function as a **countercurrent multiplier**. The kidneys use this osmotic gradient to excrete concentrated urine.

Production of concentrated urine by the kidneys occurs in the following way (see FIGURE 19.16):

1. ***Symporters in thick ascending limb cells of the loop of Henle cause a buildup of Na^+ and Cl^- in the renal medulla.*** In the thick ascending limb of the loop of Henle, the Na^+–K^+–$2Cl^-$ symporters reabsorb Na^+ and Cl^- from the tubular fluid (FIGURE 19.16a). Water is not reabsorbed in this segment, however, because the cells are impermeable to water. As a result, there is a buildup of Na^+ and Cl^- ions in the interstitial fluid of the medulla.
2. ***Countercurrent flow through the descending and ascending limbs of the loop of Henle establishes an osmotic gradient in the renal medulla.*** Because tubular fluid constantly moves from the descending limb to the thick ascending limb of the loop of Henle, the thick ascending limb is constantly reabsorbing Na^+ and Cl^- ions. Consequently, the reabsorbed Na^+ and Cl^- become increasingly concentrated in the interstitial fluid of the medulla, which results in the formation of an osmotic gradient that ranges from 300 mosmol/liter in the outer medulla to 1200 mosmol/liter deep in the inner medulla. The descending limb of the loop of Henle is very permeable to water but impermeable to solutes except urea. Because the osmolarity of the interstitial fluid outside the descending limb is higher than the tubular fluid within it, water moves out of the descending limb via osmosis. This causes the osmolarity of the tubular fluid to increase. As the fluid continues along the descending limb, its osmolarity increases even more: At the hairpin turn of the loop, the osmolarity can be as high as 1200 mosmol/liter in juxtamedullary nephrons. As you have already learned, the ascending limb of the loop is impermeable to water, but its symporters reabsorb Na^+ and Cl^- from the tubular fluid into the interstitial fluid of the renal medulla, so the osmolarity of the tubular fluid progressively decreases as it flows through the ascending limb. At the junction of the medulla and cortex, the osmolarity of the tubular fluid has fallen to about 100 mosmol/liter. Overall, tubular fluid becomes progressively more concentrated as it flows along the descending limb and progressively more dilute as it moves along the ascending limb.
3. ***Cells in the collecting ducts reabsorb more water and urea.*** When ADH increases the water permeability of the principal cells, water quickly moves via osmosis out of the collecting duct tubular fluid, into the interstitial fluid of the inner medulla, and then into the vasa recta. With loss of water, the urea left behind in the tubular fluid of the collecting duct becomes increasingly concentrated. Because duct cells deep in the medulla are permeable to it, urea diffuses from the fluid in the duct into the interstitial fluid of the medulla.
4. ***Urea recycling causes a buildup of urea in the renal medulla.*** As urea accumulates in the interstitial fluid, some of it diffuses into the tubular fluid in the descending and thin ascending limbs of the long loops of Henle, which are also permeable to urea (FIGURE 19.16a). However, while the fluid flows through the thick ascending limb, distal tubule, and cortical portion of the collecting duct, urea remains in the lumen because cells in these segments are impermeable to it. As fluid flows along the collecting duct, water reabsorption continues via osmosis because ADH is present. This water reabsorption *further increases* the concentration of urea in the tubular fluid, more urea diffuses into the interstitial fluid of the inner renal medulla, and the cycle repeats. The constant transfer of urea between segments of the renal tubule and the interstitial fluid of the medulla is termed *urea recycling*. In this way, reabsorption of water from the tubular fluid of the ducts promotes the buildup of urea in the interstitial fluid of the renal medulla, which in turn promotes water reabsorption. The solutes left behind in the lumen thus become very concentrated, and a small volume of concentrated urine is excreted.

Countercurrent Exchange

Countercurrent exchange is the process by which solutes and water are passively exchanged between the blood of the vasa recta and interstitial fluid of the renal medulla as a result of countercurrent flow. Note in FIGURE 19.16b that the vasa recta also consists of descending and ascending limbs that are parallel to each other and to the loop of Henle. Just as tubular fluid flows in opposite directions in the loop of Henle, blood flows in opposite directions in the descending and ascending parts of the vasa recta. Because countercurrent flow between the descending and ascending limbs of the vasa recta allows for exchange of solutes and water between the blood and interstitial fluid of the renal medulla, the vasa recta is said to function as a **countercurrent exchanger**.

Blood entering the vasa recta has an osmolarity of about 300 mosmol/liter. As it flows along the descending part into the renal medulla, where the interstitial fluid becomes increasingly concentrated, Na^+, Cl^-, and urea diffuse from interstitial fluid into the blood, and water diffuses from the blood into the interstitial fluid. But after its osmolarity increases, the blood flows into the ascending part of the vasa recta. Here, blood flows through a region where the interstitial fluid becomes increasingly less concentrated.

Critical Thinking

A Tale of Two Students

Brenda and Cecily are two senior college students who are just a few months from graduation. It is Saturday night and Brenda has decided that she will be staying in and studying for a major physiology exam on Monday. She has planned her evening right down to the fresh coffee she just bought. Cecily has the same physiology exam, but she considers Saturday night as her only time to have fun and reduce stress. Cecily has already consumed a couple of alcoholic beverages in her apartment as she and her friends wait to go out.

Goldmund Luki / Getty Images

Sergey Nivens / Shutterstock

Brenda finishes the barista coffee as she goes over her lecture notes. She has several chapters to get through, so she decides to brew a whole pot of coffee to keep her awake. Each cup of coffee has 100–160 mg caffeine, with a half-life of 5.7 hours, and Brenda keeps on drinking more as she studies. Thus, her blood level of caffeine is increasing. Soon Brenda feels the need to urinate. As she returns to her studies, Brenda refills her coffee and continues to sip as she reads. Within a short time, Brenda once again has to visit the bathroom. The frequency of Brenda's bathroom visits has increased to a point where it seems to Brenda that she is spending more time urinating than she is studying.

Cecily has arrived at the club and soon is dancing and having a great time. She has had several alcoholic drinks between dances in addition to those she had earlier in the evening. Like Brenda, Cecily finds herself going to the restroom at an increasing frequency. She even had to leave the dance floor before one of her favorite songs was over.

SOME THINGS TO KEEP IN MIND:

Caffeine and alcohol are both diuretics, substances that promote diuresis (an increase in urine flow rate). It takes about 250–300 mg of caffeine (two to three cups of coffee) to have a significant diuretic effect. It doesn't take that much alcohol to have a significant effect on diuresis: For every 1 gram of alcohol consumed, urine output increases 10 ml.

Is Cecily producing large amounts of urine merely because of the increase in volume from her alcoholic drinks?

Brenda is quietly studying and therefore unlikely to be losing any body water due to sweating. Could the increase in her urine output be due to simple increases in blood volume from drinking large amounts of coffee?

What is the mechanism by which alcohol causes diuresis?

Is the mechanism of caffeine-induced diuresis the same as the mechanism of alcohol-induced diuresis?

As a result, Na^+, Cl^-, and urea diffuse from the blood back into interstitial fluid, and water diffuses from interstitial fluid back into the vasa recta. The osmolarity of blood leaving the vasa recta is only slightly higher than the osmolarity of blood entering the vasa recta. Thus, the vasa recta provides oxygen and nutrients to the renal medulla without washing out or diminishing the osmotic gradient. Whereas the long loop of Henle *establishes* the osmotic gradient in the renal medulla by countercurrent multiplication, the vasa recta *maintains* the osmotic gradient in the renal medulla by countercurrent exchange.

CLINICAL CONNECTION

Diuretics

Diuretics are substances that slow renal reabsorption of water and thereby cause diuresis, an elevated urine flow rate, which in turn reduces blood volume. Diuretic drugs are often prescribed to treat hypertension (high blood pressure) because lowering blood volume usually reduces blood pressure. Naturally occurring diuretics include caffeine in coffee, tea, and sodas, which inhibits Na^+ reabsorption, and alcohol in beer, wine, and mixed drinks, which inhibits secretion of ADH. Most diuretic drugs act by interfering with a mechanism for reabsorption of filtered Na^+. For example, loop diuretics, such as furosemide (Lasix®), selectively inhibit the Na^+–K^+–$2Cl^-$ symporters in the thick ascending limb of the loop of Henle (see **FIGURE 19.10**). The thiazide diuretics, such as chlorthiazide (Diuril®), act in the distal tubule, where they promote loss of Na^+ and Cl^- in the urine by inhibiting Na^+–Cl^- symporters.

CHECKPOINT

14. How do symporters in the ascending limb of the loop of Henle and principal cells in the collecting duct contribute to the formation of concentrated urine?
15. How does ADH regulate facultative water reabsorption?
16. What are the two countercurrent mechanisms? Why are they important?

19.7 Evaluation of Kidney Function

OBJECTIVE

- Describe the various methods used to evaluate kidney function.

Routine assessment of kidney function involves evaluating both the quantity and quality of urine and the levels of wastes in the blood.

A Urinalysis Examines the Various Properties of Urine

An analysis of physical, chemical, and microscopic properties of urine, called a **urinalysis**, reveals much about the state of the body. **TABLE 19.4** summarizes the major characteristics of normal urine. The average volume of urine eliminated per day in a normal adult is about 1.5–2 liters. However, urine volume can vary depending on a variety of factors, such as fluid intake, diet, blood pressure, and blood osmolarity.

Water accounts for about 95% of the total volume of urine. The remaining 5% consists of electrolytes, solutes derived from cellular metabolism, and exogenous substances such as drugs. Normal urine is virtually free of proteins and blood cells. Typical solutes normally present in urine include filtered and secreted electrolytes that are not reabsorbed; urea (from breakdown of proteins); creatinine (from breakdown of creatine phosphate in muscle fibers); uric acid (from breakdown of nucleic acids); urobilin (from breakdown of hemoglobin); and small quantities of other substances, such as fatty acids, pigments, enzymes, and hormones.

If disease alters body metabolism or kidney function, traces of substances not normally present may appear in the urine, or normal constituents may appear in abnormal amounts. **TABLE 19.5** lists several abnormal constituents in urine that may be detected as part of a urinalysis. To learn how a urinalysis is performed, see Tool of the Trade: Performing a Urinalysis.

Blood Tests for Renal Function Include the Blood Urea Nitrogen Test and Measurement of Plasma Creatinine

Two blood-screening tests can provide information about kidney function. One is the **blood urea nitrogen (BUN)** test, which measures the blood nitrogen that is part of the urea resulting from catabolism

TABLE 19.4 Characteristics of Normal Urine

Characteristic	Description
Volume	1.5–2.0 liters in 24 hours but varies considerably.
Color	Yellow or amber but varies with urine concentration and diet. Color is due to urochrome (pigment produced from breakdown of bile) and urobilin (from breakdown of hemoglobin). Concentrated urine is darker in color. Diet (reddish-colored urine from beets), medications, and certain diseases affect color. Kidney stones may produce blood in urine.
Turbidity	Transparent when freshly voided but becomes turbid (cloudy) upon standing.
Odor	Mildly aromatic but becomes ammonia-like upon standing. Some people inherit the ability to form methylmercaptan from digested asparagus that gives urine a characteristic odor. Urine of diabetics has a fruity odor due to presence of ketone bodies.
pH	Ranges between 4.6 and 8.0; average 6.0; varies considerably with diet. High-protein diets increase acidity; vegetarian diets increase alkalinity.
Specific gravity	Specific gravity (density) is the ratio of the weight of a volume of a substance to the weight of an equal volume of distilled water. In urine, it ranges from 1.001 to 1.035. The higher the concentration of solutes, the higher the specific gravity.

TABLE 19.5 Summary of Abnormal Constituents in Urine

Abnormal Constituent	Comments
Albumin	A normal constituent of plasma, it usually appears in only very small amounts in urine because negatively charged glycoproteins in the filtration membrane oppose its filtration. The presence of excessive albumin in the urine—**albuminuria** (al′-bū-mi-NOO-rē-a)—indicates an increase in the permeability of filtration membranes due to injury or disease, increased blood pressure, or irritation of kidney cells by substances such as bacterial toxins, ether, or heavy metals.
Glucose	The presence of glucose in the urine is called **glucosuria** (gloo-kō-SOO-rē-a) and usually indicates diabetes mellitus. Occasionally it may be caused by stress, which can cause excessive amounts of epinephrine to be secreted. Epinephrine stimulates the breakdown of glycogen and liberation of glucose from the liver.
Erythrocytes	The presence of erythrocytes in the urine is called **hematuria** (hēm-a-TOO-rē-a) and generally indicates a pathological condition. One cause is acute inflammation of the urinary organs as a result of disease or irritation from kidney stones. Other causes include tumors, trauma, and kidney disease, or possible contamination of the sample by menstrual blood.
Ketone bodies	High levels of ketone bodies in the urine, called **ketonuria** (kē-tō-NOO-rē-a), may indicate diabetes mellitus, anorexia, starvation, or simply too little carbohydrate in the diet.
Bilirubin	When erythrocytes are destroyed by macrophages, the globin portion of hemoglobin is split off and the heme is converted to biliverdin. Most of the biliverdin is converted to bilirubin, which gives bile its major pigmentation. An above-normal level of bilirubin in urine is called **bilirubinuria** (bil′-ē-roo-bi-NOO-rē-a).
Urobilinogen	The presence of urobilinogen (breakdown product of hemoglobin) in urine is called **urobilinogenuria** (ū′-rō-bi-lin′-ō-je-NOO-rē-a). Trace amounts are normal, but elevated urobilinogen may be due to hemolytic or pernicious anemia, infectious hepatitis, biliary obstruction, jaundice, cirrhosis, congestive heart failure, or infectious mononucleosis.
Casts	**Casts** are tiny masses of material that have hardened and assumed the shape of the lumen of the tubule in which they formed. They are then flushed out of the tubule when filtrate builds up behind them. Casts are named after the cells or substances that compose them or based on their appearance. For example, there are leukocyte casts, erythrocyte casts, and epithelial cell casts that contain cells from the walls of the tubules.
Microbes	The number and type of bacteria vary with specific infections in the urinary tract. One of the most common is *E. coli*. The most common fungus to appear in urine is the yeast *Candida albicans*, a cause of vaginitis. The most frequent protozoan seen is *Trichomonas vaginalis*, a cause of vaginitis in females and urethritis in males.

TOOL OF THE TRADE

PERFORMING A URINALYSIS

During a urinalysis, a sample of urine is collected, and then its physical and chemical properties are examined in a variety of ways. To determine the specific gravity of urine, a floating device known as a *urinometer* is placed into the urine sample. To determine the presence of casts (tiny masses of hardened material) and cells in urine, the urine sample is centrifuged and the sediment that forms at the bottom of the centrifuge tube is observed under a microscope. To determine the presence of an abnormal solute in urine, a test strip is placed into the urine sample. *Test strips* contains small squares of reagent paper that change color when they contact specific reagents (chemicals). Test strips can be used to measure the pH of urine and the presence or absence in urine of glucose, albumin (protein), hemoglobin (blood), ketone bodies, urobilinogen, and bilirubin (see the **FIGURE**).

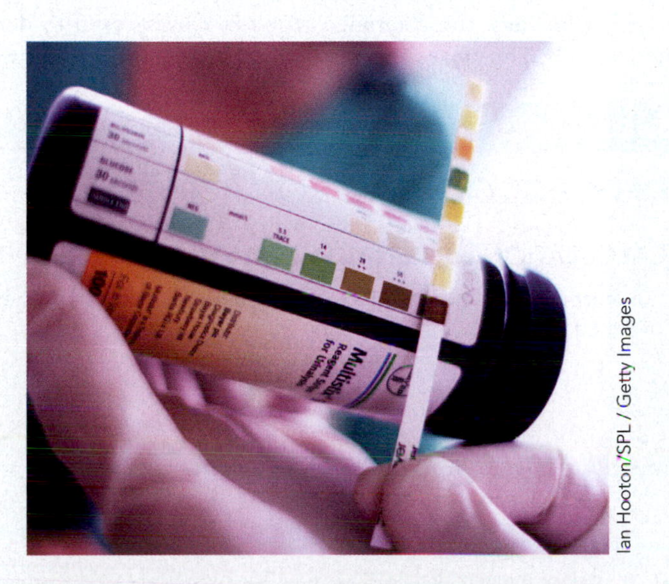

Ian Hooton/SPL / Getty Images

Test stick previously placed in urine is compared to a chart to detect abnormal constituents in urine.

and deamination of amino acids. When glomerular filtration rate decreases severely, as may occur with renal disease or obstruction of the urinary tract, BUN rises steeply. One strategy in treating such patients is to minimize their protein intake, thereby reducing the rate of urea production.

Another test often used to evaluate kidney function is measurement of **plasma creatinine**, which results from catabolism of creatine phosphate in skeletal muscle. Normally, the blood creatinine level remains steady because the rate of creatinine excretion in the urine equals its discharge from muscle. A creatinine level above 1.5 mg/dL (135 mmol/liter) is usually an indication of poor renal function. Normal values for selected blood tests are listed in Appendix C, along with situations that may cause the values to increase or decrease.

Renal Plasma Clearance Indicates How Effectively the Kidneys Are Removing a Substance from Blood Plasma

Even more useful than BUN and blood creatinine values in the diagnosis of kidney problems is an evaluation of how effectively the kidneys are removing a given substance from blood plasma and excreting it into urine. **Renal plasma clearance** is the volume of plasma that is cleared of a substance by the kidneys per unit of time, usually expressed in units of *milliliters per minute.*

The clearance of a substance (X) depends on what happens to that substance once it has been filtered. As substance X passes through the renal tubule and collecting duct, it may undergo net reabsorption, net secretion, or neither one of these processes. As a result, the clearance of substance X may be equal to the GFR, less than the GFR, or greater than the GFR.

If substance X is filtered and is neither reabsorbed nor secreted, then the clearance of substance X equals the GFR (**FIGURE 19.17a**). For example, the plant polysaccharide **inulin** is a substance that is filtered but neither reabsorbed nor secreted. (Do not confuse inulin with the hormone insulin, which is produced by the pancreas.) Typically, the clearance of inulin is about 125 mL/min, which equals the GFR. Clinically, the clearance of inulin can be used to determine the GFR. The clearance of inulin is obtained in the following way: Inulin is administered intravenously and then the concentrations of inulin in plasma and urine are measured along with the urine flow rate. Although using the clearance of inulin is an accurate method for determining the GFR, it has a few drawbacks: Inulin is not produced by the body and it must be infused continuously while clearance measurements are being determined. Measuring the creatinine clearance, however, is an easier way to assess the GFR because creatinine is a substance that is naturally produced by the body as an end product of muscle metabolism. Once creatinine is filtered, it is not reabsorbed, and it is secreted only to a very small extent. Because there is a small amount of creatinine secretion, the creatinine clearance is only a close estimate of the GFR and is not as accurate as using the inulin clearance to determine the GFR.

If substance X is filtered and undergoes net reabsorption, then the clearance of substance X is less than the GFR (**FIGURE 19.17b**). For example, glucose is filtered and then completely reabsorbed; its clearance is zero. The waste product urea is filtered and then undergoes net reabsorption; its clearance typically is less than the GFR, about 70 mL/min.

If substance X is filtered and undergoes net secretion, then the clearance of substance X is greater than the GFR (**FIGURE 19.17c**). As an example, consider the organic anion **para-aminohippuric acid (PAH)**. After PAH is administered intravenously, it is filtered and secreted in nearly a single pass through the kidneys; its clearance is greater than the GFR, about 585 mL/min.

Renal Failure Occurs Because of Inadequate Kidney Function

Renal failure refers to the inability of the kidneys to perform the excretory functions required to maintain homeostasis. In **acute renal failure (ARF)**, the kidneys abruptly stop working entirely (or almost entirely). The main feature of ARF is the suppression of urine flow, usually characterized either by *oliguria* (daily urine output between 50 mL and 250 mL), or by *anuria* (daily urine output less than 50 mL). Causes include low blood volume (for example, due to hemorrhage), decreased cardiac output, damaged renal tubules, kidney stones, and certain drugs.

PHYSIOLOGICAL EQUATION

CALCULATION OF RENAL PLASMA CLEARANCE

The **renal plasma clearance** of a substance (X) is calculated using the following equation:

$$C_X = \frac{U_X \times V}{P_X}$$

where C_X is the renal plasma clearance of substance X measured in mL/min,
U_X is the concentration of substance X in urine measured in mg/mL,
V is the urine flow rate measured in mL/min, and
P_X is the concentration of substance X in plasma measured in mg/mL.

For example, if the concentration of substance X in urine is 6 mg/mL and in plasma it is 2 mg/mL, and the urine flow rate is 1 mL/min, then the clearance of substance X is 3 mL/min:

$$C_X = \frac{6 \text{ mg/mL} \times 1 \text{ mL/min}}{2 \text{ mg/mL}}$$
$$= 3 \text{ mL/min.}$$

This means that 3 mL of plasma are cleared of substance X by the kidneys each minute.

FIGURE 19.17 Renal plasma clearance and GFR. (a) Substance X is filtered and neither reabsorbed nor secreted; therefore, the clearance of X equals the GFR. (b) Substance X is filtered and undergoes net reabsorption; as a result, the clearance of X is less than the GFR. (c) Substance X is filtered and undergoes net secretion; consequently, the clearance of X is greater than the GFR.

Renal plasma clearance is the volume of plasma that is cleared of a substance by the kidneys per unit of time.

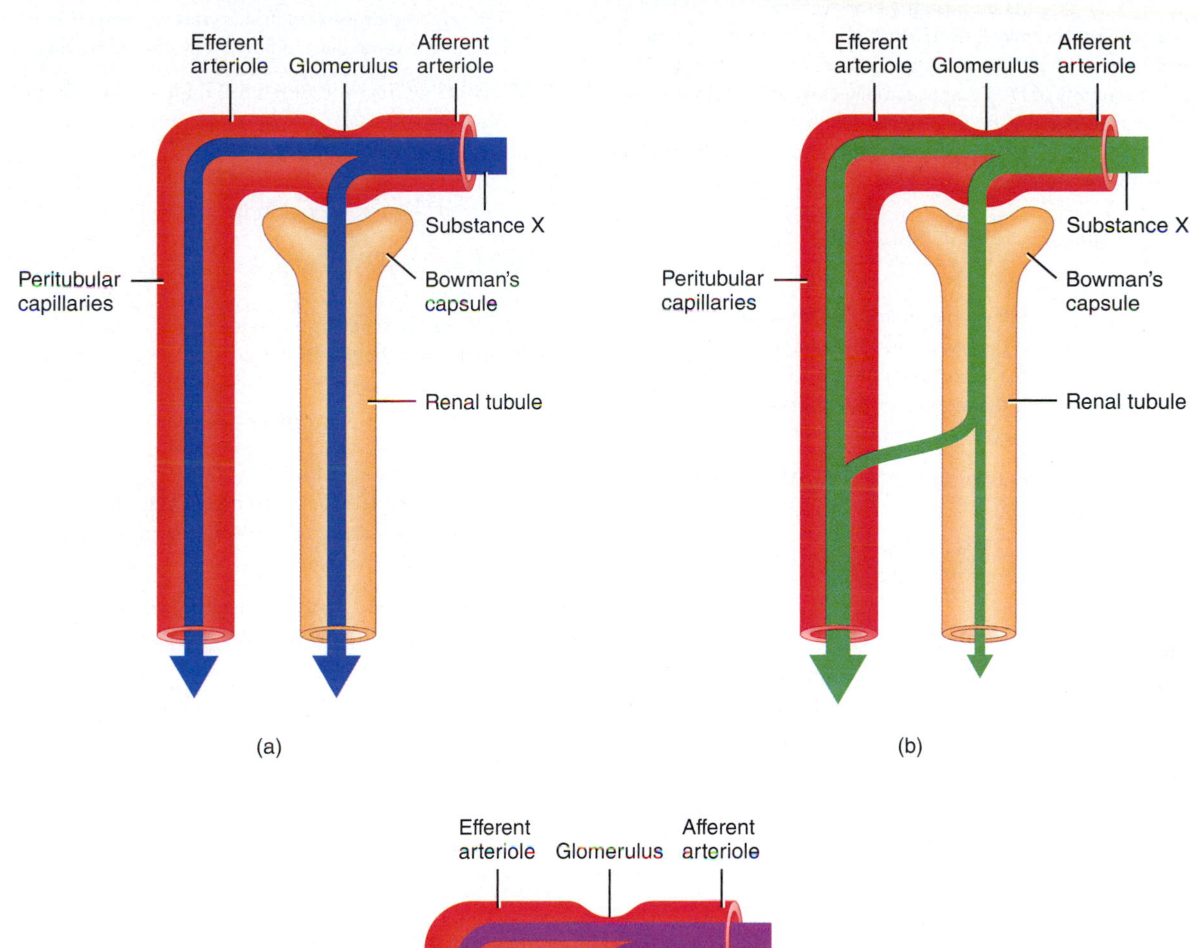

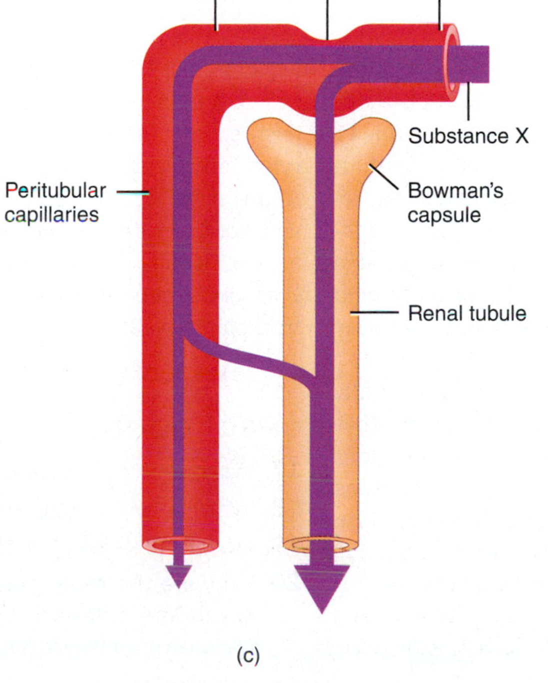

Inulin is a substance that is filtered and neither reabsorbed nor secreted. What does this mean about the clearance of inulin with respect to the GFR?

PHYSIOLOGICAL EQUATION

CALCULATION OF RENAL PLASMA FLOW, FILTRATION FRACTION, AND RENAL BLOOD FLOW

As you have just learned, the organic anion para-aminohippuric acid (PAH) is secreted in nearly a single pass through the kidneys. Thus, the clearance of PAH, which is about 585 mL/min, is used as an approximation of the **renal plasma flow (RPF)**, the amount of plasma that passes through the kidneys in one minute. The amount of PAH that is actually cleared in a single pass through the kidneys is about 90% (or 0.90), a value that is referred to as the **PAH extraction ratio**. Thus, a more accurate estimation of RPF can be made by using the following equation:

$$\text{RPF} = \frac{C_{\text{PAH}}}{\text{PAH extraction ratio}}$$

where RPF is the renal plasma flow measured in mL/min,
C_{PAH} is the renal plasma clearance of PAH measured in mL/min, and
PAH extraction ratio is the amount of PAH cleared in a single pass through the kidneys.

Because the C_{PAH} is typically about 585 mL/min and the PAH extraction ratio is 0.90, the RPF is 650 mL/min:

$$\text{RPF} = \frac{585 \text{ mL/min}}{0.90} = 650 \text{ mL/min}$$

Once GFR is determined using the inulin clearance and RPF is determined using the PAH clearance and PAH extraction ratio, then the **filtration fraction (FF)**, the amount of plasma that undergoes filtration, can be calculated in the following way:

$$\text{FF} = \frac{\text{GFR}}{\text{RPF}}$$

where FF is the filtration fraction,
GFR is the glomerular filtration rate measured in mL/min, and
RPF is the renal plasma flow measured in mL/min.

If GFR is 125 mL/min and RPF is 650 mL/min, then FF is 0.19:

$$\text{FF} = \frac{125 \text{ mL/min}}{650 \text{ mL/min}} = 0.19$$

This means that about 20% of plasma is filtered as it passes through the kidneys per minute.

Once RPF has been clinically determined, it can also be used to calculate the **renal blood flow (RBF)** by using the following equation:

$$\text{RBF} = \frac{\text{RPF}}{1 - \text{HCT}}$$

where RBF is the renal blood flow measured in mL/min,
RPF is the renal plasma flow measured in mL/min, and
HCT is the hematocrit.

If RPF is 650 mL/min and the hematocrit is 0.45 (see Section 16.1 for details about the hematocrit), then RBF is about 1200 mL/min:

$$\text{RBF} = \frac{650 \text{ mL/min}}{1 - 0.45} = 1182 \text{ mL/min}$$

Renal failure causes a multitude of problems. There is edema due to salt and water retention and acidosis due to an inability of the kidneys to excrete acidic substances. In the blood, urea builds up due to impaired renal excretion of metabolic waste products and potassium level rises, which can lead to cardiac arrest. Often, there is anemia because the kidneys no longer produce enough erythropoietin for adequate erythrocyte production. Because the kidneys are no longer able to produce calcitriol, which is needed for adequate calcium absorption from the small intestine, osteomalacia (softening of bones) also may occur.

Chronic renal failure (CRF) refers to a progressive and usually irreversible decline in kidney function. CRF may result from other kidney disorders or from traumatic loss of kidney tissue. CRF develops in three stages. In the first stage, *diminished renal reserve*, nephrons are destroyed until about 75% of the functioning nephrons are lost. At this stage, a person may have no signs or symptoms because the remaining nephrons enlarge and take over the function of those that have been lost. Once 75% of the nephrons are lost, the person enters the second stage, called *renal insufficiency*, characterized by a decrease in glomerular filtration rate (GFR) and increased blood levels of nitrogen-containing wastes and creatinine. Also, the kidneys cannot effectively concentrate or dilute the urine. The final stage, called **end-stage renal failure**, occurs when about 90% of the nephrons have been lost. At this stage, GFR diminishes to 10–15% of normal, oliguria is present, and blood levels of nitrogen-containing wastes and creatinine increase further. People with end-stage renal failure need dialysis therapy (see Clinical Connection: Dialysis) and are possible candidates for a kidney transplant operation.

CHECKPOINT

17. What are the characteristics of normal urine?
18. What chemical substances normally are present in urine?
19. How may kidney function be evaluated?
20. Why are the renal plasma clearances of glucose, urea, and creatinine different? How does each clearance compare to glomerular filtration rate?

CLINICAL CONNECTION

Dialysis

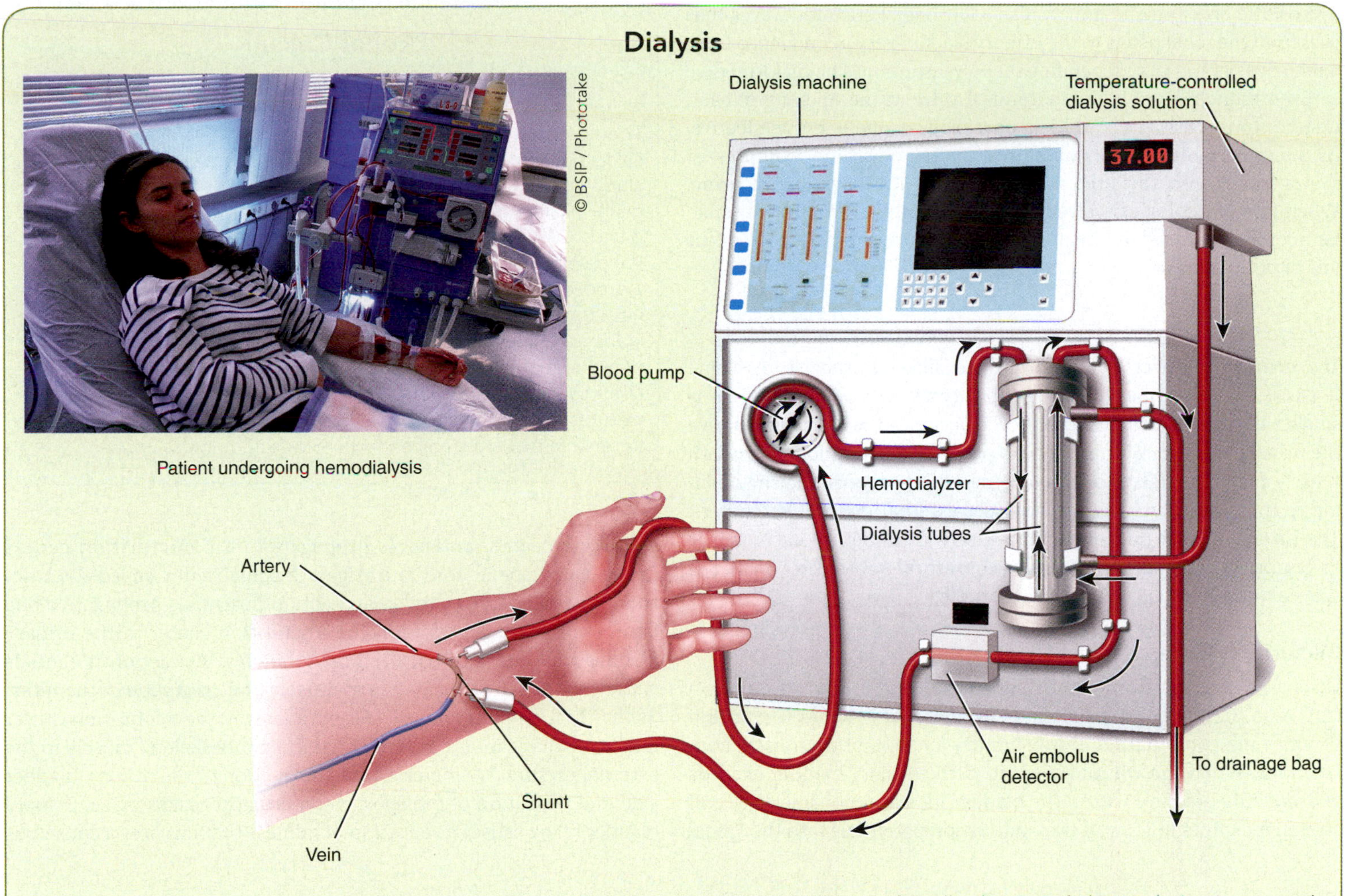

If a person's kidneys are so impaired by disease or injury that they are unable to function adequately, then blood must be cleansed artificially by **dialysis** (dī-AL-i-sis; *dialyo* = to separate), the separation of large solutes from smaller ones by diffusion through a selectively permeable membrane. One method of dialysis is **hemodialysis** (hē-mō-dī-AL-i-sis; *hemo-* = blood), which directly filters the patient's blood by removing wastes and excess electrolytes and fluid and then returning the cleansed blood to the patient. Blood removed from the body is delivered to a *hemodialyzer* (artificial kidney) (see the FIGURE). Inside the hemodialyzer, blood flows through tubing made of *dialysis membrane*, which contains pores large enough to permit the diffusion of small solutes. A dialysis solution is pumped into the hemodialyzer so that it surrounds the dialysis membrane. The dialysis solution is specially formulated to maintain diffusion gradients that remove wastes from the blood (for example, urea, creatinine, uric acid, excess phosphate, potassium, and sulfate ions) and to add needed substances (for example, glucose and bicarbonate ions) to it. The cleansed blood is passed through an air embolus detector to remove air and then returned to the body. An anticoagulant (heparin) is added to prevent blood from clotting in the hemodialyzer. As a rule, most affected people require about 6–12 hours on dialysis each week (roughly every other day).

19.8 Urine Transportation, Storage, and Elimination

OBJECTIVES

- Describe the organs that are involved in urine transportation, storage, and elimination.
- Explain how the micturition reflex operates.

Once urine is formed in the kidneys, it drains into the minor and major calyces and then into the renal pelvis (see **FIGURE 19.1**). From the renal pelvis, urine drains into the ureters and then into the urinary bladder, where it is stored until it is excreted through the urethra.

The Ureters Carry Urine to the Urinary Bladder

Each of the two **ureters** (Ū-re-ters or ū-RĒ-ters) is responsible for transporting urine from the renal pelvis of one kidney to the urinary bladder (**FIGURE 19.18**). Peristaltic contractions of the smooth muscle in the walls of the ureters push urine toward the urinary bladder, but hydrostatic pressure and gravity also contribute. Peristaltic waves that pass from the renal pelvis to the urinary bladder vary in frequency from one to five per minute, depending on how fast urine is being formed.

Even though there is no anatomical valve at the opening of each ureter into the urinary bladder, a physiological one is quite effective. As the urinary bladder fills with urine, pressure within it compresses the openings into the ureters and prevents the backflow of urine. When this physiological valve is not operating properly, it is possible for microbes to travel up the ureters from the urinary bladder to infect one or both kidneys.

The Urinary Bladder Stores Urine

The **urinary bladder** is a hollow, distensible sac that stores urine (**FIGURE 19.18**). Urinary bladder capacity averages 700–800 mL. It is smaller in females because the uterus occupies the space just above the urinary bladder. Within the wall of the urinary bladder is smooth muscle called the **detrusor muscle** (de-TROO-ser). Around the opening to the urethra are the internal and external urethral sphincters. The **internal urethral sphincter** consists of smooth muscle and is involuntarily controlled. The **external urethral sphincter** consists of skeletal muscle and is voluntarily controlled.

Micturition Reflex

Discharge of urine from the urinary bladder is called **micturition** (mik′-too-RISH-un), also known as *urination* or *voiding*. Micturition occurs via a combination of involuntary and voluntary muscle contractions. When the volume of urine in the urinary bladder exceeds 200–400 mL, pressure within the bladder increases considerably, and stretch receptors in its wall transmit action potentials into the spinal cord. These action potentials propagate to the **micturition center** in the sacral spinal cord and trigger a spinal reflex called the **micturition reflex**. In this reflex arc, action potentials propagate along parasympathetic nerves from the micturition center to the urinary bladder wall and internal urethral sphincter. The action potentials cause *contraction* of the detrusor muscle and *relaxation* of the internal urethral sphincter muscle. Simultaneously, the micturition center inhibits somatic motor neurons that innervate skeletal muscle in the external urethral sphincter. Upon contraction of the urinary bladder wall and relaxation of the sphincters, urination takes place. Urinary bladder filling causes a sensation of fullness that initiates a conscious

CLINICAL CONNECTION

Renal Calculi

The crystals of salts present in urine occasionally precipitate and solidify into insoluble stones called **renal calculi** or **kidney stones**. They commonly contain crystals of calcium oxalate, uric acid, or calcium phosphate. Conditions leading to calculus formation include the ingestion of excessive calcium, low water intake, abnormally alkaline or acidic urine, and overactivity of the parathyroid glands. When a stone lodges in a narrow passage, such as a ureter, the pain can be intense. **Shock-wave lithotripsy** (LITH-ō-trip′-sē) is a procedure that uses high-energy shock waves to disintegrate kidney stones and offers an alternative to surgical removal. Once the kidney stone is located using X-rays, a device called a lithotripter delivers brief, high-intensity sound waves through a water- or gel-filled cushion placed under the back. Over a period of 30 to 60 minutes, 1000 or more shock waves pulverize the stone, creating fragments that are small enough to wash out in the urine.

FIGURE 19.18 The ureters, urinary bladder, and urethra.

Urine is stored in the urinary bladder before being expelled via the process of micturition.

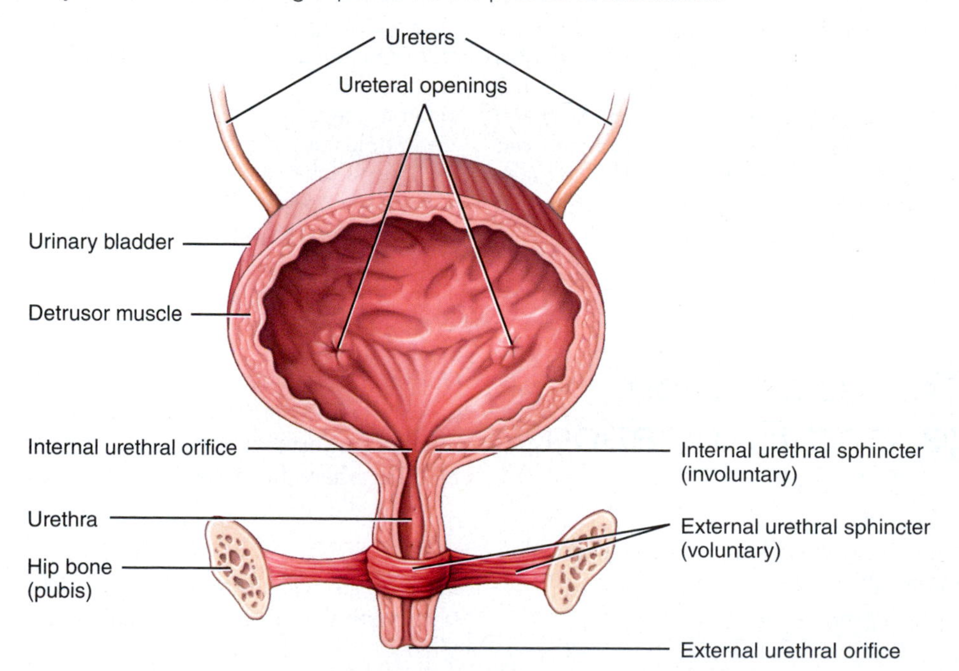

What is a lack of voluntary control over micturition called?

desire to urinate before the micturition reflex actually occurs. Although emptying of the urinary bladder is a reflex, in early childhood we learn to initiate it and stop it voluntarily. Through learned control of the external urethral sphincter muscle and certain muscles of the pelvic floor, the cerebral cortex can initiate micturition or delay its occurrence for a limited period.

The Urethra Conveys Urine to the Outside Environment

The **urethra** (ū-RĒ-thra) is a small tube leading from the floor of the urinary bladder to the exterior of the body (FIGURE 19.18). The opening of the urethra to the exterior is called the **external urethral orifice**, which is located between the clitoris and vaginal opening in a female (see FIGURE 23.8c) and at the tip of the penis in males (see FIGURE 23.2c). In both males and females, the urethra is the terminal portion of the urinary system and the passageway for discharging urine from the body. In males, it also discharges semen (fluid that contains sperm). The length of the urethra in females is about 4 cm (1.5 in.), whereas in males it is longer—about 20 cm (8 in.).

CLINICAL CONNECTION

Urinary Incontinence

A lack of voluntary control over micturition is called **urinary incontinence**. In infants and children under 2–3 years old, incontinence is normal because neurons to the external urethral sphincter muscle are not completely developed; voiding occurs whenever the urinary bladder is sufficiently distended to stimulate the micturition reflex. Urinary incontinence also occurs in adults. There are four types of urinary incontinence—stress, urge, overflow, and functional. **Stress incontinence** is the most common type of incontinence in young and middle-aged females, and results from weakness of the deep muscles of the pelvic floor. As a result, any physical stress that increases abdominal pressure, such as coughing, sneezing, laughing, exercising, straining, lifting heavy objects, and pregnancy, causes leakage of urine from the urinary bladder. **Urge incontinence** is most common in older people and is characterized by an abrupt and intense urge to urinate followed by an involuntary loss of urine. It may be caused by irritation of the urinary bladder wall by infection or kidney stones, stroke, multiple sclerosis, spinal cord injury, or anxiety. **Overflow incontinence** refers to the involuntary leakage of small amounts of urine caused by some type of blockage or weak contractions of the musculature of the urinary bladder. When urine flow is blocked (for example, from an enlarged prostate or stones) or the urinary bladder muscles can no longer contract, the urinary bladder becomes overfilled and the pressure inside increases until small amounts of urine dribble out. **Functional incontinence** is urine loss resulting from the inability to get to a toilet facility in time as a result of conditions such as stroke, severe arthritis, and Alzheimer's disease. Choosing the right treatment option depends on correct diagnosis of the type of incontinence. Treatments include Kegel exercises, urinary bladder training, medication, and possibly even surgery.

CLINICAL CONNECTION

Urinary Tract Infections

The term **urinary tract infection (UTI)** is used to describe either an infection of part of the urinary system or the presence of large numbers of microbes in urine. UTIs are more common in females because their urethras are shorter, allowing the microbes easier access to the other components of the urinary system. In addition, the urethral and anal openings are closer in females. Most first-time UTIs are caused by ***Escherichia coli (E. coli)*** bacteria that have migrated to the urethra from the anal area. Symptoms of UTIs include painful or burning urination, urgent and frequent urination, low back pain, and bed-wetting. Drinking cranberry juice can prevent the attachment of ***E.coli*** to the lining of the urinary bladder so that they are more readily flushed away during urination.

CHECKPOINT

21. What forces help propel urine from the renal pelvis to the urinary bladder?
22. What is micturition? How does the micturition reflex occur?

19.9 Waste Management in Other Body Systems

OBJECTIVE

- Describe the ways that body wastes are handled.

As you have learned, one of the many functions of the urinary system is to help rid the body of some kinds of waste materials. In addition to the kidneys, several other tissues, organs, and processes contribute to the temporary confinement of wastes, the transport of waste materials for disposal, the recycling of materials, and the excretion of excess or toxic substances in the body. These waste management systems include the following:

- ***Body buffers.*** Buffers in body fluids bind excess hydrogen ions (H^+), thereby preventing an increase in the acidity of body fluids. Buffers, like wastebaskets, have a limited capacity; eventually the H^+, like the paper in a wastebasket, must be eliminated from the body by excretion.
- ***Blood.*** The bloodstream provides pickup and delivery services for the transport of wastes, in much the same way that garbage trucks and sewer lines serve a community.
- ***Liver.*** The liver is the primary site for metabolic recycling, as occurs, for example, in the conversion of amino acids into glucose or of glucose into fatty acids. The liver also converts toxic substances into less toxic ones, such as ammonia into urea. These functions of the liver are described in Chapters 20 and 21.
- ***Lungs.*** With each exhalation, the lungs excrete CO_2, and expel heat and a little water vapor.
- ***Sweat (sudoriferous) glands.*** Especially during exercise, sweat glands in the skin help eliminate excess heat, water, and CO_2, plus small quantities of salts and urea.
- ***Gastrointestinal tract.*** Through defecation, the gastrointestinal tract excretes solid, undigested foods; wastes; some CO_2; water; salts; and heat.

CHECKPOINT

23. What roles do the liver and lungs play in the elimination of wastes?

To appreciate the many ways that the urinary system contributes to homeostasis of other body systems, examine *Focus on Homeostasis: Contributions of the Urinary System.*

FOCUS on HOMEOSTASIS

CONTRIBUTIONS OF THE URINARY SYSTEM

FOR ALL BODY SYSTEMS

- Kidneys regulate volume, composition, and pH of body fluids by removing wastes and excess substances from blood and excreting them into urine
- Ureters transport urine from the kidneys to the urinary bladder, which stores urine until it is eliminated through the urethra

INTEGUMENTARY SYSTEM

- Kidneys and skin both contribute to synthesis of calcitriol, the active form of vitamin D

SKELETAL SYSTEM

- Kidneys help adjust levels of blood calcium and phosphates, needed for building extracellular bone matrix

MUSCULAR SYSTEM

- Kidneys help adjust level of blood calcium, needed for contraction of muscle

NERVOUS SYSTEM

- Kidneys regulate blood levels of sodium and potassium, which influence neuronal excitability

ENDOCRINE SYSTEM

- Kidneys participate in synthesis of calcitriol, the active form of vitamin D
- Kidneys release erythropoietin, the hormone that stimulates production of erythrocytes

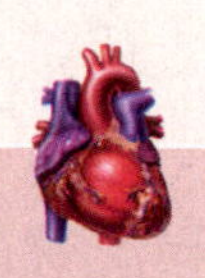

CARDIOVASCULAR SYSTEM

- By increasing or decreasing their reabsorption of water filtered from blood, kidneys help adjust blood volume and blood pressure
- Renin released by juxtaglomerular cells in kidneys raises blood pressure
- Some bilirubin from hemoglobin breakdown is converted to a yellow pigment (urobilin), which is excreted in urine

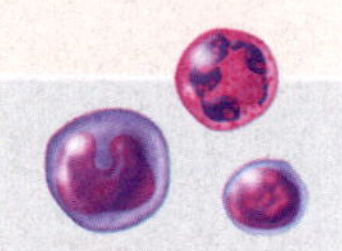

IMMUNE SYSTEM

- The flow of urine helps rid the body of microbes and their toxins

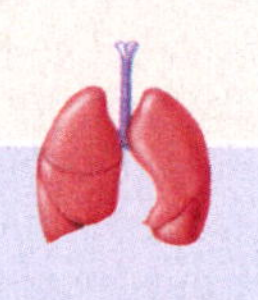

RESPIRATORY SYSTEM

- Kidneys and lungs cooperate in adjusting pH of body fluids

DIGESTIVE SYSTEM

- Kidneys help synthesize calcitriol, the active form of vitamin D, which is needed for absorption of dietary calcium

REPRODUCTIVE SYSTEMS

- In males, the portion of the urethra that extends through prostate and penis is a passageway for semen as well as urine

FROM RESEARCH TO REALITY

Carbon Monoxide and Kidney Transplants

Reference

Yoshida, J. et al. (2010). "*Ex vivo* application of carbon monoxide in UW solution prevents transplant-induced renal ischemia/reperfusion injury in pigs." *American Journal of Transplantation.* 10: 763–772.

Could kidney transplant outcomes be improved using carbon monoxide?

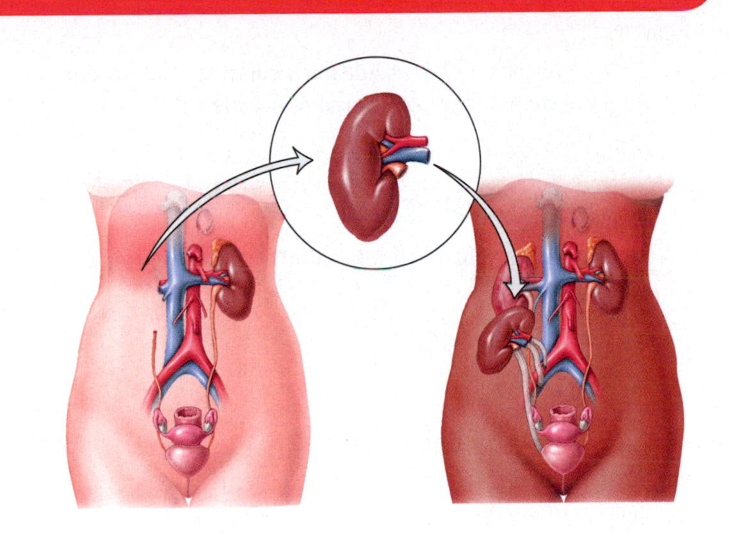

Organ transplants can fail for a number of reasons. When performing an organ transplant, a surgeon tries to take as many precautions as possible to ensure the viability of the organ and the success of the operation, especially because the need for organs is higher than the supply. Recent studies in transplantation indicate that using a solution supplemented with carbon monoxide can have positive impacts on the success of transplantations. Is that also true with kidney transplantation?

Article description:

Researchers wanted to explore the ability of carbon monoxide to enhance organ transplant success in pig kidneys. Kidneys were stored in a common solution for cold storage of organs or in the common solution supplemented with small concentrations of carbon monoxide. Various measures of kidney transplantation success were used to determine whether carbon monoxide improved outcomes.

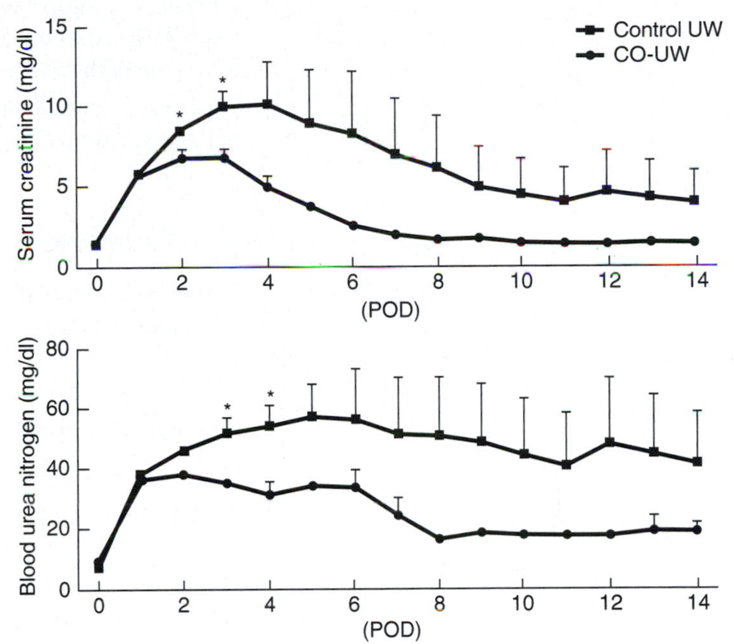

Go to WileyPLUS Learning Space and use the data from this article to answer the questions posed there and to learn more about kidney transplants.

CHAPTER REVIEW

19.1 Overview of Kidney Functions

1. The kidneys excrete wastes, produce the hormones calcitriol and erythropoietin, and maintain blood osmolarity.
2. The kidneys also regulate blood ionic composition, blood volume, blood pressure, and blood pH.

19.2 Organization of the Kidneys

1. The kidneys consist of a renal cortex, renal medulla, minor and major calyces, and a renal pelvis.

2. The nephron is the functional unit of the kidneys. A nephron consists of a renal corpuscle (glomerulus and Bowman's capsule) and a renal tubule. A renal tubule consists of a proximal tubule, a loop of Henle, and a distal tubule, which drains into a collecting duct (shared by several nephrons).
3. Blood flows into the kidney through the renal artery, which divides into smaller arteries that give rise to an afferent arteriole. From the afferent arteriole, blood flows successively into the glomerulus, efferent arteriole, peritubular capillaries, and then into small veins that give rise to the renal vein, which drains the kidney.
4. A cortical nephron has a short loop that dips only into the superficial region of the renal medulla; a juxtamedullary nephron has a long loop of Henle that extends into the deepest region of the renal medulla.
5. The juxtaglomerular apparatus (JGA) consists of the macula densa of the initial portion of the distal tubule and the juxtaglomerular cells of an afferent arteriole.

19.3 Overview of Renal Physiology

1. Nephrons perform three basic tasks: glomerular filtration, tubular reabsorption, and tubular secretion.

19.4 Glomerular Filtration

1. Fluid that enters Bowman's space is glomerular filtrate.
2. The filtration membrane consists of the glomerular endothelium, basement membrane, and filtration slits between pedicels of podocytes.
3. Most substances in blood plasma easily pass through the filtration membrane. However, blood cells and most proteins normally are not filtered.
4. Glomerular filtrate amounts to about 180 liters of fluid per day. This large amount of fluid is filtered because the filtration membrane is porous and thin, the glomerular capillaries are long, and the capillary blood pressure is high.
5. Glomerular capillary hydrostatic pressure (P_{GC}) and Bowman's space colloid osmotic pressure (π_{BS}) promote filtration; plasma colloid osmotic pressure (π_{GC}) and Bowman's space hydrostatic pressure (P_{BS}) oppose filtration. Net filtration pressure (NFP) = $(P_{GC} + \pi_{BS}) - (\pi_{GC} + P_{BS})$. NFP is about 10 mmHg.
6. Glomerular filtration rate (GFR) is the amount of filtrate formed in both kidneys per minute; it is normally 125 mL/min.
7. Glomerular filtration rate depends on renal autoregulation, neural regulation, and hormonal regulation.

19.5 Tubular Reabsorption and Tubular Secretion

1. Tubular reabsorption is a selective process that reclaims materials from tubular fluid and returns them to the bloodstream. Reabsorbed substances include water, glucose, amino acids, urea, and ions, such as sodium, chloride, potassium, bicarbonate, and phosphate.
2. Some substances not needed by the body are removed from the blood and discharged into the urine via tubular secretion. Included are ions (K^+, H^+, and $NH_4{}^+$), urea, creatinine, and certain drugs.
3. Reabsorption routes include both paracellular (between tubule cells) and transcellular (across tubule cells) routes. The maximum amount of a substance that can be reabsorbed per unit time is called the transport maximum (T_m).

4. About 80% of water reabsorption is obligatory; it occurs via osmosis, together with reabsorption of solutes, and is not hormonally regulated. The remaining 20% is facultative water reabsorption, which varies according to body needs and is regulated by ADH.
5. Na^+ ions are reabsorbed throughout the basolateral membrane via primary active transport.
6. In the proximal tubule, sodium ions are reabsorbed through the apical membranes via Na^+–glucose symporters and Na^+/H^+ antiporters; water is reabsorbed via osmosis; Cl^-, K^+, Ca^{2+}, Mg^{2+}, and urea are reabsorbed via passive diffusion; and NH_4^+ is secreted.
7. The loop of Henle reabsorbs 20–30% of the filtered Na^+, K^+, Ca^{2+}, and HCO_3^-; 35% of the filtered Cl^-; and 15% of the filtered water.
8. The distal tubule reabsorbs sodium and chloride ions via Na^+–Cl^- symporters.
9. In the late distal tubule and collecting duct, principal cells reabsorb Na^+ and secrete K^+; intercalated cells reabsorb K^+ and HCO_3^-, and secrete H^+.
10. Antidiuretic hormone, aldosterone, atrial natriuretic peptide, and parathyroid hormone regulate water and solute reabsorption.

19.6 Production of Dilute and Concentrated Urine

1. In the absence of ADH, the kidneys produce dilute urine; renal tubules absorb more solutes than water.
2. In the presence of ADH, the kidneys produce concentrated urine; large amounts of water are reabsorbed from the tubular fluid into interstitial fluid, increasing the solute concentration of the urine.
3. The countercurrent multiplier establishes an osmotic gradient in the interstitial fluid of the renal medulla that enables production of concentrated urine when ADH is present; the gradient is maintained through countercurrent exchange.

19.7 Evaluation of Kidney Function

1. A urinalysis is an analysis of the volume and physical, chemical, and microscopic properties of a urine sample.
2. Chemically, normal urine contains about 95% water and 5% solutes. The solutes normally include urea, creatinine, uric acid, urobilin, and various ions.
3. Several abnormal components can be detected in a urinalysis, including albumin, glucose, erythrocytes, leukocytes, ketone bodies, bilirubin, urobilinogen, casts, and microbes.
4. Renal clearance refers to the volume of plasma that is cleared of a substance by the kidneys per unit of time.
5. If a substance is filtered and is neither reabsorbed nor secreted, then the clearance of that substance equals the GFR; if a substance is filtered and undergoes net reabsorption, then the clearance of that substance is less than the GFR; if a substance is filtered and undergoes net secretion, then the clearance of that substance is greater than the GFR.

19.8 Urine Transportation, Storage, and Elimination

1. The ureters transport urine from the renal pelvis to the urinary bladder, primarily via peristalsis.
2. The urinary bladder is a distensible sac that stores urine.
3. The micturition reflex discharges urine from the urinary bladder.

4. The urethra is a tube leading from the floor of the urinary bladder to the exterior. In both sexes, the urethra discharges urine from the body; in males, it discharges semen as well.

19.9 Waste Management in Other Body Systems

1. In addition to the kidneys, several other tissues, organs, and processes temporarily confine wastes, transport waste materials for disposal, recycle materials, and excrete excess or toxic substances.
2. Buffers bind excess H^+, the blood transports wastes, the liver converts toxic substances into less toxic ones, the lungs exhale CO_2, sweat glands help eliminate excess heat, and the gastrointestinal tract eliminates solid wastes.

PONDER THIS

1. You administer a drug that blocks the kidneys' ability to move useful filtered substances from the nephron lumen back into the blood. Which of the major nephron functions does this drug prevent? Explain your answer.

2. You are working in an ER and you administer a bag of saline that is infused with a large amount of albumin to a patient. As a result, the patient's plasma colloid osmotic pressure has increased significantly above normal. How would this affect the glomerular filtration rate? Explain your answer.

3. Amy is curious to find out how her kidney handles a new drug that her doctor has prescribed. With the help of her uncle who is a physician, she obtains the data below. Use this data to determine if there has been net reabsorption or secretion of the drug, or neither.

[drug] in urine = 10 mg/ml
[drug] in plasma = 145 mg/ml
Urine flow rate = 1mL/min

[inulin] in urine = 300 mg/ml
[inulin] in plasma = 2.5 mg/ml

ANSWERS TO FIGURE QUESTIONS

19.1 Nephrons are the functional units of the kidneys.

19.2 Secreted penicillin is being removed from the bloodstream.

19.3 Endothelial fenestrations (pores) in glomerular capillaries are too small for erythrocytes to pass through.

19.4 Obstruction of the right ureter would increase Bowman's space hydrostatic pressure and thus decrease NFP in the right kidney; the obstruction would have no effect on the left kidney.

19.5 *Auto-* means self; tubuloglomerular feedback is an example of autoregulation because it takes place entirely within the kidneys.

19.6 The tight junctions between tubule cells form a barrier that prevents diffusion of transport proteins between the apical and basolateral membranes.

19.7 Glucose enters a proximal tubule cell via a Na^+–glucose symporter in the apical membrane and leaves via facilitated diffusion through the basolateral membrane.

19.8 The electrochemical gradient promotes movement of Na^+ into the tubule cell through the apical membrane antiporters.

19.9 Reabsorption of the solutes creates an osmotic gradient that promotes the reabsorption of water via osmosis.

19.10 This is considered secondary active transport because the symporter uses the energy stored in the electrochemical gradient of Na^+ between extracellular fluid and the cytosol. No water is reabsorbed here because the thick ascending limb of the loop of Henle is virtually impermeable to water.

19.11 In principal cells, aldosterone stimulates reabsorption of Na^+ and secretion of K^+.

19.12 When the ADH levels decline, the aquaporin-2 channels are removed from the apical membrane via endocytosis.

19.13 Obligatory water reabsorption means that water automatically follows solutes when they are reabsorbed. Facultative water reabsorption means that water reabsorption occurs independently of solute reabsorption and is hormonally regulated based on the body's needs.

19.14 Secretion occurs in the proximal tubule, the loop of Henle, and the late distal tubule and collecting duct.

19.15 Dilute urine is produced when the thick ascending limb of the loop of Henle, the distal tubule, and the collecting duct reabsorb more solutes than water.

19.16 The high osmolarity of interstitial fluid in the renal medulla is due mainly to Na^+, Cl^-, and urea.

19.17 Because inulin is filtered and neither reabsorbed nor secreted, the clearance of inulin equals the GFR.

19.18 Lack of voluntary control over micturition is termed urinary incontinence.

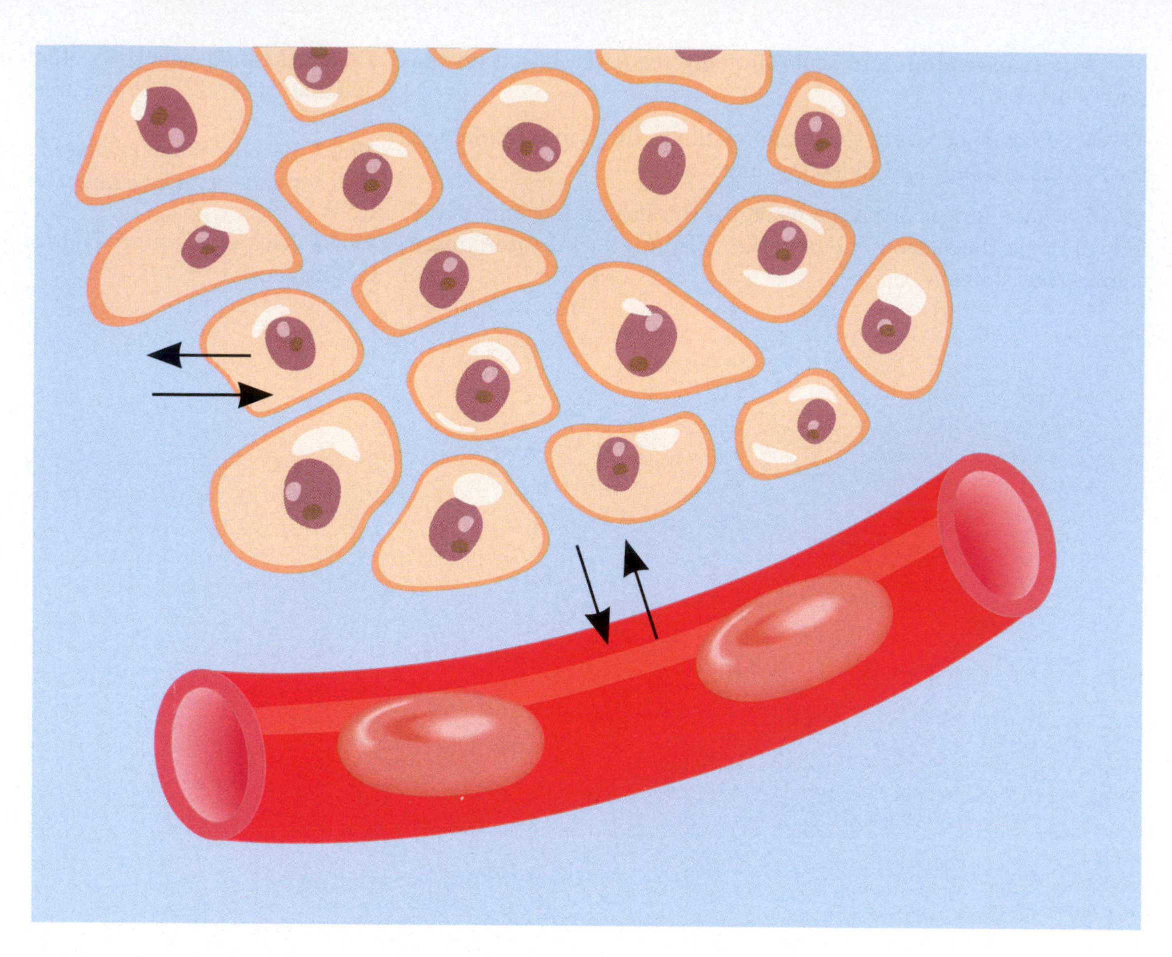

20 Fluid, Electrolyte, and Acid–Base Homeostasis

Fluid, Electrolyte, and Acid–Base Homeostasis

The regulation of the volume and composition of body fluids, their distribution throughout the body, and balancing the pH of body fluids is crucial to maintaining overall homeostasis and health.

LOOKING BACK TO MOVE AHEAD...

- A hypertonic solution causes cells to shrink, a hypotonic solution causes cells to swell, and an isotonic solution does not affect the volume or shape of cells (Section 5.3).
- In the renin–angiotensin–aldosterone pathway, renin is released from the kidneys and subsequently converts angiotensinogen into angiotensin I. Angiotensin-converting enzyme converts angiotensin I to angiotensin II, which stimulates the adrenal glands to release aldosterone (Section 13.5).
- Plasma is the liquid portion of blood; it consists of water, proteins, and other substances such as electrolytes, nutrients, gases, hormones, and wastes (Section 16.1).
- The late distal tubules and the collecting ducts of the kidneys contain two different types of cells: (1) principal cells, which have receptors for both antidiuretic hormone (ADH) and aldosterone, and (2) intercalated cells, which help regulate blood pH (Section 19.5).

In Chapter 19 you learned how the kidneys form urine. One important function of the kidneys is to help maintain fluid balance in the body. The water and dissolved solutes throughout the body constitute the **body fluids**. Regulatory mechanisms involving the kidneys and other organs normally maintain homeostasis of the body fluids. Malfunction in any or all of them may seriously endanger the functioning of organs throughout the body. In this chapter, you will explore the mechanisms that regulate the volume and distribution of body fluids and examine the factors that determine the concentrations of solutes and the pH of body fluids.

20.1 Fluid Compartments and Fluid Balance

OBJECTIVES

- Discuss the various fluid compartments of the body.
- Describe the sources and regulation of water and solute gain and loss.
- Explain how fluids move between compartments.

In lean adults, body fluids constitute between 55% and 60% of total body mass in females and males, respectively (**FIGURE 20.1**). Body fluids are present in two main "compartments"—inside cells and outside cells. About two-thirds of body fluid is **intracellular fluid (ICF)** (*intra-* = within) or **cytosol**, the fluid within cells. The other third, called **extracellular fluid (ECF)** (*extra-* = outside) is outside cells and includes all other body fluids. About 80% of the ECF is **interstitial fluid** (*inter-* = between), which occupies the microscopic spaces between cells, and 20% of the ECF is **plasma**, the liquid portion of the blood. Other extracellular fluids that are grouped with interstitial fluid include lymph in lymphatic vessels; cerebrospinal fluid in the nervous system; synovial fluid in joints; aqueous humor and vitreous humor in the eyes; endolymph and perilymph in the ears; and pleural, pericardial, and peritoneal fluids between the membranes surrounding the lungs, heart, and abdominal organs, respectively.

Two general "barriers" separate intracellular fluid, interstitial fluid, and plasma:

1. The *plasma membrane* of individual cells separates intracellular fluid from the surrounding interstitial fluid. You learned in Chapter 5 that the plasma membrane is a selectively permeable barrier: It

FIGURE 20.1 Body fluid compartments.

The term *body fluid* refers to body water and its dissolved substances.

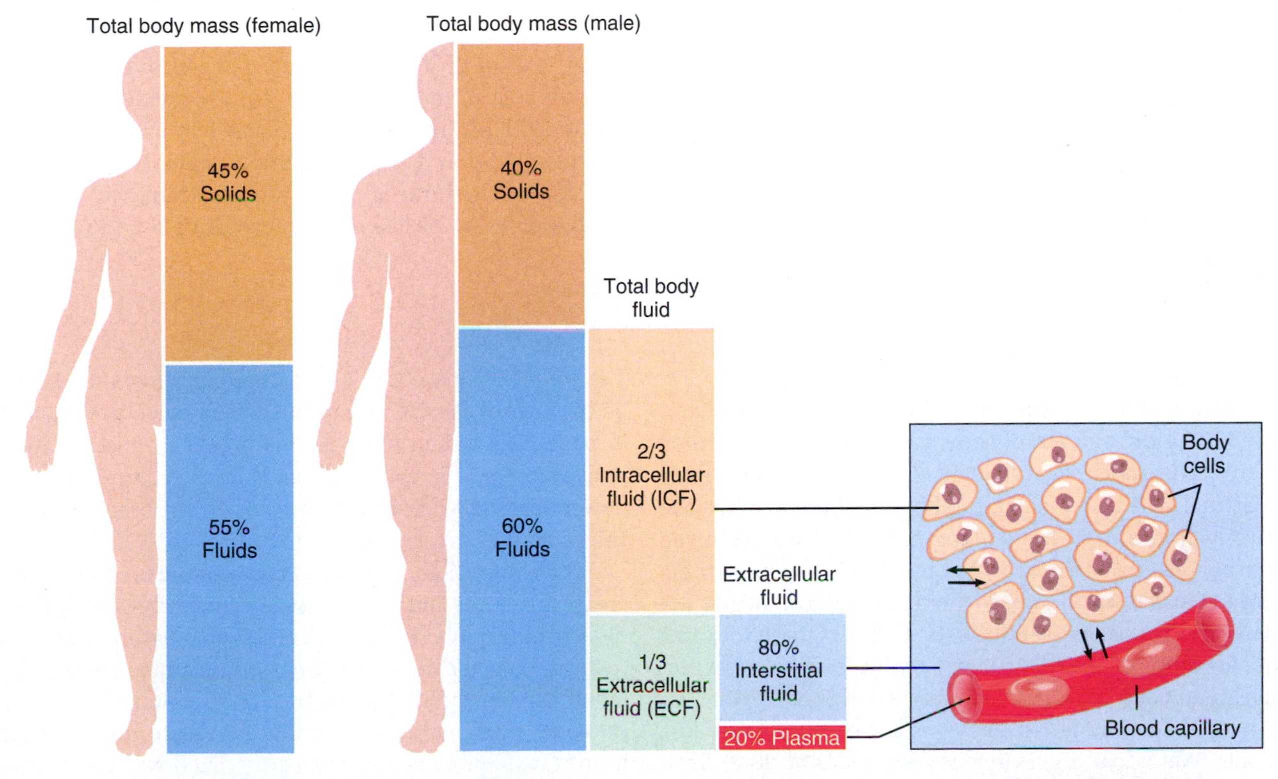

(a) Distribution of body solids and fluids in average lean adult female and male

(b) Exchange of water among body fluid compartments

What is the approximate volume of plasma in a lean 60-kg male? In a lean 60-kg female? (Note: One liter of body fluid has a mass of 1 kilogram.)

allows some substances to cross but blocks the movement of other substances. In addition, active transport pumps work continuously to maintain different concentrations of certain ions in intracellular fluid and interstitial fluid.

2. *Blood vessel walls* divide the interstitial fluid from plasma. Only in capillaries (the smallest blood vessels) and postcapillary venules are the walls thin enough and leaky enough to permit the exchange of water and solutes between plasma and interstitial fluid.

The body is in **fluid balance** when the required amounts of water and solutes are present and are correctly proportioned among the various compartments. **Water** is by far the largest single component of the body, making up 45–75% of total body mass, depending on age, gender, and the amount of adipose tissue (fat) present in the body. Obese people have proportionally less water than leaner people because water comprises less than 20% of the mass of adipose tissue. Skeletal muscle tissue, by contrast, is about 65% water. Infants have the highest percentage of water, up to 75% of body mass. The percentage of body mass that is water decreases until about 2 years of age. Until puberty, water accounts for about 60% of body mass in boys and girls. In lean adult males, water still accounts for about 60% of body mass. However, lean adult females have more subcutaneous fat than do lean adult males. Thus, their percentage of total body water is lower, accounting for about 55% of body mass.

The processes of filtration, reabsorption, diffusion, and osmosis allow continual exchange of water and solutes among body fluid compartments (**FIGURE 20.1b**). Yet the volume of fluid in each compartment remains remarkably stable. The pressures that promote filtration of fluid from blood capillaries and reabsorption of fluid back into capillaries can be reviewed in Section 14.2. Because osmosis is the primary means of water movement between intracellular fluid and interstitial fluid, the concentration of solutes in these fluids determines the *direction* of water movement. Most solutes in body fluids are **electrolytes**, inorganic compounds that dissociate into ions. Therefore, fluid balance and electrolyte balance are closely related. Because intake of water and electrolytes rarely occurs in exactly the same proportions as their presence in body fluids, the ability of the kidneys to excrete excess water by producing dilute urine, or to excrete excess electrolytes by producing concentrated urine, is of utmost importance in the maintenance of homeostasis.

The Body Can Gain or Lose Water

The body can gain water by ingestion and by metabolic synthesis (**FIGURE 20.2**). The main sources of body water are ingested liquids (about 1600 mL) and moist foods (about 700 mL) absorbed from the gastrointestinal (GI) tract, which total about 2300 mL/day. The other source of water is **metabolic water** that is produced in the body mainly when electrons are accepted by oxygen during aerobic respiration (see **FIGURE 4.13**) and, to a smaller extent, during dehydration synthesis reactions (see **FIGURE 2.13**). Metabolic water gain accounts for only 200 mL/day. Daily water gain from these two sources totals about 2500 mL.

Normally, body fluid volume remains constant because water loss equals water gain. Water loss occurs in four ways (**FIGURE 20.2**). Each day the kidneys excrete about 1500 mL in urine, the skin evaporates about 600 mL (400 mL through *insensible perspiration*, sweat that evaporates before it is perceived as moisture, and 200 mL as sweat), the lungs exhale about 300 mL as water vapor, and the gastrointestinal tract eliminates about 100 mL in feces. In women of reproductive age, additional water is lost in menstrual flow. On average, daily water loss totals about 2500 mL. The amount of water lost by a given route can vary considerably over time. For example, water may literally pour from the skin in the form of sweat during strenuous exertion. In other cases, water may be lost in diarrhea during a GI tract infection.

FIGURE 20.2 Sources of daily water gain and loss under normal conditions. Numbers are average volumes for adults.

Normally, daily water loss equals daily water gain.

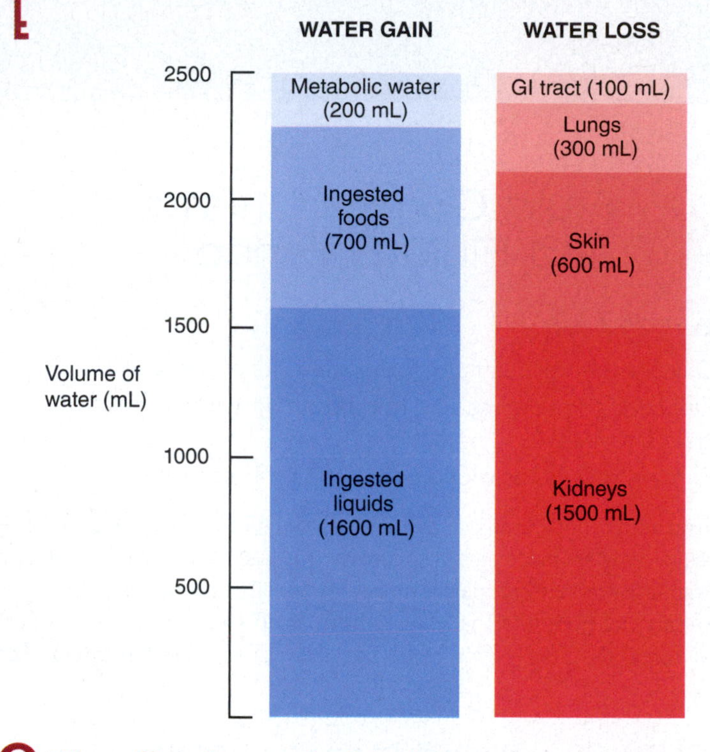

What effect does each of the following have on fluid balance: hyperventilation, vomiting, and diuretics?

Body Water Gain Is Regulated Mainly by the Volume of Water Intake

The volume of metabolic water formed in the body depends entirely on the level of aerobic respiration, which reflects the demand for ATP in body cells. When more ATP is produced, more water is formed. Body water gain is regulated mainly by the volume of water intake, or how much fluid you drink. An area in the hypothalamus known as the **thirst center** governs the urge to drink.

When water loss is greater than water gain, **dehydration**—a decrease in volume and an increase in osmolarity of body fluids—occurs. A decrease in blood volume causes blood pressure to fall. Increased activity from osmoreceptors in the hypothalamus, triggered by increased blood osmolarity, stimulates the thirst center in the hypothalamus (**FIGURE 20.3**). Other signals that stimulate the thirst center come from (1) volume receptors in the atria that detect the decrease in blood volume, (2) baroreceptors in blood vessels that detect the decrease in blood pressure, (3) angiotensin II that is formed due to activation of the renin–angiotensin–aldosterone pathway by the decrease in blood pressure, and (4) neurons in the mouth that detect dryness due to a

FIGURE 20.3 Pathways involved in the thirst response.

A major stimulus that promotes the sensation of thirst is an increase in the osmolarity of body fluids.

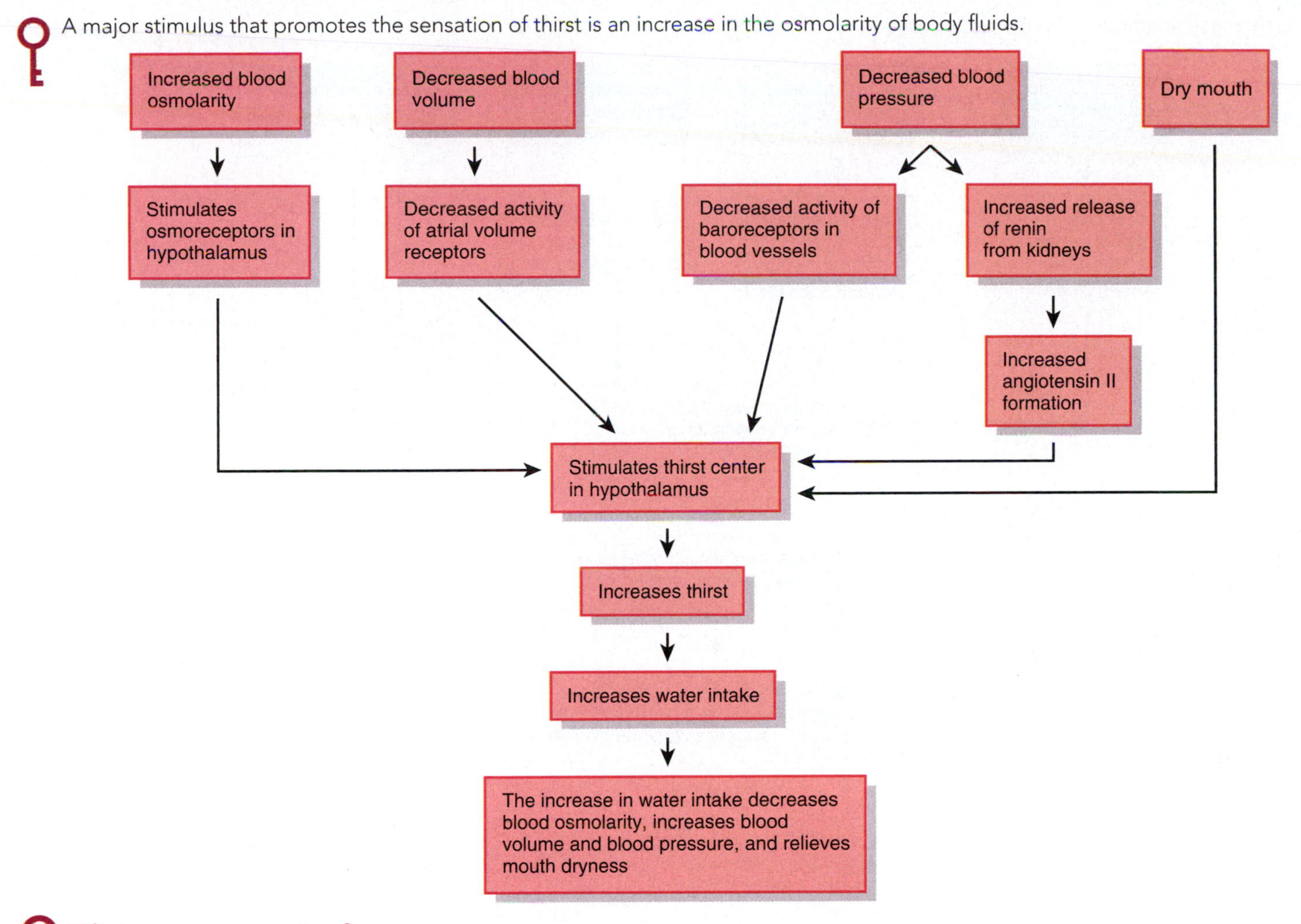

What are osmoreceptors?

decreased flow of saliva. As a result of these stimuli, the sensation of thirst increases, which usually leads to increased fluid intake (as long as fluids are available) and restoration of normal fluid volume.

Body Water or Solute Loss Is Regulated Mainly by Control of Their Loss in Urine

Even though the loss of water and solutes through sweating and exhalation increases during exercise, elimination of *excess* body water or solutes occurs mainly by control of their loss in urine. The extent of *urinary salt (NaCl) loss* is the main factor that determines body fluid *volume*. The reason for this is that "water follows solutes" by osmosis, and the two main solutes in extracellular fluid (and in urine) are sodium ions (Na^+) and chloride ions (Cl^-). In a similar way, the main factor that determines body fluid *osmolarity* is the extent of *urinary water loss*.

The major hormone that regulates water loss is **antidiuretic hormone (ADH)**. This hormone, also known as *vasopressin*, is produced by neurosecretory cells in the hypothalamus and stored in the posterior pituitary gland. When the osmolarity of body fluids increases, osmoreceptors in the hypothalamus not only stimulate thirst; they also increase the synthesis and release of ADH (FIGURE 20.4). ADH promotes the insertion of water-channel proteins (aquaporin-2) into the apical membranes of principal cells in the late distal tubules and collecting ducts of the kidneys (see FIGURE 19.12). As a result, the permeability of these cells to water increases. Water molecules move by osmosis from the renal tubular fluid into the cells and then from the cells into the bloodstream. This results in a decrease in blood osmolarity, an increase in blood volume and blood pressure, and the production of a small volume of concentrated urine. Once the body has adequate water, the ADH level in the bloodstream decreases. As the amount of ADH in the blood declines, some of the aquaporin-2 channels are removed from the apical membrane via endocytosis. Consequently, the water permeability of the principal cells decreases and more water is lost in the urine.

Factors other than blood osmolarity influence ADH secretion (FIGURE 20.4). A decrease in blood volume or blood pressure also stimulates ADH release. Atrial volume receptors detect the decrease in blood volume, and baroreceptors in blood vessels detect the decrease in blood pressure. ADH release is also stimulated by factors that are unrelated to water balance, such as pain, nausea, and stress. Secretion of ADH is inhibited by alcohol, which is why consumption of alcoholic beverages promotes diuresis (voiding large amounts of urine).

Because our daily diet contains a highly variable amount of NaCl, urinary excretion of Na^+ and Cl^- must also vary to maintain homeostasis. Hormones regulate the urinary loss of Na^+ ions. Cl^- ions usually

FIGURE 20.4 Role of antidiuretic hormone (ADH) in water balance.

ADH increases the amount of water reabsorption in the kidneys.

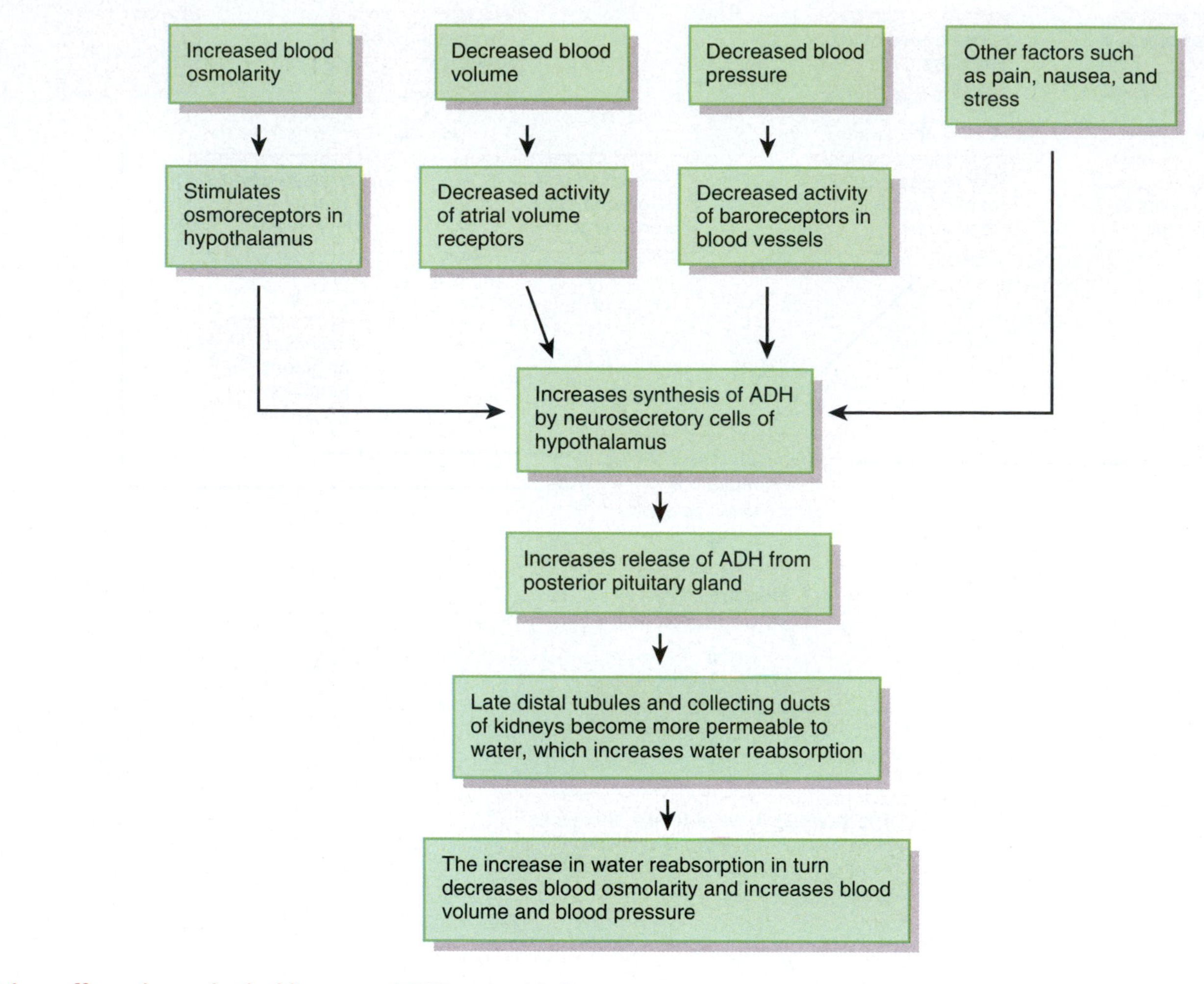

What effect does alcohol have on ADH secretion?

follow Na^+ ions because of electrical attraction or because they are transported along with Na^+ ions via symporters. The two most important hormones that regulate the extent of renal Na^+ reabsorption (and thus how much is lost in the urine) are aldosterone and atrial natriuretic peptide.

1. ***Aldosterone.*** When there is a decrease in blood pressure, which occurs in response to a decrease in blood volume, or when there is a deficiency of Na^+ in the plasma, the kidneys release renin, which activates the renin–angiotensin–aldosterone pathway (FIGURE 20.5). Once aldosterone is formed, it increases Na^+ reabsorption in the late distal tubules and collecting ducts of the kidneys, which relieves the Na^+ deficiency in the plasma. Because antidiuretic hormone (ADH) is also released when blood pressure is low, water reabsorption accompanies Na^+ reabsorption via osmosis. This conserves the volume of body fluids by reducing urinary loss of water.
2. ***Atrial natriuretic peptide.*** An increase in blood volume, as might occur after you finish one or more supersized drinks, stretches the atria of the heart and promotes release of **atrial natriuretic peptide (ANP)** (FIGURE 20.6). ANP promotes **natriuresis,** elevated excretion of Na^+ into the urine. The osmotic consequence of excreting more Na^+ is loss of more water in urine, which decreases blood volume and blood pressure. In addition to stimulating the release of ANP, an increase in blood volume also slows the release of renin from the kidneys. When the renin level declines, less aldosterone is formed, which causes reabsorption of filtered Na^+ to slow in the late distal tubules and collecting ducts of the kidneys. More filtered Na^+ and water (due to osmosis) thus remain in the tubular fluid to be excreted in the urine.

Water Can Move Between Body Fluid Compartments

Normally, the cells of the body neither shrink nor swell because the extracellular fluid that surrounds them is isotonic. This means that intracellular fluid and extracellular fluid have the same osmolarity (concentration of solutes). Changes in the osmolarity of extracellular fluid, however, cause fluid imbalances. If extracellular fluid becomes hypertonic (i.e., it has a greater concentration of solutes than intracellular fluid because its osmolarity has increased), water moves from cells

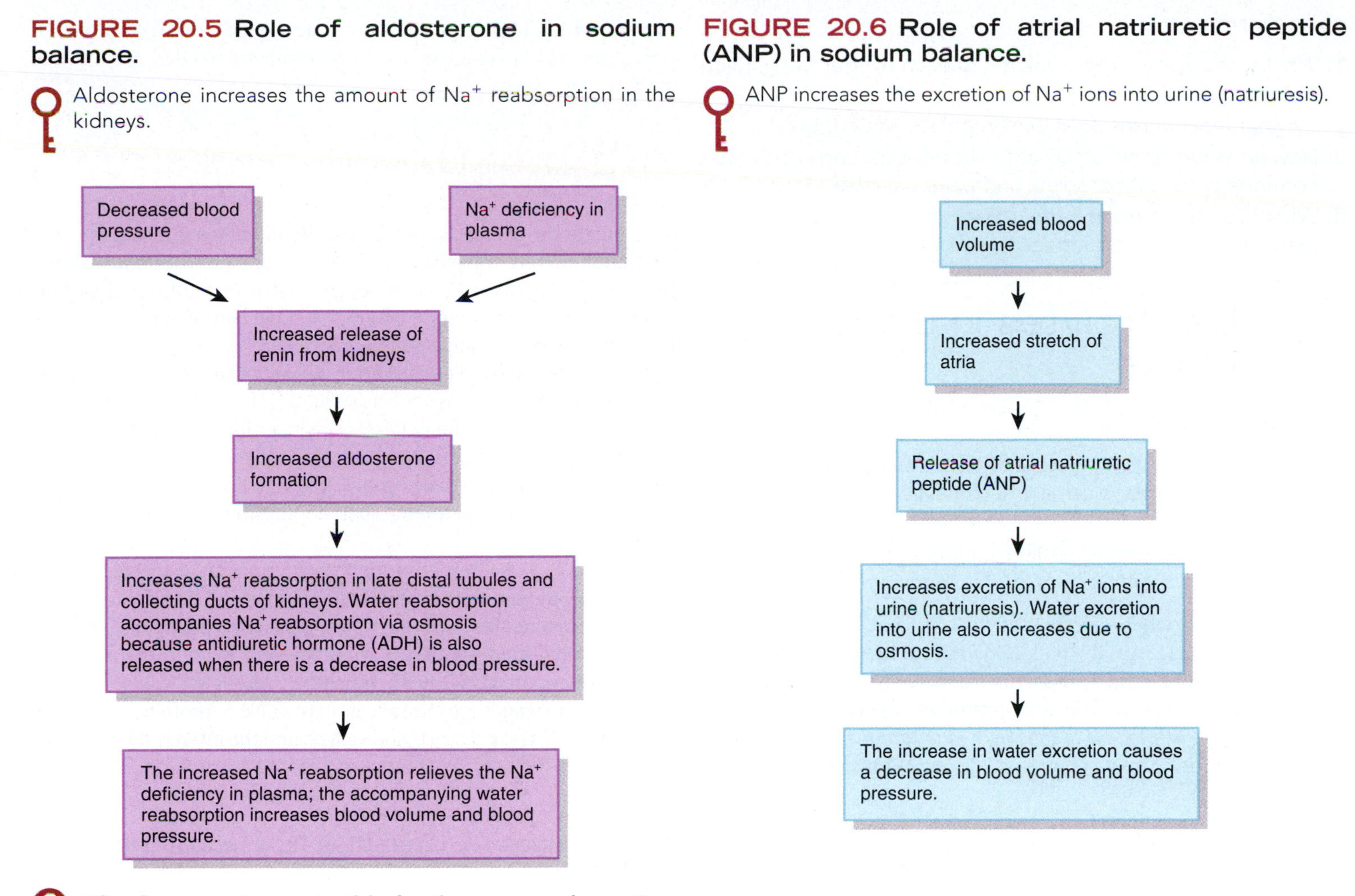

into extracellular fluid by osmosis, causing the cells to shrink. If extracellular fluid becomes hypotonic (i.e., it has a lower concentration of solutes than intracellular fluid because its osmolarity has decreased) water moves from extracellular fluid into cells by osmosis, causing the cells to swell. Changes in osmolarity most often result from changes in the concentrations of Na^+ and Cl^- (the major contributors to osmolarity of extracellular fluid).

An *increase* in the osmolarity of extracellular fluid can occur, for example, after you eat a salty meal. The increased intake of NaCl produces an increase in the levels of Na^+ and Cl^- in extracellular fluid. As a result, the osmolarity of extracellular fluid increases, which causes net movement of water from cells into extracellular fluid. Such water movement shrinks the cells of the body. If neurons of the brain remain in this state for a significant period of time, mental confusion, convulsions, coma, and even death can occur. Body cells usually shrink only slightly and only for a short duration in response to an increase in the osmolarity of extracellular fluid because corrective measures such as the thirst mechanism and secretion of antidiuretic hormone increase the amount of body water, thereby reducing the concentration of solutes in extracellular fluid back to normal levels.

A *decrease* in the osmolarity of extracellular fluid can occur, for example, after drinking a large volume of water. This dilution causes the levels of Na^+ and Cl^- in extracellular fluid to fall below the normal range. When the extracellular concentrations of Na^+ and Cl^- decrease, the osmolarity of extracellular fluid also decreases. The net result is movement of water from extracellular fluid into cells, which causes the cells to swell. Usually when the osmolarity of extracellular fluid decreases, secretion of ADH is inhibited and the kidneys excrete a large volume of dilute urine, which restores the osmolarity of body fluids back to normal. As a result, body cells swell only slightly and only for a brief period. But when a person steadily consumes water faster than the kidneys can excrete it (the maximum urine flow rate is about 15 mL/min) or when renal function is poor, the result may be **water intoxication**, a state in which excessive body water causes cells to swell dangerously. As is the case when neurons of the brain shrink, swelling of the brain's neurons can result in mental confusion, seizures, coma, and possibly death. To prevent this dire sequence of events in cases of severe electrolyte and water loss, solutions given for intravenous or oral rehydration therapy (ORT) include a small amount of table salt (NaCl).

CHECKPOINT

1. What is the approximate volume of each of your body fluid compartments?
2. By what mechanism does thirst regulate water intake?
3. How do aldosterone, atrial natriuretic peptide, and antidiuretic hormone contribute to solute and water balance?
4. What factors control the movement of water between extracellular fluid and intracellular fluid?

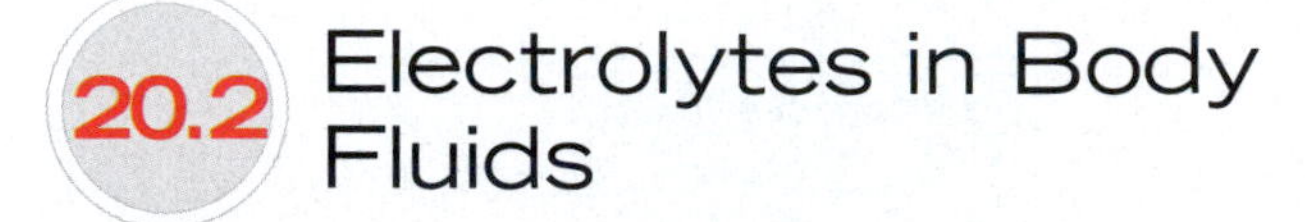

20.2 Electrolytes in Body Fluids

OBJECTIVES

- Compare the electrolyte composition of the three major fluid compartments: plasma, interstitial fluid, and intracellular fluid.
- Discuss the functions of sodium, chloride, potassium, bicarbonate, calcium, phosphate, and magnesium ions.

The ions formed when electrolytes dissolve and dissociate serve four general functions in the body: (1) Because they are largely confined to particular fluid compartments and are more numerous than nonelectrolytes, certain ions *control the osmosis of water between fluid compartments.* (2) Ions *help maintain the acid–base balance* required for normal cellular activities. (3) Ions *carry electrical current,* which allows production of action potentials and graded potentials. (4) Several ions *serve as cofactors* needed for optimal activity of enzymes.

The Electrolyte Concentrations in Body Fluids Can Be Measured in Milliequivalents per Liter

To compare the charge carried by ions in different solutions, the concentration of ions is typically expressed in units of **milliequivalents per liter (mEq/liter).** These units give the concentration of cations or anions in a given volume of solution. One equivalent is a mole of positive or negative charges; a milliequivalent is one-thousandth of an equivalent. Recall that a mole of a substance is its molecular weight expressed in grams. For ions such as sodium (Na^+), potassium (K^+), and bicarbonate (HCO_3^-), which have a single positive or negative charge, the number of mEq/liter is equal to the number of mmol/liter. For ions such as calcium (Ca^{2+}) or phosphate (HPO_4^{2-}), which have two positive or negative charges, the number of mEq/liter is twice the number of mmol/liter.

FIGURE 20.7 compares the concentrations of the main electrolytes and protein anions in plasma, interstitial fluid, and intracellular fluid. The chief difference between the two types of extracellular fluid—plasma and interstitial fluid—is that plasma contains many protein anions, in contrast to interstitial fluid, which has very few. Because normal capillary membranes are virtually impermeable to proteins, only a few plasma proteins leak out of blood vessels into the interstitial fluid. This difference in protein concentration is largely responsible for the

FIGURE 20.7 Electrolyte and protein anion concentrations in plasma, interstitial fluid, and intracellular fluid. The height of each column represents the milliequivalents per liter (mEq/liter).

The electrolytes present in extracellular fluid are different from those present in intracellular fluid.

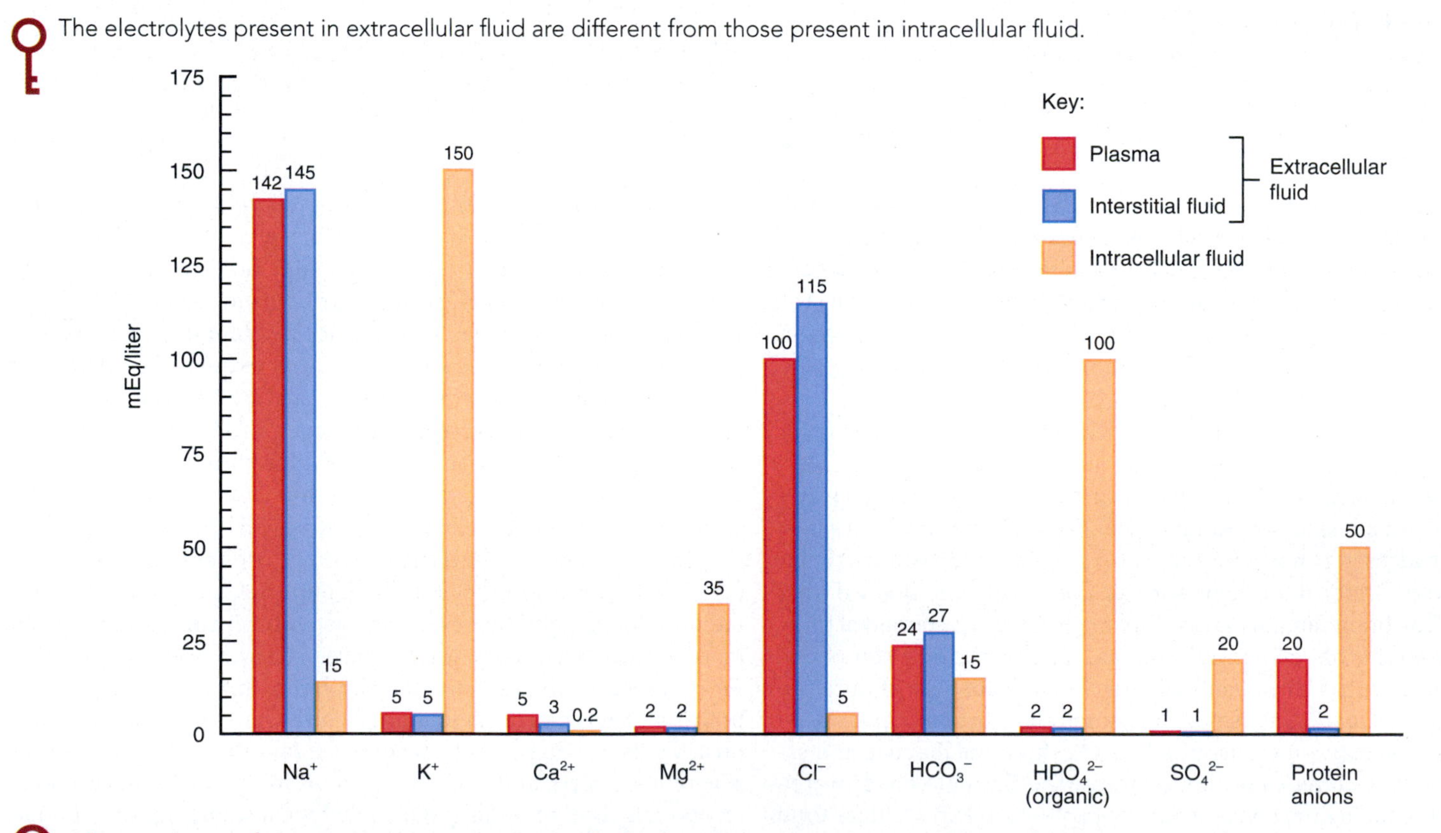

What cation and two anions are present in the highest concentrations in ECF and ICF?

colloid osmotic pressure exerted by plasma. In other respects, the two fluids are similar.

The electrolyte content of intracellular fluid differs considerably from that of extracellular fluid. In extracellular fluid, the most abundant cation is Na^+, and the most abundant anion is Cl^-. In intracellular fluid, the most abundant cation is K^+, and the most abundant anions are proteins and phosphates (HPO_4^{2-}). By actively transporting Na^+ out of cells and K^+ into cells, sodium–potassium pumps (Na^+/K^+ ATPases) play a major role in maintaining the high intracellular concentration of K^+ and high extracellular concentration of Na^+.

Electrolytes Perform Many Functions in the Body

Sodium

Sodium ions (Na^+) are the most abundant ions in extracellular fluid, accounting for 90% of the extracellular cations. The normal plasma Na^+ concentration is 136–148 mEq/liter. As you have already learned, Na^+ plays a pivotal role in fluid and electrolyte balance because it accounts for almost half of the osmolarity of extracellular fluid (142 of about 300 mosmol/liter). The flow of Na^+ through voltage-gated channels in the plasma membrane is also necessary for the generation and conduction of action potentials in neurons and muscle fibers. The typical daily intake of Na^+ in North America often far exceeds the body's normal daily requirements, due largely to excess dietary salt. The kidneys excrete excess Na^+, but they can also conserve it during periods of shortage.

The Na^+ level in the blood is controlled by aldosterone, antidiuretic hormone (ADH), and atrial natriuretic peptide (ANP). Aldosterone increases renal reabsorption of Na^+. When the plasma concentration of Na^+ drops below 135 mEq/liter, a condition called **hyponatremia**, ADH release ceases. The lack of ADH in turn permits greater excretion of water in urine and restoration of the normal Na^+ level in ECF. Atrial natriuretic peptide (ANP) increases Na^+ excretion by the kidneys when Na^+ level is above normal, a condition called **hypernatremia**.

Chloride

Chloride ions (Cl^-) are the most prevalent anions in extracellular fluid. The normal plasma Cl^- concentration is 95–105 mEq/liter. Cl^- moves relatively easily between the extracellular and intracellular compartments because most plasma membranes contain many Cl^- leakage channels and antiporters. For this reason, Cl^- can help balance the level of anions in different fluid compartments. One example is the chloride shift that occurs between erythrocytes and plasma as the blood level of carbon dioxide either increases or decreases (see **FIGURE 18.14**). In this case, the antiporter exchange of Cl^- for HCO_3^- maintains the correct balance of anions between ECF and ICF. Chloride ions are also part of the hydrochloric acid secreted into gastric juice. As mentioned earlier, processes that increase or decrease renal reabsorption of sodium ions also affect reabsorption of chloride ions either because chloride ions are electrically attracted to sodium ions or because they are transported along with sodium ions by means of symporters.

Indicators of Na^+ Imbalance

If excess sodium ions remain in the body because the kidneys fail to excrete enough of them, water is also osmotically retained. The result is increased blood volume, increased blood pressure, and **edema**, an abnormal accumulation of interstitial fluid. Renal failure and hyperaldosteronism (excessive aldosterone secretion) are two causes of Na^+ retention. Excessive urinary loss of Na^+, by contrast, causes excessive water loss, which results in **hypovolemia**, an abnormally low blood volume. Hypovolemia related to Na^+ loss is most frequently due to the inadequate secretion of aldosterone associated with adrenal insufficiency or overly vigorous therapy with diuretic drugs.

Potassium

Potassium ions (K^+) are the most abundant cations in intracellular fluid (150 mEq/liter). K^+ plays a key role in establishing the resting membrane potential and in the repolarization phase of action potentials in neurons and muscle fibers; K^+ also helps maintain normal intracellular fluid volume. When K^+ moves into or out of cells, it often is exchanged for H^+ and thereby helps regulate the pH of body fluids.

The normal plasma K^+ concentration is 3.5– 5.0 mEq/liter and is controlled mainly by aldosterone. When plasma K^+ concentration is high, more aldosterone is secreted into the blood. Aldosterone then stimulates principal cells of the late distal tubules and collecting ducts of the kidneys to secrete more K^+ so excess K^+ is lost in the urine. Conversely, when plasma K^+ concentration is low, aldosterone secretion decreases and less K^+ is excreted in urine. Because K^+ is needed during the repolarization phase of action potentials, abnormal K^+ levels can be lethal. For instance, **hyperkalemia** (above-normal concentration of K^+ in blood) can cause death due to ventricular fibrillation.

Bicarbonate

Bicarbonate ions (HCO_3^-) are the second most prevalent extracellular anions. Normal plasma HCO_3^- concentration is 22–26 mEq/liter in systemic arterial blood and 23–27 mEq/liter in systemic venous blood. HCO_3^- concentration increases as blood flows through systemic capillaries because the carbon dioxide released by metabolically active cells combines with water to form carbonic acid; the carbonic acid then dissociates into H^+ and HCO_3^-. As blood flows through pulmonary capillaries, however, the concentration of HCO_3^- decreases again as carbon dioxide is exhaled. (**FIGURE 18.14** shows these reactions.) Intracellular fluid also contains a small amount of HCO_3^-. As previously noted, the exchange of Cl^- for HCO_3^- helps maintain the correct balance of anions in extracellular fluid and intracellular fluid.

The kidneys are the main regulators of blood HCO_3^- concentration. The intercalated cells of the renal tubule can either form HCO_3^- and release it into the blood when the blood level is low (see **FIGURE 20.9**) or excrete excess HCO_3^- in the urine when the level in blood is too high. Changes in the blood level of HCO_3^- are considered later in this chapter in the section on acid–base balance.

Calcium

Because such a large amount of calcium is stored in bone, it is the most abundant mineral in the body. About 98% of the calcium in adults is located in the skeleton and teeth, where it is combined with phosphates to form a crystal lattice of mineral salts. In body fluids, calcium is mainly an extracellular cation (Ca^{2+}). The normal concentration of free or unattached Ca^{2+} in plasma is 4.5–5.5 mEq/liter. About the same amount of Ca^{2+} is attached to various plasma proteins. Besides contributing to the hardness of bones and teeth, Ca^{2+} plays important roles in blood clotting, neurotransmitter release, maintenance of muscle tone, and excitability of nervous and muscle tissue.

The most important regulator of Ca^{2+} concentration in plasma is parathyroid hormone (PTH) (see **FIGURE 13.19**). A low level of Ca^{2+} in plasma promotes release of more PTH, which stimulates osteoclasts in bone tissue to release calcium (and phosphate) from bone extracellular matrix. Thus, PTH increases bone *resorption.* Parathyroid hormone also enhances *reabsorption* of Ca^{2+} from glomerular filtrate through renal tubule cells and back into blood, and increases production of calcitriol (the form of vitamin D that acts as a hormone), which in turn increases Ca^{2+} *absorption* from food in the gastrointestinal tract. Recall that calcitonin (CT) produced by the thyroid gland inhibits the activity of osteoclasts, accelerates Ca^{2+} deposition into bones, and thus lowers blood Ca^{2+} levels.

Phosphate

About 85% of the phosphate in adults is present as calcium phosphate salts, which are structural components of bone and teeth. The remaining 15% is ionized. Three phosphate ions ($H_2PO_4^-$, HPO_4^{2-}, and PO_4^{3-}) are important intracellular anions. At the normal pH of body fluids, HPO_4^{2-} is the most prevalent form. Phosphates contribute about 100 mEq/liter of anions to intracellular fluid. HPO_4^{2-} is an important buffer of H^+, both in body fluids and in the urine. Although some are "free," most phosphate ions are covalently bound to organic molecules such as lipids (phospholipids), proteins, carbohydrates, nucleic acids (DNA and RNA), and adenosine triphosphate (ATP).

The normal plasma concentration of ionized phosphate is only 1.7–2.6 mEq/liter. The same two hormones that govern calcium homeostasis—parathyroid hormone (PTH) and calcitriol—also regulate the level of HPO_4^{2-} in plasma. PTH stimulates resorption of bone extracellular matrix by osteoclasts, which releases both phosphate and calcium ions into the bloodstream. In the kidneys, however, PTH inhibits reabsorption of phosphate ions while stimulating reabsorption of calcium ions by renal tubular cells. Thus, PTH increases urinary excretion of phosphate and lowers blood phosphate level. Calcitriol promotes absorption of both phosphates and calcium from the gastrointestinal tract.

Magnesium

In adults, about 54% of the total body magnesium is part of bone matrix as magnesium salts. The remaining 46% occurs as magnesium ions (Mg^{2+}) in intracellular fluid (45%) and extracellular fluid (1%). Mg^{2+} is the second most common intracellular cation (35 mEq/liter). Functionally, Mg^{2+} is a cofactor for certain enzymes needed for the metabolism of carbohydrates and proteins and for the sodium–potassium pump. Mg^{2+} is essential for normal neuromuscular activity, synaptic transmission, and myocardial functioning. In addition, secretion of parathyroid hormone (PTH) depends on Mg^{2+}.

Normal plasma Mg^{2+} concentration is low, only 1.3–2.1 mEq/liter. Several factors regulate the plasma level of Mg^{2+} by varying the rate at which it is excreted in the urine. The kidneys increase urinary excretion of Mg^{2+} in response to high plasma levels of Ca^{2+}, high plasma levels of Mg^{2+}, increases in extracellular fluid volume, decreases in parathyroid hormone, and acidosis. The opposite conditions decrease renal excretion of Mg^{2+}.

CHECKPOINT

5. What are the functions of electrolytes in the body?
6. Name three important extracellular electrolytes and three important intracellular electrolytes, and indicate how each is regulated.

20.3 Acid–Base Balance

OBJECTIVES

- Compare the roles of buffers, exhalation of carbon dioxide, and kidney excretion of H^+ in maintaining the pH of body fluids.
- Describe the different types of acid–base imbalances and their compensatory mechanisms.

From our discussion thus far, it should be clear that various ions play different roles that help maintain homeostasis. A major homeostatic challenge is keeping the H^+ concentration (pH) of body fluids at an appropriate level. This task—the maintenance of acid–base balance—is of critical importance to normal cellular function. For example, the three-dimensional shape of all body proteins, which enables them to perform specific functions, is very sensitive to pH changes. When the diet contains a large amount of protein, as is typical in North America, cellular metabolism produces more acids than bases, which tends to acidify the blood. Before proceeding with this section of the chapter, you may wish to review the discussion of acids, bases, and pH in Section 2.4 of the text.

In a healthy person, several mechanisms help maintain the pH of systemic arterial blood between 7.35 and 7.45 (average = 7.4). A pH of 7.4 corresponds to a H^+ concentration of 0.00004 mEq/liter = 40 nEq /liter. Because metabolic reactions often produce a huge excess of H^+, the lack of any mechanism for the disposal of H^+ would cause the H^+ level in body fluids to rise quickly to a lethal level. Homeostasis of H^+ concentration within a narrow range is thus essential to survival. The removal of H^+ from body fluids and its subsequent elimination from the body depend on the following three major mechanisms:

1. ***Buffer systems.*** Buffers act quickly to temporarily bind H^+, removing the highly reactive, excess H^+ from solution. Buffers thus raise the pH of body fluids but do not remove H^+ from the body.
2. ***Exhalation of carbon dioxide.*** By increasing the rate and depth of breathing, more carbon dioxide can be exhaled. Within minutes this reduces the level of carbonic acid in blood, which raises the blood pH (reduces blood H^+ level).
3. ***Kidney excretion of H^+.*** The slowest mechanism, but the only way to eliminate acids other than carbonic acid, is through their excretion in urine.

Let's now examine each of these mechanisms in more detail.

Buffer Systems Convert Strong Acids and Bases into Weak Acids and Bases

Most buffer systems in the body consist of a weak acid and the salt of that acid, which functions as a weak base. Buffers prevent rapid, drastic changes in the pH of body fluids by converting strong acids and bases into weak acids and weak bases within fractions of a second. Strong acids lower pH more than weak acids because strong acids release H^+ more readily and thus contribute more free hydrogen ions. Similarly, strong bases raise pH more than weak ones. The principal buffer systems of the body fluids are the carbonic acid–bicarbonate buffer system, the protein buffer system, and the phosphate buffer system.

Carbonic Acid–Bicarbonate Buffer System

The **carbonic acid–bicarbonate buffer system** is based on *carbonic acid* (H_2CO_3), which can act as a weak acid, and the *bicarbonate ion* (HCO_3^-), which can act as a weak base. HCO_3^- is a significant anion in both intracellular and extracellular fluids (see FIGURE 20.7). H_2CO_3 is produced from the combination of CO_2 and H_2O:

$$\underset{\text{Carbon dioxide}}{CO_2} + \underset{\text{Water}}{H_2O} \longrightarrow \underset{\text{Carbonic acid}}{H_2CO_3}$$

If there is a shortage of H^+, the H_2CO_3 can function as a weak acid and provide H^+ as follows:

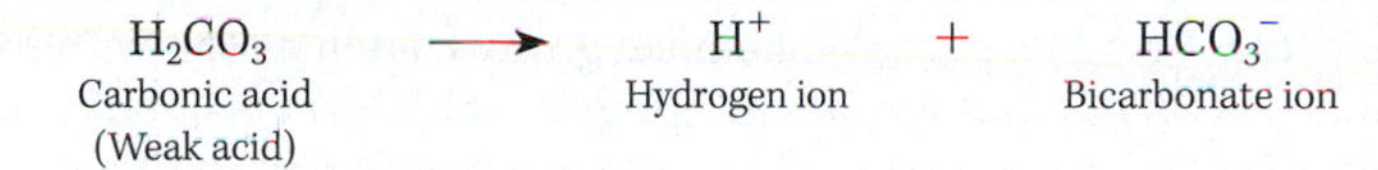

Conversely, if there is an excess of H^+, the HCO_3^- can function as a weak base and remove the excess H^+ as follows:

$$\underset{\text{Hydrogen ion}}{H^+} + \underset{\text{Bicarbonate ion (Weak base)}}{HCO_3^-} \longrightarrow \underset{\text{Carbonic acid}}{H_2CO_3}$$

Then H_2CO_3 dissociates into water and carbon dioxide, and the CO_2 is exhaled from the lungs.

The carbonic acid–bicarbonate buffer system is the most important buffer in extracellular fluid. However, because CO_2 and H_2O combine to form H_2CO_3, this buffer system cannot protect against pH changes due to respiratory problems in which there is an excess or shortage of CO_2.

PHYSIOLOGICAL EQUATION

HENDERSON–HASSELBALCH EQUATION

The relationship between pH and the concentrations of the acid and base components of a buffer system is provided by the **Henderson–Hasselbalch equation**:

$$pH = pK + \log \frac{[A^-]}{[HA]}$$

where pK is the negative log of the acid's dissociation constant,
$[A^-]$ is the concentration of the base in millimoles/liter, and
$[HA]$ is the concentration of the acid in millimoles/liter.

The Henderson–Hasselbalch equation is important because it indicates that the pH of a solution depends on two factors: (1) the pK of the acid and (2) the ratio of the amount of base to acid. Because the pK of a given acid is constant, a change in the ratio of the amount of base to acid results in a change in pH.

For the carbonic acid–bicarbonate buffer system, the acid is H_2CO_3, the base is HCO_3^-, and the pK is 6.1 (the pK is different for different acids). Therefore, for this buffer system, the Henderson–Hasselbalch equation can be written as follows:

$$pH = 6.1 + \log \frac{[HCO_3^-]}{[H_2CO_3]}$$

Normally, the HCO_3^- concentration is about 24 mmol/L and the H_2CO_3 concentration is about 1.2 mmol/liter. If you plug these values into the Henderson–Hasselbalch equation, you will see that the pH calculates to be 7.4:

$$\begin{aligned} pH &= 6.1 + \log \frac{24 \text{ mmol/L}}{1.2 \text{ mmol/L}} \\ &= 6.1 + \log \frac{20}{1} \\ &= 7.4 \end{aligned}$$

So, bicarbonate ions must outnumber carbonic acid molecules by 20 to 1 (24 mmol/L versus 1.2 mmol/L) to maintain the pH at 7.4. If there is a change in this ratio, then the pH in turn changes. For example, if the concentration of HCO_3^- increases, then the pH increases. If the concentration of H_2CO_3 increases, then the pH decreases.

Because CO_2 combines with H_2O to form H_2CO_3, the concentration of CO_2 can be used to represent the concentration of H_2CO_3. The level of CO_2 in the body is typically expressed as a partial pressure (P_{CO_2}) in mmHg. To express the P_{CO_2} level in millimoles/liter (which is the same unit used to express the bicarbonate concentration), the P_{CO_2} is multiplied by the conversion factor 0.03. Hence, the Henderson–Hasselbalch equation for the carbonic acid–bicarbonate buffer system can now be written as follows:

$$pH = 6.1 + \log \frac{[HCO_3^-]}{0.03\,(P_{CO_2})}$$

The normal P_{CO_2} in arterial blood is 40 mmHg. Thus,

$$\begin{aligned} pH &= 6.1 + \log \frac{24 \text{ mmol/L}}{0.03\,(40 \text{ mm Hg})} \\ &= 6.1 + \log \frac{24 \text{ mmol/L}}{1.2 \text{ mmol/L}} \\ &= 6.1 + \log \frac{20}{1} \\ &= 7.4 \end{aligned}$$

Again, this tells you that bicarbonate ions must outnumber CO_2 molecules by 20 to 1 in order to keep the pH at 7.4. If the concentration of HCO_3^- increases, the pH increases; if the concentration of CO_2 increases, the pH decreases.

As you will soon learn, the kidneys and lungs work together to maintain the ratio of HCO_3^- to CO_2 at 20 to 1, with the kidneys regulating the concentration of HCO_3^- and the lungs regulating the concentration of CO_2.

Protein Buffer System

The **protein buffer system** is the most abundant buffer in body cells and plasma, and the most important buffer in intracellular fluid. For example, the protein hemoglobin is an especially good buffer within erythrocytes, and albumin is the main protein buffer in plasma. Proteins contain chemical groups, such as the carboxyl group (—COOH) and the amino group ($—NH_2$); these groups are the functional components of the protein buffer system. The carboxyl group can act like an acid by releasing H^+ when pH rises; it dissociates as follows:

$$R—COOH \longrightarrow R—COO^- + H^+$$

The H^+ is then able to react with any excess OH^- (hydroxide) in the solution to form water. The amino group can act as a base by combining with H^+ when pH falls, as follows:

$$R—NH_2 + H^+ \longrightarrow R—NH_3^+$$

Hence, proteins not only have the ability to buffer acids, but they also have the ability to buffer bases.

As noted earlier, the protein hemoglobin is an important buffer of H^+ in erythrocytes (see **FIGURE 18.14**). As blood flows through the systemic capillaries, CO_2 passes from tissue cells into erythrocytes, where it combines with H_2O to form H_2CO_3. Once formed, H_2CO_3 dissociates into H^+ and HCO_3^-. Reduced hemoglobin picks up most of the H^+, which helps to prevent body fluids from becoming too acidic:

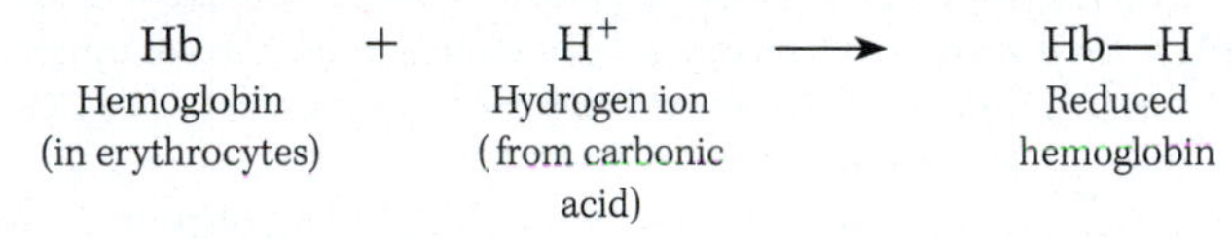

Phosphate Buffer System

The **phosphate buffer system** consists of the ions *dihydrogen phosphate* ($H_2PO_4^-$) and *monohydrogen phosphate* (HPO_4^{2-}). Recall that phosphates are major anions in intracellular fluid and minor ones in extracellular fluids (see **FIGURE 20.7**). The dihydrogen phosphate ion can act as a weak acid that buffers bases such as OH^-:

$$\underset{\text{Hydroxide ion}}{OH^-} + \underset{\text{Dihydrogen phosphate (weak acid)}}{H_2PO_4^-} \longrightarrow \underset{\text{Water}}{H_2O} + \underset{\text{Monohydrogen phosphate}}{HPO_4^{2-}}$$

The monohydrogen phosphate ion can act as a weak base that removes the H^+ released by an acid:

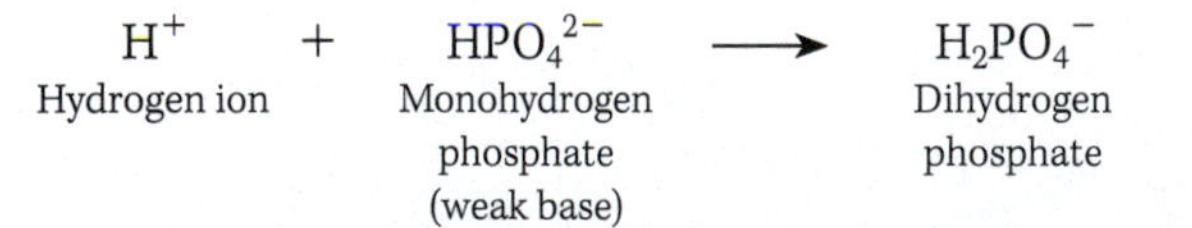

Because the concentration of phosphates is highest inside cells, the phosphate buffer system is an important regulator of pH in intracellular fluid. It also acts to a smaller degree in extracellular fluids and buffers acids in urine. $H_2PO_4^{2-}$ is formed when excess H^+ in the kidney tubule fluid combines with HPO_4^{2-} (see **FIGURE 20.9**). The H^+ that becomes part of the $H_2PO_4^-$ passes into the urine. This reaction is one way the kidneys help maintain blood pH by excreting H^+ in the urine.

Exhalation of Carbon Dioxide Helps Eliminate H^+ Ions

The simple act of breathing also plays an important role in maintaining the pH of body fluids. An increase in the CO_2 concentration in body fluids increases H^+ concentration and thus lowers the pH (makes body fluids more acidic). Because H_2CO_3 can be eliminated by exhaling CO_2, it is called a **volatile acid**. Conversely, a decrease in the CO_2 concentration of body fluids raises the pH (makes body fluids more alkaline). This chemical interaction is illustrated by the following reversible reactions:

$$\underset{\text{Carbon dioxide}}{CO_2} + \underset{\text{Water}}{H_2O} \rightleftharpoons \underset{\text{Carbonic acid}}{H_2CO_3} \rightleftharpoons \underset{\text{Hydrogen ion}}{H^+} + \underset{\text{Bicarbonate ion}}{HCO_3^-}$$

Changes in the rate and depth of breathing can alter the pH of body fluids within a couple of minutes. With increased ventilation, more CO_2 is exhaled. When CO_2 levels decrease, the reaction is driven to the left, H^+ concentration falls, and blood pH increases. Doubling the ventilation increases pH by about 0.23 units, from 7.4 to 7.63. If ventilation is slower than normal, less carbon dioxide is exhaled. When CO_2 levels increase, the reaction is driven to the right, the H^+ concentration increases, and blood pH decreases. Reducing ventilation to one-quarter of normal lowers the pH by 0.4 units, from 7.4 to 7.0. These examples show the powerful effect of alterations in breathing on the pH of body fluids.

The pH of body fluids and the rate and depth of breathing interact via a negative feedback loop (**FIGURE 20.8**). When the blood acidity increases, the decrease in pH (increase in concentration of H^+) is detected by central chemoreceptors in the medulla oblongata and peripheral chemoreceptors in the aortic and carotid bodies, both of which stimulate the dorsal respiratory group in the medulla oblongata. As a result, the diaphragm and other respiratory muscles contract more forcefully and frequently, so more CO_2 is exhaled. As less H_2CO_3 forms and fewer H^+ are present, blood pH increases. When the response brings blood pH (H^+ concentration) back to normal, there is a return to acid–base homeostasis. The same negative feedback loop operates if the blood level of CO_2 increases. Ventilation increases, which removes more CO_2, reducing the H^+ concentration and increasing the blood's pH.

By contrast, if the pH of the blood increases, the respiratory center is inhibited and the rate and depth of breathing decreases. A decrease in the CO_2 concentration of the blood has the same effect. When breathing decreases, CO_2 accumulates in the blood so its H^+ concentration increases.

The Kidneys Remove H^+ Ions by Excreting Them into Urine

Metabolic reactions produce **nonvolatile acids** such as sulfuric acid at a rate of about 1 mEq of H^+ per day for every kilogram of body mass. The only way to eliminate this huge acid load is to excrete H^+ in the urine. Given the magnitude of these contributions to acid–base balance, it's not surprising that renal failure can quickly cause death.

As you learned in Chapter 19, cells in the proximal tubules, late distal tubules, and collecting ducts of the kidneys secrete hydrogen ions into the tubular fluid. In the proximal tubules, Na^+/H^+ antiporters secrete H^+ as they reabsorb Na^+ (see **FIGURE 19.8**). Even more important for regulation of pH of body fluids, however, are the

FIGURE 20.8 Negative feedback regulation of blood pH by the respiratory system.

Exhalation of carbon dioxide lowers the H^+ concentration of blood.

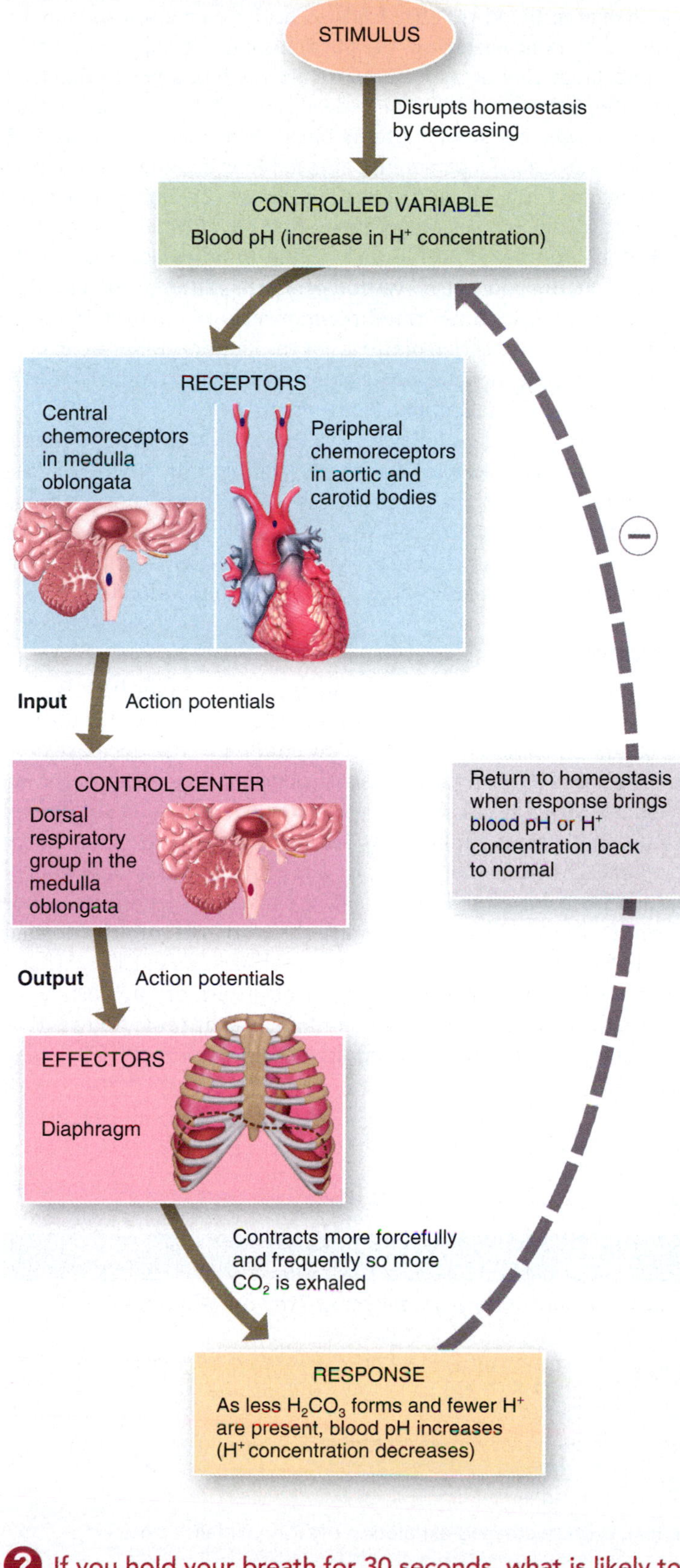

If you hold your breath for 30 seconds, what is likely to happen to your blood pH?

intercalated cells of the late distal tubule and collecting duct. The *apical* membranes of some intercalated cells include **proton pumps (H^+ ATPases)** that secrete H^+ into the tubular fluid (**FIGURE 20.9**).

FIGURE 20.9 Secretion of H^+ by intercalated cells in the late distal tubule and collecting duct. HCO_3^- = bicarbonate ion; CO_2 = carbon dioxide; H_2O = water; H_2CO_3 = carbonic acid; Cl^- = chloride ion; NH_3 = ammonia; NH_4^+ = ammonium ion; HPO_4^{2-} = monohydrogen phosphate ion; $H_2PO_4^-$ = dihydrogen phosphate ion.

Urine can be up to 1000 times more acidic than blood due to the operation of proton pumps in the late distal tubules and collecting ducts of the kidneys.

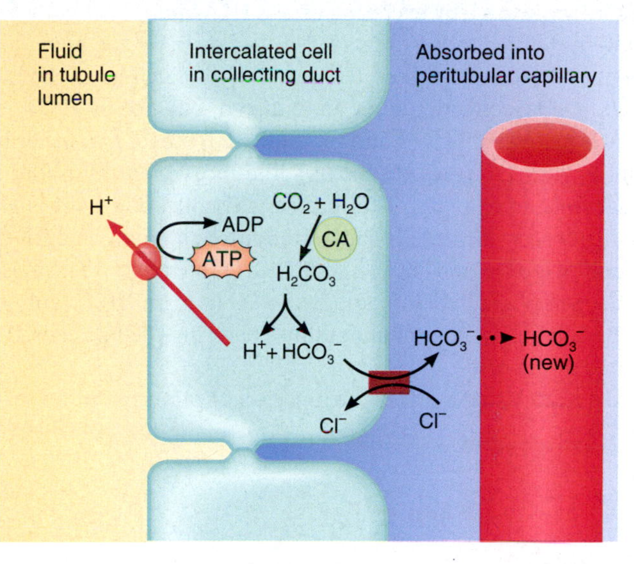

(a) Secretion of H^+

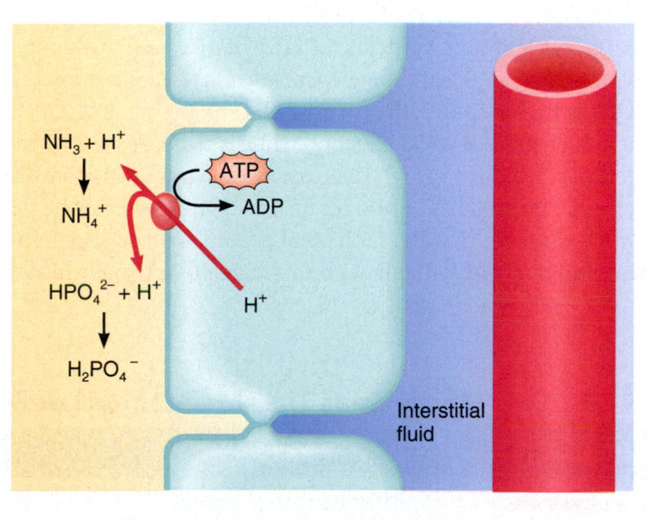

(b) Buffering of H^+ in urine

Key:
- Proton pump (H^+ ATPase) in apical membrane
- HCO_3^-–Cl^- antiporter in basolateral membrane
- Diffusion

What would be the effects of a drug that blocks the activity of carbonic anhydrase?

Intercalated cells can secrete H^+ against a concentration gradient so effectively that urine can be up to 1000 times (3 pH units) more acidic than blood. HCO_3^- produced by dissociation of H_2CO_3 inside intercalated cells crosses the basolateral membrane by means of **Cl^-/HCO_3^- antiporters** and then diffuses into peritubular capillaries (**FIGURE 20.9a**). The HCO_3^- that enters the blood in this way is *new* (not filtered). For this reason, blood leaving the kidney in the renal vein may have a higher HCO_3^- concentration than blood entering the kidney in the renal artery.

Interestingly, a second type of intercalated cell has proton pumps in its *basolateral* membrane and Cl^-/HCO_3^- antiporters in its apical membrane. These intercalated cells secrete HCO_3^- and reabsorb H^+. Thus, the two types of intercalated cells help maintain the pH of body fluids in two ways—by excreting excess H^+ when pH of body fluids is too low and by excreting excess HCO_3^- when pH is too high.

Some hydrogen ions secreted into the tubular fluid of the late distal tubule and collecting duct are buffered, but not by HCO_3^-, most of which has been filtered and reabsorbed. Instead, two other buffers combine with H^+ (**FIGURE 20.9b**). The most plentiful buffer in the tubular fluid of the late distal tubule and collecting duct is HPO_4^{2-} (monohydrogen phosphate ion). In addition, a small amount of NH_3 (ammonia) is also present. H^+ combines with HPO_4^{2-} to form $H_2PO_4^-$ (dihydrogen phosphate ion) and with NH_3 to form NH_4^+ (ammonium ion). Because these ions cannot diffuse back into tubule cells, they are excreted in the urine.

TABLE 20.1 summarizes the mechanisms that maintain the pH of body fluids.

There Are Different Types of Acid–Base Imbalances

The normal pH range of systemic arterial blood is between 7.35 (= 45 nEq of H^+/liter) and 7.45 (= 35 nEq of H^+/liter), with the average being 7.4 (= 40 nEq of H^+/liter). **Acidosis** (or *acidemia*) is a condition in which blood pH is below 7.35; **alkalosis** (or *alkalemia*) is a condition in which blood pH is higher than 7.45.

The major physiological effect of acidosis is depression of the central nervous system through depression of synaptic transmission. If the systemic arterial blood pH falls below 7, depression of the nervous system is so severe that the individual becomes disoriented, then comatose, and may die. Patients with severe acidosis usually die while in a coma. A major physiological effect of alkalosis, by contrast, is overexcitability in both the central nervous system and peripheral nerves. Neurons conduct action potentials repetitively, even when not stimulated by normal stimuli; the results are nervousness, muscle spasms, and even convulsions and death.

A change in blood pH that leads to acidosis or alkalosis may be countered by **compensation**, the physiological response to an acid–base imbalance that acts to normalize arterial blood pH. Compensation may be either *complete*, if pH indeed is brought within the normal range, or *partial*, if systemic arterial blood pH is still lower than 7.35 or higher than 7.45. If a person has altered blood pH due to metabolic causes, hyperventilation or hypoventilation can help bring blood pH back toward the normal range; this form of compensation, termed **respiratory compensation**, occurs within minutes and reaches its maximum within hours. If, however, a person has altered blood pH due to respiratory causes, then **renal compensation**—changes in secretion of H^+ and reabsorption of HCO_3^- by the kidney tubules—can help reverse the change. Renal compensation may begin in minutes, but it takes days to reach maximum effectiveness.

In the discussion that follows, note that both respiratory acidosis and respiratory alkalosis are disorders resulting from changes in the partial pressure of CO_2 (P_{CO_2}) in systemic arterial blood (normal range is 35–45 mmHg). By contrast, both metabolic acidosis and metabolic alkalosis are disorders resulting from changes in HCO_3^- concentration (normal range is 22–26 mEq/liter in systemic arterial blood).

Respiratory Acidosis

The hallmark of **respiratory acidosis** is an abnormally high P_{CO_2} in systemic arterial blood—above 45 mmHg. Inadequate exhalation of CO_2 causes the blood pH to drop. Any condition that decreases the movement of CO_2 from the blood to the alveoli of the lungs to the atmosphere causes a buildup of CO_2, H_2CO_3, and H^+. Such conditions include emphysema, pulmonary edema, injury to the respiratory center of the medulla oblongata, airway obstruction, or disorders of the muscles involved in breathing. If the respiratory problem is not too severe, the kidneys can help raise the blood pH into the normal range by increasing excretion of H^+ and reabsorption of HCO_3^- (renal compensation). The goal in treatment of respiratory acidosis is to increase the exhalation of CO_2, as, for instance, by providing ventilation therapy. In addition, intravenous administration of HCO_3^- may be helpful.

TABLE 20.1 Mechanisms That Maintain pH of Body Fluids

Mechanism	Comments
Buffer systems	Most consist of a weak acid and the salt of that acid, which functions as a weak base. They prevent drastic changes in body fluid pH.
Carbonic acid–bicarbonate	The most important buffer of extracellular fluid (ECF).
Proteins	The most abundant buffers in body cells and plasma, and the most important buffer of intracellular fluid (ICF).
Phosphates	Important buffers in intracellular fluid and in urine.
Exhalation of CO_2	With increased exhalation of CO_2, pH rises (fewer H^+). With decreased exhalation of CO_2, pH falls (more H^+).
Kidneys	Renal tubules secrete H^+ into the urine and reabsorb HCO_3^- so it is not lost in the urine.

Critical Thinking

Drinking Yourself to Death

Matthew is a 21-year-old college student who wants to become a member of a popular fraternity on campus. Matthew goes to all of the fraternity events and eventually is chosen to join the group. While hazing has been outlawed as part of the initiation, this fraternity still subjects their initiates to rigorous rites prior to their final acceptance as members. Matthew is being forced to jog in place in a hot basement and every 10 minutes he must drink a liter of water. He is sweating profusely but has not had to use the restroom since he started. Matthew is protesting that he feels sick, has a headache and can barely keep moving, but because he wants to join this fraternity, he pushes on despite feeling very ill. Matthew is able to continue for a couple of hours until he finally passes out. Matthew is very red in the face and has hot skin to the touch. His breathing is very rapid and shallow. The fraternity brothers can't revive him and finally call the ambulance. Matthew's life is in jeopardy.

SOME THINGS TO KEEP IN MIND:

When one hears of a person drinking themselves to death, the image of alcohol consumption comes to mind. However, it is possible and has been documented that people have died from drinking too much water in a short period of time. How can this be? Recall that the body's water balance is maintained through an array of homeostatically controlled mechanisms. However, when a person ingests water faster than the kidneys can excrete it (the maximum urine flow rate is about 15 mL/min), water intoxication occurs: The excessive body water causes cells (including neurons of the brain) to swell, which can ultimately lead to death. The LD_{50} (average or median lethal dose) of water is 6 liters for a person who weighs about 165 lbs.

Image Source / Getty Images

How does the rate of Matthew's water intake compare to the ability of his kidneys to excrete the water?

What effect do the overhydration and excessive sweating have on Matthew's blood sodium level?

Why did Matthew pass out?

What other problems could this cause for Matthew?

© UpperCut Images / Alamy Stock Photo

Respiratory Alkalosis

In **respiratory alkalosis**, systemic arterial blood P_{CO_2} falls below 35 mmHg. The cause of the drop in P_{CO_2} and the resulting increase in pH is hyperventilation, which occurs in conditions that stimulate the inspiratory area in the brain stem. Such conditions include oxygen deficiency due to high altitude or pulmonary disease, cerebrovascular accident (stroke), or severe anxiety. Again, renal compensation may bring blood pH into the normal range if the kidneys are able to decrease excretion of H^+ and reabsorption of HCO_3^-. Treatment of respiratory alkalosis is aimed at increasing the level of CO_2 in the body. One simple treatment is to have the person inhale and exhale into a paper bag for a short period; as a result, the person inhales air containing a higher-than-normal concentration of CO_2.

Metabolic Acidosis

In **metabolic acidosis**, the systemic arterial blood HCO_3^- level drops below 22 mEq/liter. Such a decline in this important buffer causes the blood pH to decrease. Three situations may lower the blood level of HCO_3^-: (1) actual loss of HCO_3^-, such as may occur with severe diarrhea or renal dysfunction; (2) accumulation of an acid other than carbonic acid, as may occur in ketosis (described in Section 22.1); or (3) failure of the kidneys to excrete H^+ from metabolism of dietary proteins. If the problem is not too severe, hyperventilation can help bring blood pH into the normal range (respiratory compensation). Treatment of metabolic acidosis consists of administering intravenous solutions of sodium bicarbonate and correcting the cause of the acidosis.

Metabolic Alkalosis

In **metabolic alkalosis**, the systemic arterial blood HCO_3^- concentration is above 26 mEq/liter. A nonrespiratory loss of acid or excessive intake of alkaline drugs causes the blood pH to increase above 7.45. Excessive vomiting of gastric contents, which results in a substantial loss of hydrochloric acid, is probably the most frequent cause of metabolic alkalosis. Other causes include gastric suctioning, use of certain diuretics, endocrine disorders, excessive intake of alkaline drugs (antacids), and severe dehydration. Respiratory compensation through hypoventilation may bring blood pH into the normal range. Treatment of metabolic alkalosis consists of giving fluid solutions to correct Cl^-, K^+, and other electrolyte deficiencies plus correcting the cause of alkalosis.

TABLE 20.2 summarizes respiratory and metabolic acidosis and alkalosis.

CHECKPOINT

7. Explain how each of the following buffer systems helps to maintain the pH of body fluids: carbonic acid–bicarbonate buffers, proteins, and phosphates.
8. Define acidosis and alkalosis. Distinguish among respiratory and metabolic acidosis and alkalosis.
9. What are the principal physiological effects of acidosis and alkalosis?

CLINICAL CONNECTION

Diagnosis of Acid–Base Imbalances

One can often pinpoint the cause of an acid–base imbalance by careful evaluation of three factors in a sample of systemic arterial blood: pH, concentration of HCO_3^-, and P_{CO_2}. These three blood chemistry values are examined in the following four-step sequence:

1. Note whether the pH is high (alkalosis) or low (acidosis).
2. Then decide which value—P_{CO_2} or HCO_3^-—is out of the normal range and could be the ***cause*** of the pH change. For example, ***elevated pH*** could be caused by ***low*** P_{CO_2} or high HCO_3^-.
3. If the cause is a ***change in*** P_{CO_2}, the problem is ***respiratory***; if the cause is a ***change in*** HCO_3^-, the problem is ***metabolic***.
4. Now look at the value that doesn't correspond with the observed pH change. If it is within its normal range, there is no compensation. If it is outside the normal range, compensation is occurring and partially correcting the pH imbalance.

TABLE 20.2 Summary of Acidosis and Alkalosis

Condition	Definition	Common Causes	Compensatory Mechanism
Respiratory acidosis	Increased P_{CO_2} (above 45 mmHg) and decreased pH (below 7.35) if there is no compensation.	Hypoventilation due to emphysema, pulmonary edema, trauma to respiratory center, airway obstructions, or dysfunction of muscles of respiration.	Renal: increased excretion of H^+; increased reabsorption of HCO_3^-.
Respiratory alkalosis	Decreased P_{CO_2} (below 35 mmHg) and increased pH (above 7.45) if there is no compensation.	Hyperventilation due to oxygen deficiency, pulmonary disease, cerebrovascular accident (CVA), or severe anxiety.	Renal: decreased excretion of H^+; decreased reabsorption of HCO_3^-.
Metabolic acidosis	Decreased HCO_3^- (below 22 mEq/liter) and decreased pH (below 7.35) if there is no compensation.	Loss of bicarbonate ions due to diarrhea, accumulation of acid (ketosis), renal dysfunction.	Respiratory: hyperventilation, which increases loss of CO_2.
Metabolic alkalosis	Increased HCO_3^- (above 26 mEq/liter) and increased pH (above 7.45) if there is no compensation.	Loss of acid due to vomiting, gastric suctioning, or use of certain diuretics; excessive intake of alkaline drugs.	Respiratory: hypoventilation, which slows loss of CO_2.

FROM RESEARCH TO REALITY

Rehydration Timing

Reference
Charkoudian, N. et al. (2003). Influences of hydration on post-exercise cardiovascular control in humans. *The Journal of Physiology.* 552: 635–644.

How much does dehydration affect your body after a workout?

Izf / Shutterstock

Sustained cardiovascular exercise leads to all sorts of changes in the body. A person's muscles require more oxygen and glucose in order to meet the demands of the exercise. Even an hour or longer after exercise, an increase in pulse rate is seen. Is dehydration alone responsible for the changes that are observed?

Article description:
Researchers investigated the effect of dehydration on a variety of physiological measures, from osmolarity of the blood to baroreceptor sensitivity. They were interested specifically in testing whether the timing of rehydration (during exercise or following exercise) had an impact on those values.

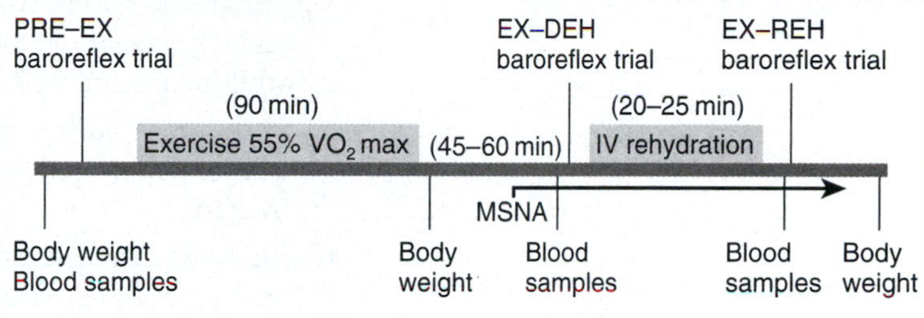

Go to WileyPLUS Learning Space and use the data from this article to answer the questions posed there and to learn more about the changes that the human body undergoes during exercise.

Blood values corresponding to each of the three baroreflex trial time points

	Trial		
Protocol 1	PRE-EX	EX-DEH	EX-REH
Hemoglobin (g dl^{-1})	13.4 ± 0.3	13.6 ± 0.4‡	12.85 ± 0.3*
Hematocrit (%)	38.9 ± 0.8	39.3 ± 0.9‡	37.2 ± 0.8*
Osmolality (mosmol kg^{-1})	286 ± 1	289 ± 1*	290 ± 1*
Na^+ ($mEq1^{-1}$)	138 ± 1	140 ± 0*	139 ± 0*
K^+ ($mEq1^{-1}$)	4.1 ± 0.1	4.6 ± 0.1*	4.4 ± 0.1*
Noradrenaline (pg ml^{-1})	194 ± 26	226 ± 22†	241 ± 25†
Adrenaline (pg ml^{-1})	20 ± 2	33 ± 4*	22 ± 2
Plasma renin activity (ng AngI ml^{-1} h^{-1})	0.79 ± 0.10	2.58 ± 0.45*	1.41 ± 0.23
Protocol 2	PRE-EX	EX-HY1	EX-HY2
Hemoglobin (g dl^{-1})	13.2 ± 0.4	12.5 ± 0.3*	12.7 ± 0.4*
Hematocrit (%)	38.2 ± 1.0	35.9 ± 1.0*	36.4 ± 1.0*
Osmolality (mosmol kg^{-1})	287 ± 0	291 ± 2	291 ± 1
Na^+ ($mEq1^{-1}$)	138 ± 1	139 ± 1	138 ± 0
K^+ ($mEq1^{-1}$)	4.0 ± 0.10	4.4 ± 0.1*	4.8 ± 0.2*
Noradrenaline (pg ml^{-1})	200 ± 34	174 ± 27	175 ± 27
Adrenaline (pg ml^{-1})	24 ± 3	27 ± 3	22 ± 3
Plasma renin activity (ng AngI ml^{-1} h^{-1})	0.69 ± 0.19	1.24 ± 0.44	1.07 ± 0.30

CHAPTER REVIEW

20.1 Fluid Compartments and Fluid Balance

1. Body fluid includes water and dissolved solutes. About two-thirds of the body's fluid is located within cells and is called intracellular fluid (ICF). The other one-third, called extracellular fluid (ECF), consists of interstitial fluid (the fluid between cells) and plasma (the liquid portion of blood).
2. Fluid balance means that the required amounts of water and solutes are present and are correctly proportioned among the various compartments.
3. An inorganic substance that dissociates into ions in solution is called an electrolyte.
4. Water is the largest single constituent in the body. It makes up 45–75% of total body mass, depending on age, gender, and the amount of adipose tissue present. Daily water gain and loss are each about 2500 mL. Sources of water gain are ingested liquids and foods, and water produced by aerobic respiration and dehydration synthesis reactions (metabolic water). Water is lost from the body via urination, evaporation from the skin surface, exhalation of water vapor, and defecation. In women, menstrual flow is an additional route for loss of body water.
5. Body water gain is regulated by adjusting the volume of water intake, mainly by drinking more or less fluid. The thirst center in the hypothalamus governs the urge to drink.
6. Although increased amounts of water and solutes are lost through sweating and exhalation during exercise, loss of excess body water or excess solutes depends mainly on regulating excretion in the urine. The extent of urinary NaCl loss is the main determinant of body fluid volume; the extent of urinary water loss is the main determinant of body fluid osmolarity.
7. The major hormone that regulates water loss from the body is antidiuretic hormone (ADH).
8. The two main hormones that regulate urinary loss of Na^+ ions are aldosterone and atrial natriuretic peptide (ANP). Aldosterone increases Na^+ reabsorption, thereby reducing urinary loss of Na^+. ANP promotes natriuresis, elevated excretion of Na^+ into the urine.
9. An increase in the osmolarity of extracellular fluid draws water out of cells, causing them to shrink. A decrease in the osmolarity of extracellular fluid brings water into cells, causing them to swell. Most often a change in osmolarity is due to a change in the concentrations of Na^+ and Cl^+ (the major contributors to osmolarity of extracellular fluid). Where there is a change in the osmolarity of extracellular fluid, body cells usually do not shrink or swell to any significant degree because compensatory mechanisms restore the extracellular fluid osmolarity back to normal.
10. When a person consumes water faster than the kidneys can excrete it or when renal function is poor, the result may be water intoxication, in which cells swell dangerously.

20.2 Electrolytes in Body Fluids

1. Ions formed when electrolytes dissolve in body fluids control the osmosis of water between fluid compartments, help maintain acid–base balance, and carry electrical current.
2. The concentrations of cations and anions are expressed in units of milliequivalents/liter (mEq/liter).
3. Plasma, interstitial fluid, and intracellular fluid contain varying types and amounts of ions.

4. Sodium ions (Na^+) are the most abundant extracellular ions. They are involved in the generation and conduction of action potentials, and play a pivotal role in fluid and electrolyte balance. Na^+ level is controlled by aldosterone, antidiuretic hormone, and atrial natriuretic peptide.
5. Chloride ions (Cl^-) are the major extracellular anions. They help balance the level of anions in different fluid compartments and are part of the HCl that is secreted into gastric juice.
6. Potassium ions (K^+) are the most abundant cations in intracellular fluid. They play a key role in the resting membrane potential and the action potentials of neurons and muscle fibers; help maintain intracellular fluid volume; and contribute to regulation of pH. K^+ level is controlled by aldosterone.
7. Bicarbonate ions (HCO_3^-) are the second most abundant anions in extracellular fluid. They are the most important buffer in blood plasma.
8. Calcium is the most abundant mineral in the body. Calcium salts are structural components of bones and teeth. Calcium ions, which are principally extracellular cations, function in blood clotting, neurotransmitter release, and contraction of muscle. Ca^{2+} level is controlled mainly by parathyroid hormone and calcitriol.
9. Phosphate ions ($H_2PO_4^-$, HPO_4^{2-}, and PO_4^{3-}) are principally intracellular anions, and their salts are structural components of bones and teeth. They are also required for the synthesis of nucleic acids and ATP and participate in buffer reactions. Their level is controlled by parathyroid hormone and calcitriol.
10. Magnesium ions (Mg^{2+}) are primarily intracellular cations. They act as cofactors in several enzyme systems.

20.3 Acid–Base Balance

1. The overall acid–base balance of the body is maintained by controlling the H^+ concentration of body fluids, especially extracellular fluid. The normal pH of systemic arterial blood is 7.35–7.45 (average is 7.4).
2. Homeostasis of pH is maintained by buffer systems, via exhalation of carbon dioxide, and via kidney excretion of H^+ and reabsorption of HCO_3^-.
3. The important buffer systems of the body include carbonic acid–bicarbonate buffers, proteins, and phosphates.
4. An increase in exhalation of carbon dioxide increases blood pH; a decrease in exhalation of CO_2 decreases blood pH.
5. In the proximal tubules of the kidneys, Na^+/H^+ antiporters secrete H^+ as they reabsorb Na^+. In the late distal tubules and collecting ducts of the kidneys, some intercalated cells reabsorb HCO_3^- and secrete H^+; other intercalated cells secrete HCO_3^-. In these ways, the kidneys can increase or decrease the pH of body fluids.
6. Acidosis is a systemic arterial blood pH below 7.35; its principal effect is depression of the central nervous system (CNS). Alkalosis is a systemic arterial blood pH above 7.45; its principal effect is overexcitability of the CNS.
7. Respiratory acidosis and alkalosis are disorders due to changes in blood P_{CO_2}; metabolic acidosis and alkalosis are disorders associated with changes in blood HCO_3^- concentration.
8. Metabolic acidosis or alkalosis can be compensated by respiratory mechanisms (respiratory compensation); respiratory acidosis or alkalosis can be compensated by renal mechanisms (renal compensation).

PONDER THIS

1. Robert is running a marathon. As he runs the race he starts to sweat profusely, which leads to a significant loss of water. In this process Robert has retained most of his solutes and instead has lost mostly water. Describe what will happen to his blood osmolarity and the mechanisms by which his body will compensate for this change.

2. Alexandra has been up all night with severe diarrhea. She is mainly losing intestinal fluid, which has a basic pH. What is this condition known as? What will happen to her blood pH and how will her body attempt to fix this change via both respiratory and renal mechanisms?

3. You are a new nurse in an ER and a patient comes in with mild dehydration. With haste in the chaos of the ER you grab and administer a saline bag that contains one liter of a hypotonic solution. Describe what would have to happen to the body's levels of ADH and aldosterone in order to compensate for this error.

ANSWERS TO FIGURE QUESTIONS

20.1 Plasma volume equals body mass × percent of body mass that is body fluid × proportion of body fluid that is ECF × proportion of ECF that is plasma × a conversion factor (1 liter/kg). For males: blood plasma volume = 60 kg × 0.60 × ⅓ × 0.20 × 1 liter/kg = 2.4 liters. Using similar calculations, female blood plasma volume is 2.2 liters.

20.2 Hyperventilation, vomiting, and diuretics all increase fluid loss.

20.3 Osmoreceptors are receptors that detect changes in the osmolarity (concentration of dissolved solutes) of body fluids.

20.4 Alcohol inhibits secretion of ADH.

20.5 ADH is responsible for the water reabsorption that accompanies aldosterone-mediated Na^+ reabsorption.

20.6 Overhydration would most likely stimulate the release of ANP.

20.7 In ECF, the major cation is Na^+, and the major anions are Cl^- and HCO_3^-. In ICF, the major cation is K^+, and the major anions are proteins and organic phosphates (for example, ATP).

20.8 Holding your breath causes blood pH to decrease slightly as CO_2 and H^+ accumulate in the blood.

20.9 A carbonic anhydrase inhibitor reduces reabsorption of Na^+ and HCO_3^- into the blood and reduces secretion of H^+. Therefore, it has a diuretic effect and can cause acidosis (lowered pH of the blood).

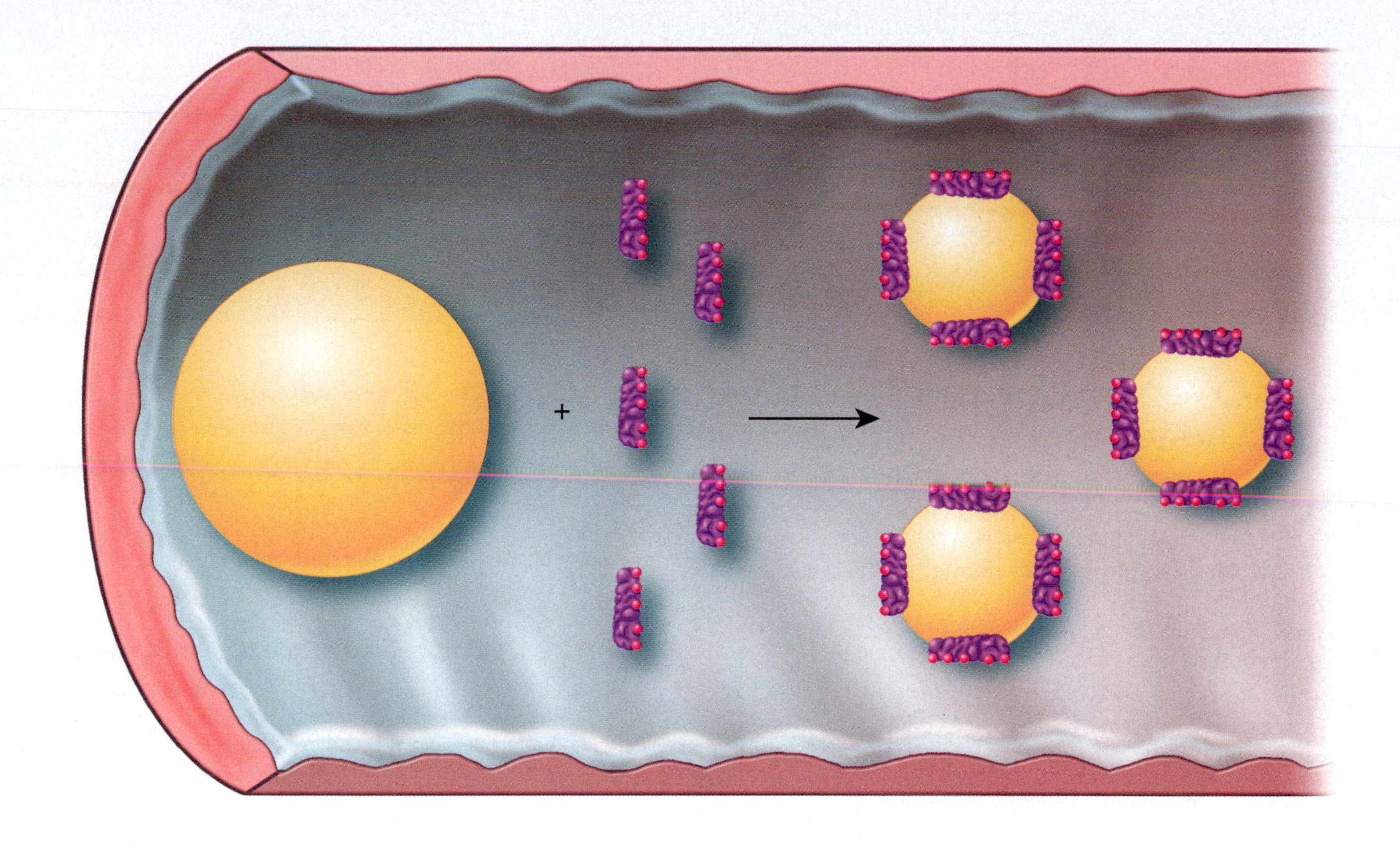

21 The Digestive System

The Digestive System and Homeostasis

The digestive system contributes to homeostasis by breaking down food into forms that can be absorbed and used by body cells. It also absorbs water, vitamins, and minerals, and eliminates wastes from the body.

LOOKING BACK TO MOVE AHEAD...

- Carbohydrates, lipids, proteins, and nucleic acids are examples of polymers, which are large molecules formed by the covalent bonding of identical or similar small building blocks called monomers (Section 2.5).
- Polymers can be broken down into their constituent monomers by hydrolysis reactions. During hydrolysis, a water molecule is split into a hydrogen atom and a hydroxyl group, and the covalent bond between the linked monomers is broken as the hydrogen atom is added to one monomer and the hydroxyl group is added to the other (Section 2.5).
- One type of smooth muscle is single-unit smooth muscle, which is so-named because it consists of fibers that contract together as a single unit. It is found in the walls of viscera (internal organs) such as the stomach and intestines (Section 11.8).
- In a portal system, blood flows from one capillary network through a portal vein and then into a second capillary network without first returning to the heart (Section 13.2).

The food we eat contains a variety of nutrients that are used for building new body tissues and repairing damaged tissues. Food is also vital to life because it is our only source of chemical energy. However, most of the food we eat consists of molecules that are too large to be used by the cells of the body. Therefore, foods must be broken down into molecules that are small enough to enter body cells, a process known as **digestion**. The organs involved in the breakdown of food—collectively called the **digestive system**—are the focus of this chapter. Like the respiratory system, the digestive system is a tubular system. It extends from the mouth to the anus, forms an extensive surface area in contact with the external environment, and is closely associated with the cardiovascular system. The combination of extensive environmental exposure and close association with blood vessels is essential for processing the food that we eat.

21.1 Overview of the Digestive System

OBJECTIVES

- Identify the organs of the digestive system.
- Describe the basic processes performed by the digestive system.
- List the functions of the layers that form the wall of the gastrointestinal tract.
- Explain the significance of the enteric nervous system.
- Describe the organization and motility patterns of gastrointestinal smooth muscle.

The Digestive System Consists of the Gastrointestinal Tract and the Accessory Digestive Organs

Two groups of organs compose the digestive system (FIGURE 21.1): the gastrointestinal (GI) tract and the accessory digestive organs. The **gastrointestinal (GI) tract**, or *alimentary canal* (*alimentary* = nourishment), is a continuous tube that extends from the mouth to the anus. Organs of the gastrointestinal tract include the mouth, pharynx, esophagus, stomach, small intestine, and large intestine. The length of the GI tract is about 4.5 meters (15 ft) in a living person. It is longer in a cadaver (about 9 meters or 30 ft) because muscles of the GI tract that are normally in a state of tonus (sustained contraction) in a living individual are no longer contracted in a cadaver. The **accessory digestive organs** include the teeth, tongue, salivary glands, pancreas, liver, and gallbladder. Teeth aid in the physical breakdown of food, and the tongue assists in chewing and swallowing. The other accessory digestive organs, however, never come into direct contact with food. They produce or store secretions that flow into the GI tract through ducts; the secretions aid in the chemical breakdown of food.

The GI tract contains food from the time it is eaten until it is digested and absorbed or eliminated. Muscular contractions in the wall of the GI tract physically break down the food by churning it and propel the food along the tract from the esophagus to the anus. The contractions also help to dissolve foods by mixing them with fluids secreted into the tract. Enzymes secreted by accessory digestive organs and cells that line the tract break down the food chemically.

There Are Six Basic Digestive Processes

Overall, the digestive system performs six basic processes (FIGURE 21.2):

1. ***Ingestion***. This process involves taking foods and liquids into the mouth (eating).
2. ***Secretion***. Each day, cells within the walls of the GI tract and accessory digestive organs secrete a total of about 7 liters of

FIGURE 21.1 Components of the digestive system.

The digestive system consists of the organs of the gastrointestinal (GI) tract (mouth, pharynx, esophagus, stomach, small intestine, and large intestine) and the accessory digestive organs (teeth, tongue, salivary glands, pancreas, liver, and gallbladder).

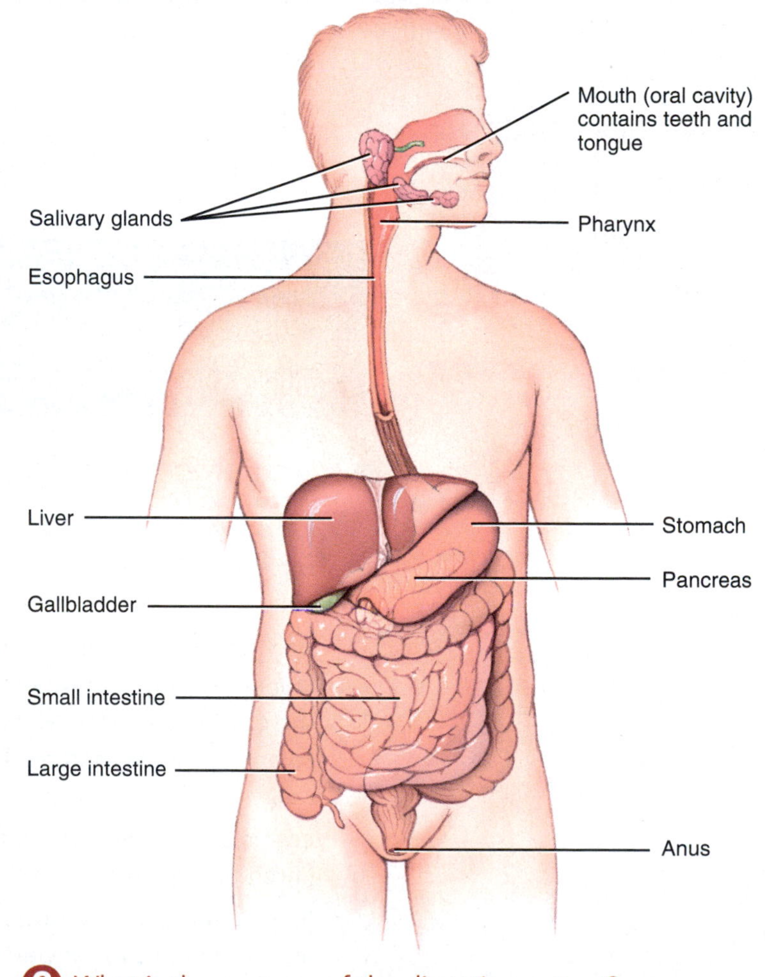

What is the purpose of the digestive system?

FIGURE 21.2 Digestive processes.

The digestive system performs six basic processes: ingestion, secretion, motility, digestion, absorption, and defecation.

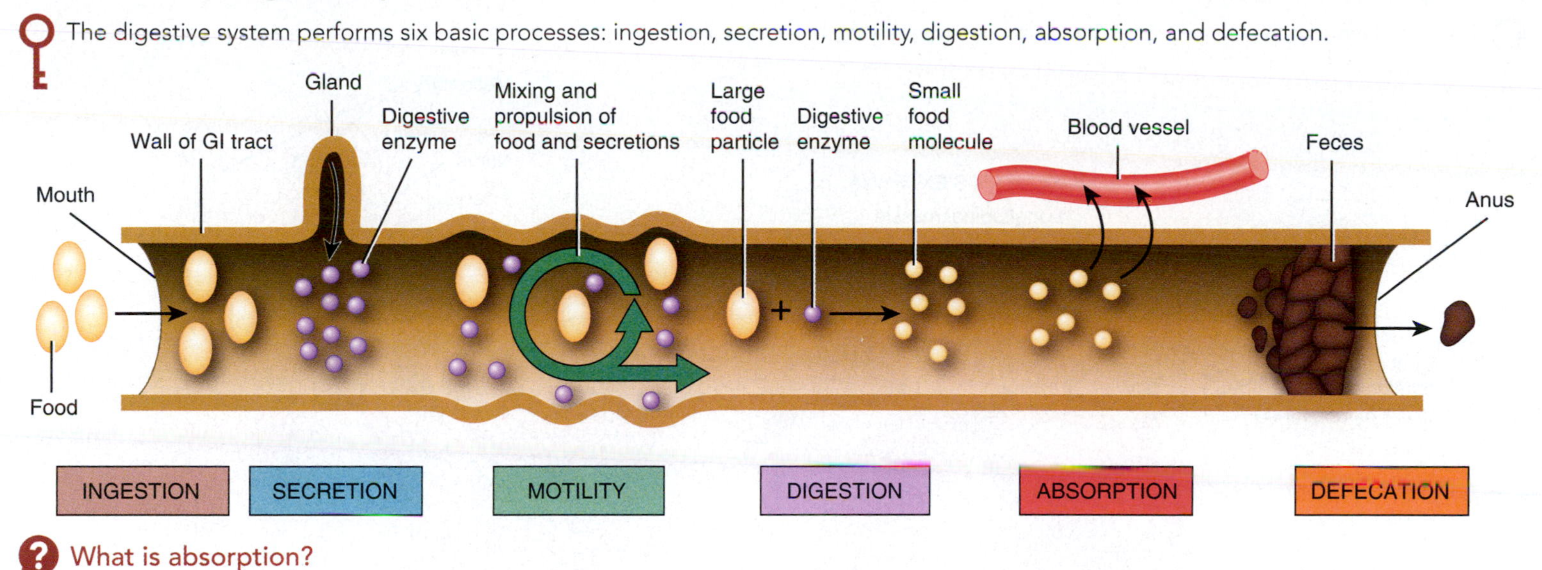

What is absorption?

water, acid, buffers, and enzymes into the lumen (interior space) of the tract.

3. ***Motility.*** Contractions of smooth muscle in the wall of the GI tract mix food and secretions, and propel them toward the anus. This capability of the GI tract to mix and move material along its length is called **motility**.
4. ***Digestion.*** Mechanical and chemical processes break down ingested food into small molecules. In **mechanical digestion**, the teeth cut and grind food before it is swallowed, and then smooth muscles of the stomach and small intestine churn the food. As a result, food molecules become dissolved and thoroughly mixed with digestive enzymes. In **chemical digestion**, the large carbohydrate, lipid, protein, and nucleic acid molecules in food are split into smaller molecules by hydrolysis (see FIGURE 2.13). Digestive enzymes produced by the salivary glands, tongue, stomach, pancreas, and small intestine catalyze these catabolic reactions.
5. ***Absorption.*** The movement of the products of digestion from the lumen of the GI tract into blood or lymph is called **absorption**. Once absorbed, these substances circulate to cells throughout the body. A few substances in food can be absorbed without undergoing digestion. These include vitamins, ions, cholesterol, and water.
6. ***Defecation.*** Wastes, indigestible substances, bacteria, cells sloughed from the lining of the GI tract, and digested materials that were not absorbed in their journey through the digestive tract leave the body through the anus in a process called **defecation**. The eliminated material is termed **feces**.

The Wall of the GI Tract Is Comprised of Four Functional Layers

The wall of the GI tract from the lower esophagus to the anus has the same basic, four-layered arrangement of tissues. These layers contribute to the various functions of the GI tract. The four layers, from the inside out, are the mucosa, submucosa, muscularis externa, and serosa (FIGURE 21.3).

Mucosa

The **mucosa**, or inner lining of the GI tract, is a mucous membrane. It is composed of (1) a layer of epithelium, (2) a lamina propria, and (3) a muscularis mucosae.

1. The **epithelium** of the mucosa is in direct contact with the contents of the GI tract. Several types of epithelial cells comprise the mucosal epithelium, and the types vary from one part of the GI tract to another. Some epithelial cells are *exocrine cells* that secrete fluids and other substances into the lumen of the GI tract. Other epithelial cells are *endocrine cells*, collectively known as **enteroendocrine cells**, that secrete hormones into the bloodstream. Still other epithelial cells are *absorptive cells* that transport nutrients from the lumen of the GI tract into the blood or lymph. Tight junctions firmly seal neighboring epithelial cells to one another to restrict leakage between the cells. The rate of renewal of GI tract epithelial cells is rapid: Every 5 to 7 days they slough off and are replaced by new cells.
2. The **lamina propria** is a layer of connective tissue that surrounds the epithelium of the mucosa. It contains small blood and lymphatic vessels, which are the sites where absorbed nutrients enter blood or lymph. Also located in the lamina propria is **gut-associated lymphoid tissue (GALT)**. These prominent lymphoid nodules contain immune system cells that protect against disease (see Chapter 16). GALT is present all along the GI tract, especially in the small intestine, appendix, and large intestine.
3. The **muscularis mucosae** is a thin layer of smooth muscle fibers that surrounds the lamina propria. Contraction of the muscularis mucosae throws the mucous membrane of the stomach and small intestine into many small folds, which increase the surface area for digestion and absorption.

Submucosa

The **submucosa** is a thick layer of connective tissue that provides the GI tract with distensibility and elasticity, allowing it to stretch as food passes through it and then to return to its original shape when food is no longer present. The submucosa contains relatively large blood and

FIGURE 21.3 Layers of the gastrointestinal tract. Different parts of the GI tract may have variations of this basic plan.

The four layers of the GI tract, from inside to outside, are the mucosa, submucosa, muscularis externa, and serosa.

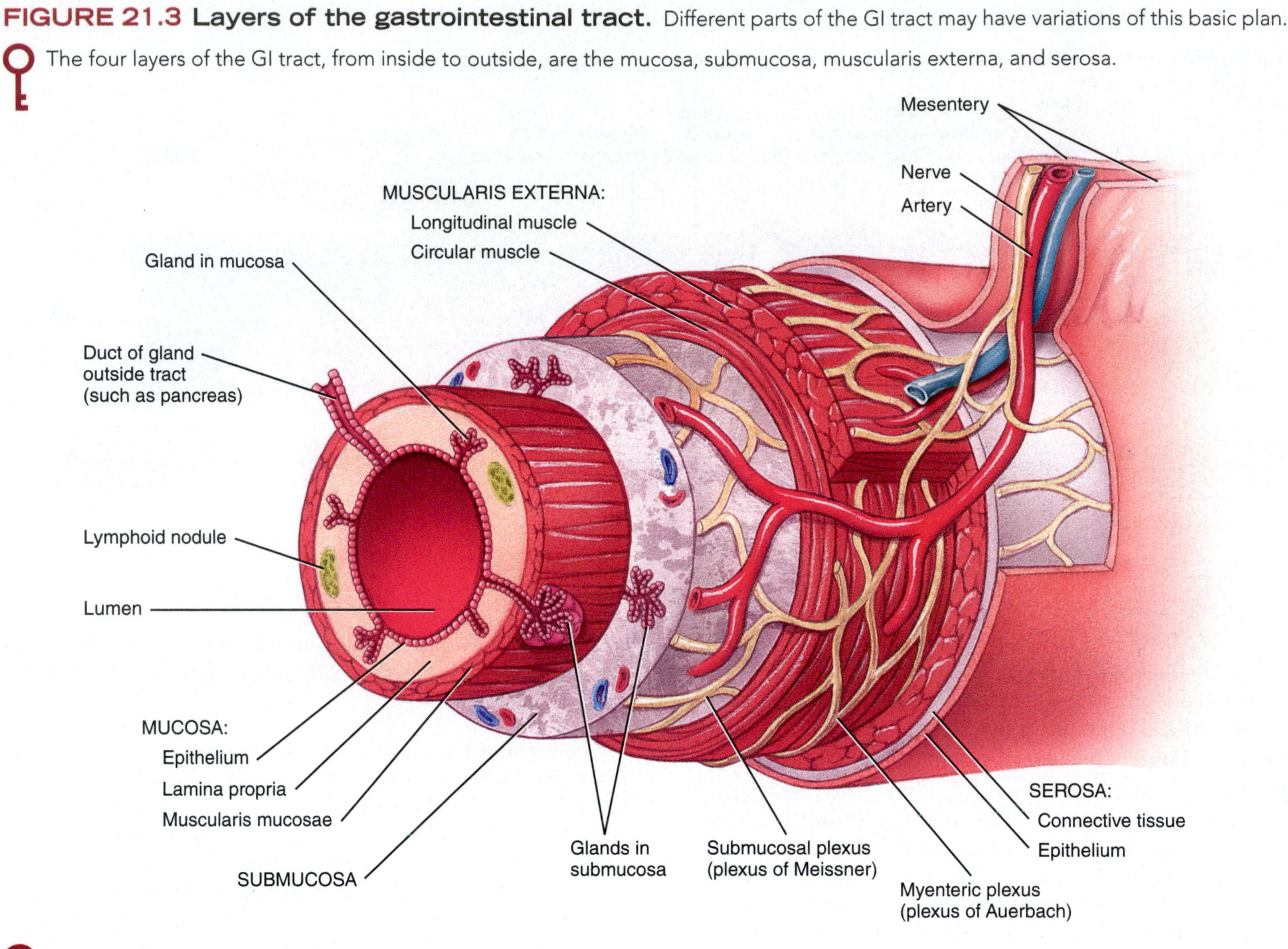

What are the functions of the lamina propria?

lymphatic vessels that receive absorbed food molecules from smaller vessels in the lamina propria. Also present in the submucosa are exocrine glands and an extensive network of neurons known as the submucosal plexus (to be described shortly).

Muscularis Externa

The **muscularis externa**, or simply the *muscularis*, is the main muscle layer of the GI tract. The muscularis externa of the mouth, pharynx, and upper to middle parts of the esophagus contains *skeletal muscle* that produces voluntary swallowing. Skeletal muscle also forms the external anal sphincter, which permits voluntary control of defecation. Throughout the rest of the GI tract, the muscularis externa consists of *smooth muscle* that is arranged in two layers: an inner layer of circular muscle and an outer layer of longitudinal muscle. Involuntary contractions of the circular and longitudinal smooth muscles propel food along the GI tract and mix food with digestive secretions. Between the two smooth muscle layers is a second plexus of neurons—the myenteric plexus (to be described shortly).

Serosa

The **serosa**, the outermost layer of the GI tract, consists of connective tissue and epithelium. It forms part of the **peritoneum**, a membrane that lines the abdominal cavity and covers the organs within that cavity. Folds of the peritoneum known as **mesenteries** bind the digestive organs to one another and to the abdominal wall. The mesenteries hold the digestive organs in place and supply the organs with blood vessels and nerves.

The GI Tract Has Neural Innervation

The gastrointestinal tract is regulated by an intrinsic set of nerves known as the enteric nervous system and by an extrinsic set of nerves that are part of the autonomic nervous system.

Enteric Nervous System

The **enteric nervous system (ENS)**, the "brain of the gut," consists of about 100 million neurons that extend from the esophagus to the anus. The neurons of the ENS are arranged into two plexuses: the myenteric plexus and the submucosal plexus (FIGURE 21.4). The **myenteric plexus**, or *plexus of Auerbach*, is located between the longitudinal and circular smooth muscle layers of the muscularis externa. The **submucosal plexus**, or *plexus of Meissner*, is found within the submucosa. The plexuses of the ENS consist of motor neurons, interneurons, and sensory neurons (FIGURE 21.4). Because the motor neurons of the myenteric plexus supply the longitudinal and circular smooth muscle layers of the muscularis externa, this plexus mostly controls GI motility (movement). The motor

FIGURE 21.4 Organization of the enteric nervous system.

The enteric nervous system consists of neurons arranged into the myenteric and submucosal plexuses.

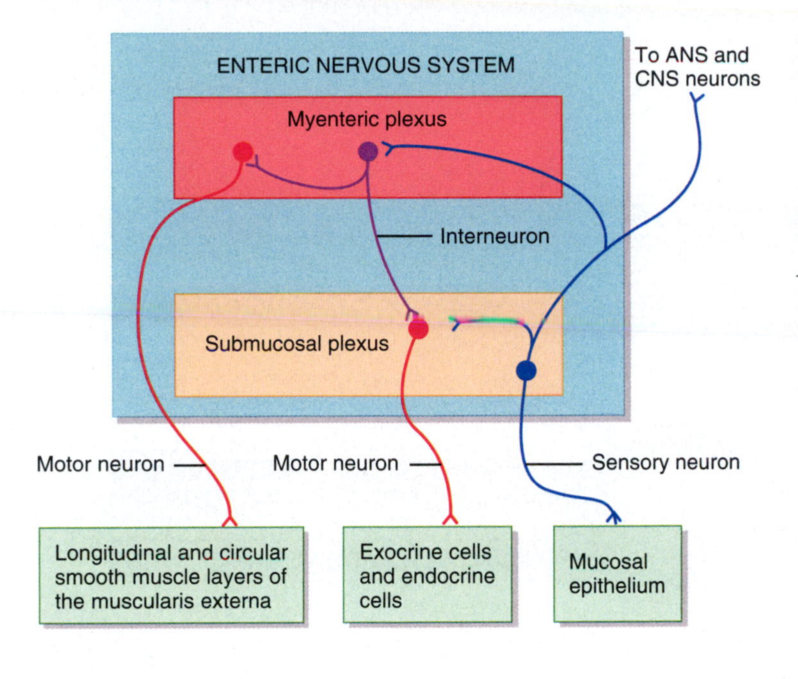

What are the functions of the myenteric and submucosal plexuses?

neurons of the submucosal plexus supply exocrine cells and endocrine cells of the GI tract, thereby controlling GI secretion. The interneurons of the ENS interconnect the neurons of the myenteric and submucosal plexuses. The sensory neurons of the ENS supply the mucosal epithelium and contain receptors that detect stimuli in the lumen of the GI tract. The wall of the GI tract contains two major types of sensory receptors: (1) *chemoreceptors*, which respond to certain chemicals in the food present in the lumen, and (2) *mechanoreceptors*, such as stretch receptors, which are activated when food distends (stretches) the wall.

Autonomic Nervous System

Although the neurons of the ENS can function independently, they are subject to regulation by the parasympathetic and sympathetic divisions of the autonomic nervous system. The vagus (X) nerves supply parasympathetic fibers to most parts of the GI tract, with the exception of the last half of the large intestine, which is supplied with parasympathetic fibers in pelvic nerves from the sacral region of the spinal cord. The parasympathetic nerves that supply the GI tract form neural connections with the ENS. Parasympathetic preganglionic neurons of the vagus (X) nerves or pelvic nerves synapse with parasympathetic postganglionic neurons located in the myenteric and submucosal plexuses. Some of the parasympathetic postganglionic neurons in turn synapse with neurons in the ENS; others directly innervate smooth muscle and glands within the wall of the GI tract. In general, stimulation of the parasympathetic nerves that innervate the GI tract causes an increase in GI secretion and motility by increasing the activity of ENS neurons.

Sympathetic nerves that supply the GI tract arise from the thoracic and upper lumbar regions of the spinal cord. Like the parasympathetic nerves, these sympathetic nerves form neural connections with the ENS. Sympathetic postganglionic neurons synapse with neurons located in the myenteric plexus and the submucosal plexus. In general, the sympathetic nerves that supply the GI tract cause a decrease in GI secretion and motility by inhibiting the neurons of the ENS. Emotions such as anger, fear, and anxiety may slow digestion because they stimulate the sympathetic nerves that supply the GI tract.

Gastrointestinal Reflex Pathways

Many neurons of the ENS are components of *gastrointestinal (GI) reflex pathways* that regulate GI secretion and motility in response to stimuli present in the lumen of the GI tract. The initial components of a typical GI reflex pathway are sensory receptors (such as chemoreceptors and mechanoreceptors) that are associated with the sensory neurons of the ENS. The axons of these sensory neurons can synapse with other neurons located in the ENS, CNS, or ANS, informing these regions about the nature of the contents and the degree of distension (stretching) of the GI tract. The neurons of the ENS, CNS, or ANS subsequently activate or inhibit GI glands and smooth muscle, altering GI secretion and motility. If the reflex pathway is confined entirely within the GI tract wall, then it is called a *short reflex*. If the reflex pathway involves not only the GI tract wall but also the CNS and autonomic nerves, then it is referred to as a *long reflex*.

GI Smooth Muscle Is Autorhythmic and Promotes Two Major Patterns of Motility

Autorhythmicity of GI Smooth Muscle

The smooth muscle in the wall of the GI tract is of the single-unit type. Recall that single-unit smooth muscle is autorhythmic (has the ability to spontaneously generate action potentials) and behaves as a functional syncytium (a single, coordinated unit) (see Section 11.8). The fibers of single-unit smooth muscle are interconnected by gap junctions, forming a network through which action potentials can spread. When an action potential is generated in one smooth muscle fiber, it rapidly spreads to all of the fibers in the syncytium and then the fibers contract in unison.

The single-unit smooth muscle of the GI tract, as a functional syncytium, consists of two types of cells: autorhythmic fibers (pacemaker cells) and contractile fibers. The autorhythmic fibers are known as the **interstitial cells of Cajal (ICCs)**, which are located in the muscularis externa between the longitudinal and circular smooth muscle layers. ICCs are noncontractile cells that undergo cycles of alternating depolarization and repolarization that do not necessarily reach threshold. These changes in membrane potential are known as **slow wave potentials** or the GI tract's **basic electrical rhythm (BER)**. The frequency of slow wave potentials varies in different parts of the GI tract, ranging from 3 to 12 per minute. If the spontaneous depolarizations of the ICCs reach threshold, action potentials are generated. The action potentials in turn are conveyed via gap junctions to contractile fibers of the smooth muscle, resulting in smooth muscle contraction. The amplitude (size) of the slow wave potentials, and therefore the ability to reach threshold, can be influenced by the ENS, ANS, hormones, and paracrine agents.

Patterns of GI Motility

As mentioned earlier, GI motility (movement) occurs because of contractions of smooth muscle in the wall of the GI tract. There are two major patterns of GI motility: peristalsis and segmentation (**FIGURE 21.5**).

1. **Peristalsis** is successive muscular contractions along the wall of a hollow muscular tube that propel the luminal contents in a forward

FIGURE 21.5 Patterns of GI motility.

The two major patterns of GI motility are peristalsis and segmentation.

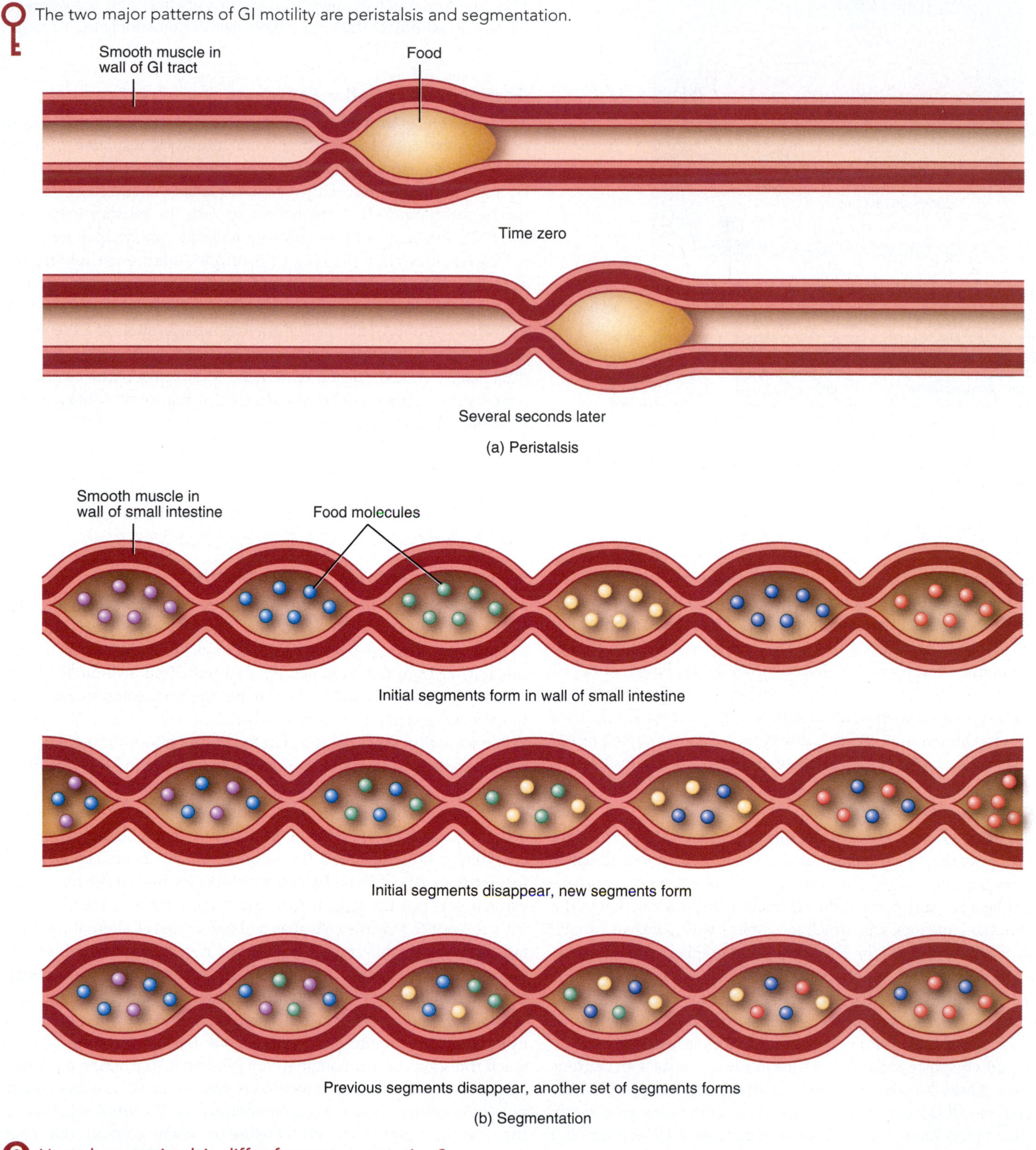

How does peristalsis differ from segmentation?

direction (FIGURE 21.5a). It occurs throughout the GI tract, from the esophagus to the anus, and in other parts of the body, including the ureters, bile ducts, and uterine (fallopian) tubes. In the GI tract, peristalsis involves successive contractions of the circular and longitudinal layers of the muscularis externa and occurs mainly in response to distension of the wall by luminal contents. In the segment of the GI tract wall just behind a mass of food, the circular layer contracts while the longitudinal layer relaxes; this shortens the wall, causing the food to move forward. Meanwhile, in the segment of the GI tract wall just in front of the food mass, the circular muscle layer relaxes while the longitudinal layer contracts; this causes the wall to push outward so that it can receive the food. As the circular and longitudinal layers undergo repeated cycles of contraction and relaxation, the food is moved along the GI tract toward the anus.

2. **Segmentation** refers to alternating muscular contractions that mix luminal contents (FIGURE 21.5b). It occurs in the small intestine in response to distension and involves contractions of the circular muscle fibers of the muscularis externa. Segmentation begins with the contraction of the circular muscle fibers at various intervals along the small intestine, an action that constricts the intestine into segments. When the circular muscle fibers relax, the segments disappear. Then circular muscle fibers at other points along the intestinal wall contract, causing new segments to form. As this sequence of events repeats, the intestinal contents slosh back and forth. Segmentation mixes food with digestive juices and brings the molecules of food into contact with the mucosa for absorption.

CHECKPOINT

1. Which components of the digestive system are GI tract organs, and which are accessory digestive organs?
2. What is the difference between mechanical digestion and chemical digestion?
3. What are some of the functions of the mucosal epithelium of the GI tract wall?
4. What general effect does the sympathetic nervous system have on the neurons of the enteric nervous system?
5. What is the functional significance of the interstitial cells of Cajal?
6. Which muscle fibers contract or relax, and in what order, during peristalsis? During segmentation?

21.2 Mouth

OBJECTIVE

- Describe the functions of the components of the mouth.

The **mouth**, also referred to as the **oral cavity**, is formed by the cheeks, lips, palate, and tongue (FIGURE 21.6). Other structures associated with the mouth are the salivary glands and teeth.

FIGURE 21.6 Components of the mouth (oral cavity).

The mouth is formed by the cheeks, lips, palate, and tongue.

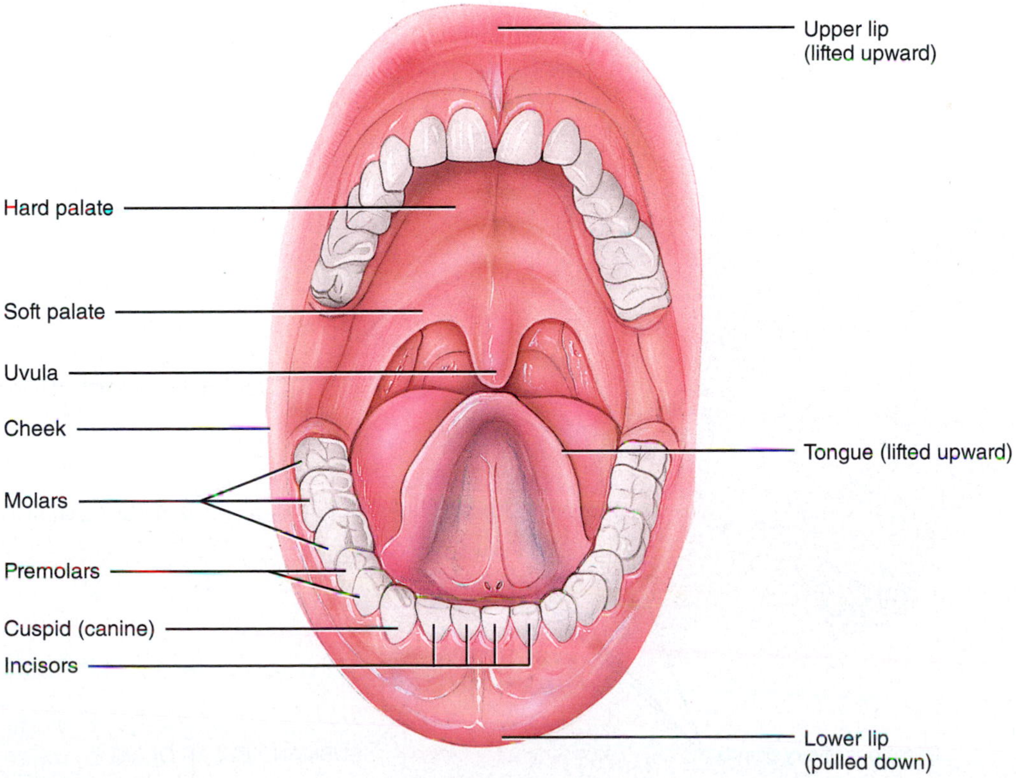

What is the function of the uvula?

The Cheeks and Lips Keep Food in the Mouth During Chewing

The **cheeks** form the side walls of the oral cavity. The **lips** are fleshy folds that surround the opening of the mouth. During chewing, the cheeks and lips keep food in the mouth between the upper and lower teeth. The lips also assist in speech.

The Palate Prevents Food from Entering the Nasal Cavity

The **palate** forms the roof of the mouth, separating the oral cavity from the nasal cavity. This important structure makes it possible to chew and breathe at the same time. Most of the roof of the mouth is a bony structure called the **hard palate**; the rest is formed by the muscular **soft palate**. Hanging from the soft palate is a projection called the **uvula**. During swallowing, the uvula and soft palate elevate to close off the upper part of the pharynx (throat). This prevents swallowed foods and liquids from entering the nasal cavity.

The Tongue Moves Food Toward the Pharynx and Produces Lingual Lipase

The **tongue** forms the floor of the oral cavity. It is an accessory digestive organ that is mainly composed of skeletal muscle. The muscles of the tongue maneuver food for chewing, shape the food into a rounded mass, force the food to the back of the mouth for swallowing, and alter the shape and size of the tongue for swallowing and speech. The tongue also contains lingual glands and taste buds. **Lingual glands** secrete mucus and a watery fluid that contains the digestive enzyme **lingual lipase**, which acts on triglycerides. The *taste buds* contain receptor cells that detect the chemicals present in food (see FIGURE 9.23).

The Salivary Glands Secrete Saliva

The **salivary glands** are accessory organs of digestion that lie outside the mouth and release their secretions into ducts emptying into the oral cavity. There are three pairs of salivary glands: the **parotid glands**, **submandibular glands**, and **sublingual glands** (FIGURE 21.7).

The fluid secreted by the salivary glands is called **saliva**. Chemically, saliva is 99.5% water and 0.5% solutes. Among the solutes are ions, including sodium, potassium, chloride, bicarbonate, and phosphate. Also present are some dissolved gases and various organic substances, including mucus, lysozyme, and salivary amylase. The water in saliva helps dissolve foods so they can be tasted and digestive reactions can begin. The bicarbonate and phosphate ions in saliva buffer acidic foods that enter the mouth, so saliva is only slightly acidic (pH 6.35–6.85). Mucus lubricates food so it can be moved around easily in the mouth, formed into a ball, and swallowed. **Lysozyme** is an enzyme that kills bacteria; however, lysozyme is not present in large enough

FIGURE 21.7 The salivary glands.

Saliva lubricates and dissolves foods and begins the chemical breakdown of carbohydrates and lipids.

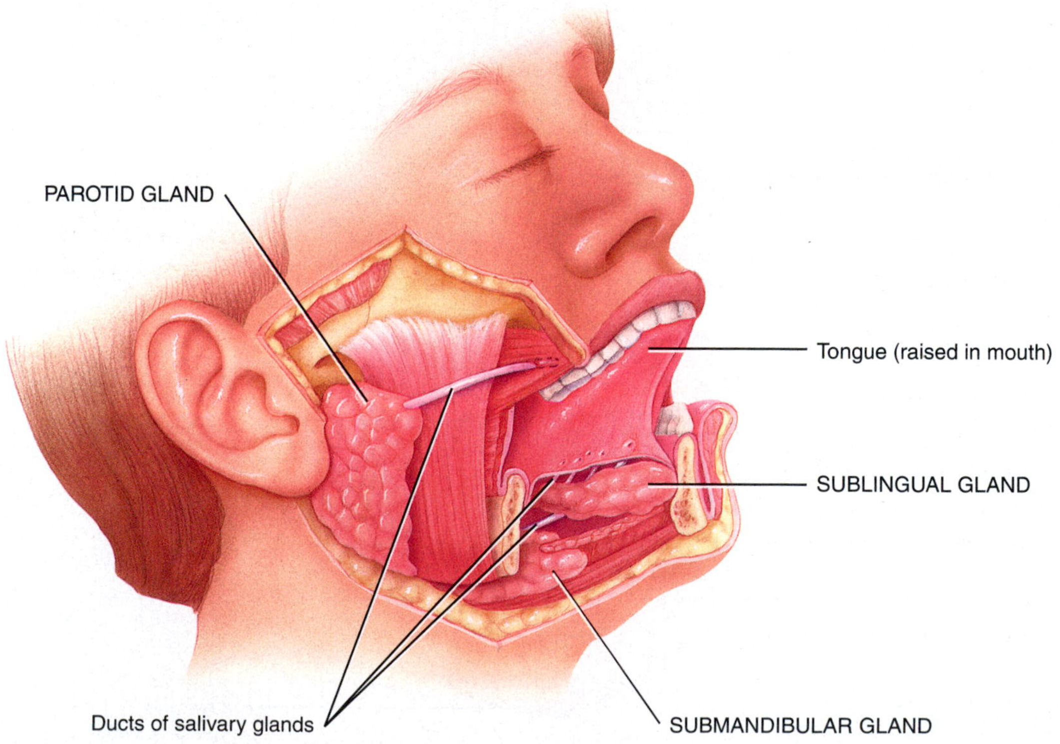

What is the purpose of salivary amylase?

quantities in saliva to eliminate all oral bacteria. **Salivary amylase** is a digestive enzyme that begins the digestion of starches; you will learn more about this enzyme shortly.

The secretion of saliva, called **salivation**, is controlled by the autonomic nervous system. Amounts of saliva secreted daily vary considerably but average about 1 liter. Normally, parasympathetic stimulation promotes continuous secretion of a moderate amount of saliva, which keeps the mouth moist and lubricates the movements of the tongue and lips during speech. The saliva is then swallowed and helps moisten the esophagus. Eventually, most components of saliva are reabsorbed, which prevents fluid loss. Sympathetic stimulation dominates during stress, resulting in dryness of the mouth. If the body becomes dehydrated, the salivary glands stop secreting saliva to conserve water; the resulting dryness in the mouth contributes to the sensation of thirst. Drinking not only restores the homeostasis of body water but also moistens the mouth.

The feel and taste of food are also potent stimulators of salivary gland secretions. Chemicals in the food stimulate receptors in taste buds on the tongue, and action potentials are conveyed from the taste buds to two salivary nuclei in the brain stem (**superior** and **inferior salivatory nuclei**). Returning action potentials along parasympathetic fibers of the facial (VII) and glossopharyngeal (IX) nerves stimulate the secretion of saliva. Saliva continues to be secreted heavily for some time after food is swallowed; this flow of saliva washes out the mouth and dilutes and buffers the remnants of irritating chemicals such as that tasty (but hot!) salsa. The sight, smell, or thought of food may also stimulate secretion of saliva.

The Teeth Physically Break Down Food

The **teeth** are accessory digestive organs located in sockets of the upper and lower jawbones. Each tooth is covered by **enamel**, which consists primarily of calcium phosphate and calcium carbonate. Enamel is harder than bone because of its higher content of calcium salts (about 95% of dry weight). In fact, enamel is the hardest substance in the body. It serves to protect the tooth from the wear and tear of chewing. It also protects against acids that can easily dissolve internal parts of the tooth.

Humans have different teeth for different functions (see **FIGURE 21.6**). *Incisors* are chisel-shaped and are adapted for cutting into food; *cuspids* (canines) have only one pointed surface (cusp) to tear and shred food; *premolars* have two cusps to crush and grind food; and *molars* have three or more blunt cusps to crush and grind food.

Mechanical Digestion in the Mouth Involves Chewing

Mechanical digestion in the mouth results from chewing, or **mastication**, in which food is manipulated by the tongue, ground by the teeth, and mixed with saliva. As a result, the food is reduced to a soft, flexible, easily swallowed mass called a **bolus**. Food molecules begin to dissolve in the water in saliva, an important activity because enzymes can react with food molecules in a liquid medium only.

Chemical Digestion in the Mouth Occurs as Food Mixes with Saliva

The enzyme salivary amylase is responsible for chemical digestion in the mouth. **Salivary amylase**, which is secreted by the salivary glands, initiates the breakdown of starch. Dietary carbohydrates are either monosaccharide and disaccharide sugars or complex polysaccharides such as starches. Most of the carbohydrates we eat are starches, but only monosaccharides can be absorbed into the bloodstream. Thus, ingested disaccharides and starches must be broken down into monosaccharides. The function of salivary amylase is to begin starch digestion by breaking down starch into smaller molecules such as the disaccharide maltose, the trisaccharide maltotriose, and short-chain glucose polymers called α-dextrins. Even though food is usually swallowed too quickly for all the starches to be broken down in the mouth, salivary amylase in the swallowed food continues to act on the starches for about another hour, at which time stomach acids inactivate it.

Saliva also contains **lingual lipase**, which is secreted by lingual glands in the tongue. This enzyme becomes activated in the acidic environment of the stomach and thus starts to work after food is swallowed. It breaks down dietary triglycerides into fatty acids and diglycerides. A diglyceride consists of a glycerol molecule that is attached to two fatty acids.

CHECKPOINT

7. What are some of the functions of the tongue?
8. How is the secretion of saliva regulated?
9. What functions do incisors, cuspids, premolars, and molars perform?
10. Which enzyme initiates starch digestion in the mouth?

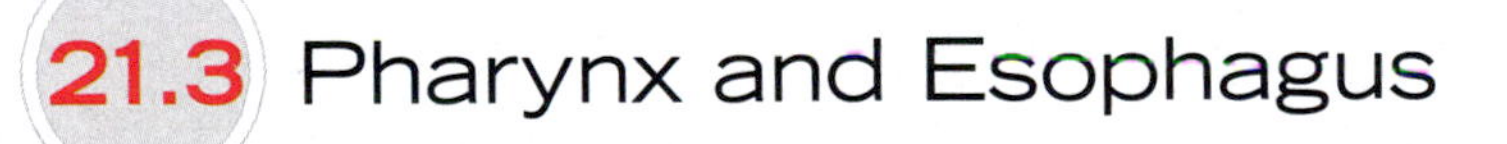

21.3 Pharynx and Esophagus

OBJECTIVES

- Describe the functions of the pharynx and the esophagus.
- Describe the three phases of deglutition.

The Pharynx Conveys Food from the Mouth to the Esophagus

When food is first swallowed, it passes from the mouth into the **pharynx** (throat), a funnel-shaped tube that extends from the nasal and oral cavities to the esophagus and to the larynx (**FIGURE 21.8**). As you learned in Chapter 18, the pharynx is part of both the digestive and respiratory systems. In terms of its digestive function, the pharynx serves as the passageway for food from the mouth into the esophagus.

The Esophagus Transports Food to the Stomach

The **esophagus** is a collapsible muscular tube that extends from the pharynx to the stomach (FIGURE 21.8). The esophagus secretes mucus and transports food into the stomach. It does not produce digestive enzymes, and it does not carry on absorption. At each end of the esophagus, the muscularis externa becomes slightly more prominent and forms two sphincters—the **upper esophageal sphincter (UES)**, which consists of skeletal muscle, and the **lower esophageal sphincter (LES)**, which consists of smooth muscle. The upper esophageal sphincter regulates the movement of food from the pharynx into the esophagus; the lower esophageal sphincter regulates the movement of food from the esophagus into the stomach.

Deglutition Is the Process by Which Food Is Swallowed

The movement of food from the mouth into the stomach is achieved by the act of swallowing, or **deglutition** (dē-gloo-TISH-un) (FIGURE 21.9). Deglutition is facilitated by the secretion of saliva and mucus and involves the mouth, pharynx, and esophagus. Swallowing occurs in three stages: (1) the voluntary stage, in which the bolus is passed from the oral cavity into the pharynx; (2) the pharyngeal stage, the involuntary passage of the bolus through the pharynx into the esophagus; and (3) the esophageal stage, the involuntary passage of the bolus through the esophagus into the stomach.

Swallowing starts when the bolus is forced to the back of the oral cavity and into the pharynx by the movement of the tongue upward and backward against the palate; these actions constitute the **voluntary stage** of swallowing. With the passage of the bolus into the pharynx, the involuntary **pharyngeal stage** of swallowing begins (FIGURE 21.9b). The bolus stimulates receptors in the pharynx, which send action potentials to the **deglutition center** in the medulla oblongata and lower pons of the brain stem. The returning action potentials cause the soft palate and uvula to elevate to close off the upper part of the pharynx, preventing swallowed foods and liquids from entering the nasal cavity. In addition, the epiglottis closes off the opening to the larynx, which prevents the bolus from entering the rest of the respiratory tract. The bolus then moves through the remainder of the pharynx. Once the upper esophageal sphincter relaxes, the bolus moves into the esophagus.

The **esophageal stage** of swallowing begins once the bolus enters the esophagus. During this phase, peristalsis propels the bolus down the esophagus toward the stomach (FIGURE 21.9c). As the bolus approaches the end of the esophagus, the lower esophageal sphincter relaxes and the bolus moves into the stomach. Mucus secreted by esophageal glands lubricates the bolus and reduces friction. The passage of solid or semisolid food from the mouth to the stomach takes 4 to 8 seconds; very soft foods and liquids pass through in about 1 second. Peristaltic contractions are so strong in the esophagus that you could do a handstand after swallowing food and the food would still reach the stomach!

FIGURE 21.8 The pharynx and esophagus.

The pharynx conveys food from the mouth to the esophagus; the esophagus in turn conveys food to the stomach.

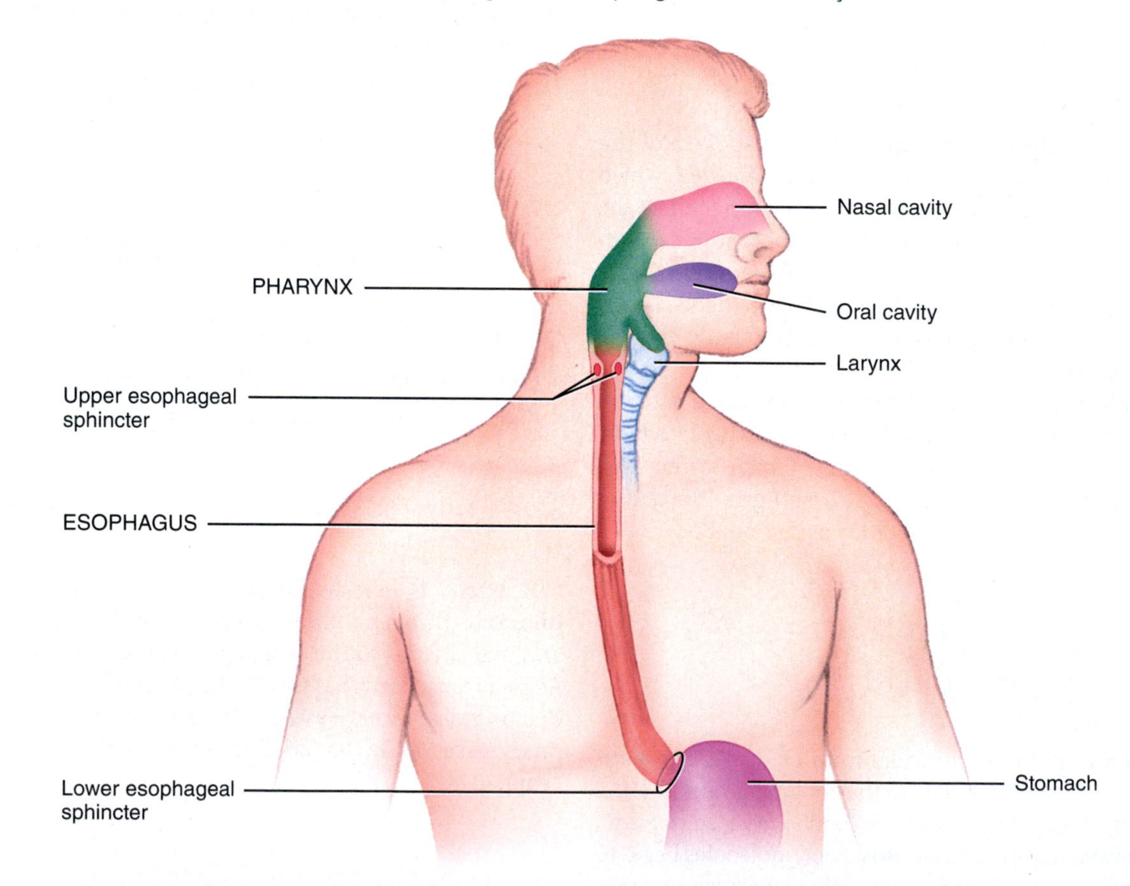

What are the functions of the upper and lower esophageal sphincters?

FIGURE 21.9 Deglutition (swallowing). During the pharyngeal stage of deglutition (b) the tongue rises against the palate, the upper part of the pharynx is closed off, the larynx rises, the epiglottis seals off the larynx, and the bolus is passed into the esophagus. During the esophageal stage of deglutition (c), food moves through the esophagus into the stomach via peristalsis.

Deglutition is the mechanism by which food moves from the mouth into the stomach.

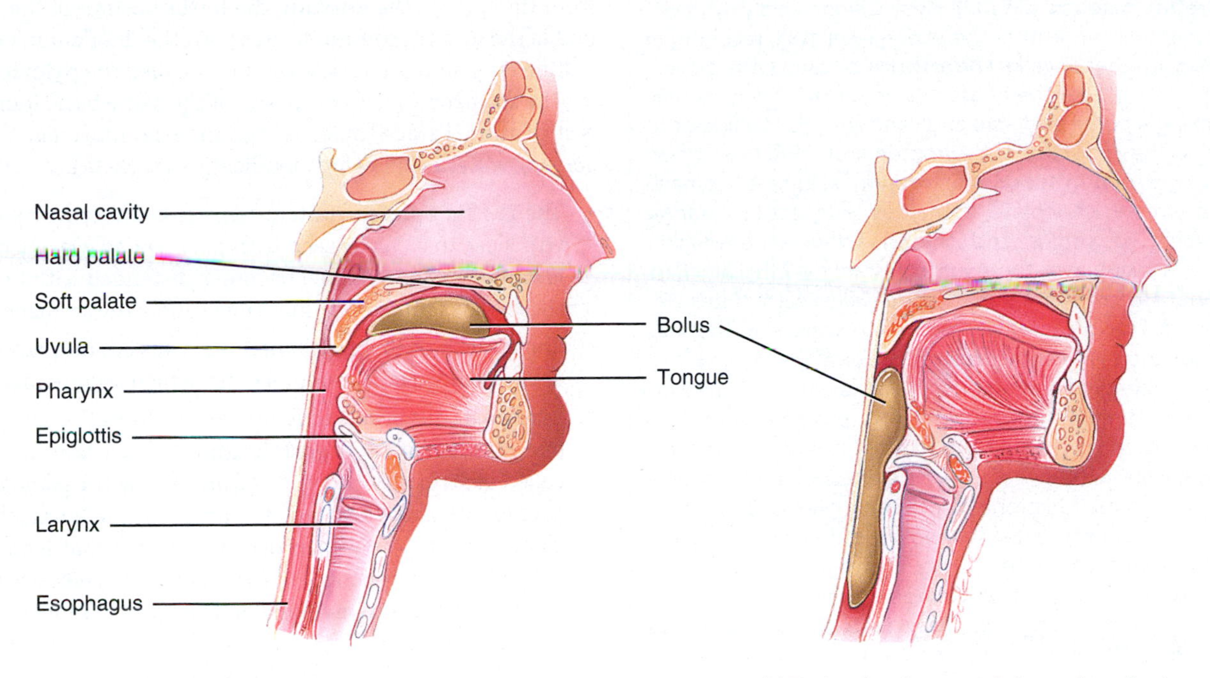

(a) Position of structures before swallowing

(b) During the pharyngeal stage of swallowing

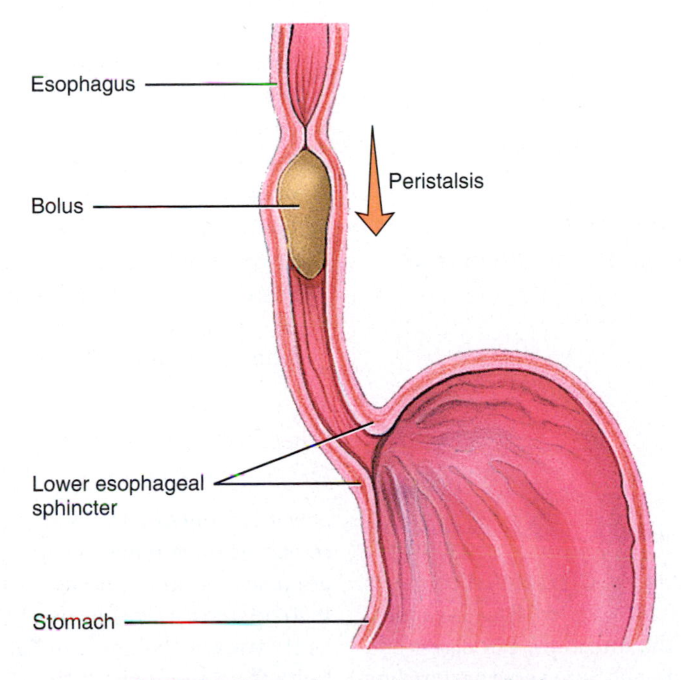

(c) During the esophageal stage of swallowing

Is swallowing a voluntary action or an involuntary action?

CLINICAL CONNECTION

Gastroesophageal Reflux Disease and Achalasia

If the lower esophageal sphincter fails to close adequately after food has entered the stomach, the stomach contents can reflux (back up) into the esophagus. This condition is known as **gastroesophageal reflux disease (GERD)**. Hydrochloric acid (HCl) from the stomach contents can irritate the esophageal wall, resulting in a burning sensation that is called **heartburn** because it is experienced in a region very near the heart; it is unrelated to any cardiac problem. Drinking alcohol and smoking can cause the sphincter to relax, worsening the problem. The symptoms of GERD can often be controlled by avoiding foods that strongly stimulate stomach acid secretion (coffee, chocolate, tomatoes, fatty foods, orange juice, peppermint, spearmint, and onions). Other acid-reducing strategies include taking over-the-counter histamine-2 (H_2) blockers such as Tagamet HB® or Pepcid AC® about 30 to 60 minutes before eating to block acid secretion, and neutralizing acid that has already been secreted with antacids such as Tums® or Maalox®.

If the lower esophageal sphincter fails to relax normally as food approaches, a whole meal may become lodged in the esophagus and enter the stomach very slowly or not at all. This condition is referred to as **achalasia** (ak′-a-LĀ-zē-a), a type of esophageal motility disorder caused by a malfunctioning myenteric plexus. Distension of the esophagus from the accumulated food causes chest pain, which is usually more severe after eating. Treatment of achalasia is directed at weakening the lower esophageal sphincter, facilitating the movement of food from the esophagus into the stomach. This can be accomplished through the use of medications that relax the sphincter, insertion of a balloon that stretches the sphincter, or surgical incision of the sphincter.

CHECKPOINT

11. To which two organ systems does the pharynx belong?

12. What are the functions of the upper and lower esophageal sphincters?

13. What are the three stages of deglutition?

14. What occurs during the voluntary phase of swallowing?

21.4 Stomach

OBJECTIVE

• Describe the functions of the stomach.

The **stomach** is a J-shaped enlargement of the GI tract that connects the esophagus to the small intestine (**FIGURE 21.10**). Because a meal can be eaten much more quickly than the intestines can digest and absorb it, one of the functions of the stomach is to serve as a mixing chamber and holding reservoir. At appropriate intervals after food is ingested, the stomach forces a small quantity of material into the duodenum, the first portion of the small intestine. Empty, the stomach it is about the size of a large sausage, but it is the most distensible part of the GI tract and can accommodate a large quantity of food, up to about 2 liters. In the stomach, digestion of starch continues, digestion of proteins and triglycerides begins, the semisolid bolus is converted to a liquid, and a few substances are absorbed.

The stomach has three main regions: the fundus, body, and antrum (**FIGURE 21.10**). The **fundus** is the upper portion of the stomach that is located above the lower esophageal sphincter. Below the fundus is the large, central portion of the stomach, called the **body**. Extending from the body is the **antrum**, the lower portion of the stomach. The end of the antrum communicates with the duodenum of the small intestine via a smooth muscle sphincter called the **pyloric sphincter**.

The stomach wall is composed of the same basic four layers as the rest of the GI tract (mucosa, submucosa, muscularis externa, and serosa), with the following modifications (**FIGURE 21.10**):

- The gastric mucosa contain large folds, called **rugae** (ROO-gē = wrinkles), that can be seen with the unaided eye when the stomach is empty. As the stomach becomes distended with food and liquid, these folds flatten out. Thus, rugae allow the stomach to expand.
- The surface of the gastric mucosa consists of a layer of epithelial cells called **surface mucous cells**, which secrete mucus.
- Epithelial cells also extend downward from the surface of the gastric mucosa to form **gastric glands** that line narrow channels called **gastric pits**. Secretions from the gastric glands flow into the gastric pits and then into the lumen of the stomach. The gastric glands contain three types of exocrine cells that secrete their products into the stomach lumen: mucous neck cells, parietal cells, and chief cells. **Mucous neck cells**, like surface mucous cells, secrete mucus. **Parietal cells** produce intrinsic factor (needed for absorption of vitamin B_{12}) and hydrochloric acid. The **chief cells** secrete pepsinogen and gastric lipase. The secretions of the mucous, parietal, and chief cells form **gastric juice**, which totals about 2 liters per day. In addition, gastric glands include a type of enteroendocrine cell, the **G cell**, which is located mainly in the antrum and secretes the hormone gastrin into the bloodstream. As you will see later in the chapter, this hormone stimulates several aspects of gastric activity.
- The muscularis externa of the stomach has three layers of smooth muscle (rather than the two found in the esophagus and the intestines): an outer longitudinal layer, a middle circular layer, and an inner oblique layer. The presence of the oblique layer enhances gastric motility. The antrum of the stomach has the largest amount of smooth muscle in its wall, whereas the fundus and body of the stomach have smaller amounts of smooth muscle.

Mechanical Digestion in the Stomach Involves Propulsion and Retropulsion

Several minutes after food enters the stomach, waves of peristalsis pass over the stomach every 15 to 25 seconds. Few peristaltic waves are observed in the fundus, which primarily has a storage function. Instead, most waves begin at the body of the stomach and intensify as they reach the antrum. Each peristaltic wave moves gastric contents from the body of the stomach down into the antrum, a process known as **propulsion**. The pyloric sphincter normally remains almost, but not completely, closed. Because most food particles in the stomach initially are too large to fit through the narrow pyloric sphincter, they are forced back into the body of the stomach, a process referred to as **retropulsion**. Another round of propulsion then

FIGURE 21.10 The stomach.

Gastric glands of the stomach produce gastric juice, which consists of hydrochloric acid, mucus, intrinsic factor, pepsin, and gastric lipase.

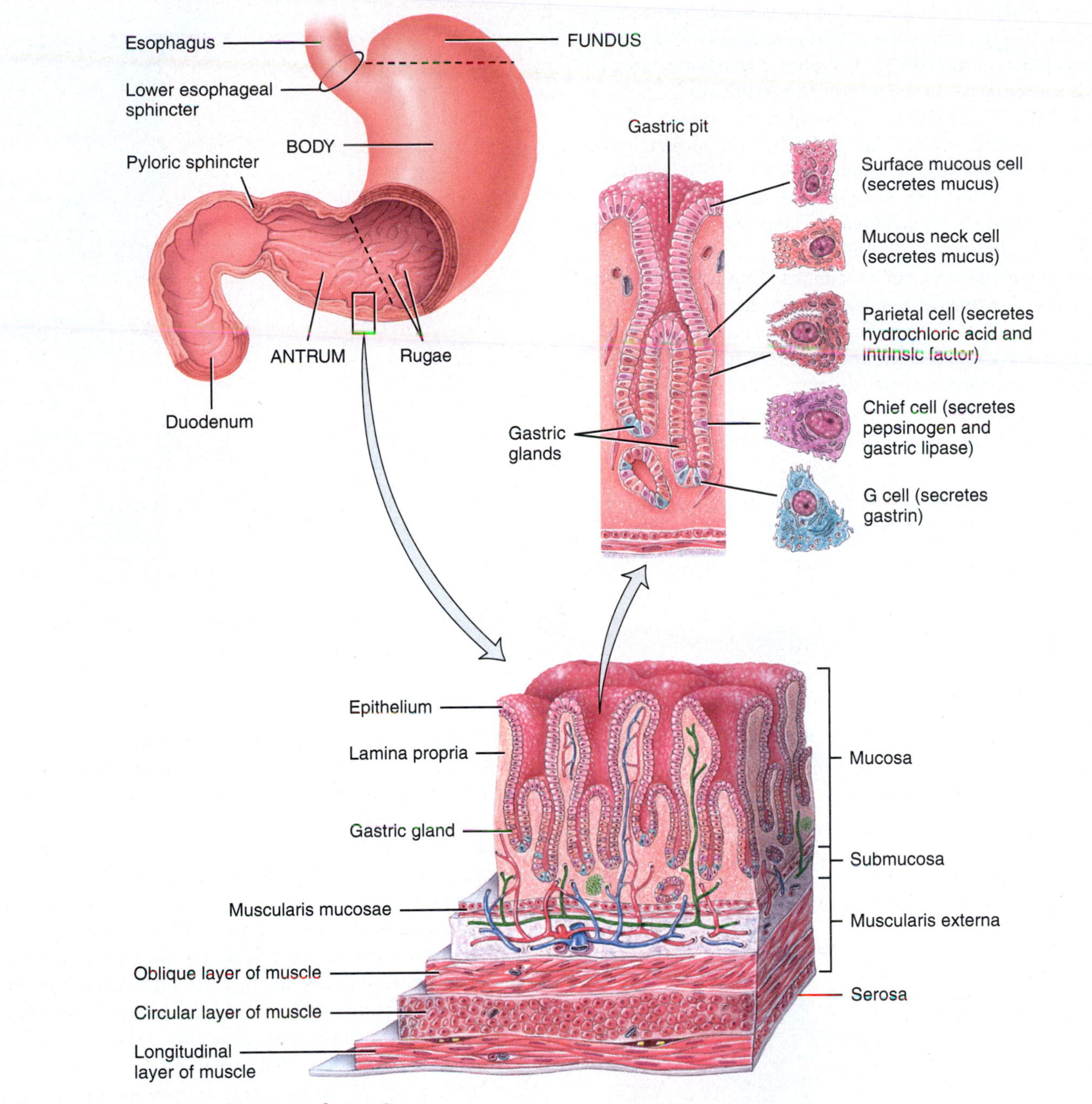

What is the purpose of intrinsic factor?

occurs, moving the food particles back down into the antrum. If the food particles are still too large to pass through the pyloric sphincter, retropulsion occurs again as the particles are squeezed back into the body of the stomach. Then yet another round of propulsion occurs and the cycle continues to repeat. The net result of these movements is that gastric contents are mixed with gastric juice, eventually becoming reduced to a soupy liquid called **chyme** (KĪM = juice). Once the food particles in chyme are small enough, they can pass through the pyloric sphincter, a phenomenon known as **gastric emptying**. Gastric emptying is a slow process: only about 3 mL of chyme move through the pyloric sphincter at a time.

Chemical Digestion in the Stomach Occurs as Food Mixes with Gastric Juice

Foods may remain in the fundus for about an hour without becoming mixed with gastric juice. During this time, digestion by salivary amylase continues. Soon, however, the churning action mixes chyme with acidic gastric juice, inactivating salivary amylase and activating lingual lipase, which starts to digest triglycerides into fatty acids and diglycerides.

Although parietal cells secrete hydrogen ions (H^+) and chloride ions (Cl^-) separately into the stomach lumen, the net effect is secretion of hydrochloric acid (HCl). **Proton pumps** powered by H^+/K^+ ATPases actively transport H^+ into the lumen while bringing potassium ions (K^+) into the cell (**FIGURE 21.11**). At the same time, Cl^- and K^+ diffuse out into the lumen through Cl^- and K^+ channels in the apical membrane. The enzyme *carbonic anhydrase*, which is especially plentiful in parietal cells, catalyzes the formation of carbonic acid (H_2CO_3) from water (H_2O) and carbon dioxide (CO_2). As carbonic acid dissociates, it provides a ready source of H^+ for the proton pumps but also generates bicarbonate ions (HCO_3^-). As HCO_3^- builds up in the cytosol, it exits the parietal cell in exchange for Cl^- via Cl^-/HCO_3^- antiporters in the basolateral membrane (next to the lamina propria). HCO_3^- diffuses into nearby blood capillaries. This "alkaline tide" of bicarbonate ions entering the bloodstream after a meal may be large enough to elevate blood pH slightly and make urine more alkaline.

HCl secretion by parietal cells can be stimulated by several sources (**FIGURE 21.12**): acetylcholine (ACh) released by parasympathetic neurons; gastrin secreted by G cells; and histamine, which is a paracrine substance released by mast cells in the nearby lamina propria. Acetylcholine and gastrin stimulate parietal cells to secrete more HCl in the presence of histamine. In other words, histamine acts synergistically, enhancing the effects of acetylcholine and gastrin. Receptors for all three substances are present in the plasma membrane of parietal cells. The histamine receptors on parietal cells are called H_2 receptors; they mediate different responses than do the H_1 receptors involved in allergic responses.

The strongly acidic fluid of the stomach kills many microbes in food. HCl partially denatures proteins in food and stimulates the secretion of hormones that promote the flow of bile and pancreatic juice. Enzymatic digestion of proteins also begins in the stomach. The only proteolytic (protein-digesting) enzyme in the stomach is **pepsin**, which is derived from pepsinogen secreted by chief cells. Pepsin severs certain peptide bonds between amino acids, breaking down a protein chain of many amino acids into smaller peptide fragments (**FIGURE 21.13**). Pepsin is most effective in the very acidic environment of the stomach (pH 2); it becomes inactive at a higher pH.

What keeps pepsin from digesting the protein in stomach cells along with the food? First, pepsin is secreted in an inactive form (pepsinogen); in this form, it cannot digest the proteins in the chief cells that produce it. Pepsinogen is not converted into active pepsin until it comes in contact with hydrochloric acid secreted by parietal cells or active pepsin molecules. Second, the stomach epithelial cells are protected from gastric juices by a 1–3 mm thick layer of alkaline mucus secreted by surface mucous cells and mucous neck cells.

Another enzyme of the stomach is **gastric lipase**, which splits the short-chain triglycerides in fat molecules (such as those found in milk) into fatty acids and monoglycerides. A monoglyceride consists of a glycerol molecule that is attached to one fatty acid molecule. This enzyme, which has a limited role in the adult stomach, operates best at a pH of 5–6. More important than either lingual lipase or gastric lipase is pancreatic lipase, an enzyme secreted by the pancreas into the small intestine.

FIGURE 21.11 Secretion of HCl (hydrochloric acid) by parietal cells in the stomach.

Proton pumps, powered by ATP, secrete H^+; Cl^- diffuses into the stomach lumen through Cl^- channels.

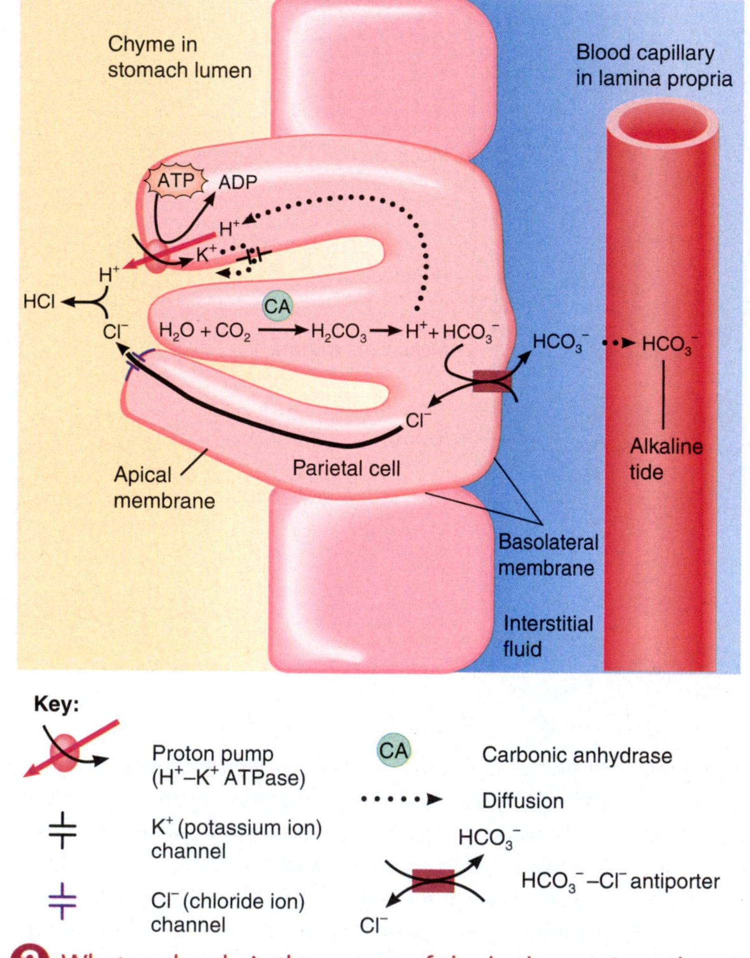

What molecule is the source of the hydrogen ions that are secreted into gastric juice?

FIGURE 21.12 Regulation of HCl secretion.

HCl secretion by parietal cells can be stimulated by acetylcholine (ACh), gastrin, and histamine.

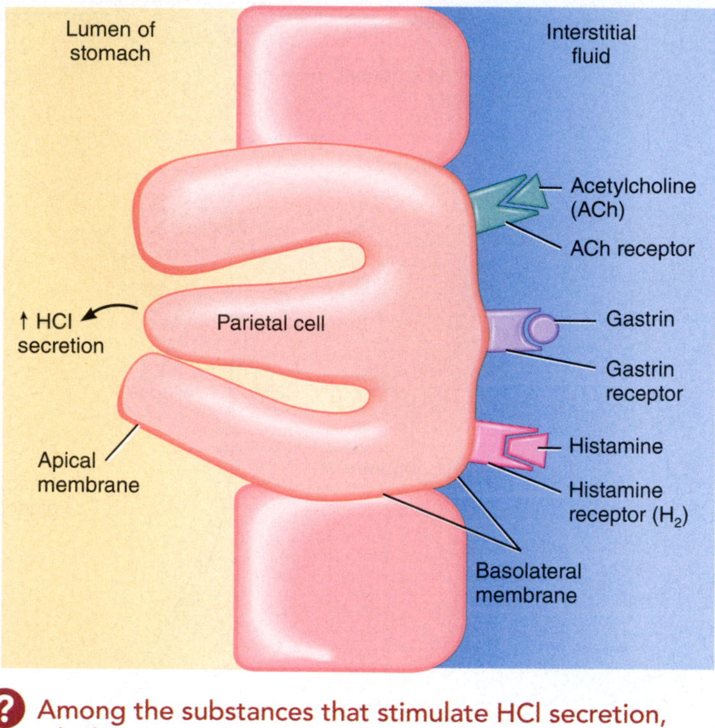

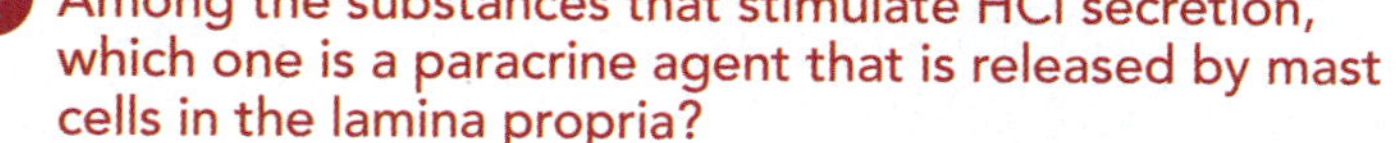

Among the substances that stimulate HCl secretion, which one is a paracrine agent that is released by mast cells in the lamina propria?

FIGURE 21.13 Enzymes involved in protein digestion. Several enzymes digest proteins: (1) pepsin, which is produced by the stomach; (2) trypsin, (3) chymotrypsin, (4) elastase, and (5) carboxypeptidase, which are produced by the pancreas; and (6) aminopeptidase, which is produced by the small intestine.

Pepsin, trypsin, chymotrypsin, and elastase break down proteins into smaller peptides; aminopeptidase and carboxypeptidase cleave off individual amino acids from peptides.

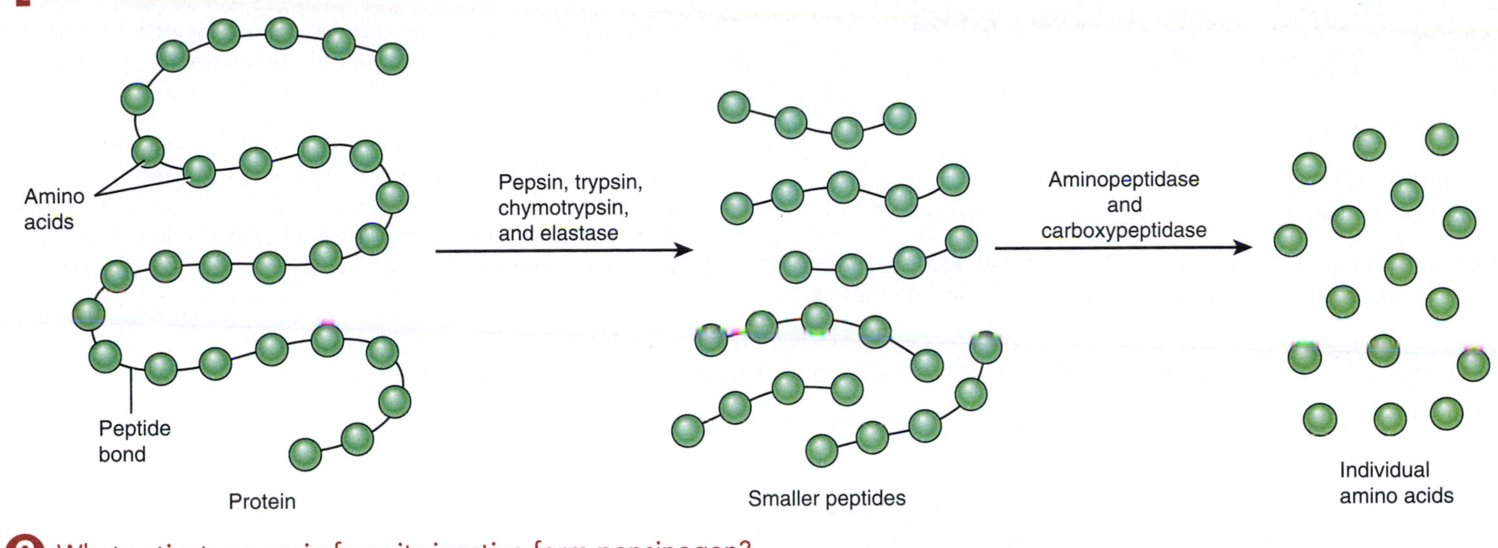

What activates pepsin from its inactive form pepsinogen?

Only a small amount of nutrients are absorbed in the stomach because its epithelial cells are impermeable to most materials. However, mucous cells of the stomach absorb some water, ions, and short-chain fatty acids, as well as certain drugs (especially aspirin) and alcohol.

Within 2 to 4 hours after eating a meal, the stomach has emptied its contents into the duodenum. Foods rich in carbohydrate spend the least time in the stomach, high-protein foods remain somewhat longer, and emptying is slowest after a fat-laden meal containing large amounts of triglycerides.

Vomiting Rapidly Expels the Contents of the GI Tract

Vomiting or *emesis* is the forcible expulsion of the contents of the GI tract (stomach and sometimes duodenum) through the mouth. The strongest stimuli for vomiting are irritation and distension of the stomach; other stimuli include unpleasant sights, general anesthesia, dizziness, and certain drugs such as morphine. Action potentials are transmitted to the **vomiting center** in the medulla oblongata, and returning action potentials propagate to the upper GI tract organs, diaphragm, and abdominal muscles. Vomiting involves squeezing the stomach between the diaphragm and abdominal muscles and expelling the contents through open esophageal sphincters. Prolonged vomiting, especially in infants and elderly people, can be serious because the loss of acidic gastric juice can lead to alkalosis (higher than normal blood pH), dehydration, and damage to the esophagus and teeth. The discomfort caused by the sensation of impending vomiting is called *nausea* (NAW-sē-a).

CHECKPOINT

15. What is the importance of rugae, mucous cells, chief cells, parietal cells, and G cells in the stomach?

16. Why do propulsion and retropulsion occur in the stomach? What is the net effect of these processes?

CLINICAL CONNECTION

Peptic Ulcer Disease

In the United States, 5–10% of the population develops **peptic ulcer disease (PUD)**. An **ulcer** is a craterlike lesion in a membrane; ulcers that develop in areas of the GI tract exposed to acidic gastric juice are called **peptic ulcers**. The most common complication of peptic ulcers is bleeding, which can lead to anemia if enough blood is lost. In acute cases, peptic ulcers can lead to shock and death. Three distinct causes of PUD are recognized: (1) the bacterium ***Helicobacter pylori***, (2) nonsteroidal anti-inflammatory drugs (NSAIDs) such as aspirin, and (3) hypersecretion of HCl.

Helicobacter pylori (previously named *Campylobacter pylori*) is the most frequent cause of PUD. The bacterium produces an enzyme called urease, which splits urea into ammonia and carbon dioxide. While shielding the bacterium from the acidity of the stomach, the ammonia also damages the protective mucous layer of the stomach and the underlying gastric cells. *H. pylori* also produces catalase, an enzyme that may protect the microbe from phagocytosis by neutrophils, plus several adhesion proteins that allow the bacterium to attach itself to gastric cells.

Several therapeutic approaches are helpful in the treatment of PUD. Cigarette smoke, alcohol, caffeine, and NSAIDs should be avoided because they can impair mucosal defensive mechanisms, which increases mucosal susceptibility to the damaging effects of HCl. In cases associated with *H. pylori*, treatment with an antibiotic drug often resolves the problem. Oral antacids such as Tums® or Maalox® can help temporarily by buffering gastric acid. When hypersecretion of HCl is the cause of PUD, histamine (H_2) blockers (such as Tagamet®) or proton pump inhibitors such as omeprazole (Prilosec®), which block secretion of H^+ from parietal cells, may be used.

17. What is the role of pepsin? Why is it secreted in an inactive form?
18. What are the functions of lingual lipase and gastric lipase in the stomach?

21.5 Pancreas, Liver, and Gallbladder

OBJECTIVES

- Describe the functions of the pancreas.
- Discuss the functions of the liver and gallbladder.

From the stomach, chyme passes into the small intestine. Because chemical digestion in the small intestine depends on activities of the pancreas, liver, and gallbladder, you will first learn about the activities of these accessory digestive organs and their contributions to digestion in the small intestine.

The Pancreas Secretes Pancreatic Juice

The **pancreas** is an elongated, tapered gland located behind the stomach (see FIGURE 21.1). It contains both an exocrine portion and an endocrine portion.

- The *exocrine portion*, which forms the bulk of the pancreas, consists of clusters of cells called **acini** (AS-i-nē; singular is *acinus*) that are attached to small ducts (FIGURE 21.14a, b). The acinar cells secrete digestive enzymes, and the duct cells secrete a fluid rich in bicarbonate. The secretions of the acinar and duct cells collectively form *pancreatic juice*. Pancreatic juice is released into the duodenum

FIGURE 21.14 Relationship of the pancreas to the liver, gallbladder, and duodenum.

Pancreatic enzymes digest starches (polysaccharides), proteins, triglycerides, and nucleic acids.

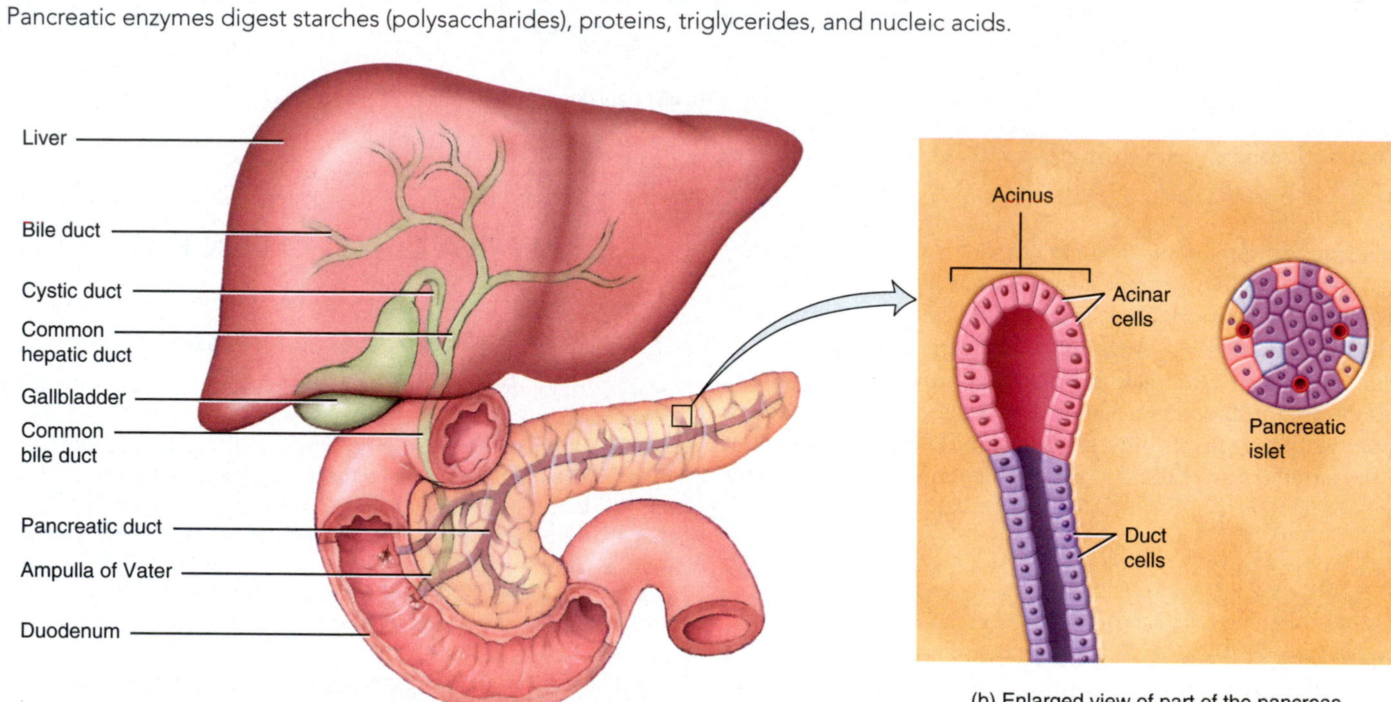

(a) Connections among the pancreas, liver, gallbladder, and duodenum

(b) Enlarged view of part of the pancreas

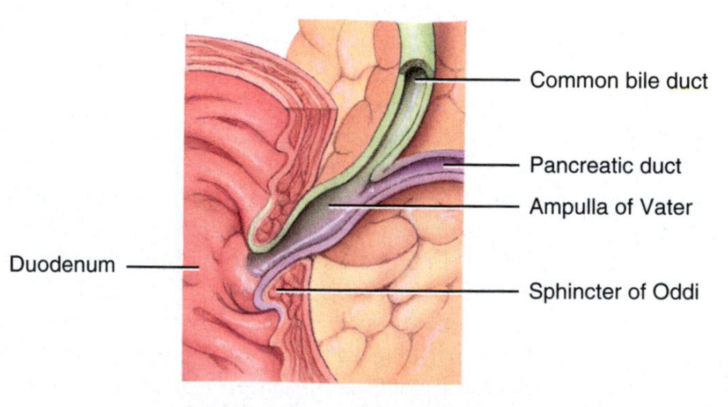

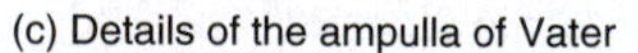
(c) Details of the ampulla of Vater

What type of fluid is found in the pancreatic duct? The common bile duct? The ampulla of Vater?

of the small intestine in the following way: The small ducts associated with the acini convey pancreatic juice into the **pancreatic duct**, which in turn unites with the common bile duct from the liver and gallbladder to form a duct called the **ampulla of Vater** *(hepatopancreatic ampulla)* that opens into the lumen of the duodenum (**FIGURE 21.14c**). The passage of pancreatic juice and bile through the ampulla of Vater into the small intestine is regulated by a mass of smooth muscle known as the **sphincter of Oddi** *(sphincter of the hepatopancreatic ampulla)*.

- The *endocrine portion* of the pancreas consists of clusters of cells called **pancreatic islets (islets of Langerhans)** that are scattered among the acini (**FIGURE 21.14b**). The cells of the pancreatic islets secrete the hormones glucagon, insulin, somatostatin, and pancreatic polypeptide. The functions of these hormones were discussed in Chapter 13.

Each day the pancreas produces about 2 liters of **pancreatic juice**, a clear, colorless liquid consisting mostly of water, some salts, sodium bicarbonate, and several enzymes. The sodium bicarbonate gives pancreatic juice a slightly alkaline pH (7.1–8.2) that buffers acidic gastric juice in chyme, stops the action of pepsin from the stomach, and creates the proper pH for the action of digestive enzymes in the small intestine. The enzymes in pancreatic juice include a starch-digesting enzyme called **pancreatic amylase**; several protein-digesting enzymes called **trypsin** (TRIP-sin), **chymotrypsin** (kī′-mō-TRIP-sin), **carboxypeptidase** (kar-bok′-sē-PEP-ti-dās), and **elastase** (ē-LAS-tās); the principal triglyceride-digesting enzyme in adults, called **pancreatic lipase**; and nucleic acid–digesting enzymes called **ribonuclease** and **deoxyribonuclease**.

The protein-digesting enzymes of the pancreas are produced in an inactive form just as pepsin is produced in the stomach as pepsinogen. Because they are inactive, the enzymes do not digest cells of the pancreas itself. Trypsin is secreted in an inactive form called **trypsinogen**. Pancreatic acinar cells also secrete a protein called **trypsin inhibitor** that combines with any trypsin formed accidentally in the pancreas or in pancreatic juice and blocks its enzymatic activity. When trypsinogen reaches the lumen of the small intestine, it encounters an activating brush-border enzyme called **enterokinase** (en′-ter-ō-KĪ-nās), which splits off part of the trypsinogen molecule to form trypsin (see **FIGURE 21.18**). In turn, trypsin acts on the inactive precursors (called **chymotrypsinogen**, **procarboxypeptidase**, and **proelastase**) to produce chymotrypsin, carboxypeptidase, and elastase, respectively.

The Liver Secretes Bile and Performs Many Other Functions

The **liver** is the largest gland of the body, weighing about 1.4 kg (about 3 lb) in an average adult. It is located below the diaphragm, mostly on the right side of the body (see **FIGURE 21.1**). The liver is organized into many functional units called **lobules** (**FIGURE 21.15a**). Each lobule is a six-sided structure (hexagon) that consists of specialized epithelial cells, called **hepatocytes**, arranged in

FIGURE 21.15 The liver.

The functional unit of the liver is the lobule, which is a six-sided structure (hexagon) of hepatocytes arranged around a central vein.

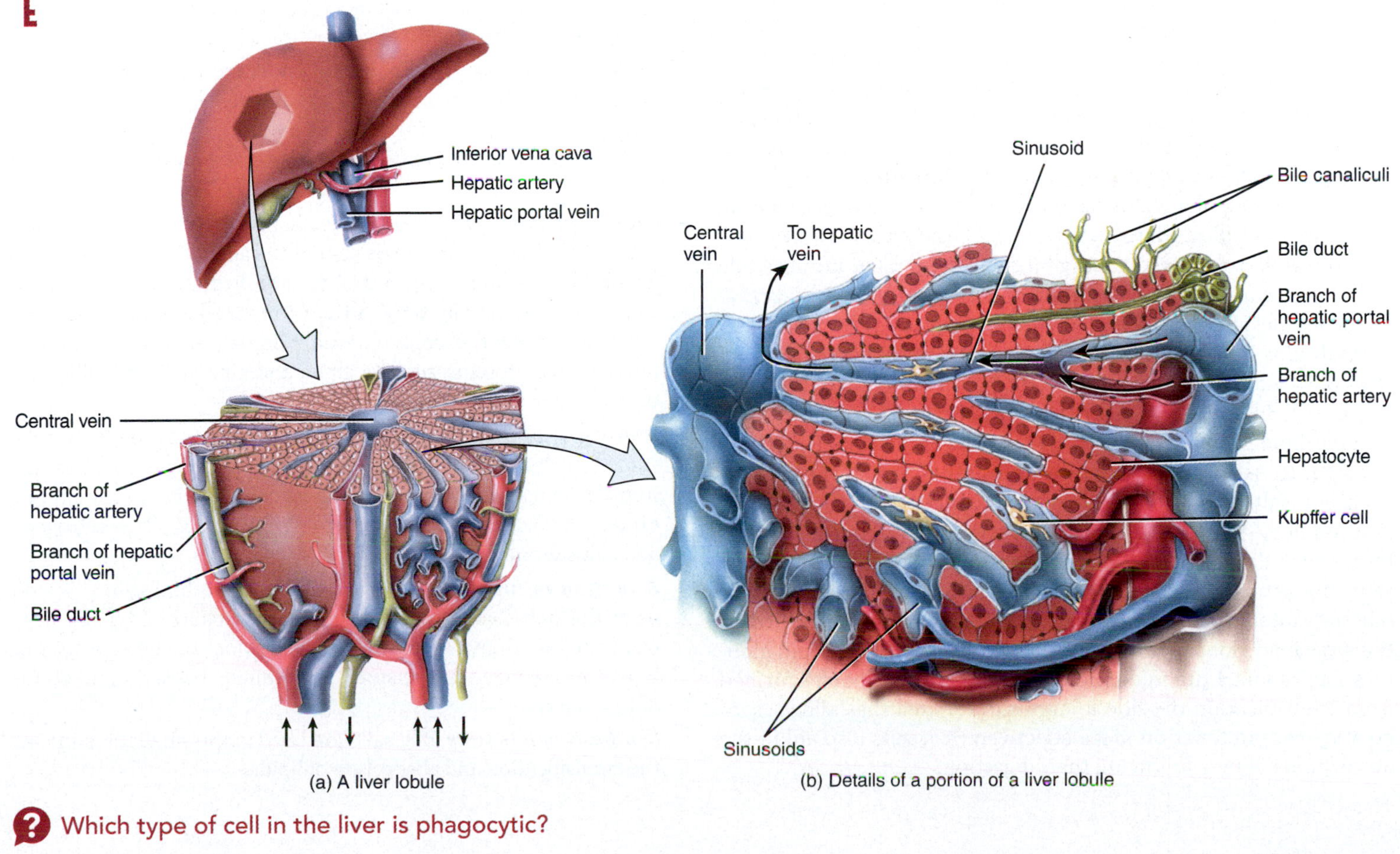

(a) A liver lobule

(b) Details of a portion of a liver lobule

Which type of cell in the liver is phagocytic?

irregular, branching, interconnecting plates around a central vein. The hepatocytes are the major functional cells of the liver and perform a wide array of metabolic, secretory, and endocrine functions. Between the hepatocytes are small ducts called **canaliculi**, which collect bile produced by the hepatocytes (FIGURE 21.15b). From the canaliculi, bile passes into *bile ducts*. The bile ducts merge to form larger ducts that eventually give rise to the *common hepatic duct*. The common hepatic duct joins the *cystic duct* from the gallbladder to form the **common bile duct** (see FIGURE 21.14a). From here, bile enters the small intestine to participate in digestion. When the small intestine is empty, the sphincter of Oddi closes, and bile backs up into the gallbladder for storage. In addition to hepatocytes and canaliculi, the liver lobule contains highly permeable capillaries called **sinusoids**, through which blood passes (FIGURE 21.15b). Also present in the sinusoids are fixed phagocytes called **Kupffer cells**, which destroy worn-out erythrocytes and leukocytes, bacteria, and other foreign matter in the venous blood draining from the gastrointestinal tract. The liver receives blood from two sources: the hepatic artery and the hepatic portal vein (FIGURE 21.15b). From the hepatic artery, the liver obtains oxygenated blood, and from the hepatic portal vein, it receives deoxygenated blood containing newly absorbed nutrients, drugs, and possibly microbes and toxins from the gastrointestinal tract. Each corner of a liver lobule contains a branch of the hepatic artery, a branch of the hepatic portal vein, and a bile duct. Branches of both the hepatic artery and the hepatic portal vein carry blood into liver sinusoids, where oxygen, most of the nutrients, and certain toxic substances are taken up by hepatocytes. Products manufactured by the hepatocytes and nutrients needed by other cells are secreted back into the blood, which then drains into the central vein and eventually passes into a hepatic vein.

Each day, hepatocytes secrete about 1 liter of **bile**, a yellow-green liquid. Bile is partially an excretory product and partially a digestive secretion. It has a pH of 7.6–8.6 and consists of water, bile pigments, bile salts, cholesterol, a phospholipid called lecithin, and several ions. The principal bile pigment is **bilirubin**, which has a yellow color. The phagocytosis of aged erythrocytes liberates iron, globin, and bilirubin (derived from heme) (see FIGURE 16.6). The iron and globin are recycled; the bilirubin is secreted into the bile and is eventually broken down in the intestine. One of its breakdown products—the brown pigment **stercobilin**—gives feces their normal color. Bile salts are also important components of bile. Two examples of bile salts are glycocholate and taurocholate (FIGURE 21.16a, b). Bile salts play a role in **emulsification**, the breakdown of large lipid globules into a suspension of small lipid globules (FIGURE 21.16c). Bile salts are **amphipathic**, which means that each bile salt has a nonpolar (hydrophobic) region and a polar (hydrophilic) region. The amphipathic nature of bile salts allows them to emulsify a large lipid globule: The nonpolar regions of bile salts interact with the large lipid globule, while the polar regions of bile salts interact with the watery intestinal chyme. Consequently, the large lipid globule is broken apart into several small lipid globules, each about 1 μm in diameter. The small lipid globules formed from emulsification provide a large surface area that allows pancreatic lipase to function more effectively. Bile salts also aid in the absorption of lipids following their digestion.

CLINICAL CONNECTION

Jaundice

Jaundice (JAWN-dis) is a yellowish coloration of the sclerae (whites of the eyes) and skin due to a buildup of the yellow pigment bilirubin. After bilirubin is formed from the breakdown of the heme pigment in aged erythrocytes, it is transported to the liver, where it is processed and eventually excreted into bile. The three main categories of jaundice are (1) ***prehepatic jaundice***, due to excess production of bilirubin; (2) ***hepatic jaundice***, due to congenital liver disease, hepatitis, or cirrhosis; and (3) ***extrahepatic jaundice***, due to blockage of bile drainage by gallstones or cancer of the bowel or the pancreas.

Because the liver of a newborn functions poorly for the first week or so, many babies experience a mild form of jaundice called ***neonatal (physiological) jaundice*** that disappears as the liver matures. Usually, it is treated by exposing the infant to blue light, which converts bilirubin into substances the kidneys can excrete.

In addition to secreting bile, which is needed for absorption of dietary fats, the liver performs many other vital functions:

- ***Carbohydrate metabolism.*** The liver is especially important in maintaining a normal blood glucose level. When blood glucose is low, the liver can break down glycogen to glucose and release the glucose into the bloodstream. The liver can also convert certain amino acids and lactic acid to glucose, and it can convert other sugars, such as fructose and galactose, into glucose. When blood glucose is high, as occurs just after eating a meal, the liver converts glucose to glycogen and triglycerides for storage.
- ***Lipid metabolism.*** Hepatocytes store some triglycerides; break down fatty acids to generate ATP; synthesize lipoproteins, which transport fatty acids, triglycerides, and cholesterol to and from body cells; synthesize cholesterol; and use cholesterol to make bile salts.
- ***Protein metabolism.*** Hepatocytes *deaminate* (remove the amino group, NH_2, from) amino acids so that the amino acids can be used for ATP production or converted to carbohydrates or fats. The resulting toxic ammonia (NH_3) is then converted into the much less toxic urea, which is excreted in urine. Hepatocytes also synthesize most plasma proteins, such as alpha and beta globulins, albumin, prothrombin, and fibrinogen.
- ***Processing of drugs and hormones.*** The liver can detoxify substances such as alcohol and excrete drugs such as penicillin, erythromycin, and sulfonamides into bile. It can also chemically alter or excrete thyroid hormones and steroid hormones such as estrogens and aldosterone.
- ***Excretion of bilirubin.*** As previously noted, bilirubin (derived from the heme of aged erythrocytes) is absorbed by the liver from the blood and secreted into bile. Most of the bilirubin in bile is metabolized in the small intestine by bacteria and eliminated in feces.
- ***Synthesis of bile salts.*** Bile salts are used in the small intestine for the emulsification and absorption of lipids.

FIGURE 21.16 Bile salts and emulsification.

Emulsification is the process by which a large lipid globule is broken down into several small lipid globules.

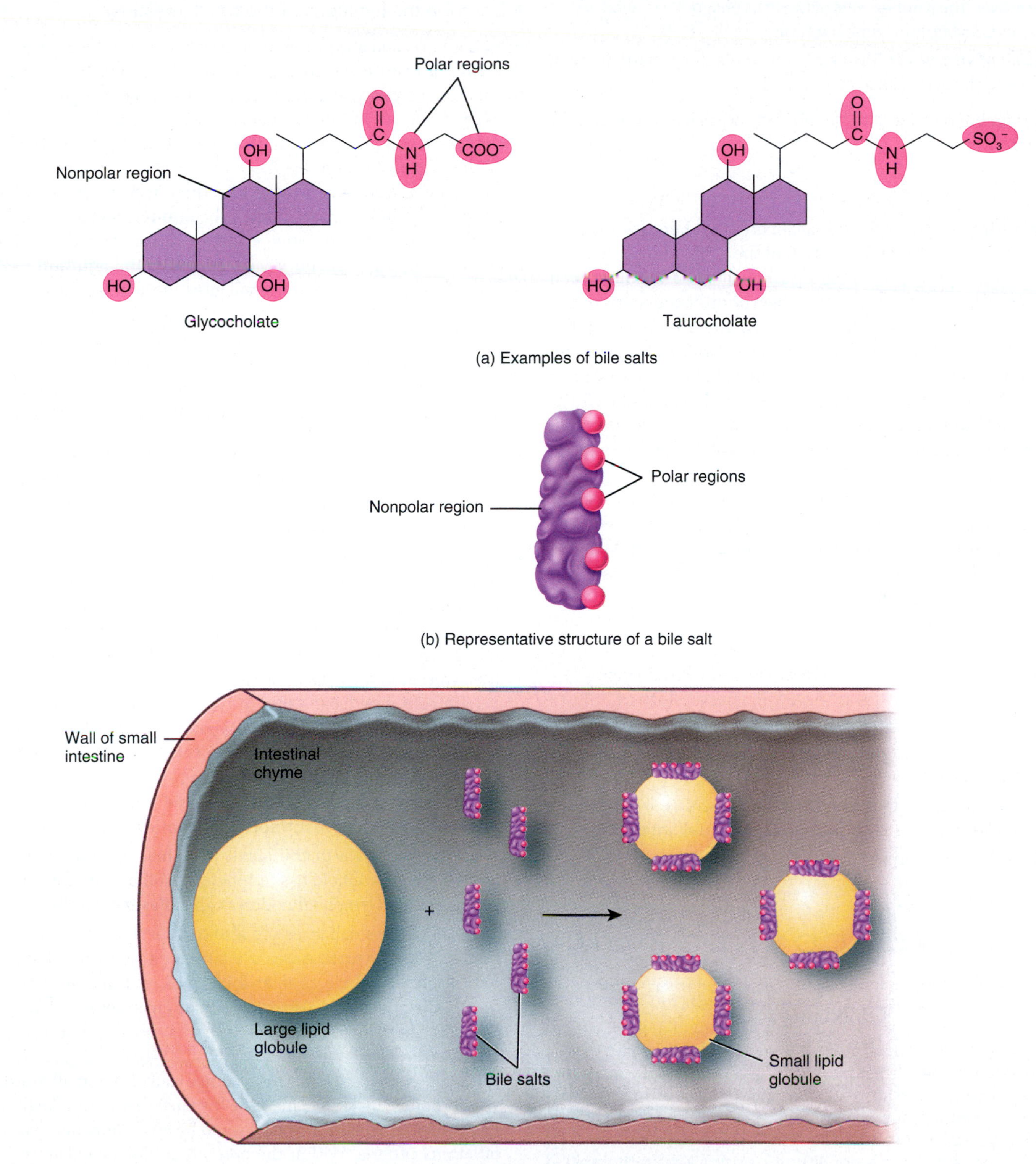

(a) Examples of bile salts

(b) Representative structure of a bile salt

(c) Emulsification of a large lipid globule

Why are bile salts able to promote emulsification?

- ***Storage.*** In addition to glycogen, the liver is a prime storage site for certain vitamins (A, B_{12}, D, E, and K) and minerals (iron and copper), which are released from the liver when needed elsewhere in the body.
- ***Phagocytosis.*** The Kupffer cells of the liver phagocytize aged erythrocytes, leukocytes, and some bacteria.
- ***Activation of vitamin D.*** The skin, liver, and kidneys participate in synthesizing the active form of vitamin D.

The liver functions related to metabolism are discussed more fully in Chapter 22.

The Gallbladder Stores and Concentrates Bile

The **gallbladder** is a pear-shaped sac located just beneath the liver (see **FIGURE 21.14a**). A major function of the gallbladder is to store bile. Bile is continuously secreted by the liver (about 1 liter each day). However, between meals after absorption has occurred, the sphincter of Oddi is closed, which prevents bile from entering the small intestine. Any bile secreted by the liver at this point backs up into the gallbladder, where it is stored. The gallbladder can store a maximum of about 50 mL of bile at any given time. Another function of the gallbladder is to concentrate bile while it is being stored. In the concentration process, some of the water and ions are removed from bile.

When the bile in the gallbladder is needed in the small intestine, contraction of smooth muscle in the wall of the gallbladder ejects the contents of the gallbladder into the cystic duct. From this duct, bile passes through the common bile duct and ampulla of Vater and then enters the lumen of the duodenum.

CHECKPOINT

19. What are pancreatic acini? How do their functions differ from those of the pancreatic islets (islets of Langerhans)?
20. What are the digestive functions of the components of pancreatic juice?
21. Describe the ducts that connect the pancreas, liver, and gallbladder with the duodenum of the small intestine.
22. What is the function of bile?

CLINICAL CONNECTION

Gallstones

If bile contains either insufficient bile salts or lecithin or excessive cholesterol, the cholesterol may crystallize to form **gallstones**. As they grow in size and number, gallstones may cause minimal, intermittent, or complete obstruction to the flow of bile from the gallbladder into the duodenum. Treatment consists of using gallstone-dissolving drugs, lithotripsy (shock-wave therapy), or surgery. For people with recurrent gallstones or for whom drugs or lithotripsy is not indicated, ***cholecystectomy***—the removal of the gallbladder and its contents—is necessary. More than half a million cholecystectomies are performed each year in the United States.

21.6 Small Intestine

OBJECTIVE

- Describe the functions of the small intestine.

Most digestion and absorption of nutrients occur in a long tube called the **small intestine**, which extends from the stomach, coils through the abdominal cavity, and eventually opens into the large intestine. It averages 2.5 cm (1 in.) in diameter; its length is about 3 m (10 ft) in a living person and about 6.5 m (21 ft) in a cadaver due to the loss of smooth muscle tone after death.

The small intestine is divided into three regions: the duodenum, jejunum, and ileum (**FIGURE 21.17**). The **duodenum** (doo-ō-DĒ-num) is the initial portion of the small intestine. It is separated from the antrum of the stomach by the *pyloric sphincter*. The **jejunum** (je-JOO-num) is the next portion of the small intestine, connecting the duodenum to the ileum. The **ileum** (IL-ē-um) is the final and longest portion of the small intestine. It joins the large intestine at a smooth muscle sphincter called the **ileocecal sphincter (valve)** (il′-ē-ō-SĒ-kal).

The wall of the small intestine is composed of the same four layers as the rest of the GI tract, with several modifications (**FIGURE 21.17**):

- The surface of the small intestinal mucosa contains two types of epithelial cells: absorptive cells and goblet cells. **Absorptive cells** of the epithelium digest and absorb nutrients in small intestinal chyme; the **goblet cells** secrete mucus.
- Epithelial cells extend downward from the surface of the small intestinal mucosa to form intestinal glands called **crypts of Lieberkühn** (LĒ-ber-koon), which secrete intestinal juice. Besides absorptive cells and goblet cells, the crypts of Lieberkühn also contain paneth cells and enteroendocrine cells. **Paneth cells** secrete lysozyme, a bactericidal enzyme, and are capable of phagocytosis. Paneth cells may have a role in regulating the microbial population in the small intestine. Three types of enteroendocrine cells are found in the crypts of Lieberkühn of the small intestine: **S cells**, **CCK cells**, and **K cells**, which secrete the hormones **secretin** (se-KRĒ-tin), **cholecystokinin** (kō-lē-sis′-tō-KĪN-in) **(CCK)**, and **glucose-dependent insulinotropic peptide (GIP)**, respectively.
- The lamina propria of the small intestinal mucosa contains an abundance of lymphoid tissue, including solitary lymphoid nodules and aggregated lymphoid nodules called **Peyer's patches**.
- The wall of the small intestine also has special structural features that increase the surface area available for absorption. These structural features include circular folds, villi, and microvilli (**FIGURE 21.17**). **Circular folds**, also called **plicae circulares** (PLĪ-kē ser-kū-LAR-ēz), are permanent folds of the mucosa and submucosa that increase the surface area by a factor of 3. **Villi** are fingerlike projections of the mucosa that give the inner wall of the small intestine a velvety appearance. They increase the surface area by a factor of 10. Each villus (singular form) is covered by epithelium and has a core of lamina propria. Within the core are a blood capillary network and a **lacteal** (LAK-tē-al), which is a lymphatic capillary. Nutrients absorbed by the epithelial cells covering the villus pass through the wall of a blood capillary or a lacteal to enter blood or lymph, respectively. In addition to circular folds and villi, the small intestine also has **microvilli**, tiny projections of the apical (free) membrane of the absorptive cells that increase the surface area by a factor of

FIGURE 21.17 The small intestine.

Most digestion and absorption occur in the small intestine; the surface area of the small intestine is increased by circular folds, villi, and microvilli.

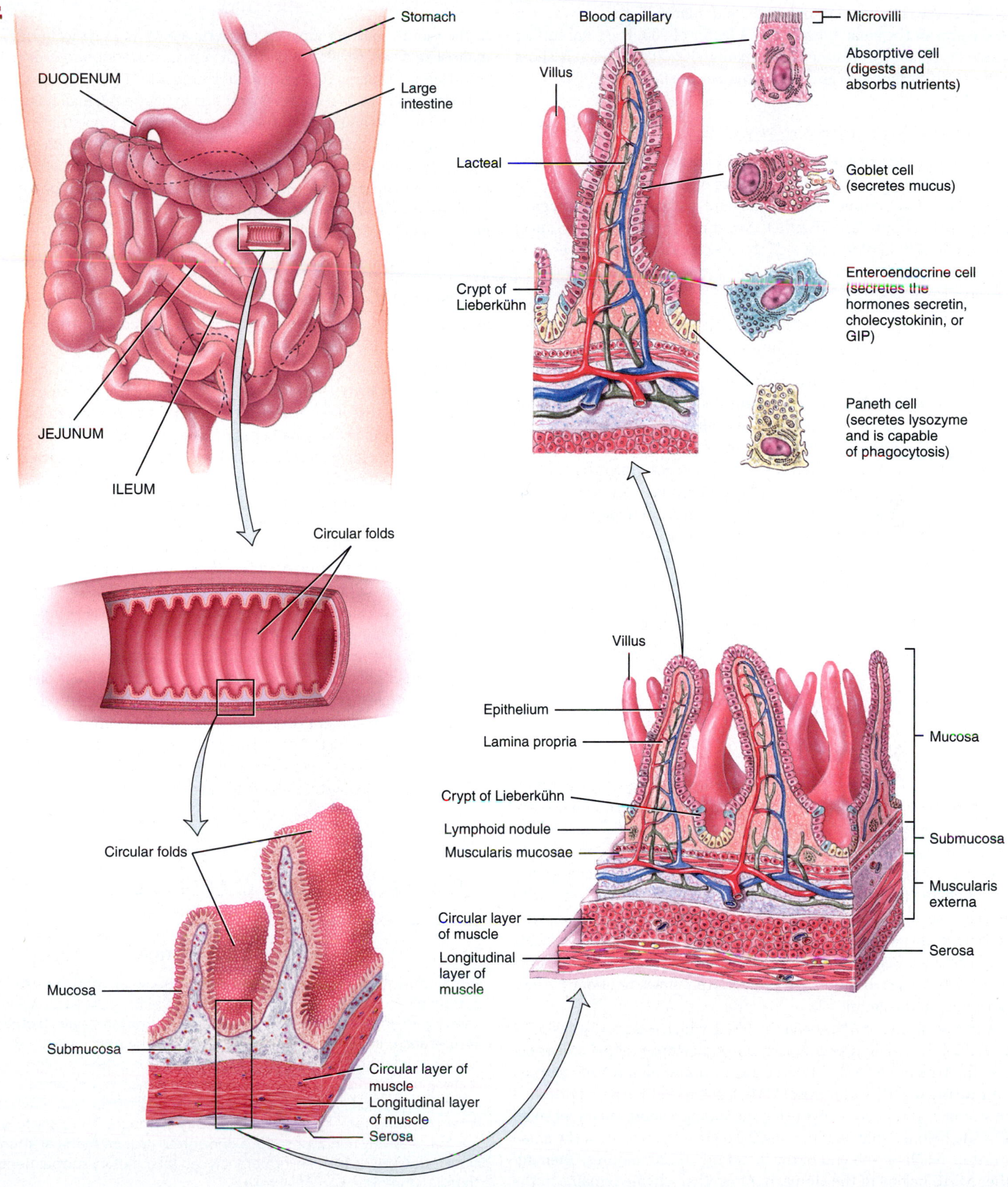

What are villi?

20. When viewed through a light microscope, the microvilli are too small to be seen individually; instead, they form a fuzzy line, called the **brush border**, extending into the lumen of the small intestine. Collectively, the circular folds, villi, and microvilli increase the surface area of the small intestine by a factor of 600. The total surface area of the small intestine is approximately 200 m^2, which is about the same surface area as that of a tennis court.

The Small Intestine Secretes Intestinal Juice

About 1 liter of **intestinal juice**, or *succus entericus* (SUK-us en-TER-i-kus), is secreted each day by the crypts of Lieberkühn. It is a clear yellow fluid that contains water, ions, and mucus and has a slightly alkaline pH (about 7.6). Together, intestinal and pancreatic juices provide a liquid medium that aids the absorption of substances from chyme in the small intestine.

Brush-Border Enzymes Are Attached to the Microvilli of Small Intestinal Absorptive Cells

The absorptive cells of the small intestine synthesize several digestive enzymes, called **brush-border enzymes**, and insert them in the plasma membrane of the microvilli. You have already learned about one brush-border enzyme, namely enterokinase, which converts trypsinogen into trypsin. Among the other brush-border enzymes are four carbohydrate-digesting enzymes called α-dextrinase, sucrase, lactase, and maltase; a protein-digesting enzyme called aminopeptidase; and two types of nucleotide-digesting enzymes, nucleosidase and phosphatase (**FIGURE 21.18**). Thus, some enzymatic digestion occurs at the surface of the absorptive cells that line the villi, rather than in the lumen exclusively, as occurs in other parts of the GI tract. Also, as absorptive cells slough off into the lumen of the small intestine, they break apart and release enzymes that help digest nutrients in the chyme.

Mechanical Digestion in the Small Intestine Involves Segmentation and the Migrating Motility Complex

The two types of movements of the small intestine—segmentation and a type of peristalsis known as the migrating motility complex—are governed mainly by the myenteric plexus. As you have learned earlier in the chapter, **segmentation** refers to localized, mixing contractions that occur in portions of the small intestine distended by a large volume of chyme (see **FIGURE 21.5b**). Segmentation mixes chyme with the digestive juices and brings the molecules of food into contact with the mucosa for absorption; it does not push the intestinal contents along the tract. Segmentation occurs most rapidly in the duodenum, about 12 times per minute, and progressively slows to about 8 times per minute in the ileum.

After most of a meal has been absorbed, which lessens distension of the wall of the small intestine, segmentation stops and peristalsis begins. The type of peristalsis that occurs in the small intestine, termed a **migrating motility complex (MMC)**, begins in the lower portion of the stomach and pushes chyme forward along a short stretch of small intestine before dying out. The MMC slowly migrates down the small intestine, reaching the end of the ileum in 90–120 minutes. Then another MMC begins in the stomach. Altogether, chyme remains in the small intestine for 3–5 hours.

Chemical Digestion in the Small Intestine Occurs as Chyme Mixes with Intestinal Juice, Pancreatic Juice, and Bile

In the mouth, salivary amylase converts starch (a polysaccharide) to maltose (a disaccharide), maltotriose (a trisaccharide), and α-dextrins (short-chain, branched fragments of starch with 5–10 glucose units). In the stomach, pepsin converts proteins to peptides (small fragments of proteins), and lingual and gastric lipases convert some triglycerides into fatty acids, diglycerides, and monoglycerides. Thus, chyme entering the small intestine contains partially digested carbohydrates, proteins, and lipids. The completion of the digestion of carbohydrates, proteins, and lipids is a collective effort of pancreatic juice, bile, and intestinal juice in the small intestine.

Digestion of Carbohydrates

Even though the action of **salivary amylase** may continue in the stomach for a while, the acidic pH of the stomach destroys salivary amylase and ends its activity. Thus, only a few starches are broken down by the time chyme leaves the stomach. Those starches not already broken down into maltose, maltotriose, and α-dextrins are cleaved by **pancreatic amylase**, an enzyme in pancreatic juice that acts in the small intestine. Although pancreatic amylase acts on both glycogen and starches, it has no effect on another polysaccharide called cellulose, an indigestible plant fiber that is commonly referred to as "roughage" as it moves through the digestive system. After amylase (either salivary or pancreatic) has split starch into smaller fragments, a brush-border enzyme called **α-dextrinase** acts on the resulting α-dextrins, clipping off one glucose unit at a time (see **FIGURE 21.18**).

Ingested molecules of sucrose, lactose, and maltose—three disaccharides—are not acted on until they reach the small intestine. Three brush-border enzymes digest the disaccharides into monosaccharides. **Sucrase** breaks sucrose into a molecule of glucose and a molecule of fructose; **lactase** digests lactose into a molecule of glucose and a molecule of galactose; and **maltase** splits maltose and maltotriose into two or three molecules of glucose, respectively (see **FIGURE 21.18**). Digestion of carbohydrates ends with the production of monosaccharides, which the digestive system is able to absorb.

Lactose Intolerance

In some people, the absorptive cells of the small intestine fail to produce enough lactase, which, as you just learned, is essential for the digestion of lactose. This results in a condition called **lactose intolerance**, in which undigested lactose in chyme causes fluid to be retained in the feces; bacterial fermentation of the undigested lactose results in the production of gases. Symptoms of lactose intolerance include diarrhea, gas, bloating, and abdominal cramps after consumption of milk and other dairy products. The symptoms can be relatively minor or serious enough to require medical attention. Persons with lactose intolerance can take dietary supplements to aid in the digestion of lactose.

FIGURE 21.18 Brush-border enzymes of the small intestine.

Brush-border enzymes complete the chemical digestion of carbohydrate, protein, and nucleic acid molecules that are present in intestinal chyme.

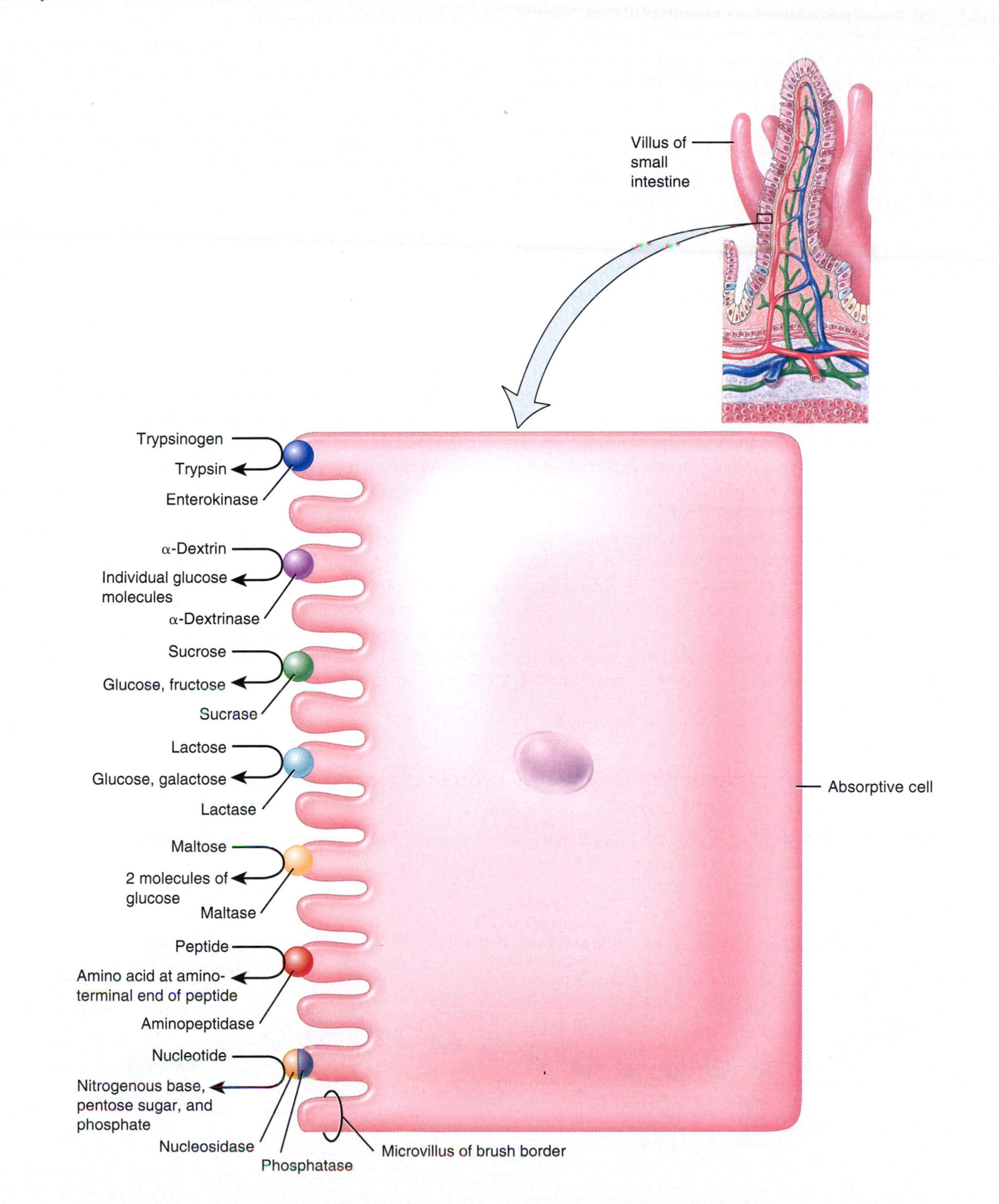

Why are brush-border enzymes so-named?

Digestion of Proteins

Recall that protein digestion starts in the stomach, where proteins are fragmented into peptides by the action of **pepsin** (see FIGURE 21.13). Three enzymes in pancreatic juice—**trypsin**, **chymotrypsin**, and **elastase**—continue to break down proteins into peptides. Pepsin, trypsin, chymotrypsin, and elastase are known as **endopeptidases** because they cleave peptide bonds within the interior of a protein to form smaller peptides (FIGURE 21.19). Although all these enzymes convert whole proteins into peptides, their actions differ somewhat because each splits peptide bonds between different amino acids. Protein digestion is completed by the enzymes aminopeptidase and carboxypeptidase. **Aminopeptidase**, which is present in the brush border (see FIGURE 21.18), cleaves off the amino acid at the amino-terminal end of a peptide. **Carboxypeptidase**, which is present in pancreatic juice, splits off the amino acid at the carboxyl-terminal end of a peptide. Aminopeptidase and carboxypeptidase are referred to as **exopeptidases** because they remove amino acids from the ends of a peptide (FIGURE 21.19). The final products of protein digestion are free amino acids, dipeptides, and tripeptides.

Digestion of Lipids

The most abundant lipids in the diet are triglycerides, which consist of a molecule of glycerol bonded to three fatty acid molecules (see FIGURE 2.17). Enzymes that split triglycerides are called **lipases**. You have already learned that there are three types of lipases that can participate in lipid digestion: **lingual lipase**, **gastric lipase**, and **pancreatic lipase**. Although some lipid digestion occurs in the stomach through the action of lingual and gastric lipases, most occurs in the small intestine through the action of pancreatic lipase. Triglycerides are broken down by pancreatic lipase into fatty acids and monoglycerides. The liberated fatty acids can be either short-chain fatty acids (with fewer than 10–12 carbons) or long-chain fatty acids.

Recall that, before a large lipid globule containing triglycerides can be digested in the small intestine, it must first undergo **emulsification**—a process in which the large lipid globule is broken down into several small lipid globules (see FIGURE 21.16c). The small lipid globules formed from emulsification provide a large surface area that allows pancreatic lipase to function more effectively.

Digestion of Nucleic Acids

Pancreatic juice contains two nucleases: **ribonuclease**, which digests RNA, and **deoxyribonuclease**, which digests DNA. The nucleotides that result from the action of the two nucleases are further digested by brush-border enzymes called **nucleosidases** and **phosphatases** into nitrogenous bases, pentose sugars, and phosphates (see FIGURE 21.18). These products are absorbed via active transport.

TABLE 21.1 summarizes the sources, substrates, and products of the digestive enzymes.

Most Nutrients and Water Are Absorbed in the Small Intestine

All of the chemical and mechanical phases of digestion, from the mouth through the small intestine, are directed toward changing food into forms that can pass through the absorptive epithelial cells lining the mucosa and into the underlying blood and lymphatic vessels. These forms are monosaccharides (glucose, fructose, and galactose) from carbohydrates; single amino acids, dipeptides, and tripeptides from proteins; and fatty acids, glycerol, and monoglycerides from triglycerides. Passage of these digested nutrients from the gastrointestinal tract into the blood or lymph is called **absorption**.

Absorption of materials occurs via simple diffusion, facilitated diffusion, osmosis, and active transport. About 90% of all absorption of nutrients occurs in the small intestine; the other 10% occurs in the stomach and large intestine. Any undigested or unabsorbed material left in the small intestine passes on to the large intestine.

FIGURE 21.19 Sites of peptide bond cleavage by protein-digesting enzymes.

Endopeptidases cleave peptide bonds within the interior of a protein; exopeptidases remove amino acids from the ends of a peptide.

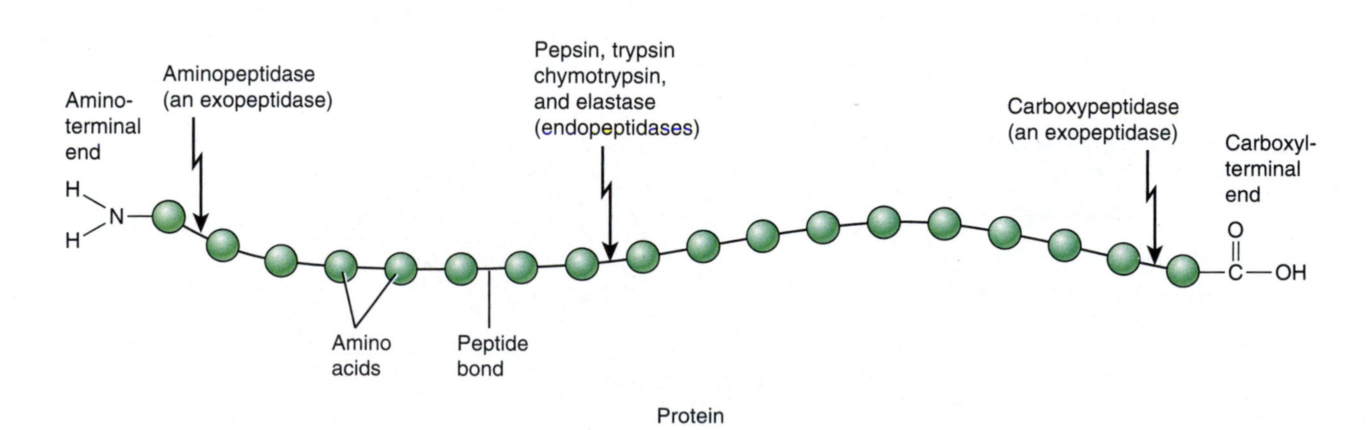

What is the function of aminopeptidase?

TABLE 21.1 Summary of Digestive Enzymes

Enzyme	Source	Substrates	Products
SALIVA			
Salivary amylase	Salivary glands.	Starches.	Maltose (disaccharide), maltotriose (trisaccharide), and α-dextrins.
Lingual lipase	Lingual glands in the tongue.	Triglycerides.	Fatty acids and diglycerides.
GASTRIC JUICE			
Pepsin	Stomach.	Proteins.	Peptides.
Gastric lipase	Stomach.	Triglycerides.	Fatty acids and monoglycerides.
PANCREATIC JUICE			
Pancreatic amylase	Pancreas.	Starches.	Maltose (disaccharide), maltotriose (trisaccharide), and α-dextrins.
Trypsin	Pancreas.	Proteins.	Peptides.
Chymotrypsin	Pancreas.	Proteins.	Peptides.
Elastase	Pancreas.	Proteins.	Peptides.
Carboxypeptidase	Pancreas.	Peptides.	Amino acid at carboxyl-terminal end of peptide.
Pancreatic lipase	Pancreas.	Triglycerides.	Fatty acids and monoglycerides.
Nucleases			
Ribonuclease	Pancreas.	Ribonucleic acid.	Nucleotides.
Deoxyribonuclease	Pancreas.	Deoxyribonucleic acid.	Nucleotides.
BRUSH BORDER			
Enterokinase	Small intestine.	Trypsinogen.	Trypsin.
α-dextrinase	Small intestine.	α-Dextrins.	Glucose.
Sucrase	Small intestine.	Sucrose.	Glucose and fructose.
Lactase	Small intestine.	Lactose.	Glucose and galactose.
Maltase	Small intestine.	Maltose.	Glucose.
Aminopeptidase	Small intestine.	Peptides.	Amino acid at amino-terminal end of peptide.
Nucleosidases and phosphatases	Small intestine.	Nucleotides.	Nitrogenous bases, pentose sugars, and phosphates.

Absorption of Monosaccharides

All carbohydrates are absorbed as monosaccharides. The capacity of the small intestine to absorb monosaccharides is huge—an estimated 120 grams per hour. As a result, all dietary carbohydrates that are digested normally are absorbed, leaving only indigestible cellulose and fibers in the feces. Monosaccharides pass from the lumen through the apical membrane via facilitated diffusion or active transport. Fructose, a monosaccharide found in fruits, is transported via facilitated diffusion; glucose and galactose are transported into absorptive cells of the villi via secondary active transport that is coupled to the active transport of Na^+ (FIGURE 21.20a). The transporter has binding sites for one glucose molecule and two sodium ions; unless all three sites are filled, neither substance is transported. Galactose competes with glucose to ride the same transporter (because both Na^+ and glucose or galactose move in the same direction, this is a *symporter*). Monosaccharides then move out of the absorptive cells through their basolateral surfaces via facilitated diffusion and enter the capillaries of the villi (FIGURE 21.20a, b).

Absorption of Amino Acids, Dipeptides, and Tripeptides

Most proteins are absorbed as amino acids via active transport processes that occur mainly in the duodenum and jejunum. About half of the absorbed amino acids are present in food; the other half come from the body itself as proteins in digestive juices and dead cells that slough off the mucosal surface! Normally, 95–98% of the protein present in the small intestine is digested and absorbed. Different transporters carry different types of amino acids. Some amino acids enter absorptive cells of the villi via Na^+-dependent secondary active transport processes that are similar to the glucose transporter; other amino acids enter by themselves via facilitated diffusion (FIGURE 21.20a). At least one symporter brings in dipeptides and tripeptides together with H^+; the peptides are then hydrolyzed to single amino acids inside the absorptive cells. Amino acids move out of the absorptive cells mainly via facilitated diffusion and enter capillaries of the villus (FIGURE 21.20a, b). Both monosaccharides and amino acids are transported in the blood to the liver by way of the hepatic portal system. If not removed by hepatocytes, they enter the general circulation.

Absorption of Lipids

All dietary lipids are absorbed via simple diffusion. Adults absorb about 95% of the lipids present in the small intestine; due to their lower production of bile, newborn infants absorb only about 85% of lipids. As a result of their emulsification and digestion, triglycerides are mainly broken down into monoglycerides and fatty acids, which can be either short-chain fatty acids or long-chain fatty acids. Although short-chain fatty acids are hydrophobic, they are very small in size. Because of their size, they can dissolve in the watery intestinal chyme, pass through the absorptive cells via simple diffusion, and follow the same route taken by monosaccharides and amino acids into a blood capillary of a villus (FIGURE 21.20a). Long-chain fatty acids and monoglycerides are large and hydrophobic and have difficulty being suspended in the watery environment of the intestinal chyme. Besides their role in emulsification, bile salts also help to make these long-chain fatty acids and monoglycerides more soluble. The bile salts in intestinal chyme surround the long-chain fatty acids and monoglycerides, forming tiny spheres called **micelles** (mī-SELZ = small morsels), each of which is 2–10 nm in diameter and includes 20–50 bile salt molecules (FIGURE 21.20a). Micelles are formed due to the amphipathic nature of bile salts: The hydrophobic regions of bile salts interact with the long-chain fatty acids and monoglycerides, and the hydrophilic regions of bile salts interact with the watery intestinal chyme. Once formed, the micelles move from the interior of the small intestinal lumen to the brush border of the absorptive cells. At that point, the long-chain fatty acids and monoglycerides diffuse out of the micelles into the absorptive cells, leaving the micelles behind in the chyme. The micelles continually repeat this ferrying function as they move from the brush border back through the chyme to the interior of the small intestinal lumen to pick up more long-chain fatty acids and monoglycerides. Micelles also solubilize other large hydrophobic molecules such as fat-soluble vitamins (A, D, E, and K) and cholesterol that may be present in intestinal chyme, and aid in their absorption. These fat-soluble vitamins and cholesterol molecules are packed in the micelles along with the long-chain fatty acids and monoglycerides.

Once inside the absorptive cells, long-chain fatty acids and monoglycerides are recombined to form triglycerides, which aggregate into globules along with phospholipids and cholesterol and become coated with proteins. These large spherical masses, about 80 nm in diameter, are called **chylomicrons** (FIGURE 21.20a). Chylomicrons leave the absorptive cell via exocytosis. Because they are so large and bulky, chylomicrons cannot enter blood capillaries—the pores in the walls of blood capillaries are too small. Instead, chylomicrons enter lacteals, which have much larger pores than blood capillaries. From lacteals, chylomicrons are transported by way of lymphatic vessels to the venous circulation (FIGURE 21.20b). The hydrophilic protein coat that surrounds each chylomicron keeps the chylomicrons suspended in blood and prevents them from sticking to each other.

Within 10 minutes after absorption, about half of the chylomicrons have already been removed from the blood as they pass through blood capillaries in the liver and adipose tissue. This removal is accomplished by an enzyme attached to the apical surface of capillary endothelial cells, called **lipoprotein lipase**, that breaks down triglycerides in chylomicrons and other lipoproteins into fatty acids and glycerol. The fatty acids and glycerol subsequently enter the hepatocytes or adipose cells and then recombine to form triglycerides. Two or three hours after a meal, few chylomicrons remain in the blood.

After participating in the emulsification and absorption of lipids, 90–95% of the bile salts are absorbed by active transport in the final segment of the small intestine (ileum) and returned by the blood to the liver through the hepatic portal system for recycling. This cycle of bile salt secretion by hepatocytes into bile, absorption by the ileum, and resecretion into bile is called the **enterohepatic circulation**. Insufficient bile salts, due either to obstruction of the bile ducts or removal of the gallbladder, can result in the loss of up to 40% of dietary lipids in feces because of diminished lipid absorption. When lipids are not absorbed properly, the fat-soluble vitamins are not adequately absorbed.

Absorption of Electrolytes

Many of the electrolytes absorbed by the small intestine come from gastrointestinal secretions, and some are part of ingested foods and liquids. Recall that electrolytes are compounds that separate into ions in water and conduct electricity. Sodium ions are actively transported out of absorptive cells by basolateral sodium–potassium pumps (Na^+/K^+ ATPase) after they have moved into absorptive cells via diffusion and secondary active transport. Thus, most of the sodium ions (Na^+) in gastrointestinal secretions are reclaimed and not lost in the feces.

FIGURE 21.20 Absorption of digested nutrients in the small intestine.

Long-chain fatty acids and monoglycerides are absorbed into lacteals; other products of digestion enter blood capillaries.

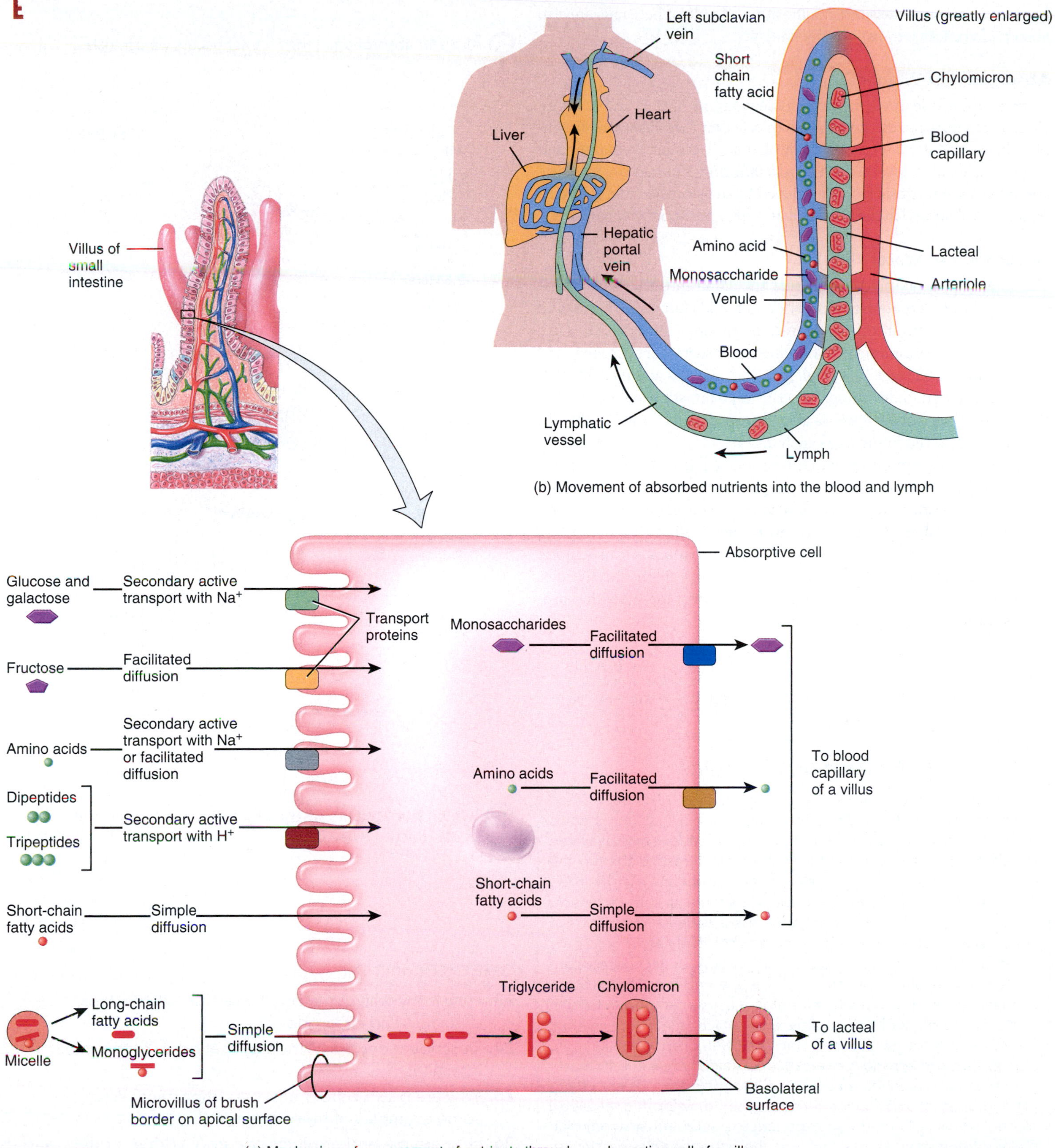

(b) Movement of absorbed nutrients into the blood and lymph

(a) Mechanisms for movement of nutrients through an absorptive cell of a villus

A monoglyceride may be larger than an amino acid. Why can monoglycerides be absorbed by simple diffusion, but amino acids cannot?

Negatively charged bicarbonate, chloride, iodide, and nitrate ions can passively follow Na^+ or be actively transported. Calcium ions are also absorbed actively in a process stimulated by calcitriol. Other electrolytes such as iron, potassium, magnesium, and phosphate ions are also absorbed via active transport mechanisms.

Absorption of Vitamins

As you have just learned, the fat-soluble vitamins A, D, E, and K are included with ingested dietary lipids in micelles and are absorbed via simple diffusion. Most water-soluble vitamins, such as most B vitamins and vitamin C, are also absorbed via simple diffusion. Vitamin B_{12}, however, combines with intrinsic factor produced by the stomach, and the combination is absorbed in the ileum via an active transport mechanism.

Absorption of Water

The total volume of fluid that enters the small intestine each day—about 9.3 liters—comes from ingestion of liquids (about 2.3 liters) and from various gastrointestinal secretions (about 7.0 liters). FIGURE 21.21 depicts the amounts of fluid ingested, secreted, absorbed, and excreted by the GI tract. The small intestine absorbs about 8.3 liters of the fluid; the remainder passes into the large intestine, where most of the rest of it—about 0.9 liter—is also absorbed. Only 0.1 liter (100 mL) of water is excreted in the feces each day.

All water absorption in the GI tract occurs via osmosis from the lumen of the intestines through absorptive cells and into blood capillaries. Because water can move across the intestinal mucosa in both directions, the absorption of water from the small intestine depends on the absorption of electrolytes and nutrients to maintain an osmotic balance with the blood. The absorbed electrolytes, monosaccharides, and amino acids establish a concentration gradient for water that promotes water absorption via osmosis.

CLINICAL CONNECTION

Absorption of Alcohol

The intoxicating and incapacitating effects of alcohol depend on the blood alcohol level. Because it is lipid-soluble, alcohol begins to be absorbed in the stomach. However, the surface area available for absorption is much greater in the small intestine than in the stomach, so when alcohol passes into the duodenum, it is absorbed more rapidly. Thus, the longer the alcohol remains in the stomach, the more slowly blood alcohol level rises. Because fatty acids in chyme slow gastric emptying, blood alcohol level rises more slowly when fat-rich foods, such as pizza, hamburgers, or nachos, are consumed with alcoholic beverages. Also, the enzyme alcohol dehydrogenase, which is present in gastric mucosa cells, breaks down some of the alcohol to acetaldehyde, which is not intoxicating. When the rate of gastric emptying is slower, proportionally more alcohol is absorbed and converted to acetaldehyde in the stomach, and thus less alcohol reaches the bloodstream. Given identical consumption of alcohol, females often develop higher blood alcohol levels (and therefore experience greater intoxication) than males of comparable size because the activity of gastric alcohol dehydrogenase is up to 60% lower in females than in males. Asian males may also have lower levels of this gastric enzyme.

FIGURE 21.21 Daily volumes of fluid ingested, secreted, absorbed, and excreted from the GI tract.

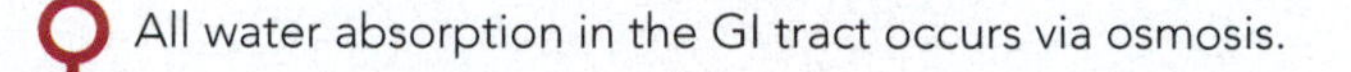

All water absorption in the GI tract occurs via osmosis.

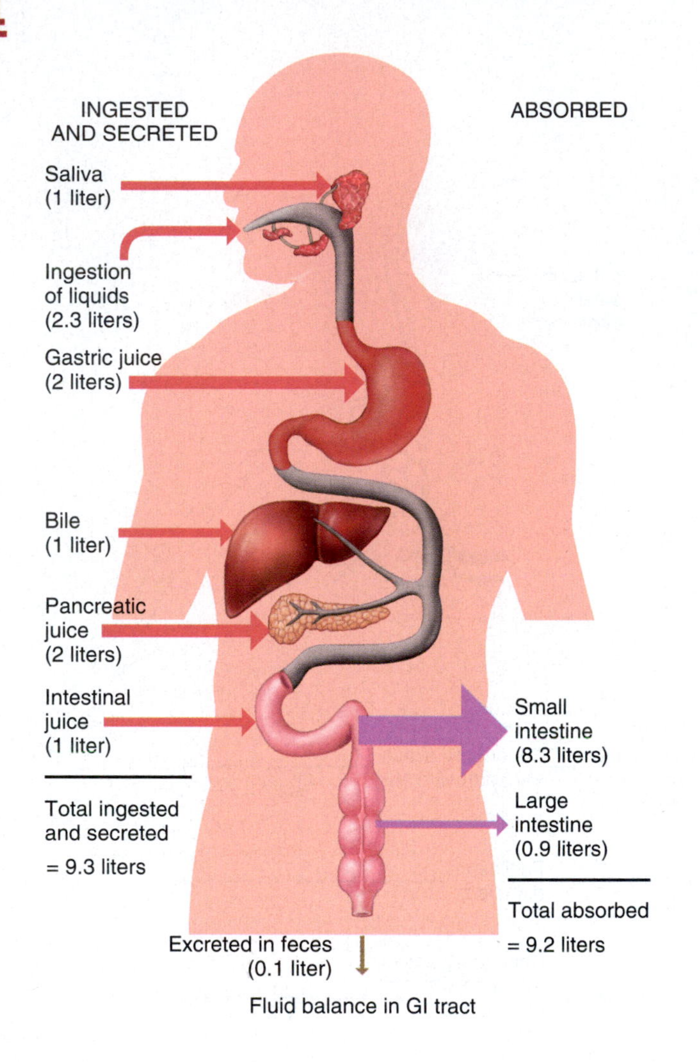

Fluid balance in GI tract

Which two organs of the digestive system secrete the most fluid?

CHECKPOINT

23. How is the wall of the small intestine adapted for digestion and absorption?
24. Describe the types of motility that occur in the small intestine.
25. Why are brush-border enzymes important? What are some examples of these enzymes?
26. How are the end products of carbohydrate, protein, and lipid digestion absorbed?
27. Describe the absorption of electrolytes, vitamins, and water by the small intestine.

Critical Thinking

Gastric Bypass Surgery and Weight Loss

Dan is a 45-year-old man who is obese. He has tried to lose weight but so far has been unsuccessful. His doctor told him that he is an ideal candidate for gastric bypass surgery. After much consideration, Dan decides to have the surgery. Within a year, Dan has lost a considerable amount of weight and is very happy with his results. He has also adopted a healthier lifestyle, with a well-balanced diet and more exercise.

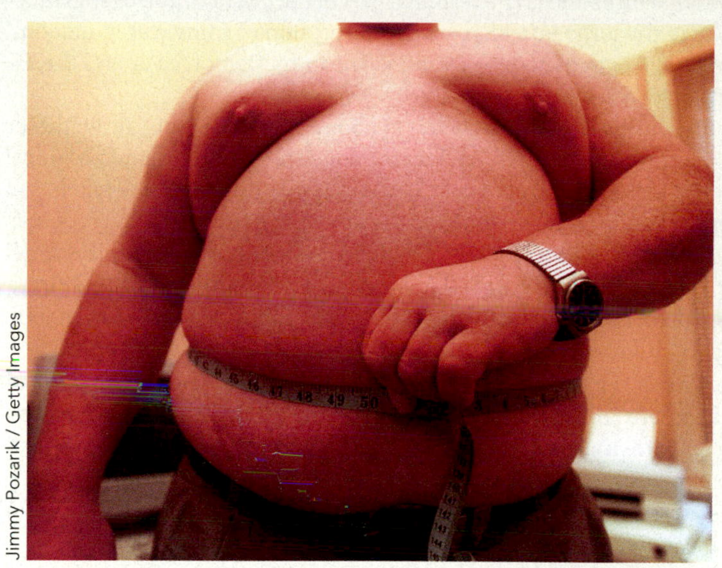
Jimmy Pozarik / Getty Images

SOME THINGS TO KEEP IN MIND:

Gastric bypass surgery is a procedure that limits the amount of food that enters the stomach and is absorbed by the small intestine in order to bring about a significant weight loss in obese individuals. The most common form of gastric bypass surgery is Roux-en-Y, in which a small pouch about the size of an egg is created at the top of the stomach. The pouch, which is only 5–10% of the stomach volume, is sealed off using surgical staples. The pouch is connected to the jejunum of the small intestine. The duodenum is reattached to the middle section of the jejunum so enzymes produced in the duodenum can still be utilized by the digestive system. This procedure results in a very small storage volume in the pouch and an enhanced rate of passage of the swallowed food into the small intestine. Patients feel full after swallowing only a small amount of food, thus mechanically reducing the number of calories that can be eaten in a single sitting. As the rate of food passage through the small intestines is increased, there is a concomitant reduction in the number of calories that can be absorbed as well. After healing, the patients should eat four to six small meals (including snacks) per day to allow the absorption of enough calories and nutrients even though each meal is very small (most people will not be able to consume more than 1.5 cups of food per meal).

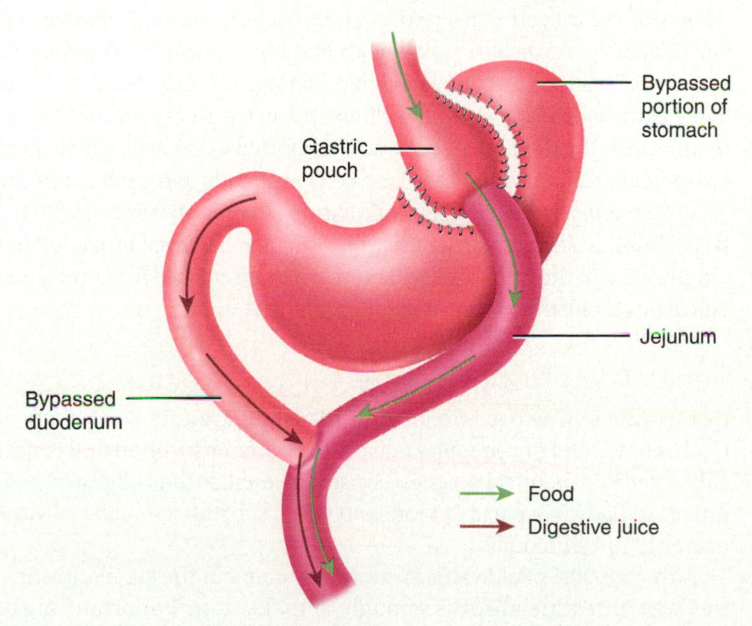

SOME INTERESTING FACTS:

More than one-third of adults (about 79 million people) in the United States are obese.

Approximately 200,000 gastric bypass surgeries are performed each year in the United States.

Patients undergoing gastric bypass surgery lose an average of about 60% of excess weight.

From a functional perspective, why isn't it problematic that most of the stomach is bypassed via gastric bypass surgery?

After having gastric bypass surgery, what would happen to Dan if he ate a meal that was too large for the gastric pouch?

Dan has noticed that after having gastric bypass surgery, he gets intoxicated very quickly when he drinks one or two alcoholic beverages. Before having the surgery, he could have a greater number of alcoholic beverages before feeling intoxicated. Why the change?

Some people who have gastric bypass surgery later begin to experience addictive behaviors (drug or alcohol dependency, compulsive shopping, or sex addiction). Why might this be the case?

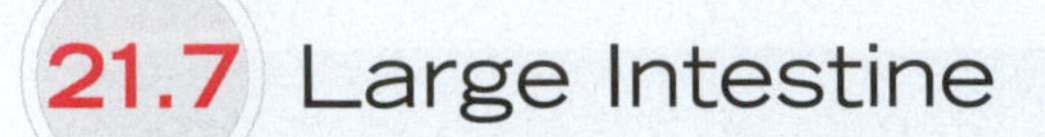

21.7 Large Intestine

OBJECTIVE

- Describe the functions of the large intestine.

The large intestine is the terminal part of the GI tract. The overall functions of the large intestine are the absorption of some water and ions, the production of certain vitamins, the formation of feces, and the expulsion of feces from the body.

The **large intestine**, which is about 6.5 cm (2.5 in.) in diameter and about 1.5 m (5 ft) in length, consists of three major regions: cecum, colon, and rectum (FIGURE 21.22). The **cecum** is the initial portion of the large intestine. At the junction of the ileum and cecum is the *ileocecal sphincter (valve)*, which allows materials from the small intestine to pass into the large intestine. The **colon** is the longest portion of the large intestine. It is further subdivided into an *ascending colon, transverse colon, descending colon*, and *sigmoid colon*. The **rectum** is the terminal portion of the large intestine. The opening of the rectum to the exterior is known as the **anus**. The anus is guarded by an **internal anal sphincter** of smooth muscle (involuntary) and an **external anal sphincter** of skeletal muscle (voluntary). Normally these sphincters keep the anus closed except during the elimination of feces.

Extending from the cecum is a small, fingerlike projection known as the **appendix** (FIGURE 21.22). This structure contains lymphoid nodules that participate in immune responses. Inflammation of the appendix, termed **appendicitis**, occurs when the lumen of the appendix becomes obstructed. If this occurs, an *appendectomy* (removal of the appendix) is recommended because an inflamed appendix is likely to rupture, releasing infectious bacteria into the abdominal cavity.

The wall of the large intestine contains the typical four layers found in the rest of the GI tract, with certain modifications (FIGURE 21.22):

- The surface of the large intestinal mucosa consists of two types of epithelial cells: absorptive cells and goblet cells. The absorptive cells absorb water and ions; the goblet cells secrete mucus that lubricates the contents of the colon.
- As in the small intestine, the epithelium of the large intestinal mucosa extends downward from the surface to form intestinal glands called **crypts of Lieberkühn**. Recall that the crypts of Lieberkühn of the small intestine contain several types of cells (see FIGURE 21.17). In the large intestine, however, the crypts of Lieberkühn contain only absorptive cells and goblet cells.
- Compared to the small intestine, the wall of the large intestine does not have as many structural features that increase surface area. There are no circular folds or villi; however, microvilli are present on the absorptive cells (FIGURE 21.22). Consequently, much more absorption occurs in the small intestine than in the large intestine.
- Unlike other parts of the GI tract, the outer longitudinal layer of smooth muscle of the muscularis externa is bundled into three bands called the **teniae coli** (TĒ-nē-ē KŌ-lī) that run most of the length of the large intestine. Contractions of these bands gather the colon into a series of pouches called **haustra** (HAWS-tra; singular is **haustrum**), which give the colon a puckered appearance.

Mechanical Digestion in the Large Intestine Involves Haustral Churning, Peristalsis, and Mass Movement

The passage of chyme from the ileum into the cecum is regulated by the action of the ileocecal sphincter. Normally, the valve remains partially closed so that the passage of chyme into the cecum usually occurs slowly. Immediately after a meal, a **gastroileal reflex** intensifies peristalsis in the ileum and forces any chyme into the cecum. The hormone gastrin also relaxes the sphincter.

Movements of the colon begin when substances pass the ileocecal sphincter. Because chyme moves through the small intestine at a fairly constant rate, the time required for a meal to pass into the colon is determined by gastric emptying time. As food passes through the ileocecal sphincter, it fills the cecum and accumulates in the ascending colon.

One movement characteristic of the large intestine is **haustral churning**. In this process, the haustra remain relaxed and become distended while they fill up. When the distension reaches a certain point, the walls contract and squeeze the contents into the next haustrum. **Peristalsis** also occurs in the large intestine but at a slower rate than in more proximal portions of the tract. A final type of movement is **mass movement**, a strong wave of contraction that begins at about the middle of the transverse colon and quickly drives the contents of the colon into the rectum. Mass movement is similar to peristalsis, except that the contraction lasts for a longer period of time. Mass movements usually take place three or four times a day, during or immediately after a meal. Hence, the presence of food in the stomach triggers mass movement in the large intestine, an event known as the **gastrocolic reflex**.

Chemical Digestion in the Large Intestine Occurs via Bacteria

The final stage of digestion occurs in the colon through the activity of bacteria that inhabit the lumen. Mucus is secreted by the glands of the large intestine, but no enzymes are secreted. Chyme is prepared for elimination by the action of bacteria, which ferment any remaining carbohydrates and release hydrogen, carbon dioxide, and methane gases. These gases contribute to **flatus** (gas) in the colon. Certain foods, such as beans, contain carbohydrates that are converted to gas by bacteria. Bacteria also convert any remaining proteins to amino acids and break down the amino acids into simpler substances: indole, skatole, hydrogen sulfide, and fatty acids. Some of the indole and skatole is eliminated in the feces and contributes to their odor; the rest is absorbed and transported to the liver, where these compounds are converted to less toxic compounds and excreted in the urine. Bacteria also decompose bilirubin to simpler pigments, including **stercobilin**, which gives feces their brown color. Bacterial products that are absorbed in the colon include several vitamins needed for normal metabolism, among them some B vitamins and vitamin K.

Feces Are Formed in the Large Intestine

By the time chyme has remained in the large intestine 3–10 hours, it has become solid or semisolid because of water absorption and is now called **feces**. Chemically, feces consist of unabsorbed digested materials; indigestible parts of food; and other substances such as water, bacteria, and stercobilin.

Although 90% of all water absorption occurs in the small intestine, the large intestine absorbs enough to make it an important organ in maintaining the body's water balance. Of the 1 liter of water that

FIGURE 21.22 The large intestine.

The large intestine absorbs some water and ions, produces certain vitamins, forms feces, and expels feces from the body.

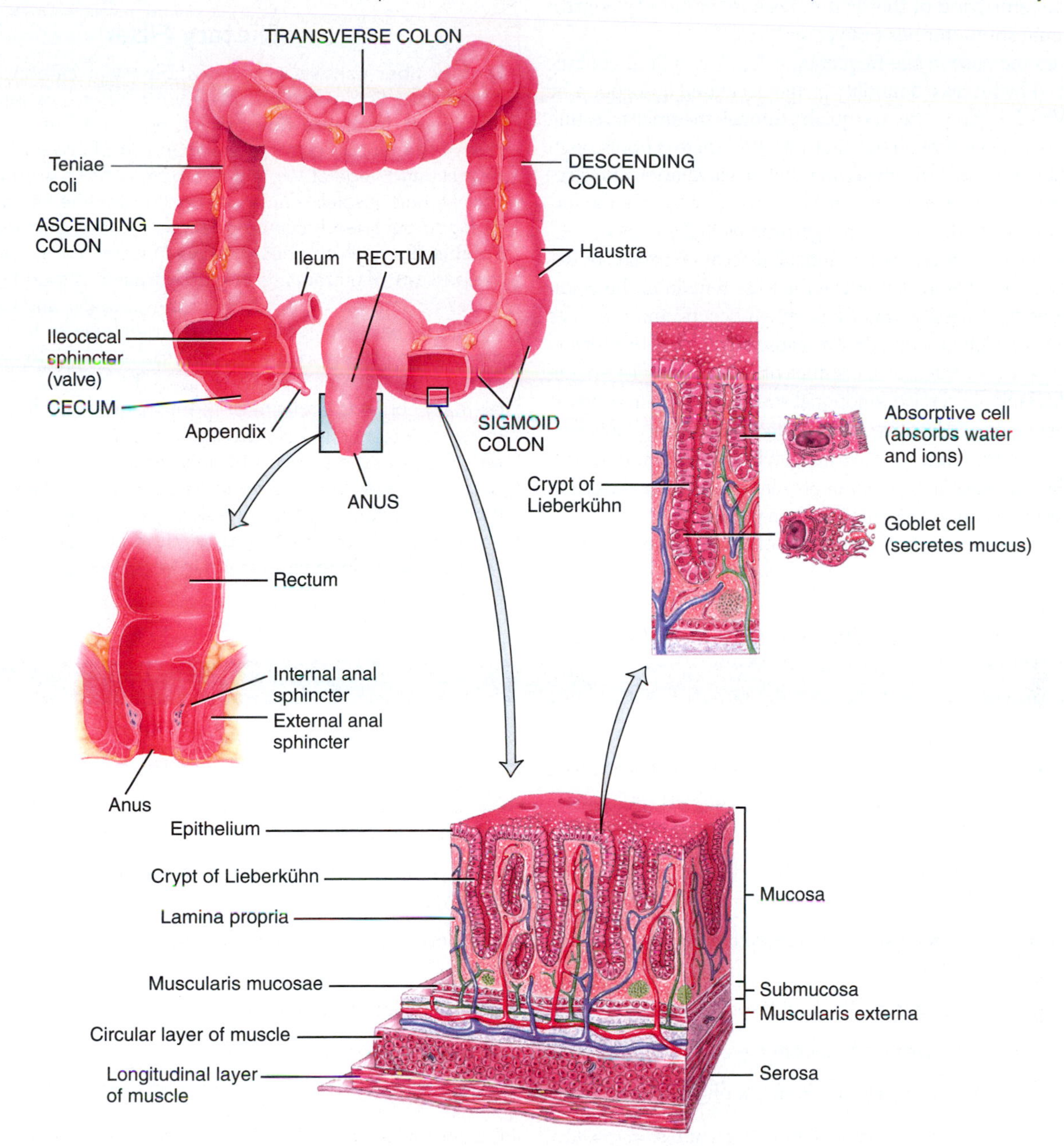

What are haustra?

enters the large intestine, all but about 100 mL is normally absorbed via osmosis. The large intestine also absorbs ions, including sodium and chloride, and some vitamins.

The Defecation Reflex Expels Feces from the Body

Mass movements push fecal material from the sigmoid colon into the rectum. The resulting distension of the rectal wall stimulates stretch receptors, which initiates a **defecation reflex** that empties the rectum. The defecation reflex occurs as follows: In response to distension of the rectal wall, the receptors send sensory input to the sacral spinal cord. Output from motor neurons in the cord travels along parasympathetic nerves back to the descending colon, sigmoid colon, rectum, and anus. The resulting contraction of the longitudinal rectal muscles shortens the rectum, thereby increasing the pressure within it. This pressure, along with parasympathetic stimulation, opens the internal anal sphincter. The external anal sphincter is voluntarily controlled. If it is voluntarily relaxed, defecation occurs and the feces are expelled through the anus. Defecation can be assisted by the **Valsalva maneuver**, which involves voluntary contractions of abdominal muscles and a forced exhalation against a closed larynx. This maneuver increases the pressure within the abdomen, which pushes the walls of the sigmoid colon and rectum inward.

A person can postpone defecation by voluntarily contracting the external anal sphincter. The feces eventually back up into the sigmoid colon until the next mass movement stimulates the stretch receptors, again creating the urge to defecate. In infants, the defecation reflex causes automatic emptying of the rectum because voluntary control of the external anal sphincter has not yet developed.

Diarrhea is an increase in the frequency, volume, and fluid content of the feces caused by increased motility of and decreased absorption by the intestines. When chyme passes too quickly through the small intestine and feces pass too quickly through the large intestine, there is not enough time for absorption. Frequent diarrhea can result in dehydration and electrolyte imbalances. Excessive motility may be caused by lactose intolerance, stress, and microbes that irritate the gastrointestinal mucosa.

Constipation refers to infrequent or difficult defecation caused by decreased motility of the intestines. Because the feces remain in the colon for prolonged periods, excessive water absorption occurs, and the feces become dry and hard. Constipation may be caused by poor habits (delaying defecation), spasms of the colon, insufficient fiber in the diet, inadequate fluid intake, lack of exercise, emotional stress, and certain drugs. A common treatment is a mild laxative, which induces defecation. However, many physicians maintain that laxatives are habit-forming, and that adding fiber to the diet, increasing the amount of exercise, and increasing fluid intake are safer ways of controlling this common problem.

TABLE 21.2 summarizes the functions of the digestive system organs.

CLINICAL CONNECTION

Dietary Fiber

Dietary fiber consists of indigestible plant carbohydrates—such as cellulose, lignin, and pectin—found in fruits, vegetables, grains, and beans. **Insoluble fiber**, which does not dissolve in water, includes the woody or structural parts of plants such as the skins of fruits and vegetables and the bran coating around wheat and corn kernels. Insoluble fiber passes through the GI tract largely unchanged but speeds up the passage of material through the tract. **Soluble fiber**, which does dissolve in water, forms a gel that slows the passage of material through the tract. It is found in abundance in beans, oats, barley, broccoli, prunes, apples, and citrus fruits.

People who choose a fiber-rich diet may reduce their risk of developing obesity, diabetes, atherosclerosis, and colorectal cancer. Soluble fiber also may help lower blood cholesterol. The liver normally converts cholesterol to bile salts, which are released into the small intestine to help fat digestion. Having accomplished their task, the bile salts are absorbed by the small intestine and recycled back to the liver. Because soluble fiber binds to bile salts to prevent their absorption, the liver makes more bile salts to replace those lost in feces. Thus, the liver uses more cholesterol to make more bile salts, and blood cholesterol level is lowered.

TABLE 21.2 Functions of Digestive System Organs

Organ	Functions
Mouth	
Cheeks and lips	Help keep food between the upper and lower teeth during chewing.
Palate	Makes it possible to chew and breathe at the same time; during swallowing, uvula and soft palate elevate to close off upper part of pharynx, preventing swallowed foods and liquids from entering the nasal cavity.
Tongue	Maneuvers food in the mouth, shapes food into a bolus, pushes food to the back of the mouth for swallowing, detects chemicals in food, and initiates digestion of triglycerides.
Salivary glands	Produce saliva, which softens, moistens, and dissolves foods; cleanses mouth and teeth; and initiates the digestion of starch.
Teeth	Cut, tear, and pulverize food to reduce solids to smaller particles for swallowing.
Pharynx	Receives a bolus from the oral cavity and passes it into the esophagus.
Esophagus	Receives a bolus from the pharynx and moves it into the stomach.
Stomach	Cycles of propulsion and retropulsion mix food with gastric juice and reduce food to chyme. The hydrochloric acid (HCl) in gastric juice activates pepsin and kills many microbes in food. Intrinsic factor aids absorption of vitamin B_{12}. The stomach serves as a reservoir for food before releasing it into the small intestine.
Pancreas	Pancreatic juice buffers acidic gastric juice in chyme (creating the proper pH for digestion in the small intestine); stops the action of pepsin from the stomach; and contains enzymes that digest carbohydrates, proteins, triglycerides, and nucleic acids.
Liver	Produces bile, which is needed for the emulsification and absorption of lipids in the small intestine.
Gallbladder	Stores and concentrates bile.
Small intestine	Segmentation mixes chyme with digestive juices; migrating motility complexes propel chyme toward the ileocecal sphincter; digestive secretions from the small intestine, pancreas, and liver complete the digestion of carbohydrates, proteins, lipids, and nucleic acids; circular folds, villi, and microvilli increase surface area for absorption; site where about 90% of nutrients and water are absorbed.
Large intestine	Haustral churning, peristalsis, and mass movements drive the contents of the colon into the rectum; bacteria produce some B vitamins and vitamin K; absorption of some water, ions, and vitamins; defecation.

CHECKPOINT

28. How does the large intestine compare to the small intestine in terms of structural features that increase the surface area for absorption?
29. Describe the mechanical movements that occur in the large intestine.
30. What is defecation and how does it occur?
31. What activities occur in the large intestine to change its contents into feces?

21.8 Phases of Digestion

OBJECTIVES

- Describe the three phases of digestion.
- Describe the major hormones that regulate digestive activities.

Digestive activities occur in three overlapping phases: the cephalic phase, the gastric phase, and the intestinal phase.

The Cephalic Phase Prepares the Mouth and Stomach for Food That Is About to Be Eaten

During the **cephalic phase** of digestion, the sight, smell, thought, or initial taste of food activates neural centers in the cerebral cortex, hypothalamus, and brain stem. The brain stem then activates the facial (VII), glossopharyngeal (IX), and vagus (X) nerves. The facial and glossopharyngeal nerves stimulate the salivary glands to secrete saliva, while the vagus nerves stimulate the gastric glands to secrete gastric juice. The purpose of the cephalic phase of digestion is to prepare the mouth and stomach for food that is about to be eaten.

The Gastric Phase Promotes Gastric Juice Secretion and Gastric Motility

Once food reaches the stomach, the **gastric phase** of digestion begins. Neural and hormonal mechanisms regulate the gastric phase of digestion to promote gastric secretion and gastric motility.

- ***Neural regulation.*** Food of any kind distends the stomach and stimulates stretch receptors in its walls. Chemoreceptors in the stomach monitor the pH of stomach chyme. When the stomach walls are distended or pH increases because proteins have entered the stomach and buffered some of the stomach acid, the stretch receptors and chemoreceptors are activated, and a neural negative feedback loop is set in motion (**FIGURE 21.23**). From the stretch receptors and chemoreceptors, action potentials propagate to the submucosal plexus, where they activate parasympathetic and enteric neurons. The resulting action potentials cause waves of peristalsis and continue to stimulate the flow of gastric juice from gastric glands. During this process, propulsion and retropulsion occur in the stomach, causing food to mix with gastric juice. Periodically, a small quantity of chyme undergoes gastric emptying into the duodenum. The pH of the stomach chyme decreases (becomes more acidic) and the distension of the stomach walls lessens because some of the chyme has passed into the small intestine, suppressing secretion of gastric juice.

FIGURE 21.23 Neural negative feedback regulation of the pH of gastric juice and gastric motility during the gastric phase of digestion.

Food entering the stomach stimulates secretion of gastric juice and causes vigorous waves of peristalsis.

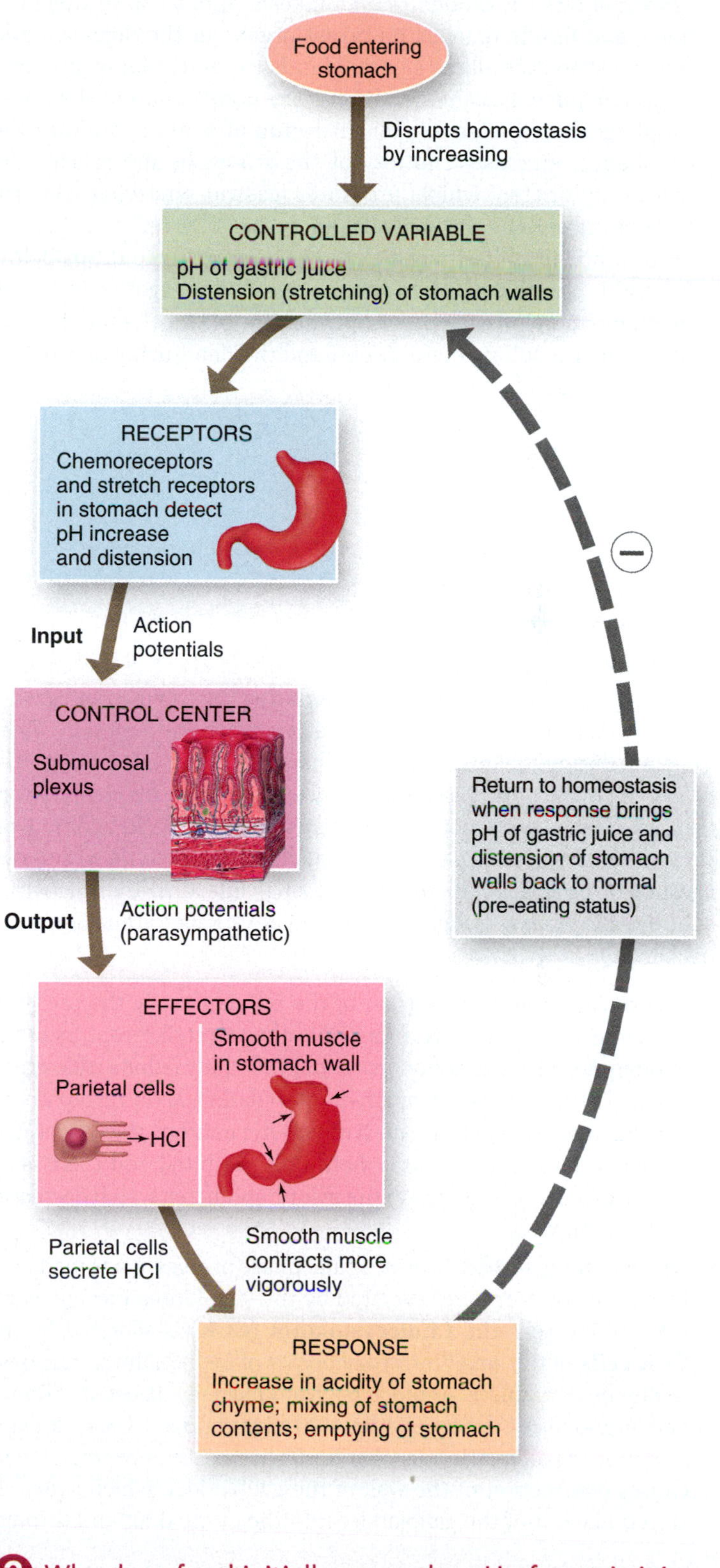

Why does food initially cause the pH of gastric juice to rise?

- ***Hormonal regulation.*** Gastric secretion during the gastric phase is also regulated by the hormone **gastrin**. Gastrin is released from the **G cells** of the gastric glands in response to several stimuli: distension of the stomach by chyme, partially digested proteins in chyme, the high pH of chyme due to the presence of food in the stomach, caffeine in gastric chyme, and acetylcholine released from parasympathetic neurons. Once it is released, gastrin enters the bloodstream, makes a round-trip through the body, and finally reaches its target organs in the digestive system. Gastrin stimulates gastric glands to secrete large amounts of gastric juice. It also strengthens the contraction of the lower esophageal sphincter to prevent reflux of acid chyme into the esophagus; increases motility of the stomach; and relaxes the pyloric sphincter, which promotes gastric emptying. Gastrin also promotes the gastroileal and gastrocolic reflexes. Gastrin secretion is inhibited when the pH of gastric juice drops below 2.0 and is stimulated when the pH rises. This negative feedback mechanism helps provide an optimal low pH for the functioning of pepsin, the killing of microbes, and the denaturing of proteins in the stomach.

The Intestinal Phase Promotes Digestion in the Small Intestine and Slows Digestion in the Stomach

The **intestinal phase** of digestion begins once food enters the small intestine. In contrast to reflexes initiated during the cephalic and gastric phases, which stimulate stomach secretory activity and motility, those occurring during the intestinal phase have inhibitory effects that slow the exit of chyme from the stomach. This prevents the duodenum from being overloaded with more chyme than it can handle. In addition, responses occurring during the intestinal phase promote the continued digestion of foods that have reached the small intestine. These activities of the intestinal phase of digestion are regulated by neural and hormonal mechanisms.

- ***Neural regulation.*** Distension of the duodenum by the presence of chyme causes the **enterogastric reflex**. Stretch receptors in the duodenal wall send action potentials to the medulla oblongata, where they inhibit parasympathetic stimulation, and to sympathetic ganglia, where they stimulate sympathetic neurons that supply the stomach. As a result, gastric motility is inhibited and there is an increase in the contraction of the pyloric sphincter, which decreases gastric emptying.
- ***Hormonal regulation.*** The intestinal phase of digestion is mediated by two major hormones secreted by the small intestine: cholecystokinin and secretin. **Cholecystokinin (CCK)** is secreted by the **CCK cells** of the small intestinal crypts of Lieberkühn in response to chyme containing amino acids from partially digested proteins and fatty acids from partially digested triglycerides. CCK stimulates secretion of pancreatic juice that is rich in digestive enzymes. It also causes contraction of the wall of the gallbladder, which squeezes stored bile out of the gallbladder into the cystic duct and through the common bile duct. In addition, CCK causes relaxation of the sphincter of Oddi, which allows pancreatic juice and bile to flow into the duodenum. CCK also slows gastric emptying by promoting contraction of the pyloric sphincter, produces satiety (a feeling of fullness) by acting on the hypothalamus in the brain, promotes normal growth and maintenance of the pancreas, and enhances the effects of secretin. Acidic chyme entering the duodenum stimulates the release of **secretin** from the **S cells** of the small intestinal crypts of Lieberkühn. In turn, secretin stimulates the flow of pancreatic juice that is rich in bicarbonate (HCO_3^-) ions to buffer the acidic chyme that enters the duodenum from the small intestine. Besides this major effect, secretin inhibits secretion of gastric juice, promotes normal growth and maintenance of the pancreas, and enhances the effects of CCK. Overall, secretin causes buffering of acid in chyme that reaches the duodenum and slows production of acid in the stomach.

There Are Many Hormones of the Digestive System

Besides gastrin, CCK, and secretin, there are many other hormones of the digestive system. For example, **ghrelin**, which is secreted by the stomach, plays a role in increasing appetite. **Glucose-dependent insulinotropic peptide (GIP)** and **glucagon-like peptide (GLP)**, which are secreted by the small intestine in response to the presence of food, stimulate the release of insulin from the pancreas, thereby increasing the blood glucose concentration. GIP and GLP are collectively referred to as **incretins**; they provide a type of feedforward control that anticipates the increase in blood glucose occurring after a typical meal. At least 10 other so-called "gut hormones" are secreted by and have effects on the GI tract. They include **motilin**, **substance P**, and **bombesin**, which stimulate motility of the intestines; **vasoactive intestinal polypeptide (VIP)**, which stimulates secretion of ions and water by the intestines and inhibits gastric acid secretion; **gastrin-releasing peptide**, which stimulates release of gastrin; and **somatostatin**, which inhibits gastrin release. Some of these hormones are thought to act as local hormones (paracrines), whereas others are secreted into the blood or even into the lumen of the GI tract. The physiological roles of these and other gut hormones are still under investigation.

TABLE 21.3 summarizes the major hormones that control digestion.

CHECKPOINT

32. What is the purpose of the cephalic phase of digestion?
33. Describe the role of gastrin in the gastric phase of digestion.
34. Outline the steps of the enterogastric reflex.
35. Explain the roles of CCK and secretin in the intestinal phase of digestion.

TABLE 21.3 Major Hormones That Control Digestion

Hormone	Stimulus and Site of Secretion	Actions
Gastrin	Distension of stomach, partially digested proteins and caffeine in stomach, and high pH of stomach chyme stimulate gastrin secretion by enteroendocrine G cells, located mainly in the mucosa of pyloric antrum of stomach.	*Major effects:* Promotes secretion of gastric juice, increases gastric motility, and promotes growth of gastric mucosa. *Minor effects:* Constricts lower esophageal sphincter, relaxes pyloric sphincter.
Secretin	Acidic (high H^+ level) chyme that enters the small intestine stimulates secretion of secretin by enteroendocrine S cells in the mucosa of the duodenum.	*Major effects:* Stimulates secretion of pancreatic juice that is rich in HCO_3^- (bicarbonate ions). *Minor effects:* Inhibits secretion of gastric juice, promotes normal growth and maintenance of the pancreas, and enhances effects of CCK.
Cholecystokinin (CCK)	Partially digested proteins (amino acids), triglycerides, and fatty acids that enter the small intestine stimulate secretion of CCK by enteroendocrine CCK cells in the mucosa of the small intestine; CCK is also released in the brain.	*Major effects:* Stimulates secretion of pancreatic juice rich in digestive enzymes, causes ejection of bile from the gallbladder and opening of the sphincter of Oddi, and induces satiety (feeling full to satisfaction). *Minor effects:* Inhibits gastric emptying, promotes normal growth and maintenance of the pancreas, and enhances effects of secretin.

21.9 Transport of Lipids by Lipoproteins

OBJECTIVE

- Describe the lipoproteins that transport lipids in the blood.

Most lipids, such as triglycerides, are nonpolar and therefore very hydrophobic molecules. They do not dissolve in water. To be transported in watery blood, such molecules first must be made more water-soluble by combining them with proteins produced by the liver and small intestine. The lipid and protein combinations thus formed are **lipoproteins**, spherical particles with an outer shell of proteins, phospholipids, and cholesterol molecules surrounding an inner core of triglycerides and other lipids (**FIGURE 21.24**). The proteins in the outer shell, called *apoproteins*, help solubilize the lipoprotein in body fluids.

Each of the several types of lipoproteins has different functions, but all essentially are transport vehicles. They provide delivery and pickup services so that lipids can be available when cells need them or removed from circulation when they are not needed. Lipoproteins are categorized and named mainly according to their density, which varies with the ratio of lipids (which have a low density) to proteins (which have a high density). From lightest to heaviest, the four major classes of lipoproteins are chylomicrons, very low-density lipoproteins, low-density lipoproteins, and high-density lipoproteins.

- **Chylomicrons** form in absorptive epithelial cells of the small intestine and transport dietary (ingested) lipids to adipose tissue for storage (see **FIGURE 21.20a, b**). Chylomicrons enter lacteals of intestinal villi and are carried by lymph into venous blood and then into the

FIGURE 21.24 A lipoprotein.

A single layer of amphipathic phospholipids, cholesterol, and proteins surrounds a core of nonpolar lipids.

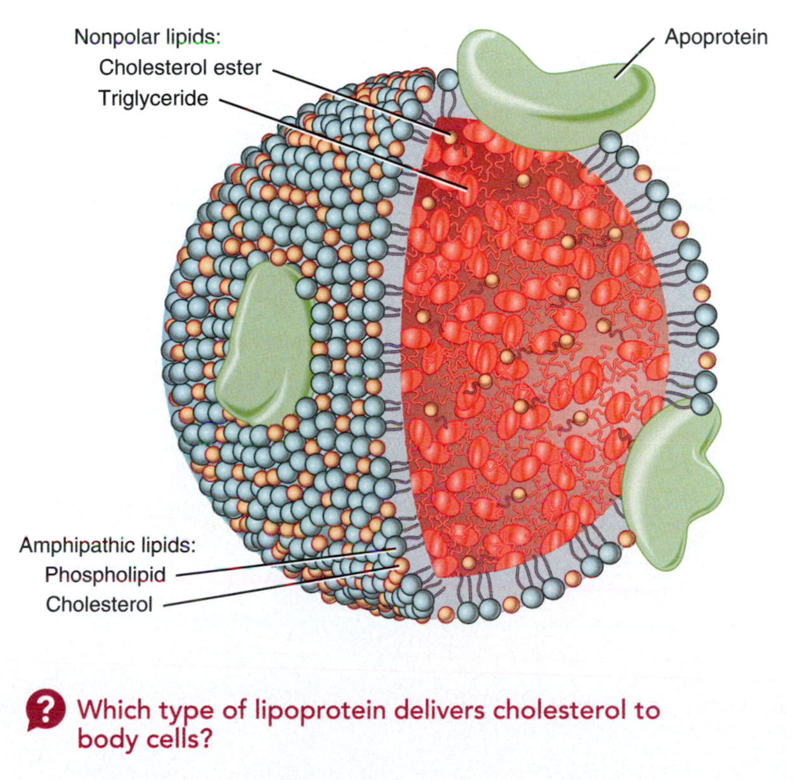

Which type of lipoprotein delivers cholesterol to body cells?

systemic circulation. Their presence gives blood a milky appearance, but they remain in the blood for only a short period of time.

- **Very low-density lipoproteins (VLDLs)** are formed in the liver and transport triglycerides made in hepatocytes to adipose tissue for storage. After depositing some of their triglycerides in adipose tissue, VLDLs are converted to low-density lipoproteins (LDLs).
- **Low-density lipoproteins (LDLs)** carry about 75% of the total cholesterol in blood and deliver it to cells throughout the body for use in repair of cell membranes and to cells in the testes, ovaries, and adrenal glands for use in the synthesis of steroid hormones. LDLs bind to LDL receptors on the plasma membrane of body cells so that the LDLs can enter the body cells via receptor-mediated endocytosis. Within a cell, an LDL is broken down, and the cholesterol is released to serve the cell's needs. Once a cell has sufficient cholesterol for its activities, a negative feedback system inhibits the cell's synthesis of new LDL receptors.
- **High-density lipoproteins (HDLs)**, which are formed in the liver, remove excess cholesterol from body cells and the blood and from atherosclerotic plaques that may be present in arterial walls. Then they transport the cholesterol to the liver, where it can be secreted into bile or used to synthesize bile salts.

CHECKPOINT

36. Which type of lipoprotein removes excess cholesterol from body cells and the blood?

Now that your exploration of the digestive system is complete, you can appreciate the many ways that this system contributes to homeostasis of other body systems by examining *Focus on Homeostasis: Contributions of the Digestive System.*

CLINICAL CONNECTION

"Bad" and "Good" Cholesterol

When present in excessive numbers, LDLs deposit cholesterol in the walls of arteries, forming atherosclerotic plaques that increase the risk of coronary artery disease (see Section 13.7). For this reason, the cholesterol in LDLs, called LDL-cholesterol, is known as **"bad" cholesterol**. Eating a high-fat diet increases the production of VLDLs, which elevates the LDL level and increases the formation of atherosclerotic plaques. Because HDLs prevent accumulation of cholesterol in blood, a high HDL level is associated with decreased risk of coronary artery disease. For this reason, HDL-cholesterol is known as **"good" cholesterol**.

Desirable levels of blood cholesterol in adults are total cholesterol under 200 mg/dL, LDL-cholesterol under 130 mg/dL, and HDL-cholesterol over 40 mg/dL. The ratio of total cholesterol to HDL-cholesterol predicts the risk of developing coronary artery disease. For example, a person with a total cholesterol of 180 mg/dL and HDL of 60 mg/dL has a risk ratio of 3. Ratios above 4 are considered undesirable; the higher the ratio, the greater the risk of developing coronary artery disease.

Among the therapies used to reduce blood cholesterol level are exercise, diet, and drugs. Regular physical activity at aerobic and nearly aerobic levels raises HDL level. Dietary changes are aimed at reducing the intake of total fat, saturated fats, and cholesterol. Drugs used to treat high blood cholesterol levels either promote excretion of bile in the feces or block the key enzyme (HMG-CoA reductase) needed for cholesterol synthesis.

FOCUS on HOMEOSTASIS

INTEGUMENTARY SYSTEM

- Excess dietary calories are stored as triglycerides in adipose cells in dermis and subcutaneous layer

SKELETAL SYSTEM

- Small intestine absorbs dietary calcium and phosphorus salts needed to build bone extracellular matrix

MUSCULAR SYSTEM

- Liver can convert lactic acid (produced by muscles during exercise) to glucose

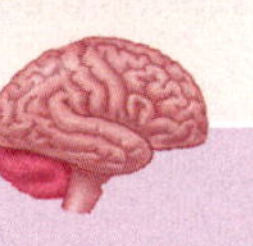

NERVOUS SYSTEM

- Gluconeogenesis (synthesis of new glucose molecules) in liver plus digestion and absorption of dietary carbohydrates provide glucose, needed for ATP production by neurons

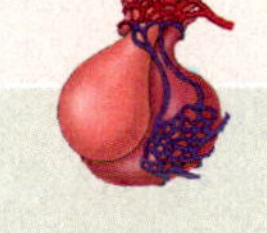

ENDOCRINE SYSTEM

- Enteroendocrine cells in stomach and small intestine release hormones that regulate digestive activities
- Pancreatic islets release insulin and glucagon

CONTRIBUTIONS OF THE DIGESTIVE SYSTEM

FOR ALL BODY SYSTEMS

- Breaks down dietary nutrients into forms that can be absorbed and used by body cells for producing ATP and building body tissues
- Absorbs water, minerals, and vitamins needed for growth and function of body tissues
- Eliminates wastes from food via feces

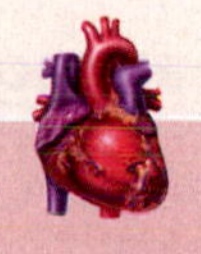

CARDIOVASCULAR SYSTEM

- GI tract absorbs water that helps maintain blood volume and iron that is needed for synthesis of hemoglobin in erythrocytes
- Bilirubin from hemoglobin breakdown is partially excreted in feces
- Liver synthesizes most plasma proteins

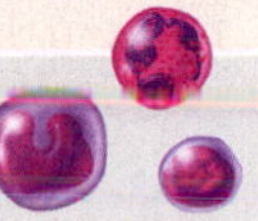

IMMUNE SYSTEM

- Acidity of gastric juice destroys bacteria and most toxins in stomach

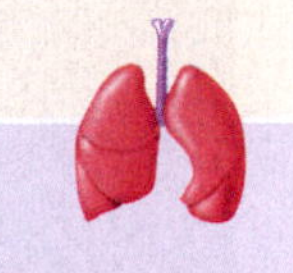

RESPIRATORY SYSTEM

- Pressure of abdominal organs against diaphragm helps expel air quickly during forced exhalation

URINARY SYSTEM

- Absorption of water by GI tract provides water needed to excrete waste products in urine

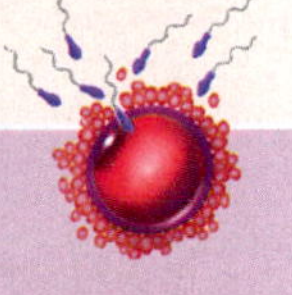

REPRODUCTIVE SYSTEMS

- Digestion and absorption provide adequate nutrients, including fats, for normal development of reproductive structures, for production of gametes (sperm and eggs), and for fetal growth and development during pregnancy

FROM RESEARCH TO REALITY

Fasting and Stomach Sensitivity

Reference

Kentish, S. et al. (2012). Diet-induced adaptation of vagal afferent function. *The Journal of Physiology.* 590: 209–221.

Are you full before you feel it?

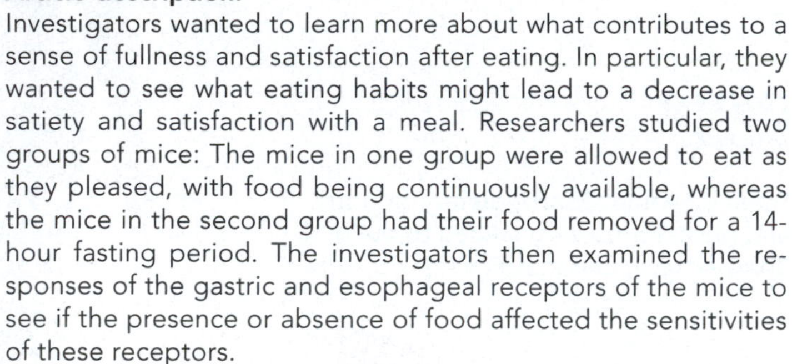

The food that you eat affects your body in myriad ways. The quantity and type of food you eat does the same. Dieting is known to cause certain changes in the body both physically and psychologically that can make staying on a diet particularly challenging. Could restricting calories actually lead your stomach to become less sensitive to the presence of food inside it?

Article description:

Investigators wanted to learn more about what contributes to a sense of fullness and satisfaction after eating. In particular, they wanted to see what eating habits might lead to a decrease in satiety and satisfaction with a meal. Researchers studied two groups of mice: The mice in one group were allowed to eat as they pleased, with food being continuously available, whereas the mice in the second group had their food removed for a 14-hour fasting period. The investigators then examined the responses of the gastric and esophageal receptors of the mice to see if the presence or absence of food affected the sensitivities of these receptors.

Go to WileyPLUS Learning Space and use the data from this article to answer the questions posed there and to discover more about fasting and stomach sensitivity.

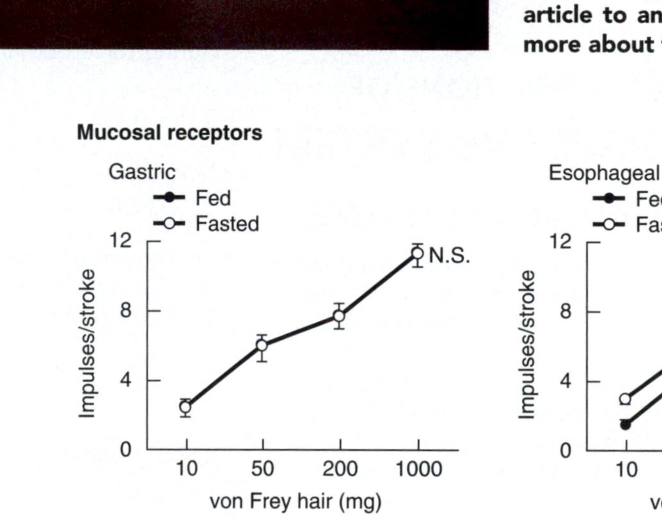

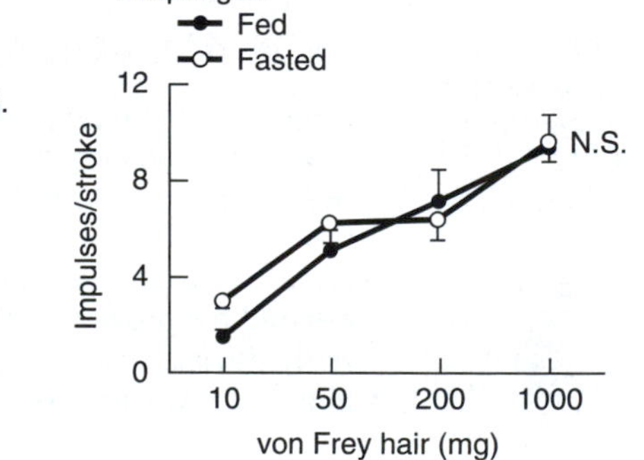

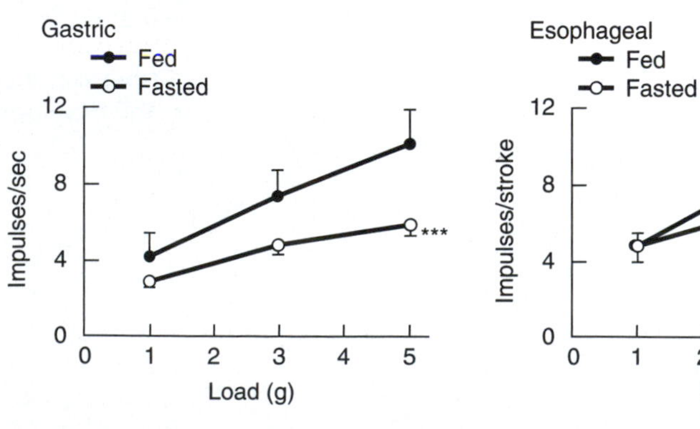

CHAPTER REVIEW

21.1 Overview of the Digestive System

1. The digestive system consists of two groups of organs: the gastrointestinal (GI) tract and the accessory digestive organs. The GI tract is a continuous tube extending from the mouth to the anus. The accessory digestive organs include the teeth, tongue, salivary glands, pancreas, liver, and gallbladder.
2. Digestion includes six basic processes: ingestion, secretion, motility, digestion, absorption, and defecation.
3. The wall of most of the GI tract consists of four layers. From the inside out, these layers are the mucosa, submucosa, muscularis externa, and serosa.
4. The GI tract is regulated by an intrinsic set of nerves known as the enteric nervous system (ENS) and by an extrinsic set of nerves that are part of the autonomic nervous system (ANS).
5. The ENS consists of neurons arranged into two plexuses: the myenteric plexus and the submucosal plexus. The myenteric plexus regulates GI tract motility (movement); the submucosal plexus regulates GI secretion.
6. Although the neurons of the ENS can function independently, they are subject to regulation by the neurons of the ANS. Parasympathetic nerves increase GI tract secretion and motility by increasing the activity of ENS neurons. Sympathetic nerves decrease GI tract secretion and motility by inhibiting ENS neurons.
7. Gastrointestinal reflex pathways regulate GI secretion and motility in response to stimuli present in the lumen of the GI tract.
8. The single-unit smooth muscle within the wall of the GI tract is autorhythmic and behaves as a functional syncytium.
9. The autorhythmicity of GI smooth muscle is due to the interstitial cells of Cajal (ICCs) located in the muscularis externa. ICCs produce slow waves (basic electrical rhythm) that can give rise to action potentials if threshold is reached. The action potentials in turn cause the contractile fibers of smooth muscle to contract.
10. There are two main patterns of GI motility: peristalsis and segmentation. Peristalsis is successive contractions of smooth muscle that propel GI contents in a forward direction. Segmentation refers to alternating contractions of smooth muscle that mix the luminal contents of the small intestine.

21.2 Mouth

1. The mouth is formed by the cheeks, lips, palate, and tongue. Also associated with the mouth are the salivary glands and the teeth.
2. The cheeks and lips help keep food between the upper and lower teeth during chewing; the lips also assist in speech.
3. The palate separates the oral cavity from the nasal cavity, allowing you to chew and breathe at the same time. The palate consists of two parts: a bony hard palate and a muscular soft palate.
4. The tongue forms the floor of the oral cavity. It contains muscles; lingual glands that secrete the enzyme lingual lipase; and taste buds, which detect chemicals present in food.
5. The salivary glands (parotid, submandibular, and sublingual) secrete saliva, a fluid that lubricates food and starts the chemical digestion of carbohydrates.
6. Salivation is controlled by the salivary nuclei in the brain stem.
7. The teeth project into the mouth and are adapted for mechanical digestion.
8. Through mastication, food is mixed with saliva and shaped into a soft, flexible mass called a bolus.
9. Salivary amylase begins the digestion of starches, and lingual lipase acts on triglycerides.

CHAPTER REVIEW

21.3 Pharynx and Esophagus

1. The pharynx extends from the nasal cavity to the esophagus and to the larynx. It has both respiratory and digestive functions. For the digestive system, the pharynx serves as the passageway for food from the mouth into the esophagus.
2. The esophagus connects the pharynx to the stomach; it secretes mucus and transports food to the stomach.
3. Deglutition, or swallowing, is the process by which food moves from the mouth into the stomach.
4. Deglutition consists of a voluntary stage, a pharyngeal stage (involuntary), and an esophageal stage (involuntary).

21.4 Stomach

1. The stomach connects the esophagus to the small intestine.
2. The three main regions of the stomach are the fundus, body, and antrum.
3. Adaptations of the stomach for digestion include rugae; gastric glands that produce mucus, hydrochloric acid, pepsinogen (which is converted to pepsin), gastric lipase, and intrinsic factor; and a three-layered muscularis.
4. Mechanical digestion in the stomach involves cycles of propulsion (movement of gastric contents from the body of the stomach down into the antrum by peristalsis) and retropulsion (forced movement of gastric contents back into the antrum due to a narrow pyloric sphincter). The net effect of these processes is that food particles are mixed with gastric juice, eventually forming a soupy mixture called chyme.
5. Chemical digestion in the stomach consists mostly of the conversion of proteins into peptides by pepsin.
6. Secretion of hydrochloric acid by parietal cells involves proton pumps and chloride (Cl^-) channels.
7. Secretion of hydrochloric acid by parietal cells is regulated by acetylcholine, gastrin, and histamine.
8. The stomach can absorb a few substances, such as water, certain ions, drugs, and alcohol.
9. Vomiting is the forcible expulsion of the contents of the upper GI tract through the mouth.

21.5 Pancreas, Liver, and Gallbladder

1. The pancreas is a gland that contains an exocrine portion (acini that secrete pancreatic juice into ducts) and an endocrine portion (pancreatic islets that secrete hormones).
2. Pancreatic juice contains enzymes that digest starch (pancreatic amylase), proteins (trypsin, chymotrypsin, carboxypeptidase, and elastase), triglycerides (pancreatic lipase), and nucleic acids (ribonuclease and deoxyribonuclease).
3. The liver is the largest gland of the body; it consists of lobules that contain hepatocytes (liver cells), sinusoids, Kupffer cells, and a central vein.
4. Hepatocytes produce bile, a substance that emulsifies dietary lipids.
5. The liver also functions in carbohydrate, lipid, and protein metabolism; processing of drugs and hormones; excretion of bilirubin; synthesis of bile salts; storage of vitamins and minerals; phagocytosis; and activation of vitamin D.
6. The gallbladder stores and concentrates bile.

CHAPTER REVIEW

21.6 Small Intestine

1. The small intestine connects the stomach to the large intestine; it is divided into a duodenum, jejunum, and ileum.
2. The wall of the small intestine contains crypts of Lieberkühn (intestinal glands) that secrete intestinal juice and mucus.
3. The small intestine also contains circular folds, villi, and microvilli, which provide a large surface area for digestion and absorption.
4. Brush-border enzymes break down α-dextrins into glucose (α-dextrinase); sucrose to glucose and fructose (sucrase); lactose to glucose and galactose (lactase); maltose to glucose (maltase); and nucleotides to nitrogenous bases, pentose sugars, and phosphates (nucleosidases and phosphatases). In addition, the brush-border enzyme aminopeptidase breaks off amino acids at the amino-terminal end of peptides.
5. Mechanical digestion in the small intestine involves segmentation and migrating motility complexes.
6. Most absorption along the GI tract occurs in the small intestine. Absorption in the small intestine occurs via simple diffusion, facilitated diffusion, osmosis, and active transport.
7. Monosaccharides, amino acids, and short-chain fatty acids pass into the blood capillaries of a villus.
8. Long-chain fatty acids and monoglycerides are absorbed from micelles, resynthesized to triglycerides, and formed into chylomicrons.
9. Chylomicrons move into lymph in the lacteal of a villus.
10. The small intestine also absorbs electrolytes, vitamins, and water.

21.7 Large Intestine

1. The large intestine is the terminal part of the GI tract; it consists of three regions: the cecum, colon, and rectum.
2. The mucosa contains many goblet cells, and the muscularis externa consists of teniae coli that contract to form haustra.
3. Mechanical movements of the large intestine include haustral churning, peristalsis, and mass movements.
4. The last stages of chemical digestion occur in the large intestine through bacterial action. Substances are further broken down, and some vitamins are synthesized.
5. The large intestine absorbs water, ions, and vitamins.
6. The elimination of feces from the rectum is called defecation.

21.8 Phases of Digestion

1. Digestive activities occur in three overlapping phases: cephalic phase, gastric phase, and intestinal phase.
2. During the cephalic phase of digestion, salivary glands secrete saliva, and gastric glands secrete gastric juice in order to prepare the mouth and stomach for food that is about to be eaten.
3. The presence of food in the stomach causes the gastric phase of digestion, which promotes gastric juice secretion and gastric motility.
4. During the intestinal phase of digestion, food is digested in the small intestine. In addition, gastric motility and gastric secretion decrease in order to slow the exit of chyme from the stomach, which prevents the small intestine from being overloaded with more chyme than it can handle.

5. The activities that occur during the various phases of digestion are coordinated by neural pathways and by hormones.

21.9 Transport of Lipids by Lipoproteins

1. Lipoproteins transport lipids in the bloodstream; they are spherical particles that contain an outer shell of protein, phospholipids, and cholesterol surrounding an inner core of triglycerides and other lipids.
2. Types of lipoproteins include chylomicrons, which carry dietary lipids to adipose tissue; very low-density lipoproteins (VLDLs), which carry triglycerides from the liver to adipose tissue; low density lipoproteins (LDLs), which deliver cholesterol to body cells; and high-density lipoproteins (HDLs), which remove excess cholesterol from body cells and the blood, and transport it to the liver for elimination.

PONDER THIS

1. You just ate a burrito that contained the following: tortilla and rice—carbohydrates; beans and chicken—protein; sour cream—fat. Trace this food through the digestive tract. Along the way discuss one major component that is digested in each compartment and name the enzyme that does this.

2. Bariatric surgery is a surgical procedure in which changes are made to the gastrointestinal tract. One type of surgery is referred to as gastric bypass and is done as a weight loss intervention. This procedure takes several inches of the small intestine as well as part of the stomach and bypasses it as an avenue for food to travel. From the standpoint of the 6 functions of digestion, why would this be a weight loss aid?

3. Cystic fibrosis is often associated with respiratory infections and other effects on the lungs. However, it has additional effects. This condition affects the production of saline in many parts of the body. In this sense, cystic fibrosis can be detrimental to digestive function, particularly in the small intestine, as a result of pancreatic problems. Explain what is happening in this case.

ANSWERS TO FIGURE QUESTIONS

21.1 The digestive system breaks down food into molecules that are small enough to enter body cells.

21.2 In the context of the digestive system, absorption is the movement of the products of digestion from the lumen of the GI tract into blood or lymph.

21.3 The lamina propria has the following functions: (1) It contains blood vessels and lymphatic vessels, which are the routes by which nutrients are absorbed from the GI tract; (2) it supports the mucosal epithelium and binds it to the muscularis mucosae; and (3) it contains gut-associated lymphoid tissue (GALT), which helps protect against disease.

21.4 The neurons of the myenteric plexus regulate GI motility, and the neurons of the submucosal plexus regulate GI secretion.

21.5 Peristalsis is propulsive, causing net forward movement of luminal contents; segmentation causes mixing of luminal contents with no net movement in any particular direction.

21.6 During swallowing, the uvula elevates to close off the upper part of the pharynx, thereby preventing swallowed foods and liquids from entering the nasal cavity.

21.7 Salivary amylase breaks down starch into smaller molecules such as maltose, maltotriose, and α-dextrins.

21.8 The upper and lower esophageal sphincters regulate the movement of food into and out of the esophagus.

21.9 Both. Initiation of swallowing is voluntary and the action is carried out by skeletal muscles. Completion of swallowing—moving a bolus along the esophagus and into the stomach—is involuntary and involves peristalsis by smooth muscle.

21.10 Intrinsic factor is needed for absorption of vitamin B_{12}.

21.11 Hydrogen ions secreted into gastric juice are derived from carbonic acid (H_2CO_3).

21.12 Histamine is a paracrine agent released by mast cells in the lamina propria.

21.13 Hydrochloric acid (HCl) activates pepsin from its inactive form pepsinogen.

21.14 The pancreatic duct contains pancreatic juice, the common bile duct contains bile, and the ampulla of Vater contains pancreatic juice and bile.

21.15 The phagocytic cell in the liver is the Kupffer cell.

21.16 Bile salts are amphipathic (they have nonpolar regions and polar regions). The nonpolar regions of bile salts interact with the large lipid globule, while the polar regions of bile salts interact with the watery intestinal chyme. As a result, the large lipid globule is broken apart into several small lipid globules.

21.17 Villi are fingerlike projections of the mucosa that increase the surface area of the small intestine by a factor of 10.

21.18 Brush-border enzymes are so-named because they are located in the microvilli (brush border) of absorptive cells of the small intestine.

21.19 Aminopeptidase removes the amino acid from the amino-terminal end of a peptide.

21.20 Because monoglycerides are hydrophobic (nonpolar) molecules, they can diffuse through the lipid bilayer of the plasma membrane.

21.21 The stomach and pancreas are the two digestive system organs that secrete the largest volumes of fluid.

21.22 Haustra are pouches of the large intestine that are formed as a result of the contraction of smooth muscle bands (teniae coli).

21.23 The pH of gastric juice rises due to the buffering action of some amino acids in food proteins.

21.24 LDLs deliver cholesterol to body cells.

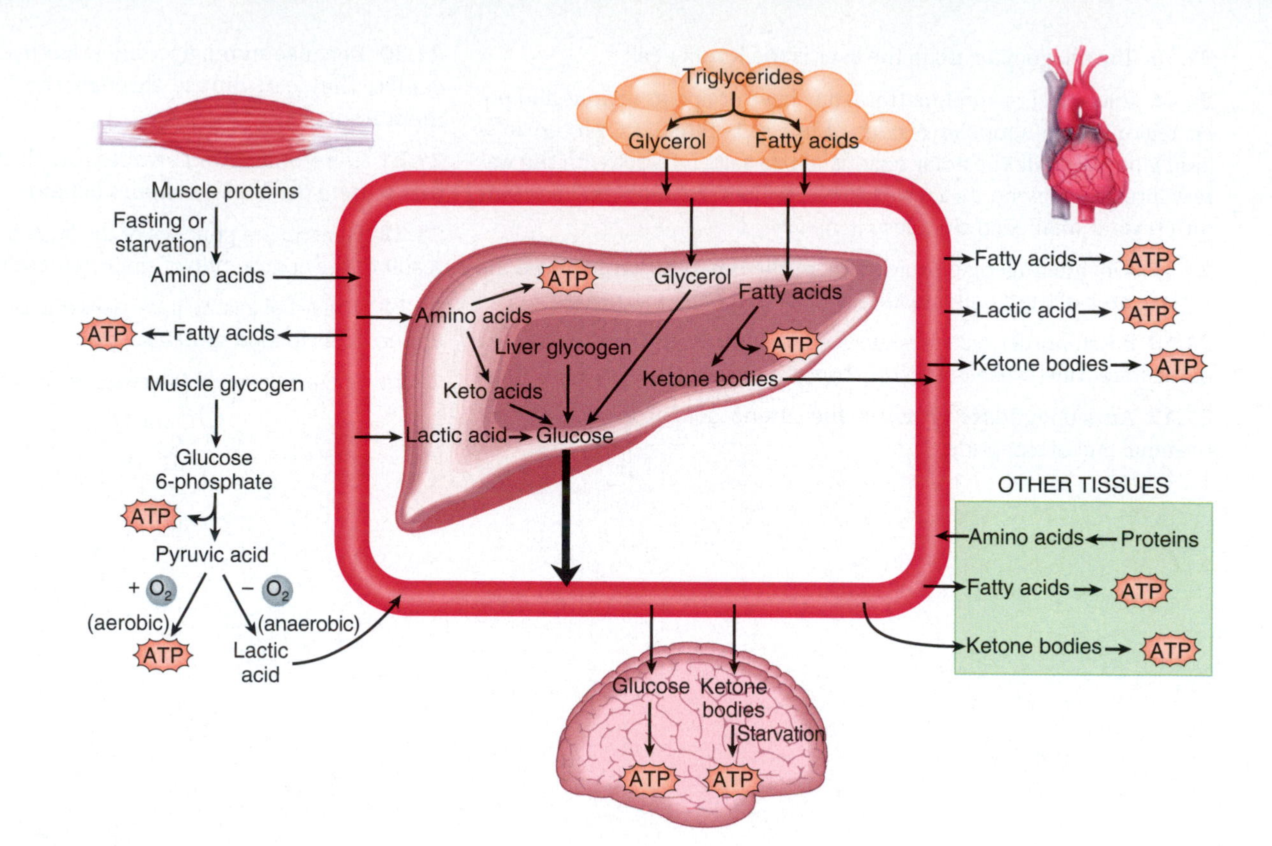

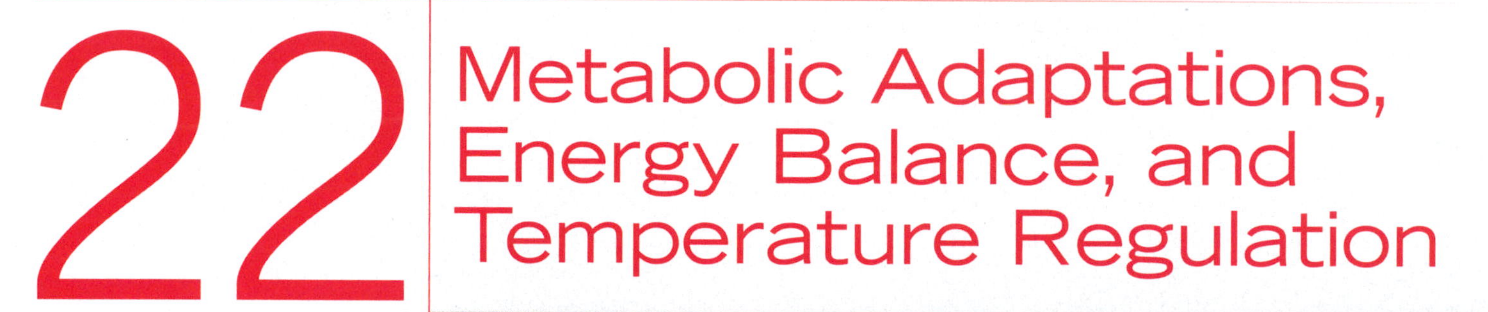

22 Metabolic Adaptations, Energy Balance, and Temperature Regulation

Metabolic Adaptations, Energy Balance, Temperature Regulation, and Homeostasis

Metabolic adaptations contribute to homeostasis by causing the appropriate reactions to occur based on the body's energy needs. Energy balance contributes to homeostasis by matching energy input to energy output to keep the body's weight stable. Temperature regulation contributes to homeostasis by promoting responses that prevent the body from becoming too hot or too cold.

LOOKING BACK TO MOVE AHEAD...

- Cellular respiration is the process by which a nutrient molecule (glucose, fatty acid, or amino acid) is broken down in the presence of oxygen to form carbon dioxide, water, and energy (ATP and heat) (Section 4.4).
- Cellular respiration of glucose involves four sets of reactions: glycolysis, formation of acetyl CoA, the Krebs cycle, and the electron transport chain. Fatty acids enter cellular respiration after being converted into acetyl CoA, whereas amino acids enter cellular respiration after being modified to form either pyruvic acid, acetyl CoA, or an intermediate of the Krebs cycle (Section 4.4).
- The hypothalamus functions as the body's thermostat; it regulates body temperature by stimulating activities that promote heat loss or heat gain (Section 8.2).
- Lipoproteins transport lipids through the bloodstream; the four major classes of lipoproteins are chylomicrons, very low-density lipoproteins (VLDLs), low-density lipoproteins (LDLs), and high-density lipoproteins (HDLs) (Section 21.9).

In Chapter 21, you learned how the digestive system breaks down food into units small enough to be absorbed into the bloodstream. In this chapter, you will discover how these absorbed nutrients are channeled into specific metabolic reactions to adapt to the energy demands of the body. You will also examine how energy balance—the precise matching of energy input to energy output—occurs in order to stabilize body weight. This chapter concludes with a discussion of the various mechanisms that help keep the core body temperature constant.

22.1 Metabolic Adaptations

OBJECTIVE

- Compare metabolism during the absorptive and postabsorptive states.

Some aspects of metabolism depend on how much time has passed since the last meal. During the **absorptive state**, ingested nutrients enter the bloodstream from the gastrointestinal (GI) tract to provide energy for the body. During the **postabsorptive state**, the GI tract lacks nutrients and energy is supplied by the breakdown of the body's own nutrient stores. A typical meal requires about 4 hours for complete absorption; given three meals a day, the absorptive state exists for about 12 hours each day. Assuming no between-meal snacks, the other 12 hours—typically late morning, late afternoon, and most of the night—are spent in the postabsorptive state.

The Absorptive State Promotes Reactions That Catabolize Nutrients, Synthesize Proteins, and Form Nutrient Stores

Soon after a meal, nutrients start to enter the blood after being absorbed from the GI tract. Recall that ingested food reaches the bloodstream mainly as glucose, amino acids, and triglycerides (in chylomicrons).

Absorptive State Reactions

During the absorptive state, some of the absorbed nutrients are catabolized for the body's energy needs or are used to synthesize proteins. The following reactions of the absorptive state reflect this function (FIGURE 22.1):

1. ***Catabolism of glucose.*** Most cells of the body produce the majority of their ATP by catabolizing glucose via cellular respiration.

FIGURE 22.1 Principal metabolic pathways during the absorptive state.

During the absorptive state, most body cells produce ATP by catabolizing glucose to CO_2 and H_2O.

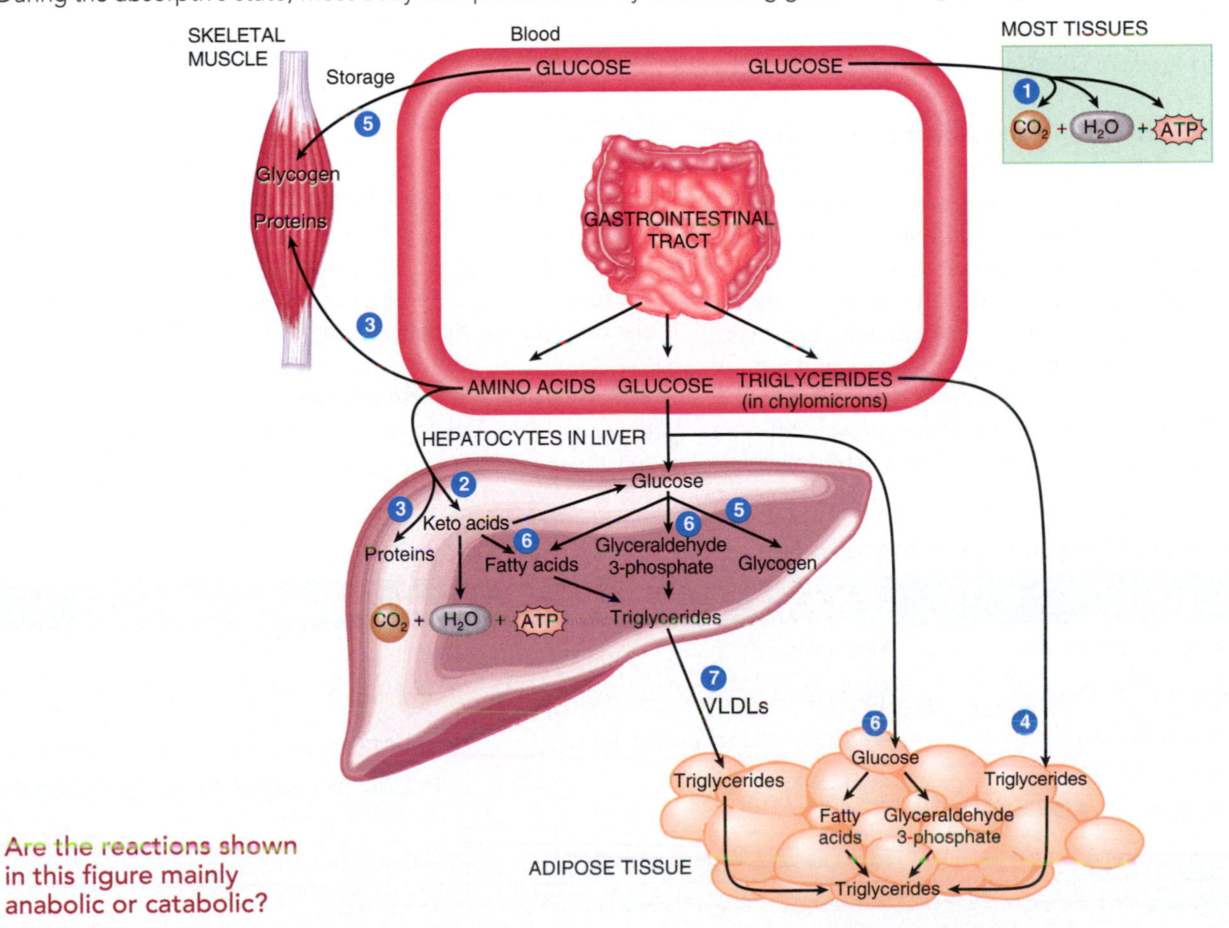

Are the reactions shown in this figure mainly anabolic or catabolic?

Hence glucose is the body's main energy source during the absorptive state. About 50% of the glucose absorbed from a typical meal is catabolized by cells throughout the body to produce ATP.

2. ***Catabolism of amino acids.*** Some amino acids enter hepatocytes (liver cells), where they are deaminated to keto acids. The keto acids in turn can either enter the Krebs cycle for ATP production or be used to synthesize glucose or fatty acids.
3. ***Protein synthesis.*** Many amino acids enter body cells, such as muscle cells and hepatocytes, for synthesis of proteins.
4. ***Catabolism of few dietary lipids.*** During the absorptive state, only a small portion of dietary lipids are catabolized for energy; most dietary lipids are stored in adipose tissue.

Another key event of the absorptive state is that absorbed nutrients in excess of the body's energy needs are converted into **nutrient stores**—namely glycogen and fat. This function is reflected by the following absorptive state reactions (FIGURE 22.1):

5. ***Glycogenesis.*** Some of the glucose that may be in excess of the body's needs is taken up by the liver and skeletal muscle and then converted into glycogen (glycogenesis).
6. ***Lipogenesis.*** The liver can also convert excess glucose or amino acids to fatty acids for use in the synthesis of triglycerides (lipogenesis). Adipocytes also take up glucose not picked up by the liver and convert it into triglycerides for storage. Overall, about 40% of the glucose absorbed from a meal is converted to triglycerides, and about 10% is stored as glycogen in skeletal muscles and the liver.
7. ***Transport of triglycerides from liver to adipose tissue.*** Some fatty acids and triglycerides synthesized in the liver remain there, but hepatocytes package most into very low-density lipoproteins (VLDLs), which carry lipids to adipose tissue for storage.

Regulation of Metabolism During the Absorptive State

Soon after a meal, glucose-dependent insulinotropic peptide (GIP), plus the rising blood levels of glucose and certain amino acids, stimulate pancreatic beta cells to release the hormone insulin. In general, insulin increases the activity of enzymes needed for anabolism and the synthesis of storage molecules; at the same time, it decreases the activity of enzymes needed for catabolic or breakdown reactions. Insulin promotes the entry of glucose and amino acids into cells of many tissues, and it stimulates the conversion of glucose to glycogen (glycogenesis) in both liver and muscle cells. In liver and adipose tissue, insulin enhances the synthesis of triglycerides (lipogenesis), and in cells throughout the body, insulin stimulates protein synthesis. (See Section 13.7 to review the effects of insulin.) Insulin-like growth factors and the thyroid hormones (T_3 and T_4) also stimulate protein synthesis.

Before glucose can be used by body cells, it must first pass through the plasma membrane and enter the cytosol. Glucose entry into most body cells occurs via **glucose transporter (GLUT)** molecules, a family of transporters that bring glucose into cells via facilitated diffusion (see Section 5.4). A high level of insulin increases the insertion of one type of GLUT, called **GLUT4**, into the plasma membranes of most body cells (especially muscle fibers and adipocytes), increasing the rate of facilitated diffusion of glucose into cells. In neurons and hepatocytes, however, other types of GLUTs are always present in the plasma membrane, so glucose entry is always "turned on." Upon entering a cell, glucose becomes phosphorylated. Because GLUT cannot transport phosphorylated glucose, this reaction traps glucose within the cell.

TABLE 22.1 summarizes the hormonal regulation of metabolism in the absorptive state.

The Postabsorptive State Promotes Reactions That Maintain the Normal Blood Glucose Level When the GI Tract Lacks Nutrients

About 4 hours after the last meal, absorption of nutrients from the small intestine is complete, and the blood glucose level starts to fall because glucose continues to leave the bloodstream and enter body cells while none is being absorbed from the GI tract. Thus, the main metabolic challenge during the postabsorptive state is to maintain the normal blood glucose level of 70–110 mg/100 mL (3.9–6.1 mmol/liter). Homeostasis of blood glucose concentration is especially important for the nervous system and for red blood cells for the following reasons:

- The dominant fuel molecule for ATP production in the nervous system is glucose because fatty acids are unable to pass the blood–brain barrier.
- Red blood cells derive all of their ATP from glycolysis of glucose because they have no mitochondria, so the Krebs cycle and the electron transport chain are not available to them.

Postabsorptive State Reactions

A key feature of the postabsorptive state is that the blood glucose concentration is maintained at a normal level due to the breakdown of the body's

TABLE 22.1 Hormonal Regulation of Metabolism in the Absorptive State

Process	Location(s)	Main Stimulating Hormone(s)
Facilitated diffusion of glucose into cells	Most cells.	Insulin.*
Active transport of amino acids into cells	Most cells.	Insulin.
Glycogenesis (glycogen synthesis)	Hepatocytes and muscle fibers.	Insulin.
Protein synthesis	All body cells.	Insulin, thyroid hormones, and insulin-like growth factors.
Lipogenesis (triglyceride synthesis)	Adipose cells and hepatocytes.	Insulin.

* Facilitated diffusion of glucose into hepatocytes (liver cells) and neurons is always "turned on" and does not require insulin.

nutrient stores (glycogen and fat) and the formation of new glucose from noncarbohydrate sources (gluconeogenesis). The reactions of the postabsorptive state that produce glucose are as follows (FIGURE 22.2):

1. ***Glycogenolysis in the liver.*** During the postabsorptive state, a major source of blood glucose is liver glycogenolysis, which can provide about a 4-hour supply of glucose. Once glycogenolysis occurs in the liver, the glucose is released into the blood.
2. ***Glycogenolysis in muscle.*** Glycogenolysis can also occur in skeletal muscle. However, in skeletal muscle, the glucose that is formed from glycogenolysis is catabolized to provide ATP for muscle contraction: Glycogen is broken down to glucose 6-phosphate, which undergoes glycolysis. If anaerobic conditions exist in the skeletal muscle, the pyruvic acid is converted to lactic acid, which is released into the blood. The liver takes up the lactic acid, converts it back to glucose, and then releases glucose into the blood.
3. ***Lipolysis.*** In adipose tissue, triglycerides are broken down into fatty acids and glycerol, which are released into the blood. The glycerol is taken up by the liver and then converted into glucose, which in turn is released into the bloodstream.
4. ***Protein catabolism.*** Modest breakdown of proteins in skeletal muscle and other tissues releases amino acids, which then can be

FIGURE 22.2 Principal metabolic pathways during the postabsorptive state.

The principal function of the postabsorptive state is to maintain a normal blood glucose level.

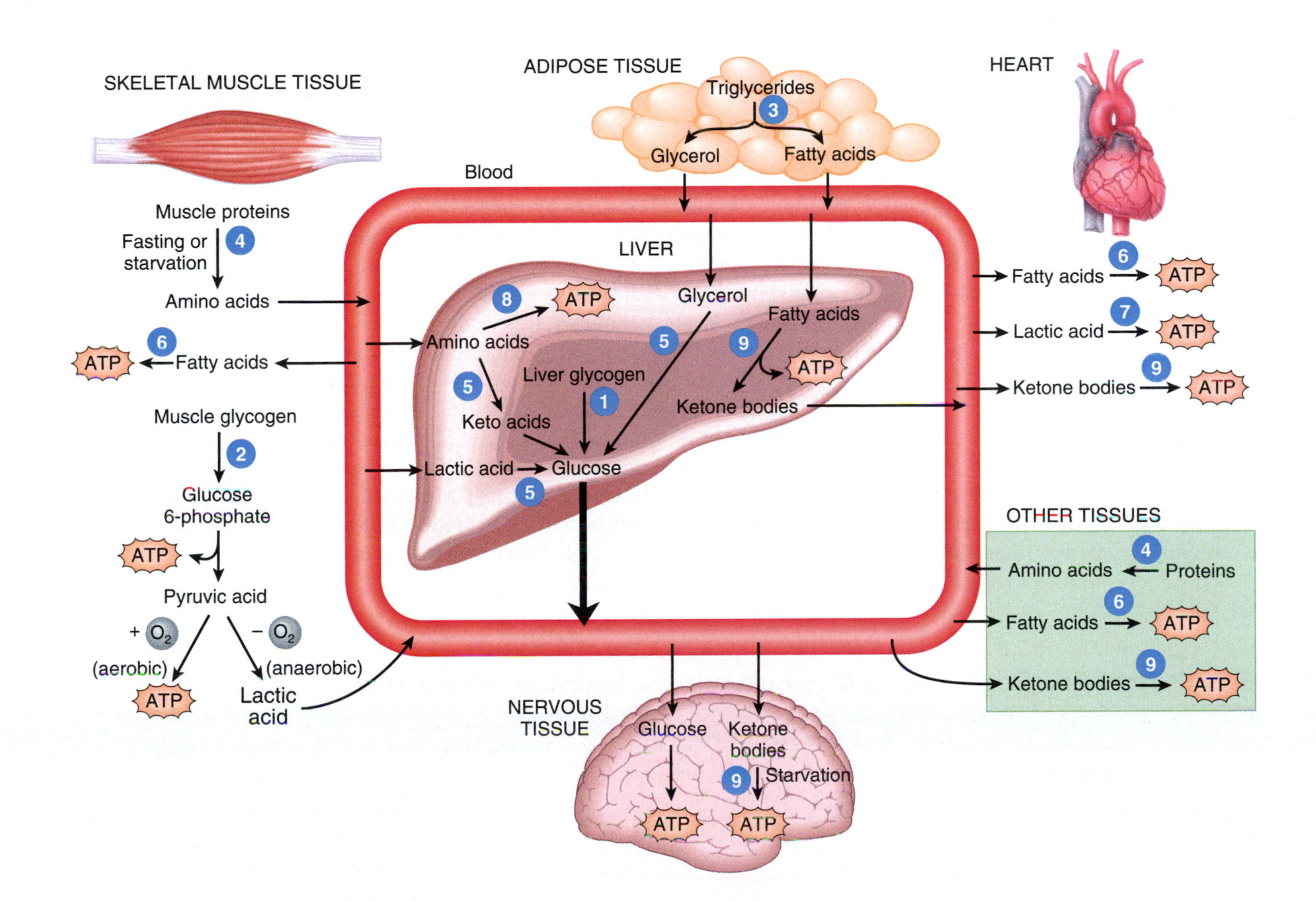

What processes directly elevate blood glucose during the postabsorptive state, and where does each occur?

converted to glucose by the liver. The glucose in turn is released into the bloodstream.

5. ***Gluconeogenesis.*** During the postabsorptive state, new glucose is formed from noncarbohydrate sources. Examples of gluconeogenesis include the formation of glucose from lactic acid, glycerol, or an amino acid.

Another hallmark feature of the postabsorptive state is that glucose sparing occurs. **Glucose sparing** means that most body cells switch to other fuels besides glucose as their main source of energy, leaving more glucose in the blood for the brain and red blood cells. The following reactions produce ATP without using glucose (FIGURE 22.2):

6. ***Catabolism of fatty acids.*** The fatty acids released by lipolysis of triglycerides cannot be used for glucose production because acetyl CoA cannot be readily converted to pyruvic acid. But most cells can catabolize the fatty acids directly, feed them into the Krebs cycle as acetyl CoA, and produce ATP through the electron transport chain.
7. ***Catabolism of lactic acid.*** Cardiac muscle can produce ATP aerobically from lactic acid.
8. ***Catabolism of amino acids.*** In hepatocytes, amino acids may be catabolized directly to produce ATP.
9. ***Catabolism of ketone bodies.*** Hepatocytes also convert fatty acids to ketone bodies (acetoacetic acid, beta-hydroxybutyric acid, and acetone), which can be used by the heart, kidneys, and other tissues for ATP production.

Regulation of Metabolism During the Postabsorptive State

Both hormones and the sympathetic division of the autonomic nervous system (ANS) regulate metabolism during the postabsorptive state. The hormones that regulate postabsorptive state metabolism sometimes are called anti-insulin hormones because they counter the effects of insulin during the absorptive state. As blood glucose level declines, the secretion of insulin falls and the release of anti-insulin hormones rises.

When blood glucose concentration starts to drop, the pancreatic alpha cells release the hormone glucagon. The primary target tissue of glucagon is the liver; the major effect is increased release of glucose into the bloodstream due to gluconeogenesis and glycogenolysis.

Low blood glucose also activates the sympathetic branch of the ANS. Glucose-sensitive neurons in the hypothalamus detect low blood glucose and increase sympathetic output. As a result, sympathetic nerve endings release the neurotransmitter norepinephrine, and the adrenal medulla releases two catecholamine hormones—epinephrine and norepinephrine—into the bloodstream. Like glucagon, epinephrine stimulates glycogen breakdown. Epinephrine and norepinephrine are both potent stimulators of lipolysis. These actions of the catecholamines help to increase glucose and free fatty acid levels in the blood. As a result, muscle uses more fatty acids for ATP production, and more glucose is available to the nervous system.

Stressful situations such as low blood glucose, hot or cold temperatures, fear, or trauma ultimately cause the release of the hormone cortisol from the adrenal gland. Cortisol in turn promotes gluconeogenesis, lipolysis, and protein catabolism.

TABLE 22.2 summarizes the hormonal regulation of metabolism in the postabsorptive state.

Metabolism During Fasting and Starvation Results in an Increase in Ketone Body Formation

The term **fasting** means going without food for many hours or a few days; **starvation** implies weeks or months of food deprivation or inadequate food intake. During fasting and starvation, glycogen stores are quickly depleted. However, there is an increase in the catabolism of lipids and proteins, which allows a person to survive without food for two months or more as long as enough water is consumed to prevent dehydration. The amount of adipose tissue the body contains determines the amount of time possible without food.

The most dramatic metabolic change that occurs with fasting and starvation is the increase in the formation of ketone bodies by hepatocytes. Ketone body production increases as catabolism of fatty acids rises. Lipid-soluble ketone bodies can diffuse through plasma membranes and across the blood–brain barrier, and be used as an alternative fuel for ATP production, especially by cardiac and skeletal muscle fibers and neurons. Normally, only a trace level of ketone bodies (0.01 mmol/liter) is present in the blood, so ketone bodies are a negligible fuel source. After two days of fasting, however, the level of ketones is 100–300 times higher and supplies roughly a third of the brain's fuel for ATP production. By 40 days of starvation, ketones provide up to two-thirds of the brain's energy needs. The presence of ketones actually reduces the use of glucose for ATP production, which in turn decreases the demand for gluconeogenesis and slows the catabolism of muscle proteins.

TABLE 22.2 Hormonal Regulation of Metabolism in the Postabsorptive State

Process	Location(s)	Main Stimulating Hormone(s)
Glycogenolysis (glycogen breakdown)	Hepatocytes and skeletal muscle fibers.	Glucagon and epinephrine.
Lipolysis (triglyceride breakdown)	Adipocytes.	Epinephrine, norepinephrine, cortisol, insulin-like growth factors, thyroid hormones, and others.
Protein breakdown	Most body cells, but especially skeletal muscle fibers.	Cortisol.
Gluconeogenesis (synthesis of glucose from noncarbohydrate sources)	Hepatocytes and kidney cortex cells.	Glucagon and cortisol.

CLINICAL CONNECTION

Ketosis

The level of ketone bodies in the blood normally is very low because other tissues use them for ATP production as fast as they are generated from the breakdown of fatty acids in the liver. During periods of excessive fatty acid catabolism, however, the production of ketone bodies exceeds their uptake and use by body cells. This might occur after a meal rich in triglycerides, or during fasting or starvation, because few carbohydrates are available for catabolism. Excessive fatty acid catabolism may also occur in poorly controlled or untreated diabetes mellitus for two reasons: (1) Adequate glucose cannot get into cells, so triglycerides are used for ATP production, and (2) insulin normally inhibits lipolysis, so a lack of insulin accelerates the pace of lipolysis. When the concentration of ketone bodies in the blood rises above normal—a condition called **ketosis**—the ketone bodies, most of which are acids, must be buffered. If too many accumulate, they decrease the concentration of buffers such as bicarbonate ions, and blood pH falls. Extreme or prolonged ketosis can lead to **ketoacidosis**, an abnormally low blood pH due to a high concentration of ketone bodies. The decreased blood pH in turn causes depression of the central nervous system, which can result in disorientation, coma, and even death if the condition is not treated. When a diabetic becomes seriously insulin-deficient, one of the telltale signs is the sweet smell on the breath from the ketone body acetone.

CHECKPOINT

1. Define the following terms: glycogenesis, glycogenolysis, gluconeogenesis, lipogenesis, and lipolysis.
2. Which of the processes listed in checkpoint question 1 occurs during the absorptive state? During the postabsorptive state?
3. Why is glucose sparing necessary during the postabsorptive state and not during the absorptive state?
4. What are the roles of insulin, glucagon, epinephrine, insulin-like growth factors, thyroid hormones, and cortisol in the regulation of metabolism?
5. Why is ketogenesis more significant during fasting or starvation than during normal absorptive and postabsorptive states?

22.2 Energy Balance

OBJECTIVES

- Explain what is meant by the term *energy balance*.
- Discuss the various factors that affect metabolic rate.
- Describe the role of the hypothalamus in the regulation of food intake.

Energy balance refers to the precise matching of energy intake (in food) to energy expenditure over time. When the energy content of food balances the energy used by all cells of the body, body weight remains constant (unless there is a gain or loss of water). In many people, weight stability persists despite large day-to-day variations in activity and food intake. In the more affluent nations, however, a large fraction of the population is overweight. Easy access to tasty, high-calorie foods and a "couch potato" lifestyle both promote weight gain. Being overweight increases the risk of dying from a variety of cardiovascular and metabolic disorders, including hypertension, varicose veins, diabetes mellitus, arthritis, and certain cancers.

The Energy Content in Food Is Expressed in Units Called Calories

As you learned in Chapter 4, when catabolic reactions occur, energy is released. About 40% of this energy is used to perform biological work, such as active transport and muscle contraction. The remaining 60% is converted to heat, some of which helps maintain normal body temperature. Excess heat is lost to the environment. When the body catabolizes the organic compounds in food, the heat energy released can be measured in units called calories. A **calorie (cal)** is defined as the amount of energy in the form of heat required to raise the temperature of 1 gram of water 1°C. Because the calorie is a relatively small unit, the **kilocalorie (kcal)** or **Calorie (Cal)** (always spelled with an uppercase C) is often used to express the energy content of foods. A kilocalorie equals 1000 calories. Thus, when we say that a particular food item contains 500 Calories, we are actually referring to kilocalories.

Essentially all of the kilocalories in our food come from the catabolism of carbohydrates, proteins, and fats. The catabolism of carbohydrates or proteins yields about the same amount of energy—about 4 kcal/g. The catabolism of fat yields much more energy—about 9 kcal/g. Some foods or beverages may contain alcohol, and the catabolism of alcohol also yields energy—about 7 kcal/g. The energy content of carbohydrates, proteins, fats, and alcohol is summarized in TABLE 22.3.

The number of kilocalories from a component in a particular food can be calculated by multiplying the number of grams of that component by its energy content. For example, suppose that one slice of pizza contains 27 g of carbohydrate, 14 g of fat, and 12 g of protein. To calculate the number of kcal from carbohydrate in this slice of pizza, multiply the number of grams of carbohydrate in the pizza by the energy content of carbohydrates: 27 g carbohydrate × 4 kcal/g = 108 kcal. To calculate the number of kcal from fat in the slice of pizza, multiple the number of grams of fat in the pizza by the energy content of fat: 14 g fat × 9 kcal/g = 126 kcal. To calculate the number of kcal from protein in the slice of pizza, multiply the number of grams of protein in the pizza by the energy content of protein: 12 g protein × 4 kcal/g = 48 kcal. Finally, to calculate the total kcal in the slice of pizza, add together all of the kcal from carbohydrate, fat, and protein: 108 kcal + 126 kcal + 48 kcal = 282 kcal.

TABLE 22.3 Energy Content of Various Nutrients and Alcohol

Nutrient	Energy Content
Carbohydrate	4 kcal/g
Protein	4 kcal/g
Fat	9 kcal/g
Alcohol	7 kcal/g

TABLE 22.4 lists the caloric content of several familiar foods. The higher the caloric content of a particular food, the greater the amount of energy released as it is catabolized. For example, the energy content of one medium apple is 80 kcal; this means that 80 kcal is the amount of energy released as the apple is catabolized. The energy content of a slice of chocolate cake is 247 kcal; this means that 247 kcal is the amount of energy released as the chocolate cake is catabolized. Suppose that you eat the apple or the chocolate cake. Based on the caloric content of these foods, your body will have to work harder (via exercise, for example) to release more energy in order to catabolize the chocolate cake compared to the apple.

Beverages can also be a source of calories. For example, a cola soft drink (12 ounces) contains 40 g of carbohydrate, 0 g of protein, and 0 g of fat, so the energy content of this soda is 160 kcal (40 g carbohydrate × 4 kcal/g). A typical serving of vodka (1.5 ounces) contains 0 g of carbohydrate, 0 g of protein, 0 g of fat, and 14 g of alcohol, so the energy content of this drink is 98 calories (14 g × 7 kcal/g). If juice, soda, or cocktail mix is added to the vodka, these solutions usually contain carbohydrates that contribute additional calories. **TABLE 22.5** lists the caloric content of several beverages.

It is not known with certainty what levels and types of carbohydrate, fat, and protein are optimal in the diet. Different populations around the world eat radically different diets that are adapted to their particular lifestyles. However, many experts recommend the following distribution of calories: 50–60% from carbohydrates, with less than 15% from simple sugars; less than 30% from fats (triglycerides are the main type of dietary fat), with no more than 10% as saturated fats; and about 12–15% from proteins.

The guidelines for healthy eating are as follows:

- Eat a variety of foods.
- Maintain a healthy weight.
- Choose foods low in fat, saturated fat, and cholesterol.
- Eat plenty of vegetables, fruits, and grain products.
- Use sugars in moderation only.

The Metabolic Rate Reflects the Amount of Energy Used by Metabolic Reactions over Time

The overall rate at which metabolic reactions use energy is termed the **metabolic rate**. As you have already learned, some of the energy is used to produce ATP, and some is released as heat. Thus, the higher the metabolic rate, the higher the rate of heat production.

Factors That Affect Metabolic Rate

Several factors affect the metabolic rate:

- ***Hormones.*** Thyroid hormones (thyroxine and triiodothyronine) are the main regulators of basal metabolic rate (BMR), the metabolic rate under basal conditions (described shortly). BMR increases as

TABLE 22.4 Caloric Content of Various Foods

Food	Serving Size	Energy (kcal)	Carbohydrate (g)	Fat (g)	Protein (g)
Apple	1	80	19	0	1
Broccoli (raw)	½ cup	16	3	0	1
Baked potato (plain)	1	160	35	0	5
Wheat bread	1 slice	65	12	1	2
Vegetable soup	1 cup	100	20	0	5
Baked chicken	3 ounces	158	0	6	26
Lean ground beef (10% fat)	3 ounces	178	0	10	22
Baked trout	3 ounces	101	0	1	23
McDonald's® Big Mac	1	541	45	29	25
Wendy's® Biggie Fry	1	530	68	25	6
Chick-fil-A® chicken sandwich (fried)	1	408	38	16	28
Burger King Whopper®	1	710	52	42	31
Pizza Hut® super supreme pizza	1 slice	282	27	14	12
Cinnabon® roll	1	808	115	32	15
Chocolate cake	1 slice	247	35	11	2
Butter	1 tablespoon	108	0	12	0
Sour cream	2 tablespoons	62	1	6	1
Mayonnaise	1 tablespoon	99	0	11	0

TABLE 22.5 Caloric Content of Various Beverages

Beverage	Serving Size	Energy (kcal)	Carbohydrate (g)	Fat (g)	Protein (g)	Alcohol (g)
Cola soft drink	12 ounces	160	40	0	0	0
Whole milk	1 cup	148	11	8	8	0
Orange juice	1 cup	108	25	0	2	0
White wine	5 ounces	102	1	0	0	14
Red wine	5 ounces	110	3	0	0	14
Beer	12 ounces	143	13	0	0	13
Vodka	1.5 ounces	98	0	0	0	14
Whiskey	1.5 ounces	98	0	0	0	14
Bourbon	1.5 ounces	98	0	0	0	14

the blood levels of thyroid hormones rise. The response to changing levels of thyroid hormones is slow, however, taking several days to appear. Thyroid hormones increase BMR in part by stimulating cellular respiration. As cells use more oxygen to produce ATP, more heat is given off, and body temperature rises. This effect of thyroid hormones on BMR is called the **calorigenic effect**. Other hormones have minor effects on BMR. Testosterone, insulin, and human growth hormone can increase the metabolic rate by 5–15%.

- ***Exercise.*** During strenuous exercise, the metabolic rate may increase to as much as 15 times the basal rate. In well-trained athletes, the rate may increase up to 20 times.
- ***Nervous system.*** During exercise or in a stressful situation, the sympathetic division of the autonomic nervous system is stimulated. Its postganglionic neurons release norepinephrine (NE), and it also stimulates release of the hormones epinephrine and norepinephrine by the adrenal medulla. Both epinephrine and norepinephrine increase the metabolic rate of body cells.
- ***Body temperature.*** The higher the body temperature, the higher the metabolic rate. Each 1°C rise in core temperature increases the rate of biochemical reactions by about 10%. As a result, metabolic rate may be increased substantially during a fever.
- ***Ingestion of food.*** The ingestion of food raises the metabolic rate 10–20% due to the energy "costs" of digesting, absorbing, and storing nutrients. This effect, **food-induced thermogenesis**, is greatest after eating a high-protein meal and is less after eating carbohydrates and lipids.
- ***Age.*** The metabolic rate of a child, in relation to its size, is about double that of an elderly person due to the high rates of reactions related to growth.
- ***Other factors.*** Other factors that affect metabolic rate include gender (lower in females, except during pregnancy and lactation), climate (lower in tropical regions), sleep (lower), and malnutrition (lower).

Basal Metabolic Rate

Because many factors affect metabolic rate, it is measured under standard conditions, with the body in a quiet, resting, and fasting condition called the **basal state**. The measurement obtained under these conditions is the **basal metabolic rate (BMR)**. The most common way to determine BMR is by measuring the amount of oxygen used per kilocalorie of food metabolized. When the body uses 1 liter of oxygen to catabolize a typical dietary mixture of triglycerides, carbohydrates, and proteins, about 4.8 kcal of energy is released. BMR is 1200–1800 kcal/day in adults, or about 24 kcal/kg of body mass in adult males and 22 kcal/kg in adult females. The added calories needed to support daily activities, such as digestion and walking, range from 500 kcal for a small, relatively sedentary person to over 3000 kcal for a person in training for Olympic-level competitions or mountain climbing.

Total Metabolic Rate

The **total metabolic rate (TMR)** is the total energy expenditure by the body per unit of time. Three components contribute to the TMR:

1. ***Basal metabolic rate.*** The basal metabolic rate accounts for about 60% of the TMR.
2. ***Physical activity.*** Physical activity typically adds 30–35% but can be lower in sedentary people. The energy expenditure is partly from voluntary exercise, such as walking, and partly from **nonexercise activity thermogenesis (NEAT)**, the energy costs for maintaining muscle tone, posture while sitting or standing, and involuntary fidgeting movements. TABLE 22.6 lists various activities and the calories that they burn per hour.
3. ***Food-induced thermogenesis.*** Food-induced thermogenesis—the heat produced while food is being digested, absorbed, and stored—represents 5–10% of the TMR.

Adipose Tissue Is the Main Site of Stored Chemical Energy

The major site of stored chemical energy in the body is adipose tissue. When energy use exceeds energy input, triglycerides in adipose tissue are catabolized to provide the extra energy, and when energy input exceeds energy expenditure, triglycerides are stored. Over time, the amount of stored triglycerides indicates the excess of energy intake over energy expenditure. Even small differences add up over time. A gain of 20 lb (9 kg) between ages 25 and 55 represents only a tiny imbalance, about 0.3% more energy intake in food than energy expenditure.

TABLE 22.6 Various Activities and the Calories Released

Activity	Energy Expenditure (kcal/hr)
Aerobics	419
Canoeing	248
Dancing	332
House cleaning	202
Office work	105
Playing the piano	170
Reading	86
Walking (3 mph)	250
Running (5 mph)	570
Sitting	102
Standing	132
Studying at desk	128
Swimming	572
Talking on phone	71
Weightlifting	224
Writing	122
Texting	40

Food Intake Is Regulated by Many Factors

Negative feedback mechanisms regulate both our energy intake and our energy expenditure. But no sensory receptors exist to monitor our weight or size. How then is food intake regulated? The answer to this question is incomplete, but important advances in understanding regulation of food intake have occurred in the past decade. It depends on many factors, including neural and endocrine signals, levels of certain nutrients in the blood, psychological elements such as stress or depression, signals from the GI tract and the special senses, and neural connections between the hypothalamus and other parts of the brain.

Within the hypothalamus are clusters of neurons that play key roles in regulating food intake. **Satiety** is a feeling of fullness accompanied by lack of desire to eat. Two hypothalamic areas involved in regulation of food intake are the *arcuate nucleus* and the *paraventricular nucleus*. In 1994, the first experiments were reported on a mouse gene, named *obese*, that causes overeating and severe obesity in its mutated form. The product of this gene is the hormone **leptin**. In both mice and humans, leptin helps decrease **adiposity**, total body-fat mass. Leptin is synthesized and secreted by adipocytes in proportion to adiposity; as more triglycerides are stored, more leptin is secreted into the bloodstream. Leptin acts on the hypothalamus to inhibit circuits that stimulate eating while also activating circuits that increase energy expenditure. The hormone insulin has a similar but smaller effect. Both leptin and insulin are able to pass through the blood–brain barrier.

When leptin and insulin levels are *low*, neurons that extend from the arcuate nucleus to the paraventricular nucleus release a neurotransmitter called **neuropeptide Y** that stimulates food intake. Other neurons that extend between the arcuate and paraventricular nuclei release a neurotransmitter called **melanocortin**, which is similar to melanocyte-stimulating hormone (MSH). Leptin stimulates release of melanocortin, which acts to inhibit food intake. Another hormone involved in the regulation of food intake is **ghrelin**, which is produced by endocrine cells of the stomach. Ghrelin plays a role in increasing appetite. It is thought that ghrelin performs this function by stimulating the release of neuropeptide Y from hypothalamic neurons. Although leptin, neuropeptide Y, melanocortin, and ghrelin are key signaling molecules for maintaining energy balance, several other hormones and neurotransmitters also contribute. Other areas of the hypothalamus plus nuclei in the brain stem, limbic system, and cerebral cortex take part. An understanding of the brain circuits involved is still far from complete.

Achieving energy balance requires regulation of energy intake. Most increases and decreases in food intake are due to changes in meal size rather than changes in number of meals. Many experiments have demonstrated the presence of satiety signals, chemical or neural changes that help terminate eating when "fullness" is attained. For example, an increase in blood glucose level, as occurs after a meal, decreases appetite. Several hormones, such as glucagon, cholecystokinin, estrogens, and epinephrine (acting via beta receptors) act to signal satiety and to increase energy expenditure. Distension of the GI tract, particularly the stomach and duodenum, also contributes to termination of food intake. Other hormones increase appetite and decrease energy expenditure. These include growth hormone–releasing hormone (GHRH), androgens, glucocorticoids, epinephrine (acting via alpha receptors), and progesterone.

Obesity Occurs When Too Much Adipose Tissue Accumulates in the Body

Obesity is body weight more than 20% above a desirable standard due to an excessive accumulation of adipose tissue. About one-third of the adult population in the United States is obese. (An athlete may be *overweight* due to higher-than-normal amounts of muscle tissue without being obese.) Even moderate obesity is hazardous to health; it is a risk factor in cardiovascular disease, hypertension, pulmonary disease, non-insulin-dependent diabetes mellitus, arthritis, and certain cancers (breast, uterus, and colon).

In a few cases, obesity may result from trauma of or tumors in the food-regulating centers in the hypothalamus. In most cases of obesity, no specific cause can be identified. Contributing factors include genetic factors, eating habits taught early in life, overeating to relieve tension, and social customs. Although leptin suppresses appetite and produces satiety in experimental animals, it is not deficient in most obese people.

Most surplus calories in the diet are converted to triglycerides and stored in adipose cells. Initially, the adipocytes increase in size, but at a maximal size, they divide. As a result, proliferation of adipocytes occurs in extreme obesity. The enzyme endothelial lipoprotein lipase regulates triglyceride storage. The enzyme is very active in abdominal fat but less active in hip fat. Accumulation of fat in the abdomen is associated with higher blood cholesterol level and other cardiac risk factors because adipose cells in this area appear to be more metabolically active.

Treatment of obesity is difficult because most people who are successful at losing weight gain it back within two years. Yet even modest

CLINICAL CONNECTION

Eating Disorders

Two types of **eating disorders** are bulimia and anorexia nervosa. These disorders typically occur in women but can also occur in men. **Bulimia**, or *binge–purge syndrome,* is a disorder characterized by overeating at least twice a week followed by purging via self-induced vomiting, strict dieting or fasting, vigorous exercise, or use of laxatives or diuretics. It occurs in response to fears of being overweight or to stress, depression, and physiological disorders such as hypothalamic tumors. **Anorexia nervosa** is a chronic disorder characterized by self-induced weight loss, negative perception of body image, and physiological changes that result from nutritional depletion. Patients with anorexia nervosa have a fixation on weight control and often insist on having a bowel movement every day despite inadequate food intake. They often abuse laxatives, which worsens the fluid and electrolyte imbalances and nutrient deficiencies. This disorder may be inherited. Individuals may become emaciated and may ultimately die of starvation or one of its complications. Also associated with the disorder are osteoporosis, depression, and brain abnormalities coupled with impaired mental performance. Treatment consists of psychotherapy and dietary regulation.

weight loss is associated with health benefits. Possible treatments for obesity include behavior modification programs, very-low-calorie diets, drugs, and surgery. Behavior modification programs, offered at many hospitals, strive to alter eating behaviors and increase exercise activity. Very-low-calorie (VLC) diets include 400 to 800 kcal/day in a commercially made liquid mixture. Drugs that treat obesity either inhibit reuptake of serotonin and norepinephrine in brain areas that govern eating behavior, or they inhibit the lipases released into the lumen of the GI tract. With less lipase activity, fewer dietary triglycerides are absorbed. For those with extreme obesity who have not responded to other treatments, a surgical procedure may be considered. The two operations most commonly performed—gastric bypass and gastroplasty—both greatly reduce the stomach size so that it can hold just a tiny quantity of food.

CHECKPOINT

6. What is a calorie? Why is the kilocalorie often used more than the calorie to express the energy content of food?
7. What are the three components that contribute to the total metabolic rate?
8. What are the functions of leptin, neuropeptide Y, melanocortin, and ghrelin?

22.3 Regulation of Body Temperature

OBJECTIVES

- Describe the various mechanisms of heat transfer.
- Explain how normal body temperature is maintained by negative feedback loops involving the hypothalamic thermostat.

Your body produces more or less heat depending on the rates of its metabolic reactions. Because homeostasis of body temperature can be maintained only if the rate of heat loss from the body equals the rate of heat production by metabolism, it is important to understand the ways in which heat can be lost, gained, or conserved. **Heat** is a form of energy that can be measured as **temperature**. Despite wide fluctuations in environmental temperature, homeostatic mechanisms can maintain a normal range for internal body temperature. If the rate of body heat production equals the rate of heat loss, the body maintains a constant core temperature near 37°C (98.6°F). **Core temperature** is the temperature in body structures deep to the skin and subcutaneous layer. **Shell temperature** is the temperature near the body surface—in the skin and subcutaneous layer. Depending on the environmental temperature, shell temperature is 1–6°C lower than core temperature. A core temperature that is too high kills by denaturing body proteins; one that is too low causes cardiac arrhythmias that result in death.

There Are Four Mechanisms of Heat Transfer

Maintaining normal body temperature depends on the ability to lose heat to the environment at the same rate as it is produced by metabolic reactions. Heat can be transferred between the body and its surroundings in four ways (**FIGURE 22.3**):

1. **Conduction** is the heat exchange that occurs between molecules of two materials that are in direct contact with each other. At rest,

FIGURE 22.3 Mechanisms of heat transfer.

Heat can be transferred between the body and its surroundings via conduction, convection, radiation, and evaporation.

What is conduction?

about 3% of body heat is lost via conduction to cooler, solid materials in contact with the body, such as a chair, clothing, and jewelry. Heat can also be gained via conduction—for example, while soaking in a hot tub. Because water conducts heat 20 times more effectively than air, heat loss or heat gain via conduction is much greater when the body is submerged in cold or hot water.

2. **Convection** is the transfer of heat by the movement of air or water between areas of different temperatures. The contact of air or water with your body results in heat transfer by both conduction and convection. When cool air makes contact with the body, the air becomes warmed and therefore less dense and is carried away by convection currents created as the less dense air rises. The faster the air moves—for example, by a breeze or a fan—the faster the rate of convection. At rest, about 15% of body heat is lost to the air via conduction and convection.
3. **Radiation** is the transfer of heat in the form of infrared rays between a warmer object and a cooler one without physical contact. Your body loses heat by radiating more infrared waves than it absorbs from cooler objects. If surrounding objects are warmer than you are, you absorb more heat than you lose by radiation. In a room at 21°C (70°F), about 60% of heat loss occurs via radiation in a resting person.
4. **Evaporation** is the conversion of a liquid to a vapor. Every milliliter of evaporating water takes with it a great deal of heat—about 0.58 kcal/mL. Under typical resting conditions, about 22% of heat loss occurs through evaporation of about 700 mL of water per day—300 mL in exhaled air and 400 mL from the skin surface. Because we are not normally aware of this water loss through the skin and mucous membranes of the mouth and respiratory system, it is termed **insensible water loss**. The rate of evaporation is inversely related to relative humidity, the ratio of the actual amount of moisture in the air to the maximum amount it can hold at a given temperature. The higher the relative humidity, the lower the rate of evaporation. At 100% humidity, heat is gained via condensation of water on the skin surface as fast as heat is lost via evaporation. Evaporation provides the main defense against overheating during exercise. Under extreme conditions, a maximum of about 3 liters of sweat can be produced each hour, removing more than 1700 kcal of heat if all of it evaporates. (Note: Sweat that drips off the body rather than evaporating removes very little heat.)

The Hypothalamus Contains the Body's Thermostat

The control center that functions as the body's thermostat is a group of neurons in the anterior part of the hypothalamus, the **preoptic area**. This area receives input from thermoreceptors in the skin (**peripheral thermoreceptors**) and in the hypothalamus itself (**central thermoreceptors**). Neurons of the preoptic area generate action potentials at a higher frequency when blood temperature increases and at a lower frequency when blood temperature decreases.

Action potentials from the preoptic area propagate to two other parts of the hypothalamus known as the **heat-losing center** and the **heat-promoting center**, which, when stimulated by the preoptic area, set into operation a series of responses that lower body temperature and raise body temperature, respectively.

Thermoregulation Maintains the Body's Temperature

If core temperature declines, mechanisms that help conserve heat and increase heat production act via negative feedback to raise the body temperature to normal (**FIGURE 22.4**). Peripheral thermoreceptors and central thermoreceptors send input to the preoptic area of the hypothalamus, which in turn activates the heat-promoting center. In response, the hypothalamus discharges action potentials and secretes thyrotropin-releasing hormone (TRH), which in turn stimulates thyrotrophs in the anterior pituitary gland to release thyroid-stimulating hormone (TSH). Action potentials from the hypothalamus and TSH then activate several effectors, which respond in the following ways to increase the core temperature to the normal value:

- ***Vasoconstriction.*** Action potentials from the heat-promoting center stimulate sympathetic nerves that cause blood vessels of the skin to constrict. Vasoconstriction decreases the flow of warm blood, and thus the transfer of heat, from the internal organs to the skin. Slowing the rate of heat loss allows the internal body temperature to increase as metabolic reactions continue to produce heat.
- ***Release of epinephrine and norepinephrine.*** Action potentials in sympathetic nerves leading to the adrenal medulla stimulate the release of epinephrine and norepinephrine into the blood. The hormones in turn bring about an increase in cellular metabolism, which increases heat production.
- ***Shivering.*** The heat-promoting center stimulates parts of the brain that increase muscle tone and hence heat production. As muscle tone increases in one muscle (the agonist), the small contractions stretch muscle spindles in its antagonist, initiating a stretch reflex. The resulting contraction in the antagonist stretches muscle spindles in the agonist, and it too develops a stretch reflex. This repetitive cycle—called **shivering**—greatly increases the rate of heat production. During maximal shivering, body heat production can rise to about four times the basal rate in just a few minutes.
- ***Release of thyroid hormones.*** The thyroid gland responds to TSH by releasing more thyroid hormones into the blood. As increased levels of thyroid hormones slowly increase the metabolic rate, body temperature rises.

If core body temperature rises above normal, a negative feedback loop opposite to the one depicted in **FIGURE 22.4** goes into action. The higher temperature of the blood stimulates peripheral and central thermoreceptors that send input to the preoptic area, which in turn stimulates the heat-losing center and inhibits the heat-promoting center. Action potentials from the heat-losing center cause dilation of blood vessels in the skin. The skin becomes warm, and the excess heat is lost to the environment via radiation and conduction as an increased volume of blood flows from the warmer core of the body into the cooler skin. At the same time, metabolic rate decreases, and shivering does not occur. The high temperature of the blood stimulates sweat glands of the skin via hypothalamic activation of sympathetic nerves. As the water in perspiration evaporates from the surface of the skin, the skin is cooled. All of these responses counteract heat-promoting effects and help return body temperature to normal.

FIGURE 22.4 Negative feedback mechanisms that conserve heat and increase heat production.

When stimulated, the heat-promoting center in the hypothalamus raises body temperature.

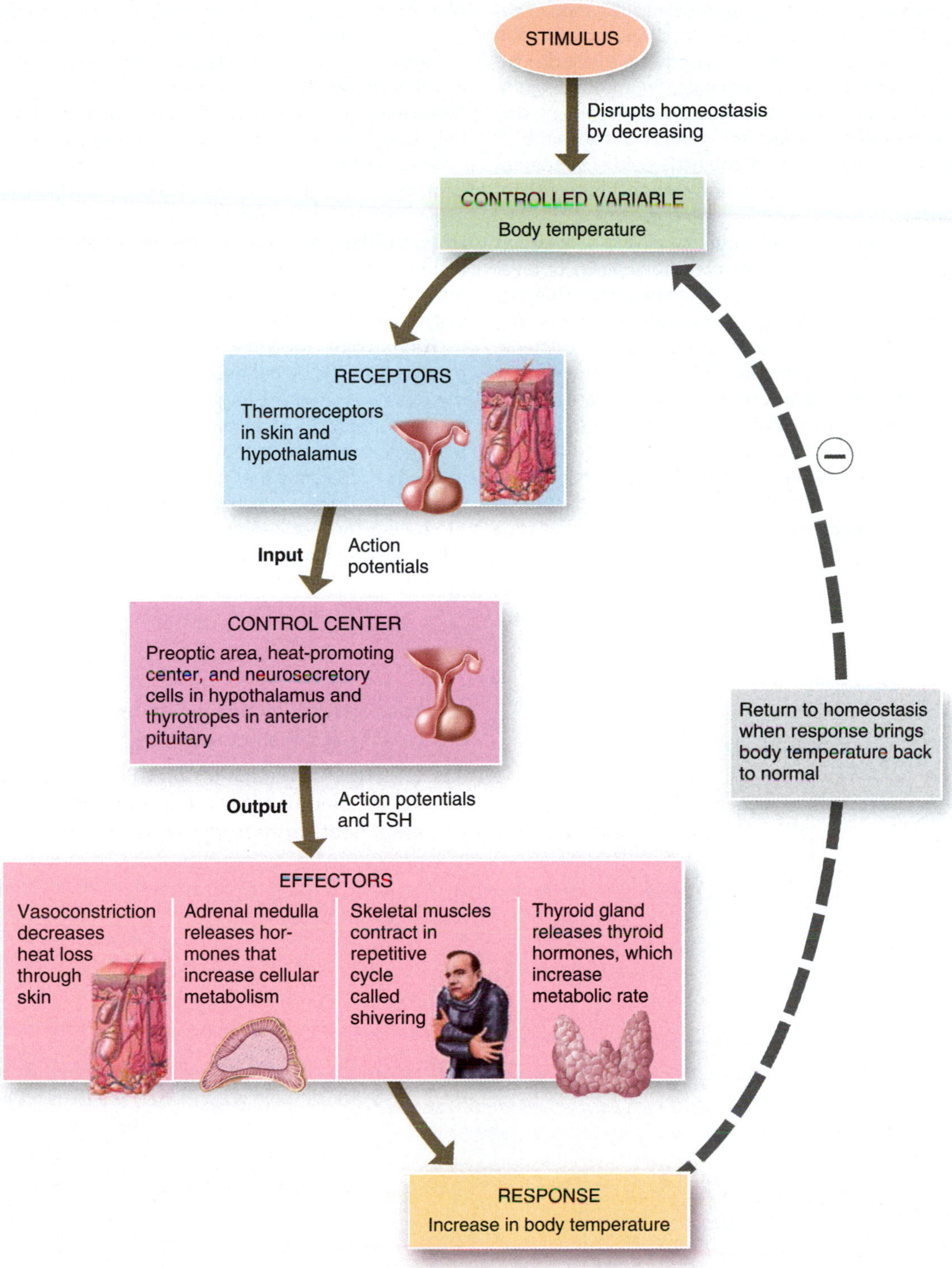

STIMULUS

Disrupts homeostasis by decreasing

CONTROLLED VARIABLE
Body temperature

RECEPTORS
Thermoreceptors in skin and hypothalamus

Input
Action potentials

CONTROL CENTER
Preoptic area, heat-promoting center, and neurosecretory cells in hypothalamus and thyrotropes in anterior pituitary

Output
Action potentials and TSH

EFFECTORS

Vasoconstriction decreases heat loss through skin

Adrenal medulla releases hormones that increase cellular metabolism

Skeletal muscles contract in repetitive cycle called shivering

Thyroid gland releases thyroid hormones, which increase metabolic rate

RESPONSE
Increase in body temperature

Return to homeostasis when response brings body temperature back to normal

−

What is the difference between peripheral and central thermoreceptors?

CLINICAL CONNECTION

Disorders of Body Temperature Regulation

There are several **disorders of body temperature regulation**: hypothermia, fever, heat exhaustion, and heatstroke. **Hypothermia** is a lowering of core body temperature to 35°C (95°F) or below. Causes of hypothermia include an overwhelming cold stress (immersion in icy water), metabolic diseases (hypoglycemia, adrenal insufficiency, or hypothyroidism), drugs (alcohol, antidepressants, sedatives, or tranquilizers), burns, and malnutrition. Symptoms of hypothermia include sensation of cold, shivering, confusion, vasoconstriction, muscle rigidity, slow heart rate, hypoventilation, loss of spontaneous movement, and coma. Death is usually caused by cardiac arrhythmias. Because the elderly have reduced metabolic protection against a cold environment coupled with a reduced perception of cold, they are at greater risk for developing hypothermia.

A **fever** is an elevation of core temperature caused by a resetting of the hypothalamic thermostat. The most common causes of fever are viral or bacterial infections and bacterial toxins; other causes are ovulation, excessive secretion of thyroid hormones, tumors, and reactions to vaccines. When phagocytes ingest certain bacteria, they are stimulated to secrete a **pyrogen** (PĪ-rō-gen), a fever-producing substance. One pyrogen is interleukin-1. It circulates to the hypothalamus and induces neurons of the preoptic area to secrete prostaglandins. Some prostaglandins can reset the hypothalamic thermostat at a higher temperature, and temperature-regulating reflex mechanisms then act to bring the core body temperature up to this new setting. ***Antipyretics*** are agents that relieve or reduce fever. Examples include aspirin, acetaminophen (Tylenol®), and ibuprofen (Advil®), all of which reduce fever by inhibiting synthesis of certain prostaglandins. Although death results if core temperature rises above 44–46°C (112–114°F), up to a point, fever is beneficial. A higher temperature intensifies the effects of interferons and the phagocytic activities of macrophages while hindering replication of some pathogens. Because fever increases heart rate, infection-fighting white blood cells are delivered to sites of infection more rapidly. In addition, antibody production and T cell proliferation increase. Moreover, heat speeds up the rate of chemical reactions, which may help body cells repair themselves more quickly.

Heat exhaustion is a condition in which the core temperature is generally normal, or a little below, and the skin is cool and moist due to profuse perspiration. Heat exhaustion is usually characterized by loss of fluid and electrolytes, especially salt (NaCl). The salt loss results in muscle cramps, dizziness, vomiting, and fainting; fluid loss may cause low blood pressure. Complete rest, rehydration, and electrolyte replacement are recommended.

Heatstroke, also known as a sunstroke, is a severe and often fatal disorder caused by exposure to high temperatures, especially when the relative humidity is high, which makes it difficult for the body to lose heat. Blood flow to the skin is decreased, perspiration is greatly reduced, and body temperature rises sharply because of failure of the hypothalamic thermostat. Body temperature may reach 43°C (110°F). Treatment, which must be undertaken immediately, consists of cooling the body by immersing the victim in cool water and by administering fluids and electrolytes.

CHECKPOINT

9. Distinguish between core temperature and shell temperature.

10. In what ways can a person lose heat to or gain heat from the surroundings? How is it possible for a person to lose heat on a sunny beach when the temperature is 40°C (104°F) and the humidity is 85%?

11. Describe how each of the following parts of the hypothalamus plays a role in thermoregulation: preoptic area, heat-promoting center, and heat-losing center.

Critical Thinking

It's Cool to Adapt

Connor is halfway through his final practice for the upcoming Iditarod Trail Race. The weather is getting colder with gray skies, and it looks like a storm is approaching. As Connor is led by his dogs along a trail, his sled is upset and Connor ends up falling through the snow and ice into frigid water. Connor was lucky that the lanyard he attached from his waist to the sled did not break and he was able to pull himself out of the water, but he is soaked to the skin. It is very unlikely, however, that he will be able to make it back to camp with the sled in its condition and him wet and cold. The storm is approaching as the winds pick up and the temperature falls rapidly. With a potential for temperatures in the low teens to single digits, Connor has to work fast to get prepared for spending the night in the wilderness. All of Connor's supplies are wet, so he will not be able to start a fire. He has already begun shivering and his breathing is labored. Connor begins to dig into the snow and hollow out a hole.

Rebecca Grabill / Getty Images

SOME THINGS TO KEEP IN MIND:

Temperatures on Earth can vary drastically. The, lowest recorded temperature of −128.5°F (−89.2°C) occurred on July 21, 1983, in Vostock, Antarctica, while the highest recorded temperature of 134°F (56.7°C), occurred on July 10, 1913, in Furnace Creek Ranch, California, in the United States. Because humans populate about 43% of the Earth's land surface, they have methods of adapting to large fluctuations in environmental temperatures, such as shivering and sweating. Human beings have an average core body temperature of 98.6°F (37.0°C), which must be maintained within a few °F because hypothermia begins when the core temperature drops to 95°F (35°C). With core temperatures below 85°F (29.4°C) the temperature regulating system found in the hypothalamus usually fails. Hyperthermia starts as the core temperature reaches 105–107°F (40.6–41.7°C), with internal organ failure occurring after only a few days at these temperatures.

© blickwinkel / Alamy Stock Photo

Why would Connor need to hollow a hole in the snow?

Connor's shivering is increasing to a point where he can barely dig. Why is Connor shivering?

Connor notices that his hands have a pale color and are especially cold. Is this a good event or a bad event for Connor's potential survival?

Connor is a fit man who trains hard to be prepared for the aerobic demands of participating in dog-sledding events. As a result, he has a lean body with little fat. It is often assumed that in cold environments, energy is acquired from the metabolism of fat stores (the primary function of brown fat in human babies is thermal regulation). But Connor has little fat; is he therefore doomed?

Connor has finished digging the hole and has pulled two of his sled dogs into the hole with him. Snow is falling hard and the wind is blowing steadily. Why would Connor pull the dogs into the hole with him?

FROM RESEARCH TO REALITY

Calorie Restriction

Reference

Pasiakos, S. et al. (2011). Appetite and endocrine regulators of energy balance after 2 days of energy restriction: Insulin, leptin, ghrelin, and DHEA-S. *Obesity*. 19(6): 1124–1130.

How does calorie restriction really affect your body?

Visual Field/ Getty Images, Inc.

Weight loss efforts typically rely on calorie restriction for success. People who restrict themselves calorically tend to feel hungrier than those who do not. Could there be a chemical reason for the increase in hunger? Could a person experience changes that increase hunger after only two days of caloric restriction?

Article description:

Researchers had two groups of subjects, those that were fed more normal meals and those that were fed meals drastically reduced in calories. Researchers monitored the concentration of a variety of hormones and chemicals in the blood of the calorically restricted and well-fed groups to identify any differences between the two groups.

Go to WileyPLUS Learning Space and use the data from this article to answer the questions posed there and to learn more about the effects of calorie restriction.

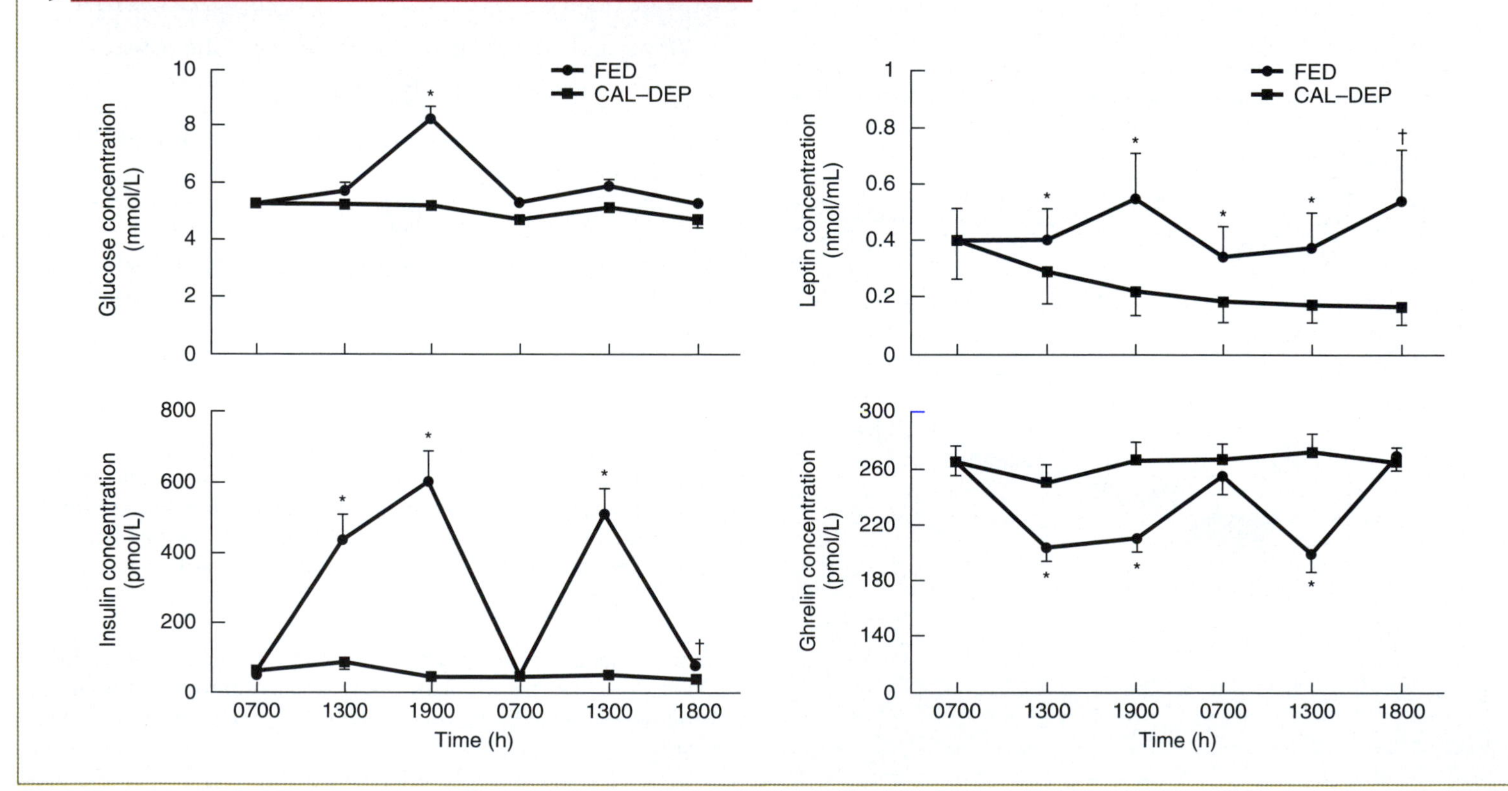

22.1 Metabolic Adaptations

1. During the absorptive state, ingested nutrients enter the bloodstream from the GI tract. Blood glucose is catabolized to form ATP, and glucose transported to the liver is converted to glycogen or triglycerides. Most triglycerides are stored in adipose tissue. Amino acids in hepatocytes are converted to carbohydrates, fats, and proteins. The main hormone regulator of the absorptive state is insulin.
2. During the postabsorptive state, absorption is complete and the ATP needs of the body are satisfied by nutrients already present in the body. The major task is to maintain normal blood glucose level by converting glycogen into glucose, converting glycerol into glucose, and converting amino acids into glucose. Fatty acids, ketone bodies, and amino acids are catabolized to supply ATP. There are several hormone regulators of the postabsorptive state: glucagon, epinephrine, norepinephrine, and cortisol.
3. Fasting is going without food for a few days; starvation implies weeks or months of inadequate food intake. During fasting and starvation, fatty acids and ketone bodies are increasingly utilized for ATP production.

22.2 Energy Balance

1. Energy balance is the precise matching of energy intake to energy expenditure over time.
2. A calorie (cal) is the amount of energy required to raise the temperature of 1 g of water 1°C. Because the calorie is a relatively small unit, the kilocalorie (kcal) or Calorie (cal) is often used to measure the body's metabolic rate and to express the energy content of foods; a kilocalorie equals 1000 calories.
3. Metabolic rate is the overall rate at which metabolic reactions use energy. Factors that affect metabolic rate include hormones, exercise, the nervous system, body temperature, ingestion of food, age, gender, climate, sleep, and malnutrition.
4. Measurement of the metabolic rate under basal conditions is called the basal metabolic rate (BMR).
5. Total metabolic rate (TMR) is the total energy expenditure by the body per unit of time. Three components contribute to the TMR: (1) BMR, (2) physical activity, and (3) food-induced thermogenesis.
6. Adipose tissue is the major site of stored chemical energy.
7. Two nuclei in the hypothalamus that help regulate food intake are the arcuate and paraventricular nuclei. The hormone leptin, released by adipocytes, inhibits release of neuropeptide Y from the arcuate nucleus and thereby decreases food intake. Melanocortin also decreases food intake. Ghrelin, released by the stomach, increases appetite by stimulating the release of neuropeptide Y.
8. Obesity refers to body weight due to an excessive accumulation of adipose tissue.
9. Two disorders of eating include bulimia and anorexia nervosa.

22.3 Regulation of Body Temperature

1. Normal core temperature is maintained by a delicate balance between heat-producing and heat-losing mechanisms.
2. Mechanisms of heat transfer include conduction, convection, radiation, and evaporation. Conduction is the transfer of heat between two substances or objects in contact with each other. Convection is the transfer of heat by movement of air or water between areas of different temperatures. Radiation is the transfer of heat from a warmer object to a cooler object without physical contact. Evaporation is the conversion of a liquid to a vapor; in the process, heat is lost.

CHAPTER REVIEW

3. The hypothalamic thermostat is in the preoptic area.
4. Responses that produce, conserve, or retain heat when core temperature falls include vasoconstriction, release of epinephrine and norepinephrine, shivering, and release of thyroid hormones.
5. Responses that increase heat loss when core temperature increases include vasodilation, decreased metabolic rate, and evaporation of perspiration.
6. Disorders of body temperature regulation include hypothermia, fever, heat exhaustion, and heatstroke.

PONDER THIS

1. Jonathan goes into his doctor's office because, despite how much food he eats, he is unable to gain any weight. His doctor discovers that he is breaking down muscle mass, his body isn't able to store fat, and his muscles are very low in glycogen stores. His doctor realizes that his condition is mimicking an extreme body state with regard to food intake. Is his body primarily in the absorptive or the postabsorptive state? Explain your answer using the terms catabolic and anabolic.

2. You wake up and realize you are late for class. You do not plan adequately and run out of your home without a jacket. Later that day it begins to snow and it is about 25°F. Explain what would happen to your body temperature and how your body would compensate under these conditions.

3. Lisa wants to get pool ready for the summer and in doing so decides that she wants to lose weight. Lisa realizes that she needs to expend more calories on a daily basis than she consumes through food. In order to accomplish this, Lisa reduces her caloric intake by 500 calories a day. How will her body maintain normal physiological function during the times she does not eat and with this caloric deficit? How will this effort help her to lose weight?

ANSWERS TO FIGURE QUESTIONS

22.1 Reactions of the absorptive state are mainly catabolic.

22.2 Processes that directly elevate blood glucose during the postabsorptive state include lipolysis (in adipocytes and hepatocytes), gluconeogenesis (in hepatocytes), and glycogenolysis (in hepatocytes).

22.3 Conduction is the heat exchange that occurs between molecules of two materials that are in direct contact with each other.

22.4 Peripheral thermoreceptors are located in the skin; central thermoreceptors are located in the hypothalamus.

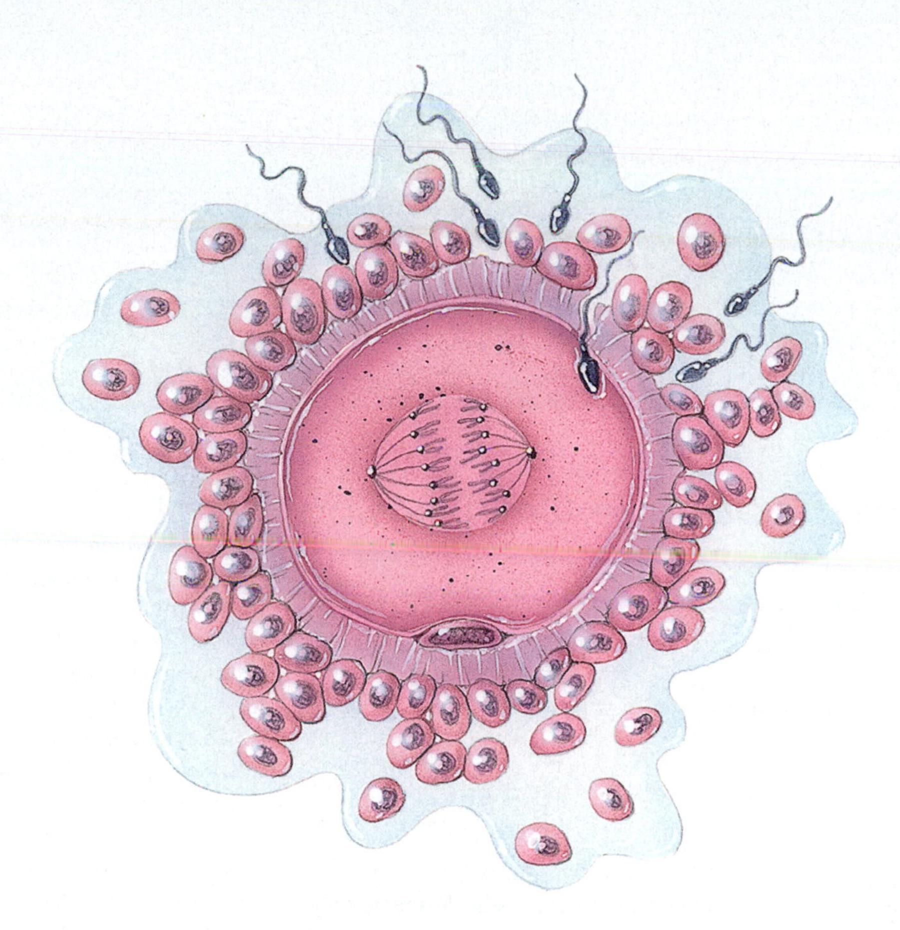

23 The Reproductive Systems

The Reproductive Systems and Homeostasis

The male and female reproductive systems are not essential for homeostasis, but they do work together to produce offspring, thereby ensuring the perpetuation of the human species. In addition, the female reproductive system sustains the growth of the embryo and fetus during pregnancy.

LOOKING BACK TO MOVE AHEAD...

- Steroids are lipids that contain four interconnected hydrocarbon rings; examples include cholesterol, testosterone, estrogens, and cortisol (Section 2.5).
- There are two types of cell division: (1) somatic cell division, which replaces dead or injured cells and adds new ones during tissue growth and (2) reproductive cell division, which produces gametes (sperm or eggs) (Section 3.6).
- Nitric oxide (NO) is a gas neurotransmitter that has widespread effects throughout the body; one of its functions is to promote erection of the penis in males (Section 6.2).
- Androgens are hormones that have masculinizing effects; examples include testosterone (produced by the testes), dihydrotestosterone (a substance derived from testosterone), and dehydroepiandrosterone (produced by the adrenal glands) (Section 13.5).

Sexual reproduction is the process by which organisms produce offspring from the union of germ cells called **gametes**. After the male gamete (sperm cell) unites with the female gamete (egg or oocyte)—an event called **fertilization**—the resulting cell contains one set of chromosomes from each parent. Males and females have distinct reproductive organs that are adapted for producing gametes; facilitating fertilization; and, in females, sustaining the growth of the embryo and fetus during pregnancy.

The organs of the male and female reproductive systems can be grouped by function. The **gonads**—testes in males and ovaries in females—produce gametes and secrete sex hormones. Various **ducts** then store and transport the gametes, and **accessory sex glands** produce substances that protect the gametes and facilitate their movement. Finally, **supporting structures**, such as the penis in males and the vagina and uterus in females, assist the delivery of gametes. The uterus is also the site for the development of the embryo and fetus.

23.1 Reproductive Cell Division

OBJECTIVES

- Compare the numbers of chromosomes in somatic cells and gametes.
- Outline the steps of the two stages of meiosis.

Somatic Cells and Gametes Have Different Numbers of Chromosomes

Before learning about reproductive cell division, you must first understand the distribution of genetic material in cells. **Somatic cells**, which include any cell in the body other than a gamete, contain 23 pairs of chromosomes, for a total of 46 chromosomes. One member of each pair is inherited from each parent. The two chromosomes that make up each pair are called **homologous chromosomes** or **homologs**; they contain similar genes arranged in the same (or almost the same) order. Because somatic cells contain two sets of chromosomes, they are called **diploid cells**, symbolized **$2n$**. As you learned in Chapter 3, somatic cell division involves a nuclear division called **mitosis** (which consists of four phases: prophase, metaphase, anaphase, and telophase) and a cytoplasmic division called **cytokinesis**. The effect of somatic cell division is to provide two identical diploid cells, each with the same number and kind of chromosomes as the original cell.

In sexual reproduction, each new organism is the result of the union of two different gametes, one produced by each parent. If gametes had the same number of chromosomes as somatic cells, the number of chromosomes would double at fertilization. Reproductive cell division consists of a special type of nuclear division called **meiosis**, in which the number of chromosomes is reduced by half, and cytokinesis. As a result, gametes contain a single set of 23 chromosomes and thus are **haploid (n) cells**. Fertilization restores the diploid number of chromosomes.

There Are Two Stages of Meiosis

Meiosis occurs in two successive stages: **meiosis I** and **meiosis II**. During the interphase that precedes meiosis I, the chromosomes of the diploid cell start to replicate. As a result of replication, each chromosome consists of two genetically identical **sister chromatids**, which are attached at their centromeres. This replication of chromosomes is similar to the one that precedes mitosis in somatic cell division.

Meiosis I

Meiosis I, which begins once chromosomal replication is complete, consists of four phases: prophase I, metaphase I, anaphase I, and telophase I (FIGURE 23.1a). Prophase I is an extended phase in which the chromosomes shorten and thicken, the nuclear envelope and nucleoli disappear, and the meiotic spindle forms. In addition, the sister chromatids of each pair of homologous chromosomes pair off, an event called **synapsis** (FIGURE 23.1a, b). The resulting four chromatids form a structure called a **tetrad**. Parts of the chromatids of the two homologous chromosomes may be exchanged with one another. Such an exchange between parts of nonsister (genetically different) chromatids is termed **crossing-over** (FIGURE 23.1b). This process, among others, permits an exchange of genes between chromatids of homologous chromosomes. Due to crossing-over, the resulting cells are genetically unlike each other and genetically unlike the starting cell that produced them. Crossing-over results in **genetic recombination**—that is, the formation of new combinations of genes—and accounts for part of the great genetic variation among humans and other organisms that form gametes via meiosis. Recall from Chapter 3 that synapsis, tetrad formation, and crossing-over do not occur during mitotic prophase of somatic cell division.

In metaphase I, the tetrads formed by the homologous pairs of chromosomes line up along the metaphase plate of the cell, with homologous chromosomes side by side (FIGURE 23.1a). During anaphase I, the members of each homologous pair of chromosomes separate as they are pulled to opposite poles of the cell by the microtubules attached to the centromeres. The paired sister chromatids, held by a centromere, remain together. (Recall that during mitotic anaphase, the centromeres split and the sister chromatids separate.) Telophase I and cytokinesis of meiosis are similar to telophase and cytokinesis of mitosis. The net effect of meiosis I is that each resulting cell contains the haploid number of chromosomes because it contains only one member of each pair of the homologous chromosomes present in the starting cell.

Meiosis II

The second stage of meiosis, meiosis II, also consists of four phases: prophase II, metaphase II, anaphase II, and telophase II (FIGURE 23.1a). These phases are similar to those that occur during mitosis; the centromeres split, and the sister chromatids separate and move toward opposite poles of the cell.

In summary, meiosis I begins with a diploid starting cell and ends with two cells, each with the haploid number of chromosomes. During meiosis II, each of the two haploid cells formed during meiosis

FIGURE 23.1 Reproductive cell division. Details of the events are discussed in the text.

In reproductive cell division, a single diploid cell undergoes meiosis I and meiosis II to produce four haploid gametes.

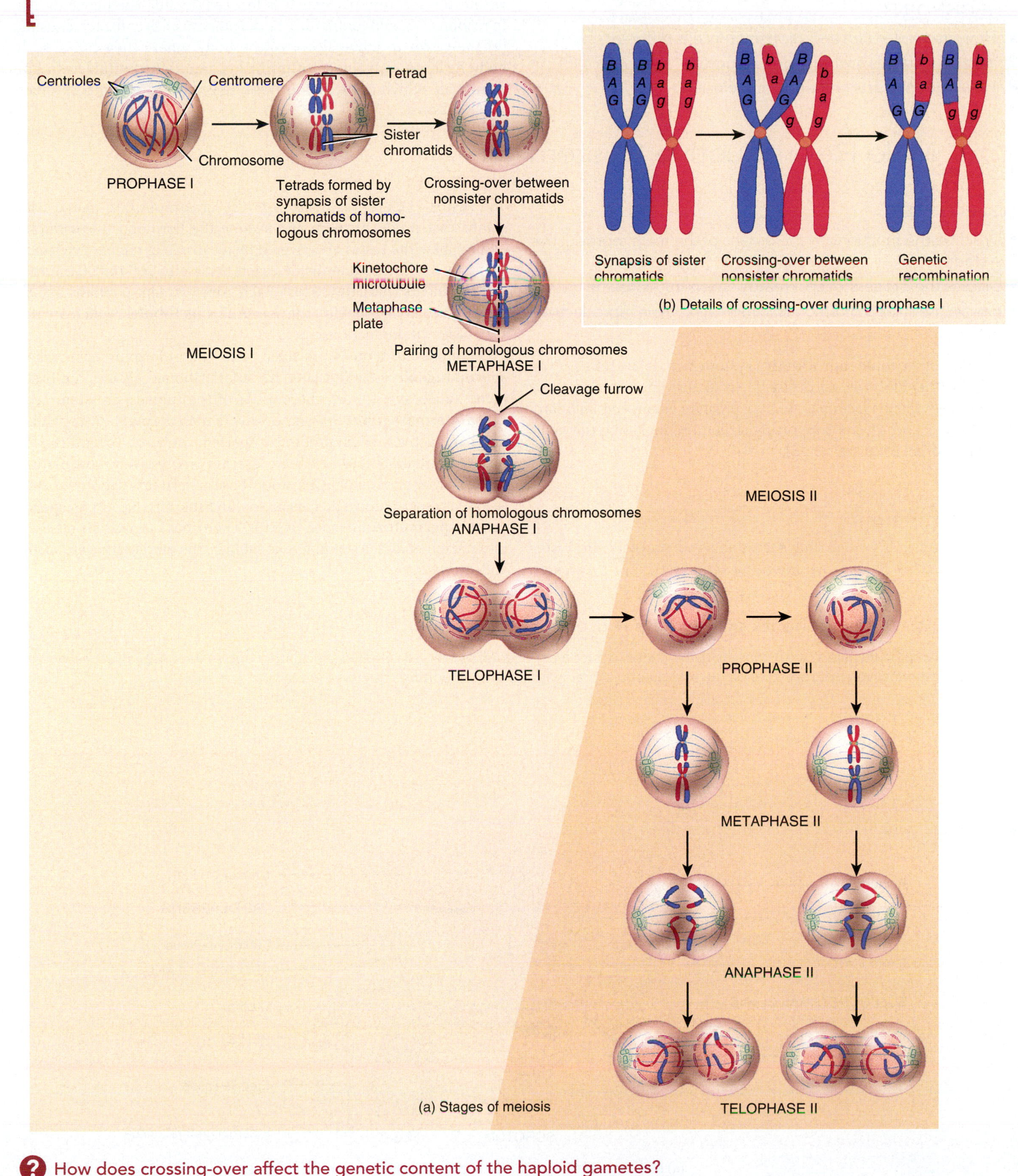

(a) Stages of meiosis

(b) Details of crossing-over during prophase I

How does crossing-over affect the genetic content of the haploid gametes?

I divides; the net result is four haploid gametes that are genetically different from the original diploid starting cell.

CHECKPOINT

1. How are haploid (n) and diploid ($2n$) cells different?
2. What is synapsis and why does it occur?
3. How does meiosis I differ from meiosis II?

23.2 Male Reproductive System

OBJECTIVES

- Describe the functions of the organs of the male reproductive system.
- Discuss the process of spermatogenesis in the testes.
- Explain the roles of hormones in regulating male reproductive function.

The organs of the **male reproductive system** include the testes; a system of ducts (epididymis, vas deferens, ejaculatory ducts, and urethra); accessory sex glands (seminal vesicles, prostate gland, and bulbourethral glands); and several supporting structures, including the penis and the scrotum (FIGURE 23.2a).

The Scrotum Protects the Testes and Regulates Their Temperature

The **scrotum** is a pouch that surrounds and protects the testes (FIGURE 23.2a). It also regulates the temperature of the testes. Normal sperm production requires a temperature about 3°C below core body temperature. This lowered temperature is maintained within the scrotum because it is located outside the pelvic cavity. In response to cold temperatures, scrotal muscles contract, causing (1) the testes to move closer to the body, where they can absorb body heat, and (2) the scrotum to become tight (wrinkled in appearance), which reduces heat loss. Exposure to warmth reverses these actions.

The Testes Produce Sperm and Secrete Hormones

The **testes** (TES-tēz; singular is *testis*), or *testicles*, are paired oval glands that produce sperm and secrete the hormones testosterone and inhibin. The testes are covered by a capsule of connective tissue that extends inward and divides each testis into 200 to 300 internal compartments (FIGURE 23.2b). Each compartment contains one to three tightly coiled tubules, the **seminiferous tubules**, which are the sites of sperm production.

Seminiferous tubules are lined with sperm-forming cells called **spermatogenic cells** (FIGURE 23.2b). Positioned against the basement membrane, toward the outside of the tubules, are stem cells called **spermatogonia**. Spermatogonia remain dormant during childhood and actively begin producing sperm at puberty. Toward the lumen of the seminiferous tubule are layers of progressively more mature cells. In order of advancing maturity, these are primary spermatocytes, secondary spermatocytes, spermatids, and sperm cells. After a **sperm cell**, or **spermatozoon** (sper′-ma-tō-ZŌ-on), has formed, it is released into the lumen of the seminiferous tubule. (The plural terms are *sperm* and *spermatozoa*.)

FIGURE 23.2 Components of the male reproductive system.

The male reproductive system consists of the testes, a duct system (epididymis, vas deferens, ejaculatory ducts, and urethra), accessory sex glands (seminal vesicles, prostate gland, and bulbourethral glands), and supporting structures (penis and scrotum).

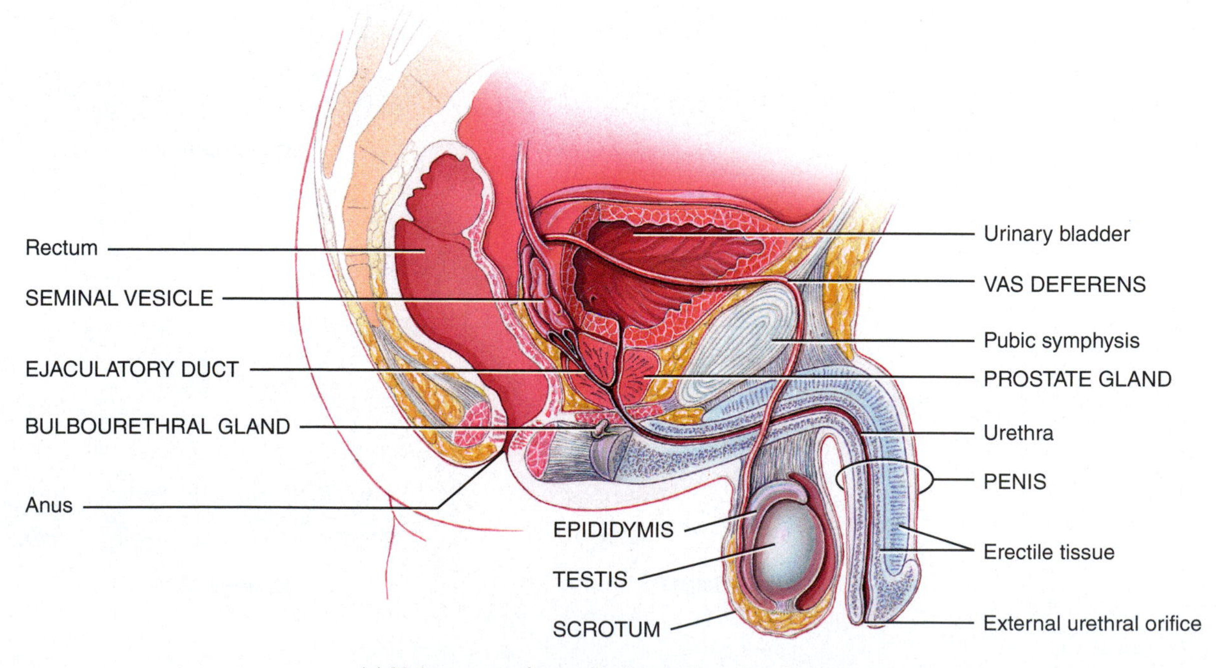

(a) Male organs of reproduction and surrounding structures

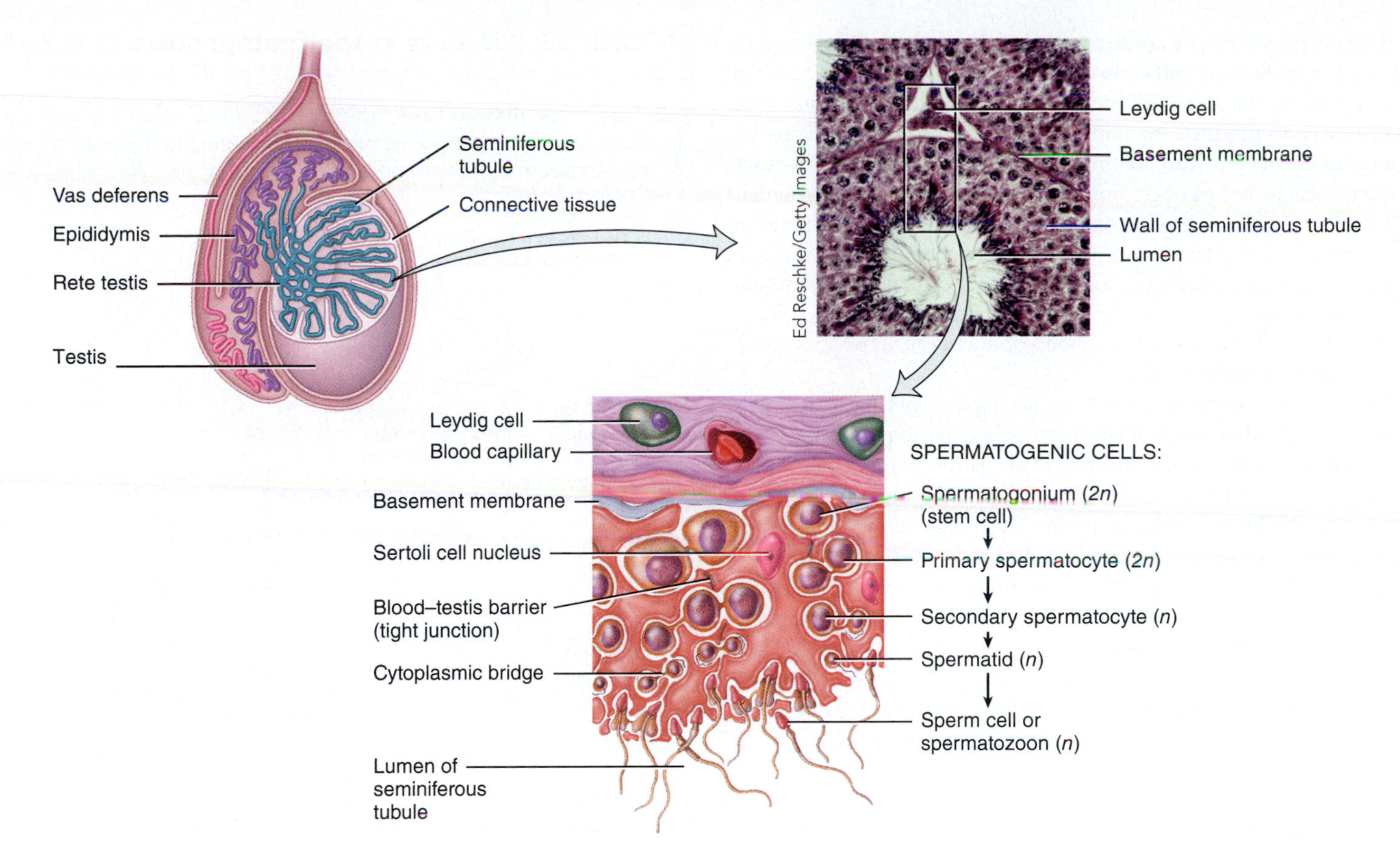

(b) Organization of the testis

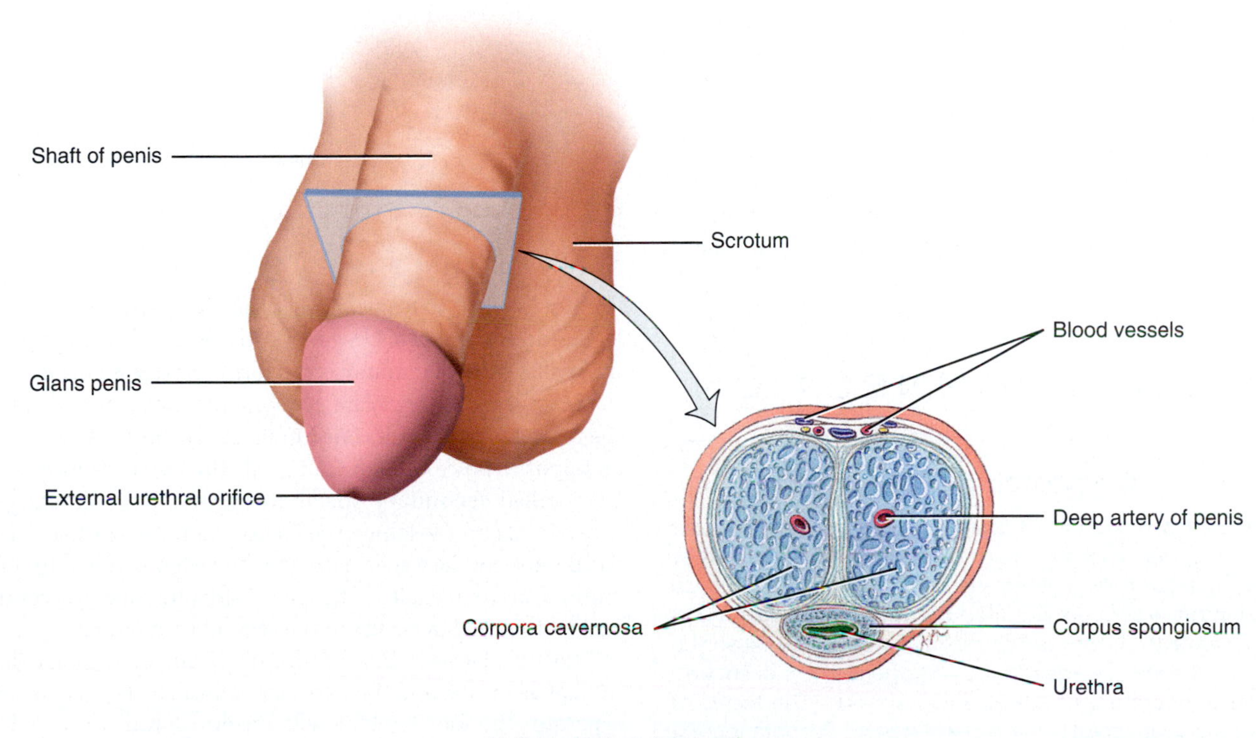

(c) Organization of the penis

What is the functional significance of the seminiferous tubules of the testes?

Embedded among the spermatogenic cells in the seminiferous tubules are large **Sertoli cells**, also known as *sustentacular cells*, which extend from the basement membrane to the lumen of the tubule (FIGURE 23.2b). Tight junctions join neighboring Sertoli cells to one another. These junctions form an obstruction known as the **blood–testis barrier** because substances must first pass through the Sertoli cells before they can reach the developing sperm. By isolating the developing gametes from the blood, the blood–testis barrier prevents an immune response against the spermatogenic cell's surface antigens, which are recognized as "foreign" by the immune system. The blood–testis barrier does not include spermatogonia because spermatogonia are located outside the barrier.

Sertoli cells support and protect developing spermatogenic cells in several ways. They nourish spermatocytes, spermatids, and sperm; phagocytize excess spermatid cytoplasm as development proceeds; and control movements of spermatogenic cells and the release of sperm into the lumen of the seminiferous tubule. They also produce fluid for sperm transport, secrete the hormone inhibin, and regulate the effects of testosterone and follicle-stimulating hormone (FSH).

In the spaces between adjacent seminiferous tubules are clusters of cells called **Leydig cells** or *interstitial cells* (FIGURE 23.2b). These cells secrete testosterone, the most prevalent androgen. An **androgen** is a hormone that promotes the development of masculine characteristics. Testosterone also promotes a man's *libido* (sexual drive).

Spermatogenesis

The process by which sperm are produced in the seminiferous tubules of the testes is called **spermatogenesis**. In humans, spermatogenesis takes about 75 days. It begins with the spermatogonia, which contain the diploid ($2n$) number of chromosomes (FIGURE 23.3). Spermatogonia are types of stem cells; when they undergo mitosis, some spermatogonia remain near the basement membrane of the seminiferous tubule in an undifferentiated state to serve as a reservoir of cells for future cell division and subsequent sperm production. The rest of the spermatogonia lose contact with the basement membrane, squeeze through the tight junctions of the blood–testis barrier, undergo developmental changes, and differentiate into **primary spermatocytes**. Primary spermatocytes, like spermatogonia, are diploid ($2n$); that is, they have 46 chromosomes.

Cryptorchidism

Before birth, the testes are initially present in the abdominal cavity and then, during the seventh month of fetal development, they descend into the scrotum. The condition in which the testes do not descend into the scrotum is called **cryptorchidism** (krip-TOR-ki-dizm). Nondescent of both testes results in sterility because the cells involved in the initial stages of spermatogenesis are destroyed by the higher temperature of the abdominal cavity. The testes of about 80% of boys with cryptorchidism will descend spontaneously during the first year of life. When the testes remain undescended, the condition can be corrected surgically, ideally before 18 months of age.

FIGURE 23.3 Events in spermatogenesis. Diploid cells ($2n$) have 46 chromosomes; haploid cells (n) have 23 chromosomes.

During spermatogenesis, spermatogonia differentiate into primary spermatocytes, which in turn sequentially give rise to secondary spermatocytes, spermatids, and spermatozoa.

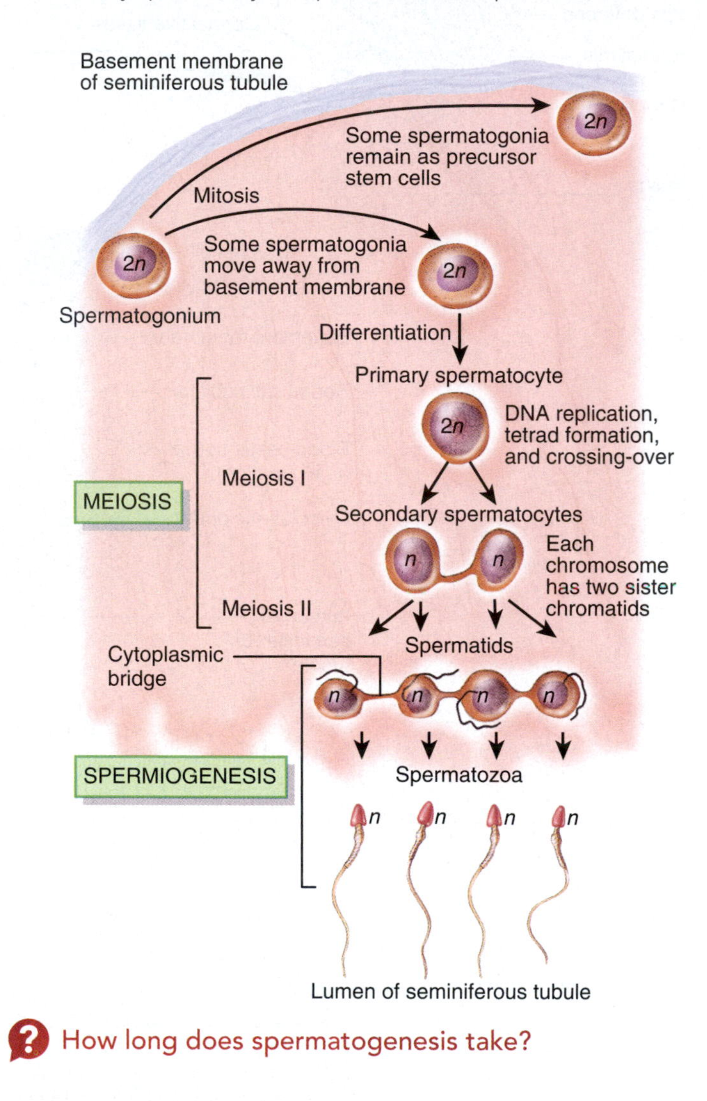

How long does spermatogenesis take?

Shortly after it forms, each primary spermatocyte replicates its DNA and then meiosis begins (FIGURE 23.3). During meiosis I, homologous pairs of chromosomes form tetrads, allowing crossing-over to occur, and then they line up along the metaphase plate. Afterward, the meiotic spindle pulls one (duplicated) chromosome of each pair to an opposite pole of the dividing cell. The two cells formed by meiosis I are called **secondary spermatocytes**. Each secondary spermatocyte has 23 chromosomes, the haploid number (n). Each chromosome within a secondary spermatocyte, however, is made up of two sister chromatids (two copies of the DNA) still attached by a centromere. No replication of DNA occurs in the secondary spermatocytes.

During meiosis II, the chromosomes line up in single file along the metaphase plate, and the two sister chromatids of each chromosome separate. The four haploid cells resulting from meiosis II are called **spermatids**. A single primary spermatocyte therefore produces four spermatids via two rounds of cell division (meiosis I and meiosis II).

A unique process occurs during spermatogenesis. As spermatogenic cells proliferate, they fail to complete cytoplasmic separation (cytokinesis). The cells remain in contact via cytoplasmic bridges throughout their entire development (see FIGURES 23.2b and 23.3). This pattern of development most likely accounts for the synchronized production of sperm in any given area of a seminiferous tubule. It also has survival value, in that half of the sperm contain an X chromosome and half contain a Y chromosome. The larger X chromosome carries genes needed for spermatogenesis that are lacking on the smaller Y chromosome.

The final stage of spermatogenesis, **spermiogenesis**, is the development of haploid spermatids into sperm. No cell division occurs in spermiogenesis; each spermatid becomes a single **sperm cell**. During this process, spherical spermatids transform into elongated, slender sperm. Sertoli cells dispose of the excess cytoplasm that sloughs off. Finally, sperm are released from their connections to Sertoli cells, an event known as **spermiation**. Sperm then enter the lumen of the seminiferous tubule. At this point, sperm are not yet able to swim. Fluid secreted by Sertoli cells pushes sperm along their way, first into a network of ducts in the testes known as the **rete testis** (RĒ-tē) (see FIGURE 23.2b) and then into the epididymis.

Sperm

Each day about 300 million sperm complete the process of spermatogenesis. A sperm consists of three structures highly adapted for reaching and penetrating an egg (oocyte): a head, a midpiece, and a tail (FIGURE 23.4). The **head** of the sperm contains a nucleus and an acrosome. The nucleus has the haploid number of chromosomes (23). The **acrosome** is a caplike vesicle filled with enzymes that help a sperm penetrate an egg to bring about fertilization. In the **midpiece** of a sperm are many mitochondria, which provide ATP for locomotion. The **tail** of a sperm, a typical flagellum, propels the sperm along its way. Once ejaculated, most sperm do not survive more than three to five days within the female reproductive tract.

FIGURE 23.4 Parts of a sperm cell.

About 300 million sperm are produced each day.

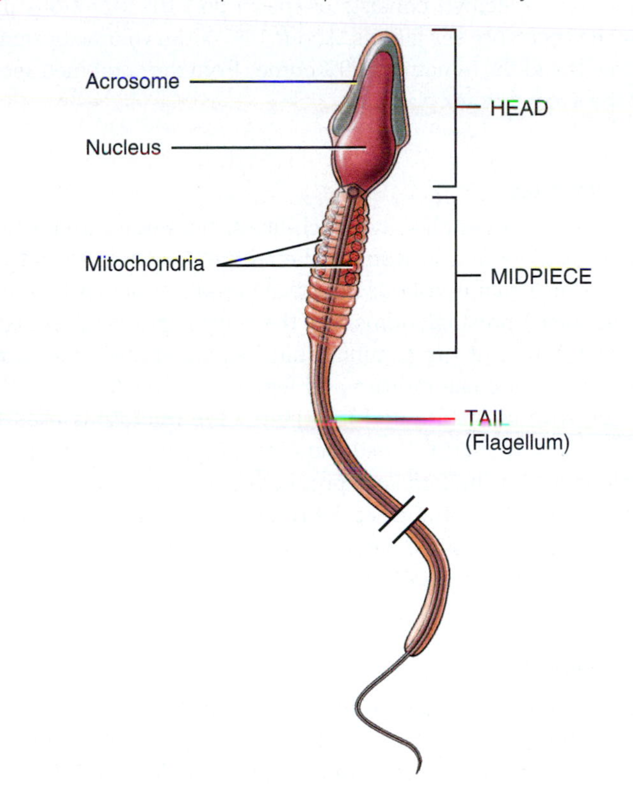

What are the functions of each part of the sperm cell?

The Male Duct System Aids in the Maturation, Storage, and Transport of Sperm

The male duct system is a series of continuous tubes within the male reproductive tract. It sequentially consists of the epididymis, vas deferens, ejaculatory ducts, and urethra. The male duct system transports and stores sperm, assists in their maturation, and conveys them to the exterior.

Epididymis

The **epididymis** (ep-i-DID-i-mis; plural is *epididymides*, ep-i-di-DIM-i-dēz) is a tightly coiled tube that connects the testis to the vas deferens (see FIGURE 23.2b). The epididymis is the site of **sperm maturation**, the process by which sperm acquire motility and the ability to fertilize an egg. This occurs over a period of about 14 days. The epididymis also helps propel sperm into the vas deferens during sexual arousal by peristaltic contraction of its smooth muscle. In addition, the epididymis stores sperm, which remain viable here for up to several months. Any stored sperm that are not ejaculated by that time are eventually phagocytized by cells in the epididymis.

Vas Deferens

The terminal portion of the epididymis gives rise to the **vas deferens**, also known as the *ductus deferens* (see FIGURE 23.2b) From the epididymis, the vas deferens ascends out of the scrotum and enters the pelvic cavity, where it loops over the side and down the back surface of the urinary bladder (see FIGURE 23.2a). The vas deferens conveys sperm during sexual arousal from the epididymis toward the urethra by peristaltic contractions of its smooth muscle. Like the epididymis, the vas deferens can also can store sperm for several months. Any stored sperm that are not ejaculated by that time are eventually phagocytized by cells in the vas deferens.

Ejaculatory Ducts

Each **ejaculatory duct** is formed by the union of the vas deferens and the duct from the seminal vesicle (see FIGURE 23.2a). The ejaculatory ducts enter the prostate gland and terminate in the urethra, where they eject sperm and seminal vesicle secretions just before the release of semen from the urethra to the exterior.

Urethra

In males, the **urethra** is the shared terminal duct of the reproductive and urinary systems; it serves as a passageway for both semen and urine. The urethra originates in the floor of the urinary bladder, passes through the prostate gland, and extends through the penis, where it opens to the exterior as the *external urethral orifice* (see FIGURE 23.2a).

The Accessory Sex Glands Add Secretions to Sperm to Form Semen

The **accessory sex glands** of the male reproductive system include the seminal vesicles, the prostate gland, and the bulbourethral glands. During sexual arousal, secretions from the seminal vesicles enter the ejaculatory ducts, secretions from the prostate gland enter the ejaculatory ducts and urethra, and secretions from the bulbourethral glands enter the urethra. Semen consists of sperm plus the secretions provided by the accessory sex glands. About 10% of the volume of semen is sperm cells and the remaining 90% comes from the combined secretions of the accessory sex glands.

Seminal Vesicles

The paired **seminal vesicles** are pouch-like structures located behind the urinary bladder and in front of the rectum (see **FIGURE 23.2a**). They secrete an alkaline, viscous fluid that contains fructose (a monosaccharide sugar), prostaglandins, and the clotting protein fibrinogen. The alkaline nature of the seminal fluid helps neutralize the acidic environment of the male urethra and female reproductive tract that would otherwise inactivate and kill sperm. The fructose is used for ATP production by sperm. Prostaglandins stimulate smooth muscle contractions within the female reproductive tract, facilitating movement of sperm cells up the tract. Fibrinogen helps semen coagulate (clot) after ejaculation. Fluid secreted by the seminal vesicles constitutes about 60% of the volume of semen.

Prostate Gland

The **prostate gland**, roughly the size and shape of a golf ball, is below the urinary bladder and surrounds the upper portion of the urethra (see **FIGURE 23.2a**). It secretes a milky, slightly acidic fluid (pH about 6.5) that contains several substances: (1) *Citric acid* in prostatic fluid is used by sperm for ATP production via the Krebs cycle; (2) *clotting enzymes* secreted by the prostate act on fibrinogen from the seminal vesicles to clot semen after ejaculation; (3) *proteolytic enzymes*, such as *prostate-specific antigen (PSA)*, fibrinolysin, and pepsinogen, eventually break down the clot; (4) *seminalplasmin* in prostatic fluid is an antibiotic that destroys bacteria that may be present in semen or in the female reproductive tract, and (5) *acid phosphatase* is also found in prostatic fluid, but its function is unknown. Prostatic secretions make up about 25% of the volume of semen and contribute to sperm motility and viability.

Bulbourethral Glands

The paired **bulbourethral glands**, or *Cowper's glands*, are pea-sized structures located below the prostate gland on either side of the urethra (see **FIGURE 23.2a**). The bulbourethral glands secrete an alkaline fluid into the urethra that protects the passing sperm by neutralizing acids from urine in the urethra. They also secrete mucus that lubricates the end of the penis and the lining of the urethra, decreasing the number of sperm damaged during ejaculation. Secretions from the bulbourethral glands constitute about 5% of the volume of semen.

Semen

Semen is a mixture of sperm and the secretions of the seminal vesicles, prostate gland, and bulbourethral glands. Despite the slight acidity of prostatic fluid, semen has a slightly alkaline pH of 7.2–7.7 due to the higher pH and larger volume of fluid from the seminal vesicles. The prostatic secretion gives semen a milky appearance, and fluids from the seminal vesicles and bulbourethral glands give it a sticky consistency. Seminal fluid provides sperm with a transportation medium, nutrients, and protection from the hostile acidic environment of the male's urethra and the female's vagina.

Once ejaculated, liquid semen coagulates within 5 minutes due to the presence of the clotting protein fibrinogen from the seminal vesicles and clotting enzymes from the prostate gland. Coagulation of semen occurs in the following way: The clotting enzymes act on fibrinogen to form the protein fibrin, which clots the semen, trapping mobile sperm cells within the fibrin meshwork. It is thought that semen coagulation occurs in order to help keep sperm cells from leaking out of the vagina. After about 10 to 20 minutes, semen reliquefies because prostate-specific antigen (PSA) and other proteolytic enzymes produced by the prostate break down the clot. As the clot dissolves, sperm cells are released and begin their journey to the egg. Abnormal or delayed liquefaction of clotted semen may cause complete or partial immobilization of sperm, thereby inhibiting their movement through the female reproductive tract.

The Penis Is the Male Organ of Copulation

The **penis** is the male organ of copulation (sexual intercourse), depositing semen into the vagina of the female reproductive tract. It is a cylindrical structure that consists of two main regions: (1) the **shaft (body) of the penis**, the longest portion of the penis, and (2) the glans penis, the terminal end of the penis (see **FIGURE 23.2c**). The penis contains **erectile tissue**, which is composed of numerous blood sinuses (vascular spaces) lined by endothelial cells and surrounded by elastic connective tissue and smooth muscle. In the shaft of the penis there are three masses of erectile tissue: two upper masses called the **corpora cavernosa** (singular is *corpus cavernosum*) and one lower mass known as the **corpus spongiosum**, which surrounds the urethra (see **FIGURE 23.2c**).

Upon sexual stimulation, the penis becomes enlarged and stiff, a phenomenon known as **erection**. Erection is a parasympathetic reflex coordinated by the spinal cord (**FIGURE 23.5a**). During an erection, parasympathetic nerves that extend from the sacral portion of the spinal cord to the penis release nitric oxide (NO). The NO causes smooth muscle in the walls of arterioles supplying the penis to relax (dilate), allowing large amounts of blood to enter the blood sinuses of the erectile tissue. As the blood sinuses expand, the veins that drain the penis are compressed, which slows blood outflow. The combination of the dilation of penile arterioles, the expansion of blood sinuses, and the compression of penile veins results in an erection. A major stimulus for erection is mechanical stimulation of the penis. Mechanoreceptors provide direct input to the erection integrating center in the spinal cord. Erotic sights, sounds, smells, and thoughts can also stimulate erection. This involves descending inputs from the brain (hypothalamus and limbic system) to the spinal cord. Negative stimuli (a bad mood, depression, anxiety, etc.) can also inhibit erection through these descending pathways.

The nitric oxide (NO) that promotes erection can be released as a neurotransmitter by parasympathetic axons or as a local mediator by nearby endothelial cells in penile arterioles. If NO is released as a local mediator, it involves the release of the neurotransmitter acetylcholine from the same parasympathetic axons that release NO as a neurotransmitter. The molecular mechanism by which nitric oxide

FIGURE 23.5 Erection.

During an erection, penile arterioles dilate, blood sinuses expand, and penile veins are compressed.

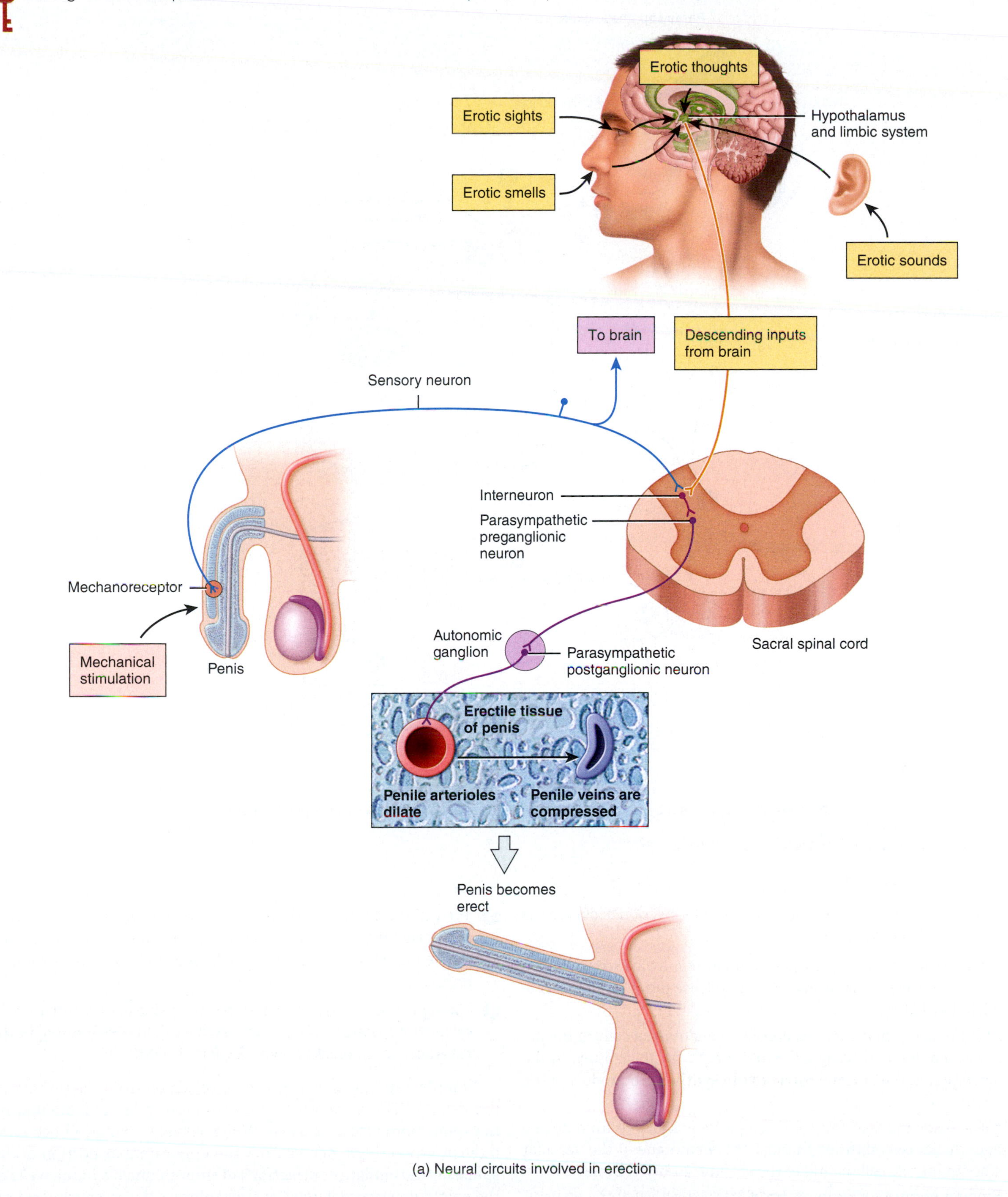

(a) Neural circuits involved in erection

Figure 23.5 (continues)

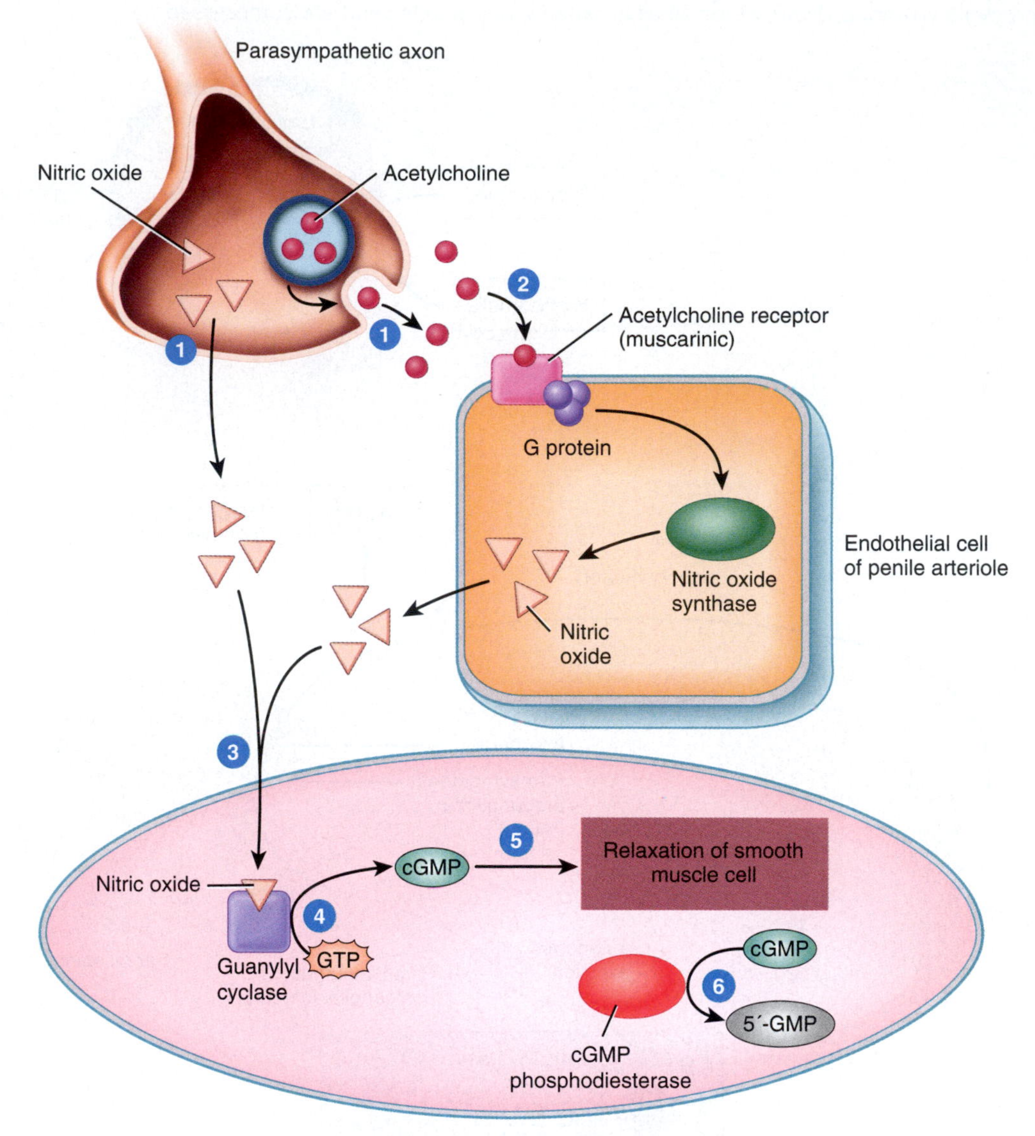

(b) Molecular mechanism by which nitric oxide relaxes a smooth muscle cell in a penile arteriole

What is the role of nitric oxide in erection?

released from parasympathetic axons promotes erection occurs in the following way (**FIGURE 23.5b**):

1. Sexual stimulation causes parasympathetic axons extending to smooth muscle in penile arterioles to release nitric oxide and acetylcholine (ACh).
2. ACh binds to muscarinic receptors on endothelial cells in the penile arterioles, activating a G protein pathway that results in the activation of the enzyme **nitric oxide synthase** to produce nitric oxide.
3. Nitric oxide released from parasympathetic axons or from nearby endothelial cells diffuses through the membrane of the vascular smooth muscle cell to bind to the enzyme **guanylyl cyclase**.
4. Binding of nitric oxide to guanylyl cyclase activates the enzyme, which produces cGMP from GTP.
5. The cGMP, in turn, activates a pathway that results in relaxation of the smooth muscle cell. Relaxation of smooth muscle causes the penile arteriole to dilate, which ultimately causes erection to occur.
6. The signaling pathway triggered by nitric oxide in a smooth muscle cell is terminated, in part, by an enzyme called **cGMP phosphodiesterase**, which breaks down cGMP to 5′-GMP.

Ejaculation, the powerful release of semen from the urethra to the exterior, is a sympathetic and somatic reflex coordinated by the spinal cord (**FIGURE 23.6**). When sexual stimulation becomes intense, sympathetic nerves from the upper lumbar portion of the spinal cord stimulate contractions of smooth muscle in the walls of the epididymis, vas deferens, and ejaculatory ducts, causing sperm to be propelled into the urethra. In addition, the sympathetic nerves

FIGURE 23.6 Ejaculation.

Ejaculation is the expulsion of semen from the penis.

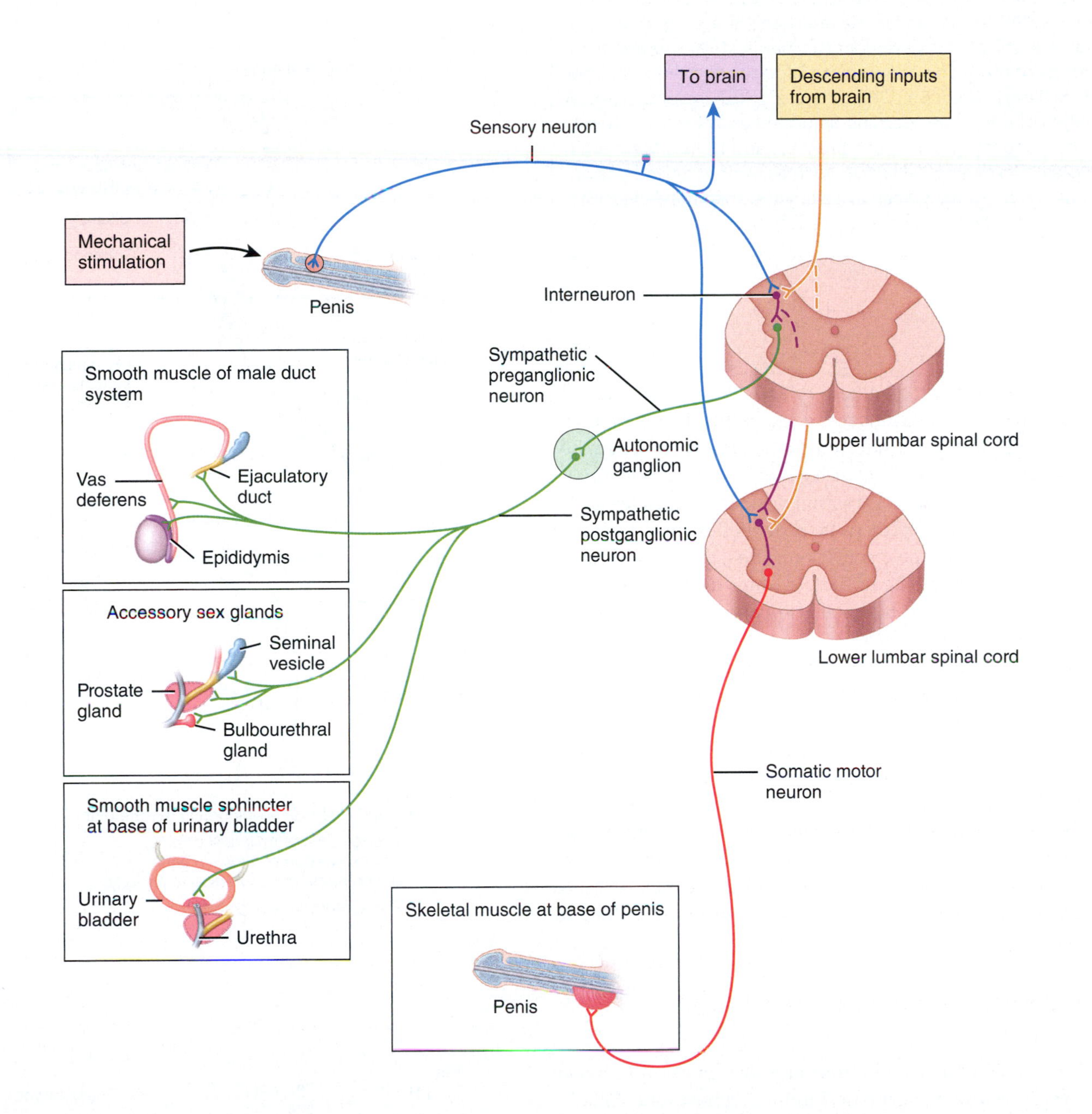

Why is ejaculation both a sympathetic and a somatic reflex?

CLINICAL CONNECTION

Erectile Dysfunction

Erectile dysfunction (ED), previously termed ***impotence***, is the consistent inability of an adult male to attain or hold an erection long enough for sexual intercourse. Many cases of ED are caused by insufficient release of nitric oxide (NO), which relaxes the smooth muscle of penile arterioles. The drugs Viagra® (sildenafil), Levitra® (verndenafil), and Cialis® (tadalafil) enhance erection by inhibiting the enzyme cGMP phosphodiesterase, which breaks down cGMP. As a result, cGMP levels remain elevated and the smooth muscle within the walls of blood vessels in the penis stays relaxed for a longer period of time. This allows more blood to remain in the penis, which prolongs erection. Other causes of ED include physical abnormalities of the penis, vascular disturbances (arterial or venous obstructions), and drugs (alcohol, narcotics, and tranquilizers are a few examples). Psychological factors such as anxiety or depression, fear of causing pregnancy, fear of sexually transmitted diseases, and religious inhibitions may also cause ED.

stimulate the accessory sex glands to release their secretions. Semen is formed as these secretions mix with sperm. The movement of semen into the urethra is referred to as **emission**. Then, somatic nerves from the lower lumbar and upper sacral portions of the spinal cord stimulate the contraction of skeletal muscles at the base of the penis, causing semen to be propelled from the urethra to the exterior. As part of the ejaculation reflex, the smooth muscle sphincter at the base of the urinary bladder closes, preventing urine from being expelled during ejaculation and preventing semen from entering the urinary bladder.

The volume of semen in a typical ejaculation is about 3 mL, with an average of 100 million sperm per mL (300 million total). When the number of sperm falls below 20 million/mL, the male is likely to be infertile. A very large number of sperm is required for successful fertilization because only a tiny fraction ever reaches the egg.

Once sexual stimulation of the penis has ended, the arterioles supplying the penis constrict and the smooth muscle within erectile tissue contracts, making the blood sinuses smaller. This relieves pressure on the veins supplying the penis and allows the blood to drain through them. Consequently, the penis returns to its **flaccid** (relaxed) state.

Male Reproductive Function Is Regulated by Several Hormones

The main hormones that regulate reproductive function in males are gonadotropin-releasing hormone, luteinizing hormone, follicle-stimulating hormone, testosterone, dihydrotestosterone, and inhibin (FIGURE 23.7). At puberty, certain hypothalamic neurosecretory cells increase their secretion of **gonadotropin-releasing hormone (GnRH)**. This hormone in turn stimulates gonadotrophs in the anterior pituitary to increase their secretion of the two gonadotropins, **luteinizing hormone (LH)** and **follicle-stimulating hormone (FSH)**.

FIGURE 23.7 Hormonal regulation of male reproductive function. Dashed lines indicate negative feedback inhibition.

The hormones that affect reproductive function in males are gonadotropin-releasing hormone, follicle-stimulating hormone, luteinizing hormone, testosterone, dihydrotestosterone, and inhibin.

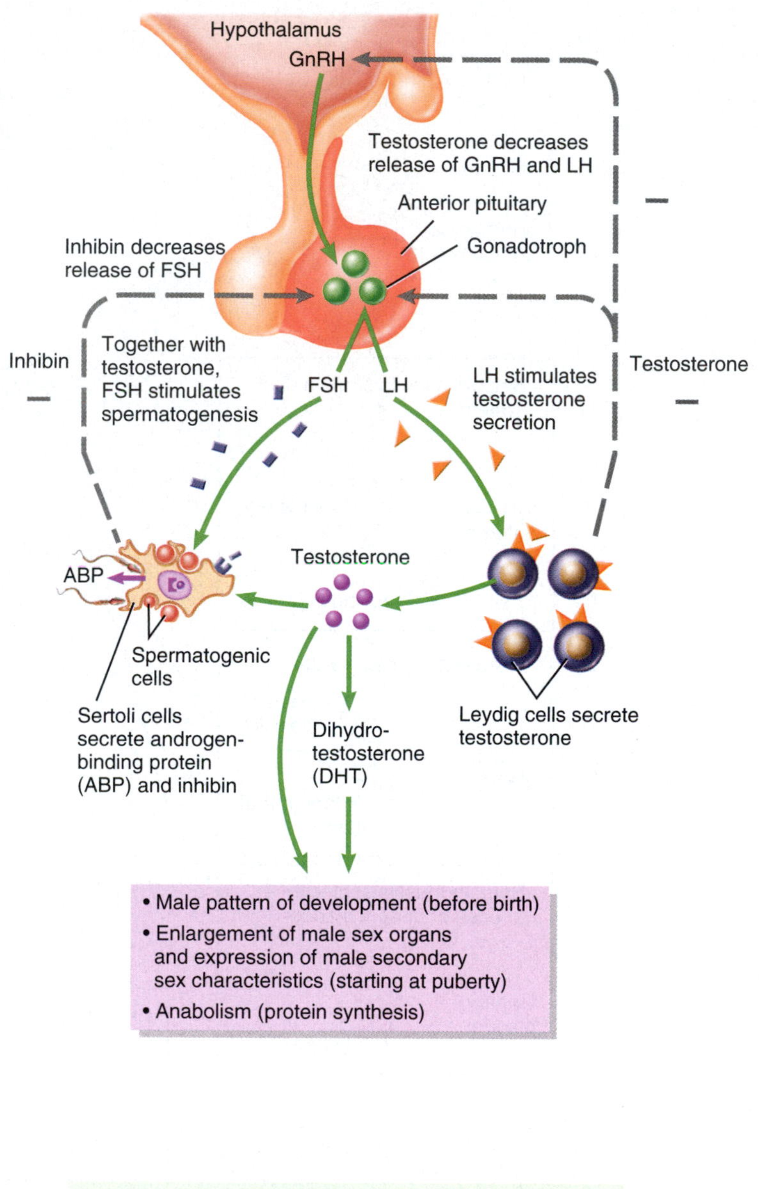

Which cells secrete inhibin?

CLINICAL CONNECTION

Anabolic Steroids

The use of **anabolic steroids** by athletes has received widespread attention. These steroid hormones, similar to testosterone, are taken to increase muscle size and thus strength during athletic contests. However, the large doses needed to produce an effect have damaging, sometimes even devastating side effects, including kidney damage, liver cancer, increased risk of heart disease, stunted growth, wide mood swings, and increased irritability and aggression (*'roid rage*). In addition, males have diminished testosterone secretion and may experience atrophy of the testes, sterility, and baldness. Females who take anabolic steroids may experience atrophy of the breasts and uterus, menstrual irregularities, sterility, facial hair growth, and deepening of the voice.

LH stimulates Leydig cells to secrete the hormone **testosterone**. This steroid hormone is synthesized from cholesterol in the testes and is the principal androgen. It is lipid-soluble and readily diffuses out of Leydig cells into the interstitial fluid and then into blood. In some target cells, such as those in the external genitalia and prostate gland, the enzyme **5 α-reductase** converts testosterone to another androgen called **dihydrotestosterone (DHT)**.

FSH acts indirectly to stimulate spermatogenesis (FIGURE 23.7). FSH and testosterone act synergistically on the Sertoli cells to stimulate secretion of **androgen-binding protein (ABP)** into the lumen of the seminiferous tubules and into the interstitial fluid around the spermatogenic cells. ABP binds to testosterone, keeping its concentration high. Testosterone stimulates the final steps of spermatogenesis in the seminiferous tubules. Once the degree of spermatogenesis required for male reproductive functions has been achieved, Sertoli cells release **inhibin**, a protein hormone named for its role in inhibiting FSH secretion by the anterior pituitary (FIGURE 23.7). If spermatogenesis is proceeding too slowly, less inhibin is released, which permits more FSH secretion and an increased rate of spermatogenesis.

Testosterone and dihydrotestosterone both bind to the same androgen receptors, which are found within the nuclei of target cells. The hormone–receptor complex regulates gene expression, turning some genes on and others off. Because of these changes, the androgens produce several effects:

- ***Prenatal development.*** Before birth, testosterone stimulates the male pattern of development of reproductive system ducts and the descent of the testes. Dihydrotestosterone stimulates development of the external genitalia (described in Section 23.5). Testosterone is also converted in the brain to estrogens, which may play a role in the development of certain regions of the brain in males.
- ***Development of male sex characteristics.*** At puberty, testosterone and dihydrotestosterone bring about development and enlargement of the male sex organs and the development of male secondary sex characteristics. **Secondary sex characteristics** are traits that distinguish males and females but do not have a direct role in reproduction. Male secondary sex characteristics include muscular and skeletal growth that results in wide shoulders and narrow hips; facial and chest hair (within hereditary limits) and more hair on other parts of the body; thickening of the skin; increased sebaceous (oil) gland secretion; and enlargement of the larynx and consequent deepening of the voice.
- ***Development of sexual function.*** Androgens contribute to male sexual behavior and spermatogenesis and to sex drive (libido) in both males and females. Recall that the adrenal cortex is the main source of androgens in females.
- ***Stimulation of anabolism.*** Androgens are anabolic hormones; that is, they stimulate protein synthesis. This effect is obvious in the heavier muscle and bone mass of most men compared to women.

A negative feedback system regulates testosterone production (FIGURE 23.7). When the testosterone concentration in the blood increases to a certain level, it inhibits the release of GnRH by cells in the hypothalamus and the release of LH from gonadotrophs in the anterior pituitary. As a result, the concentration of LH in blood decreases. With less stimulation by LH, the Leydig cells in the testes secrete less testosterone and there is a return to homeostasis. If the testosterone concentration in the blood falls too low, however, GnRH is again released by the hypothalamus and stimulates secretion of LH by the anterior pituitary. LH in turn stimulates testosterone production by the testes.

CHECKPOINT

4. Describe the function of the scrotum in protecting the testes from temperature fluctuations.
5. What is meant by the term *spermiation*?
6. Describe the functions of the epididymis, vas deferens, ejaculatory ducts, and urethra.
7. How do the seminal vesicles, prostate gland, and bulbourethral glands contribute to the composition of semen?
8. Explain the physiological processes involved in erection and ejaculation.
9. What are the roles of FSH, LH, testosterone, and inhibin in the male reproductive system? How is secretion of these hormones controlled?
10. How is dihydrotestosterone formed?

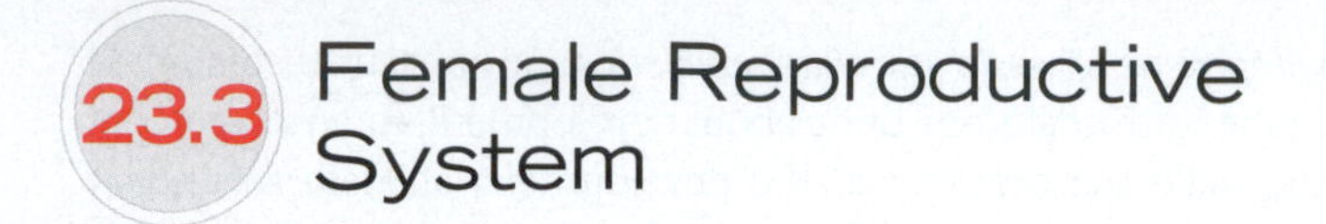

23.3 Female Reproductive System

OBJECTIVES

- Describe the functions of the organs of the female reproductive system.
- Discuss the process of oogenesis in the ovaries.
- Explain the functions of the various female reproductive hormones.
- Compare the major events of the female reproductive cycle.

The organs of the **female reproductive system** include the ovaries, fallopian tubes, uterus, vagina, vulva (external genitalia), and mammary glands (FIGURE 23.8).

The Ovaries Produce Eggs and Secrete Hormones

The **ovaries** are paired glands located in the pelvic cavity on either side of the uterus (FIGURE 23.8b). They are homologous to the testes. (Here, *homologous* means that the two organs have the same embryonic origin.) The ovaries produce eggs (oocytes) and several hormones, including estrogens, progesterone, inhibin, and relaxin.

FIGURE 23.8 Components of the female reproductive system. In the right side of the drawing of part (b), the fallopian tube and uterus have been sectioned to show internal structures.

The organs of reproduction in females include the ovaries, fallopian tubes, uterus, vagina, vulva, and mammary glands. After ovulation, an egg is swept into a fallopian tube by fimbriae.

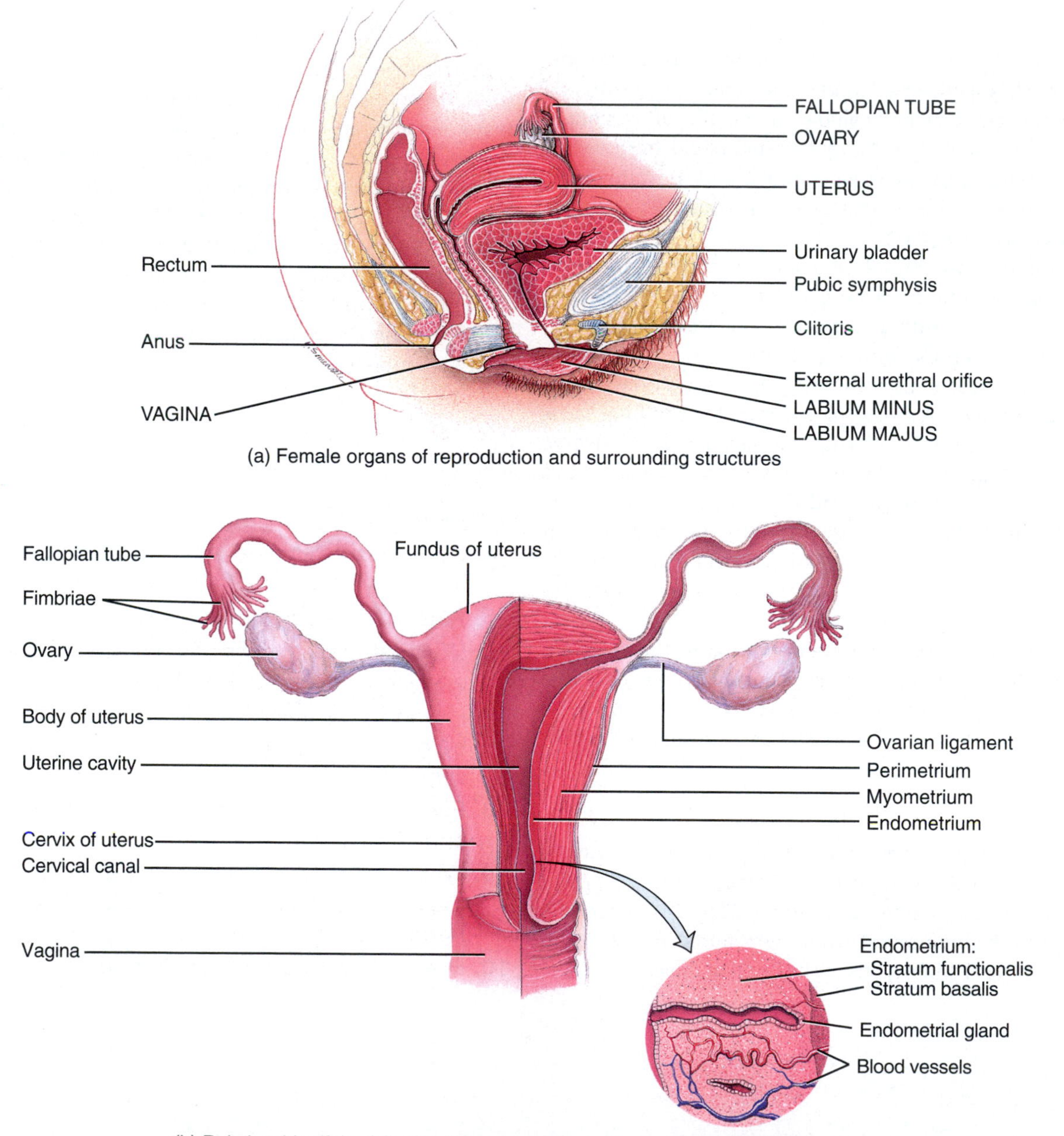

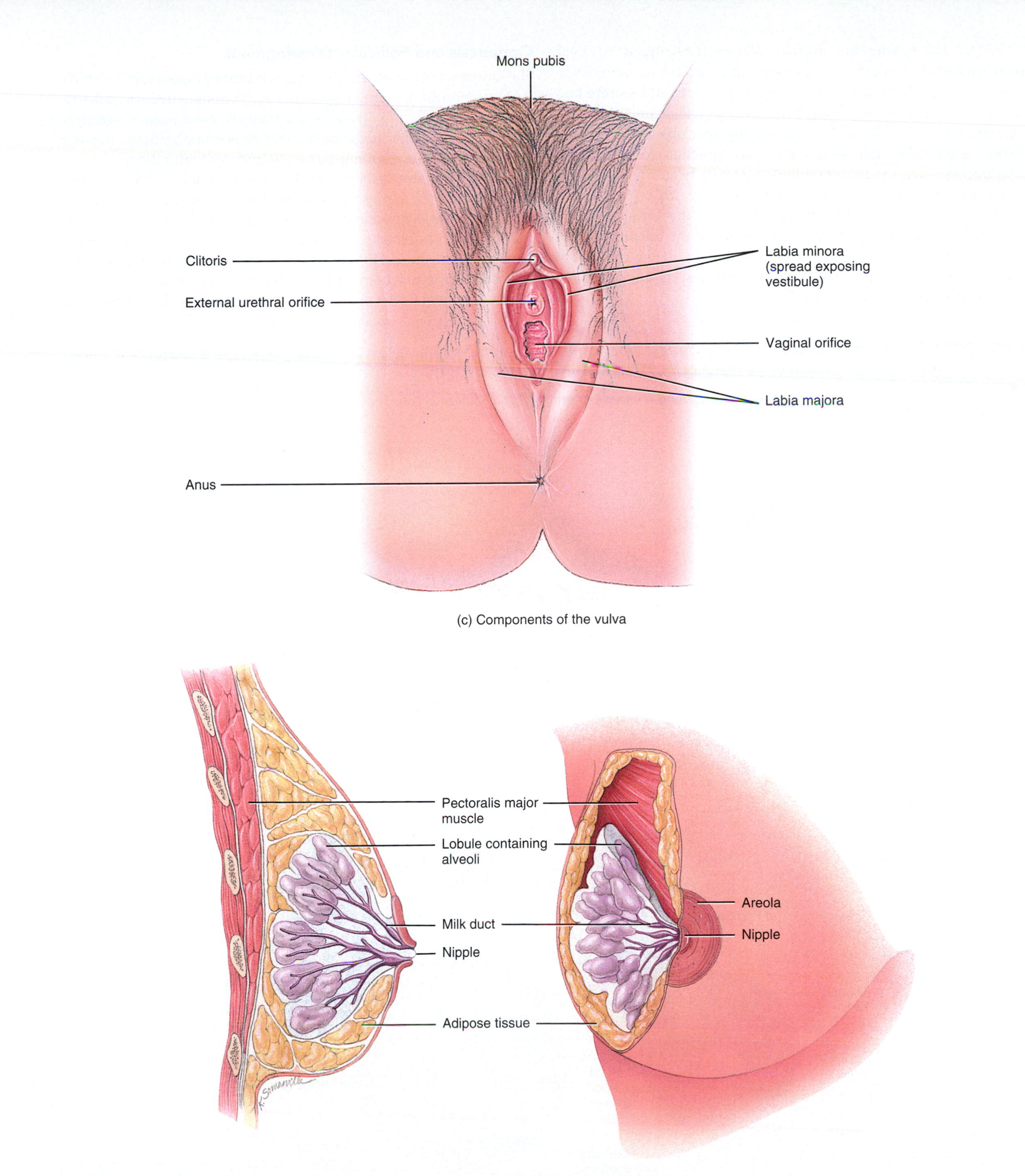

(c) Components of the vulva

(d) The mammary gland within the breast

Which part of the uterine lining sheds during menstruation?

Within the ovaries are ovarian follicles (FIGURE 23.9). Each **ovarian follicle** consists of an **oocyte** and a variable number of surrounding cells that nourish the developing oocyte and secrete hormones as the follicle grows larger. The ovarian follicle initially begins as a primordial follicle, which sequentially develops into a primary follicle, a secondary follicle, and a mature (graafian) follicle. Ovulation occurs when the mature follicle ruptures, releasing the oocyte into the pelvic cavity. From the pelvic cavity, the oocyte is normally swept into the fallopian tube. Within the ovary, the remnants of the ruptured follicle develop into a structure called the **corpus luteum** (= yellow body). If the released oocyte is not fertilized, the corpus luteum eventually degenerates into fibrous scar tissue called the **corpus albicans** (= white body).

Oogenesis and Follicular Development

The formation of gametes in the ovaries is termed **oogenesis** (ō-ō-JEN-e-sis). In contrast to spermatogenesis, which begins in males at puberty, oogenesis begins in females before they are even born. Oogenesis occurs in essentially the same manner as spermatogenesis; meiosis takes place and the resulting germ cells undergo maturation.

During fetal development, cells in the ovaries differentiate into **oogonia** (ō-ō-GŌ-nē-a; singular is **oogonium**). Oogonia are diploid ($2n$) stem cells that divide mitotically to produce millions of germ cells. Even before birth, most of these germ cells degenerate in a process known as **atresia** (a-TRĒ-zē-a). A few, however, develop into larger cells called **primary oocytes** (Ō-ō-sītz) that enter prophase of meiosis I during fetal development but do not complete that phase until

FIGURE 23.9 Organization of the ovary and ovarian follicles. The arrows in the ovary indicate the sequence of developmental stages that occur as part of the maturation of an egg during the ovarian cycle.

An ovarian follicle consists of an oocyte (egg) and a variable number of surrounding cells that nourish the developing oocyte and secrete hormones as the follicle grows larger.

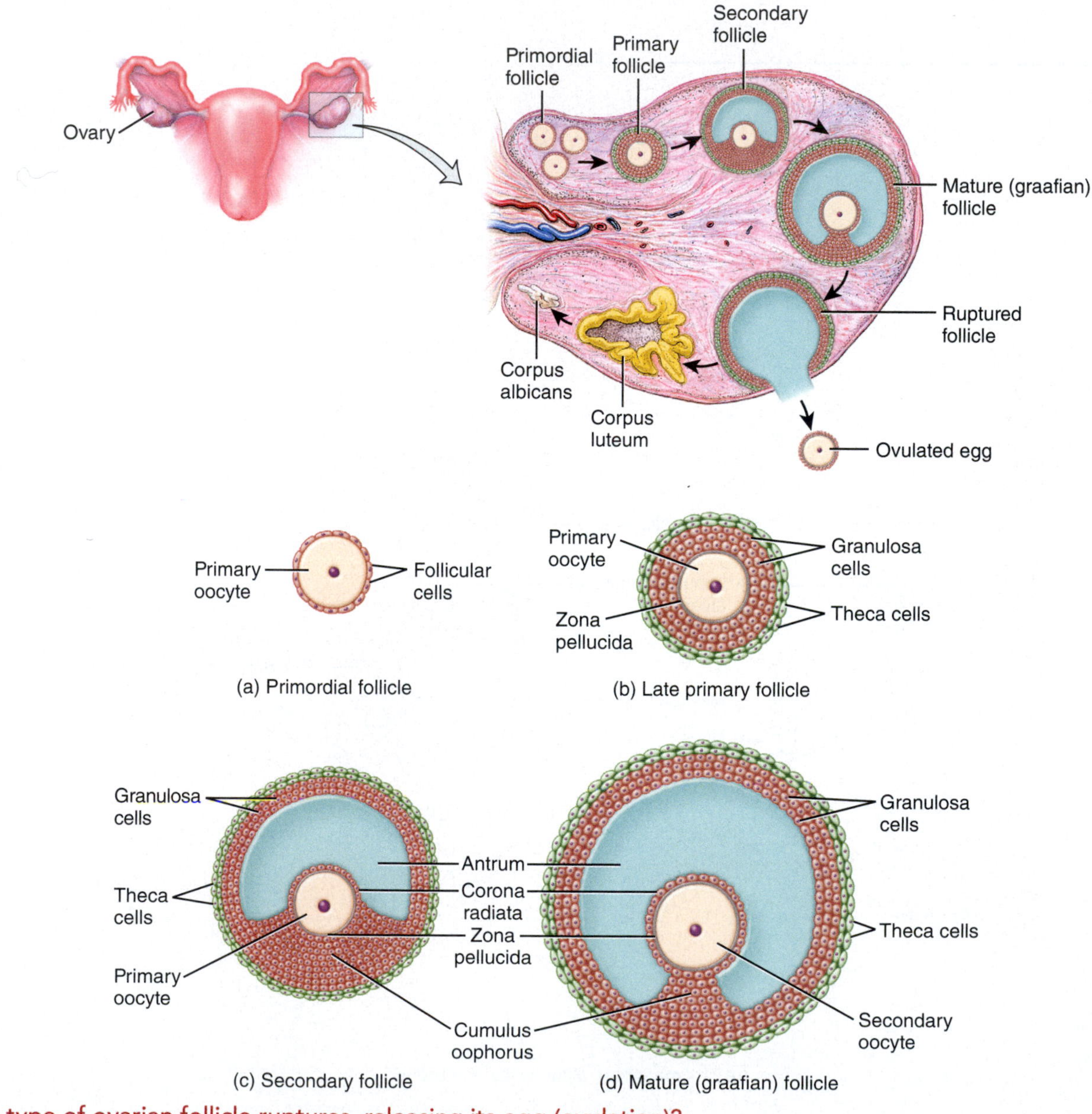

Which type of ovarian follicle ruptures, releasing its egg (ovulation)?

after puberty. During this arrested stage of development, each primary oocyte is surrounded by a single layer of **follicular cells**, and the entire structure is called a **primordial follicle** (FIGURE 23.9a). At birth, approximately 200,000 to 2,000,000 primary oocytes remain in each ovary. Of these, about 40,000 are still present at puberty, and around 400 will mature and ovulate during a woman's reproductive lifetime. The remainder undergo atresia.

Each month after puberty until menopause, gonadotropins (FSH and LH) secreted by the anterior pituitary further stimulate the development of several primordial follicles, although only one typically reaches the maturity needed for ovulation. A few primordial follicles start to grow, developing into **primary follicles** (FIGURE 23.9b). Each primary follicle consists of a primary oocyte that is surrounded in a later stage of development by several layers of follicular cells, which are now called **granulosa cells**. As the primary follicle grows, a clear glycoprotein layer called the **zona pellucida** forms between the primary oocyte and the granulosa cells. In addition, two layers of **theca cells** develop along the outer wall of the follicle.

With continuing maturation, a primary follicle develops into a secondary follicle (FIGURE 23.9c). In a **secondary follicle**, the granulosa cells begin to secrete follicular fluid, which builds up in a cavity called the **antrum** in the center of the follicle. Some of the granulosa cells form a mound called the **cumulus oophorus** that projects into the antrum and encloses the primary oocyte. The layer of granulosa cells that immediately surrounds the oocyte is referred to as the **corona radiata**.

The secondary follicle eventually becomes larger, turning into a **mature (graafian) follicle** (FIGURE 23.9d). While in this follicle, and just before ovulation, the diploid primary oocyte completes meiosis I, producing two haploid (n) cells of unequal size—each with 23 chromosomes (FIGURE 23.10). The smaller cell produced by meiosis I, called the **first polar body**, is essentially a packet of discarded nuclear material. The larger cell, known as the **secondary oocyte**, receives most of the cytoplasm. Once a secondary oocyte is formed, it begins meiosis II but then stops in metaphase. The mature (graafian) follicle soon ruptures and releases its secondary oocyte, a process known as **ovulation**.

At ovulation, the secondary oocyte, which is surrounded by the corona radiata and zona pellucida, is expelled into the pelvic cavity along with the first polar body. These cells are usually swept into the fallopian tube. If fertilization does not occur, the cells degenerate. If sperm are present in the fallopian tube and one penetrates the secondary oocyte, however, meiosis II resumes (FIGURE 23.10). The secondary oocyte splits into two haploid cells, again of unequal size. The larger cell is the **ovum**, or mature egg; the smaller one is the **second polar body**. The nuclei of the sperm cell and the ovum then unite, forming a diploid **zygote**, the beginning of the embryonic period of development. If the first polar body undergoes another division to produce two polar bodies, then the primary oocyte ultimately gives rise to three haploid polar bodies, which all degenerate, and a single haploid ovum. Thus, one primary oocyte gives rise to a single gamete (an ovum). By contrast, recall that in males one primary spermatocyte produces four gametes (sperm).

TABLE 23.1 summarizes the events of oogenesis and follicular development.

The Fallopian Tube Transports an Egg from the Ovary to the Uterus

Females have two **fallopian tubes**, also known as *uterine tubes* or *oviducts*, that extend from the upper part of the uterus (see FIGURE 23.8b).

FIGURE 23.10 Oogenesis. Diploid cells ($2n$) have 46 chromosomes; haploid cells (n) have 23 chromosomes.

Oogenesis refers to the formation of eggs (oocytes) in the ovaries.

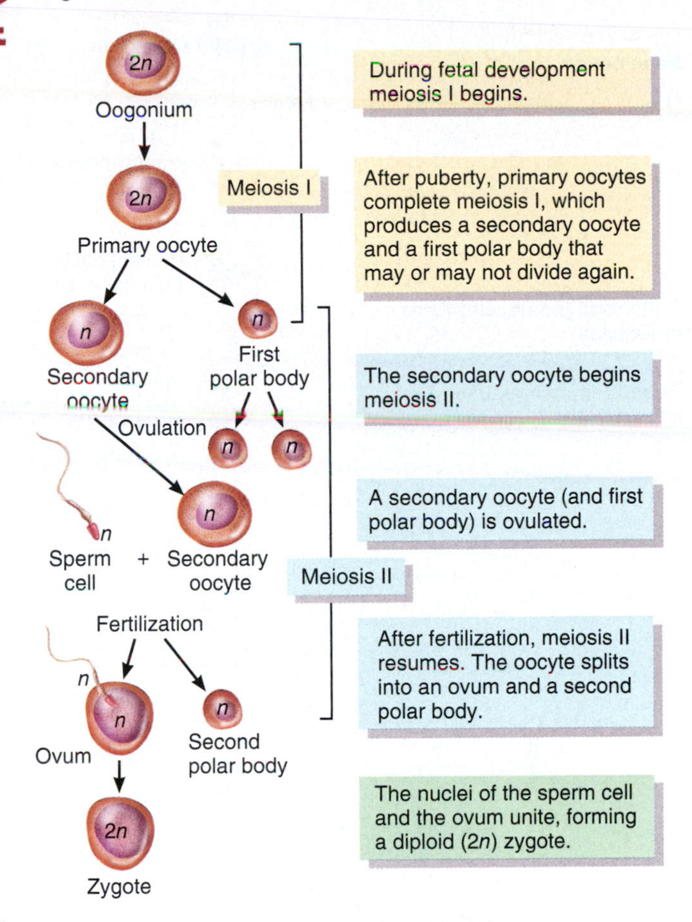

How does the age of a primary oocyte in a female compare with the age of a primary spermatocyte in a male?

The funnel-shaped end of each tube is close to an ovary and contains fingerlike projections called **fimbriae** (FIM-brē-ē). Functionally, the fallopian tube transports an egg (oocyte) from the ovary to the uterus. This occurs in the following way: After ovulation, local currents produced by movements of the fimbriae sweep the egg into the fallopian tube. The egg is then moved along the tube by cilia in the tube's lining and peristaltic contractions of the tube's smooth muscle layer.

The fallopian tube also serves as the site where fertilization normally takes place. After semen is deposited into the vagina by the penis during sexual intercourse, sperm move into the uterus and then into the fallopian tubes. If an egg is present in one of the tubes, it may be fertilized by a sperm cell. The cilia and peristaltic contractions of the fallopian tube move the fertilized egg into the uterus, where it implants. Fertilization can occur at any time up to about 24 hours after ovulation. If an egg is not fertilized by that time, it degenerates.

The Uterus Has Many Reproductive Functions

The **uterus**, or *womb*, is a hollow, pear-shaped organ situated between the urinary bladder and the rectum. Parts of the uterus include a dome-shaped portion above the fallopian tubes called the **fundus**, a tapering central portion called the **body**, and a lower narrow portion

TABLE 23.1 Summary of Oogenesis and Follicular Development

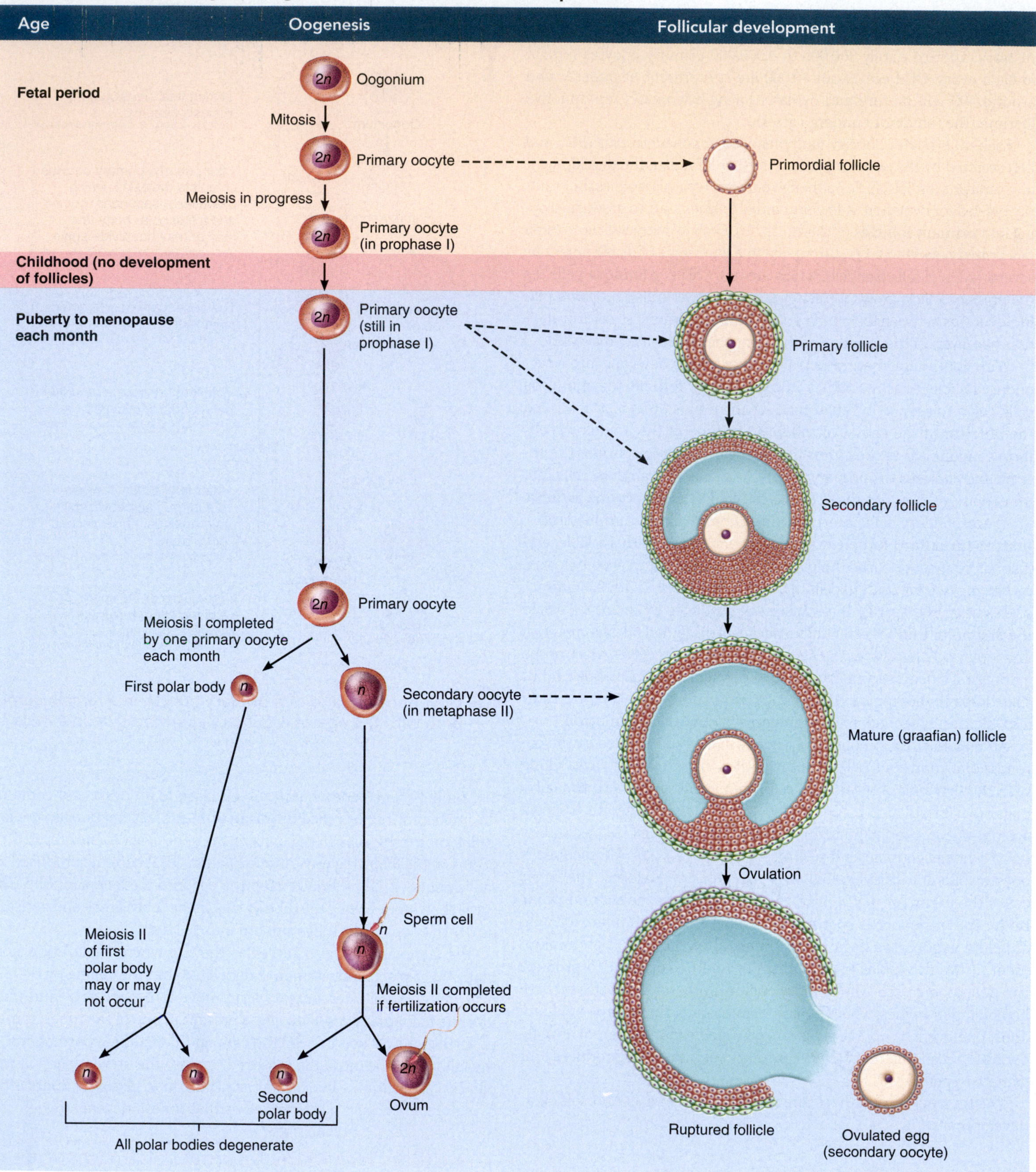

opening into the vagina called the **cervix** (see FIGURE 23.8b). Sperm cells deposited in the vagina during intercourse reach the fallopian tubes by passing through the cervix and the body of the uterus. The lining of the cervix contain glands that secrete **cervical mucus**. Most of the time, the mucous is viscous, forming a cervical plug that impedes the spread of bacteria and the passage of sperm cells. However, at or near the time of ovulation, cervical mucus is less viscous, making it easier for sperm cells to pass through.

The wall of the uterus is composed of three layers: endometrium, myometrium, and perimetrium (see FIGURE 23.8b). The inner layer—the **endometrium**—contains large numbers of blood vessels and consists of epithelial tissue, connective tissue, and endometrial glands. The endometrium is further divided into two layers: the stratum functionalis and the stratum basalis. The **stratum functionalis** (*functional layer*) lines the uterine cavity. In a nonpregnant woman of reproductive age, the stratum functionalis sloughs off every month, a process known as *menstruation*. The subsequent release of blood and tissue from the body is referred to as *menstrual flow*. In a pregnant woman, the stratum functionalis remains attached to the rest of the uterus and serves as the site where the fertilized egg has implanted and develops into a fetus. The **stratum basalis** (*basal layer*) is the deeper part of the endometrium. It is a permanent layer that gives rise to a new stratum functionalis after each menstruation. The middle layer of the uterus, the **myometrium**, consists of smooth muscle and forms the bulk of the uterine wall. During labor and childbirth, coordinated contractions of the myometrium in response to oxytocin from the posterior pituitary help expel the fetus from the uterus. The outer layer of the uterus is the **perimetrium**. It consists of epithelium and connective tissue and provides support to the uterus.

In summary, the uterus performs several functions. It serves as part of the pathway for sperm deposited in the vagina to reach the fallopian tubes. It is also the site of implantation of a fertilized egg, the development of the fetus during pregnancy, and the contractions that push the fetus out of a woman's body during childbirth. In addition, the uterus is the source of menstrual flow.

The Vagina Is the Female Organ of Copulation

The **vagina** is a tubular canal that extends from the uterine cervix to the exterior of the body (see FIGURE 23.8a, b). It is the female organ of copulation, serving as the receptacle for the penis during intercourse. It also functions as the outlet for menstrual flow and as the *birth canal*, the passageway through which the fetus moves from the uterus to the outside environment during delivery. The opening of the vagina to the exterior is known as the *vaginal orifice* (see FIGURE 23.8c). In young females, the vaginal orifice is partially covered by the *hymen*, a thin membrane containing many blood vessels. The hymen is usually torn by the first sexual intercourse or by strenuous exercise. The lining of the vagina is bathed by acidic **vaginal fluid**. The resulting acidic environment retards microbial growth, but it is also harmful to sperm. Alkaline components of semen, mainly from the seminal vesicles, neutralize the acidity of the vagina and increase the viability of sperm.

The Vulva Refers to the Female External Genitalia

The term **vulva**, or *pudendum* (pū-DEN-dum), collectively refers to the external genitalia of the female. It includes the mons pubis, labia majora, labia minora, clitoris, and vestibule (see FIGURE 23.8c). The **mons pubis** is a rounded, fatty area that covers the joint between the pubic bones. Extending down and back from the mons pubis are two prominent folds of skin called the **labia majora** (singular is *labium majus*), which are homologous to the scrotum of the male. The labia majora enclose the **labia minora** (singular is *labium minus*), two smaller folds of skin that are homologous to the shaft of the penis. The **clitoris** is a small cylindrical mass of erectile tissue located at the front end of the vulva. It is homologous to the glans penis. Like its male counterpart, the clitoris fills with blood and enlarges during sexual arousal. The **vestibule** is the space between the labia minora. Within the vestibule are the *external urethral orifice* (the opening of the urethra to the exterior) and the *vaginal orifice* (the opening of the vagina to the exterior). Near the external urethral orifice are the **paraurethral** (*Skene's*) **glands**, which secrete mucus into the urethra. These glands are homologous to the male prostate gland. On either side of the vaginal orifice are the **greater vestibular** (*Bartholin's*) **glands**, which secrete a lubricating mucus into the vestibule during sexual arousal. The greater vestibular glands are homologous to the bulbourethral glands in males.

TABLE 23.2 summarizes the homologous structures of the female and male reproductive systems.

The Mammary Glands Function in Lactation

The **mammary glands**, one located in each breast, are modified sweat glands that produce and eject milk. Together, milk production and ejection constitute **lactation**. A mammary gland consists of 15 to 20 lobes separated by a variable amount of adipose tissue. In each lobe are several smaller *lobules* composed of clusters of milk-secreting glands termed **alveoli** (see FIGURE 23.8d). When milk is being produced, it passes from the alveoli into ducts that convey it to the *nipple*, a pigmented projection on the surface of the breast. Contraction of **myoepithelial cells** surrounding the alveoli helps propel milk toward the nipple. Milk production is stimulated by the hormone prolactin from the anterior pituitary. The ejection of milk is stimulated by oxytocin, which is released from the posterior pituitary in response to the suckling action of an infant on the mother's nipple. Lactation is described in detail later in this chapter.

Female Reproductive Function Is Regulated by Many Hormones

Several hormones affect reproductive function in females: gonadotropin-releasing hormone, follicle-stimulating hormone, luteinizing

TABLE 23.2 Summary of Homologous Structures of the Female and Male Reproductive Systems

Female Structures	Male Structures
Ovaries	Testes
Egg	Sperm cells
Labia majora	Scrotum
Labia minora	Shaft of penis
Clitoris	Glans penis
Paraurethral glands	Prostate gland
Greater vestibular glands	Bulbourethral glands

hormone, estrogens, progesterone, relaxin, and inhibin (**FIGURE 23.11**). Beginning at puberty, the hypothalamus increases secretion of **gonadotropin-releasing hormone (GnRH)**. GnRH in turn stimulates the release of **follicle-stimulating hormone (FSH)** and **luteinizing hormone (LH)** from the anterior pituitary. FSH initiates follicular growth, while LH stimulates further development of the ovarian follicles. In addition, both FSH and LH stimulate the ovarian follicles to secrete estrogens. LH stimulates the theca cells of a developing follicle to produce androgens (the main thecal androgen is **androstenedione**). Under the influence of FSH, the androgens are taken up by the granulosa cells of the follicle and are then converted into estrogens. The enzyme **aromatase** catalyzes this reaction. LH also triggers ovulation and then promotes formation of the corpus luteum, the reason for the name *luteinizing hormone*. Stimulated by LH, the corpus luteum produces and secretes estrogens, progesterone, relaxin, and inhibin.

At least six different estrogens have been isolated from the plasma of human females, but only three are present in significant quantities: *beta (β)-estradiol*, *estrone*, and *estriol*. In a nonpregnant woman, the most abundant estrogen is β-estradiol, which is synthesized from cholesterol in the ovaries.

Estrogens secreted by ovarian follicles and the corpus luteum have several important functions (**FIGURE 23.11**):

- Estrogens promote the development and maintenance of female reproductive structures and secondary sex characteristics. Female secondary sex characteristics include distribution of adipose tissue in the breasts, abdomen, mons pubis, and hips; a higher voice pitch; a broad pelvis; and a female pattern of hair growth on the head and body.
- Estrogens increase protein anabolism, including the building of strong bones.

FIGURE 23.11 Hormonal regulation of female reproductive function. Dashed lines indicate negative feedback inhibition.

The hormones that affect reproductive function in females are gonadotropin-releasing hormone, follicle-stimulating hormone, luteinizing hormone, estrogens, progesterone, relaxin, and inhibin.

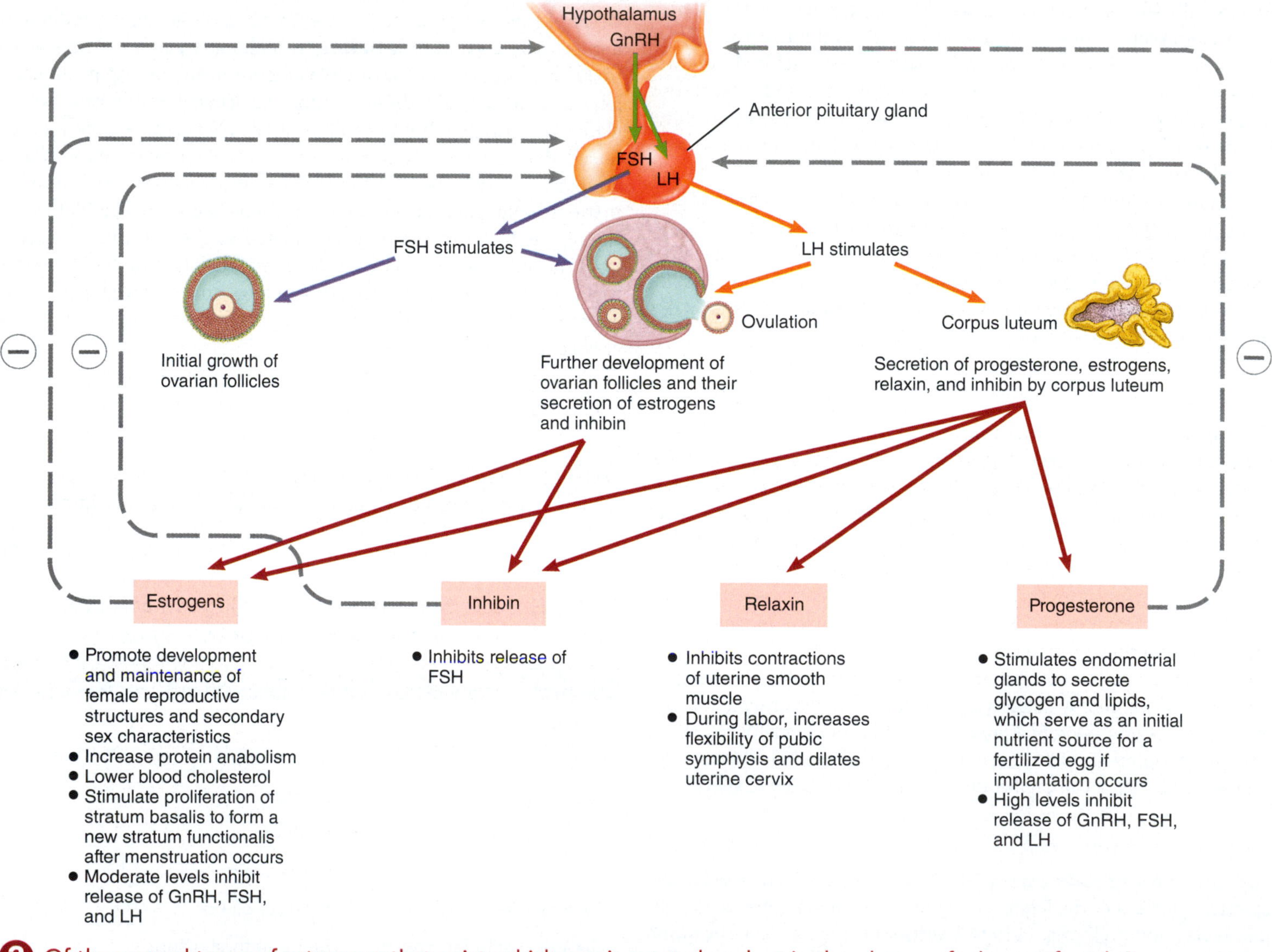

Of the several types of estrogens that exist, which one is most abundant in the plasma of a human female?

- Estrogens lower blood cholesterol level, which is probably the reason that women under age 50 have a much lower risk of coronary artery disease than do men of comparable age.
- Every month, after menstruation occurs, estrogens stimulate proliferation of the stratum basalis to form a new stratum functionalis that replaces the one that has sloughed off.
- Moderate levels of estrogens in the blood inhibit both the release of GnRH by the hypothalamus and secretion of LH and FSH by the anterior pituitary.

Progesterone, produced by cells of the corpus luteum, prepares the endometrium of the uterus for the possible implantation of a fertilized egg by stimulating the endometrial glands to secrete glycogen and lipids. These secretions serve as a nutrient source for the embryo during the early stages of pregnancy until the placenta develops. High levels of progesterone also inhibit secretion of GnRH, FSH, and LH.

The small quantity of **relaxin** produced by the corpus luteum during each monthly cycle relaxes the uterus by inhibiting contractions of the myometrium. Presumably, implantation of a fertilized egg occurs more readily in a "quiet" uterus. During pregnancy, the placenta produces much more relaxin, and it continues to relax uterine smooth muscle. At the end of pregnancy, relaxin also increases the flexibility of the pubic symphysis and may help dilate the uterine cervix, both of which ease delivery of the baby.

Inhibin is secreted by granulosa cells of growing follicles and by the corpus luteum after ovulation. It inhibits secretion of FSH.

The Female Reproductive Cycle Refers to the Cyclical Changes in the Ovaries and Uterus

During their reproductive years, nonpregnant females normally exhibit cyclical changes in the ovaries and uterus. Each cycle takes about a month and involves both oogenesis and preparation of the uterus to receive a fertilized egg. Hormones secreted by the hypothalamus, anterior pituitary, and ovaries control the main events. The **ovarian cycle** is a series of events in the ovaries that occur during and after the maturation of an egg. The **uterine (menstrual) cycle** is a concurrent series of changes in the endometrium of the uterus to prepare it for the arrival of a fertilized egg that will develop there until birth. If fertilization does not occur, ovarian hormones wane, which causes the stratum functionalis of the endometrium to slough off. The general term **female reproductive cycle** encompasses the ovarian and uterine cycles, the hormonal changes that regulate them, and the related cyclical changes in the breasts and cervix. The duration of the female reproductive cycle typically ranges from 24 to 35 days. For this discussion, it is assumed that the cycle has a duration of 28 days and is divided into four phases: the menstrual phase, the preovulatory phase, ovulation, and the postovulatory phase (FIGURE 23.12).

Menstrual Phase

The **menstrual phase** (MEN-stroo-al), also called **menstruation** (men′-stroo-Ā-shun) or *menses* (MEN-sēz = month), lasts for roughly the first 5 days of the cycle. By convention, the first day of menstruation is day 1 of a new cycle.

EVENTS IN THE OVARIES Under the influence of FSH, several primordial follicles develop into primary follicles and then into secondary follicles. This developmental process may take several months to occur. Therefore, a follicle that begins to develop at the beginning of a particular menstrual cycle may not reach maturity and ovulate until several menstrual cycles later.

EVENTS IN THE UTERUS Menstrual flow from the uterus consists of 50–150 mL of blood, tissue fluid, mucus, and epithelial cells shed from the stratum functionalis of the endometrium. This discharge occurs because the declining levels of estrogens and progesterone stimulate release of prostaglandins that cause the uterine arteries to constrict. As a result, the cells they supply become oxygen-deprived and start to die. Eventually, the entire stratum functionalis sloughs off. At this time the endometrium is very thin, about 2–5 mm, because only the stratum basalis remains. The menstrual flow passes from the uterine cavity through the cervix and vagina to the exterior.

Preovulatory Phase

The **preovulatory phase** is the time between the end of menstruation and the beginning of ovulation. The preovulatory phase of the cycle is more variable in length than the other phases and accounts for most of the differences in length of the cycle. It lasts from days 6 to 13 days in a 28-day cycle.

EVENTS IN THE OVARIES Some of the secondary follicles in the ovaries begin to secrete estrogens and inhibin. By about day 6, a single secondary follicle in one of the two ovaries has outgrown all of the others to become the **dominant follicle**. Estrogens and inhibin secreted by the dominant follicle decrease the secretion of FSH, which causes other, less well-developed follicles to stop growing and undergo atresia. Fraternal (nonidentical) twins or triplets result when two or three secondary follicles become codominant and later are ovulated and fertilized at about the same time.

Normally, the one dominant secondary follicle becomes the **mature (graafian) follicle**, which continues to enlarge until it is more than 20 mm in diameter and ready for ovulation (see FIGURE 23.9). This follicle forms a blisterlike bulge due to the swelling antrum on the surface of the ovary. During the final maturation process, the mature follicle continues to increase its production of estrogens (FIGURE 23.12).

With reference to the ovarian cycle, the menstrual and preovulatory phases together are termed the **follicular phase** because ovarian follicles are growing and developing.

EVENTS IN THE UTERUS Estrogens liberated into the blood by growing ovarian follicles stimulate the repair of the endometrium; cells of the stratum basalis undergo mitosis and produce a new stratum functionalis. As the endometrium thickens, the endometrial glands develop, and blood vessels coil and lengthen as they penetrate the stratum functionalis. The thickness of the endometrium approximately doubles, to about 4–10 mm. With reference to the uterine cycle, the preovulatory phase is also termed the **proliferative phase** because the endometrium is proliferating.

Ovulation

Ovulation, the rupture of the mature (graafian) follicle and the release of the egg (secondary oocyte) into the pelvic cavity, usually occurs on day 14 in a 28-day cycle. During ovulation, the secondary oocyte remains surrounded by its zona pellucida and corona radiata.

The *high levels of estrogens* during the last part of the preovulatory phase exert a *positive* feedback effect on the cells that secrete LH and

FIGURE 23.12 The female reproductive cycle. The length of the female reproductive cycle typically is 24 to 36 days; the preovulatory phase is more variable in length than the other phases. (a) Events in the ovarian and uterine cycles and the release of anterior pituitary hormones are correlated with the sequence of the cycle's four phases. In the cycle shown, fertilization and implantation have not occurred. (b) Relative concentrations of anterior pituitary hormones (FSH and LH) and ovarian hormones (estrogens and progesterone) during the phases of a normal female reproductive cycle.

Estrogens are the primary ovarian hormones before ovulation; after ovulation, both progesterone and estrogens are secreted by the corpus luteum.

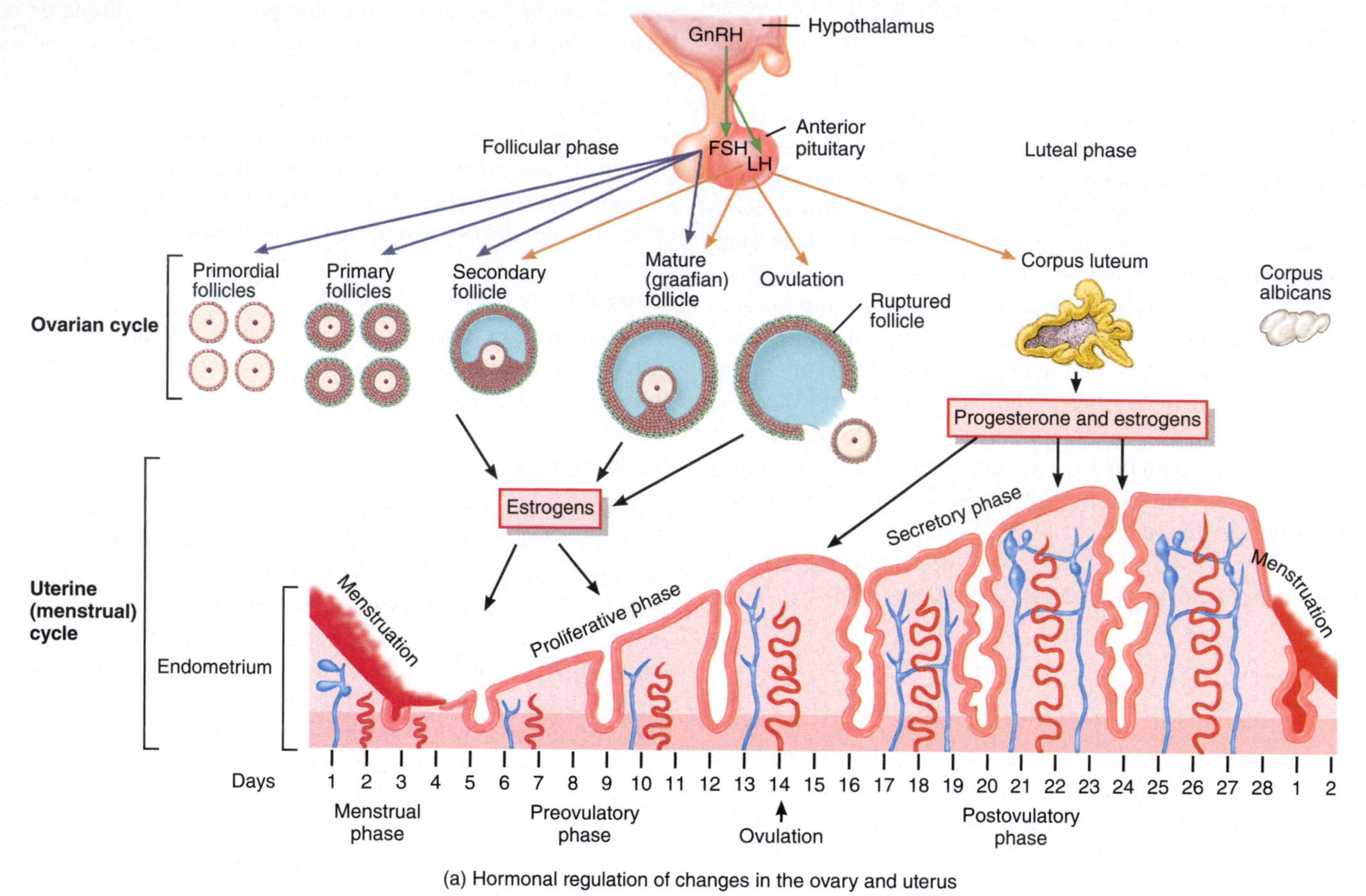

(a) Hormonal regulation of changes in the ovary and uterus

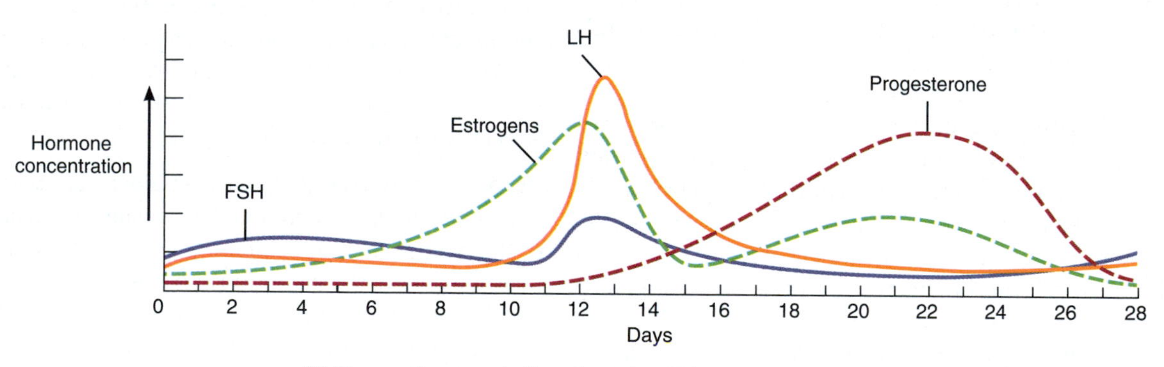

(b) Changes in concentration of anterior pituitary and ovarian hormones

Which hormone is responsible for ovulation?

gonadotropin-releasing hormone (GnRH) and cause ovulation, as follows (FIGURE 23.13):

1. A high concentration of estrogens stimulates more frequent release of GnRH from the hypothalamus. It also directly stimulates gonadotrophs in the anterior pituitary to secrete LH.
2. GnRH promotes the release of FSH and additional LH by the anterior pituitary.
3. LH causes rupture of the mature (graafian) follicle and expulsion of a secondary oocyte about 9 hours after the peak of the **LH surge**. The ovulated oocyte and its zona pellucida and corona radiata cells are usually swept into the fallopian tube.

From time to time, an oocyte is lost in the pelvic cavity, where it later disintegrates. The small amount of blood that sometimes leaks into the pelvic cavity from the ruptured follicle can cause pain, known

FIGURE 23.13 High levels of estrogens exert a positive feedback effect (green arrows) on the hypothalamus and anterior pituitary, thereby increasing secretion of GnRH and LH.

At midcycle, a surge of LH triggers ovulation.

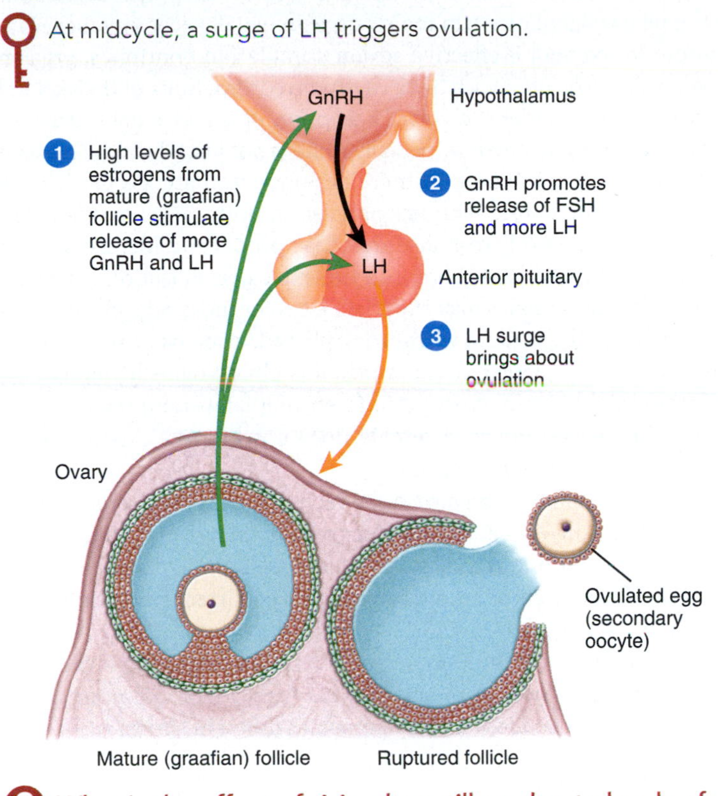

What is the effect of rising but still moderate levels of estrogens on the secretion of GnRH, LH, and FSH?

as **mittelschmerz** (MIT-el-shmärts = pain in the middle), at the time of ovulation.

An over-the-counter home test that detects a rising level of LH can be used to predict ovulation a day in advance.

Postovulatory Phase

The **postovulatory phase** of the female reproductive cycle is the time between the end of ovulation and the onset of the next menses. In duration, it is the most constant part of the female reproductive cycle. It lasts for 14 days in a 28-day cycle, from day 15 to day 28 (see FIGURE 23.12).

Events in One Ovary After ovulation, the ruptured follicle is transformed into the corpus luteum under the influence of LH. Stimulated by LH, the corpus luteum secretes progesterone, estrogens, relaxin, and inhibin. With reference to the ovarian cycle, this phase is also called the **luteal phase**.

Later events in an ovary that has ovulated an oocyte depend on whether the oocyte is fertilized. If the oocyte *is not fertilized*, the corpus luteum has a lifespan of only 2 weeks. After that, its secretory activity declines, and it degenerates into a corpus albicans (see FIGURES 23.9 and 23.12). As the levels of estrogens, progesterone, and inhibin decrease, release of GnRH, FSH, and LH rises due to loss of negative feedback suppression by the ovarian hormones. Follicular growth resumes and a new ovarian cycle begins.

If the oocyte *is fertilized* and begins to divide, the corpus luteum persists past its normal 2-week lifespan. It is "rescued" from degeneration by **human chorionic gonadotropin (hCG)**. This hormone is produced by the chorion of the embryo beginning about 8 days after fertilization. Like LH, hCG stimulates the secretory activity of the corpus luteum. The presence of hCG in maternal blood or urine is an indicator of pregnancy and is the hormone detected by home pregnancy tests.

Events in the Uterus Estrogens secreted by the corpus luteum cause continued growth of the stratum functionalis. As a result, the endometrium thickens to about 12–18 mm (0.48–0.72 in.). Progesterone produced by the corpus luteum promotes further growth and coiling of the endometrial glands and vascularization of the superficial endometrium. In addition, progesterone stimulates the endometrial glands to secrete glycogen and lipids, which serve as an initial nutrient source for a fertilized egg if implantation occurs. Because of the secretory activity of the endometrial glands, this period is called the **secretory phase** of the uterine cycle. These preparatory changes peak about 1 week after ovulation, at the time a fertilized egg might arrive in the uterus. If a woman is not pregnant, the levels of estrogens and progesterone decline due to degeneration of the corpus luteum. Withdrawal of estrogens and progesterone causes menstruation.

CHECKPOINT

11. Describe the principal events of oogenesis.
12. Why are the fimbriae of the fallopian tubes important?

CLINICAL CONNECTION

Female Athlete Triad: Disordered Eating, Amenorrhea, and Premature Osteoporosis

The female reproductive cycle can be disrupted by many factors, including weight loss, low body weight, disordered eating, and vigorous physical activity. The observation that three conditions—disordered eating, amenorrhea, and osteoporosis—often occur together in female athletes led researchers to coin the term **female athlete triad**.

Many athletes experience intense pressure from coaches, parents, peers, and themselves to lose weight to improve performance. Hence, they may develop disordered eating behaviors and engage in other harmful weight-loss practices in a struggle to maintain a very low body weight. **Amenorrhea** (ā-men′-ō-RĒ-a) is the absence of menstruation. The most common causes of amenorrhea are pregnancy and menopause. In female athletes, amenorrhea results from reduced secretion of gonadotropin-releasing hormone, which decreases the release of LH and FSH. As a result, ovarian follicles fail to develop, ovulation does not occur, synthesis of estrogens and progesterone wanes, and monthly menstrual bleeding ceases. Most cases of the female athlete triad occur in young women with very low amounts of body fat. Low levels of the hormone leptin, secreted by adipose cells, may be a contributing factor.

Because estrogens help bones retain calcium and other minerals, chronically low levels of estrogens are associated with loss of bone mineral density. The female athlete triad causes "old bones in young women." In one study, amenorrheic runners in their twenties had low bone mineral densities similar to those of postmenopausal women 50 to 70 years old! Short periods of amenorrhea in young athletes may cause no lasting harm. However, long-term cessation of the reproductive cycle may be accompanied by a loss of bone mass, and adolescent athletes may fail to achieve an adequate bone mass; both situations can lead to premature osteoporosis and irreversible bone damage.

13. What is the functional significance of the stratum functionalis of the endometrium?
14. What is the purpose of the myoepithelial cells surrounding the alveoli of mammary glands?
15. Briefly outline the major events of each phase of the female reproductive cycle.
16. Diagram the major hormonal changes that occur during the female reproductive cycle.

23.4 The Human Sexual Response

OBJECTIVE

- Compare the sexual responses of males and females.

During heterosexual **sexual intercourse**, also called **copulation** or **coitus** (KŌ-i-tus), the penis is inserted into the vagina. The similar sequence of physiological and emotional changes experienced by both males and females before, during, and after intercourse is termed the **human sexual response**. William Masters and Virginia Johnson, who began their pioneering research on human sexuality in the late 1950s, divided the human sexual response into four phases: excitement, plateau, orgasm, and resolution.

During the **excitement** phase, there is **vasocongestion**—engorgement with blood—of genital tissues, resulting in erection of the penis in men and erection of the clitoris and swelling of the labia and vagina in women. In addition, vasocongestion causes the breasts to swell and the nipples to become erect. The excitement phase is also associated with an increase in the secretion of fluid that lubricates the walls of the vagina. When the connective tissue of the vagina becomes engorged with blood, lubricating fluid oozes from the capillaries and seeps through the epithelial lining via a process called **transudation**. Glands within the cervical mucosa and the greater vestibular (Bartholin's) glands contribute a small quantity of lubricating mucus. Without satisfactory lubrication, sexual intercourse is difficult and painful for both partners and inhibits orgasm. Other changes that occur during the excitement phase include increased heart rate and blood pressure, increased skeletal muscle tone throughout the body, and hyperventilation. Direct physical contact (as in kissing or touching), especially of the penis, clitoris, nipples of the breasts, and earlobes is a potent initiator of excitement. However, anticipation or fear; memories; visual, olfactory, and auditory sensations; and fantasies can enhance or diminish the likelihood that excitement will occur.

The changes that begin during excitement are sustained at an intense level in the **plateau** phase, which may last for only a few seconds or for many minutes. During this phase, many females and some males display a **sex flush**, a rashlike redness of the face and chest due to vasodilation of blood vessels in those parts of the body. The head of the penis increases in diameter and the testes swell. Late in the plateau phase, pronounced vasocongestion of the lower third of the vagina swells the tissue and narrows the opening. Because of this response, the vagina grips the penis more firmly.

Generally, the briefest phase is **orgasm** (*climax*), during which both sexes experience several rhythmic muscular contractions about 0.8 sec apart, accompanied by intense, pleasurable sensations and a further increase in blood pressure, heart rate, and respiratory rate. The sex flush is also most prominent at this time. In males, contraction of smooth muscle in the walls of the epididymis, vas deferens, and ejaculatory ducts as well as secretion of fluid by the accessory sex glands cause semen to move into the urethra (emission). Then, rhythmic contractions of skeletal muscles at the base of the penis propel semen out of the penis (ejaculation). In males, orgasm usually accompanies ejaculation. In women, if effective sexual stimulation continues, orgasm may occur, associated with 3–12 rhythmic contractions of the skeletal muscles that underlie the vulva. Reception of the ejaculate provides little stimulus for a female, especially if she is not already at the plateau phase; this is why a female partner does not automatically experience orgasm simultaneously with her partner. In both males and females, orgasm is a total body response that may produce milder sensations on some occasions and more intense, explosive sensations at other times. Whereas females may experience two or more orgasms in rapid succession, males enter a **refractory period**, a recovery time during which a second ejaculation and orgasm is physiologically impossible. In some males, the refractory period lasts only a few minutes; in others it lasts for several hours. A female does not have to experience an orgasm for fertilization to occur.

In the final phase—**resolution**, which begins with a sense of profound relaxation—genital tissues, heart rate, blood pressure, breathing, and muscle tone return to the unaroused state. If sexual excitement has been intense but orgasm has not occurred, resolution takes place more slowly.

The four phases of the human sexual response are not always clearly separated from one another and may vary considerably among different people, and even in the same person at different times.

CHECKPOINT

17. What happens during each of the four phases of the human sexual response?

23.5 Sex Determination and Sex Differentiation

OBJECTIVES

- Explain how genetic sex is determined.
- Describe how parts of an embryo differentiate into male or female reproductive organs.

Sex Determination Is Based on the Types of Sex Chromosomes That Are Present

As mentioned earlier in the chapter, somatic cells are diploid ($2n$): They contain 23 pairs of homologous chromosomes, for a total of 46 chromosomes. Of these chromosomes, there are 22 pairs of autosomes and one pair of sex chromosomes. **Autosomes** code for the overall form of the human body and for specific traits such as eye color and height. The two sex chromosomes—a large **X chromosome** and a smaller **Y chromosome**—determine the genetic sex of an individual. In a genetic female, somatic cells contain two X chromosomes. In a genetic male, somatic cells contain one X and one Y chromosome. Determination of genetic sex by the sex chromosomes is known as **sex determination**.

In gametes (sperm or eggs), which are haploid (n), there are only 23 total chromosomes. Of these chromosomes, there are 22 autosomes and 1 sex chromosome. In sperm cells, the sex chromosome is either

X or Y—approximately half of the sperm cells produced by meiosis contain an X and the other half a Y. In an egg, the sex chromosome is always an X. Genetic sex is established at the moment of conception by the type of sperm cell (X-bearing or Y-bearing) that fertilizes the egg. If an X-bearing sperm fertilizes the egg, the embryo formed will be a genetic female (XX). If a Y-bearing sperm fertilizes the egg, the embryo formed will be a genetic male (XY).

Sex Differentiation Involves Sex Chromosomes and the Presence or Absence of Certain Hormones

The early embryo is **bipotential**, which means that it has the ability to form either male or female reproductive organs. The first step in the development of the reproductive organs occurs in response to the genetic sex of the embryo. If the embryo is genetically male, testes develop; if the embryo is genetically female, ovaries develop. Once testes form in a male embryo, they begin to secrete androgens (masculinizing hormones), which cause a male reproductive tract and male external genitalia to develop. Female embryos, which contain ovaries instead of testes, do not produce testicular androgens. The lack of testicular androgens in a female embryo causes a female reproductive tract and female external genitalia to develop by default. Such a default pathway is ideal because both male and female embryos are exposed to high levels of estrogens and progesterone from the mother's placenta and ovaries during pregnancy. If female sex hormones played a role in sex differentiation, then all embryos (whether genetically male or female) would develop female reproductive organs. **Sex differentiation** is the process by which reproductive organs develop along male or female lines. To understand the steps involved in sex differentiation, you will first examine how the internal reproductive organs are formed and then you will discover how the external genitalia are developed.

Differentiation of Internal Reproductive Organs

Toward the end of the fifth week of development, the embryo contains several undifferentiated internal reproductive structures: primitive gonads, Wolffian ducts, and Müllerian ducts (**FIGURE 23.14**).

FIGURE 23.14 Development of the internal reproductive organs.

Primitive gonads develop into testes or ovaries, Wolffian ducts develop into the male reproductive tract and certain glands, and the Müllerian ducts develop into the female reproductive tract.

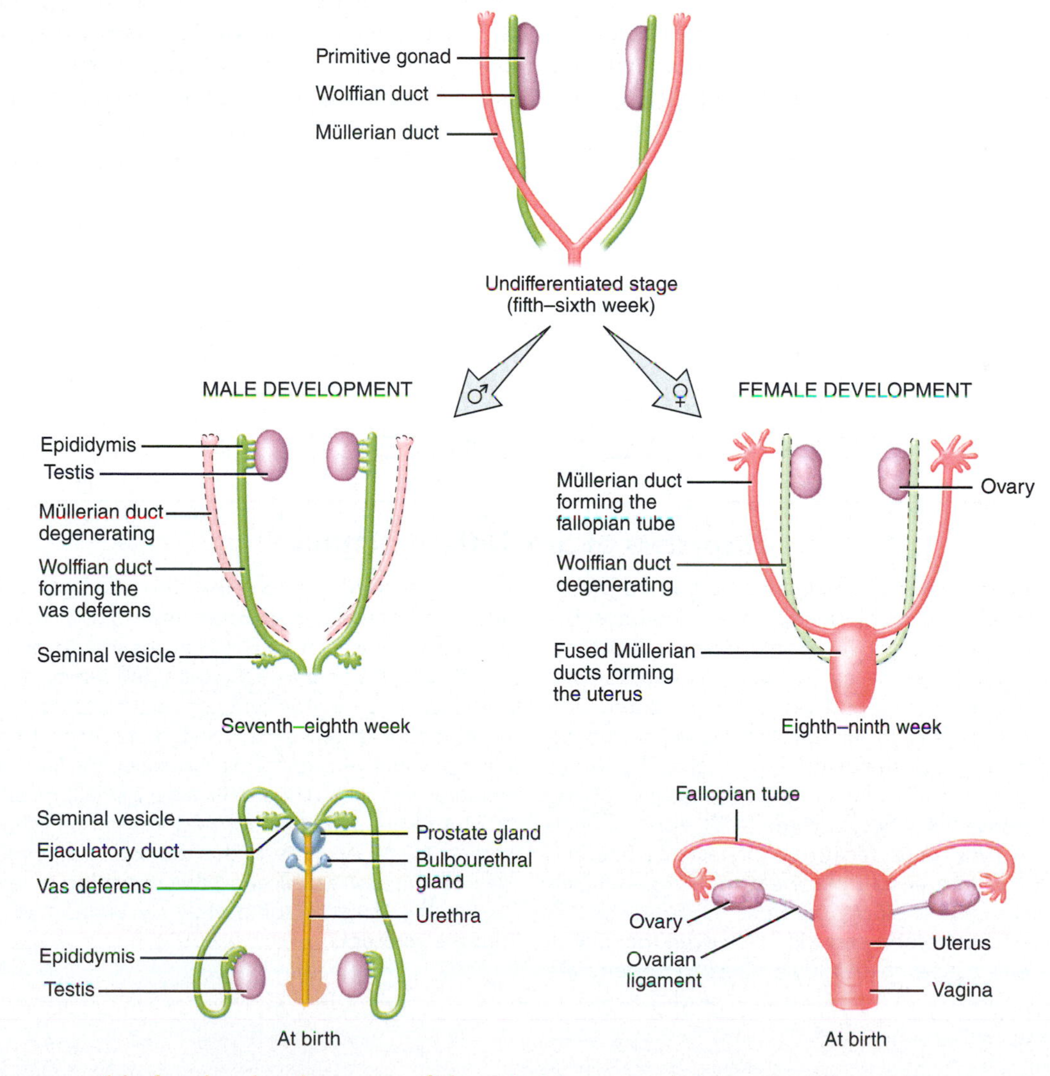

Which gene is responsible for the development of the primitive gonads into testes?

The primitive **gonads** have the ability to develop into testes or ovaries; the **Wolffian ducts**, also known as *mesonephric ducts*, have the ability to develop into the male reproductive tract and certain glands; and the **Müllerian ducts**, also referred to as *paramesonephric ducts*, have the ability to develop into the female reproductive tract. Thus, an early embryo has the potential to follow either the male or the female pattern of development because it contains both sets of ducts and primitive gonads that can differentiate into either testes or ovaries.

If an embryo is genetically male, it develops testes. The male pattern of development is initiated by a "master switch" gene on the Y chromosome named ***SRY***, which stands for *S*ex-determining *R*egion of the *Y* chromosome. When the *SRY* gene is expressed during development, its protein product causes the primitive gonads to differentiate into testes around the seventh week of development. Then around the eighth week of development, the primitive Leydig cells in the embryonic testes begin to secrete the androgen **testosterone**. Testosterone stimulates development of the Wolffian duct on each side into the epididymis, vas deferens, ejaculatory duct, and seminal vesicle (FIGURE 23.14). In addition, Sertoli cells in the embryonic testes secrete a hormone called **Müllerian-inhibiting substance (MIS)**, which causes degeneration of the Müllerian ducts, thereby preventing female internal structures from forming in the male embryo.

If an embryo is genetically female, it develops ovaries because a female embryo lacks a Y chromosome and therefore does not have the SRY gene. The lack of the SRY gene causes the primitive gonads to begin differentiating into ovaries around the ninth week of development. Because the female embryo does not have testes, neither MIS nor testosterone is secreted. The lack of MIS allows the Müllerian ducts to develop into the fallopian tubes, uterus, and part of the vagina (FIGURE 23.14). The lack of testosterone causes the Wolffian ducts to degenerate, preventing male internal reproductive structures from forming in the female embryo.

Differentiation of External Genitalia

The external genitalia of both male and female embryos remain undifferentiated until about the eighth week of development. Before differentiation, all embryos have a swelling on the outer surface that consists of a **genital tubercle**, **urethral folds**, **urethral groove**, and **labioscrotal swellings** (FIGURE 23.15). These structures have the potential to develop into male or female external genitalia.

The presence or absence of the hormone **dihydrotestosterone (DHT)** determines which set of external genitalia the embryo will develop. If DHT is present, male external genitalia develop (FIGURE 23.15). In a male embryo, some of the testosterone released from the testes is converted into DHT by the enzyme **5 α-reductase**. The DHT produced causes the genital tubercle to develop into the glans penis, the urethral folds and urethral groove to develop into the shaft of the penis, and the labioscrotal swellings to develop into a scrotum.

In the absence of DHT, female external genitalia develop (FIGURE 23.15). In a female embryo, DHT is not present because testosterone is absent. The lack of DHT causes the genital tubercle to develop into a clitoris, the urethral folds to develop into the labia minora, the urethral groove to become the vestibule, and the labioscrotal swellings to develop into labia majora.

CLINICAL CONNECTION

Disorders of Sex Differentiation

When disorders of sex differentiation occur, an embryo of a particular genetic sex may develop some of the reproductive organs that are characteristic of the opposite sex. In **adrenogenital syndrome**, the adrenal glands of a genetic female secrete excessive amounts of adrenal androgen during development. Normal female internal organs (ovaries and reproductive tract) still develop. However, the excess androgen levels masculinize the external genitalia: The clitoris is enlarged, resembling a penis, and the labia are fused, resembling a scrotum. In **androgen-insensitivity syndrome (AIS)**, also known as **testicular feminization syndrome**, receptors for testosterone and dihydrotestosterone (DHT) are absent or defective in a genetically male embryo. The testes still develop (but remain undescended at birth) and they secrete testosterone and MIS. However, the lack of sensitivity to testosterone causes the Wolffian ducts to degenerate and, because MIS is secreted, the Mullerian ducts regress as well. Thus, an internal reproductive tract does not develop. In addition, the lack of sensitivity to DHT causes the development of female external genitalia. In **5 α-reductase deficiency**, a genetically male embryo lacks the enzyme that converts testosterone to DHT. The testes develop (but remain undescended at birth) and a normal male internal reproductive tract forms. However, the lack of DHT feminizes the external genitalia: The penis is very small, resembling a clitoris; the scrotum is incompletely formed, resembling the labia; and a partial vagina exists. At puberty the testosterone level rises and the DHT deficiency resolves, allowing this condition to reverse: The penis enlarges, the scrotum completely develops, and the testes descend into the scrotum.

FIGURE 23.15 Development of the external genitalia.

The presence or absence of the hormone dihydrotestosterone (DHT) determines whether male or female external genitalia develop.

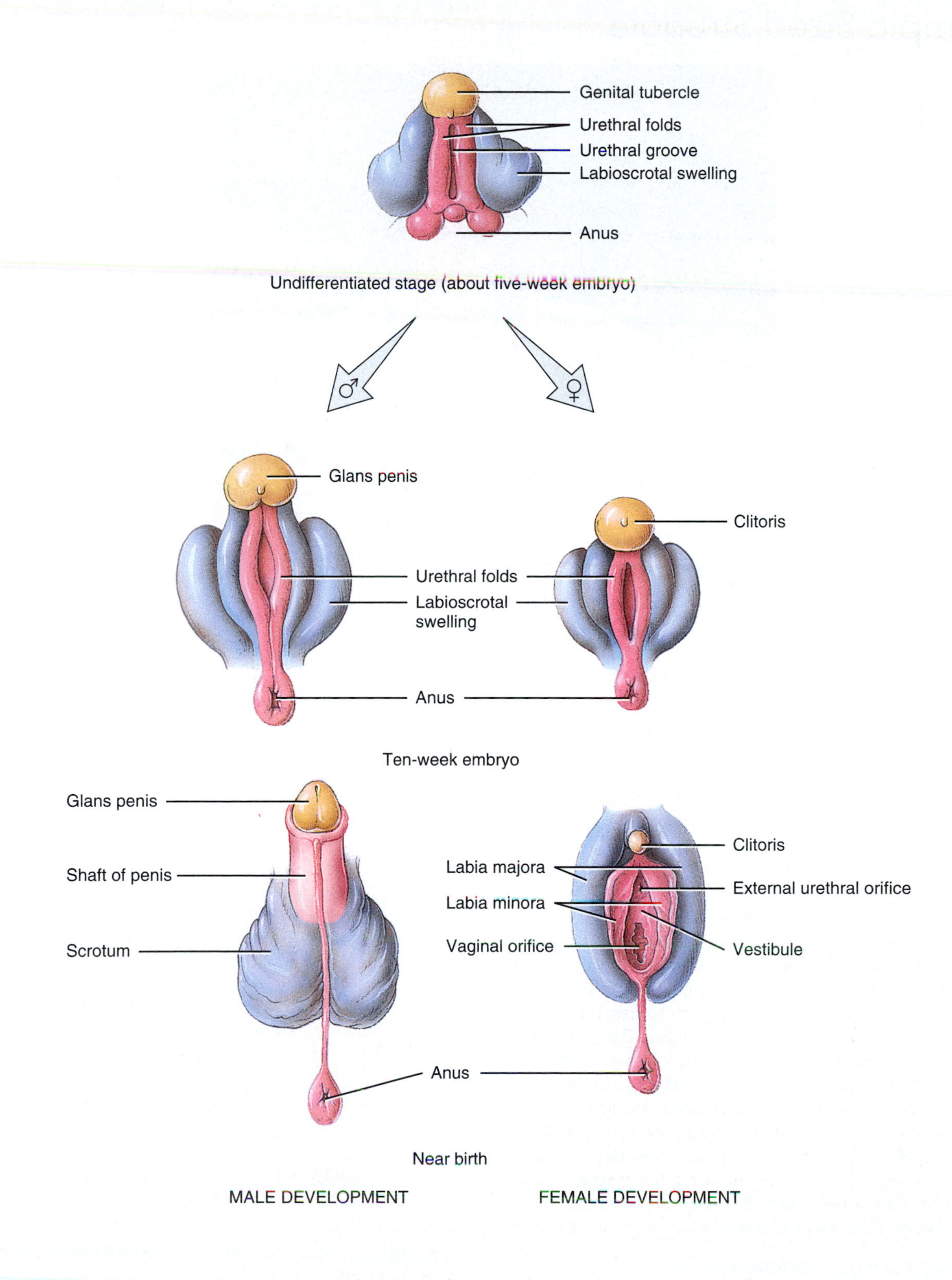

What enzyme converts testosterone into dihydrotestosterone?

Critical Thinking

An Olympic-Sized Struggle

María José Martínez Patiño

María José Martínez Patiño was a hurdler for the Spanish national team. Because of suspected cheating in women's sports, all female athletes had to be "confirmed" female at high-level competitions. In the late 1960s, this confirmation meant that the women had to be visually observed by a panel of gynecologists. If they had the correct anatomy, by visual inspection, they were allowed to compete. Obviously, this was demeaning and undignified, so an alternate method of determining sex was devised. This was a relatively simple yet somewhat unreliable genetic test. María had "passed" the test and had received her "certificate of femininity" in 1983. However, she forgot her certificate when she went to compete at the 1985 World University Games in Kobe, Japan. Prior to this competition, skin cells from the lining of the mouth were collected. This buccal smear was Giemsa-stained so the technician could visualize the chromosomes. In each somatic cell, genetic females have two X chromosomes, but one undergoes inactivation and becomes what is called a Barr body—the "definitive" test of femininity. María was told by the team doctor that there was a problem with her test results and that she could not compete in the race. A devastated María returned to Spain and underwent more testing, including the generation of a karyotype. It was discovered that María's cells had no Barr bodies, but they did display XY chromosomes, making her a genetic male. Further testing showed that María had no uterus but did have undescended testes, which were making androgens such as testosterone and androstenedione.

SOME THINGS TO KEEP IN MIND:

It was concluded that María had androgen-insensitivity syndrome, a condition caused by the absence or deficiency of receptors for testosterone and dihydrotestosterone in a genetically male embryo. Undescended testes are present at birth, no internal reproductive tract forms, and the external genitalia look female. When María was born, she looked like a normal baby girl and continued to develop as a female, including breast development; however, she never had menstrual periods.

Because sports officials thought that the presence of testosterone gave María an advantage over other female athletes, she was banned from competing in any games. María fought to have her right to compete in sports restored and eventually it was, but by that time her momentum was lost and she retired from running. Now she is a college professor, with a PhD in sports sciences. She writes about her experience in an effort to change regulations about the perceived advantages of female athletes with congenital differences.

Why are testosterone and dihydrotestosterone considered androgens?

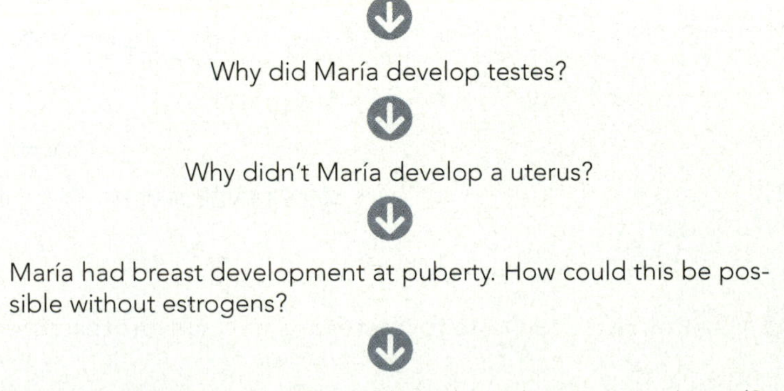

CHECKPOINT

18. How does the type of sperm cell (X-bearing or Y-bearing) determine the genetic sex of the embryo?
19. Describe the role of hormones in the differentiation of the Wolffian ducts, the Müllerian ducts, and the external genitalia.

23.6 Aging and the Reproductive Systems

OBJECTIVE

- Describe the effects of aging on the reproductive systems.

During the first decade of life, the reproductive system is in a juvenile state. At about age 10, hormone-directed changes start to occur in both sexes. **Puberty** is the period when secondary sex characteristics begin to develop and the potential for sexual reproduction is reached. The onset of puberty is marked by pulses or bursts of LH and FSH secretion, each triggered by a pulse of GnRH. Most pulses occur during sleep. As puberty advances, the hormone pulses occur during the day as well as at night. The pulses increase in frequency during a three- to four-year period until the adult pattern is established. The stimuli that cause the GnRH pulses are still unclear, but a role for the hormone **leptin**, which is secreted by adipose tissue, is starting to unfold. Just before puberty, leptin levels rise in proportion to adipose tissue mass. Interestingly, leptin receptors are present in both the hypothalamus and anterior pituitary. Mice that lack a functional leptin gene from birth are sterile and remain in a prepubertal state. Giving leptin to such mice elicits secretion of gonadotropins, and they become fertile. Leptin may signal the hypothalamus that long-term energy stores (triglycerides in adipose tissue) are adequate for reproductive functions to begin.

In females, the reproductive cycle normally occurs once each month from **menarche** (me-NAR-kē), the first menses, to **menopause**, the permanent cessation of menses. Thus, the female reproductive system has a time-limited span of fertility between menarche and menopause. For the first 1 to 2 years after menarche, ovulation occurs in only about 10% of the cycles and the luteal phase is short. Gradually, the percentage of ovulatory cycles increases, and the luteal phase reaches its normal duration of 14 days. With age, fertility declines. Between the ages of 40 and 50, the pool of remaining ovarian follicles becomes exhausted. As a result, the ovaries become less responsive to hormonal stimulation. The production of estrogens declines, despite copious secretion of FSH and LH by the anterior pituitary. Many women experience hot flashes and heavy sweating, which coincide with bursts of GnRH release. Other symptoms of menopause are headache, hair loss, muscular pains, vaginal dryness, insomnia, depression, weight gain, and mood swings. Some atrophy of the ovaries, fallopian tubes, uterus, vagina, external genitalia, and breasts occurs in postmenopausal women. Due to loss of estrogens, most women experience a decline in bone mineral density after menopause. Sexual desire (libido) does not show a parallel decline; it may be maintained by adrenal sex steroids.

In males, declining reproductive function is much more subtle than in females. Healthy men often retain reproductive capacity into their eighties or nineties. At about age 55 a decline in testosterone synthesis leads to reduced muscle strength, fewer viable sperm, and decreased sexual desire. Although sperm production decreases 50–70% between ages 60 and 80, abundant sperm may still be present even in old age.

CHECKPOINT

20. What changes occur in males and females at puberty?
21. What is the difference between menarche and menopause?

23.7 Pregnancy and Labor

OBJECTIVES

- Describe the major events that occur during the embryonic and fetal periods of development.
- Discuss the three stages of labor.
- Explain the hormonal control of lactation.

Once sperm and an egg have developed through meiosis and maturation, and the sperm have been deposited in the vagina, pregnancy can occur. **Pregnancy** refers to the period of time when a fertilized egg develops within a female. It begins at the moment of fertilization and normally ends with birth about 38 weeks later, or 40 weeks after the last menstrual period. From fertilization through the eighth week (second month) of development, a stage called the **embryonic period**, the developing human is called an **embryo**. The **fetal period** begins at week 9 (third month) and continues until birth. During this time, the developing human is called a **fetus**.

The Embryonic Period Involves Several Important Events

The embryonic period is characterized by several significant events, including fertilization, cleavage of the zygote, blastocyst formation, implantation, gastrulation, and organogenesis.

Fertilization

During **fertilization**, also referred to as **conception**, the genetic materials from a haploid sperm cell and a haploid egg (secondary oocyte) merge into a single diploid nucleus. Of the 300 million sperm introduced into the vagina, fewer than 2 million (less than 1%) reach the cervix of the uterus and only about 200 reach the secondary oocyte. Fertilization normally occurs in the fallopian tube within 12 to 24 hours after ovulation. Sperm can survive up to five days in the female reproductive tract, although a secondary oocyte is viable for only about 24 hours after ovulation. Thus, pregnancy is *most likely* to occur if intercourse takes place during a 5-day window—from 4 days before ovulation to 1 day after ovulation.

Sperm swim from the vagina into the cervical canal by the whiplike movements of their tails (flagella). The passage of sperm through the rest of the uterus and then into the fallopian tube results mainly from contractions of the walls of these organs. Prostaglandins in semen are believed to stimulate uterine motility at the time of intercourse and to aid in the movement of sperm through the uterus and into the fallopian tube. The ability of sperm to fertilize an egg is enhanced by **capacitation**, a series of functional changes that occur when sperm are acted upon by the secretions of the female reproductive tract, mostly in the fallopian tube. Capacitation causes the sperm's tail to beat even more vigorously and prepare its plasma membrane to fuse with the oocyte's plasma membrane.

For fertilization to occur, a sperm cell first must penetrate two layers: the **corona radiata**, the granulosa cells that surround the secondary oocyte, and the **zona pellucida**, the clear glycoprotein layer between the corona radiata and the oocyte's plasma membrane (FIGURE 23.16a). The acrosome, a helmetlike structure that covers the head of a sperm, contains several enzymes. Acrosomal enzymes and strong tail movements by the sperm help it penetrate the cells of the corona radiata and come in contact with the zona pellucida. One of the glycoproteins in the zona pellucida called ZP3 acts as a sperm receptor. Its binding to specific membrane proteins in the sperm head triggers the **acrosomal reaction**, the release of the contents of the acrosome. The acrosomal enzymes digest a path through the zona pellucida as the lashing sperm tail pushes the sperm cell onward. Although many sperm bind to ZP3 molecules and undergo acrosomal reactions, only the first sperm cell to penetrate the entire zona pellucida and reach the oocyte's plasma membrane fuses with the oocyte.

FIGURE 23.16 Selected structures and events in fertilization.

During fertilization, genetic material from a sperm cell and a secondary oocyte merge to form a single diploid nucleus.

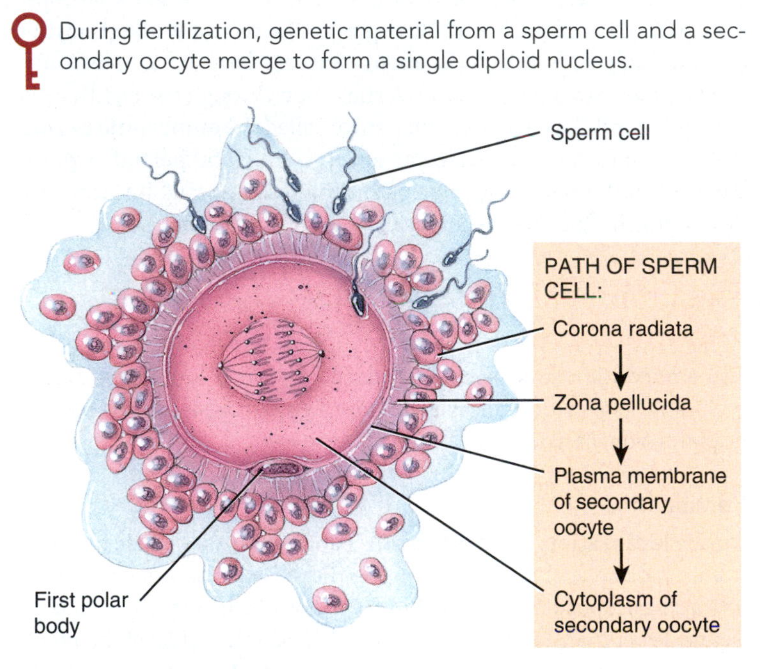

(a) Sperm cell penetrating secondary oocyte

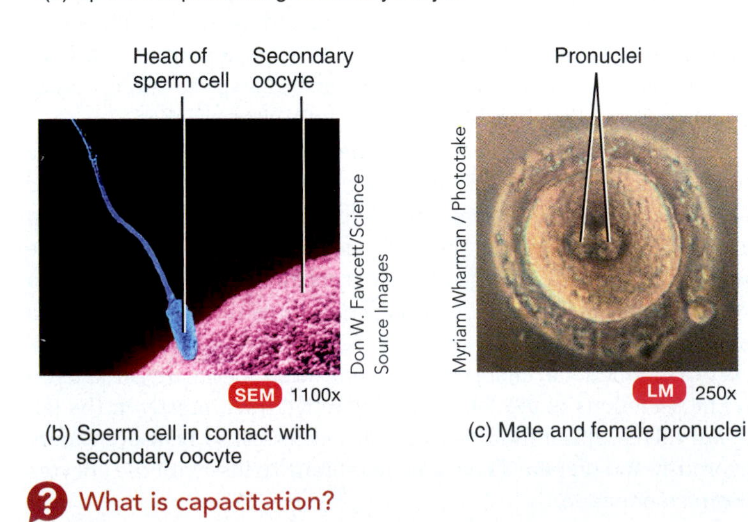

(b) Sperm cell in contact with secondary oocyte

(c) Male and female pronuclei

What is capacitation?

The fusion of a sperm cell with a secondary oocyte sets in motion events that block **polyspermy**, fertilization by more than one sperm cell. Within a few seconds, the cell membrane of the oocyte depolarizes, which acts as a *fast block to polyspermy*—a depolarized oocyte cannot fuse with another sperm. Depolarization also triggers the intracellular release of calcium ions, which stimulate exocytosis of secretory vesicles from the oocyte. The molecules released by exocytosis inactivate ZP3 and harden the entire zona pellucida, events called the *slow block to polyspermy*.

Once a sperm cell enters a secondary oocyte, the oocyte first must complete meiosis II. It divides into a larger ovum (mature egg) and a smaller second polar body that fragments and disintegrates (see FIGURE 23.10). The nucleus in the head of the sperm develops into the **male pronucleus**, and the nucleus of the fertilized ovum develops into the **female pronucleus** (FIGURE 23.16c). After the male and female pronuclei form, they fuse, producing a single diploid nucleus that contains 23 chromosomes from each pronucleus. Thus, the fusion of the haploid (n) pronuclei restores the diploid number ($2n$) of 46 chromosomes. The fertilized ovum is now called a **zygote**.

Dizygotic (fraternal) twins are produced from the independent release of two secondary oocytes and the subsequent fertilization of each by different sperm. They are the same age and in the uterus at the same time, but genetically they are as dissimilar as any other siblings. Dizygotic twins may or may not be the same sex. Because **monozygotic (identical) twins** develop from a single fertilized ovum, they contain exactly the same genetic material and are always the same sex. Monozygotic twins arise from separation of the developing cells into two embryos, which in 99% of the cases occurs before 8 days have passed. Separations that occur later than 8 days are likely to produce **conjoined twins**, a situation in which the twins are joined together and share some body structures.

Cleavage of the Zygote

After fertilization, rapid mitotic cell divisions of the zygote called **cleavage** take place (FIGURE 23.17). The first division of the zygote begins about 24 hours after fertilization and is completed about 6 hours later (FIGURE 23.17a). Each succeeding division takes slightly less time. By the second day after fertilization, the second cleavage is completed and there are four cells (FIGURE 23.17b). By the end of the third day, there are 16 cells. The progressively smaller cells produced by cleavage are called **blastomeres**. Successive cleavages eventually produce a solid ball of cells called the **morula**. The morula is still surrounded by the zona pellucida and is about the same size as the original zygote (FIGURE 23.17c).

Blastocyst Formation

By the end of the fourth day, the number of cells in the morula increases as it continues to move through the fallopian tube toward the uterine cavity. When the morula enters the uterine cavity on day 4 or 5, a glycogen-rich secretion from the glands of the endometrium of the uterus penetrates the morula; collects between the blastomeres; and reorganizes them around a large fluid-filled cavity called the **blastocyst cavity**, also called the *blastocoel* (FIGURE 23.17e). Once the blastocyst cavity is formed, the developing hollow ball of cells is called the **blastocyst**. Though it now has hundreds of cells, the blastocyst is still about the same size as the original zygote. During the formation of the blastocyst, two distinct cell populations arise: the trophoblast and inner cell mass

FIGURE 23.17 Cleavage and the formation of the morula and blastocyst.

Cleavage refers to the early, rapid mitotic divisions of a zygote.

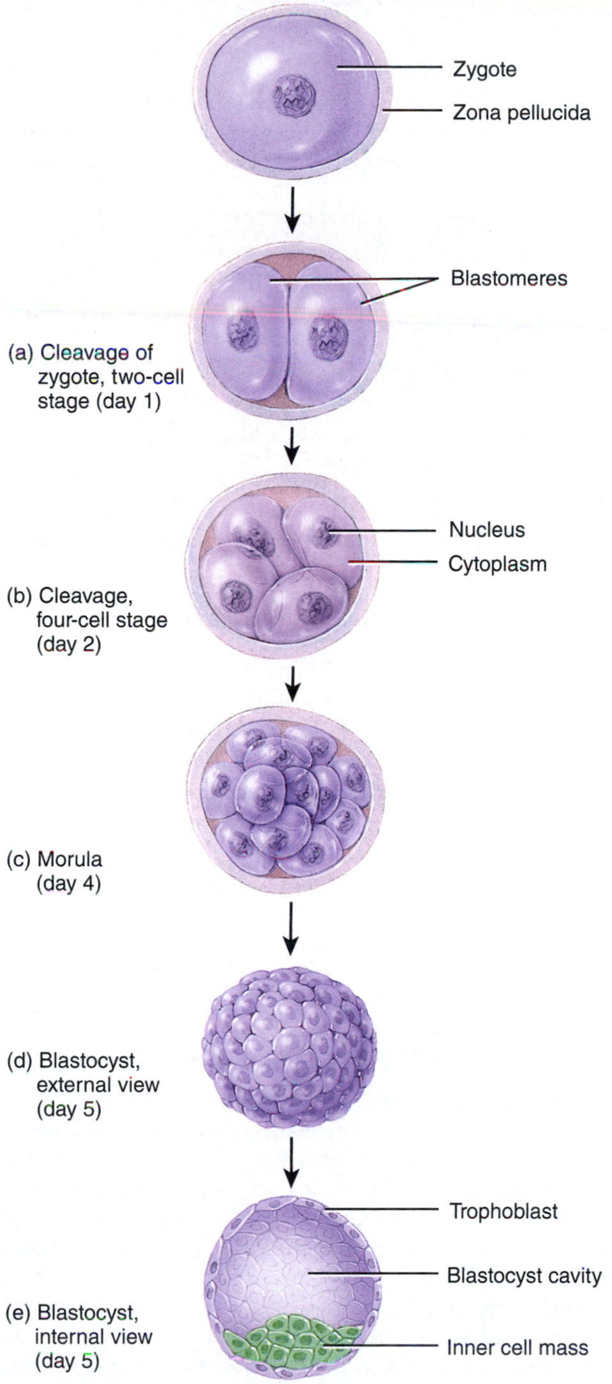

Zygote
Zona pellucida
Blastomeres
(a) Cleavage of zygote, two-cell stage (day 1)
Nucleus
Cytoplasm
(b) Cleavage, four-cell stage (day 2)
(c) Morula (day 4)
(d) Blastocyst, external view (day 5)
Trophoblast
Blastocyst cavity
Inner cell mass
(e) Blastocyst, internal view (day 5)

What is the difference between a morula and a blastocyst?

(FIGURE 23.17e). The **trophoblast** is the outer superficial layer of cells that forms the spherelike wall of the blastocyst. It will ultimately develop into the chorion, an extraembryonic membrane that forms the fetal portion of the placenta. The **inner cell mass** is located internally and eventually develops into the embryo proper and the other extraembryonic membranes (amnion, allantois, and yolk sac). On about the fifth day after fertilization, the blastocyst "hatches" from the zona pellucida by digesting a hole in it with an enzyme and then squeezing through the hole. This shedding of the zona pellucida is necessary in order to permit the next step, implantation (attachment) into the vascular, glandular endometrial lining of the uterus.

Implantation

The blastocyst remains free within the uterine cavity for about 2 days before it attaches to the uterine wall. At this time the endometrium is in its secretory phase. About 7 days after fertilization, the blastocyst attaches to the endometrium in a process called **implantation**. Initially,

CLINICAL CONNECTION

Stem-Cell Research

Stem cells are unspecialized cells that have the ability to divide for indefinite periods and give rise to specialized cells. In the context of human development, a zygote (fertilized egg) is a stem cell. Because it has the potential to form an entire organism, a zygote is known as a ***totipotent stem cell*** (tō-TIP-ō-tent; ***totus-*** = whole; ***-potentia*** = power). As development continues, embryonic cells become ***pluripotent stem cells*** (ploo-RIP-ō-tent; ***pluri-*** = several), each of which can give rise to many (but not all) different types of cells. Later, pluripotent stem cells can undergo further specialization into ***multipotent stem cells*** (mul-TIP-ō-tent), stem cells that have the ability to develop into a few different types of cells.

Because pluripotent stem cells can give rise to many cell types in the body, they are extremely important in research and health care. For example, they might be used to generate cells and tissues for transplantation to treat conditions such as cancer, Parkinson's disease, Alzheimer's disease, and spinal cord injury. Pluripotent stem cells currently used in research are derived from (1) extra embryos that were destined to be used for infertility treatments but were not needed and (2) embryos terminated during the first trimester of pregnancy. Pluripotent stem cells that are obtained from an embryo are referred to as **embryonic stem cells**. Embryonic stem-cell research is quite controversial because it raises ethical questions about the use of an embryo as a stem-cell source.

Because of this controversy, scientists are investigating the clinical applications of **adult stem cells**—stem cells that remain in the body throughout adulthood. Adult stem cells have been found in bone marrow, adipose tissue, muscle, the brain, and other parts of the body. These stem cells can potentially be harvested from a patient and then used to repair other tissues and organs in that patient's body without having to use stem cells from embryos. Unlike embryonic stem cells, most adult stem cells are multipotent, producing only a few types of cells. However, scientists are discovering ways to coax adult stem cells to develop into a wider variety of cell types. Currently, adult stem-cell therapy has been used to treat certain cancers, blood disorders, and immune system disorders. Clinical trials are underway to determine what other disorders adult stem-cell therapy might be able to treat.

FIGURE 23.18 Implantation.

Implantation, the attachment of a blastocyst to the endometrium, occurs about 7 days after fertilization.

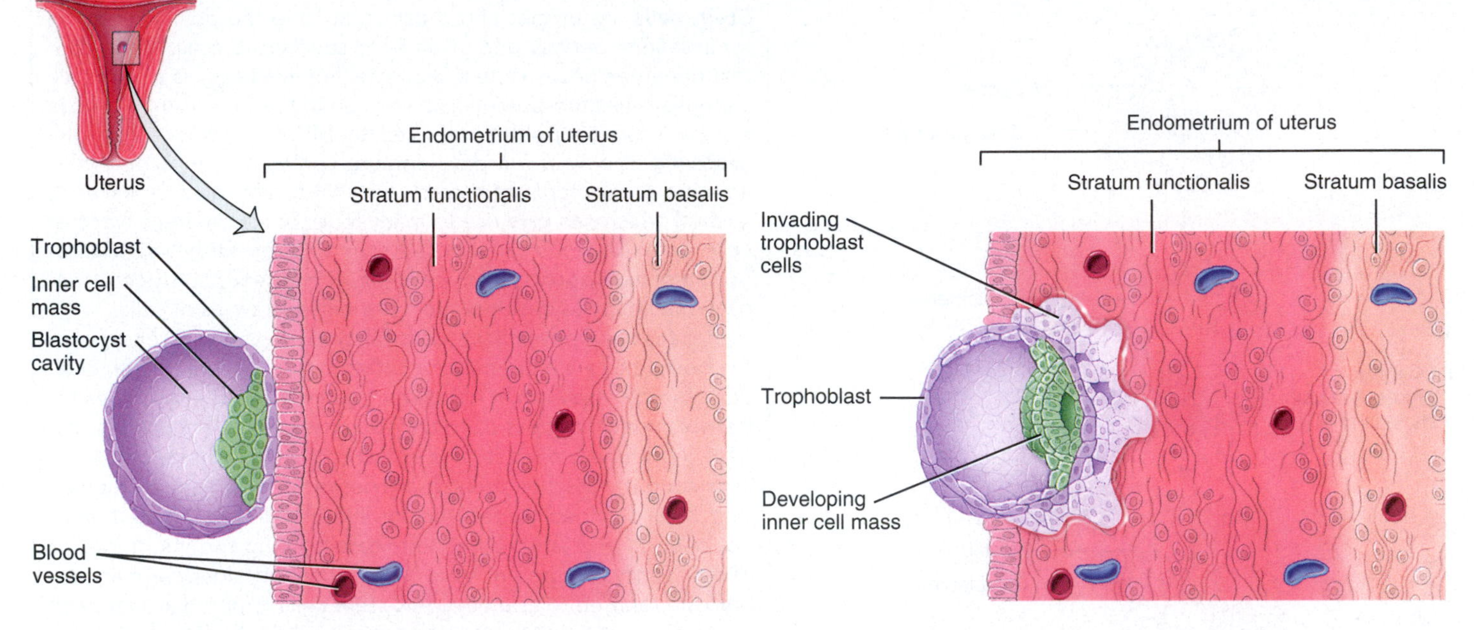

(a) Blastocyst loosely attached to the endometrium

(b) Blastocyst burrowing into the endometrium

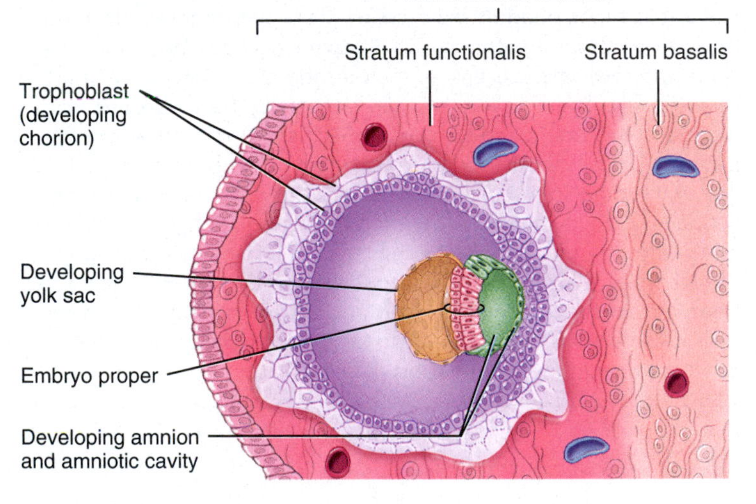

(c) Blastocyst completely embedded within the endometrium

? How does a blastocyst merge with and burrow into the endometrium?

the blastocyst only loosely attaches to the endometrium (FIGURE 23.18a). As the blastocyst implants, it orients with the inner cell mass toward the stratum functionalis of the endometrium. Soon thereafter, the blastocyst secretes enzymes, burrows into the stratum functionalis, and eventually becomes surrounded by it (FIGURE 23.18b, c).

The major events associated with the first week of development are summarized in FIGURE 23.19.

Gastrulation

The implanted embryo continues to develop in the uterine wall. By the third week after fertilization, some of the cells of the embryo have rearranged themselves into three primary germ layers: the ectoderm, mesoderm, and endoderm (FIGURE 23.20). The process by which the three germ layers form is known as **gastrulation**. The primary germ layers are the major embryonic tissues from which the various tissues and organs of the body develop. As the embryo develops, the **endoderm** ultimately becomes the epithelial lining of the gastrointestinal tract, respiratory tract, and several other organs. The **mesoderm** gives

CLINICAL CONNECTION

Ectopic Pregnancy

Ectopic pregnancy is the development of an embryo or fetus outside the uterine cavity. An ectopic pregnancy usually occurs when movement of the fertilized egg through the fallopian tube is impaired by scarring due to a prior tubal infection, decreased movement of the fallopian tube smooth muscle, or abnormal tubal anatomy. Although the most common site of ectopic pregnancy is the fallopian tube, ectopic pregnancies may also occur in the ovary, abdominal cavity, or uterine cervix. Women who smoke are twice as likely to have an ectopic pregnancy because nicotine in cigarette smoke paralyzes the cilia in the lining of the fallopian tube (as it does in the respiratory airways). Scars from pelvic inflammatory disease, previous fallopian tube surgery, and previous ectopic pregnancy may also hinder movement of the fertilized egg. The signs and symptoms of ectopic pregnancy include one or two missed menstrual cycles followed by bleeding and acute abdominal and pelvic pain. Unless removed, the developing embryo can rupture the fallopian tube, often resulting in death of the mother. Treatment options include surgery or the use of a cancer drug called methotrexate, which causes embryonic cells to stop dividing and eventually disappear.

FIGURE 23.19 Summary of events associated with the first week of development.

During the first week of development, fertilization, cleavage, blastocyst formation, and implantation occur.

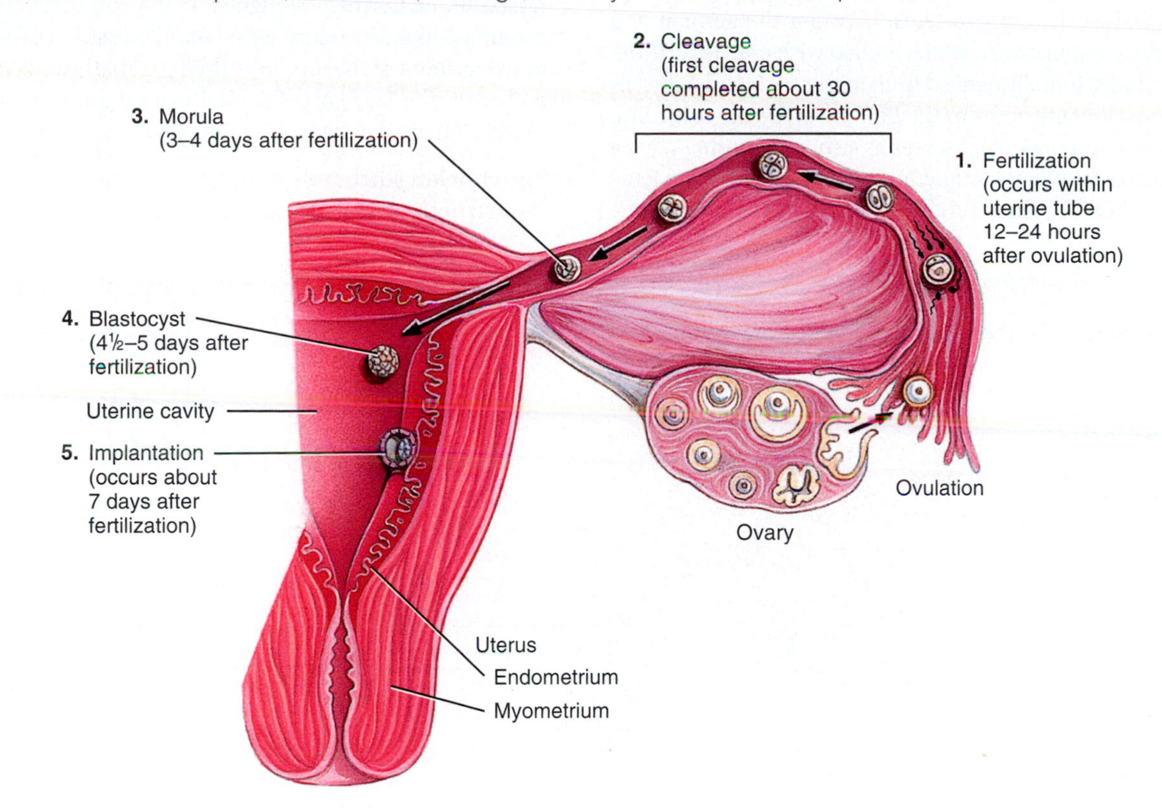

In which phase of the uterine cycle does implantation occur?

FIGURE 23.20 Gastrulation.

During gastrulation, cells of the embryo form three primary germ layers: the ectoderm, mesoderm, and endoderm.

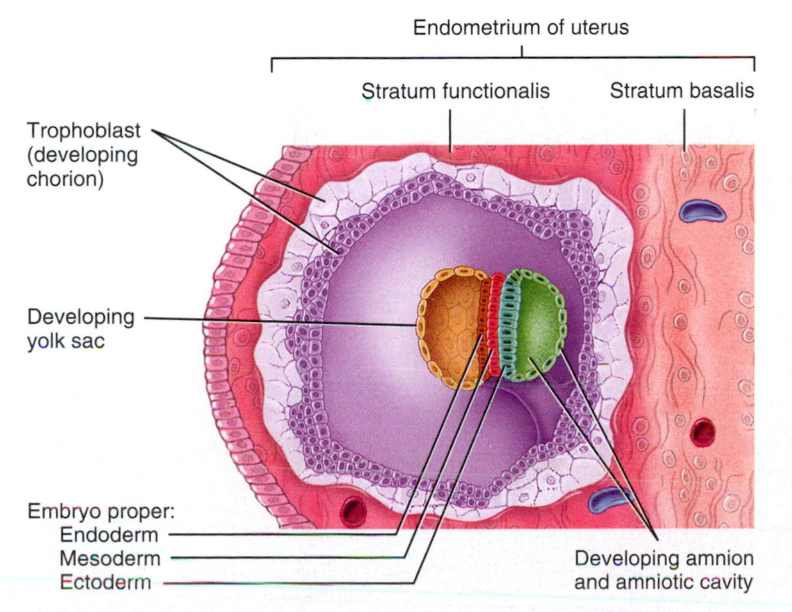

What body structures develop from the ectoderm?

rise to muscles, bones, and other connective tissues. The **ectoderm** develops into the epidermis of the skin and the nervous system.

Organogenesis

The fourth through eighth weeks of development are very significant in embryonic development because all major organs appear during this time. The term **organogenesis** refers to the formation of body organs and systems. By the end of the eighth week, all the major body systems have begun to develop, although their functions for the most part are minimal.

The Fetal Period Is a Time of Growth and Differentiation

During the fetal period, tissues and organs that developed during the embryonic period grow and differentiate. Very few new structures appear during the fetal period, but the rate of body growth is remarkable, especially during the second half of intrauterine life. For example, during the last two and one-half months of intrauterine life, half of the full-term weight is added. At the beginning of the fetal period, the head is half the length of the body. By the end of the fetal period, the head size is only one-quarter the length of the body. During the same period, the limbs also increase in size from one-eighth to one-half the fetal length. The fetus is also less vulnerable to the damaging effects of drugs, radiation, and microbes than it was as an embryo.

The Extraembryonic Membranes Have Ancillary Developmental Roles

There are four **extraembryonic membranes** associated with the embryo and later the fetus during pregnancy: the amnion, chorion, yolk sac, and allantois (FIGURE 23.21). These membranes are present by the third week after fertilization.

1. The **amnion** encloses the embryo/fetus. Between the amnion and the embryo is the *amniotic cavity*, which is filled with **amniotic fluid**. Most amniotic fluid is initially derived from maternal blood. Later, as the fetus develops, it contributes to the fluid by excreting urine into the amniotic cavity. Amniotic fluid serves as a shock absorber for the fetus, helps regulate fetal body temperature, helps prevent the fetus from drying out, and prevents adhesions between the skin of the fetus and surrounding tissues. The amnion usually ruptures just before birth; it and its fluid constitute the "bag of waters." Embryonic or fetal cells are normally sloughed off into amniotic fluid. They can be examined in a procedure called *amniocentesis*, which involves withdrawing some of the amniotic fluid that bathes the developing fetus and analyzing the fetal cells and dissolved substances (see Tool of the Trade: Prenatal Diagnostic Testing presented later in this section).
2. The **chorion** surrounds the amnion. Part of the chorion develops into structures called *chorionic villi*, which function as the fetal

FIGURE 23.21 Relationship of the extraembryonic membranes to the decidua and the placenta.

The placenta is formed by the decidua basalis of the mother and the chorionic villi of the fetus.

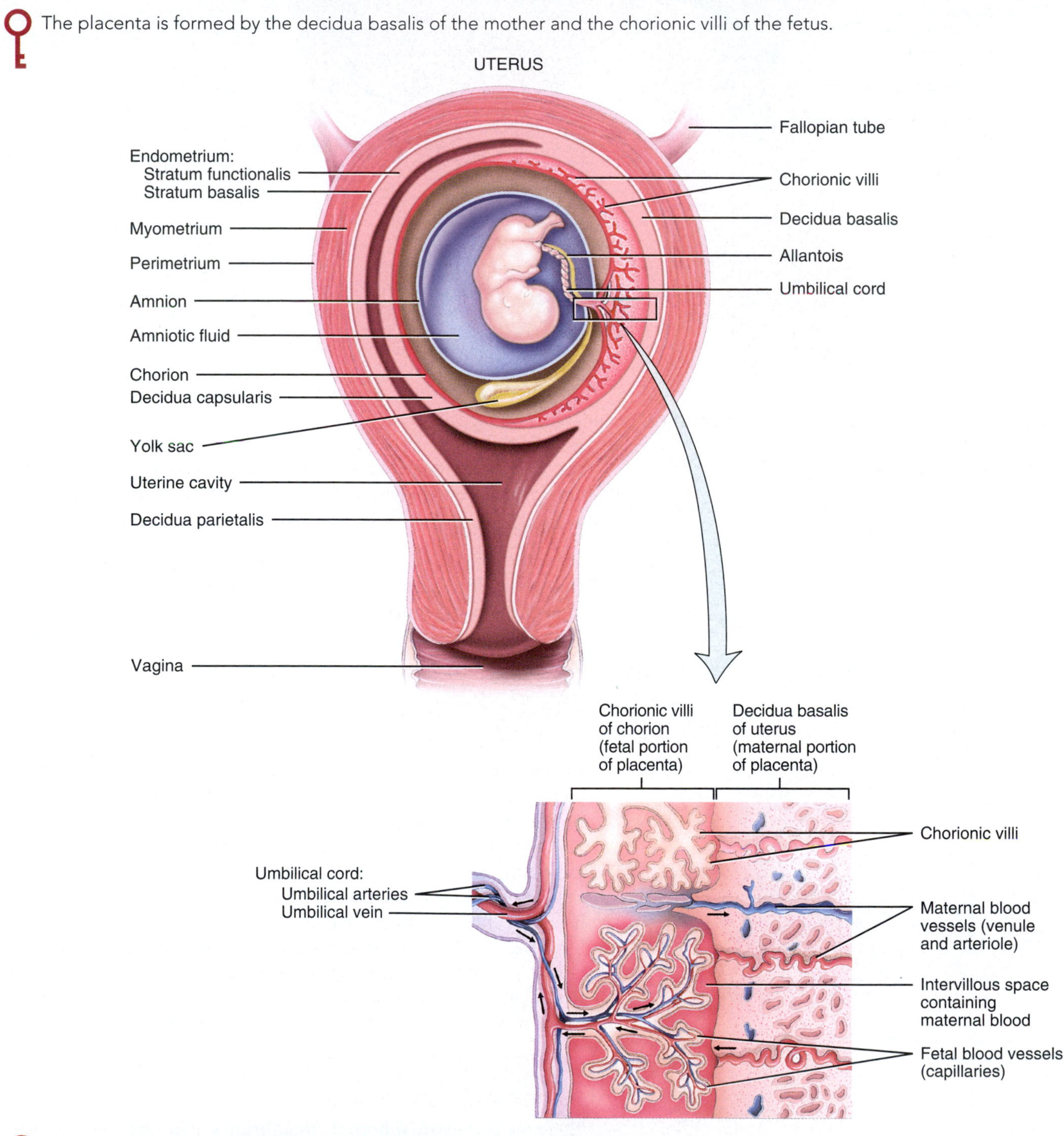

What is the decidua basalis?

CLINICAL CONNECTION

Early Pregnancy Tests

Early pregnancy tests detect the tiny amounts of human chorionic gonadotropin (hCG) in the urine that begin to be excreted about 8 days after fertilization. The test kits can detect pregnancy as early as the first day of a missed menstrual period—that is, at about 14 days after fertilization. Chemicals in the kits produce a color change if a reaction occurs between hCG in the urine and hCG antibodies included in the kit.

part of the placenta (described shortly). In addition, the chorion produces the hormone **human chorionic gonadotropin (hCG)**. Recall that hCG rescues the corpus luteum from degeneration and sustains its secretion of estrogens and progesterone—an activity required to prevent menstruation and for the continued attachment of the embryo or fetus to the lining of the uterus. Secretion of hCG occurs as early as the eighth day after fertilization. By the third month of pregnancy, the placenta is established and produces the estrogens and progesterone that continue to sustain the pregnancy. Because the corpus luteum is no longer essential at this point, hCG levels significantly decrease and the corpus luteum degrades.

3. Because human embryos receive their nutrients from the endometrium, the **yolk sac** is relatively empty and small, and it decreases in size as development progresses. Nevertheless, the yolk sac has several important functions in humans: It supplies nutrients to the embryo during the second and third weeks of development; is the source of blood cells from the third through sixth weeks; forms part of the gut (gastrointestinal tract); and contains the first cells (primordial germ cells) that will eventually migrate into the developing gonads, differentiate into the primitive germ cells, and form gametes.
4. The **allantois** is a small outpouching that forms from the wall of the yolk sac. In most other mammals, the allantois is used for gas exchange and waste removal. Because of the role of the human placenta in these activities, the allantois is not a prominent structure in humans. Nevertheless, it does function in the early formation of blood and blood vessels, and it is associated with the development of the urinary bladder.

The Decidua Refers to the Stratum Functionalis of the Uterus in a Pregnant Woman

During pregnancy, the stratum functionalis of the endometrium is known as the **decidua** (dē-SID-ū-a = falling off). The decidua separates from the rest of the uterus after the fetus is delivered, much as it does in normal menstruation. Different regions of the decidua are named based on their positions relative to the site of the implanted embryo (**FIGURE 23.21**). The **decidua basalis** is the portion of the stratum functionalis between the embryo and the stratum basalis of the endometrium; it provides large amounts of glycogen and lipids for the developing embryo and fetus, and later becomes the maternal part of the placenta. The **decidua capsularis** is the portion of the stratum functionalis located between the embryo and the uterine cavity. The **decidua parietalis** is the remaining part of the stratum functionalis that lines the noninvolved areas of the rest of the uterus. As the embryo and later the fetus enlarges, the decidua capsularis bulges into the uterine cavity and fuses with the decidua parietalis, thereby obliterating the uterine cavity. By about 27 weeks, the decidua capsularis degenerates and disappears.

The Placenta Is the Site of Nutrient and Waste Exchange Between the Mother and Fetus and Has Other Important Functional Roles

The **placenta** is a pancake-shaped structure that develops around the third month of pregnancy. It consists of two distinct parts: (1) the maternal portion formed by the **decidua basalis** of the endometrium and (2) the fetal portion formed by the **chorionic villi**, fingerlike projections of the chorion that are rich in capillaries (**FIGURE 23.21**). Surrounding the chorionic villi are **intervillous spaces**, cavities filled with maternal blood. The placenta is unique because it develops from two separate individuals, the mother and the fetus. The actual connection between the placenta and fetus is through the **umbilical cord**. The umbilical cord consists of two *umbilical arteries* that carry deoxygenated fetal blood to the placenta and one *umbilical vein* that carries oxygen and nutrients acquired from maternal blood into the fetus.

The placenta is the site of nutrient and waste exchange between the mother and the fetus. This function of the placenta is carried out without the maternal or fetal blood mixing. Instead, oxygen and nutrients in the maternal blood of the intervillous spaces diffuse across the walls of the chorionic villi capillaries into fetal blood. Waste products such as carbon dioxide diffuse in the opposite direction.

In addition to serving as the site of exchange of nutrients and wastes between the mother and fetus, the placenta stores nutrients such as carbohydrates, proteins, calcium, and iron, which are released into fetal circulation as required. The placenta is also a protective barrier because most microorganisms cannot pass through it. However, certain viruses, such as those that cause AIDS, German measles, chickenpox, measles, encephalitis, and poliomyelitis, can cross the placenta. Many drugs, alcohol, and some substances that can cause birth defects also pass freely. Maternal antibodies can cross the placenta, providing protection against certain harmful microbes.

A final function of the placenta is the production of five hormones that have significant roles in pregnancy: estrogens, progesterone, relaxin, human chorionic somatomammotropin, and corticotropin-releasing hormone. As noted earlier, during the first two months of pregnancy, the corpus luteum in the ovary continues to secrete high levels of **progesterone** and **estrogens**, which maintain the lining of the uterus during pregnancy. From the third month through the remainder of the pregnancy, the placenta itself provides the high levels of progesterone and estrogens required to sustain the pregnancy. After delivery, estrogens and progesterone in the blood decrease to normal levels.

Relaxin, a hormone produced first by the corpus luteum of the ovary and later by the placenta, increases the flexibility of the pubic symphysis and helps dilate the uterine cervix during labor. Both of these actions ease delivery of the baby.

A third hormone produced by the placenta is **human chorionic somatomammotropin (hCS)**, also known as **human placental lactogen (hPL)**. It is thought to help prepare the mammary glands for lactation, enhance maternal growth by increasing protein synthesis, and regulate certain aspects of metabolism in both mother and fetus. For example, hCS decreases the use of glucose by the mother and promotes the release of fatty acids from her adipose tissue, making more glucose available to the fetus.

The hormone most recently found to be produced by the placenta is **corticotropin-releasing hormone (CRH)**, which in nonpregnant people is secreted only by neurosecretory cells in the hypothalamus. CRH is now thought to be part of the "clock" that establishes the timing of birth. Secretion of CRH by the placenta begins at about 12 weeks and increases enormously toward the end of pregnancy. Women who have higher levels of CRH earlier in pregnancy are more likely to deliver prematurely; those who have low levels are more likely to deliver after their due date. CRH from the placenta has a second important effect: It increases secretion of cortisol, which is needed for maturation of the fetal lungs and the production of surfactant (see Section 18.2).

Teratogens Cause Developmental Defects in the Embryo or Fetus

Exposure of a developing embryo or fetus to certain environmental factors can damage the developing organism or even cause death. A **teratogen** (TER-a-tō-jen) is any agent or influence that causes developmental defects in the embryo or fetus. Several examples are discussed in the following sections.

Chemicals and Drugs

Because the placenta is not an absolute barrier between the maternal and fetal circulations, any drug or chemical should be considered potentially dangerous to the fetus when given to the mother. Alcohol is by far the number-one fetal teratogen. Intrauterine exposure to even a small amount of alcohol may result in **fetal alcohol syndrome (FAS)**, one of the most common causes of mental retardation and the most common preventable cause of birth defects in the United States. The symptoms of FAS may include slow growth before and after birth, defective heart and other organs, malformed limbs, genital abnormalities, and central nervous system damage. Behavioral problems, such as hyperactivity, extreme nervousness, reduced ability to concentrate, and an inability to appreciate cause-and-effect relationships, are common.

Other teratogens include certain viruses (hepatitis B and C and certain papilloma viruses that cause sexually transmitted diseases); pesticides; industrial chemicals; some antibiotics; thalidomide, and numerous other prescription drugs; LSD; and cocaine. A pregnant woman who uses cocaine, for example, subjects the fetus to higher risk of retarded growth, attention and orientation problems, hyperirritability, a tendency to stop breathing, malformed or missing organs, strokes, and seizures. The risks of spontaneous abortion, premature birth, and stillbirth also increase with fetal exposure to cocaine.

Cigarette Smoking

Strong evidence implicates cigarette smoking during pregnancy as a cause of low infant birth weight; there is also a strong association between smoking and a higher fetal and infant mortality rate. Women who smoke have a much higher risk of an ectopic pregnancy. Cigarette smoke may be teratogenic and may cause cardiac and brain abnormalities. Maternal smoking has also been linked with sudden infant death syndrome (SIDS). Even a mother's exposure to secondhand cigarette smoke (breathing air containing tobacco smoke) during pregnancy or while nursing predisposes her baby to increased incidence of respiratory problems, including bronchitis and pneumonia, during the first year of life.

Irradiation

Ionizing radiation of various kinds is a potent teratogen. Exposure of pregnant mothers to X-rays or radioactive isotopes during the embryo's susceptible period of development may cause microcephaly (small head size relative to the rest of the body), mental retardation, and skeletal malformations. Caution is advised, especially during the first trimester of pregnancy.

Labor Involves Complex Hormonal Interactions and Occurs in Three Stages

Labor, also referred to as **parturition** or **birth**, is the process by which the fetus is expelled from the uterus through the vagina. The onset of labor is determined by complex interactions of several placental and fetal hormones. The placental hormone progesterone inhibits uterine contractions; therefore, labor cannot take place until its effects are diminished. Toward the end of gestation, the levels of estrogens in the mother's blood rise sharply, producing changes that overcome the inhibiting effects of progesterone. The rise in estrogens results from increasing secretion by the placenta of corticotropin-releasing hormone, which stimulates the anterior pituitary gland of the fetus to secrete adrenocorticotropic hormone (ACTH). In turn, ACTH stimulates the fetal adrenal gland to secrete cortisol and dehydroepiandrosterone (DHEA), the major adrenal androgen. The placenta then converts DHEA into estrogens. High levels of estrogens cause the number of receptors for oxytocin on uterine muscle fibers to increase and cause uterine muscle fibers to form gap junctions with one another. Oxytocin released by the posterior pituitary stimulates uterine contractions, and relaxin from the placenta assists by increasing the flexibility of the pubic symphysis and helping dilate the uterine cervix. Estrogens also stimulate the placenta to release prostaglandins, which induce production of enzymes that digest collagen fibers in the cervix, causing it to soften.

Control of labor contractions during parturition occurs via a positive feedback cycle (see Chapter 1, **FIGURE 1.6**). Contractions of the uterine myometrium force the baby's head or body into the cervix, distending (stretching) the cervix. Stretch receptors in the cervix send action potentials to neurosecretory cells in the hypothalamus, causing them to release oxytocin into blood capillaries of the posterior pituitary gland. Oxytocin then is carried by the blood to the uterus, where it stimulates the myometrium to contract more forcefully. As the contractions intensify, the baby's body stretches the cervix still more, and the resulting action potentials stimulate the secretion of yet more oxytocin. With the birth of the infant, the positive feedback cycle is broken because cervical distension suddenly lessens.

Uterine contractions occur in waves (quite similar to the peristaltic waves of the gastrointestinal tract) that start at the top of the uterus and move downward, eventually expelling the fetus. Labor can be divided into three stages (**FIGURE 23.22**):

1. ***Stage of dilation.*** The time from the onset of labor to the complete dilation of the cervix is the **stage of dilation**. This stage, which typically lasts 6–12 hours, features regular contractions of the uterus, usually a rupturing of the amniotic sac, and complete dilation (to 10 cm) of the cervix. If the amniotic sac does not rupture spontaneously, it is ruptured intentionally.
2. ***Stage of expulsion.*** The time (10 minutes to several hours) from complete cervical dilation to delivery of the baby is the **stage of expulsion**. Once the baby has been born, the umbilical cord is tied

FIGURE 23.22 Stages of labor.

Labor is the process by which the fetus is expelled from the uterus through the vagina.

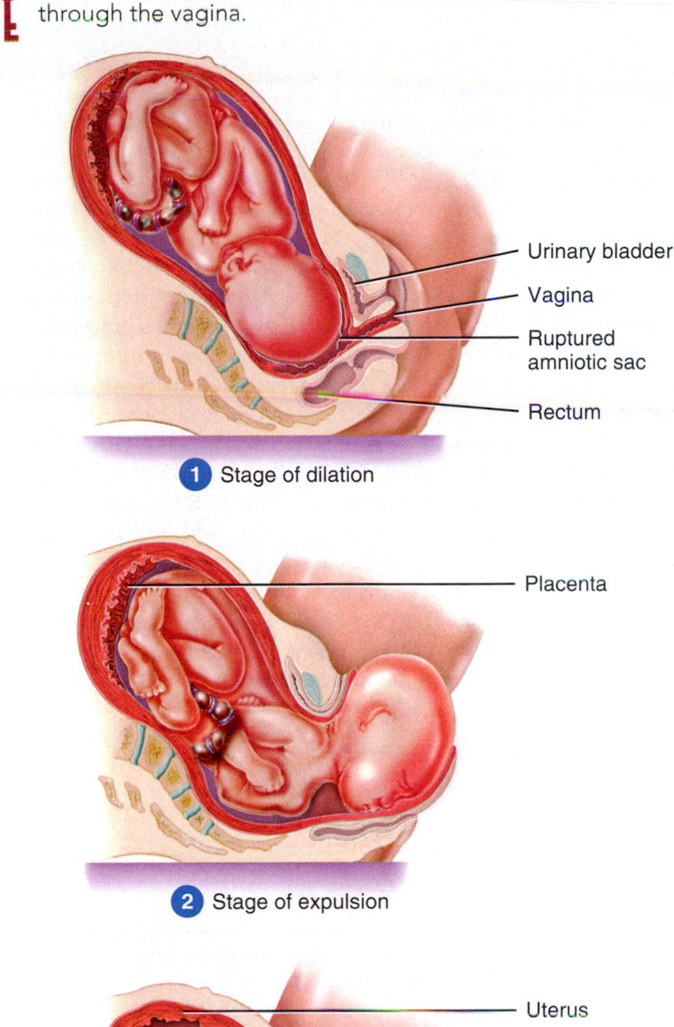

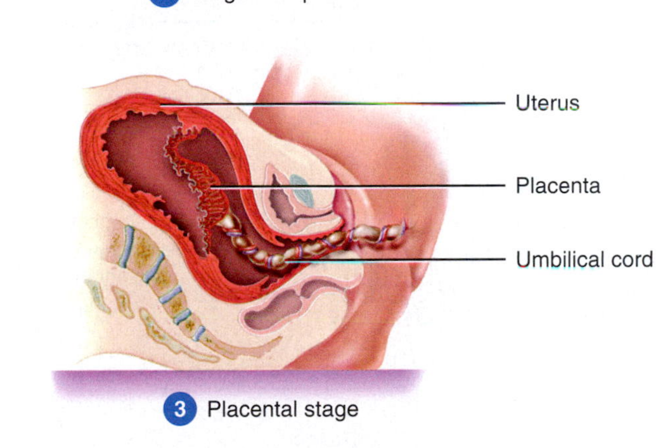

What event marks the beginning of the stage of expulsion?

off and then severed. The area on the infant where the umbilical cord was attached eventually forms the **umbilicus** (navel).

3. ***Placental stage.*** The **placental stage** is the time (5–30 minutes or more) after delivery until uterine contractions detach the placenta from the uterine wall and expel it from the mother's body as the **afterbirth**. The uterine contractions that occur during this stage also constrict blood vessels that were torn during delivery, reducing the likelihood of hemorrhage.

As a rule, labor lasts longer with first babies, typically about 14 hours. For women who have previously given birth, the average duration of labor is about 8 hours—although the time varies enormously among births.

About 7% of pregnant women do not deliver by 2 weeks after their due date. Such cases carry an increased risk of brain damage to the fetus, and even fetal death, due to inadequate supplies of oxygen and nutrients from an aging placenta. Post-term deliveries may be facilitated by inducing labor, which is initiated by administration of oxytocin (Pitocin®), or by surgical delivery (cesarean section).

Following the delivery of the baby and placenta is a 6-week period during which the maternal reproductive organs and physiology return to the prepregnancy state. This period is called the **puerperium** (pū-er-PER-ē-um). Through a process of tissue catabolism, the uterus undergoes a remarkable reduction in size, called **involution**. In addition, the cervix loses its elasticity and regains its prepregnancy firmness. For 2–4 weeks after delivery, women have a uterine discharge called **lochia** (LŌ-kē-a), which consists initially of blood and later of serous fluid derived from the former site of the placenta.

Lactation Is the Process by Which the Mammary Glands Produce and Eject Milk

Lactation is the production and ejection of milk from the mammary glands. A principal hormone in promoting milk production is **prolactin (PRL)**, which is secreted from the anterior pituitary gland. Even though prolactin levels increase as the pregnancy progresses, no milk production occurs because progesterone inhibits the effects of prolactin. After delivery, the levels of estrogens and progesterone in the mother's blood decrease, and the inhibition is removed. The principal stimulus in maintaining prolactin secretion during lactation is the suckling action of the infant during **breast feeding**. Suckling initiates action potentials from stretch receptors in the nipples to the hypothalamus; the action potentials decrease hypothalamic release of prolactin-inhibiting hormone (PIH) and increase release of prolactin-releasing hormone (PRH), so more prolactin is released by the anterior pituitary.

CLINICAL CONNECTION

Dystocia and Cesarean Section

Dystocia (dis-TŌ-sē-a), or difficult labor, may result either from an abnormal position (presentation) of the fetus or a birth canal of inadequate size to permit vaginal delivery. In a **breech presentation**, for example, the fetal buttocks or lower limbs, rather than the head, enter the birth canal first; this occurs most often in premature births. If fetal or maternal distress prevents a vaginal birth, the baby may be delivered surgically through an abdominal incision. A low, horizontal cut is made through the abdominal wall and lower portion of the uterus, through which the baby and placenta are removed. Even though it is popularly associated with the birth of Julius Caesar, the true reason this procedure is termed a **cesarean section (C-section)** is because it was described in Roman law, *lex cesarea*, about 600 years before Julius Caesar was born. Even a history of multiple C-sections need not exclude a pregnant woman from attempting a vaginal delivery.

TOOL OF THE TRADE

PRENATAL DIAGNOSTIC TESTING

Several tests are available to detect genetic disorders and assess fetal well-being during pregnancy. Examples of **prenatal diagnostic testing** include fetal ultrasonography, amniocentesis, and chorionic villi sampling (CVS).

If there is a question about the normal progress of a pregnancy, **fetal ultrasonography** may be performed. By far the most common use of diagnostic ultrasound is to determine a more accurate fetal age when the date of conception is unclear. It is also used to confirm pregnancy, evaluate fetal viability and growth, determine fetal position, identify multiple pregnancies, identify fetal–maternal abnormalities, and serve as an adjunct to special procedures such as amniocentesis. During fetal ultrasonography, a transducer, an instrument that emits high-frequency sound waves, is passed back and forth over the abdomen. The reflected sound waves from the developing fetus are picked up by the transducer and converted to an on-screen image called a **sonogram**.

Amniocentesis (am-nē-ō-sen-TĒ-sis) involves withdrawing some of the amniotic fluid that bathes the developing fetus and analyzing the fetal cells and dissolved substances. It is used to test for the presence of certain genetic disorders, such as Down syndrome (DS), Tay-Sachs disease, and sickle-cell disease. It is also used to help determine survivability of the fetus. The test is usually done at 14–16 weeks of gestation. All gross chromosomal abnormalities and over 50 biochemical defects can be detected through amniocentesis. During amniocentesis, the position of the fetus and placenta is first identified using ultrasound and palpation. After the skin is prepared with an antiseptic and a local anesthetic is given, a hypodermic needle is inserted through the mother's abdominal wall and into the amniotic cavity within the uterus. Then, 10 to 30 mL of fluid and suspended cells are aspirated (FIGURE A) for microscopic examination and biochemical testing. Chromosome studies, which require growing the cells for 2–4 weeks in a culture medium, may reveal rearranged, missing, or extra chromosomes. Amniocentesis is performed only when a risk for genetic defects is suspected because there is about a 0.5% chance of spontaneous abortion after the procedure.

In **chorionic villi sampling (CVS)**, a catheter is guided through the vagina and cervix of the uterus and then advanced to the chorionic villi under ultrasound guidance (FIGURE B). About 30 milligrams of tissue are suctioned out and prepared for chromosomal analysis. Alternatively, the chorionic villi can be sampled by inserting a needle through the abdominal cavity, as performed in amniocentesis. CVS can identify the same defects as amniocentesis because chorion cells and fetal cells contain the same genome. CVS offers several advantages over amniocentesis: It can be performed as early as eight weeks of gestation, and test results are available in only a few days, permitting an earlier decision about whether to continue the pregnancy. However, CVS is slightly riskier than amniocentesis; after the procedure there is a 1–2% chance of spontaneous abortion.

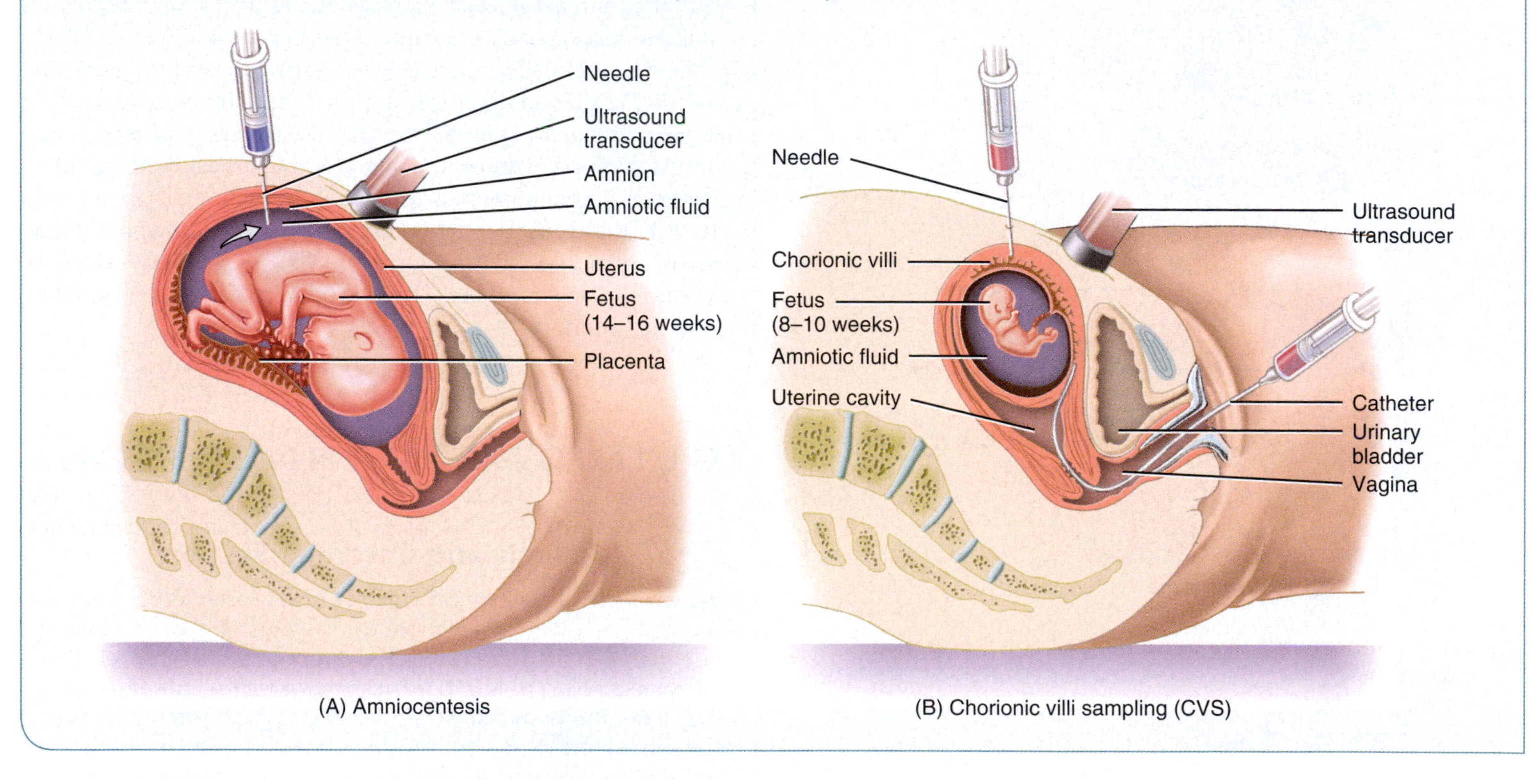

(A) Amniocentesis

(B) Chorionic villi sampling (CVS)

Oxytocin causes release of milk into the mammary ducts via the **milk ejection reflex** (FIGURE 23.23). Milk formed by the glandular cells of the breasts is stored until the baby begins active suckling. Stimulation of touch receptors in the nipple initiates sensory action potentials that are relayed to the hypothalamus. In response, secretion of oxytocin from the posterior pituitary increases. Carried by the bloodstream to the mammary glands, oxytocin stimulates contraction of myoepithelial (smooth-muscle-like) cells surrounding the glandular cells and ducts. The resulting compression moves the milk from the alveoli of the mammary glands into the mammary ducts, where it can be suckled. This process is termed **milk ejection** or **milk let-down**. Stimuli other than suckling, such as hearing a baby's cry, also can trigger oxytocin release and milk ejection. The suckling stimulation that produces the release of oxytocin also inhibits the release of PIH; this results in increased secretion of prolactin, which maintains lactation.

FIGURE 23.23 The milk ejection reflex, a positive feedback cycle.

Oxytocin stimulates contraction of myoepithelial cells in the breasts, which squeeze the glandular and duct cells and causes milk ejection.

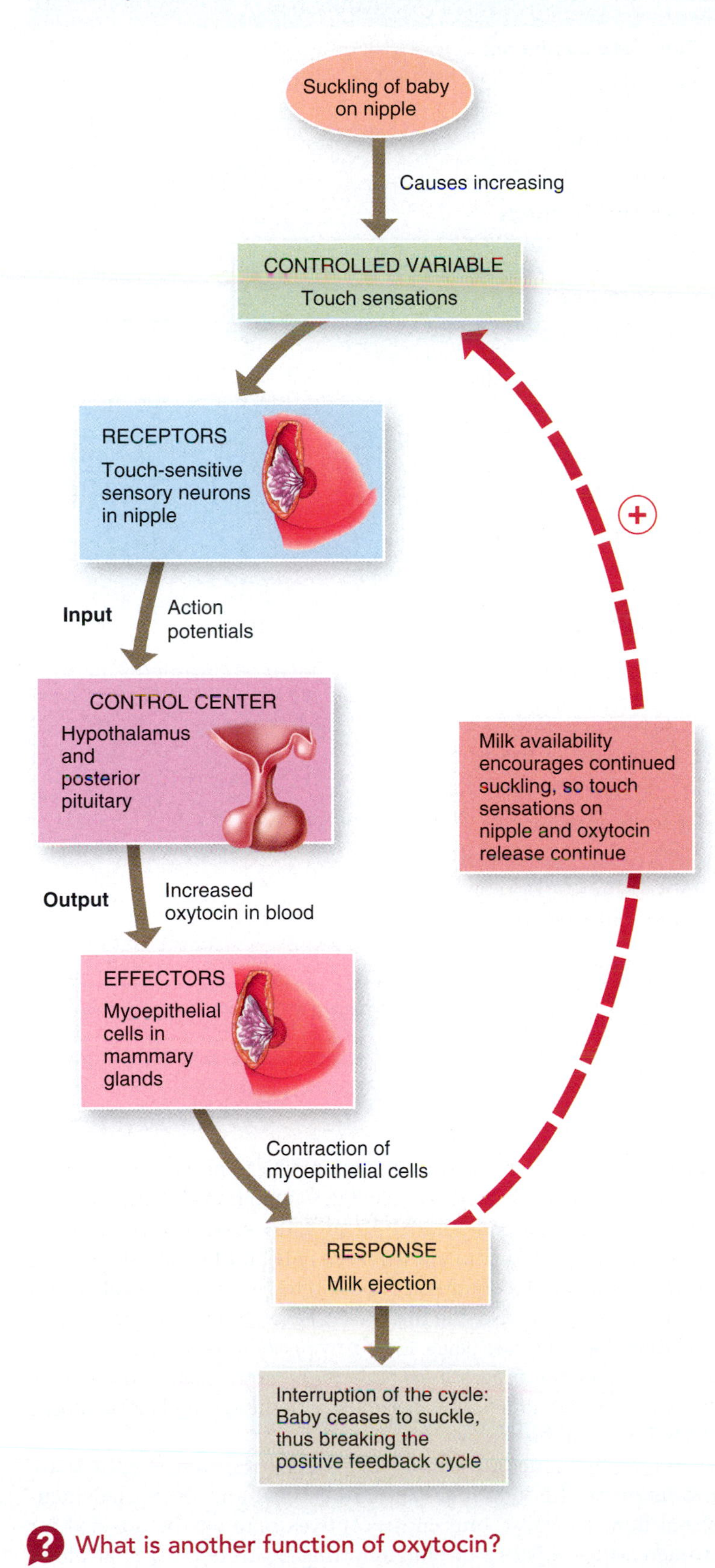

What is another function of oxytocin?

During late pregnancy and the first few days after birth, the mammary glands secrete a cloudy fluid called **colostrum**. Although it is not as nutritious as milk—it contains less lactose and almost no fat—colostrum serves adequately until the appearance of true milk on about the fourth day. Colostrum and maternal milk contain important antibodies that protect the infant during the first few months of life.

Following birth of the infant, the prolactin level starts to return to the nonpregnant level. However, each time the mother nurses the infant, action potentials from the nipples to the hypothalamus increase the release of PRH (and decrease the release of PIH), resulting in a tenfold increase in prolactin secretion by the anterior pituitary that lasts about an hour. Prolactin acts on the mammary glands to provide milk for the next nursing period. If this surge of prolactin is blocked by injury or disease, or if nursing is discontinued, the mammary glands lose their ability to produce milk in only a few days. Even though milk production normally decreases considerably within 7–9 months after birth, it can continue for several years if breast feeding continues.

Lactation often blocks ovarian cycles for the first few months following delivery if the frequency of suckling is about 8–10 times a day. This effect is inconsistent, however, and ovulation commonly precedes the first menstrual period after delivery of a baby. As a result, the mother can never be certain she is not fertile. Breast feeding is therefore an unreliable birth control measure. The suppression of ovulation during lactation is believed to occur as follows: During breast feeding, neural input from the nipple reaches the hypothalamus and causes it to produce neurotransmitters that suppress the release of gonadotropin-releasing hormone (GnRH). As a result, production of LH and FSH decreases, and ovulation is inhibited.

A primary benefit of breast feeding is nutritional: Human milk is a sterile solution that contains amounts of fatty acids, lactose, amino acids, minerals, vitamins, and water that are ideal for the baby's digestion, brain development, and growth. Breast milk also contains maternal antibodies (IgA) and white blood cells that help protect the baby from microbes and other foreign substances.

Years before oxytocin was discovered, it was common practice in midwifery to let a firstborn twin nurse at the mother's breast to speed the birth of the second child. Now we know why this practice is helpful—it stimulates the release of oxytocin. Even after a single birth, nursing promotes expulsion of the placenta (afterbirth) and helps the uterus return to its normal size. Synthetic oxytocin (Pitocin) is often given to induce labor or to increase uterine tone and control hemorrhage just after parturition.

CHECKPOINT

22. How is polyspermy prevented?
23. When, where, and how does implantation occur?
24. What is the significance of gastrulation?
25. What are the functions of the extraembryonic membranes?
26. List the hormones secreted by the placenta and describe the functions of each.
27. What happens during the stage of dilation, the stage of expulsion, and the placental stage of labor?
28. Which hormones contribute to lactation? What is the function of each?

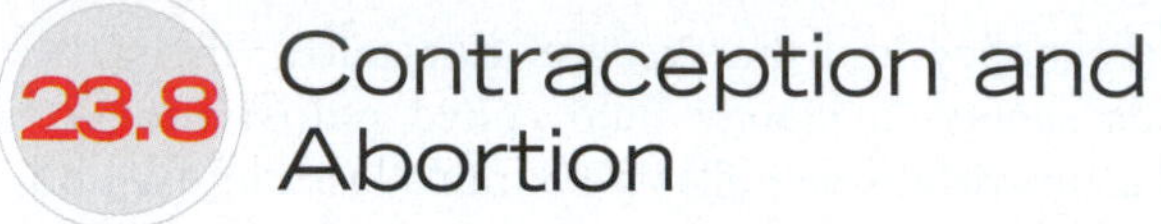

23.8 Contraception and Abortion

OBJECTIVES

- Explain the differences among the various methods of contraception and compare their effectiveness.
- Distinguish between a spontaneous abortion and an induced abortion.

Various Methods of Contraception Control Fertility and Prevent Conception

Contraception, or *birth control*, refers to the avoidance of pregnancy by various methods that are designed to control fertility and prevent conception. No single, ideal method of birth control exists. The only method of preventing pregnancy that is 100% reliable is **complete abstinence**, the practice of refraining from sexual intercourse. Several other methods of contraception are available; each has its advantages and disadvantages. These include sterilization, hormonal methods, intrauterine devices, spermicides, barrier methods, periodic abstinence, and coitus interruptus. TABLE 23.3 provides the failure rates for different types of contraceptive methods.

TABLE 23.3 Failure Rates for Several Contraceptive Methods

Method	Failure Rates (%)*	
	Perfect Use†	Typical Use
Complete abstinence	0	0
Sterilization		
Vasectomy	0.10	0.15
Tubal ligation	0.5	0.5
Essure®	0.2	0.2
Hormonal Methods		
Oral contraceptives		
Combined pill	0.3	1–2
Seasonale®	0.3	1–2
Minipill	0.5	2
Emergency contraception pill	25	25
Nonoral contraceptives		
Contraceptive skin patch	0.1	1–2
Vaginal contraceptive ring	0.1	1–2
Hormone injections	0.3	1–2
Hormone implant (Implanon®)	0.05	0.05
Intrauterine devices		
Copper IUD (Paragard®)	0.6	0.8
Hormonal IUD (Mirena®)	0.2	0.2
Spermicides (alone)	15	29
Barrier methods		
Male condom	2	15
Vaginal pouch	5	21
Diaphragm (with spermicide)	6	16
Cervical cap (with spermicide)	9	16
Periodic abstinence		
Rhythm	9	25
Sympto-thermal	2	20
Coitus interruptus	4	18
No method	85	85

* Defined as the percentage of women having an unintended pregnancy during the first year of use.
† Failure rate when the method is used correctly and consistently.

Sterilization

Sterilization is a procedure that renders an individual incapable of further reproduction. The principal method for sterilization of males is a **vasectomy**, in which a portion of each vas deferens is removed. To gain access to the vas deferens, an incision is made with a scalpel (conventional procedure) or a puncture is made with special forceps (nonscalpel vasectomy). Then a small section of each vas deferens is cut out and the remaining free ends are tied off or cauterized. Although sperm production continues in the testes, sperm can no longer reach the exterior. The sperm degenerate and are destroyed by phagocytosis. Because blood vessels are not cut, testosterone levels in the blood remain normal, so vasectomy has no effect on sexual desire, sexual performance, or ejaculation.

Sterilization in females is achieved most often by performing a **tubal ligation**, in which both fallopian tubes are closed off. This can be achieved in a few different ways. "Clips" or "clamps" can be placed on the fallopian tubes, the tubes can be tied off and/or cut, and sometimes they are cauterized. In any case the result is that the egg cannot pass through the fallopian tube, and sperm cannot reach the egg.

An alternative to tubal ligation is a type of nonincisional sterilization known as Essure®. In the Essure® procedure, a soft, tiny coil made of polyester fibers and nickel-titanium alloy is inserted with a catheter into the vagina, through the uterus, and into each fallopian tube. Over a three-month period, the coil stimulates growth of scar tissue in and around itself, blocking the fallopian tubes. As with tubal ligation, the egg cannot pass through the fallopian tubes, and sperm cannot reach the egg.

Hormonal Methods

Aside from complete abstinence or sterilization, hormonal methods are the most effective means of birth control. **Oral contraceptives**, or *birth control pills*, contain various mixtures of synthetic estrogens and progestin (a chemical with actions similar to progesterone). They prevent pregnancy by negative feedback inhibition of GnRH release from the hypothalamus, and FSH and LH release from the anterior pituitary gland. The low levels of FSH and LH usually prevent the development of a dominant follicle in the ovary. As a result, levels of estrogens do not rise, the midcycle LH surge does not occur, and ovulation does not take place. Even if ovulation does occur, as it does in some cases, the progestin in oral contraceptives thickens cervical mucus, making it difficult for sperm to reach the egg, and it also blocks implantation in the uterus.

Among the noncontraceptive benefits of oral contraceptives are regulation of the length of menstrual cycle and decreased menstrual flow. However, oral contraceptives may not be advised for women with a history of blood-clotting disorders, cerebral blood vessel damage, migraine headaches, hypertension, liver malfunction, or heart disease. Women who take the pill and smoke face far higher odds of having a heart attack or stroke than do nonsmoking pill users. Smokers should quit smoking or use an alternative method of contraception.

There are several variations of *oral* hormonal methods of contraception:

- **Combined pill.** Contains both estrogens and progestin and is typically taken once a day for 3 weeks. The pills taken during the fourth week are inactive (do not contain hormones) and permit menstruation to occur.
- **Seasonale®.** Contains both estrogens and progestin and is taken once a day in 3-month cycles of 12 weeks of hormone-containing pills followed by 1 week of inactive pills. Menstruation occurs during the thirteenth week.
- **Minipill.** Contains progestin only and is taken every day of the month.
- **Emergency contraception (EC) pill.** Consists of one pill containing progestin that is used to prevent pregnancy following unprotected sexual intercourse. The relatively high level of progestin in an EC pill suppresses ovulation, thickens cervical mucus to prevent sperm from reaching the egg, and blocks implantation. An EC pill is also known as *the morning after pill*. An example is *Plan B®*, which is available without a prescription. The pill should be taken as soon as possible but within 72 hours of unprotected intercourse.

Nonoral hormonal methods of contraception are also available, including:

- **Contraceptive skin patch (Ortho Evra®).** Contains both estrogens and progestin delivered in a skin patch placed on the skin (upper outer arm, back, lower abdomen, or buttocks) once a week for 3 weeks. After 1 week, the patch is removed from one location and then a new one is placed elsewhere. During the fourth week no patch is used, allowing menstruation to occur.
- **Vaginal contraceptive ring (NuvaRing®).** A flexible doughnut-shaped ring about 5 cm (2 in.) in diameter that contains estrogens and progestin and is inserted by the female herself into the vagina. It is left in the vagina for three weeks to prevent conception and then removed for one week to permit menstruation.
- **Hormone injections (Depo-provera®).** An injectable progestin given intramuscularly by a health-care practitioner once every 3 months.
- **Hormone implant (Implanon®).** A matchstick-sized plastic rod containing progestin that is surgically implanted under the skin of the arm using local anesthesia. It slowly and continuously releases progestin, which inhibits ovulation and thickens cervical mucus. The effect lasts for 3 years and is even more reliable than sterilization. Removing the implant restores fertility.

Intrauterine Devices

An **intrauterine device (IUD)** is a small, T-shaped object that is inserted by a health-care professional into the cavity of the uterus. Two types of IUDs are available in the United States: the copper IUD and the hormonal IUD. The **copper IUD** (ParaGard®) contains a plastic frame that is covered with a copper wire. The copper causes changes in the uterine lining that prevent implantation of the fertilized egg. ParaGard is approved for up to 10 years of use and has long-term effectiveness comparable to that of tubal ligation. The **hormonal IUD** (Mirena®) has a plastic frame that surrounds a reservoir containing progestin. The progestin is slowly released from the IUD and functions like the other progestin-containing contraceptives: It suppresses ovulation, thickens cervical mucus, and blocks implantation. Mirena is effective for up to 5 years. Some women cannot use IUDs because of expulsion, bleeding, or discomfort.

Spermicides

Various foams, creams, jellies, and suppositories that contain sperm-killing agents, or **spermicides**, make the vagina and cervix unfavorable for sperm survival and are available without prescription. They are placed in the vagina before sexual intercourse. The most widely used spermicide is *nonoxynol-9*, which kills sperm by disrupting their plasma membranes. A spermicide is more effective when used with a barrier method such as a male condom, vaginal pouch, diaphragm, or cervical cap.

Barrier Methods

Barrier methods use a physical barrier and are designed to prevent sperm from gaining access to the uterine cavity and fallopian tubes. In addition to preventing pregnancy, certain barrier methods (male condom and vaginal pouch) may also provide some protection against sexually transmitted diseases (STDs) such as AIDS. In contrast, oral contraceptives and IUDs confer no such protection. Among the barrier methods are the male condom, vaginal pouch, diaphragm, and cervical cap.

A **male condom** is a covering of latex, polyurethane, or animal membrane that is placed over the penis to prevent deposition of sperm in the female reproductive tract. A **vaginal pouch**, sometimes called a **female condom**, is designed to prevent sperm from entering the uterus. It is made of two flexible rings connected by a polyurethane sheath. One ring lies inside the sheath and is inserted to fit over the cervix; the other ring remains outside the vagina and covers the female external genitalia. A **diaphragm** is a rubber, dome-shaped structure that covers the cervix and must be fitted by a health-care professional. It is used in conjunction with a spermicide and can be inserted by the female up to 6 hours before intercourse. The diaphragm stops most sperm from passing into the cervix and the spermicide kills most sperm that do get by. Although diaphragm use does decrease the risk of some STDs, it does not fully protect against HIV infection because the vagina is still exposed. **A cervical cap** resembles a diaphragm but is smaller and more rigid. It fits snugly over the cervix and also must be fitted by a health-care professional. Spermicides should be used with the cervical cap.

Periodic Abstinence

A couple can use their knowledge of the physiological changes that occur during the female reproductive cycle to decide either to abstain from intercourse on those days when pregnancy is a likely result or to plan intercourse on those days if they wish to conceive a child. In females with normal and regular menstrual cycles, these physiological events help to predict the day on which ovulation is likely to occur.

The first physiologically based method, developed in the 1930s, is known as the **rhythm method**. It takes advantage of the fact that the egg is only viable for up to 24 hours and that sperm can survive 3 to 5 days in the female reproductive tract. Thus, couples using this contraceptive method should avoid sexual intercourse for several days before ovulation, the day of ovulation, and several days after ovulation (just in case ovulation occurs a few days after day 14). The effectiveness of the rhythm method for birth control is poor in many women due to the irregularity of the female reproductive cycle.

Another system is the **sympto-thermal method**, in which couples are instructed to know and understand certain signs of fertility. The signs of ovulation include a slight decrease in basal body temperature prior to ovulation and then a slight increase in basal body temperature just after ovulation; the production of abundant clear, stretchy cervical mucus; and pain associated with ovulation (*mittelschmerz*). If a couple abstains from sexual intercourse when the signs of ovulation are present, the chance of pregnancy is decreased. A big problem with this method is that fertilization can still occur if intercourse takes place a few days *before* ovulation.

Coitus Interruptus

Coitus interruptus is withdrawal of the penis from the vagina just before ejaculation. Failures with this method are due either to failure to withdraw before ejaculation or to preejaculatory emission of sperm-containing fluid from the urethra. In addition, this method offers no protection against transmission of STDs.

Abortion Results in the Termination of a Pregnancy

Abortion refers to the premature expulsion of the products of conception from the uterus, usually before the twentieth week of pregnancy. An abortion may be *spontaneous* (naturally occurring, also called a *miscarriage*) or *induced* (intentionally performed).

There are several types of induced abortions. One involves **mifepristone**, called **miniprex** in the United States and **RU 486** in Europe. It is a hormone approved only for pregnancies 9 weeks or less when taken with misoprostol (a prostaglandin). Mifepristone is an antiprogestin; it blocks the action of progesterone by binding to and blocking progesterone receptors. Progesterone prepares the uterine endometrium for implantation and then maintains the uterine lining after implantation. If the level of progesterone falls during pregnancy or if the action of the hormone is blocked, menstruation occurs, and the embryo sloughs off along with the uterine lining. Within 12 hours after taking mifepristone, the endometrium starts to degenerate, and within 72 hours it begins to slough off. Misoprostol stimulates uterine contractions and is given after mifepristone to aid in expulsion of the endometrium.

Another type of induced abortion is called **vacuum aspiration** (suction) and can be performed up to the sixteenth week of pregnancy. A small, flexible tube attached to a vacuum source is inserted into the uterus through the vagina. The embryo or fetus, placenta, and lining of the uterus are then removed by suction. For pregnancies between 13 and 16 weeks, a technique called **dilation and evacuation** is commonly used. After the cervix is dilated, suction and forceps are used to remove the fetus, placenta, and uterine lining. From the 16th to 24th week, a **late-stage abortion** may be employed using surgical methods similar to dilation and evacuation or through nonsurgical methods using a saline solution or medications to induce abortion. Labor may be induced by using vaginal suppositories, intravenous infusion, or injections into the amniotic fluid through the uterus.

CHECKPOINT

29. How do oral contraceptives reduce the likelihood of pregnancy?
30. How do some methods of birth control protect against sexually transmitted diseases?
31. What effect does mifepristone have on the endometrium?

23.9 Infertility

OBJECTIVE

- Explain the different types of techniques used to help infertile couples have a baby.

Infertility is the inability to produce offspring. Female infertility may be caused by ovarian disease, obstruction of the fallopian tubes, or conditions in which the uterus is not adequately prepared to receive a fertilized egg. Male infertility may be caused by obstruction of male reproductive ducts or by inadequate production of viable, normal sperm. The seminiferous tubules of the testes are sensitive to many factors—X-rays, infections, toxins, malnutrition, and higher-than-normal scrotal temperatures—that may cause degenerative changes and produce male infertility.

Another cause of infertility in females is inadequate body fat. To begin and maintain a normal reproductive cycle, a female must have a minimum amount of body fat. Even a moderate deficiency of fat (10–15% below normal weight for height) may delay the onset of menstruation (menarche), inhibit ovulation during the reproductive cycle, or cause amenorrhea (cessation of menstruation). Both dieting and intensive exercise may reduce body fat below the minimum amount and lead to infertility that is reversible if weight gain or reduction of intensive exercise, or both, occurs. Studies of very obese women indicate that they, like very lean ones, experience problems with amenorrhea and infertility. Males also experience reproductive problems in response to undernutrition and weight loss. For example, they produce less prostatic fluid and reduced numbers of motile sperm.

Many fertility-expanding techniques now exist for assisting infertile couples in having a baby. To achieve **in vitro fertilization (IVF)**—fertilization in a laboratory dish—the mother-to-be is given follicle-stimulating hormone (FSH) soon after menstruation so that several eggs, rather than the typical single egg, will be produced (superovulation). When several follicles have reached the appropriate size, a small incision is made near the umbilicus, and the eggs are aspirated from the stimulated follicles and transferred to a solution containing sperm, where the eggs undergo fertilization. Alternatively, an egg may be fertilized in vitro by suctioning a sperm or even a spermatid obtained from the testis into a tiny pipette and then injecting it into the egg's cytoplasm. This procedure, termed **intracytoplasmic sperm injection (ICSI)**, has been used when infertility is caused by impairments in sperm motility or to the failure of spermatids to develop into spermatozoa. When the zygote achieved by IVF or ICSI reaches the 8-cell or 16-cell stage, it is introduced into the uterus for implantation and subsequent growth.

In **embryo transfer**, a man's semen is used to artificially inseminate a fertile egg donor. After fertilization in the donor's fallopian tube, the morula or blastocyst is transferred from the donor to the infertile woman, who then carries it (and subsequently the fetus) to term. Embryo transfer is indicated for women who are infertile or who do not want to pass on their own genes because they are carriers of a serious genetic disorder.

In **gamete intrafallopian transfer (GIFT)** the goal is to mimic the normal process of conception by uniting sperm and egg in the prospective mother's fallopian tubes. It is an attempt to bypass conditions in the female reproductive tract that might prevent fertilization, such as high acidity or inappropriate amounts of mucus. In this procedure, a woman is given FSH and LH to stimulate the production of several eggs, which are aspirated from the mature follicles, mixed outside the body with a solution containing sperm, and then immediately inserted into the fallopian tubes.

CHECKPOINT

32. What are some causes of infertility in men and women?

33. How does embryo transfer differ from gamete intrafallopian transfer (GIFT)?

To appreciate the many ways that the reproductive systems contribute to homeostasis of other body systems, examine *Focus on Homeostasis: Contributions of the Reproductive Systems.*

FOCUS on HOMEOSTASIS

CONTRIBUTIONS OF THE REPRODUCTIVE SYSTEMS

INTEGUMENTARY SYSTEM

- Androgens promote the growth of body hair
- Estrogens stimulate the deposition of fat in the breasts, abdomen, and hips
- Mammary glands produce milk
- Skin stretches during pregnancy as the fetus enlarges

SKELETAL SYSTEM

- Androgens and estrogens stimulate the growth and maintenance of bones of the skeletal system

MUSCULAR SYSTEM

- Androgens stimulate the growth of skeletal muscles

NERVOUS SYSTEM

- Androgens influence libido (sex drive)
- Estrogens may play a role in the development of certain regions of the brain in males

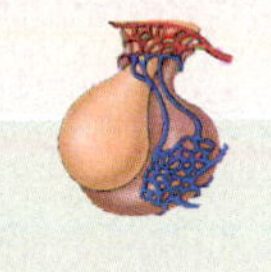

ENDOCRINE SYSTEM

- Testosterone and estrogens exert feedback effects on the hypothalamus and anterior pituitary gland

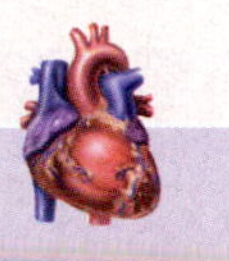

CARDIOVASCULAR SYSTEM

- Estrogens lower blood cholesterol level and may reduce the risk of coronary artery disease in women under age 50

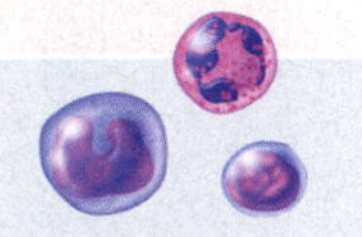

IMMUNE SYSTEM

- The presence of an antibiotic-like chemical in semen and the acidic pH of vaginal fluid provide innate immunity against microbes in the reproductive tract

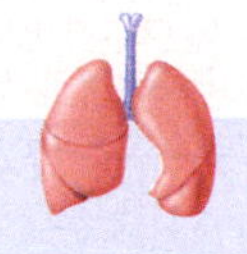

RESPIRATORY SYSTEM

- Sexual arousal increases the rate and depth of breathing

DIGESTIVE SYSTEM

- The presence of the fetus during pregnancy crowds the digestive organs, which leads to heartburn and constipation

URINARY SYSTEM

- In males, the portion of the urethra that extends through the prostate and penis is a passageway for urine as well as semen

FOR ALL BODY SYSTEMS

- The male and female reproductive systems produce gametes (sperm and eggs) that unite to form embryos and fetuses, which contain cells that divide and differentiate to form all of the organ systems of the body

FROM RESEARCH TO REALITY

Pregnancy and Age

Reference
Khalil, A. et al. (2013). Maternal age and adverse pregnancy outcome: A cohort study. *Ultrasound Obstetrics Gynecology*. 42: 634–643.

Does being older make pregnancy more risky?

Jean-Philippe WALLET / Shutterstock

Although considered a safe biological process, pregnancy is fraught with the possibility of complications. Certain factors can be used to predict whether a woman is at higher or lower risk for complications, which allows physicians to act accordingly. Is age a risk factor for pregnancy complications?

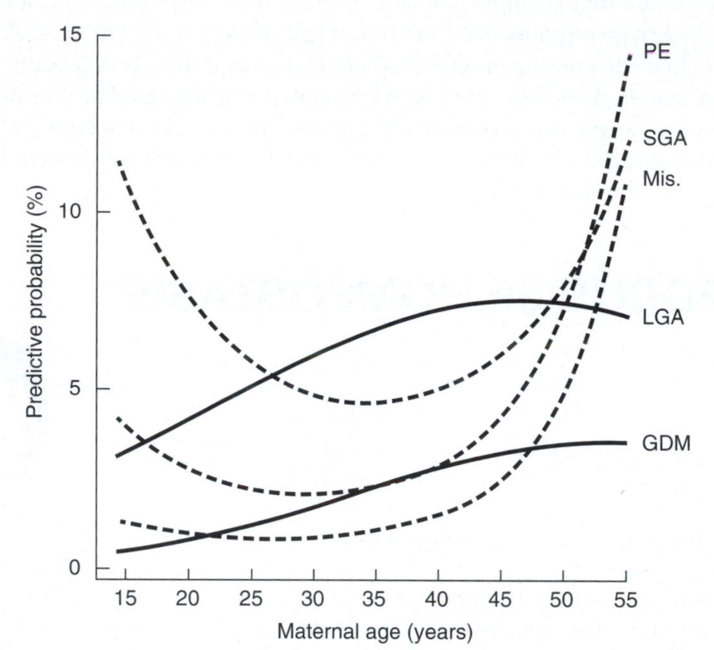

Article description:
Researchers ran a large retrospective cohort study following over 70,000 women during and after their pregnancy. They looked at the women's age and their risk of developing certain complications of pregnancy. Possible confounding factors were adjusted for, and the data were analyzed.

Go to WileyPLUS Learning Space and use the data from this article to answer the questions posed there and to discover more about pregnancy complications at any age.

CHAPTER REVIEW

23.1 Reproductive Cell Division

1. Gametes contain the haploid (n) chromosome number (23), and somatic cells contain the diploid ($2n$) chromosome number (46).
2. Meiosis is the process that produces haploid gametes; it consists of two successive nuclear divisions called meiosis I and meiosis II.
3. During meiosis I, homologous chromosomes undergo synapsis (pairing) and crossing-over; the net result is two haploid cells that are genetically unlike each other and unlike the original cell that produced them.
4. During meiosis II, two haploid daughter cells divide to form four haploid cells.

23.2 Male Reproductive System

1. The male organs of reproduction include the testes, a system of ducts (epididymis, vas deferens, ejaculatory ducts, and urethra), accessory sex glands (seminal vesicles, prostate gland, and bulbourethral glands), and supporting structures (penis and scrotum).
2. The scrotum is a sac that supports the testes and regulates their temperature.

3. The testes are the male gonads. They contain seminiferous tubules, in which sperm cells are made; Sertoli cells (sustentacular cells), which nourish sperm cells and secrete inhibin; and Leydig (interstitial) cells, which produce testosterone. Spermatogenesis is the process by which sperm are produced in the seminiferous tubules of the testes. The spermatogenesis sequence, which includes meiosis I, meiosis II, and spermiogenesis, results in the formation of four haploid sperm (spermatozoa) from each primary spermatocyte. A sperm cell consists of three parts: a head, midpiece, and tail (flagellum).
4. The epididymis is the site of sperm maturation and storage. The vas deferens also stores sperm and, during ejaculation, it propels sperm toward the urethra. Each ejaculatory duct is formed by the union of the vas deferens and the duct from the seminal vesicle. It is the passageway for ejection of sperm and secretions of the seminal vesicles into the urethra. The urethra is the shared terminal duct of the male reproductive and urinary systems; it serves as a passageway for both semen and urine.
5. The seminal vesicles secrete an alkaline, viscous fluid that contains fructose (used by sperm for ATP production). Seminal fluid constitutes about 60% of the volume of semen and contributes to sperm viability. The prostate secretes a slightly acidic fluid that constitutes about 25% of the volume of semen and contributes to sperm motility. The bulbourethral (Cowper's) glands secrete mucus for lubrication and an alkaline substance that neutralizes acid. Semen is a mixture of sperm and the secretions of the accessory sex glands; it provides the fluid in which sperm are transported, supplies nutrients, and neutralizes the acidity of the male urethra and the vagina.
6. The penis consists of erectile tissue that contains blood sinuses (vascular spaces). During sexual arousal, the tissue of the penis fills with blood (erection).
7. At puberty, gonadotropin-releasing hormone (GnRH) stimulates anterior pituitary secretion of FSH and LH. LH stimulates production of testosterone; FSH and testosterone stimulate spermatogenesis. Sertoli cells secrete androgen-binding protein (ABP), which binds to testosterone and keeps the testosterone concentration high in the seminiferous tubule.
8. Testosterone controls the growth, development, and maintenance of sex organs; stimulates bone growth, protein anabolism, and sperm maturation; and stimulates development of masculine secondary sex characteristics.
9. Inhibin is produced by Sertoli cells; its inhibition of FSH helps regulate the rate of spermatogenesis.

23.3 Female Reproductive System

1. The female organs of reproduction include the ovaries, fallopian tubes, uterus, vagina, vulva, and mammary glands.
2. The ovaries produce eggs (oocytes) and hormones (estrogens, progesterone, inhibin, and relaxin). Within the ovaries are ovarian follicles. An ovarian follicle initially begins as a primordial follicle, which sequentially develops into a primary follicle, secondary follicle, and then a mature (graafian) follicle. The mature follicle ruptures, releasing an egg (secondary oocyte), a process referred to as ovulation. The remnants of the ruptured follicle in the ovary develop into the corpus luteum, which eventually degrades into the corpus albicans if fertilization does not occur. Oogenesis is the production of eggs by the ovaries. The oogenesis sequence includes meiosis I and meiosis II, which goes to completion only after an ovulated secondary oocyte is fertilized by a sperm cell.
3. The fallopian tubes transport eggs from the ovaries to the uterus and are the normal sites of fertilization.
4. The uterus consists of three layers: (1) endometrium, which is further divided into two layers (stratum functionalis and stratum basalis); (2) myometrium, which consists of smooth muscle that contracts during labor to help expel the fetus; and (3) perimetrium

(connective tissue). The uterus functions in menstruation, implantation of a fertilized egg, development of a fetus during pregnancy, and labor. It is also part of the pathway for sperm to reach the fallopian tube to fertilize an egg.

5. The vagina is the receptacle of the penis during sexual intercourse and also functions as the outlet for menstrual flow and as the birth canal. The vulva is a collective term for the external genitalia of the female. It consists of the mons pubis, labia majora, labia minora, clitoris, and vestibule. The mammary glands are modified sweat glands (one in each breast) that produce and eject milk (lactation).
6. At puberty, gonadotropin-releasing hormone (GnRH) stimulates anterior pituitary secretion of FSH and LH. FSH and LH stimulate development of ovarian follicles and secretion of estrogens by the follicles. LH also stimulates ovulation, formation of the corpus luteum, and the secretion of progesterone and estrogens by the corpus luteum.
7. Estrogens promote the development of female reproductive structures and secondary sex characteristics; increase protein anabolism; lower blood cholesterol level; and stimulate proliferation of the stratum basalis to form a new stratum functionalis each month after menstruation occurs.
8. Progesterone stimulates endometrial glands to secrete glycogen and lipids, which serve as a nutrient source for a fertilized egg if implantation occurs.
9. Relaxin relaxes the myometrium at the time of possible implantation. At the end of a pregnancy, relaxin increases the flexibility of the pubic symphysis and helps dilate the uterine cervix to facilitate delivery.
10. Inhibin, which is secreted by the growing follicles and by the corpus luteum, inhibits secretion of FSH and, to a lesser extent, LH.
11. The female reproductive cycle refers to the cyclical changes in the ovaries and uterus. It usually lasts 28 days and consists of four phases: menstrual phase, preovulatory phase, ovulation, and postovulatory phase. During the menstrual phase, the stratum functionalis of the endometrium is shed, discharging blood, tissue fluid, mucus, and epithelial cells. During the preovulatory phase, a group of follicles in the ovaries begins to undergo final maturation. One follicle outgrows the others and becomes dominant, while the others degenerate. At the same time, endometrial repair occurs in the uterus. Estrogens are the dominant ovarian hormone during the preovulatory phase. Ovulation is the rupture of the mature (graafian) follicle and the release of an egg (secondary oocyte) into the pelvic cavity. It is brought about by a surge of LH. During the postovulatory phase, both progesterone and estrogens are secreted in large quantities by the corpus luteum of the ovary, and the uterine endometrium prepares for implantation.
12. If fertilization and implantation do not occur, the corpus luteum degenerates, and the resulting low levels of progesterone and estrogens allow shedding of the stratum functionalis of the endometrium followed by the initiation of another reproductive cycle. If fertilization and implantation do occur, the corpus luteum is maintained by hCG. The corpus luteum and later the placenta secrete progesterone and estrogens to support pregnancy.

23.4 The Human Sexual Response

1. The similar sequence of changes experienced by both males and females before, during, and after intercourse is termed the human sexual response. It occurs in four phases: excitement, plateau, orgasm, and resolution.
2. During excitement, there is vasocongestion (engorgement with blood) of genital tissues. Other changes that occur during this phase include increased heart rate and blood pressure, increased skeletal muscle tone throughout the body, and hyperventilation.
3. During the plateau phase, the changes that began during the excitement phase are sustained at an intense level.

4. During orgasm, there are several rhythmic muscular contractions, accompanied by pleasurable sensations and a further increase in blood pressure, heart rate, and respiration rate.
5. During the resolution phase, genital tissues, heart rate, blood pressure, breathing, and muscle tone return to the unaroused state.

23.5 Sex Determination and Sex Differentiation

1. The sex chromosomes (X and Y) determine the genetic sex of an individual. In a genetic female, somatic cells contain two X chromosomes; in a genetic male, somatic cells contain one X and one Y chromosome.
2. The early embryo is bipotential (has the ability to form either male or female reproductive organs).
3. In males, the SRY gene on the Y chromosome causes the gonads to develop into testes. In females, the gonads differentiate into ovaries because the Y chromosome (and therefore the SRY gene) is not present.
4. In males, testosterone stimulates development of each Wolffian duct into an epididymis, vas deferens, ejaculatory duct, and seminal vesicle, and Müllerian-inhibiting substance (MIS) causes the Müllerian ducts to degenerate. In females, testosterone and MIS are absent; the Müllerian ducts develop into the fallopian tubes, uterus, and part of the vagina, and the Wolffian ducts degenerate.
5. The external genitalia are stimulated to develop into male structures (penis and scrotum) by the hormone dihydrotestosterone (DHT). The external genitalia develop into female structures (clitoris, labia minora, vestibule, and labia majora) when DHT is not produced, the normal situation in female embryos.

23.6 Aging and the Reproductive Systems

1. Puberty is the period when secondary sex characteristics begin to develop and the potential for sexual reproduction is reached.
2. The onset of puberty is marked by pulses or bursts of LH and FSH secretion, each triggered by a pulse of GnRH. The hormone leptin, released by adipose tissue, may signal the hypothalamus that long-term energy stores (triglycerides in adipose tissue) are adequate for reproductive functions to begin.
3. In females, the reproductive cycle normally occurs once each month from menarche, the first menses, to menopause, the permanent cessation of menses.
4. Between the ages of 40 and 50, the pool of remaining ovarian follicles becomes exhausted and levels of progesterone and estrogens decline. Most women experience a decline in bone mineral density after menopause, together with some atrophy of the ovaries, fallopian tubes, uterus, vagina, external genitalia, and breasts.
5. In older males, decreased levels of testosterone are associated with decreased muscle strength, waning sexual desire, and fewer viable sperm.

23.7 Pregnancy and Labor

1. Pregnancy refers to the period of time when a fertilized egg is developing within a female. It begins at fertilization and normally ends in birth. The embryonic period is from fertilization through the eighth week of development; the fetal period begins at week 9 and continues until birth.
2. During fertilization, a sperm cell penetrates an egg (secondary oocyte) and their pronuclei unite. Penetration of the zona pellucida is facilitated by enzymes in the sperm's acrosome. The resulting cell is a zygote. Normally, only one sperm cell fertilizes a secondary oocyte because of the fast and slow blocks to polyspermy.

3. Early rapid cell division of a zygote is called cleavage. The solid sphere of cells produced by cleavage is a morula. The morula develops into a blastocyst, a hollow ball of cells differentiated into a trophoblast and an inner cell mass. The attachment of a blastocyst to the endometrium is termed implantation; it occurs as a result of enzymatic degradation of the endometrium.
4. After implantation, the blastocyst continues to develop. By the third week after fertilization, cells in the embryo have formed three primary germ layers (gastrulation): ectoderm, mesoderm, and endoderm. These layers give rise to the various organs and tissues of the body. During the fourth through the eighth weeks of development, all major organs appear (organogenesis). The fetal period is primarily concerned with the growth and differentiation of tissues and organs that developed during the embryonic period.
5. There are four extraembryonic membranes that are associated with the embryo/fetus: amnion, chorion, yolk sac, and allantois.
6. The decidua refers to the stratum functionalis of the endometrium when a woman is pregnant. It consists of three regions: decidua basalis, decidua capsularis, and decidua parietalis. The placenta develops around the third month of pregnancy. It consists of chorionic villi (fetal portion) and decidua basalis (maternal portion). The placenta is the site of exchange of nutrients and wastes between the mother and fetus. It also functions as a protective barrier, stores nutrients, and produces several hormones to maintain pregnancy (estrogens, progesterone, relaxin, human chorionic somatomammotropin, and corticotropin-releasing hormone).
7. Teratogens are agents that cause physical defects in developing embryos. Among the more important teratogens are alcohol, pesticides, industrial chemicals, some prescription drugs, cocaine, LSD, nicotine, and ionizing radiation.
8. Several prenatal diagnostic tests are used to detect genetic disorders and to assess fetal well-being. These include fetal ultrasonography, in which an image of a fetus is displayed on a screen; amniocentesis, the withdrawal and analysis of amniotic fluid and the fetal cells within it; and chorionic villi sampling (CVS), which involves withdrawal of chorionic villi tissue for chromosomal analysis. CVS can be done earlier than amniocentesis, and the results are available more quickly, but it is slightly riskier than amniocentesis.
9. Labor is the process by which the fetus is expelled from the uterus through the vagina to the outside. It involves dilation of the cervix, expulsion of the fetus, and delivery of the placenta. Oxytocin stimulates uterine contractions via a positive feedback cycle.
10. Lactation refers to the production and ejection of milk by the mammary glands. Milk production is promoted by prolactin (PRL); milk ejection is stimulated by oxytocin.

23.8 Contraception and Abortion

1. Methods of contraception include complete abstinence, surgical sterilization (vasectomy, tubal ligation), hormonal methods (combined pill, minipill, contraceptive skin patch, vaginal contraceptive ring, emergency contraception, hormonal injections), intrauterine devices, spermicides, barrier methods (male condom, vaginal pouch, diaphragm, cervical cap), and periodic abstinence (rhythm and sympto-thermal methods).
2. An abortion is the premature expulsion from the uterus of the products of conception; it may be spontaneous or induced.

23.9 Infertility

1. Infertility is the inability to produce offspring.
2. Methods for assisting infertile couples in having a baby include in vitro fertilization, intracytoplasmic sperm injection, embryo transfer, and gamete intrafallopian transfer.

PONDER THIS

1. Lisa goes in to her obstetrician to discuss her plans for childbirth. She has decided that she would like parturition to be induced by using some form of medication. Her obstetrician tells her that they will use a drug called dinoprostone. When Lisa asks about the drug, her obstetrician tells her that it is a form of a synthetic prostaglandin that will carry out the same functions as prostaglandins when released in the body. Explain why this drug would help Lisa in the process of childbirth.

2. Alex and Stephanie want to have a baby but have been encountering some difficulty. After trying for a year they decide to see a fertility expert. After running many different tests, the fertility expert discovers that Alex has a very low sperm count but normal levels of testosterone in his blood. The fertility expert also discovers that, during a one month period, Stephanie's levels of estrogen do not change or fluctuate. What are the problems with Alex and Stephanie, respectively?

3. What would occur if a drug that functioned as a competitive inhibitor of progesterone were introduced during days 26–28 of the menstrual cycle without fertilization? With fertilization?

ANSWERS TO FIGURE QUESTIONS

23.1 As a result of crossing-over, the haploid gametes are genetically unlike each other and genetically unlike the starting cell that produced them.

23.2 Spermatogenesis occurs in the seminiferous tubules of the testes.

23.3 Spermatogenesis takes about 75 days.

23.4 The sperm head contains the nucleus with 23 highly condensed chromosomes and an acrosome that contains enzymes for penetration of an egg; the midpiece contains mitochondria for ATP production for locomotion and metabolism; the tail is a flagellum, which provides motility.

23.5 Nitric oxide activates a signaling pathway that ultimately causes the smooth muscle of penile arterioles to relax (dilate).

23.6 Ejaculation is a sympathetic reflex because sympathetic neurons stimulate contraction of smooth muscle in the walls of the epididymis, vas deferens, ejaculatory duct, and urinary sphincter and stimulate secretion of the accessory sex glands; ejaculation is also a somatic reflex because somatic motor neurons stimulate the contraction of skeletal muscles at the base of the penis.

23.7 Sertoli cells secrete inhibin.

23.8 The stratum functionalis of the endometrium sheds during menstruation.

23.9 The mature (graafian) follicle ruptures, releasing its egg into the pelvic cavity, from which it is normally swept into the fallopian tube.

23.10 Primary oocytes are present in the ovary at birth, so they are as old as the woman. In males, primary spermatocytes are continually being formed from stem cells (spermatogonia) and thus are only a few days old.

23.11 The principal estrogen in female plasma is β-estradiol.

23.12 The hormone responsible for ovulation is LH.

23.13 The effect of rising but moderate levels of estrogens is negative feedback inhibition of the secretion of GnRH, LH, and FSH.

23.14 The SRY gene on the Y chromosome is responsible for the development of the primitive gonads into testes.

23.15 The enzyme 5 α-reductase converts testosterone into dihydrotestosterone.

23.16 Capacitation is a series of functional changes that enhances the ability of sperm to fertilize the egg. It occurs after sperm have been deposited in the female reproductive tract.

23.17 A morula is a solid ball of cells; a blastocyst consists of a rim of cells (trophoblast) surrounding a cavity (blastocyst cavity) and an inner cell mass.

23.18 The blastocyst secretes enzymes that eat away the endometrial lining at the site of implantation.

23.19 Implantation occurs during the secretory phase of the uterine cycle.

23.20 The ectoderm gives rise to the epidermis of the skin and the structures of the nervous system.

23.21 The decidua basalis is the portion of the stratum functionalis between the embryo/fetus and the stratum basalis of the endometrium.

23.22 Complete dilation of the cervix marks the onset of the stage of expulsion.

23.23 Oxytocin also stimulates contraction of the uterus during delivery of a baby.

Appendix A: Measurements

U.S. CUSTOMARY SYSTEM

Parameter	Unit	Relation to Other U.S. Units	SI (Metric) Equivalent
Length	inch	1/12 foot	2.54 centimeters
	foot	12 inches	0.305 meter
	yard	36 inches	0.914 meter
	mile	5280 feet	1.609 kilometers
Mass	grain	1/1,000 pound	64.799 milligrams
	dram	1/16 ounce	1.772 grams
	ounce	16 drams	28.350 grams
	pound	16 ounces	453.6 grams
	ton	2000 pounds	907.18 kilograms
Volume (Liquid)	ounce	1/16 pint	29.574 milliliters
	pint	16 ounces	0.473 liter
	quart	2 pints	0.946 liter
	gallon	4 quarts	3.785 liters
Volume (Dry)	pint	1/2 quart	0.551 liter
	quart	2 pints	1.101 liters
	peck	8 quarts	8.810 liters
	bushel	4 pecks	35.239 liters

INTERNATIONAL SYSTEM (SI)

Base Units			Prefixes		
Unit	Quantity	Symbol	Prefix	Multiplier	Symbol
meter	length	m	tera-	$10^{12} = 1{,}000{,}000{,}000{,}000$	T
kilogram	mass	kg	giga-	$10^{9} = 1{,}000{,}000{,}000$	G
second	time	s	mega-	$10^{6} = 1{,}000{,}000$	M
liter	volume	L	kilo-	$10^{3} = 1000$	k
mole	amount of matter	mol	hecto-	$10^{2} = 100$	h
			deca-	$10^{1} = 10$	da
			deci-	$10^{-1} = 0.1$	d
			centi-	$10^{-2} = 0.01$	c
			milli-	$10^{-3} = 0.001$	m
			micro-	$10^{-6} = 0.000{,}001$	μ
			nano-	$10^{-9} = 0.000{,}000{,}001$	n
			pico-	$10^{-12} = 0.000{,}000{,}000{,}001$	p

TEMPERATURE CONVERSION

Fahrenheit (F) To Celsius (C)
°C = (°F − 32) ÷ 1.8

Celsius (C) To Fahrenheit (F)
°F = (°C × 1.8) + 32

U.S. TO SI (METRIC) CONVERSION

When You Know	Multiply by	To Find
inches	2.54	centimeters
feet	30.48	centimeters
yards	0.91	meters
miles	1.61	kilometers
ounces	28.35	grams
pounds	0.45	kilograms
tons	0.91	metric tons
fluid ounces	29.57	milliliters
pints	0.47	liters
quarts	0.95	liters
gallons	3.79	liters

SI (METRIC) TO U.S. CONVERSION

When You Know	Multiply by	To Find
millimeters	0.04	inches
centimeters	0.39	inches
meters	3.28	feet
kilometers	0.62	miles
liters	1.06	quarts
cubic meters	35.31	cubic feet
grams	0.035	ounces
kilograms	2.21	pounds

Appendix B: Periodic Table

The periodic table lists the known **chemical elements**, the basic units of matter. The elements in the table are arranged left to right in rows in order of their **atomic number**, the number of protons in the nucleus. Each horizontal row, numbered from 1 to 7, is a **period**. All elements in a given period have the same number of electron shells as their period number. For example, an atom of hydrogen or helium each has one electron shell, while an atom of potassium or calcium each has four electron shells. The elements in each column, or **group**, share chemical properties. For example, the elements in column IA are very chemically reactive, whereas the elements in column VIIIA have full electron shells and thus are chemically inert.

Scientists now recognize 118 different elements; 92 occur naturally on Earth, and the rest are produced from the natural elements using particle accelerators or nuclear reactors. Elements are designated by **chemical symbols**, which are the first one or two letters of the element's name in English, Latin, or another language.

Twenty-six of the 92 naturally occurring elements normally are present in your body. Of these, just four elements—oxygen (O), carbon (C), hydrogen (H), and nitrogen (N) (coded blue)—constitute about 96% of the body's mass. Eight others—calcium (Ca), phosphorus (P), potassium (K), sulfur (S), sodium (Na), chlorine (Cl), magnesium (Mg), and iron (Fe) (coded pink)—contribute 3.8% of the body's mass. An additional 14 elements, called **trace elements** because they are present in tiny amounts, account for the remaining 0.2% of the body's mass. The trace elements are aluminum, boron, chromium, cobalt, copper, fluorine, iodine, manganese, molybdenum, selenium, silicon, tin, vanadium, and zinc (coded yellow). Table 2.1 in Chapter 2 provides information about the main chemical elements in the body.

Key: 23 ← Atomic number; V ← Chemical symbol; 50.942 ← Atomic mass (weight)

Percentage of body mass
- 96% (4 elements) (blue)
- 3.8% (8 elements) (pink)
- 0.2% (14 elements) (yellow)

Period	IA	IIA	IIIB	IVB	VB	VIB	VIIB	VIIIB	VIIIB	VIIIB	IB	IIB	IIIA	IVA	VA	VIA	VIIA	VIIIA
1	1 Hydrogen H 1.0079																	2 Helium He 4.003
2	3 Lithium Li 6.941	4 Beryllium Be 9.012											5 Boron B 10.811	6 Carbon C 12.011	7 Nitrogen N 14.007	8 Oxygen O 15.999	9 Fluorine F 18.998	10 Neon Ne 20.180
3	11 Sodium Na 22.989	12 Magnesium Mg 24.305											13 Aluminum Al 26.9815	14 Silicon Si 28.086	15 Phosphorus P 30.974	16 Sulfur S 32.066	17 Chlorine Cl 35.453	18 Argon Ar 39.948
4	19 Potassium K 39.098	20 Calcium Ca 40.08	21 Scandium Sc 44.956	22 Titanium Ti 47.87	23 Vanadium V 50.942	24 Chromium Cr 51.996	25 Manganese Mn 54.938	26 Iron Fe 55.845	27 Cobalt Co 58.933	28 Nickel Ni 58.69	29 Copper Cu 63.546	30 Zinc Zn 65.38	31 Gallium Ga 69.723	32 Germanium Ge 72.59	33 Arsenic As 74.992	34 Selenium Se 78.96	35 Bromine Br 79.904	36 Krypton Kr 83.80
5	37 Rubidium Rb 85.468	38 Strontium Sr 87.62	39 Yttrium Y 88.905	40 Zirconium Zr 91.22	41 Niobium Nb 92.906	42 Molybdenum Mo 95.94	43 Technetium Tc (99)	44 Ruthenium Ru 101.07	45 Rhodium Rh 102.905	46 Palladium Pd 106.42	47 Silver Ag 107.868	48 Cadmium Cd 112.40	49 Indium In 114.82	50 Tin Sn 118.69	51 Antimony Sb 121.75	52 Tellurium Te 127.60	53 Iodine I 126.904	54 Xenon Xe 131.30
6	55 Cesium Cs 132.905	56 Barium Ba 137.33		72 Hafnium Hf 178.49	73 Tantalum Ta 180.948	74 Tungsten W 183.85	75 Rhenium Re 186.2	76 Osmium Os 190.2	77 Iridium Ir 192.22	78 Platinum Pt 195.08	79 Gold Au 196.967	80 Mercury Hg 200.59	81 Thallium Tl 204.38	82 Lead Pb 207.19	83 Bismuth Bi 208.980	84 Polonium Po (209)	85 Astatine At (210)	86 Radon Rn (222)
7	87 Francium Fr (223)	88 Radium Ra (226)		104 Rutherfordium Rf (267)	105 Dubnium Db (268)	106 Seaborgium Sg (271)	107 Bohrium Bh (272)	108 Hassium Hs (270)	109 Meitnerium Mt (276)	110 Ds (281)	111 Rg (280)	112 Uub (285)	113 Uut (284)	114 Uuq (289)	115 Uup (288)	116 Uuh (293)		118 Uuo (294)

57–71, Lanthanides

57 Lanthanum La 138.91	58 Cerium Ce 140.12	59 Praseodymium Pr 140.907	60 Neodymium Nd 144.24	61 Promethium Pm 144.913	62 Samarium Sm 150.35	63 Europium Eu 151.96	64 Gadolinium Gd 157.25	65 Terbium Tb 158.925	66 Dysprosium Dy 162.50	67 Holmium Ho 164.930	68 Erbium Er 167.26	69 Thulium Tm 168.934	70 Ytterbium Yb 173.04	71 Lutetium Lu 174.97

89–103, Actinides

89 Actinium Ac (227)	90 Thorium Th 232.038	91 Protactinium Pa (231)	92 Uranium U 238.03	93 Neptunium Np (237)	94 Plutonium Pu 244.064	95 Americium Am (243)	96 Curium Cm (247)	97 Berkelium Bk (247)	98 Californium Cf 242.058	99 Einsteinium Es (254)	100 Fermium Fm 257.095	101 Mendelevium Md 258.10	102 Nobelium No 259.10	103 Lawrencium Lr 260.105

Appendix C: Normal Values for Selected Blood Tests

The system of international (SI) units (Système Internationale d'Unités) is used in most countries and in many medical and scientific journals. Clinical laboratories in the United States, by contrast, usually report values for blood and urine tests in conventional units. The laboratory values in this Appendix give conventional units first, followed by SI equivalents in parentheses. Values listed for various blood tests should be viewed as reference values rather than absolute "normal" values for all well people. Values may vary due to age, gender, diet, and environment of the subject or the equipment, methods, and standards of the lab performing the measurement.

KEY TO SYMBOLS

g = gram	mL = milliliter
mg = milligram = 10^{-3} gram	μL = microliter
μg = microgram = 10^{-6} gram	mEq/L = milliequivalents per liter
U = units	mmol/L = millimoles per liter
L = liter	μmol/L = micromoles per liter
dL = deciliter	> = greater than; < = less than

BLOOD TESTS

Test (Specimen)	U.S. Reference Values (SI Units)	Values Increase In	Values Decrease In
Aminotransferases (serum)			
Alanine aminotransferase (ALT)	0–35 U/L (same)	Liver disease or liver damage due to toxic drugs.	
Aspartate aminotransferase (AST)	0–35 U/L (same)	Myocardial infarction, liver disease, trauma to skeletal muscles, severe burns.	Beriberi, uncontrolled diabetes mellitus with acidosis, pregnancy.
Ammonia (plasma)	20–120 μg/dL (12–55 μmol/L)	Liver disease, heart failure, emphysema, pneumonia, hemolytic disease of the newborn.	Hypertension.
Bilirubin (serum)	Conjugated: <0.5 mg/dL (<5.0 μmol/L) Unconjugated: 0.2–1.0 mg/dL (18–20 μmol/L) Newborn: 1.0–12.0 mg/dL (<200 μmol/L)	Conjugated bilirubin: liver dysfunction or gallstones. Unconjugated bilirubin: excessive hemolysis of red blood cells.	
Blood urea nitrogen (BUN) (serum)	8–26 mg/dL (2.9–9.3 mmol/L)	Kidney disease, urinary tract obstruction, shock, diabetes, burns, dehydration, myocardial infarction.	Liver failure, malnutrition, overhydration, pregnancy.
Carbon dioxide content (bicarbonate 1 dissolved CO_2) (whole blood)	Arterial: 19–24 mEq/L (19–24 mmol/L) Venous: 22–26 mEq/L (22–26 mmol/L)	Severe diarrhea, severe vomiting, starvation, emphysema, aldosteronism.	Renal failure, diabetic ketoacidosis, shock.
Cholesterol, total (plasma) **HDL cholesterol** (plasma) **LDL cholesterol** (plasma)	<200 mg/dL (<5.2 mmol/L) is desirable >40 mg/dL (>1.0 mmol/L) is desirable <130 mg/dL (<3.2 mmol/L) is desirable	Hypercholesterolemia, uncontrolled diabetes mellitus, hypothyroidism, hypertension, atherosclerosis, nephrosis.	Liver disease, hyperthyroidism, fat malabsorption, pernicious or hemolytic anemia, severe infections.

BLOOD TESTS Continued

Test (Specimen)	U.S. Reference Values (SI Units)	Values Increase In	Values Decrease In
Creatine (serum)	Males: 0.15–0.5 mg/dL (10–40 μmol/L) Females: 0.35–0.9 mg/dL (30–70 μmol/L)	Muscular dystrophy, damage to muscle tissue, electric shock, chronic alcoholism.	
Creatine kinase (CK), also known as **creatine phosphokinase (CPK)** (serum)	0–130 U/L (same)	Myocardial infarction, progressive muscular dystrophy, hypothyroidism, pulmonary edema.	
Creatinine (serum)	0.5–1.2 mg/dL (45–105 μmol/L)	Impaired renal function, urinary tract obstruction, giantism, acromegaly.	Decreased muscle mass, as occurs in muscular dystrophy or myasthenia gravis.
Electrolytes (plasma)	See Figure 20.7 in Chapter 20		
Gamma-glutamyl transferase (GGT) (serum)	0–30 U/L (same)	Bile duct obstruction, cirrhosis, alcoholism, metastatic liver cancer, congestive heart failure.	
Glucose (plasma)	70–110 mg/dL (3.9–6.1 mmol/L)	Diabetes mellitus, acute stress, hyperthyroidism, chronic liver disease, Cushing's syndrome.	Addison's disease, hypothyroidism, hyperinsulinism.
Hemoglobin (whole blood)	Males: 14–18 g/100 mL (140–180 g/L) Females: 12–16 g/100 mL (120–160 g/L) Newborns: 14–20 g/100 mL (140–200 g/L)	Polycythemia, congestive heart failure, chronic obstructive pulmonary disease, living at high altitude.	Anemia, severe hemorrhage, cancer, hemolysis, Hodgkin disease, nutritional deficiency of vitamin B_{12}, systemic lupus erythematosus, kidney disease.
Iron, total (serum)	Males: 80–180 mg/dL (14–32 μmol/L) Females: 60–160 mg/dL (11–29 μmol/L)	Liver disease, hemolytic anemia, iron poisoning.	Iron-deficiency anemia, chronic blood loss, pregnancy (late), chronic heavy menstruation.
Lactic dehydrogenase (LDH) (serum)	71–207 U/L (same)	Myocardial infarction, liver disease, skeletal muscle necrosis, extensive cancer.	
Lipids (serum) **Total** **Triglycerides**	 400–850 mg/dL (4.0–8.5 g/L) 10–190 mg/dL (0.1–1.9 g/L)	Hyperlipidemia, diabetes mellitus.	Fat malabsorption, hypothyroidism.
Platelet count (whole blood)	150,000–400,000/μL	Cancer, trauma, leukemia, cirrhosis.	Anemias, allergic conditions, hemorrhage.
Protein (serum) **Total** **Albumin** **Globulin**	 6–8 g/dL (60–80 g/L) 4–6 g/dL (40–60 g/L) 2.3–3.5 g/dL (23–35 g/L)	Dehydration, shock, chronic infections.	Liver disease, poor protein intake, hemorrhage, diarrhea, malabsorption, chronic renal failure, severe burns.
Erythrocyte (red blood cell) count (whole blood)	Males: 4.5–6.5 million/μL Females: 3.9–5.6 million/μL	Polycythemia, dehydration, living at high altitude.	Hemorrhage, hemolysis, anemias, cancer, overhydration.
Uric acid (urate) (serum)	2.0–7.0 mg/dL (120–420 μmol/L)	Impaired renal function, gout, metastatic cancer, shock,starvation.	
Leukocyte (white blood cell) count, total (whole blood)	5000–10,000/μL (See Figure 16.1 in Chapter 16 for relative percentages of different types of leukocytes.)	Acute infections, trauma, malignant diseases, cardiovascular diseases.	Diabetes mellitus, anemia.

Appendix D: Normal Values for Selected Urine Tests

URINE TESTS

Test (Specimen)	U.S. Reference Values (SI Units)	Clinical Implications
Amylase (2 hour)	35–260 Somogyi units/hr (6.5–48.1 units/hr)	Values increase in inflammation of the pancreas (pancreatitis) or salivary glands, obstruction of the pancreatic duct, and perforated peptic ulcer.
Bilirubin* (random)	Negative	Values increase in liver disease and obstructive biliary disease.
Blood* (random)	Negative	Values increase in renal disease, extensive burns, transfusion reactions, and hemolytic anemia.
Calcium (Ca^{2+}) (random)	10 mg/dL (2.5 mmol/L); up to 300 mg/24 hr (7.5 mmol/24 hr)	Amount depends on dietary intake; values increase in hyperparathyroidism, metastatic malignancies, and primary cancer of breasts and lungs; values decrease in hypoparathyroidism and vitamin D deficiency.
Casts (24 hour)		
Epithelial	Occasional	Values increase in nephrosis and heavy metal poisoning.
Granular	Occasional	Values increase in nephritis and pyelonephritis.
Hyaline	Occasional	Values increase in kidney infections.
Erythrocyte	Occasional	Values increase in glomerular membrane damage and fever.
Leukocyte	Occasional	Values increase in pyelonephritis, kidney stones, and cystitis.
Chloride (Cl^-) (24 hour)	140–250 mEq/24 hr (140–250 mmol/24 hr)	Amount depends on dietary salt intake; values increase in Addison's disease, dehydration, and starvation; values decrease in pyloric obstruction, diarrhea, and emphysema.
Color (random)	Yellow, straw, amber	Varies with many disease states, hydration, and diet.
Creatinine (24 hour)	Males: 1.0–2.0 g/24 hr (9–18 mmol/24 hr) Females: 0.8–1.8 g/24 hr (7–16 mmol/24 hr)	Values increase in infections; values decrease in muscular atrophy, anemia, and kidney diseases.
Glucose*	Negative	Values increase in diabetes mellitus, brain injury, and myocardial infarction.
Hydroxycorticosteroids (17-hydroxysteroids) (24 hour)	Males: 5–15 mg/24 hr (13–41 μmol/24 hr) Females: 2–13 mg/24 hr (5–36 mmol/24 hr)	Values increase in Cushing's syndrome, burns, and infections; values decrease in Addison's disease.
Ketone bodies* (random)	Negative	Values increase in diabetic acidosis, fever, anorexia, fasting, and starvation.
17-Ketosteroids (24 hour)	Males: 8–25 mg/24 hr (28–87 μmol/24 hr) Females: 5–15 mg/24 hr (17–53 μmol/24 hr)	Values decrease in surgery, burns, infections, adrenogenital syndrome, and Cushing's syndrome.

URINE TESTS Continued

Test (Specimen)	U.S. Reference Values (SI Units)	Clinical Implications
Odor (random)	Aromatic	Becomes acetonelike in diabetic ketosis.
Osmolality (24 hour)	500–800 mosmol/kg water (500–800 mmol/kg water)	Values increase in cirrhosis, congestive heart failure (CHF), and high-protein diets; values decrease in aldosteronism, diabetes insipidus, and hypokalemia.
pH* (random)	4.6–8.0	Values increase in urinary tract infections and severe alkalosis; values decrease in acidosis, emphysema, starvation, and dehydration.
Phenylpyruvic acid (random)	Negative	Values increase in phenylketonuria (PKU).
Potassium (K^+) (24 hour)	40–80 mEq/24 hr (40–80 mmol/24 hr)	Values increase in chronic renal failure, dehydration, starvation, and Cushing's syndrome; values decrease in diarrhea, malabsorption syndrome, and adrenal cortical insufficiency.
Protein* (albumin) (random)	Negative	Values increase in nephritis, fever, severe anemias, trauma, and hyperthyroidism.
Sodium (Na^+) (24 hour)	75–200 mEq/24 hr (75–200 mmol/24 hr)	Amount depends on dietary salt intake; values increase in dehydration, starvation, and diabetic acidosis; values decrease in diarrhea, acute renal failure, emphysema, and Cushing's syndrome.
Specific gravity* (random)	1.001–1.035 (same)	Values increase in diabetes mellitus and excessive water loss; values decrease in absence of antidiuretic hormone (ADH) and severe renal damage.
Urea (24 hour)	25–35 g/24 hr (420–580 mmol/24 hr)	Values increase in response to increased protein intake; values decrease in impaired renal function.
Uric acid (24 hour)	0.4–1.0 g/24 hr (1.5–4.0 mmol/24 hr)	Values increase in gout, leukemia, and liver disease; values decrease in kidney disease.
Urobilinogen* (24 hour)	1.7–6.0 μmol/24 hr	Values increase in anemias, hepatitis A (infectious), biliary disease, and cirrhosis; values decrease in cholelithiasis and renal insufficiency.
Volume, total (24 hour)	1000–2000 mL/24 hr (1.0–2.0 L/24 hr)	Varies with many factors.

* Test often performed using a **dipstick**, a plastic strip impregnated with chemicals that is dipped into a urine specimen to detect particular substances. Certain colors indicate the presence or absence of a substance and sometimes give a rough estimate of the amount(s) present.

Glossary

PRONUNCIATION KEY

1. The most strongly accented syllable appears in capital letters, for example, bilateral (bī-LAT-er-al) and diagnosis (dī-ag-NŌ-sis).

2. If there is a secondary accent, it is noted by a prime (′), for example, constitution (kon′-sti-TOO-shun) and physiology (fiz′-ē-OL-ō-jē). Any additional secondary accents are also noted by a prime, for example, decarboxylation (dē′-kar-bok′-si-LĀ-shun).

3. Vowels marked by a line above the letter are pronounced with the long sound, as in the following common words:

ā as in *māke*	*ō* as in *pōle*
ē as in *bē*	*ū* as in *cūte*
ī as in *īvy*	

4. Vowels not marked by a line above the letter are pronounced with the short sound, as in the following words:

a as in *above* or *at*	*o* as in *not*
e as in *bet*	*u* as in *bud*
i as in *sip*	

5. Other vowel sounds are indicated as follows:

oy as in *oil*
oo as in *root*

6. Consonant sounds are pronounced as in the following words:

b as in *bat*	*m* as in *mother*
ch as in *chair*	*n* as in *no*
d as in *dog*	*p* as in *pick*
f as in *father*	*r* as in *rib*
g as in *get*	*s* as in *so*
h as in *hat*	*t* as in *tea*
j as in *jump*	*v* as in *very*
k as in *can*	*w* as in *welcome*
ks as in *tax*	*z* as in *zero*
kw as in *quit*	*zh* as in *lesion*
l as in *let*	

A

Absorption (ab-SORP-shun) Intake of fluids or other substances by cells of the skin or mucous membranes; the passage of digested foods from the gastrointestinal tract into blood or lymph.

Acetylcholine (as′-ē-til-KŌ-lēn) **(ACh)** A neurotransmitter liberated by many peripheral nervous system neurons and some central nervous system neurons. It is excitatory at neuromuscular junctions and excitatory or inhibitory at other synapses.

Acquired immunodeficiency syndrome (AIDS) A disease caused by the human immunodeficiency virus (HIV). Characterized by a positive HIV-antibody test, low helper T cell count, and certain indicator diseases (for example Kaposi's sarcoma, pneumocystis carinii pneumonia, tuberculosis, fungal diseases). Other symptoms include fever or night sweats, coughing, sore throat, fatigue, body aches, weight loss, and enlarged lymph nodes.

Acrosome (AK-rō-sōm) A lysosomelike organelle in the head of a sperm cell containing enzymes that facilitate the penetration of a sperm cell into a secondary oocyte.

Actin (AK-tin) A contractile protein that is part of thin filaments in muscle fibers.

Action potential (AP) An electrical signal that propagates along the membrane of a neuron or muscle fiber (cell); a rapid change in membrane potential that involves a depolarization followed by a repolarization. Also called a nerve action potential or nerve impulse as it relates to a neuron, and a muscle action potential or muscle impulse as it relates to a muscle fiber.

Activation (ak′-ti-VĀ-shun) **energy** The minimum amount of energy required for a chemical reaction to occur.

Adaptation (ad′-ap-TĀ-shun) A decrease in the response of a sensory receptor during a maintained, constant stimulus; the adjustment of the pupil of the eye to changes in light intensity.

Adenosine triphosphate (a-DEN-ō-sēn trī-FOS-fāt) **(ATP)** The main energy currency in living cells; used to transfer the chemical energy needed for metabolic reactions. ATP consists of the purine base adenine and the five-carbon sugar ribose, to which are added, in linear array, three phosphate groups.

Adrenal cortex (a-DRĒ-nal KOR-teks) The outer portion of an adrenal gland, divided into three zones: the zona glomerulosa secretes mineralocorticoids, the zona fasciculata secretes glucocorticoids, and the zona reticularis secretes androgens.

Adrenal glands Two glands located above each kidney. Also called the suprarenal (soo′-pra-RĒ-nal) glands.

Adrenal medulla (me-DUL-la) The inner part of an adrenal gland, consisting of cells that secrete epinephrine, norepinephrine, and a small amount of dopamine in response to stimulation by sympathetic preganglionic neurons.

Adrenergic (ad′-ren-ER-jik) **neuron** A neuron that releases norepinephrine (noradrenaline) or epinephrine (adrenaline) as its neurotransmitter.

Adrenocorticotropic (ad-rē-nō-kor-ti-kō-TRŌP-ik) **hormone (ACTH)** A hormone produced by the anterior pituitary that influences the production and secretion of cortisol by the adrenal cortex.

Aerobic (ār-Ō-bik) Requiring molecular oxygen.

Agonist A substance that binds to and activates a receptor, mimicking the effect of the natural neurotransmitter or hormone.

Agglutination (a-gloo-ti-NĀ-shun) Clumping of microorganisms or blood cells, typically due to an antigen–antibody reaction.

Aldosterone (al-DOS-ter-ōn) A mineralocorticoid produced by the adrenal cortex that promotes sodium and water reabsorption by the kidneys and potassium excretion in urine.

Allergen (AL-er-jen) An antigen that evokes a hypersensitivity reaction.

All-or-none-principle If a stimulus depolarizes a neuron to threshold, the neuron fires at its maximum voltage (all); if threshold is not reached, the neuron does not fire at all (none). Above threshold, stronger stimuli do not produce stronger action potentials.

Alpha (AL-fa) **cell** A type of cell in the pancreatic islets (islets of Langerhans) in the pancreas that secretes the hormone glucagon.

Alpha (α) receptor A type of adrenergic receptor for norepinephrine and epinephrine; present on many visceral effectors innervated by sympathetic postganglionic neurons.

Alveolar macrophage Highly phagocytic cell found in the alveolar walls of the lungs.

Alveolus (al-VĒ-ō-lus) A small hollow or cavity; an air sac in the lungs; milk-secreting portion of a mammary gland. Plural is alveoli (al-VĒ-ō-lī).

Alzheimer's (ALTZ-hī-mers) **disease (AD)** Disabling neurological disorder characterized by dysfunction and death of specific cerebral neurons, resulting in widespread intellectual impairment, personality changes, and fluctuations in alertness.

Amenorrhea (ā-men-ō-RĒ-a) Absence of menstruation.

Ampulla of Vater (VA-ter) A small, raised area in the duodenum where the combined common bile duct and main pancreatic duct empty into the duodenum. Also called the hepatopancreatic (hep′-a-tō-pan′-krē-A-tik) ampulla.

Amnion (AM-nē-on) A thin, protective membrane that holds the fetus suspended in amniotic fluid. Also called the "bag of waters."

Anabolism (a-NAB-ō-lizm) The formation of complex chemical substances from smaller, simpler components.

Anaerobic (an-ar-Ō-bik) Not requiring oxygen.

Analgesia (an-an-JĒ-zē-a) Pain relief; absence of the sensation of pain.

Anaphase (AN-a-fāz) The third stage of mitosis in which the sister chromatids that have separated at the centromeres move to opposite poles of the cell.

Anatomic dead space The volume of the conducting zone (nose to the terminal bronchioles), totaling about 150 mL of the 500 mL in a quiet breath (tidal volume). Air in the anatomic dead space does not reach the alveoli to participate in gas exchange.

Androgens (AN-drō-jenz) Masculinizing sex hormones produced by the testes in males and the adrenal cortex in both sexes; responsible for libido (sexual desire). The two main androgens are testosterone and dihydrotestosterone.

Anemia (a-NĒ-mē-a) Condition of the blood in which the number of functional erythrocytes or their hemoglobin content is below normal.

Angina pectoris A pain in the chest related to reduced coronary circulation due to coronary artery disease (CAD) or spasms of vascular smooth muscle in coronary arteries.

Angiogenesis (an′-jē-ō-JEN-e-sis) The growth of new networks of blood vessels.

Antagonist (an-TAG-ō-nist) A substance that binds to and blocks a receptor, preventing the natural neurotransmitter or hormone from exerting Its effect.

Anterior pituitary Anterior lobe of the pituitary gland. Also called the adenohypophysis (ad′-e-nō-hī-POF-i-sis).

Antibody (AN-ti-bod′-ē) **(Ab)** A protein produced by plasma cells in response to a specific antigen. The antibody combines with that antigen to neutralize, inhibit, or destroy it. Also called an immunoglobulin (im-ū-nō-GLOB-ū-lin) or Ig.

Anticoagulant (an-tī-cō-AG-ū-lant) A substance that can delay, suppress, or prevent the clotting of blood.

Antidiuretic hormone (ADH) Hormone produced by neurosecretory cells in the paraventricular and supraoptic nuclei of the hypothalamus that stimulates water reabsorption from kidney tubule cells into the blood and vasoconstriction of arterioles. Also called vasopressin (vāz-ō-PRES-in).

Antigen (AN-ti-jen) **(Ag)** A substance that has immunogenicity (the ability to provoke an immune response) and reactivity (the ability to react with the antibodies or cells that result from the immune response); derived from the term *antibody generator*. Also termed a complete antigen.

Antigen-presenting cell (APC) Special class of migratory cell that processes and presents antigens to T cells during an immune response. APCs include macrophages, B cells, and dendritic cells, which are present in the skin, mucous membranes, and lymph nodes.

Anus (Ā-nus) The outlet of the rectum.

Aortic (ā-OR-tik) **body** Cluster of chemoreceptors on or near the arch of the aorta that respond to changes in blood levels of oxygen, carbon dioxide, and hydrogen ions (H^+).

Apnea (AP-nē-a) Temporary cessation of breathing.

Apoptosis (ap′-ōp-TŌ-sis *or* ap-ō-TŌ-sis) Programmed cell death; a normal type of cell death that removes unneeded cells during embryological development, regulates the number of cells in tissues, and eliminates many potentially dangerous cells such as cancer cells.

Aqueous humor (ĀK-wē-us HŪ-mer) The watery fluid, similar in composition to cerebrospinal fluid, that fills the anterior cavity of the eye.

Arousal (a-ROW-zal) Awakening from sleep, a response due to stimulation of the reticular activating system (RAS).

Arrhythmia An irregular heart rhythm. Also called a dysrhythmia.

Arteriole (ar-TĒ-rē-ōl) A small microscopic artery that delivers blood to a capillary.

Arteriosclerosis (ar-tē-rē-ō-skle-RŌ-sis) Group of diseases characterized by thickening of the walls of arteries and loss of elasticity.

Artery (AR-ter-ē) A blood vessel that carries blood away from the heart.

Asthma (AZ-ma) Usually an allergic reaction characterized by smooth muscle spasms in bronchi resulting in wheezing and difficult breathing.

Astigmatism (a-STIG-ma-tizm) An irregular curvature of the lens or cornea of the eye causing the image to be out of focus and producing faulty vision.

Astrocyte (AS-trō-sīt) A neuroglial cell having a star shape that participates in brain development and the metabolism of neurotransmitters, helps form the blood–brain barrier, helps maintain the proper balance of K^+ for generation of action potentials, and provides a link between neurons and blood vessels.

Atherosclerosis (ath-er-ō-skle-RŌ-sis) A progressive disease characterized by the formation in the walls of large and medium-sized arteries of lesions called atherosclerotic plaques.

Atherosclerotic (ath-er-ō-skle-RO-tik) **plaque** A lesion that results from accumulated cholesterol and smooth muscle fibers (cells) of an artery; may become obstructive.

Atom Unit of matter that makes up a chemical element; consists of a nucleus (containing positively charged protons and uncharged neutrons) and negatively charged electrons that orbit the nucleus.

Atresia (a-TRĒ-zē-a) Degeneration and reabsorption of an ovarian follicle before it fully matures and ruptures.

Atrial natriuretic peptide (ANP) Peptide hormone, produced by the atria of the heart, that causes natriuresis (urinary excretion of sodium).

Atrioventricular (ā′-trē-ō-ven-TRIK-ū-lar) **(AV) bundle** The part of the conduction system of the heart that begins at the atrioventricular (AV) node, passes through the cardiac skeleton separating the atria and the ventricles, then extends a short distance down the interventricular septum before splitting into right and left bundle branches. Also called the bundle of His (HIZ).

Atrioventricular (AV) node The part of the conduction system of the heart made up of a compact mass of conducting cells located in the septum between the two atria.

Atrioventricular (AV) valve A heart valve, made up of membranous flaps or cusps, that allows blood to flow in one direction only, from an atrium into a ventricle.

Atrium (Ā-trē-um) An upper chamber of the heart. Plural is atria.

Auditory ossicle (Aw-di-tō-rē OS-si-kul) One of the three small bones of the middle ear called the malleus, incus, and stapes.

Auscultation (aws-kul-TĀ-shun) Examination by listening to sounds in the body.

Autocrine A type of local mediator that acts on the same cell that secreted it.

Autoimmunity An immunological response against a person's own tissues.

Autolysis (aw-TOL-i-sis) Self-destruction of cells by their own lysosomal digestive enzymes after death or in a pathological process.

Autonomic ganglion (aw′-tō-NOM-ik GANG-lē-on) A cluster of cell bodies of sympathetic or parasympathetic neurons located in the peripheral nervous system.

Autonomic nervous system (ANS) The portion of the peripheral nervous system that conveys output to cardiac muscle, smooth muscle, and glands.

Autophagy (aw-TOF-a-jē) Process by which worn-out organelles are digested within lysosomes.

Autopsy The examination of the body after death.

Autorhythmicity (aw′-tō-rith-MIS-i-tē) The ability to repeatedly generate spontaneous action potentials.

Axon (AK-son) The usually single, long process of a neuron that propagates an action potential toward the axon terminals.

B

Baroreceptor (bar′-ō-rē-SEP-tor) Sensory receptor that responds to changes in blood pressure.

Basal nuclei Paired clusters of gray matter deep in each cerebral hemisphere including the globus pallidus, putamen, and caudate nucleus.

Basilar (BĀS-i-lar) **membrane** A membrane in the cochlea of the internal ear that separates the cochlear duct from the scala tympani and on which the organ of Corti rests.

Basophil (BĀ-sō-fil) A type of leukocyte characterized by a pale nucleus and large granules that stain blue-purple with basic dyes.

Beta (BĀ-ta) **cell** A type of cell in the pancreatic islets (islets of Langerhans) in the pancreas that secretes the hormone insulin. Also called a B cell.

Beta (β) receptor A type of adrenergic receptor for norepinephrine and epinephrine; found on many visceral effectors innervated by sympathetic postganglionic neurons.

Bicuspid (mitral) valve Atrioventricular (AV) valve on the left side of the heart. Also called the mitral valve or left atrioventricular valve.

Bile (BĪL) A secretion of the liver consisting of water, bile salts, bile pigments, cholesterol, lecithin, and several ions that emulsifies lipids prior to their digestion.

Bilirubin (bil-ē-ROO-bin) A yellow-orange pigment that is one of the end products of hemoglobin breakdown in hepatocytes and is excreted as a waste material in bile.

Blastocyst (BLAS-tō-sist) In the development of an embryo, a hollow ball of cells that consists of a blastocoel (the internal cavity), trophoblast (outer cells), and inner cell mass.

Blood The fluid that circulates through the heart and blood vessels and that constitutes the chief means of transport within the body.

Blood pressure (BP) Force exerted by blood against the walls of blood vessels due to contraction of the heart and influenced by the elasticity of the vessel walls; clinically, a measure of the pressure in arteries during ventricular systole and ventricular diastole.

Blood reservoir (REZ-er-vwar) Systemic veins and venules that contain large amounts of blood that can be moved quickly to parts of the body requiring the blood.

Blood–brain barrier (BBB) A barrier consisting of specialized brain capillaries and astrocytes that prevents the passage of materials from the blood to the cerebrospinal fluid and brain.

Blood–testis barrier A barrier formed by Sertoli cells that prevents an immune response against antigens produced by spermatogenic cells by isolating the cells from the blood.

Body fluids Dilute, watery solutions containing dissolved chemicals that are found inside cells as well as surrounding them.

Bolus (BŌ-lus) A soft, rounded mass, usually food, that is swallowed.

Bony labyrinth (LAB-i-rinth) A series of cavities within the temporal bone of the cranium forming the vestibule, cochlea, and semicircular canals of the inner ear.

Bowman's (BŌ-manz) **capsule** A double-walled epithelial cup at the beginning of a nephron that encloses the glomerular capillaries. Also called the glomerular capsule.

Bradycardia (brād′-i-KAR-dē-a) A slow resting heart or pulse rate (under 60 beats per minute).

Brain The part of the central nervous system contained within the cranium.

Bronchial (BRON-kē-al) **tree** The trachea, bronchi, and their branching structures.

Bronchiole (BRONG-kē-ōl) Branch of a tertiary bronchus further dividing into terminal bronchioles, which divide into respiratory bronchioles.

Buffer system A weak acid and the salt of that acid (which functions as a weak base). Buffers prevent drastic changes in pH by converting strong acids and bases to weak acids and bases.

Bulbourethral (bul′-bō-ū-RĒ-thral) **gland** One of a pair of glands located below the prostate that secretes an alkaline fluid into the urethra. Also called a Cowper's (KOW-perz) gland.

Bulimia (boo-LĒ-mē-a) A disorder characterized by overeating at least twice a week followed by purging by self-induced vomiting, strict dieting or fasting, vigorous exercise, or use of laxatives or diuretics. Also called binge–purge syndrome.

C

Calcitonin (kal-si-TŌ-nin) **(CT)** A hormone produced by the parafollicular cells of the thyroid gland that can lower the amount of blood calcium and phosphates by inhibiting bone resorption (breakdown of bone extracellular matrix) and by accelerating uptake of calcium and phosphates into bone matrix.

Cancer A group of diseases characterized by uncontrolled or abnormal cell division.

Capacitation (ka-pas′-i-TĀ-shun) The functional changes that sperm undergo in the female reproductive tract that allow them to fertilize a secondary oocyte.

Carbohydrate Organic compound consisting of carbon, hydrogen, and oxygen; the ratio of hydrogen to oxygen atoms is usually 2:1. Examples include sugars, glycogen, starches, and glucose.

Cardiac cycle A complete heartbeat consisting of systole (contraction) and diastole (relaxation) of both atria plus systole and diastole of both ventricles.

Cardiac output (CO) Volume of blood ejected from the left ventricle (or the right ventricle) into the aorta (or pulmonary trunk) each minute.

Cardiovascular (kar-dē-ō-VAS-kū-lar) **(CV) center** Groups of neurons scattered within the medulla oblongata that regulate heart rate, force of contraction, and blood vessel diameter.

Cardiovascular (kar-dē-ō-VAS-kū-lar) **system** Body system that consists of blood, the heart, and blood vessels.

Carotid (ka-ROT-id) **body** Cluster of chemoreceptors on or near the carotid sinus that respond to changes in blood levels of oxygen, carbon dioxide, and hydrogen ions.

Carotid sinus (SĪ-nus) A dilated region of the internal carotid artery just above where it branches from the common carotid artery; it contains baroreceptors that monitor blood pressure.

Carrier protein A protein that undergoes a conformational change to move a solute across the plasma membrane. Also called a transporter.

Catabolism (ka-TAB-ō-lizm) The breakdown of complex chemical substances into simpler components.

Catalyst Chemical compounds that speed up chemical reactions by lowering the activation energy needed for a reaction to occur.

Cataract (KAT-a-rakt) Loss of transparency of the lens of the eye or its capsule or both.

Cecum (SĒ-kum) A blind pouch at the beginning of the large intestine that attaches to the ileum.

Cell The basic structural and functional unit of all organisms; the smallest structure capable of performing all activities vital to life.

Cell cycle Growth and division of a single cell into two identical cells; consists of interphase and cell division.

Cell division Process by which a cell reproduces itself that consists of a nuclear division (mitosis) and a cytoplasmic division (cytokinesis); types include somatic and reproductive cell division.

Cellular respiration The process by which a nutrient molecule such as glucose, a fatty acid, or an amino acid is broken down in the presence of oxygen to form carbon dioxide (CO_2), water, and energy (ATP and heat).

Central nervous system (CNS) That portion of the nervous system that consists of the brain and spinal cord.

Centrioles (SEN-trē-ōlz) Paired, cylindrical structures of a centrosome, each consisting of a ring of microtubules and arranged at right angles to each other.

Centromere (SEN-trō-mēr) The constricted portion of a chromosome where the two sister chromatids are joined; serves as the point of attachment for the microtubules that pull chromatids during anaphase of cell division.

Centrosome (SEN-trō-sōm) A dense network of small protein fibers near the nucleus of a cell, containing a pair of centrioles and pericentriolar material.

Cerebellum (ser′-e-BEL-um) The part of the brain lying posterior to the medulla oblongata and pons; governs balance and coordinates skilled movements.

Cerebrospinal (se-rē′-brō-SPĪ-nal) **fluid (CSF)** A fluid produced by choroid plexuses in the ventricles of the brain. The fluid circulates in the ventricles, the central canal, and the subarachnoid space around the brain and spinal cord.

Cerumen (se-ROO-men) Waxlike secretion produced by glands in the external auditory canal. Also termed earwax.

Cervix (SER-viks) Neck; any constricted portion of an organ, such as lower cylindrical part of the uterus.

Chemical reaction The formation of new chemical bonds or the breaking of old chemical bonds between atoms.

Chemistry (KEM-is-trē) The science of the structure and interactions of matter.

Chemoreceptor (kē′-mō-rē-SEP-tor) Sensory receptor that detects the presence of a specific chemical.

Chief cell The secreting cell of a gastric gland that produces pepsinogen, the precursor of the enzyme pepsin, and the enzyme gastric lipase. Also called a zymogenic (zī′-mō-JEN-ik) cell. Cell in the parathyroid glands that secretes parathyroid hormone (PTH).

Cholesterol (kō-LES-te-rol) Classified as a lipid, the most abundant steroid in animal tissues; located in cell membranes and used for the synthesis of steroid hormones and bile salts.

Cholinergic (kō-lin-ER-jik) **neuron** A neuron that liberates acetylcholine as its neurotransmitter.

Chordae tendineae (KOR-dē TEN-din-ē-ē) Tendonlike, fibrous cords that connect atrioventricular valves of the heart with papillary muscles.

Chorion (KŌ-rē-on) The most superficial fetal membrane that becomes the principal embryonic portion of the placenta; serves a protective and nutritive function.

Choroid (KŌ-royd) A dark, vascular component of the middle layer of the eye.

Choroid plexus (PLEK-sus) A network of capillaries and ependymal cells in the ventricles of the brain that produces cerebrospinal fluid.

Chromaffin (KRŌ-maf-in) **cell** Cell that has an affinity for chromium salts due in part to the presence of the precursors of the neurotransmitter epinephrine; found, among other places, in the adrenal medulla.

Chromatin (KRŌ-ma-tin) The threadlike mass of genetic material, consisting of DNA and histone proteins, that is present in the nucleus of a nondividing or interphase cell.

Chromosome (KRŌ-mō-sōm) One of the small, threadlike structures in the nucleus of a cell, normally 46 in a human diploid cell, that bears the genetic material; composed of DNA and proteins (histones) that form a delicate chromatin thread during interphase; becomes packaged into compact rodlike structures that are visible under the light microscope during cell division.

Chyme (KĪM) The semifluid mixture of partly digested food and digestive secretions found in the stomach and small intestine during digestion of a meal.

Cilia (SIL-ē-a) A hair or hairlike process projecting from a cell that may be used to move the entire cell or to move substances along the surface of the cell. Singular is cilium.

Ciliary (SIL-ē-ar′-ē) **body** The part of the middle layer of the eye that secretes aqueous humor and promotes accommodation.

Circadian (ser-KĀ-dē-an) **rhythm** The pattern of biological activity on a 24-hour cycle, such as the sleep–wake cycle.

Circular folds Permanent folds of the mucosa and submucosa of the small intestine that increase the surface area for absorption. Also called plicae circulares (PLĪ-kē SER-kū-lar-ēs).

Circulation time Time required for a drop of blood to pass through the pulmonary and systemic circulations; normally about 1 minute.

Cleavage (KLĒV-ij) The rapid mitotic divisions following the fertilization of a secondary oocyte, resulting in an increased number of progressively smaller cells, called blastomeres.

Clitoris (KLI-to-ris) An erectile organ of the female, located at the front end of the vulva, that is homologous to the male penis.

Clone (KLŌN) A population of identical cells.

Cochlea (KOK-lē-a) A winding, cone-shaped tube forming a portion of the inner ear and containing the organ of Corti.

Cochlear duct The membranous cochlea consisting of a spirally arranged tube enclosed in the bony cochlea and lying along its outer wall. Also called the scala media (SCA-la MĒ-dē-a).

Colon The portion of the large intestine consisting of ascending, transverse, descending, and sigmoid portions.

Colony-stimulating factor (CSF) One of a group of molecules that stimulates development of leukocytes.

Common bile duct A tube formed by the union of the common hepatic duct and the cystic duct that empties bile into the duodenum at the ampulla of Vater (hepatopancreatic ampulla).

Cone The type of photoreceptor in the retina that is specialized for highly acute color vision in bright light.

Concentration gradient A difference in the concentration of a chemical from one place to another.

Connective tissue One of the most abundant of the four basic tissue types in the body, performing the functions of binding and supporting; consists of relatively few cells in a generous matrix (the ground substance and fibers between the cells).

Consciousness (KON-shus-nes) A state of wakefulness in which an individual is fully alert, aware, and oriented, partly as a result of feedback between the cerebral cortex and reticular activating system.

Contractility (kon′-trak-TIL-i-tē) The ability of cells or parts of cells to generate force actively to undergo shortening for movements. Muscle fibers (cells) exhibit a high degree of contractility.

Control center Part of a feedback system that sets the range of values within which a controlled variable should be maintained, evaluates input from receptors, and generates output commands.

Convergence (con-VER-jens) A synaptic arrangement in which the synaptic end bulbs of several presynaptic neurons terminate on one postsynaptic neuron. The medial movement of the two eyeballs so that both are directed toward a near object being viewed in order to produce a single image.

Cornea (KOR-nē-a) The part of the outer layer of the eye that is transparent.

Corona radiata (kō-RŌ-na rā-dē-A-ta) The innermost layer of granulosa cells that is firmly attached to the zona pellucida around a secondary oocyte.

Coronary artery disease (CAD) A condition such as atherosclerosis that causes narrowing of coronary arteries so that blood flow to the heart is reduced. The result is coronary heart disease, in which the heart muscle receives inadequate blood flow due to an interruption of its blood supply.

Corpus albicans (KOR-pus AL-bi-kanz) A mass of white fibrous scar tissue in the ovary that forms after the corpus luteum regresses.

Corpus luteum (LOO-tē-um) A yellowish body in the ovary formed when a follicle has discharged its secondary oocyte; secretes estrogens, progesterone, relaxin, and inhibin.

Corticobulbar pathway Motor pathway that conveys information for voluntary control of skeletal muscles of the head.

Corticobulbar tract Motor (descending) tract that conveys information from the motor cortex to the brain stem for voluntary control of skeletal muscles of the head.

Corticospinal pathway Motor pathway that conveys information for voluntary control of skeletal muscles of the limbs and trunk.

Cranial nerve One of 12 pairs of nerves that connect the brain to sensory receptors and effectors in the head, neck, and many organs in the thoracic and abdominal cavities. Each is designated by a Roman numeral and a name.

Cushing's syndrome Condition caused by a hypersecretion of glucocorticoids characterized by spindly legs, "moon face," "buffalo hump," pendulous abdomen, flushed facial skin, poor wound healing, hyperglycemia, osteoporosis, hypertension, and increased susceptibility to disease.

Cytokinesis (sī′-tō-ki-NĒ-sis) Distribution of the cytoplasm into two separate cells during cell division; coordinated with nuclear division (mitosis).

Cytolysis (sī-TOL-i-sis) The rupture of living cells in which the contents leak out.

Cytoplasm (SĪ-tō-plasm) Cytosol plus all organelles except the nucleus.

Cytoskeleton Complex internal structure of cytoplasm consisting of microfilaments, microtubules, and intermediate filaments.

Cytosol (SĪ-tō-sol) Fluid portion of cytoplasm in which solutes are dissolved and organelles are suspended. Also called intracellular fluid.

D

Defecation (def-e-KĀ-shun) The discharge of feces from the rectum.

Deglutition (dē-gloo-TISH-un) The act of swallowing.

Dehydration (dē-hī-DRĀ-shun) Excessive loss of water from the body or its parts.

Dendrite (DEN-drīt) A neuronal process that carries electrical signals, usually graded potentials, toward the cell body.

Deoxyribonucleic (dē-ok′-sē-rī-bō-nū-KLĒ-ik) **acid (DNA)** A nucleic acid constructed of nucleotides consisting of one of four bases (adenine, cytosine, guanine, or thymine), deoxyribose, and a phosphate group. Encoded in the nucleotides is genetic information.

Diabetes mellitus (dī-a-BĒ-tēz MEL-i-tus) An endocrine disorder caused by an inability to produce or use insulin. It is characterized by the three "polys": polyuria (excessive urine production), polydipsia (excessive thirst), and polyphagia (excessive eating).

Diagnosis Distinguishing one disease from another or determining the nature of a disease from signs and symptoms by inspection, palpation, laboratory tests, and other means.

Dialysis The removal of waste products from blood by diffusion through a selectively permeable membrane.

Diaphragm (DĪ-a-fram) Any partition that separates one area from another, especially the dome-shaped skeletal muscle between the thoracic and abdominal cavities; a dome-shaped device that is placed over the cervix, usually with a spermicide, to prevent conception.

Diarrhea (dī-a-RĒ-a) Frequent defecation of liquid caused by increased motility of the intestines.

Diastole (dī-AS-tō-lē) In the cardiac cycle, the phase of relaxation or dilation of the heart muscle, especially of the ventricles.

Diastolic (dī-as-TOL-ik) **pressure (DP)** The force exerted by blood on arterial walls during ventricular relaxation; the lowest blood pressure measured in the large arteries, normally about 70 mmHg in a young adult.

Diencephalon (dī-en-SEF-a-lon) A part of the brain consisting of the thalamus, hypothalamus, and pineal gland.

Diffusion (di-FŪ-zhun) The random mixing of particles from one location to another because of the particles' kinetic energy.

Digestion (dī-JES-chun) The mechanical and chemical breakdown of food to simple molecules that can be absorbed and used by body cells.

Digestive system Body system that ingests food, breaks it down, processes it, and eliminates wastes from the body.

Direct motor pathways Motor tracts that convey information from the motor cortex to cause voluntary movements of skeletal muscles. Also called the pyramidal pathways.

Disease An illness characterized by a recognizable set of signs and symptoms.

Disorder Any abnormality of structure or function.

Dorsal column Sensory (ascending) tract that conveys information up the spinal cord to the brain for sensations of touch, pressure, vibration, and proprioception.

Dorsal column pathway Sensory pathway that conveys information for touch, pressure, vibration, and proprioception.

Dorsal root The structure composed of axons of sensory (afferent) neurons that joins a ventral root to form a spinal nerve. Also called a posterior root.

Down-regulation Phenomenon in which there is a decrease in the number of receptors in response to an excess of a hormone or neurotransmitter.

Dual innervation Innervation of most organs of the body by both sympathetic and parasympathetic neurons.

Duodenum (doo′-ō-DĒ-num or doo-OD-e-num) The first part of the small intestine.

E

Edema (e-DĒ-ma) An abnormal accumulation of interstitial fluid.

Effector (e-FEK-tor) An organ of the body, either a muscle or a gland, that is innervated by somatic or autonomic motor neurons.

Ejaculation (ē-jak-ū-LĀ-shun) The reflex ejection or expulsion of semen from the penis.

Ejaculatory (ē-JAK-ū-la-tō-rē) **duct** A tube that transports sperm from the vas deferens to the urethra.

Electrical excitability (ek-sīt′-a-BIL-i-tē) Ability to respond to certain stimuli by producing action potentials.

Electrical gradient A difference in electrical charges between two regions.

Electrocardiogram (e-lek′-trō-KAR-dē-ō-gram) A recording of the electrical changes that accompany the cardiac cycle that can be detected at the surface of the body; may be resting, stress, or ambulatory.

Electrochemical gradient The combined influence of the concentration gradient and the electrical gradient on the movement of a particular ion.

Embolus (EM-bō-lus) A blood clot, bubble of air or fat from broken bones, mass of bacteria, or other debris or foreign material transported by the blood.

Embryo (EM-brē-ō) The young of any organism in an early stage of development; in humans, the developing organism from fertilization to the end of the eighth week of development.

Emergent properties New properties that exist at a higher level of the body's organizational plan that are not present at lower levels.

Emission (ē-MISH-un) Propulsion of sperm into the urethra due to peristaltic contractions of the ducts of the testes, epididymides, and vas deferens as a result of sympathetic stimulation.

Emphysema (em-fi-SĒ-ma) A lung disorder in which alveolar walls disintegrate, producing abnormally large air spaces and loss of elasticity in the lungs; typically caused by exposure to cigarette smoke.

Emulsification (e-mul-si-fi-KĀ-shun) The dispersion of large lipid globules into smaller, uniformly distributed particles in the presence of bile.

Enamel (e-NAM-el) The hard, white substance covering the surface of a tooth.

Endocrine (EN-dō-krin) **gland** A gland that secretes hormones into interstitial fluid and then the blood; a ductless gland.

Endocrine (EN-dō-krin) **system** All endocrine glands and hormone-secreting cells.

Endocytosis (en′-dō-sī-TŌ-sis) The uptake into a cell of large molecules and particles by vesicles formed from the plasma membrane.

Endolymph (EN-dō-limf′) The fluid within the membranous labyrinth of the internal ear.

Endometrium The inner layer of the uterus.

Endoplasmic reticulum (en′-dō PLAS-mik re-TIK-ū-lum) **(ER)** A network of channels running through the cytoplasm of a cell that serves in intracellular transportation, support, storage, synthesis, and packaging of molecules. Portions of ER where ribosomes are attached to the outer surface are called rough ER; portions that have no ribosomes are called smooth ER.

Energy The capacity to do work.

Enteric (en-TER-ik) **nervous system (ENS)** One of the subdivisions of the autonomic nervous system that is embedded in the wall of the gastrointestinal (GI) tract; it governs motility and secretions of the GI tract.

Enteroendocrine (en-ter-ō-EN-dō-krin) **cell** A cell of the gastrointestinal tract that secretes a hormone.

Enzyme (EN-zīm) A substance that accelerates chemical reactions; an organic catalyst, usually a protein.

Eosinophil (ē-ō-SIN-ō-fil) A type of leukocyte characterized by granules that stain red or orange with acid dyes.

Ependymal (ep-EN-de-mal) **cells** Neuroglial cells that cover choroid plexuses and produce cerebrospinal fluid (CSF); they also line the ventricles of the brain and probably assist in the circulation of CSF.

Epididymis (ep′-i-DID-i-mis) A comma-shaped organ that lies along the border of the testis. Plural is epididymides (ep′-i-di-DIM-i-dēz).

Epiglottis (ep′-i-GLOT-is) A large, leaf-shaped piece of cartilage lying on top of the larynx; its unattached portion is free to move up and down to cover the rest of the larynx during swallowing.

Epinephrine (ep-ē-NEF-rin) Hormone secreted by the adrenal medulla that produces actions similar to those that result from sympathetic stimulation. Also called adrenaline (a-DREN-a-lin).

Epithelial (ep-i-THĒ-lē-al) **tissue** The tissue that covers body surfaces, lines hollow organs and ducts, and forms glands. Also called epithelium.

Equilibrium (ē-kwi-LIB-rē-um) The state of being balanced.

Equilibrium (ē-kwi-LIB-rē-um) **potential** The membrane potential at which the concentration gradient and electrical gradient for a particular ion are equal in magnitude but opposite in direction and there is no net movement of that ion across the plasma membrane.

Erection (ē-REK-shun) The enlarged and stiff state of the penis or clitoris resulting from the engorgement of the spongy erectile tissue with blood.

Erythrocytes Blood cells that contain the oxygen-carrying protein hemoglobin; responsible for oxygen transport throughout the body. Also called red blood cells (RBCs).

Erythropoietin (e-rith′-rō-POY-ē-tin) **(EPO)** A hormone released by the kidneys that stimulates erythrocyte production.

Esophagus (e-SOF-a-gus) The hollow muscular tube that connects the pharynx and the stomach.

Estrogens (ES-trō-jenz) Feminizing sex hormones produced by the ovaries; govern development of oocytes, maintenance of female reproductive structures, and appearance of secondary sex characteristics; also affect fluid and electrolyte balance, and protein anabolism.

Eupnea (ŪP-nē-a) Normal quiet breathing.

Eustachian (ū-STĀ-shun or ū-STĀ-kē-an) **tube** The tube that connects the middle ear with the pharynx. Also called the auditory tube or pharyngotympanic tube.

Exocytosis (ek-sō-sī-TŌ-sis) A process in which membrane-enclosed secretory vesicles form inside the cell, fuse with the plasma membrane, and release their contents into the interstitial fluid; achieves secretion of materials from a cell.

Expiration Breathing out; expelling air from the lungs into the atmosphere. Also called exhalation.

External auditory canal A curved tube in the temporal bone that leads to the middle ear.

Exteroceptor (EKS-ter-ō-sep′-tor) A sensory receptor that detects external stimuli.

Extracellular chemical messenger A molecule that is released by a cell, enters extracellular fluid, and then binds to a receptor on or in its target cell to cause a response.

Extracellular fluid (ECF) Fluid outside body cells, such as interstitial fluid and plasma.

F

Facilitated diffusion Process in which an integral membrane protein assists in the movement of a specific substance across the membrane.

Fallopian (fal-LŌ-pē-an) **tube** Duct that transports ova from the ovary to the uterus. Also called the uterine (Ū-ter-in) tube or oviduct.

Fat A triglyceride that is a solid at room temperature.

Feces (FĒ-sēz) Material discharged from the rectum and made up of bacteria, excretions, and food residue. Also called stool.

Feedback system Cycle of events in which the status of a controlled variable is monitored, evaluated, changed, remonitored, and reevaluated.

Feedforward control Events that occur in anticipation of a change in a controlled variable.

Female reproductive cycle General term for the ovarian and uterine cycles, the hormonal changes that accompany them, and cyclic changes in the breasts and cervix; includes changes in the endometrium of a nonpregnant female that prepares the lining of the uterus to receive a fertilized ovum.

Fertilization (fer-til-i-ZĀ-shun) Penetration of a secondary oocyte by a sperm cell, meiotic division of secondary oocyte to form an ovum, and subsequent union of the nuclei of the gametes.

Fetus (FĒ-tus) In humans, the developing organism in utero from the beginning of the third month to birth.

Fever Abnormally high body temperature due to a resetting of the hypothalamic thermostat.

Fight-or-flight response The effects produced upon stimulation of the sympathetic division of the autonomic nervous system. First of three stages of the stress response.

Filtration (fil-TRĀ-shun) The flow of a liquid through a filter (or membrane that acts like a filter) due to hydrostatic pressure; occurs in capillaries due to blood pressure.

Flagella (fla-JEL-a) Hairlike, motile processes on the extremity of a bacterium, protozoan, or sperm cell. Singular is flagellum.

Follicle-stimulating hormone (FSH) Hormone secreted by the anterior pituitary; initiates development of ova and stimulates the ovaries to secrete estrogens in females, and initiates sperm production in males.

Fovea (FŌ-vē-a) A depression in the center of the macula lutea of the retina, containing cones only. It is the area of highest visual acuity (sharpness of vision).

Free radical An atom or molecule with an unpaired electron in the outermost shell. It is unstable, highly reactive, and destroys nearby molecules.

G

G protein A membrane protein that binds guanosine nucleotides, either guanosine diphosphate (GDP) or guanosine triphosphate (GTP).

Gallbladder A small pouch, located below the liver, that stores bile and empties by means of the cystic duct.

Gamete (GAM-ēt) A male or female reproductive cell; a sperm cell or secondary oocyte.

Ganglion (GANG-glē-on) A group of neuronal cell bodies located in the peripheral nervous system (PNS). Plural is ganglia (GANG-glē-a).

Gastric (GAS-trik) **glands** Glands in the mucosa of the stomach composed of cells that empty their secretions into narrow channels called gastric pits.

Gastrointestinal (gas-trō-in-TES-tin-al) **(GI) tract** A continuous tube extending from the mouth to the anus. Also called the alimentary (al′-i-MEN-tar-ē) canal.

Gene (JĒN) Biological unit of heredity; a segment of DNA located in a definite position on a particular chromosome; a sequence of DNA that codes for a particular mRNA, rRNA, or tRNA.

Genome (JĒ-nōm) The complete set of genes of an organism.

Glomerular filtrate The fluid produced when blood is filtered by the filtration membrane in the glomeruli of the kidneys.

Glomerulus A rounded mass of nerves or blood vessels, especially the microscopic tuft of capillaries that is surrounded by Bowman's capsule of each kidney tubule. Plural is glomeruli.

Glucagon (GLOO-ka-gon) A hormone produced by the alpha cells of the pancreatic islets (islets of Langerhans) that increases blood glucose level.

Glucocorticoids (gloo′-kō-KOR-ti-koyds) Hormones secreted by the cortex of the adrenal gland, especially cortisol, that influence glucose metabolism.

Glucosuria The presence of glucose in the urine; may be temporary or pathological. Also called glycosuria.

Glycogen (GLĪ-kō-jen) A highly branched polymer of glucose containing thousands of subunits; functions as a compact store of glucose molecules in liver and muscle fibers (cells).

Goiter (GOY-ter) An enlarged thyroid gland.

Golgi (GOL-jē) **complex** An organelle in the cytoplasm of cells consisting of four to six flattened sacs (cisternae), stacked on one another, with expanded areas at their ends; functions in processing, sorting, packaging, and delivering proteins and lipids to the plasma membrane, lysosomes, and secretory vesicles.

Gonad (GŌ-nad) A gland that produces gametes and hormones; the ovary in the female and the testis in the male.

Graded potential A small deviation from the membrane potential that makes the membrane either less polarized (inside less negative) or more polarized (inside more negative).

Gray matter Aggregations of unmyelinated nervous tissue, including neuronal cell bodies, dendrites, unmyelinated axons, axon terminals, and neuroglia.

Growth An increase in size due to an increase in (1) the number of cells, (2) the size of existing cells as internal components increase in size, or (3) the size of intercellular substances.

Growth hormone (GH) Hormone secreted by the anterior pituitary that stimulates growth of body tissues, especially bone and muscle. Also known as somatotropin.

Gustation (gus-TĀ-shun) The sense of taste.

H

Hair root plexus (PLEK-sus) A type of sensory receptor that detects movements on the skin surface that disturb hairs.

Haustra (HAWS-tra) A series of pouches that characterize the colon; caused by tonic contractions of the teniae coli. Singular is haustrum.

Heart Organ of the cardiovascular system responsible for pumping blood throughout the body.

Heart block An arrhythmia (dysrhythmia) of the heart in which the atria and ventricles contract independently because of a blocking of electrical impulses through the heart at some point in the conduction system.

Heart murmur An abnormal sound that consists of a flow noise that is heard before, between, or after the normal heart sounds, or that may mask normal heart sounds.

Hematocrit (he-MAT-ō-krit) **(Hct)** The percentage of blood made up of erythrocytes. Usually measured by centrifuging a blood sample in a graduated tube and dividing the volume of erythrocytes by the total volume of blood in the sample.

Hematopoiesis (hem′-a-tō-poy-E-sis) Blood cell production, which occurs in red bone marrow after birth. Also called hemopoiesis (hēm-ō-poy-Ē-sis).

Hemodynamics (hē-mō-dī-NAM-iks) The forces involved in circulating blood throughout the body.

Hemoglobin (hē-mō-GLŌ-bin) A substance in erythrocytes consisting of the protein globin and the iron-containing red pigment heme that transports most of the oxygen and some carbon dioxide in blood.

Hemolysis (hē-MOL-i-sis) The escape of hemoglobin from the interior of an erythrocyte into the surrounding medium; results from disruption of the cell membrane by toxins or drugs, freezing or thawing, or hypotonic solutions.

Hemolytic disease of the newborn (HDN) A hemolytic anemia of a newborn child that results from the destruction of the infant's erythrocytes by antibodies produced by the mother; usually the antibodies are due to an Rh blood type incompatibility.

Hemophilia (hē′-mō-FIL-ē-a) A hereditary blood disorder in which a deficient production of certain factors involved in blood clotting results in excessive bleeding into joints, deep tissues, and elsewhere.

Hemorrhage (HEM-o-rij) Bleeding; the escape of blood from blood vessels, especially when the loss is profuse.

Hepatocyte (he-PAT-ō-cīt) A liver cell.

Homeostasis The condition in which the body's internal environment remains relatively constant within physiological limits.

Homologous (hō-MOL-ō-gus) **chromosomes** Two chromosomes that belong to a pair. Also called homologs.

Hormone (HOR-mōn) A secretion of an endocrine cell that is carried by the blood to a distant target cell.

Human chorionic gonadotropin (kō-rē-ON-ik gō-nad-ō-TRŌ-pin) **(hCG)** A hormone produced by the developing placenta that maintains the corpus luteum.

Hyperplasia (hī-per-PLĀ-zē-a) An increase in the number of normal cells in a tissue or organ, increasing its size.

Hypersecretion (hī′-per-se-KRĒ-shun) Overactivity of glands resulting in excessive secretion.

Hypertension (hī′-per-TEN-shun) High blood pressure.

Hypertonia (hī′-per-TŌ-nē-a) Increased muscle tone that is expressed as spasticity or rigidity.

Hypertonic (hī′-per-TON-ik) **solution** Solution that causes cells to shrink due to loss of water by osmosis.

Hypertrophy (hī-PER-trō-fē) An excessive enlargement or overgrowth of tissue without cell division.

Hyperventilation (hī′-per-ven-til-LĀ-shun) A rate of inspiration and expiration higher than that required to maintain a normal partial pressure of carbon dioxide in the blood.

Hyposecretion (hī′-pō-se-KRĒ-shun) Underactivity of glands resulting in diminished secretion.

Hypothalamus (hī′-pō-THAL-a-mus) The portion of the diencephalon beneath the thalamus.

Hypothermia (hī′-pō-THER-mē-a) Lowering of body temperature below 35°C (95°F); in surgical procedures, it refers to deliberate cooling of the body to slow down metabolism and reduce oxygen needs of tissues.

Hypotonia (hī′-pō-TŌ-nē-a) Decreased or lost muscle tone in which muscles appear flaccid.

Hypotonic (hī′-pō-TON-ik) **solution** Solution that causes cells to swell and perhaps rupture due to gain of water by osmosis.

Hypoxia (hī-POKS-ē-a) Lack of adequate oxygen at the tissue level.

I

Ileum (IL-ē-um) The terminal part of the small intestine.

Immunity (i-MŪ-ni-tē) The state of being resistant to damage or disease, particularly by invading pathogens and foreign proteins. Also called resistance.

Immune system System that provides the body with resistance to disease.

Immunoglobulin (im-ū-nō-GLOB-ū-lin) **(Ig)** A protein synthesized by plasma cells derived from B lymphocytes in response to a specific antigen. Also called an antibody.

Indirect motor pathways Motor tracts that convey information from the brain stem to cause involuntary movements that regulate posture, balance, and muscle tone. Also known as the extrapyramidal pathways.

Inhibin A hormone secreted by the gonads that inhibits release of follicle-stimulating hormone (FSH) by the anterior pituitary.

Inhibiting hormone Hormone secreted by the hypothalamus that can suppress secretion of hormones by the anterior pituitary.

Inspiration The act of drawing air into the lungs. Also called inhalation.

Insulin (IN-soo-lin) A hormone produced by the beta cells of a pancreatic islet (islet of Langerhans) that decreases the blood glucose level.

Integration The process by which several components work together for a common, unified purpose.

Integrins (IN-te-grinz) A family of transmembrane glycoproteins in plasma membranes that function in cell adhesion; they are present in hemidesmosomes, which anchor cells to a basement membrane, and they mediate adhesion of neutrophils to endothelial cells during emigration.

Intercalated (in-TER-ka-lāt-ed) **disc** An irregular transverse thickening of sarcolemma that contains desmosomes, which hold cardiac muscle fibers (cells) together, and gap junctions, which aid in conduction of muscle action potentials from one fiber to the next.

Intermediate filament Protein filament, ranging from 8 to 12 nm in diameter, that may provide structural reinforcement, hold organelles in place, and give shape to a cell.

Interneuron Neuron located entirely within the central nervous system; it integrates sensory input and motor output. Also called an association neuron.

Interoceptor (IN-ter-ō-sep′-tor) A sensory receptor that detects internal stimuli.

Interphase (IN-ter-fāz) The period of the cell cycle between cell divisions, consisting of the G_1 (gap or growth) phase, when the cell is engaged in growth, metabolism, and production of substances required for division; S (synthesis) phase, during which chromosomes are replicated; and G_2 phase.

Intracellular (in′-tra-SEL-ū-lar) **fluid (ICF)** Fluid located within cells. Also called cytosol.

Intrafusal (in′-tra-FŪ-sal) **fibers** Three to ten specialized muscle fibers (cells), partially enclosed in a spindle-shaped connective tissue capsule, that make up a muscle spindle.

Intraocular (in′-tra-OK-ū-lar) **pressure** Pressure in the eyeball, produced mainly by aqueous humor.

Ion channel An integral transmembrane protein containing a pore that allows passage of small, inorganic ions that are too hydrophilic to penetrate the nonpolar interior of the lipid bilayer.

Interstitial (in′-ter-STISH-al) **fluid** The portion of extracellular fluid that fills the microscopic spaces between the cells of tissues.

Iris The colored portion of the middle layer of the eye that contains circular and radial smooth muscle. The hole in the center of the iris is the pupil.

Ischemia (is-KĒ-mē-a) A lack of sufficient blood to a body part due to obstruction or constriction of a blood vessel.

Isotonic (ī′-sō-TON-ik) **solution** A solution having the same concentration of nonpenetrating solutes as cytosol.

J

Jaundice (JON-dis) A condition characterized by yellowness of the skin, the white portions of the eyes, mucous membranes, and body fluids because of a buildup of bilirubin.

Jejunum (je-JOO-num) The middle part of the small intestine.

Joint kinesthetic (kin′-es-THET-ik) **receptor** A proprioceptive receptor located in a joint, stimulated by joint movement.

Juxtaglomerular apparatus (JGA) Consists of the macula densa (cells of the distal tubule adjacent to the afferent and efferent arterioles) and juxtaglomerular cells (modified cells of the afferent and sometimes efferent arterioles); secretes renin when blood pressure starts to fall.

K

Kidney One of the paired reddish organs located near the lower back that regulates the composition, volume, and pressure of blood and produces urine.

Kinesthesia (kin′-es-THĒ-zē-a) The perception of the extent and direction of movement of body parts; this sense is possible due to action potentials generated by proprioceptors.

Kinetic energy The energy associated with matter in motion.

Kinetochore (ki-NET-ō-kor) Protein complex attached to the outside of a centromere to which kinetochore microtubules attach.

L

Labia majora (LĀ-bē-a ma-JŌ-ra) Two folds of skin extending downward and backward from the mons pubis of the female.

Labia minora (min-OR-a) Two folds of skin that are inward to the labia majora of the female.

Lactation (lak-TĀ-shun) The secretion and ejection of milk by the mammary glands.

Lacteal (LAK-tē-al) One of many lymphatic vessels in villi of the intestines that absorb triglycerides and other lipids from digested food.

Lamina propria (PRŌ-prē-a) Areolar connective tissue with elastic fibers and a plexus of veins; part of the mucosa of the organs such as the ureters, urinary bladder, and urethra.

Large intestine The portion of the gastrointestinal tract extending from the ileum of the small intestine to the anus, divided structurally into the cecum, colon, and rectum.

Larynx (LAR-ingks) The voice box, a short passageway that connects the pharynx with the trachea.

Lateral corticospinal tract Motor (descending) tract that conveys information from the motor cortex to the spinal cord for voluntary control of skeletal muscles in the distal parts of the limb.

Lens A transparent organ constructed of proteins (crystallins) located behind the pupil and iris of the eye and in front of the vitreous humor.

Leukemia (loo-KĒ-mē-a) A malignant disease of the blood-forming tissues characterized by either uncontrolled production and accumulation of immature leukocytes in which many cells fail to reach maturity (acute) or an accumulation of mature leukocytes in the blood because they do not die at the end of their normal life span (chronic).

Leukocytes (LOO-kō-sīt) Blood cells that are responsible for protecting the body from foreign substances via phagocytosis or immune reactions. Also called white blood cells (WBCs).

Ligand (LĪ-gand) A chemical substance that binds to a specific receptor.

Limbic system The part of the brain concerned with various aspects of emotion and behavior. It includes the cingulate gyrus, amygdala, hippocampus, dentate gyrus, and parahippocampal gyrus of the cerebrum; portions of the thalamus and hypothalamus; and the olfactory bulbs.

Lipid (LIP-id) An organic compound composed of carbon, hydrogen, and oxygen that is usually insoluble in water but soluble in alcohol, ether, and chloroform; examples include triglycerides (fats and oils), phospholipids, steroids, and eicosanoids.

Lipid bilayer Arrangement of phospholipid, glycolipid, and cholesterol molecules in two parallel sheets in which the hydrophilic "heads" face outward and the hydrophobic "tails" face inward; found in cellular membranes.

Lipoprotein (lip′-ō-PRŌ-tēn) One of several types of particles containing lipids (cholesterol and triglycerides) and proteins that make it water soluble for transport in the blood; high levels of low-density lipoproteins (LDLs) are associated with increased risk of atherosclerosis, whereas high levels of high-density lipoproteins (HDLs) are associated with decreased risk of atherosclerosis.

Liver Large organ that produces bile; synthesizes most plasma proteins; interconverts nutrients; detoxifies substances; stores glycogen, iron, and vitamins; carries on phagocytosis of worn-out blood cells and bacteria; and helps synthesize the active form of vitamin D.

Local mediator A molecule that acts on a target cell without entering the bloodstream.

Long-term potentiation (pō-ten′-shē-Ā-shun) **(LTP)** Prolonged, enhanced synaptic transmission that occurs at certain synapses within the hippocampus of the brain; believed to underlie some aspects of memory.

Loop of Henle The part of the renal tubule that receives fluid from the proximal tubule and transmits it to the distal tubule. Also called the nephron loop.

Lungs Main organs of respiration that lie on either side of the heart in the thoracic cavity.

Luteinizing (LOO-tē-in′-īz-ing) **hormone (LH)** A hormone secreted by the anterior pituitary that stimulates ovulation and formation of the corpus luteum in females; stimulates testosterone secretion by the testes in males.

Lymph (LIMF) Fluid confined in lymphatic vessels and flowing through the lymphatic system until it is returned to the blood.

Lymph node An oval or bean-shaped structure located along a lymphatic vessel.

Lymphocyte (LIM-fō-sīt) A type of leukocyte that helps carry out cell-mediated and antibody-mediated immune responses; found in blood and in lymphoid tissues.

Lysosome (LĪ-sō-sōm) An organelle in the cytoplasm of a cell, enclosed by a single membrane and containing powerful digestive enzymes.

Lysozyme (LĪ-sō-zīm) A bactericidal enzyme found in tears, saliva, and perspiration.

M

Macrophage (MAK-rō-fāj) Phagocytic cell derived from a monocyte; may be fixed or wandering.

Macula (MAK-ū-la) A discolored spot or a colored area; a small, thickened region on the wall of the utricle and saccule that contains receptors for static equilibrium.

Macula lutea (LOO-tē-a) The yellow spot in the center of the retina.

Major histocompatibility complex (MHC) antigens Surface proteins on leukocytes and other nucleated cells that are unique for each person (except for identical siblings); used to type tissues and help prevent rejection of transplanted tissues. Also known as human leukocyte antigens (HLA).

Mammary (MAM-ar-ē) **gland** Modified sweat gland of the female that produces milk for the nourishment of the young.

Mastication (mas′-ti-KĀ-shun) Chewing.

Mechanoreceptor (me-KAN-ō-rē-sep-tor) Sensory receptor that detects mechanical deformation of the receptor itself or adjacent cells; stimuli so detected include those related to touch, pressure, vibration, proprioception, hearing, equilibrium, and blood pressure.

Medulla oblongata (me-DOOL-la ob′-long-GA-ta) The most inferior part of the brain stem. Also termed the medulla.

Medullary respiratory center The portion of the respiratory center in the medulla that helps regulate breathing.

Meiosis (mī-Ō-sis) A type of cell division that occurs during production of gametes, involving two successive nuclear divisions that result in cells with the haploid (*n*) number of chromosomes.

Meissner (MĪS-ner) **corpuscle** A sensory receptor that detects the onset of touch and low-frequency vibrations. Also called a corpuscle of touch.

Melatonin (me-a-TŌN-in) A hormone secreted by the pineal gland that helps set the timing of the body's biological clock.

Membrane potential The voltage (electrical potential difference) that exists across the plasma membrane of a cell.

Membranous labyrinth (MEM-bra-nus LAB-i-rinth) The part of the labyrinth of the internal ear that is located inside the bony labyrinth and separated from it by the perilymph; made up of the semicircular ducts, the saccule and utricle, and the cochlear duct.

Menarche (me-NAR-kē) The first menses (menstrual flow) and beginning of the ovarian and uterine cycles.

Meninges (me-NIN-jēz) Three membranes covering the brain and spinal cord, called the dura mater, arachnoid mater, and pia mater. Singular is meninx (MEN-inks).

Menopause (MEN-ō-pawz) The termination of the menstrual cycles.

Menstruation (men′-stroo-Ā-shun) Periodic discharge of blood, tissue fluid, mucus, and epithelial cells that usually lasts for 5 days; caused by a sudden reduction in estrogens and progesterone. Also called the menstrual phase or menses.

Merkel disc A type of sensory receptor that responds to continuous touch and to pressure. Also called a type I cutaneous mechanoreceptor.

Metabolism (me-TAB-ō-lizm) All of the chemical reactions that occur within an organism, including the catabolic (decomposition) reactions and the anabolic (synthesis) reactions.

Metaphase (MET-a-fāz) The second stage of mitosis, in which chromatid pairs line up on the metaphase plate of the cell.

Metarteriole (met′-ar-TĒ-rē-ōl) A blood vessel that emerges from an arteriole, traverses a capillary network, and empties into a venule.

Microglia (mī-KROG-lē-a) Neuroglial cells that carry on phagocytosis.

Microtubule (mī-krō-TOO-būl) Cylindrical protein filament, from 18 to 30 nm in diameter, consisting of the protein tubulin; provides support, structure, and transportation.

Microvilli (mī-krō-VIL-ī) Microscopic, fingerlike projections of the plasma membranes of cells that increase surface area for absorption, especially in the small intestine and proximal tubules of the kidneys.

Micturition The act of expelling urine from the urinary bladder. Also called urination (ū-ri-NĀ-shun).

Midbrain The part of the brain between the pons and the diencephalon. Also called the mesencephalon (mes′-en-SEF-a-lon).

Mineralocorticoids (min′-er-al-ō-KORT-ti-koyds) A group of hormones of the adrenal cortex that help regulate sodium and potassium balance.

Mitochondrion (mī-tō-KON-drē-on) A double-membraned organelle that plays a central role in the production of ATP; known as the "powerhouse" of the cell. Plural is mitochondria.

Mitosis (mī-TŌ-sis) The orderly division of the nucleus of a cell that ensures that each new nucleus has the same number and kind of chromosomes as the original nucleus. The process includes the replication of chromosomes and the distribution of the two sets of chromosomes into two separate and equal nuclei.

Mitotic spindle Collective term for a football-shaped assembly of microtubules (nonkinetochore, kinetochore, and aster) that is responsible for the movement of chromosomes during cell division.

Molecule (mol′-e-KŪL) A combination of two or more atoms that share electrons.

Monocyte (MON-ō-sīt′) The largest type of leukocyte, characterized by a granular cytoplasm.

Monounsaturated fat A fatty acid that contains one double covalent bond between its carbon atoms; not completely saturated with hydrogen atoms. Plentiful in triglycerides of olive and peanut oils.

Mons pubis (MONZ PŪ-bis) The rounded, fatty prominence over the joint between the pubic bones, covered by coarse pubic hair.

Motor end plate Region of the sarcolemma of a muscle fiber (cell) that includes acetylcholine (ACh) receptors, which bind ACh released by synaptic end bulbs of somatic motor neurons.

Motor neuron (NOO-ron) Neuron that conveys action potentials away from the central nervous system to effectors in the periphery. Also called an efferent neuron.

Motor unit A motor neuron together with the muscle fibers (cells) it stimulates.

Mucosa (mū-KŌ-sa) A membrane that lines a body cavity that opens to the exterior. Also called the mucous membrane.

Muscarinic (mus′-ka-RIN-ik) **receptor** A type of cholinergic receptor for acetylcholine found on all visceral effectors innervated by parasympathetic postganglionic axons and on sweat glands innervated by cholinergic sympathetic postganglionic axons.

Muscle fatigue (fa-TĒG) Inability of a muscle to maintain its strength of contraction or tension; may be related to insufficient oxygen, depletion of glycogen, and/or lactic acid buildup.

Muscle spindle An encapsulated proprioceptor in a skeletal muscle, consisting of specialized intrafusal muscle fibers and nerve endings; detects static muscle length and changes in muscle length.

Muscle tissue A tissue that is specialized to contract. Types include skeletal, cardiac, and smooth.

Muscle tone A sustained, partial contraction of portions of a skeletal or smooth muscle in response to activation of stretch receptors or a baseline level of action potentials in the innervating motor neurons.

Muscular dystrophy (DIS-trō-fē) Inherited muscle-destroying diseases, characterized by degeneration of muscle fibers (cells) that causes progressive atrophy of the skeletal muscle.

Muscular system System that consists of all of the skeletal muscles of the body.

Mutation (mū-TĀ-shun) Any change in the sequence of bases in a DNA molecule resulting in a permanent alteration in some inheritable trait.

Myelin (MĪ-e-lin) **sheath** Multilayered lipid and protein covering, formed by Schwann cells and oligodendrocytes, around axons of many peripheral and central nervous system neurons.

Myenteric (mī-en-TER-ik) **plexus** A network of neurons located in the muscularis externa of the gastrointestinal tract. Also called the plexus of Auerbach (OW-er-bak).

Myocardial infarction (mī′-ō-KAR-dē-al in-FARK-shun) Gross necrosis of myocardial tissue due to interrupted blood supply. Also called a heart attack.

Myocardium (mī′-ō-KAR-dē-um) The middle layer of the heart wall, made up of cardiac muscle tissue, lying between the epicardium and the endocardium and constituting the bulk of the heart.

Myoglobin (mī-ō-GLŌB-in) The oxygen-binding, iron-containing protein present in the sarcoplasm of muscle fibers (cells); contributes the red color to muscle.

Myogram (MĪ-ō-gram) The record or tracing produced by a myograph, an apparatus that measures and records the force of muscular contractions.

Myometrium (mī′-ō-MĒ-trē-um) The middle layer of the uterus, composed of smooth muscle.

Myopia (mī-Ō-pē-a) Defect in vision in which objects can be seen distinctly only when very close to the eyes; nearsightedness.

Myosin (MĪ-ō-sin) The contractile protein that makes up the thick filaments of muscle fibers.

N

Necrosis A pathological type of cell death that results from disease, injury, or lack of blood supply in which many adjacent cells swell, burst, and spill their contents into the interstitial fluid, triggering an inflammatory response.

Negative feedback system A feedback system that reverses a change in a controlled variable.

Neoplasm A new growth that may be benign or malignant.

Nephron The functional unit of the kidney.

Nerve A bundle of neuronal axons in the peripheral nervous system.

Nervous system A network of billions of neurons and even more neuroglia that is organized into two main divisions: central nervous system (brain and spinal cord) and peripheral nervous system (nervous tissue outside the central nervous system).

Nervous tissue Tissue that can detect and respond to changes in the environment.

Neuroglia (noo-RŌG-lē-a) Cells of the nervous system that perform various supportive functions. The neuroglia of the central nervous system are the astrocytes, oligodendrocytes, microglia, and ependymal cells; neuroglia of the peripheral nervous system include Schwann cells and satellite cells. Also called glia (GLĒ-a).

Neuromuscular (noo-rō-MUS-kū-lar) **junction (NMJ)** A synapse between the axon terminals of a motor neuron and the sarcolemma of a muscle fiber (cell).

Neuron (NOO-ron) A nerve cell, consisting of a cell body, dendrites, and an axon.

Neurosecretory (noo-rō-SĒK-re-tō-rē) **cell** A neuron that secretes a hypothalamic releasing hormone or inhibiting hormone into blood capillaries of the hypothalamus; a neuron that secretes oxytocin or antidiuretic hormone into blood capillaries of the posterior pituitary.

Neurotransmitter (noo′-rō-trans-MIT-er) A molecule released from a neuron into the synaptic cleft in response to an action potential.

Neutrophil (NOO-trō-fil) A type of leukocyte characterized by granules that stain pale lilac with a combination of acidic and basic dyes.

Nicotinic (nik′-ō-TIN-ik) **receptor** A type of cholinergic receptor for acetylcholine found on both parasympathetic and sympathetic postganglionic neurons and on skeletal muscle at the motor end plate.

Nociceptor (nō′-sē-SEP-tor) A free nerve ending that detects painful stimuli.

Norepinephrine (nor′-ep-ē-NEF-rin) **(NE)** A hormone secreted by most sympathetic postganglionic neurons and by the adrenal medulla. Also called noradrenaline (nor-a-DREN-a-lin).

Nucleic (noo-KLĒ-ik) **acid** An organic compound that is a long polymer of nucleotides, with each nucleotide containing a pentose sugar, a phosphate group, and one of four possible nitrogenous bases (adenine, cytosine, guanine, and thymine or uracil).

Nucleoli Spherical bodies within a cell nucleus composed of protein, DNA, and RNA that are the sites of the assembly of small and large ribosomal subunits. Singular is nucleolus.

Nucleosome (NOO-klē-ō-sōm) Structural subunit of a chromosome consisting of histones and DNA.

Nucleus (NOO-klē-us) A spherical or oval organelle of a cell that contains the hereditary factors of the cell, called genes; a cluster of neuronal cell bodies in the central nervous system; the central part of an atom made up of protons and neutrons.

Nutrient A chemical substance in food that provides energy, forms new body components, or assists in various body functions.

O

Obesity (ō-BĒS-i-tē) Body weight more than 20% above a desirable standard due to excessive accumulation of fat.

Olfaction (ōl-FAK-shun) The sense of smell.

Olfactory bulb A mass of gray matter containing cell bodies of neurons that form synapses with neurons of the olfactory (I) nerve.

Olfactory receptor cell A sensory neuron that detects olfactory stimuli.

Oligodendrocyte (OL-i-gō-den′-drō-sīt) A neuroglial cell that supports neurons and produces a myelin sheath around axons of neurons of the central nervous system.

Oncogene (ON-kō-jēn) Cancer-causing gene; it derives from a normal gene, termed a protooncogene, that encodes proteins involved in cell growth or cell regulation but has the ability to transform a normal cell into a cancerous cell when it is mutated or inappropriately activated.

Oogenesis (ō-ō-JEN-e-sis) Formation and development of female gametes (oocytes).

Optic chiasm (kī-AZM) A crossing point of the two branches of the optic (II) nerve.

Optic disc A small area of the retina containing openings through which the axons of the ganglion cells emerge as the optic (II) nerve. Also called the blind spot.

Organ A structure composed of two or more different kinds of tissues with a specific function and usually a recognizable shape.

Organ of Corti (KOR-tē) The organ of hearing, consisting of hair cells and supporting cells that rest on the basilar membrane and extend into the endolymph of the cochlear duct. Also called the spiral organ.

Organelle (or-ga-NEL) A permanent structure within a cell with characteristic morphology that is specialized to serve a specific function in cellular activities.

Organism A total living form; one individual.

Osmoreceptor (oz′-mō-rē-SEP-tor) Receptor in the hypothalamus that is sensitive to changes in blood osmolarity and, in response to high osmolarity (low water concentration), stimulates synthesis and release of antidiuretic hormone (ADH).

Osmosis (oz-MŌ-sis) The net movement of water molecules through a selectively permeable membrane from an area of higher water concentration to an area of lower water concentration until equilibrium is reached.

Otolith (Ō-tō-lith) A particle of calcium carbonate embedded in the otolithic membrane.

Otolithic (ō-tō-LITH-ik) **membrane** Thick, gelatinous, glycoprotein layer located directly over hair cells of the macula in the saccule and utricle of the internal ear.

Oval window A small, membrane-covered opening between the middle ear and inner ear into which the footplate of the stapes fits.

Ovarian (ō-VAR-ē-an) **cycle** A monthly series of events in the ovary associated with the maturation of a secondary oocyte.

Ovarian follicle (FOL-i-kul) A general name for oocytes (immature ova) in any stage of development, along with their surrounding epithelial cells.

Ovary (Ō-var-ē) Female gonad that produces oocytes and the hormones estrogens, progesterone, inhibin, and relaxin.

Ovulation (ov′-ū-LĀ-shun) The rupture of a mature ovarian (Graafian) follicle with discharge of a secondary oocyte into the pelvic cavity.

Ovum (Ō-vum) The female reproductive or germ cell; an egg cell; arises through completion of meiosis in a secondary oocyte after penetration by a sperm.

Oxyhemoglobin (ok′-sē-HĒ-mō-glō-bin) Hemoglobin combined with oxygen.

Oxytocin (ok-sē-TŌ-sin) **(OT)** A hormone secreted by neurosecretory cells in the paraventricular and supraoptic nuclei of the hypothalamus that stimulates contraction of smooth muscle in the pregnant uterus and myoepithelial cells around the ducts of mammary glands.

P

P wave The deflection wave of an electrocardiogram that signifies atrial depolarization.

Pacinian (pa-SIN-ē-an) **corpuscle** A type of sensory receptor that detects high-frequency vibrations. Also called a lamellated corpuscle.

Pancreas (PAN-krē-as) An elongated, tapered gland located behind the stomach. It is both an exocrine gland (secreting pancreatic juice) and an endocrine gland (secreting insulin, glucagon, somatostatin, and pancreatic polypeptide).

Pancreatic (pan′-krē-AT-ik) **duct** A single large tube that unites with the common bile duct from the liver and gallbladder and drains pancreatic juice into the duodenum at the ampulla of Vater (hepatopancreatic ampulla).

Pancreatic islet (ī-let) A cluster of endocrine gland cells in the pancreas that secretes insulin, glucagon, somatostatin, and pancreatic polypeptide. Also called an islet of Langerhans (LANG-er-hanz).

Paracrine A type of local mediator that acts on a neighboring target cell.

Parasympathetic (par′-a-sim-pa-THET-ik) **nervous system** One of the subdivisions of the autonomic nervous system; primarily concerned with activities that conserve and restore body energy.

Parathyroid (par′-a-THĪ-royd) **gland** One of four small endocrine glands partially embedded in the back surface of the thyroid gland.

Parathyroid hormone (PTH) A hormone secreted by the chief (principal) cells of the parathyroid glands that increases blood calcium level and decreases blood phosphate level.

Parietal cell A type of secretory cell in gastric glands that produces hydrochloric acid and intrinsic factor.

Pathogen (PATH-ō-jen) A disease-producing microbe.

Pedicel Footlike structure, as on podocytes of a glomerulus.

Penis (PĒ-nis) The organ of copulation and urination in males; used to deposit semen into the female vagina.

Pepsin Protein-digesting enzyme secreted by chief cells of the stomach in the inactive form pepsinogen, which is converted to active pepsin by hydrochloric acid.

Pericardial (per′-i-KAR-dē-al) **cavity** Small potential space between the visceral and parietal layers of the pericardium that contains pericardial fluid.

Pericardium (per′-i-KAR-dē-um) A loose-fitting membrane that encloses the heart.

Perilymph (PER-i-limf) The fluid contained between the bony and membranous labyrinths of the inner ear.

Perimetrium (per′-i-MĒ-trē-um) The outer layer of the uterus.

Peripheral nervous system (PNS) The part of the nervous system that lies outside the central nervous system; it includes nerves and sensory receptors.

Peristalsis (per′-i-STAL-sis) Successive muscular contractions along the wall of a hollow muscular structure.

Peroxisome (pe-ROKS-i-sōm) Organelle similar in structure to a lysosome; contains enzymes that use molecular oxygen to oxidize various organic compounds to produce hydrogen peroxide; abundant in liver cells.

pH A measure of the concentration of hydrogen ions (H^+) in a solution.

Phagocytosis (fag′-ō-sī-TŌ-sis) The process by which phagocytes ingest and destroy microbes, cell debris, and other foreign matter.

Pharynx (FAR-inks) The throat; a tube that extends from the nasal and oral cavities to the larynx and esophagus.

Photopigment A substance that can absorb light and undergo structural changes, which can lead to the development of a receptor potential in photoreceptor of the eye. Also called a visual pigment.

Photoreceptor Receptor that detects light shining on the retina of the eye.

Physiology The study of the functions of an organism and its constituent parts.

Pineal (PĪN-ē-al) **gland** A cone-shaped gland of the diencephalon that secretes melatonin.

Pinna (PIN-na) The projecting part of the external ear that collects sound waves and channels them into the external auditory canal.

Pituitary (pi-TOO-i-tār-ē) **gland** A small endocrine gland attached to the hypothalamus by the infundibulum. Also called the hypophysis (hī-POF-i-sis).

Plasma (PLAZ-ma) The extracellular fluid found in blood vessels; blood minus the cellular elements.

Plasma cell Cell that develops from a B cell (lymphocyte) and produces antibodies.

Plasma membrane Outer, limiting membrane that separates the cell's internal parts from extracellular fluid or the external environment.

Platelet (PLĀT-let) A fragment of cytoplasm enclosed in a cell membrane and lacking a nucleus; found in the circulating blood; plays a role in hemostasis.

Platelet plug Aggregation of platelets at a site where a blood vessel is damaged that helps stop or slow blood loss.

Polycythemia (pol′-ē-sī-THĒ-mē-a) Disorder characterized by an above-normal hematocrit (above 55%); can cause hypertension, thrombosis, and hemorrhage.

Polyunsaturated fat A fatty acid that contains more than one double covalent bond between its carbon atoms; abundant in triglycerides of corn oil, sunflower oil, and fish oils.

Pons (PONZ) The part of the brain stem that forms a "bridge" between the medulla oblongata and the midbrain.

Pontine respiratory center The portion of the respiratory center in the pons that helps regulate breathing.

Positive feedback system Feedback system that strengthens a change in a controlled variable.

Posterior pituitary Posterior lobe of the pituitary gland. Also called the neurohypophysis (noo-rō-hī-POF-i-sis).

Postganglionic neuron (post′-gang-lē-ON-ik NOO-ron) The second autonomic motor neuron in an autonomic pathway, having its cell body and dendrites located in an autonomic ganglion and its axon extending to cardiac muscle, smooth muscle, or a gland.

Potential energy The energy stored by matter due to its position.

Preganglionic (prē-gang-lē-ON-ik) **neuron** The first autonomic motor neuron in an autonomic pathway, with its cell body and dendrites in the brain or spinal cord and its axon extending to an autonomic ganglion, where it synapses with a postganglionic neuron.

Presbyopia (prez-bē-Ō-pē-a) A loss of elasticity of the lens of the eye due to advancing age, with resulting inability to focus clearly on near objects.

Primary motor cortex A region of the cerebral cortex in the precentral gyrus of the frontal lobe of the cerebrum that controls specific muscles or groups of muscles.

Primary somatosensory cortex A region of the cerebral cortex in the postcentral gyrus of the parietal lobe of the cerebrum that localizes exactly the points of the body where somatic sensations originate.

Primary active transport The process in which a solute moves across the membrane against its concentration or electrochemical gradient by carriers that use energy supplied by hydrolysis of ATP.

Progesterone (prō-JES-te-rōn) A female sex hormone produced by the ovaries that helps prepare the endometrium of the uterus for implantation of a fertilized ovum and the mammary glands for milk secretion.

Prolactin (prō-LAK-tin) **(PRL)** A hormone secreted by the anterior pituitary that initiates and maintains milk secretion by the mammary glands.

Prophase The first stage of mitosis, during which chromatid pairs are formed and aggregate around the metaphase plate of the cell.

Proprioceptor (PRO-prē-ō-sep′-tor) A receptor located in muscles, tendons, or joints (muscle spindles, tendon organs, and joint kinesthetic receptors) that provides information about body position and movements.

Prostaglandins (pros′-ta-GLAN-dins) **(PG)** Lipids released by damaged cells that intensify the effects of histamine and kinins.

Prostate (PROS-tāt) A doughnut-shaped gland below the urinary bladder that surrounds the upper portion of the male urethra; secretes a slightly acidic solution that contributes to sperm motility and viability.

Proteasome (PRŌ-tē-a-sōm) Tiny cellular organelle in cytosol and nucleus containing proteases that destroy unneeded, damaged, or faulty proteins.

Protein An organic compound consisting of carbon, hydrogen, oxygen, nitrogen, and sometimes sulfur and phosphorus; synthesized on ribosomes and made up of amino acids linked by peptide bonds.

Proto-oncogene Gene responsible for some aspect of normal growth and development; it may transform into an oncogene, a gene capable of causing cancer.

Puberty (PŪ-ber-tē) The time of life during which the secondary sex characteristics begin to appear and the capability for sexual reproduction is possible; usually occurs between the ages of 10 and 17.

Pulmonary edema (e-DĒ-ma) An abnormal accumulation of interstitial fluid in the tissue spaces and alveoli of the lungs due to increased pulmonary capillary permeability or increased pulmonary capillary pressure.

Pulmonary embolism (EM-bō-lizm) The presence of a blood clot or a foreign substance in a pulmonary arterial blood vessel that obstructs circulation to lung tissue.

Pulse (PULS) The rhythmic expansion and elastic recoil of a systemic artery after each contraction of the left ventricle.

Pupil The hole in the center of the iris; the area through which light enters the posterior cavity of the eyeball.

Purkinje (pur-KIN-jē) **fiber** Muscle fiber (cell) in the ventricular tissue of the heart specialized for conducting an action potential to the myocardium; part of the conduction system of the heart.

Pus The liquid product of inflammation containing leukocytes or their remains and debris of dead cells.

Q

QRS complex The deflection waves of an electrocardiogram that represent onset of ventricular depolarization.

R

Receptor A specialized cell or a distal portion of a neuron that responds to a specific sensory modality, such as touch, pressure, cold, light, or sound, and converts it to an electrical signal (receptor potential); a specific molecule or cluster of molecules that recognizes and is bound by a particular ligand.

Rectum (REK-tum) The terminal part of the gastrointestinal tract, from the sigmoid colon to the anus.

Red nucleus A cluster of cell bodies in the midbrain that looks reddish due to an iron-containing pigment and a rich blood supply.

Referred pain Pain that is felt at a site remote from the place of origin.

Reflex Fast response to a change (stimulus) in the internal or external environment that attempts to restore homeostasis.

Reflex arc The most basic conduction pathway through the nervous system, connecting a receptor and an effector; consists of a receptor, a sensory neuron, an integrating center in the central nervous system, a motor neuron, and an effector. Also called a reflex circuit.

Relaxin (RLX) A female hormone produced by the ovaries and placenta that increases flexibility of the pubic symphysis and helps dilate the uterine cervix to ease delivery of a baby.

Releasing hormone Hormone secreted by the hypothalamus that can stimulate secretion of hormones of the anterior pituitary.

Renal corpuscle Bowman's capsule and its enclosed glomerulus.

Reproduction The formation of new cells for growth, repair, or replacement; the production of a new individual.

Reproductive cell division Type of cell division in which gametes (sperm and oocytes) are produced; consists of meiosis and cytokinesis.

Reproductive system Body system composed of organs involved in the production of a new individual.

Respiration (res-pi-RĀ-shun) The process of supplying the body with O_2 and removing CO_2.

Respiratory center Neurons in the pons and medulla oblongata of the brain stem that regulate breathing; divided into the medullary respiratory center and the pontine respiratory center.

Respiratory (RES-pi-ra-tō-rē) **system** Body system consisting of the nose, pharynx, larynx, trachea, bronchi, and lungs.

Reticular (re-TIK-ū-lar) **activating system (RAS)** A portion of the reticular formation that has many ascending connections with the cerebral cortex; when this area of the brain stem is active, action potentials pass to the thalamus and widespread areas of the cerebral cortex, resulting in generalized alertness or arousal from sleep.

Reticular formation A network of small groups of neuronal cell bodies scattered among bundles of axons (mixed gray and white matter) beginning in the medulla oblongata and extending upward through the central part of the brain stem.

Retina (RET-i-na) The inner layer of the eye that converts light into action potentials.

Ribonucleic (rī-bō-noo-KLĒ-ik) **acid (RNA)** A single-stranded nucleic acid made up of nucleotides, each consisting of a nitrogenous base (adenine, cytosine, guanine, or uracil), ribose, and a phosphate group; three types are messenger RNA (mRNA), transfer RNA (tRNA), and ribosomal RNA (rRNA), each of which has a specific role during protein synthesis.

Ribosome (RĪ-bō-sōm) A cellular structure in the cytoplasm of cells, composed of a small subunit and a large subunit that contain ribosomal RNA and ribosomal proteins; the site of protein synthesis.

Rigidity (ri-JID-i-tē) Hypertonia characterized by increased muscle tone; does not affect reflexes.

Rigor mortis State of partial contraction of muscles after death due to lack of ATP; myosin heads (cross-bridges) remain attached to actin, thus preventing relaxation.

Rod One of two types of photoreceptor in the retina of the eye; specialized for vision in dim light.

Round window A small opening between the middle and internal ear, directly below the oval window, covered by membrane.

Ruffini corpuscle A type of sensory receptor that responds to skin stretching and to pressure. Also called a type II cutaneous mechanoreceptor.

Rugae (ROO-gē) Large folds in the mucosa of an empty hollow organ, such as the stomach and vagina.

S

Saccule (SAK-ūl) The smaller of the two chambers in the membranous labyrinth inside the vestibule of the internal ear; contains a receptor organ that detects linear acceleration or deceleration in a vertical direction.

Saliva (sa-LĪ -va) A clear, alkaline, somewhat viscous secretion produced mostly by the three pairs of salivary glands; contains various salts, mucin, lysozyme, salivary amylase, and lingual lipase (produced by glands in the tongue).

Salivary amylase (SAL-i-ver-ē AM-i-lās) An enzyme in saliva that initiates the chemical breakdown of starch.

Salivary gland One of three pairs of glands (parotid, submandibular, and sublingual) that lie external to the mouth and pour their secretory product (saliva) into ducts that empty into the oral cavity.

Sarcolemma (sar′-kō-LEM-ma) The cell membrane of a muscle fiber (cell), especially of a skeletal muscle fiber.

Sarcomere (SAR-kō-mēr) A contractile unit in a striated muscle fiber (cell) extending from one Z disc to the next Z disc.

Sarcoplasm (SAR-kō-plazm) The cytoplasm of a muscle fiber (cell).

Sarcoplasmic reticulum (sar′-kō-PLAZ-mik re-TIK-ū-lum) **(SR)** A network of sacs and tubes surrounding myofibrils of a muscle fiber (cell), comparable to endoplasmic reticulum; functions to reabsorb calcium ions during relaxation and to release them to cause contraction.

Satellite (SAT-i-līt) **cells** Flat neuroglial cells that surround cell bodies of peripheral nervous system ganglia to provide structural support and regulate the exchange of material between a neuronal cell body and interstitial fluid.

Saturated fat A fatty acid that contains only single bonds (no double bonds) between its carbon atoms; all carbon atoms are bonded to the maximum number of hydrogen atoms; prevalent in triglycerides of animal products such as meat, milk, milk products, and eggs.

Scala tympani (SKA-la TIM-pan-ē) The lower spiral-shaped channel of the bony cochlea, filled with perilymph.

Scala vestibuli (ves-TIB-ū-lē) The upper spiral-shaped channel of the bony cochlea, filled with perilymph.

Schwann cell (SCHVON or SCHWON) A neuroglial cell that produces a myelin sheath around axons of neurons of the peripheral nervous system.

Sclera (SKLE-ra) The portion of the outer layer of the eye that is comprised of tough connective tissue.

Scrotum (SKRŌ-tum) A skin-covered pouch that contains the testes and their accessory structures.

Sebum (SĒ-bum) Secretion of sebaceous (oil) glands.

Secondary sex characteristics Traits that distinguish males and females but do not have a direct role in reproduction.

Secondary active transport Process in which the energy supplied by an ionic electrochemical gradient is used to move a solute against its concentration or electrochemical gradient.

Semen (SĒ-men) A fluid discharged at ejaculation by a male that consists of a mixture of sperm and the secretions of the seminiferous tubules, seminal vesicles, prostate, and bulbourethral (Cowper's) glands.

Semicircular canals Three bony channels filled with perilymph, in which lie the semicircular ducts.

Semicircular ducts Membranous semicircular canals filled with endolymph that float in the perilymph of the bony semicircular canals; contain cristae that are concerned with rotational acceleration or deceleration.

Semilunar (sem′-ē-LOO-nar) **(SL) valve** A valve between the aorta or the pulmonary trunk and a ventricle of the heart.

Seminal vesicle (SEM-i-nal VES-i-kul) One of a pair of convoluted, pouchlike structures, lying behind and below the urinary bladder and in front of the rectum, that secrete a component of semen into the ejaculatory ducts.

Seminiferous tubule (sem′-i-NI-fer us TOO-būl) A tightly coiled duct, located in the testis, where sperm are produced.

Sensation A state of awareness of external or internal conditions of the body.

Sensory neuron (NOO-ron) Neuron that carries action potentials into the central nervous system. Also called an afferent neuron.

Serum Blood plasma minus its clotting proteins.

Set point The narrow range within which a controlled variable should be maintained.

Sex chromosomes The twenty-third pair of chromosomes, designated X and Y, which determine the genetic sex of an individual; in males, the pair is XY; in females, XX.

Shock Failure of the cardiovascular system to deliver adequate amounts of oxygen and nutrients to meet the metabolic needs of the body due to inadequate cardiac output.

Sign Any objective evidence of disease that can be observed or measured, such as a lesion, swelling, or fever.

Simple diffusion Process in which solutes move freely through the lipid bilayer of the plasma membranes of cells without the help of membrane transport proteins.

Signal transduction The process by which a signal molecule (extracellular chemical messenger) is transduced (converted) into a cellular response.

Sinoatrial (si-nō-Ā-trē-al) **(SA) node** A small mass of cardiac muscle fibers (cells) located in the right atrium that serves as the natural pacemaker of the heart.

Sinusoid (SĪ-nū-soyd) A large, thin-walled, and leaky type of capillary with large intercellular clefts that may allow proteins and blood cells to pass from a tissue into the bloodstream.

Sister chromatid (KRŌ-ma-tid) One of a pair of identical connected nucleoprotein strands that are joined at the centromere and separate during cell division, each becoming a chromosome of one of the two daughter cells.

Sleep A state of partial unconsciousness from which a person can be aroused; associated with a low level of activity in the reticular activating system.

Small intestine A long tube of the gastrointestinal tract that extends from the stomach to the large intestine; it is divided into three segments: duodenum, jejunum, and ileum.

Somatic (sō-MAT-ik) **cell division** Type of cell division in which a single starting cell duplicates itself to produce two identical cells; consists of mitosis and cytokinesis.

Somatic nervous system The portion of the peripheral nervous system that conveys output from the central nervous system to skeletal muscles.

Spasticity (spas-TIS-i-tē) Hypertonia characterized by increased muscle tone, increased tendon reflexes, and pathological reflexes (Babinski sign).

Sperm cell A mature male gamete. Also called a spermatozoon (sper′-ma-tō-ZŌ-on).

Spermatogenesis (sper′-ma-tō-JEN-e-sis) The formation and development of sperm in the seminiferous tubules of the testes.

Spinal (SPĪ-nal) **cord** A cylindrical mass of nerve tissue that extends from the brain.

Spinal nerve One of the 31 pairs of nerves that connect the spinal cord to sensory receptors and effectors in most parts of the body.

Spinothalamic (spī-nō-tha-LAM-ik) **pathway** Sensory pathway that conveys information for pain, temperature, itch, and tickle. Also called the anterolateral (an′-ter-ō-LAT-er-al) pathway.

Spinothalamic (spī-nō-tha-LAM-ik) **tract** Sensory (ascending) tract that conveys information up the spinal cord to the brain for sensations of pain, temperature, itch, and tickle.

Spleen (SPLĒN) Large lymphoid organ that filters microbes and aged or defective erythrocytes from blood.

Starvation (star-VĀ-shun) The loss of energy stores in the form of glycogen, triglycerides, and proteins due to inadequate intake of nutrients or inability to digest, absorb, or metabolize ingested nutrients.

Stem cell An unspecialized cell that has the ability to divide for indefinite periods and give rise to a specialized cell.

Stenosis (sten-Ō-sis) An abnormal narrowing or constriction of a duct or opening.

Stimulus Any stress that changes a controlled variable; any change in the internal or external environment that excites a sensory receptor, a neuron, or a muscle fiber.

Stomach The J-shaped enlargement of the gastrointestinal tract between the esophagus and small intestine.

Stratum basalis The layer of the endometrium next to the myometrium that is maintained during menstruation and gestation and produces a new stratum functionalis following menstruation or parturition.

Stratum functionalis (funk′-shun-AL-is) The layer of the endometrium next to the uterine cavity that is shed during menstruation or forms the maternal portion of the placenta during gestation.

Stroke Destruction of brain tissue (infarction) resulting from obstruction or rupture of blood vessels that supply the brain. Also called a cerebrovascular (se-rē-brō-VAS-kū-lar) accident (CVA) or brain attack.

Subarachnoid (sub′-a-RAK-noyd) **space** A space between the arachnoid mater and the pia mater that surrounds the brain and spinal cord and through which cerebrospinal fluid circulates.

Subatomic particles Components of an atom.

Submucosal plexus A network of neurons located in the submucosa of the gastrointestinal tract. Also called the plexus of Meissner (MĪZ-ner).

Surfactant (sur-FAK-tant) Complex mixture of phospholipids and lipoproteins, produced by type II alveolar (septal) cells in the lungs, that decreases surface tension.

Sympathetic (sim′-pa-THET-ik) **nervous system** One of the subdivisions of the autonomic nervous system; primarily concerned with processes involving the expenditure of energy.

Symptoms Subjective changes in body functions that are not apparent to an observer.

Synapse (SIN-aps) The functional junction between two neurons or between a neuron and an effector, such as a muscle or gland; may be electrical or chemical.

Synaptic (sin-AP-tik) **cleft** The narrow gap at a chemical synapse that separates the axon terminal of one neuron from another neuron or muscle fiber (cell) and across which a neurotransmitter diffuses to affect the postsynaptic cell.

Synaptic end bulb Expanded distal end of an axon terminal that contains synaptic vesicles.

Synaptic vesicle Membrane-enclosed sac in a synaptic end bulb that stores neurotransmitters.

System An association of organs that have a common function.

Systemic circulation The routes through which oxygenated blood flows from the left ventricle through the aorta to all the organs of the body and deoxygenated blood returns to the right atrium.

Systole (SIS-tō-lē) In the cardiac cycle, the phase of contraction of the heart muscle, especially of the ventricles.

Systolic (sis-TOL-ik) **pressure (SP)** The force exerted by blood on arterial walls during ventricular contraction; the highest pressure measured in the large arteries, about 110 mmHg under normal conditions for a young adult.

T

T wave The deflection wave of an electrocardiogram that represents ventricular repolarization.

Tachycardia (tak′-i-KAR-dē-a) An abnormally rapid resting heartbeat or pulse rate (over 100 beats per minute).

Target cell A cell that can respond to an extracellular chemical messenger.

Tectorial (tek-TŌ-rē-al) **membrane** A gelatinous membrane projecting over and in contact with the hair cells of the organ of Corti in the cochlear duct.

Teeth (TĒTH) Accessory structures of digestion that cut, shred, crush, and grind food.

Telophase (TEL-ō-fāz) The final stage of mitosis.

Tendon (TEN-don) A white fibrous cord of connective tissue that attaches muscle to bone.

Tendon organ A proprioceptive receptor, sensitive to changes in muscle tension and force of contraction, found chiefly near the junctions of tendons and muscles. Also called a Golgi (GOL-jē) tendon organ.

Tendon reflex A polysynaptic, ipsilateral reflex that protects tendons and their associated muscles from damage that might be brought about by excessive tension. The receptors involved are called tendon organs.

Teniae coli (TĒ-nē-ē KŌ-lī) The three flat bands of thickened, longitudinal smooth muscle running the length of the large intestine, except in the rectum. Singular is tenia coli.

Testis (TES-tis) Male gonad that produces sperm and the hormones testosterone and inhibin. Plural is testes. Also called a testicle.

Testosterone (tes-TOS-te-rōn) A male sex hormone (androgen) secreted by Leydig cells of a mature testis; needed for development of sperm; together with a second androgen termed dihydrotestosterone (DHT), controls the growth and development of male reproductive organs, secondary sex characteristics, and body growth.

Thalamus (THAL-amus) The main component of the diencephalon of the brain.

Thermoreceptor (THER-mō-rē-sep-tor) Sensory receptor that detects changes in temperature.

Thrombopoietin (TPO) Hormone produced by the liver that stimulates formation of platelets from megakaryocytes.

Thrombosis (throm-BŌ-sis) The formation of a clot in an unbroken blood vessel, usually a vein.

Thrombus (THROM-bus) A stationary clot formed in an unbroken blood vessel, usually a vein.

Thymus (THĪ-mus) A bilobed organ in which T cells develop immunocompetence.

Thyroid gland An endocrine gland that secretes thyroxine (T_4), triiodothyronine (T_3), and calcitonin.

Thyroid-stimulating hormone (TSH) A hormone secreted by the anterior pituitary that stimulates the synthesis and secretion of thyroxine (T_4) and triiodothyronine (T_3). Also known as thyrotropin.

Thyroxine (thī-ROK-sēn) **(T_4)** A hormone secreted by the thyroid gland that regulates metabolism, growth and development, and the activity of the nervous system. Also called tetraiodothyronine.

Tissue A group of similar cells that work together to perform a specific function.

Tongue A large muscular structure located on the floor of the oral cavity.

Trachea (TRĀ-kē-a) Tubular air passageway that extends from the larynx to the primary bronchi. Also called the windpipe.

Tract A bundle of neuronal axons in the central nervous system.

Transcription The process of copying the information represented by the sequence of base triplets in DNA into a complementary sequence of codons.

Translation Process in which the nucleotide sequence in an mRNA molecule specifies the amino acid sequence of a protein.

Transverse (T) tubules (TOO-būls) Small, cylindrical invaginations of the sarcolemma of striated muscle fibers (cells) that conduct muscle action potentials toward the center of the muscle fiber.

Triad (TRĪ-ad) A complex of three units in a muscle fiber composed of a transverse tubule and the sarcoplasmic reticulum terminal cisternae on both sides of it.

Tricuspid (trī-KUS-pid) **valve** Atrioventricular (AV) valve on the right side of the heart.

Triglyceride (trī-GLI-ser-īd) A lipid formed from one molecule of glycerol and three molecules of fatty acids that may be either solid (fats) or liquid (oils) at room temperature; the body's most highly concentrated source of chemical potential energy; found mainly within adipocytes. Also called a neutral fat or a triacylglycerol.

Triiodothyronine (trī-ī-ō-dō-THĪ-rō-nēn) **(T_3)** A hormone produced by the thyroid gland that regulates metabolism, growth and development, and the activity of the nervous system.

Trophoblast (TRŌF-ō-blast) The superficial covering of cells of the blastocyst.

Tropic (TRŌ-pik) **hormone** A hormone whose target is another endocrine gland.

Tubal ligation (lī-GĀ-shun) A sterilization procedure in which the fallopian tubes are tied and cut.

U

Ureter One of two tubes that connect the kidney with the urinary bladder.

Urethra (ū-RĒ-thra) The duct from the urinary bladder to the exterior of the body that conveys urine in females and urine and semen in males.

Urinalysis An analysis of the volume and physical, chemical, and microscopic properties of urine.

Urinary bladder A hollow, muscular organ that stores urine until it is excreted through the urethra.

Urinary system The body system consisting of the kidneys, ureters, urinary bladder, and urethra.

Urine The fluid produced by the kidneys that contains wastes and excess materials; excreted from the body through the urethra.

Uterus (Ū-te-rus) The hollow, muscular organ in females that is the site of menstruation, implantation, development of the fetus, and labor. Also called the womb.

Utricle (Ū-tri-kul) The larger of the two divisions of the membranous labyrinth located inside the vestibule of the inner ear, containing a receptor organ that detects head tilt and linear acceleration or deceleration that occurs in a horizontal direction.

V

Vagina (va-JĪ-na) A muscular, tubular organ that functions as the female organ of copulation, the passageway for menstrual flow, and the birth canal.

Varicose (VAR-i-kōs) Pertaining to an unnatural swelling, as in the case of a varicose vein.

Vas deferens (DEF-er-ens) The duct that carries sperm from the epididymis to the ejaculatory duct. Also called the ductus deferens.

Vascular spasm Contraction of the smooth muscle in the wall of a damaged blood vessel to prevent blood loss.

Vasectomy (va-SEK-tō-mē) A means of sterilization of males in which a portion of each vas deferens is removed.

Vasoconstriction (vāz-ō-kon-STRIK-shun) A decrease in the size of the lumen of a blood vessel caused by contraction of the smooth muscle in the wall of the vessel.

Vasodilation (vāz-ō-dī-LĀ-shun) An increase in the size of the lumen of a blood vessel caused by relaxation of the smooth muscle in the wall of the vessel.

Vein A blood vessel that conveys blood from tissues back to the heart.

Ventricle (VEN-tri-kul) A cavity in the brain filled with cerebrospinal fluid. A lower chamber of the heart.

Venule (VEN-ūl) A small vein that collects blood from capillaries and delivers it to a vein.

Ventral corticospinal tract Motor (descending) tract that conveys information from the motor cortex to the spinal cord for voluntary control of skeletal muscles in the trunk and proximal parts of the limbs.

Ventral root The structure composed of axons of motor (efferent) neurons that joins a dorsal root to form a spinal nerve. Also called an anterior root.

Vesicle (VES-i-kul) A small bladder or sac containing liquid.

Vestibular (ves-TIB-ū-lar) **apparatus** Collective term for the organs of equilibrium; includes the saccule, utricle, and semicircular ducts.

Vestibular membrane The membrane that separates the cochlear duct from the scala vestibuli.

Villi (VIL-ī) Fingerlike folds of the epithelium and lamina propria of the small intestinal mucosa that increase the surface area for absorption. Singular is villus (VIL-lus).

Vision The act of seeing.

Vitamin An organic molecule necessary in trace amounts that acts as a catalyst in normal metabolic processes in the body.

Vitreous (VIT-rē-us) **humor** A soft, jellylike substance that fills the posterior cavity of the eye.

Vulva (VUL-va) Collective designation for the external genitalia of the female.

W

White matter Aggregations of myelinated axons located in the brain and spinal cord.

X

X chromosome One of the two sex chromosomes. Sex chromosomes determine the genetic sex of an individual.

Y

Y chromosome One of the two sex chromosomes. Sex chromosomes determine the genetic sex of an individual.

Z

Zona fasciculata (ZŌ-na fa-sik′-ū-LA-ta) The middle zone of the adrenal cortex; secretes glucocorticoid hormones, mainly cortisol.

Zona glomerulosa (glo-mer′-ū-LŌ-sa) The outer zone of the adrenal cortex; secretes mineralocorticoid hormones, mainly aldosterone.

Zona pellucida (ZŌ-na pe-LOO-si-da) Clear glycoprotein layer between a secondary oocyte and the surrounding granulosa cells of the corona radiata.

Zona reticularis (ret-ik′-ū-LAR-is) The inner zone of the adrenal cortex; secretes androgens.

Zygote (ZĪ-gōt) The single cell resulting from the union of male and female gametes; the fertilized ovum.

Index

NOTE: Figures and tables are indicated by italic *f* and *t*, respectively, following the page reference.

M